MOTOR AIR CONDITIONER & HEATER MANUAL

5th Edition

First Printing

Louis C. Forier, SAE
Editor-in-Chief

Larry Solnik, SAE
Executive Editor

Michael J. Kromida, SAE
Managing Editor

Dan Irizarry, SAE • Warren Schildknecht, SAE
Associate Editors

Katherine Keen • William Lilieholm, SAE • Keith D. Rasmussen
Paul Sylvester • John J. Keebler
Assistant Editors

Patricia S. Walsh
Editorial Assistant

Published by

MOTOR

224 West 57th St., New York, N.Y. 10019

The Automotive Business Magazine

Library of Congress Catalog Number: 80-81643

ISBN 0-87851-529-1

DECIMAL & MILLIMETER EQUIVALENTS

INCH	INCH	MM
1/64	.015625	.397
1/32	.03125	.794
3/64	.046875	1.191
1/16	.0625	1.587
5/64	.078125	1.984
3/32	.09375	2.381
7/64	.109375	2.778
1/8	.125	3.175
9/64	.140625	3.572
5/32	.15625	3.969
11/64	.171875	4.366
3/16	.1875	4.762
13/64	.203125	5.159
7/32	.21875	5.556
15/64	.234375	5.953
1/4	.25	6.350
17/64	.265625	6.747
9/32	.28125	7.144
19/64	.296875	7.541
5/16	.3125	7.937
21/64	.328125	8.334
11/32	.34375	8.731

INCH	INCH	MM
23/64	.359375	9.128
3/8	.375	9.525
25/64	.390625	9.922
13/32	.40625	10.319
27/64	.421875	10.716
7/16	.4375	11.113
29/64	.453125	11.509
15/32	.46875	11.906
31/64	.484375	12.303
1/2	.5	12.700
33/64	.515625	13.097
17/32	.53125	13.494
35/64	.546875	13.890
9/16	.5625	14.287
37/64	.578125	14.684
19/32	.59375	15.081
39/64	.609375	15.478
5/8	.625	15.875
41/64	.640625	16.272
21/32	.65625	16.669
43/64	.671875	17.065

INCH	INCH	MM
11/16	.6875	17.462
45/64	.703125	17.859
23/32	.71875	18.265
47/64	.734375	18.653
3/4	.75	19.050
49/64	.765625	19.447
25/32	.78125	19.884
51/64	.796875	20.240
13/16	.8125	20.637
53/64	.828125	21.034
27/32	.84375	21.431
55/64	.859375	21.828
7/8	.875	22.225
57/64	.890625	22.622
29/32	.90625	23.019
59/64	.921875	23.415
15/16	.9375	23.812
61/64	.953125	24.209
31/32	.96875	24.606
63/64	.984375	25.003
1		25.400

Special Service Tools

Throughout this manual references are made and illustrations may depict the use of special tools required to perform certain jobs. These special tools can generally be ordered through the dealers of the make vehicle being serviced. It is also suggested that you check with local automotive supply firms as they also supply tools manufactured by other firms that will assist in the performance of these jobs. The vehicle manufacturers special tools are supplied by:

American Motors & General Motors .... Service Tool Division
Kent-Moore Corporation
1501 South Jackson Street
Jackson, Michigan 49203

Chrysler Corp. .... Miller Special Tools
A Division of Utica Tool Co.
32615 Park Lane
Garden City, Michigan 48135

Ford Motor Co. .... Owatonna Tool Company
Owatonna, Minnesota 55060

Table of Contents

Data reported in this manual is subject to change. To report current additions or revisions between editions of the manual, a "Manual Update" page is published every month for your guidance in MOTOR Magazine.

Air Conditioning Specifications: 1-315

I. Theory

II. Automatic Temperature Control Systems

Chrysler Corporation

Ford Motor Company

General Motors Corporation

Buick

Cadillac

Chevrolet

Oldsmobile

Pontiac

III. Maintenance • Diagnosis • Testing

Maintenance & Refrigerant Component Diagnosis

TABLE OF CONTENTS

TABLE OF CONTENTS

IV. Unit Replacement

Part 1 — Servicing Refrigeration System Components

TABLE OF CONTENTS

Part 2 — Servicing Electrical, Manual, Vacuum System Components, Evaporator & Heater Core

American Motors & Jeep

Chrysler Corp. (Cars & Trucks)

TABLE OF CONTENTS

TABLE OF CONTENTS

General Motors Corp.

BUICK

CADILLAC

CHEVROLET & GMC

TABLE OF CONTENTS

OLDSMOBILE

PONTIAC

TABLE OF CONTENTS

GM AUTO TEMPERATURE CONTROL CALIBRATIONS

V. VIR Service

VI. Compressor Service

VII. Aftermarket Units

VIII. Truck & Cargo Coolers

TABLE OF CONTENTS

IX. Glossary

X. Vacuum & Wiring Diagrams

American Motors

Chrysler Corporation

Chrysler & Imperial

Dodge & Plymouth

TABLE OF CONTENTS

TABLE OF CONTENTS

Buick-Full Size

Cadillac

Chevrolet-Compact & Intermediate

1980 Chevrolet Citation, Buick Skylark, Oldsmobile Omega & Pontiac Phoenix

Chevrolet-Full Size

Chevrolet Corvette

TABLE OF CONTENTS

I. THEORY

Basic Concept

We all know what air conditioning does for us but few of us understand how or why it works. An air conditioner is functionally very similar to a refrigerator and refrigeration is accomplished by making practical use of three laws of nature. These laws of nature and their practical application are discussed in the following paragraphs.

HEAT TRANSFER

When two substances of different temperature are placed near each other, the heat in the warmer substance always travels to the colder substance until both are of equal temperature. For example, a piece of ice in an ice box does not transmit its coldness to a bottle of milk standing nearby. Rather, in obedience to a law of nature, the heat in the warm milk automatically flows into the ice which has a lesser degree of heat.

In order to determine the amount of heat that transfers from one substance to another, science has established a definite standard of measurement called the British Thermal Unit or BTU. One BTU is the amount of heat required to raise the temperature of one pound of water 1 degree Fahrenheit. For example, to raise the temperature of one pound of water from 32 degrees F to 212 degrees F, one BTU of heat must be added for each degree rise in temperature or a total of 180 BTU's of heat. Conversely, in order to lower the temperature of one pound of water from 212 degrees F to 32 degrees F, 180 BTU's of heat must be removed from the water.

LATENT HEAT OF VAPORIZATION

Every substance exists in one of the three states of natural matter—solid, liquid or gas. The specific state being dependent on the amount of heat contained by the substance. Also, every substance has two points on the temperature scale where it absorbs heat. For example, water absorbs heat at 32 degrees F. when it changes from a solid (ice) to a liquid, and again absorbs heat at 212 degrees F. when it boils and changes from a liquid to a vapor. Therefore, the quantity of heat required to change the state of a substance is called latent heat or hidden heat. This heat which cannot be directly measured with a thermometer, or felt, is the basis of all refrigeration systems.

When a liquid boils (changes from a liquid to a gas) it absorbs heat without raising the temperature of the resulting gas. When the gas condenses (changes back to a liquid), it gives off heat without lowering the temperature of the resulting liquid.

For example, place one pound of water at 32 degrees F in an open container over a flame. With each BTU of heat that the water absorbs from the flame, its temperature rises 1 degree F. Thus, after it has absorbed 180 BTU's of heat, the water reaches a temperature of 212 degrees F. Here the law of nature is again encountered. Even though the flame continues to give its heat to the water, the temperature of the water remains at 212 degrees F. No matter how large or hot you make the flame, you cannot make water any hotter than 212 degrees F at sea level.

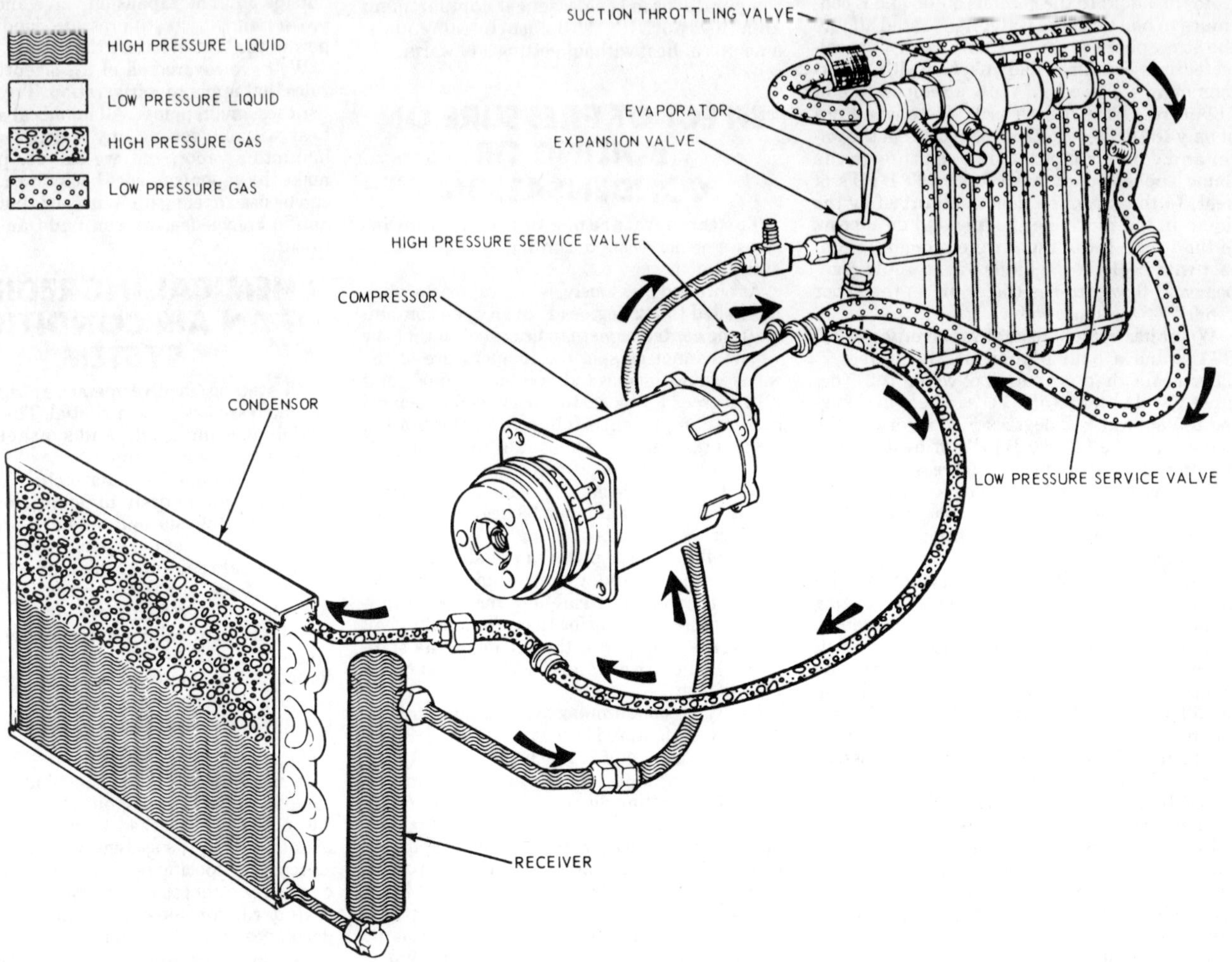

Fig. 1 Air conditioning cycle

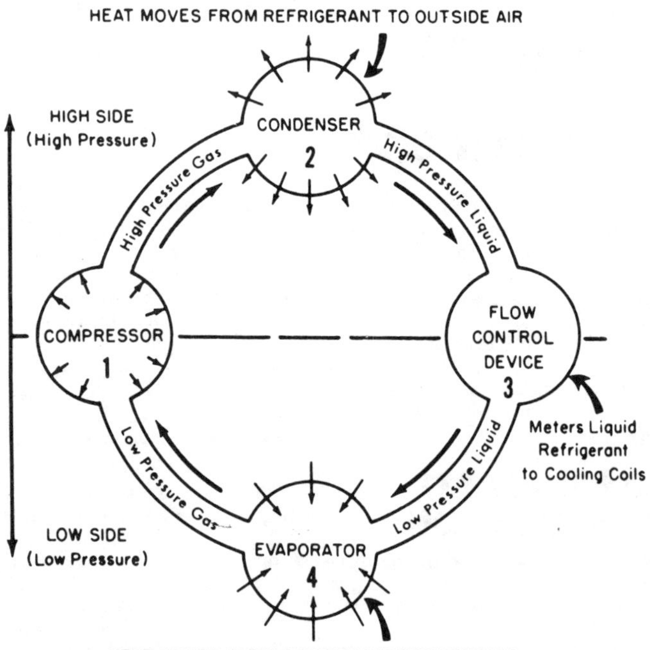

Fig. 2 Air conditioning theory of operation

The water, however, starts to boil or change from the liquid to the gaseous state, and it continues to boil until the water has passed off into the atmosphere as vapor. If this vapor were collected in a container and checked with a thermometer, it also would show a temperature of 212 degrees F. In other words, there was a rise of only 180 degrees F (from 32 to 212) in the water and vapor temperature even though the flame applied many more than 180 BTU's of heat. In this case, the heat is absorbed by the liquid in the process of boiling and disappears in the vapor. If the vapor were brought in contact with cool air, the hidden heat would reappear and flow into the cooler air as the vapor condensed back into water.

Water has a latent heat of vaporization of 970 BTU's and a boiling point of 212 degrees F. This means that one pound of water at 212 degrees F will absorb 970 BTU's of heat in changing to vapor at 212 degrees F. Conversely, the vapor will give off 970 BTU's of heat in condensing back to water at 212 degrees F.

This tremendous heat transfer, that occurs when a liquid boils or a vapor condenses, forms the basic principle of all conventional refrigeration systems.

For a liquid to be a good refrigerant, the amount of heat that it absorbs when vaporizing is not the only factor. It must also have a low boiling point. That is, the temperature at which it boils must be lower than the substance to be cooled. To illustrate with water, place a bottle of milk at room temperature (70 degrees F) next to boiling water (212 degrees F). The heat would flow from the water (higher temperature) to the milk (lower temperature). The milk would be heated rather than cooled, because the boiling point of water is too high.

In order to make practical use of the heat transfer that takes place when a liquid boils, we must choose a liquid with a low boiling point. Refrigerant 12 is the liquid most commonly used in automotive air conditioning systems because it boils at 21.7 degrees F. below zero. Here is a liquid that boils or vaporizes well below temperatures in the passenger compartment and, in vaporizing, will absorb tremendous amounts of heat without getting any warmer itself.

EFFECT OF PRESSURE ON BOILING OR CONDENSATION

The temperature of a liquid or vapor increases or decreases according to the pressure exerted on it.

As outlined previously, water vaporizes when it is boiled to 212 degrees F. in an open container. If the container were to be sealed so that the vapors cannot escape, the temperature of the water would increase above the boiling point (212 degrees F.) by 2½ to 3 degrees for each psi increase in pressure. Therefore, the boiling point of the water is increased by increasing the pressure. This is the principle used in the pressurized cooling system. By using a pressurized radiator cap, the water can be heated far above 212 degrees. F. without boiling.

A similar thing happens in the air conditioning system, but under controlled conditions. Liquid refrigerant is carefully metered into a low pressure area causing the refrigerant to boil within the evaporator, thus changing its state from a liquid to a gas and absorbing great quantities of heat.

In any air conditioning system, liquid refrigerant (R-12) is stored in the receiver under high pressure, Figs. 1 and 2. When the liquid R-12 is released into the evaporator by the expansion valve the resulting decrease in pressure lowers its temperature to —21.7 degrees F; its boiling point. As the R-12 flows through the evaporator coils, the passenger compartment air passes over the outside surface of the coils. As it boils, the R-12 absorbs heat from the air and thus cools the air in the passenger compartment. The heat from the passenger compartment is absorbed by the boiling refrigerant and hidden in the vapor. The refrigeration cycle is now under way. To completethe cycle, the following remains to be done:

1. Dispose of the heat in the vapor.
2. Convert the vapor back to liquid for re-use.
3. Return the liquid to the starting point in the refrigeration cycle.

The compressor performs these functions. It pumps the refrigerant (containing the hidden heat) out of the evaporator and forces it under high pressure into the condenser which is located in the outside air stream at the front of the car. The increased pressure in the condenser raises the R-12 vapor temperature to a point higher than that of the outside air. As the heat transfers from the hot vapor to the cooler air, the R-12 condenses back to a liquid. The liquid under high pressure now returns to the reservoir where it is stored for re-use.

It may seem somewhat difficult to understand how heat is transferred from a relatively cooler passenger compartment to the hotter outside air. The answer is in the difference between the refrigerant pressure that exists in the evaporator and the pressure that exists in the condenser. In the evaporator, the expansion valve reduces the pressure and thereby reduces the boiling point below the temperature in the passenger compartment. Thus, heat transfers from the passenger compartment to the boiling refrigerant. In the condensor, the compressor raises the condensation point above the temperature of the outside air. Thus, the heat transfers from the condensing refrigerant to the outside air. The expansion valve and the compressor simply create pressure conditions that permit the laws of nature to function.

We have covered all of the scientific ground rules that apply to refrigeration. Try to remember these main points. All liquids absorb lots of heat without getting any warmer when they boil into a vapor, and, we can use pressure to make the vapor condense back into a liquid so it can be used over again. With just that much scientific knowledge, we can build an air conditioner.

CHEMICAL INGREDIENTS OF AN AIR CONDITIONER SYSTEM

All systems involve metals, refrigerant and oil which are basic and essential. The desiccant, or dehydrating agent, and synthetic rubber make it even more complex.

All of the ingredients have chemical properties which are entirely different from each of the others. Only one undesirable element added or allowed to enter the system can start a chain of chemical reactions which upsets stability and interferes with the operation of the system. A sealed refrigerating system is a complex physical-chemical combination which is designed for stability within certain operating limits. If these limits are exceeded, many physical reactions occur to the system which will eventually cause it to fail.

Metals

In most cases, metals contribute to the decomposition of R-12 and oil in varying amounts. All metals are attacked by acids. Each of the metals used in the system have been carefully selected for specific reasons; heat conductivity, durability, strength and chemical composition.

Under favorable conditions the amount of decomposition of R-12 and oil is negligible. If undesirable substances are added and the temperature is increased, the rate of decomposition and the production of harmful acids increases proportionally.

Refrigerant

Refrigerant-12 is transparent and colorless. It has a boiling point of —21.7 degrees F. and under normal temperatures and pressures, it will be a vapor. The vapor is heavier than air, noninflammable, nonexplosive, non-poisonous (except when in contact with an open flame) and noncorrosive (except when in contact with water).

The chemical properties of refrigerants are very important factors in the stability of the system since the refrigerant penetrates to every part of the system.

Among the many desirable properties of R-12, is its stability under operating conditions. While more stable than the other refrigerants under the same conditions, it can be caused to form harmful acids which will eventually cause the system to fail.

Oil

Oil is the most complex of all organic chemicals. Its stability in a refrigeration system is dependent upon the source of crude oil and its method of refining. A good refrigerating oil must be free of sludge gum-forming substances and impurities such as sulphur. It must be stabilized to resist oxidation and must have a high degree of resistance to carbonization.

The chemical properties of the lubricating oil form another very important consideration in the chemical stability within the system. Like the refrigerant, it travels to every part of the system. Leaving a container uncapped for even a few minutes allows the oil to absorb moisture from the air.

Many system failures have been caused by chemical reactions which were started by servicemen adding contaminated oil.

Desiccant (Dehydrating Agent)

An ideal desiccant must have the followinng characteristics:

1. High capacity.
2. High efficiency.
3. Low tendency to powder.
4. Absorb moisture without reacting chemically with it.
5. Allow refrigerant to flow through it with minimum restriction.
6. Retain moisture at high temperature.

PRIMARY CAUSES OF SYSTEM FAILURE

Leaks

A shortage of refrigerant causes oil to be trapped in the evaporator. Oil may be lost with the refrigerant at point of leakage. Both of these conditions can cause compressor seizure.

Oil circulates in the system with the refrigerant; in solution with the liquid and in globules with the vapor. It leaves the compressor by the action of the pistons and mixes with the refrigerant liquid in the condenser. The oil then enters the evaporator with the liquid, and with the evaporator properly flooded, is returned to the compressor through the low pressure line. Some of the oil returns as globules in the vapor but more importantly, it is swept as a liquid along the walls of the tubing by the velocity of the vapor. If the evaporator is starved, the oil cannot return in sufficient quantity to keep the compressor properly lubricated.

High Temperature And Pressure

An increase in temperature causes an increase in pressure. This accelerates chemical instability due to existing contaminants in the system, and initiates chemical instability in clean systems. Other results are brittle hoses, "O" rings and bypass valve diaphragms with possible decomposition, broken compressor discharge reeds and seized compressor bearings.

High temperature and pressure is also caused by air in the system.

Air In The System

Air results from a discharged system or careless servicing procedures. This reduces system capacity and efficiency and causes oxidation of oil into gum and varnish.

When a leak causes a system to be discharged, the resulting vacuum will cause air to be drawn in. Air in a system is a non-condensable gas and will build up in the condenser as it would in an air tank. The resultant heat produced will contribute to the conditions previously discussed.

Many systems are contaminated by servicemen who are either careless or do not know proper servicing methods. Too frequently, systems are left open during servicing and then are not properly purged or evacuated. Often, as the result of a collision, a system will be opened because of ruptured hoses or damaged condenser, etc. In these cases, the dehydrator should be replaced and the system purged, evacuated and recharged. Remember, any air in the system, is too much.

Poor Connections

Hose clamps must be properly located and tightened. Hose should be installed over the sealing flanges with the end of the hose at the stop flange. The hose should never extend beyond the stop flange.

When compression fittings are used, over tightening can cause damage to the "O" ring gasket and will result in leaks.

Restrictions

These may be caused by powdered desiccant or dirt and foreign matter. This may result in a starved evaporator and loss of cooling, high temperature at the bypass hose, or a seized compressor.

When the amount of moisture in a system exceeds the capacity of the dessicant, it can break down the dessicant and cause it to powder. The powder passes through the system with the refrigerant liquid to the expansion value screen. Some of it may pass through into the evaporator where it may quickly build up to cause a restriction. Sufficient oil then cannot be returned to the compressor and it may seize.

Dirt

Dirt, which is any foreign material, may come from cleaner residues, cutting, machining or preserving oils, metal dust or chips, lint or dust, loose rust, soldering or brazing fluxes, paint or loose oxide scale. These can also cause seized bearings, discharge or expansion valve failure, decomposition of refrigerant and oil or corrosion of metal parts.

Corrosion

Corrosion and its by-products can restrict valve and drier screens, roughen bearings surfaces or hasten fatigue of discharge reeds. This can result in high temperature and pressure, decomposition or leaks.

The major cause of corrosion is moisture in the system.

Moisture

Moisture is the greatest enemy of refrigeration systems. Combined with metal, it produces oxide, Iron Hydroxide, and Aluminum Hydroxide. Combined with R-12, it produces Carbonic acid, Hydrochloric acid and Hydrofluoric acid. Moisture can also cause freeze-up of expansion valve and powdered desiccant.

AIR CONDITIONER COMPONENTS

Receiver-Dehydrator Unit

The air conditioning system stores the liquid refrigerant (R-12) under pressure in a combination receiver and dehydrator. The pressure in the receiver normally varies from about 80 to 300 psi, depending upon the surrounding air temperature and compressor speed.

The dehydrator serves the purpose of removing any moisture that may have accumulated in the system. Even small amounts of moisture will cause an air cooling unit to malfunction.

Evaporator Unit

When a cooling system is operating, the liquid refrigerant (R-12) flows from the receiver-dehydrator unit through a hose to the evaporator where it is allowed to evaporate at a reduced pressure, to cool the evaporator. Air is blown through the evaporator fins and is thus cooled by the evaporator.

Expansion Valve

The rate of refrigerant evaporation is controlled by an expansion valve which allows only enough refrigerant to flow into the evaporator to keep it operating efficiently, depending on its heat load.

The expansion valve consists of the valve and a temperature sensing capillary tube and bulb. The bulb is clamped to the outlet pipe of the evaporator. By use of internal passages leading to and from the underside of the valve diaphragm, the diaphragm senses the refrigerant temperature and pressure as it leaves the evaporator core. Thus the valve is controlled by evaporator outlet temperature.

The restricting effect of the expansion valve at the evaporator causes a low pressure on the low pressure side of the system of 12 to 50 psi, depending on the surrounding air temperature and compressor speed.

Valves in Receiver (VIR)

This assembly contains the thermostatic expansion valve (TX), POA suction throttling valve, receiver-dehydrator and sight glass into one unit. This assembly eliminates the external line between the expansion valve and outlet of the POA valve. The thermobulb and capillary line are also eliminated as the power element or diaphragm end of the expansion valve capsule is exposed directly to the refrigerant vapor entering the VIR unit. The EE (EEVIR) model of the VIR eliminates the equalizer port between the POA valve and TX valve in the housing.

Compressor Unit

The evaporated refrigerant leaving the evaporator at a pressure of 12 to 50 psi is pumped by the compressor, located on the engine, into the top of the condenser which is in front of the radiator.

The compressor maintains a pressure on its high side of from 80 to 300 psi, again depending on the surrounding air temperature and compressor speed.

Condenser

The condenser, located in front of the radiator, receives heated and compressed refrigerant gas from the compressor.

As the refrigerant gas flows down through the condenser, it is cooled by air passing between the sections of the condenser. The cooled, compressed gas condenses to a liquid refrigerant which then flows into the receiver.

Sight Glass

A liquid sight glass is mounted in the high pressure refrigerant line between the receiver and the expansion valve. The glass is used to determine whether there is enough liquid refrigerant in the system.

Magnetic Clutch

The compressor is cut in and out of operation by an electrically operated magnetic clutch mounted on the compressor crankshaft. The clutch is controlled by the blower switch, a clutch switch and a de-icing switch or suction throttling valve.

Thermostatic De-Icing Switch

This switch is connected in series with the blower switch and the clutch switch and controls the operation of the compressor by controlling the magnetic clutch. The temperature sensing tube of the switch is placed in contact with the evaporator fins. When the temperature of the evaporator becomes too cold, the switch opens the clutch electrical circuit, thereby disconnecting the compressor from the engine. When the temperature of the evaporator rises, the switch closes and energizes the clutch. This connects the compressor to the engine and cooling action begins again.

When the blower switch is off or the control lever is not in an A/C position, the clutch cannot be energized for compressor operation.

When the blower switch is on and the control lever is in the cooling range, the clutch is controlled by the thermostatic switch which is controlled by evaporator temperature.

Suction Throttling Valve

This valve performs the same function as the de-icing switch explained previously. Since pressure affects temperature, the main function of the valve is to maintain the refrigerant gas in the evaporator at a pressure which will maintain the temperature of the evaporator core surface above 32 degrees F. This will prevent evaporator icing.

To make the expansion valve work effectively with the suction throttling valve, an equalizer line is connected between the suction side of the suction throttling valve and the expansion valve diaphragm.

POA Evaporator Pressure Control Valve

The function of the POA (Pilot Operated Absolute) valve is to control evaporator pressure. This is done by throttling or restricting the evaporator outlet, the end result being the same as with the suction throttling valve used on other models. There is no similarity in the operation of the two valves. The valve is factory set for 29.5 pounds per square inch gauge pressure, at 900 ft. and is not to be tampered with. If it is determined that the valve is malfunctioning it must be replaced.

Ambient Temperature Switch

A system that does not use a de-icing switch will use an ambient temperature switch. This switch operates the compressor switch. When the temperature at the switch is above 40 degrees F, the compressor must be on. When the temperature at the switch is below 25 degrees F, the compressor must be off.

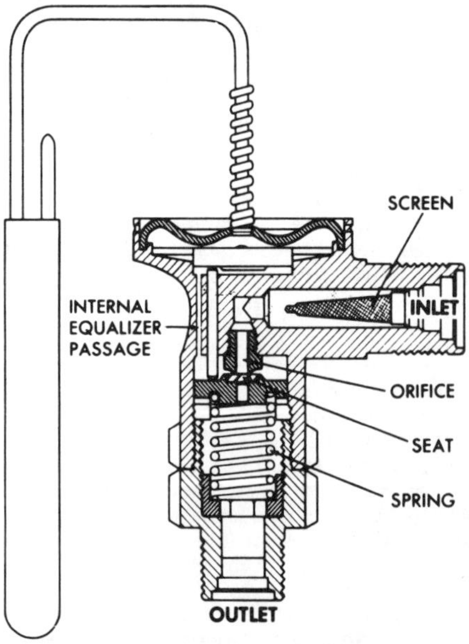

Fig. 3 Internally equalized expansion valve.

High Pressure Relief Valve

This valve is used to release excess refrigerant pressure in the system, by opening and allowing refrigerant vapor to escape under extreme pressure conditions. This way, excess pressure that may damage the compressor is released.

Low Pressure Cut-Off Switch

This switch senses low refrigerant charge conditions and protects the compressor from damage by creating an open circuit in the compressor clutch circuit. On some systems, this switch also serves as an ambient temperature switch, to prevent compressor operation at low temperatures.

METHODS OF TEMPERATURE CONTROL

Expansion Valves

Internally Equalized Type

Referring to Fig. 3, the refrigerant enters the inlet and screen as a high pressure liquid. The refrigerant flow is restricted by a metered orifice through which it must pass. As the refrigerant passes through this metered orifice, it changes from a high pressure liquid to a low pressure liquid or from the high side to the low side.

The refrigerant flow through the metered orifice is of the utmost importance. Anything that interferes with this flow seriously affects operation of the entire system.

A sufficient drop in temperature of the area cooled by the evaporator changes the heat transfer requirement of the evaporator. To continue the same flow of refrigerant into the evaporator will result in extreme cooling of the evaporator fins and coils to such an extent that they will freeze over with ice and stop air flow over them.

The flow of refrigerant is controlled by varying the amount of refrigerant allowed to pass through the metered orifice. This is accomplished by internal balancing pressures moving a valve or a seat to control the refrigerant flow.

A thermal bulb connected to a diaphragm by a small line and filled with refrigerant gas such as Freon is secured firmly by a clamp to the evaporator tail pipe. The thermal bulb is sensitive to tail pipe temperatures. If the tail pipe becomes warm, the gas inside the bulb will begin to expand, exerting pressure against the diaphragm on the top plate that is connected to the seat or valve by a pin or pins (whichever construction is used). This expansion will then move the seat away from the orifice, allowing an increased refrigerant flow. As the tail pipe temperature drops, the pressure in the thermal bulb also drops, allowing the valve to restrict flow as required by the evaporator.

The pressure of the refrigerant entering the evaporator is fed back to the underside of the diaphragm through the internal equalizing passage. Expansion of the gas in the thermal bulb must overcome the internal balancing pressure before the valve will open to increase refrigerant flow.

A spring is installed against the valve and adjusted to a pre-determined setting at the time of manufacture. This is the super-heat spring, the purpose of which is to prevent slugging of the evaporator with excessive liquid. The adjusted tension of this spring is the determinating factor in the opening and closing of the expansion valve. During opening or closing, the superheat spring tension will retard or assist valve operation as required.

NOTE: For all practical purposes we should not be concerned with the adjustment of this spring. Tension is adjusted from 4 to 16 pounds as required for the unit on which it is to be installed. This original setting is sufficient for the life of the valve, and tampering with the adjustment will only lead to an unsatisfactory operating unit. Special equipment is required in most cases to accurately calibrate this adjustment.

Externally Equalized Type

Operation of this type valve, Fig. 4, is the same as the internally equalized valve except that the evaporator pressure is fed against the underside of the diaphragm from the tail pipe of the evaporator by an equalizer line. This balances the temperature of the tail pipe through the expansion valve thermal bulb against evaporator pressure taken from the tail pipe.

Service Precautions

From the foregoing description of the thermostatic expansion valve, it can readily be seen that the valve is more sensitive to foreign materials than any other unit in the A/C system. Therefore, the following points should be observed to protect the valve while in service and while servicing the system.

1. The AC/C system should be operated at least for short periods throughout the winter season to prevent the internal moving parts from corroding and sticking, which will result in the valve becoming inoperative and necessitating replacement. Periodic operation will also lubricate the entire system and especially the compressor seal, keeping it soft and in condition properly to seal the refrigerant.
2. Clean and replace all accessible screens when servicing.
3. Install a filter if system has excessive foreign materials present.
4. Follow correct evacuation procedure to re-

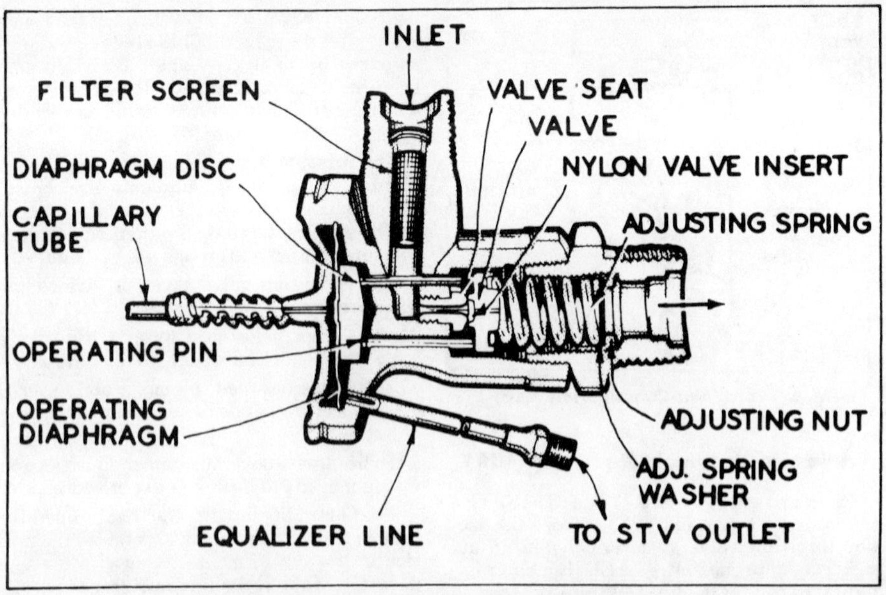

Fig. 4 Externally equalized expansion valve

move all possible moisture from system.

5. Cap or cover any lines opened for service to prevent entrance of moisture or dirt.
6. Replace dehydrator as soon as excess moisture content is evident. Any system should have the dehydrator replaced at least upon opening the system the third time for service. Also, any time the system has been opened for a prolonged period from accident or rupture, replace the dehydrator.
7. Handle the thermal bulb and line with extreme care, as excessive bending and rough handling can cause a break that will release the gas, rendering the valve inoperative.
8. Use a back-up wrench when removing any connection to prevent twisting of a line, which may result in weakening or breaking it.
9. Replace expansion valve only with a comparable valve. A numbering system is used to designate the orifice size. Replacing the valve with one with either too small or too large an orifice can seriously affect system operation.
10. Maintain corrosion-free positive contact between thermal bulb and tail pipe. Over a period of time corrosion will form between the two contact surfaces and insulate the two surfaces to such an extent that the operation of the expansion valve will be affected. The clamp may also work loose, preventing positive contact between the two and again affect operation of the expansion valve.

Frigidaire Expansion Tube (Orifice)

The expansion tube is located in the evaporator inlet line, and provides resistance to the high pressure liquid, thereby controlling the flow of refrigerant to the evaporator and allowing it to enter the evaporator as a low pressure liquid.

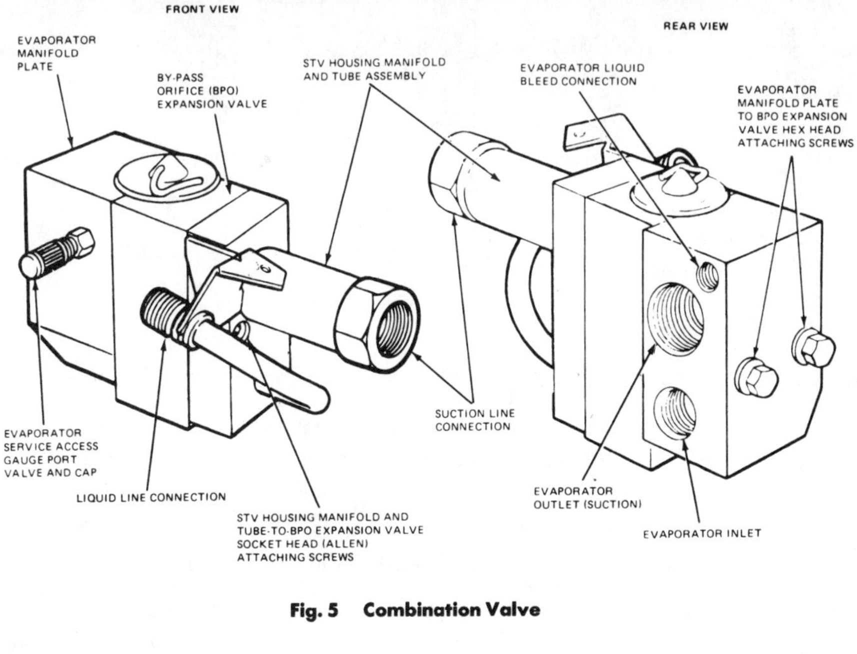

Fig. 5 Combination Valve

Chrysler Air Temp & Ford "H" Block Valve

This valve is a "block type" standard expansion valve that is internally equalized through passages and does not require a sensing bulb. On some Air Temp units, the low pressure cut-off switch is part of this assembly.

SUCTION THROTTLING REGULATORS

Suction throttling regulated systems include any type control used to regulate the flow of refrigerant from the evaporator to the compressor. This control will be located at some point between the tail pipe of the evaporator and the compressor in the low side or suction line, as it is called.

Sometimes it may be difficult to determine exactly what type of control system is used on a particular installation. However, if underhood inspection of the compressor and lines does not show a control device, it is safe to assume that a thermostat-controlled recycling clutch is used. *Generally, then, all suction throttling regulator controls will be located under the hood. The thermostat will not be used, and the clutch will be energized during total operation of this type system.*

Constant heat load, ram air, and compressor speed will allow the expansion valve to meter an even flow of refrigerant into the evaporator. Change any of these conditions, however, and the refrigerant flow through the expansion valve will change. Without some means provided to regulate refrigerant flow, the evaporator cooling will become excessive and freeze the moisture condensing on the coils. This results in evaporator icing or freeze-up and requires that the ice on the evaporator be melted before it can resume cooling action.

Installation of a suction throttling control device will compensate for these varying conditions. As the pressure and temperature within the evaporator drop, spring pressure forces a valve within the regulating device toward its seat, retarding the flow of refrigerant. This action increases the pressure back into the evaporator, raising both pressure and temperature. The temperature will not be allowed to drop low enough to freeze over the evaporator coils. A point of balance will soon be reached whereby the refrigerant flow past the valve in the regulator will reach a degree of pressure and temperature sufficient to maintain efficient cooling without evaporator freezing.

Chrysler EPR Valve

This suction control device is used on many Chrysler units. The EPR valve (Evaporator Pressure Regulator) is calibrated to produce maximum "cooling" without causing frost or ice on the evaporator fins and tubing. If for any reason the factory calibration has been disturbed, the EPR valve may restrict the flow of refrigerant at an evaporator pressure which is either too high for maximum performance or too low to prevent coil freeze-up.

Chrysler Air-Temp ETR Valve

The ETR (Evaporator Temperature Valve) was chosen for the Auto-Temp system because the EPR valve used on standard A/C systems would not prevent coil freeze-up at the lower ambient temperatures at which the compressor is required to operate. The ETR valve consists of an electrical solenoid which is actuated by the temperature switch which senses coil fin temperature.

Ford Combination Valve

This valve, Fig. 5, used on some 1976-80 Ford Motor Co. vehicles combines the function of the thermal expansion valve and suction throttling valve. Much of the tubing used with these valves has been eliminated.

The expansion valve regulates the liquid refrigerant flow while the interior of the vehicle and the evaporator core are being cooled. After this cool-down time, the suction throttling valve controls refrigerant flow.

The suction throttling valve maintains evaporator pressure above 28.5 psi to prevent evaporator freeze-up. Compressor suction pressure will always be less than evaporator pressure, the amount determined by compressor speed and evaporator heat load.

The liquid bleed valve is a Shrader valve calibrated to open at 10-20 psi. Its purpose is to bypass oil laden refrigerant around the restriction formed when the STV is closed, thereby providing oil return to the compressor.

CAUTION: Do not substitute the Shrader valve installed in the expansion valve with one of the gage port valves. Substitution could result in compressor damage due to oil starvation because a gage port Shrader valve will not open.

Ford Mini-Combination Valve

This valve, Fig. 6, is used on some 1980 Ford Motor Co. vehicles and is mounted directly to the evaporator core tubes. This unit combines the function of the expansion valve and suction throttling valve and performs the same function as the combination valve explained previously.

The suction throttling valve is mounted on the expansion valve. The expansion valve part of this valve is an internally equalized H-type valve with a slotted ball seat.

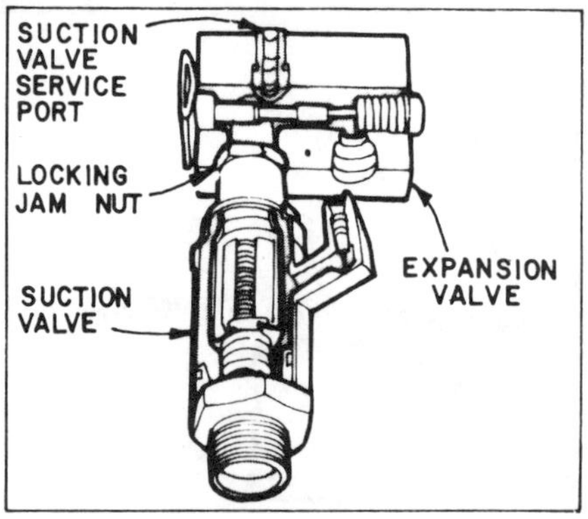

Fig. 6 Ford Mini-Combination valve

Frigidaire Valves In Receiver (VIR)

This assembly, Fig. 7, contains the thermostatic expansion (TX) valve, POA suction throttling valve, receiver-dehydrator and sight glass into one unit. The EE (Equalizer Expansion) model of the VIR, eliminates the equalizer port between the POA and TX valve cavities in the valve housing. By omitting the O-ring from the upper groove of the TX valve, the equalizer function is accomplished by allowing pressure within the inlet connector shell assembly to be transmitted to the TX valve equalizer port, Fig. 8.

Some VIR units incorporate a moisture indicator as a servicing aid and means of visual moisture check to assist in proper diagnosis of a malfunction resulting from moisture in the air conditioning system. These units can be identified by a color guide on the data label affixed to the receiver shell. The moisture indicator consists of a sleeve with a moisture sensitive coating assembled onto the lower end of the expansion capsule, Fig. 9. This sleeve is visible through the sight glass and changes in color from blue (dry system) to pink (wet system) when the air conditioning system has absorbed about 10 grams of water.

The pressure and temperature sensitive expansion valve, Fig. 10, automatically controls the amount of refrigerant entering the evaporator, by sensing the refrigerant temperature and pressure as it passes through the VIR unit.

The expansion valve cavity is divided into three pressure zones:

1. The area below the lower O-ring corresponds to discharge or high pressure side.
2. The area between the upper and lower O-rings corresponds to evaporator inlet pressure.
3. The area above the upper O-ring corresponds to evaporator outlet pressure.

The power diaphragm, which is the operating part of the expansion valve, controls the position of the valve seat in response to the temperature of the returning refrigerant passing across the top and around the sides of the expansion valve, the pressure of the refrigerant acting on the bottom of the diaphragm, head pressure acting upon the bottom of the valve and spring pressure holding the valve closed. The power diaphragm is partially filled with charcoal granules and charged with a specific amount of R-22 refrigerant. As the returning refrigerant from the evaporator passes across the TX valve diaphragm cover and around the sides of the valve above the O-ring, the refrigerant temperature is transmitted to the valve. Slight temperature changes will change the pressure of the R-22 gas, thus, causing the diaphragm operating pin and valve seat to open

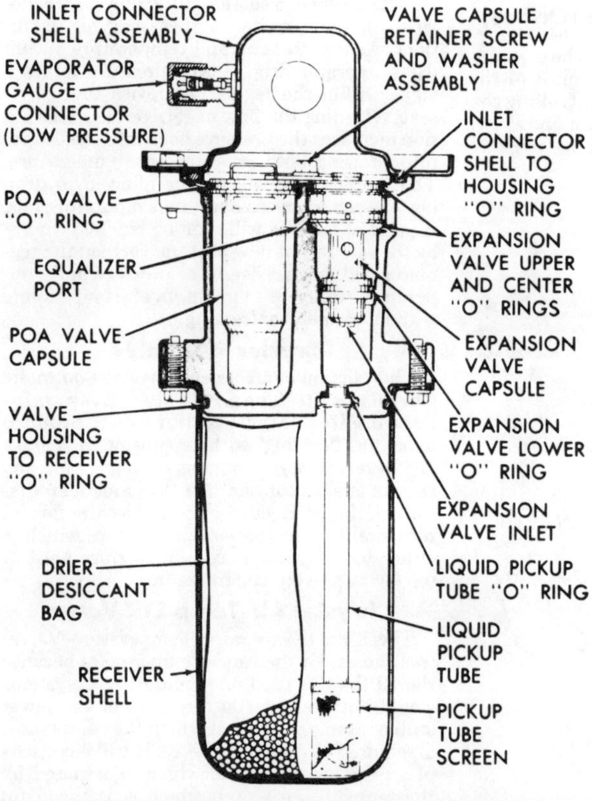

Fig. 7 Valves in receiver (V.I.R.)

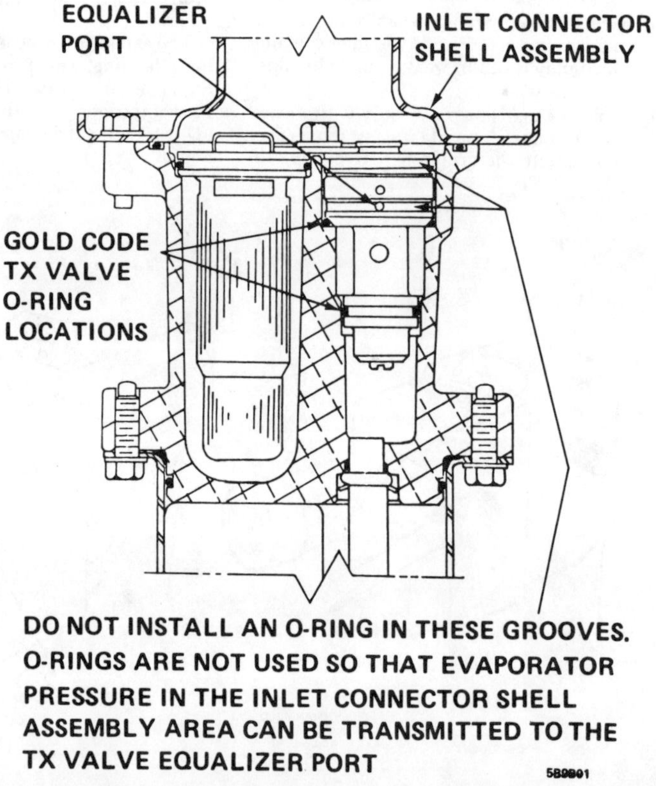

Fig. 8 VIR-EE TX valve capsule "O" ring locations

the TX valve. Evaporator outlet pressure applied to the bottom of the diaphragm through the equalizer port opposes the pressure of the diaphragm. As head pressure and spring pressure balance out, the valve seat opens enough to provide a constant evaporator temperature. If the evaporator runs short of liquid refrigerant, the refrigerant will become too warm when it reaches the expansion valve, increasing the R-22 gas pressure against the top of the diaphragm. Acting through the operating pin, this opens the expansion valve allowing more liquid refrigerant to the evaporator. If the evaporator has too much liquid refrigerant, the refrigerant passing over the expansion valve will become too cold, therefore relieving some of the pressure on the diaphragm, allowing it to move toward the closed position and reducing the flow of liquid refrigerant.

The equalizer port is used to speed up the opening of the expansion valve under certain conditions. If the Suction Throttling Valve (STV) is closing to slow down the cooling, the upper end of the evaporator core will eventually become warm, thus opening the expansion valve due to the pressure of the R-22 refrigerant. At higher vehicle speeds, the length of time for valve opening would not provide an even outlet temperature. To eliminate this fluctuation of temperature, the equalizer feature is used to help the expansion valve power diaphragm move the valve to the open position without waiting for the expansion valve to warm up.

As pressure drops in the equalizer port and under the diaphragm, the expansion valve will operate as the TX valve was being heated and exerting increased pressure on top of the diaphragm. The expansion valve will open and allow refrigerant flow into the evaporator core because of the reduced pressure under the diaphragm.

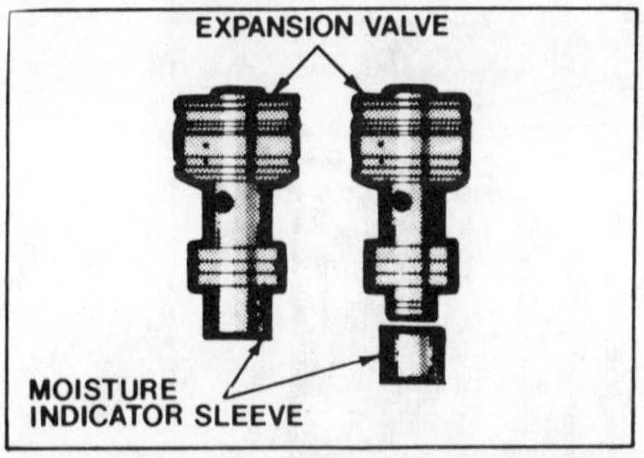

Fig. 9 Expansion valve & moisture indicator

The equalizer port in the previous VIR assemblies is drilled between the STV cavity of the VIR and the bottom of the TX valve diaphragm, but does not necessarily have low compressor pressure available at the line since if pressure drops in the low side of the system, the oil bleed check valve opens and raises the equalizer pressure above what it would be with the oil bleed valve closed. This allows some pressure from the evaporator into the equalizer through the STV cavity, leaving the pressure in the equalizer between the evaporator pressure of 29.5 psi and low pressure as low as 15 psi.

The equalizer port in the EE VIR is a drilled hole in the TX valve below the power diaphragm. This area is exposed to evaporator pressure by eliminating the top O-ring used in previous VIR assemblies. This allows evaporator pressure to enter the equalizer port which eliminates the need for a drilled passage between the STV and TX valve cavities.

Frigidaire STV

The function of the STV (Suction Throttling Valve), Fig. 11, is to regulate the flow of refrigerant from the evaporator so that a constant pressure of about 29.5 psi is maintained in the evaporator core. At this pressure the evaporator core tubes and fins are maintained at about 32 degrees F. If the pressure in the core were to drop much below 29.5 psi, ice would form on the core and block air flow. The ability of the STV to maintain 29.5 psi is limited by compressor capacity. At low compressor speed and high evaporator heat load, the evaporator pressure may be greater than 29.5 psi and the valve will be wide open.

During operation, the valve is controlled by opposing forces on the valve piston. Evaporator pressure on one side of the piston is opposed by spring pressure and control pressure on the other side.

Evaporator pressure above 29.5 psi exerts pressure on the piston, and through a bleed hole in the piston, into the bellows chamber where it is reduced to control pressure by the bellows operated pilot needle valve. This pressure overcomes spring tension and control pressure, causing the piston to open until a balanced condition is reached, returning evaporator pressure to the desired setting.

When evaporator pressure drops below 29.5 psi, spring tension overcomes evaporator pressure, causing the piston to close until the evaporator pressure returns to the desired setting.

Frigidaire POA Valve

The POA valve (Pilot Operated Absolute),

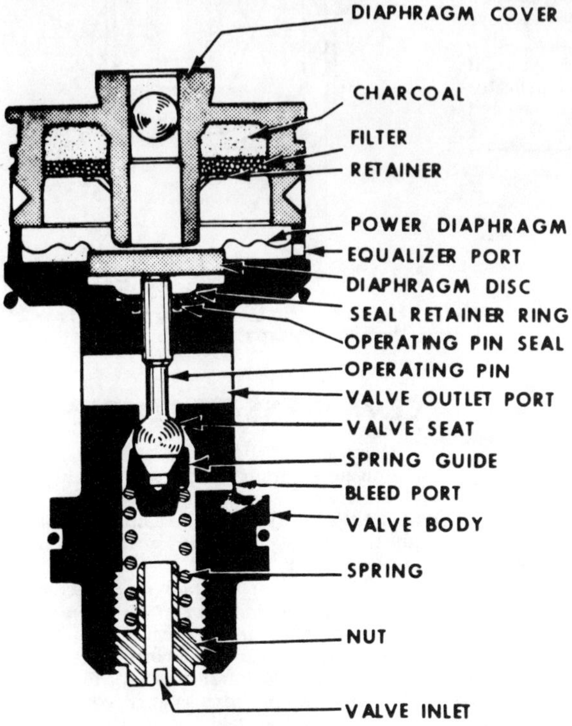

Fig. 10 Expansion (TX) valve

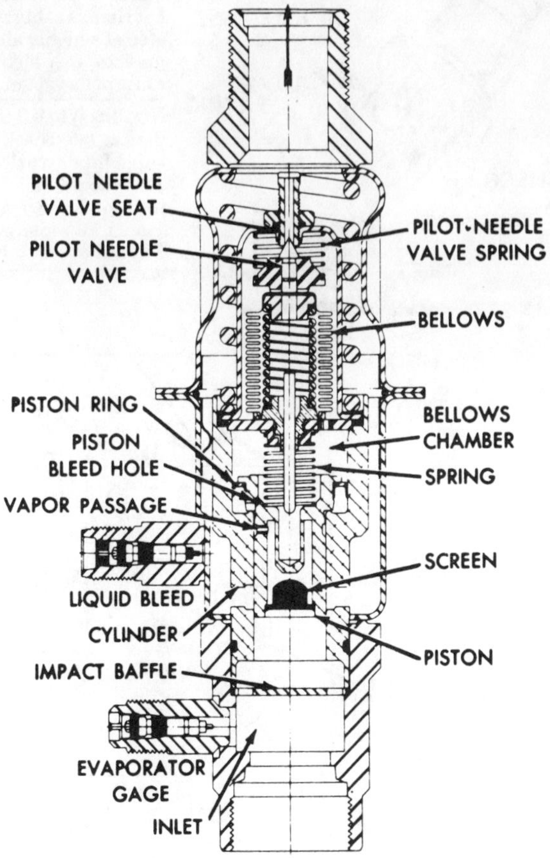

Fig. 11 Sectional view of STV/POA (suction throttling valve/pilot operated absolute)

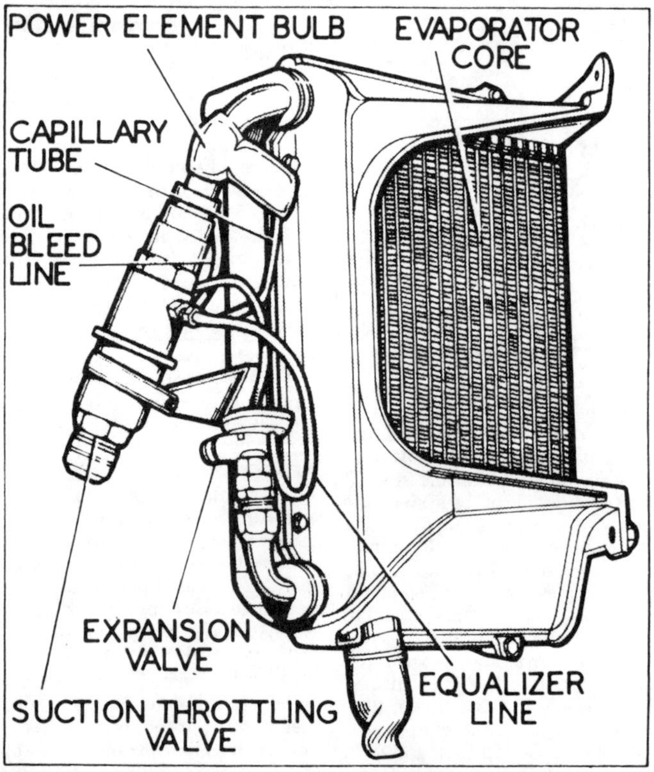

Fig. 12 Typical installation of GM POA valve and associated parts

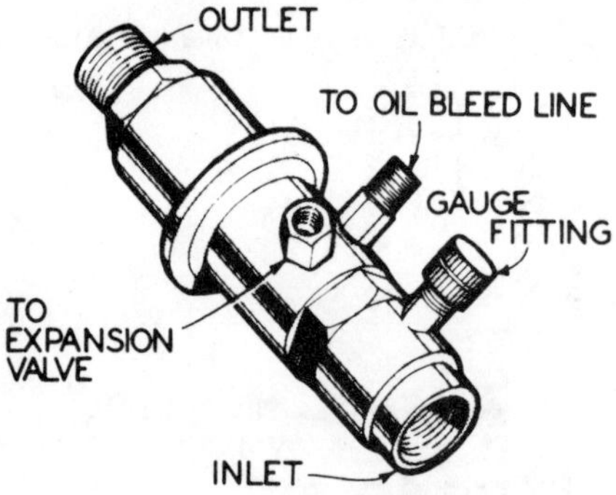

Fig. 13 Exterior view of POA valve

Figs. 12 and 13, differs from the STV in that it has no vacuum diaphragm. The POA valve will not change calibration when the system is operated at a higher altitude due to the atmospheric pressure to which the STV was subjected because of its vacuum diaphragm.

NOTE: When it is found necessary to replace a POA valve, check the interior of the defective valve for corrosion or crystalization of salts. This should indicate moisture in the system. If this condition exists, the receiver-dehydrator should be replaced and the system evacuated for one hour.

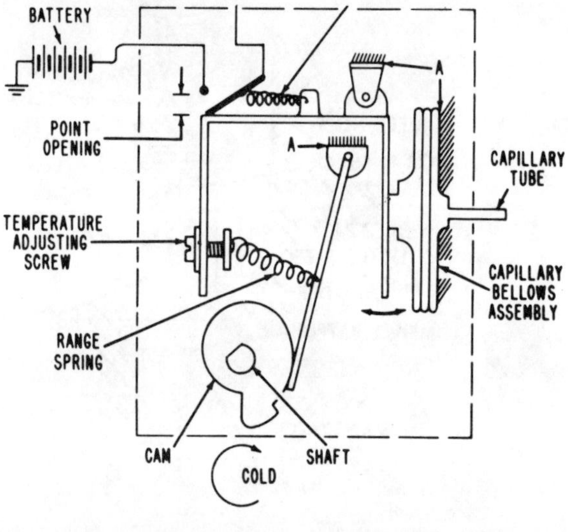

Fig. 15 Wiring diagram of a typical adjustable thermostatic switch

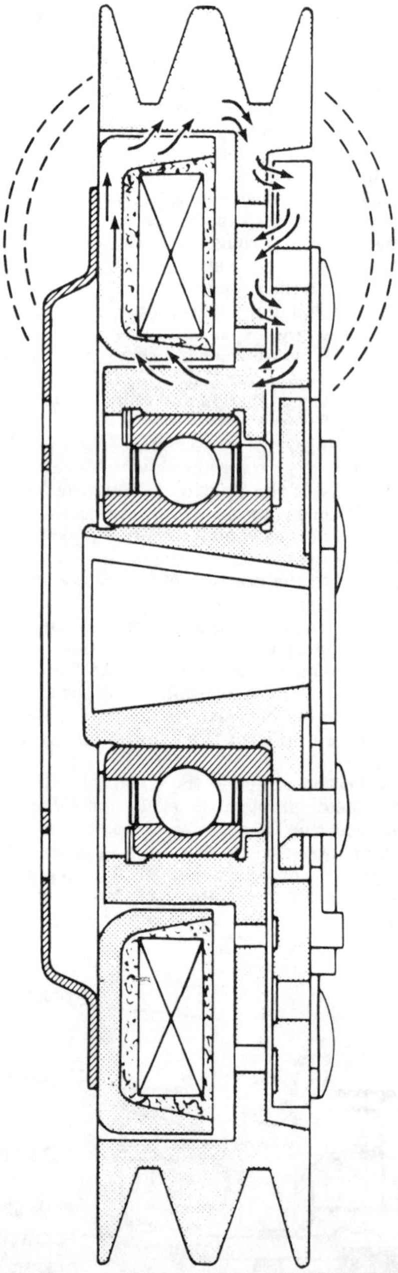

Fig. 14 Path of magnetic flow through compressor clutch

COMPRESSOR CLUTCHES

The purpose of the compressor clutch is to transmit power from the compressor and provide a means of engaging and disengaging the air conditioning system. The clutches used are of the rotor plate and friction surface type with a stationary coil mounted on the compressor frame. This coil being stationary, correct spacing becomes important to prevent the rotating pulley from contacting the coil and still bring the hub and armature into position to obtain maximum attraction of the magnetic force.

The compressor clutch is engaged by magnetic force and disengaged by spring tension. A wound core within a metal cup acts in a similar manner as a horseshoe magnet when the coil is energized, Fig. 14.

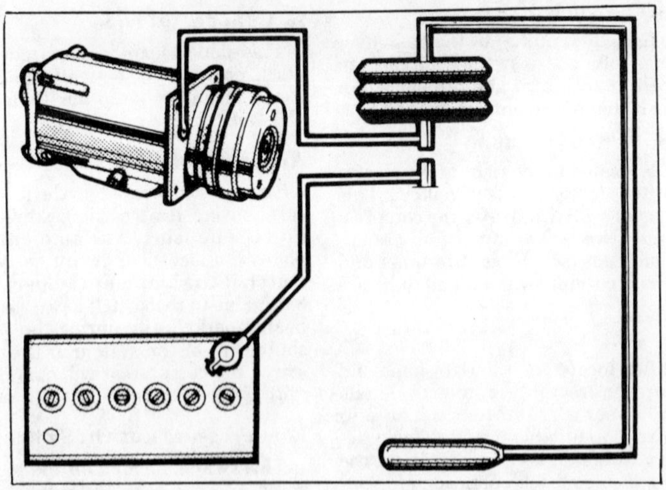

Fig. 16 Schematic layout of a thermostatic controlled clutch circuit

THERMOSTAT CONTROL SWITCH

When an A/C system is not provided with a suction throttling valve (STV, POA or EPR), the pumping action of the compressor is stopped by a thermostatically controlled switch, Figs. 15 and 16.

The opening and closing of the electrical contacts are controlled by a movement of a temperature sensitive diaphragm or bellows. The bellows has a capillary tube connected to it which is filled with Freon or CO_2. The capillary tube is positioned so that it may have either the cold air from the evaporator pass over it or it may be connected to the tail pipe of the evaporator. In either position, evaporator temperature will affect the temperature-sensitive compound in the capillary tube by causing it to contract as the evaporator becomes colder. The contraction of the gas will cause the bellows to contract. This action separates the electrical contacts and breaks the electrical circuit to the compressor clutch, which stops compressor operation.

The evaporator begins to warm which, in turn, causes the gas in the capillary tube to expand. The bellows will also begin to expand, moving the electrical contacts closer to each other. At a predetermined point, the expansion of the bellows will bring the contacts together, closing the circuit to the compressor clutch, energizing it and bringing the compressor into operation again. This cycling action will be repeated as long as air conditioning operation is required.

II. AUTOMATIC TEMPERATURE CONTROL SYSTEMS

This section will provide a description of the various Automatic Temperature Control (ATC) and Automatic Climate Control (ACC) systems used.

Chrysler Corporation

1974-78 AUTO TEMP II

This Auto-Temp II system automatically controls the heating and air conditioning operations of the vehicle to maintain a selected comfort level. The system provides heat or cool dehumidified air depending upon climatic conditions.

Operating Controls

The Automatic Temperature Control assembly is mounted in the center of the instrument panel and contains a comfort control potentiometer and lever and a push button switch.

Comfort Control Lever

The front face of the push button control assembly has five numbers, 65-70-75-80-85, to permit the operator to select a desired comfort level using the comfort control lever.

Pushbutton Switch

Off

The system will not operate. The servo will be automatically positioned in Lo-Heat (Park) or M1 A/C (Park), the blower will not operate and the outside air door will be closed.

Vent

Same as "Lo-Auto," however, the blower will operate at M2 speed only. The system will be in the A/C mode with clutch disengaged.

Lo-Auto

The blower will remain off and the outside air door will close, if heat is required, until water temperature reaches 125°F. Temperature controlled air will then be discharged from the heater slots, with normal defrost bleed, or the air conditioning outlets as required by the system. There are five blower speeds (M4, M3, M2, M1 and Lo) available in the air conditioning mode and the 1976-78 heat mode. Four blower speeds (M3, M2, M1 and Lo) are available in the 1974-75 heat mode.

Hi-Auto

Same as "Lo-Auto," however, the blower will operate at the higher speeds only. There are four blower speeds (M2, M3, M4 and Hi) available in the 1974-75 air conditioning mode. Three blower speeds (M2, M3 and Hi) are available in the heat mode and the 1976-78 air conditioning mode. All speeds are higher in the "Hi-Auto" position than in "Lo-Auto." This "Hi-Auto" position permits the operator to select the higher blower speeds that are required to maintain comfort during city traffic operation where air distribution may be inadequate, satisfy rear seat passengers under extreme conditions and provide adequate smoke removal under extreme conditions.

Lo-Def

The blower will immediately activate and air will be discharged through the defroster outlets, with normal heat bleed. The system is then controlled the same as in "Hi-Auto." The blower will operate at higher speeds, however, the "Hi" speed is not used. With the system controls in the "Lo-Def" position, maximum air conditioned air may be discharged onto the windshield, also the higher blower speed that is available in the air conditioning mode is obtained. However, it is probable that the system would not be operated in the "Lo-Def" position while full cooling of the vehicle interior is required.

Hi-Def

Same as "Lo-Def," however, the blower will operate at the highest (Hi) speed used in the heat mode only and discharge air will be maximum temperature.

Both "Lo-Def" and "Hi-Def" positions override a vacuum circuit to start blower operation, regardless of water temperature. Therefore, if the operator returns the control level to either the "Lo-Auto" or "Hi-Auto" position, the blower will continue to function even though the water temperature is less than 125° F. However, the blower may be deactivated by depressing the "Off" button.

In positions except "Off," the compressor will operate if the ambient temperature is above 32°F.

System Components

In-Car Sensor

This sensor, located behind a grille in the instrument panel, senses the car interior temperature and automatically signals the amplifier to compensate for any variation from the selected temperature setting.

Aspirator/Air Scoop

The aspirator, located on the rear surface of the evaporator assembly, is a tube shaped device connected to the air scoop with a piece of flexible hose. The air scoop is mounted inside the evaporator assembly so the air flowing through the evaporator is routed to the aspirator. The air creates a slight vacuum in the aspirator which then pulls the air over the in-car sensor.

Ambient Sensor

The ambient sensor, located in the plenum panel above the evaporator assembly, senses outside air temperature and sends an electrical signal to the amplifier.

Comfort Control Potentiometer & Lever

The lever, located above the ATC control assembly push buttons, permits the operator to select any interior temperature setting from 65 to 85. The lever controls a linear potentiometer which creates a resistance to signal the amplifier to control the system at the selected temperature setting.

Master On-Off Switch

This switch, located on top of the passenger side housing, is color coded green. This vacuum operated switch activates the blower when eight or more inches of vacuum are applied.

Compressor Switch (Vacuum)

This switch, located to the right of the master "On-Off" switch, is color coded yelllow. This vacuum operated switch activates the compressor clutch when two inches of vacuum are applied if the ambient switch on 1974 units and the low pressure cut out switch on all units are closed.

Amplifier

The amplifier, located on the right hand end of the evaporator assembly, converts signals from the in-car sensor, ambient sensor and comfort control potentiometer into a single signal to the servo motor. This signal positions the servo motor and a feedback potentiometer creates a return signal which deactivates the amplifier until a new signal is generated.

The amplifier also utilizes a battery voltage signal to "Park" the system, positioned in "Lo-Heat" or M1 A/C depending on ambient temperature, whenever the push button and/or the ignition is "Off."

Check Valves

The check valves are located in the vacuum harness on the passenger side near the top of the evaporator case. For identification, the valves are referred to as Number 1 and Number 2, with Number 1 check valve being nearest the servo assembly.

Vent Relay

This relay, located on the rear surface of the evaporator assembly above the aspirator, is activated in all modes except when the "Vent" push button is depressed. When the relay is deactivated it substitutes the sensor string with a fixed resistor so the amplifier can position the system in the "Vent" mode.

Hi-Blower Relay, 1974-75

This relay, located in the relay bank insulator to the left of the steering column, is activated only in the "Hi-Def" or "Hi-Auto" modes when the system requires "Hi" blower speed for heating or cooling.

Resistor Block

The resistor block, located on the top right hand corner of the passenger side housing, consists of resistance coils which provide the various blower motor speeds.

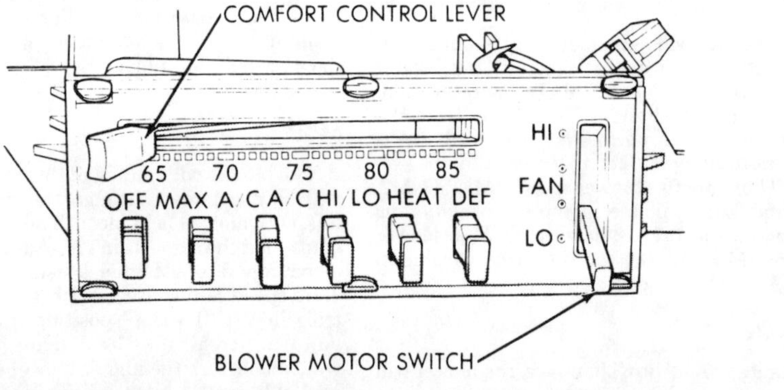

Fig. 1 Control panel assembly. Except 1980 Cordoba, Diplomat, LeBaron & Mirada

ATC Diode, 1974-75

This diode, located to the right of the resistor block, prevents the activation of the Hi-Blower Relay in any mode other than "Hi-Auto" or "Hi-Def."

Ambient Switch, 1974

This switch, located on the face of the engine side housing, has a sensing capillary protruding into the housing. The ambient switch completes the electrical circuit to the compressor clutch if the ambient temperature is above 32°F. Due to the switch location, it will become heated periodically during the winter by brief shutdowns of the system or the engine. On restarts, the compressor will operate briefly, permitting the compressor seals to be lubricated.

Low-Pressure Cut-Off Switch

This switch, located on the receiver drier, is wired in series with the compressor clutch. It disconnects the electrical power to the clutch when refrigerant pressure drops below the control point of the switch.

Whenever the system is deactivated by this switch due to loss of refrigerant, refrigerant oil may also have been lost. To prevent damage to the compressor due to operation without sufficient lubrication, the leak must be repaired and the compressor oil level checked prior to final charge of the system.

The switch is a sealed calibrated unit and cannot be adjusted or repaired. Replace if found defective.

Servo Assembly (Blend Air Servo)

The servo assembly's internal components are positioned by a signal from the amplifier through a small electric motor and gear train assembly. The electric motor actuates a rotary vacuum switch and a wiper arm which moves over a printed circuit board to change the blower speeds.

The blend air door is positioned automatically by the servo arm which is directly connected to the servo gear train. The door is connected to the servo arm with a link terminated in cable wire.

Water Valve (On-Off)

This vacuum operated valve, located above the right wheel housing in the engine compartment, is normally open in all modes of operation except when the "Off" push button is depressed.

Cold Engine Lockout Switch

This switch, mounted on the water valve, prevents the blower motor from starting, when heat is required, when the water temperature is below 125° F. and the "Lo-Auto" or "Hi-Auto" push-buttons are depressed. This function is accomplished by preventing vacuum from reaching the master switch.

Vacuum Transfer Switch

This switch, located in the outside/recirculating air door housing, positions the door in one of three positions: "Off," 20% outside air or 100% outside air. The door is operated by a spring loaded vacuum actuator and the transfer switch.

Vacuum Reservoir

The vacuum reservoir, mounted in the engine compartment, maintains sufficient vacuum in the vacuum circuit during periods when

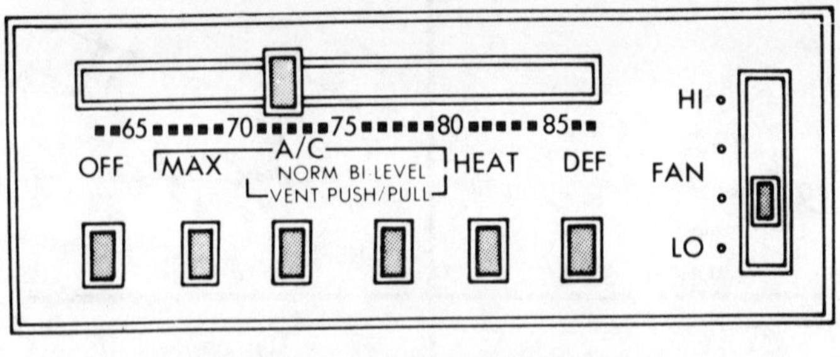

Fig. 2 Control panel assembly. 1980 Cordoba, Diplomat, LeBaron & Mirada

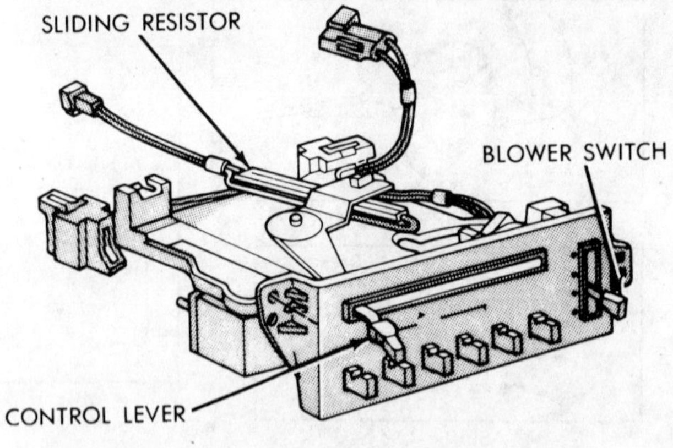

Fig. 3 Control head assembly. 1980 Cordoba, Diplomat, LeBaron & Mirada

manifold vacuum is reduced. A check valve is located at the inlet port which closes when vacuum supply diminishes.

Blower Motor

The blower motor, located on the engine side housing under the right front fender, operates automatically when any push-button other than "Off" is depressed and the master switch is closed.

1979-80 SEMI-AUTOMATIC TEMPERATURE CONTROL (S.A.T.C.)

This system maintains the selected temperature level within the vehicle while permitting the selection of various modes of operation. In cold weather, the system automatically moves the blend air door to provide heat and in warm weather, the door movement provides cool and/or dehumidified air at the preset temperature.

After the desired temperature setting and pushbutton mode is selected, the system will cool or heat in varying degrees, according to interior temperature, ambient temperature and temperature control setting. Temperature control is not obtainable in Max A/C since the water valve is closed and no water flows through the heater core while air is recirculated on the inside of the vehicle.

Operating Controls

The control assembly, Figs. 1 and 2, is located on the instrument panel and consists of a temperature control lever, a pushbutton mode switch and a four speed blower switch. The pushbutton and blower switches are identical to those used on standard A/C systems.

Comfort Control

The face of the comfort control has five numbers, 65-70-75-80-85, to permit the operator to select the desired comfort level

Comfort Control Lever

This lever, which is located above the pushbuttons in the control assembly, allows the operator to select the desired comfort level. On 1979 models, movement of the lever operates an arm on the S.A.T.C. bi-metal sensor through a flexible control cable. On 1980 models, movement of the lever operates a sliding resistor in the control head, Fig. 3.

Pushbutton Switch

The vacuum portion of this switch controls the operation of the water valve and positions all doors in the unit except the blend air door. The electrical portion of the switch controls the A/C compressor, blower motor and on 1980 Cordoba, Diplomat, LeBaron and Mirada models the electronic servo-motor operation.

When the switch is in the off position, the inlet air door is closed to outside air, the mode door is in the A/C position, the Heat-Defrost door is in the heat position, and the heater core coolant flow is off. No air flows through the unit and the compressor and S.A.T.C. electronics is off, since the blower and compressor clutch circuits are open.

When the switch is in the Max A/C position, all of the doors are in the same position as in the Off position. The electrical circuits to the lower motor, compressor clutch and on 1980 Cordoba, Diplomat, LeBaron and Mirada models the S.A.T.C. electronics are closed. This mode is recommended for initial cool down, extreme outside humidity, or high ambient temperatures. Only high and medium 2 (M2) blower speeds will operate with no automatic temperature control in this mode.

The A/C on 1979 models, or Norm on 1980 models position can be used when the vehicle has been cooled to the desired temperature. The vacuum application at the outside recirculating air door actuator is transferred to the rod side, moving the door away from the outside-air inlet and closing the recirculating inlet. The water valve is open and all other vacuum applications and door positions are the same as the Max A/C position. The blower motor and compressor are on. In this position, a Vent mode can be obtained by first pushing the Norm button and then pulling it back out. This will open the clutch circuit and allow the blower motor to force outside air into the passenger compartment.

In the Hi-Lo on 1979 models, or Bi-Level (A/C Vent) on 1980 models position, outside air enters the vehicle and is directed through 2 levels of air outlets. The upper level (A/C outlets and lap coolers) and the lower level (heater outlets and defroster bleed). All four blower speeds are operational and the compressor is on. On 1980 Cordoba, Diplomat LeBaron and Mirada models, when the button is pulled out all the way, the compressor goes off, providing a vent position. Interior comfort is obtained with a split in airflow and temperatures, meaning that a temperature difference exists between the two levels. Since the upper level is cooler than the lower level, increased comfort is available during a cold and sunny day.

In the Heat position, outside air door is open as in the A/C position. Vacuum is applied to the Hi-Lo door actuator on 1979 models, and to the Bi-Level door actuator on 1980 models, closing off the passage to air conditioning distribution duct and opening the passage to the heater-defroster duct. Since the heater-defroster door is in the heat position, the full flow of heated air is routed through the heater outlets, except for a small amount that bleeds through the defroster outlets. The water valve is open, the blower switch is activated and the compressor is off.

In the Defrost position, all conditions are the same as for heater operation, except that no vacuum is applied to defroster door actuator which is spring loaded toward the defrost position. The recirculating air door will be open to outside air and the defroster door will be open. There will be a small amount of heat through the heater outlets, and the compressor will be operating. On 1980 models, compressor operation can be eliminated in the Defrost position by pulling the Defrost button fully out. This procedure should be used only in cold weather below 32° F. All other defroster operations will continue with button in either position.

Blower Motor Switch

This switch can be used to select four different blower motor speeds. The blower motor will operate at the speed selected in all push button positions except Off. In Max A/C position, only the two highest blower motor speeds can be selected.

NOTE: On 1979 models, activation of heated back light, prevents operation of the high blower speed.

Major System Components Except 1980 Cordoba, Diplomat, LeBaron & Mirada, Fig. 4

Ambient Compensator

This compensator is located inside the evaporator heater assembly on the rear surface. It is a molded chamber connected to the in-car air inlet grille with a flexible hose. In-vehicle air is drawn through the compensator and either is cooled or heated by outside air passing over the

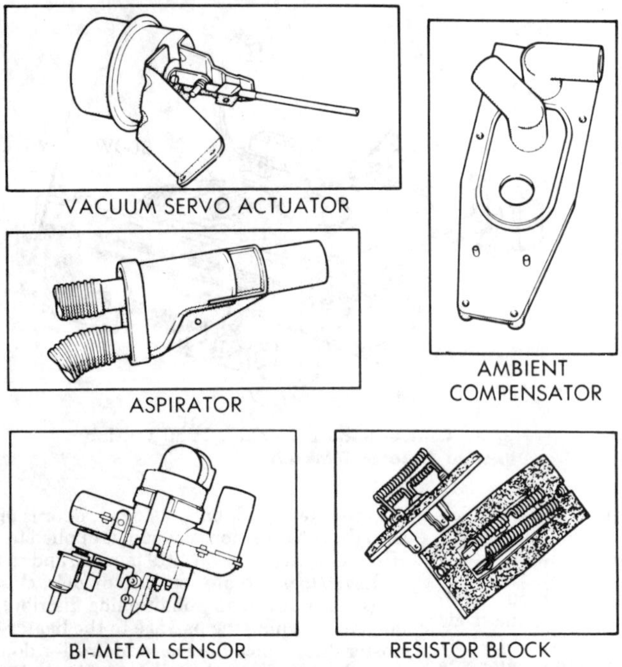

Fig. 4 System components. Except 1980 Cordoba, Diplomat, LeBaron & Mirada

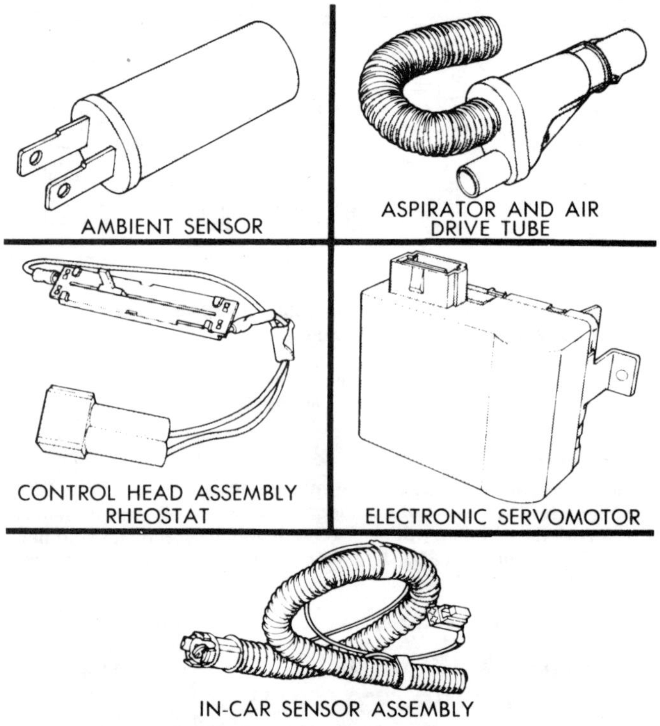

Fig. 5 System components. 1980 Cordoba, Diplomat, LeBaron & Mirada

outer surface of the chamber. An additional port, located at the top of the compensator is used to power the aspirator.

Bi-Metal Sensor

This sensor is located on the rear surface of the evaporator heater assembly. The bi-metal sensor responds to in-vehicle and outside-vehicle temperature changes in order to maintain the desired temperature. Inside the sensor is a bi-metal spring which controls vacuum supplied to the vacuum servo actuator in response to varying temperatures.

Vacuum Servo Actuator

This actuator is located on the left front of the evaporator heater assembly. The servo actuator is a vacuum operated, linear actuator, which responds to vacuum from the bi-metal sensor and positions the blend air door through a control cable. The vacuum is delivered from the bi-metal through a portion of the vacuum harness.

Aspirator

The aspirator is located on the rear surface of the blower motor housing. It is tubular shaped and is connected to the air drive source (part of the compensator) by a flexible hose. The air creates a slight vacuum in the aspirator, which pulls air through the in-car air grille, the compensator and across the bi-metal sensor. Another flexible hose connects the bi-metal sensor with the aspirator and both air sources are dispersed under the instrument panel.

In-Car-Air Inlet Grille

This grille is located on the instrument panel. On Newport and New Yorker models it is located directly above the comfort control and on Charger and Magnum models, it is located at the right side of the instrument cluster. The grille permits air to be drawn from the vehicle's interior and signal the bi-metal sensor to compensate for a variation in the selected comfort level.

Resistor Block

The resistor block is located in the plenum panel above the evaporator heater assembly. It consists of resistance coils which allow various blower motor speeds to be selected.

Vacuum Reservoir

The vacuum reservoir is mounted in the engine compartment on the right wheel housing. It is used to maintain sufficient vacuum levels in the vacuum circuit during low manifold vacuum conditions. The reservoir has a check valve at the inlet port which closes when engine manifold vacuum level drops below that of the reservoir.

Water Valve

The water valve is located on the right side of the engine compartment. It is normally open (on) and is vacuum operated. It is open in all modes of operation except when the Off or Max A/C buttons are pushed.

Major System Components, 1980 Cordoba, Diplomat, LeBaron & Mirada, Fig. 5

In-Car-Sensor

The in-car-sensor assembly is connected to the aspirator suction nipple and the instrument panel. Air from the passenger compartment is drawn through the instrument panel opening past a temperature sensing thermistor located in the sensor assembly. Electrical resistance of the thermistor increases when in-vehicle air temperature decreases and decreases when the temperature increases. This action permits the system to detect in-vehicle air temperature changes.

Ambient (Outside Air) Sensor

This sensor is located inside the rear surface of the evaporator-heater assembly. Ambient air, which is drawn into the ventilation system plenum, is blown across the ambient sensor. The sensor's electrical resistance increases when ambient air temperature decreases and decreases when temperature increases. This action permits the system to detect outside air temperature changes.

Comfort Control Sliding Resistor

This resistor is mounted on the control head assembly and is mechanically linked to the control lever. The resistor's electrical resistance increases when the lever is moved to the right or high temperature settings. Moving the lever to the left decreases resistance. This control permits the operator to select a desired temperature level.

Electronic Servomotor

The electronic servomotor is located on the evaporator-heater assembly above the blend-air door shaft. The servomotor measures the resistance of the sensors and the control head resistor and moves the blend-air door to a position which corresponds to a certain electrical resistance. High sensor or control head resistor resistances cause the servomotor to move the blend-air door to a higher (more heat) position.

Aspirator & Air Drive Tube

The aspirator is located on the rear surface of the blower motor housing. Pressurized air from the housing is fed through the air drive tube to the aspirator and the evaporator-heater assembly. The air passing through the aspirator creates a slight vacuum which draws air from the passenger compartment through the in-car sensor assembly.

Ford Motor Co.

1974-77 LINCOLN

Air Flow

During operation, outside air is drawn from the cowl vent just below the windshield, except at maximum cooling when recirculated air is used.

The system is called a reheat system to provide conditioned air to the vehicle interior. With this type of system all air flow from the blower passes through the evaporator core. Temperature is then regulated by reheating the cooler air to the desired temperature. Temperature of the outlet air is varied by the Temperature Blend Door (5) which controls the amount of cooled air that flows through and/or around the heater core, and then mixes in the distribution plenum chamber, Fig. 1. From here it is diverted to the heater outlets, the defroster nozzles, or the air conditioning registers.

Control Operation

The automatic temperature control (ATC) box is located in the left cowl side panel, Fig. 1. The controls in the instrument panel are operated by moving the functional control lever to any one of the six control positions and setting the temperature control lever to any desired temperature.

High—Cooling

As the passenger compartment temperature approaches the temperature control setting, the Outside Air Door (1) is opened (no vacuum) admitting outside air and the Recirculating Air Door (2) is closed (no vacuum).

When the passenger compartment temperature and the control setting temperature are nearly balanced, the Temperature Blend Door (5) is actuated. The door moves to a position so that part of the air leaving the evaporator core is directed through the heater core where it is warmed. Blend air (the mixture of cool and warm air) is then directed into the passenger compartment through the air conditioning ducts and registers.

The position of the Temperature Blend Door (5) changes automatically to maintain the temperature selected on the control lever; at the same time, blower speeds are reduced. If the passenger compartment temperature rises above the temperature control setting, the Temperature Blend Door (5) moves to admit more cool air and less warm air. If the passenger compartment temperature drops below the temperature control setting, the Temperature Blend Door (5) moves to admit more warm air and less cool air. As maximum heating or cooling operation is approached, blower speeds are automatically increased.

Low—Cooling

Same as *HIGH* maximum cooling except for lower blower speeds, and fresh air is used at all times in LOW.

Low or High Heating

When passenger compartment temperature is well below the temperature control lever setting (cold weather), an engine thermostat switch will delay the operation of the blower until the engine coolant reaches 130 degrees F. This will prevent a sudden cold air blast from the heater outlets. The system will then heat the passenger compartment to the desired temperature, and level out to maintain the pre-set temperature.

When the passenger compartment temperature and the temperature control lever setting are nearly balanced, the Temperature Blend Door (5) is actuated. The door moves to a position so that only part of the air leaving the evaporator core is directed through the heater core where it is warmed. Blend air (the mixture of cool and warm air) then is directed into the passenger compartment.

The position of the Temperature Blend Door (5) changes automatically to maintain the temperature selected on the control lever; at the same time blower speeds are reduced. If the passenger compartment temperature rises above the temperature control lever setting, the Temperature Blend Door (5) moves to admit more cool air and less warm air. If the passenger compartment temperature drops below the temperature control lever setting, the Temperature Blend Door (5) moves to admit more warm air and less cool air. As maximum heating or cooling operation is approved, blower speeds are automatically increased.

Approximately 10 percent of this air flow is diverted to the defroster nozzles.

Vent or Economy Position

In *vent* or economy position the (1) door is open to outside air (no vacuum) and the (2) door is closed to inside air (no vacuum). The heater water valve (4) is closed (vacuum) to water through heater core (no heating). The (5)

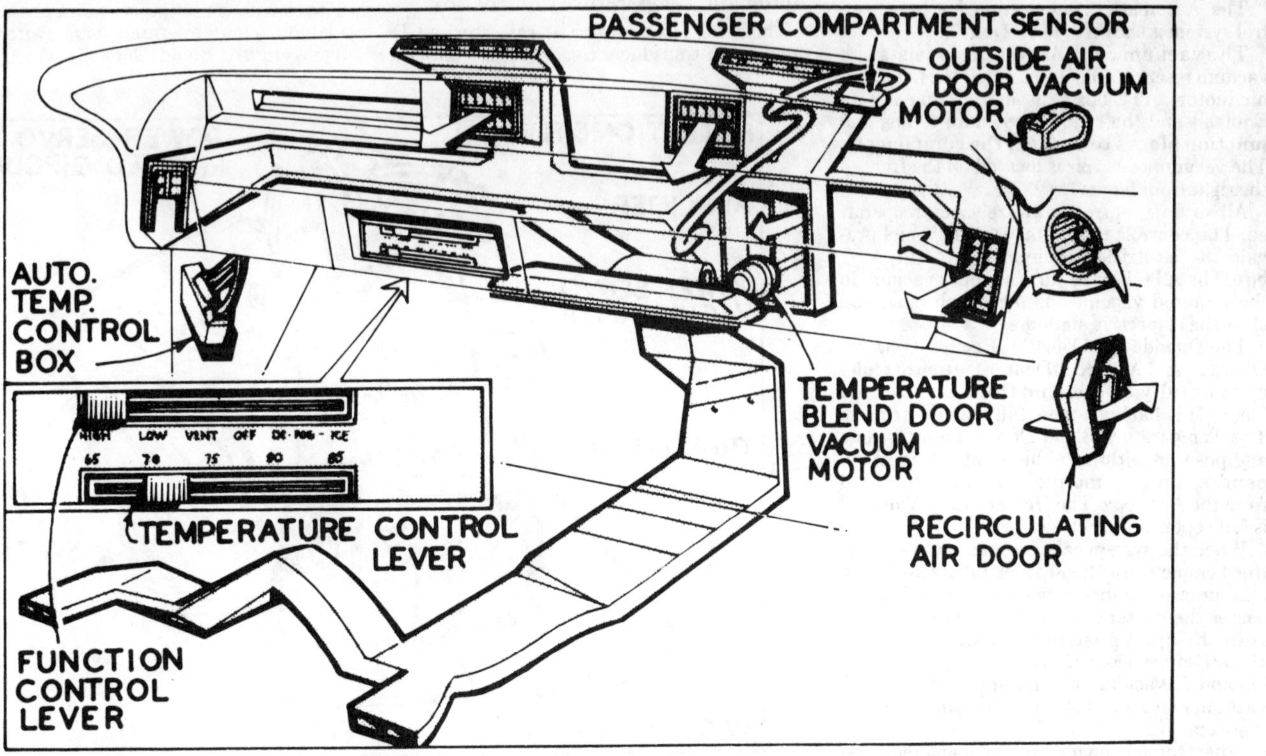

Fig. 1 Automatic Temperature Control System. 1974-77 Lincoln Continental (typical)

door is fully closed (full vacuum) by-passing all air around the heater core. The (6) door is open to the A/C registers (vacuum). The A/C compressor clutch is disengaged (no refrigerant through the evaporator — no cooling). The blower speed is at a predetermined minimum. Therefore, fresh outside air is forced through the A/C registers.

Off Position

In *OFF* position the Outside Air Door (1) is closed and the Recirculating Air Door (2) is open (vacuum). The AC/Heat Door (6) is closed (no vacuum), and the Heat/Defrost Door (7) is in the closed (no vacuum) position. The Heater Water Valve (4) is open (no vacuum) the blower is off, the A/C clutch is off and the Temperature Blend Door (5) remains in the last operational position.

Defog Position

In *DEFOG,* operation is the same as in *HIGH HEATING,* except that the Heat/Defrost Door (7) is in mid-position (7a vacuum). Forty percent of the air flow is to the defroster nozzles with the remainder directed to the heat ducts. There is no system operational delay in *DEFOG* due to engine coolant temperatures.

De-Ice or Defrost Position

When the functional control lever is in the *DE-ICE* or *DEFROST* position, the Temperature Blend Door (5) is fully open and all incoming air is diverted through the heater core to be warmed, and then directed to the defroster nozzles (Heat/Defrost Door 7a and 7b no vacuum). The system operates on high blower only in *DE-ICE. Ten percent of this airflow is bled to the heater floor outlets.*

Vacuum Control System

The Automatic Temperature Vacuum control system is shown in Figs. 2 and 3.

The vacuum control system consists of a vacuum reserve tank, check valve (4), five vacuum motors, ATC box (Automatic Temperature Control) and the required vacuum tubing and junction blocks to connect the components. The vacuum reservoir is located on the front of the right front fender.

All air flow control doors are vacuum operated. The controls on the instrument panel provide the electrical requirements to the ATC box. The ATC box, in turn, supplies vacuum to the required vacuum motors which open and close the respective air doors.

The Outside Air Door (1), Recirculating Air Door (2), and AC/Heat Door (6) are either fully open or fully closed, and the Heat/Defrost Door (7) is halfway open, fully open or closed. The Temperature Blend Door (5) may assume any position within the limits of its travel depending on the amount of vacuum supplied from the ATC box. The Heater Water Valve (4) is fully open or fully closed.

When the system calls for maximum cooling, the Temperature Blend Door (5) is closed at a minimum of 10 inches vacuum and all air bypasses the heater core. The AC/Heat Door (6) is in the open position (vacuum) and the Heat/Defrost Door (7) is in the open position (7a and 7b vacuum) at a minimum of 10 inches vacuum and air is distributed through the A/C registers.

When the passenger compartment temperature approaches the desired temperature setting, the AC/Heat Door (6) may change from the A/C register (vacuum) to the floor heat position (no vacuum), or from heat to air conditioning if the Temperature Blend Door (5) reaches approximately mid-position. The AC/Heat Door (6) is positioned by a switch contact change at the power servo switch. This energizes the AC/Heat Door (6) solenoid to direct air to the registers when the system calls for A/C or de-energizes the solenoid and directs the blended air to the heater floor outlets when the system calls for heat.

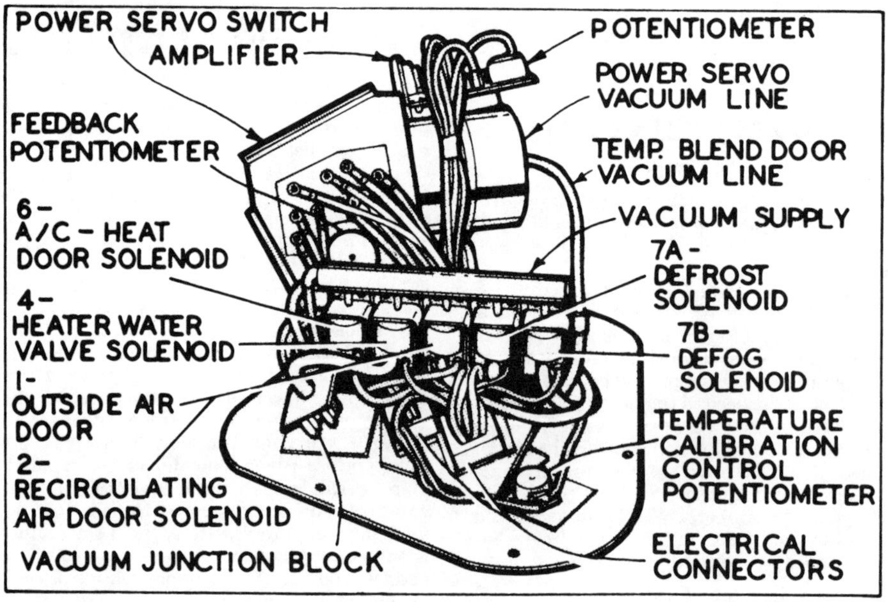

Fig. 2 A.T.C. box door control solenoids

Electrical System

Automatic Temperature Control Box

The ATC box contains a transistorized DC amplifier, a transducer to convert the amplifier electrical output to a variable vacuum supply, one relay, power servo, and five vacuum control solenoids, Figs. 2 and 3.

The input signal to the amplifier is dependent on the series resistance of the temperature control lever rheostat, the fixed valve of a calibration rheostat (located in the control unit), and the combined resistance of the two sensors.

The amplifier output is used to control a transducer, thereby converting electrical energy to a variable vacuum supply to accomplish the required automatic operating sequences. The transducer output vacuum controls a modulated vacuum operated servo switch, and the Temperature Blend Door (5) which regulates

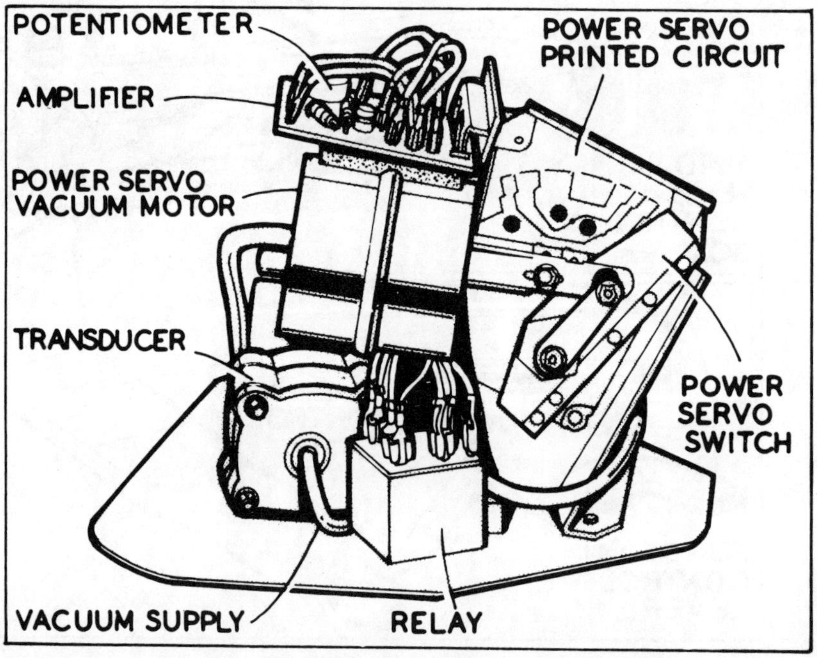

Fig. 3 A.T.C. box power servo components

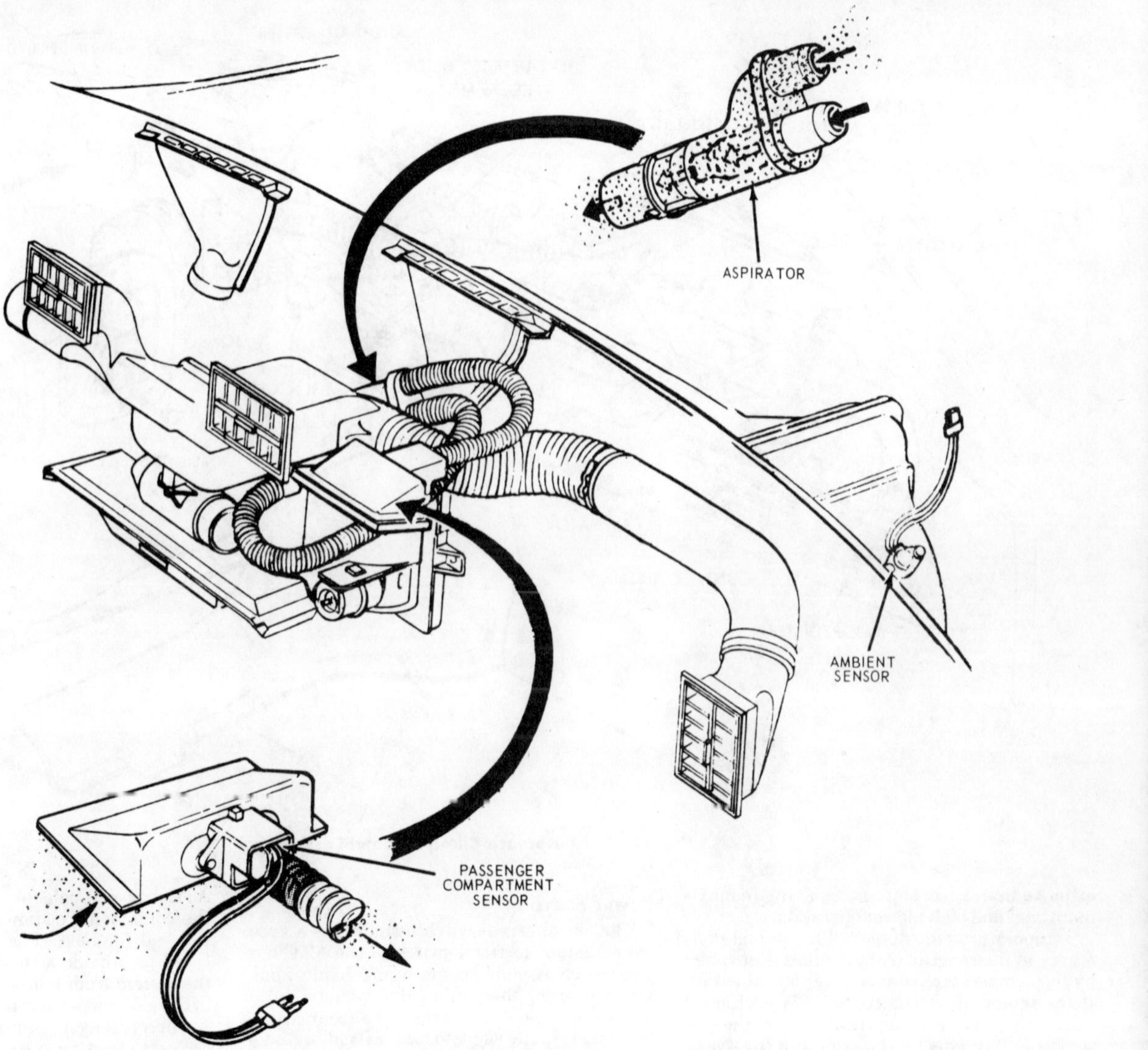

Fig. 4 A.T.C. system sensing components

the hot and cold air flow. The vacuum operated servo switch located in the ATC box selects blower speeds and operates the vacuum solenoids which control air distribution through the heater outlets of air conditioning registers, and also controls the Outside Air Door (1) and the Recirculating Air Door (2).

Cold Weather Operation

Assuming an initial cold weather operation, the system functions as follows: Set the functional control lever for *HIGH* if extremely cold, and set the temperature control lever to a comfortable setting (approximately 75 degrees F.). The heating system will be automatically positioned for maximum heat and high blower operation but will not turn on until the engine coolant temperature reaches approximately 130 degrees F. (This normally takes 3-5 minutes of engine operations).

In cold temperatures, the sensors will have a relatively high series resistance. This controls the amplifier and transducer in a manner to produce a minimum outlet vacuum and calls for maximum heater operation.

As the vehicle interior warms up, the passenger compartment sensor resistance value decreases resulting in an increased transducer vacuum output. As vacuum output increases, the Temperature Blend Door (5) position will change reducing discharge air temperature, and the vacuum operated servo switch will cause the blower speed to drop off. As the passenger compartment temperature approaches the temperature control lever setting, the transducer vacuum will hold at the level required to balance the heat loss or gain from the vehicle to the outside air.

Hot Weather Operation

Now assume that the vehicle is to be operated in hot weather. With the same control lever settings as used for cold weather operation, the system will come on when the engine is started.

In hot temperatures, the sensors will have a low series resistance. This controls the amplifier and transducer and produces a high vacuum. Under these conditions the transducer vacuum is at its highest value (at or above 10 inches) calling for maximum air conditioning (high blower and recirculated air). As the passenger compartment cools down the passenger compartment sensor resistance values increase causing decreased transducer vacuum output. This results in the system changing from recirculated air to outside air and also in reduced blower speeds. The Temperature Blend Door (5) position will change causing some air to flow through the heater core for temperature regulation.

For mild ambient conditions, the temperature control lever should be set at 75 degrees F. and the functional control lever in *LOW. Automatic control in LOW* is similar to *HIGH* except that the system is locked on outside air, and there are five blower speeds instead of three.

A disturbance, such as opening the windows on a hot day, causes the sensor string resistance to drop, resulting in increased transducer vacuum output. With *HIGH transducer vacuum output, more air conditioning is called for to counteract the hot air coming in the vehicle windows.*

In *DEFOG,* system operation is the same as in *HIGH* except that outside air is used at all times and the Heat/Defrost Door (7) is in the mid-position. In *DE/ICE,* an electrical signal is supplied directly to the amplifier, overriding all

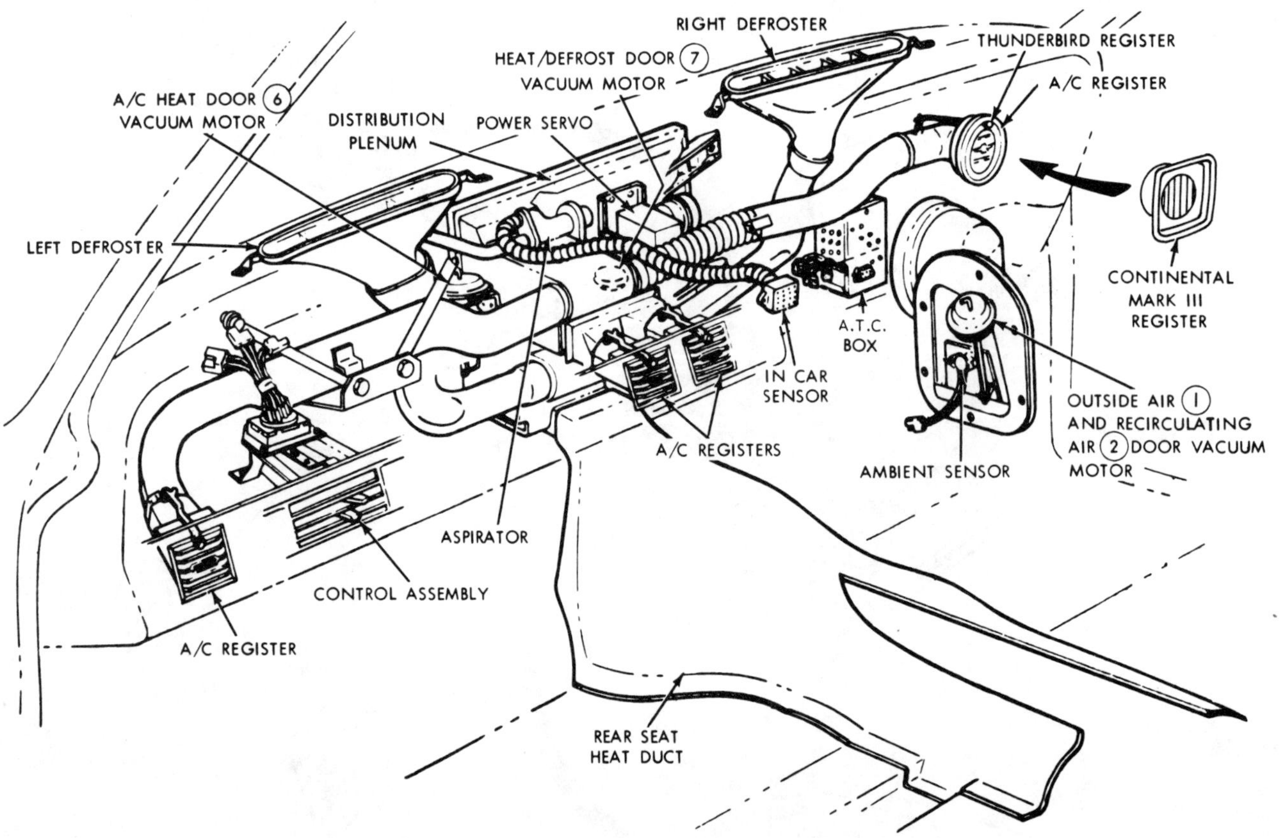

Fig. 5 Automatic Climate Control assembly

other temperature signals resulting in maximum heat and high blower operation.

Temperature adjustment of individual preference in the range of 65-85 degrees is provided by the temperature control lever and rheostat. If the temperature control lever is positioned warmer, the rheostat adds resistance to the sensor string. The effect is the same as if the sensor thermistors were cooled and had more resistance. This causes an amplifier and transducer response that decreases the vacuum output. This in turn increases heater output until the sensor thermistors have become warmer and the sensor string resistance has changed to compensate for the temperature control lever adjustment. The system then regulates at the new temperature that is indicated by the temperature control lever position. A similar but opposite reaction occurs when the temperature control lever is set cooler.

Blower Motor Speed

Blower motor speed is controlled by the vacuum-operated power servo switch. With either high or low vacuum from the transducer, the switch cuts out blower resistors to provide high blower speed. As the transducer vacuum regulates between the two extremes and passenger compartment temperature approaches the temperature control lever setting, the blower voltage and speed are reduced by cutting in resistors. There are five power servo switch positions to provide five blower speeds in *LOW. In HIGH the instrument panel control eliminates the two lowest switch steps and activates a relay to cut out an additional range resistor, causing three higher blower voltages and corresponding speeds.*

Power Servo

The power servo switch assembly is a vacuum actuated electrical switch in the ATC box. The switch assembly contains mechanical linkage connecting the power servo vacuum motor to the electrical switch arm and a feedback potentiometer. The vacuum motor is calibrated so that a known vacuum is required to move it to any corresponding position. The feedback potentiometer signals the amplifier the related position of the servo switch arm and the same vacuum is also provided to the Temperature Blend Door (5) vacuum motor.

The switch assembly is divided into three sections:

1. Function—Outside Air Door (1), Recirculating Air Door (2), Heater Water Valve (4), AC/Heat Door (6), Heat/Defrost Door (7a) and Heat/Defrost Door (7b).
2. Blower motor control during air delivery through the floor outlets (warm air).
3. Blower motor control during air delivery through the registers (cool air).

Temperature Sensors

The Automatic Temperature Control System makes use of two temperature sensing devices, called thermistors (sensors) for its operation, Fig. 4. An ambient sensor is located in the evaporator case to sense temperature of the incoming air. The second sensor is located behind a grille in the instrument panel to sense passenger compartment air temperature. The resistance of these sensors changes with temperature and provides varying electrical values to the ATC box. The sensors are wired in series with the temperature control rheostat.

When the combined resistance is low, the ATC box will call for air conditioning. When the combined resistance is high, the ATC box will call for heat. The ambient sensor, is encased to provide a thermal delay and prevent the system from following momentary changes in the ambient air temperature.

The passenger compartment sensor has the greatest effect on system operation. It samples the passenger compartment air drawn past it by an aspirator.

Sensor Aspirator

A passenger compartment sensor aspirator is built into the air distribution system to draw a sample of passenger compartment air through an opening in the instrument panel and past the sensor, Fig. 5. This system insures minimum sensor response time to the passenger compartment temperature changes, and samples the passenger compartment temperature at a location as close to breath-level as practical. The sensor location also provides the best balance between ambient and passenger compartment temperature.

1974 MARK IV & THUNDERBIRD

The automatic temperature control system will automatically control the temperature, mode of air delivery, blower speed, and reduce the relative humidity of air inside the vehicle. The operator need only move the control lever to the desired temperature and select a control position and the A.T.C. system will automatically select the mode and temperature of air delivery as required to obtain the car interior temperature selected. The system will then maintain the set comfort level automatically regardless of the operating conditions and requires little or no change in the setting to compensate for changes due to outside weather conditions.

Automatic control of temperature is provided year around. In hot weather, it will cool the car to the pre-set comfort level and then modulate cooling to whatever degree is required to maintain the desired temperature.

In mild weather, the interior of the vehicle remains comfortable without resetting the controls. In cold weather, provision is made to delay the operation of the system until engine coolant has warmed enough to provide warm air flow at initial heater system start-up. The system will then heat the car quickly to the desired temperature and level out to maintain the pre-set temperature.

Most of the time the system uses outside air except at maximum cooling when recirculated air is used.

The integral A/C system operates on the reheat principal to provide conditioned air to the car interior. With this type of system all air flow from the blower passes through the evaporator core. Temperature is then regulated by reheating the cooled air to the desired temperature. Temperature of the outlet air is varied by the temperature blend door which controls the percent of cooler air that flows through and/or around the heater core. The air is then mixed and directed into the distribution chamber. From there it is diverted to the heater ducts, the defroster nozzles, or the air conditioning registers.

When warmer air is required to maintain the desired level, the air is distributed through openings at the bottom of the unit for the front passenger compartment. When cooler air is required to maintain the desired level, the air is distributed through the four adjustable registers on the instrument panel.

Air for defroster operation is distributed through two defroster ducts onto the windshield at maximum heat only. Air for defog operation is split between floor outlets and defroster nozzles and is automatically blended to maintain the control temperature.

At outside air temperature above 40 degrees F, the compressor is in operation and the system air flow is cooled and dehumidified as it passes through the evaporator core. The removal of excess moisture from the air increases passenger comfort, particularly when traveling in humid weather. Also window fogging and interior condensation is reduced considerably.

The automatic temperature control utilizes both electrical and vacuum control. The controls on the instrument panel provide electrical signals to an automatic temperature control (A.T.C.) box. The A.T.C. box then supplies vacuum signals to the various vacuum motors which open and close the respective air doors and operates a blower switch to provide blower motor speed control.

Control Operation

The automatic temperature control includes a functional control slide lever for AUTO HIGH, AUTO LOW, VENT, OFF, DEFOG and DEFROST and a temperature control slide lever that determines the temperature at which the system will operate. The controls are located to the left of the steering column.

Temperature selection, in the range of 65 to 85 degrees, is made by sliding the temperature lever to the desired temperature on the control dial. Each position of the functional control lever is detented for positive engagement. With the lever in the OFF position, the blower motor is off, the outside air door is closed and the system is inoperative.

With the lever in the AUTO LOW position, the system selects fresh air for heating, or cooling, and adjusts air flow with any one of the five blower speeds. If the interior temperature is below the dial setting, warmed air is delivered through the heater outlets. Provision is made to delay the operation of the system until the engine coolant has warmed enough to provide warm air flow at initial heater system start-up.

If the interior temperature of the car is above the temperature setting on the dial, cooled air is delivered immediately without a water temperature delay to the air conditioning ducts and registers. This is done to provide a more rapid cool down rate.

System operation with the slide lever in AUTO HIGH position is similar to LOW except for blower motor speed and the use of recirculated air when maximum air conditioning is required. With the system in HIGH, blower motor speed is selected from three speeds which are somewhat higher than the five speeds used in LOW to provide optimum system performance in temperature extremes. The highest speed is automatically selected and as the temperature approaches the pre-set level, the system switches to fresh air operation, and the blower motor speed and the intensity of the heated or cooled air are lowered to maintain the pre-set temperature.

With the slide lever in VENT position, outside air is delivered out of the registers. Air temperature out of the registers will be approximately the same as outside ambient air, since the water supply to the heater is turned off and the compressor clutch is off. A temperature rise does occur because of engine heat radiation and/or sun load. The single (VENT) blower speed is predetermined by the blower motor resistor in the vent circuit.

With the slide lever in DEFOG position, the system operation is the same as in HIGH except that the air is split with 50 percent discharged from the defroster nozzles and 50 percent discharged from the floor outlets rather than air conditioning registers. There is no water warm-up blower delay in DEFOG or DEFROST; and in DEFOG, the temperature is controlled automatically. With the slide lever in DEFROST position, the system selects the maximum heat available at the highest blower speed and the air is discharged from the defroster nozzles. In DEFROST, approximately 10 percent of the air flow is directed to the heat ducts. In AUTO LOW and AUTO HIGH, with heater type operation, approximately 10 percent of the heater air is diverted from the heat duct to the defroster outlet to keep the windshield clear. No air is directed to the windshield in air conditioning type operation.

In DEFOG, as in AUTO HIGH and AUTO LOW, both the temperature and blower speed are generally at maximum initially, but as the actual interior car temperature moves toward the preselected temperature, both the blower motor speed and the heat intensity are gradually reduced. In DEFROST, maximum blower speed and maximum heat are in effect at all times. In order to return to normal in-car air distribution, it will be necessary to slide the lever to the AUTO HIGH or to the AUTO LOW position. There is no cold-engine blower delay in either DEFROST or DEFOG.

The register at each air conditioner outlet is adjustable for air flow direction by positioning louvers in the register.

1974-77 FORD & MERCURY FULL SIZE, COUGAR, ELITE, MONTEGO & TORINO 1975-77 MARK IV & THUNDERBIRD

This automatic temperature system automatically controls temperature blower speed and reduces the relative humidity inside the vehicle. The control lever is used to select the desired temperature setting. The system will then deliver cooled or heated air to maintain the selected temperature automatically, regardless of outside weather conditions.

In hot weather, the system will cool the air, then modulate the cooling to maintain the selected temperature.

In cold weather, a provision is made to delay system operation until engine coolant has warmed enough to reduce the cold air blast from the heater outlets. This system will heat the vehicle interior, then maintain the selected temperature.

During system operation, outside air is drawn from the cowl vent, located just below the windshield, except at maximum cooling when recirculated inside air is utilized.

This ATC system is of the re-heat type in which all air flow from the blower is passed through the evaporator core. The air temperature is varied by the temperature blend door which controls the amount of the cooled air that flows through and/or around the heater core. This air is then mixed and directed into the distribution plenum chamber where the blended air is diverted to the heater outlets, defroster nozzles or air conditioning registers.

When warmer air is required to maintain the selected temperature, the air is distributed through the floor outlet openings at the bottom of the distribution plenum chamber. When cooler air is required, the air is distributed through ducts to the adjustable air conditioning registers on the instrument panel.

On Ford and Mercury full size models, the shift from Heat or floor discharge to A/C or panel discharge involves a mid-way or split-shift position. In this intermediate position, moderate air temperature is discharged through the floor outlets and the panel registers.

Air for defroster operation is distributed through two defroster ducts at maximum blower speed. Air for defog operation is divided between the floor outlets and the defrost nozzles and the air temperature is maintained at the selected setting.

When the A/C compressor is operating at temperatures above 40° F. on Ford and Mercury full size models, 45° F. on 1974 models except Ford and Mercury full size or 55° F. on

1975-76 models except Ford and Mercury full size, excess moisture is removed from the air. This limits window fogging and interior condensation.

CONTROL OPERATION

The ATC system incorporates a functional control lever that determines which modes the system will operate in, a temperature control lever and a blower switch.

The blower switch provides two automatic blower speed sequences, high and low, and one fixed high blower speed. The switch is operative in Vent or Economy, Auto and Defog. The switch is inoperative in Defrost since the highest blower speed is available only.

In the Off position, the blower motor is off, the outside air door is closed and the system is inoperative.

In the Vent or Economy position, outside air is delivered through the A/C registers and on 1976 Ford and Mercury full size models, the floor ducts. Air flow is adjusted with any of the four blower speeds in the low and high blower switch settings. Air temperature is regulated to maintain the selected temperature level, however, no cooled air is supplied since the A/C compressor is inoperative.

In the Auto position, fresh air is selected for heating or cooling and recirculated air for maximum cooling. Air flow is automatically adjusted with any of the four blower speeds in the low or high Auto blower setting. If the in-car temperature is below the selected level, warm air is delivered through the floor outlets with a portion of the air directed to the defroster nozzles. System operation is delayed until engine coolant has warmed sufficiently to minimize the cold air blast from the heater outlets.

If the in-car temperature is above the selected level, cooled air is delivered through the A/C registers. No system delay is utilized since this provides a rapid cool-down rate.

In the Defog position, approximately half the air flow is delivered through the defroster nozzles and the remaining half through the floor outlets. Discharge air temperature is maintained at the selected level. Also, no system delay is used.

In the Defrost position, all air flow is discharged from the defroster nozzles at the maximum blower speed.

AIR FLOW

1974-77 Ford & Mercury Full Size

Refer to Figs. 6 and 7 for operation of the vacuum motors and doors which control the air flow in the various control settings.

Off

In this position, the outside air door is closed, the recirculating air door is open and the heat-defrost door is in the heat position. The heater water valve is open, the blower is off and the temperature blend door and the A/C-heat door assume whatever position the sensor vacuum indicates.

Vent or Economy

In this position, outside air is delivered through the panel registers and on 1975-76 models, the floor outlets. The temperature blend door is in the same position as in Auto.

Auto, Maximum Cooling (Recirculating Air)

In this position, with the in-car temperature above the selected level and the ambient temperature above 40° F., the outside air door is closed and the recirculating air door is open. The A/C-heat door is in the A/C position, the heat-defrost door is in the heat position and the temperature blend door is positioned to permit the air flow to bypass the heater core. The water valve is closed. Warm air, drawn from the passenger compartment, is passed through the evaporator core where it is cooled and returned to the passenger compartment through the A/C registers.

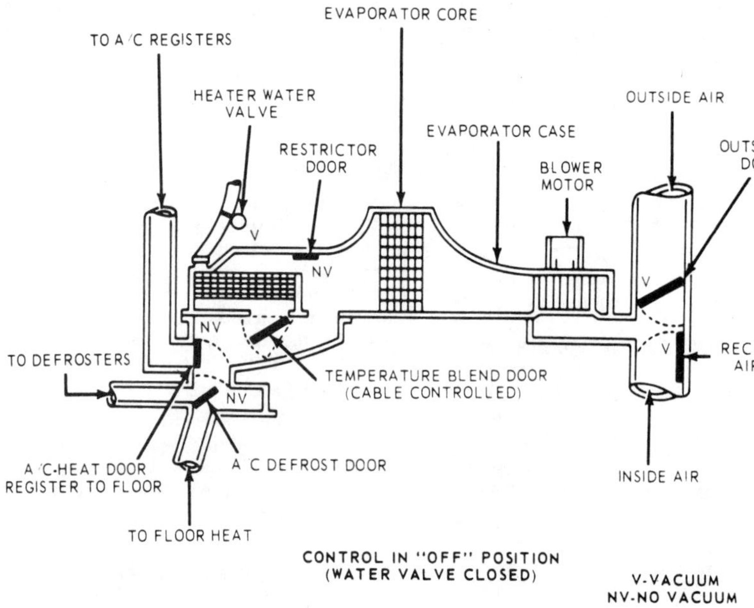

Fig. 6 System airflow schematic. 1974 Ford & Mercury full size

Auto, Cooling (Outside Air)

As the in-car temperature reaches the desired setting, the outside air door opens, the recirculating air door closes and the water valve opens.

Auto, Modulated Cooling (Outside Air)

When the in-car temperature is balanced with the desired setting, the temperature blend door is moved to a position so a portion of the air is routed through the heater core. The blend air is directed through the A/C ducts and registers. As more warm air is blended, the system automatically proportions some air through the floor outlets. The position of the blend door automatically changes to maintain the selected temperature and the blower speeds are reduced in the low or high blower switch position.

If the in-car temperature falls below the desired setting, the blend door moves to admit more warm air and less cool air. As the maximum cooling or heating operation is approached, blower speeds are automatically increased when the blower switch is in Auto.

Auto, Maximum Heating

In this position, with the in-car temperature below the desired setting, the outside air door is open and the recirculating air door is closed. The A/C-heat door and the heat-defrost door are in the heat position and the temperature blend door is positioned so all air flow is directed through the heater core. The heated air

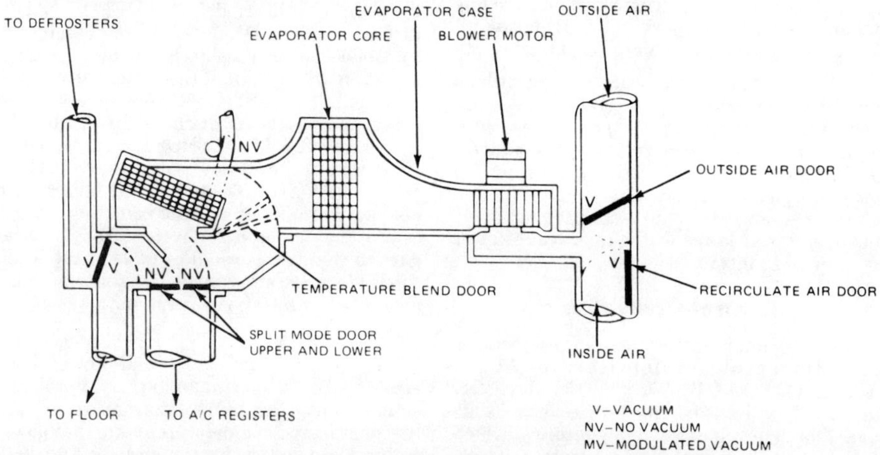

Fig. 7 System airflow schematic. 1975-77 Ford & Mercury full size

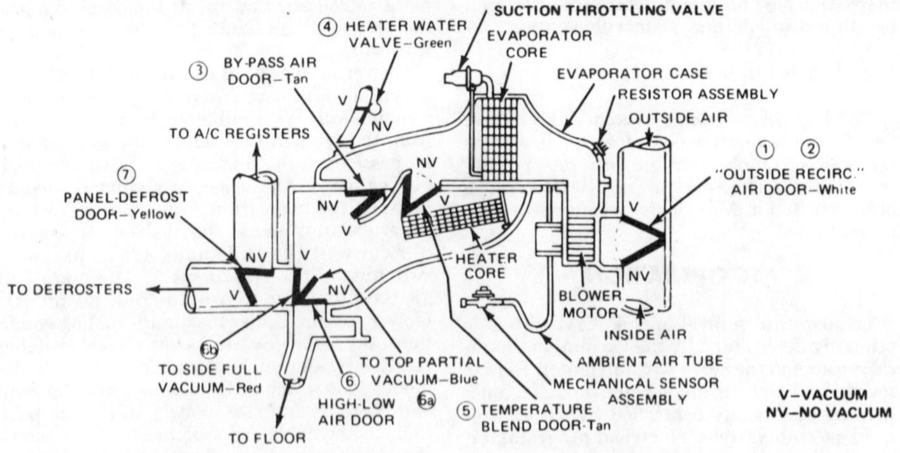

Fig. 8 System airflow schematic. 1974-77 Cougar, Elite, Montego, Torino & 1975-77 Mark IV & Thunderbird

flows through the heater ducts and the floor outlets. A portion of the heated air is directed to the defroster nozzles.

Auto, Modulated Heating

As the in-car temperature reaches the desired setting, the blend door moves to a position so a portion of the air is directed through the heater core. The position of the blend door is automatically changed to maintain the selected temperature and the blower speeds are reduced when the blower switch is in the Auto position. As maximum heating or cooling operation is approached, blower speeds are automatically increased.

Defog

In this position, system operation is the same as in Auto Heating except the heat-defrost is in the mid-position. About half the air is directed to the defroster nozzles and the remaining half to the floor outlets.

Defrost

In this position, the blend door modulates and the heat-defrost door is in the full defrost position. All air flow is directed to the windshield at the maximum, blower speed.

1974-77 Cougar, Elite, Montego & Torino; 1975-77 Mark IV & Thunderbird

Refer to Fig. 8 for operation of the vacuum motors and air doors which control the air flow in the various control settings.

Off

In this position, the Outside-Recirculating air door is in the recirculating position, the Mode and Panel-Defrost air doors are in the defrost position. The blower motor is off and the Blend and By-Pass air doors and the water valve assume whatever position the sensor vacuum indicates.

Vent or Economy

In this position, outside air is delivered through the panel A/C registers. The Blend and By-Pass air doors and the water valve assume whatever position the sensor vacuum indicates.

Auto, Maximum Cooling (Recirculating Air)

In this position, with the in-car temperature above the selected level, the Outside-Recirculating air door is in the recirculating position, the Mode door is in the through flow position and the Panel-Defrost air door is in the Panel position. The water valve and the Blend air door are closed and the By-Pass air door is open. Warm air, drawn from the passenger compartment, is directed through the evaporator core and is delivered through the A/C ducts and registers.

Auto, Modulated Cooling (Outside Air)

When the in-car temperature is balanced with the desired setting, the temperature blend door is moved to a position so a portion of the air is routed through the heater core. The blend air is directed through the A/C ducts and registers. As more warm air is blended, the system automatically proportions some air through the floor outlets. The position of the blend door automatically changes to maintain the selected temperature and the blower speeds are reduced in the low or high blower switch position.

If the in-car temperature falls below the desired setting, the blend door moves to admit more warm air and less cool air. As the maximum cooling or heating operation is approached, blower speeds are automatically increased when the blower switch is in Auto.

Auto, Maximum Heating

In this position, with the in-car temperature below the selected level, the Outside-Recirculating air door is in the outside position, the Mode air door is in the Heat position and the Panel-Defrost air door is in the defrost position. The water valve and the Blend air door are opened and the By-Pass air door is closed. Air from the evaporator is passed through the heater core and is distributed through the heater ducts and floor outlets. A portion of this air is diverted to the defroster nozzles.

Auto, Modulated Heating

As the in-car temperature reaches the desired setting, the blend door moves to a position so a portion of the air is directed through the heater core. The position of the blend door is automatically changed to maintain the selected temperature and the blower speeds are reduced when the blower switch is in the Auto position. As maximum heating or cooling operation is approached, blower speeds are automatically increased.

Defog

In this position, system operation is the same as in Auto Heating except the Mode air door is in the mid-position. About two-thirds of the air flow is directed to the defroster nozzles and the remaining third to the floor outlets.

Defrost

In this position, the Blend air door modulates and the Mode and Panel-Defrost air doors are in the defrost position. All air flow is directed to the windshield at maximum blower speed with slight bleed to the floor outlets.

SYSTEM COMPONENTS

In-Car Sensor

The in-car sensor assembly contains a bi-metallic sensor strip. The sensor output controls a vacuum modulator in such a manner that higher temperatures cause greater vacuum levels. A bias to the vacuum output levels is accomplished by a control arm. Therefore, modulated vacuum is the result of the effects of the control bias and the in-car temperature.

The sensor assembly contains an aspirator which provides a very small amount of vacuum to draw a continuous sampling of in-car air across the in-car sensor. Ambient air is directed to the assembly by an insulated tube from the blower air stream to operate the venturi aspirator.

Water Valve Solenoid

The water valve solenoid is an electrically operated vacuum switch. The vacuum switch connects the water valve vacuum motor to the supply vacuum when voltage is applied to the solenoid. When the solenoid is de-energized, the vacuum switch vents the water valve vacuum motor to the atmosphere.

Electro Vacuum Relay (E.V.R.)

The E.V.R. is a combination solenoid vacuum valve and a single pole, single throw, normally closed electrical relay. The vacuum motor and the relay are actuated by the solenoid. The E.V.R. performs two functions in this ATC system. First, during cold engine operation when heating is required, the E.V.R. locks out blower operation and closes the outside air door. Second, during maximum A/C cooling the E.V.R. switches the system to the recirculating air mode.

Servo

The servo is a calibrated vacuum motor with an attached electrical switch. The vacuum motor is used to position the temperature blend door for specific vacuum levels. The electrical switch provides specific blower speeds, two panel door operations, water valve operation and the recirculation mode for specific vacuum levels. The electrical switch is a conventional printed circuit board and slider which was designed so it can be replaced without altering the electrical/vacuum relationship.

On models except Ford and Mercury full size, an attached servo vacuum diverter valve controls the operation of the water valve and the

By-Pass, High-Low and Panel-Defrost air doors.

Servo Vacuum Diverter Valve, Exc. Ford & Mercury Full Size

The servo vacuum diverter valve is operated by the servo vacuum motor. This valve is also controlled by the modulating vacuum levels in the in-car sensor. The diverter valve controls vacuum directly to the by-pass air door and water valve vacuum motors. When cooled air is needed, the diverter valve closes the water valve and opens the by-pass door. When heated air is needed, the diverter valve opens the water valve and closes the by-pass door.

This valve also controls vacuum through the control head to the High-Low air door and Panel-Defrost air door in the Auto position. When cooling air is needed, the diverter valve closes the High-Low air door and opens the Panel-Defrost door. When heated air is needed, the diverter valve closes the Panel-Defrost door and opens the High-Low air door.

Resistor Pack

The blower motor resistor pack is mounted on the evaporator housing. There are four resistor elements on the resistor board and, depending upon the blower speed determined by the servo, series resistance is increased or decreased to the blower motor circuit to vary the blower speed.

Water Temperature Switch

In Auto, a circuit from the engine coolant temperature switch to the E.V.R. prevents blower operation during heating conditions until engine coolant temperature reaches 130° F.

Ambient Switch

This double pole switch prevents A/C compressor operation and the recirculation air mode when ambient temperature is below 40° F.

Control Head

The control head contains the mode selector, the blower switch and the temperature control lever.

The mode selector is a vacuum selector switch combined with a four cam operated electrical switch. The vacuum selector directs source vacuum to the various vacuum motors and also interrupts the modulated vacuum in the Vent, Defog and Defrost modes. The selector connects the vacuum signal from the E.V.R. to the recirculating air door and the outside air door in the Auto mode. Two double pole cam operated switches are also operated by the vacuum selector switch. The combination of these switches is called the electrical selector switch which controls the electrical supply to the compressor clutch, servo switch board, manual blower switch and the high blower relay.

The blower switch provides three modes of blower operation. In Max, the switch activates the high blower relay, in turn supplying direct battery voltage to the blower motor. In High-Auto, the switch directs the blower feed from the four step blower switch of the servo, through the resistor pack, to the blower motor. In Lo-Auto, the switch directs the blower feed from the servo through the range resistor to reduce each of the four blower Auto speeds.

The temperature control lever is connected to the sensor assembly by a Bowden cable. Movement of the control lever causes a corresponding movement on the pre-load portion of the sensor assembly and determines the temperature at which the system will maintain.

High Blower Relay

The high blower relay switches the blower feed from the resistor output to the low resistance direct supply from the battery. The control signal to the relay is supplied from the blower switch in Max or the selector switch in Defrost.

ATC OPERATION

Temperature control of this ATC system is primarily determined by the vacuum mechanical sensor and the servo vacuum motor. Blower speed, discharge air mode and overtravel function are electrically controlled by the control head assembly, servo electrical programmer, water valve solenoid on Ford and Mercury full size models, the E.V.R. and the vacuum diverter valve on 1974-77 Cougar, Elite, Montego and Torino and 1975-77 Mark IV and Thunderbird models. The temperature setting control is accomplished by an adjustable control cable between the control head and sensor assemblies.

The sensor assembly is a bimetal operated vacuum modulator which converts air temperature values into corresponding values of modulated vacuum. The bi-metallic element senses the in-car temperature and is mechanically linked to a vacuum modulator to directly control and regulate a vacuum signal which is proportional to the in-car temperature. The bimetal sensor is placed within a plastic aspirator. A small amount of ambient air is diverted from the blower outlet through an insulated tube to the sensor assembly. This ambient air then flows through the aspirator which draws in-car air over the bimetal element. The combined air is exhausted from the sensor.

The modulated vacuum signal from the sensor is transmitted to the calibrated vacuum motor of the servo assembly which positions the temperature blend door. The blend door mixes various proportions of the heated air and cooled air to maintain the selected setting. High in-car temperatures produce high sensor output vacuum which positions the blend door to deliver more cooled air. Low in-car temperatures produce low system vacuum levels delivering more warm air.

Blower motor speed is controlled by the servo vacuum motor and attached electrical switchboard. With either high or low vacuum from the temperature sensor, the switchboard by-passes the resistor pack to provide a high blower speed. As the sensor regulated vacuum between the two extremes and the in-car temperature approaches the desired setting, the blower voltage and speed are reduced by switching in the resistor pack. There are four switch positions to provide four automatic blower speeds. Under maximum cooling conditions, the servo electrical switchboard switches the E.V.R. and on Ford and Mercury full size models, the water valve solenoid. Also, on Ford and Mercury full size models, in its mid-position, the switchboard switches, in increments, the heat-to-panel and floor solenoid switches. On models except Ford and Mercury full size, the vacuum diverter valve, part of the servo assembly, closes the water valve under maximum cooling operations and also controls the mode shift.

1978-80 ALL MODELS

This automatic temperature control system is of the split case design with an independent blower motor, and automatically controls the temperature and relative humidity inside the vehicle. The two control levers provided maintain the desired temperature and system function, Figs. 9, 10 and 11. The system will deliver cooled or heated air to the inside of the vehicle and will automatically maintain the selected comfort level. The blower motor speed can be adjusted for more or less air as desired.

Automatic temperature control is maintained in both hot and cold weather. In hot weather, it will cool the inside of the vehicle to the selected comfort level and then modulate cooling as necessary to maintain the desired temperature. In mild weather, the interior of the vehicle is heated or cooled automatically as required to maintain comfort. In cold weather, a provision is made in the HEAT position

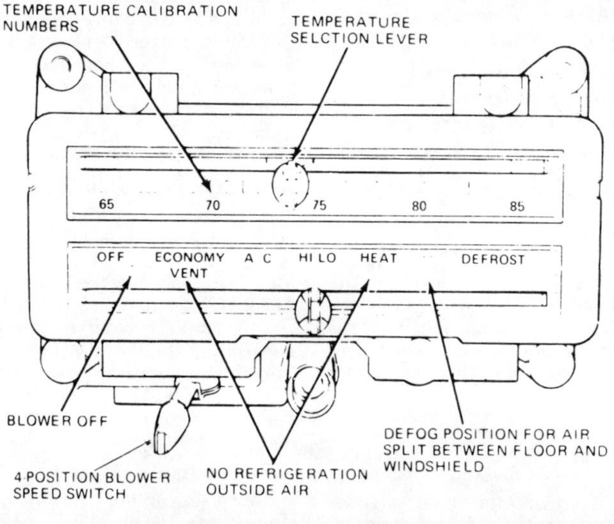

Fig. 9 Control panel. 1978-79 Cougar, LTD II, Mark V and Thunderbird

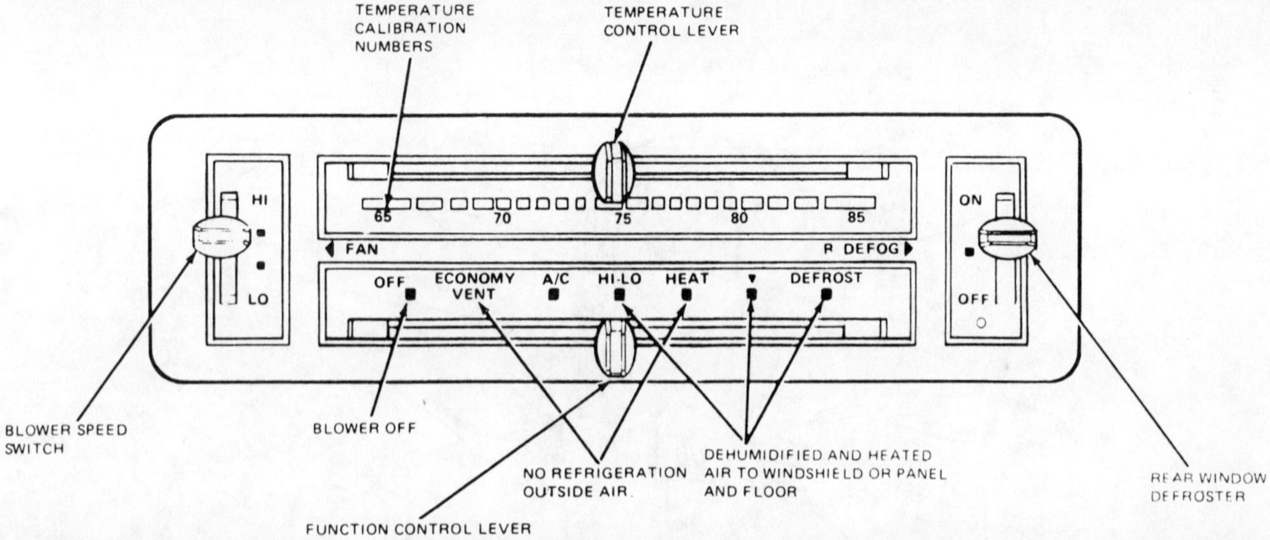

Fig. 10 Control panel. 1978 Ford and Mercury full size, 1978-79 Lincoln Continental and 1978-80 Granada, Monarch and Versailles

(FLOOR position on some models), of the function lever to delay system operation until the engine coolant temperature is warm enough to minimize the discharge of cold air from the heater outlet.

During operation, outside air is drawn from the cowl vent just below the windshield at all times except during maximum cooling when recirculated air is used.

This system uses a "reheat" method to provide conditioned air. With this method all air from the blower passes through the evaporator core, where it is cooled and dehumidified. Temperature is then controlled by reheating a portion of the cooled dry air and blending it with the remaining cool air to the desired temperature. Temperature blending is automatically controlled by the temperature blend door which controls the amount of cooled air flowing through or around the heater core where it is mixed and directed into the distribution plenum. The air is finally directed to the heater ducts, defrost nozzles, or instrument panel registers, according to the selected function lever position.

System Operation

Automatic temperature control of this system is determined by the modulated vacuum produced by the sensor assembly, Fig. 12, that controls the servo vacuum motor, Figs. 13 and 14. Temperature level changes are accomplished by means of a control cable between the control assembly and sensor assembly. System air flow is manually controlled by the control assembly. A vacuum selector valve, controlled by the function control lever, distributes vacuum to the various door vacuum motors which in turn direct the air flow through the system, Figs. 15 thru 22.

A bi-metal operated vacuum modulator, within the sensor assembly, Fig. 12, converts air temperature into corresponding values of modulated vacuum. The bi-metal senses in-vehicle temperatures and is mechanically connected to an internal vacuum modulator to control and regulate a vacuum which is proportional to temperature. The bi-metal and vacuum modulators are placed within a plastic aspirator. Air is diverted by an air scoop near the high pressure side of the blower (before the evaporator) through a tube to the sensor assembly. The air flows through the aspirator which draws air from the vehicle's interior over the bi-metal. This combined air is then exhausted from the sensor underneath the instrument panel.

Modulated vacuum from the sensor is directly connected to the large calibrated vacuum motor portion of the servo assembly. The modulated vacuum controls and positions the blend door attached to the arm of the servo motor assembly. The calibrated vacuum motor moves the blend door to mix various proportions of hot air from the heater core and cold air that has bypassed the heater core. The vehicle's interior temperature produces compensating discharge air temperatures. High temperatures produce high sensor output vacuum levels which position the blend door to deliver cool discharge air. Low temperatures produce low sensor vacuum levels which cause warm discharge air.

A vacuum diverter valve (part of the power servo assembly) turns off the water valve under maximum cooling conditions and controls the outside air/recirc door for maximum cooling (recirculating air) conditions when the function control lever is in the A/C position (PANEL position on some models).

Function Control Lever Operation

This system has a function control lever for OFF, ECONOMY-VENT, A/C, HI-LO, HEAT, DEFOG and DEFROST on Granada, Monarch, Versailles, 1978 Ford and Mercury full size and 1978-79 Cougar, Lincoln Continental,

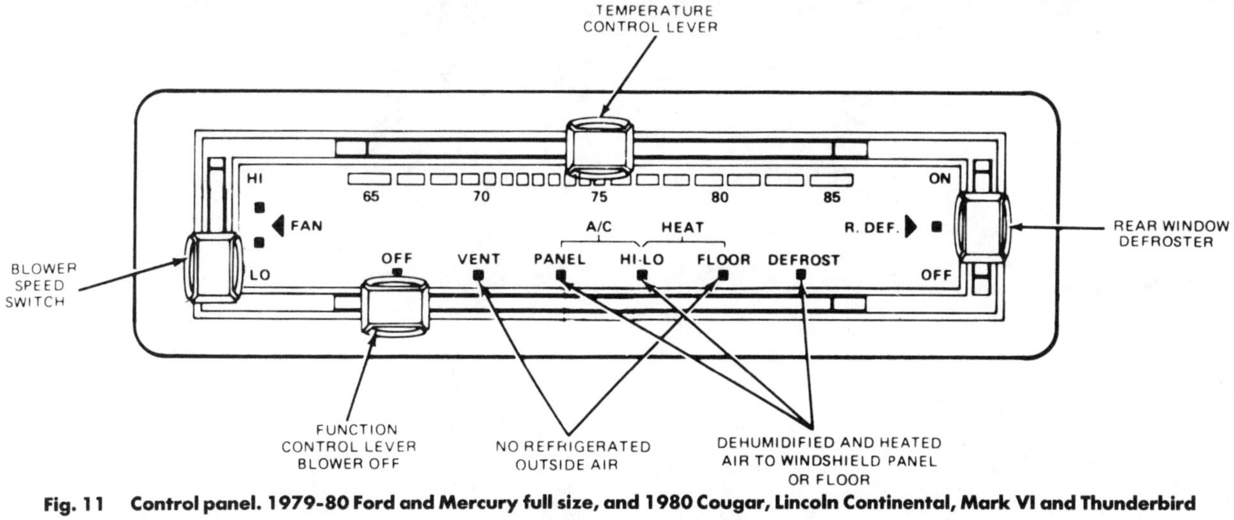

Fig. 11 Control panel. 1979-80 Ford and Mercury full size, and 1980 Cougar, Lincoln Continental, Mark VI and Thunderbird

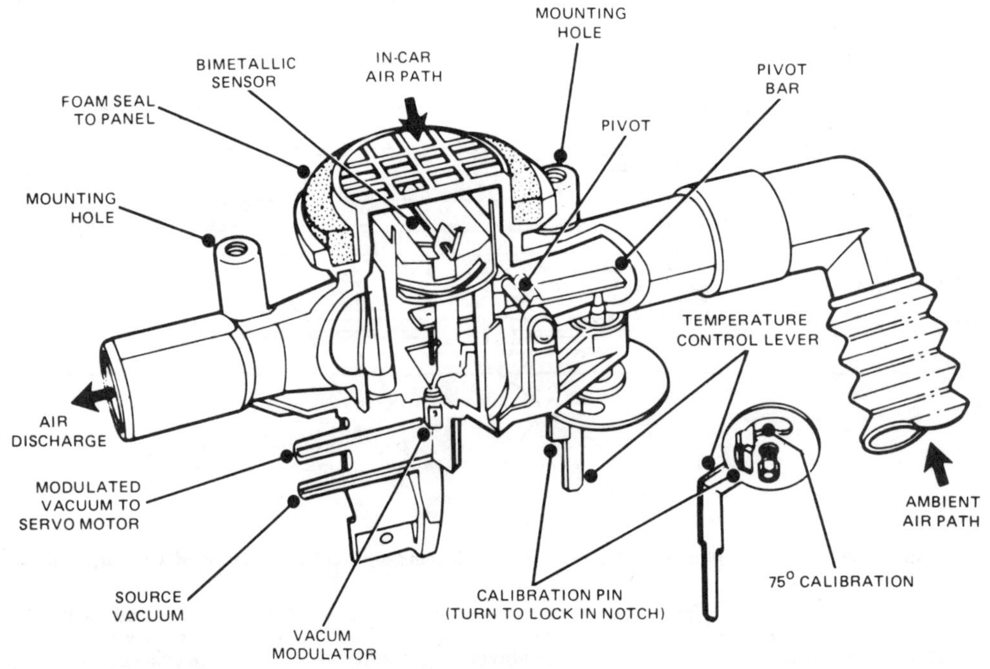

Fig. 12 Sensor assembly

LTD II, Mark V and Thunderbird, Figs. 15 and 16, and a function control lever for OFF, VENT, PANEL, HI-LO, FLOOR and DEFROST on 1979-80 Ford and Mercury full size models and 1980 Cougar, Lincoln Continental, Mark VI and Thunderbird, Fig. 11.

The function control lever determines the manner in which the system will operate a calibrated temperature control lever for setting the desired comfort temperature and a blower switch to control the volume of air movement.

System Air Flow, Figs. 17 thru 22

OFF Position

In this position, the outside-recirc door is in the recirculating position. On Granada, Monarch and Versailles models, the A/C-DEFROST door is in the defrost position and the Heat-Defrost door is in the full heat position. On Cougar, Ford and Mercury full size Lincoln Continental, LTD II, Mark V and VI and Thunderbird models, the A/C-HEAT (PANEL on 1979-80 Ford and Mercury full size and 1980

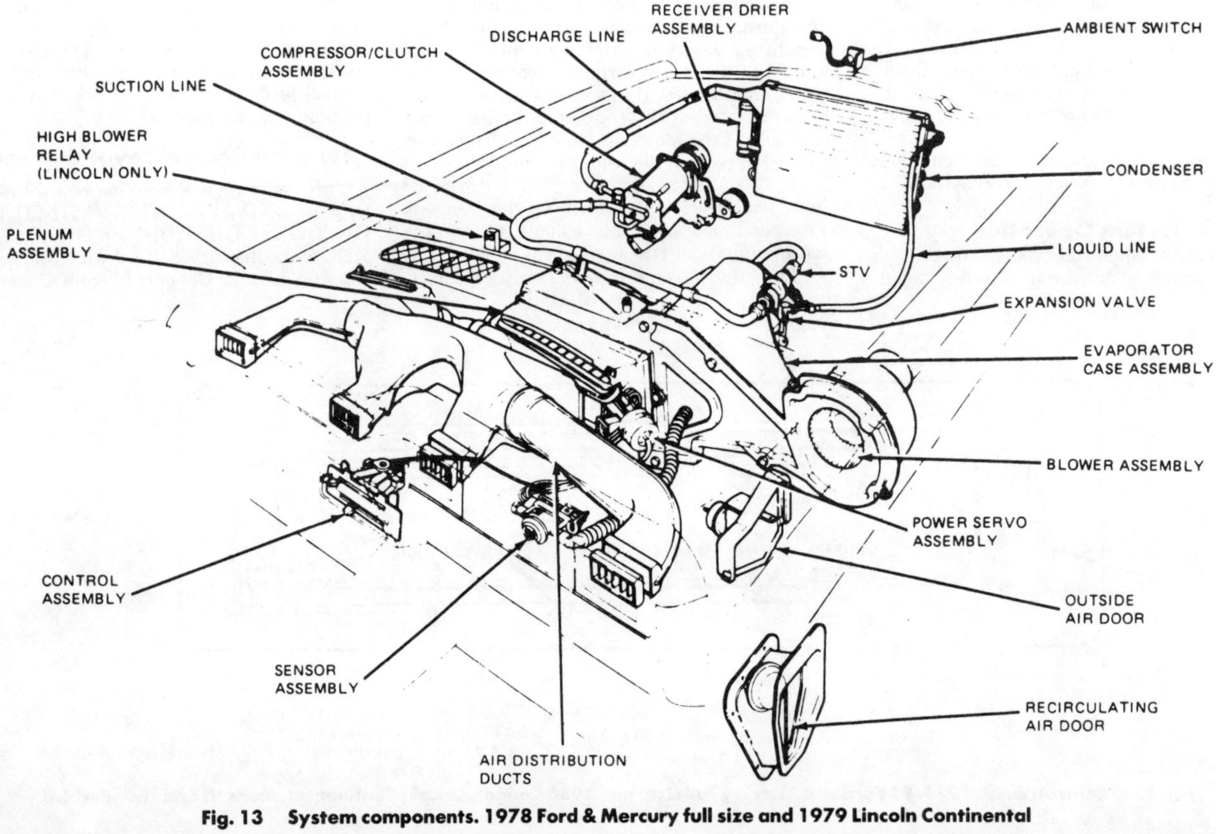

Fig. 13 System components. 1978 Ford & Mercury full size and 1979 Lincoln Continental

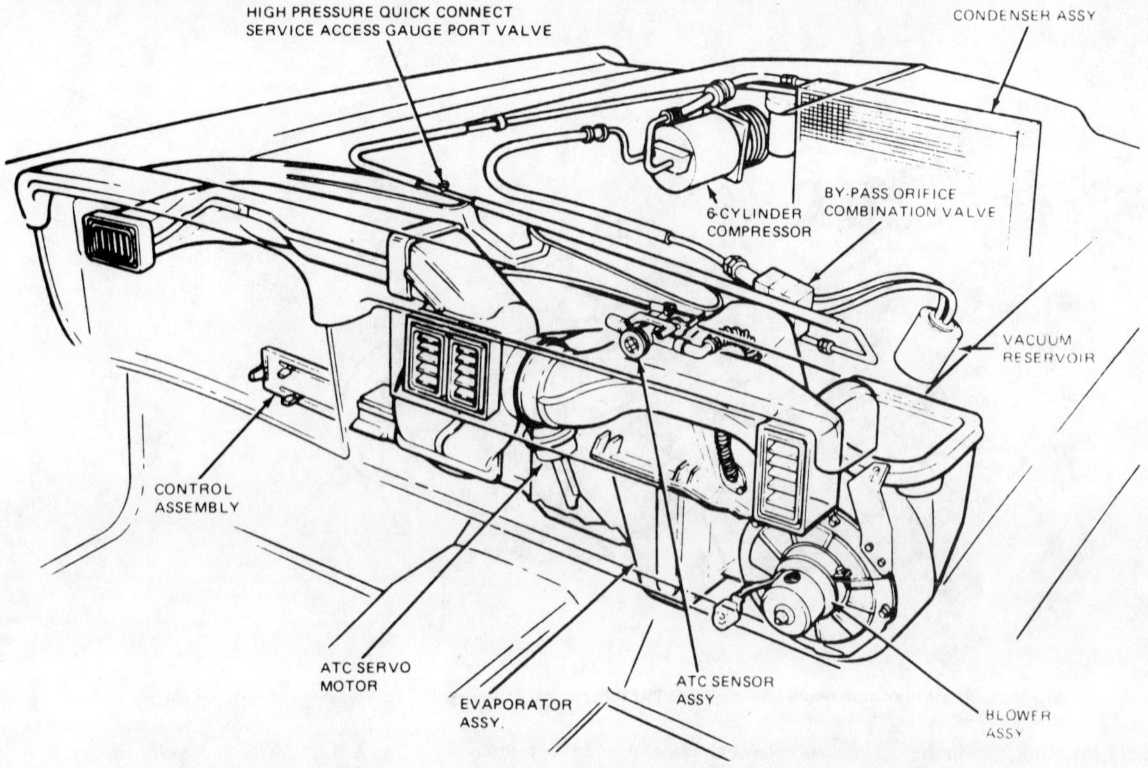

Fig. 14 System components. 1978-80 Granada, Monarch and Versailles, 1979 Lincoln Continental

Cougar, Lincoln Continental, Mark VI and Thunderbird models) door is in the heat position and the HEAT-DEFROST (FLOOR-DEFROST on 1979-80 Ford and Mercury full size and 1980 Cougar, Lincoln Continental, Mark VI and Thunderbird models) door is in the heat position. The blower motor is off and the temperature blend door and water valve assume a position dependent upon the modulated vacuum from the sensor to the power servo.

NOTE: A modulated vacuum is produced in all function control lever positions and is totally dependent upon the temperature lever setting and existing in-vehicle temperature conditions.

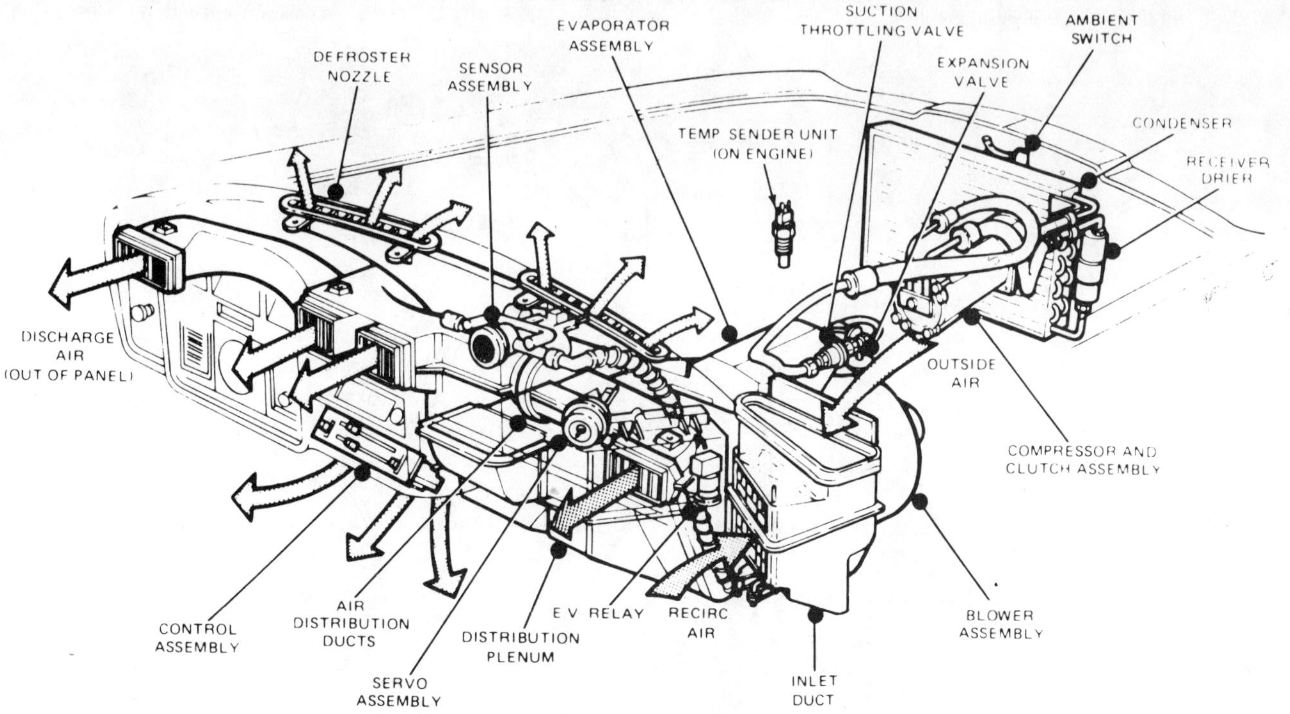

Fig. 15 System Components and airflow. 1979 Ford and Mercury full size

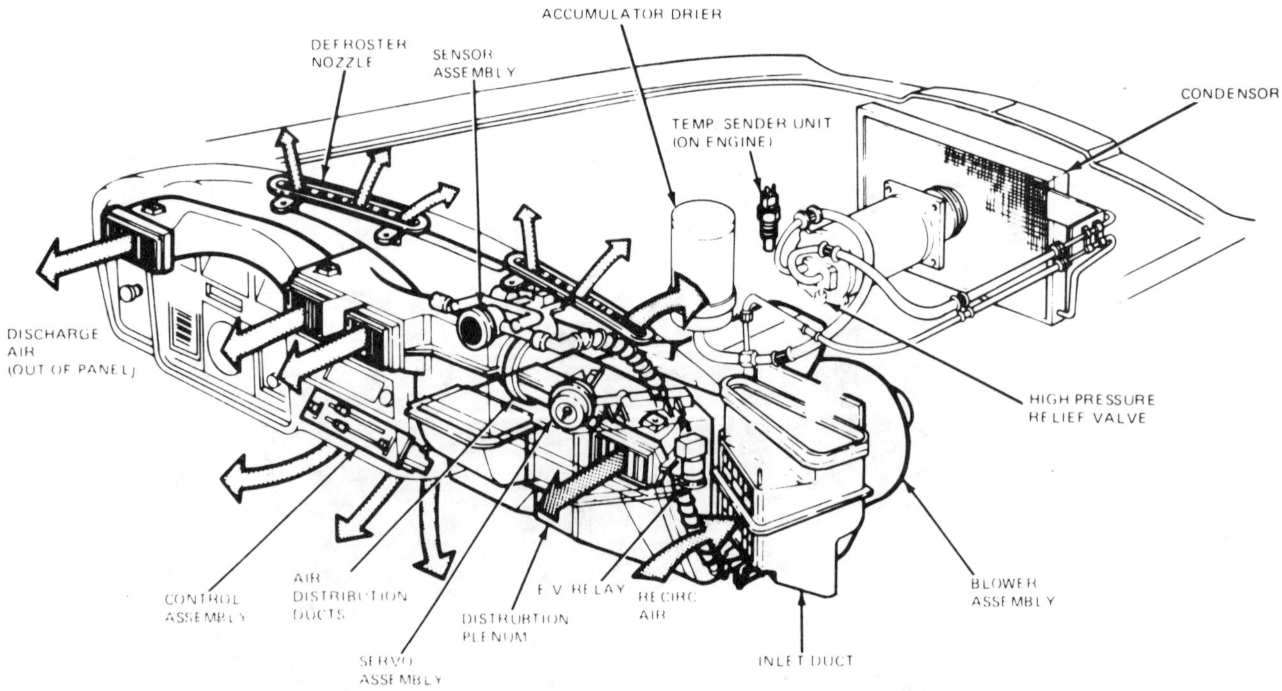

Fig. 16 System components and airflow. 1980 Ford and Mercury full size & Lincoln Continental

ECONOMY-VENT Position, Granada, Monarch & Versailles, 1978 Ford & Mercury Full Size & 1978-79 Cougar, Lincoln Continental, LTD II, Mark V & Thunderbird.

In this position, outside air is directed through the A/C registers, with automatic control of discharge air temperatures and water valve actuation. Refrigerated cooling is not available. ECONOMY-VENT and HEAT positions provide the most economical operations since the compressor does not operate.

VENT Position, 1979-80 Ford & Mercury Full Size, 1980 Cougar, Lincoln Continental, Mark VI & Thunderbird

In this position, outside air is directed through the A/C registers, with automatic control of discharge air temperature, however, refrigerated cooling is not available. VENT and FLOOR positions provide the most economical operations since the compressor does not operate.

A/C Position—Maximum Cooling (Recirculate), Granada, Monarch & Versailles, 1978 Ford & Mercury Full Size & 1978-79 Cougar, Lincoln Continental, Mark V & Thunderbird. PANEL Position—Maximum Cooling (Recirculate), 1979-80 Ford & Mercury Full Size, 1980 Cougar, Lincoln Continental, Mark VI & Thunderbird

In this position and with the vehicle's interior

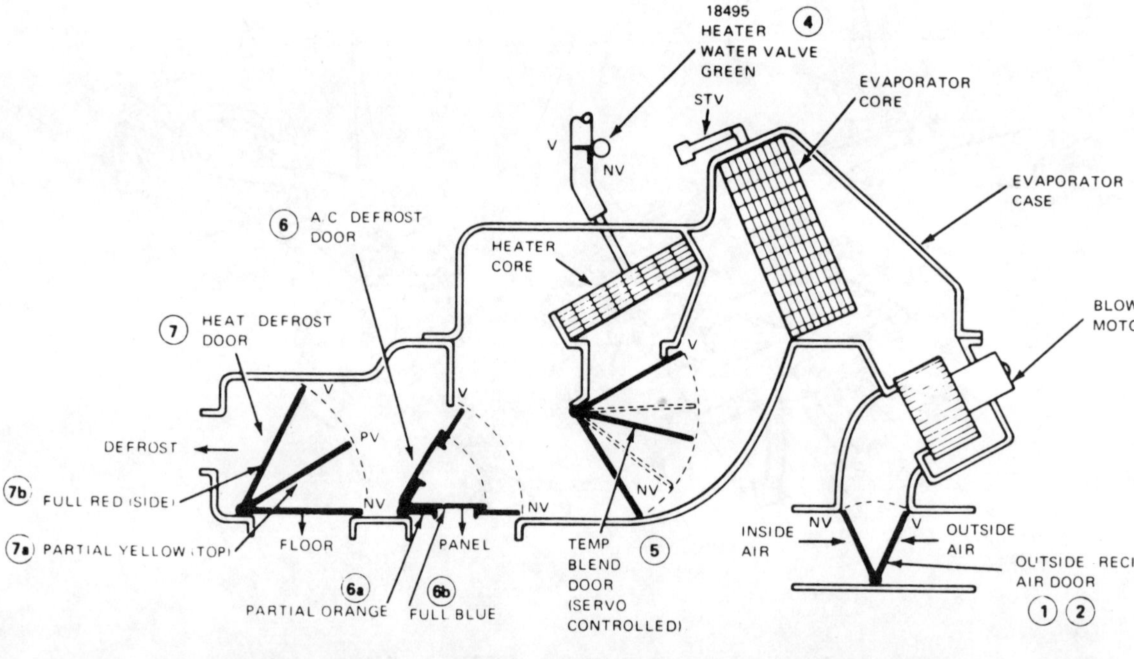

Fig. 17 System airflow schematic. 1978-79 Cougar, LTD II, Mark V and Thunderbird

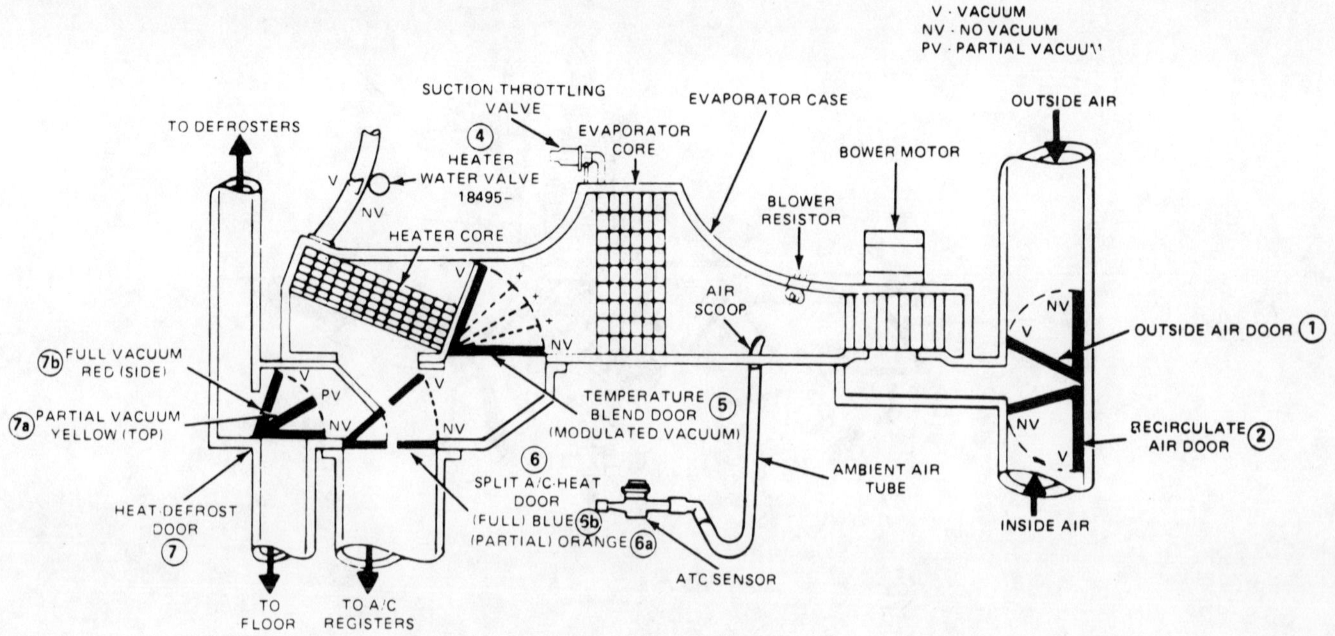

Fig. 18 System airflow schematic. 1978 Ford and Mercury full size and 1978-79 Lincoln Continental

considerably above the temperature setting, the modulated vacuum is high at the servo motor. This causes the following to occur:

1. The outside recirc door to go to the recirc position
2. The water valve to close on Granada, Monarch, Versailles, 1978 Ford & Mercury Full Size and 1978-79 Cougar, Lincoln Continental, Mark V and Thunderbird.
3. The temperature blend door to close, allowing all air to by-pass the heater core. Air is drawn from the passenger compartment, passed through the evaporator core where it is cooled and returned to the vehicle's interior through the A/C ducts and registers.

A/C Position—Modulated Cooling (Outside Air), Granada, Monarch & Versailles, 1978 Ford & Mercury Full Size & 1978-79 Cougar, Lincoln Continental, Mark V & Thunderbird. PANEL Position—Modulated Cooling (Outside Air), 1979-80 Ford & Mercury Full Size, 1980 Cougar, Lincoln Continental, Mark VI & Thunderbird

In this position, when less cooling is required to maintain the temperature setting, the temperature blend door is actuated to a mid-position where part of the air leaving the evaporator core is passed through the heater core. Blend air (combination of cooled and heated air) is then mixed in the plenum and directed to the vehicle's interior through the A/C ducts and registers. The temperature blend door position changes automatically to maintain the selected temperature. If the in-vehicle temperature rises above the selected temperature, the blend door moves to allow more cooled air and less heated air. If the in-vehicle temperature drops below the selected temperature, the blend door moves to allow more heated air and less cooled air.

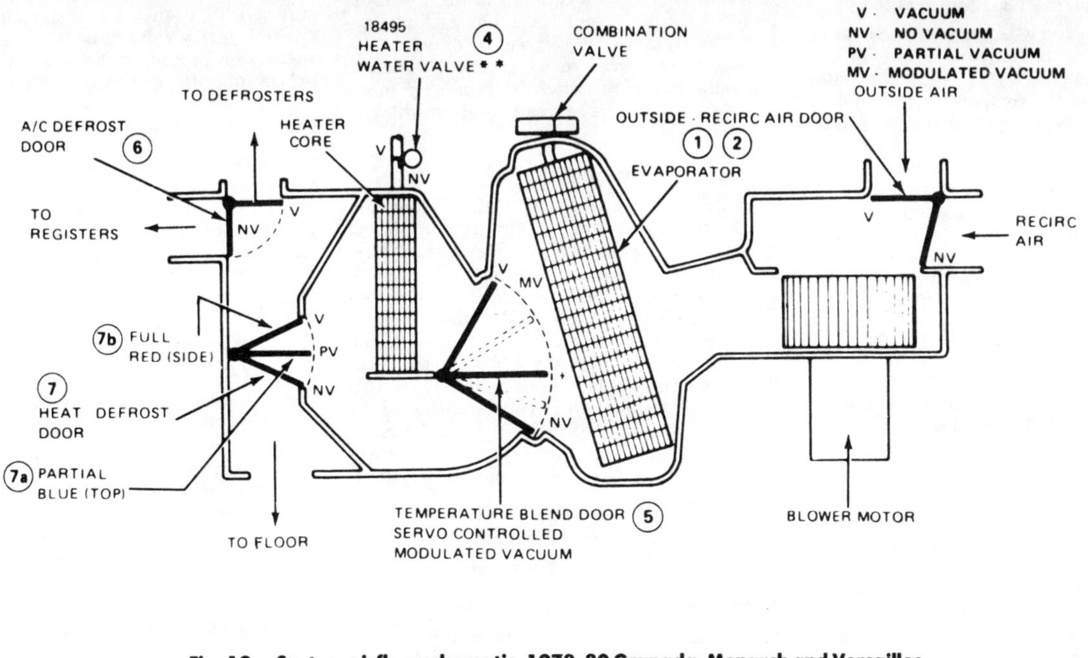

Fig. 19 System airflow schematic. 1978-80 Granada, Monarch and Versailles

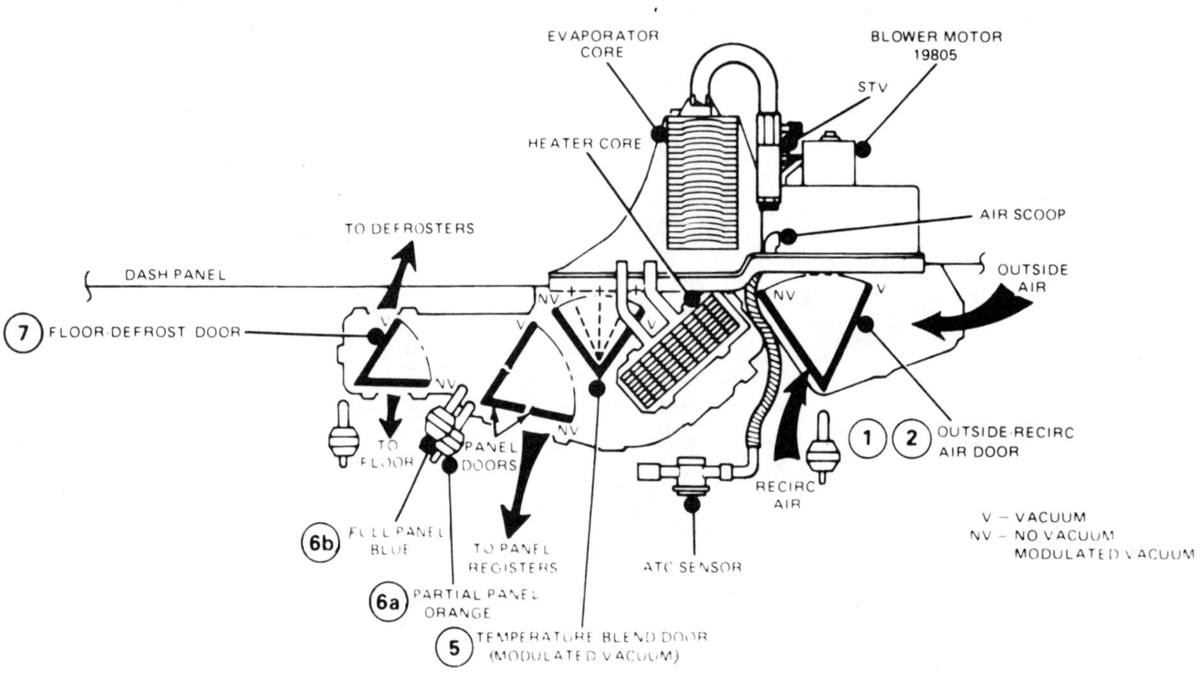

Fig. 20 System airflow schematic. 1979 Ford and Mercury full size.

HI-LO Position

Operation is similar to that described in the A/C or PANEL positions above, except that air flow is divided between the panel registers and floor ducts. In this position, recirculated air will not be used for maximum cooling.

HEAT Position—Maximum Heating, Granada, Monarch & Versailles, 1978 Ford & Mercury Full Size & 1978-79 Cougar, Lincoln Continental, Mark V & Thunderbird. FLOOR Position—Maximum Heating, 1979-80 Ford & Mercury Full Size, 1980 Cougar, Lincoln Continental, Mark VI & Thunderbird

In this position and with the vehicle's interior temperature below the temperature setting, the modulated vacuum is low, causing the water valve to open and the temperature blend door to fully open. Blower motor operation will be delayed until the engine coolant temperature reaches approximately 120° F, then the air is forced by the blower motor through the evaporator core and all of the air passes through the heater core before being distributed through the heater floor ducts into the vehicle's interior. A portion of this air is directed to the defroster nozzles.

HEAT Position—Modulated Heating, Granada, Monarch & Versailles, 1978 Ford & Mercury Full Size & 1978-79 Cougar, Lincoln Continental, Mark V & Thunderbird. FLOOR Position—Maximum Heating, 1979-80 Ford & Mercury Full Size, 1980 Cougar, Lincoln Continental, Mark VI & Thunderbird

When less heating is required to maintain the selected temperature, the temperature blend door moves to a mid-position so that only part of the air leaving the evaporator core is directed through the heater core. Blend air is then directed into the vehicle. The compressor does not operate in this position.

DEFOG Position

In this position, operation is the same as HEAT position, except that the A/C-Defrost door is in the defrost position on Granada, Monarch & Versailles, the A/C-Heat door is in the heat position on 1978 Cougar, Ford & Mercury full size, LTD II, Mark V & Thunderbird and 1978-79 Lincoln Continental, and the Defrost door on 1978 Cougar, LTD II, Mark V and Thunderbird and the Heat-Defrost door on Granada, Monarch & Versailles, 1978 Ford and Mercury Full size, 1978-79 Lincoln Continental and 1979 Cougar, LTD, Mark V and Thunderbird is in the partial position. The air flow is directed to the defroster nozzles and the floor outlets. The A/C compressor will be on when the outside air temperature is high enough to close the ambient cut-off switch. This helps in minimizing windshield fogging by dehumidifying the incoming air.

DEFROST Position

In this position, the A/C-Defrost door and the Heat-Defrost door on Granada, Monarch and Versailles, 1978 Ford and Mercury full size, 1978-79 Cougar, Lincoln Continental, Mark V and Thunderbird, and the PANEL doors and FLOOR-Defrost doors are in the full defrost positions so that all incoming air is directed to the defroster nozzles with a slight bleed to the floor. The temperature blend door modulates in this position. Maximum heat can be obtained by moving the temperature lever to the highest temperature position. The A/C compressor will be on to dehumidify the air and reduce windshield fogging.

System Components, Figs. 12 thru 16

Sensor Assembly

The sensor assembly has a bi-metallic sensor which senses the passenger compartment air temperature. A small opening in the instrument panel allows passenger compartment air to enter the sensor. The output controls a vacuum modulator and higher temperatures cause higher vacuum gauge readings. A bias to the vacuum output level is accomplished by means of a control arm attached by a cable assembly to the temperature control lever. The modulated vacuum output is the result of the effects of the control bias and the temperature of the air across the bi-metallic sensor. The sensor also has an aspirator which provides a very small vacuum to draw a continuous sampling of passenger compartment air across the air temperature sensing bi-metal. Primary air is ducted to the sensor through a tube from the blower motor stream providing the power to operate the aspirator for the in-vehicle sensor.

Servo Assembly

The servo assembly is a calibrated vacuum motor and diverter valve assembly. The calibrated vacuum motor provides precise temperature blend door positions for a specific vacuum level. The vacuum diverter valve controls water valve operation and recirc or outside air operation.

Electro-Vacuum Relay (E. V. Relay)

This relay is a combination single pole single throw normally closed electrical relay and a normally closed vacuum solenoid valve. Both are actuated by one electrical coil. The E.V. relay is used during cold engine operation and when the function control lever is in the HEAT or FLOOR position, to lock-out blower motor operation, and to switch the system to recirculating air until the engine coolant temperature warms up to about 120° F.

Control Assembly

The control assembly consists of a vacuum

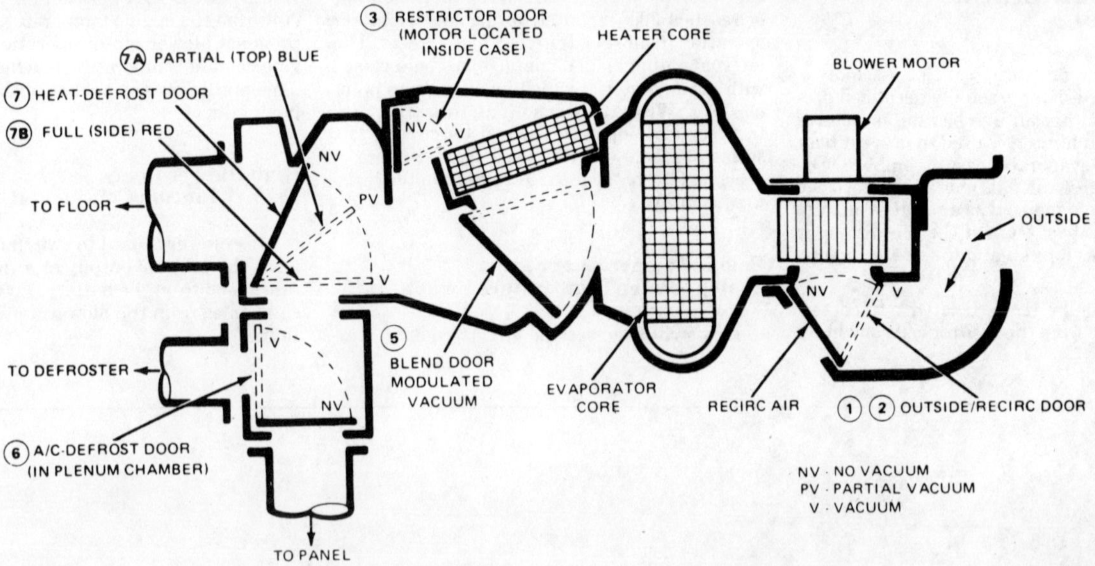

Fig. 21 System airflow schematic. 1980 Cougar and Thunderbird

selector valve, controlled by the function control lever that distributes vacuum to the various door vacuum motors which in turn direct air flow through the system, a blower switch that provides four speeds of blower motor operating, and a temperature control lever which connects through a control cable to the sensor assembly.

The vacuum selector valve directs source vacuum to various vacuum motors and connects the vacuum from the electrovacuum relay to the outside air-recirc door in HEAT or FLOOR positions. Three electrical switches are controlled by the selector. The combination of these switches controls the electrical supply to the A/C clutch and blower motor switch.

The blower switch provides a current path for by-passing the resistors in the resistor assembly, one at a time, until full blower speed is obtained in the Hi blower switch position.

The temperature control lever is connected to the sensor assembly by a control cable. Movement of the control lever from 65° F to 85° F causes a corresponding movement on the preload (bias) portion of the sensor assembly, and determines the temperature that the system will maintain.

Resistor Assembly, 1978 Cougar, LTD II, Mark V & Thunderbird

The blower motor resistor assembly has three resistance elements to provide four blower speeds. Depending on the blower switch position, series resistance is added or shorted out of the blower motor circuit to decrease or increase speed.

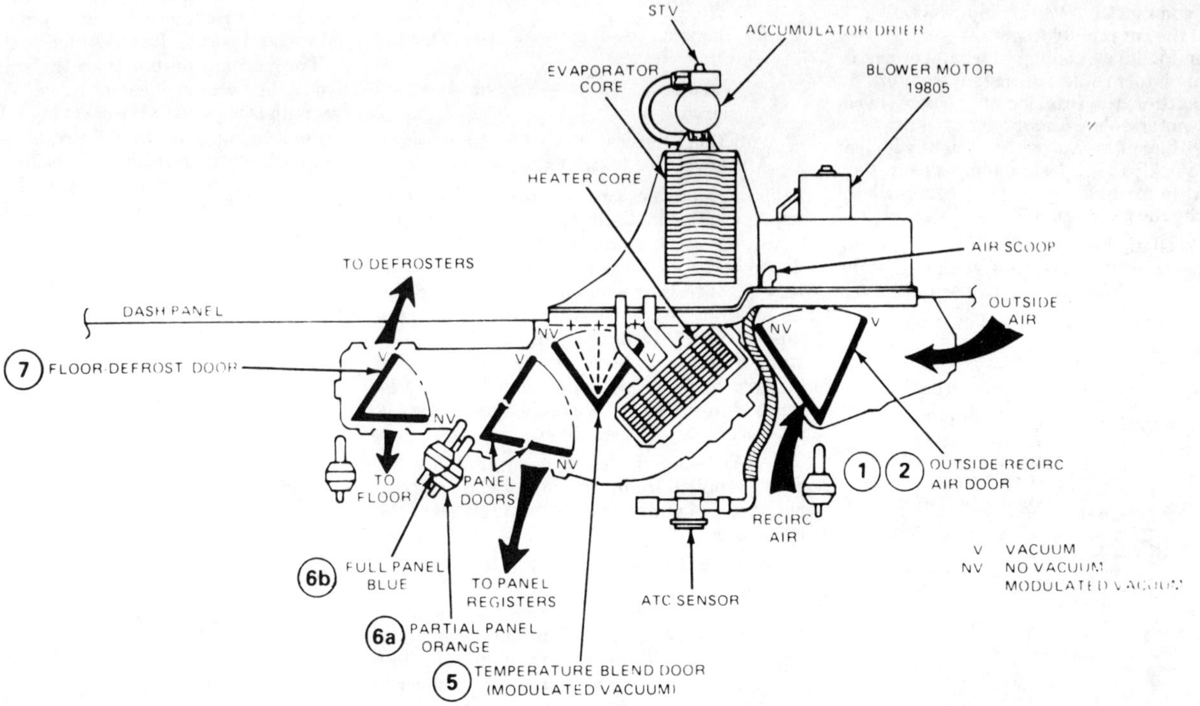

Fig. 22 System airflow schematic. 1980 Ford and Mercury full size, Lincoln Continental and Mark IV

Thermal Limiter Resistor Assembly, Exc. 1978 Cougar, LTD II, Mark V & Thunderbird

The operation and design of this assembly is similar to the Resistor assembly described previously, except that an overheating protective device (thermal limiter) is used to prevent heat damage to the evaporator case assembly. Overheating of the resistor coils may occur when the system air flow is reduced as a result of an inlet or outlet air restriction or if the blower motor wheel becomes locked and/or does not turn freely.

The thermal limiter is located a predetermined distance from the resistor coils and in series with the coil circuit. It will open the coil circuit when the temperature of the thermal limiter reaches 300° F, interrupting blower motor operation in all speeds except high blower. The internal spring loaded contacts are held closed with a wax material which has a melting point of 300° F. When the wax softens, the spring contacts separate, opening the resistor circuit. The spring contacts will not close again, making it necessary to replace the thermal limiter resistor assembly.

Engine Temperature Sender Switch/Water Temperature Switch

This switch senses the engine coolant temperature near the engine thermostat. In the HEAT or FLOOR position only, a ground circuit from the engine temperature sender switch prevents blower motor operation until the engine coolant temperature reaches about 120° F. This minimizes any cold air discharge from the heater ducts.

High Blower Relay, 1978-79 Lincoln Continental

This blower is used to switch the blower feed from the resistor output to a direct low resistance feed from the battery. Power to the relay coil comes from the blower switch "high" position.

General Motors Corporation

1974-76

A.C.C. or Comfortron system is designed to automatically control the heating and air conditioning components in the automobile so that a constant interior temperature is maintained, regardless of varying ambient conditions. The Automatic Climate Control system is beneficial in both summer and winter. In hot weather, it will cool the car rapidly to pre-set comfort level and then modulate cooling to whatever degree is required to maintain constant comfort. In mild weather, the interior of the vehicle remains comfortable without having to reset the controls. In cold weather, the system will heat the car quickly to the desired temperature, then level out to maintain the pre-set comfort level desired by the passengers.

The existing heater and air conditioning components provide a series system so that the primary control function is to position an air mix door and a mode door in order to release properly heated or cooled air from the proper duct outlets. Components of the control system are:

1. Sensing Devices—Thermistors
2. Programmer
   (a) Signal Amplifier—DC Transistor Amplifier
   (b) Transducer—Electrical to Vacuum Converter
   (c) Vacuum Motor—Vacuum-Powered Servo
3. Function Devices—Door Diaphragms, Air Mix Door, Hot Water Valve, etc.

This system uses two or three electrical sensors, the ambient, discharge air and in-car units to regulate the amount of vacuum supplied to the power servo. Each sensor is a semi-conductor type of resistor whose resistance value varies inversely to temperature. As the temperature increases, resistance through the sensor string decreases. These sensors are connected in series with the potentiometer. When resistance of any of these devices changes, a new signal is sent to the DC amplifier.

The DC amplifier enlarges or amplifies the signal supplied from the sensors and the temperature dial and feeds this strong signal to the transducer.

The transducer is actually a solenoid with a vacuum valve connected to the end of the plunger which is located inside the windings. The plunger is positioned inside the winding, depending upon the amount of current flow from the amplifier. The transducer plunger is spring-loaded so that when no current flows through the windings, the plunger is pulled out of the windings. When current flows through the windings, the plunger is pulled into the winding, causing the transducer vacuum valve to have a slight leak to outside air. This results in a lower vacuum level from the regulated vacuum port of the transducer, Fig. 1.

If a decrease in current flow from the DC amplifier occurs, the plunger moves out of the winding, opening the transducer vacuum valve and resulting in more vacuum from the regulated vacuum port of the transducer.

The vacuum output from the transducer is fed to the vacuum checking relay. When engine vacuum is applied to the checking relay, the relay opens, Fig. 2, allowing the regulated vacuum output from the transducer to be fed directly to the vacuum motor, positioning it. If engine vacuum to the checking relay is not present, the relay closes, Fig. 3, sealing the vacuum applied to the vacuum motor. This locks the vacuum motor in place until engine vacuum is again applied to the vacuum relay.

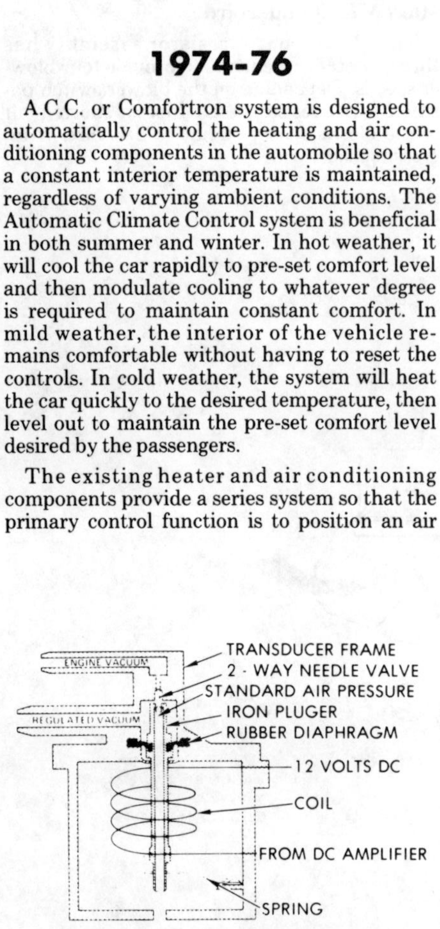

Fig. 1 Crossectional view of transducer

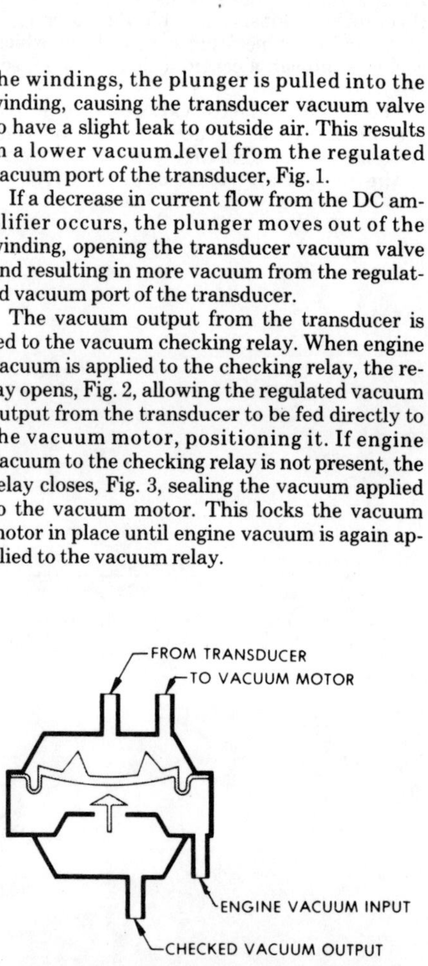

Fig. 2 Checking relay open

The vacuum checking relay also has another section which checks the input vacuum to the door diaphragms whenever engine vacuum is not present. This holds the doors in place until the engine vacuum is again present.

The vacuum motor mechanism is positioned as the result of the vacuum from the transducer. The vacuum motor mechanism is directly connected to electrical wiper contacts that control the various blower speeds. The mechanism is also connected to a rotary vacuum valve in the programmer and also to an output shaft.

The rotary vacuum valve channels vacuum to various vacuum diaphragms and to the rotary vacuum valve on the control head. The output shaft controls the position of the air-mix door, controlling the temperature of the air which is distributed into the passenger compartment.

As the vacuum motor in the programmer moves, a gear meshed with the mechanism rotates the amplifier feedback potentiometer. This potentiometer indicates when the vacuum motor has reached the proper position and cancels out the change in resistance of the sensor string or the temperature dial. This signals the amplifier to stop the movement of the vacuum motor.

Whenever an appreciable amount of variation in resistance occurs in the sensor string of the temperature dial, the vacuum motor moves to supply either warmer or cooler conditioned air to counteract the resistance change.

1975-76 Buick, Cadillac and Chevrolet models are equipped with a revised design ATC programmer, designated the MKII. This new programmer has a revised electrical-to-vacuum-transducer mechanism. The supply vacuum is metered through a .016 inch orifice in the vacuum relay and the transducer provides a variable vent action which regulates the vacuum signal to the power diaphragm. Since the transducer has a high vent capacity, a porous metal restriction is located in the power diaphragm vacuum feed line to maintain position of the programmer output until the vacuum relay can lock at vehicle shutdown. The power diaphragm incorporates a vacuum level valve which is actuated by the diaphragm plate at the maximum heat end of travel to limit the power diaphragm level to approximately 6.0 inches Hg. This limits the programmer vacuum level, thereby preventing the possibility of locking in the maximum heater position under low engine vacuum conditions.

BASIC SYSTEM OPERATION

The numbers on the temperature dial thumb-wheel or temperature lever located on the control head indicate the various temperatures which can be obtained in the passenger compartment of the car. This temperature dial serves the same function as the thermostat in your home. The temperature dial controls the in-car temperature when the selector lever is in the "Off," "Lo," "Auto," "Hi," "Bi-Level," or "DEF" position.

All Exc. 1974-76 Oldsmobile, 1975-76 Pontiac & 1976 Buick Century, Regal, LeSabre

Off Position

The Automatic Climate Control system is completely "shut off" when the ignition switch

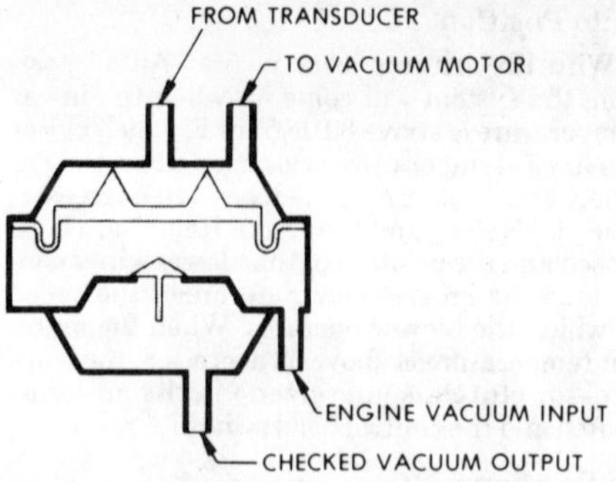

Fig. 3 Checking relay closed

of the car is in the "Off" position. When the car ignition switch is turned on, the electrical circuit to the A.C.C. system is accomplished. With the control head selector lever in the "Off" position, the system will come on when the engine water temperature reaches about 120 degrees F, or if the inside car temperature reaches about 80 degrees F. The fan will run at "Lo Blower" speed and conditioned air will flow from the heater outlets. The temperature of the air depends on the temperature dial setting.

Vent or Economy Position

When the selector lever is in the "Vent" position, the system comes on immediately whenever the ignition switch is in the "Run" position. The blower runs at "Lo Blower" speed and outside air is distributed into the passenger compartment of the car through the A/C outlets.

Lo Position

With the control head selector lever in the "Lo" position, the system will not come on until either the inside car temperature reaches about 80 degrees F, or the engine water temperature reaches about 120 degrees F. The blower will then operate at "Lo Blower" speed only. Conditioned air will flow from the A/C outlets, the heater outlets, or both when the system is "Bi-Level" operation.

Auto Position

The operation of the A.C.C. system when the selector lever is in the "Auto" position is the same as the "Lo" position, except that the blower is no longer locked in "Lo Blower" operation. The programmer will select any of the four blower speeds ("Lo," "Med. 1," "Med. 2," and "Hi") in order to maintain the proper comfort level in the car.

Hi Position

When the control head selector lever is placed in the "Hi" position, the blower operates only at "Hi Blower" speed. The temperature selected on the temperature dial will be maintained inside the passenger compartment. If maximum A/C is required, the inside air will be recirculated through the A/C system for maximum cooling. When the system operation has stabilized and the blower is operating at lower speeds, the conditioned air will be distributed from both the A/C and heater outlets. This is "Bi-Level" operation.

Bi-Level Position

The blower operates at "Lo," "Med. 1," "Med. 2," or "Hi" speeds when the control head selector lever is in the "BI-LEVEL" position. The desired temperature level inside the car will be maintained. Conditioned air is distributed from the heater, A/C, and defroster outlets, resulting in "tri-level" operation. The system will come on immediately when the engine thermal switch or the in-car temperature switch is closed.

Def Position

With the control head selector lever in the "DEF" position, the system comes on immediately when the ignition switch is turned on. The fan is locked on "Hi Blower" speed and all of the air is distributed from the defroster outlets onto the windshield. Some air is bled to floor on "DEF." The comfort level selected on the temperature dial will be maintained in the car.

1974-76 Oldsmobile, 1975-76 Pontiac & 1976 Buick Century, Regal, LeSabre

Off Position, All

The system is inoperative when the ignition switch is in the off position. With the panel selector lever in the Off position, the blower is inoperative until engine coolant temperature reaches 120° F., after which the blower will run at low speed, regardless of blower switch position. The discharge air temperature will be maintained at the desired level, however, since the A/C compressor is inoperative in this mode, cold air is not available.

Air Cond Position, 1974 Oldsmobile

Blower motor speed is controlled by the blower switch unless the temperature is in the 65° position. If the temperature lever is in the 65° position, the blower will operate at high speed regardless of blower switch position. Cold air is discharged from the A/C registers.

Max Position, Exc. 1974 Oldsmobile

In this mode, the A/C compressor is operative when the ambient temperature is above 35° F. The water control valve is closed and only cold air is available. The blower runs in high speed and air is discharged from the A/C outlets only. The system uses recirculated air from the vehicle interior for maximum cooling.

Norm Position, Exc. 1974 Oldsmobile

Air is discharged from the A/C outlets and the blower will operate at the speed determined by blower switch position. Also, the A/C compressor will operate when the ambient temperature is above 35° F.

Bi-Level Position, All

In this position, conditioned air is distributed from the A/C and heater outlets. Also, a small amount of air is discharged from the defroster outlets.

Vent Position, All

The A/C compressor is inoperative, therefore cold air is not available. The blower operates at a speed as determined by blower switch position. Outside air is delivered through the A/C outlets.

Heater Position, All

Blower operation is delayed until engine coolant temperature reaches 120° F., after which the blower runs at a speed determined by the blower switch position. The A/C compressor is inoperative and heater air is discharged through the heater outlets with a slight bleed through the defroster outlets.

Defrost Position, All

The blower operates at a speed determined by blower switch position. Heated air is discharged through the defroster outlets with a slight bleed through the heater outlets. Also, the A/C compressor operates when the ambient temperature is above 35° F.

ELECTRICAL SYSTEM OPERATION

All Exc. 1974-76 Oldsmobile, 1975-76 Pontiac & 1976 Buick Century, Regal, LeSabre

Off Position

When the control head selector lever is in the "OFF" position, the system is locked in "Lo Blower" operation. When the ignition switch is in the "Run" position and the engine thermal switch is open, the heater delay relay is closed and the "Lo Relay" is closed. The engine thermal switch opens whenever the engine water temperature is above 120 degrees F. When the engine thermal switch opens, the heater delay relay springs close. If the temperature inside the car is above 80 degrees F, the in-car temperature switch is closed, which also causes the "Lo Relay" to close, resulting in "Lo Blower" operation.

With ignition switch in the "Run" position, power is supplied to the programmer, allowing it to operate. The in-car sensor, the ambient sensor, and the temperature dial setting on the control head cause the programmer to move the air-mix door. The proper hot-cold air mixture is then distributed from the heater outlets, maintaining the in-car temperature at the comfort level selected on the temperature dial.

Vent or Economy Position

With the ignition switch in "Run" and the selector lever in the "Vent" position, the "Lo Relay" will be closed and the system will come on immediately with the fan locked on "Lo Blower" speed. The vacuum motor in the programmer is in the A/C position, since the in-car sensor and the ambient sensor are shorted out by the control head switch. The ambient switch is closed when the ambient temperature is above 32 degrees F. However, the compressor will not run, since no voltage is applied to it through the control head switch. The engine thermal switch is closed when the engine water temperature is above 120 degrees F. The in-car temperature switch is closed if the in-car temperature is above 80 degrees F. In "Vent" immediate start-up occurs, irrespective of temperature delay settings. The "Auto Relay" is always open in this mode. The engine thermal switch is open when the engine water temperature is above 120 degrees F, which closes the heater delay relay. This causes the "Lo Relay" to be energized.

Lo Position

With the selector lever in the "Lo" position, the system will come on when the in-car temperature is above 80 degrees F, which closes the in-car temperature switch, or when the engine water temperature reaches 120 degrees F. The "Lo Relay" is then closed, allowing the blower to be locked on "Lo Blower" speed. The "Auto Relay" remains open.

When the ambient temperature is above 32 degrees F, the ambient switch is closed, supplying voltage to energize the compressor clutch via the control head switch.

Auto Position

With the selector lever in the "Auto" position, the system will come on when the in-car temperature is above 80 degrees F, which closes the in-car temperature switch, or when the engine water temperature reaches 120 degrees F. The "Lo Relay" and the "Auto Relay" are both closed and the position of the blower wiper contacts in the programmer determines the speed at which the blower operates. When the ambient temperature is above 32 degrees F, the compressor clutch is energized via the ambient switch and the control head switch.

Hi Position

When the selector lever is in the "Hi" position, the system turn-on is the same as described in the "Auto" position section. Both the "Lo" and "Auto" relays are closed when the system is operating and a direct circuit path via the control head switch is connected from the battery circuit to the blower motor. This locks the blower on "Hi Blower" operation. When the ambient temperature is above 32 degrees F, the ambient switch is closed and the compressor clutch is energized. If the system is calling for maximum cooling, the system will operate in the recirculation mode; however, there is no change in the electrical function when this mode is achieved.

Bi-Level Position

With the selector lever in the "BI-LEVEL" position, the system turn-on is the same as described in the "Auto" position section. The "Lo" and "Auto" relays are both closed and the position of the blower wiper contacts in the programmer determines the speed at which the blower operates. When the ambient temperature is above 32 degrees F, the ambient switch is closed and the compressor clutch is energized.

Def Position

With the selector lever in the "DEF" position, the system comes on immediately and the blower is locked on "Hi" speed. The "Lo Relay" and the "Auto Relay" always remain energized.

The temperature dial must be used to obtain maximum heat when it is desired to override the automatic control.

1974-76 Oldsmobile, 1975-76 Pontiac & 1976 Buick Century, Regal & LeSabre

Off Position, All

Current from the fuse block is directed to the thermal relay, then to the control when engine coolant temperature reaches approximately 120° F. The blower operates at low speed regardless of blower switch position. The A/C compressor is inoperative.

Air Cond Position, 1974 Oldsmobile

With the blower switch in the LO position, current is directed through the resistor block, by-passing the thermal relay, through the Hi Blower Relay to the blower motor. In the M1 position, current is directed through two resistors to the blower motor. In the M2 position, current is directed through one resistor to the blower motor. In the Hi position, current from the blower switch energizes the High Blower Relay and current from the junction block is directed through the relay to the blower motor.

If the ambient temperature is above approximately 32° F, the ambient switch is closed and the A/C compressor clutch is energized. Also, current is directed from the control to the Recirculating Override Switch. With the temperature lever in the 65° position, the Recirculating Override Switch is closed and current is directed to the Hi terminal on the blower switch, in turn operating the blower at high speed regardless of blower switch position.

Max Position, Exc. 1974 Oldsmobile

In this position, current is directed to the Hi terminal on the blower speed switch, in turn to the Hi blower relay which is then energized. Current from the junction block flows through the energized relay and runs the blower at high speed regardless of blower switch position.

Current from the control energizes the A/C compressor clutch through the ambient switch and thermal limiter. The ambient switch closes when the ambient temperature is above 32° F. The thermal limiter protects the compressor if the refrigerant charge is lost.

Norm Position, Exc. 1974 Oldsmobile

Current is directed to the blower switch and operates the blower motor immediately when the switch is in M1, M2 or Hi. When the blower switch is in the Lo position, blower operation is delayed by the thermal relay until engine coolant temperature reaches approximately 120° F. Hi blower speed and A/C compressor activation is the same as in the Max Position.

Bi-Level Position, All

Same as the Norm position.

Vent & Heater Position, All

Same as Norm position except the A/C compressor is inoperative.

Defrost Position, All

Same as the Norm position.

VACUUM SYSTEM OPERATION

All Exc. 1974-76 Oldsmobile, 1975-76 Pontiac & 1976 Buick Century, Regal, LeSabre

Off Position

With the selector lever in the "Off" position and the engine running, the system is turned on whenever the engine thermal switch or the in-car temperature switch is closed. The vacuum motor in the programmer moves to a position that will moderate the in-car temperatures. Air flows from the heater outlets at "Lo Blower" speed.

Vent or Economy Position

Air is drawn through the outside air door and is distributed from the A/C and defroster outlets at "Lo Blower" speed. The programmer is in the maximum A/C position.

Lo Position

When maximum cooling is required, cold air is distributed from the A/C outlets. The blower is locked on "Lo Blower" speed. Recirculation of air is not possible in the "Lo" selector lever position. A small amount of this cold air is blown onto the windshield.

When the vacuum motor moves from maximum A/C, the porting in the programmer vacuum valve changes.

As the air temperature from the outlets reaches a moderate temperature, the system goes into "Bi-Level" operation and air flows from the A/C and the heater outlets. Some moderate temperature dry air is also released from the defroster outlets onto the windshield.

When heating is required, hot air is distributed from the heater outlets. A small amount of hot, dry air is also released from the defroster outlets.

Auto Position

Even though maximum A/C is required, the system cannot go into recirculation operation. Air is released from the A/C outlets at "Hi Blower" speed and some cold, dry air is also blown onto the windshield.

As the in-car temperature begins to lower, the programmer moves out of the maximum A/C position.

When the outlet air temperature reaches a moderate temperature, the system goes into "Bi-Level" operation. Conditioned air flows from both the heater and A/C outlets. Also, a small amount of moderate, dry air is distributed from the defroster outlets.

When maximum heating is required from the system, air is distributed from the heater outlets at "Hi Blower" speed. Some hot, dry air is also blown onto the windshield.

Hi Position

The passenger compartment air is recirculated through the A/C system when maximum cooling is required. Cold air is distributed from the A/C outlets at "Hi Blower" speed and a small amount of cold, dry air is also blown onto the windshield.

As the programmer moves from the maximum A/C position, air is drawn in from outside the car to be conditioned. This air is distributed from the A/C and defroster outlets at "Hi Blower" speed.

As the outlet air reaches moderate temperature, the system begins to operate in the "Bi-Level" mode. Some cool, dry air is also released from the defroster outlets.

When the system goes into heater operation, the warm air is released from the heater outlets at "Hi Blower" speed. Some warm, dry air is also released from the defroster outlets.

Bi-Level Position

If maximum cooling is required, the outside air door is closed for recirculation operation and the blower is operating at "Hi Blower" speed. Cold, dry air is distributed from the heater, A/C, and defroster outlets, resulting in "Tri-Level" operation.

When maximum cooling is no longer required, the outside air door opens and the blower speed decreases. Complete blower programming is used in the "BI-LEVEL" position. When heating is required from the system, the blower speed increases and reaches "Hi Blower" speed at maximum heating.

Def Position

With the selector lever in the "DEF" position, the outside air door is closed when maximum cooling is required. All of the air is directed onto the windshield from the defroster outlets at "Hi Blower" speed. When maximum cooling is no longer required, the outside air door opens but the fan remains on "Hi Blower" speed.

1974-76 Oldsmobile, 1975-76 Pontiac & 1976 Buick Century, Regal, LeSabre

Off Position, All

In this position, vacuum is directed to the control vacuum valve. However, all ports at the valve are either vented or sealed.

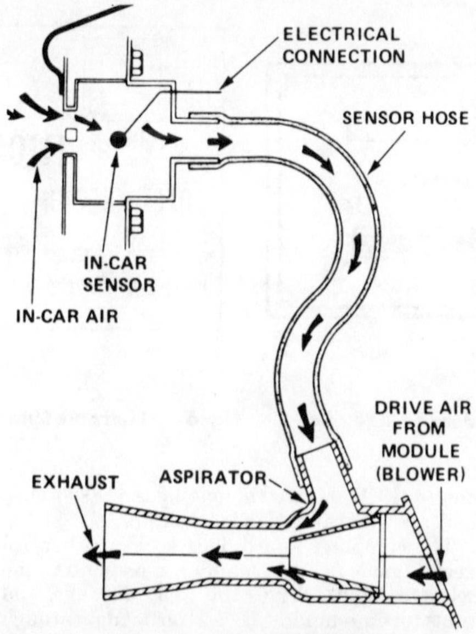

Fig. 4 In-car sensor and aspirator

Air Cond. Position, 1974 Oldsmobile

Vacuum system operation with the temperature lever at the 65° position is the same as 1975-76 Max position. When the temperature lever is moved approximately ¾ inch toward the 85° position, vacuum system operation is the same as 1975-76 Norm position.

Max Position, Exc. 1974 Oldsmobile

Vacuum is directed from the vacuum reserve tank to the control vacuum valve. Vacuum is then applied to all diaphragms except the defroster door diaphragms.

Norm Position, Exc. 1974 Oldsmobile

Vacuum is applied to the upper and lower mode door diaphragms, in turn opening the doors to the A/C mode. Also, the water valve is opened.

Bi-Level Position, All

Vacuum from the vacuum reserve tank is applied to the lower mode door diaphragm to open the door to the A/C mode. Engine vacuum is directed through a restrictor to the defroster door to partially open the door. Also, engine vacuum is used to close the defroster door when the engine is shut off. When the engine is restarted, the restrictor in the defroster door vacuum line causes a time delay before the defroster door reopens. This reduces window fogging.

Vent Position, All

Vacuum system operation in this mode is the same as in Bi-Level except the upper mode door is also opened by vacuum from the vacuum tank.

Heat Position, All

Engine vacuum is directed through a restrictor to partially open the defroster door. Also, engine vacuum is used to close the defroster door when the engine is shut off. When the engine is restarted, the restrictor in the defroster door vacuum line causes a time delay before the defroster door reopens.

Defrost Position, All

Engine vacuum is applied to the off center port on the defroster door diaphragm to partially open the defroster door. Vacuum from the vacuum reserve tank is applied to the center port on the defroster door diaphragm to fully open the defroster door.

1977-80 BUICK AUTOMATIC TEMPERATURE CONTROL, CADILLAC CLIMATE CONTROL, CHEVROLET COMFORTRON; 1979-80 PONTIAC AUTOMATIC TEMPERATURE CONTROL; 1980 CADILLAC ELECTRONIC CLIMATE CONTROL

This system provides automatically controlled interior temperatures. With this system the desired comfort level can be set with a dial on the control panel. The system will then maintain the comfort level and will require little or no change in setting to compensate for changes in outside weather conditions.

The system will remove excess moisture from the air when the compressor is engaged and in operation. This removal of moisture from the air increases passenger comfort, and limits window fogging and interior condensation.

Two temperature sensing devices called thermistors are used in this system for its operation. The in-car sensor, Fig. 4, located in the vehicle, senses vehicle interior temperature. The ambient sensor, located in the air passage in the cowl vent opening, senses temperature of the incoming air. These sensors are wired in series with the temperature dial (variable resistor) on the control panel. As the temperature in the area of the sensor varies, the total resistance changes, supplying a variable input signal to the amplifier in the programmer. The amplifier through the transducer changes the position of the vacuum motor link which controls the temperature door, the rotary vacuum valve, and the blower speed.

System Components

Control Panel Except 1980 Eldorado & Seville

The control panel temperature dial is graduated in 5° F increments from 65° F to 85° F and is color coded red for hot, blue for cold and white for moderate temperatures, Fig. 5. The dial varies the resistance of a wire-wound rheostat.

The circuit board electrical switch is mounted on the underside of the baseplate. The control lever has wiper contacts that ride on the circuit board pads to provide the required electrical connections.

The in-car turn on switch is a thermostatic switch located in the control panel. It senses car interior temperature and will turn on the system if the interior temperature is above approximately 75° F, even if the blower delay (turn-on) switch, located in the right cylinder head is cold.

On 1979 Riviera and all 1980 Buick models, there is an override feature that prevents the heater from being turned on when ambient temperatures are above 85° F.

The control vacuum valve, located at the forward edge of the baseplate, is a nine-port rotary valve which interconnects or vents the vacuum hoses attached to it to operate the mode at the proper times.

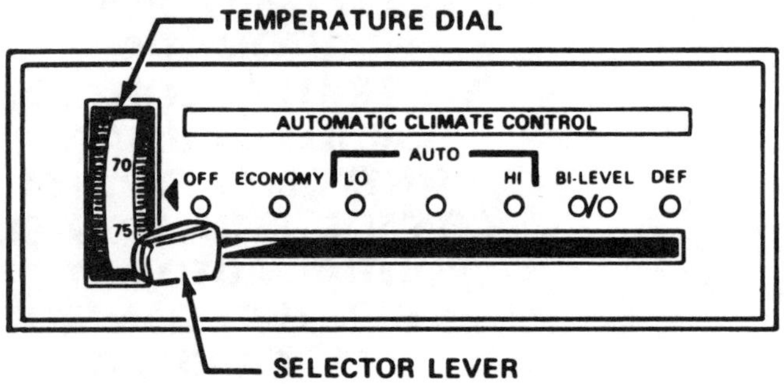

Fig. 5 Typical control panel. Except 1980 Eldorado & Seville

Fig. 6 Electronic Climate Control control panel. 1980 Eldorado & Seville

Control Panel, 1980 Eldorado & Seville

The control panel on the Electronic Climate Control consists of a keyboard and circuit board assembly. The red and blue push buttons serve to increase and decrease the temperature in one degree increments between 65° F and 85° F, Fig. 6. A continuous touch of the button will cause a change in the selected temperature at the rate of two degrees per second until the button is released or the end of the scale is reached.

Temperature settings of 60° F and 90° F are available, representing maximum cooling or maximum heating respectively. These two settings override all in-vehicle temperature control circuits. Depressing the cooler and warmer pushbuttons simultaneously will display a temperature setting of 60° F immediately. Otherwise, the 60° F and 90° F settings are obtained in one degree or two degree steps.

The two digit readout displays the temperature setting. However, vehicles with Digital Electronic Fuel Injection will display trouble codes for that system when the "Off" mode pushbutton and the "warmer" pushbutton are depressed simultaneously for a short period of time.

The six mode pushbuttons allow the selection of the appropriate control function. When equipped with a rear window defogger, a seventh mode pushbutton is incorporated into the control panel for that function. Light emitting diodes (LED) located above the mode switches indicate which function is selected.

The computer board, located in the control head, signals the programmer to respond to the selected inputs, drives the LED displays, and transmits a temperature setting signal through the resistor string to the programmer. When the ignition switch is turned off, the control head will remember the previous setting and return to it when the vehicle is restarted. The control head will remember the information as long as the battery remains connected. If the battery is disconnected, the control will automatically reset to the AUTO mode and to a temperature setting of 75° F.

Power Module, 1980 Eldorado & Seville

This module controls the operation of the blower motor and is located on top of the evaporator and blower assembly. The power module receives the blower drive signal from the programmer and amplifies the signal to a strong output signal proportional to the input. The power module is transistorized and provides variable blower speeds. If the vehicle's interior is hotter or colder than the set temperature, the blower will start at high speed and gradually decrease to low speed as the pre-selected temperature is selected. A heat sink, attached to the power module, protrudes into the evaporator air stream as a means of dissipating heat which the power module generates. The power module, which receives a signal from the programmer, also cycles the A/C compressor.

Temperature Sensors

The temperature sensors are thermistors which vary resistance inversely with temperature changes, Fig. 7. For example, an increase in temperature causes a decrease in resistance, and a decrease in temperature causes an increase in resistance.

Aspirator

To provide accurate sensing of the interior temperature, a small amount of air is drawn into the in-car-sensor housing and passed over the thermistor. This air movement is accomplished by the use of an aspirator, Fig. 4, mounted on the programmer chassis, and a rubber hose that connects the in-car sensor housing to the aspirator. The aspirator is a tube-within-a-tube arrangement. Air from the blower case is discharged into the inner tube, causing a suction at the end of the inner tube so that air is pulled from the rubber hose.

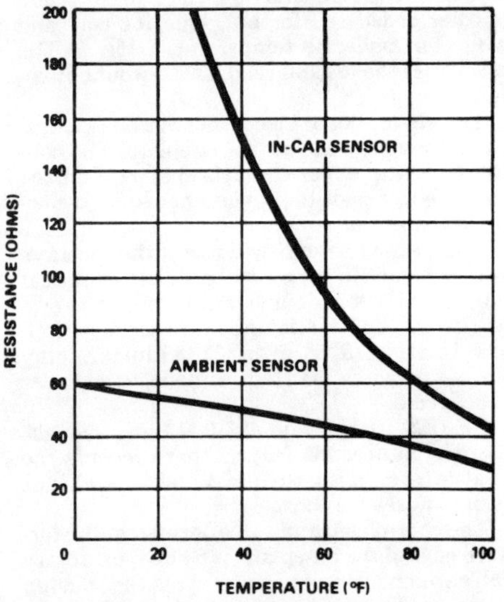

Fig. 7 Sensor resistance chart

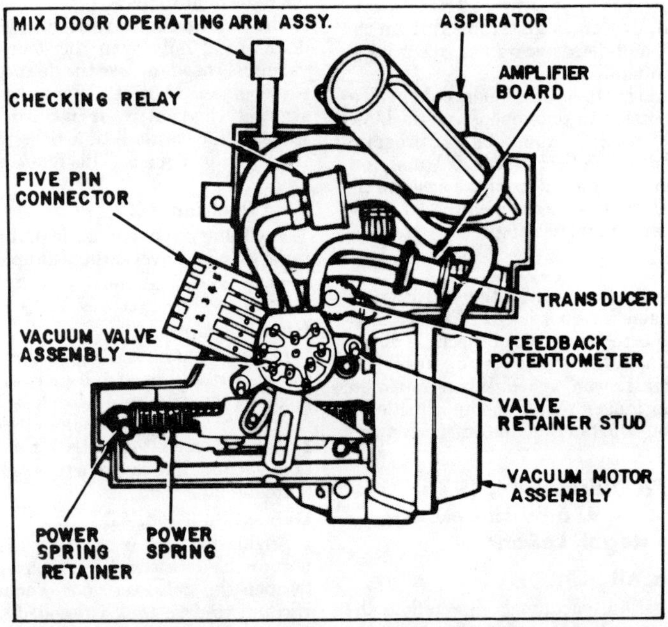

Fig. 8 Programmer assembly. Except Eldorado & Seville Mk II

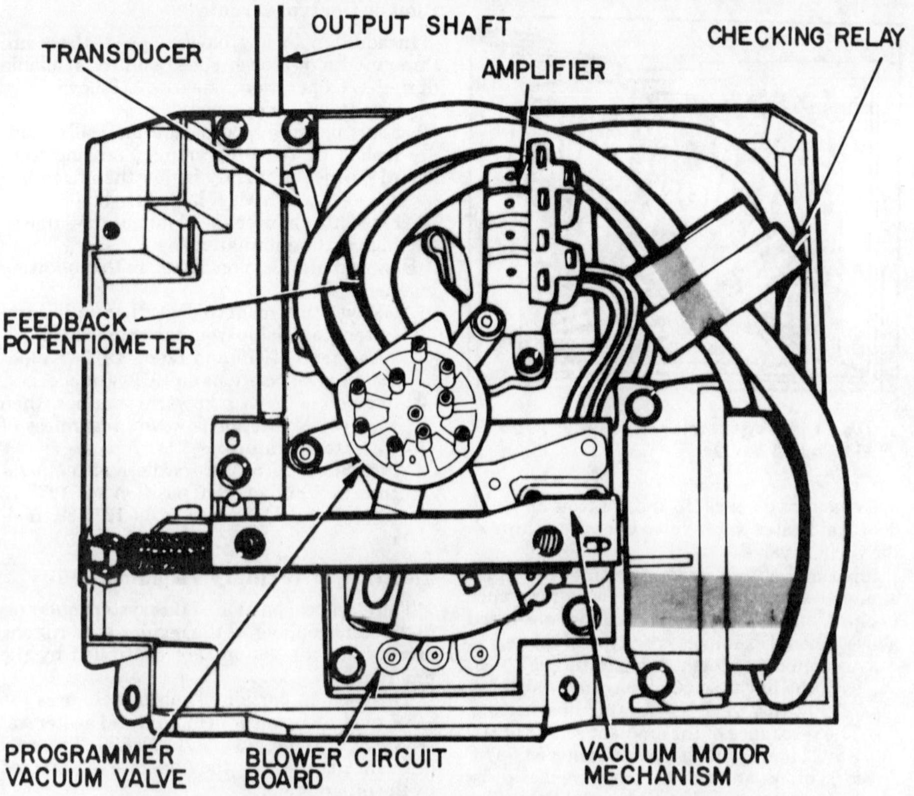

Fig. 9 Programmer assembly. 1977-79 Eldorado & Seville Mk II

Programmer Except 1980 Eldorado & Seville

The programmer, Figs. 8 and 9, contains most of the temperature control components and is located in the passenger compartment on the right-hand end of the module. It consists of a plastic chassis, a removable plastic cover, and the functional components described below.

The 1979-80 Riviera programmer is a modified 1978 Electra and LeSabre unit. A separate adapter plate is used for mounting it to the heater assembly.

The amplifier accepts a weak electrical signal from the sensors and the temperature dial and provides a strong output signal. The amplifier should be serviced only as a unit. No attempt should be made to repair an amplifier except by a qualified technician.

The transducer is mounted directly to the amplifier board and is basically a throttling device. The vacuum level is set by balancing an air force against a ball with a magnetic attraction between the armature and pole piece. Output current from the amplifier flows through the solenoid coil to control the vacuum level of the transducer.

The vacuum checking relay, Fig. 10, provides two separate vacuum checking functions. In the upper part of the relay is a check valve which is opened by engine vacuum applied at port 5. This allows port 3, which supplies vacuum to the rotary vacuum valve on the programmer and control head, to be evacuated. If engine vacuum drops upon acceleration or steep grades, the check valve closes, maintaining the vacuum level at port 3.

In the lower part of the relay is a spring-loaded diaphragm which blocks or passes regulated vacuum from the transducer, which is connected at port 2, to the vacuum motor which is connected to port 1. When engine vacuum at port 5 is high, the diaphragm is pulled upward allowing regulated vacuum to pass to the vacuum motor. When engine vacuum drops, the diaphragm moves downward blocking passage and maintaining vacuum in the vacuum motor.

The vacuum motor produces all mechanical motion within the programmer. It positions the vacuum motor mechanism which actuates the rotary shaft that drives the air-mix door link. This same mechanism moves the blower circuit switch, the vacuum valve operating arm and the feedback potentiometer. The vacuum motor mechanism is spring loaded. With no vacuum, it is fully extended out of the vacuum motor. This is the Max A/C position. When vacuum is applied to the vacuum motor, the mechanism is pulled into the vacuum motor to the Max-Heat position. To avoid loss of position during stop/restart operations, a restriction is placed between the vacuum checking relay and the transducer.

The blower switch is located in the center part of the programmer. Spring-loaded contacts move across the board making the required blower circuit connections. The purpose of the switch places the resistors into the blower circuit at the required points in the program to control blower speed. A low blower resistor is also provided.

The vacuum valve, located in the center of the assembly, vents or connects the various ports as required, to move the mode doors and operate the heater valve.

The feedback potentiometer is located above the vacuum valve and limits movement of the vacuum motor, so that a small signal from the sensors does not drive the motor its full travel. The feedback potentiometer is located, electrically, in the sensor circuit and operates in oppo-

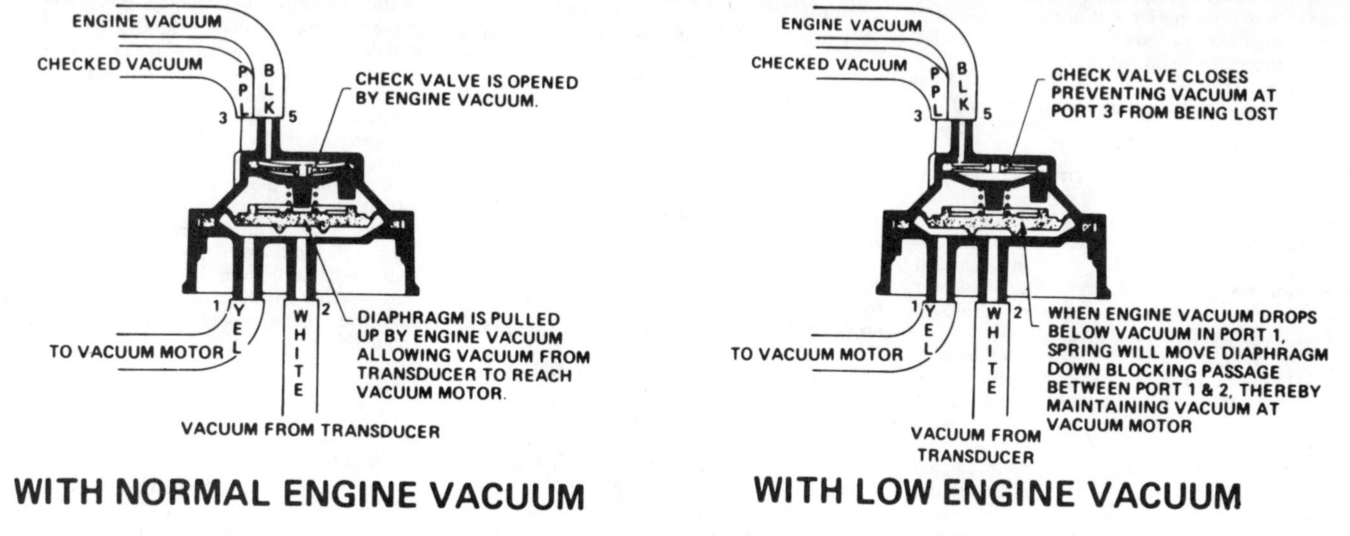

Fig. 10 Vacuum checking relay

sition to the sensors. If a sensor drives the vacuum motor to Max-Heat, the potentiometer will rotate so that its resistance change tends to offset the resistance change of the sensor. This unit is adjustable.

Programmer, 1980 Eldorado & Seville

The programmer, Fig. 11, contains temperature control components and is located on the right end of the heater and air conditioner assembly. The programmer can be identified by the electrical and vacuum connections attached to it. The programmer consists of a plastic case and the components listed below.

The circuit board is an amplifier and logic board that accepts electrical signals from the temperature sensors and mode pushbuttons in the control head and provides output control signals. The circuit board transmits the blower control signal, the A/C compressor drive signal, the DC motor control signal which adjusts the air mix door position, and the regulated 8 volt signal which powers the control head and the power regulator. The potentiometer is adjustable for programmer calibration changes. However, no attempt should be made to repair a circuit board except by a qualified technician with the proper service equipment.

The DC motor, located in the plastic programmer case, actuates the rotary shaft which drives the air mix door link, and is reversible.

Vacuum solenoids, inside the plastic programmer case, are actuated by signals originating within the circuit board. The solenoids control the vacuum to the vacuum actuators which control the hot water valve, the recirculating door, the mode door and the defroster door.

Blower Relay, 1977-78

A single-pole, single throw relay is used to supply power to the blower motor. It is located on the blower case in the engine compartment. The relay supplies power in all positions of the control lever except Off and Lo.

Blower Turn On Switch/Blower Delay Switch

This switch is a bi-metal, snap action switch that senses engine temperature. It is closed above about 120° F and open below that temperature.

System Operation

The automatic temperature control system is made up of three major sections which perform the following operations:

1. Temperature control
2. Blower control
3. Electrical and auxiliary vacuum control

Temperature Control Circuit

The temperature control circuit consists of: 1) inputs to the programmer from the in-car sensor, ambient sensor, temperature dial and feedback potentiometer; 2) programmer control actions of the amplifier, the transducer and vacuum relay; and 3) output of the transducer to the vacuum motor, which moves the link to control the air temperature door.

The resistance of the two sensors varies with their temperature. Their resistance plus that of the temperature dial, form the input signal to the programmer. This weak electrical signal is amplified to produce a current flow through the transducer. The transducer converts this signal to a vacuum signal which is routed through the vacuum relay to the vacuum motor. The vacuum motor positions the air-mix door through the link, controls the blower speed through the movement of the switch, and controls all mode door and heater water valve operation through the vacuum valve.

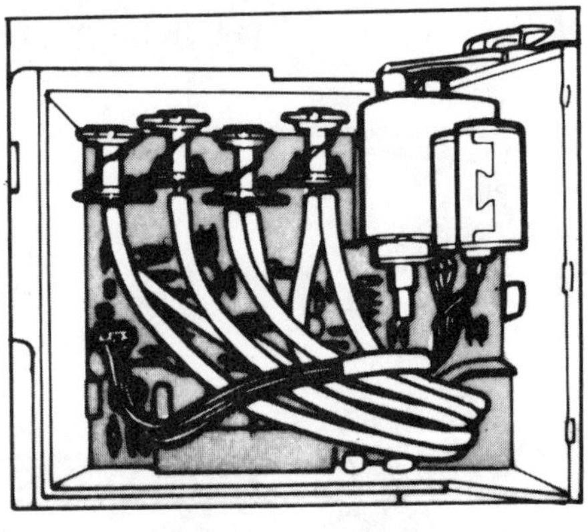

Fig.11 Programmer assembly. 1980 Eldorado & Seville

Input and output characteristics of the programmer are shown above. Input is resistance in ohms, output consists of the functions listed above. Several examples are outlined below:

1. Assume the ambient temperature is 80° F, with the in-car sensor being maintained at 75° F with a 75° F temperature dial setting. The resistance of the ambient sensor at 80° F is 32 ohms, the in-car resistance at 75° F is 68 ohms, and the temperature dial resistance at the 75° F setting is 36 ohms. This resistance positions the programmer to produce M1 blower speed, and intermediate discharge air temperature (about 60° F), A/C mode, an open heater water valve, and outside air.
2. If the ambient temperature were to increase to 90° F, the ambient sensor resistance would decrease to 26 ohms, causing the amplifier output current to decrease, which would cause the transducer output vacuum to the vacuum motor to decrease. The new point of operation would be at the 130 ohm point, producing M2 blower speed and colder discharge air. The system would still be in the A/C mode, with the water valve open, using outside air. In this way, the in-car temperature is maintained at 75° F despite the change in ambient temperature.
3. If the heat load on the vehicle were to be decreased, the in-car sensor temperature would tend to decrease. This would cause its resistance to increase so that the total resistance to the programmer would increase. This would cause the programmer to move back to the right to a warmer position.

In the above examples, a decrease in amplifier output current causes a decrease in transducer output. Hot sensors result in low sensor string resistance, causing the programmer to position the vacuum motor in the cool or cold area of operation. Cold sensors cause high resistance, causing the programmer to position the vacuum motor in the warm or hot area of operation.

When diagnosing a system, it is important to note that an open sensor (disconnected or open wire) would indicate a high resistance (as if the sensor were very cold) which would drive the programmer to maximum heat, while a shorted sensor would produce low resistance and drive the system to Max A/C.

Blower Control Circuit

In addition to low blower speed, there are three modes of blower speed control available during A/C operation. They are as follows:

1. Fixed low blower speed.
2. Four blower speeds automatically controlled by the programmer, ranging from M1, which is slightly higher than fixed low speed, through M2, M3 and High.
3. Fixed high blower speed (about the same as high in the automatic range).

Blower motor turn on occurs in the following manner:

1. Delayed turn on in cold weather until warm water is available from the engine in all positions except Off and Lo on 1977-78 models, and Def positions on 1979-80 models.
2. Immediate turn on in warm weather when the vehicle interior is warm, regardless of water temperature.
3. Immediate turn on regardless of temperature in Vent or Def position on 1977-78 models and Def position on 1979-80 models.

Electrical & Auxiliary Vacuum Circuits

The electrical portion of this system operates in the same manner as the regular A/C system, except that the actions are controlled by the programmer.

The vacuum portion of this system uses the same mode door vacuum motors and heater water valve as the regular A/C system. The only difference is the way they operate as directed by the programmer.

Mode Lever Settings

Off

At this setting, outside air is circulated at a fixed low blower speed through the A/C module and is discharged into the passenger compartment from the heater outlet when the ignition is On and turn on requirements are met (by the blower delay switch/blower turn on switch and in-car sensor). The temperature of the air is regulated to maintain the temperature on the dial. The system will not cool in warm weather, since the compressor is not operating.

Economy

At this setting, outside air is circulated at a selected blower through the evaporator and heater assembly and is then discharged into the passenger compartment from the A/C or heater outlets. Since the compressor is not operating, the ability of the system to maintain the dial setting is restricted. It may be necessary to set the dial about 5° F cooler than normal to maintain comfort at 50° F to 70° F ambient temperatures.

In the upper part of the usable range of this setting, air is delivered from the air conditioning. Below 50° F, air is delivered in the BI-LEVEL mode, from both the air conditioner and heater outlets or from the heater outlet only in cold weather. System turn on will be delayed until the engine coolant temperature reaches about 130° F.

Lo

At this setting, operation is the same as that provided at "Auto" except that the blower operates at a fixed low speed.

Auto

At this setting, regulation of blower speed, discharge air temperature, air delivery mode,

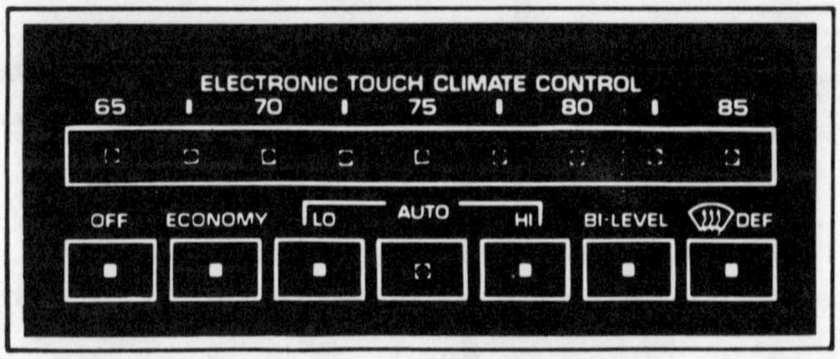

Fig. 12 Electronic Touch Climate Control control panel. 1980 Electra & LeSabre

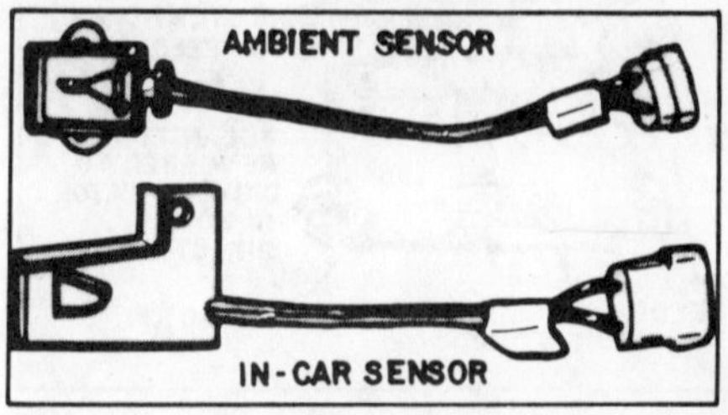

Fig. 13 In-car sensor and ambient sensor

system turn-on and compressor operation is controlled automatically to maintain the selected temperature on the dial.

Discharge air temperature will be controlled automatically between 45° F and 150° F. If discharge air temperature is above 90° F, the air will be delivered from the heater outlet. Below 75° F, air will be delivered from the A/C outlets. Between these temperatures, air will be delivered in the BI-LEVEL mode, from both the heater and A/C outlets.

If the vehicle interior is above 75° F, the system will turn on immediately when the vehicle is operated. If the vehicle interior is below 75° F, system turn-on will be delayed until the engine coolant temperature is about 130° F.

Hi

At this setting, operation is the same as that provided at "Auto" except that blower motor operation will be at a fixed maximum speed.

Bi-Level

At this setting, operation is the same as that provided at "Auto" except that bi-level air delivery will be provided. About half the air will be delivered through the heater outlet and half through the A/C outlet.

Def

At this setting, all the air is delivered to the windshield and the blower motor operates at a fixed high speed. Temperature regulation is maintained, and blower motor turn-on will be immediate, even with a cold engine. On 1979-80 Electra and LeSabre, there will be some air bleed to the heater outlets.

Temperature Dial

On 1979-80 models, a manual override is provided at full heat. Placing the dial at the highest temperature above 85° F will lock the system in "Max" heat. Air will be delivered from the outlets as selected by the mode lever setting.

1980 BUICK ELECTRA & LE SABRE ELECTRONIC TOUCH CLIMATE CONTROL

This system is a completely electrically controlled system that automatically controls in-vehicle temperatures regardless of outside temperature changes.

Temperature changes are accomplished by touching the control panel surface at the desired mode or temperature position, Fig. 12. An LED (Light Emitting Diode) illuminates at each position, indicating the mode and temperature in operation.

Electric motor driven actuators, controlled by the control panel, move each mode door and temperature door to the correct position, releasing properly heated or cooled air from the proper outlets.

If a mode door or temperature door does not move to its proper position, the temperature LED (for the temperature door or mode door) will flash off and on continuously. Touching the panel either at the position where the LED is flashing or at any other position will cause the control center to try again to position the mode and temperature doors into their proper positions. If this cannot be achieved within 20-30 seconds, the temperature LED's will again begin to flash, and the control center will shut off current to the affected motor(s) to prevent damage.

The extreme temperature position of 65° F will drive the temperature door to maximum cold, and the extreme temperature position of 85° F will drive the temperature door to maximum heat.

System Components

Control Panel

The control panel, Fig. 12, consists of a face plate and a control box which contains the electronic components. Neither component is serviceable, and should either require replacement, the control assembly must be replaced as a unit.

Low current switches under the surface of the face plate relay control selections to a microprocessor inside the control box which selects the desired mode and temperature control through power transistors. The microprocessor also stores the last control setting, and displays that setting each time the ignition switch is turned on.

Three connectors are attached to the control panel. The 22-wire connector supplies power and feed-back circuits to the 5 electric actuators, the 6-wire connector supplies battery voltage, in-car sensor circuits and backlighting power, the third connector supplies circuits to all A/C components in the engine compartment.

Engine Coolant Temperature Switch

This switch delays blower motor turn-on until the coolant is warm enough. This delay occurs only in the heater or bi-level mode. To provide immediate cooling in a hot vehicle, the delay switch is by-passed.

The coolant temperature switch is threaded into the right rear cylinder head.

Sensors

An in-car sensor and an ambient (outside temperature) sensor are used, Fig. 13. The in-

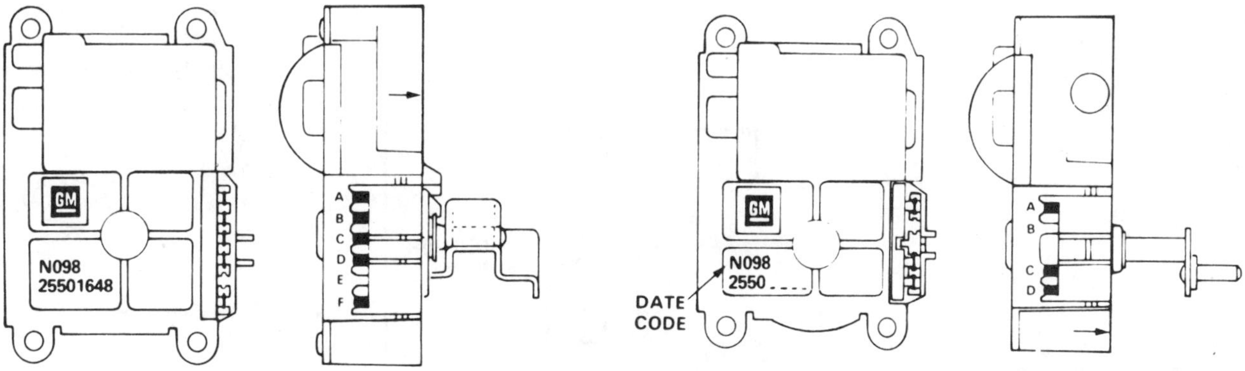

Fig. 14 Temperature door actuator (left) & mode actuator (right)

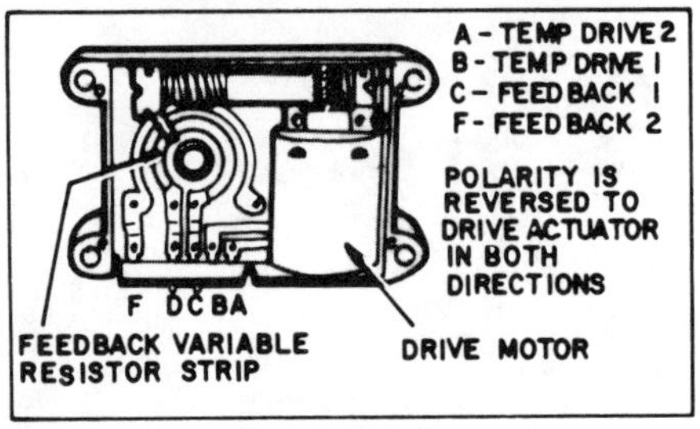

Fig. 15 Temperature door actuator resistor strip

car sensor is located on the instrument panel carrier and the ambient sensor is located under the air inlet screen. Both sensors are thermistors (temperature sensitive resistors) which provide the basic control inputs for the system.

These sensors are specific to this system and are identified with paper tape marked C-68, and violet wire leads.

Aspirator

The aspirator causes a small amount of air to flow from the cold air portion of the heater case. This air flow causes a slight vacuum effect at the end of the aspirator to which the in-car sensor hose is connected, resulting in air from the passenger compartment to be pulled through the hose and aspirator. In this manner, the in-car sensor will be sensing true in-car air and adjusting the signal to the microprocessor to compensate for in-vehicle temperature changes.

Temperature Door Actuator

This actuator, Fig. 14, is mounted on the right side of the A/C module, and contains a drive motor and a feedback variable resistor strip, Fig. 15. The resistor strip provides temperature door position information in terms of ohms back to the control. The control is programmed to provide a specific mode and blower speed for each temperature door position based on the resistance at each position.

If the temperature door actuator is misadjusted, or if the actuator arm is binding or obstructed, the control senses an incorrect actuator arm position and will try to reposition the door. If the actuator does not move, current to the actuator will be shut off, and the control panel temperature LED will flash.

Temperature door adjustment can be checked by touching 65° F, then waiting 60 seconds and touch 85° F and waiting 60 seconds. If the LEDs do not flash, door adjustment is

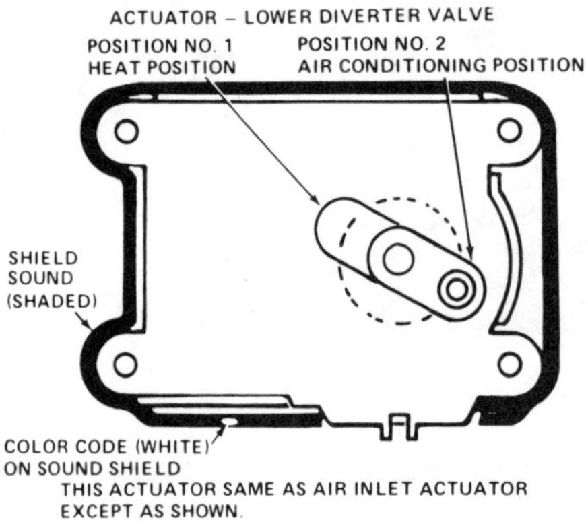

Fig. 16 Lower actuator

correct unless the door link is missing or disconnected. This procedure drives the temperature door actuator to extreme cold and then hot positions. An incorrectly adjusted door will bottom out and stop the actuator arm at an incorrect position.

Mode Actuators & Water Valve

The mode and water valve actuators, Figs. 16 thru 19, contain drive and feedback circuitry permitting them to be stopped at specific positions. Contacts in each actuator close a circuit through a fixed resistor located at each actuator arm position. To change positions, the control center supplies current and runs the actuator until the feedback circuit is closed through the appropriate resistor, then current is shut off. Through this feedback, the control center knows the exact position of each actuator.

If the feedback circuit becomes open through faulty connectors, dirty contacts in the actuator, broken wiring or the resistor valve is incorrect, the actuator will run for 20 seconds until shut off by the control panel timer. This condition will be indicated by a flashing LED on the control panel.

Defroster Actuator

This actuator, Fig. 20, is always in the sealed defrost position when the ignition is switched on. A timer circuit in the control center moves the door to bleed position after the ignition switch has been on for 30-60 seconds. The actuators can be identified by a stamped number on the arm. The recirc is stamped with a 2, the lower mode with a 7, the upper mode with a 6 and the def with a 3.

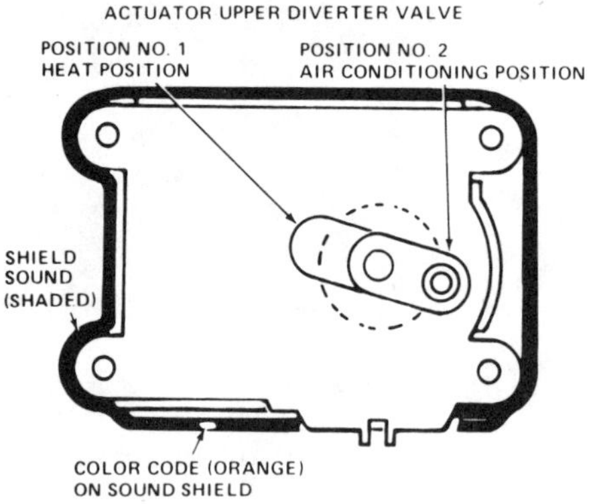

Fig. 17 Upper actuator

Blower & A/C Clutch Control

This control is located in the blower motor air stream in the engine compartment and converts low current signals from the control center to high current feed to the blower motor. Blower motor speeds are infinitely variable, controlled through a variable resistor strip on the temperature door actuator. The variable resistor strip acts as a rheostat feeding into the control panel which relays the proper blower motor speed signal to the blower control. A delay function provides for gradual increase or decrease of blower motor speeds including upon start-up. Also contained in the blower motor control unit is a power transistor circuit which engages the compressor clutch.

1977 BUICK CUSTOM-AIRE; 1977-78 PONTIAC AUTOMATIC TEMPERATURE CONTROL; 1977-80 OLDSMOBILE TEMPMATIC

This system provides semi-regulated controlled interior temperatures, maintains the set comfort level and requires little or no change in setting to compensate for changes in ambient conditions.

This system uses both in-car temperature

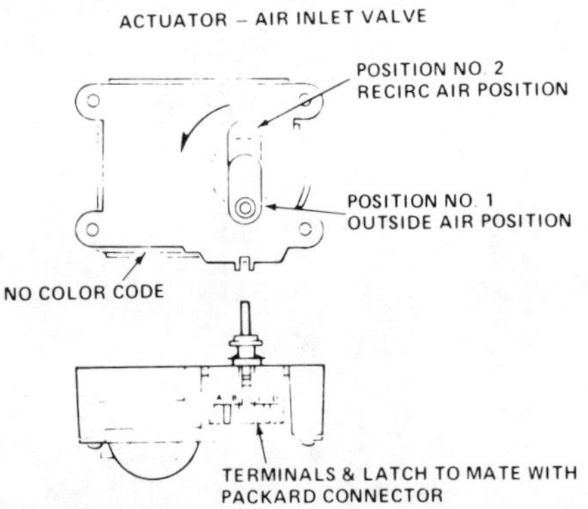

Fig. 18 Air inlet actuator

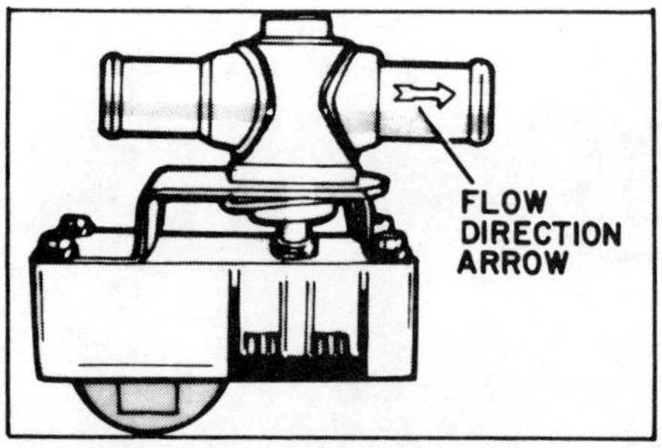

Fig. 19 Water valve and actuator

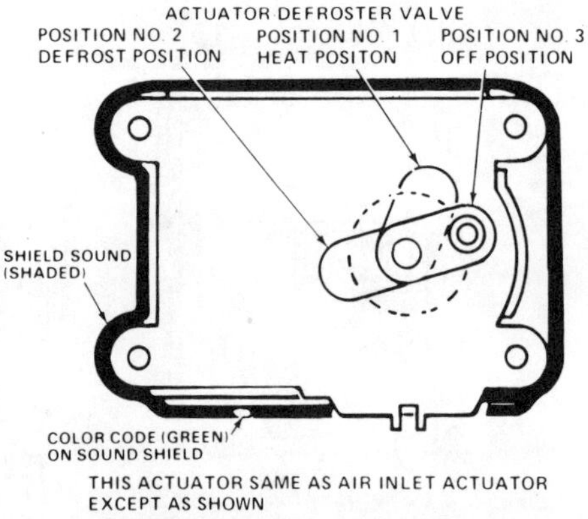

Fig. 20 Defroster actuator

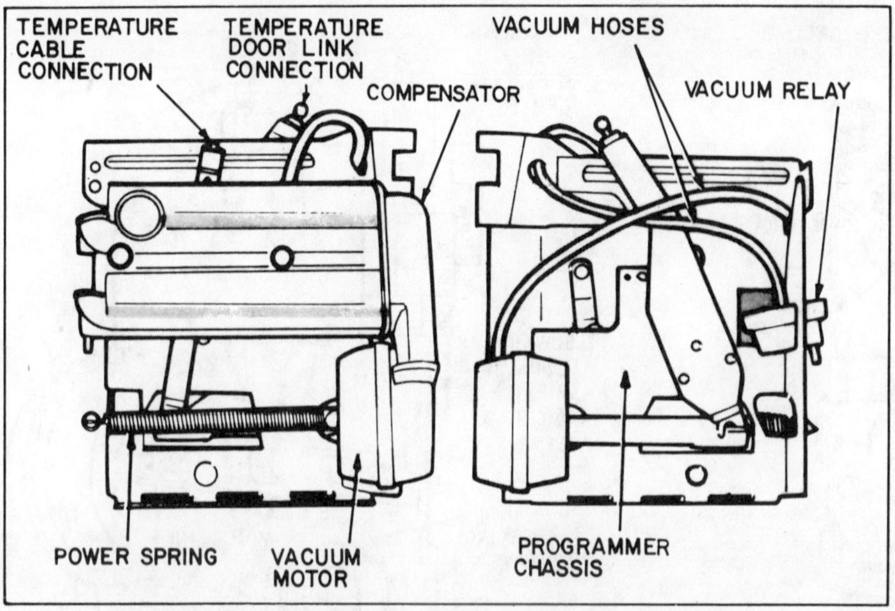

Fig. 21 Programmer assembly. 1977 Buick Century & Regal

and the temperature lever settings to signal the programmer. An air inlet on the instrument panel allows air from inside the vehicle to sense the temperature. The air sample, along with the temperature lever position signals the programmer. The programmer responds by controlling the temperature door which in turn regulates the air outlet temperature.

System Operation

Operation of this system is controlled by the settings on the control panel. The fan switch controls the amount of air flow by varying the blower motor speed. Interior comfort level is controlled by the position of the temperature lever. Overall operation of the system, including fan speed and comfort level control, is determined by the selector lever on the control panel. The selector lever has seven modes of system operation.

System Components

Control Panel

The control panel permits system control by means of a selector lever, a temperature control lever and a fan switch.

The selector panel permits selection of mode of operation (OFF, MAX, NORM, BI-LEVEL, VENT, HEATER, DEFROST) by positioning a master (select) electrical switch and a vacuum select valve. The vacuum select valve controls flow of hot water to the heater core and applies vacuum to or vents the vacuum diaphragms which control the movement of the air valves in the heater case and the air inlet assembly. The select (electrical) switch position controls the compressor and also provides power to the fan switch.

The temperature lever selects the desired comfort level (65-70-75-80-85) by positioning a cam on the control panel. The cam operates a bowden cable which references the programmer assembly. The programmer assembly then controls the heater case temperature valve to regulate air outlet temperature to maintain the selected comfort level.

The fan switch permits selection of four blower speeds.

Programmer

The programmer, Figs. 21 thru 25, consists of a compensator assembly, a vacuum motor, a vacuum checking relay, the output linkage and control linkage. The programmer is mounted on the A/C heater case and is connected to the temperature door by means of a threaded rod from the heater case. There are three input connections to the programmer.

1. The vacuum connection from the vacuum harness supplies engine vacuum relay.
2. The temperature control cable from the control panel attaches to the control lever of the programmer to position the programmer for the desired comfort level.
3. The in-car air source hose is connected to the compensator assembly. This provides a sample of in-car air temperature to the progammer, allowing regulation of air outlet temperature to compensate for temperature changes inside the vehicle.

Vacuum Motor

The vacuum motor is spring loaded to hold

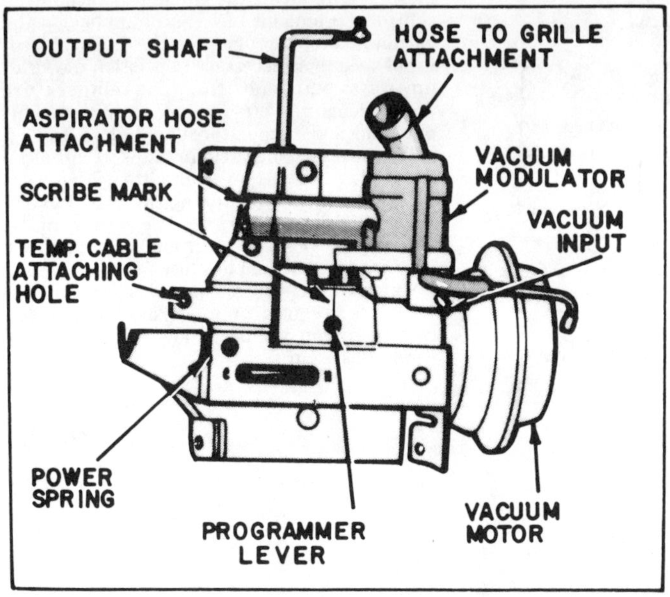

Fig. 22 Programmer assembly. 1977 Olds Cutlass & 1977-78 Toronado

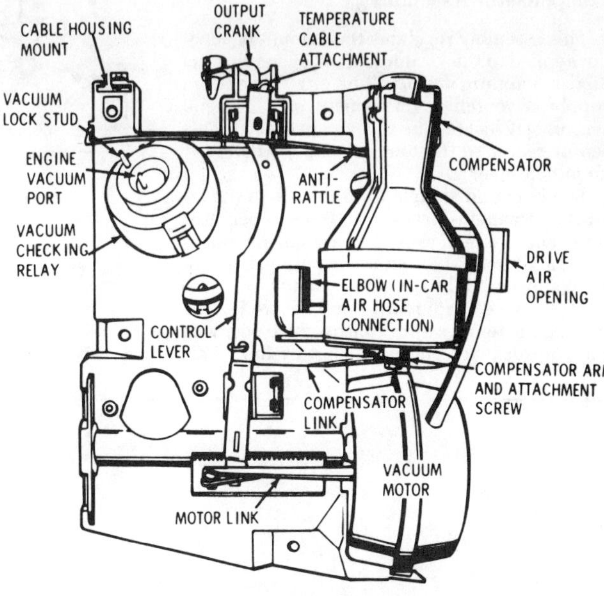

Fig. 23 Programmer assembly, front view. 1977 Buick Full Size, 1977-78 Pontiac & 1977-80 Olds 88, 98

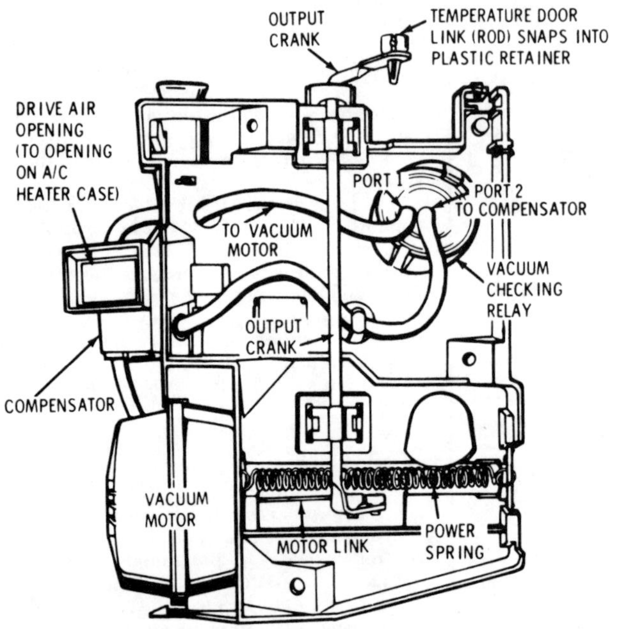

Fig. 24 Programmer assembly, rear view. 1977 Buick Full Size, 1977-78 Pontiac & 1977-80 Olds 88, 98

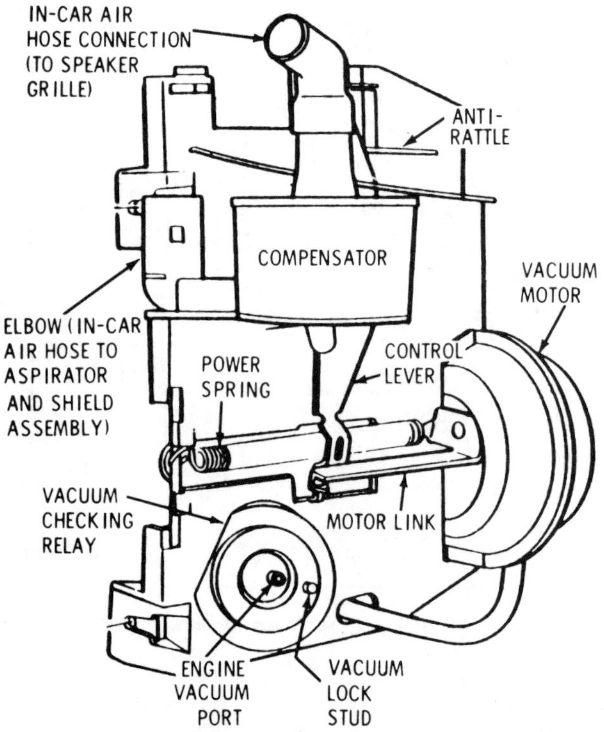

Fig. 25 Programmer assembly. 1979-80 Olds Toronado

the motor link out to the full cold (65° F) position.

As vacuum is applied to the motor, the link is pulled into the motor. An increase in vacuum causes the motor link to be pulled further into the motor, then the programmer output crank moves toward full heat and the temperature door will be positioned for warmer outlet air flow. Approximately 7½ inches of vacuum is required to pull the motor link all the way into the motor for full heat (85° F) operation. A venting valve in the vacuum motor prevents the motor from locking in the full heat position.

Compensator Assembly

This assembly regulates the amount of vacuum applied to the vacuum motor by venting the engine vacuum supply. The engine vacuum supply is vented by movement of a bi-metal sensor coil located in the compensator. This sensor reacts to the temperature of the in-car air passing over it.

The in-car air sample from the top of the instrument panel is drawn into the compensator by a venturi effect, caused by a small amount of air being forced through a separate passage in the compensator by the blower motor.

When the control panel temperature lever is moved, the temperature control cable positions the control lever and compensator link. This changes the sensor coil tension to provide control of comfort level at the selected temperature lever position.

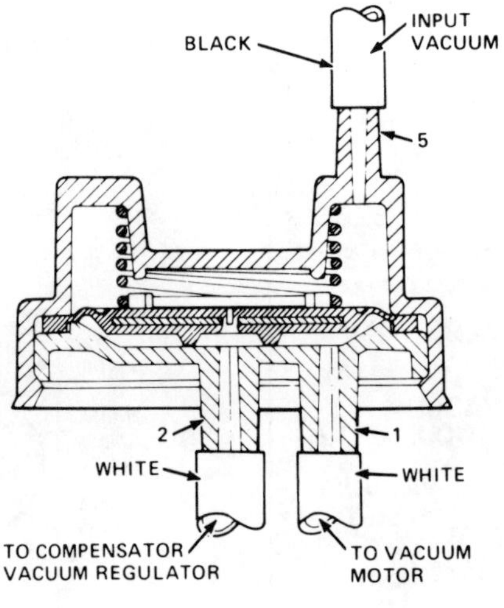

Fig. 26 Vacuum checking relay

Vacuum Checking Relay

Engine vacuum is supplied to a port (on front side) of the vacuum checking relay, Fig. 26, through the vacuum hose harness. The engine vacuum port has a locking feature to maintain vacuum. Port No. 2 is constantly bleeding to atmosphere (reducing the amount of vacuum) to control the amount of vacuum applied to the vacuum motor through port No. 1. This bleeding of vacuum is necessary to position the vacuum motor which positions the temperature door through the programmer motor link, output crank and temperature door link.

Whenever engine vacuum drops to within 1½ inch or less of control vacuum, the spring-loaded diaphragm is pushed against the base of ports 1 and 2, sealing port 1. The vacuum motor is locked in position until engine is restored. When engine vacuum increases, the vacuum relay opens and system control resumes. This prevents the programmer from going to full cold operation whenever low engine vacuum conditions occur.

Maintenance & Refrigerant Component Diagnosis

GENERAL PRECAUTIONS

SAFETY

The Freon refrigerant used in car air conditioners is also known as R-12 or F-12. It is colorless and odorless both as a gas and a liquid. Since it boils (vaporizes) at —21.7 deg. F., it will usually be in a vapor state when being handled in a repair shop. But if a portion of the liquid coolant should come in contact with the hands or face, remember that its temperature momentarily will be at least 22 deg. below zero!

Protective goggles should be worn when opening any refrigerant lines. If liquid coolant does touch the eyes, bathe the eyes quickly in cold water. Then apply a bland disinfectant oil to the eyes. See an eye doctor.

When checking a system for leaks with a torch type leak detector, do not breathe the vapors coming from the flame. Do not discharge refrigerant in the area of a live flame. A poisonous phosgene gas is produced when R-12 or F-12 is burned. While the small amount of this gas produced by a leak detector is not harmful unless inhaled directly at the flame, the quantity of refrigerant released into the air when a system is purged can be extremely dangerous if allowed to come in contact with an open flame. Thus, when purging a system, be sure that the discharge hose is routed to a well ventilated place where no flame is present. Under these conditions the refrigerant will be quickly dissipated into the surrounding air.

Never allow the temperature of refrigerant drums to exceed 125° F. The resultant increase in temperature will cause a corresponding increase in pressure which may cause the safety plug to release or the drum to burst.

If it is necessary to heat a drum of refrigerant when charging a system, the drum should be placed in water that is no hotter than 125° F. Never use a blowtorch, or other open flame. If possible, a pressure release mechanism should be attached before the drum is heated.

CLEANLINESS

Air conditioning systems are extremely sensitive to moisture and dirt. The importance of clean working conditions cannot be overemphasized as the smallest particle of foreign matter in an air conditioning system will contaminate the refrigerant, causing rust, ice or damage to the compressor. For this reason, all replacement parts are sold in vacuum sealed containers and should not be opened until they are to be installed in the system. If, for any reason, a part has been removed from its container for any length of time, the part must be completely flushed using only R-12 or F-12 to remove any dust or moisture that may have accumulated during storage. In cases as collision repairs, where the system has been open for any length of time, the entire system must be purged completely and a new receiver-dehydrator must be installed because the element of the existing unit will have become saturated and will be unable to remove any moisture from the system once the system is recharged.

When making gauge connections, purge the gauge lines first by cracking the charging valve and allowing a small amount of refrigerant to flow through the lines, then connect the lines immediately.

Cleanliness is especially important when servicing compressors because of the very close tolerances used in these units. Consequently, repairs to the compressor itself should not be attempted unless all proper tools are at hand and a virtually spotless work area is provided.

General Service

Use care when disconnecting or connecting refrigerant lines; always use a back-up wrench and be careful not to over-tighten any connection. Over-tightening will result in line and flare seat distortion and a system leak.

When making pressure checks on systems having service valves, be sure valve is in the intermediate position. If turned in too far, the hose connection will be closed, a position used for isolating the compressor. When closing the gauge port, do not over-tighten the valve or damage to the seat will result.

After disconnecting gauge lines, check the valve areas to be sure service valves are correctly seated and Schraeder valves, if used, are not leaking.

EXERCISE SYSTEM

An important fact most car owners ignore is that car coolers must be used periodically. Car manufacturers are emphatic on this point. They caution that when the air conditioner is not used regularly, particularly during cold months, it should be turned on for a few minutes once every two or three weeks while the engine is running. This keeps the system in good operating condition.

Checking out the system for the effects of disuse before the onset of summer is one of the most important aspects of car cooler servicing.

First on the list is to clean out the condenser core, mounted in all cases at the front of the car's radiator. All obstructions, such as leaves, bugs, dirt, and the like, must be removed, as these will reduce heat transfer and impair the efficiency of the system. Make sure the space between the condenser and the radiator also is free of foreign matter.

Make certain the evaporator water drain is open. Certain systems have two evaporators, one in the engine compartment and one in the trunk. The evaporator cools and dehumidifies the air before it enters the car; there the refrigerant is changed from a liquid to a vapor. As the core cools the air, moisture condenses on it but the moisture is prevented from collecting in the evaporator by means of the water drain.

OIL LEVEL CHECK

Chrysler RV-2

1. Slowly discharge refrigerant system, Fig. 1, then near completion of discharge, flush oil dipstick with freon. This will ensure dipstick is clean and approximately the same temperature as oil in compressor sump.

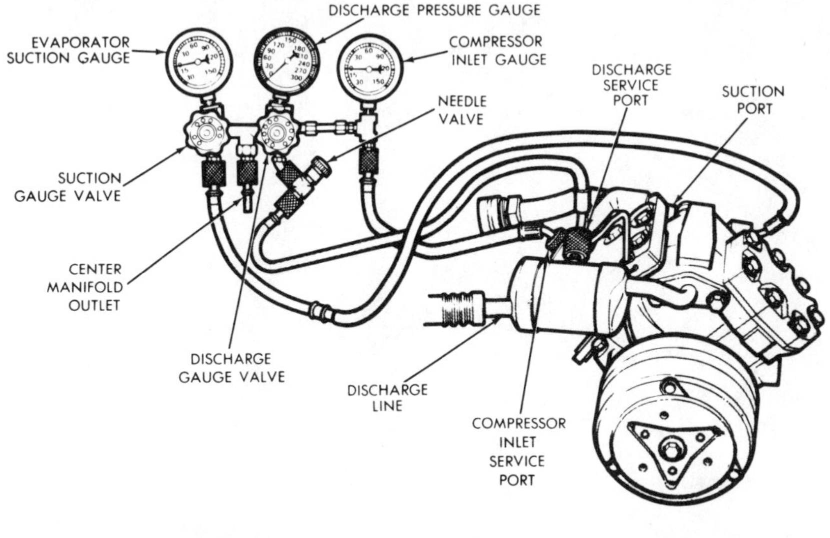

Fig. 1 Discharging system using manifold gauge set. RV-2 compressor

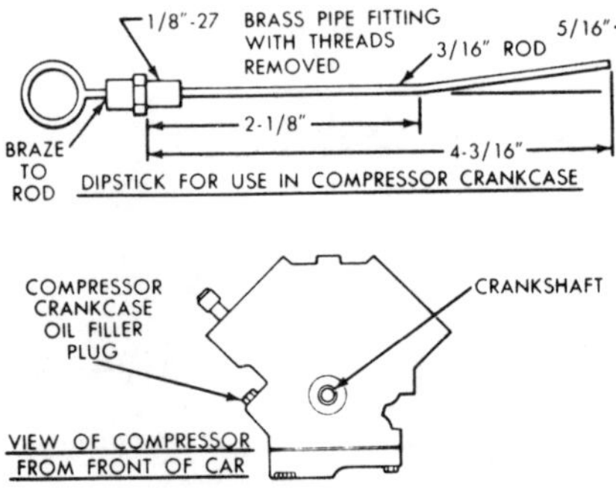

Fig. 2 RV-2 compressor oil level dipstick fabrication & filler plug location

2. Carefully remove compressor sump oil filler plug, then insert dipstick into hole until it bottoms in sump, Fig. 2.

NOTE: When removing compressor oil sump plug a face shield should be worn. Refrigerant dissolved in compressor oil could cause oil to percolate out through filler plug opening. Oil level should be checked only after refrigerant has boiled off and oil surface has stabilized.

3. Remove dipstick and measure oil level. Refer to "A/C Specifications" and add refrigerant as necessary to bring oil level within limits.
4. If any of the following components are to be replaced, it will be necessary to add additional oil to the system.
   Evaporator .... 2 ounces
   Condenser .... 1 ounce
   Filter-Drier .... 1 ounce
5. Install oil filler plug, then evacuate and recharge system.

Tecumseh & York

NOTE: Compressor oil level should be checked only if a refrigerant component is to be replaced or if system has developed a leak. Oil level should be checked with compressor installed on vehicle.

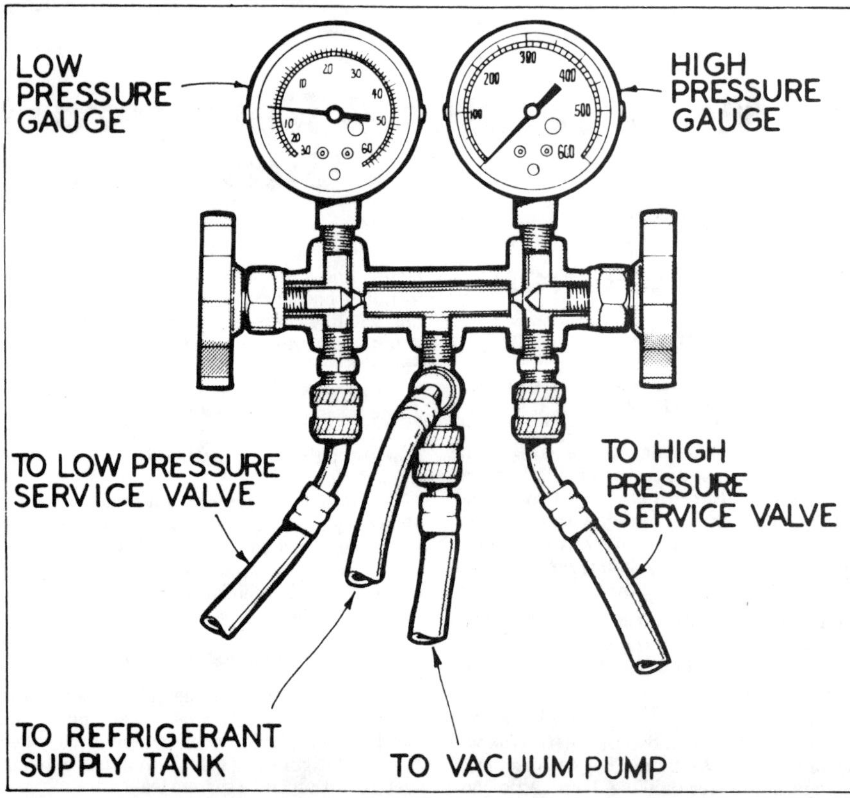

Fig. 3 Manifold gauge set connections. Frigidaire, Nippondenso, Tecumseh and York compressors

1. Connect manifold gauge set, Fig. 3, then operate system for approximately 10 minutes or until system pressures stabilize. This will allow oil in the system to return to compressor sump.
2. On American Motors models, isolate compressor as outlined later.
3. On Ford Motor Co. models, discharge entire refrigerant system.
4. On all models, slowly loosen compressor oil filler plug to relieve any internal pressures in compressor.

NOTE: A face shield should be worn when loosening or removing oil filler plug.

5. Insert clean dipstick into filler plug hole until it bottoms in sump, Figs. 4 and 5.

NOTE: On York compressors ensure keyway in shaft faces head of compressor before checking oil level.

6. Remove dipstick and measure oil level. Refer to "A/C Data Table" and add refrigerant oil as necessary. Install filler plug and "O" ring.

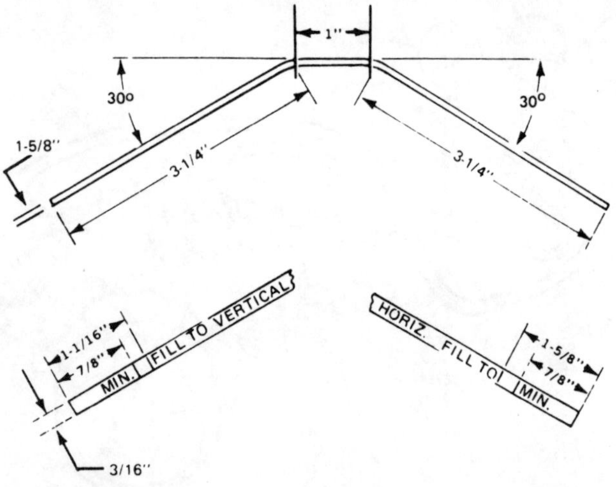

Fig. 4 Tecumseh compressor oil level dipstick fabrication

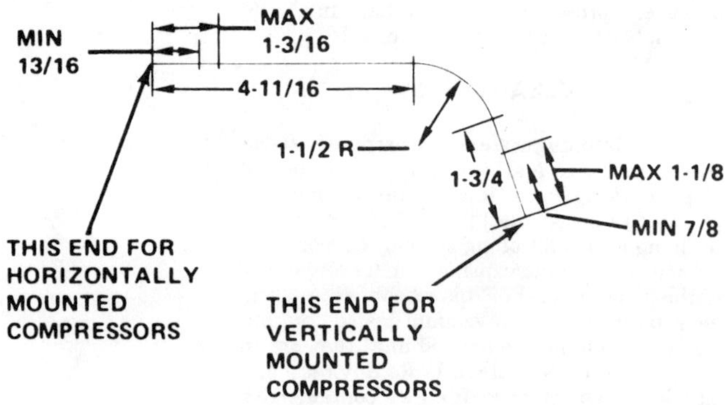

Fig. 5 York compressor oil level dipstick fabrication

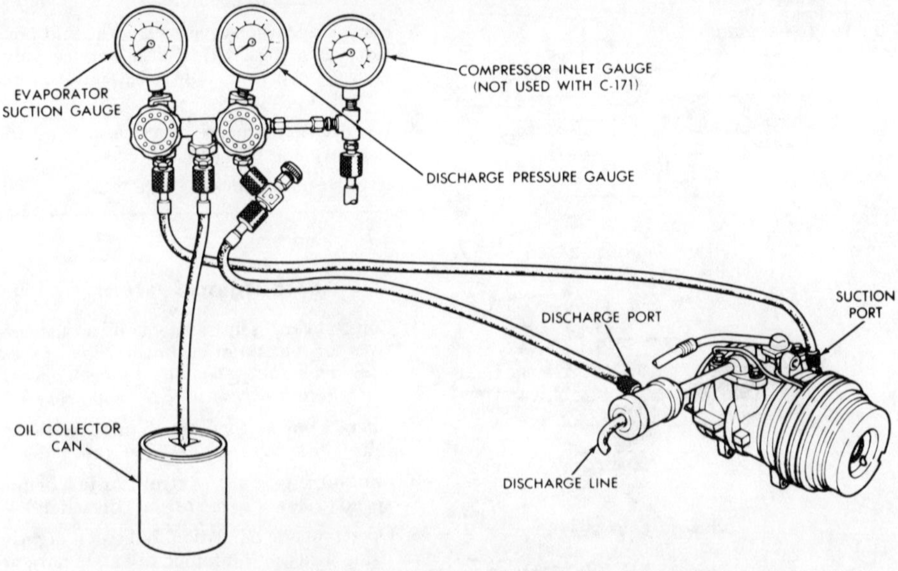

Fig. 6 Discharging system using manifold gauge set. C-171 compressor

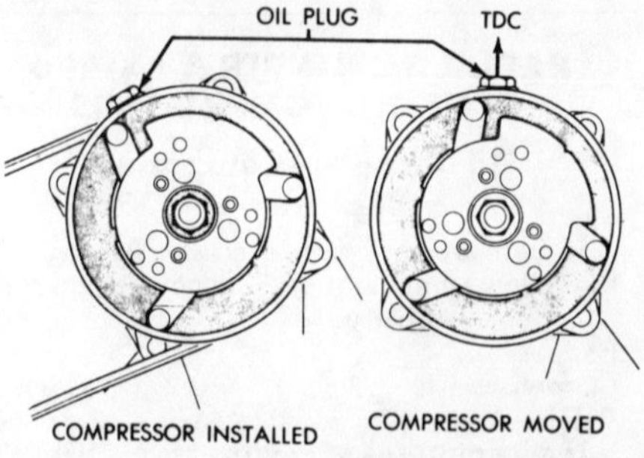

Fig. 7 Positioning compressor oil filler hole. Sankyo compressor

7. On American Motors models, purge air from compressor.
8. On Ford Motor Co. models, evacuate and recharge system.

C-171

1. Discharge system, Fig. 6, and remove suction and discharge lines from compressor.
2. Remove compressor assembly from vehicle then drain compressor oil from suction and discharge ports.
3. Add 9 oz. of refrigerant oil to compressor through suction port.
4. Install compressor assembly and using new gaskets, attach suction and discharge lines.
5. Evacuate and charge system.

   If any of the following components are to be replaced, it will be necessary to add oil to the system.

Evaporator	2 ounces
Condenser	1 ounce
Filter-Drier	1 ounce

Sankyo

1. After loosening or removing compressor mounting bolts, rotate compressor so that oil filler hole is at 12 o'clock, Fig. 7.
2. Clean filler plug and surrounding area of contaminants then slowly remove plug.
3. Rotate the front plate hub so that the notch in the center of the lobe is indexed 110 degrees from 6 o'clock on the compressor, Fig. 8. At this point the ball end of the top piston rod lines up with the oil fill port, Fig. 9.
4. Insert dipstick tool C-4504 from upper right to lower left until stop contacts filler port surface, Fig. 10.
5. The dipstick is marked in one ounce increments. Add oil as necessary to bring level to 3-4 ounces of oil.

CAUTION: When removing compressor oil sump plug a face shield should be worn. Refrigerant dissolved in compressor oil could cause oil to percolate out through filler plug opening. Oil level should be checked only after refrigerant has boiled off and oil surface has stabilized.

Nippondenso

1. Operate compressor for ten to fifteen minutes at maximum cooling to stabilize system.
2. Discharge system and remove compressor from vehicle. Remove drain plug and measure the amount of oil drained from the compressor.
3. If less than three ounces of oil were drained from compressor, add six ounces of refrigerant oil to drained service replacement compressor.
4. If between three and six ounces of oil were drained from compressor, the oil is properly distributed throughout the system. Add an equal amount of refrigerant oil to drained service replacement compressor.
5. Never add more than six ounces of refrigerant oil to a replacement compressor. Never install a replacement compressor containing more than six ounces of refrigerant oil.

   Clean refrigerant oil should be added to the following replacement components:

Evaporator	3 ounces
Condenser	1 ounce
Accumulator	1 ounce

FRIGIDAIRE COMPRESSORS

NOTE: These compressors have no provision to accurately check oil level on the car. Periodic inspection of oil level has, therefore, been eliminated from installations using these compressors. Oil level check performed with the compressor removed from the vehicle is to be made only under conditions of severe oil loss caused by a compressor oil leak, broken refrigerant hose, or a rupture from crash damage. The system is designed to hold the quantity of refrigeration oil listed in the *A/C Data Table.* The affinity of Freon 12 and refrigeration oil and the

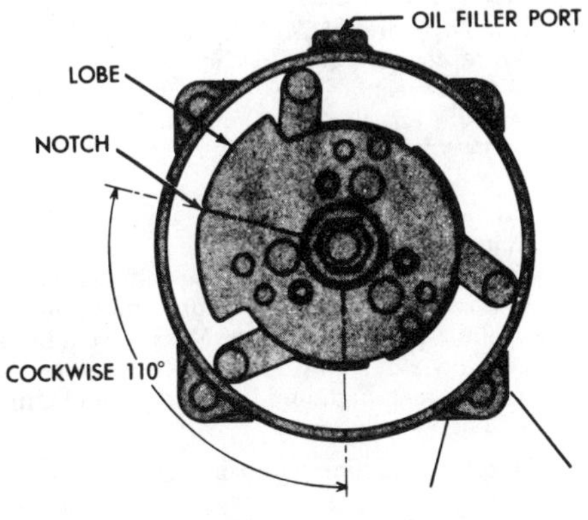

Fig. 8 Front plate hub aligned for oil level check. Sankyo compressor

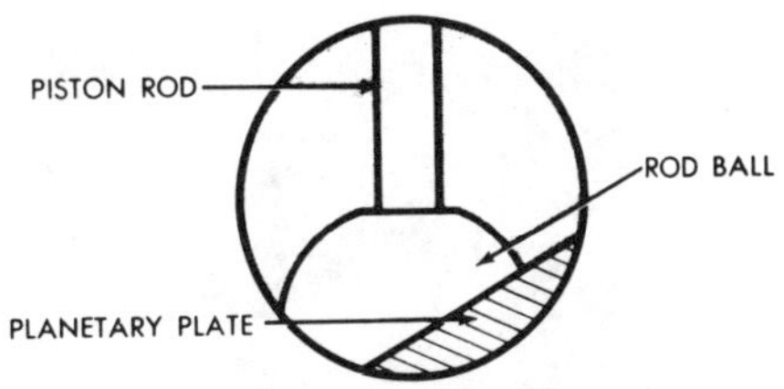

Fig. 9 Piston rod correctly aligned with oil filler port. Sankyo compressor

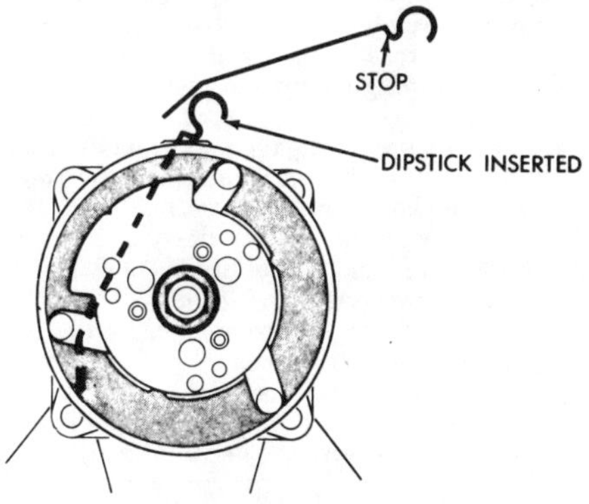

Fig. 10 Dipstick properly inserted. Sankyo compressor

design of the compressor will prevent a full oil charge being contained in the compressor.

RADIAL 4 CYLINDER & AXIAL 6 CYLINDER COMPRESSORS

Oil Charge—Component Replacement

If there are no signs of excessive leakage, add the following amount of oil depending on component to be replaced:

Evaporator .... 3 ounces
Condenser .... 1 ounce
VIR .... 1 ounce

If accumulator or compressor is to be replaced, drain oil from component to be replaced and measure; then add same amount of new oil to replacement component plus one additional ounce.

NOTE: The radial 4 cylinder compressor does not have an oil sump.

Oil Charge—Leak Condition

On models with axial 6 cylinder compressor, both accumulator and compressor must be removed, oil drained and measured in cases of excessive oil leakage. If oil recovered is 4 ounces or more, add same amount of new refrigerant oil to system. If amount of oil recovered is less than 4 ounces, add 6 ounces of new refrigerant oil to system.

On models with radial 4 cylinder compressor, it will only be necessary to drain and measure oil from accumulator assembly in cases of excessive oil leakage. The radial 4 cylinder compressor does not have an oil sump, therefore it is not necessary to remove compressor in cases of oil leakage. If amount of oil drained is less than 2 ounces, add 2 ounces of new refrigerant oil to system.

NOTE: On all models, if accumulator is replaced, one additional ounce of oil must be added to replace amount captured by desiccant in old accumulator.

SYSTEM DISCHARGING & EVACUATION

Service Note: American Motors

Isolating Compressor From System

It is not necessary to discharge system for compressor removal. Compressor can be isolated from system, eliminating need for recharging when performing compressor service. Proceed as follows:

1. Connect pressure gauge and manifold set, then close both gauge hand valves and mid-position both service valves, Fig. 11.
2. Start engine and operate air conditioning.
3. Slowly turn suction service valve clockwise toward front-seated position. When suction pressure drops to zero or less, stop engine and compressor and quickly finish front-seating suction service valve.
4. Slowly loosen oil sump filler plug to relieve any internal pressures in compressor.

NOTE: A face shield should be worn when loosening oil filler plug.

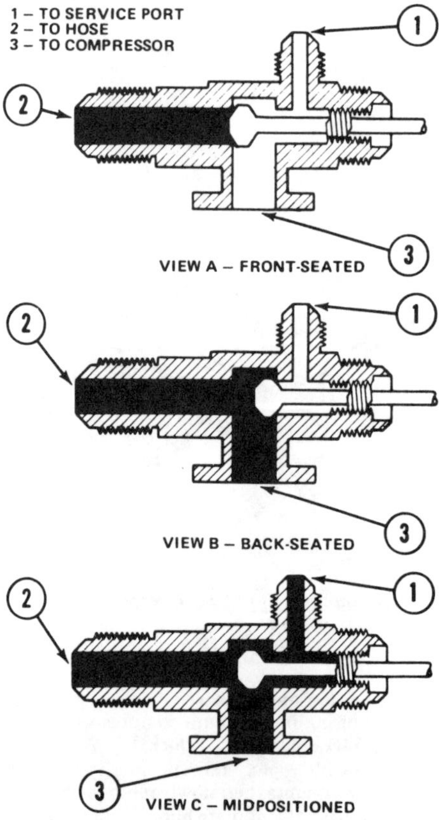

Fig. 11 Service valve positions. (Typical)

5. Compressor is now isolated from system, service valve can now be removed from compressor.

NOTE: After checking oil level or installing compressor refer to following procedure to purge compressor of air.

Purging Compressor Of Air

Compressor must be purged of air whenever it has been isolated from system for oil level check or compressor service. Proceed as follows:

1. Connect service valve and lines to compressor, then cap service gauge ports on both service valves.
2. Back-seat suction valve to allow system refrigerant to enter compressor.
3. Place discharge service valve in mid-position, then loosen discharge service valve gauge port cap to allow refrigerant to force air out of compressor.
4. Back-seat discharge service valve and tighten gauge port cap
5. Remove manifold and gauge set.

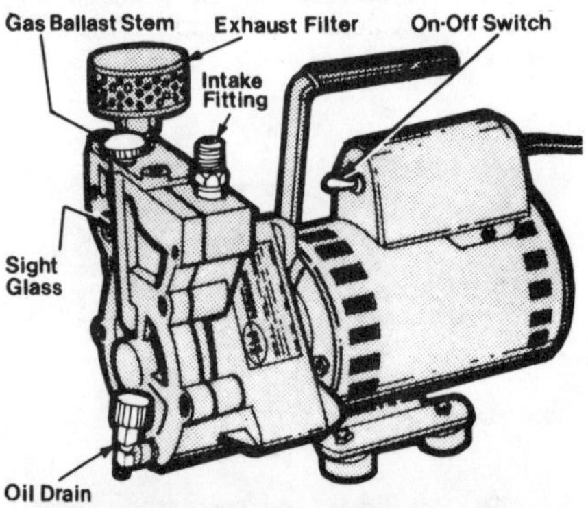

Fig. 12 Robinaire vacuum pump

Discharging System

1. Connect gauges into system and adjust controls for maximum cooling. *This is necessary, when the system has not been operating to return excess oil to the compressor.*
2. Operate engine for 10 to 15 minutes to stabilize the system at 1500-1750 rpm.
3. Shut off engine and controls. Adjust engine speed to slow idle to prevent "dieseling"
4. Open low side hand manifold valve slightly, using a rag or container to catch refrigerant. *Do not discharge the refrigerant near an open flame as a toxic gas (phosgene) can result.*
5. Open high side manifold hand valve. *Open hand valves only enough to bleed refrigerant from system. Too rapid purging will draw excessive oil from compressor and system.*
6. Close gauge manifold shut-off valves when refrigerant ceases to bleed from hose. Both gauges will read zero psi.

Evacuate System With Vacuum Pump

Vacuum pumps suitable for removing air and moisture from A/C systems are commercially available such as the Robinair pump, Fig. 12. A specification for system pump-down used here is 28-29½" vacuum. This reading can be attained at or near sea level only. For each 1000 feet of altitude this operation is being performed, the reading will be 1" vacuum higher. As an example, at 5000 feet evaluation, only 23-24½" of vacuum can be obtained.

NOTE: The system must be completely discharged before it can be evacuated. Damage to the vacuum pump will result if pressurized refrigerant is allowed to enter.

NOTE: On General Motors vehicles, disconnect wire connector from thermal limiter assembly. Connect a jumper wire between feed wire and clutch coil wire, if equipped.

1. Connect vacuum pump to gauge manifold. With gauges connected into system, remove cap from vacuum hose connector. Install center hose from gauge manifold to vacuum pump connector. Midposition high and low side compressor service valves (if used). Open high and low side gauge manifold hand valves.
2. Operate vacuum pump a minimum of 30 minutes for air and moisture removal. Watch compound gauge that system pumps down into a vacuum. System will reach 28-29½" vacuum in not over 5 min-

utes. If system does not pump down, check all connections and leak-test if necessary.
3. Close gauge manifold hand valves and shut off vacuum pump.
4. Check ability of system to hold vacuum. Watch compound gauge to see that gauge does not rise at a faster rate than 1" vacuum every 4 or 5 minutes. If compound gauge rises at too rapid a rate, install partial charge and leak-test. Then discharge system as outlined above.
5. If system holds vacuum, charge system with refrigerant.

Evacuate System Using Charging Station

A vacuum pump is built into the charging station and is constructed to withstand repeated and prolonged use without damage. Complete moisture removal from the system if possible only with a vacuum pump constructed for the purpose.

NOTE: The system must be completely discharged before it can be evacuated. Damage to the vacuum pump will result if pressurized refrigerant is allowed to enter.

NOTE: On General Motors vehicles, disconnect wire connector from thermal limiter assembly. Connect a jumper wire between feed wire and compressor clutch coil wire, if equipped.

1. Operate vacuum pump. Connect hose to vacuum pump if system was purged through charging station.
2. Open high and low side gauge valves of charging station.
3. Connect station into 110-volt current.
4. Engage "Off-On" switch to vacuum pump according to directions of specific station being used.
5. System should pump down into a 28-29½" vacuum in not more than 5 minutes. If system fails to meet this specification, a leak is present and must be repaired before continuing.
6. Operate pump a minimum of 30 minutes to remove all air and moisture.

NOTE: Do not use alcohol in the system to remove moisture.

7. Close high and low side gauge valves and turn off pump.
8. Check ability of system to hold vacuum by watching compound gauge to see that it does not rise at a rate higher than 1" of vacuum every 4 or 5 minutes: If rise rate is not within specifications, repair system as necessary. If rise rate is within specifications, charge system with refrigerant.

CHARGING THE SYSTEM

Using 15-Ounce Containers

To prevent waste and to aid in more accurate charging, refrigerant manufacturers package refrigerant in 15-ounce cans. The small containers are advantageous to small shops doing only a limited amount of A/C service work. The small containers are handled in the same manner as the larger drums except care must be taken not to overheat the cans because of the danger of explosion.

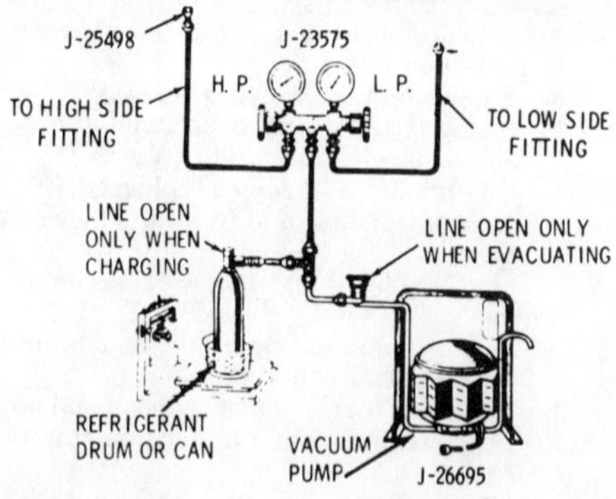

Fig.13 Connections for evacuating & charging refrigerant system. (Typical)

NOTE: Do not use 15 ounce cans to charge into high pressure side of system or into a system that is at high temperature. High system pressures could cause charging can to explode.

NOTE: On General Motors vehicles, disconnect wire connector from thermal limiter assembly. Connect a jumper wire between feed wire and compressor clutch coil wire, if equipped.

1. Install "Fitz-All" valve to container(s). The Fitz-All valve is available for single can or multi cans. Whichever is used, preliminary installation to the can(s) is the same.
2. Close shut-off valve of Fitz-All valve. Pierce can with mechanism which is part of Fitz-All valve.
3. With the system pumped-down, install charging hose to Fitz-All valve. Loosen charging hose at center connector on gauge manifold.
4. Crack Fitz-All shut-off valve to purge air from charging hose. Tighten charging hose connection on gauge manifold and close shut-off valve.
5. Fully open manifold valves and place cans of refrigerant into a pan containing water heated to 125° F.

NOTE: Cooling the condenser with a large fan will speed the charging process.

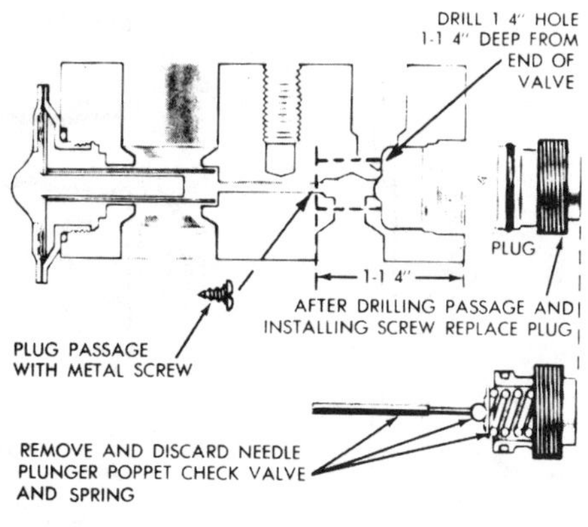

Fig.14 Reworking discarded H valve

6. Refer to the A/C specifications to determine capacity of system being serviced. Do not overfill. Charge system as follows: Close high side gauge manifold hand valve. Start engine and adjust idle to 1500-1750 rpm. Set A/C controls for maximum cooling. Open low side gauge manifold hand valve to allow refrigerant to be drawn into system. If single containers are used, replace as necessary.
7. After system is fully charged, test system for leaks and check operation. Refer to "System Testing" section for specifications and pressure of particular units.

Using Charging Station

NOTE: On General Motors vehicles, disconnect wire connector from thermal limiter assembly. Connect a jumper wire between feed wire and compressor clutch coil wire, if equipped.

Most stations contain a charging cylinder into which the exact amount of refrigerant required for the system being serviced may be placed while system pump-down is being performed. The refrigerant charging cylinder in the station is heated to the correct temperature to insure proper refrigerant flow to all parts of the system as a gas during the charging operations, Fig. 13. Following correct evacuation procedure as to length of time for complete moisture removal, the vacuum pump will so efficiently pump-down the system that opening the correct valves will completely charge the system from the high side, and the use of the compressor in the charging operation will not be required.

1. Prepare charging cylinder for filling: Open storage drum valve. Close all valves on station. Read storage tank gauge pressure. Rotate dial shroud on charging cylinder to correlate with pressure on gauge. Open cylinder fill valve.
2. Fill charging cylinder: Determine system capacity from *A/C Data Table.* Intermittently open and close pressure relief valve. *When pressure relief valve opens, Freon will enter cylinder and boil. Closing the valve will increase pressure on the Freon, changing it to a liquid to stabilize the refrigerant in the sight glass. Fill to specified level in sight glass and close pressure relief valve.*
3. Charge system: With gauges connected into system, open refrigerant control valve. Open high pressure valve. After full charge has entered system, close refrigerant control valve and high pressure valve. Reseat suction and discharge service valves and remove charging station hoses.
4. Test system for leaks and check operation. Refer to "System Testing" section for specifications and pressures of particular units.

FLUSHING A/C SYSTEM W/H TYPE EXPANSION VALVE

Chrysler Corp.

1. Rework discarded H valve as shown in Fig. 14.
2. Discharge system, then remove H valve and install reworked H valve.

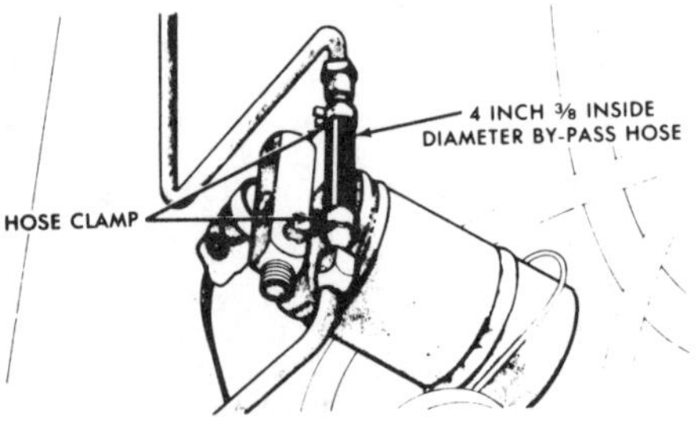

Fig. 15 Filter drier by-pass hose installation

3. Disconnect lines from filter drier, then install bypass as shown in Fig. 15.
4. Disconnect suction line from compressor and line over fender into collector can.
5. Using tool No. C-4089, remove discharge line valve core, then connect charging hose to tool, Fig. 16.
6. Slowly open refrigerant container valve and allow 5 pounds minimum of refrigerant through system and out open end of suction line.
7. After system is completely flushed, change oil in compressor sump.
8. Connect compressor suction line, then remove expansion valve bypass and install H valve, using new "O" rings.
9. Replace discharge line muffler service valve core.
10. Remove filter-drier bypass and install filter-drier using new "O" rings, then recharge system.

ACCUMULATOR

General Motors

1. Accumulator assembly, Fig. 17, should only be replaced when:
   a. Expansion tube screen is plugged.
   b. Physical perforation of accumulator resulting in a leak is found.
   c. Compressor inlet screen is plugged.
   d. Evaporator fails due to internal corrosion.
2. Do not replace accumulator assembly when:
   a. A dent not resulting in a leak is found on outer shell of accumulator.
   b. Vehicle is involved in a collision and no physical perforation of accumulator is found.

RECEIVER-DRIER (DEHYDRATOR)

American Motors & Ford Motor Co.

The receiver drier can not be repaired; it must be replaced if it leaks, is restricted or clogged, or has been punctured in a collision. In addition to filtering and storing refrigerant, the receiver drier also collects moisture; therefore, if contamination of system components is found, the drier should be replaced.

THERMOSTATIC SWITCH TEST

General Motors

1. Install gauge set, then position control lever in Norm, Lo blower and operate engine at 1500 rpm on all models except Oldsmobile.

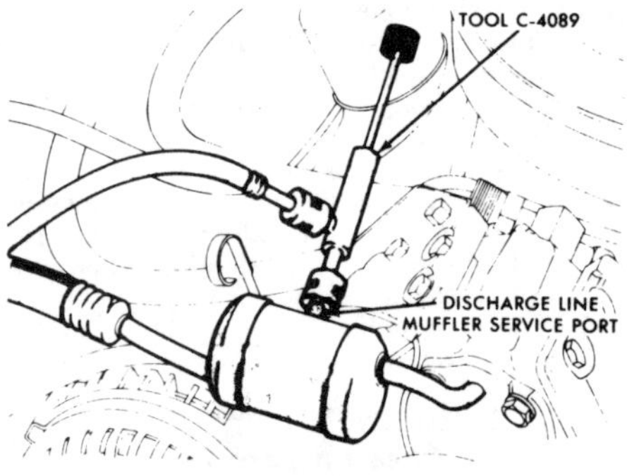

Fig. 16 Tool C-4089 installed on discharge line muffler service port

On Oldsmobile models, operate engine at 2000 rpm.

a. Thermostatic switch, Fig. 18, should cycle compressor off in 5 minutes or less. If compressor continues to operate and accumulator becomes frosted, check for loose clamps around capillary tube to evaporator inlet pipe or a defective switch.
b. If compressor does not operate and battery voltage is available at one switch terminal with A/C on, thermostatic switch is defective and should be replaced.

NOTE: If after completing the above procedure switch seems to operate properly but outlet air temperature is abnormal, some switches may be adjusted. Refer to Part 2 of the "Unit Replacement" section for correct adjustment procedure on 1974-76 Vega and 1976 Chevette.

THERMOSTATIC DE-ICING SWITCH TEST

1974-77 Maverick & Comet

1. Connect an ohmmeter to thermostatic switch terminals to ensure switch contacts are closed. Placed capillary tube into a solution of ice, salt and water. Solution temperature should be 20°F or lower.

NOTE: Ensure solution does not enter the control. Test can also be performed by spraying capillary tube with Refrigerant-12.

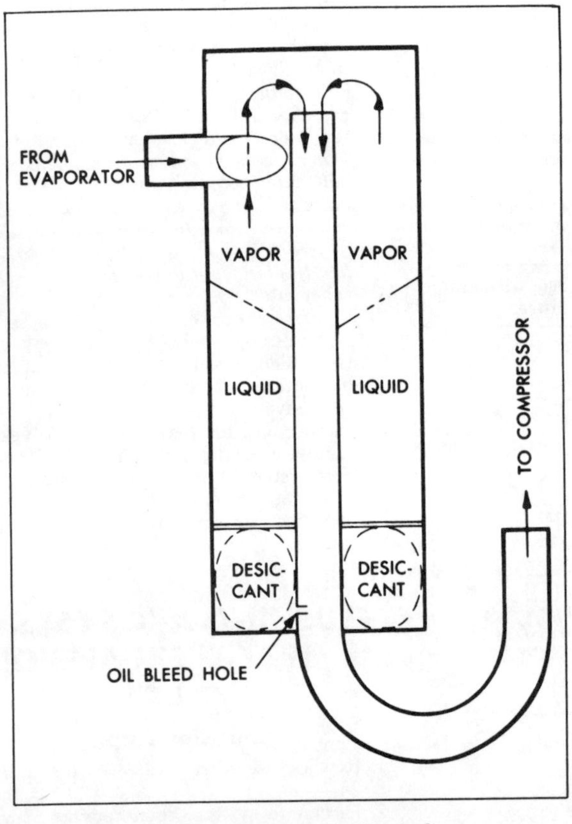

Fig. 17 Accumulator (Typical)

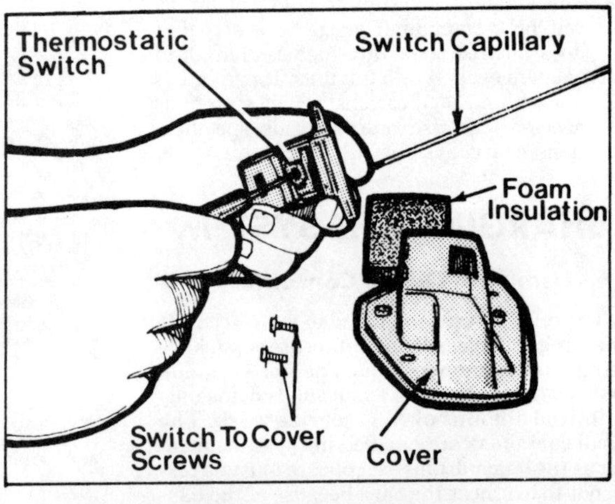

Fig. 18 Thermostatic switch (Typical)

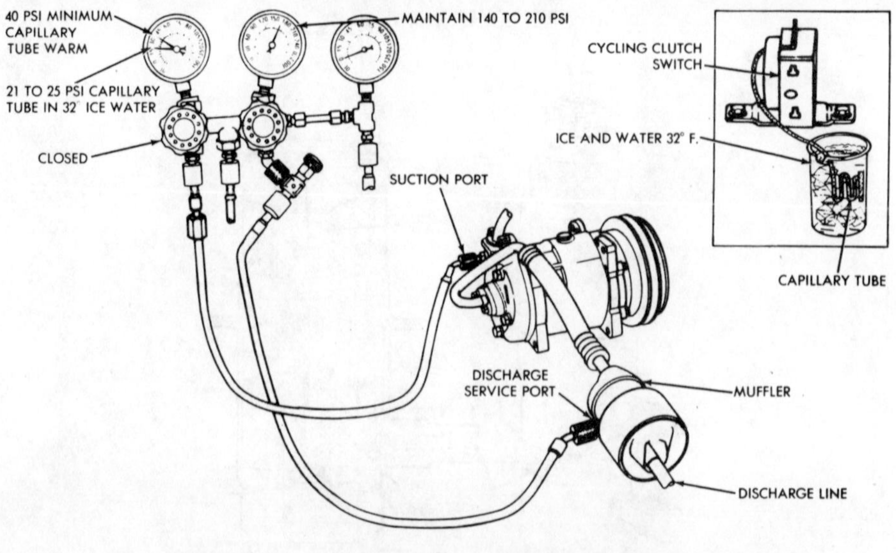

Fig. 19 Testing cycling clutch switch. Sankyo compressor

2. While tube is in solution the switch contacts should remain open.
3. Remove tube from solution and allow to warm up, switch contacts should close.
4. If switch contacts do not open and close as indicated, replace switch.
5. Using an ohmmeter check switch contact resistance. Resistance should be 1 ohm or less. If resistance is above 1 ohm, replace switch.

LOW PRESSURE CUT-OFF SWITCH

Chrysler Corp.

1. Remove rubber boot from cut-off switch and connect a jumper wire between connector lead terminals.
2. Turn ignition switch on and listen for compressor clutch to engage.
3. If compressor clutch does not engage, clutch, clutch fuse, cycling clutch switch (if equipped), or wiring may be defective.
4. If compressor clutch engages, connect a manifold gauge set and note discharge pressure reading on 1974-78 vehicles. On 1979-80 vehicles, note suction pressure.

NOTE: The low pressure cut-off switch is located on the filter-drier on 1974-78 vehicles; on 1979-80 vehicles it is located on the H valve assembly.

5. If the discharge pressure is below 30 psi on 1974 models or 24 psi on 1975-78 models, the system is low on refrigerant and should be inspected for leaks and recharged.
6. On 1979-80 models, except 1980 Omni/Horizon, suction pressure of 12 psi or more must cause switch to energize clutch. Suction pressure of 16 psi or more will cause switch to energize clutch on 1980 Omni/Horizon models. If evaporator suction pressure is 25 psi or less, the system is low on refrigerant and must be leak tested and recharged.
7. Remove jumper wire and reinstall switch connector. Repeat step 2. If the clutch does not engage, discharge system, replace switch and recharge system.

CYCLING CLUTCH SWITCH TEST

Chrysler Corp.

The cycling clutch switch is wired in series with the compressor clutch coil and is secured to the suction line next to the H valve assembly. The CC switch utilizes a capillary tube to monitor the temperature of the suction line and turns the compressor on and off as necessary. The switch is calibrated and sealed; if defective it must be replaced. To test the CC switch proceed as follows:

a. Disconnect wires from switch.
b. With A/C system OFF, check the continuity of the switch. At temperatures above 45 degrees, the switch is normally closed. If the contacts are open indicated by no continuity, replace switch.
c. If the contacts are closed, switch calibration may be incorrect or capillary tube may be insulated from suction tube by dirt or a loose connection. Ensure that capillary tube is properly secured to suction tube and that contacting surfaces are clean. Then, remove switch and immerse capillary tube in a container of ice water, Fig. 19.
d. Using an ohmmeter or continuity tester, check switch for continuity at a temperature of 38 to 40 degrees F. If switch contacts are not open at this temperature, replace switch.
e. Raise temperature slowly to 52 to 54 degrees F. If contacts do not close at this temperature, replace switch.
f. A properly operating switch will pass tests outlined in steps d and e. Remove and install new switch, if necessary, as outlined in "Unit Replacement" section.

LOW PRESSURE CUT-OUT SWITCH TEST

1979 Lincoln Continental

The low pressure cut-out switch prevents the compressor from operating if the system becomes discharged. In addition, it prevents compressor operation during low ambient temperature conditions. To test for open ground circuit in switch, proceed as follows:

a. Disconnect clutch ground wire from cut-out switch.
b. Using a self-powered test light, check switch operation.
c. If test light does not glow, attach a manifold gauge set to refrigerant system and observe system pressure.
d. If pressure is above 45 psi, discharge system and replace low pressure cut-out switch.
e. If pressure is between 35 and 45 psi, connect a jumper wire from the wire disconnected from switch to ground.
f. Operate system and use test light to check switch operation. If test light does not glow when pressure is above 45 psi, discharge system and replace cut-out switch.

To test for shorted switch, proceed as follows:

a. Attach a manifold gauge set to refrigerant system and observe system pressure. If the high side pressure is 35 psi or below, self-powered test light connected to switch terminal and ground should not glow. If test light glows, switch is shorted and must be replaced.
b. If high side pressure is 35 psi or more, reduce pressure either by moving the vehicle outside in low ambient temperature conditions or by discharging system.
c. Test light should not glow when pressure falls below 35 psi; if it does, switch is shorted and must be replaced.

DISCHARGE PRESSURE SWITCH

This switch is located in the discharge side of the system and is wired in series with the compressor clutch to protect the compressor in the event of loss of refrigerant or low ambient temperature.

NOTE: On some models the discharge pressure switch is located in the rear head of late model GM compressors and should not be confused with the superheat switch which, on earlier compressors, was located in the same spot.

If the switch is faulty, it must be replaced; it cannot be adjusted or repaired.

SUPERHEAT SHUTOFF SWITCH

The superheat switch is located in the suction passage at the rear head of the compressor. Its contacts will close if the refrigerant in the system should get down to 2 lbs. Current will then pass through the superheat switch and the thermal limiter to ground causing the thermal limiter fuse to blow. Test the switch as follows:

1. Disconnect wire from switch.
2. Attach a test light to 12 volt source and center terminal of switch.
3. Test light should be off if system is charged; test light should be on if system is discharged.

PRESSURE CYCLING SWITCH

1979-80

The pressure cycling switch is used in place of the thermostatic switch on some 1979 and all 1980 General Motors models to cycle the compressor off and on thus maintaining proper cooling and preventing evaporator freeze-up. In addition, if the accumulator pressure drops below a predetermined pressure, the switch cycles the compressor off to prevent compressor damage. The pressure cycling switch can be replaced without discharging the refrigerant system as the accumulator incorporates a Schrader-type valve.

SUCTION THROTTLING REGULATORS

Suction throttling regulated systems include any type control used to regulate the flow of refrigerant from the evaporator to the compressor. This control will be located at some point between the tail pipe of the evaporator and the compressor in the low side or suction line.

Sometimes it may be difficult to determine exactly what type of control system is used on a particular installation. However, if under-hood inspection of the compressor and lines does not show a control device, it is safe to assume that a thermostat-controlled cycling clutch is used. *Generally, then, all suction throttling regulator controls will be located under the hood. The thermostat will not be used, and the clutch will be energized during total operation of this type system.*

Constant head load, ram air, and compressor speed will allow the expansion valve to meter an even flow of refrigerant into the evaporator. Change any of these conditions, however, and the refrigerant flow through the expansion valve will change. Without some means provided to regulate refrigerant flow, the evaporator cooling will become excessive and freeze the moisture condensing on the coils. This results in evaporator icing or freeze-up and requires that the ice on the evaporator be melted before it can resume cooling action.

Installation of a suction throttling control device will compensate for these varying conditions. As the pressure and temperature within the evaporator drop, spring pressure forces a valve within the regulating device toward its seat, retarding the flow of refrigerant. This action increases the pressure back into the evaporator, raising both pressure and temperature. The temperature will not be allowed to drop low enough to freeze over the evaporator coils. A point of balance will soon be reached whereby the refrigerant flow past the valve in the regulator will reach a degree of pressure and temperature sufficient to maintain efficient cooling without evaporator freezing.

Ford STV Test

1974-79 & 1980 Less Cycling Clutch

NOTE: 1980 vehicles equipped with cycling clutch system incorporate a suction accumulator/drier. Refer to Performance Test in the "System Testing" section.

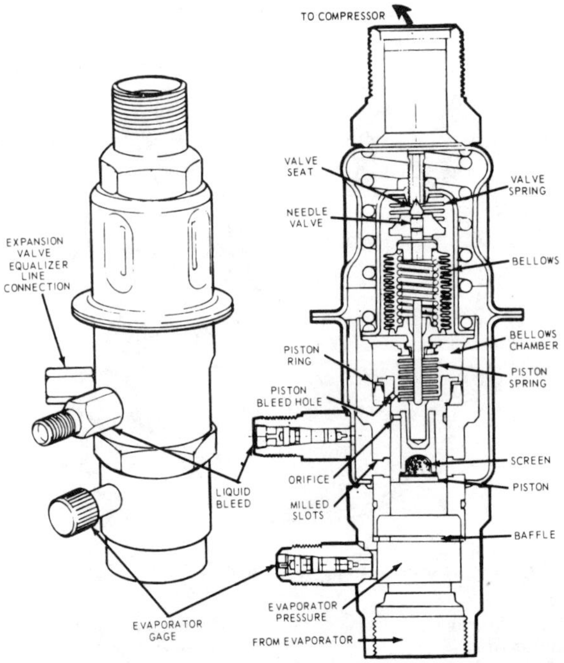

Fig. 20. Sectional view of Ford STV valve & GM POA valve

1. Connect the manifold gauge set into the system, and connect a separate pressure gauge to the evaporator connection on the suction throttling valve, Fig. 20.
2. Perform the air conditioner performance test as outlined in the foregoing procedure. Make any repairs indicated by the test.
3. After operating the system for 10 to 15 minutes, be sure that normal operating pressures are indicated on the manifold gauge set. Check the reading on the evaporator gauge.
4. Any gauge used to check the evaporator pressure at the suction throttling valve will be affected by atmospheric pressure, whereas the pressure within the valve remains unaffected. To compensate for this, use the STV pressure/altitude chart to determine the required gauge reading for correct evaporator pressure at a given altitude on 1974-78 models. In addition, a change in ambient temperature will affect high side system pressures. Observe the following conditions before beginning STV system tests on 1979-80 models:
   a. System pressures stabilized.
   b. In-car temperature stabilized at 70 to 80 degrees F.
   c. Controls set to Maximum A/C (Recirculating Air).
   d. Blower speed set to Medium.
   e. Engine speed at 1500 rpm.
   f. Compressor clutch engaged.

Refer to "System Testing" section for pressure readings versus temperature and altitude; if readings are not within specifications, the refrigerant system pressure evaluation chart details probable causes.

STV PRESSURE/ALTITUDE CHART

Altitude	Required Gauge Reading × ±1 psi
Sea Level	28.5
1000 ft.	29.0
2000 ft.	29.5
3000 ft.	30.0
4000 ft.	30.5
5000 ft.	31.0
6000 ft.	31.4
7000 ft.	31.8
8000 ft.	32.3
9000 ft.	32.7
10000 ft.	33.2

Ford Combination Valve Test

The combination valve combines the functions of the thermal expansion and suction throttling valve. Refer to Ford STV Test for testing procedure. The system pressures of models using the STV/BPO (suction throttling valve/bypass orifice) should be compared with the normal pressures shown in the System Testing section.

NOTE: Units equipped with a combination valve of the STV/BPO type differ from the earlier system in that they do not have a suction pressure service access gauge port for testing. They do have a high pressure service access gauge port and an evaporator (low) pressure service access gauge port. To connect a manifold gauge set into the high pressure side requires a special adapter as the high side port does not have the familiar threaded stem. The adapter snaps over the port to provide a leak-free connection.

Chrysler EPR-2 & EPR Valve

The EPR (Evaporator Pressure Regulator) valve is calibrated to produce maximum cooling without causing frost or ice to form on the evaporator. If for any reason the factory calibration has been disturbed, the EPR valve may restrict the flow of refrigerant at an evaporator pressure which is either too high for maximum performance or too low to prevent coil freeze-up. The following test determines whether or not the valve is functioning properly.

The EPR valve supersedes the EPR-2 valve but performs the same function despite some outward visual differences. System test figures will be the same for the EPR and EPR-2.

EPR-2 Valve Test Less H-Valve

1. Close windows and set A/C controls for "Max A/C", high blower and temperature control to full reheat position.
2. Run V8 engines at 800 RPM and 6 cylinder engines at 900 RPM.
3. Operate for a minimum of five minutes to obtain partial stabilization and sufficient reheat to load system.
4. Pressure at the discharge port should be 140 to 210 psi at this time. If not, check pressure drop across EPR valve which should be 1 to 6 psi on 1974-76 units. If pressure drop is greater than the above limits, the EPR valve and nylon filter suction screen should be replaced.
5. If the EPR valve was found satisfactory, remove the expansion valve thermo bulb from the suction line and immerse bulb into a container filled with ice and water. The compressor inlet pressure should now be 15 psi or less on 1974-76 units with the evaporator suction pressure as specified in Fig. 8. If the compressor inlet pressure is more than specified the EPR valve and nylon filter suction screen is faulty and must be replaced.

Chrysler EPR Valve, H-Valve & Compressor Test

NOTE: Test must be performed with ambient temperature between 70 and 85 degrees F.

1. Close windows and operate engine at 800 rpm.
2. On models with manual A/C, set A/C controls for Max. A/C, high blower. On Intermediate models except Aspen and Volare, place temperature lever in full reheat position. On Aspen, Volare and Full Size models, disconnect and plug water valve vacuum hose in engine compartment. On the Semi-Automatic Temperature Control system, remove the sensor vacuum connector and connect the large port to the small port with a ⅛ to 3/16 inch jumper hose.
3. On models with Auto-Temp II, install A.T.C. tester C-4144 or 4135. Set tester at "Heat," "Blower Volts," then depress "Hi-Auto" push button. .Air should flow from heater outlets. After system has been operating approximately 15 minutes, disconnect and plug recirculating-outside air door actuator vacuum hose (light green).
4. Operate system for at least 5 minutes, then note pressure reading at discharge service port. If head pressure of 140-240 psi cannot be obtained, check refrigerant level and pressure drop across EPR valve. Recharge system or replace EPR valve as necessary.
5. Spray H valve control head with liquid freon for 10 to 15 seconds. Evaporator suction pressure should drop below 15" of vacuum; if not, H valve is stuck open and should be replaced. Discontinue spraying H valve with freon; suction pressure should increase to a minimum of 38 psi and then stabilize at a pressure of 25 to 35 psi. If H valve does not perform in this manner, it should be replaced.

NOTE: Before handling freon, refer to General Precautions—Safety, located in this chapter.

6. EPR valve should now be wide open and pressure drop across valve should be no greater than 6 psi. Pressure drop is the difference between evaporator suction and compressor inlet pressures. If pressure drop is greater than 6 psi, EPR valve is defective or there is a restriction in system. Replace EPR valve or restricted component.
7. On Intermediate models except Aspen and Volare, place temperature lever in full cool position. On Aspen and Volare and Full Size models less Auto Temp II, reconnect vacuum hose to water valves. On models with Auto Temp II, place tester switch in A/C position. On models equipped with SATC, remove the jumper hose and plug the small port. Operate engine on all models for 15 minutes to cool evaporator. Then, place blower switch in low speed position on models with manual A/C or mid-position on models with Auto Temp II and operate engine at 1100 rpm for 5 minutes to stabilize system.
8. Evaporator pressure should read 22 to 28 psi and pressure drop between evaporator and compressor inlet should be greater than 6 psi if system is operating properly. If evaporator suction pressure does not drop to 28 psi or below and pressure drop between evaporator suction and compressor inlet is greater than 6 psi, the EPR valve is stuck closed and should be replaced. If after replacing EPR valve pressure drop is below 6 psi, remove compressor heads and check for defective head gaskets and reed valves.
9. Set engine speed at 800 rpm and blower switch on Hi position; evaporator suction should be between 22 and 30 psi and compressor inlet pressure between 15 and 24 psi. If compressor discharge pressure is above 240 psi, check for restricted discharge line, radiator overheating, air in system or defective fan drive. If compressor discharge pressure is below 140 psi, check for defective compressor gaskets or discharge reed.
10. On models with Auto-Temp II, reconnect recirculating air door vacuum hose, remove A.T.C. tester and connect wiring harness. On models with SATC, reconnect the sensor vacuum connector.

Chrysler H Valve Test Models W/Cycling Clutch

NOTE: Test must be performed with ambient temperature between 70 and 85 degrees F.

1. Close windows and operate engine at 800 RPM. Set controls for Max. A/C, high blower and temperature control lever in full reheat position. Use jumper wires to by-pass cycling clutch switch, and low-pressure cut off switch, if equipped, then disconnect and plug the water valve vacuum hose in the engine compartment.
2. Operate the system for 5 minutes and check the pressure at the discharge service port. Pressure should reach 140 to 240 psi; if not check system charge level.
3. Spray H valve control head with liquid freon for 10 to 15 seconds. Evaporator suction pressure must drop to below 15" of vacuum; if not, H valve is stuck open and should be replaced. Discontinue spraying H valve with freon and check suction pressure. Pressure should increase to a minimum of 38 psi and then stabilize at a pressure of 25 to 35 psi. If H valve does not perform in this manner, it should be replaced.

NOTE: Before handling freon, refer to General Precautions—Safety, located in this chapter.

4. Reconnect A/C water valve vacuum hose and set temperature control to cool position.
5. With engine speed at 800 RPM and blower on HI, the evaporator suction pressure should be between 20 and 30 psi. If compressor discharge is higher than 240 psi, check for a restricted discharge line, radiator overheating, air in system or faulty viscous fan drive. If discharge pressure is less than 140 psi, check compressor head gaskets and discharge reeds.
6. Reconnect the wires to the cycling clutch switch.

Frigidaire POA Valve

General Motors

The POA valve (Pilot Operated Absolute), Fig. 20, differs from the STV in that it has no vacuum diaphragm. The POA valve will not change calibration when the system is operated at a higher altitude due to the atmospheric pressure to which the STV was subjected because of its vacuum diaphragm.

NOTE: When it is found necessary to replace a POA valve, check the interior of the defective valve for corrosion or crystalization of salts. This should indicate moisture in the system. If this condition exists, the receiver-dehydrator should be replaced and the system evacuated for one hour.

When leak-testing the POA valve, it is necessary to check only the hose coupling ends. When using a Propane Torch Leak Detector, no evidence of refrigerant should be present at the POA valve.

The POA valve will provide different gauge readings based on the altitude where the readings are being taken. Correct gauge reading at sea level is 29.5 psi. Gauge readings will be one

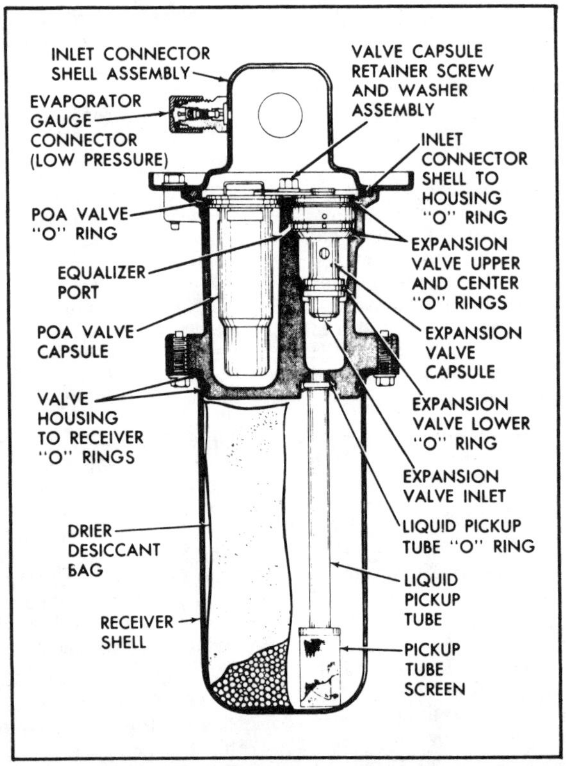

Fig. 21 Sectional view of Valve-in-Receiver assembly

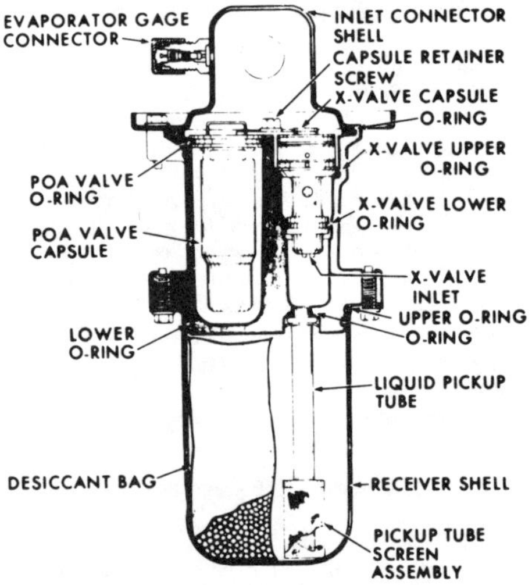

Fig. 22 Sectional view of Evaporator Equalized Valves-in-Receiver assembly

half psi higher for each additional 1000 feet of elevation.

The only test for POA valve operation is to check the suction pressure at the valve when making a performance test. If the valve gives improper gauge readings, it must be replaced since it is not repairable or adjustable.

VIR Control

Valves-In-Receiver, General Motors

The VIR assembly, shown in Fig. 21, incorporates the thermostatic expansion valve. POA suction throttling valve, receiver-dehydrator and sight glass. The construction of this unit eliminates the need for an external equalizer line between the thermostatic expansion valve and POA suction throttling valve. Also, the thermobulb and capillary line has been eliminated.

The equilizer function is affected by a port in the wall separating the POA suction throttling valve and thermostatic expansion valve cavities in the VIR unit. The functions of the thermobulb and capillary line are now accomplished directly by the thermostatic expansion valve. The diaphragm end of the thermostatic expansion valve is exposed to refrigerant vapor entering the VIR from the outlet of the evaporator.

NOTE: A revised design VIR assembly, the Evaporator Equalized Valves In Receiver (EEVIR), Fig. 22, eliminates the equalizer port between the POA and thermostatic expansion valve cavities in the valve housing. The upper "O" ring is omitted from the thermostatic expansion valve permitting pressure within the inlet connector shell to be applied to the thermostatic expansion valve equalizer port.

The sight glass is located at the inlet of the thermostatic expansion valve and the evaporator gauge fitting is located in the inlet connector shell. The VIR unit is mounted adjacent to the evaporator.

The liquid bleed valve, located in the VIR valve housing, opens when a pressure differential greater than 10-20 PSI occurs between evaporator and suction pressures. The valve opening returns oil to the compressor under low refrigerant charge or low evaporator load conditions. The valve closes when the pressure differential is below 10 PSI to prevent loss of capacity. Refer to VIR Diagnosis chart, Fig. 23.

NOTE: The filter screen, drier desiccant, thermostatic expansion valve and POA suction throttling valve are serviceable.

EXPANSION VALVES

Internally Equalized Type

Referring to Fig. 24, the refrigerant enters the inlet and screen as a high pressure liquid. The refrigerant flow is restricted by a metered orifice through which it must pass. As the refrigerant passes through this metered orifice, it changes from a high pressure liquid to a low pressure liquid or from the high side to the low side.

The refrigerant flow through the metered orifice is of the utmost importance. Anything that interferes with this flow seriously affects operation of the entire system.

A sufficient drop in temperature of the area cooled by the evaporator changes the heat transfer requirement of the evaporator. To continue the same flow of refrigerant into the evaporator will result in extreme cooling of the evaporator fins and coils to such an extent that they will freeze over with ice and stop air flow over them.

The flow of refrigerant is controlled by varying the amount of refrigerant allowed to pass through the metered orifice. This is accomplished by internal balancing pressures moving a valve or a seat to control the refrigerant flow.

A thermal bulb connected to a diaphragm by a small line and filled with refrigerant gas such as Freon is secured firmly by a clamp to the evaporator tail pipe. The thermal bulb is sensitive to tail pipe temperatures. If the tail pipe becomes warm, the gas inside the bulb will begin to expand, exerting pressure against the diaphragm on the top plate that is connected to the seat or valve by a pin or pins (whichever construction is used). This expansion will then move the seat away from the orifice, allowing an increased refrigerant flow. As the tail pipe temperature drops, the pressure in the thermal bulb also drops, allowing the valve to restrict flow as required by the evaporator.

The pressure of the refrigerant entering the evaporator is fed back to the underside of the diaphragm through the internal equalizing passage. Expansion of the gas in the thermal bulb must overcome the internal balancing pressure before the valve will open to increase refrigerant flow.

A spring is installed against the valve and adjusted to a pre-determined setting at the time of manufacture. This is the super-heat spring, the purpose of which is to prevent slugging of the evaporator with excessive liquid. The adjusted tension of this spring is the determining factor in the opening and closing of the expansion valve. During opening or closing, the superheat spring tension will retard or assist valve operation as required.

NOTE: For all practical purposes we should not be concerned with the adjustment of this spring. Tension is adjusted from 4 to 16 pounds as required for the unit on which it is to be installed. This original setting is sufficient for the life of the valve, and tampering with the adjustment will only lead to an unsatisfactory operating unit. Special equipment is required in most cases to accurately calibrate this adjustment.

Externally Equalized Type

Operation of this type valve, Fig. 25, is the same as the internally equalized valve except that the evaporator pressure is fed against the underside of the diaphragm from the tail pipe of

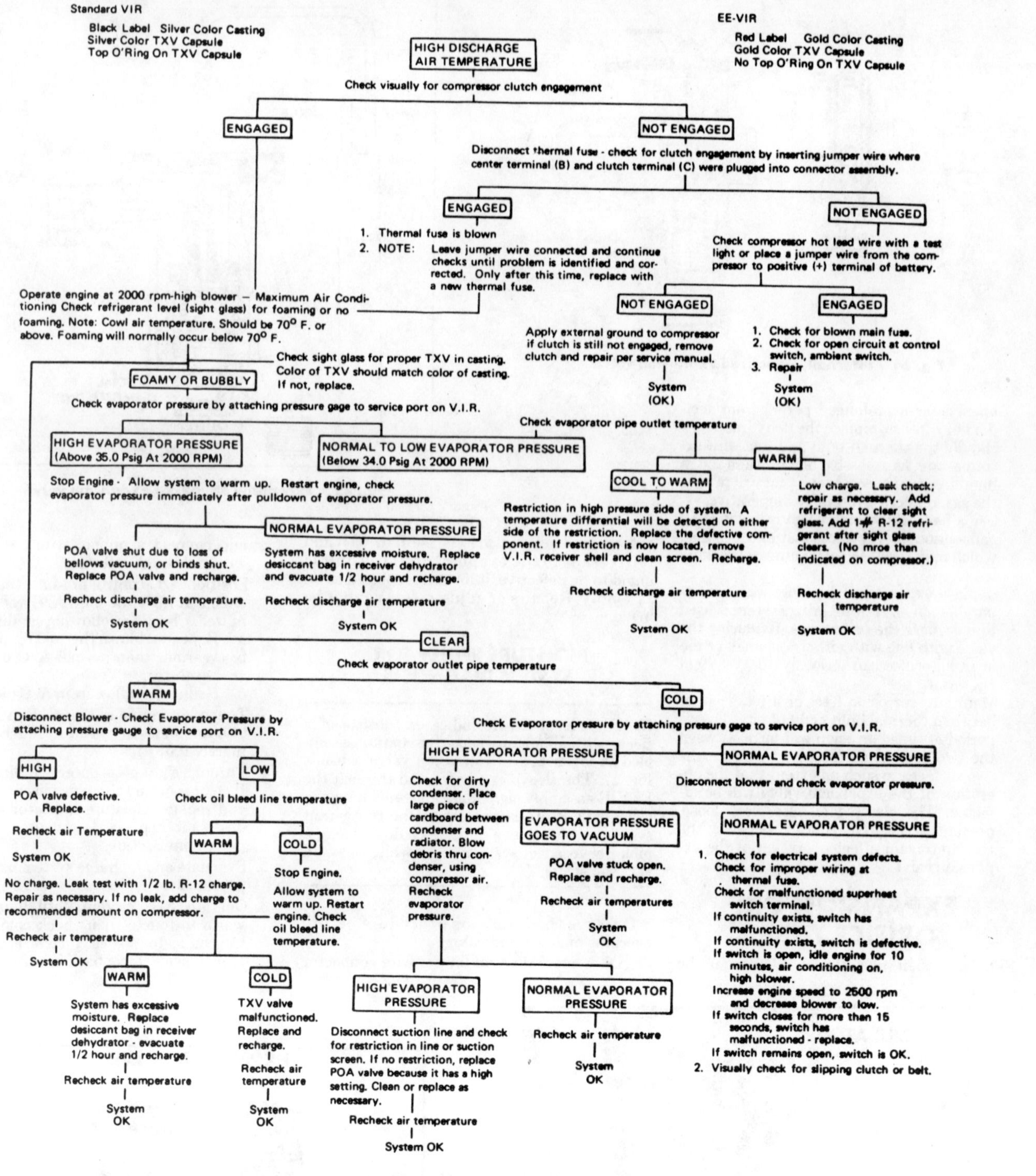

Fig. 23 VIR Diagnosis Chart

the evaporator by an equalizer line. This balances the temperature of the tail pipe through the expansion valve thermal bulb against evaporator pressure taken from the tail pipe.

Service Precautions

From the foregoing description of the thermostatic expansion valve, it can readily be seen that the valve is more sensitive to foreign materials than any other unit in the A/C system. Therefore, the following points should be observed to protect the valve while in service and while servicing the system.

1. The A/C system should be operated at least for short periods throughout the winter season to prevent the internal moving parts from corroding and sticking, which will result in the valve becoming inoperative and necessitating replacement. Periodic operation will also lubricate the entire system and especially the compressor seal, keeping it soft and in condition properly to seal the refrigerant.
2. Clean and replace all accessible screens when servicing.
3. Install a filter if system has excessive foreign materials present.
4. Follow correct evacuation procedure to remove all possible moisture from system.
5. Cap or cover any lines opened for service to prevent entrance of moisture or dirt.
6. Replace dehydrator as soon as excess moisture content is evident. Any system should have the dehydrator replaced at least upon opening the system the third time for service. Also, any time the system has been

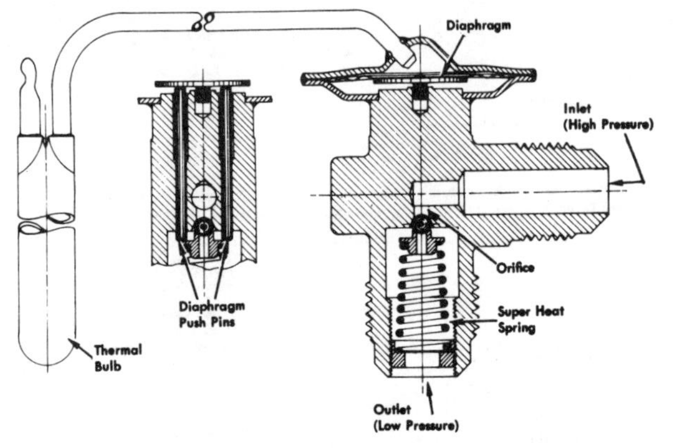

Fig. 24 Internally equalized expansion valve

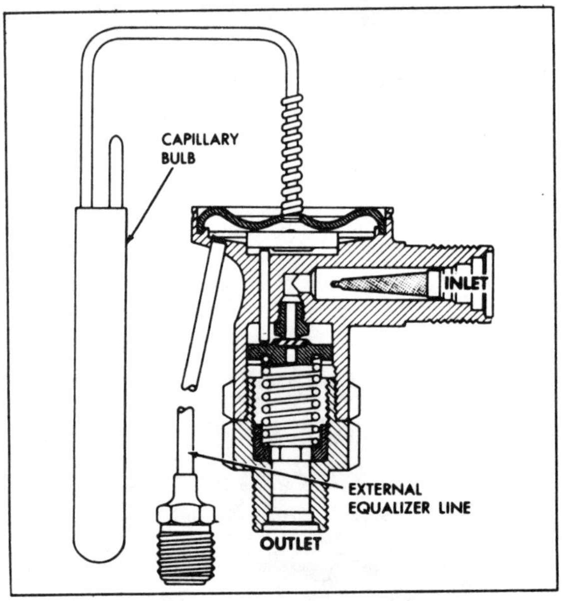

Fig. 25 Externally equalized expansion valve

opened for a prolonged period from accident or rupture, replace the dehydrator.

7. Handle the thermal bulb and line with extreme care, as excessive bending and rough handling can cause a break that will release the gas, rendering the valve inoperative.
8. Use a back-up wrench when removing any connection to prevent twisting of a line, which may result in weakening or breaking it.
9. Replace expansion valve only with a comparable valve. A numbering system is used to designate the orifice size. Replacing the valve with one with either too small or too large an orifice can seriously affect system operation.
10. Maintain corrosion-free positive contact between thermal bulb and tail pipe. Over a period of time corrosion will form between the two contact surfaces and insulate the two surfaces to such an extent that the operation of the expansion valve will be affected. The clamp may also work loose, preventing positive contact between the two and again affect operation of the expansion valve.

EXPANSION OR ORIFICE TUBE

The expansion tube, Fig. 26, is located in the evaporator inlet and meters the flow of liquid refrigerant to the evaporator. If the unit is found to be defective, it must be replaced as an assembly. Repairs or adjustments cannot be made.

MOISTURE INDICATOR SYSTEM DIAGNOSIS

NOTE: The moisture indicator consists of a sleeve with a moisture sensitive coating assembled to lower end of expansion valve capsule, Fig. 27. The sleeve can be viewed through the EEVIR assembly sight glass and will change in color if moisture is present in the refrigerant system. A blue sleeve indicates a dry system; a pink sleeve, a wet system; and a white sleeve, a very wet system.

Observe the following procedure before checking moisture indicator:

1. Disconnect thermal limiter wire connector and connect a jumper wire between feed wire and clutch coil wire.
2. Position A/C control in Max., then operate engine at high idle until A/C system is stabilized. Check for following conditions:
   a. Repeated blown thermal limiter.
   b. Warmer than normal A/C outlet temperatures.
   c. Reduced air flow from A/C outlets.
3. Touch liquid line in area of EEVIR assembly, line must be warm to touch (approximately 120° F).
4. Through sight glass observe indicator color. If color is other than blue, discharge system and replace moisture indicator sleeve and desiccant bag, as described in Unit Replacement Section.
5. Evacuate and recharge system as described previously.
6. Operate A/C system until liquid line is warm to touch, then check color of moisture indicator sleeve. If color is other than blue, place vehicle back in service for a pe-

Fig. 26 Expansion tube assembly

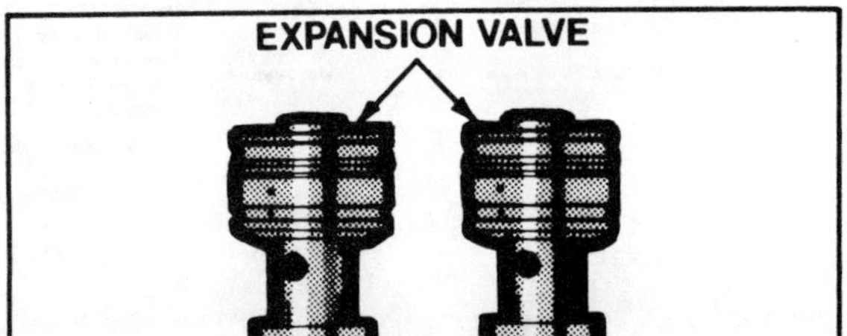

Fig. 27 Moisture indicator sleeve installed on expansion valve capsule

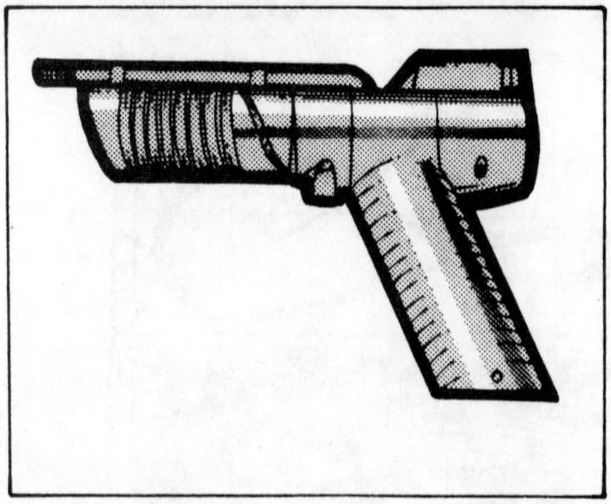

Fig. 28 Robinaire battery powered electronic leak detector

Fig. 29 General Electric (Tracker) battery powered electronic leak detector

riod of 2 days and recheck indicator sleeve color.

TESTING CLUTCH COIL

Refer to "System Testing" section for proper magnetic clutch coil diagnosis.

LEAK TESTS

Testing the refrigerant system for leaks is one of the most important phases of trouble shooting. One or more of the methods outlined will prove useful in detecting suspected leaks or checking connections if service work is performed. Before beginning any leak test, attach a manifold gauge set and note pressure. If very little or no pressure is indicated, it will be necessary to install a partial charge. Check all connections, compressor head gasket, oil filler plug and compressor shaft seal for leaks.

Electronic Leak Detectors

There are a number of electronic leak detectors available to perform leak tests, Figs. 28 through 32. Refer to operating instructions for the particular unit being used and observe these general procedures:

a. Move the detector probe at a rate of one inch per second in areas of suspected leaks.
b. Position the probe below the point being tested as refrigerant gas is heavier than air.
c. Be sure to check service access gauge port valve fittings, particularly when valve caps are missing, as dirt accumulations can destroy the sealing area of valve core when manifold gauge set is attached. Replace missing valve caps after cleaning valve core area.

NOTE: Valve caps should only be finger tightened. Using pliers to tighten valve caps may distort sealing surface of valve.

d. Check manifold gauge set and hoses for leaks as well as the rest of the system.

Flame-Type (Halide) Leak Detectors

CAUTION: When using flame-type detectors, avoid inhaling fumes produced by burning refrigerant. Do not use this type detector where concentrations of combustible or explosive gases, dusts or vapors may exist.

1. Adjust detector flame as small as possible to obtain maximum sensitivity. Be sure copper element is cherry red and not burned away. The flame will be almost colorless.
2. Slowly move detector along areas of suspected leaks. A slight leak will cause the flame to change to a bright yellow-green color. A large leak will be indicated by a brilliant blue flame. Position detector under areas being tested as refrigerant gas is heavier than air.

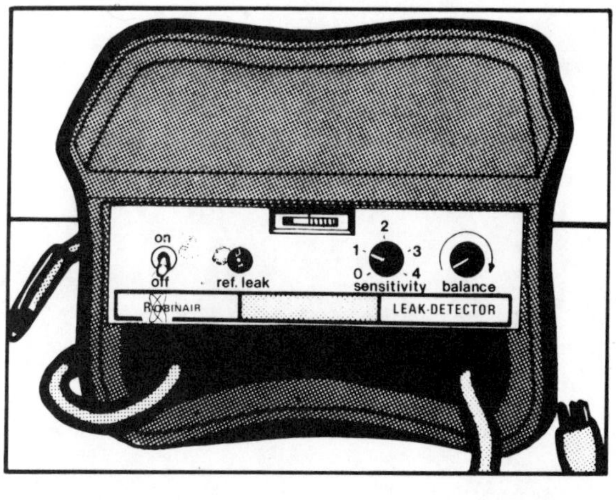

Fig. 30 Robinaire Electronic leak detector

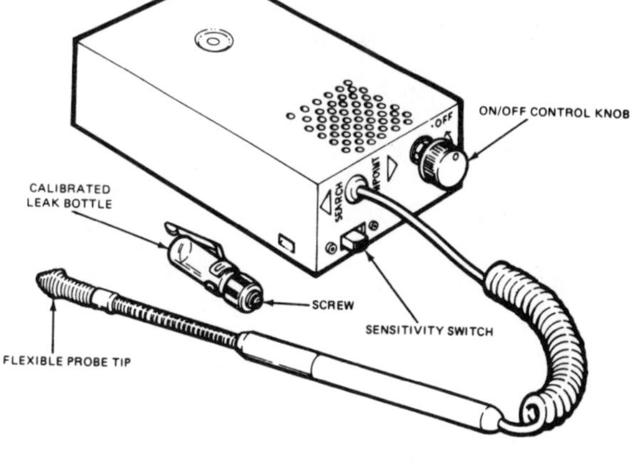

Fig. 31 Ford Electronic leak detector

NOTE: The presence of dust in the pick-up hose may cause a change in the color of the flame. If not recognized, a false diagnosis could be made. Store leak detector in a clean place and be sure the hose is free of dust before leak testing.

3. Check the manifold gauge set and hoses for leaks as well as the rest of the system.
4. Use a small fan to ventilate areas where the leak detector indicates refrigerant all the time. These areas are contaminated with refrigerant and must be ventilated before leak can be pinpointed.

Fluid Leak Detectors

Apply leak detector solution around joints to be tested. A cluster of bubbles will form immediately if there is a leak. A white foam that forms after a short while will indicate an extremely small leak. In some confined areas such as sections of the evaporator and condenser, electronic leak detectors will be more useful.

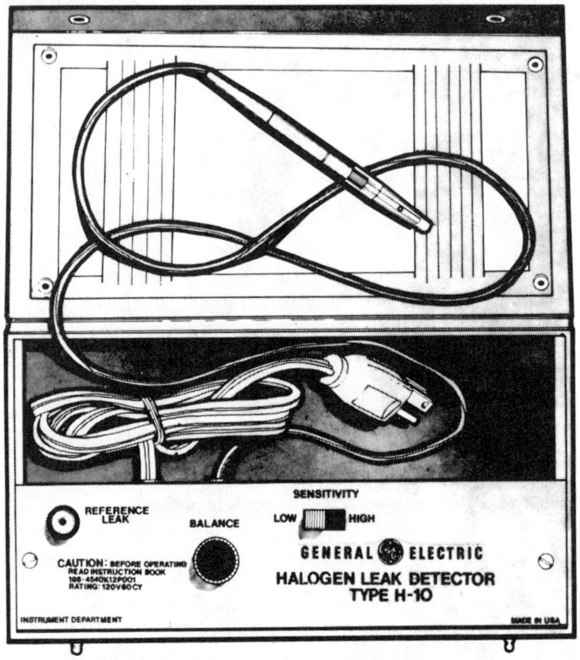

Fig. 32 General Electric AC powered electronic leak detector

Trouble-Shooting

GENERAL DIAGNOSIS & TROUBLE-SHOOTING

While a detailed diagnostic procedure for all air conditioning systems would be impractical, due to the many variations in construction and operation, there are three fundamental divisions of a total diagnosis:

1. Refrigerant systems must have an adequate, but not excessive charge.
2. Determine whether the refrigerant system is governed by a cycling clutch compressor or by an evaporator pressure control valve.
3. The air distribution system (blower motor, switches, vacuum lines and air ducts) must be operational before checking the refrigerant system.

Check the blower; if inoperative, examine switches, fuses, connections, wiring and the blower motor. In the event the blower is operating but the air output is low, check for loose wire connections or shorts, undercharged battery, dirty or loose switch contacts, or a faulty blower motor. Inspect the air distribution system for obstructions and ensure proper door operation.

If the blower is circulating the air but there is no cooling, check the compressor drive belt; ensure it is not broken or slipping. If the pulley is turning but the compressor shaft is not, check the magnetic clutch. On GM models equipped with a cycling clutch, the following hand check method will quickly reveal whether the problem lies in the refrigerant system or further testing of the distribution system is required.

1. With engine warmed up and at normal idle, set selector lever to Norm, temperature lever to Cold and blower on Hi.
2. Place one hand on the evaporator inlet pipe and the other on the accumulator surface with the compressor engaged.
3. If both surfaces are the same temperature and some degree colder than ambient, refrigerant system is normal. Refer to the "System Testing" section if the vehicle is not being sufficiently cooled.
4. If the inlet pipe is cooler than the accumulator surface, refrigerant system is low on charge. Add small amounts of refrigerant until both feel the same temperature. Then add 14 oz. (one can) of additional refrigerant.
5. If inlet pipe is frosted over and accumulator surface is warmer, proceed as in step 4.

The individual components of the refrigerant system will often give clear signs of their malfunctioning. Use the following general descriptions to pinpoint faulty components.

Compressor

A faulty compressor will display one or more of the following symptoms: noise, seizure, leakage or low inlet and discharge pressure. A steady, resonant noise from the compressor should not cause alarm, but irregular metallic rattling may indicate broken parts and should be investigated. A thumping noise from the compressor and a cool, sweating suction line into it may indicate an overcharged system. Check seizure by disengaging the magnetic clutch and rotating the driven plate. If the compressor is seized, the driven plate will not rotate.

NOTE: False compressor seizure may occur after an extended period of disuse or storage. Lubricating oil drains away from the polished surfaces of ball seats and axial plate and the compressor appears to be seized. Use a clutch hub holding tool to turn the compressor in the opposite direction of rotation at least three revolutions. Check for false compressor seizure if compressor has not been used in a month or longer.

If compressor is not seized but will not rotate, check for current at magnetic coil. Low discharge pressure may be caused by faulty seals within the compressor, a restriction in the compressor or elsewhere, or by a low refrigerant charge. The compressor must have the correct amount of the proper viscosity oil; excess oil will restrict refrigerant circulation and reduce compressor outlet pressure.

Condenser

The condenser may malfunction either by leaking or by being restricted. If restricted, compressor discharge pressure will be excessive. Icy or frosty spots on the condenser will indicate a partial restriction within the condenser. Be sure all foreign matter is removed from the front of the condenser; similarly, bent cooling fins will block air flow through the condenser and result in high discharge pressures.

Evaporator

A faulty evaporator will provide insufficient cooling to the vehicle. The core may be restricted with dirt, the case may be cracked, or a seal may be leaking sufficiently to prevent cooling. If evaporator restriction is due to icing, the expansion valve, capillary tube or suction throttling valve, if equipped, may be at fault and should be investigated.

Since there is a constant condensation of atmospheric moisture on the outside of the evaporator coils, ensure that the draining system is unobstructed and clean.

NOTE: Some vehicles have an auxiliary evaporator in the trunk or between the headliner and the roof.

Receiver-Dehydrator

A restriction inside the receiver-dehydrator will result in high head pressures if the restriction is on the inlet side of the unit. A restriction at the outlet side will cause low head pressures and little or no cooling. An exceedingly cold receiver-dehydrator may be restricted.

If the system has been in service for a considerable amount of time, the dessicant element may have lost its moisture-absorbing ability. This condition is indicated by the constant presence of small bubbles in the sight glass if equipped and a wide difference in temperature between the inlet and outlet receiver-dehydrator lines.

Thermostatic Expansion Valve

Faulty expansion valves will be indicated by low suction and discharge pressures on the manifold gauge set. In most cases the power element fails and the valve closes. In rare instances, the inlet screen becomes clogged with contamination or dessicant beads loose in the system.

POA or Suction Throttling Valve

The failure of these type valves will result in evaporator pressure being too high or too low depending on the nature of the defect. Moisture in the system can cause the thermal expansion valve or the POA valve to freeze up. To determine if the valve is malfunctioning due to ice, turn the compressor off and allow the system to defrost. Then repeat the performance test.

Refrigerant Line Restrictions

1. A restricted suction line will be indicated by low suction pressure at the compressor, low discharge pressure and little or no cooling.
2. A restricted discharge line will usually cause the pressure relief valve to open.
3. A restricted liquid line will cause low suction and discharge pressures and little or no cooling.

Thermal Limiter Diagnosis

Refer to Figs. 1 through 3, for blown thermal limiter diagnosis.

AMERICAN MOTORS

1974-80

PERFORMANCE DIAGNOSIS

Compressor Noise

1. Valves broken.
2. Overcharged.
3. Oil level incorrect.
4. Piston Slap.
5. Broken rings.

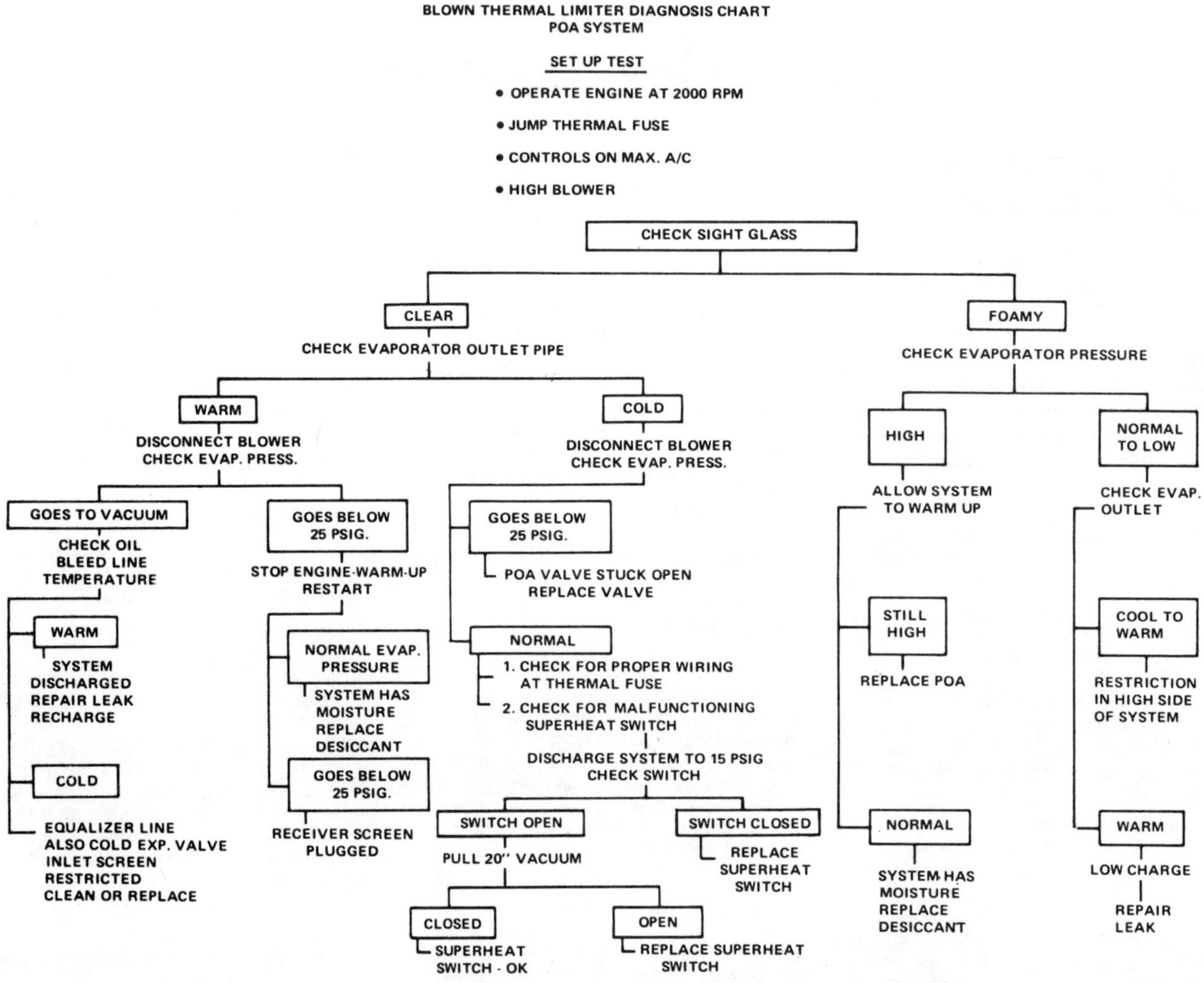

Fig. 1 Blown thermal limiter diagnosis chart. Models w/POA system

Excessive Vibration

1. Belt tension incorrect.
2. Compressor clutch loose.
3. Overcharged.
4. Misaligned pulley.
5. Compressor mounting bracket loose.

Condensation Dripping In Passenger Compartment

1. Plugged or improperly installed drain hose.
2. Improperly installed insulation.

A/C Airflow Stops On Acceleration

1. Vacuum storage tank defective.
2. Defective vacuum line or switch.
3. Vacuum leak.

Frozen Evaporator Coil

1. Thermostat faulty.
2. Improperly installed thermostat capillary tube.

CHRYSLER CORP.

1974-76

System Operates Only On Maximum Cooling

1. Short circuit in ambient or in-car sensor.
2. Short circuit in comfort control potentiometer.
3. Defective amplifier.
4. Defective servo.
5. Shorted sensor circuit.

System Operates On A/C Mode At All Times With Controlled Air Temperature

1. No vacuum supply to pot side of mode actuator.

System Operates Only On Maximum Heating & Hi-Blower

1. Open circuit in ambient or in-car sensor.
2. Open circuit in comfort control potentiometer.
3. Defective servo assembly.
4. Defective amplifier.
5. Sensor circuit open.

System Controls In Heat Mode For Most Ambient Temperature

1. Defective ambient sensor.
2. Defective amplifier.

System Operates On Heat Mode At All Times

1. No vacuum supply to rod side of mode actuator.
2. Defective ambient sensor.

System Puts Out Only Hot Air For All Models At All Times

1. Blend air door link disconnected.

System Puts Out Only Cold Air In All Modes At All Times

1. Blend air door link disconnected.
2. Blocked heater hose.
3. Defective water valve.
4. Crossed vacuum line or defective push button switch.

Blower & Compressor Will Not Operate & System Will Not Go Into Selected Mode

1. Open circuit.
2. Vacuum leak at engine fitting or reservoir tank.
3. Loose connection at push button switch.
4. Vacuum harness at push button switch leaking.
5. Defective push button switch.
6. Leak in vacuum circuit.
7. Vacuum lines improperly installed at vacuum reservoir.
8. Pinched or blocked vacuum line.
9. Defective actuator.

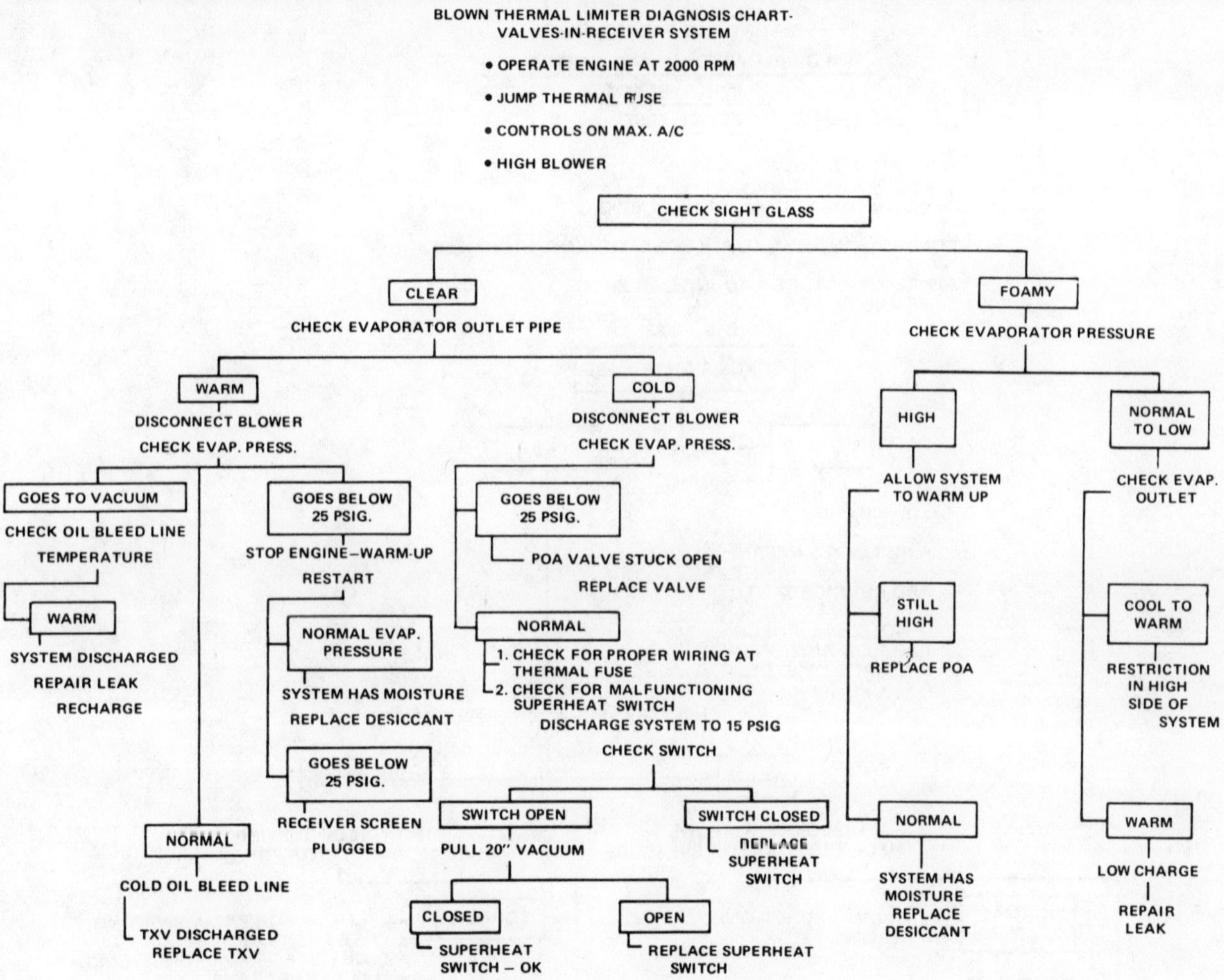

Fig. 2 Blown thermal limiter diagnosis chart. Models w/VIR system

Blower Will Not Operate In Any Mode

1. Blower motor wiring harness disconnected.
2. Blown fuse.
3. Defective blower motor.
4. Defective control switch.
5. Leak in vacuum circuit.
6. Defective resistor block.
7. Pinched or blocked vacuum line.
8. Improperly grounded blower motor.
9. Defective wiring at Hi blower relay insulator.
10. Defective Hi blower relay.

Compressor Will Not Operate At Any Time

1. Disconnected clutch wire.
2. Defective vacuum operated compressor switch.
3. Pinched or blocked vacuum line.
4. Defective low pressure cut-off switch.
5. Refrigerant system low.
6. Defective fuse.
7. Defective push button switch.
8. Defective compressor clutch wiring.

Hot Air From A/C Outlets & Cold Air From Heater Outlets

1. Reversed A/C and heater mode actuator vacuum lines.

Air Coming Out Of A/C Outlets When System Is In Defrost or Hi-Defrost Position

1. Vacuum lines to mode actuator and defrost actuator reversed.
2. Defective control switch.

Most Of Air Coming Out Of Defroster When System Should Be On Heat Mode

1. Vacuum Hoses to defroster actuator reversed.

Air Comes Out Of Defroster When System Should Be In Heat Mode & Out Of Heater Outlets When System Is In A/C

1. Vacuum lines to mode and defroster actuators crossed.

Air Comes Out Of All Outlets When System Is In Defrost Or A/C & Out Of A/C Outlets When System Is In Heat Mode.

1. Vacuum lines to mode and defroster actuators crossed.

System Will Not Change Modes When In Auto Or Hi-Auto

1. Doors binding.
2. Disconnected vacuum line.
3. Pinched or blocked vacuum line to actuator.

System Does Not Achieve Comfortable Condition—Cycling Of Temperature & Possibly Blower Speed

1. Defective servo assembly.
2. Defective aspirator.
3. Defective engine thermostat.
4. Aspirator scoop missing.

System Does Not Achieve Comfortable Condition—Temperature Drifts Up & Down

1. Pinched vacuum line to aspirator.
2. Defective sensor leads.
3. Defective amplifier.
4. Defective radiator cap.
5. Defective engine thermostat.

Blower Does Not Shut Off When Control Switch Is In The Off Position

1. Check valve defective or improperly installed.
2. Defective control switch.
3. Pinched vacuum line.

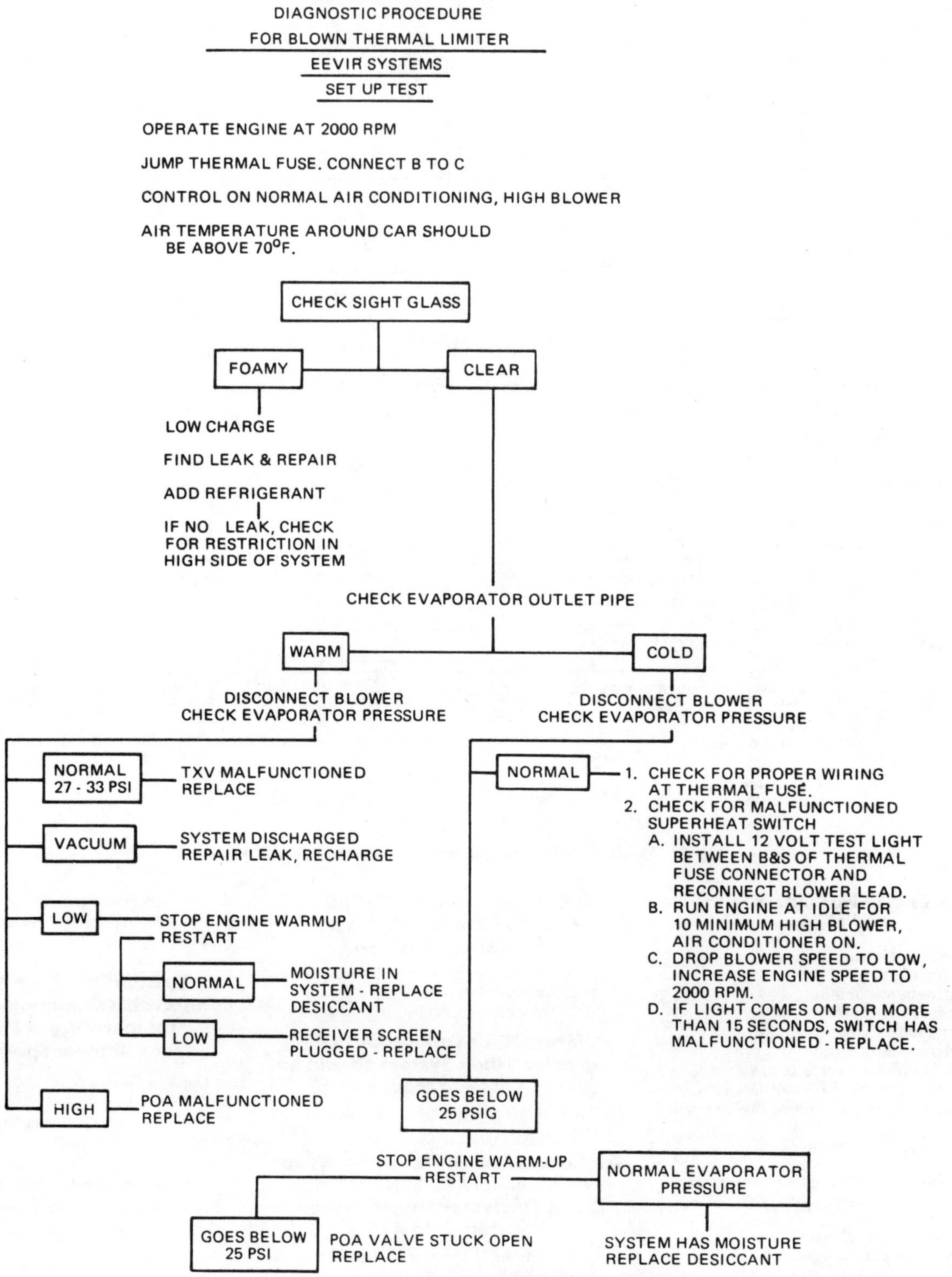

Fig. 3 Blown thermal limiter diagnosis chart. Models w/EEVIR system

Blower Does Not Come On In Defrost or Hi-Defrost When Engine Is Cold

1. Defective or improperly installed check valve.
2. Defective blower relay.
3. Defective master switch.

Blower Changes Speed With Changes In Car Voltage

1. Defective amplifier.

Air Coming Out Of Defroster When Should Be In A/C Mode

1. A/C-heat and heater-defroster mode actuator vacuum lines reversed.

Insufficient Air Flow

1. Air distribution ducts disconnected.
2. Defective blower motor.
3. Defective resistor block.
4. Freeze-up condition.

Blower Speeds Always Too Great On Lo-Auto

1. Defective resistor block.
2. Loose connection at resistor block.

System Blows Cold Air At The Start During Cold Weather Operation

1. Defective cold engine lockout switch.
2. Defective blower master switch.
3. Plugged vacuum bleed.
4. Defective or improperly installed check valve #2.

Insufficient Heat When Maximum Heat Is Required

1. Defective servo assembly.
2. Defective or clogged heater core.
3. Defective engine thermostat.
4. Improperly adjusted or binding blend air door.

Erratic Temperature Control

1. Heater hoses improperly installed.
2. Low coolant level.
3. Defective radiator cap.
4. Vacuum or electrical connection loose at control unit assembly.
5. Defective engine thermostat.
6. Defective servo assembly.
7. Intermittent sensor or temperature control potentiometer.
8. Erratic amplifier.

Insufficient Cooling When Maximum Cooling Is Required

1. Defective EPR valve.
2. System low on refrigerant.
3. Defective H-valve.
4. Defective control switch.
5. Compressor clutch not running.
6. Outside air door stuck.

Discharge Air Too Cold At Times

1. Defective EPR valve.

After Continuous Operation Air Flow Drops Off

1. Evaporator coil freeze-up.

Air Comes Out Of A/C or Heater Outlets When Driving In Off Position

1. Vacuum hoses on recirculating door actuator reversed.
2. Defective blower master switch.
3. Recirculating air door not sealing.
4. Defective or improperly installed check valve #2.

Odors Being Discharged Through A/C Or Heater Outlets

1. Vacuum hoses on recirculating door improperly installed.

System Quits On Acceleration

1. Defective vacuum reservoir tank check valve.
2. Vacuum leak.
3. Defective or improperly installed check valve #1.

System Will Not Give Maximum Heat On Hi-Defrost

1. Defective defrost override circuit in amplifier.
2. Disconnected lead from Hi-Defrost switch to amplifier.
3. Defective Hi-Defrost control switch.
4. Defective servo.

System Does Not Achieve Comfortable Condition

1. Miscalibrated comfort control potentiometer.
2. Miscalibrated amplifier.
3. A/C outlets improperly directed.
4. Pinched or disconnected line to aspirator.

Erratic Changes In Temperature Or Blower Speeds

1. Loose connection at sensors, amplifier or servo assembly.
2. Defective amplifier.
3. Defective servo assembly.
4. Defective aspirator or scoop assembly.
5. Blend air door link disconnected.

Blower Speeds Not In Sequence

1. Wiring harness improperly installed at servo or resistor block.

System Will Not Center

1. Defective amplifier.
2. Defective servo.
3. Defective inline fuse.
4. Disconnected battery feed.
5. System battery feed not connected to fuse block battery terminal.

System Comes On In M1-A/C Or Low Heat & Remains

1. Partially connected main harness.
2. Defective fuse.
3. Defective compressor switch.
4. Vacuum hose to compressor switch pinched or disconnected.

Hi Blower Inoperative In A/C & Heat Modes

1. Defective blower motor.
2. Defective Hi blower relay.
3. Defective relay switch in servo assembly.
4. Defective wiring.

Compressor Clutch Engaged In Vent Mode

1. Defective control switch.
2. Defective wiring.

System Operates In Hi Blower Speed In Lo Auto Mode

1. Defective ATC diode.
2. Defective wiring.

System Operates In A/C Mode & M2 Speed Only

1. Defective vent relay.
2. Defective control switch.

System Goes To Max. A/C Or Max. Heat While In Vent Position

1. Defective vent resistor in servo.

System Operates Only In Hi Blower & Hi Auto With Temperature Control

1. Defective Hi blower relay.
2. Defective relay switch in servo.

System Operates In M2 & Hi Speed While In A/C Or Heat Mode & Hi Auto

1. Defective wiring.
2. Open diode.
3. Faulty servo.

1977-80

For trouble-shooting and diagnosis procedures, Refer to System Testing Section.

FORD MOTOR CO.

Insufficient, Erratic or No Cooling

1. Overheated engine.
2. Faulty blower switch.
3. Faulty blower.
4. Leaking or obstructed ducts.
5. Blend door not closing properly.
6. Clutch not operating properly.
7. Compressor drive belt slipping.
8. Compressor service valves not set properly.
9. Leak in refrigerant system, loss of charge.
10. System pressures not normal.
11. Suction pressure extremely low.
12. Suction pressure high, outlet pressure normal.
13. Suction pressure high, outlet pressure low.

Blower Misses Speed Range or Does Not Turn Off

1. Loose or broken connections at blower.
2. Improperly grounded motor.
3. Excessive current draw.
4. Faulty blower switch.
5. Resistor to blower switch lead(s) disconnected or damaged.
6. Faulty resistor.
7. Faulty blower motor.

Inoperative Magnetic Clutch or Erratic Clutch Operation

1. Foreign material on commutator ring.
2. Worn or broken brushes.
3. No continuity at following points:
   (a) Clutch lead.
   (b) Clutch control switch.
   (c) Thermostatic switch-to-clutch control switch.
   (d) Thermostatic switch.
   (e) Ignition switch to thermostatic switch.
   (f) Blower switch to ignition switch.

Excessive Compressor Noise or Vibration

1. Incorrect brackets or braces.
2. Broken or cracked brackets, braces, idlers or pulleys.
3. Drive belts slipping or misaligned.
4. Condenser surface has accumulation of leaves, mud or other foreign material which can cause high system pressure.
5. Clutch rotor rubbing clutch field.
6. Loose clutch on compressor shaft.
7. Damaged clutch bearing.
8. Foreign material or damaged parts in compressor.
9. Maximum allowable clutch runout exceeds specifications.
10. System pressures are abnormal.

Temperature Control (Upper) Lever Door Does Not Provide Adequate Heating or Cooling

1. Rod between heater water control valve and thermostatic switch improperly installed.
2. Temperature control bowden cable not adjusted properly.
3. Insufficient cooling.

Functional Control (Lower) Lever Does Not Operate A/C System Doors Properly

1. Vacuum leak.
2. Faulty control assembly.
3. A/C doors binding.

GENERAL MOTORS

1974 Buick

Blower Inoperative

1. Disconnected loose or corroded blower ground wire.
2. Disconnected feed wire.
3. Defective blower.
4. Defective fuse.
5. Defective blower switch in programmer. Test programmer with Tester J-23678.

Blower, Programmer, and Compressor Inoperative

1. Connection broken at:
   (a) Engine harness connector plug (no. 10 red wire and green wire).
   (b) Cluster extension harness to I/P harness connector plug.
   (c) Cluster extension harness to body extension harness connector plug.

Hi Blower Only MAX HEAT Position

1. Disconnected plug at programmer.

Lo Blower Only

1. Disconnected plug at control head to body extension harness.
2. Disconnected plug at auto relay.

No LO Blower in OFF, VENT, and LO Positions

1. Disconnected plug at the lo relay.
2. Defective relay.

Immediate Lo Blower—Car Start-Up—In-Car Temperature Below 80 Degrees

1. Disconnected plug at engine thermal switch.
2. Defective engine thermal switch.

Immediate Lo Blower—Car Start-Up—In-Car Temperature Above 80 Degrees

1. Check preceding causes.
2. Disconnected plug at heater delay relay.
3. Wrong or defective heater delay relay.

Lo Blower Inoperative All Positions

1. Grounded wire or defective engine thermal switch.

Compressor Inoperative Above 32 Degrees Ambient

1. Disconnected ambient switch.
2. Defective ambient switch.
3. Disconnected plug at compressor.
4. Defective compressor coil.
5. Missed terminal in engine harness connection.

Compressor Operates Below 32 Degrees

1. Shorted ambient switch.
2. Shorted wire in compressor circuit.
3. Defective compressor clutch.

Maximum Heat Mode AUTO RANGE Positions—No Temperature Control—No Automatic Blower Changes

1. Disconnected ambient sensor.
2. Disconnected in-car sensor. Buzzing of programmer transducer may occur when selector lever is changed to "VENT" position.
3. Disconnected vacuum hose in programmer—vacuum valve to checking relay (purple).

Maximum Heat Mode All Positions—No Temperature Control—No Automatic Blower Changes

1. Disconnected vacuum plug at:
   (a) Control Heat (Vacuum)
   (b) Programmer (Vacuum)
   (c) Main vacuum harness connector.
   (d) Vacuum manifold defective.

Maximum Heat Mode and Cold Air From Heater Outlet

1. Disconnected vacuum hose in programmer—vacuum valve to checking relay (black).
2. Disconnected vacuum hose at vacuum source line (black at firewall).
3. Defective vacuum manifold.

Max. A/C All Positions

1. Shorted ambient sensor.
2. Shorted in-car sensor.
3. Disconnected vacuum hose in programmer at:
   (a) Source to transducer (short black hose).
   (b) Vacuum motor to transducer (black and white hose).

Max. A/C Only in AUTO RANGE Positions—Cold Air Out of Outlets

1. Disconnected vacuum hose in programmer—vacuum motor to checking relay (yellow)—shorted sensor string.

Erratic Temperature Control

1. Kinked or disconnected aspirator hose.
2. Defective programmer.
3. Defective control head. Test programmer with Tester J-23678.

Temperature of Discharge Air Too Hot or Too Cool at Mode Change

1. Misadjusted temperature door linkage.

Insufficient Heat

1. Misadjusted temperature door linkage.
2. Defective water valve.
3. Defective engine thermostat.
4. Low coolant.

Modes Will Not Change in AUTO RANGE Positions from A/C to Heat

1. Black and grey vacuum source lines switched.
2. Disconnected source vacuum hose to reservoir.
3. Leaking vacuum reservoir.

Excessive Temperature Difference at Outlets Bi-Level Operation (Left Outlet Warmer Than Center)

1. Vacuum lines switched at upper and lower mode door diaphragms.

Partial Air Flow to Windshield in DEF and No Air Flow to Windshield in BI-LEVEL Position

1. Vacuum lines to defroster (dual) diaphragm switched.
2. Normal BI-LEVEL operation has a delay before door opens.

No Air Flow to Windshield in Either BI-LEVEL or DEF Position

1. Either vacuum line to defroster diaphragm disconnected.
2. Leaking dual diaphragm.

1975-76 Buick Exc. A.T.C.

Blower Inoperative

1. Defective blower motor ground wire.
2. Disconnected feed wire.
3. Defective blower motor.
4. Defective fuse.
5. Defective high blower relay.

Lo Blower Only

1. Disconnected plug at control head to instrument harness.
2. Defective engine thermoswitch.

No Lo Blower In Off

1. Engine thermoswitch disconnected.
2. Open circuit in engine thermoswitch.

Immediate Lo Blower—Car Start-Up

1. Defective engine thermoswitch.
2. Warm engine—normal operation.

Temperature Of Discharge Air Too Hot Or Too Cold

1. Disconnected or misadjusted temperature door cable.
2. Defective water valve.
3. Defective engine thermostat.
4. Low coolant level.

No Full A/C or Heater (Bi-Level Only)

1. Vacuum lines improperly installed at upper and lower mode door diaphragms.

Partial Air Flow To Windshield In DEF & No Air Flow To Windshield In Bi-Level Position

1. Vacuum lines improperly installed at defroster diaphragm.
2. Normal Bi-Level operation has a delay before door opens.

No Air Flow to Windshield In Either Bi-Level or DEF Position

1. Vacuum line to defroster diaphragm disconnected.
2. Defroster diaphragm leaking.

1975-76 Buick w/A.T.C.

System Operates Only At Maximum Air Conditioning

1. With system operating, disconnect programmer wiring harness and note programmer movement through slot in bottom cover. If programmer remains in max. A/C position, remove vacuum connector and check for no or low vacuum supply. If proper vacuum supply is obtained, check voltage to programmer. If correct voltage is obtained, the programmer must be removed and checked using tester J-23678.
2. Other external causes are:
   (a) In-car sensor or ambient sensor shorted.
   (b) Temperature dial miscalibrated or shorted.
   (c) Short in sensor circuit.

System Operates Only At Maximum Heating

1. With system operating, set temperature dial to 65 degrees and control lever to Economy position. Note programmer movement through slot in lower half of cover. If programmer moves to the maximum A/C position, check for the following:
   (a) Disconnected or faulty in-car sensor.

   (b) Disconnected or faulty ambient sensor.
   (c) Loose connection or open in sensor circuit.
   (d) Loose connection at three-way connector.
2. If programmer did not move when the control lever was put in the Economy position, check for the following:
   (a) Disconnected or faulty temperature dial.
   (b) Disconnected or open in ground circuit between control head and programmer.
   (c) Faulty programmer, check with tester J-23678.

System Fluctuates Blower Speeds And/Or Mode Shifts During Acceleration

1. Faulty programmer vacuum check relay.
2. Programmer vacuum motor leaking.

Dial Setting Does Not Provide Comfort

1. Use tester J-23678 to check and reset calibration on the following components:
   (a) Temperature dial.
   (b) Feed back potentiometer in programmer.
   (c) Temperature door link.
   (d) Sensor string.

Blower Cycling & Mode Cycling

1. Normal if temperature dial is moved in large increments.
2. Check system using tester J-23678.

Insufficient Heating Or Cooling

1. Check compressor clutch.
2. Low refrigerant charge.
3. Low engine coolant level.
4. Heater hose or core blocked.
5. Faulty blower relay, blocked air passage or disconnected air hose.
6. Miscalibrated temperature dial.
7. Miscalibrated programmer feedback potentiometer.

Blower Speeds & Modes Shift Occurs Without Temperature Change

1. Check air mix door link connection to programmer shaft and air door crankarm.

System Operates at Low Blower Only

1. Auto blower relay disconnected.
2. Faulty auto blower relay.
3. Loose connection at six-way connector, at programmer or at control head.
4. Open circuit in auto blower circuit.

No Blower Operation In Any Lever Setting

1. Loose connection at blower motor or radio capacitor.
2. Faulty fuse.
3. Faulty blower motor.
4. Open circuit in blower motor wiring harness.
5. Loose connection at six-way connector near programmer.
6. Faulty inline fuse from alternator.

No Blower In Off or Low Settings

1. Loose connection at low blower relay.
2. Faulty low blower relay.
3. Open low speed resistor on programmer.
4. Loose connection at six-way connector or at programmer.
5. Open in low blower circuit.

Operates At Low Blower Except In Hi & DEF

1. Faulty programmer wiper contact or board.
2. Loose connection at programmer.
3. Open circuit in programmer internal wiring.

Blower Operates Only In Hi & Defrost

1. Loose connection at programmer.

Blower Speeds Skipped In Auto & Economy Settings

1. Open resistor on programmer circuit board.

No Heater Turn On In Cold Weather Exc. In Defrost

1. Wiring harness disconnected at thermoswitch, also check wiring continuity of circuit.
2. Defective thermoswitch.

NOTE: Disconnect control head electrical connector, if blower turns off with engine warmed up, ground thermoswitch feed wire. If blower now comes on at low speed, the thermoswitch is defective.

Immediate Heater Turn On In Cold Weather

1. Defective thermoswitch.
2. Defective in-car switch.
3. Check wiring continuity of circuit.

A/C Delayed In Hot Weather Until Engine Warms Up

1. Check wiring continuity of in-car switch circuit.

1977-79 Buick Century, Regal, Skyhawk & Skylark

Blower Motor Inoperative

1. Blower motor lead disconnected, corroded or loose.
2. Feed wire disconnected.
3. Defective blower motor.
4. Defective high blower relay.
5. Blown fuse.

Low Blower Only

1. Disconnected wire connector at control head to instrument panel wiring harness.
2. Defective thermoswitch (Century & Regal).

No Low Blower In Off Position

1. Defective wiring.
2. Disconnected or defective thermoswitch. (Century or Regal).

Discharge Air Temperature Too Cool or Hot

1. Disconnected or improperly adjusted temperature door.

Insufficient Heat

1. Disconnected or improperly adjusted temperature door cable.
2. Defective engine thermostat.
3. Low coolant level.
4. Defective water valve (Century or Regal).

No Full Heat Or A/C (Bi-Level Only)

1. Vacuum lines improperly installed at upper and lower mode door diaphragm.

Intermediate Low Blower During Start Up (Engine Cold)

1. Defective engine thermoswitch.

Partial Air Flow To Windshield In Defroster & No Air Flow In Bi-Level Position

1. Vacuum lines to defroster diaphragm improperly installed.

No Air Flow To Windshield In Either Bi-Level Or Defrost Position

1. Vacuum line to defroster disconnected.
2. Defroster diaphragm leaking.

1977-79 Buick Full Size & 1980 All

For trouble-shooting and diagnosis procedures, refer to System Testing Section.

1974-76 Cadillac

System Operates In Maximum A/C Only

1. With system operating, disconnect programmer wiring harness and note programmer movement through slot in bottom cover. If programmer remains in Max. A/C position, remove vacuum connector and check for no or low vacuum supply. If proper vacuum is obtained, check voltage to programmer. If correct voltage is obtained, the programmer must be removed and checked with tester J-23678.
2. If programmer moved to Max. heating when electrical connector was removed, check for the following:
   1973-76 models:
   (a) Shorted ambient or in-car sensor.
   (b) Shorted or miscalibrated temperature dial.
   (c) Short in sensor wiring harness.
   1976 models:
   (d) Disconnected or open in ground circuit between control head and programmer.
   (e) No voltage being delivered to programmer.
   (f) No ground at amplifier.
   (g) Programmer assembly malfunctioning.

System Operates In Maximum Heating

1. On 1974-75 models, check for the following:
   (a) Disconnected or improperly installed 3 way connector near right side of programmer.
   (b) Disconnected or open in ground circuit between control head and programmer.
   (c) No voltage at programmer.
   (d) No ground at amplifier.
   (e) Malfunctioning programmer assembly.
2. On 1974-76 models, check for the following:
   (a) Disconnected or malfunctioning ambi-

ent or in-car sensor.
   (b) Open circuit or improperly installed sensor string wiring harness.
   (c) Disconnected or malfunctioning temperature dial.

System Fluctuates Blower Speeds and/or Mode Shifts During Acceleration

1. Defective vacuum checking relay.
2. Programmer vacuum motor leaking.

Install vacuum and electrical connectors, then disconnect ambient sensor connector to force programmer to maximum heat. Remove vacuum hose connector from programmer. Programmer should remain in maximum heat. If programmer does not stay in maximum heat, install ambient sensor connector and pinch vacuum motor supply hose with a pair of needle nose pliers. If programmer still moves the vacuum motor is leaking. If the programmer does not move the checking relay is malfunctioning. If mode shifts without blower change, the check valve portion of the checking relay is malfunctioning.

Dial Setting Does Not Provide Comfort

1. Use tester J-23678 to check and reset calibration on the following components.
   (a) Temperature dial.
   (b) Feedback potentiometer in programmer.
   (c) Temperature door link.
   (d) Sensor string.

Blower & Mode Cycling

1. Poor electrical connections.
2. Malfunctioning sensor or amplifier.

Insufficient Heating or Cooling

1. Improper compressor clutch operation. Check low refrigerant detection system.
2. Low refrigerant charge.
3. Low engine coolant level.
4. Heater hose or core blocked.
5. Faulty blower relay, blocked air passage or disconnected air hose.
6. Miscalibrated temperature dial.
7. Air mix door improperly adjusted.
8. Miscalibrated programmer feedback potentiometer.

Blower Speeds & Mode Shift Occurs W/O Temperature Change

1. Check air mix door link connection to programmer shaft and air door crank arm.

System Operates At Low Blower

1. Auto blower relay disconnected.
2. Defective auto blower relay.
3. Loose connection at six-way connector, at programmer or control head.
4. Open circuit in auto blower wiring.

No Blower Operation In Any Lever Setting

1. Loose connection at blower motor or radio capacitor.
2. Defective fuse.
3. Defective blower motor.
4. Open circuit in blower motor wiring harness.
5. Loose connection at six-way connector near programmer.

No Blower Operation In Off or Lo Settings

1. Wiper contacts on control head circuit board malfunctioning.
2. Open low speed resistor on programmer.
3. Open circuit in blower wiring.

Operates At Low Blower Only Exc. In Hi & Def

1. Programmer wiper contact or board malfunctioning.
2. Loose or disconnected electrical connector at programmer.
3. Open circuit in programmer wiring.

Blower Operation In Hi or Def Only

1. Disconnected electrical connector at programmer. System operates at Max A/C only.

Blower Speed Varies In Hi or Def

1. Control head circuit board switch malfunctioning.
2. Open circuit in control head wiring harness.
3. Open in by-pass wiring harness.

Blower Operates With Ignition Off

1. Auto blower relay malfunctioning.

Blower Speeds Skipped In Economy Or Auto Settings

1. Open resistor on programmer blower circuit board.

No Heater Turn-On In Cold Weather Exc. In Off, Lo & Def

1. Heater turn-on switch disconnected.
2. Defective heater turn-on switch.
3. Faulty wiring.

Immediate Heater Turn-On In Cold Weather

1. Defective heater turn-on switch.
2. Defective in-car sensor.
3. Faulty wiring.

A/C Delayed In Hot Weather Until Engine Warms Up

1. Defective in-car sensor.
2. Open circuit in control head wiring harness.
3. Faulty wiring in the in-car sensor.

1974-76 Chevrolet

Blower Inoperative

1. Defective fuse (fuse block or in-line).
2. Disconnected, loose, or corroded blower ground wire.
3. Disconnected feed wire.
4. Defective blower.
5. Defective blower switch in programmer. Except De-Ice and Off.

NOTE: Test programmer with Tester J-23678. Defective programmer blower resistor or wiper assembly.

Blower, Programmer, and Compressor Inoperative

1. Electrical connection broken at compressor or blower.
2. Check 25 AMP A/C fuse.
3. Check programmer wiring harness.

Hi Blower Only, HI, AUTO, or DE-ICE—Max. Heat

1. Disconnected electrical plug at programmer.

LO Blower Only With Max. Heat

1. Disconnect electrical plug at control head.

LO Blower Only A/C or Heat

1. Disconnected electrical plug at auto relay.
2. Defective auto relay.

No LO Blower in OFF, VENT and LO Positions

1. Disconnected electrical plug at the lo relay.
2. Defective lo relay.

Immediate Blower-Cold Eng.-In-Car Temperature Below 70 Degrees

1. Defective control head thermal switch.
2. Defective engine thermal switch.
3. ETS shorted to ground.

Hi Blower DE-ICE Only In-Car Temperature Below 70 Degrees

1. Engine thermal switch or connector.

Compressor Inoperative Above 40 Degrees Ambient

1. Disconnected ambient switch.
2. Defective ambient switch.
3. Disconnected electrical plug at compressor
4. Defective compressor coil.
5. Control head connector disconnected.
6. Control head switch defective.

Compressor Operates Below 25 Degrees

1. Shorted ambient switch.
2. Shorted wire in compressor circuit.
3. Defective compressor clutch.

Maximum Heat Mode LO, AUTO, Hi—No Temperature Control—No Automatic Blower Changes

1. Disconnected ambient sensor.
2. Disconnected in-car sensor.

NOTE: Buzzing of programmer transducer may occur when selector lever is changed to "VENT" position.

AUTO Range—No Temperature Control—No Auto. Blower Change

1. Disconnected vacuum hose in programmer—vacuum valve to checking relay (purple).

Control in AUTO Head Mode all Positions—No Temperature Control—No Automatic Blower Changes

1. Pinched vacuum harness or disconnected vacuum plug at:
   (a) Control head (vacuum).
   (b) Programmer (vacuum).
   (c) In-Line connector.
2. Vacuum tank defective.

Maximum Heat Mode and Cold Air from Heater Outlet

1. Disconnected or pinched vacuum hose in programmer—vacuum valve to checking relay (black).
2. Disconnected or pinched vacuum hose at vacuum source line (black at firewall).
3. Defective vacuum tank.
4. Temperature door link adjustment loose or misadjusted.
5. Defective water valve.

Max. A/C All Positions

1. Shorted ambient sensor.
2. Shorted in-car sensor.
3. Disconnected or pinched vacuum hose in programmer at:
   (a) Source to transducer (short black hose).
   (b) Vacuum motor to transducer (black and white hose).

Max. Heat All Positions

1. Defective checking relay.

Max. A/C Only in Auto Range Positions—Cold Air Out of Outlets

1. Disconnected or pinched vacuum hose in programmer—vacuum motor to checking relay (yellow)
2. Shorted sensor string.

Erratic Temperature Control

1. Kinked or disconnected aspirator hose.
2. Defective programmer.
3. Defective control head.

NOTE: Test programmer with Tester J-23678.

Temperature of Discharge Air Too Hot or Too Cool at Mode Change

1. Misadjusted temperature door linkage.

Insufficient Heat

1. Misadjusted temperature door linkage.
2. Defective water valve.
3. Defective engine thermostat.
4. Low coolant.

Modes Will Not Change in Auto Range Positions from A/C to Heat

1. Black and gray vacuum source lines switched, or pinched.
2. Disconnected or pinched source vacuum hose to reservoir.
3. Leaking vacuum reservoir.

Excessive Temperature Difference at Outlets Bi-Level Operation (Left Outlet Warmer than Center)

1. Vacuum lines switched or pinched at upper and lower mode door diaphragms.

Partial Air Flow to Windshield in De-Ice and No Air Flow to Windshield in De-Fog Position

1. Vacuum lines to defroster (dual) diaphragm switched, pinched or disconnected.

NOTE: Normal de-fog operation has a delay before door opens.

No Air Flow to Windshield in Either De-Fog or De-Ice Position

1. Either vacuum line to defroster diaphragm disconnected or pinched.
2. Leaking dual diaphragm.

1977-80 Chevrolet

For trouble-shooting and diagnosis procedures, refer to System Testing Section.

1974-76 Oldsmobile

Evaporator Pressure Too High

1. Defective suction throttling valve.
2. Restriction in suction line.
3. Loose compressor drive belt.
4. Defective clutch or coil.
5. Defective expansion valve.
6. Expansion valve capillary tube not too tight to evaporator suction line.

High Pressure Side of System Too High

1. Engine overheated.
2. Restricted flow through condenser.
3. Air in system or overcharge of refrigerant. Momentarily discharge system on discharge side with engine not running; then, operate system and recheck pressure. Repeat as necessary. Check sight glass with system underload.
4. Restriction in condenser, dehydrator receiver assembly, or any discharge or liquid line.
5. Too much oil in the compressor.

Nozzle Discharge Air Too Warm (All Other Readings OK)

1. Air hoses not properly connected.
2. Defective or mispositioned evaporator drain hoses.
3. Poor Seal—Evaporator to cowl.

Clutch Slippage

1. Head pressure too high.
   Discharge system until bubbles appear in sight glass and then add one pound of refrigerant.
2. Pulley wobbles.
   Check and replace, if necessary, the pulley bearing. If pulley has been worn by the bearing, replace pulley.

Velocity of Air Discharge Nozzles Too Low

1. Restricted evaporator core in evaporator assembly.
2. Restricted air hoses.
3. Defective blower motor.
4. Defective switches.
5. Poor wiring connection (Low voltage at blower).

Evaporator Pressure Too Low

1. Insufficient refrigerant charge.
2. Restricted air flow through evaporator.

Water Blowing Out Air Discharge Nozzle

1. Plugged or kinked evaporator drain hose.

Inoperative Controls

1. Inadequate vacuum.
   Check vacuum. All controls should move with 10 in. of vacuum. Check hoses.
2. Control cables improperly adjusted.

Compressor Not Engaged

1. No voltage at compressor coil.
   Trace system to power source, determine place of loss and correct.
2. Compressor coil inoperative although there is proper voltage to coil.

Compressor Engaged But Not Operational

1. Clutch slipping.
2. Belt slipping.
3. High torque compressor (Seized).
4. Compressor noisy.
   Clutch not engaged, remove belt to determine location, check for interference between coil housing and pulley hub; if exists, repair compressor.

Blower Motor Inoperative in All Speeds

1. Blown fuse.
2. Defective HI blower relay.
3. Defective motor.
4. Break in wiring or loose connections.

Blower Motor Inoperative in High Selector Position

1. Blown fuse.
2. Defective high blower relay.
3. Defective fan switch.

Blower Motor Inoperative in LO, M1 or M2 Speeds

1. Open circuit between fan switch and resistor.
2. Defective resistor.

Automatic Climate Control System Operates Only at Maximum Cooling and High Blower

1. Defective amplifier.
2. Defective transducer.
3. Power servo stuck.
4. Defective vacuum motor or vacuum relay.
5. Temperature door link disconnected.
6. Shorted in-car sensor.

Automatic Climate Control System Operates Only at Maximum Heating and High Blower

1. Defective amplifier.
2. Amplifier connector loose or disconnected.
3. Defective transducer.
4. Poor ground at transducer case.
5. In-car sensor wire disconnected.
6. Defective in-car sensor.
7. Control panel not grounded.
8. Defective temperature dial rheostat.
9. Short to ground in sensor string.

Insufficient Cooling Air Not Cold

1. No refrigerant in system.
2. Feedback potentiometer shaft misadjusted.
3. Defective transducer.
4. Defective amplifier.
5. Ambient duct or in-car sensor touching cold object.
6. Duct sensor hose kinked, cracked or disconnected.
7. Ambient switch contacts dirty.
8. Compressor inoperative. No voltage at clutch coil.
9. Compressor clutch slipping.
10. Compressor belt slipping.

1977-80 Oldsmobile

For trouble-shooting and diagnosis procedures, refer to System Testing Section.

1974-76 Pontiac

Insufficient Heating

1. Check for proper engine thermostat.
2. Check blower operation.
3. Check temperature lever and control cable for proper operation.
4. Check water valve for proper operation, if equipped.

Insufficient Cooling

1. Check blower operation.
2. Check for obstructions, proper routing and proper connection of the vacuum hoses and check valve.
3. Insufficient air flow.
4. Air leaks in air system.
5. Leak in refrigerant system.
6. Low refrigerant charge.
7. Restriction in high pressure side of system.
8. Defective POA valve.
9. Moisture in refrigerant system.

Compressor Discharge Pressure Too High

1. Engine overheated.
2. Overcharge of refrigerant or air in system. Systems with excess discharge pressures should be slowly depressurized at the receiver inlet connection, observing behavior of high pressure gauge indicator.

NOTE: Disconnect thermal limiter before depressurizing system.

3. Overcharge of Freon or air in system. Install gauge set and bleed off Freon from discharge side for 20 seconds. After 20 seconds close valves and recheck operating pressures. Repeat until discharge pressure is normal. Check sight glass. If bubbles appear it indicates that air was in system.
4. Restriction in condenser, receiver-dehydrator or any high pressure line.
5. Condenser air flow blocked.
6. P.O.A. valve inlet pressure too high.

Compressor Discharge Pressure Too Low

1. Insufficient Freon. Check for presence of bubbles or foam. If bubbles or foam are noted, charge with Freon.

NOTE: It is not unusual for bubbling to occur on minimum cooling and "LO" blower in mild weather even with fully charged system.

2. Defective compressor.
3. Plug in Freon system.
   - *(a)* Disconnect fittings assembly and detach hoses from the compressor; disconnect receiver-dehydrator inlet and outlet tubes. Seal the compressor ports and receiver fittings.
   - *(b)* Check ends of lines for slipping plugs or torn-off pieces of these plugs left in at assembly.
   - *(c)* Blow dry nitrogen, Freon or dry air thru lines to determine if lines or condensor is plugged.

   CAUTION: Bleed air hose of all moisture.

   - *(d)* If plug in the system has not been found, disconnect P.O.A. valve from the evaporator.
   - *(e)* Blow thru expansion valve and evaporator, to check for plugged evaporator.
4. P.O.A. valve inlet pressure too low.

P.O.A. Valve Inlet Pressure Too High

1. P.O.A. Valve Schrader stuck open.

CAUTION: Use only Schrader valve prescribed.

2. Expansion valve capillary tube to evaporator tube. Remove insulation and inspect for clearance between tube and bulb. If gap exists, move bulb to establish contact, reclamp and reinsulate.
3. Expansion valve inoperative. Remove expansion valve and inspect screen for foreign objects. If present, there is a possibility seat is being held open. Install new expansion valve; if condition is corrected, discard valve removed.

P.O.A. Valve Inlet Pressure Too Low

1. P.O.A. Valve stuck open. Shut off engine. If inlet pressure does not rise valve is stuck open. Also indicated by less than 3 to 4 psi pressure differential between suction pressure and P.O.A. Valve inlet pressure.
2. Expansion valve capillary tube broken, inlet screen plugged or valve otherwise fails.
3. Restriction in system hoses or tubes.

Nozzle Outlet Temperature Too Cold

1. P.O.A. valve faulty.

Nozzle Outlet Temperature Too Warm

1. Poor seal—evaporator core to evaporator inlet case or evaporator to heater case.
2. Defective or missing evaporator drain hose.
3. Air hoses not properly connected.
4. Vacuum control hoses not connected properly.
5. Insufficient Freon.
6. P.O.A. Valve faulty.
7. Expansion valve faulty.

Blown Thermal Limiter

1. Low or totally discharged refrigerant system.
2. Expansion valve stuck closed.
3. Restricted oil bleed line.
4. POA valve stuck open.
5. Incorrect valve core in oil bleed line fitting on POA or VIR.
6. Blown during charging of system.
7. Faulty super heat switch.
8. Thermal limiter installed in improper location.

1977-80 Pontiac

For trouble-shooting and diagnosis procedures, refer to System Testing Section.

System Testing

The following material deals with the performance and testing of the various systems used. It is presented by car manufacturers because of the differences in the units.

AMERICAN MOTORS

System Pressure Check

The chart, Fig. 1, listing temperatures and pressures is based on in-shop or service stall operating conditions and may be used as a guide when diagnosing system operation.

Should the existing ambient temperature vary from those listed on the chart, averages may be calculated.

A thermometer should be inserted in the air discharge outlet to measure the discharge air temperature.

Connect the pressure gauge and manifold assembly to the compressor service valves and "crack" both valves. Operate the system with the engine at 1500 RPM and the thermostat turned fully clockwise but not into the "Desert Only" detent or Max. or Cold detent. Insert a thermometer in the discharge air outlet and observe the air temperature. If pressures are abnormal, refer to Trouble-Shooting American Motors Pressure Diagnosis.

Pressure Diagnosis

Low Side Low, High Side Low

1. System low on refrigerant.
2. Expansion valve clogged.

Low Side High, High Side Low

1. Compressor has internal leak.
2. Compressor head gasket leaking.
3. Expansion valve defective.
4. Compressor drive belt slipping.

Low Side High, High Side High

1. Condensor fin clogged.
2. Air in refrigerant system.
3. Expansion valve defective.
4. Worn or loose fan belts.

Low Side Low, High Side High

1. Expansion valve defective.
2. Restricted refrigerant liquid line.
3. Restricted receiver or condensor.

Low Side & High Side Normal But Inadequate Cooling

1. Air in refrigerant system.
2. Moisture in refrigerant system.

A/C System Diagnosis

When diagnosing A/C problems, refer to Figs. 2 through 6, for vacuum motor airflow door position and control system functions. Refer to Figs. 7 through 9 for A/C system diagnosis procedures.

Magnetic Clutch Diagnosis

Refer to Figs. 10 through 18, for magnetic clutch diagnosis procedures.

Vacuum System Check

Matador & Pacer

1. Check and note amount of vacuum available at intake manifold source.
2. Connect a vacuum gauge to vacuum storage tank outlet fitting and note reading.

Relative Humidity (percent)	Surrounding Air Temperature (°F)	Engine Speed (RPM)	Maximum Desirable Center Register Discharge Air Temp. (°F)	Suction Pressure PSI (REF)	Head Pressure PSI (+25 PSI)
	70		40	11	177
20	80	1500	41	15	208
	90		42	20	226
	100		43	23	255
	70		40	12	181
30	80	1500	41	16	214
	90		42	22	234
	100		44	26	267
	70		40	13	185
40	80	1500	42	18	220
	90		43	23	243
	100		44	26	278
	70		40	14	189
50	80	1500	42	19	226
	90		44	25	251
	100		46	27	289
	70		41	15	193
60	80	1500	43	21	233
	90		45	25	259
	100		46	28	300
	70		41	16	198
70	80	1500	43	22	238
	90		45	26	267
	100		46	29	312
	70		42	18	202
80	80	1500	44	23	244
	90		47	27	277
	100		—	—	—
	70		42	19	206
90	80	1500	47	24	250
	90		48	28	284
	100		—	—	—

Fig. 1 Temperature — pressure chart

3. Turn off ignition and note if there is a vacuum drop. If a vacuum drop occurs, the check valve in the vacuum storage tank is defective and the vacuum tank should be replaced.
4. If no vacuum drop occurs, start engine, then connect vacuum gauge to each inlet fitting of the vacuum control switch and note reading.
5. Vacuum reading should be the same at each inlet fitting as that noted at intake manifold and storage tank.
6. If vacuum reading is low, check vacuum hose for leaks.
7. If the above vacuum checks, do not indicate a vacuum leak, connect vacuum gauge to suspect hose at airflow door vacuum motor and check for reading on gauge.
8. If vacuum reading drops, this indicates either vacuum hose is leaking or vacuum control switch is defective.
9. If vacuum reading does not drop, disconnect vacuum gauge and reconnect vacuum hose.
10. Place control switch in desired position and check vacuum motor for proper operation and leaks.
11. If vacuum motor does not operate and a vacuum leak is indicated, replace vacuum motor.
12. If no vacuum leak is indicated, check airflow door for binding and lubricate all pivot points with Lubriplate.
13. On 1975-77 Pacer models, check floor door motor by depressing Heat push button. Door should be fully opened and manifold vacuum applied to both motor fittings. Do

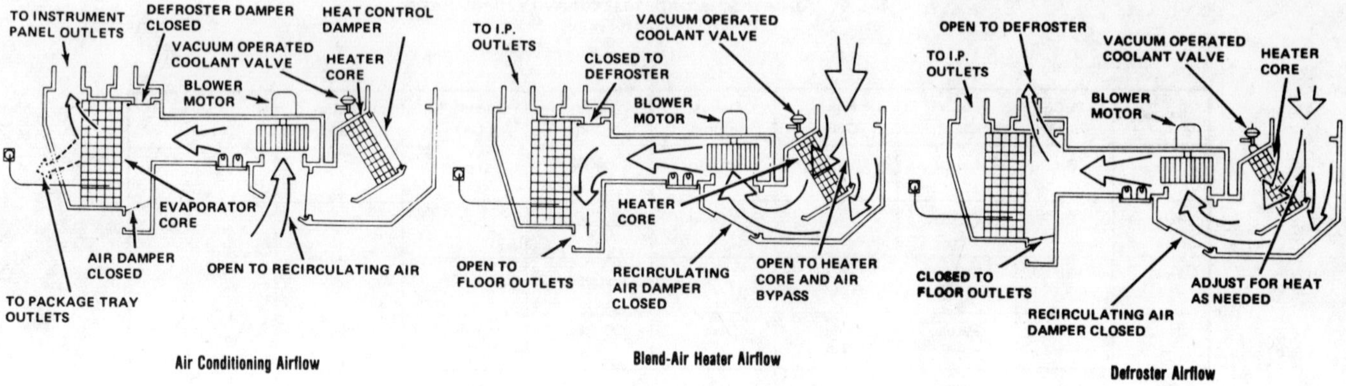

Fig. 2 Airflow schematic test chart. 1974-80 Except Ambassador, Matador & Pacer

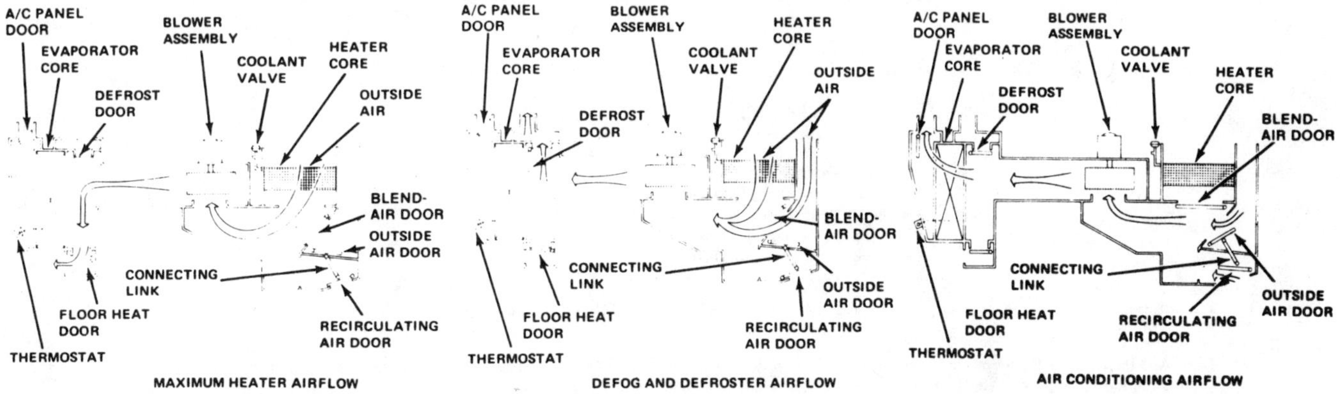

Fig. 3 Airflow schematic test chart. 1974 Ambassador & 1974-78 Matador

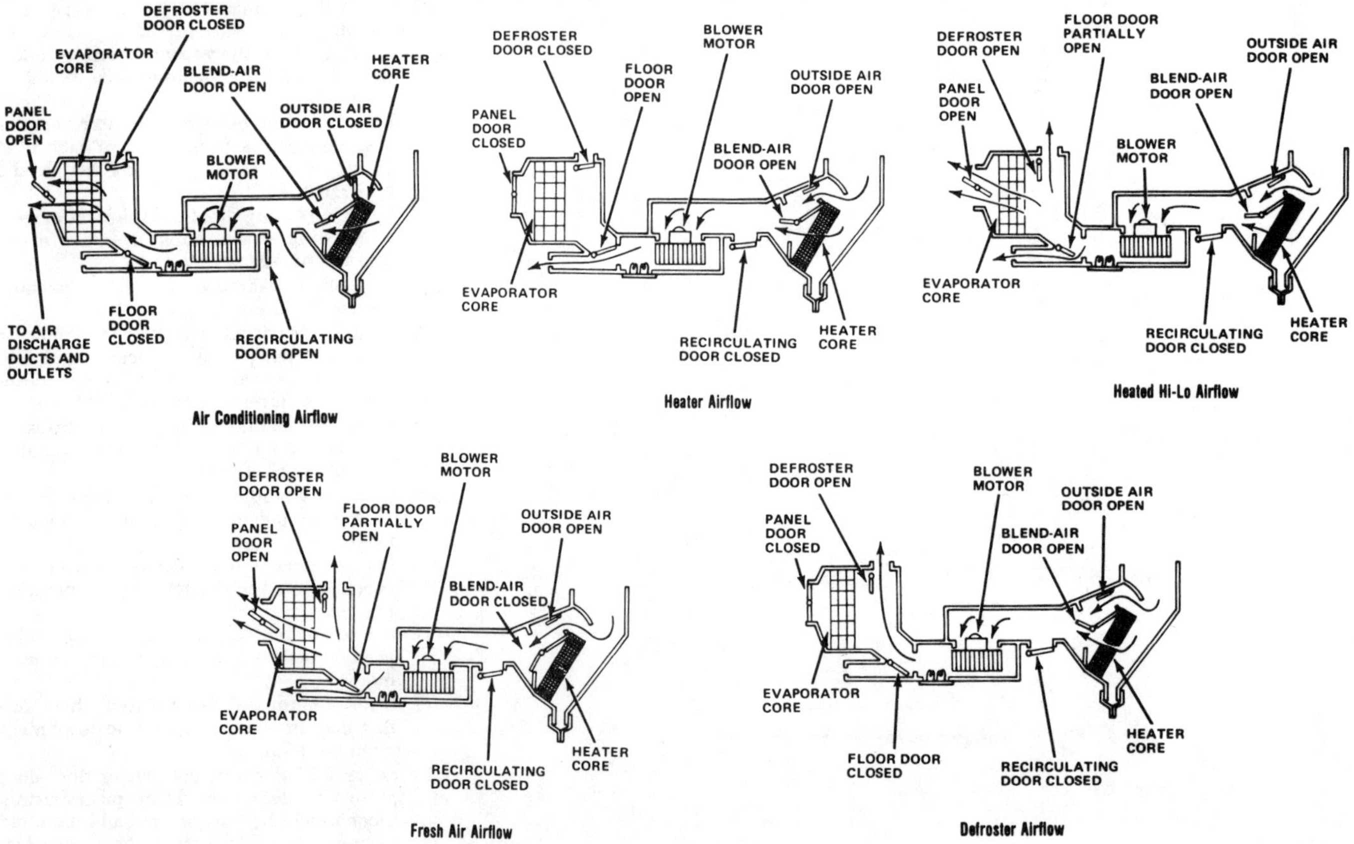

Fig. 4 Airflow schematic test chart. 1975-80 Pacer

Mode	A/C Clutch	Floor Door	Panel Door	Defrost Door	Recirc Door	Outside Air Door	Water Valve	Blower Fan
Off	Off	Closed	Closed	Closed	Closed	Closed	Closed	——
A/C	On	Closed	Open	Closed	Open	Closed	Closed	Operators Choice 4 Speeds With Off
Heat	Off	Open	Closed	Closed	Closed	Open	Open	
Hi-Lo	Off	Partial Open	Open	Open	Closed	Open	Open	
Def	Off	Closed	Closed	Open	Closed	Open	Open	

1. Outside air 100% in heat, hi-lo and defrost modes and 0 to 30 % in A/C mode dependent on temperature control lever position from maximum cold to warm.

2. A/C clutch on selection by mode pushbutton dependent on fan switch being energized.

Floor Door – Vacuum to Open — 3 position vacuum motor; 1 port partial movement; 2 ports full movement
Panel Door – Vacuum to Open
Defrost Door – Vacuum to Close
Outside Air Door – Vacuum to Close
Water Valve – Vacuum to Close
Recirc. Door – Vacuum to Open

3. In heat, hi-lo, defrost modes the water valve is shut off at maximum cold setting of temperature control lever.

4. In A/C mode, the A/C thermostat is adjusted by the temperature control lever. The cold detent position of temperature control lever adjusts A/C thermostat to provide coldest temperature without danger of icing evaporator core.

Fig. 5 Control system functional test chart. 1975-80 Pacer

Control System Mode Selection

Mode Lever	A/C Clutch	Floor Door	Panel Door	Defrost Door	Outside Air*			Water Valve
					0%	Partial	100%	
Off	Off	Closed	Closed	Closed	●			Closed
A/C	On	Closed	Open	Closed	●**	●		Closed
Bi-Lvl	Off	Open	Open	Open			●	Open ***
Heat	Off	Open	Closed	Closed			●	Open ***
Defog	On	Closed	Closed	Open			●	Open ***
Def	Off	Closed	Closed	Open			●	Open ***

Temperature Lever – Operates the A/C thermostat to left of center and the heater temperature control door to right of center.
Fan Switch – Is independent of mode setting except must be energized in A/C and DEFOG modes to activate A/C compressor clutch.
* Outside Air Door linked to recirculating door (outside door open/recirculating door closed).
** Outside Air Door – Set for 100% recirculating at full cool setting of Temperature Lever.
*** Water Valve – Closed at full cool setting of Temperature Lever.

Vacuum Motor Operation:
Floor Door – Vacuum to Open
Panel Door – Vacuum to Open
Defrost Door – Vacuum to Close
Recirculating Door – Vacuum to Open – 2 Stage: End Nipple Partial, Both Nipples Full
Water Valve – Vacuum to Close

Fig. 6 Control system functional test chart. 1976-78 Matador

press Hi-Low push button, door should be partially open and manifold vacuum applied to end motor fitting and side fitting should be vented.

Water Valve Test

1974-75

1. Position temperature control lever in off position on all models except Ambassador and Matador. On Ambassador and Matador, position temperature control lever in Max. Cool.
2. Start engine and hold water valve while removing vacuum line.
3. Movement should be felt as water valve opens. If no movement is felt, check vacuum line for vacuum. If vacuum is present, replace water valve. If no vacuum is present, check for ruptured or plugged line.

1976-77 All & 1978 Matador

1. Disconnect vacuum hose from water valve and tee-in vacuum gauge between hose and water valve fitting, then start engine.
2. On Gremlin and Hornet models, position temperature control lever in high. On Matador and Pacer, position temperature control lever on warm and selector in heat mode. No vacuum reading should be indicated on gauge.
3. On Gremlin and Hornet models, position temperature lever in off. On Matador and Pacer models, position selector lever in off or A/C. Water valve should close and manifold vacuum should be indicated on gauge.
4. On Matador and Pacer models, position selector lever in heat, water valve should open and no vacuum reading should be indicated on gauge. Move temperature lever to Max. cold position, water valve should close and manifold vacuum should be indicated on gauge.
5. On all models, if water valve does not function as described above, refer to Water Valve Diagnosis.

1078-80 AMX, Concord, Eagle, Gremlin & Spirit

1. Disconnect vacuum hose from water valve and tee-in a vacuum gauge between hose and water valve fitting, then start engine.
2. Position temperature control lever on warm and selector lever on heat. No vacuum should be indicated on gauge.
3. While moving temperature lever to cool and/or selector lever to A/C, observe water valve. Water valve should close and vacuum gauge should indicate manifold vacuum.
4. If water valve does not function as described above, refer to Water Valve Diagnosis.

1978-80 Pacer

1. Disconnect vacuum line from water valve and tee-in a vacuum gauge between hose and water valve fitting, then start engine.
2. Press heat or defrost push button and note vacuum gauge reading. No vacuum should be indicated on gauge.
3. While pressing off A/C or vent push button, observe water valve. Water valve should close and vacuum gauge should indicate manifold vacuum.
4. While pressing heat or defrost push button observe water valve. Water valve should close and no vacuum should be indicated.
5. If water valve does not function as described above, refer to Water Valve Diagnosis.

Water Valve Diagnosis

1976-77 Gremlin & Hornet

1. Water valve does not close with vacuum applied.
   a. Defective water valve.
2. No vacuum applied to water valve when temperature control lever is in off position.
   a. Vacuum hose disconnected from intake manifold.
   b. Defective check valve.
   c. Leaking or plugged vacuum hose from check valve to switch.
   d. Vacuum hose disconnected from switch.
   e. Improperly adjusted or defective switch.
   f. Leaking or plugged vacuum hose from switch to water valve.

1976-77 Pacer & 1976-78 Matador

1. Water valve does not close with vacuum applied.
   a. Defective water valve.
2. No vacuum applied to water valve with selector lever in off or A/C position (operates properly with temperature control lever).
   a. Improper vacuum hose routing.
   b. Defective vacuum control switch (push button switch on Pacer models).
   c. Leaking vacuum hose.
3. No vacuum applied to water valve with temperature control lever in Max. Cold position (operates properly in off or A/C modes).
   a. Temperature control lever not depressing vacuum switch button.
   b. Defective vacuum switch.
   c. Leaking vacuum hose.
4. Vacuum applied to water valve with selector in Heat, Bi-Level, Hi-Lo or Defrost mode (temperature control lever in warm position).
   a. Improper vacuum hose routing.
   b. Defective vacuum control switch (push button switch on Pacer models).
   c. Water valve vacuum switch plunger sticking in depressed position.

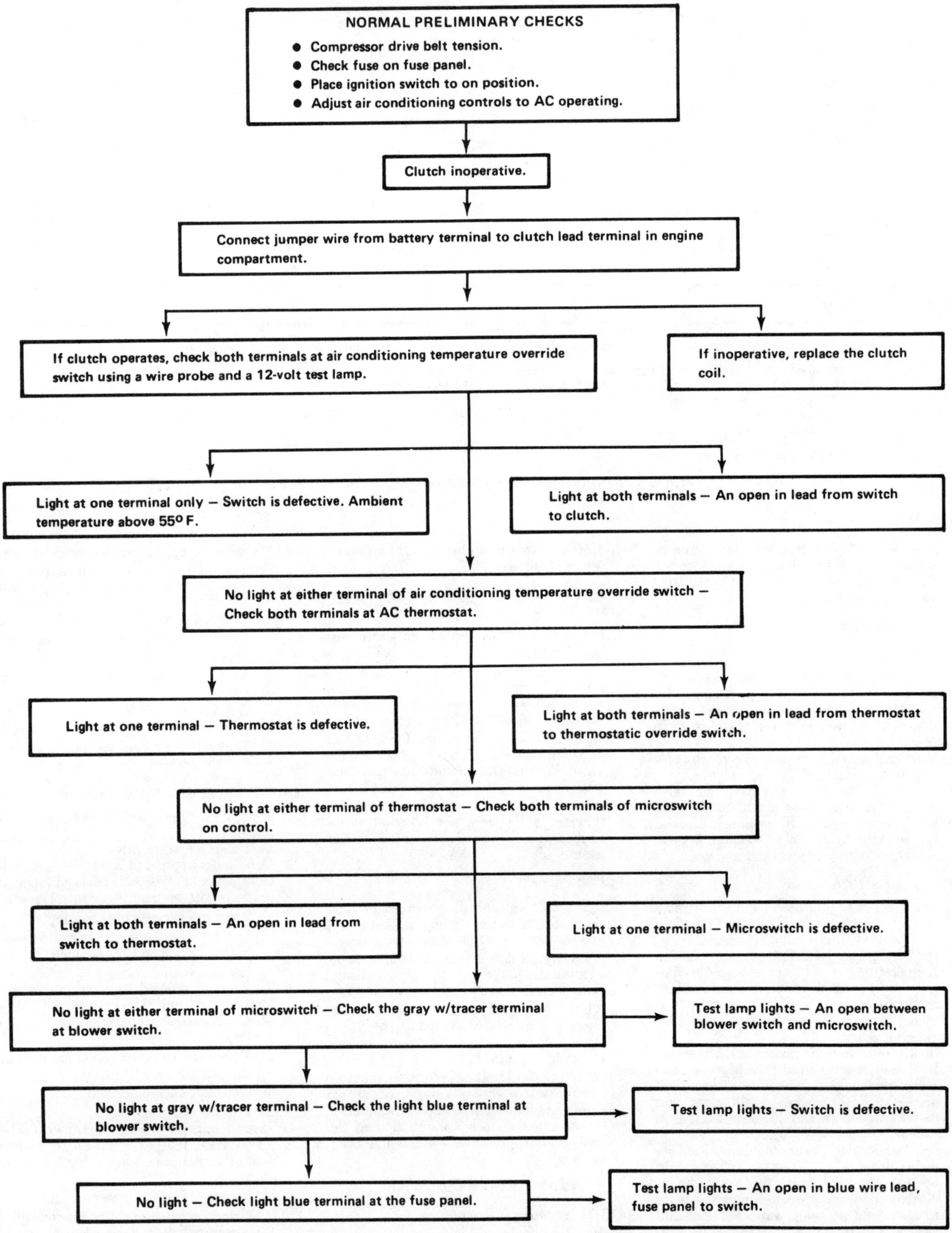

Fig. 7 A/C diagnosis (chart 1). 1974 models

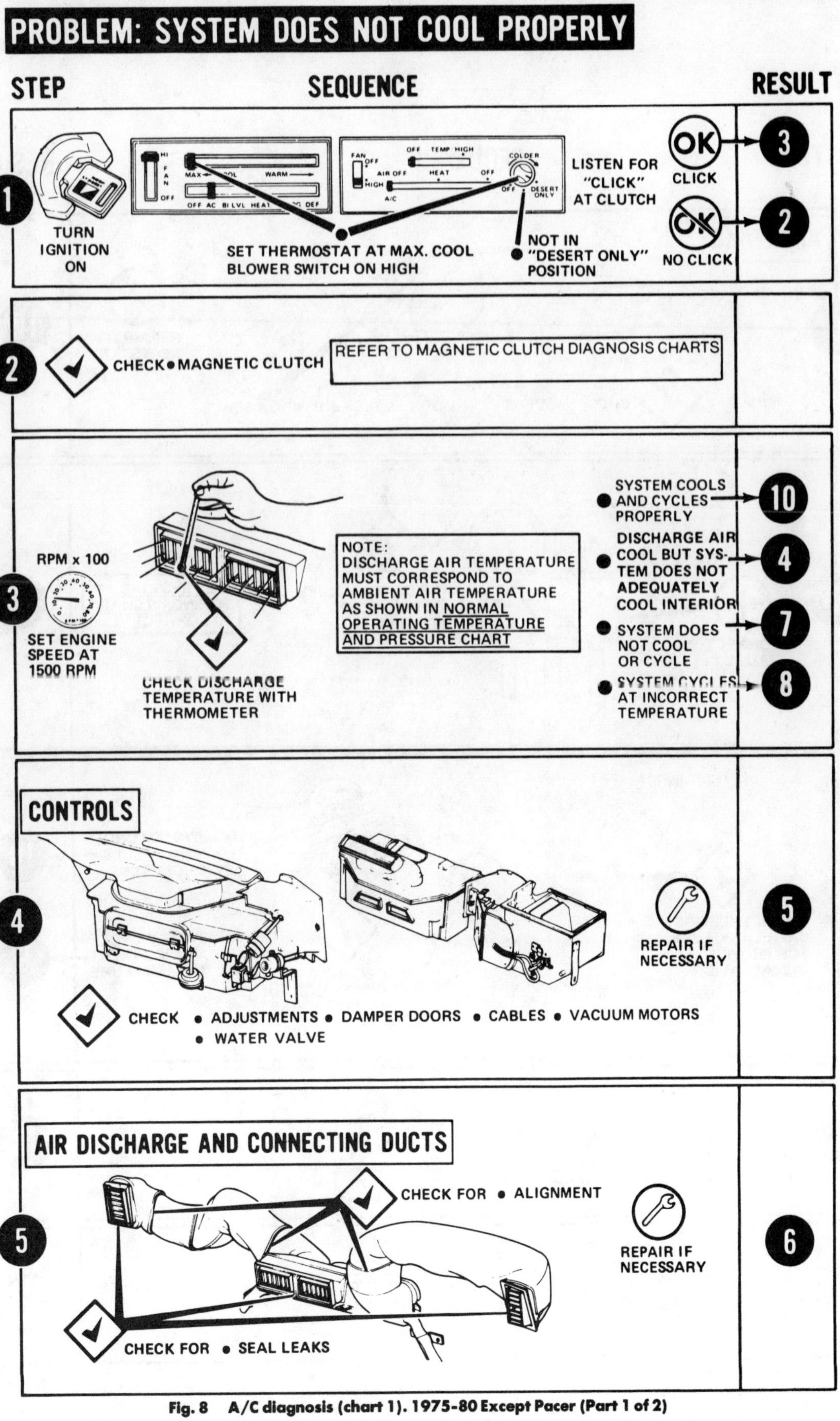

Fig. 8 A/C diagnosis (chart 1). 1975-80 Except Pacer (Part 1 of 2)

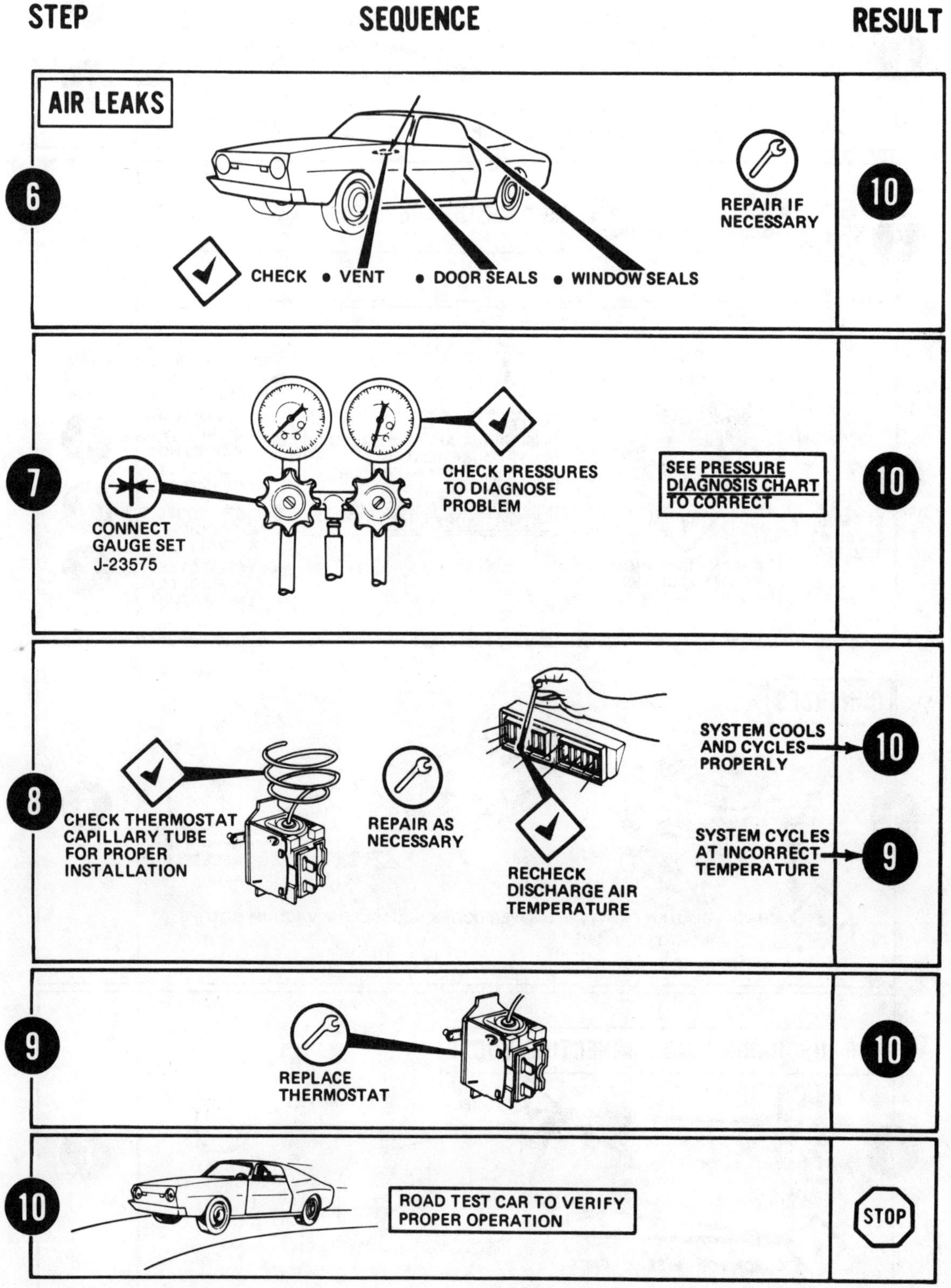

Fig. 8 A/C diagnosis (chart 1). 1975-80 Except Pacer (Part 2 of 2)

PROBLEM: SYSTEM DOES NOT COOL PROPERLY

STEP	SEQUENCE	RESULT
1	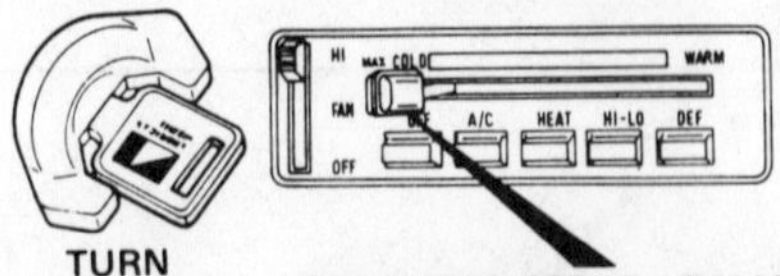 TURN IGNITION ON; SET THERMOSTAT AT MAX. COLD, BLOWER SWITCH ON HIGH, PRESS A/C PUSHBUTTON; LISTEN FOR "CLICK" AT CLUTCH — OK CLICK → 3; NO CLICK → 2	3 / 2
2	CHECK • MAGNETIC CLUTCH — REFER TO MAGNETIC CLUTCH DIAGNOSIS CHARTS	
3	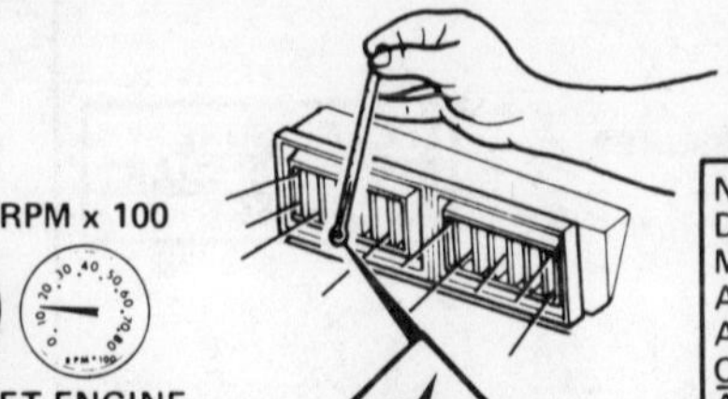 RPM x 100; SET ENGINE SPEED AT 1500 RPM; CHECK DISCHARGE TEMPERATURE WITH THERMOMETER. NOTE: DISCHARGE AIR TEMPERATURE MUST CORRESPOND TO AMBIENT AIR TEMPERATURE AS SHOWN IN NORMAL OPERATING TEMPERATURE AND PRESSURE CHART. • SYSTEM COOLS AND CYCLES PROPERLY → 10 • DISCHARGE AIR COOL BUT SYSTEM DOES NOT ADEQUATELY COOL INTERIOR → 4 • SYSTEM DOES NOT COOL OR CYCLE → 7 • SYSTEM CYCLES AT INCORRECT TEMPERATURE → 8	10 / 4 / 7 / 8
4	CONTROLS 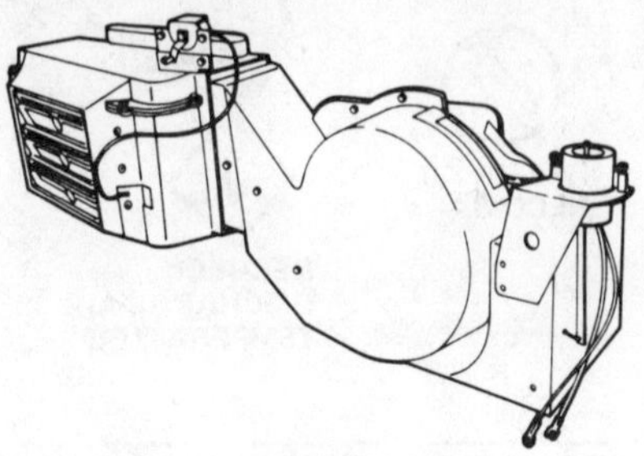 CHECK • ADJUSTMENTS • DAMPER DOORS • CABLES • VACUUM MOTORS • HOT WATER SHUT OFF VALVE; REPAIR IF NECESSARY	5
5	AIR DISCHARGE AND CONNECTING DUCTS 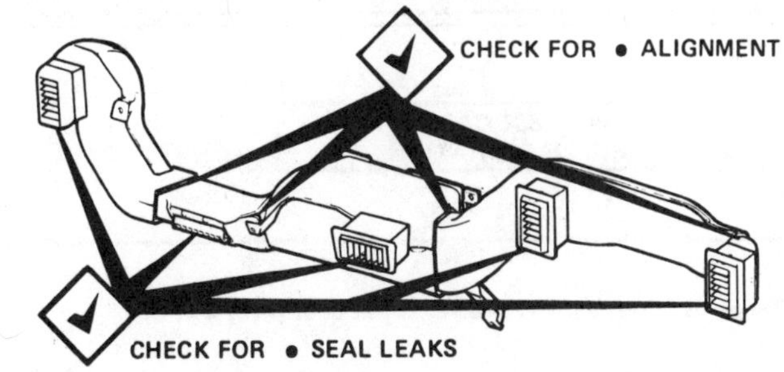 CHECK FOR • ALIGNMENT; CHECK FOR • SEAL LEAKS; REPAIR IF NECESSARY	6

Fig. 9 A/C diagnosis (chart 1). 1975-80 Except Pacer (Part 1 of 2)

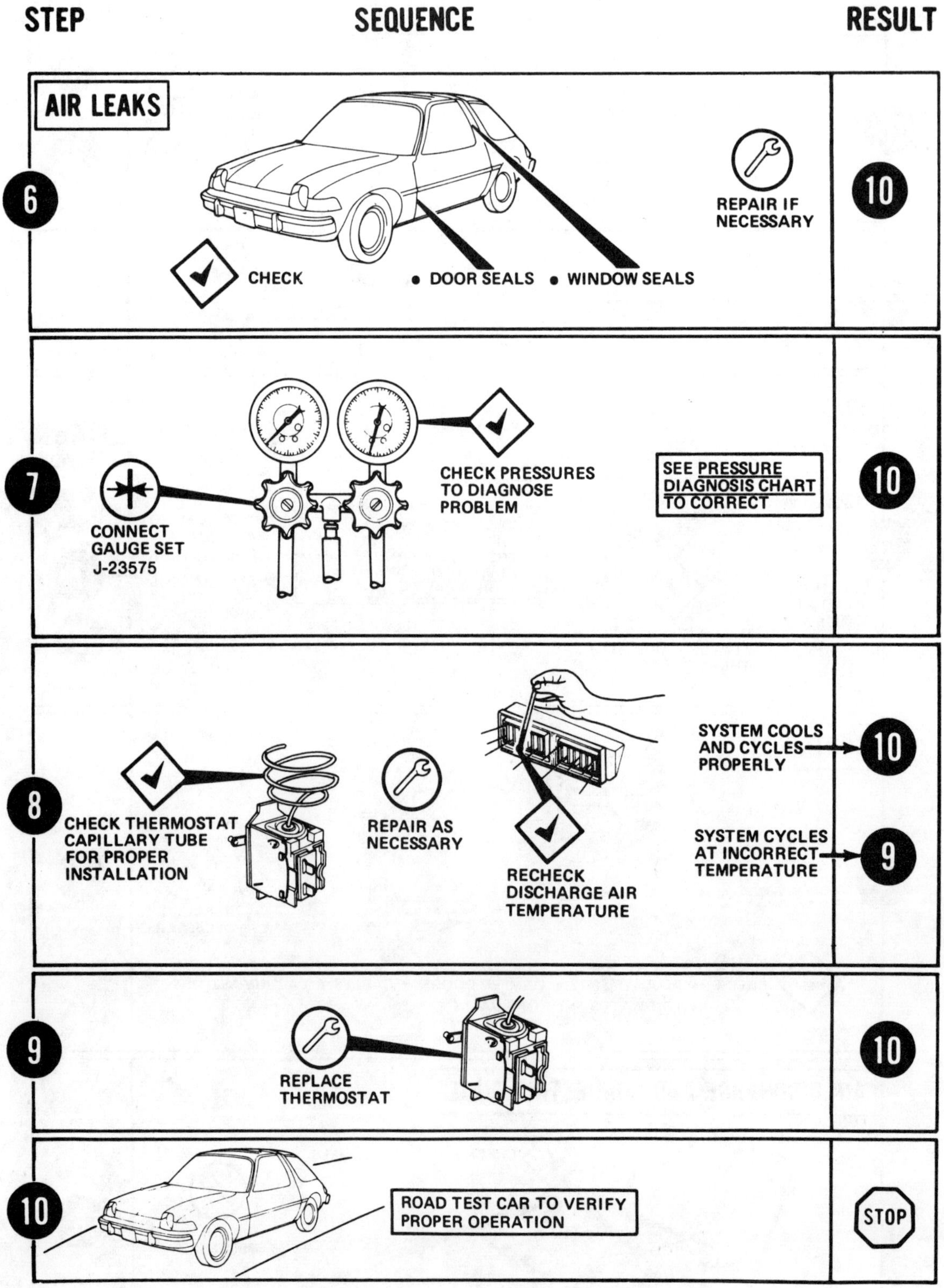

Fig. 9 A/C diagnosis (chart 1). 1975-80 Except Pacer (Part 2 of 2)

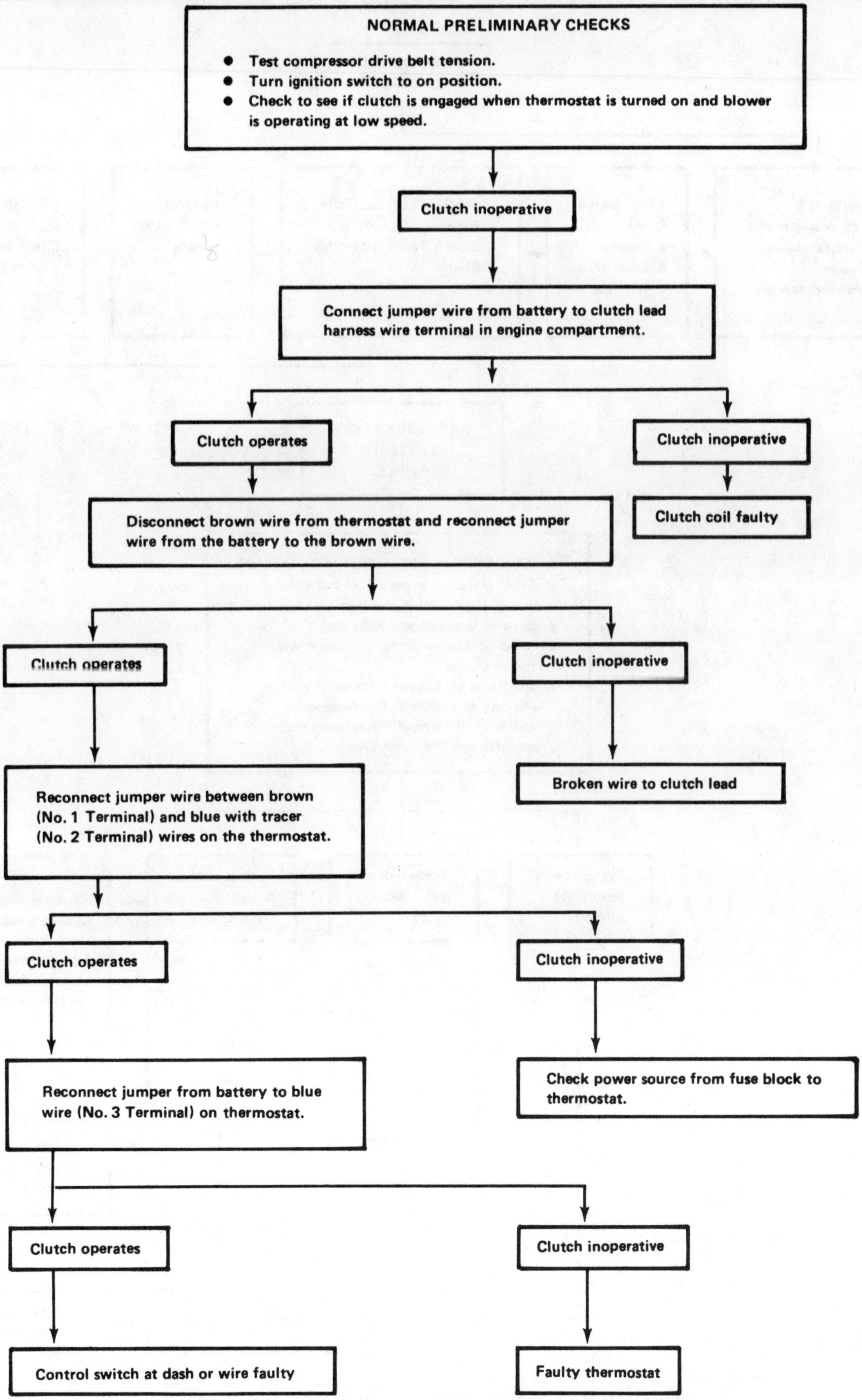

Fig. 10 Magnetic clutch diagnosis (chart 2). 1974 Except Ambassador & Matador

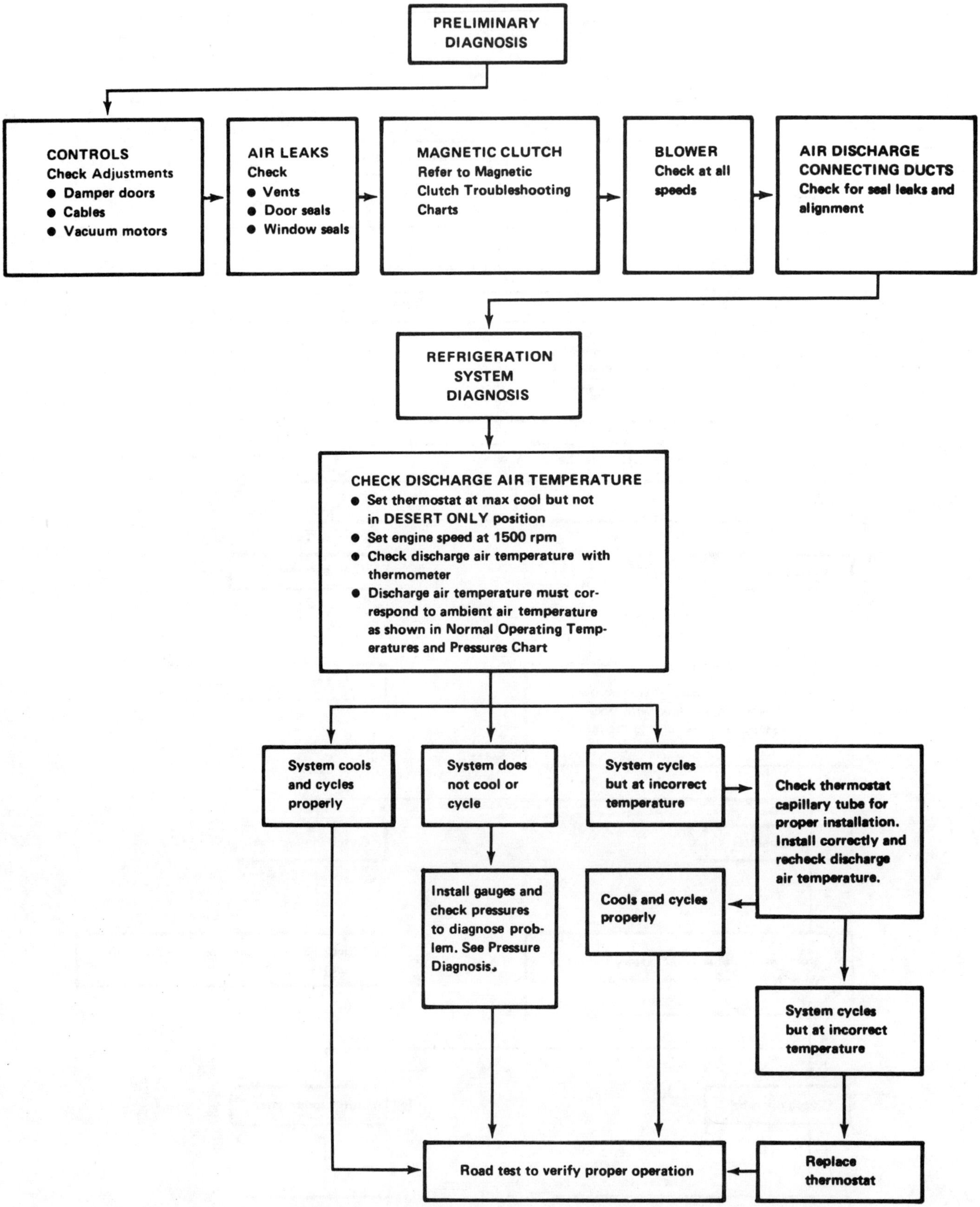

Fig. 11 Magnetic clutch diagnosis (chart 2). 1974 Ambassador & Matador

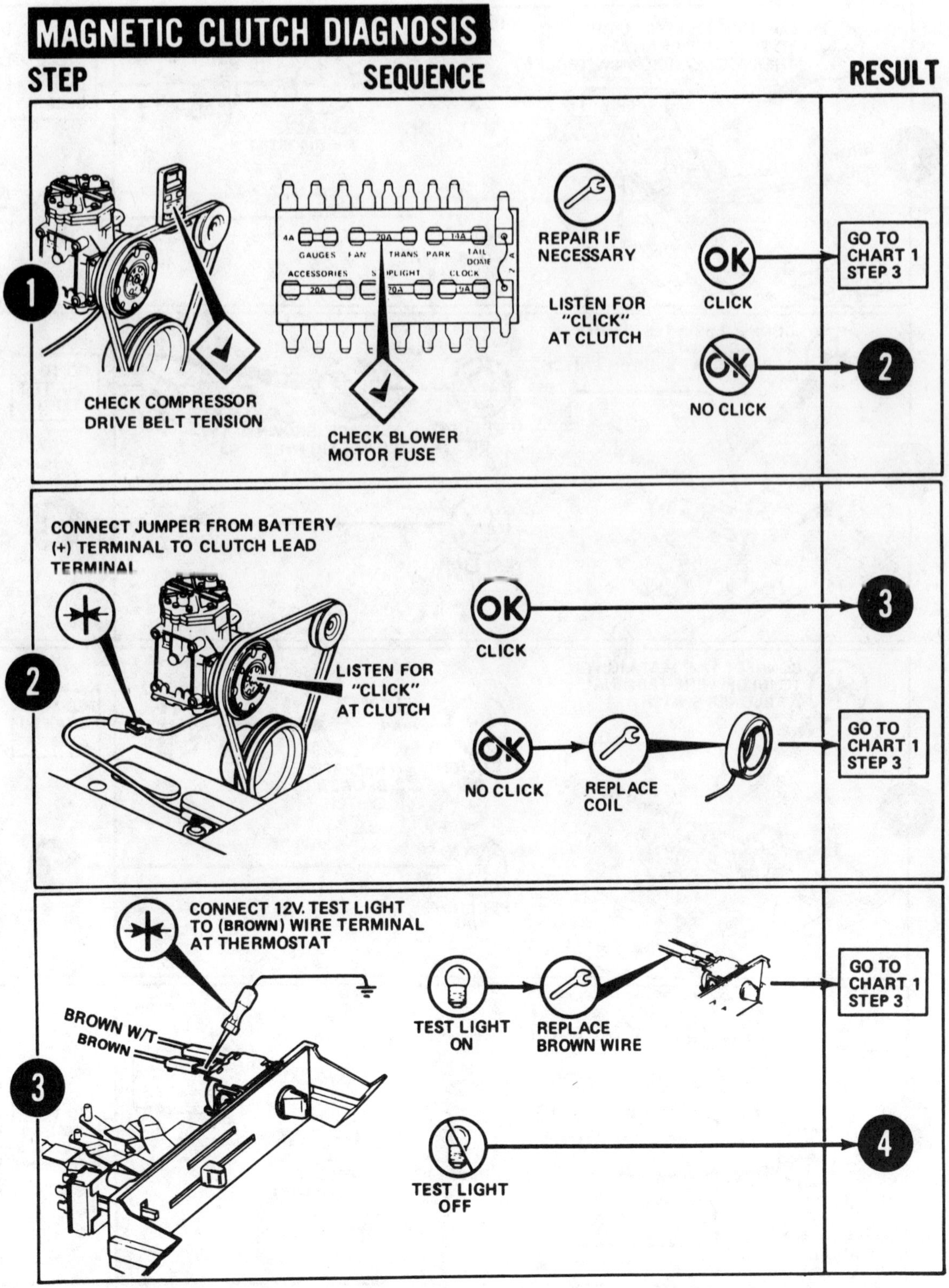

Fig. 12 Magnetic clutch diagnosis (chart 2). 1975-77 Except Matador & Pacer (Part 1 of 2)

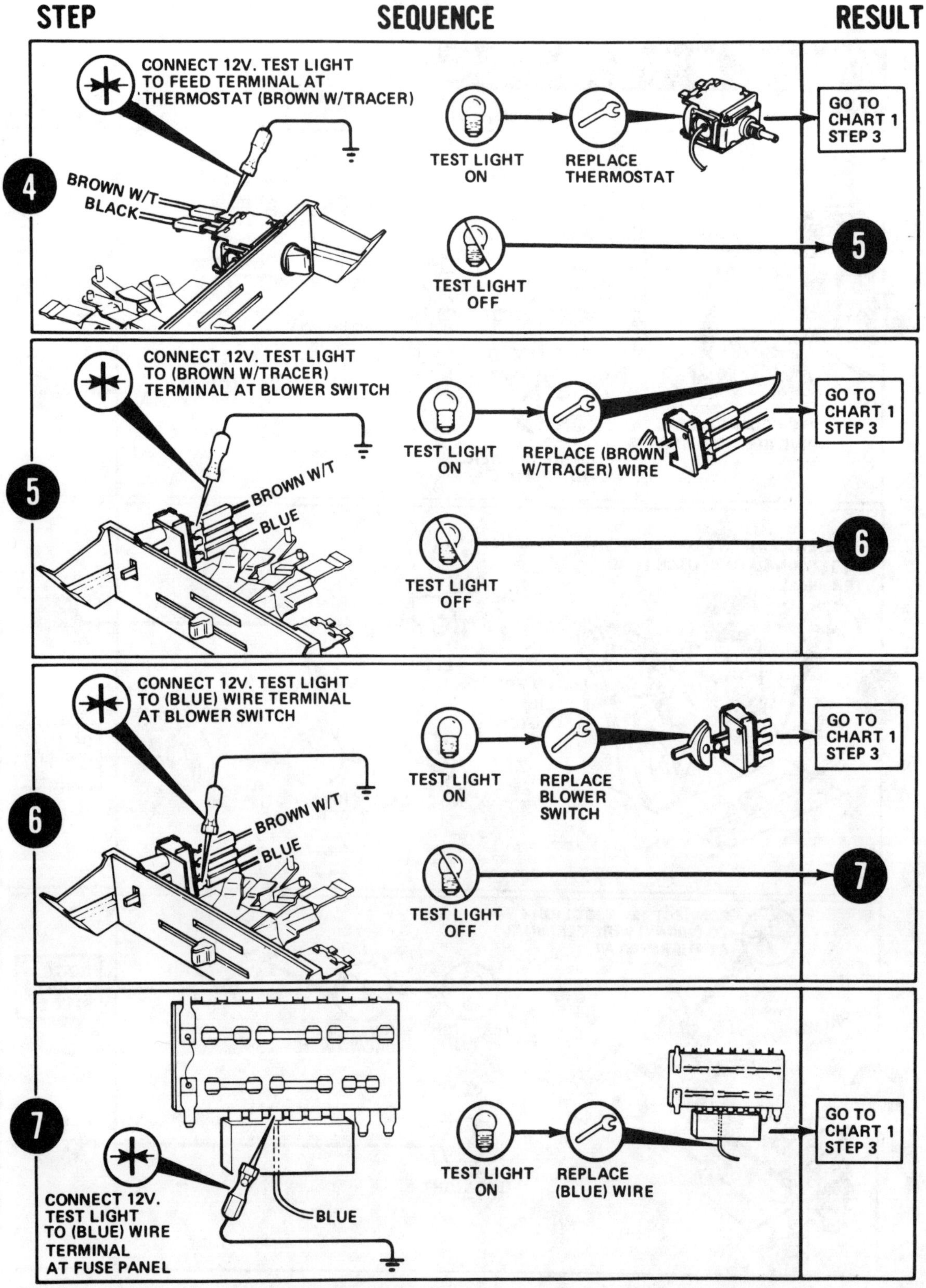

Fig. 12 Magnetic clutch diagnosis (chart 2). 1975-77 Except Matador & Pacer (Part 2 of 2)

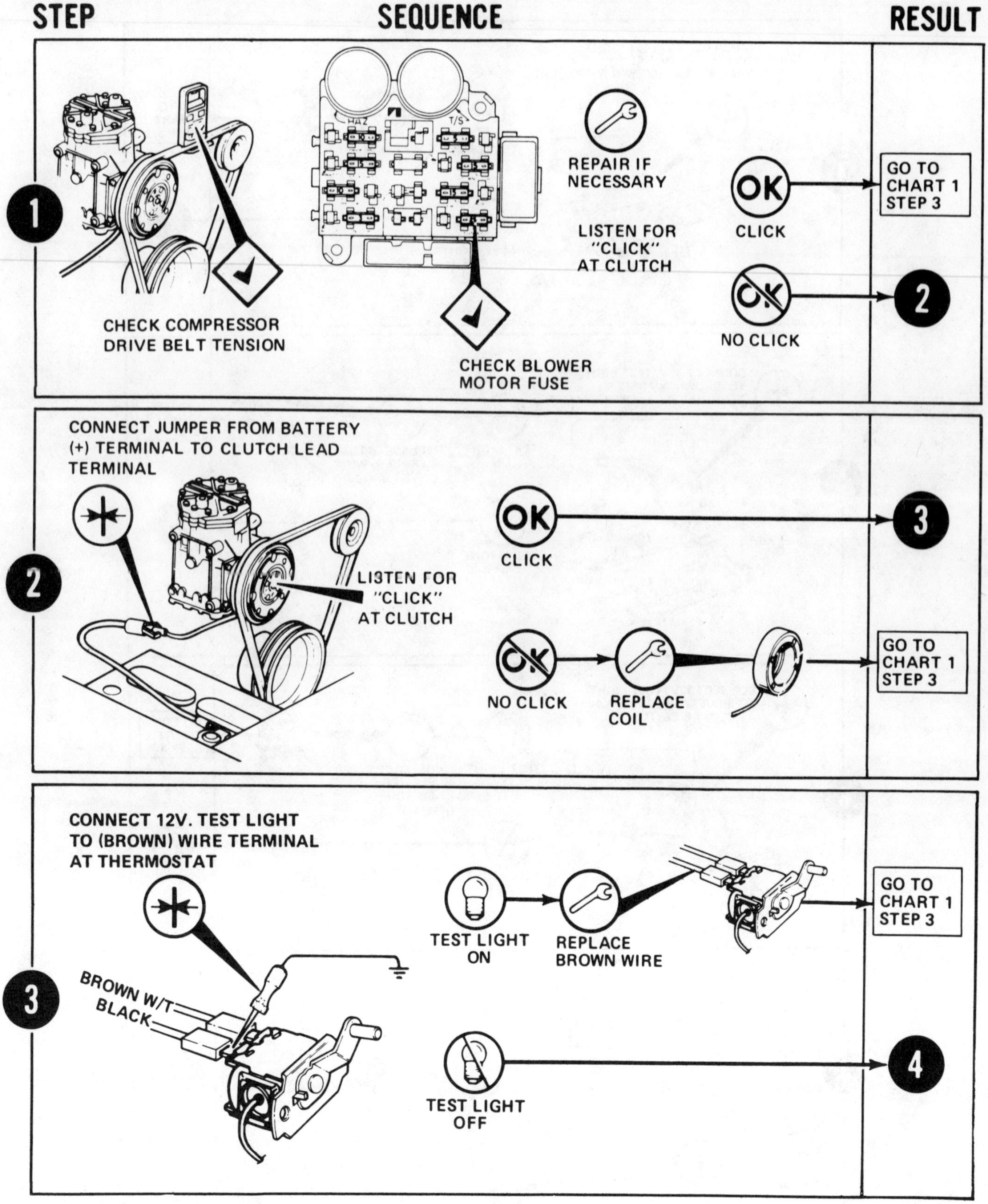

Fig. 13 Magnetic clutch diagnosis (chart 2). 1978-80 Except Matador & Pacer & 1980 models w/4 cyl. Engine (Part 1 of 2)

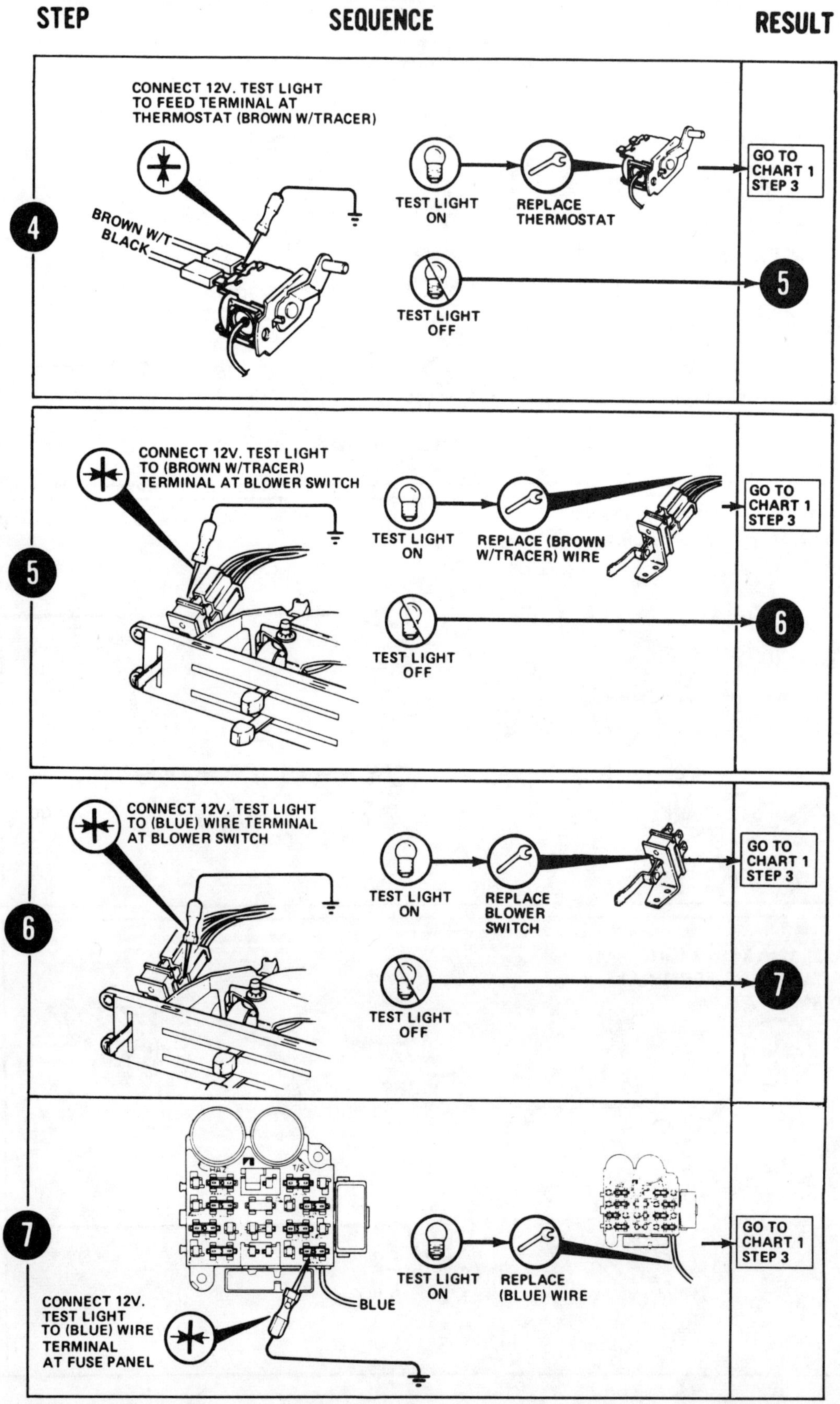

Fig. 13 Magnetic clutch diagnosis (chart 2). 1978-80 Except Matador & Pacer & 1980 models w/4 cyl. Engine (Part 2 of 2)

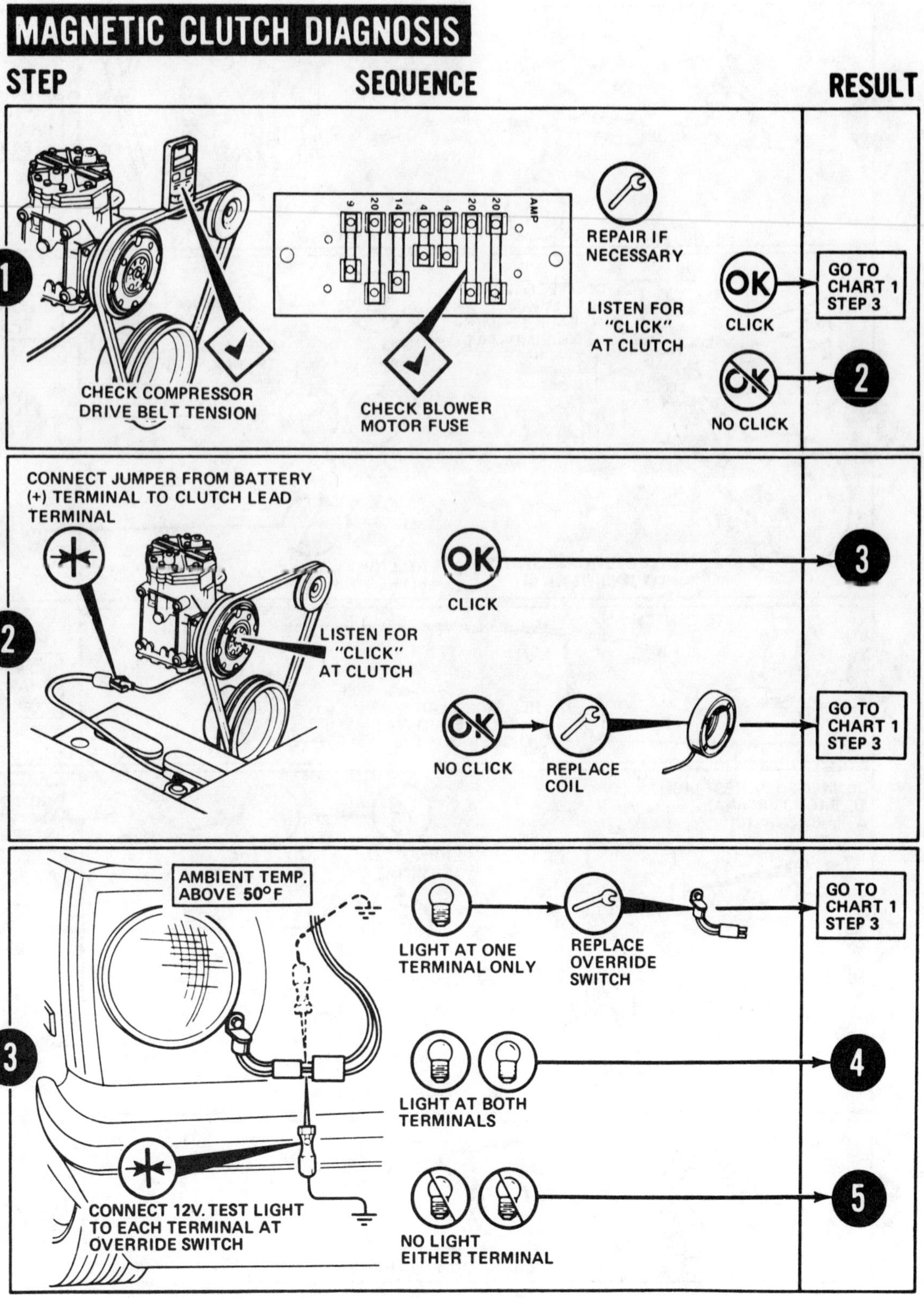

Fig. 14 Magnetic clutch diagnosis (chart 2). 1975-78 Matador (Part 1 of 3)

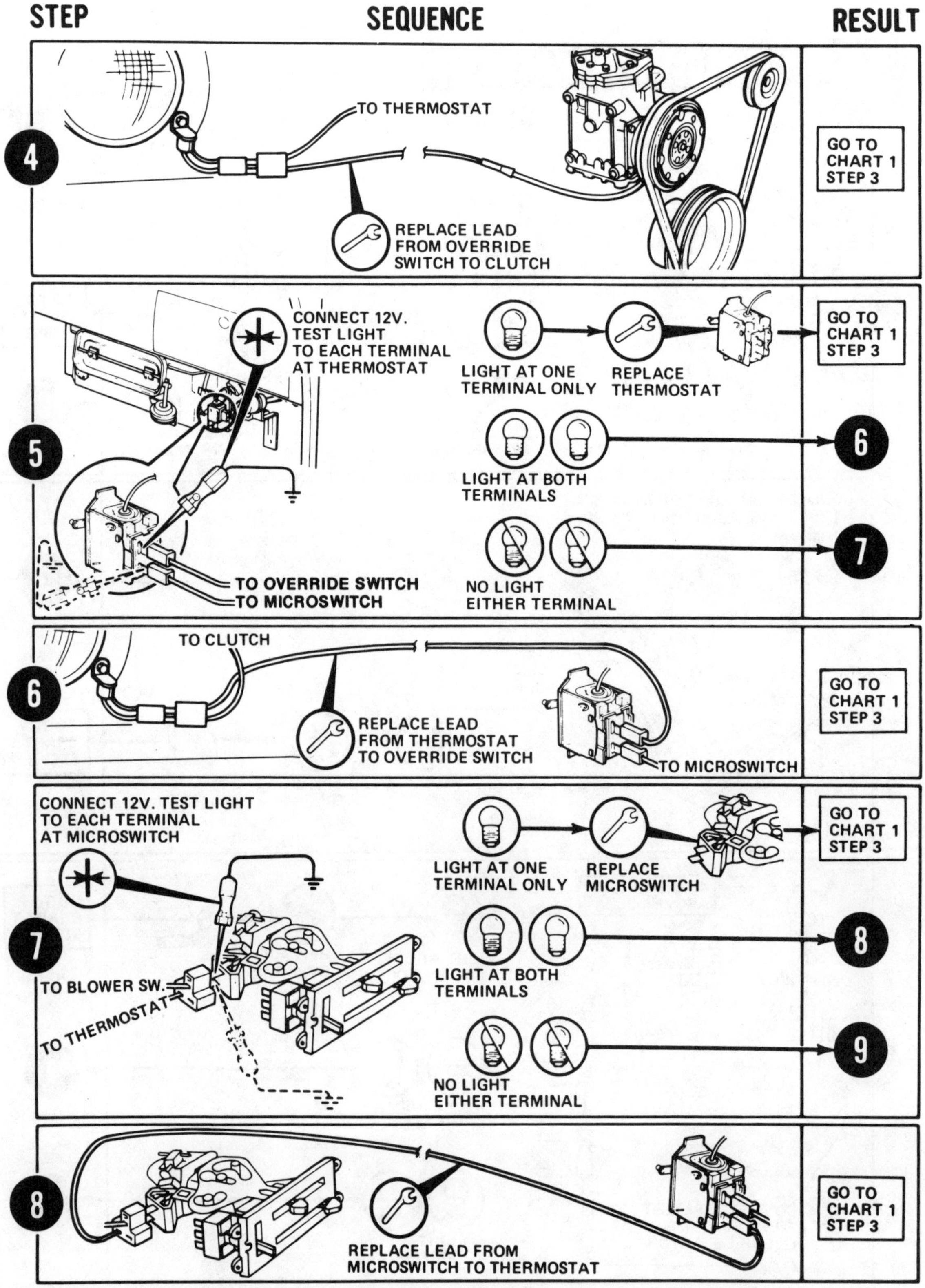

Fig. 14 Magnetic clutch diagnosis (chart 2). 1975-78 Matador (Part 2 of 3)

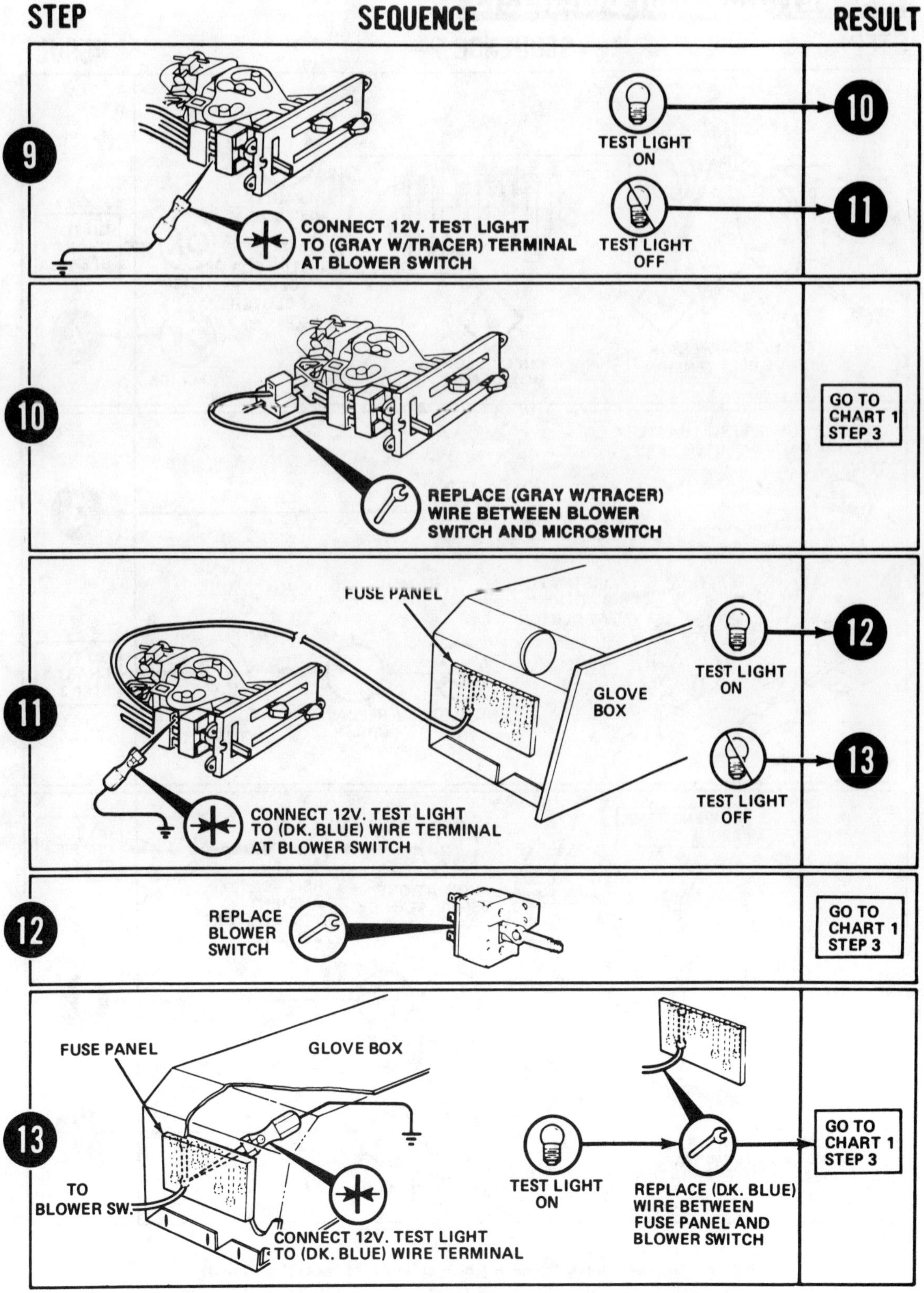

Fig. 14 Magnetic clutch diagnosis (chart 2). 1975-78 Matador (Part 3 of 3)

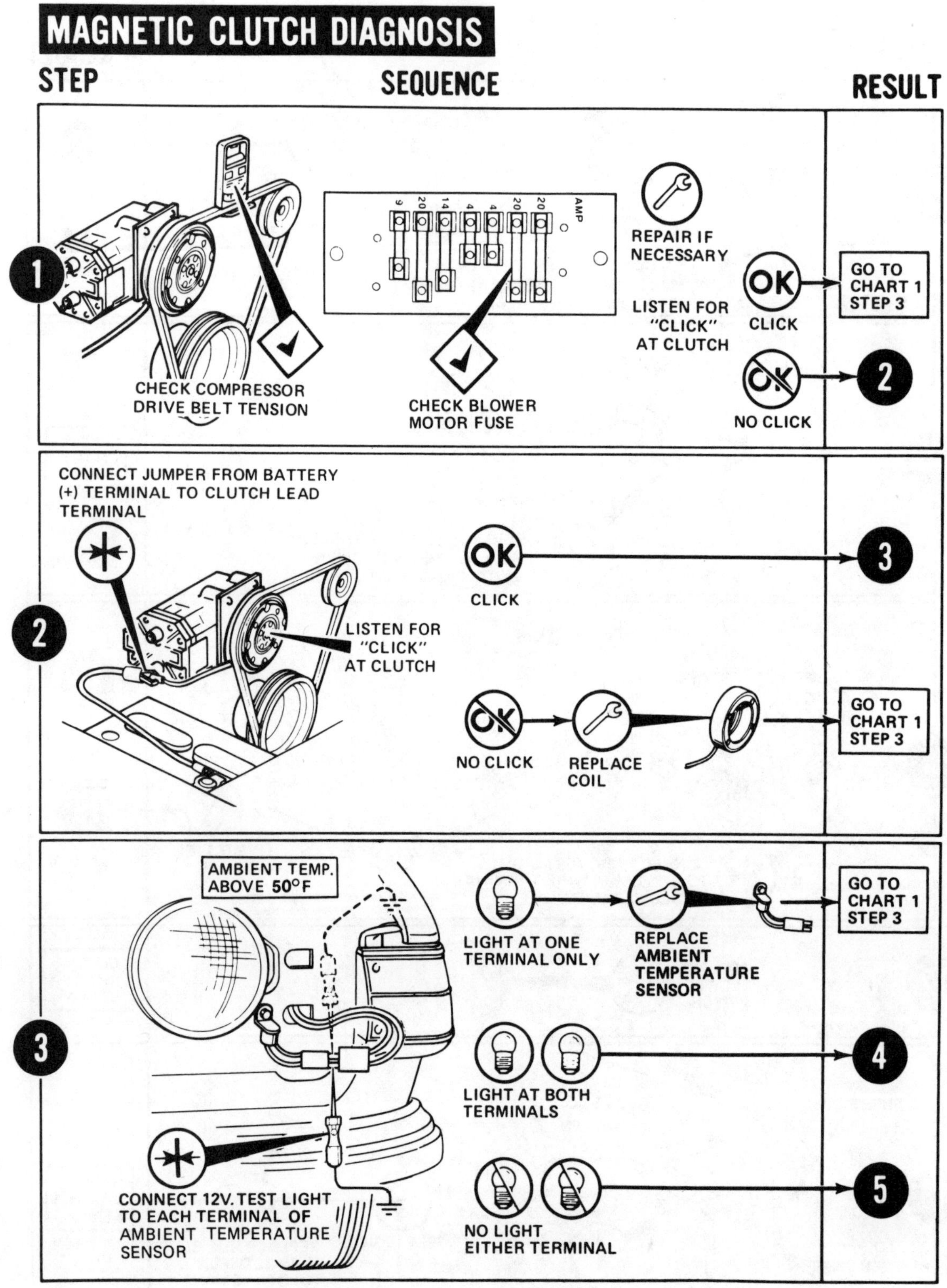

Fig. 15 Magnetic clutch diagnosis (chart 2). 1975-76 Pacer (Part 1 of 3)

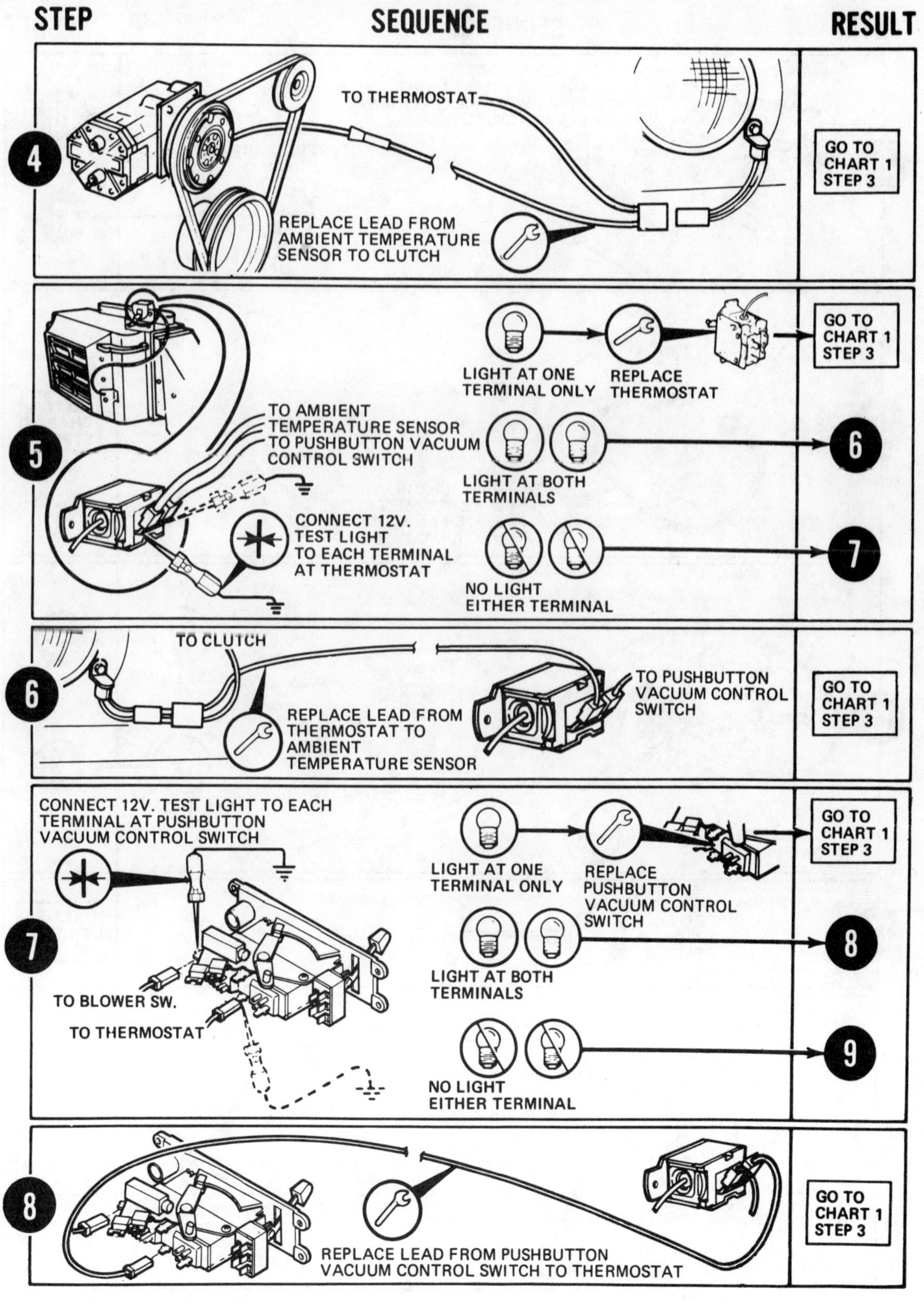

Fig. 15 Magnetic clutch diagnosis (chart 2). 1975-76 Pacer (Part 2 of 3)

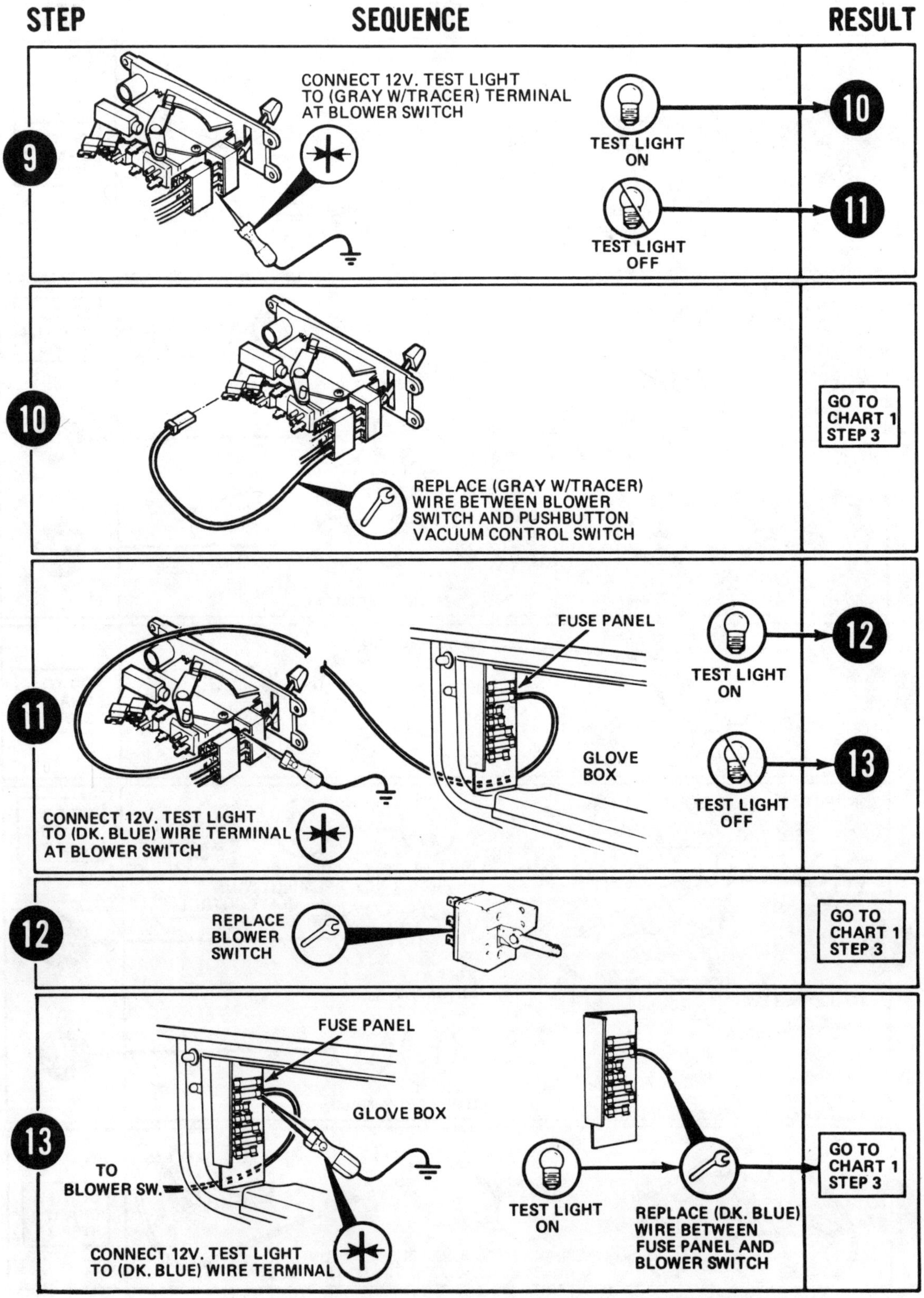

Fig. 15 Magnetic clutch diagnosis (chart 2). 1975-76 Pacer (Part 3 of 3)

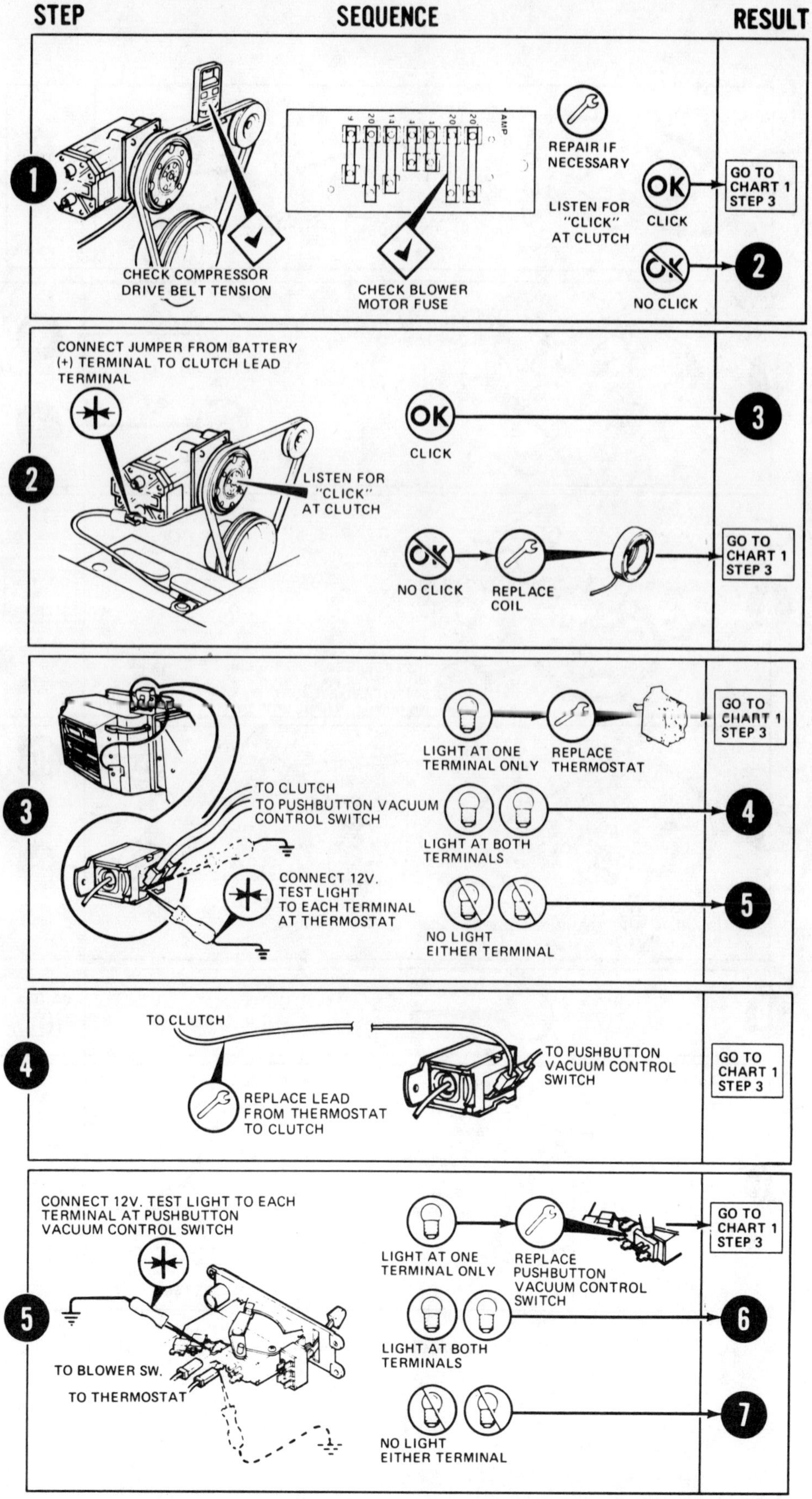

Fig. 16 Magnetic clutch diagnosis (chart 2). 1977-78 Pacer (Part 1 of 2)

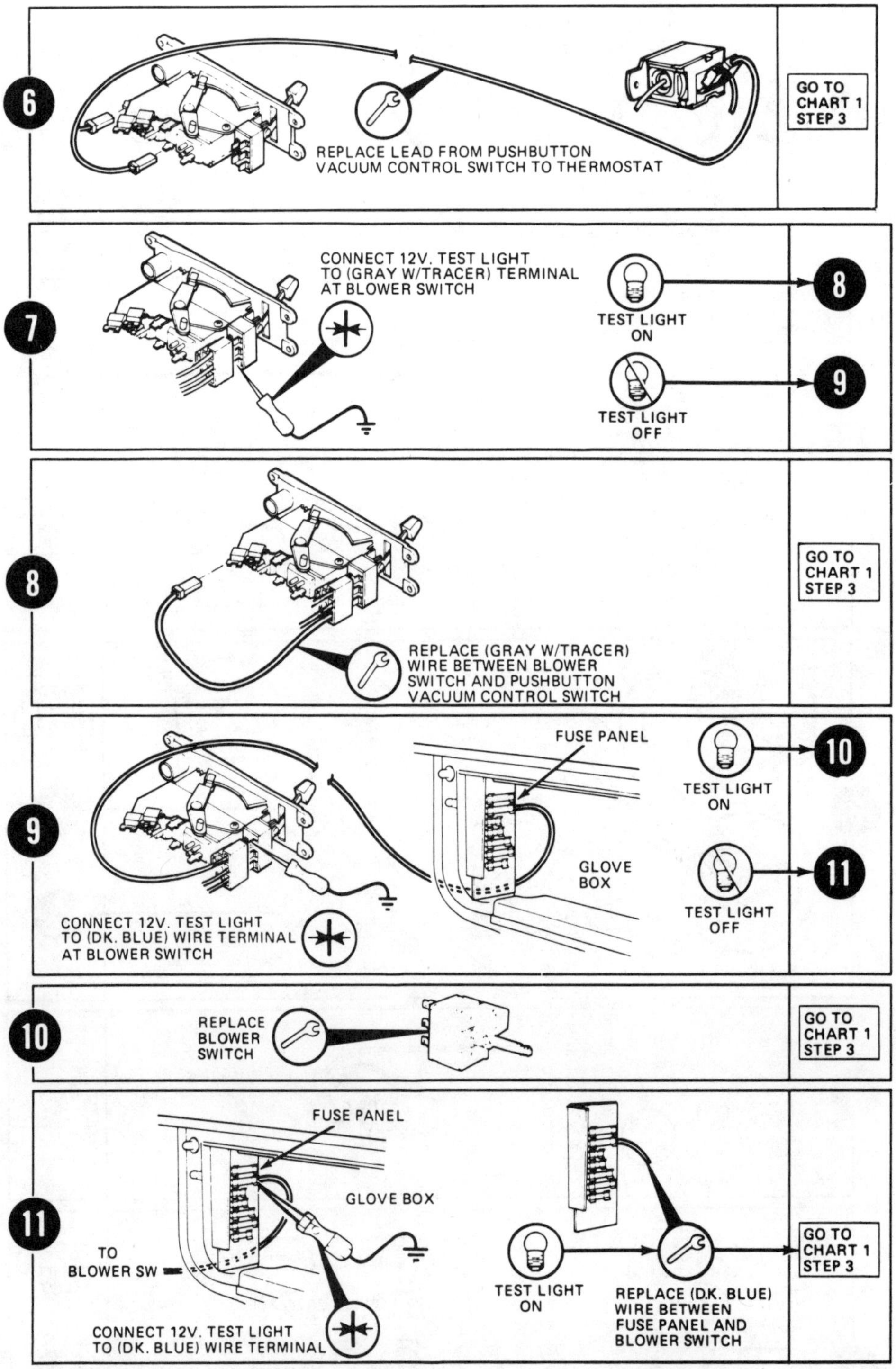

Fig. 16 Magnetic clutch diagnosis (chart 2). 1977-78 Pacer (Part 2 of 2)

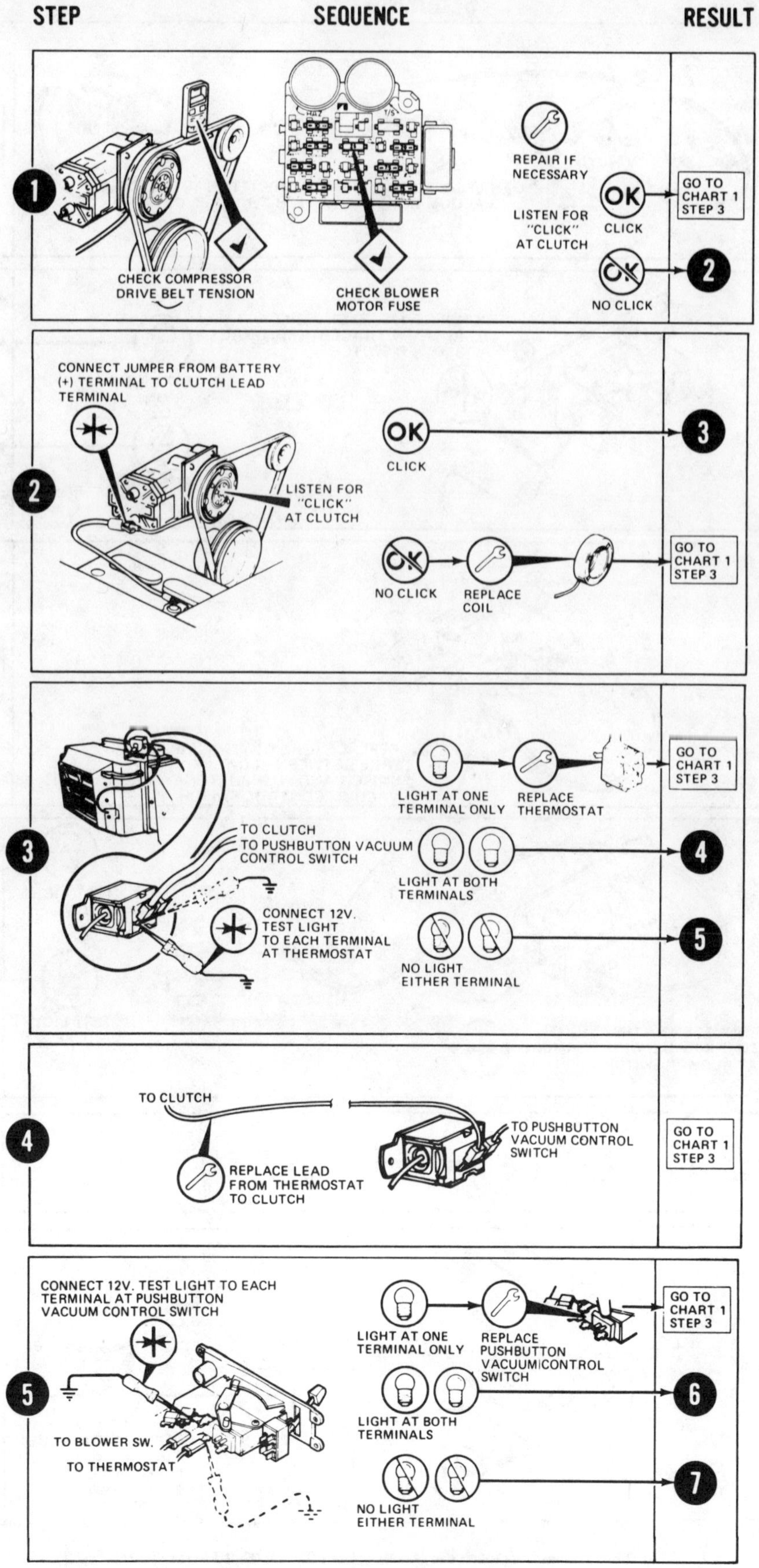

Fig. 17 Magnetic clutch diagnosis (chart 2). 1979-80 Pacer (Part 1 of 2)

STEP	SEQUENCE	RESULT

6

GO TO CHART 1 STEP 3

7

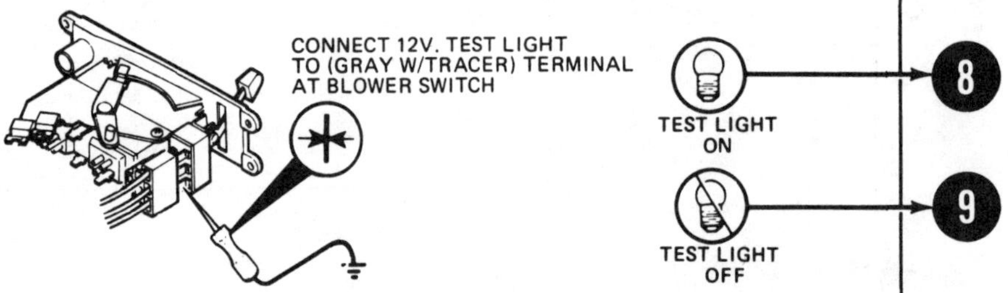

8

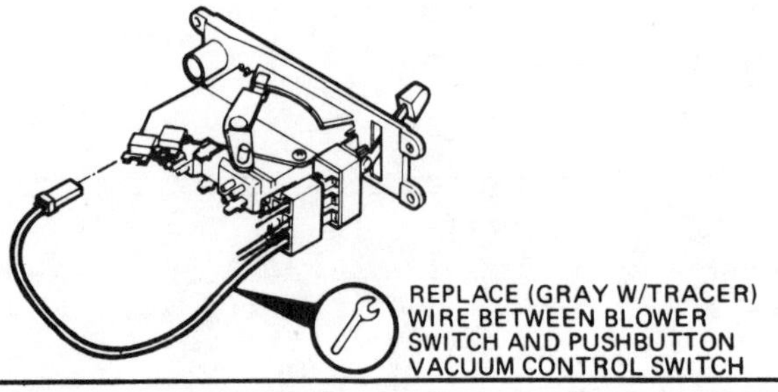

GO TO CHART 1 STEP 3

9

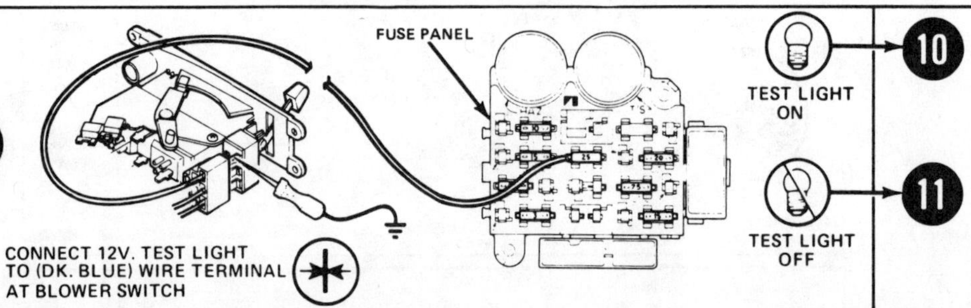

10

GO TO CHART 1 STEP 3

11

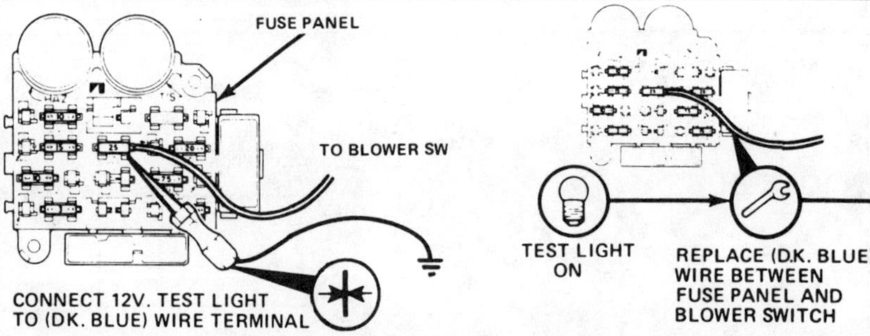

GO TO CHART 1 STEP 3

Fig. 17 Magnetic clutch diagnosis (chart 2). 1979-80 Pacer (Part 2 of 2)

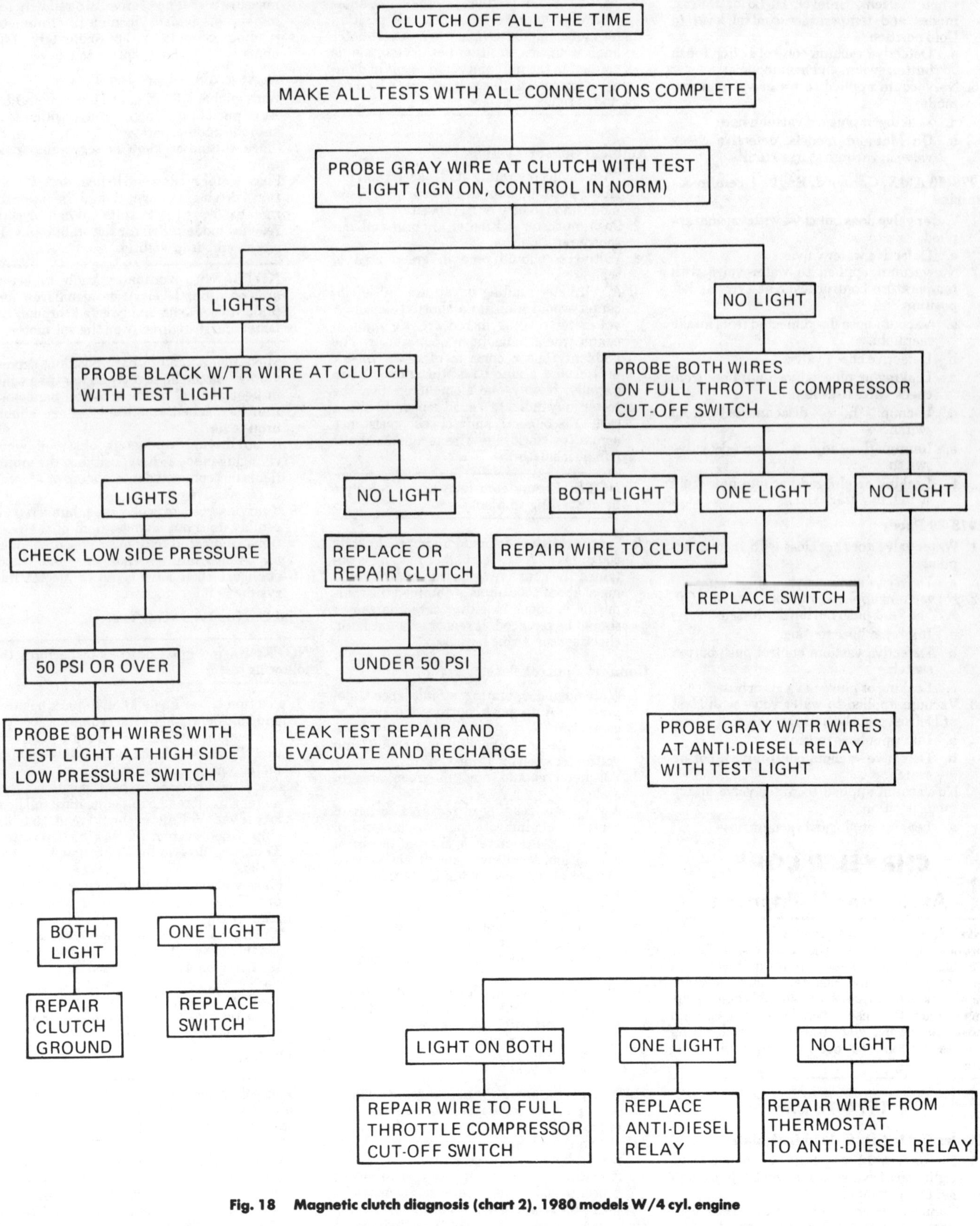

Fig. 18 Magnetic clutch diagnosis (chart 2). 1980 models W/4 cyl. engine

5. Vacuum leak at vacuum control switch when in Heat, Bi-level, Hi-Lo or Defrost modes and temperature control lever in Cold position.
   a. Defective vacuum control switch (push button switch on Pacer models)
6. No vacuum applied to water valve in any mode.
   a. Leaking or plugged vacuum hose.
   b. On Matador models, defective check valve in vacuum storage tank.

1978-80 AMX, Concord, Eagle, Gremlin & Spirit

1. Water valve does not close with vacuum applied.
   a. Defective water valve.
2. No vacuum applied to water valve with temperature control lever at extreme left position.
   a. Vacuum hose disconnected from intake manifold.
   b. Defective check valve.
   c. Leaking or plugged vacuum hose from check valve to switch.
   d. Vacuum hose disconnected from switch.
   e. Improperly adjusted or defective switch.
   f. Leaking or plugged vacuum hose from switch to water valve.

1978-80 Pacer

1. Water valve does not close with vacuum applied.
   a. Defective water valve.
2. No vacuum applied to water valve with Off, A/C or Vent push button depressed.
   a. Improper hose routing.
   b. Defective vacuum control push button switch.
   c. Leaking or plugged vacuum hose.
3. Vacuum applied to water valve with Heat or Defrost push button depressed.
   a. Improper hose routing.
   b. Defective vacuum control push button switch.
4. No vacuum applied to water valve in any mode position.
   a. Leaking or plugged vacuum hose.

CHRYSLER CORP.

Auto Temp II Diagnosis

NOTE: Before testing ensure check valves are properly installed and that park circuit fuse is not blown. Check bleed wire for correct length and proper routing. Ensure vacuum lines are not blocked or improperly routed. When repairing vacuum lines use only nylon tape or vacuum hose, do not use electrical tape. Check wiring harness for loose connections and proper routing.

1974-76 Units

Tester C-4144 Or C-4315 Installation

1. Disconnect 10 way service connector at right end of evaporator assembly in passenger compartment.
2. Connect tester in series with service connector and attach tester ground wire to a good body ground.
3. Tee vacuum gauge into vacuum transfer switch hose.

NOTE: When testing system do not allow voltmeter knob to be in any test position for more than 30 seconds. When voltmeter knob is in test position battery voltage is applied to the unit and could result in damage if exposed to this voltage over a long period of time.

Ambient Sensor Test

1. With engine operating at idle speed and Lo-Auto push button depressed, set tester mode knob to heat or A/C position.
2. Position tester voltmeter knob in ambient sensor position.
3. Voltmeter should read in green area of scale.
4. A voltmeter reading in red area to left of center would indicate a shorted sensor. A voltmeter reading in red area to right of center would indicate an open sensor. The ambient sensor consists of a thermistor button and a shunt resistor connected in parallel. If only one element is open the tester may indicate a good sensor. In either case disconnect and check resistance across two dark blue wire terminals at the female insulator.

NOTE: Ensure vent relay is not defective before testing ambient sensor.

5. Sensor resistance should be approximately 550 ohms at 75°F. If a higher reading is obtained the shunt resistor is open. If a reading of about 1500 ohms is obtained the thermistor is open. In either case the sensor should be replaced. If sensor is not at fault, check sensor wiring harness.

Comfort Control Test

1. With engine operating at idle speed, depress Lo-Auto push button and position control on 75.
2. Place tester mode knob in heat or A/C and voltmeter knob on Temp Control.
3. Voltmeter should read in green area of scale.
4. A voltmeter reading in red area to left of center would indicate that comfort control dial or potentiometer is shorted or out of calibration. A voltmeter reading in red area to right of center would indicate that comfort control dial or potentiometer is open or out of calibration.
5. To test calibration of dial or potentiometer, disconnect tester and check resistance across two violet wire terminals of female insulator.
6. Ohmmeter reading should be 1000 ohms when comfort control is on 75. If reading is not 1000 ohms, the comfort control potentiometer must be replaced.

In-Car Sensor Test

1. With engine operating at idle speed and Lo-Auto push button depressed, position tester mode knob in heat or A/C.
2. Set voltmeter knob or in-car sensor. Voltmeter should read in green area.
3. A voltmeter reading in red area to left of center would indicate sensor or sensor wiring harness is shorted. A voltmeter reading in red area to right of center would indicate sensor or sensor wiring harness is opened. When testing sensor, temprature should be approximately 75°F.
4. On 1975-76 models, disconnect tester and measure resistance across two red wire terminals of female insulator. Ohmmeter reading should be approximately 1400 ohms at 75°F, if not replace sensor.

Servo Assembly-Amplifier Test

1. With engine operating at 1100 rpm and Lo-Auto push button depressed, position tester mode knob in park.
2. Place voltmeter knob in servo-amp position.
3. Place tester mode knob in heat or A/C position, driving voltage should be approximately 3½ to 4½ volts. When system reaches mode called for the stabilized voltage should drop slightly.

NOTE: With voltmeter knob in servo-amp position, the electrical signal from amplifier to servo assembly is fed through the tester and read directly on the voltmeter.

4. Move mode knob to park position; driving voltage should be slightly higher than when in heat or A/C position. When park position is reached stabilized voltage should drop to zero.
5. If no driving voltage is observed when changing mode switch positions, the amplifier is not connected into system or amplifier is defective.
6. If driving voltage is observed, but servo assembly does not change to all of the positions (indicated by blower speed changes, air temperature changes and door changes), then servo is not connected into system or it is defective.

Cold Engine Lock-Out Test

NOTE: Engine must be cold to perform the following test.

1. Position tester mode knob in park position and voltmeter knob on servo-amp.
2. Start engine and depress Lo-Auto push button. There should be no air flow or voltmeter reading at this time.
3. Place mode knob in heat position, voltmeter should read 3 to 4 volts until full heat is reached. Voltmeter should read 2.5 to 3.5 volts when system reaches heat position. There should still be no air flow from any outlets.
4. Place voltmeter knob in blower volts position. There should be no air flow or voltage reading.
5. When engine coolant temperature reaches 125°F, blower should come on M3 speed on 1974-75 models and Hi speed on 1976 models. System will be in full heat position with air flow from heater outlets with normal defroster bleed. This should occur within approximately 3 minutes.

Park Test

NOTE: Park test is a check of where the system positions itself when the ignition or push button switches are turned off.

1. With engine operating, position tester mode knob in heat and voltmeter on blower volts. System will stabilize in full heat M3 blower on 1974-75 models and full heat Hi blower on 1976 models.
2. Place mode knob in park. Blower speeds should decrease in steps to low blower and hold.

3. Place mode knob in A/C. System will shift to A/C and blower speeds will increase to M4 blower on 1974-75 models and to Hi blower on 1976 models.
4. Place mode knob in park. Blower speeds will drop in steps to M1 A/C and hold.
5. Place voltmeter knob in servo-amp position and mode knob in heat position. Wait 40 seconds, then press off push button switch. Voltmeter should read approximately 5 volts until park position is reached.
6. Place voltmeter knob in blower volts position and mode knob in A/C, press Lo-Auto push button. System will come on in low blower heat mode, then shift to A/C mode and increase in steps to M4 A/C on 1974-75 models and Hi blower A/C on 1976 models.
7. Place voltmeter knob in servo-amp position, then turn ignition switch off. Voltmeter should read approximately 5 volts until park position is reached, then drop to zero.

NOTE: When system is operating in heat mode and ignition or push button switches are turned off, system will park in Lo blower. When system is operating in A/C mode it will park in M1 A/C.

Calibration Check

NOTE: This check is performed with engine operating at 1100 rpm, Lo-Auto push button depressed and comfort control set at 75. Position mode knob in A/C on 1974-76 models and voltmeter knob on servo-amp.

1. To perform servo check, place mode knob in park position. When voltmeter reading drops to zero, disconnect 8 way servo assembly connector on 1974-75 models and 6 way servo assembly male connector on 1976 models. Check resistance across violet and black wire terminals of electrical connector. Resistance should be 1100 to 1350 ohms on 1974-76 models. If resistance is not within limits, the servo is out of calibration and should be replaced.
2. To perform amplifier check, position tester mode knob in heat. When voltmeter reads full heat M3 blower speed on 1974-75 models, full heat Hi blower on 1976 models, position mode knob in mid. After voltmeter drops to zero, disconnect 8 way servo assembly connector on 1974-75 models and 6 way servo assembly male connector on 1976 models. Check resistance across violet and black wire terminals of electrical connector. Resistance should be 1000-1200 ohms. If resistance is not within limits, the amplifier is out of calibration and should be replaced.

Sequence Test

The sequence test is an operational check of the system through all its functions. Operate engine at 1100 rpm. Position push buttons, tester mode knob and voltmeter knob as shown in Sequence Test Chart, Figs. 19 and 20.

NOTE: Allow approximately 1 minute for system to travel from A/C high blower to heat high blower.

High Blower Operation 1974-75 Auto Temp II

1. Operate engine at 1100 rpm and position voltmeter knob on blower volts and tester mode knob on heat.
2. Depress Lo-Auto push button and wait until servo assembly stops driving. Blower voltage should stabilize at 8 to 8.5 volts.
3. Depress Hi-Auto push button switch. Airflow should increase and voltmeter should read above 10 volts. If not system is defective.
4. If voltmeter reading is above 10 volts while in Lo-Auto position and there is no change in blower volts when Hi-Auto push button is depressed, the system is defective.
5. If system is found defective in high blower operation, check wiring harness, resistor block, diode, push button switch, relay switch and high blower relay.

Vent Mode Check 1974-76 Auto Temp II

NOTE: Remove tester from system.

1. Operate engine at 1100 rpm and depress Hi-Defrost push button switch. When system reaches max. blower speed, depress Vent push button switch. Compressor clutch should be disengaged and air should be delivered from heater outlets.
2. Blower speeds should decrease and system should change from heat mode to A/C mode. The servo should stop at M2 A/C.
3. If system does not operate as described, check A/C mode operation, push button switch vent contacts, vent relay and vent resistor.

Diode Test 1974-76 Auto Temp II

1. Connect one ohmmeter lead to each diode terminal and note reading.
2. Reverse ohmmeter leads and note reading.
3. Resistance reading should be high in one direction and low in the other.

Relay Switch Test 1974-75 Auto Temp II

1. With engine operating depress Hi-Defrost push button.
2. Wait until system drives to max. heat position.
3. Using test light, check No. 14 wire (gray switch feed wire) at 8 way servo connector. If test lamp does not light check system wiring harness.
4. If test lamp lights check No. 13 wire (gray with tracer). Test lamp should light.
5. Place comfort lever on 65 and depress Hi-Auto push button.

NOTE: Ambient temperature must be above 65°F.

6. Test light should go off within 1 minute. If test lamp does not go off within 1 minute after Hi-Auto push button is pushed, the relay switch is defective and the servo assembly should be replaced.

Vent Resistor Test 1974-76 Auto Temp II

1. Disconnect 8 way servo electrical connector on 1974-75 models, 6 way servo male electrical connector on 1976 models.
2. Connect an ohmmeter between light blue and violet wire terminals of servo electrical connector.
3. Resistance should be approximately 2500 ohms, if not servo assembly should be replaced.

Vent Relay Test 1974-76 Auto Temp II

1. Disconnect 3 way connector from relay.
2. Depress Lo-Auto push button, then using a test light check for continuity between center terminal and the other two terminals. One terminal should be opened and the other closed.
3. Depress Vent push button, then using test light check for continuity between center terminal and the other two terminals. Terminals should check opposite of when Lo-Auto push button was depressed.

High Blower Relay Test 1974-75 Auto Temp II

1. Remove relay from relay bank.
2. Check continuity between center terminal and the other two terminals. One should be opened and the other closed.
3. Using a 12 volt power supply energize the relay. Check continuity between terminals. The terminal that was opened should now be closed and the terminal that was closed should now be opened.

Control Unit Assembly

If the control unit assembly is found defective, it is to be replaced as an assembly, and no attempts should be made to service it.

It should be noted here that other component failures in the system can create the same symptoms as a defective control unit assembly. Therefore, a thorough check of other components should be made as directed in the service diagnosis charts before the control unit is replaced.

However, if a complaint is made that system blows cold air immediately at the start during cold weather, it is more than likely the cause is the thermostat vacuum delay valve. It could also be caused by the blower master switch if it failed in the closed position, or by a failed check valve.

When it has been definitely determined that vacuum delay valve is faulty, the whole water valve assembly should be replaced.

System Vacuum Leak Test

1. On 1974-76 models, connect a 0 to 30 inch vacuum gauge at in-line vacuum connector located in orange vacuum hose near vacuum transfer switch. This hose is connected with the master switch. On all models with the "Lo-Auto" button, place system on Heat mode by rotating tester vacuum control dial to maximum vacuum.
2. Allow vacuum at master switch gauge to build up to at least 18 inches and stabilize.
3. Turn engine off and observe master switch gauge vacuum decay for one minute.

If the vacuum gauge did not build up to at least 18 inches vacuum with the engine running, or if the vacuum decay rate with the engine off exceeded 5 inches per minute, the system will fail the performance test. Check for disconnected hoses and leaks, repair as necessary.

Check Valves

Check valves are used in the Auto-Temp system to prevent the venting of certain vacuum circuits, when the vacuum supply is lost or removed.

The check valve(s), which is an integral part of the reservoir tank(s), is discussed in the vacuum reservoir tank check-out procedure. There are two other check valves in the Auto-Temp II system. These valves are a part of the vacuum harness assembly.

The valves are spaced apart on the harness, and for identification, they will be referred to in sequence as numbers 1, 2, 3, and 4; where number 1 is the closest to the control unit assembly.

Each check valve also serves as a "tee" connector, thereby eliminating extra connections. In the Auto-Temp II system valve number 1

AUTO-TEMP II SEQUENCE CHART

Push Button Switch	Tester Controls Voltmeter Switch	Tester Controls Mode Switch	Outside Air Door Position	AC Door Position	Heat Door Position	Defrost Door Position	Blower Motor Speed
OFF	Blower Volts	A/C	0%	Open or Closed	Closed or Open	Closed or Bleed	Off
LO-AUTO	Blower Volts	Park	0%	Open or Closed	Closed or Open	Closed or Bleed	On
		A/C	100%	Open	Closed	Closed	M1
			100%	Open	Closed	Closed	M1 to M2
			100%	Open	Closed	Closed	M2 to M3
			20%	Open	Closed	Closed	M3
			20%	Open	Closed	Closed	M3 to M4
			20%	Open	Closed	Closed	M4
		Heat	20%	Open	Closed	Closed	M4
			20%	Open	Closed	Closed	M4 to M3
			100%	Open	Closed	Closed	M3
			100%	Open	Closed	Closed	M3 to M2
			100%	Open	Closed	Closed	M2 to M1
			100%	Open	Closed	Closed	M1 to Lo
			100%	Closed	Open	Bleed	Lo
			100%	Closed	Open	Bleed	Lo to M1
			100%	Closed	Open	Bleed	M1 to M2
			100%	Closed	Open	Bleed	M2 to M3
			100%	Closed	Open	Bleed	M3
		A/C	100%	Closed	Open	Bleed	M3
			100%	Closed	Open	Bleed	M3 to M2
			100%	Closed	Open	Bleed	M2 to M1
			100%	Closed	Open	Bleed	M1 to Lo
			100%	Closed	Open	Bleed	Lo
			100%	Open	Closed	Closed	Lo to M1
			100%	Open	Closed	Closed	M1 to M2
			100%	Open	Closed	Closed	M2 to M3
			20%	Open	Closed	Closed	M3
			20%	Open	Closed	Closed	M3 to M4
			20%	Open	Closed	Closed	M4
HI-AUTO	Blower Volts	A/C	20%	Open	Closed	Closed	Hi
		Heat	20%	Open	Closed	Closed	Hi
			20%	Open	Closed	Closed	Hi to M4
			20%	Open	Closed	Closed	M4 to M3
			100%	Open	Closed	Closed	M3
			100%	Open	Closed	Closed	M3 to M2
			100%	Closed	Open	Bleed	M2
			100%	Closed	Open	Bleed	M2 to M3
			100%	Closed	Open	Bleed	M3 to Hi
		A/C	100%	Closed	Open	Bleed	Hi
			100%	Closed	Open	Bleed	Hi to M3
			100%	Closed	Open	Bleed	M3 to M2
			100%	Open	Closed	Closed	M2
			100%	Open	Closed	Closed	M2 to M3
			20%	Open	Closed	Closed	M3
			20%	Open	Closed	Closed	M3 to M4
			20%	Open	Closed	Closed	M4 to Hi
			20%	Open	Closed	Closed	Hi
LO-DEFROST	Blower Volts	A/C	20%	Closed	Bleed	Open	M4
		Heat	20%	Closed	Bleed	Open	M4
			20%	Closed	Bleed	Open	M4 to M3
			100%	Closed	Bleed	Open	M3
			100%	Closed	Bleed	Open	M3 to M2
			100%	Closed	Bleed	Open	M2
			100%	Closed	Bleed	Open	M2 to M3
			100%	Closed	Bleed	Open	M3
		A/C	100%	Closed	Bleed	Open	M3
			100%	Closed	Bleed	Open	M3 to M2
			100%	Closed	Bleed	Open	M2
			100%	Closed	Bleed	Open	M2 to M3
			20%	Closed	Bleed	Open	M3
			20%	Closed	Bleed	Open	M3 to M4
			20%	Closed	Bleed	Open	M4
HI-DEFROST	Blower Volts	Heat	100%	Closed	Bleed	Open	Hi

Blower Motor Speeds

Hi—Maximum Speed	M3—One speed down from M4	M1—One speed down from M2
M4—One speed down from Hi	M2—One speed down from M3	Lo—One speed down from M1

Fig. 19 Sequence test chart. 1974-75 Auto Temp II

AUTO-TEMP II SEQUENCE CHART

Push Button Switch	Tester Controls Voltmeter Switch	Mode Switch	Outside Air Door Position	AC Door Position	Heat Door Position	Defrost Door Position	Blower Motor Speed
OFF	Blower Volts	A/C	0%	Open or Closed	Closed or Open	Closed or Bleed	Off
LO-AUTO	Blower Volts	Park	0%	Open or Closed	Closed or Open	Closed or Bleed	On
		A/C	100%	Open	Closed	Closed	M1
			100%	Open	Closed	Closed	M1 to M2
			100%	Open	Closed	Closed	M2 to M3
			20%	Open	Closed	Closed	M3
			20%	Open	Closed	Closed	M3 to Hi
			20%	Open	Closed	Closed	Hi
		Heat	20%	Open	Closed	Closed	Hi
			20%	Open	Closed	Closed	Hi to M3
			100%	Open	Closed	Closed	M3
			100%	Open	Closed	Closed	M3 to M2
			100%	Open	Closed	Closed	M2 to M1
			100%	Open	Closed	Closed	M1 to Lo
			100%	Closed	Open	Bleed	Lo
			100%	Closed	Open	Bleed	Lo to M1
			100%	Closed	Open	Bleed	M1 to M2
			100%	Closed	Open	Bleed	M2 to M3
			100%	Closed	Open	Bleed	M3 to Hi
			100%	Closed	Open	Bleed	Hi
		A/C	100%	Closed	Open	Bleed	Hi to M3
			100%	Closed	Open	Bleed	M3 to M2
			100%	Closed	Open	Bleed	M2 to M1
			100%	Closed	Open	Bleed	M1 to Lo
			100%	Closed	Open	Bleed	Lo
			100%	Open	Closed	Closed	Lo to M1
			100%	Open	Closed	Closed	M1 to M2
			100%	Open	Closed	Closed	M2 to M3
			20%	Open	Closed	Closed	M3
			20%	Open	Closed	Closed	M3 to Hi
			20%	Open	Closed	Closed	Hi
HI-AUTO	Blower Volts	A/C	20%	Open	Closed	Closed	Hi
		Heat	20%	Open	Closed	Closed	Hi
			20%	Open	Closed	Closed	Hi to M3
			100%	Open	Closed	Closed	M3
			100%	Open	Closed	Closed	M3 to M2
			100%	Closed	Open	Bleed	M2
			100%	Closed	Open	Bleed	M2 to M3
			100%	Closed	Open	Bleed	M3 to Hi
		A/C	100%	Closed	Open	Bleed	Hi
			100%	Closed	Open	Bleed	Hi to M3
			100%	Closed	Open	Bleed	M3 to M2
			100%	Open	Closed	Closed	M2
			100%	Open	Closed	Closed	M2 to M3
			20%	Open	Closed	Closed	M3
			20%	Open	Closed	Closed	M3 to Hi
			20%	Open	Closed	Closed	Hi
LO-DEFROST	Blower Volts	A/C	20%	Closed	Bleed	Open	Hi
		Heat	20%	Closed	Bleed	Open	Hi
			20%	Closed	Bleed	Open	Hi to M3
			100%	Closed	Bleed	Open	M3
			100%	Closed	Bleed	Open	M3 to M2
			100%	Closed	Bleed	Open	M2
			100%	Closed	Bleed	Open	M2 to M3
			100%	Closed	Bleed	Open	M3 to Hi
			100%	Closed	Bleed	Open	Hi
		A/C	100%	Closed	Bleed	Open	Hi to M3
			100%	Closed	Bleed	Open	M3 to M2
			100%	Closed	Bleed	Open	M2
			100%	Closed	Bleed	Open	M2 to M3
			20%	Closed	Bleed	Open	M3
			20%	Closed	Bleed	Open	M3 to Hi
			20%	Closed	Bleed	Open	Hi
HI-DEFROST	Blower Volts	Heat	100%	Closed	Bleed	Open	Hi

Blower Motor Speeds

Hi—Maximum Speed
M3—One speed down from Hi
M2—One speed down from M3
M1—One speed down from M2
Lo—One speed down from M1

Fig. 20 Sequence test chart. 1976 Auto Temp II

serves as one "tee", while each of the others serves as two "tees". On the body of each valve has an arrow which points to the port or ports being checked, Fig. 21.

Check Out Procedure for Valves

The following is presented as an aid for diagnosing check valve failures:

1. The most probable type of failure with the check valves is leakage, and it will be the only one discussed here. This type of failure will not in any case make the system totally inoperative. There will be certain malfunctions created by failed check valves. The valves and their malfunction symptoms are listed:

Check Valve Number 1 Leaking

System stays on 20% outside air and will not go to 100% outside air. Also, system shuts off when vacuum supply is lost.

Check Valve Number 1 Installed Backwards

System malfunctions the same as with a leaking valve. Also, with a cold engine the blower will not come on when the "Defrost" or "Hi-Defrost" button is pushed.

Check Valve Number 2 Leaking

Blower starts immediately and blows cold air during cold weather.

Check Valve Number 2 Installed Backwards

Blower starts immediately and blows cold air during cold weather. With the engine cold the blower will not shut off when the "Off" button is pushed.

1. When check valve failure is suspected, the following should be done:
   (a) Inspect harness to see that valves are installed correctly.
   (b) Remove any valves which are suspected to be faulty, and check as follows:

Connect a vacuum gauge and supply vacuum to valve, Fig. 21.

Apply full vacuum attainable from car or from an external source. Note vacuum at gauge and remove vacuum supply hose. The reading should have remained approximately the same. If vacuum is dropping off at the gauge, the check valve is faulty. Change any valves which are found faulty and check out system to see if malfunction has been corrected. It should be noted that the most severe test on a check valve is below freezing weather. It is possible that a defective valve will not work in such weather, but work at room ambient temperature.

CAUTION: Do not use a lubricant of any type to attempt to increase the efficiency of faulty check valves as serious damage will result to the entire vacuum system.

Blower Master ON-OFF Switch

The master switch is vacuum operated and turns on the blower. A magnet mounted inside the switch holds the electrical contacts open. When sufficient vacuum is applied an internal diaphragm with attached electrical contact is pulled away from the magnet and the electrical circuit is completed.

This switch can be checked by applying 8 ± 1 inch Hg. vacuum to it with an external vacuum source and checking for electrical continuity, and by removing the vacuum source and checking to see that it opens at less than 5 inches Hg. vacuum. The resistance of this switch can be checked by measuring the voltage at the motor when the system is in Hi-Blower airconditioning. If the motor voltage is low, and it can be corrected by replacing the master switch with a jumper, the switch should be replaced.

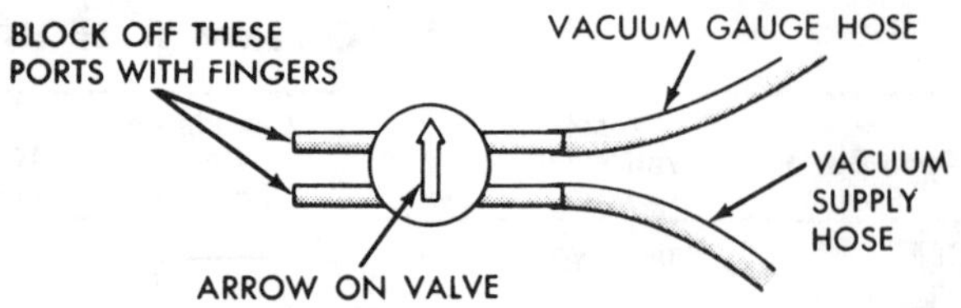

Fig. 21 Check valve vacuum test

Compressor Vacuum Switch

This switch is vacuum operated and turns on the compressor and amplifier (if the ambient switch is closed). It is identical to the blower master switch except that it closes at 2 ± 1 inch Hg. vacuum and completes the electrical circuit to the compressor clutch. The switch is actuated immediately in any push-button position except off.

The performance of the switch can be checked by applying 2 ± 1 inch Hg. vacuum with an external vacuum source and checking for electrical continuity then removing the vacuum source and checking to see that it opens. The resistance of the switch can be checked by measuring the voltage at the clutch any time the compressor should be running. If the clutch voltage is low, and can be corrected by replacing the switch with a jumper, the switch should be replaced.

Ambient Switch 1974 & Early 1975 Auto Temp II

The ambient switch can be checked by removing it and checking for continuity at a temperature slightly above 32°F and for being open at temperatures slightly below 32 degrees F.

Aspirator and Tube Test

With air flowing through the passenger side housing the aspirator should pull air into the in-car sensor grille. This can be checked by holding a smoking object (cigarette) next to the grille with the system on Hi blower in heat mode. If smoke is not pulled into the grille the aspirator is not working properly. The aspirator and air scoop must be inspected for incomplete connection or kinks.

Vacuum Reservoir Tank

The reservoir tank is used for the purpose of maintaining sufficient vacuum in the system in cases where source vacuum is lost momentarily. The tank has a check valve at the inlet port which closes when the vacuum supply drops off.

Operation of the system can be checked in the following manner:

1. Connect a 0-29 inch Hg. vacuum gauge to outlet port of a tank.
2. Start engine and allow to run at idle speed.
3. Vacuum at gauge should build up to 16 inch Hg. or above in the single tank system and 20 inch Hg. or above in the dual tank system, within one minute.
4. Shut engine off, and observe gauge.
   (a) If vacuum does not drop off, the check valve and tank are operating satisfactorily.
   (b) If vacuum drops off, there is a leak and tank assembly should be replaced.

1977-78 Units

Refer to diagnosis charts, Fig. 22 (Parts 1 through 8), and the following test procedures when diagnosing problems on 1977-78 Auto Temp II.

TESTER C-4144 OR C-4315 INSTALLATION

1. Disconnect 10 way service connector located inside passenger compartment behind lower skirt near lower right hand corner of ATC unit.
2. Connect tester in series with service connectors, then connect tester ground wire to a good body ground.
3. Tee-in tester vacuum hose at orange hose near vacuum transfer switch.
4. Position comfort control lever on instrument panel to 75 degrees before beginning any tests.

SYSTEM TEST PROCEDURE

1. With system in off position, place a thermometer into right center outlet and start engine.
2. With engine operating at 1000 rpm, place voltmeter on blower volts. There should be no reading on voltmeter.
3. Depress High-Auto push button and observe the following:
   a. Compressor should be operating and system should be in A/C mode driving toward Hi blower A/C with outside air door open to 20 percent outside air, if ambient temperature is above 32 degrees F.
   b. Blower motor should be operating at high speed and tester voltmeter should read approximately the same voltage as voltage supplied by the voltage regulator.
   c. Air should be discharged from A/C upper outlets and temperature should be dropping toward 40 degrees F.
   d. If system is operating as described above, system is operating properly in this position.

Lo-Auto Test

1. Depress Lo-Auto push button, voltmeter should indicate a blower voltage reduction.
2. Rotate tester mode knob to Heat and observe the following:
   a. Blower speed should drop one step as outside air door goes from 20 percent to 100 percent outside air with no change in outlet air temperature.
   b. Blower speed should drop a second step and an increase of approximately 10 degrees F. in air outlet temperature should be noted.

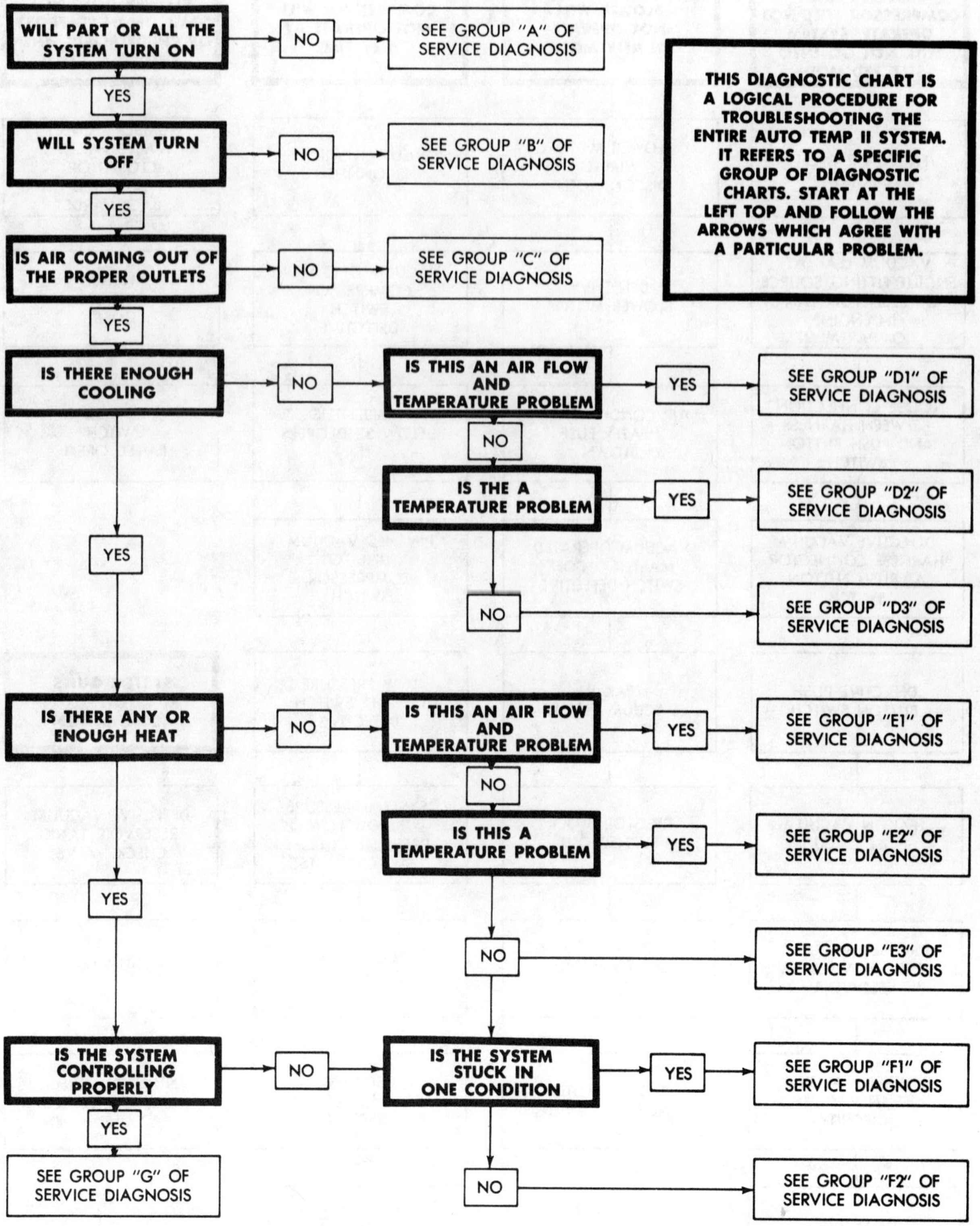

Fig. 22 Auto Temp II diagnosis chart. 1977-78 Units (Part 1 of 8)

GROUP A — PART OR ALL OF SYSTEM WILL NOT TURN ON

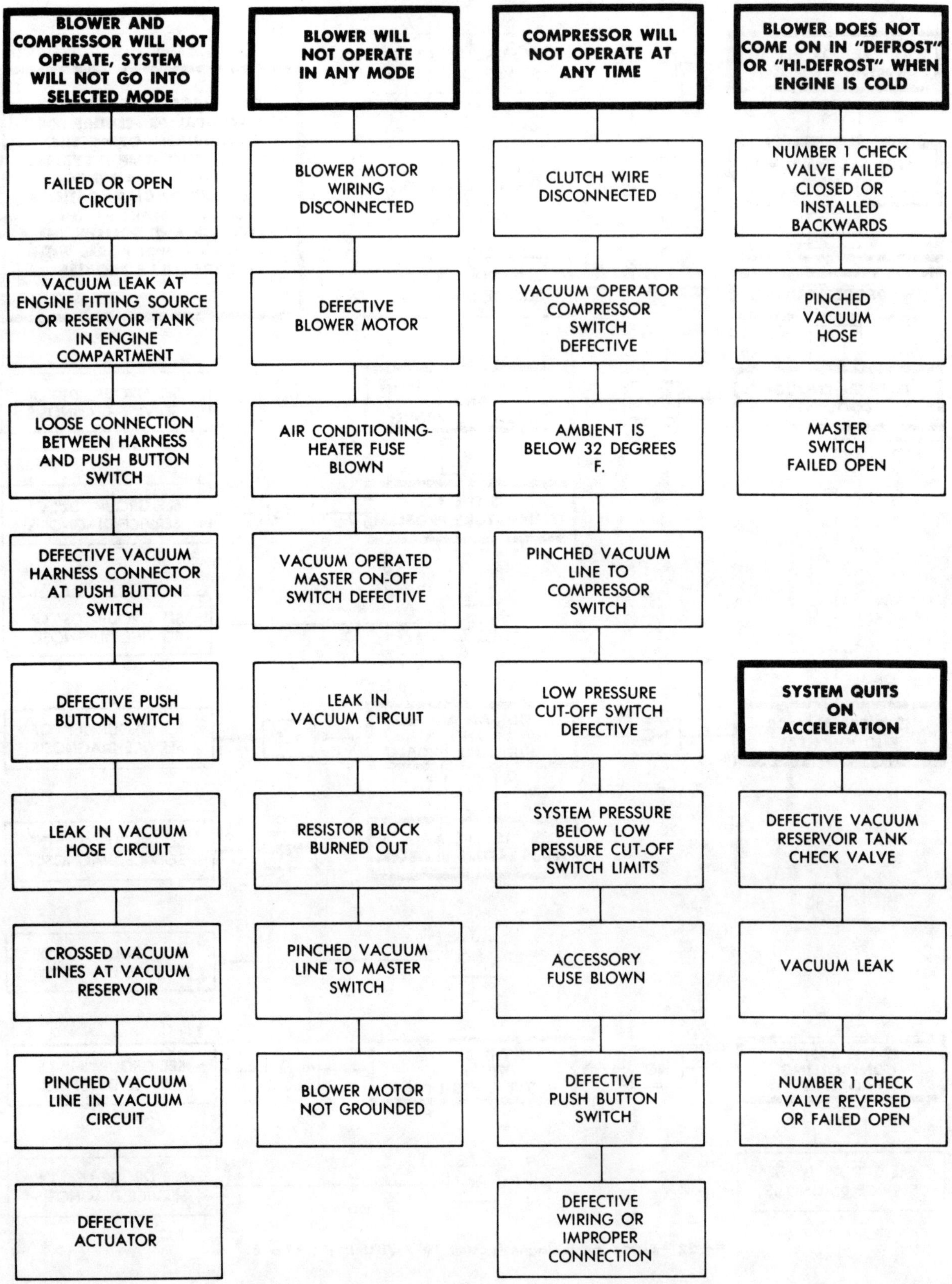

Fig. 22 Auto Temp II diagnosis chart. 1977-78 Units (Part 2 of 8)

GROUP B — SYSTEM WILL NOT TURN OFF

BLOWER DOES NOT SHUT OFF WHEN OFF BUTTON IS PUSHED
- NUMBER 2 CHECK VALVE FAILED CLOSED OR INSTALLED BACKWARD—ONLY WHEN ENGINE IS COLD
- DEFECTIVE PUSH BUTTON SWITCH
- PINCHED VACUUM HOSE

GROUP C — AIR COMES OUT OF THE WRONG OUTLETS

A/C AND HEAT MODES REVERSED — HOT AIR FROM A/C OUTLETS AND COLD AIR FROM HEAT OUTLETS
- A/C AND HEAT MODE ACTUATOR VACUUM HOSES REVERSED

AIR COMING OUT OF A/C OUTLETS WHEN SYSTEM IS IN THE DEFROST OR HI-DEFROST MODE
- VACUUM HOSES TO MODE ACTUATOR AND DEFROST ACTUATOR REVERSED
- DEFECTIVE PUSHBUTTON SWITCH ASSEMBLY

SYSTEM WILL NOT CHANGE MODE IN AUTO OR HI-AUTO — EVEN WITH TESTER OPERATING CONTROL UNIT
- DOORS BINDING
- VACUUM HOSE OFF
- PINCHED VACUUM HOSE TO ACTUATOR

AIR COMES OUT OF A/C OUTLETS OR HEAT OUTLETS WHILE DRIVING IN THE OFF POSITION
- RECIRCULATION DOOR IN OUTSIDE AIR POSITION—VACUUM HOSES ON RECIRCULATION DOOR ACTUATOR REVERSED
- BLOWER MASTER SWITCH—FAILED IN CLOSED POSITION
- RECIRCULATING DOOR NOT SEALING PROPERLY
- NUMBER 2 CHECK VALVE REVERSED OR FAILED CLOSED

MOST OF THE AIR COMING OUT OF DEFROSTERS WHEN SYSTEM SHOULD BE IN HEAT MODE
- VACUUM HOSES TO SHAFT AND POT SIDE OF DEFROST ACTUATOR REVERSED

AIR COMES OUT OF THE DEFROSTER WHEN SYSTEM SHOULD BE IN HEAT MODE, AIR COMES OUT HEATER OUTLET WHEN SYSTEM SHOULD BE IN A/C
- ALL VACUUM LINES TO MODE AND DEFROST ACTUATORS COMPLETELY CROSSED
- MODE DOOR STUCK IN HEAT POSITION

AIR COMING OUT OF DEFROSTER WHEN SYSTEM SHOULD BE IN THE A/C MODE
- A/C HEAT AND HEAT DEFROSTER MODE ACTUATOR HOSES REVERSED ON ACTUATORS
- DEFROST BUTTON MAY BE DEPRESSED

AIR COMES OUT OF ALL OUTLETS WHEN SYSTEM SHOULD BE IN DEFROST OR A/C. AIR COMES OUT A/C OUTLETS WHEN SYSTEM SHOULD BE IN HEAT MODE
- ALL VACUUM LINES TO MODE AND DEFROST ACTUATORS COMPLETELY CROSSED

Fig. 22 Auto Temp II diagnosis chart. 1977-78 Units (Part 3 of 8)

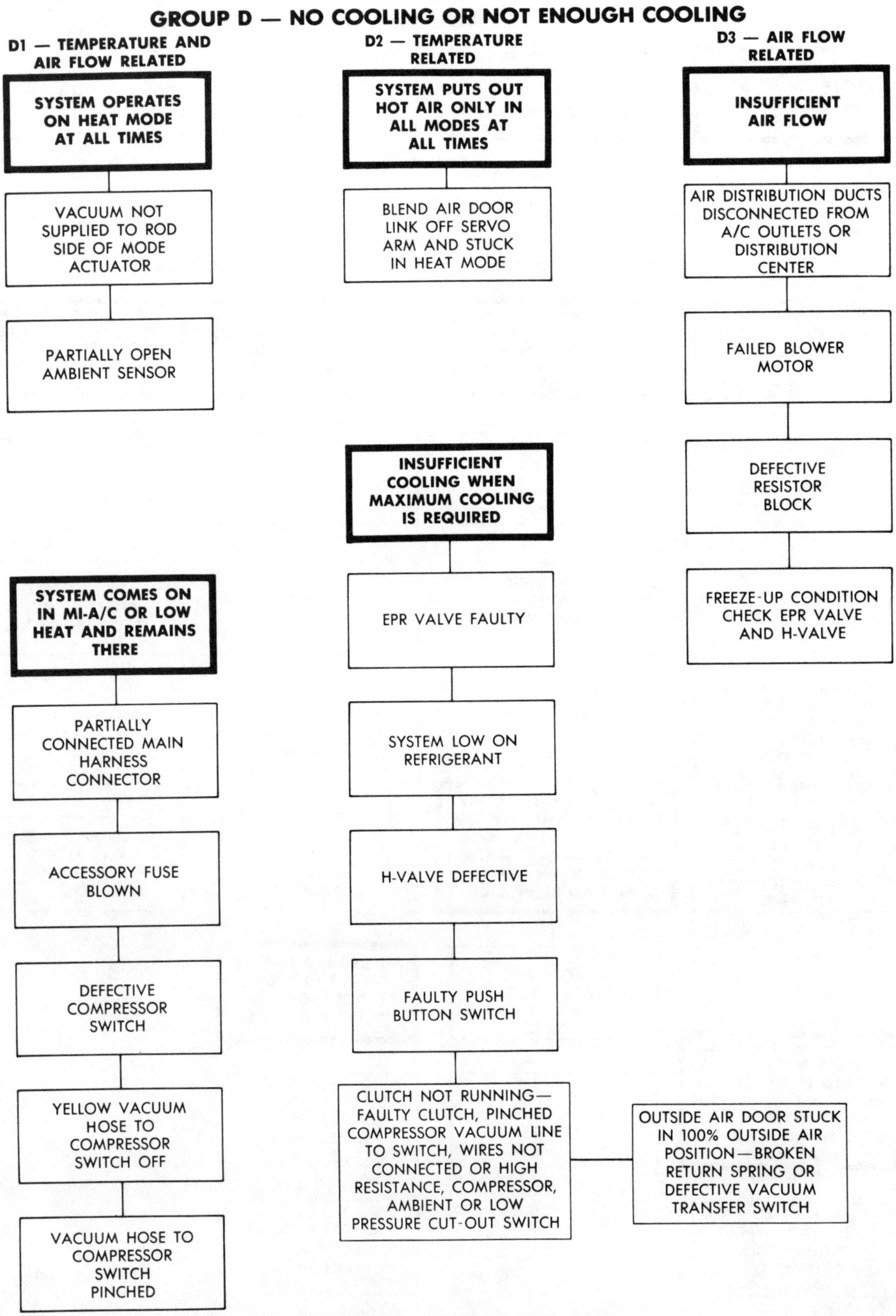

Fig. 22 Auto Temp II diagnosis chart. 1977-78 Units (Part 4 of 8)

GROUP E — NO HEATING OR NOT ENOUGH HEATING

E1 — TEMPERATURE AND AIR FLOW RELATED

SYSTEM WILL NOT GO TO MAXIMUM HEAT ON HI-DEFROST

- DEFROST OVERRIDE CIRCUIT IN AMPLIFIER
- HI-DEFROST FEED FROM PUSH BUTTON TO AMPLIFIER NOT CONNECTED AT SWITCH
- PUSH BUTTON ACTUATED HI-DEFROST FEED SWITCH IN CONTROL HEAD FAULTY
- DEFECTIVE SERVO

SYSTEM COMES ON IN MI-A/C OR LOW HEAT AND REMAINS THERE

- PARTIALLY CONNECTED MAIN HARNESS CONNECTOR
- ACCESSORY FUSE BLOWN
- DEFECTIVE COMPRESSOR SWITCH
- YELLOW VACUUM HOSE TO COMPRESSOR SWITCH OFF OR PINCHED

E2 — TEMPERATURE RELATED

SYSTEM BLOWS COLD AIR AT THE START DURING COLD WEATHER OPERATION

- COLD ENGINE LOCKOUT SWITCH FAULTY
- BLOWER MOTOR MASTER SWITCH INOPERATIVE IN CLOSED POSITION
- VACUUM BLEED PLUGGED
- NUMBER 2 CHECK VALVE FAILED OPEN OR REVERSED

INSUFFICIENT HEAT WHEN MAXIMUM HEATING IS REQUIRED

- DEFECTIVE SERVO ASSEMBLY
- HEATER CORE FAULTY OR PLUGGED
- DEFECTIVE ENGINE THERMOSTAT
- BLEND AIR DOOR NOT ADJUSTED PROPERLY OR BINDING

DISCHARGE AIR TOO COLD AT TIMES

- EPR VALVE FAILED OPEN

SYSTEM PUTS OUT ONLY COLD AIR IN ALL MODES AT ALL TIMES

- BLEND AIR DOOR LINK OFF SERVO ARM AND STUCK IN A/C
- KINKED WATER HOSE
- DEFECTIVE WATER VALVE
- DEFECTIVE PUSH BUTTON SWITCH
- CROSSED VACUUM LINE TO PUSH BUTTON SWITCH

E3 — AIR FLOW RELATED

INSUFFICIENT AIR FLOW

- AIR DISTRIBUTION DUCTS DISCONNECTED FROM A/C OUTLETS
- AIR DISTRIBUTION DUCTS DISCONNECTED FROM DISTRIBUTION CENTER
- FAILING BLOWER MOTOR
- DEFECTIVE RESISTOR BLOCK
- FREEZE-UP CONDITION
- STUCK EPR VALVE
- STUCK H-VALVE

Fig. 22 Auto Temp II diagnosis chart. 1977-78 Units (Part 5 of 8)

GROUP F — SYSTEM NOT CONTROLLING PROPERLY

F1 — SYSTEM IS STUCK IN ONE CONDITION

SYSTEM OPERATES ONLY ON MAXIMUM COOLING

1. A SHORT CIRCUIT IN THE AMBIENT OR IN-CAR SENSOR
2. A SHORT CIRCUIT IN THE COMFORT CONTROL POTENTIOMETER
3. DEFECTIVE AMPLIFIER
4. DEFECTIVE SERVO
5. SENSOR CIRCUIT SHORTED TO GROUND

SYSTEM OPERATES ONLY ON MAXIMUM HEATING AND HI-BLOWER

1. AN OPEN CIRCUIT IN THE AMBIENT OR IN-CAR SENSOR
2. AN OPEN CIRCUIT IN THE COMFORT CONTROL POTENTIOMETER
3. DEFECTIVE SERVO ASSEMBLY
4. DEFECTIVE AMPLIFIER
5. SENSOR CIRCUIT OPEN

SYSTEM OPERATES ON A/C MODE AT ALL TIMES WITH CONTROLLED AIR TEMPERATURE

1. VACUUM NOT SUPPLIED TO POT SIDE OF MODE ACTUATOR

BLOWER SPEED ALWAYS TOO GREAT ON LOW-AUTO

1. RESISTOR BLOCK FAULTY OR SHORTED
2. CONNECTIONS AT RESISTOR BLOCK LOOSE OR DISCONNECTED

SYSTEM WILL NOT CHANGE MODE IN AUTO OR HI-AUTO EVEN WITH TESTER OPERATING CONTROL UNIT

1. DOORS BINDING
2. VACUUM HOSE OFF
3. PINCHED HOSE TO ACTUATOR

SYSTEM COMES ON IN MI-A/C OR LOW HEAT AND REMAINS THERE

1. PARTIALLY CONNECTED MAIN HARNESS CONNECTOR
2. ACCESSORY FUSE BLOWN
3. DEFECTIVE COMPRESSOR SWITCH
4. YELLOW VACUUM HOSE TO COMPRESSOR SWITCH OFF OR PINCHED

SYSTEM CONTROLS IN HEAT MODE FOR MOST AMBIENT TEMPERATURES

1. PARTIALLY OPEN AMBIENT SENSOR
2. DEFECTIVE AMPLIFIER

BLOWER SPEEDS ALWAYS TOO GREAT ON HI-AUTO

1. NO LOWER SPEED PROVIDED FOR HI-AUTO—SYSTEM NORMAL

Fig. 22 Auto Temp II diagnosis chart. 1977-78 Units (Part 6 of 8)

F2 — SYSTEM NOT CONTROLLING PROPERLY

ERRATIC TEMPERATURE CONTROL
- HEATER HOSES REVERSED AT CORE INLET AND OUTLET TUBES
- RADIATOR WATER LEVEL LOW
- DEFECTIVE RADIATOR CAP
- LOOSE VACUUM OR ELECTRICAL CONNECTION AT CONTROL UNIT ASSEMBLY
- DEFECTIVE ENGINE THERMOSTAT
- FAULTY SERVO ASSEMBLY
- INTERMITTENT SENSOR OR TEMPERATURE CONTROL POTENTIOMETER
- ERRATIC AMPLIFIER
- ASPIRATOR DISCONNECTED

SYSTEM WILL NOT PARK
- DEFECTIVE AMPLIFIER
- DEFECTIVE SERVO
- BLOWN FUSE
- SYSTEM BATTERY FEED DISCONNECTED
- SYSTEM BATTERY FEED CONNECTED TO FUSE BLOCK ACCESSORY TERMINAL

SYSTEM DOES NOT ACHIEVE A COMFORTABLE LEVEL — ERRATIC CHANGES OF TEMPERATURE AND POSSIBLE BLOWER SPEED
- LOOSE ELECTRICAL CONNECTION AT SENSORS, AMPLIFIER OR SERVO ASSEMBLY
- DEFECTIVE AMPLIFIER
- DEFECTIVE SERVO ASSEMBLY
- ASPIRATOR AND SCOOP ASSEMBLY
- BLEND AIR DOOR LINK NOT CONNECTED TO SERVO

SYSTEM DOES NOT ACHIEVE A COMFORTABLE LEVEL — TEMPERATURE SLOWLY DRIFTS UP AND DOWN
- ASPIRATOR NOT OPERATING PROPERLY—PINCHED OR DISCONNECTED TUBE
- FRAYED OR BARE SECTIONS IN SENSOR LEADS
- DEFECTIVE AMPLIFIER
- DEFECTIVE RADIATOR PRESSURE CAP
- DEFECTIVE ENGINE THERMOSTAT

DISCHARGE AIR TOO COLD AT TIMES
- EPR VALVE FAILED OPEN

BLOWER SPEEDS NOT IN SEQUENCE
- WIRING HARNESS ERROR AT SERVO OR RESISTOR BLOCK

CLUTCH ENGAGED IN VENT MODE
- FAULTY PUSH BUTTON SWITCH
- FAULTY WIRING

Fig. 22 Auto Temp II diagnosis chart. 1977-78 Units (Part 7 of 8)

F2 — SYSTEM NOT CONTROLLING PROPERLY

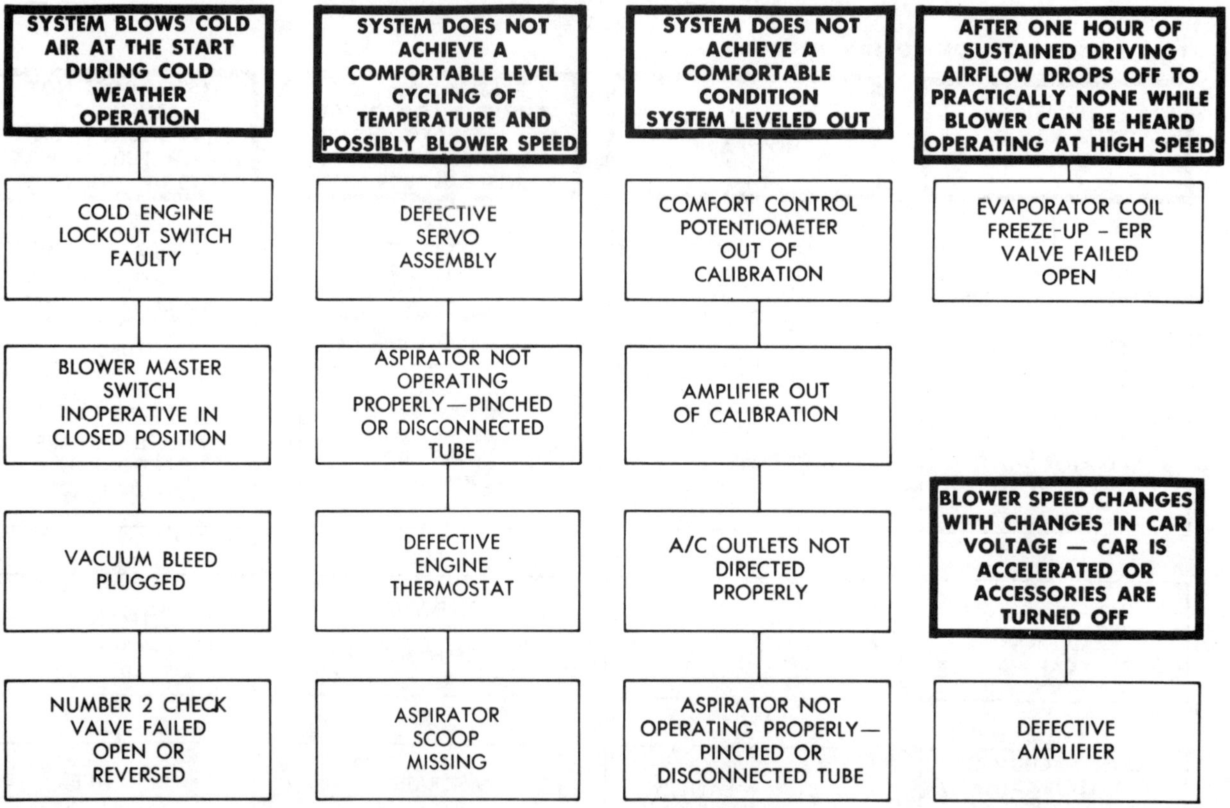

GROUP G — MISCELLANEOUS

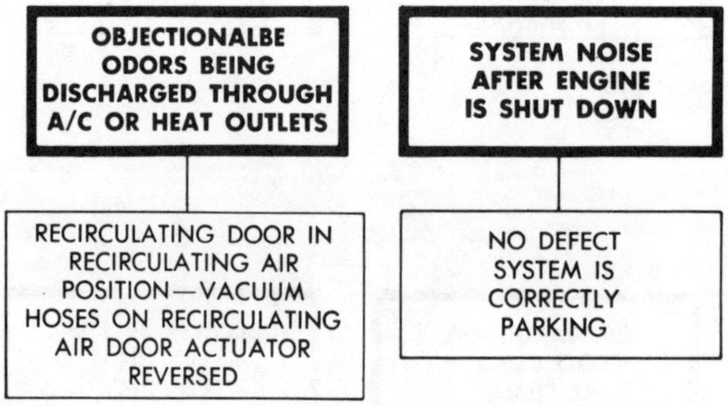

Fig. 22 Auto Temp II diagnosis chart. 1977-78 Units (Part 8 of 8)

c. Blower motor should then drop a third and fourth additional step, with a slight increase in outlet temperature at each step drop.

d. As system goes from A/C mode to heat mode, move thermometer from A/C outlet to heater outlet.

e. After system is in heat mode, blower motor should increase one step with an additional rise in heater outlet temperature.

f. Blower motor should increase a second and third additional step, with an additional rise in outlet temperature at each step.

g. Blower motor should increase a fourth step and outlet air temperature should be at maximum, approximately 140 degrees F.

Hi-Auto

1. Place tester mode knob on A/C, system should change from heat mode to A/C and drive to Hi blower.
2. Push Hi-Auto push button and observe the following:
   a. Air velocity should increase.
   b. Voltmeter reading should increase.
   c. There should be no change in vacuum gauge reading and little or no change in outlet air temperature.

Lo-Defrost

1. Depress Lo-Defrost push button and observe the following:
   a. Air should be discharged from defroster outlets, with bleed air coming from heater outlets. No air should be coming out of A/C outlets.
   b. Outlet air temperature should remain approximately the same as in Lo-Auto.
   c. Blower speed should drop one speed.

Hi-Defrost

1. Depress Hi-Defrost push button and observe the following:

a. Defroster air outlet temperature should rise to maximum heat.
b. Blower speed should increase in steps to high blower.
c. Tester mode knob is no longer effective except for Park mode.

Vent Test

1. Place tester mode knob on A/C.
2. Depress Vent push button and note that compressor clutch disengages.
3. When system has reached the A/C mode, perform test as described under Lo-Auto. Blower speed and air circulation will change in the same sequence as Lo-Auto, except that compressor will not be operating and outlet air will not be cooled.

Sensor Test

1. With engine operating at idle speed and Lo-Auto push button depressed, rotate tester voltmeter knob from off to ambient sensor position.

NOTE: To prevent damage to sensors, do not leave voltmeter in any sensor position for more than 30 seconds.

2. If voltmeter indicator is located to left in the red area, sensor is shorted.
3. If voltmeter indicator is located to right in the red area, sensor is opened or disconnected.
4. In either case, sensor should be connected or replaced as necessary.
5. Use the above procedure to check comfort control at 75 degrees and in-car sensor.

Sequence Test

The sequence test is a means of checking the system through all its functions. Operate engine at 1100 rpm. Position push buttons, tester mode knob and voltmeter knob as indicated in Sequence Test Chart, Fig. 23.

Park Test

1. With engine operating and system in Lo-Auto position, turn tester voltmeter knob to blower volts and mode knob to Heat. After system has stabilized in maximum heat position and high blower, place mode knob in park position. Blower motor speed should drop in steps and stop. This park position is low blower heat mode.

NOTE: When tester is installed and in Park, blower motor will remain on unless off button is depressed to vent system.

2. Place mode knob on A/C, blower speed should rise in steps to high blower. Rotate mode knob to Park, system should return Ml-A/C and stop.
3. Place voltmeter to Servo Amps and mode knob on Heat. Wait approximately 40 seconds, then depress off push button. The voltmeter should show a reading of approximately 5 volts until Park position is reached. When Park position is reached voltmeter should read zero.
4. Place voltmeter knob on Blower Volts and mode knob on A/C, then depress Lo-Auto push button. System should start in low blower heat mode then shift to A/C mode and blower speed increases in steps to High blower A/C.
5. Place voltmeter knob to Servo Amp and turn ignition switch off. The system should function as described in step 3.

NOTE: This system has two Park positions. When system is in heat mode, when ignition switch is turned off or off push button is depressed, system should Park in low blower heat mode. When system is in A/C mode, system will park in Ml-A/C.

Servo Assembly-Amplifier Test

The amplifier receives electrical signals from the sensors and temperature control and positions the servo assembly with an electrical signal to give the required amount of cooling or heating. Position tester voltmeter on Servo Amps and note voltmeter reading. The driving voltage from the amplifier to the servo assembly should be 3 to 5 volts in either A/C or Heat modes. When system stabilizes, voltage should drop slightly. When servo assembly is driving to the Park position, voltage will be slightly higher. When the servo assembly stabilizes in the Park position, voltage will drop to zero. If no driving voltage is available, check for the following:

a. Ground wire to wiring harness connector disconnected.
b. Amplifier disconnected.
c. Blown fuse.
d. Defective amplifier.

The servo assembly should change position when driving voltage is at least 1.5 volts. If no change in airflow, temperature of air or mode shift, the servo assembly is disconnected or defective and should be connected or replaced.

COMPONENT TEST PROCEDURES

Vacuum Transfer Switch Test

The three port vacuum transfer switch and spring loaded vacuum actuator are used to position the outside recirculating air door in off, 20 percent outside air and 100 percent outside air. To test vacuum transfer switch, proceed as follows:

1. Depress off push button, then start engine and observe the following:
   a. Outside air door should be in the zero percent outside air position.
   b. Switch plunger should be retracted and touching lever.
2. Depress Lo-Auto push button, then position mode knob in Park position and observe the following:
   a. Outside air door should be in the 100 percent outside air position.
   b. Switch plunger should move outward and not be contacting lever.
3. With Lo-Auto push depressed, rotate mode knob to Park position and observe the following:
   a. Outside air door should be in 20 percent outside air position.
   b. Switch plunger should be part way out and in contact with lever.
4. If system operated as described above, the vacuum transfer switch is operating properly. If system did not function as described above, proceed as follows:
   a. With engine operating, depress off push button.
   b. Disconnect vacuum plug from transfer switch and connect a vacuum source to middle port of switch.
   c. Depress switch plunger into switch and check for vacuum to be present at switch upper port. Also check for an absence of vacuum at switch lower port.
   d. Release plunger and check for vacuum to be present at switch lower and an absence of vacuum at switch upper port.
   e. If the vacuum transfer switch did not operate as described above, the switch is defective and should be replaced.

Blower Master Switch Test

Using an external vacuum source, apply approximately 8 in. Hg. of vacuum to switch and check for continuity. When 3 in. Hg. or less of vacuum is applied to switch, switch should indicate an open circuit. Check resistance of switch, by measuring voltage of blower motor with system in Hi-Blower A/C. If blower motor voltage seems low, the switch should be replaced.

Compressor Vacuum Switch

Using an external vacuum source, apply approximately 2 in. Hg. to switch and check for continuity. After removing vacuum source, check to ensure that switch indicated an opened circuit. Check resistance of switch, by measuring compressor clutch voltage. If voltage is low, replace switch.

Comfort Control Potentiometer Test

1. Disconnect electrical leads from potentiometer, then connect a suitable ohmmeter.
2. If ohmmeter indicated infinite reading, potentiometer is opened and should be replaced.
3. If ohmmeter indicates a zero reading, potentiometer is shorted and should be replaced.
4. Potentiometer should indicated a resistance of approximately 200 to 1800 ohms as lever is moved from 65 to 85, if not replace potentiometer.
5. Check potentiometer calibration, position control lever at 75. Ohmmeter should read 1000 ohms, if not control lever is out of calibration and should be replaced.

In-Car Sensor Test

Disconnect wire connector from service connector and check resistance across the two red wire terminals using a suitable ohmmeter. Maximum resistance should be approximately 1850 ohms at 65 degrees F, 1400 ohms at 75 degrees F and 1070 ohms at 85 degrees F. If sensor is defective, replace as necessary.

Ambient Sensor Test

Disconnect service connector and check resistance across two dark blue wire terminals on female connector using a suitable ohmmeter. Maximum resistance should be 650 ohms at 65 degrees F, 550 ohms at 75 degrees F and 450 ohms at 85 degrees F. If resistance is constantly 1500 ohms, the sensor is opened and should be replaced. If resistance is constantly above those listed above, the shunt resistor is opened and sensor should be replaced.

Aspirator & Tube Test

The aspirator and tube assembly is used to pull air into the in-car sensor grille. The aspirator and tube assembly can be checked by holding a lit cigarette next to the grille with system operating at high blower heat mode. If smoke is not drawn into the grille, the as-

AUTO-TEMP II SEQUENCE CHART

Push Button Switch	Tester Controls Voltmeter Switch	Tester Controls Mode Switch	Outside Air Door Position	AC Door Position	Heat Door Position	Defrost Door Position	Blower Motor Speed
OFF	Blower Volts	A/C	0%	Open or Closed	Closed or Open	Closed or Bleed	Off
LO-AUTO AND VENT	Blower Volts	Park	100%	Open or Closed	Closed or Open	Closed or Bleed	On
		A/C	100%	Open	Closed	Closed	M1
			100%	Open	Closed	Closed	M1 to M2
			100%	Open	Closed	Closed	M2 to M3
			20%	Open	Closed	Closed	M3
			20%	Open	Closed	Closed	M3 to Hi
			20%	Open	Closed	Closed	Hi
		Heat	20%	Open	Closed	Closed	Hi
			20%	Open	Closed	Closed	Hi to M3
			100%	Open	Closed	Closed	M3
			100%	Open	Closed	Closed	M3 to M2
			100%	Open	Closed	Closed	M2 to M1
			100%	Open	Closed	Closed	M1 to Lo
			100%	Closed	Open	Bleed	Lo
			100%	Closed	Open	Bleed	Lo to M1
			100%	Closed	Open	Bleed	M1 to M2
			100%	Closed	Open	Bleed	M2 to M3
			100%	Closed	Open	Bleed	M3 to Hi
			100%	Closed	Open	Bleed	Hi
		A/C	100%	Closed	Open	Bleed	Hi
			100%	Closed	Open	Bleed	Hi to M3
			100%	Closed	Open	Bleed	M3 to M2
			100%	Closed	Open	Bleed	M2 to M1
			100%	Closed	Open	Bleed	M1 to Lo
			100%	Closed	Open	Bleed	Lo
			100%	Open	Closed	Closed	Lo to M1
			100%	Open	Closed	Closed	M1 to M2
			100%	Open	Closed	Closed	M2 to M3
			20%	Open	Closed	Closed	M3
			20%	Open	Closed	Closed	M3 to Hi
			20%	Open	Closed	Closed	Hi
HI-AUTO	Blower Volts	A/C	20%	Open	Closed	Closed	Hi
		Heat	20%	Open	Closed	Closed	Hi
			20%	Open	Closed	Closed	Hi to M3
			100%	Open	Closed	Closed	M3
			100%	Open	Closed	Closed	M3 to M2
			100%	Closed	Open	Bleed	M2
			100%	Closed	Open	Bleed	M2 to M3
			100%	Closed	Open	Bleed	M3 to Hi
		A/C	100%	Closed	Open	Bleed	Hi
			100%	Closed	Open	Bleed	Hi to M3
			100%	Closed	Open	Bleed	M3 to M2
			100%	Open	Closed	Closed	M2
			100%	Open	Closed	Closed	M2 to M3
			20%	Open	Closed	Closed	M3
			20%	Open	Closed	Closed	M3 to Hi
			20%	Open	Closed	Closed	Hi
LO-DEFROST	Blower Volts	A/C	20%	Closed	Bleed	Open	Hi
		Heat	20%	Closed	Bleed	Open	Hi
			20%	Closed	Bleed	Open	Hi to M3
			100%	Closed	Bleed	Open	M3
			100%	Closed	Bleed	Open	M3 to M2
			100%	Closed	Bleed	Open	M2
			100%	Closed	Bleed	Open	M2 to M3
			100%	Closed	Bleed	Open	M3 to Hi
			100%	Closed	Bleed	Open	Hi
		A/C	100%	Closed	Bleed	Open	Hi to M3
			100%	Closed	Bleed	Open	M3 to M2
			100%	Closed	Bleed	Open	M2
			100%	Closed	Bleed	Open	M2 to M3
			20%	Closed	Bleed	Open	M3
			20%	Closed	Bleed	Open	M3 to Hi
			20%	Closed	Bleed	Open	Hi
HI-DEFROST	Blower Volts	Heat	100%	Closed	Bleed	Open	Hi

Blower Motor Speeds

Hi—Maximum Speed
M3—One speed down from Hi
M2—One speed down from M3
M1—One speed down from M2
Lo—One speed down from M1

Fig. 23 Sequence test chart. 1977-78 Auto Temp II

pirator is improperly connected or kinked.

Vacuum Reservoir Test

1. Connect a 0 to 29 in. Hg. vacuum gauge to vacuum tank outlet port, then start engine and operate at idle speed.
2. Vacuum gauge should read 16 in. Hg. or more within 1 minute.
3. Turn engine off and observe vacuum gauge. Vacuum should not drop off. If vacuum drops off, the tank assembly or check valve is leaking and the tank assembly should be replaced.

Check Valve Test

NOTE: In the following procedure the check valves are referred to as check valve No. 1 and No. 2, the check valve nearest the servo assembly is check valve No. 1.

1. Check valves that are leaking or improperly installed may cause the following symptoms:
   a. If check valve No. 1 is leaking, system will stay on 20 percent outside air and will not go to 100 percent outside air. System will shut off when vacuum supply is exhausted.
   b. If check valve No. 1 is installed backwards, system will function as described in step a. When engine is cold, blower motor will not come on when system in Defrost or Hi-Defrost.
   c. If check valve No. 2 is leaking, blower motor will start immediately and discharge cold air during cold weather.
   d. If check valve No. 2 is installed backwards, blower motor will operate as described in step c. When engine is cold, blower motor will turn off when Off push button is depressed.
2. If check valve failure is suspected, proceed as follows:
   a. Check to ensure that check valves are properly installed in vacuum harness.
   b. Remove suspected check valve and connect an external vacuum source and a vacuum gauge, Fig. 21. Place fingers over the two opened check valve ports and note vacuum gauge reading, then remove vacuum supply hose. Vacuum reading should remain approximately the same. If vacuum drops off, the check valve is defective and should be replaced.

   NOTE: It is possible for a defective check valve to work at room temperature but not at temperatures below freezing. Also do not use any type of lubricant to increase the efficiency of a defective check valve.

SYSTEM VACUUM LEAK TEST

1. Connect a 0 to 30 in. Hg. vacuum gauge to inline vacuum connector at orange vacuum hose near vacuum transfer switch.
2. Depress Lo-Auto push button, then place system in heat mode by rotating tester mode knob.
3. Allow vacuum to build up to 18 in. Hg. and stabilize, then shut off engine and observe vacuum gauge for vacuum drop.
4. If vacuum does not build up to 18 in. Hg. or if vacuum drops at a rate of more than 5 in. Hg. per minute, check system for leaks or disconnected vacuum hoses and repair as necessary.

COLD ENGINE LOCKOUT TEST

1. With engine cold, set tester voltmeter to Servo Amps and mode knob to Park, then start engine and depress Lo-Auto push button. There should be no air flow or voltmeter reading.
2. Set tester mode knob to Heat, voltmeter should now read 3 to 4½ volts until full heat is reached, then voltage should drop to ½ volt. There should be no air flow out of any outlets.
3. Set tester voltmeter knob to Voltage position, there should be no voltage reading or air flow. When engine coolant temperature reaches 125 degrees F, which should occur in approximately 3 minutes, blower motor should come on at high blower speed and system should be in full heat position. Air flow should be from heater ducts with a normal defroster bleed.

CALIBRATION CHECK-OUT PROCEDURE

NOTE: This procedure is used to check the calibration of the servo assembly and amplifier.

1. With tester connected to service connector, set voltmeter to off and mode knob to A/C.
2. Connect tachometer to engine and operate engine at 1100 RPM.
3. Set Auto Temp II controls to 75 and depress Lo-Auto push button.
4. Check ambient sensor, in-car sensor and comfort control potentiometer and lever as described previously.
5. Position tester voltmeter to Servo Amps.
6. With system in on position and in A/C mode, place tester mode knob on Park. When voltage on tester voltmeter drops to zero, proceed as follows:
   a. Disconnect servo assembly wire connector, then using a suitable ohmmeter, check resistance between servo black and violet wires.
   b. If resistance is 1200 to 1500 ohms, reconnect servo wire connector and proceed to next step. If resistance is not within limits, replace servo assembly.
7. Position tester mode knob to Heat.
8. When voltage indicated in voltmeter drops to zero, disconnect servo assembly wire connector and connect an ohmmeter between black and violet wires, then proceed as follows:
   a. Resistance should be between 1000 and 1200 ohms. If resistance is within limits, reconnect servo wire connector. If resistance is not within limits, replace amplifier assembly.

Semi-Automatic Temperature Control (S.A.T.C.)

1979 Cordoba, Magnum, 1979-80 Newport, New Yorker, St. Regis & 1980 Gran Fury

Refer to diagnosis charts, Fig. 24 (Parts 1 through 9) and the following test procedures when diagnosing problems on the S.A.T.C. system.

System Inspection

1. Check system components for the following before performing diagnosis procedures on system:
   a. Check all vacuum hoses for proper connection.
   b. Temperature control cable for proper attachment and adjustment.
   c. Blend air door and vacuum servo actuator cable for proper attachment and adjustment.
   d. Aspirator and in-car sensor hoses for proper connections.
   e. Bi-metal sensor for proper attachment to evaporator/heater assembly.
   f. Compensator hose for loose or damaged seals.

System Test

1. Tee-in a suitable vacuum gauge to bi-metal sensor ports, Fig. 25. If a buzzing noise is encountered when installing vacuum gauge, this does not indicate that sensor is defective or that system will not operate properly.
2. Position instrument panel control lever to 65, then start engine and operate at 1000 rpm.
3. Position a thermometer in right center A/C outlet duct, then depress A/C push button, Max. A/C push button if ambient temperature is above 80 degrees F, and place blower switch on high. Allow approximately 5 to 10 minutes for system to stabilize with doors and windows closed.
4. If cold airflow does not come out of A/C outlets, refer to Semi-Automatic Temperature Control Diagnosis Charts, Fig. 24 (Parts 1 through 9) and Fig. 26, then proceed to following step.
5. If system has stabilized in Max. A/C, depress A/C push button.
6. Move comfort control lever from 65 to 85, then after approximately 1 minute, return lever to 65 and note vacuum gauge and temperature reading. Air outlet temperature and vacuum reading should increase when lever is moved from 65 to 85 and decrease when lever is moved back to 65.
7. If air outlet temperature does not vary, check for the following:
   a. Check to ensure that bi-metal sensor control cable is properly attached to sensor.
   b. As comfort control lever is being moved, check to ensure that bi-metal sensor arm is actuated. If arm is not actuated, check cable for proper attachment at S.A.T.C. control and repair or replace cable as necessary.
   c. If cable is attached properly, disconnect cable from bi-metal sensor and check to ensure that arm moves freely. If arm does not move freely, replace sensor.
8. If bi-metal sensor arm moves freely and outlet air temperature does not vary, proceed as follows:
   a. Remove vacuum hose connector from bi-metal sensor and connect a vacuum gauge to small port on connector, Fig. 27.
   b. With engine operating, vacuum should be at least 15 in. Hg. If vacuum is less than 15 in. Hg., there may be a leak in the A/C vacuum harness, push button switch, actuators or defective component. If there is a loss of temperature control or no heat in modes other than A/C, the above problems may also apply.

AIR CONDITIONING BLOWER MOTOR AND CONTROL SYSTEM DIAGNOSIS (ELECTRICAL)

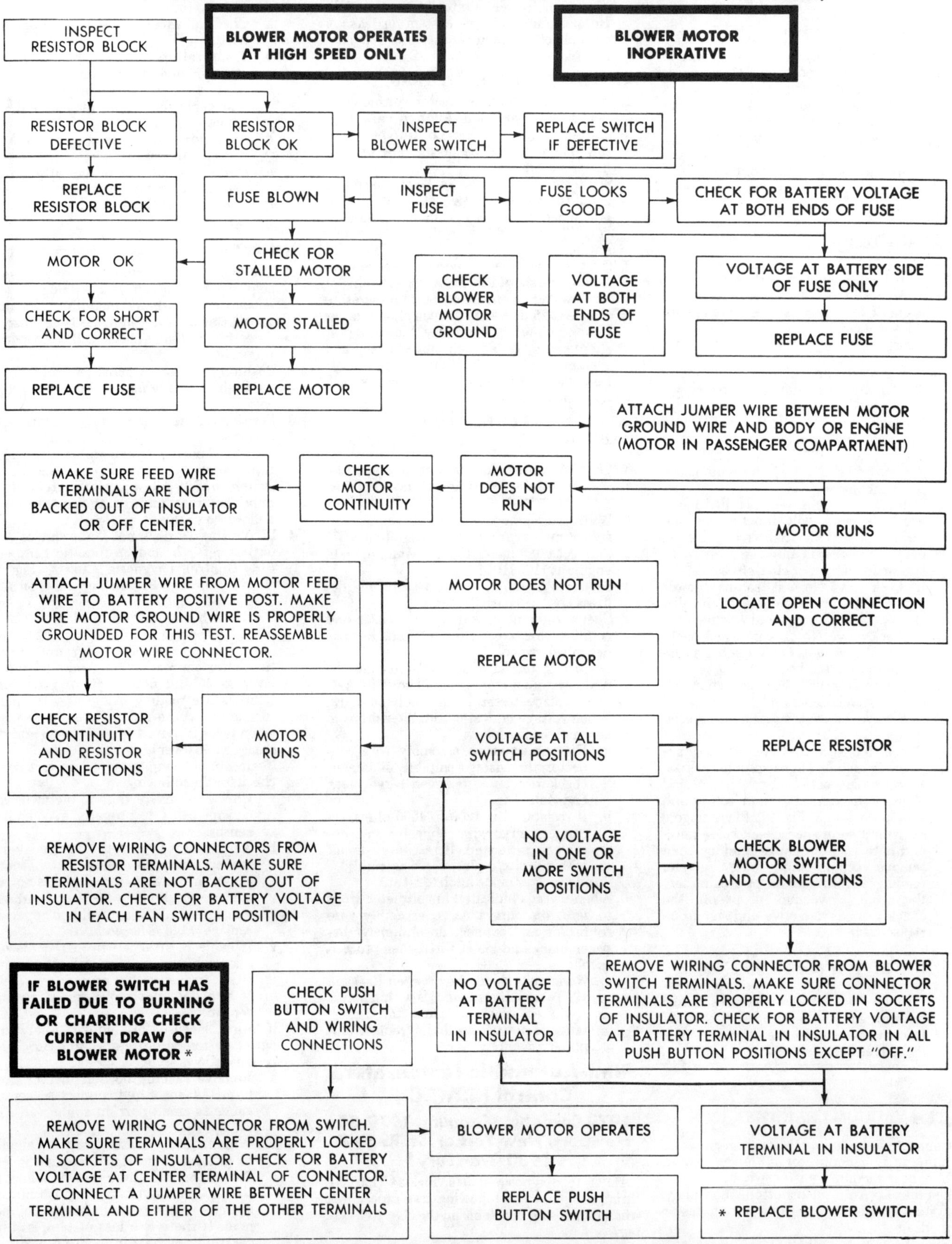

Fig. 24 Semi-Automatic Temperature Control (S.A.T.C.) diagnosis chart. 1979 Cordoba, Magnum, 1979-80 Newport, New Yorker, St. Regis & 1980 Gran Fury (Part 1 of 9)

SEMI-AUTOMATIC TEMPERATURE CONTROL (S.A.T.C.) DIAGNOSIS

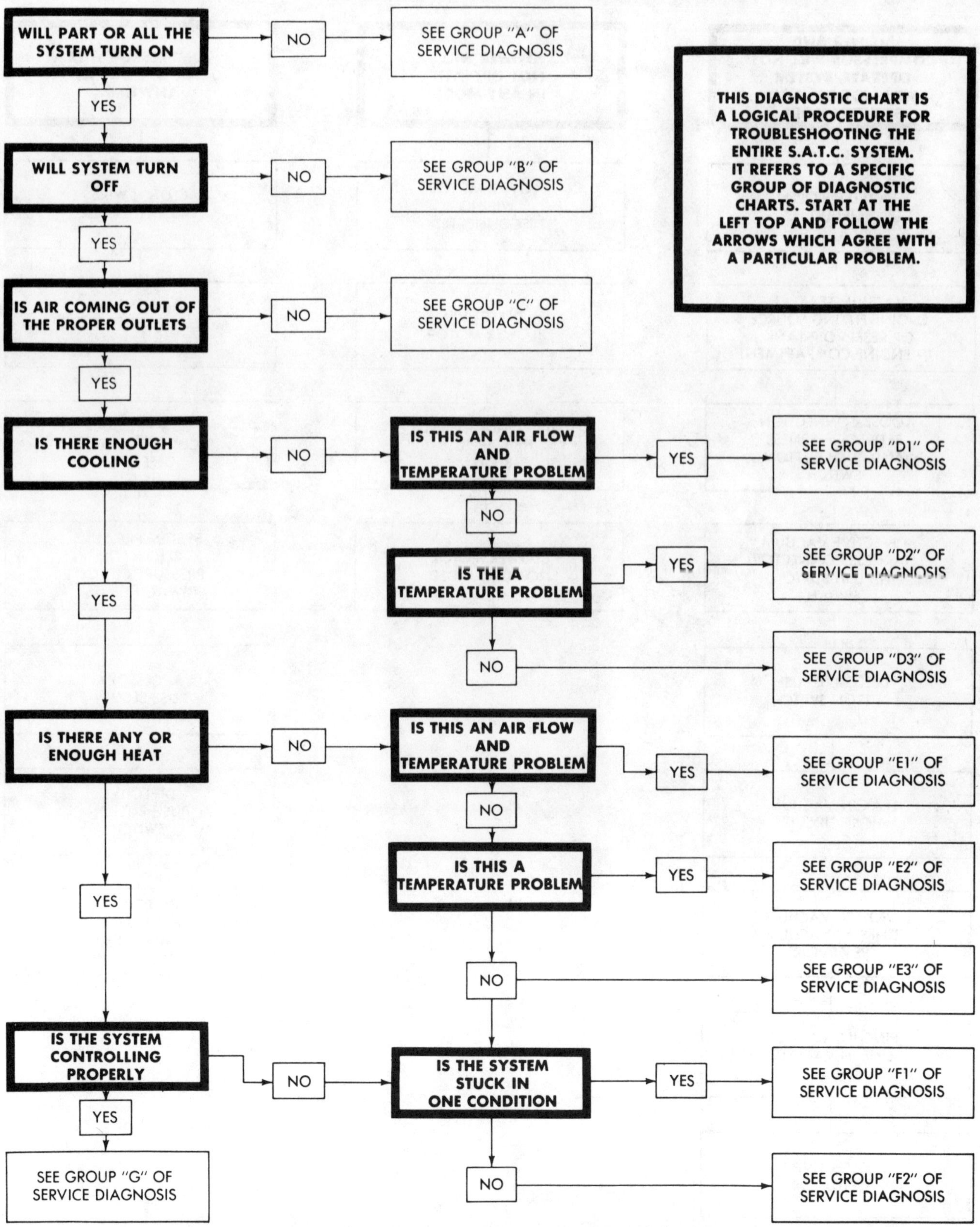

Fig. 24 **Semi-Automatic Temperature Control (S.A.T.C.) diagnosis chart. 1979 Cordoba, Magnum, 1979-80 Newport, New Yorker, St. Regis & 1980 Gran Fury (Part 2 of 9)**

GROUP A—PART OR ALL OF SYSTEM WILL NOT TURN ON

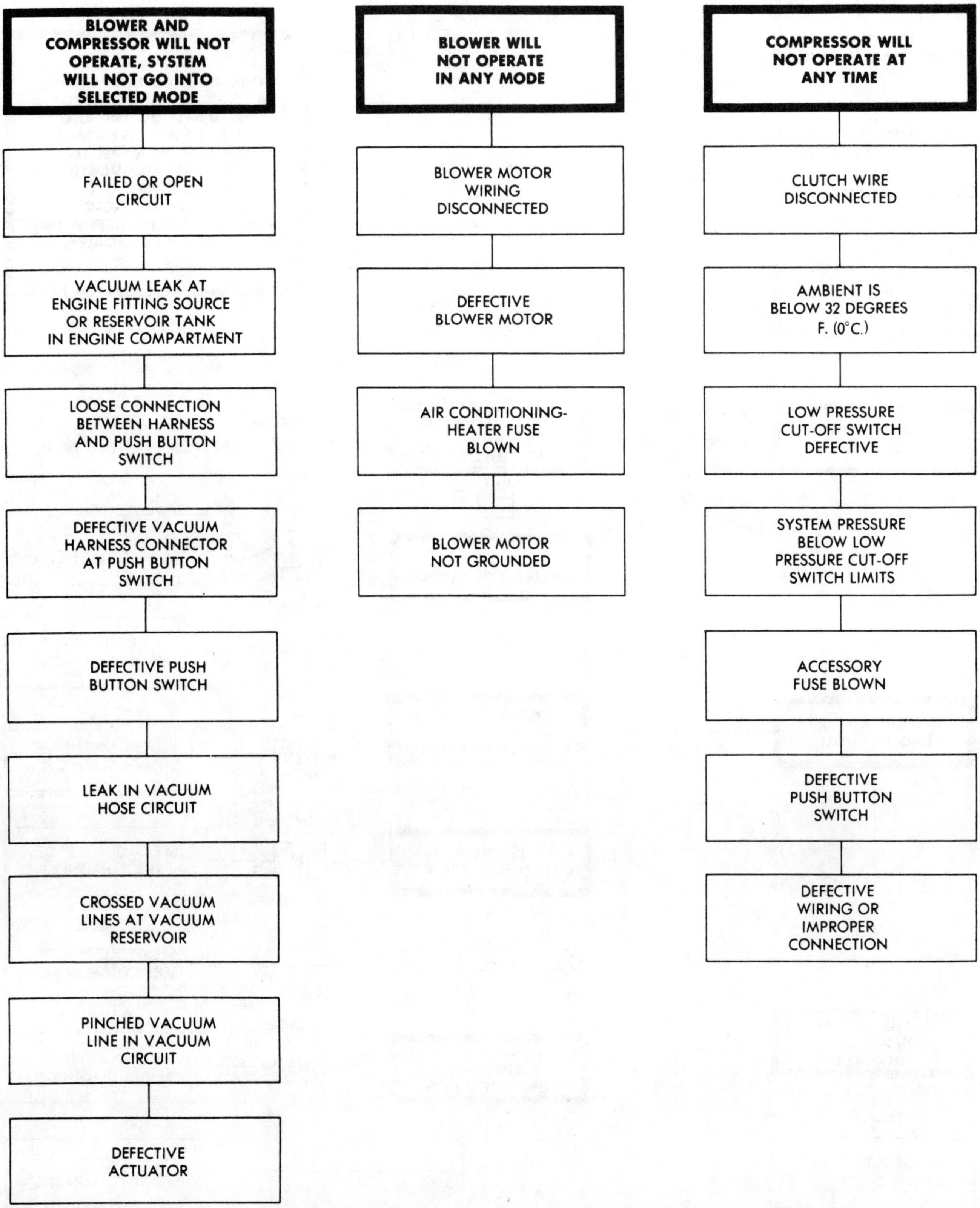

Fig. 24 Semi-Automatic Temperature Control (S.A.T.C.) diagnosis chart. 1979 Cordoba, Magnum, 1979-80 Newport, New Yorker, St. Regis & 1980 Gran Fury (Part 3 of 9)

GROUP B—SYSTEM WILL NOT TURN OFF

BLOWER DOES NOT SHUT OFF WHEN OFF BUTTON IS PUSHED
- DEFECTIVE PUSH BUTTON SWITCH

GROUP C—AIR COMES OUT OF THE WRONG OUTLETS

A/C AND HEAT MODES REVERSED — HOT AIR FROM A/C OUTLETS AND COLD AIR FROM HEAT OUTLETS
- A/C AND HEAT MODE ACTUATOR VACUUM HOSES REVERSED

MOST OF THE AIR COMING OUT OF DEFROSTERS WHEN SYSTEM SHOULD BE IN HEAT MODE
- HEAT — DEFROST VACUUM LINES CROSSED

AIR COMES OUT OF ALL OUTLETS WHEN SYSTEM SHOULD BE IN DEFROST OR A/C. AIR COMES OUT A/C OUTLETS WHEN SYSTEM SHOULD BE IN HEAT MODE
- ALL VACUUM LINES TO MODE AND DEFROST ACTUATORS COMPLETELY CROSSED
- MODE DOOR STUCK IN NEUTRAL POSITION

AIR COMING OUT OF A/C OUTLETS WHEN SYSTEM IS IN THE DEFROST OR HEAT MODE
- VACUUM HOSES TO MODE ACTUATOR AND DEFROST ACTUATOR REVERSED
- DEFECTIVE PUSHBUTTON SWITCH ASSEMBLY

AIR COMES OUT OF THE DEFROSTER WHEN SYSTEM SHOULD BE IN HEAT MODE, AIR COMES OUT HEATER OUTLET WHEN SYSTEM SHOULD BE IN A/C
- ALL VACUUM LINES TO MODE AND DEFROST ACTUATORS COMPLETELY CROSSED
- MODE DOOR STUCK IN HEAT POSITION
- VACUUM LEAK OR FAILED MODE ACTUATOR

AIR COMES OUT OF A/C OUTLETS OR HEAT OUTLETS WHILE DRIVING IN THE OFF POSITION
- RECIRCULATION DOOR IN OUTSIDE AIR POSITION—VACUUM HOSES ON RECIRCULATION DOOR ACTUATOR REVERSED
- VACUUM HOSE OFF
- RECIRCULATING DOOR NOT SEALING PROPERLY

AIR COMING OUT OF DEFROSTER WHEN SYSTEM SHOULD BE IN THE A/C MODE
- A/C HEAT AND HEAT DEFROSTER MODE ACTUATOR HOSES REVERSED ON ACTUATORS
- MODE DOOR NOT SEALING COMPLETELY

Fig. 24 **Semi-Automatic Temperature Control (S.A.T.C.) diagnosis chart. 1979 Cordoba, Magnum, 1979-80 Newport, New Yorker, St. Regis & 1980 Gran Fury (Part 4 of 9)**

GROUP D—NO COOLING OR NOT ENOUGH COOLING

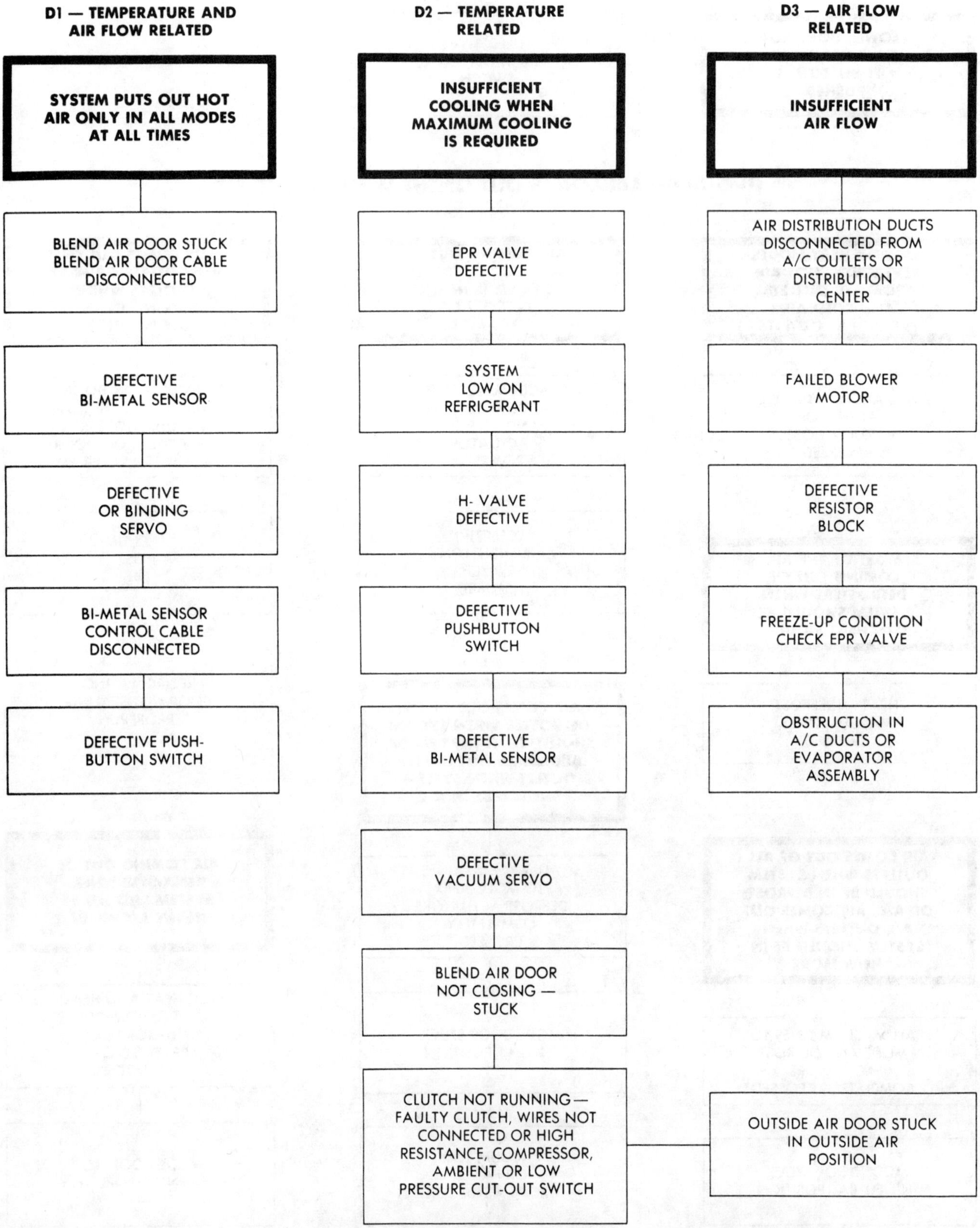

Fig. 24 Semi-Automatic Temperature Control (S.A.T.C.) diagnosis chart. 1979 Cordoba, Magnum, 1979-80 Newport, New Yorker, St. Regis & 1980 Gran Fury (Part 5 of 9)

GROUP E— NO HEATING OR NOT ENOUGH HEATING

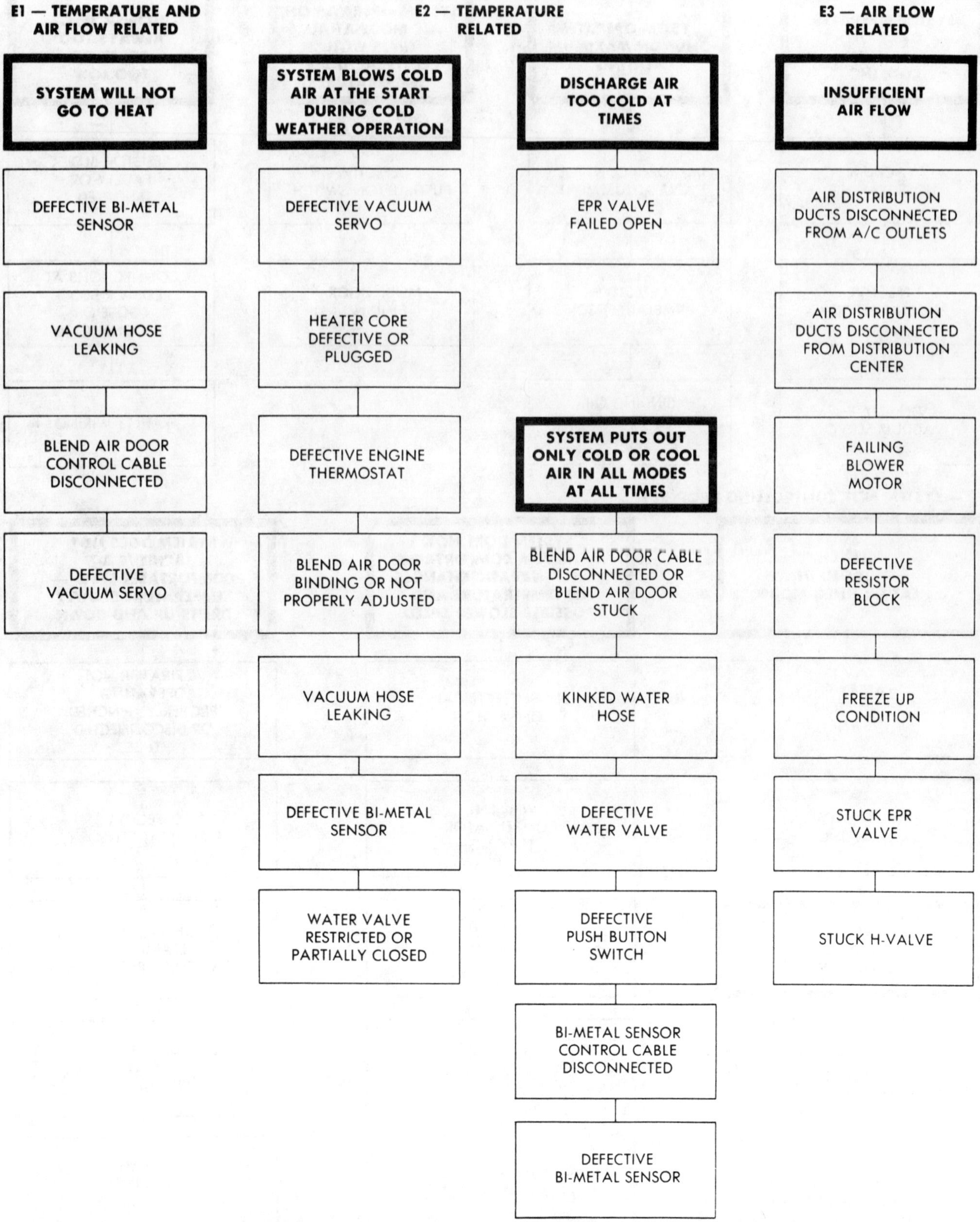

Fig. 24 Semi-Automatic Temperature Control (S.A.T.C.) diagnosis chart. 1979 Cordoba, Magnum, 1979-80 Newport, New Yorker, St. Regis & 1980 Gran Fury (Part 6 of 9)

GROUP F—SYSTEM NOT CONTROLLING PROPERLY

F1 — SYSTEM IS STUCK IN ONE CONDITION

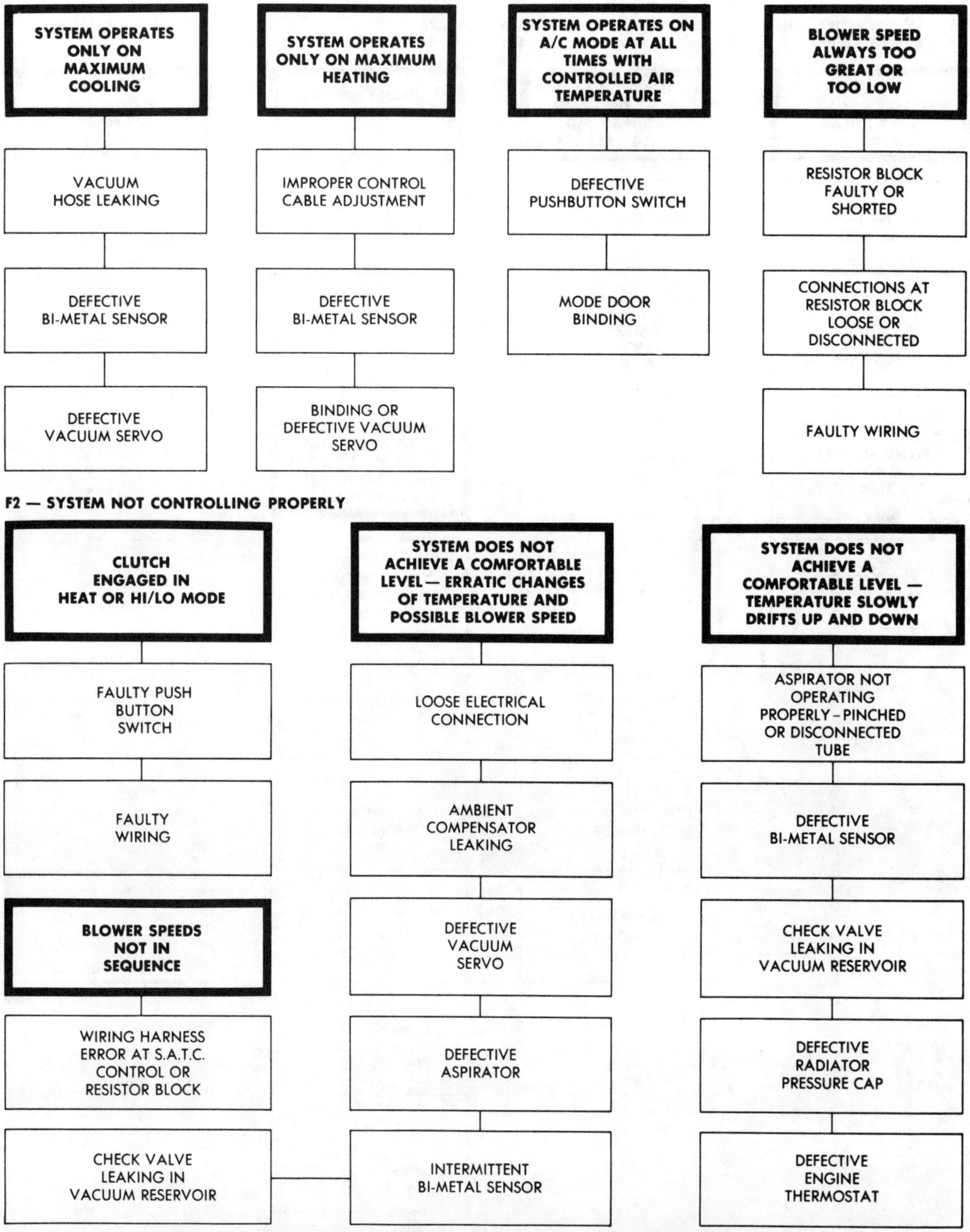

Fig. 24 Semi-Automatic Temperature Control (S.A.T.C.) diagnosis chart. 1979 Cordoba, Magnum, 1979-80 Newport, New Yorker, St. Regis & 1980 Gran Fury (Part 7 of 9)

F-2 SYSTEM NOT CONTROLLING PROPERLY

ERRATIC TEMPERATURE CONTROL

- HEATER HOSES REVERSED AT CORE INLET AND OUTLET TUBES
- RADIATOR WATER LEVEL LOW
- DEFECTIVE RADIATOR CAP
- VACUUM HOSE LOOSE
- DEFECTIVE ENGINE THERMOSTAT
- FAULTY SERVO ASSEMBLY
- INTERMITTENT BI-METAL SENSOR
- BLEND AIR DOOR BINDING
- ASPIRATOR DISCONNECTED

DISCHARGE AIR TOO COLD AT TIMES

- EPR VALVE FAILED OPEN

SYSTEM DOES NOT ACHIEVE A COMFORTABLE TEMPERATURE WITHIN WHITE RANGE OF COMFORT CONTROL LEVER

- DEFECTIVE OR IMPROPERLY INSTALLED BI-METAL SENSOR
- BLEND AIR DOOR CABLE IMPROPERLY ADJUSTED
- BI-METAL SENSOR CONTROL CABLE IMPROPERLY ADJUSTED
- ASPIRATOR NOT OPERATING PROPERLY — PINCHED OR DISCONNECTED TUBE
- DEFECTIVE VACUUM SERVO
- VACUUM LEAK IN SYSTEM

SYSTEM DOES NOT ACHIEVE A COMFORTABLE LEVEL — CYCLING OF TEMPERATURE

- DEFECTIVE SERVO ASSEMBLY
- ASPIRATOR NOT OPERATING PROPERLY — PINCHED OR DISCONNECTED TUBE
- DEFECTIVE ENGINE THERMOSTAT
- IMPROPERLY INSTALLED BI-METAL SENSOR
- SEAL LEAKING AT AMBIENT COMPENSATOR
- DEFECTIVE BI-METAL SENSOR
- VACUUM LEAK IN SYSTEM

SYSTEM DOES NOT ACHIEVE A COMFORTABLE CONDITION SYSTEM LEVELED OUT

- DEFECTIVE BI-METAL SENSOR
- DEFECTIVE VACUUM SERVO
- A/C OUTLETS NOT DIRECTED PROPERLY
- ASPIRATOR NOT OPERATING PROPERLY PINCHED OR DISCONNECTED TUBE
- IMPROPERLY ADJUSTED BI-METAL SENSOR CONTROL CABLE
- IMPROPERLY ADJUSTED BLEND AIR DOOR CONTROL CABLE
- VACUUM LEAK IN SYSTEM

AFTER ONE HOUR OF SUSTAINED DRIVING AIRFLOW DROPS OFF TO PRACTICALLY NONE WHILE BLOWER CAN BE HEARD OPERATING AT HIGH SPEED

- EVAPORATOR COIL FREEZE-UP — EPR VALVE FAILED OPEN
- FAILED CYCLING CLUTCH SWITCH

Fig. 24 Semi-Automatic Temperature Control (S.A.T.C.) diagnosis chart. 1979 Cordoba, Magnum, 1979-80 Newport, New Yorker, St. Regis & 1980 Gran Fury (Part 8 of 9)

GROUP G—MISCELLANEOUS

DISCHARGE AIR CYCLING AT A/C OUTLETS — SYSTEM STABILIZED A/C OR MAX A/C MODE

NO DEFECT — SEE VACUUM SERVO TEST PROCEDURE

BI-METAL SENSOR MAKES BUZZING NOISE BUT OPERATES PROPERLY

BI-METAL SENSOR NOT DEFECTIVE

Fig. 24 Semi-Automatic Temperature Control (S.A.T.C.) diagnosis chart. 1979 Cordoba, Magnum, 1979-80 Newport, New Yorker, St. Regis & 1980 Gran Fury Part 9 of 9

PUSH BUTTON	OFF	A/C MAX	A/C NORM	A/C BI-LEVEL	HEAT	DEFROST
Inlet Air Door (Open to)	Inside	Inside	Outside	Outside	Outside	Outside
Heat-Defrost Door (Open to)	Heat	Heat	Heat	Heat	Heat	Defrost
Upper Bi-Lev Door (Open to)	Bi-Lev-A/C	Bi-Lev-A/C	Bi-Lev-A/C	Bi-Lev-A/C	Heat	Heat
Lower Bi-Lev Door (Open to)	A/C	A/C	A/C	Bi-Lev/Heat	Bi-Lev/Heat	Bi-Lev-Heat
Compressor Clutch	Off	On	On*	On*	Off	On***
Blower	Off	On**	On	On	On	On
Water Valve	Off	Off	On	On	On	On

* Pulling out NORM button provides VENT mode with refrigeration off. BI-LEVEL button pull-out works same way.
** Only high and medium 2 blower speeds operational.
*** Pulling out DEF button provides DEFROST mode with refrigeration off.

Fig. 26 Control system chart. 1979 Cordoba, Magnum, 1979-80 Newport, New Yorker, St. Regis & 1980 Gran Fury S.A.T.C.

9. If no vacuum leaks are indicated, remove vacuum gauge from vacuum hose connector small port, then plug small port opening and connect a hand vacuum pump to large port on connector and apply 15 in. Hg. of vacuum. A/C outlet air temperature should increase and vacuum reading should remain constant, Fig. 28. If vacuum reading remains constant and temperature increases, replace bi-metal sensor.

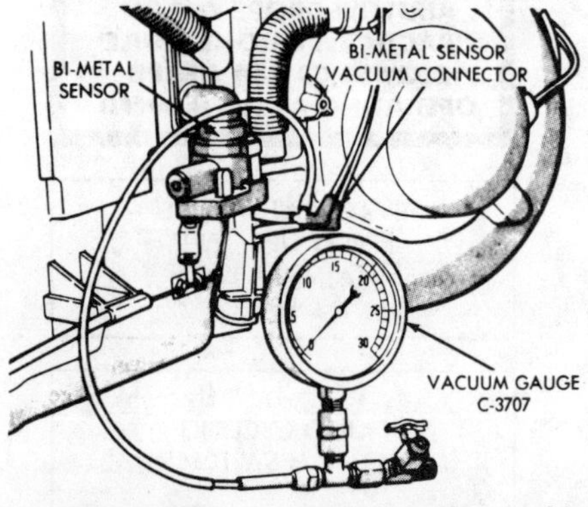

Fig. 25 Vacuum gauge connection to bi-metal sensor. 1979 Cordoba, Magnum, 1979-80 Newport, New Yorker, St. Regis & 1980 Gran Fury S.A.T.C.

10. If a vacuum leak is indicated at vacuum connector larger port, check for a loose vacuum connection at vacuum servo actuator.
11. If leak is still present, plug vacuum servo actuator connector end and repeat step 9, if vacuum leak is still indicated, replace vacuum hose.
12. If no vacuum leak was indicated in step 9 and outlet air temperature does not increase, check to ensure that blend air door is properly attached to vacuum servo and blend air door crank, also check to ensure that cable is not defective.
13. If cable is properly connected but temperature does not increase, disconnect cable and check to ensure that blend air door moves freely in housing and repair as necessary. If door moves freely, and temperature increases to approximately 150 degrees F, replace servo actuator.
14. If blend air door moves freely but only cold air is discharged from air outlet, check to ensure that both heater hoses are warm at heater core. If one or both heater hoses are cool at heater core, check engine thermostat and water valve for proper operation. Vacuum should only be applied to water valve when system is in the Off or Max. A/C positions. When no vacuum is applied, water valve should be opened allowing coolant to flow. If vacuum is present in any other mode than Off or Max. A/C, check push button switch and vacuum harness for proper hose routing.

Temperature Control Test

1. Connect vacuum gauge to bi-metal sensor ports, then position temperature control

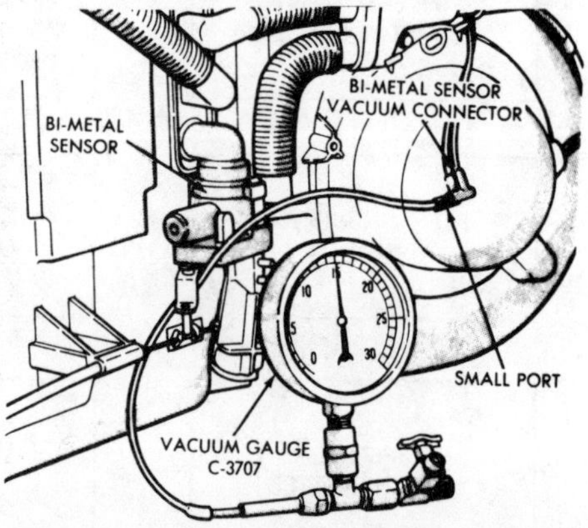

Fig. 27 Vacuum gauge connection to bi-metal sensor connector small port. 1979 Cordoba, Magnum, 1979-80 Newport, New Yorker, St. Regis & 1980 Gran Fury S.A.T.C.

lever to 75. With engine operating, A/C outlet temperature should either increase or decrease depending on ambient temperature.

2. Allow system vacuum reading to stabilize on Hi Blower, then position psychrometer C-3704 on middle of rear seat and turn psychrometer on.
3. With windows closed and system stabilized, interior dry bulb temperaure should be between 70 and 80 degrees F.
4. If interior dry bulb temperature does not stabilize between 70 and 80 degrees F, move temperature control lever to bring temperature within limits. System should stabilize within the 70 to 80 setting on the temperature control lever.
5. If dry bulb temperature cannot be stabilized within the 70 to 80 temperature control lever setting, proceed as follows:
   a. Check to ensure that bi-metal sensor control cable is properly adjusted and attached to bi-metal sensor and temperature control lever.
   b. Check to ensure that blend air door control cable is properly attached to blend air door and blend air door crank attaching nut is tight.
   c. Check to ensure that aspirator hoses and in-car sensor hoses for proper connection and damage.
   d. Check to ensure that ambient sensor seal is not damaged and that sensor is properly installed on evaporator/-heater assembly.
   e. Check to ensure bi-metal sensor is properly installed and sealed at compensator.
   f. Check aspirator for proper operation by placing system in heat mode at Hi-blower. Position a piece of paper in front of in-car sensor air inlet grille, the paper should be drawn against the grille if aspirator is operating properly. If paper is not drawn against in-car sensor grille, remove aspirator and check to ensure that aspirator is not blocked. Also check to ensure that aspirator hoses are properly attached and not damaged.
   g. If temperature control is still operating improperly, replace bi-metal sensor.

NOTE: The cycling clutch switch may also cause A/C outlet air temperature to cycle. Also refrigerant system freeze-up over long periods of driving may effect system comfort and control.

1980 Cordoba, Diplomat, LeBaron & Mirada

Refer to diagnosis charts, Figs. 29 Parts 1 through 6, and the following test procedures when diagnosing problems on the S.A.T.C. system.

Positioning Blend Air Door For System Diagnosis

After removing glove compartment & right hand cowl panel, set control to low blower Bi-Level position when locking blend air door in full or minimum heat position. To position blend air door in full heat position, disconnect in-car sensor then turn system on. To position blend air door in the minimum heat position, connect a jumper wire between point E of red wiring harness connector and point B harness ground connection, Fig. 30, then turn system on. To position blend air door between minimum and full heat positions, disconnect battery ground cable, then disconnect S.A.T.C. harness and blower motor ground connections, when connect blower motor ground connection. Connect a jumper wire between S.A.T.C. ground wire and a good body ground, then reconnect battery ground cable and move temperature control lever to bring blend air door to the desired position. When blend air door has reached the desired position, disconnect jumper wire connecting S.A.T.C. ground wire to body ground.

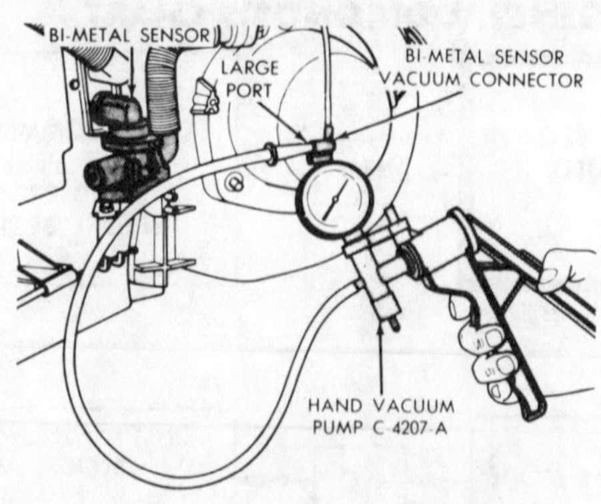

Fig. 28 Vacuum gauge connection to bi-metal sensor connector small port. 1979 Cordoba, Magnum, 1979-80 Newport, New Yorker, St. Regis & 1980 Gran Fury S.A.T.C.

NOTE: Do not apply power to system with blower motor ground wire disconnected.

Electrical Continuity Test

The glove compartment and right hand cowl trim panel must be removed before performing the continuity test. Check to ensure that at least 11 volts is available at test points shown in Fig. 30. Before testing system for continuity, disconnect power feed connector and in-car sensor. Refer to Fig. 31 Parts 1 and 2, for continuity test procedures.

Ambient Sensor Test

1. Remove ambient sensor and check resistance using a suitable ohmmeter.

NOTE: Do not hold sensor in hand or apply ohmmeter current for more than 5 seconds otherwise sensor resistance and resistance reading accuracy will be affected.

2. With an ambient temperature of 70 to 80 degrees F, sensor resistance should be 255 to 335 ohms.

Aspirator Test

1. Place controls on high blower heat mode.
2. Place a piece of paper over in-car sensor air inlet grille.
3. Suction from aspirator should hold paper against grille. If aspirator does not hold paper against grille, refer to Fig. 29 Part 5 of 6.

Control Unit Resistor Test

1. Connect ohmmeter leads to resistor wire terminals.
2. With control lever at 65, resistance should be 390 ohms.
3. Slowly move temperature control lever to right until ohmmeter reading is 930 ohms. Ohmmeter reading should increase smoothly and temperature lever should be near the 75 setting.
4. Move temperature control lever to 85. Ohmmeter reading should smoothly increase to at least 1500 ohms.

Electronic Servomotor Test

1. Remove servomotor assembly from vehicle. After removing servomotor, check to ensure blend air door is operating properly.
2. To perform this test, a fully charged 12 volt battery, an ambient sensor, a series connected instrument panel bulb and socket and a test lead. Connect series connected instrument panel lamp to battery positive post and test lead to battery negative post.
3. Connect test leads as shown in Figs. 32 and 33, to check servomotor operation. The servomotor should run smoothly as it moves to internal stops. The test lamp glows dimly as motor moves to stop and brighten when stop is reached. The test lamp will glow brightly if motor binds between stops.
4. If servomotor shaft fails to move, recheck electrical test connections and condition of 12 volt battery. If motor moves only in the counterclockwise direction check ambient sensor. The motor must move to the full clockwise and counterclockwise positions to be considered acceptable.

In-Car Sensor Test

1. Remove glove compartment, then disconnect in-car sensor from wiring harness. Do not disconnect aspirator assembly or remove sensor from instrument panel.
2. Check aspirator assembly as described under Aspirator Test.
3. Position a thermometer into air inlet grille near in-car sensor, then set blower motor speed to M2. Depress Max. A/C push button, then pull push button out to turn A/C compressor off and close water valve.
4. With blower motor operating, connect ohmmeter to in-car sensor wire terminals. Resistance of the in-car sensor should be between 1100 and 1800 ohms at a temperature of 70 to 80 degrees F.

NOTE: Do not leave ohmmeter connected to in-car sensor for a period of time longer than 5 seconds, otherwise an inaccurate resistance reading may be obtained.

All Systems

Performance Tests

Over-All Performance Test

Humidity (the amount of moisture in the air) has an important bearing on the temperature of the air delivered to the vehicle's interior. This is true of all air-conditioned systems whether in the home, office, or vehicle. It is important to understand the effect humidity has on the performance of the system. When humidity is high, the evaporator has to perform a double duty. It must lower the air temperature and the temperature of the moisture carried in the air. Condensing the moisture in the air transfers a great deal of heat energy into the evaporator fins and tubing. This reduces the amount of heat the evaporator can absorb from the air. In other words, high humidity greatly reduces the evaporator's ability

GENERAL DIAGNOSIS CHART

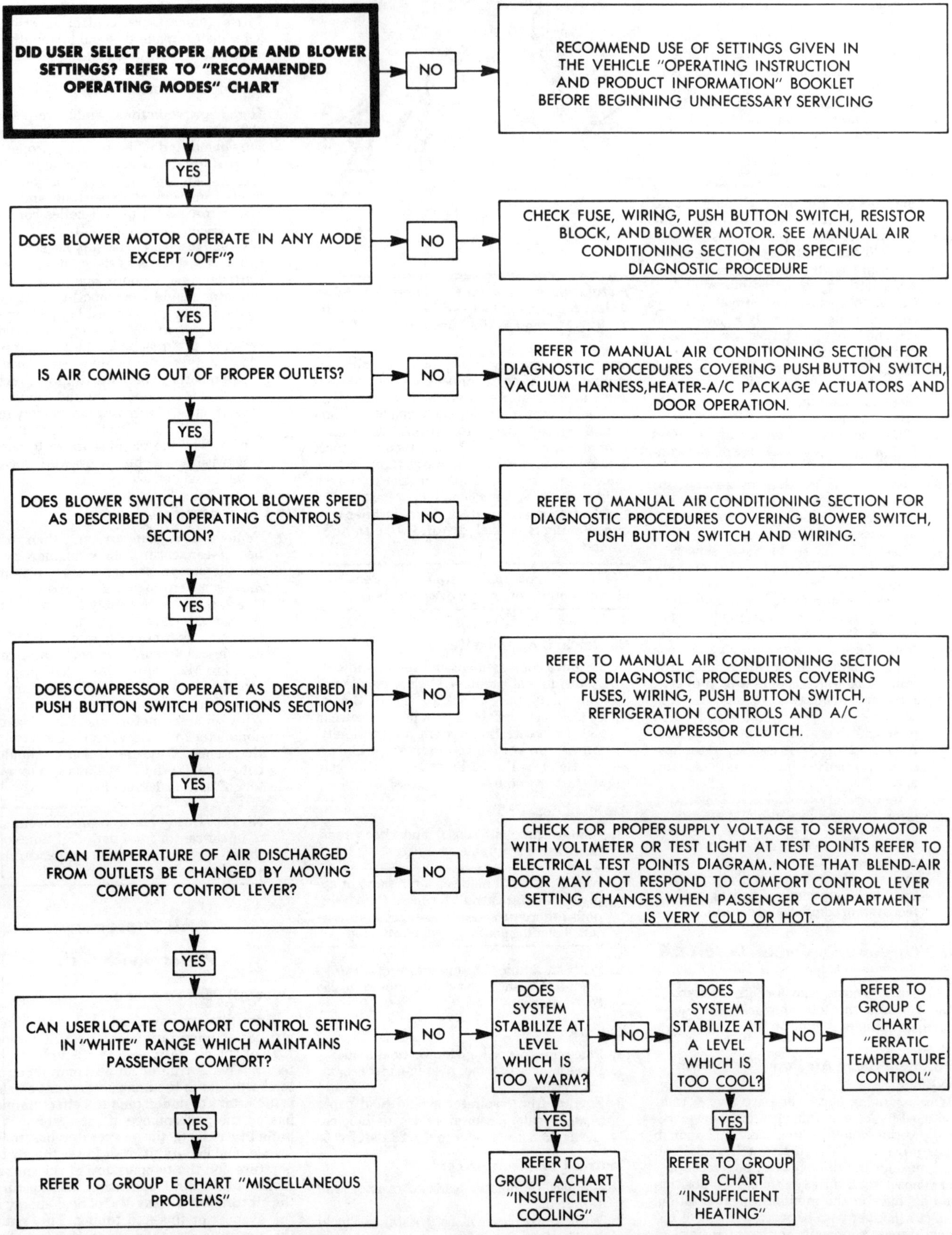

Fig. 29 Semi-Automatic Temperature Control (S.A.T.C.) diagnosis chart. 1980 Cordoba, Diplomat, LeBaron & Mirada Part 1 of 6

GROUP A—INSUFFICIENT COOLING

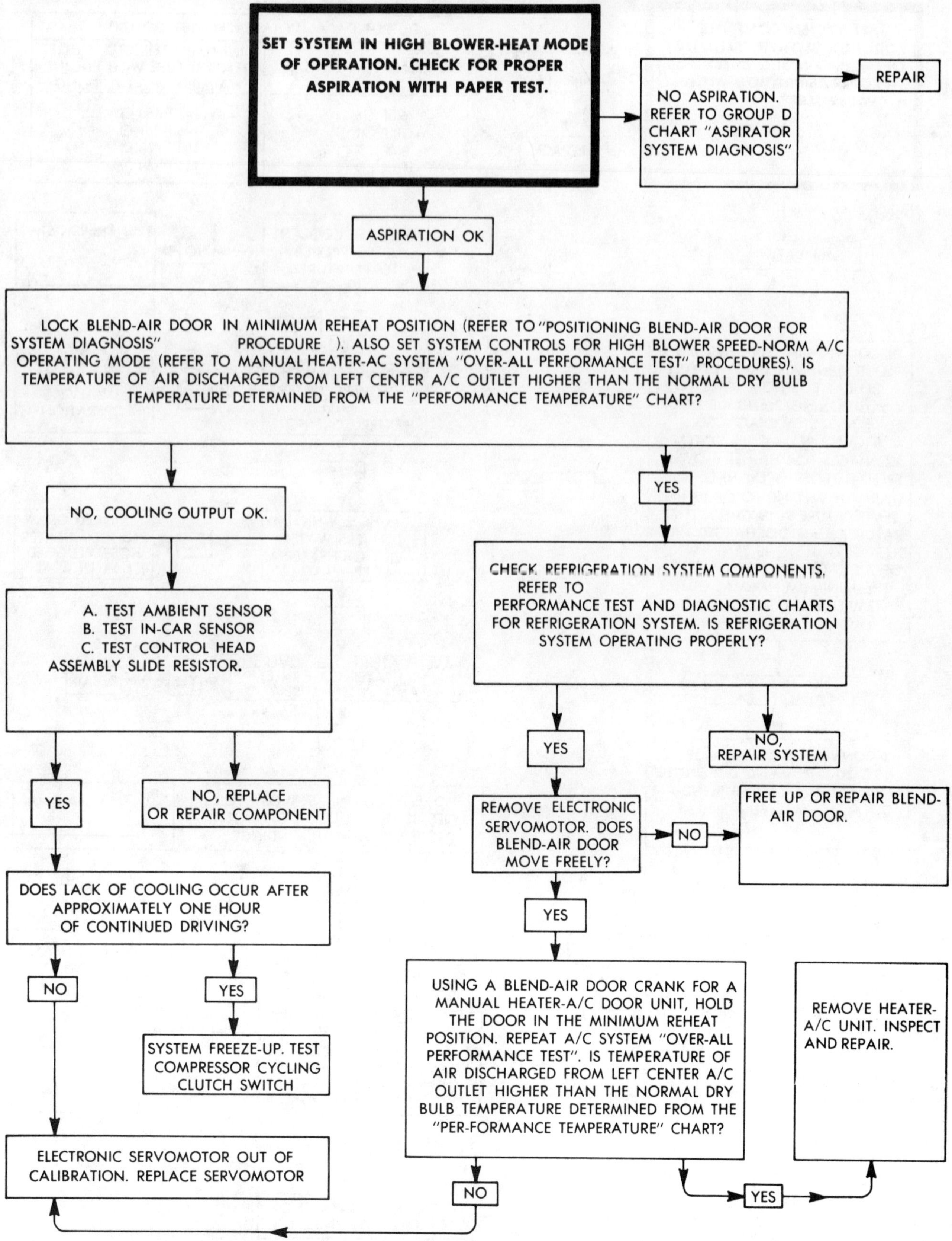

Fig. 29 Semi-Automatic Temperature Control (S.A.T.C.) diagnosis chart. 1980 Cordoba, Diplomat, LeBaron & Mirada (Part 2 of 6)

GROUP B — INSUFFICIENT HEATING

OUTPUT CAPACITY (AT DEFROSTER OUTLET)	
AMBIENT TEMPERATURE	MINIMUM DEFROSTER OUTLET TEMPERATURE WITH ENGINE WARMED UP AND IDLING
60°F (16°C)	142°F (60°C)
70°F (21°C)	146°F (61°C)
80°F (27°C)	150°F (63°C)
90°F (32°C)	154°F (64°C)

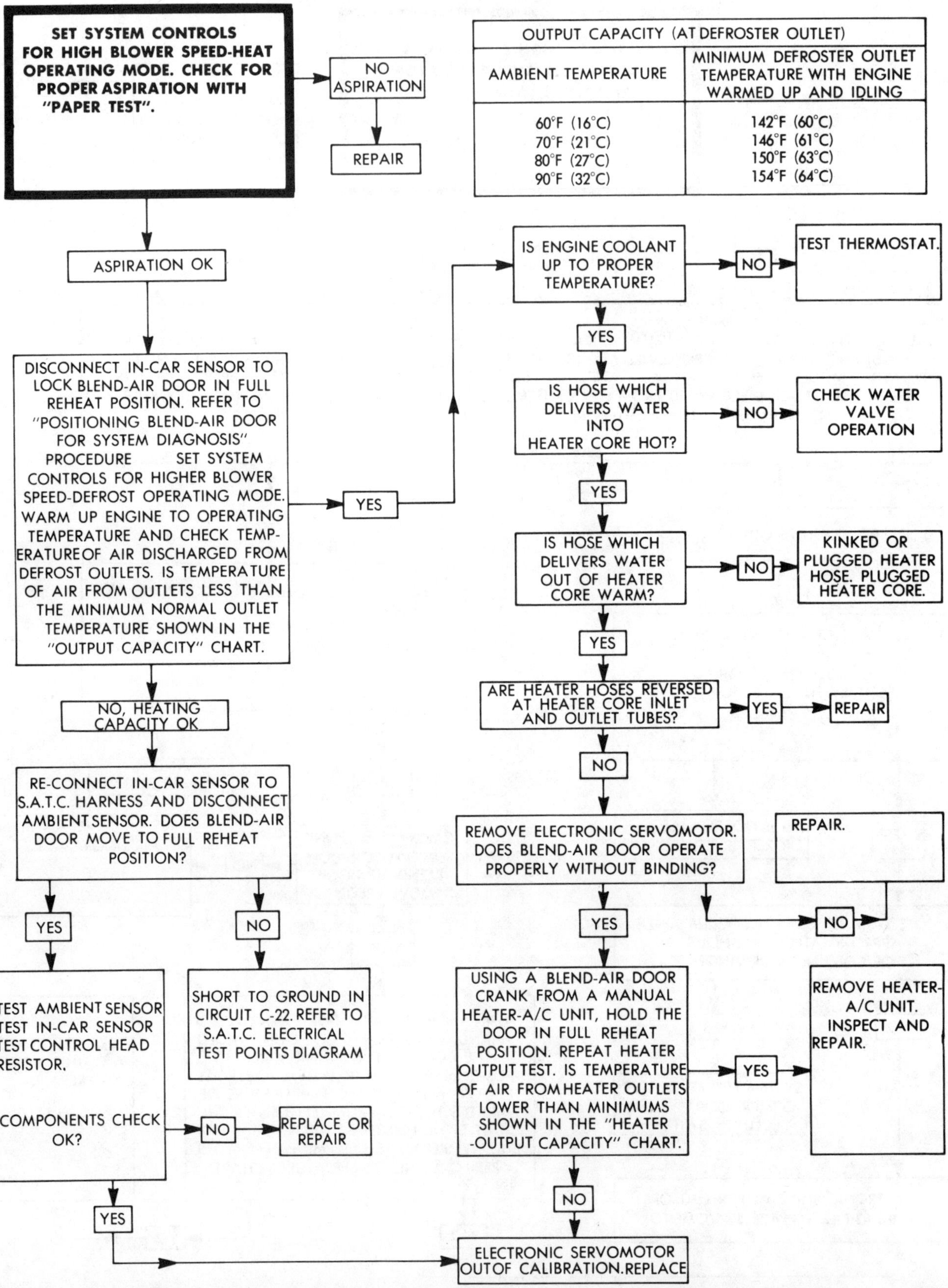

Fig. 29 Semi-Automatic Temperature Control (S.A.T.C.) diagnosis chart. 1980 Cordoba, Diplomat, LeBaron & Mirada (Part 3 of 6)

GROUP C — ERRATIC TEMPERATURE CONTROL

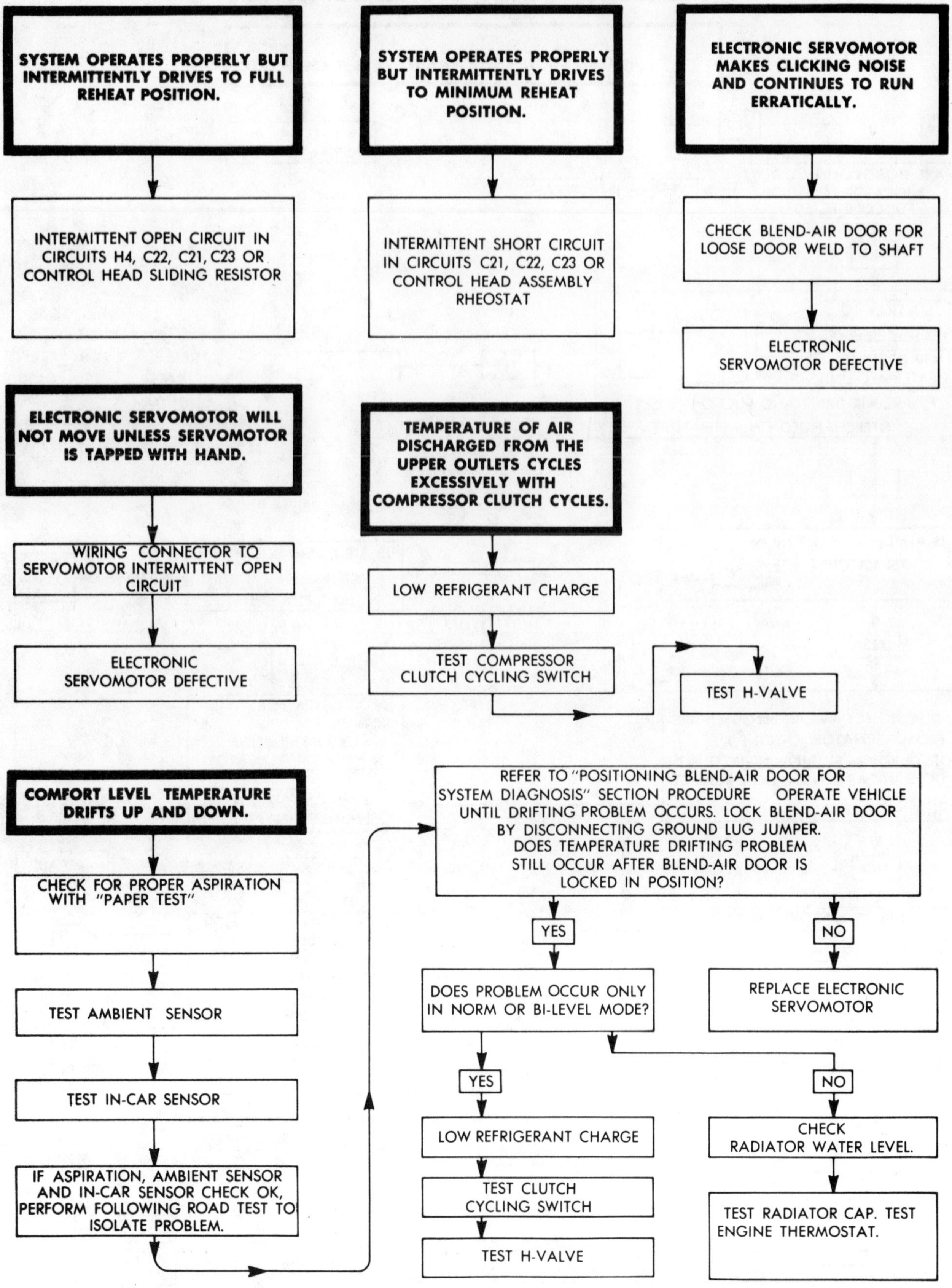

Fig. 29 **Semi-Automatic Temperature Control (S.A.T.C.) diagnosis chart. 1980 Cordoba, Diplomat, LeBaron & Mirada (Part 4 of 6)**

GROUP D — ASPIRATOR SYSTEM DIAGNOSIS

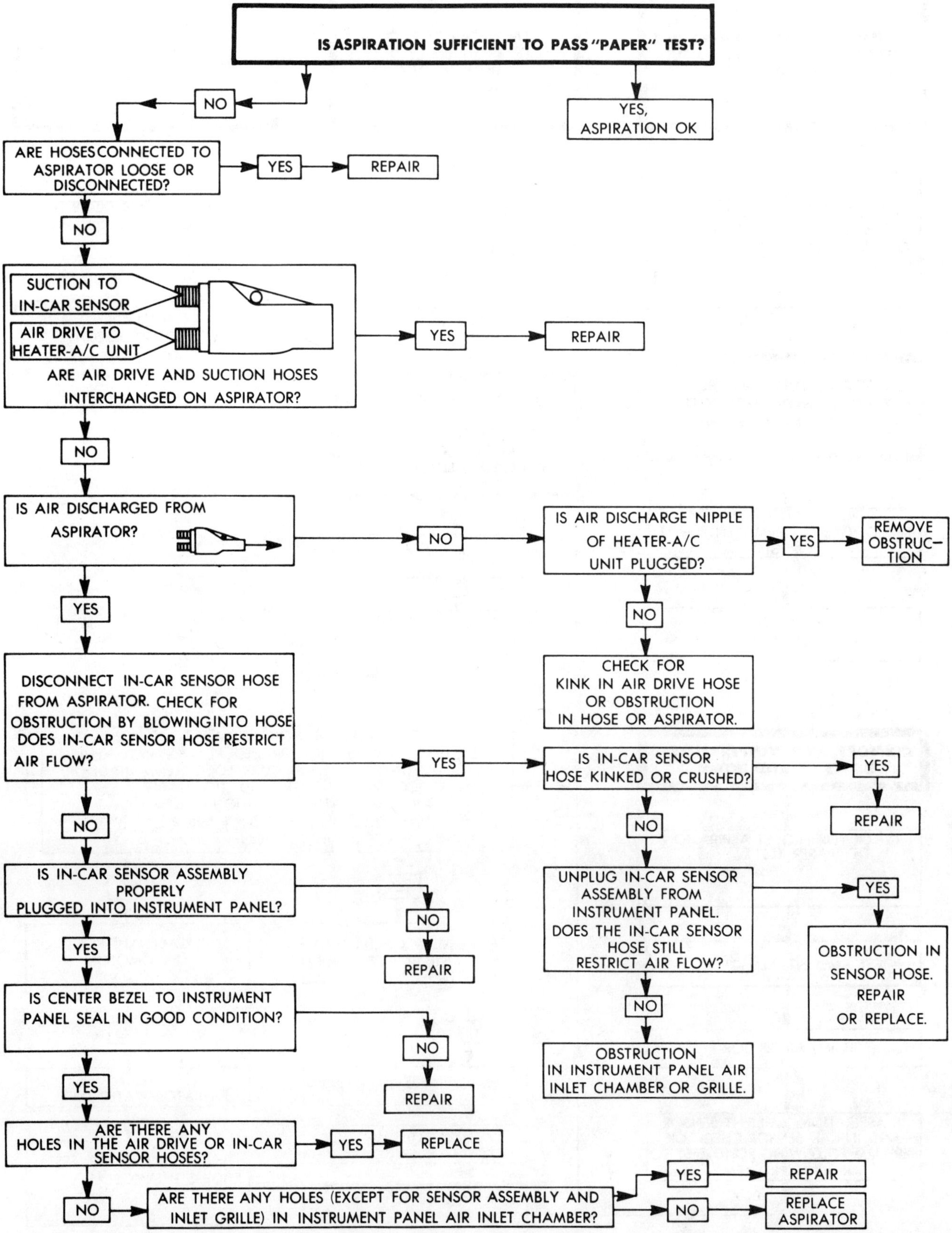

Fig. 29 Semi-Automatic Temperature Control (S.A.T.C.) diagnosis chart. 1980 Cordoba, Diplomat, LeBaron & Mirada Part (5 of 6)

GROUP E—MISCELLANEOUS PROBLEMS

SYSTEM INITIALLY DISCHARGES HOT AIR FROM UPPER OUTLETS IN COLD WEATHER USE.

↓

USER SHOULD SELECT HEATER MODE FOR INITIAL COLD WEATHER WARM UP. SEE "RECOMMENDED OPERATING MODES". DO NOT USE LOW BLOWER SPEED FOR INITIALLY HEATING VEHICLE.

SYSTEM INITIALLY DISCHARGES COLD AIR IN COLD WEATHER.

↓

ALLOW ENGINE COOLANT TEMPERATURE TO WARM UP BEFORE USING HIGH BLOWER SPEED.

SERVOMOTOR MAKES EXCESSIVE NOISE. SOUND IS LOUDER THAN BLOWER WHEN OPERATED IN HIGH BLOWER-HEATER MODE.

↓

SERVOMOTOR DEFECTIVE

Fig. 29 Semi-Automatic Temperature Control (S.A.T.C.) diagnosis chart. 1980 Cordoba, Diplomat, LeBaron & Mirada (Part 6 of 6)

to lower the temperature of the air delivered to the vehicle interior.

Evaporator capacity used to reduce the amount of moisture in the air is not wasted. Wringing some of the moisture out of the air entering the vehicle adds materially to the comfort of the passengers. However, an owner may expect too much from his air-conditioning system on humid days. A performance test is the best way to determine whether or not the system is performing up to standard. This test also provides valuable clues to the possible cause of trouble.

Air temperature in test room must be 75°F.

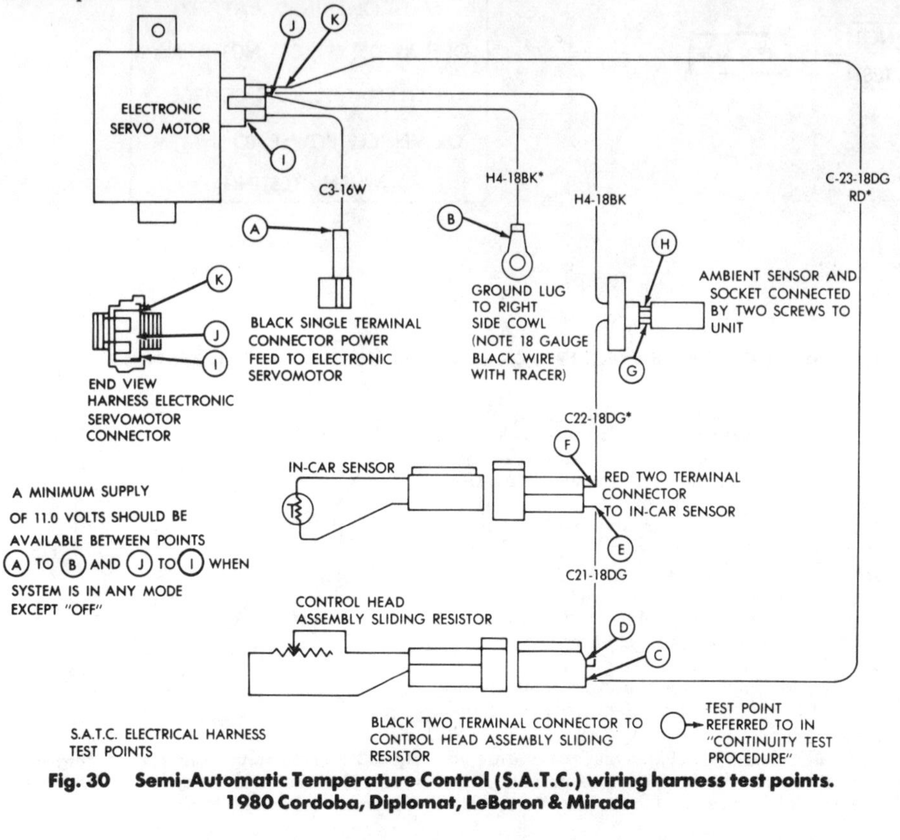

Fig. 30 Semi-Automatic Temperature Control (S.A.T.C.) wiring harness test points. 1980 Cordoba, Diplomat, LeBaron & Mirada

minimum for this test.

1974-76 Front Unit Performance Test

1. Start the engine and open the windows.
2. On 1974-76 models, position temperature control lever in cool. Push in "A/C" button, place fan switch on high. Open all grille outlets.
3. Adjust engine RPM as outlined in the following chart:

Model	Engine	Engine RPM
1974-76 All	6 Cyl.	1600
	V8	1300

4. Arrange gauge set manifold hoses and tachometer leads to allow hood to be lowered, then close hood.
5. Place motor-driven psychrometer Tool C-3704 at cowl inlet opening. Distilled water should be used with this meter to prevent drying out and hardening the wet sock.
6. Place thermometer Tool C-3623 fully into center grille opening on 1974-76 models. All outlets should be fully extended and directed towards rear of vehicle.
7. Operate the air-conditioning system until a stabilized condition on the gauges and thermometers has been established. One of the most important steps in making the overall performance test is that the engine must be operated for approximately five minutes to allow all the under-hood components of the system to reach their operating temperature.
8. *Partially* close the needle valve, located below the discharge pressure gauge, to minimize oscillation of the pointer. Do not close the needle valve completely since this would prevent the discharge pressure gauge from registering pressure.
   This test should be performed with the discharge pressure from 190 to 210 psi. The 190 to 210 pound pressure is for *test purposes only.* To increase pressure restrict the air flow across the condenser using cardboard, paper, etc. to decrease pressure, increase air flow across condenser with external floor fans.
9. Observe and record both the "Inlet Dry Bulb Temperature" and "Inlet Wet Bulb Temperature" as registered on the psychrometer.

Observe and record "Discharge Air Temperature" registered by the thermometer at right hand grille outlet.

From the appropriate "Performance Temperature Chart," for vehicle and type installation being tested, Figs. 34 through 36, determine the maximum allowable discharge air

S.A.T.C. SYSTEM CONTINUITY TEST PROCEDURE—I

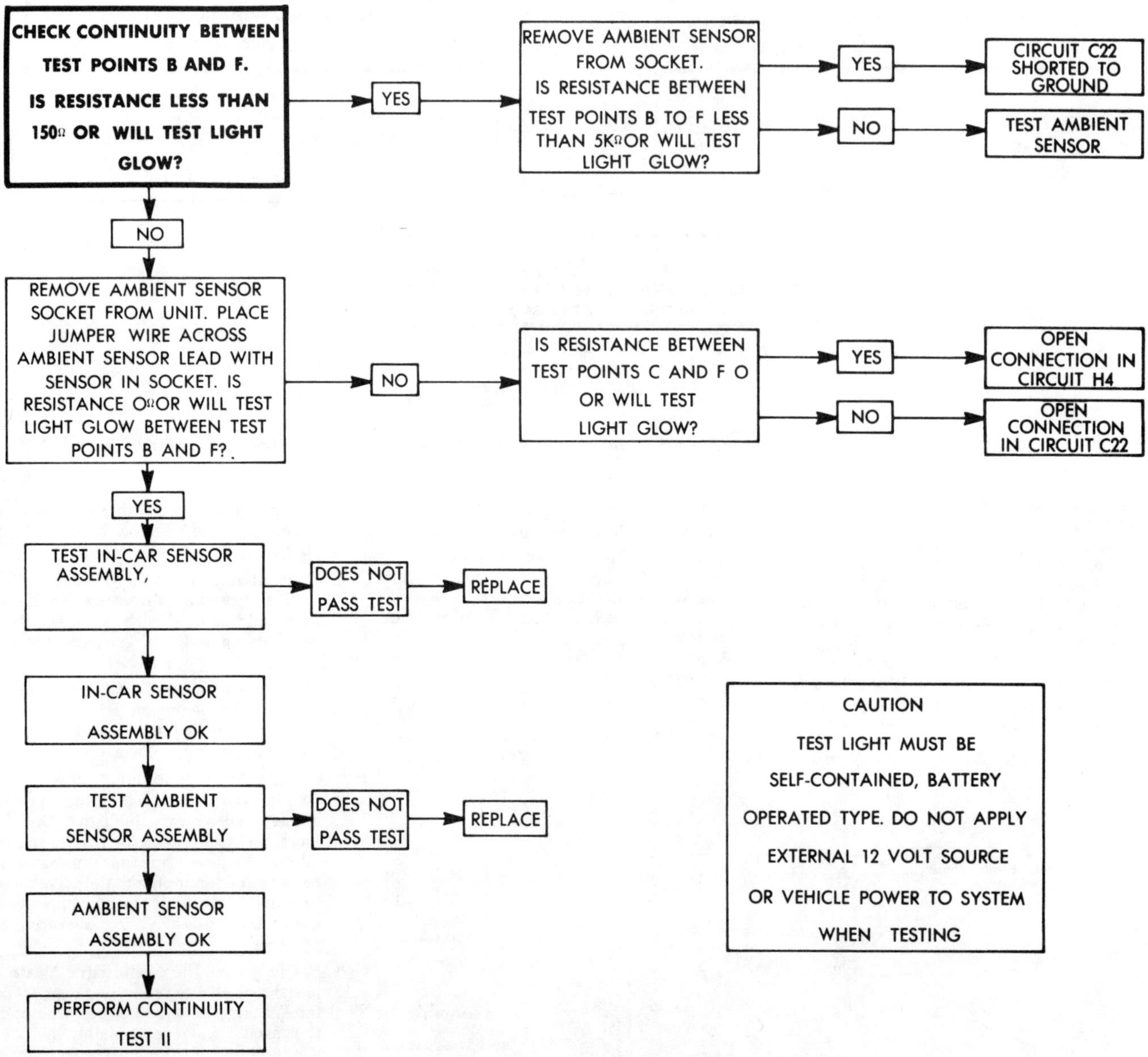

Fig. 31 Semi-Automatic Temperature Control (S.A.T.C.) continuity test. 1980 Cordoba, Diplomat, LeBaron & Mirada Part 1 of 2

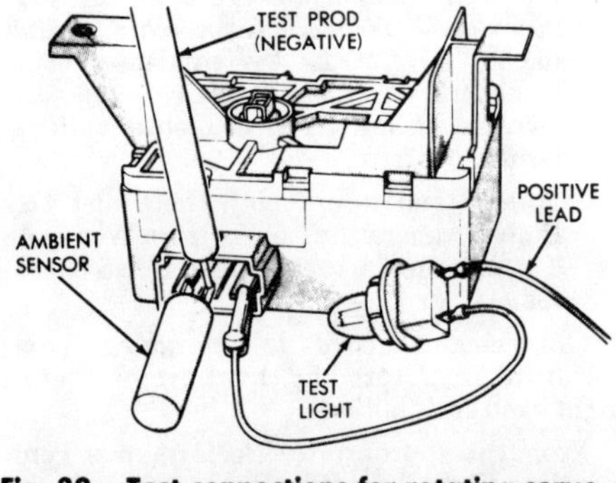

Fig. 32 Test connections for rotating servo-motor shaft clockwise. 1980 Cordoba, Diplomat, LeBaron & Mirada

perature for the prevailing "Dry" and "Wet" bulb temperatures recorded. If the vehicle discharge air temperature is at or below the temperature given on the Performance Chart, the air-conditioning is delivering its cooling capacity. However, to assure trouble-free operation, continue with the "Expansion Valve and Evaporator Pressure Regulator Valve Test."

On 1976 models, if discharge air temperature is above maximum allowable on "Performance Chart", check temperature cable adjustment or attachment at water valve on intermediate models. On compact and full size models, check water valve vacuum and blend air door operation.

On all models if discharge air temperature at the outlet grilles is above the maximum allowable on Performance Chart, perform the

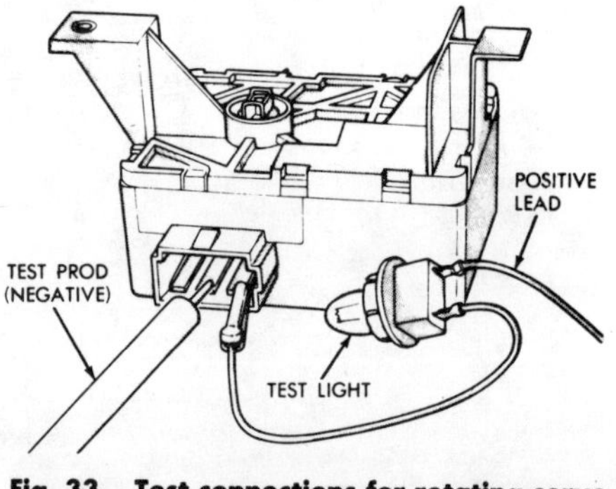

Fig. 33 Test connections for rotating servo-motor shaft counterclockwise. 1980 Cordoba, Diplomat, LeBaron & Mirada

S.A.T.C. SYSTEM CONTINUITY TEST PROCEDURE—II

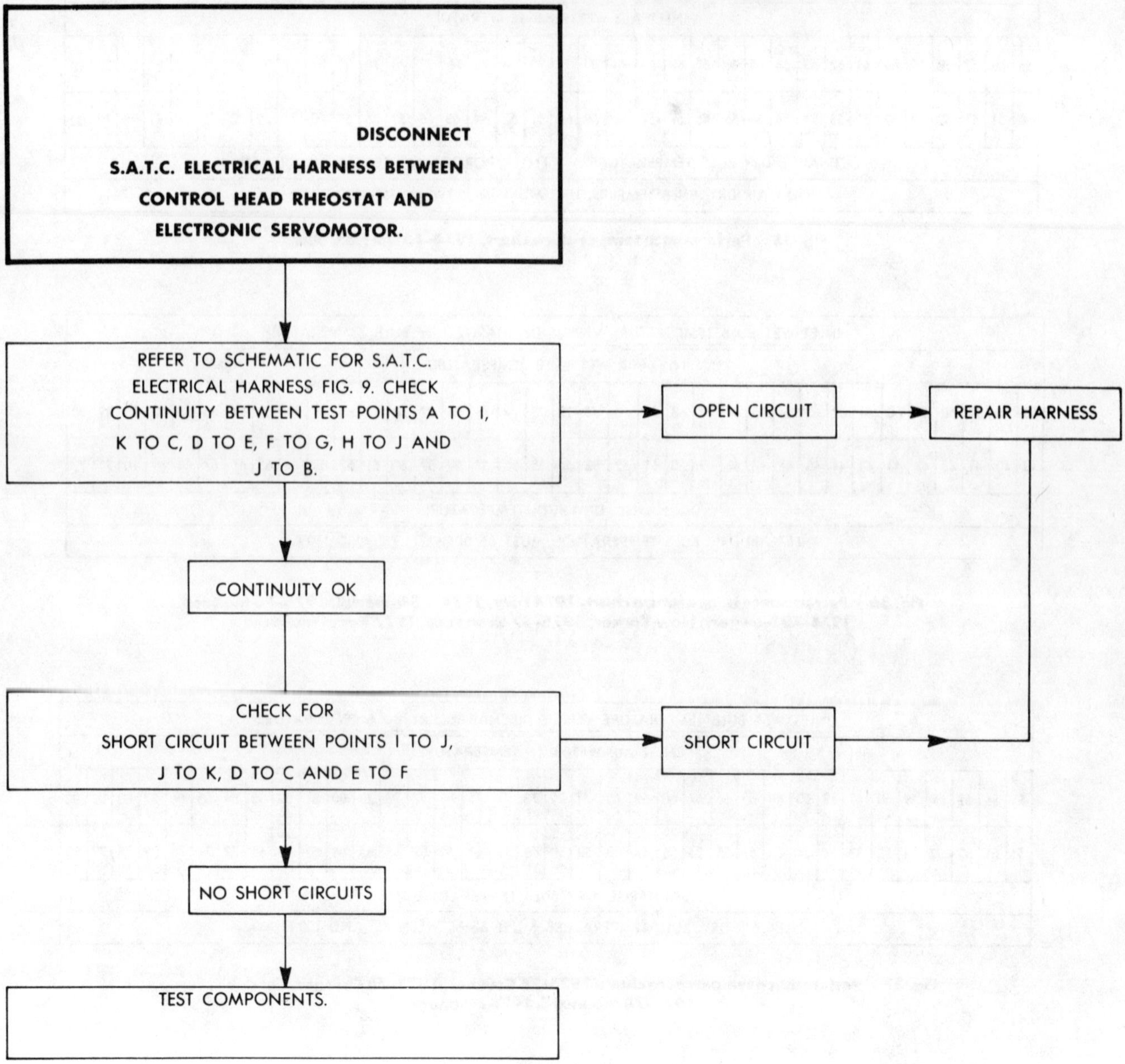

Fig. 31 Semi-Automatic Temperature Control (S.A.T.C.) continuity test. 1980 Cordoba, Diplomat, LeBaron & Mirada (Part 2 of 2)

INLET WET BULB TEMPERATURE VERSUS DISCHARGE DRY BULB TEMPERATURE																																			
INLET AIR WET BULB TEMPERATURE																																			
55	56	57	58	59	60	61	62	63	64	65	66	67	68	69	70	71	72	73	74	75	76	77	78	79	80	81	82	83	84	85	86	87	88	89	90
51	52	52	53	54	54	55	55	56	56	57	57	58	58	59	59	60	60	61	62	64	65	66	67	68	69	70	71	72	73	75	76	77	78	79	80
DISCHARGE DRY BULB TEMPERATURE																																			
INLET AIR DRY BULB TEMPERATURE MUST BE BETWEEN 75° AND 100°F.																																			

Fig. 34 Performance temperature chart. 1974 Barracuda, Challenger, Charger, Coronet & Satellite

INLET WET BULB TEMPERATURE VERSUS DISCHARGE DRY BULB TEMPERATURE

INLET AIR WET BULB TEMPERATURE

55	56	57	58	59	60	61	62	63	64	65	66	67	68	69	70	71	72	73	74	75	76	77	78	79	80	81	82	83	84	85	86	87	88	89	90
50	51	51	52	52	52	53	53	54	54	55	55	56	57	57	58	59	60	61	62	64	65	66	67	68	69	70	71	72	73	75	76	77	78	79	80

DISCHARGE DRY BULB TEMPERATURE AT CENTER PORTION OF CENTER AIR OUTLET

INLET AIR DRY BULB TEMPERATURE MUST BE BETWEEN 75° AND 100°F.

Fig. 35 Performance temperature chart. 1974-76 Dart & Valiant

INLET WET BULB TEMPERATURE VERSUS DISCHARGE DRY BULB TEMPERATURE

INLET AIR WET BULB TEMPERATURE

55	56	57	58	59	60	61	62	63	64	65	66	67	68	69	70	71	72	73	74	75	76	77	78	79	80	81	82	83	84	85	86	87	88	89	90
40	40	41	42	42	43	43	44	45	46	47	48	49	50	51	52	53	54	55	56	57	58	59	60	61	62	63	64	65	66	67	68	69	70	71	72

DISCHARGE DRY BULB TEMPERATURE

INLET AIR DRY BULB TEMPERATURE MUST BE BETWEEN 75° AND 110°F.

Fig. 36 Performance temperature chart. 1974 Fury, 1974-75 Imperial, 1974-76 Monaco, 1974-78 Newport, New Yorker, 1975-77 Gran Fury, 1977 Royal Monaco

INLET WET BULB TEMPERATURE VERSUS DISCHARGE DRY BULB TEMPERATURE

INLET AIR WET BULB TEMPERATURE

55	56	57	58	59	60	61	62	63	64	65	66	67	68	69	70	71	72	73	74	75	76	77	78	79	80	81	82	83	84	85	86	87	88	89	90
51	51	52	52	53	53	53	54	54	55	55	55	56	56	57	58	58	59	60	61	61	62	63	64	65	66	67	68	69	72	74	75	77	78	79	80

DISCHARGE DRY BULB TEMPERATURE

INLET AIR DRY BULB TEMPERATURE MUST BE BETWEEN 75° AND 100°F.

Fig. 37 Performance temperature chart. 1975-76 Coronet, 1975-78 Charger, Cordoba, Fury, 1977-78 Monaco & 1978 Magnum

INLET WET BULB TEMPERATURE VERSUS DISCHARGE DRY BULB TEMPERATURE

(upper box) ← CELSIUS

(lower box) ← FAHRENHEIT

INLET AIR WET BULB TEMPERATURE

12.5	13	14	14.5	15	15.5	16	16.5	17	17.5	18	19	19.5	20	20.5	21	21.5	22	22.5	23	24	24.5	25	25.5	26	26.5	27	27.5	28	29	29.5	30	30.5	31	31.5	32
55	56	57	58	59	60	61	62	63	64	65	66	67	68	69	70	71	72	73	74	75	76	77	78	79	80	81	82	83	84	85	86	87	88	89	90

DISCHARGE AIR DRY BULB TEMPERATURE AT LEFT CENTER OUTLET

6	6	6.5	7	7	8	8	8.5	9	9.5	10	10.5	11	11.5	12	12.5	13	13.5	14.5	15	15.5	16	16.5	17	18	18.5	19	19.5	20	20.5	21	21.5	22	23	23.5	24
43	43	44	45	45	46	46	47	48	49	50	51	52	53	54	55	56	57	58	59	60	61	62	63	64	65	66	67	68	69	70	71	72	73	74	75

INLET AIR DRY BULB TEMPERATURE MUST BE BETWEEN 24° AND 38°C (75° AND 100°F)

Fig. 38 Performance temperature chart. 1976-80 Aspen, Volare, 1977-80 Diplomat, LeBaron, 1980 Cordoba & Mirada

INLET WET BULB TEMPERATURE VERSUS DISCHARGE DRY BULB TEMPERATURE

(CELSIUS / FAHRENHEIT)

INLET AIR WET BULB TEMPERATURE																																				
CELSIUS	12.5	13	14	14.5	15	15.5	16	16.5	17	17.5	18	19	19.5	20	20.5	21	21.5	22	22.5	23	24	24.5	25	25.5	26	26.5	27	27.5	28	29	29.5	30	30.5	31	31.5	32
FAHRENHEIT	55	56	57	58	59	60	61	62	63	64	65	66	67	68	69	70	71	72	73	74	75	76	77	78	79	80	81	82	83	84	85	86	87	88	89	90
DISCHARGE DRY BULB TEMPERATURE AT LEFT CENTER OUTLET																																				
CELSIUS	12.5	12.5	12.5	13	14	14.5	15	15.5	15.5	15.5	15.5	16	16.5	17	17	17	17.5	18	19	19.5	20	20	20.5	21	21.5	22	22.5	23	24	24.5	25	25.5	26	26.5	27	27.5
FAHRENHEIT	55	55	55	56	57	58	59	60	60	60	60	61	62	63	63	63	64	65	66	67	68	68	69	70	71	72	73	74	75	76	77	78	79	80	81	82
INLET AIR DRY BULB TEMPERATURE MUST BE BETWEEN 24° AND 38°C (75°-100°F.)																																				

Fig. 39 Performance temperature chart. 1978-80 Horizon & Omni

"Expansion Valve and Evaporator Pressure Regulator Valve Test."

1977-80 Front Unit Performance Test

1. Start engine and open windows.
2. On models less Semi-Automatic Temperature Control (S.A.T.C.), place temperature lever in cool position.
3. On 1979-80 models equipped with Semi-Automatic Temperature Control (S.A.T.C.), the blend air door must be locked in the minimum heat position. On 1979 Cordoba, Magnum, 1979-80 Newport, New Yorker, St. Regis and 1980 Gran Fury with S.A.T.C., disconnect and plug bi-metal sensor vacuum hose. On 1980 Cordoba, Diplomat, LeBaron and Mirada with S.A.T.C., disconnect in-car sensor and ground sensor wire connector terminal to which wiring harness connector, point E Fig. 30, is connected.
4. Position A/C control in Norm A/C and blower switch on high, then open all discharge air outlets and position outlets so that air flow is towards rear of vehicle.
5. On Aspen and Volare models, close floor air outlets.
6. On models equipped with cycling clutch switch, bypass clutch switch terminals using a jumper wire to prevent system from cycling during test procedure.
7. Adjust engine speed to 1400 rpm on 4 cylinder models, 1600 rpm on 6 cylinder models or 1300 rpm on 8 cylinder models.
8. Position refrigerant gauge set and tachometer so that hood can be lowered and closed.
9. On all models except Horizon and Omni, position motor driven psychrometer C-3704 near cowl inlet opening. On Horizon and Omni models, position motor driven psychrometer C-3668 near cowl inlet opening. On all models, distilled water should be used with meter to prevent dry sock from hardening.
10. On 1977-78 models except Aspen, Diplomat, Horizon, LeBaron, Omni and Volare, position thermometer C-3623 into right center air outlet. On 1977-78 Aspen, Diplomat, Horizon, LeBaron, Omni, Volare and all 1979-80 models, position thermometer C-3623 into left center air outlet.
11. Operate A/C system until pressure reading on gauge and temperature reading on thermometer have stabilized.

NOTE: Engine should be operated at specified rpm for approximately 5 minutes to allow underhood components of system to reach operating temperature.

12. On gauge sets equipped with a needle valve, partially close needle valve to minimize gauge pointer oscillation. Do not completely close needle valve, as this will prevent discharge pressure from registering on gauge.
13. For test purposes only, a discharge pressure of 190 to 210 psig. should be obtained. To increase pressure, use a piece of cardboard to restrict air flow across condenser. To decrease pressure, increase air flow across condenser using an external floor fan.
14. Observe and note inlet wet bulb temperature as indicated on psychrometer and discharge air temperature as indicated on thermometer located in air outlet duct. Referring to Figs. 36 through 40, determine maximum allowable discharge air temperature for dry and wet bulb temperatures noted previously. If discharge air temperature is at or below the maximum allowable temperature shown in the chart, the A/C system is cooling properly. If temperature is above the maximum allowable, check temperature control cable adjustment, blend air door operation and water valve vacuum. If discharge air temperature is still above the maximum limit, refer to A/C system diagnosis charts, Figs. 41 through 46.

Vacuum Control System Tests 1974-80

The test of the push-button operation determines whether or not the vacuum and elec-

INLET WET BULB TEMPERATURE VERSUS DISCHARGE DRY BULB TEMPERATURE

(CELSIUS / FAHRENHEIT)

INLET AIR WET BULB TEMPERATURE																																				
CELSIUS	12.5	13	14	14.5	15	15.5	16	16.5	17	17.5	18	19	19.5	20	20.5	21	21.5	22	22.5	23	24	24.5	25	25.5	26	26.5	27	27.5	28	29	29.5	30	30.5	31	31.5	32
FAHRENHEIT	55	56	57	58	59	60	61	62	63	64	65	66	67	68	69	70	71	72	73	74	75	76	77	78	79	80	81	82	83	84	85	86	87	88	89	90
DISCHARGE AIR DRY BULB TEMPERATURE AT LEFT CENTER OUTLET																																				
CELSIUS	6	6.5	7	8	8.5	9	9.5	10	10.5	11	11.5	12	12.5	13	13.5	14.5	15	15.5	16	16.5	17	18	18.5	19	19.5	20	20.5	21	21.5	22	23	23.5	24	24.5	25	25.5
FAHRENHEIT	43	44	45	46	47	48	49	50	51	52	53	54	55	56	57	58	59	60	61	62	63	64	65	66	67	68	69	70	71	72	73	74	75	76	77	78
INLET AIR DRY BULB TEMPERATURE MUST BE BETWEEN 24° AND 38°C (75° AND 100°F)																																				

Fig. 40 Performance temperature chart. 1979 Cordoba, Magnum, 1979-80 Newport, New Yorker, & 1980 Gran Fury

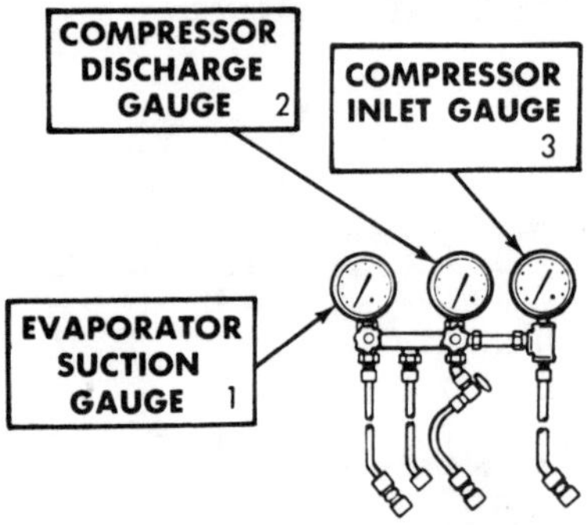

REFRIGERANT SYSTEM DIAGNOSIS WITH "H" TYPE EXPANSION AND EPR VALVE

- CHECK REFRIGERANT LEVEL
- INSTALL GAUGE SET
- PROCEED WITH OVERALL PERFORMANCE TEST
- TEST CONDITIONS AMBIENT – 70° TO 85° F. (21° TO 29° C) MODE – MAX A/C WINDOWS – CLOSED SET VEHICLE IN NEUTRAL OR PARK APPLY PARKING BRAKE
- USING GAUGE SET FOLLOW DIAGNOSIS SEQUENCE AS INDICATED

ENGINE RPM	BLOWER MOTOR SPEED	EVAPORATOR SUCTION GAUGE PSIG (kPag)1	COMPRESSOR DISCHARGE GAUGE PSIG (kPag)2	COMPRESSOR INLET GAUGE PSIG (kPag)3	DIAGNOSIS

DISCONNECT AND PLUG WATER VALVE VACUUM HOSE SET TEMPERATURE CONTROL IN FULL REHEAT

PRESSURE DROP NOT GREATER THAN 6 PSIG (42 kPag) — NORMAL

800	HI	25-35 PSIG (172 TO 240 kPag)	140-240 PSIG (966 TO 1656 kPag)	19-29 PSIG (133 TO 202 kPag)	

PRESSURE DROP GREATER THAN 6 PSIG (42 kPag) — EPR-2 STUCK CLOSED OR LOW SIDE RESTRICTION

SPRAY "H" VALVE HEAD WITH LIQ-UID REFRIGERANT FOR 15 SECONDS — 15 INCHES OF VACUUM (–50 kPag) — "H" VALVE STUCK OPEN IF CONDITION CANNOT BE MET

DISCONTINUE SPRAY — 38 PSI (262 kPag) OR MORE THEN DROPS TO 25 TO 35 PSIG (172 TO 240 kPag) — "H" VALVE STUCK CLOSED IF CONDITIONS CANNOT BE MET

RECONNECT WATER VACUUM HOSE SET TEMPERATURE CONTROL IN COOL POSITION

PRESSURE DROP GREATER THAN 6 PSIG (42 kPag) — NORMAL

1100	LOW	22-28 PSIG (152 TO 193 kPag)	140-240 PSIG (966 TO 1656 kPag)	15 PSIG OR LESS (104 kPag)	EPR STUCK OPEN REPLACE EPR

PRESSURE DROP NOT GREATER THAN 6 PSIG (42 kPag) — IF PRESSURE DROP REMAINS LOW CHECK COMPRESSOR HEAD GASKETS AND REEDS

ENGINE RPM	BLOWER MOTOR SPEED	EVAPORATOR SUCTION	COMPRESSOR DISCHARGE	COMPRESSOR INLET	DIAGNOSIS
800	HI	22-30 PSIG (152 TO 207 kPag)	240 PSIG (1656 kPag) OR HIGHER	15-24 PSIG (104 TO 166 kPag)	RESTRICTED CONDENSER — RESTRICTED DISCHARGE LINE — RADIATOR OVERHEATING — AIR IN SYSTEM — OVERCHARGED SYSTEM — FAULTY VISCOUS FAN DRIVE
		22-30 PSIG (152 TO 207 kPag)	140 PSIG (966 kPag) OR LESS	30 PSIG (207 kPag) OR LESS	DEFECTIVE COMPRESSOR HEAD GASKETS OR DISCHARGE REEDS

Fig. 41 Refrigerant system diagnosis chart. 1977-79 models with Airtemp RV2 compressor

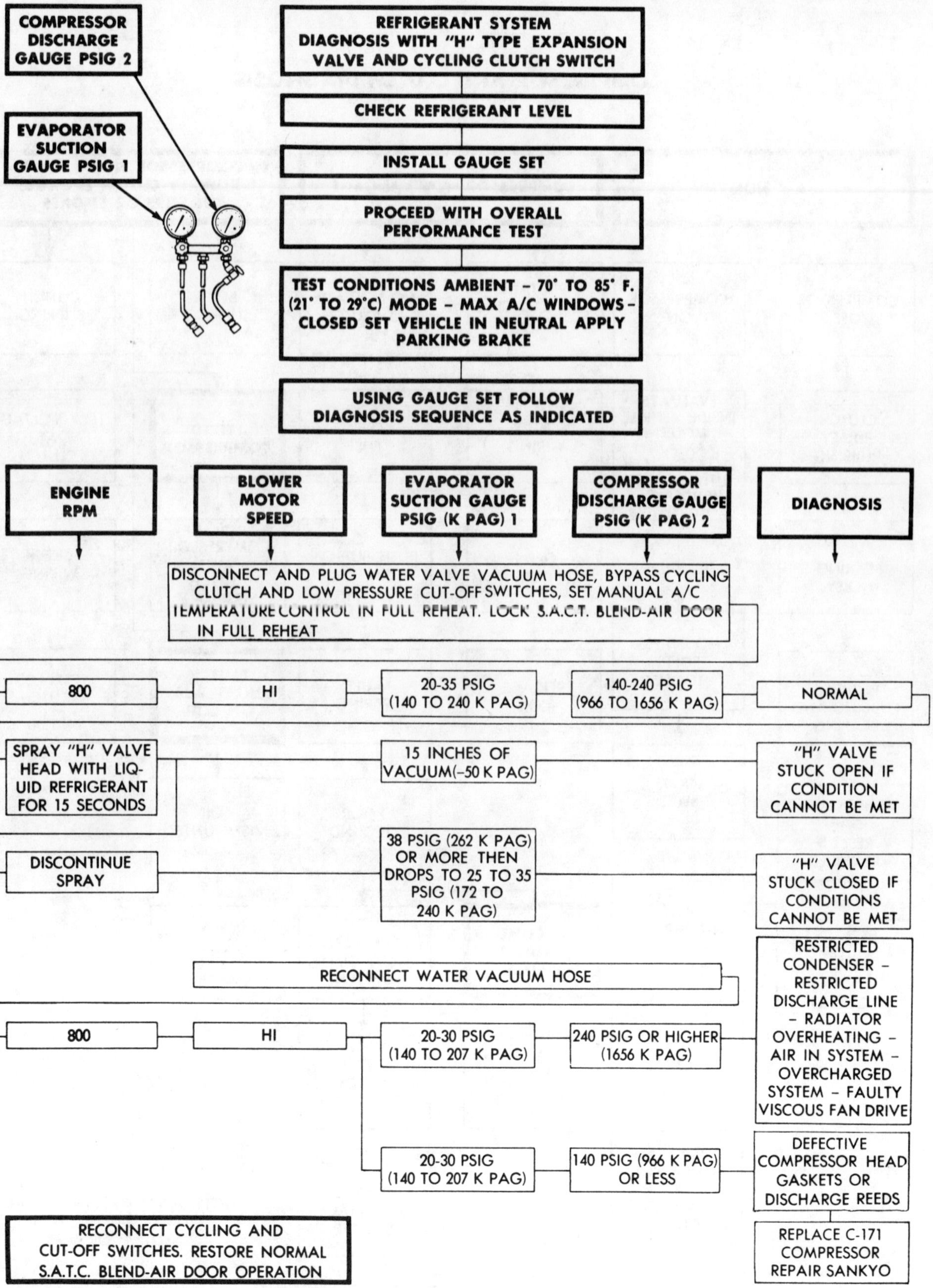

Fig. 42 Refrigerant system diagnosis chart. 1978-80 models with C-171 or Sankyo compressor

COMPRESSOR AND CLUTCH DIAGNOSIS

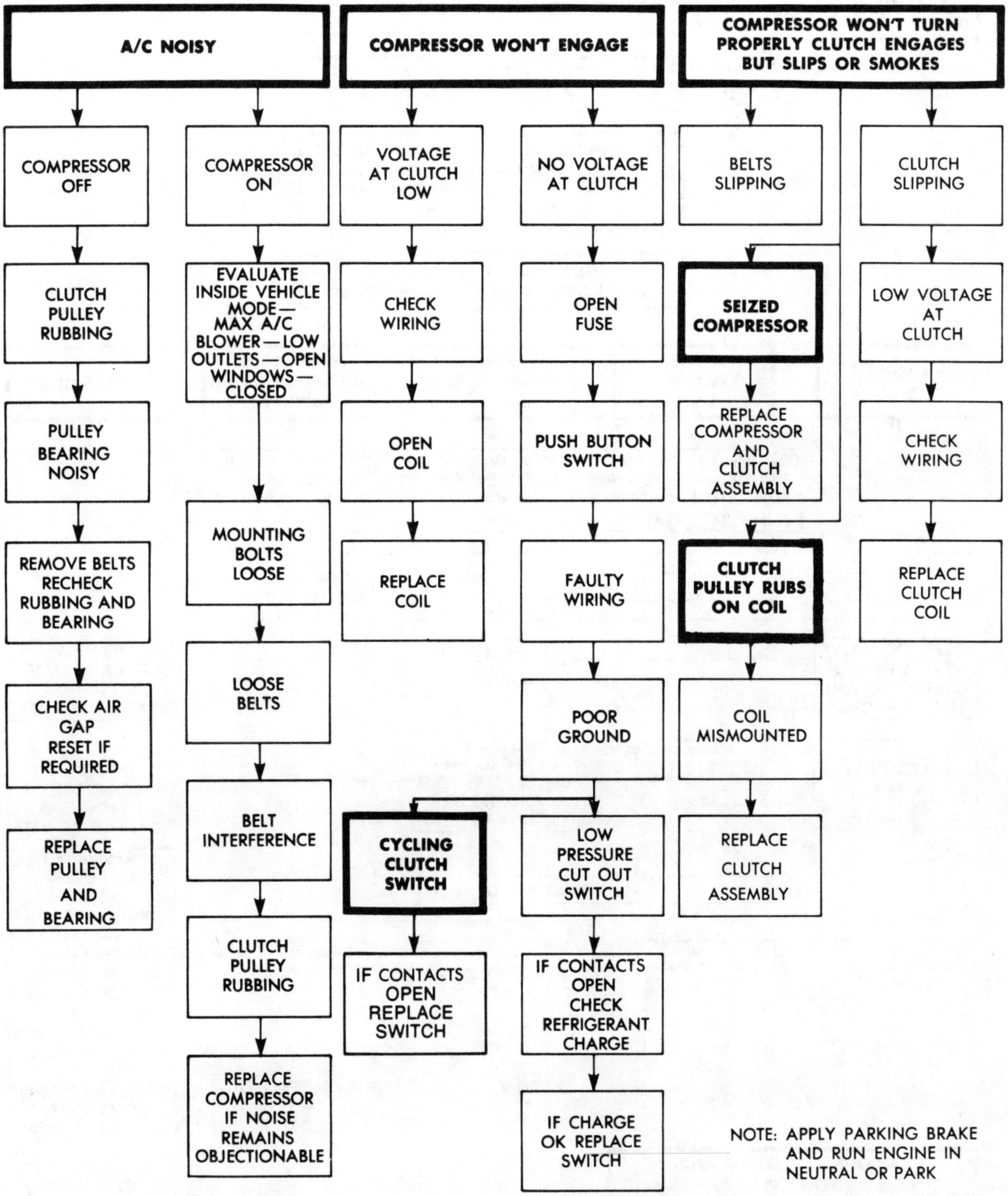

NOTE: APPLY PARKING BRAKE AND RUN ENGINE IN NEUTRAL OR PARK

Fig. 43 Compressor & clutch diagnosis chart. 1977-80 models

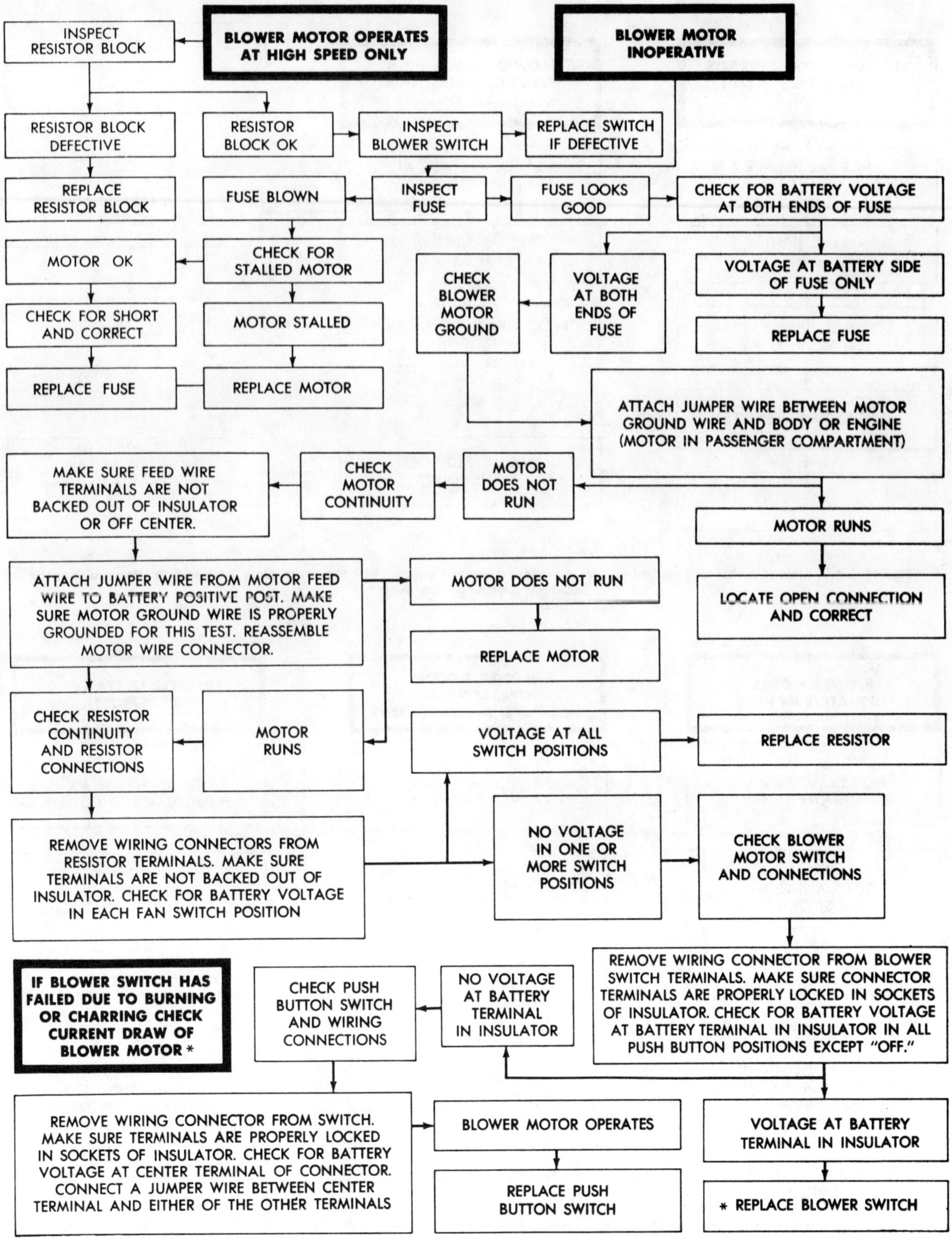

Fig. 44 A/C blower motor & control system electrical diagnosis chart. 1977-80 models except 1979 Newport, New Yorker & St. Regis w/heated back light & 65 amp. alternator

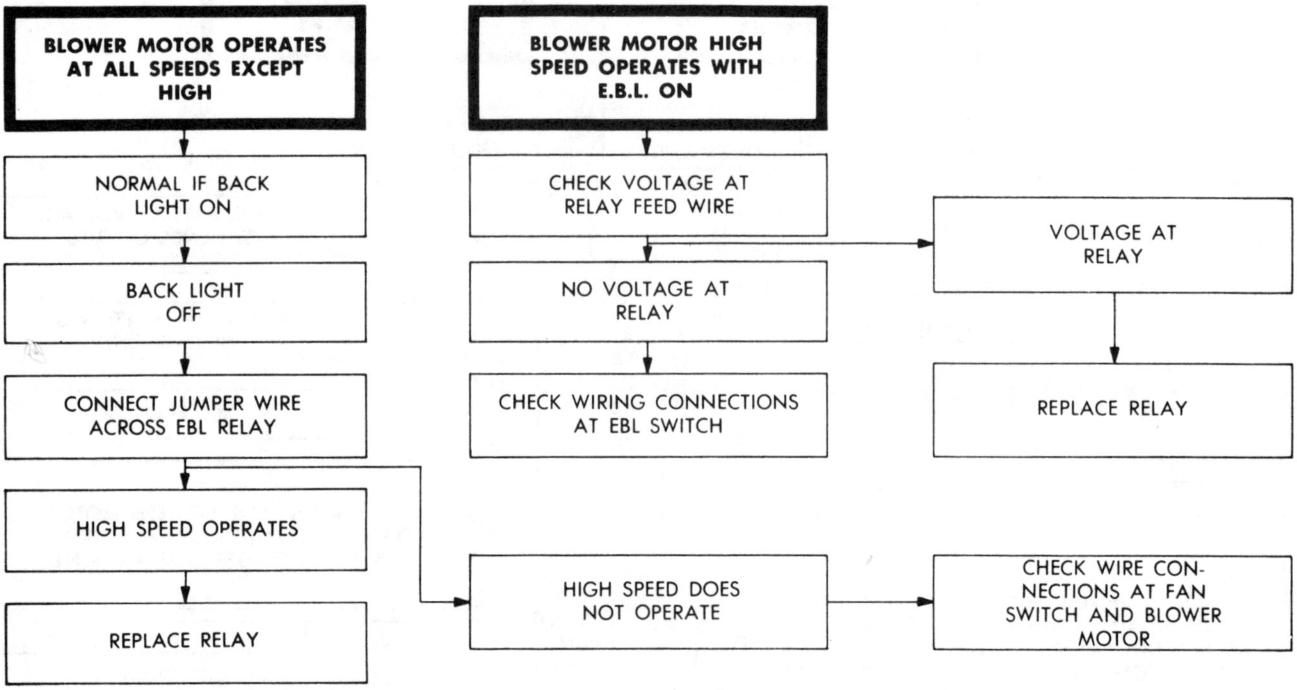

Fig. 45 A/C blower motor and control system electrical diagnosis chart. 1979 Newport, New Yorker & St. Regis w/heated backlight & 65 amp. alternator.

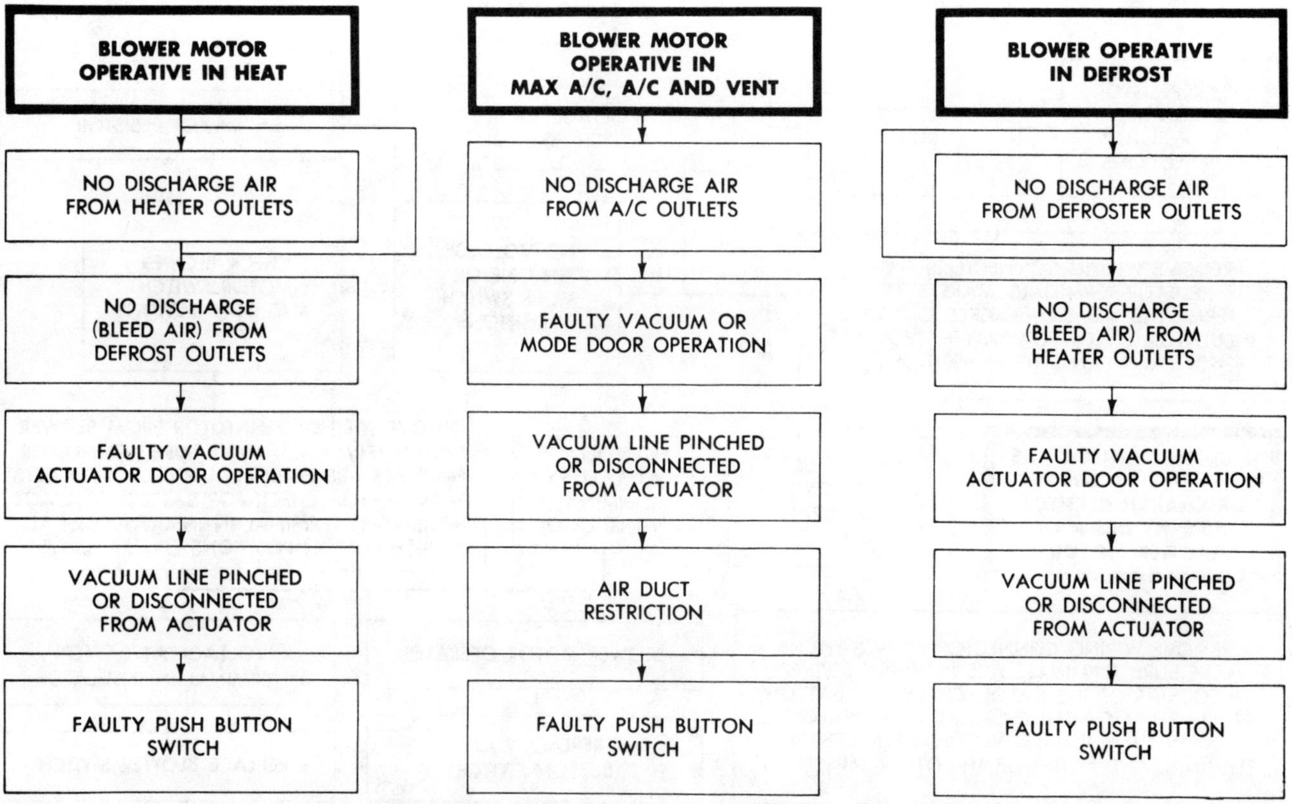

Fig. 46 A/C blower motor and control system mechanical diagnosis chart. 1977-80 models

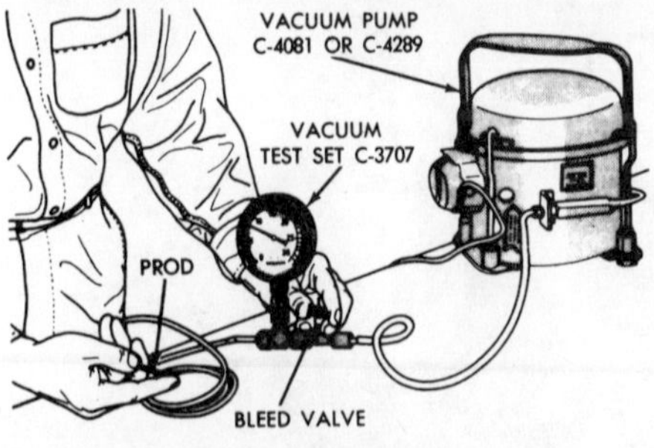

Fig.47 Vacuum tester bleed valve adjustment. 1974-80 Chrysler Corp.

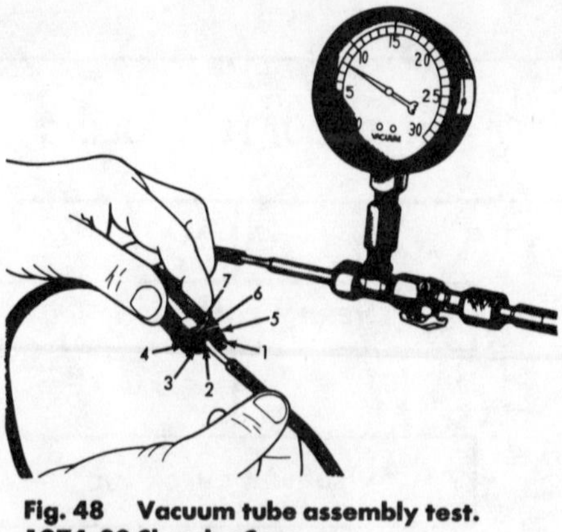

Fig. 48 Vacuum tube assembly test. 1974-80 Chrysler Corp.

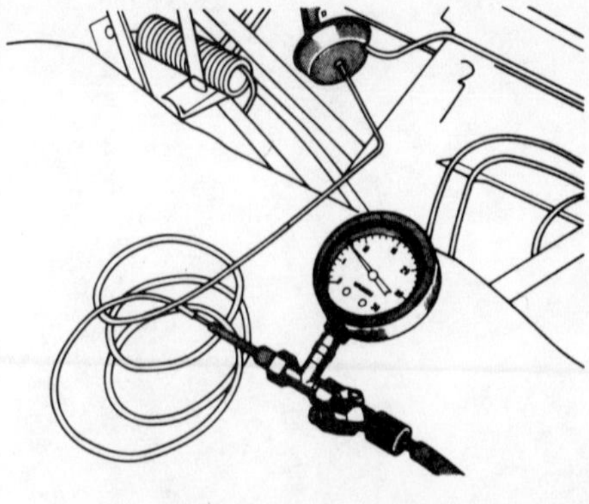

Fig. 49 Vacuum actuator. 1974-80 Chrysler Corp.

trical circuits are properly connected and the controls are functioning properly. However, it is possible that a vacuum control system that operates perfectly at the high vacuum provided at engine idle speed may not function properly at high engine speeds. Before starting this test, stop engine and make certain the vacuum source hose at engine intake manifold is tight on its connector.

Start vacuum pump (Tool C-3652) and connect to the vacuum test set (Tool C-3707). Adjust bleed valve on test set to obtain exactly 8 inches of vacuum with a finger blocking the prod on end of test hose, Fig. 47.

It is essential that the bleed valve be adjusted so the vacuum gauge pointer will return to exactly 8 inches when the prod is covered by a finger. Otherwise a false reading will be obtained when the control circuit is tested.

1. Disconnect engine vacuum source hose at intake manifold, then connect tester hose prod to vacuum source hose leading to control switch. Position vacuum gauge on cowl so that it can be viewed from inside vehicle.
2. Depress heat push button and note vacuum gauge reading. Vacuum gauge reading should drop, then after vacuum actuator has operated, return to 7¼ to 8 inches of vacuum. Depress remaining push buttons, allowing time for vacuum actuators to actuate and vacuum to stabilize at each push button position. The maximum allowable vacuum drop below 8 inches after each operation is ¾ inch of vacuum.

NOTE: Bypass vacuum storage tank, if equipped.

3. If vacuum drop is more than ¾ inch, recheck tester for a reading of exactly 8 inches of vacuum. If tester reading is correct, check to ensure that the seven port vacuum connector is properly connected to the control switch.

NOTE: Do not use any type of lubricant to properly position connector on control switch. If it is impossible to properly seat connector all the way on switch, place a drop or two of clean water into connector plug ports, so that plug will slide completely on switch.

4. If vacuum drop is still in excess of ¾ inch, remove 7 port connector from control switch and place finger over port No. 6 to check vacuum source hose. Remove tester prod from vacuum source hose and insert prod alternately into each port of the 7 port connector except No. 6, Fig. 48. After each actuator has operated, note the amount of vacuum drop. If vacuum test gauge returns to 8 inches of vacuum at each vacuum port, the control switch is defective and should be replaced. If an excessive vacuum drop is indicated at one or more switch ports, locate defective vacuum hose or actuator. A leaking actuator can be located by attaching tester vacuum line directly to the actuator, Fig. 49. A leaking line may be detected by attaching tester and running fingers along line while noting vacuum gauge reading.
5. A leaking vacuum line can be repaired by cutting out the portion of hose leaking and ends of vacuum line into a ⅛ in. inside diameter rubber tube. It may be necessary to wet tube to aid insertion of vacuum lines.

FORD

Insufficient Cooling Diagnosis

Refer to Fig. 50 parts 1 through 6 for insufficient cooling diagnosis procedure.

A/C Refrigerant System Diagnosis

Refer to Fig. 51 parts 1 through 5 for A/C refrigerant system diagnosis procedure.

Magnetic Clutch Diagnosis

Refer to Fig. 52 for magnetic clutch diagnosis procedure.

Compressor Diagnosis

Refer to Fig. 53 parts 1 and 2 for compressor diagnosis procedure.

Blower Motor Test

1974-76 Current Draw Test

1. Disconnect blower motor wire at connector.
2. Connect negative lead of ammeter to blower motor wire and positive lead to battery positive terminal, Fig. 54.
3. Blower motor should operate and ammeter should read as indicated in Blower Motor Test Chart, Fig. 55.

1977-80 Voltage Test

1. Fabricate two jumper wires with a male spade connector on one end and a female spade connector on the other end. Strip insulation from female connector on the end of one jumper wire.
2. Disconnect blower motor wire connector located near blower motor. Connect jumper wire with insulation removed, to blower motor feed wire terminals (orange wire with black strip) on each wire connector. Connect other jumper wire to blower motor ground wire terminals (black wires) at each connector. If difficulty is encountered in determining blower motor feed and ground wire terminals, refer to wiring diagrams located elsewhere in this manual.
3. Connect positive lead of voltmeter to jumper wire where insulation has been removed from female connector. Connect negative lead of voltmeter to a good body ground.
4. Position temperature control lever at midway point to allow average air flow around and through heater core.
5. Position mode lever on Heat and place blower switch to low position, if necessary, then turn ignition switch to on position.
6. With a fully charged battery, blower motor voltage should be as specified, Fig. 56, for each blower switch position.

1977-80 Current Draw Test

1. If necessary, fabricate jumper wires as described in step 1, 1977-80 Blower Motor Voltage Test.
2. Disconnect blower motor wire connector located near blower motor. Connect jumper wire with insulation removed to blower motor feed terminal (orange wire with black stripe) of female wire connector. Connect ammeter between jumper wire female connector and blower motor feed terminal (orange wire with black stripe) male wire connector. Connect remaining jumper wire between blower motor ground terminals (black wire) at each wire connector. If difficulty is encountered in identifying blower motor feed and ground wire terminals, refer to wiring diagrams located elsewhere in this manual.
3. Position temperature control lever at midway point and mode control lever to Heat. Place blower switch on low, if necessary, then turn ignition switch to on position.
4. With a fully charged battery, blower motor current draw should be as specified, Fig. 56, at each blower switch position.

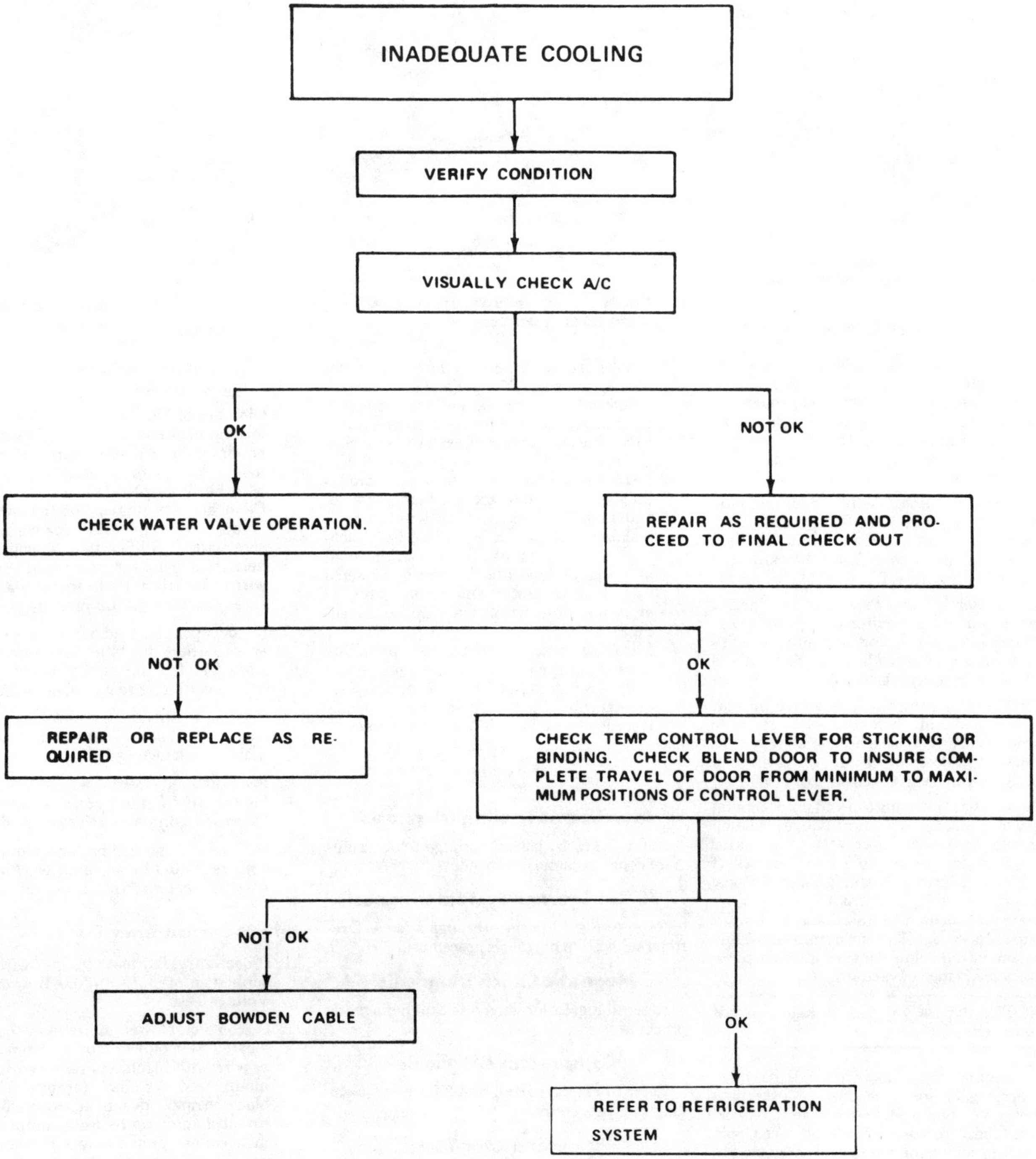

Fig. 50 Insufficient cooling diagnosis chart. Ford Motor Co. (Part 1 of 6)

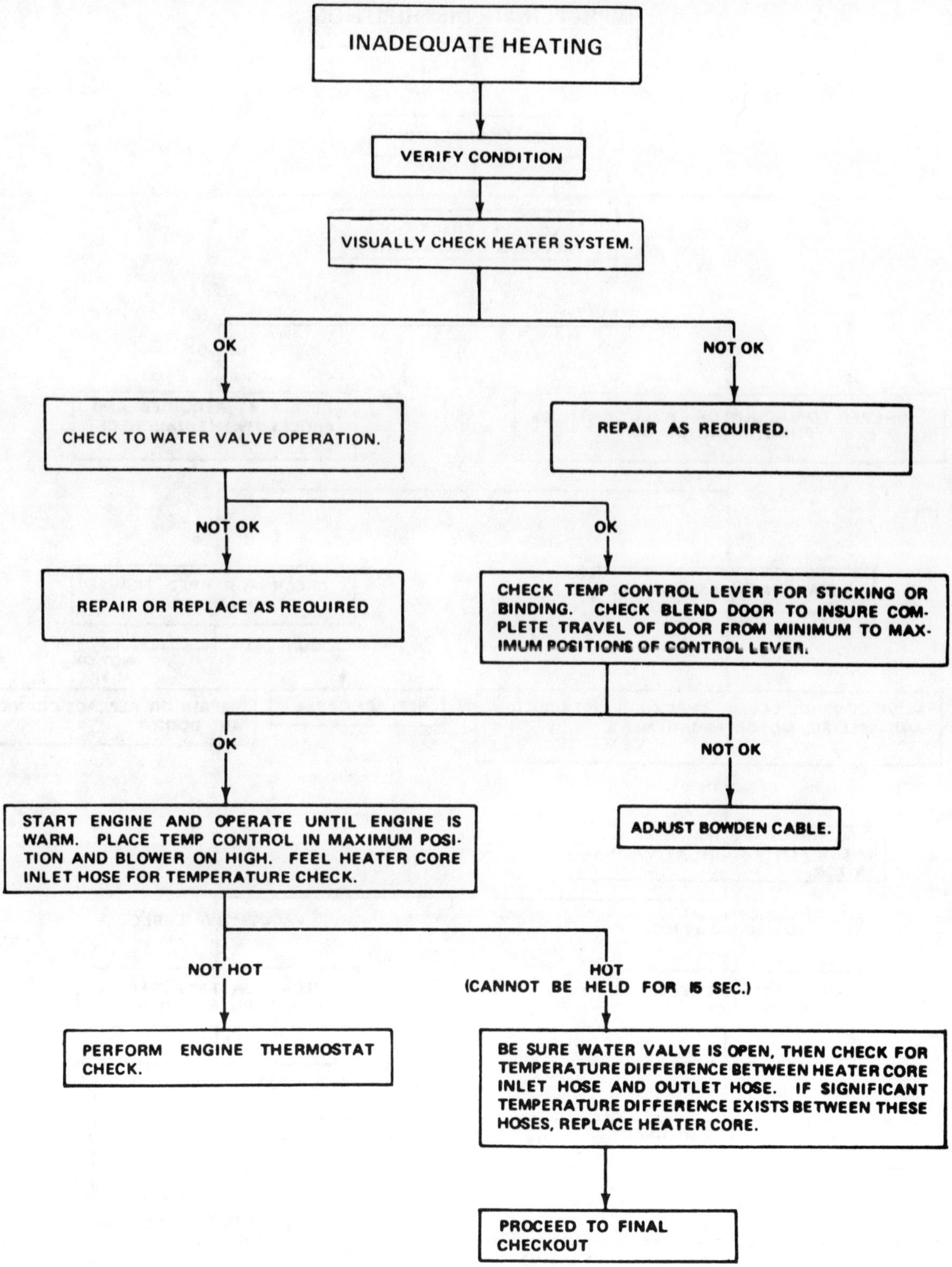

Fig. 50 Insufficient cooling diagnosis chart. Ford Motor Co. (Part 2 of 6)

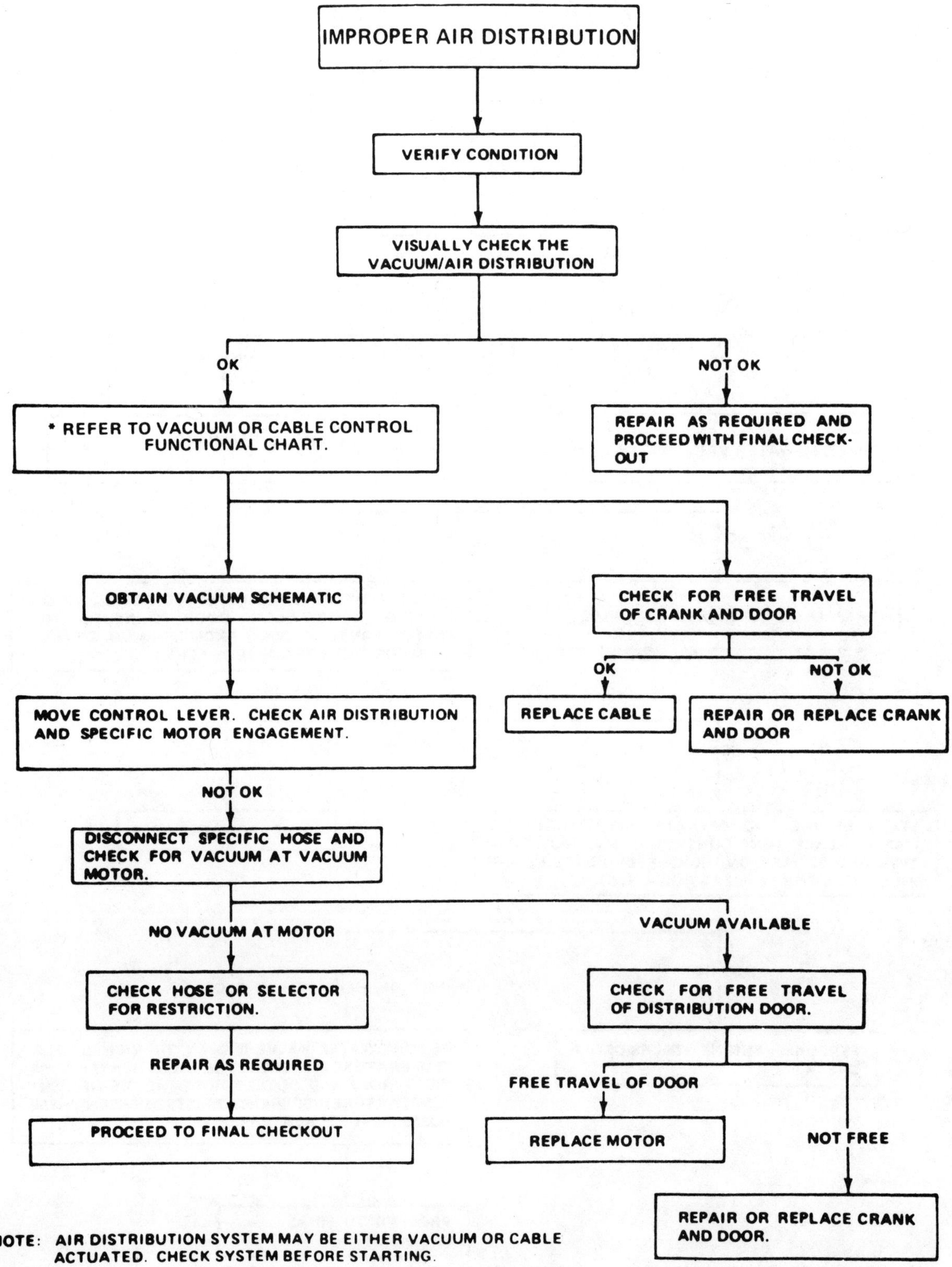

• NOTE: AIR DISTRIBUTION SYSTEM MAY BE EITHER VACUUM OR CABLE ACTUATED. CHECK SYSTEM BEFORE STARTING.

NOTE: IF ANY REPAIR IS MADE THAT IS JUDGED TO CORRECT CONDITION, THEN PROCEED TO FINAL CHECK OUT

Fig. 50 Insufficient cooling diagnosis chart. Ford Motor Co. (Part 3 of 6)

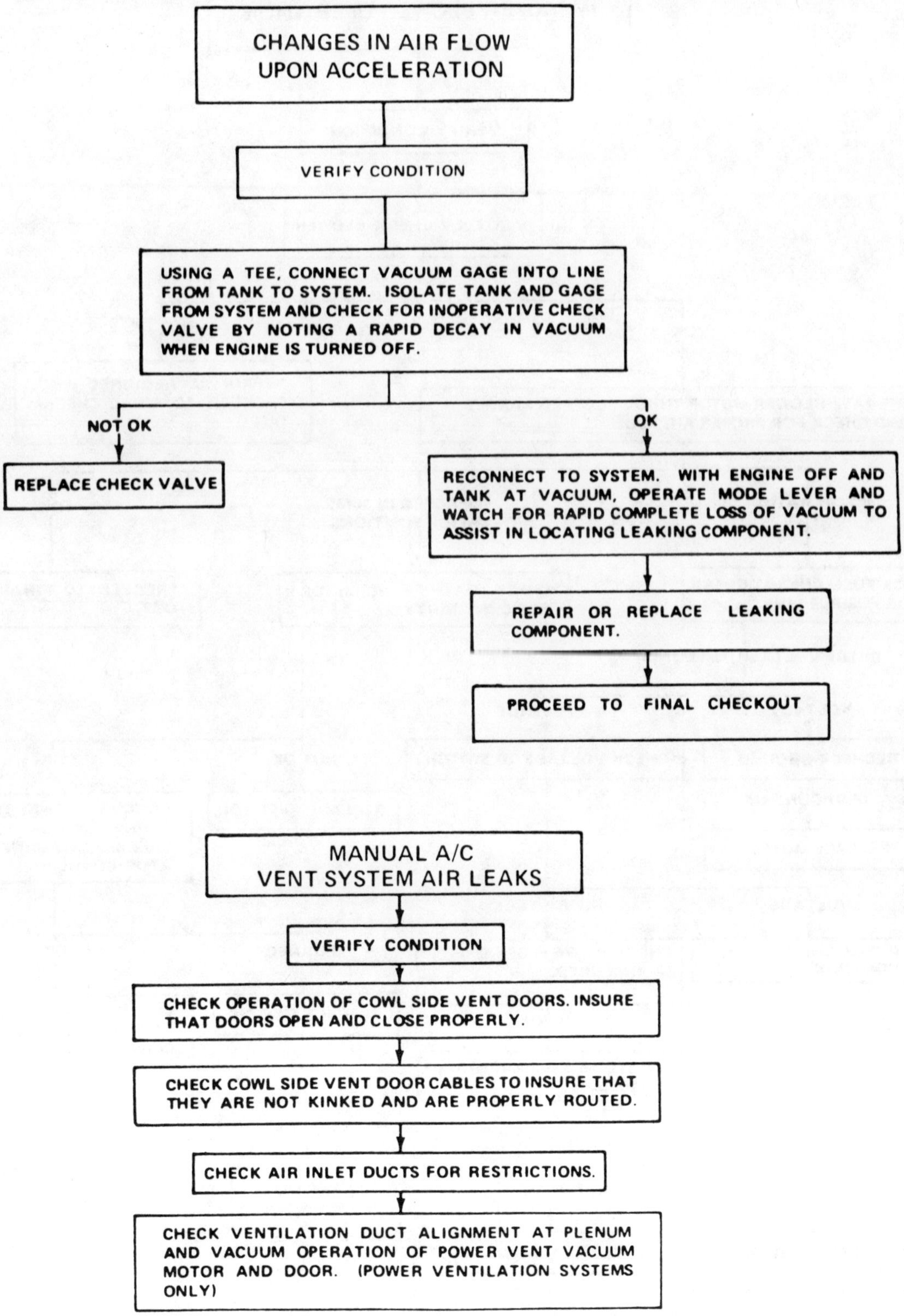

Fig. 50 Insufficient cooling diagnosis chart. Ford Motor Co. (Part 4 of 6)

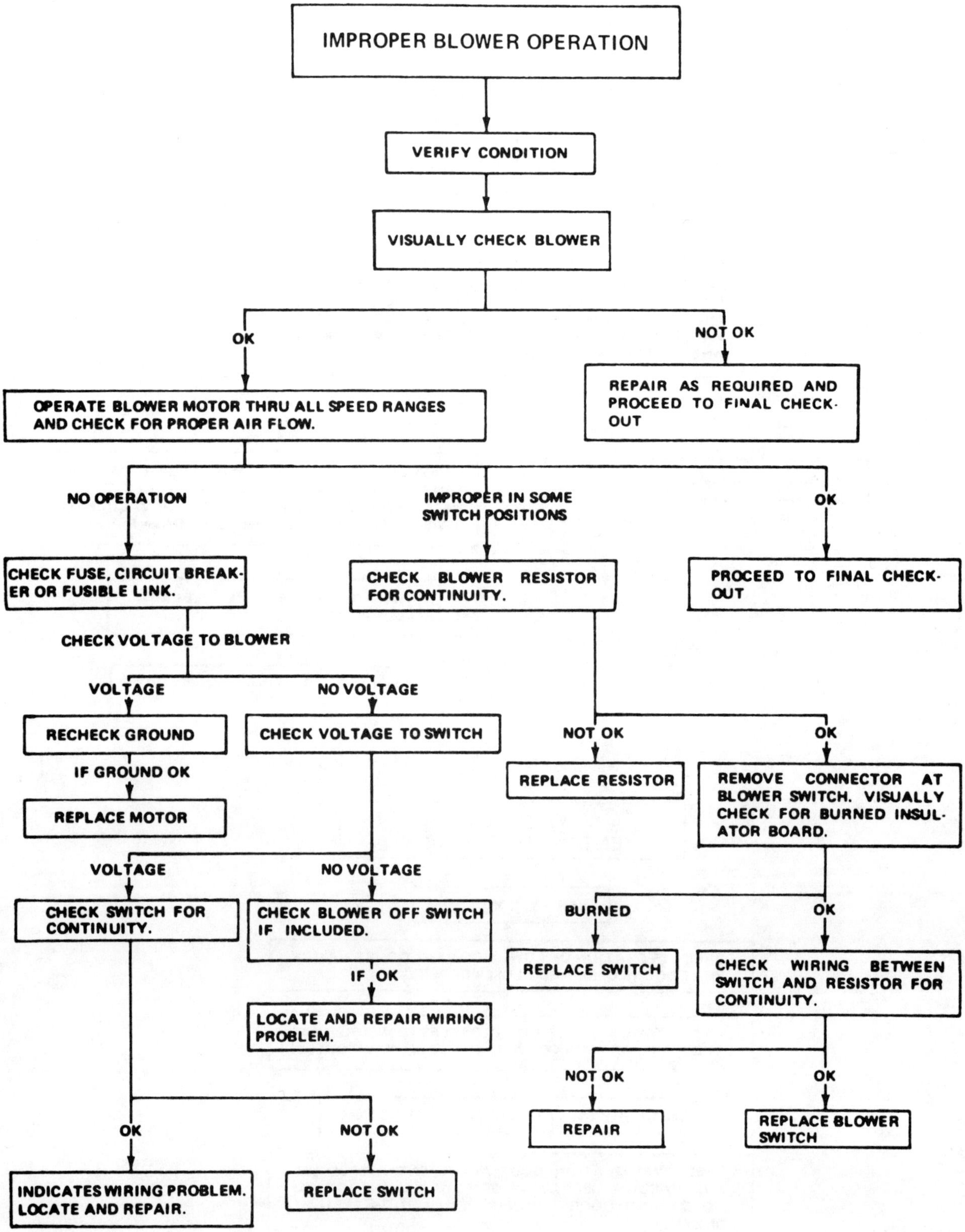

NOTE: IF ANY REPAIR IS MADE THAT IS JUDGED TO CORRECT CONDITION, THEN PROCEED TO FINAL CHECKOUT

Fig. 50 Insufficient cooling diagnosis chart. Ford Motor Co. (Part 5 of 6)

FINAL CHECK OUT MANUAL A/C

1. FUNCTION CONTROL LEVER THROUGH ALL MODES OF OPERATION ON HIGH BLOWER AND CHECK FOR PROPER AIR DISTRIBUTION. VISUALLY CHECK WATER VALVE OPERATION AND FRESH/RECIRC DOOR OPERATION BY SOUND.

2. SET CONTROL TO HEAT MODE AND CHECK FOR PROPER BLOWER SPEED OPERATION AT ALL SWITCH POSITIONS.

3. SET CONTROL TO HEAT MODE AND OPERATE THE TEMPERATURE CONTROL LEVER FROM THE COOL TO WARM POSITION TO ESTABLISH PROPER OPERATION OF THE BLEND DOOR. THE DISCHARGE AIR TEMPERATURE SHOULD RANGE FROM COOL TO HOT.

   NOTE: THE TEMPERATURE CONTROL LEVER SHOULD NOT STAY BOTTOMED OUT AT THE EDGE OF THE SLOT ON EITHER END OF THE CONTROL.

4. MEASURE THE DISCHARGE AIR TEMPERATURE AND COMPARE TO TABLE OF NORMAL OPERATING CONDITIONS ON PAGE 6 FOR SUCTION THROTTLING VALVE SYSTEMS AND 40-50° FOR CLUTCH CYCLING SYSTEMS.

5. SYSTEM IS OKAY, RELEASE.

Fig. 50 Insufficient cooling diagnosis chart. Ford Motor Co. (Part 6 of 6)

Performance Test

NOTE: Pressure developed on high pressure and low pressure side of the compressor indicate whether the system is operating properly.

1. Connect a manifold gauge set to service valves, ensure gauge valves are closed.
2. Refer to Figs. 57 through 66 and observe conditional requirements for performance testing.

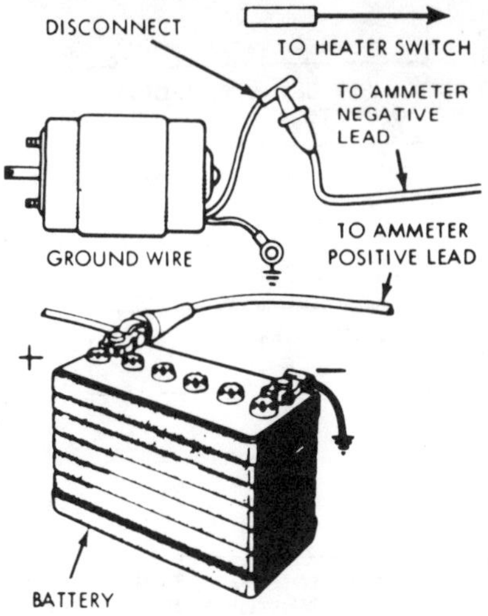

Fig. 54 Blower motor current draw test connections. 1974-76 Ford Motor Co.

3. Ensure front of vehicle is at least 5 ft. from any wall.

NOTE: The actual pressure indicated on the gauge will depend upon surrounding air temperature and humidity. Higher air temperature and humidity will give higher gauge readings. At idle with surrounding air temperatures at 110°F, head pressure may be as high as 300 psi. If system is tested under these conditions, a fan must be placed in front of the radiator and condenser.

4. System pressures are detailed and evaluated for each type refrigerant system in Figs. 57 through 66.

Vacuum Control System Test

1974-76 Cougar, Elite, Mark IV, Montego, Thunderbird, Torino & 1974-77 Lincoln

1. Connect a vacuum gauge into vacuum supply line between check valve and vacuum selector.
2. Position functional lever on Defrost and temperature lever on warm.
3. Start engine and allow to warm up. Turn ignition switch to accessory position and note vacuum gauge reading. If gauge indicates a loss of vacuum, there is a leak in the vacuum supply line or at the check valve. If vacuum is maintained, system is operating properly up to the selector valve.
4. Refer to Figs. 67 through 70 for control lever positions and vacuum motor operation. Perform test as described above for each control lever setting. If a vacuum loss is indicated, the vacuum motor(s) or vacuum line(s) for the prescribed control lever setting is leaking. If vacuum is maintained the system is operating properly. If no vaccum leaks are indicated throughout all control lever positions, check the motor or door that could be causing the complaint.

Models W/Manual Control Except 1974-76 Cougar, Elite, Montego, Thunderbird & Torino

With engine operating, move control lever through all mode positions. A momentary hissing noise should be heard as control lever is moved from one position to another, indicating that vacuum is being supplied to the control head. A continuous hissing noise would indicate a leak in the vacuum system. Refer to Fig. 71, when diagnosing the vacuum system. Also refer to airflow schematic and vacuum control chart, Figs. 72 through 85. These charts indicate mode door positions and whether or not vacuum is applied to vacuum motor. An inoperative vacuum motor will result in incorrect air flow. When checking air flow position blower switch on

BLOWER MOTOR TEST CHART

Year	Model	Amps.
1974-76	Comet & Maverick	16-20
	Lincoln	21-24
1974-76	Montego & Torino	23
	Mark IV & Thunderbird	23
1974	Ford & Mercury	22
1974-76	Bobcat, Mustang II & Pinto	25 max.
	Cougar & Elite	23
1975-76	Granada & Monarch	23
	Ford & Mercury	24

Fig. 55 Blower motor current draw test specifications. 1974-76 Ford Motor Co.

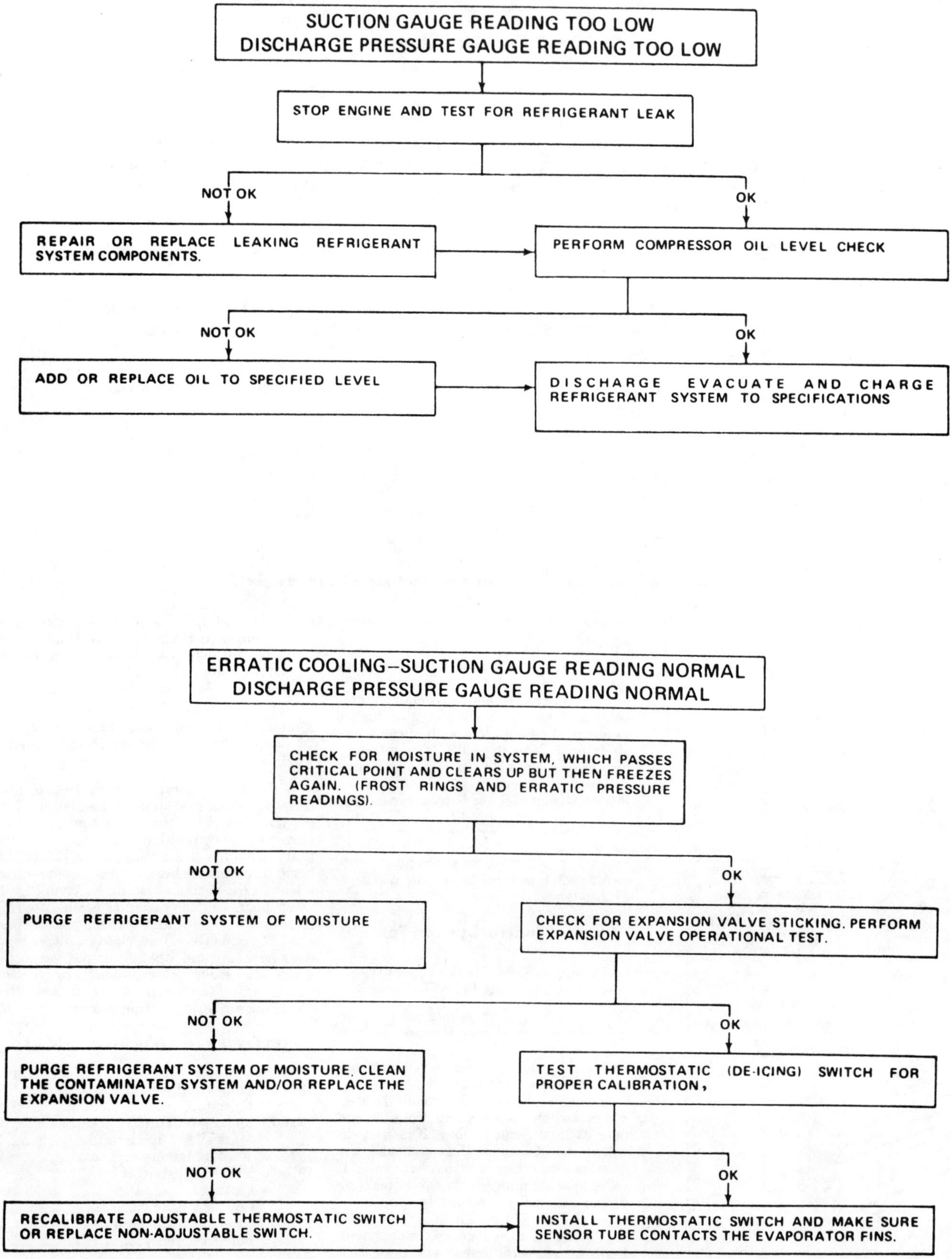

Fig. 51 A/C refrigerant system diagnosis chart. Ford Motor Co. (Part 1 of 5)

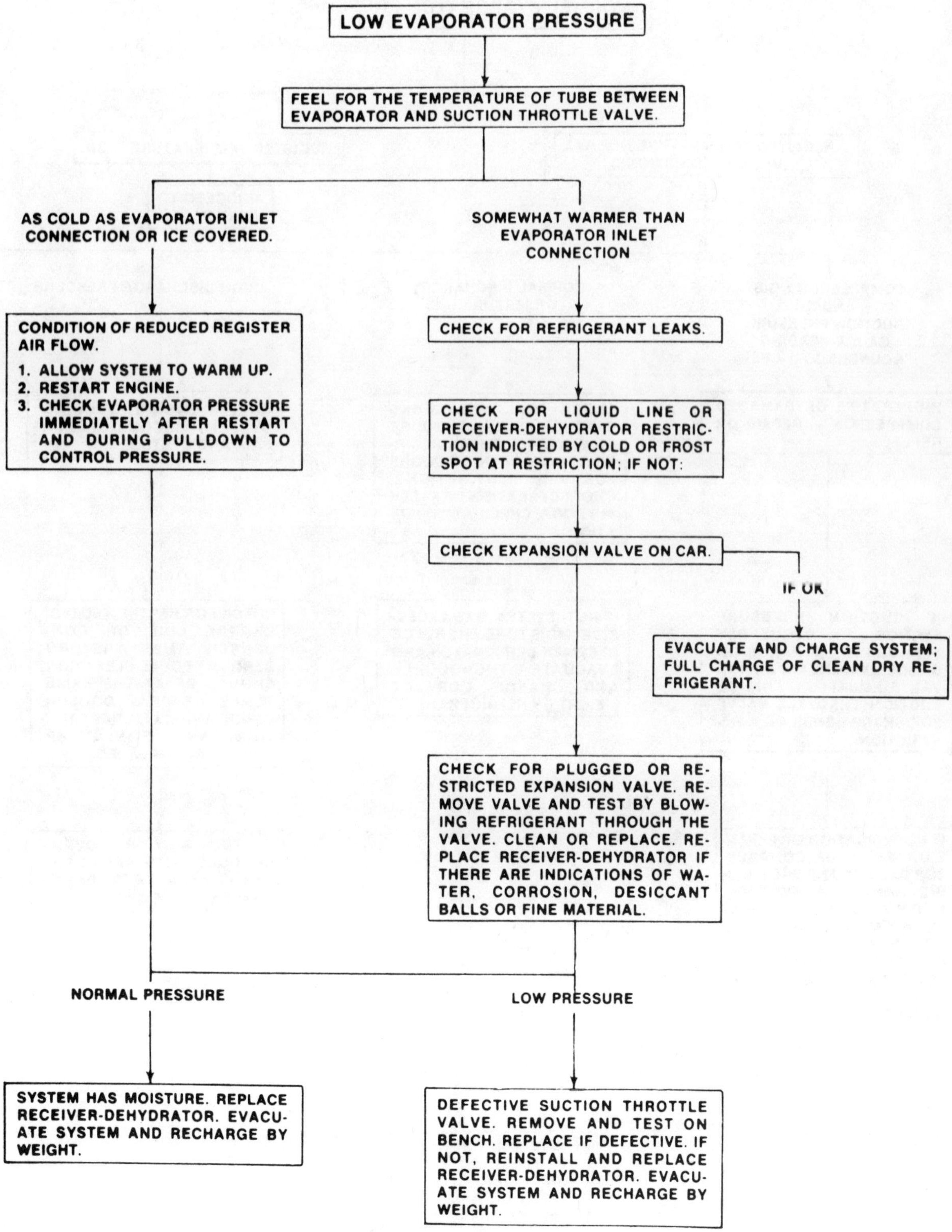

Fig. 51 A/C refrigerant system diagnosis chart. Ford Motor Co. (Part 2 of 5)

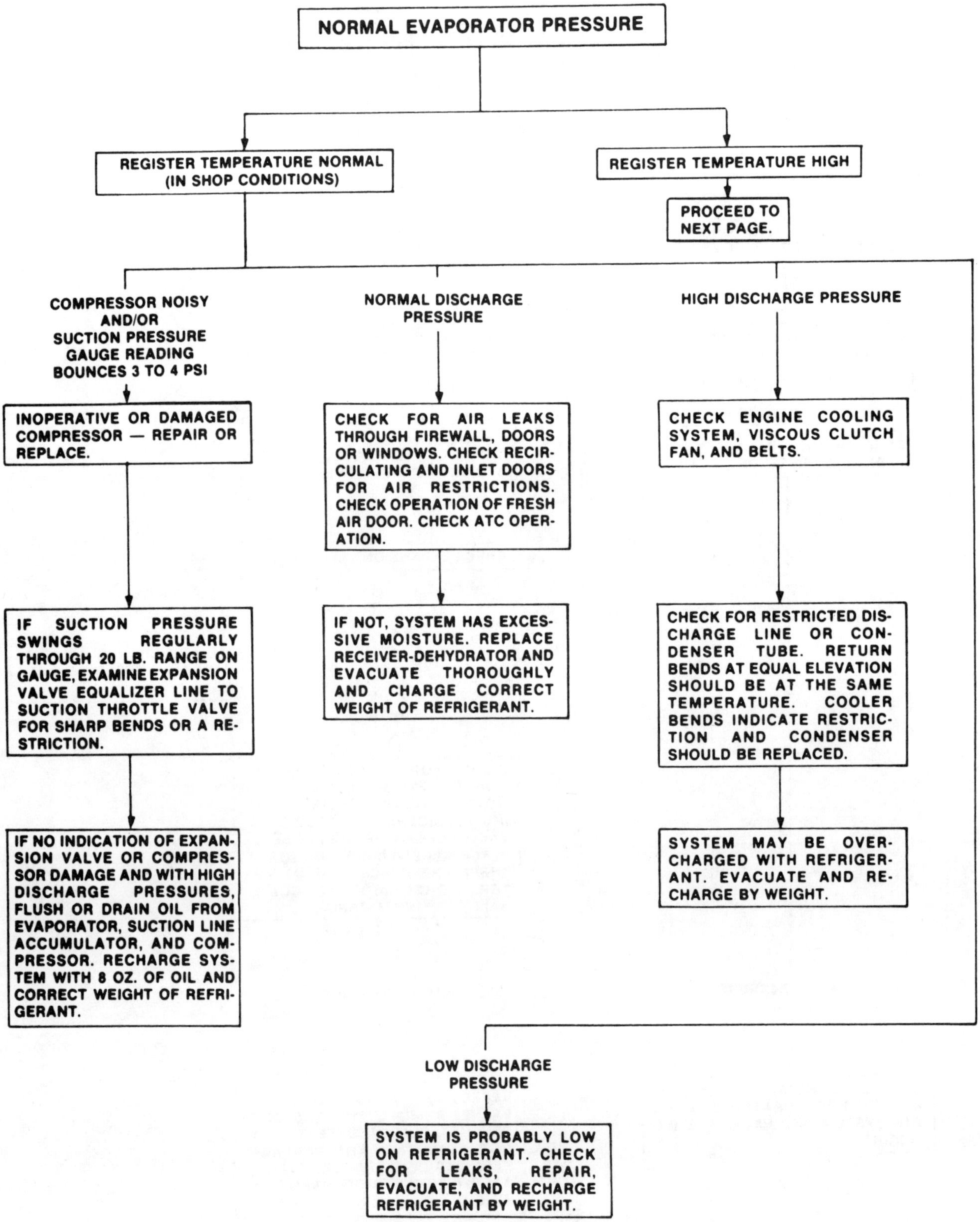

Fig. 51 A/C refrigerant system diagnosis chart. Ford Motor Co. (Part 3 of 5)

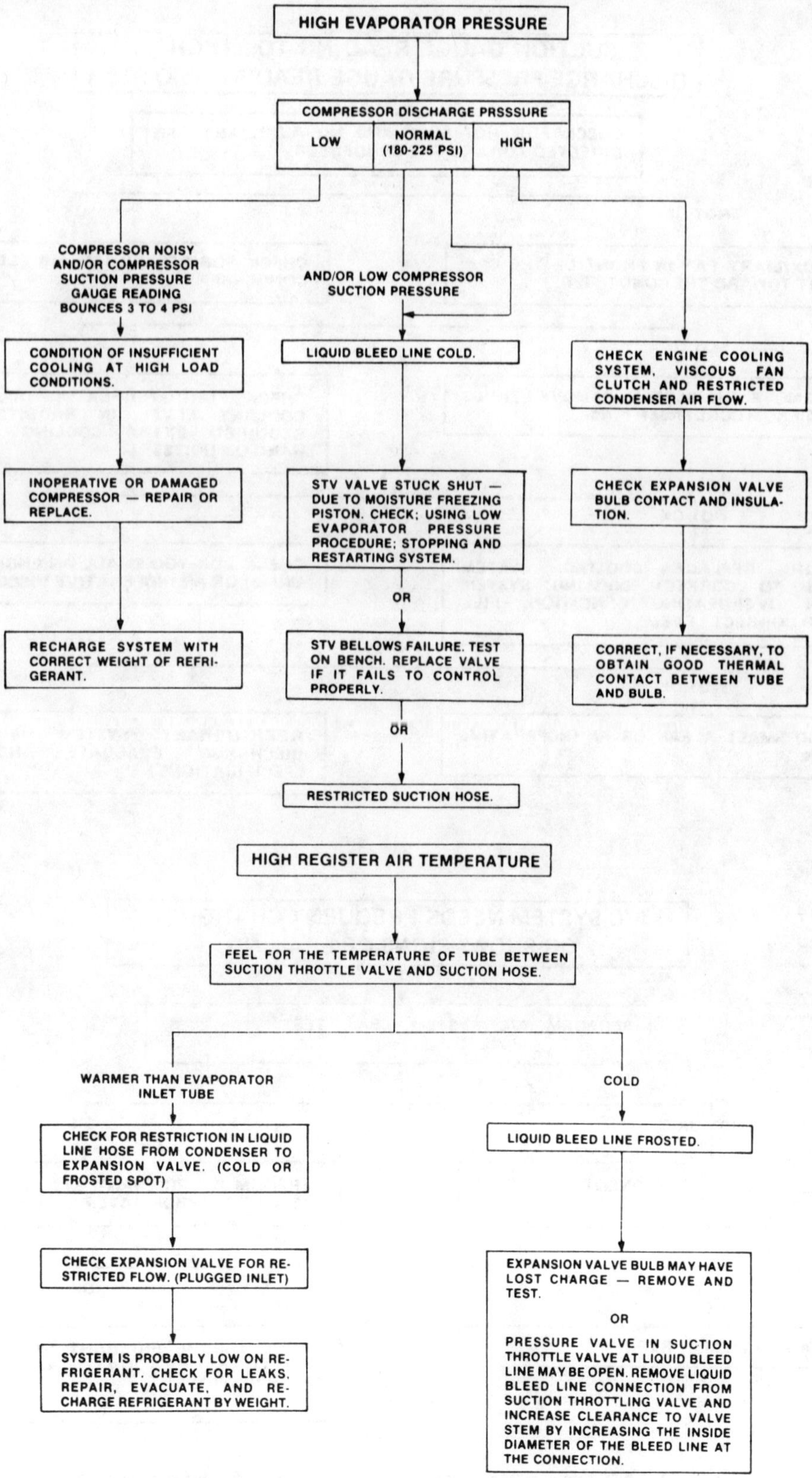

Fig. 51 A/C refrigerant system diagnosis chart. Ford Motor Co. (Part 4 of 5)

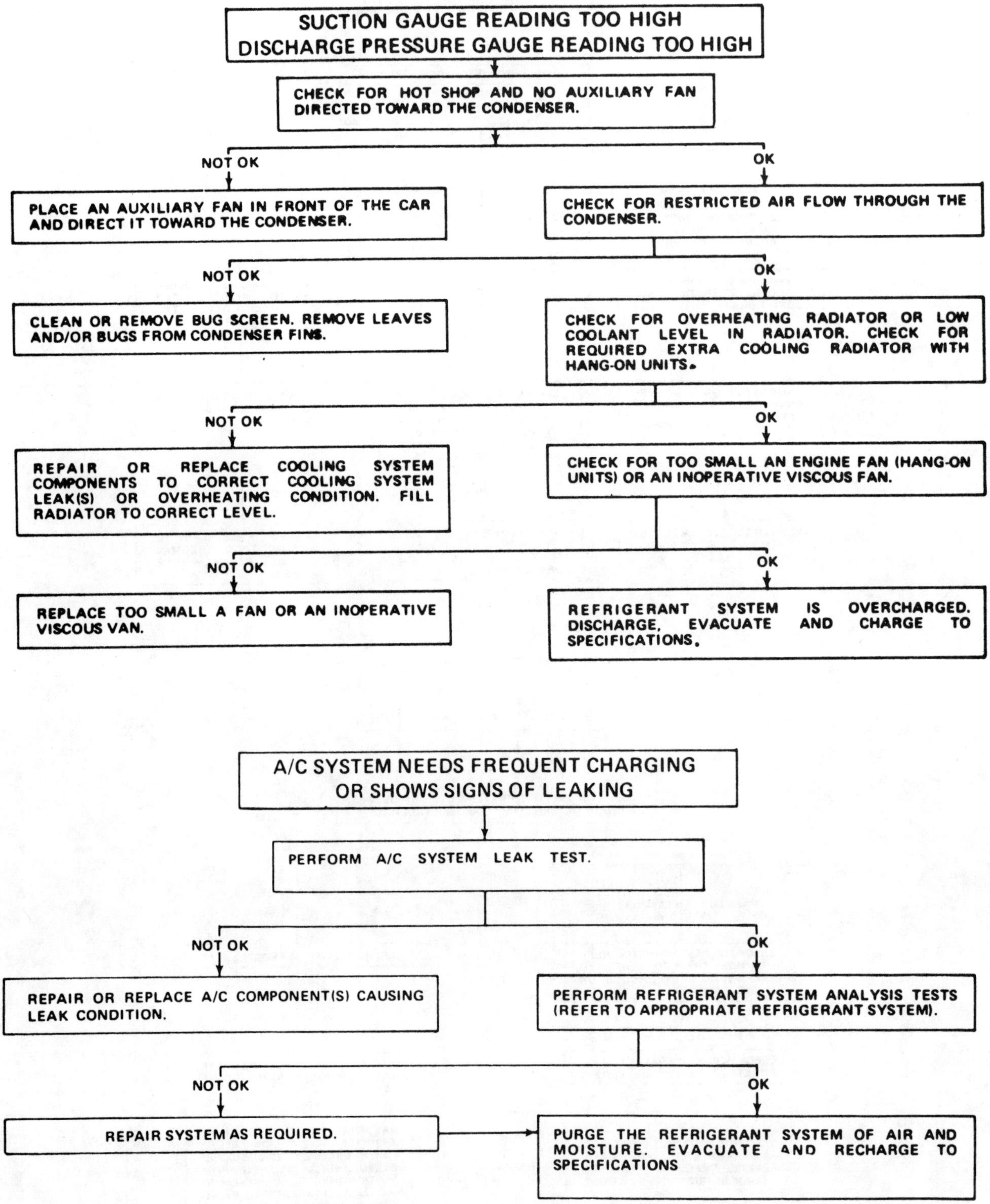

Fig. 51 A/C refrigerant system diagnosis chart. Ford Motor Co. (Part 5 of 5)

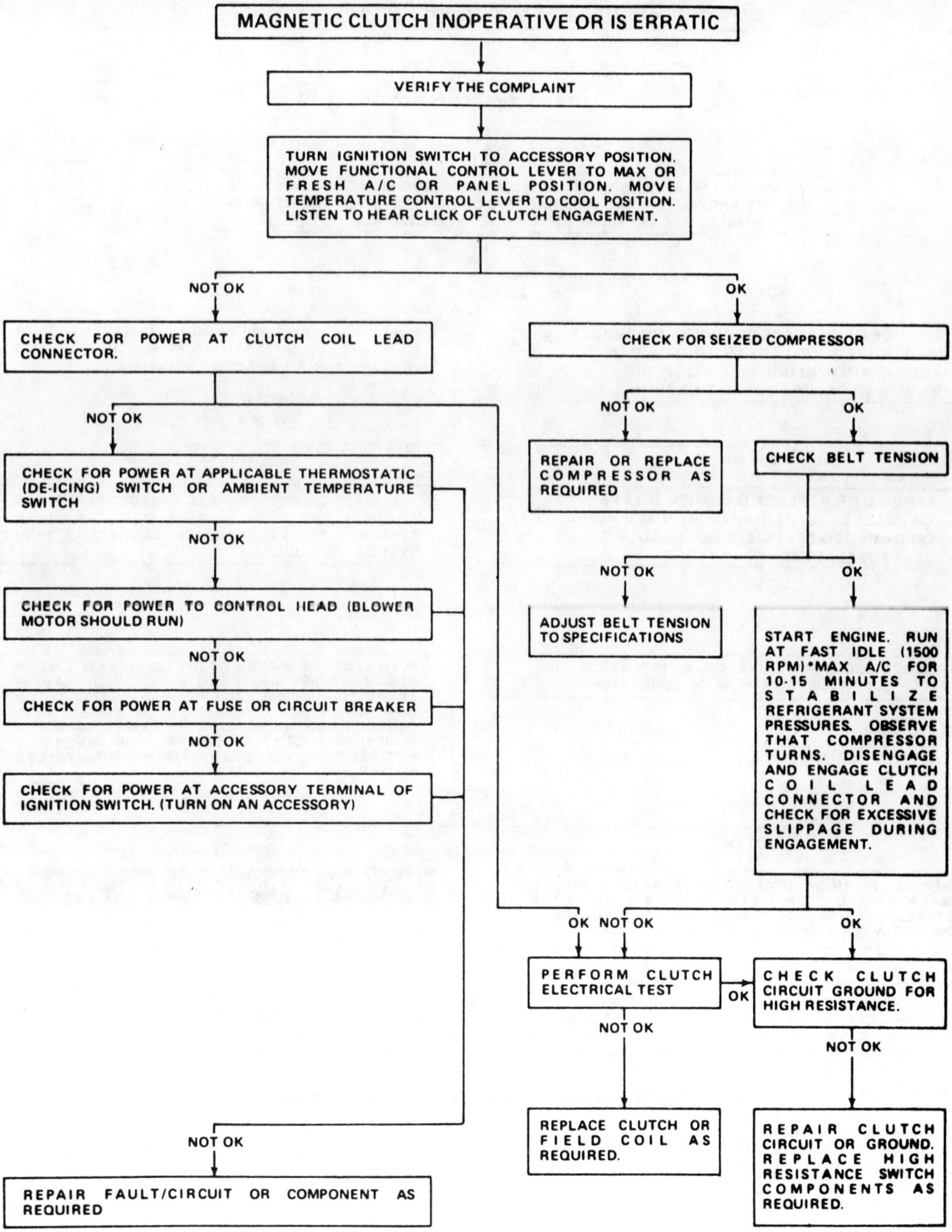

Fig. 52 Magnetic clutch diagnosis chart. Ford Motor Co.

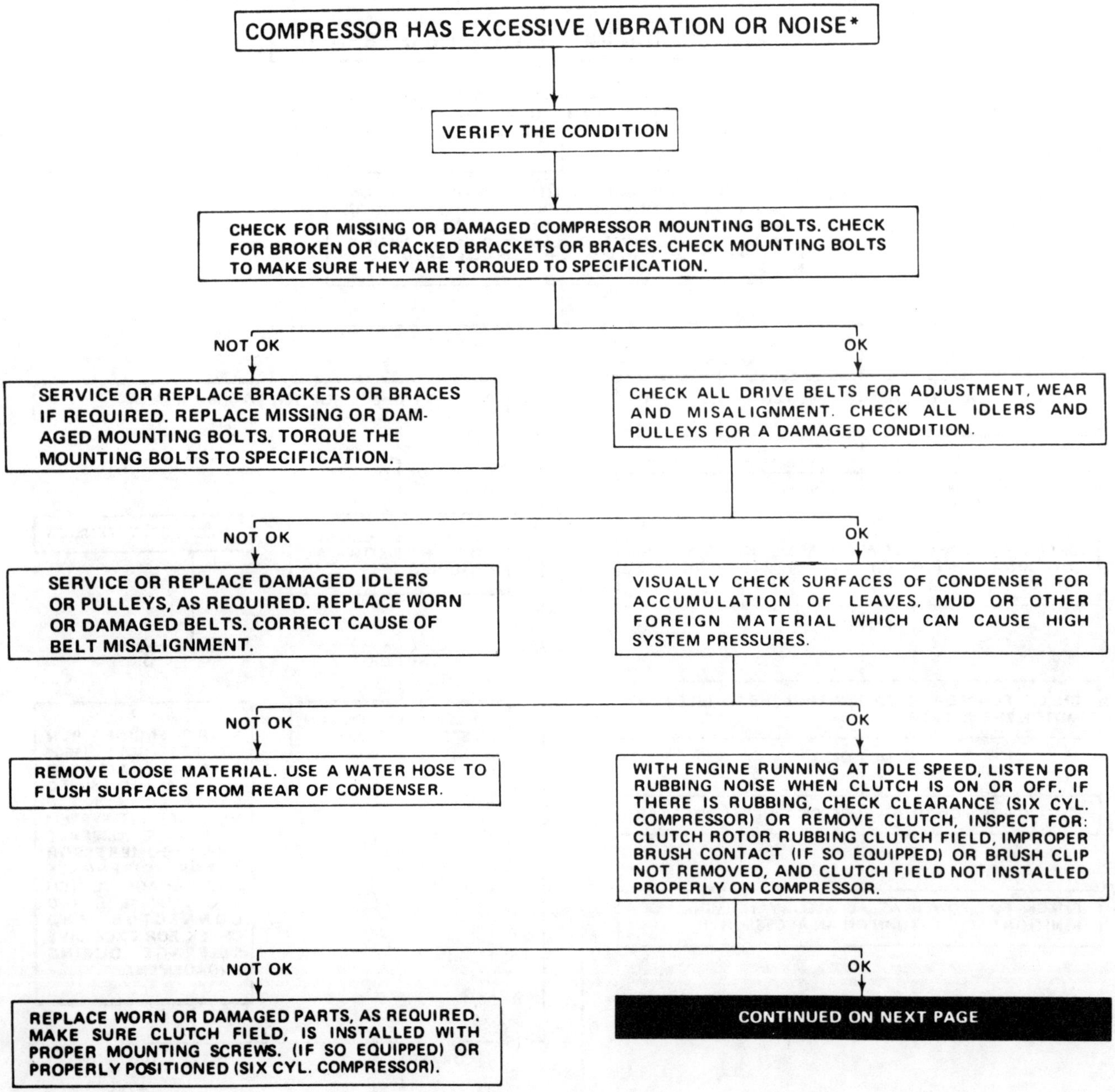

*SOME NOISE AND VIBRATION IS NORMAL IN ALL AIR CONDITIONING SYSTEMS. HIGH TEMPERATURES AND PROGRESSIVELY INCREASED SYSTEM PRESSURES WILL INCREASE THE NOISE AND VIBRATION LEVEL. THIS IS A NORMAL CONDITION. COMPARISON AND EXPERIENCE WITH AIR CONDITIONING SYSTEMS UNDER VARIOUS OPERATING CONDITIONS WILL TELL THE TECHNICIAN WHICH SYSTEMS ARE NOT CONFORMING TO THE NORMAL PUMPING SOUND AND VIBRATION PATTERNS MADE BY A PROPERLY FUNCTIONING COMPRESSOR.

Fig. 53 Compressor diagnosis chart. Ford Motor Co. (Part 1 of 2)

COMPRESSOR HAS EXCESSIVE VIBRATION OR NOISE – Continued

CONTINUED FROM PREVIOUS PAGE

CHECK CLUTCH FOR LOOSENESS OR PLAY ON COMPRESSOR SHAFT. REMOVE CLUTCH AND INSPECT TAPER (2 CYL. COMPRESSOR) AND KEYWAY IF LOOSENESS IS SUSPECTED.

- NOT OK: REPLACE KEY IF WORN. REMOVE MINOR IMPERFECTIONS FROM TAPER WITH A CROCUS CLOTH OR FINE HONING STONE.
- OK: OPERATE ENGINE AT IDLE SPEED WITH A/C CLUTCH DISENGAGED. IF CLUTCH BEARING IS DAMAGED, THE NOISE WILL STOP WHEN THE CLUTCH IS ENGAGED (2 CYL. COMPRESSOR ONLY). THE BEARING NOISE MAY BECOME LOUDER WITH SIX CYL. COMPRESSOR.

- NOT OK: REPLACE THE DAMAGED CLUTCH BEARING
- OK: WITH ENGINE OPERATING AT IDLE SPEED. CHECK A/C CLUTCH FOR SCRAPING NOISE WHEN CLUTCH IS DISENGAGED. ENGAGE CLUTCH TO SEE IF NOISE DISAPPEARS.

- NOT OK: CYCLE CLUTCH A FEW TIMES TO BURNISH AND SEAT THE CLUTCH FACE WHEN ACTIVATED. REPLACE CLUTCH, IF REQUIRED.
- OK: RUN ENGINE AT IDLE SPEED. CHECK CLUTCH MAXIMUM RUNOUT WITH A DIAL INDICATOR.

- NOT OK: IF MAXIMUM ALLOWABLE RUNOUT EXCEEDS SPECIFICATIONS, REPLACE CLUTCH.
- OK: PERFORM REFRIGERANT SYSTEM PRESSURE CHECK (REFER TO APPROPRIATE ANALYSIS CHART) EXCESSIVE PRESSURES CAN CAUSE VIBRATION OR NOISE. *

- NOT OK: SERVICE OR REPLACE THE A/C COMPONENTS AS INDICATED.
- OK: PERFORM COMPRESSOR TEST

- NOT OK: REPLACE THE VALVE PLATE AND/OR HEAD GASKET, AS REQUIRED. REPLACE THE COMPRESSOR IF THE CYLINDER WALLS ARE SCORED OR PIECES OF METAL ARE IMBEDDED IN THE PISTON.
- OK: RUN THE ENGINE AT 1500 RPM, A/C ON MAXIMUM COOLING AND HIGH BLOWER. LISTEN TO THE COMPRESSOR FOR AN EXCESSIVE RUMBLE OR KNOCK THAT DISAPPEARS WHEN CLUTCH IS DEACTIVATED.*

- NOT OK: SERVICE OR REPLACE COMPRESSOR IF RUMBLE OR KNOCK DISAPPEARS WHEN CLUTCH IS DEACTIVATED.

*SOME NOISE AND VIBRATION IS NORMAL IN ALL AIR CONDITIONING SYSTEMS. HIGH TEMPERATURES AND PROGRESSIVELY INCREASED SYSTEM PRESSURES WILL INCREASE THE NOISE AND VIBRATION LEVEL. THIS IS A NORMAL CONDITION. COMPARISON AND EXPERIENCE WITH AIR CONDITIONING SYSTEMS UNDER VARIOUS OPERATING CONDITIONS WILL TELL THE TECHNICIAN WHICH SYSTEMS ARE NOT CONFORMING TO THE NORMAL PUMPING SOUND AND VIBRATION PATTERNS MADE BY A PROPERLY FUNCTIONING COMPRESSOR.

Fig. 53 Compressor diagnosis chart. Ford Motor Co. (Part 2 of 2)

Blower Motor Test Chart

Model	Year	System	Blower Switch Position							
			Voltage Test				Current Draw Test			
			Lo (Volts)	Med. Lo (Volts)	Med. Hi (Volts)	Hi (Volts)	Lo (Amps.)	Med. Lo (Amps.)	Med. Hi (Amps.)	Hi (Amps.)
Bobcat &	1977	All	—	—	—	—	3.5	6	10	20
Pinto	1978-80	All	7-9	9-11.5	11.5-13	13-14.8	5	8	15	22
Mustang &	1977-79	All	7-9	9-11.5	11.5-13	13-14.8	5	8	15	22
1979-80 Capri	1980	All	3.9	6.2	8.9	12.8	4	8	13	23
Comet & Maverick	1977	All	—	—	—	—	3.5	6	10	20
Fairmont & Zephyr	1978-80	All	3.9	6.2	8.9	12.8	4	8	13	23
Granada	1978-80	①	5	7	10	14	8	11	20	25
Monarch & Versailles	1980	Manual	7-8.5	8.5-10	10-12.5	12.5-14.8	6	10	17	28
Cougar, LTD II,	1977-79	Manual	5	7	10	12.4	6	8	15	22
Mark V, VI &	1977-79	A.T.C.	5	7	10	12.4	6	10	17	25
Thunderbird	1980	All	3.9	6.2	8.9	12.8	4	8	13	23
Ford &	1977	A.T.C.	—	5-10②	5.5-12③	12-14.7④	—	22②	28③	—
Mercury	1977-78	Manual	7-8.5	8.5-10	10-12.5	12.5-14.8	6	10	17	28
Full Size	1978	A.T.C.	5	7	10	12.4	6	10	17	25
	1979-80	Manual	5	7	10	12.4	6	8	15	25
	1979-80	A.T.C.	5	7	10	12.4	8	10	17	25
Lincoln	1977	A.T.C.	—	—	—	12	—	—	—	21-24
Continental	1978	A.T.C.	5	7	10	12.4	6	10	17	25
	1979-80	A.T.C.	5	7	10	12.4	8	10	17	25

① —1977-79 all & 1980 models w/A.T.C.
② —Auto. Lo.
③ —Auto. Hi.
④ —Max.

Fig. 56 Blower motor voltage & current draw test specifications. 1977-80 Ford Motor Co.

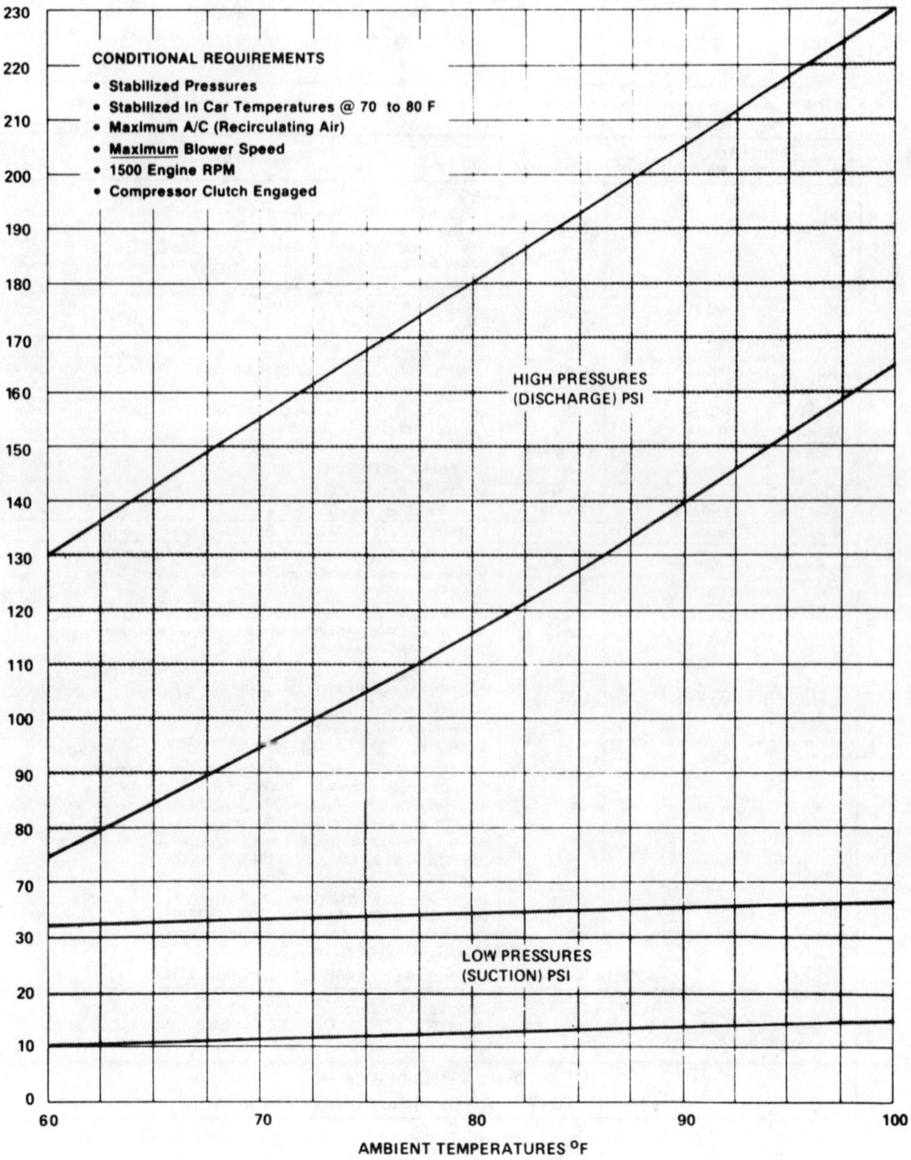

Fig. 57 Discharge & suction pressures vs. ambient temperatures. 1974-77 Cycling clutch systems w/de-icing switch

NOTE: Normal system conditional requirements must be maintained to properly evaluate refrigerant system pressures. Refer to chart applicable to system under test.

HIGH PRESSURE		LOW PRESSURE	POSSIBLE CAUSES
•LIQUID LINE	•COMPRESSOR		
HIGH	HIGH	HIGH	ENGINE OVERHEATING EXPANSION VALVE SENSING BULB — Poor Contact, Inadequate Insulation HEATER WATER VALVE — Insufficient Closure/ EVAPORATOR — Too Much Heat. Observe Conditional Requirements. EXPANSION VALVE — Sticking Open
HIGH	HIGH	NORMAL TO HIGH	AIR IN REFRIGERANT SYSTEM
HIGH	HIGH	NORMAL	REFRIGERANT — Overcharge
HIGH	HIGH	LOW	COMPRESSOR-TO-CONDENSER A/C LINE — Restriction: After Service Access Valve
NORMAL TO HIGH	NORMAL TO HIGH	HIGH	CONDENSER — Inadequate Air Flow (Bent Fins, Debris, Loose Radiator Fan Belt, Slipping Fan Clutch)
NORMAL TO HIGH	NORMAL TO HIGH	NORMAL TO HIGH	HUMIDITY WELL ABOVE NORMAL EXCESSIVE REFRIGERANT OIL
NORMAL TO ERRATIC	NORMAL TO ERRATIC	NORMAL TO ERRATIC	MOISTURE IN REFRIGERANT SYSTEM — Affects Expansion Valve Function
ERRATIC	ERRATIC	ERRATIC	EXPANSION VALVE SENSING BULB — Kinked Capillary Tube
NORMAL TO LOW	NORMAL TO LOW	NORMAL TO LOW	EVAPORATOR — Restriction, Inadequate Air Flow — Reset Blower to Conditional Requirement and/or Confirm Blower Condition
LOW	HIGH	LOW	RECEIVER/DRIER — Restriction A/C LIQUID LINE — Restriction; Before Service Access Valve CONDENSER — Restriction
LOW	LOW	HIGH	A/C SUCTION LINE — Restriction; After Service Access Valve SUCTION ACCUMULATOR — Restriction; After Service Access Valve COMPRESSOR — Damaged Valves — (Inadequate compression)
LOW	LOW	LOW	A/C SUCTION LINE — Restriction; Before Service Access Valve SUCTION ACCUMULATOR — Restriction; Before Service Access Valve EVAPORATOR — Iced-up/CONDENSER — Plugged/ RECEIVER/DRIER — Plugged A/C LIQUID LINE — Restriction, After Service Access Valve/EXPANSION VALVE SENSING BULB — Perforated Bulb/EXPANSION VALVE — Sticking Closed REFRIGERANT LEAK AND/OR UNDERCHARGE COMPRESSOR-TO-CONDENSER A/C LINE — Restriction, Before Service Access Valve DE-ICING SWITCH — Sticking Closed/COMPRESSOR CLUTCH — Seized

ADDITIONAL POSSIBLE CAUSES ASSOCIATED WITH INADEQUATE COMPRESSOR OPERATION

COMPRESSOR CLUTCH — Slipping/COMPRESSOR DRIVE BELT — Loose/ CLUTCH COIL — Shorted coils; Slipping/CLUTCH COIL — Open Circuit/Blown Clutch Circuit Fuse/ DE/ICING SWITCH — Sticking Open/Improper Clutch Circuit Contact in Control Assembly Switch

ADDITION POSSIBLE CAUSES ASSOCIATED WITH A NON-FUNCTIONAL COMPRESSOR

DE-ICING SWITCH — Sticking Closed/ COMPRESSOR CLUTCH — Seized REFRIGERANT LEAK

Fig. 58 Refrigeration system analysis chart. 1974-77 Cycling clutch systems w/de-icing switch

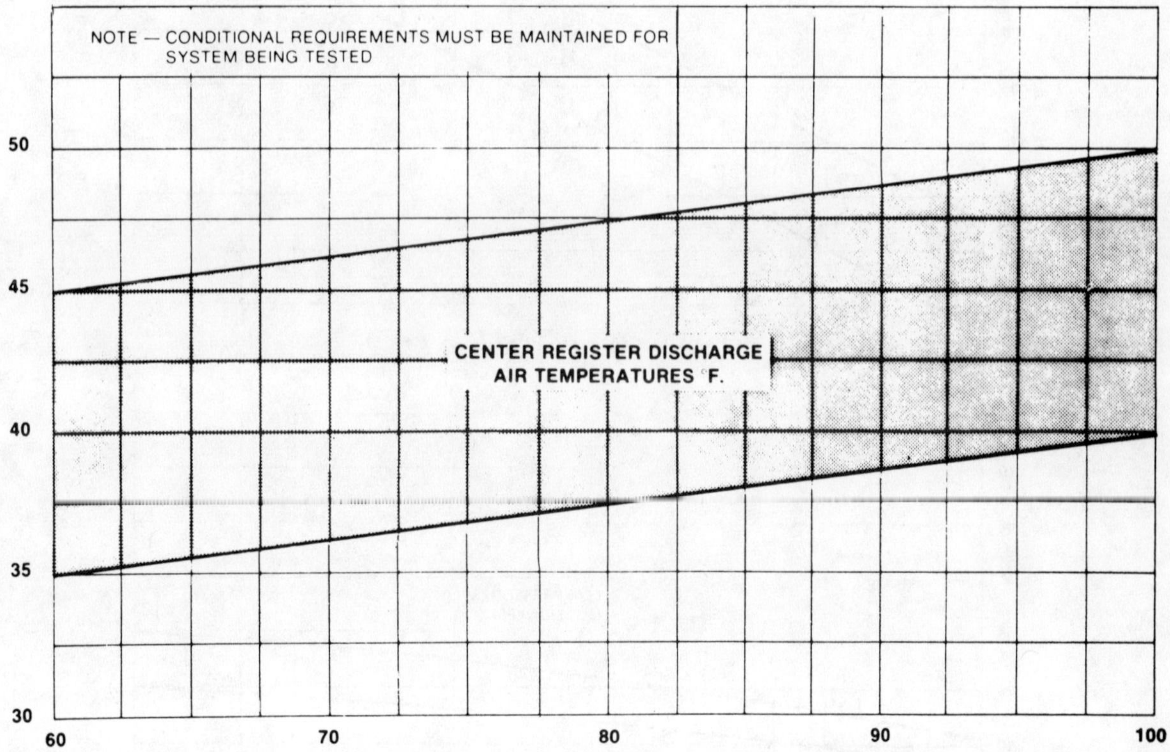

Fig. 59 **Normal center register discharge air temperatures vs. ambient temperatures. All models except 1980 cycling clutch systems**

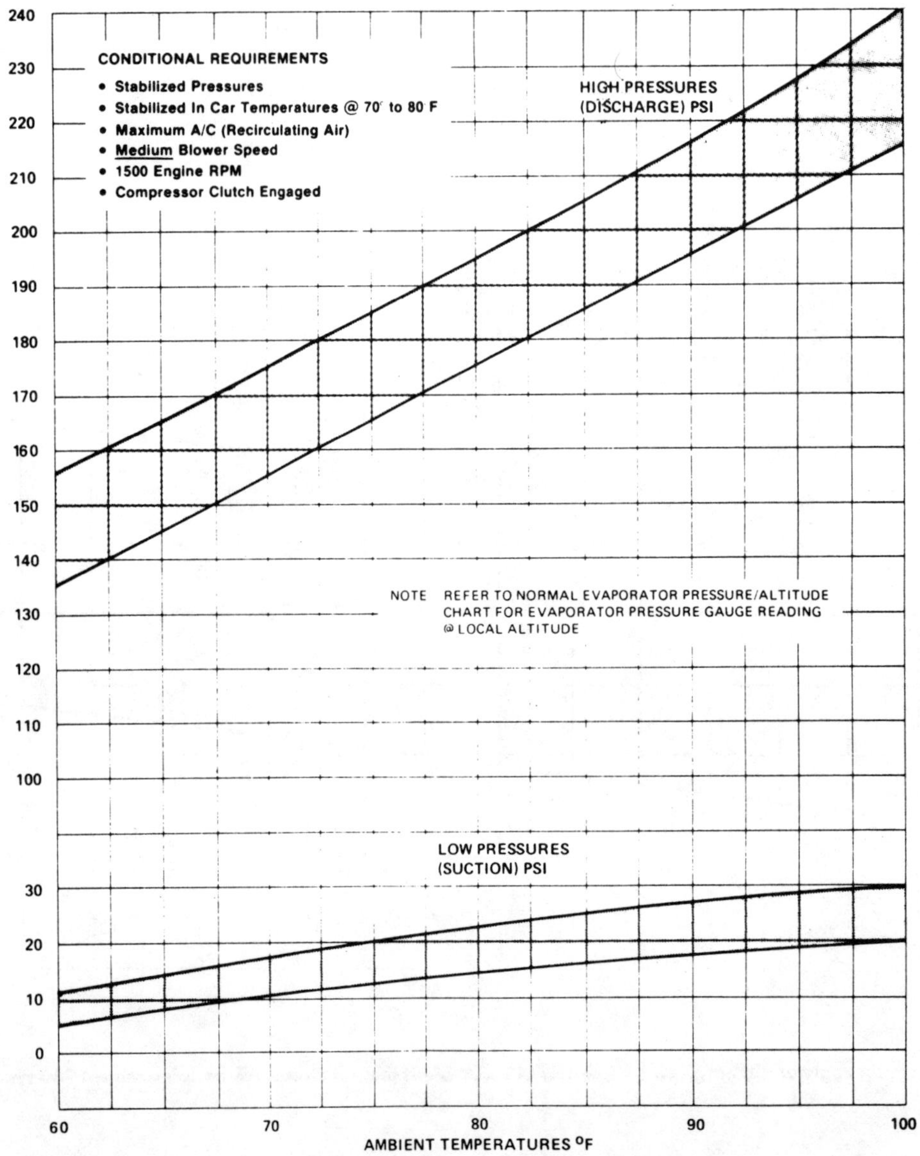

Fig. 60 Discharge & suction pressures vs. ambient temperatures. STV systems w/o BPO

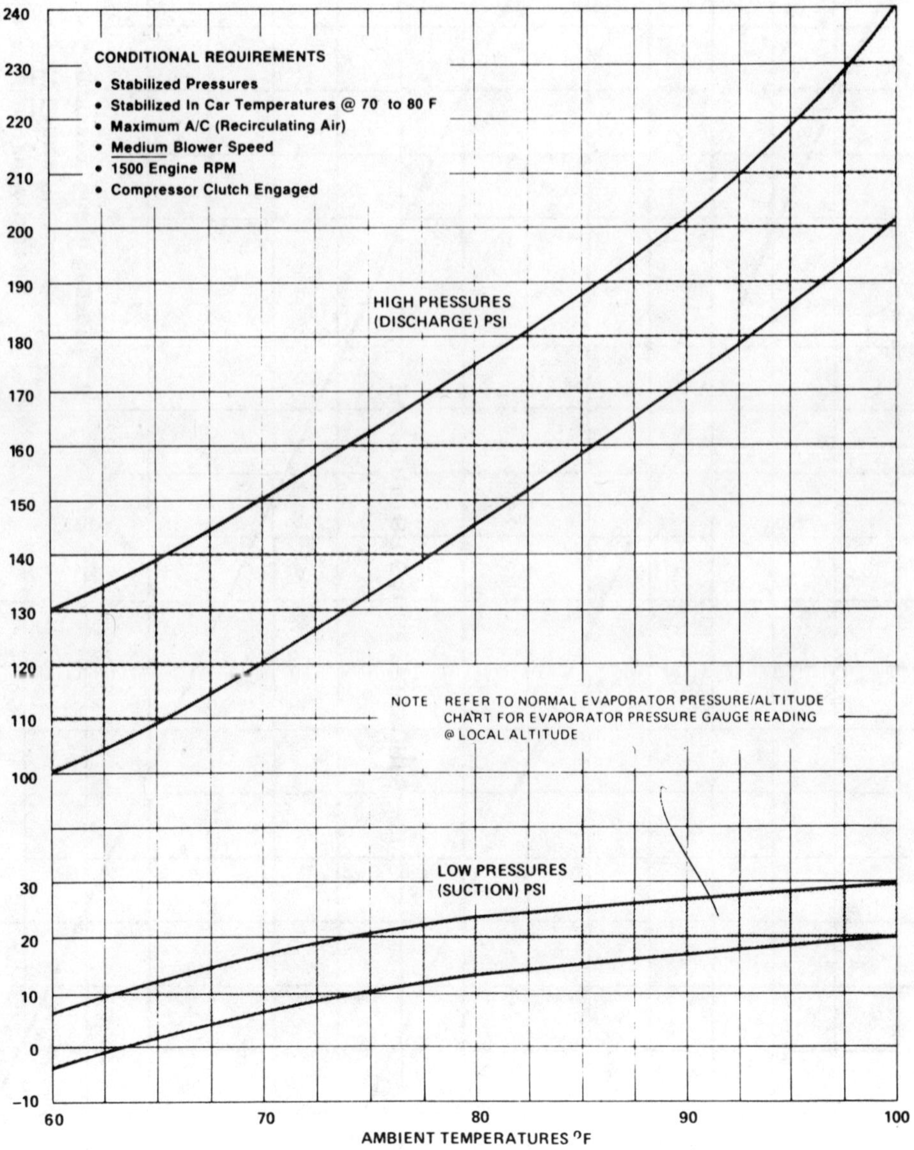

Fig. 61 Discharge & suction pressures vs. ambient temperatures. STV systems w/BPO

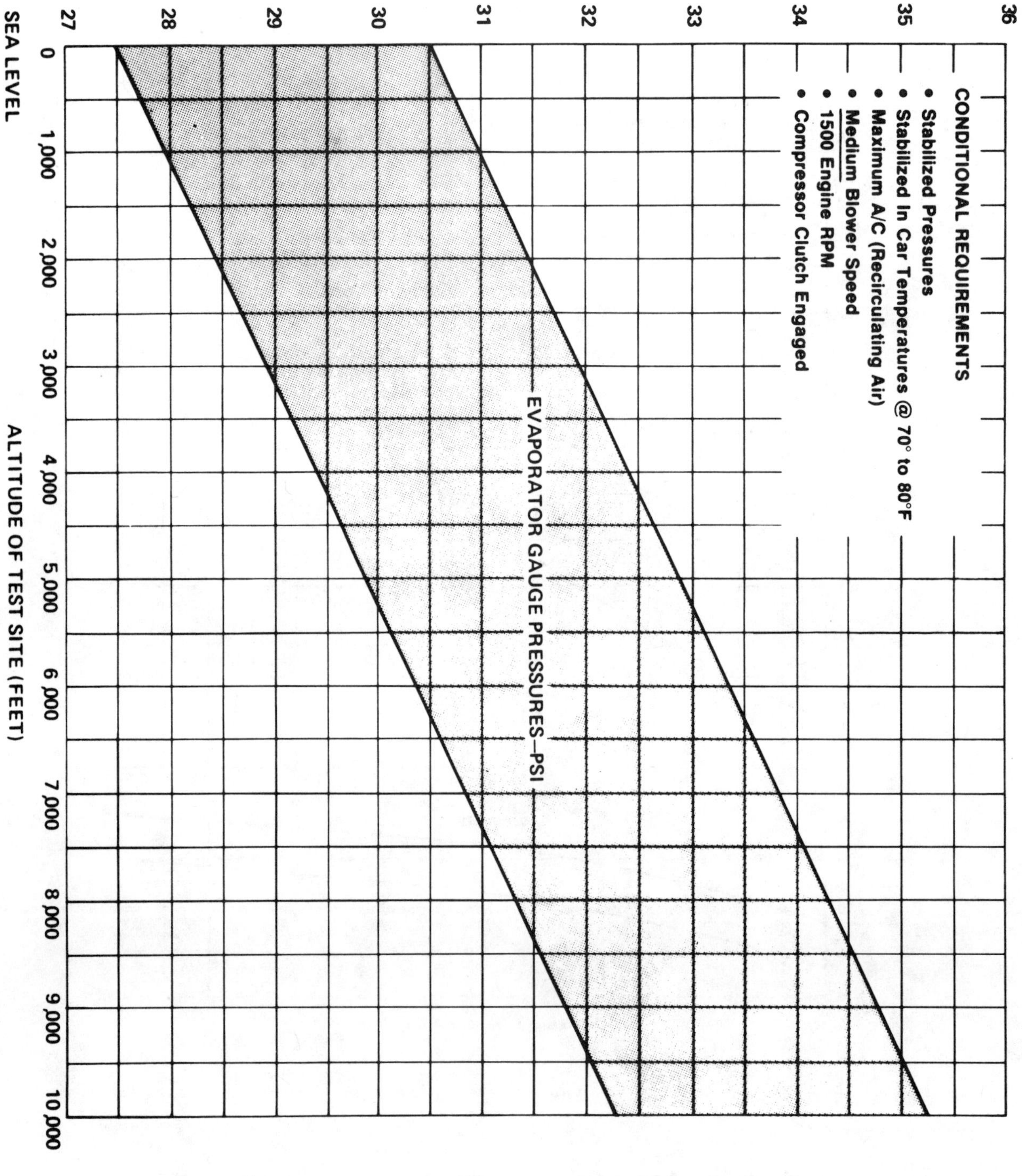

Fig. 62 Evaporator pressures vs. altitude. All models except 1980 cycling clutch systems

NOTE: Normal system conditional requirements must be maintained to properly evaluate refrigerant system pressures. Refer to chart applicable to system under test.

HIGH PRESSURE		LOW PRESSURE	EVAPORATOR PRESSURE	POSSIBLE CAUSES
● LIQUID LINE	@ COMPRESSOR			
HIGH	HIGH	HIGH	HIGH	ENGINE OVERHEATING / EXPANSION VALVE SENSING BULB Poor Contact, Inadequate Insulation
HIGH	HIGH	HIGH	NORMAL TO HIGH	HEATER WATER VALVE Insufficient Closure / EVAPORATOR Too Much Air Flow · Reset Blower to Conditional Requirement / EXPANSION VALVE Sticking Open
HIGH	HIGH	NORMAL TO HIGH	NORMAL	AIR IN REFRIGERANT SYSTEM
HIGH	HIGH	NORMAL	NORMAL	REFRIGERANT Overcharge
HIGH	HIGH	LOW	LOW	COMPRESSOR-TO-CONDENSER A/C LINE – Restriction; After Service Access Valve
NORMAL TO HIGH	NORMAL TO HIGH	HIGH	HIGH	CONDENSER Inadequate Air Flow (Bent Fins, Debris, Loose Radiator Fan Belt, Slipping Fan Clutch)
NORMAL TO HIGH	NORMAL TO HIGH	NORMAL TO HIGH	NORMAL TO HIGH	HUMIDITY WELL ABOVE NORMAL
NORMAL TO HIGH	NORMAL TO HIGH	NORMAL TO HIGH	NORMAL	EXCESSIVE REFRIGERANT OIL
NORMAL	NORMAL	NORMAL	LOW	SUCTION THROTTLING VALVE – Sticking Open – May Cause Iced-Up Evaporator, May Be Related to Moisture.
NORMAL TO ERRATIC	NORMAL TO ERRATIC	NORMAL TO ERRATIC	NORMAL TO ERRATIC	MOISTURE IN REFRIGERANT SYSTEM – Affects Expansion Valve and STV Functions.
ERRATIC	ERRATIC	ERRATIC	ERRATIC	EXPANSION VALVE SENSING BULB – Kinked Capillary Tube
NORMAL TO LOW	NORMAL TO LOW	NORMAL TO LOW	NORMAL	EVAPORATOR · Restriction, Inadequate Air Flow · Reset Blower to Conditional Requirement and/or Confirm Blower Condition
LOW	HIGH	LOW	LOW	RECEIVER/DRIER – Restriction / A/C LIQUID LINE – Restriction, Before Service Access Valve / CONDENSER – Restriction
LOW	LOW	HIGH	NORMAL TO HIGH	A/C SUCTION LINE · Restriction, After Service Access Valve / SUCTION ACCUMULATOR · Restriction, After Service Access Valve
LOW	LOW	HIGH	HIGH	COMPRESSOR · Damaged Valves (Inadequate compression)
LOW	LOW	LOW	HIGH	SUCTION THROTTLING VALVE – Sticking Closed · May Be Related to Moisture
LOW	LOW	LOW	NORMAL TO HIGH	A/C SUCTION LINE – Restriction, Before Service Access Valve / SUCTION ACCUMULATOR – Restriction, Before Service Access Valve
LOW	LOW	LOW	LOW	EVAPORATOR – Iced-up / CONDENSER – Plugged / RECEIVER/DRIER – Plugged / A/C LIQUID LINE – Restriction; After Service Access Valve / EXPANSION VALVE SENSING BULB – Perforated Bulb / EXPANSION VALVE – Sticking Closed / REFRIGERANT LEAK AND/OR UNDERCHARGE COMPRESSOR-TO-CONDENSER A/C LINE – Restriction, Before Service Access Valve

ADDITIONAL POSSIBLE CAUSES ASSOCIATED WITH INADEQUATE COMPRESSOR OPERATION

COMPRESSOR CLUTCH – Slipping / COMPRESSOR DRIVE BELT – Loose / CLUTCH COIL – Shorted Coils; Slipping / CLUTCH COIL – Open Circuit / AMBIENT CUT-OFF SWITCH – Sticking Open / BLOWN CLUTCH CIRCUIT FUSE – Improper Clutch Circuit Contact in Control Assembly Switch.

ADDITION POSSIBLE CAUSES ASSOCIATED WITH A NON-FUNCTIONAL COMPRESSOR

PRESSURE DIFFERENTIAL VALVE (In Liquid Bleed Line) – Sticking Closed / AMBIENT CUT-OFF SWITCH – Sticking Closed / SUCTION THROTTLING VALVE – Sticking Open / REFRIGERANT LEAK

CCL 1236-B

Fig. 63 Refrigeration system analysis chart. STV & STV/BPO systems w/low pressure access gauge port valve

NOTE: Normal system conditional requirements must be maintained to properly evaluate refrigerant system pressures. Refer to chart applicable to system under test.

HIGH PRESSURE		EVAPORATOR PRESSURE	POSSIBLE CAUSES
● LIQUID LINE	@ COMPRESSOR		
HIGH	HIGH	HIGH	ENGINE OVERHEATING / EXPANSION VALVE SENSING BULB – Poor Contact, Inadequate Insulation
HIGH	HIGH	NORMAL TO HIGH	HEATER WATER VALVE – Insufficient Closure / EVAPORATOR – Too Much Air Flow · Reset Blower to Conditional Requirement / EXPANSION VALVE – Sticking Open
HIGH	HIGH	NORMAL	AIR IN REFRIGERANT SYSTEM REFRIGERANT – Overcharge
HIGH	HIGH	LOW	COMPRESSOR-TO-CONDENSER A/C LINE – Restriction; After Service Access Valve
NORMAL TO HIGH	NORMAL TO HIGH	HIGH	CONDENSER – Inadequate Air Flow (Bent Fins, Debris, Loose Radiator Fan Belt, Slipping Fan Clutch)
NORMAL TO HIGH	NORMAL TO HIGH	NORMAL TO HIGH	HUMIDITY WELL ABOVE NORMAL
NORMAL TO HIGH	NORMAL TO HIGH	NORMAL	EXCESSIVE REFRIGERANT OIL
NORMAL	NORMAL	LOW	SUCTION THROTTLING VALVE – Sticking Open – May Cause Iced-Up Evaporator, May Be Related to Moisture.
NORMAL TO ERRATIC	NORMAL TO ERRATIC	NORMAL TO ERRATIC	MOISTURE IN REFRIGERANT SYSTEM – Affects Expansion Valve and STV Functions.
ERRATIC	ERRATIC	ERRATIC	EXPANSION VALVE SENSING BULB – Kinked Capillary Tube
NORMAL TO LOW	NORMAL TO LOW	NORMAL	EVAPORATOR – Restriction, Inadequate Air Flow · Reset Blower to Conditional Requirement and/or Confirm Blower Condition
LOW	HIGH	LOW	RECEIVER/DRIER – Restriction / A/C LIQUID LINE – Restriction; Before Service Access Valve / CONDENSER – Restriction
LOW	LOW	NORMAL TO HIGH	A/C SUCTION LINE – Restriction; SUCTION ACCUMULATOR – Restriction
LOW	LOW	HIGH	COMPRESSOR – Damaged Valves · (Inadequate compression) SUCTION THROTTLING VALVE – Sticking Closed – May Be Related to Moisture.
LOW	LOW	LOW	EVAPORATOR – Iced-up / CONDENSER – Plugged / RECEIVER/DRIER – Plugged / A/C LIQUID LINE – Restriction; After Service Access Valve / EXPANSION VALVE SENSING BULB – Perforated Bulb / EXPANSION VALVE – Sticking Closed / REFRIGERANT LEAK AND/OR UNDERCHARGE COMPRESSOR-TO-CONDENSER A/C LINE – Restriction; Before Service Access Valve

ADDITIONAL POSSIBLE CAUSES ASSOCIATED WITH A NON-FUNCTIONAL COMPRESSOR

COMPRESSOR CLUTCH – Slipping / COMPRESSOR DRIVE BELT – Loose / CLUTCH COIL – Shorted Coils; Slipping / CLUTCH COIL – Open Circuit / AMBIENT CUT-OFF SWITCH – Sticking Open / BLOWN CLUTCH CIRCUIT FUSE / Improper Clutch Circuit Contact in Control Assembly Switch

ADDITIONAL POSSIBLE CAUSES ASSOCIATED WITH INADEQUATE COMPRESSOR OPERATION

PRESSURE DIFFERENTIAL VALVE (In Liquid Bleed Line) – Sticking Closed / AMBIENT CUT-OFF SWITCH – Sticking Closed / SUCTION THROTTLING VALVE – Sticking Open / REFRIGERANT LEAK

Fig. 64 Refrigeration system analysis chart. STV & STV/BPO systems w/o pressure access gauge port valve

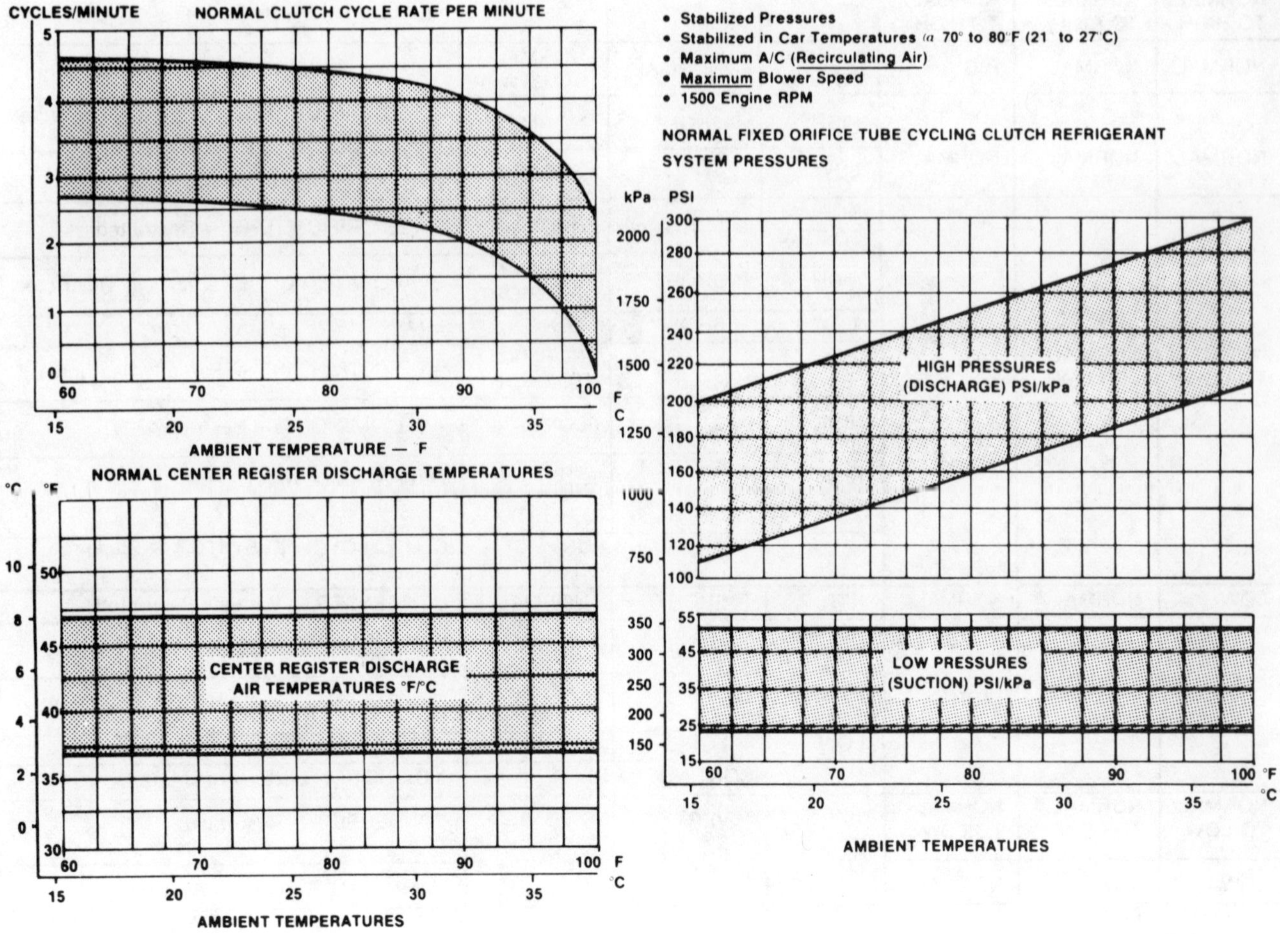

Fig. 65 Refrigeration system conditional requirements. 1980 Cycling clutch systems w/fixed orifice tube

NOTE: Normal system conditional requirements must be maintained to properly evaluate refrigerant system pressures. Refer to chart applicable to system under test.

HIGH PRESSURE @ LIQUID LINE	HIGH PRESSURE @ COMPRESSOR	LOW PRESSURE	CLUTCH CYCLE TIME RATE	CLUTCH CYCLE TIME ON	CLUTCH CYCLE TIME OFF	COMPONENT — CAUSES
HIGH	HIGH	HIGH	LOW	NORMAL TO LOW	NORMAL	ENGINE OVERHEATING
HIGH	HIGH	NORMAL TO HIGH	LOW	LONG	SHORT	CONDENSER — Inadequate Airflow
HIGH	HIGH	NORMAL	LOW	LONG	SHORT	AIR IN REFRIGERANT REFRIGERANT OVERCHARGE (a)
NORMAL TO HIGH	NORMAL TO HIGH	NORMAL TO HIGH	LOW	LONG	SHORT	HUMIDITY VERY HIGH (b)
NORMAL	NORMAL	HIGH	NORMAL TO LOW	NORMAL	NORMAL TO LOW	CLUTCH CYCLING SWITCH — High Setting
NORMAL	NORMAL	NORMAL	HIGH	SHORT	NORMAL	FIXED ORIFICE TUBE — Partially Restricted A/C LIQUID LINE — Partially Restricted
NORMAL	NORMAL	NORMAL	LOW	LONG	SHORT	MOISTURE IN REFRIGERANT SYSTEM EXCESSIVE REFRIGERANT OIL
NORMAL TO LOW	NORMAL TO LOW	NORMAL	NORMAL TO LOW	LONG	SHORT	FIXED ORIFICE TUBE — Restricted
NORMAL TO LOW	NORMAL TO LOW	NORMAL	HIGH	SHORT	NORMAL TO SHORT	EVAPORATOR CORE — Partially Restricted
NORMAL TO LOW	NORMAL TO LOW	NORMAL	LOW	SHORT	LONG	EVAPORATOR — Low Airflow
NORMAL TO LOW	NORMAL TO LOW	NORMAL	HIGH	SHORT	NORMAL TO SHORT	A/C SUCTION LINE — Partially Restricted
NORMAL TO LOW	NORMAL TO LOW	NORMAL	HIGH	VERY SHORT	LONG	A/C LIQUID LINE — Restricted
NORMAL TO LOW	NORMAL TO LOW	NORMAL	HIGH	SHORT TO VERY SHORT	SHORT TO VERY SHORT	LOW REFRIGERANT CHARGE
NORMAL TO LOW	NORMAL TO LOW	NORMAL	CONTINUOUS RUN			CLUTCH CYCLING SWITCH — Sticking Closed
LOW	NORMAL	NORMAL	HIGH	SHORT	NORMAL	CONDENSER — Partially Restricted
LOW	LOW	HIGH	NORMAL TO LOW	LONG	SHORT	FIXED ORIFICE TUBE — O-Ring Leaking
LOW	LOW	HIGH	RAPID	VERY SHORT	VERY SHORT	CLUTCH CYCLING SWITCH — Low Setting
LOW	LOW	HIGH	LOW	LONG	SHORT	COMPRESSOR — Low Performance
LOW	LOW	HIGH	CONTINUOUS RUN			A/C SUCTION LINE — Restricted
NORMAL TO LOW	NORMAL TO LOW	NORMAL TO LOW	HIGH	VERY SHORT	LONG	CONDENSER — Restricted
LOW	LOW	NORMAL	HIGH	VERY SHORT	LONG	EVAPORATOR CORE — Restricted
ERRATIC OPERATON OR COMPRESSOR NOT RUNNING			—	—	—	CLUTCH CYCLING SWITCH — Dirty Contacts or Sticking Open.

ADDITIONAL POSSIBLE CAUSE COMPONENTS ASSOCIATED WITH INADEQUATE COMPRESSION OPERATION

- COMPRESSOR CLUTCH Slipping • LOOSE DRIVE BELT
- CLUTCH COIL With Open, Blown Circuit, Blown Fuse or Loose Mounting
- CONTROL ASSEMBLY SWITCH — Dirty Contacts or Sticking Open

ADDITIONAL POSSIBLE CAUSE COMPONENTS ASSOCIATED WITH A DAMAGED COMPRESSOR

- CLUTCH CYCLING PRESSURE SWITCH — Sticking Closed or Compressor Clutch Seized
- SUCTION ACCUMULATOR/DRIER — Refrigerant Oil Bleed Hole Plugged
- REFRIGERANT LEAKS

(a) Compressor may make noise on initial run. This is a slugging condition caused by excessive liquid refrigerant.
(b) Compressor clutch may not cycle during very high humidity condition.

Fig. 66 Refrigeration system analysis chart. 1980 Cycling clutch systems w/fixed orifice tube

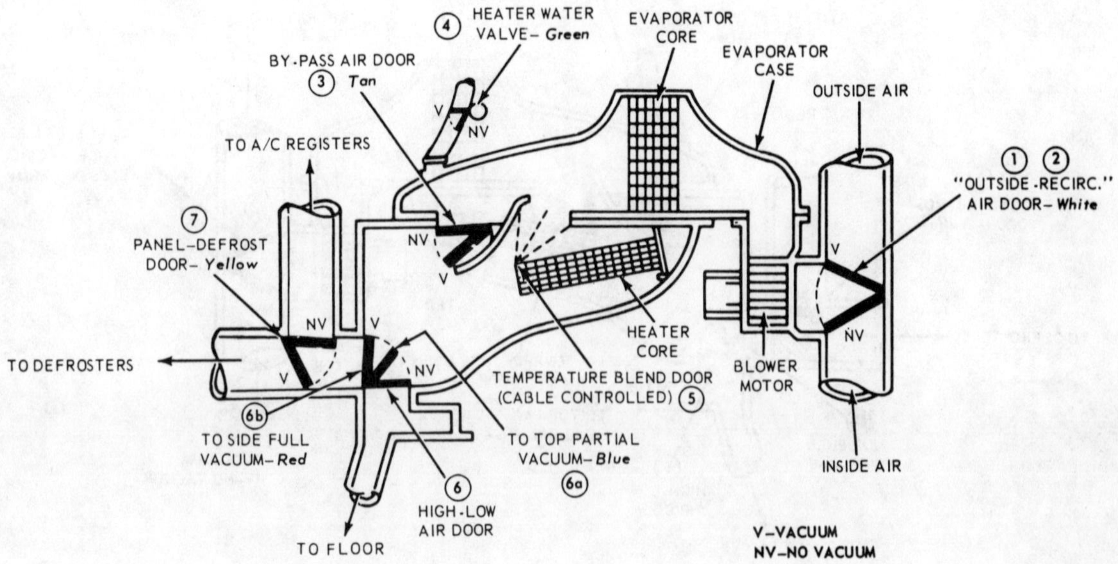

MANUAL A/C-HEATER VACUUM MOTOR TEST CHART

Functional Control Lever Position	Temperature Control Lever Position	Vacuum Motors Applied With Vacuum						Air Flow
OFF	ANY POSITION	1–2	–	4	6a	6b	–	NONE
VENT/ECONOMY	ANY POSITION	–	3	4	–	–	7	OUTSIDE AIR THRU REGISTERS
A/C	COOL–1	1–2	3	4	–	–	7	RECIRC AIR THRU REGISTERS
	COOL–2	1–2	3	–	–	–	7	
	COOL–3, TEMP–4, WARM–5	–	3	–	–	–	7	OUTSIDE AIR THRU REGISTERS
HI–LO	COOL–1			4	6a	–	7	RECIRC AIR THRU REGISTERS AND TO FLOOR
	COOL–2			–	6a	–	7	
	COOL–3, TEMP–4 WARM–5	–		–	6a	–	7	OUTSIDE AIR THRU REGISTERS AND TO FLOOR
HEAT	COOL–1	–	–	4	6a	6b	–	OUTSIDE AIR TO FLOOR
	COOL–2 AND 3, TEMP–4, WARM–5	–	–	–	6a	6b	–	
HEAT ▽ DEFROST	COOL–1	–	–	4	6a	–	–	OUTSIDE AIR TO FLOOR AND AND THRU DEFROSTER OUTLETS
	COOL–2 AND 3, TEMP–4, WARM–5	–	–	–	6a	–	–	
DEFROST	COOL–1	–	–	4	–	–	–	OUTSIDE AIR THRU DEFROSTER OUTLETS
	COOL–2 AND 3, TEMP–4, WARM–5	–	–	–	–	–	–	
VACUUM LINE COLOR CODE		WHITE	TAN	GREEN	BLUE (TOP)	RED (SIDE)	YELLOW	

Fig. 67 Airflow schematic & vacuum control test chart. 1974-76 Cougar, Elite, Montego, Thunderbird & Torino w/manual control

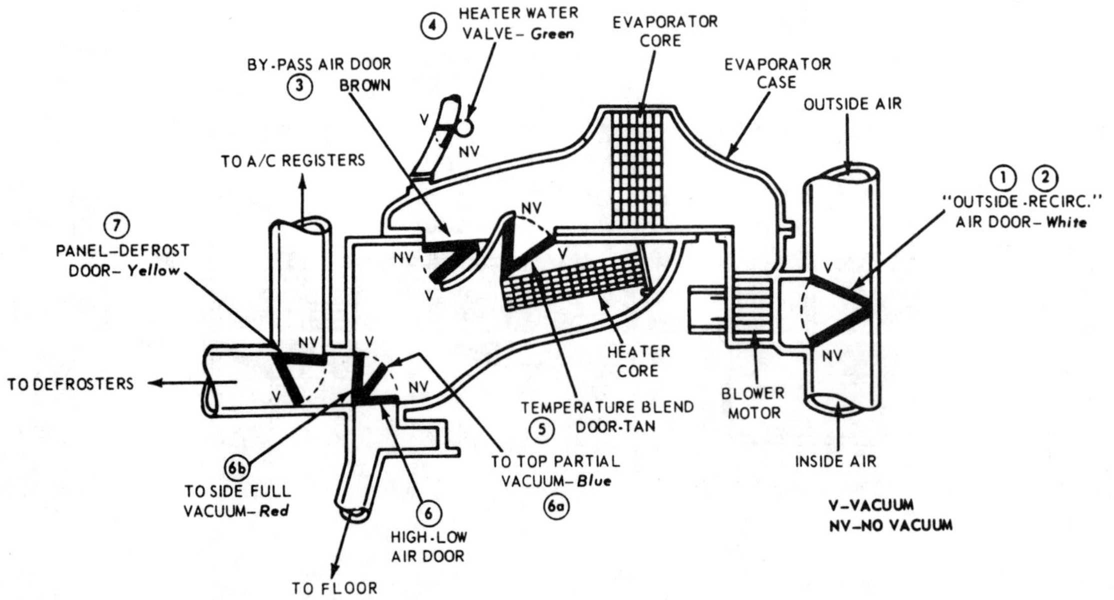

THUNDERBIRD AND MARK IV AUTOMATIC TEMPERATURE CONTROL VACUUM MOTOR TEST CHART										
Functional Control Lever Position		A.T.C. Requirement	Vacuum Motors Applied With Vacuum							Air Flow
AUTO	HIGH	A/C*	1–2	3	4	5	–	–	7	COLD AIR THROUGH REGISTERS
AUTO	HIGH	HEAT**	–	–	–	5†	6a	6b	–	HEATED AIR TO FLOOR
AUTO	LOW	A/C*	–	3	4	5†	–	–	7	COOL AIR THROUGH REGISTERS
AUTO	LOW	HEAT**	–	–	–	5†	6a	6b	–	HEATED AIR TO FLOOR
VENT		FRESH AIR	–	3	4	5	–	–	7	OUTSIDE AIR TEMP THROUGH REGISTERS
OFF		NO AIR FLOW	1–2	–	–	–	–	–	–	————
AUTO	DEFOG	HEAT	–	–	–	5†	6a	–	–	HEATED AIR TO FLOOR AND DEFROSTERS
DEFROST		HEAT	–	–	–	–	–	–	–	MAXIMUM HEATED AIR TO DEFROSTERS
VACUUM LINE COLOR CODE			WHITE	BROWN	GREEN	TAN	BLUE (TOP)	RED (SIDE)	YELLOW	

*Temperature Lever Set for Lower Than In-Car Temperature
**Temperature Lever Set for Higher Than In-Car Temperature
†Vacuum Modulates Between NO Vacuum and Full Vacuum

Fig. 68 Airflow schematic & vacuum control test chart. 1974 Mark IV & Thunderbird w/A.T.C.

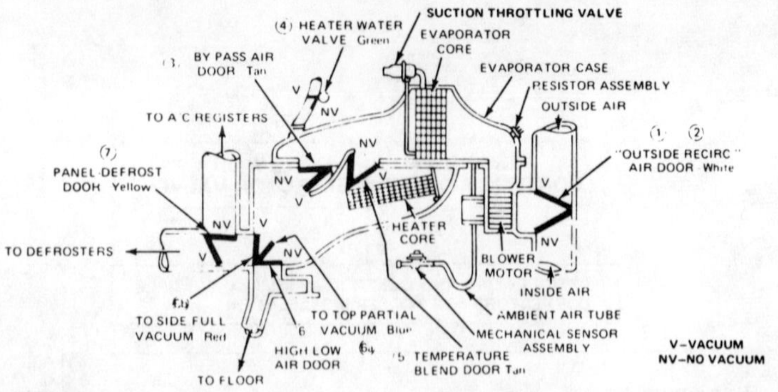

Functional Control Lever Position	A.T.C. Requirement	Vacuum Motors Applied with Vacuum						
OFF	A/C†*	1-2	3	4	5+	6a	6b	–
	HEAT**	1-2	–	–	5+	6a	6b	–
VENT/ECONOMY	A/C†	–	3	4	5+	–	–	7
	HEAT**	–	–	–	5+	–	–	7
AUTO	A/C*	1-2	3	4	5+	–	–	7
	HEAT**	–	–	–	5+	6a	6b	–
DEFOG	A/C*	–	3	4	5+	6a	–	–
	HEAT**	–	–	–	5+	6a	–	–
DEFROST	A/C*	–	3	4	5+	–	–	–
	HEAT**	–	–	–	5+	–	–	–
Vacuum Line Color Code		White	Tan 2 Stripes	Green	Tan	Blue (Top)	Red (Side)	Yellow

* A/C – with Temperature Lever Setting lower than In-Car Temperature
** HEAT – with Temperature Lever Setting higher than In-Car Temperature
‡ Modulated Vacuum somewhere between No Vacuum and Full Vacuum
– NO Vacuum
† A/C Clutch not engaged

Fig. 69 Airflow schematic & vacuum control test chart. 1974-76 Cougar, Elite, Montego, Torino, 1975-76 Mark IV & Thunderbird w/A.T.C.

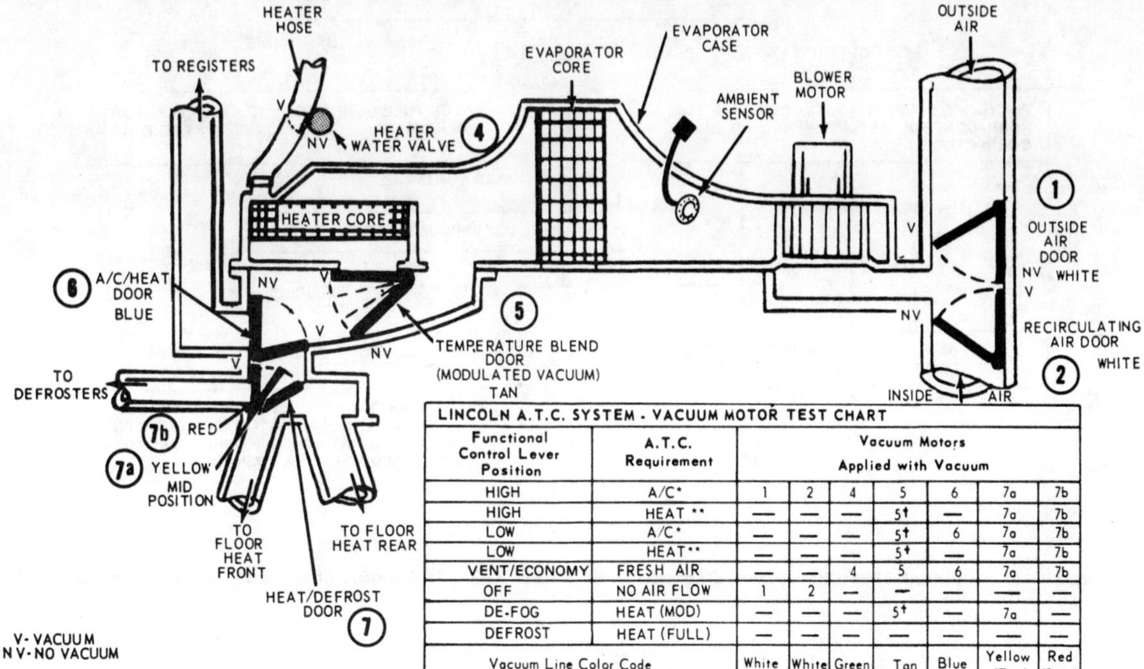

LINCOLN A.T.C. SYSTEM - VACUUM MOTOR TEST CHART

Functional Control Lever Position	A.T.C. Requirement	Vacuum Motors Applied with Vacuum						
HIGH	A/C*	1	2	4	5	6	7a	7b
HIGH	HEAT **	—	—	—	5†	—	7a	7b
LOW	A/C*	—	—	—	5†	6	7a	7b
LOW	HEAT**	—	—	—	5†	—	7a	7b
VENT/ECONOMY	FRESH AIR	—	—	4	5	6	7a	7b
OFF	NO AIR FLOW	1	2	—	—	—	—	—
DE-FOG	HEAT (MOD)	—	—	—	5†	—	7a	—
DEFROST	HEAT (FULL)	—	—	—	—	—	—	—
Vacuum Line Color Code		White	White	Green	Tan	Blue	Yellow (Top)	Red (Side)

*TEMPERATURE LEVER SET FOR LOWER THAN IN-CAR TEMPERATURE
** TEMPERATURE LEVER SET FOR HIGHER THAN IN CAR TEMPERATURE
†VACUUM MODULATES BETWEEN NO VACUUM AND FULL VACUUM

Fig. 70 Airflow schematic & vacuum control test chart. 1974-77 Lincoln Continental w/A.T.C.

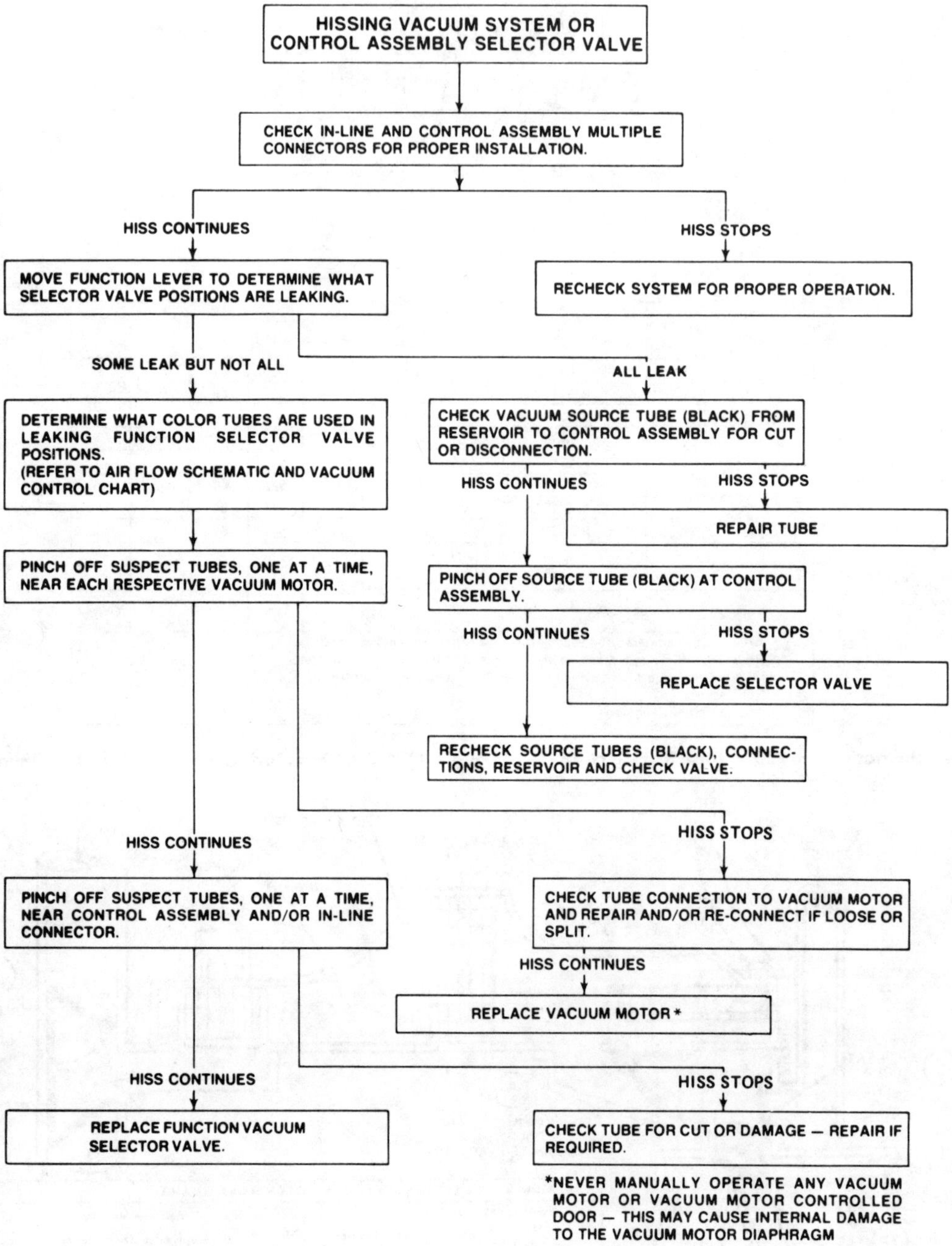

Fig. 71 Manual control vacuum system diagnosis chart. Except 1974-76 Cougar, Elite, Montego, Thunderbird & Torino

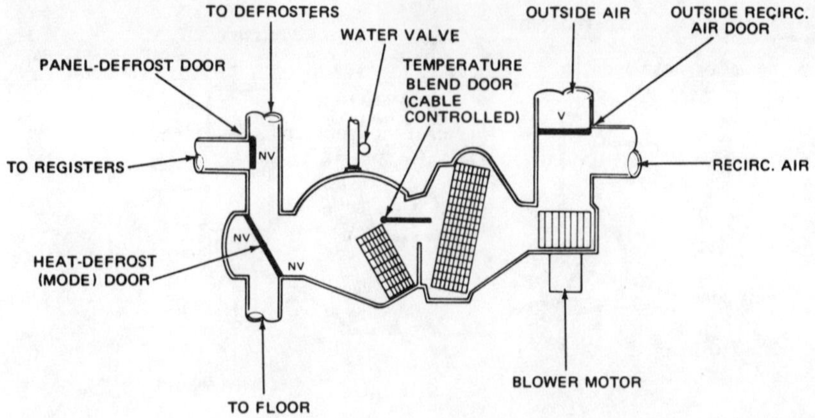

A/C-HEATER CONTROL SETTING

COMPONENT			HOSE COLOR	MODE LEVER POSITION: A/C MAX	A/C FRESH	OFF	VENT•• HI/LO	HEAT	–	DEFROST
VACUUM MOTORS	HEAT DEFROST DOOR @ (MODE)	PARTIAL	BLUE (TOP)	NV	NV	NV	V	V	V	NV
		FULL	RED (SIDE)	NV	NV	NV	NV	V	NV	NV
	A/C-DEFROST DOOR		YELLOW	V	V	NV	V	NV	NV	NV
	OUTSIDE-RECIRC DOOR		WHITE	V	NV	V	NV	NV	NV	NV
	WATER VALVE		BROWN	V	V	NV	NV	NV	NV	NV
BLOWER MOTOR				ON $L-M_1$ M_2-H	ON $L-M_1$ M_2-H OFF – RAM AIR	OFF *	ON – $L-M_1-M_2-H$ OFF – RAM AIR			
CLUTCH #			ABOVE 40° F	ON		OFF/ON				
			BELOW 40° F	OFF						
WATER VALVE VACUUM SWITCH △			COOL	OPEN		SEALED				
			MOD	SEALED						
			WARM							
TEMPERATURE DOOR CABLE CONTROLLED			COOL	ALL COLD AIR BY-PASSES HEATER CORE			OUTSIDE AIR BY-PASSES HEATER CORE			
			MOD	COLD AIR PASSES THROUGH AND AROUND HEATER CORE			OUTSIDE AIR PASSES THROUGH AND AROUND HEATER CORE – THEN MIXED			
			WARM	ALL COLD AIR PASSES THROUGH HEATER CORE			OUTSIDE AIR PASSES THROUGH HEATER CORE			

@ – 3 POSITION MOTOR
* – ON – RECIRC DEFROST AIR
\# – 40° F AMBIENT TEMPERATURE SWITCH
△ – CONTROLLED BY TEMPERATURE LEVER
V – VACUUM
NV – NO VACUUM
L – LOW
M_1 – MEDIUM LOW
M_2 – MEDIUM HIGH
H – HIGH
•• – ECONOMY FOR 1976

Fig. 72 Airflow schematic & vacuum control test chart. 1974-76 Bobcat & Pinto

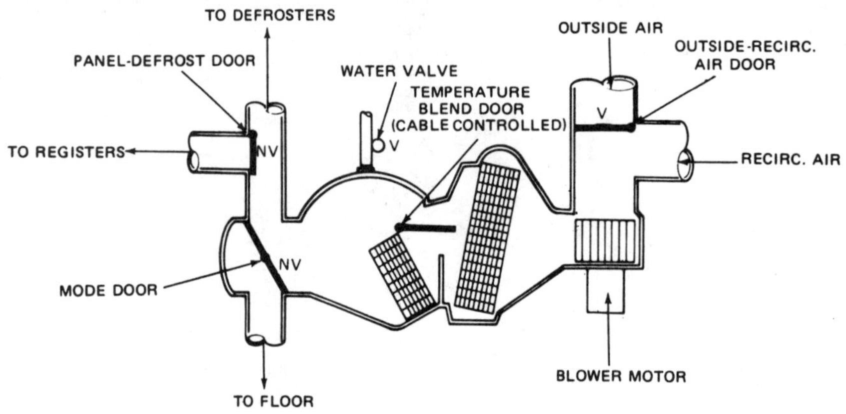

MODE LEVER POSITION

COMPONENT			HOSE COLOR	OFF	VENT••	A/C	HI/LO	HEAT	∇	DEFROST
VACUUM MOTORS	HEAT-DEFROST DOOR @	PARTIAL	Blue (Top)	V	NV	NV	V	V	V	NV
		FULL	Red (Side)	V	NV	NV	NV	V	NV	NV
	A/C-DEFROST DOOR		Yellow	NV	V	V	V	NV	NV	NV
	OUTSIDE-RECIRC DOOR		White	V Open To Recirc Air	NV Open To Outside Air	①	NV Open To Outside Air	NV Open To Outside Air	NV Open To Outside Air	NV Open To Outside Air
	WATER VALVE		Green	V Closed	V Closed	①	①	①	①	①
BLOWER MOTOR				OFF	ON Low – Medium 1 – Medium 2 – High					
CLUTCH *			Above 40° F	OFF		ON	OFF/ON			
			Below 40° F	OFF						

TEMPERATURE LEVER POSITION

COMPONENT		HOSE COLOR 3 Port Connector	TEMPERATURE COOL ⟷ WARM			
VACUUM MOTORS	Water Valve ①	Turquoise	V Closed	V Closed	NV Open	NV Open
	Recirc. Outside ①	Purple	V Open To Recirc Air	NV Open To Outside Air	NV Open To Outside Air	NV Open To Outside Air
TEMPERATURE BLEND DOOR (CABLE CONTROLLED)			Dwell Closed – All Air By-Passes Heater Core		BLEND Combination of Air Which By-Passes Heater Core and that Which Has Passed Thru Heater Core	Open All Air Passes Thru Heater Core

@ – 3 Position Motor
① – Dependent on Temperature Lever
* – 40° F Ambient Temperature Switch
•• – ECONOMY for 1976

RECIRC – Inside Air
V – Vacuum
NV – No Vacuum

Fig. 73 Airflow schematic & vacuum control test chart. 1974-76 Mustang

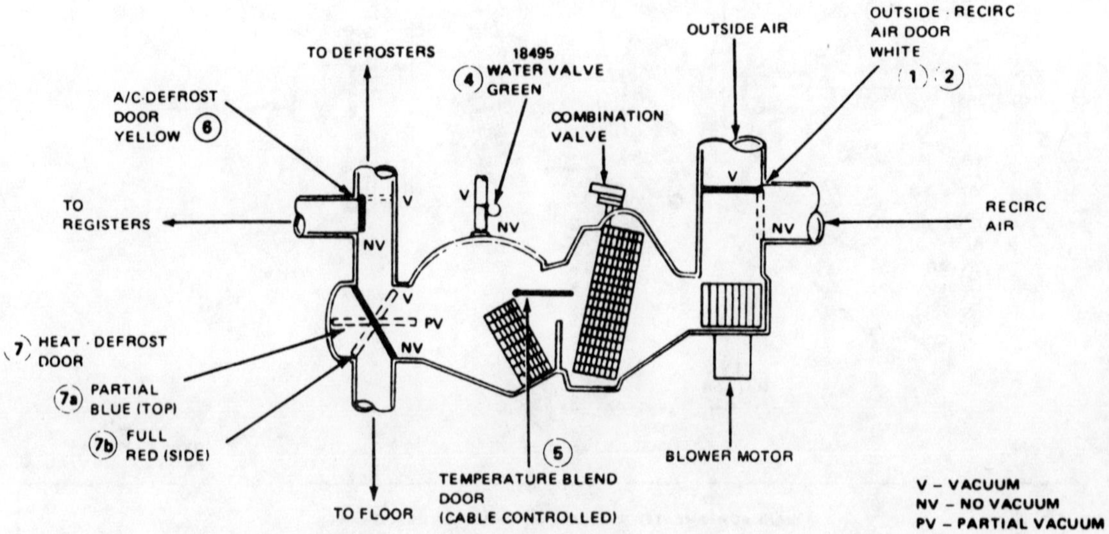

1977 PINTO AND BOBCAT
MANUAL A/C-HEATER SYSTEM VACUUM MOTOR TEST CHART

FUNCTION CONTROL LEVER POSITION		TEMPERATURE CONTROL LEVER POSITION	VACUUM MOTORS APPLIED WITH VACUUM				
			OUTSIDE-RECIRC.	WATER VALVE **	A/C DEFROST	HEAT · DEFROST PARTIAL	HEAT · DEFROST FULL
A/C	MAX *	Cool · 1	1 · 2	4	6	–	–
		Cool · 2	1 · 2	–	6	–	–
		Cool · 3 Temp · 4 Warm · 5	1 · 2	–	6	–	–
	FRESH *	Cool · 1	–	4	6	–	–
		Cool · 2	–	–	6	–	–
		Cool · 3 Temp · 4 Warm · 5	–	–	6	–	–
OFF		Any Position	1 · 2	–	–	–	–
ECONOMY VENT		Any Position	–	–	6	7a	–
HEAT		Cool · 1	–	–	–	7a	7b
		Cool · 2	–	–	–	7a	7b
		Cool · 3 Temp · 4 Warm · 5	–	–	–	7a	7b
– (DEFOG) *		Cool · 1	–	–	–	7a	–
		Cool · 2	–	–	–	7a	–
		Cool 3 · Temp · 4 Warm · 5	–	–	–	7a	–
DEFROST *		Cool · 1	–	–	–	–	–
		Cool · 2	–	–	–	–	–
		Cool · 3 Temp · 4 Warm · 5	–	–	–	–	–
VACUUM LINE COLOR CODE			WHITE	GREEN	YELLOW	BLUE (TOP)	RED (SIDE)

* COMPRESSOR CLUTCH ENGAGED WHEN BLOWER IS ON AND AMBIENT CUT-OFF SWITCH IS CLOSED
** CONTROLLED BY VACUUM SWITCH ON CONTROL HEAD VIA TEMP. LEVER
– NO VACUUM

Fig. 74 Airflow schematic & vacuum control test chart. 1977 Bobcat & Pinto

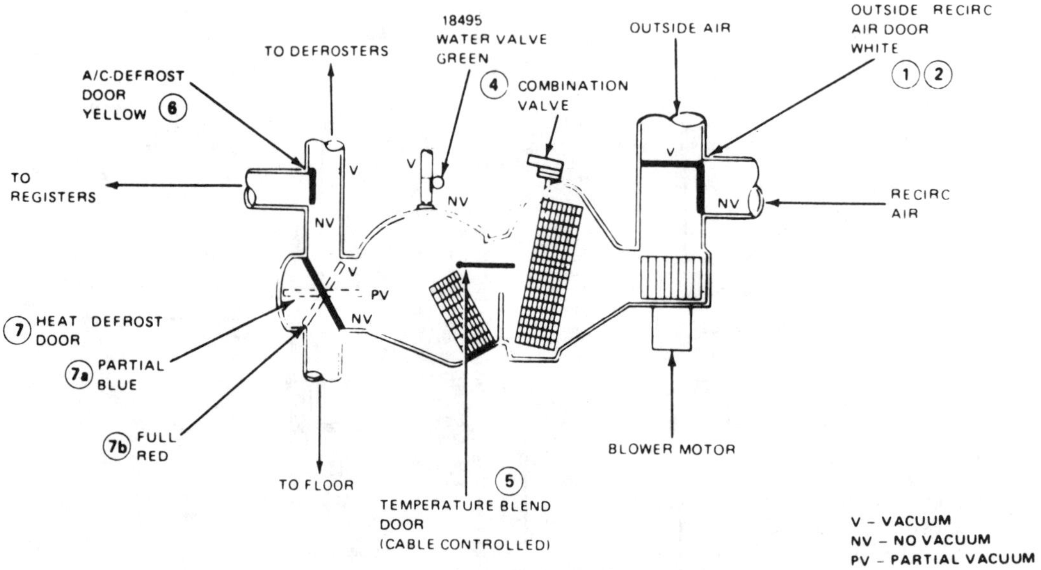

MANUAL A/C-HEATER SYSTEM VACUUM MOTOR TEST CHART

FUNCTION CONTROL LEVER POSITION	TEMPERATURE CONTROL LEVER POSITION	VACUUM MOTORS APPLIED WITH VACUUM				
		OUTSIDE-RECIRC.	WATER VALVE	A/C DEFROST	HEAT-DEFROST PARTIAL	HEAT-DEFROST FULL
OFF	Any Position	1-2	4		7a	7b
ECONOMY VENT	Any Position	–	4	6	–	–
A/C *	Cool-1	1-2	4	6	–	–
	Cool-2	–	4	6	–	–
	Cool-3 Temp-4 Warm-5	–	–	6		–
HI-LO	Cool-1	–	4	6	7a	–
	Cool-2	–	4	6	7a	–
	Cool-3 Temp-4 Warm-5	–	–	6	7a	–
HEAT	Cool-1	–	4	-	7a	7b
	Cool-2	–	4	-	7a	7b
	Cool-3 Temp-4 Warm-5	–	–		7a	7b
▽ **(DEFOG)** *	Cool-1	–	4		7a	
	Cool-2	–	4		7a	
	Cool 3-.Temp-4 Warm-5	–	-		7a	
DEFROST *	Cool-1	–	4			
	Cool-2	–	4			
	Cool-3 Temp-4 Warm-5	–	-			
VACUUM LINE COLOR CODE		WHITE	GREEN	YELLOW	BLUE (TOP)	RED (SIDE)

* COMPRESSOR CLUTCH ENGAGED WHEN AMBIENT CUT-OFF SWITCH IS CLOSED
– NO VACUUM

Fig. 75 Airflow schematic & vacuum control test chart. 1977-78 Mustang, 1978-80 Bobcat, Fairmont, Pinto & Zephyr

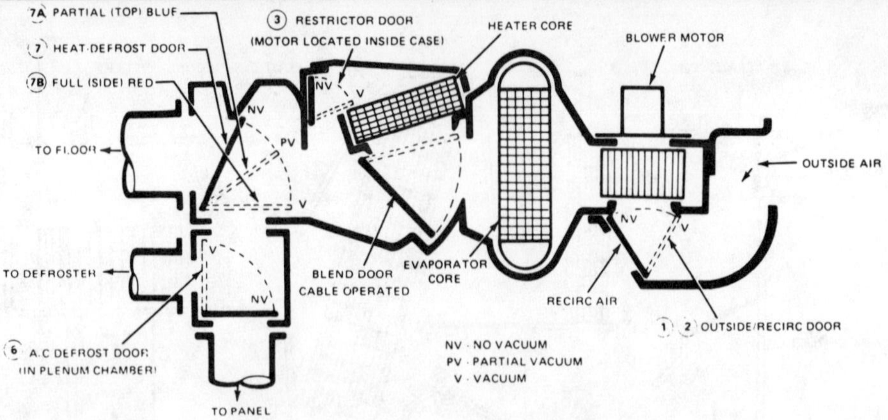

MANUAL A/C-HEATER SYSTEM VACUUM MOTOR TEST CHART

FUNCTION SELECTOR LEVER POSITION		TEMPERATURE SELECTOR LEVER POSITION	VACUUM MOTORS APPLIED WITH VACUUM				
			OUTSIDE RECIRC. DOOR	RESTRICTOR DOOR	HEAT DEFROST DOOR		A/C-DEFROST DOOR
					PARTIAL	FULL	
OFF		ANY POSITION	1-2	—	7A	7B	6
A/C	MAX	RANGE 1	1-2	3	—	—	6
		RANGE 2	1-2	—	—	—	6
	NORM	RANGE 1	—	3	—	—	6
		RANGE 2	—	—	—	—	6
VENT		RANGE 1	—	3	—	—	6
		RANGE 2	—	—	—	—	6
HI-LO		ANY POSITION	—	—	7A	—	6
HEAT		ANY POSITION	—	—	7A	7B	—
*DEFROST		ANY POSITION	—	—	—	—	—
VACUUM HOSE COLOR CODE			White	Green	Blue	Red	Yellow

*Compressor clutch engaged when ambient cut-off switch is closed
— No Vacuum (atmosphere)

Fig. 76 Airflow schematic & vacuum control test chart. 1979-80 Capri & Mustang

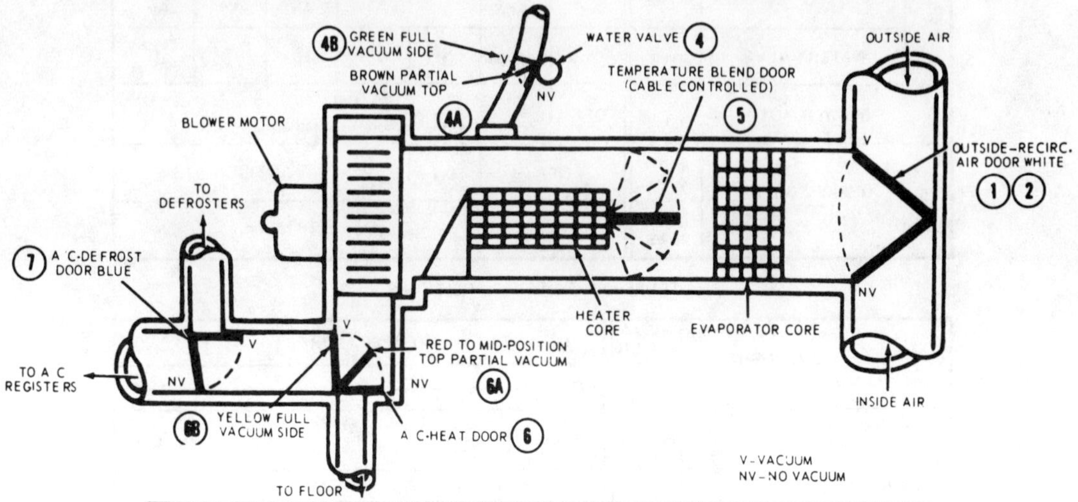

MAVERICK AND COMET A C-HEATER SYSTEM VACUUM MOTOR TEST CHART							
Functional Control Lever Position	Temperature Control Lever Position	Vacuum Motors Applied with Vacuum					
		OUTSIDE PECIRC.	WATER VALVE		A-C-HEAT		A C DEFROST
MAX. A C	COOL	1–2	4a	4b	–	–	7
	WARM	1–2	4a	–	–	–	7
FRESH A C	COOL	–	4a	4b	–	–	7
	WARM	–	4a	–	–	–	7
OFF	WARM	1–2	–	–	–	–	–
HI-HEAT	WARM	–	–	–	6a	–	7
LO-HEAT	WARM	–	–	–	6a	6b	
FOG-DEFROST	WARM	–	–	–	6a	–	–
ICE-DEFROST	WARM	–	–	–	–	–	–
VACUUM LINE COLOR CODE		WHITE	BROWN (TOP)	GREEN (SIDE)	RED (TOP)	YELLOW (SIDE)	BLUE
VACUUM MOTOR CONTROLLED BY LEVER		FUNCTIONAL CONTROL	FUNCTIONAL CONTROL	TEMPERATURE CONTROL	FUNCTIONAL CONTROL	FUNCTIONAL CONTROL	FUNCTIONAL CONTROL

Fig. 77 Airflow schematic & vacuum control test chart. 1974-77 Comet & Maverick

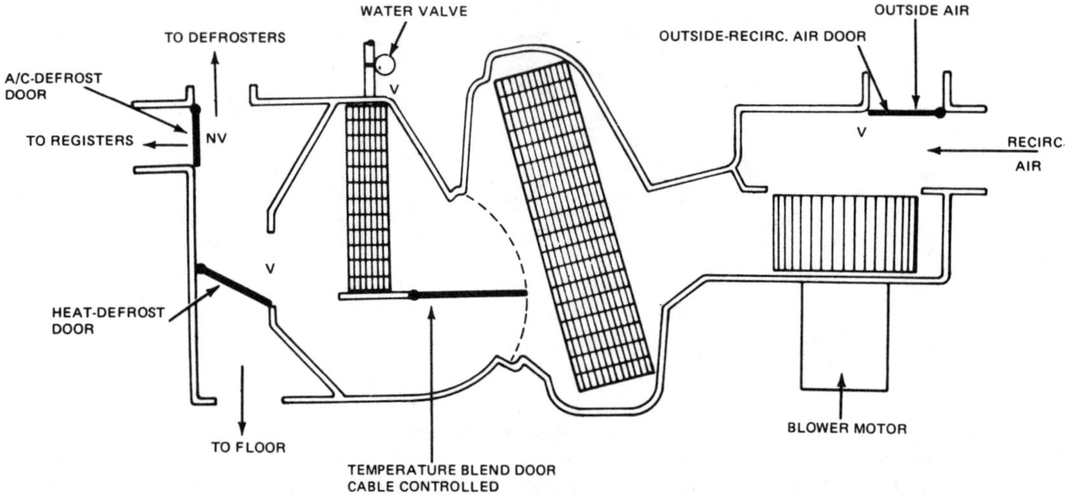

COMPONENT			HOSE COLOR	OFF	VENT/ ECON.	A/C	HI/LO	HEAT	∇	DEFROST
VACUUM MOTORS	HEAT-DEFROST DOOR @	PARTIAL FLOOR	Blue (Top)	V	NV	NV	V	V	V	NV
		FULL FLOOR	Red (Side)	V	NV	NV	NV	V	NV	NV
	A/C-DEFROST DOOR		Yellow	NV	V	V	V	NV	NV	NV
	OUTSIDE-RECIRC DOOR		White	V Open To Recirc Air	NV Open To Outside Air	①	NV Open To Outside Air	NV Open To Outside Air	NV Open To Outside Air	NV Open To Outside Air
	WATER VALVE		Green	V Closed	V Closed	①	①	①	①	①
BLOWER MOTOR				OFF	ON Low — Medium 1 — Medium 2 — High					
COMPRESSOR CLUTCH			Above 40° F	OFF		ON	OFF/ON			
			Below 40° F	OFF						

TEMPERATURE LEVER POSITION

COMPONENT		HOSE COLOR 3 Port Con-nector	TEMPERATURE		
			COOL ⟵⟶ WARM		
VACUUM MOTORS	Water Valve ①	Turquoise	V Closed	NV Open	NV Open
	Recirc. Outside ①	Purple	V Open To Recirc Air	NV Open To Outside Air	NV Open To Outside Air
TEMPERATURE BLEND DOOR (CABLE CONTROLLED)			Dwell Closed — All Air By-Passes Heater Core	BLEND Combination of Air Which By-Passes Heater Core and that Which Has Passed Thru Heater Core	Open All Air Passes Thru Heater Core

@ — 3 Position Motor
① — Dependent on Temperature Lever
* — 40° F Ambient Temperature Switch

RECIRC — Inside Air
V — Vacuum
NV — No Vacuum

Fig. 78 Airflow schematic & vacuum control test chart. 1975-76 Granada & Monarch

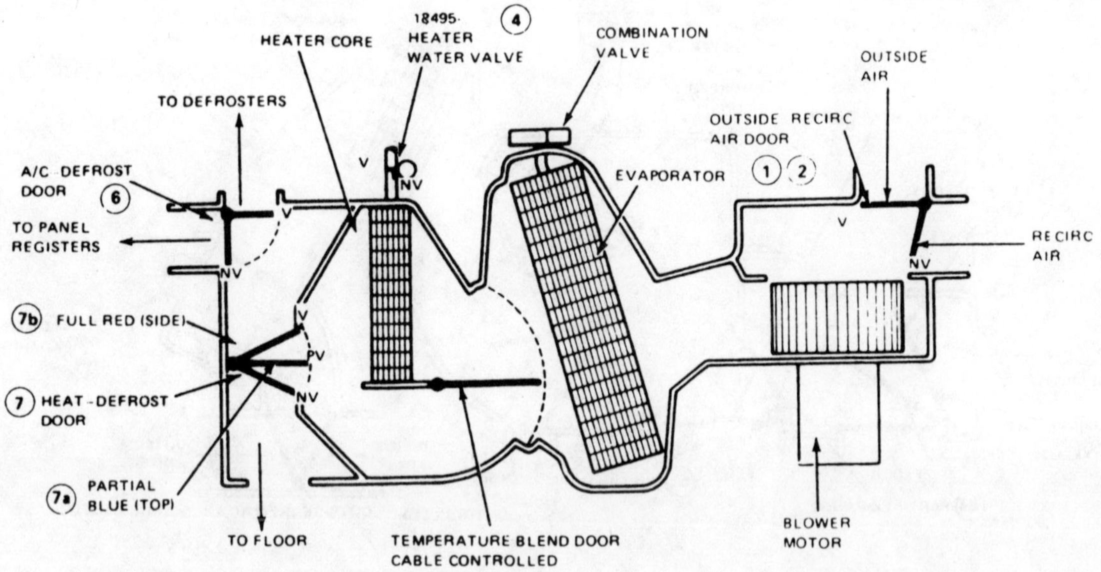

MANUAL A/C-HEATER SYSTEM VACUUM MOTOR TEST CHART

FUNCTION CONTROL LEVER POSITION	TEMPERATURE CONTROL LEVER POSITION	VACUUM MOTORS APPLIED WITH VACUUM				
		OUTSIDE RECIRC	WATER VALVE	A/C DEFROST	HEAT DEFROST PARTIAL	HEAT DEFROST FULL
OFF	Any Position	1 · 2	4		7a	7b
ECONOMY VENT	Any Position	–	4	6		
A/C *	Cool · 1	1 · 2	4	6		
	Cool · 2	1 · 2		6		
	Cool · 3 Temp · 4 Warm · 5	–	–	6		
HI · LO	Cool · 1	–	4	6	7a	
	Cool · 2		-	6	7a	
	Cool · 3 Temp · 4 Warm · 5	–	-	6	7a	
HEAT	Cool · 1	–	4		7a	7b
	Cool · 2		-		7a	7b
	Cool · 3 Temp · 4 Warm · 5	–			7a	7b
∇ **(DEFOG)** *	Cool · 1	-	4		7a	
	Cool · 2	-			7a	
	Cool 3 · Temp · 4 Warm · 5	-			7a	
DEFROST *	Cool · 1	–	4			
	Cool · 2	–				
	Cool · 3 Temp · 4 Warm · 5					
VACUUM LINE COLOR CODE		WHITE	GREEN	YELLOW	BLUE (TOP)	RED (SIDE)

* COMPRESSOR CLUTCH ENGAGED WHEN AMBIENT CUT-OFF SWITCH IS CLOSED
NO VACUUM

Fig. 79 Airflow schematic & vacuum control test chart. 1977-80 Granada & Monarch w/manual control

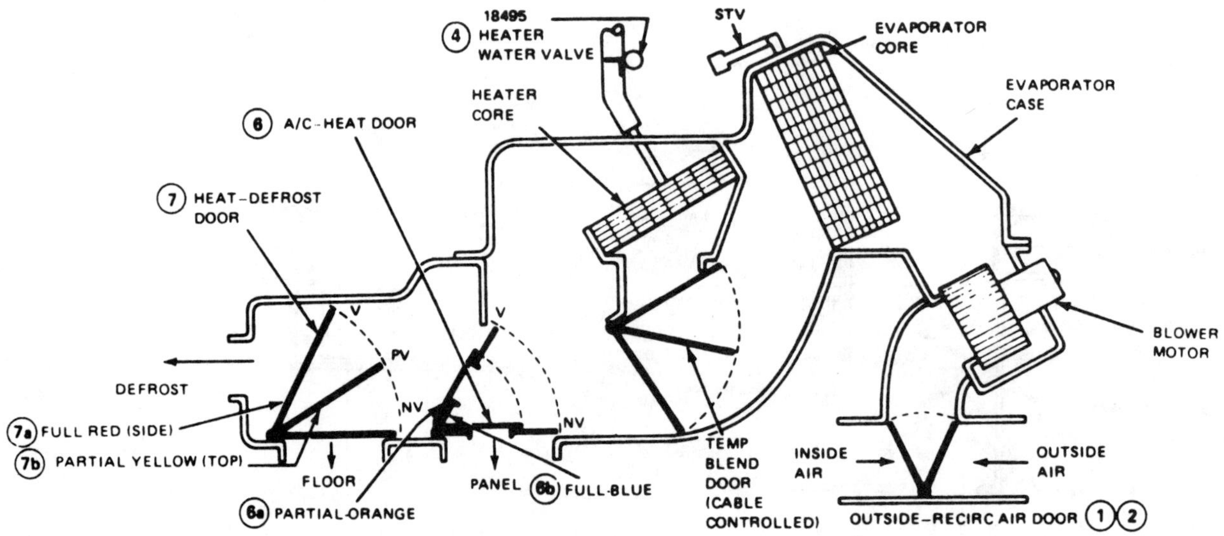

MANUAL A/C-HEATER SYSTEM VACUUM MOTOR TEST CHART							
		VACUUM MOTORS APPLIED WITH VACUUM					
FUNCTION CONTROL LEVER POSITION	TEMPERATURE CONTROL LEVER POSITION	OUTSIDE-RECIRC.	WATER VALVE	A/C-HEAT PARTIAL	A/C-HEAT FULL	HEAT-DEFROST PARTIAL	HEAT-DEFROST FULL
OFF	Any Position	1 · 2	4	–	–	7a	7b
ECONOMY VENT	Any Position	–	4	6a	6b	7a	7b
A/C *	Cool · 1	1 · 2	4	6a	6b	7a	7b
	Cool · 2	1 · 2	–	6a	6b	7a	7b
	Cool · 3 Temp · 4 Warm · 5	–	–	6a	6b	7a	7b
HI · LO	Cool · 1	–	4	6a	–	7a	7b
	Cool · 2	–	–	6a	–	7a	7b
	Cool · 3 Temp · 4 Warm · 5	–	–	6a	–	7a	7b
HEAT	Cool · 1	–	4	–	–	7a	7b
	Cool · 2	–	–	–	–	7a	7b
	Cool · 3 Temp · 4 Warm · 5	–	–	–	–	7a	7b
▽ (DEFOG) *	Cool · 1	–	4	–	–	7a	–
	Cool · 2	–	–	–	–	7a	–
	Cool · 3 Temp · 4 Warm · 5	–	–	–	–	7a	-
DEFROST *	Cool · 1	–	4	–	–	–	·
	Cool · 2	–	--	–	–	–	–
	Cool · 3 Temp · 4 Warm · 5	–	–	–	–	–	
VACUUM LINE COLOR CODE		WHITE	GREEN	ORANGE	BLUE	YELLOW (TOP)	RED (SIDE)

* COMPRESSOR CLUTCH ENGAGED WHEN AMBIENT CUT-OFF SWITCH IS CLOSED.
NO VACUUM

Fig. 80 Airflow schematic & vacuum control test chart. 1977-79 Cougar, LTD II & Thunderbird w/manual control

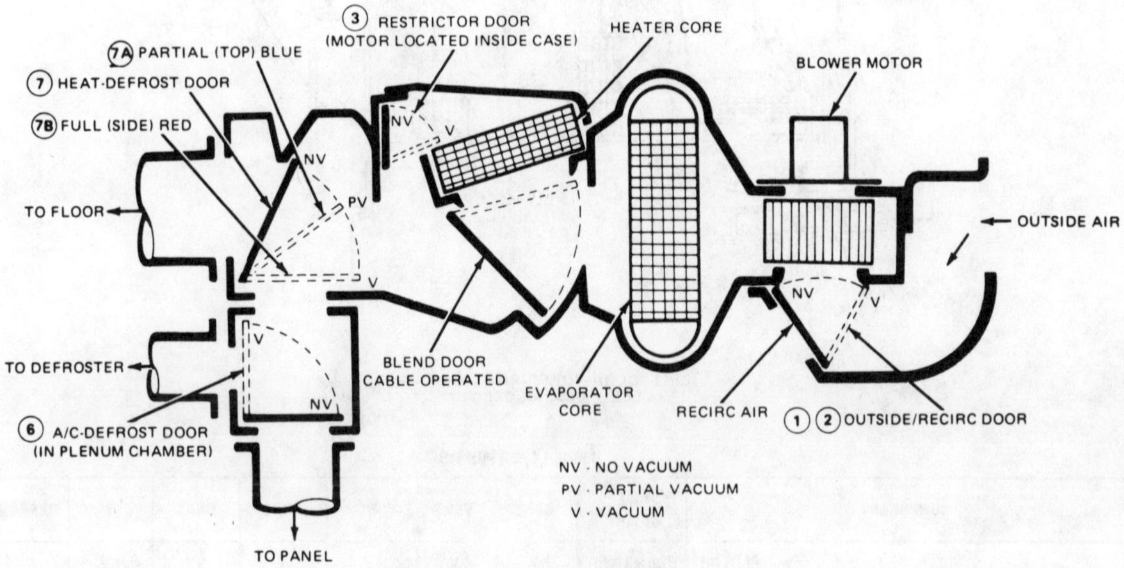

THUNDERBIRD AND COUGAR
MANUAL A/C-HEATER SYSTEM VACUUM MOTOR TEST CHART

FUNCTION SELECTOR LEVER POSITION		TEMPERATURE SELECTOR LEVER POSITION	VACUUM MOTORS APPLIED WITH VACUUM: OUTSIDE RECIRC. DOOR	RESTRICTOR DOOR	HEAT DEFROST DOOR: PARTIAL	HEAT DEFROST DOOR: FULL	A/C-DEFROST DOOR
OFF		ANY POSITION	1-2	—	7A	7B	6
A/C-	MAX	RANGE 1	1-2	3	—	—	6
		RANGE 2	1-2	—	—	—	6
	NORM	RANGE 1	—	3	—	—	6
		RANGE 2	—	—	—	—	6
VENT		RANGE 1	—	3	—	—	6
		RANGE 2	—	—	—	—	6
HEAT	HI-LO	ANY POSITION	—	—	7A	—	6
	FLOOR	ANY POSITION	—	—	7A	7B	—
*DEFROST		ANY POSITION	—	—	—	—	—
VACUUM HOSE COLOR CODE			White	Green	Blue	Red	Yellow

*Compressor clutch engaged when ambient cut-off switch is closed.
— No Vacuum (atmosphere).

Fig. 81 Airflow schematic & vacuum control test chart. 1980 Cougar & Thunderbird w/manual control

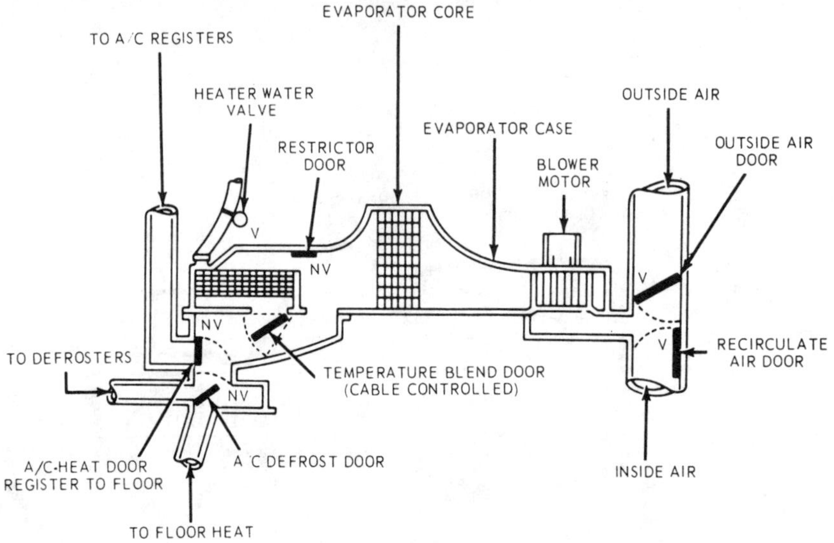

CONTROL IN "OFF" POSITION
(WATER VALVE CLOSED)

V-VACUUM
NV-NO VACUUM

MODE LEVER POSITION

COMPONENT			HOSE COLOR	OFF	VENT	A/C	HI-LO	HEAT	▼	DEFROST
VACUUM MOTORS	A/C HEAT DOOR (MODE) @	PARTIAL	Orange (Top)	NV	V	V	V	NV	NV	NV
		FULL	Blue (Side)	NV	V	V	NV	NV	NV	NV
	HEAT DEFROST DOOR @	PARTIAL	Yellow (Top)	NV	V	V	V	V	V	NV
		FULL	Red (Side)	NV	V	V	V	V	NV	NV
	OUTSIDE AIR DOOR		White	V Closed	NV Open	②	NV Open	NV Open	NV Open	NV Open
	RECIRC DOOR		White	V Open	NV Closed	②	NV Closed	NV Closed	NV Closed	NV Closed
	RESTRICTOR AIR DOOR		Tan	NV Open	V Closed	V Closed	NV Open	NV Open	NV Open	NV Open
	WATER VALVE		Green	V Closed	V Closed	①	①	①	①	①
BLOWER MOTOR				OFF	ON Low – Medium 1 – Medium 2 – High					
CLUTCH*			Above 40° F	OFF		ON	OFF			
			Below 40° F	OFF						

TEMPERATURE LEVER POSITION

COMPONENT		HOSE COLOR 3 Port Connector	TEMPERATURE COOL ←→ WARM			
VACUUM MOTORS	Water Valve ①	Turquoise	V Closed	NV Open	NV Open	NV Open
	Recirc. ②	Purple	V Open	V Open	NV Closed	NV Closed
	Outside ②		V Closed	V Closed	NV Open	NV Open
TEMPERATURE BLEND DOOR (CABLE CONTROLLED)			Dwell closed – All air by-passes heater core		BLEND Combination of air which by-passes heater core and that which has passed through heater core	Open – All air passes through heater core

@ – 3 position motor
① ② – Dependent on temperature lever
* – 40° F ambient temperature switch

RECIRC – Inside air
V – Vacuum
NV – No vacuum.

Fig. 82 Airflow schematic & vacuum control test chart. 1974 Ford & Mercury w/manual control

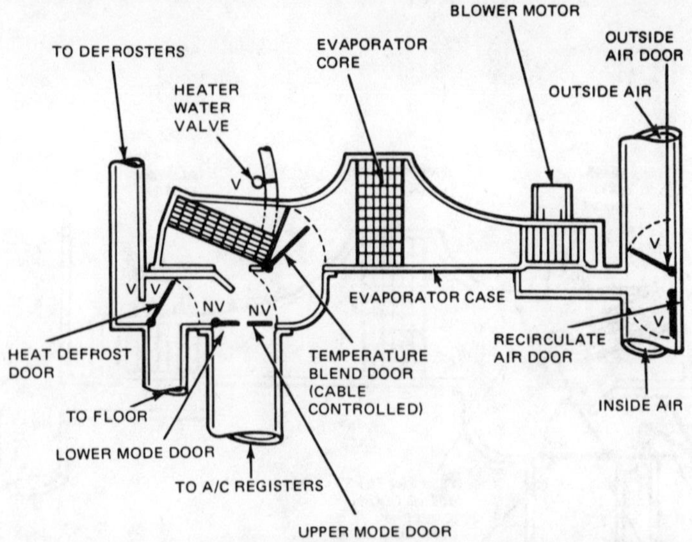

MODE LEVER POSITION

	COMPONENT		HOSE COLOR	OFF	VENT	A/C	HI-LO	HEAT	▼	DEFROST
VACUUM MOTORS	LOWER MODE (HI-LO) DOOR *		Orange	NV	V	V	V	NV	NV	NV
	UPPER MODE DOOR		Blue	NV	V	V	NV	NV	NV	NV
	HEAT DEFROST DOOR @	PARTIAL	Yellow (Top)	V	V	V	V	V	V	NV
		FULL	Red (Side)	V	V	V	V	V	NV	NV
	OUTSIDE AIR DOOR		White	V Closed	NV Open	②	NV Open	NV Open	NV Open	NV Open
	RECIRC DOOR		White	V Open	NV Closed	②	NV Closed	NV Closed	NV Closed	NV Closed
	WATER VALVE		Green	V Closed	V Closed	①	①	①	①	①
BLOWER MOTOR				OFF	ON Low – Medium 1 – Medium 2 – High					
CLUTCH *			Above 45° F	OFF		ON	OFF		ON	
			Below 30° F	OFF						

TEMPERATURE LEVER POSITION

COMPONENT		HOSE COLOR 3 Port Connector	TEMPERATURE COOL ⟵			⟶ WARM
VACUUM MOTORS	Water Valve ①	Turquoise	V Closed	NV Open	NV Open	NV Open
	Recirc. ②	Purple	V Open	V Open	NV Closed	NV Closed
	Outside ②	Purple	V Closed	V Closed	NV Open	NV Open
TEMPERATURE BLEND DOOR (CABLE CONTROLLED)			Dwell Closed – All Air By-Passes Heater Core		BLEND Combination of Air Which By-Passes Htr. Core and that Which Has Passed Thru Heater Core	Open All Air Passes Thru Htr. Core

@ – 3 Position Motor
①② – Dependent on Temp. Lever
* – 30°F Ambient Temperature Switch

RECIRC. – Inside Air
V – Vacuum
NV – No Vacuum

Fig. 83 Airflow schematic & vacuum control test chart. 1975-76 Ford & Mercury w/manual control

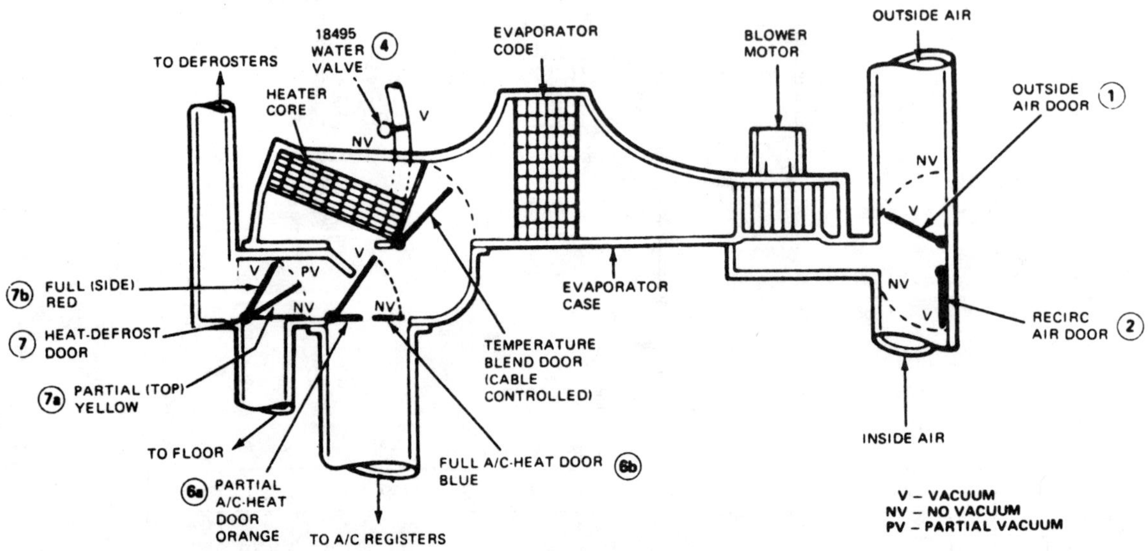

MANUAL A/C-HEATER SYSTEM VACUUM MOTOR TEST CHART

FUNCTION CONTROL LEVER POSITION	TEMPERATURE CONTROL LEVER POSITION	VACUUM MOTORS APPLIED WITH VACUUM						
		OUTSIDE	RECIRC	WATER VALVE	A/C HEAT		HEAT DEFROST	
					PARTIAL	FULL	PARTIAL	FULL
OFF	Any Position	1	2	4			7a	7b
ECONOMY VENT	Any Position	–	–	4	6a	6b	7a	7b
A/C •	Cool 1	1	2	4	6a	6b	7a	7b
	Cool 2	1	2	-	6a	6b	7a	7b
	Cool 3 Temp 4 Warm 5				6a	6b	7a	7b
HI - LO	Cool 1			4	6a		7a	7b
	Cool 2				6a		7a	7b
	Cool 3 Temp 4 Warm 5				6a		7a	7b
HEAT	Cool 1			4			7a	7b
	Cool 2						7a	7b
	Cool 3 Temp 4 Warm 5						7a	7b
▽ (DEFOG) •	Cool 1			4			7a	
	Cool 2						7a	
	Cool 3 Temp 4 Warm 5						7a	
DEFROST •	Cool 1			4				
	Cool 2							
	Cool 3 Temp 4 Warm 5							
VACUUM LINE COLOR CODE		WHITE	WHITE	GREEN	ORANGE	BLUE	YELLOW (TOP)	RED (SIDE)

• COMPRESSOR CLUTCH ENGAGED WHEN AMBIENT CUT-OFF SWITCH IS CLOSED.
—NO VACUUM

Fig. 84 Airflow schematic & vacuum control test chart. 1977-78 Ford & Mercury w/manual control

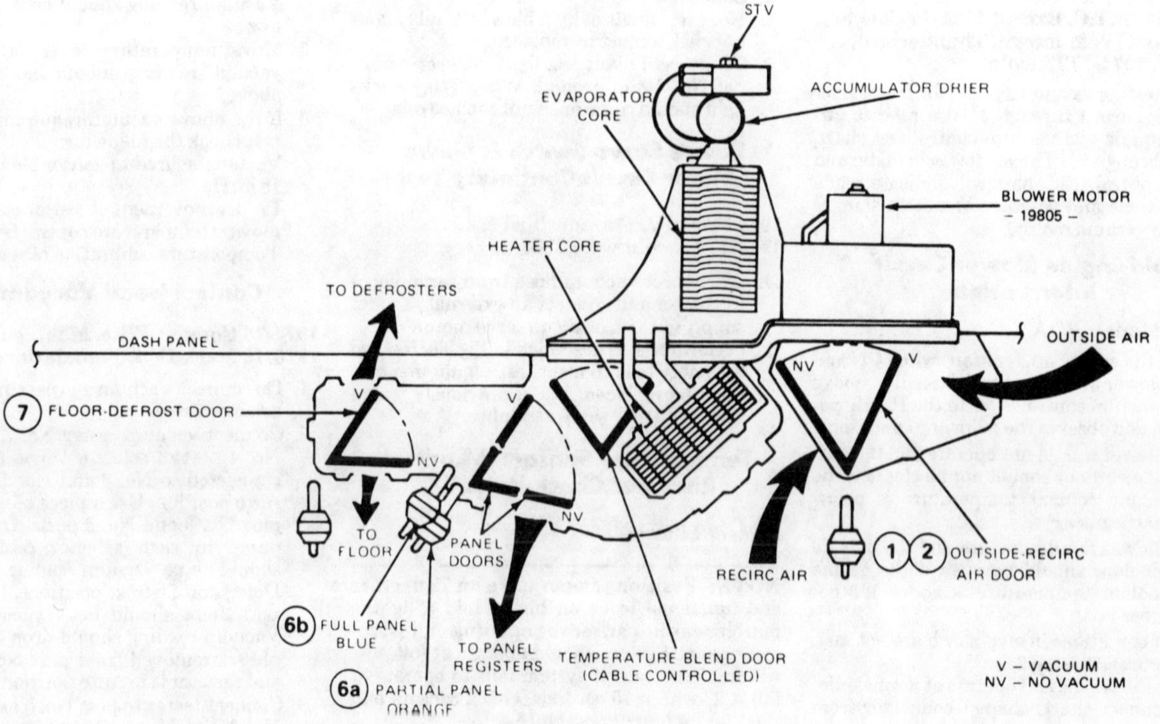

FORD AND MERCURY
MANUAL A/C — HEATER SYSTEM VACUUM MOTOR TEST CHART

FUNCTION CONTROL LEVER POSITION		VACUUM MOTORS APPLIED WITH VACUUM			
		OUTSIDE — RECIRC.	PANEL PARTIAL	PANEL FULL	FLOOR-DEFROST
	OFF	1·2	—	—	7
A/C	MAX *	1·2	6a	6b	7
A/C	NORM *	—	6a	6b	7
	VENT	—	6a	6b	7
HEAT	HI-LO	—	6a	—	7
HEAT	FLOOR	—	—	—	7
	DEFROST *	—	—	—	—
VACUUM LINE COLOR CODE		WHITE	ORANGE	BLUE	YELLOW

— No Vacuum

*Compressor clutch engaged when ambient cut-off switch is closed

Fig. 85 Airflow schematic & vacuum control test chart. 1979-80 Ford & Mercury w/manual control

high. If a vacuum motor is found to be defective, check motor using tester 21-0014 or equivalent. If vacuum motor does not operate properly, check for disconnected, kinked or faulty vacuum hose leading to motor.

Models W/A.T.C. Except 1974-76 Cougar, Elite, Mark IV, Montego, Thunderbird, Torino & 1974-77 Lincoln

When testing vacuum system for leaks, refer to Fig. 86, parts 1 through 4. Also refer to airflow schematic and vacuum control test chart, Figs. 87 through 95. The airflow schematic and vacuum control test chart will indicate mode door positions and whether or not vacuum is applied to vacuum motor.

Cold Engine Blower Circuit Interruption

1978-80 Models W/A.T.C.

1. With the engine off, ignition switch ON and the blower switch in the HI position, move the function control lever to the HEAT position and observe the following conditions:
   a. Blower should not operate and the outside air door should not be closed if the engine coolant temperature is below 100 degrees F.
   b. Blower should operate and the outside air door should be open if the engine coolant temperature is above 120 degrees F.
   c. If conditions in step a or b are not met, proceed to step 2.
2. Start the engine and operate at normal idle.
3. Disconnect the D-shaped connector from the engine temperature sender unit. If the blower does not turn on, check for continuity between Electro-vacuum relay terminals S1 and S2, Fig. 96. Below 100 degrees F the contacts should not have more than one ohm resistance. Replace the EVR if more than one ohm resistance is present. If the EVR is functioning properly, check for a poor ground in the 244 (yellow with white hash) circuit of the wire harness.
4. Connect a jumper wire from suitable ground to the 244 circuit in the D-shaped temperature sender connector. The blower should turn off.

CAUTION: Do not accidentally ground the other terminal (I) of the D connector if the vehicle is equipped with a temperature gauge as damage to the gauge will result.

If the blower does not turn off, check for continuity between EVR terminals C1 and C2. There should be approximately 31 ohms resistance. If there is no continuity, replace the EVR.

5. If steps 1a and 1b do not agree with steps 3 and 4, test the engine temperature sender for improper operation as follows:
   a. The terminal marked R (vehicles equipped with a temperature gauge) or C (vehicles equipped with a temperature warning light) should have an engine ground circuit with engine coolant temperature below 100 degrees F. and an open circuit above 120 degrees F.
   b. If terminal resistance is one ohm or more with coolant temperature below 100 degrees F. or if the circuit does not open at coolant temperatures above 120 degrees F., replace the temperature sender.

High Blower Relay Test

1974-76 Cougar, Elite, Montego, Torino, 1975-76 Mark IV & Thunderbird w/A.T.C.

1. Connect a jumper wire between No. 3 and No. 4 terminals of high blower. If relay does not click it must be replaced.
2. Connect a 12 volt test light between terminal No. 2 and ground. When relay clicks light should come on, if not replace relay.

Power Servo Switch & Blower Motor Circuit Continuity Test

1974 Mark IV, Thunderbird & 1974-77 Lincoln w/A.T.C.

1. Disconnect vacuum hose from servo vacuum motor and connect an external vacuum supply of 13 in. of Hg. to servo motor.
2. Perform test as described in Fig. 89. If servo assembly fails to meet test requirements it must be replaced. Use a previously tested relay with new servo assembly.

Temperature Sensors & Control Rheostat Check W/A.T.C.

1974-77 Lincoln

NOTE: Position temperature on 75 degrees and functional lever on high. Hold a lighted match near in-car sensor opening on instrument panel. System should operate at full A/C within 15 seconds. If system fails to operate at full A/C within 15 seconds, check sensors and rheostat for proper resistance values.

1. Disconnect battery ground cable.
2. Position functional lever on off and temperature lever at 75 degrees.
3. Disconnect black wire connector from A.T.C. box.
4. Connect an ohmmeter between green wire of wiring harness and ground.
5. With vehicle and sensor temperature at approximately 70° F to 80° F, sensor string and rheostat resistance should be 1200 to 1300 ohms. If resistance is not within limits proceed to next step.
6. Position temperature lever at 85 degrees and note ohmmeter reading, then move lever to 65 degrees. The difference in resistance between 65 degrees and 85 degrees temperature lever settings should be 480 to 590 ohms. If resistance is not within limits, the rheostat must be replaced.
7. Disconnect wire connector from ambient sensor and connect an ohmmeter between the two sensor terminals. With ambient temperature at approximately 70° F to 80° F, sensor resistance should be 165 to 185 ohms.
8. Disconnect ambient sensor wire connector. Connect an ohmmeter between blue-yellow wire and ground. Resistance should be 750 to 900 ohms.

Sensor Test

1974-76 Cougar, Elite, Montego, Torino, 1975-76 Mark IV & Thunderbird W/A.T.C.

1. Remove glove box and check ambient inlet hose for proper connection. Ensure ambient hose is not pinched or blocked.
2. Tee a vacuum gauge into servo vacuum line (tan hose).
3. Position control lever in Auto, then bring in-car temperature to approximately 75 degrees.
4. Position temperature lever on 85 degrees and allow time for vacuum to stabilize. Vacuum reading should be 2 in. Hg. or below
5. Move temperature lever to 65 degrees, vacuum reading should be 10 in. Hg. or above.
6. If the above vacuum requirements are not met check the following:
   Vacuum source to sensor should be above 18 in. Hg.
   Proper movement of sensor control arm by moving temperature control lever.
   Temperature calibration of sensor.

Control Head Vacuum Test

1974-76 Cougar, Elite, Montego, Torino, 1975-76 Mark IV & Thunderbird W/A.T.C.

1. Disconnect vacuum harness from vacuum selector valve.
2. Connect vacuum tester No. ZRE-10662 to No. 4 port of selector valve. No. 4 port is connected to No. 3 and No. 5 ports in the Auto position. Use a piece of vacuum hose, plug No. 3 and No. 5 ports. Operate tester pump in each selector position. There should be no vacuum reading in Off, Vent, Defog and Defrost positions. In Auto position there should be a vacuum reading. Vacuum reading should drop to zero when plug is removed from port No. 3 or No. 5 and selector is in Auto position.
3. Connect tester to port No. 5 and plug ports No. 3 and No. 4. Operate tester pump at each selector position. A vacuum reading should be obtained in all selector positions. Vacuum reading should drop to zero when plug is removed from port No. 3 or No. 4 and selector is in Auto position.
4. Connect tester to port No. 3 and plug ports No. 4 and No. 5. A vacuum reading should be obtained in Defog and Auto. With selector in Auto remove plug from port No. 4 or No. 5, vacuum reading should drop to zero. In Off, Vent and Defrost positions there should be no vacuum reading.
5. Connect tester to port No. 9 and plug port No. 3. In Defog position a vacuum reading should be obtained and when plug is removed from No. 3 port vacuum should drop to zero. Perform test with port No. 1 plugged and selector in Off position, then with No. 7 port plugged and selector in Vent position. In each instance a vacuum reading should be obtained and drop to zero when port is unplugged.
6. Perform test to remaining ports referring to Fig. 29. If any test differs from that indicated in Fig. 97, replace the selector valve.

ATC Control Box Test

NOTE: The ATC box should be removed from vehicle and tested on a bench. When testing do not move servo switch by hand, use vacuum only.

1974 Mark IV, Thunderbird & 1974-77 Lincoln Continental

When diagnosing A.T.C. control box problems, refer to Fig. 99 for 1974 Mark IV and Thunderbird test procedures and Fig. 100 for 1974-77 Lincoln Continental test procedures.

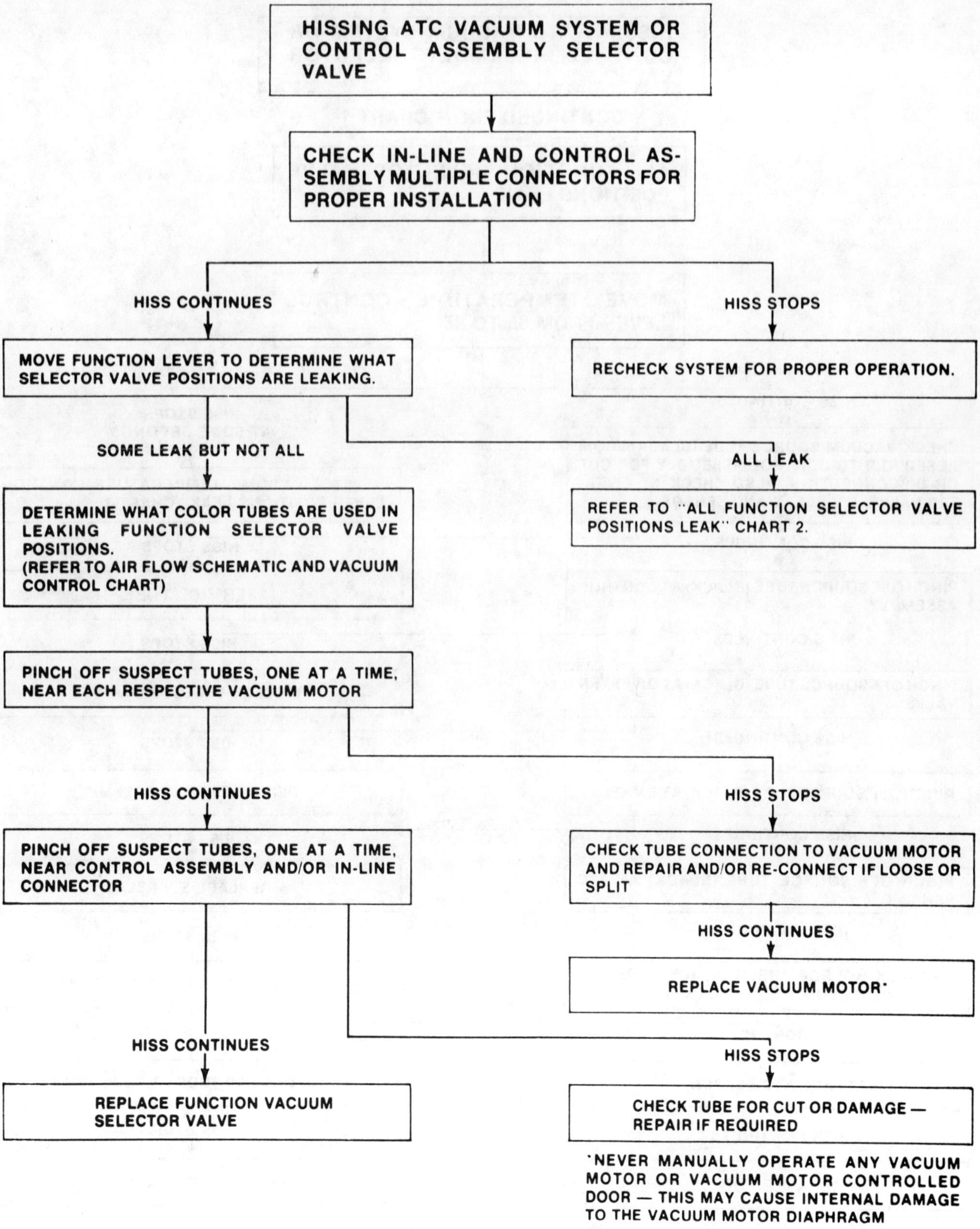

Fig. 86 A.T.C. vacuum system diagnosis chart. Except 1974-76 Cougar, Elite, Mark IV, Montego, Thunderbird, Torino & 1974-77 Lincoln Continental (Part 1 of 4)

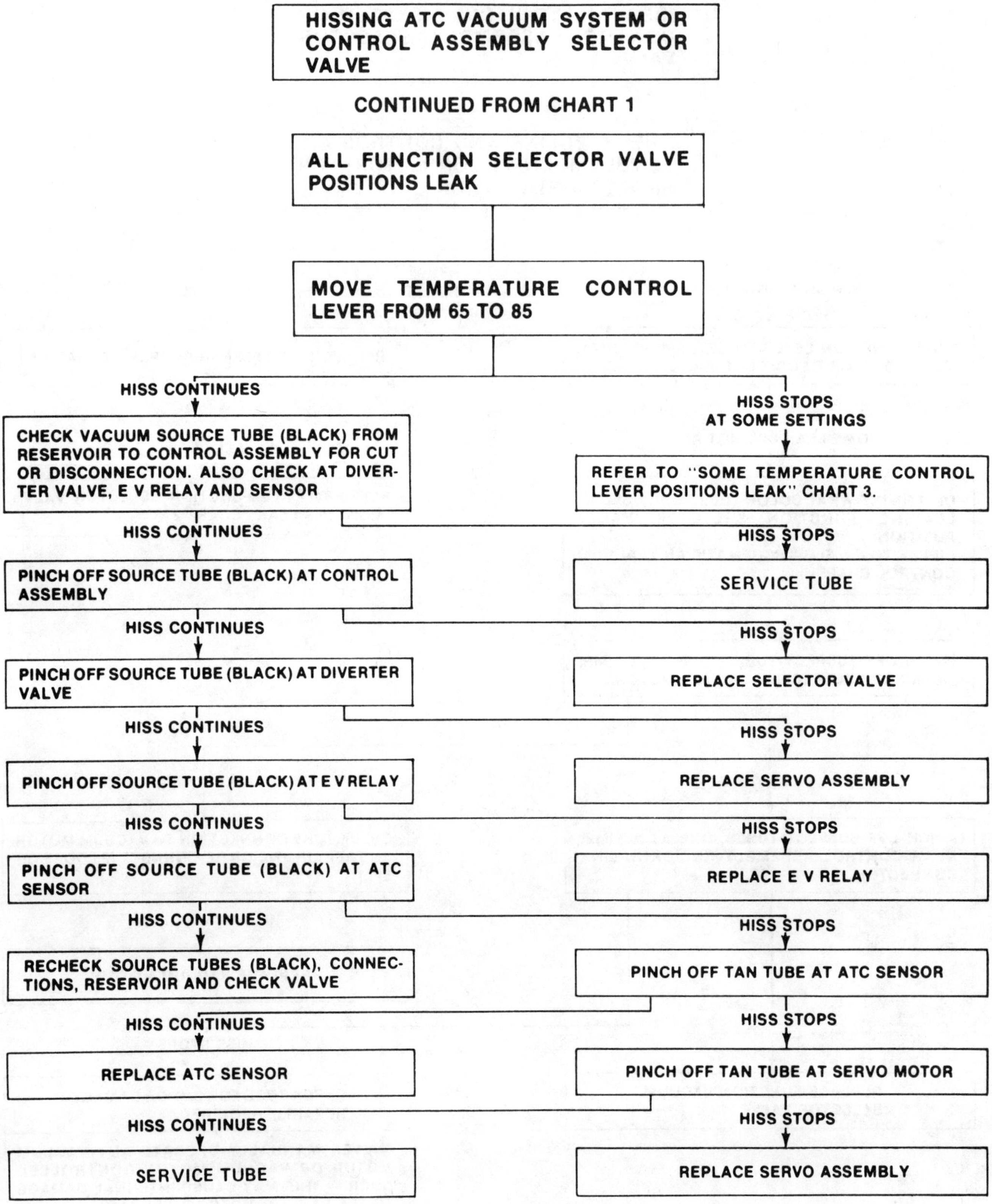

Fig. 86 A.T.C. vacuum system diagnosis chart. Except 1974-76 Cougar, Elite, Mark IV, Montego, Thunderbird, Torino & 1974-77 Lincoln Continental (Part 2 of 4)

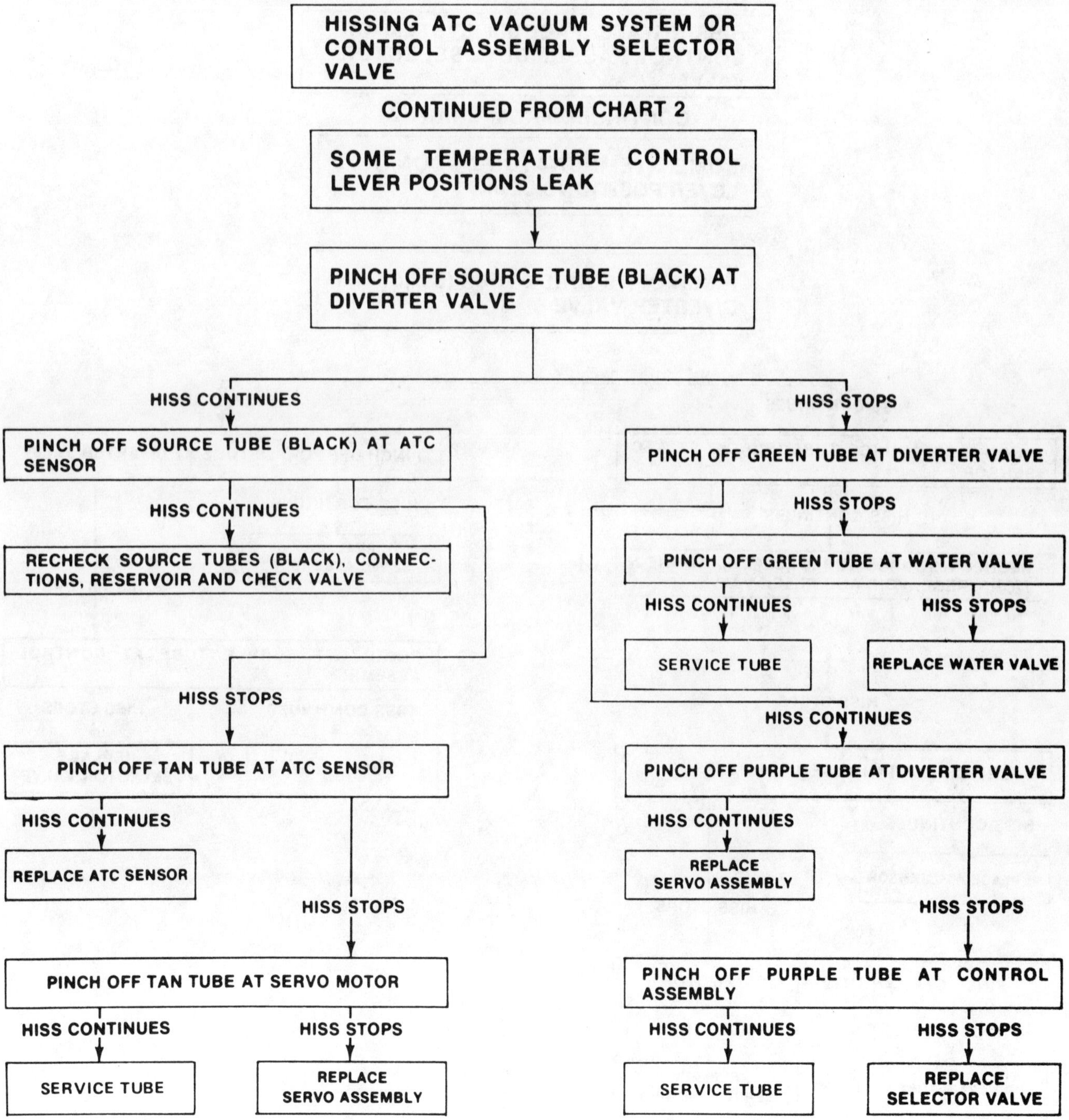

Fig. 86 A.T.C. vacuum system diagnosis chart. Models w/water valve except 1974-76 Cougar, Elite, Mark IV, Montego, Thunderbird, Torino & 1974-77 Lincoln Continental (Part 3 of 4)

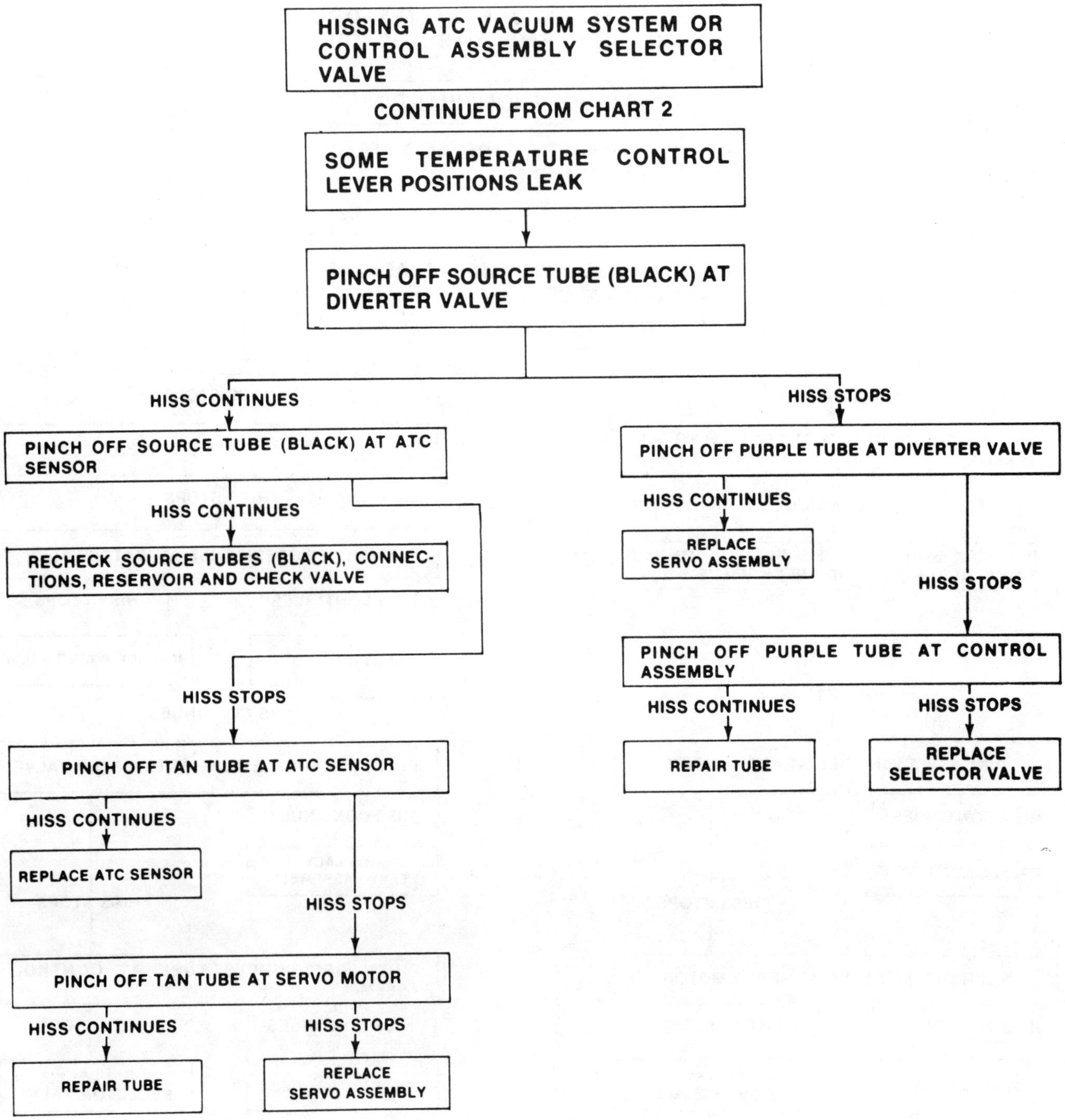

Fig. 86 A.T.C. vacuum system diagnosis chart. Models less water valve except 1974-76 Cougar, Elite, Mark IV, Montego, Thunderbird, Torino & 1974-77 Lincoln Continental (Part 4 of 4)

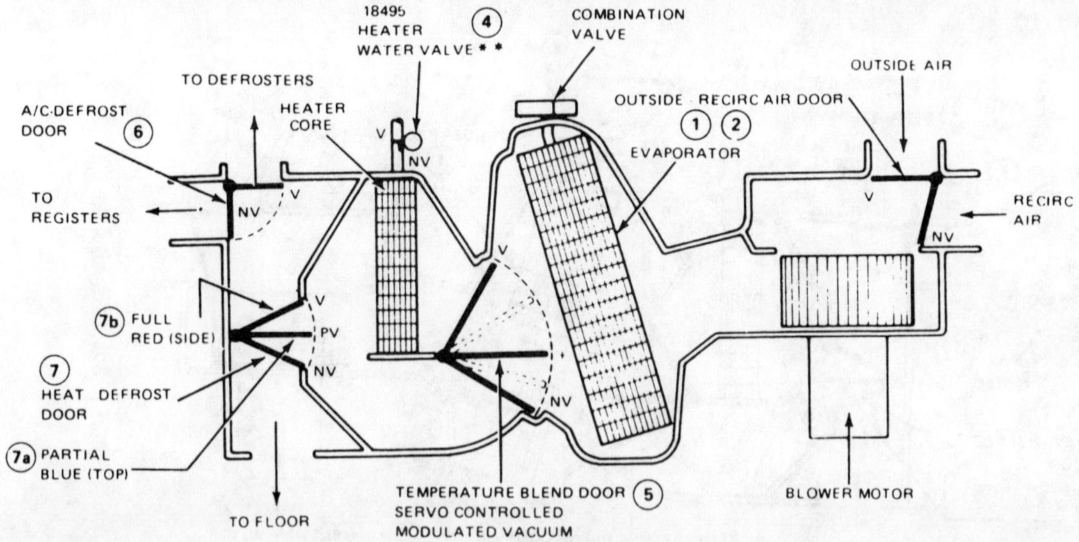

AUTOMATIC TEMPERATURE CONTROL SYSTEM
VACUUM MOTOR TEST CHART

FUNCTION CONTROL LEVER POSITION	ATC REQUIREMENT	OUTSIDE - RECIRC	VACUUM MOTORS APPLIED WITH VACUUM			HEAT - DEFROST	
			WATER VALVE * *	TEMP BLEND +	A/C DEFROST	PARTIAL	FULL
OFF	–	1 · 2	* *	+		7a	7b
ECONOMY VENT	Cool	–	4	5	6		–
	Mid	–	–	+	6	–	
	Warm	–	–		6	–	–
A/C *	Cool	1 · 2 * *	4	5	6		
	Mid	–	–	+	6		
	Warm	–	–	–	6	–	–
HI - LO *	Cool	–	4	5	6	7a	–
	Mid	–	–	+	6	7a	–
	Warm	–	–	–	6	7a	–
HEAT	Cool	–	4	5		7a	7b
	Mid	–	–	+		7a	7b
	Warm	1 · 2 †	–	–		7a	7b
▽ (DEFOG) *	Cool	–	4	5		7a	–
	Mid	–	–	+	–	7a	–
	Warm	–	–	–		7a	
DEFROST *	Cool	–	4	5	–	–	
	Mid	–	–	+	–		
	Warm	–	–	–	–		
VACUUM LINE COLOR CODE		WHITE	GREEN	TAN	YELLOW	BLUE (TOP)	RED (SIDE)

* COMPRESSOR CLUTCH ENGAGED WHEN AMBIENT CUT-OFF SWITCH IS CLOSED
† DELAYED BY ENGINE TEMP. SENDER AND E.V. RELAY UNTIL 130°F ENGINE COOLANT TEMP
* * DEPENDENT ON HEATING/COOLING REQUIREMENT OF SERVO DIVERTER VALVE
\+ MODULATED VACUUM
– NO VACUUM

Fig. 87 Airflow schematic & vacuum control test chart. 1977 Versailles, 1977-80 Granada & Monarch w/A.T.C.

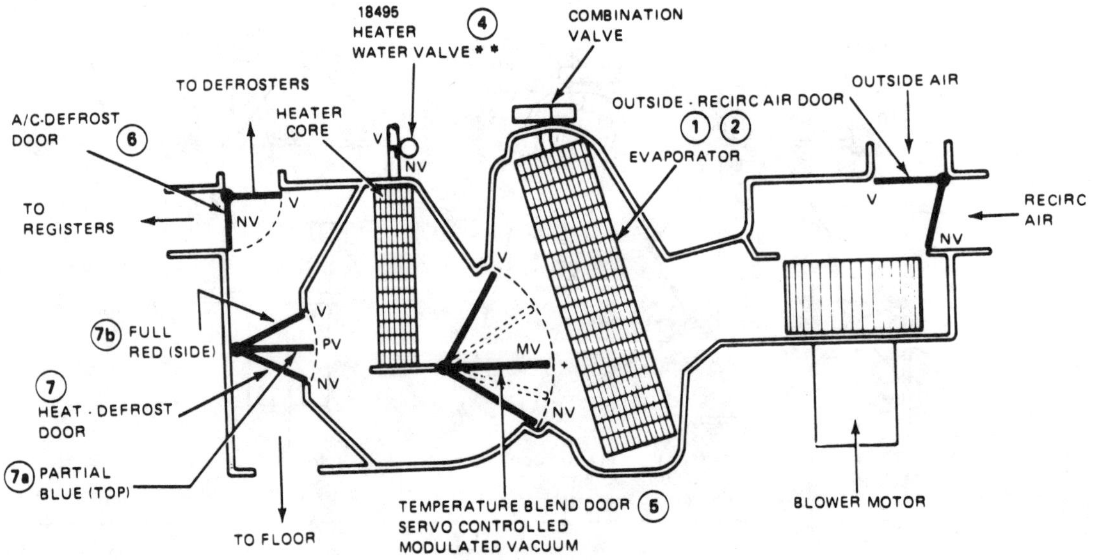

AUTOMATIC TEMPERATURE CONTROL SYSTEM
VACUUM MOTOR TEST CHART

FUNCTION CONTROL LEVER POSITION	ATC REQUIREMENT	OUTSIDE–RECIRC	VACUUM MOTORS APPLIED WITH VACUUM				
			WATER VALVE * *	TEMP BLEND +	A/C–DEFROST	HEAT–DEFROST PARTIAL	HEAT–DEFROST FULL
OFF	–	1 · 2	* *	+	–	7a	7b
ECONOMY VENT	Cool	–	4	5	6	–	–
	Mid	–	–	+	6	–	–
	Warm	–	–	–	6	–	–
A/C *	Cool	1 · 2 * *	4	5	6	–	–
	Mid	–	–	+	6	–	–
	Warm	–	–	–	6	–	–
HI · LO *	Cool	–	4	5	6	7a	–
	Mid	–	–	+	6	7a	–
	Warm	–	–	–	6	7a	–
HEAT	Cool	–	4	5	–	7a	7b
	Mid	–	–	+	–	7a	7b
	Warm	1 · 2 †	–	–	–	7a	7b
▽ (DEFOG) *	Cool	–	4	5	–	7a	–
	Mid	–	–	+	–	7a	–
	Warm	–	–	–	–	7a	–
DEFROST *	Cool	–	4	5	–	–	–
	Mid	–	–	+	–	–	–
	Warm	–	–	–	–	–	–
VACUUM LINE COLOR CODE		WHITE	GREEN	TAN	YELLOW	BLUE (TOP)	RED (SIDE)

* COMPRESSOR CLUTCH ENGAGED WHEN AMBIENT CUT-OFF SWITCH IS CLOSED.
† DELAYED BY ENGINE TEMP. SENDER AND E.V. RELAY UNTIL 130°F. ENGINE COOLANT TEMP.
* * DEPENDENT ON HEATING/COOLING REQUIREMENT OF SERVO DIVERTER VALVE
\+ MODULATED VACUUM

Fig. 88 Airflow schematic & vacuum control test chart. 1978-80 Versailles w/A.T.C.

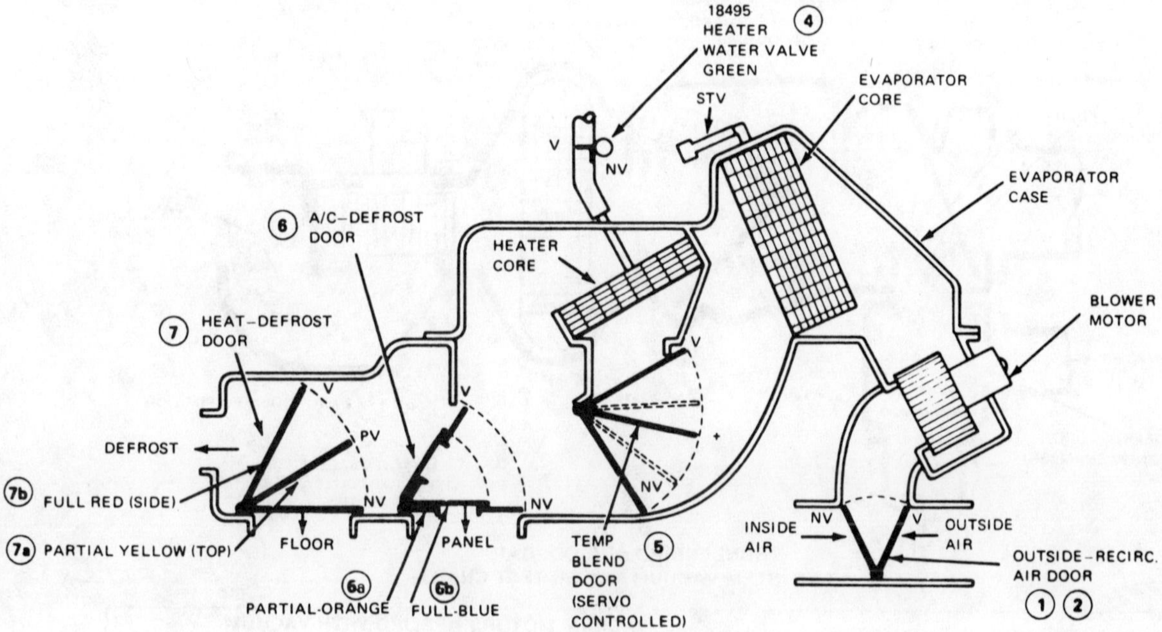

LTD II, COUGAR, THUNDERBIRD AND MARK V
AUTOMATIC TEMPERATURE CONTROL SYSTEM
VACUUM MOTOR TEST CHART

FUNCTION CONTROL LEVER POSITION	ATC REQUIREMENT	VACUUM MOTORS APPLIED WITH VACUUM						
		OUTSIDE – RECIRC.	WATER VALVE **	TEMP. BLEND +	A/C–HEAT		HEAT–DEFROST	
					PARTIAL	FULL	PARTIAL	FULL
OFF	–	1 · 2	**	+	–	–	7a	7b
ECONOMY VENT	Cool	–	4	5	6a	6b	7a	7b
	Mid	–	–	+	6a	6b	7a	7b
	Warm	–	–	–	6a	6b	7a	7b
A/C *	Cool	1 · 2 **	4	5	6a	6b	7a	7b
	Mid	–	–	+	6a	6b	7a	7b
	Warm	–	–	–	6a	6b	7a	7b
HI · LO *	Cool	–	4	5	6a	–	7a	7b
	Mid	–	–	+	6a	–	7a	7b
	Warm	–	–	–	6a	–	7a	7b
HEAT	Cool	–	4	5	–	–	7a	7b
	Mid	–	–	+	–	–	7a	7b
	Warm	1 · 2 †	–	–	–	–	7a	7b
▽ (DEFOG) *	Cool	–	4	5	–	–	7a	–
	Mid	–	–	+	–	–	7a	–
	Warm	–	–	–	–	–	7a	–
DEFROST *	Cool	–	4	5	–	–	–	–
	Mid	–	–	+	–	–	–	–
	Warm	–	–	–	–	–	–	–
VACUUM LINE COLOR CODE		WHITE	GREEN	TAN	ORANGE	BLUE	YELLOW (TOP)	RED (SIDE)

* COMPRESSOR CLUTCH ENGAGED WHEN AMBIENT CUT-OFF SWITCH IS CLOSED
† DELAYED BY ENGINE TEMP. SENDER AND E.V. RELAY UNTIL 130°F. ENGINE COOLANT TEMP.
** DEPENDENT ON HEATING/COOLING REQUIREMENT OF SERVO DIVERTER VALVE
\+ MODULATED VACUUM
– NO VACUUM

Fig. 89 Airflow schematic & vacuum control test chart. 1977-79 Cougar, LTD II, Mark V & Thunderbird w/A.T.C.

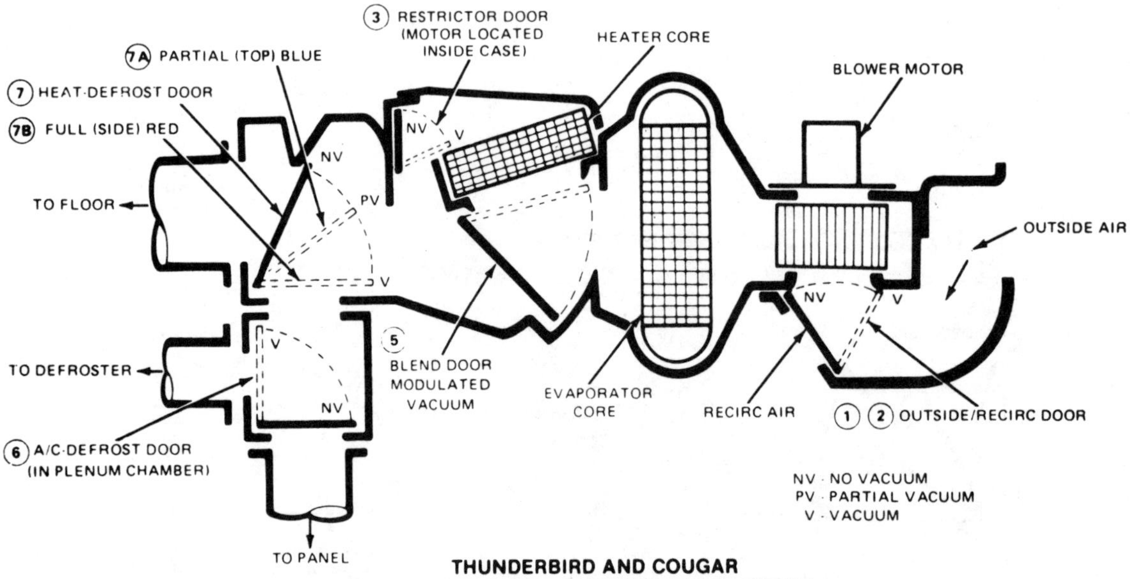

THUNDERBIRD AND COUGAR
ATC SYSTEM VACUUM MOTOR TEST CHART

FUNCTION CONTROL LEVER POSITION		ATC REQUIREMENT	VACUUM MOTORS APPLIED WITH VACUUM: OUTSIDE RECIRC	RESTRICTOR DOOR	TEMP BLEND	HEAT-DEFROST: PARITAL	HEAT-DEFROST: FULL	A/C DEFROST
OFF			1-2	—		7A	7B	—
VENT		COOL	—	—	5	—	—	6
		MID	—	—	+	—	—	6
		WARM	—	—	—	—	—	6
A/C	PANEL*	COOL	1-2**	3	5	—	—	6
		MID	—	—	+	—	—	6
		WARM	—	—	—	—	—	6
	HI-LO*	COOL	—	—	5	7A	—	6
		MID	—	—	+	7A	—	6
HEAT		WARM	—	—	—	7A	—	6
	FLOOR	COOL	—	—	5	7A	7B	—
		MID	—	—	+	7A	7B	—
		WARM	1-2†	—	—	7A	7B	—
DEFROST*		COOL	—	—	5	—	—	—
		MID	—	—	+	—	—	—
		WARM	—	—	—	—	—	—
VACUUM LINE COLOR CODE			WHITE	GREEN	TAN	BLUE	RED	ORANGE

* COMPRESSOR CLUTCH ENGAGED WHEN AMBIENT CUT OFF SWITCH IS CLOSED.
† DELAYED BY ENGINE TEMP SENDER AND EV RELAY UNTIL 120°F ENGINE COOLANT TEMP.
** DEPENDENT ON HEATING COOLING REQUIREMENT OF SERVO DIVERTER VALVE.
\+ MODULATED VACUUM.
— NO VACUUM.

Fig. 90 Airflow schematic & vacuum control test chart. 1980 Cougar & Thunderbird w/A.T.C.

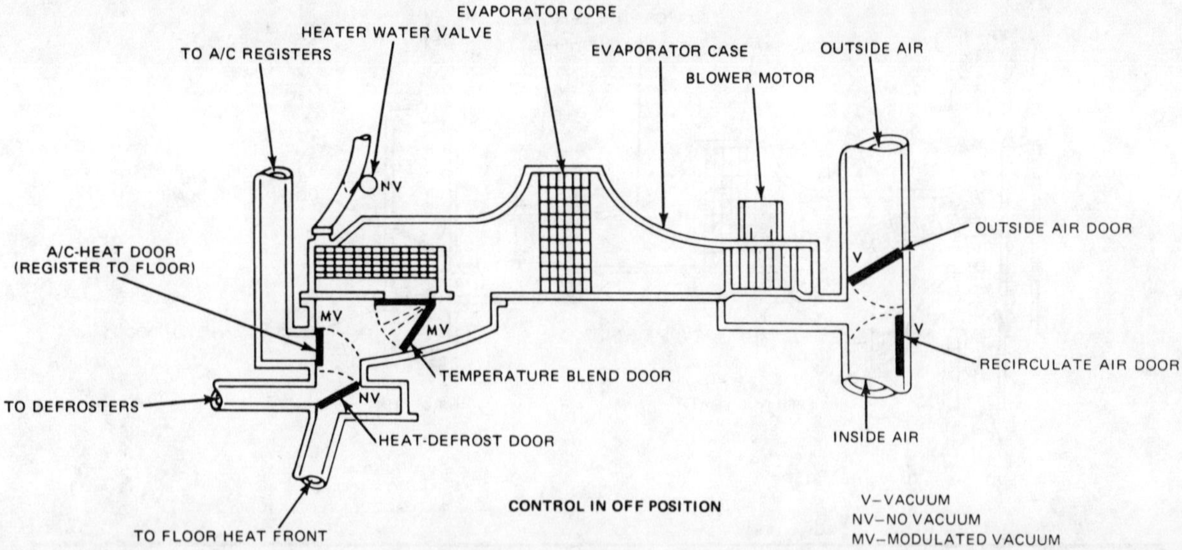

		VACUUM COLOR CODE	TEMPERATURE CONTROL LEVER (Automatic Vacuum Control)	FUNCTIONAL CONTROL LEVER (VACUUM CONTROLLED) OFF	VENT	AUTO	DEFOG	DEFROST
AIR DOOR VACUUM MOTORS	OUTSIDE AIR DOOR	White		Closed V	Open NV	Closed Max A/C V / NV Open		OPEN NV
	RECIRC DOOR	White		Open V	Closed NV	Open * Max A/C V / NV Closed		CLOSED NV
	A/C REGISTER – FLOOR HEAT DOOR (MODE)	Blue	Max Heat	Heat Position Floor NV	A/C Register Position V	Heat Position Floor No Vacuum		
			Mod	Floor – Registers Modulated V	A/C Register Position V	Floor – Registers Modulated V	Heat Position Floor NV	
			Max Cool	A/C Register Position Vacuum			Heat Position Floor NV	
	HEAT-DEFROST DOOR	Red (Side)		NV	V		NV	NV
				Defrost Position	Heat Position		Heat-Defrost Mid Position	Defrost Position
		Yellow (Top)		NV	V		V	NV
	TEMPERATURE BLEND DOOR	Tan	Max Heat	All Air Passes Thru Heater Core No Vacuum				
				Air Passes Thru and Around Heater Core Modulated Vacuum				
			Max Cool	All Air By-Passes Heater Core Vacuum				
WATER VALVE		Green		Open NV	Open No Vacuum / Closed Vacuum *			

V – Vacuum NV – No Vacuum Mod – Modulated * Above 40° Outside Air Temperature

Fig. 91 Airflow schematic & vacuum control test chart. 1974 Ford & Mercury w/A.T.C.

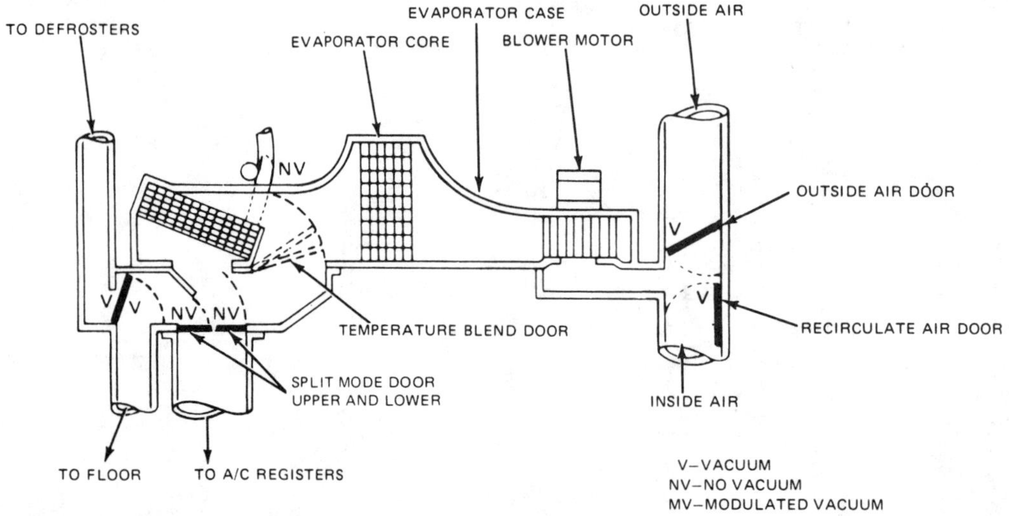

			VACUUM COLOR CODE	TEMP. CONTROL LEVER (Automatic Vacuum Control)	FUNCTIONAL CONTROL LEVER (VACUUM CONTROLLED) OFF	VENT	AUTO	DEFOG	DEFROST
AIR DOOR	OUTSIDE AIR DOOR		White		Closed V	Open NV	Closed Max A/C V / NV Open	Open NV	Open NV
AIR DOOR	RECIRC DOOR		White		Open V	Closed NV	Open Max A/C V / NV Closed	Closed NV	Closed NV
VACUUM MOTORS	MODE DOOR *	Small Door	Orange	COOL	NO VACUUM	V	V	NO VACUUM	NO VACUUM
VACUUM MOTORS	MODE DOOR *	Large Door	Blue	COOL	NO VACUUM	V	V	NO VACUUM	NO VACUUM
VACUUM MOTORS	MODE DOOR *	Small Door	Orange	MOD	NO VACUUM	NV	NV	NO VACUUM	NO VACUUM
VACUUM MOTORS	MODE DOOR *	Large Door	Blue	MOD	NO VACUUM	NV	NV	NO VACUUM	NO VACUUM
VACUUM MOTORS	MODE DOOR *	Small Door	Orange	WARM	NO VACUUM	V	NV	NO VACUUM	NO VACUUM
VACUUM MOTORS	MODE DOOR *	Large Door	Blue	WARM	NO VACUUM	NV	NV	NO VACUUM	NO VACUUM
VACUUM MOTORS	HEAT-DEFROST DOOR		Red (Side) / Yellow (Top)		*	V Heat Position V	V Heat Position V	NV Heat-Defrost Mid Position V	NV Defrost Position NV
VACUUM MOTORS	TEMPERATURE BLEND DOOR		Tan	Max Heat	All Air Passes Thru Heater Core No Vacuum				
VACUUM MOTORS	TEMPERATURE BLEND DOOR (continued)		Tan		Air Passes Thru and Around Heater Core Modulated Vacuum				
VACUUM MOTORS	TEMPERATURE BLEND DOOR (continued)		Tan	Max Cool	All Air By-Passes Heater Core Vacuum				
V – Vacuum NV – No Vacuum MOD – Modulated	WATER VALVE		Green		Open NV	Open No Vacuum Closed Vacuum			

Fig. 92 Airflow schematic & vacuum control test chart. 1975-76 Ford & Mercury w/A.T.C.

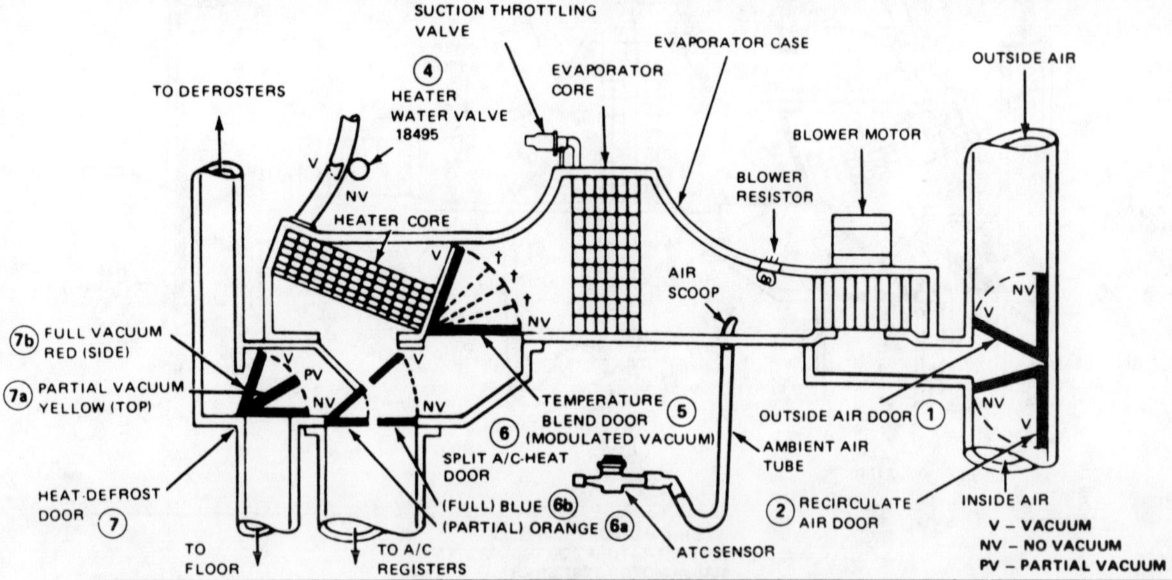

1977 FORD AND MERCURY AUTOMATIC TEMPERATURE CONTROL SYSTEM VACUUM MOTOR TEST CHART

FUNCTION CONTROL LEVER POSITION	ATC REQUIREMENT		VACUUM MOTORS APPLIED WITH VACUUM							
							A/C · HEAT		HEAT · DEFROST	
	TEMP	BLOWER SPEED	OUTSIDE	RECIRC.	WATER VALVE	TEMP. BLEND+	PARTIAL	FULL	PARTIAL	FULL
OFF			1	2	–	5	–	–	7a	7b
ECONOMY VENT	COOL	4	–	–	4	5	6a	6b	7a	7b
		3	–	–	–	+	6a	6b	7a	7b
		2	–	–	–	+	6a	6b	7a	7b
		1	–	–	–	+ Mid	6a	–	7a	7b
	HEAT	2	–	–	–	+	6a	–	7a	7b
		3	–	–	–	+	6a	–	7a	7b
		4	–	–	–	–	6a	–	7a	7b
AUTO *	A/C	4	1	2	4	5	6a	6b	7a	7b
		3	1	2	–	+	6a	6b	7a	7b
		2	–	–	–	+	6a	6b	7a	7b
		1	–	–	–	+ Mid	6a	–	7a	7b
	HEAT †	2	–	–	–	+	–	–	7a	7b
		3	–	–	–	+	–	–	7a	7b
		4 †	– †	–	–	–	–	–	7a	7b
(DEFOG) *	A/C	4	–	–	4	5	–	–	7a	–
		3	–	–	–	+	–	–	7a	–
		2	–	–	–	+	–	–	7a	–
		1	–	–	–	+ Mid	–	–	7a	–
	HEAT	2	–	–	–	+	–	–	7a	–
		3	–	–	–	+	–	–	7a	–
		4	–	–	–	–	–	–	7a	–
DEFROST *	A/C	4	–	–	4	5	–	–	–	–
		4	–	–	–	+	–	–	–	–
		4	–	–	–	+	–	–	–	–
		4	–	–	–	+ Mid	–	–	–	–
	HEAT	4	–	–	–	+	–	–	–	–
		4	–	–	–	+	–	–	–	–
		4	–	–	–	–	–	–	–	–
VACUUM LINE COLOR CODE			WHITE	WHITE	GREEN	TAN	ORANGE	BLUE	YELLOW (TOP)	RED (SIDE)

\+ MODULATED VACUUM
– NO VACUUM
• COMPRESSOR CLUTCH ENGAGED WHEN AMBIENT CUT-OFF SWITCH IS CLOSED.
† DELAYED BY ENGINE TEMP. SENDER AND E.V. RELAY UNTIL 130°F. ENGINE COOLANT TEMP.

Fig. 93 Airflow schematic & vacuum control test chart. 1977 Ford & Mercury w/A.T.C.

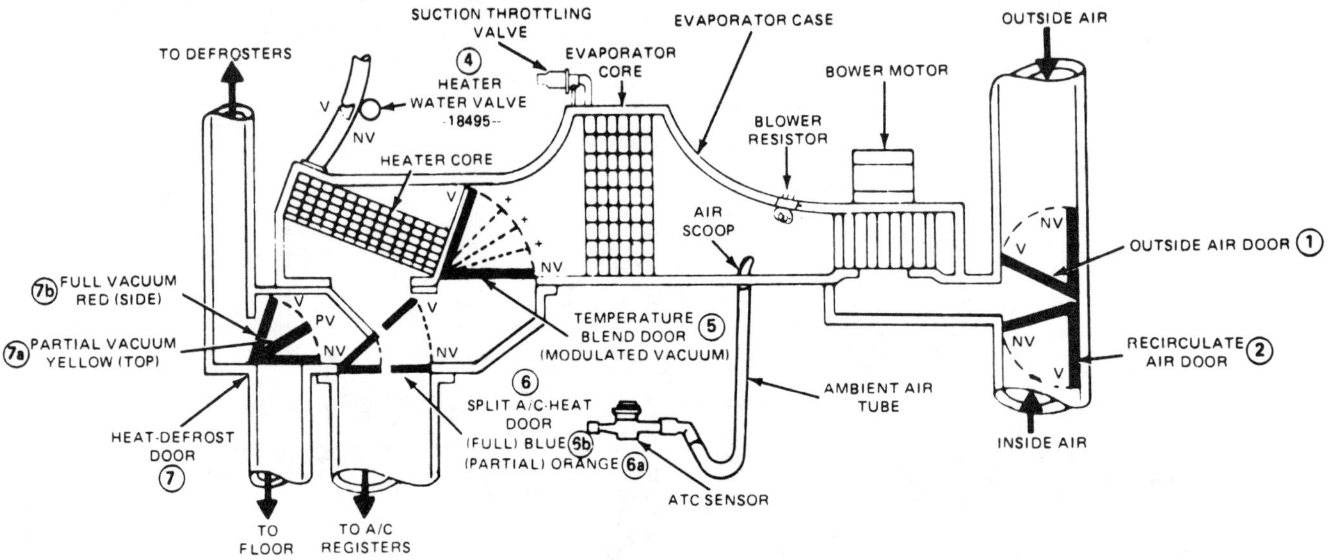

FORD, MERCURY, LINCOLN AUTOMATIC TEMPERATURE CONTROL SYSTEM VACUUM MOTOR TEST CHART									
		VACUUM MOTORS APPLIED WITH VACUUM							
						A/C HEAT		HEAT-DEFROST	
FUNCTION CONTROL LEVER POSITION	ATC REQUIREMENT	OUT SIDE	RE CIRC.	WATER VALVE **	TEMP BLEND ·	PARTIAL	FULL	PARTIAL	FULL
OFF		1	2	**	·		–	7a	7b
ECONOMY VENT	Cool	–	–	4	5	6a	6b	7a	7b
	Mid	–	–		·	6a	6b	7a	7b
	Warm	–	–			6a	6b	7a	7b
A/C *	Cool	1**	2**	4	5	6a	6b	7a	7b
	Mid	–	–		·	6a	6b	7a	7b
	Warm	–	–			6a	6b	7a	7b
HI · LO *	Cool	–	–	4	5	6a	--	7a	7b
	Mid	–	–		·	6a	–	7a	7b
	Warm	–	–			6a	–	7a	7b
HEAT	Cool	–	–	4	5		–	7a	7b
	Mid	–	–		·		–	7a	7b
	Warm	1†	2†				–	7a	7b
▽ (DEFOG) *	Cool	–	–	4	5		–	7a	–
	Mid	–	–		+		–	7a	–
	Warm	–	–			–	–	7a	–
DEFROST *	Cool	–	–	4	5		–	–	–
	Mid	–	–		+		–	–	–
	Warm	··	–		–	–	–	–	–
VACUUM LINE COLOR CODE		WHITE	WHITE	GREEN	TAN	ORANGE	BLUE	YELLOW (TOP)	RED (SIDE)

* COMPRESSOR CLUTCH ENGAGED WHEN AMBIENT CUT-OFF SWITCH IS CLOSED
† DELAYED BY ENGINE TEMP. SENDER AND E.V. RELAY UNTIL 120°F. ENGINE COOLANT TEMP.
** DEPENDENT ON HEATING/COOLING REQUIREMENT OF SERVO DIVERTER VALVE
\+ MODULATED VACUUM
– NO VACUUM

Fig. 94 Airflow schematic & vacuum control test chart. 1978 Ford, Mercury & 1978-79 Lincoln Continental w/A.T.C.

CONTROL HEAD VACUUM TEST CHART

VACUUM PORT	HOSE COLOR	VACUUM SELECTOR POSITION					FUNCTION
		OFF	VENT	AUTO	DEFOG	DEFROST	
4	RED	–	–	TO 3 AND 5	–	–	TO HIGH-LOW DOOR ⑥b SIDE
5	CHARTREUSE	SEAL	SEAL	TO 3 AND 4	SEAL	SEAL	FROM SERVO DIVERTER VALVE // 1 PORT
6	NOT CONNECTED	SEAL	SEAL	SEAL	SEAL	SEAL	NOT CONNECTED
3	BLUE	–	–	TO 4 AND 5	V	–	TO HIGH-LOW DOOR ⑥a TOP
9	BLACK	S	S	S	S	S	SOURCE VACUUM
1	WHITE	V	–	TO 2	–	–	TO OUTSIDE-RECIRC DOOR ①②
2	TURQUOISE	SEAL	SEAL	TO 1	SEAL	SEAL	FROM E.V. RELAY
7	YELLOW	–	V	TO 8	–	–	TO PANEL-DEFORST DOOR ⑦
8	PURPLE	SEAL	SEAL	TO 7	SEAL	SEAL	FROM SERVO DIVERTER VALVE# 3 PORT

V = VACUUM. ⊟ = NO VACUUM (ATMOSPHERE) S = SYSTEM SOURCE VACUUM

Fig. 98 Control head vacuum test. 1974-76 Cougar, Elite, Montego, Torino, 1975-76 Mark IV & Thunderbird

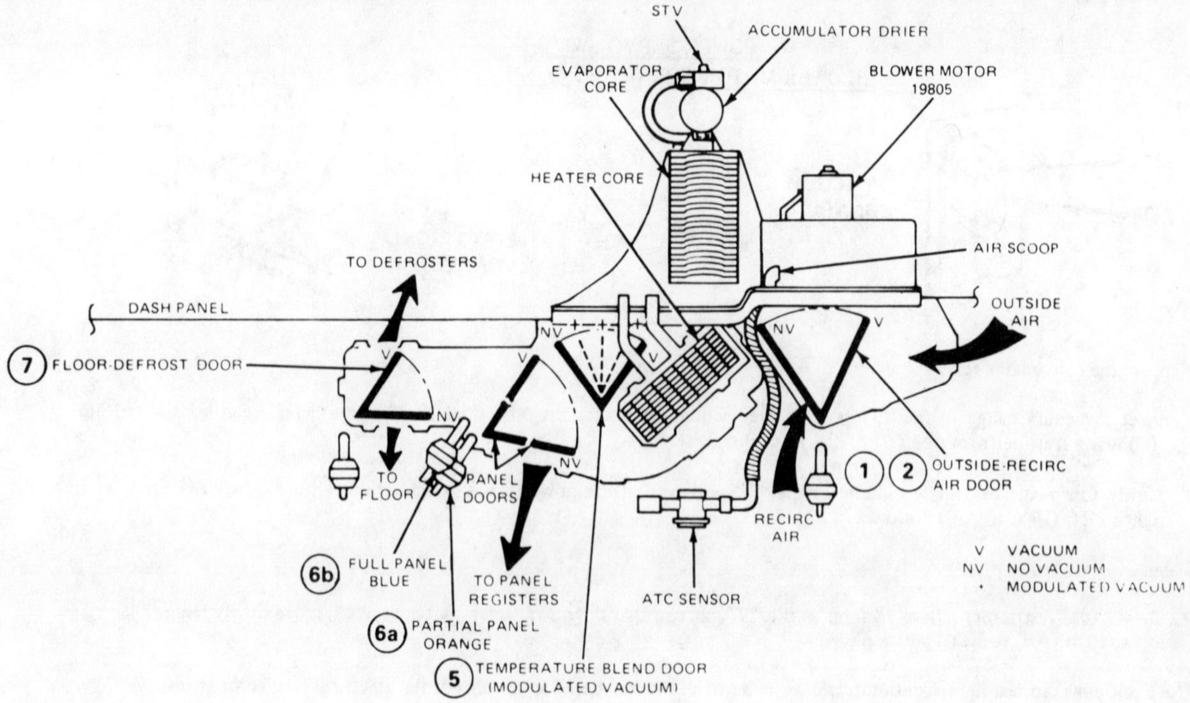

AUTOMATIC TEMPERATURE CONTROL SYSTEM VACUUM MOTOR TEST CHART

FUNCTION CONTROL LEVER POSITION	ATC REQUIREMENT	VACUUM MOTORS APPLIED WITH VACUUM				
		OUTSIDE-RECIRC	TEMP BLEND	PANEL PARTIAL	PANEL FULL	FLOOR-DEFROST
OFF		1-2	•			7
VENT	COOL		5	6a	6b	7
VENT	MID		•	6a	6b	7
VENT	WARM			6a	6b	7
A/C PANEL*	COOL	1-2**	5	6a	6b	7
A/C PANEL*	MID		•	6a	6b	7
A/C PANEL*	WARM			6a	6b	7
HEAT HI LO*	COOL		5	6a		7
HEAT HI LO*	MID		•	6a		7
HEAT HI LO*	WARM			6a		7
HEAT FLOOR	COOL		5			7
HEAT FLOOR	MID		•			7
HEAT FLOOR	WARM	1-2 †				7
DEFROST*	COOL		5			
DEFROST*	MID		•			
DEFROST*	WARM					
VACUUM LINE COLOR CODE		WHITE	TAN	ORANGE	BLUE	YELLOW

* COMPRESSOR CLUTCH ENGAGED WHEN AMBIENT CUT OFF SWITCH IS CLOSED
† DELAYED BY ENGINE TEMP. SENDER AND E.V. RELAY UNTIL 120 F. ENGINE COOLANT TEMP
** DEPENDENT ON HEATING COOLING REQUIREMENT OF SERVO DIVERTER VALVE
\+ MODULATED VACUUM
— NO VACUUM

Fig. 95 Airflow schematic vacuum control test chart. 1979-80 Ford, Mercury, 1980 Lincoln Continental & Mark VI w/A.T.C.

Fig. 96 Electro-vacuum relay terminal locations. 1978-80 models w/A.T.C.

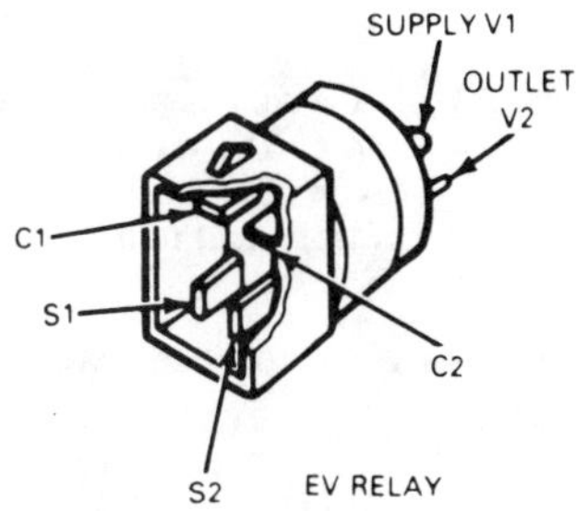

POWER SERVO SWITCH
BLOWER MOTOR CIRCUIT CONTINUITY TEST

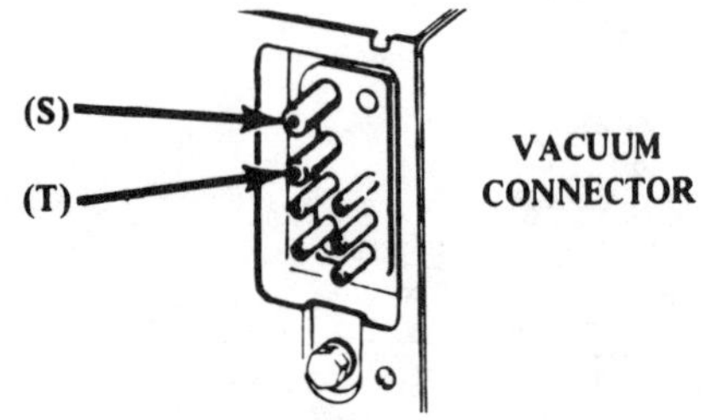

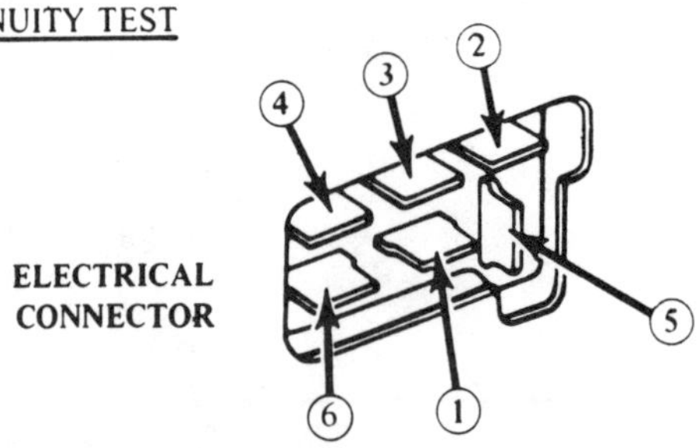

Remove the cold engine relay*

Connect a vacuum gauge to the "S" port of the vacuum connector on ATC Box. If a vacuum gauge reading is noted during the following test, perform step (6) of the ATC Box Test.

1. **Slowly** (use restrictor) apply vacuum to the "T" port so that the power servo switch arm will move **slowly** from the extreme right to left (Heat to A/C) position.

Then:

2. **Slowly** (use restrictor) release vacuum to the "T" port so that the power servo switch arm will move **slowly** from the extreme left to right (A/C to Heat) position.

Use a self powered test light. Continuity will exist between the numbered spade pins of the electrical connector as follows:

Heater Blower Switch Position	**Electrical Connector Spade Pin Numbers**	**Visual Check**
High	5-1	Light will remain on when arm is in maximum HEAT position (no vacuum). Then go off when vacuum is applied.*
2 3 4	5-2 5-3 5-4	Light will blink once* (when on the right side of mid-position) as arm moves from extreme right to left or from extreme left to right.
NO CONTINUITY IN MID-POSITION		
A/C Blower Switch Position	**Electrical Connector Spade Pin Numbers**	**Visual Check**
High	6-1	Light will remain on when arm is in maximum A/C position (maximum vacuum). Then go off when vacuum is removed.*
2 3 4	6-2 6-3 6-4	Light will blink once* (when on the left side of mid-position) as arm moves from extreme right to left or from extreme left to right.

*NOTE: If the cold engine relay is not removed before performing the test, continuity will exist between spade pin numbers 5 and 6. The test light will remain ON in both the maximum HEAT (extreme right, no vacuum) and maximum A/C (extreme left, maximum vacuum) positions. The test light will also blink twice as the arm travels either from left to right or from right to left.

Fig. 97 Power servo switch & blower motor circuit continuity test chart. 1974 Mark IV, Thunderbird & 1974-77 Lincoln Continental

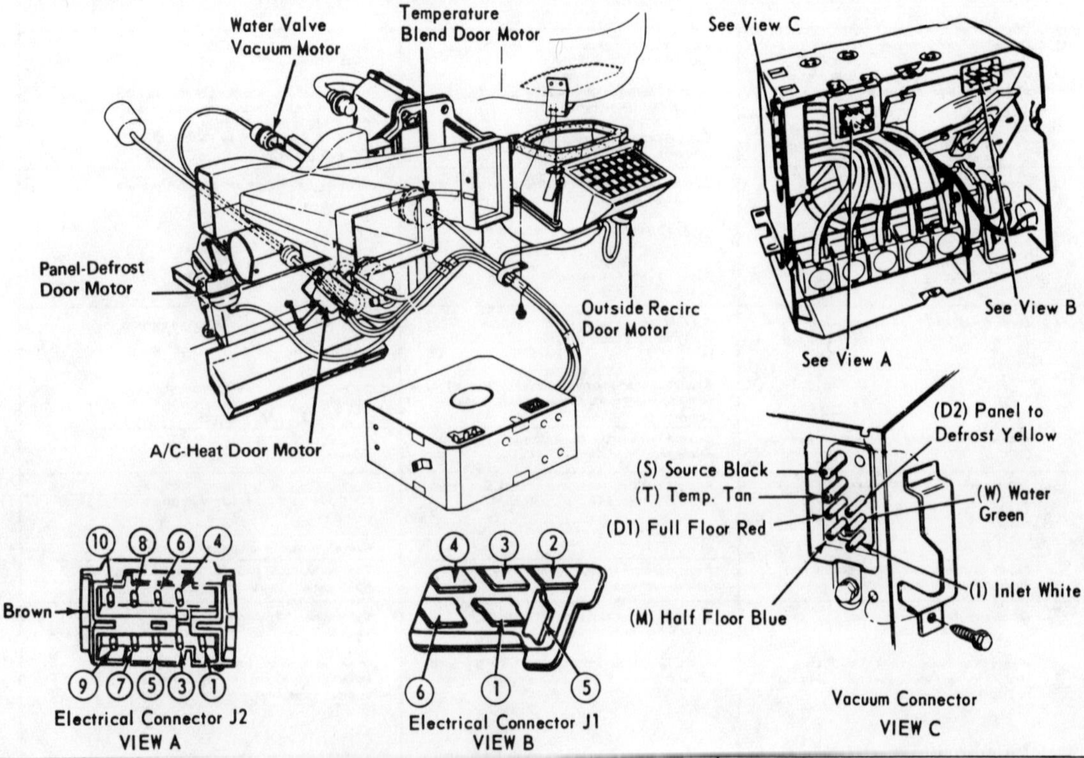

TEST EQUIPMENT REQUIRED		PRELIMINARY STEPS
1 – Jumper wire (14–16 gauge) 3' to 6' long with alligator clip on one end, two C7AB-14488-B female pin terminals and one B9A-14294-L female spade terminal connectors on other end.  1 – Jumper wire (14–16 gauge) 3' long with alligator clip on one end and one C7AB-14488-B female pin terminal connector on other end.  1 – Jumper wire for ground, 3' to 6' long with alligator clip on each end. 	4 – Vacuum gauges or 1 – Vacuum gauge and 3 – Vacuum motors 1 – Ohmmeter or 1 – Self-powered test light 1 – 12 volt test light 3 – 1/8" x 1' vacuum hose 1 – 3/16" x 1' vacuum hose 1 – 1/4" x 1' vacuum hose 1 – 383001-S vacuum connector (3/16" to 1/8") 1 – 383005-S vacuum connector (1/4" to 1/4")	1. Disconnect the vacuum harness connector and the two electrical connector plugs from the A.T.C. box. 2. Remove the box from the vehicle and set it on a work bench. 3. Connect a vacuum supply to the "S" port on the box. 4. Make sure that the A.T.C. box is grounded. 5. Perform each test with a minimum of 13 inches of vacuum. **NOTE: When performing any one of the tests, make all indicated vacuum connections before energizing the electrical circuits.**

TEST	PROCEDURE	RESULT REQUIRED
① Check half floor solenoid (M) Ability of box to operate half floor portion of A/C-Heat door vacuum motor.	Connect Vacuum gauge to "M" port on box. Apply 12 volts to terminal pin No. 10 on box (Brown connector). or Apply 12 volts to terminal of half floor (Blue) solenoid.	Vacuum Port "M" should go to full vacuum. No other port should show vacuum. A shorted diode could cause D1 to be activated too.

Fig. 99 A.T.C. control box test chart. 1974 Mark IV & Thunderbird (Part 1 of 2)

② Check full floor Solenoid (D1) Ability of box to operate full floor portion of A/C-Heat door vacuum motor.	Connect vacuum gauge to "D1" and "M" ports on box. Apply 12 volts to terminal pin No. 5 on box (Brown connector). or Apply 12 volts to terminal of full floor (Red solenoid).	Vacuum ports "D1" and "M" should both go to full vacuum. Anytime vacuum is applied to the full floor position (Red) it should also be applied to the half floor position (Blue). Failure Can Be Caused By Faulty "Diode" Or "Full Floor Solenoid"
③ Check "inlet" door solenoid (I) Ability of box to operate Outside / Recirc. Door Vacuum motor.	Connect vacuum gauge to "I" port vacuum connector on box. Apply 12 volts to terminal pin No. 3 on box (Brown connector) or Apply 12 volts to terminal of inlet (White solenoid).	Vacuum port "I" should go to full vacuum.
④ Check operation of "cold engine water" blower cut off relay (Inlet solenoid circuit)	Connect vacuum gauge to "D1" and gauge or motor to "I" port of vacuum connector on box. Apply 12 volts to terminal pin No. 5 (Brown) connector) and spade terminal No. 6 (Black connector).	Vacuum port D1 should go to full vacuum. Vacuum port "I" should remain at zero vacuum.
Ability of cold engine relay in box to override Step 3.	Ground Terminal pin No. 4 (Brown connector).	Vacuum port "I" should go to full vacuum.
⑤ Check cold engine relay (blower circuit).	Apply 12 volts to terminal pin No. 5. Ground terminal pin No. 4 (Brown connector)	Electrical continuity should cease between No. 5 and No. 6 spade terminals (Black connector). (Use ohmmeter or self-powered test light.)
Ability of cold engine relay in box to interrupt heater blower circuit	Remove ground from terminal pin No. 4 (Brown connector).	Continuity should return between No. 5 and No. 6 spade terminals (Black connector).
⑥ Check amplifier and transducer. Ability of amplifier in box to call for full vacuum through transducer. If OK, proceed to Step 7.	Connect vacuum gauge to "T" port of vacuum connector on box. Apply 12 volts to terminal pin No. 7 and ground terminal pin No. 8 (Brown Connector).	Vacuum port "T" should go to full vacuum. Power Servo switch arm must move from extreme right to extreme left (full vacuum) in about 10 seconds.
⑥a Check transducer. Ability of transducer to control vacuum supply to temperature blend door and power servo modulating vacuum motors.	Connect vacuum gauge to "T" port of vacuum connector on box. Disconnect wire harness No. 1, 2, and 3 pin connectors from transducer. Connect fused (1/4 amp) +12 volts to No. 2 pin terminal on transducer. Ground clip of 12 volt test light to frame of box and touch probe to No. 1 pin terminal of transducer.	Vacuum port "T" should go to full vacuum. Power servo switch arm must move from extreme right (zero vacuum) to extreme left (full vacuum) in about 10 seconds. When probe of 12 volt test light is removed, vacuum must maintain for an indefinite period of time (no leakage).
	Touch probeof 12 volt test light to No. 3 pin terminal of transducer.	Vacuum port "T" should go to zero vacuum. Power servo switch arm must move from extreme left (full vacuum) to extreme right (zero vacuum) in about 10 seconds.
⑦ Check Panel-Defrost door "D2" and water valve (W) solenoids.	Install vacuum gauge on "T" port of vacuum connector and motors or gauges on "D1", "D2" M & W ports. Apply 12 volts to terminal pins No. 7 and No. 5 and then ground terminal pin No. 8 (Brown connector).	Vacuum ports (T), (D2), and (W) should go to full vacuum. Vacuum ports (D1) and (M) should go to zero vacuum. "T" should always move relatively slowly compared to the solenoids.
Ability of box to operate Panel-Defrost door, Heater, Water Valve, and By-Pass Air door Vacuum motors.	Remove ground from terminal pin No. 8 (Brown connector).	Vacuum ports (T, (D2) and (W) should go to zero vacuum; and (D1) and (M) should go to full vacuum.
If the A.T.C. Box fails to pass any one of the above tests, the box should be repaired or replaced. If the box passes all the tests, there is probably trouble elsewhere in the system.	Connect vacuum harness connector to box. Connect both Brown (pin) and Black (spade) electrical connectors to box.	

Fig. 99 A.T.C. control box test chart. 1974 Mark IV & Thunderbird (Part 2 of 2)

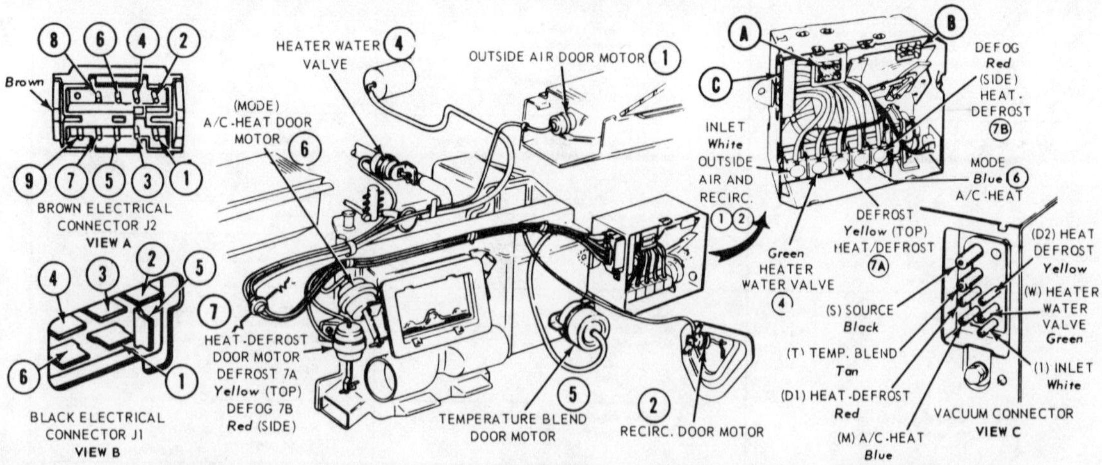

TEST	PROCEDURE	RESULT REQUIRED
EQUIPMENT 1 – Jumper wire (14–16 gauge) 3' to 6' long with alligator clip on one end, two C7AB-14488-B female pin terminals and one B9A-14294-L female spade terminal connectors on other end.  1 – Jumper wire (14–16 gauge) 3' long with alligator clip on one end and one C7AB-14488-B female pin terminal connector on other end.  1 – Jumper wire for ground, 3' to 6' long with alligator clip on each end.  4 – Vacuum gauges or 1 – Vacuum gauge and 3 – Vacuum motors 1 – Ohmmeter or 1 – Self-powered test light 1 – 12 volt test light 3 – 1/8" x 1' vacuum hose 1 – 3/16" x 1' vacuum hose 1 – 1/4" x 1' vacuum hose 1 – 383001-S vacuum connector (3/16" to 1/8") 1 – 383005-S vacuum connector (1/4" to 1/4")	**PRELIMINARY STEPS** 1. Disconnect the vacuum harness connector and the two electrical connector plugs from the A.T.C. box. 2. Remove the box from the vehicle and set it on a work bench. 3. Connect a vacuum supply to the "S" port on the box. 4. Make sure that the A.T.C. box is grounded. 5. Perform each test with a minimum of 13 inches of vacuum. **NOTE: When performing any one of the tests, make all indicated vacuum connections before energizing the electrical circuits.**	INDICATES SATISFACTORY PERFORMANCE
① Check "Defrost" Solenoid (D2) Ability of box to operate defrost portion of Heat-Defrost vacuum motor (7B)	Connect vacuum gauge to "D2" port on box. Apply 12 volts to terminal pin No. 2 on box (Brown connector). or Apply 12 volts to terminal of Defrost (yellow) solenoid).	Vacuum port "D2" should go to full vacuum.
② Check "Defog" solenoid (D1) Ability of box to operate defog portion of Heat-Defrost vacuum motor (7A)	Connect vacuum gauge to "D1" port on box. Apply 12 volts to terminal pin No. 5 on box (Brown connector). or Apply 12 volts to terminal of Defog (red) solenoid.	Vacuum port "D1" should go to full vacuum.

Fig. 100 A.T.C. control box test chart. 1974-77 Lincoln Continental (Part 1 of 2)

③ Check "inlet" door solenoid (I). Ability of box to operate Outside Air Door Vacuum motor ① and Recirc. Door Vacuum motor ②	Connect vacuum gauge to "I" port vacuum connector on box. Apply 12 volts to terminal pin No. 3 on box (Brown connector). or Apply 12 volts to terminal of Inlet (white) solenoid.	Vacuum port "I" should go to full vacuum.
④ Check operation of "cold engine water" blower cut off relay (Inlet solenoid circuit).	Connect vacuum gauge to "D1" and gauge or motor to "I" port of vacuum connector on box. Apply 12 volts to terminal pin No. 5 (Brown connector) and spade terminal No. 6 (Black connector).	Vacuum port "D1" should go to full vacuum. Vacuum port "I" should go to zero vacuum.
Ability of cold engine relay in box to override Step 3.	Ground terminal pin No. 4 (Brown connector).	Port "I" should go to full vacuum.
⑤ Check cold engine relay (blower circuit).	Apply 12 volts to terminal pin No. 5, ground terminal pin No. 4 (Brown connector).	Electrical continuity should cease between No. 5 and No. 6 spade terminals (Black connector). (Use ohmmeter or self-powered test light.)
Ability of cold engine relay in box to interrupt heater blower circuit.	Remove ground from terminal pin No. 4 (Brown connector).	Continuity should exist between No. 5 and No. 6 spade terminals (Black connector).
⑥ Check amplifier and transducer. Ability of amplifier in box to call for full vacuum through transducer. If OK, proceed to step 7.	Connect vacuum gauge to "T" port of vacuum connector on box. Apply 12 volts to terminal pin No. 7 and ground terminal pin No. 8 (Brown Connector).	Vacuum port "T" should go to full vacuum. Power servo switch arm must move from extreme right to extreme left (full vacuum) in about **10 seconds.**
⑥a Check transducer. Ability of transducer to control vacuum supply to temperature blend door ⑤ and power servo modulating vacuum motors.	Connect vacuum gauge to "T" port of vacuum connector on box. Disconnect wire harness No. 1, 2 and 3 pin connectors from transducer. Connect fused (1/4 amp.) + 12 volts to No. 2 pin terminal on transducer. Ground clip of 12 volt test light to frame of box and touch probe to No. 1 pin terminal of transducer.	Vacuum port "T" should to to full vacuum. Power servo switch arm must move from extreme right (zero vacuum) to extreme left (full vacuum) in about **10 seconds.** When probe of 12 volt test light is removed, vacuum must maintain for an indefinite period of time (no leakage).
	Touch probe of 12 volt test light to No. 3 pin terminal of transducer.	Vacuum port "T" should go to zero vacuum. Power servo switch arm must move from extreme left (full vacuum) to extreme right (zero vacuum) in about **10 seconds.**
⑦ **Check A/C-Heat door (M) and water valve (W) solenoids.**	Install vacuum gauge on "T" port of vacuum connector and motors or gauges on "D1, M & W" ports. Apply 12 volts to terminal pins No. 7 and No. 5, and then ground terminal pin No. 8 (Brown connector).	Vacuum ports (T), (D1), (M), and (W) should go to full vacuum.
Ability of box to operate A/C-Heat door ⑥ and Heater Water Valve ④ vacuum motors.	Remove ground from terminal pin No. 8 (Brown connector).	Vacuum ports (T), (M) and (W) should go to zero vacuum and (D1) remain at full vacuum.
If the A.T.C. box fails to pass any one of the above tests, the box is bad and should be repaired or replaced. If the box passes all the tests, there is probably trouble elsewhere in the system.	Connect vacuum harness connector to box. Connect both Brown (pin) and Black (spade) electrical connectors to box.	

Fig. 100 A.T.C. control box test chart. 1974-77 Lincoln Continental (Part 2 of 2)

Vacuum	Blower Speed	Water Valve	Fresh/Recirc. Air
High (Above 14.0 in.) Cold Air Out of A/C Registers	Hi M.H. Med Low	Closed * Above 12.0 in.	Recirc. Air * Above 11.0 in.
Low (Below 2.0 in.) Hot Air Out of Floor	Med M.H. Hi	Open Below 9.0 in.	Fresh Air Below 8.0 in.

* Unless outside air temperature is below 40° F.

Fig. 101 Water valve & air function diagnosis chart. 1974-77 Ford & Mercury w/A.T.C.

1974-77 Ford & Mercury A.T.C. System Diagnosis

NOTE: Engine must be at operating temperature. Operate engine at 1500 rpm with vehicle interior at approximately 75°F.

1. Connect a vacuum gauge into servo motor vacuum line.
2. Connect a voltmeter into blower motor line.
3. Position control lever on Auto and blower switch on Auto Lo or Hi.
4. Move temperature control lever from 85 degrees to 65 degrees. Vacuum should cycle from low to high vacuum. Note blower motor voltage, water valve operation, fresh air door and recirculating air door operation, refer to Fig. 101.
5. After performing the above test procedure, refer to chart 1, Fig. 102, to further diagnose malfunction in A.T.C. system.

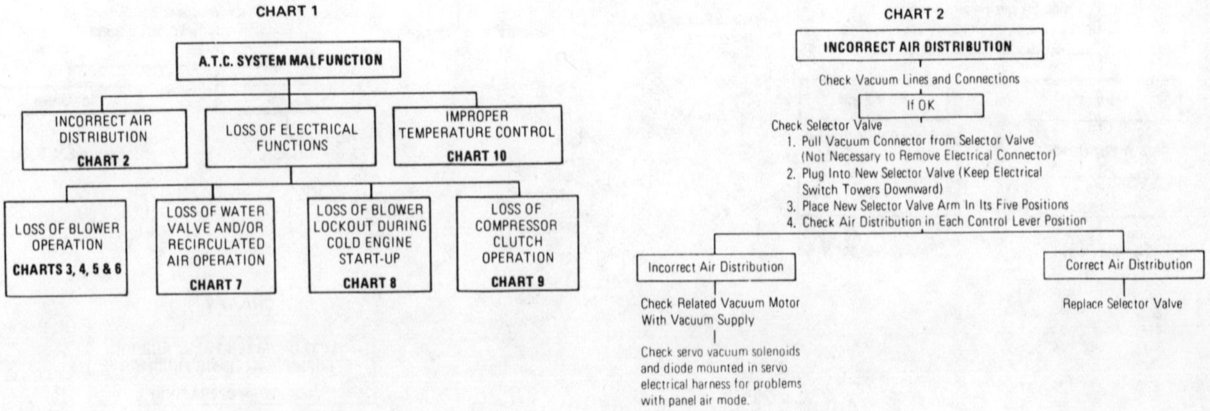

Fig. 102 A.T.C. system diagnosis chart. 1974-77 Ford & Mercury (Part 1 of 4)

CHART 3

INSTRUCTIONS: Find the Appropriate Row That Describes The Blower Problem. Problem Must Be Present In Only The Control Lever And Blower Switch Position As Indicated.

BLOWER PROBLEM DESCRIPTION	FUNCTIONAL CONTROL LEVER POSITION(S)	BLOWER SWITCH POSITION(S)	POSSIBLE INOPERATIVE COMPONENTS	DIRECTIONS
No Blower Operation In Heat and A/C (85° F and 65° F Setting)	All Positions	All Positions	Circuit Breaker, Selector Valve, High Blower Relay, Blower Motor	See Checkout Chart No. 4
	Vent, Auto, and Defog Only	Max Only	Blower Switch, High Blower Relay	Set in Max. Measure Voltage at High Blower Relay Wire – Orange/Black Dot. Voltage – Replace Relay. No Voltage – Replace Switch.
		Hi & Lo Auto	Selector Valve, Servo, Blower Switch, High Blower Relay	See Chart No. 5
	Vent, Auto, or Defog Only	Hi & Lo Auto	Selector Valve	Replace Selector Valve
No Blower Operation for One, Two, or Three Automatic Speeds in Heat and A/C (85° F and 65° F Setting)	Vent, Auto and Defog Only	Hi & Lo Auto	Servo, Resistor Pack	Check Continuity of Resistor Pack Per Schematic – If OK – Replace Servo Switchboard Assembly
Now Blower Operation In Heat Only (85° F Setting)	Auto Only	Hi & Lo Auto	E.V. Relay, Servo, Engine Temperature Switch	See Chart No. 6
No Blower Operation In A/C Only (65° F Setting)	Vent, Auto, and Defog Only	Hi & Lo Auto	Servo	Replace Servo Switch Board Assembly
No High Automatic Blower Speed in Heat or A/C (85° F or 65° F Setting)	Vent, Auto, and Defog Only	Hi & Lo Auto	Servo	Calibrate Servo

Fig. 102 A.T.C. system diagnosis chart. 1974-77 Ford & Mercury (Part 2 of 4)

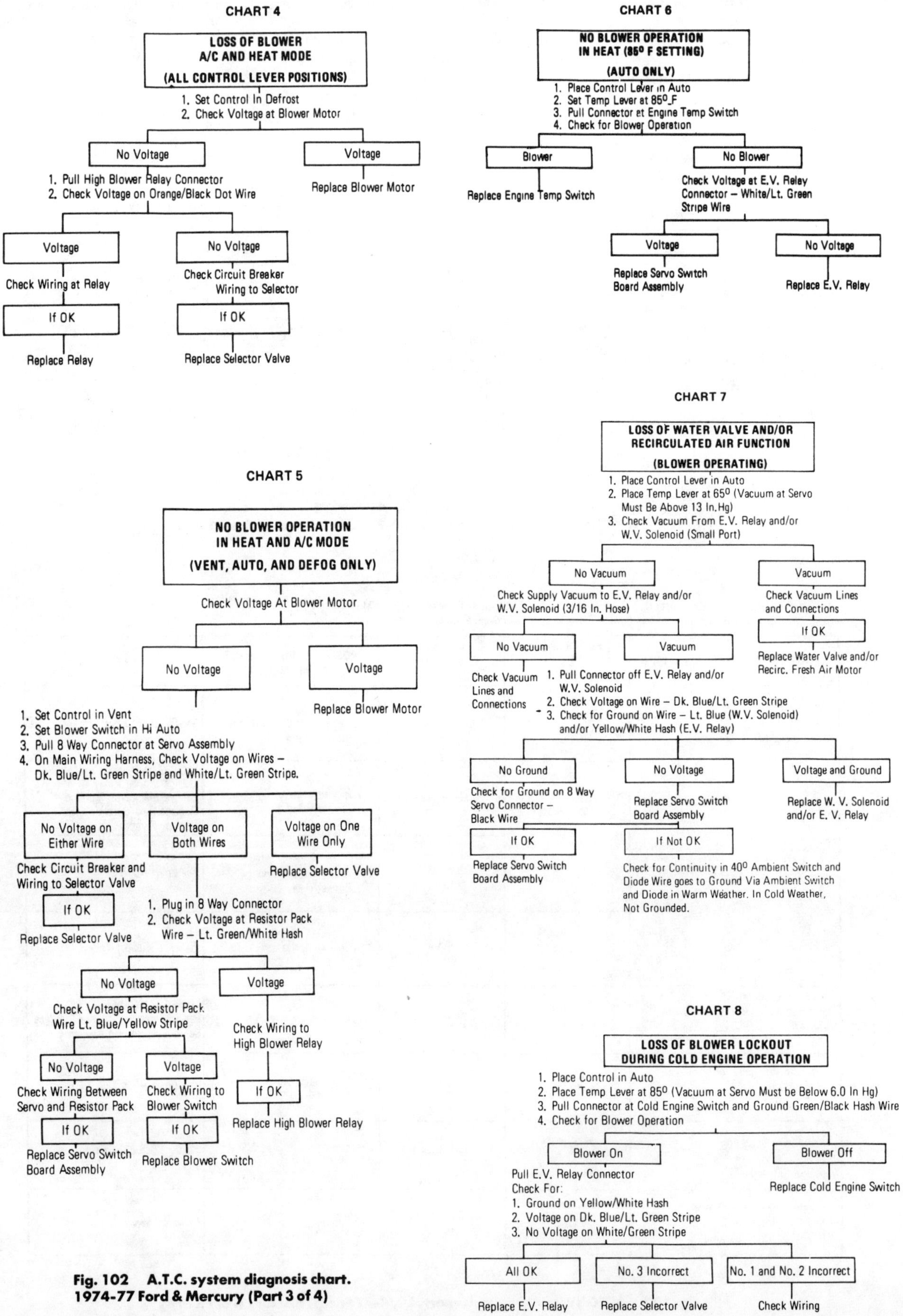

Fig. 102 A.T.C. system diagnosis chart. 1974-77 Ford & Mercury (Part 3 of 4)

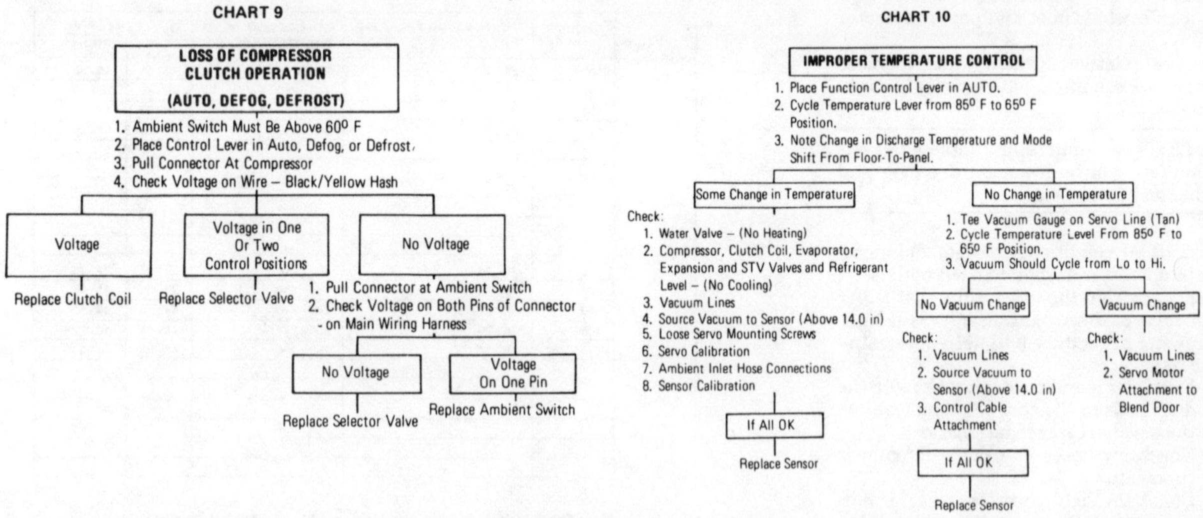

Fig. 102 A.T.C. system diagnosis chart. 1974-77 Ford & Mercury (Part 4 of 4)

GENERAL MOTORS

1974-80 Buick

PERFORMANCE TEST

1974-76

Performance testing is a measurement of the air conditioner system performance to determine if discharge air temperature, pressure in suction line, and pressure in discharge line are within specific limitations, Figs. 103 to 106.

To perform test proceed as follows:

1. Remove protective caps from Schrader valve located on suction throttle valve and Schrader valve located on compressor discharge port.
2. Interconnect manifold and gage set, gage charging lines and gage adapters to air conditioning system.
3. Open doors and hood of car.
4. Set temperature lever to COLD position and fan lever to HI. Air control lever in A/C position.
5. Idle engine at 2000 RPM in neutral.
6. Place a high volume industrial type fan in front of radiator grille to insure minimum differential between temperature of air passing through radiator grille and con

Ambient Air in Degrees F. (In Auxiliary Fan Air Blast Ahead of Condenser)	70°		80°		90°		100°		110°	
Air Quality	Arid	Humid	Arid	Humid	Arid	Humid	Arid	Humid	Arid	Humid
*Average Compressor Head Pressure in PSI	163 to 187	193 to 217	193 to 217	223 to 247	223 to 247	263 to 287	258 to 282	313 to 337	298 to 322	373 to 397
Average Evaporator Pressure **PSI AT SEA LEVEL	27.5 to 29.5	28.5 to 30.5	28 to 30	29 to 31	28 to 30	30 to 32	28 to 30	33 to 35	28.5 to 30.5	37.5 to 39.5
Center Outlet Temperature in Degrees F.	38° to 42°	41° to 45°	39° to 43°	43° to 47°	40° to 44°	47° to 51°	41° to 45°	52° to 56°	43° to 47°	59° to 63°

*NOTE: These pressures are for engine with engine fan clutch engaged. With fan clutch disengaged, pressures generally are 25-35 psi higher than shown here.

**NOTE: Interior pressure of the evaporator is isolated from exterior atmospheric pressure. As a result, the controlling element (vacuum bellows) of the P.O.A. valve is able to operate independently of the effect of atmospheric pressure. However, any gauge used to check the pressure will not be free from the effect of atmospheric pressure. This altitude effect on the gauge must be taken into account when interpreting a reading. As the altitude increases and atmospheric pressure goes down, the pressure reading on the gauge will go up.

The increase noted in the above readings will be approximately .5 psi per 1,000 feet above sea level. For example at 90°F, in an arid climate at 2,000 feet above sea level, evaporator pressures would be 29 to 31 psi.

Fig. 103 Performance test chart. 1974 Apollo

denser, and temperature of air flow through cowl air inlet and past evaporator core.

7. Measure relative humidity and ambient temperature in immediate vicinity of car to be tested.

NOTE: The temperature obtained at the air outlets will be lower on dry days and higher on humid days.

8. Open all air conditioner outlets and measure temperature at right and left outlets.

If it appears from the test results that either the POA valve or the expansion valve is at fault, the following procedure will help determine which to replace.

(a) Check temperature door, make sure the door seals in the cool position, readjust the bowden cable if necessary.
(b) Check air hoses and ducts for proper connection.
(c) Check the sight glass for "clear" condition and make sure compressor clutch is engaged.

SERIES	AMBIENT TEMP. °F.	EVAPORATOR PRESSURE P.S.I.G. AT VIR	COMPRESSOR HEAD PRESSURE P.S.I.G.	OUTLET TEMP. °F. LEFT	RIGHT
CENTURY & REGAL	80	28 - 30	200 - 245	40 - 45	40 - 45
	90	28 - 30	240 - 280	45 - 49	45 - 49
	100	28 - 30	260 - 320	48 - 52	48 - 52
	110	28 - 30	300 - 340	51 - 55	51 - 55
ELECTRA, LESABRE & RIVIERA	70	28 - 30	160 - 190	37 - 41	37 - 41
	80	28 - 30	180 - 230	41 - 45	41 - 45
	90	28 - 31	200 - 270	44 - 50	44 - 50
	100	28 - 30	235 - 275	46 - 52	46 - 52
	110	29 - 32	280 - 340	48 - 54	48 - 54

Fig. 104 Performance test chart. 1974-76 Century, Electra, LeSabre, Regal & Riviera

AMBIENT TEMP. °F.	EVAPORATOR PRESSURE P.S.I.G. AT VIR	COMPRESSOR HEAD PRESSURE P.S.I.G.	OUTLET LEFT	TEMP. °F. RIGHT
70	28.5 – 30	180 – 210	39 – 40	40 – 43
80	29 – 31	205 – 230	41 – 45	43 – 47
90	30 – 32	235 – 270	45 – 52	47 – 54
100	30 – 35	275 – 310	48 – 54	50 – 56

Fig. 105 Performance test chart. 1975-76 Skylark

Ambient Air in Degrees F. (In Auxiliary Fan Air Blast Ahead of Condenser)	70°		80°		90°		100°		110°	
Air Quality	Arid	Humid	Arid	Humid	Arid	Humid	Arid	Humid	Arid	Humid
*Average Compressor Head Pressure in PSI	140	180	180	220	200	240	230	260	280	300
Average Evaporator Pressure **PSI AT SEA LEVEL	28	30	24	30	24	30	24	30	24	30
Center Outlet Temperature in Degrees F.	40°	42°	42°	44°	44°	47°	45°	48°	47°	51°

*NOTE: These pressures are for engine with engine fan clutch engaged. With fan clutch disengaged, pressures generally are 25-35 psi higher than shown here.

**NOTE: Interior pressure of the evaporator is isolated from exterior atmospheric pressure. As a result, the controlling element (vacuum bellows) of the P.O.A. valve is able to operate independently of the effect of atmospheric pressure. However, any gauge used to check the pressure will not be free from the effect of atmospheric pressure. This altitude effect on the gauge must be taken into account when interpreting a reading. As the altitude increases and atmospheric pressure goes down, the pressure reading on the gauge will go up.

The increase noted in the above readings will be approximately .5 psi per 1,000 feet above sea level. For example at 90°F, in an arid climate at 2,000 feet above sea level, evaporator pressures would be 29 to 31 psi.

Fig. 106 Performance test chart. 1975 Apollo & 1975-76 Skylark

After these basic visual checks, install evaporator and head pressure gages. Operate the engine at 1500 RPM, "Recirc" control setting and "Lo" blower.

(d) If evaporator pressure is 30 psi or less (and discharge air temperatures are too warm) replace the expansion valve.

(e) If evaporator pressure is above 30 psi, even with blower wire disconnected, make sure the expansion valve feeler bulb is clamped tightly to the evaporator outlet pipe and the feeler bulb insulation is in place. If the bulb and insulation are OK, replace the POA valve.

(f) If evaporator pressure is 30 psi ±1 psi (and discharge air temperatures are normal), partially cover the condenser to obtain head pressure from 325 psi to 379 psi maximum. If evaporator pressure rises above 30 psi, change the expansion valve. If expansion pressure remains at 30 psi, install a new receiver dehydrator.

1977-79 All & 1980 Except Skylark

1. On 1977-78 Skylark and 1977-79 Century and Regal, perform test with hood and windows opened and doors closed. On 1977-80

AMBIENT TEMP. AT GRILL	70°F	80°F	90°F	100°F
AVERAGE COMPRESSOR HEAD PRESSURE IN PSI	150 TO 190	180 TO 220	240 TO 280	300 TO 340
AVERAGE EVAPORATOR PRESSURE-PSI AT SEA LEVEL	28 TO 31	28 TO 31	28 TO 31	28 TO 31
LEFT CENTER OUTLET TEMP.	36 TO 40	36 TO 40	38 TO 42	41 TO 45

Fig. 107 Performance test chart. 1977-78 Skyhawk

AMBIENT TEMP. AT GRILL	70°F	80°F	90°F	100°F	110°F
AVERAGE COMPRESSOR HEAD PRESSURE IN PSI	125 TO 170	140 TO 190	165 TO 230	200 TO 260	220 TO 300
AVERAGE EVAPORATOR PRESSURE-PSI AT SEA LEVEL	24 TO 27	24 TO 27	24 TO 27	24 TO 35	34 TO 45
LEFT CENTER OUTLET TEMP.	40 TO 44	42 TO 45	43 TO 47	45 TO 52	45 TO 60

Fig. 108 Performance test chart. 1977-78 Skylark

AMBIENT TEMP. AT GRILL	70°F	80°F	90°F	100°F	110°F
AVERAGE COMPRESSOR HEAD PRESSURE IN PSI	N/A TO	140 TO 195	170 TO 230	210 TO 270	235 TO 330
AVERAGE EVAPORATOR PRESSURE-PSI AT SEA LEVEL	N/A TO	24 TO 28	24 TO 32	24 TO 40	30 TO 48
LEFT CENTER OUTLET TEMP.	N/A TO	41 TO 46	41 TO 49	41 TO 58	46 TO 67

Fig. 109 Performance test chart. 1977-78 Century & Regal

AMBIENT TEMP. AT GRILL	70°F	80°F	90°F	100°F	110°F
AVERAGE COMPRESSOR HEAD PRESSURE IN PSI	140 TO 180	170 TO 210	200 TO 240	215 TO 300	265 TO 305
AVERAGE EVAPORATOR PRESSURE-PSI AT SEA LEVEL	24 TO 27	24 TO 27	24 TO 27	24 TO 32	24 TO 33
LEFT CENTER OUTLET TEMP.	35 TO 43	35 TO 44	35 TO 44	35 TO 45	35 TO 45

Fig. 110 Performance test chart. 1977-78 Electra, LeSabre & Riviera

Temperature of Air Entering Condenser		70°F (21°C)	80°F (27°C)	90°F (32°C)	100°F (38°C)	110°F (43°C)
Compressor Out Pressure (Before Orifice)	CENTURY & REGAL	110-150 psi 758-1034 kPa	130-170 psi 896-1172 kPa	160-200 psi 1103-1379 kPa	195-235 psi 1344-1620 kPa	230-25- psi 1586-1724 kPa
	ELECTRA & LESABRE	120-160 psi 827-1103 kPa	150-190 psi 1034-1310 kPa	185-225 psi 1276-1551 kPa	215-255 psi 1482-1758 kPa	270-310 psi 1861-2137 kPa
	SKYLARK	105-145 psi 724-1000 kPa	130-170 psi 896-1172 kPa	170-210 psi 1172-1448 kPa	200-24- psi 1379-1655 kPa	235-275 psi 1620-1896 kPa
	SKYHAWK	130-170 psi 896-1172 kPa	180-220 psi 1241-1517 kPa	215-255 psi 1482-1758 kPa	240-280 psi 1655-1931 kPa	255-295 psi 1758-2134 kPa
Evaporator Pressure (At Accumulator)	CENTURY & REGAL	24-30 psi 165-207 kPa	24-30 psi 165-207 kPa	24-30 psi 165-207 kPa	24-30 psi 165-207 kPa	26-30 psi 179-207 kPa
	ELECTRA & LESABRE	25-30 psi 173-207 kPa	25-30 psi 173-207 kPa	25-30 psi 173-207 kPa	26-32 psi 180-220 kPa	26-32 psi 180-220 kPa
	SKYLARK	26-32 psi 165-220 kPa	26-32 psi 165-220 kPa	26-32 psi 165-220 kPa	26-32 psi 165-220 kPa	26-32 psi 165-220 kPa
	SKYHAWK	24-30 psi 166-207 kPa	24-30 psi 166-207 kPa	24-30 psi 166-207 kPa	24-32 psi 166-220 kPa	26-32 psi 186-220 kPa
Discharge Air Temperature-Left Center Outlet	CENTURY & REGAL	34-38°F 1.1 - 3.3°C	34-38°F 1.1 - 3.3°C	35-39°F 1.7 - 3.9°C	38-42°F 3.3 - 5.6°C	40-44°F 4.4 - 6.7°C
	ELECTRA & LESABRE	38-42°F (3-6°C)	38-42°F (3-6°C)	38-42°F (3-6°C)	38-42°F (3-6°C)	42-48°F (6-9°C)
	SKYLARK	38-42°F 3.3 - 5.6°C	38-42°F 3.3 - 5.6°C	38-42°F 3.3 - 5.6°C	38-42°F 3.3 - 5.6°C	38-42°F 3.3 - 5.6°C
	SKYHAWK	38-42°F (3-6°C)	38-42°F (3-6°C)	38-42°F (3-6°C)	38-42°F (3-6°C)	38-42°F (3-6°C)

Fig. 111 Performance test chart. 1979 Buick all models

TEMPERATURE OF AIR ENTERING CONDENSER		70°F. (21°C.)	80°F. (27°C.)	90°F. (32°C.)	100°F. (38°C.)	110°F. (43°C.)
*COMPRESSOR OUT PRESSURE (BEFORE ORIFICE)	CENTURY & REGAL	110-150 psi 758-1034 kPa	130-170 psi 896-1172 kPa	160-200 psi 1103-1379 kPa	195-235 psi 1344-1620 kPa	230-250 psi 1586-1724 kPa
	ELECTRA & LESABRE	120-160 psi 827-1103 kPa	150-190 psi 1034-1310 kPa	185-225 psi 1276-1551 kPa	215-255 psi 1482-1758 kPa	270-310 1861-2137 kPa
	SKYHAWK	130-170 psi 896-1172 kPa	180-220 psi 1241-1517 kPa	215-255 psi 1482-1758 kPa	240-280 psi 1655-1931 kPa	255-295 psi 1758-2034 kPa
*EVAPORATOR PRESSURE (AT ACCUMULATOR)**	CENTURY & REGAL	24-30 psi 165-207 kPa	24-30 psi 165-207 kPa	24-30 psi 165-207 kPa	24-30 psi 165-207 kPa	26-30 psi 179-207 kPa
	ELECTRA & LESABRE	25-30 psi 173-207 kPa	25-30 psi 173-207 kPa	25-30 psi 173-207 kPa	26-32 psi 180-220 kPa	26-32 psi 180-220 kPa
	SKYHAWK	24-30 psi 166-207 kPa	24-30 psi 166-207 kPa	24-30 psi 166-207 kPa	24-32 psi 166-220 kPa	26-32 psi 186-220 kPa
*DISCHARGE AIR TEMPERATURE—LEFT CENTER OUTLET	CENTURY & REGAL	34-38°F. 1.1-3.3°C.	34-38°F. 1.1-3.3°C.	35-39°F. 1.7-3.9°C.	38-42°F. 3.3-5.6°C.	40-44°F. 4.4-6.7°C.
	ELECTRA & LESABRE	38-42°F. (3-6°C.)	38-42°F. (3-6°C.)	38-42°F. (3-6°C)	38-42°F. (3-6°)	42-48°F. (6-9°C.)
	SKYHAWK	38-42°F. (3-6°C.)	38-42°F. (3-6°C.)	38-42°F. (3-6°C.)	38-42°F. (3-6°C.)	38-42°F. (3-6°C.)

Fig. 112 Performance test chart. 1980 Buick all models

Electra, LeSabre, Riviera, Skyhawk, 1978-80 Skylark and 1980 Century and Regal, perform test with hood opened and doors and windows closed.

2. On all models, manually disengage fan clutch by tightening spring on fan clutch to first notch past normal setting.
3. Position a 2 ft. diameter auxiliary cooling fan ahead of front bumper.
4. Connect manifold gauge set to A/C system.
5. Position suitable thermometers at front of grille and in instrument panel left center A/C air outlet.
6. Set A/C controls to the maximum full cold position, with A/C outlets in the fully opened position.
7. Check to ensure that vehicle interior temperature is below 90 degrees F.
8. Operate A/C system for a minimum of 10 minutes prior to noting test readings.
9. With engine operating at 2000 RPM, note test readings and compare with those listed in Figs. 107 through 112.

NOTE: Obtain reading just before compressor clutch releases, as clutch will cycle throughout test procedure. The temperature obtained at the A/C outlets will be lower than those listed on dry days and higher than those listed on humid days.

10. If outlet temperature is within or below limits, place blower motor switch on low and observe accumulator and evaporator pipes for frost. If no frost is present, system is operating properly. If frost is present on 1977-78 models, the thermostatic switch is defective and should be replaced. If frost is present on 1979-80 models, the pressure cycling switch is defective and should be replaced.
11. If outlet temperature is above the specified limit, check accumulator pressure and refer to Figs. 107 through 112. If accumulator pressure is low, check for plugged expansion tube and repair and replace as necessary, then replace accumulator and evacuate and recharge system. If accumulator pressure is normal or high, check compressor clutch cycling. If compressor clutch cycles on and off, replace thermostatic switch on 1977-78 models. On 1979-80 models, replace pressure cycling switch. If compressor operates continuously, check for missing expansion tube. If expansion tube is in place, check compressor inlet screen. If inlet screen is plugged, repair or replace as necessary, then replace accumulator and evacuate and recharge system. If compressor screen is not plugged, system is overcharged and must be evacuated and recharged.

1980 Skylark

Refer to Insufficient Cooling Diagnosis Procedure, 1980 Skylark when performing the performance test.

INSUFFICIENT COOLING DIAGNOSIS

1977-78 All & 1979 Except Skyhawk

When diagnosing insufficient cooling problems, refer to Fig. 113 for diagnostic procedure.

1979 Skyhawk & 1980 Models Except Skylark

When diagnosing insufficient cooling problems, refer to Fig. 114 parts 1 and 2 for diagnostic procedure.

1980 Skylark

When diagnosing insufficient cooling problems, refer to Fig. 115 parts 1 through 3 for diagnostic procedure.

1980 SKYLARK COMPRESSOR CLUTCH DIAGNOSIS

Refer to Compressor clutch diagnosis chart, Fig. 116, for diagnostic procedure.

1980 SKYLARK ENGINE COOLING FAN DIAGNOSIS

Refer to engine cooling fan diagnosis chart, Fig. 117, for diagnostic procedure.

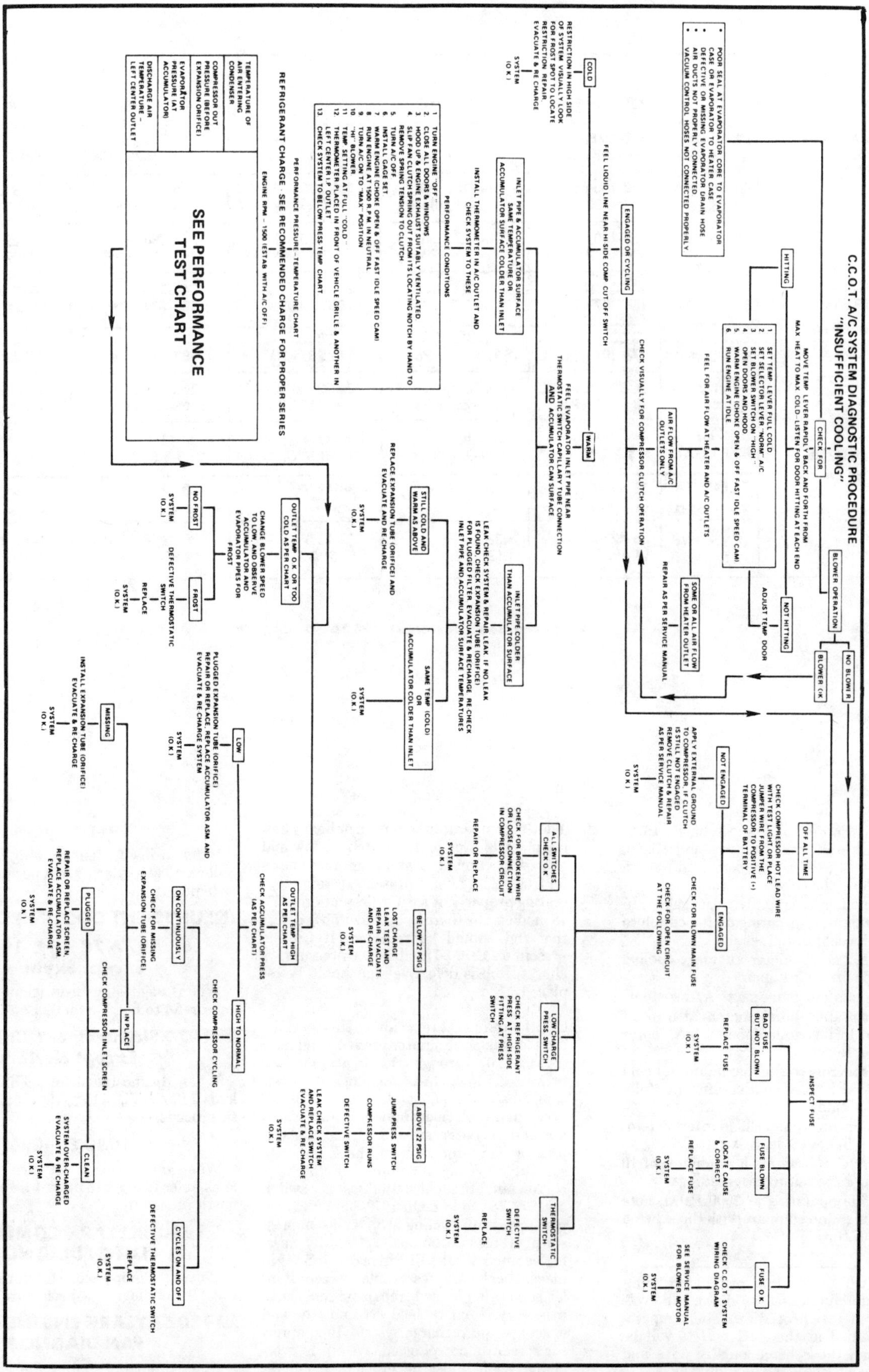

Fig. 113 Insufficient cooling diagnosis chart. 1977-78 Buick all & 1979 Except Skyhawk

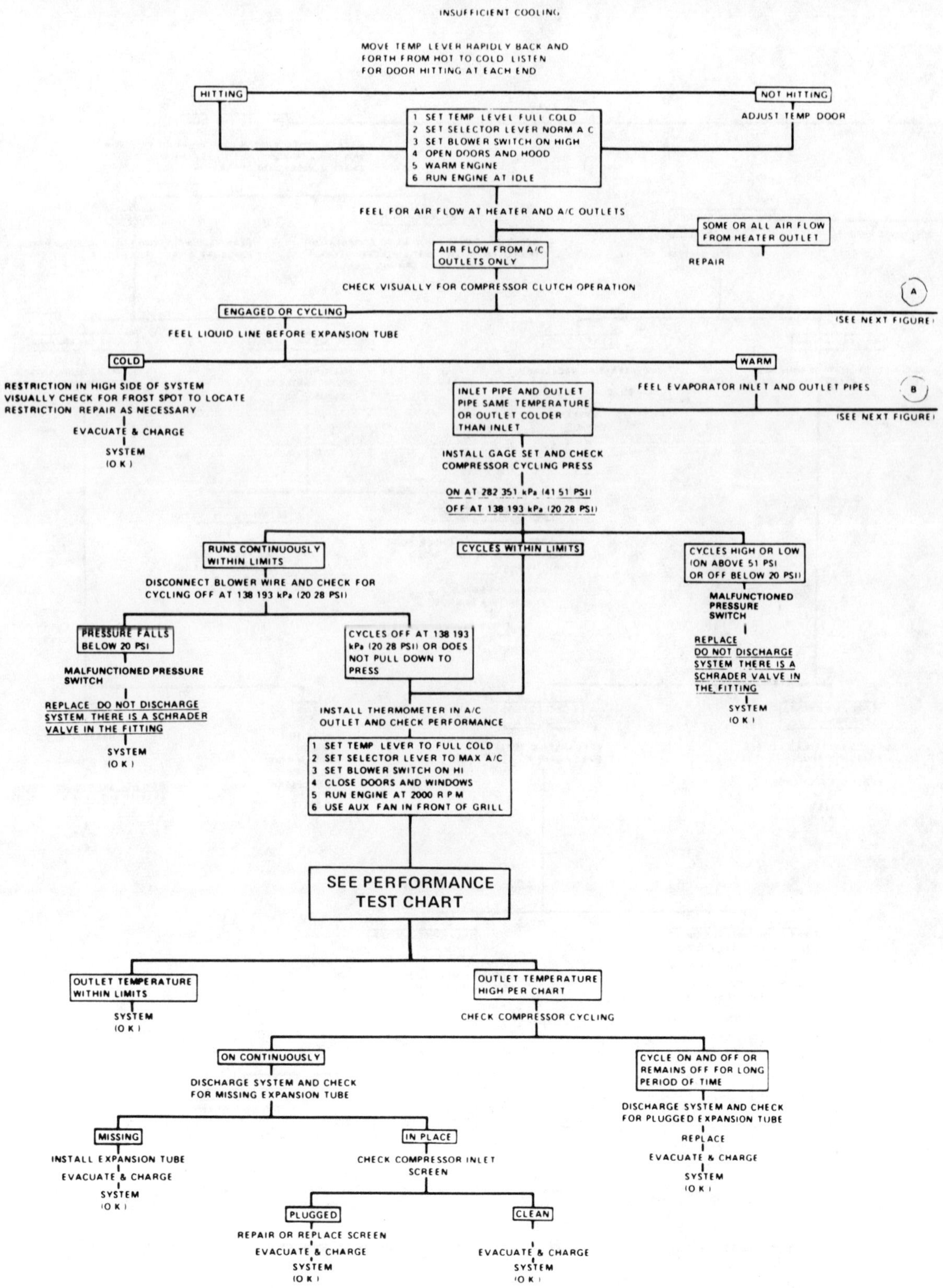

Fig. 114 Insufficient cooling diagnosis chart. 1979-80 G.M. models w/pressure cycling switch (Part 1 of 2)

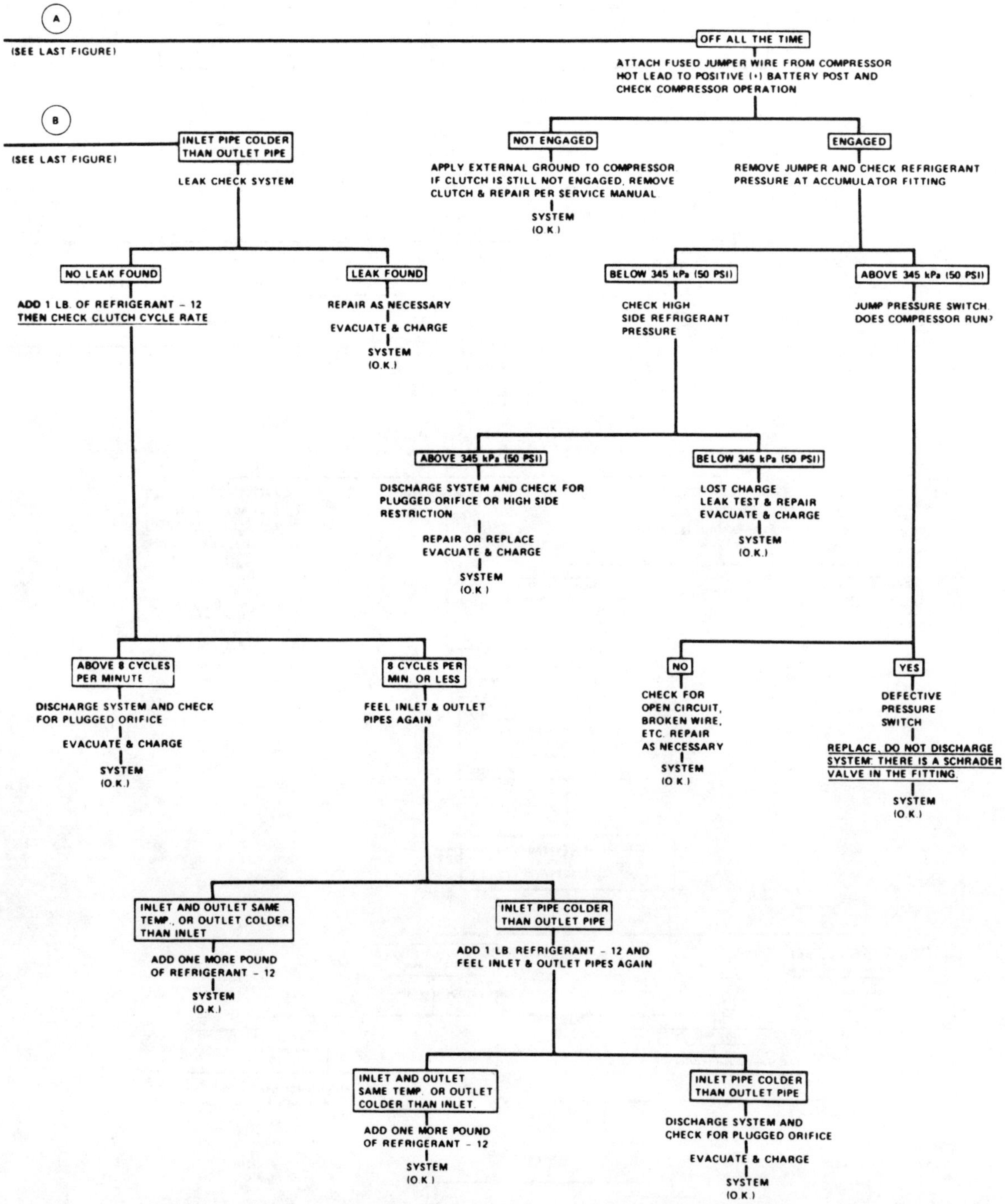

Fig. 114 Insufficient cooling diagnosis chart. 1979-80 G.M. models w/pressure cycling switch (Part 2 of 2)

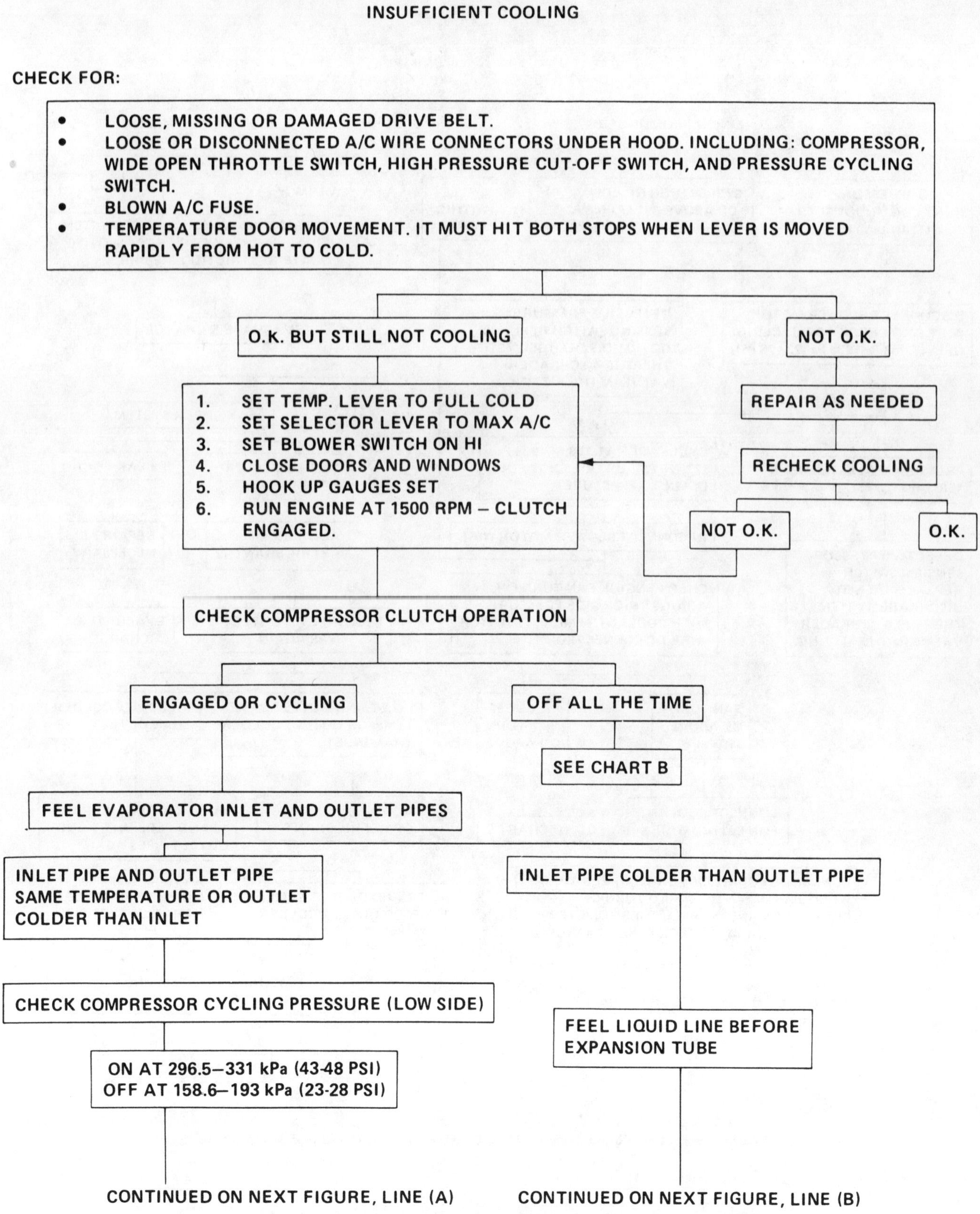

Fig. 115 Insufficient cooling diagnosis chart. 1980 Citation, Omega, Phoenix & Skylark (Part 1 of 3)

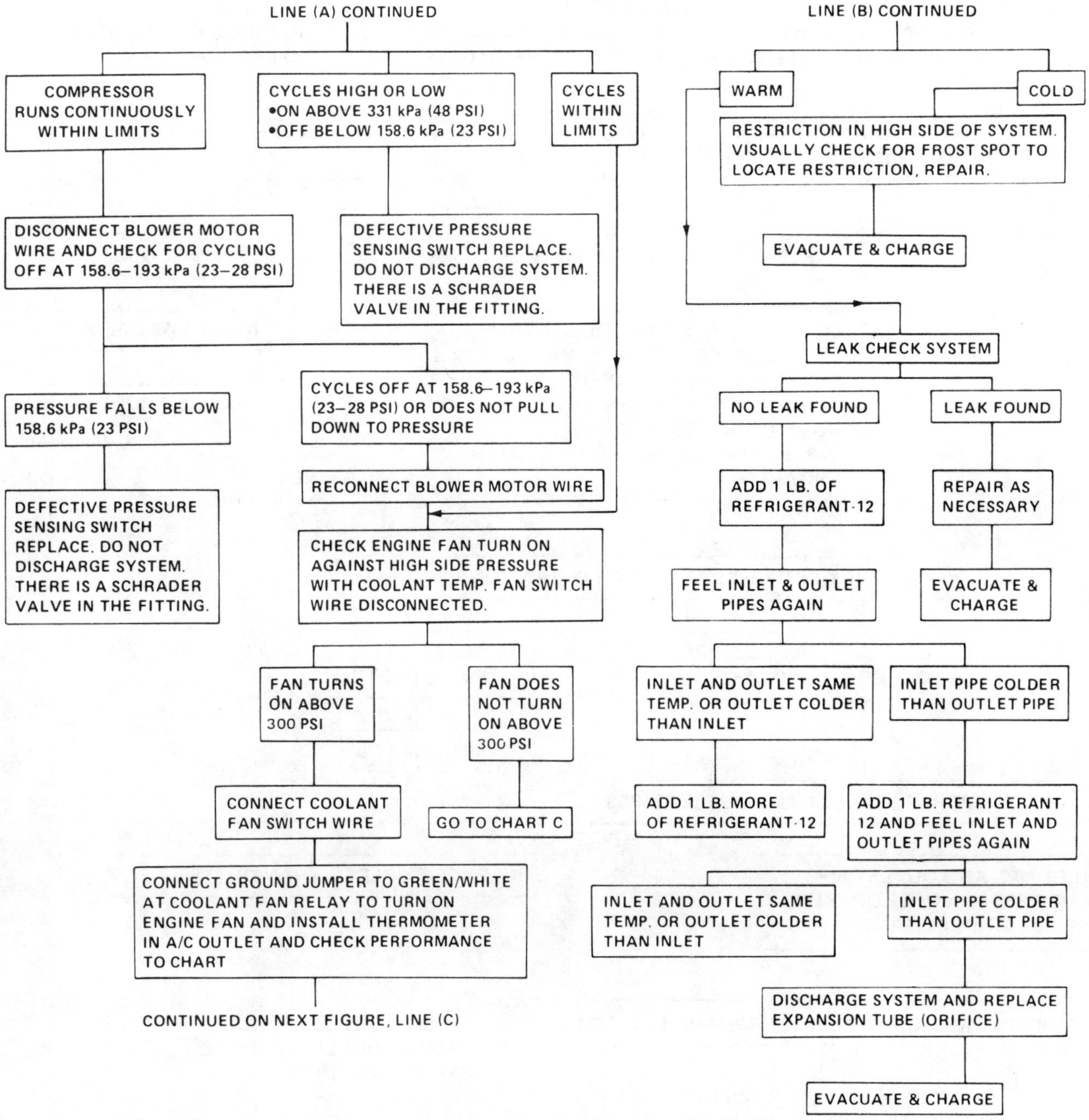

Fig. 115 Insufficient cooling diagnosis chart. 1980 Citation, Omega, Phoenix & Skylark (Part 2 of 3)

LINE (C) CONTINUED

TEMP. OF AIR ENTERING THE CONDENSER	70°F.	80°F.	90°F.	100°F.	110°F.
*COMPRESSOR OUT PRESSURE	135-170 psi 930-1170 kPa	170-205 psi 1170-1410 kPa	205-245 psi 1410-1690 kPa	245-285 psi 1690-1960 kPa	285-320 psi 1960-2100 kPa
*ACCUMULATOR PRESSURE	23-28 psi 158-193 kPa	23-29 psi 158-200 kPa	26-35 psi 185-240 kPa	31-40 psi 215-275 kPa	35-44 psi 185-305 kPa
DISCHARGE AIR TEMP.—LEFT CENTER OUTLET	36-43°F. 1-6°C.	36-43°F. 1-6°C.	36-43°F. 1-6°C.	42-48°F. 5-9°C.	48-53°F. 9-11°C.

*JUST BEFORE COMPRESSOR ENGAGES, IF CYCLING.

- OUTLET TEMPERATURE WITHIN LIMITS
  - COMPARE COMPRESSOR OUT PRESSURE (HIGH SIDE) TO CHART
    - WITHIN LIMITS
      - DISCONNECT JUMPER FROM FAN RELAY
      - SYSTEM OK
    - NOT IN LIMITS
      - SYSTEM OVERCHARGED
      - DISCHARGE REFRIGERANT UNTIL CLUTCH CYCLES
      - DISCONNECT JUMPER FROM FAN RELAY
- OUTLET TEMPERATURE HIGH
  - CHECK FOR OVERCHARGE BY DISCHARGING REFRIGERANT 30 SECONDS AT A TIME UP TO 4 TIMES CHECK OUTLET TEMPERATURE AFTER EACH DISCHARGE.
    - TEMPERATURE O.K.
      - DISCONNECT JUMPER FROM FAN RELAY
    - TEMPERATURE NOT O.K.
      - DISCHARGE AND REPLACE EXPANSION TUBE (ORIFICE)
      - EVACUATE AND RECHARGE
      - DISCONNECT JUMPER FROM FAN RELAY

Fig. 115 Insufficient cooling diagnosis chart. 1980 Citation, Omega, Phoenix & Skylark (Part 3 of 3)

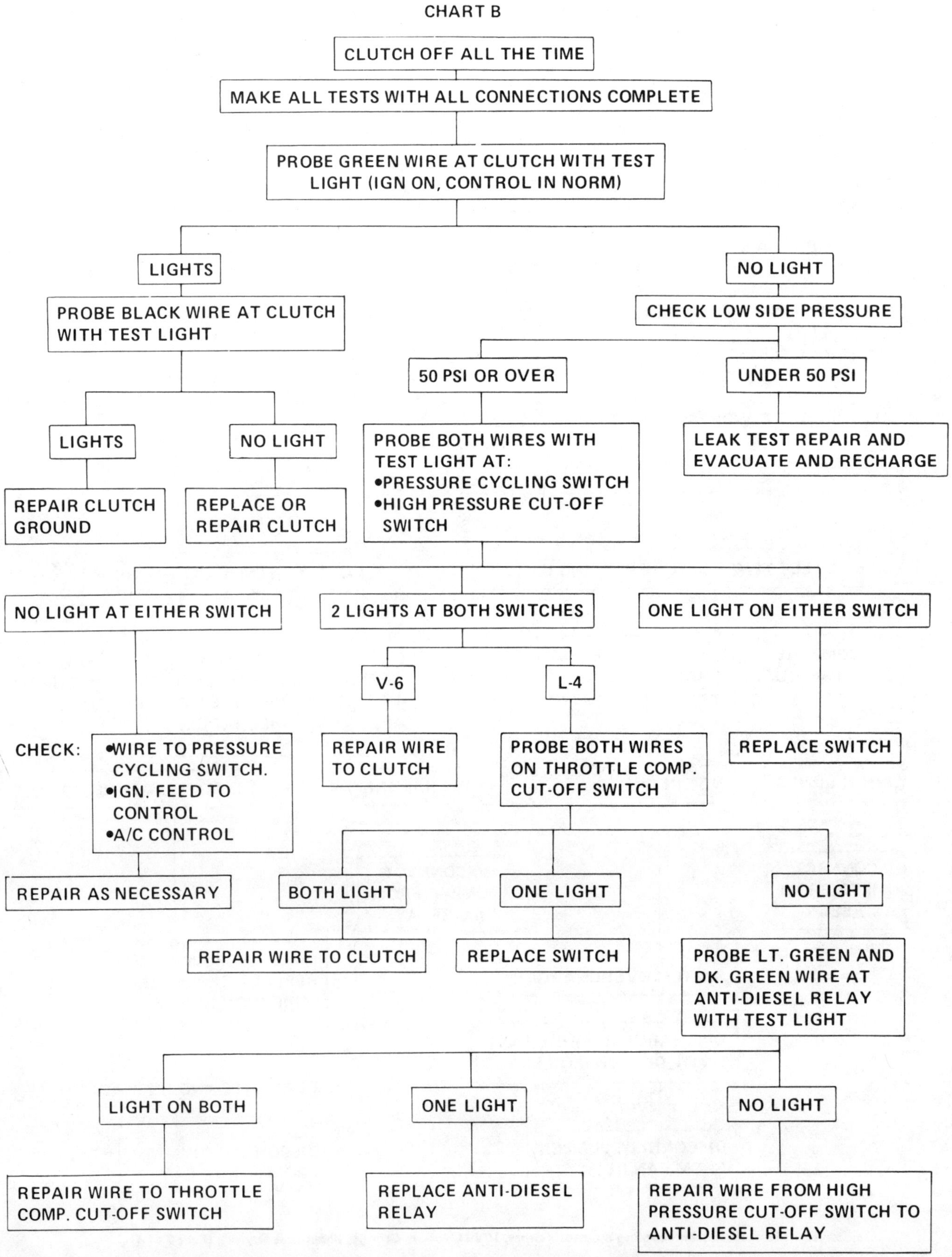

Fig. 116 Compressor clutch diagnosis chart. 1980 Citation, Omega, Phoenix & Skylark

CHART C

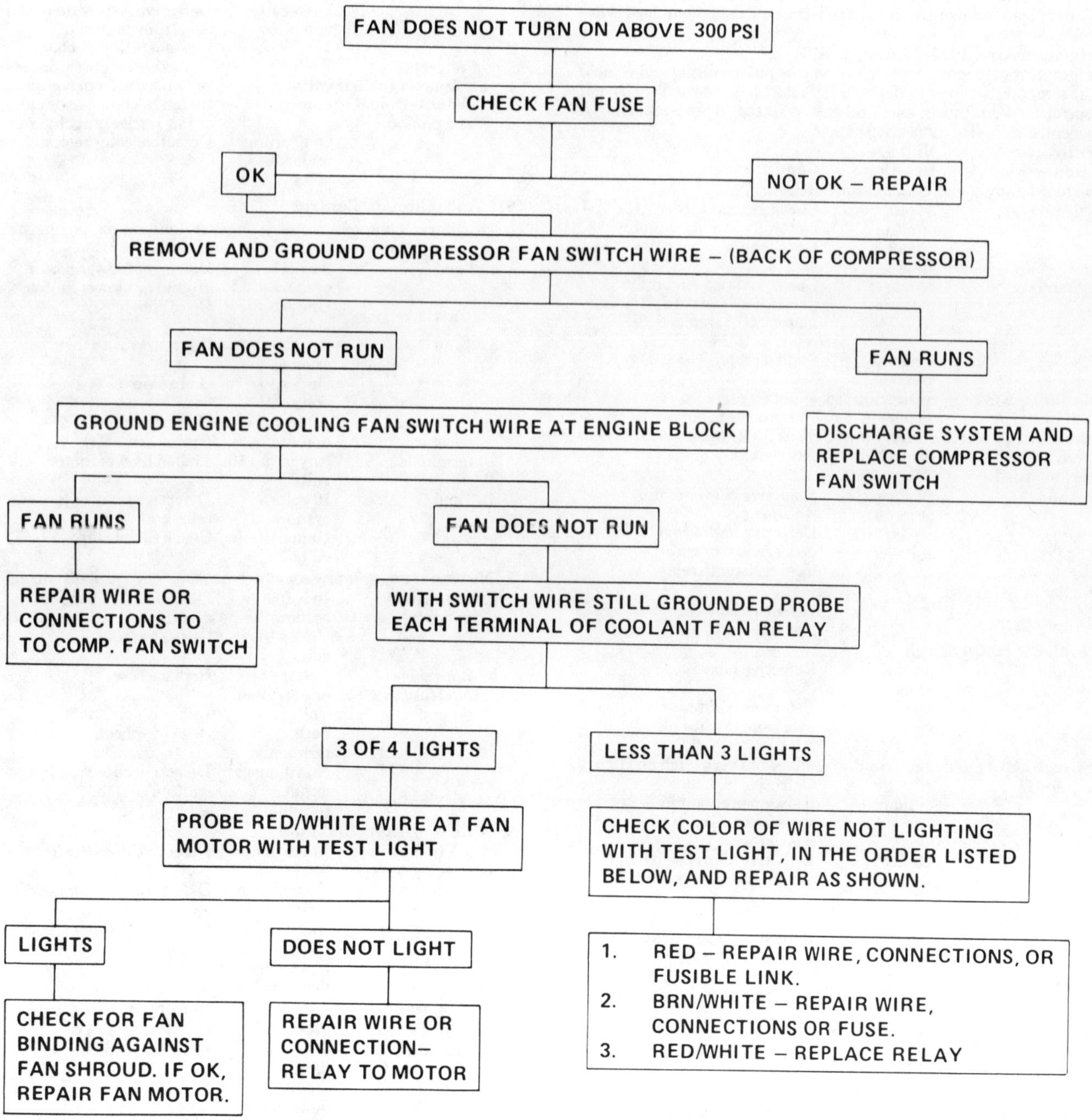

Fig. 117 Engine cooling fan diagnosis chart. 1980 Citation, Omega, Phoenix & Skylark

1977-78 Full Size A/C Electrical Diagnosis Connector

These models are equipped with two electrical diagnosis connectors, Figs. 118 through 120. To perform the following diagnosis procedures, a voltmeter and jumper wire are required. If a switch or relay is suspected of being defective, connect a jumper wire across switch or relay. If system operates, the switch or relay is defective. If system does not operate, the wire or connector is defective. With jumper wire connected, connect voltmeter from jumper wire to ground. If no voltage is present, the positive lead is defective. If voltage is present the ground lead is defective.

1977 Custom Aire, 1977-78 Manual A/C

Perform test with ignition switch on, engine not operating and vehicle interior temperature over 55 degrees F. Place blower switch on Lo and set control for Vent mode unless otherwise stated. Before replacing a component, check wire connection at unit.

Voltmeter Connections	Voltage Reading	Condition
1. No blower any speed, compressor clutch engages.		
a. 1 and G	3 volts or more	Check ground wire to motor, if ok, replace motor
	Under 3 volts	Go to step b.
b. 2 and G	3 volts or more	Check for open circuit in purple wire (relay to motor). If ok, replace relay.
	Under 3 volts	Stalled motor.
2. No blower and compressor clutch does not engage. If heater A/C fuse is blown, check for ground in circuit, or excessive motor or clutch current. If not, check for an open circuit in lead from ignition switch to low blower delay switch.		
3. No low speed.		
a. 3 and G	10 volts or more	Defective blower motor resistor.
	Under 10 volts	Defective low blower delay switch or wire from resistor to relay.
4. No low or M1 speed.		Defective blower motor resistor.
5. No blower except in high.		
a. 2 and G	3 volts or more	Defective relay.
	Under 3 volts	Defective resistor or wire from resistor to relay.
6. No high blower under any condition. Place blower switch on High.		
a. 9 and G	10 volts or more	Defective high blower relay, open circuit in red wire to relay or poor relay ground.
	Under 10 volts	Defective blower switch or wire from blower switch to relay coil.
7. No M1 speed.		Defective blower switch or wire to resistor.
8. No M2 speed.		Defective blower switch or wire to resistor.
9. No M1, M2, or high with switch. Place blower switch in High.		
a. Voltage at blower switch feed terminal from A/C control	10 volts or more	Defective blower switch.
	Under 10 volts	Defective A/C control or wire to blower switch
10. Low blower only, compressor clutch inoperative.		
a. 4 and G	10 volts or more	Defective A/C control.
	Under 10 volts	Defective wire from low blower delay switch to A/C control.
11. Compressor clutch and blower motor inoperative. See step 2.		
12. Compressor clutch operates in Norm., Max., Bi-Level or Defrost, but not in all.		Defective A/C control.
13. Compressor clutch inoperative, blower motor operates, A/C control in Norm.		
a. 6 and G	10 volts or more	Go to step c.
	Under 10 volts	Go to step b.
b. 8 and G	10 volts or more	Defective thermostatic switch or lead to clutch
	Under 10 volts	Defective A/C control or lead from control to thermostatic switch.
c. Connect voltmeter to each clutch terminal and ground.	Both 10 volts or more	Defective compressor cut-off switch, wire from compressor switch to compressor clutch or low refrigerant charge.
	One terminal under 10 volts	Defective compressor clutch.

1977 Auto Climate Control

Voltmeter Connections	Voltage Reading	Condition
1. No blower, compressor clutch engages, ignition on, engine stopped and over 120 degrees F, Control in off position.		
a. 1 and G	3 volts or more	Defective blower motor ground or blower motor.
	Under 3 volts	Go to step b.
b. 3 and G	10 volts or more	Go to step c.
	Under 10 volts	Go to step d.
c. Volts at all resistor terminals	All 10 volts or more	Defective circuit 65 (splice to motor).
	Any under 10 volts	Stalled blower motor.
d. 2 and G	10 volts or more	Defective relay or leads to relay coil.
	Under 10 volts	Check fusible link.
2. No blower and compressor clutch does not engage, ignition on.		
a. 4 to engine ground	10 volts or more	Defective A/C harness ground.
	Under 10 volts	Go to step b.
b. Both sides of A/C fuse to ground	Voltage at one terminal	Replace fuse.
	Voltage at both terminals	Defective circuit 50 (fuse to splice).
	No voltage at either terminal	Defective circuit 300 or ignition switch.
3. No blower except in Defrost.		
Check in-car and engine temperature	In-car above 75° F	Defective in-car switch.
	Engine above 120° F	Defective cylinder head switch
	In-car below 75° F and engine below 120° F	System operating properly.
4. No low blower.		Defective resistor.
5. Low blower only, ignition, engine temperature over 120 degrees F.		
Voltage at circuit 42 at control (E)	10 volts or more	Defective control.
	Under 10 volts	Defective circuit 42.
6. Low blower only except in High or Defrost. Engine operating with engine temperature over 120 degrees F and A/C control set at Economy position.		
a. 5 and G	10 volts or more	Defective blower circuit in programmer.
	Under 10 volts	Go to step b.
b. Voltage at circuit 64 at control D	10 volts or more	Defective circuit 64, control to programmer.
	Under 10 volts	Defective control.

Voltmeter Connections	Voltage Reading	Condition
7. Blower speed does not vary in Auto. Engine operating with engine temperature above 120° F. Check vacuum hoses at engine and vacuum reservoir for damage and proper connection.		
a. Stays in High and Max. heat or Max. cold	Open sensor string	Go to step b.
b. Ground circuit 62 at programmer 3 terminal connector K and hold for 10 seconds	No speed change	Defective programmer.
	Speed changes	Go to step c.
c. Ground circuit 198 at control J	No speed change	Faulty temperature dial.
	Speed changes	Defective in-car sensor.
d. Remove circuit 62 (K) at programmer for 10 seconds	No speed change	Defective programmer.
	Speed changes	Short in sensor or grounded sensor string, go to step e.
e. Reconnect circuit 62 (K) at programmer and connect voltmeter from that wire to ground. Move temp. dial from 65 to 85 and watch voltmeter	Varies about .5 volt	Shorted in-car sensor or sensor lead.
	Does not vary	Shorted temperature dial or ground wire from temperature dial to programmer.
8. Skipping any speed(s) in Auto. Ignition on with engine temperature over 120° F.		
a. Control on Defrost check voltage at each resistor terminal	Any under 10 volts	Defective resistor.
	All under 10 volts	Go to step b.
b. Control on Economy blower switch connector disconnected. Check voltage at each connector terminal	All 3 volts or more	Defective blower switch.
	Any under 3 volts	Faulty lead to terminal with low voltage.
9. No high blower.		Faulty circuit 65.
10. Blower operates with ignition off.		Defective relay.
11. Compressor clutch does not engage, blower motor operates satisfactorily. Engine operating, with engine temperature above 120° F.		
a. 6 and G	10 volts or more	Faulty clutch ground, clutch or compressor wiring harness.
	Under 10 volts	Go to step b.
b. 8 and G	10 volts or more	Defective temperature switch or circuit 59 to clutch.
	Under 10 volts	Go to step c.
c. 7 and G	10 volts or more	Low refrigerant charge, defective refrigerant pressure switch or circuit 67 to temperature switch
	Under 10 volts	Go to step d.
d. Circuit 66 (A) and 50 (B) to ground at control	10 volts or more	Defective control.
	Under 10 volts	Faulty circuit 50 splice to control.
12. Compressor does not engage, blower motor does not operate.		Refer to step 2.
13. A/C does not turn on until engine warms-up.		Defective in-car switch.

1978 Auto Climate Control

Voltmeter Connections	Voltage Reading	Condition
1. No blower, compressor clutch engages, ignition on, engine stopped and over 120 degrees F and control in off position.		
a. 1 and G	3 volts or more	Faulty blower motor ground or motor.
	Under 3 volts	Go to step b.
b. 3 and G	10 volts or more	Go to step c.
	Under 10 volts	Go to step d.
c. Voltage at all resistor terminals	All 10 volts or more	Faulty splice (circuit 65) to blower motor.
	Any under 10 volts	Stalled blower motor.
d. 2 and G	10 volts or more	Defective relay or leads to relay coil.
	Under 10 volts	Check fusible link.
2. Blower motor and compressor clutch do not operate, ignition on.		
a. 4 and engine ground	10 volts or more	Refer to steps 1 and 11.
	Under 10 volts	Go to step b.
b. Both sides of A/C fuse to ground	Voltage at one terminal	Replace fuse.
	Voltage at both terminals	Faulty circuit 50, fuse to splice.
3. No blower motor operation except in Defrost.		
a. Check in-car and engine temperature	In-car above 75° F	Defective in-car switch.
	Engine above 120°F	Defective cylinder head switch.
	In-car below 75° F & engine below 120° F	System operating properly.
4. No low blower.		Defective resistor.
5. Low blower only, ignition on, engine temperature above 120° F.		
a. Voltage at circuit 42 (E) at control (relay to control)	10 volts or more	Defective control.
	Under 10 volts	Faulty circuit 42.
6. Low blower only except in high or Defrost. Engine operating with engine temperature over 120° F and control on Economy.		
a. 5 and G	10 volts or more	Faulty blower circuit in programmer.
	Under 10 volts	Go to step b.
b. Voltage at circuit 64 at control (D)	10 volts or more	Faulty circuit 64, control to programmer.
	Under 10 volts	Defective control.
7. Blower speed does not vary in Economy, Auto. and Bi-Level, engine operating with engine temperature over 120 degrees F.		
a. Disconnect temp. door rod and recheck	Blower speed varies	Temperature door sticking.
	Blower speed does not vary	Check condition of vacuum hoses. If satisfactory, proceed to step b.
b. Voltage at circuit 62 (K) at programmer to ground while cycling temp. dial	Varies approx. .5 to 1.5 volts	Defective programmer.
	Does not vary .5 to 1.5 volt	Go to step c.
c. Blower operates at high speed only	System remains in Max. Heat	Go to step d.
	System remains in Max. cooling	Check circuits B and H and ground J at programmer. If satisfactory, go to step f.

Voltmeter Connections	Voltage Reading	Condition
d. Ground circuit 62 at programmer 3 terminal connector (K) and hold for 10 seconds	No speed change	Defective programmer.
	Speed changes	Go to step e.
e. Ground circuit 198 at control (J)	No speed change	Defective temp. dial.
	Speed changes	Defective in-car sensor.
f. Remove circuit 62 at programmer for 10 seconds (K)	No speed change	Defective programmer.
	Speed changes	Go to step g.
g. Reconnect circuit 62 (K) at programmer and connect voltmeter from that wire to ground and cycle temp. dial from 65 to 85	Voltage varies about .5 volts	Faulty in-car sensor or sensor lead.
	Voltage does not vary	Faulty temp. dial or grounded wire from dial to programmer.

8. Skipping any speed(s) in Auto. Ignition on and engine temperature above 120° F.

Voltmeter Connections	Voltage Reading	Condition
a. Set control at Defrost volts at each resistor terminal	Any under 10 volts	Defective resistor.
	All 10 volts or more	Go to step b.
b. Set control at Economy	All 3 volts or more	Defective blower switch.
Blower switch connector removed. Check volts at each terminal	Any under 3 volts	Faulty lead to one with low voltage.

9. No high blower.		Faulty circuit 65.
10. Blower operates with ignition off.		Defective relay.

11. Compressor does not engage, blower motor operates properly. Engine operating, with engine temperature over 120° F and control in Auto.

Voltmeter Connections	Voltage Reading	Condition
a. 6 and G	10 volts or more	Go to step c.
	Under 10 volts	Go to step b.
b. 8 and G	10 volts or more	Faulty thermostatic switch or circuit 59 to clutch.
	Under 10 volts	Go to step d.
c. Voltage at both clutch terminals	10 volts or more at both	Low refrigerant pressure or faulty compressor cutoff switch or circuit to clutch.
	10 volts or more at one terminal	Defective compressor clutch.
d. Circuit 66 (A) and 50 (B) to ground at control	10 volts or more	Faulty circuit 66, control switch to thermostatic switch
	10 volts or more to one terminal	Defective control.
	Under 10 volts at both	Faulty circuit 50 splice to control.

12. Compressor clutch does not engage, blower operation.		Refer to step 2.
13. A/C does not turn on until engine warms up.		Defective in-car switch.

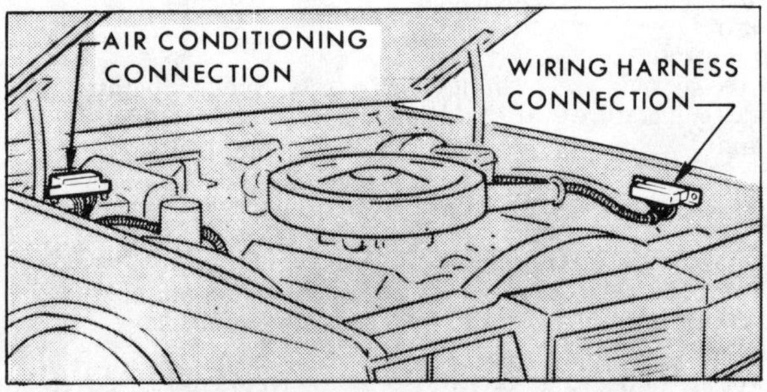

Fig. 118 Electrical diagnosis connector locations. 1977-80 Buick (except 1979-80 Riviera), Chevrolet, Oldsmobile (except Toronado) & Pontiac Full Size Models

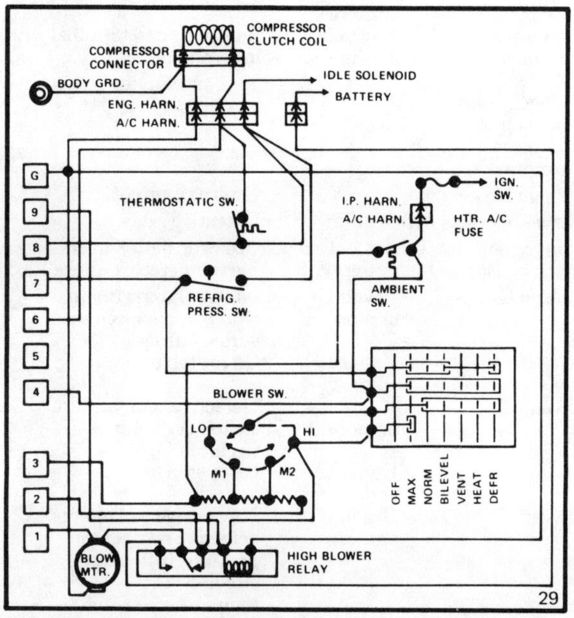

Fig. 119 Diagnosis connector schematic. 1977-78 Buick & Pontiac Full Size with manual A/C

MANUAL A/C

A/C Vacuum System Diagnosis

1978-80 Century & Regal

Refer to Fig. 121 parts 1 through 5, for A/C system diagnostic procedures.

1980 Skylark

Refer to Fig. 122 parts 1 and 2, for A/C vacuum system diagnostic procedures.

Control System Functional Test

1976-80 Skyhawk, Skylark, 1977-80 Electra, LeSabre, Riviera & 1978-80 Century & Regal

Refer to functional test charts, Figs. 123 to 129, when testing control system operation.

1974 AUTO CLIMATE CONTROL

Testing Procedure

1. Start the car and place the control head selector lever in the "auto" position and the temperature dial at 75. Allow enough time for the car engine to warm up and A.C.C. system to come on if the system were acting

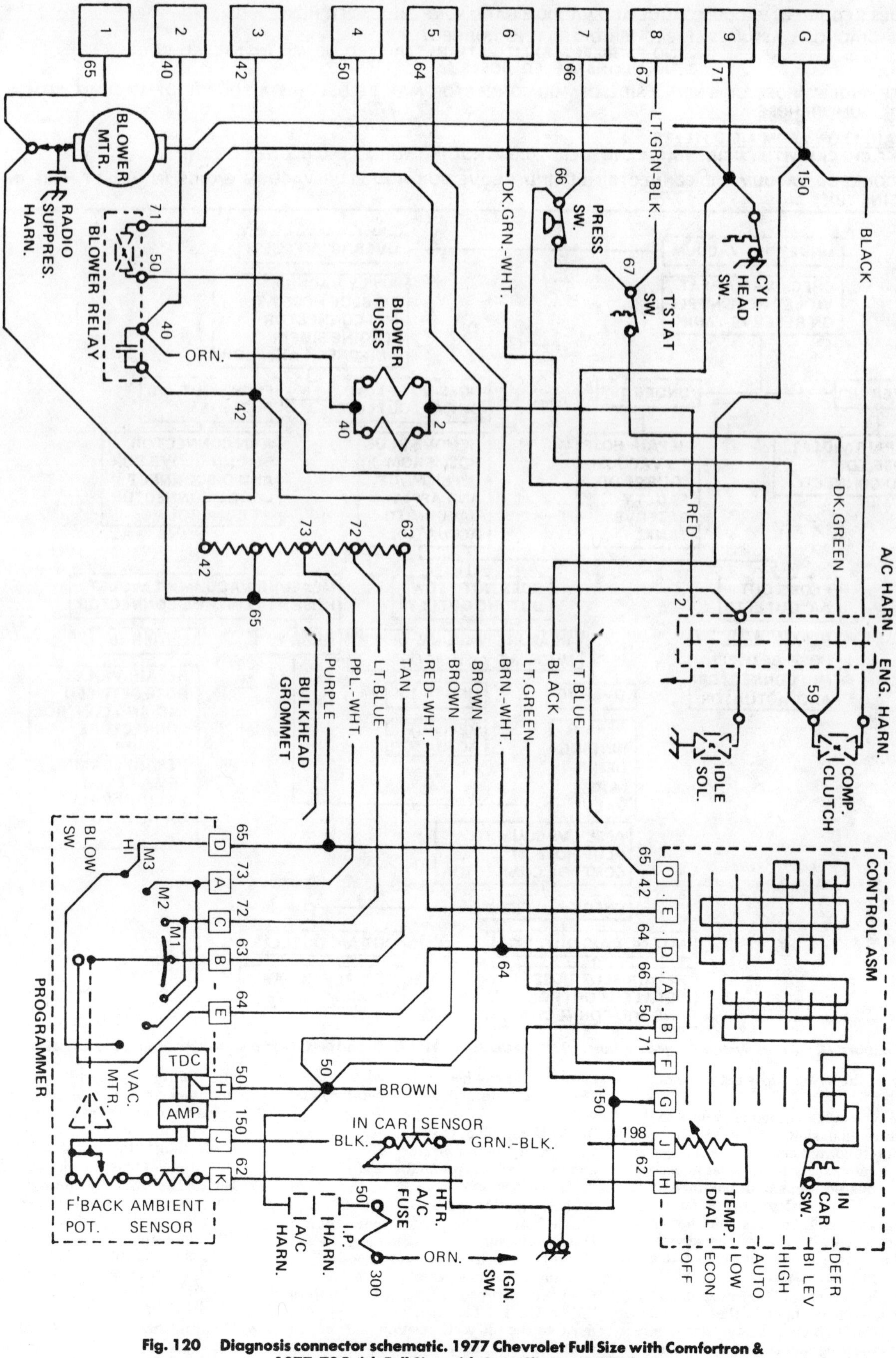

Fig. 120 Diagnosis connector schematic. 1977 Chevrolet Full Size with Comfortron & 1977-78 Buick Full Size with Auto Climate Control

AIR FLOW DIAGNOSIS

TOOLS REQUIRED: VACUUM GAUGE AND VACUUM HAND PUMP OR JUMPER HOSE.

THIS DIAGNOSIS ASSUMES:
1. BLOWER OPERATING CORRECTLY.
2. NO PROBLEMS EXIST WITH RESTRICTED OR MIS-INSTALLED DUCTS.
3. NO MIS-INSTALLED HOSES.

NOTE: VIOLET HOSE ON ENGINE SIDE OF MID-CONNECTOR MAY BE USED AS A SOURCE OF VACUUM WITH A JUMPER HOSE.

NO AIR FLOW FROM A/C OUTLETS

MAKE CHECKS WITH ENGINE WARM AND IDLING, CONTROL IN "NORM" AND BLOWER ON "HI".

DISCONNECT VACUUM MID-CONNECTOR BEHIND GLOVE BOX AND PLUG VACUUM GAUGE IN VIOLET HOSE IN ENGINE SIDE.

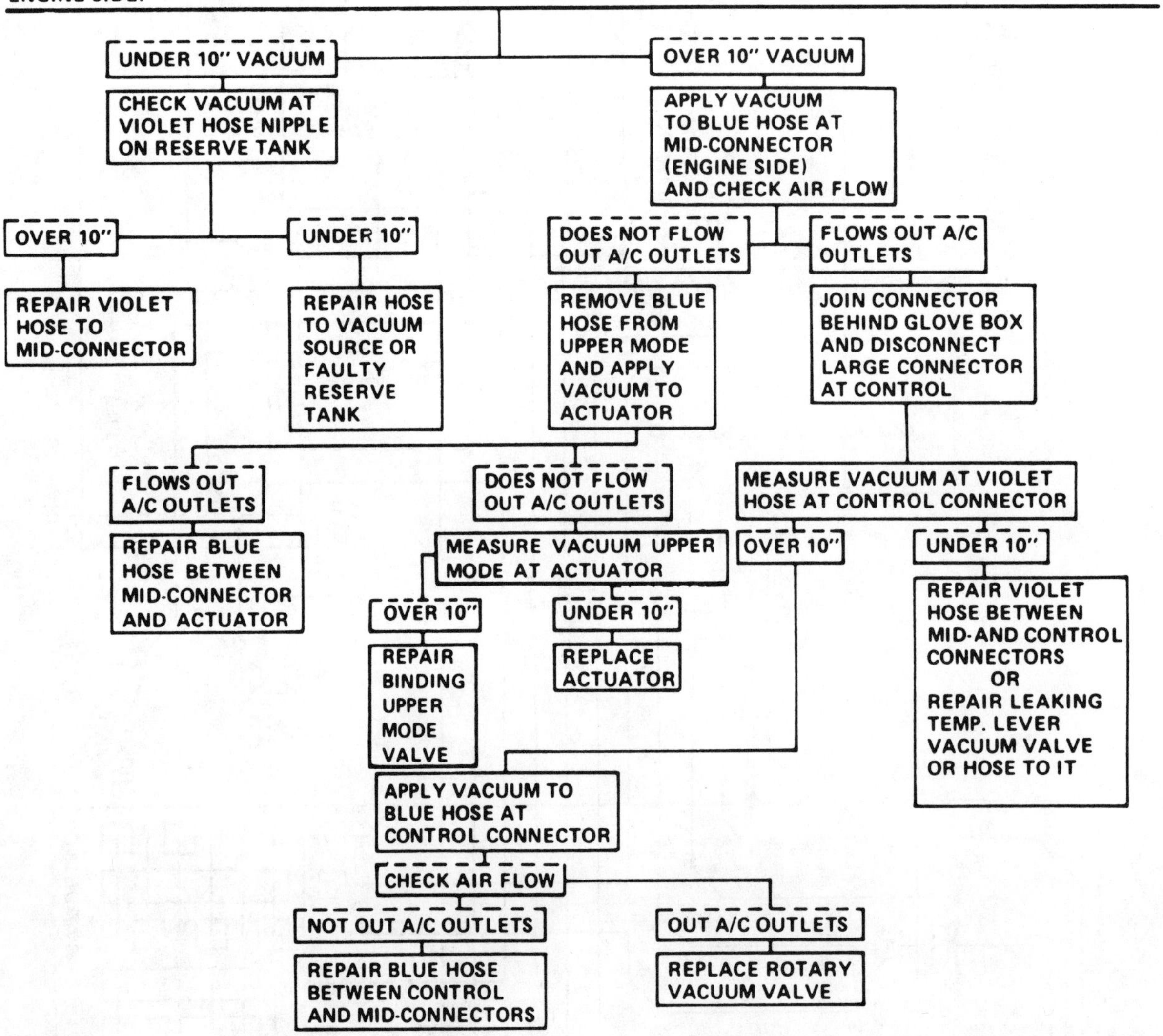

Fig. 121 Manual A/C vacuum system diagnosis chart. 1978-80 Century, Cutlass, Grand Prix, LeMans, Malibu, Monte Carlo & Regal (Part 1 of 5)

normally. If the system does not come on, skip to Step 3.

2. After the system has turned on, rotate the temperature dial slowly back and forth looking for abnormal operation of the system. Then move the selector lever to each of the various positions and look for any malfunction in the system's operation. By observing any malfunctions and noting when they occur, oftentimes the serviceman will be able to isolate the problem to a certain area of the system.
3. Remove the cover of the programmer (while it is still mounted in the car) and connect J-23678 A.C.C. Tester electrical harness to the programmer and the car harness.
4. Place the control head selector lever in the "Auto" position and the temperature dial on 75.
5. On the tester, place the "Temperature Dial Calibrator" switch in the "Off" position and the "Manual-Automatic" switch in the "Manual" position.
6. Rotate the "Manual Control" knob to "Max. Heat." The programmer should move to the full heat position and the fan should run at high-blower speed. (The vacuum motor mechanism will move into the vacuum motor.) Rotate the knob to "Max. Cold." The programmer should move to the full A/C position and the fan should run at high-blower speed. (The vacuum motor mechanism will move out of the vacuum motor.) If only partial programmer movement occurs, or "Hi Blower" is not obtained at both extremes, make the air mix door link adjustment *before* proceeding to the next step. If the programmer does *not* move at all, skip to Step 9.
7. Rotate the "Manual Control" slowly counterclockwise. The vacuum motor mechanism should first start to move at precisely 180 ohms (plus or minus 1 ohm) on the "Manual Control" knob. If this first movement occurs before or after 180 ohms, make the programmer amplifier calibration (feedback "pot" adjustment).
8. Rotating the "Manual Control" knob to the "Max. Cold" position should cause the system to shift to full A/C operation and the

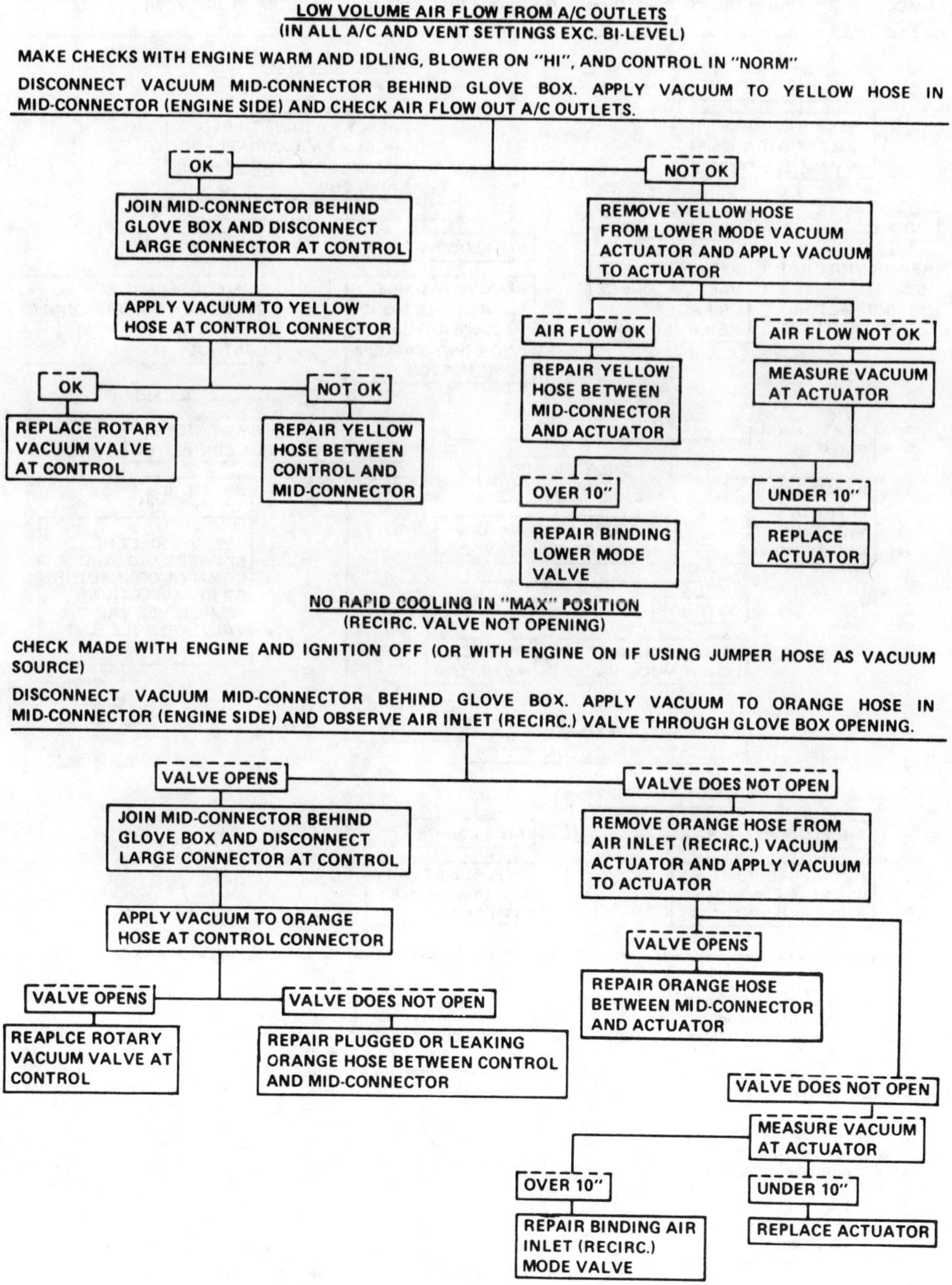

Fig. 121 Manual A/C vacuum system diagnosis chart. 1978-80 Century, Cutlass, Grand Prix, LeMans, Malibu, Monte Carlo & Regal (Part 2 of 5)

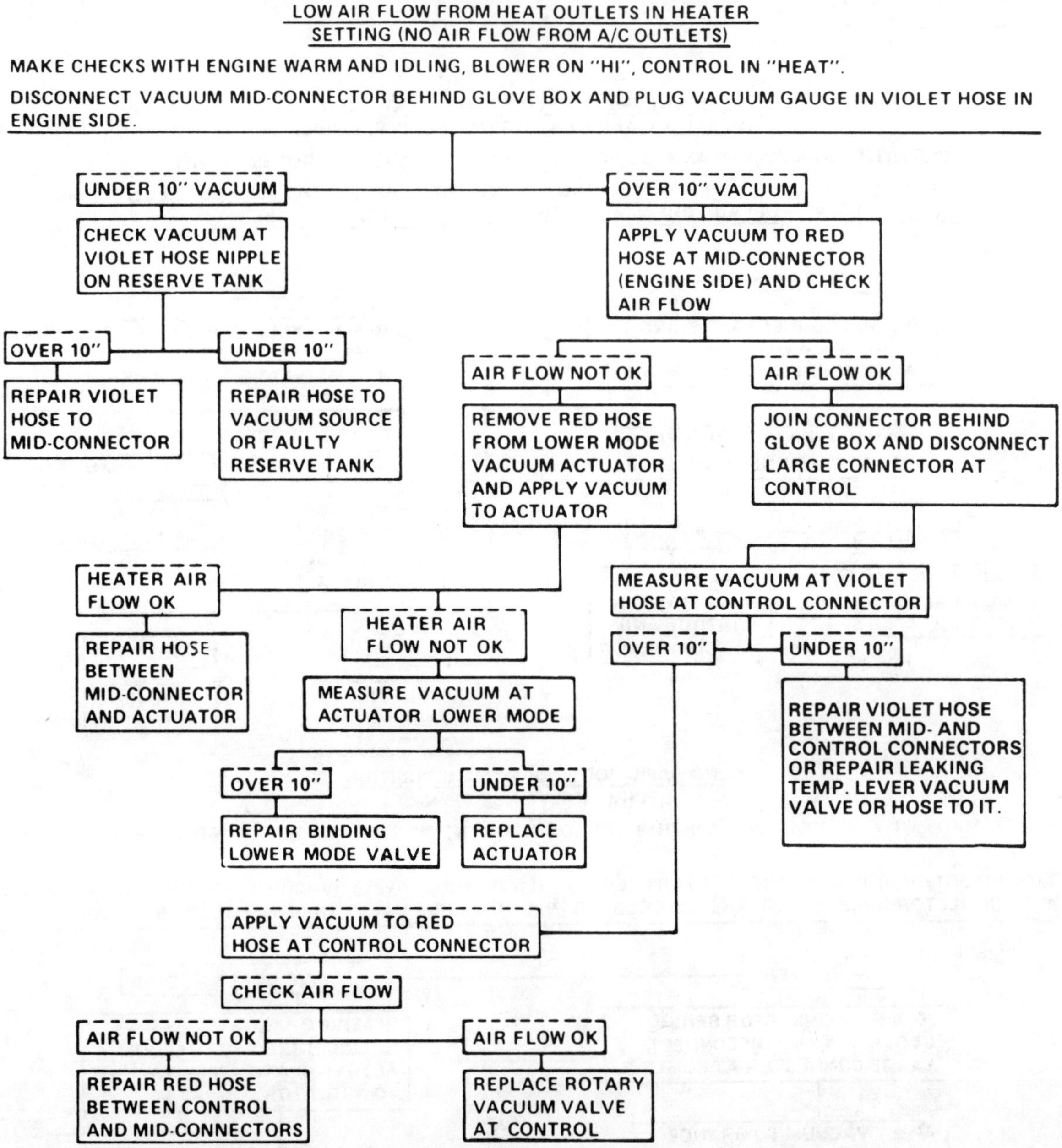

Fig. 121 Manual A/C vacuum system diagnosis chart. 1978-80 Century, Cutlass, Grand Prix, LeMans, Malibu, Monte Carlo & Regal (Part 3 of 5)

vacuum motor mechanism will move out of the vacuum motor. If the programmer moves normally when rotating the "Manual Control," skip to Step 11.

Programmer Does Not Move

9. On the A.C.C. Tester place the "Temperature Dial Calibration" switch in the "Off" position and the "Voltmeter" control in the "Programmer 12-V Supply" position. Battery voltage should appear on meter. No voltage indicates the *lack* of a ground on terminal No. 1 of the programmer or the *lack* of battery supply to terminal No. 2 of the programmer.

10. Plug the tester's dummy vacuum plug on the programmer vacuum valve. Connect the dummy hose to the vacuum supply hose (Port No. 2) in the car vacuum harness. Make sure vacuum is present. The programmer now has vacuum supply with no car vacuum system components connected. If the programmer functions using the "Manual Control" on the tester, troubleshoot the A.C.C. vacuum system. If the programmer does not function, the defect is in programmer. Remove the dummy plug and reconnect the vacuum harness after making this test.

Sensor String Test

11. Place the control head selector lever in the "Auto" position and the temperature dial on 75.
12. On the J-23678 A.C.C. Tester, place the "Manual-Automatic" switch in the "Auto" position.
13. Observe the position of the vacuum motor mechanism.
14. On the A.C.C. Tester, switch the "Manual Automatic" switch to the "Manual" position and then adjust the "Manual Control" knob until the vacuum motor mechanism assumes the same position as it had in Step 13. When the "Manual Control" knob is properly adjusted, switching the "Manual-Automatic" switch back and forth will result in no movement of the vacuum motor mechanism.
15. Read the setting of the "Manual Control." This resistance reading is the same as the resistance of the two sensors and the temperature dial combined and should be 120 to 150 ohms at 70 to 75 degrees F. room

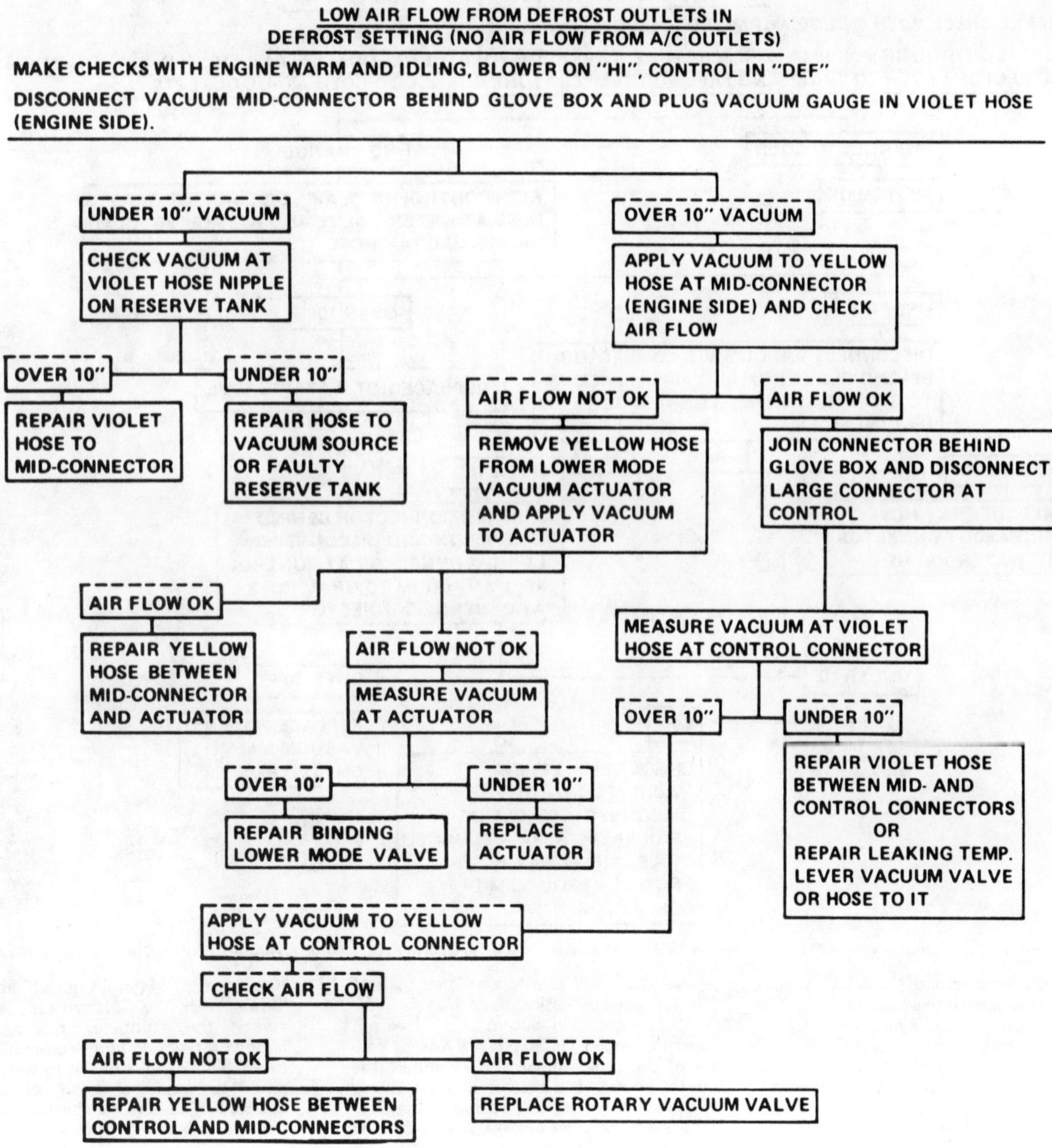

Fig. 121 Manual A/C vacuum system diagnosis chart. 1978-80 Century, Cutlass, Grand Prix, LeMans, Malibu, Monte Carlo & Regal (Part 4 of 5)

temperature. If this reading is incorrect, try calibrating the temperature dial. If the calibration can be accomplished, then the temperature dial is "good." Visually check the ambient and in-car sensors for shorts or bad connections.

Lo Relay Test

16. Place the A.C.C. Tester's "Temperature Dial Calibration" switch in the "Off" position and the "Voltmeter" knob in the "Lo Relay" position.
17. Place the control head selector lever in the "Vent" or "De-Ice" position. If the "Lo Relay" is energized, the voltmeter will read battery voltage. The "Lo Relay" is operated by: 1) Hot engine thermal switch; 2) Hot in-car temperature switch; 3) Control head in "Vent" or "De-Ice" position. If the relay does *not* energize in the control head "Vent" or "De-Ice" position, the relay or the control head switch is defective.

Programmer Blower Switch Test

18. Place the A.C.C. Tester's "Temperature Dial Calibrator" switch in the "Off" position, the "Voltmeter" knob in the "Blower" position, and the "Manual-Automatic" switch in the "Manual" position.
19. Using the "Manual Control" move the programmer from full heat to the full A/C position. The voltage at the blower (coming from programmer terminal No. 5) should be battery voltage in full heat. As the programmer moves from full heat, the voltage should drop in steps, indicating different blower speeds, and then increase in steps to battery voltage in full A/C. If the voltage steps are not present or battery voltage is not indicated on the tester voltmeter when in full heat and full A/C operation, then the programmer is defective.

On-The-Bench Testing Procedure For Programmer

Before attempting to repair or adjust a programmer, locate the "output shaft," the "feedback potentiometer," the "vacuum motor," the "blower switch," the "vacuum checking relay," the "transducer," and the "rotary vacuum valve."

When following the troubleshooting procedure, do not skip any steps unless instructed to do so.

The A.C.C. tester "Manual-Automatic" switch should always be in the "Manual" position and the "Temperature Dial Calibration" switch should always be in the "Off" position when troubleshooting the programmer on the bench.

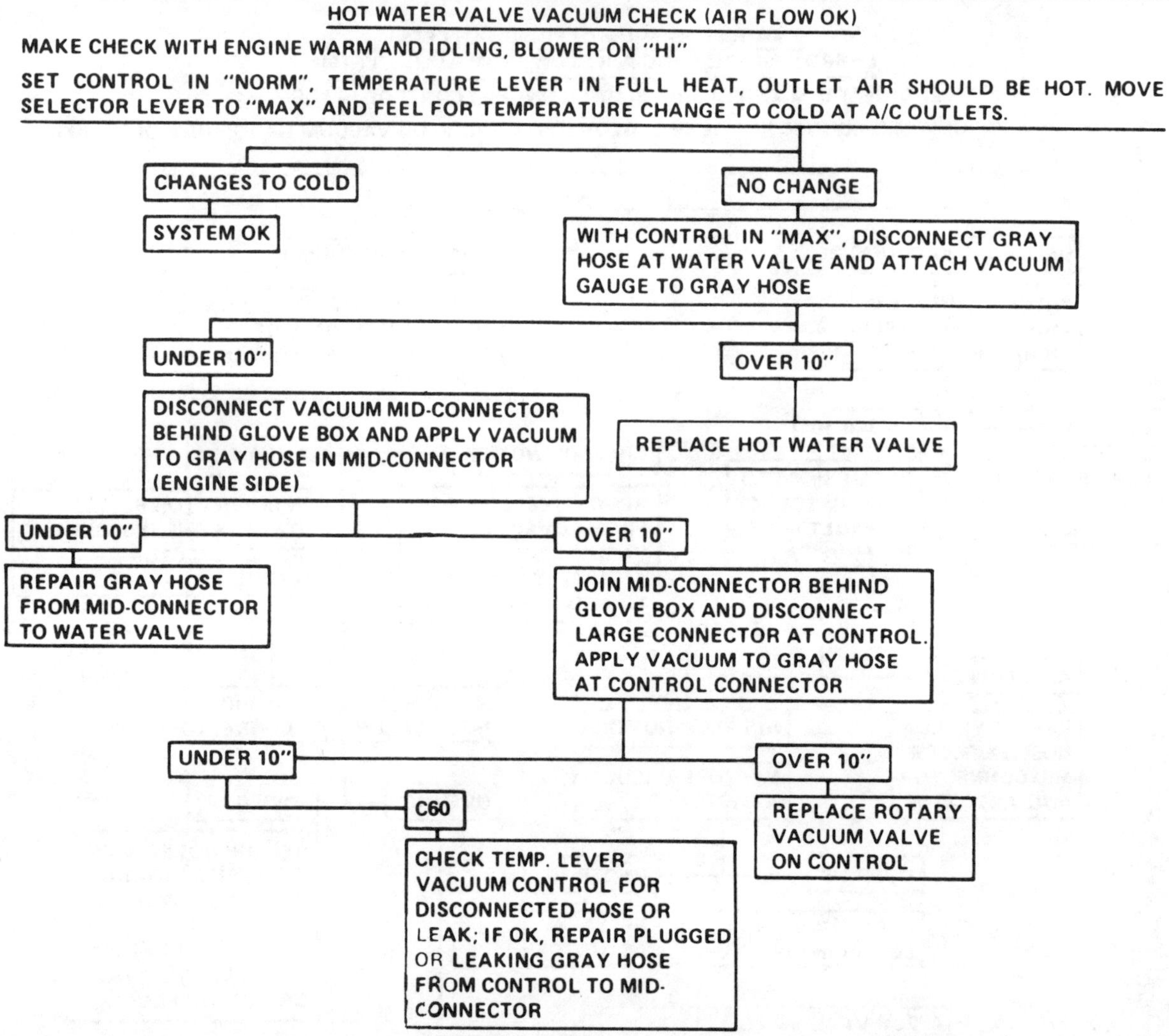

Fig. 121 Manual A/C vacuum system diagnosis chart. 1978-80 Century, Cutlass, Grand Prix, LeMans, Malibu, Monte Carlo & Regal (Part 5 of 5)

1. Plug the male connector of the A.C.C. Tester electrical harness into programmer.
2. Plug the electrical test connector No. J-23713 into female connector of A.C.C. tester electrical harness.
3. Plug the hard rubber dummy plug from the tester onto the programmer rotary valve and then connect it to the vacuum supply. Turn on the vacuum supply to 20 inches Hg. or more.
4. Connect the positive lead of a 12-volt DC power supply altered to the yellow wire from electrical test connector and connect the negative lead of the supply to the black wire of the test connector. The "Voltage" knob of the tester should read 12 volts when in the "No. 2 Feed" position, the "No. 7 Lo Relay" position, and the "No. 8 Auto Relay" position. If it doesn't, check the voltage supply and hook-up.
5. On the A.C.C. tester, place the "Manual-Automatic" switch in the "Manual" position and the "Temperature Dial Calibration" switch in the "Off" position.
6. Rotate the A.C.C. tester "Manual Control" knob from "Max. Heat" to "Max. Cold." (Make sure the power supply is adjusted for 12 volts DC and vacuum supply is at 20 inches Hg. or more.) The programmer output shaft should rotate 120 degrees (⅓ rotation of shaft) from maximum heat to maximum cold. If it does rotate normally, adjust the feedback potentiometer.
7. If the programmer output shaft does not operate properly in Step 6, remove the cover of the programmer and reconnect the tester electrical connector and the rotary vacuum valve rubber dummy plug. Recheck programmer operation as described in Step 6. If the programmer output shaft now operates normally, look for pinched vacuum hose, loose electrical connection, pinched or shorted wires, shorted Darlington Amplifier heat sink, or a mechanical bind with the programmer cover. Be sure to tap the programmer components with the handle of a screwdriver to check for intermittent problems.

Amplifier Test

8. If the programmer output shaft still does not operate properly, place the "Voltage" knob of the tester in the "Probe and Clip" position. Connect the alligator clip lead to the transducer terminal with the gray wire.

CAUTION: Do not short transducer terminals together or amplifier will be damaged.

Push the probe into the programmer connector body making contact with terminal No. 2 which connects to a yellow wire.

9. Rotating the "Manual Control" knob from "Max. Heat" to "Max. Cold" (or vice-versa), the voltmeter reading should *change 5 volts or more* (after it is stabilized). This change indicates that the amplifier is "good." If the 5 volts or more change is *not* obtained, check the amplifier heat sink insulator for proper positioning.

NOTE: If the heat sink is shorted to the programmer chassis, the programmer will go to the full AC position.

If the voltage change cannot be obtained and the insulator is okay, check the calibration of the amplifier according to the instructions. If the programmer still does not function properly, replace the amplifier circuit board and be sure to *install the new heat sink insulator properly.*

The new amplifier *must be calibrated* accordingly.

Transducer Test

10. If the 5 volts change was present in the amplifier check in Step 9, but the programmer output shaft does not operate normally, leave the "Probe and Clip" connected, as instructed in Step 8, *throughout Transducer Test Procedure.*
11. Disconnect vacuum hose from the small diameter port of the transducer and connect

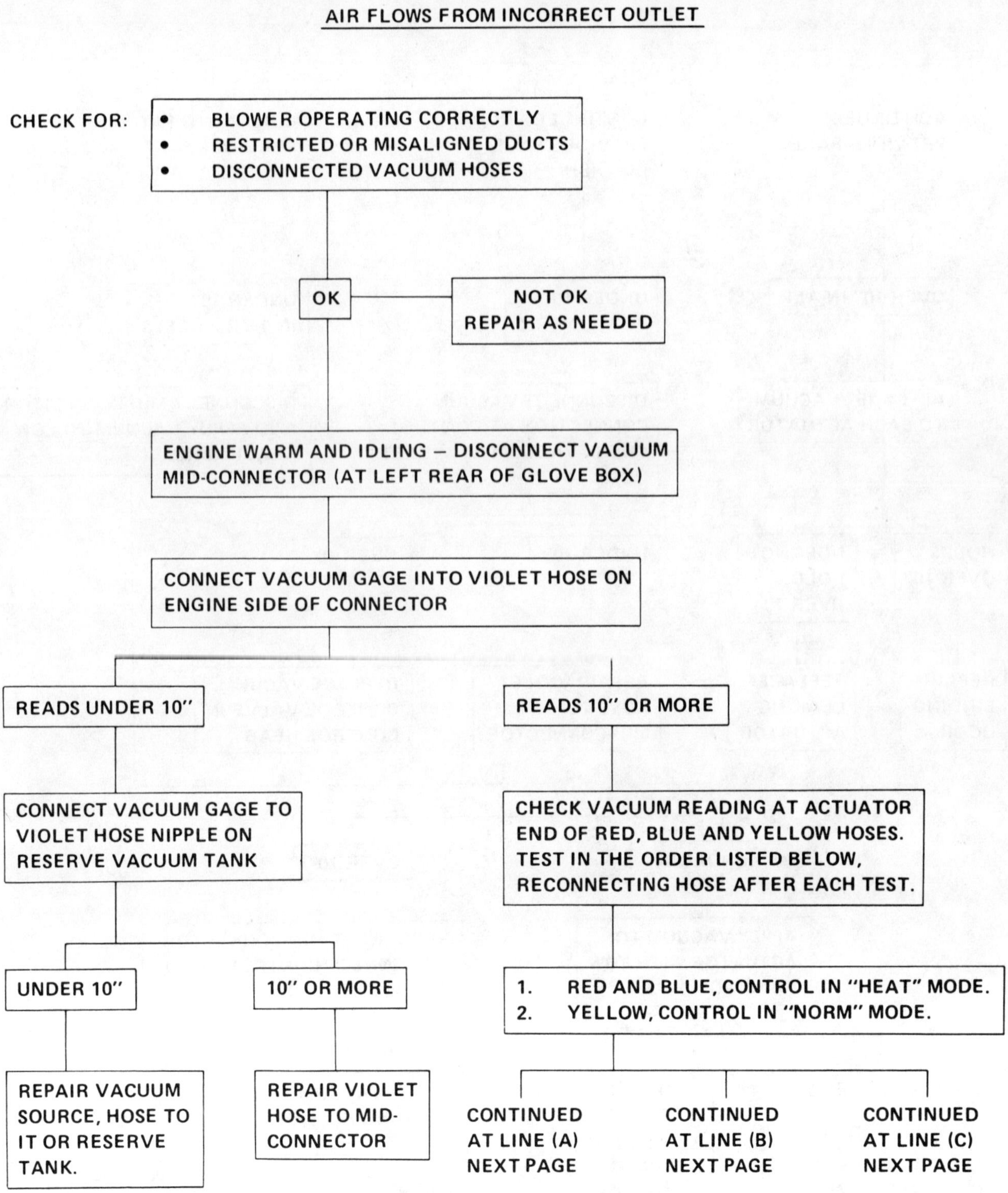

Fig. 122 Manual A/C vacuum system diagnosis chart. 1980 Citation, Omega, Phoenix & Skylark (Part 1 of 2)

A.C.C. tester's hose to the vacuum gauge directly to this hose. About 20 inches Hg. or more vacuum should be present on this transducer vacuum supply hose. If vacuum is proper, restore hose connection. If 20 inches Hg. or more is *not* present, check the vacuum supply vacuum level and then go to Steps 14, 15, 16 and 17 to check for vacuum leaks in the vacuum checking relay and the rotary vacuum valve.

12. To check for proper regulated vacuum output from the transducer disconnect the long hose from the transducer to the vacuum checking *relay at the relay.*

NOTE: This long hose must be at least 15 inches long, or the transducer will make a buzzing noise.

Connect the A.T.C. tester's vacuum gauge directly to the long hose at the vacuum checking relay.

13. Position the programmer in an upright position so that the output shaft points straight up. With the "Manual Control" knob in the "Max. Heat" position, 0 volts should be read on the voltmeter and the vacuum should be 9 to 11 inches Hg. Turn-

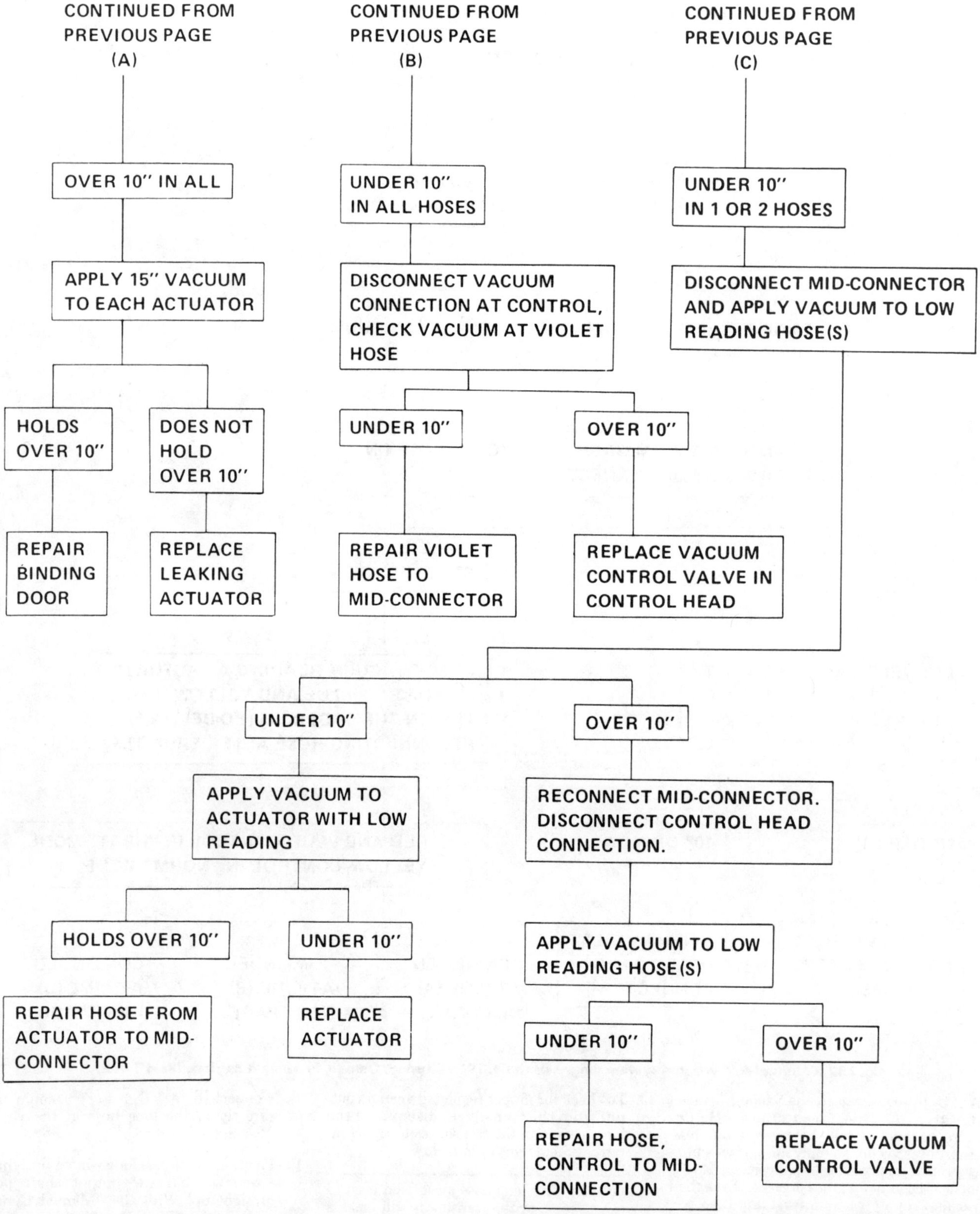

Fig. 122 Manual A/C vacuum system diagnosis chart. 1980 Citation, Omega, Phoenix & Skylark (Part 2 of 2)

SELECTOR LEVER POSITION	BLOWER SPEEDS AVAILABLE	AIR SOURCE	DIVERTER DOOR		A/C-DEF DOOR	COMPRESSOR	AIR ENTERS VEHICLE
			# 1	# 2			
OFF	LO	OUTSIDE	UP	UP	OPEN TO DEF	OFF	*FLOOR OUTLETS
MAX	HI	INSIDE %	DOWN	DOWN	OPEN TO A/C	ON &	DASH OUTLETS
NORM	ALL	OUTSIDE	DOWN	DOWN	OPEN TO A/C	ON &	DASH OUTLETS
BI-LEVEL	ALL	OUTSIDE	UP	DOWN	OPEN TO A/C	ON &	FLOOR & DASH OUTLETS
VENT	ALL	OUTSIDE	DOWN	DOWN	OPEN TO A/C	OFF	DASH OUTLETS
HTR	ALL	OUTSIDE	UP	UP	OPEN TO DEF	OFF	*FLOOR OUTLETS
DEF	ALL	OUTSIDE	MID	DOWN	OPEN TO DEF	ON &	**DEFROSTER OUTLETS

NOTE: *Some bleed-by to defroster outlets. **Some bleed-by to floor outlets.
%100% Inside air is not available, some bleed-through of outside air is allowed.
&Providing the ambient temperature is above 37° F at the ambient temperature switch.

Fig. 123 Control system functional test chart. 1976-78 Skyhawk

ing the "Manual Control" knob to the "Max. Cold" position should make the vacuum drop to 0 inches Hg. and the voltage should increase 5 volts or more. If the above indications *do not occur,* the transducer is defective. After the check is completed, restore hose connection. Disconnect probe and clip from the programmer.

Rotary Vacuum Valve and Vacuum Checking Relay Test

14. Connect the tester's vacuum gauge into the vacuum feed line to the rubber dummy plug on the programmer rotary vacuum valve. Disconnect the center hose from the side port of the rotary vacuum valve and seal off the port by placing your finger over the port. The tester's vacuum gauge should read 20 inches Hg. or more. If it does not, either the vacuum checking relay or the rotary vacuum valve is leaky. If 20 inches Hg. or more vacuum is read, skip to Step 16 and restore the hose connection.
15. Remove all three hoses from the side ports of the rotary vacuum valve. Remove the short hose from the transducer and connect it across the two outer ports on the side of the rotary valve. Place finger over center side port of the rotary vacuum valve. If the tester's vacuum gauge now reads 20 inches Hg. or more, the *vacuum checking relay* is defective. If it does *not,* the rotary valve is defective and leaking. After repair of programmer, restore all hose connections.

Vacuum Motor and Vacuum Checking Relay Test

16. Disconnect electrical connector from programmer. *Do not disconnect* the vacuum input to programmer. The programmer should go to the full heat position (the vacuum motor mechanism will move into the vacuum motor). If it does not, then either the vacuum checking relay or the vacuum motor has a severe leak. If it does go to full heat, a slight leak may be present and the vacuum motor will move toward "Max. Cold" after removing the rubber dummy plug from the programmer.
17. If the vacuum motor mechanism moves, the vacuum motor or vacuum checking relay is leaking. Disconnect the short hose at vacuum motor. Apply raw vacuum to the vacuum motor input port, then pinch the hose at the vacuum motor with long nose pliers and hold for 30 seconds. If the mechanism moves, the vacuum motor is defective. If it does *not* move, the vacuum checking relay is defective.

Programmer Blower Switch Test

18. Place the "Voltage" knob in the "No. 6 Blower" position. Rotate the "Manual Control" knob to the "Max. Heat" position. The voltmeter reading should be 12 volts DC (supply voltage). As the "Manual Control" knob is slowly rotated toward "Max. Cold", at about "175", the voltage will drop slightly. At about "160" the voltage will again drop slightly and at about "155", the voltage will drop slightly once more. At about "130", the voltage will increase slightly. At about "120", the voltage will again increase slightly and at about "110", a slight increase will again be noted. If these steps in blower voltage are not present, the blower switch in the programmer is defective.
19. Tap the components of the programmer while rotating the "Manual Control" knob and watch for any erratic operation due to intermittent problems. Replace programmer cover and rotate "Manual Control" knob, checking for normal rotation of the output shaft.

1975-76 AUTO CLIMATE CONTROL

Tester-J-23678 Procedure

Programmer Functional Test

1. Connect tester electrical harness to programmer and car harness to tester.
2. Operate car at idle, approx. 1000 RPM.
3. Set control head lever at auto.
4. Turn temp. dial calibration switch off.
5. Place Manual automatic switch in manual position.

NOTE: This substitutes variable resistor for sensor string. At all control head settings.

6. Place the manual control at max. heat and then rotate slowly counter-clockwise. Vacuum motor mechanism which can be seen through the slot in the programmer cover should start moving at 175 ohms. Rotating the knob to max. cold should move the programmer to its full A/C position. Check air-mix door for full travel, and no obstructions. If the programmer operates in this manner, the components governing programmer movement — amplifier, transducer and vacuum motor — are ok. Proceed to sensor string test.
7. If the programmer moves but the movement does not start at 175 ohms adjust feedback pot.
8. If the programmer does not move, check as follows for supply voltage and supply vacuum:
   (a) Place voltmeter switch in 2 position. Battery voltage should appear on meter. No voltage indicates lack of ground on term. 1 or battery supply to term. 2 of programmer.
   (b) Plug tester's dummy vacuum plug on programmer valve. Connect dummy hose to vacuum supply (Port 2 hose in car vacuum harness). Make sure vacuum is present (11 in. hg. or more). Programmer now has vacuum supply with no car vacuum system components connected. If programmer functions using manual control, troubleshoot car vacuum system. If programmer does not function, defect is in programmer.

Sensor String Test (Do Programmer Functional Test First)

1. Set control lever at auto and temperature dial at 75.
2. Place temp. dial calibration switch at off.
3. Place manual-automatic switch at automatic position.
4. Observe position of vacuum motor mechanism through slot in programmer cover.
5. Switch to manual and adjust manual control for same mechanism position.
6. When manual control is properly adjusted, moving manual-automatic switch back and forth will not move vacuum motor.
7. Read manual control. This indicates the resistance of the sensors and the temperature dial combined and should be 120 to 150 ohms, at 70 to 75 room temp.
8. If reading is not 120 to 150 ohms perform the temperature dial calibration test to check out the temp. dial. If temp. dial checks O.K., the in-car or ambient sensor or connecting wiring is defective.

Temperature Dial Calibration

1. Place control lever at Econ. position.
2. Place manual-automatic switch in manual position.
3. Place calibration switch in cal. position.
4. Disconnect in-car sensor wire connector

COMFORT PREFERENCE		POSITION SELECTOR LEVER	POSITION TEMPERATURE LEVER	POSITION FAN BLOWER LEVER	COMMENTS
NO HEAT		"OFF"	"COLD"	"LO"	Air Conditioning Compressor is off. Fixed LO blower speed (after engine warm-up). Outside air enters thru heater outlet.
VENTILATION	ECONOMY	"VENT"	"COLD" (Outside air temperature) to "HOT" (Warm incoming air)	"LO" to "HI"	Air Conditioning compressor is off. Outside air enters thru air conditioning outlets.
HEAT	ECONOMY	"HEATER"	"HOT" (For maximum heat) or toward "COLD" to lower temperature	"LO" to "HI"	Air Conditioning compressor is off. Most air enters thru heater outlet - some out defroster (delayed until moist air is removed from system).
WINDSHIELD DEFROSTING		"BI-LEVEL"	"HOT" (For maximum heat) or toward "COLD" to lower temperature	"LO" to "HI"	Air Conditioning compressor is on above 45°F. (7°C.) to dehamidify. Air enters thru Air Conditioning, heater and defroster outlets to defog windshield and side windows. Air to windshield is delayed until moist air is removed from system. Smallest portion of air to windshield.
		"DEFROSTER"			Air Conditioning compressor on above 45°F. (7°C.) to dehumidify. Most air directed to windshield - some out heater.
TEMPERED — MAXIMUM COOL	AIR CONDITIONING	"MAXIMUM"	"COLD"	Blower fixed on "HI" regardless of setting	FOR MAXIMUM COOLING - Air Conditioning compressor on to cool and dehumidify. For rapid cool-down or quick hamidity removal, incoming air is cooled and 80% recirculated thru system for additional cooling. Air thru Air Conditioning outlets.
TEMPERED — NORMAL COOL	AIR CONDITIONING	"NORMAL"	Between "COLD" and "HOT"	"LO" to "HI"	FOR NORMAL COOLING - Air Conditioning compressor on to cool and dehumidify. Air cooled but not recirculated out Air Conditioning outlets.
TEMPERED — BI-LEVEL	AIR CONDITIONING	"BI-LEVEL"			FOR BI-LEVEL COOLING - Air Conditioning compressor on to cool and dehumidify. Air enters thru Air Conditioning heater and defroster outlets.

Fig. 124 Control system functional test chart. 1976-78 Skylark

and plug into tester wiring harness adapter. (Ground both sensor wires).

5. Read voltmeter.
6. Press pushbutton and adjust temp. dial on control head to get same voltmeter reading as above.
7. Temp. dial should read as indicated on tester panel near pushbutton. If it does not, slip temp. dial clutch to correct.

NOTE: If same voltmeter reading cannot be obtained, the temp. dial is defective.

Blower Circuit Tests

1. Position calibration switch in off and manual-automatic switch in manual.
2. Position control lever in Lo. Using voltmeter knob, there should be voltage at 7 and approximately half battery voltage at 6. If there is no voltage at 7, there is a malfunction in the Lo relay circuits. If there is voltage at 7, but no voltage at 6 the programmer is malfunctioning.
3. Position control lever on auto, there should be battery voltage at 8 and voltage at 6 should decrease in steps and increase in steps as control knob is rotated from max. heat to max. cold. If voltage at 6 does not change in steps, the programmer is malfunctioning. Ensure there is a good connection between 6 and blower motor. If blower wire is disconnected 6 will read battery voltage only.

Programmer Vacuum Circuit

1. Position temperature dial calibration switch in off and manual-automatic switch in manual. Place manual control in max. heat.
2. If programmer moves to max. heat, skip to Vacuum Motor and Checking Relay. If programmer does not move to max. heat proceed to step 3.
3. Remove programmer cover, then install dummy plug with vacuum gauge teed-in to port 2 of vehicle harness plug.
4. If vacuum drops below supply vacuum found in Programmer Function Test, using long nose pliers pinch two black vacuum hoses leading from side of rotary valve. If vacuum readings improve, check programmer vacuum hoses for leaks, then proceed to Vacuum Motor and Checking Relay.

STEP	CONTROL SETTINGS			SYSTEM RESPONSE					
	MODE CONTROL	TEMPERATURE CONTROL	FAN SWITCH	BLOWER SPEED	COMPRESSOR	AIR VALVE	HEATER OUTLETS	A/C OUTLETS	DEFROSTER OUTLETS
1	OFF	COLD	LO	LOW	OFF	O/S AIR	AIR FLOW	NO AIR FLOW	MINIMUM AIR FLOW
2	MAX	COLD	LO	HIGH	ON	I/S AIR	NO AIR FLOW	COLD AIR FLOW	NO AIR FLOW
3	NORM	COLD	LO	LOW	ON	O/S AIR	NO AIR FLOW	COLD AIR FLOW	NO AIR FLOW
4	NORM	COLD	LO TO HI	LOW TO HIGH	ON	O/S AIR	NO AIR FLOW	COLD AIR FLOW	NO AIR FLOW
5	BI-LEVEL	COLD	HI	HIGH	ON	O/S AIR	COLD AIR FLOW	COLD AIR FLOW	MINIMUM AIR FLOW
6	VENT	COLD	HI	HIGH	OFF	O/S AIR	*NO AIR FLOW	AMBIENT AIR	NO AIR FLOW
7	HEATER	HOT	HI	HIGH	OFF	O/S AIR	HOT AIR	NO AIR FLOW	MINIMUM AIR FLOW
8	DEF	HOT	HI	HIGH	ON	O/S AIR	MINIMAL FLOW	NO AIR FLOW	MAXIMUM AIR FLOW

*CHEVY C60 HAS AMBIENT AIR FROM HEATER & A/C OUTLETS.

Fig. 125 Control system functional test chart. 1977 Century, Electra, LeSabre, Regal & Riviera manual A/C

5. If reading does not improve, the rotary valve is leaking.

Vacuum Motor & Checking Relay

1. With tester installed as described in Programmer Vacuum Circuit, programmer should go to full heat.
2. Disconnect dummy plug, vacuum checking relay should hold vacuum motor in full heat. If mechanism moves, vacuum motor or checking relay is leaking. Apply vacuum to motor, then pinch hose at motor to see if leak is in motor. A good motor will hold position.

Amplifier & Transducer

1. Connect tester harness into system, then position manual-automatic switch in manual and temperature dial calibration switch in off.
2. Remove small vacuum hose from transducer and connect tester vacuum gauge to hose. A reading of 11 in. Hg. or more of vacuum should be obtained. If there is a delay in obtaining 11 in. Hg. or more, the restrictor located in the small transducer hose may be plugged. Reinstall hose on transducer.
3. Remove large vacuum hose from transducer and connect tester vacuum gauge to large port on transducer.
4. Rotate manual control, vacuum readings of 1 to 11 in. Hg. should be obtained. If the

ENGINE WARM AND IDLING

	CONTROL SETTINGS			SYSTEM RESPONSE					
	MODE LEVER	TEMP. LEVER	BLOWER SWITCH	BLOWER SPEED	COMP. CLUTCH	AIR VALVE	HEATER OUTLET	A/C OUTLETS	DEFROST OUTLETS
1	OFF	COLD	LO TO HI	LO ONLY	OFF	OUTSIDE AIR	AMBIENT AIR FLOW	NO AIR FLOW	SLIGHT AIR FLOW
2	MAX.	COLD	LO TO HI	SAME AS SWITCH	ON	INSIDE AIR	SLIGHT AIR FLOW	COLD AIR FLOW	NO AIR FLOW
3	NORM.	COLD	LO TO HI	SAME AS SWITCH	ON	OUTSIDE AIR	SLIGHT AIR FLOW	COLD AIR FLOW	NO AIR FLOW
4	BI-LEVEL	COLD	LO TO HI	SAME AS SWITCH	ON	OUTSIDE AIR	COLD AIR FLOW	COLD AIR FLOW	NO AIR FLOW
5	VENT	COLD	LO TO HI	SAME AS SWITCH	OFF	OUTSIDE AIR	SLIGHT AIR FLOW	AMBIENT AIR FLOW	NO AIR FLOW
6	HEATER	HOT	LO TO HI	SAME AS SWITCH	OFF	OUTSIDE AIR	HOT AIR FLOW	NO AIR FLOW	SLIGHT AIR FLOW
7	HEATER	COLD	LO TO HI	SAME AS SWITCH	OFF	OUTSIDE AIR	AMBIENT AIR FLOW	NO AIR FLOW	SLIGHT AIR FLOW
8	DEFROST	HOT	LO TO HI	SAME AS SWITCH	ON	OUTSIDE AIR	SLIGHT AIR FLOW	NO AIR FLOW	HOT AIR FLOW

Fig. 126 Control system functional test chart. 1978 Riviera, 1978-79 Century, Electra, LeSabre & 1979 Skyhawk manual A/C

	CONTROL SETTINGS			SYSTEM RESPONSE					
STEP	MODE CONTROL	TEMPERATURE CONTROL	FAN SWITCH	BLOWER SPEED	COMPRESSOR	AIR VALVE	HEATER OUTLETS	A/C OUTLETS	DEFROSTER OUTLETS
1	OFF	COLD	LO	LOW	OFF	O/S AIR	AIR FLOW	NO AIR FLOW	MINIMUM AIR FLOW
2	MAX	COLD	LO	LOW	ON	I/S AIR	NO AIR FLOW	COLD AIR FLOW	NO AIR FLOW
3	MAX	COLD	LO TO HI	LOW TO HIGH	ON	I/S AIR	NO AIR FLOW	COLD AIR FLOW	NO AIR FLOW
4	NORM	COLD	HI	HIGH	ON	O/S AIR	NO AIR FLOW	COLD AIR FLOW	NO AIR FLOW
5	BI-LEVEL	COLD	HI	HIGH	ON	O/S AIR	COLD AIR FLOW	COLD AIR FLOW	MINIMUM AIR FLOW
6	VENT	COLD	HI	HIGH	OFF	O/S AIR	NO AIR FLOW	AMBIENT AIR	NO AIR FLOW
7	HEATER	HOT	HI	HIGH	OFF	O/S AIR	HOT AIR FLOW	NO AIR FLOW	MINIMUM AIR FLOW
8	DEF	HOT	HI	HIGH	ON	O/S AIR	MINIMUM AIR FLOW	NO AIR FLOW	MAXIMUM AIR FLOW

Fig. 127 Control system functional test chart. 1979 Riviera

A/C FUNCTIONAL TEST									
	CONTROL SETTINGS			SYSTEM RESPONSE					
STEP	MODE CONTROL	TEMPERATURE CONTROL	FAN SWITCH	BLOWER SPEED	COMPRESSOR	AIR VALVE	HEATER OUTLETS	A/C OUTLETS	DEFROSTER OUTLETS
1	OFF	COLD	LO	OFF	OFF	O/S AIR	RAM AIR FLOW	NO AIR FLOW	RAM AIR FLOW
2	MAX	COLD	LO	LO	ON	I/S AIR	NO AIR FLOW	COLD AIR FLOW	NO AIR FLOW
3	MAX	COLD	LO TO HI	LO TO HI	ON	I/S AIR	NO AIR FLOW	COLD AIR FLOW	NO AIR FLOW
4	NORM	COLD	HI	HI	ON	O/S AIR	NO AIR FLOW	COLD AIR FLOW	NO AIR FLOW
5	BI-LEVEL	COLD	HI	HI	ON	O/S AIR	COLD AIR FLOW	COLD AIR FLOW	MINIMUM AIR FLOW
6	VENT	COLD	HI	HI	OFF	O/S AIR	NO AIR FLOW	AMBIENT AIR	NO AIR FLOW
7	HEATER	HOT	HI	HI	OFF	O/S AIR	HOT AIR FLOW	NO AIR FLOW	MINIMUM AIR FLOW
8	DEF	HOT	HI	HI	ON	O/S AIR	MINIMUM AIR FLOW	NO AIR FLOW	MAXIMUM AIR FLOW

Fig. 128 Control system functional test chart. 1980 Citation, Omega, Phoenix & Skylark

ALL SERIES A/C FUNCTIONAL TEST ENGINE WARM, CHOKE OFF AND IDLING OFF FAST IDLE SPEED CAM									
	CONTROL SETTINGS			SYSTEM RESPONSE					
	MODE LEVER	TEMP. LEVER	BLOWER SWITCH	BLOWER SPEED	COMP. CLUTCH	AIR VALVE POSITION	HEATER OUTLET AIR FLOW?	A/C OUTLET AIR FLOW?	DEFROST OUTLET AIR FLOW?
1	OFF	COLD	LO TO HI	LO ONLY	OFF	OUTSIDE AIR	AMBIENT AIR FLOW	NONE	SLIGHT
2	MAX.	COLD	LO TO HI	SAME AS SWITCH	ON	INSIDE AIR	SLIGHT FLOW	COLD FLOW	SLIGHT NONE (A)
3A	NORM.	COLD	LO TO HI	SAME AS SWITCH	ON	OUTSIDE AIR	SLIGHT FLOW	COLD FLOW	SLIGHT NONE (A)
3B	NORM. (A,B SERIES)	HOT	LO TO HI	SAME AS SWITCH	ON	OUTSIDE AIR	SLIGHT FLOW	COLD FLOW	SLIGHT NONE (A)
4	BI-LEVEL	COLD	LO TO HI	SAME AS SWITCH	ON	OUTSIDE AIR	COLD FLOW	COLD FLOW	SLIGHT NONE (A)
5	VENT	COLD	LO TO HI	SAME AS SWITCH	OFF	OUTSIDE AIR	SLIGHT FLOW	AMBIENT AIR FLOW	NONE
6	HEATER	HOT	LO TO HI	SAME AS SWITCH	OFF	OUTSIDE AIR	HOT FLOW	NONE	SLIGHT AIR FLOW
7	HEATER	COLD	LO TO HI	SAME AS SWITCH	OFF	OUTSIDE AIR	AMBIENT AIR FLOW	NONE	SLIGHT AIR FLOW
8	DEFROST	HOT	LO TO HI	SAME AS SWITCH	ON	OUTSIDE AIR	SLIGHT FLOW	NONE	HOT AIR FLOW

Fig. 129 Control system functional test chart. 1980 Buick except Skylark

amplifier-transducer is functional any value from 1 to 11 in. Hg. is obtainable by careful adjustment of the manual control knob.

Feedback Potentiometer Adjustment

1. Using screwdriver, slip feedback pot shaft counter-clockwise to stop (gear does move). Programmer should move to full heat.
2. Place manual-automatic switch in manual position. Rotate manual control to max. heat.
3. Move manual control to 175. Do not overtravel.
4. Slip feedback pot shaft slowly clockwise looking for signs of vacuum motor mechanism movement.
5. Check adjustment with manual control. Travel should start when 175 is reached as knob is turned counter-clockwise from max. heat.
6. If feedback pot cannot be adjusted, replace amplifier assembly.

Blower Switch Bench Test. (Using J-23713 Bench Test Harness)

1. Place voltmeter knob at 6.
2. With manual-automatic switch in manual, rotate manual control back and forth. Voltage should drop then increase in steps. Steps will be smaller than those seen in car.

Programmer Moves But Vacuum Operated Door(s) Malfunctions

If one of the doors does not operate, malfunction could be caused by plugged or pinched harness or plugged valve on programmer. To check programmer valve, cut off sealed hose on dummy plug for circuit malfunctioning and connect the vacuum gauge to the plug. Plug the dummy plug on the programmer vacuum valve. Connect dummy hose to vacuum supply hose (Port 2) in car vacuum harness. Monitor vacuum while operating programmer with manual control. When completed, seal dummy plug hose with rivet. Do not use a screw.

Temperature Door Check And Adjustment

A maladjusted temperature door link may not allow the programmer to travel fully to max. heat or to max. cold and could result in poor heating or poor cooling complaints. To check adjustment, attach tester electrical harness between programmer and car harness. Put manual-automatic switch in manual and control lever in auto. Put voltmeter knob in 6. Rotate manual control from max. heat to max. cold; high blower (Battery Voltage) should be obtained in both extremes. With control lever in high, recirc. operation should be obtained at max. cold (3 min. delay on some cars). With these conditions met, temperature door adjustment is correct. To readjust:

1. Start engine and position control lever in Def.
2. Loosen programmer shaft screw.
3. Position manual-automatic switch in manual and manual control at max. heat. Programmer should go to full heat.
4. Push door link away from programmer to place temperature door in full heat position. Blower air will hold door in position.
5. Tighten programmer shaft screw and recheck adjustment.

1977-80 AUTO CLIMATE CONTROL

Functional Test

With engine at operating temperature, run engine at 1000 RPM or higher, and perform functional test as described in Figs. 130 and 131. During the functional test, note the following:

a. Check to ensure that air flow is not coming from both A/C and heater outlets when only one mode is selected. If air delivery is from both outlets when only one mode is selected, this indicates a vacuum leak or a sticking airflow door.
b. When programmer changes mode, blower speed, air delivery mode etc., discharge air temperature should change. If temperature does not change, check air mix door to programmer link connection.
c. If neither programmer functions nor air delivery temperature changes, an incorrect electrical or vacuum signal to the programmer is indicated or the programmer may be faulty.
d. If a malfunction of a particular vacuum door is indicated, check for a disconnected vacuum line at door vacuum actuator.
e. Using information obtained during the functional test, isolate problem to one of the following areas: temperature control problem, blower control problem, auxiliary vacuum problem or refrigeration system problem and refer to the appropriate problem diagnosis listed below. After system problem has been diagnosed and corrected, perform functional check to ensure that system is now operating properly.

TEMPERATURE CONTROL PROBLEM DIAGNOSIS

The function of the temperature control circuit is to determine the temperature of air to be discharged into the passenger compartment by positioning the air mix door. If the problem is poor or no heating or cooling, but the air mix door and programmer function properly, the problem is the refrigerant circuit or heater water system. Blower cycling or mode cycling may be caused by a temperature control problem.

If a temperature control problem is noted, check for the following:

a. Disconnected sensor or temperature dial. This will interrupt the electrical signal and drive programmer to the max. heat position.
b. Improper sensor connection. An improper sensor connection will increase resistance to the sensor string and drive system to a higher temperature than desired.
c. Disconnected vacuum supply hose at vacuum checking relay. This will lock vacuum motor in a fixed position.
d. Disconnected vacuum hose in transducer-programmer vacuum motor line. This will cause system to drive to max. cooling.
e. Leak in auxiliary vacuum circuit. This may

STEP	CONTROL SETTINGS		SYSTEM RESPONSE					
	MODE CONTROL	TEMPERATRUE DIAL	BLOWER SPEED	COMPRESSOR	AIR VALVE	HEATER OUTLETS	A/C OUTLETS	DEFROSTER OUTLETS
1	OFF	85	LOW	OFF	O/S AIR	NO AIR FLOW	NO AIR FLOW	MINIMUM AIR FLOW
2	ECONOMY	85	AUTO	OFF	O/S AIR	NO AIR FLOW	NO AIR FLOW	MINIMUM AIR FLOW
3	LO	85	LOW	ON	O/S AIR	NO AIR FLOW	HOT AIR FLOW	NO AIR FLOW
4	AUTO	85 TO 65	AUTO	ON	O/S AIR	AIR FLOW	H TO C AIR FLOW	NO AIR FLOW
5	HI	65 TO 85	HIGH	ON	I/S AIR	AIR FLOW	C TO H AIR FLOW	NO AIR FLOW
6	BI-LEVEL	85	AUTO	ON	O/S AIR	AIR FLOW	AIR FLOW	AIR FLOW
7	DEF	85	HIGH	ON	O/S AIR	MINIMUM AIR FLOW	NO AIR FLOW	MAXIMUM AIR FLOW

Fig. 130 Control system functional test chart. 1977-78 Buick with Auto Climate Control

reduce the vacuum supply to the transducer below control requirements, causing off calibration or a poor heating condition.

f. A loss of vacuum supply will usually result in cold air flow from floor outlets. Diagnosis of problems in the temperature control circuit can be accomplished using tester J-23678, refer to Auto Climate Control System & Programmer Diagnosis. Symptoms and causes of temperature control problems are as follows:

System Operates At Max. A/C Only

With system operating, disconnect in-car sensor wire connector and observe programmer movement through hole in bottom of programmer cover. Programmer should move full in to the right (max. heat position). If programmer remains at the full out to left position (max. cooling), remove vacuum connector and check for vacuum at black hose. If no or low vacuum is indicated, check for leaking or disconnected vacuum hose. If adequate vacuum supply is indicated, check voltage at programmer. If voltage is indicated at programmer, remove cover and check for disconnected wiring or vacuum hoses. If problem in programmer cannot be located, use tester J-23678 to test programmer, refer to Auto Climate Control System and Programmer Diagnosis.

If programmer moves to the full heat position when sensor wire connector is disconnected, the programmer is not at fault and the following areas should be checked:

a. In-car sensor shorted.
b. Miscalibrated or shorted temperature dial.
c. A shorted sensor circuit wiring harness.
d. Disconnected or open ground circuit between programmer and chassis ground.
e. No voltage supply at programmer.
f. Amplifier not grounded.

System Operates At Max. Heat Only

Position temperature control dial at 65 and control lever to any position except off. Check to see if programmer air mix door arm moves to the max. A/C position. If programmer does not move, check for the following:

a. Disconnected or defective in-car or ambient sensor.
b. Improperly connected or open circuit in sensor string wiring harness.

Place temperature control on 85 and check to see if programmer moves. If programmer does not move, check for the following:

a. Disconnected or defective temperature dial.
b. Open circuit or disconnected wire between programmer and chassis ground.
c. Defective programmer. Use tester J-23678 to test programmer.

System Changes Blower Speeds And/Or Modes During Acceleration

Remove programmer cover, then reconnect electrical and vacuum connections. With system operating, remove in-car sensor wire connector to drive programmer to the max. heat, full vacuum position. Disconnect vacuum hose connector from programmer; programmer should remain in the full heat position. If programmer does not remain in the full heat position, repeat procedure, then pinch programmer vacuum supply hose using needle nose pliers. If programmer still does not remain in the full heat position, the programmer vacuum motor is leaking. If programmer remains in the full heat position with vacuum supply hose pinched, the checking relay is defective.

d. If problem is mode changes without change in blower speed, check valve portion of the checking relay may be defective.

Temperature Dial Setting Does Not Provide Comfort

Using tester J-23678, refer to Auto Climate Control System and Programmer Diagnosis to check and reset calibration on temperature dial, feedback potentiometer in programmer, sensor string and temperature door link adjustment. Also check to ensure that in-car sensor aspirator hose is connected to sensor and aspirator and is not kinked.

Mode & Blower Cycling

NOTE: Some cycling of system may occur if temperature control dial is moved in large increments.

This problem may be caused by improper electrical connection, poor solder joints or a defective sensor or amplifier. If problem is intermittent, tap areas around sensor, control head, and programmer. Also shake electrical connections attached to these components. If problem cannot be located by the transducer buzzing, remove programmer cover and reconnect electrical and vacuum connectors, then connect a voltmeter to transducer output terminals. While observing voltmeter, continue to tap areas described above and shake electrical connections. If voltmeter indicates erratic voltage fluctuations, replace component or resolder connections as necessary.

Insufficient Cooling Or Heating

This condition may be caused by a temperature control refrigerant system or heater system problem, to determine which is at fault proceed as follows:

a. No cooling, check compressor clutch operation. If compressor clutch is not operating, problem is in refrigerant system. Also check for cold suction line.
b. No heating, check engine coolant level. Also feel heater hoses to see if engine coolant is entering the heater core.
c. Check for proper air flow at Auto. and high settings. Improper air flow may be caused by a blower-related problem, blocked air passage or disconnected air hose.
d. Check calibration of temperature control dial.
e. Check calibration of programmer feedback potentiometer.

If problem seems to be in the temperature control system, use tester J-23678 to locate and correct problem, refer to Auto Climate Control System and Programmer Diagnosis.

Blower Speed & Mode Shifts Without Temperature Change

Check air mix door link to programmer shaft and door crankarm connections.

BLOWER CONTROL CIRCUIT PROBLEM DIAGNOSIS

Blower control circuit problems are divided into two categories, blower speed control and blower turn-on. Blower circuit problems and probable causes are as follows:

System Operates At Low Blower Only

1. Check for improper connection at five-way connector connected at programmer and control head.
2. Check for an open circuit in Auto. blower wiring circuit.

Blower Does not Operate In Any Lever Position

1. Check for improperly connected or open circuit in blower wiring harness.
2. Check for a blown A/C fuse.

	CONTROL SETTINGS		SYSTEM RESPONSE						
STEP	MODE CONTROL	TEMPERATURE DIAL	BLOWER SPEED	COMPRESSOR	AIR VALVE	HEATER OUTLETS	A/C OUTLETS	DEFROSTER OUTLETS	AIR TEMPERATURE
1	OFF	85	LOW	OFF	O/S AIR	AIR FLOW	NO AIR FLOW	NO AIR FLOW	HOT
2	ECONOMY	85	AUTO	OFF	O/S AIR	AIR FLOW	NO AIR FLOW	MINIMUM AIR FLOW	
3	LO	85	LOW	ON	O/S AIR	AIR FLOW	NO AIR FLOW		
4	AUTO	85 TO 65	AUTO	ON	O/S AIR I/S AIR	AIR FLOW NO AIR FLOW	NO AIR FLOW AIR FLOW	NO AIR FLOW	HOT TO COLD
5	HI	65 TO 85	HIGH	ON	I/S AIR O/S AIR	NO AIR FLOW AIR FLOW	AIR FLOW NO AIR FLOW	NO AIR FLOW	COLD TO HOT
6	BI-LEVEL	85	AUTO	ON	O/S AIR	AIR FLOW	AIR FLOW	MINIMUM AIR FLOW	HOT
7	DEF	85	HIGH	ON	O/S AIR	MINIMUM AIR FLOW	NO AIR FLOW	MAXIMUM AIR FLOW	

Fig. 131 Control system functional test chart. 1979-80 Buick with Auto Climate Control

3. Check for stalled blower motor.
4. Check for an improperly connected five-way wire connector at control head or programmer.

No Blower In Off or Lo Positions

1. Check for a defective blower relay.
2. Open circuit in low speed resistor.
3. Open circuit in blower wiring circuit.

Blower Will Operate In Hi Or Defrost Only

Check electrical connections at programmer.

Blower Speed Varies In Hi & Defrost

1. Check for defect in control head circuit board switch.
2. Open circuit in control head wiring circuit.
3. Open circuit in bypass wiring circuit.

Blower Operates With Ignition Switch In Off Position

1. Check for a defective blower relay.

Heater Will Not Turn On In Cold Weather Except In Defrost

1. Check for disconnected wire connector at blower delay switch located on engine cylinder head.
2. Check for defective blower delay switch by grounding switch feed wire, system should turn on. If problem does not occur with vehicle in garage, disconnect control head wire connector to open in-car sensor circuit, if blower motor turns off with engine at operating temperature, ground blower delay switch feed wire. If blower is now operating, the blower delay switch is defective.
3. If problem is still encountered, refer to wiring diagrams located elsewhere in this manual and check circuit for wiring problem.

Heater Will Turn On Immediately In Cold Weather In Any Control Setting Except Defrost

1. Check for defective (closed) blower delay switch.
2. Check for defective (closed) in-car sensor.
3. If problem is still encountered, refer to wiring diagrams located elsewhere in this manual and check circuit for wiring problem.

A/C Delayed In Warm Weather Until Engine Reaches Operating Temperature

1. Check for defective (opened) in-car switch.
2. Check for open circuit in control head wiring harness (in-car sensor circuit).
3. If problem is still encountered, refer to wiring circuit diagrams located elsewhere in this manual and check in-car sensor circuit for wiring problem.

AUXILIARY VACUUM PROBLEM DIAGNOSIS

Auxiliary vacuum circuits in the Auto Climate Control system are circuits which do not affect the temperature of discharge air. The auxiliary vacuum circuits are as follows:

1. Upper and lower mode door vacuum circuits on Electra, LeSabre and Riviera, heater-A/C mode door on Century and Regal.
2. Defroster bleed vacuum circuit on Electra, LeSabre and Riviera.
3. Defroster door vacuum circuit.
4. Recirculating air door vacuum circuit.
5. Heater water valve vacuum circuit.
6. Auxiliary vacuum circuit check valve.

It should be noted that a leak in one vacuum circuit may reduce the vacuum supply to another vacuum circuit.

A/C Mode Door Actuation Problem

This condition may be indicated by the following symptoms, no A/C operation and split mode operation. The system requires vacuum to move door(s) to the A/C position. Check vacuum hose connections at mode door actuators. Split modes may occur due to problems in the door(s) circuits or because of leaks in other auxiliary vacuum circuits. Refer to procedure described under Auxiliary Vacuum Circuit Test.

Defroster Bleed Problems (Electra, LeSabre & Riviera)

1. If there is no defroster bleed, check vacuum hose connection at defroster vacuum actuator. If vacuum connections are satisfactory, the porous flow plug may have to be replaced. Before installing a replacement porous plug, check for vacuum in yellow line and connect hose to actuator without the plug to check bleed door operation.
2. If rapid fogging of windshield is encountered when heat is turned on, check for improperly installed or missing porous plug.

Defroster Door Actuation Problem

This condition is indicated by no air flow from defroster outlets with selector lever in Defrost position. On Electra, LeSabre and Riviera models, vacuum to defroster actuator is required for defroster operation. On Century and Regal models, no vacuum is required for defroster operation. Defroster door actuation problems may be caused by the following:

1. Electra, LeSabre & Riviera.
   a. Improper hose connection at defroster actuator.
   b. Check for vacuum at control head vacuum supply hose.
   c. If system remains in A/C mode with control lever in Defrost position, check vacuum circuit to determine why mode override did not occur.
2. Century and Regal.
   a. Check to ensure that defroster door is not sticking.
   b. If vacuum door is not sticking, refer to Auxiliary Vacuum Circuit Test procedure.

Recirculating Air Door Problem

This condition is indicated by no recirculating air flow at low temperature setting. Check to ensure vacuum hose is properly connected to recirculating air door actuator. Disconnect three wire connector at programmer to drive system to recirculating air mode. If system does not go to recirculating air mode, check adjustment of rod to temperature door.

Heater Water Valve Problem

The water valve is normally open and requires vacuum to stop water flow. A leak in this vacuum circuit may result in a reduced vacuum supply for other vacuum functions. The most probable complaint would be a split mode delivery. Refer to Auxiliary Vacuum Circuit Test.

Auxiliary Vacuum Circuit Check Valve

The check valve for the auxiliary vacuum circuit is located in the programmer checking relay assembly. A malfunctioning check valve will result in a loss of a specific function during acceleration or under a heavy engine load. With system operating, disconnect and plug black vacuum hose supplying vacuum to programmer checking relay. System vacuum function should hold, if not reconnect black hose and pinch purple hose. If system function now holds, the check valve is defective. If system continues to leak down, the vacuum circuit leading to the check valve is leaking.

Auxiliary Vacuum Circuit Test

Disconnect vehicle vacuum harness connector from programmer. This should cause air to flow to floor with defroster bleed on Electra, LeSabre and Riviera. On Century and Regal, air will flow from both heater and defroster outlets. Connect a tee with short length of vacuum hose and union to tester J-23678 vacuum gauge hose, inserting tee into port No. 2 of vacuum harness plug and probing each of the other ports in turn, Fig. 132. Position control lever in specified position and cover ports 3 and 9 as specified, Fig. 133. At each lever position, full vacuum should be read after actuator travel is completed. A low vacuum reading would indicate a leak in the control head valve, vacuum hoses or components listed for the particular lever setting. If a function does not occur even though full vacuum is available, reconnect vacuum harness plug to programmer and check for vacuum at actuator end of hose indicated. If vacuum is obtained at actuator end of hose, check for binding in actuator or door.

COMPRESSOR ELECTRICAL DIAGNOSIS

Compressor electrical problems would be indicated by no cooling or insufficient cooling. Check for the following:

1. Blown A/C-heater fuse.
2. Improper electrical connection at compressor clutch or clutch coil ground.
3. Improperly connected or defective pressure cut-off switch or thermostatic switch.
4. On models equipped with V8-301 engine, check for disconnected or defective wide open throttle switch.
5. Improperly connected control head wire connector.
6. Open circuit in wiring harness.

REFRIGERANT SYSTEM DIAGNOSIS

Use the following procedure to separate refrigerant system problems from temperature control problems.

1. Place control lever at Off, Economy, Lo, Auto., Hi, Bi-Level and Defrost and check for proper system operation, Figs. 130 and 131.

2. If system operates properly, place control lever on Auto. and temperature dial on 65.
3. Position a suitable thermometer in A/C center outlet, then wait approximately 1 minute and note thermometer reading. Thermometer should read approximately 40 to 50° F, unless humidity is excessively high.
4. If discharge air temperature is significantly higher than 50° F, disconnect programmer amplifier wire connector at programmer, then check to ensure that air mix door link arm is in the max. A/C position and temperature door rod is properly connected and adjusted.
5. If when amplifier wire connector is disconnected, air temperature drops to 50° F, the problem is in the temperature control system, refer to Temperature Control Diagnosis.
6. If temperature does not drop and is significantly higher than 50° F, the problem is in the refrigerant system, refer to Insufficient Cooling Diagnosis.

Auto Climate Control Programmer & System Diagnosis

Using tester J-23678 perform test procedure described in Fig. 134 parts 1 and 2.

1976-77 CUSTOM AIRE

Functional Test

NOTE: Test should be performed with engine running, coolant warm and ambient temperature at approximately 75°F.

1. Position selector lever in Off, fan switch in Hi and temperature lever at 85.
   (a) Warm air is delivered from heater outlet.
   (b) Blower motor runs at low speed.
   (c) Compressor is not operating.
2. Position selector lever in Defrost.
   (a) Blower motor changes to high speed.
   (b) Warm air is delivered from defroster outlet with small amount of air flow from heater outlet.
   (c) Compressor is operating.
3. Position selector in Heater.
   (a) Warm air is delivered from heater outlet with small amount of air flow from defroster outlet.
   (b) Blower motor remains at high speed.
   (c) Compressor is not operating.

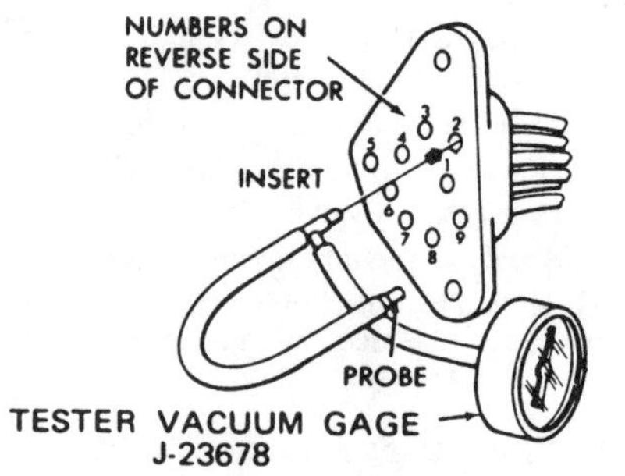

Fig. 132 Auxilliary vacuum circuit test connections. 1977-80 Buick, Cadillac, 1979-80 Pontiac Auto Climate Control & 1977-80 Chevrolet Comfortron

4. Position selector lever in Vent.
   (a) Warm air flows from A/C outlets.
   (b) Blower motor remains at high speed.
   (c) Compressor is not operating.
5. Position selector lever in Bi-Level.
   (a) Warm air is delivered from heater and A/C outlets with small amount of air flow from defroster outlets.
   (b) Blower motor remains at high speed.
   (c) Compressor is operating.
6. Position selector lever in Norm and temperature lever to 65.
   (a) Cold air is delivered from A/C outlets.
   (b) Blower motor remains at high speed.
   (c) Compressor is operating.
7. Pausing at each detent, move fan switch to Low.
   (a) Blower motor speed and air flow should decrease.
8. Position selector lever in Max.
   (a) Blower motor changes from low to high speed.
   (b) System goes to recirculation air flow. Air noise can be heard from right front corner of passenger compartment.
   (c) Cold air is delivered from A/C outlets.
   (d) Compressor is operating.

NOTE: During the functional test the following items may be noted.
1. A change in temperature lever positions will effect outlet air temperature in all selector lever positions except Max. Engine coolant is not allowed to flow through the heater core when selector lever is in the Max position. Only maximum cooling is available in Max position.
2. When coming from any selector position except Bi-Level or Defrost to Heater, a restrictor in the vacuum line will delay air flow to the defroster outlets.
3. The blower motor will not operate in the Off or Heater positions until the engine is warmed up to approximately 120°F.
4. The compressor will not operate if ambient temperature is below 35°F. At temperatures above 35°F the compressor will operate in the following selector positions: Max, Norm, Bi-Level and Defrost.

Temperature Control Diagnosis

NOTE: A malfunction in the temperature control may cause poor regulation of outlet air temperature, warm air flow only or cold air flow only.

1. Remove glove box so that programmer, linkage and temperature door crank arm can be viewed.
2. With engine running and at operating temperature, position selector lever in Norm and fan switch in Hi.
3. Move temperature lever from full cold to full warm and back to full cold. When lever is moved from full cold to full warm, outlet air temperature should increase then decrease when lever is moved back to full cold. Note movement of programmer linkage and temperature door crank.
4. If outlet air temperature does not vary and programmer link does not move when temperature lever is moved, check control cable to ensure it is not disconnected or broken.
5. If control cable is found satisfactory, but programmer link does not move and outlet air temperature does not vary, check for disconnected programmer linkage and temperature door for binding.
6. If temperature door moves freely but only cold air is delivered, check for vacuum at water control valve. When selector lever

Control Lever	Probe Port:	Hose Color	Cover Up	Other Hose In Circuit	Action Expected
Bi-Level	None	Black	Probe	Yellow	Defroster door moves to "bleed" position
Bi-Level	1	Gray	None	None	Hot water valve closes - Check air in "max. heat"
Bi-Level		White	Port 9	Violet, Tan	Lower mode door moves to A/C position
Bi-Level	4	Orange	None	None	Recirc, door opens - 1-2 min. delay
Lo	8	Dark Green	Port 9	Violet	Checks control head vacuum leakage
Lo	6 or 8	Red	None	None	Upper mode door moves to A/C position
Hi	6	Light Green	None	Tan	Lower mode door moves to A/C position
Def	9	Violet	Port 3	White, Blue	Defroster door moves to full "DEF" position

Fig. 133 Auxiliary vacuum circuit test chart. 1977-80 Buick with Auto Climate Control

DIAGNOSING DEFECTS IN PROGRAMMER. THE FOLLOWING TESTS SHOULD BE PERFORMED TO DETERMINE A DEFECTIVE INTERNAL PART. TESTS SHOULD BE MADE IN THE NUMBERED SEQUENCE.

TEST 1 — PROGRAMMER VACUUM CIRCUIT

WITH ADAPTER HARNESS J 23678-77 INSTALLED, PROCEED AS FOLLOWS

1. DISCONNECT SINGLE WIRE IN LIGHT GREEN WIRE OF HARNESS J 23678-77. IF PROGRAMMER GOES TO MAX. HEAT, PROCEED TO TEST 2.
2. REMOVE PROGRAMMER COVER AND INSTALL DUMMY VACUUM PLUG WHICH HAS BEEN MODIFIED AS SHOWN.
3. TEE THE TESTERS VACUUM GAGE INTO THE DUMMY PLUG AND PORT 2 OF CAR VACUUM HARNESS.
4. USE LONG NEEDLE NOSE PLIERS TO PINCH THE CENTER AND RIGHT SIDE HOSES FROM THE VACUUM VALVE.
5. IF VACUUM READING IMPROVES OVER VACUUM READING TAKEN DURING PROGRAMMER FUNCTIONAL TEST, CHECK PROGRAMMER VACUUM HOSES FOR LEAKS.
6. IF VACUUM READING DOES NOT IMPROVE, LEAK IS IN VACUUM VALVE. REPLACE VALVE.

TEST 2 — VACUUM MOTOR AND CHECKING RELAY

1. WITH SINGLE CONNECTOR IN LIGHT GREEN WIRE OF HARNESS J 23678-77 DISCONNECTED, PROGRAMMER SHOULD GO TO FULL HEAT.
2. DISCONNECT DUMMY VACUUM PLUG FROM PROGRAMMER. VACUUM CHECKING RELAY SHOULD HOLD VACUUM MOTOR IN FULL HEAT.
3. IF MECHANISM MOVES, VACUUM MOTOR OR CHECKING RELAY IS LEAKING.
4. REINSTALL CAR VACUUM HARNESS, THEN PINCH HOSE AT MOTOR TO SEE IF LEAK IS IN MOTOR. REPAIR AS REQUIRED.

TEST 3 — AMPLIFIER AND TRANSDUCER

WITH ADAPTER HARNESS J 23678-77 INSTALLED, PROCEED AS FOLLOWS

1. WITH MANUAL-AUTOMATIC SWITCH IN MANUAL AND SINGLE CONNECTOR ATTACHED TO PROGRAMMER TEST POINT, PLACE TEMPERATURE DIAL CALIBRATION SWITCH IN OFF POSITION.
2. PLACE VOLTMETER SWITCH IN PROBE AND CLIP POSITION CONNECT CLIP TO BOTTOM TRANSDUCER WIRE.

   CAUTION: TRANSDUCER WIRES ARE VERY DELICATE. USE APPROPRIATE CAUTION WHEN MAKING OR REMOVING THIS CONNECTION. DO NOT SHORT TRANSDUCER TERMINALS TOGETHER OR AMPLIFIER WILL BE DAMAGED.
3. PUSH PROBE INTO PROGRAMMER CONNECTOR BODY MAKING CONTACT WITH TERM. 1 (YELLOW WIRE).
4. ROTATING THE MANUAL CONTROL FROM MAX. HEAT TO MAX. COLD, THE VOLTAGE READING SHOULD CHANGE 5 VOLTS OR MORE FOR A GOOD AMPLIFIER.
5. IF READING DOES NOT CHANGE, REPLACE AMPLIFIER.

TEST 4 — TRANSDUCER TEST

WITH TESTER AND LEADS SAME AS IN AMPLIFIER TEST PROCEED AS FOLLOWS

1. 11 IN. HG. OR MORE VACUUM SHOULD BE PRESENT ON SMALL TRANSDUCER SUPPLY HOSE.
2. TO CHECK OUTPUT REGULATED VACUUM, PULL CHECKING RELAY FROM POSITION AND REMOVE HOSE FROM TRANSDUCER. CONNECT VACUUM GAGE TO HOSE. USE LARGE TEE TO MAKE CONNECTION AND SEAL EXTRA PORT WITH THUMB.
3. ROTATE MANUAL CONTROL TO 10 VOLTS ON VOLTMETER. THE VACUUM SHOULD BE 9 TO 11 IN. HG. TURNING MANUAL CONTROL TO MAX COLD SHOULD MAKE VACUUM DROP, TOO, 0.5 IN. HG FOR A GOOD TRANSDUCER. RECONNECT HOSES.

TEST 5 — FEEDBACK POTENTIOMETER ADJUSTMENTS

WITH ADAPTER HARNESS J 23678-77 INSTALLED AND SINGLE CONNECTOR UNHOOKED FROM PROGRAMMER, PROCEED AS FOLLOWS:

1. USING SCREWDRIVER, SLIP FEEDBACK POT SHAFT COUNTERCLOCKWISE TO STOP. PROGRAMMER SHOULD MOVE TO FULL HEAT.
2. PLACE MANUAL-AUTOMATIC SWITCH IN MANUAL POSITION. ROTATE MANUAL CONTROL TO MAX HEAT.
3. MOVE MANUAL CONTROL TO 180. DO NOT OVERTRAVEL.
4. SLIP FEEDBACK POT SHAFT SLOWLY CLOCKWISE LOOKING FOR SIGNS OF VACUUM MOTOR MECHANISM MOVEMENT AND STOP.
5. CHECK ADJUSTMENT WITH MANUAL CONTROL TRAVEL SHOULD START WHEN 180 IS REACHED AS KNOB IS TURNED COUNTERCLOCKWISE FROM MAX. HEAT.
6. IF FEEDBACK POT CANNOT BE ADJUSTED, REPLACE AMPLIFIER ASSEMBLY.

PROGRAMMER MOVES BUT VACUUM DOORS DO NOT

IF ONE OF THE DOORS DOES NOT OPERATE, MALFUNCTION COULD BE CAUSED BY PLUGGED OR PINCHED HARNESS OR PLUGGED VALVE ON PROGRAMMER.

1. TO CHECK PROGRAMMER VALVE, CUT OFF SEALED HOSE OF DUMMY PLUG FOR CIRCUIT MALFUNCTIONING AND CONNECT THE VACUUM GAGE TO THE PLUG.
2. PLUG THE DUMMY PLUG ON THE PROGRAMMER VACUUM VALVE CONNECT DUMMY HOSE TO VACUUM SUPPLY HOSE (PORT 2) IN CAR VACUUM HARNESS.
3. MONITOR VACUUM WHILE OPERATING PROGRAMMER WITH MANUAL CONTROL.
4. WHEN COMPLETED, SEAL DUMMY PLUG HOSE WITH RIVET. DO NOT USE A SCREW. REFER TO SHOP MANUAL FOR PROPER VACUUM VALVE OPERATION.

TEMPERATURE DOOR CHECK AND ADJUSTMENT

A MIS-ADJUSTED TEMPERATURE DOOR LINK MAY NOT ALLOW THE PROGRAMMER TO TRAVEL FULLY TO MAX. HEAT OR TO MAX. COLD AND COULD RESULT IN POOR HEATING OR POOR COOLING COMPLAINTS. TO CHECK ADJUSTMENT, ATTACH TESTER ELECTRICAL HARNESS BETWEEN PROGRAMMER AND CAR HARNESS USING ADAPTER HARNESS J 23678-77. CONNECT SINGLE CONNECTOR TO PROGRAMMER TEST POINT. PUT MANUAL-AUTOMATIC SWITCH IN MANUAL AND CONTROL LEVER IN "AUTO". PUT VOLTMETER KNOB IN 6. ROTATE MANUAL CONTROL FROM MAX HEAT TO MAX. COLD. HIGH BLOWER (BATTERY VOLTAGE) SHOULD BE OBTAINED IN BOTH EXTREMES WITH CONTROL LEVER IN "HIGH" RECIRC. OPERATION SHOULD BE OBTAINED AT MAX. COLD (1 MIN. DELAY). WITH THESE CONDITIONS MET, TEMPERATURE DOOR ADJUSTMENT IS CORRECT TO READJUST.

1. START ENGINE AND PUT CONTROL LEVER IN "DEF".
2. REMOVE THREADED DOOR LINK FROM PROGRAMMER ARM.
3. DISCONNECT SINGLE CONNECTOR IN LIGHT GREEN WIRE OF ADAPTER HARNESS J 23678-77. PROGRAMMER SHOULD GO TO FULL HEAT.
4. MOVE TEMPERATURE DOOR TO FULL HEAT POSITION BY PULLING DOOR LINK.
5. REINSTALL THREADED DOOR LINK TO PROGRAMMER ARM AND CHECK AS ABOVE.

Fig. 134 Programmer & system diagnosis chart. 1977-80 Buick, Cadillac, 1979-80 Pontiac Auto Climate Control & 1977-80 Chevrolet Comfortron (Part 1 of 2)

SYSTEM FUNCTIONAL TEST — ALWAYS START DIAGNOSIS WITH THIS TEST

SET CONTROL LEVER AT:		SYSTEM SHOULD OPERATE AS FOLLOWS (ENGINE RUNNING – 70°-80° AMBIENT):
DEF	85°	FIXED HIGH BLOWER. MOST OF AIR DELIVERED FROM DEFROSTER OUTLETS. SOME AIR FROM HEATER OUTLET. IMMEDIATE BLOWER TURN ON
BI LEVEL	75°	LOWER BLOWER SPEED. COOLER AIR FROM BOTH A/C AND HEATER OUTLETS. VERY SMALL AMOUNT OF AIR OUT OF DEFROSTER
HI	65°	FIXED HIGH BLOWER. COLD AIR FROM A/C OUTLETS. RECIRC SHOULD OCCUR
AUTO	65°	RECIRC. MAINTAINED. BLOWER SPEED MAY DROP
AUTO	85°	BLOWER SPEED SHOULD DROP THEN INCREASE. AIR SHOULD BECOME WARMER. AIR SHOULD CHANGE FROM A/C TO BI LEVEL TO HEATER
LO	85°	BLOWER SHOULD DROP TO FIXED LOW SPEED
ECONOMY	85°	SAME AS AUTO EXCEPT NO COMPRESSOR OPERATION
OFF		FIXED LOW BLOWER. MILD AIR OUT HEATER OUTLETS. NO AIR OUT A/C OR DEFROSTER DUCTS. NO COMPRESSOR OPERATION

PROGRAMMER FUNCTIONAL TEST

1. CONNECT TESTER ELECTRICAL HARNESS TO PROGRAMMER AND CAR HARNESS TO TESTER USING ADAPTER HARNESS J 23678-77. LEAVE SINGLE CONNECTOR UNHOOKED.
2. OPERATE CAR AT IDLE (APPROX. 1000 RPM).
3. SET CONTROL HEAD LEVER AT "AUTO".
4. TURN TEMP. DIAL CALIBRATION SWITCH OFF.
5. PLACE MANUAL-AUTOMATIC SWITCH IN MANUAL POSITION.
   NOTE: THIS SUBSTITUTES VARIABLE RESISTOR (MANUAL CONTROL SWITCH IN TESTER) FOR SENSOR STRING AT ALL CONTROL HEAD SETTINGS.
6. PLACE THE MANUAL CONTROL AT MAX. HEAT AND THEN ROTATE SLOWLY COUNTERCLOCKWISE. THE VACUUM MOTOR MECHANISM WHICH CAN BE SEEN THROUGH THE SLOT IN THE PROGRAMMER COVER SHOULD START MOVING AT 180 OHMS. ROTATING THE KNOB TO MAX. COLD SHOULD MOVE THE PROGRAMMER TO ITS FULL A/C POSITION. CHECK AIR-MIX DOOR FOR FULL TRAVEL AND NO OBSTRUCTIONS. IF THE PROGRAMMER OPERATES IN THIS MANNER, THE COMPONENTS GOVERNING PROGRAMMER MOVEMENT - AMPLIFIER, TRANSDUCER AND VACUUM MOTOR - ARE OK. PROCEED TO SENSOR STRING TEST.
7. IF THE PROGRAMMER MOVES BUT THE MOVEMENT DOES NOT START AT 180 OHMS ± 3OHMS, ADJUST FEEDBACK POT AS DESCRIBED IN TESTS.
8. IF THE PROGRAMMER DOES NOT MOVE, CHECK AS FOLLOWS FOR SUPPLY VOLTAGE AND SUPPLY VACUUM.
   a. PLACE VOLTMETER SWITCH IN POSITION 2. BATTERY VOLTAGE SHOULD BE INDICATED ON METER. NO VOLTAGE INDICATES LACK OF GROUND ON TERMINAL 2 OR NO BATTERY SUPPLY TO TERMINAL 1 OF PROGRAMMER.
   b. DUMMY VACUUM PLUG MUST BE MODIFIED AS SHOWN. PLUG TESTER'S DUMMY VACUUM PLUG ON PROGRAMMER VALVE. CONNECT DUMMY HOSE TO VACUUM SUPPLY (PORT 2 HOSE IN CAR VACUUM HARNESS). MAKE SURE VACUUM SUPPLY IS PRESENT (11 IN. HG OR MORE). PROGRAMMER NOW HAS VACUUM SUPPLY WITH NO CAR VACUUM SYSTEM COMPONENTS CONNECTED. IF PROGRAMMER FUNCTIONS USING MANUAL CONTROL, TROUBLESHOOT CAR VACUUM SYSTEM. IF PROGRAMMER DOES NOT FUNCTION, DEFECT IS IN PROGRAMMER. SEE TESTS 1 AND 2.

SENSOR STRING TEST
(DO PROGRAMMER FUNCTIONAL TEST FIRST)

1. SET CONTROL LEVER AT "AUTO" AND TEMPERATURE DIAL AT 75.
2. PLACE TEMPERATURE DIAL CALIBRATION SWITCH AT OFF.
3. PLACE MANUAL-AUTOMATIC SWITCH AT AUTOMATIC POSITION AND REMOVE SINGLE CONNECTOR FROM PROGRAMMER TEST POINT.
4. OBSERVE POSITION OF VACUUM MOTOR MECHANISM THROUGH SLOT IN PROGRAMMER COVER. MARK POSITION ON COVER.
5. TURN TESTER MANUAL-AUTOMATIC SWITCH TO MANUAL AND CONNECT SINGLE CONNECTOR TO PROGRAMMER TEST POINT. ADJUST MANUAL CONTROL UNTIL VACUUM MOTOR IS IN SAME POSITION AS STEP 4.
6. READ MANUAL CONTROL. THIS INDICATES THE RESISTANCE OF THE SENSORS AND THE TEMPERATURE DIAL COMBINED AND SHOULD BE 120 TO 150 OHMS AT 70°F TO 75°F ROOM TEMPERATURE.
7. IF READING IS NOT 120 TO 150 OHMS, PERFORM THE TEMPERATURE DIAL CALIBRATION TEST BELOW TO CHECK THE TEMPERATURE DIAL. IF TEMPERATURE DIAL CHECKS OK, THE IN-CAR OR AMBIENT SENSOR OR CONNECTING WIRING IS MALFUNCTIONING.

TEMPERATURE DIAL CALIBRATION

TO CALIBRATE TEMPERATURE DIAL, REMOVE CONTROL HEAD AND REMOVE TWO GREEN WIRES FROM TEMPERATURE DIAL POT. CONNECT AN OHMMETER TO TERMINALS AND SET TEMPERATURE DIAL AT 75. OHMMETER SHOULD READ 36 OHMS. IF NOT, SLIP TEMPERATURE DIAL CLUTCH TO ADJUST.

BLOWER CIRCUIT TESTS

1. PLACE CALIBRATION SWITCH IN OFF POSITION.
2. PLACE MANUAL-AUTOMATIC SWITCH IN MANUAL POSITION AND CONNECT SINGLE CONNECTOR TO PROGRAMMER TEST POINT.
3. SET CONTROL LEVER AT "LO". WITH VOLTMETER KNOB AT 6, THERE SHOULD BE 3.5-4.5 VOLTS ON VOLTMETER. IF THERE IS NO VOLTAGE AT 6, THERE IS A DEFECT IN THE BLOWER CIRCUIT.
4. SET CONTROL LEVER AT "AUTO". THERE SHOULD BE BATTERY VOLTAGE AT 8 AND THE VOLTAGE AT 6 SHOULD STAIR STEP DOWN AND THEN BACK UP AS THE MANUAL CONTROL IS ROTATED FROM MAX. HEAT TO MAX. COLD. IF THERE IS NO VOLTAGE AT 8, THERE IS A DEFECT IN THE BLOWER CIRCUIT. IF THE VOLTAGE AT 6 DOES NOT CHANGE AS DESCRIBED, THE PROGRAMMER IS DEFECTIVE. NOTE: A GOOD CONNECTION BETWEEN 6 AND THE BLOWER MOTOR IS ASSUMED. IF THE BLOWER WIRE IS DISCONNECTED, ONLY BATTERY VOLTAGE WILL APPEAR AT 6.

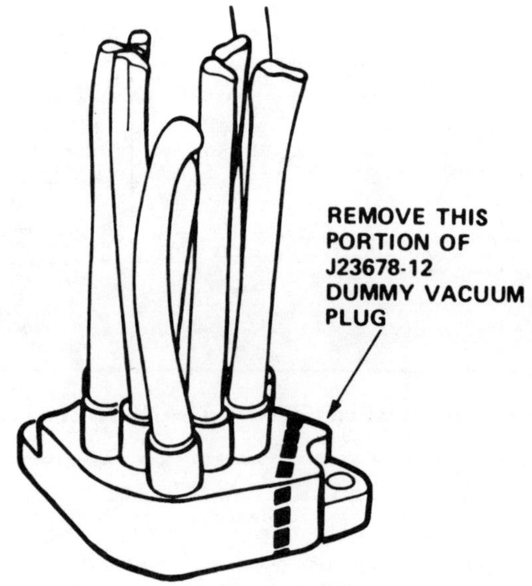

MODIFIED VACUUM PLUG

Fig. 134 Programmer & system diagnosis chart. 1977-80 Buick, Cadillac, 1979-80 Pontiac Auto Climate Control & 1977-80 Chevrolet Comfortron (Part 2 of 2)

is in the Norm position, no vacuum should be applied to the water control valve. If vacuum is present check system for source of vacuum. If vacuum is not present but heater hose from water control valve to heater core is not hot, replace the water control valve. If heater hose is hot check for vacuum at programmer. If vacuum is not present at programmer check vacuum for leaks. If vacuum is present, disconnect hose from vacuum motor and apply vacuum directly to motor. If vacuum motor mechanism does not move replace vacuum motor. If mechanism does move check vacuum circuit from vacuum checking relay to vacuum motor and compensator assembly for leaks. If no leakage found replace the vacuum checking relay.

7. If temperature door moves freely but only warm air is delivered check vacuum motor for proper operation. Vacuum motor mechanism should be fully drawn into vacuum motor when in full heat operation. Disconnect vacuum hose, if mechanism moves out of motor replace the compensator assembly. If vacuum motor mechanism does not move, replace vacuum motor.
8. If poor temperature control is noted and outlet air temperature varies, check for the following:
   (a) Improperly adjusted temperature control cable.
   (b) Improperly installed or leaking air distribution hoses.
   (c) Faulty compensator assembly in programmer.

1980 ELECTRONIC TOUCH CLIMATE CONTROL SYSTEM

General Diagnosis

Observe the following conditions to aid in diagnosing system malfunctions:

1. A 65 degree temperature selection drives system to maximum cold.
2. An 85 degree temperature selection drives system to maximum heat.
3. Mode L.E.D.s will flash if a mode is malfunctioning.
4. Temperature L.E.D.s will flash if the temperature control actuator malfunctions.
5. A mode door actuator with a disconnected wire will cause mode L.E.D.s to flash.
6. A mode door which is inoperative but stopped in a position which is correct for a certain mode will not cause mode panel L.E.D.s to flash until the control panel calls for the actuator to move to another position.
7. Blower delay affects only heater and bi-level modes. Switch can be readily checked by touching 65 degrees on the control panel. Blower operation at 65 degrees and no blower when 85 degrees is touched indicates faulty coolant switch circuit or cold engine.
8. Control must have constant B+ (battery voltage) first and then switched ignition power supply to operate correctly.
9. An open sensor string circuit will drive the system to maximum heat except at 65 degree selection.
10. A shorted sensor string circuit will drive the system to maximum cold except at 85 degree selection.

Control Panel Function Test

Test

1. Turn ignition on, touch 75 degrees, auto mode. Observe if L.E.D.s come on.
2. Turn ignition on and off several times and observe whether L.E.D.s always light 75 degrees, auto mode.

Conclusion

No lights indicate ignition power is not reaching control panel. Check A/C fuse or brown feed wire to control panel harness in instrument panel.

A mode or temperature L.E.D. that reads differently each time the ignition is turned on indicates the orange B+ feed wire is disconnected from fuse block or is not supplying current all the time or the fuse may be blown.

Check the center connector on the rear of control panel for proper engagement, proper wire locations and for current. Pin B should have battery power at all times even with the ignition off.

Since L.E.D.s may fail, check other modes and temperature positions to determine if a control head L.E.D. has failed or is malfunctioning. Replace control panel if panel L.E.D.s fail to light and fail to function as described in steps 1 and 2.

Mode Function Tests

NOTE: Perform these tests with engine running and at operating temperature.

1. Touch 85 degrees, Auto mode. As soon as ignition switch is turned on, blower speed should gradually increase to maximum with all air coming from heater outlets. In 30 to 60 seconds defroster actuator should mode to bleed position allowing a small amount of air to be directed to the windshield. Discharge air should be hot. Control panel L.E.D.s will flash in 20 to 30 seconds if temperature or mode actuators have not reached heater and maximum heat positions. Compressor will cycle when ambient temperature is above 50 degrees F.
2. Touch 65 degrees, Auto mode. Blower speed will decrease to low and gradually increase to maximum blower. Air will recirculate and water valve will close. Air should become cool and move to A/C outlets. Compressor will cycle when ambient temperature is above 50 degrees F. Panel L.E.D.s will flash if temperature door does not reach maximum cool and if mode doors are not in A/C and recirculation or water valve is not closed.
3. Touch Bi-Level. Blower speed will not change and air will be split between A/C heater outlets.
4. Touch Lo-Blower. Blower speed will gradually decrease to low blower.
5. Touch 85 degrees, Defrost mode. Blower will gradually increase to Hi and temperature will become hot. Compressor will cycle above 50 degrees F. ambient. Defrost door will move from bleed to open.
6. Touch 85 degrees, Economy mode. Compressor should turn off and air should come from the heater outlets.

System Diagnosis

System Operates Only at Maximum A/C

1. Touch 85 degree panel. If system remains in maximum A/C, proceed with diagnosis of temperature door actuator and wiring.
2. If system goes to heat condition, check the following items:
   a. Shorted in-car sensor, ambient sensor or incorrect sensors.
   b. Miscalibrated temperature panel.
   c. Faulty temperature door actuator.

System Operates Only at Maximum Heat

1. Touch 65 degree panel. If system remains in maximum heat, proceed with diagnosis of temperature door actuator and wiring.
2. If system goes to maximum cool, check the following items:
   a. Open sensor string.
   b. Incorrect sensors.

System Does Not Provide Comfort at Control Panel Settings

1. Check the following components:
   a. Aspirator System. Airflow should hold a small piece of paper against sensor air inlet grille.
   b. Temperature door rod adjustment.
   c. Panel and temperature actuator calibration.
   d. In-car sensor box cover.

Mode Actuator Function Tests

Flashing Mode L.E.D.s

1. Perform functional test and check for airflow from correct outlets, Fig. 135.
   a. If airflow is correct, proceed to step 2.
   b. If airflow is incorrect, proceed with actuator diagnosis.
2. Touch 85 degrees, Auto Hi mode and turn ignition switch off and back on. Check for defroster door sealing. Defroster should run to bleed position after 45 to 60 seconds.
   a. If defroster seal and bleed are OK, proceed to step 3.
   b. If defroster seal and bleed are not OK. proceed with defroster actuator diagnosis.
3. Touch valve panel between 65 and 85 degrees, Auto mode and observe water valve operation.
   a. Water valve is functioning properly if it starts and stops within 10 seconds.
   b. If water valve is inoperative or runs 15 seconds or longer, diagnose water valve actuator and circuit.

Mode Actuator Diagnosis

1. Use the airflow chart, Fig. 135, to determine exactly which mode actuators are malfunctioning.
2. Check electrical connection to actuator. Then touch mode again and observe mode change.

NOTE: A flashing mode light indicates a power shut-off to the affected mode. Touching the mode panel supplies power for an additional 20 seconds.

3. Observe the suspected faulty actuator through a mode change to ensure correct actuator is being diagnosed. Determine whether:
   a. Actuator is inoperative.
   b. Actuator stops in incorrect position.
   c. Actuator stops in correct position but mode door does not seal.

<table>
<tr><th colspan="2"></th><th>BLOWER SPEED</th><th>AIR INLET DOOR POSITION</th><th>TEMPERATURE DOOR POSITION</th><th>MODE DOORS POSITION</th><th>DEFROSTER DOOR POSITION</th><th>COMPRESSOR</th><th>HTR WATER SHUT OFF VALVE</th><th>IS BLOWER DELAYED FOR ENGINE WATER WARMUP?</th></tr>
<tr><td rowspan="7">MODE SELEC-TION</td><td>OFF</td><td>FIXED LOW</td><td rowspan="3">OUTSIDE AIR</td><td rowspan="7">VARIES — RESPONSIVE TO SENSORS</td><td>FORCED TO HEATER</td><td>SEALED</td><td rowspan="2">DOESN'T OPERATE</td><td rowspan="5">USUALLY OPEN-EXC. CLOSED AT MAX. A/C POSITION OF TEMPER-ATURE DOOR</td><td rowspan="6">YES EXCEPT IN THE A/C MODE WHEN CAR IN-TERIOR IS WARM IN WHICH CASE BLOWER TURNS ON IMMEDIATELY EVEN IF ENGINE IS COLD.</td></tr>
<tr><td>ECCN.</td><td>VARIABLE BLOWER PROGRAM</td><td rowspan="4">EITHER HEATER, BI-LEVEL OR A/C DEPENDS ON TEMPERATURE DOOR</td><td rowspan="5">DOOR WILL ASSUME "BLEED" POSITION AFTER 45 SECOND DELAY AND WILL DELIV-ER BLEED AIR TO W/S IF IN HEATER OR BI-LEVEL MODE</td></tr>
<tr><td>LO</td><td>FIXED LOW</td><td rowspan="5">OPERATES ABOVE APPROX. 37°F (3°C)</td></tr>
<tr><td>AUTO</td><td>VARIABLE BLOWER PROGRAM</td><td rowspan="4">RECIRCULATE AIR WHEN TEMPERATURE DOOR IS AT MAX. A/C OTHERWISE OUTSIDE AIR</td></tr>
<tr><td>HI</td><td>FIXED HI</td></tr>
<tr><td>BI-LEVEL</td><td>VARIABLE BLOWER PROGRAM</td><td>BI-LEVEL</td><td rowspan="2">OPEN</td></tr>
<tr><td>DEF.</td><td>FIXED HI</td><td>FORCED TO HEATER</td><td>FULL OPEN TO W/S</td><td>NO-BLOWER IS TURNED ON</td></tr>
</table>

Fig. 135 Control system functional test chart. 1980 Electronic Touch Climate Control

Once actuator malfunction has been pinpointed, proceed as follows:

Actuator Inoperative, Ignition On & Mode Lights Flashing

1. Connect a voltmeter between Pin C of actuator and ground. If 8 or more volts are observed, proceed to step 2. If less than 2 volts are observed to step 3.
2. Connect a voltmeter between Pins A and C and touch flashing mode light. Mode light should stay on for 20 seconds. If 8 volts or more are observed, check for obstruction to links or mode doors or stripped actuator gears. If no obstruction is found, replace actuator. If less than 2 volts are observed, proceed to step 4.
3. Connect a voltmeter between Pin C of control panel and ground. If 8 volts or more are observed, repair wire to actuator pin. If less than 2 volts are observed, replace control panel assembly.
4. Connect a voltmeter between Pin C of control panel and pin corresponding to Pin A of faulty actuator. Touch flashing mode light and observe voltage. If 8 volts or more are observed, check wire continuity. If less than 2 volts are measured, replace control panel assembly.

NOTE: Never connect an ohmmeter to a powered circuit as damage to the instrument will result.

Actuator Stops At Incorrect Position

1. Connect leads of an auxiliary 12 volt DC power source to terminals A and C of actuator to drive actuator 180 degrees from its starting position. Disconnect power source. Use an ohmmeter to check resistance between terminals C and D of actuator at that position. Note ohmmeter reading. Reconnect auxiliary power source and drive actuator back to its original position. Check resistance between terminals C and D at this position. If one ohmmeter reading is infinity and the other corresponds to the resistance values listed in Fig. 135A, check continuity of wire between Pin C of actuator and 22 pin connector at control. If continuity exists, replace control head. If there is no change in resistance between the two actuator positions, replace actuator.
2. Ensure proper actuator is installed in vehicle by checking part number. Due to differences in crank arm geometry, actuators are not interchangeable.

NOTE: For testing purposes only, a suspected faulty actuator may be interchanged with a known operational one. Any mode actuator can be used to test any actuator connector circuit except defroster door which requires a special actuator.

Actuator Stops in Correct Position, Mode Door Does Not Seal

1. Ensure proper door link connection and assembly.

Temperature Actuator Function Tests

Flashing Temperature L.E.D.'s

1. Touch 65 degrees and observe panel L.E.D. for 30 seconds.
   a. If flashing stops, temperature rod is maladjusted; refer to "Unit Replacement Section" for adjustment procedure.
   b. If flashing continues, proceed to step 2.
2. Touch 85 degrees and observe panel L.E.D. for 30 seconds.
   a. If flashing stops, refer to "Unit Replacement Section" and correctly adjust temperature rod.
   b. If flashing continues, proceed to step 3.
3. Check system circuitry. Refer to Temperature Actuator Diagnosis.

Temperature Actuator Diagnosis

Temperature Actuator Inoperative

Ensure temperature actuator arm is not obstructed, then check temperature door rod operation and adjustment. Inspect electrical connection to actuator. If a visual check fails to reveal the malfunction, proceed to step 1.

1. Connect a 12 volt DC power source across terminals A and B of actuator for 2 to 3 seconds then reverse test leads. Actuator should operate, arm should rotate and reverse direction when test leads are reversed. If so, proceed to step 2; if not, replace inoperative actuator. If actuator runs but arm does not move, actuator gears are stripped. Replace actuator.
2. Using an ohmmeter, check the resistance between actuator terminals C and D. There should be 5000 ohms resistance at 70 to 85 degrees F. ambient. If so, proceed to step 3. If reading is infinity, actuator has a broken resistor strip and the actuator should be replaced.
3. Connect a 12 volt DC power source across terminals A and B to drive actuator arm to position 3 (calibration). Using an ohmmeter, check resistance between terminals C and F. There should be 1230 ohms resistance; if so, proceed to step 4. If not, actuator feedback circuit is faulty and actuator must be replaced.
4. Using an ohmmeter or self-powered test light, check continuity of wire harness from actuator to control panel. Refer to the following when checking continuity:
   a. An open circuit in wire A (temp drive 2) will result in a stalled actuator.
   b. An open circuit in wire B (temp drive 1) will result in a stalled actuator.
   c. An open circuit in wire C (feedback 1) will result in actuator driving only toward cold position.
   d. An open circuit in wire D (feedback 3) will result in actuator driving only toward heat position.
   e. An open circuit in wire F (feedback 2) will result in actuator driving only toward cold position.

RESISTANCE BETWEEN TERMINALS C & D			
ACTUATOR ASSEMBLY	ARM POSITION		
	1	2	3
AIR INLET	.150 Ω	6.8K Ω	
LWR DIVERTER	6.8K Ω	.150 Ω	
UPR DIVERTER	6.8K Ω	.150 Ω	
DEFROST	6.8K Ω	.150 Ω	27.0K Ω

Fig. 135A Mode actuator resistance chart. 1980 Electronic Touch Climate Control

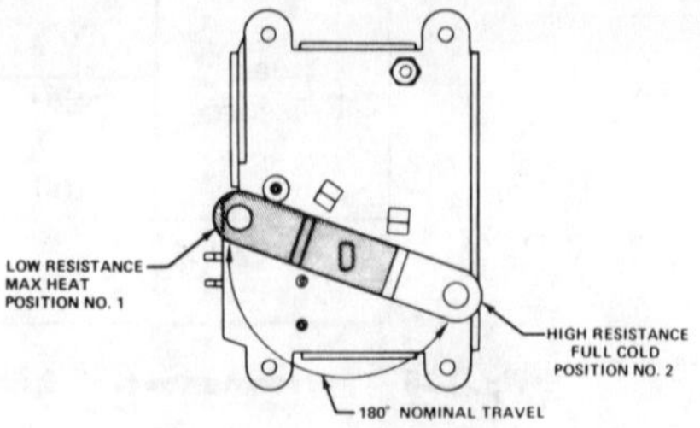

Fig. 136 Temperature door actuator travel. 1980 Electronic Touch Climate Control

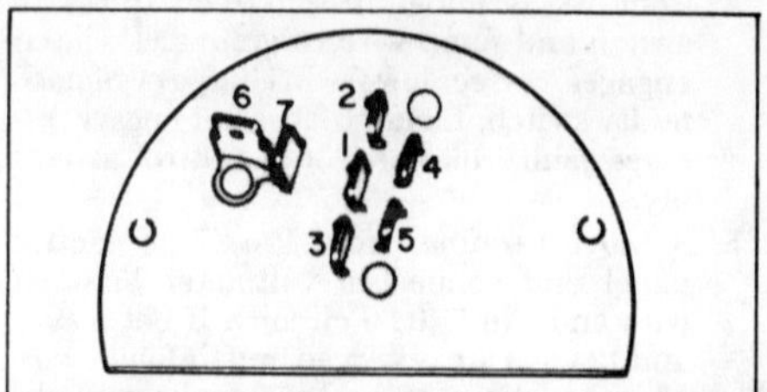

TERMINAL NUMBER	FUNCTION
1	BLOWER SPEED INPUT SIGNAL FROM CONTROL HEAD
2	B+(12 VOLTS) DIRECT FROM FUSE BLOCK
3	IGNITION SWITCHED 12 VOLTS
4	COMPRESSOR CLUTCH SIGNAL FROM CONTROL HEAD
5	COMPRESSOR CLUTCH ENGAGEMENT FEED
6	BLOWER MOTOR FEED
7	SYSTEM GROUND

Fig. 137 Blower control unit power transistor circuit terminal identification. 1980 Electronic Touch Climate Control

Repair open circuits as necessary.

5. If wire harness was found satisfactory in step 4 and temperature actuator is still inoperative, replace control panel.

Actuator Does Not Move Through Full Travel

If actuator moves less than 180 degrees from maximum heat to maximum cold position, Fig. 136, perform actuator resistor tests outlined previously in steps 2, 3, 4, and 5. Replace actuator or control panel as necessary.

Blower Diagnosis

The blower and compressor clutch control is located in the blower motor air stream in the engine compartment. It converts low current signals from the control panel into high current voltage feed to the blower motor. The resistor strip on the temperature door actuator acts as a rheostat to control blower motor speed. A power transistor circuit in the control engages the compressor clutch.

Blower malfunctions can be divided into five problem areas; observe the following conditions when testing blower control circuits:

1. No Blower — Engine Off, Engine Temperature over 120 degrees F., Control Off, Ignition On, Panel L.E.D.s On, 65 degree Panel Setting.
   a. Connect a voltmeter between blower feed and ground, Pin #6 (2-way) Circuit #65 E, Fig. 137. If 3 volts or more are observed, repair faulty ground or motor. If less than 3 volts are observed, proceed to step b.
   b. Connect a voltmeter between B+ and ground, Pin #2 (5-way) Circuit #2. If less than 3 volts are observed, repair faulty B+ connection or fusible link. If battery voltage is observed, proceed to step c.
   c. Connect a voltmeter between ignition On and ground, Pin #3 (5-way) Circuit #901. If less than 3 volts are observed, check wire continuity and repair if necessary. If battery voltage is observed, proceed to step d.
   d. Connect a voltmeter between blower control ground and ground, Pin #7 (2-way) Circuit #151A. If 1 volt or more is observed, repair open ground wire. If no voltage is observed, proceed to step e.
   e. Connect a voltmeter between blower control and ground, Pin #1 (5-way) Circuit 900. If 3 volts or more are observed, replace blower control. If less than 3 volts are observed, proceed to step f.
   f. Connect a voltmeter between panel 8-way connector, Pin B, and ground (Circuit #900). If fluctating voltage of 3 volts or more is observed, check blower control wire continuity. If less than 3 volts are observed, replace control panel.
2. No Blower — Panel Set at 85 degrees, Ignition On, Engine Temperature over 120°F., Panel L.E.D.s On.
   a. Touch 65 degrees; if blower operates, check for faulty engine temperature switch or attaching wire. Then proceed to step b. If blower does not operate, refer to test 1.
   b. Touch 85 degrees, ground engine temperature switch wire, turn ignition switch off and then on (71 circuit). If blower operates, replace engine temperature switch. If blower does not operate, proceed to step c.
   c. Ensure wire terminal connections on control panel are secure at 8-way connector, then ground Pin A (71 circuit). If blower operates, clean or replace terminals. If blower does not operate, replace control panel.
3. High Blower Only — No Low Blower in OFF and LO Positions.
   a. Connect a voltmeter between blower control, Pin #1, and ground (Circuit #900). If fluctuating voltage of between 3 and 7 volts is observed, replace blower control. If fluctuating voltage of between 7 and 14 volts is observed, replace control panel.
4. No Blower in HI or DEF — L.E.D.s On.
   a. Connect a voltmeter between blower control, Pin #1, and ground (Circuit #900). If fluctuating voltage of between 3 and 7 volts is observed, replace control panel. If fluctuating voltage of between 7 and 14 volts is observed, proceed to step b.
   b. Connect a voltmeter between blower feed, Pin #6, and ground (Circuit #65E). If between 11 and 14 volts are observed, replace blower motor. If less than 10 volts are observed, replace blower control.
5. Blower Speed Does Not Vary in ECON, AUTO, or BI-LEV Mode.
   a. Touch 65 degrees, then touch 85 degrees allowing 30 seconds between changes. Observe L.E.D.s; if they flash, proceed with temperature door actuator diagnosis. If L.E.D.s remain on, proceed to step b.
   b. Connect a voltmeter between blower feed, Pin #6, and ground (Circuit #65E). Touch 65 and 85 degrees alternately. If voltage changes from 12 to 3 volts and back to 12 volts, replace blower motor. If there is no change in voltage, proceed to step c.
   c. Connect a voltmeter between blower control, Pin #1, and ground (Circuit #900). Touch 65 and 85 degrees alternately. If voltage changes between 3 and 12 volts, replace blower control. If there is no change in voltage, proceed to step d.
   d. Connect a voltmeter between control panel, Pin B (8-way), and ground (Circuit #900). Touch 65 and 85 degrees alternately. If there is a voltage change, repair wire (Circuit #900). If there is no voltage change, replace control panel.

Compressor Clutch Diagnosis

If clutch does not engage when ambient temperature is above 55°F., ensure engine is running, panel mode and temperature L.E.D.s are on and proceed as follows:

1. Connect a voltmeter between clutch input and ground. If between 8 and 12 volts are observed, repair ground or replace faulty clutch. If no voltage is observed, proceed to step 2.
2. Connect a voltmeter between clutch control input and ground, Pin #4 (5-way) Circuit #966, Fig. 137. If between 8 and 12 volts are observed, proceed to step 3. If no voltage is observed, proceed to step 5.
3. Connect a voltmeter between ignition On and ground, Pin #3 (5-way) Circuit #901. If between 8 and 12 volts are observed, proceed to step 4. If below 8 volts are observed, check continuity of B "run" line.

4. Remove terminal from freon pressure switch and jump wire to ground. If clutch engages, correct low freon charge or replace faulty switch. If clutch does not engage, replace faulty blower-clutch control assembly.
5. Remove terminal from Pin C on control panel and connect a voltmeter between wire and Pin C (966 circuit). If between 8 and 12 volts are observed, repair faulty wire or connection, If below 8 volts are observed, replace faulty control panel.

	70°F	80°F	90°F	100°F
Head Pressure	120 TO 160	150 TO 190	195 TO 235	230 TO 270
Evaporator Pressure	27-45 PSIG			

Fig. 138 Performance test chart. 1977-79 DeVille & Fleetwood Brougham

Cadillac

1974-76 AUTO CLIMATE CONTROL

HOW TO ISOLATE THE PROBLEM

Step 1

To confirm the system defect make a brief check of system operation by sitting in the car and operating the controls, with the engine warmed up and running at 1000 RPM or higher, in the following sequence:

1. Operate engine at 1000 rpm or better.
2. Place A/C outlets in full open position.
3. Place a thermometer in center A/C outlet.
4. Position control lever in Def on 1974-76 models and temperature dial at 85.
   (a) Air should be delivered from defroster outlets.
   (b) Blower should operate at high speed.
   (c) Some air should come from heater outlets.
5. Position control lever in Bi-Level on 1974-76 models and temperature dial at 75.
   (a) Air should be delivered from A/C and heater outlets.
   (b) Blower should operate at lower speed.
   (c) Recirculating air door should open.
6. Position lever at Hi and temperature dial at 65.
   (a) Cool to cold air should be delivered from A/C outlets.
   (b) Blower should operate at high speed.
   (c) Recirculating air door should open.
7. Position lever in Auto and temperature dial at 65.
   (a) Recirculating air door should close.
   (b) After 45 seconds discharge temperature should be 50°F or lower. Discharge air temperature may be higher if ambient temperature is 90 or above.
8. Position lever in Auto and temperature dial at 85.
   (a) Blower speeds should drop.
   (b) Recirculating air door should close.
   (c) Discharge air temperature should increase.
   (d) Air delivery may change from A/C outlets to bi-level to heater outlets.
9. Position lever at Lo and temperature dial at 85.
   (a) Air should be delivered at low blower speed.
10. On 1974 models, position lever in Vent and temperature dial at 65.
    (a) Air delivery should be from A/C outlets (or bi-level).
    (b) Blower should operate at low speed.
    (c) Compressor should not operate.
11. On 1975-76 models, position lever in Economy and temperature dial at 65.
    (a) Air should be delivered from A/C outlet, bi-level or heater outlets depending on ambient temperature.
    (b) Blower speed may vary.
    (c) Compressor should not operate.
12. On all models, position lever in off and temperature dial at 65.
    (a) Air should be delivered from heater outlets.
    (b) Blower should operate at low speed.
    (c) Compressor should not operate.

NOTE: During the above tests, be sure to note the following:

(a) Check to insure that air delivery is *not* coming from both the A/C and heater outlets, when only one mode is indicated. A split air delivery is indicative of a vacuum leak.

(b) Note whether program events (air delivery mode change, blower speed change, recirc air, etc.) occur without change in discharge air temperature. This would indicate the programmer is operating without moving the temperature air door. Check the air mix door link to programmer connection.

(c) If neither the program events nor the discharge air temperature change, a faulty vacuum or electrical signal to the programmer or a programmer malfunction is indicated.

(d) Failure of a specific vacuum operated door function could indicate a vacuum disconnect of the vacuum diaphragm at that door.

UNDERHOOD AREA

Step 2

A simple, visual inspection of the easily accessible underhood and instrument panel electrical and vacuum connections will, in many instances, reveal the defect on the spot. The following checks can be made without removing any car parts:

(a) Heater water valve vacuum connection (center of car behind engine).
(b) Low and auto blower relay electrical connections (located on evaporator case).
(c) Ambient switch and sensor electrical connection (located on evaporator case).
(d) A/C compressor electrical connection and ground wire attachment.

AMBIENT AIR TEMPERATURE	21° C (70° F)	27° C (80° F)	32° C (90° F)	38° C (100° F)
COMPRESSOR HEAD PRESSURE* ± 138 KPA (± 20 PSI)	965 KPA (140 PSI)	1171 KPA (170 PSI)	1481 KPA (215 PSI)	1723 KPA (250 PSI)
ACCUMULATOR PRESSURE * ± 55 KPA (± 8 PSI)	241 KPA (35 PSI)	241 KPA (35 PSI)	241 KPA (35 PSI)	241 KPA (35 PSI)
CENTER OUTLET TEMPERATURE ± 1° C (± 2° F)	3° C (37° F)	3° C (38° F)	4° C (40° F)	6° C (42° F)
*JUST BEFORE COMPRESSOR DISENGAGES				

Fig. 139 Performance test chart. 1980 Cadillac all models

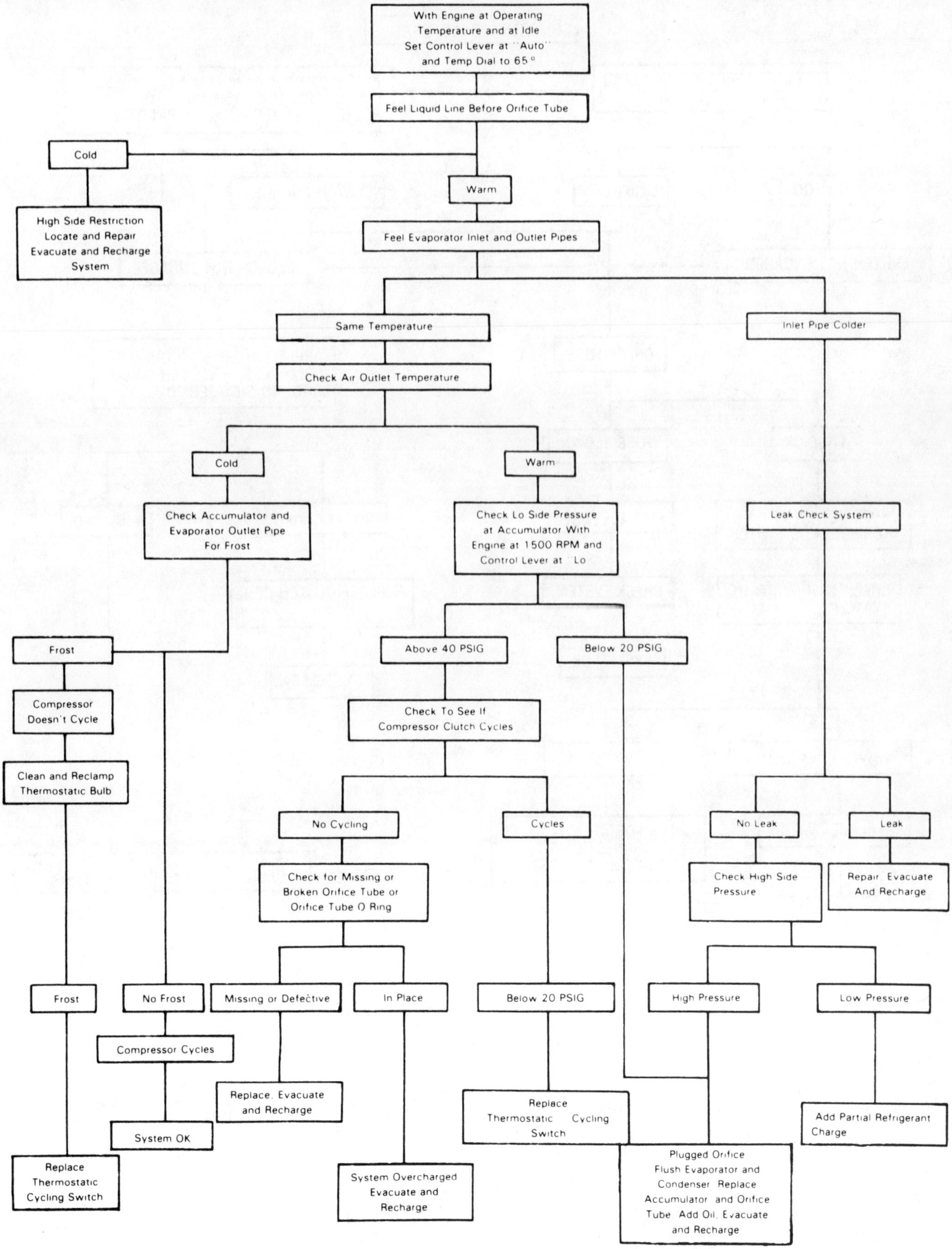

Fig. 140 Insufficient cooling diagnosis chart. 1977-80 Cadillac all models

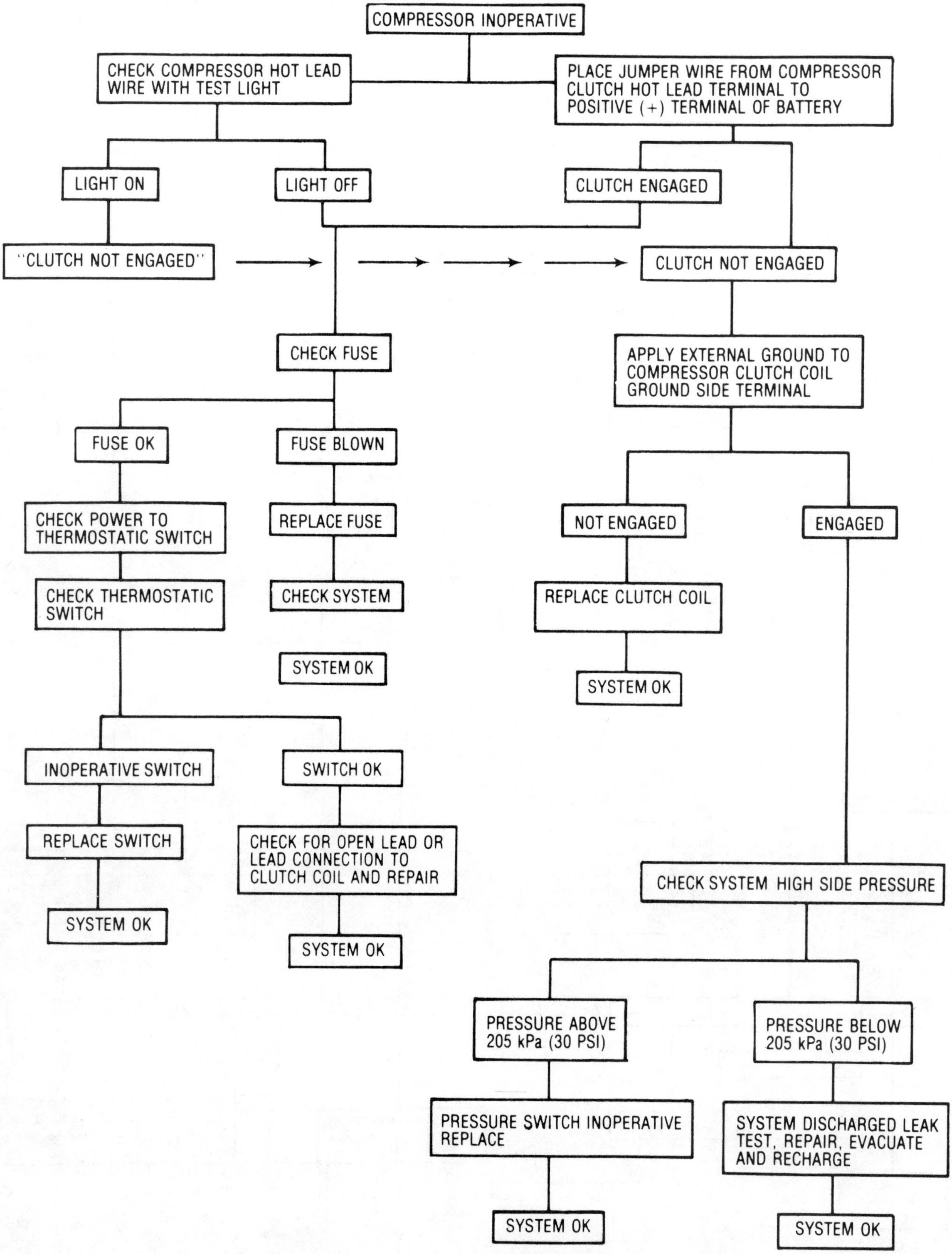

Fig. 141 Low pressure switch diagnosis chart. 1977-80 Cadillac all models

CONTROL SETTING	SYSTEM SHOULD OPERATE AS FOLLOWS
"DEF" DIAL @ 85°	Air should be delivered out of defroster outlets at a fixed high blower speed. Some floor bleed out heater.
"BI-LEVEL" DIAL @ 85°	Air should be delivered from **both** the A/C and heater outlets. Only a small portion of air will come out the defroster outlet.
"HI" DIAL @ 65°	Air should be delivered out the A/C outlets at a fixed high blower speed and cool to cold temperature. The recirculating air door should open (blower noise increases). Door movement will be slow because of vacuum delay porous plug.
"AUTO" DIAL @ 85°	Blower speeds should drop (Hi-M_3-M_2-M_1-etc.). Recirc air door should close (noise level will drop). Discharge air temperature should increase. Depending on the temperature in the work area, the air delivery mode should change from the A/C outlets to the heater outlets.
"LO" DIAL @ 85°	Air should be delivered at a fixed low blower speed.
"ECONOMY" DIAL @ 65°	Blower speeds should increase (M_1-M_2-M_3-etc.) Depending on the temperature in the work area, the air delivery mode should change from the heater outlets to both heater and A/C outlets or the A/C outlets only. The A/C compressor should not operate.
"OFF" DIAL @ 65°	Air should be discharged out the heater outlet at a fixed low blower speed. No A/C compressor operation.

Fig. 142 Auto Climate Control functional test chart. 1977-79 Eldorado, 1977-80 DeVille & Fleetwood Brougham & 1977-80 Chevrolet Comfortron

CONTROL SETTING	SYSTEM SHOULD OPERATE AS FOLLOWS:
"DEF" DIAL @ 85°F	Air should be delivered out of defroster outlets at a fixed high blower speed. Some floor bleed out heater.
"BI-LEVEL" DIAL @ 75°F	Air should be delivered from *both* the A/C and heater outlets at a reduced blower speed. Only a small portion of air will come out the defroster outlet.
"HI" DIAL @ 65°F	Air should be delivered out the A/C outlets at a fixed high blower speed and cool to cold temperature. The recirculating air door should open (blower noise increases).
"AUTO" DIAL @ 85°F	Blower speeds should drop (Hi - M_3 - M_2 - M_1 etc.). Recirc air door should close (noise level will drop). Discharge air temperature should increase. Depending on the temperature in the work area, the air delivery mode should change from the A/C outlets to the heater outlets.
"LO" DIAL @ 85°F	Air should be delivered at a fixed low blower speed with immediate turn on regardless of engine coolant temperature.
"ECONOMY" DIAL @ 65°F	Blower speeds should increase (M_1 - M_2 - M_3 etc.) Depending on the temperature in the work area, the air delivery mode should change from the heater outlets to both heater and A/C outlets or the A/C outlets only. The A/C compressor should not operate.
"OFF" DIAL @ 65°F	Air should be discharged out the heater outlet at a fixed low blower speed. No A/C compressor operation. Blower will turn on immediately regardless of engine coolant temperature.

Fig. 143 Auto Climate Control functional test chart. 1977-79 Seville

(e) Heater turn-on switch electrical connection (located at front of R.H. cylinder head behind generator). Switch color should be red.

(f) Blower motor electrical feed connection and ground wire attachment.

INSTRUMENT PANEL AREA

(a) Control head vacuum and electrical connections.

(b) Defroster door and defroster bleed vacuum connections (L.H. side of heater case under instrument panel).

(c) Vacuum source connection and vacuum tee connections (at centerline of car under instrument panel).

(d) Upper and lower mode door vacuum connections (located between the cowl and heater case toward left side of heater case).

(e) 15 amp A/C-heater fuse in fuse block.

(f) A/C programmer electrical and vacuum connections (located at R.H. side of heater case under instrument panel).

(g) A/C programmer output shaft crank arm attachment.

(h) Recirc air door vacuum hose connection (orange stripe) (at R.H. side of car under instrument panel).

(i) Water valve vacuum hose (white stripe) in-line connection (located at R.H. side of car under instrument panel).

(j) Six-way and three-way electrical connections which feed underhood harness (located at R.H. side of car under instrument panel).

In addition to the above readily available checks, the following areas require removal of the glove box and need be made only if further diagnosis indicates the necessity:

(a) In-car sensor electrical and air aspirator hose connections (located in map light area).

(b) Hose connection to the in-car sensor air aspirator (mounted on top surface of heater case).

(c) Air mix door link connection to air mix door crank arm (located on top surface of heater case).

Step 3

Based on information gained during the functional test performed in Step 1, try to relate the problem to one of the following areas:

1. Temperature Control Problems
2. Blower Control Problems
3. Auxiliary Vacuum Problems
4. Refrigeration System Problems

Once a problem has been isolated to one of the areas listed above, refer to the diagnosis section. *It is of special importance to be familiar with the general description of the major sections of the system as presented in the Theory of Operation Section.*

Step 4

After the problem has been properly diagnosed and the repair made, it is important to run through the brief check listed in Step 2 in order to insure that the system is now performing correctly.

Temperature Control Problem Diagnosis

The primary function of the temperature control circuit is to determine the correct temperature of the air to be discharged into the passenger compartment and to set the air mix door position to accomplish that function. The signal used for this purpose (vacuum to position the programmer vacuum motor) is also used in the blower speed control circuit and the auxiliary vacuum function circuit. Those uses may be disregarded when dealing with problems which relate only to the temperature control circuit.

Examples of temperature control problems are: "System operates at maximum air conditioning (no heat) or only at maximum heating (no air conditioning)", "Temperature dial does not provide comfort", "Poor heating or poor cooling". Blower cycling and mode cycling are also caused by a malfunction in the temperature control circuit.

It is important to separate temperature control circuit problems from refrigeration or heater circuit problems. If the complaint is "poor" or "no" heating or cooling and the programmer moves to both extremes of travel with the air mix door and program functions known to follow, the problem probably lies in the refrigerant or heater water circuits.

Many temperature control problems result from poor electrical or vacuum line connections. The following relationships may aid diagnosis:

1. A *disconnected sensor or temperature dial* interrupts the electrical signal and drives the programmer to maximum heating.
2. A *poor sensor connection* adds resistance to the sensor string driving the system hotter.
3. An *open amplifier power feed* eliminates the output signal and drives the programmer to maximum heating.
4. A *disconnected transducer lead* drives the programmer to maximum heating.
5. A *disconnected vacuum hose* supplying the vacuum checking relay actuating nipple will lock the relay in an intermediate position.
6. A *disconnected vacuum hose* in the transducer-programmer vacuum motor line will drive the system to maximum cooling.
7. A *leak* in the auxiliary vacuum circuit may reduce the transducer vacuum supply level below control requirements, causing an "off-calibration" or "poor heating" type of complaint.
8. A *loss* of supply vacuum usually results in cold air flow on the floor.

Diagnosis of most of the problems encountered in the temperature control circuit will be greatly aided by use of tester, J-23678.

Specific problems and probable causes are as follows:

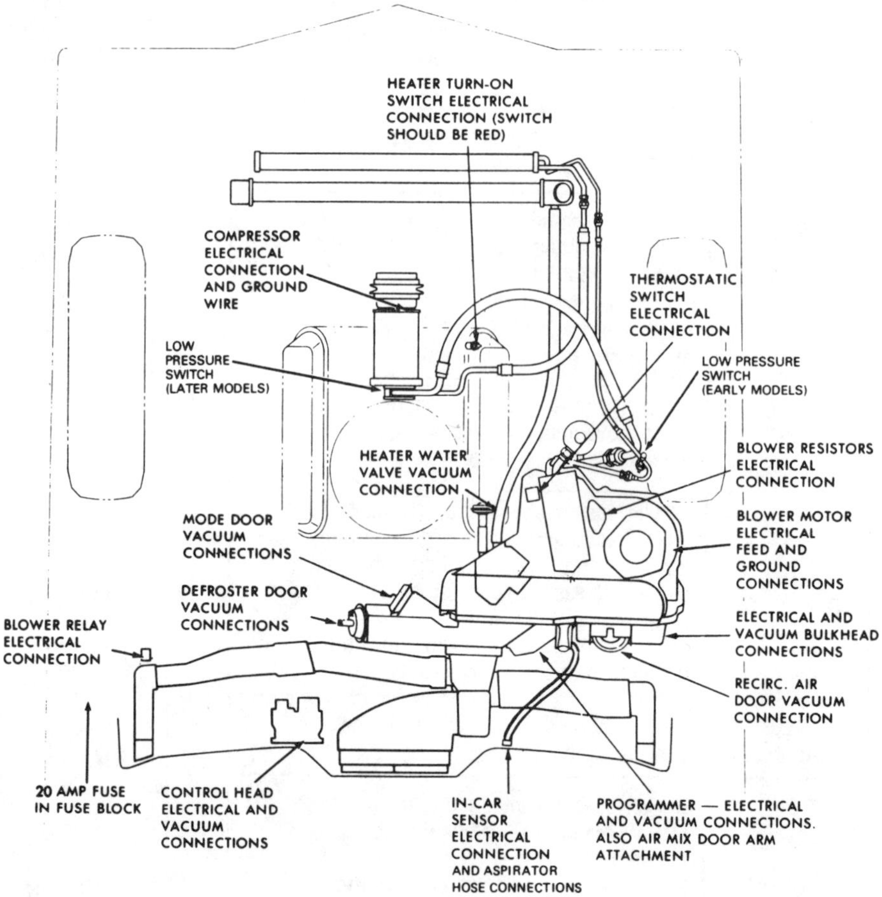

Fig. 144 In-car & underhood diagnosis areas. 1977-80 DeVille & Fleetwood Brougham

TESTER J-23678 (1974-76) PROCEDURE

PROGRAMMER FUNCTIONAL TEST

1. Connect tester electrical harness to programmer and car harness to tester.
2. Operate car at idle. Approx. 1000 RPM.
3. Set control head lever at auto.
4. Turn temp. dial calibration switch off.
5. Place Manual automatic switch in manual position.

NOTE: This substitutes variable resistor for sensor string at all control head settings.

6. Place the manual control at max. heat and then rotate slowly counter-clockwise. Vacuum motor mechanism which can be seen through the slot in the programmer cover should start moving at 180 ohms on 1974 models and 175 ohms on 1975-76 models. Rotating the knob to max. cold should move the programmer to its full A/C position. Check air-mix door for full travel, and no obstructions. If the programmer operates in this manner, the components governing programmer movement — amplifier, transducer and vacuum motor — are ok. Proceed to sensor string test.
7. If the programmer moves but the movement does not start at 180 ohms on 1974 models and 175 ohms on 1975-76 models, adjust feedback pot.
8. If the programmer does not move, check as follows for supply voltage and supply vacuum:
   (a) Place voltmeter switch in 2 position. Battery voltage should appear on meter. No voltage indicates lack of ground on term. 1 or battery supply to term. 2 of programmer.
   (b) Plug tester's dummy vacuum plug on programmer valve. Connect dummy hose to vacuum supply (Port 2 hose in car vacuum harness). Make sure vacuum is present (11 in. Hg. or more). Programmer now has vacuum supply with no car vacuum system components connected. If programmer functions using manual control, troubleshoot car vacuum system. If programmer does not function, defects are in programmer. See Programmer Vacuum Circuit.

Sensor String Test (Do Programmer Functional Test First)

1. Set control lever at auto and temperature dial at 75.
2. Place temp. dial calibration switch at off.
3. Place manual-automatic switch at automatic position.
4. Observe position of vacuum motor mechanism through slot in programmer cover.
5. Switch to manual and adjust manual control for same mechanism position.
6. When manual control is properly adjusted, moving manual automatic switch back

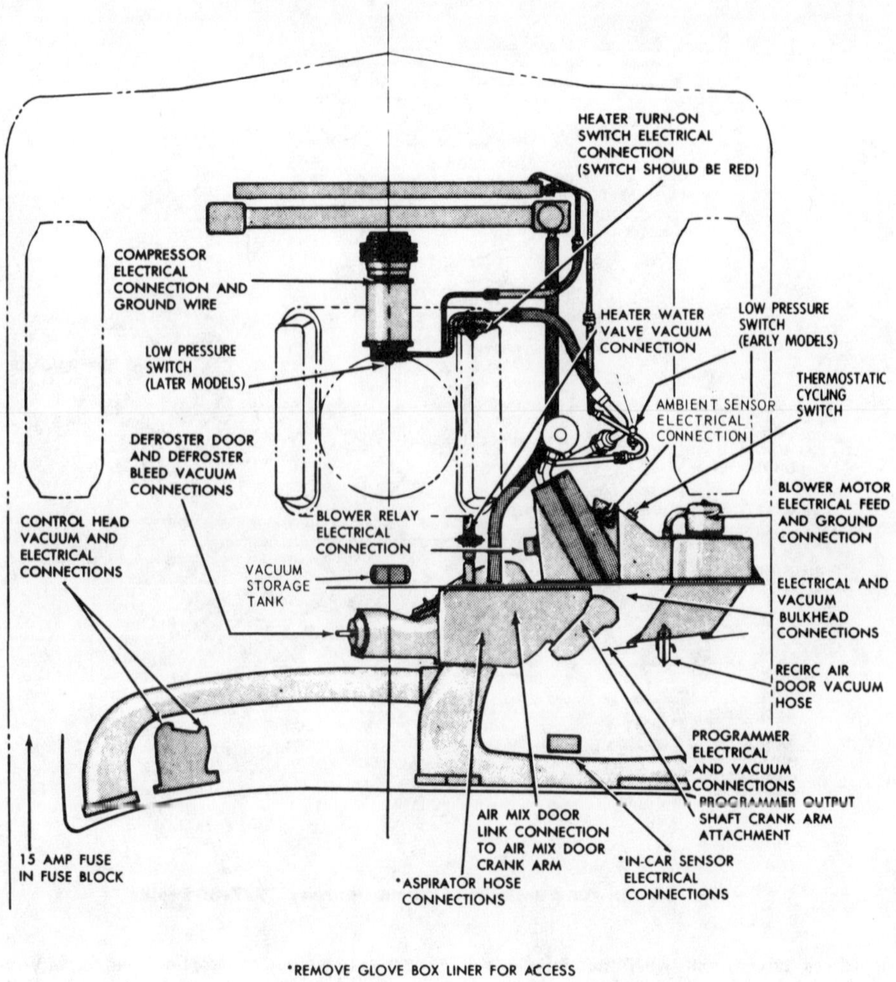

Fig. 145 In-car & underhood diagnosis areas. 1977-80 Eldorado

and forth will not move vacuum motor.

7. Read manual control. This indicates the resistance of the sensors and the temperature dial combined and should be 100 to 150 ohms on 1974 models and 120-150 ohms on 1975-76 models, at 70 to 75 room temp.
8. If reading is not 100 to 150 ohms on 1974 models and 120-150 ohms on 1975-76 models, perform the temperature dial calibration test to check out the temp. dial. If temp. dial checks ok, the in-car or ambient sensor or connecting wiring is defective.

TEMPERATURE DIAL CALIBRATION

1. Place control lever at Vent position on 1974 models and Economy position on 1975-76 models.
2. Place manual-automatic switch in manual position.
3. Place calibration switch in cal. position.
4. On 1975-76 models, disconnect in-car sensor and plug connector into tester harness, ground both sensor wires.
5. On all models read voltmeter.
6. Press pushbutton and adjust temp. dial on control head to get same voltmeter reading as above.
7. Temp. dial should read as indicated on tester panel near pushbutton. If it does not, slip temp. dial clutch to correct.

NOTE: If same voltmeter reading cannot be obtained, the temp. dial is defective.

8. On 1975-76 models, reconnect in-car sensor.

BLOWER CIRCUIT TESTS

1. Place calibration switch in off position.
2. Place manual-automatic switch in manual position.
3. Set control lever at Lo, using voltmeter knob, there should be battery voltage in brn. wire at programmer (from Lo relay) and approx. half battery voltage in tan wire at programmer (from blower). If there is no voltage in brn. wire, there is a defect in the Lo relay circuit. If there is voltage in brn. wire but no voltage in tan wire, the programmer is defective.
4. Set control lever at auto. There should be battery voltage in or/bl wire at programmer dial (from Auto relay) and the voltage in tan wire should stairstep down and then back up as the manual control is rotated from max. heat to max. cold. If full battery voltage is not obtained in or/bl wire, there is a defect in the auto relay circuit. If the voltage in tan wire does not change as described, the programmer is defective.

NOTE: A good connection between programmer and the blower motor is assumed; if the blower wire is disconnected, only battery voltage will appear in tan wire.

Programmer Internal Tests

PROGRAMMER VACUUM CIRCUIT

NOTE: In diagnosing defects in programmer, the following tests should be performed to determine a defective internal part. Tests should be made in the following sequence.

On 1974 models, remove electrical connector from programmer. If programmer moves to max. heat position, skip to Vacuum Motor and Checking Relays, otherwise, perform following test:

On 1975-76 models, turn temperature dial calibration switch off. Position automatic-manual switch in manual and manual control in max. heat. If programmer moves to max. heat position, skip to Vacuum Motor and Checking Relay, otherwise perform the following test:

1. Remove programmer cover, install dummy vacuum plug with vacuum gauge lead-in to port 2 of car harness plug.
2. If vacuum drops below supply vacuum found in programmer functional test, use long-nose pliers to pinch the two black hoses leading from side of rotary valve. If vacuum reading improves, check programmer hoses for leaks, then go to Vacuum Motor and Checking Relay.
3. If reading does not improve, leak is in rotary valve.

VACUUM MOTOR AND CHECKING RELAY

1. Disconnect electrical connector from programmer. Programmer should go to full heat.
2. Disconnect vacuum dummy plug or car vacuum harness from programmer. Vacuum checking relay should hold vacuum motor in full heat. If mechanism moves, vacuum motor or checking relay is leaking. Apply raw vacuum to vacuum motor then pinch hose at motor to see if leak is in motor.

1974 AMPLIFIER & TRANSDUCER

1. Connect tester harness to system. Place manual automatic switch in manual and temp. dial calibration switch in off position.
2. Place voltmeter switch in probe and clip position connect clip to transducer terminal with grey wire.

CAUTION: Do not short transducer terminals together or amplifier will be damaged.

Push probe into programmer connector body making contact with terminal 2 (Yellow Wire).

3. Rotating the manual control from max. heat to max. cold, the voltage reading should change 5 volts or more for a good

amplifier. If amplifier checks bad, check position of mylar insulator on side of amplifier.

1975-76

1. Connect tester harness into system, then position manual-automatic switch in manual and temperature dial calibration switch in off.
2. Remove small vacuum hose from transducer and connect tester vacuum gauge to hose. A reading of 11 in. Hg. or more of vacuum should be obtained. If there is a delay in obtaining 11 in. Hg. or more, the restrictor located in the small transducer hose may be plugged. Reinstall hose on transducer.
3. Remove large vacuum hose from transducer and connect tester vacuum gauge to large port on transducer.
4. Rotate manual control; vacuum readings of 1 to 5 in. Hg. in max. A/C and 5 to 22 in. Hg. in max. heat should be obtained. Any reading from 2 to 17 in. Hg. is obtainable by adjusting the manual control if the amplifier-transducer is functional.

1974 TRANSDUCER TEST

1. Connect tester and leads same as in amplifier test.
2. 11 in. Hg. or more vacuum should be present on small transducer supply hose.
3. To check output regulated vacuum, connect vacuum gauge to white striped hose from transducer at vacuum checking relay. Use large tee to make connection: seal extra port with thumb.
4. Rotate manual control to get 0 volts on voltmeter. Turning manual control to max. cold should make vacuum drop to 0 in. Hg. Reconnect hoses.

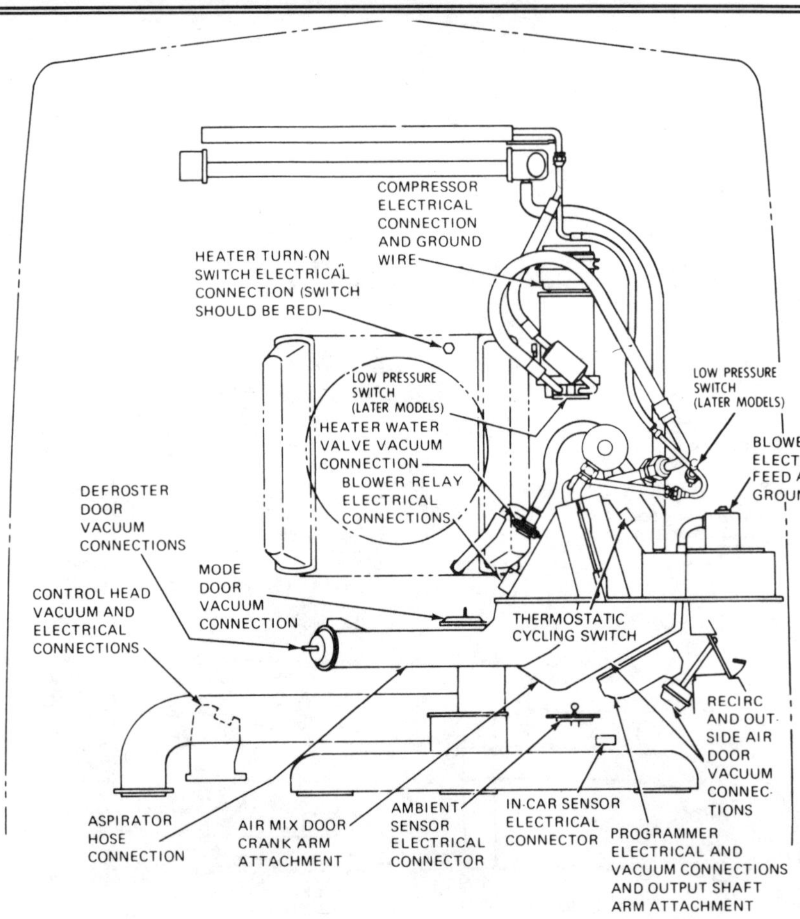

Fig. 146 In-car & underhood diagnosis areas. 1977-80 Seville

FEEDBACK POTENTIOMETER ADJUSTMENT

1. Using screwdriver, slip feedback pot shaft counter-clockwise to stop (gear does move). Programmer should move to full heat.
2. Place manual-automatic switch in manual position. Rotate manual control to max. heat.
3. Move manual control to 180 on 1974 models and 175 on 1975-76 models. Do not overtravel.
4. Slip feedback pot shaft slowly clockwise looking for signs of vacuum motor mechanism movement.
5. Check adjustment with manual control. Travel should start when 180 on 1974 models and 175 on 1975-76 models is reached as knob is turned counter-clockwise from max. heat.
6. If feedback pot cannot be adjusted, replace amplifier assembly.

BLOWER SWITCH BENCH TEST. (USING J-23713 BENCH TEST HARNESS)

1. Place voltmeter knob at 6.
2. With manual-automatic switch in manual, rotate manual control back and forth. Voltage should drop then increase in steps. Steps will be smaller than those seen in car.

PROGRAMMER MOVES BUT VACUUM OPERATED DOOR(S) MALFUNCTIONS

If one of the doors does not operate, malfunction could be caused by plugged or pinched harness or plugged valve. Cut off sealed hose on dummy plug for circuit malfunctioning and connect the vacuum gauge to the plug. Plug the dummy plug on the programmer vacuum valve. Connect dummy hose to vacuum supply hose

Control Lever	Probe Port:	Hose Color	Cover Up	Other Hoses In Circuit	Action Expected
Lo	None	Black	Probe	Yellow	Defroster door moves to "bleed" position
	1	White	None	None	Hot water valve closes - Check air in "max. heat"
	4	Orange	None	None	Recirc. door opens - 1-2 min. delay
	5	Dk. Brown	Port 9	Purple	None - Checks control head vacuum leakage
	7	Chartreuse	None	Green	Upper mode door moves to A/C position
	8	Tan	None	None	Upper and lower mode doors move to A/C position
Def	9	Purple	None	Dk. Blue	Defroster door moves to full "DEF" position

A low vacuum reading in any step indicates a leak in control head valve or in the hoses and components listed in that step. If some function is missing even though full vacuum is available, reconnect harness plug to programmer and check vacuum at extreme end of hoses involved, to distinguish between a pinched hose and a mechanical bind in actuator door.

Fig. 147 Auxiliary vacuum circuit test chart. 1977-80 DeVille & Fleetwood Brougham

(Port 2) in car vacuum harness. Monitor vacuum while operating programmer with manual control. When completed, seal dummy plug hose with rivet. Do not use a screw.

TEMPERATURE DOOR CHECK & ADJUSTMENT

A mis-adjusted temperature door link may not allow the programmer to travel fully to max. heat or to max. cold and could result in poor heating or poor cooling complaints. To check adjustment, attach tester electrical harness between programmer and car harness. Put manual-automatic switch in manual and control lever in auto. Put voltmeter knob in 6. Rotate manual control from max. heat to max. cold; high blower (Battery Voltage) should be obtained in both extremes. With control lever in high, recirc. operation should be obtained at max. cold (3 min. delay on some cars). With these conditions met, temperature door adjustment is correct. To readjust:

1. Start engine and put control lever in defrost.
2. Loosen screw on programmer shaft.
3. On 1974 models, remove electrical connector from programmer. Programmer should go to full heat.
4. On 1975-76 models, position manual-automatic switch in manual and manual control in max. heat, programmer should go to full heat.
5. On all models, move temperature door to full heat position by pushing door link away from programmer. The blower air will hold door.
6. Tighten screw and re-check as above.

1977-80 AUTO CLIMATE CONTROL

Performance Tests

Before repairs are attempted on systems found to provide insufficient cooling, perform an A/C performance test and compare results with specifications listed in Figs. 138 and 139 which detail performance readings on properly operating systems. Observe the following conditions before beginning test:

1. Operate engine at 2000 RPM and allow A/C system to stabilize.
2. Place a thermometer in the center outlet and open all A/C outlets.
3. Close all doors and windows, set controls to Hi and 65 degrees.
4. Connect manifold gauge set or charging station.
5. Place an auxiliary fan in front of radiator.

Insufficient Cooling Diagnosis

Refer to Figs. 140 and 141 which detail proper diagnosis procedure to pinpoint faulty refrigeration system components.

How To Isolate The Problem

Step 1

To confirm the system defect, make a brief check of system operation by sitting in the car and operating the controls with the engine warmed up and running at 1000 RPM or higher. Refer to Figs. 142 and 143 and perform control system functional test in sequence.

Step 2

A simple, visual inspection of the easily accessible underhood and instrument panel electrical and vacuum connections will, in most cases, reveal the defect on the spot. Refer to Figs. 144, 145 and 146 for diagnosis areas on each model.

Step 3

Based on information gained during the functional test performed in step 1, try to relate the problem to one of the following areas:

1. Temperature Control Problems
2. Blower Control Problems
3. Auxiliary Vacuum Problems
4. Refrigeration System Problems

Once the problem has been isolated, refer to the following diagnosis.

Step 4

After the problem has been properly diagnosed and repaired, it is important to run through the brief check listed in step 1 to insure that the system is performing correctly.

TEMPERATURE CONTROL PROBLEM DIAGNOSIS

The function of the temperature control circuit is to determine the temperature of air to be discharged into the passenger compartment by positioning the air mix door. If the problem is poor or no heating or cooling, but the air mix door and programmer function properly, the problem is in the refrigerant circuit or heater water system. Blower cycling or mode cycling may be caused by a temperature control problem.

If a temperature control problem is noted, check for the following:

a. Disconnected sensor or temperature dial. This will interrupt the electrical signal and drive programmer to the max. heat position.
b. Improper sensor connection. An improper sensor connection will increase resistance to the sensor string and drive system to a higher temperature than desired.
c. Disconnected vacuum supply hose at vacuum checking relay. This will lock vacuum motor in a fixed position.
d. Disconnected vacuum hose in transducer-programmer vacuum motor line. This will cause system to drive to max. cooling.
e. Leak in auxiliary vacuum circuit. This may reduce the vacuum supply to the transducer below control requirement, causing an off calibration or poor heating condition.
f. A loss of vacuum supply will usually result in cold air flow from floor outlets.
g. An open amplifier power feed eliminates the outlet signal and drives the programmer to max. A/C.

Diagnosis of problems in the temperature control circuit can be accomplished using tester J-23678, refer to Auto Climate Control Programmer System Diagnosis. Symtoms and causes of temperature control are as follows:

System Operates At Max. A/C Only

To determine whether this is a vacuum or electrical problem, install test harness J-23678-77 with single connector open and 8 pin connectors fastened together, between amplifier connector and amplifier terminals. While operating the system, observe the vacuum motor position through the slot at the bottom of the programmer cover. If the programmer remains in the maximum air conditioner position, remove the vacuum harness and check for supply vacuum at the black hose. If there is low or no vacuum, check for leaks or disconnects in vacuum hose assembly. If vacuum exists at the black hose, programmer is faulty. Remove programmer cover and inspect for obvious disconnects. If no programmer problem is obvious, use Tester J-23678, Adapter harness J-23678-77 and instructions J-23678-51 to analyze and correct problem.

If the programmer moved to maximum heater position with test harness installed, the problem is outside the programmer. Check for the following conditions:

a. Shorted in-car sensor.
b. Shorted or miscalibrated temperature dial.
c. A short in the sensor circuit of the wiring harness.
d. Disconnect or open in ground circuit between control head and programmer.
e. No electrical power to programmer.

System Operates At Max. Heat Only

Position temperature control dial at 65 and control lever to any position except off. Check

CONTROL LEVER	PROBE PORT	HOSE COLOR	COVER UP	OTHER HOSES IN CIRCUIT	ACTION EXPECTED
Bi-Level	None	Black	Probe	Yellow	Defroster door moves to "bleed" position
Bi-Level	1	White	None	None	Hot water valve closes — Check air in "max. heat"
Bi-Level	3	Red	Port 9	Purple, Tan	Lower mode door moves to A/C position
Bi-Level	4	Orange	None	None	Recirc. door opens — 1-2 min. delay
Lo	6	Chartreuse	Port 9	Purple	Checks control head vacuum leakage
Lo	6 or 8	Green	None	None	Upper mode door moves to A/C position
Hi	7	Dk. Brown	None	Tan	Lower mode door moves to A/C position
Def	9	Purple	Port 3	Red, Dk. Blue	Defroster door moves to full "DEF" position

Fig. 148 Auxiliary vacuum circuit test chart. 1977-79 Eldorado

SET CONTROL AT:	SYSTEM SHOULD OPERATE AS FOLLOWS (ENGINE RUNNING – 70° - 80° GARAGE
DEF . . . . 85°	Fixed high blower – most of air delivered from defroster outlets - some air from heater outlet – instant blower turn on.
BI-LEVEL . . . . 75°	Lower blower speed – cooler air from both A/C and HTR outlets – only very small amount of air out defroster.
HI . . . . 65°	Fixed high blower – cold air from A/C outlets – recirc should occur (will take approx. 3 min. on some cars because of restrictor in hose.
AUTO . . . . 65°	No recirc – door will approx. take 5 min. on some cars
AUTO . . . . 85°	before outside air enters.
LO . . . . 85°	Blower should drop to fixed low speed.
ECON . . . . 85°	No compressor operation, otherwise same as AUTO.
OFF . . . . 85°	Fixed low blower – warm air out heater outlets – no air out A/C or defroster ducts – no compressor operation.

PROGRAMMER FUNCTIONAL TEST

1. Connect tester electrical harness to programmer and car harness to tester. Use harness adapter No. J-24774.
2. Operate car at idle – approx. 1000 RPM.
3. Set control head lever at auto.
4. Turn temp dial calibration switch off.
5. Place manual-automatic switch in manual position.

   (NOTE: This substitutes variable resistor for sensor string at all control head settings.)

6. Place the manual control at max. heat and then rotate slowly counter-clockwise. The vacuum motor mechanism which can be seen through the slot in the programmer cover should start moving at 175 OHMS. Rotating the knob to max. cold should move the programmer to its full A/C position. Check air mix door for full travel and no obstructions. If the programmer operates in this manner, the components governing programmer movement – amplifier, transducer and vacuum motor – are OK. Proceed to sensor string test.
7. If the programmer moves but the movement does not start at 175 OHMS, adjust feedback pot per procedure on the next page.
8. If the programmer does not move check as follows for supply voltage and supply vacuum:
   a. Place voltmeter switch in 2 position. Battery voltage should appear on meter. No voltage indicates lack of ground on Term. 1 or battery supply to Term. 2 of programmer.
   b. Plug testers dummy vac plug on programmer valve. Connect dummy hose to vac supply (port 2 hose in car vacuum harness). Make sure vac is present (11 in. hg. or more). Programmer now has vac supply with no car vac system components connected. If programmer functions using manual control, troubleshoot car vac system. If programmer does not function problem is in programmer. See next page.

SENSOR STRING TEST (MUST PASS PROGRAMMER FUNCTIONAL TEST FIRST)

(NOTE: Do not perform sensor string test on car which has been run at idle for a long period of time in a heated building. This condition produces sufficient engine heat to heat the ambient sensor and cause lower resistance readings than specified in the procedure even though no problem exists in the sensor string.)

1. Set control lever at auto and temperature dial at 75.
2. Place temp dial calibration switch at off.
3. Place manual-automatic switch at automatic position.
4. Observe position of vacuum motor mechanism through slot in programmer cover.
5. Switch to manual and adjust manual control for same mechanism position.
6. When manual control is properly adjusted moving manual-automatic switch back and forth will not move vacuum motor.
7. Read manual control. This indicates the resistance of the sensors and the temperature dial combined and should be 120 to 150 OHMS at 70° to 75° room temp.
8. If reading is not 120 to 150 OHMS perform the temperature dial calibration test to check out the temp dial. If temp dial checks OK, the in-car or ambient sensor or connecting wiring is the problem.

TEMPERATURE DIAL CALIBRATION (CONTROL LEVER IN ECON.)

1. Place control lever at economy.
2. Place manual-automatic switch in manual position.
3. Place calibration switch in cal. position.
4. Disconnect in-car sensor connector and plug connector into mating connector on tester harness adaptor (ground both sensor wires).
5. Read voltmeter.
6. Press pushbutton and adjust temp dial on control head to get same voltmeter reading as above.
7. Temp dial should read as indicated on tester panel near pushbutton. If it does not, slip temp dial clutch to correct.

   (NOTE: If same voltmeter reading cannot be obtained the temp dial is malfunctioning.)

8. Reconnect in-car sensor.

BLOWER CIRCUIT TESTS

1. Place calibration switch in off position.
2. Place manual-automatic switch in manual position.
3. Set control lever at LO using voltmeter knob. There should be battery voltage at 7 (approx. 11 volts) and approx. half-battery voltage at 6 (4.0 to 4.5 volts). If there is no voltage at 7, there is a problem in the LO blower circuit. If there is voltage at 7 but no voltage at 6 the programmer is malfunctioning.
4. Set control lever at auto. There should be battery voltage at 8 (approx. 11 volts) and the voltage at 6 should stairstep down and then back up as the manual control is rotated from max. heat to max. cold. If there is no battery voltage at 8, there is a problem in the auto relay circuit. If the voltage at 6 does not change as described, the programmer is malfunctioning.

   (NOTE: A good connection between 6 and the blower motor is assumed. If the blower wire is disconnected, only battery voltage will appear at 6.)

Fig. 149 Programmer & system diagnosis chart. 1977-79 Eldorado & Seville (Part 1 of 2)

DIAGNOSING MALFUNCTIONS IN PROGRAMMER. THE FOLLOWING TESTS SHOULD BE PERFORMED TO DETERMINE MALFUNCTIONING INTERNAL PART. TESTS SHOULD BE MADE IN THE NUMBERED SEQUENCE.

TEST 1 PROGRAMMER VACUUM CIRCUIT

Turn temp dial calibration switch off. Place manual-automatic switch in manual. Position manual control–max. heat. If programmer moves to max. heat position skip to Test 2, otherwise perform following test.

1. Remove programmer cover. Install dummy vacuum plug with vacuum gauge feed-in to port 2 of car harness plug.
2. If vacuum drops below supply vacuum found in programmer functional test, use long-nose pliers to pinch the two black hoses leading from side of rotary valve. If vacuum reading improves, check programmer hoses for leaks, then go to Test 2.
3. If reading does not improve, leak is in rotary valve.

TEST 2 VACUUM MOTOR AND CHECKING RELAY

1. Leave tester set as in Test 1 above. Programmer should go to full heat.
2. Disconnect vacuum dummy plug or car vacuum harness from programmer. Vacuum checking relay should hold vacuum motor in full heat. If mechanism moves, vacuum motor or checking relay is leaking. Apply raw vacuum to vacuum motor then pinch hose at motor to see if leak is in motor. A good motor holds position.

TEST 3 AMPLIFIER AND TRANSDUCER

1. Connect tester harness to system. Place manual-automatic switch in manual and temp dial calibration switch in off position.
2. Remove small transducer vacuum hose and connect tester vacuum gauge to hose. 11 in. hg. or more vac. should be present on small transducer supply hose. Delay in obtaining 11 in. hg. or more may be due to a plugged restrictor located in the small transducer hose. Re-install hose on transducer.
3. Remove large vacuum hose from transducer. Connect tester vacuum gage to large port on transducer.
4. By rotating the manual control, vacuum readings of 1 to 5 in. hg. in max. A/C and 5 to 22 in. hg. in max. heat should be obtainable. Any value from 2 to 17 in. hg. can be obtained by careful adjustment of the manual control if the amplifier-transducer is functional.

TEST 4 FEEDBACK POTENTIOMETER ADJUSTMENT

1. Using screwdriver, slip feedback pot shaft counter-clockwise with respect to gear. Programmer should move to full heat.
2. Place manual-automatic switch in manual position. rotate manual control to max. heat.
3. Move manual control to 175.
4. Slip feedback pot shaft slowly clockwise looking for signs of vacuum motor mechanism movement.
5. Check adjustment with manual control. Travel should start when 175 is reached as knob is turned counter-clockwise from max. heat.
6. If feedback pot cannot be adjusted, replace amplifier-transducer assembly.

BLOWER SWITCH BENCH TEST – (USING J-23713 BENCH TEST HARNESS)

1. Place voltmeter knob at 6.
2. With manual-automatic switch in manual, rotate manual control back and forth. Voltage should drop then increase in steps. Steps will be smaller than those seen in car.

PROGRAMMER MOVES BUT VACUUM OPERATED DOOR(S) MALFUNCTION

If one of the doors does not operate, malfunction could be caused by plugged or pinched harness or plugged valve on programmer. To check programmer valve, cut off sealed hose on dummy plug for circuit malfunctioning and connect the vacuum gauge to the plug. Plug the dummy plug on the programmer vacuum valve. Connect dummy hose to vacuum supply hose (port 2) in car vacuum harness. Monitor vacuum while operating programmer with manual control. When completed seal dummy plug hose with rivet. **Do not use a screw.** Refer to Shop Manual for proper vac valve operation.

TEMPERATURE DOOR CHECK AND ADJUSTMENT

A mis-adjusted temperature door link may not allow the programmer to travel fully to max. heat or to max. cold and could result in poor heating or poor cooling complaints. To check adjustment, attach tester electrical harness between programmer and car harness. Put manual-automatic switch in manual and control lever in aut. Put voltmeter knob in 6. Rotate manual control from max. heat to max. cold. High blower should be obtained in A/C and M_3 in max. heat with control lever in high. Recirc operation should be obtained at max. cold. A 3 minute delay on some cars with these conditions met. Temperature door adjustment is correct to readjust.

1. Start engine and put control lever in defrost.
2. Loosen screw on programmer shaft.
3. Place manual-automatic switch in manual. Manual control at max. heat programmer should go to full heat.
4. Move temperature door to full heat position by pushing door link away from programmer. The blower air hold door.
5. Tighten screw and re-check as above.

Fig. 149 Programmer & system diagnosis chart. 1977-79 Eldorado & Seville (Part 2 of 2)

to see if programmer air mix door arm moves to the max. A/C position. If programmer does not move, check for the following:

a. Disconnected or defective in-car or ambient sensor.
b. Improperly connected or open circuit in sensor string wiring harness.

Place temperature control on 85 and check to see if programmer moves. If programmer does not move, check for the following:

a. Disconnected or defective temperature dial.
b. Open circuit or disconnected wire between programmer and chassis ground.
c. Defective programmer. Use tester J-23678 to test programmer.

System Changes Blower And/Or Modes During Acceleration

a. Install test harness J-23678-77 with single connector open and 8 pin connectors fastened together between amplifier connector and amplifier terminals, forcing the programmer to max. heat, full vacuum position.
b. Remove programmer cover, then reconnect electrical and vacuum connections. With system operating, remove in-car sensor wire connector to drive programmer to the max. heat, full vacuum position. Disconnect vacuum hose connector from programmer; programmer should remain in the full heat position. If programmer does not remain in the full heat position, repeat step a, then pinch programmer vacuum supply hose using needle nose pliers. If programmer still does not remain in the full heat position, the programmer vacuum motor is leaking. If programmer remains in the full heat position with vacuum supply hose pinched, the checking relay is defective.
c. If problem is mode changes without change in blower speed, the check valve portion of the checking relay may be defective.

Control Settings	Temp.	System Should Operate as Follows
FRT DEF	90°	Warm air should be delivered from defroster outlets at a fixed high blower speed. Some floor bleed from heater outlet.
HI	90°	Warm air should be delivered from the heater outlet with only a small amount from the defroster outlets. The blower should stay at high speed.
AUTO	75°	After a short delay, the blower speed should gradually decrease. Discharged air should become cooler. Depending on the temperature in the work area, the air delivery mode should change to A/C outlets from the heater outlets.
AUTO	60°	Blower speed should increase to a fixed high value. Discharged air should become maximum cold. The recirculating air door should slowly open (signalled by blower noise increase).
LO	60°	After a short delay, the blower speed should gradually decrease to a very low value. The discharged air should remain cold and come from the A/C outlets. The recirc. door should slowly close.
ECON	60°	The A/C compressor should be off. Blower speed should gradually increase to a fixed high value. Discharged air should continue to come from A/C outlets and should be warmer than the temperature in the work area.
OFF	60°	After a short delay, the blower speed should gradually decrease to a very low value. The discharged air should be delivered from the heater outlet and should be somewhat warmer than the temperature in the work area. The A/C compressor should be off.
RR DEF	NONE	LED above button should light up. Heat should be applied to rear window.

Fig. 150 Electronic Climate Control system functional test. 1980 Eldorado & Seville

Temperature Dial Setting Does Not Provide Comfort

Using tester J-23678, refer to Auto Climate Control System and Programmer Diagnosis to check and reset calibration on temperature dial, feedback potentiometer in programmer, sensor string and temperature door link adjustment. Also check to ensure that in-car sensor aspirator hose is connected to sensor and aspirator and is not kinked.

Mode & Blower Cycling

NOTE: Some cycling of system may occur if temperature control dial is moved in large increments.

This problem may be caused by improper electrical connection, cold solder joints or a defective sensor or amplifier. If problem is intermittent, tap areas around sensor, control head, and programmer. Also shake electrical connections attached to these components. If problem can not be located by the transducer buzzing, remove programmer cover and reconnect electrical and vacuum connectors, then connect a voltmeter to transducer output terminals. While observing voltmeter, continue to tap areas described above and shake electrical connections. If voltmeter indicates erratic voltage fluctuations, replace component or resolder connections as necessary.

Insufficent Cooling Or Heating

This condition may be caused by a temperature control, refrigerant system or heater system problem to determine which is at fault. Proceed as follows:

a. No cooling, check compressor clutch operation. If compressor clutch is not operating, problem is in refrigerant system. Also check for cold suction line.
b. No heating, check engine coolant level. Also feel heater hoses to see if engine coolant is entering the heater core.
c. Check for proper air flow at Auto. and high settings. Improper air flow may be caused by a blower related problem, blocked air passage or disconnected air hose.
d. Check calibration of temperature control dial.
e. Check calibration of programmer feedback potentiometer.

If problem seems to be in the temperature control system, use tester J-23678 to locate and correct problem, refer to Auto Climate Control Programmer and System Diagnosis.

Blower Speed & Mode Shifts W/O Temperature Change

Check air mix door link to programmer shaft and air door crankarm connections.

BLOWER CONTROL CIRCUIT PROBLEM DIAGNOSIS

Blower control circuit problems are divided into two categories, blower speed control and blower turn-on. Blower circuit problems and probable causes are as follows:

System Operates Only at Low Blower

1. Check for electrical disconnection at blower relay. Also check for faulty blower relay.
2. Check for poor or no connection at programmer or at control head. Refer to the appropriate wiring diagram located elsewhere in this manual and trace blower circuit.

<table>
<tr><th colspan="10">SYSTEM RESPONSE</th></tr>
<tr><th></th><th></th><th>BLOWER SPEED</th><th>AIR INLET DOOR POSITION</th><th>TEMPERATURE DOOR POSITION</th><th>MODE DOORS POSITION</th><th>DEFROSTER DOOR POSITION</th><th>COMPRESSOR</th><th>HTR-WATER SHUT-OFF VALVE</th><th>IS BLOWER DELAYED FOR ENGINE WATER WARMUP?</th></tr>
<tr><td rowspan="6">CONTROL LEVER SETTINGS</td><td>OFF</td><td>FIXED LOW</td><td rowspan="3">OUTSIDE AIR</td><td rowspan="6">VARIES — RESPONSIVE TO SENSORS</td><td>FORCED TO HEATER</td><td rowspan="5">DOOR WILL ASSUME "BLEED" POSITION AFTER 45 SECOND DELAY AND WILL DELIVER BLEED AIR TO W/S IF IN HEATER MODE</td><td rowspan="2">DOESN'T OPERATE</td><td rowspan="5">USUALLY OPEN — EXC. CLOSED AT MAX. A/C POSITION OF PROGRAMMER</td><td rowspan="5">YES — EXCEPT IN A/C MODE IN WHICH CASE THE CONTROL HEAD TURNS ON THE BLOWER IMMEDIATELY EVEN IF ENGINE IS COLD</td></tr>
<tr><td>ECON.</td><td>VARIABLE BLOWER PROGRAM</td><td rowspan="4">EITHER HEATER OR A/C MODE DEPENDING ON PROGRAMMER POSITION</td></tr>
<tr><td>LO</td><td>FIXED LOW</td><td rowspan="4">CYCLES ON AND OFF TO MAINTAIN EVAPORATOR BETWEEN MINUS 1°C AND 71/2°C (31°F TO 34°F)</td></tr>
<tr><td>AUTO</td><td>VARIABLE BLOWER PROGRAM</td><td rowspan="2">RECIRCULATE AIR WHEN PROGRAMMER IS AT MAX. A/C OTHERWISE OUTSIDE</td></tr>
<tr><td>HI</td><td>FIXED HI</td></tr>
<tr><td>DEF</td><td>FIXED HI</td><td>OUTSIDE AIR</td><td>FORCED TO HEATER</td><td>FULL OPEN TO W/S</td><td>OPEN</td><td>NO DELAY</td></tr>
</table>

Fig. 151 Electronic Climate Control system operational chart. 1980 Eldorado & Seville

No Blower Operation in Any Lever Setting

1. Check for electrical disconnection at blower motor, blower relay or at radio capacitor in series with blower feed, if equipped. Also check for faulty blower motor or relay.
2. Check for blown A/C-heater fuse.
3. Check for open circuit in wiring harness to blower motor.
4. Check for electrical disconnection at programmer connector.

No Blower Operation in Off or Low Settings

1. Check for open low speed resistor on blower resistor board.
2. Check for poor or no connection at programmer.
3. Check for open in low blower circuit wiring.
4. Check for faulty blower relay.

System Operations Only at Low Blower Except in Hi & Def.

1. Check for faulty programmer wiper contacts or board.
2. Check for poor or no connection at programmer. Also check for open circuit in programmer internal wiring.
3. Check for poor or no connection at blower resistor board. Also check for open resistor.
4. Check for disconnection at control head to programmer feed terminal.
5. Check for open in blower circuit wiring.

Blower Operates Only in Hi & Def.

1. Check for electrical disconnection at blower resistor assembly.
2. Check for open circuit in wiring harness between resistor assembly and blower motor.

Blower Operates at Low Speed in Hi & Def.

1. Check for faulty control head wipers or circuit board.
2. Check for open in by-pass circuit in wiring harness.
3. Check for open circuit in control head wiring harness.

Blower Operates with Ignition Off

1. Check for faulty blower relay.

Blower Speeds Skipped in Economy, Auto or Bi-Level

1. Check for open resistor in blower resistor circuit board.

No Heater Turn-On in Cold Weather Except in Def, Off and Lo

1. Check for electrical disconnection at heater turn-on switch. Also check for faulty switch.
2. Check for open in wiring circuit.

Immediate Heater Turn-On in Cold Weather

1. Check for faulty heater turn-on switch.
2. Check for faulty in-car switch.

NOTE: This is a normal condition in Def setting.

A/C Delayed in hot Weather Until Engine Warms Up

1. Check for faulty in-car switch.
2. Check for open circuit in control head wiring harness.
3. Check continuity of in-car switch circuit.

A LOGICAL SERVICE PROCEDURE IS AVAILABLE TO ISOLATE AN ECC PROBLEM. ALL SYSTEM DIAGNOSIS SHOULD BEGIN BY CONDUCTING THE FUNCTIONAL TEST IN THE EXACT ORDER IN WHICH IT IS PRESENTED. IF THE ANSWER TO ANY FUNCTIONAL TEST QUESTION IS NO, THEN TURN TO THE SPECIFIC TROUBLE TREE FOR ANALYSIS OF THE MALFUNCTION. IT IS ESSENTIAL THAT THE FUNCTIONAL TEST BE PERFORMED WITHOUT OMITTING STEPS BECAUSE OMISSIONS MAY RESULT IN THE INABILITY TO ISOLATE SPECIFIC MALFUNCTIONS.

IT IS NOT NECESSARY TO PERFORM THE FUNCTIONAL TEST AT THE BEGINNING OF EACH TROUBLE TREE IF THE FUNCTIONAL TEST HAS BEEN UTILIZED TO SELECT THE PROPER TROUBLE TREE. WHEN A MALFUNCTION HAS BEEN ISOLATED AND REPAIRED, THE TECHNICIAN SHOULD EXIT FROM THE TROUBLE TREE AT THE POINT OF REPAIR.

PROCEDE AS FOLLOWS:

1. CHECK FUSES (CHECK STOP HAZARD OPERATION TO VERIFY STOP HAZARD FUSE)
2. WARM ENGINE BEFORE TESTS
3. CHECK LED ABOVE EACH BUTTON AS TEST IS PERFORMED.

SYSTEM CHECKS	CONTROL SETTING	TROUBLE TREES
1. DO THE MPG AND CONTROL HEAD DISPLAY? (MPG IS OPTIONAL)	ALL	1
2. DO COOLER AND WARMER PUSHBUTTONS OPERATE?	ALL	1
3. SET AT 60°		
A. DOES THE BLOWER OPERATE?	LO, AUTO, HI	2A
B. IS THERE LO BLOWER?	LO	2B
C. IS THERE HI BLOWER?	HI	2C
D. IS THERE AIR FROM A/C OUTLETS?	ECON, LO, AUTO, HI	3
E. DOES COMP CLUTCH ENGAGE?	LO, AUTO, HI	4
F. DOES COMP CLUTCH DISENGAGE?	OFF, ECON	5
G. IS THERE COLD AIR FROM A/C OUTLETS? IF NOT:	LO, AUTO, HI	6
1) IS THERE HEAT ONLY?		
2) IS THERE INSUFFICIENT COOLING		7
H. *DOES RECIRC DOOR OPEN FULLY?	AUTO, HI	8
4. *SET AT 90°		
A. IS THERE SUFFICIENT HEAT FROM HEATER OUTLETS?	AUTO	9
5. *SET AT 85°		
A. IS THERE WARM OR HOT AIR AT HEATER OUTLET?	LO, AUTO	10
6. DOES FRONT DEFROSTER OPERATE?	FRT DEF	11
7. DOES REAR DEFROSTER OPERATE?	RR DEF	12
8. DOES REAR DEFROSTER TURN OFF?	RR DEF OFF	13

*WAIT ONE OR TWO MINUTES FOR SYSTEM TO CHANGE.

Fig. 152 Electronic Climate Control system diagnosis chart. 1980 Eldorado & Seville (Part 1 of 16)

AUXILIARY VACUUM PROBLEM DIAGNOSIS

Auxiliary vacuum circuits in the Auto Climate Control system, are circuits which do not affect the temperature of discharge air. The auxiliary vacuum circuits are as follows:

1. Upper and lower mode door vacuum circuits.
2. Defroster bleed vacuum circuit.
3. Defroster door vacuum circuit.
4. Recirculating air door vacuum circuit.
5. Heater water valve vacuum circuit.
6. Auxiliary vacuum circuit check valve.

It should be noted that a leak in one vacuum circuit may reduce the vacuum supply to another vacuum circuit.

A/C Mode Door Actuation Problem

This condition may be indicated by the following symptions, no A/C operation and split mode operation. The system requires vacuum to move door(s) to the A/C position. Check vacuum hose connections at mode door actuators. Split modes may occur due to problems in the door(s) circuits or because of leaks in other auxiliary vacuum circuits. Refer to procedure described under Auxiliary Vacuum Circuit Test.

Defroster Bleed Problems

1. If there is no defroster bleed, check vacuum hose connection at defroster vacuum actuator. If vacuum connections are here satis-

1 NO DISPLAY

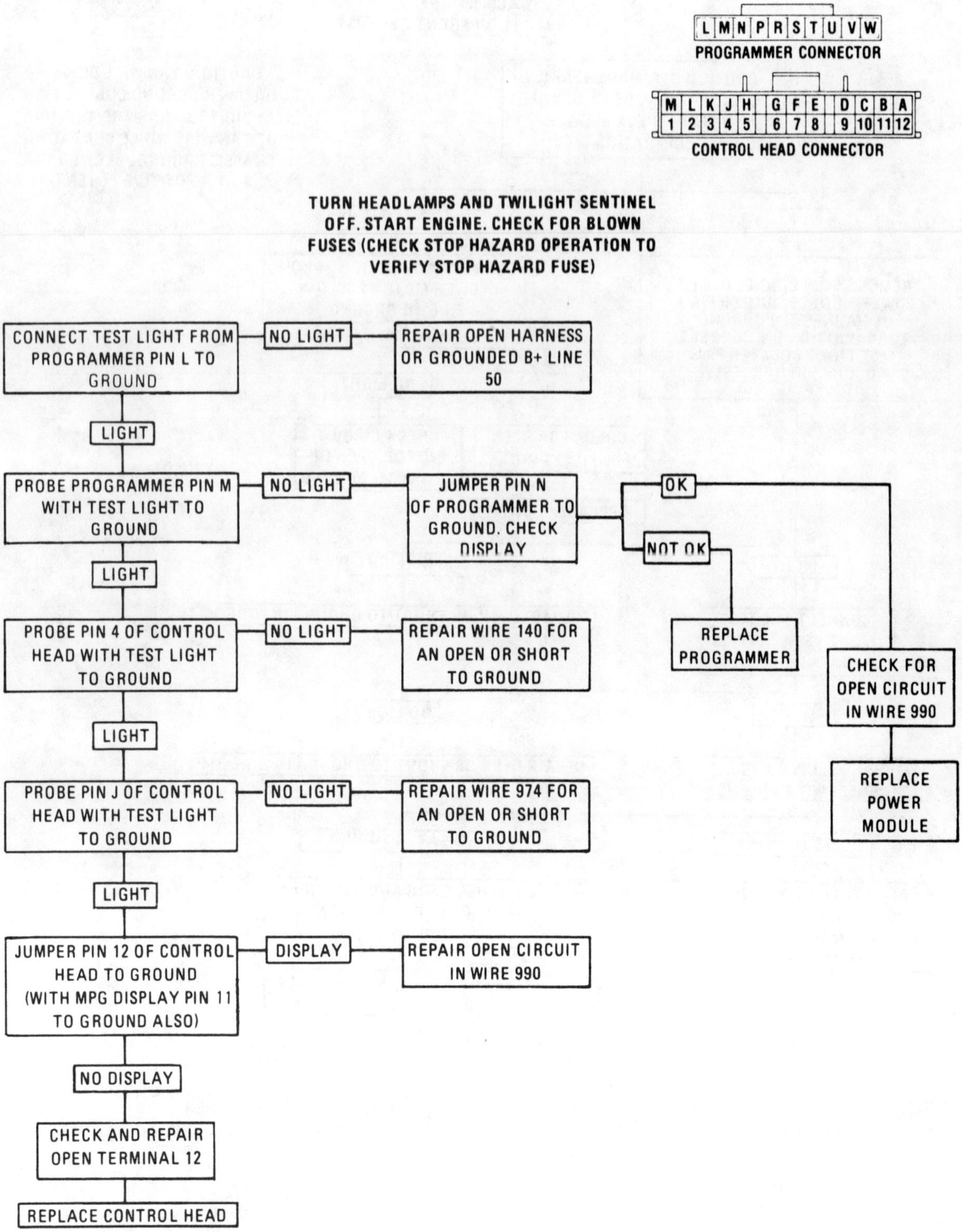

Fig. 152 Electronic Climate Control system diagnosis chart. 1980 Eldorado & Seville (Part 2 of 16)

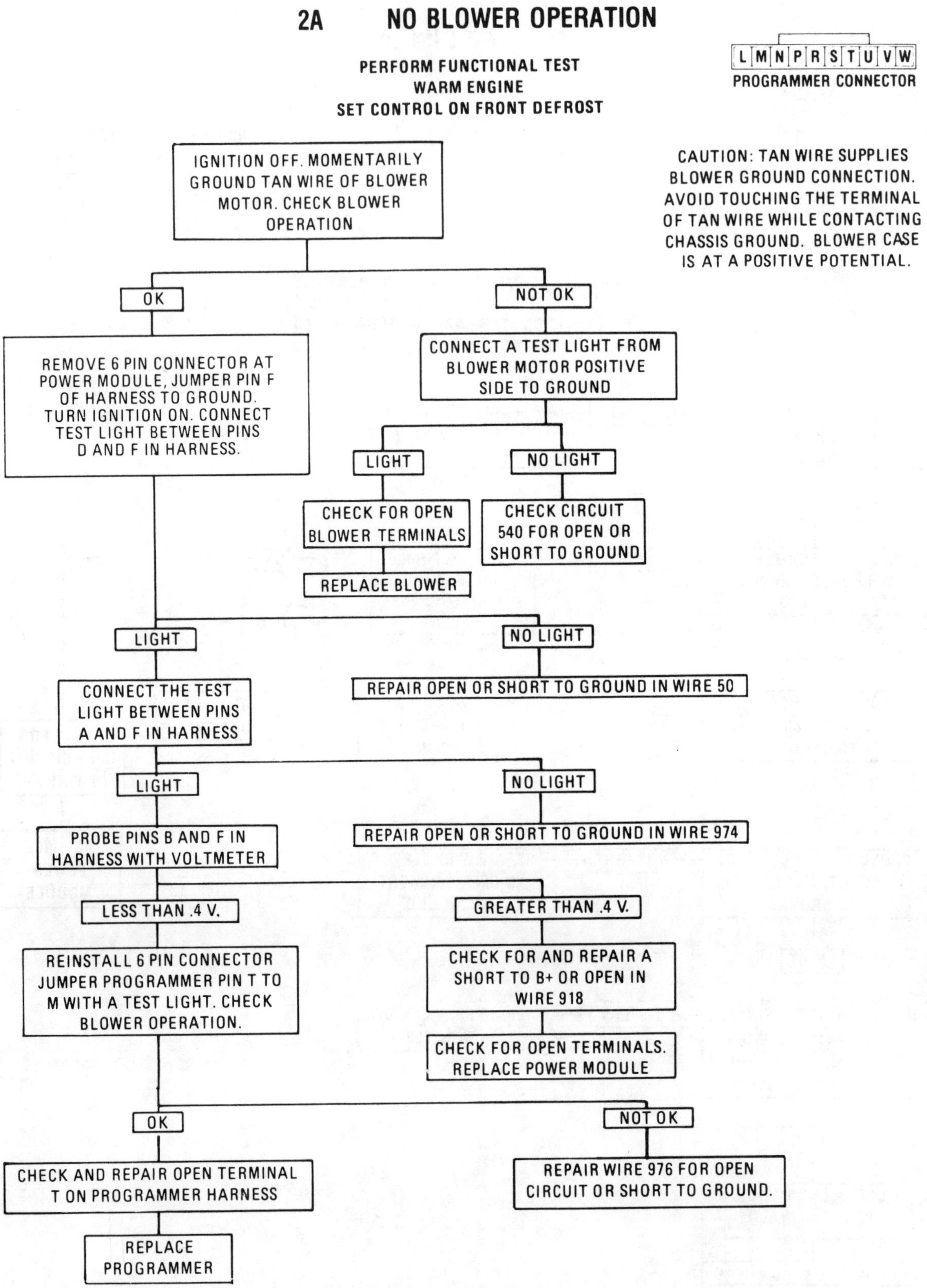

Fig. 152 Electronic Climate Control system diagnosis chart. 1980 Eldorado & Seville (Part 3 of 16)

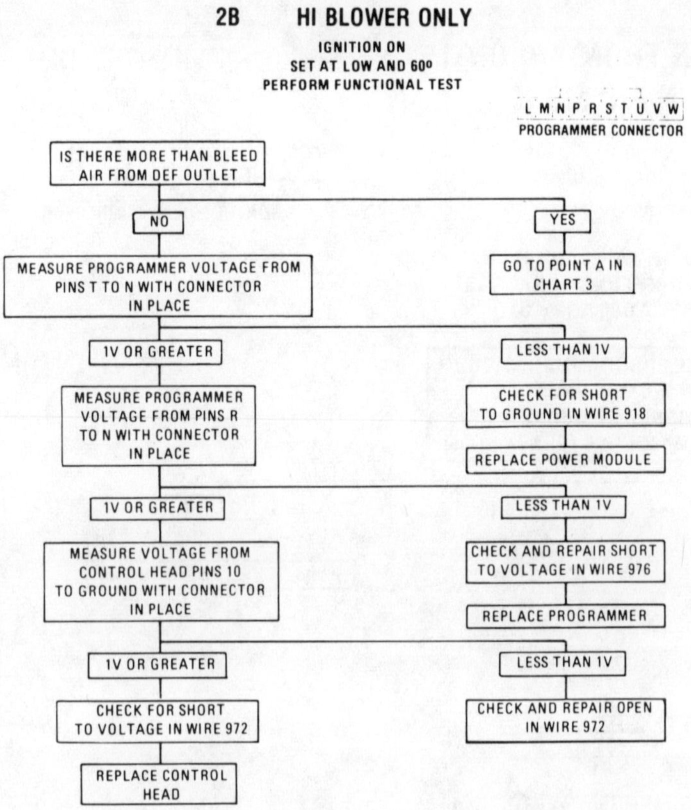

Fig. 152 Electronic Climate Control system diagnosis chart. 1980 Eldorado & Seville (Part 4 of 16)

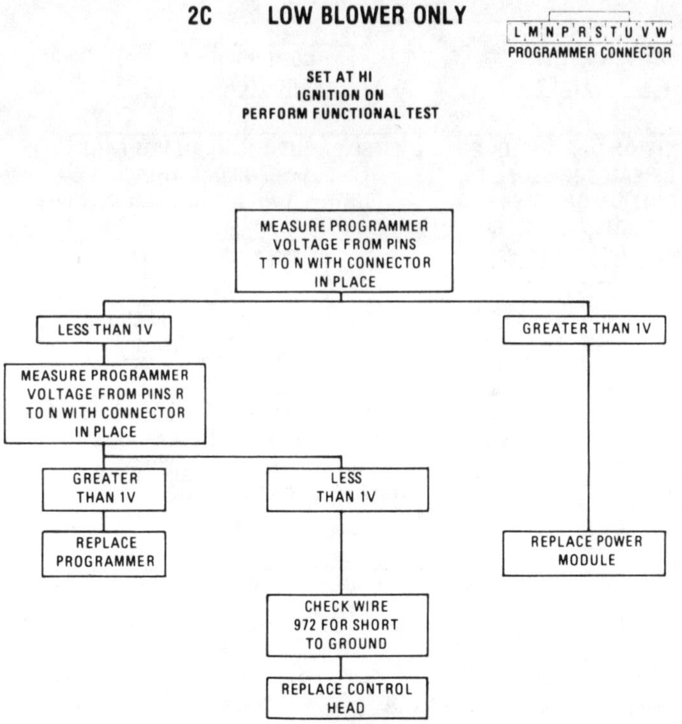

Fig. 152 Electronic Climate Control system diagnosis chart. 1980 Eldorado & Seville (Part 5 of 16)

factory, the porous flow plug may have to be replaced. Before installing a replacement porous plug, check for vacuum in yellow line and connect hose to actuator without the plug to check bleed door operation.

2. If rapid fogging of windshield is encountered when heat is turned on, check for improperly installed or missing porous plug.

Defroster Door Actuation Problem

This condition is indicated by no air flow from defroster outlets with selector lever in Defrost position. Vacuum to defroster actuator is required for defroster operation. Defroster door actuation problems may be caused by the following:

a. Improper hose connection at defroster actuator.
b. Check for vacuum at control head vacuum supply hose.
c. If system remains in A/C mode with control lever in Defrost position, check vacuum circuit to determine why mode override did not occur.
d. Check to ensure that defroster door is not sticking.
e. If vacuum door is not sticking, refer to Auxiliary Vacuum Circuit Test procedure.

Recirculating Air Door Problem

As vacuum is required to obtain recirculating air, check hose connection at recirc. air door. Remove vacuum delay plug to check for adequate vacuum at Hi and Auto settings and temperature dial at 65 degrees F. If vacuum is available but door does not move with plug in circuit after approximately three minutes, replace plug.

Heater Water Valve Problem

The water valve is normally open and requires vacuum to close valve to stop water flow. A leak in this vacuum circuit may result in a reduced vacuum supply for other vacuum functions. The most probable complaint would be a split mode delivery. Refer to Auxiliary Vacuum Circuit Test.

Auxiliary Vacuum Circuit Check Valve

The check valve for the auxiliary vacuum circuit is located in the programmer checking relay assembly. A malfunctioning check valve will result in a loss of a specific function during acceleration or under a heavy engine load. With system operating, disconnect and plug black vacuum hose supplying vacuum to programmer checking relay. System vacuum function should hold, if not reconnect black hose and pinch purple hose. If system function now holds, the check valve is defective. If system continues to leak down, the vacuum circuit leading to the check valve is leaking.

Auxiliary Vacuum Circuit Test, Except Seville

Disconnect vacuum harness plug from programmer valve; this should direct all airflow to

3 NO AIR FROM A/C OUTLET IN A/C MODE

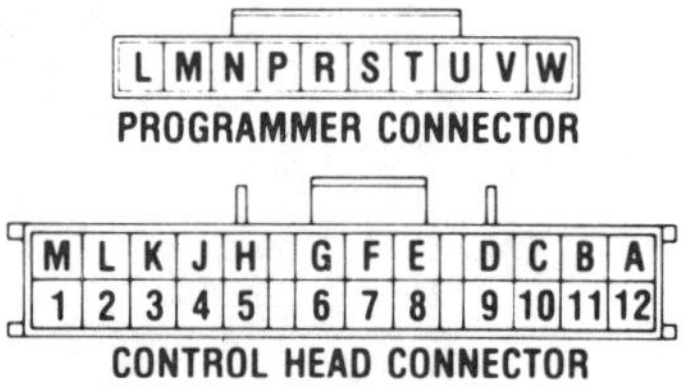

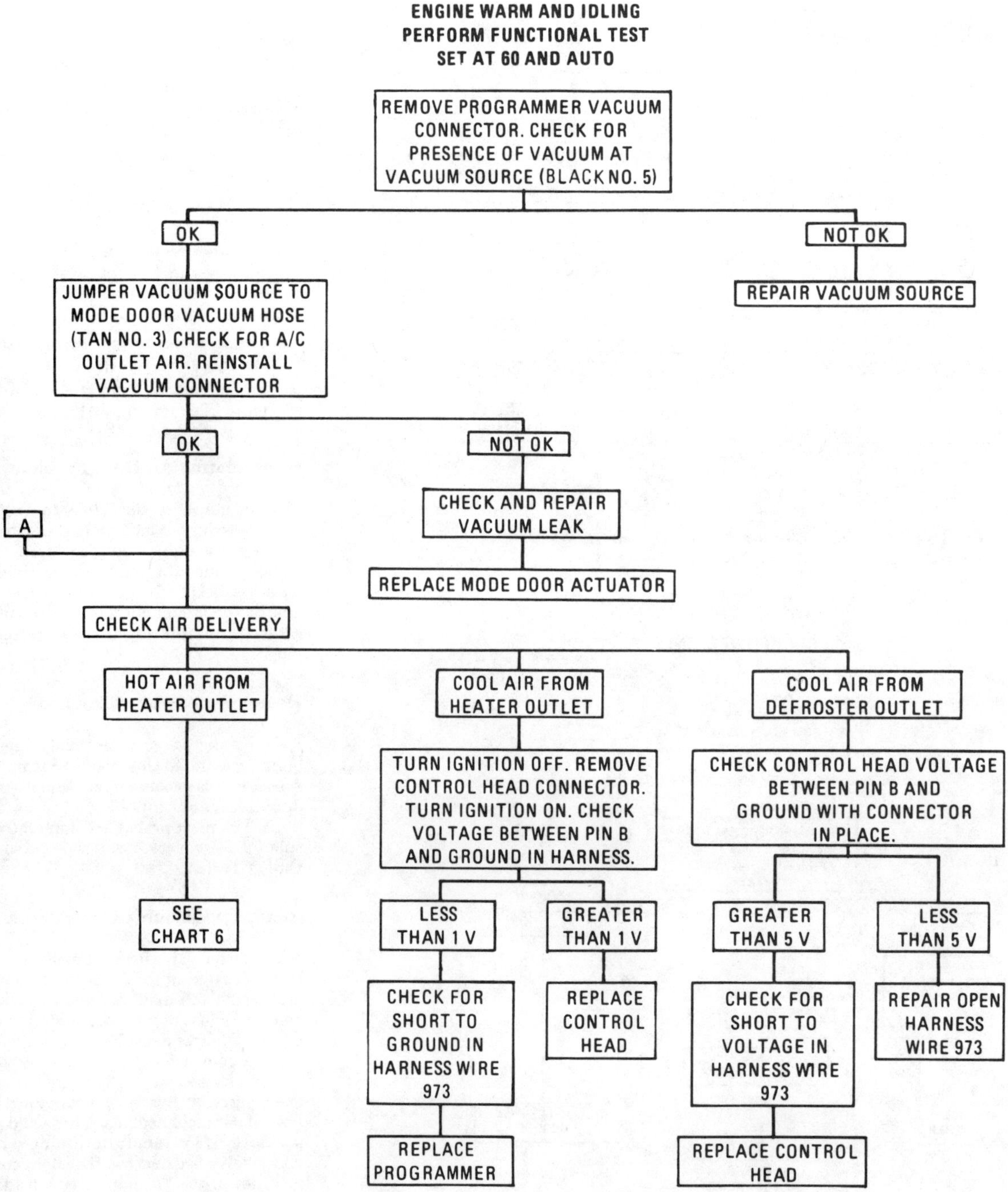

Fig. 152 Electronic Climate Control system diagnosis chart. 1980 Eldorado & Seville (Part 6 of 16)

4 COMPRESSOR NOT WORKING

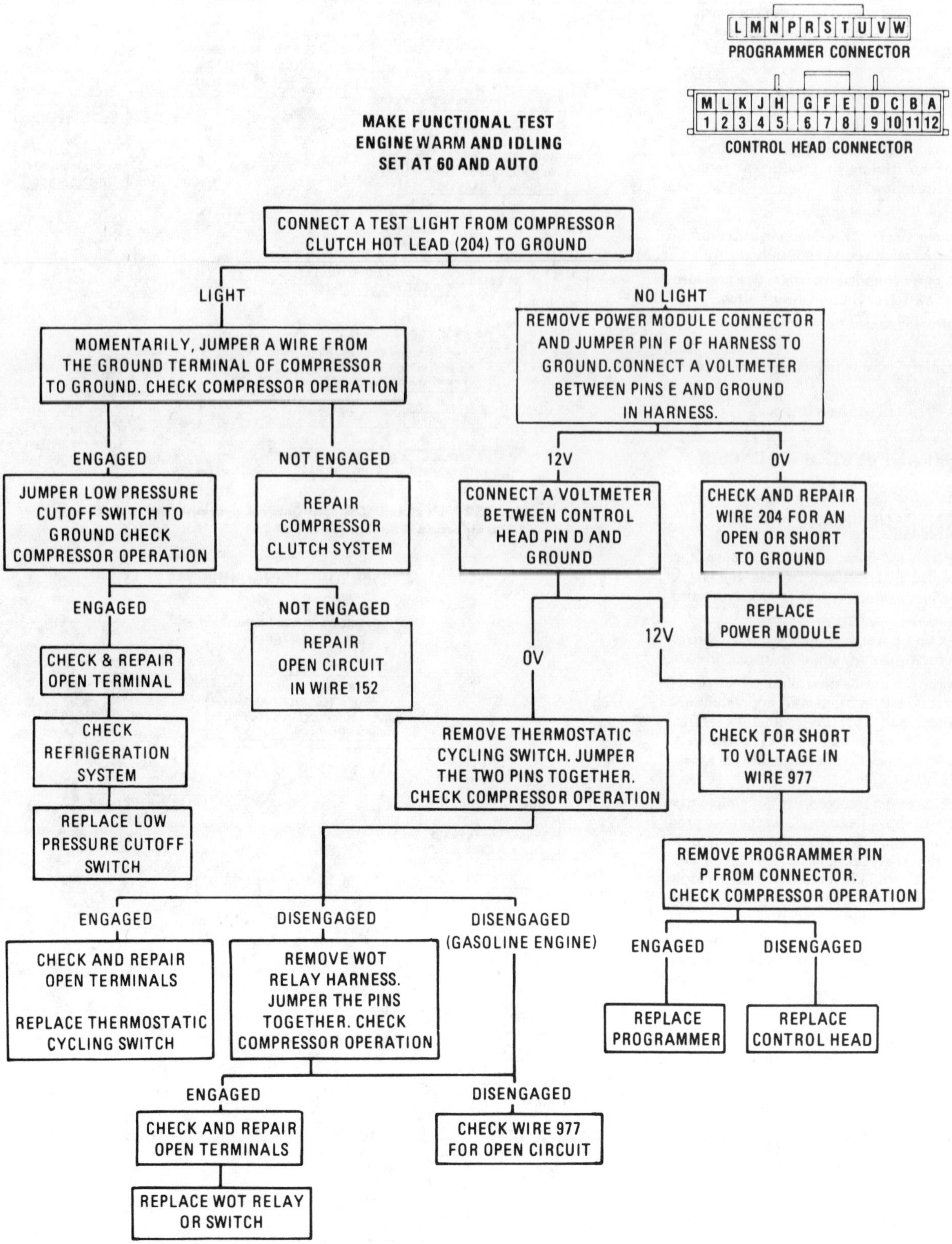

Fig. 152 Electronic Climate Control system diagnosis chart. 1980 Eldorado & Seville (Part 7 of 16)

floor outlets except for defroster bleed air. Insert tee of tester J-23678 vacuum gauge into port No. 2 vacuum harness plug, Fig. 132, and probe each of the other ports as outlined in Figs. 147 and 148. Ensure control lever is in proper position for step of test being performed. For each step, full vacuum should be read after actuator travel is completed.

COMPRESSOR ELECTRICAL DIAGNOSIS

Compressor electrical problems would be indicated by no cooling or insufficient cooling. Check for the following:

1. Blown A/C — heater fuse.
2. Improper electrical connection at compressor clutch or clutch coil ground.
3. Improperly connected or defective pressure cut-off switch or thermostatic switch.
4. Compressor seized or compressor belt slipping.
5. Improperly connected control head wire connector.
6. Open circuit in wiring harness.

REFRIGERANT SYSTEM DIAGNOSIS

Use the following procedure to separate refrigerant system problems from temperature control problems.

1. Place control lever at Off, Economy, Lo, Auto., Hi, Bi-Level and Defrost and check for proper system operation, Figs. 142 and 143.
2. If system operates properly, place control lever on Auto. and temperature dial on 65.
3. Position a suitable thermometer in right-hand A/C outlet, then wait approximately 1 minute and note thermometer reading. Thermometer should read approximately 40 to 50 degrees F, unless humidity is excessively high.
4. If discharge air temperature is significantly higher than 50 degrees F, disconnect programmer amplifier wire connector at programmer, then check to ensure that air mix door link arm is in the max. A/C position and temperature door rod is properly connected and adjusted.
5. If when amplifier wire connector is disconnected, air temperature drops to 50 degrees F, the problem is in the temperature control system, refer to Temperature Control Diagnosis.
6. If temperature does not drop and is significantly higher than 50 degrees F, the problem is in the refrigerant system, refer to Insufficient Cooling Diagnosis.

Auto Climate Control System & Programmer Diagnosis

Use instructions in Fig. 134, Parts 1 and 2, to evaluate control system performance with tester J-23678 on 1977-80 Cadillac models except Eldorado and Seville. Refer to instructions in Fig. 149, Parts 1 and 2, to evaluate control system performance with tester J-23678 on 1977-80 Eldorado and Seville models.

5 ECONOMY DOES NOT OPERATE

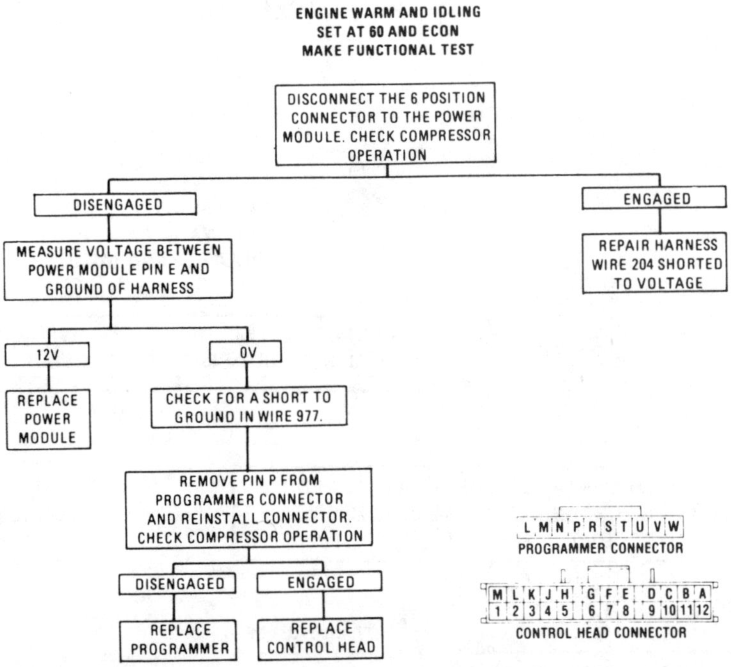

Fig. 152 Electronic Climate Control system diagnosis chart. 1980 Eldorado & Seville (Part 8 of 16)

6 SENSOR STRING PROBLEM

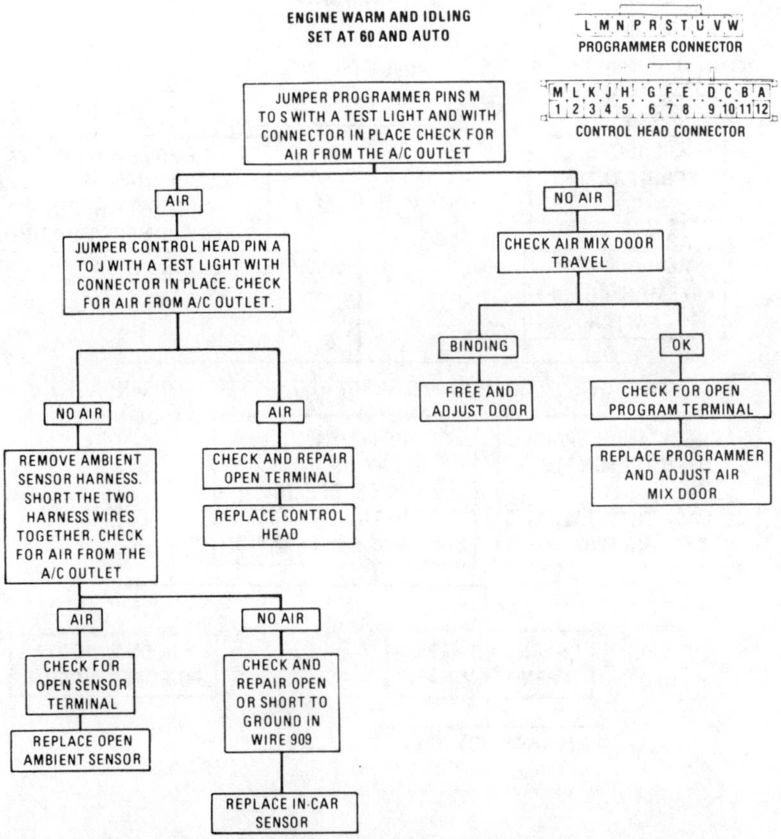

Fig. 152 Electronic Climate Control system diagnosis chart. 1980 Eldorado & Seville (Part 9 of 16)

7 INSUFFICIENT COOLING

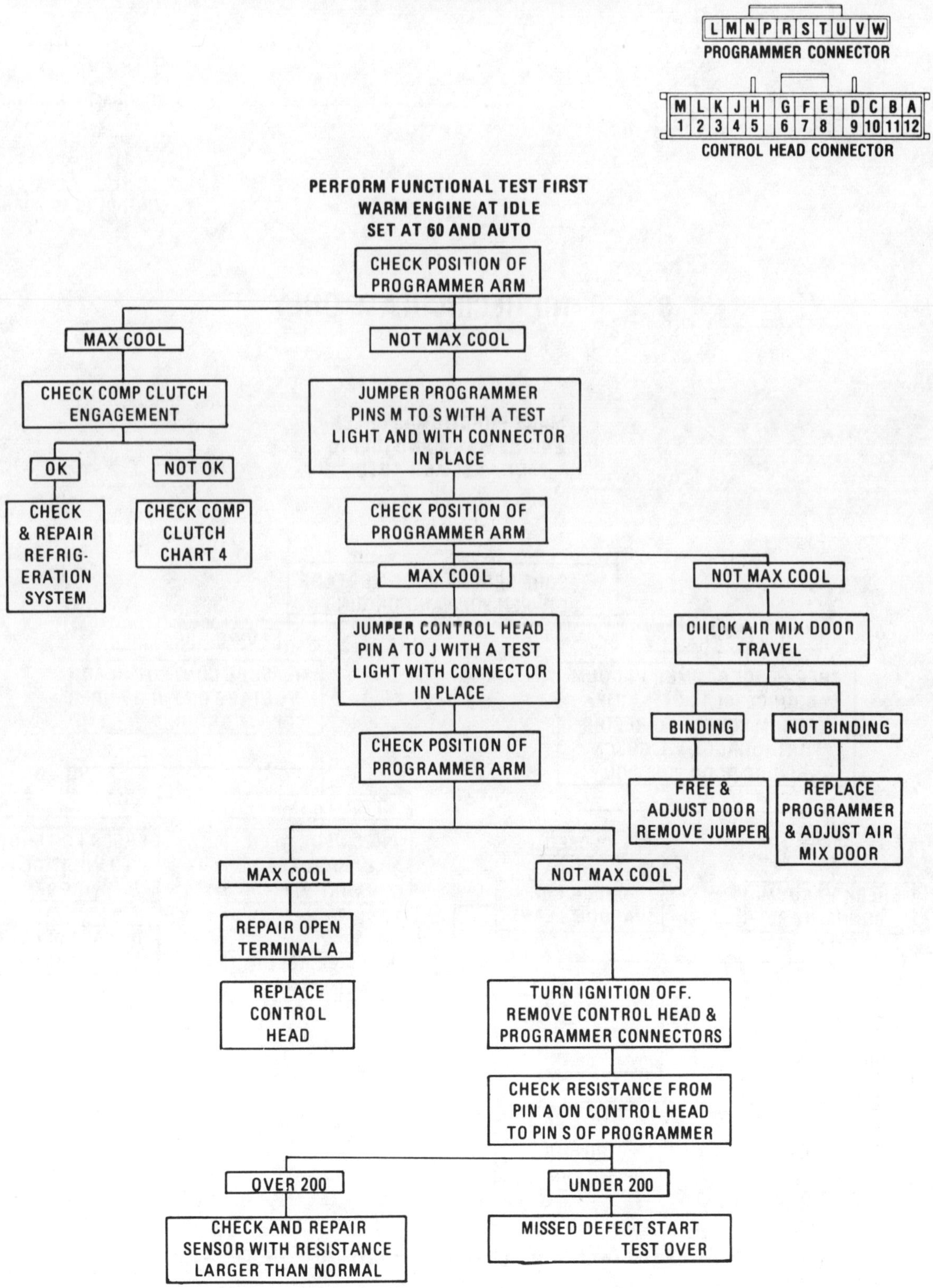

Fig. 152 Electronic Climate Control system diagnosis chart. 1980 Eldorado & Seville (Part 10 of 16)

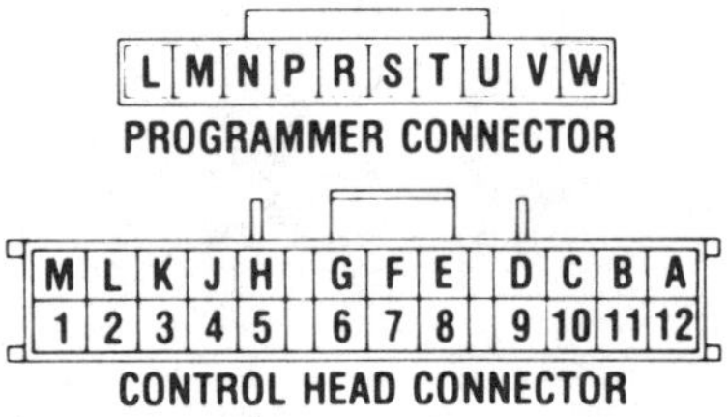

8 NO RECIRCULATE ONLY

MAKE FUNCTIONAL CHECK
ENGINE WARM AND IDLING
SET AT 60 AND AUTO

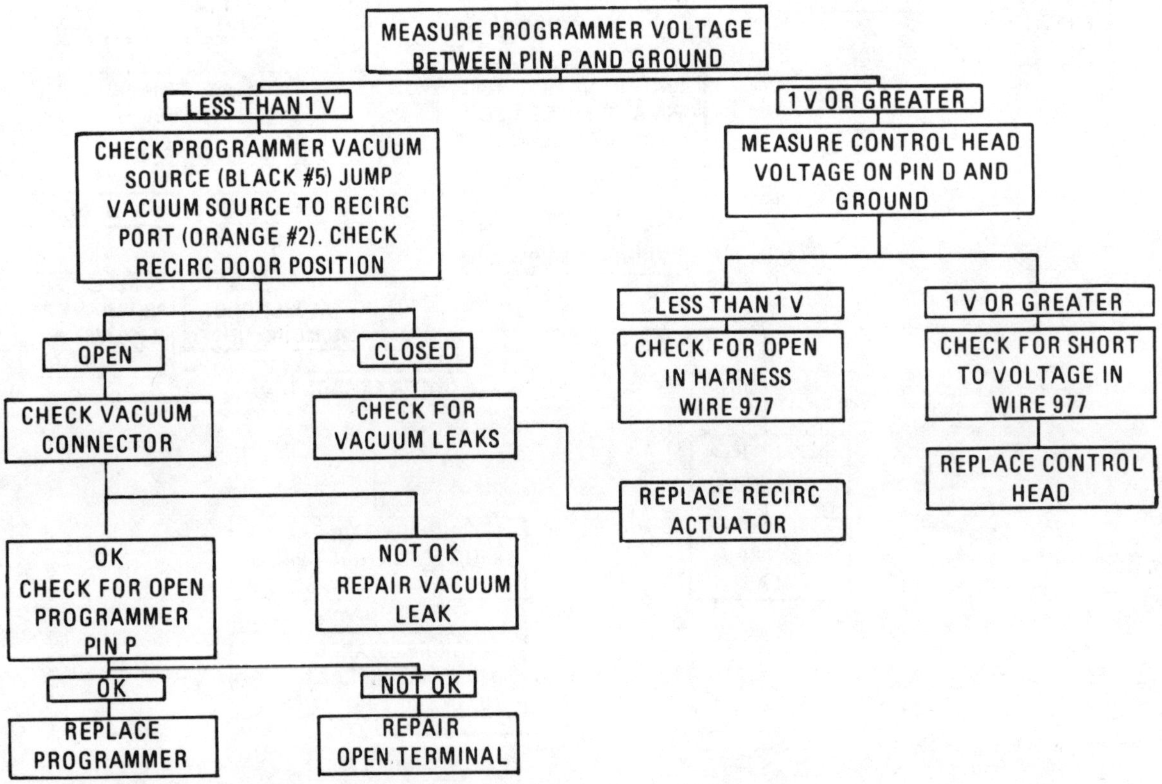

Fig. 152 Electronic Climate Control system diagnosis chart. 1980 Eldorado & Seville (Part 11 of 16)

9 INSUFFICIENT HEAT

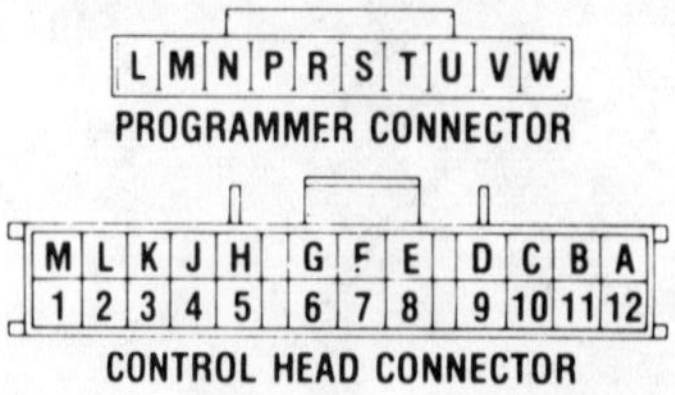

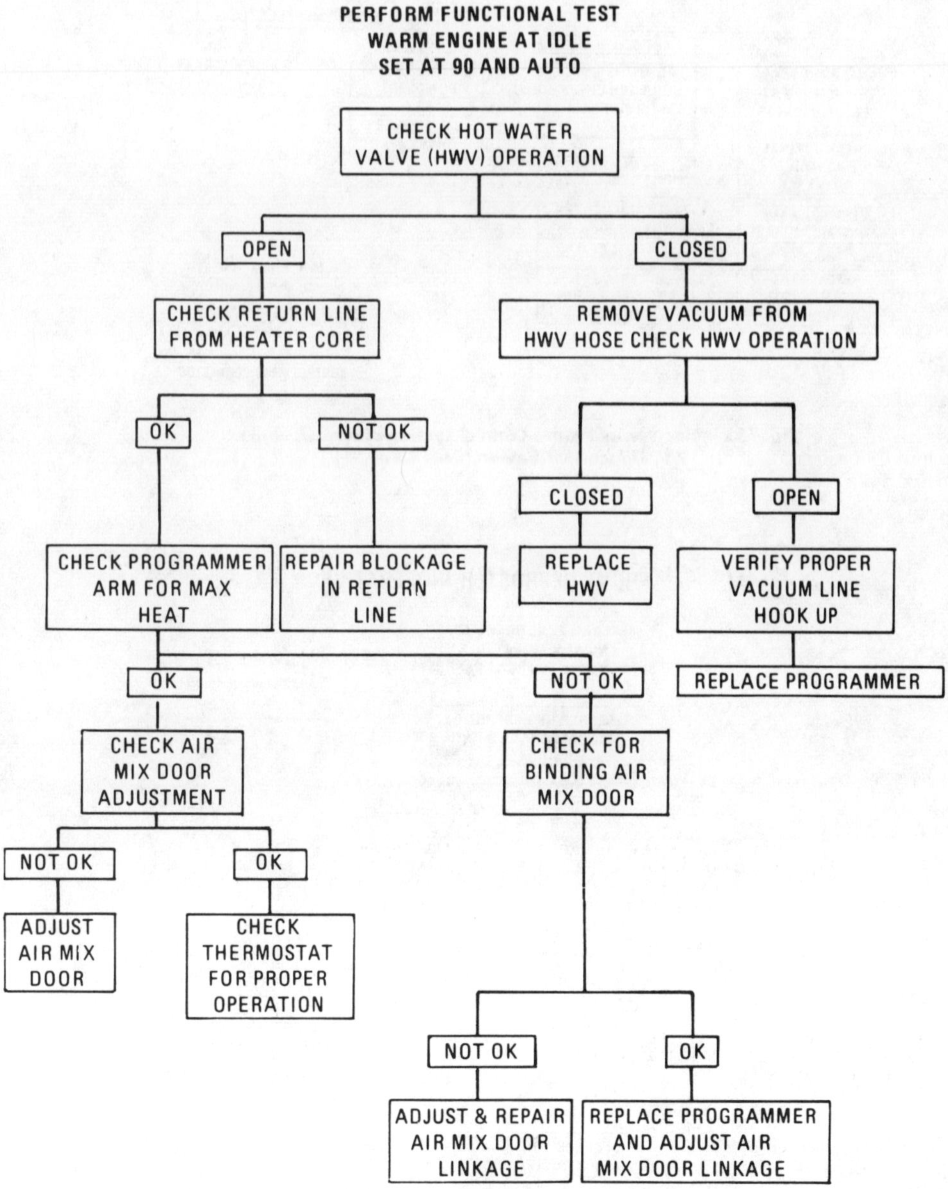

Fig. 152 Electronic Climate Control system diagnosis chart. 1980 Eldorado & Seville (Part 12 of 16)

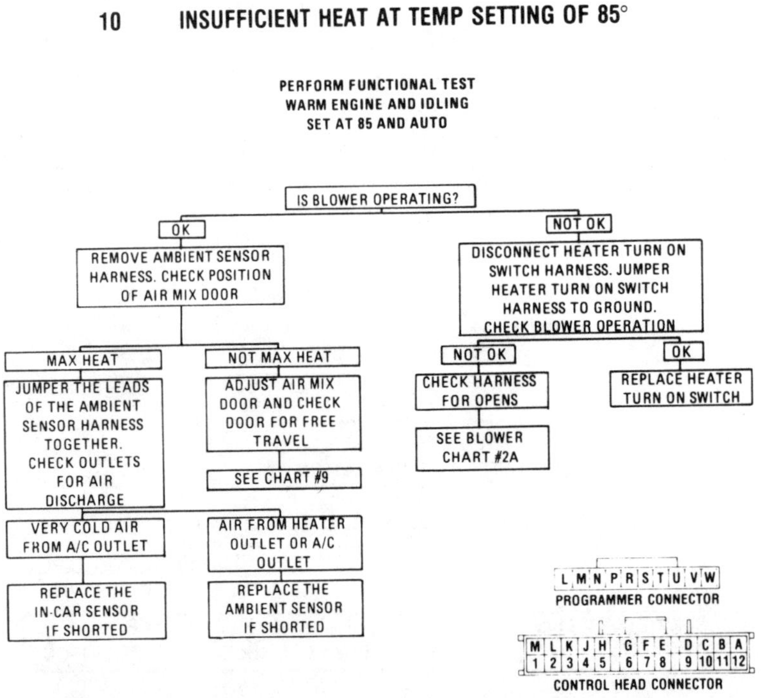

Fig. 152 Electronic Climate Control system diagnosis chart. 1980 Eldorado & Seville (Part 13 of 16)

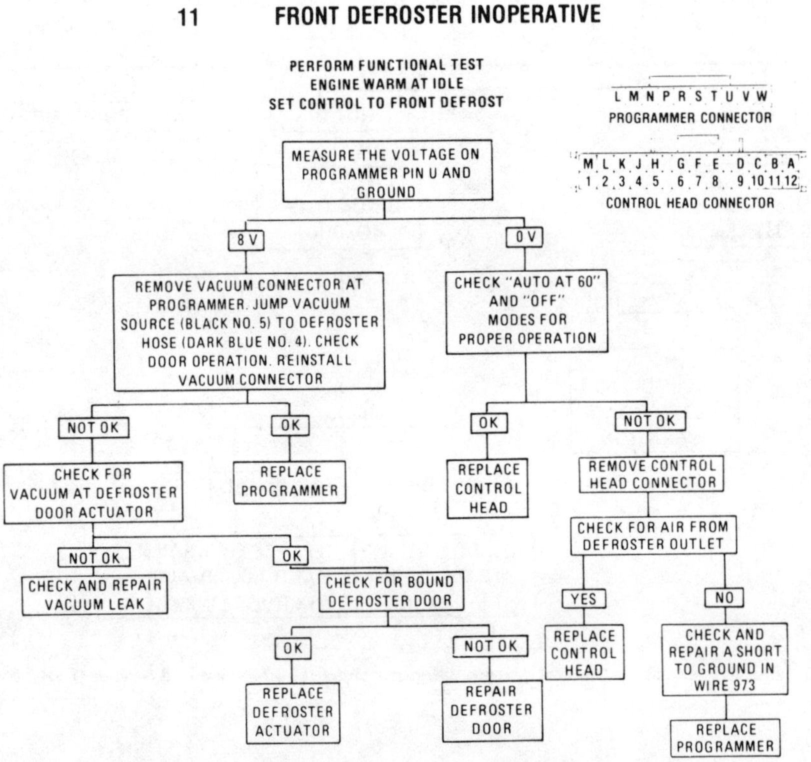

Fig. 152 Electronic Climate Control system diagnosis chart. 1980 Eldorado & Seville (Part 14 of 16)

12 REAR DEFROST INOPERATIVE

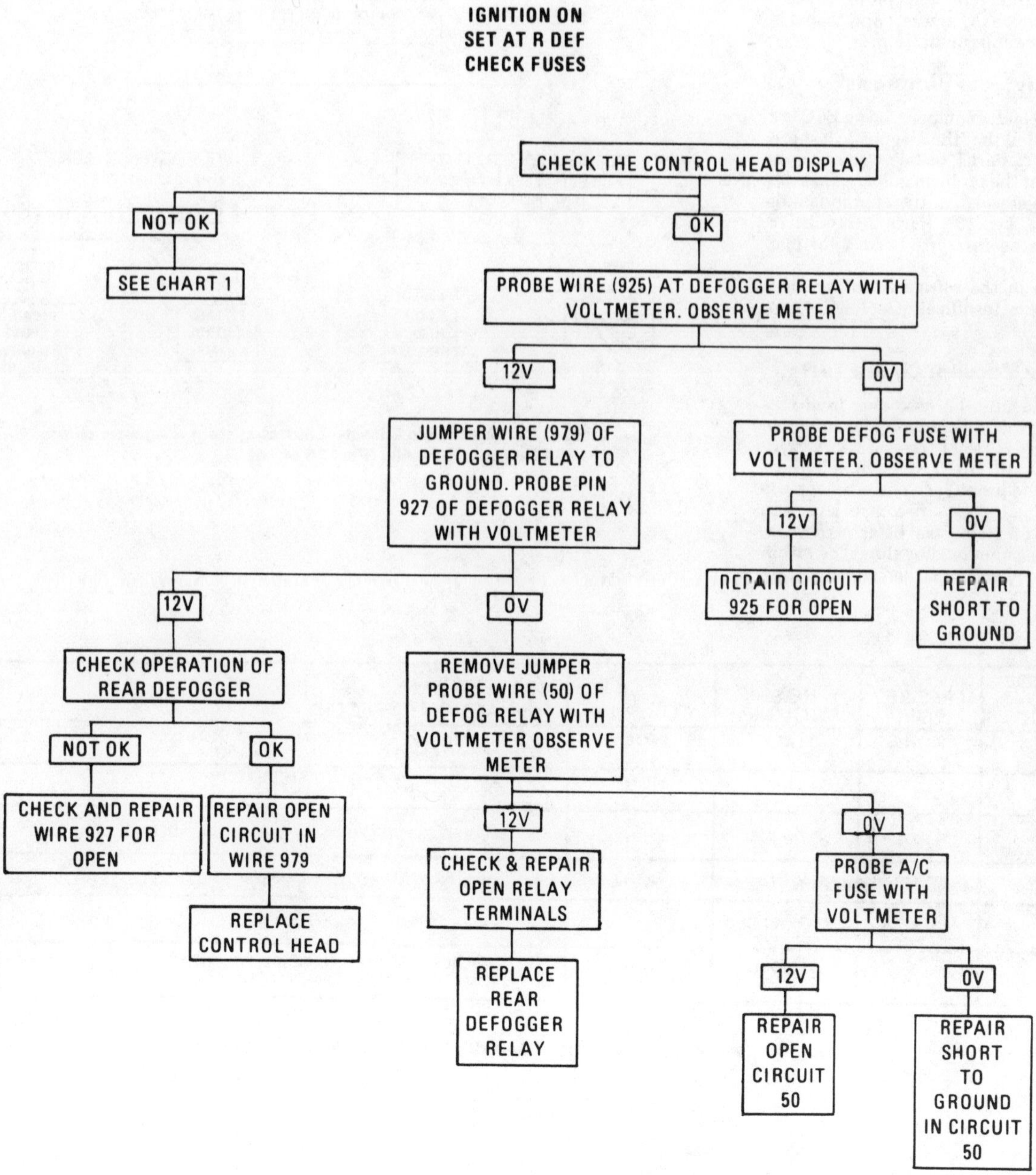

Fig. 152 Electronic Climate Control system diagnosis chart. 1980 Eldorado & Seville (Part 15 of 16)

1980 ELECTRONIC CLIMATE CONTROL (ECC)

Functional Test

Before beginning control system functional test, ensure engine is warm and idling. Refer to Fig. 150 to perform test. Fig. 151 details system response to various control lever settings. These charts outline functional performance of properly operating ECC systems and should be used for reference and comparison.

ECC System Diagnosis

In diagnosing faulty components on ECC systems, it is essential that the diagnostic test procedure in Fig. 152, Part 1, be followed exactly in order to pinpoint the problem quickly. Once the problem has been isolated, the appropriate diagnosis chart(s), Fig. 152, Parts 2 through 16, detail(s) repairs necessary to correct the malfunction.

If a problem in the refrigeration system is suspected, refer to Insufficent Cooling Diagnosis.

Auxiliary Vacuum Circuit Test

Disconnect vacuum harness plug from programmer valve; this should direct all airflow to floor outlets. Insert tee of tester J-23678 vacuum gauge into port No. 5 of vacuum harness plug, Fig. 132, and probe each of the other ports as outlined in Fig. 153. Ensure control lever is in proper position for step of test being performed. For each step, vacuum reading should be within 2 or 3 inches of full vacuum when actuator travel is completed.

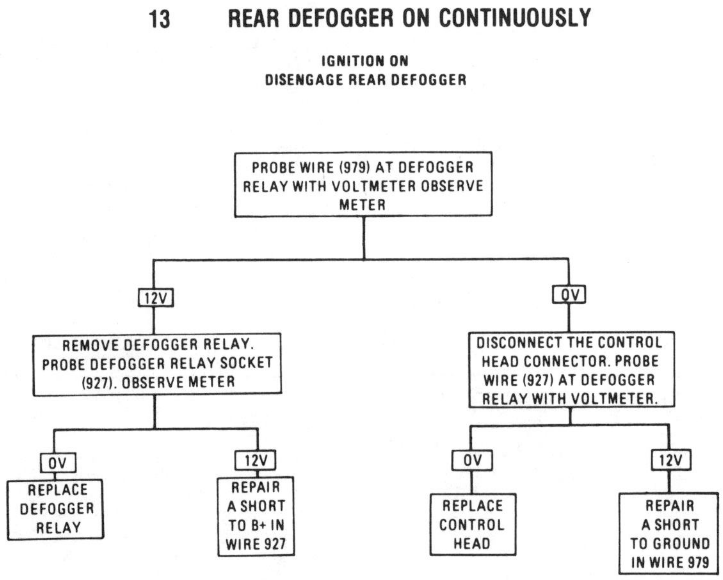

Fig. 152 Electronic Climate Control system diagnosis chart. 1980 Eldorado & Seville (Part 16 of 16)

CONTROL HEAD SETTING	PROBE PORT	HOSE COLOR	COVER UP	ACTION EXPECTED
HI	NONE	BLACK	PROBE	VACUUM READING WITHIN 2-3 IN HG AS COMPARED TO MANIFOLD VACUUM
HI	3	TAN	NONE	AIR DELIVERY SHOULD SHIFT FROM HEATER TO A C OUTLETS.
HI	2	ORANGE	NONE	RECIRC DOOR OPENS — 1-2 MINUTE DELAY.
HI	1	WHITE	NONE	HEATER WATER VALVE CLOSES.
DEF	4	DARK BLUE	6	MOST OF THE AIR SHOULD SWITCH TO DEF OUTLET.
DEF	6	YELLOW	NONE	DEF DOOR MOVES TO BLEED POSITION. SOME AIR EXITS FROM DEF OUTLETS.

Fig. 153 Auxiliary vacuum circuit test chart. 1980 Eldorado & Seville

Temperature of Air Entering Condenser	70°	80°	90°	100°	110°
Engine rpm	2000				
Compressor Head Pressure	155-165	180-190	200-210	230-240	265-275
Evaporator Pressure at VIR	28.5 - 31.5 psi*				
Discharge Air Temp. @ Right Hand Outlet	35-39	36-40	36-40	37-41	39-43

Fig. 154 Performance test chart. 1974-76 Nova

Temperature of Air Entering Condenser	70°	80°	90°	100°	110°
Engine rpm	2000				
Compressor Head Pressure	150-160	180-190	210-220	250-260	285-295
Evaporator Pressure @ VIR	29.5 - 32.5 psi*				
Discharge Air Temp. @ Right Hand Outlet	38-42	39-43	40-44	44-48	48-52

Fig. 155 Performance test chart. 1974-76 Camaro

Chevrolet Line

PERFORMANCE DATA & TESTS

The following Performance Data define normal operation of the system under the above conditions. Relative humidity does not appear in the tables because after running the prescribed length of time on recirculated air and maximum cooling, the relative humidity of the air passing over the evaporator core will remain at approximately 35% to 40% regardless of the ambient temperature or humidity.

Should excessive head pressures be encountered at higher ambient temperatures, an 18" fan placed in front of the vehicle and blowing into the condenser will provide the extra circulation of air needed to bring the pressures to within the limits specified.

Performance Test

This test may be conducted to determine if the system is performing in a satisfactory manner and should be used as a guide by the serviceman in diagnosing trouble within the system. The following fixed conditions must be adhered to in order to make it possible to compare the performance of the system being tested with the standards given in Figs. 154 to 166.

Except 1980 Citation

1. Doors and windows closed. (Car inside or in shade.)
2. Hood up and engine exhaust suitably ventilated.
3. Vehicle in NEUTRAL with engine running at 2000 rpm.

NOTE: On 1974-77 Vega and 1978-79 Monza "S" coupe and station wagon, operate engine at 1500 rpm.

Temperature of Air Entering Condenser	70°	80°	90°	100°	110°
Engine rpm	2000				
Compressor Head Pressure	135 145	170 180	190 200	220 230	265 275
Evaporator Pressure @ VIR	29.0 - 31.0 psi*				
Discharge Air Temp. @ Right Hand Outlet	33-37	35-39	35-39	37-41	38-42

Fig. 156 Performance test chart. 1974-76 Chevelle & Monte Carlo

Temperature of Air Entering Condenser	70°	80°	90°	100°
Engine rpm	2000			
Compressor Head Pressure	115 125	145 155	175 185	210 220
Evaporator Pressure @ VIR	29-31 psi*			
Discharge Air Temp. @ Right Hand Outlet	36-40	38-42	39-43	40-44

Fig. 157 Performance test chart. 1974-76 Chevrolet full size

Temperature of Air Entering Condenser	70°	80°	90°	100°	110°
Engine rpm	1500				
Compressor out Pressure (before expansion tube)	150 205	185 235	215 270	250 300	280 335
Evaporator Pressure - at accumulator (psi)	20-26	20-28	21-29	25-33	30-39
Discharge Air Temp. — Right Hand Center Outlet	35° 41°	36°-43°	36°-44°	42°-52°	47°-57°

Fig. 158 Performance test chart. 1974-77 Vega

Temperature of Air Entering Condenser	70°	80°	90°	100°	110°
Engine rpm	2000				
Compressor Head Pressure	180-190	245-255	290-300	315-325	350-360
Evaporator Pressure at VIR	28.5 - 31.5 psi*				
Discharge Air Temp. @ Right Hand Outlet	48-52	48-52	49-53	50-54	52-56

Fig. 159 Performance test chart. 1974-77 Corvette

Temperature of Air Entering Condenser	70°	80°	90°	100°	110°
Engine rpm	2000				
Compressor Head Pressure	140 160	170 190	200 220	240 260	270 290
Evaporator Pressure @ VIR	31 34	30 33	28.5 31.5		
Discharge Air Temp. @ Right Hand Outlet	49 53	48 52	44 48	48 52	48 52

Fig. 160 Performance test chart. 1975-76 Monza

Temperature of air entering condenser	70°F	80°F	90°F	100°F	110°F
Engine RPM	2000				
Compressor out pressure before expansion tube (psig)	110- 140	145- 175	185- 215	220- 250	265- 295
Evaporator pressure at accumulator (psig)	24- 30	25- 31	25- 31	26- 32	27- 33-
Discharge air temp @ right hand outlet*	38°F 44°F	39°F 45°F	39°F 45°F	40°F 46°F	41°F 47°F

Fig. 161 Performance test chart. 1976-80 Chevette

TEMPERATURE OF AIR ENTERING CONDENSER	70 F	80 F	90 F	100 F	110 F
COMPRESSOR OUT PRESSURE (BEFORE EXPANSION ORIFICE)	125–175 psi	145–195 psi	170–220 psi	195–245 psi	225–275 psi
EVAPORATOR PRESSURE (AT ACCUMULATOR)	25–32 psi	25–32 psi	25–32 psi	25–32 psi	25–32 psi
DISCHARGE AIR TEMPERATURE – LEFT CENTER OUTLET	39–44 F	40–45 F	41–46 F	43–48 F	45–50 F

Fig. 162 Performance test chart. 1977 Chevrolet

Model	Grille Temp. (°F)	High Side Pressure (PSIG.)
Camaro	80	215
	100	245
Chevelle & Monte Carlo	80	180
	100	230
Nova	80	180
	100	235

Fig. 163 Performance test chart (high side pressure). 1977 Except Chevrolet full size

TEMPERATURE OF AIR ENTERING CONDENSER		70°F (21°C)	80°F (27°C)	90°F (32°C)	100°F (38°C)	110°F (43°C)
*COMPRESSOR OUT PRESSURE (BEFORE EXPANSION ORIFICE	MALIBU & MONTE CARLO	125-175 psi 862-1207 kPa	145-195 psi 1000-1345 kPa	170-220 psi 1172-1517 kPa	195-245 psi 1345-1689 kPa	225-275 psi 1551-1896 kPa
	FULL SIZE	140-180 psi 965-1241 kPa	170-210 psi 1172-1448 kPa	200-240 psi 1379-1655 kPa	235-275 psi 1620-1896 kPa	265-305 psi 1827-2103 kPa
	CAMARO		195-235 psi 1345-1620 kPa		225-265 psi 1551-1827 kPa	
	NOVA		160-200 psi 1103-1379 kPa		215-255 psi 1482-1758 kPa	
*EVAPORATOR PRESSURE (AT ACCUMULATOR)**	MALIBU & MONTE CARLO	25-32 psi 172-221 kPa	25-32 psi 172-221 kPa	25-32 psi 172-221 kPa	25-32 psi 172-221 kPa	25-32 psi 172-221 kPa
	FULL SIZE	24-27 psi 165-186 kPa	24-27 psi 165-186 kPa	24-27 psi 165-186 kPa	24-27 psi 165-186 kPa	24-27 psi 165-186 kPa
	CAMARO		24-27 psi 165-186 kPa		24-27 psi 165-186 kPa	
	NOVA		24-27 psi 165-186 kPa		24-27 psi 165-186 kPa	
*DISCHARGE AIR TEMPERATURE— LEFT CENTER OUTLET	MALIBU & MONTE CARLO	39-44°F	40-45°F	41-46°F	43-48°F	45-50°F
	FULL SIZE	41-43°F	42-44°F	42-44°F	43-45°F	43-45°F
	CAMARO		38-45°F		38-45°F	
	NOVA		38-45°F		38-45°F	

*Just before compressor disengages **At sea level

Fig. 164 Performance test chart. 1978 Models Except Chevette & Monza

AMBIENT TEMP.	21 C (70 F)	27 C (80 F)	32 C (90 F)	38 C (100 F)	43 C (110 F)
A C OUTLET TEMP.	4 7 (40 45)	4 7 (40 45)	6 8 (42 46)	6 8 (42 46)	7 10 (44 50)

Fig. 165 Performance test chart. 1979-80 Camaro

AMBIENT TEMP.		21°C (70°F)	27°C (80°F)	32°C (90°F)	38°C (100°F)	43°C (110°F)
A/C OUTLET TEMP.	WAGON	7-9°C (44-48)	7-9°C (44-48)	7-9°C (44-48)	8-10°C (46-50)	10-12°C (50-54)
	COUPE	3-6°C (38-42)	3-6°C (38-42)	3-6°C (38-42)	3-6°C (38-42)	3-6°C (38-42)

Fig. 166 Performance test chart. 1979-80 Monza

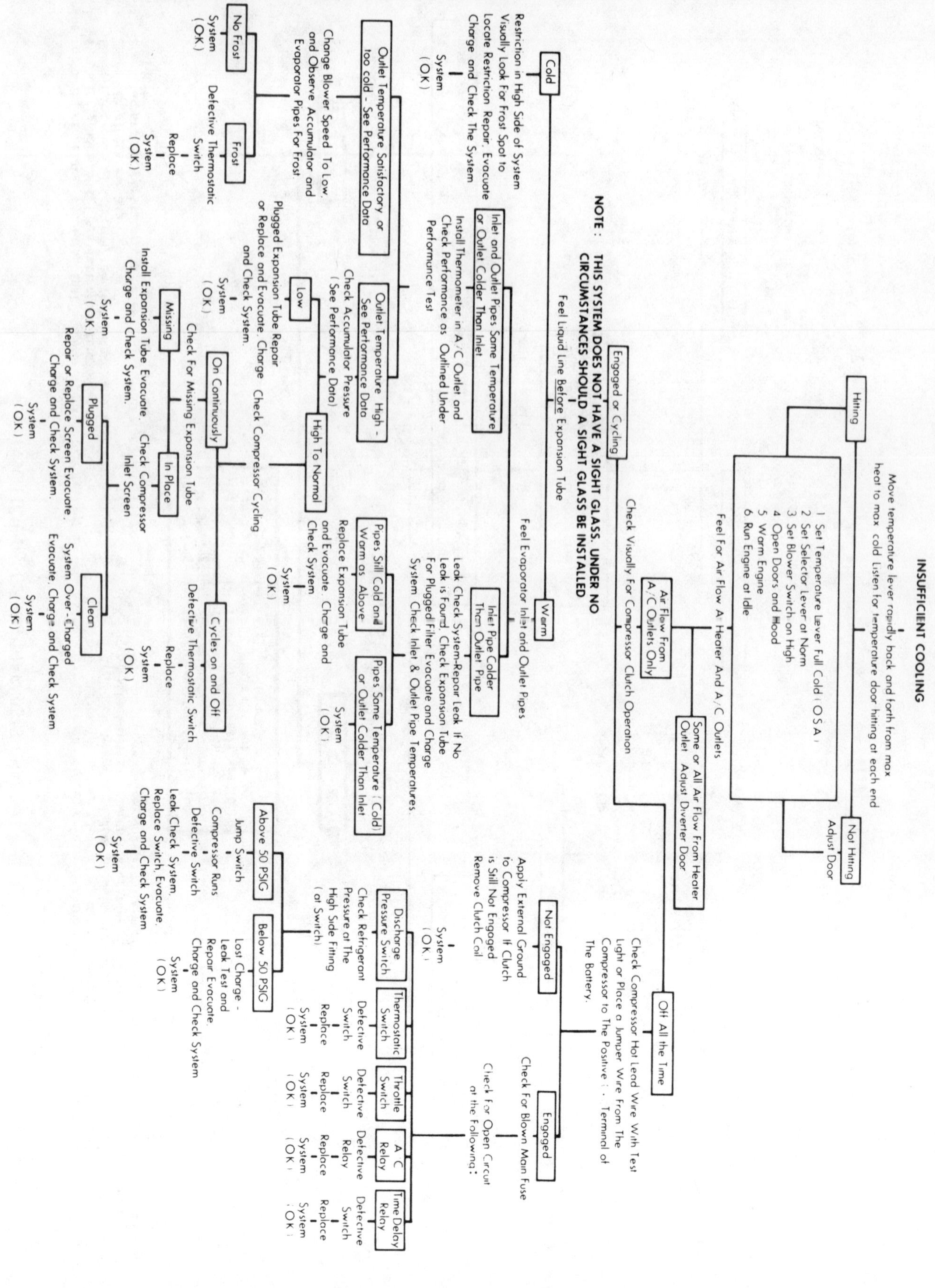

Fig. 167 Insufficient cooling diagnosis chart. 1974-76 Chevrolet Vega & Pontiac Astre

INSUFFICIENT COOLING DIAGNOSIS CHART #1

The following procedures should be applied before performance testing an A/C System.

1. Check for proper belt installation and tension with J-23600.
2. Check for proper clutch coil terminal connector installation.
3. Check for clutch air Gap
4. Check for broken, burst, or cut hoses. Also check for loose fittings on all components.
5. Check for condenser air blockage due to foreign material.
6. Check for proper air ducting hose connections.
7. Check temperature door adjustment, adjust if incorrect.
8. Check gasket between evaporator and firewall for air leak, repair if leaking.
9. Install pressure gages and thermometer and make performance test.

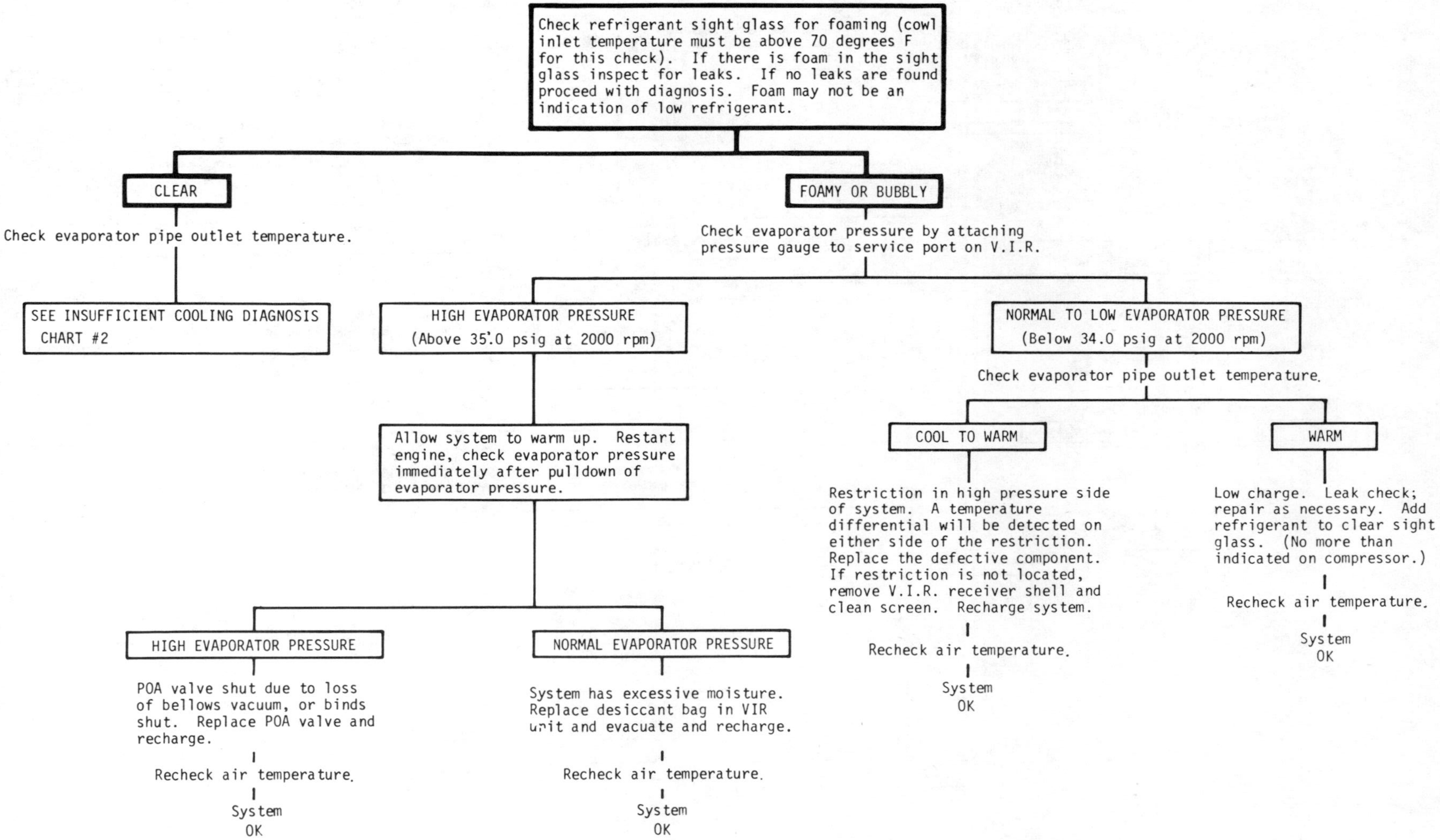

Fig. 168 Insufficient cooling diagnosis chart. 1974-76 Models (Except Chevette & Vega) & 1977 Corvette (Part 1 of 2)

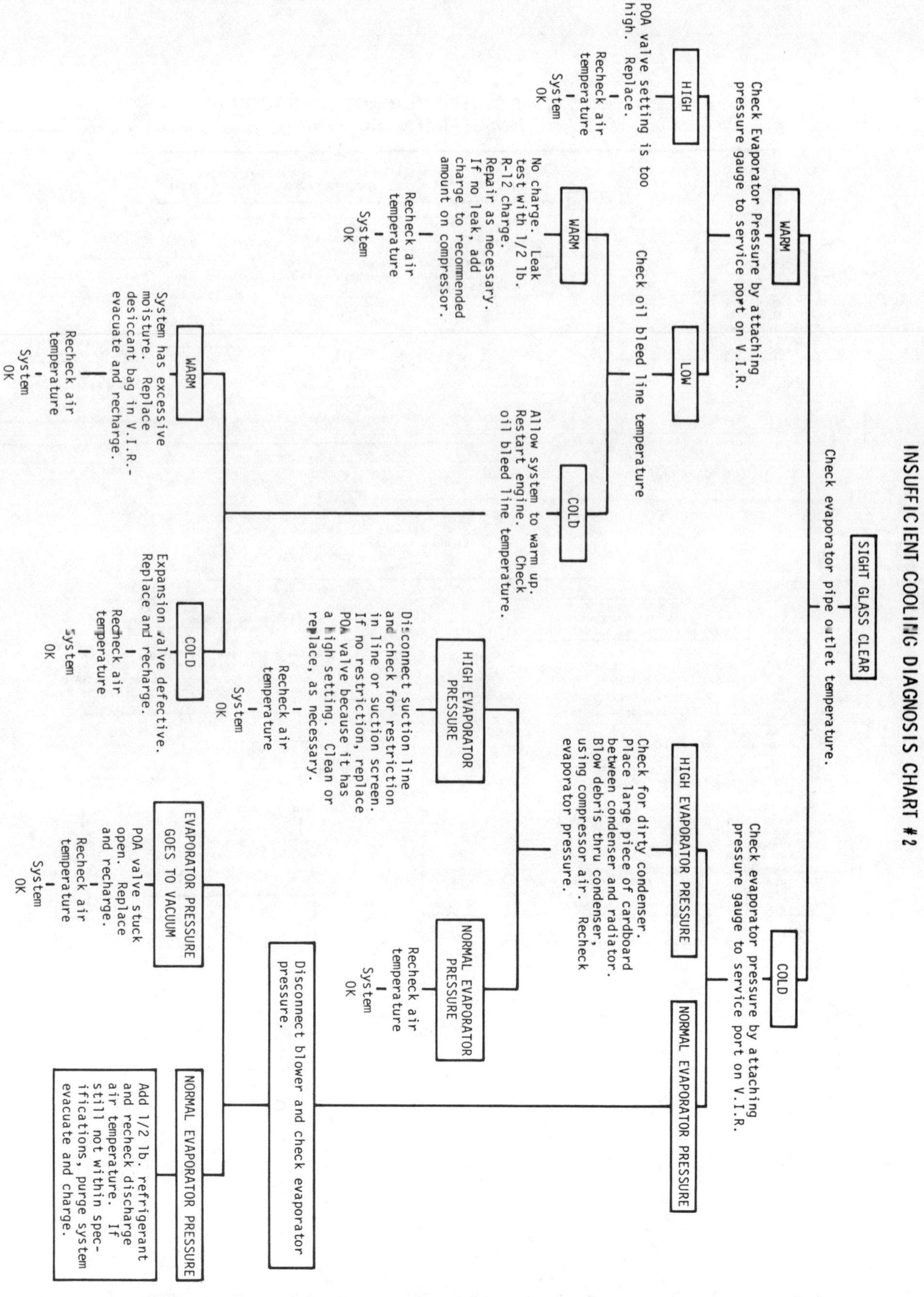

Fig. 168 Insufficient cooling diagnosis chart. 1974-76 Models (Except Chevette & Vega) & 1977 Corvette (Part 2 of 2)

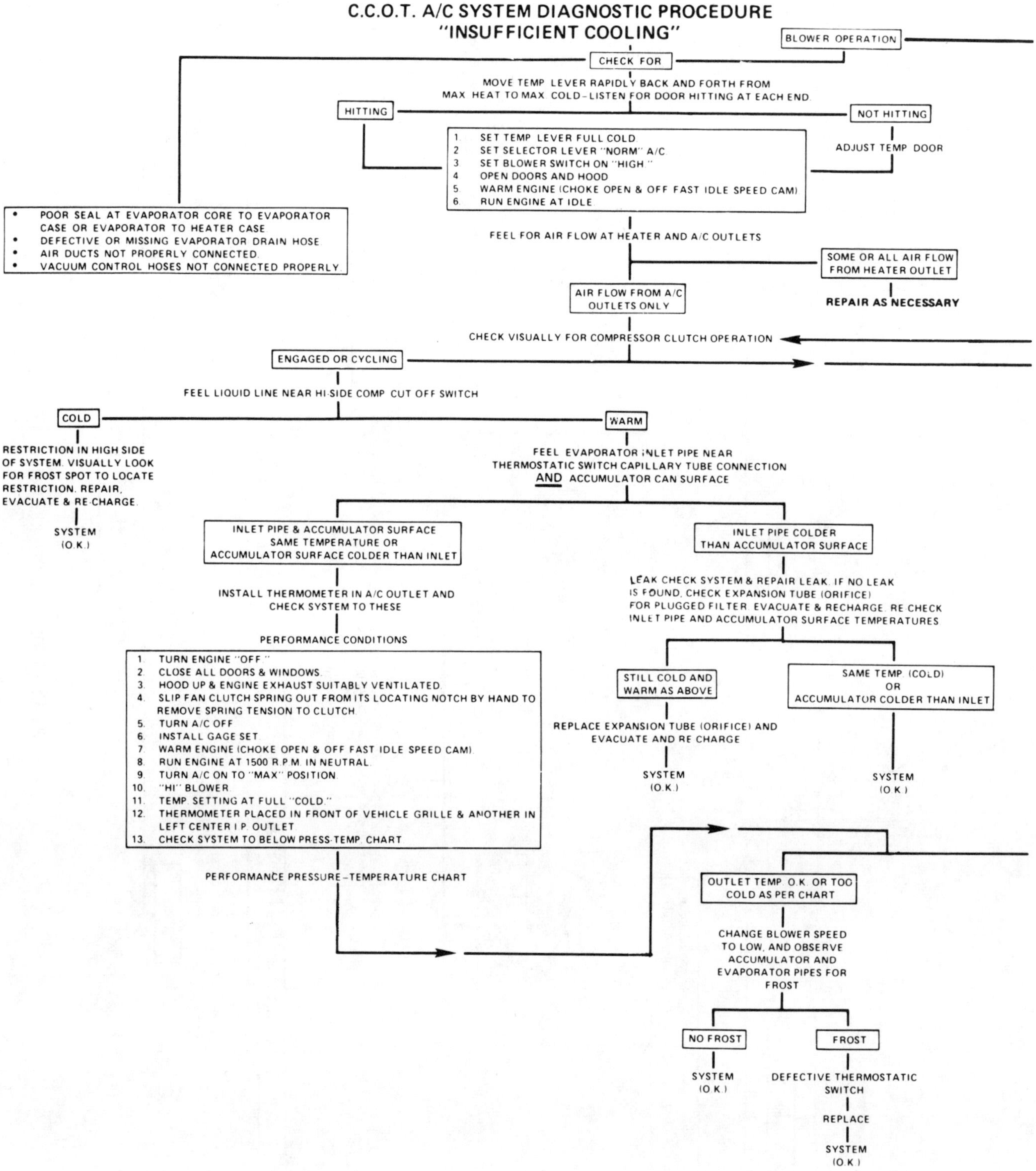

Fig. 169 Insufficient cooling diagnosis chart. 1976-78 Chevette, 1977 models (except Corvette) & 1978 Monza (Part 1 of 2)

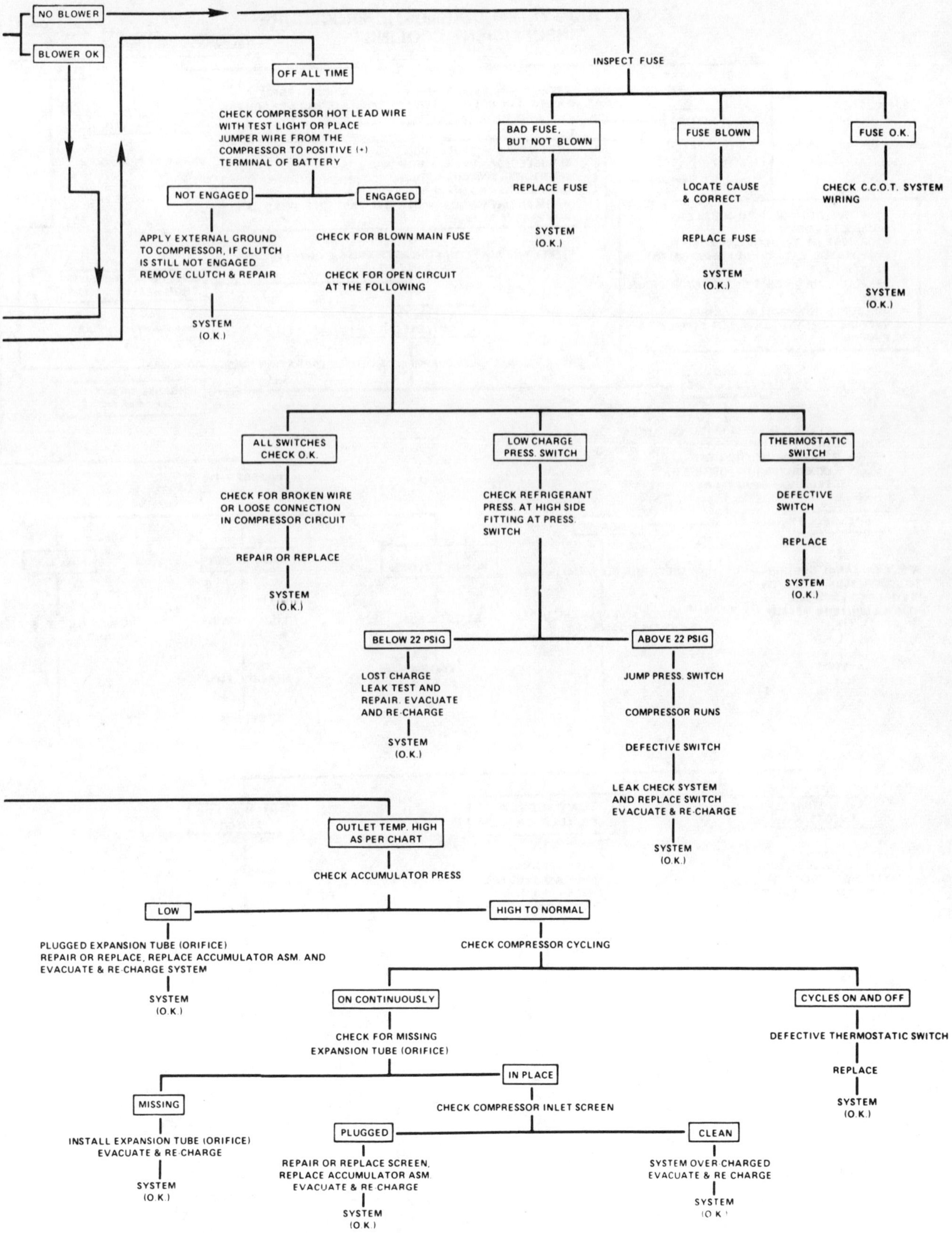

Fig. 169 Insufficient cooling diagnosis chart. 1976-78 Chevette, 1977 models (except Corvette) & 1978 Monza (Part 2 of 2)

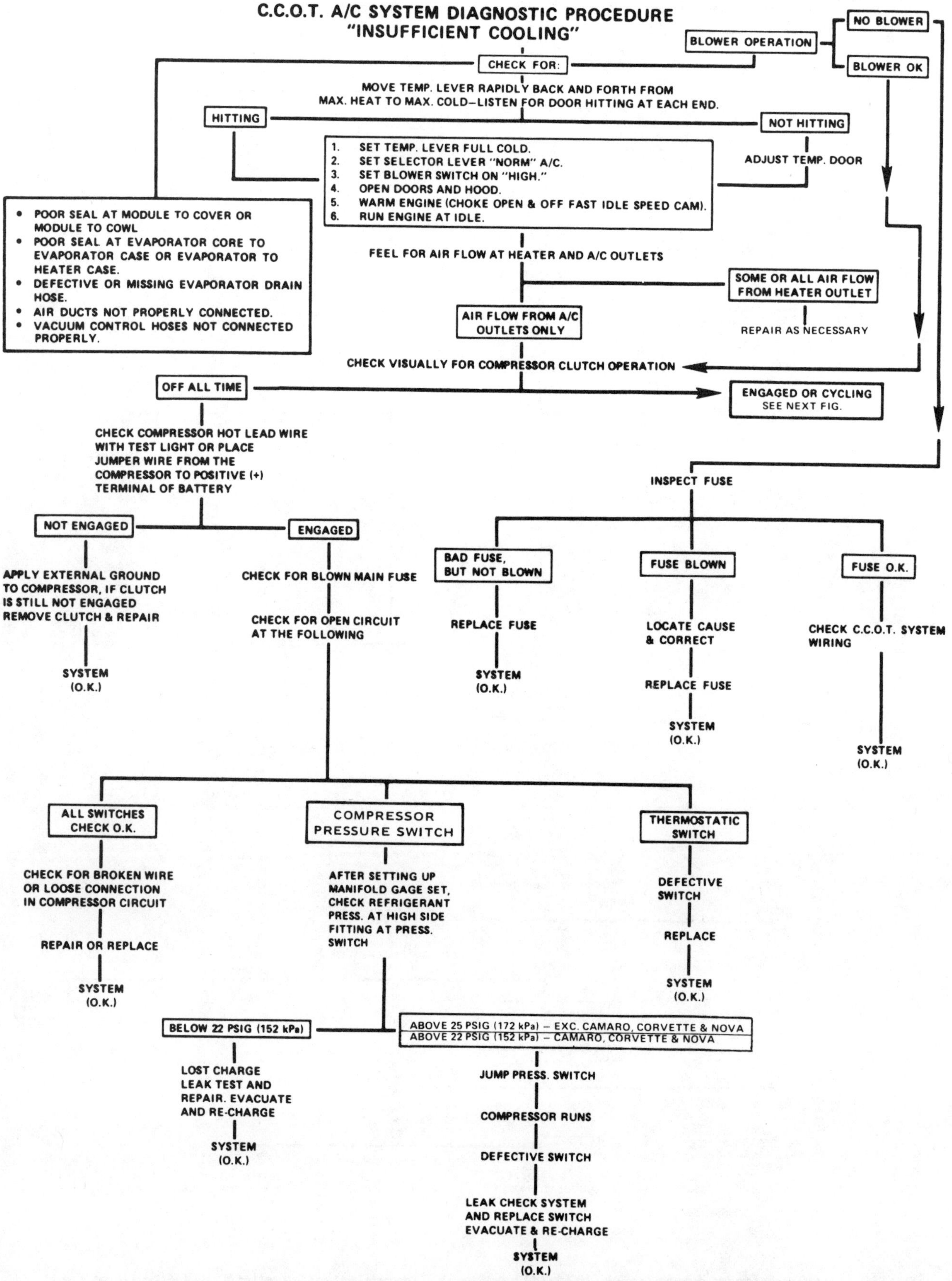

Fig. 170 Insufficient cooling diagnosis chart. 1978 models except Chevette & Monza (Part 1 of 2)

C.C.O.T. A/C SYSTEM DIAGNOSTIC PROCEDURE "INSUFFICIENT COOLING"

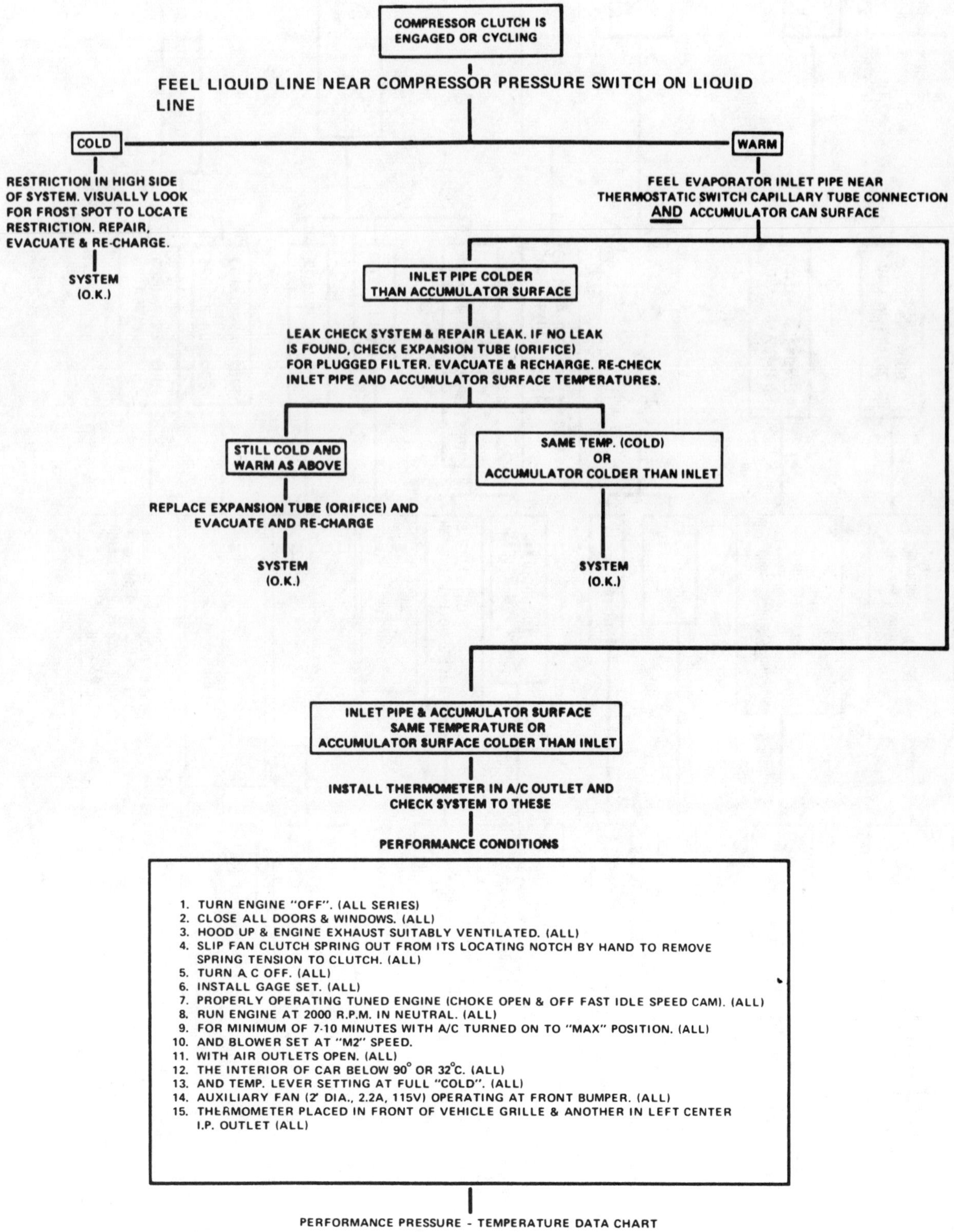

Fig. 170 Insufficient cooling diagnosis chart. 1978 models except Chevette & Monza (Part 2 of 2)

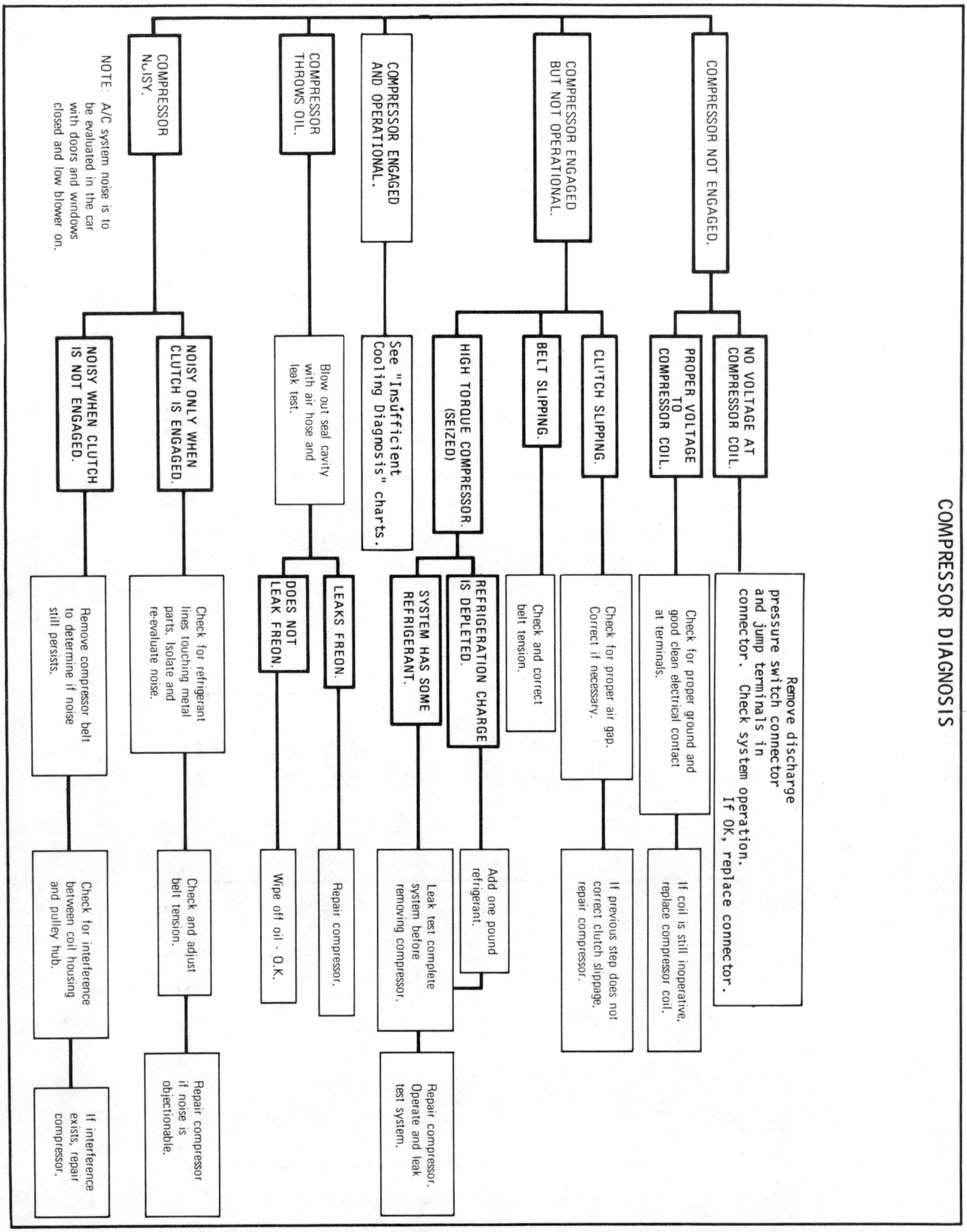

Fig. 171 Compressor clutch diagnosis chart. 1974-76 models (except Chevette & Vega) & 1977 Corvette

4. Air Conditioning controls set for—
   (a) Maximum cooling.
   (b) High blower speed.
5. TEMP control set at "COLD" and all air conditioning outlets open.
6. Gauge set installed.
7. System settled out (run-in approximately 10 minutes).
8. A thermometer placed in front of vehicle grille and another in the left center A/C outlet.

   NOTE: On 1974-76 Vega, install thermometer in right hand A/C outlet.

NOTE: Higher temperatures and pressures will occur at higher ambient temperatures. In areas of high humidity it is possible to have thermometer and gauge readings approach but not reach the figures listed in the performance tables and still have a satisfactory operating unit. However, it is important to remember that low pressure has a direct relationship to nozzle outlet temperature. If pressure is too low, ice will gradually form on the evaporator fins, restricting airflow into the passenger area and resulting in insufficient or no cooling.

1980 Citation

Refer to Insufficient Cooling Diagnosis Chart, Fig. 115 parts 1 through 3, when performing the performance test.

Insufficient Cooling Diagnosis

1974-76 Vega

These units incorporate a thermostatic switch in the refrigerant system. When diagnosing insufficient cooling problems, refer to Fig. 167 for diagnostic procedure.

1974-76 Models (Except Chevette & Vega) & 1977 Corvette

These units incorporate the use of a V.I.R. or E.E.V.I.R. unit in the refrigerant system. When diagnosing insufficient cooling problems, refer to Fig. 168 parts 1 and 2, for insufficient cooling diagnostic procedure.

1976 Chevette, 1977 Models Except Corvette, 1978 All & 1979 Models Except Camaro, Chevette & Monza

These models are equipped with the C.C.O.T. A/C system, which incorporates the use of a thermostatic switch. When diagnosing insufficient cooling problems, refer to Figs. 169 and 170 parts 1 and 2, for insufficient cooling diagnostic procedures.

1979 Camaro, Chevette, Monza & 1980 Models Except Citation

These models are equipped with the C.C.O.T. A/C system, which incorporates the use of a pressure cycling switch. When diagnosing insufficient cooling problems, refer to Fig. 114 parts 1 and 2, for insufficient cooling diagnostic procedure.

1980 Citation

When diagnosing insufficient cooling problems on these models, refer to Fig. 115 parts 1 through 3, for diagnostic procedure.

Compressor Clutch Diagnosis

NOTE: For models not listed below, refer to Insufficient Cooling Diagnosis, when problems with compressor clutch are encountered.

1974-76 Models (Except Chevette & Vega) & 1977 Corvette

Refer to compressor clutch diagnosis chart, Fig. 171, for diagnostic procedure.

1980 Citation

Refer to Compressor Clutch Diagnosis chart, Fig. 116, for diagnostic procedure.

1980 Citation Engine Cooling Fan Diagnosis

Refer to engine cooling fan diagnosis chart, Fig. 117, for diagnostic procedure.

POA Valve Capsule

The only check for proper POA valve operation is to check the suction pressure at the valve as during a performance test. The POA valve is an absolute valve and will provide different gauge readings based on the altitude where the readings are being taken. Correct gauge reading at sea level is 29.5 psi. Gauge readings will be one-half psi higher for each additional 1000 feet of elevation. If a valve gives improper gauge readings, it must be replaced since it is not repairable or adjustable.

Diagnosis For Defective VIR Unit

A defective VIR unit may be caused by a defective POA or expansion valve capsule. A restriction in the VIR or evaporator inlet will be indicated by low head pressures and little or no cooling. An excessively cold receiver shell may be indicative of a restriction.

1976-78 Chevette A/C Electrical Diagnosis Connector

Before starting diagnosis procedure, check to ensure that battery is fully charged and that circuit No. 2 in passenger compartment is operational. When performing test refer to Figs. 172, 173 and 174. To perform this test, a voltmeter is required.

Voltmeter Connections	Voltage Reading	Condition
1. Blower motor inoperative, compressor operates properly. Perform test with ignition switch on, A/C control off and blower switch in low.		
a. 1 and G (V1)	Over 3 volts	Loose blower motor wire connector or defective blower motor.
	Under 3 volts	Go to step b.
b. 2 and G (V2)	Over 3 volts	Disconnected or defective high blower relay.
	Under 3 volts	Go to step c.
c. 3 and G (V3)	Over 10 volts	Open circuit in blower resistor or wire from resistor to high blower relay.
	Under 10 volts	Open circuit in wire from time-temperature relay to blower resistor.
2. Blower and compressor inoperative. Perform test with ignition switch on, A/C control off and blower switch on low.		
a. 4 and G (V4)	Over 10 volts	Temperature switch or wiring grounded.
	Under 10 volts	Blown fuse or open circuit between fuse and time temperature relay.
3. No high blower in either high blower or Max. A/C. Perform test with ignition on, A/C control on Vent and blower switch on high.		
a. 9 and G (V9)	Over 10 volts	Blown high blower fuse, defective high blower relay or disconnected feed wire.
	Under 10 volts	Open circuit between A/C control and high blower relay coil.
4. No low blower speed or high blower speed only. Perform test with ignition on, A/C control in Vent and blower switch on low.		
a. 3 and G (V3)	Over 10 volts	If blower runs in any other speed but low, resistor. If blower runs only on high speed, resistor or wiring from resistor to high blower relay is defective.
	Under 10 volts	Open circuit in wire between time-temperature relay and resistor.

Voltmeter Connections	Voltage Reading	Condition
5. No medium blower speeds.		
		Defective blower switch or open wire from blower switch to resistor.

6. Blower operates at high blower speed in all switch positions. Perform test with ignition on, A/C control in Vent and blower switch in low.

a. 9 and G (V9)	Any voltage reading	Short circuit in high blower relay or feed wire, blower switch or A/C control. Zero voltage reading in Norm, Heat or Defrost positions indicates problem in A/C control. Zero voltage in Off position indicates defective wire between A/C control and blower switch or defective A/C control. If voltage drops to zero as blower switch is moved, check for defective blower switch.
	Zero voltage reading	Defective high blower relay or short circuit in wire from relay to blower motor.

7. Compressor inoperative, blower motor operates properly. Perform test with ignition on, A/C control in Norm and blower switch in Low.

a. 6 and G (V6)	Over 10 volts	Check for defective compressor clutch or clutch wiring.
	Under 10 volts	Go to step b.
b. 7 and G (V7)	Over 10 volts	Open circuit in compressor clutch feed wire.
	Under 10 volts	Open circuit in A/C control.

1977-78 Chevrolet Full Size

These models are equipped with two electrical diagnosis connectors, Figs. 118, 120, 175, 176 and 177. To perform the following diagnosis, a voltmeter and jumper wire are required. If a switch or relay are suspected of being defective, connect a jumper wire across switch or relay. If system operates, the switch or relay is defective. If system does not operate, the wire or connector is defective. With jumper wire connected, connect voltmeter from jumper wire to ground. If no voltage is present, the positive lead is defective. If voltage is present, the ground lead is defective. Perform test with ignition switch on, engine stopped and inside car temperature above 55°F, place blower switch on low and A/C control on Vent unless otherwise noted.

1977 Four Season

Voltmeter Connections	Voltage Reading	Condition
1. No blower any speed, compressor clutch engages.		
a. 1 and G	3 volts or more	Defective blower motor, ground or capacitor lead.
	Under 3 volts	Go to step b.
b. 2 and G	3 volts or more	Defective relay or lead from relay to blower motor.
	Under 3 volts	Stalled blower motor.

2. No blower and compressor clutch does not engage. Check A/C fuse and if blown, check for ground in circuit or excessive motor or clutch current. Also check for open circuit in wire from ignition switch to ambient switch or an improperly grounded A/C harness.

3. No low blower speed.

a. 3 and G	10 volts or more	Defective blower motor resistor.
	Under 10 volts	Defective ambient switch or lead to resistor.

4. No low or M1 blower speed. — Defective resistor.

5. No blower except in high or Max. A/C.

a. 2 and G	3 volts or more	Defective relay.
	Under 3 volts	Defective resistor or lead wire from resistor to relay.

6. No high blower in any switch or control position. Place blower switch on high.

a. 9 and G	10 volts or more	Defective high blower relay, open circuit in feed wire to relay or poor relay ground.
	Under 10 volts	Faulty wire from blower switch to relay coil.

7. No high blower speed except in Max. A/C. — Defective blower switch.

8. No high blower speed in Max. A/C. — Defective A/C control or wire from A/C control to blower switch.

9. No M1 blower speed. — Defective blower switch or wire to resistor.

10. No M2 blower speed. — Defective blower switch or wire to resistor.

11. No M1, M2 or high blower using blower switch, but high blower speed is obtained in Max. A/C. Place blower motor switch in high.

Voltage at blower switch feed terminal from A/C control	10 volts or more	Defective blower switch.
	Under 10 volts	Defective A/C control or wire to blower switch.

12. Compressor clutch and blower motor inoperative. — Refer to step 2.

13. Compressor clutch operates in Norm. or Max. but not in both. — Defective A/C control.

14. Compressor clutch inoperative, blower motor on. Place A/C control in Norm.

a. 6 and G	10 volts or more	Defective blower switch.
	Under 10 volts	Go to step b.

Voltmeter Connections	Voltage Reading	Condition
b. 8 and G	10 volts or more	Defective temperature switch or lead wire to compressor clutch.
	Under 10 volts	Go to step c.
c. 7 and G	10 volts or more	Low refrigerant charge or defective refrigerant low pressure switch or lead from low pressure switch to temperature switch.
	Under 10 volts	Defective A/C control or lead from control to refrigerant low pressure switch.

15. Low blower only, compressor clutch inoperative.

a. 4 and G	10 volts or more	Defective A/C control.
	Under 10 volts	Faulty connection from ambient switch to A/C control.

1978 Four Season

1. No blower, any speed compressor clutch engages.

a. 1 and G	3 volts or more	Defective blower motor, ground or capacitor lead.
	Under 3 volts	Go to step b.
b. 2 and G	3 volts or more	Defective relay or lead from relay to motor.
	Under 3 volts	Stalled blower motor.

2. No blower operation and compressor clutch does not engage. Check A/C fuse and if blown, check for ground in circuit or excessive motor or clutch current. Also check for an open circuit in wire from ignition switch to ambient switch or poor A/C harness ground.
3. No low blower speed.

a. 3 and G	10 volts or more	Defective resistor.
	Under 10 volts	Defective ambient switch or lead to resistor.

4. No low or M1 blower speed. — Defective resistor.

5. No blower except in high.

a. 2 and G	3 volts or more	Defective relay.
	Under 3 volts	Defective resistor or wire from resistor to relay.

6. No high blower under any condition. Place blower switch on high.

a. 9 and G	10 volts or more	Defective high blower relay, open circuit in feed wire to relay or poor relay ground.
	Under 10 volts	Faulty wire from blower switch to relay coil.

7. No M1 blower speed. — Defective blower switch or wire to relay.

Voltmeter Connections	Voltage Reading	Condition

8. No M2 blower speed. — Defective blower switch or wire to relay.

9. No M1, M2 or high blower using blower switch, but high blower is obtained in Max. A/C. Place blower switch in high.

a. Voltage at blower switch feed terminal from A/C control	10 volts or more	Defective blower switch.
	Under 10 volts	Defective A/C control or wire to blower switch.

10. Compressor clutch and blower motor inoperative. — See step 2.

11. Compressor clutch operates in Norm. or Max., but not in both. — Defective A/C control.

12. Compressor clutch inoperative, blower on. Place A/C control in Norm.

a. 6 and G	10 volts or more	Defective compressor clutch or pressure switch.
	Under 10 volts	Go to step b.
b. 8 and G	10 volts or more	Defective thermostatic switch or lead to clutch.
	Under 10 volts	Go to step c.
c. 7 and G	4 volts or more	Low refrigerant charge or defective refrigerant low pressure switch or wire from pressure switch to compressor.
	Under 4 volts	Defective A/C control or lead from control to low pressure switch.

13. Low blower only, compressor clutch inoperative.

a. 4 and G	10 volts or more	Defective A/C control.
	Under 10 volts	Faulty connection from ambient switch to A/C control.

1977-78 Comfortron

1. No blower, compressor clutch engages. Ignition switch on, engine stopped, engine temperature above 120°F and A/C control on Off.

a. 1 and G	3 volts or more	Faulty blower motor ground or defective blower motor.
	Under 3 volts	Go to step b.
b. 3 and G	10 volts or more	Go to step c.
	Under 10 volts	Go to step d.
c. Voltage at all resistor terminals	All 10 volts or more	Faulty circuit 65, splice to blower motor.
	Any under 10 volts	Stalled blower motor.

Voltmeter Connections	Voltage Reading	Condition
d. 2 and G	10 volts or more	Defective relay or leads to relay coil.
	Under 10 volts	Go to step e.
e. Check blower fuses (2 in parallel)	Both blown	Grounded circuit or defective blower motor.
	One blown or both ok	Go to step f.
f. Voltage at red wires at fuse holders to G	Both 10 volts or more	Faulty circuit 40, fuse to relay.
	Both under 10 volts	Faulty circuit 2, engine wiring harness to fuse.
	One under 10 volts	Faulty lead to that fuse holder.

2. No blower and compressor clutch does not engage. Ignition switch on.

Voltmeter Connections	Voltage Reading	Condition
a. 4 to engine ground	10 volts or more	Refer to steps 1 and 11.
	Under 10 volts	Go to step b.
b. Both sides of A/C fuse to ground	Voltage at one terminal	Replace fuse.
	Voltage at both terminals	Faulty circuit 50, fuse to splice.
	No voltage at either terminal	Faulty circuit 300 or ignition switch.

3. No blower except in Defrost. Check in-car and engine temperature. If in-car temperature is above 75°F, the in-car switch is defective. If engine temperature is above 120°F the cylinder head switch is defective. If in-car temperature is below 75°F and engine temperature is below 120°F, system is operating properly.

4. No low blower. Defective resistor.

5. Low blower only. Ignition on and engine temperature over 120°F.

Voltmeter Connections	Voltage Reading	Condition
a. Voltage at circuit 42 at control (E)	10 volts or more	Defective control or programmer.
	Under 10 volts	Faulty circuit 42.

6. Low blower only except in High or Defrost. Engine operating, engine temperature above 120°F and A/C control set at Economy.

Voltmeter Connections	Voltage Reading	Condition
a. 5 and G	10 volts or more	Faulty blower circuit in programmer.
	Under 10 volts	Go to step b.
b. Voltage at circuit 64 at control (D)	10 volts or more	Faulty circuit 64, control to programmer.
	Under 10 volts	Defective control.

7. Blower speed does not vary in Auto. With engine operating and engine temperature above 120°F, check hoses at engine and vacuum reservoir for condition and proper connection.

a. If system stays in high and max. heat, there is an opened circuit in the sensor string, proceed to step b. If system remains in high and max. cold, go to step d.

Voltmeter Connections	Voltage Reading	Condition
b. Ground circuit 62 at programmer 3 wire connector (K) and hold for 10 seconds.	No speed change	Defective programmer.
	Speed changes	Go to step c.
c. Ground circuit 198 at control (J)	No speed change	Defective temperature dial.
	Speed changes	Defective in-car sensor.
d. Disconnect circuit 62 (K) at programmer for 10 seconds	No speed change	Defective programmer.
	Speed changes	Shorted sensor or grounded sensor string, proceed to step e.
e. Reconnect circuit 62 (K) at programmer and connect a voltmeter from that wire to ground and cycle temp. dial between 65 and 85	Varies about .5 volt	Shorted in-car sensor or lead wire or sensor.
	Does not vary	Shorted temperature dial or wire from temperature dial to programmer.

8. Skipping any speed in Auto. Ignition on, engine temperature above 120°F.

Voltmeter Connections	Voltage Reading	Condition
a. Set control at Defrost and check voltage at each resistor terminal	Any under 10 volts	Defective resistor.
	All 10 volts or more	Go to step b.
b. Set control at Ecomony, blower switch wire connector removed, check voltage at each connector terminal	All 3 volts or more	Defective blower switch.
	Any under 3 volts	Faulty lead to terminal with low voltage.

9. No high blower. Faulty circuit 65.

10. Blower operates with ignition switch in off position.

Defective relay.

11. Compressor clutch does not engage, blower motor operates. Engine operating with engine temperature above 120°F.

Voltmeter Connections	Voltage Reading	Condition
a. 6 and G	10 volts or more	Defective pressure switch or compressor clutch.
	Under 10 volts	Go to step b.
b. 8 and G	10 volts or more	Defective thermostatic switch or circuit 69 to clutch.
	Under 10 volts	Go to step c.

Voltmeter Connections	Voltage Reading	Condition
c. 7 and G	10 volts or more	Low refrigerant charge, defective low refrigerant pressure switch or circuit (67 on 1977 models, 202 on 1978 models) to pressure switch.
	Under 10 volts	Go to step d.
d. Circuit 66 (A) and 50 (B) to	10 volts or more at both	Faulty circuit 66, A/C control to thermostatic switch.

Voltmeter Connections	Voltage Reading	Condition
Ground at control	10 volts or more at one	Defective A/C control.
	Under 10 volt at both	Faulty circuit 50, splice to control.

11A. Compressor clutch does not engage, blower inoperative.
Refer to step 2.

12. A/C does not operate until engine warms up.
Defective in-car sensor.

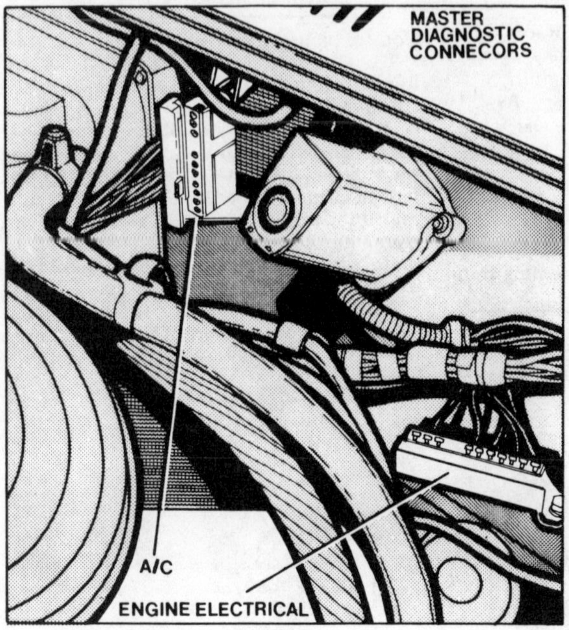

Fig. 172 A/C diagnosis connector location. 1976-78 Chevette

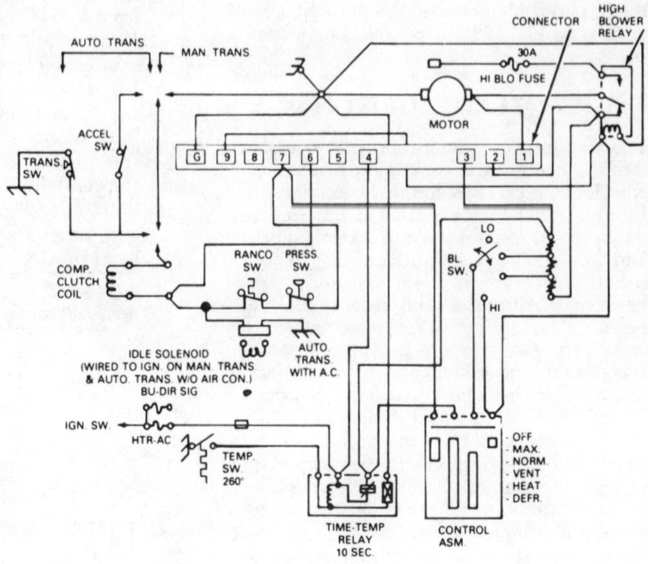

Fig. 173 A/C diagnosis connector schematic. 1976-78 Chevette

FOUR SEASON VACUUM SYSTEM DIAGNOSIS

1974-80 EXCEPT 1978-80 MALIBU, MONTE CARLO & 1980 CITATION

Start the engine and allow it to idle—move the selector lever to each position and refer to the *vacuum diagrams* for proper airflow, air door functioning and vacuum circuits. If airflow is not out of the proper outlets at each selector lever position, then proceed as follows:

Check for Good Hose Connections

At the vacuum actuators, control head valve, reservoir, tees, etc.

Check the vacuum source circuit as follows

Install vacuum tee and gauge (with restrictor) at the vacuum tank outlet if equipped. Idle the engine and read the vacuum (a normal vacuum is equivalent to manifold vacuum) at all selector lever positions. If there is no vacuum tank read manifold vacuum.

(a) *Vacuum Less Than Normal At All Positions*

Remove the tee and connect the vacuum gauge line directly to the tank or if not equipped to the feed circuit — read the vacuum. If still low, then the problem lies in the feed circuit, the feed circuit to the tank, or in the tank itself. If vacuum is now normal, then the problem lies downstream.

(b) *Vacuum Less Than Normal at Some Positions*

If vacuum was low at one or several of the selector lever positions, a leak is indicated in these circuits.

(c) *Vacuum Normal at All Positions*

If vacuum was normal and even at all positions, then the malfunction is probably caused by improperly connected or plugged lines or a defective vacuum valve or valves.

Specific Vacuum Circuit Check

Place the selector lever in the malfunctioning position and check for vacuum at the pertinent vacuum actuators. If vacuum exists at the actuator but the door does not move, then the actuator is defective or the door is mechanically bound. If low or no vacuum exists at the actuator, then the next step is to determine whether the cause is the vacuum harness or the vacuum valve. Check the vacuum harness first.

1978-80 MALIBU & MONTE CARLO

When diagnosing vacuum system problems, refer to Fig. 121 parts 1 through 5 for diagnostic procedure.

1980 CITATION

When diagnosing A/C vacuum system problems, refer to Fig. 122 parts 1 and 2 for diagnostic procedures.

Control System Functional Test

1974-77 Vega, 1975-80 Monza, 1976-79 Nova, 1976-80 Camaro, Chevelle, Chevette, Malibu, Monte Carlo, 1977-80 Chevrolet & Corvette

The control system functional test should be performed with engine at operating temperature, and ambient temperature should be between 45 and 90° F. Refer to Figs. 178 through 184 for control system operation.

1980 Citation

When performing control system functional test, refer to Fig. 128 for control system operation.

1974-76 COMFORTRON

The Comfortron System normally operates within the heating and cooling ranges that are needed for comfort considering the existing ambient and in-car temperatures. Temperature dial movement will not provide maximum heat in hot weather or maximum cooling in cool weather. With warm ambient and in-car temperatures, turning the temperature dial between 65°F and 85°F will cause the system to operate between maximum cooling and partial heating. With cool ambient and in-car temperatures, turning the temperature dial between 65°F and 85°F will cause the system to operate between partial cooling and maximum heating.

The Comfortron System must be checked for the proper functioning of the complete installation before and after all repair operations are initiated and completed, Fig. 185.

This check must be made with the engine operating at minimum of 2000 rpm and coolant warm (above 95°F). All air conditioning outlets must be open.

The recommended control operation sequence and the corresponding normal system responses are as follows:

1. With control lever in "off" position and temperature dial at 65:
   (a) Blower — low speed
   (b) Air flows out the heater outlet
2. Lever in "vent" position, dial set at 65:
   (a) Blower — low speed
   (b) Cool air flows out A/C outlets
3. Lever in "low" position, dial set at 65:
   (a) Blower — low speed
   (b) Cool air flows from A/C outlets
4. Lever in "auto" position, dial set at 65:
   (a) Blower speed will increase
   (b) Cool air flows from A/C outlets
   (c) Rotate dial to 85
   (d) Blower will decrease to low speed, then increase to medium or high
   (e) Air temperature changes from cold to hot and comes out heater outlet
5. Lever in "hi" position, dial set at 65:
   (a) Blower goes to high speed and maintains high air flow.
   (b) All air flows from A/C outlets
   (c) Rotate dial to 85
   (d) Air changes from cold to hot and shifts from A/C to the heater outlets (approx. 15 sec. required to complete shift)
6. Lever in "defog" position, dial set at 65:
   (a) Blower — medium or high speed
   (b) Air flows from defroster, heater, and A/C outlets
7. Lever in "de-ice" position, dial set at 65:
   (a) Blower — high speed
   (b) Most air comes from defroster nozzle with small amount of air coming from heater outlet.

NOTE: The only time the recirc. air door will move to inside air is when the dial is set at max. cold (65) and the lever is in "hi" after a 1.5 min. delay, providing the programmer moves to the *max. cold* position.

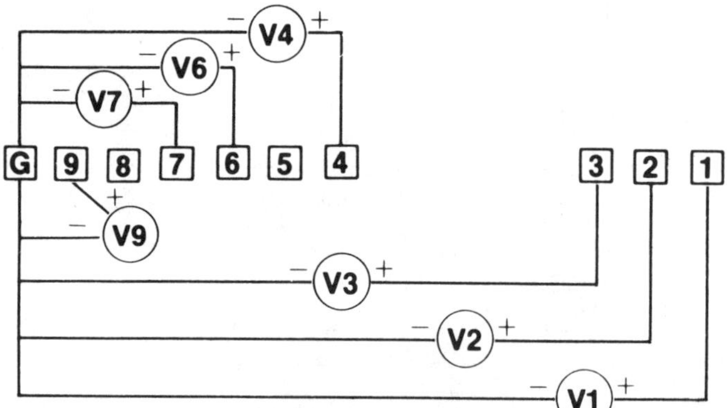

Fig. 174 A/C diagnosis connector test connections. 1976-78 Chevette

Comfortron Tester J-23678

The Comfortron System Tester J-23678 has been designed and developed to be used in troubleshooting the Comfortron System on Chevrolet and 125" W.B. Station Wagons. The tester can be used to isolate a malfunction to the control panel, the sensor string, the Comfortron vacuum system or the programmer. If a malfunction is determined to be in the programmer, the tester can also be used to completely troubleshoot each component in the programmer.

A hard rubber dummy plug, two sizes of vacuum tees and a supply of hose unions are located in the tester's storage compartment.

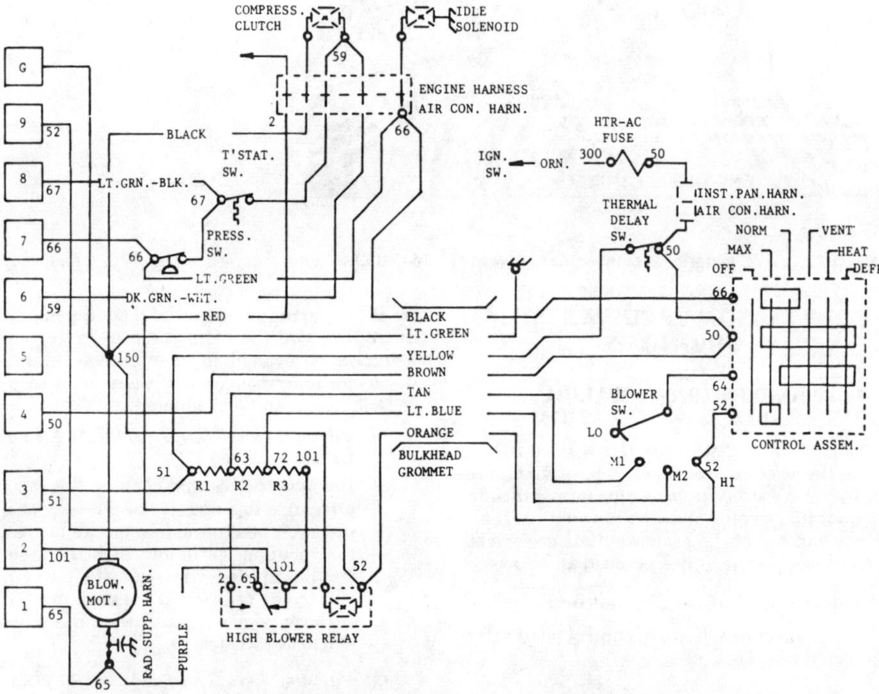

Fig. 175 A/C diagnosis connector schematic. 1977 Chevrolet full size w/Four Season A/C

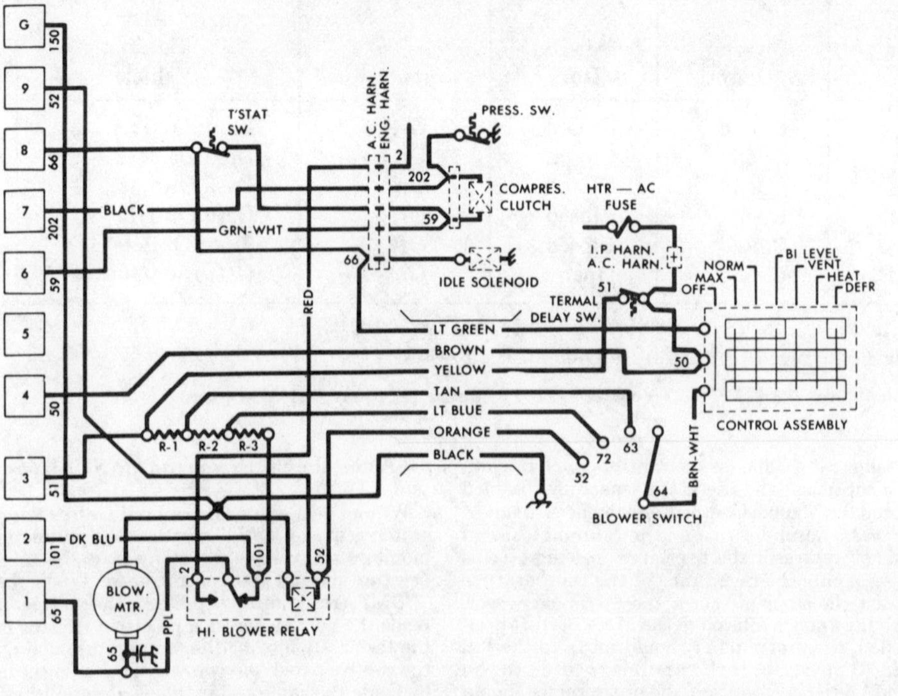

Fig. 176 A/C diagnosis connector schematic. 1978 Chevrolet full size w/Four Season A/C

The rubber dummy plug is used to plug all of the vacuum ports on the programmer except the raw vacuum input for isolating a vacuum problem between the programmer and the vacuum system. The tees and unions are used to connect the rubber hose from the tester's vacuum gauge into the system to make various vacuum checks.

The wiring harness from the tester is actually a patchcable which can be plugged into the programmer and into the car electrical harness.

The voltmeter on the tester panel has a range of 0 to 15 volts D/C. The voltage monitored by the meter is controlled by the knob just below the meter at all times except when the Temperature Dial Calibration switch is in the "CAL." (calibrate) position. The "#2 Programmer" position allows the tester to directly monitor the input voltage to the programmer. The "#6 Blower" position allows the tester to directly monitor the voltage applied to the blower motor. The "#7 LO-Relay" position indicates that the Lo-relay has energized and that the relay contacts have closed if battery voltage is read on the meter. The "#8 Auto-Relay" position indicates that the auto-relay has energized and that the relay contacts have closed if battery voltage is read on the meter. The "Probe and Clip" position connects the probe and clip

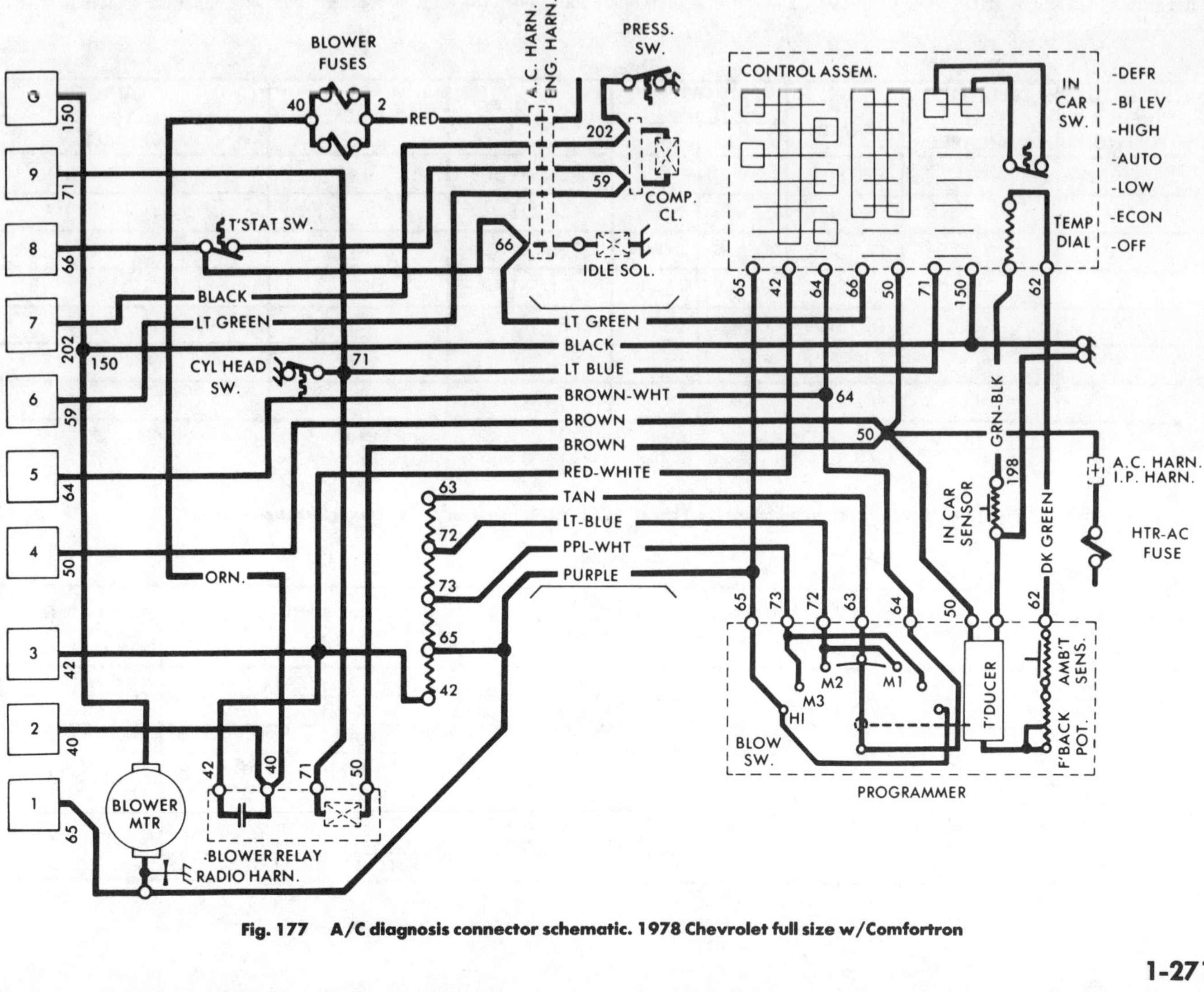

Fig. 177 A/C diagnosis connector schematic. 1978 Chevrolet full size w/Comfortron

Selector Lever Position	Blower Speeds Available	Air Source	Diverter Door	Air-Def Door	Compressor	Air Enters Vehicle
Off	All	Outside	Closed	Closed	Off	Floor Outlets
Max	Hi Only	Inside%	Open	Closed	On	Dash Outlets
Norm	All	Outside	Open	Closed	On	Dash Outlets
Vent	All	Outside	Open	Closed	Off	Dash Outlets
Htr	All	Outside	Closed	Closed	Off	Floor Outlets*
Def	All	Outside	Closed	Open	On	Defroster Outlets**

NOTE: *Some bleed-by to defroster outlets. **Some bleed-by to floor outlets.
%100% Inside air is not available, some bleed-through of outside air is allowed.

Fig. 178 Control system functional test chart. 1974-77 Vega & 1978-79 Monza "S" coupe & station wagon

in the tester's storage compartment directly to the voltmeter. This makes it possible to use the voltmeter to check various other voltages in the system. The red probe should always be connected to positive and the alligator clip lead should always be connected to the more negative terminal.

When the Manual-Automatic switch is in the AUTOMATIC position and the Temperature Dial Calibration switch is OFF, the tester monitors voltages on the voltmeter according to the voltage knob position. When the Manual-Automatic switch is in the MANUAL position, the temperature dial on the control panel, the ambient sensor and the in-car sensor are disabled and the Manual Control knob replaces them.

The numbers around the Manual Control knob represent the resistance in ohms that is replacing the resistance of the temperature dial, the ambient sensor and the in-car sensor. If the knob is rotated to the MAX. COLD position, the programmer should move to the full A/C position. If the knob is rotated to the MAX. HEAT position, the programmer should move to the maximum heat position. The Manual Control knob is operational when the Manual-Automatic switch is in the MANUAL position.

When the Temperature Dial Calibration switch is in the "CAL." position, the voltmeter monitors the voltage directly across the temperature dial on the control head. When the "COMPARE" button is pressed, the voltmeter reads the voltage across a precision resistor in the tester. By moving the temperature dial on the control panel, the two voltage readings can be made the same. After this is accomplished, the temperature dial on the control head should read the correct temperature as indicated on

SELECTOR LEVER POSITION	BLOWER SPEEDS AVAILABLE	AIR SOURCE	DIVERTER DOOR		A/C-DEF DOOR	COMPRESSOR	AIR ENTERS VEHICLE
			# 1	#2			
OFF	LO	OUTSIDE	UP	UP	OPEN TO DEF	OFF	*FLOOR OUTLETS
MAX	HI	INSIDE %	DOWN	DOWN	OPEN TO A/C	ON	DASH OUTLETS
NORM	ALL	OUTSIDE	DOWN	DOWN	OPEN TO A/C	ON	DASH OUTLETS
BI-LEVEL	ALL	OUTSIDE	DOWN	UP	OPEN TO A/C	ON	FLOOR & DASH OUTLETS
VENT	ALL	OUTSIDE	DOWN	DOWN	OPEN TO A/C	OFF	DASH OUTLETS
HTR	ALL	OUTSIDE	UP	UP	OPEN TO DEF	OFF	*FLOOR OUTLETS
DEF	ALL	OUTSIDE	DOWN	UP	OPEN TO DEF	ON	**DEFROSTER OUTLETS

NOTE: *Some bleed-by to defroster outlets. **Some bleed-by to floor outlets.
%100% Inside air is not available, some bleed-through of outside air is allowed.

Fig. 179 Control system functional test chart. 1975-79 Monza except "S" coupe & station wagon

SELECTOR LEVER POSITION	COMPRESSOR	BLOWER SPEEDS AVAIL	AIR SOURCE	AIR ENTERS VEHICLE	HEATER A/C DOOR-OPEN TO:	HEATER DEFROSTER DOOR-OPEN TO:
OFF	OFF	LOW	OUTSIDE	FLOOR OUTLETS	HEATER and DEF	HEATER
MAX A/C	ON	HI	*RECIRC AIR	DASH OUTLETS	A/C	HEATER and DEF
NORM A/C	ON	ALL	OUTSIDE	DASH OUTLETS	A/C	HEATER and DEF
BI-LEVEL	ON	ALL	OUTSIDE	FLOOR, DEF and DASH OUTLETS	A/C, HEATER and DEFROST	HEATER and DEFROST
VENT	OFF	ALL	OUTSIDE	DASH OUTLETS	A/C	HEATER
HTR	OFF	ALL	OUTSIDE	FLOOR OUTLETS	HEATER and DEF	HEATER
DEF	ON	ALL	OUTSIDE	DEFROST OUTLETS	HEATER and DEF	DEFROST

Fig. 180 Control system functional test chart. 1976-77 Camaro & Nova

SELECTOR LEVER POSITION	COMPRESSOR	BLOWER SPEEDS AVAIL	AIR SOURCE	WATER VALVE	AIR ENTERS VEHICLE	UPPER DIVERTER DOOR–OPEN TO:	LOWER DIVERTER DOOR–OPEN TO:	HEATER DEFROSTER DOOR OPEN TO:
OFF	OFF	LOW	OUTSIDE	OPEN	FLOOR OUTLET	HEATER	HEATER	HEATER
MAX A/C	ON &	HIGH	INSIDE*	CLOSED	A/C OUTLETS	A/C	A/C	HEATER
NORM A/C	ON &	ALL	OUTSIDE	OPEN	A/C OUTLETS	A/C	A/C	HEATER
BI-LEVEL	ON &	ALL	OUTSIDE	CLOSED	FLOOR - DEF and A/C OUTLETS	A/C	HEATER	HEATER and DEFROST
VENT	OFF	ALL	OUTSIDE	CLOSED	A/C OUTLETS	A/C	A/C	HEATER
HTR	OFF	ALL	OUTSIDE	CLOSED	DEFROST and FLOOR OUTLETS	HEATER	HEATER	HEATER and DEFROST
DEF	ON &	ALL	OUTSIDE	CLOSED	DEFROST OUTLETS	HEATER	HEATER	DEFROST

NOTE: * 100% inside air is not available, some bleed through of outside air is allowed.
& Providing the ambient temperature is above 37° at the ambient temperature switch

Fig. 181 Control system functional test chart. 1976-77 Chevelle & Monte Carlo Four Season A/C

the tester panel. If it does not, the temperature dial clutch should be held and the dial slipped to read the correct temperature (see Temperature Dial Calibration Procedure).

The Temperature Dial Calibration switch should always be in the OFF position unless the temperature dial on the control head is being calibrated.

On Car Diagnosis

NOTE: Do not skip any steps in the Troubleshooting Procedure unless instructed to do so.

1. Start car and place control head selector lever in the AUTO position and the temperature dial at 75°. Allow enough time for the car engine to warm up and the system to come on if the system were acting normally. If the system does not come on, skip to Step 3.
2. After the system has turned on, rotate the temperature dial slowly back and forth looking for abnormal operation of the system. Then move the selector lever to each of the various positions and look for any malfunction in system operation. By ob-

SELECTOR LEVER POSITION	BLOWER SPEEDS AVAILABLE	AIR SOURCE	HEATER DEFROSTER DOOR	A/C DOOR	AIR ENTERS VEHICLE	COMPRESSOR
OFF	ALL	OUTSIDE	OPEN	CLOSED	HEATER AND DEFROSTER OUTLETS	OFF
MAX	HI ONLY	INSIDE %	CLOSED	OPEN	DASH OUTLETS	ON &
NORM	ALL	OUTSIDE	CLOSED	OPEN	DASH OUTLETS	ON &
VENT	ALL	OUTSIDE	CLOSED	OPEN	DASH OUTLETS	OFF
HEATER	ALL	OUTSIDE	OPEN	CLOSED	HEATER OUTLETS*	OFF
DEF	ALL	OUTSIDE	OPEN (TO DEFROST ONLY)	CLOSED	DEFROSTER OUTLETS**	OFF

NOTE: *SOME BLEED-BY TO DEFROSTER OUTLETS. **SOME BLEED-BY TO FLOOR OUTLETS.
% 100% INSIDE AIR IS NOT AVAILABLE, SOME BLEED-THROUGH OF OUTSIDE AIR IS ALLOWED.
& PROVIDING THE DISCHARGE PRESSURE SWITCH PRESSURE IS ABOVE APPROX. 42 PSI.

Fig. 182 Control system functional test chart. 1976-80 Chevette & 1977 Corvette

SELECTOR LEVER POSITION	COMPRESSOR	BLOWER SPEEDS AVAIL	AIR SOURCE	AIR ENTERS VEHICLE	HEATER A/C DOOR-OPEN TO:	DEFROSTER DOOR-OPEN TO:
OFF	OFF	LOW	OUTSIDE	FLOOR OUTLETS	HEATER and DEF	
MAX A/C	ON	HI	*RECIRC AIR	DASH OUTLETS	A/C	
NORM A/C	ON	ALL	OUTSIDE	DASH OUTLETS	A/C	
BI-LEVEL	ON	ALL	OUTSIDE	FLOOR, DEF and DASH OUTLETS	A/C, HEATER and DEFROST	DEFROST
VENT	OFF	ALL	OUTSIDE	DASH OUTLETS	A/C	
HTR	OFF	ALL	OUTSIDE	FLOOR OUTLETS	HEATER and DEF	
DEF	ON	ALL	OUTSIDE	DEFROST OUTLETS	HEATER and DEF	DEFROST

Fig. 183 Control system functional test chart. 1977 Chevrolet full size Four Season A/C

A/C FUNCTIONAL TEST ENGINE WARM AND IDLING									
	CONTROL SETTINGS			SYSTEM RESPONSE					
	MODE LEVER	TEMP. LEVER	BLOWER SWITCH	BLOWER SPEED	COMP. CLUTCH	AIR VALVE	HEATER OUTLET	A/C OUTLETS	DEFROST OUTLETS
1	OFF	COLD	LO TO HI	LO ONLY	OFF	OUTSIDE AIR	AMBIENT AIR FLOW	NO AIR FLOW	SLIGHT AIR FLOW
2	MAX.	COLD	LO TO HI	SAME AS SWITCH	ON	INSIDE AIR	SLIGHT AIR FLOW	COLD AIR FLOW	NO AIR FLOW
3	NORM.	COLD	LO TO HI	SAME AS SWITCH	ON	OUTSIDE AIR	SLIGHT AIR FLOW	COLD AIR FLOW	NO AIR FLOW
5	VENT	COLD	LO TO HI	SAME AS SWITCH	OFF	OUTSIDE AIR	SLIGHT AIR FLOW	AMBIENT AIR FLOW	NO AIR FLOW
6	HEATER	HOT	LO TO HI	SAME AS SWITCH	OFF	OUTSIDE AIR	HOT AIR FLOW	NO AIR FLOW	SLIGHT AIR FLOW
7	HEATER	COLD	LO TO HI	SAME AS SWITCH	OFF	OUTSIDE AIR	AMBIENT AIR FLOW	NO AIR FLOW	SLIGHT AIR FLOW
8	DEFROST	HOT	LO TO HI	SAME AS SWITCH	ON	OUTSIDE AIR	SLIGHT AIR FLOW	NO AIR FLOW	HOT AIR FLOW

Fig. 184 Control system functional test chart. 1978-80 except Chevette, Citation & Monza Four Season

serving any malfunctions and noting when they occur, it is often possible to isolate the problem to a certain area of the system.

3. Remove the cover of the programmer (while it is still mounted in the car) and connect the J-23678 Comfortron Tester electrical harness to the programmer and the car harness, reconnect the vacuum connector.
4. Place the control head selector lever in the AUTO position and the temperature dial at 75°.
5. On the tester, place the Temperature Dial Calibration switch in the OFF position and the Manual-Automatic switch in the MANUAL position.
6. Rotate the Manual Control knob to MAX. HEAT. The programmer should move to the full heat position and the fan should run at high blower speed. (The vacuum

LEVER POSITION	DISCHARGE AIR TEMPERATURE	OUTSIDE AIR DOOR POSITION	PROGRAMMER AND TEMP. DOOR	POSITION OF MODE DOORS	DEFROSTER DOOR POSITION	COMPRESSOR CLUTCH	BLOWER RELAYS	BLOWER SPEEDS
OFF	OUTSIDE AIR TEMPERATURE OR WARMER DEPENDING UPON DIAL SETTING	OUTSIDE AIR (NO VACUUM AT DIAPHRAGM)	VARIES	HEATER (OUTSIDE AIR FROM HEATER OUTLETS-NO VACUUM AT DIAPHRAGM)	CLOSED (NO VACUUM AT EITHER PORT)	DE-ENERGIZED	LOW RELAY IS CLOSED BY THERMO SWITCHES	FIXED LOW
VENT	OUTSIDE AIR TEMPERATURE		FORCED TO MAXIMUM A/C	A/C (OUTSIDE AIR FROM A/C OUTLETS-VACUUM AT DIAPHRAGMS)	BLEED POSITION IF SELECTOR LEVER IS MOVED FROM RIGHT TO LEFT TO VENT POSITION		LOW RELAY IS CLOSED BY CONTROL SWITCH	
LO	CONTROLLED TEMPERATURE DEPENDING UPON DIAL SETTING		VARIES FROM MAXIMUM HEAT TO MAXIMUM A/C DEPENDING ON AMBIENT TEMPERATURE, IN-CAR TEMPERATURE AND DIAL SETTING	EITHER HEATER TRI-LEVEL OR A/C DEPENDING ON PROGRAMMER POSITION (CONDITIONED AIR FROM HEATER, A/C OUTLETS, OR ALL OUTLETS DEPENDING UPON DIAL SETTING AND IN-CAR TEMPERATURE. A SMALL AMOUNT WILL COME OUT DEFROSTER OUTLETS)	BLEED POSITION AND WILL DELIVER SMALL AMOUNT OF AIR TO WINDSHIELD IF IN HEATER OR TRI-LEVEL MODE (VACUUM AT OFF CENTER PORT. NO VACUUM AT CENTER PORT)	ENERGIZED IF ABOVE FREEZING	LOW RELAY IS CLOSED BY THERMO SWITCHES	
AUTO							LOW AND AUTO RELAYS ARE CLOSED IF EITHER ENGINE OR IN-CAR SWITCHES ARE CLOSED	VARIABLE BLOWER PROGRAM
HI	CONTROLLED TEMPERATURE DEPENDING UPON DIAL SETTING (WILL RECIRCULATE FOR MAXIMUM A/C WHEN REQUIRED)	RECIRCULATE AIR WHEN PROGRAMMER IS IN MAXIMUM A/C OTHERWISE OUTSIDE AIR (VACUUM AT DIAPHRAGM IN RECIRCULATE)						FIXED HIGH
BI-LEVEL				TRI-LEVEL				VARIABLE BLOWER PROGRAM
DEF				HEATER (HOT OR COLD AIR FROM DEFROSTER AND HEATER OUTLETS DEPENDING ON DIAL SETTING)	FULL OPEN TO WINDSHIELD (VACUUM AT BOTH PORTS)		LOW AND AUTO RELAYS CLOSED BY CONTROL SWITCH	FIXED HIGH

Fig. 185 Comfortron functional test chart. 1974-76 models

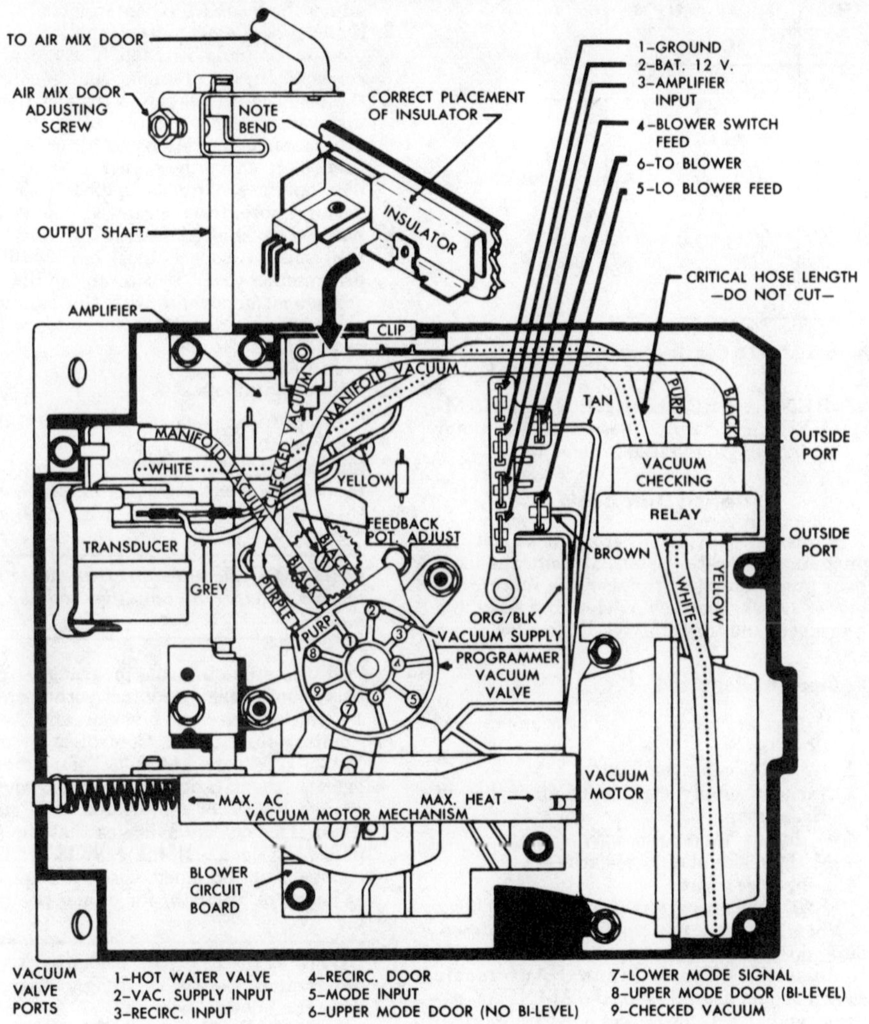

Fig. 186 Delco Radio Programmer schematic. 1974-76 Comfortron

motor mechanism will move into the vacuum motor.) Rotate the knob to MAX. COLD. The programmer should move to the full A/C position and the fan should run at high blower speed. (The vacuum motor mechanism will move out of the vacuum motor.) If only partial programmer movement occurs or "Hi blower" is not obtained at both extremes, check the air mix (temperature) door link adjustment before proceeding to the next step. If the programmer does not move at all, skip to Step 9.

7. With the tester on MAX. HEAT, rotate the Manual Control knob slowly counterclockwise. The vacuum motor mechanism should first start to move between 180 to 175 ohms on the Manual Control knob. If this first movement occurs before or after this range, make the Programmer Amplifier Calibration (Feedback "Pot" Adjustment) described under BENCH PROCEDURE-PROGRAMMER, Step #18.
8. Rotating the Manual Control knob to the MAX. COLD position should cause the system to shift to full A/C operation and the vacuum motor mechanism will move out of the vacuum motor. If the programmer moves normally when rotating the Manual Control knob, skip to Step 11.

Programmer Does Not Move

9. On the Comfortron Tester, place the Temperature Dial Calibration switch in the OFF position and the VOLTMETER control in the "PROGRAMMER 12-V SUPPLY" position. Battery voltage should appear on meter. No voltage indicates the lack of a ground on terminal #1, Fig. 186, of the programmer or the lack of battery supply to terminal #2 of the programmer.
   (*a*) If battery voltage appears on voltmeter, proceed to Step 10.
   (*b*) If no voltage appears on voltmeter, place voltmeter knob in the "Probe and Clip" position. Determine if programmer terminal #1 is ungrounded and/or terminal #2 has no battery voltage applied to it. Trace appropriate wire(s) to find malfunction in car electrical system by probing points 2, 9, 10, 11 on wiring diagram for battery voltage.
10. Plug the tester's dummy vacuum plug on the programmer vacuum valve. Connect the dummy plug hose to the vacuum supply hose (Port #2) in the car vacuum harness. Make sure vacuum is present. The programmer now has vacuum supply with no car vacuum system components connected. If the programmer functions using the MANUAL CONTROL on the tester, troubleshoot the vacuum system. If the programmer does not function, the defect is in the programmer. Remove the dummy plug and reconnect the vacuum harness after making this test. If defect is in programmer, remove the programmer and follow procedure under BENCH PROCEDURE-PROGRAMMER.

NOTE: Programmer functioning allows movement of mix door only during this test.

Sensor String Diagnosis

11. Place the control head selector lever in the AUTO position and the temperature dial on 75°.
12. On the J-23678 Tester, place the Manual-Automatic switch in the AUTOMATIC position.
13. Observe the position of the vacuum motor mechanism.
14. On the Tester, switch the Manual-Automatic switch to the MANUAL position and adjust the Manual Control knob until the vacuum motor mechanism assumes the same position it had in Step 13. When the Manual Control Knob is properly adjusted, switching the Manual-Automatic switch back and forth will result in no movement of the vacuum motor mechanism.
15. Read the setting of the MANUAL CONTROL. This resistance reading is the same as the resistance of the two sensors and the temperature dial combined and should be 120 to 150 ohms at 70° to 75°F. room temperature. If this reading is incorrect, try calibrating the temperature dial (See TEMPERATURE DIAL CALIBRATION). If the calibration can be accomplished, the temperature dial is "good". Visually check the ambient and in-car sensors for shorts or bad connections.

NOTE: If the vehicle is in a warm garage and the engine is operated for a long period of time, the ambient sensor may be affected by the heat radiated from the engine. A good ambient sensor may read as low as 100 ohms under these conditions. If a reading of 100-120 ohms is obtained, perform a temperature dial calibration test. If calibration is correct, check ambient and in-car sensor for shorts, opens and proper resistance values. Refer to Fig. 187.

To Check Sensor Resistance

(*a*) Disconnect connector, use thermometer to obtain temperature near sensor.
(*b*) Connect ohmmeter leads to sensor terminals and read resistance (Rx1) scale. Refer to Fig. 187.

Lo Relay Diagnosis

16. Place the Tester's Temperature Dial Calibration switch in the OFF position and the Voltmeter knob in the "LO RELAY" position.
17. Place the control head selector lever in the VENT or DE-ICE position. If the Lo Relay is energized, the voltmeter will read battery voltage. The Lo Relay is operated by:

TEMP.	AMBIENT SENSOR	IN-CAR SENSOR
	OHMS	OHMS
60°	36	95
65°	34	85
70°	32	76
75°	31	69
80°	29	60
85°	27	55
90°	26	50
95°	24	45
100°	23	40

Fig. 187 Sensor resistance chart. 1974-76 Comfortron

(a) Hot engine thermal switch

(b) Hot in-car temperature switch

(c) Control head in VENT or DE-ICE position.

If the relay does not energize in the control head VENT or DE-ICE position, the relay, control head switch, or electrical harness is defective. Using another voltmeter, probe #12 for battery voltage, connecting probe #13 to ground, if relay energizes the control head switch or the electrical harness is defective. If the relay does not energize the relay is defective.

Auto Relay Diagnosis

18. Place the Tester's Temperature Dial Calibration switch in the OFF position and the Voltmeter knob in the "AUTO RELAY" position.
19. Place the control selector lever in the DE-ICE position if the Auto Relay is energized, the voltmeter will read battery voltage. The Auto Relay is operated by:
    (a) Hot engine thermal switch
    (b) Hot in-car temperature switch
    (c) Control head in DE-ICE.

If the relay does not energize in control head DE-ICE position, the relay, electrical harness or control head switch is defective. With another voltmeter, probe #14 for battery voltage, connecting probe #15 to ground, if relay energizes the control head switch or the electrical harness is defective. If the relay does not energize the relay is defective.

Blower Switch Diagnosis

20. Place Tester Temperature Dial Calibration switch in the OFF position, the Voltage knob in the "BLOWER" position, and the Manual Automatic switch in the MANUAL position.
21. Using the MANUAL CONTROL, move the programmer from max. heat to the max. cold position. The voltage at the blower (coming from programmer terminal #6) should be battery voltage in max. heat. As the programmer moves from max. heat, the voltage should drop in steps, indicating different blower speeds, and then increase in steps to battery voltage in max. cold. If the voltage steps are not present or battery voltage is not indicated on the tester voltmeter when in max. heat and max. cold operation, the programmer is defective. Remove and troubleshoot the programmer (See BENCH PROCEDURE PROGRAMMER). If voltage is present blower motor or blower ground lead is defective.

Bench Diagnosis

Before attempting to repair or adjust the programmer, locate the output shaft, the feedback potentiometer, the vacuum motor, the blower switch, the vacuum checking relay, the transducer and the rotary vacuum valve.

Equipment Required

1. D/C power supply rated at 5 amperes at 12 volts.
2. J-23678 Comfortron Tester.
3. Vacuum supply capable of 20 inches of Hg. or more.
4. ¼" hollow socket and drive.
5. Medium size blade screwdriver.
6. Long-nose pliers.
7. J-23713 electrical test connector.

When following the troubleshooting procedure, do not skip any steps unless instructed to do so. The Tester Manual-Automatic switch should always be in the MANUAL position and the Temperature Dial Calibration switch should always be in the OFF position when troubleshooting the programmer on the bench.

1. Plug male connector of Tester electrical harness into programmer
2. Plug J-23713 electrical test connector into female connector of Tester electrical harness.
3. Plug dummy plug from tester onto the programmer rotary valve and connect it to the vacuum supply. Turn on the vacuum supply to 20 inches Hg. or more.
4. Connect positive lead of 12 volt D/C power supply to yellow wire from electrical test connector. Connect negative lead of power supply to black wire of test connector. Voltmeter on tester should read 12 volts when in "#2 Feed" position, "#7 Lo-Relay" position and "#8 Auto Relay" position. If it does not, check the voltage supply and hook-up.
5. On the Tester, place the Manual-Automatic switch in the MANUAL position and the Temperature Dial Calibration switch in the OFF position.
6. Rotate the Tester Manual Control knob from MAX. HEAT to MAX. COLD. (Make sure the power supply is adjusted for 12 volts D/C and vacuum supply is at 20 inches hg. or more.) The programmer output shaft should rotate 120 degrees (⅓ rotation of shaft) from maximum heat to maximum cold. If it does rotate normally, adjust the feedback potentiometer.
7. If the programmer output shaft does not operate properly in Step 6, remove the cover of the programmer and reconnect the tester electrical connector and the rotary vacuum valve dummy plug. Recheck programmer operation as described in Step 6. If the programmer output shaft now operates normally, look for pinched vacuum hose, loose electrical connection, pinched or shorted Darlington Amplifier heat sink or a mechanical bind with the programmer cover. Be sure to tap the programmer components with the handle of a screwdriver to check for intermittent problems.

Amplifier Diagnosis

8. If the programmer output shaft still does not operate properly, place the Voltage knob of the tester in the "Probe and Clip" position. Connect the alligator clip lead to the transducer terminal with the gray wire.

CAUTION: *Do not short transducer terminals together or amplifier will be damaged.*

Push the probe into the programmer connector body making contact with terminal #2 which connects to a yellow wire.

9. Rotating the Manual Control knob from MAX. HEAT to MAX. COLD (or viceversa), the voltmeter reading should change 5 volts or more (after it is stabilized). This change indicates that the amplifier is "good". If the 5 volts or more change is not obtained, check the amplifier heat sink insulator for proper positioning.

NOTE: If the heat sink is shorted to the programmer chassis, the programmer will go to the full A/C position.

If the voltage change cannot be obtained and the insulator is okay, check the calibration of the amplifier according to the instructions in Step 18. If the programmer still does not function properly, replace the amplifier circuit board and be sure to install the new heat sink insulator properly. The new amplifier must be calibrated accordingly.

Transducer Diagnosis

10. If the 5 volts change was present in the amplifier check in Step 9 but the programmer output shaft does not operate normally, leave the "Probe and Clip" connected as instructed in Step 8 throughout transducer test procedure.
11. Disconnect vacuum hose from the small diameter port of the transducer and connect Tester's hose to the vacuum gauge directly to this hose. About 20 inches Hg. or more vacuum should be present on this transducer vacuum supply hose. If vacuum is proper, restore hose connection. If 20 inches Hg. or more vacuum is not present, check the vacuum supply vacuum level and then go to Steps 14, 15, 16 and 17 to check for vacuum leaks in the vacuum checking relay and the rotary vacuum valve.
12. To check for proper regulated vacuum output from the transducer, disconnect the long hose from the transducer to the vacuum checking relay at the relay.

NOTE: This long hose must be at least 15 inches long or the transducer will make a buzzing noise. Connect the Tester's vacuum gauge directly to the long hose at the vacuum checking relay.

13. Position the programmer in an upright position so that the output shaft points straight up. With the Manual Control knob in the maximum heat position, "0"volts should be read on the voltmeter and the vacuum should be 9 to 11 inches. Turning the Manual Control knob to the MAX. COLD position should make the vacuum drop to "0" inches and the voltage should increase 5 volts or more. If these indications do not occur, the transducer is defective. After the check is completed, restore hose connection. Disconnect probe and clip from the programmer.

Rotary Vacuum Valve and Vacuum Checking Relay Diagnosis

14. Connect the tester's vacuum gauge into the vacuum feed line to the dummy plug on the programmer rotary vacuum valve. Disconnect the center hose from the side port of the rotary vacuum valve and seal off the port by placing your finger over the port. The tester's vacuum gauge should read 20 inches. Hg. or more. If it does not, either the vacuum checking relay or the rotary vacuum valve is leaky. If 20 inches Hg. or more vacuum is read, skip to Step 16 and restore the hose connection.
15. Remove all three hoses from the side ports of the rotary vacuum valve. Remove the short hose from the transducer and connect it across the two outer ports on the side of the rotary valve. Place finger over center side port of the rotary vacuum valve. If the tester's vacuum gauge now reads 20 inches Hg. or more, the vacuum checking relay is defective and leaking. After repair of programmer, restore all hose connections.

Vacuum Motor and Vacuum Checking Relay Diagnosis

16. Disconnect electrical connector from programmer. Do not disconnect vacuum input to programmer. The programmer should go to the full heat position (the vacuum motor mechanism will move into the vacuum motor). If it does not, either the vacuum checking relay or the vacuum motor has a severe leak. If it does go to full heat, a slight leak may be present and the vacuum motor will move toward maximum cold after removing the rubber dummy plug from the programmer.
17. If the vacuum motor mechanism moves, the vacuum motor or vacuum checking relay is leaking. Disconnect the short hose at vacuum motor. Apply raw vacuum to the vacuum motor input port then pinch the hose at the vacuum motor with long nose pliers and hold for 30 seconds. If the mechanism moves, the vacuum motor is defective. If it does not move, the vacuum checking relay is defective.

Feedback Potentiometer Adjustment

18. With programmer connected to tester, 12 volt supply and vacuum supply as instructed in Steps 1 thru 4 under PROGRAMMER DIAGNOSIS-OFF CAR, and with cover removed use a screwdriver to turn the feedback potentiometer fully counterclockwise. The vacuum motor will now be in the full heat position. (Mechanism moves into motor.)
    (a) With MANUAL AUTOMATIC switch in MANUAL position rotate MANUAL CONTROL to MAX. HEAT.
    (b) Carefully adjust MANUAL CONTROL to 180, do not go beyond, then back.
    (c) Turn the feedback potentiometer very slowly clockwise while watching the vacuum motor mechanism, not the output shaft. Stop immediately when movement of the mechanism occurs. Check adjustment by rotating the MANUAL CONTROL to MAX. HEAT then slowly rotating MANUAL CONTROL counterclockwise while watching the vacuum motor mechanism. When the mechanism just starts to move the MANUAL CONTROL should be at exactly 180. Slight readjustment is required if MANUAL CONTROL is not at 180.

Blower Switch Diagnosis

19. Place the Voltage knob in the "#6 Blower" position. Rotate the Manual Control knob to the MAX. HEAT position. The voltmeter reading should be 12 volts D/C (supply voltage). As the Manual Control knob is slowly rotated toward MAX COLD at about "175" the voltage will drop slightly. At about "160" the voltage will again drop slightly, and again at about "155". At about "130" the voltage will increase slightly. At about "120", the voltage will again increase slightly and again at about "110" a slight increase should be noted. If these steps in blower voltage are not present, the blower switch in the programmer is defective.
20. Tap components of the programmer while rotating the Manual Control knob and watch for any erratic operation due to intermittent problems. Replace programmer cover and rotate Manual Control knob checking for normal rotation of the output shaft.

Compressor Clutch Diagnosis

Set the control head on "auto". If the compressor ambient switch is above 40°F the compressor must operate on "lo" "auto", "hi", "de-fog", or "bi-level", and "de-ice", if the operation is erratic in some positions the control head switch is defective. If compressor does not operate at all, probe for battery voltage at points 16, 17, 18, and 19 on Comfortron wiring diagrams. Visually check compressor clutch connector and ground lead.

1977-80 COMFORTRON

Functional Test

With engine at operating temperature, run engine at 1000 rpm or higher and perform the functional test as described in Fig. 142. During the functional test, note the following:

a. Check to ensure that air flow is not coming from both A/C and heater outlets when only one mode is selected. If air delivery is from both outlets when only one mode is selected, this indicates a vacuum leak or a sticking airflow door.
b. When programmer changes mode, blower speed, air delivery mode, etc., discharge air temperature should change. If temperature does not change, check for proper air mix door to programmer link connection.
c. If neither, programmer functions nor air delivery temperature changes, an incorrect electrical or vacuum signal to the programmer is indicated or the programmer may be faulty.
d. If a malfunction of a particular vacuum door is indicated, check for a disconnected vacuum line at door actuator.
e. Using information obtained during the functional test, isolate problem to one of the following areas, temperature control problem, blower control problem, auxiliary vacuum circuit problem or refrigeration system problem and refer to the appropriate problem diagnosis listed below. After system problem has been diagnosed and corrected, perform functional check to ensure that system is now operating properly.

TEMPERATURE CONTROL PROBLEM DIAGNOSIS

The function of the temperature control circuit is to determine the temperature of air to be discharged into the passenger compartment by positioning the air mix door. If problem is poor or no heating or cooling, but the air mix door and programmer function properly, the problem is in the refrigerant circuit or heater water system. Blower cycling or mode cycling may be caused by a temperature control problem.

If a temperature control problem is noted, check for the following:

a. Disconnected sensor or temperature dial. This will interrupt the electrical signal and drive programmer to the max. heat position.
b. Improper sensor connection. An improper sensor connection will increase resistance to the sensor string and drive system to a higher temperature than desired.
c. Disconnected vacuum supply hose at vacuum checking relay. This will lock vacuum motor in a fixed position.
d. Disconnected vacuum hose in transducer-programmer vacuum motor line. This will cause system to drive to max. cooling.
e. Leak in auxiliary vacuum circuit. This may reduce the vacuum supply to the transducer below control requirement, causing an off calibration or poor heating condition.
f. A loss of vacuum supply will usually result in cold air flow from floor outlets. Diagnosis of problems in the temperature control circuit can be accomplished using tester J-23678, refer to Comfortron Programmer System Diagnosis. Symptoms and causes of temperature control are as follows:

System Operates At Max. A/C Only

With system operating, disconnect in-car sensor wire connector and observe programmer movement through hole in bottom of programmer cover. Programmer should move full in the right (max. heat position). If programmer remains at the full out to left position (max. cooling), remove vacuum connector and check for vacuum at black hose. If no or low vacuum is indicated, check for leaking or disconnected vacuum hose. If adequate vacuum supply is indicated, check voltage at programmer. If voltage is indicated at programmer, remove cover and check for disconnected wiring or vacuum hoses. If problem in programmer cannot be located, use tester J-23678 to test programmer, refer to Comfortron Programmer and System Diagnosis.

If programmer moves to the full heat position

when sensor wire connector is disconnected, the programmer is not at fault and the following areas should be checked:

a. In-car sensor shorted.
b. Miscalibrated or shorted temperature dial.
c. A shorted sensor circuit wiring harness.
d. Disconnected or open ground circuit between programmer and chassis ground.
e. No voltage supply at programmer.
f. Amplifier not grounded.

System Operates At Max. Heat Only

Position temperature control dial at 65 and control lever to any position except off. Check to see if programmer air mix door arm moves to the max. A/C position. If programmer does not move, check for the following:

a. Disconnected or defective in-car or ambient sensor.
b. Improperly connected or open circuit in sensor string wiring harness.

Place temperature control on 85 and check to see if programmer moves. If programmer does not move, check for the following:

a. Disconnected or defective temperature dial.
b. Open circuit or disconnected wire between programmer and chassis ground.
c. Defective programmer. Use tester J-23678 to test programmer.

System Changes Blower Speeds &/Or Modes During Acceleration

Remove programmer cover, then reconnect electrical and vacuum connections. With system operating, remove in-car sensor wire connector to drive programmer to the max. heat, full vacuum position. Disconnect vacuum hose connector from programmer, programmer should remain in the full heat position. If programmer does not remain in the full heat position, repeat procedure, then pinch programmer vacuum supply hose using needle nose pliers. If programmer still does not remain in the full heat position, the programmer vacuum motor is leaking. If programmer remains in the full heat position with vacuum supply hose pinched, the checking relay is defective.

c. If problem is mode changes without change in blower speed, the check valve portion of the checking relay may be defective.

Temperature Dial Setting Does Not Provide Comfort

Using tester J-23678, refer to Comfortron Programmer and System Diagnosis. To check and reset calibration on temperature dial, feedback potentiometer in programmer, sensor string and temperature door link adjustment. Also check to ensure that in-car sensor aspirator hose is connected to sensor and aspirator and is kinked.

Mode & Blower Cycling

NOTE: Some cycling of system may occur if temperature control dial is moved in large increments.

This problem may be caused by improper electrical connection, poor solder joints or a defective sensor or amplifier. If problem is intermittent, tap areas around sensor, control head, and programmer. Also shake electrical connections attached to these components. If problem cannot be located by the transducer buzzing, remove programmer cover and reconnect electrical and vacuum connectors, then connect a voltmeter to transducer output terminals. While observing voltmeter, continue to tap areas described above and shake electrical connections. If voltmeter indicates erratic voltage fluctuations, replace component or resolder connections as necessary.

Insufficient Cooling Or Heating

This condition may be caused by a temperature control or refrigerant system or heater system is at fault, proceed as follows:

a. No cooling, check compressor clutch operation. If compressor clutch is not operating, problem is in refrigerant system. Also check for cold suction line.
b. No heating, check engine coolant level. Also feel heater hoses to see if engine coolant is entering the heater core.
c. Check for proper air flow at Auto. and high settings. Improper air flow may be caused by a blower related problem, blocked air passage or disconnected air hose.
d. Check calibration of temperature control dial.
e. Check calibration of programmer feedback potentiometer.

If problems seems to be in the temperature control system, use tester J-23678 to locate and correct problem, refer to Comfortron Programmer and System Diagnosis.

Blower Speed & Mode Shifts W/O Temperature Change

Check air mix door link to programmer shaft and air door crankarm connections.

BLOWER CONTROL CIRCUIT PROBLEM DIAGNOSIS

Blower control circuit problems are divided into two categories, blower speed control and blower turn-on. Blower circuit problems and probable causes are as follows:

System Operates At Low Blower Only

1. Check for improper connection at five-way connector connected at programmer and control head.
2. Check for an open circuit in Auto. blower wiring circuit.

Blower Does Not Operate In Any Lever Position

1. Check for improperly connected or open circuit in blower wiring harness.
2. Check for a blown A/C fuse.
3. Check for stalled blower motor,
4. Check for an improperly connected five-way wire connector at control head or programmer.

No Blower In Off or Lo Positions

1. Check for a defective blower relay.
2. Open circuit in low speed resistor.
3. Open circuit in blower wiring circuit.

Blower Will Operate In Hi or Defrost Only

Check electrical connections at programmer.

Blower Speed Varies In Hi & Defrost

1. Check for defect in control head circuit board switch.
2. Open circuit in control head wiring circuit.
3. Open circuit in bypass wiring circuit.

Blower Operates With Ignition Switch In Off Position

Check for a defective blower relay.

Heater Will Not Turn On In Cold Weather Except In Defrost

1. Check for disconnected wire connector at blower delay switch located on engine cylinder head.
2. Check for defective blower delay switch by grounding switch feed wire, system should turn on. If problem does not occur with vehicle in garage, disconnect control head wire connector to open in-car sensor circuit, if blower motor turns off with engine at operating temperature, ground blower delay switch feed wire. If blower is now operating, the blower delay switch is defective.
3. If problem is still encountered, refer to wiring diagram located elsewhere in this manual and check circuit for wiring problem.

Heater Will Turn-on Immediately In Cold Weather In Any Control Setting Except Defrost

1. Check for defective (closed) blower delay switch.
2. Check for defective (closed) in-car sensor.
3. If problem is still encountered, refer to wiring diagrams located elsewhere in this manual and check circuit for wiring problem.

A/C Delayed In Warm Weather Until Engine Reaches Operating Temperature

1. Check for defective (opened) in-car switch.
2. Check for open circuit in control head wiring harness (in-car sensor circuit).
3. If problem is still encountered, refer to wiring circuit diagrams located elsewhere in this manual and check in-car sensor circuit for wiring problem.

AUXILIARY VACUUM PROBLEM DIAGNOSIS

Auxiliary vacuum circuits in the Comfortron system, are circuits which do not affect the temperature of discharge air. The auxiliary vacuum circuits are as follows:

1. Upper and lower mode door vacuum circuits on Chevrolet Full Size, heater-A/C mode door on Chevelle, Malibu and Monte Carlo.
2. Defroster bleed circuit on Chevrolet Full Size.
3. Defroster vacuum door vacuum circuit.
4. Recirculating air door vacuum circuit.
5. Heater water valve vacuum circuit.
6. Auxiliary vacuum circuit check valve.

It should be noted that a leak in one vacuum circuit may reduce the vacuum supply to another vacuum circuit.

A/C Mode Door Actuation Problem

This condition may be indicated by no A/C operation and split mode operation. The system requires vacuum to move door(s) to the A/C position. Check vacuum hose connections at mode door actuators. Split modes may occur due to problems in the door(s) circuits or because of leaks in other auxiliary vacuum circuits.

Defroster Bleed Problems (Chevrolet Full Size)

1. If there is no defroster bleed, check vacuum

CONTROL LEVER POSITION ➡	OFF	A/C	VENT	HEAT	BI-LEVEL	DEFROST
BLOWER	LOW SPEED IF WATER TEMPERATURE IS ABOVE 120^0	LOW OR SET BY BLOWER SWITCH UNLESS TEMPERATURE LEVER IN "65°" POSITION THEN OPERATES ON "HI" BLOWER	★ LOW SPEED OR SET BY BLOWER SWITCH UNLESS TEMPERATURE LEVER IS SET AT "65°". THEN OPERATES ON "HI" BLOWER. ★ DELAYED IN ALL BLOWER POSITIONS UNTIL ENGINE COOLANT REACHES 120° F.	★		
COMPRESSOR CLUTCH	DE-ENERGIZED	ENERGIZED IF ABOVE FREEZING	DE-ENERGIZED		ENERGIZED IF ABOVE FREEZING	
OUTSIDE AIR DOOR POSITION	OPEN TO OUTSIDE AIR (NO VACUUM AT DIAPHRAGM)	OPEN TO RECIRCULATE AIR WHEN IN "MAX." COLD POSITION OTHERWISE OUTSIDE AIR (VACUUM AT DIAPHRAGM FOR RECIRCULATE)	OPEN TO OUTSIDE AIR (NO VACUUM AT DIAPHRAGM)			
TEMPERATURE DOOR POSITION	VARIES WITH TEMPERATURE LEVER POSITION AND IN CAR TEMPERATURE					
UPPER MODE DOOR POSITION	HEATER POSITION (AIR OUT HEATER OUTLETS AND NO VACUUM AT DIAPHRAGM)	A/C POSITION (AIR OUT A/C OUTLETS AND VACUUM AT DIAPHRAGM)	HEATER POSITION (AIR OUT HEATER OUTLETS AND NO VACUUM AT DIAPHRAGM)			
LOWER MODE DOOR POSITION	HEATER POSITION (AIR OUT HEATER OUTLETS AND NO VACUUM AT DIAPHRAGM)	A/C POSITION (AIR OUT A/C OUTLETS AND VACUUM AT DIAPHRAGM		HEATER POSITION (AIR OUT HEATER OUTLETS AND NO VACUUM AT DIAPHRAGM)	A/C POSITION (AIR OUT A/C OUTLETS AND VACUUM AT DIAPHRAGM)	HEATER POSITION (AIR OUT HEATER OUTLETS AND NO VACUUM AT DIAPHRAGM)
DEFROSTER DOOR POSITION	CLOSED POSITION (NO VACUUM AT DIAPHRAGM)			PARTIALLY OPEN (SMALL AMOUNT OF AIR OUT DEFROSTER OUTLETS AND VACUUM AT OUTER PORT OF DIAPHRAGM)		FULL OPEN (AIR OUT DEFROSTER OUTLETS AND VACUUM AT BOTH PORTS OF DIAPHRAGM)
WATER VALVE	CLOSED ONLY AT FULL COLD TEMPERATURE LEVER POSITION (VACUUM CLOSES WATER VALVE)					

Fig. 188 Functional test chart. 1974 Cutlass, 88, 98, & Toronado

hose connection at defroster vacuum actuator. If vacuum connections are satisfactory, the porous flow plug may have to be replaced. Before installing a replacement porous plug, check for vacuum in yellow line and connect hose to actuator without the plug to check bleed door operation.

2. If rapid fogging of windshield is encountered when heater is turned on, check for improperly installed or missing porous plug.

Defroster Door Actuation Problem

This condition is indicated by no air flow from defroster outlets with selector lever in Defrost position. On Chevrolet full size models, vacuum to defroster actuator is required for defroster operation. On Chevelle, Malibu and Monte Carlo, no vacuum is required for defroster operation. Defroster door actuation problems may be caused by the following:

1. Chevrolet Full Size.
   a. Improper hose connection at defroster actuator.
   b. Check for vacuum at control head vacuum supply hose.
   c. If system remains in A/C mode with control lever in Defrost position, check vacuum circuit to determine why mode override did not occur.
2. Chevelle, Malibu and Monte Carlo.
   a. Check to ensure that defroster door is not sticking.
   b. If vacuum door is not sticking, check vacuum circuit.

Recirculating Air Door Problem

This condition is indicated by no recirculating air flow at low temperature setting. Check to ensure vacuum hose is properly connected to recirculating air door actuator. Disconnect three wire connector at programmer to drive system to recirculating air mode. If system does not go to recirculating air mode, check adjustment of rod to temperature door.

Heater Water Valve Problem

The water valve is normally open and requires vacuum to close valve to stop water flow. A leak in this vacuum circuit may result in a reduced vacuum supply to other vacuum circuits. The most probable complaint would be a split mode delivery. Refer to Auxiliary Vacuum Circuit Test.

Auxiliary Vacuum Circuit Check Valve

The check valve for the auxiliary vacuum circuit is located in the programmer checking relay assembly. A malfunction check valve will result in a loss of a specific function during acceleration or under a heavy engine load. With system operating, disconnect and plug black

CONTROL LEVER POSITION	OFF	MAX	NORM	BI-LEVEL	VENT	HEAT	DEFROST
BLOWER	LOW SPEED IF WATER TEMPERATURE IS ABOVE 120° / DELAYED	AUTOMATIC HI BLOWER	SPEED SET BY BLOWER SWITCH / DELAYED ON "LO" UNTIL COOLANT IS 120°	SPEED SET BY BLOWER SWITCH / DELAYED ON "LO" UNTIL COOLANT IS 120°	SPEED SET BY BLOWER SWITCH. / DELAYED ON "LO" UNTIL COOLANT IS 120°	BLOWER SPEEDS DELAYED IN ALL POSITIONS UNTIL COOLANT IS 120°	SPEED SET BY BLOWER SWITCH / DELAYED ON "LO" UNTIL COOLANT IS 120°
COMPRESSOR CLUTCH	DE-ENERGIZED	ENERGIZED IF ABOVE FREEZING	ENERGIZED IF ABOVE FREEZING	ENERGIZED IF ABOVE FREEZING	DE-ENERGIZED	DE-ENERGIZED	ENERGIZED IF ABOVE FREEZING
OUTSIDE AIR DOOR POSITION	OPEN TO OUTSIDE AIR (NO VACUUM AT DIAPHRAGM)	OPEN TO RECIRCULATE POSITION (VACUUM AT DIAPHRAGM FOR RECIRCULATE)	OPEN TO OUTSIDE AIR (NO VACUUM AT DIAPHRAGM)	OPEN TO OUTSIDE AIR (NO VACUUM AT DIAPHRAGM)	OPEN TO OUTSIDE AIR (NO VACUUM AT DIAPHRAGM)	OPEN TO OUTSIDE AIR (NO VACUUM AT DIAPHRAGM)	OPEN TO OUTSIDE AIR (NO VACUUM AT DIAPHRAGM)
TEMPERATURE DOOR POSITION	VARIES WITH TEMPERATURE LEVER POSITION AND IN CAR TEMPERATURE	VARIES WITH TEMPERATURE LEVER POSITION AND IN CAR TEMPERATURE	VARIES WITH TEMPERATURE LEVER POSITION AND IN CAR TEMPERATURE	VARIES WITH TEMPERATURE LEVER POSITION AND IN CAR TEMPERATURE	VARIES WITH TEMPERATURE LEVER POSITION AND IN CAR TEMPERATURE	VARIES WITH TEMPERATURE LEVER POSITION AND IN CAR TEMPERATURE	VARIES WITH TEMPERATURE LEVER POSITION AND IN CAR TEMPERATURE
UPPER MODE DOOR POSITION	HEATER POSITION (AIR OUT HEATER OUTLETS AND NO VACUUM AT DIAPHRAGM)	A/C POSITION (AIR OUT A/C OUTLETS AND VACUUM AT DIAPHRAGM)	A/C POSITION (AIR OUT A/C OUTLETS AND VACUUM AT DIAPHRAGM)	HEATER POSITION (AIR OUT HEATER OUTLETS AND NO VACUUM AT DIAPHRAGM)	A/C POSITION AIR OUT A/C OUTLETS AND VACUUM AT DIAPHRAGM)	HEATER POSITION (AIR OUT HEATER OUTLETS AND NO DIAPHRAGM)	HEATER POSITION (AIR OUT HEATER OUTLETS AND NO DIAPHRAGM)
LOWER MODE DOOR POSITION	HEATER POSITION (AIR OUT HEATER OUTLETS AND NO VACUUM AT DIAPHRAGM)	A/C POSITION (AIR OUT A/C OUTLETS AND VACUUM AT DIAPHRAGM)	A/C POSITION (AIR OUT A/C OUTLETS AND VACUUM AT DIAPHRAGM)	A/C POSITION (AIR OUT A/C OUTLETS AND VACUUM AT DIAPHRAGM)	A/C POSITION (AIR OUT A/C OUTLETS AND VACUUM AT DIAPHRAGM)	HEATER POSITION (AIR OUT HEATER OUTLETS AND NO VACUUM AT DIAPHRAGM)	HEATER POSITION (AIR OUT HEATER OUTLETS AND NO DIAPHRAGM)
DEFROSTER DOOR POSITION	CLOSED POSITION (NO VACUUM AT DIAPHRAGM)	CLOSED POSITION (NO VACUUM AT DIAPHRAGM)	CLOSED POSITION (NO VACUUM AT DIAPHRAGM)	PARTIALLY OPEN (SMALL AMOUNT OF AIR OUT DEFROSTER OUTLETS & VACUUM AT OUTER PORT OF DIAPHRAGM)	PARTIALLY OPEN VACUUM AT OUTER PORT OF DIAPHRAGM	PARTIALLY OPEN (SMALL AMOUNT OF AIR OUT DEFROSTER OUTLETS & VACUUM AT OUTER PORT OF DIAPHRAGM)	FULLY OPEN (AIR OUT DEFROSTER OUTLETS & VACUUM DIAPHRAGM)
WATER VALVE	OPEN	CLOSED (VACUUM AT DIAPHRAGM)	OPEN	OPEN	OPEN	OPEN	OPEN

Fig. 189 Functional test chart. 1975 Cutlass, 88, 98, & Toronado

SELECTOR LEVER POSITION	COMPRESSOR	BLOWER SPEEDS AVAIL	AIR SOURCE	AIR ENTERS CAR	HEATER A/C DOOR-OPEN TO:	HEATER DEFROSTER DOOR-OPEN TO:
OFF	OFF	LOW	OUTSIDE	FLOOR OUTLETS	HEATER and DEF	HEATER
MAX A/C	ON	ALL	RECIR. AIR 80%	UPPER OUTLETS	A/C	HEATER and DEF
NORM A/C	ON	ALL	OUTSIDE	UPPER OUTLETS	A/C	HEATER and DEF
BI-LEVEL	ON	ALL	OUTSIDE	FLOOR, DEF and UPPER OUTLETS	A/C, HEATER and DEFROST	HEATER and DEFROST
VENT	OFF	ALL	OUTSIDE	UPPER OUTLETS	A/C	HEATER
HTR	OFF	ALL	OUTSIDE	FLOOR OUTLETS	HEATER and DEF	HEATER
DEF	ON	ALL	OUTSIDE	DEFROST OUTLETS	HEATER and DEF	DEFROST

Fig. 190 Functional test chart. 1975-79 Omega

vacuum hose supplying vacuum to programmer checking relay. System vacuum function should hold, if not reconnect black hose and pinch purple hose. If system function now holds, the check valve is defective. If system continues to leak down, the vacuum circuit leading to the check valve is leaking.

COMPRESSOR ELECTRICAL DIAGNOSIS

Compressor electrical problems would be indicated by no cooling or insufficient cooling. Check for the following:

1. Blown A/C-heater fuse.
2. Improper electrical connection at compressor clutch or clutch coil ground.
3. Improperly connected or defective pressure cut-off switch or thermostatic switch.
4. Improperly connected control head wire connector.
5. Open circuit in wiring harness.

Comfortron Programmer & System Diagnosis

When diagnosing Comfortron programmer and system problems, use tester J-23678 and refer to Fig. 134 parts 1 and 2.

OLDSMOBILE

1974-80 FUNCTIONAL TEST

Except 1980 Omega

Refer to Figs. 188 through 196 which outline functional test results for properly operating A/C systems. Perform test with engine warm and idling and move the control levers to the various positions indicated. Door position or discharge airflow should be as outlined in chart; also note blower speed, compressor engagement and water valve operation.

After the problem has been diagnosed and repaired, repeat the functional test to insure satisfactory system performance.

1980 Omega

Refer to Fig. 128 which outlines functional test results for a properly operating A/C system.

1974-76 Custom & Tempmatic Systems Operational Test

Install air conditioning gauge set, operate the engine at 1500 rpm in Neutral. Set control on Recirculate with blower at low speed.

1. Check evaporator pressure. If pressure is 30 psi or less and outlet air temperature is too warm, check to see that the expansion valve is clamped tightly to the evaporator outlet pipe and that the insulation is in place.

If the feeler bulb installation and insulation are satisfactory, a leak in the bulb is indicated and the expansion valve should be replaced.

2. If the evaporator pressure is above 30 psi, even with blower motor wire disconnected, replace the P.O.A. valve.
3. If evaporator pressure is 29 psi ±1 psi and the outlet air temperature is normal, partially cover the condenser to obtain head pressure of 300 psi. If evaporator pressure rises above 30 psi replace the expansion valve. If evaporator pressure remains at 29 psi *during test,* there is a possibility of water or moisture in the system which could freeze under certain conditions, and a new receiver dehydrator may be required.

Performance Test

The Performance Test should be made with the car doors open, the temperature control lever fully to the left, A.C. control set at NORMAL, blower speed switch on "HI," all air outlets open, an auxiliary fan in front of the radiator, and the car hood lowered.

1. Remove Schrader valve fitting cap at the suction throttling valve or V.I.R.
2. Install Adapter J-5420 on the low pressure gauge hose, and connect the adapter to the Schrader valve fitting on the suction throttling valve or V.I.R., then momentarily open low pressure gauge valve to purge gauge hose.
3. Remove the compressor high pressure Schrader valve protective cap and install high pressure gauge hose with Adapter J-5420. Be sure high pressure gauge valve is closed.
4. Momentarily open high pressure gauge to purge the gauge and hose.
5. With transmission in PARK or NEUTRAL and parking brake applied, adjust engine speed to 2000 rpm.
6. After temperature and humidity have been determined, compare test results with the PERFORMANCE CHART, Figs. 197 through 208.
7. When test is completed, disconnect gauge hoses, and install protective caps.
8. Install Schrader valve fitting cap on suction throttling valve or V.I.R.

Tempmatic Programmer Diagnosis

Refer to charts, Fig. 209, parts 1 and 2 for Programmer Diagnosis.

1977-80 CUSTOM & TEMPMATIC SYSTEMS

Performance Test

Except 1980 Omega

Observe the following conditions before beginning performance testing:

1. Temperature lever set to full cold.
2. Selector lever set to Max A/C.
3. Blower switch set to Hi,
4. Doors and windows closed.
5. Engine speed set to 2000 rpm.
6. Fan clutch manually disengaged.
7. Auxiliary fan placed in front of grille.
8. Manifold gauge set connected.
9. Refer to Figs. 210 through 227 which outline performance test results for properly operating A/C systems.

NOTE: System pressures and outlet temperatures indicated are attained just before compressor clutch releases.

1980 Omega

Refer to Insufficient Cooling Diagnosis, Fig. 115, Parts 1 through 3, which outlines performance test results for a properly operating A/C system.

Insufficient Cooling Diagnosis

1977-78 TORONADO

1. Refer to A/C functional test and check blower motor operation. If blower is not operating properly, refer to blower motor diagnosis. If blower is operating properly, check discharge airflow by referring to A/C functional test.
2. If discharge airflow is not satisfactory, inspect air door operation and repair as necessary.
3. If discharge airflow is normal, check temperature door operation.
4. If temperature door does not operate properly:
    a. Inspect Bowden cable and temperature door for binding on CUSTOM A/C units. Adjust cable as required.
    b. On Tempmatic units, refer to Tempmatic Temperature Control Diagnosis.
5. If temperature door operates properly, inspect compressor clutch. With engine running and control on A/C, compressor clutch should engine if ambient temperature is above freezing. If compressor drive belt is broken or slipping, replace or tension as necessary.
6. If compressor does not engage, proceed to Step 13.
7. If compressor does engage, ensure compressor shaft is turning. If it is not and clutch is slipping, ensure air gap between clutch faces is approximately 1/32 inch.

CONTROL LEVER POSITION	OFF	MAX	NORM	BI-LEVEL	VENT	HEAT	DEFROST
BLOWER	LOW SPEED	AUTOMATIC HI BLOWER	SPEED SET BY BLOWER SWITCH	SPEED SET BY BLOWER SWITCH	SPEED SET BY BLOWER SWITCH	SPEED SET BY BLOWER SWITCH	SPEED SET BY BLOWER SWITCH
COMPRESSOR CLUTCH	DE-ENERGIZED	ENERGIZED IF ABOVE FREEZING	ENERGIZED IF ABOVE FREEZING	ENERGIZED IF ABOVE FREEZING	DE-ENERGIZED	DE-ENERGIZED	ENERGIZED IF ABOVE FREEZING
OUTSIDE AIR DOOR POSITION	OPEN TO OUTSIDE AIR (NO VACUUM AT DIAPHRAGM)	OPEN TO RECIRCULATE POSITION (VACUUM AT DIAPHRAGM FOR RECIRCULATE)	OPEN TO OUTSIDE AIR (NO VACUUM AT DIAPHRAGM)	OPEN TO OUTSIDE AIR (NO VACUUM AT DIAPHRAGM)	OPEN TO OUTSIDE AIR (NO VACUUM AT DIAPHRAGM)	OPEN TO OUTSIDE AIR (NO VACUUM AT DIAPHRAGM)	OPEN TO OUTSIDE AIR (NO VACUUM AT DIAPHRAGM)
TEMPERATURE DOOR POSITION	VARIES WITH TEMPERATURE LEVER POSITION AND IN CAR TEMPERATURE	VARIES WITH TEMPERATURE LEVER POSITION AND IN CAR TEMPERATURE	VARIES WITH TEMPERATURE LEVER POSITION AND IN CAR TEMPERATURE	VARIES WITH TEMPERATURE LEVER POSITION AND IN CAR TEMPERATURE	VARIES WITH TEMPERATURE LEVER POSITION AND IN CAR TEMPERATURE	VARIES WITH TEMPERATURE LEVER POSITION AND IN CAR TEMPERATURE	VARIES WITH TEMPERATURE LEVER POSITION AND IN CAR TEMPERATURE
UPPER MODE DOOR POSITION	HEATER POSITION (AIR OUT HEATER OUTLETS AND NO VACUUM AT DIAPHRAGM)	A/C POSITION (AIR OUT A/C OUTLETS AND VACUUM AT DIAPHRAGM)	A/C POSITION (AIR OUT A/C OUTLETS AND VACUUM AT DIAPHRAGM)	HEATER POSITION (AIR OUT HEATER OUTLETS AND NO VACUUM AT DIAPHRAGM)	A/C POSITION AIR OUT A/C OUTLETS AND VACUUM AT DIAPHRAGM)	HEATER POSITION (AIR OUT HEATER OUTLETS AND NO DIAPHRAGM)	HEATER POSITION (AIR OUT HEATER OUTLETS AND NO DIAPHRAGM)
LOWER MODE DOOR POSITION	HEATER POSITION (AIR OUT HEATER OUTLETS AND NO VACUUM AT DIAPHRAGM)	A/C POSITION (AIR OUT A/C OUTLETS AND VACUUM AT DIAPHRAGM)	A/C POSITION (AIR OUT A/C OUTLETS AND VACUUM AT DIAPHRAGM)	A/C POSITION (AIR OUT A/C OUTLETS AND VACUUM AT DIAPHRAGM)	A/C POSITION (AIR OUT A/C OUTLETS AND VACUUM AT DIAPHRAGM)	HEATER POSITION (AIR OUT HEATER OUTLETS AND NO VACUUM AT DIAPHRAGM)	HEATER POSITION (AIR OUT HEATER OUTLETS AND NO DIAPHRAGM)
DEFROSTER DOOR POSITION	CLOSED POSITION (NO VACUUM AT DIAPHRAGM)	CLOSED POSITION (NO VACUUM AT DIAPHRAGM)	CLOSED POSITION (NO VACUUM AT DIAPHRAGM)	PARTIALLY OPEN (SMALL AMOUNT OF AIR OUT DEFROSTER OUTLETS & VACUUM AT OUTER PORT OF DIAPHRAGM)	PARTIALLY OPEN VACUUM AT OUTER PORT OF DIAPHRAGM	PARTIALLY OPEN (SMALL AMOUNT OF AIR OUT DEFROSTER OUTLETS & VACUUM AT OUTER PORT OF DIAPHRAGM)	FULLY OPEN (AIR OUT DEFROSTER OUTLETS & VACUUM DIAPHRAGM)
WATER VALVE	OPEN	CLOSED (VACUUM AT DIAPHRAGM)	OPEN	OPEN	OPEN	OPEN	OPEN

Fig. 191 Functional test chart. 1976 Cutlass, 88, 98 & Toronado

8. If air gap is satisfactory but shaft does not turn, inspect compressor for broken reed valve or other internal damage.
9. If compressor shaft does turn and compressor is quiet, inspect under hood for oil traces near compressor and refrigerant lines.
10. If there are no leaks, proceed to Refrigeration System Diagnosis. If leaks are found, repair as required.
11. If the compressor is noisy when clutch is not engaged, remove compressor belt to determine if the noise is still present. If so, problem is not in the compressor. If the noise disappears, inspect for noisy pulley bearing or coil housing to pulley interference.
12. If the compressor is noisy only when clutch is engaged, inspect compressor mounting for loose of missing parts and ensure A/C hoses do not bind against each other or

SELECTOR LEVER POSITION	BLOWER SPEEDS AVAILABLE	AIR SOURCE	DIVERTER DOOR #1	DIVERTER DOOR #2	A/C-DEF DOOR	COMPRESSOR	AIR ENTERS VEHICLE
OFF	LO	OUTSIDE	UP	UP	OPEN TO DEF	OFF	FLOOR OUTLETS
MAX	ALL	INSIDE	DOWN	DOWN	OPEN TO A/C	ON	DASH OUTLETS
NORM	ALL	OUTSIDE	DOWN	DOWN	OPEN TO A/C	ON	DASH OUTLETS
BI-LEVEL	ALL	OUTSIDE	DOWN	UP	OPEN TO A/C	ON	FLOOR & DASH OUTLETS
VENT	ALL	OUTSIDE	DOWN	DOWN	OPEN TO A/C	OFF	DASH OUTLETS
HTR	ALL	OUTSIDE	UP	UP	OPEN TO DEF	OFF	FLOOR OUTLETS
DEF	ALL	OUTSIDE	DOWN	MID	OPEN TO DEF	ON	DEFROSTER OUTLETS

Fig. 192 Functional test chart. 1976-80 Starfire

CONTROL LEVER POSITION	OFF	MAX	NORM	BI-LEVEL	VENT	HEATER	DEFROST
BLOWER SPEED	LOW	AS SELECTED BY SWITCH					
COMPRESSOR CLUTCH	DE-ENERGIZED	ON ABOVE FREEZING			OFF		ON ABOVE FREEZING
OUTSIDE AIR DOOR	OPEN-NO	INSIDE AIR	OPEN TO OUTSIDE AIR - NO VACUUM				
TEMPERATURE DOOR POSITION	SELECTED BY TEMP. LEVER AND PROGRAMMER						
UPPER MODE DOOR POSITION	OPEN TO HEATER	OPEN TO AC OUTLETS		OPEN TO HEATER	OPEN TO AC OUTLET	OPEN TO HEATER OUTLET	
LOWER MODE DOOR POSITION	OPEN TO HEATER	OPEN TO AC OUTLETS				OPEN TO HEATER OUTLET	
DEFROSTER DOOR	CLOSED	CLOSED		PARTIALLY OPEN		FULL OPEN	
WATER VALVE	OPEN	CLOSED	OPEN				

Fig. 193 Functional test chart. 1977 Cutlass & 1977-78 Toronado

against other components. If noise still exists, compressor must be removed for repair.

13. If compressor does not engage, inspect heater-A/C fuse; repair short circuit if fuse is blown.
14. If the fuse is not blown, turn ignition switch On, set control to A/C and ensure ambient temperature is above freezing. Connect a test light to ground and to green wire terminal at compressor.
15. If the test light is on, inspect terminal connection and clutch ground to engine. If satisfactory, replace clutch coil.
16. If the test light is off, touch test light to brown wire at center terminal of thermal limiter. If the test light goes on, proceed to step 18.
17. If the test light remains off, touch test light to both terminals of ambient switch.
    a. If test light is on at both terminals, repair open circuit in brown wire, ambient switch to thermal limiter.
    b. If test light is on at one terminal, replace ambient switch.
    c. If test light is off at both terminals, probe green wire at control. If test light goes on, repair open circuit in green wire from control to ambient switch. If test light remains off repair control as required.
18. If test lamp in step 16 goes on, probe both outside terminals of the thermal limiter.
19. If test light is on at both terminals, repair open circuit in green wire from control to ambient switch.
20. If test light remains off at both terminals, thermal limiter fuse is blown. Disconnect thermal limiter and insert a jumper wire across center terminal and clutch terminal (B and C) of connector.
21. If compressor clutch engages, proceed to step 24.
22. If compressor clutch does not engage, use a jumper wire from positive terminal of battery to compressor clutch. If compressor still does not engage and compressor is properly grounded, remove and repair clutch.
23. If jumper wire from battery causes compressor clutch to engage, check heater-A/C fuse, open circuit at control switch and ambient switch, and repair or replace as necessary.
24. If compressor clutch engaged in step 21, use an ammeter to test compressor clutch coil current draw. If current draw is more than 4 amps, replace clutch coil. If current draw is less than 4 amps, connect a manifold gauge set and proceed to Refrigeration System Diagnosis.

Refrigeration System Diagnosis

1. Inspect the liquid line at the V.I.R. inlet; it should be warm (120 degrees F. or more). Then inspect the moisture indicator in the sight glass. If the indicator is blue, indicating a dry system, proceed to step 3. If it is pink, indicating some moisture in the system, or white, indicating a very wet system, proceed to the next step.
2. If desiccant must be replaced if the refrigerant system has developed moisture. The system must be evacuated and recharged. Then recheck the moisture indicator. If it is blue, the system is moisture-free. If it is pink, return the vehicle to service for 1 or 2 days with the A/C in operation; recheck the moisture indicator. If it is not blue, the system must again be evacuated and recharged.
3. Check the evaporator outlet pipe temperature, if it is cold, proceed to step 5. If it is warm, disconnect the blower and check the evaporator pressure. If vacuum exists at the evaporator, the system is discharged. Repair leak(s) as necessary and recharge. Refer to Performance Test chart; if evaporator pressure is abnormally high, replace POA valve. If evaporator pressure is abnormally low, a restriction in the high pressure side of the system is indicated. A temperature difference will be noted on either side of the restriction. Replace defective component. If a restriction is not noted, remove V.I.R. receiver shell and clean pick-up tube assembly screen. Evacuate and recharge system. If evaporator pressure is normal, proceed to next step.

	CONTROL SETTINGS			SYSTEM RESPONSE					
	MODE LEVER	TEMP. LEVER	BLOWER SWITCH	BLOWER SPEED	COMP. CLUTCH	AIR VALVE	HEATER OUTLET	A/C OUTLETS	DEFROST OUTLETS
1	OFF	COLD	LO TO HI	LO ONLY	OFF	OUTSIDE AIR	AMBIENT AIR FLOW	NO AIR FLOW	SLIGHT AIR FLOW
2	MAX.	COLD	LO TO HI	SAME AS SWITCH	ON	INSIDE AIR	SLIGHT AIR FLOW	COLD AIR FLOW	NO AIR FLOW
3	NORM.	COLD	LO TO HI	SAME AS SWITCH	ON	OUTSIDE AIR	SLIGHT AIR FLOW	COLD AIR FLOW	NO AIR FLOW
4	BI-LEVEL	COLD	LO TO HI	SAME AS SWITCH	ON	OUTSIDE AIR	COLD AIR FLOW	COLD AIR FLOW	NO AIR FLOW
5	VENT	COLD	LO TO HI	SAME AS SWITCH	OFF	OUTSIDE AIR	SLIGHT AIR FLOW	AMBIENT AIR FLOW	NO AIR FLOW
6	HEATER	HOT	LO TO HI	SAME AS SWITCH	OFF	OUTSIDE AIR	HOT AIR FLOW	NO AIR FLOW	SLIGHT AIR FLOW
7	HEATER	COLD	LO TO HI	SAME AS SWITCH	OFF	OUTSIDE AIR	AMBIENT AIR FLOW	NO AIR FLOW	SLIGHT AIR FLOW
8	DEFROST	HOT	LO TO HI	SAME AS SWITCH	ON	OUTSIDE AIR	SLIGHT AIR FLOW	NO AIR FLOW	HOT AIR FLOW

Fig. 194 Functional test chart. 1978-80 Cutlass

CONTROL SETTINGS		AIR FLOW					
SELECTOR LEVER	TEMP. LEVER	BLOWER SPEED	COMPRESSOR	AIR VALVE	HEATER OUTLETS	A/C OUTLETS	DEFROSTER OUTLETS
OFF	COLD	LOW	OFF	OUTSIDE AIR	AIR FLOW	NO AIR FLOW	MINIMUM AIR FLOW
MAX	COLD	HIGH	ON	INSIDE AIR	NO AIR FLOW	COLD AIR FLOW	NO AIR FLOW
NORM	COLD	LOW	ON	OUTSIDE AIR	SLIGHT AIR FLOW	COLD AIR FLOW	NO AIR FLOW
BI-LEVEL	COLD	HIGH	ON	OUTSIDE AIR	AIR FLOW	COLD AIR FLOW	MINIMUM AIR FLOW
VENT	COLD	HIGH	OFF	OUTSIDE AIR	SLIGHT AIR FLOW	AMBIENT AIR	NO AIR FLOW
HEATER	HOT	HIGH	OFF	OUTSIDE AIR	HOT AIR FLOW	NO AIR FLOW	MINIMUM AIR FLOW
DEF	HOT	HIGH	ON	OUTSIDE AIR	MINIMUM AIR FLOW	NO AIR FLOW	MAXIMUM AIR FLOW

Fig. 195 Functional test chart. 1977-80 88 & 98

4. Remove expansion valve and check by blowing on the end of the valve. If air comes out of the small hole only, replace expansion valve. If air comes out of the small and large hole, remove and clean pick-up tube assembly screen. Install new desiccant bag and reinstall original expansion valve with new moisture indicator. Evacuate and recharge system.
5. If evaporator outlet pipe temperature was found to be cold in step 3, disconnect blower and check evaporator pressure. Referring to the Performance Test chart, if the pressure goes below normal immediately, inspect POA valve for binding or dirt in the screen. Clean or replace as necessary. Evacuate and recharge system, after installing a new desiccant bag. If the evapora-

CONTROL SETTINGS		AIR FLOW					
SELECTOR LEVER	TEMP. LEVER	BLOWER SPEED	COMPRESSOR	AIR VALVE	HEATER OUTLETS	A/C OUTLETS	DEFROSTER OUTLETS
OFF	COLD	LOW	OFF	OUTSIDE AIR	AIR FLOW	NO AIR FLOW	MINIMUM AIR FLOW
MAX	COLD	HIGH	ON	INSIDE AIR	NO AIR FLOW	COLD AIR FLOW	NO AIR FLOW
NORM	COLD	LOW	ON	OUTSIDE AIR	NO AIR FLOW	COLD AIR FLOW	NO AIR FLOW
BI-LEVEL	COLD	HIGH	ON	OUTSIDE AIR	COLD AIR FLOW	COLD AIR FLOW	MINIMUM AIR FLOW
VENT	COLD	HIGH	OFF	OUTSIDE AIR	NO AIR FLOW	AMBIENT AIR	NO AIR FLOW
HEATER	HOT	HIGH	OFF	OUTSIDE AIR	HOT AIR FLOW	NO AIR FLOW	MINIMUM AIR FLOW
DEF	HOT	HIGH	ON	OUTSIDE AIR	MINIMUM AIR FLOW	NO AIR FLOW	MAXIMUM AIR FLOW

Fig. 196 Functional test chart. 1979-80 Toronado

IN FRONT OF CONDENSER RELATIVE* HUMIDITY	IN FRONT OF CONDENSER AIR TEMP. °F.	EVAPORATOR PRESSURE AT SUCTION THROTTLING VALVE ± 1 PSI	ENGINE RPM	DISCHARGE AIR TEMP. LEFT CENTER OUTLET ± 2/F.	HEAD PRESSURE (DISCHARGE) ± 20 PSI
20	70	29.0	2000	42	200
	80	29.0		45	242
	90	29.5		49	282
	100	36.5		58	336
30	70	29.0	2000	42	203
	80	29.0		46	244
	90	31.0		52	291
	100	39.0		61	342
40	70	29.0	2000	43	206
	80	29.0		48	247
	90	33.0		55	296
	100	43.0		65	358
50	70	29.0	2000	44	212
	80	29.0		49	259
	90	36.0		58	307
	100	46.5		69	372
60	70	29.0	2000	45	220
	80	30.0		51	258
	90	39.0		61	315
	100	50.5		74	389
70	70	29.0	2000	45	224
	80	37.0		54	266
	90	42.0		65	333
	100	54.0		77	389
80	70	29.0	2000	47	224
	80	34.0		56	273
	90	45.0		69	342
	100	57.5		80	408
90	70	29.0	2000	48	231
	80	36.1		59	282
	90	47.0		71	345

Fig. 197 Performance test chart. 1974 Omega

IN FRONT OF CONDENSER RELATIVE HUMIDITY	IN FRONT OF CONDENSER AIR TEMP °F	EVAPORATOR PRESSURE AT V.I.R. GAGE CONNECTION ±1 PSI	ENGINE RPM	DISCHARGE AIR TEMP. R.H. NOZZLE ±2°F.	HEAD PRESSURE (DISCHARGE) ±20 PSI
20	70	29.5	2000	40	210
	80	29.5		43	250
	90	30.0		45	280
	100	30.0		53	325
30	70	29.5	2000	40	210
	80	29.5		43	260
	90	30.0		48	300
	100	31.0		60	335
40	70	29.5	2000	41	220
	80	29.5		44	270
	90	30.0		51	305
	100	32.0		63	355
50	70	29.5	2000	42	235
	80	30.0		46	280
	90	30.0		57	320
	100	33.0		67	375
60	70	29.5	2000	42	245
	80	30.0		48	300
	90	30.0		58	325
	100	38.0		70	380
70	70	30.0	2000	44	260
	80	30.0		50	310
	90	32.0		63	330
	100	42.0		74	395
80	70	30.0	2000	45	275
	80	30.0		54	320
	90	32.0		66	340
	100	44.0		77	400
90	70	30.0	2000	46	285
	80	31.0		56	320
	90	33.0		69	350
	100	46.0		80	404

Fig. 198 Performance test chart. 1974 Cutlass

tor pressure was found to be normal, proceed to next step.

6. Ensure superheat switch is correctly wired, correct if necessary. If wiring is in order, check for defective superheat switch as outlined elsewhere in this chapter.
7. If evaporator pressure was found to be normal to low, leak test system. Repair leaks as necessary. If no leaks are found, check for restriction as outlined in step 3. If no restriction is found, install a partial charge and inspect sight glass. Sight glass should be clear, if not, recheck for restriction in system.

1977-78 EXCEPT TORONADO & 1979-80 ALL EXCEPT 1980 OMEGA

Models W/Thermostatic Switch

1. Before beginning diagnostic procedure, make the following preliminary checks:
   a. Blower operation.
   b. Damaged or missing evaporator drain hose.
   c. Functional test of system.
   d. Proper sealing of evaporator or module case.
2. With engine running, switch A/C on and observe compressor clutch operation. If compressor engages or cycles, proceed to step 4. If compressor does not engage, use a test light to check for current at compressor feed wire. If there is no current, proceed to step 3. If there is current at compressor feed wire, disconnect wire at low pressure cut-off switch (except Omega models) and ground that wire. On Omega models, round black wire at clutch coil connection. If clutch is still not engaged, remove and repair clutch.
3. If there is no current at compressor feed wire, check for blown main fuse and repair short circuits as necessary. If fuse is not

IN FRONT OF CONDENSER RELATIVE HUMIDITY	IN FRONT OF CONDENSER AIR TEMP. °F.	EVAPORATOR PRESSURE AT SUCTION THROTTLING VALVE ± 1 PSI	ENGINE RPM	DISCHARGE AIR TEMP. LEFT CENTER OUTLET ± 2°F.	PRESSURE HIGH (DISCHARGE) ± 20 PSI
30	70	29.5	2000	39	200
	80	29.5		42	220
	90	30.0		48	260
	100	32.0		55	300
40	70	29.5	2000	39	205
	80	29.5		45	222
	90	31.0		51	260
	100	33.0		59	305
50	70	29.5	2000	43	210
	80	30.0		46	235
	90	31.0		55	295
	100	34.0		64	330
60	70	29.5	2000	45	215
	80	30.0		51	255
	90	36.0		60	305
	100	40.0		67	350
70	70	30.0	2000	45	235
	80	33.0		56	260
	90	35.0		67	360
80	70	30.0	2000	45	240
	80	31.0		55	290
	90	37.0		70	380
90	70	30.0	2000	50	250
	80	32.0		62	340
	90	38.0		74	400

Fig. 199 Performance test chart. 1974 88 & 98

IN FRONT OF CONDENSER RELATIVE′ HUMIDITY	IN FRONT OF CONDENSER AIR TEMP. °F.	EVAPORATOR PRESSURE AT V.I.R. GAGE CONNECTION ± 1 PSI	ENGINE RPM	DISCHARGE LEFT CENTER OUTLET ± 2°F.	PRESSURE HIGH (DISCHARGE) ± 20 PSI
20	70	29.5	2000	41	195
	80	29.5		43	220
	90	29.5		44	265
	100	30.0		52	325
30	70	29.5	2000	41	200
	80	29.5		44	230
	90	29.5		47	280
	100	31.0		56	340
40	70	29.5	2000	42	210
	80	29.5		45	240
	90	29.5		51	295
	100	33.0		60	350
50	70	29.5	2000	42	210
	80	29.5		46	250
	90	30.5		55	310
	100	36.0		62	375
60	70	29.5	2000	43	220
	80	29.5		47	260
	90	32.0		57	320
	100	38.0		66	390
70	70	29.5	2000	43	225
	80	30.0		50	270
	90	33.0		59	330
80	70	29.5	2000	44	230
	80	30.5		52	285
	90	35.0		63	345
90	70	29.5	2000	45	237
	80	31.0		55	295
	90	37.5		66	362

Fig. 200 Performance test chart. 1974 Toronado

Ambient Air in Degrees F. (In Auxiliary Fan Air Blast Ahead of Condenser)	70°		80°		90°		100°		110°	
Air Quality	Arid	Humid	Arid	Humid	Arid	Humid	Arid	Humid	Arid	Humid
Average Compressor Head Pressure in PSI	145	185	185	225	205	245	240	270	260	310
Average Evaporator Pressure PSI AT SEA LEVEL	29 to 32									
Center Outlet Temperature Degrees F.	42	44	44	47	45	48	47	51	53	55

Fig. 201 Performance test chart. 1975 Starfire

Ambient Air in Degrees F. (In Auxiliary Fan Air Blast Ahead of Condenser)	70°		80°		90°		100°		110°	
Air Quality	Arid	Humid	Arid	Humid	Arid	Humid	Arid	Humid	Arid	Humid
Average Compressor Head Pressure in PSI	140	180	180	220	200	240	230	260	250	300
Average Evaporator Pressure PSI AT SEA LEVEL	28 to 31									
Center Outlet Temperature Degrees F.	40	42	42	44	44	47	45	48	47	51

Fig. 202 Performance test chart. 1975-76 Omega w/4 cyl. radial compressor

blown check for open circuits in the thermostatic switch and the compressor pressure switch. If both switches are operational, check for broken wire or loose connection in compressor circuit; repair as necessary. If thermostatic switch is faulty, it must be replaced. Check the compressor pressure switch by observing the refrigerant pressure at the high side fitting. If high side pressure is below 22 psi, perform leak test and repair leak(s). Evacuate and recharge system. If high side pressure is above 22 psi, connect a jumper wire across pressure switch. If the compressure runs, replace switch, then evacuate and recharge system.

4. If compressor engages or cycles, feel the liquid line ahead of expansion tube (orifice). If the liquid line is cold, check for a restriction in the high side of system. Perform repairs as necessary, then evacuate and recharge system. If liquid line feels warm, feel evaporator inlet pipe near thermostatic switch capillary tube connection and accumulator can surface. If the inlet pipe is colder than accumulator surface, leak test system and repair leak(s) as necessary. If no leaks are found, check expansion tube (orifice) for plugged filter. Repair, evacuate and recharge. Recheck inlet pipe and accumulator surface temperatures. If normal system operation is not indicated by both being the same cold temperature or accumulator being somewhat cooler than inlet pipe, expansion tube must be replaced. Again evacuate and recharge system.

IN FRONT OF CONDENSER RELATIVE HUMIDITY	IN FRONT OF CONDENSER AIR TEMP. °F.	EVAPORATOR PRESSURE AT V.I.R. GAUGE CONNECTION	ENGINE RPM	DISCHARGE AIR TEMP. LEFT CENTER OUTLET ± 2°F.	PRESSURE HIGH (DISCHARGE) ± 20 PSI
20	70	29.5	2000	44	203
	80	29.5		46	248
	90	29.5		49	300
	100	34.0		55	350
30	70	29.5	2000	44	206
	80	29.5		47	250
	90	32.0		52	309
	100	35.0		57	360
40	70	29.5	2000	45	210
	80	29.5		49	255
	90	32.0		55	314
	100	37.0		61	372
50	70	29.5	2000	46	215
	80	29.5		50	265
	90	35.0		58	325
	100	40.0		65	385
60	70	29.5	2000	48	223
	80	29.5		53	269
	90	37.0		61	345
	100	45.0		68	400
70	70	29.5	2000	48	228
	80	37.0		57	275
	90	42.0		65	353
	100	49.0		70	408
80	70	29.5	2000	49	230
	80	34.0		59	280
	90	45.0		69	360
	100	52.0		72	420
90	70	29.5	2000	50	234
	80	35.0		61	300
	90	44.0		71	360

Fig. 203 Performance test chart. 1975-76 Omega w/6 cyl. compressor

Ambient Air in Degrees F. (In Auxiliary Fan Air Blast Ahead of Condenser)	70°		80°		90°		100°		110°	
Air Quality	Arid	Humid	Arid	Humid	Arid	Humid	Arid	Humid	Arid	Humid
Average Compressor Head Pressure in PSI	145	185	185	225	205	245	240	270	260	310
Average Evaporator Pressure PSI AT SEA LEVEL	28 to 31									
Center Outlet Temperature in Degrees F.	40	42	42	44	44	47	45	48	47	51

Fig. 204 Performance test chart. 1975-76 Cutlass w/4 cyl. radial compressor

5. Insert a thermometer in the left center instrument panel outlet and install manifold gauge set. Refer to Performance Testing and observe conditions outlined, then compare performance test results with values given for properly operating system. If outlet temperature is satisfactory or too cold, change blower speed to low and observe accumulator and evaporator pipes for frost. If frost is present, replace thermostatic switch.
6. If outlet temperature is too high, check accumulator pressure. If pressure is too low, inspect for plugged expansion tube (orifice) and repair or replace as necessary. Then replace accumulator assembly, evacuate and recharge system. If accumulator pressure is high to normal, check compressor cycling. If the compressor cycles on and off more than 4 times per minute, replace the thermostatic switch. If the compressor runs continuously, ensure expansion tube is not missing. If it is, install new expansion tube (orifice) and evacuate and recharge system. If expansion tube is not missing, inspect compressor inlet screen on 6 cylinder compressors. If plugged, repair or replace screen, replace accumulator assembly then evacuate and recharge system. If compressor inlet screen is clean, ensure system is not overcharged. If so, evacuate and recharge.

Models w/Pressure Cycling Switch

Refer to Fig. 228, Parts 1 and 2, for diagnosis.

Tempmatic Visual Inspection

Often a visual inspection of system components will reveal the cause of a specific malfunction. Before beginning diagnostic procedure, inspect the following:

1. Vacuum hose routing and connection.
2. Electrical harness connections.
3. In-car air hose connections at programmer and instrument panel grille. Ensure hose is not kinked.
4. Module case to programmer drive air seal.
5. Temperature control adjusment and cable connection.
6. Compressor clutch for proper engagement.
7. Compressor drive belt for proper tension.
8. All programmer linkages for binding or loose connections.
9. Sight glass for "clear" condition. (1977-78 Toronado only.)
10. Moisture indicator for blue color. (1977-78 Toronado only.)

Tempmatic Temperature Control Diagnosis

Temperature control malfunctions may be indicated by warm airflow only, cold airflow only, or poor regulation through the entire range of outlet air temperature. Proceed as follows to diagnose problems in temperature control:

1. Remove glove box to gain access to programmer and temperature door threaded rod.
2. With engine running and at operating temperature, move the control panel lever to NORM and the fan switch to HI.
3. Move the temperature lever to full cold, then to full hot and then back to full cold.

IN FRONT OF CONDENSER: RELATIVE HUMIDITY	IN FRONT OF CONDENSER: AIR TEMP. °F.	EVAPORATOR PRESSURE AT E.E.V.I.R. GAUGE CONNECTION ± 1 PSI	ENGINE RPM	DISCHARGE AIR TEMP. R.H. NOZZLE ± 2°F.	PRESSURE HIGH (DISCHARGE) ± 20 PSI
20	70	29.5	2000	40	150
	80	29.5		44	190
	90	30.0		48	245
	100	31.0		57	305
30	70	29.5	2000	42	150
	80	30.0		47	205
	90	31.0		51	265
	100	32.0		61	325
40	70	29.5	2000	45	165
	80	30.0		49	215
	90	32.0		55	280
	100	39.0		65	345
50	70	30.0	2000	47	180
	80	32.0		53	235
	90	34.0		59	295
	100	40.0		69	350
60	70	30.0	2000	48	180
	80	33.0		56	240
	90	36.0		63	300
	100	43.0		73	360
70	70	30.0	2000	50	185
	80	34.0		58	245
	90	38.0		65	305
	100	44.0		75	365
80	70	30.0	2000	50	190
	80	34.0		59	250
	90	39.0		67	310
90	70	30.0	2000	50	200
	80	36.0		62	265
	90	42.0		71	330

Fig. 205 Performance test chart. 1975-76 Cutlass w/6 cyl. compressor

IN FRONT OF CONDENSER: RELATIVE HUMIDITY	IN FRONT OF CONDENSER: AIR TEMP. °F.	EVAPORATOR PRESSURE AT E.E.V.I.R. GAUGE CONNECTION ± 1 PSI	ENGINE RPM	DISCHARGE AIR TEMP. LEFT CENTER OUTLET ± 2°F.	PRESSURE HIGH (DISCHARGE) ± 20 PSI
30	70	29.5	2000	42	190
	80	30.5		48	230
	90	32.5		54	295
	100	39.0		64	370
40	70	29.5	2000	45	175
	80	31.5		52	240
	90	34.0		59	310
	100	42.0		69	380
50	70	29.5	2000	46	185
	80	32.5		54	250
	90	36.0		62	320
	100	42.5		71	385
60	70	29.5	2000	47	190
	80	33.5		56	250
	90	38.0		64	325
	100	43.5		72	390
70	70	29.5	2000	48	200
	80	34.5		57	265
	90	40.0		67	335
80	70	29.5	2000	49	210
	80	35.0		58	280
	90	43.0		71	350
90	70	29.5	2000	50	220
	80	36.5		59	290
	90	44.0		72	360

Fig. 206 Performance test chart. 1975-76 88 & 98

IN FRONT OF CONDENSER: RELATIVE HUMIDITY	IN FRONT OF CONDENSER: AIR TEMP. °F. ± 1 PSI	EVAPORATOR PRESSURE AT E.E.V.I.R. GAUGE CONNECTION	ENGINE RPM	DISCHARGE AIR TEMP. LEFT CENTER OUTLET ± 2°F.	PRESSURE HIGH (DISCHARGE) ± 20 PSI
20	70	29.5	2000	42	195
	80	29.5		44	240
	90	30.0		46	290
	100	32.0		54	335
30	70	29.5	2000	43	200
	80	29.5		47	250
	90	30.0		50	300
	100	35.0		59	355
40	70	29.5	2000	44	210
	80	30.0		47	260
	90	33.0		54	315
	100	39.0		64	375
50	70	29.5	2000	44	210
	80	30.0		49	265
	90	35.0		58	325
	100	42.0		67	380
60	70	29.5	2000	45	215
	80	30.0		51	275
	90	37.0		61	330
	100	45.0		72	400
70	70	29.5	2000	46	225
	80	32.0		53	280
	90	39.0		64	335
80	70	29.5	2000	47	230
	80	34.0		56	295
	90	41.0		67	350
90	70	29.5	2000	48	235
	80	36.0		58	300
	90	43.0		70	365

Fig. 207 Performance test chart. 1975-76 Toronado

PERFORMANCE TEST CONDITIONS

Hood	Raised	Fan Switch	HI
Front Doors	Closed	Temperature Lever	Full Cold
Front Windows	Open	Nozzles and Air Outlets	Open
Select Lever	Max. (Recirc.)	Engine Speed	2000 RPM

TEST READINGS: Ambient Air in Degrees Fahrenheit / Celsius (In Auxiliary Fan Air Blast Ahead of Condenser)	70° / 21°		80° / 27°		90° / 32°		100° / 38°		110° / 43°	
Air Quality	Arid	Humid	Arid	Humid	Arid	Humid	Arid	Humid	Arid	Humid
Average Compressor Head Pressure in PSI	145	185	185	225	205	245	240	270	260	310
Average Evaporator Pressure PSI AT SEA LEVEL	29 to 32									
Right Hand Outlet Temperature Fahrenheit / Celsius	49° / 9°	53° / 12°	48° / 9°	52° / 11°	44° / 7°	48° / 9°	48° / 9°	52° / 11°	48° / 9°	52° / 11°

Fig. 208 Performance test chart. 1976 Starfire

Viewed through the glove box opening, the programmer link and temperature door threaded rod should move. Air outlet temperature should decrease as lever is moved to full cold, and increase as it is moved to full hot.

4. If air outlet temperature does not vary and the programmer link does not move, ensure that the temperature control cable is properly attached to both control panel lever and programmer arm. Replace as necessary.
5. If the air outlet temperature does not vary and the programmer link does move, proceed as follows:
   a. Ensure vacuum motor linkage is properly connected at programmer.
   b. Disconnect linkage at temperature door crank and move crank to both extremes to ensure door is not binding within the heater case. Repair door as necessary.
   c. If temperature door moves freely but only cold air is distributed, check vacuum hose connections at water control valve in the heater water inlet hose. If vacuum exists and the valve is closed, check A/C control for source of vacuum to the valve. Vacuum should only be

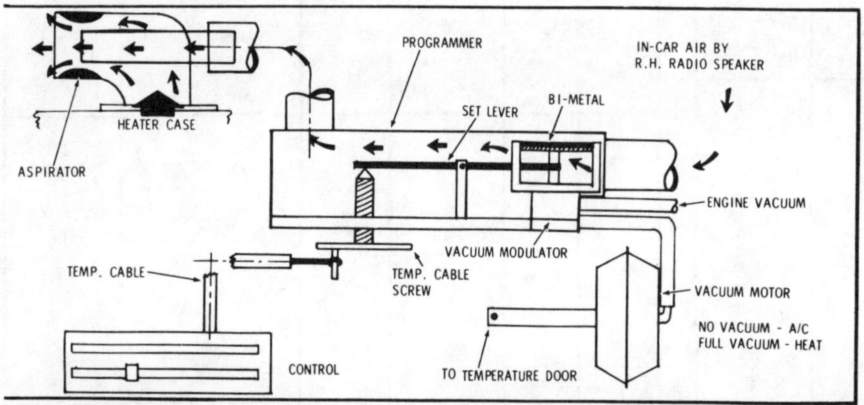

The temperature lever on the control rotates a screw which positions the vacuum valve through the set lever. The vacuum valve modulates engine vacuum to the vacuum motor to set temperature door position. The position of the temperature door determines heater-A/C discharge air temperature.

A small portion of the air flow through the heater case goes through the aspirator creating a low pressure at the small tube. Thus in-car air by the right hand radio speaker is drawn through the aspirator and hoses over the bi-metal sensing element in the programmer. The bi-metal acts directly on the vacuum valve to maintain the set temperature in response to changes in in-car temperature.

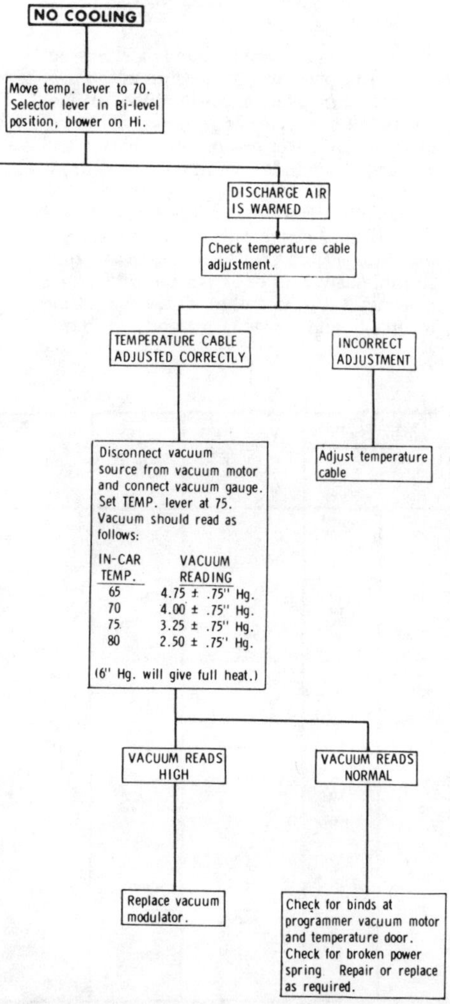

Fig. 209 Tempmatic programmer diagnosis chart. 1974-76 models (Part 1 of 2)

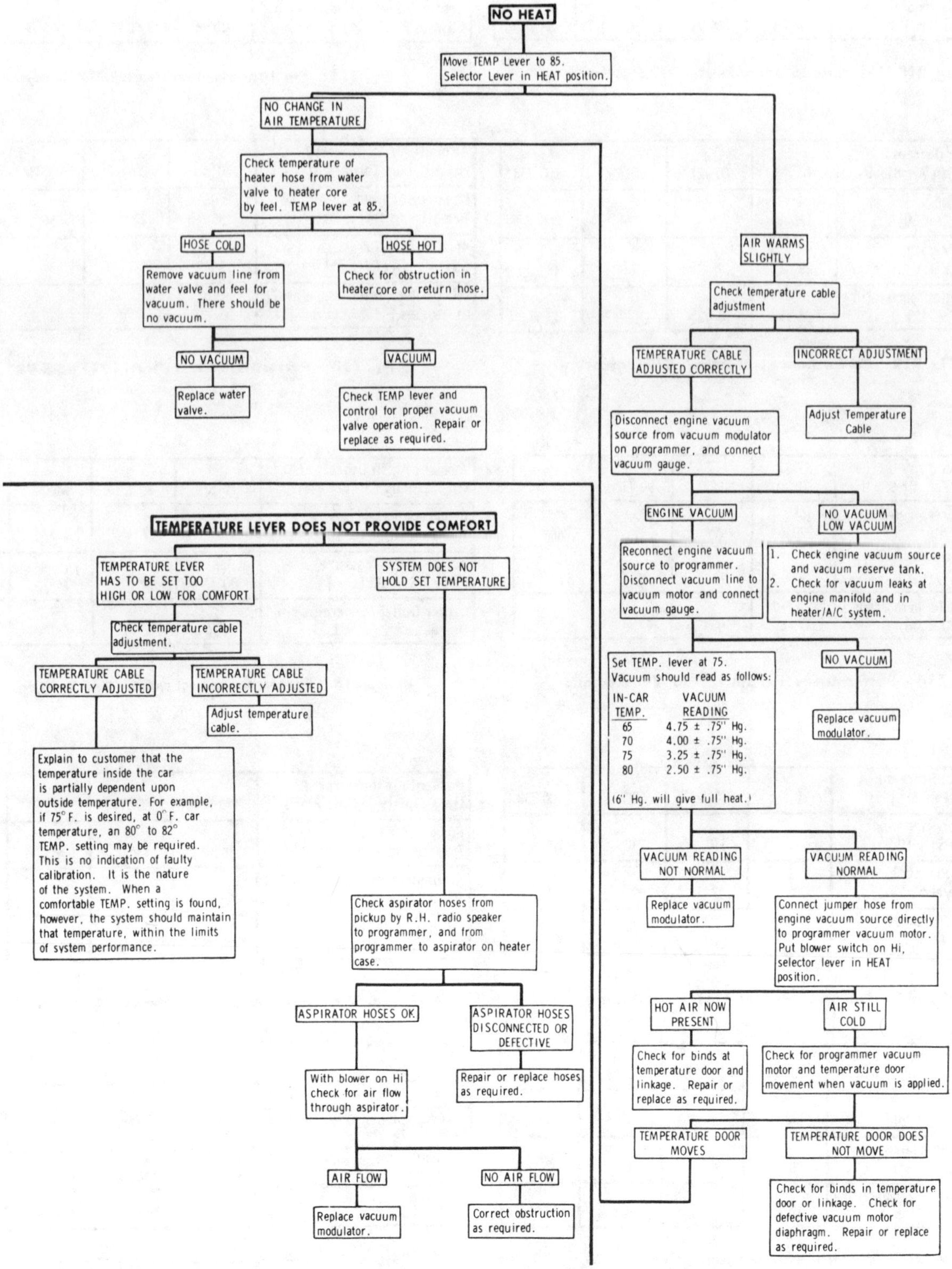

Fig. 209 Tempmatic programmer diagnosis chart. 1974-76 models (Part 2 of 2)

Ambient Air in Degrees F°/C° (In Auxiliary Fan Air Blast)	70°/21°	80°/27°	90°/32°	100°/38°
Compressor Head Pressure in PSI ± 20	170	200	260	320
Accumulator Pressure PSI AT SEA LEVEL ± 1	29.5	29	29.5	29.5
Center Outlet Temperature in Degrees F ± 2 / C ± 1	38°/3°	38°/3°	30°/-1°	43°/6°

Fig. 210 Performance test chart. 1977 Starfire.

Ambient Air in Degrees F°/C° (In Auxiliary Fan Air Blast)	70°/21°	80°/27°	90°/32°	100°/38°
Average Compressor Head Pressure in PSI ± 20	175	200	250	270
Accumulator Pressure PSI at Sea Level ± 1	24	27	30	33
Center Outlet Temperature in Degrees F ± 2 / C ± 1	40°/4°	41°/5°	45°/7°	49°/9°

Fig. 211 Performance test chart. 1977 Omega

Ambient Air in Degrees F°/C°. (In Auxiliary Fan Air Blast)	70°/21°	80°/27°	90°/32°	100°/38°
Compressor Head Pressure in PSI ± 20	180	200	225	260
Accumulator Pressure PSI AT SEA LEVEL ± 1	25	27	29	33
Left Center Outlet Temperature in Degrees F ± 2 / C ± 1	38°/3°	36°/2°	40°/4°	45°/7°

Fig. 212 Performance test chart. 1977 Cutlass

Ambient Air in Degrees F°/C°. (In Auxiliary Fan Air Blast)	70°/21°	80°/27°	90°/32°	100°/38°
Compressor Head Pressure in PSI ± 20	185	220	250	270
Accumulator Pressure PSI AT SEA LEVEL ± 1	23	27	31	36
Left Center Outlet Temperature in Degrees F ± 2 / C ± 1	36°/2°	42°/6°	48°/9°	54°/12°

Fig. 213 Performance test chart. 1977 88 & 98

Ambient Air in Degrees F°/C° (In Auxiliary Fan Air Blast)	70°/21°	80°/27°	90°/32°	100°/38°
Average Compressor Head*	165 220	200 265	225 295	325 350
Average Evaporator Pressure PSI AT SEA LEVEL ± 1	29	29	30	32
Center Outlet Temperature in Degrees F ± 2 / C ± 1	35°/2°	37°/3°	40°/4°	42°/6°

Fig. 214 Performance test chart. 1977-78 Toronado

Ambient Air in Degrees F°/C°. (In Auxiliary Fan Air Blast)	70°/21°	80°/27°	90°/32°	100°/38°
Compressor Head Pressure in PSI ± 20	170	220	255	280
Accumulator Pressure PSI AT SEA LEVEL ± 1	30	30	30	32
Center Outlet Temperature in Degrees F ± 2 / C ± 1	42°/6°	42°/6°	42°/6°	42°/6°

Fig. 215 Performance test chart. 1978 Starfire

Ambient Air in Degrees F°/C°. (In Auxiliary Fan Air Blast)	70°/21°	80°/27°	90°/32°	100°/38°
Average Compressor Head Pressure in PSI ± 20	145	170	210	240
Accumulator Pressure PSI at Sea Level ± 1	32	32	32	32
Center Outlet Temperature in Degrees F ± 2 / C ± 1	42°/6°	42°/6°	42°/6°	42°/6°

Fig. 216 Performance test chart. 1978 Omega

Ambient Air in Degrees F°/C°. (In Auxiliary Fan Air Blast)	70°/21°	80°/27°	90°/32°	100°/38°
Compressor Head Pressure in PSI ± 20	150	170	200	235
Accumulator Pressure PSI AT SEA LEVEL ± 1	30	30	30	30
Left Center Outlet Temperature in Degrees F ± 2 / C ± 1	38°/3°	38°/3°	39°/4°	42°/6°

Fig. 217 Performance test chart. 1978 Cutlass

Ambient Air in Degrees F°/C°. (In Auxiliary Fan Air Blast)	70°/21°	80°/27°	90°/32°	100°/38°
Compressor Head Pressure in PSI ± 20	160	190	225	255
Accumulator Pressure PSI AT SEA LEVEL ± 1	30	30	30	32
Left Center Outlet Temperature in Degrees F ± 2 / C ± 1	42°/6°	42°/6°	42°/6°	42°/6°

Fig. 218 Performance test chart. 1978 88 & 98

AMBIENT AIR TEMPERATURE	70° F (21° C)	80° F (27° C)	90° F (32° C)	100° F (38° C)
COMPRESSOR HEAD PRESSURE* ± 20 PSI (± 138 kPa)	170 PSI (1172 kPa)	200 PSI (1379 kPa)	260 PSI (1792 kPa)	320 PSI (2205 kPa)
ACCUMULATOR PRESSURE* ± 1 PSI (± 6.89 kPa)	29 PSI (200 kPa)	29 PSI (200 kPa)	29 PSI (200 kPa)	29 PSI (200 kPa)
CENTER OUTLET TEMPERATURE ± 2° F (± 1° C)	38° F (3° C)	38° F (3° C)	40° F (4° C)	43° F (6° C)

Fig. 219 Performance test chart. 1979 Starfire

Ambient Air In Degrees F°/C°. (In Auxiliary Fan Air Blast)	70°/21°	80°/27°	90°/32°	100°/38°
Average Compressor Head Pressure± 20 PSI (± 138 kPa)	175 PSI (1206 kPa)	200 PSI (1397 kPa)	250 PSI (1723 kPa)	270 PSI (1861 kPa)
Accumulator Pressure ± 1 PSI (± 6. 89 kPa)	24 PSI (165 kPa)	27 PSI (186 kPa)	30 PSI (207 kPa)	33 PSI (228 kPa)
Center Outlet Temperature In Degrees ± 2° F (± 1° C)	40° F (4° C)	41° F (5° C)	45° F (7° C)	49° F (9° C)

Fig. 220 Performance test chart. 1979 Omega

Ambient Air In Degrees F°/C°. (In Auxiliary Fan Air Blast)	70°/21°	80°/27°	90°/32°	100°/38°
Compressor Head Pressure ± 20 PSI (± 138 kPa)	160 PSI (1102 kPa)	175 PSI (1206 kPa)	210 PSI (1466 kPa)	230 PSI (1604 kPa)
Accumulator Pressure ± 1 PSI (± 6. 89 kPa)	26 PSI (179 kPa)	28 PSI (193 kPa)	30 PSI (207 kPa)	31 PSI (213 kPa)
Left Center Outlet Temperature In Degrees ± 2° F (± 1° C)	38° F (3° C)	38° F (3° C)	39° F (4° C)	42° F (6° C)

Fig. 221 Performance test chart. 1979 Cutlass

Ambient Air In Degrees F°/C°. (In Auxiliary Fan Air Blast)	70°/21°	80°/27°	90°/32°	100°/38°
Compressor Head Pressure ± 20 PSI (± 138 kPa)	165 PSI (1137 kPa)	185 PSI (1275 kPa)	210 PSI (1466 kPa)	235 PSI (1638 kPa)
Accumulator Pressure ± 1 PSI (± 6. 89 kPa)	25 PSI (172 kPa)	27 PSI (186 kPa)	30 PSI (207 kPa)	32 PSI (214 kPa)
Left Center Outlet Temperature In Degrees ± 2° F (± 1° C)	38° F (3° C)	38° F (3° C)	40° F (4° C)	42° F (6° C)

Fig. 222 Performance test chart. 1979 88 & 98

Ambient Air In Degrees F°/C°. (In Auxiliary Fan Air Blast)	70°/21°	80°/27°	90°/32°	100°/38°
Compressor Head Pressure ± 20 PSI (± 138 kPa)	160 PSI (1102 kPa)	175 PSI (1206 kPa)	195 PSI (1344 kPa)	225 PSI (1579 kPa)
Accumulator Pressure ± 1 PSI (± 6. 89 kPa)	29 PSI (200 kPa)	29 PSI (200 kPa)	30 PSI (207 kPa)	32 PSI (214 kPa)
Center Outlet Temperature In Degrees ± 2° F (± 1° C)	37° F (3° C)	38° F (3° C)	40° F (4° C)	42° F (6° C)

Fig. 223 Performance test chart. 1979 Toronado

Ambient Air in Degrees F°/C°. (In Auxiliary Fan Air Blast)	70° F (21° C)	80° F (27° C)	90° F (32° C)	100° F (38° C)
Compressor Head Pressure ± 20 PSI (± 140 kPa)	173 PSI (1190 kPa)	203 PSI (1400 kPa)	264 PSI (1820 kPa)	325 PSI (2240 kPa)
Accumulator Pressure ± 1 PSI (± 7 kPa)	30 PSI (207 kPa)	29 PSI (203 kPa)	30 PSI (207 kPa)	30 PSI (207 kPa)
Left Center Outlet Temperature in Degrees ± 2° F (±1° C)	37° F (3° C)	37° F (3° C)	39° F (4° C)	43° F (6° C)

Fig. 224 Performance test chart. 1980 Starfire

	CUTLASS				CUTLASS SUPREME			
Ambient Air in Degrees F°/C°. (In Auxiliary Fan Air Blast)	70°F (21° C)	80° F (27° C)	90° F (32° C)	100° F (38° C)	70° F (21° C)	80° F (27° C)	90° F (32° C)	100° F (38° C)
Compressor Head Pressure ± 20 PSI (± 140 kPa)	162 PSI (1120 kPa)	192 PSI (1330 kPa)	223 PSI (1540 kPa)	263 PSI (1820 kPa)	152 PSI (1085 kPa)	178 PSI (1225 kPa)	208 PSI (1435 kPa)	248 PSI (1715 kPa)
Accumulator Pressure ± 1 PSI (± 7 kPa)	25 PSI (175 kPa)	29 PSI (203 kPa)	34 PSI (231 kPa)	39 PSI (266 kPa)	24 PSI (168 kPa)	25 PSI (175 kPa)	30 PSI (210 kPa)	36 PSI (245 kPa)
Left Center Outlet Temperature in Degrees ± 2° F (±1° C)	34° F (1° C)	37° F (3° C)	45° F (7° C)	50° F (10° C)	34° F (1° C)	34° F (1° C)	41° F (5° C)	46° F (8° C)

Fig. 225 Performance test chart. 1980 Cutlass

Ambient Air in Degrees F°/C°. (In Auxiliary Fan Air Blast)	70° F (21° C)	80° F (27° C)	90° F (32° C)	100° F (38° C)
Compressor Head Pressure ± 20 PSI (± 140 kPa)	147 PSI (1015 kPa)	166 PSI (1148 kPa)	188 PSI (1295 kPa)	218 PSI (1505 kPa)
Accumulator Pressure ± 1 PSI (± 7 kPa)	25 PSI (175 kPa)	25 PSI (175 kPa)	25 PSI (175 kPa)	30 PSI (210 kPa)
Left Center Outlet Temperature in Degrees ± 2°F (±1° C)	37° F (3° C)	37° F (3° C)	37° F (3° C)	39° F (4° C)

Fig. 226 Performance test chart. 1980 88 & 98

Ambient Air in Degrees F°/C°. (In Auxiliary Fan Air Blast)	70° F (21° C)	80° F (27° C)	90° F (32° C)	100° F (38° C)
Compressor Head Pressure ± 20 PSI (± 140 kPa)	162 PSI (1120 kPa)	178 PSI (1225 kPa)	208 PSI (1435 kPa)	239 PSI (1645 kPa)
Accumulator Pressure ± 1 PSI (± 7 kPa)	24 PSI (168 kPa)	24 PSI (168 kPa)	27 PSI (189 kPa)	31 PSI (217 kPa)
Left Center Outlet Temperature in Degrees ± 2° F (± 1° C)	34° F (1° C)	34° F (1° C)	37° F (2° C)	41° F (5° C)

Fig. 227 Performance test chart. 1980 Toronado

applied to the valve when control is in Max. A/C. If vacuum exists at the water control valve and the heater hose from the valve to the heater core is hot, replace faulty water control valve.

d. If the temperature door moves freely and hot water is available to the heater core but only cold air is distributed, check vacuum hose connections at the programmer. If vacuum is not present, inspect vacuum circuitry. If vacuum is present at the programmer, disconnect hose at the vacuum motor and apply vacuum directly to the motor. If vacuum motor link does not move, replace faulty motor. If vacuum motor link does move, inspect hoses from the vacuum checking relay to the vacuum motor and the compensator assembly. Repair leaks or disconnects as necessary. If no leaks or disconnects are found, replace the vacuum checking relay.

e. If the temperature door moves freely but only warm air is distributed, inspect programmer vacuum motor link. In full heat operation, motor link should be fully drawn into the vacuum motor. Disconnect vacuum hose from vacuum motor. If motor link moves out from motor, replace the compensator assembly. If not, replace vacuum motor.

6. If air outlet temperature varies in step 3 but poor temperature control is observed, proceed as follows:
   a. Inspect temperature control adjustment and adjust if necessary.
   b. Ensure in-car air hose and ambient drive air source are not kinked or loosely connected. Repair leaks or disconnects as necessary.
   c. If temperature control cable is properly connected, the in-car and ambient air hoses are securely connected and routed but poor temperature control is still observed, replace the compensator assembly in the programmer.

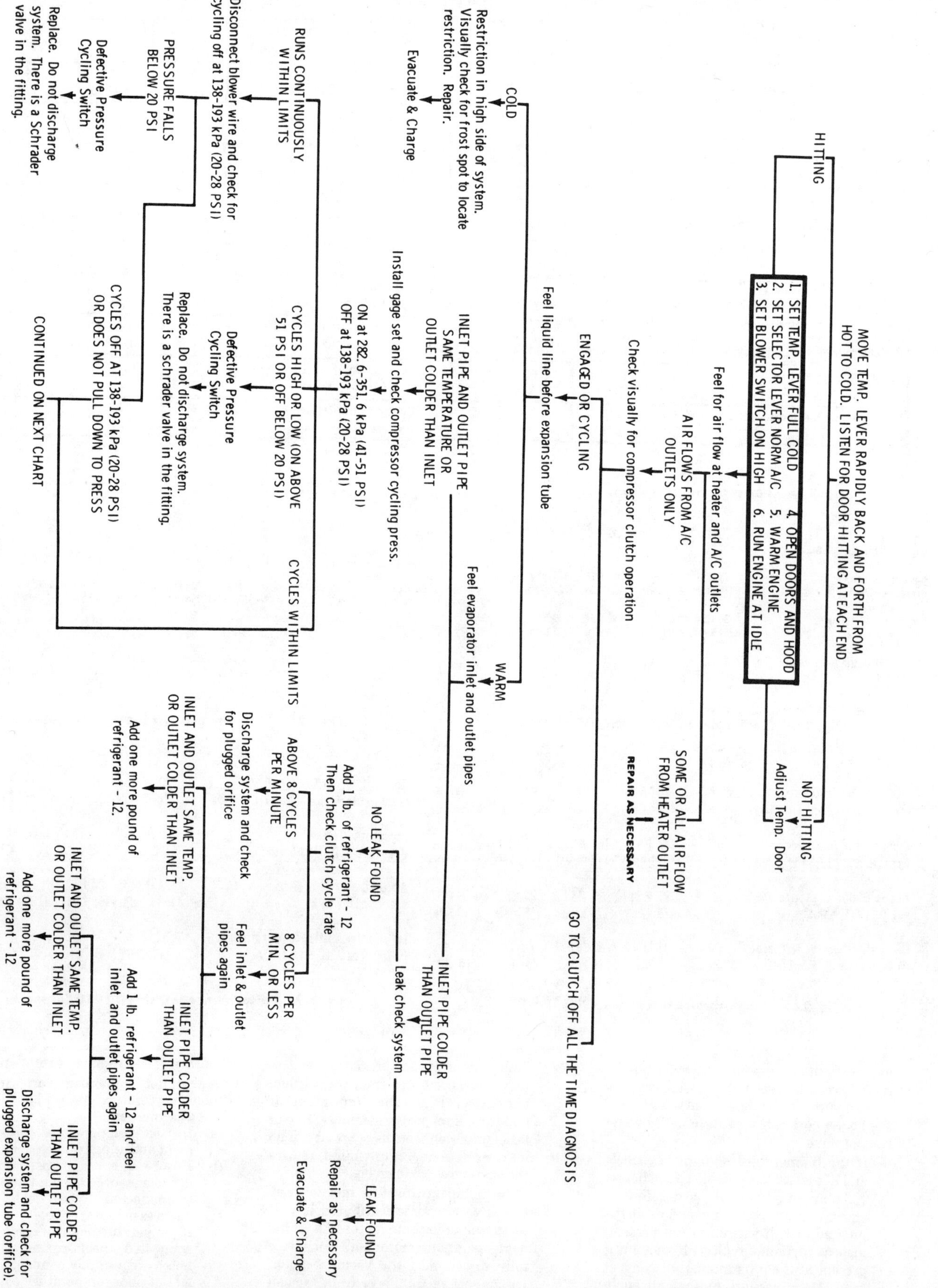

Fig. 228 Insufficient cooling diagnosis chart. Models w/pressure cycling switch (Part 1 of 2)

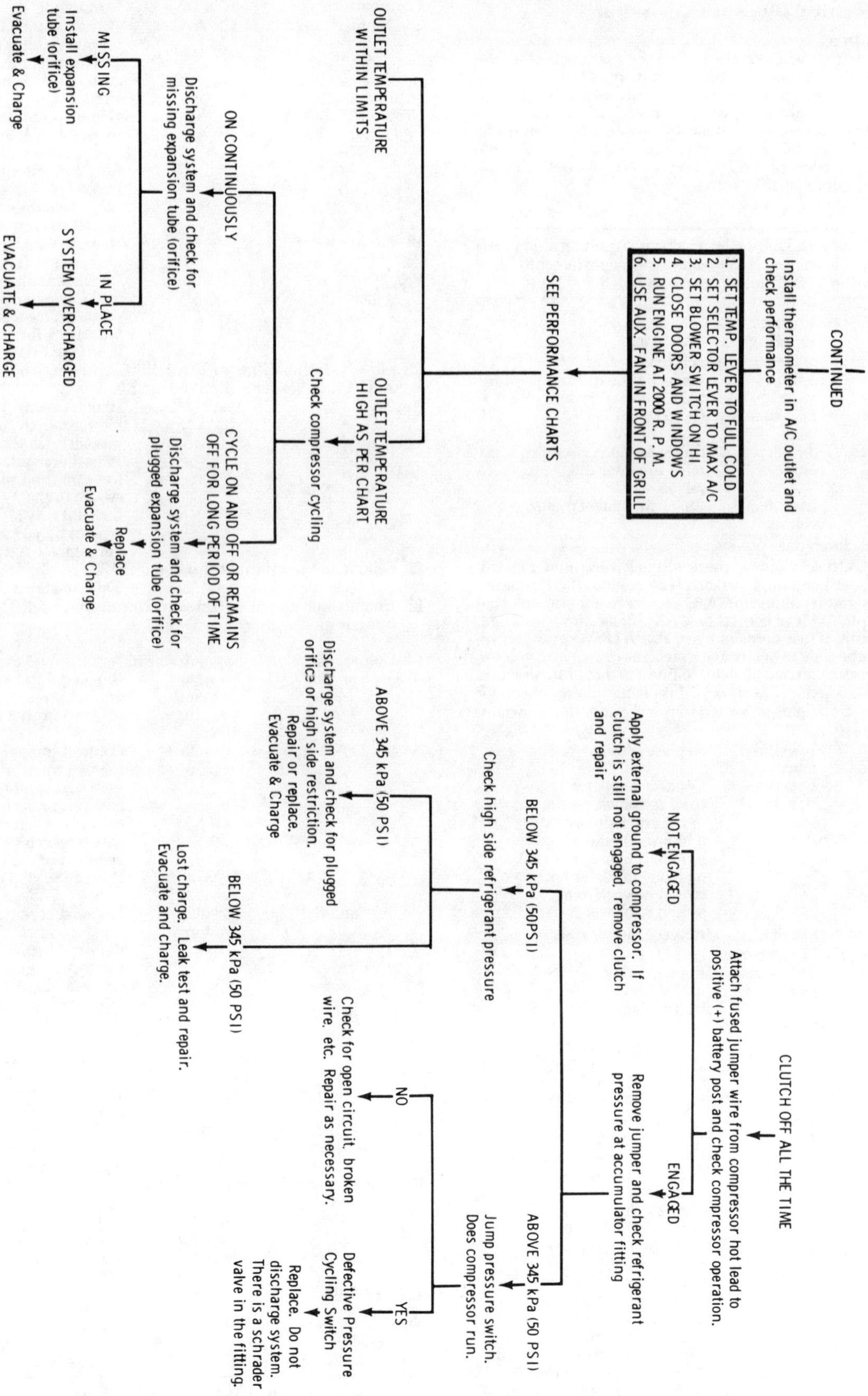

Fig. 228 Insufficient cooling diagnosis chart. Models w/pressure cycling switch (Part 2 of 2)

1977-78 88 & 98 A/C
Electrical Diagnosis Connector

Use a voltmeter to make the indicated checks across the diagnostic connector, Fig. 118. This will reveal whether the malfunction is in a component or in the wire connecting that component in the circuit. If a jumper wire across a suspected faulty component makes the system operational, replace the faulty component. If not, check for current at the feed wire or repair poor ground. Refer to diagnosis connector, Fig. 119, which details the similar Buick and Pontiac schematic before beginning diagnostic procedure on 1977 vehicles and Fig. 229 before beginning diagnostic procedure on 1978 vehicles.

1977 88 & 98

NOTE: Perform tests with ignition on, engine not running, and inside car temperature over 55 degrees F. Blower switch should be set to LOW and A/C control set to VENT unless otherwise noted.

Voltmeter Connections	Voltage Reading	Condition
1. No blower at any speed, compressor clutch engages.		
a. 1 and G	3 volts or more	Inspect ground wire to motor; if OK, replace motor.
	Less than 3 volts	Go to step b.
b. 2 and G	3 volts or more	Check for open circuit in relay to motor. If OK, replace relay.
	Less than 3 volts	Check for stalled motor.
2. No blower and clutch does not engage.		

Inspect heater A/C fuse; if blown, check wiring harness under instrument panel and under hood for short circuit to ground. If OK, disconnect blower and 3 terminal connector (A/C harness to idle stop solenoid and compressor clutch). Move control to MAX. If fuse blows again, recheck for short circuit. If fuse does not blow, attach blower and 3 terminal connector one at a time to determine which one causes fuse to blow. Disconnect compressor clutch connector to find out if malfunction is in clutch or idle stop solenoid. If heater/AC fuse is not blown, check for open circuit in wire from ignition switch to the in-car ambient switch.

Voltmeter Connections	Voltage Reading	Condition
3. No low blower speed.		
a. 3 and G	10 volts or more	Replace blower resistor
	Less than 10 volts	Connect jumper wire to the 2 terminals on the in-car ambient switch; if blower operates, replace switch. If blower does not operate, check for open circuit in wire from switch to blower resistor.
4. No low or M1 blower speed.		Replace blower resistor.
5. No blower except in High or Max. A/C.		
a. 2 and G	3 volts or more	Replace relay.
	Less than 3 volts	Observe voltage at all terminals on resistor; if any are less than 3 volts, replace resistor. If all are OK, check wire from resistor to relay for open circuit.
6. No high blower under any condition. Place blower switch in High.		
a. 9 and G	10 volts or more	If relay does not click when control is moved from OFF to MAX and relay ground is OK, replace relay. If relay clicks, check for 12 volts at hi blower feed wire at relay. If OK, replace relay. If less than 12 volts, check for disconnect at feed wire connector or open circuit in the wire.
	Less than 10 volts	Check for open circuit in wire from blower switch to high blower relay coil.
7. No high speed except when control is in MAX. A/C.		Replace blower switch.
8. No high speed in MAX. A/C.		Remove control and inspect for open circuit in wire from control to blower switch. If OK, replace control.
9. No M1 speed.		Remove control and unplug blower switch connector. Check for voltage at back feed through resistor. If OK, replace blower switch. If not, check for open circuit in wire to blower resistor.
10. No M2 speed.		Remove control and unplug blower switch connector. Check for voltage at back feed through resistor. If OK, replace blower switch. If not, check for open circuit in wire to blower resistor.
11. No M1, M2 or High with switch. High blower operates in MAX. A/C. Place blower switch in High.		Remove control and ensure blower switch connector is secure. Check for 10 or more volts in the wire from the blower switch to the A/C control (brn dbl wht). If OK, replace blower switch. If not, replace A/C control.
12. Clutch and blower motor inoperative.		Refer to step 2.
13. Compressor clutch operates in Norm., Bi-Level, Defrost or Max. but not in all.		Replace A/C control.
14. Compressor clutch inoperative; blower OK. A/C control in Norm.		
a. 6 and G	10 volts or more	Replace clutch or repair ground wire.
	Less than 10 volts	Proceed to step b.
b. 8 and G	10 volts or more	Connect jumper wires across the two wires on temperature switch. If compressor clutch engages, replace switch. If not, check for open circuit in wire between switch and compressor clutch connector.
	Less than 10 volts	Proceed to step c.
c. 7 and G	10 volts or more	Disconnect refrigerant pressure switch. Connect a jumper wire across the two connector terminals. If clutch engages, check for low refrigerant charge. If OK, replace pressure switch. If clutch did not engage, check for open circuit in wire from pressure switch to temperature switch.
	Less than 10 volts	Check for open circuit in wire from refrigerant pressure switch to control. If OK, replace control.
15. Low blower only; clutch inoperative.		
a. 4 and G	10 volts or more	Replace A/C control.
	Less than 10 volts	Check for open circuit in wire from ambient switch to control.

1978 88 & 98

NOTE: Perform tests with ignition on, engine not running, and inside car temperature over 55 degrees F. Blower switch should be set to LOW and A/C control set to VENT unless otherwise noted.

Voltmeter Connections	Voltage Reading	Condition
1. No blower at any speed, compressor clutch engages.		
a. 1 and G	3 volts or more	Inspect ground wire to motor; if OK, replace motor.

Voltmeter Connections	Voltage Reading	Condition
	Less than 3 volts	Go to step b.
b. 2 and G	3 volts or more	Check for open circuit in relay to motor. If OK, replace relay.
	Less than 3 volts	If O volts are observed, check A/C harness ground; check for stalled motor.

2. No blower and clutch does not engage.

Inspect heater A/C fuse; if blown, check wiring harness under instrument panel and under hood for short circuit to ground. If OK, disconnect blower and 3 terminal connector (A/C harness to idle stop solenoid and compressor clutch). Move control to MAX. If fuse blows again, recheck for short circuit. If fuse does not blow, attach blower and 3 terminal connector one at a time to determine which one causes fuse to blow. Disconnect compressor clutch connector to find out if malfunction is in clutch or idle stop solenoid. If heater/AC fuse is not blown, check for open circuit in wire from ignition switch to the low blower ambient switch.

Voltmeter Connections	Voltage Reading	Condition
3. No low blower speed above 55 degrees in-car ambient.		
a. 3 and G	10 volts or more	Replace blower resistor.
	Less than 10 volts	Connect jumper wire to the 2 terminals on the low blower ambient switch; if blower operates, replace switch. If blower does not operate, check for open circuit in wire from switch to blower resistor.
4. No low or M1 blower speed.		Replace blower resistor.
5. No blower except in High or Max. A/C.		
a. 2 and G	3 volts or more	Replace hi-blower relay.
	Less than 3 volts	Observe voltage at all terminals on resistor; if any are less than 3 volts, replace resistor. If all are OK, check wire from resistor to relay for open circuit.
6. No high blower under any condition. Place blower switch in High.		
a. 9 and G	10 volts or more	If relay does not click when control is moved from OFF to MAX. and relay ground is OK, replace relay. If relay clicks, check for 12 volts at hi blower feed wire at relay. If OK, replace relay. If less than 12 volts, check for disconnect at feed wire connector or open circuit in the wire.
	Less than 10 volts	Check for open circuit in wire from blower switch to high blower relay coil.
7. No M1 speed.		Remove control, disconnect blower switch connector and check for voltage at M1 circuit from blower switch to resistor. If OK, replace blower switch. If no voltage is observed, check for open circuit in the wire to blower resistor.
8. No M2 speed.		Remove control, disconnect blower switch connector. Check for voltage at M2 circuit from blower switch to resistor. If OK, replace blower switch. If no voltage is observed, check for open circuit in the wire to blower resistor.
9. No M1, M2 or High with switch (OK in Max. A/C). Place blower switch in High.		Remove control, check for loose connector at blower switch. If 10 or more volts are observed in the wire from the blower switch to the A/C control (brn dbl wht), replace blower switch. If less than 10 volts are observed, replace A/C control switch.
10. Low blower only. Compressor clutch inoperative.		
a. 4 and G	10 volts or more	Replace A/C control switch.
	Less than 10 volts	Check for open circuit in wire from low blower ambient switch to control switch.
11. Compressor clutch and blower motor inoperative. Refer to step 2.		
12. Compressor clutch operates in Norm., Bi-Level, Defrost, or Max. but not in all.		Replace A/C control switch.
13. Compressor clutch inoperative; blower OK.		
a. 6 and G	10 volts or more	Proceed to step c.
	Less than 10 volts	Proceed to step b.
b. 8 and G	10 volts or more	Connect a jumper wire across thermostatic switch terminals; if clutch engages, replace switch. If not, proceed to step c.
	Less than 10 volts	Check for open circuit in wire from A/C control to thermostatic switch. If OK, replace A/C control switch.
c. Connect a voltmeter between ground and each clutch terminal.		
	Both terminals 10 volts or more	Disconnect compressor cut-off switch and touch to ground. If clutch engages, check for low refrigerant charge. If OK, replace cut-off switch.
	One terminal less than 10 volts	Replace clutch coil.

Blower Diagnosis

1974 Omega; 1975-77 Starfire & Omega

1. Before beginning blower diagnosis, determine whether:
   a. All speeds are inoperative; if so, proceed to step 2.
   b. High speed only is inoperative; if so, proceed to step 3.
   c. High speed is functional, but one or all of the other speeds are inoperative; if so, proceed to step 4.
2. If all speeds are inoperative, inspect heater-A/C fuse in fuse panel. If fuse is not blown, check for open circuit in purple wire from resistor to blower motor. If there is no open circuit and blower motor is properly grounded, replace blower motor. If heater-A/C fuse is blown, locate and repair short circuit to ground in one or more of the following wires:
   a. Brown wire from fuse panel to low blower switch.
   b. Light blue, light green or yellow wires from fan switch to resistor or orange wire from switch to relay.
   c. Purple wire from resistor to relay or to blower motor.
3. If high speed only is inoperative, inspect in-line fuse from engine harness to A/C relay. If fuse is blown, locate and repair short circuit to ground in feed wire from in-line fuse to relay or in wire from relay to blower motor. If fuse is not blown, switch ignition On and move fan switch to Hi. Then using a test light, probe smaller gauge orange wire at relay. If test light does not go on, locate and repair open circuit in orange wire from relay to fan switch. If test light goes on, replace relay.
4. If high speed is functional, but one or all of the other speeds are inoperative, check for

open circuit in the following fan switch to resistor wires:
a. Lo — Yellow wire
b. M1 — Light green wire
c. M2 — Light blue wire

If no open circuits are found, replace resistor.

1974-76 88 & 98; 1974-78 Cutlass & Toronado

1. Before beginning blower diagnosis, determine whether:
   a. All speeds are inoperative; if so, proceed to step 2.
   b. High speed only is inoperative; if so, proceed to step 3.
   c. High speed is functional, but all other speeds are inoperative; if so, proceed to step 4.
   d. High speed is functional, but Lo, M1 or M2 speeds are inoperative; if so, proceed to step 5.
2. If all speeds are inoperative, ensure engine is at operating temperature and ignition On. Then move blower switch to Lo position. Inspect heater-A/C fuse in fuse block. If the fuse is blown, locate and repair short circuit in one or more of the following wires:
   a. Brown wire from fuse block to control or from control to resistor block or thermal relay.
   b. Brown wire from thermal relay to blower resistor. Refer to wiring diagram located elsewhere in this manual.
   c. Light blue and yellow wires from blower resistor to fan switch.
   d. Black/white or orange wires to high blower relay.
   e. Purple wire to blower motor.

   If the fuse is not blown, connect a test light to blower motor connector. If the test light goes on and the blower motor is properly grounded, replace blower motor. If the test light does not go on, probe each wire connector on high blower connector with the test light after moving fan switch to HI. If the test light goes on at all terminals, repair open circuit in purple wire to blower. If the test light goes on at all terminals except purple, replace high blower relay.
3. If high speed only is inoperative and car is equipped with heated back glass, refer to Hi Blower Diagnosis chart. If not, check in-line fuse, junction block to high blower relay. If fuse is blown, locate and repair short circuit in wire to high blower relay. If fuse is not blown, move fan switch to HI and temporarily connect a 12 volt power source to orange wire (single connector) at high blower relay. If blower does not run at high speed, ensure high blower relay is properly grounded. If it is, replace high blower relay. If blower does run at high speed after connecting a 12 volt power source, use a test light to probe orange wire on relay and orange wire at fan switch. If test light is off at both positions, replace fan switch. If test light is on at fan switch only, locate and repair open circuit in orange wire from fan switch to relay.
4. If high speed is functional but all other speeds are inoperative, connect a test light to black/white wire connector at high blower relay. If the test light goes on, replace high blower relay. If not, probe black/white wire connector at resistor. If test light remains off, replace resistor. If test light comes on, locate and repair open circuit in black/white wire between resistor and high blower relay.

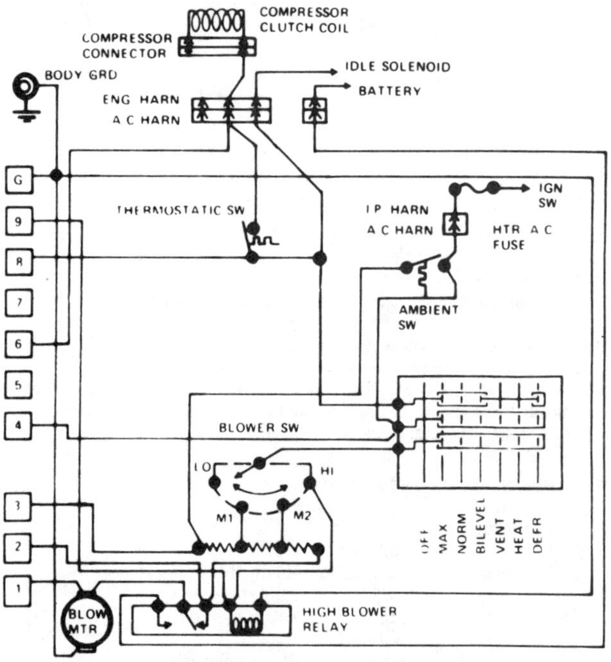

Fig. 229 Diagnosis connector schematic. 1978 88 & 98

5. If high speed is functional but Lo, M1 or M2 speeds are inoperative, determine whether blower is inoperative in Lo only. If so, on models without a thermal relay, connect a test light to brown wire terminal at resistor. If test light goes on, replace resistor. If not, locate and repair open circuit in brown wire between resistor and control. On models equipped with a thermal relay, ensure engine is at operating temperature to close thermal relay. Turn ignition on, A/C off and move fan switch to Lo. Connect a test light and probe each wire terminal at thermal relay. If test light is off at one terminal, (brown/white stripe wire on Tempmatic) replace thermal relay. If test light is off at both terminals, locate and repair open circuit in brown wire between thermal relay and control. If test light is on at both terminals, probe brown wire terminal at resistor. If test light goes on, replace resistor. If not, locate and repair open circuit in brown wire between resistor and thermal relay. If M1 or M2 speeds are inoperative, on models with or without thermal relay, switch ignition on, and move blower speed switch to position which fails to operate. Using a test light, probe yellow M1 wire or light blue M2 wire at resistor. If the test light goes on and the blower is also inoperative in Lo, replace resistor. However, if blower functions in Lo, clean wire connections to resistor. If the test light does not go on, probe same color wire at blower speed switch. If test light remains off, replace blower speed switch. If test light goes on, locate and repair open circuit in wire between blower speed switch and resistor.

1978 Starfire & Omega; 1979-80 All

1. Before beginning blower diagnosis, determine whether:
   a. All speeds are inoperative; if so, proceed to step 2.
   b. High speed only is inoperative; if so, proceed to step 3.
   c. High speed is functional, but one or all of the other speeds are inoperative; if so, proceed to step 4.
2. If all speeds are inoperative, inspect heater-A/C fuse in fuse panel. If fuse is not blown, check for open circuit in purple wire from relay to blower motor. If there is no open circuit and blower motor is properly grounded, replace blower motor. If heater-A/C fuse is blown, locate and repair short circuit to ground in one or more of the follown wires:
   a. Brown wire from fuse panel to control or resistor.
   b. Light blue or tan wires from fan switch to resistor or orange wire from switch to relay.
   c. Dark blue wire from resistor to relay or purple wire from relay to blower motor.
3. If high speed only is inoperative, inspect red wire from engine harness to A/C relay for open circuit. If open circuit in red wire is found, repair as necessary. Ensure there is not an open circuit in purple wire from relay to blower motor. If open circuit in red wire is not found, switch ignition On and move fan switch to Hi. Then using a test light, probe smaller gauge orange wire at relay. If test light does not go on, locate and repair open circuit in orange wire from relay to fan switch. If test light goes on, replace relay.
4. If high speed is functional, but one or all of the other speeds are inoperative, check for open circuit in the following fan switch to resistor wires:
   a. Lo — Brown wire
   b. M1 — Tan wire
   c. M2 — Light blue wire

   If no open circuits are found, replace resistor.

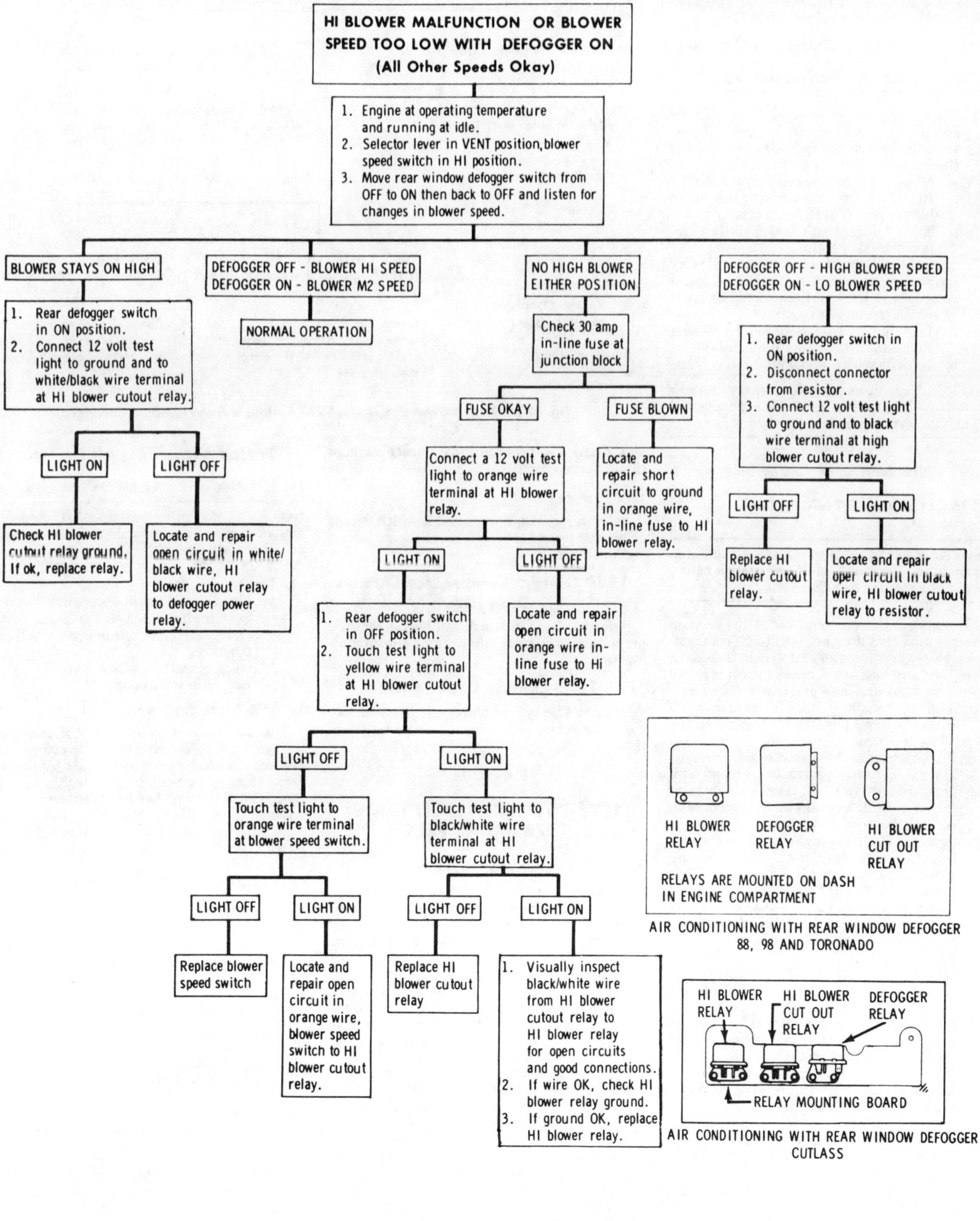

Fig. 230 High blower diagnosis. 1974-76 Cutlass, 88, 98 & Toronado w/rear window defogger

High Blower Diagnosis

Refer to Figs. 230 and 231 which outline high blower diagnostic procedure for vehicles equipped with a rear window defogger.

A/C & Defogger Relay On-Car Test

NOTE: Perform test with engine idling.

1. Place A/C control lever in Norm, blower speed switch in Low, and defogger Off. Listen or feel for Hi Blower Relay operation as the following controls are moved:
   a. Move the blower speed switch to Hi; Hi Blower Relay should operate. Keep blower switch in Hi for next test.
   b. Turn defogger switch On; all three relays should operate.
2. Using a test light, check for current at relay terminals. Switchers should be positioned as in 1a and 1b. Test light should be on at all switch terminals except:
   a. Light green wire between Hi Blower Relay and Hi Blower Cutout Relay.
   b. Dark blue wire at Defogger Relay.

NOTE: Test light may be slightly dimmer when black/double white stripe, purple and light blue wire terminals are probed.

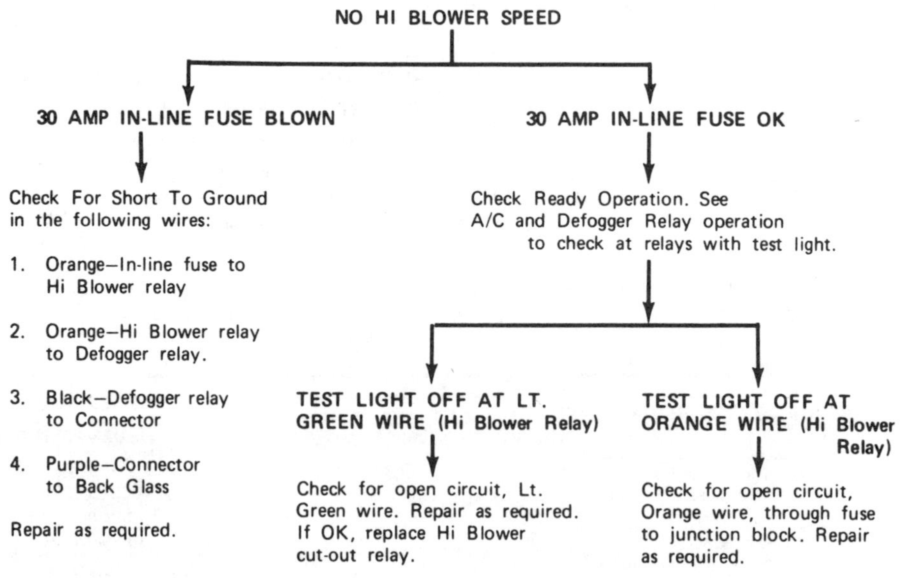

Fig. 231 High blower diagnosis. 1977 Cutlass w/rear window defogger

Vacuum Leak Diagnosis

1976-77 Cutlass; 1976-78 88, 98 & Toronado

NOTE: Refer to the appropriate vacuum diagram in the Vacuum and Wiring section before beginning vacuum leak diagnosis.

Disconnect the vacuum hose at the vacuum reserve tank that does not lead to intake manifold. Unplug compressor clutch coil, then use a "tee" and two hoses to connect a vacuum gauge to reserve tank and hose previously disconnected. With engine idling, check for vacuum at all A/C-heater control positions. Vacuum reading should be the same as manifold vacuum reading. Perform tests in the following sequence:

1. Move A/C-heater control to OFF and temperature lever to HOT. If low vacuum is indicated, check for leaks at vacuum hose to control (hose into which vacuum gauge is tapped). On Tempmatic units also check hose to programmer, programmer vacuum motor and modulator.
2. Move control to HEAT. If low vacuum is indicated, inspect vacuum hose to defroster door vacuum motor. Models with two hoses to defroster door, inspect hose with restrictor.
3. Move control to DEFROST. If low vacuum is indicated, inspect dark blue hose to defroster door vacuum motor.
4. Move control to BI-LEVEL. If low vacuum is indicated, inspect vacuum hose to lower mode door vacuum motor.
5. Move control to NORM. If low vacuum is indicated, inspect vacuum hose to upper mode door vacuum motor.
6. Move control to MAX. If low vacuum is indicated, move temperature lever to HOT. If discharge air is hot, inspect vacuum hose to water valve. If discharge air is cool, inspect vacuum hose to outside air door vacuum motor.

1978-80 Cutlass

Refer to Fig. 121, Parts 1 through 5, which details diagnostic procedure to correct vacuum circuit malfunctions.

1980 Omega

Refer to Fig. 122, Parts 1 and 2, which details diagnostic procedure to correct vacuum circuit malfunctions.

1980 Omega Compressor Clutch Diagnosis

Refer to Fig. 116 which outlines diagnostic procedure to correct compressor clutch malfunction.

1980 Omega Cooling Fan

Refer to Fig. 117 for diagnostic procedure to correct electric engine cooling fan malfunctions.

Pontiac

1974-76 MODELS EXCEPT LATE 1976 GRAND PRIX

Preliminary Checks

1. Check compressor belt for proper tension.
2. Check all refrigeration lines for leaks, kinks, or other restrictions.
3. Check all air hoses for leaks or restrictions. Air restriction may indicate a plugged (or partially plugged) evaporator core.
4. Check outer surfaces of radiator and condenser cores to be sure they are not plugged with dirt, leaves or other foreign material. Be sure to check between the condenser and radiator as well as the outer surfaces.
5. Start engine and operate at 2000 rpm with system in A/C mode, temperature control lever set for maximum cooling and blower speed on HI. After at least five minutes of engine operation, observe for bubbling at the sight glass (above 70°F ambient). If the system is low on refrigerant, add refrigerant until receiver-dehydrator just shows clear and add an additional one (1) pound of refrigerant.
6. Decrease blower speed and observe decreases in air flow.

Performance Test All Systems

1974-76 Models Exc. Astre & Sunbird

NOTE: To determine whether A/C electrical, air, vacuum and refrigeration systems are operating properly refer to Fig. 232

1. Test conditions:
   (a) Hood closed and doors open.
   (b) Selector lever in A/C mode, fan switch on Hi and temperature lever at full cold position.
   (c) Nozzles and air outlets open.
   (d) Engine at low idle speed.

1975-76 Astre & 1976 Sunbird

This test may be conducted to determine if the system is performing in a satisfactory manner and should be used as a guide by the serviceman in diagnosing trouble within the system. The following fixed conditions must be adhered to in order to make it possible to compare the performance of the system being tested with the standards given in Figs. 233 through 235

1. Doors and windows closed. (Car inside or in shade.)
2. Hood up and engine exhaust suitably ventilated.
3. Operate engine at 1500 rpm on Astre models and 2000 rpm on Sunbird models.
4. Air conditioning controls set for—
   (a) Maximum cooling.
   (b) High blower speed.
5. TEMP control set at "COLD" and all air conditioning outlets open.
6. Gauge set installed.
7. System settled out (run-in approximately 10 minutes).
8. A thermometer placed in front of vehicle grille and another in the right hand A/C outlet.

NOTE: High temperatures and pressures will occur at higher ambient temperatures. In areas of high humidity it is possible to have thermometer and gauge readings approach but not reach the figures listed in the performance tables and still have a

Ambient Air in Degrees F. (In Auxiliary Fan Air Blast Ahead of Condenser)	70°		80°		90°		100°		110°	
Air Quality	Arid	Humid	Arid	Humid	Arid	Humid	Arid	Humid	Arid	Humid
*Average Compressor Head Pressure in PSI	163 to 187	193 to 217	193 to 217	223 to 247	223 to 247	263 to 287	258 to 282	313 to 337	298 to 322	373 to 397
Average Evaporator Pressure **PSI AT SEA LEVEL	27.5 to 29.5	28.5 to 30.5	28 to 30	29 to 31	28 to 30	30 to 32	28 to 30	33 to 35	28.5 to 30.5	37.5 to 39.5
Center Outlet Temperature in Degrees F.	38° to 42°	41° to 45°	39° to 43°	43° to 47°	40° to 44°	47° to 51°	41° to 45°	52° to 56°	43° to 47°	59° to 63°

*NOTE: These pressures are for engine with engine fan clutch engaged. With fan clutch disengaged, pressures generally are 25-35 psi higher than shown here

**NOTE: Interior pressure of the evaporator is isolated from exterior atmospheric pressure. As a result, the controlling element of the P.O.A. valve is able to operate independently of the effect of atmospheric pressure. However, any gauge used to check the pressure will not be free from the effect of atmospheric pressure. This altitude effect on the gauge must be taken into account when interpreting a reading. As the altitude increases and atmospheric pressure goes down, the pressure reading on the gauge will go up.

The increase noted in the above readings will be approximately .5 psi per 1,000 feet above sea level. For example at 90°F in an arid climate at 2,000 feet above sea level, evaporator pressures would be 29 to 31 psi.

Fig. 232 Performance test chart. 1974-76 Models except Astre & Sunbird & Late 1976 Grand Prix

satisfactory operating unit. However, it is important to remember that low pressure has a direct relationship to nozzle outlet temperature. If pressure is too low, ice will gradually form on the evaporator fins, restricting airflow into the passenger area and resulting in insufficient or no cooling.

Insufficient Cooling Diagnosis, All Systems

1974-76 Except Astre & Sunbird

When diagnosing insufficient cooling problems, with cowl inlet temperature above 70° F, check refrigerant sight glass for foaming. If foaming is present, refer to insufficient cooling diagnosis chart, Fig. 236 part 1 of 2. If sight glass is clear refer to insufficient cooling diagnosis chart, Fig. 236 part 2 of 2.

1975-76 Astre & Sunbird

When diagnosing insufficient cooling problems, refer to insufficient cooling diagnosis chart, Fig. 167.

1974 CUSTOM A/C

FUNCTIONAL TEST

To determine if the air conditioning system is performing as it should, the following functional test should be used to evaluate system operation:

NOTE: *This test must be performed on complete car with engine running, coolant warm and environment above 45 degrees F.*

1974 Full Size, LeMans, Grand Am & Grand Prix

1. Move selector lever to OFF, set blower switch to HI blower (full up) and move temperature lever to full HOT (all the way to the right.)
   (a) The blower will run on Lo speed with air flow out the heater outlet only.
   (b) The compressor is not running.
2. Move selector lever to Defrost on 1974 models.

Temperature of Air Entering Condenser	70°	80°	90°	100°	110°
Engine rpm	1500				
Compressor out Pressure (before expansion tube)	150 205	185 235	215 270	250 300	280 335
Evaporator Pressure - at accumulator (psi)	20-26	20-28	21-29	25-33	30-39
Discharge Air Temp. — Right Hand Center Outlet	35° 41°	36°-43°	36°-44°	42°-52°	47°-57°

Fig. 233 Performance test chart. 1975-76 Astre

TEMPERATURE OF AIR ENTERING CONDENSER	70°	80°	90°	100°	110°
ENGINE RPM	2000				
COMPRESSOR HEAD PRESSURE	140 160	170 190	200 220	240 260	270 290
EVAPORATOR PRESSURE @ VIR	31 34	30 33	28.5 31.5		
DISCHARGE AIR TEMP. @ RIGHT HAND OUTLET	49 53	48 52	44 48	48 52	48 52

Fig. 234 Performance test chart. 1976 Sunbird w/4 cyl. engine

AMBIENT TEMP. DEGREES F.	EVAPORATOR PRESSURE P.S.I.G. AT EE–VIR	COMPRESSOR HEAD PRESSURE P.S.I.G.	OUTLET TEMP. DEGREES F. LEFT	RIGHT
70	28.5 – 30	180 – 210	39–40	40–43
80	29 – 31	205 – 230	41–45	43–47
90	30 – 32	235 – 270	45–52	47–54
100	30 – 35	275 – 310	48–54	50–56

Fig. 235 Performance test chart. 1976 Sunbird w/ V6-231 engine

---

NOTE: *Each lever position has a detent to lock the lever in place.*

(a) Blower comes on high speed.
(b) Hot air comes chiefly from defroster nozzles with some air from heater outlet.
(c) Compressor is running.

3. Move selector lever to Bi-Level on 1974 models.
   (a) Hot air comes from defroster nozzles, heater outlet and A/C outlets.
   (b) The compressor continues to run.
4. Move selector lever to HEATER
   (a) Hot air comes chiefly from heater outlet with some air from defroster nozzles.
   (b) Compressor is not running.
5. Move selector lever to VENT.
   (a) Hot air comes from heater outlet and A/C outlets. Air may come from defroster nozzle for first 20 to 30 seconds and then stop.
   (b) Compressor is not running.
6. Move selector lever to A/C (air conditioning) and temperature lever to a point below the word "A/C" (most of the way to the left).
   (a) Cold air comes from the air conditioning outlets.
   (b) Compressor is running.
7. Move the blower switch to full down pausing at each detent on way.
   (a) Blower speed and air flow decrease a step at each position.
8. Move temperature lever to full COLD (rest of the way to the left).
   (a) Blower comes on high speed.
   (b) Air noise is heard from R.H. front corner of passenger compartment as system goes on inside air.

NOTE: *There is a restrictor in the vacuum line which will delay obtaining air from the defroster nozzle in the heater mode when coming from any position except DEFOG or DE-ICE. This delay should be 30 to 50 seconds.*

1974 Firebird

1. Move selector to OFF, set blower switch to HI, and move temperature lever to full WARM.
   (a) No air flow from any outlets.
   (b) Compressor is not running.
   (c) On Firebird models, blower motor is running at low speed to achieve evaporator purging but no air enters the car.
2. Move selector lever to DE-ICE.

NOTE: *Each lever position has a detent to lock the lever in place.*

   (a) Blower comes on high speed.
   (b) Hot air comes chiefly from defroster nozzles with some air from heater outlet.
   (c) Compressor is running.
3. Move selector lever to HEAT.
   (a) Hot air comes chiefly from heater outlet with some air from defroster nozzles.
   (b) Compressor is running.
4. Move selector lever to BI-LEVEL and temperature lever to mid-travel.
   (a) Air flows from both heater and air conditioning outlets.
   (b) Air conditioning outlet air slightly cooler than heater outlet air.
   (c) Compressor is running.
5. Move selector lever to A/C (air conditioning) and temperature lever to full COOL.
   (a) Cold air comes from the air conditioning outlets.

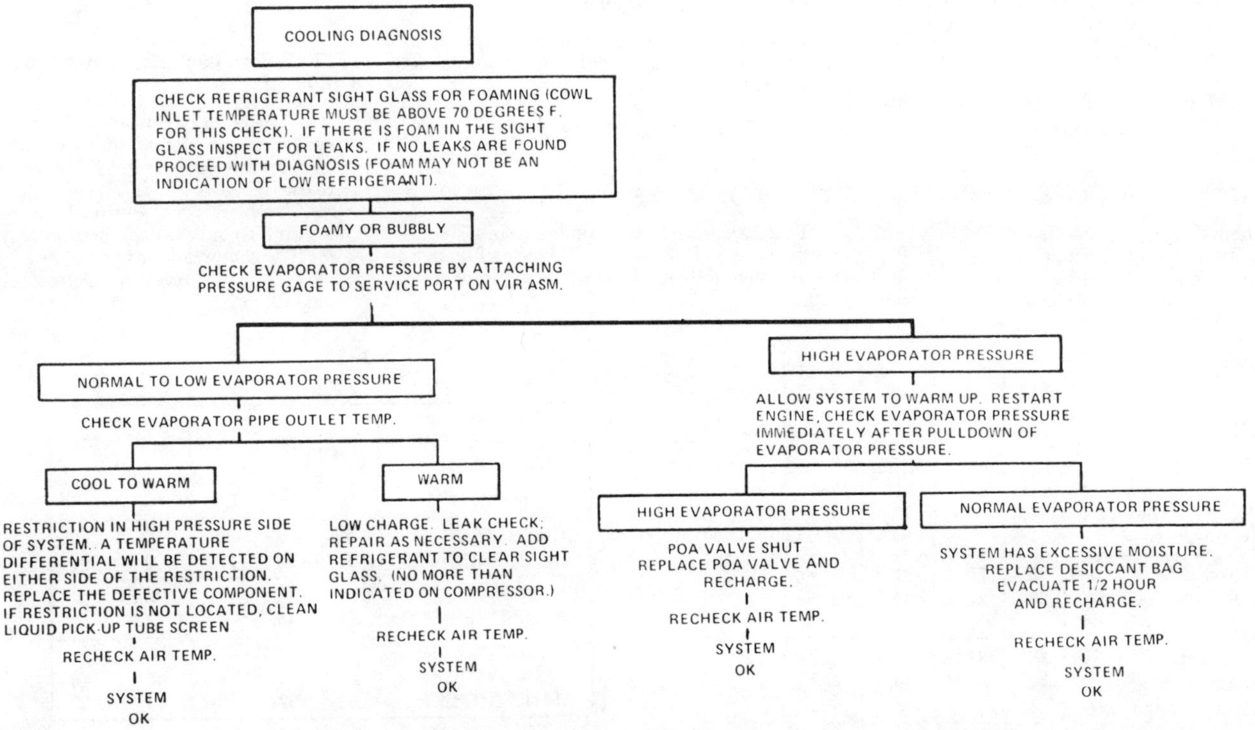

Fig. 236 Insufficient cooling diagnosis chart (sight glass foamy). 1974 all & 1975-76 except Astre & Late 1976 Grand Prix (Part 1 of 2)

(b) Compressor is running.

6. Move compressor (VENT-NORM) switch to VENT.
   (a) Outside air comes directly through air conditioning outlets with no heating or cooling.
   (b) Compressor does not run.
7. Move selector lever to INSIDE.
   (a) Blower noise level increases, also a slight increase in air flow.
   (b) Compressor switch returns to the NORM position.
   (c) Compressor is running.
8. Move the blower switch to LO pausing at each detent on way.
   (a) Blower speed and air flow decrease a step at each position.

1974 Auto Temp Control

FUNCTIONAL TEST

This functional test is to be made on car with engine running, coolant warm, hood closed and in approximately 75°F. ambient temperature.

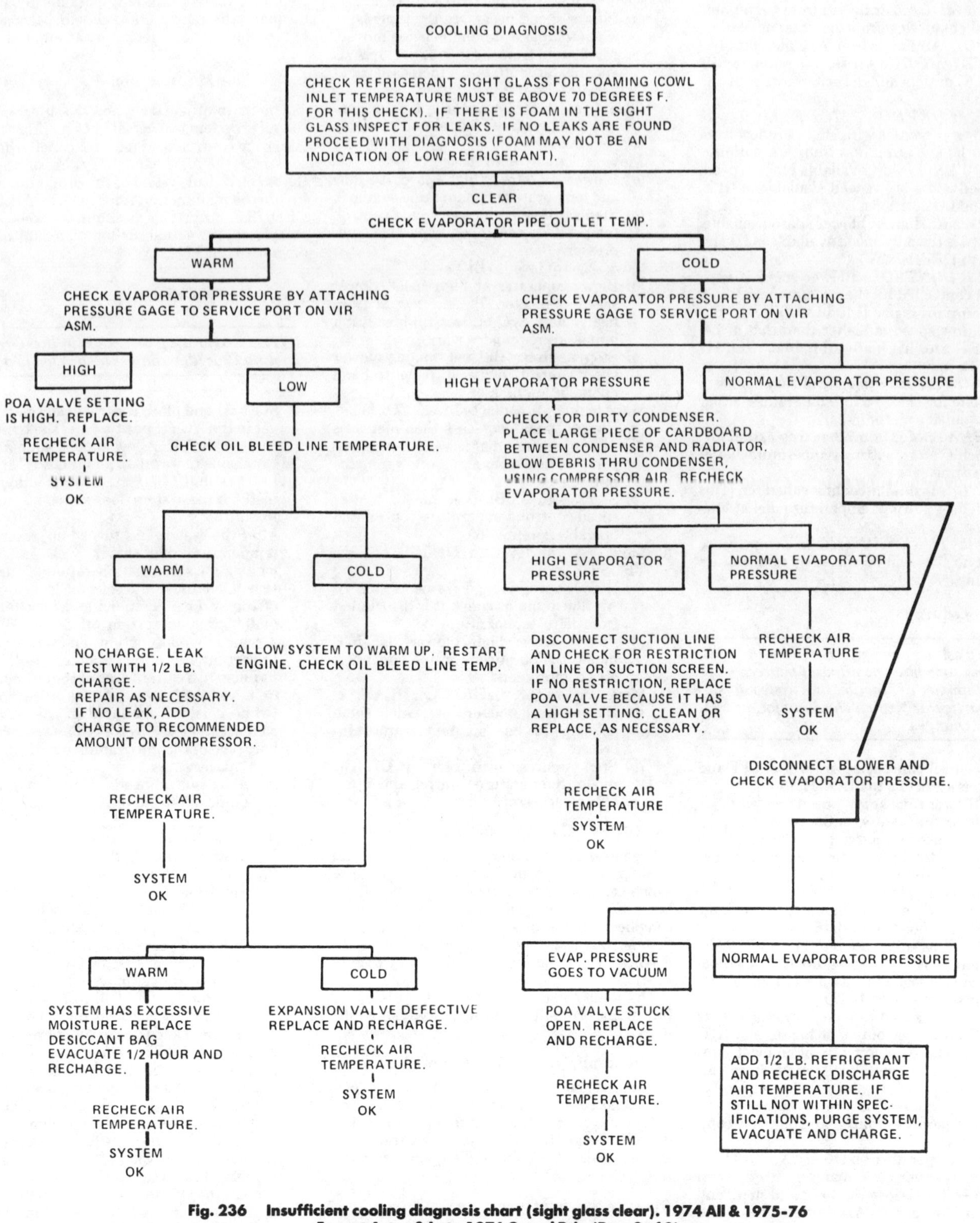

Fig. 236 Insufficient cooling diagnosis chart (sight glass clear). 1974 All & 1975-76 Except Astre & Late 1976 Grand Prix (Part 2 of 2)

Each position of the control panel selector lever has a detent to retain lever in that position. Be sure lever engages in detent.

Variations in ambient and in-car temperatures will affect system response and cause one or two mode shifts with a given setting of the control panel.

Terms used in the functional test are defined as follows:

HEATER MODE: Air flow is from heater outlet with slight air flow from defroster nozzles.
A/C MODE: Air flow is from A/C outlets only.
BI-LEVEL MODE: Air flow is from heater outlet, A/C outlets and defroster nozzles at the same time.
DE-ICE or DEFROST MODE: Air flow is from defroster nozzles with slight air flow from heater outlet and no air flow from A/C outlets.
LO BLOWER: Lowest available blower speed. This speed is the only speed available in OFF, VENT and LO positions.
HI BLOWER: Highest blower speed available. This speed is the only speed available in HI and DE-ICE or DEFROST.
BLOWER-CONTROL: Blower speed is variable and controlled by the programmer. When blower control is provided (AUTO and BI-LEVEL), low speed is higher than that in LO BLOWER and high speed is less than HI BLOWER.
MODE CONTROL: System can vary modes to compensate for changes in temperature, either ambient (outside air) or in-car.
MAX. HEAT: Maximum heating called for. This is achieved by setting temperature dial at highest setting.
MAX. A/C: Maximum cooling called for. This is achieved by setting temperature dial at lowest setting.
TEMPERATURE CONTROL: Temperature of outlet air flow can be varied by temperature dial setting.

Test Procedure

NOTE: *Test is to be performed in order indicated and time must be allowed between steps for programmer to function and system to stabilize, particularly when checking for air temperature changes.*

1. Set control panel selector lever to OFF and temperature dial to MAX. HEAT.
   (a) Blower runs at low speed and warm air flows from heater outlet only.
   (b) Step assures: purge (Lo) blower, sealed defroster, open water control valve and inoperable compressor.
2. Move selector lever to VENT.
   (a) Blower speed remains low. Cool air flows from A/C outlets.
   (b) Step assures: Lo blower, A/C mode, minimum temperature, inoperable compressor and closed water valve.
3. Move selector lever to LO.
   (a) Blower speed remains low. Air flow is from heater outlet or heater and A/C outlets (depending upon ambient temperature) and air temperature increases from that in step 2.
   (b) Step assures: Lo blower, mode control, temperature control and operable compressor.
4. Move temperature dial to MAX. COLD.
   (a) Blower speed remains low. Air temperature is lower than that in step 3 and air flow is from A/C outlets or A/C and heater outlets.
   (b) Step assures: Lo blower, mode control, temperature control and operable compressor.
5. Move selector lever to AUTO.
   (a) Blower speed increases. Air flow and air temperature remain same as in step 4.
   (b) Step assures: blower control, temperature control, mode control and operable compressor.
6. Move temperature dial to MAX. HEAT.
   (a) Blower speed decreases then increases when temperature dial is being moved. Air flow is from heater outlet or A/C and heater outlets at increased temperature from step 5.
   (b) Step assures: blower control, mode control, temperature control and operable compressor.
7. Move selector lever to HI.
   (a) Blower operates at high speed. Air flow and temperature remain same as step 6.
   (b) Step assures: Hi blower, mode control, temperature control and operable compressor.
8. Move selector lever to Bi-Level.
   (a) Blower operates at high speed (lower speed than step 7). Air flow is from heater and A/C outlets (bi-level) and temperature remains warm.
   (b) Step assures: Bi-level mode, blower speed control, temperature control and operable compressor.
9. Move temperature dial to MAX. COLD.
   (a) Blower speed decreases then increases as temperature dial is varied. Air temperature decreases and air flow remains bi-level.
   (b) Step assures: Bi-level mode, blower speed control, temperature control and operable compressor.
10. Move selector lever to DE-ICE or DEFROST.
    (a) Blower speed is high (same as step 7). Air flow remains cool but is distributed from defroster nozzles.
    (b) Step assures: high blower, DE-ICE mode, temperature control and operable compressor.
11. Move temperature dial to MAX. HEAT.
    (a) Blower speed and air flow remain same as step 10 but air temperature increases.
    (b) Step assures: high blower, DE-ICE mode, temperature control and operable compressor.

DIAGNOSTIC PROCEDURE

Diagnosis of a malfunction in the system can be simplified by a logical, systematic procedure which can minimize diagnosis time and effort. Lack of a logical, systematic procedure could complicate diagnosis of a relatively minor malfunction or waste time and effort attempting to diagnose a malfunction which does not actually exist.

To simplify diagnosis of the ATC system, the following procedure should always be employed.

Understand System Operation

Before a malfunction in the ATC system can be diagnosed and corrected, an understanding of how the system operates under normal conditions is essential. System operation is described under "Automatic Temperature Control Systems."

Observe System Operation

The most important part of diagnosing a problem is to determine exactly what the complaint is and whether this complaint actually stems from a malfunction in the system. (Without a thorough understanding of how the system is supposed to operate, a complaint stemming from misunderstanding of system operation or improper setting of the control panel lever could result in the search for a malfunction that does not exist). If complaint is of a condition that occurs only periodically (intermittent), the malfunction should be observed before diagnosis and repair are attempted.

Isolate The Malfunction

The main objective when troubleshooting the ATC system is to isolate the malfunction to either the control head (control panel), the sensor string, the ATC vacuum system on the car, the car electrical system or the programmer. To isolate the malfunction refer to ON CAR PROCEDURE. After this preliminary isolation is completed, the actual malfunction can then be diagnosed quickly.

ON CAR PROCEDURE

NOTE: *Do not skip any steps in the Troubleshooting Procedure unless instructed to do so.*

1. Start car and place control head selector lever in the AUTO position and the temperature dial at 75°. Allow enough time for the car engine to warm up and the ATC system to come on if the system were acting normally. If the system does not come on, skip to Step 3.
2. After the system has turned on, rotate the temperature dial slowly back and forth looking for abnormal operation of the system. Then move the selector lever to each of the various positions and look for any malfunction in system operation. By observing any malfunctions and noting when they occur, it is often possible to isolate the problem to a certain area of the system.
3. Remove the cover of the programmer (while it is still mounted in the car) and connect the J-23678 ATC Tester electrical harness to the programmer and the car ATC harness.
4. Place the control head selector lever in the AUTO position and the temperature dial at 75°.
5. On the tester, place the Temperature Dial Calibration switch in the OFF position and the Manual Automatic switch in the MANUAL position.
6. Rotate the Manual Control knob to MAX. HEAT. The programmer should move to the full heat position and the fan should run at high blower speed. (The vacuum motor mechanism will move into the vacuum motor.) Rotate the knob to MAX. COLD. The programmer should move to the full A/C position and the fan should run at high blower speed. (The vacuum motor mechanism will move out of the vacuum motor.) If only partial programmer movement occurs or "Hi blower" is not obtained at both extremes, make the air mix (temperature) door link adjustment before proceeding to the next step. If the programmer does not move at all, skip to Step 9.
7. Rotate the Manual Control knob slowly counter-clockwise. The vacuum motor mechanism should first start to move at

precisely 180 ohms (± 1 ohm) on the Manual Control knob. If this first movement occurs before or after 180 ohms, make the Programmer Amplifier Calibration (Feedback "Pot" Adjustment) described under ON BENCH PROCEDURE-PROGRAMMER.

8. Rotating the Manual Control knob to the MAX. COLD. position should cause the system to shift to full A/C operation and the vacuum motor mechanism will move out of the vacuum motor. If the programmer moves normally when rotating the Manual Control knob, skip to Step 11.

Programmer Does Not Move

9. On the ATC Tester, place the Temperature Dial Calibration switch in the OFF position and the VOLTMETER control in the "PROGRAMMER 12-V SUPPLY" position. Battery voltage should appear on meter. No voltage indicates the lack of a ground on terminal #1 of the programmer or the lack of battery supply to terminal #2 of the programmer.
   (a) If battery voltage appears on voltmeter, proceed to step 10.
   (b) If no voltage appears on voltmeter, place voltmeter knob in the "Probe and Clip" position. Determine if programmer terminal #1 is ungrounded and/or terminal #2 has no battery voltage applied to it. Trace appropriate wire(s) to find malfunction in car ATC electrical system.
10. Plug the tester's dummy vacuum plug on the programmer vacuum valve. Connect the dummy plug hose to the vacuum supply hose (Port #2) in the car vacuum harness. Make sure vacuum is present. The programmer now has vacuum supply with no car vacuum system components connected. If the programmer functions using the MANUAL CONTROL on the tester, troubleshoot the ATC vacuum system. If the programmer does not function, the defect is in the programmer. Remove the dummy plug and reconnect the vacuum harness after making this test. If defect is in programmer, remove the programmer and follow procedure under ON BENCH PROCEDURE-PROGRAMMER.

Sensor String Test

NOTE: *Do not perform sensor string test on car which has been run at idle for a long period of time in a heated building. This condition produces sufficient engine heat to heat the ambient sensor and cause lower resistance readings than specified in following procedure even though no problem exists in the sensor string.*

11. Place the control head selector lever in the AUTO position and the temperature dial on 75°
12. On the J-23678 ATC Test, place the Manual Automatic switch in the AUTOMATIC position.
13. Observe the position of the vacuum motor mechanism.
14. On the ATC Test, switch the Manual-Automatic switch to the MANUAL position and adjust the Manual Control knob until the vacuum motor mechanism assumes the same position it had in Step 13. When the Manual Control knob is properly adjusted, switching the Manual-Automatic switch back and forth will result in no movement of the vacuum motor mechanism.
15. Read the setting of the MANUAL CONTROL. This resistance reading is the same as the resistance of the two sensors and the temperature dial combined and should be 120 to 150 ohms at 70° to 75°F room temperature. If this reading is incorrect, try calibrating the temperature dial. If the calibration can be accomplished, the temperature dial is "good." Visually check the ambient and in-car sensors for shorts or bad connections.

Lo Relay Test

16. Place the ATC Tester's Temperature Dial Calibration switch in the OFF position and the Voltmeter knob in the "LO RELAY" position.
17. Place the control head selector lever in the VENT or DE-ICE position. If the Lo Relay is energized, the voltmeter will read battery voltage. The Lo Relay is operated by:
   (a) Hot engine thermal switch.
   (b) Hot in-car temperature switch.
   (c) Control head in VENT or DE-ICE position.

If the relay does not energize in the control head VENT or DE-ICE position, the relay or the control head switch is defective.

Auto Relay Test

18. Place the ATC Tester's Temperature Dial Calibration switch in the OFF position and the Voltmeter knob in the "AUTO RELAY" position.
19. Place the control selector lever in the DE-ICE position. If the Auto Relay is energized, the Voltmeter will read battery voltage. The Auto Relay is operated by:
   (a) Hot engine thermal switch.
   (b) Hot in-car temperature switch.
   (c) Control head in DE-ICE.

If the relay does not energize in control head DE-ICE position, the relay or the control head switch is defective.

Programmer Blower Switch Test

20. Place ATC Tester Temperature Dial Calibration switch in the OFF position, the Voltage knob in the "BLOWER" position, and the Manual-Automatic switch in the MANUAL position.
21. Using the MANUAL CONTROL, move the programmer from full heat to the full A/C position. The voltage at the blower (coming from programmer terminal #6) should be battery voltage in full heat. As the programmer moves from full heat, the voltage should drop in steps, indicating different blower speeds, and then increase in steps to battery voltage in full A/C. If the voltage steps are not present or battery voltage is not indicated on the tester voltmeter when in full heat and full A/C operation, the programmer is defective. Remove and troubleshoot the programmer (See ON BENCH PROCEDURE-PROGRAMMER).

ON BENCH PROCEDURE-PROGRAMMER

Before attempting to repair or adjust the programmer, locate the output shaft, the feedback potentiometer, the vacuum motor, the blower switch, the vacuum checking relay, the transducer and the rotary vacuum valve. Refer to Fig. 95 for identification of all programmer components.

Equipment Required

1. D.C. power supply rated at 5 amperes at 12 volts.
2. J-23678 Automatic Temperature Control Tester.
3. Vacuum supply capable of 20 inches of Hg. or more.
4. ¼" hollow socket and drive.
5. Medium size blade screwdriver.
6. Long-nose pliers.
7. J-23713 electrical test connector.

When following the troubleshooting procedure, do not skip any steps unless instructed to do so. The ATC Tester Manual-Automatic switch should always be in the MANUAL position and the Temperature Dial Calibration switch should always be in the OFF position when troubleshooting the programmer on the bench.

1. Plug male connector of ATC Tester electrical harness into programmer.
2. Plug J-23713 electrical test connector into female connector of ATC Tester electrical harness.
3. Plug dummy plug from tester onto the programmer rotary valve and connect it to the vacuum supply. Turn on the vacuum supply to 20 inches Hg. or more.
4. Connect positive lead of 12 volt DC power supply to yellow wire from electrical test connector. Connect negative lead of power supply to black wire of test connector. Voltmeter on tester should read 12 volts when in "#2 Feed" position, "#7 Lo Relay" position and "#8 Auto Relay" position. If it does not, check the voltage supply and hook-up.
5. On the ATC Tester, place the Manual-Automatic switch in the MANUAL position and the temperature Dial Calibration switch in the OFF position.
6. Rotate the ATC Tester Manual Control knob from MAX. HEAT to MAX. COLD. (Make sure the power supply is adjusted for 12 volts DC and vacuum supply is at 20 inches Hg. or more.) The programmer output shaft should rotate 120 degrees (⅓ rotation of shaft) from maximum heat to maximum cold. If it does not rotate normally, adjust the feedback potentiometer.
7. If the programmer output shaft does not operate properly in Step 6, remove the cover of the programmer and reconnect the tester electrical connector and the rotary vacuum valve dummy plug. Recheck programmer operation as described in Step 6. If the programmer output shaft now operates normally, look for pinched vacuum hose, loose electrical connection, pinched or shorted Darlington Amplifer heat sink or a mechanical bind with the programmer cover. Be sure to tap the programmer components with the handle of a screwdriver to check for intermittent problems.

Amplifier Test

8. If the programmer output shaft still does not operate properly, place the Voltage knob of the tester in the "Probe and Clip" position. Connect the alligator clip lead to the transducer terminal with the gray wire. CAUTION: DO NOT SHORT TRANSDUCER TERMINALS TOGETHER OR AMPLIFIER WILL BE DAMAGED. Push the probe into the programmer connector body making contact with terminal #2 which connects to a yellow wire.

9. Rotating the Manual Control knob from MAX. HEAT to MAX. COLD (or vice versa), the voltmeter reading should change 5 volts or more (after it is stabilized). This change indicates that the amplifier is "good." If the 5 volts or more change is not obtained, check the amplifier heat sink insulator for proper positioning. (NOTE: If the heat sink is shorted to the programmer chassis, the programmer will go to the full A/C position.) If the voltage change cannot be obtained and the insulator is okay, check the calibration of the amplifier accordingly. If the programmer still does not function properly, replace the amplifier circuit board and be sure to install the new heat sink insulator properly. The new amplifier must be calibrated.

Transducer Test

10. If the 5 volts change was present in the amplifier check in Step 9 but the programmer output shaft does not operate normally, leave the "Probe and Clip" connected as instructed in Step 8 throughout transducer test procedure.
11. Disconnect vacuum hose from the small diameter port of the transducer and connect ATC Tester's hose to the vacuum gauge directly to this hose. About 20 inches Hg. or more vacuum should be present on this transducer vacuum supply hose. If vacuum is proper, restore hose connection. If 20 inches Hg. or more vacuum is not present, check the vacuum supply vacuum level and then go to Steps 14, 15, 16 and 17 to check for vacuum leaks in the vacuum checking relay and the rotary vacuum valve.
12. To check for proper regulated vacuum output from the transducer, disconnect the long hose from the transducer to the vacuum checking relay at the relay. (NOTE: This long hose must be at least 15 inches long or the transducer will make a buzzing noise.) Connect the ATC Tester's vacuum gauge directly to the long hose at the vacuum checking relay.
13. Position the programmer in an upright position so that the output shaft points straight up. With the Manual Control knob in the maximum heat position, "0" volts should be read on the voltmeter and the vacuum should be 9 to 11 inches. Turning the Manual Control knob to the MAX. COLD position should make the vacuum drop to "0" inches and the voltage should increase 5 volts or more. If these indications do not occur, the transducer is defective. After the check is completed, restore hose connection. Disconnect probe and clip from the programmer.

Rotary Vacuum Valve and Vacuum Checking Relay Test

14. Connect the tester's vacuum gauge into the vacuum feed line to the dummy plug on the programmer rotary vacuum valve. Disconnect the center hose from the side port of the rotary vacuum valve and seal off the port by placing your finger over the port. The tester's vacuum gauge should read 20 inches Hg. or more. If it does not, either the vacuum checking relay or the rotary vacuum valve is leaky. If 20 inches Hg. or more vacuum is read, skip to Step 16 and restore the hose connection.
15. Remove all three hoses from the side ports of the rotary vacuum valve. Remove the short hose from the transducer and connect it across the two outer ports on the side of the rotary valve. Place finger over center side port of the rotary vacuum valve. If the tester's vacuum gauge now reads 20 inches Hg. or more, the vacuum checking relay is defective. If it does not the rotary valve is defective and leaking. After repair of programmer, restore all hose connections.

Vacuum Motor and Vacuum Checking Relay Test

16. Disconnect electrical connector from programmer. Do not disconnect vacuum input to programmer. The programmer should go to the full heat position (the vacuum motor mechanism will move into the vacuum motor). If it does not, either the vacuum checking relay or the vacuum motor has a severe leak. If it does go to full heat, a slight leak may be present and the vacuum motor will move toward maximum cold after removing the rubber dummy plug from the programmer.
17. If the vacuum motor mechanism moves, the vacuum motor or vacuum checking relay is leaking. Disconnect the short hose at vacuum motor. Apply raw vacuum to the vacuum motor input port then pinch the hose at the vacuum motor with long nose pliers and hold for 30 seconds. If the mechanism moves, the vacuum motor is defective. If it does not move, the vacuum checking relay is defective.

Feedback Potentiometer Adjustment

18. With programmer connected to tester, 12 volt supply and vacuum supply as instructed in Steps 1 thru 4 under PROGRAMMER DIAGNOSIS-OFF CAR, and with cover removed use a screwdriver to turn the feedback potentiometer fully counterclockwise. The vacuum motor will now be in the full heat position. (Mechanism moves into motor.)
    - *(a)* With MANUAL AUTOMATIC switch in MANUAL position rotate MANUAL CONTROL to MAX. HEAT.
    - *(b)* Carefully adjust MANUAL CONTROL to 180, do not go beyond, then back.
    - *(c)* Turn the feedback potentiometer very slowly clockwise while watching the vacuum motor mechanism, not the output shaft. Stop immediately when movement of the mechanism occurs.
    - *(d)* Check adjustment by rotating the MANUAL CONTROL to MAX. HEAT then slowly rotating MANUAL CONTROL counter-clockwise while watching the vacuum motor mechanism. When the mechanism just starts to move the MANUAL CONTROL should be at exactly 180. Slight readjustment is required if MANUAL CONTROL is not at 180.

Programmer Blower Switch Test

19. Place the Voltage knob in the "#6 Blower" position. Rotate the Manual Control knob to the MAX. HEAT position. The voltmeter reading should be 12 volts D.C. (supply voltage). As the Manual Control knob is slowly rotated toward MAX. COLD at about "175" the voltage will drop slightly. At about "160" the voltage will again drop slightly, and again at about "155." At about "130," the voltage will increase slightly. At about "120," the voltage will again increase slightly and again at about "110" a slight increase should be noted. If these steps in blower voltage are not present, the blower switch in the programmer is defective.
20. Tap components of the programmer while rotating the Manual Control knob and watch for any erratic operation due to intermittent problems. Replace programmer cover and rotate Manual Control knob checking for normal rotation of the output shaft.

1975-76 Custom A/C

FUNCTIONAL TEST

NOTE: Test must be performed with engine running, coolant warm and environment above 45°F. Each selector lever position has a detent to lock lever in place.

1975-76 Full Size, Grand Prix (Except Late 1976) & LeMans

1. Position selector lever in Off, fan switch in Hi and temperature lever in full Hot.
    - *(a)* Blower will run at low speed with air flow from heater outlet only. If temperature at control panel is below 40°F blower will not operate.
    - *(b)* Compressor is not operating.
2. Position selector lever in Defrost.
    - *(a)* Blower comes on high speed.
    - *(b)* Hot air is delivered from defroster outlet with some air from heater outlet.
    - *(c)* Compresser is operating.
3. Position selector lever in Heater.
    - *(a)* Hot air is delivered from heater outlet with some air from defroster outlet.
    - *(b)* Blower remains at high speed.
    - *(c)* Compressor is not operating.
4. Position selector lever in Vent.
    - *(a)* Hot air is delivered from A/C outlets.
    - *(b)* Blower remains at high speed.
    - *(c)* Compressor is not operating.
5. Position selector lever in Bi-Level.
    - *(a)* Hot air is delivered from heater and A/C outlets with some air from defroster outlet.
    - *(b)* Blower remains at high speed.
    - *(c)* Compressor is operating.
6. Position selector lever in Normal and temperature lever in full cold.
    - *(a)* Cold air is delivered from A/C outlets.
    - *(b)* Blower operates at high speed.
    - *(c)* Compressor is operating.
7. Pausing at each detent, move fan switch to full down position.
    - *(a)* Blower speed and airflow should decrease.
8. Position selector lever in Max.
    - *(a)* Blower operates at high speed.
    - *(b)* As system goes on inside air noise can be heard from right hand side of passenger compartment.

NOTE: A restrictor in the vacuum line will delay obtaining air from defroster outlet in Heater when coming from any position except Defrost or Bi-Level.

1975-76 Firebird

1. Position selector lever in Off, blower switch in Hi and temperature lever in full warm.

(a) Air flow from heater outlet.
(b) Compressor not operating.
(c) Blower operating at low speed.

2. Position selector lever in Defrost.
   (a) Blower comes on high speed.
   (b) Hot air is delivered from defroster outlet with some air from heater outlet.
   (c) Compressor is operating.
3. Position selector lever in Heat.
   (a) Hot air is delivered from heater outlet with some air from defroster outlet.
   (b) Blower remains at high speed.
   (c) Compressor is operating.
4. Position selector lever in Bi-Level.
   (a) Air flow from heater and A/C outlets with a slight flow of air from defroster outlet.
   (b) Blower remains at high speed.
   (c) Compressor is operating.
5. Position selector lever in A/C and temperature lever in full cool.
   (a) Cold air is delivered from A/C outlets.
   (b) Blower remains at high speed.
   (c) Compressor is operating.
6. Place Norm-Economy switch in Economy.
   (a) Outside air is delivered through A/C outlets.
   (b) Blower remains at high speed.
   (c) Compressor is not operating.
7. Move selector lever to Inside.
   (a) Air flow and blower noise will increase.
   (b) Blower remains at high speed.
   (c) Norm-Economy switch will return to Norm.
   (d) Compressor is operating.
8. Pausing at each detent, move fan switch to low.
   (a) Blower speed and air flow will decrease.

1975-76 Auto Temp. Control

Functional Test

NOTE: Test should be performed with engine running, coolant warm and ambient temperature at approximately 75°F.

1. Position selector lever on Off, fan switch on Hi and temperature lever at 85.
   (a) Warm air is delivered from heater outlet.
   (b) Blower motor runs at low speed.
   (c) Compressor is not operating.
2. Position selector lever in Defrost.
   (a) Blower motor changes to high speed.
   (b) Warm air is delivered from defroster outlet with small amount of air flow from heater outlet.
   (c) Compressor is operating.
3. Position selector in Heat.
   (a) Warm air is delivered from heater outlet with small amount of air flow from defroster outlet.
   (b) Blower motor remains at high speed.
   (c) Compressor is not operating.
4. Position selector lever in Vent.
   (a) Warm air flows from A/C outlets.
   (b) Blower motor remains at high speed.
   (c) Compressor is not operating.
5. Position selector lever in Bi-Level.
   (a) Warm air is delivered from heater and A/C outlets with small amount of air flow from defroster outlets.
   (b) Blower motor remains at high speed.
   (c) Compressor is operating.
6. Position selector lever in Norm and temperature lever to 65.
   (a) Cold air is delivered from A/C outlets.
   (b) Blower operates at high speed.
   (c) Compressor is operating.
7. Pausing at each detent, move fan switch to Low.
   (a) Blower motor speed and air flow should decrease.
8. Position selector lever in Max.
   (a) Blower motor changes from low to high speed.
   (b) System goes to recirculation air flow. Air noise can be heard from right corner of passenger compartment.
   (c) Cold air is delivered from A/C outlets.
   (d) Compressor is operating.

NOTE: During the functional test the following items may be noted.

1. A change in temperature lever position will affect outlet air temperature in all selector lever positions except Max. Engine coolant is not allowed to flow through the heater core when selector lever is in the Max position. Only maximum cooling is available in Max. position.
2. When coming from any selector position except Bi-Level or Defrost to Heater, a restrictor in the vacuum line will delay air flow to the defroster outlets.
3. The blower motor will not operate in the Off or Heater positions until the engine is warmed up to approximately 120°F.
4. The compressor will not operate if ambient temperature is below 35°F. At temperatures above 35°F the compressor will operate in the following selector positions: Max, Norm, Bi-Level and Defrost.

Temperature Control Diagnosis

NOTE: A malfunction in the temperature control may cause poor regulation of outlet air temperature, warm air flow only or cold air flow only.

1. Remove glove box so that programmer, linkage and temperature door crank arm can be viewed.
2. With engine running and at operating temperature, position selector lever in Norm and fan switch in Hi.
3. Move temperature lever from full cold to full warm and back to full cold. When lever is moved from full cold to full warm, outlet air temperature should increase then decrease when lever is moved back to full cold. Note movement of programmer linkage and temperature door crank.
4. If outlet air temperature does not vary and programmer link does not move when temperature lever is moved, check control cable to ensure it is not disconnected or broken.
5. If control cable is found satisfactory, but programmer link does not move and outlet air temperature does not vary, check for disconnected programmer linkage and temperature door for binding.
6. If temperature door moves freely but only cold air is delivered, check for vacuum at water control valve. When selector lever is in the Norm position, no vacuum should be applied to the water control valve. If vacuum is present check system for source of vacuum. If vacuum is not present but heater hose from water control valve to heater core is not hot, replace the water control valve. If heater hose is hot check for vacuum at programmer. If vacuum is not present at programmer check vacuum system for leaks. If vacuum is present, disconnect hose from vacuum motor and apply vacuum directly to motor. If vacuum motor mechanism does not move replace vacuum motor. If mechanism does move check vacuum circuit from vacuum checking relay to vacuum motor and compensator assembly for leaks. If no leakage is found, replace the vacuum checking relay.
7. If temperature door moves freely but only warm is delivered, check vacuum motor for proper operation. Vacuum motor mechanism should be fully drawn into vacuum motor when in full heat operation. Disconnect vacuum hose, if mechanism moves out of motor replace the compensator assembly. If vacuum motor mechanism does not move, replace vacuum motor.
8. If poor temperature control is noted and outlet air temperature varies, check for the following:
   (a) Improperly adjusted temperature control cable.
   (b) Improperly installed or leaking air distribution hoses.
   (c) Faulty compensator assembly in programmer.

LATE 1976 GRAND PRIX & 1977-80 ALL MODELS

Performance Test, All Systems

Except 1980 Phoenix

1. Connect manifold and gauge set to A/C system.
2. On 1977 models except Grand Prix and LeMans, slip fan clutch spring from locating notch and tighten to first notch past normal setting. On late 1976 Grand Prix, 1977 Grand Prix and LeMans and all 1978-80 models, slip fan clutch spring from locating notch to relieve spring tension from clutch.
3. Perform test with hood opened, windows and doors closed.
4. Open all A/C outlets, then position a thermometer in left center instrument panel A/C outlet and another thermometer in front of vehicle grille.
5. Start engine and allow to reach operating temperature.
6. Place a 2 ft. diameter auxiliary fan at vehicle front bumper.
7. Place control lever on Max. A/C and temperature lever on full cold. On 1978 models, place blower switch on M2, on 1979-80 models, place blower switch on High.
8. On late 1976 Grand Prix and 1977 Grand Prix and LeMans, operate engine at 1500 rpm. On all other models operate engine at 2000 rpm.
9. Note pressure and temperature readings, then refer to performance test charts, Figs. 237 through 242.
10. On Late 1976 Grand Prix, 1977-78 all models and 1979 models except Firebird and Sunbird, if outlet temperature is within limits or too cold, change blower speed to low and check accumulator and evaporator pipes from frost. If no frost is present, system is operating properly. If frost is present, the thermostatic switch is defective

TEMPERATURE OF AIR ENTERING CONDENSER		70 F (21 C)	80 F (27 C)	90 F (32 C)	100 F (38 C)	110 F (43 C)
*COMPRESSOR OUT PRESSURE (BEFORE EXPANSION ORIFICE	LEMANS GRAND PRIX	125-175 psi 862-1207 kPa	145-195 psi 1000-1345 kPa	170-220 psi 1172-1517 kPa	195-245 psi 1345-1689 kPa	225-275 psi 1551-1896 kPa
	FULL SIZE	140-180 psi 965-1241 kPa	170-210 psi 1172-1448 kPa	200-240 psi 1379-1655 kPa	235-275 psi 1620-1896 kPa	265-305 psi 1827-2103 kPa
	FIREBIRD		195-235 psi 1345-1620 kPa		225-265 psi 1551-1827 kPa	
	PHOENIX VENTURA		160-200 psi 1103-1379 kPa		215-255 psi 1482-1758 kPa	
	ASTRE		160-200 psi 1103-1379 kPa		230-270 psi 1586-1862 kPa	
	SUNBIRD		160-200 psi 1103-1379 kPa		240-280 psi 1655-1931 kPa	
*EVAPORATOR PRESSURE (AT ACCUMULATOR)**	LEMANS GRAND PRIX	25-32 psi 172-221 kPa	25-32 psi 172-221 kPa	25-32 psi 172-221 kPa	25-32 psi 172-221 kPa	25-32 psi 172-221 kPa
	FULL SIZE	24-27 psi 165-186 kPa	24-27 psi 165-186 kPa	24-27 psi 165-186 kPa	24-27 psi 165-186 kPa	24-27 psi 165-186 kPa
	FIREBIRD		24-27 psi 165-186 kPa		24-27 psi 165-186 kPa	
	PHOENIX VENTURA		24-27 psi 165-186 kPa		24-27 psi 165-186 kPa	
	ASTRE		24-27 psi 165-186 kPa		24-27 psi 165-186 kPa	
	SUNBIRD		24-27 psi 165-186 kPa		24-27 psi 165-186 kPa	
*DISCHARGE AIR TEMPERATURE—LEFT CENTER OUTLET	LEMANS GRAND PRIX	39-44 F	40-45 F	41-46 F	43-48 F	45-50 F
	FULL SIZE	41-43 F	42-44 F	42-44 F	43-45 F	43-45 F
	FIREBIRD		38-45 F		38-45 F	
	PHOENIX VENTURA		38-45 F		38-45 F	
	ASTRE		38-45 F		38-45 F	
	SUNBIRD		38-45 F		38-45 F	

*Just before compressor disengages **At sea level

Fig. 237 Performance test chart. 1977 all models

and should be replaced. If outlet temperatures are not within limits, check the following:

a. Check accumulator pressure. If pressure is low, the expansion tube orifice is plugged and should be replaced. If pressure is high to normal, check compressor cycling.
b. If compressor operates continuously, check for missing expansion tube orifice and replace if missing. If expansion tube orifice is not missing, check compressor inlet screen (axial 6 cyl. compressor). If inlet screen is plugged, remove screen and repair or replace as necessary. If screen is not plugged, system is overcharged, evacuate and recharge system.
c. If compressor cycles on and off, replace thermostatic switch.

11. On 1979 Firebird, Sunbird and 1980 all models, if outlet temperature is within limits, system is operating properly. If outlet temperature is not within limits, check the following:
    a. Check compressor cycling.
    b. If compressor operates continuously, check for missing expansion tube orifice and replace if missing. If expansion tube orifice is not missing, check compressor inlet screen (axial 6 cyl. compressor). If compressor inlet screen is plugged, repair or replace as necessary. If inlet screen is not plugged, evacuate and recharge system.
    c. If compressor cycles on and off or remains off for a long period of time, check for a plugged expansion tube and replace as necessary.

TEMPERATURE OF AIR ENTERING CONDENSER		70°F. (21°C.)	80°F. (27°C.)	90°F. (32°C.)	100°F. (38°C.)	110°F. (43°C.)
*COMPRESSOR OUT PRESSURE (BEFORE ORIFICE)	LEMANS GRAND PRIX	110-150 psi 758-1034 kPa	130-170 psi 896-1172 kPa	160-200 psi 1103-1379 kPa	195-235 psi 1344-1620 kPa	230-250 psi 1586-1724 kPa
	FULL SIZE	120-160 psi 827-1103 kPa	150-190 psi 1034-1310 kPa	185-225 psi 1276-1551 kPa	215-255 psi 1482-1758 kPa	270-310 psi 1861-2137 kPa
	FIREBIRD	155-175 psi 1069-1344 kPa	205-245 psi 1413-1689 kPa	220-260 psi 1517-1793 kPa	255-275 psi 1758-2034 kPa	295-335 psi 2034-2310 kPa
	PHOENIX	105-145 psi 724-1000 kPa	130-170 psi 896-1172 kPa	170-210 psi 1172-1448 kPa	200-240 psi 1379-1655 kPa	235-275 psi 1620-1896 kPa
	SUNBIRD WAGON	135-175 psi 931-1207 kPa	170-210 psi 1172 1448 kPa	215-255 psi 1487-1758 kPa	265-305 psi 1827-2103 kPa	295-335 psi 2034-2310 kPa
	SUNBIRD EXC. WAGON	130-170 psi 896-1172 kPa	180-220 psi 1241-1517 kPa	215-255 psi 1482-1758 kPa	240-280 psi 1655-1931 kPa	255-295 psi 1758-2034 kPa
*EVAPORATOR PRESSURE (AT ACCUMULATOR)**	LEMANS GRAND PRIX	24-30 psi 165-207 kPa	24-30 psi 165-207 kPa	24-30 psi 165-207 kPa	24-30 psi 165-207 kPa	26-30 psi 179-207 kPa
	FULL SIZE	25-30 psi 173-207 kPa	25-30 psi 173-207 kPa	25-30 psi 173-207 kPa	26-32 psi 180-220 kPa	26-32 psi 180-220 kPa
	FIREBIRD	24-30 psi 166-207 kPa	24-30 psi 166-207 kPa	25-30 psi 173-207 kPa	27-32 psi 186-220 kPa	27-32 psi 186-220 kPa
	PHOENIX	26-32 psi 165-220 kPa	26-32 psi 165-220 kPa	26-32 psi 165-220 kPa	26-32 psi 165-220 kPa	26-32 psi 165-220 kPa
	SUNBIRD WAGON	24-30 psi 166-207 kPa	24-30 psi 166-207 kPa	25-30 psi 173-207 kPa	26-32 psi 186-220 kPa	26-32 psi 186-220 kPa
	SUNBIRD EXC. WAGON	24-30 psi 166-207 kPa	24-30 psi 166-207 kPa	24-30 psi 166-207 kPa	24-32 psi 166-220 kPa	26-32 psi 186-220 kPa
*DISCHARGE AIR TEMPERATURE—LEFT CENTER OUTLET	LEMANS GRAND PRIX	34-38°F. 1.1-3.3°C.	34-38°F. 1.1-3.3°C.	35-39°F. 1.7-3.9°C.	38-42°F. 3.3-5.6°C.	40-44°F. 4.4-6.7°C.
	FULL SIZE	38-42°F. (3-6°C.)	38-42°F. (3-6°C.)	38-42°F. (3-6°C.)	38-42°F. (3-6°C.)	42-48°F. (6-9°C.)
	FIREBIRD	40-45°F. (4-7°C.)	40-45°F. (4-7°C.)	42-46°F. (6-8°C.)	42-46°F. (6-8°C.)	44-50°F. (7-10°C.)
	PHOENIX	38-42°F. 3.3-5.6°C.	38-42°F. 3.3-5.6°C.	38-42°F. 3.3-5.6°C.	38-42°F. 3.3-5.6°C.	38-42°F. 3.3-5.6°C.
	SUNBIRD WAGON	44-48°F. (7-9°C.)	44-48°F. (7-9°C.)	44-48°F. (7-9°C.)	46-50°F. (8-10°C.)	50-54°F. (10-12°C.)
	SUNBIRD EXC. WAGON	38-42°F. (3-6°C.)	38-42°F. (3-6°C.)	38-42°F. (3-6°C.)	38-42°F. (3-6°C.)	38-42°F. (3-6°C.)

*JUST BEFORE COMPRESSOR DISENGAGES
**AT SEA LEVEL

Fig. 238 Performance test chart. 1978 all models & 1979 models except Firebird & Sunbird

AMBIENT TEMP.	21°C (70°F)	27°C (80°F)	32°C (90°F)	38°C (100°F)	43°C (110°F)
A/C OUTLET TEMP.	4-7 (40-45)	4-7 (40-45)	6-8 (42-46)	6-8 (42-46)	7-10 (44-50)

Fig. 239. Performance test chart. 1979-80 Firebird

AMBIENT TEMP.		21°C (70°F)	27°C (80°F)	32°C (90°F)	38°C (100°F)	43°C (110°F)
A/C OUTLET TEMP.	WAGON	7-9°C (44-48)	7-9°C (44-48)	7-9°C (44-48)	8-10°C (46-50)	10-12°C (50-54)
	COUPE	3-6°C (38-42)	3-6°C (38-42)	3-6°C (38-42)	3-6°C (38-42)	3-6°C (38-42)

Fig. 240. Performance test chart. 1979-80 Sunbird

1980 Phoenix

Refer to insufficient cooling diagnosis chart, Fig. 115 parts 1 through 3, when performing the performance test.

Insufficient Cooling Diagnosis, All Systems

Late 1976 Grand Prix, 1977-78 All 1979 Except Firebird & Sunbird

These models are equipped with the C.C.O.T. A/C system, which incorporates the use of a thermostatic switch. When diagnosing insufficient cooling problems, refer to Fig. 243 parts 1 and 2, for diagnosis procedure.

1979 Firebird, Sunbird & 1980 Models Except Phoenix

These models are equipped with the C.C.O.T. A/C system which incorporates the use of a pressure cycling switch. When diagnosing insufficient cooling problems, refer to Fig. 114 parts l and 2, for diagnosis procedure.

1980 Phoenix

When diagnosing insufficient cooling problems, refer to Fig. 115 parts 1 through 3, for diagnostic procedure.

1980 Phoenix Compressor Clutch Diagnosis

Refer to compressor clutch diagnosis chart, Fig. 116, for diagnostic procedure.

1980 Phoenix Engine Cooling Fan Diagnosis

Refer to engine cooling fan diagnostic chart, Fig. 117, for diagnostic procedure.

Ambient Temp. (°F)	70	80	90	100	110
A/C Outlet Temp. (°F)	34-38	34-38	35-39	38-42	40-44

Fig. 241 Performance test chart. 1980 Grand Prix & LeMans

Ambient Temp. (°F)	70	80	90	100	110
A/C Outlet Temp. (°F)	38-42	38-42	38-42	38-42	42-48

Fig. 242 Performance test chart. 1980 Pontiac full size

A/C Electrical Diagnosis Connector, All Systems

1977-78 FULL SIZE MODELS

These models are equipped with two electrical diagnosis connectors, Figs. 118 and 119. To perform the following diagnosis procedures, a voltmeter and jumper wire are required. If a switch or relay is suspected of being defective, connect a jumper wire across switch or relay. If system operates, the switch or relay is defective. If system does not operate, the wire or connector is defective. With jumper wire connected, connect voltmeter from jumper wire to ground. If no voltage is present, the positive lead is defective. If voltage is present, the ground lead is defective.

Early 1977 Custom A/C & Auto Temp. Control

NOTE: This procedure is for 1977 models built before July 1977. These models can be identified by the following serial numbers, from serial No. 7P100001 to 7P421425 (models built at Pontiac Plant) and serial No. 7X100001 to 7X144805 (models built at Fairfax Plant). On these models, when A/C control is on Max. A/C, blower motor will operate at high speed only.

Perform test with ignition switch on, engine not operating and vehicle interior temperature above 55° F. Place blower switch on low and set A/C control for Vent mode unless otherwise stated. Before replacing a component, check wire connection at unit.

Voltmeter Connections	Voltage Reading	Condition.
1. No blower any speed, compressor clutch engages.		
a. 1 and G	3 volts or more	Defective blower motor or ground or capacitor lead.
	Under 3 volts	Go To Step b.
b. 2 and G.	3 volts or more	Defective relay or wire from relay to blower motor
	Under 3 volts	Stalled blower motor.

2. Blower motor inoperative and compressor clutch does not engage. If A/C-heater fuse is blown, check for grounded circuit or excessive blower motor or compressor clutch draw. If not, check for open circuit between ignition switch and low blower delay switch or poor A/C harness ground.

Voltmeter Connections	Voltage Reading	Condition
3. No low blower speed.		
a. 3 and G	10 volts or more	Defective blower motor resistor.
	Under 10 volts	Defective low blower delay switch or wire to blower motor resistor.
4. No low or M1 blower speed.		Defective blower motor resistor.
5. No blower except in High or Max. A/C.		
a. 2 and G	3 volts or more	Defective high blower relay.
	Under 3 volts	Defective resistor or wire from resistor to relay.
6. No high blower under any condition. Place blower switch in High.		
a. 9 and G	10 volts or more	Defective high blower relay, open circuit in red wire to relay or poor relay ground.
	Under 10 volts	Faulty wire from blower switch to relay coil.
7. No High blower speed except in Max. A/C.		Defective blower switch
8. No high blower speed in Max. A/C.		Defective A/C control or wire from A/C control to blower switch.
9. No M1 blower speed.		Defective blower switch or wire to resistor.
10. No M2 blower speed.		Defective blower switch or wire to resistor.
11. No M1, M2 or high blower speed using switch, but high speed is obtained in MAX A/C. Place blower switch in High.		
a. Voltage at control terminal No. 2	10 volts or more	Faulty blower switch or wire to blower switch.
	Under 10 volts	Defective A/C control.
12. Low blower speed only, compressor clutch does not change.		
a. 4 and G	10 volts or more	Defective A/C control.
	Under 10 volts	Faulty wire from low blower delay switch to A/C control.
13. Compressor clutch does not engage and blower motor inoperative.		Refer to step 2.
14. Compressor clutch engages in Norm., Max., Bi-Level or Defrost, but not in all.		Defective A/C control.

15. Gompressor clutch does not engage, blower motor operates. Place A/C control in Norm.

Voltmeter Connections	Voltage Reading	Condition
a. 6 and G	10 volts or more	Defective clutch, ground or compressor clutch wiring harness.
	Under 10 volts	Go to step b.
b. 8 and G	10 volts or more	Defective thermostatic wire to compressor clutch.
	Under 10 volts	Go to step c.
c. 7 and G	10 volts or more	Low refrigerant charge, defective low refrigerant pressure switch or wire from low pressure switch.
	Under 10 volts	Defective A/C control or wire from control to low pressure switch.

Late 1977 & 1978 Custom A/C & Auto Temp. Control

NOTE: This procedure is for 1977 models built from July 1977 on and all 1978 models. The late 1977 models can be identified by the following serial numbers, from serial No. 7P421426 on (models built at Pontiac Plant) and serial No. 7X144806 on (models built at Fairfax Plant). On these models, when A/C control is on Max. A/C, blower motor will operate at speed selected on blower switch.

Perform test with ignition switch on, engine not operating and vehicle interior temperature above 55° F. Place blower switch on low and set A/C control for Vent mode unless otherwise stated. Before replacing a component, check wire connection at unit.

Voltmeter Connections	Voltage Reading	Condition
1. No blower any speed, compressor clutch engages.		
a. 1 and G	3 volts or more	Defective blower motor or poor motor ground.
	Under 3 volts	Go to step b.
b. 2 and G	3 volts or more	Defective relay or lead from relay to blower motor
	Under 3 volts	Stalled blower motor, If voltage is zero, check A/C wiring harness ground.
2. No blower and compressor clutch does not engage. If A/C-heater fuse is blown, check for grounded circuit or excessive blower motor or compressor clutch current. If not, check for open circuit in wire from ignition switch to low blower delay switch.		
3. No low blower speed.		
a. 3 and G	10 volts or more	Defective blower motor resistor.
	Under 10 volts	Defective low blower delay switch or wire to resistor.
4. No low or M1 blower speed.		Defective blower motor resistor.
5. No blower operation except in High.		
a. 2 and G	3 volts or more	Defective high blower relay.
	Under 3 volts	Defective resistor or wire from resistor to relay.
6. No high blower speed under any condition. Place blower switch in High.		
a. 9 and G	10 volts or more	Defective high blower relay; open circuit in red wire to relay or poor relay ground.
	Under 10 volts	Defective blower switch or wire from blower switch to relay coil.
7. No M1 blower speed.	Defective Blower	Defective blower switch or wire to resistor.
8. No M2 blower speed.		Defective blower switch or wire to resistor.
9. No M1, M2 or High blower speed using switch. Place blower switch on High.		
a. Voltage at blower switch feed terminal from A/C Control	10 volts or more	Defective blower switch.
	Under 10 volts	Defective A/C control or wire from control to blower switch.
10. Low blower only, compressor clutch does not engage.		
a. 4 and G	10 volts or more	Defective A/C control
	Under 10 volts	Faulty wire from low blower delay switch to A/C control.
11. Compressor does not engage and blower motor inoperative.		Refer to step 2.
12. Compressor clutch engages in Norm, Max., Bi-Level or Defrost, but not in all.		Defective A/C control.
13. Compressor clutch engages, blower motor operates. Place A/C control in Norm.		
a. 6 and G	10 volts or more	Go to step c.
	Under 10 volts	Go to step b.
b. 8 and G	10 volts or more	Defective thermostatic switch or lead wire to clutch.
	Under 10 volts	Defective A/C control or lead wire from control to thermostatic switch.
c. Connect voltmeter to each clutch terminal and ground.	Both 10 volts or more	Defective compressor cutoff wire from cut-off switch to comp. clutch or low refrigerant charge.
	On terminal under 10 volts	Defective clutch.

Late 1976 Grand Prix & 1977-80 Custom A/C

A/C VACUUM SYSTEM DIAGNOSIS
1978-80 Grand Prix & LeMans

Refer to Fig. 121 parts 1 through 5, for A/C vacuum system diagnostic procedures.

1980 Phoenix

Refer to Fig. 122 parts 1 and 2 for A/C vacuum system diagnostic procedure.

CONTROL SYSTEM FUNCTIONAL TEST
Late 1976 Grand Prix & 1977 Grand Prix & LeMans

NOTE: Before performing test, make sure that engine is at normal operating temperature and that outside temperature is above 45° F when checking A/C system.

1. Move mode lever to off, set blower switch to HI and move temperature lever to full cold.
   a. The blower motor should be running at low speed.
   b. Air should flow out of the heater outlet only.
   c. The A/C compressor should be off.
2. Move mode lever to defrost.
   a. Blower motor should run at high speed.
   b. Air should flow primarily from the defroster nozzles with some air flow from the heater outlet.
   c. Compressor should be running.
3. Move mode lever to heater.
   a. Air should flow primarily from the heater outlet with some air flow from defroster nozzles.
   b. Compressor should be off.
4. Move temperature lever to maximum heat.
   a. Air flow from heater outlet should increase in speed.

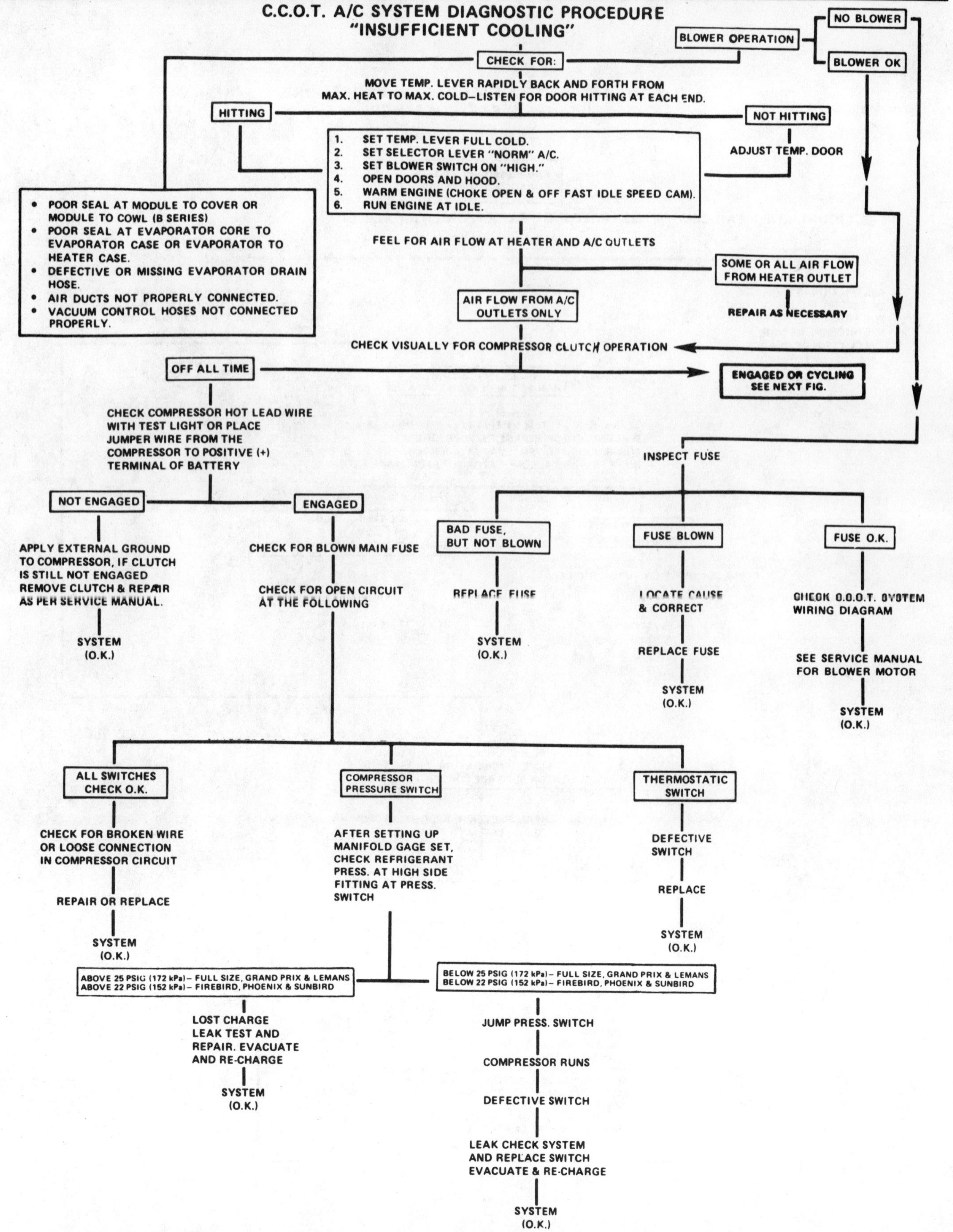

Fig. 243 Insufficient cooling diagnosis chart. 1977-78 all models & 1979 models except Firebird & Sunbird (Part 1 of 2)

C.C.O.T. A/C SYSTEM DIAGNOSTIC PROCEDURE "INSUFFICIENT COOLING"

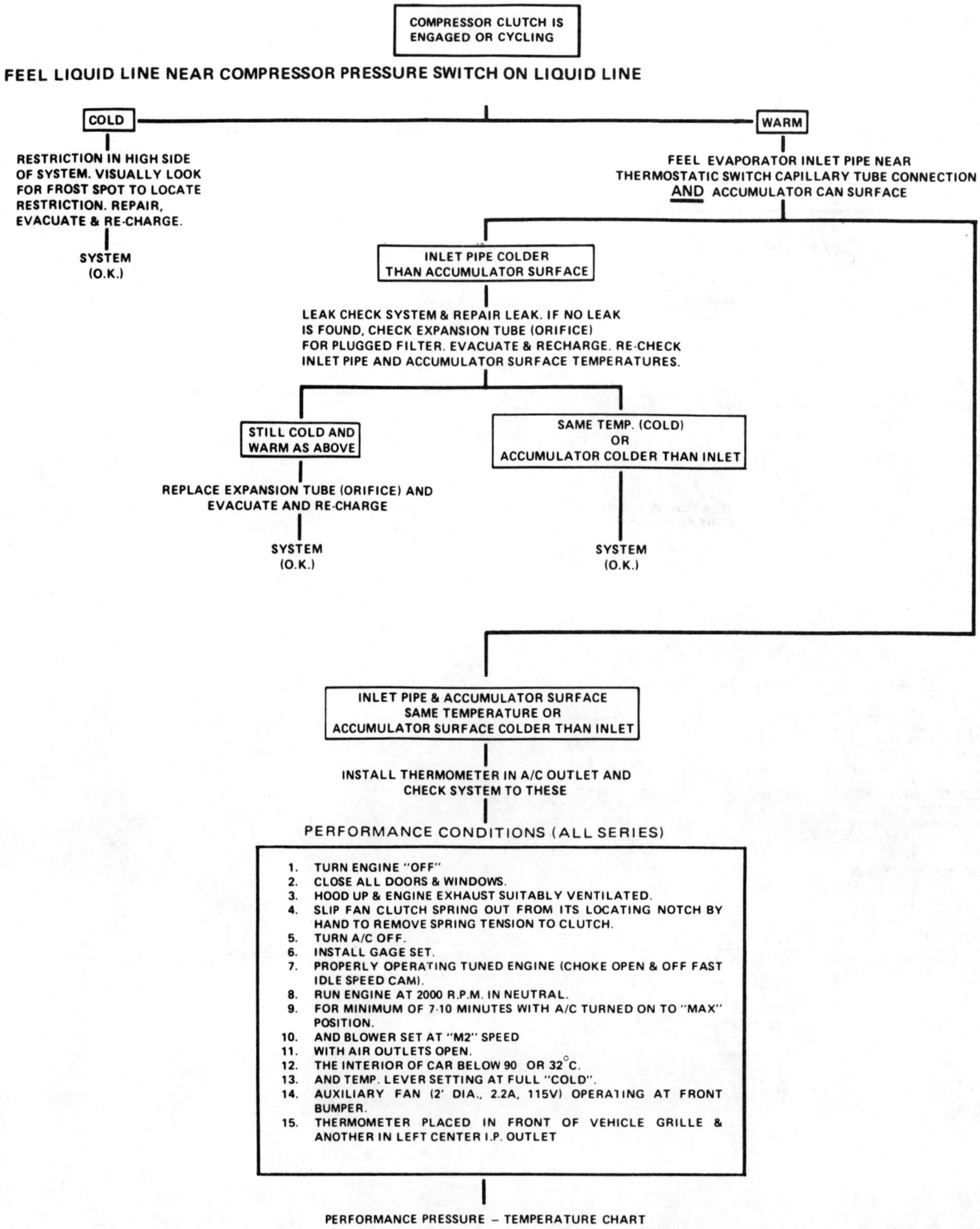

Fig. 243 Insufficient cooling diagnosis chart. 1977-78 all models & 1979 models except Firebird & Sunbird (Part 2 of 2)

5. Move mode lever to vent.
   a. Air should flow from A/C outlets.
   b. The compressor should not be running.
6. Move mode lever to Bi-Level.
   a. Air should flow from the defroster outlets, heater outlet and A/C outlet.
   b. The compressor should be running.
7. Move mode lever to normal.
   a. Air should flow from the A/C outlets.
   b. The compressor should be running.
8. Move the blower switch to full down, pausing at each detent.
   a. Blower speed and air flow should decrease at each detent position.
9. Move mode lever to max.
   a. Blower motor should come on high speed.
   b. Air noise should be heard from the right front corner passenger compartment as the system goes on inside air.
   c. A/C compressor should be running.

NOTE: There is a restrictor in the line which will delay obtaining air from the defroster in heater and bi-level mode when coming from any position except defrost.

1977 Bonneville & Catalina

NOTE: Before performing test, idle engine off the fast idle speed cam for about 20 minutes with temperature lever set to maximum cold. Ambient temperatures should be about 75° F, but not less than 45° F or greater than 90° F. Check effort required to move the control lever using a pull scale. The effort should be between 2-6.5 pounds; if not, repair or adjustments should be made.

1. Slide mode lever to off, set blower switch to High and leave temperature lever in maximum cold position. Air, the same temperature as that outside of the vehicle, should flow from the heater outlets. The blower motor should be running at low speed and the A/C compressor should be off.
2. Slide mode lever to norm, noting the detent at each mode position and ensure that lever efforts are within limits described previously. Blower motor speed should increase to high speed and air should flow from the A/C outlets. The compressor should come on and may cycle off and on. Air temperature should be about 45° F at the A/C outlets.
3. Slide mode lever to Bi-Level. Air should flow from the A/C and heater outlets. The compressor should come on and may cycle off and on. Air temperature should be about 45° F from both outlets.
4. Slide mode lever to vent. Air should flow from the A/C outlets only. Air flow should be greater than in step 2. The compressor should be off and the air temperature from the outlets should be the same as that outside the vehicle.
5. Slide mode lever to Heater. Air should flow primarily from the heater outlet with some air flow from the defroster outlet. The compressor should be off, and air temperature should be about the same as that outside of the vehicle. Blower speed should still be high.
6. Slide the temperature lever from maximum Cold to maximum Hot, noting that the effort required to slide the lever is as specified above. The temperature of air from the heater outlet should rise rapidly from outside ambient temperature to about 140° F. Blower speed should still be high.
7. Slide the mode lever to Defrost. Air should flow primarily from the defroster nozzle with some air flow from the heater outlet. The compressor should come on and may cycle off and on. Air temperature should be 140° F and blower speed should be High.
8. Move blower switch to Low, pausing at each of the two intermediate positions. Blower speed and air flow should decrease accordingly at each position.
9. Slide temperature lever to maximum Cold and mode selector lever to Max. Blower should come on High speed and air should flow from the A/C outlets only. Air noise should be heard from the right front corner of the passenger compartment as the system goes on "inside" air. The compressor should come on and may cycle off and on. Air temperature should be about 45° F at the A/C outlets.

1978-79 All & 1980 Except Phoenix

Refer to control system functional test chart, Fig. 244, when testing control system operation.

1980 Phoenix

Refer to control system functional test chart, Fig. 128, when testing control system operation.

Late 1976 Grand Prix & 1977-78 Auto Temp. Control

CONTROL SYSTEM FUNCTIONAL TEST

Late 1976 Grand Prix & 1977 Grand Prix & LeMans

NOTE: Before performing test, start engine and allow to reach operating temperature. When performing test ambient temperature should be approximately 75° F.

1. Place mode lever in Off, blower switch on High and temperature control lever at 65.
   a. Blower motor should be operating at low speed.
   b. Air flow should be from heater outlets only.
   c. Compressor should be off.
2. Place mode lever on Defrost. Note that

	CONTROL SETTINGS			SYSTEM RESPONSE					
	MODE LEVER	TEMP. LEVER	BLOWER SWITCH	BLOWER SPEED	COMP. CLUTCH	AIR VALVE POSITION	HEATER OUTLET AIR FLOW?	A/C OUTLET AIR FLOW?	DEFROST OUTLET AIR FLOW?
1	OFF	COLD	LO TO HI	LO ONLY	OFF	OUTSIDE AIR	AMBIENT AIR FLOW	NONE	SLIGHT
2	MAX.	COLD	LO TO HI	SAME AS SWITCH	ON	INSIDE AIR	SLIGHT FLOW	COLD FLOW	SLIGHT NONE (A&G)
3A	NORM.	COLD	LO TO HI	SAME AS SWITCH	ON	OUTSIDE AIR	SLIGHT FLOW	COLD FLOW	SLIGHT NONE (A&G)
3B	NORM. (A,B,G SERIES)	HOT	LO TO HI	SAME AS SWITCH	ON	OUTSIDE AIR	SLIGHT FLOW	HOT FLOW	SLIGHT NONE (A&G)
4	BI-LEVEL	COLD	LO TO HI	SAME AS SWITCH	ON	OUTSIDE AIR	COLD FLOW	COLD FLOW	SLIGHT NONE (A&G)
5	VENT	COLD	LO TO HI	SAME AS SWITCH	OFF	OUTSIDE AIR	SLIGHT FLOW	AMBIENT AIR FLOW	NONE
6	HEATER	HOT	LO TO HI	SAME AS SWITCH	OFF	OUTSIDE AIR	HOT FLOW	NONE	SLIGHT AIR FLOW
7	HEATER	COLD	LO TO HI	SAME AS SWITCH	OFF	OUTSIDE AIR	AMBIENT AIR FLOW	NONE	SLIGHT AIR FLOW
8	DEFROST	HOT	LO TO HI	SAME AS SWITCH	ON	OUTSIDE AIR	SLIGHT FLOW	NONE	HOT AIR FLOW

Fig. 244 Control system functional test chart. 1978 all systems & 1979-80 Custom A/C

STEP	CONTROL SETTINGS		SYSTEM RESPONSE						
	MODE CONTROL	TEMPERATURE DIAL	BLOWER SPEED	COMPRESSOR	AIR VALVE	HEATER OUTLETS	A/C OUTLETS	DEFROSTER OUTLETS	AIR TEMPERATURE
1	OFF	85	LOW	OFF	O/S AIR	AIR FLOW	NO AIR FLOW	NO AIR FLOW	HOT
2	ECONOMY	85	AUTO	OFF	O/S AIR	AIR FLOW	NO AIR FLOW	MINIMUM AIR FLOW	
3	LO	85	LOW	ON	O/S AIR	AIR FLOW	NO AIR FLOW		
4	AUTO	85 TO 65	AUTO	ON	O/S AIR I/S AIR	AIR FLOW NO AIR FLOW	NO AIR FLOW AIR FLOW	 NO AIR FLOW	HOT TO COLD
5	HI	65 TO 85	HIGH	ON	I/S AIR O/S AIR	NO AIR FLOW AIR FLOW	AIR FLOW NO AIR FLOW	NO AIR FLOW MINIMUM	COLD TO HOT
6	BI-LEVEL	85	AUTO	ON	O/S AIR	AIR FLOW	AIR FLOW	AIR FLOW	HOT
7	DEF	85	HIGH	ON	O/S AIR	MINIMUM AIR FLOW	NO AIR FLOW	MAXIMUM AIR FLOW	

Fig. 245 Control system functional test chart. 1979-80 Auto Temp. Control

each lever position has a detent lock to hold lever in position.

a. Blower motor should be operating at High speed.
b. Air flow should be from defroster outlets with slight bleed air from heater outlets.
c. Compressor should be operating.

3. Place mode lever on Heater.
   a. Air should flow from heater outlets, with a slight bleed air flow from defroster outlets.
   b. Compressor should not be operating.
4. Move temperature control lever to 85.
   a. Air flow from heater outlets should increase in speed.
5. Place mode lever on Vent.
   a. Air flow should be from A/C outlets.
   b. Compressor should not be operating.
6. Place mode lever on Bi-Level.
   a. Air flow should be from defroster, heater and A/C outlets.
   b. Compressor should be operating.
7. Place mode lever on Normal.
   a. Air flow should be from A/C outlets.
   b. Compressor should be operating.
8. Move blower switch to low speed, pausing at each switch detent.
   a. Blower speed and air flow should decrease at each detent position.
9. Place mode lever on Max.
   a. Blower motor should come on at high speed.
   b. Air noise should be heard from right corner of passenger compartment as system goes to inside air.
   c. Compressor should be operating.

NOTE: A restrictor located in the line will delay obtaining airflow from defroster outlets in Heater and Bi-Level mode when coming from any position except defrost.

1977 Bonneville & Catalina

NOTE: Before performing test, idle engine off fast idle speed cam for about 20 minutes with A/C control blower switch set at Hi blower, the temperature lever at 65° F, the mode lever in the Norm position and doors closed for interior stabilization. Ambient temperature should be about 75° F. Check the effort required to move the control lever using a pull scale. The effort should be between 2-6.5 pounds, if not, repair or adjustments should be made.

1. Slide mode lever to off, set blower switch to Hi and leave temperature lever at 65. Air temperature equivalent to that outside of the vehicle should be discharged from the heater outlet. The blower motor should run at low speed and the A/C compressor should be off.
2. Slide the mode lever to Norm. Note detent at each position and insure that lever efforts are as described previously. Blower motor speed should increase to high speed and air should flow from the A/C outlets. The compressor should come on and may cycle off and on. Air flow should be cool.
3. Slide the mode lever to Bi-Level. Air should flow from the A/C outlets and heater outlets. Air flow should decrease from that noted in step 2. The compressor should come on and may cycle off and on.
4. Slide the mode lever to Vent. Air should flow from the A/C outlets only, and the speed should be higher than that noted in step 2.
5. Slide the mode lever to Heater. Air should flow primarily from the heater outlet with some air flow from the defroster nozzle. The A/C compressor should be off and the air temperature should be at least the same temperature as that outside the vehicle. Blower speed should still be high.
6. Slide the temperature lever from 65 to 85, noting the lever effort requirements. The air temperature from the heater outlet should increase a noticeable amount, but no greater than 115° F. Blower speed should still be high.
7. Slide the mode lever to Defrost. Air should flow primarily from the defroster nozzle with some air flow from the heater outlet. The compressor should come on and may cycle off and on. Air temperature should be about 100° F to 115° F. Blower speed should still be high.
8. Move blower switch to Low, pausing at each of the two intermediate positions. Blower speed and air flow should decrease as the switch lever is moved through each position.
9. Slide mode lever to Max. Blower should come on high and air should flow from the A/C outlets only. Air noise should be heard from the right front corner of the passenger compartment as the system goes on "inside" air. The A/C compressor should come on and may cycle off and on. Air temperature should be about 60° F at the A/C outlets.

1978 All Models

Refer to control system functional test chart, Fig. 244, when testing control operation.

1979-80 Auto Temp. Control

CONTROL SYSTEM FUNCTIONAL TEST

1. To diagnose an air conditioning problem in the least amount of time and effort, the following recommended procedure should be followed:
   a. Obtain an accurate, detailed description of the problem.
   b. Perform functional test of system while in the vehicle and operating the controls with engine at normal operating temperature and running at 1000 rpm or greater. Refer to Fig. 245, for functional test chart.
2. During the functional test, note the following:
   a. Check that air is not flowing from both the A/C and heater outlets when only one mode is selected. A split air delivery indicates a vacuum leak or stuck door.
   b. Note if program events (air delivery mode change, blower speed change, recirc. air, etc.) occur without change in discharge air temperature. This would indicate that the programmer is operating without moving the temperature air door. Check air mix door link to programmer connection.
   c. If neither the program events nor discharge air temperature change, an incorrect vacuum or electrical signal to the programmer or a programmer problem is indicated.
   d. Malfunction of a specific vacuum operated door function could indicate a vacuum line disconnected at the vacuum actuator of that door. Refer to Vacuum Circuits located elsewhere in this manual.
3. Perform the easiest checks first. A visual inspection of the easily accessible underhood and instrument panel electrical and vacuum connections can reveal the defect.
4. Based on the functional test performed in step 1A, try to relate the problem into one of the following areas then, refer to that diagnosis further on:
   a. Temperature control problem.
   b. Blower control problem.
   c. Auxiliary vacuum problem.
   d. Refrigeration system problem.
5. After the problem has been diagnosed and repaired, perform the functional test to assure that the system is operating properly. Refer to Fig. 245, for functional test chart.

TEMPERATURE CONTROL PROBLEM DIAGNOSIS

The primary function of the temperature control circuit is to determine the correct temperature of the air flow into the passenger compartment by regulating the air mix door position.

Temperature control problems can be divided into the following:

1. System operates at maximum A/C (no heat) or maximum heat (no A/C.)
2. Temperature dial does not provide comfort.
3. Poor heating or poor cooling.
4. Blower cycling and mode cycling may also be caused by a problem in the temperature control circuit.

Temperature control circuit problems should be separated from refrigeration or heater problems. If the symptom is poor or no heating or cooling, and the air mix door and program events function properly, the problem is in the refrigeration or heater water systems.

Since many of the temperature control problems are a result of poor electrical or vacuum connections, the following suggestions may aid diagnosis:

1. A disconnected sensor or temperature dial interrupts the electrical signal and drives the programmer to maximum heating.
2. A poor sensor connection adds resistance to the sensor string, driving the system hotter.
3. An open amplifier power feed does not allow the transducer to operate, and the vacuum motor moves to MAX A/C.
4. A disconnected vacuum hose supplying the vacuum checking relay will lock the vacuum motor in a fixed position.
5. A disconnected vacuum hose in the transducer-programmer vacuum motor line will drive the system to maximum cooling.
6. A leak in the auxiliary vacuum circuit may reduce transducer vacuum supply level below control requirements, causing an "off calibration" or poor heating type of complaint.
7. A loss of supply vacuum will usually result in cold air flow on the floor.

If problem in programmer cannot be located, use tester J-23678 to test programmer, refer to Auto. Temp. Control Programmer and System Diagnosis.

System Operates Only At Max. A/C

This can be either an electrical or vacuum problem, the wrong input to the programmer or internal programmer defects. With the system operating, disconnect the in-car sensor connector and observe the programmer movement through the slot at the bottom of the programmer cover. The programmer should move to the bottom of the programmer cover to the max heat position (full in to the right). If the programmer remains in the max A/C position (full out, to the left), remove the multiple vacuum connector and check for supply vacuum at the black hose. If no or low vacuum supply, check for leaks or disconnects in vacuum hose assembly. If vacuum supply is okay, check voltage to the programmer. If okay, problem is in the programmer. Remove programmer cover and inspect for obvious problems. If there are no obvious problems, use tester J-23678 to diagnose.

If the programmer moves to the max heat position when the electrical connector is removed, the problem is external to the programmer, the following should checked:

1. Shorted in-car sensor.
2. Shorted or miscalibrated temperature dial.
3. Short in sensor circuit of the wiring harness.
4. Disconnected or opened ground circuit (black wire) between programmer and ground.
5. No electrical power to programmer.
6. Amplifier not grounded.

System Operates Only At Max Heating

This problem is usually caused by an electrical malfunction. Problems outside the programmer can usually be separated from internal programmer problems. To check, set temperature dial to 65° and the control lever to any position other than "Off" position. Observe the programmer mix door operating arm to check if programmer now moves to the maximum air conditioning position. If the programmer does not move, check the following:

1. Disconnected or defective in-car sensor.
2. Disconnected or defective ambient sensor.
3. Open circuit or backed-out terminal in the wiring harness sensor string circuit.

If the programmer does not move when the control is moved from 65° to 85°, check the following:

1. Disconnected or defective temperature dial.
2. Disconnected wire, or open in ground circuit (black wire) between programmer and ground.
3. Internal programmer problem. To check, use tester J-23678.

System Changes Blower Speeds And/Or Shifts Modes During Acceleration

With system operating and programmer cover removed, reconnect the vacuum and electrical connectors and proceed as follows:

1. Disconnect the in-car sensor connector to force the programmer to maximum heat (full vacuum).
2. Disconnect the vacuum hose assembly connector. The programmer should remain in the maximum heat position. If it does not, repeat step 1 and then pinch the programmer vacuum motor supply hose with needle nose pliers. If the programmer still moves, the vacuum motor is leaking. If the programmer does not move, the checking relay is defective.
3. If the original problem was mode shifts without blower change, check valve portion of the checking relay is probably defective.

Dial Setting Does Not Provide Comfort

Use tester J-23678 to check and reset calibration of the following components:

1. Feedback potentiometer in the programmer.
2. Temperature door link adjustment.
3. The sensor string.

Check that in-car sensor aspirator hose is attached at both the in-car sensor and aspirator and is not kinked.

Blower Cycling & Mode Cycling

Erratic system operation may be caused by poor electrical connections, poor solder joints or a defective sensor or amplifier. The problem may be intermittent and occur as the vehicle moves. To isolate the problem, tap the areas around the sensors, control head and programmer and shake the wires attached to those areas. If the problem cannot be isolated by the tranducer buzzing, remove programmer cover, reconnect electrical and vacuum connectors and connect a voltmeter to the transducer output terminals. Repeat the tapping and wire shaking and observe voltmeter for erratic voltage fluctuations. Replacement of the components or resoldering of terminals may be necessary. Some cycling may also occur and is normal if the temperature dial is moved in large increments.

Inadequate Cooling Or Heating

This condition may be caused by a control system malfunction or by the refrigerant or heater system. To check, proceed as follows:

1. If a no cooling condition exists, check for compressor clutch actuation. Refer to Insufficient Cooling Diagnosis.
2. If a no cooling condition exists, check for cold suction line.
3. If a no heating condition exists, check engine coolant level.
4. With engine running hot, check temperature of heater hoses by feeling them.
5. Check for air flow at "Auto" and "Hi" lever settings. Improper air flow may be a blower relay problem, a blocked air passage or a disconnected air hose.
6. Check calibration of programmer feedback potentiometer.
7. Check calibration of control head temperature dial.

If a refrigeration problem exists, refer to appropriate diagnosis chart. If the problem is caused by the temperature control system, use tester J-23678 to diagnose the system.

Blower Speeds & Mode Shifts Occur Without Temperature Change

Check connection of air mix door link to programmer shaft and to air door crank arm.

BLOWER CONTROL CIRCUIT PROBLEM DIAGNOSIS

Blower control can be divided into the blower turn-on circuit and blower speed control circuit. Blower turn-on is obtained when a ground circuit is complete for the relay coil. The relay coil ground circuit is completed as follows:

1. The blower delay switch provides delay for heater water warm-up in cold weather operation (except in "Def" mode).
2. The in-car switch, that provides immediate turn-on in summer operation (except in Def).
3. Immediate turn-on in "Def" mode. Blower speed control is accomplished by actuating only the low blower relay in the "Off" and "Lo" positions, by selective use of the resistors on the blower circuit board in the "Auto", "Bi-Level" and "Economy" lever settings or by a high blower override circuit in the "Hi" and "Defrost" lever positions. Note that the system air doors are always positioned to permit air delivery into the passenger compartment.

System Operates Only At Low Blower

1. Backed-out terminal or power connection at five-way connector at programmer, or control head.
2. Open in "Auto" blower circuit wiring.

No Blower Operation In Any Lever Setting

1. Connector disconnected at blower motor.
2. Blown or Open A/C-heater fuse.
3. Stalled blower motor.
4. Open circuit in wiring harness to blower motor.

5. Disconnected five-way connector at programmer.

No Blower in Off Or Lo Setting

1. Inoperative blower relay.
2. Open low speed resistor.
3. Open in blower circuit wiring.

Blower Operation in "Hi" or "Defrost"

1. Connector disconnected at programmer.

Blower Speed Varies in "Hi" & "Defrost"

1. Defective wipers or circuit board in control head circuit board switch.
2. Open in control head wiring harness.
3. Open in by-pass circuit in wiring harness.

Blower Operates With Ignition Off

1. Defective blower relay.

Blower Speeds Skipped in "Auto", "Bi-Level", & "Economy" Settings

1. Open blower resistor on programmer blower switch.

No Heater Turn-On In Cold Weather Except in "Defrost"

1. Disconnected connector at blower delay switch (located at rear of right hand cylinder head).
2. Defective blower delay switch. Check by grounding switch feed wire. System should go on. If the problem does not occur indoors, it is possible that the system is turned on by the in-car switch. Disconnect control head electrical connector to open the in-car switch. If the blower turns off (with engine warmed up), ground the blower delay switch feed wire. If the blower now turns on (in low blower speed), the blower delay switch is defective. If problem still exists, refer to electrical circuit diagrams for wiring problem.

Immediate Heater Turn-On In Cold Weather

1. Normal condition in "Defrost" control heat lever settings.
2. Defective (closed) blower delay switch. Remove switch and check when cold.
3. Defective (closed) in-car switch.
4. If problem still exists, refer to electrical circuit diagrams for wiring problem.

A/C Delayed In Hot Weather Until Engine Is Warm Unless Control Lever Is In "Def"

1. Defective (open) in-car switch.
2. Open in control head wiring harness (in-car switch circuit).
3. If problem persists, check wiring in the in-car switch circuit.

AUXILIARY VACUUM PROBLEM DIAGNOSIS

Auxiliary vacuum functions are those vacuum circuits which do not affect the temperature of the discharge air. This includes all vacuum functions except the transducer, the relay portion of the vacuum checking relay and the program vacuum motor. The auxiliary circuits are as follows:

1. Upper mode door vacuum circuit. Grand Am, Grand LeMans, Grand Prix and LeMans models use a heater A/C mode door.
2. Lower mode door vacuum circuit on Bonneville and Catalina models.
3. Defroster bleed vacuum circuit on Bonneville and Catalina models.
4. Defroster door vacuum circuit.
5. Recirc. door vacuum circuit.
6. Heater water valve vacuum circuit.
7. Auxiliary vacuum circuit check valve.

Since many of the vacuum lines are interconnected, a leak in one circuit may reduce the vacuum supply to another circuit, causing a malfunction of a second circuit. For example, a disconnected water valve vacuum hose may prevent the mode doors from moving their full travel causing a split mode condition.

A/C Mode Door Actuating Problems

Grand Am, Grand LeMans, Grand Prix and LeMans models use a single mode door while Bonneville and Catalina models use upper and lower mode doors. The most likely problem in this circuit would be "no vacuum operation" and "split mode operation." Vacuum is required to move the door(s) to the A/C mode. Check the hose connections at the mode door vacuum actuator(s). Split modes may occur because of problems in the door circuit(s), or because of leaks in other auxiliary vacuum circuits. The external circuit test procedure listed below can be used to locate mode door vacuum problems.

Defroster Bleed Problems, Bonneville & Catalina Models

Movement of the side (bleed) port of the defroster diaphragm is delayed by a porous flow plug inserted in the vacuum hose. Problems relating to defroster bleed operation will usually be as specified:

1. No Defroster Bleed-Check hose connection at defroster vacuum actuator. If connections are okay, porous plug is probably too restrictive. Before replacing plug, check for vacuum in yellow line and apply hose to actuator, without the plug, to check bleed door operation.
2. Rapid fogging of windshield when heater turns on — Either there is no porous plug or plug is incorrectly installed.

Defroster Door Actuation Problem

The most likely problem to be expected in this circuit would be "no air out of defroster outlets in Defrost." Vacuum to the defroster actuator is required for defroster operation on Bonneville and Catalina models, and no vacuum is required on all other models.

Problems on Bonneville and Catalina models would most likely be caused by a bad hose connection at defroster actuator. Check as follows:

1. Check for vacuum at control head vacuum supply (purple) hose.
2. If system stays in A/C mode at "DEFROST," check vacuum circuit to determine why mode override did not occur.

Problems on Grand Am, Grand LeMans, Grand Prix and LeMans models would most likely be caused by a stuck door. If problem is not the door, check vacuum circuit.

Recirculating Air Door Actuation Problems

The most likely problem to be expected in this circuit would be "no recirc. at low thumbwheel setting." Since vacuum is required to obtain recirc. air, check hose connection at recirc. air door. Remove 3-wire connector at programmer to drive programmer to "recirc." mode. If "recirc." does occur, check setting of rod to temp. door.

Heater Water Valve Control Problems

The water valve is normally open and requires vacuum to close water flow. A vacuum leak in this circuit may result in reduced vacuum levels available in other vacuum functions. The most likely result of this problem would be a split mode delivery, check vacuum circuit.

Auxiliary Vacuum Circuit Check Valve Problems

The check valve for the auxiliary vacuum circuit is located in the programmer checking relay assembly. A malfunctioning check valve will probably result in loss of a specific function, such as A/C shutting off during acceleration or under heavy engine load. The check valve can be checked with the system operating by disconnecting and plugging the black hose supplying the programmer checking relay. The vacuum functions should hold, if not, reconnect the black hose and pinch the purple hose. If the functions hold now, the check valve is defective. If the functions continue to leak down, there is a circuit leak further upstream in the vacuum circuit.

REFRIGERATION SYSTEM PROBLEM DIAGNOSIS

The purpose of the refrigeration system is to maintain the evaporator core surfaces at 32° F when refrigeration is required. A properly operating refrigeration system will maintain this temperature if engine speed is sufficient and the heat load on the system is not excessive. In most cases, a problem will result in higher-than-normal core temperature and loss of effective cooling. In a few cases, the evaporator core will operate at a temperature lower than 32° F and condensation in the core will freeze and block air flow, again causing loss of cooling.

It is important to separate refrigeration problems from temperature control problems. This should be done during the functional check, as follows:

1. Set lever at all modes of operation and check for proper system operation according to Fig. 245.
2. If system operates properly, return lever to AUTO and set dial at 65° F.
3. Install thermometer in center A/C outlet and wait about 1 minute. Thermometer should read 40-50° F unless temperature or humidity of test area are excessively high.
4. If discharge temperature is significantly higher than 50° F, disconnect programmer amplifier connector attached to programmer. Check to make sure that the air-mix door link is pivoted fully to the "Max" A/C position, and that temperature door rod is properly connected and adjusted.
5. If the discharge air temperature drops to or below 50° F as a result of disconnecting the programmer amplifier connector, the problem is in the temperature control circuit.
6. If the discharge air temperature does not change and is still significantly higher than 50° F, a refrigeration problem exists.

AUTO TEMP. CONTROL PROGRAMMER & SYSTEM DIAGNOSIS

Use tester J-23678 to diagnose A.T.C. programmer and system problems. When using tester follow procedure described in diagnosis charts, Fig. 134 parts 1 and 2.

Air Conditioning Specifications

Year	Model	Refrigerant Capacity Lbs.	Refrigeration Oil		
			Viscosity	Total System Capacity, Ounces	Compressor Oil Level Check Inches
AMERICAN MOTORS					
1974	Ambassador & Matador	3¼	300	7	①
	Others	2	300	7	①
1975-76	Ambassador & Matador	3¼	300	7	①
	Others Exc. Pacer	2	300	7	①
	Pacer	2⅛	300	7	①
1977	Matador	3	②	7	①
	Gremlin & Hornet	2	②	7	①
	Pacer	2⅛	②	7	①
1978	Matador	3	②	7	①
	AMX, Concord & Gremlin	2	②	7	①
	Pacer	2⅛	②	7	①
1979-80	Exc. Pacer	2	②	7	①
	Pacer	2⅛	②	7	①
BUICK — FULL SIZE					
1974-79	All	3¾	525	④	③
1980	Electra & LeSabre	3¾	525	④	③
	Riviera	3½	525	④	③
BUICK — COMPACT & INTERMEDIATE (EXC. SKYHAWK & 1980 SKYLARK)					
1974	All	3¾	525	10½	③
1975-76	Apollo & Skylark	3½	525	④	③
	Century & Regal	3¾	525	④	③
1977-79	Skylark	3½	525	④	③
	Century & Regal	3¾	525	④	③
1980	Century & Regal	3½	525	④	③
CADILLAC					
1974	All	3¾	525	10½	③
1975-79	Exc. Seville	⑤	525	⑥	③
1976	Seville	3½	525	10½	③
1977-79	Seville	3½	525	10	③
1980	All	3¾	525	6	③
CAMARO, CHEVROLET, CHEVELLE, MALIBU, MONTE CARLO & NOVA					
1974	All	3¾	525	11	③
1975-76	4 Cyl. Comp.	3¾	525	5½-6½	③
	6 Cyl. Comp.	3¾	525	⑦	③
1977-79	Nova	3½	525	④	③
1977-80	Exc. Camaro & Nova	3¾	525	④	③
	Camaro	3¼	525	④	③
CHEVROLET CHEVETTE					
1976-80	All	2¼	525	5½-6½	③
1980 CHEVROLET CITATION, BUICK SKYLARK, OLDS OMEGA & PONTIAC PHOENIX					
1980	All	2¾	525	6	③

AIR CONDITIONING SPECIFICATIONS

Year	Model	Refrigerant Capacity Lbs.	Refrigeration Oil		
			Viscosity	Total System Capacity, Ounces	Compressor Oil Level Check Inches
CHEVROLET MONZA, BUICK SKYHAWK, OLDS STARFIRE & PONTIAC SUNBIRD					
1975-76	Skyhawk	3	25	5½-6½	③
	Starfire	4	525	5½-6½	③
1975-80	Monza	2 13/16	525	④	③
1976	Sunbird	2 13/16	525	5½-6½	③
1977	Skyhawk	3	525	6	③
	Sunbird	3¼	525	6	③
1977-78	Starfire	2½	525	6	③
1978-80	Skyhawk & Sunbird	2½	525	6	③
1979-80	Starfire	⑧	525	6	③
CHEVROLET VEGA & PONTIAC ASTRE					
1974-75	All	2 13/16	525	11	③
1976	Astre	2 13/16	525	5½-6½	③
1976-77	Vega	2 13/16	525	5½-6½	③
1977	Astre	2½	525	5½-6½	③
CHEVROLET CORVETTE					
1974-80	All	3	525	⑦	③
CHRYSLER & IMPERIAL					
1974-75	Exc. Cordoba	4½	300	10-12	1⅝-2⅜⑨
1975	Cordoba	3¾	300	10-12	1⅝-2⅜⑨
1976-77	Newport & New Yorker	2¾	⑩	10-12	2⅜⑨
	Cordoba & LeBaron	2⅝	⑩	10-12	2⅜⑨
1978-79	All	2⅝	500	⑪	⑫
1980	Newport & New Yorker	2⅞	500	⑪	⑫
	Cordoba & LeBaron	2⅝	500	⑪	⑫
DODGE — FULL SIZE					
1974-75	All	4¼	300	10-12	1⅝-2⅜⑨
1976-77	All	2¾	⑩	10-12	2⅜⑨
1978-79	All	2⅝	500	⑪	⑫
1980	All	2⅞	500	⑪	⑫
DODGE — COMPACT & INTERMEDIATE					
1974	Exc. Dart	2¾	300	10-12	⑨⑬
	Dart	3¼	300	10-12	⑨⑬
1975	Exc. Dart	3¾	300	10-12	⑨⑬
	Dart	3	300	10-12	⑨⑬
1976-77	All	2⅝	⑩	10-12	2⅜⑨
1978-80	All	2⅝	500	⑪	⑫
DODGE OMNI & PLYMOUTH HORIZON					
1978-79	All	2⅛	500	7-8	⑭
1980	All	2⅛	500	⑪	⑮
FORD — FULL SIZE					
1974-79	All	4¼	⑯	⑰	③
1980	All	3¼	⑯	⑰	③

AIR CONDITIONING SPECIFICATIONS

Year	Model	Refrigerant Capacity Lbs.	Refrigeration Oil: Viscosity	Refrigeration Oil: Total System Capacity, Ounces	Refrigeration Oil: Compressor Oil Level Check Inches
FORD — COMPACT & INTERMEDIATE (EXC. MUSTANG & PINTO)					
1974-76	York Comp.	⑱	②	10	①⑨
	Tecumseh Comp.	⑱	②	11	①⑨
1976-78	Granada	4¼	525	10½	③
1977-79	York Comp.	⑲	②	10	①⑨
	Tecumseh Comp.	⑲	②	11	①⑨
1979-80	Granada	4	525	10½	③
1980	York Comp.	3½	②	10	①⑨
	Tecumseh Comp.	3½	②	11	①⑨
FORD MUSTANG & MERCURY CAPRI					
1974-78	York Comp.	3¼	②	10	①⑨
	Tecumseh Comp.	3¼	②	11	①⑨
1979-80	York Comp.	3½	②	10	①⑨
	Tecumseh Comp.	3½	②	11	①⑨
FORD PINTO & MERCURY BOBCAT					
1975-80	York Comp.	2¼	②	10	①⑨
	Tecumseh Comp.	2¼	②	11	①⑨
LINCOLN CONTINENTAL, MARK IV, V, VI, & VERSAILLES					
1974-78	Mark IV & V	4½	⑯	10½	③
	Lincoln	4¼	⑯	10½	③
1977-78	Versailles	4¼	⑯	10½	③
1979-80	Exc. Versailles	4¼	⑯	⑰	③
	Versailles	4	⑯	⑰	③
MERCURY — FULL SIZE					
1974-79	All	4¼	⑯	⑰	③
1980	All	3¼	⑯	⑰	③
MERCURY — COMPACT & INTERMEDIATE (EXC. BOBCAT & CAPRI)					
1974-76	York Comp.	⑳	②	10	①⑨
	Tecumseh Comp.	⑳	②	11	①⑨
1976-78	Monarch	4¼	525	10½	③
1977-79	York Comp.	㉑	②	10	①⑨
	Tecumseh Comp.	㉑	②	11	①⑨
1979-80	Monarch	4	525	10½	③
1980	York Comp.	3½	②	10	①⑨
	Tecumseh Comp.	3½	②	11	①⑨

AIR CONDITIONING SPECIFICATIONS

Year	Model	Refrigerant Capacity Lbs.	Refrigeration Oil		
			Viscosity	Total System Capacity, Ounces	Compressor Oil Level Check Inches
OLDSMOBILE—FULL SIZE					
1974-76	All	4	525	10½	③
1977-80	All	3¾	525	④	③
OLDSMOBILE—COMPACT & INTERMEDIATE (EXC. STARFIRE & 1980 OMEGA)					
1974-76	All	4	525	④	③
1977-79	Omega	3½	525	④	③
1977-80	Cutlass	3¾	525	④	③
PLYMOUTH—FULL SIZE					
1974-75	All	4¼	300	10-12	1⅝-2⅜⑨
1976-77	All	2¾	⑩	10-12	2⅜⑨
1978-79	All	2⅝	500	⑪	⑫
1980	All	2⅞	500	⑪	⑫
PLYMOUTH—COMPACT & INTERMEDIATE					
1974	Exc. Valiant	2¾	300	10-12	⑨⑬
	Valiant	3¼	300	10-12	⑨⑬
1975	Exc. Valiant	3¾	300	10-12	⑨⑬
	Valiant	3	300	10-12	⑨⑬
1976-77	All	2⅝	⑩	10-12	2⅜⑨
1978-80	All	2⅝	500	⑪	⑫
PONTIAC—FULL SIZE					
1974-75	All	㉒	525	11	③
1976	All	3¾	525	10½	③
1977-79	All	3¾	525	④	③
1980	All	3½	525	④	③
PONTIAC—COMPACT & INTERMEDIATE (EXC. ASTRE, SUNBIRD & 1980 PHOENIX)					
1974-75	All	㉒	525	④	③
1976	Exc. Firebird	3¾	525	④	③
	Firebird	3¼	525	④	③
1977-80	All	㉓	525	④	③
THUNDERBIRD					
1974-79	All	㉔	525	10½	③
1980	York Comp.	3½	②	10	①⑨
	Tecumseh Comp.	3½	②	11	①⑨

① — York Compressor-Vertical mount, ⅞-1⅛ inch, horizontal mount, 13/16-1 3/16 inch. Tecumseh Compressor-Vertical mount, ⅞-1⅜ inch; horizontal mount,⅞-1 ⅝ inch.
② — Suniso 5G or Capella E.
③ — Note that "Oil level inches" cannot be checked. Refer to total capacity in ounces, and to text for procedure.
④ — Axial compressor, 10 oz.; radial compressor, 6 oz.
⑤ — With one evaporator, 3¾ lbs. With two evaporators, 5 lbs.
⑥ — With one evaporator, 10½ oz. With two evaporators, 13½ oz.
⑦ — 1975, 11 oz.; 1976-79, 10½ oz.; 1980, with axial compressor, 10½ oz.; with radial compressor, 6 oz.
⑧ — With axial compressor, 3 lbs.; with radial compressor, 2½ lbs.
⑨ — Dipstick reading with compressor installed.
⑩ — 1976, 300 viscosity; 1977, 500 viscosity.
⑪ — With RV2 compressor, 10-12 oz. With Sankyo compressor, 3-4 oz. With C-171 compressor, 9-10 oz.
⑫ — RV2 compressor, 2⅜ inches. On Sankyo compressor, the oil level should be at the third to fourth increment on the dipstick which represents 3 to 4 ounces of oil. Refer to text for procedure. On C-171 compressor, refer to text.
⑬ — Six cylinder engines, 1¾-2⅜ inch; V8 engines, 1⅝-2⅜ inch.
⑭ — Oil level should be at the third to fourth increment on the dipstick which represents 3-4 oz. of oil. Refer to text for procedure.
⑮ — Refer to text for C-171 compressor.
⑯ — Frigidaire compressor, 525 viscosity; Nippondenso compressor, 500 viscosity.
⑰ — Frigidaire compressor, 10½ oz.; Nippondenso compressor, 13 oz.
⑱ — Maverick, 1⅞ lbs.; Elite & Torino, 4½ lbs.; 1975 Granada, 4 lbs.
⑲ — Maverick, 1⅞ lbs.; LTD II, 4¼ lbs.; Fairmont, 3½ lbs.
⑳ — Comet, 1⅞ lbs.; Cougar & Montego, 4½lbs.; 1975 Monarch, 4 lbs.
㉑ — Comet, 1⅞ lbs.; Cougar 4¼ lbs.; Zephyr, 3½ lbs.
㉒ — Refer to compressor decal when servicing these systems.
㉓ — Firebird, 3¼ lbs.; Ventura, 3½ lbs.; Grand Prix & Le Mans, 3¾ lbs.
㉔ — 1975-76, 4½ lbs.; 1977-79, 4¼ lbs.

CHARGING VALVE LOCATIONS

Year & Model	High Pressure	Low Pressure
AMERICAN MOTORS		
1974-79	Compressor	Compressor
1980	①	①
BUICK—FULL SIZE		
1974-1976	Compressor	VIR
1977-1980	②	Accumulator
BUICK—COMPACT & INTERMEDIATE EXC. SKYHAWK & 1980 SKYLARK		
1974-1976	Compressor	VIR
1977-1980	Muffler	Accumulator
CADILLAC		
1974-1976		
Exc. Seville	Vapor Line	VIR
Seville	Compressor	VIR
1977-1979	④	Accumulator
1980	②	Accumulator
CAMARO, CHEVELLE, CHEVROLET, MALIBU, MONTE CARLO & NOVA		
1974-1976	⑤	VIR
1977-1980	④	Accumulator
CHEVROLET CHEVETTE		
1976-1979	Dis. Press. Switch	Accumulator
1980	Muffler	Accumulator
1980 CHEVROLET CITATION, BUICK SKYLARK, OLDSMOBILE OMEGA & PONTIAC PHOENIX		
1980	Muffler	Accumulator
CHEVROLET CORVETTE		
1974-1976	VIR	VIR
1977-1980	②	Accumulator
CHEVROLET VEGA & PONTIAC ASTRE		
1974-1977	Dis. Press. Switch	Accumulator
CHRYSLER & IMPERIAL		
1974-1977	Compressor	Compressor
1978-1980	⑥	Compressor
DODGE (ALL)		
1974-1977	Compressor	Compressor
1978-1980	⑥	Compressor
FORD (ALL)		
1974-1979⑦	Compressor	Compressor
1977-1979⑧	⑨	STV
1980⑦	⑩	⑪
1980⑧	⑨	Accumulator
LINCOLN CONTINENTAL, MARK IV, V, VI & VERSAILLES		
1974-1976	Compressor	Compressor
1977-1980	⑨	Accumulator
MERCURY (ALL)		
1974-1979⑦	Compressor	Compressor
1977-1979⑧	⑨	STV
1980⑦	⑩	⑪
1980⑧	⑨	Accumulator
MONZA, SKYHAWK, STARFIRE & SUNBIRD		
1974-1976		
Exc. Monza	Compressor	VIR
Monza	⑤	VIR
1977-1979	⑫	⑬
1980	②	Accumulator
OLDSMOBILE—EXC. STARFIRE & 1980 OMEGA		
1974-1977③	Compressor	VIR
1977-1980⑭	Compressor	Accumulator
PLYMOUTH (ALL)		
1974-1977	Compressor	Compressor
1978-1980	⑥	Compressor
PONTIAC—EXC. ASTRE & SUNBIRD & 1980 PHOENIX		
1974⑭	Compressor	STV
1975-1976③	Compressor	VIR
1977-1980	②	Accumulator
THUNDERBIRD		
1974-1979	Compressor	⑮
1980⑧	⑨	Accumulator

① — 6 cyl. engine, high & low pressure service ports, on comp.; 4 cyl. engine, high & low pressure service ports, on service valve adapter.
② — Liquid line or muffler.
③ — With VIR system.
④ — On high pressure line from compressor.
⑤ — Axial comp., VIR inlet line; radial comp., located on compressor.
⑥ — With RV2 compressor, on compressor. With Sankyo or C-171 compressor, on muffler.
⑦ — 2 cyl. compressor.
⑧ — 6 cyl. compressor.
⑨ — On comp. discharge line.
⑩ — On liquid line at combination valve.
⑪ — On combination valve.
⑫ — 4 cyl. engine on line from condenser; 6 cyl. & 8 cyl. engines, on Dis. Press. Sw.
⑬ — 4 cyl. & 6 cyl. engines on VIR; 8 cyl. engine on accumulator.
⑭ — With cycling clutch system.
⑮ — On low pressure line from compressor.

AIR CONDITIONING SPECIFICATIONS

A/C COMPRESSOR BELT TENSION

Year & Model	New Lbs.	Used Lbs.
AMERICAN MOTORS		
1974-75	125-155	105-130
1976-80 Exc. 4-151	125-155	90-115
1980 4-151	90-115	90-115
BUICK — EXC. SKYHAWK & 1980 SKYLARK		
1974	90	90
1975	150	100
1976-79	155	90
1980 5/16" belt	80	50
⅜" belt	140	70
15/32" belt	165	90
CADILLAC		
1974-76 Exc. Seville	170	120
1976 Seville	125	90
1977-80	110-140	60-120
CHEVROLET — EXC. CHEVETTE, CITATION, MONZA & VEGA		
1974	135-145	95
1975-80	135-145	90
CHEVROLET CHEVETTE		
1976-80	135-145	90-100
1980 CHEVROLET CITATION, BUICK SKYLARK, OLDS OMEGA & PONTIAC PHOENIX		
1980	150	80
CHEVROLET MONZA, BUICK SKYHAWK, OLDS STARFIRE & PONTIAC SUNBIRD		
1975 V6-231	140-160	100
V8-350	120-130	75
1975-76 V8-262	120-130	75
1975-77 4-140	120-130	75
1976-77 V6-231	145-160	90
1976-79 V8-305	120-130	75
1977 4-151	110-140	70
1978 V6-196,231	145-160	90
1978-79 4-151	120-130	75
1979 V6-196,231	168	90
1980 5/16" belt	80	50
⅜" Exc. cog belt	140	70
⅜" cog belt	140	60
15/32" belt	165	90
CHEVROLET VEGA & PONTIAC ASTRE		
1974-77	135-145	95
CHRYSLER, DODGE, IMPERIAL & PLYMOUTH EXC. OMNI & HORIZON		
1974-1979	120	70
1980	120	65

Year & Model	New Lbs.	Used Lbs.
DODGE OMNI & PLYMOUTH HORIZON		
1978-80	55	40
FORD & MERCURY COMPACT & INTERMEDIATE		
1974-80 Exc. ¼" belt	140	110
¼" belt	80	60
FORD & MERCURY FULL SIZE		
1974-1979 Exc. ¼" belt	140	110
¼" belt	65	50
1980 All	120-160	75-120
FORD MUSTANG & PINTO — MERCURY BOBCAT & CAPRI		
1974 2000 c.c. eng.	140	110
1974-78 2800 cc. eng. Exc. ¼"	140	110
2800 cc. eng. ¼"	65	50
1974-79 2300 cc. eng. Exc. ¼"	140	110
2300 cc. eng. ¼"	65	50
1980 All	140	110
LINCOLN CONTINENTAL, MARK IV, V, VI & VERSAILLES		
1974-1979 Exc. ¼" belt	140	110
1975-1979 ¼" belt	65	50
1980	140	110
OLDSMOBILE — EXC. STARFIRE & 1980 OMEGA		
1974-79	110-140	70
1980 5/16" belt	80	50
⅜" Exc. cog belt	140	70
⅜" cog belt	140	60
15/32" belt	165	90
PONTIAC — EXC. ASTRE, SUNBIRD & 1980 PHOENIX		
1974 6 cyl.	135-145	95
1974 V8	135-165	100-105
1975-76 6-250	135-145	95
Exc. V8-260	135-165	100-105
V8-260	100-140	70
1977 Exc. 4-151 & V8 305, 350①	135-165	100-105
4-151	110-140	75-80
1978-80 Exc. 4-151	135-145	90-100
4-151	120-130	70-80
THUNDERBIRD		
1974	140	110
1975-79 Exc. ¼" belt	140	105
1975-79 ¼" belt	65	50
1980	140	110

① — Distributor located at rear of engine, clockwise distributor rotation

IV. UNIT REPLACEMENT

Part 1

Servicing Refrigeration System Components

COMPRESSOR CLUTCH, REPLACE

Air-Temp. Tecumseh & York

After determining that the clutch coil is defective, replace the coil as outlined below, referring to Figs. 1 and 2.

Remove Compressor Pulley

1. Loosen and remove drive belt.
2. If possible, energize clutch and remove clutch mounting bolt.
3. Thread a ⅝ inch-11 x 2½ inch bolt into hub center.
4. Tighten bolt against compressor shaft and lift off pulley when loosened.

Replace Coil

1. Disconnect coil lead(s).
2. Remove coil retainer screws from seal plate and remove coil from compressor.
3. Install new coil into seal plate and tighten bolts finger tight.
4. Install plastic aligner ring (if used) on coil.
5. Alternately tighten coil retainer bolts gradually and evenly until all are tightened securely.
5. Connect coil wire(s).

Install pulley

1. Align keyway in hub with key in shaft and press on pulley.
2. Install retainer bolt. Install and adjust drive belt.
3. Momentarily engage clutch coil and torque bolt to 20 ft. lbs.

C-171 Compressor

Removal, Fig. 3

1. Remove drive belt from pulley. Disconnect field coil wiring.
2. Use a suitable spanner wrench to hold clutch hub plate and remove shaft nut and lock washer.
3. Install front plate puller C-4561 or suitable puller and remove front plate, key and shims.
4. Remove pulley retaining snap ring and slide pulley off compressor.
5. Remove field coil retaining snap ring and slide off housing.

Installation

1. Align hole in back of field coil with pin on compressor housing and position coil into place making sure the field coil wires are properly routed. Install field coil wire retaining clip and then field coil retaining snap ring, bevel side outward.
2. Install pulley assembly onto compressor then, install snap ring with bevel side outward.
3. Install shims and shaft key in place.
4. Install front plate assembly onto compressor shaft.
5. With front plate fully seated, check air gap between pulley and front plate assembly. Air gap should be 020-.035 inch. If not, change shims to correct air gap.
6. Install lockwasher and nut. Torque nut to 155 in. lbs. Recheck air gap and adjust if necessary.

Frigidaire & Motorcraft Axial Compressor

NOTE: The compressor clutch may be removed with the compressor on the car if no other operations are to be performed. Do not clean pulley bearing with any type of solvent. Bearing is supplied with correct lubricant when assem-

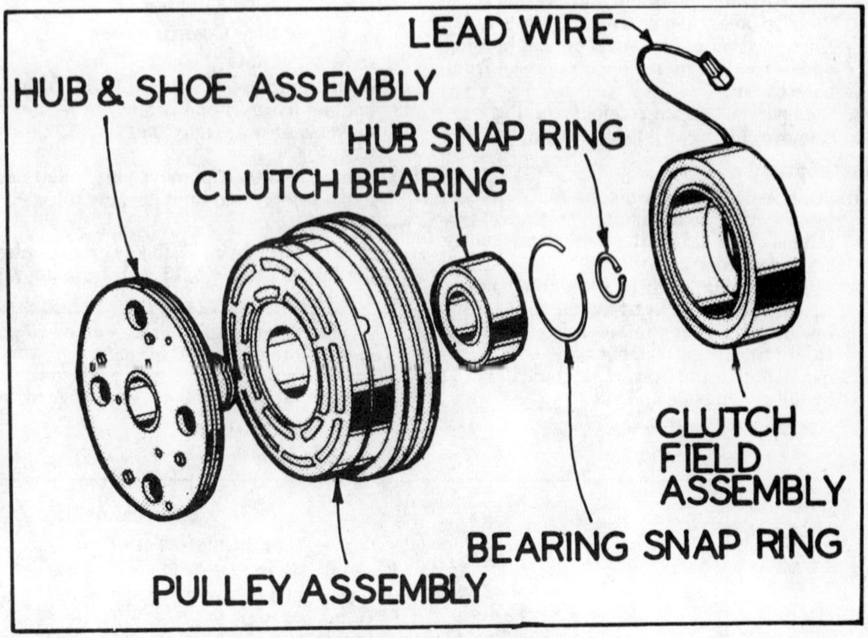

Fig. 1 Clutch assembly. Air-Temp compressors

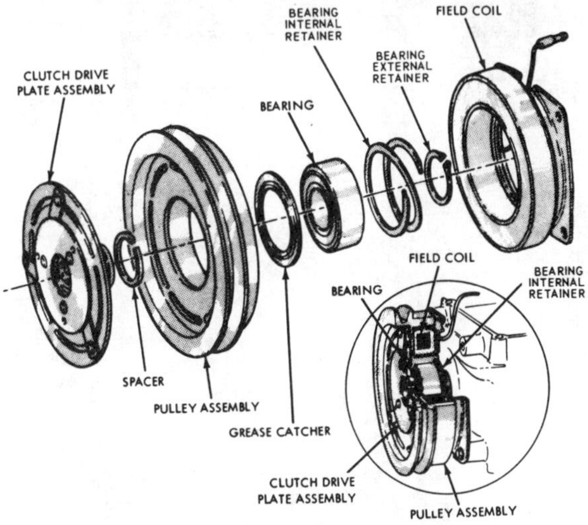

Fig. 2 Clutch assembly. Tecumseh and York compressors

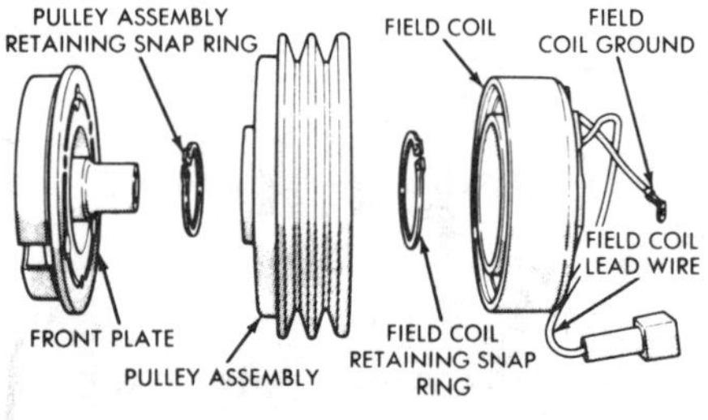

Fig. 3 Clutch assembly. C-171 compressors

bled and requires no other lubricant at any time.

Removal, Fig. 4

1. Remove clutch plate and hub assembly.
2. Remove pulley and bearing, Figs. 5, 6, 7.
3. Note position of terminals on coil housing and scribe location on compressor front head casting.
4. Remove coil housing retainer ring, Fig. 8.
5. Remove clutch coil housing assembly.

Installation

1. Position coil and housing assembly on compressor front head casting so electrical terminals line up with marks previously scribed on compressor.
2. Align locating extrusions on coil housing with holes in front head casting.
3. Install coil housing retainer rings with flat side of ring facing coil, Fig. 8.
4. Install pulley and bearing assembly.
5. Install clutch plate and hub.

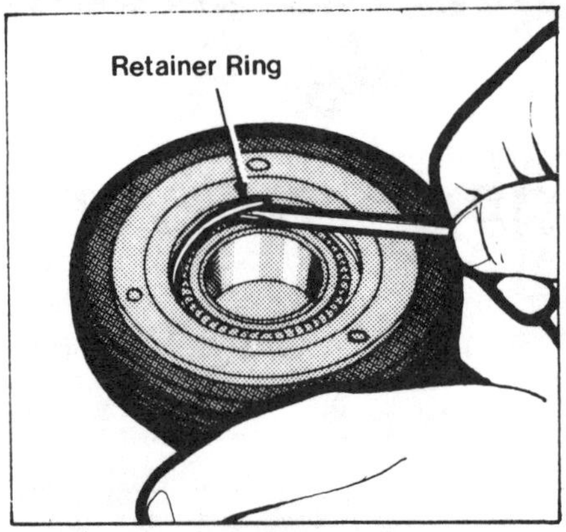

Fig. 5 Removing pulley and bearing retainer ring

Frigidaire Radial Compressor

Removal

1. With compressor secured to a suitable holding fixture and clamped in a vise, remove shaft nut using tools J-25030 and J-9399, Fig. 9.
2. Using tool J-9401, remove clutch plate and hub assembly and shaft key, Fig. 10.

Installation

1. Install shaft key into hub key groove, allowing key to project 3/16 inch from keyway, Fig. 11.

NOTE: The key is slightly curved to provide an interference fit in shaft.

2. With shaft key and keyway aligned, assemble clutch plate and hub on shaft.

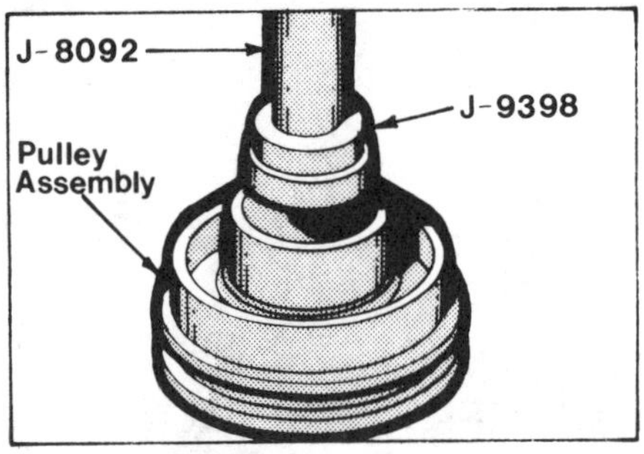

Fig. 6 Removing bearing from pulley

NOTE: Clean clutch plate and clutch rotor frictional surfaces before installation. Also, when installing the assembly, do not apply pressure to clutch hub and shaft.

3. Using tools J-9480-1 and J-9480-2, press hub onto shaft until approximately .020 to .040 inch clearance exists between clutch plate and rotor frictional surfaces, Fig. 12.
4. Install new shaft nut with small diameter boss facing toward crankshaft shoulder, using removal tools. Torque shaft nut to 8-12 ft. lbs. Ensure specified clearance is maintained between clutch plate and rotor.

Nippondenso Six Cylinder Compressor

Clutch Hub & Pulley, Replace

1. Partially drain coolant and disconnect upper hose from radiator.
2. Energize clutch and remove clutch hub re-

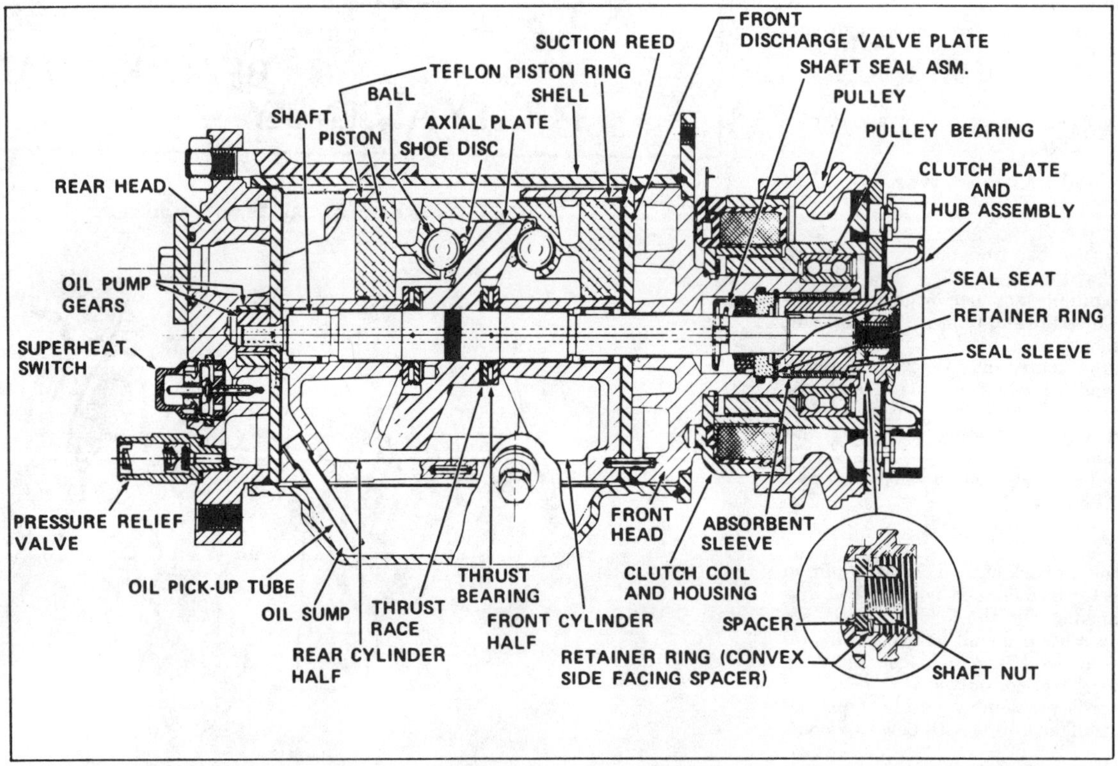

Fig. 4 Frigidaire & Motorcraft axial compressor

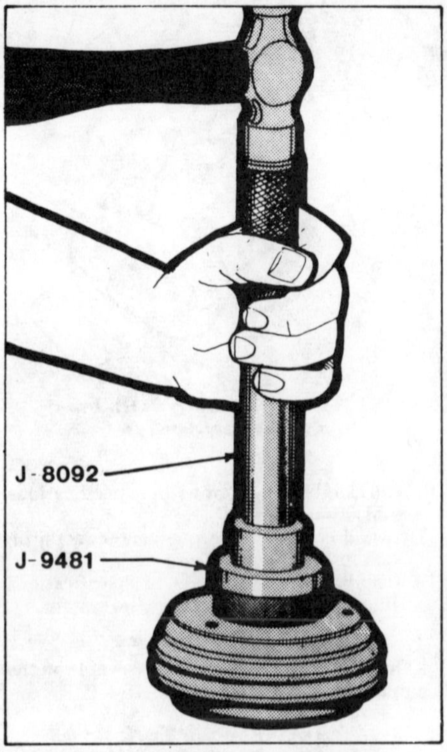

Fig. 7 Installing bearing on pulley

taining nut.

3. Remove drive belt from the pulley.
4. Remove clutch hub and shim(s) from compressor shaft using tool T80L-19703-B or equivalent, Fig. 13.
5. Remove pulley retaining snap ring. Then remove pulley and bearing assembly using tools T71P-19703-B, D80L-19703-AJ, T80L-19703-K and T80L-19703-G or equivalent puller, if pulley cannot be removed by hand.
6. Reverse procedure to install. Pulley snap ring must be installed with bevel side of snap ring out. Install hub with two thickest shims between compressor shaft and hub. Use tool T80L-19703-F or suitable puller to install hub onto compressor shaft if necessary. Torque clutch hub nut to 10-14 ft. lbs. Measure air gap between pulley and hub which should be .021-.036 inch. Adjust by changing shims between hub and compressor shaft.

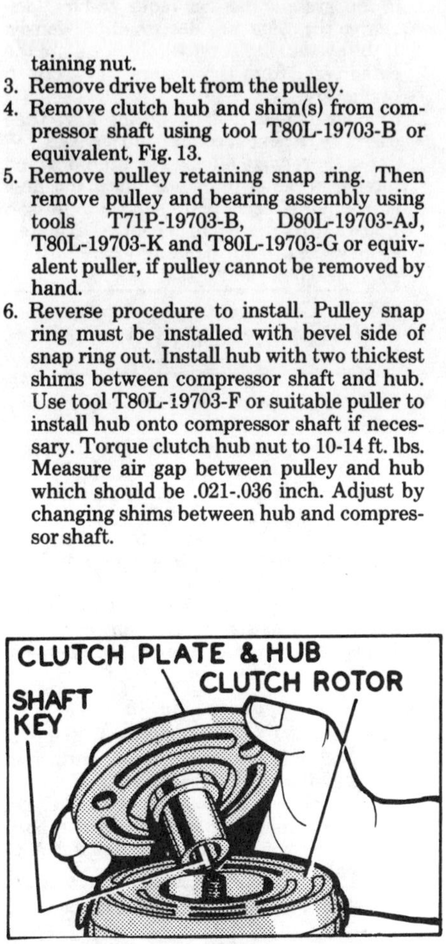

Fig. 11 Installing shaft key

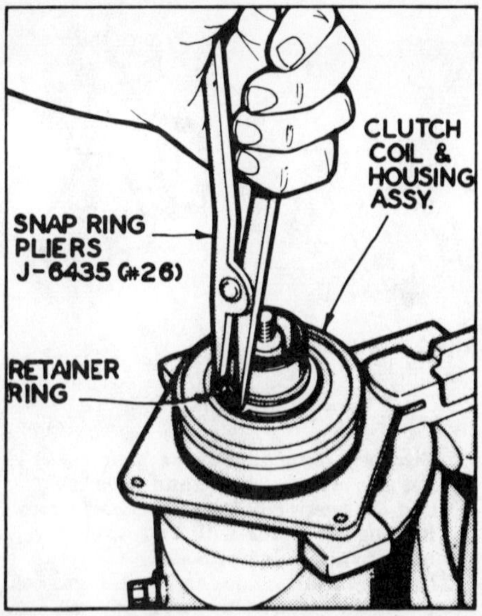

Fig. 8 Replacing coil housing retaining ring

NOTE: Do not drive hub into position. Damage to compressor will result. Use only a suitable puller.

Clutch Field Coil, Replace

1. Remove clutch hub and pulley as described in "Clutch Hub & Pulley, Replace".
2. Disconnect coil wiring.
3. Remove field coil retaining snap ring and pull coil from compressor.
4. Reverse procedure to install. Install coil retaining snap ring with bevel side facing out.

NOTE: Ensure field coil locator hole aligns with pin on compressor housing during installation of snap ring.

Sankyo Compressor

Removal, Fig. 14.

1. Remove drive belt from pulley and disconnect field coil wiring.

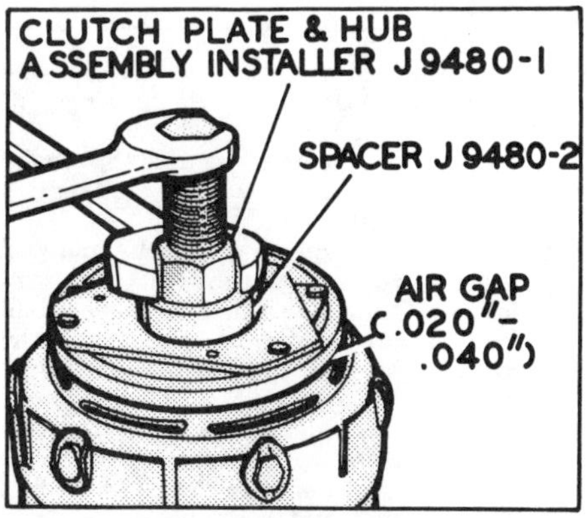

Fig. 12 Installing clutch plate & hub assembly

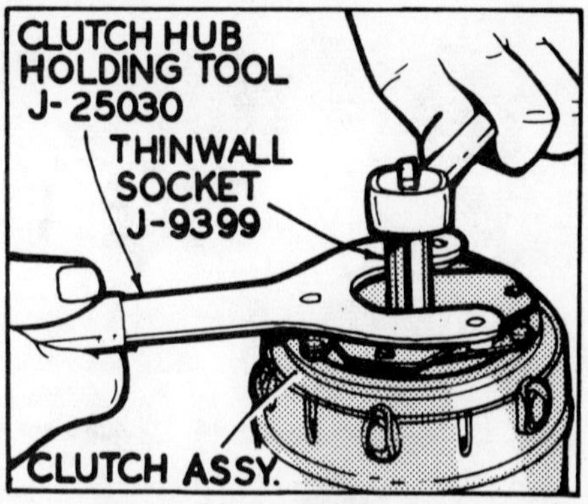

Fig. 9 Replacing compressor shaft nut

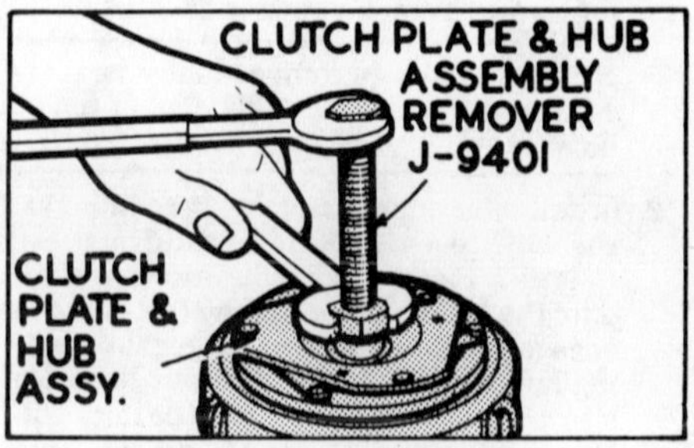

Fig. 10 Removing clutch plate & hub assembly

2. Use a suitable spanner wrench to hold clutch hub and remove retaining nut.
3. Place tool C-4490 or suitable puller over end of shaft and install three (3) bolts into the ¼-20 holes in clutch hub and pull hub from shaft.
4. Remove shaft key and shims from compressor shaft.
5. Remove pulley bearing internal and external snap rings. Then install rotor pulley puller, tool C-4495-15 into the rotor cavity so that the lips on the puller jaws engage the snap ring groove adjacent to the outer race of bearing. Remove rotor pulley from compressor.
6. Remove field coil retaining screws and pull coil assembly from compressor housing.

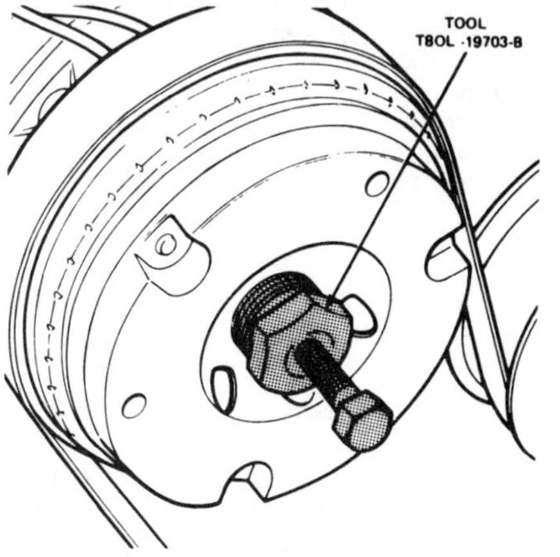

Fig. 13 Removing compressor clutch hub. Nippondenso

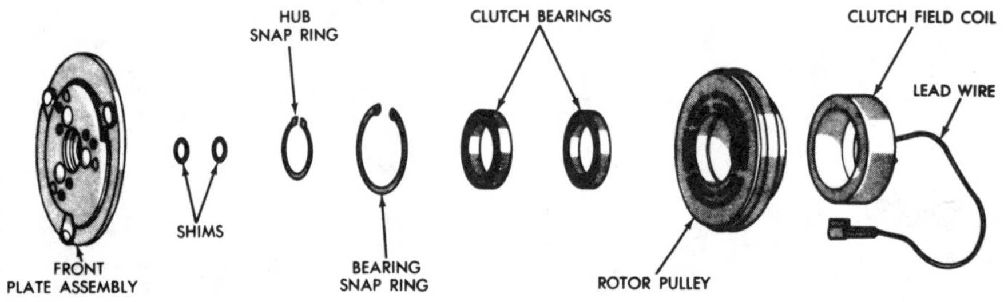

Fig. 14 Compressor clutch assembly. Sankyo

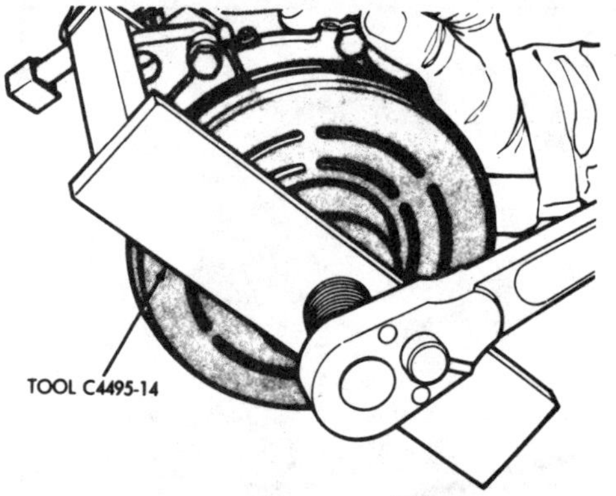

Fig. 15 Installing clutch front plate assembly. Sankyo

Installation

1. Align tab on back of field coil with recess in compressor housing and install coil retaining screws. Torque screws to 25-30 in. lbs.

NOTE: Lead wire from coil must be routed directly to the retaining clip on top of housing.

2. Install rotor pulley assembly onto compressor with tool C-4495-14 and drive sleeve C-4495-2. Place clutch pulley over the shaft onto the hub and position the tool frame to engage the pins with the diagonal mounting lugs (unused), Fig. 15. Insert the drive sleeve between the frame and the rotor pulley with the conical end against the inner race of the bearing. Rotate screw in center of frame clockwise to engage the drive sleeve. Continue turning the screw until the rotor pulley is bottomed against the shoulder of compressor.
3. Install both bearing snap rings.
4. Install shims and key previously removed.
5. Install the front clutch plate assembly onto shaft with lands facing clutch pulley. Position the frame of tool C-4495-14 to engage compressor mounting lugs as in step (2) and place C-4495-1 drive sleeve between the frame and the front plate hub. Rotate the screw to contact the drive sleeve. Continue until the armature hub bottoms.
6. Check the air between the front plate and rotor pulley. Air gap should be .016-.032 inch. If not, change shims with shims of a different thickness as required.
7. Install compressor shaft nut and torque to 25-30 ft. lbs. Recheck air gap.

Clutch Bearing, Replace, Fig. 14.

1. Remove bearing retaining ring.
2. Remove bearings using suitable press.
3. Reverse procedure to install. Bearings must be installed with black shields facing each other and the colored shields facing outward. Install until fully seated.

COMPRESSOR SEAL, REPLACE

Air-Temp

NOTE: The seal may be replaced with the compressor installed in the vehicle or removed from the vehicle. Special care should be used when installing the seal on a compressor installed in the vehicle, that the carbon ring does not fall out of its housing. Adequate lubrication of the rotating seal assembly prior to installation will prevent the carbon ring from falling out. If the compressor is removed, the suction opening on the back must be sealed to prevent loss of oil.

Removal, Fig. 16

1. Discharge refrigerant from system.
2. Remove belt, clutch, coil and drive key.
3. Remove bearing housing bolts and remove housing by prying with two screwdrivers inserted in the slots provided.
4. Remove bearing housing oil seal, gas seal plate and O-ring from bearing housing.

NOTE: The gas seal plate and O-ring are part of the replacement package and must be replaced.

Installation

1. Immerse the new seal seat in clean refrigerant oil, then install seal into bearing housing with the smooth side facing up. Using a sleeve with a minimum inside diameter of 1⅜ inch, carefully drive seal into housing until it is fully seated.
2. Prior to installing the cartridge seal assembly, inspect the assembly to assure that the tangs of the carbon seal, scallops and retainer are indexed in the slots of its corresponding steel part, Fig. 17.
3. Immerse the seal assembly with the carbon ring facing up in clean refrigerant oil, then while holding the seal assembly at the outside edge to prevent the ring from coming out, install the seal assembly.

NOTE: Do not touch the polished sealing surface of the carbon ring.

4. After seal assembly bottoms against crankshaft bearing inspect the index alignment.
5. Using clean refrigerant oil, lubricate the new bearing housing oil seal and install it.

NOTE: Assure that seal is evenly stretched into place and that it is not rolled over.

6. Wipe the seal seat clean with a lint-free cloth and lubricate with clean refrigerant oil.
7. Install bearing housing assuring that end of crankshaft does not contact seal seat in bearing housing.
8. Install the bearing housing bolts and pull bearing housing evenly into place by turning bolts ½ turn at a time to prevent ball bearing outer race from jamming in the housing. Torque bolts to 95-135 inch lbs.
9. Install drive key, then assemble clutch coil onto compressor and check torque required to rotate crankshaft. If torque required to turn crankshaft is more than 10 inch lbs., remove clutch and loosen bearing housing bolts until shaft loosens up, then retorque bolts evenly to 95-135 inch lbs.
10. Check oil level. If not within specifications, add as necessary.
11. Install clutch and torque center mounting bolt to 20 ft. lbs.
12. Install and adjust belts to specifications, then evacuate and recharge the system.

C-171 Compressor

The compressor seal is not serviceable on this compressor.

Tecumseh & York

Removal, Figs. 18 & 19

1. If the system has Schrader Valves, discharge the system as described previously. If the system has service valves, isolate the compressor from the system as described previously.
2. Remove clutch and Woodruff key, then carefully remove secondary dust shield.
3. Carefully remove all accumulated dirt and foreign material from seal plate and area surrounding seal plate on compressor. Place a small drain pan beneath seal plate.
4. Remove seal plate bolts and carefully remove seal plate and gasket.

NOTE: Use care to avoid marring or scratching the sealing surfaces or polished surfaces.

5. Remove carbon seal ring and seal housing assembly.
6. Remove all gasket material from seal plate and compressor. Assure that shaft, seal plate and compressor gasket surfaces are completely clean.
7. Check face of crankshaft front bearing journal in the seal housing to assure that there are no nicks or burrs. Check crankshaft surface and all parts of the seal assembly to make sure that they are not damaged.
8. Inspect compressor internal components for damage.

Installation, Tecumseh, Fig. 18

1. Lubricate new seal assembly in clean refrigerant oil and lubricate exposed surface of crankshaft with clean refrigerant oil.
2. Place seal over crankshaft with the end that fits carbon ring facing out, then position the carbon ring over the shaft and onto the seal.

NOTE: The raised rim of the carbon ring must face out.

3. Insert O-ring in crankcase mating surface

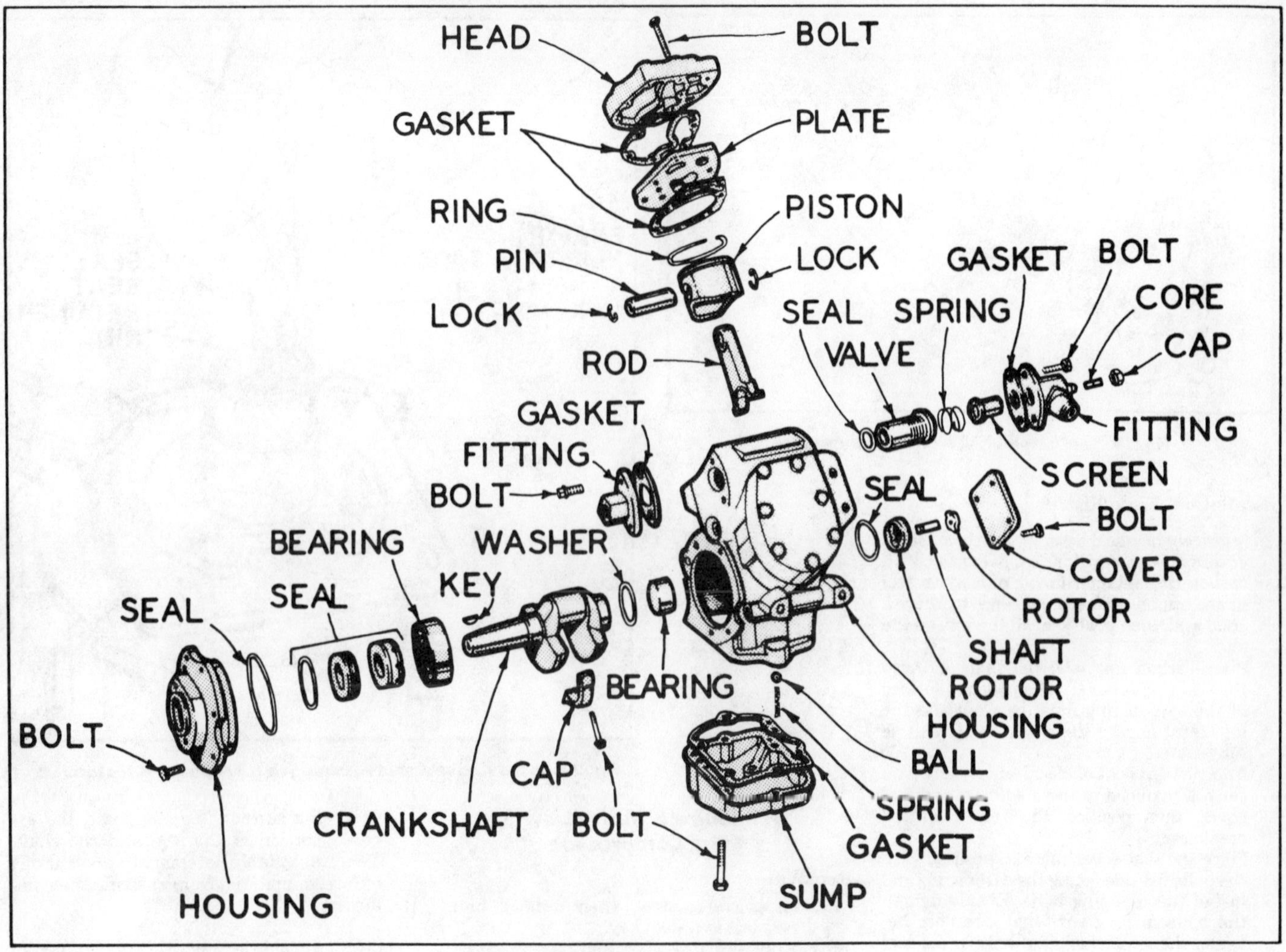

Fig. 16 Chrysler Air-Temp compressor

of seal plate, then position seal plate, shield and seal plate retainer onto the compressor aligning the bolt holes. Push seal plate retainer against compressor and install and evenly torque the retaining bolts to 54-78 inch lbs.

4. Rotate shaft by hand 15-20 times to seat the seal.
5. Assuring that there are no burrs or dirt on crankshaft, install the key and magnetic clutch onto the shaft.
6. Install belt and adjust tension to specifications.
7. Check oil level. If not within specifications, add as necessary.
8. Evacuate and recharge system.

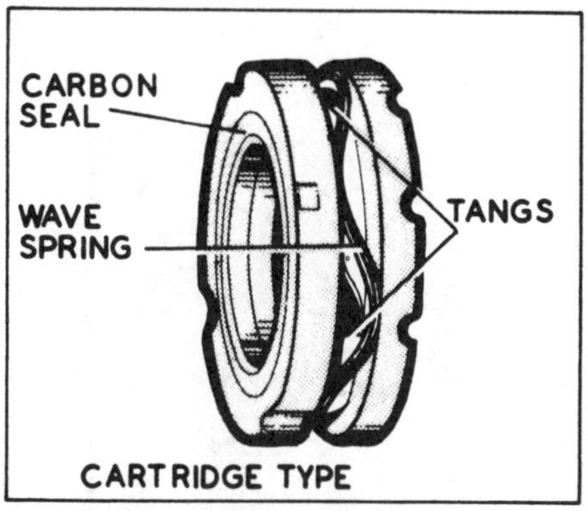

Fig. 17 Cartridge type seal assembly. Chrysler Air-Temp

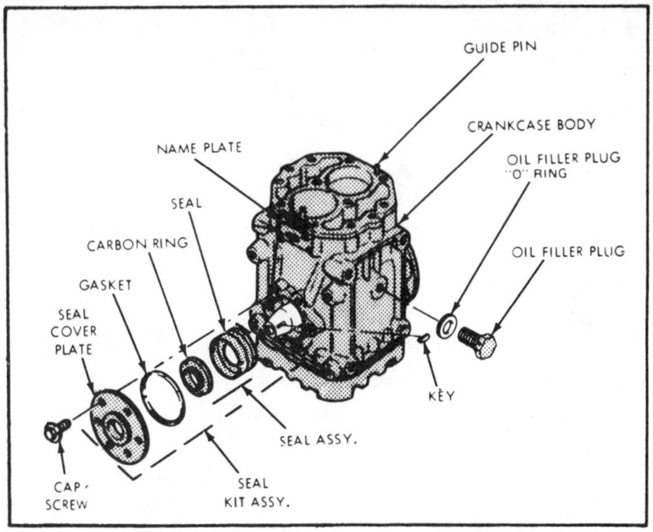

Fig. 18 Seal assembly. York compressor

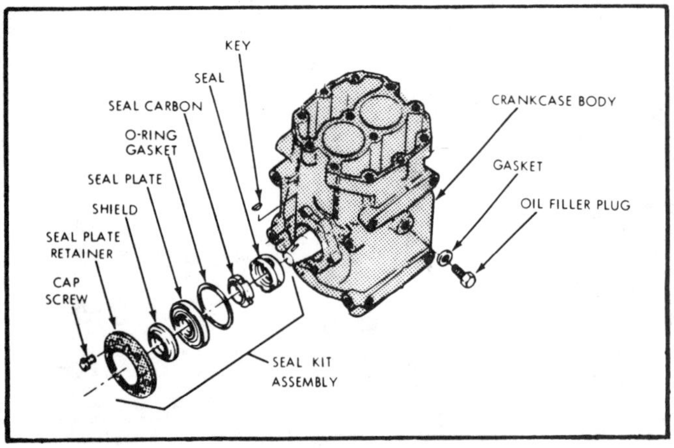

Fig. 19 Seal assembly. Tecumseh compressor

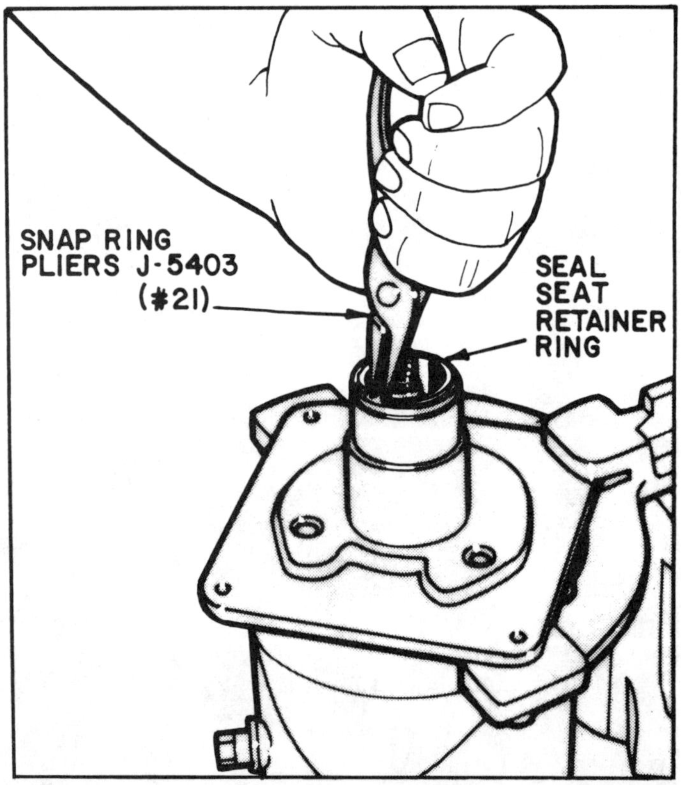

Fig. 20 Removing seal seat retaining ring. Frigidaire & Motorcraft

Installation, York, Fig. 19

1. Lubricate new seal assembly in clean refrigerant oil, then position seal over shaft with carbon ring retainer facing out. Move seal in and out on shaft several times to insure a good seal, then push seal all the way on the shaft.
2. Place carbon ring over the shaft and into the seal ring retainer. The polished surface of the carbon ring must face out and the lugs must engage the ring retainer and be fully seated.
3. Apply a light coat of clean refrigerant oil on the mating faces of the crankcase and seal cover, then position the gasket on the crankcase.
4. Place the seal cover plate in position with the polished side facing the carbon ring and install the retaining bolts. Evenly tighten the bolts making sure that clearance between the crankshaft and hole in the seal cover plate is even all around the shaft. If clearance is not equal, gently tap seal face into position until the clearance is equal. Torque the bolts evenly to 5-7 ft. lbs. (small seal) and 7-13 ft. lbs. (large seal).
5. Install belt and adjust tension to specifications.
6. Check oil level. If not within specifications, add as necessary.
7. Evacuate and recharge system.

Frigidaire & Motorcraft Axial Compressor

Removal

1. Depressurize system, then remove hub, drive plate and shaft key.
2. Pry out sleeve retainer and remove absorbent sleeve.
3. Using snap ring pliers J-5403 (No. 21), remove shaft seal retaining ring, Fig. 20.
4. Thoroughly clean inside of compressor neck area surrounding the shaft, the exposed portion of the seal and the shaft. This is absolutely necessary to prevent dirt or foreign material from entering the compressor.
5. Position seal protector J-22974 over end of shaft to prevent damaging ceramic seat, Fig. 21. Engage tangs of tool J-32128 into recessed portion of seal seat by turning handle clockwise and then remove seat

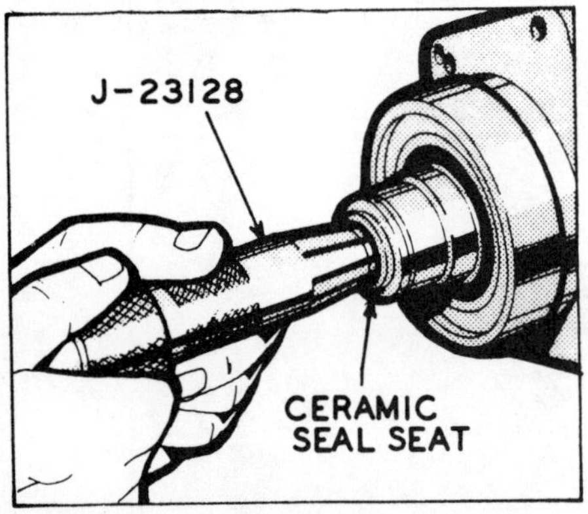

Fig. 21 Removing seal seat. Frigidaire & Motorcraft

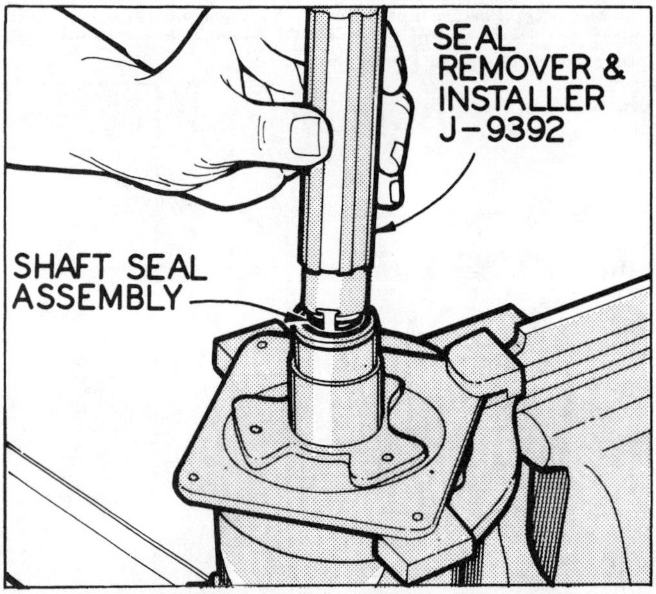

Fig. 22 Removing shaft seal. Frigidaire & Motorcraft

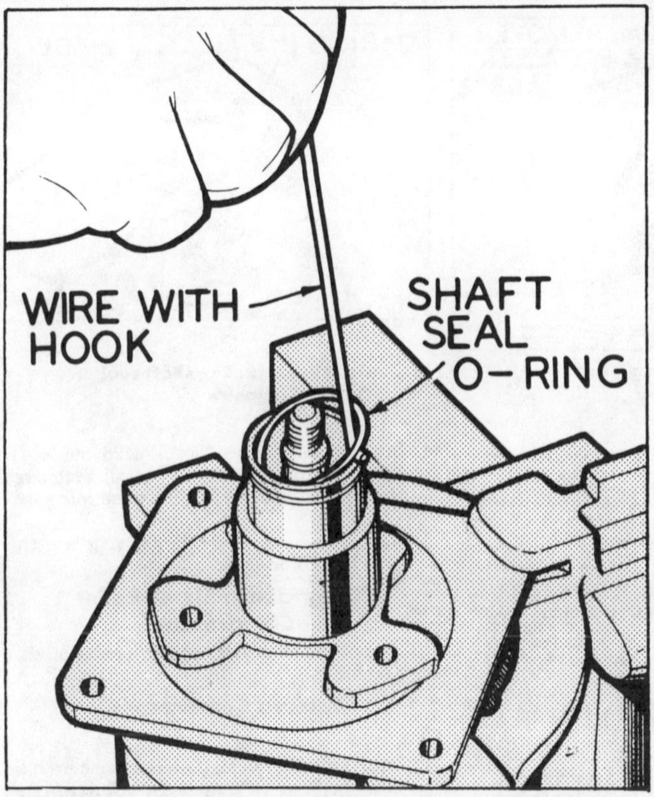

Fig. 23 Removing shaft seal O-ring. Frigidaire & Motorcraft

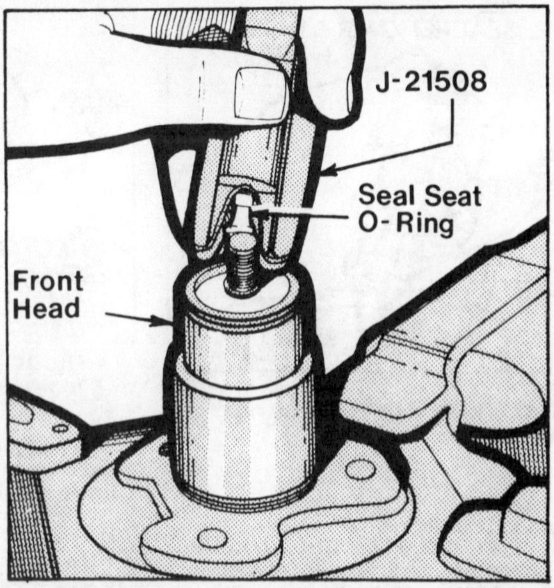

Fig. 24 Installing seal seat O-ring. Frigidaire & Motorcraft

from compressor with a rotary motion.

NOTE: The tool handle must be hand-tightened only. Do not use tools or pliers to tighten tool handle.

6. Using tool J-9392, remove seal assembly. Press tool downward while turning it clockwise to engage the tabs of the seal assembly, then gently but firmly, pull tool straight out, Fig. 22.
7. Using a hooked wire, remove seal seat O-ring, Fig. 23.
8. Recheck shaft and inside of compressor neck. Make sure that these areas are perfectly clean before installing the new parts.

Installation

1. Coat the new seal seat O-ring with clean refrigerant oil, then using tool J-21508, Fig. 24, install O-ring in bottom groove of compressor neck, Fig. 25.
2. Coat O-ring and seal face of new seal with clean refrigerant oil. Install seal assembly onto tool J-9392 by engaging tabs of the seal with tangs of the tool.
3. Place seal protector J-22974 over end of shaft and carefully slide the new seal assembly and tool onto the shaft. Gently twist tool clockwise while pushing seal assembly down the shaft until the seal engages the flats on the shaft and is seated

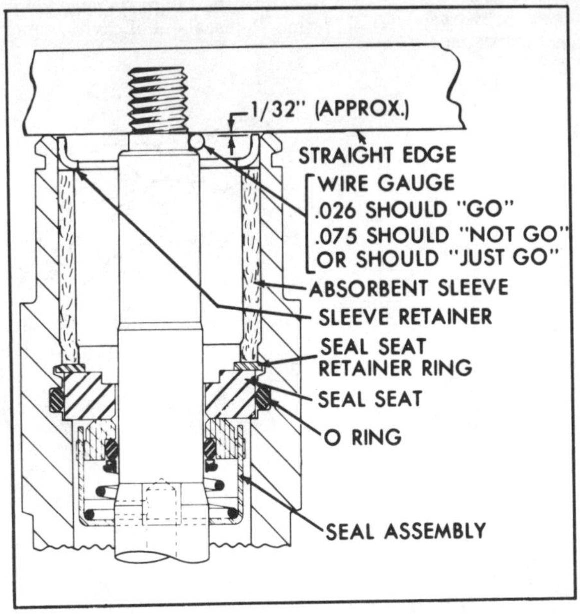

Fig. 25 Gauging seal retainer. Frigidaire & Motorcraft

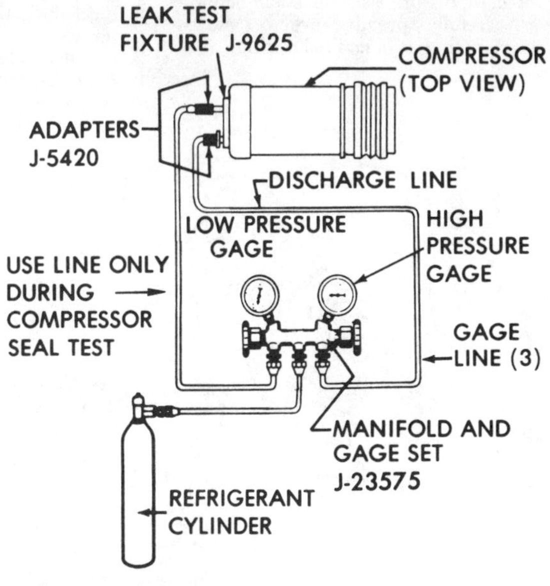

Fig. 26 Leak testing compressor seal. Frigidaire & Motorcraft

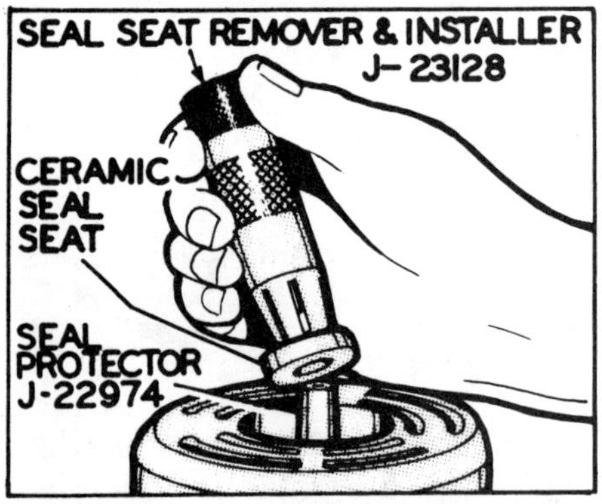

Fig. 27 Replacing shaft seal seat. Frigidaire

in place. Disengage tool by pressing downward and twisting counter-clockwise.

4. Coat the seal face of new seal seat with clean refrigerant oil, then install seal seat on tool J-23128 and install into compressor neck using care to avoid dislodging the seal seat O-ring and assuring that seal seat makes a good seal with O-ring.
5. Using snap ring pliers J-5403 (No. 21), install seal seat retainer ring, Fig. 20. Use sleeve from tool J-23128 to press on seal seat retainer so that it snaps into place. Remove seal protector J-22974.
6. Install Compressor Leak Test Fixture J-9625 and connect gage charging lines, Fig. 26. Pressurize suction side of compressor to drum pressure. Temporarily install the shaft nut and with compressor horizontal and sump down, rotate compressor shaft several times in normal direction of rotation and check for leaks around the seal area using a torch type leak detector. Correct any leaks found and remove and discard shaft nut.
7. Remove any excess oil from the shaft and inside the compressor neck area.
8. Install the new absorbent sleeve by rolling the material into a cylinder, then slipping sleeve into compressor neck with overlap at top of compressor. Using a small screwdriver, carefully spread sleeve to remove the overlap, so that in its final position, the ends butt together at the top vertical centerline.
9. Install new sleeve retainer so that its flange face will be against the front end of the sleeve, then using sleeve from tool J-23128 and a mallet, tap sleeve retainer until outer edge of the sleeve retainer is recessed about 1/32 inch from the face of the compressor neck, Fig. 25.
10. Install key, hub and drive plate.
11. Recharge refrigeration system.

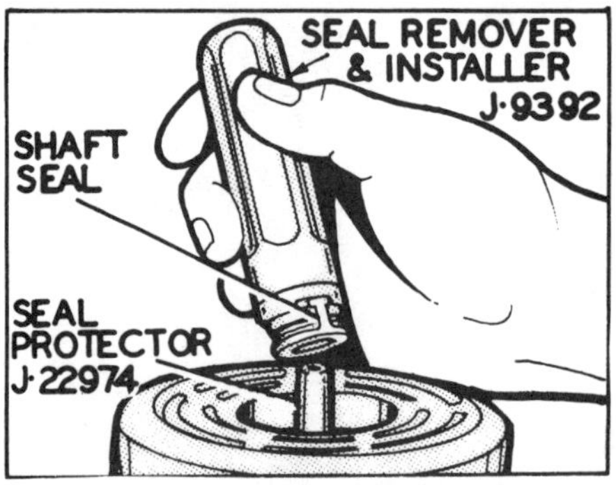

Fig. 28 Replacing shaft seal. Frigidaire

Frigidaire Radial Compressor

Removal

1. Discharge refrigerant and remove clutch plate and hub outlined previously.
2. Using tool J-5403, remove shaft seal seat retainer ring.
3. Remove shaft seal seat with tool J-23128, Fig. 27, then the shaft seal with tool J-9392, Fig. 28.
4. Remove seal seat "O" ring with tool J-9553.

Installation

NOTE: Ensure compressor neck and shaft area is free from lint, dirt, or other foreign material prior to installing new seal parts.

1. Lubricate seal seat "O" ring in refrigerant oil and assemble onto tool J-21508, Fig. 29.
2. Insert tool J-21508 into compressor neck until bottomed and install seal in proper position.
3. Lubricate shaft seal "O" ring with refrigerant oil and install on tool J-9392. Slide seal protector, tool J-22974, over shaft and using tool J-9392, install shaft seal, Fig. 28.

Fig. 29 Installing shaft seal O-ring. Frigidaire

4. Install seal seat onto tool J-23128 and lubricate with refrigerant oil. Install seat over shaft and tool J-22974, then push into position, Fig. 27.
5. Install new seal seat retainer ring and the clutch hub and rotor.

Nippondenso Six Cylinder Compressor

The compressor seal is not serviceable on this compressor.

Sankyo Compressor

Removal

1. Depressurize system, then remove clutch as described in "Compressor Clutch, Replace".
2. Remove the key and shims from compressor shaft.
3. Remove the seal seat retaining snap ring, Fig. 30. Then remove the seal seat using tool C-4492.

NOTE: Engage tool jaws onto groove of the seal seat, Fig. 31.

4. Using tool C-4493, remove seal cartridge from compressor shaft. Insert tool into cavity until it contacts the seal frame. Rotate tool until tangs on tool enter slots in the frame. Push against the seal and continue to rotate until the tangs are engaged with seal frame. Pull seal out of cavity, continu-

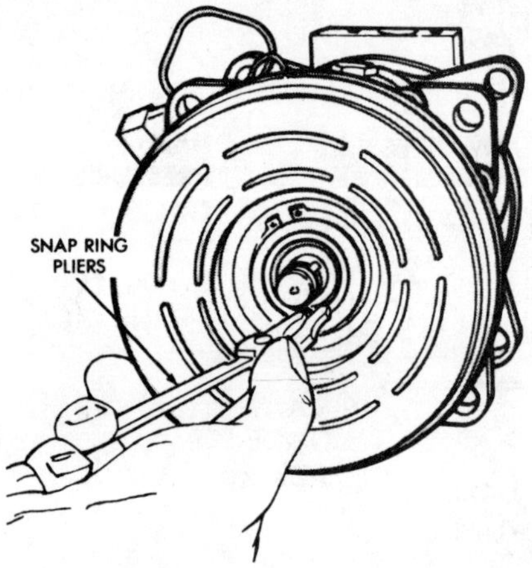

Fig. 30 Removing seal seat snap ring. Sankyo

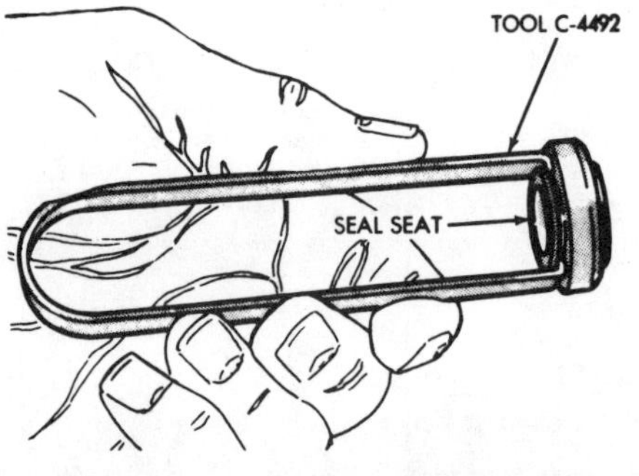

Fig. 31 Removing seal seat. Sankyo

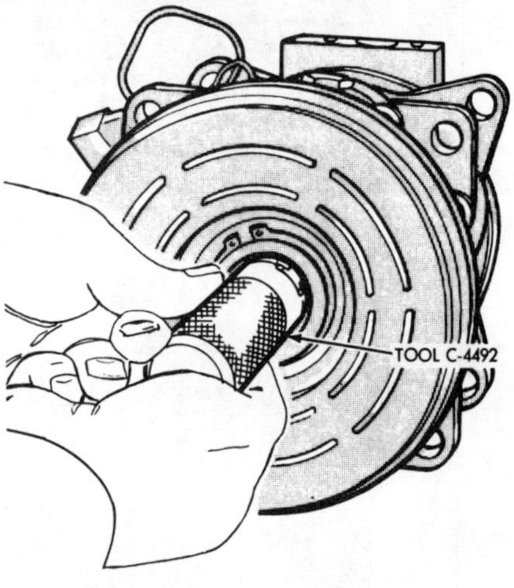

Fig. 32 Removing seal cartridge. Sankyo

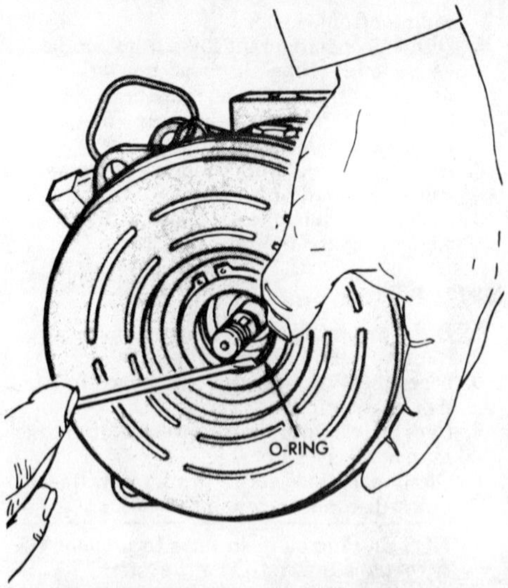

Fig. 33 Removing O-ring. Sankyo

ing to apply pressure in the same direction of rotation used to engage the tool with seal frame, Fig. 32.

5. Remove seal seat O-ring from its groove, Fig. 33.

Installation

NOTE: Ensure compressor neck and shaft area is free of lint, dirt, or other foreign material prior to installing new seal parts.

1. Install seal seat O-ring into its groove, Fig. 33.
2. Coat surface of seal with clean refrigerant oil. Insert tool C-4493 over the cartridge (carbon face towards tool), ensuring not to touch lapped face of seal with tool. Engage tabs of tool into slots in seal frame by pushing and rotating tool, Fig. 32.
3. Coat surface of seal protector C-4494 with refrigerant oil. Insert seal protector into seal cartridge from side opposite lapped carbon face until protector is nearly through, Fig. 34. Place open end of seal protector onto compressor shaft and slide seal onto shaft by pushing installation tool. Push seal until it contacts shoulder on the shaft. Then, with a rotating motion, maintaining pressure, rotate the seal frame to engage the flats on the shaft. When the frame is aligned with flats, the seal can be pushed an additional ⅛ inch on the shaft, Fig. 32.
4. Lubricate seal seat with refrigerant oil, taking care not to touch the lapped face.
5. Engage the jaws of seat removal tool C-4492 into groove on back of seal seat, Fig. 31. Place the seat over compressor shaft and push seat into the housing bore, Fig. 32. As the seat contacts O-ring, use a slight rocking and rotating motion while pushing seat into place. Continue pushing until snap ring groove in housing bore is visible. If the groove is not visible, remove the tool and place seal ring installation tool C-4496 onto the shaft and engage the threads. Tighten the tool until snap ring groove is visible.
6. Install seat retaining ring into the groove with the beveled edge facing away from the compressor. Ensure the snap ring is fully seated with a gap of 5/32 inch between lugs of the ring.

NOTE: Seal seat installation tool C-4496 can also be used to seat snap ring.

7. Connect a refrigerant source to system and apply a holding charge. Check for evidence of leakage.
8. Install shims, shaft key and clutch assembly.
9. Check refrigerant oil level and add if necessary.
10. Evacuate and recharge system.

RECEIVER-DEHYDRATOR, REPLACE, FIG. 35

NOTE: Whenever the system has been open for any amount of time as in the case of a collision, the receiver-dehydrator must be replaced because the dehyrator element will have become saturated with moisture. A new unit is required to remove excess moisture from the rest of the system.

AMERICAN MOTORS & JEEP

1979-80 Jeep

1. Discharge refrigerant from system.
2. On Cherokee, Wagoneer and Truck models, remove headlamp trim ring, headlamp and grille. Then remove bolt from top of inner panel. Pull panel back and block.
3. On all models, disconnect refrigerant lines from receiver-drier. Plug all open lines to prevent entry of dirt and moisture into system.

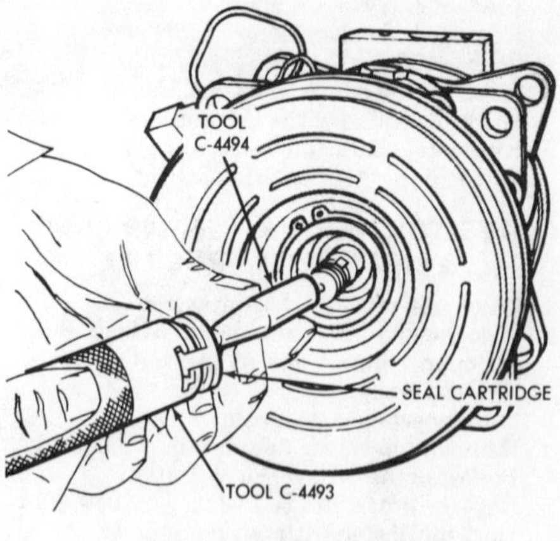

Fig. 34 Installing seal protector. Sankyo

4. Remove receiver-drier retaining bolts and receiver-drier.
5. Reverse procedure to install. Evacuate, charge and leak test system. Check system for proper operation.

1974-80 Exc. 1979-80 Jeep

1. Remove condenser assembly from vehicle as outlined in "CONDENSER, REPLACE" procedure.
2. Disconnect refrigerant lines from receiver-dehydrator. Plug all open connections to

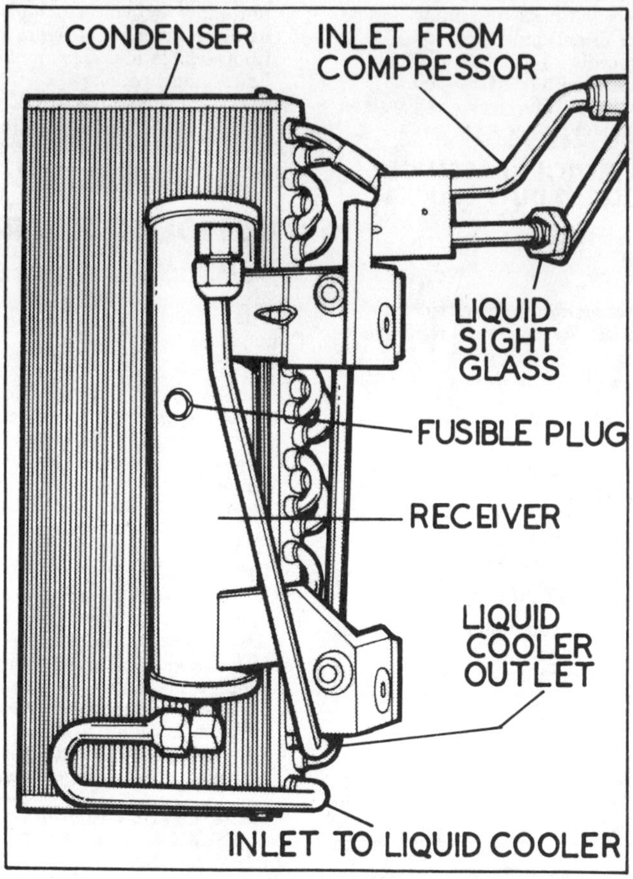

Fig. 35 Receiver-dehydrator. Typical

prevent entry of dirt and moisture.
3. Remove receiver-dehydrator mounting screws and the receiver-dehydrator.
4. Reverse procedure to install. Use new "O" rings lubricated with refrigerant oil.
5. Evacuate, charge and leak test system and check for proper operation.

CHRYSLER CORP. PASSENGER CARS & LIGHT DUTY TRUCKS

1. Discharge refrigerant from system.
2. Disconnect refrigerant lines at fittings. Plug all open connections to prevent entry of dirt and moisture.
3. Disconnect electrical wiring.
4. Remove receiver-dehydrator mounting bolts and the receiver-dehydrator.
5. Reverse procedure to install. Use new "O" rings lubricated with refrigerant oil.
6. Evacuate, charge and leak test system and check for proper operation.

FORD MOTOR CO. PASSENGER CARS & LIGHT DUTY TRUCKS

NOTE: On some models, the receiver-drier is permanently attached to the condenser assembly, requiring replacement of both units.

1. On Light Duty Truck models, refer to "Condenser, Replace" procedure for receiver-drier replacement.
2. On all models, discharge refrigerant from system.
3. Disconnect refrigerant lines at fittings. Plug all open connections to prevent entry of dirt and moisture into system.
4. Remove receiver-drier mounting screws and receiver-drier from vehicle.
5. Reverse procedure to install. Use new O-rings lubricated with refrigerant oil.
6. Evacuate, charge and leak test system. Check system for proper operation.

GENERAL MOTORS PASSENGER CARS & LIGHT DUTY TRUCKS

Buick

1974 Apollo

1. Discharge refrigerant from system.
2. Disconnect battery cables and remove battery.
3. Remove receiver assembly mounting screws.
4. Disconnect refrigerant line connections at receiver.
5. Separate receiver from brackets.
6. Reverse procedure to install. Use new "O" rings lubricated with refrigerant oil.
7. Evacuate, charge and leak test system and check for proper operation.

1974-77 Chevrolet & G.M.C. Van Models

1. Disconnect battery ground cable and compressor clutch connector.
2. Discharge refrigerant from system. Plug all open lines to prevent entry of dirt and moisture into system.
3. Remove front headlamp bezels, parking lamp lenses and grille.

NOTE: Place a piece of protective tape on upper surface of bumper.

4. Disconnect refrigerant lines from receiver-drier, then remove receiver-drier from vehicle.
5. Reverse procedure to install. Add one ounce of refrigerant oil to new receiver-drier.
6. Evacuate, charge and leak test system. Check system for proper operation.

Oldsmobile

1974 Omega

1. Disconnect thermal limiter then discharge system.
2. Disconnect battery cables and remove battery.
3. Remove receiver bracket screws, then disconnect the inlet and outlet fittings and remove receiver-dehydrator assembly.

NOTE: Plug all lines to prevent entry of moisture and dirt.

4. Separate receiver-dehydrator from brackets.
5. Add one ounce of refrigerant oil to new receiver-dehydrator, then reverse removal procedure to install.
6. Evacuate, charge and leak test system and check for proper operation.

Pontiac

1974 Ventura

1. Discharge refrigerant from system.
2. Disconnect battery cables and remove battery from vehicle.
3. Remove receiver-dehydrator bracket mounting screws.
4. Disconnect refrigerant lines from receiver-dehydrator. Plug all open connections to prevent entry of dirt and moisture.
5. Remove receiver-dehydrator assembly from vehicle and separate receiver-dehydrator from brackets.
6. Reverse procedure to install. Use new "O" rings lubricated with refrigerant oil.
7. Evacuate, charge and leak test system and check for proper operation.

CONDENSER, REPLACE

AMERICAN MOTORS & JEEP

1974 Javelin & Matador, 1975-78 Matador Coupe Models

1. Discharge refrigerant from system.
2. AMX and Javelin models, remove upper air baffle to fender tie braces. On all models, remove upper air baffle.
3. Disconnect compressor discharge to condenser coupling. Then disconnect receiver to evaporator hose at disconnect coupling. Plug all open connections to prevent entry of dirt and moisture.
4. On 1975-78 Matador models, remove radiator mounting bolts and reposition radiator close to engine to gain access to condenser assembly mounting bolts.
5. On all models, remove condenser assembly mounting bolts.
6. Separate receiver from condenser.
7. Reverse procedure to install.
8. Evacuate, charge and leak test system. Check system for proper operation.

1974-80 Concord, Gremlin, Hornet, Spirit; 1977-80 AMX & Eagle with 6 Cyl. & V8 Engine

1. Discharge refrigerant from system.
2. Drain coolant from radiator and remove radiator from vehicle.
3. Disconnect refrigerant lines from condenser assembly. Plug all open connections to prevent entry of dirt and moisture.
4. Remove condenser assembly mounting bolts and condenser assembly.
5. Remove receiver from condenser assembly.
6. Reverse procedure to install.
7. Evacuate, charge and leak test system. Check system for proper operation.

1977-80 Pacer

1. Discharge refrigerant from system.
2. Remove radiator from vehicle.
3. Disconnect compressor discharge-to-condenser coupling.
4. Remove clamp from compressor discharge hose at receiver-drier.
5. Disconnect receiver-to-evaporator line at quick-disconnect coupling.

NOTE: Plug all open lines to prevent entry of moisture and dirt into system.

6. Remove condenser attaching bolts and condenser assembly with receiver-drier.
7. Reverse procedure to install.
8. Evacuate, charge and leak test system. Check for proper operation.

1977-80 AMX, Concord, Gremlin, Hornet & Spirit With Four-Cylinder Engine

1. Drain radiator coolant.
2. Disconnect upper and lower radiator hoses from radiator. Then disconnect heater hose from radiator.
3. Disconnect and plug automatic transmission oil cooler lines, if equipped.
4. Remove charcoal canister, mounting bracket, inlet air duct, fan shroud and radiator attaching bolts. Them remove radiator from vehicle.
5. Discharge refrigerant from system.
6. Disconnect compressor discharge-to-condenser line.
7. Disconnect receiver/drier-to-evaporator line.

NOTE: Plug all open lines to prevent entry of dirt and moisture into system.

8. Remove condenser attaching screws and condenser with receiver-drier assembly.
9. Reverse procedure to install.
10. Evacuate, charge and leak test system. Check for proper operation.

1975-76 Pacer

1. Discharge refrigerant from system.
2. Disconnect high pressure hose at receiver disconnect coupling.
3. Disconnect discharge service valve and hose assembly from compressor. Plug all open connections to prevent entry of dirt and moisture.
4. Remove fan shroud upper attaching screws.
5. Remove radiator grille reinforcement. Push radiator rearward to gain access to condenser mounting screws. Insert a block of wood to hold radiator in position.
6. Remove condenser assembly mounting screws and condenser.
7. Reverse procedure to install. Ensure headlight wiring harness is not pinched behind condenser.
8. Evacuate, charge and leak test system. Check system for proper operation.

1974-80 Jeep Models

NOTE: Condenser is removed together with receiver-drier as an assembly.

1. Discharge refrigerant from system.
2. Drain coolant and remove fan shroud and radiator from vehicle.
3. Disconnect pressure line from condenser. Plug all open lines to prevent entry of dirt and moisture into system.
4. Remove condenser attaching screws and tilt bottom of condenser toward engine.
5. From underneath vehicle, disconnect refrigerant line from receiver-drier assembly.
6. Remove condenser and receiver-drier assembly from vehicle.
7. Remove receiver-drier assembly from condenser if necessary.
8. Reverse procedure to install.
9. Evacuate, charge and leak test system. Check system for proper operation.

Ford Motor Co. Passenger Cars & Light Duty Trucks

1974-77 Comet & Maverick

1. Discharge refrigerant from system.
2. Remove both grille upper supports and the hood latch and support assembly.
3. Disconnect evaporator to condenser hose at receiver-dehyrator. Disconnect compressor to condenser hose at condenser. Remove clip securing hoses at radiator support. Plug open connections to prevent entry of dirt and moisture.
4. Remove condenser assembly mounting bolts and condenser assembly.
5. Remove receiver-dehydrator from condenser.
6. Reverse procedure to install. Use new "O" rings lubricated with refrigerant oil.
7. Evacuate, charge and leak test system. Check system for proper operation.

1979-80 Bobcat & Pinto

1. Discharge refrigerant from system.
2. Remove radiator grille assembly.
3. Disconnect ambient cut-off switch wiring connector.
4. Remove hood latch support, support brace and hood latch from between grille and radiator. Then position the assembly aside.
5. Remove power steering oil cooler and cooler line bracket bolts from radiator support and position cooler with lines aside without disconnecting them.
6. Disconnect refrigerant hoses from condenser and receiver-drier assembly. Plug all openings to prevent entry of dirt and moisture into system.
7. Remove condenser attaching screws and condenser.
8. Reverse procedure to install. Use new O-rings lubricated with refrigerant oil. Add two ounces of refrigerant oil to condenser and receiver-drier assembly.
9. Evacuate, charge and leak test system. Check system for proper operation.

1974-78 Pinto & 1975-78 Bobcat

1. Discharge refrigerant from system.
2. On 1974-76 models, remove lower hood latch bracket.
3. On all models, disconnect refrigerant hoses from condenser. Plug all open connections to prevent entry of dirt and moisture.
4. Remove condenser assembly mounting screws and condenser assembly.
5. Remove receiver from condenser.
6. Reverse procedure to install. Use new "O" rings lubricated with refrigerant oil.
7. Evacuate, charge and leak test system. Check system for proper operation.

1979-80 Ford, Mercury; 1980 Continental Mark VI & Lincoln Continental

1. Remove battery from vehicle.
2. Drain coolant and disconnect upper hose from radiator.
3. Discharge refrigerant from system and disconnect refrigerant hoses from condenser. Plug all open connections to prevent entry of dirt and moisture into system.
4. Remove air inlet duct retaining screws and position duct aside.
5. Remove fan shroud retaining screws and position shroud to the rear.
6. Remove radiator upper bracket screws and tilt radiator rearward.
7. Remove condenser retaining screws and condenser.

NOTE: Retain all four condenser rubber mounts.

8. Reverse procedure to install. Use new O-rings lubricated with refrigerant oil. Add two ounces of refrigerant oil to new condenser.
9. Evacuate, charge and leak test system. Check system for proper operation.

1978-80 Fairmont, Zephyr; 1979-80 Capri, Mustang; 1980 Cougar & Thunderbird

1. Discharge refrigerant system.
2. Remove the radiator grille. Move ambient cut-off switch away from front of condenser.
3. Disconnect battery and remove from vehicle.
4. Disconnect refrigerant lines from condenser. Plug all open connections to prevent entry of dirt and moisture into system.
5. Remove condenser retaining bolts and condenser assembly from vehicle.
6. Reverse procedure to install. Use new O-rings lubricated with refrigerant oil.

NOTE: If condenser is being replaced, replace receiver-drier also.

7. Evacuate, charge and leak test system. Check system for proper operation.

1977-79 Continental Mark V, Cougar, LTD II & Thunderbird

1. Discharge refrigerant from system.
2. On Continental Mark V models, remove upper plastic shield, upper radiator mounts and upper condenser mounts from radiator support.
3. On all models, remove condenser isolation mounting pad nuts, and the liquid line to condenser bracket screw.
4. Disconnect refrigerant lines from condenser assembly. Plug all openings to prevent entry of dirt and moisture into system.
5. Remove condenser assembly from vehicle.
6. Reverse procedure to install. Use new O-rings lubricated with refrigerant oil.
7. Evacuate, charge and leak test system. Check system for proper operation.

1974-76 Continental Mark IV, Montego, Torino & Thunderbird, 1974-76 Cougar & Elite

1. Discharge refrigerant from system.
2. Remove radiator upper support brackets from vehicle. Tilt radiator rearward to gain access to condenser assembly.
3. On Cougar, Elite, Montego and Torino models, remove condenser upper right isolation mount. Then, remove nuts attaching remaining isolation mounts to auxiliary brackets. On Thunderbird models, remove nuts attaching each lower isolation mount to its auxiliary bracket. Then, remove studs securing upper auxiliary brackets to radiator support.
4. On all models, disconnect refrigerant lines. Plug all open connections to prevent entry of dirt and moisture.
5. Remove condenser assembly from vehicle.
6. Remove receiver-dehydrator from condenser.
7. Reverse procedure to install. Use new "O" rings lubricated with refrigerant oil.
8. Evacuate, charge and leak test system. Check system for proper operation.

1974-78 Ford & Mercury

1. Discharge refrigerant from system and drain engine coolant from radiator.
2. Remove radiator from vehicle.
3. Disconnect refrigerant lines at fittings. Plug all open connections to prevent entry of dirt and moisture.
4. Remove condenser mounting bolts and nuts, then the condenser assembly from vehicle by sliding assembly rearward through radiator support.
5. Remove receiver from condenser.
6. Reverse procedure to install. Use new "O" rings lubricated with refrigerant oil.
7. Evacuate, charge and leak test system. Check system for proper operation.

1974-78 Mustang II, 1975-80 Granada, Monarch & 1978-80 Versailles

1. Discharge refrigerant from system.
2. Remove lower hood lock and vertical support brace.
3. Disconnect ambient temperature switch wiring.
4. On Mustang II models equipped with power steering, remove power steering cooler tube mounting screws and, without disconnecting lines, reposition cooler tube upward and outward to permit condenser removal.
5. Disconnect refrigerant lines at fittings. Plug all open connections to prevent entry of dirt and moisture.
6. Remove condenser assembly mounting screws and condenser assembly.
7. Remove receiver-dehydrator from condenser.
8. Reverse procedure to install. Use new "O" rings lubricated with refrigerant oil.
9. Evacuate, charge and leak test system. Check system for proper operation.

1979-80 Econoline

NOTE: The condenser is mounted on the engine side of radiator support.

1. Discharge refrigerant from system.
2. Disconnect refrigerant lines from condenser. Plug all open lines to prevent entry of dirt and moisture into system.
3. Remove hood latch attaching screws and position latch aside.

4. Remove screws attaching top edge of grille to radiator support. Then remove screw attaching center of grille to grille center support.
5. From under vehicle, reposition splash shields and remove condenser lower mounting nuts.
6. Remove bolts attaching condenser to radiator upper support.
7. Remove upper radiator support attaching bolts. Pull top of grille forward and remove radiator upper support.
8. Remove condenser from vehicle. Remove receiver-drier from condenser assembly.
9. Reverse procedure to install. Use new O-rings lubricated with refrigerant oil.
10. Evacuate, charge and leak test system. Check system for proper operation.

1975-78 Econoline

1. Remove upper radiator support.
2. Discharge refrigerant from system.
3. Disconnect refrigerant lines from condenser. Plug all open lines to prevent entry of dirt and moisture into system.
4. Remove condenser lower attaching nuts and condenser.
5. Remove receiver-drier from condenser assembly.
6. Reverse procedure to install. Use new O-rings lubricated with refrigerant oil.
7. Evacuate, charge and leak test system. Check for proper operation.

1974-78 Light Duty Truck, Exc. 1975-78 Econoline

1. Remove hood lock assembly mounting bracket.
2. Discharge refrigerant from system.
3. Disconnect refrigerant lines. Plug all open lines to prevent entry of dirt and moisture.
4. Remove condenser attaching bolts and condenser from vehicle.
5. Remove receiver-drier from condenser.
6. Reverse procedure to install. Use new O-rings lubricated with refrigerant oil.
7. Evacuate, charge and leak test system. Check for proper operation.

1980 Light Duty Truck Exc. Econoline

1. Discharge refrigerant from system.
2. Disconnect refrigerant lines from condenser. Plug all open lines to prevent entry of dirt and moisture into system.
3. Partially drain coolant from radiator and disconnect upper radiator hose.
4. From under vehicle, remove lower condenser attaching bracket screws from radiator support.
5. Remove upper condenser bracket screws. Then tilt radiator towards engine and lift condenser from vehicle.
6. Reverse procedure to install. Use new O-rings lubricated with refrigerant oil.
7. Evacuate, charge and leak test system. Check for proper operation.

1979 Light Duty Truck Exc. Econoline

1. Discharge refrigerant from system.
2. Remove grille molding to grille upper support screws and upper support.
3. Remove grille heat shield, if equipped.
4. Remove screw attaching lower center of grille to mounting bracket.
5. Remove hood latch support brace.
6. Disconnect refrigerant lines from condenser. Plug all open lines to prevent entry of dirt and moisture into system.
7. Remove oil cooler attaching bolts and position cooler aside, if equipped.
8. Remove condenser attaching screws and condenser.
9. Reverse procedure to install. Use new O-rings lubricated with refrigerant oil.
10. Evacuate, charge and leak test system. Check for proper operation.

General Motors Corp. Passenger Cars & Light Duty Trucks

Buick

1980 Skylark

1. Discharge refrigerant from system.
2. Disconnect refrigerant lines from condenser. Plug open connections to prevent entry of dirt and moisture into system.
3. Remove grille center support, condenser mounting screws and condenser.
4. Reverse procedure to install. Use new O-rings lubricated with refrigerant oil.
5. Evacuate, charge and leak test system. Check system for proper operation.

1974-75 Apollo & 1975-79 Skylark

1. Discharge refrigerant from system.
2. Remove left and right front end panel fillers.
3. Disconnect hood latch cable and remove hood latch.
4. Remove condenser mounting screws and condenser.
5. Disconnect refrigerant lines. Plug open connections to prevent entry of dirt and moisture.
6. Remove condenser mounting screws and condenser.
7. Reverse procedure to install. Use new "O" rings lubricated with refrigerant oil.
8. Evaculate, charge and leak test system. Check system for proper operations.

1977-78 Riviera, 1977-80 Electra, Estate Wagon & LeSabre

1. Discharge refrigerant from system.
2. Disconnect refrigerant lines from condenser. Plug open lines to prevent entry of dirt and moisture.
3. Remove cold air duct from core support. Then remove upper half of fan shroud.
4. Remove condenser top mount brackets and condenser.
5. Reverse procedure to install. Use new O-rings lubricated with refrigerant oil.
6. Evacuate, charge and leak test system. Check system for proper operation.

1979-80 Riviera

1. Disconnect the negative battery cable and discharge refrigerant from system.
2. Disconnect refrigerant lines from condenser. Plug open lines to prevent entry of dirt and moisture into system.
3. Remove radiator upper support to radiator support screws and upper fan shroud. Remove lower fan shroud.
4. Remove upper condenser brackets and condenser from vehicle.
5. Reverse procedure to install. Add one ounce of refrigerant oil to system. Use new O-rings lubricated with refrigerant oil.
6. Evacuate, charge and leak test system. Check system for proper operation.

1974-76 Full Size Models

1. Discharge refrigerant from system.
2. Remove upper radiator mounting screws and panel.
3. Remove upper condenser mounting bolts and the bracket on each side of upper tie bar.
4. Disconnect refrigerant lines from fittings. Plug open connections to prevent entry of dirt and moisture.
5. Remove condenser from vehicle.
6. Reverse procedure to install. Use new "O" rings lubricated, with refrigerant oil
7. Evacuate,charge and leak test system. Check system for proper operation.

1980 Skyhawk

1. Discharge refrigerant from system.
2. Disconnect refrigerant lines from condenser. Plug all open connections to prevent entry of dirt and moisture into system.
3. Remove upper radiator shroud. Then remove condenser bolts and condenser.
4. Reverse procedure to install. Use new O-rings lubricated with refrigerant oil.
5. Evacuate, charge and leak test system. Check system for proper operation.

1975-79 Skyhawk

1. Disconnect battery ground cable and discharge refrigerant from system.
2. Remove valance panel and grille center support.
3. Disconnect refrigerant lines. Plug open connections to prevent entry of dirt and moisture.
4. Remove condenser bracket to radiator support retaining screws.
5. Carefully lower condenser from the vehicle.
6. Reverse procedure to install. Use new "O" rings lubricated with refrigerant oil.
7. Evacuate, charge and leak test system. Check system for proper operation.

Cadillac

1977-80

1. Remove radiator from vehicle.
2. Discharge refrigerant from system.
3. Disconnect refrigerant lines from condenser. Plug all open lines to prevent entry of dirt and moisture into system.
4. Remove condenser mounting attachments and condenser.
5. Reverse procedure to install. Use new O-rings lubricated with refrigerant oil.
6. Evacuate, charge and leak test system. Check system for proper operation.

1974-76 Models Exc. Seville

1. Discharge refrigerant from system.
2. Remove screws attaching upper radiator mount bracket at each side of radiator cradle, then the screws attaching wheelhouse brace to bracket. Remove bracket from vehicle.
3. Disconnect high pressure vapor line.
4. Remove clip securing liquid line to condenser. Disconnect line at condenser fitting. Plug open connections to prevent entry of dirt and moisture.
5. Remove condenser bracket to rubber mount attaching nuts and clips attaching acromat to condenser frame.
6. Remove condenser from vehicle.
7. Reverse procedure to install. Use new "O" rings lubricated with refrigerant oil.
8. Evacuate, charge and leak test system. Check system for proper operation.

1976 Seville

1. Discharge refrigerant from system.
2. Drain coolant from radiator.

3. Loosen hose clamps and disconnect coolant hoses from radiator.
4. Disconnect transmission fluid cooler lines. Plug all open connections to prevent loss of fluid.
5. Remove radiator cover panel screws and panel. Then, loosen fan shroud and tilt shroud rearward.
6. Remove radiator.
7. Disconnect refrigerant lines at fittings. Plug all open connections to prevent entry of dirt and moisture.
8. Remove condenser mounting bolts and condenser.
9. Reverse procedure to install. Use new "O" rings lubricated with refrigerant oil.
10. Evacuate, charge and leak test system. Check system for proper operation.

Chevrolet

1980 Citation

1. Discharge refrigerant from system.
2. Disconnect refrigerant lines from condenser. Plug all open connections to prevent entry of dirt and moisture into system.
3. Remove grille center support. Then remove condenser mounting screws and condenser.
4. Reverse procedure to install. Use new O-rings lubricated with refrigerant oil.
5. Evacuate, charge and leak test system. Check system for proper operation.

1974-80 Corvette

1. Disconnect battery ground cable and idscharge refrigerant from system.
2. Mark hood hinge attachment location to aid in reassembly. Remove hood from vehicle.
3. Disconnect refrigerant lines from condenser. Plug all open connections to prevent entry of dirt and moisture.
4. Remove condenser attaching screws and condenser.
5. Reverse procedure to install. Use new "O" rings lubricated with refrigerant oil.
6. Evacuate, charge and leak test system. Check system for proper operation.

1974-79 Nova

1. Disconnect battery ground cable discharge refrigerant from system.
2. Remove hood catch mounting screws and remove assembly.
3. Remove hood latch support screws and remove support.
4. Disconnect refrigerant lines from condenser. Plug all open connections to prevent entry of dirt and moisture. On 1974-77 models, remove inlet line clamp and screw.
5. On all models, remove condenser mounting screws and condenser from vehicle.
6. Reverse procedure to install. Use new O-rings lubricated with refrigerant oil.
7. Evacuate, charge and leak test system. Check system for proper operation.

1974-77 Vega; 1978 Monza "S" Hatchback Coupe & 1978-79 Monza Wagon

1. Disconnect battery ground cable and discharge refrigerant from system.
2. Remove valance panel and grille center support.
3. Disconnect refrigerant lines from condenser. Plug all open connections to prevent entry of dirt and moisture into system.
4. Remove condenser mounting bolts and condenser.
5. Reverse procedure to install. Use new O-rings lubricated with refrigerant oil.
6. Evacuate, charge and leak test system. Check system for proper operation.

1977-80 Chevrolet, Impala & Caprice

1. Discharge refrigerant from system.
2. Disconnect refrigerant lines.
3. Remove cold air duct from core support.
4. Remove upper half of fan shroud.
5. Remove condenser mounting brackets and condenser.
6. Reverse procedure to install. Use new O-rings lubricated with refrigerant oil.
7. Evacuate, charge and leak test system. Check system for proper operation.

1978-80 Malibu & Monte Carlo

1. Discharge refrigerant from system.
2. Disconnect refrigerant lines.
3. Remove upper fan shroud.
4. Remove condenser mounting brackets and condenser.
5. Reverse procedure to install. Use new O-rings lubricated with refrigerant oil.
6. Evacuate, charge and leak test system. Check system for proper operation.

1974-76 Chevrolet; 1974-77 Chevelle, Monte Carlo & 1974-80 Camaro

1. Disconnect battery ground cable and discharge refrigerant from system.
2. Remove radiator shroud upper screws.
3. Remove radiator upper retaining bracket screws and retaining bracket.
4. Push top of radiator rearward and place a wood block between radiator upper right corner and radiator support.
5. Remove condenser mounting screws and nuts.
6. Disconnect refrigerant lines from condenser. Plug all open connections to prevent entry of dirt and moisture.
7. Lift right side of condenser until condenser upper bracket clears radiator support. Remove condenser assembly from vehicle by lifting diagonally upward while guiding left end of condenser past radiator.
8. Reverse procedure to install. Use new "O" rings lubricated with refrigerant oil.
9. Evacuate, charge and leak test system. Check system for proper operation.

1975-76 Monza 2 + 2

1. Disconnect battery cables and remove battery from vehicle.
2. Discharge refrigerant from system.
3. Remove valance panel lower screw and grille center support.
4. Disconnect refrigerant lines from condenser. Plug all open connections to prevent entry of dirt and moisture.
5. Move radiator rearward for access. Remove condenser bracket mounting screws and carefully lower condenser from vehicle.
6. Reverse procedure to install. Use new "O" rings lubricated with refrigerant oil.
7. Evacuate, charge and leak test system. Check system for proper operation.

1975-80 Monza Exc. 1975 2+2; 1978 "S" Hatchback Coupe & 1978-79 Wagon

1. Disconnect battery ground cable and discharge refrigerant from system.
2. Remove grille face plate, front bumper valance panel and grille center support from vehicle.
3. Remove radiator upper support screws and push radiator assembly rearward to gain access to condenser upper mounting screws.
4. Disconnect refrigerant lines from condenser. Plug all open connections to prevent entry of dirt and moisture.
5. Remove condenser mounting screws and lower condenser from vehicle.
6. Reverse procedure to install. Use new "O" rings lubricated with refrigerant oil.
7. Evacuate, charge and leak test system. Check system for proper operation.

1976-80 Chevette

1. Disconnect battery ground cable and discharge refrigerant from system.
2. Remove radiator upper support brackets. Then, remove condenser bracket to radiator support nuts.
3. Disconnect refrigerant lines from condenser. Plug all open connections to prevent entry of dirt and moisture.
4. Reposition radiator rearward and lift condenser from vehicle.
5. Reverse procedure to install. Use new "O" rings lubricated with refrigerant oil.
6. Evacuate, charge and leak test system. Check system for proper operation.

Oldsmobile

NOTE: It may be necessary to remove radiator or other components interfering with condenser replacement.

1. Discharge refrigerant from system.
2. Disconnect refrigerant lines. Plug all open connections to prevent entry of dirt and moisture.
3. Remove condenser mounting bolts and condenser.
4. Reverse procedure to install. Use new "O" rings lubricated with refrigerant oil.
5. Evacuate, charge and leak test system. Check system for proper operation.

Pontiac

1980 Phoenix

1. Discharge refrigerant from system.
2. Disconnect refrigerant lines from condenser.
3. Remove grille center support. Remove condenser mounting screws and condenser.
4. Reverse procedure to install. Use new "O" rings lubricated with refrigerant oil.
5. Evacuate, charge and leak test system. Check system for proper operation.

1977 Ventura & 1977-79 Phoenix

1. Discharge refrigerant from system.
2. Disconnect refrigerant lines from condenser.
3. Remove condenser mounting screws and condenser.
4. Reverse procedure to install. Use new "O" rings lubricated with refrigerant oil.
5. Evacuate, charge and leak test system. Check system for proper operation.

1974-76 Ventura

1. Discharge refrigerant from system.
2. On 1974 models, remove front bumper and hood latch support from vehicle.

3. On all models, disconnect refrigerant lines. Plug all open connections to prevent entry of dirt and moisture.
4. On 1975-76 models, remove center hood latch support assembly.
5. On all models, remove condenser mounting screws and condenser.

NOTE: On 1975-76 models, condenser is removed from underneath vehicle.

6. Reverse procedure to install. Use new "O" rings lubricated with refrigerant oil.
7. Evacuate, charge and leak test system. Check system for proper operation.

1977-80 Firebird

1. Discharge refrigerant from system.
2. Disconnect refrigerant lines from condenser.
3. Remove condenser attaching bolts and condenser.
4. Reverse procedure to install. Use new "O" rings lubricated with refrigerant oil.
5. Evacuate, charge and leak test system. Check system for proper operation.

1974-76 Firebird

1. Discharge refrigerant from system.
2. Disconnect upper refrigerant hose. Plug all open connections to prevent entry of dirt and moisture.
3. Remove upper hood latch support to radiator support screws.
4. Disconnect parking lights from valance panel and remove valance panel from vehicle.
5. Remove screws attaching center hood latch support and reposition support aside.
6. Disconnect condenser lower refrigerant line. Plug all open connections.
7. Remove condenser mounting nuts and the condenser from beneath vehicle.
8. Reverse procedure to install. Use new "O" rings lubricated with refrigerant oil.
9. Evacuate, charge and leak test system. Check system for proper operation.

1977-80 Bonneville, Catalina, Grand Safari & Safari

1. Discharge refrigerant from system.
2. Disconnect refrigerant lines.
3. Remove cold air duct from core support.
4. Remove upper half of fan shroud.
5. Remove condenser mounting brackets and condenser.
6. Reverse procedure to install. Use new "O" rings lubricated with refrigerant oil.
7. Evacuate, charge and leak test system. Check system for proper operation.

1978-80 Grand Am, Grand Prix & LeMans

1. Discharge refrigerant from system.
2. Disconnect refrigerant lines.
3. Remove upper fan shroud.
4. Remove condenser mounting brackets and condenser.
5. Reverse procedure to install. Use new "O" rings lubricated with refrigerant oil.
6. Evacuate, charge and leak test system. Check system for proper operation.

1974-76 Full Size Models & 1974-77 Intermediate Size Models

1. Discharge refrigerant from system.
2. Disconnect refrigerant hose at upper condenser fitting. Plug all open connections to prevent entry of dirt and moisture.
3. On models except Grand Prix, remove upper hood latch support screw at radiator support. On Grand Prix models, remove upper air baffle at radiator support. On all models, remove lower air baffle.
4. Disconnect refrigerant line at lower fitting. Plug all open connections.
5. Remove condenser assembly mounting screws and condenser assembly from beneath vehicle.
6. Remove brackets from condenser.
7. Reverse procedure to install. Use new "O" rings lubricated with refrigerant oil.
8. Evacuate, charge and leak test system. Check system for proper operation.

1977 Astre & 1977-80 Sunbird

1. Discharge refrigerant from system.
2. Disconnect refrigerant lines from condenser.
3. Remove condenser attaching screws and condenser.
4. Reverse procedure to install. Use new "O" rings lubricated with refrigerant oil.
5. Evacuate, charge and leak test system. Check system for proper operation.

1976 Astre & Sunbird

1. Disconnect battery ground cable and discharge refrigerant from system.
2. Remove grille face plate, front bumper valance panel and grille center support.
3. Remove radiator upper support screws. Push radiator, support and shroud rearward to gain access to condenser upper mounting bolts.
4. Disconnect refrigerant lines from condenser. Plug all open connections to prevent entry of dirt and moisture.
5. Remove condenser mounting screws and lower condenser from vehicle.
6. Reverse procedure to install. Use new "O" rings lubricated with refrigerant oil.
7. Evacuate, charge and leak test system. Check system for proper operation.

1975 Astre

1. Disconnect battery ground cable and discharge refrigerant from system.
2. Remove valance panel and grille center support.
3. Disconnect refrigerant lines from condenser. Plug all open connections to prevent entry of dirt and moisture.
4. Remove condenser mounting bolts and condenser.
5. Reverse procedure to install. Use new "O" rings lubricated with refrigerant oil.
6. Evacuate, charge and leak test system. Check system for proper operation.

1978-80 Chevrolet & G.M.C. Vans

1. Disconnect battery ground cable. Discharge refrigerant from system.
2. Remove grille, hood lock and support as an assembly.
3. Disconnect refrigerant lines from condenser. Plug all open lines to prevent entry of dirt and moisture into system.
4. Remove condenser attaching screws and condenser.
5. Reverse procedure to install. Use new "O" rings lubricated with refrigerant oil. And one ounce of refrigerant oil to new condenser.
6. Evacuate, charge and leak test system. Check for proper operation.

1974-77 Chevrolet & G.M.C. Vans

1. Disconnect battery ground cable and compressor clutch connector.
2. Discharge refrigerant from system.
3. Remove headlamp bezels and parking lens.
4. Place a protective strip of tape on upper surface of bumper. Then remove front grille.
5. Remove radiator center brace and upper supports.
6. Move top of radiator rearward to gain access to condenser.
7. Remove condenser attaching bolts. Disconnect refrigerant lines from condenser. Plug all open lines to prevent entry of dirt and moisture into system.
8. Remove condenser from vehicle.
9. Reverse procedure to install. Use new "O" rings lubricated with refrigerant oil. Add one ounce of refrigerant oil to new condenser.
10. Evacuate, charge and leak test system. Check system for proper operation.

1974-80 Chevrolet & G.M.C. Light Duty Trucks Exc. Vans & Motor Home

1. Disconnect battery ground cable.
2. Discharge refrigerant from system.
3. Remove grille and grille center support. Remove left grille support to upper fender support screws.
4. Disconnect refrigerant lines from condenser. Plug all open lines to prevent entry of dirt and moisture into system.
5. Remove condenser attaching screws, bend left grille support outboard, pull condenser forward and lower from vehicle.
6. Reverse procedure to install. Use new "O" rings lubricated with refrigerant oil. Add one ounce of refrigerant oil to new condenser.
7. Evacuate, charge and leak test system. Check system for proper operation.

1974-80 Chevrolet & G.M.C. Light Duty Motor Home

1. Disconnect battery ground cable.
2. Discharge refrigerant from system. Disconnect refrigerant lines from condensor. Plug all open lines to prevent entry of dirt and moisture into system.
3. Remove condenser retaining screws and condenser.
4. Reverse procedure to install. Use new "O" rings lubricated with refrigerant oil. Add one ounce of refrigerant oil to new condenser.
5. Evacuate, charge and leak test system. Check system for proper operation.

THERMOSTATIC SWITCH, REPLACE

American Motors & Jeep

1978-80 Exc. Jeep, Matador & Pacer

1. Disconnect battery negative cable.
2. Remove package tray outlet assembly and instrument panel center housing.
3. Remove screw and clip attaching thermostat capillary tube to evaporator housing.
4. Remove capillary tube from tube guide.
5. Remove radio, control panel attaching screws and center discharge duct.
6. Lower control panel and disconnect thermostat wiring.
7. Remove thermostat.
8. Install thermostat. Then connect wiring and cable.

9. Install center discharge duct, control panel attaching screws and radio.
10. Route capillary tube to evaporator housing. Make a 90° bend (at colored tape) in capillary tube and insert end into tube guide mounted of front of evaporator housing between the core fins.
11. Secure capillary tube to housing with clip and screw.
12. Install instrument panel center panel, package outlet assembly and battery negative cable.
13. Check system operation.
14. Refer to "Control Cable, Adjust" for adjustment.

1974-77 Hornet & 1974-77 Gremlin

1. Disconnect battery ground cable.
2. Remove package tray outlet assembly and instrument panel center housing.
3. Remove thermostatic switch capillary tube from evaporator housing.
4. On 1975-77 models, remove radio, control attaching screws, and center discharge duct. Lower control. On all models, disconnect wiring from switch, noting the location of wiring connectors.
5. Remove thermostatic switch knob, retaining nut and the switch.
6. Install replacement switch and retaining nut; place knob on shaft.
7. Connect wiring to switch.
8. On 1975-77 models, position switch in place. Install center discharge duct, control attaching screws and radio.

NOTE: Be careful not to kink the capillary tube.

9. Make a 90° bend 3½ inches from the tube end and insert the tube between the evaporator fins. On all models, install the capillary tube retaining screw.
10. Install instrument panel center housing and package tray outlet assembly.
11. Connect battery ground cable.

1974 Javelin

1. Disconnect battery ground cable.
2. Remove the crash pad, lower right moldings, assist handle, finish panel and the right side cowl trim panel. Remove instrument panel reinforcement.
3. Remove the capillary tube retaining screw and pull the tube from the housing.
4. Disconnect wiring from thermostatic switch and remove switch through glove compartment opening.
5. Install replacement switch.
6. Route capillary tube to the evaporator housing. Do not kink the tubing. Make a 90° bend in the tube 3¼ inches from the tube end. Insert the end into the housing between the evaporator fins. Install tube retaining screw.
7. Connect electrical wiring to switch.
8. Install instrument panel reinforcement, cowl trim panel, finish panel, assist handle, moldings and crash pad.
9. Connect battery ground cable.

1974 Ambassador & 1974-78 Matador

1. Disconnect battery ground cable.
2. Remove thermostatic switch control cable and remove cable.
3. Disconnect wiring from switch.
4. Remove thermostatic switch attaching screw and capillary tube attaching screw.
5. Pull insulation away from evaporator housing and remove thermostatic switch.
6. Make a 90° bend in capillary tube 4⅞ inches from tube end and insert into evaporator housing between core fins. Install capillary tube attaching screw.
7. Position thermostatic switch and install attaching screw.
8. Place insulation in its original position on evaporator housing.
9. Connect switch wiring and control cable. Install control cable attaching screw. Adjust control cable, as described elsewhere in this manual.
10. Connect battery ground cable.

1975-80 Pacer

1. Disconnect battery ground cable.
2. Remove radio overlay and radio.
3. Remove instrument panel crash pad. Remove left corner and windshield side finish moldings.
4. Remove upper instrument panel to lower instrument panel attaching screws. Remove right and left scuff plates and cowl trim panels.
5. Tape steering column to prevent damage to paint. Loosen steering column support nuts.
6. Remove screws attaching lower instrument panel to "A" pillars and pull instrument panel to rear.
7. Remove insulation covering capillary tube and evaporator housing.
8. Disconnect wiring from thermostatic switch and disconnect control cable.
9. Remove capillary tube to evaporator housing attaching screw. Remove adjusting screws securing thermostatic switch bracket to the evaporator housing. Remove capillary tube from evaporator housing.
10. Remove thermostatic switch and bracket from vehicle. Remove switch from bracket.
11. Position replacement switch on bracket and install mounting screws.
12. Make a 90° bend (at colored tape) in capillary tube 4½ inches from tube end on 1975-76 models and 4⅛ inches from tube end on 1977-80 models. Then install end of capillary tube into tube guide mounted on the front of evaporator housing between evaporator core fins. Install capillary tube retaining screw.
13. Position switch and bracket on evaporator housing and install, but do not tighten adjusting screws. Connect control cable to thermostatic switch.
14. Move temperature control lever to MAX COLD position and tighten adjusting screws. Connect wiring to switch.
15. Replace insulation in its original position on evaporator housing.
16. Return lower instrument panel to its original position and install mounting screws to "A" pillars. Install upper instrument panel to lower instrument panel attaching screws.
17. Tighten steering column support nuts to 10 foot pounds torque. Remove protective tape from steering column.
18. Install left corner and windshield side finish moldings. Install instrument panel crash pad.
19. Install radio and radio overlay.
20. Install right and left cowl trim panels and scuff plates.
21. Connect battery ground cable.
22. Check operation of thermostatic switch.

1974-80 Jeep

1. Disassemble evaporator core housing to gain access to switch.
2. Remove switch.
3. Reverse procedure to install. Insert capillary tube into coil 3 inches on 1974-76; and 2 inches on 1977-80 models.

CHRYSLER CORP. PASSENGER CARS & LIGHT DUTY TRUCKS

1979-80 Light Duty Trucks & 1980 Passenger Cars (Cycling Clutch Switch)

1. Disconnect switch wiring.
2. Remove switch retaining screws, wrap a rag around the capillary tube at the well opening and pull tube out of well while holding rag firmly around tube so that grease stays in well. Do not bend the capillary tube.

NOTE: For proper A/C operation, grease must be retained in well.

3. Reverse procedure to install. Check operation.

Exc. 1979-80 Light Duty Trucks & 1980 Passenger Cars (Cycling Clutch Switch)

1. Disconnect switch wiring.
2. Remove switch retaining screws.
3. Remove capillary tube from suction line by first removing the insulation material, then loosen the clamp enough to slide the wound portion of capillary tube from under clamp.
4. Remove any dirt or insulation from capillary tube position on suction line. mark capillary tube location to ensure proper location of replacement.
5. Reverse procedure to install. Torque switch mounting screws and clamp to 15-20 in. lbs.

NOTE: Flat portion of clamp must cover capillary tube wound end.

6. Install insulation material.

FORD MOTOR CO. PASSENGER CARS & LIGHT DUTY TRUCKS

1974 Comet & Maverick

1. Disconnect battery ground cable.
2. Remove instrument cluster then remove ash tray assembly and disconnect light.
3. Remove left duct assembly, center register and right duct assembly.
4. Disconnect vacuum hose from heat-defrost door at plenum. Push plenum assembly to the left for access to thermostatic switch.
5. Disconnect wiring connectors from switch and remove switch retaining screws. Then, working through the center grille opening and between package tray and the instrument panel, remove thermostatic switch and pull sensing tube out of evaporator core.
6. To install replacement switch, feed sensing tube carefully into evaporator core fins and position the switch onto the evaporator. Install switch mounting screws and connect wiring connectors to switch.
7. Position plenum in place and install right duct assembly. Connect heat/defrost door vacuum hose to the plenum.
8. Install center register assembly and left

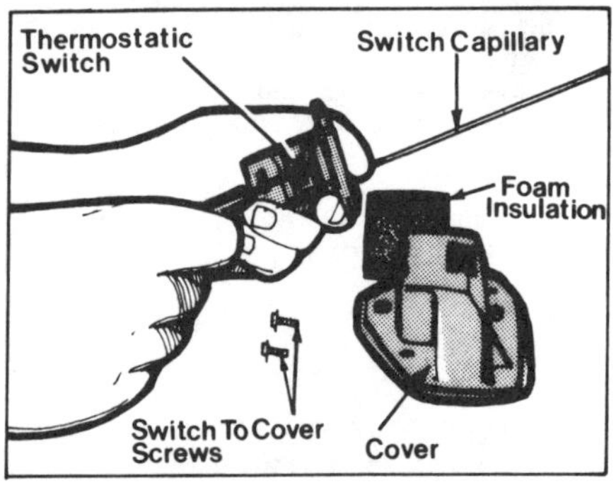

Fig. 36 Thermostatic switch adjustment. 1974-76 Vega & 1976 Chevette

air duct.
9. Install ash tray assembly and instrument cluster.
10. Connect battery ground cable.

1975-77 Comet & Maverick

1. Disconnect battery ground cable.
2. Remove lower instrument panel extension, center register and right duct assembly.
3. Disconnect wiring connectors from thermostatic switch and remove switch retaining screws.
4. From under instrument panel, remove switch and pull sensing tube from evaporator core.
5. To install replacement switch, insert sensing tube into the core fins. Position switch onto evaporator and install switch retaining screws. Connect wiring to switch.
6. Install right duct assembly and lower instrument panel extension. On 1975 models, connect the left and right ducts to the register. On 1976-77 models, connect the left duct to the panel register.
7. On all models, connect battery ground cable.

1974-79 Light Duty Truck Exc. Econoline

1. Disconnect battery ground cable.
2. Remove glove box liner and right A/C duct.
3. Disconnect vacuum line from fresh air door motor.
4. Remove evaporator rear housing from dash panel and the fresh air inlet boot.

NOTE: Install one upper nut to retain front housing to dash after rear housing is removed.

5. Remove switch from evaporator assembly.
6. Reverse procedure to install. Be sure sensing tube is in notch of housing.

1975-80 Econoline

1. Remove battery from vehicle.
2. Disconnect wiring from switch.
3. Remove switch cover screws and switch with cover.

NOTE: Pull switch with cover straight out of evaporator assembly.

4. Remove switch from cover.
5. Reverse procedure to install.

1974 Econoline

1. Position carpet or floor mat aside.
2. Remove evaporator register from floor.
3. Position sensor tube out of core.
4. Remove control housing cover and switch lever.
5. Disconnect switch wiring and remove switch and tube assembly.
6. Reverse procedure to install.

GENERAL MOTORS CORP. PASSENGER CARS & LIGHT DUTY TRUCKS

1976 Astre, Chevette & Vega

1. Disconnect electrical harness at switch connector.
2. Remove screw attaching sensing element clamp and remove clamp. Remove insulation from around sensing element.
3. Remove screws attaching switch to evaporator case and remove switch with sensing element.
4. Reverse procedure to install.

NOTE: Insulation removed from sensing element must be replaced with a suitable insulating compound. When replacing sensing element clamp, tighten clamp screw securely enough to ensure a clean tight fit between sensing element and inlet tube. Do not overtighten.

1974-75 Vega & 1975 Astre

1. Disconnect electrical harness at switch connector.
2. Remove screws attaching thermostatic switch to evaporator case cover and remove switch. Be careful not to damage sensing line. Mark location of the sensing line in the evaporator core to aid in assembly.
3. Reverse removal procedure to install.

NOTE: Be sure that the sensing line is making good contact with the core.

1976-79 Passenger Cars & 1977-79 Light Duty Trucks Exc. Motor Home

1. Disconnect electrical connectors from switch.
2. Remove switch to evaporator case retaining screws.
3. Carefully remove insulation and loosen clamps enough to pull end of capillary tube out from under clamps at evaporator inlet pipe.
4. Using solvent, clean pipe area to remove remaining insulation material and dirt. Mark capillary tube location on inlet pipe.
5. Reverse procedure to install. Mount formed end of capillary tube in same position as original tube.

NOTE: Do not install capillary tube more than ½ inch in either direction of previously marked location.

6. Adjust each clamp before tightening so that width of clamp covers capillary tube's formed end. Torque clamps to 15-20 inch lbs.
7. Recover tube and pipe with insulation material. If insulation is damaged, replace with same type material.

1974-76 Light Duty Trucks Exc. Van & Motor Home

1. Disconnect battery ground strap and switch wiring.
2. Remove switch attaching screws and switch.

NOTE: Mark position of sensing element in core to aid in assembly.

3. Reverse procedure to install.

1974-80 Motor Home

1. Remove evaporator case cover plate.

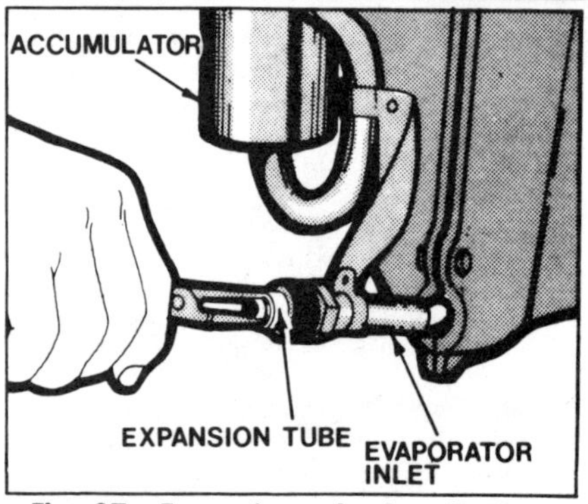

Fig. 37 Expansion tube installation. Models with accumulator

2. Remove switch retaining screws and switch.
3. Reverse procedure to install. Position capillary tube in same location as original.

THERMOSTATIC SWITCH, ADJUST

1974-76 Vega, Light Duty Truck & 1976 Chevette

1. Install gauge set on vehicle.
2. Start and run engine at approximately 2000 RPM on all models exc. Vega; 1500 RPM on Vega.
3. Set temperature control for maximum cooling and place blower switch in high blower position. Close vehicle doors and windows and open all A/C outlets.
4. Permit system to operate for 10 minutes before performing any adjustments.
5. Suction pressure should be as listed in the Performance Test chart in the "Maintenance, Diagnosis & Testing" section when the compressor clutch disengages.
6. If adjustment is required, remove switch, then the fiber end plate for access to adjusting screw.
7. If outlet temperature is less than specified at end of cooling cycle, turn adjusting screw a partial turn counter-clockwise, Fig. 36. If outlet temperature is greater than specified temperature, turn screw clockwise.

NOTE: One turn of screw will change outlet temperature approximately 4 degrees.

8. Install fiber end plate, then the switch.
9. Check system performance and readjust switch if necessary.

ACCUMULATOR, REPLACE

FORD MOTOR CO. PASSENGER CARS & LIGHT DUTY TRUCKS

1980 Ford, Lincoln Continental, Mark VI, Mercury & Light Duty Trucks Exc. Econoline

1. Discharge refrigerant from system.
2. Disconnect refrigerant lines from accumulator. Plug all open lines to prevent entry of dirt and moisture.
3. Disconnect the pressure control de-ice switch wiring plug.
4. Remove retaining bracket screw and remove accumulator.
5. Reverse procedure to install. Use new O-rings lubricated with refrigerant oil.
6. Evacuate, charge and leak test system. Check system for proper operation.

GENERAL MOTORS CORP. PASSENGER CARS & LIGHT DUTY TRUCKS

1. Disconnect battery ground cable and discharge refrigerant from system.
2. Disconnect inlet and outlet lines from accumulator. Plug open connections at once to prevent entry of moisture.
3. Disconnect cycling clutch switch wiring plug if equipped.
4. Remove accumulator bracket mounting screws and then the accumulator.

NOTE: If replacing accumulator, drain and measure amount of oil in old accumulator, then add an equal amount of new oil to new accumulator.

5. Reverse removal procedure to install. When connecting lines, use new "O" ring seals coated with clean refrigerant oil.

NOTE: Do not uncap accumulator until the last minute before installing lines.

6. Evacuate, charge and leak test system and check for proper operation.

SUPERHEAT SHUTOFF SWITCH, REPLACE

GENERAL MOTORS CORP. PASSENGER CARS & LIGHT DUTY TRUCKS

NOTE: The low refrigerant charge protection system consists of a superheat shutoff switch located in the rear head of the compressor, connected in series with a thermal limiter. A blown thermal limiter indicates that the system is either low or completely out of refrigerant charge.

1. Disconnect superheat shutoff switch electrical connector. Discharge refrigerant from system. Then remove superheat switch retainer ring, using internal snap ring pliers.
2. Remove switch from compressor rear head by pulling at terminal housing groove with seal seat remover J-9393 or other suitable tool. Remove "O" ring from the switch cavity.
3. To install, first check superheat switch cavity and "O" ring groove for dirt or foreign material. Install a new "O" ring lubricated with refrigerant oil.
4. Lubricate superheat switch housing with refrigerant oil. Insert switch carefully into cavity until switch bottoms.
5. Using internal snap ring pliers, install switch retaining ring with high point of curved sides adjacent to switch housing. Be sure snap ring is properly seated in groove.
6. Evacuate, charge and leak test system. Connect switch electrical connector and check for proper system operation.

COMPRESSOR DISCHARGE PRESSURE/LOW PRESSURE CUT-OFF SWITCH, REPLACE

NOTE: The switch is located on compressor or high pressure line of GM vehicles; receiver-drier or H-valve assembly on Chrysler vehicles and on the compressor of 1979 Lincoln Continental vehicles.

1. Discharge refrigerant from system and disconnect electrical connector from switch.
2. Remove switch.
3. Install replacement switch using a new "O" ring lubricated with refrigerant oil.
4. Evacuate, charge and leak test system and check for proper operation.

CYCLING CLUTCH SWITCH, REPLACE

CHRYSLER CORP. PASSENGER CARS & LIGHT DUTY TRUCKS

Refer to "Thermostatic Switch, Replace" for replacement procedure.

FORD MOTOR CO. PASSENGER CARS & LIGHT DUTY TRUCKS

1980 Ford, Lincoln Continental, Mark VI, Mercury & Light Duty Trucks Exc. Econoline

NOTE: The clutch cycling pressure switch is mounted on a schrader type fitting on the accumulator. Replacement of switch does not require discharge of refrigerant from system.

GENERAL MOTORS CORP. PASSENGER CARS & LIGHT DUTY TRUCKS

1. Disconnect wiring from switch.
2. Remove switch from accumulator.

NOTE: The switch is mounted on a schrader valve type fitting. Refrigerant discharge is not required.

3. Reverse procedure to install. Use a new O-ring lubricated with refrigerant oil.
4. Check system for proper operation.

EXPANSION TUBE/FIXED ORIFICE TUBE, REPLACE

FORD MOTOR CO. PASSENGER CARS & LIGHT DUTY TRUCKS

1980 Ford, Lincoln Continental, Mark VI, Mercury & Light Duty Trucks Exc. Econoline

1. Discharge refrigerant from system.
2. Disconnect liquid line from evaporator core inlet tube connection and cap the line.
3. Insert fixed orifice removal tool into evaporator inlet tube (over fixed orifice tube). Turn the tool slightly to the right, engaging tangs of orifice tube into slots of tool.

CAUTION: Do not turn tool more than slightly to the right or risk breaking orifice tube in half. If tube is broken, an extractor tool will remove broken off portion from the evaporator.

4. Install tool nut onto tool and remove orifice tube.
5. Reverse procedure to install. Use new "O" rings lubricated with refrigerant oil.
6. Evacuate, charge and leak test system. Check system for proper operation.

GENERAL MOTORS CORP. PASSENGER CARS & LIGHT DUTY TRUCKS

NOTE: The expansion tube is located in the evaporator core inlet line on models equipped with an accumulator.

1. Discharge refrigerant from system.
2. Disconnect liquid line from evaporator inlet pipe.
3. Using needle nose pliers, remove expansion tube and screen assembly from evaporator inlet, Fig. 37.
4. Reverse procedure to install. Use new O-rings lubricated with refrigerant oil. Install expansion tube with small screen end in first.
5. Evacuate, charge and leak test system. Check system for proper operation.

EXPANSION VALVE, REPLACE

AMERICAN MOTORS & JEEP

1974-80 Exc. Jeep

1. Discharge refrigerant from system.
2. On 1974 Ambassador and 1974-78 Matador, remove lower instrument finish panel. On 1974-80 Exc. Ambassador, Matador and Pacer, remove package tray and steering column finish panel. On 1975-80 Pacer, remove steering tube cover and left intermediate duct.
3. Remove insulation wrapped around suction line and thermal bulb. Mark the bulb location on the suction line.
4. Disconnect inlet and outlet refrigerant lines.
5. Reverse procedure to install. Clamp replacement bulb in exactly the same position as the old unit and be sure it is securely clamped to a clean surface. The insulation must be wrapped around the bulb and line after the bulb has been installed on the line.

1974-80 Jeep

1. Discharge refrigerant from system.
2. Disconnect inlet (suction) line from compressor.
3. Disconnect receiver-drier to evaporator hose at quick-disconnect coupling if equipped.
4. Remove hose clamps and dash grommet retaining screws.
5. Remove evaporator housing attaching screws and lower housing. Pull hoses with grommet through dash panel.
6. Remove insulation from expansion valve. Mark position of capillary tube.
7. Remove expansion valve.
8. Reverse procedure to install. Clean suction line to assure positive contact of new capillary tube.
9. Evacuate, charge and leak test system. Check for proper operation.

CHRYSLER CORP. PASSENGER CARS & LIGHT DUTY TRUCKS

1974 Passenger Cars & 1975-76 Dart & Valiant, Fig. 38

1. Discharge the system.
2. Disconnect the equalizer line from the suction line fitting. Cap all openings to prevent dirt and moisture from entering the system.
3. Disconnect the expansion valve from the liquid line and the evaporator and cap openings.
4. Carefully pull out the capillary tube from the well in the suction line.
5. Remove the rubber seal from the capillary tube. Inspect condition of inlet screen.

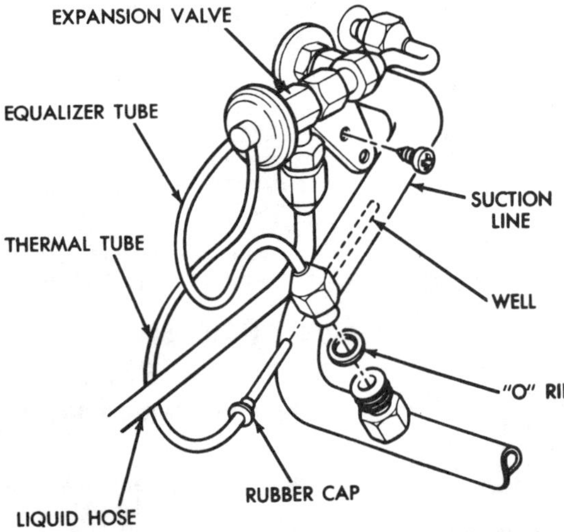

Fig. 38 Expansion valve details, except H-Type. Chrysler Corp.

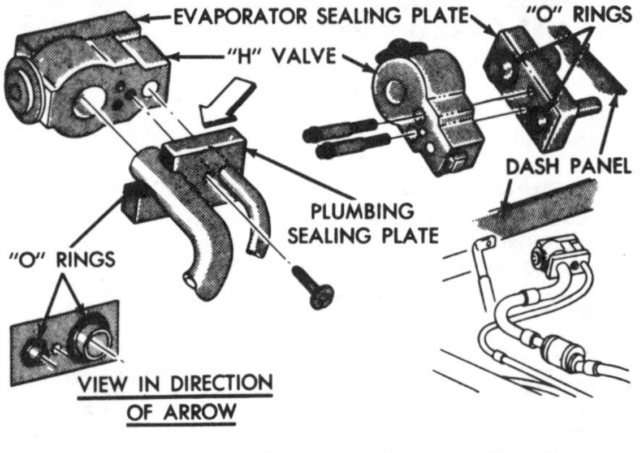

Fig. 39 Expansion valve details, H-Type. Chrysler Corp.

6. Reverse procedure to install and test system.

1975-80 Passenger Cars Exc. Dart & Valiant & 1979-80 Vans, Fig. 39

1. Discharge refrigerant from system.
2. Remove center bolt from sealing plate and carefully pull refrigerant lines toward front of vehicle.

NOTE: Use care to avoid scratching the valve sealing surfaces with the tube pilots.

3. Hold valve, then remove the Allen-head retaining bolts and remove valve.
4. Reverse procedure to install.
5. Evacuate, charge and leak test system. Check for proper operation.

1974-79 Light Duty Truck Exc. 1979 Van Models

1. Discharge refrigerant from system.
2. Remove equalizer line from suction tube. Plug all open lines to prevent entry of dirt and moisture into system.
3. Disconnect expansion valve from refrigerant line and plug open ends.
4. Carefully pull capillary tube from well in suction line.
5. Remove rubber seal from capillary tube and inspect inlet screen.
6. Reverse procedure to install.
7. Evacuate, charge and leak test system. Check for proper operation.

FORD MOTOR CO. PASSENGER CARS & LIGHT DUTY TRUCKS

1974-80 Exc. Models With Combination Valve, Mini-Combination Valve & Fixed Orifice Tube

1. Discharge refrigerant from system.
2. Remove insulation from expansion valve and hose ends.
3. Loosen temperature bulb clamp(s) retaining screws and pull bulb from clamp.
4. Disconnect high pressure hose from expansion valve and disconnect valve from evaporator core at tube fitting.
5. Reverse procedure to install. Use new O-rings lubricated with refrigerant oil.
6. Evacuate, charge and leak test system. Check for proper operation.

GENERAL MOTORS CORP. PASSENGER CARS & LIGHT DUTY TRUCKS EXC. VIR SYSTEMS

NOTE: Refer to "VIR Service" section for service procedures on VIR systems.

1974 Apollo, Omega & Ventura

1. Discharge system.
2. Disconnect oil bleed line, equalizer fitting and inlet and outlet fittings. Cap all opened lines to prevent entry of dirt.
3. Remove screw retaining valve to bracket and remove valve.
4. To install, reverse removal procedure using new O-ring coated with refrigerant oil.

1974-80 Light Duty Truck Overhead System Exc. 1974-77 Vans

1. Disconnect battery ground cable.
2. Discharge refrigerant from system.
3. Remove rear duct.
4. Disconnect wiring from blower motor.
5. Remove lower evaporator case screws and lower case and motor assembly.

CAUTION: Before removing case screws, support lower case and motor assembly.

6. Remove sensing bulb clamps, and remove expansion valve from refrigerant lines. Plug all open lines to prevent entry of dirt and moisture into system.
7. Reverse procedure to install. Use new O-rings lubricated with refrigerant oil.
8. Evacuate, charge and leak test system. Check for proper operation.

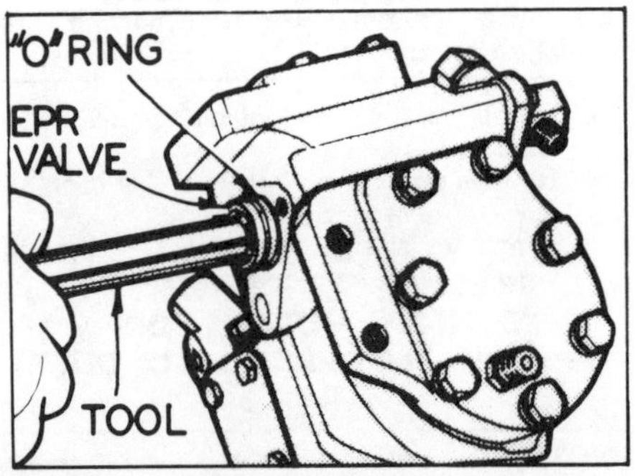

Fig. 40 Evaporator pressure regulator. Chrysler Air-Temp compressors

1974-80 Motor Home Chassis

1. Discharge refrigerant from system.
2. Remove cover plate and separate upper and lower evaporator case halves.
3. Disconnect refrigerant lines from expansion valve. Plug all open lines to prevent entry of dirt and moisture into system.
4. Remove sensing bulb from evaporator outlet manifold. Then remove expansion valve.
5. Reverse procedure to install. Use new O-rings lubricated with refrigerant oil.
6. Evacuate, charge and leak test system. Check for proper operation.

1974-77 Light Duty Van Models

1. Discharge refrigerant from system.
2. Disconnect battery ground cable.
3. Remove blower-evaporator shield, shield bracket and heater intermediate duct.
4. Unsnap engine cover latches, remove lower tab mounting bolts and remove engine cover.
5. Loosen both steering column instrument panel reinforcement screws; remove one screw.
6. Remove left foot cooler bracket to panel screws, disconnect outlet from duct and remove outlet and bracket assembly.
7. Disconnect speedometer cable. Remove instrument panel to lower reinforcement screws and rotate panel upwards for access.
8. Remove sensing bulb clamps, and remove expansion valve from refrigerant lines. Plug all open lines to prevent entry of dirt and moisture into system.
9. Reverse procedure to install. Use new O-rings lubricated with refrigerant oil.
10. Evacuate, charge and leak test system. Check system for proper operation.

1974-77 Light Duty Van Overhead System

1. Disconnect battery ground cable. Discharge refrigerant from system.
2. Disconnect drain tubes from blower-evaporator shroud and remove shroud.
3. Disconnect blower motor ground and lead wires.
4. For removal of inner expansion valve, proceed as follows:
   a. Support lower case, remove lower case attaching screws and lower case assembly.
   b. Remove sensing bulb clamps and re-

move expansion valve from refrigerant lines. Plug all lines to prevent entry of dirt and moisture into system.

5. For removal of outer expansion valve, proceed as follows:
   a. Disconnect refrigerant lines at rear of blower-evaporator assembly. Plug all open lines to prevent entry of dirt and moisture into system.
   b. Remove blower-evaporator assembly from vehicle and place on work bench upside down.
   c. Remove lower case from evaporator assembly.
   d. Remove upper shroud from upper case and then remove upper case from evaporator core.
   e. Remove sensing bulb clamps and expansion valve from refrigerant lines. Plug all open lines to prevent entry of dirt and moisture into system.
6. Reverse procedure to install. Use new O-rings lubricated with refrigerant oil.
7. Evacuate, charge and leak test system. Check system for proper operation.

EVAPORATOR PRESSURE REGULATOR, REPLACE, FIG. 40

CHRYSLER CORP. PASSENGER CARS & LIGHT DUTY TRUCKS

1. Discharge the system.
2. Remove the two EPR/EPR2 valve suction line fitting bolts and the fitting which also contains the compressor suction screen, spring and the gasket.
3. To remove EPR valve, tool C3822 is required. To remove EPR2, tool C3822 can be modified to tool C3822A by enlarging the slots to 1/16 inch, Fig. 41. Remove the EPR valve and "O" ring from the compressor

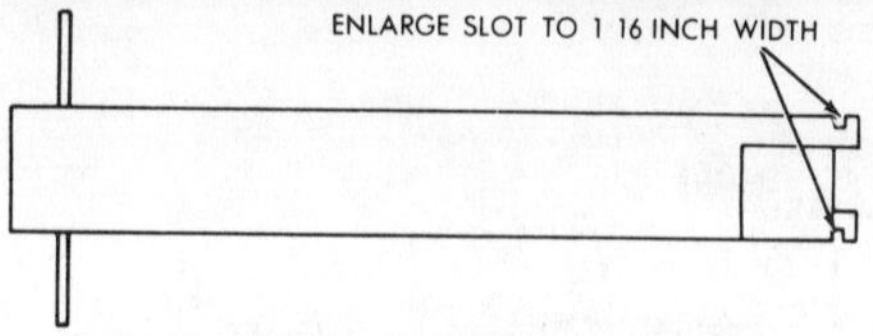

Fig. 41 EPR Valve tool modification. Chrysler Air-Temp compressors

with special tool by rotating the valve counterclockerwise slightly.

NOTE: Do not handle the EPR valve more than necessary. The valve should be inspected externally and wiped clean with a lint free cloth. Place the valve in a plastic bag until ready to be installed.

---

4. Reverse procedure to install, using new "O" rings.

PILOT OPERATED ABSOLUTE/SUCTION THROTTLING VALVE

FORD MOTOR CO. PASSENGER CARS & LIGHT DUTY TRUCKS

1974-79 Exc. Models With Combination Valve

1. Discharge refrigerant from system.
2. Disconnect equalizer line and refrigerant lines from valve.
3. Retaining screws (if used) and remove valve.
4. Reverse procedure to install. Use new O-rings lubricated with refrigerant oil.
5. Evacuate, charge and leak test system. Check system for proper operation.

1976-80 Models With Combination Valve Or Mini-Combination Valve

Refer to "Combination Valve, Replace" or "Mini-Combination Valve, Replace" for procedure.

COMBINATION VALVE, REPLACE

FORD MOTOR CO. PASSENGER CARS

1976-80 Granada, Monarch & 1977-80 Versailles, Fig. 42; 1977-78 Mustang, 1977-80 Bobcat & Pinto, Fig. 43

1. Discharge refrigerant from system.
2. On Bobcat, Mustang and Pinto model, remove battery cables, battery and heat shield.
3. On all models, disconnect refrigerant lines from STV housing manifold.

NOTE: Use a backup wrench on STV housing manifold to prevent damage.

---

4. Remove retaining bolts from side of manifold plate and remove expansion valve and STV housing manifold from evaporator manifold plate.
5. Remove allen screws and separate STV housing manifold from expansion valve body.
6. Using curved jaw channel lock pliers, grasp STV body on surface nearest tapered end, then turn in a counter-clockwise direction and remove the STV.
7. Reverse procedure to install. Use new "O" rings lubricated with refrigerant oil. Install STV into expansion valve body hand tight.

NOTE: To avoid STV damage, grasp body of STV on end nearest taper, opposite from slotted end.

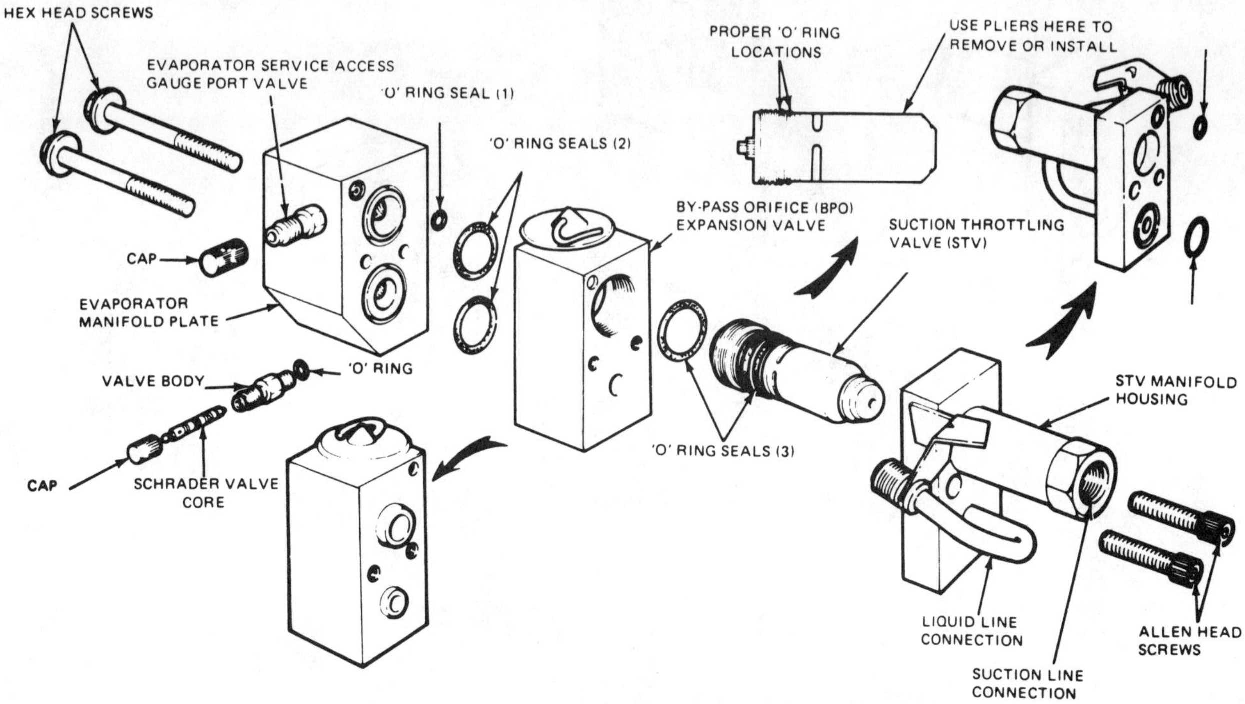

Fig. 42 Combination valve assembly. 1976-80 Granada, Monarch & Versailles

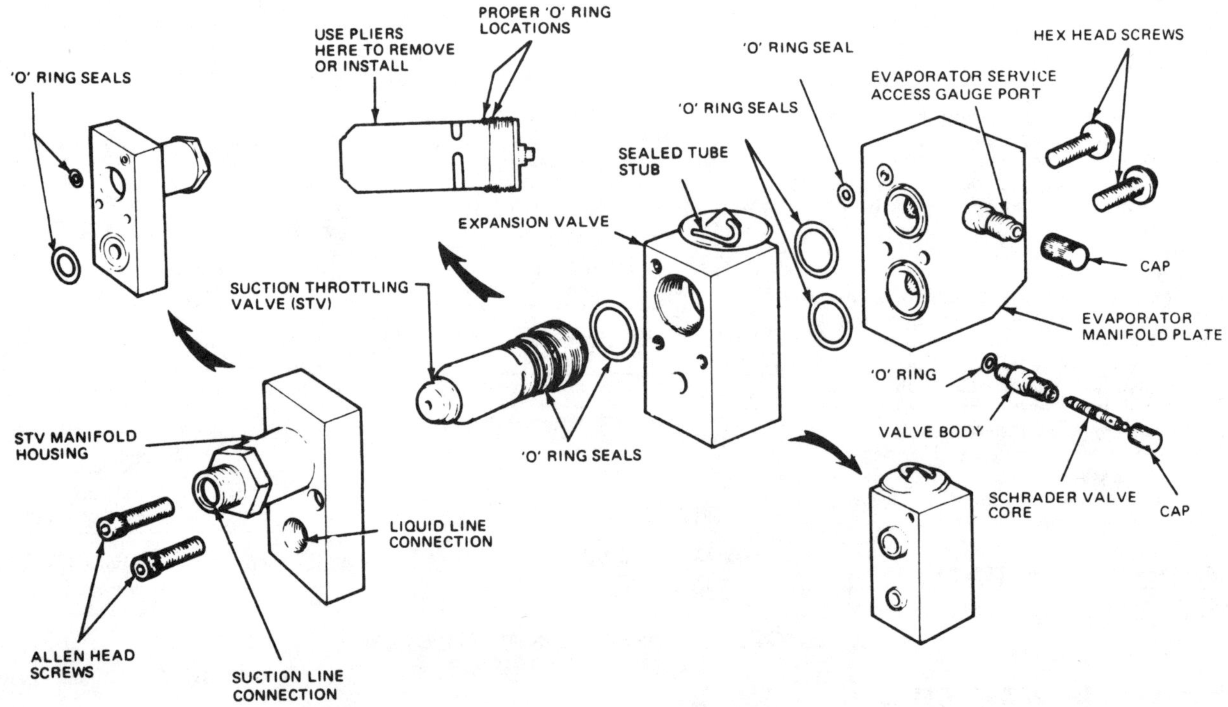

Fig. 43 Combination valve assembly. 1977-78 Mustang; 1977-80 Bobcat & Pinto

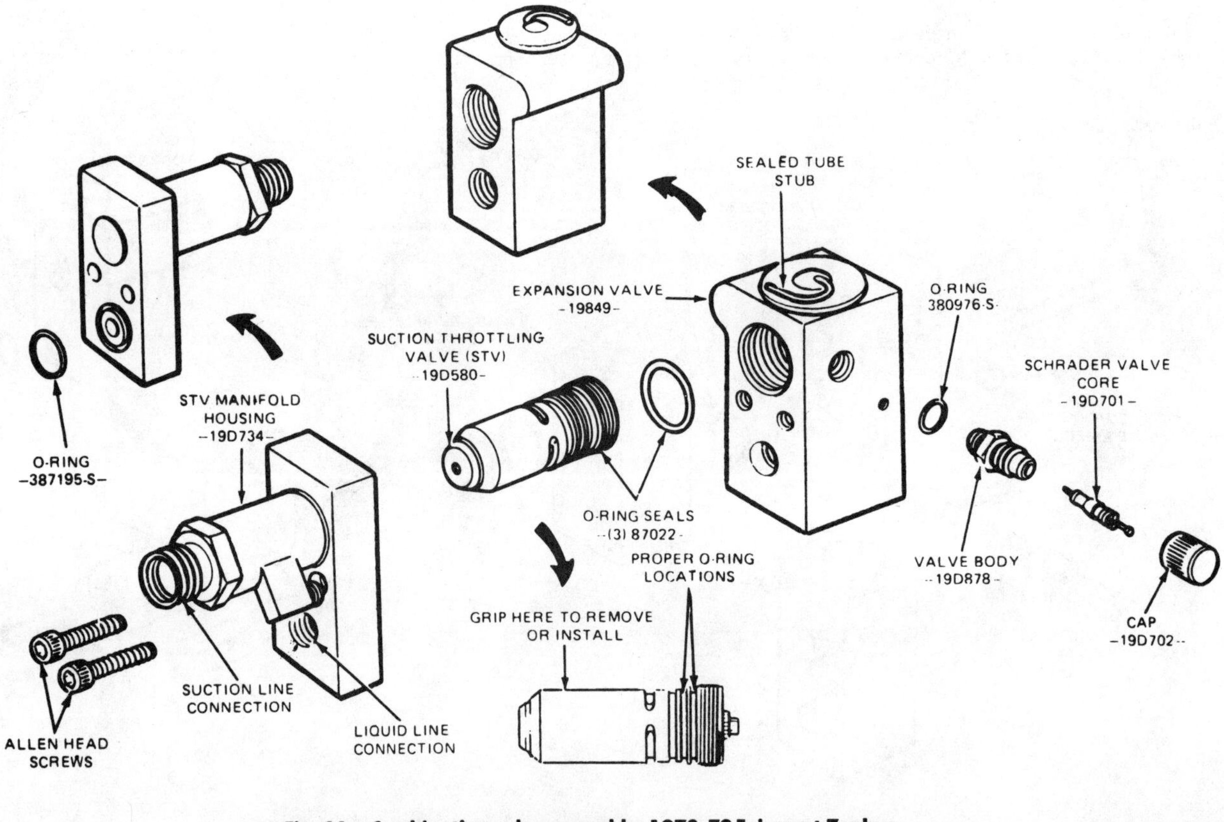

Fig. 44 Combination valve assembly. 1978-79 Fairmont Zephyr

8. Evacuate, charge and leak test system. Check system for proper operation.

1978-79 Fairmont, Zephyr; 1979 Capri & Mustang, Fig. 44

1. Discharge refrigerant from system.
2. Disconnect the high and low pressure lines and the liquid bleed line from the STV housing manifold of the combination valve.

NOTE: Use a back-up wrench on STV housing manifold to prevent damage.

3. To remove STV (suction throttling valve):
   a. Remove two allen screws and separate STV housing manifold from expansion valve body.
   b. Using a curved jaw channel lock pliers, remove STV from expansion valve body.

   NOTE: To avoid STV damage, grasp valve on end near taper, Fig. 44.

4. To remove combination valve:
   a. Disconnect the evaporator fittings from combination valve housing.

   NOTE: Use a back-up wrench on housing to prevent damage.

   b. Remove screw attaching housing to bracket and remove expansion valve.

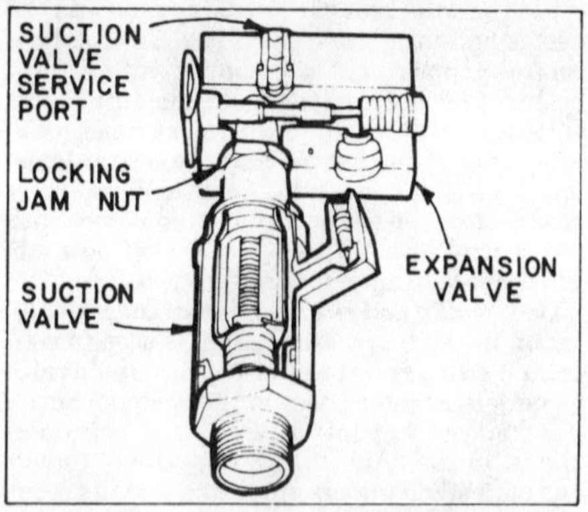

Fig. 45 Mini-combination valve. 1980 Cougar, Fairmont, Thunderbird & Zephyr

5. Reverse procedure to install. Use new "O" rings lubricated with refrigerant oil. Install STV into expansion valve body hand tight.

NOTE: To avoid STV damage, grasp body of STV on end nearest taper, opposite from slotted end.

6. Evacuate, charge and leak test system. Check system for proper operation.

MINI-COMBINATION VALVE, REPLACE

FORD MOTOR CO. PASSENGER CARS

1980 Capri, Cougar, Fairmont, Mustang, Thunderbird & Zephyr, Fig. 45

1. Discharge refrigerant from system.
2. Disconnect high and low pressure refrigerant lines, then remove the liquid bleed line from the STV.
3. To remove the STV (suction throttling valve), unscrew from expansion valve body.
4. To remove expansion valve:
   a. Disconnect the evaporator fittings from combination valve body.
   b. Remove screw attaching housing to bracket and remove housing.
5. Reverse procedure to install. Use new "O" rings lubricated with refrigerant oil.
6. Evacuate, charge and leak test system. Check system for proper operation.

General Motors Corp. Passenger Cars & Light Duty Trucks

NOTE: When replacing POA valve, check interior of old valve for corrosion or crystallation of salts, which indicate excessive moisture in the system. If such is the case, the receiver-dehydrator should be replaced and the system evacuated for one hour.

Part 2

Servicing Electrical, Manual, Vacuum System Components, Evaporator & Heater Core

AMERICAN MOTORS PASSENGER CARS & JEEP

CONTROL CABLE ADJUSTMENTS

1977-80 With A/C Exc. Matador & Pacer

Blend-Air Door Cable

1. Hold heater temperature control lever in High on 1977 models or Warm on all other models, and position blend air door crank arm in extreme right end of travel (full heat).
2. Slip self-adjusting clip ½ inch along cable wire moving blend-air door crank arm toward control panel.

NOTE: Clip is located at blend-air door crank arm at front of heater housing.

3. Move temperature control lever to "OFF" position on 1977 models or "COLD" on all other models, causing cable to slip through clip for proper adjustment. Recheck adjustment.
4. Adjust water valve vacuum switch if necessary.

Defroster Door and Floor Heat Door

NOTE: Air control lever controls door operation for Defrost and Heat.

1. Remove cable retaining clips from evaporator housing.
2. Hold air control lever in "OFF" position on 1977 models or "MAX A/C" on all others.
3. Hold defroster door firmly closed and install cable retaining clip. Repeat procedure for floor heat door.

Outside/Recirculating Air Door

NOTE: Clip is located at door crank arm on the top right of heater/blower housing.

1. Hold air control lever in "OFF" position on 1977 models or "MAX A/C" on all other models.
2. Slip self-adjusting clip ½ inch along wire, moving door crank arm away from control panel.
3. Move air control lever from "OFF" or "MAX A/C" to "HEAT" position, causing control wire to slip through clip for proper adjustment.

1977-80 Less A/C Exc. Matador & Pacer

Blend-Air Door Cable

Refer to "Blend-Air Door Cable" adjustment procedure on models with A/C.

Heater and Defroster Door Cable

NOTE: The cable is equipped with a self-adjust clip located on outside air door crank arm. The clip is accessible after removal of glove box.

1. Position air control lever to "OFF" and heater-defroster door crank arm to lowest position (full heat).
2. Slip self-adjusting clip ½ inch along control cable wire, moving crank arm towards cable clamp.
3. Move air control lever from "OFF" to "DEF" position to adjust cable.

Outside Air Door Cable

Refer to "Outside/Recirculating Air Door" adjustment procedure for models with A/C.

1974 Ambassador & Matador With A/C

Heater Valve & A/C Thermostat Cable

To adjust either cable, slide the upper control lever to the mid-dot position. Rotate turnbuckle at least 1½ turns to increase cable length. Place a slight pressure on the lever and at the same time rotate the turnbuckle in the opposite direction. When the lever begins to move, to the right when adjusting heater valve cable or to the left when adjusting the A/C thermostat cable, rotate the respective cable turnbuckle another ¼ turn.

Recirculating Air Damper Cable

NOTE: On 1974 models the recirculating air damper is vacuum actuated and no adjustment is required.

Loosen the two attaching screws. Manually hold damper door closed and tighten screws.

Heater Air Damper Cable

NOTE: On 1974 models, the heater air damper is vacuum actuated and no adjustment is required.

Loosen the two attaching screws. Manually hold damper closed and tighten screws.

Defroster Air Damper Cable

NOTE: On 1974 models, the defroster air damper is vacuum actuated and no adjustment is required.

Loosen the two attaching screws. Manually hold damper door open and tighten screws.

A/C Panel Damper Air Door

The A/C panel damper vacuum motor has two elongated slots on the mounting bracket to permit adjustment of the damper door. To adjust, ensure that no vacuum is applied to the vacuum motor and loosen the vacuum motor attaching screws. Manually close the damper and tighten the motor attaching screws.

1974-78 Matador Less A/C

Blend-Air Door Cable

Rotate the adjustment turnbuckle as required to close the blend-air damper door when the temperature control lever is in the far left position.

NOTE: The adjustment turnbuckle is located midway between the control panel and heater housing.

Heater Air and Defroster Damper Door Cable

1. Loosen cable retaining screws at heater housing.
2. Place air control lever in AIR OFF position.
3. Hold damper door levers closed and slide cable housings until cables are taut.
4. Tighten cable retaining screws and check air control lever operation.

1974-76 Gremlin, Javelin, & Hornet With A/C

The upper lever on the system control panel positions the heat control damper door in the heater housing. The control cable to the heat control damper door is equipped with a turnbuckle to facilitate adjustment. The turnbuckle is located beneath the instrument panel midway between the control and the heater housing. To adjust, rotate the turnbuckle to completely close the heat control damper door with the control lever in the "Heat, Off" position. Then, check damper for proper operation.

On Gremlin and Hornet models, the lower lever on the system control panel is used to control all damper door operations. On Javelin models, the center lever on the system control panel is used to control all damper door operations. In the "Air Off, A/C" position, the air and defroster dampers are closed and the recirculating air door is fully open. In the "Heat" position, the air damper is open and the defroster damper and recirculating air door are closed. In the "Def" position, the defroster damper is open and the air damper and recirculating air door are closed. To adjust the damper door, remove the cable retaining clips. Then, set the control lever for the damper door being adjusted to be in the closed position. Hold the damper door in the closed position and install the cable retaining clip.

1974-76 Gremlin, Javelin & Hornet Less A/C

Blend-Air Door Cable

Rotate the adjustment turnbuckle as required to close the blend-air damper door when the temperature control lever is in the far left position.

NOTE: The adjustment turnbuckle is located midway between the control panel and heater housing on Gremlin and Hornet models, and through the glove box on Javelin models.

Heater Air and Defroster Damper Door Cable

1. Remove cable housing retaining clips.
2. Position air control lever in AIR OFF and hold the damper door levers closed.
3. Install cable retaining clips and check air control lever operation.

1975-78 Matador

Blend Air Damper Cable

1. Remove lower instrument finish panel, glove box door and liner.
2. Place temperature control lever in center detent position.
3. Rotate turnbuckle counter-clockwise to raise blend air door crank arm approximately ⅛ inch.
4. Rotate turnbuckle clockwise until the crank arm is in the full downward position, then rotate turnbuckle an additional ¼ turn.
5. Check temperature control lever operation for full travel of blend air door.
6. Install glove box liner, door and lower instrument finish panel.

A/C Thermostat Cable

1. Remove instrument cluster bezel, clock, clock opening cover or fuel economy gauge.
2. Remove control panel attaching screws and pull panel rearward.
3. Place the temperature control in the "Off" position.
4. Rotate turnbuckle clockwise to lower thermostat crank arm downward approximately ⅛ inch.
5. Rotate turnbuckle counter-clockwise until crank arm is in the full upward position, then rotate turnbuckle an additional ¼ turn counter-clockwise.

NOTE: The temperature control may not travel to the stop in the center detent position when properly adjusted since overtravel is built into the thermostat.

6. Install control panel, clock, clock opening cover, fuel economy gauge, if equipped, and the instrument cluster bezel.

A/C Panel Damper Air Door

The A/C panel damper vacuum motor has two elongated slots on the mounting bracket to permit adjustment of the damper door. To adjust, ensure that vacuum is not applied to the door vacuum motor and loosen vacuum motor attaching screws. Manually close the damper and tighten the screws.

1977-80 Pacer Less A/C

Blend-Air Door

1. Position temperature control lever to far left (COOL).
2. Position blend-air door crank arm in full cool position (forward).

NOTE: Blend-air crank arm is located in engine compartment.

3. Move self-adjusting clip ½ inch along control cable wire, moving crank arm toward front of vehicle.
4. Move temperature control lever from "COOL" to "WARM" to adjust cable.

Floor Door

1. Position mode control lever in "OFF" and slide adjusting clip ½ inch up cable toward top of instrument panel.

NOTE: The crank arm is located on blower housing, accessible through glove box opening.

2. Move mode control lever from "OFF" to "DEF" to adjust cable.
3. Check operation of floor door.

Defroster Door

NOTE: The defroster door crank arm is located on heater housing, under instrument panel to left of heater control panel.

1. Hold mode control lever in "OFF" position.
2. Slip self-adjusting clip ½ inch along cable wire, moving defroster door crank arm toward rear of vehicle.
3. Move mode control lever from OFF to DEF to adjust.

NOTE: It may be necessary to slide crank arm adjusting clip slightly (1/16 inch) toward front of vehicle to ensure proper arm movement.

4. Check operation of defroster door.

1975-76 Pacer Less A/C

Blend-Air Door

Refer to "1975-80 Pacer with A/C" for blend-air door cable adjustment.

1975-80 Pacer With A/C

Blend Air Door Cable

1. Remove radio speaker grille from instrument panel.
2. Place the temperature in the full heat position.
3. Working through the speaker grille opening, rotate cable turnbuckle to fully close the blend air.
4. Check door for proper operation and install the radio speaker grille.

Temperature Control Thermostat Cable

1. Disconnect battery ground cable.
2. Remove radio overlay and radio.
3. Loosen thermostat adjusting screws.
4. Place temperature control lever in "Max Cold" position and tighten adjusting screws with the thermostat lever at the extreme end of travel.
5. Check temperature control thermostat for proper operation.
6. Install radio and overlay.
7. Connect battery ground cable.

1978-80 Jeep CJ models

Air Inlet Door

This cable has a turnbuckle adjuster located about 6 inches above the cable adjustment to right side of heater. Adjust until the passenger vent door is closed when the air control knob is pushed in.

NOTE: If cable is adjusted past closed door position, it may prevent the driver's vent door from closing.

1974-80 Jeep Exc. CJ models

Blend-Air Door

This cable has a turnbuckle adjuster located under the instrument panel to the right of the steering column. Adjust turnbuckle until blend-air door closes when the temperature control lever is in "COOL" position (far left).

1974-77 Jeep CJ models

1. Remove control cable housing retaining clip screw.
2. Move control knob to "OFF" position (Full in).
3. Hold door tightly closed and install cable housing retaining clip screw.
4. Check operation of controls.

COOLANT CONTROL VALVE, REPLACE

1974-80

1. Drain approximately two quarts of coolant from radiator.
2. Disconnect vacuum line from coolant control valve.
3. On all engines except 1974-75 V8's, disconnect heater hose connections at coolant control valve and remove valve. On 1974-75 V8 engines, disconnect heater hose connections from coolant control valve and rotate valve counter-clockwise to remove valve from manifold.
4. Reverse procedure to install.

CONTROL PANEL, REPLACE

1974 Javelin

1. Disconnect battery ground cable.
2. Remove lower right hand mouldings, assist handle, finish panel and lower center housing.
3. Remove control panel attaching nuts, then disconnect the control cables and electrical connectors from control panel.
4. Remove thermostatic switch attaching screws.
5. Remove control panel from vehicle.
6. Reverse procedure to install.

1974 Ambassador & 1974-78 Matador

1. Disconnect battery ground cable.
2. Remove instrument cluster bezel screws and the radio attaching nuts. Disconnect remote mirror control from bezel. Then, remove the cluster bezel.
3. Remove clock, clock opening cover or fuel economy gauge, if equipped.
4. Remove control panel attaching screws, then disconnect control cables and electrical connectors from control panel. On 1975-78 models, disconnect vacuum hose connector from air outlet damper vacuum switch and water valve vacuum switch.
5. On all models, remove control panel from vehicle.
6. Reverse procedure to install.

1974-80 Exc. Ambassador, Javelin, Matador & Pacer

1. Disconnect battery ground cable.
2. Remove instrument panel center housing.
3. Remove A/C thermostat control knob and attaching nut.
4. Remove control panel attaching screws.
5. On 1975-80 models, remove center upper discharge air duct and the radio.
6. On all models, disconnect control cables and electrical connectors from control panel, then remove the panel.
7. Reverse procedure to install.

1975-80 Pacer

1. Disconnect battery ground cable.
2. Remove radio overlay and radio.
3. Remove control panel attaching screws, then disconnect the control cables and electrical connectors from control panel.
4. Disconnect vacuum hose connectors from vacuum switches and remove control panel from vehicle.
5. Reverse procedure to install.

1975-80 Jeep Exc. CJ Models

1. Disconnect battery ground cable.
2. Remove instrument cluster retaining screws.
3. Disconnect speedometer cable, electrical connectors and vacuum hoses from cluster.

NOTE: Mark hoses according to their location.

4. Remove cluster lights and disconnect cable from temperature control lever.
5. Remove cluster from vehicle.
6. Remove heater-defroster control from cluster.
7. Remove fan control switch.
8. Reverse procedure to install.

1974 Jeep Exc. CJ Models

1. Disconnect vacuum lines and control cable from heater control assembly.
2. Unsnap lamp bulb from control and disconnect terminal connector located in wiring.
3. Remove two nuts and mounting bracket from bottom of control, then remove by pushing out on bottom.
4. Reverse procedure to install.

1975 Jeep CJ

1. Remove cable to panel retaining clips.
2. Disconnect fan control switch wiring.
3. Disconnect control cables at damper doors.
4. Remove control panel attaching screws and panel.
5. Reverse procedure to install.

VACUUM MOTORS, REPLACE

1975-80 Pacer

Floor Door Vacuum Motor

1. Remove instrument panel crash pad and the upper to lower instrument panel attaching screws.
2. Disconnect vacuum hose from motor.
3. Remove vacuum motor attaching screws and clip, then the vacuum motor.
4. Reverse procedure to install.

Panel Door Vacuum Motor

NOTE: The panel door vacuum motor is located on the front of the evaporator housing.

1. Remove radio.
2. Disconnect vacuum hose from vacuum motor.
3. Remove vacuum motor attaching screws and clip, then the vacuum motor.
4. Reverse procedure to install.

Defroster Door Vacuum Motor

NOTE: The defroster door vacuum motor is located at the left end of the evaporator housing.

1. Disconnect vacuum hose from motor.
2. Working beneath instrument panel, remove vacuum motor attaching screws and clip, then the vacuum motor.
3. Reverse procedure to install.

Recirculating Air Door Vacuum Motor

NOTE: The recirculating air door vacuum motor is located on the right end of the blower motor housing.

1. Remove instrument panel crash pad, upper to lower instrument panel attaching screws and the recirculating air grille.
2. Disconnect vacuum hose from vacuum motor.
3. Remove vacuum motor attaching screws and clip, then the vacuum motor.
4. Reverse procedure to install.

EVAPORATOR CORE, REPLACE

1974-80 Exc. Ambassador, Javelin, Matador & Pacer

1. Disconnect battery ground cable and discharge refrigerant from system.
2. Remove package tray and duct assembly.
3. Remove instrument panel center housing, ash tray, radio air outlets and duct.
4. Remove insulation and disconnect refrigerant lines from evaporator.
5. Disconnect control cables and remove thermostat capillary tube at evaporator housing.
6. Remove evaporator housing, then remove housing cover and evaporator core.
7. Reverse procedure to install.

1974 Javelin

1. Disconnect battery ground cable and discharge refrigerant system.
2. Remove hose retaining clips and dash panel hose grommet.
3. Disconnect suction service valve from compressor and the refrigerant hose from the disconnect coupling.
4. Remove assist handle, moldings and instrument panel top cover.
5. Remove lower right instrument panel finish panel and the right hand instrument panel support brace.
6. Remove thermostat capillary tube from evaporator.
7. Remove A/C control assembly attaching nuts and the lower center housing. Do not disconnect control cables.
8. Remove evaporator housing to dash panel screws and loosen the heater core and

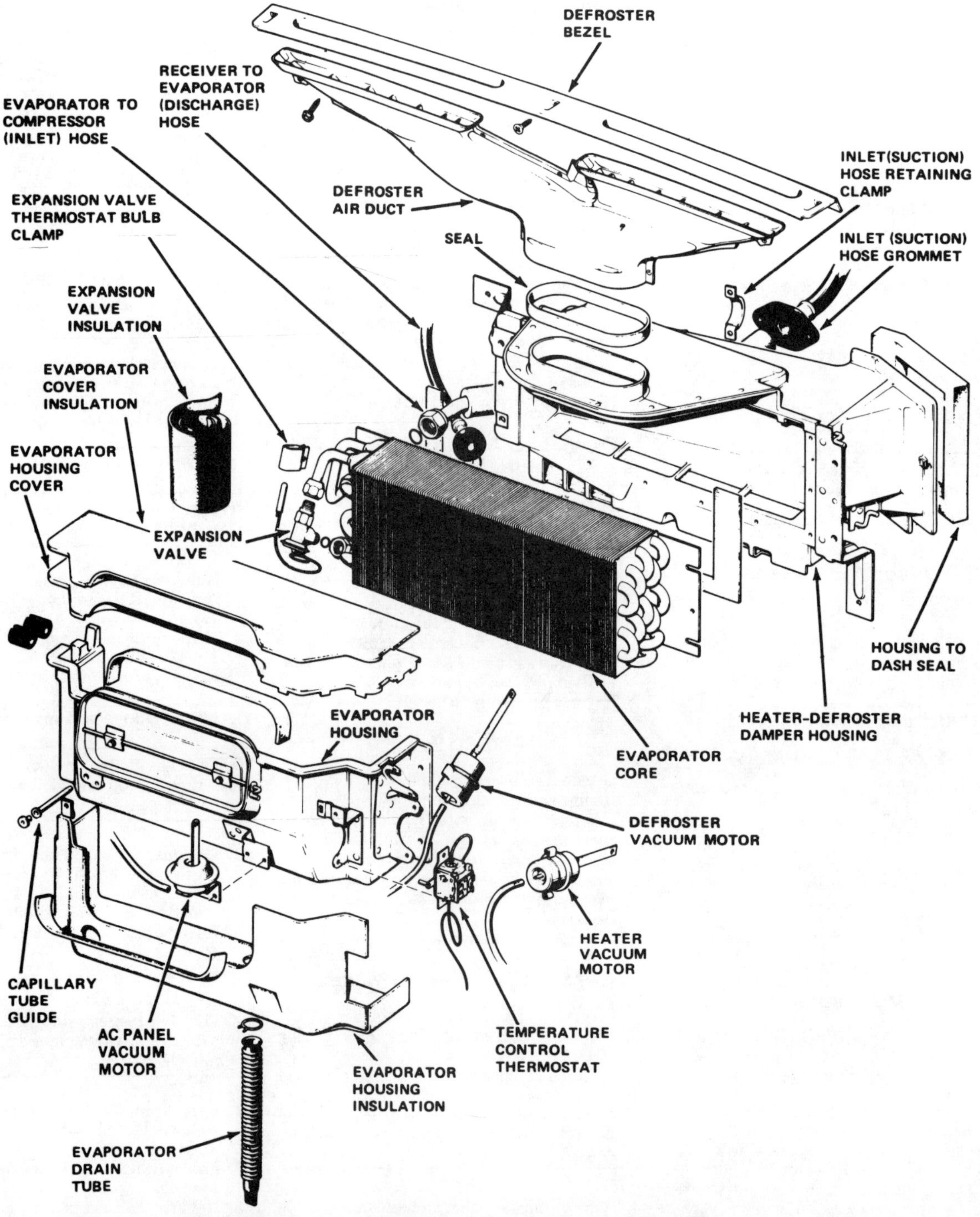

Fig. 1 Evaporator core assembly. 1974-78 Ambassador & Matador

blower housing attaching screws.

9. Slide evaporator housing to the left and off the heater core and blower housing.
10. Loosen instrument panel attaching nuts at steering column to permit panel movement while the evaporator assembly is being removed.
11. Reverse procedure to install.

1974 Ambassador & 1974-78 Matador

1. Disconnect battery ground cable and discharge refrigerant system.
2. Remove lower instrument finish panel.
3. On 1975-76 models, remove instrument cluster bezel, clock, fuse panel, glove box and radio.
4. On all models, disconnect inlet and discharge hoses, then the A/C panel damper motor vacuum hoses, Fig. 1.
5. Remove drain tube.
6. Disconnect thermostat control cable and wire, then remove the defroster, floor and recirculating damper vacuum motors.
7. Remove evaporator housing attaching screws and disconnect A/C ducts from evaporator housing.
8. Move evaporator housing downward and toward the right, then remove housing from vehicle.
9. Reverse procedure to install.

1975-80 Pacer

1. Remove heater core housing as outlined under "Heater Core, Replace", 1975-80 Pacer procedure.
2. Remove evaporator housing to dash panel attaching nuts and loosen blower housing to dash panel attaching screw, Fig. 2.
3. Remove crash pad, then the left corner and windshield side finish moulding.
4. Remove radio overlay panel and radio.
5. Disconnect speedometer cable, then the control cable and electrical wiring from temperature control thermostat.
6. Disengage instrument panel wiring harness from bracket on evaporator housing and from under thermostat.
7. Disconnect vacuum hose from panel door vacuum motor.
8. Remove instrument panel attaching screws, then the right and left scuff plates and cowl trim panels.
9. Remove steering column tube cover and the left intermediate air discharge duct.
10. Loosen steering column bracket nuts.

NOTE: Tape steering column to prevent paint damage.

11. Remove hood release cable bracket attaching screws.
12. Remove insulation from hose fittings, then disconnect and cap the suction and high pressure hoses.
13. Disconnect vacuum hose from defroster door vacuum motor.
14. Remove lower instrument panel to right "A" pillar attaching screw, then the lower instrument panel to left "A" pillar attaching screw and loosen the upper screw.
15. Pull lower instrument panel outward and loosely install lower instrument panel to right "A" pillar screw.
16. Remove floor outlet duct and drain hose.
17. Pull evaporator housing from dash and remove from vehicle.
18. Remove evaporator core from housing.
19. Reverse procedure to install.

1974-80 Jeep Models

1. Discharge refrigerant from system.
2. Disconnect inlet line at compressor. Plug all open lines.
3. Disconnect evaporator to receiver line.

NOTE: On all except CJ models, disconnect refrigerant line at quick-disconnect coupling.

4. Remove hose clamps and dash grommet retaining screws.
5. Remove evaporator housing to dash attaching screws and bracket.
6. Lower housing and pull hoses through dash opening. Remove housing from vehicle.
7. Disassemble housing to remove evaporator core, Figs. 3 & 4.
8. Reverse procedure to install.

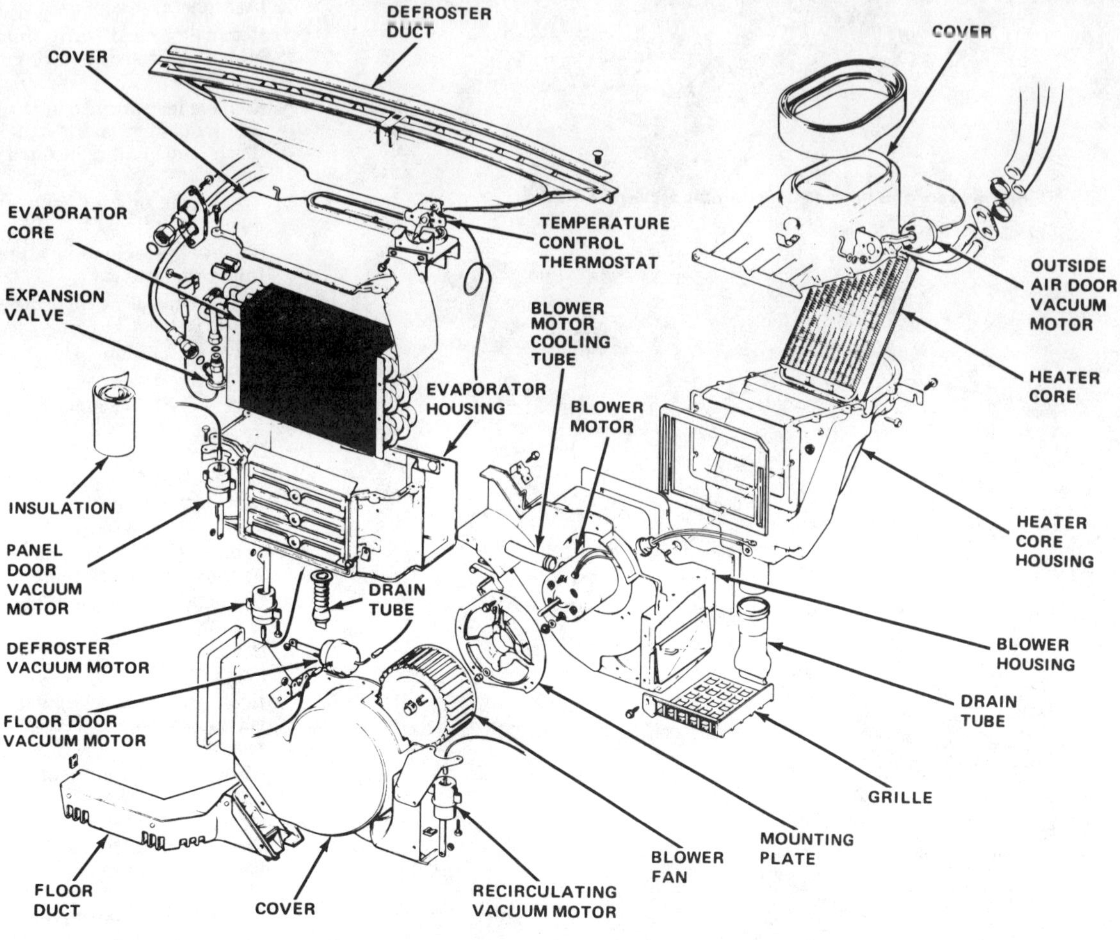

Fig. 2 Evaporator core assembly. 1975-80 Pacer

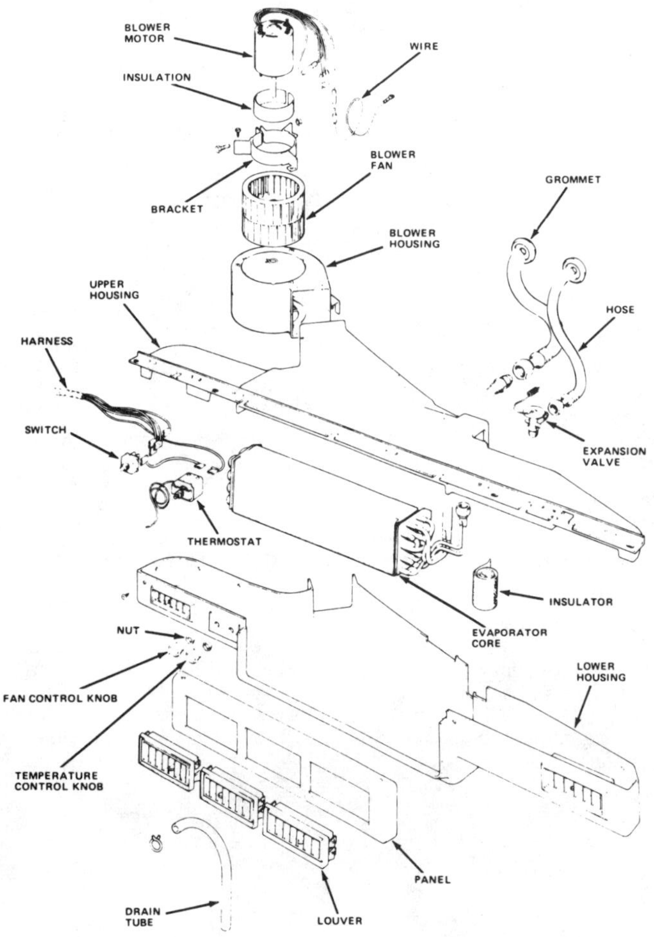

Fig. 3 Evaporator housing disassembled. Jeep CJ models

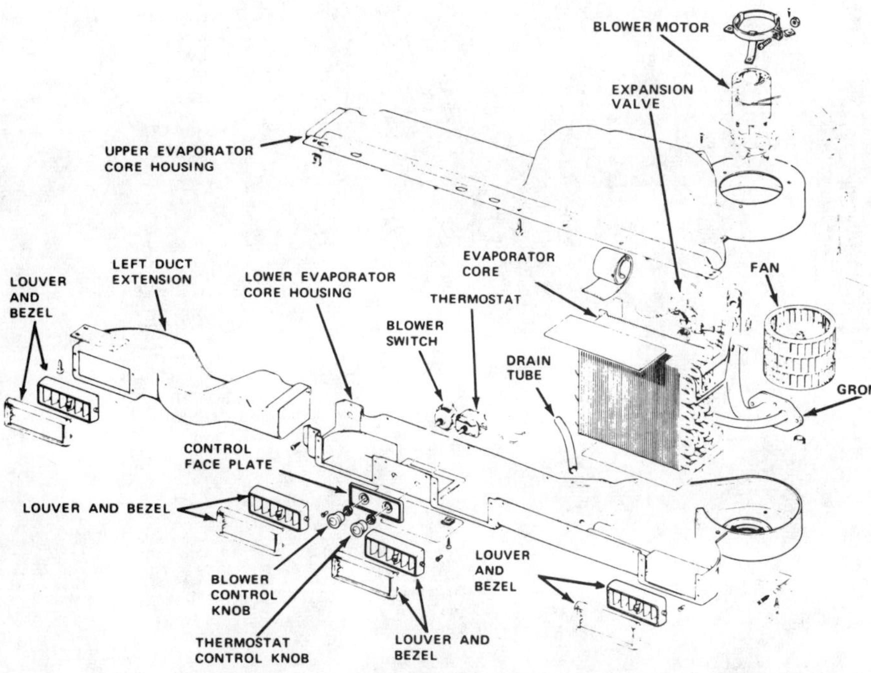

Fig. 4 Evaporator housing disassembled. Jeep Exc. CJ models

HEATER CORE, REPLACE

1974-80 Exc. Ambassador, Javelin, Jeep, Matador & Pacer

1. Open heater valve and drain about 2 qts. from cooling system.
2. Disconnect heater hoses and plug heater core tubes.
3. Remove blower motor and fan.
4. Remove package tray if so equipped.
5. Disconnect wire connector at resistor.
6. Remove instrument panel bezel, outlet and duct.
7. Disconnect control cables from damper levers.
8. Remove right side windshield pillar moulding and the instrument panel upper attaching screws and right side cap screw at the door hinge post.
9. Remove right side kick panel and heater housing attaching screws.
10. Pull the right side of the instrument panel slightly rearward and remove the housing.
11. Reverse procedure to install.

1974 Javelin

1. Drain 2 qts. of water from cooling system and disconnect hoses from core tubes in engine compartment. Install corks in hoses and core tubes.
2. Disconnect blower motor wires.
3. Remove ground wire from dash panel ground stud and remove ground stud nut and washer.
4. Remove instrument panel top cover, right side mouldings, assist handle, lower right finish panel and right hand panel support brace.
5. Disconnect air vent cable at heater housing.
6. Remove heater housing attaching screws in front compartment.
7. After removal of housing assembly the heater core can be removed.
8. Reverse procedure to install.

1977-78 Matador

1. Disconnect battery ground cable.
2. Drain about 2 qts. of coolant then disconnect heater hoses and plug ends of hoses and core openings.
3. Remove glove box to gain access to heater housing attaching screws.
4. Disconnect blend-air cable at heater core housing.
5. Remove fuse panel.
6. Remove lower instrument finish panel, right side windshield pillar and corner finish mouldings for access to upper right mounting screws.
7. Disconnect vacuum motor hoses. Remove remaining heater core housing attaching screws.
8. Remove capscrew attaching instrument panel to right A-pillar. Then pull right side of instrument panel slightly rearward and remove heater core housing assembly.
9. Remove heater core from housing, Fig. 5.
10. Reverse procedure to install.

1974 Ambassador & 1974-76 Matador

1. Disconnect battery ground cable.
2. Drain about 2 quarts of coolant then disconnect heater hoses and plug ends of hoses and core openings.
3. Remove lower instrument finish panel and glove box.
4. Disconnect control cables and vacuum motor hoses.
5. Remove retaining screws and remove heater core housing assembly.
6. Reverse procedure to install.

1975-80 Pacer

1. Drain approximately 2 qts. from cooling system.
2. Remove heater hoses and install plugs in hoses and core openings.
3. Position vacuum hoses away from core housing.
4. Remove core housing cover screws, disconnect overcenter spring and remove cover.
5. Remove screws attaching core to housing and remove core.
6. Reverse procedure to install.

1978-80 Jeep CJ

1. Disconnect battery ground cable and drain cooling system.
2. Disconnect heater core hoses.
3. Disconnect damper door control cables.
4. Disconnect blower motor wire.
5. Disconnect water drain hose and defroster hose.
6. Remove nuts retaining heater housing assembly to dash panel, then remove heater housing assembly by tilting down to disengage it from air inlet duct and pulling towards rear of vehicle.
7. Remove heater core from housing, Fig. 6.
8. Reverse procedure to install.

1975-77 Jeep CJ

Remove battery and battery box.
2. Drain cooling system and disconnect hoses from heater core.
3. Disconnect control cables, then disconnect water drain hose and defroster hose.
4. Remove glove box, then disconnect water drain hose and defroster hose.
5. Disconnect heater to air deflector duct at heater housing.
6. From engine compartment, remove heater housing stud nuts, then the heater assembly.
7. Remove heater core cover retaining screws, then the cover and heater core, Fig. 7.
8. Reverse procedure to install.

1974 Jeep CJ

1. Disconnect battery ground cable.
2. Drain cooling system and disconnect hoses from heater core.
3. Mark duct halves to assure proper reassembly, then remove screws retaining duct halves together.
4. Remove core retaining screws and the heater core, Fig. 7.
5. Reverse procedure to install.

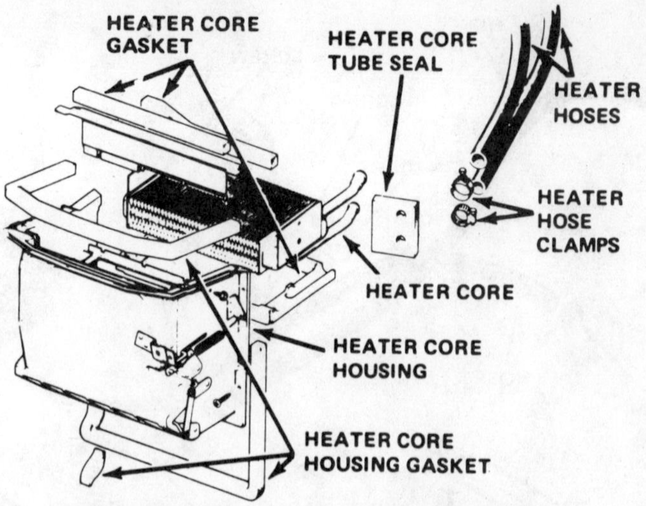

Fig. 5 Heater core and housing. 1977-78 Matador

1974-80 Jeep Exc. CJ models

1. Disconnect battery ground cable.
2. Drain cooling system and disconnect heater hoses from heater core.
3. Disconnect control cable and resistor harness connector.
4. On 1974 models:
   a. Remove the four heater core and duct-to-firewall retaining nuts.

   NOTE: Two of the nuts are located inside of the vehicle just to the right of the transition duct.

   b. Remove heater and duct assembly.
   c. Mark duct halves to assure proper reassembly, then remove screws and separate duct halves, Fig. 8.
5. On 1975-80 models:
   a. Remove heater core housing to dash panel retaining nuts and the heater core housing assembly.
   b. Remove screws retaining housing halves together and separate housing, Figs. 8 & 9.
6. Remove heater core retaining screws and the heater core.
7. Reverse procedure to install.

BLOWER MOTOR, REPLACE

1974-80 Exc. Ambassador, Jeep, Matador & Pacer models

1. Working in engine compartment, disconnect blower motor wire.

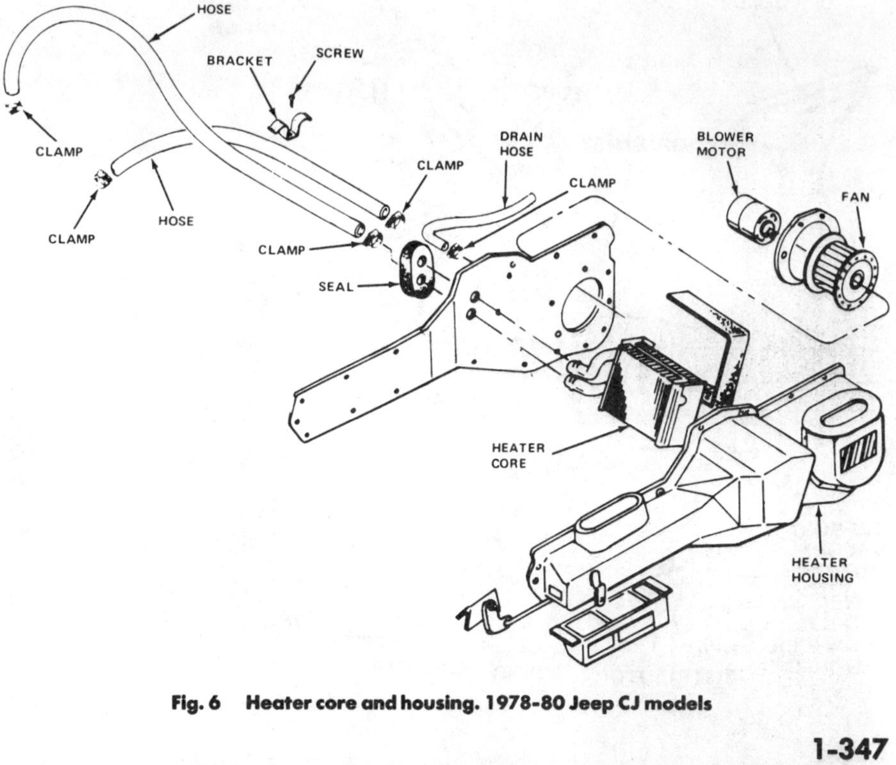

Fig. 6 Heater core and housing. 1978-80 Jeep CJ models

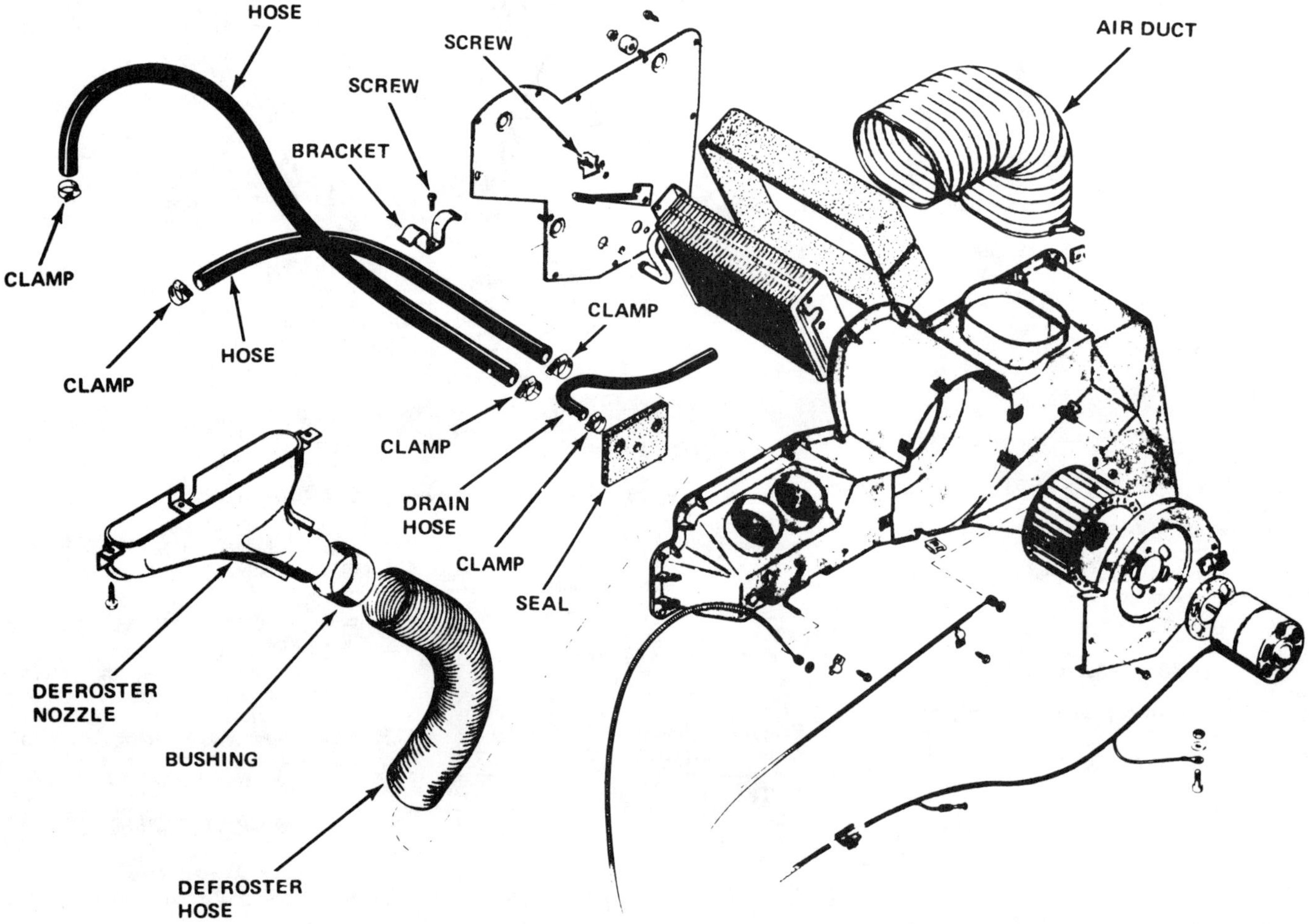

Fig. 7 Heater core and housing. 1974-77 Jeep CJ models

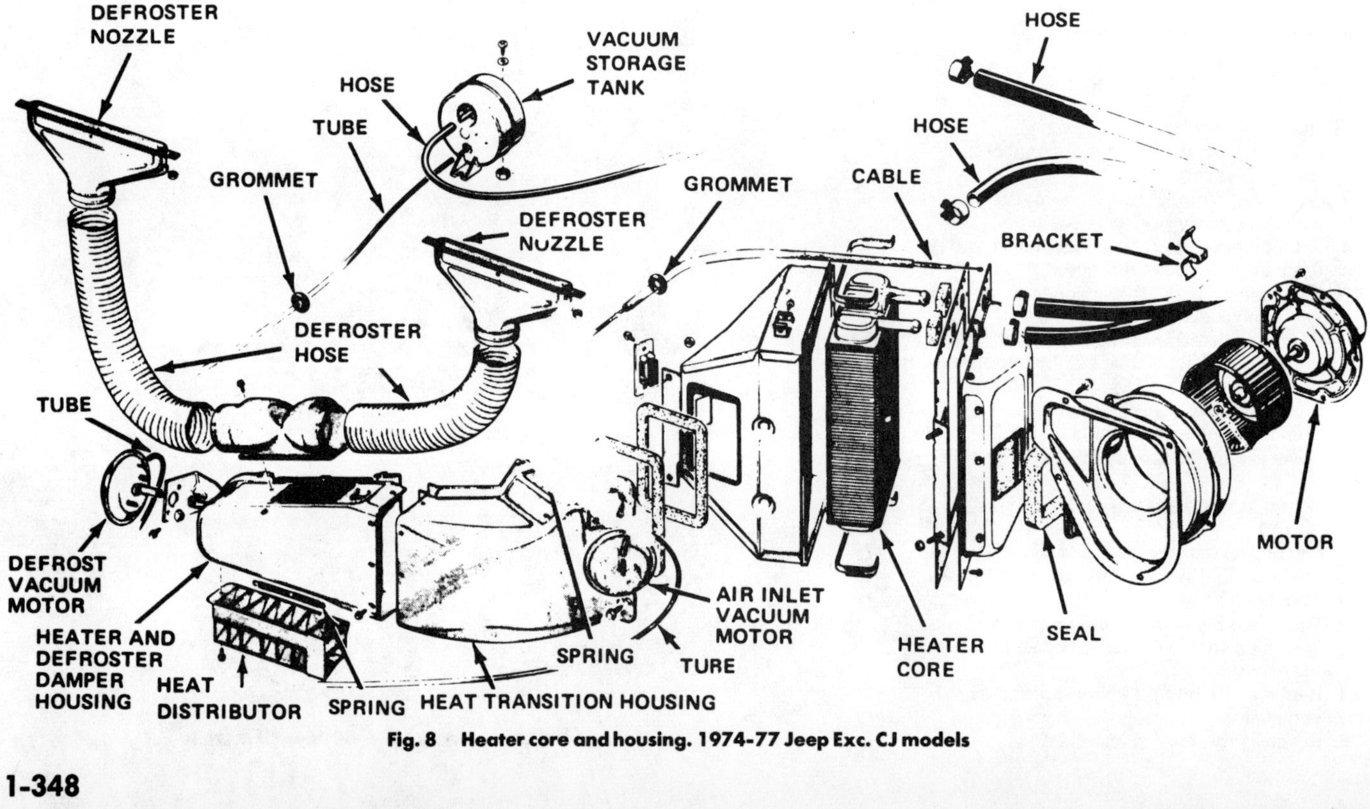

Fig. 8 Heater core and housing. 1974-77 Jeep Exc. CJ models

2. Remove three retaining nuts for blower scroll cover to which blower assembly is attached and remove blower motor and fan.
3. Reverse procedure to install.

1974 Ambassador & 1974-78 Matador

1. Disconnect blower motor wiring.
2. Remove blower motor attaching screws and remove blower motor.
3. Reverse procedure to install.

1975-80 Pacer

Less Air Conditioning

1. Disconnect battery ground cable.
2. Remove right side windshield molding, then the instrument panel crash pad.
3. Remove right scuff plate and cowl panel.
4. Remove lower instrument panel to pillar attaching screws, then pull instrument panel rearward and reinstall attaching screws, allowing panel to rest on screws.
5. Remove heater core housing attaching nuts and screw, position vacuum lines aside, disconnect air door, then pull housing forward and position on upper control arm.
6. Remove blower motor attaching screws, disconnect relay ground wire and resistor wiring.
7. Remove blower motor brace.
8. Loosen heater housing to instrument panel attaching nuts.
9. Pull blower motor housing rearward and downward, disconnect vacuum hoses from motors and remove blower housing.
10. Remove housing cover, disconnect wire and remove blower to housing attaching screws.
11. Remove blower from housing, then separate fan and mounting bracket from motor.
12. Reverse procedure to install.

With Air Conditioning

1. Disconnect battery ground cable.
2. Remove right scuff plate and cowl panel.
3. Remove radio overlay, then the instrument panel crash pad.
4. Remove instrument panel to right pillar attaching screws, then the upper to lower instrument panel attaching screws, located above glove box.
5. Disconnect air door cable from heater core housing and remove housing to floor pan brace screw.
6. Disconnect wires from blower motor resistor and vacuum lines from vacuum motors.
7. Remove heater core housing attaching nuts and screw, position vacuum lines aside, then pull housing forward and position on upper control arm.
8. Remove floor duct.
9. Disconnect wires from blower motor relay and remove blower motor housing attaching screw from firewall.
10. Loosen evaporator housing to dash panel attaching nuts, remove blower motor housing attaching screw from dash panel, then pull blower housing rearward and downward.
11. Pull right side of instrument panel rearward and remove blower motor.
12. Remove blower housing cover, Fig. 2.
13. Remove blower motor from housing, the separate fan and mounting bracket from motor.
14. Reverse procedure to install.

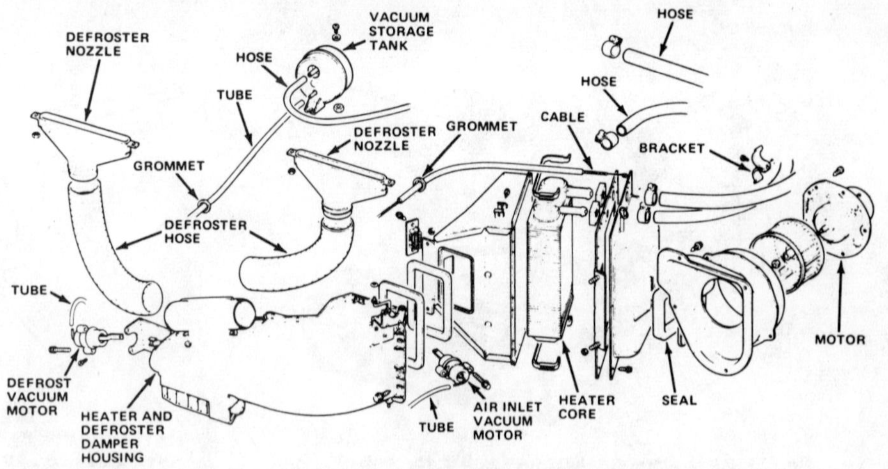

Fig. 9 Heater core and housing. 1978-80 Jeep Exc. CJ models

1978-80 Jeep CJ

Less Air Conditioning

1. Refer to "Heater Core, Replace" under 1978-80 Jeep CJ models and perform steps 1 thru 6 to remove heater housing assembly.
2. Remove blower motor assembly retaining screws and remove blower motor.
3. Reverse procedure to install.

With Air Conditioning

NOTE: It is not necessary to discharge and remove the evaporator core to service the blower motor. The evaporator housing may be lowered from the instrument panel to gain access to blower motor retaining screws. Refer to "Evaporator, Replace" procedure.

1974-80 Jeep Exc. CJ

Less Air Conditioning

1. Disconnect blower motor wiring.
2. Remove blower motor to housing attaching screws and blower motor.
3. Reverse procedure to install.

With Air Conditioning

NOTE: It is not necessary to discharge and remove the evaporator core to service the blower motor. The evaporator housing may be lowered from the instrument panel to gain access to blower motor retaining screws. Refer to "Evaporator, Replace" procedure.

1974-77 Jeep CJ

Less Air Conditioning

1. Disconnect battery ground cable.
2. Disconnect blend-air cable.
3. Disconnect blower motor wiring.
4. Remove blower motor housing retaining screws and blower motor housing.
5. Remove blower motor from housing.
6. Reverse procedure to install.

With Air Conditioning

NOTE: It is not necessary to discharge and remove the evaporator core to service the blower motor. The evaporator housing may be lowered from the instrument panel to gain access to blower motor retaining screws. Refer to "Evap-

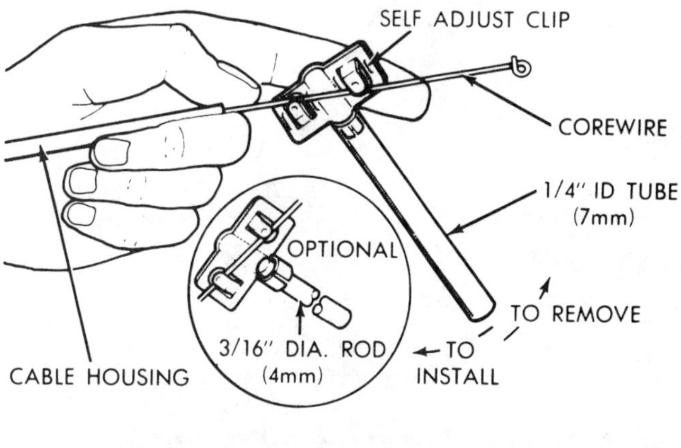

Fig. 10. Installing self-adjust clip. 1979-80 Typical

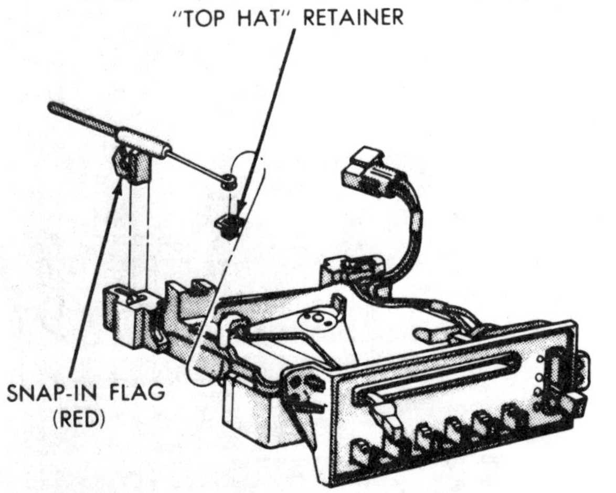

Fig. 11 Installing temperature cable to control assembly. 1979-80 Typical

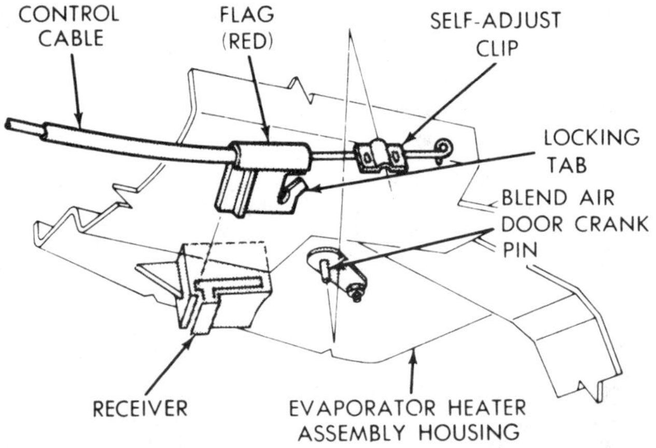

Fig. 12 Installing temperature cable to heater or heater-evaporator assembly. 1979-80 Cordoba, Gran Fury, Magnum, Mirada, Newport, New Yorker & St. Regis.

orator, Replace" procedure.

CHRYSLER CORP.

Control Cable Adjustments

PASSENGER CARS

1979-80 All Models

Position self-adjusting clip on wire cord 2 inches from loop at heater or heater-evaporator assembly. Pry clip into locked position using a ¼ inch tube, Fig. 10. Connect cable to control panel lever and heater or heater-evaporator assembly, Figs. 11 & 12. Adjust cable by moving lever all the way to the right.

CAUTION: Ensure that cable is properly routed under relief notch in defroster duct on Gran Fury, Newport, New Yorker & St. Regis models. Failure to do so could cause interference with accelerator cable.

1978 Charger, Cordoba, Fury, Magnum & Monaco Models With A/C

Place temperature control lever in "COOL" position. Attach cable core wire to control lever pin and snap cable housing into retaining clip so housing extends ⅜ inch beyond clip. Install control housing into instrument panel. Pull cable from engine compartment until tape butts against grommet. Cable must be routed over brake assembly and starter relay. Clip cable to plenum mounting clip. Position self-adjusting clip 1½ inch from loop end and pry into place, Fig. 10. Move control lever to "WARM" position to adjust cable.

1978 Horizon & Omni

Position self-adjusting clip 1½ inches from small loop end. Align crank arm with clip and install flag into receiver on heater or evaporator-heater housing, Fig. 13. Place temperature control or mode control lever in extreme left position. Attach cable core end to lever and snap cable housing into retaining clip. The tape marker must be against retainer on side facing away from lever. If cable housing has no marking tape, install cable so housing extends ⅜ inch from retaining clip. Install control assembly into panel. Adjust cable by moving lever all the way to the right.

1974-78 Intermediate Models Less A/C Exc. Barracuda, Challenger, Dart & Valiant

Position control lever all the way to the right. Disconnect clip from crank arm and slide along wire core until against cable housing. Position control lever in "COOL" or "OFF" position and install clip on crank arm. Adjust by moving control lever all the way to the right, until it contacts edge of bezel slot.

1974-77 Intermediate Models With A/C Exc. Aspen, Dart, Diplomat, LeBaron, Valiant & Volare

Place temperature control lever in the Max. Heat position. Place self-adjusting clip on cable 2 inches on 1974 models or 1½ inches on

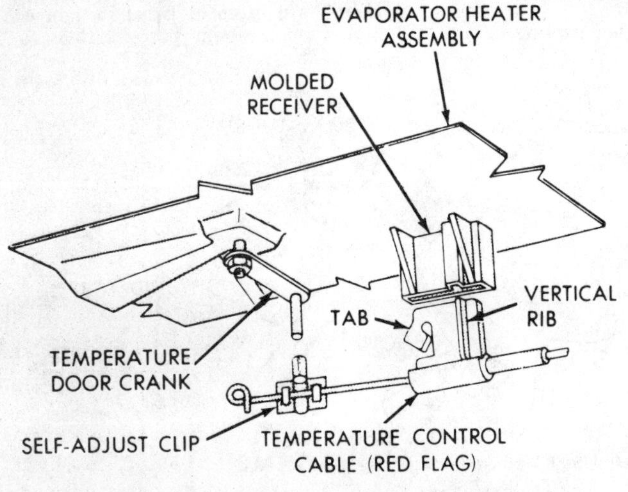

Fig. 13 Temperature control cable to evaporator housing. Typical Horizon & Omni

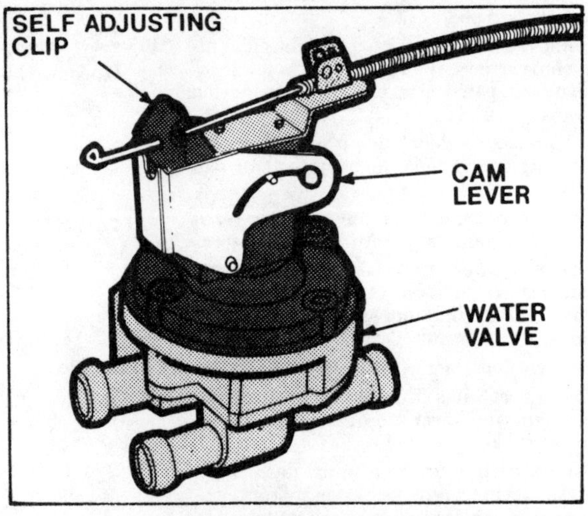

Fig. 14 Chrysler temperature cable adjustment (Typical). 1974-78 Intermediates exc. Aspen, Dart, Diplomat, LeBaron, Valiant & Volare

1975-77 models from the water valve end loop, Fig. 14. Install clip with cable on water valve cam and so the housing projects ⅝ inch on 1974-76 models, ½ inch on 1977 models, beyond clip. To adjust cable, move temperature control lever to the cool position on 1974 models or to the warm position on 1975-77 models.

1974-76 Dart & Valiant

Place temperature control lever in the warm position and remove the self-adjust clip from door crank. Then, move the temperature control lever to the cool position and slide the self-adjust clip to the cable housing. Place the temperature control lever in the warm position and install the self-adjust clip on door crank, avoiding damage to the core wire. Move temperature control lever toward the cool position until the lever bottoms against the side of the bezel.

1974-78 Chrysler, Dodge & Plymouth Full Size Models Less A/C

Position self-adjust clip on cable core 1½ inches from loop end. Position control lever to extreme left end of travel. Attach the self-adjust clip onto the proper door crank arm and snap cable housing into retaining clip so housing extends ½ inch from end of clip. Attach opposite end of cable to proper control lever. Snap control cable housing into retaining clip so housing extends ½ inch beyond clip. Install control assembly into instrument panel. Adjust cable by moving control lever to the extreme right end of travel.

1974-78 Chrysler, Dodge & Plymouth Full Size With A/C

Install a self-adjust clip on cable core wire approximately two inches from the temperature door end loop, Fig. 15. Place temperature control lever in the extreme cool position and connect the cable self-adjust clip to the proper door crank arm, then snap the cable housing to the retaining clip so the cable housing projects approximately 2⅛ inch from the end of the clip. Connect the opposite end of the cable core wire to the corresponding control lever pin and snap the cable housing into the retainer clip so the housing extends approximately ⅜ inch beyond end of clip. Move temperature control lever to the fully warm position.

1976-78 Aspen, Diplomat, LeBaron & Volare With A/C

Place self-adjust clip on cable core 1½ inches from small loop on A/C unit end, then install cable and clip on temperature control door crank arm, Fig. 16. Place the temperature control lever in the fully cool position and attach the cable core wire to the control lever pin, then snap the cable housing into the retaining clip so colored marking tape is next to retaining clip on side facing away from control lever pin, Fig. 17. If cable has no marking tape, position cable housing to extend ⅜ inch beyond retaining clip.

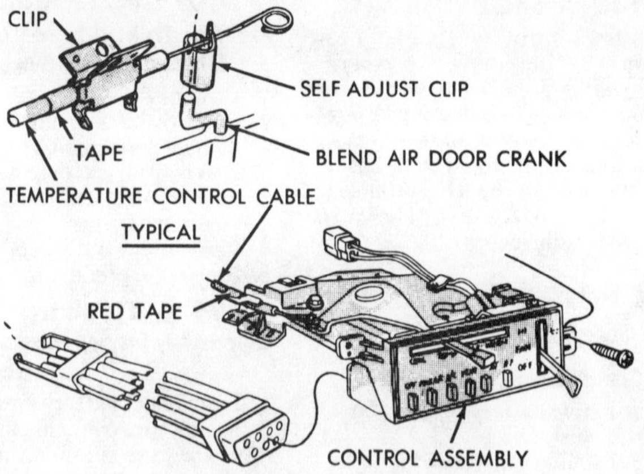

Fig. 15 Chrysler temperature cable adjustment. 1974-78 Full Size

1974 Barracuda & Challenger Less A/C

Temperature Control Cable

Remove retaining clip from cable at temperature control door. Place control lever in "LOW" position. Rotate control door crank arm to extreme clockwise position and reinstall cable retaining clip. Be sure control lever remains in "LOW" position during adjustment.

Heat-Defrost Cable

Place control lever in "OFF" position, loosen cable retaining screw at control assembly. Move cable housing towards instrument panel to shorten travel or away from panel to lengthen travel of cable. Tighten cable retaining screws.

TRUCKS & VANS

1979-80 Van Models

Temperature Control Cable

Position self-adjusting clip on wire core 2 inches from loop at heater or heater-evaporator housing end. Pry clip into locked position using a ¼ inch tube, Fig. 10. Connect cable to temperature control lever, Fig. 11. Place self-adjusting clip over blend-air door crank and snap cable housing into retaining clip so that locating tape is located next to side of clip away from crank arm. Adjust by moving control lever to the extreme right position of travel.

Mode Control Cable

Position self-adjust clip 2 inches from loop at distribution duct end. Pry clip into a locked position using a ¼ inch tube, Fig. 10. Slide cable core wire loop on mode control lever pin and snap flag into receiver on heater control. Install heater control assembly. Place mode control lever in extreme left position (OFF). Partially insert flag into receiver on distribution duct then slide self-adjust clip over door crank arm and snap flag into place. Adjust cable by moving mode control lever to extreme right (DEFROST).

1979 Truck Models Less A/C

Temperature Control Cable

Attach cable to heater control assembly and install into dash panel. Pull cable housing through dash panel into engine compartment until grommet is seated. Place temperature lever in "COOL" position. Place end loop on water valve crank pin and hold crank pin in the fully extended position. While pulling on cable, tighten cable housing clamp.

Model Control Cable

Attach cable to heater control assembly and install into dash panel. Move control lever to "OFF" position. Place cable end loop on mode door crank arm in its fully extended position, pull cable and snap it into retaining clip.

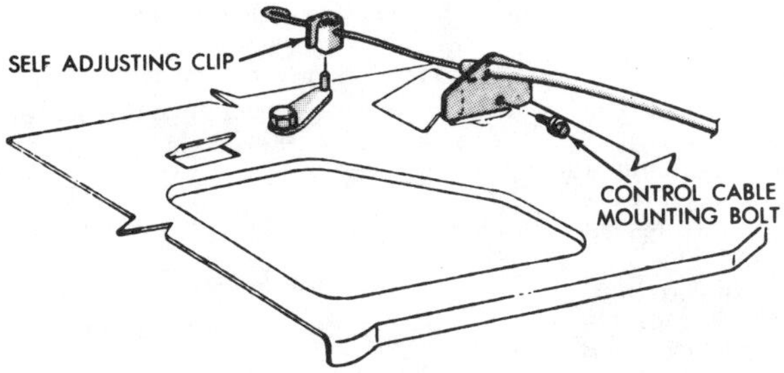

Fig. 16 Chrysler temperature cable adjustment at door end. 1976-78 Aspen, Diplomat, LeBaron & Volare

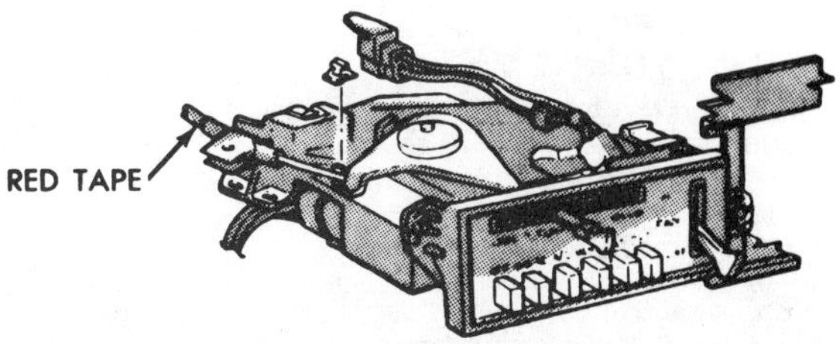

Fig. 17 Chrysler temperature control cable adjustment at control end. 1976-78 Aspen, Diplomat, LeBaron & Volare

1974-79 Truck Models With A/C

Place temperature control lever in Max. heat position. Position self-adjust clip on control cable 2 inches from end loop. Place clip on water valve crank pin and snap cable housing into retaining clip so that the locating tape is next to clip on side away from crank pin. If cable has no locating tape, install cable so housing extends ½ inch away from clip. Move control lever to "WARM" position to adjust clip.

Control Panel, Replace

PASSENGER CARS

1980 Cordoba & Mirada

1. Disconnect negative battery cable.
2. Remove center cluster bezel.
3. Disconnect ATC slide control wiring.

NOTE: The ATC slide control is located to the right of the wire retaining clip at the far left face of the ATC voltage control.

4. Remove control mounting screws.
5. Pull control from panel and disconnect all control cables, electrical leads and vacuum harness.
6. Remove control assembly from vehicle.
7. Reverse procedure to install.

1978-80 Horizon & Omni

1. Reach under instrument panel to disengage headlamp switch knob release button on switch and pull knob and shaft assembly from switch.
2. Remove lower left bezel attaching screws and bezel.
3. Remove control assembly mounting screws and pull control assembly rearward.
4. Disconnect wiring, cables and vacuum lines, then remove control assembly.
5. Reverse procedure to install.

1974 Barracuda & Challenger

1. Disconnect battery ground cable.
2. Remove control panel mounting bracket and the control assembly mounting screws.
3. Disconnect vacuum harness, electrical leads and control cables from control assembly.
4. Remove control panel from vehicle.
5. Reverse procedure to install.

1974 Charger, Coronet & Satellite

1. Disconnect battery ground cable.
2. Disconnect vacuum lines and electrical leads from control panel by reaching under instrument panel.
3. Remove support bracket mounting screw from rear of control panel.
4. Remove control panel mounting nuts from rear of panel.
5. Lower control panel from rear of instrument panel and disconnect control cables.
6. Remove control assembly from vehicle.
7. Reverse procedure to install.

1974-76 Dart & Valiant

1. Disconnect battery ground cable.
2. Remove radio.
3. Remove nuts securing control assembly from beneath instrument panel.
4. Remove control panel support bracket to radio speaker mounting stud nut.
5. Lower the control panel and disconnect electrical leads, control cables and vacuum hoses.
6. Remove control assembly from vehicle.
7. Reverse procedure to install.

1974-80 Chrysler, Dodge & Plymouth Full Size; 1976-80 Aspen, Diplomat, LeBaron & Volare

1. Disconnect battery ground cable.
2. Remove instrument cluster bezel and the control panel mounting screws.
3. Pull control panel from cluster and disconnect control cables, electrical leads and vacuum harness.
4. Remove control assembly from vehicle.
5. Reverse procedure to install.

1975-79 Charger, Cordoba, Coronet, Fury, Magnum & 1977-78 Monaco

1. Disconnect battery ground cable.
2. Remove instrument cluster lower bezel and the screws retaining the control panel to the cluster housing.
3. Pull control panel from cluster housing and disconnect control cables, electrical leads and vacuum harness. Disconnect harness from beneath panel.
4. Remove control assembly from vehicle.
5. Reverse procedure to install.

TRUCKS & VANS

1974-79 Truck Models

1. Disconnect fusible link in engine compartment.
2. Remove left air conditioner duct, if equipped.
3. Working under instrument panel, remove control assembly bracket mounting screw.
4. Remove control panel mounting screws, then lower control panel from dash far enough to disconnect cables, wiring and vacuum hose from control assembly.
5. Remove control assembly from under instrument panel.
6. Reverse procedure to install.

1978-80 Van Models

1. Disconnect battery.
2. Remove control panel bezel.
3. Remove control panel attaching screws.
4. Pull control assembly from dash and disconnect temperature control cable, wiring, vacuum lines and remove control assembly.
5. Reverse procedure to install.

1974-77 Van Models Less A/C

1. Disconnect fusible link located in engine compartment.
2. Remove instrument cluster screws and pull cluster out of dash panel far enough to disconnect speedometer cable and wiring. Remove cluster.
3. Loosen allen screws on cable knobs and remove knobs.
4. Working through cluster opening lower cables, remove panel mounting nuts and remove panel.
5. Reverse procedure to install.

1974-77 Van Models With A/C

1. Disconnect fusible link located in engine compartment.
2. Remove instrument cluster screws and pull cluster out of dash panel far enough to disconnect speedometer cable and wiring. Remove cluster.
3. Working through cluster opening, remove control assembly mounting nuts, lower assembly from under dash and disconnect wiring and vacuum hoses. Remove control assembly.
4. Reverse procedure to install.

Servo, Replace

1974-78 Auto-Temp

Removal

1. Disconnect battery ground cable.
2. Remove instrument panel lower assembly.
3. Disconnect servo electrical and vacuum connections.
4. Remove servo retaining screws, then the blend air door link and clip assembly from servo arm.
5. Remove screws from unit.

Installation

1. Slide self-adjust clip on wire toward the blend air door shaft until clip bottoms against link.
2. Position new servo in place and install self-adjust clip and wire assembly on servo arm pin.
3. Push servo assembly forward and install retaining screws.
4. Connect vacuum and electrical connections to servo, then the battery ground cable.
5. Start engine and push "Hi-Def" button. Observe servo arm movement while clip is self-adjusting.
6. When servo stops in the Max Heat position, turn off engine.
7. Ensure all connections are properly secured and install instrument panel lower assembly.

In-Car Sensor, Replace

1980 Cordoba, Diplomat, LeBaron & Mirada With S.A.T.C.

1. On Cordoba and Mirada models, remove center instrument panel bezel and radio.
2. On all models, remove glove box.
3. Disconnect in-car sensor assembly by pulling and slowly rotating the sensor assembly plastic housing to unsnap it from panel.

NOTE: Do not pull on sensor tube assembly.

4. Disconnect sensor assembly from aspirator tube, then separate wiring connector and remove sensor.
5. Reverse procedure to install.

1974-78 Auto-Temp

1. Disconnect battery ground cable.
2. Remove instrument panel lower assembly and the center air distribution duct.
3. Disconnect in-car sensor hose from aspirator, then the sensor electrical connector and remove hose from retaining clip.
4. Remove in-car sensor mounting screw, then the sensor.

S.A.T.C. Bi-Metal Sensor, Replace

1979 All & 1980 Full Size

1. Disconnect bi-metal control cable.
2. Remove retaining screws, vacuum connector, aspirator hose and sensor.

NOTE: Do not tear the compensator seal when removing sensor.

3. Reverse procedure to install. Position self-adjusting clip 1 inch from end of sensor cable.

Evaporator Core, Replace

1979 Magnum, 1979-80 Cordoba & Full Size, 1980 Gran Fury & Mirada

1. Disconnect battery ground cable and drain cooling system.
2. Discharge air conditioning refrigerant system.
3. Disconnect heater hoses from heater core.
4. Remove "H" valve and cap refrigerant lines. Remove condensate drain tube.
5. Disconnect vacuum lines from engine compartment and push grommet and vacuum lines through dash panel.

NOTE: On manual air conditioning systems, the vacuum lines are connected to the manifold vacuum tee and water valve. On semi-automatic air conditioning systems, the vacuum lines are connected to the vacuum reservoir and water valve.

6. Remove four nuts attaching evaporator heater assembly to dash panel.
7. Move front seat rearward and remove console, if equipped.
8. On models except Cordoba, Magnum and Mirada, remove air conditioning switch control from dash panel and disconnect vacuum harness from harness extension.
9. On all models, remove ash tray and housing and the glove box assembly.
10. Disconnect right hand lap cooler tube from lap cooler and remove the right hand trim panel or bezel.
11. Remove right hand cowl trim pad and disconnect blower motor electrical connections.
12. On Cordoba, Magnum and Mirada, disconnect vacuum harness from harness extension, then remove heater distribution duct.
13. On all models, disconnect temperature control cable from evaporator heater housing.
14. On models except Cordoba, Magnum and Mirada, remove heater distribution housing and the center distribution duct.
15. On all models, remove the hi/lo door actuator, Figs. 18 & 19.
16. On models with semi-automatic air conditioning system, remove vacuum servo actuator from housing and the in-car air hose from compensator.
17. On all models, support housing and remove brace between housing and plenum. Then, pull housing rearward and move toward right side to remove.
18. Remove housing top cover and the evaporator core retaining screw.
19. Lift evaporator core from housing.
20. Reverse procedure to install.

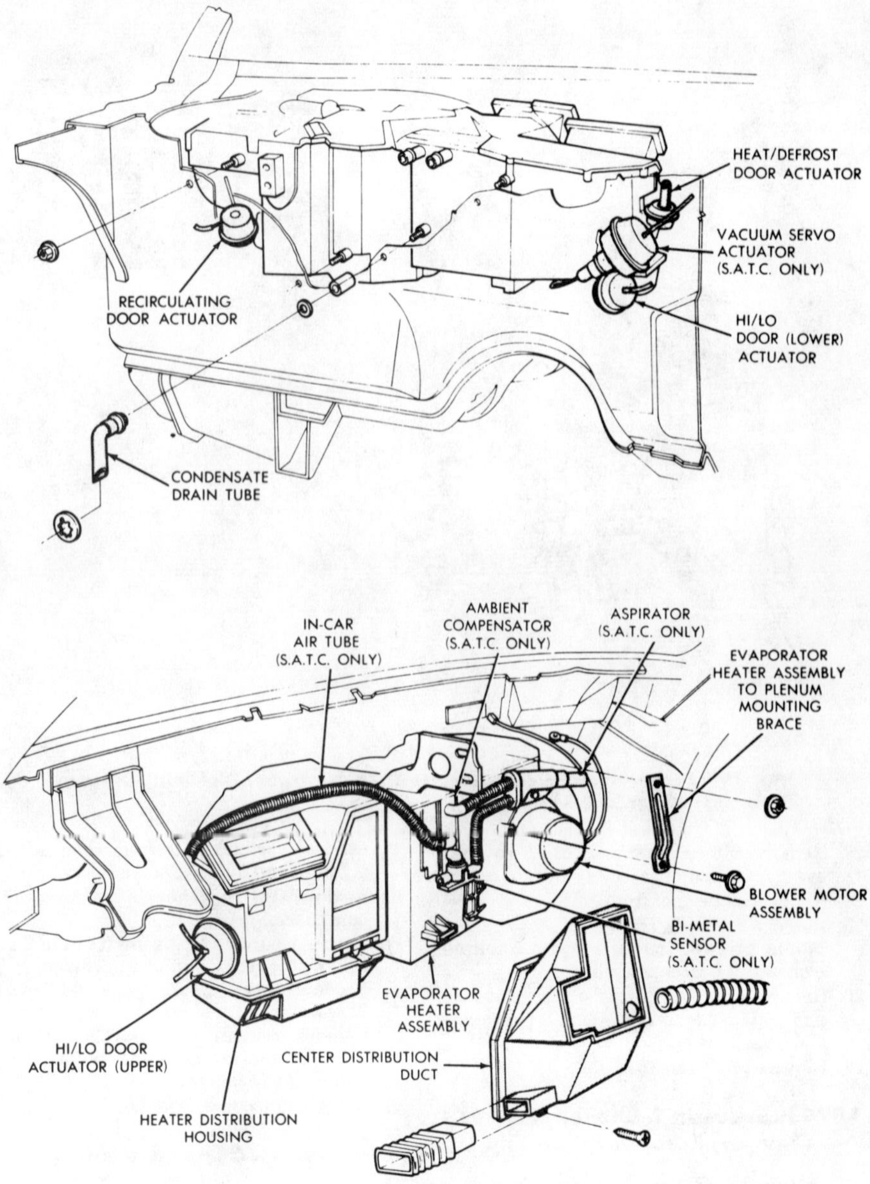

Fig. 18 Evaporator-heater assembly. 1979 Cordoba, Magnum; 1979-80 Gran Fury, Newport, New Yorker & St. Regis

1978-80 Horizon & Omni

1. Disconnect ground cable, drain radiator and discharge refrigerant system.
2. Disconnect temperature door cable from evaporator heater assembly.
3. Remove glove box.
4. Disconnect vacuum harness from control panel and the blower motor feed wire and anti-diesel relay wires.
5. Remove right hand bezel.
6. Remove central air duct cover from central air distribution duct.
7. Remove three screws retaining central air distribution duct, then the duct.
8. Remove defroster duct adapter.
9. Remove "H" valve and disconnect heater hoses. Plug heater core tube openings.
10. Remove condensate drain tube from evaporator heater unit.
11. Disconnect vacuum lines at intake manifold and water valve.
12. Remove evaporator heater unit to dash panel retaining nuts, Fig. 20.
13. Remove panel support bracket.
14. Remove right side cowl lower bracket.
15. Remove instrument panel pivot bracket screw from right side.
16. Remove screws securing lower instrument panel at steering column.
17. Pull back carpet from under evaporator heater assembly as far as possible.
18. Remove nut from A/C to plenum mounting brace and blower motor ground cable.
19. Support unit and remove brace from stud.
20. Lift evaporator heater unit and pull rearward to clear dash panel and liner. Also, the lower instrument panel will have to be pulled rearward to provide adequate clearance for unit removal. Slowly lower the evaporator heater unit, slide rearward and remove from vehicle.
21. Remove ¼-20 nut from mode door actuator arm on top cover and two retaining clips from front edge of cover.
22. Remove mode door actuator attaching screws and the actuator.

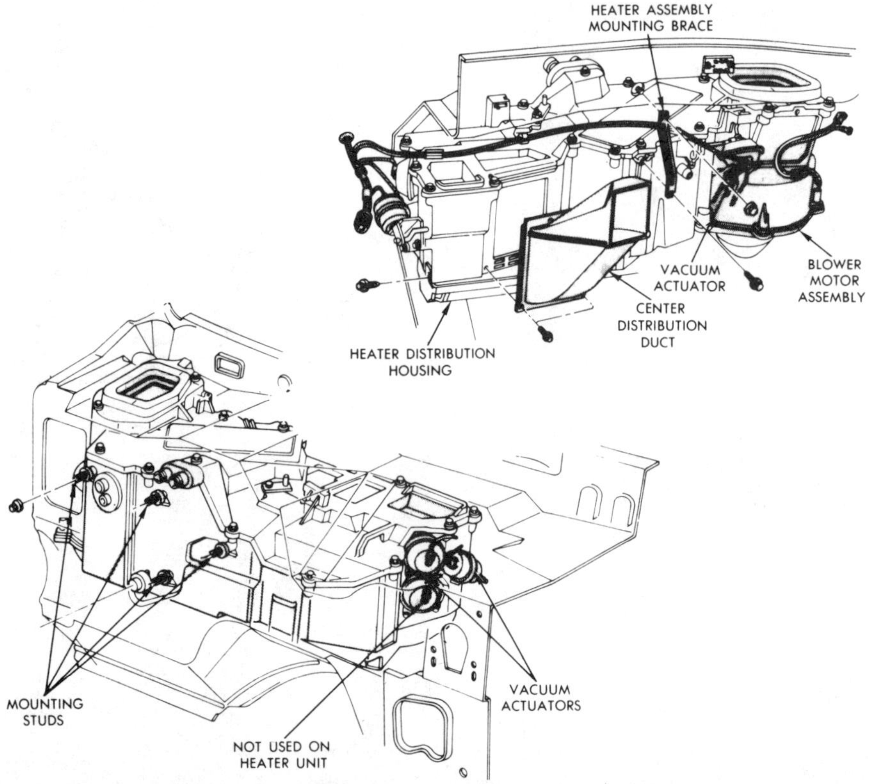

Fig. 19 Evaporator-heater assembly. 1976-80 Aspen, Diplomat, LeBaron, Volare; 1980 Cordoba & Mirada

23. Remove fifteen screws securing cover to evaporator heater assembly and the cover.
24. Remove mode door from unit.
25. Remove screw from heater core tube retaining bracket, then the heater core from evaporator heater assembly.
26. Remove expansion valve sealing plate seal and seal bracket screw, then lift evaporator coil from housing.
27. Reverse procedure to install.

1974 Barracuda & Challenger; 1974 Charger, Coronet & Satellite

1. Disconnect battery ground cable and remove air cleaner.
2. Drain cooling system and disconnect heater hoses. Plug heater core tubes to prevent spilling coolant when unit is removed.
3. Discharge refrigerant system and disconnect refrigerant lines at dash panel. Use two wrenches to loosen each connection and plug all lines to prevent entry of dirt.
4. Disconnect blower motor wires, then remove blower motor cooling tube and blower motor.
5. Remove glove box, appearance shield from lower edge of instrument panel, left spot cooler and air distribution duct.
6. Disconnect wires from resistor block, then disconnect antenna lead and remove radio.
7. Disconnect vacuum hose harness from control switch and disconnect water valve cable from bracket on left end of housing.
8. Remove stud nuts from engine compartment side, then remove support bracket from housing to plenum panel and remove unit.
9. Carefully remove plenum air seal and disconnect vacuum hose from inlet air door actuator and bypass door actuator.
10. Remove seal from heater and evaporator core tubes.
11. Remove screws retaining front and rear halves together and one screw from between evaporator core tubes and separate housings.
12. Remove evaporator access plate, then remove the four evaporator retaining screws and remove evaporator.
13. Reverse procedure to install.

1974-76 Dart & Valiant

1. Discharge refrigerant from system and disconnect battery ground cable.
2. Disconnect liquid and suction line from evaporator. Use two wrenches to loosen each connection. Cap all open lines to prevent entry of dirt.
3. Remove distribution duct and glove box, then disconnect support bracket from rear of housing plenum.
4. If used, remove courtesy light and relay from right bottom edge of instrument panel.
5. Remove right kick-pad and pivot bolt from right lower corner of panel and pull panel slightly rearward.
6. Disconnect vacuum lines from actuator and electrical connector from resistor block.
7. Disconnect drain tube and leave connected to firewall, then remove the four retaining nuts and remove unit by pulling rearward.
8. Remove fresh air inlet duct retaining screws, then remove the nine screws retaining cover and remove cover.
9. Reverse procedure to install using new O-rings coated with refrigerant oil on all connections.

1975-78 Charger, Cordoba, Coronet, Fury, Magnum & 1977-78 Monaco

1. Remove air cleaner and disconnect battery ground cable.
2. Drain cooling system and disconnect heater hoses. Plug heater core tubes to prevent spilling coolant when unit is removed.
3. Discharge refrigerant from system.
4. Disconnect line from H-valve and cover sealing plate, then remove expansion valve and cover both sealing surfaces of valve.
5. Disconnect blower motor wires and remove blower motor cooling tube.
6. Remove glove box, ash tray and housing, appearance shield, right spot cooler, right cowl panel and air distribution duct.
7. Remove A/C mode door vacuum actuator from mounting bracket and shift actuator 90° forward on top of unit.
8. Disconnect wiring from seat belt interlock unit and from resistor block, then disconnect antenna and remove radio.
9. Disconnect vacuum harness from extension to control, then from engine compartment, remove the housing stud nuts and remove drain tube.
10. Remove bracket from rear unit to plenum, then pull unit so that tubing clears dash panel and rotate so that right end of unit comes out first from under instrument panel.
11. Remove air seal and disconnect vacuum from air door actuator.
12. Remove seal from core tubes, then remove clamps and screws holding housing together and separate halves.
13. Remove access plate, then remove evaporator core retaining screws and remove evaporator core.
14. Reverse procedure to install.

1974-78 Chrysler, Dodge & Plymouth Full Size Models, Fig. 21

1. Discharge refrigerant from system.
2. Disconnect battery ground cable, drain cooling system and remove air cleaner.
3. Disconnect heater hoses. Plug heater core tubes to prevent spilling coolant when unit is removed.
4. Remove H-valve. Refer to "UNIT REPLACEMENT" Section 1.
5. Remove distribution duct, disconnect antenna lead, wires and vacuum hoses from unit and remove drain tube.
6. On Auto-Temp units:
   (a) Disconnect wires and vacuum connector from servo.
   (b) Disconnect wires from amplifier.
   (c) Disconnect wires and vacuum hoses from master and compressor switches.
   (d) Disconnect aspirator tube.
7. Disconnect control cable from clip on unit and remove retaining nut from support bracket.
8. Through engine compartment, remove the six retaining stud nuts, then roll or tip housing out from under instrument panel.
9. Remove air door levers from shafts, then remove retaining screws and lift off top cover.
10. Remove evaporator core retaining screws and remove evaporator core.
11. Reverse procedure to install.

1976-80 Aspen, Diplomat, LeBaron & Volare, Fig. 19

1. Discharge refrigerant from system.
2. Disconnect battery ground cable, drain cooling system and remove air cleaner.
3. Disconnect heater hoses. Plug both heater core tubes to prevent spilling coolant when unit is removed.
4. Remove "H" valve. Refer to "UNIT REPLACEMENT" Section 1.
5. Remove cluster bezel assembly, instrument panel upper cover, steering column cover, right intermediate side cowl trim, lower instrument panel and instrument panel center to lower reinforcement.
6. If used, remove floor console.
7. Remove right center air distribution duct and locking tab on defroster duct.
8. Disconnect control cable from evaporator housing, blower motor resistor block wiring and vacuum lines from water valve and vacuum source "Tee".
9. Remove wiring from evaporator housing and vacuum lines from inlet air housing and disconnect vacuum harness coupling.
10. Remove rubber drain tube and housing mounting stud nuts from engine compartment side.
11. Remove hanger strap from evaporator plenum stud.
12. Roll evaporator so that tubing clears firewall and remove unit from vehicle.
13. Remove blend air door lever from shaft, then remove screws around top of unit and remove top cover.
14. Remove evaporator core by lifting out.
15. Reverse procedure to install.

TRUCKS & VANS

1974-79 Truck Models, Fig. 22

1. Discharge refrigerant from system.
2. Disconnect battery ground cable.
3. Disconnect refrigerant lines from evaporator at dash panel. Plug all open lines to prevent entry of dirt and moisture into system.
4. Remove glove box, ash tray, left A/C duct, right A/C duct, distribution duct and center outlet duct.
5. Disconnect blower wiring from resistor.
6. Remove vacuum lines from rear half of housing, then remove housing drain tubes from engine side of dash.
7. Remove 22 screws from rear housing. Remove 2 plugs from housing and remove 2 screws located inside using an extension.
8. Holding defroster door to heat position, remove rear half of housing.
9. Remove evaporator core retaining screws from engine compartment and each side of core.
10. Remove evaporator core from housing.
11. Reverse procedure to install. Use new O-rings lubricated with refrigerant oil.
12. Evacuate, charge and leak test system. Check system for proper operation.

1979-80 Van Models, Fig. 23

1. Discharge refrigerant from system and disconnect battery ground cable.
2. Disconnect heater hoses from heater core.

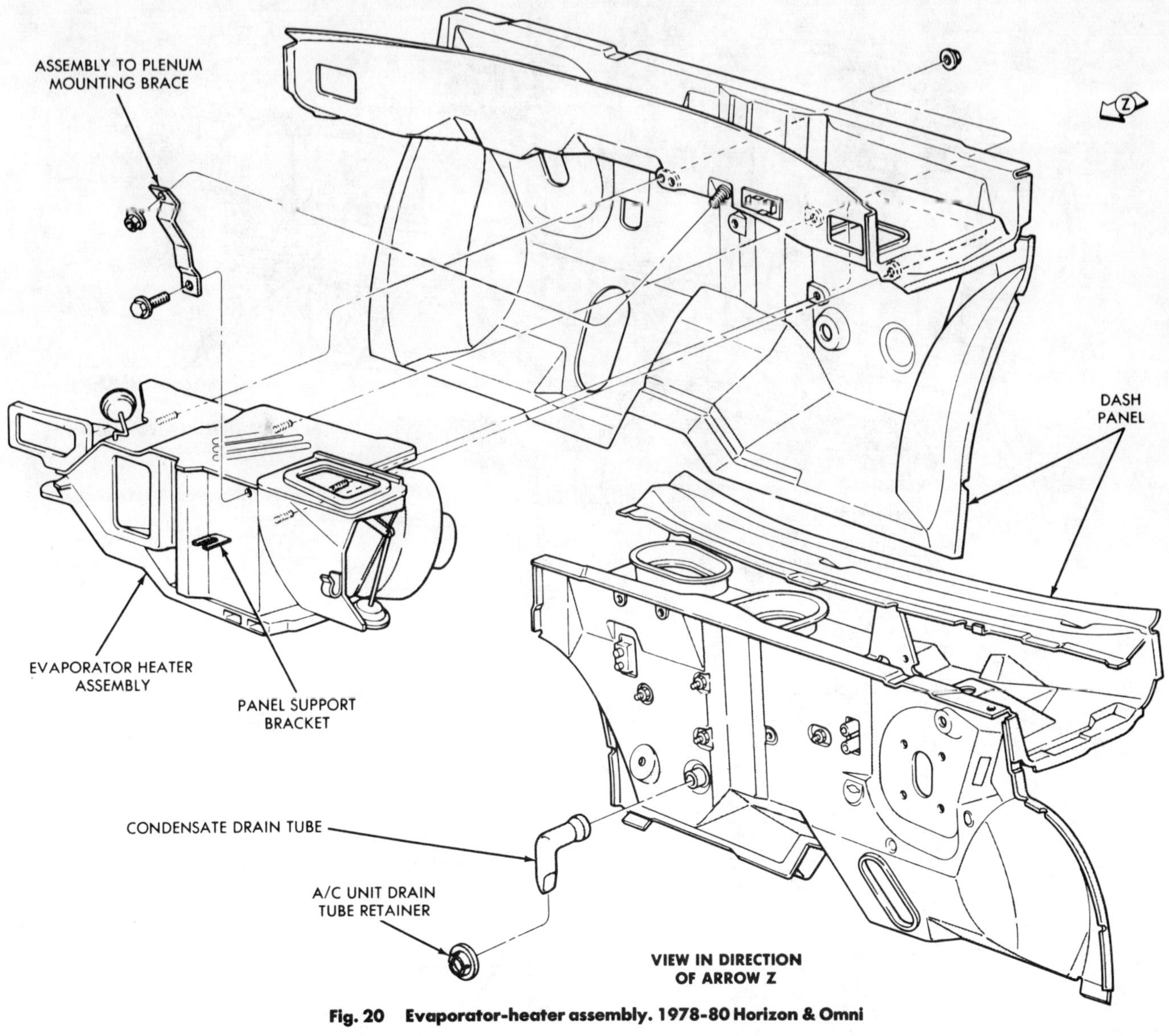

Fig. 20 Evaporator-heater assembly. 1978-80 Horizon & Omni

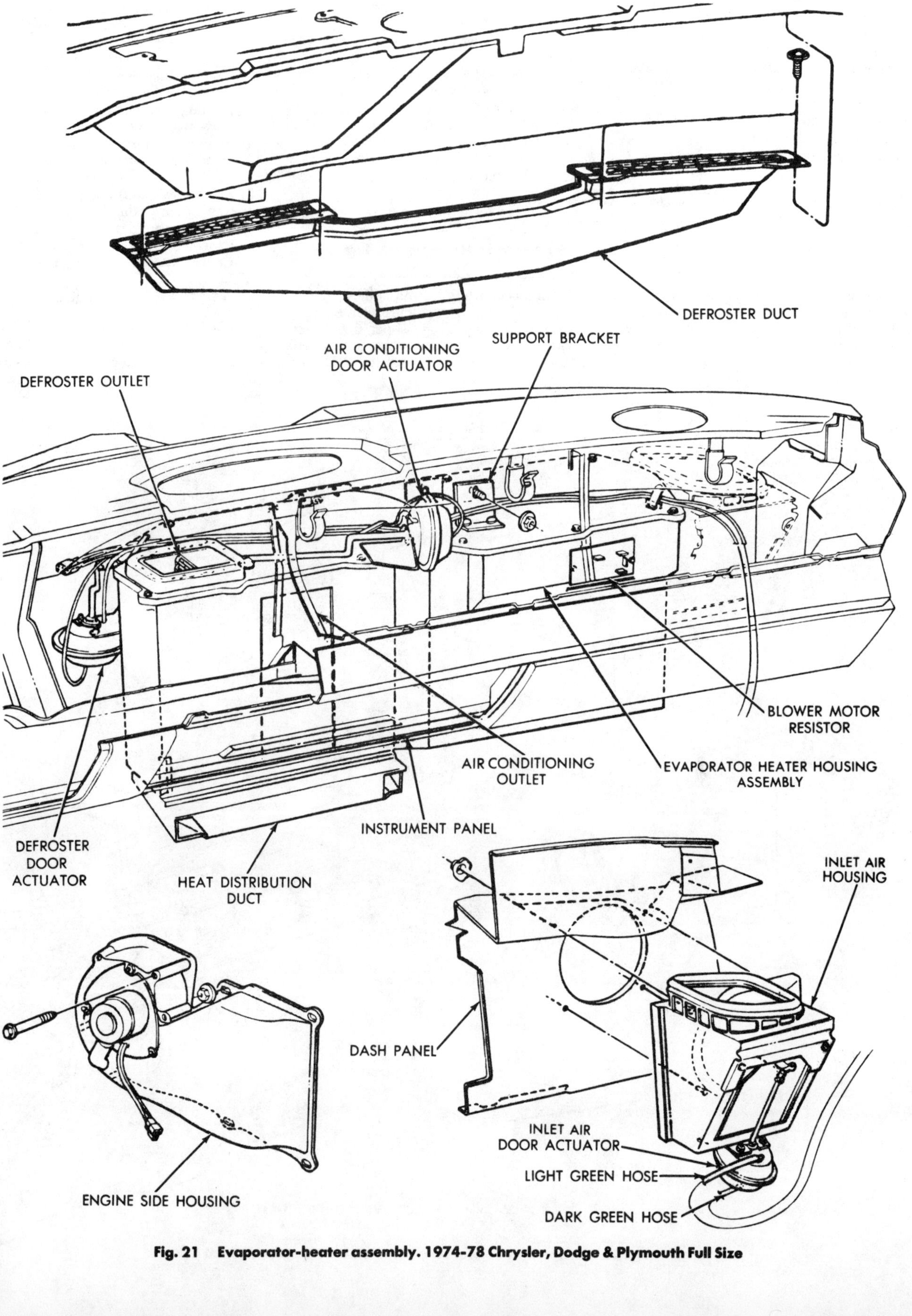

Fig. 21 Evaporator-heater assembly. 1974-78 Chrysler, Dodge & Plymouth Full Size

NOTE: Prevent coolant from spilling onto alternator.

3. Remove H-valve, then remove receiver-drier bracket screws and swing refrigerant plumbing aside. Plug all refrigerant lines to prevent entry of dirt and moisture into system.
4. Remove temperature cable from cover.
5. Remove glove box, spot cooler bezel and appearance shield.
6. Working through glove box opening, and under instrument panel, remove evaporator housing mounting screws and nuts.
7. Remove 2 screws attaching blower housing to evaporator housing flange connection and remove evaporator housing from vehicle.
8. Place housing on bench and remove cover to gain access to evaporator core.
9. Remove one screw from strap on heater core tubes and pull core out of housing.
10. Remove screw from under plumbing attachment plate and pull evaporator from housing.
11. Reverse procedure to install. Evacuate, charge and leak test system. Check system for proper operation.

1974-78 Van Models, Fig. 24

1. Discharge refrigerant from system and disconnect battery ground cable.
2. Remove grille, condenser and radiator from vehicle as an assembly.
3. Disconnect heater hoses and refrigerant lines from evaporator-heater housing. Plug all refrigerant lines to prevent entry of dirt and moisture into system.

NOTE: Prevent coolant from spilling onto alternator.

4. Remove glove box, spot cooler bezel and appearance shield.
5. Working through glove box opening, and under instrument panel, remove evaporator housing to dash panel mounting nuts and bolts.
6. Remove the wiper motor. Disconnect all hoses and wiring from evaporator housing.
7. Remove the crossbar bolts and front seal plate bolts. Separate evaporator housing from blower housing and remove from vehicle.
8. Place housing on bench. Carefully pry heater core out of housing.

NOTE: Do not remove air seal from heater core.

9. Remove evaporator mounting screws and housing attaching screws and separate housing.
10. Remove evaporator core from housing.
11. Reverse procedure to install. Use new O-rings lubricated with refrigerant oil.
12. Evacuate, charge and leak test system. Check for proper operation.

1976-80 Van Auxiliary Evaporator

1. Discharge refrigerant from system.
2. Working under vehicle, disconnect refrigerant lines from evaporator. Plug all open lines to prevent entry of dirt and moisture into system.
3. Remove housing lower cover, refrigerant line seal and cover plate.
4. Remove evaporator coil.
5. Reverse procedure to install. Evacuate, charge and leak test system. Check for proper operation.

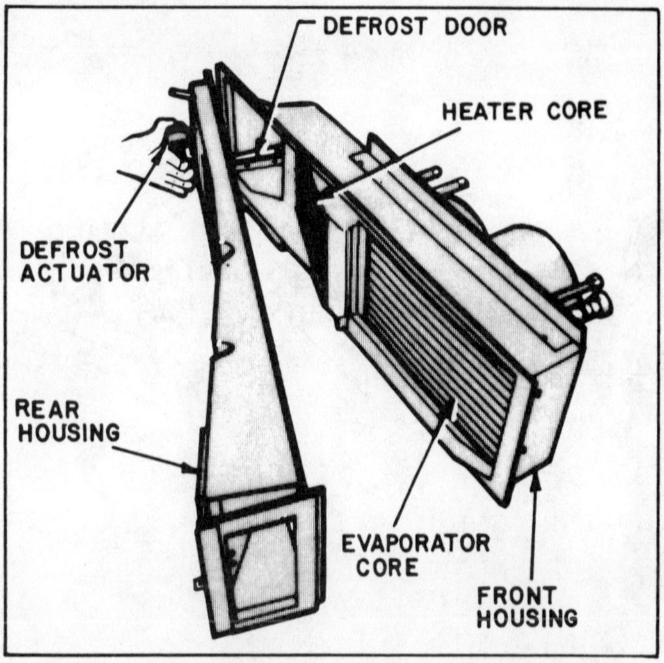

Fig. 22 Evaporator-heater assembly. 1974-79 Truck models

Heater Core, Replace

PASSENGER CARS

1979 Magnum; 1979-80 Cordoba & Full Size, 1980 Gran Fury & Mirada

Less Air Conditioning, Fig. 25

1. Disconnect battery ground cable and drain cooling system.
2. Disconnect heater hoses from heater core.
3. Disconnect vacuum hoses from water valve and manifold tee, then push hoses and grommet through dash panel.
4. Remove four nuts securing heater assembly to dash panel.
5. Move front seat rearward and remove console, if equipped.

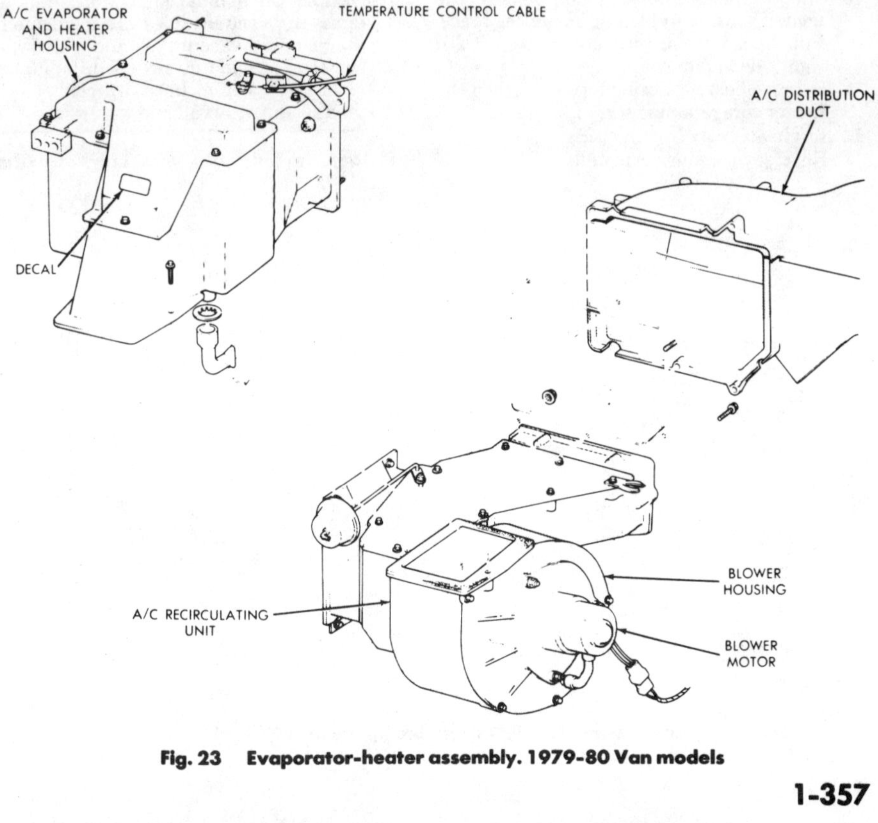

Fig. 23 Evaporator-heater assembly. 1979-80 Van models

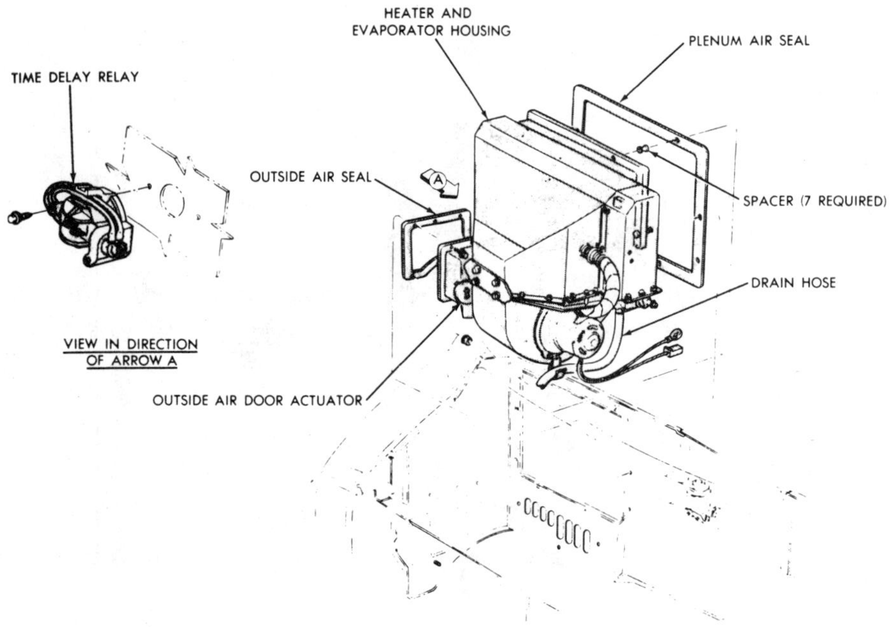

Fig. 24 Evaporator-heater assembly. 1974-78 Van models

6. On models except Cordoba and Magnum, remove heater control and disconnect vacuum harness from extension harness.
7. On all models, remove glove box, ash tray and housing, right hand lap cooler duct, lower right hand trim panel or bezel, and the right hand cowl trim pad.
8. Disconnect blower motor electrical connections.
9. Disconnect temperature control cable from heater housing, then remove heater distribution housing.
10. Support heater housing and remove the heater housing to plenum mounting brace. Pull housing rearward and move toward right side to remove.
11. Remove heater housing top cover, then the heater core retaining screw.
12. Lift heater core from housing.
13. Reverse procedure to install.

With Air Conditioning, Fig. 18

1. Disconnect battery ground cable and drain cooling system.
2. Discharge air conditioning refrigerant system.
3. Disconnect heater hoses from heater core.
4. Remove "H" valve and cap refrigerant lines. Remove condensate drain tube.
5. Disconnect vacuum lines from engine compartment and push grommet and vacuum lines through dash panel.

NOTE: On manual air conditioning systems, the vacuum lines are connected to the manifold vacuum tee and water valve. On semi-automatic air conditioning systems, the vacuum lines are connected to the vacuum reservoir and water valve.

6. Remove four nuts attaching evaporator heater assembly to dash panel.
7. Move front seat rearward and remove console, if equipped.
8. On models except Cordoba and Magnum, remove air conditioning switch control from dash panel and disconnect vacuum harness from harness extension.
9. On all models, remove ash tray and housing and the glove box assembly.
10. Disconnect right hand lap cooler tube from lap cooler and remove the right hand trim panel or bezel.
11. Remove right hand cowl trim pad and disconnect blower motor electrical connections.
12. On Cordoba and Magnum, disconnect vacuum harness from harness extension, then remove heater distribution duct.
13. On all models, disconnect temperature control cable from evaporator heater housing.
14. On models except Cordoba and Magnum, remove heater distribution housing and the center distribution duct.
15. On all models, remove the hi/lo door actuator.
16. On models with semi-automatic air conditioning system, remove vacuum servo actuator from housing and the in-car air hose from compensator.
17. On all models, support housing and remove brace between housing and plenum. Then, pull housing rearward and move toward right side to remove.
18. Remove housing top cover and the heater core retaining screw.
19. Lift heater core from housing.
20. Reverse procedure to install.

1978-80 Horizon & Omni

Less Air Conditioning, Fig. 26

1. Disconnect battery ground cable and drain cooling system.
2. Remove center outside air floor vent housing.
3. Remove ash tray.
4. Remove two screws retaining defroster duct adapter to heater unit.

NOTE: The left hand screw is accessible through the ash tray housing access hole.

5. Remove the adapter and push the flexible connector upward and position aside.
6. Disconnect temperature control cable.
7. Remove glove box and door assembly.
8. Disconnect blower motor wiring connector.
9. Disconnect heater hoses and plug heater core tube openings.
10. Remove two nuts retaining heater unit to dash panel.
11. Remove screw attaching heater brace bracket to instrument panel.
12. Remove heater unit support strap nut. Disconnect strap from plenum stud and lower heater unit from panel.
13. Disconnect mode control cable and remove heater unit from vehicle.
14. Remove left heater outlet duct.
15. Remove four screws from blower motor mounting plate, then the blower motor from heater unit.
16. Remove defroster duct adapter.
17. Remove outside air and defroster door cover.
18. Remove screw from center of defroster door, then the door, screw and screw re-

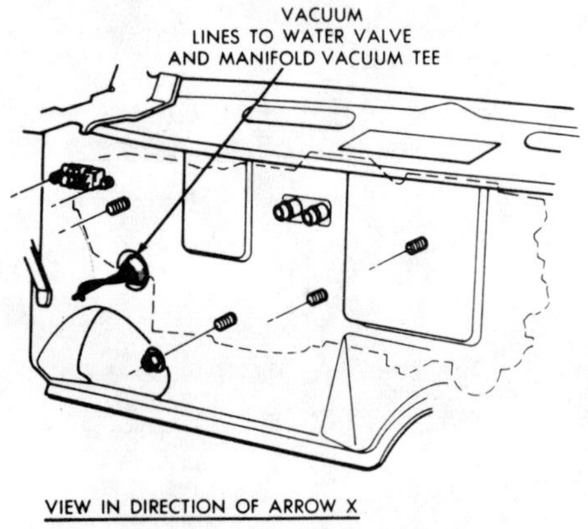

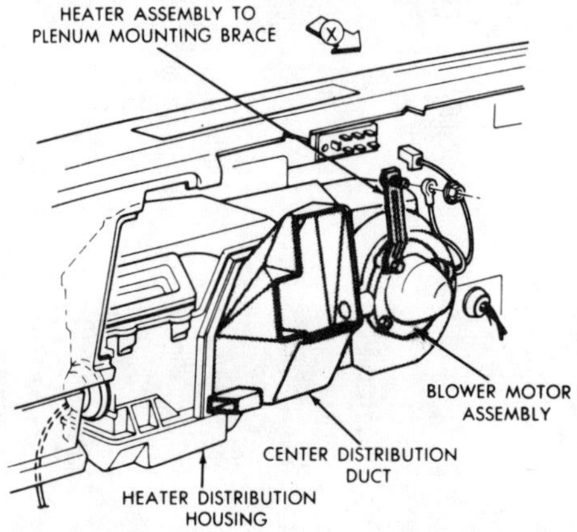

Fig. 25 Heater assembly. 1979 Cordoba, Magnum; 1979-80 Gran Fury, Newport, New Yorker & St. Regis

taining plate.
19. Remove defroster door control rod from heater unit.
20. Remove heater core cover.
21. Slide heater core from heater unit.
22. Reverse procedure to install.

With Air Conditioning, Fig. 20

1. Disconnect ground cable, drain radiator and discharge refrigerant system.
2. Disconnect temperature door cable from evaporator heater assembly.
3. Remove glove box.
4. Disconnect vacuum harness from control panel and the blower motor feed wire and anti-diesel relay wires.
5. Remove right hand bezel.
6. Remove central air duct cover from central air distribution duct.
7. Remove three screws retaining central air distribution duct, then the duct.
8. Remove defroster duct adapter.
9. Remove "H" valve and disconnect heater hoses. Plug heater core tube openings.
10. Remove condensate drain tube from evaporator heater unit.
11. Disconnect vacuum lines at intake manifold and water valve.
12. Remove evaporator heater unit to dash panel retaining nuts.
13. Remove panel support bracket.
14. Remove right side cowl lower bracket.
15. Remove instrument panel pivot bracket screw from right side.
16. Remove screws securing lower instrument panel at steering column.
17. Pull back carpet from under evaporator heater assembly as far as possible.
18. Remove nut from A/C to plenum mounting brace and blower motor ground cable.
19. Support unit and remove brace from stud.
20. Lift evaporator heater unit and pull rearward to clear dash panel and liner. Also, the lower instrument panel will have to be pulled rearward to provide adequate clearance for unit removal. Slowly lower the evaporator heater unit, slide rearward and remove from vehicle.
21. Remove ¼-20 nut from mode door actuator arm on top cover and two retaining clips from front edge of cover.
22. Remove mode door actuator attaching screws and the actuator.
23. Remove fifteen screws securing cover to evaporator heater assembly and the cover.
24. Remove mode door from unit.
25. Remove screw from heater core tube retaining bracket, then heater core from evaporator heater assembly.
26. Reverse procedure to install.

1974 Barracuda, Challenger, Charger, Coronet & Satellite

Less Air Conditioning

1. Remove three mounting nuts from studs around blower motor and remove flange and air seal.
2. Unplug antenna from radio and place to one side.
3. Remove screw from housing to plenum support rod on right side of housing above fresh air opening.
4. Disconnect three air door cables.
5. Disconnect wires from blower resistor.
6. Tip unit down and out from under panel.

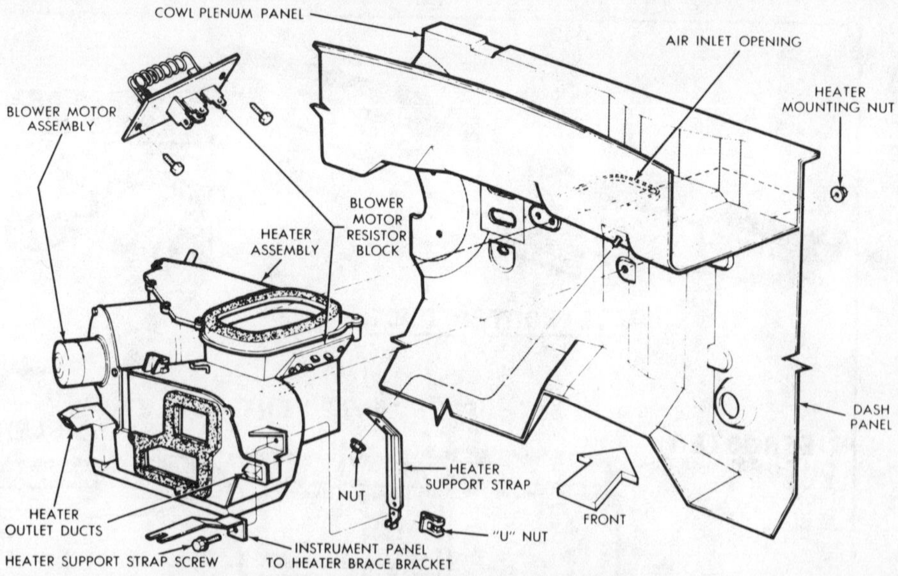

Fig. 26 Heater assembly. 1978-80 Horizon & Omni

With Air Conditioning

1. Remove air cleaner and slowly discharge refrigerant from system.
2. Disconnect refrigerant lines at dash panel. Leave expansion valve attached to line. Cap all lines.
3. Disconnect blower motor wires and remove motor cooling tube and motor.
4. Remove glove box and appearance shield from lower edge of dash panel.
5. Remove left spot cooler duct and air distribution duct.
6. Disconnect wires from blower resistor and antenna wire from radio.
7. Remove radio.
8. Disconnect vacuum harness from back of control switch.
9. Remove water valve cable from bracket on left end of housing.
10. Remove nuts from housing mounting studs in engine compartment.
11. Remove rubber drain tube.
12. Remove support bracket from housing plenum panel.
13. Carefully remove plenum air seal.
14. Disconnect vacuum hose from inlet air door actuator and by-pass door actuator.
15. Remove air seal from heater and evaporator core tubes.
16. Remove 18 screws holding front and rear covers together and one screw from between core tubes. Separate housings.
17. Remove 3 screws from evaporator core access plate, remove plate, this provides access to two evaporator core mounting screws.
18. Remove four screws holding evaporator core to front cover and remove core. Carefully lift left half of housing seal from rear cover. Do not remove entire seal as bottom portion is a water seal.
19. Remove two core retaining screws from mounting plates and one from between core tubes in back of rear cover. Lift core out of housing.

1974-76 Dart & Valiant

Less Air Conditioning

1. Remove heater hoses to dash panel seal and retainer plate.
2. Remove heater motor seal retainer plate and seal from dash panel.
3. Disconnect control cables from heater.
4. Remove heater motor resistor wire from resistor.
5. Remove defroster tubes from heater.
6. Disconnect heater housing support rod from outside air duct.
7. Remove heater assembly.

With Air Conditioning

1. Remove battery, drain radiator and disconnect heater from unit.
2. Remove core tube seal nut, bracket and seal.
3. Remove A/C duct, ash tray and housing and radio.
4. Remove heat-defrost vacuum actuator pot; let hang by rod.
5. To remove heat distribution duct, remove three screws on front cover, two on each end and work housing out of lip and remove to the left side.
6. Remove left defroster duct. Remove right defroster duct and let it hang from top.
7. To remove the rear distribution housing, reach through radio opening and remove three screws top and bottom and one at left end.
8. With housing off, core will be settling loose. Separate seal and lift out.

1974-78 Chrysler, Dodge, Plymouth Full Size

Less Air Conditioning, Fig. 27

1. Disconnect battery and drain radiator.
2. Disconnect heater hoses and plug hose fittings on heater core.
3. Move front seat rearward and remove instrument panel lower cover. Unplug antenna lead from radio and disconnect upper level vent vacuum line.
4. Disconnect the upper level vent from heater housing and the mounting bracket

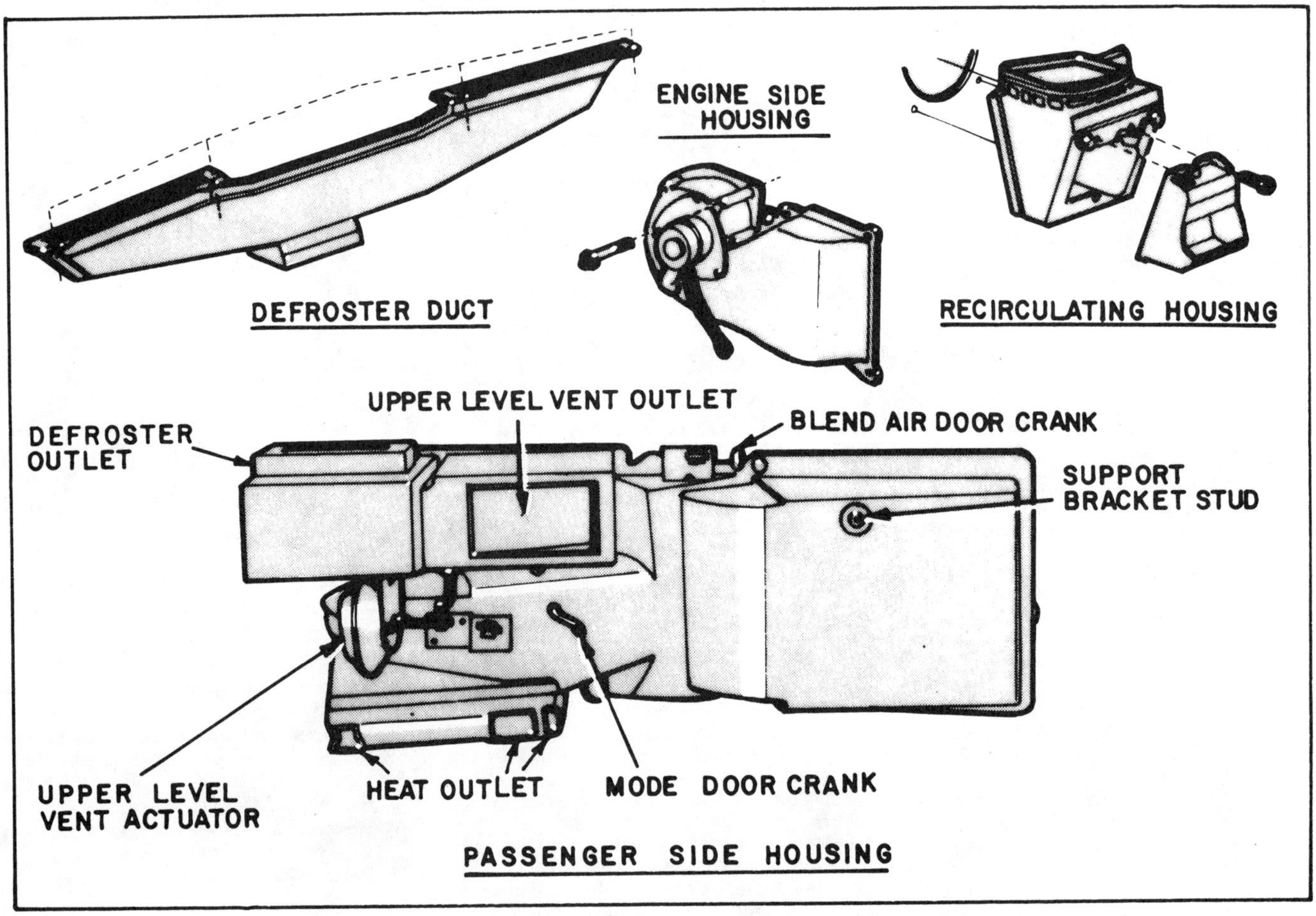

Fig. 27 Heater assembly. 1974-78 Chrysler, Dodge & Plymouth Full Size

from dash, then swing duct back.

5. Disconnect electrical connectors from blower motor resistor and control cable from clip and crank.
6. Disconnect support bracket and swing bracket out of way.
7. From engine compartment, remove five retaining nuts from studs.
8. Roll or tip housing out from under instrument panel and disconnect control cable from clip and the blend air door crank.
9. Remove core tube locating screw from between core tubes, remove housing retaining nuts, separate housing and remove core.

With Air Conditioning, Fig. 21

1. Discharge refrigerant system.
2. Disconnect fusible link, drain cooling system, remove air cleaner and disconnect heater hoses at heater core. Install plugs in heater core tube to prevent coolant spillage when removing heater core.
3. Remove "H" valve, then cap refrigerant lines to prevent entry of dirt and moisture.
4. Move front seat rearward to provide room to remove unit.
5. Remove lap cooler and instrument panel lower cover.
6. Remove A/C distribution duct.
7. Disconnect antenna lead from radio and wires and vacuum lines from unit.
8. Remove drain tube. On models with ATC, disconnect wire connectors and vacuum connector from servo, amplifier and compressor switches. Disconnect aspirator tube.
9. On all models, remove temperature control cable from clip on unit.
10. Remove retaining nut from support bracket.
11. From engine side of dash panel, remove retaining nuts from studs.
12. Tilt housing and move from under instrument panel.
13. Remove mode door and blend air door levers.
14. Remove attaching screws and lift off top cover.
15. Remove four heater core retaining screws and three screws from core tube seal, then lift heater core from housing.
16. Reverse procedure to install.

1975-78 Charger, Cordoba, Coronet, Fury, Magnum & 1977-78 Monaco

Less Air Conditioning

1. Disconnect fusible link and drain coolant.
2. Disconnect heater hoses from core tubes and plug tubes to prevent coolant from spilling into interior of vehicle.
3. Remove nuts from around blower motor and one from center of dash panel, then remove lower instrument panel bezel, glove box and door.
4. Disconnect antenna lead from radio, then remove screw from housing to plenum support rod on right side of housing above outside air opening.
5. Disconnect air door control cables and wire from blower motor resistor.
6. Tip heater assembly down and out from under instrument panel.
7. Remove screws retaining front cover to heater assembly, then cut rubber seal in two places where front cover separates cover from housing.
8. Remove core tube retaining screw located behind housing and between core tubes.
9. Remove sponge rubber gaskets from core tubes and remove heater core from housing.

With Air Conditioning

1. Disconnect fusible link and drain coolant.
2. Disconnect heater hoses from core tubes and plug tubes to prevent coolant from spilling into interior of vehicle.
3. Discharge A/C refrigerant and disconnect line from H-valve and cover tubing sealing plate. Remove expansion valve from evaporator and cover evaporator sealing plate and both sealing surfaces of expansion valve.
4. Disconnect blower motor wires and remove blower motor cooling tube.
5. Remove glove box, ash tray and housing

bezel, right lap cooler, right side cowl panel and air distribution duct.
6. Remove A/C mode door vacuum actuator from bracket and shift actuator forward on top of unit.
7. Disconnect connector from seat belt interlock control module and disconnect wires from blower motor resistor.
8. Disconnect antenna lead from radio and remove radio.
9. Remove housing retaining nuts through engine compartment, then remove rubber drain tube.
10. Remove support bracket from rear of unit to plenum.
11. Pull unit back so that tubes clear dash panel, then rotate so that right end of unit comes out from under the instrument panel first.
12. Carefully remove plenum air seal, then disconnect vacuum hose from inlet air door actuator.
13. Remove the 14 clamps and 4 screws and separate housings.
14. Carefully lift left half of housing seal from rear cover. Do not remove entire seal as the lower portion is a water seal.
15. Remove two retaining screws from mounting plate and one from between heater core tubes and lift heater core from housing.

1976-80 Aspen, Diplomat, LeBaron & Volare

Less Air conditioning, Fig. 28

1. Disconnect battery ground cable and drain cooling system.
2. Disconnect and plug heater hoses from dash panel.
3. Remove heater core tube dash panel seals and retainer.
4. Remove instrument cluster bezel assembly, upper cover, steering column cover, right intermediate side cowl trim panel and the lower instrument panel.
5. Remove instrument panel center to lower reinforcement.
6. Remove right vent control cable from unit.
7. Disconnect temperature and mode door control cables from unit, then the blower motor resistor block wiring.
8. Remove heater assembly mounting nuts in engine compartment.
9. Remove heater support to plenum bracket and pull heater unit from dash panel.
10. Separate heater housing by removing retainer clips.
11. Remove heater core tube support clamp and slide heater core from housing.
12. Reverse procedure to install.

With Air Conditioning, Fig. 19

1. Remove evaporator assembly as outlined under "Evaporator Core, Replace".
2. Remove blend air door lever from shaft.
3. Remove top cover screws and the cover.
4. Remove heater core from housing.

TRUCKS & VANS

1979-80 Van Models, Fig. 23

With Air Conditioning

1. Discharge refrigerant from system.
2. Disconnect battery ground cable and drain cooling system.
3. Place a waterproof cover over alternator to prevent coolant from spilling over it, then disconnect and plug heater core hoses.
4. Disconnect refrigerant lines at H-valve, then remove two screws from the filter drier bracket and position lines aside. Cap all refrigerant lines. Disconnect temperature control cable from cover.
5. Front inside vehicle, remove glove box, spot cooler bezel and appearance shield.
6. Through glove box opening and under instrument panel, remove screws and nuts retaining evaporator housing to dash panel.
7. Remove the two screws from flange connection to blower housing. Separate evaporator housing from blower housing and carefully remove evaporator housing from vehicle.
8. Remove cover from housing, then remove the heater core strap retaining screw and remove heater core.
9. Reverse procedure to install.

1974-78 Van Models, Fig. 24

With Air Conditioning

1. Discharge refrigerant from A/C system.
2. Disconnect battery ground cable.
3. Drain cooling system and disconnect hoses from heater core.
4. Remove grille, condenser and radiator.
5. Disconnect refrigerant lines.

NOTE: Use two wrenches when disconnecting refrigerant lines to prevent twisting refrigerant lines out of evaporator core.

6. Remove glove box, spot cooler and appearance shield.
7. Remove screws and nuts retaining evaporator housing to dash panel.
8. Remove windshield wiper motor.
9. Disconnect vacuum hoses and electrical connections at evaporator housing, then disconnect blower motor, cooler hose and drain hoses.
10. Remove the two bolts from crossbar and the four screws from sealplate on front of housing, then separate evaporator housing from blower motor housing and remove evaporator housing from vehicle.
11. Remove receiver-drier and cap openings, then carefully pry heater core out of housing.

NOTE: Do not remove air seal from front of core unless seal is damaged and is to be replaced.

12. Reverse procedure to install.

1978-80 Van Models, Fig. 29

Less Air Conditioning

1. Disconnect battery ground cable and drain cooling system.
2. Disconnect heater core hoses.
3. Disconnect temperature control cable from heater core cover and blend air door crank.
4. Disconnect blower motor feed wire from resistor block located on distribution duct in cab.
5. Set coolant overflow tank aside.
6. Remove the three nuts retaining heater assembly to side cowl and the two nuts retaining heater assembly to cowl panel, then remove heater assembly from vehicle.
7. Remove the heater core cover, then remove

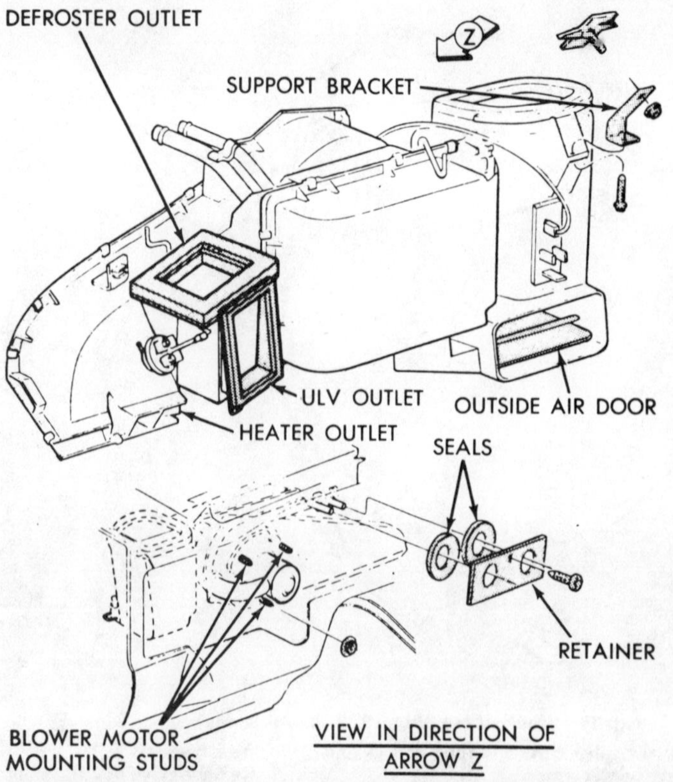

Fig. 28 Heater assembly. 1976-80 Aspen, Diplomat, LeBaron & Volare

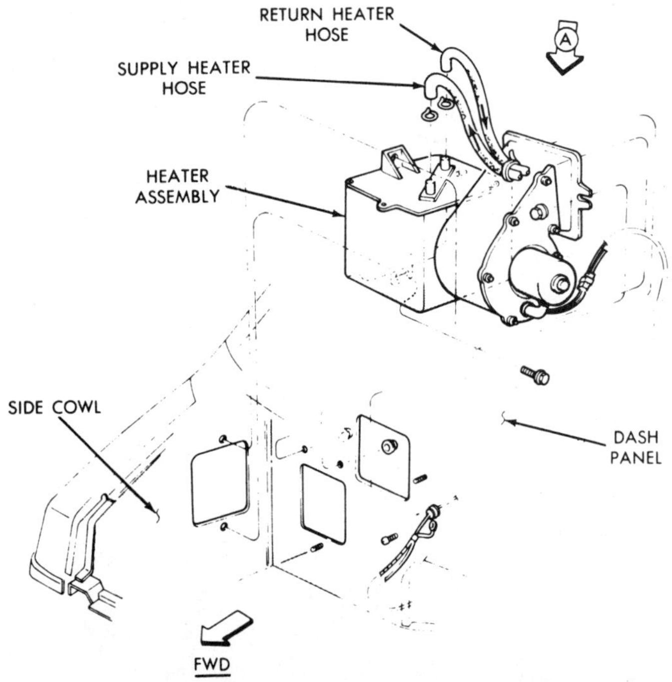

Fig. 29 Heater assembly. 1978-80 Van models

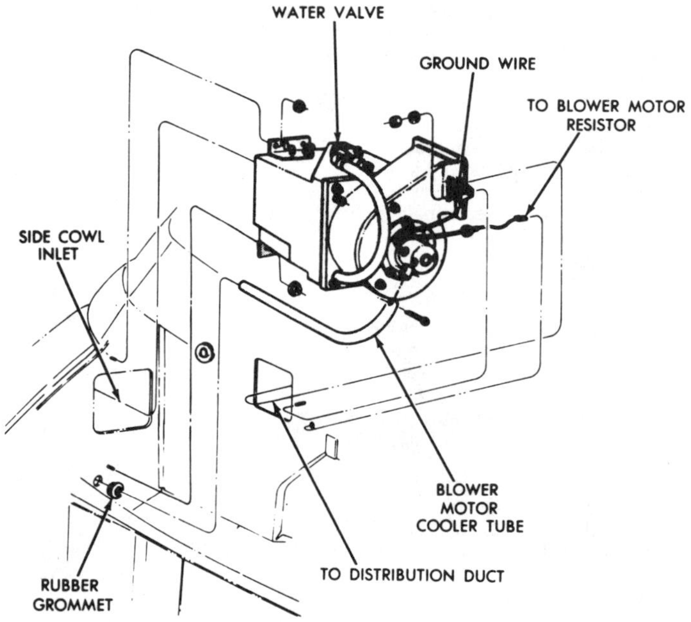

Fig. 30 Heater assembly. 1974-77 Van models

the two heater core retaining screws and remove heater core.
8. Reverse procedure to install.

1974-77 Van Models, Fig. 30

Less Air Conditioning

1. Disconnect battery ground cable.
2. Drain cooling system and disconnect hoses from heater core.
3. Disconnect ground wire and blower motor resistor wires.
4. Disconnect control cable and remove water valve retaining screws.
5. Remove blower motor cooling tube.
6. Remove heater assembly mounting nuts and the heater assembly.
7. Remove heater cover screws and the cover, then the heater core from cover.
8. Reverse procedure to install.

1974-79 Truck Models

Less Air Conditioning

1. Disconnect battery ground cable.
2. Drain cooling system and disconnect hoses from heater core.
3. Disconnect electrical connector from resistor.
4. Disconnect control cables and remove defroster ducts.
5. Disconnect ground wire and cooling tube from blower motor.
6. Remove bracket from right end of instrument panel and pull toward rear of cab.
7. Remove the seven retaining nuts from engine side of firewall and from inside of cab near right hand kick panel.
8. Remove heater from vehicle.
9. Remove screws and separate heater housing.
10. Remove core retaining screws and carefully slide core out of heater housing.
11. Reverse procedure to install.

With Air Conditioning, Fig. 22

1. Disconnect battery ground cable.
2. Drain cooling system and disconnect hoses from heater core.
3. Remove glove box and ash tray.
4. Remove right and left A/C ducts, distribution duct and center air outlet and duct.
5. Disconnect wiring harness from resistor and vacuum lines from rear housing.
6. Remove the 24 retaining screws, then while holding defroster toward heat position, separate the units.
7. Remove one screw from engine side and two screws from each end of core and slide core out.
8. Reverse procedure to install.

1974-80 Van Auxiliary Heater Core

Less Air Conditioning

1. Disconnect battery ground cable and drain cooling system.
2. Disconnect heater hoses from heater assembly.
3. Disconnect blower wiring and remove heater assembly from vehicle.
4. Remove housing cover and heater core from cover.
5. Reverse procedure to install. Bleed heater hoses.

With Air Conditioning

1. Drain coolant from vehicle.
2. Working under vehicle, disconnect heater hoses and remove lower cover plate.
3. Remove heater core support bracket and heater core.
4. Reverse procedure to install. Bleed heater hoses.

Blower Motor, Replace

PASSENGER CARS

1979 Magnum; 1979-80 Cordoba & Full Size, 1980 Gran Fury & Mirada

NOTE: The blower motor is accessible from under right side of instrument panel.

1. Disconnect battery ground cable.
2. Remove glove box assembly.
3. Disconnect blower motor electrical connections.
4. Remove heater housing or evaporator heater housing to plenum brace.
5. Remove blower motor mounting screws and the blower motor, Figs. 18, 19 & 25.
6. Reverse procedure to install.

1978-80 Horizon & Omni

Less Air Conditioning, Fig. 26

1. Disconnect battery ground cable.
2. Disconnect blower motor wiring connector.
3. Remove left heater outlet duct.
4. Remove four screws retaining blower motor mounting plate to heater unit.
5. Remove blower motor assembly, Fig. 20.
6. Reverse procedure to install.

With Air Conditioning, Fig. 31

1. Disconnect battery ground cable.
2. Remove three screws securing glove box to instrument panel and the glove box.
3. Disconnect blower motor feed and ground wires. Remove wires from retaining clip on recirculating housing.
4. Disconnect blower motor vent tube from A/C unit.
5. Loosen recirculation door actuator from bracket and remove actuator from housing. Do not disconnect vacuum lines.
6. Remove seven screws securing recirculating housing to A/C unit, then the housing.
7. Remove three blower motor mounting flange nuts and the blower motor.
8. Reverse procedure to install.

1974-76 Dart & Valiant

Less Air Conditioning

The heater assembly must be removed to

service the blower motor. See procedure for Heater Core Removal to remove assembly, then proceed to Remove Blower Motor as follows:

1. Disconnect lead wire from the blower motor to the heater assembly.
2. Remove motor cooler tube.
3. Remove heater backplate assembly from heater.
4. Remove fan from motor shaft. Remove blower motor from backplate.

With Air Conditioning

NOTE: All service to the blower motor is made from the engine compartment side.

1. Disconnect the feed wire at the connector and the ground wire. Remove the air tube.
2. Remove three screws located on the outer surface of the mounting plate.
3. Remove the mounting plate and blower motor as an assembly.

1974-78 Intermediate Exc. Aspen, Dart, Diplomat, Lebaron, Valiant & Volare

Less Air Conditioning

The heater assembly must be removed to service the blower motor. Refer procedure for Heater Core Removal to remove assembly, then proceed to Remove Blower Motor as follows:

1. Disconnect blower motor lead from resistor block. Disconnect the ground wire from the mounting plate.
2. Remove six screws and retaining clips holding blower motor mounting plate to housing. Separate the blower motor and mounting plate from the housing.
3. Remove the blower wheel from the motor shaft.
4. Remove the two retaining nuts and separate the motor from the mounting plate. Remove the motor.

With Air Conditioning

NOTE: All service to the blower motor is made from the engine compartment side.

1. Disconnect the feed wire at the connector and the ground wire. Remove the air tube.
2. Remove three screws located on the outer surface of the mounting plate.
3. Remove the mounting plate and blower motor as an assembly.

1974-78 Chrysler, Dodge & Plymouth Full Size

The blower motor is mounted to the engine side housing under the right front fender between the inner fender shield and the fender. The inner fender shield must be removed to service the blower motor.

1976-80 Aspen, Diplomat, LeBaron & Volare

Less Air Conditioning

1. Remove heater assembly and heater core as outlined under "Heater Core, Replace".
2. Remove blower motor vent tube.
3. Remove blower motor mounting nuts and the blower motor.

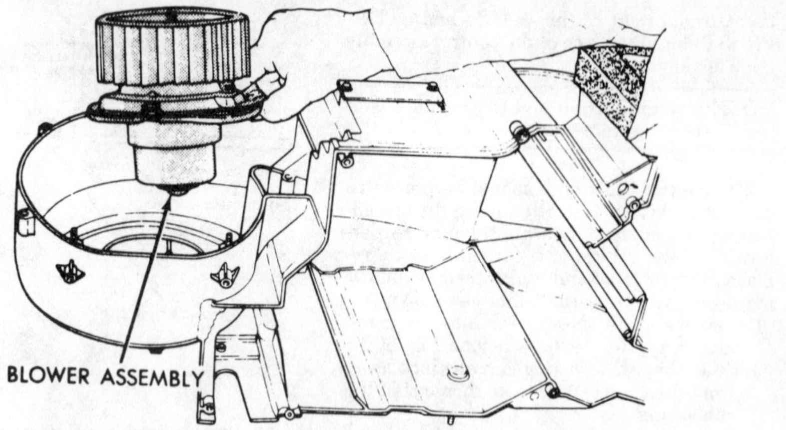

Fig. 31 Blower motor assembly. 1978-80 Horizon & Omni with air conditioning

With Air Conditioning, Fig. 19

NOTE: All service to the blower motor is performed under the right side of the instrument panel.

1. Disconnect battery ground cable and the blower motor feed wire.
2. Remove blower motor mounting nuts from bottom of recirculation housing.
3. Separate upper and lower blower motor housing, then remove blower motor mounting plate screws.
4. Remove wire grommet, mounting plate and blower motor.

TRUCKS & VANS

1979-80 Van Models, Fig. 23

With Air Conditioning

1. Disconnect battery ground cable.
2. Remove top half of shroud by removing 2 screws from shroud to radiator support and two screws holding the halves of the shroud together. Top right screw on vehicles with six cylinder engine is hidden behind discharge line muffler. Move top half of shroud out of the way.
3. Disconnect blower motor electrical connector and remove blower motor cooling tube.
4. Remove the three retaining nuts from studs holding blower motor.
5. Pull A/C suction and discharge lines inboard and upward while pulling blower motor assembly from housing.
6. Remove blower motor.
7. Reverse procedure to install.

1974-78 Van Models, Fig. 24

With Air Conditioning

1. Disconnect battery ground cable.
2. Remove radiator.
3. Remove blower motor cooler tube.
4. Disconnect lead wire at connector and ground wire from retaining screw.
5. Remove the three retaining nuts and washers and remove blower motor from housing.
6. Reverse procedure to install.

1978-80 Van Models, Fig. 29

Less Air Conditioning

1. Disconnect battery ground cable and blower motor wiring.
2. On 1978-79 models, remove heater housing as described in "Heater Core, Replace".
3. On all models, remove blower motor backing plate from heater housing.
4. Remove blower wheel and motor from backing plate.
5. Reverse procedure to install.

1974-77 Van Models, Fig. 29

Less Air Conditioning

1. Disconnect battery ground cable.
2. Remove blower motor cooler tube.
3. Disconnect blower motor resistor wire and the ground wire.
4. Remove three retaining nuts from mounting plate, then the blower motor.
5. Reverse procedure to install.

1974-79 Truck Models

1. Disconnect battery ground cable and blower motor wiring.
2. Disconnect blower motor vent tube.
3. Remove blower assembly mounting screws and blower assembly.
4. Separate wheel from motor noting position of wheel on shaft.
5. Remove blower mounting plate from motor.
6. Reverse procedure to install.

1974-80 Van Models Auxiliary Heater

1. Remove auxiliary heater assembly as described in "Heater Core, Replace".
2. Remove blower motor mounting screws and blower motor.
3. Reverse procedure to install.

1976-80 Van Models Auxiliary Evaporator-Heater

1. Disconnect battery ground cable and blower motor wiring.
2. Remove blower motor mounting screws and motor.
3. Reverse procedure to install.

FORD MOTOR CO.

Control Cables, Adjust

PASSENGER CARS

1978 Fairmont, Zephyr & 1979-80 All Models

The function and/or temperature control cables self-adjust when the levers are moved to the extreme right of the WARM and/or DEFROST slot in the face of the control assembly upon installation.

NOTE: Some function and temperature control levers move from top to bottom.

The control cable wire should be pre-set to prevent kinking before attempting the self-adjusting procedure. Cable pre-set may be performed in the vehicle or before installing a replacement cable assembly. Proceed as follows to pre-set cable before installing assembly:

1. Insert a small screwdriver into the crank arm end of the control wire loop, Fig. 32.
2. Slide the self-adjusting clip one inch away from the control wire loop, then install the cable assembly.
3. Position the self-adjusting clip by moving the control lever to the extreme right of the WARM and/or DEFROST slot.
4. Ensure proper control operation.

To pre-set cable after installation, move control lever to the COOL and/or OFF position and proceed as follows:

1. Firmly secure the crank arm, then insert a small screwdriver into the control wire loop, Fig. 32.
2. Pull the control wire loop through the self-adjusting clip until there is a one inch gap between the loop and the clip.
3. Position the self-adjusting clip by moving the control lever to the extreme right of the WARM and/or DEFROST slot.
4. Ensure proper control operation and full door travel.

1974-77 Comet & Maverick

Less Air Conditioning

1. Loosen the temperature control cable mounting screw near the door crank arm.
2. Slide the temperature control lever to the full WARM position at the lower end of the control slot, then back it off ⅛ inch.
3. Move the door crank arm to the full counterclockwise position.
4. With the door crank arm and the control lever properly positioned, tighten the retaining clip screw.

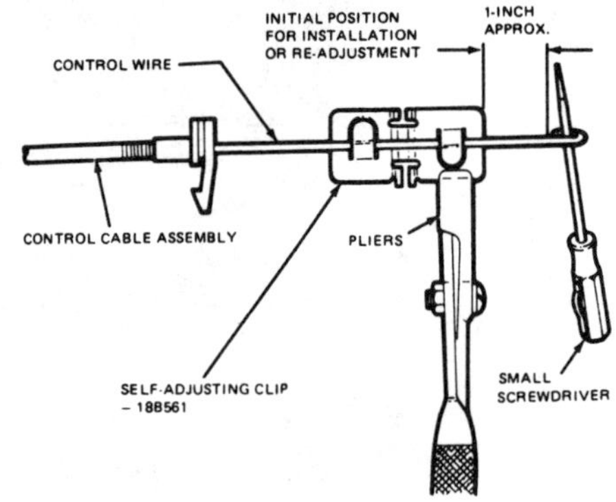

Fig. 32 Control cable adjustment. 1978 Fairmont, Zephyr & 1979-80 Passenger Cars

With Air Conditioning

1. Move temperature control lever to the extreme warm and cool positions.
2. If temperature control lever does not contact either edge of slot, tighten retaining screw on cable mounting clip near temperature door crank arm on the evaporator assembly.
3. If temperature control lever contacts the upper edge of the slot only, loosen retaining screw and pull cable downward until temperature control lever does not contact the edge of the slot. Tighten retaining screw.
4. If the temperature control lever contacts the lower edge of the slot only, loosen retaining screw and pull cable upward until the temperature control lever does not contact edge of slot. Tighten retaining screw.

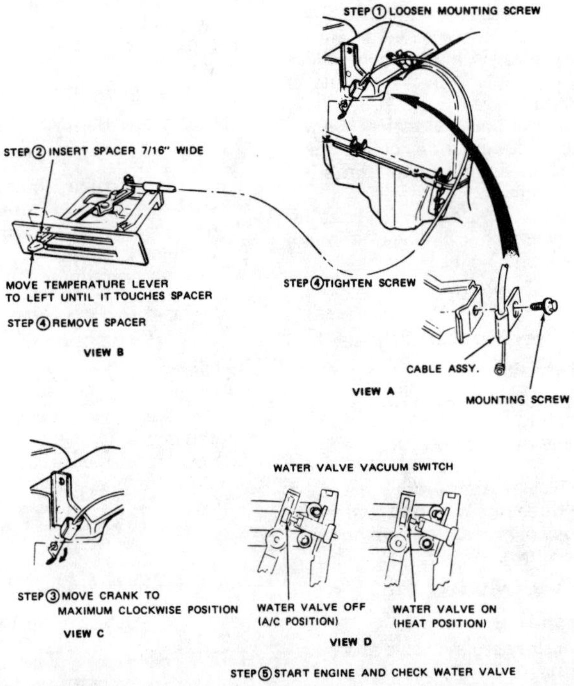

Fig. 33 Temperature cable adjustment. 1974-78 Bobcat, Mustang II & Pinto With A/C

1974-78 Mustang II, Pinto & 1975-78 Bobcat

Less Air Conditioning

1. Loosen the temperature control cable mounting screw.
2. Slide the temperature control lever to the full WARM position at the right end of the control slot. Then, back it off ⅛ inch on Mustang II or ¼ inch on Pinto and Bobcat.
3. Move the door crank arm toward the mounting bracket and tighten the mounting screw.

With Air Conditioning

1. Loosen temperature control cable mounting screw.
2. Insert a 7/16 inch spacer into the temperature lever slot between the lever and left end of slot, then move the temperature control lever to the left until lever contacts spacer, Fig. 33.
3. Move the temperature blend door crank arm clockwise to the maximum cool position.
4. Tighten temperature control cable mounting screw and remove 7/16 inch spacer.
5. Start engine and ensure water valve closes when the temperature control lever is in the maximum cool position.

1974-76 Montego, Torino, 1974-78 Cougar, 1975-76 Elite, 1975-78 Granada, Monarch, 1977-78 LTD II & Thunderbird

Less Air Conditioning

1. Slide the temperature control lever to the WARM end of the slot.
2. Loosen the cable retaining screw and rotate the door crank arm to the full clockwise position.
3. Move the control lever to provide 3/16 inch clearance between the lever and the edge of the slot. Ensure the door does not move.
4. Tighten the cable retaining screw.

1974-76 Cougar, Elite, Montego, Thunderbird & Torino

With Air Conditioning

There should be a ⅛ inch clearance between the temperature control lever and the edge of the slot, Fig. 34, with the temperature control lever in the warm position. Adjust turnbuckle as follows:

1. Rotate turnbuckle while applying pressure against lever and noting direction of lever travel. If lever is moving in wrong direction, rotate turnbuckle in opposite direction.
2. After correct clearance is obtained, check adjustment by moving temperature lever from the warm to cool position, then back to the warm position.

1974-78 Ford & Mercury

Less Air Conditioning

1. Slide the temperature control lever to maximum WARM position.
2. Perform cable adjustment at the bracket on the bottom of the plenum by sliding cable flag in bracket groove until the temperature door crank is in the full heat position.
3. Ensure ⅛ inch clearance between the temperature control lever and the right end of the lever slot; this establishes maximum heat position.
4. Tighten cable screw at plenum bracket.

1977-78 Cougar, LTD II & Thunderbird

With Air Conditioning

Refer to "1974-78 Ford & Mercury, Less Air Conditioning" for adjustment procedure.

1974-78 Ford & Mercury Full Size

With Air Conditioning

1. Place temperature control lever in the fully cool position.
2. Adjust cable at bracket on plenum by sliding cable flag in bracket groove until temperature door crank arm is in the full right position.
3. Tighten cable screw at plenum bracket.

1974-77 Mustang II & 1975-78 Granada & Monarch

With Air Conditioning

1. Loosen temperature control cable mounting screw.
2. Place temperature control lever in the full left hand position, Fig. 33.
3. Move temperature blend door crank arm clockwise to the maximum cool position.
4. Tighten cable mounting screw.

TRUCKS & VANS

1980 Light Duty Pickup & Bronco

Less Air Conditioning

Refer to "1978 Fairmont, Zephyr & 1979-80 Passenger Cars" cable adjustment procedure.

With Air Conditioning

1. Remove glove box liner.
2. Loosen cable attaching screw at bracket located on top of plenum.
3. Position control lever ⅛ inch from left end of slot in control assembly.
4. Rotate temperature door crank arm to extreme left and tighten screw.
5. Move lever to the left and right to be sure the lever bounces back lightly at each end of slot.

1975-80 Econoline With Standard Heater

Function Control Cable

1. Rotate door crank arm to full clockwise position.

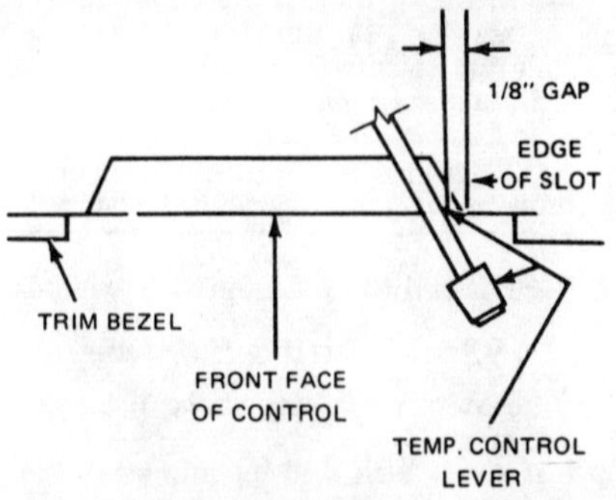

Fig. 34 Temperature cable adjustment. 1974-76 Cougar, Elite, Montego, Thunderbird & Torino With A/C

2. Move control lever to "DEFROST" position allowing a gap of .30 inch between lever and edge of slot.
3. With the door crank and control lever in position, tighten cable screw and nut.
4. Re-adjust cable as necessary.

Temperature Control Cable

1. Rotate temperature door crank to full clockwise position.
2. Place temperature control lever to "WARM" position allowing a gap of .35 inch between lever and edge of slot.
3. With door crank arm and control lever in position, tighten cable screw and nut.
4. Re-adjust cable as necessary.

1974-79 Light Duty Pickup & 1979 Bronco

Function Control Cable, Less Air Conditioning

Loosen cable clamp screw at plenum. Move control lever to DEFROST position. Then, push air door crank arm to the extreme right position and tighten cable clamp screw.

Temperature Control Cable

Loosen cable clamp screw at heater housing. Move control lever to WARM position. Then, push the temperature door crank arm to the extreme right position and tighten cable clamp screw.

1978-79 Econoline With Hi-Lo Heater

Temperature Control Cable

Move temperature control lever to WARM position. Rotate cable turnbuckle to obtain a 0.35 inch gap between the slot edge and lever edge.

NOTE: The turnbuckle is located under the instrument panel to the right of the engine cover. It may be necessary to remove the right heater duct to obtain access to the turnbuckle.

1978 Light Duty Pickup & Bronco With Hi-Lo Heater

Air Blend Door

Loosen cable clamp screw. Move temperature control lever to extreme left position (COOL). Hold the control lever in position and move the door crank arm to the extreme forward position and tighten cable clamp.

1974-77 Bronco

Less Air Conditioning

The heat/defrost, outside air door and water valve control cables must be adjusted so that the doors and valve are at the end of their travel when the control knobs are 1/16 to ⅛ inch from the end of travel (pushed in). Loosen the cable retaining screw at the air door of water valve crank arm and move cable housing back and forth to adjust. Then, tighten retaining screw.

1975-80 Econoline

With Air Conditioning

Refer to "1974-80 Econoline With Standard Heater" for temperature cable adjustment.

1974-79 Light Pickup & 1978-79 Bronco

With Air Conditioning

1. Loosen cable clamp screw.
2. Position temperature control lever in "COOL" (left).
3. Hold control lever and move air blend door crank arm to the extreme forward position and tighten cable clamp screw.

1974 Econoline

Less Air Conditioning

The defrost door and air door must be adjusted so that each door is at the end of its travel when the control knob is 1/16 to ⅛ inch from end of its travel (up position). Remove glove box and radio to gain access to heater control. Loosen the cable retaining screw. Move heat control lever to "OFF" position. Then, adjust door crank to closed position by rotating crank towards rear of vehicle and tighten retaining screw.

To adjust defroster door, disconnect defrost door control cable from ball-joint on control assembly. With control lever in "OFF" position, push link up to close defroster door and adjust the link length so that it will snap back on the ball-joint.

ATC Sensor Calibration

1. On all models except Mark V, remove instrument panel pad. On Mark V, remove right hand applique.
2. On Mark V, loosen sensor and bracket from instrument panel.
3. On all models, while moving the temperature control lever, rotate the calibration locking pin into the 75 degree calibration notch of the sensor control arm, Fig. 35.
4. Check position of temperature control lever; it should be at the 75 degree location. If not, loosen the bowden cable housing to sensor attaching screw and move the temperature control lever to the 75 degree position while the sensor control arm is locked in the calibration position. Then tighten bowden cable housing to sensor attaching screw, ensuring temperature control lever remains at the 75 degree position with the sensor arm locked.

Readjusting Calibration Point

The following procedure should be used when it is necessary to change the calibration point to suit individual preferences.

1. Determine difference between present and desired temperature level settings. Also determine direction of change.
2. To change setting, adjust temperature control lever position with the sensor arm locked in the 75 degree calibration position. Loosen the bowden cable housing to sensor attaching screw and move the temperature control lever the desired amount, then retighten attaching screw. If equipped with turnbuckle, adjust as necessary.

NOTE: Move the temperature control lever to a higher temperature setting to decrease the desired temperature. Move the temperature control lever to a lower temperature setting to increase the desired temperature.

CAUTION: If a readjustment of the temperature control lever of more than 4 degrees is required and the cable adjustment is found to be correct, replace the sensor assembly.

Servo Motor, Adjust

1978 Ford & Mercury & 1978-79 Lincoln Continental

1. Remove vacuum from the servo motor by disconnecting vacuum hose.
2. Loosen motor retaining screws and slide motor forward until the blend door just contacts its stop (full Heat). Continue to slide motor forward another ⅛ inch and tighten the retaining screws.

NOTE: The motor diaphragm should move about ⅛ inch inside the housing.

3. Reconnect the tan vacuum hose to motor.

1974-77 Ford & Mercury

1. Evacuate servo motor above 19 inches of vacuum.
2. Insert a ⅛ inch drill bit into servo motor and arm, Fig. 36.
3. Remove vacuum supply from motor and loosen motor mounting screws.
4. Slide motor rearward until blend door is closed.

NOTE: Do not exert a force greater than 5 pounds while sliding motor rearward.

5. Tighten mounting screws while maintaining a 5-10 pound force on motor.
6. Evacuate motor and remove drill bit.

1974-76 Cougar, Elite, Mark IV, Montego, Thunderbird & Torino

The servo motor must be adjusted to both the blend door and the servo switchboard.

NOTE: On Mark IV and Thunderbird, it is necessary to remove and lower the ATC plenum to perform the following adjustments.

Servo Motor To Blend Door

1. Apply 12 or more inches of vacuum to servo motor.
2. Insert a ⅛ inch twist drill into servo link, Fig. 36.

NOTE: A number 5 machine screw or pin, ⅛ inch diameter by 2 inches long, may also be used.

3. Remove vacuum supply from servo motor. Ensure the sled pin is engaged in the servo linkage slot.
4. Loosen servo motor mounting screws and slide the servo motor rearward until the blend door closes against the inner frame of the plenum chamber in the full A/C position. Do not apply more than 5 pounds of axial force while sliding servo rearward.
5. Tighten servo mounting screws. There should be sufficient rearward lead to cause a slight compression of the foam seal, approximately 1/16 inch, on the blend door.
6. Apply sufficient vacuum to servo motor and remove adjustment pin.

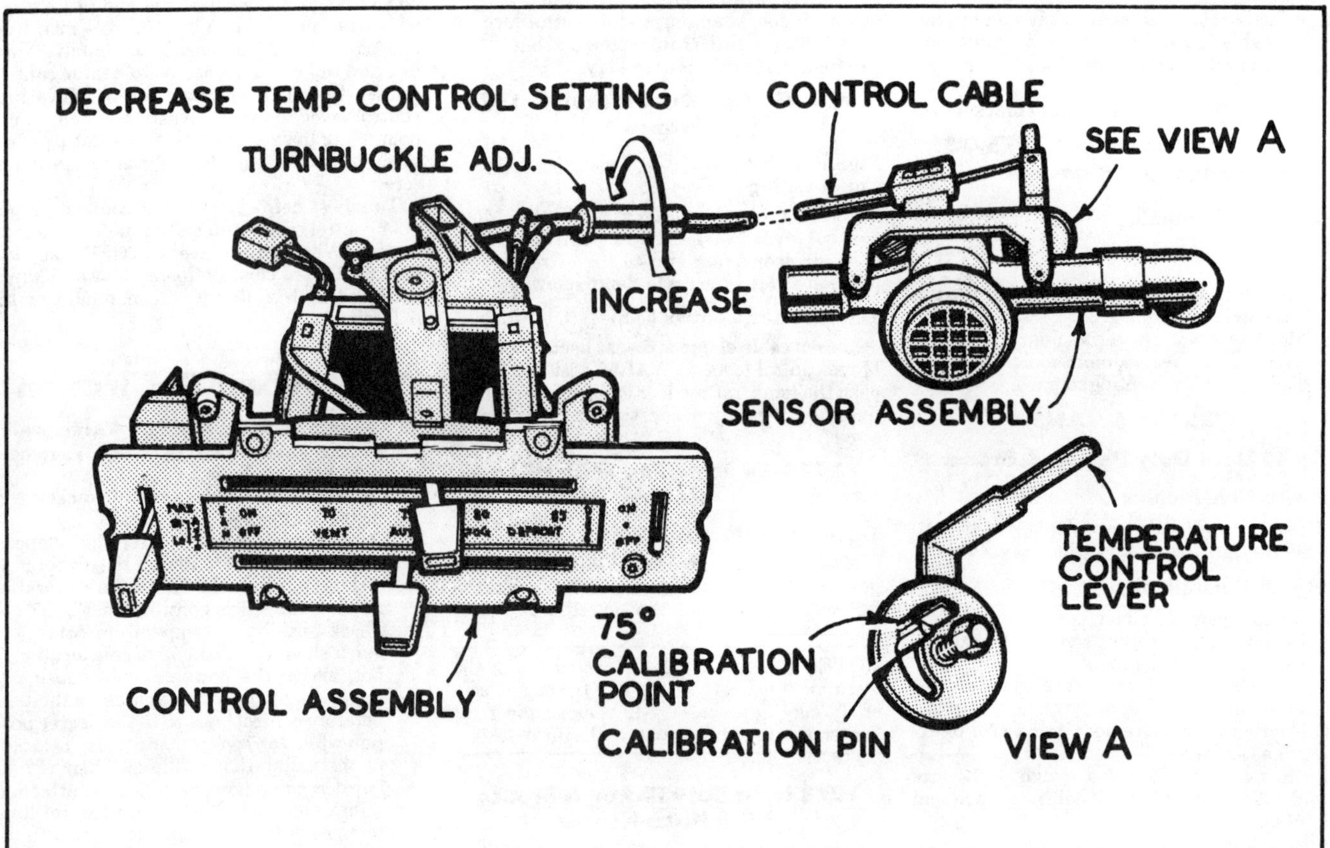

Fig. 35 ATC Sensor Calibration. Ford Motor Co.

Servo Motor To Servo Switchboard

There are two alternate procedures used to perform this calibration. In the first procedure, the servo motor vacuum supply is controlled by the sensor. If 15 or more inches of vacuum cannot be obtained in "Step b, Procedure 1," a malfunctioning sensor is indicated. Use "Procedure 2," in which the vacuum supply from the sensor is disconnected from the servo and an outside vacuum source is used.

Procedure 1:

a. Tee a vacuum gauge into the servo modulated vacuum line.
b. Start engine to build vacuum supply and place blower programmed switch in Auto-Hi position. Move temperature control lever to 65° so the vacuum level is above 15 inches and the blower is operating at high speed.
c. Slowly move the temperature control lever and note the vacuum level at which blower speed changes from high speed to the next lowest speed.
d. Adjust the blower speed shift point by sliding the servo assembly fore or aft on the bracket slot so the shift occurs between 12 and 13 inches Hg. Sliding the servo rearward raises the shift point and sliding it forward lowers the shift point. Tighten the servo mounting screws while holding the servo in the proper position.
e. Check system operation by removing the servo modulated vacuum line, thereby reducing the vacuum to zero. This should result in maximum blower speed and heat.

Procedure 2:

a. Disconnect and plug servo modulated vacuum line, then connect a vacuum gauge and an outside vacuum source to the servo motor.
b. Start engine to build vacuum supply to diverter valve and place blower programmer switch in Auto-Hi position. Apply vacuum to the servo motor until the vacuum level is above 15 inches and the blower is operating at high speed.
c. Slowly reduce vacuum to the servo and note vacuum level at which the blower speed changes to the next lowest speed.
d. Adjust the blower speed shift point by sliding the servo assembly fore and aft on the bracket slot so the shift point occurs between 12 and 13 inches Hg. Sliding the servo rearward raises the shift point and sliding it forward lowers the shift point. Tighten servo mounting screws while holding the servo in proper position.
e. Check system for proper operation by removing vacuum source from servo. This should result in maximum blower speed and heat.

Temp. Control Rheostat Calibration, Adjust

1974-77 Lincoln Continental

1. Set temperature control lever at 75°.
2. Loosen screws securing linear rheostat.
3. Connect an ohmmeter between the white and black leads with connector disconnected from wiring harness.
4. Move rheostat until 268 ohms are obtained on the ohmmeter.
5. Tighten screws and check adjustment.

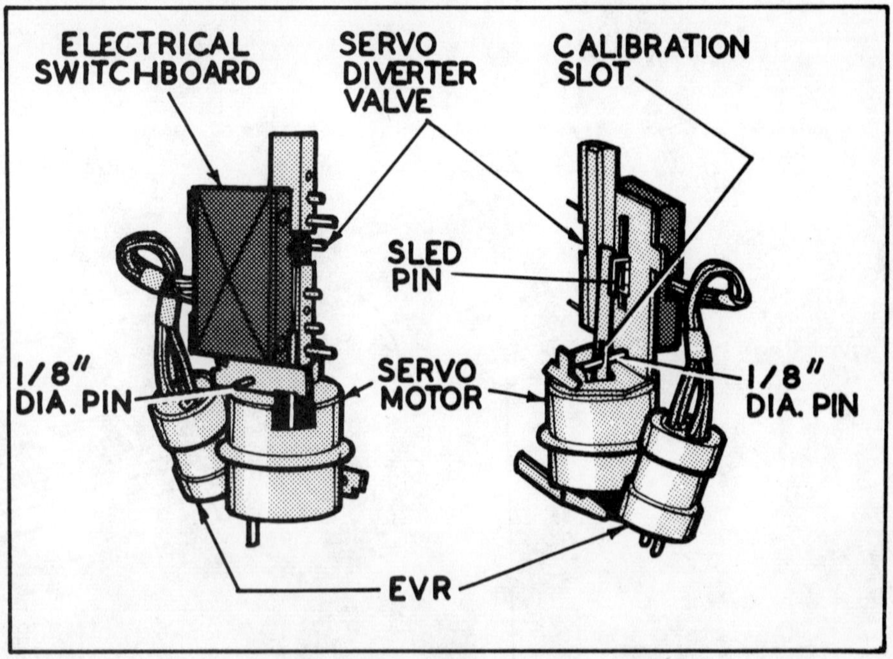

Fig. 36 ATC Servo Motor Calibration. 1974-76 Cougar, Elite, Mark IV, Montego, Thunderbird, Torino; 1974-77 Ford & Mercury

ATC Box Temperature Calibration, Adjust

1974-77 Lincoln Continental

This adjustment should not be performed if the ATC system is not performing properly. This calibration adjustment is used only to compensate for variations in the ATC box temperature sensors and control assembly rheostat. However, the calibration may be changed to suit individual preferences. The temperature calibration control is located on the face of the ATC box. Each calibration division corresponds to approximately 2 degrees F. To raise the temperature level, rotate the calibration control clockwise as indicated by the curved arrow and the warmer notation. To lower temperature level, rotate control counter-clockwise.

Vacuum Motor, Adjust

1974-77 Maverick & Comet

Use the adjusting screw and slot on the two-piece actuating shaft assembly to adjust the vacuum motor. With the door in its NV (no vacuum applied) position, proceed as follows:

1. Loosen adjusting screw.
2. Hold the door in its NV position and adjust the shaft so that the preload indicator notch is flush with the motor body.
3. Tighten adjusting screw.

1974-77 Lincoln Continental

Outside Air Door

1. Remove the cowl air inlet screen.
2. Disconnect vacuum hose and ensure door opens and closes without sticking or binding.
3. Loosen vacuum motor mounting screws, Fig. 37 and move the motor outboard until the door contacts the limit stop. Then, move the motor approximately ⅛ inch further outboard to preload the motor and tighten the mounting screws.
4. Connect vacuum hose and install cowl air inlet screen.

Recirculating Air Door

1. Remove cowl trim panel.
2. Disconnect vacuum hose and ensure door opens and closes without sticking or binding.
3. Loosen vacuum motor mounting screws, Fig. 38, and with the air door in the closed position, move vacuum motor outboard.
4. Align preload notch on motor shaft with base of motor and tighten mounting screws.
5. Connect vacuum hose and install cowl trim panel.

Temperature Blend Door

1. Loosen vacuum motor mounting screws.
2. Position the motor until the preload notch nearest the motor is ¼ inch from the motor body with the door in its normal position and tighten the mounting screws.

1974 Mark IV & Thunderbird

Temperature Blend Door

1. Loosen vacuum motor mounting screws.
2. Position vacuum motor and bracket assembly on the heater case bracket so the forward edges of the two brackets are aligned and tighten mounting screws.

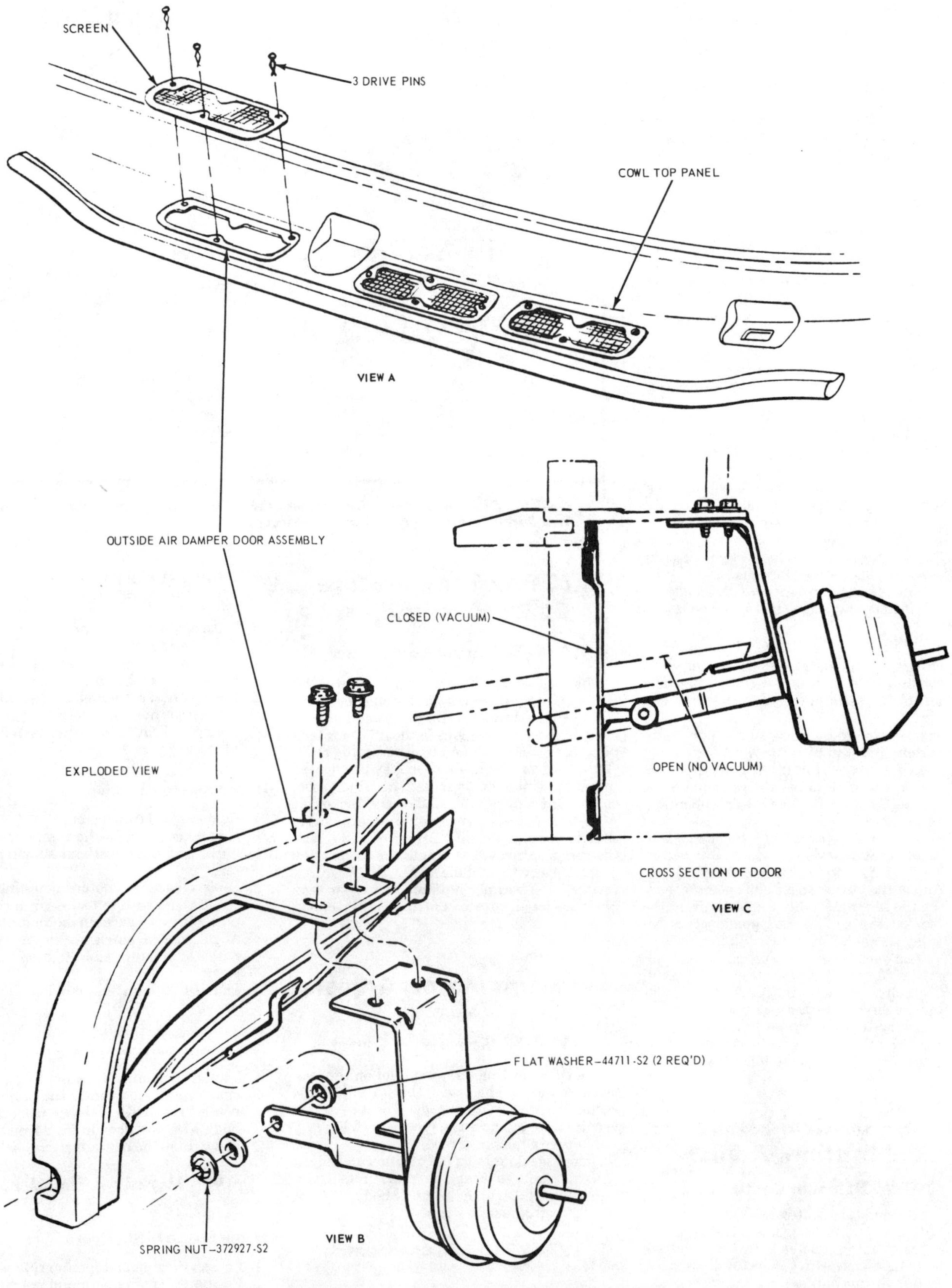

Fig. 37 Outside door vacuum motor. 1974-77 Lincoln Continental

Heater Water Valve Vacuum Switch Adjustment

1974-77 Comet & Maverick

1. Place temperature control lever in the full cool position. If water valve vacuum switch is not fully depressed, loosen adjusting screw and move switch door crank arm until plunger is fully depressed. Tighten adjusting screw.

Control Assembly, Replace

PASSENGER CARS

1974-77 Comet & Maverick

To remove control, remove instrument cluster. Pull knobs off control levers and switches. Carefully remove the face plate by prying it off at top and bottom alternately. If this is not done, the mounting pins may be broken. Pry off the lens plate from control assembly, then remove screws retaining control assembly to instrument panel. Move control assembly out through cluster opening and disconnect all switches and cables.

1980 Lincoln Continental

1. Disconnect battery ground cable.
2. Remove radio control knobs, steering column cover, center finish panels and right lower center finish panel.

NOTE: Two right lower center finish panel screws are located in the ashtray opening.

3. Remove control assembly attaching screws, pull control out of instrument panel opening and disconnect wiring, cable and vacuum hoses.
4. Reverse procedure to install.

1978 Ford & Mercury Full Size & 1978-79 Lincoln Continental With ATC

1. Remove radio knobs and right instrument cluster trim panel retaining screws.
2. Pull lower edge of right instrument cluster trim panel away from instrument panel and disengage the top right instrument cluster trim panel tabs (2) from instrument panel and remove trim panel.
3. Remove ashtray and ashtray support bracket.
4. Disconnect wiring and vacuum hoses from control assembly.
5. Remove control assembly screws and pull assembly out of instrument panel for access to temperature control cable.
6. Disconnect control cable and remove control assembly from panel.
7. Reverse procedure to install.

1974-77 Lincoln Continental

1. Disconnect battery ground cable.
2. Remove steering column lower support panel and lower steering column.
3. Remove lower instrument panel pad.
4. Remove knobs from control assembly. Control lever knobs are retained with set screws.
5. Remove bezel assembly and retainers.
6. Remove control assembly screws and lower assembly from rear of instrument panel.
7. Disconnect vacuum hoses and electrical connections from control assembly, Fig. 39, then remove the assembly.
8. Reverse procedure to install.

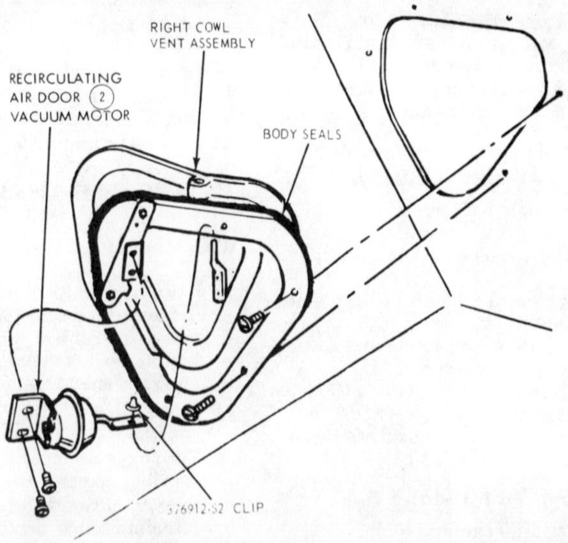

Fig. 38 Recirculating air door assembly. 1974-77 Lincoln Continental

1974-79 Cougar, Elite, LTD II, Mark IV, V, Montego, Thunderbird & Torino

Less Air Conditioning

1. Disconnect battery ground cable.
2. Remove knobs and cables from control levers.
3. Remove control assembly screws and push through instrument panel towards engine compartment.
4. Disconnect wiring and vacuum hoses from control assembly and remove from vehicle.
5. Reverse procedure to install.

With Air Conditioning

1. Disconnect battery ground cable.
2. Remove knobs from control assembly levers.
3. On 1974 Mark IV and Thunderbird, remove control assembly applique.
4. On all models, remove control assembly mounting screws.
5. Disconnect temperature control cable, wiring, vacuum harness from control assembly, then remove control from instrument panel.
6. Reverse procedure to install.

1974-77 & 1978 Less ATC Ford & Mercury Full Size

Less Air Conditioning

1. Disconnect battery ground cable.
2. Remove knobs from control assembly levers and switches.
3. Remove control assembly applique, then disconnect control cables from assembly.
4. Remove control assembly mounting nuts and slide assembly forward out of engine side of instrument panel.
5. Disconnect wiring and vacuum connectors and remove control from vehicle.
6. Reverse procedure to install.

With Air Conditioning

1. Remove knobs from control assembly levers and switches.
2. On 1977-78 models, remove ashtray and ashtray bracket.
3. On all models, remove applique from covering control assembly.
4. Remove control assembly screws.
5. Disconnect control cable, vacuum harness and wiring from control assembly.
6. Remove control from rear of instrument panel.
7. Reverse procedure to install.

1978-80 Bobcat & Pinto Less A/C

1. Remove glove compartment.
2. Disconnect control cables from heater assembly.
3. Remove control assembly attaching screws and pull control out and away from instrument panel.
4. Disconnect wiring and cables from control assembly and remove from vehicle.
5. Reverse procedure to install.

1974-77 Bobcat & Pinto Less A/C

1. Remove control assembly mounting screws and pull control out of instrument panel.
2. Disconnect wiring and control cables from control assembly and remove control.
3. Reverse procedure to install.

1978-80 Bobcat & Pinto With A/C

1. Remove consolette to instrument panel-to-floor brace screws and pull consolette rearward only enough to unsnap and remove control trim cover.
2. Remove control to consolette attaching screws and remove consolette from control.
3. Disconnect wiring, vacuum hoses and control cables and remove control.
4. Reverse procedure to install.

1974-77 Bobcat & Pinto With A/C

1. Disconnect battery ground cable.
2. Remove control assembly retaining screws.

3. Disconnect temperature control cable from evaporator case.
4. Pull control assembly from console.
5. Disconnect vacuum harness and electrical leads from control assembly.
6. Remove control assembly from vehicle.
7. Reverse procedure to install.

1978-80 Fairmont & Zephyr; 1979-80 Capri & Mustang

1. Remove instrument cluster trim panel.
2. Remove control assembly attaching screws and pull control rearward out of dash panel.
3. Disconnect control cables from evaporator-heater assembly.
4. Disconnect wiring connectors and vacuum hoses from control assembly.
5. Remove control assembly with cables from instrument panel.
6. Disconnect cables from control assembly.
7. Reverse procedure to install.

1974-78 Mustang II

1. Disconnect battery ground cable.
2. Remove control assembly mounting screws from the rear of the instrument panel. Then, remove the control support bracket screw from the instrument panel lower lip.
3. Lower control assembly rear of panel.
4. Disconnect temperature control cable, vacuum harness and electrical leads from control assembly.
5. Remove control assembly from vehicle.
6. Reverse procedure to install.

1974-80 Granada & Monarch

Less Air Conditioning

1. Remove windshield wiper, light switch and radio control knobs. Remove light switch control bezel.
2. Remove screws from top of cluster bezel and pull lower edge of bezel to disengage the snap-in retainers.
3. Disconnect control cables from control assembly.
4. Remove control assembly attaching screws, disconnect wiring and remove control from rear of instrument panel.
5. Reverse procedure to install.

1979-80 Granada & Monarch

With Manual Air Conditioning

1. Disconnect battery negative cable.
2. Remove steering column cover, then reach under instrument panel, depress release button on headlamp switch and remove knob and shaft assembly from switch.
3. Remove headlight switch bezel, radio control knobs and attaching nuts.
4. Turn on ignition switch and shift transmission into manual low position.
5. Remove cluster bezel screws and disengage retaining studs using a thin screwdriver, then tilt top of cluster out and remove from instrument panel.
6. Remove control assembly screws and pull control out of dash.
7. Disconnect control cables, wiring and vacuum connectors from control assembly.
8. Remove control assembly from vehicle.
9. Reverse procedure to install.

1975-78 Granada & Monarch

With Manual Air Conditioning

1. Disconnect battery ground cable.
2. On 1975 models, remove seat belt interlock module and mounting bracket from brake pedal support.
3. On all models, disconnect vacuum harness at inline connector, then the vacuum connector from the 9 port selector valve, Fig. 40.
4. Disconnect control cable and electrical wiring from control assembly.
5. Disconnect wiring from defogger switch and indicator lamp, if equipped.
6. Remove light switch knob and bezel, then the wiper switch and radio knobs.
7. Remove cluster applique upper screws and pull applique rearward to disengage snap-in clips along the lower edge.
8. Remove control assembly mounting screws and pull assembly from instrument panel.

1979-80 Granada, Monarch & Versailles

With ATC Air Conditioning

1. Disconnect battery ground cable.
2. Remove headlamp switch knob, headlight switch bezel and radio knobs.
3. Remove instrument panel finish panel.
4. Remove control assembly screws and lower control assembly from under instrument panel.
5. Disconnect control cable, vacuum harness and wiring, then remove control assembly.
6. Reverse procedure to install.

1977-78 Granada, Monarch & Versailles

With ATC Air Conditioning

1. Remove control assembly knobs and accessory switch knob.
2. Remove control assembly retaining screws.
3. Disconnect control cable, vacuum harness and wiring from control and remove control from vehicle.
4. Reverse procedure to install.

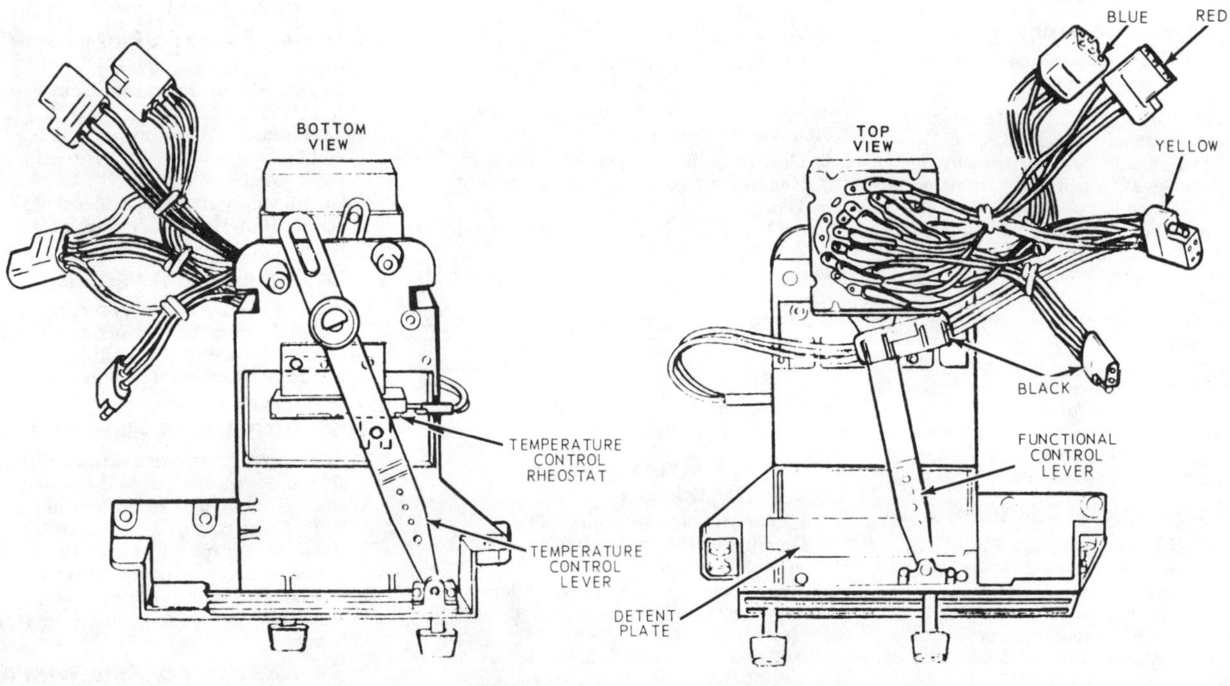

Fig. 39 Control assembly. 1974-77 Lincoln Continental

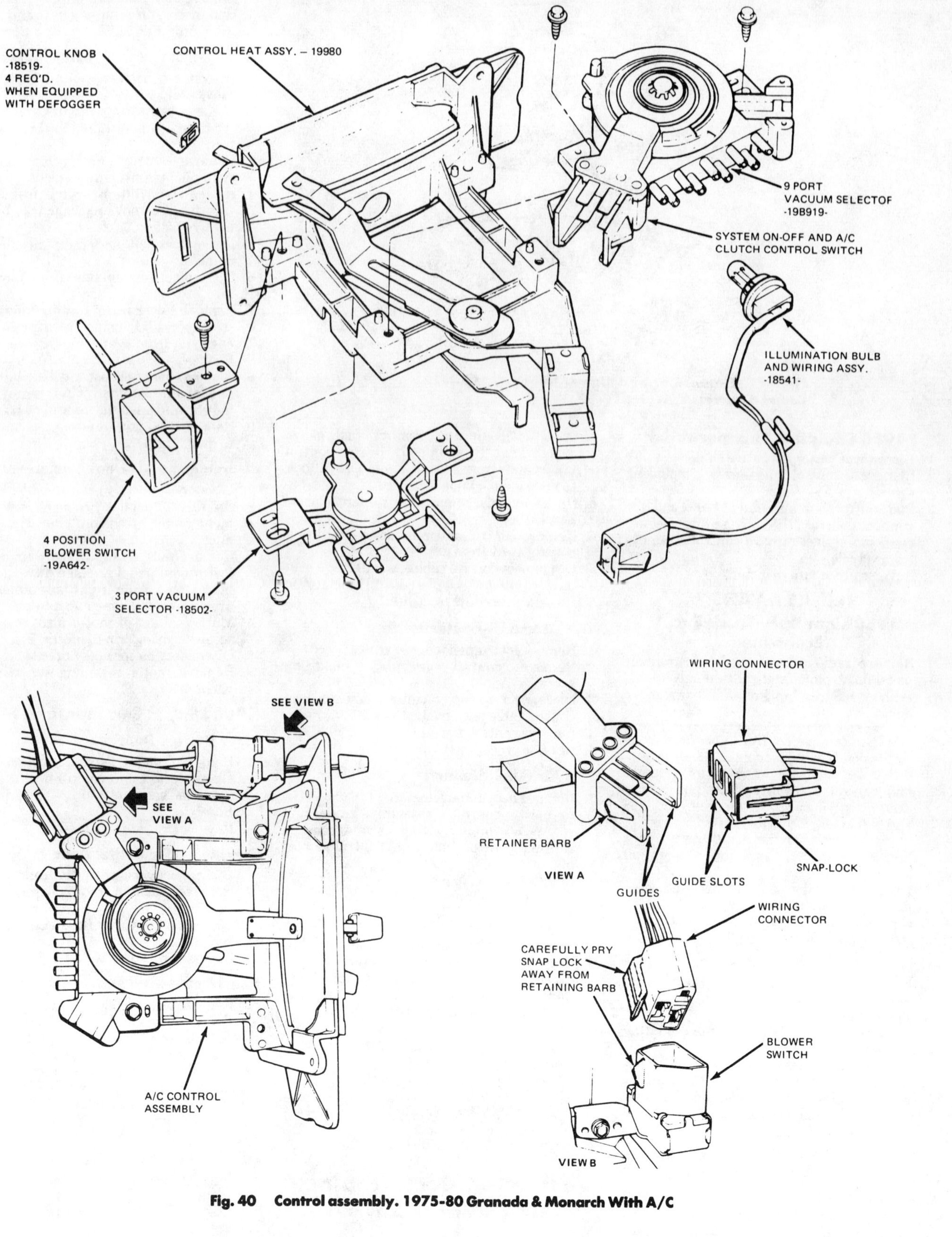

Fig. 40 Control assembly. 1975-80 Granada & Monarch With A/C

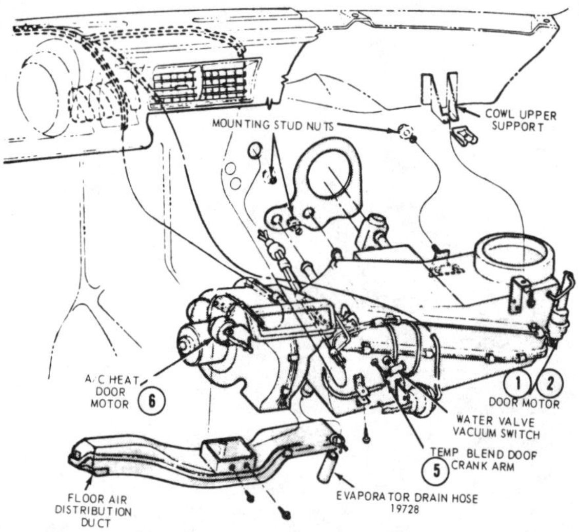

Fig. 41 Evaporator and blower housing. 1974-77 Comet & Maverick (Typical)

1980 Cougar & Thunderbird

1. Disconnect battery negative cable.
2. Remove control assembly mounting screws.
3. Pull control from instrument panel and disconnect control cable, wiring and vacuum connectors, then remove control assembly from vehicle.
4. Reverse procedure to install.

TRUCKS & VANS

1980 Light Duty Truck Exc. Econoline

1. Remove radio control knobs, instrument panel finish panel, control assembly knobs, ashtray, ashtray bracket and floor register duct.

NOTE: Remove control knobs from control assembly by placing a small screwdriver between the knob spring retainer and control assembly. Then pull screwdriver, applying pressure on spring retainer and pull knob off control shaft.

2. Remove control assembly mounting screws.
3. Disconnect wiring, control cables and vacuum harness from control assembly.
4. Remove control assembly from vehicle.
5. Reverse procedure to install.

1975-80 Econoline

1. Remove knobs from control assembly levers.
2. Remove applique.
3. Remove control assembly from mounting bracket.
4. Carefully pull control assembly from opening in mounting bracket and disconnect control cables, wiring connectors and vacuum harness from control assembly.
5. Reverse procedure to install.

1974-79 Light Duty Truck Exc. Econoline & 1974-77 Bronco

With Standard Heater

1. Disconnect battery negative cable.
2. Remove radio control knobs, headlight switch knob and shaft.
3. Remove nut and washer from each radio shaft and remove radio bezel.
4. On 1974-78 models, remove fuel gauge switch knob (if equipped), heater control knobs and wiper washer knob. Use a hook to release each knob lock tab.
5. On all models, remove instrument cluster bezel.
6. Remove control assembly screws and pull control away from dash.
7. Disconnect control cables, wiring and vacuum harness and remove control assembly.
8. Reverse procedure to install.

With Deluxe Hi-Lo Heater

1. Remove instrument cluster trim.
2. Remove control assembly mounting screws.
3. Disconnect wiring, vacuum connector and control cable from control assembly.
4. Remove control assembly.
5. Reverse procedure to install.

With Air Conditioning

1. Remove instrument cluster.
2. Remove control assembly mounting screws. Disconnect wiring, vacuum connectors and control cable from control assembly.
3. Remove control assembly.
4. Reverse procedure to install.

1974 Econoline Heater Control

1. Remove control assembly knobs.
2. Remove glove box or radio if equipped.
3. Remove control front finish panel by removing retaining clip at bottom of panel (behind instrument panel).
4. Remove engine cover and lower control assembly screw retaining control to lower edge of instrument panel.
5. Remove upper control assembly retaining screw and control assembly through glove box opening after disconnecting control cables and switch wiring.
6. Reverse procedure to install.

Vacuum Motors, Replace

1974-77 Comet & Maverick

A/C-Heat Door Vacuum Motor

1. Disconnect battery ground cable and remove radio.
2. Disconnect vacuum hoses from motor and remove hose retaining clip.
3. Remove motor retaining nuts and the vacuum motor, Fig. 41.

Heat/Defrost Door Vacuum Motor

1. Disconnect battery ground cable and remove radio.
2. Remove two screws retaining center register to instrument panel and remove register.
3. Remove ashtray and disconnect left and right ducts from center register.
4. Disconnect ignition switch from package tray. Then remove package tray by removing five bolts.
5. Disconnect left and right air ducts from outside registers.
6. Disconnect vacuum hose from blower housing.
7. Remove lower nut retaining blower motor to evaporator housing. Rotate motor to disengage it from evaporator. Lower motor to floor.
8. Remove heat/defrost air distribution duct or plenum chamber. Then remove retaining clip and open the door in duct.
9. Remove two nuts retaining vacuum motor to duct, Fig. 41.

Recirculating Air Door Vacuum Motor

1. Disconnect battery ground cable.
2. On 1975-77 models, remove lower instrument panel extension, then the right air duct.
3. Remove nuts retaining right air register to instrument panel package tray. Remove inboard bolt from right instrument panel bracket, then the register assembly.
4. Mark position of motor arm, then remove screw securing arm to motor, Fig. 41.
5. Disconnect motor vacuum hose.
6. Remove motor retaining screws and the vacuum motor.

1980 Lincoln Continental & Mark VI

Floor Defrost Door

1. Remove floor air distribution duct.
2. Disconnect vacuum motor from crank arm.
3. Remove vacuum motor and disconnect hose.
4. Reverse procedure to install.

Outside Recirculating Door

1. Disconnect vacuum motor from crank arm.
2. Remove vacuum motor from mounting bracket.
3. Disconnect vacuum hose and remove motor.
4. Reverse procedure to install.

Panel Door—Full and/or Partial

1. Remove plenum from vehicle.
2. Reach through defroster nozzle opening and remove sleeve nut attaching vacuum motor to door.
3. Remove vacuum motor from plenum and disconnect hose from motor.
4. Reverse procedure to install.

1978-79 Lincoln Continental

A/C Heat Door

1. Remove floor air distribution duct from plenum.
2. Disconnect vacuum motor arm from door crank.
3. Remove vacuum motor and disconnect hose.
4. Reverse procedure to install.

Heat-Defrost Door

1. Disconnect vacuum hoses from motor.
2. Disconnect vacuum motor from door crank.
3. Remove vacuum motor from bracket.
4. Reverse procedure to install.

Outside Air Door

1. Remove right cowl air intake screen located under hood near base of windshield.
2. Remove outside air door assembly from cowl panel.
3. Disconnect vacuum hose from motor.
4. Disconnect vacuum motor from door and remove motor from air door assembly.
5. Reverse procedure to install.

Recirculating Air Door

1. Remove sill plate from right front door opening.
2. Remove right cowl side trim panel.
3. Disconnect vacuum motor from cowl vent assembly.
4. Disconnect vacuum motor vacuum hose and remove motor from vehicle.
5. Reverse procedure to install.

1974-77 Lincoln Continental

Temperature Blend Door Vacuum Motor

1. Disconnect battery ground cable.
2. Disconnect vacuum motor operating arm from shaft of blend door, Fig. 42.
3. Disconnect vacuum hose, unfasten and remove motor.

A/C-Heat Door Vacuum Motor

1. Disconnect battery.
2. Remove lower instrument panel cover and glove box.
3. Disconnect defroster hose located in front of motor, Fig. 42.
4. Remove clip from arm. Remove two nuts retaining motor to right side of plenum chamber. Disconnect hose and remove motor.

Heat/Defrost Door Vacuum Motor

1. Disconnect battery ground cable.
2. Remove lower instrument panel cover.
3. Remove clip from motor arm and remove retaining nuts, Fig. 42. Disconnect hose and remove vacuum motor.

1979-80 Ford & Mercury Full Size

Refer to "1980 Lincoln Continental" procedure for vacuum motor replacement.

1974-78 Ford & Mercury Full Size

The fresh air damper door vacuum is serviced by removing the cowl air intake screen under the hood, Fig. 35. The recirculating air door motor is accessible by removing the right cowl trim panel. The air restrictor door motor and water valve are serviceable from the engine compartment. The mode door motor and the A/C-defrost door motor are located on the plenum left side and require instrument panel pad removal to gain access to these motors.

1978-80 Fairmont Zephyr; 1979-80 Capri, Mustang; 1980 Cougar & Thunderbird

A/C-Defrost Door OR Heat-Defrost Door

1. Disconnect battery negative cable. Remove radio.
2. Remove floor air distribution duct from evaporator case.
3. Remove A/C defrost distribution duct to center register duct attaching screws, then pull A/C defrost distribution duct down from its installed position.
4. Disconnect vacuum motor from A/C defrost door.
5. Remove vacuum motor from bracket and disconnect vacuum hose from motor.
6. Reverse procedure to install.

Outside Recirculation Door

1. Remove glove compartment and disconnect vacuum hose from motor.
2. Disconnect motor arm from door shaft.
3. Remove motor from bracket.
4. Reverse procedure to install.

Restrictor Door

1. Remove heater core.
2. Remove vacuum motor from bracket and restrictor door.
3. Disconnect vacuum hose from motor and remove motor through heat-defrost door opening.
4. Reverse procedure to install.

1974-78 Mustang; 1974-80 Bobcat & Pinto

The outside air door vacuum motor is accessible by removing the glove box. The panel-defrost and the A/C-heat vacuum motors can be serviced from beneath the instrument panel, Fig. 43.

1975-80 Granada, Monarch & Versailles

Recirculating Air Door Vacuum Motor

This vacuum motor is located forward of the recirculating air inlet duct.

1. Remove vacuum motor mounting screws.
2. Remove clip attaching motor arm to air door crank arm.
3. Remove vacuum motor from vehicle.

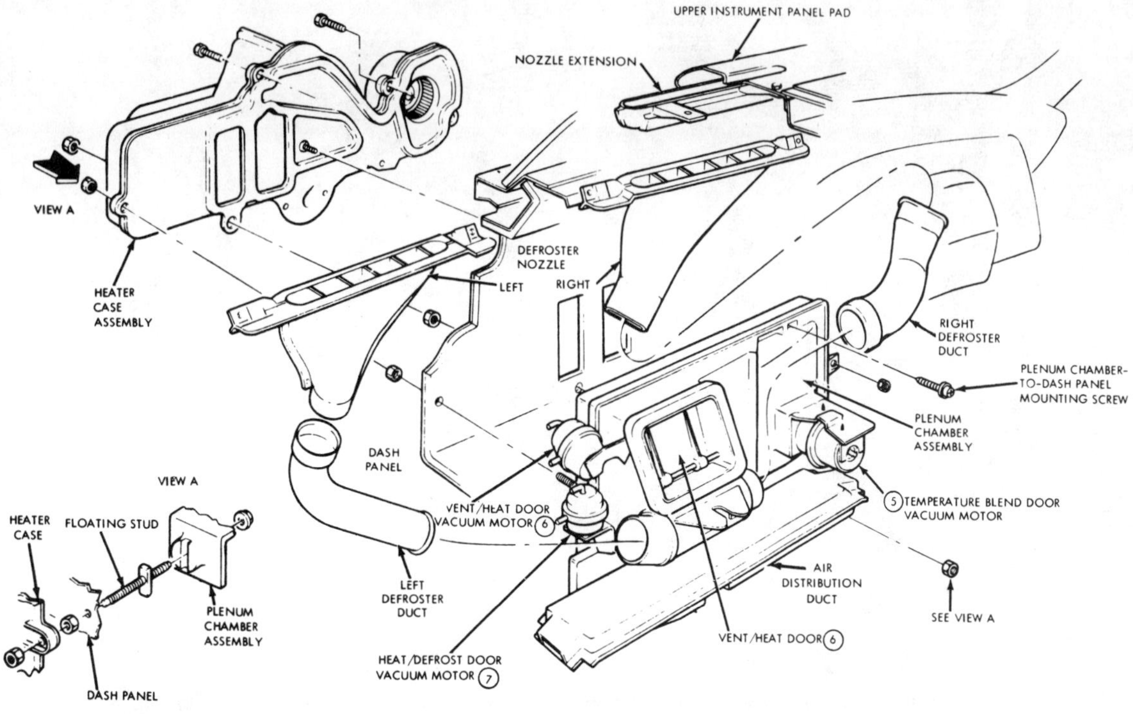

Fig. 42 Vacuum motor locations. 1974-77 Lincoln Continental

Blower Resistor, Replace

1. Disconnect wiring connector from resistor assembly.
2. Remove resistor assembly screws.
3. Remove resistor assembly.
4. Reverse procedure to install.

Ambient Sensor, Replace

1974-77 Lincoln Continental

The ambient sensor is secured to the evaporator case with two sleeve rivets and is accessible from the engine compartment. It is necessary to drill out the rivets for sensor replacement.

1974 Mark IV & Thunderbird

1. Disconnect battery ground cable and remove glove box liner.
2. Remove recirculating air duct from blower assembly.
3. Disconnect sensor wiring harness connector and remove harness retaining clip.
4. Remove sensor from recirculating air door.

Outside Temperature Cut-Off Switch, Replace

1974 Cougar Elite, Mark IV, Montego, Thunderbird & Torino

1. Disconnect battery ground cable and remove glove box liner.
2. Remove recirculating air duct from blower assembly.
3. Disconnect switch wiring harness connector and remove harness retaining clip.
4. Remove switch from recirculating air door.

In-Car Sensor, Replace

1979-80 Ford, Mercury; 1980 Cougar, Fairmont, Lincoln Continental, Mark VI & Thunderbird

1. Disconnect battery negative cable.
2. Remove instrument panel pad.
3. Remove sensor mounting screws, cable housing attaching screw and disconnect control cable from sensor.

NOTE: Secure control cable with tape or service wire to prevent it from falling from sight.

4. Disconnect air and vacuum hoses from sensor and remove from vehicle.
5. Reverse procedure to install. Adjust cable as described under "ATC Sensor Calibration".

1977-80 Granada, Monarch & Versailles

1. Remove instrument panel pad.
2. Carefully disconnect air inlet tube and elbow from sensor.
3. Disconnect control cable and vacuum connector from sensor assembly.
4. Remove sensor attaching screws and sensor.
5. Reverse procedure to install. Adjust cable as described under "ATC Sensor Calibration"

1974 Mark IV & Thunderbird

1. Disconnect battery ground cable and remove instrument panel upper finish panel.
2. Remove sensor to instrument panel screws.
3. Disconnect electrical connector and aspirator hose from sensor and remove sensor from vehicle.

1974-78 Ford, Mercury; 1974-79 Cougar, Elite, LTD II, Montego, Torino; 1975-79 Lincoln Continental, Mark IV, Mark V & Thunderbird

1. Disconnect battery ground cable and remove instrument panel pad.
2. Remove foam insulated ambient air hose, then disconnect vacuum connector and control cable.

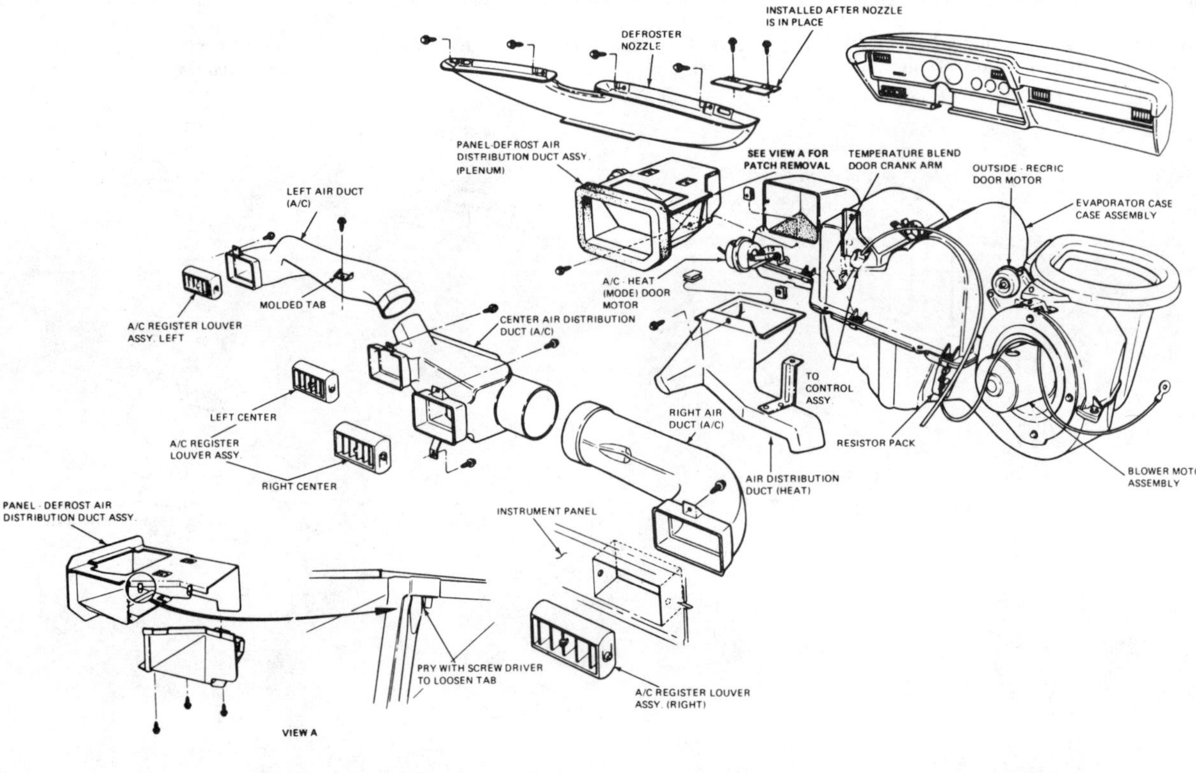

Fig. 43 Evaporator-heater assembly. 1974-78 Mustang II; 1974-80 Bobcat & Pinto (Typical)

3. Remove sensor mounting screws and the sensor.

Servo Assembly, Replace

1979-80 Ford, Mercury; 1980 Cougar, Fairmont, Lincoln Continental, Mark VI & Thunderbird

1. Disconnect battery negative cable.
2. Disconnect glove compartment door stop and allow door to hang.
3. Disconnect vacuum hose from servo and remove mounting screws.
4. Remove vacuum harness from vacuum diverter valve, servo arm protective cover and nut that connects servo motor to blend door.
5. Remove servo motor from vehicle.
6. Reverse procedure to install.

1978 Ford, Mercury; 1978-79 Lincoln Continental

1. Remove center A/C distribution duct to gain access to servo arm attachment.
2. Disconnect vacuum hoses from both A/C-Heat door vacuum motors. Using a hand vacuum pump connected to both motors, evacuate motors to gain access to servo motor arm-to-blend door connecting pin.
3. Loosen servo motor attaching screws to relieve pressure on connecting pin and remove the pin to separate the door arm from servo link and overtravel spring.
4. Remove servo motor attaching screws and disconnect vacuum hose. Pull motor away from the plenum to clear arm and lower for access to vacuum diverter valve.
5. Disconnect vacuum harness from vacuum diverter valve and remove vacuum servo.
6. Reverse procedure to install.

1977-79 Cougar, LTD II, Mark V & Thunderbird

1. Remove ATC plenum, and disconnect hose from servo.
2. Remove vacuum harness from diverter valve.
3. Remove plate covering blend door crank arm.
4. Remove retainer from blend door crank arm, servo mounting screws and servo.
5. Reverse procedure to install.

1974-77 Ford, Mercury; 1974-76 Cougar, Elite, Mark IV, Montego, Thunderbird & Torino

1. Disconnect battery ground cable and remove glove box.
2. Mark position of servo motor on mounting bracket, Fig. 44.
3. Disconnect vacuum hoses from servo motor and electro vacuum relay (EVR).
4. Disconnect electrical connector from main wiring harness.
5. Disconnect vacuum connector from servo diverter valve.
6. Remove distribution plenum retaining screws, then the plenum from dash panel. Remove blend door arm to servo motor link pin and separate door arm from overtravel spring on motor link end.
7. Remove servo assembly to mounting bracket screws, then the servo assembly from distribution plenum chamber.
8. Reverse procedure to install.

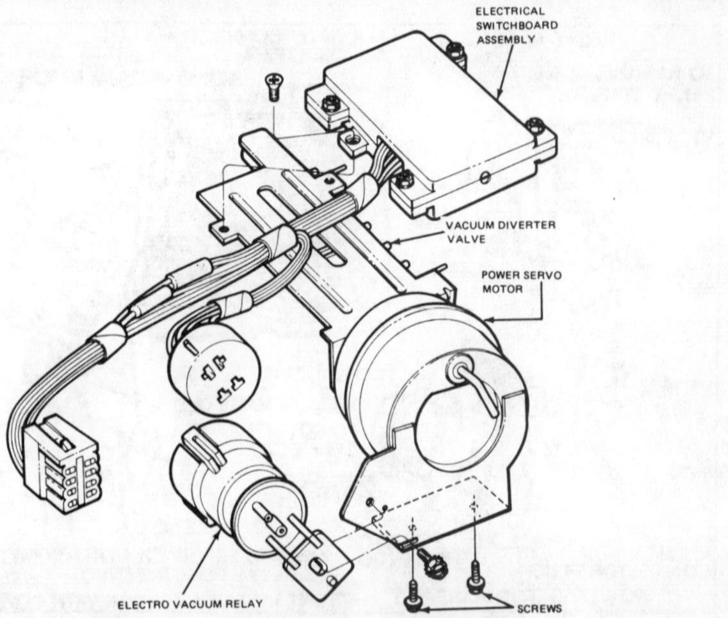

Fig. 44 ATC servo assembly. 1974-76 Cougar, Elite, Mark IV, Montego, Thunderbird, Torino; 1974-77 Ford & Mercury with ATC

1977-80 Granada, Monarch & Versailles

1. Remove glove compartment liner.
2. Disconnect vacuum hose from servo (tan).
3. Remove vacuum harness from servo diverter valve.
4. Disconnect servo arm and spring from blend door crank arm and remove servo.
5. Reverse procedure to install.

Servo Electrical Switchboard, Replace

1974-77 Ford, Mercury Full Size; 1974-76 Cougar, Elite, Montego, Torino; 1975-76 Mark IV & Thunderbird

1. Remove servo assembly as outlined under "Servo Assembly" procedure.
2. Remove switchboard attaching screws and the switchboard from servo assembly, Fig. 44.

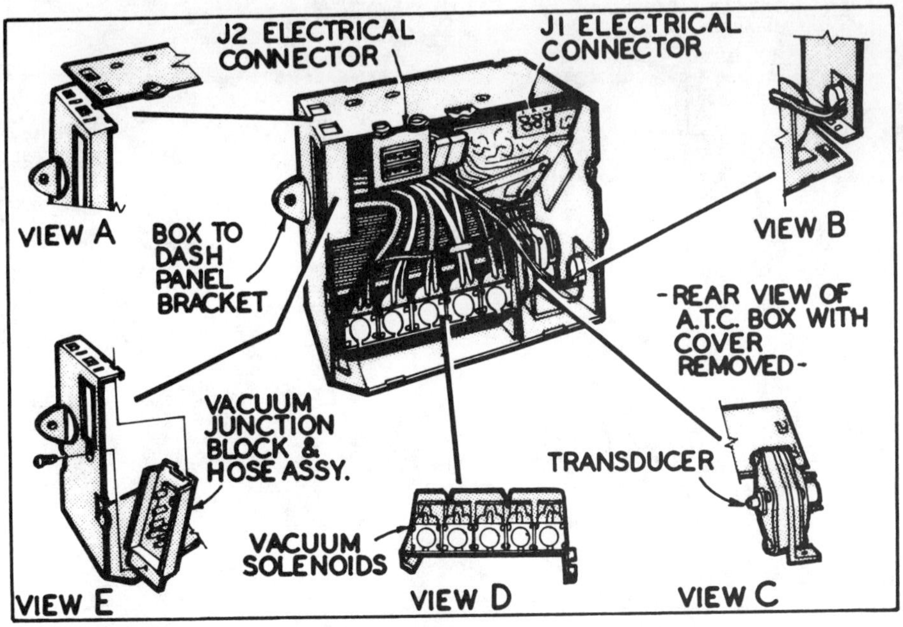

Fig. 45 ATC Box, rear view. 1974 Mark IV, Thunderbird & 1974-77 Lincoln Continental with ATC

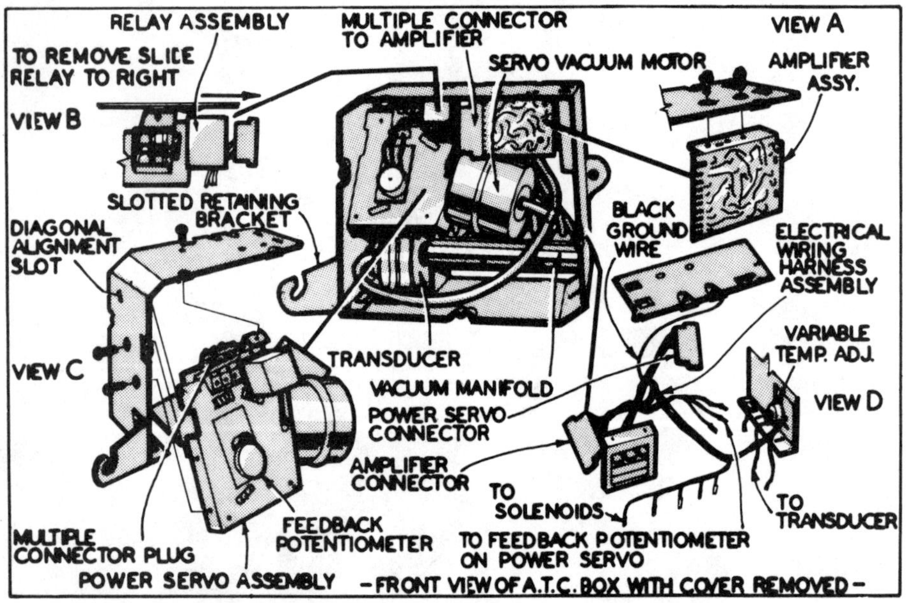

Fig. 46 ATC Box, front view. 1974 Mark IV, Thunderbird & 1974-77 Lincoln Continental with ATC

ATC Box, Replace & Service

1974 Mark IV & Thunderbird

1. Disconnect battery ground cable.
2. Remove instrument panel applique located above glove box.
3. Remove ATC box mounting screws and slide ATC box rearward through opening in instrumental panel.

1974-77 Lincoln Continental

1. Remove glove box liner screws, disconnect glove box light and remove glove box liner.
2. Remove ATC box retaining nuts.
3. Disconnect electrical wiring from box.
4. Remove vacuum hose retainer, disconnect vacuum hoses and remove ATC box from vehicle.

Service

Transducer, Replace

1. Remove front and rear cover plates.
2. Slide locking tabs from retaining slots and separate sides of ATC box at upper left and lower right corners.
3. Disconnect three wires from transducer and remove transducer mounting screws from bottom of box, Fig. 45.
4. Disconnect vacuum hose and the manifold hose from transducer.
5. Remove transducer from box.

Vacuum Solenoids, Replace

1. Remove transducer as outlined previously.
2. Remove one mounting plate attaching screw from lower left corner of the box.
3. Disconnect vacuum hoses and electrical leads from solenoids, Fig. 45.
4. Remove solenoids and mounting plate from bottom of box as an assembly.

Amplifier, Replace

1. Remove front cover plate.
2. Remove electrical connector from amplifier, Fig. 46.
3. Remove amplifier mounting screws from top of ATC box, then the amplifier.

Relay Assembly, Replace

1. Remove front cover plate.
2. Disconnect electrical connector from amplifier and position amplifier aside, Fig. 46.
3. Slide relay on guide rails toward amplifier and lift relay from box.

Power Servo Assembly, Replace

1. Remove front cover plate.
2. Disconnect multiple electrical connector from power servo plug, then the vacuum

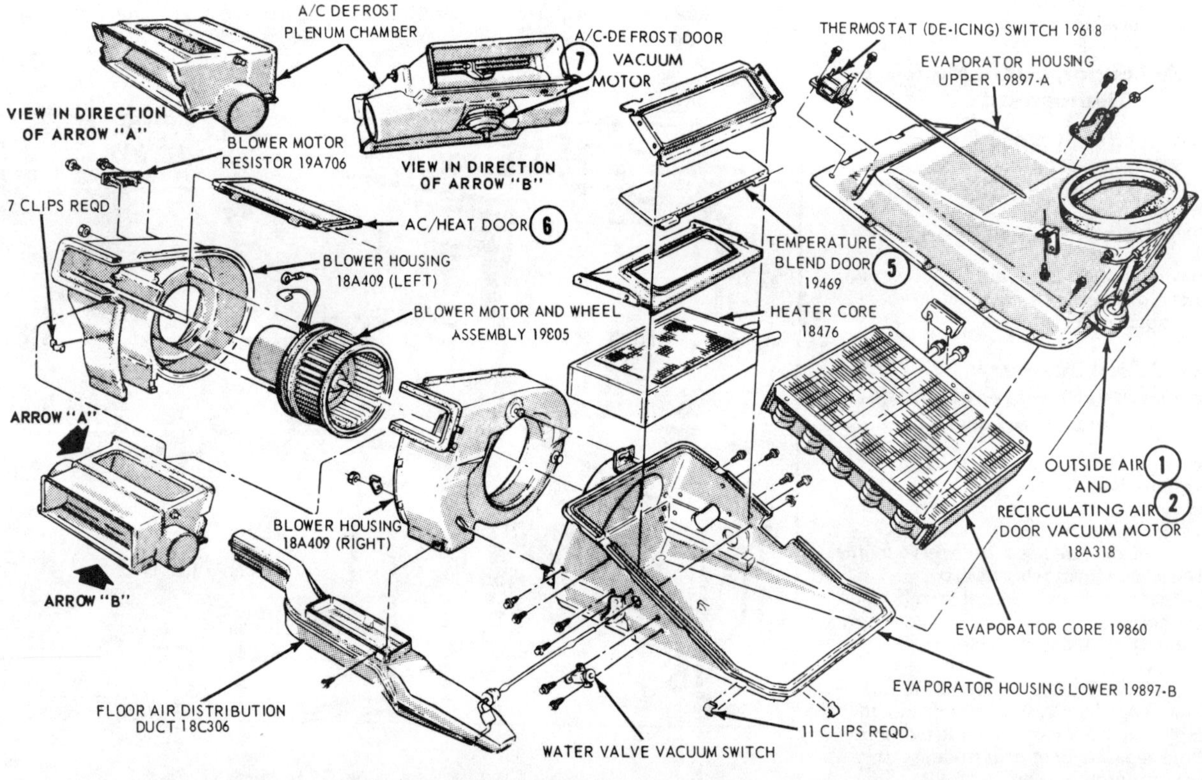

Fig. 47 Evaporator and blower housing. 1974-77 Comet & Maverick (Typical)

hose from the servo motor, Fig. 46.
3. Remove assembly mounting screws from top and side of ATC box, then the power servo assembly.

Electrical Wiring Harness, Replace

1. Remove front and rear cover plates.
2. Disconnect solenoid and transducer wiring. Fig. 46.
3. Disconnect electrical connectors from amplifier board and power servo circuit board.
4. Disconnect wiring from feedback potentiometer on power servo.
5. Remove J2 electrical connector mounting plate screws.
6. Unsolder the black ground wire from inside top surface of box.
7. Unsolder purple and grey wires from potentiometer in the lower right corner of the box, then remove the harness assembly.

Evaporator Core, Replace

PASSENGER CARS

1974-77 Comet & Maverick

1. Disconnect battery ground cable, remove air cleaner, drain cooling system and discharge refrigerant from system.
2. Disconnect expansion valve from evaporator and heater hoses from heater core.
3. From engine side, remove the three A/C housing stud nuts, Fig. 47.
4. Remove lower instrument panel and disconnect right and left register ducts from plenum.
5. On 1974-77 models, remove floor distribution duct and center register, then pull plenum chamber to disengage it from blower housing.
6. Disconnect vacuum hoses from actuators and from water vacuum switch and disconnect control cable from door crank arm.
7. Remove any tape or clips retaining vacuum hoses to unit, then disconnect wires from A/C thermostat and resistor.
8. On 1974-75 models, disconnect blower motor ground wire from windshield wiper mounting bracket.
9. On 1976-77 models, disconnect blower motor connector and vacuum source connector and remove blower motor.
10. Remove evaporator housing to cowl upper support retaining screw and move assembly rearward to clear mounting studs.
11. Pull drain tube from floor pan, then lower assembly and remove from vehicle.
12. Remove flange clips from housing and separate housing.
13. Remove thermostatic switch and carefully pull capillary tube out.
14. Remove retaining screws and evaporator.
15. Reverse procedure to install.

1977-79 Cougar, LTD II, Mark V & Thunderbird

1. Discharge refrigerant from system.
2. Raise and support vehicle. Remove the right front wheel.
3. Drill a 1.5 inch hole in fender apron, Fig. 48. A drill dimple is provided for proper location of hole.
4. Drain engine coolant and disconnect heater hoses at heater core.
5. Disconnect refrigerant hoses from evaporator. Plug all open hoses to prevent entry of dirt and moisture into system. Remove wiring connector from blower resistor.
6. Remove evaporator housing to dash panel fasteners. One is removed from inside of vehicle and is located on the lower flange of the evaporator housing on the inboard of evaporator core, Fig. 49.
7. Remove the self-tapping screws (¼-20) located at top of evaporator assembly.
8. Remove nuts from inboard side of evaporator (these nuts also retain A/C plenum).
9. Remove top nut on outboard side of evaporator (this nut also retains blower assembly).
10. Through service hole in fender apron, loosen lower nut on outboard side and run the nut to end of stud.

NOTE: Do not remove nut from stud.

11. On inside of vehicle, remove instrument panel pad and right cowl trim panel. Then, remove two fasteners retaining the instrument panel to the body (one is located on cowl inner panel, the other is located on cowl side panel).
12. Remove radio speaker, support rod, defroster nozzle, right side duct, floor ducts and center duct.
13. Remove the two outboard nuts retaining the plenum. Then, remove screw retaining blower assembly support bracket to inner cowl.
14. Move plenum and blower assemblies rearward to reduce the length of stud protruding through the dash panel.

NOTE: The plenum and blower assemblies do not have to be removed from vehicle.

15. Remove evaporator-heater housing from vehicle.
16. Separate evaporator case halves and remove evaporator core.
17. Reverse procedure to install. Use new case seals. Use new O-rings lubricated with refrigerant oil. Install a plug button in service hole in fender apron.
18. Evacuate, charge and leak test system. Check for proper operation.

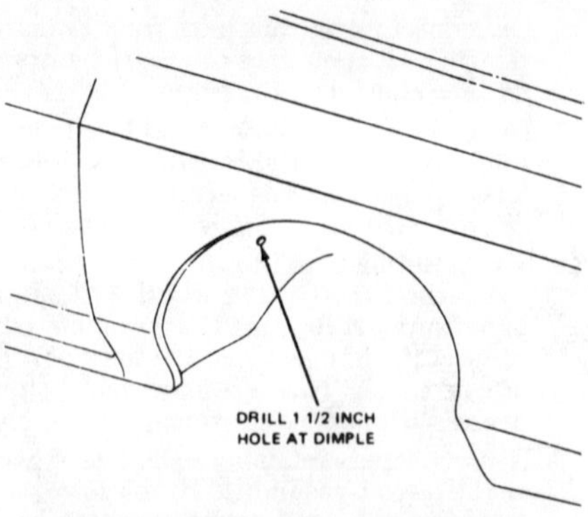

Fig. 48 Location for cutting access hole. 1977-79 Cougar, LTD II, Mark V & Thunderbird

1974-76 Cougar Elite, Mark IV, Montego, Thunderbird, Torino

1. Disconnect battery ground cable, discharge refrigerant from system, drain cooling system and disconnect heater hoses.
2. Remove expansion valve and suction throttling valve.
3. From engine compartment, remove two nuts and three screws from evaporator case

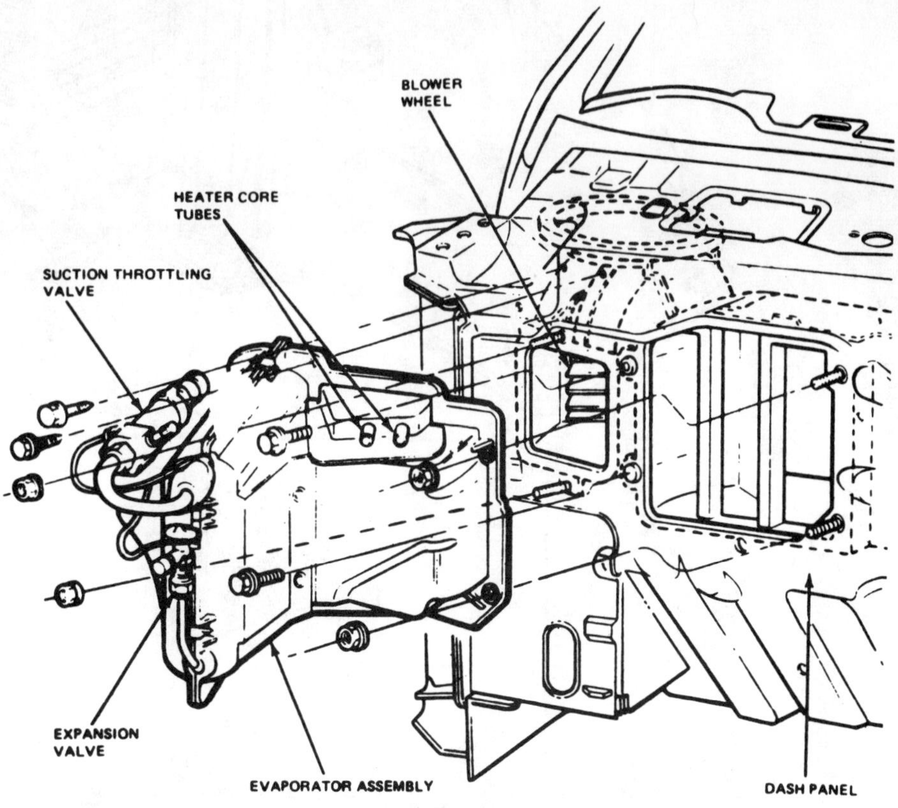

Fig. 49 Evaporator assembly. 1977-79 Cougar, LTD II, Mark V & Thunderbird

mounting flange, then from under instrument panel, remove two plenum right lower screws and one blower housing to dash to evaporator screw, Fig. 50.

NOTE: On all except Thunderbird models, remove the plenum mounting screws and move plenum rearward to obtain clearance.

4. Remove evaporator case by lifting it upward to clear engine.
5. Cut flange gasket at points shown, Fig. 50, then remove the seven retaining screws, separate case and lift evaporator out.
6. Reverse procedure to install.

1979-80 Ford, Mercury, 1980 Lincoln Continental & Mark VI

1. Disconnect negative battery cable. Discharge refrigerant from system.
2. Disconnect refrigerant lines from evaporator. Plug all open lines to prevent entry of dirt and moisture into system.
3. Disconnect heater hoses from heater core. Plug hoses and heater core tubes to prevent loss of coolant. Disconnect pressure control de-ice switch wiring connector, if equipped.
4. Turn steering wheel to the left to position right wheel so that the wheel well splash panel support bracket is accessible. Disconnect the bracket from the back portion of wheel well. This will allow panel to be depressed for added clearance.
5. Remove screws retaining right side of hood seal bracket assembly. Remove copper hood ground clip from underneath hood seal and fold the seal assembly towards the left side of vehicle.
6. Position all vacuum hoses and movable wires away from evaporator assembly.
7. Disconnect blower motor ground wire and accompanying engine ground wire from body frame.
8. Disconnect blower motor wiring harness from main harness. Then, disconnect the two large wiring harnesses (which pass over evaporator housing) at the various hard shell connecting points and position aside.
9. From under instrument panel, fold right side carpsting back. Remove the bottom left screw that supports the inlet recirculation air duct assembly.
10. From engine compartment, remove the three nuts from the evaporator assembly mounting studs. Remove the two screws lo-

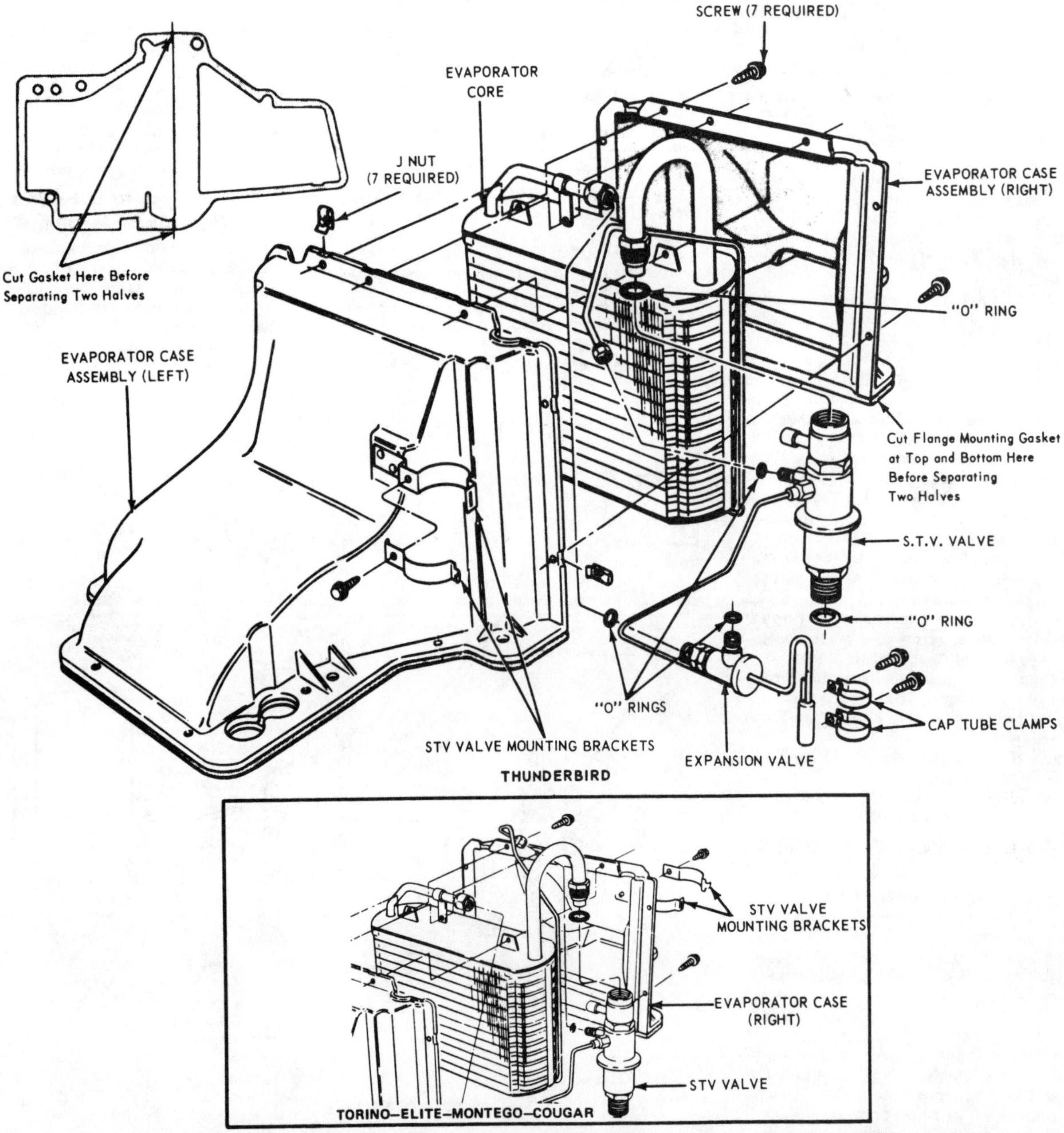

Fig. 50 Evaporator assembly. 1974-76 Cougar, Elite, Mark IV, Montego, Thunderbird & Torino

cated at the blower motor housing portion of evaporator housing, Fig. 51.

11. Pull bottom of case away from dash panel. Then move top of case away and remove evaporator case assembly from vehicle.
12. Separate evaporator case assembly and remove evaporator core.
13. Reverse procedure to install. Use new seals and heat shield on case assembly. Use new O-rings lubricated with refrigerant oil.
14. Evacuate, charge and leak test system. Check for proper operation.

1974-78 Ford, Mercury & 1974-79 Lincoln Continental

1. Discharge refrigerant from system.
2. Disconnect suction line from STV and liquid line from expansion valve. Cap all open lines to prevent entry of dirt and moisture.
3. Remove insulation from expansion valve bulb and disconnect bulb from line.
4. Remove suction throttling valve clamp and refrigerant line support bracket clamp.
5. Remove evaporator core cover flange screws and liquid line support bracket, Fig. 52.
6. Remove left and right cover attaching screws and remove left cover, then pull evaporator core and right cover diagonally forward out of housing.
7. Disconnect oil return line and equalizer line from suction throttling valve.
8. Disconnect suction throttling valve and expansion valve from evaporator core, then remove right cover and gasket.
9. Reverse procedure to install.

1978-80 Fairmont, Zephyr; 1979-80 Capri, Mustang; 1980 Cougar & Thunderbird

1. Disconnect battery negative cable.
2. Remove instrument panel pad, steering column lower opening cover and steering column trim shrouds.
3. On vehicles equipped with automatic transmission, remove four nuts attaching steering column to brake pedal support and carefully lower steering column only enough to gain access to the transmission gear shift selector and cable assembly.

CAUTION: If steering column is lowered too far, damage to shift linkage or cable will result.

4. Reach between steering column and instrument panel and gently lift the selector lever cable off the lever. Then, remove cable clamp from steering column tube and lay steering column on seat.
5. On all models, remove screw attaching instrument panel to brake pedal support at steering column opening. Remove screw attaching lower brace to the lower edge of the instrument panel below the radio. Remove screw attaching the brace to lower edge of instrument panel.
6. Disconnect vacuum, control cable(s) and wiring connectors from evaporator housing.
7. Disconnect in-line connector to blower motor.
8. Support instrument panel and remove the three remaining screws attaching panel to cowl.
9. Remove one screw at each end of panel to cowl side panels.
10. Move instrument panel rearward and disconnect speedometer cable and any wires that will not allow panel to lay on front seat.
11. Discharge refrigerant from system. Drain coolant from vehicle.
12. Disconnect refrigerant hoses from combination valve. Plug all open connections to prevent entry of dirt and moisture into system.
13. Disconnect heater hoses from heater core and plug the core and hoses.
14. Disconnect blower housing support brace from cowl top panel.
15. Disconnect vacuum supply hose (black) from in-line check valve in engine compartment.
16. Disconnect blower motor wiring.
17. Working under hood, remove two nuts retaining evaporator housing to dash panel.
18. Working inside vehicle, remove screw attaching evaporator housing support bracket to cowl top panel. Remove nut attaching bracket at left end of evaporator housing to dash panel, and nut retaining the bracket below the evaporator case to dash panel.
19. Carefully pull evaporator housing away from dash panel and remove from vehicle, Fig. 53.
20. Remove heater core access cover, heater core and restrictor door vacuum motor.

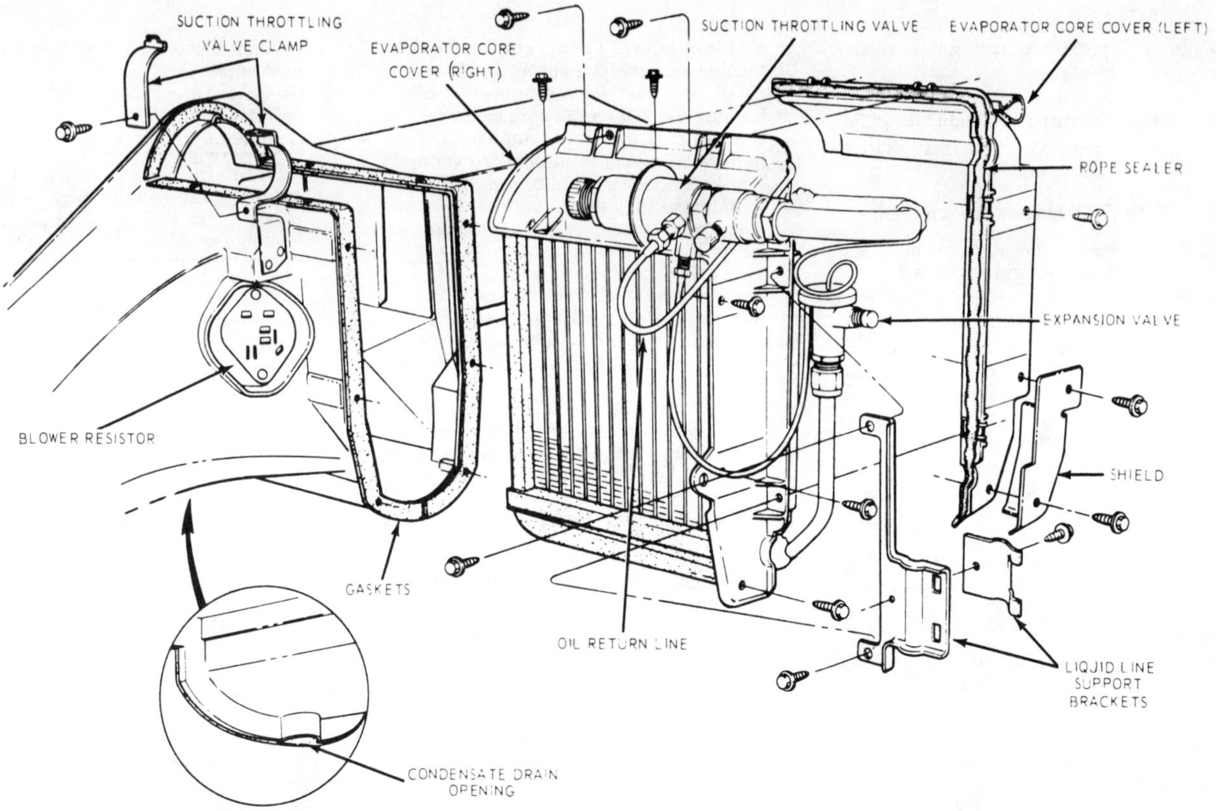

Fig. 51 Evaporator assembly. 1979-80 Ford, Mercury; 1980 Lincoln Continental & Mark VI

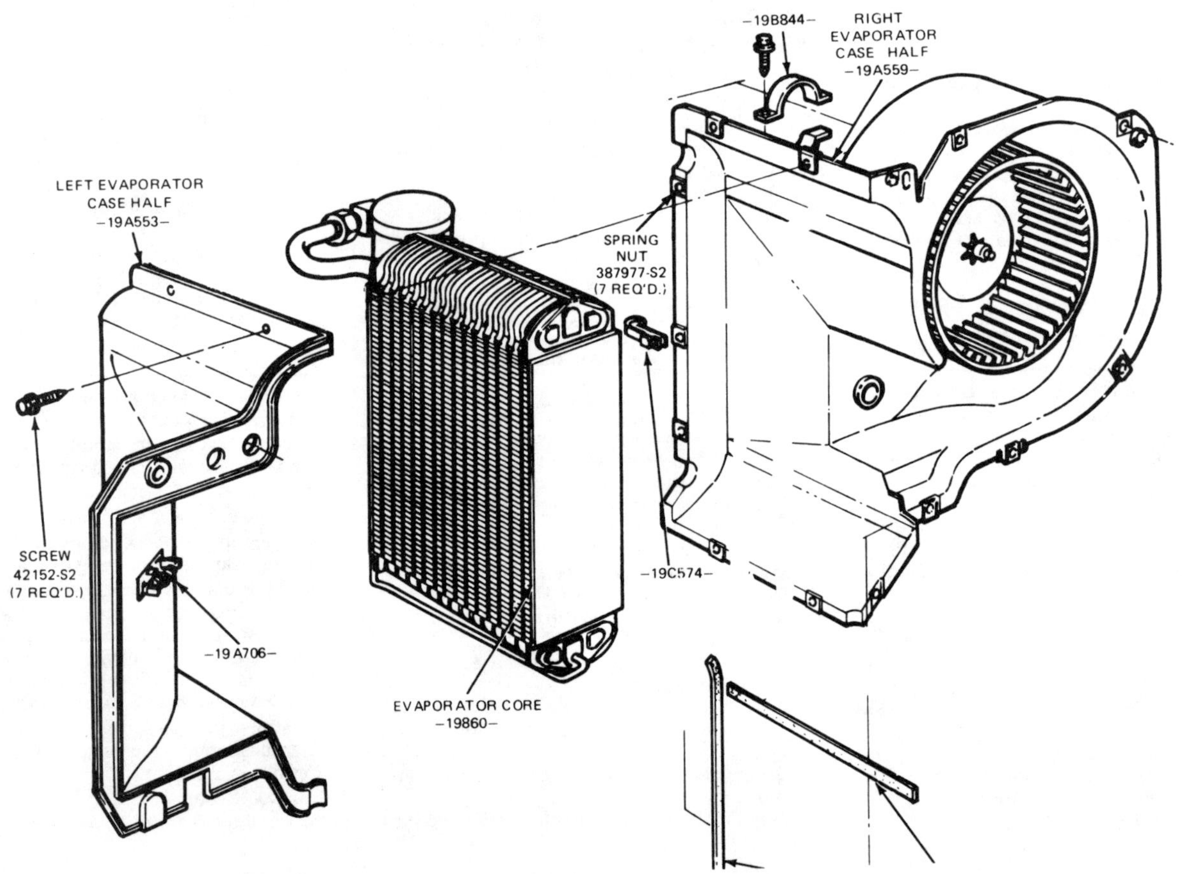

Fig. 52 Evaporator assembly. 1974-78 Ford, Mercury; 1974-79 Lincoln Continental

21. Separate evaporator housing halves and remove evaporator core.
22. Reverse procedure to install. Use new O-rings lubricated with refrigerant oil.
23. Evacuate, charge and leak test system. Check for proper operation.

1979-80 Bobcat & Pinto

1. Disconnect battery negative ground cable. Discharge refrigerant from system.
2. Drain coolant from system. Disconnect heater hoses from heater core.
3. Remove the hex head screws attaching the evaporator manifold plate to the expansion valve body and separate the parts. Plug all open connections to prevent the entry of dirt and moisture into system.
4. Remove glove compartment liner and sight shield along lower edge of instrument panel.
5. Disconnect control assembly vacuum harness at connector near evaporator housing.
6. Disconnect control cable from evaporator housing.
7. Remove floor air distribution duct, A/C defrost adapter duct, right register duct and A/C defrost distribution duct.
8. Disconnect blower motor wiring and remove blower motor.
9. Install a ¼-20 bolt through the flange tab on the outside recirculating duct to cowl upper bracket to hold the duct in place, Fig. 54.
10. Remove three screws attaching air inlet duct to evaporator housing through blower motor opening. Then, remove screw attaching evaporator housing to outside recirculating duct (located between blower housing and evaporator core below the outside recirculating door vacuum motor).
11. Remove two screws attaching evaporator housing upper brackets to cowl top panel.
12. Working in engine compartment, remove evaporator housing to dash panel nuts and rotate evaporator housing down and out from under instrument panel.
13. Remove evaporator housing upper case attaching bolts and upper case half.
14. Move evaporator core tube seal forward to clear case mounting stud and lift core from case.
15. Reverse procedure to install. Use new O-rings lubricated with refrigerant oil.
16. Evacuate, charge and leak test system. Check system for proper operation.

1974 Mustang & 1974-78 Bobcat & Pinto

1. Drain cooling system, discharge refrigerant from system and disconnect battery ground cable.
2. Disconnect refrigerant lines and front half of refrigerant manifold. Remove manifold stud to provide clearance when removing evaporator case.
3. Remove heater hoses and A/C drain hose.
4. Remove glove box then disconnect vacuum hoses and temperature control cable from evaporator case.
5. Remove heat distribution duct, Fig. 54.
6. On Mustang II models, remove mode door vacuum motor and lower section of A/C-defrost plenum.
7. On Bobcat and Pinto models, remove A/C-defrost plenum.
8. Remove blower motor assembly.

NOTE: Install ¼-20 hex head screw into mounting tab on inlet duct to hold duct in place. Leave this screw in place when installing assembly.

9. Remove three inlet duct to evaporator case screws through blower motor opening and remove one upper case to inlet duct screw located under outside recirc motor mounting bracket.
10. Remove two evaporator to upper cowl bracket screws and four evaporator to dash panel stud nuts.
11. Rotate unit down and away from dash panel and out from under instrument panel.
12. Reverse procedure to install.

1975-80 Granada, Monarch, & Versailles

1. Disconnect battery ground cable, discharge refrigerant from system and drain cooling system.
2. Remove front half of refrigerant manifold and cover manifold plates to prevent entry of dirt and moisture.
3. Disconnect heater hoses and disconnect vacuum supply hose from water valve and system supply hose from supply tank, then push vacuum line grommet through dash panel.

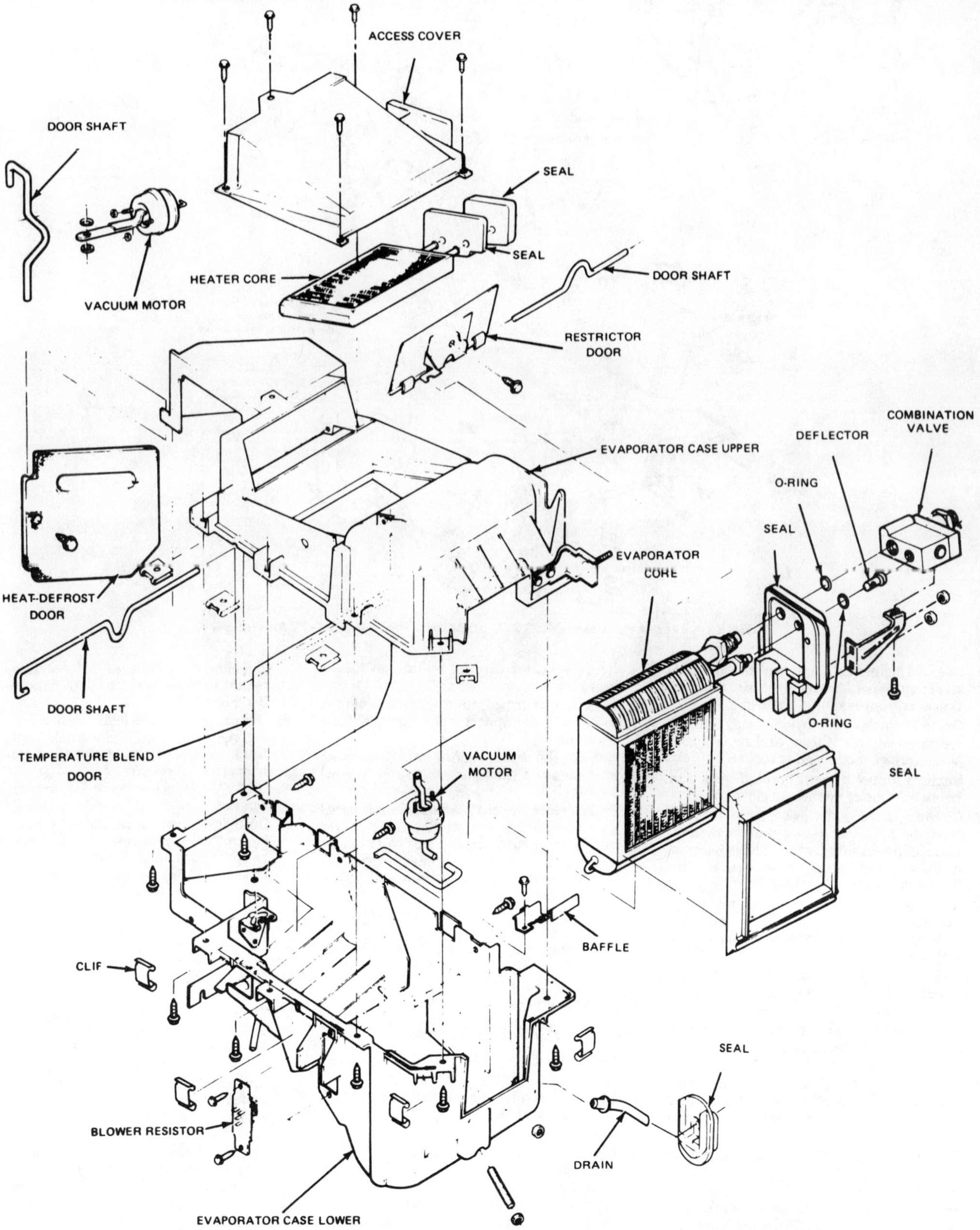

Fig. 53 Evaporator assembly. 1978-80 Fairmont, Zephyr; 1979-80 Capri, Mustang; 1980 Cougar & Thunderbird

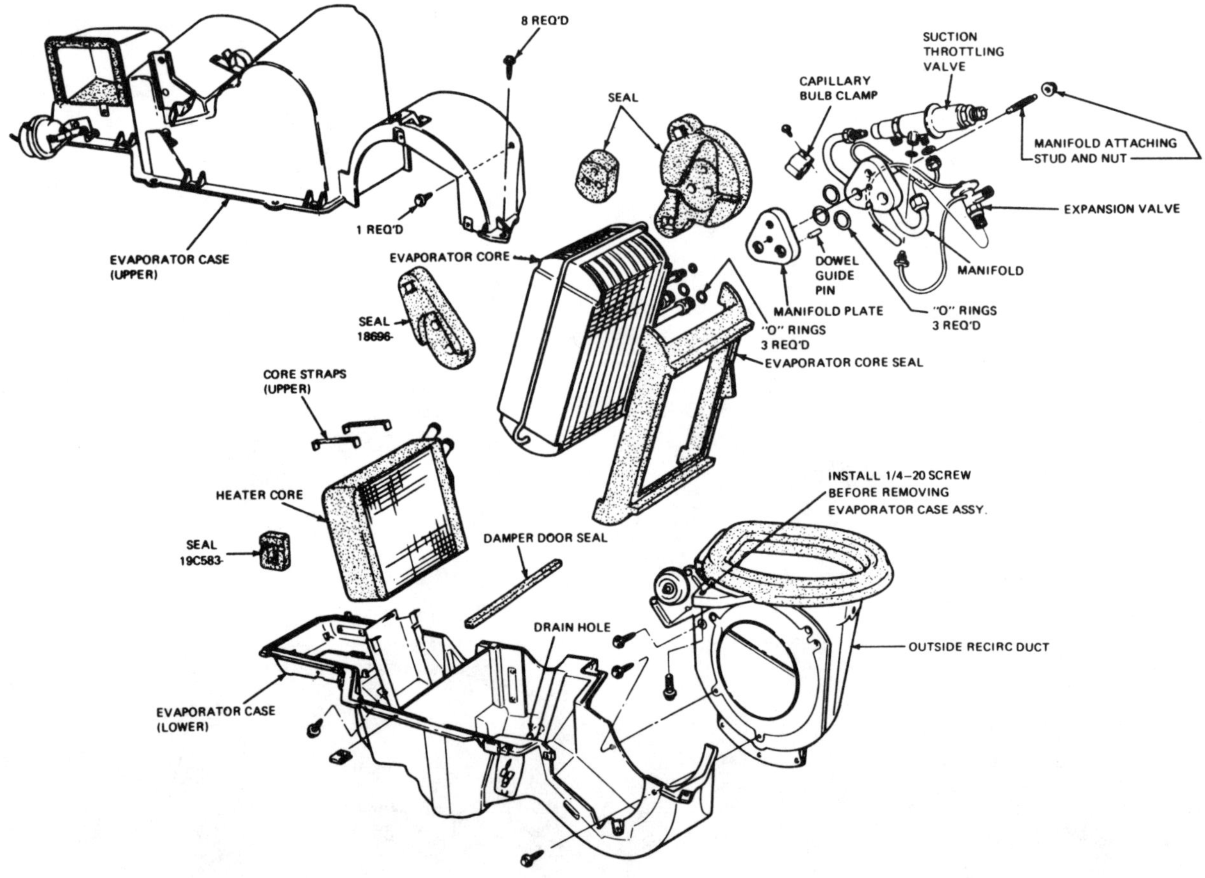

Fig. 54 Evaporator assembly. 1974-80 Bobcat, Pinto & 1974-78 Mustang II

4. Remove sight shield and brackets from instrument panel and remove heat distribution duct from evaporator housing, Fig. 55.
5. On 1975 models, remove steering column vibration damper if used, and remove seat belt interlock module and bracket.
6. Remove tunnel to cowl brace if used, and remove defroster nozzle and glove box.
7. Disconnect vacuum hoses from A/C-defrost and A/C-heat door actuators, then remove clip retaining vacuum hose harness, and disconnect vacuum hose harness at inline connector to control switch.
8. Remove nuts retaining heat/defrost vacuum actuator and swing vacuum actuator rearward on crank arm and disconnect control cable.
9. Loosen right door sill scuff plate and right A-pillar trim panel and remove right cowl side trim panel.
10. Remove right lower instrument panel to cowl side bolt and loosen right upper instrument panel to cowl upper bolt.
11. Remove cowl to lower instrument panel brace and disconnect evaporator mid-case to cowl mounting strap, leaving strap loosely attached to evaporator.
12. Remove plenum to left mounting bracket screws and plenum to evaporator case mounting screws and clips.
13. Swing bottom of plenum away from evaporator case to disengage S-clip and raise plenum to clear tabs on top of evaporator case. Move plenum to the left and place it on floor tunnel.
14. Remove heater core by pulling on tab or rear gasket.
15. Disconnect wiring from resistor and blower motor and disconnect right courtesy light from clip.
16. Remove the three evaporator to dash panel stud nuts and the right evaporator to cowl bracket bolt.
17. Pull evaporator case away from dash panel and raise plenum end while pulling blower end down. Swing plenum end to the right and pull case rearward under instrument panel rail, then swing plenum end to the left and remove unit from vehicle.
18. Remove blower motor assembly and recirculating air inlet duct.
19. Remove screws, clamps and clips, then separate case halves and remove evaporator core from upper case.
20. Reverse procedure to install.

1975-78 Mustang

1. Remove battery, drain cooling system and discharge refrigerant from system.
2. Remove instrument panel pad assembly.
3. On 1975 models, remove cluster finish panel and cluster assembly.
4. Remove radio speaker, both A-pillar mouldings and both side kick panel pad assemblies.
5. Remove lower steering column cover.
6. On 1975 models, remove right vent control assembly.
7. Remove steering column to cowl panel shake brace, then remove accelerator pedal assembly and bottom bolt retaining center shake brace to instrument panel assembly, Fig. 54.
8. Disconnect antenna lead, connectors at left and right cowl panel, windshield wiper plug from wiper switch and connectors from blower motor resistor.
9. On 1975 models, remove left vent control screws.
10. On 1976-78 models, disconnect temperature control cable and remove two evaporator to upper cowl bracket screws.
11. Disconnect main wiring harness and all remaining connectors in engine compartment and push harness and connectors into passenger compartment.
12. Disconnect turn signal indicator switch assembly, then remove the four steering column mounting nuts and lower steering column.
13. Disconnect connectors at ignition switch and stop light switch.
14. Remove instrument panel to steering column center support, then remove instrument panel bolts and remove instrument panel.
15. Disconnect heater hoses and lines at evaporator manifold valve assembly.
16. Remove screw, loosen nut and remove refrigerant manifold from bracket, then remove stud nut and remove outer manifold.
17. Remove heater to dash panel nuts and remove heater assembly.
18. Remove defroster and A/C duct assembly.
19. Remove mounting brackets from evaporator case and remove assembly.

20. Remove evaporator case retaining screws, separate case and remove evaporator core.
21. Reverse procedure to install.

TRUCKS & VANS

1980 F Series & Bronco

1. Disconnect battery ground cable. Discharge refrigerant from system.
2. Disconnect pressure switch and remove from suction accumulator, Fig. 56.
3. Disconnect refrigerant lines from accumulator and evaporator. Plug all open lines to prevent entry of dirt and moisture into system.
4. Working in passenger compartment, remove the screw attaching bottom of plenum to dash panel.
5. Working under the hood, remove the nut retaining the upper left corner of evaporator case to dash panel.
6. Remove six screws attaching left evaporator housing to evaporator case.
7. Remove two spring clips holding left side of housing cover plate and left housing together at dash panel.
8. Remove evaporator housing from case.
9. Remove evaporator core and suction accumulator from evaporator case.
10. Reverse procedure to install. Use new O-rings lubricated with refrigerant oil.
11. Evacuate, charge and leak test system. Check for proper operation.

1974-80 Econoline Models

1. Disconnect blower resistor wiring and vacuum line from outside recirculated door vacuum motor, Fig. 57.
2. Remove A/C blower cover. Then, remove push nut from outside recirculated door shaft.
3. Disconnect control cable from evaporator housing. Remove A/C blower housing.
4. Partially drain engine coolant and disconnect heater hoses from heater core.
5. Remove battery from vehicle. Discharge refrigerant from system.
6. Disconnect refrigerant lines from evaporator core. Plug all open connections to prevent the entry of dirt and moisture into system.
7. Remove drain tube from bottom of evaporator case assembly. Then, remove evaporator assembly from dash, Fig. 57.
8. Remove blend door and housing. Then remove insulation tape from expansion valve bulb.
9. Remove evaporator tube retainer, seal and evaporator core.
10. Reverse procedure to install. Use new O-rings lubricated with refrigerant oil.
11. Evacuate, charge and leak test system. Check for proper operation.

1974-80 Econoline Auxiliary Evaporator

1. Remove first bench seat (if equipped).
2. Remove cover assembly, Fig. 58.
3. Discharge refrigerant from system.
4. Disconnect refrigerant lines from evaporator core. Plug all open connections to prevent entry of dirt and moisture into system.
5. Remove evaporator core and seal assembly with expansion valve attached.
6. Reverse procedure to install. Use new O-rings lubricated with refrigerant oil.
7. Evacuate, charge and leak test system. Check for proper operation.

1974-78 C Series

1. Discharge refrigerant from system.
2. Remove A/C unit cover, screen and evaporator cover.
3. Disconnect expansion valve and refrigerant line from evaporator core. Plug all open lines to prevent entry of dirt and moisture into system.
4. Remove evaporator attaching screws, pull thermostatic switch sensing tube from core and remove evaporator from vehicle.

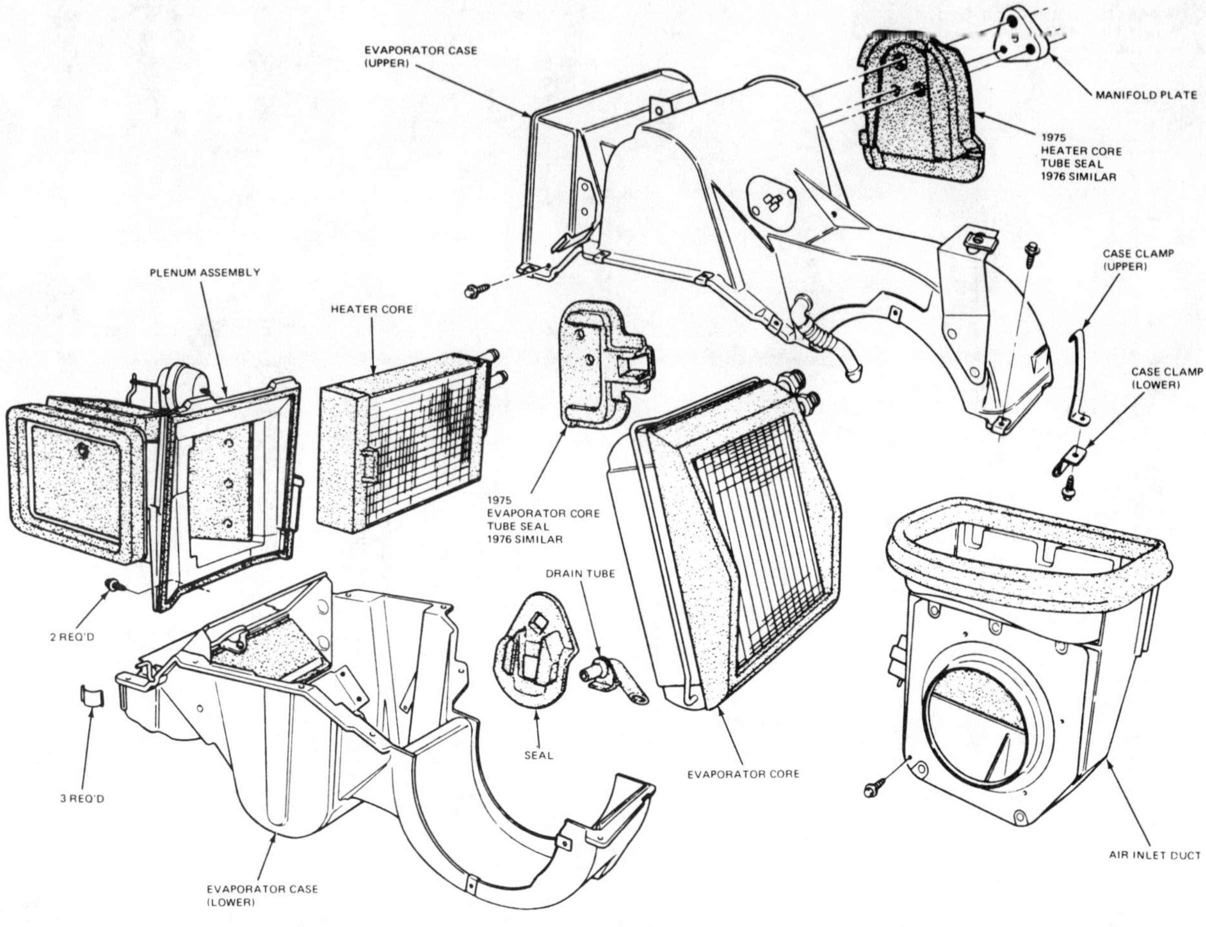

Fig. 55 Evaporator assembly. 1975-80 Granada, Monarch & Versailles

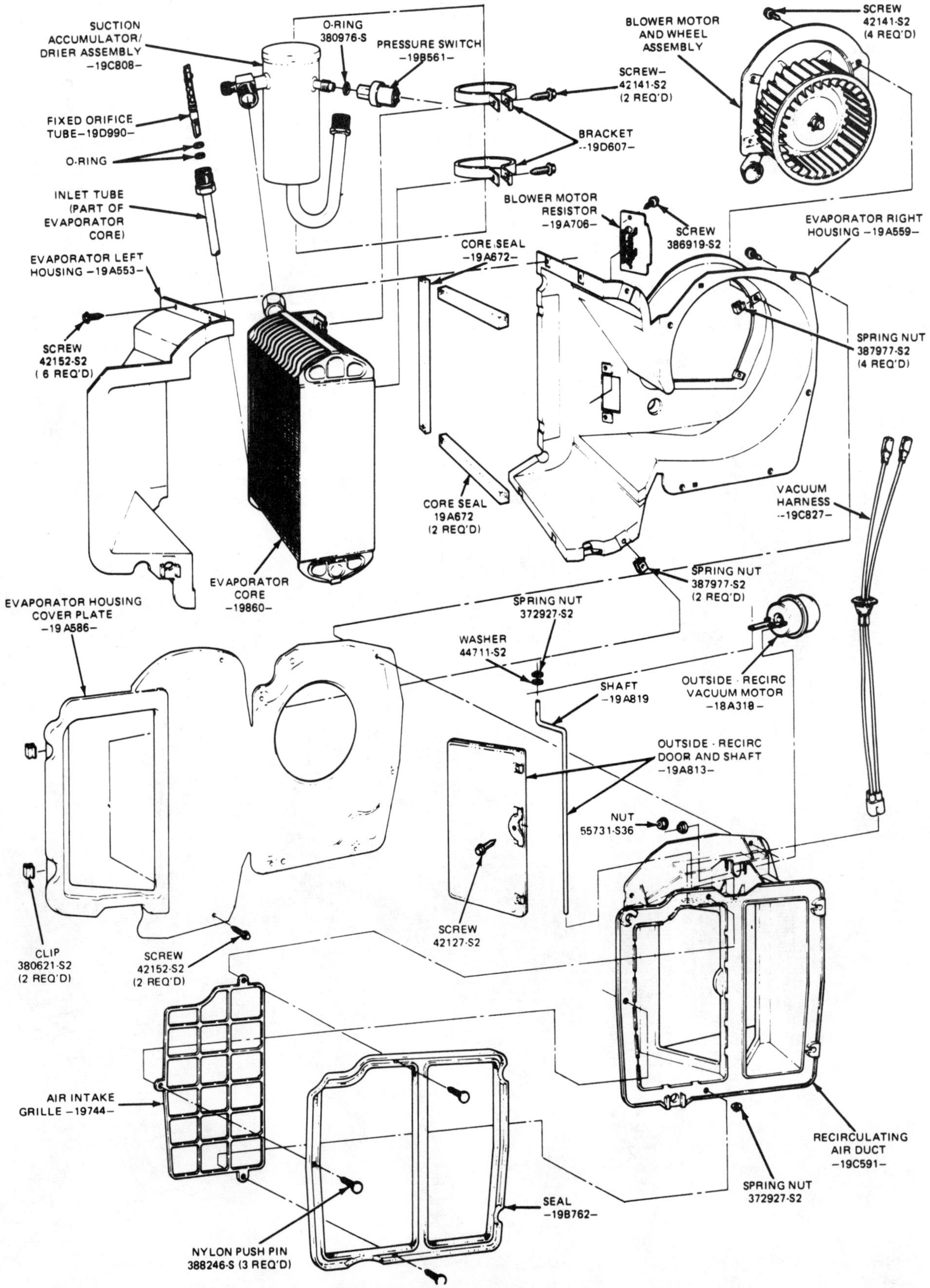

Fig. 56 Evaporator assembly. 1980 'F' Series and Bronco

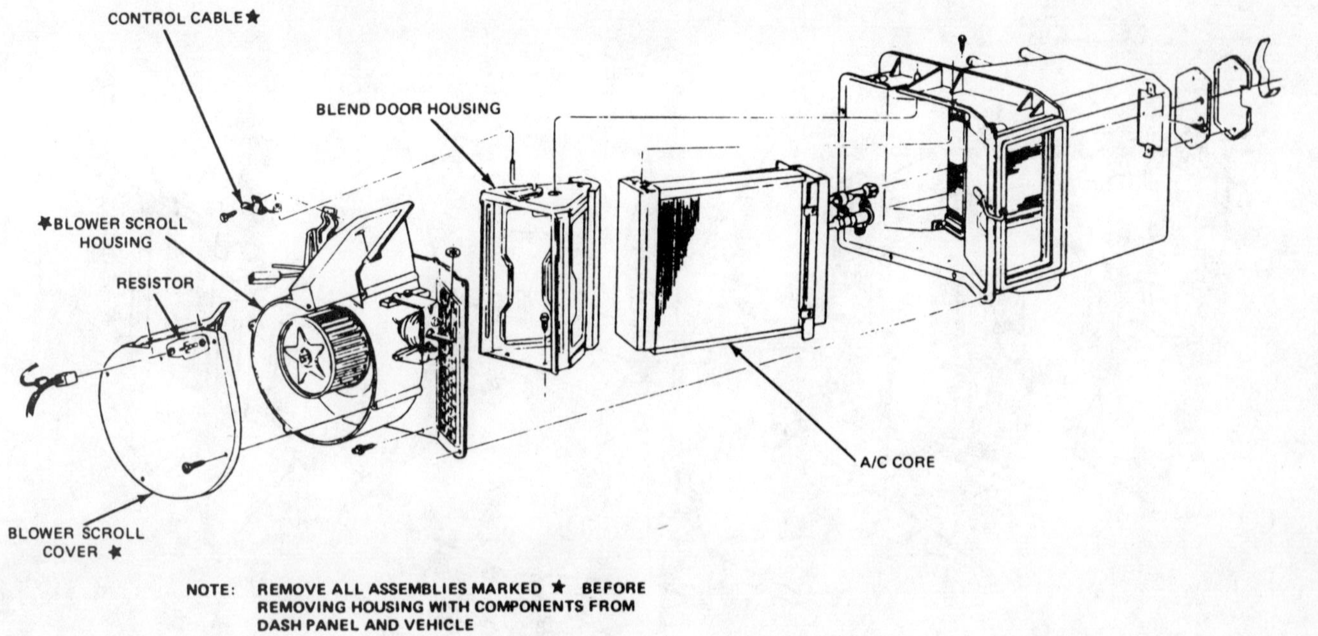

Fig. 57 Evaporator assembly. 1974-80 Econoline

5. Reverse procedure to install. Use new O-rings lubricated with refrigerant oil.
6. Evacuate, charge and leak test system. Check for proper operation.

1974-79 F Series & 1978-79 Bronco

1. Disconnect battery ground cable and remove carburetor air cleaner.
2. Discharge refrigerant from system.
3. Remove A/C refrigerant line support strap from cowl.
4. Remove insulation from expansion valve and sensing bulb. Then, remove cover plate and seal from evaporator housing at the expansion valve.
5. Remove glove box liner and right A/C duct.
6. Disconnect vacuum hose from outside recirculating door vacuum motor.
7. Remove rear evaporator housing from dash panel and fresh air inlet boot, Fig. 59. Install one upper nut to retain evaporator assembly to dash.
8. Remove de-icing switch and mounting plate, Fig. 59. Push wire grommet out of mounting plate.
9. Remove plenum from evaporator assembly.
10. Disconnect refrigerant lines from evaporator. Plug all open lines to prevent entry of dirt and moisture into system.
11. Remove evaporator core attaching screws and evaporator core.
12. Reverse procedure to install. Use new O-rings lubricated with refrigerant oil.
13. Evacuate, charge and leak test system. Check for proper operation.

Heater Core, Replace

PASSENGER CARS

1978-80 Fairmont, Zephyr; 1979-80 Capri & Mustang

Less Air Conditioning

1. Drain cooling system and disconnect battery ground cable.
2. Disconnect heater hoses from heater core and plug core openings.
3. Remove glove box liner.
4. Remove instrument panel to cowl brace retaining screws and the brace.
5. Move temperature control lever to warm position.
6. Remove the four heater core cover retaining screws, then the cover through the glove box opening.
7. Remove heater core assembly mounting stud nuts from engine compartment.
8. Push core tubes and seal toward passenger compartment to loosen core from case assembly.
9. Remove heater core from case through the glove box opening, Fig. 60.
10. Reverse procedure to install.

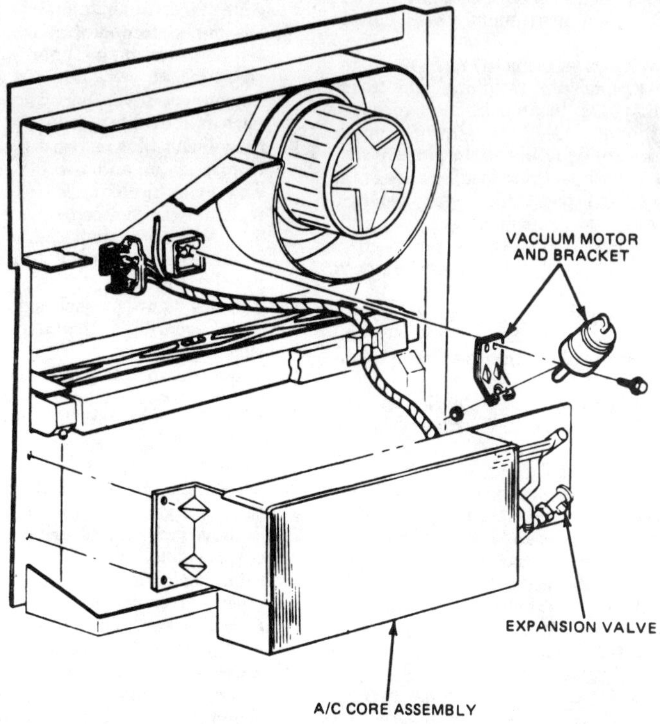

Fig. 58 Auxiliary Evaporator assembly. Econoline

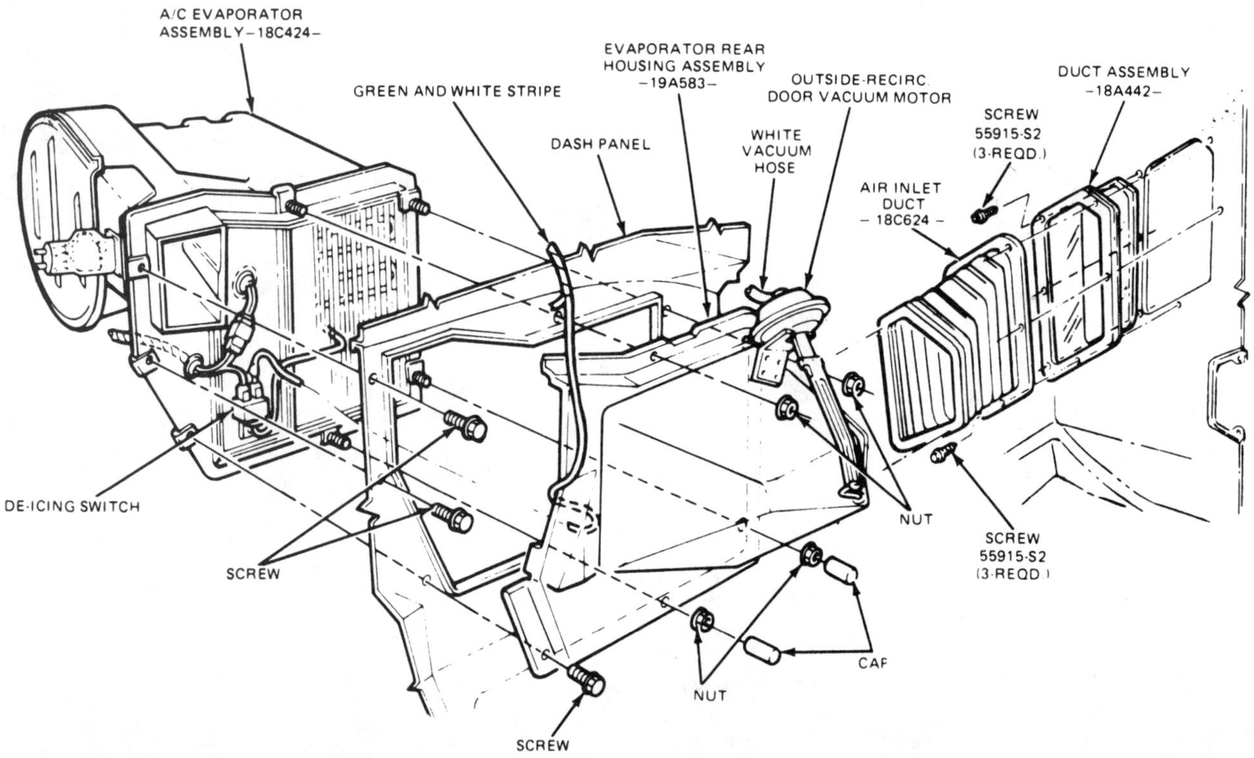

Fig. 59 Evaporator assembly. 1974-79 'F' Series and 1978-79 Bronco

With Air Conditioning

1. Disconnect battery ground cable.
2. Remove screws securing the left side of instrument panel pad retaining tabs to the instrument panel, in the upper right and left corners of the instrument panel cluster area.
3. Remove screws securing the right side of instrument panel pad retaining tabs to instrument panel, located in the two openings at the top edge of the right instrument panel trim applique, above the glove box.
4. Remove screws securing leading edge of instrument panel pad to defroster openings. Use a magnetic or locking tang type phillips screwdriver. Do not let screws drop into defroster openings since plenum door damage may result.
5. Raise the overhanging edge of the instrument panel pad to clear retaining tabs and pull pad rearward to remove from top of instrument panel.
6. Remove screws attaching steering column opening lower cover to instrument panel, then the cover.
7. Remove steering column trim shrouds.
8. On models with automatic transmission, remove nuts securing steering column to brake pedal support and lower steering column carefully for access to gearshift selector lever and cable assembly.
9. Reach between steering column and instrument panel and lift selector lever cable off selector lever. Then, remove cable clamp from steering column tube.
10. On all models, rest steering column on front seat.
11. Remove screw attaching instrument panel to brake pedal support at steering column opening.
12. Remove screw attaching lower brace to lower edge of instrument panel, below the radio.
13. Remove screw attaching brace to lower edge of instrument panel.
14. Disconnect temperature control cable from temperature blend door and evaporator case bracket.
15. Disconnect seven-port vacuum hose connectors at evaporator case.
16. Disconnect blower resistor wire connector from resistor and the blower motor feed wire at in-line connector near the blower resistor wire connector.
17. Support instrument panel and remove three screws securing top of instrument panel to the cowl.
18. Remove screw at each side of instrument panel securing instrument panel to cowl side panels.
19. Move instrument panel rearward and disconnect speedometer cable and any wiring that will not permit the panel to lay on the front seat.
20. Drain cooling system.
21. Disconnect heater hoses from heater core tubes. Plug heater core tubes.
22. Remove two nuts securing evaporator case to dash panel from engine compartment.
23. From the passenger compartment, remove screws attaching the evaporator case support bracket and air inlet duct support bracket to cowl top panel.
24. Remove one nut retaining the bracket at left end of evaporator case to the dash panel and the one nut securing the brackets below the case to the dash panel.
25. Pull evaporator case assembly from dash panel to gain access to the screws retaining heater core access cover to the evaporator case.
26. Remove five heater core access cover attaching screws and the access cover, Fig. 53.
27. Remove heater core and seals from evaporator case.
28. Remove the two seals from the heater core tubes.
29. Reverse procedure to install.

1974-78 Mustang II; 1974-80 Bobcat & Pinto

Less Air Conditioning

1. Drain coolant and disconnect battery.
2. Disconnect blower motor wire at engine side of dash.
3. Disconnect heater hoses at engine block.
4. Remove four heater assembly-to-dash mounting nuts from the engine side of the dash.
5. Remove the glove box.
6. Disconnect control cables from heater. Remove mounting bracket clips and disconnect cables from door crank arms.
7. Remove radio as outlined previously.
8. Working inside car, remove snap rivet that attaches the forward side of the defroster air duct to the plenum chamber. Move the air duct back into the defroster nozzle to disengage it from the tabs on the plenum chamber. Now, tilt the forward edge of the duct up and forward to disengage it from the nozzle and remove it from the left side of the heater assembly.
9. Remove heater case-to-instrument panel support bracket mounting screw and re-

move the heater case. At the same time, pull the two heater hoses in through the dash panel. Then disconnect the hoses from the heater core in the case.
10. Remove compression gasket from cowl air inlet.
11. Remove eleven clips from around the front and rear case flanges and separate the front and rear halves of the case, Fig. 61.
12. Lift heater core from front half of case.

With Air Conditioning

1. Disconnect battery ground cable, drain radiator and discharge refrigerant from A/C system.
2. Remove refrigerant lines and the front half of refrigerant manifold.

NOTE: Remove manifold mounting stud to ensure clearance for removal of evaporator case.

3. Disconnect heater hoses from core tubes and remove condensation drain hose in engine compartment.
4. Remove glove box.
5. Disconnect vacuum hoses from evaporator case and temperature control cable from blend door crank arm.
6. Remove heat distribution duct, Fig. 54.

NOTE: On Mustang II, remove mode door vacuum motor.

7. On Mustang II, remove lower section of the air conditioning defrost plenum.
8. On 1974 Pinto, remove staples retaining fold down door on plenum, Fig. 62. Bend fold down door from locating tabs on plenum and remove adapter duct.

NOTE: During installation, position fold down door between locating tabs and tape in place with two pieces of black tape, Fig. 62.

9. On all models, remove blower motor and wheel from blower scroll.
10. Install a ¼-20 hex washer head screw in mounting tab of inlet duct to upper cowl bracket, holding inlet duct in place.

NOTE: Leave screw in position during in-

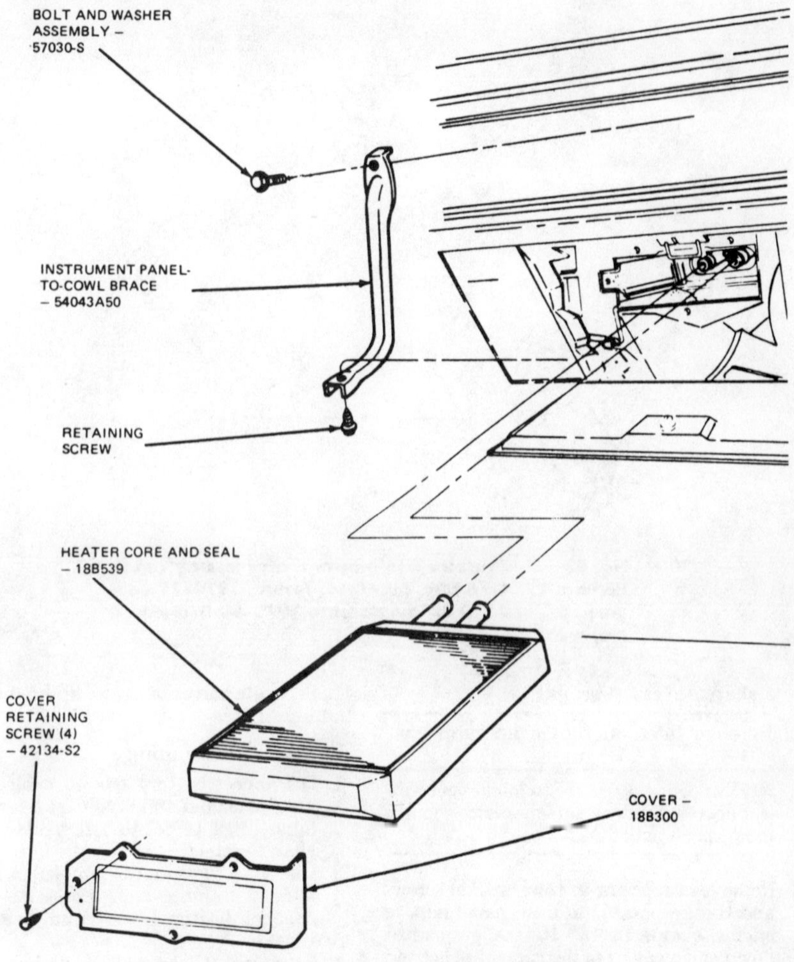

Fig. 60 Heater core removal. 1978-80 Fairmont, Zephyr; 1979-80 Capri & Mustang Less A/C

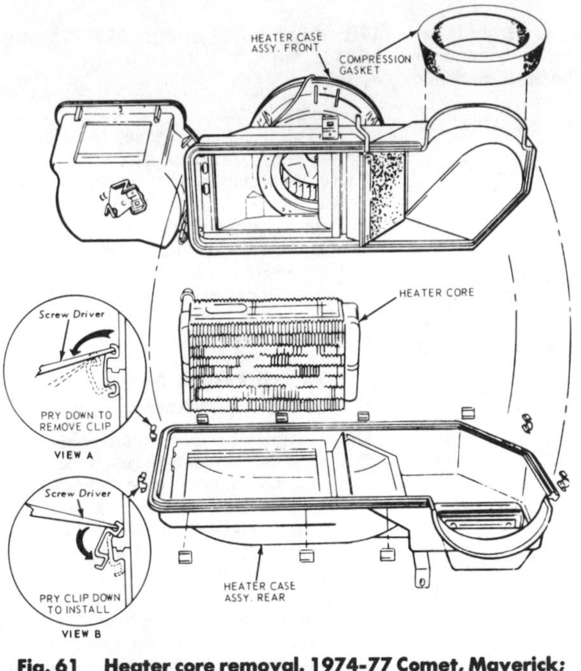

Fig. 61 Heater core removal. 1974-77 Comet, Maverick; 1974-78 Mustang II; 1974-80 Bobcat & Pinto Less A/C

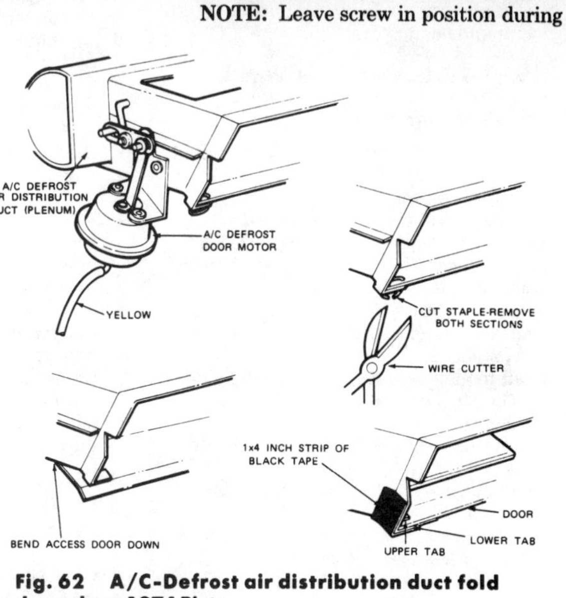

Fig. 62 A/C-Defrost air distribution duct fold down door. 1974 Pinto

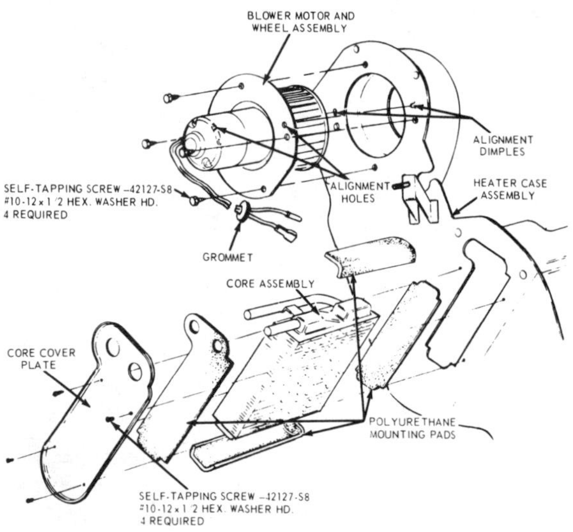

Fig. 63 Heater core and blower motor less air conditioning. 1974-76 Elite, Montego, Torino; 1974-79 Cougar; 1977-79 LTD II, Thunderbird; 1975-80 Granada & Monarch

stallation of case assembly.

11. Remove inlet duct to evaporator case screws.

NOTE: One upper case to inlet duct screw is located under outside-recirc. motor mounting bracket.

12. Remove evaporator to cowl bracket screws and the evaporator to dash panel nuts in engine compartment. Rotate evaporator down and away from instrument panel and remove from under panel.
13. Remove upper to lower case screws and rubber seal from heater core tubes.
14. Remove upper half of evaporator case and move rubber seal on evaporator core forward and pull evaporator core from lower case.
15. Remove heater core upper straps, air deflector mounting screw, then remove air deflector and heater core.

1974-77 Comet & Maverick

Less Air Conditioning

1. Disconnect battery ground cable and drain engine cooling system.
2. Disconnect blower motor ground wire from fender apron at engine side of dash.
3. Disconnect heater hoses at engine block.
4. Remove heater assembly to dash panel mounting nuts.
5. On 1974-75 units, remove glove compartment and on all models remove right cowl trim panel.
6. On all models remove cable retaining clips and the push nuts at door crank arms. Disconnect control cables from crank arms.
7. Remove defroster air duct from left side of heater assembly.
8. Disconnect motor lead from resistor assembly on bottom of heater.
9. Remove heater case to instrument panel support bracket mounting screw.
10. Pull heater hoses through dash panel and disconnect hoses from heater core.
11. Separate halves of heater and remove core, Fig. 61.

With Air Conditioning

1. Disconnect battery ground cable, remove air cleaner and drain cooling system.
2. Connect gauge set to compressor and discharge system.
3. Disconnect evaporator core tubes and tape ends to keep out foreign material and disconnect heater hoses from heater core tubes.
4. Remove A/C assembly-to-dash panel stud nuts.
5. On 1974 units, remove right cowl trim panel and remove glove compartment.
6. On all models disconnect right and left A/C air ducts.
7. Remove floor distribution duct from blower. Disconnect vacuum hoses from door motor and the vacuum harness multiple connector.
8. Disconnect temperature control cable from door crank arm on evaporator.
9. Remove screw that retains evaporator to cowl upper support and remove A/C rearward and away from dash panel.
10. Disconnect wire connectors from A/C thermostat switch on the evaporator housing and also the resistor on blower housing.
11. Disconnect blower motor ground wire from wiper mounting bracket and lower the blower and evaporator housing removing it from vehicle.
12. Separate halves of evaporator housing and remove water valve vacuum switch, Fig. 47.
13. Remove temperature blend door shaft, frames and door from lower half of evaporator housing.
14. Lift heater core out of lower housing and remove pads from core.

1974-76 Elite, Montego & Torino; 1975-79 Cougar; 1977-79 Mark V & Thunderbird

Less Air Conditioning

1. Drain cooling system and disconnect both heater hoses at dash.
2. Unfasten heater from dash.
3. Disconnect temperature and defroster cables at heater.
4. Disconnect wires from resistor, and blower motor wires and clip retaining heater to defroster nozzle.
5. Remove glove box.
6. Remove bolt and nut retaining right air duct control to instrument panel.
7. Remove nuts retaining right air duct and remove duct.
8. Take heater assembly to bench. Then remove heater core cover and pad, Fig. 63, and lift out core.

With Air Conditioning, 1974-76 Cougar Elite, Montego & Torino

1. Drain engine coolant and disconnect heater hoses from core.
2. Remove glove box.
3. Remove heater air outlet register from plenum assembly (2 snap clips).
4. Remove temperature control cable assembly mounting screw, and disconnect the end of the cable from the blend door crank arm (1 spring nut).
5. Remove the blue and red vacuum hoses from the high-low door vacuum motor; the yellow hose from the panel-defrost door motor, and the brown hose at the inline tee connector to the temperature by-pass door motor.
6. Disconnect wiring connector from resistor.
7. Remove ten screws from around flange of plenum case and remove rear case half of the plenum, Fig.64.
8. Remove mounting nut from heater core tube support bracket.
9. Remove heater core.

NOTE: During installation of heater core, be sure to apply body sealer around the case flanges to provide a positive seal. Also make sure core mounting gasket is properly installed.

10. Reverse procedure to install.

With Air Conditioning: 1977-79 Cougar, LTD II, Mark V & Thunderbird

1. Disconnect battery ground cable and drain cooling system.
2. Disconnect heater hoses from heater core and plug core openings.
3. Remove heater core cover plate retaining screws and the plate, Fig 65.
4. Press downward on heater core and tip toward front of vehicle to release the heater core seal from housing.
5. Lift heater core from case and remove from vehicle.
6. Reverse procedure to install.

1979-80 Ford, Mercury; 1980 Lincoln Continental & Mark VI

1. Disconnect battery ground cable.
2. Disconnect heater hoses from core. Plug hoses and heater core tubes to prevent coolant loss during core removal.
3. Remove one bolt located below the windshield wiper motor retaining left end of plenum to dash panel.
4. Remove one nut retaining the upper left corner of evaporator case to dash panel.
5. Disconnect vacuum supply hose from vacuum source, then push grommet and hose into passenger compartment.

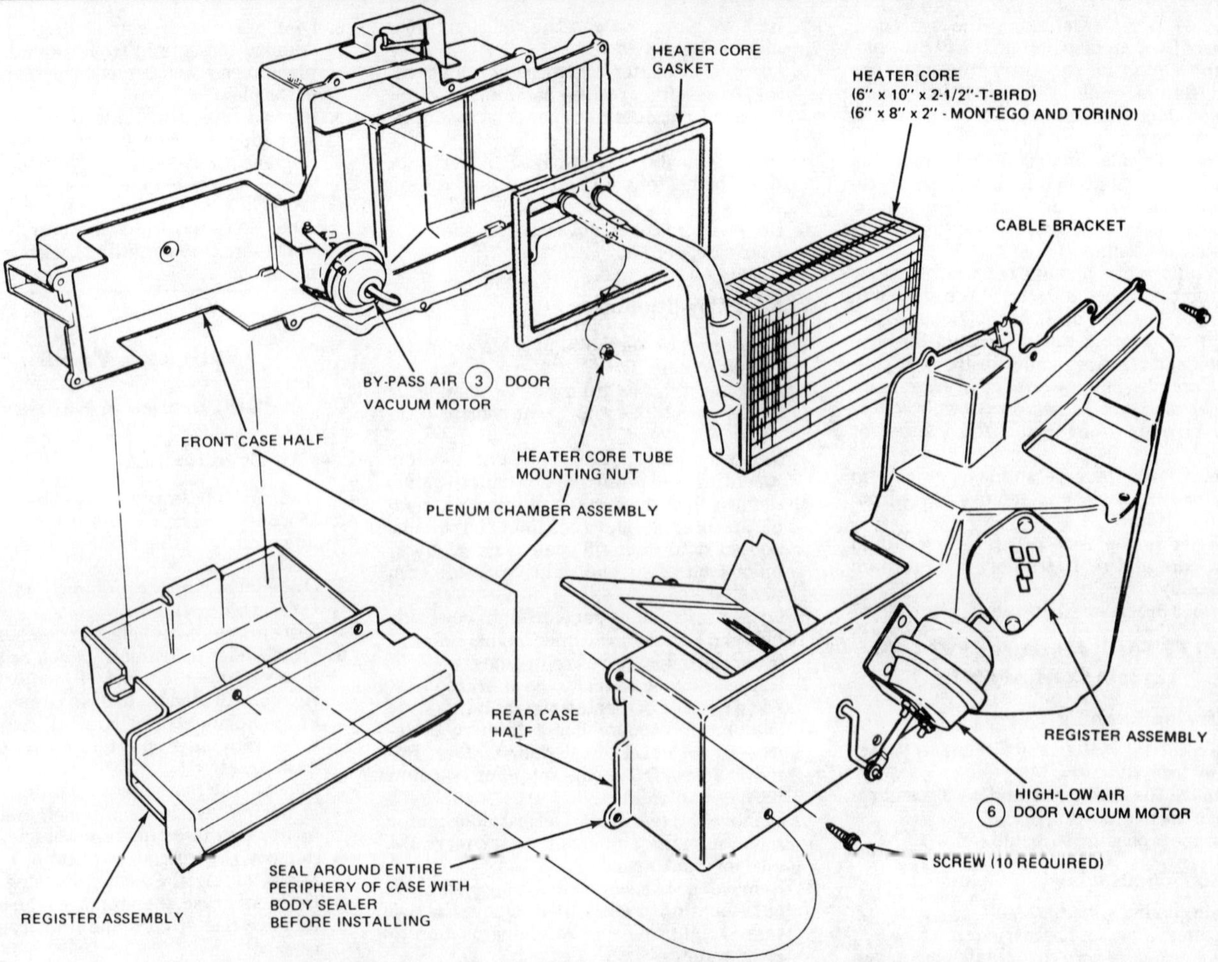

Fig. 64 Heater core assembly. 1974-76 Cougar, Elite, Montego & Torino With A/C

6. Remove glove compartment, then loosen right door sill plate and remove right side cowl trim panel.
7. Remove bolt retaining lower right end of instrument panel to side cowl, then remove instrument panel pad as follows:
   a. Remove screws retaining instrument panel pad to instrument panel at each defroster opening.
   b. Remove the one screw retaining each outboard end of pad to instrument panel.
   c. Remove the five screws retaining lower edge of instrument panel pad, then pull instrument panel pad rearward and remove it.
8. Disconnect temperature control cable housing from bracket at top of plenum, then disconnect cable from temperature blend door crank arm.
9. Remove push clip retaining the center register duct bracket to the plenum and rotate bracket to the right.
10. Disconnect vacuum jumper harness at multiple vacuum connector near the floor air distribution duct, then disconnect white vacuum hose from the outside-recirculating door vacuum motor.

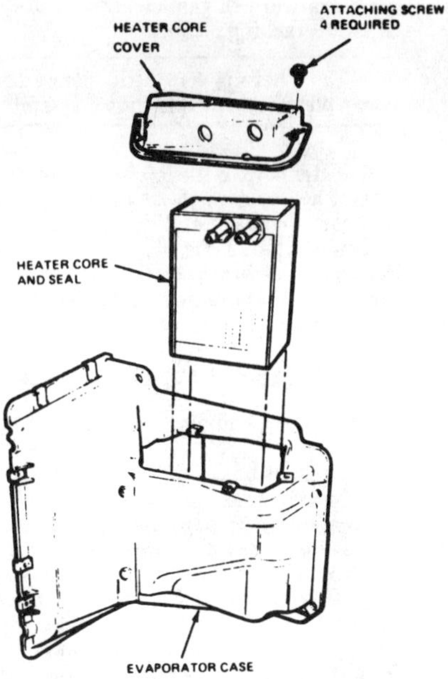

Fig. 65 Heater core assembly. 1977- 79 Cougar, LTD II & Thunderbird With A/C

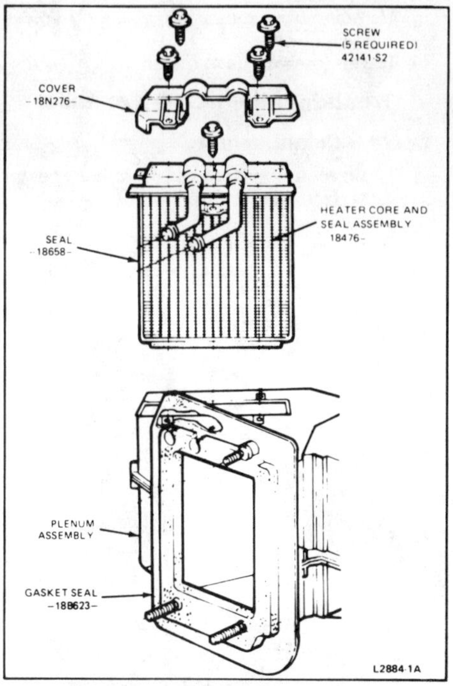

Fig. 66 Heater core assembly. 1979-80 Ford, Mercury; 1980 Lincoln Continental & Mark VI

11. Remove screws retaining the passenger side of floor air distribution duct to the plenum. It may be necessary to remove the two screws retaining the partial (lower) panel door vacuum motor to mounting bracket to gain access to right screw.
12. Remove the plastic push fastener retaining floor air distribution duct to left end of plenum and remove floor air distribution duct.
13. Remove nuts from the two studs along lower edge of plenum.
14. Carefully move plenum rearward to allow heater core tubes and stud at top of plenum to clear holes in dash panel. Remove plenum by rotating top of plenum forward, down and out from under instrument panel. Carefully pull lower edge of instrument panel rearward as necessary while rolling the plenum from behind the instrument panel.
15. Remove the four retaining screws from heater core cover and remove cover from plenum, Fig. 66.
16. Remove heater core retaining screw then pull core and seal assembly from plenum assembly.
17. Reverse procedure to install.

1974-78 Ford, Mercury; 1974-79 Lincoln Continental

Less Air Conditioning

1. Drain cooling system and disconnect heater hoses from the core.
2. Remove the core cover and gasket and remove the heater core, Fig. 67.
3. Reverse procedure to install.

With Air Conditioning

1. Drain cooling system.
2. Disconnect heater hoses from core tubes.
3. Remove heater core cover plate and gasket, Fig. 68.
4. Pull heater core and mounting gasket up out of case.
5. Remove core mounting gasket. Remove heater core.
6. Reverse procedure to install.

Granada, Monarch & Versailles

Less Air Conditioning

1. Drain cooling system and disconnect heater hoses from heater core.

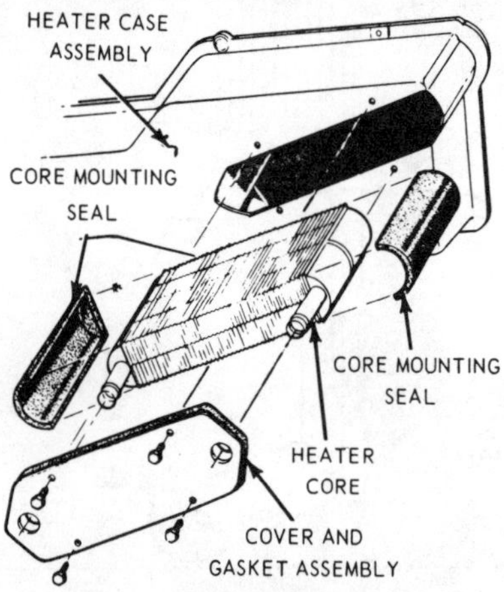

Fig. 67 Heater core assembly. 1974-78 Ford & Mercury

2. Remove glove box and the right and floor air distribution ducts.
3. Disconnect heater control cables and wiring harness from resistor assembly.
4. Remove vent duct to upper cowl mounting bolt.
5. Remove heater case to dash panel nuts, then the heater case and vent duct assembly.
6. Remove core cover and seal, then slide core from case, Fig. 63.

With Air Conditioning

1. Disconnect battery ground cable and drain cooling system.
2. Disconnect hoses from heater core. Plug heater core tubes to prevent coolant leakage.
3. Remove washer nut from plenum assembly mounting stud on engine side of dash panel.
4. Remove floor duct, seat belt interlock module and bracket, glove box liner and shields.
5. Loosen right door sill scuff plate, right A-pillar trim cover and right cowl side trim panel.
6. Loosen instrument panel to right cowl side bolt and remove instrument panel brace at lower rail below glove box opening.
7. If used, remove tunnel to cowl brace located on left side of plenum assembly.
8. Disconnect vacuum hoses from A/C-Defrost and Heat/Defrost door motors. Remove screw from clip retaining vacuum harness to plenum.
9. Remove the two Heat/Defrost door motor mounting nuts and swing motor rearward on door crank arm.
10. Remove the two screws retaining plenum to left mounting bracket, then remove the two screws and three clips retaining plenum to evaporator case.
11. Swing bottom of plenum away from evaporator case to disengage S-clip on forward flange of plenum, then raise plenum to clear tabs on top of evaporator case.
12. Move plenum to left while pulling instrument rearward to gain clearance. Use care to avoid cracking plastic.

NOTE: There is very little clearance between plenum and wiper motor assembly.

13. Pull heater core to the left, then as rear surface of heater core clears evaporator case, pull core rearward and downward to clear instrument panel, Fig. 55.
14. Reverse procedure to install. Before installing core, make sure that heater core tube to dash panel seal is in place.

1974-76 Mark IV & Thunderbird

1. Drain coolant and disconnect heater hoses from core.
2. Remove glove box.
3. Remove heater air outlet register from plenum assembly (2 snap clips).
4. Remove temperature control cable assembly mounting screw and disconnect end of cable from blend door crank arm (1 spring nut).
5. Remove the blue and red vacuum hoses from the high-low door vacuum motor; the yellow hose from the panel-defrost door motor, and the brown hose at the inline tee connector to the temperature by-pass door motor, Fig. 64.
6. Disconnect wiring connector from resistor.
7. Remove ten screws from around flange of plenum case and remove the rear case half of the plenum.
8. Remove mounting nut from heater core tube support bracket.
9. Remove core.

NOTE: When installing, apply body sealer between front and rear halves and make sure core mounting gasket is properly installed.

TRUCKS & VANS

1980 Bronco & F Series

Less Air Conditioning

1. Disconnect battery ground cable.
2. Disconnect temperature cable from blend door and mounting bracket on top of heater case.
3. Disconnect wires from blower motor resistor and blower motor.
4. Disconnect and plug heater core hoses.
5. From under instrument panel, remove two nuts retaining left end of heater case and right end of plenum to dash panel.
6. From engine compartment, remove one screw retaining top center of heater case to dash panel.
7. Remove the two screws retaining right end of heater case to dash panel, and remove heater case assembly from vehicle.
8. Remove the nine screws and one bolt and nut retaining the heater housing plate to the heater case, then remove housing plate.
9. Remove heater core and seal from heater case.
10. Reverse procedure to install.

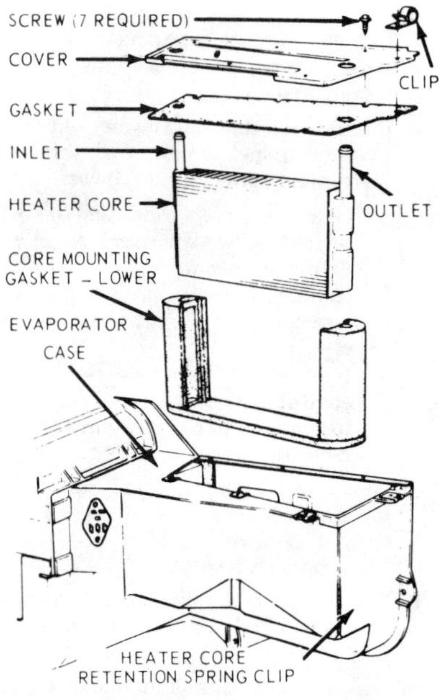

Fig. 68 Heater core removal. 1974-78 Ford, Mercury; 1974-79 Lincoln Continental

With Air Conditioning

1. Disconnect battery ground cable.
2. Disconnect and plug heater core hoses.
3. Remove glove compartment liner.
4. Remove screws retaining heater core cover to plenum and remove cover.
5. Remove heater core.
6. Reverse procedure to install.

1979 Bronco & F Series

Less Air Conditioning

1. Disconnect battery ground cable.
2. Disconnect temperature and function control bowden cables from heater housing.
3. Disconnect wires from blower motor resistor and remove the five screws retaining the air inlet (vent) duct to heater housing.
4. Disconnect blower motor leads.
5. Drain cooling system and disconnect heater core hoses.
6. Remove the three heater stud retaining nuts and remove heater assembly.
7. Remove heater core cover and gasket.
8. Pull heater core and lower support from heater.
9. Reverse procedure to install.

1974-79 Ranchero

For service procedures, refer to the appropriate section under 1974-76 Torino and 1977-79 LTD II.

1974-79 F Series; 1978-79 Bronco

With Air Conditioning

1. Disconnect battery ground cable and remove air cleaner housing.
2. Drain cooling system and disconnect hoses from heater core.
3. Remove A/C hose support from cowl, then the tape from expansion valve and remove cover plate and seal from evaporator housing at expansion valve.
4. Remove glove box liner and right A/C duct.
5. Disconnect vacuum hose from fresh air door vacuum motor.
6. Remove evaporator rear housing from dash panel and fresh air inlet boot, then reinstall one upper nut to retain evaporator front housing to dash after rear housing is removed.
7. Remove icing switch and mounting plate. Push wire grommet out of mounting plate.
8. Remove plenum from evaporator front housing and lower plenum.
9. On 1975-79 vehicles:
   a. Apply protective tape on A-pillar inner cowl panel at right corner on instrument panel.
   b. Remove lower right instrument panel to A-pillar bolt and lower the center instrument panel brace, bolt and nut.
   c. Position instrument panel rearward and install the A-pillar bolt to hold panel in rearward position.
10. Remove the four evaporator core to housing retaining screws, then pull core away from housing and position it upward and rearward from housing.
11. Remove the three screws and two plates, then the heater core from housing.
12. Reverse procedure to install.

1974-78 Bronco

Less Air Conditioning

1. Disconnect battery ground cable.
2. Drain cooling system and disconnect hoses from heater core.
3. Remove nuts and washers retaining heater to dash.
4. Disconnect right and left defroster hoses from plenum.
5. Disconnect air inlet cowl and position heater on floor.
6. Disconnect control cables from crank arms and the electrical wires and connectors.
7. Remove heater from vehicle, Fig. 69.
8. Remove rear cover retaining screws, then the core retaining screws and core.
9. Reverse procedure to install.

1974-80 F Series except 350-500 Series

Less Air Conditioning

1. Disconnect battery ground cable.
2. Drain cooling system and disconnect hoses from heater core.
3. Remove heater to dash retaining nuts and washers, Fig. 70.
4. Remove glove box.
5. Remove screws retaining air inlet duct to cowl, then pull heater assembly from dash and disconnect both defroster nozzles from heater.
6. Disconnect blower motor and resistor electrical connectors and control cables from heater and remove heater from vehicle.
7. Remove core cover stop bracket, then the retaining clip and heater core from case.
8. Reverse procedure to install.

1974-80 F350-500 Series

Less Air Conditioning

1. Disconnect battery ground cable.
2. Disconnect control cables and electrical connectors from blower motor resistor.
3. Remove screws retaining air inlet vent duct to heater housing and disconnect blower motor electrical connector.
4. Drain cooling system and disconnect hoses from heater core.
5. Remove the heater retaining stud nuts, then the heater from vehicle.
6. Remove heater core cover and gasket, then the heater core and lower support from heater case, Fig. 70.
7. Reverse procedure to install.

1975-80 Econoline

With Air Conditioning

1. Disconnect battery ground cable.
2. Drain cooling system and disconnect hoses from heater core.
3. Disconnect electrical leads from resistor on front face of A/C blower scroll cover.
4. Remove blower cover screws and remove cover.
5. Remove push nut and washer from fresh air-recirc door shaft.
6. Disconnect control cable from bracket and slide it over bracket, then remove cable wire loop from blend door shaft.
7. Remove blend door housing and blower motor housing.
8. Remove heater core retaining brackets and the heater core.
9. Reverse procedure to install.

Less Air Conditioning

1. Remove battery.
2. Drain cooling system and disconnect hoses from heater core.
3. Disconnect resistor and blower motor electrical leads.
4. Remove heater retaining screws and heater from vehicle.

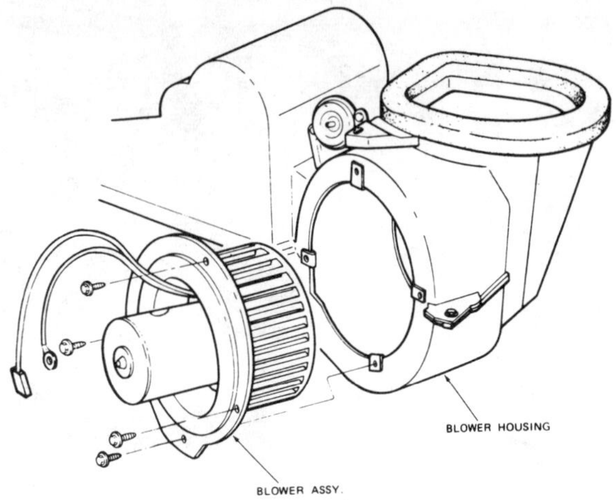

Fig. 70 Heater assembly. 1974-80 'F' Series Exc. 350-500 Series

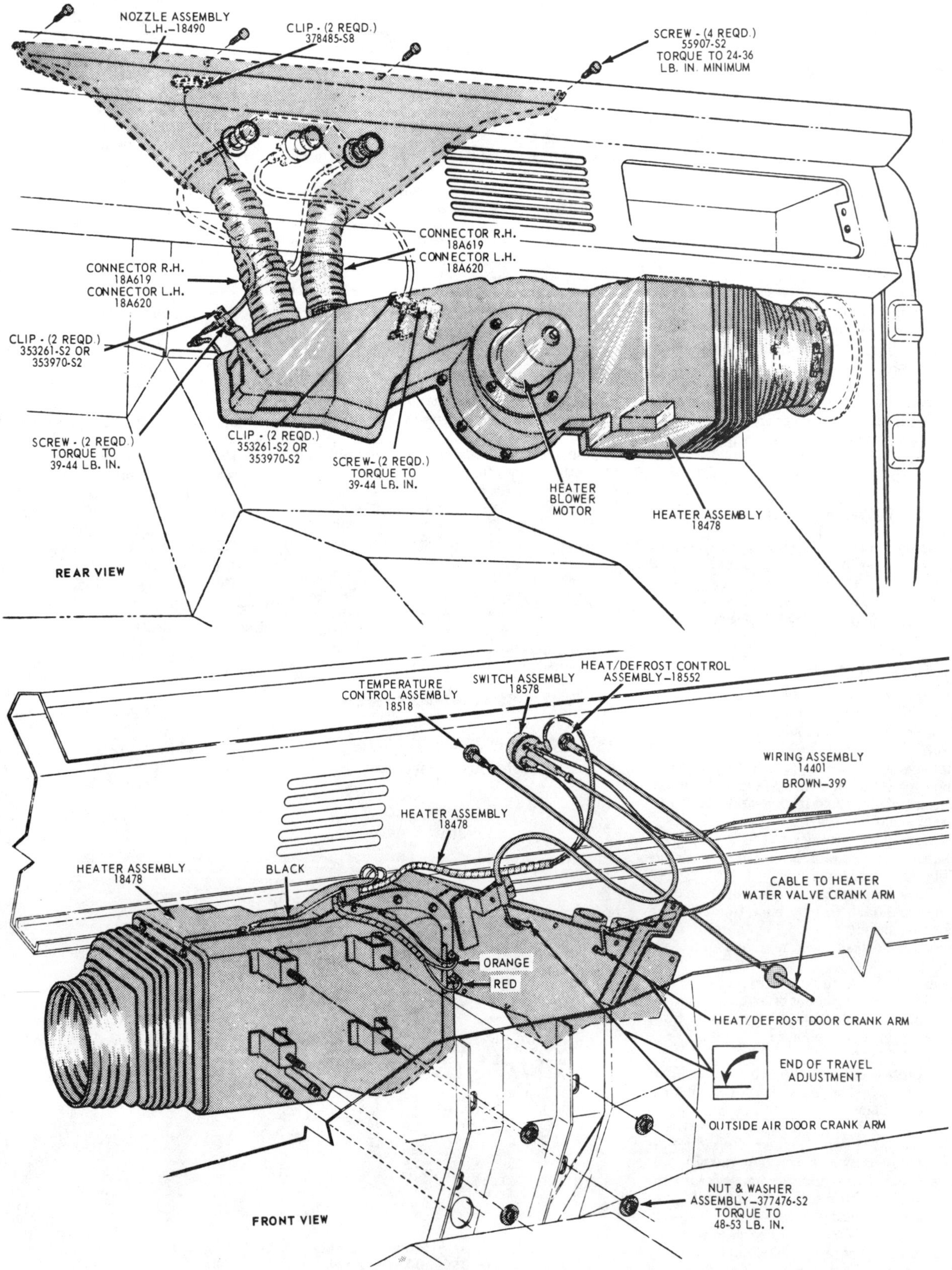

Fig. 69 Heater core and blower motor. 1974-77 Bronco

5. Cut seal at top and bottom edge of heater core retainer and remove retainer.
6. Slide seal and core assembly from case.
7. Reverse procedure to install.

1974 Econoline

Less Air Conditioning

1. Remove battery.
2. Drain cooling system and disconnect hoses from heater core.
3. Disconnect resistor and blower motor electrical connectors.
4. Remove heater to dash retaining bolts, then move heater out, disconnect control cable and remove heater from vehicle.
5. Separate heater case halves and remove heater core.
6. Reverse procedure to install.

With Air Conditioning

Refer to the "Heater Core, Replace," Less Air Conditioning, 1974 Econoline procedure.

1974-80 L Series

Less Air Conditioning

1. Disconnect battery ground cable.
2. Drain cooling system and disconnect hoses from heater core.
3. Disconnect control cable and blower motor electrical leads, then remove heater front cover.
4. Pull heater core from heater case.
5. Reverse procedure to install.

With Air Conditioning

1. Disconnect battery ground cable.
2. Drain cooling system and disconnect hoses from heater core.
3. Discharge refrigerant from A/C system.
4. Disconnect low pressure hose at dash panel and remove refrigerant low pressure fitting retaining nut and lockwasher.
5. Remove heater-air conditioner assembly to dash panel retaining bolts and disconnect drain tube from bottom of housing.
6. Disconnect high pressure hose at side of housing assembly.
7. Lower assembly to floor of vehicle and disconnect control cables from assembly.
8. Disconnect wires from blower switch and the ground wire.
9. Remove heater-air conditioner assembly.
10. Remove wire harness and blower resistor from housing.
11. Remove screws retaining evaporator core air deflector to air inlet end of housing.
12. Remove housing cover retaining screws and cover from housing.
13. Remove evaporator core retaining screws and the evaporator core and hoses.
14. Remove blower and adaptor plate retaining screws and the blower and adapter plate.
15. Remove screws retaining A/C door lever retainers to door, then the door from bellcrank.
16. Remove screws retaining evaporator core air deflector to housing, then the assembly from housing.
17. Remove heater baffle from housing.
18. Remove heater core retainer attaching screws and the heater core.
19. Reverse procedure to install.

1974-80 C Series

1. Disconnect battery ground cable.
2. Drain cooling system and disconnect hoses from heater core.
3. Disconnect defroster hoses from plenum chamber.
4. Remove glove box and disconnect blower motor electrical connectors.
5. Remove heater retaining nuts and washers and separate heater from dash, then dis-

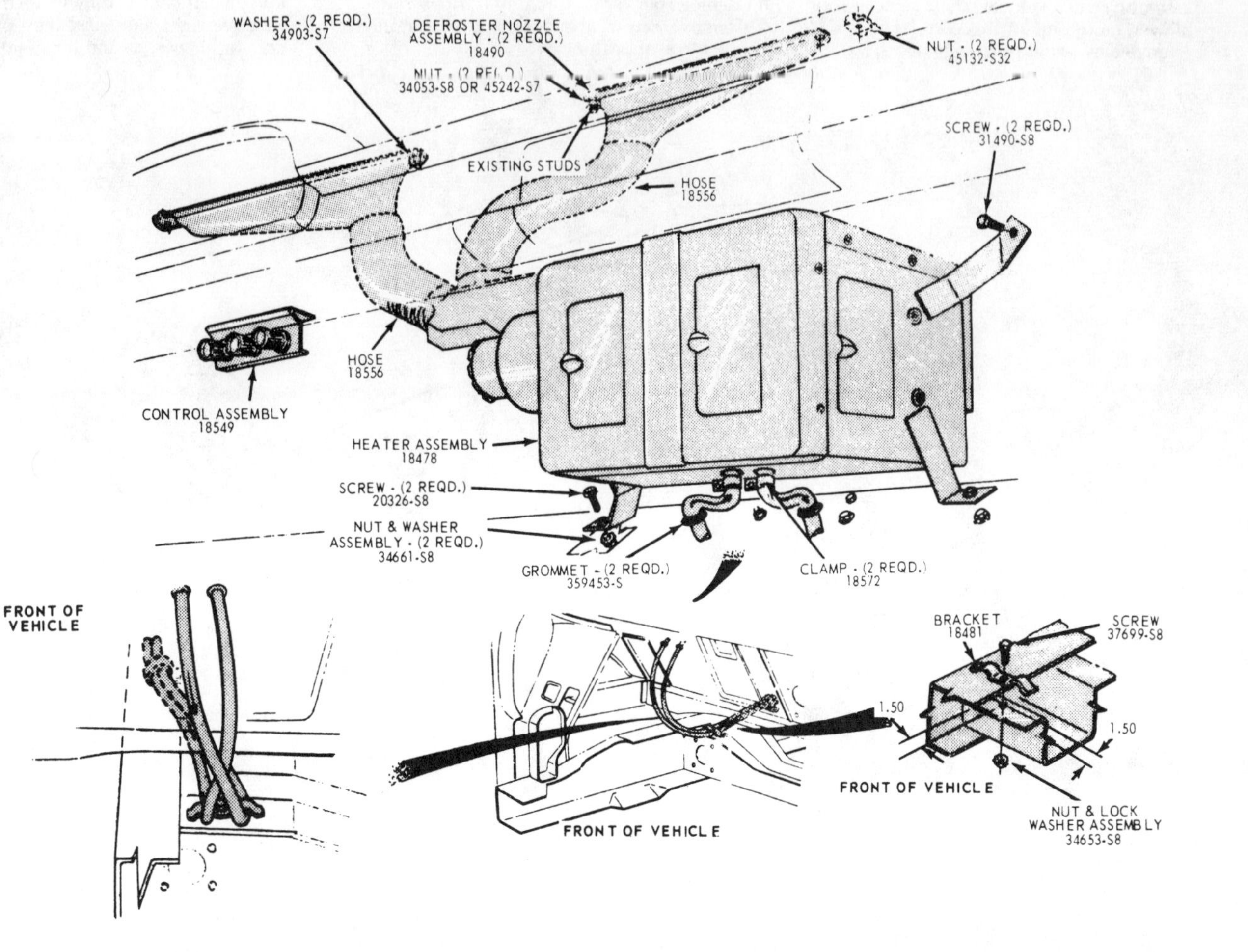

Fig. 71 Heater assembly. 1974-80 'C' Series

connect control cables from heater and remove heater from vehicle, Fig. 71.
6. Remove heater case top cover, then the heater core retaining screws and the heater core from case.
7. Reverse procedure to install.

Blower Motor, Replace

PASSENGER CARS

1980 Cougar & Thunderbird

1. Disconnect battery ground cable.
2. Remove glove box, then disconnect vacuum hose from outside-recirculating air door vacuum motor.
3. Remove screw attaching support brace to top of air inlet duct.
4. Disconnect blower motor power lead at wire connector.
5. Remove blower motor housing lower support bracket to heater/evaporator attaching nut.
6. Remove side cowl trim panel.
7. Remove blower motor ground wire attaching screw.
8. Remove screw attaching top of air inlet duct to heater/evaporator case.
9. Remove air inlet duct and blower housing assembly down and away from heater/evaporator case assembly.
10. Remove air inlet duct and blower housing assembly from vehicle, then remove four blower motor mounting plate screws and remove blower motor from housing, Fig. 72.
11. Reverse procedure to install.

1979-80 Capri & Mustang

Less Air Conditioning

1. Disconnect battery ground cable.
2. Remove right ventilator assembly.
3. Disconnect blower lead wire connector from spade terminal of resistor assembly and push back through hole in case.
4. Remove right side trim panel for access to blower ground terminal lug, then remove the ground terminal lug retaining screw.
5. Remove blower motor flange retaining screws from inside the blower housing.
6. Remove blower motor from housing.
7. Reverse procedure to install.

With Air Conditioning

1. Disconnect battery ground cable.
2. Remove glove box and disconnect hose from outside-recirculation door vacuum motor.
3. Remove instrument panel lower right to side cowl attaching bolt.
4. Remove screw attaching support brace to top of air inlet duct.
5. Disconnect blower motor feed wire at connector.
6. Remove nut retaining blower housing lower support bracket to evaporator case.
7. Remove side cowl trim panel.
8. Remove blower motor ground wire screw.
9. Remove screw attaching top of air inlet duct to evaporator case.
10. Pull air inlet duct and blower housing assembly downward and away from evaporator case.
11. Remove four blower motor mounting plate screws, then the blower motor assembly from blower housing.
12. Reverse procedure to install.

1979-80 Ford & Mercury Full Size; 1980 Continental & Mark VI

1. Disconnect battery ground cable.
2. Disconnect blower motor ground wire and the engine ground wire and position wiring aside.
3. Disconnect blower motor lead connector from wiring harness hard shell connector.
4. Remove blower motor cooling tube from blower motor.
5. Remove blower motor retaining screws.
6. Rotate blower motor slightly to the right so the bottom edge of the mounting plate follows the contour of the wheelwell splash panel. Then, lift blower motor up and out of housing assembly.
7. Reverse procedure to install.

1979-80 Fairmont & Zephyr

Less Air Conditioning

1. Disconnect battery ground cable.
2. Remove right ventilator assembly, then disconnect blower electrical connector from spade terminal of resistor assembly and push it back through hole in case.
3. Remove right side cowl trim panel for access to blower ground terminal lug (black)

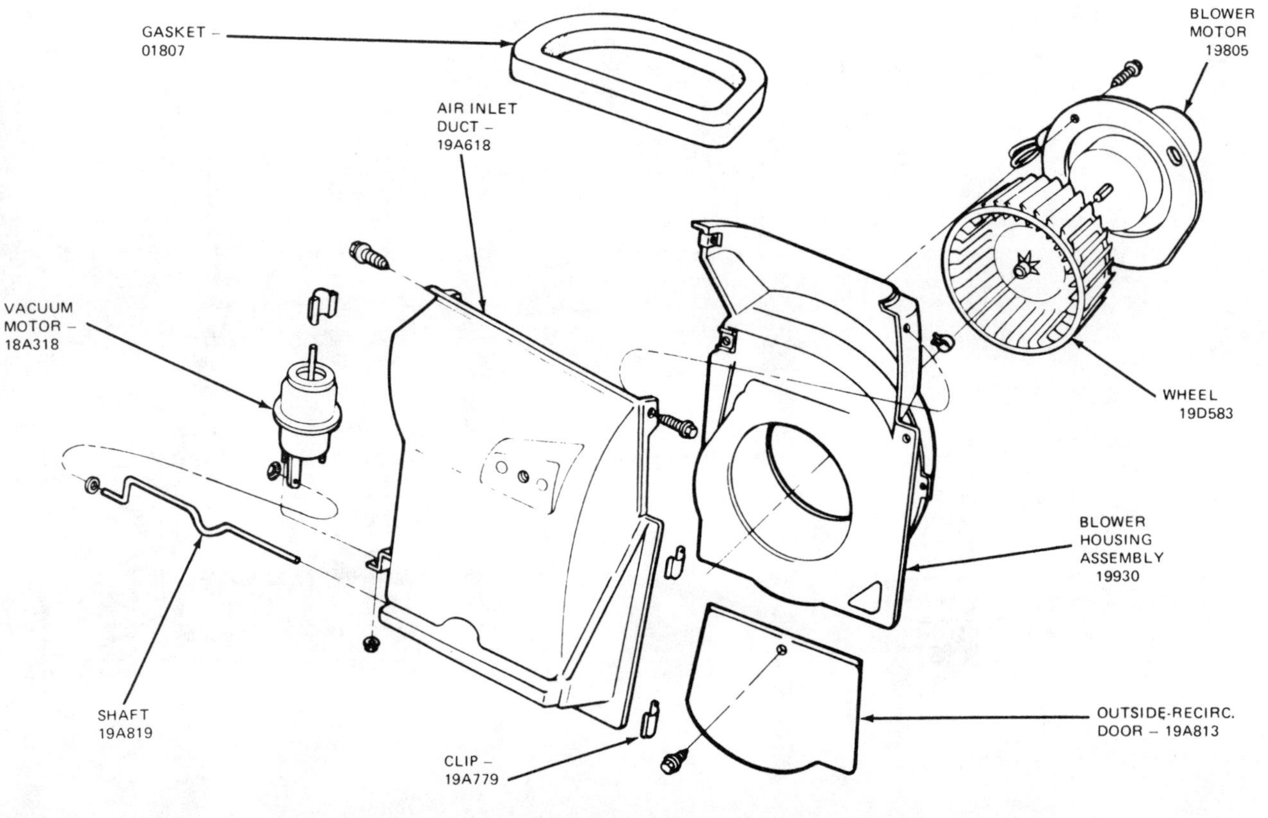

Fig. 72 Blower motor assembly. 1980 Cougar & Thunderbird

and remove ground terminal lug retaining screw.

4. Remove blower flange retaining screw from inside blower housing.

NOTE: Blower wheel can be removed for greater access.

5. Remove blower motor from housing.
6. Reverse procedure to install.

With Air Conditioning

1. Disconnect battery ground cable.
2. Remove four blower motor mounting plate screws and remove blower motor and wheel assembly from housing.

NOTE: Do not remove mounting plate from motor.

3. Reverse procedure to install.

1978 Fairmont & Zephyr

Less Air Conditioning

1. Disconnect battery ground cable.
2. Remove screw securing right register duct mounting bracket to lower edge of instrument panel.
3. Remove screws securing ventilator control cable lever assembly to lower edge of instrument panel.
4. Remove glove box liner.
5. Remove plastic rivets securing grille to ventilator floor outlet, then the grille from the bottom of the ventilator assembly.
6. Remove right register duct and register assembly.
7. Remove screws securing ventilator assembly to blower housing portion of the heater case assembly.
8. Slide ventilator assembly toward the right, then downward to remove from under instrument panel.
9. Remove push nut from door crank arm.
10. Remove control cable housing retaining screw, then control cable assembly from ventilator assembly.
11. Disconnect blower lead wire from resistor assembly, push through hole in case.
12. Remove right side cowl trim panel and the ground terminal lug retaining screw.
13. Remove blower motor retaining screws and the blower motor.
14. Reverse procedure to install.

With Air Conditioning

1. Disconnect battery ground cable.
2. Remove glove box and disconnect vacuum hose from outside-recirc air door motor.
3. Remove instrument panel lower right to side attaching bolt.
4. Remove screw attaching support brace to top of air inlet duct.
5. Disconnect blower motor lead wire.
6. Remove nut securing blower motor housing lower bracket to evaporator case.
7. Remove side cowl trim panel and the blower ground wire screw.
8. Remove screw securing top of air inlet duct to evaporator case.
9. Move air inlet duct and blower housing.

1977-80 Versailles

1. Disconnect battery ground cable.
2. Remove glove box.
3. Loosen right door sill scuff plate, right A-pillar trim cover and remove right cowl side trim panel.
4. Remove right lower instrument panel to cowl side bolt.

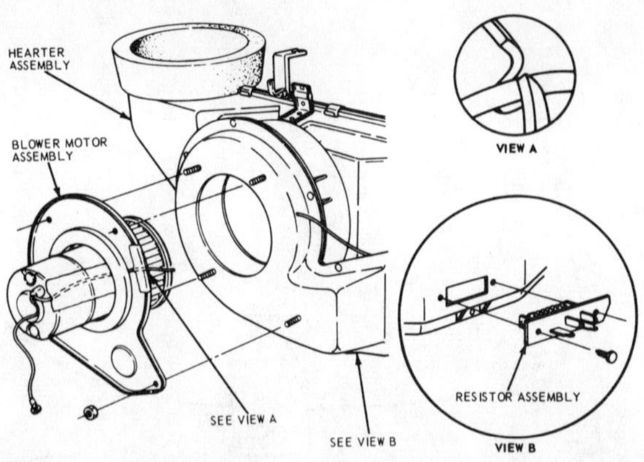

Fig. 74 Blower motor assembly. 1974-77 Comet & Maverick; 1974-78 Mustang II; 1975-80 Bobcat & Pinto

5. Remove cowl to lower instrument panel brace bolt.
6. Disconnect electrical connector from motor.
7. Remove blower motor assembly retaining screws and then blower motor. Pull rearward on lower edge of instrument panel to gain clearance.

NOTE: Do not remove mounting plate from blower motor, as the plate location is critical and should not be changed.

8. Reverse procedure to install.

1975-76 Elite, Montego & Torino; 1975-79 Cougar; 1977-79 LTD II & Thunderbird

Less Air Conditioning

1. Follow the procedure to remove the heater core as described previously.
2. Remove the heater assembly from the vehicle, and place it on a bench.
3. Remove the four mounting screws and remove the blower motor and wheel assembly from the blower, Fig. 63.
4. Reverse procedure to install.

With Air Conditioning: 1975-76 Cougar, Elite, Montego & Torino

1. Remove the glove box to gain access.
2. Remove the recirculation air register and duct assembly from the blower assembly.
3. Remove the two screws that attach the blower lower housing to the dash panel.
4. Disconnect the white hose from the outside recirculation air door vacuum motor, and remove the vacuum motor from the blower lower housing (2 screws). Leave the motor actuator connector to the door crank arm.
5. Disconnect the blower motor lead wire (orange) from the harness connector, and disconnect the motor ground wire (black).
6. Remove six upper to lower blower housing flange screws.
7. Separate the blower lower housing and motor assembly and remove it from under the instrument pad.
8. Remove the blower motor and wheel assembly from the lower housing (4 screws), Fig. 73.
9. The upper flange of the recirculating duct is retained to the blower upper housing with two S-clips that remained on the housing during removal. Make sure that the duct is installed properly in the two clips during installation.
10. Reverse procedure to install.

With Air Conditioning: 1977-79 Cougar, LTD II & Thunderbird

1. Remove instrument panel pad, glove box and side cowl trim panel.
2. Remove instrument panel attachment on right side.
3. Remove one nut attaching blower motor housing to engine side of dash panel.
4. Remove one nut attaching blower motor to passenger compartment side of dash panel.
5. Remove one blower housing mounting bracket and cowl top inner screw.
6. Disconnect vacuum line from outside-recirculating air door vacuum motor.
7. Disconnect blower motor lead wire from connector and blower motor ground wire.
8. Remove blower housing assembly.
9. Remove blower motor and wheel as an assembly from blower housing.
10. Reverse procedure to install.

1974-78 Mustang II; 1974-80 Bobcat & Pinto

Less Air Conditioning

1. Follow procedure to remove the heater core as described previously.
2. Disconnect the blower motor lead wire (orange) from the resistor.
3. Remove the four blower mounting plate nuts and remove the motor and wheel assembly, Fig. 74.
4. Reverse procedure to install.

With Air Conditioning

1. Remove glove box.
2. Remove four screws retaining blower motor and wheel to blower scroll, Fig. 70.
3. Remove blower motor and wheel assembly.

1974-77 Comet & Maverick

Less Air Conditioning

1. Remove the heater case, Fig. 75, as described previously.
2. Disconnect the blower motor lead wire (orange) from the resistor.
3. Remove the four blower motor mounting plate nuts and remove the motor and wheel assembly from the heater assembly, Fig. 74.

With Air Conditioning

The blower housing has to be removed to gain access to the blower motor.

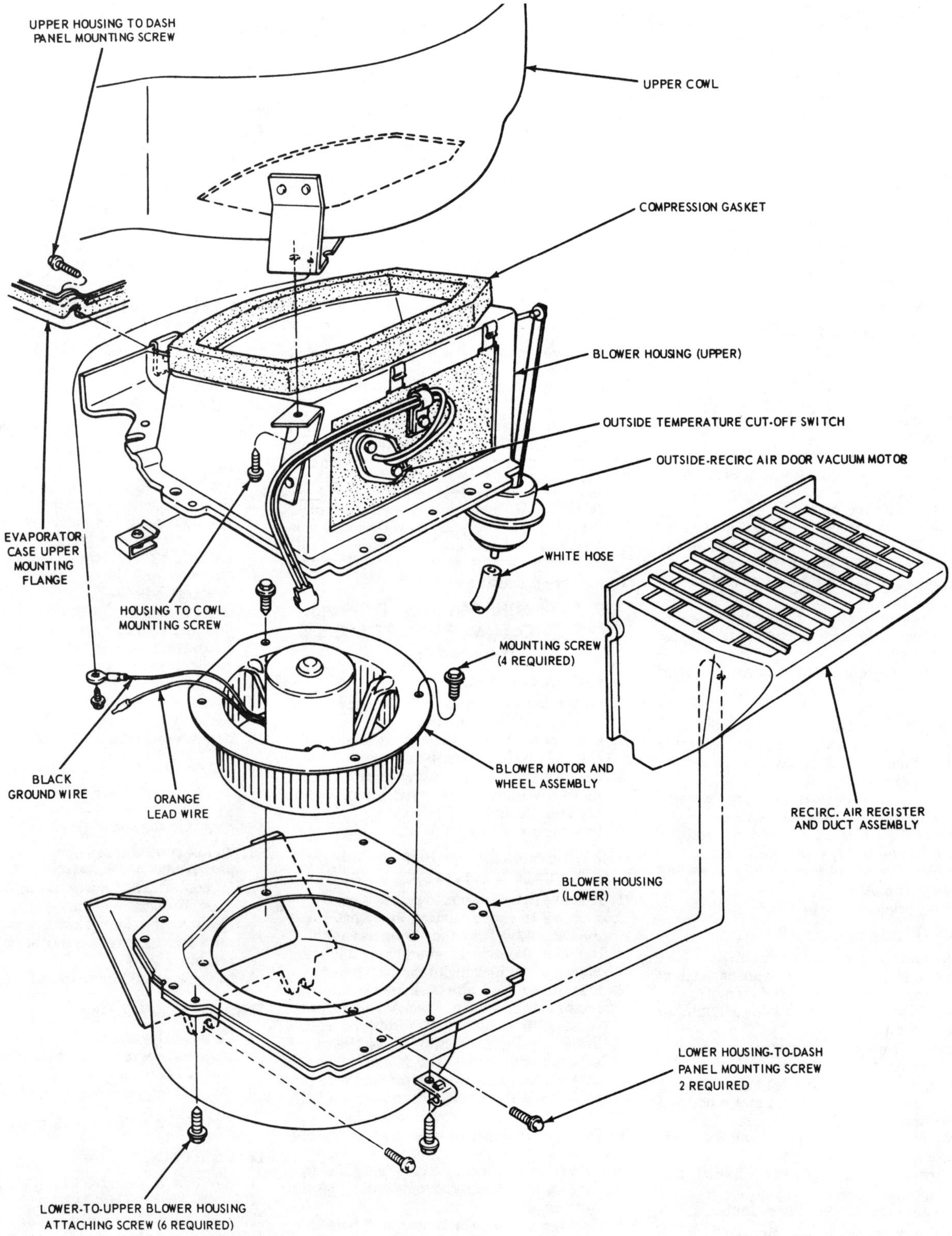

Fig. 73 Blower motor assembly. 1974-76 Cougar, Elite, Mark IV, Montego, Thunderbird & Torino

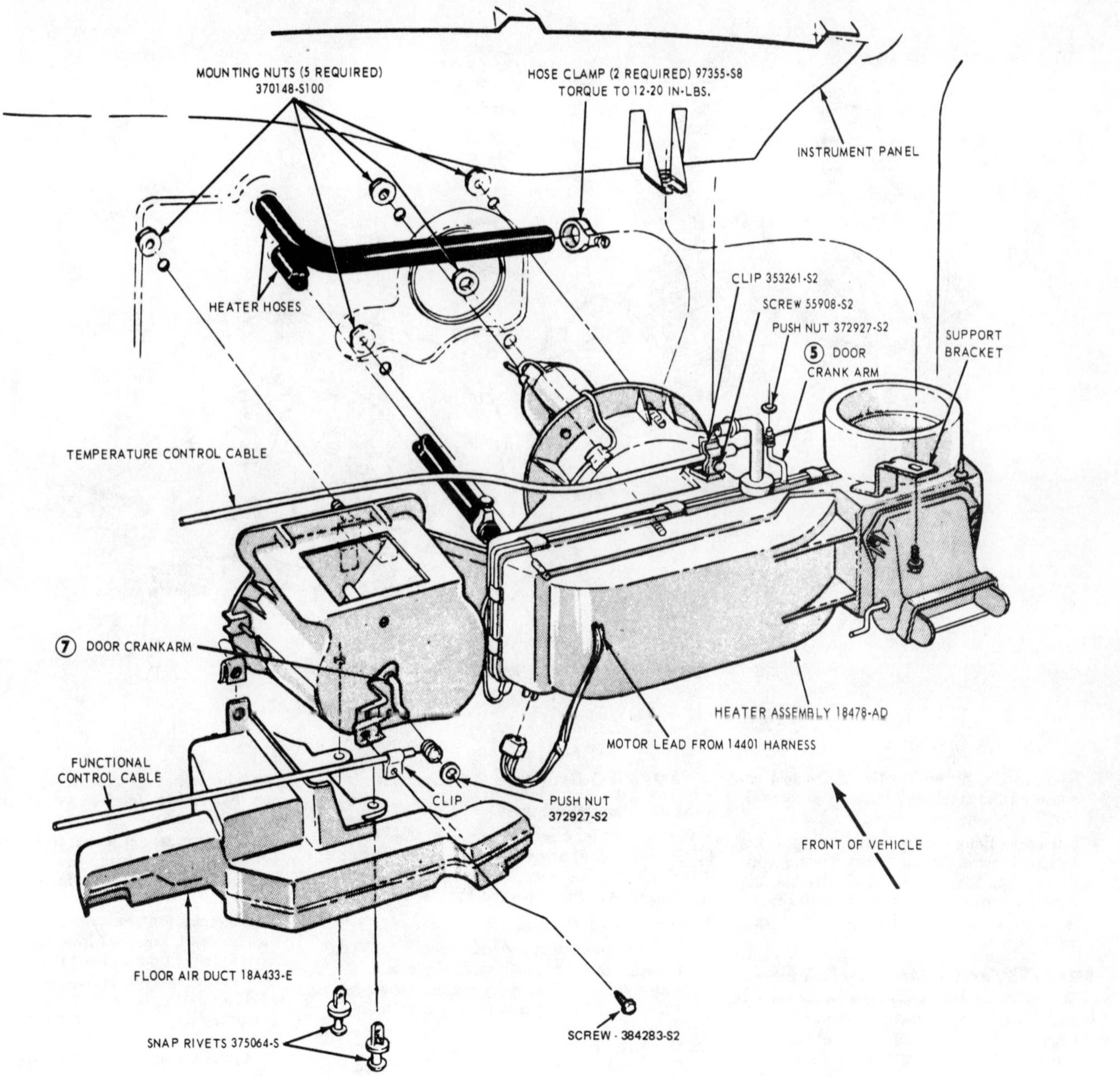

Fig. 75 Heater assembly. 1974-77 Comet & Maverick

1. Remove the radio. Remove the utility shelf (5 retaining bolts), and disconnect both right and left A/C register air duct assemblies from the plenum chamber.
2. Remove the floor air distribution duct from the bottom of the blower housing. Remove the blower housing mounting stud nut and lock plate.
3. Rotate the blower housing to unlock the slotted tabs on the blower housing from their lock pins on the evaporator housing. There are two tabs and two pins. Disconnect the red and yellow hoses at the vacuum motor on the blower housing. Disconnect the resistor and ground wires, and remove the blower housing.
4. Cut the gaskets around the A/C outlets at the break line before removing seven clips that separate both right and left halves of the blower housing.
5. Remove blower motor mounting plate retaining nuts, and remove the motor and wheel assembly from the housing, Fig. 76.
6. When installing be sure that the A/C Heat-Door is positioned properly before clipping the right and left housings together.

1974-76 Cougar, Elite, Montego & Torino

Less Air Conditioning

1. Follow the procedure to remove the heater core as described previously.
2. Remove the heater assembly from the vehicle, and place it on a bench.
3. Remove the four mounting screws and remove the blower motor and wheel assembly from the blower, Fig. 63.
4. Reverse the above procedure to install.

With Air Conditioning

1. Remove the glove box to gain access.
2. Remove the recirculation air register and duct assembly from the blower assembly.
3. Remove the two screws that attach the blower lower housing to the dash panel.
4. Disconnect the white hose from the outside-recirculation air door vacuum motor, and remove the vacuum motor from the blower lower housing (2 screws). Leave the motor actuator connected to the door crank arm.
5. Disconnect the blower motor lead wire (orange) from the harness connector, and disconnect the motor ground wire (black).
6. Remove six upper to lower blower housing flange screws.
7. Separate the blower lower housing and motor assembly from the upper housing and remove it from under the instrument pad.

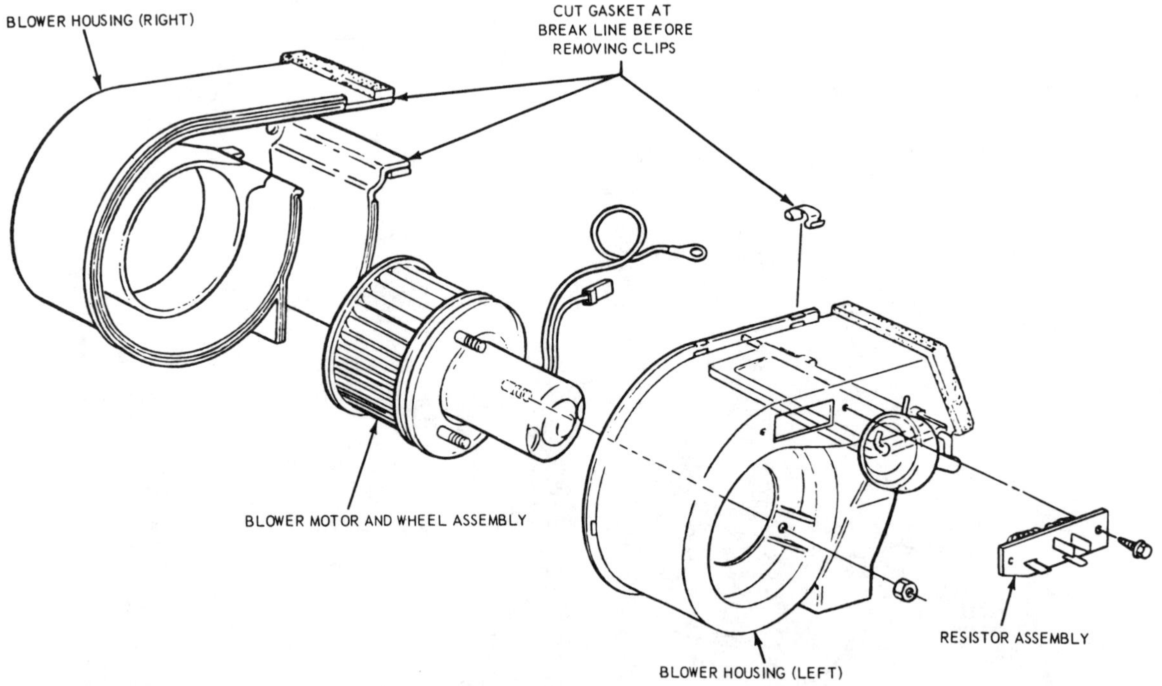

Fig. 76 Blower motor assembly. 1974-77 Comet & Maverick

8. Remove the blower motor and wheel assembly from the lower housing (4 screws), Fig. 73.
9. The upper flange of the recirculating duct is retained to the blower upper housing with two S-clips that remained on the housing during removal. Make sure that the duct is installed properly in the two clips during installation.

1974-78 Ford & Mercury Full Size

NOTE: The blower motor is located in the right side of the case to the right of the hood hinge under the right front fender. It is necessary to cut an opening in the right front fender apron to gain access to the blower motor assembly.

1. Disconnect the blower motor lead wire (orange) at the rear of the right hood hinge.
2. Remove the ground wire (black) from the upper cowl.
3. Remove the right front tire and wheel assembly.
4. Locate and cut opening in fender apron, Fig. 77. Care must be taken to avoid damage to the heater case by drill push-through or over-travel.
5. Remove the four blower motor mounting screws and disconnect the cooler tube from the motor.
6. Carefully move the motor and wheel assembly forward out of the heater case through the access hole.
7. A replacement cover plate is available from Ford part No. #18A475.

1975-80 Granada & Monarch

Less Air Conditioning

Remove heater case as outlined in the "1975-80 Granada & Monarch" procedure under "Heater Core, Removal," then remove screws securing blower motor to case, Fig. 63.

With Air Conditioning

1. Disconnect battery ground cable.
2. Loosen right door sill scuff plate and remove right cowl side trim panel, then remove right lower instrument panel to cowl side bolt.
3. Remove cowl to loosen instrument panel brace bolt, then disconnect wiring harness connectors from blower motor and remove cooling tube from motor.
4. Remove the four blower motor attaching screws and remove blower motor from scroll by pulling on lower edge of instrument panel to gain clearance.

NOTE: Use extreme care when pulling back on instrument panel to avoid cracking plastic panel.

5. Reverse procedure to install.

1974-79 Lincoln Continental

1. Remove the hood.
2. Remove the right hood hinge and right fender inner support brace as an assembly.
3. Disconnect the blower motor air cooling tube from the motor.
4. Disconnect the motor lead wire from the harness and the ground wire from the dash panel.
5. Disconnect the rear section of the right front fender apron from the fender around the wheel opening (7 screws) and remove the two lower fender-to-cowl mounting screws.
6. Separate the fender apron from the fender wheel opening so that the apron can be pushed downward away from the blower motor.
7. Remove the four blower motor mounting plate screws, Fig. 78. Move the motor and wheel forward out of the blower scroll and remove the assembly through the opening while applying pressure to the fender apron to enlarge the opening at the hinge area.

1974-76 Mark IV & Thunderbird; 1977-79 Mark V

1. Disconnect the battery ground cable.
2. Remove the glove box for access.
3. Remove the recirc air register and duct assembly from the blower lower housing to the dash panel.
4. Disconnect the white hose from the outside-recirc air door vacuum motor and remove the vacuum motor from the blower lower housing. Leave the motor actuator connected to the door crank arm, Fig. 73.
5. Disconnect the blower lead wire (orange) from the harness connector, disconnect the motor ground wire (black).
6. Remove the six upper to lower blower housing flange screws, Fig. 73.
7. Separate the blower lower housing and motor assembly from the upper housing and remove it from under the instrument panel.
8. Remove the blower motor and wheel assembly from the lower housing.
9. Reverse the above to install.

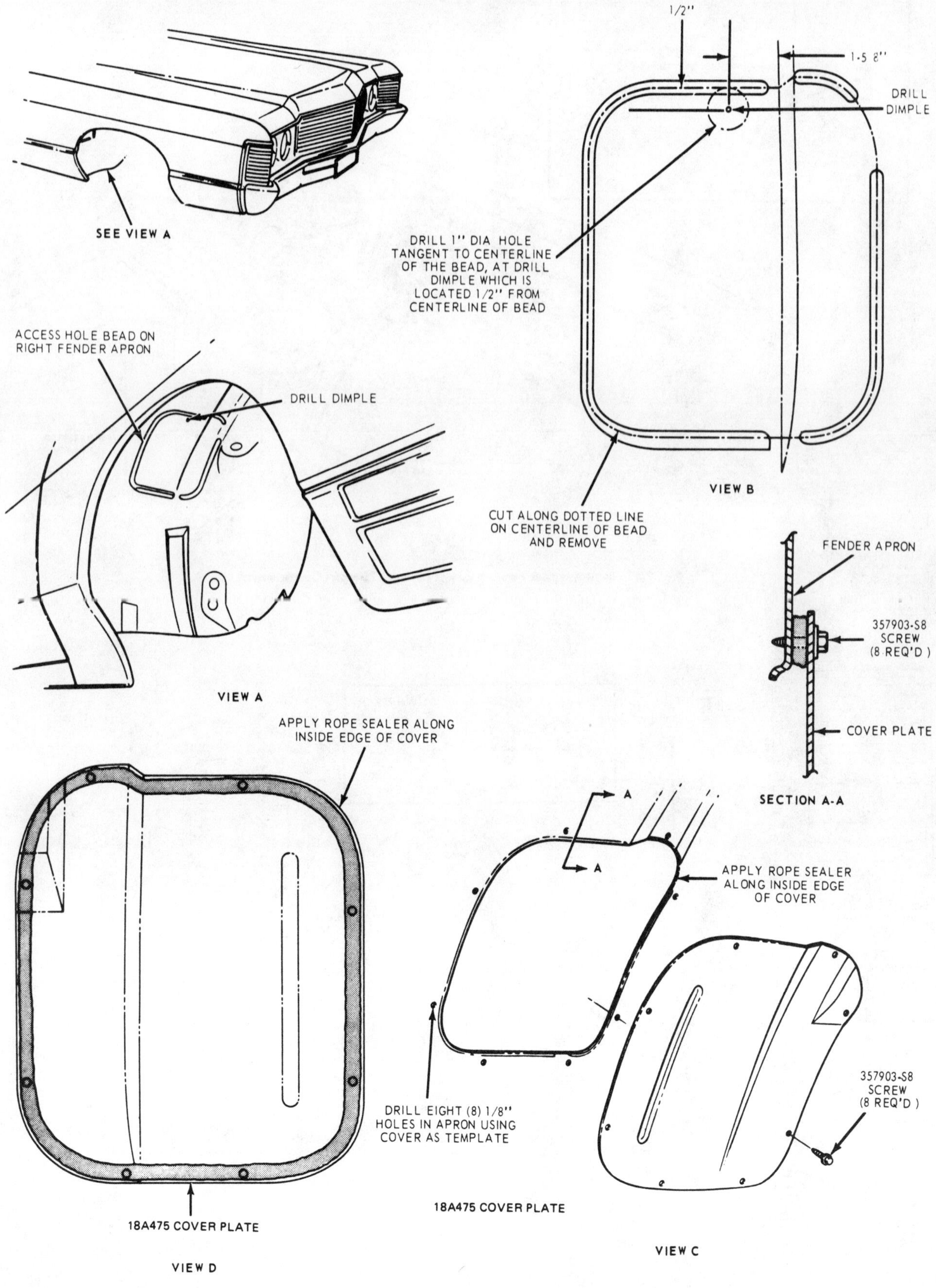

Fig. 77 Blower motor access hole. 1974-78 Ford & Mercury Full Size

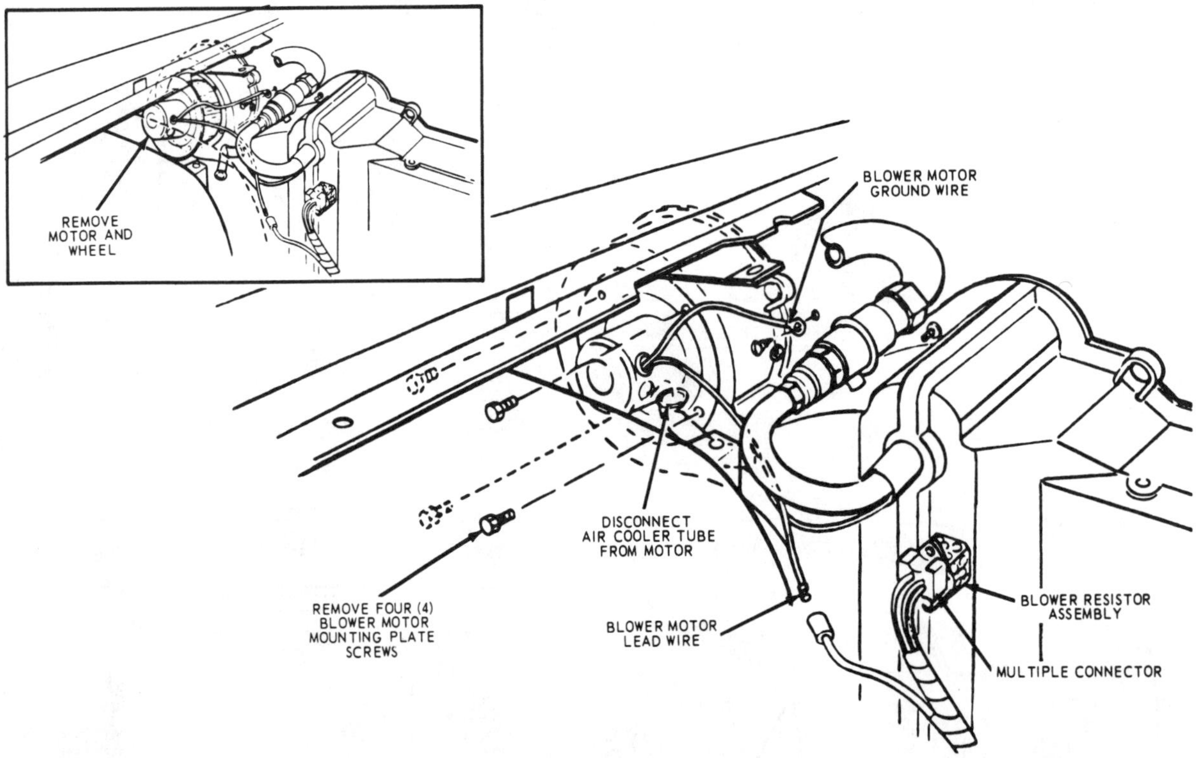

Fig. 78 Blower motor assembly. 1974-79 Lincoln Continental

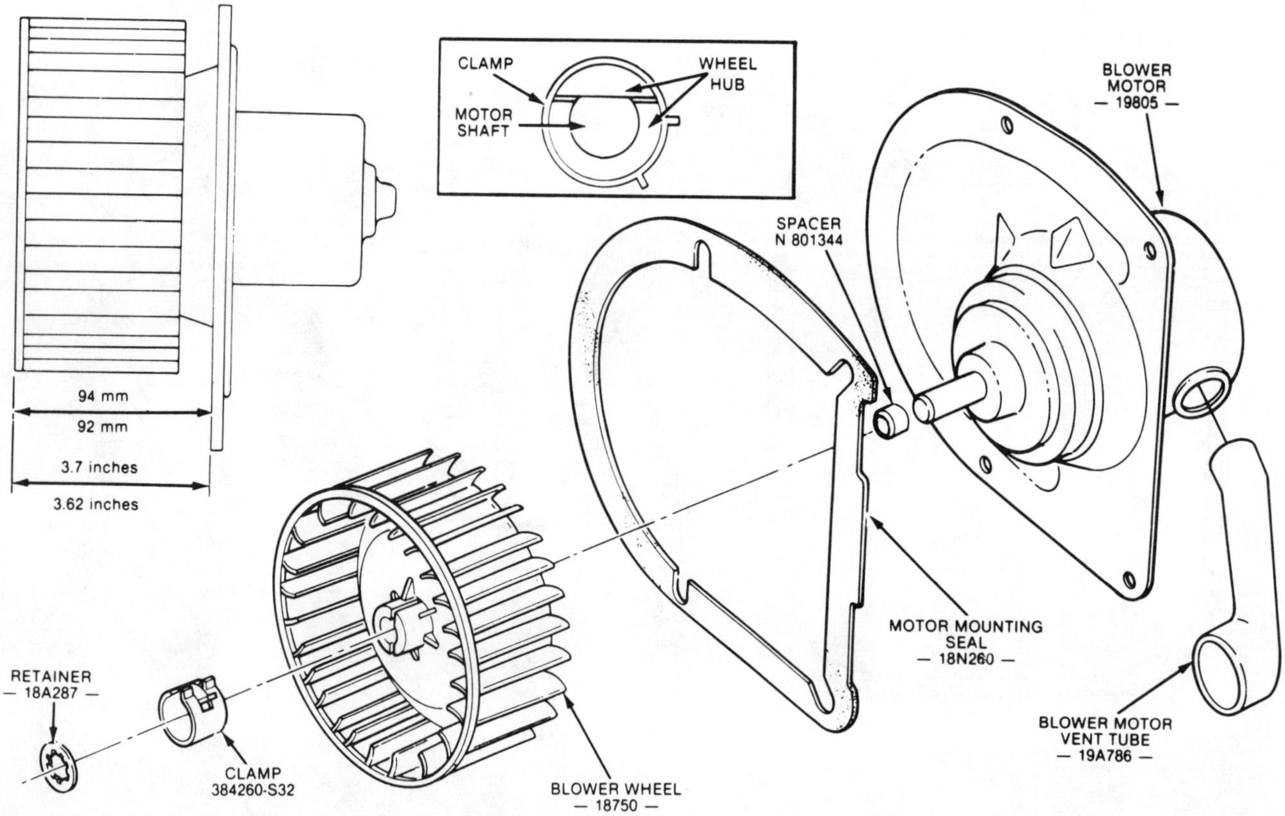

Fig. 79 Blower motor assembly. 1980 Bronco & F Series

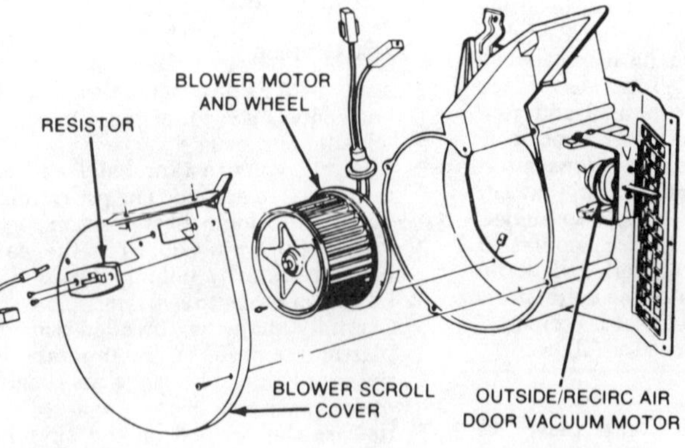

Fig. 80 Blower Motor Assembly With Air Conditioning. 1974-80 Econoline

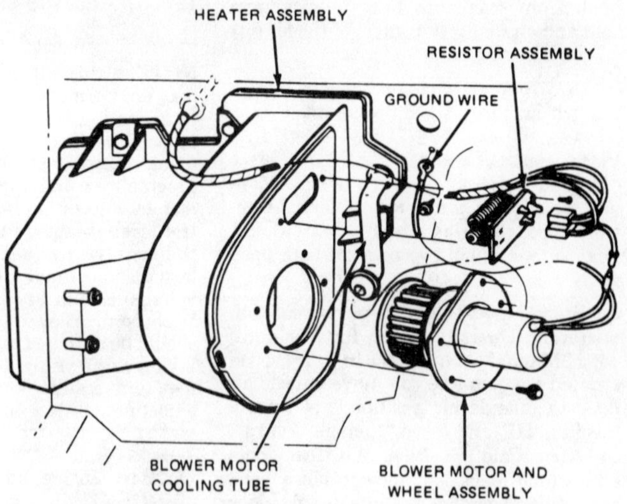

Fig. 81 Blower Motor Assembly Less Air Conditioning. 1974-80 Econoline

TRUCKS & VANS

1980 Bronco & F Series

1. Disconnect battery ground cable.
2. Disconnect electrical leads at connector and at ground screw.
3. Remove the four blower motor retaining screws, then remove blower motor assembly, Fig. 79.
4. Reverse procedure to install.

1979 Bronco & F Series

Less Air Conditioning

1. Refer to "Heater Core, Replace" and perform steps 1 thru 6 to remove heater assembly.
2. Remove the two screws and nuts retaining blower motor to heater assembly.
3. Reverse procedure to install.

1975-80 Econoline

Less Air Conditioning

1. Disconnect blower motor electrical leads.
2. Disconnect blower motor cooling tube from motor, Fig. 80.
3. Remove blower motor assembly retaining screws and remove blower motor.
4. Reverse procedure to install.

With Air Conditioning

1. Disconnect battery ground cable.
2. Disconnect electrical leads from resistor, Fig. 81.
3. Remove blower scroll cover.
4. Push wiring grommet forward out of hole in blower motor housing.
5. Remove the four blower mounting plate retaining screws and remove blower motor.
6. Reverse procedure to install.

1974-80 C Series

1. Disconnect battery ground cable.
2. Remove blower housing front cover.
3. Disconnect electrical connector and ground wire.
4. Remove bracket retaining screws and remove motor and blower assembly.
5. Remove blower fan from motor and motor from mounting plate.
6. Reverse procedure to install.

1974-80 L Series

Less Air Conditioning

1. Disconnect battery ground cable.
2. Disconnect control cable from top of heater, then remove the spring nut, clip and screw.
3. Disconnect blower motor electrical leads and remove heater front cover.
4. Remove motor to mounting plate retaining screws, then rotate motor to unlock it and remove motor and wheel assembly.
5. Loosen set screw and remove blower wheel from motor.
6. Reverse procedure to install.

With Air Conditioning

Refer to the "Heater Core, Replace," with air conditioning, L Series procedure. Perform steps 1 through 14 to remove blower motor, then reverse procedure to install

1974-79 Ranchero

For service procedures, refer to the appropriate section under 1974-76 Torino and 1977-79 LTD II.

1974-78 Bronco

Less Air Conditioning

Refer to "Heater Core, Replace," procedure.

1974-78 F Series

Less Air Conditioning

1. Remove heater assembly as described previously under "Heater Core."
2. Remove the two screws and nuts retaining blower motor to heater.
3. Remove blower fan from motor and motor from mounting plate.
4. Reverse procedure to install

With Air Conditioning

1. Refer to "Heater Core, Replace," with air conditioning, 1974-79 F Series procedure. Perform steps 1 through 10.
2. Unsnap temperature door arm and remove the arm support.
3. Remove temperature door arm pivot retainer.
4. Remove blower motor from housing.
5. Reverse procedure to install.

1974 Econoline

Less Air Conditioning

1. Remove heater assembly as described under "Heater Core" for 1974 models.
2. Remove the two retaining screws and lift off blower motor and wheel assembly from front half of heater housing.
3. Remove blower wheel and motor mounting bracket.
4. Reverse procedure to install

With Air Conditioning

Refer to the "Blower Motor, Replace," less air conditioning, 1974 Econoline procedure.

GENERAL MOTORS CORP.

Buick

ATC PROGRAMMER, LINK & TEMPERATURE DOOR, ADJUST

1974-76

1. Loosen door link hex screw at programmer output shaft, Fig. 82.
2. Place control head selector lever in DEF position.
3. Disconnect electrical connector from programmer. This results in proper positioning of programmer output shaft.
4. Make sure that air mix door is in full heat position. Blower air flow will hold air mix door in proper position.
5. Without disturbing door link or output shaft position, tighten link hex screw.
6. On 1974 models, connect tester J-23678 to wiring harness and the programmer. With "Manual-Automatic" switch in the "Manual" position, rotate "Manual Control"

knob from maximum heat to maximum cold and check for full travel of the air mix door and programmer.

7. On 1975-76 models, connect tester J-23678 into wiring harness and programmer using harness adapter J-24774 and J-24774-75. Place control head in "AUTO." Place "Manual-Automatic" switch in the "Manual" position. Then, rotate "Manual Control" through "Max. Heat" "Max. Cold." High blower should respond in both positions.
8. On 1975-76 models, check for recirculation operation. Operate system for 5 minutes with "Manual Control" set to "150" so restricted vacuum line can move outside air door to outside air position. Set control head to "HI" and move "Manual Control" to "Max. Cold" position. With car doors and windows closed, blower noise level should increase when outside air door moves to the recirculated air position after a three minute delay.

1977-78 All; 1979-80 Except Riviera

NOTE A mis-adjusted temperature door link may not permit the programmer to travel fully to maximum heat or maximum cold, resulting in poor heating or cooling. To check adjustment, connect tester electrical harness between programmer and the vehicle's harness. Place manual-automatic switch in manual and control lever in auto. Place voltmeter knob in 6, then rotate manual control from maximum heat to maximum cold. High blower (battery voltage) should be obtained in both extremes. With control lever in high, recirc., operation should be obtained at max. cold (after slight delay on some vehicles) with these conditions met, temperature door adjustment is correct. If adjustment is necessary, proceed as follows:

1. Start engine and place control lever in "DEF."
2. Remove threaded door rod from programmer arm.
3. Remove electrical connector from in-car sensor. Programmer should go to full heat.
4. Move temperature door to full heat position by pulling door link.
5. Reconnect threaded door rod to programmer, and recheck.

1979-80 Riviera

1. Apply 9-15 inches vacuum to ports 2 and 3 and connect a vacuum gauge to port 4. Seal all other ports, Fig. 83.
2. Apply 12 volts to terminal 1 and connect terminal 2 to ground. The programmer will move to maximum heat position.
3. Preload the air mix door 1 in. the maximum heat position by pulling threaded rod (6 pounds of force) to seal door.
4. Carefully snap the threaded rod into the plastic retainer E on the programmer output arm F. Do not change the programmer or door positions.
5. Release the preload on the air mix door. Apply 12 volts to terminal 1 and connect terminal 3 to ground through a 33 ohm resistor. The programmer will index to maxi-

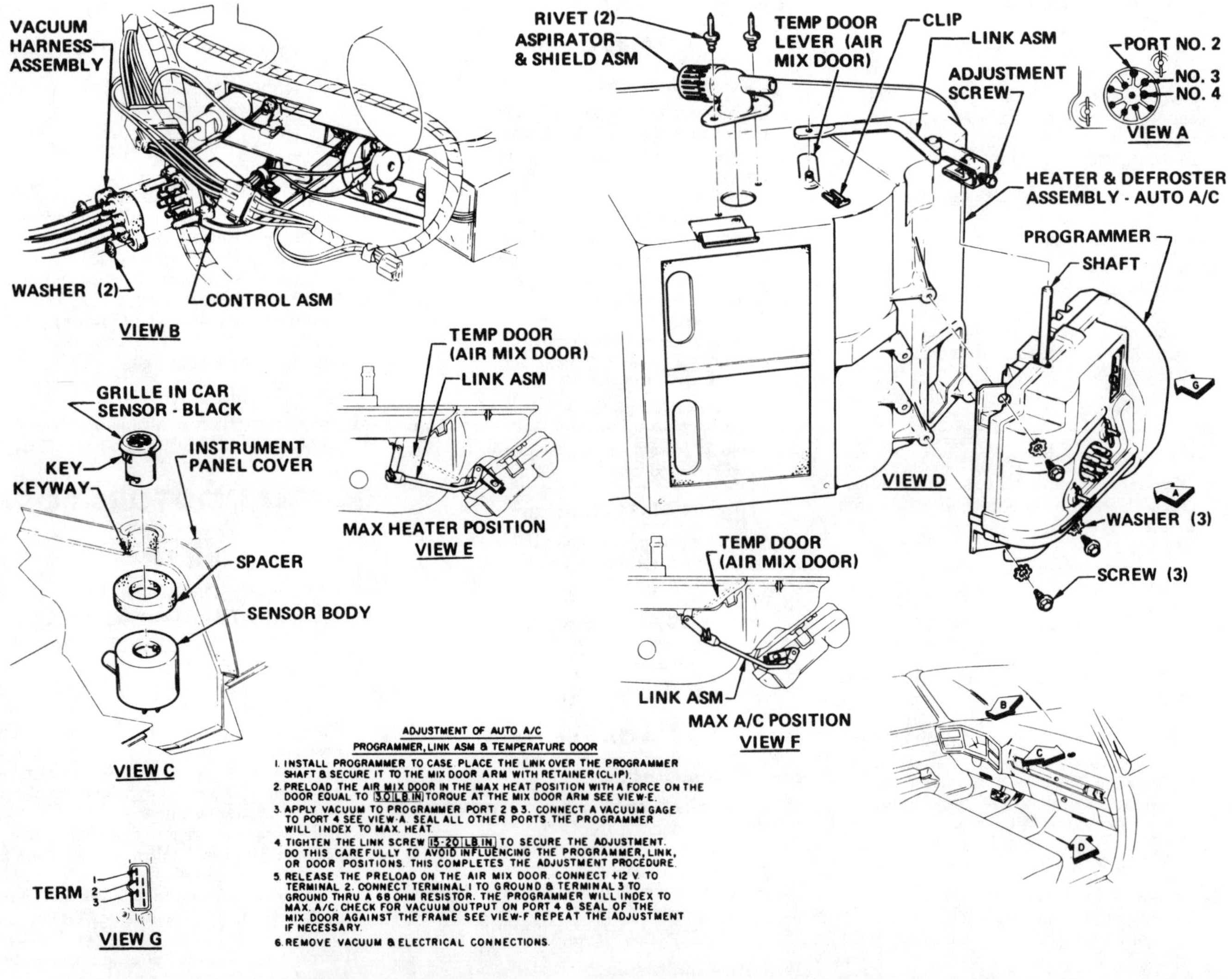

Fig. 82 Programmer link & Temperature door adjustment. 1974-76 All with Auto Temp Control

mum A/C. Check for vacuum output at port 4 and sealing of the air mix door against the frame. Repeat adjustment procedure if necessary.

TEMPERATURE DOOR ACTUATOR, ADJUST

1980 Electra & LeSabre With Electronic Touch Climate Control

Remove right instrument panel trim plate, then turn ignition switch on and touch 65° button while observing actuator arm travel and position. Arm should rotate toward front of vehicle. Touch 85° button while observing arm travel. Arm should rotate toward rear of vehicle with at least 180° angular travel. If door is improperly adjusted, temperature L.E.D.'s will flash at either 65 or 85 control panel position. To adjust temperature door, touch 85°; when system goes to maximum heat, disconnect temperature door rod from actuator retainer, then pull rod and snap into retainer.

A/C CONTROL CABLE, ADJUST

Temperature Control Cable

1974-76 (Less ATC or Custom Aire) Electra, LeSabre & Riviera, 1975 Skyhawk, 1974-79 (Less ATC Or Custom Aire) Century & Regal

1. Assemble control cable to instrument panel control.
2. Secure control cable to temperature control valve.
3. Adjust control cable to obtain 1/16" to ⅛" springback when control is placed in "Hot" position, Fig. 84.

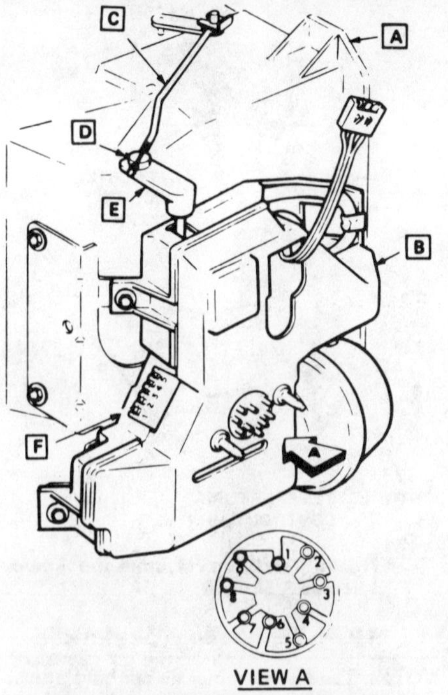

Fig. 83 Programmer link & Temperature door adjustment. 1977-78 All; 1979-80 Except Riviera with Auto Temp Control

1976-77 All With Custom Aire

1. Remove glove box.
2. Set control panel temperature lever to 75.
3. Check if hole in programmer control lever aligns with hole in programmer chassis, Fig. 85. If holes do not align disconnect cable at programmer.
4. Align the holes and adjust cable turnbuckle so cable fits on programmer control lever.
5. Reconnect cable and check adjustment.

1977-80 (Less ATC Or Custom Aire) Electra & LeSabre, 1980 (Less ATC Or Custom Aire) Century & Regal

1. Remove push-on retainer and disconnect temperature cable sliding clip from bellcrank post.
2. Snap the temperature door's threaded operating rod into the bellcrank's white nylon retaining button groove so that .4 inch of threaded rod end is exposed beyond the outer edge of the white nylon button.
3. Reconnect temperature cable retaining clip to bellcrank's vertical post.
4. Move temperature control lever its full travel in both directions to automatically allow cable adjustment. The temperature door should be striking its seat as the lever is moved.

1980 Skyhawk

1. Make sure that cable and temperature lever are positioned at their OFF position.
2. Loosen screw at slotted cable flag.
3. Perform adjustment at slotted cable flag, then tighten screw at flag.

1980 Skylark

1. Connect cable assembly to control head, Fig. 86.
2. Place temperature lever in "OFF" position.

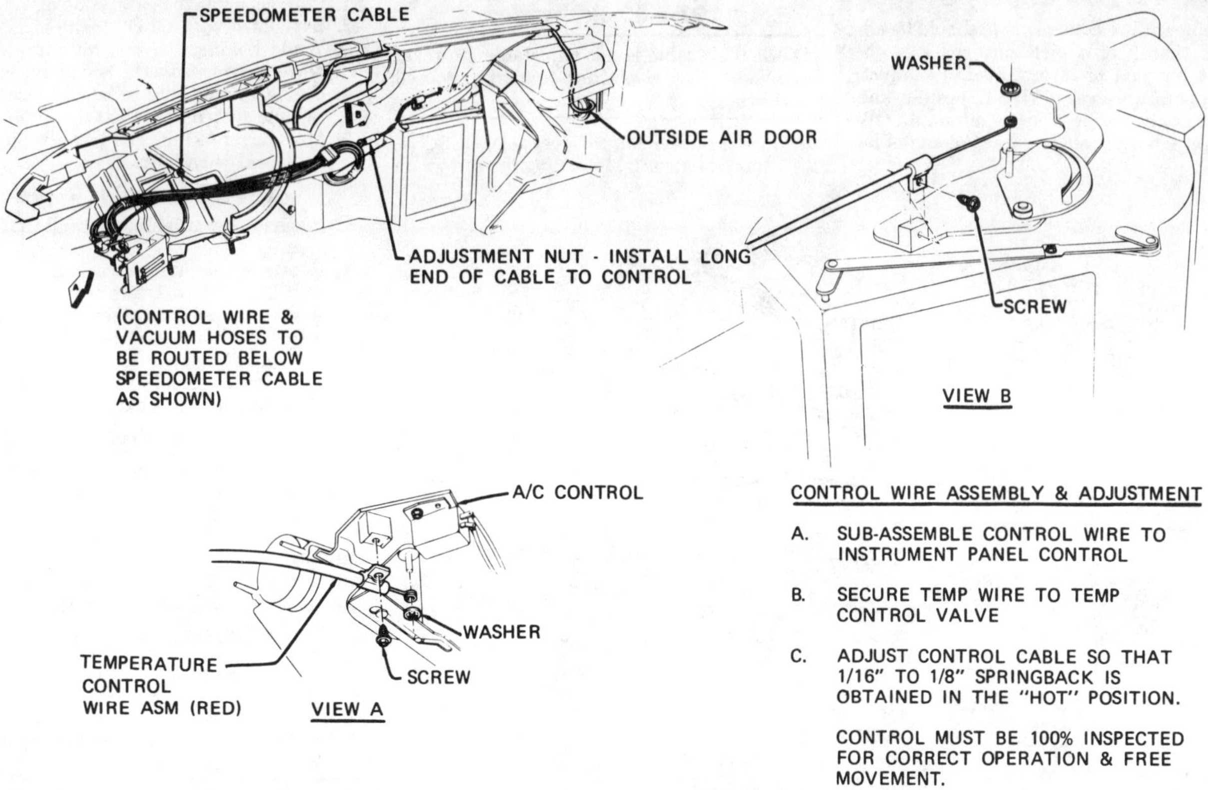

Fig. 84 Temperature cable adjustment (Less Auto Temp Control). 1974-76 Buick Electra, LeSabre & Riviera; 1975 Skyhawk; 1974-79 Century & Regal

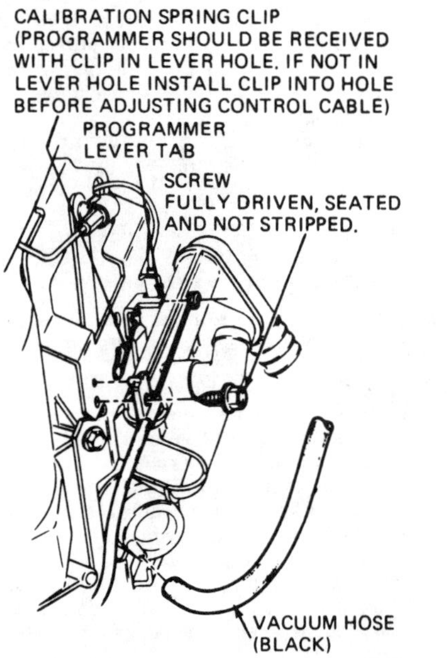

Fig. 85 Temperature cable adjustment. 1976 Buick with Custom Aire

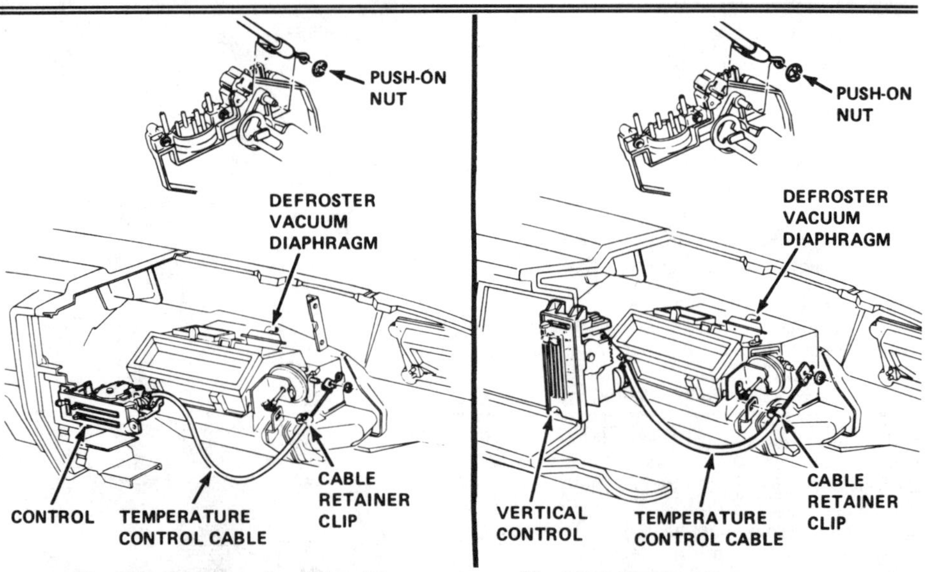

Fig. 86 A/C control panel and temperature cable. 1980 Citation, Omega, Phoenix & Skylark

3. Connect cable to temperature door post and push sheath toward lever until door seats and all lash is removed from cable and control head lever.
4. Tighten cable securing screw.

HEATER TEMPERATURE CONTROL CABLE, ADJUST

1974-76 Except Apollo, Skyhawk & Skylark, 1977-79 Century & Regal

The temperature control cable should be adjusted so that 1/16-⅛ inch springback is obtained at hot end of lever travel. To adjust, move temperature lever to "HOT" position and rotate the control cable adjuster nut until 1/16-⅛ inch springback is obtained at hot end of lever travel.

1974-79 Apollo & Skylark

NOTE: The heater cables do not have adjustable turnbuckles. Cable adjustment is obtained in the positioning of the cable attaching bracket at the heater case.

1. Connect cable at heater control.
2. Set control lever to closed (full up) position.
3. Move the corresponding heater case door to the OFF (full closed) position.
4. Connect cable wire to heater case door lever and tighten cable attaching bracket screws.
5. Check for proper cable operation.

1976-80 Skyhawk

NOTE: This cable is not adjustable, however if the door does not position properly, proceed as follows:

1. Place the lower control lever at "COLD."
2. Adjust the door by loosening the screws on top of the distributor duct. Position door so that it seals at the forward side of the duct, then tighten screws.
3. Recheck door operation. If it does not seal at the forward side of the duct with temperature lever at "COLD," perform readjustment.

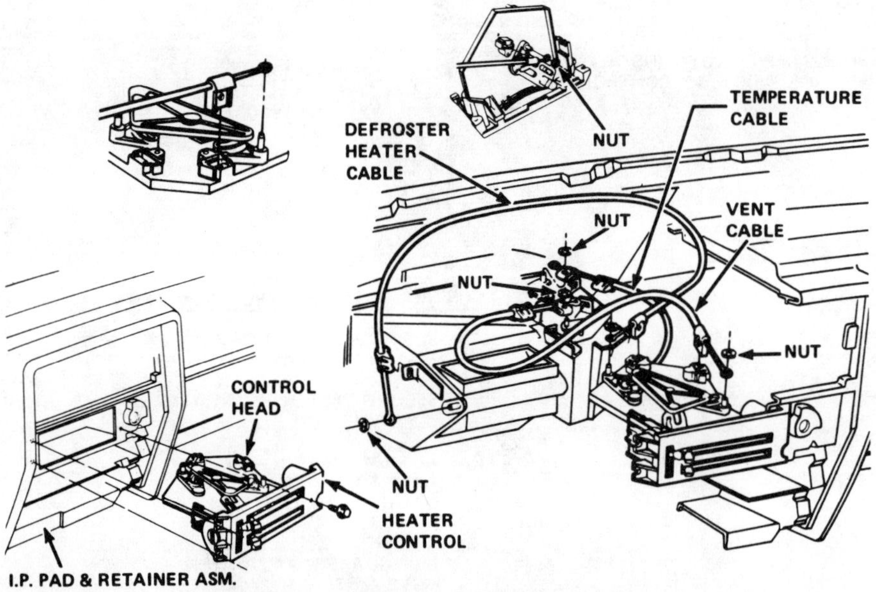

Fig. 87 Heater control panel and temperature cable (horizontal mount). 1980 Citation, Omega, Phoenix & Skylark

1977-80 Electra & LeSabre

NOTE: The temperature cables do not have adjustable turnbuckles. Cable adjustment is in the positioning of the slip clip at the heater case end of the control cable.

1. Connect cable to heater control.
2. Set control to "COLD" position.
3. While holding the control wire, slide the slip clip away from the end of the wire.
4. Connect cable wire to heater case door lever and snap in cable attaching tab.
5. Move lever to full "HOT" position and the cable will self adjust.

1980 Century & Regal

1. Place temperature lever at full COLD position.
2. Rotate temperature door bell crank to full COLD position (crank rotated full clockwise as viewed from above.)
3. Adjust turnbuckle until cable loop aligns with pin on bell crank.
4. Move lever to full HOT. If door is not heard striking its seat, repeat adjustment.

1980 Skylark

The temperature cable incorporates a slider type self-adjust feature A/C control cable adjust, Figs. 87 and 88. As the control head temperature lever is moved through its full range of travel, the cable clip will assume a position assuring that the temperature valve will seat in the full hot and full cold positions.

AIR-DEFROST DOOR-CABLE

1976-79 Skyhawk

NOTE: This cable is not adjustable; however, if the doors do not position properly, proceed as follows:

1. Remove floor outlet duct.
2. Place the upper control lever at "HEAT-

ER." The air defrost door should be parallel with the front and rear sides of the duct.
3. Adjust the door by loosening the two screws on top of the distributor duct and repositioning the door. Tighten screws.
4. Check door operation. In the "DEF" position, the door should seat at the rearward side of the duct and have a ⅛-¼ inch gap at the forward side. Readjust if necessary.
5. Install the floor outlet duct.

SHUT OFF VALVE (PURGE DOOR) CONTROL CABLE, ADJUST

1980 Electra & LeSabre

1. Rotate bell crank to pull valve (door) closed against module case.
2. Adjust turnbuckle to allow installation of cable end to bell crank post.

A/C CONTROL ASSEMBLY, REPLACE

1974 Full Size Less Air Cushion Restraint System

1. Disconnect battery.
2. Unscrew headlight escutcheon and remove headlight switch.
3. Remove lower dash trim.
4. Remove two see-lights from trim plate.
5. Remove four screws from control face.
6. Remove one screw from under dash which connects heater control to instrument panel forward support.
7. Disconnect vacuum and electrical connectors.
8. Remove control assembly.
9. Reverse procedure to install.

1974-76 Intermediates Exc. Apollo & Skylark

1. Disconnect battery ground cable.
2. Remove instrument panel trim plate by pulling rearward and unsnapping from instrument panel.
3. Remove control attaching screws. Then, pull control from dash panel and disconnect vacuum and electrical connectors and Bowden cable.
4. Reverse procedure to install.

1974-76 Apollo & Skylark

1. Disconnect battery ground cable and remove radio.
2. Remove control panel attaching screws and lower control panel slightly from dashboard.
3. Disconnect vacuum hoses, electrical connectors and heater cables. Lower control panel from dashboard, Figs. 139 and 140.
4. Reverse procedure to install.

1974 Full Size With Air Cushion Restraint System

1. Disconnect battery ground cable and tape end.
2. Remove both lower instrument panel cover trim plates by prying underneath and pulling out.
3. Disconnect emergency brake release cable and trip set control cable, if equipped. Remove lower left instrument panel cover.
4. Remove headlight switch control shaft and escutcheon. Then, pull trim plate off carefully. Ensure see-lights pull out of receptacles.
5. Remove screws from control face and the screw connecting control to forward support.
6. Disconnect vacuum and electrical connectors and the Bowden cable, then, remove control assembly.
7. Reverse procedure to install.

1975-76 Full Size

1. On models equipped with air cushion restraint system, turn ignition switch to "Lock."
2. On all models, disconnect battery ground cable and tape end.
3. Remove left lower instrument panel trim by carefully prying and pulling trim out.
4. Remove three screws from control face.
5. Disconnect vacuum hoses, electrical connectors and the Bowden cable.
6. Remove control assembly from dashboard.
7. Reverse removal procedure to install.

1975 Skyhawk

1. Disconnect battery ground cable.
2. Remove clock knob and wood grain trim panel, if equipped.
3. Remove instrument panel cover attaching screws, glove box and four attaching nuts from dash inner side above glove box opening, then, the instrument panel cover.
4. Remove nuts securing steering column to upper bracket guide and lower the steering column.
5. Remove screw attaching cluster assembly to dash support and disconnect speedometer cable from speedometer head.
6. Remove cluster assembly screws and pull control panel assembly from dash panel, Fig. 89.
7. Remove Bowden cable, disconnect vacuum hoses and electrical connectors.
8. Reverse removal procedure to install. Torque steering column to dash bracket nuts to 15-25 ft. lbs.

1977-79 Skylark

1. Disconnect battery ground cable.
2. Remove radio.
3. Remove control panel to instrument panel retaining screws and lower control panel slightly from instrument panel.
4. Disconnect vacuum, electrical connections and heater cables from control panel, then lower the control from dash, Fig. 140.
5. Reverse procedure to install.

1977-80 Century & Regal

1. Disconnect battery ground cable.
2. Remove instrument trim plate by pulling rearward and unsnapping from instrument panel.
3. Pull control out and disconnect vacuum, electrical connectors and release the Bowden cable and pull down.
4. Reverse procedure to install. Adjust control cable so that 1/16 to 1/8 inch springback is obtained with temperature lever at hot position.

1977-78 Riviera & 1977-80 Electra & LeSabre

1. Disconnect battery ground cable.
2. Remove left and right instrument panel trim and the center instrument panel trim.

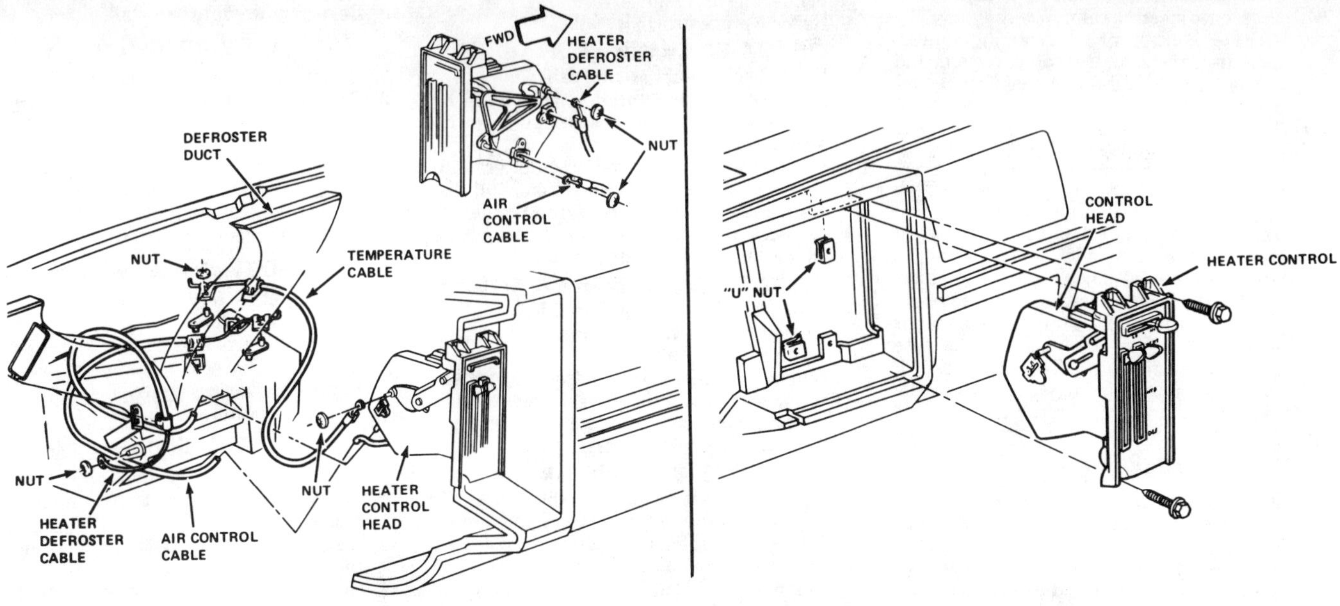

Fig. 88 Heater control panel and temperature cable (vertical mount). 1980 Citation, Omega, Phoenix & Skylark

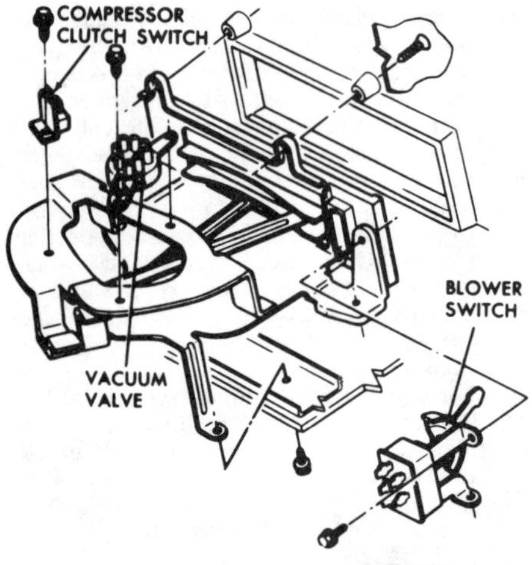

Fig. 89 Control panel installation. 1974-77 Astre & Vega; 1975-80 Monza, Starfire & Sunbird.

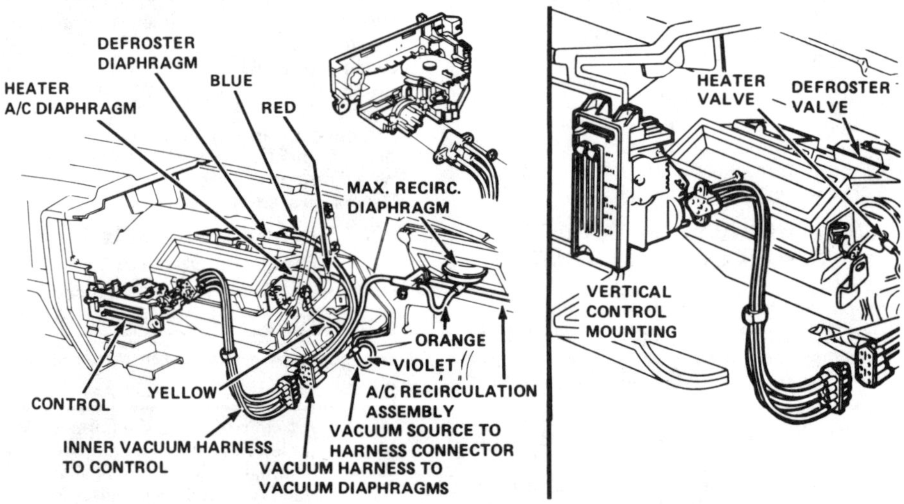

Fig. 90 A/C Control Panel and vacuum connections. 1980 Citation, Omega, Phoenix & Skylark.

3. Remove glove box.
4. Disconnect control cable from A/C module.
5. Remove screws from control face.
6. Disconnect vacuum and electrical connectors and cable.
7. Remove control assembly.
8. Reverse procedure to install.

1976-80 Skyhawk

1. Disconnect battery ground cable.
2. Remove left and right hand dash outlets using a flat tool to disengage them.
3. Remove the 11 screws from around edge of instrument panel pad.
4. Remove left A/C duct.
5. Using a prying action raise pad so that clips at rear of pad are allowed to disengage, then remove pad.
6. Remove steering column retaining nuts and lower assembly, resting the steering wheel on the driver's seat.
7. Remove instrument panel to dash retaining screws, then place a protective covering over steering column and lower instrument panel onto steering column.
8. Remove control panel retaining screws, then disconnect control cable, vacuum and wiring harness and remove control assembly.
9. Reverse procedure to install. Torque steering column retaining nuts to 15-25 ft. lbs.

CAUTION: The steering column to dash bracket fastener is an important retention part as it could affect the performance of vital components and systems and/or could result in a major repair. If replacement of this part is necessary, it must be replaced with one of the same part number or equivalent. Do not use a part of substitute design or lesser quality. Torque values (15-25 ft. lbs.) must be adhered to during reassembly to assure proper retention.

1980 Skylark

1. Disconnect battery ground cable.
2. Disconnect temperature control cable, Fig. 90.
3. Remove control panel retaining screws, then pull control out and disconnect vacuum and electrical connections.
4. Reverse procedure to install.

HEATER CONTROL ASSEMBLY, REPLACE

1974 Electra, LeSabre & Riviera

1. Disconnect battery ground cable.
2. Remove head light switch and dash panel trim.
3. Remove the two see-lights from trim plate and the four screws from control assembly face.
4. Remove 1 screw from under dash which connects control assembly to instrument panel forward support.
5. Disconnect vacuum harness, electrical connectors and control cables, then remove control assembly.
6. Reverse procedure to install.

1974-79 Apollo & Skylark

1. Disconnect battery ground cable.
2. Remove radio.
3. Remove the three control assembly retaining screws and lower the assembly.
4. Disconnect control cables and electrical connectors and remove assembly.
5. Reverse procedure to install.

1974-80 Century & Regal

1. Disconnect battery ground cable.
2. Remove instrumentation trim plates by pulling rearward and unsnapping from instrument panel.
3. Remove the four control assembly retaining screws, then pull control out and disconnect electrical connectors and cables.
4. Reverse procedure to install. Make sure that cables are connected to correct levers.

1975-76 Electra, LeSabre & Riviera

1. Disconnect battery ground cable.
2. Remove left lower instrument panel trim by prying out carefully.
3. Remove the three control assembly retaining screws, then disconnect vacuum and eletrical connectors and remove assembly.
4. Reverse procedure to install.

1975-80 Skyhawk

1. Disconnect battery ground cable.
2. Remove instrument panel bezel and radio.
3. Remove control assembly retaining screws, then push control assembly forward and disconnect control cables and electrical connectors.
4. Remove control assembly.
5. Reverse procedure to install.

1977-78 Electra & LeSabre

1. Disconnect battery ground cable.
2. Remove left hand, right hand and center instrument panel trim plates.
3. Disconnect the three control cables from heater assembly.
4. Remove control assembly retaining screws, then disconnect electrical connector and control cable and remove control assembly.
5. Reverse procedure to install.

1979 Electra & LeSabre

1. Disconnect battery ground cable.
2. Remove the four control assembly retaining screws and the flexible hose outlet to instrument panel duct.
3. Disconnect temperature control and mode selector cables at heater assembly.
4. Pull control assembly outward and then disconnect electrical connectors and cables from assembly and remove control assembly.
5. Reverse procedure to install.

1979-80 Riviera

1. Disconnect battery ground cable.
2. Remove radio knobs and center instrument panel trim plate.
3. Remove control assembly retaining screws.
4. Disconnect electrical connector and vacuum connector from control assembly, then remove control assembly.
5. Reverse procedure to install.

1980 Electra & LeSabre

1. Disconnect battery ground cable.
2. Remove trim plates.
3. Remove control assembly retaining screws, disconnect harness and control cable and remove control assembly.
4. Reverse procedure to install.

1980 Skylark

1. Disconnect battery ground cable.
2. Remove control panel screws and pull control out.
3. Disconnect electrical connection and cables from control.
4. Remove blower switch from control panel, if necessary.
5. Reverse procedure to install.

PROGRAMMER, REPLACE

1974-75 All; 1976 Electra & Riviera

Removal

1. Remove glove box.
2. Loosen adjusting screw on link assembly.
3. Disconnect vacuum hoses and electrical connections.
4. Remove programmer mounting screws and the programmer.

Installation

1. Position programmer on heater defroster assembly and install mounting screws.
2. Install link assembly on output shaft, leaving the attaching screw loose.
3. Set control panel selector lever to "Def" and install vacuum harness assembly.
4. Ensure air mix door is in full heat position. Blower air flow will then hold air mix door in the proper position.
5. Tighten securing screw on door link. Use caution not to disturb door link or output shaft position when tightening screw.
6. Connect tester J-23678 to wiring harness and programmer. Set tester "Manual Automatic" switch to "Manual" and rotate "Manual Control" knob from maximum heat to maximum cold. Check for full travel of air mix door and programmer.
7. Connect electrical connections and install glove box.

1976 LeSabre; 1976-77 Century & Regal

Removal

1. Remove glove box.
2. Disconnect programmer temperature control cable and the vacuum harness.
3. Disconnect in-car air source hose at programmer and ambient air source hose at cowl.
4. Disconnect temperature door link from programmer.
5. Remove programmer mounting screws and the programmer.

Installation, Except 1977 LeSabre

1. If original programmer is to be installed, move programmer temperature control lever to the full right position as viewed from the front of the programmer. Lift the control lever locating spring out from the alignment hole in programmer chassis, then move control to align with hole in chassis, allowing the locating spring to rest in control lever hole.
2. Install the programmer in reverse order of removal.
3. Set control panel temperature lever at 75, then attach mounting tab of temperature control cable to programmer. Adjust cable turnbuckle until cable loop aligns with programmer control lever and attach cable loop to lever. Move control panel temperature lever to 65 to release locating spring.

CAUTION: Do not move control panel temperature lever toward 85 while locating spring is engaged as the programmer control lever will be damaged.

4. If a new programmer is to be installed, install programmer in reverse order of removal, then set control panel temperature lever at 75. Adjust cable turnbuckle until cable loop aligns with programmer control lever and attach loop to lever. Move control panel temperature lever to 65 to release the locating spring.

CAUTION: Do not move control panel temperature lever towards 85 while locating spring is engaged as the programmer control lever will be damaged.

5. Install glove box.

Installation, 1977 LeSabre

1. Install programmer in reverse order of removal.
2. Set control panel temperature lever at 75.
3. Attach mounting tab of temperature control cable to programmer after connecting cable wire.
4. Install glove box.

1977-78 Electra & Riviera, 1978 LeSabre

Removal

1. Disconnect battery ground cable.
2. Remove glove box.
3. Disconnect temperature door link threaded rod at programmer.
4. Disconnect vacuum and electrical connectors.
5. Remove hose and adapter from programmer.
6. Remove programmer retaining screws and programmer.

Installation, 1977

1. Position programmer onto module and install retaining screws.
2. Check to make sure that air mix door is in the full heat position. The blower motor air flow will now hold the mix door in the proper position.
3. Without disturbing the door rod or the output shaft position, reconnect door rod to programmer.
4. Reconnect battery ground cable. Check programmer operation using tester J-23678.

Installation, 1978

1. Position programmer onto module and install retaining screws.
2. Connect 3-wire connector to programmer.
3. Place control head selector lever in DEF position and connect vacuum harness assembly.
4. Disconnect in-car sensor.
5. Check to make sure that air mix door is in the full heat position. The blower motor air flow will now hold the mix door in the proper position.
6. Without disturbing the door rod or the output shaft position, reconnect door rod to programmer.
7. Check programmer operation using tester J-23678. With "Manual-Automatic" switch in the "Manual" position, rotate the "Manual Control" knob from maximum heat to maximum cold and check for full travel of air mix door and programmer.
8. Reconnect in-car sensor and battery.

1978-79 Century & Regal, 1979-80 Electra & LeSabre

Removal

1. Disconnect battery ground cable.
2. Remove glove box. On Electra and LeSabre models, remove right hand hush panel.
3. Disconnect temperature door link threaded rod at programmer.
4. Disconnect vacuum and electrical connections.
5. Remove hose from programmer.
6. Remove programmer retaining screws, then remove programmer.

Installation

1. Install programmer onto heater-air conditioner module.
2. Place control head thumbwheel switch to highest temperature position and install vacuum harness assembly and electrical connections and reconnect battery.
3. Make sure that air mix door is in the full heat position. The blower air flow will now hold the air mix door in the proper position.
4. Without disturbing door rod or output shaft position, reconnect door rod to programmer.
5. Remove 3-way connector from programmer and check for full travel of air mix door.

1979-80 Riviera

Removal

1. Disconnect battery ground cable.
2. Remove right hand hush panel and glove box.
3. Remove the three programmer mounting screws.
4. Disconnect electrical and vacuum connectors from programmer.
5. Disconnect rod from actuator link and remove programmer.

Installation

1. Install programmer and connect the rod onto the actuator link.
2. Connect all electrical and vacuum connectors to programmer.
3. Install the three programmer mounting screws.
4. Install glove box and hush panel.
5. Test programmer operation using tester J-23678.

VACUUM MOTORS & ACTUATORS, REPLACE

1974-79 Apollo & Skylark

Air Inlet Door Vacuum Motor

1. Remove kick panel vacuum motor cover.
2. Disconnect vacuum hose at motor and remove vacuum motor to kick panel retaining screws.
3. Remove link and spring at recirculating air door, then the vacuum motor with bracket.
4. Separate vacuum motor from bracket.
5. Reverse procedure to install.

Defroster Door Vacuum Motor

1. Remove heater outlet duct.
2. Remove vacuum motor retaining screws and actuator link retaining screw.
3. Disconnect vacuum hose and remove vacuum motor.
4. Reverse procedure to install.

Plenum Air Inlet Door Vacuum Motor

1. Remove windshield wiper arms and cowl vent grille.
2. Remove plenum vacuum motor attaching screws, Fig. 91. Disconnect vacuum hose from motor.
3. Position air inlet door in closed position and lift vacuum motor assembly from vehicle.
4. Remove vacuum motor to bracket and link screws and the vacuum motor.
5. Reverse procedure to install.

1980 Electra & LeSabre

Bi-Level Actuators

1. Remove battery ground cable.
2. Remove left instrument panel sound absorber.
3. Remove cross duct and heater floor outlet.

4. Remove heater floor baffle by removing screw through opening on right of heater outlet opening.
5. Remove transmission kickdown switch, then working through heater outlet opening, remove door link retainers.
6. Remove 2 screws on forward surface of heater outlet opening and insert punch and push straight forward to free actuators and mounting brackets from locating studs.
7. Remove assembly by pulling to left, then disconnect wiring.
8. Reverse procedure to install.

Defroster Actuator

1. Disconnect battery ground cable.
2. Remove the left instrument panel sound absorber panels.
3. Remove the cross car duct.
4. Remove heater floor outlet.
5. Disconnect transmission kickdown switch connector.
6. Remove defroster door link retainer and spring.
7. Remove screws retaining actuator and bracket to case and remove assembly.
8. Disconnect electrical connector, remove actuator from bracket, transfer parts.
9. Reverse procedure to install.

Recirc Door Actuator

1. Disconnect battery ground cable.
2. Remove the right insrument panel sound absorber panel.
3. Remove glove box door and compartment.
4. Remove trimplate above glove box.
5. Remove right A/C grille and duct.
6. Remove screws attaching actuator and bracket to recirc housing and unplug electrical connector.
7. Lift up and work actuator linkage off of recirc door.
8. Transfer linkage, mounting bracket and sound deadener cover.
9. Reverse procedure to install.

Temperature Door Actuator

1. Disconnect battery ground cable.
2. Remove left instrument panel sound absorber panel.
3. Remove glove box door and compartment.
4. Remove trimplate above glove box.
5. Remove right A/C grille and duct.
6. Disconnect aspirator hose from assembly.
7. Unsnap temp door rod from actuator and disconnect electrical connector.
8. Remove screws attaching bracket assembly to case assembly.
9. Remove actuator from mounting bracket.
10. To install, transfer sound deadener cover and reverse removal procedures, except before installing the temp door rod, connect the neg. battery cable and touch 85° on the control panel. This will drive the system to max. heat. Pull the rod before snapping it into place.

MASTER SWITCH, REPLACE

1977-80 Skyhawk

1. Disconnect battery ground cable.
2. Remove control panel retaining screws and pull out control panel. Do not disconnect cables.
3. Disconnect wiring and vacuum hoses at switch.
4. Remove switch retaining screws and switch.
5. Reverse procedure to install.

AMBIENT SWITCH, REPLACE

1974-79 Apollo & Skylark

The underhood ambient switch is located in the evaporator case. To remove switch, disconnect electrical connector. Then remove switch retaining screws and the switch.

1980 All With Auto Temp Control & Electronic Touch Climate Control

1. Disconnect battery ground cable.
2. Remove hood seal from lip of A/C module.
3. Remove screw from screen nearest sensor.
4. Disconnect connector from sensor and remove sensor.
5. Reverse procedure to install.

IN-CAR SENSOR, REPLACE

1974 All & 1975-77 Exc. Full Size

1. Open glove box door, reach up through opening and grasp sensor body. Twist ¼ turn clockwise and pull down through opening.
2. Disconnect wire connector from sensor and aspirator hose.
3. Reverse procedure to install. Ensure sensor body spacer is located between body and sensor grille.

1975-77 Full Size

1. Through opening in glove compartment, remove sensor retaining screws.
2. Disconnect wiring connector and aspirator hose from sensor. Remove sensor through glove compartment opening.
3. Reverse procedure to install.

1978 All & 1979-80 Electra & LeSabre

1. Disconnect battery ground cable.
2. Remove instrument panel pad.
3. Remove in-car sensor retaining screw, then disconnect connector and hose and remove sensor.
4. Reverse procedure to install.

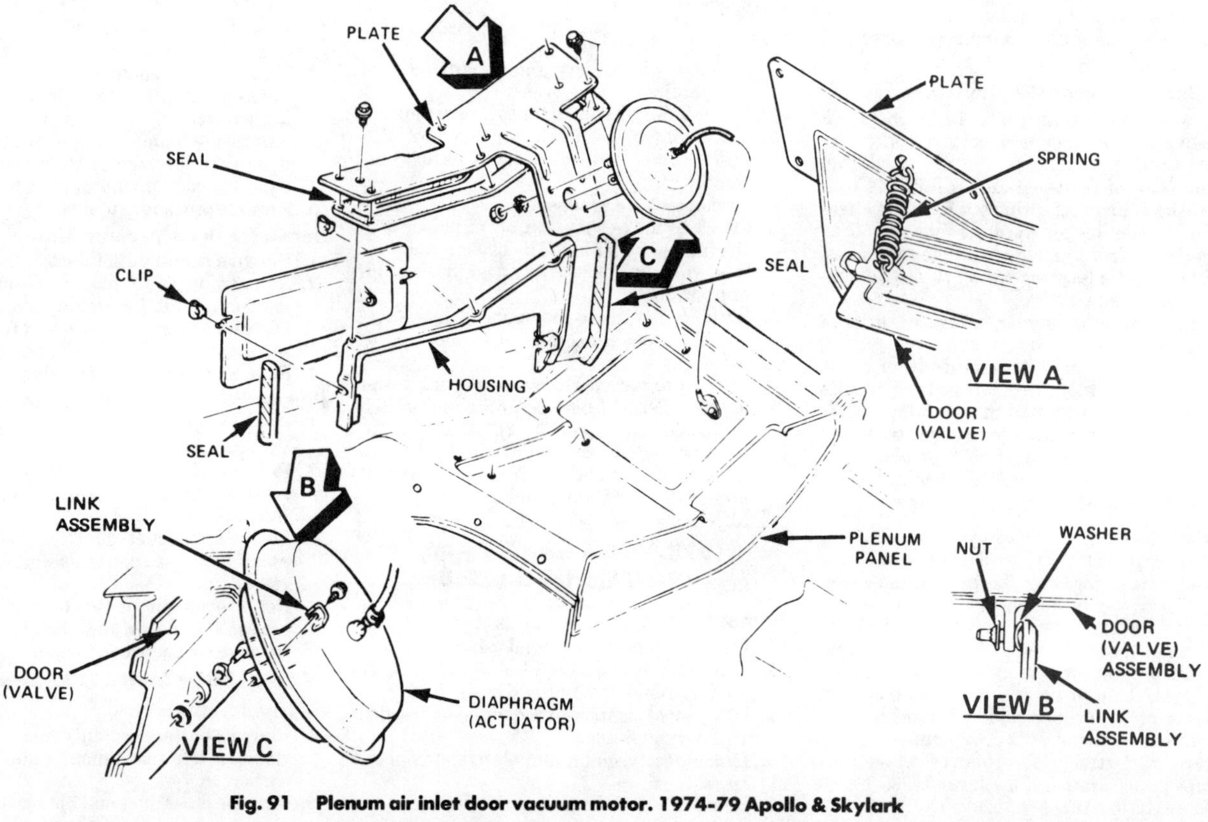

Fig. 91 Plenum air inlet door vacuum motor. 1974-79 Apollo & Skylark

1979-80 Century & Regal

1. Disconnect battery ground cable.
2. Release glove box side tabs and lower the glove box.
3. Release the sensor snap tab, then disconnect wire and hose and remove sensor.
4. Reverse procedure to install.

EVAPORATOR CORE, REPLACE

1974-79 Apollo & Skylark

1. Discharge refrigerant from system.
2. On 1974-76 models, remove VIR assembly as described under "VIR SERVICE" section. On 1977-79 models, remove accumulator assembly.
3. Disconnect all wiring from evaporator case, then remove evaporator case halves retaining screws and nuts, Figs. 92, 93 and 94.
4. Carefully pull left half of evaporator case out and up, then remove evaporator core retaining screws and remove core.
5. Reverse procedure to install using new O-rings coated with refrigerant oil on all line connections.

1974-76 Full Size

1. Discharge refrigerant from system.
2. Disconnect blower motor, resistor and ambient switch connectors.
3. Remove VIR assembly as described under "VIR SERVICE" section.
4. Loosen bolts retaining case to bulkhead so that case can be rotated to clear studs through bulkhead, Fig. 95.
5. Remove screws retaining case halves, then remove inboard half and remove evaporator core.
6. Reverse procedure to install using new O-rings coated with refrigerant oil on all line fittings.

1974-79 Century & Regal

1. Discharge refrigerant from system and disconnect battery ground cable.
2. Disconnect all wiring from evaporator-blower case and disconnect blower motor ground wire.
3. On 1974-76 models, remove VIR assembly as described under "VIR SERVICE" section. On 1977-79 models, remove accumulator assembly.
4. Remove all retaining screws and nuts and remove evaporator-blower case assembly.
5. Remove evaporator-blower case retaining screws and remove evaporator core, Figs. 96, 97 and 98.
6. Reverse procedure to install using new O-rings coated with refrigerant oil on all line fittings.

1975-80 Skyhawk

1. Disconnect battery ground cable.
2. Discharge refrigerant from system, drain cooling system and disconnect heater hoses.
3. Remove glove box, right air outlet duct, deflector, instrument panel bezel and pad and left air outlet duct and deflector and feed duct.
4. Lower steering column, then remove instrument panel assembly and remove control assembly from instrument panel.
5. Remove radio, defroster duct and center distribution duct.
6. On 1975-76 models, remove VIR assembly as described under "VIR SERVICE" section. On 1977-80 models, remove accumulator.
7. Disconnect electrical connectors at control assembly and from plenum and disconnect lines from vacuum tank.
8. Remove cowl to heater-distributor case and stud nuts and remove assembly, Fig. 99.
9. Separate case halves and remove evaporator core.
10. Reverse procedure to install.

1977-79 Riviera; 1977-80 Electra & LeSabre

1. Disconnect battery ground cable.

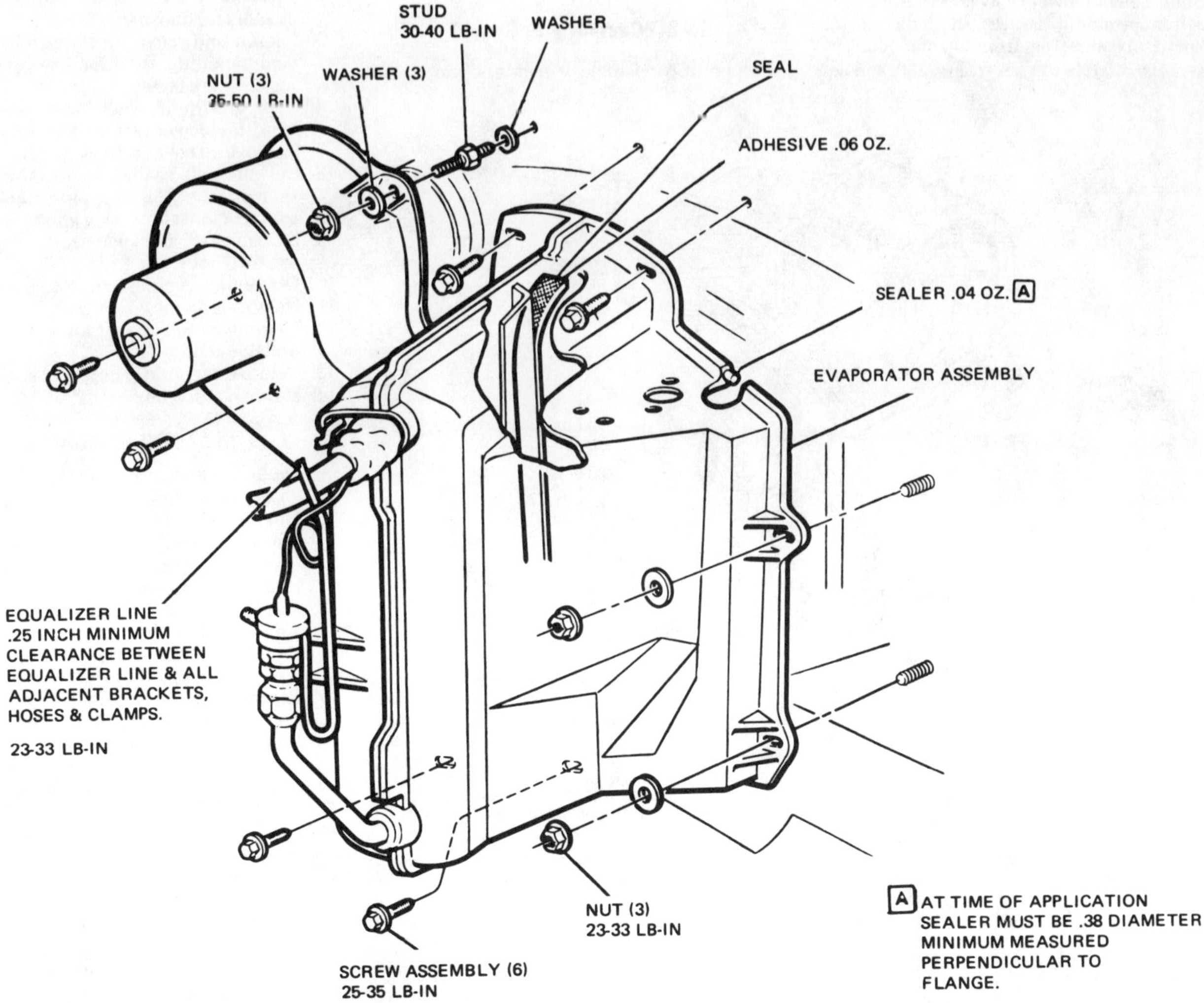

Fig. 92 Evaporator & blower motor assembly (With POA). 1974 Apollo, Omega & Ventura

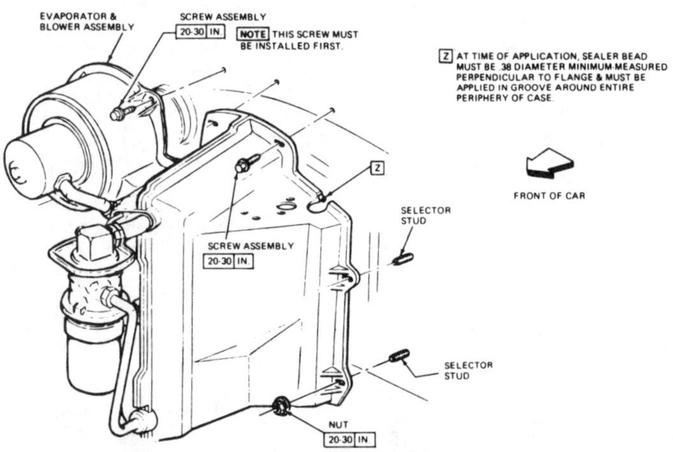

Fig. 93 Evaporator & blower motor assembly (With VIR). 1974-76 Nova; 1975-76 Apollo, Omega & Ventura

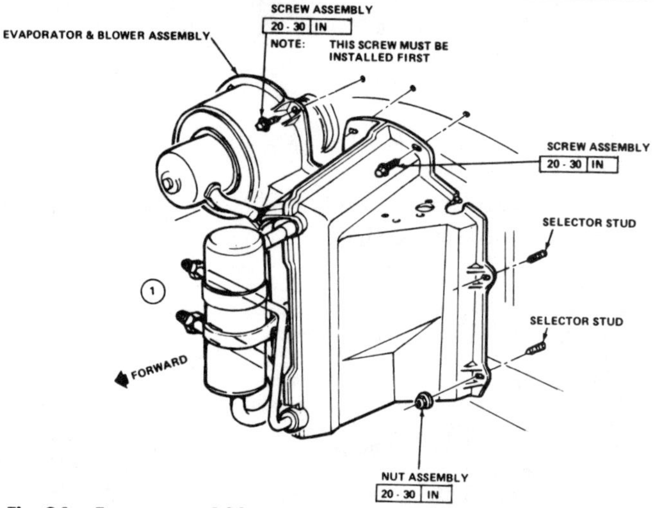

Fig. 94 Evaporator & blower motor assembly. 1977-79 Apollo, Omega, Nova, Phoenix & Ventura

2. Discharge refrigerant from system.
3. Remove module seal, screen and right hand windshield wiper arm.
4. Remove diagnostic connector on 1977-78 models, high blower relay and thermostatic switch retaining screws, then disconnect electrical connectors from top of module.
5. Remove module top cover, Fig. 100, and accumulator bracket screws.
6. Disconnect refrigerant lines and remove evaporator core.
7. Reverse procedure to install.

1980 Century & Regal

1. Operate windshield wipers and stop when right hand wiper is at up position by turning ignition key off.
2. Disconnect battery ground cable.
3. Disconnect seal and remove screens.
4. Disconnect all electrical connectors from evaporator housing.
5. Loosen and move up lower windshield reveal moulding, then remove reveal moulding cowl brackets.
6. Tape a strip of wood below lower edge of windshield over module for protection.
7. Remove all cover screws.
8. Cut through sealing material along cowl, then pry cover off from side (not down from top) to avoid contacting glass.
9. Lift cover off and away from flange of fender-cowl brace, Fig. 98.
10. Disconnect cycling switch connector.
11. Remove accumulator bracket screws.
12. Disconnect refrigerant lines at accumulator and liquid line.
13. Remove evaporator core and seal.
14. Reverse procedure to install.

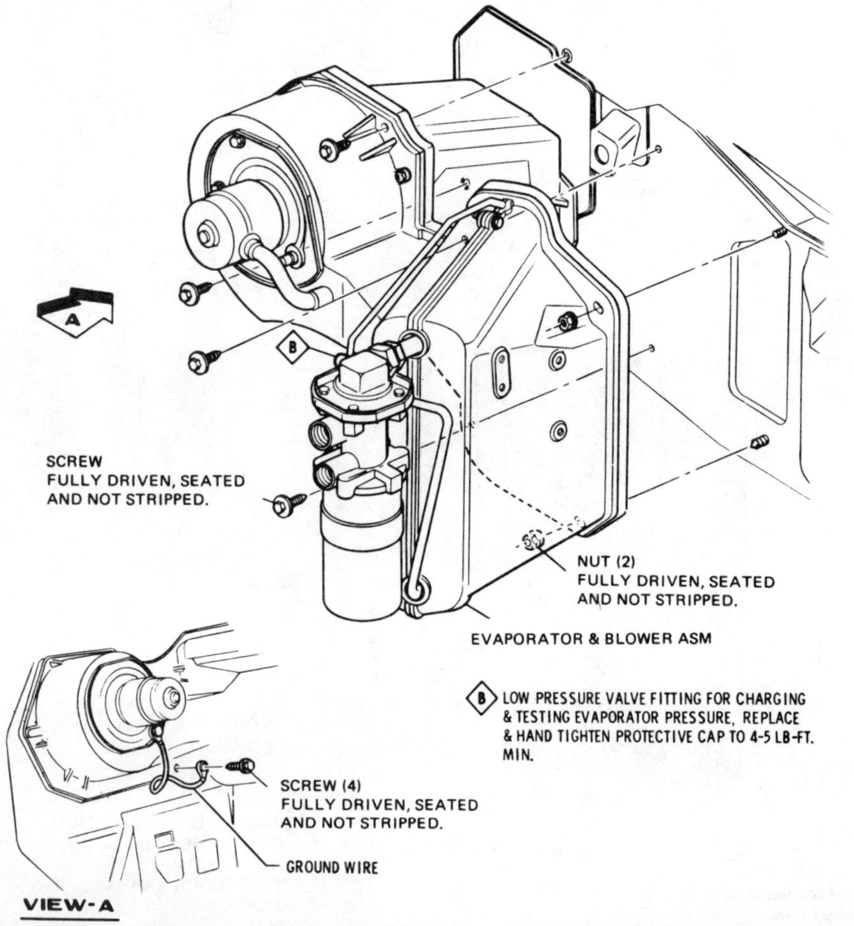

Fig. 95 Evaporator & blower motor assembly. 1974-76 Buick, Chevrolet, Oldsmobile & Pontiac full size models

1980 Riviera

1. Disconnect battery ground cable.
2. Discharge refrigerant from system.
3. Disconnect electrical connectors from blower resistor, blower relay and thermostatic switch, and disconnect ground wire.
4. Remove vacuum tank.
5. Disconnect line from evaporator and line from accumulator.
6. Remove blower motor as described elsewhere in this section.
7. Remove the two screws and one nut from bottom of evaporator and blower assembly.
8. Remove the two screws and one nut from top of evaporator and blower assembly.
9. Remove evaporator and blower assembly, Fig. 101.
10. Disconnect accumulator line from evaporator.
11. Remove accumulator from evaporator case, then remove evaporator.
12. Reverse procedure to install.

1980 Skylark

1. Disconnect battery ground cable and discharge refrigerant system.
2. Remove air cleaner.

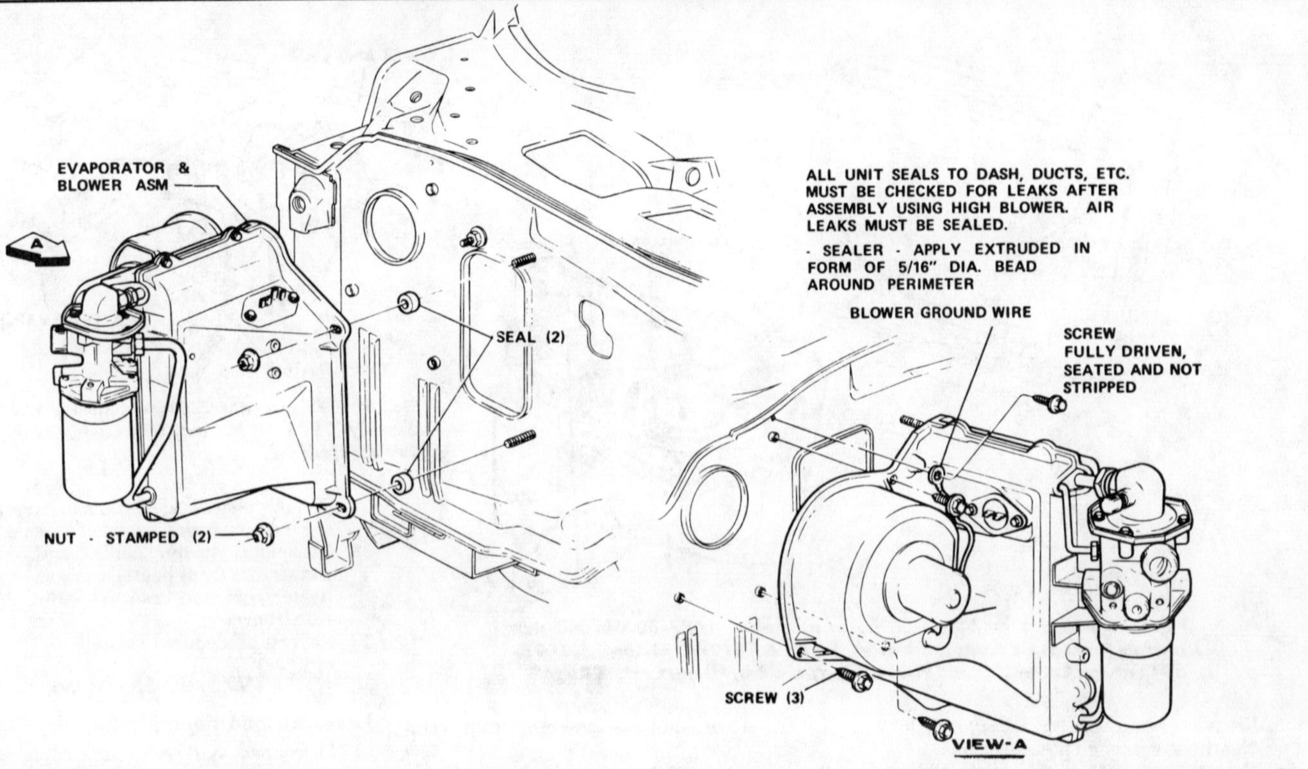

Fig. 96 Evaporator & blower motor assembly. 1974-76 Century, Chevelle, Cutlass, Grand Prix, LeMans, Monte Carlo & Regal.

3. Disconnect air conditioning electrical connections along the cowl and position wiring harness aside.
4. Remove vacuum tank.
5. Disconnect liquid line at evaporator inlet and the low pressure line at accumulator, Fig. 102.
6. Remove outer blower case to cowl screws, then the blower and case assembly.
7. Remove evaporator from vehicle.
8. Reverse procedure to install.

HEATER CORE, REPLACE

1974-76 Full Size

CAUTION: On vehicles equipped with an Air Cushion Restraint System, turn ignition switch to "Lock," disconnect battery ground cable and tape end, thereby deactivating system.

1. Disconnect vacuum hoses from defroster door and outside air inlet door actuator dia phragms and control cable from temperature door lever.
2. Unfasten connector from blower motor resistor.
3. Remove nuts securing heater to dash, Fig. 103.
4. Remove screws securing defroster outlet adapter to heater and raise adapter away from heater.
5. Work heater rearward until studs clear dash and remove.
6. Reverse procedure to install.

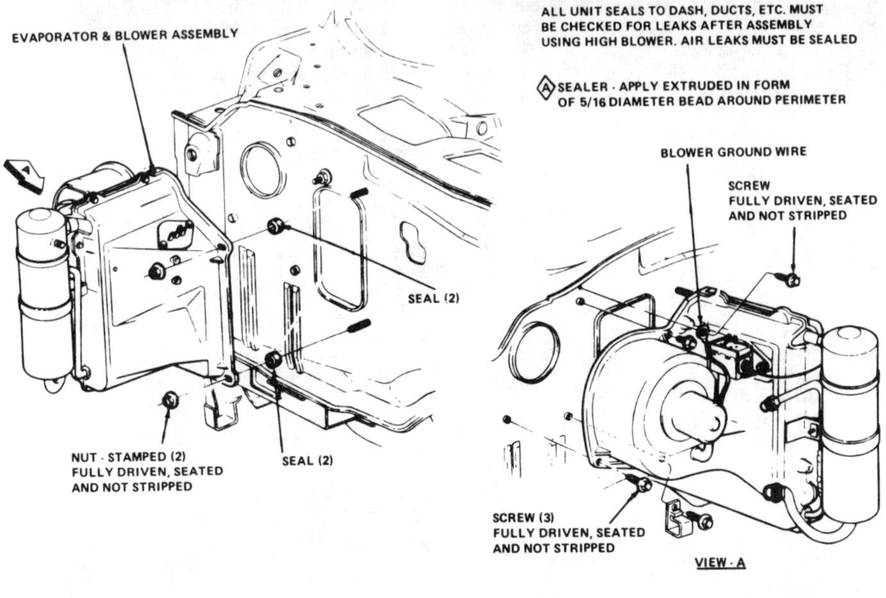

Fig. 97 Evaporator & blower motor assembly. 1977 Century & Regal

1974-77 Century & Regal

Without Air Conditioning

1. Disconnect battery ground cable.
2. Drain cooling system and disconnect heater hoses from heater core.
3. Disconnect control cables from door levers.
4. Remove four nuts securing heater to dash, Fig. 104.
5. Remove screw securing defroster outlet to heater.
6. Work heater assembly rearward until studs clear dash.
7. Reverse procedure to install

With Air Conditioning

1. Disconnect battery ground cable.
2. Drain cooling system and disconnect heater hoses from heater core.
3. Disconnect control cables from temperature door guides and vacuum hoses from actuator diaphragms.
4. Remove resistor assembly and reach through opening and remove one attaching nut to dash. Remove one attaching nut to dash directly over transmission and two attaching nuts to upper and lower inboard evaporator case half.

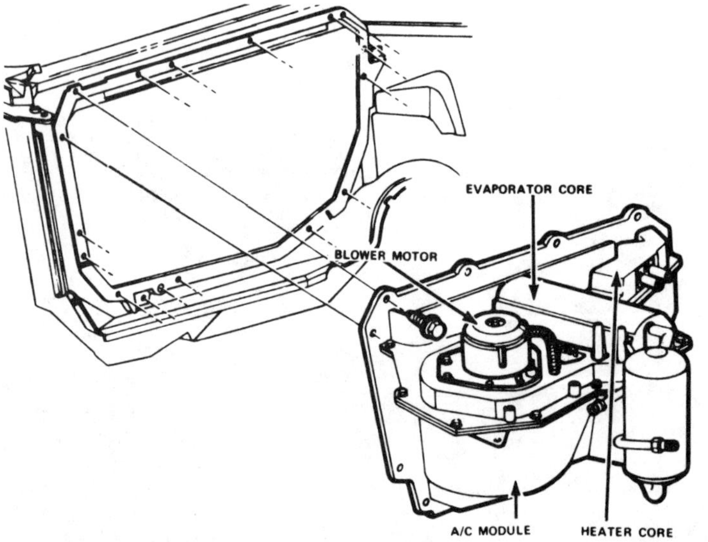

Fig. 98 Evaporator & blower motor assembly. 1977-80 GM full size models (except Eldorado, Toronado, Seville & 1979-80 Riviera) & 1978-80 Century, Cutlass, Grand Prix, LeMans, Malibu, Monte Carlo & Regal

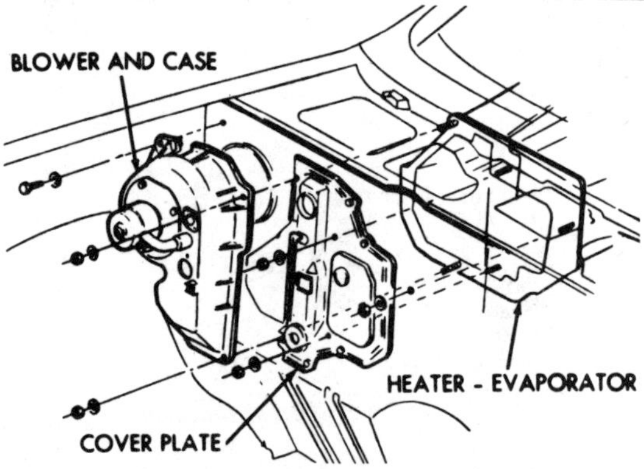

Fig. 99 Evaporator & blower motor assembly. 1975-80 Monza, Skyhawk, Starfire & Sunbird

5. From inside car remove one screw in lower righthand corner on passenger side.
6. Remove lower attaching outlets and work assembly rearward until studs clear dash.
7. Reverse procedure to install.

1974-79 Apollo & Skylark

Less Air Conditioning

1. Disconnect battery ground cable and drain radiator.
2. Disconnect heater hoses from core and plug hoses and core openings to prevent coolant spillage.
3. Remove core case retaining nuts from engine side of dash, Fig. 105.
4. Remove glove compartment and door.
5. Using a ¼-inch twist drill, drill out lower right hand heater case stud from inside vehicle.
6. Pull heater core and case assembly from dash.
7. Disconnect heater cables and blower motor resistor connector from case, then remove heater core and case assembly from vehicle.
8. Remove core tube seal and retaining strips, then heater core from case.
9. Reverse procedure to install.

With Air Conditioning

1. Disconnect battery ground cable and drain radiator.
2. Disconnect upper heater hose and plug hose and core openings. Remove accessible heater core and case attaching nuts.
3. Remove right front fender skirt bolts, lower skirt and disconnect lower heater hose from core tube. Plug hose and core openings.
4. Remove lower right hand heater core and case attaching nuts, Fig. 105.
5. Remove glove compartment and door.
6. Remove right hand kick panel recirculation vacuum diaphragm.
7. Remove heater outlet from bottom of heater case, then cold air distributor duct.
8. Remove heater case extension screws, then separate extension from heater case.
9. Disconnect heater cables and electrical connectors from heater case, then remove heater core and case assembly. Separate core from case.
10. Reverse procedure to install.

1975-80 Skyhawk

Less Air Conditioning

1. Disconnect battery ground cable and disconnect blower motor electrical lead.
2. Place a pan under vehicle, then disconnect heater hoses from core and secure hoses in a raised position.
3. Remove blower inlet to dash panel screws and nuts, then remove blower inlet and blower motor and wheel assembly.
4. Remove core retaining strap screws and remove core.
5. Reverse procedure to install making certain that blower inlet sealer is intact, replace as necessary.

1975-76 With Air Conditioning

1. Disconnect battery ground cable.
2. Remove glove box and right air duct.
3. Remove insrument panel bezel and pad.
4. Remove left air duct and feed duct.
5. Lower steering column and remove instrument panel.
6. Remove control assembly, radio, defroster duct and large distribution duct.

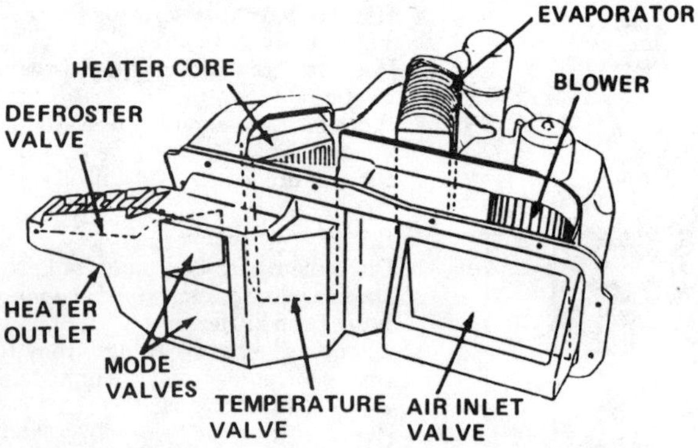

Fig. 100 Evaporator core & blower motor assembly. 1977-80 GM Full Size Except Eldorado, Toronado, Seville & 1979-80 Riviera

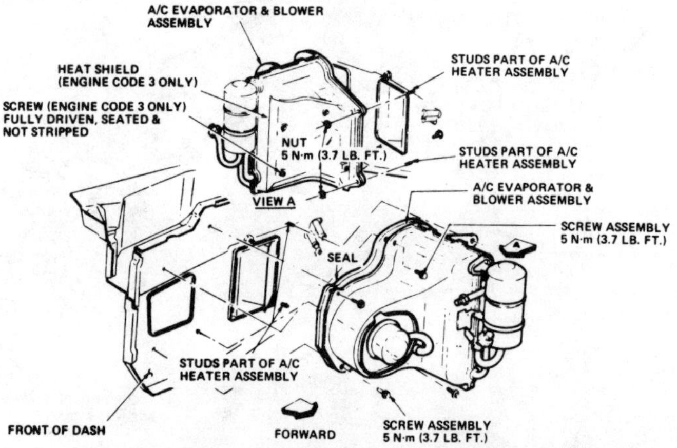

Fig. 101 Evaporator core & blower motor assembly. 1979-80 Riviera

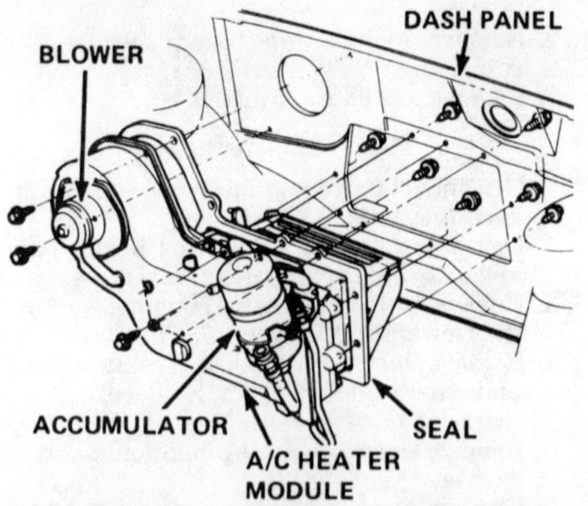

Fig. 102 Module assembly. 1980 Citation, Omega, Phoenix & Skylark

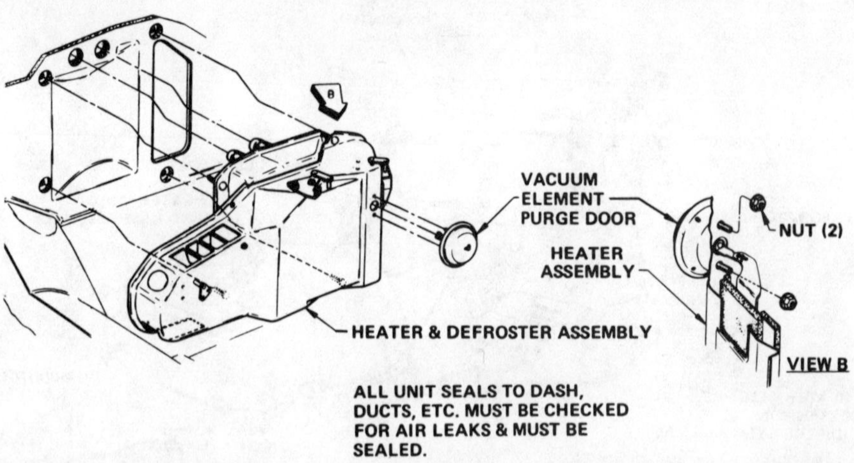

Fig. 103 Heater & defroster assembly. 1974-76 Buick & Oldsmobile Full Size (Typical)

7. Purge system of refrigerant.
8. Disconnect heater hoses at core.
9. Clean VIR exterior surface and line connections, then disconnect compressor inlet line, oil bleed line and condenser outlet line from VIR assembly.

NOTE: Cap all line and VIR ports.

10. Loosen evaporator inlet and outlet lines and remove VIR mounting clamp, then slide VIR assembly from evaporator lines.
11. Remove heater-distributor case to cowl attaching nuts and remove case, then disconnect wiring harness and vacuum lines.
12. Separate heater case from distributor, then remove heater core from case.
13. Reverse procedure to install.

1977-80 With Air Conditioning

NOTE: The heater core can be removed without purging the A/C refrigerant system or removing the evaporator case half of the assembly by using the following procedure.

1. Disconnect battery ground cable.
2. Remove floor outlet duct, then remove glove box and door as an assembly.
3. Remove left and right hand dash outlets using a suitable tool.
4. Remove instrument panel pad.
5. Disconnect vacuum hoses at valves on left end of heater and evaporator unit.
6. Remove insulation tray located below instrument cluster, then loosen console and slide console rearward.
7. Remove instrument panel to dash attaching screws. Place a protective covering over steering column, then lower instrument panel to steering column. Disconnect speedometer cable, radio leads and control head connectors.
8. Remove right hand instrument panel and lap cooler as an assembly.
9. Remove modular duct to heater and evaporator case screw, then remove duct assembly.
10. Disconnect temperature door cable and wiring harness.
11. Disconnect heater hoses from heater core. Plug core tubes to prevent spillage when removing heater core.
12. Remove three heater case stud nuts located in engine compartment.
13. Remove heater core case to evaporator core case attaching screws.
14. Drive on studs to break case loose from dash panel.
15. Remove heater core case assembly.
16. Remove heater core to case attaching screws, then remove heater core.
17. Reverse procedure to install.

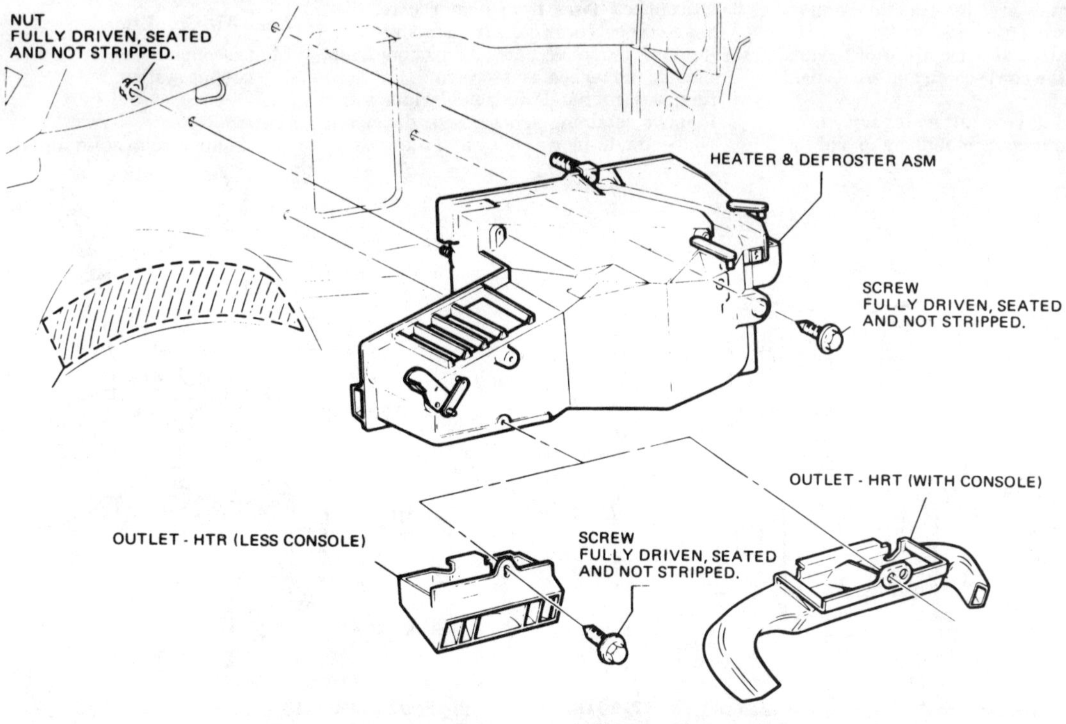

Fig. 104 Heater core & defroster assembly. 1974-77 Century & Regal

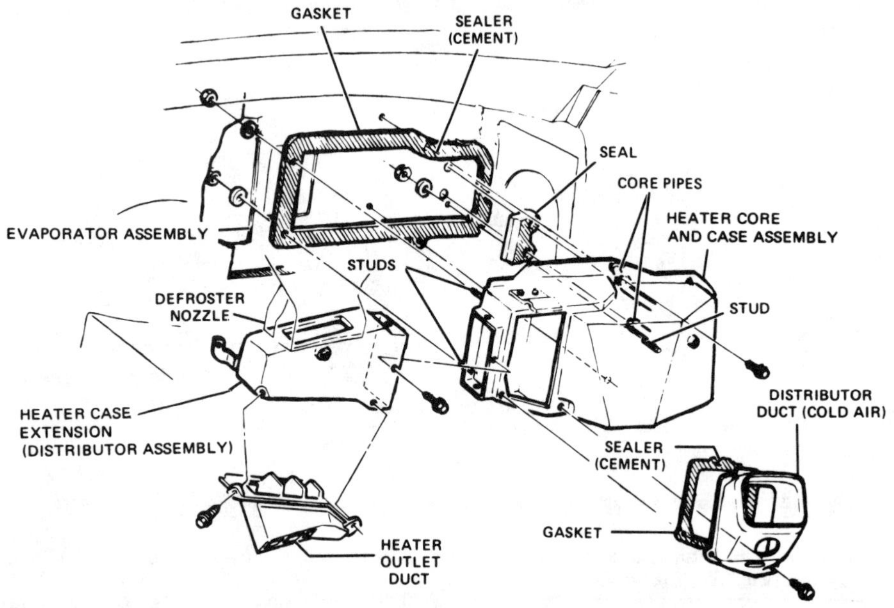

Fig. 105 Heater core. 1974-79 Apollo, Nova & Skylark (Typical)

1977-80 FULL SIZE EXCEPT 1979-80 RIVIERA

Without Air Conditioning

1. Disconnect battery ground cable and drain radiator.
2. Disconnect heater hoses from module, then the electrical connections from module front case.
3. Remove module front case attaching screws, then the heater core, Fig. 106.
4. Reverse procedure to install.

With Air Conditioning

1. Disconnect battery ground cable and drain radiator.
2. Disconnect heater hoses from module.
3. Remove diagnostic connector and thermostatic switch from module.
4. Remove weather seal from top of module cover, then the cowl screen and windshield washer nozzle.
5. Remove module cover attaching screws.
6. Remove heater core retaining clip and the heater core from vehicle, Fig. 100.
7. Reverse procedure to install.

1978-80 Century & Regal

Without Air Conditioning

1. Disconnect battery ground cable and partially drain cooling system.
2. Disconnect hoses from heater core, then the electrical connectors.
3. Remove front module retaining screws and module cover, Fig. 107.
4. Remove heater core.
5. Reverse procedure to install.

With Air Conditioning

1. Disconnect battery ground cable and partially drain cooling system.
2. Disconnect hoses from heater core, then the electrical connectors from module.
3. Remove retaining bracket, ground strap, module rubber seal and screen.
4. Remove right hand windshield wiper arm.
5. Remove retaining screws from diagnostic connector, hi-blower relay and thermostatic switch.
6. Remove module top cover, then heater core, Fig. 98.
7. Reverse procedure to install.

1979-80 Riviera

1. Disconnect battery ground cable and drain cooling system.
2. Disconnect heater hoses from heater core and install plugs in heater core outlets.
3. Remove instrument panel sound absorbers, then lower steering column.
4. Remove instrument cluster as described under Instrument Cluster, Replace.
5. Remove radio front speakers.
6. Remove screws attaching manifold to heater case.
7. Remove upper and lower instrument panel attaching screws.
8. Disconnect parking brake release cable.
9. Disconnect instrument panel wiring harness from dash wiring harness.
10. Disconnect right hand remote control mirror cable from instrument panel pad.
11. Disconnect speedometer cable and temperate control cable at heater case.
12. Disconnect radio, A/C wiring and vacuum lines, and all wiring necessary to remove instrument panel assembly. If equipped with pulse wiper, remove wiper switch, unlock wire connector from cluster carrier and separate pulse wiper jumper harness from wiper switch wire connector.
13. Remove instrument panel and harness assembly.
14. Remove defroster ducts, then remove blower motor resistor.
15. Remove A/C-heater housing to dash panel nuts, Fig. 108.
16. Remove housing to dash screw and clip from inside vehicle, then remove housing assembly from dash.
17. Remove heater core from housing assembly.

1980 Skylark

Less Air Conditioning

1. Disconnect battery ground cable and drain cooling system.
2. Disconnect the heater hoses from heater core.
3. Remove radio noise suppression strap.

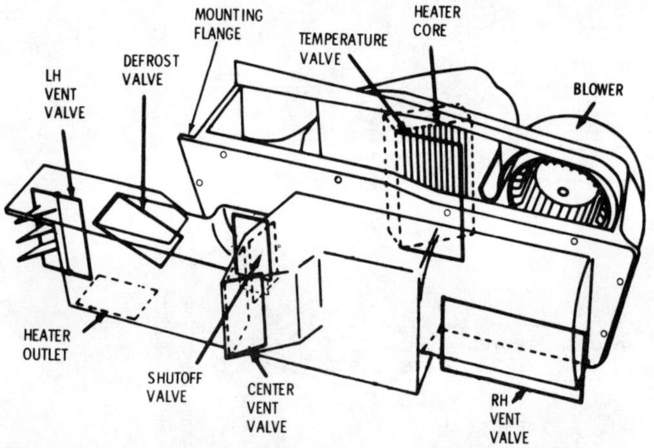

Fig. 106 Heater core & blower motor assembly. 1977-80 GM Full Size Except Eldorado, Toronado, Seville & 1979-80 Riviera

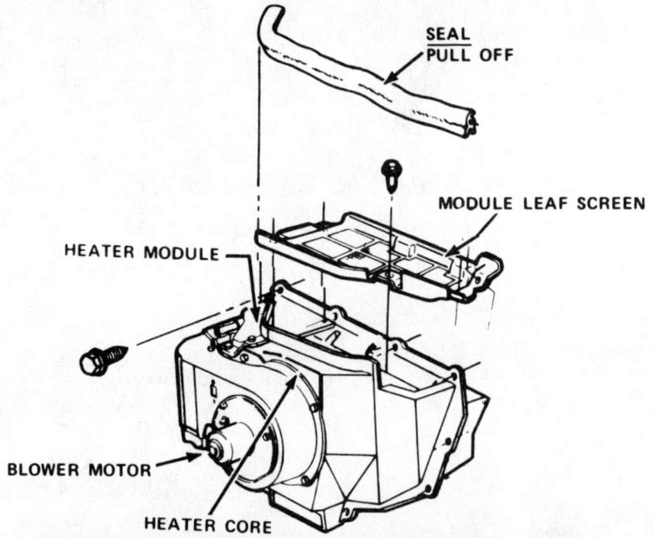

Fig. 107 Module assembly. 1978-80 Century, Cutlass, Grand Am, Grand Prix, LeMans, Malibu, Monte Carlo & Regal

4. Remove heater core cover retaining screws and the cover, Figs. 109 and 110.
5. Remove heater core from vehicle.
6. Reverse procedure to install.

With Air Conditioning

1. Disconnect battery ground cable and drain cooling system.
2. Disconnect heater hoses from heater core.
3. Remove right side hush panel and open the glove box.
4. Remove heater duct retaining screw and the duct.
5. Remove instrument panel support bracket.
6. Remove heater case side cover retaining screws and the cover.
7. Remove heater core retaining clamps and the inlet and outlet tube support clamps.
8. Remove heater core from case.
9. Reverse procedure to install.

BLOWER MOTOR, REPLACE

1974-76 Electra, LeSabre, Riviera

CAUTION: On vehicles equipped with an Air Cushion Restraint System, turn ignition switch to "Lock," disconnect battery ground cable and tape end, thereby deactivating system.

1. On 1975-76 Riviera, support hood and loosen hood hinge from extension. Remove extension and plate assembly.
2. Disconnect all wires, unfasten and remove motor. Fig. 103.

1974-77 Century & Regal

1. Disconnect blower motor wire.
2. Remove screws securing motor to air inlet and remove blower motor, Fig. 104.

1975-80 Skyhawk

1. Disconnect battery ground cable and disconnect blower motor electrical leads. On some models, it may be necessary to remove coolant recovery tank attaching screws and position tank aside. Do not disconnect hoses from tank.
2. Scribe blower motor flange to case position.
3. Remove blower motor retaining screws and remove blower motor and wheel assembly. Pry flange gently, if sealer acts as an adhesive.
4. Reverse removal procedure to install, aligning scribe marks made during removal.

1974-79 Apollo & Skylark

1. Disconnect battery ground cable and raise vehicle on hoist.
2. Remove fender skirt attaching bolts except those attaching skirt to radiator support.
3. Pull skirt out and down, then place a block of wood between skirt and fender to provide clearance for motor removal.
4. Disconnect blower motor electrical connections and remove motor attaching screws and blower motor.
5. Remove blower motor retaining nut and separate impeller from motor.
6. Reverse procedure to install.

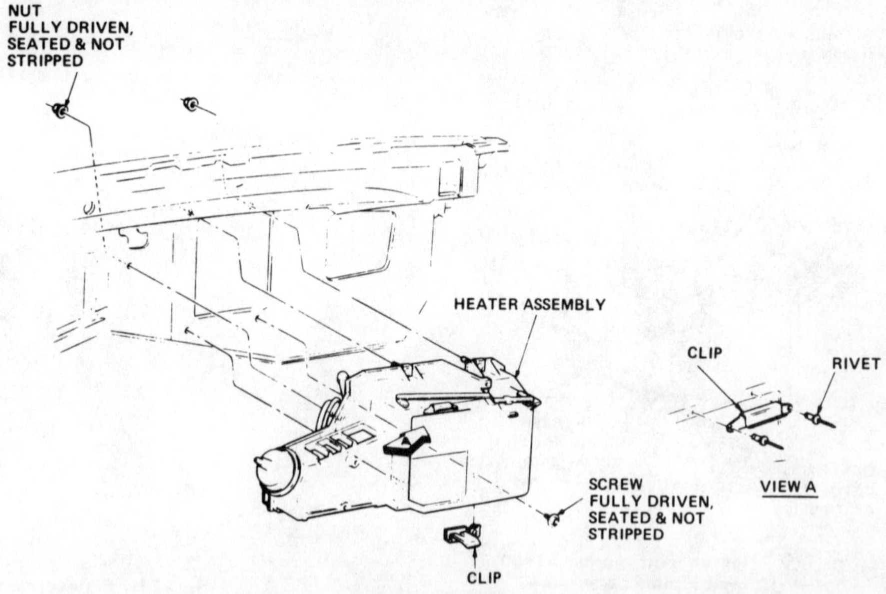

Fig. 108 Heater core. 1979-80 Eldorado, Riviera & Toronado

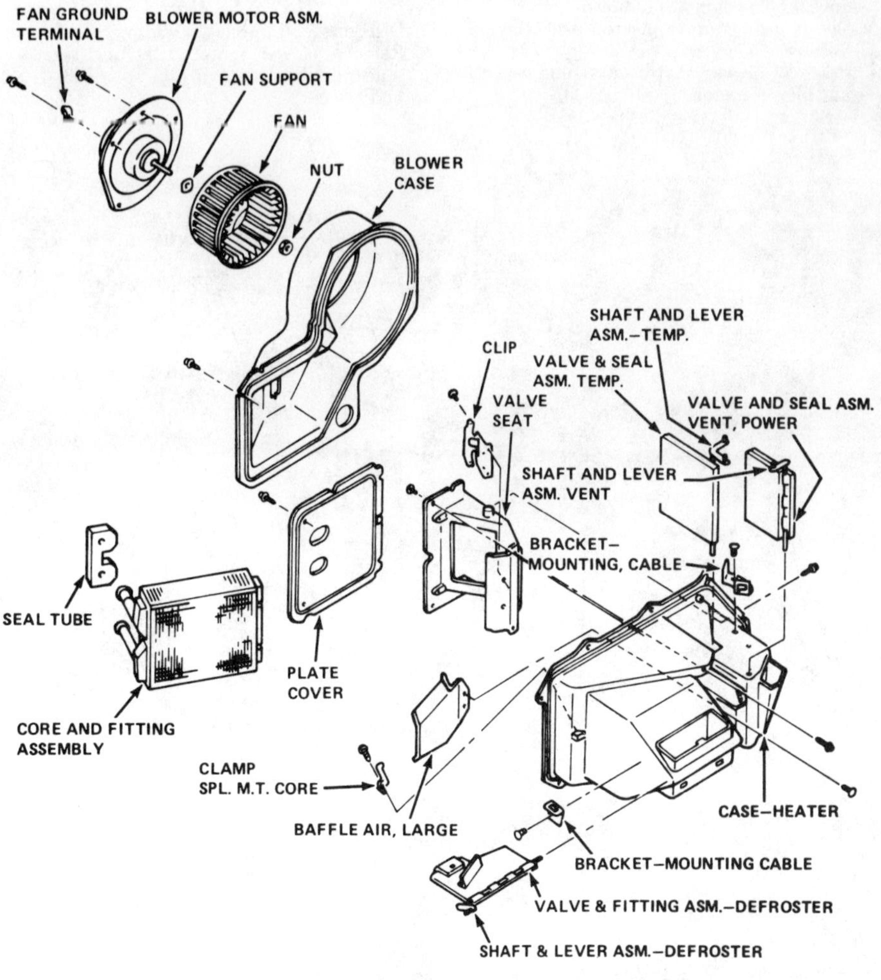

Fig. 109 Heater core & case assembly. 1980 Citation, Omega, Phoenix & Skylark

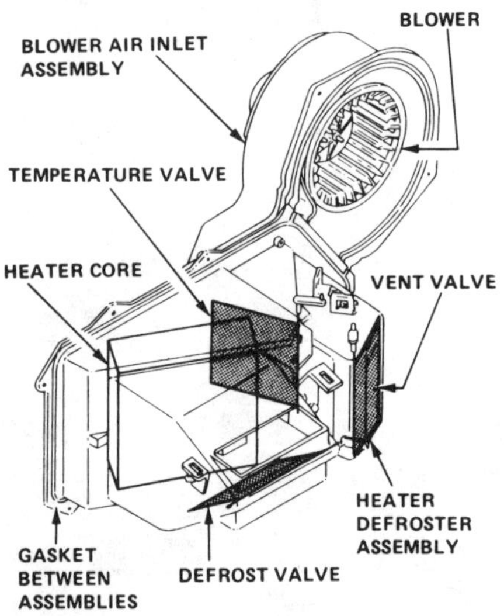

Fig. 110 Heater vent module. 1980 Citation, Omega, Phoenix & Skylark

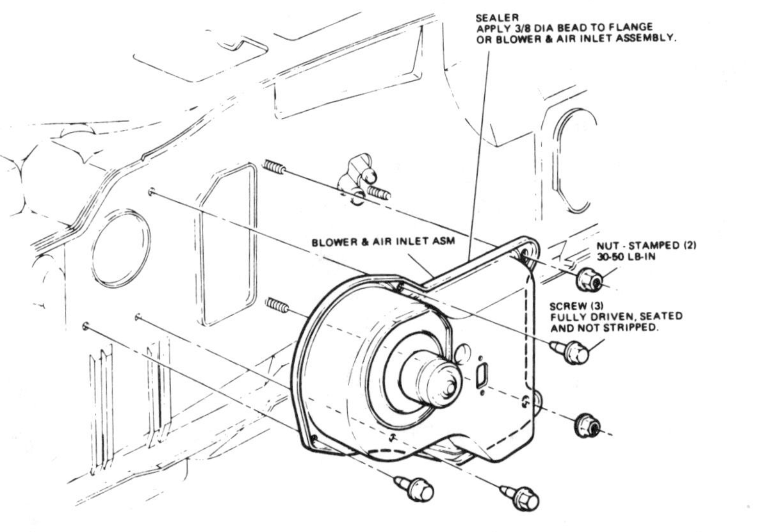

Fig. 111 Blower motor. 1978-80 Century Regal

1979-80 Riviera

1. Disconnect battery ground cable.
2. Remove right fender skirt to provide clearance for blower motor removal.
3. Disconnect wire connector and cooling hose.
4. Remove blower motor attaching screws and blower motor.

1978-80 Century & Regal, 1977-78 Riviera, 1977-80 Electra & LeSabre & 1980 Skylark

1. Disconnect blower motor wiring.
2. Remove blower motor attaching screws and the blower motor, Figs. 102, 106, 110 and 111.

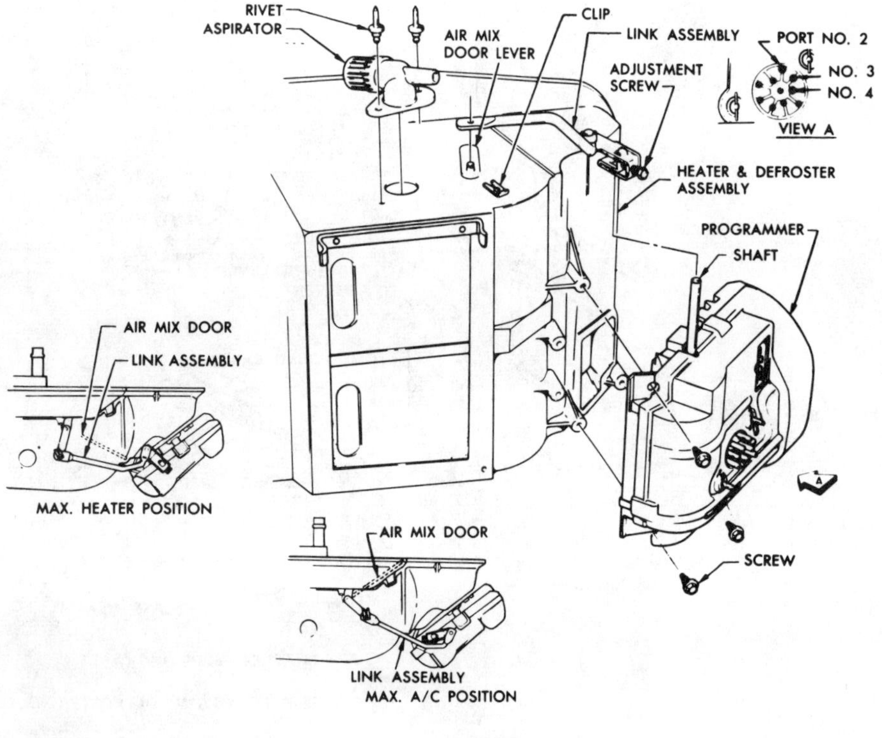

Fig. 112 Air-mix door link adjustment & programmer installation. 1974-76 models (except Seville) & 1977-78 Eldorado

Cadillac

AIR-MIX DOOR LINK ADJUSTMENT

1974-76 All, 1977-78 Eldorado & 1977-79 Seville Auto Climate Control

1. Loosen hex screw of door link at programmer output shaft, Figs. 112 and 113.
2. On 1974-79 models, set lever to "Def."
3. On 1974-79 models, remove ambient sensor electrical connector. On all models, programmer output shaft will go to the full heat position. Blower air flow will now hold the door in the proper position.
4. Without disturbing door link or output shaft position, tighten door link hex screw.
5. Connect tester J-23678 into wiring harness and programmer. On 1974-79 models, use harness adapter J-24774. On all models, set control head to "Auto" and set tester "Manual-Automatic" switch to "Manual." Using tester "Manual Control," swing programmer to "Max. Heat" and then to "Max. Cold." High blower should operate in both positions.
6. Check recirculation operation. Set tester "Manual Control" to 150 and operate system for five minutes so that the restricted vacuum line can move the outside air door to the outside air position. With control head in "Hi," move "Manual Control" to "Max. Cold." With all vehicle doors and windows closed, blower noise level should increase when recirculation occurs, after approximately a three minute delay.

1977-80 Auto Climate Control Except Eldorado & Seville

To check air-mix door for proper operation or to position door for adjustment, connect tester J-23678 between programmer and vehicle wiring harness using adapter J-23678-77. Connect single connector to programmer test point. Place manual-automatic switch on manual, control lever on Auto. and voltmeter on 6. Rotate manual control from max. heat to max. cold, high blower speed should be obtained at both extremes with control lever in high. Recirculating air operation should be obtained at

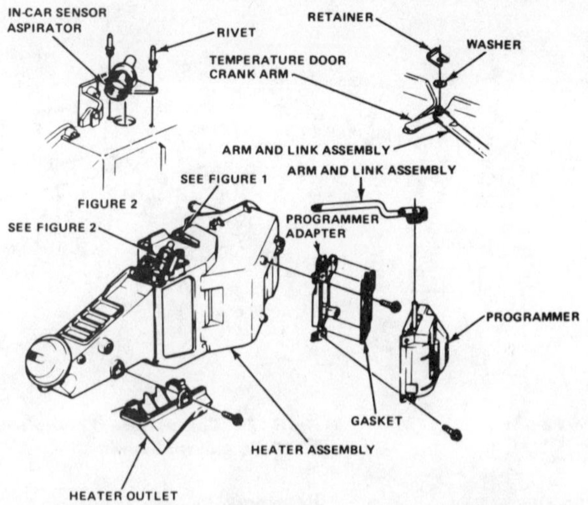

Fig. 113 Air-mix door link adjustment & programmer installation. 1976-79 Seville

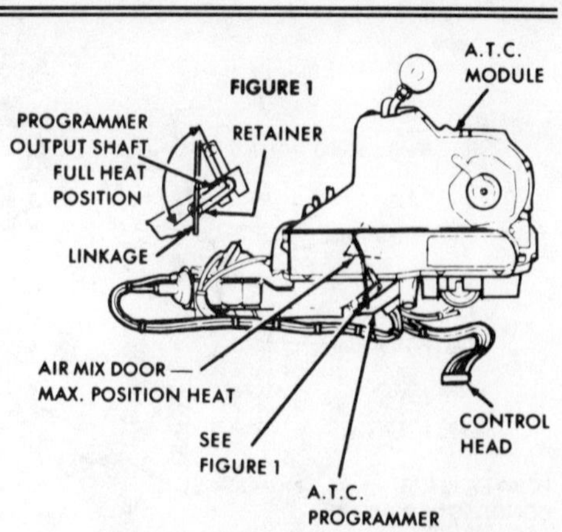

Fig. 114 Air-mix door link adjustment. 1977-80 except Eldorado & Seville

max. cold. With these conditions met, air-mix door can now be adjusted.

1. Disconnect air-mix door linkage from plastic retainer on programmer output shaft.
2. Place A/C control lever on Defrost, then start engine.
3. Disconnect single wire connector in light green wire of adapter harness J-23678-77. This will properly position the programmer output shaft in the full heat position, Fig. 114.
4. Move temperature door to the full heat position by pulling on linkage, then seat linkage into plastic retainer on programmer output arm.
5. Recheck air-mix door for proper operation as described above.

1979 Eldorado Auto Climate Control

1. Apply 9-15 inches vacuum to ports 2 and 3 and connect a vacuum gauge to port 4. Seal all other ports, Fig. 115.
2. Apply 12 volts to terminal 1 and connect terminal 2 to ground. The programmer will move to maximum heat position.
3. Preload the air mix door link the maximum heat position by pulling threaded rod (6 pounds of force) to seal door.
4. Carefully snap the threaded rod into the plastic retainer E on the programmer output arm F. Do not change the programmer or door positions.
5. Release the preload on the air mix door. Apply 12 volts to terminal 1 and connect terminal 3 to ground through a 33 ohm resistor. The programmer will index to maximum A/C. Check for vacuum output at port 4 and sealing of the air mix door against the frame. Repeat adjustment procedure if necessary.

1980 Eldorado & Seville Electronic Climate Control

1. Remove glove box.
2. Set control panel temperature for 90° F, allow approximately two minutes for programmer arm to travel to the max. heat position, Fig. 116.
3. Disconnect threaded rod from plastic retainer on programmer output arm, then check to see air-mix door is not binding.
4. Preload air-mix door in the max. heat position by pulling on threaded rod.
5. Check to ensure that programmer output arm is aligned with arrow on programmer case, then carefully snap threaded rod into plastic retainer on programmer arm.

NOTE: When snapping threaded rod into programmer arm plastic retainer, use care not to disturb position of air-mix door or programmer output arm.

6. Set control panel temperature for 60° F and check to ensure that programmer output arm and air-mix door travel to the max. cooling position, Fig. 116.

CONTROL PANEL, REPLACE

1974-76 Models (Except Seville) & 1977-78 Eldorado

1. Disconnect battery ground cable.
2. Remove four screws securing lower cover to lower steering column cover reinforcement. Then, remove four screws securing lower cover to instrument panel horizontal support. Remove steering column lower cover.
3. Remove two control to support mounting screws. From left rear corner of control, remove control to bracket mounting screw, Fig. 117.
4. Remove control and disconnect vacuum and electrical connectors and the bulb.
5. Reverse procedure to install.

1976-79 Seville

1. Disconnect battery ground cable.
2. Remove two screws from lower edge of instrument panel. Then remove two screws from upper sides of steering column cover.
3. Lower steering column cover and disconnect rear defogger circuit breaker.
4. Remove lower steering column cover and reinforcement.
5. Remove four screws securing speedometer cluster bezel to lower instrument panel and cluster assembly.
6. Press bezel downward slightly, rotate top outward and carefully remove speedometer cluster bezel.

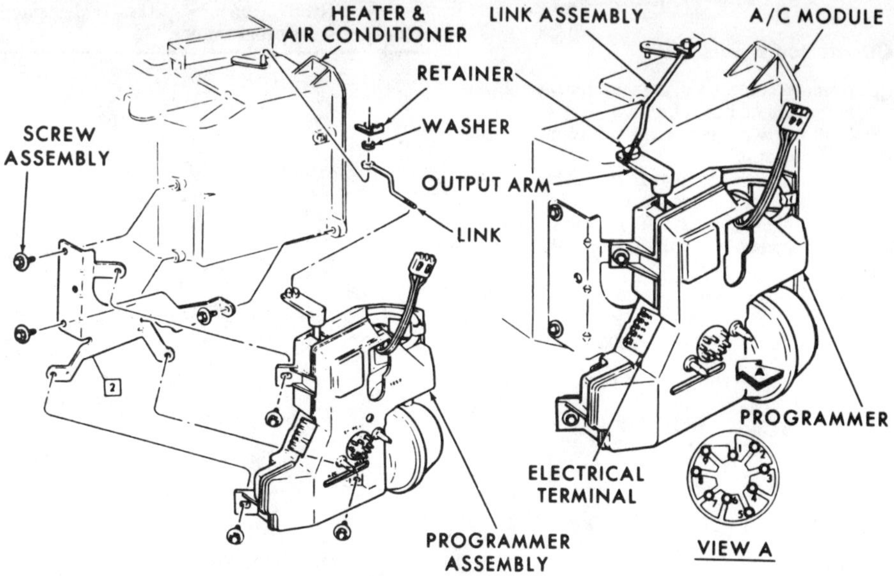

Fig. 115 Air-mix door link adjustment & programmer installation. 1979 Seville

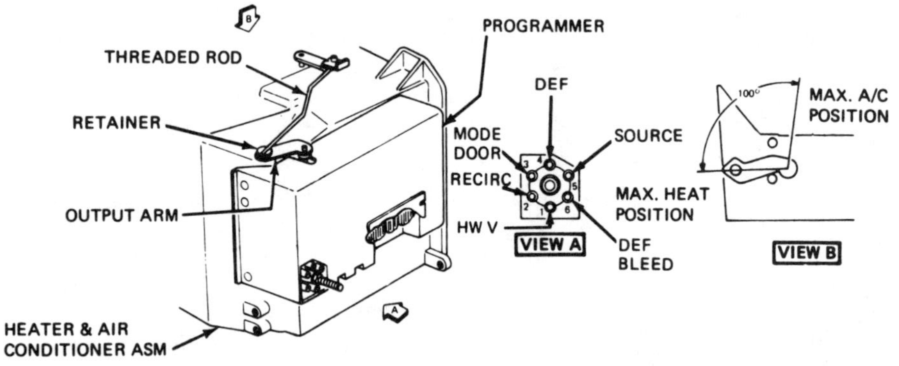

Fig. 116 Air-mix door link adjustment. 1980 Eldorado & Seville

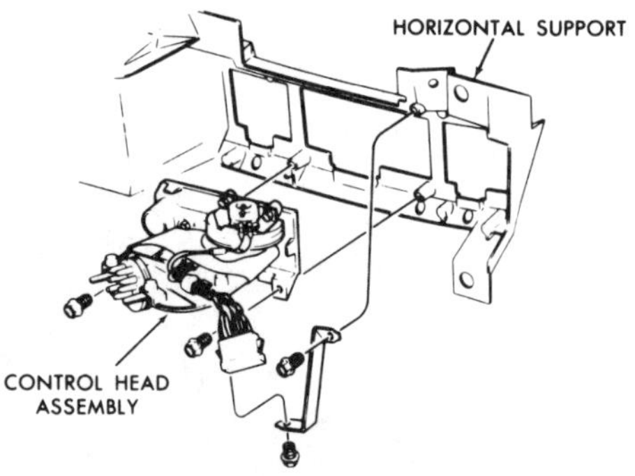

Fig. 117 Control panel installation. 1974-76 except Seville

7. Remove trim screw from left side of lower panel, then the two screws securing left lower instrument panel to top cover, Fig.117.
8. Loosen screw securing lower panel to reinforcement.
9. Pull left lower panel out and disconnect control panel electrical and vacuum connectors. Remove control panel mounting screws and remove control panel from vehicle.
10. Reverse procedure to install.

1977-80 Brougham, DeVille, 1979-80 Eldorado & 1980 Seville

1. Disconnect battery ground cable.
2. Remove instrument panel center bezel.
3. Remove the two control assembly to instrument panel retaining screws.
4. Remove control assembly from instrument panel, then disconnect vacuum and electrical connectors and remove assembly from vehicle, Fig. 119.
5. Reverse procedure to install.

CONTROL PANEL SERVICE

1974-76 Models

Circuit Board Removal

1. Remove screw securing wire harness clip to control head base plate.
2. Remove two green wires from temperature dial rheostat.
3. Remove circuit board mounting screws from bottom of base plate.
4. Set control lever to "Off" position. Slide circuit board out from under control panel.

Circuit Board Installation

1. On 1974-76 models, set lever to "Def." On all models, feed circuit board connector and wiring through base plate from bottom.
2. Position circuit board and install mounting screws.
3. From top of control head, connect two green wires to temperature dial rheostat. Secure harness clip to base plate with screw.

Temperature Control Rheostat, Replace

1. Disconnect two green wires from rheostat.
2. Remove rheostat bracket mounting screw and the rheostat with bracket.
3. Reverse procedure to install. After control panel installation, adjust temperature dial as previously outlined.

Control Head Vacuum Valve, Replace

1. Remove vacuum valve mounting screws, then the vacuum control valve.
2. Position vacuum control valve on control head with pin engaging lever arm.
3. Install valve mounting screws.

PROGRAMMER, REPLACE

1974-76 Models Exc. Seville

NOTE: On models equipped with the Theft Deterrent System, ensure system is disarmed before servicing programmer.

Removal

1. Remove Theft Deterrent System switch or Track Master Controller from top of glove compartment liner, if equipped. On all models, remove glove compartment liner.
2. Remove clips from vacuum valve studs and disconnect vacuum harness from programmer vacuum valve. Disconnect wiring harness from programmer.
3. Disconnect door link from programmer output shaft, Fig. 112. Remove programmer mounting screws and programmer.

Installation

1. Position programmer to heater case and install mounting screws.
2. Connect electrical connector to programmer and connect vacuum harness to programmer vacuum valve. Secure harness with clips on pins of valve.
3. Install door link and adjust as previously outlined.
4. Insert glove compartment liner into glove compartment opening. Install mounting screws except for upper right screw.
5. Position door stop cable and install upper right screw.
6. Install Theft Deterrent System switch or Track Master Controller, if equipped.

1976-79 Seville

1. Remove litter receptacle.
2. Remove screws securing recirculating air

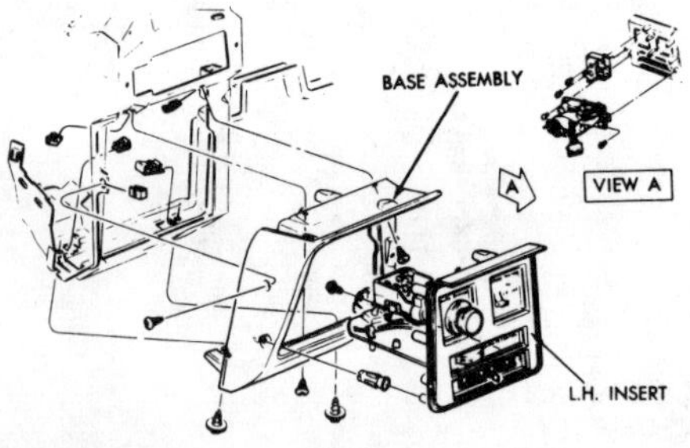

Fig. 118 Control panel installation. 1976-79 Seville

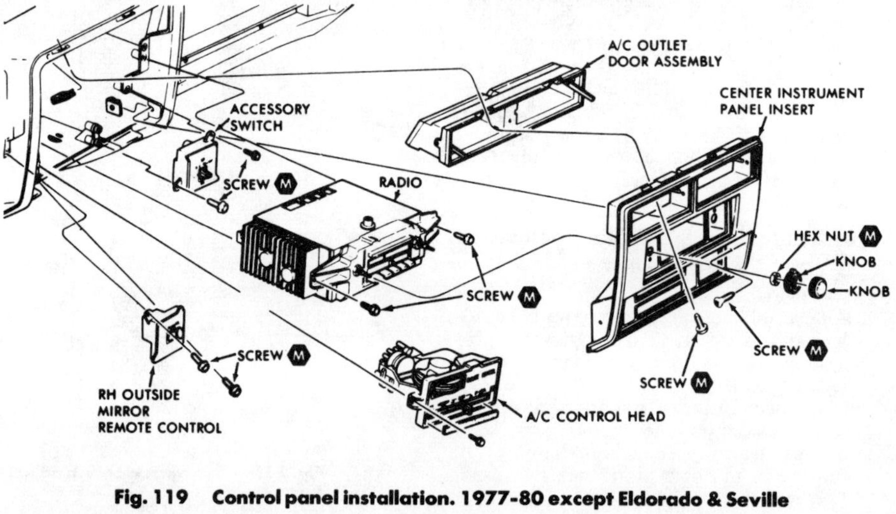

Fig. 119 Control panel installation. 1977-80 except Eldorado & Seville

door vacuum motor shroud to right kick panel.
3. Disconnect vacuum hoses and electrical connectors from programmer.
4. Remove programmer mounting screws, Fig. 113.
5. Lower programmer and loosen air mix door linkage set screw and disconnect linkage.
6. Remove programmer from vehicle.
7. Reverse procedure to install. Adjust mix door linkage as outlined previously.

1977-79 Eldorado & 1977-80 Brougham & DeVille

1. Disconnect battery ground cable.
2. Remove glove compartment. On 1980 models, remove programmer insulating cover by removing two retaining screws.
3. Disconnect vacuum harness and electrical connectors from programmer.
4. Disconnect door link from programmer output shaft by prying rod from retainer.
5. Remove the three programmer retaining screws, then remove programmer, Figs. 115 and 120.
6. Reverse procedure to install. Adjust air-mix door as described elsewhere in this section.

1980 Eldorado & Seville

1. Disconnect battery ground cable.
2. Remove glove box.
3. Remove right hand hush panel.
4. Remove two screws retaining ECM (DEFI) or ECU (EFI) to instrument panel, then disconnect wiring connectors and remove unit.
5. Remove left hand hush panel. If equipped with CB radio, remove the two retaining screws, then disconnect wiring and remove CB unit.
6. Remove programmer arm from programmer by snapping out of clip, then disconnect electrical and vacuum connectors. Remove the two retaining screws, then remove programmer.
7. Reverse procedure to install. Adjust programmer air-mix door as described elsewhere in this section.

IN-CAR SENSOR, REPLACE

1974-76 Models (Except Seville) & 1977-78 Eldorado

1. Disconnect battery ground cable.
2. Remove right, left and right center air outlet grilles using tool J-21612 to compress release tabs. On 1976 models, use tool J-24612.
3. Working through air outlet openings, remove three fasteners securing instrument panel pad to dash panel support. Remove screws securing pad to dash panel horizontal support.
4. Pull pad outward and disconnect electrical connector from windshield wiper switch, then remove pad.
5. Remove screws securing instrument panel top cover to left and right side of instrument panel, Fig. 121. Then, remove two top cover to horizontal support studs. Loosen, but do not remove the nut securing top cover to instrument panel brace on firewall.

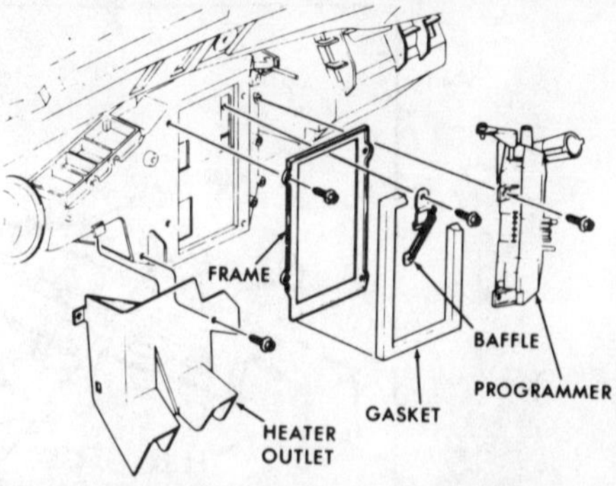

Fig. 120 Programmer installation. 1977-80 except Eldorado & Seville

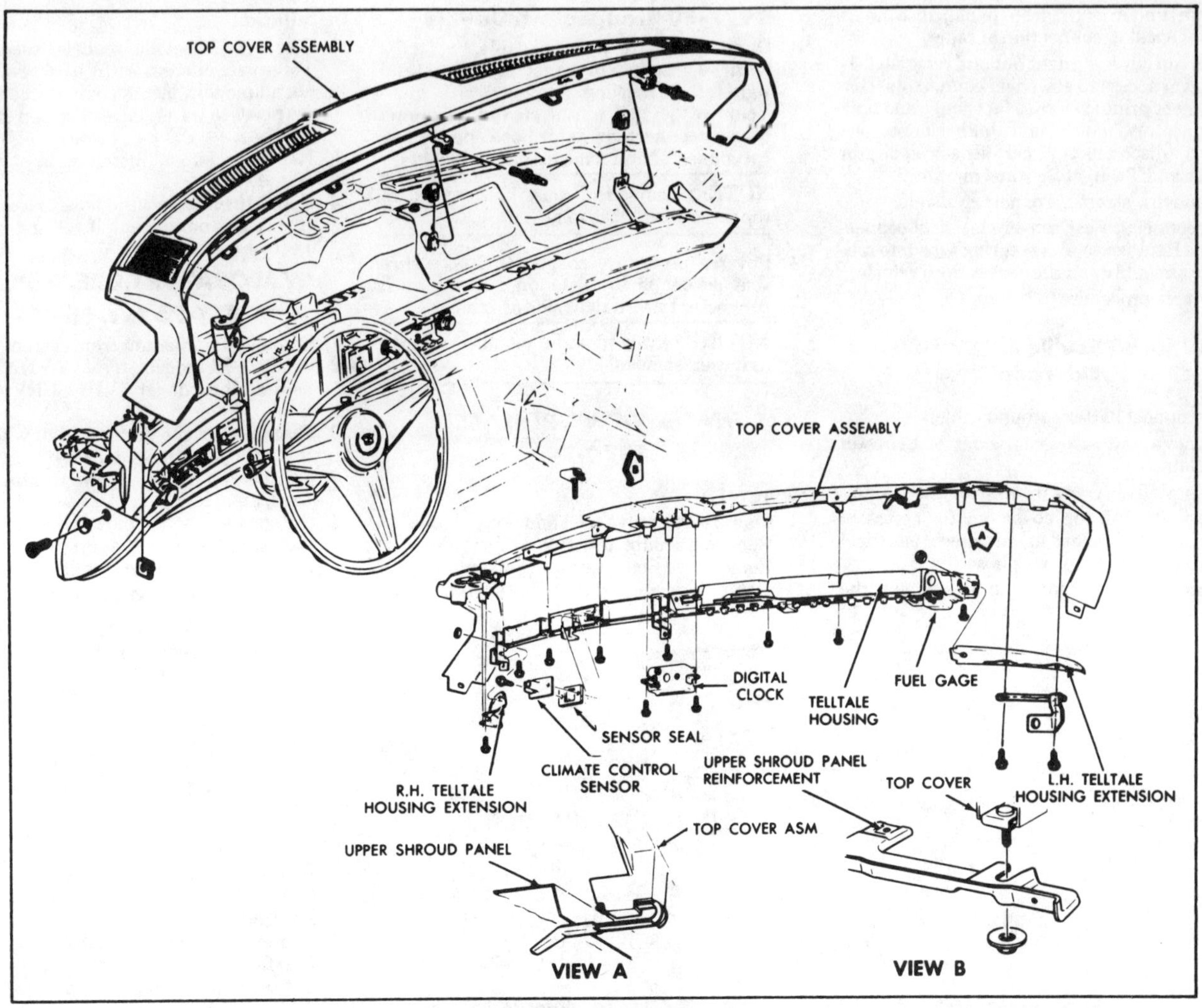

Fig. 121 In-car sensor installation. 1974-76 models (except Seville) & 1977-78 Eldorado

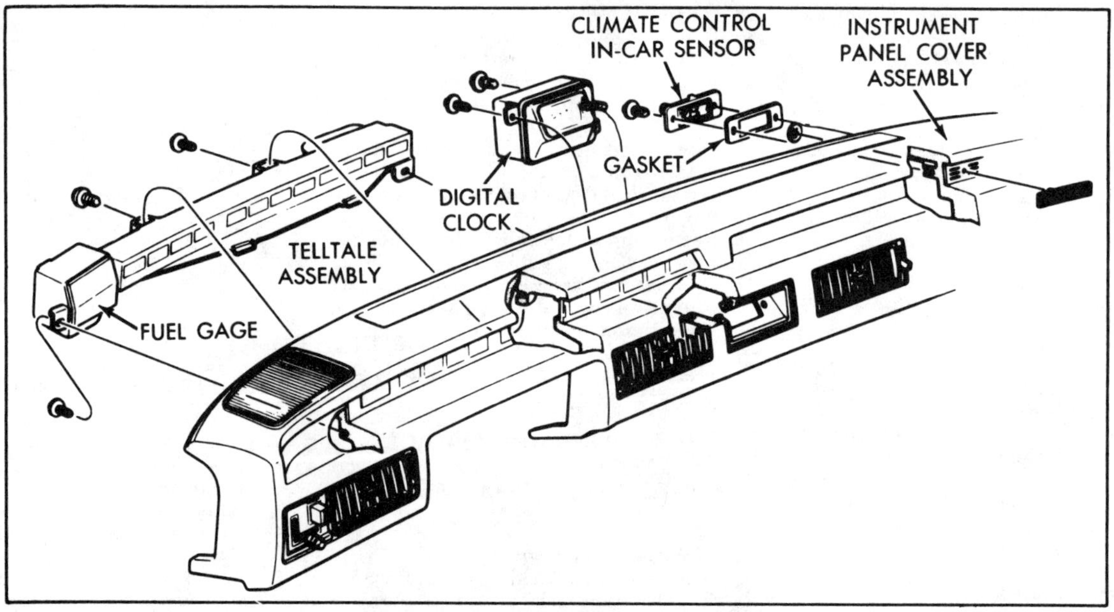

Fig. 122 In-car sensor installation. 1976-79 Seville

The nut is located forward and to the left of the clock.

6. From underneath dash panel, disconnect front speaker connection at radio.
7. Pull top cover straight out and raise slightly to gain access to electrical connectors. Disconnect printed circuit, fuel gauge and bulb connectors, clock and bulb connectors. Also, disconnect in-car sensor aspirator hose and Twilight Sentinel photocell.
8. Remove instrument panel top cover.
9. Disconnect in-car sensor electrical connectors. Remove screws securing sensor to telltale assembly and the sensor from vehicle.
10. Reverse procedure to install.

1976-80 Seville & 1979-80 Eldorado

1. Disconnect battery ground cable.
2. Remove four screws on sides of defroster outlets.
3. Remove five screws from cover lower edge.
4. Carefully pull top cover out for access to wiring. Disconnect in-car sensor electrical connector and aspirator hose.
5. Remove sensor mounting screws and the sensor, Figs. 122 and 123.
6. Reverse procedure to install.

1977-80 Brougham & DeVille

1. Disconnect battery ground cable.
2. Remove instrument panel top cover and right telltale housing.
3. Remove the two sensor retaining screws, then disconnect aspirator hose and electrical connector and remove sensor, Fig. 124.

NOTE: The sensor must be installed in the same position as removed.

4. Reverse procedure to install. Make sure that sensor is installed on housing with connector toward right side of vehicle.

NOTE: Sensor must be installed exactly as it was removed.

WATER VALVE, REPLACE

1974-79

Removal

1. Clamp water valve inlet and outlet hoses.
2. Remove vacuum hose from water valve.
3. Remove clamps securing coolant hoses to water valve and separate hoses from valve
4. On 1976-79 Seville models, remove water valve mounting screw and the water valve.

Installation

1. On 1976-79 Seville models, secure water valve to condenser with one screw.
2. On all models, install coolant inlet and outlet hoses to water valve. Tighten clamps securely.
3. Remove clamps installed in "Removal" Step 1.
4. Connect white vacuum hose to water valve.
5. Check engine coolant level and adjust as necessary.

EVAPORATOR CORE, REPLACE

1974-76 Exc. Seville

1. Discharge refrigerant from system.
2. On 1974-76 models, remove VIR assembly as described under "VIR SERVICE" section.
3. Disconnect right tie strut from wheel housing and place it aside, then remove nuts and screws retaining evaporator case to cowl, Fig. 125.
4. Cut gasket, then separate evaporator case halves and remove evaporator.
5. Reverse procedure to install.

1976-79 Seville

1. Discharge refrigerant from system, drain radiator and remove right hand wheelhousing strut.
2. Support front of hood, then tape padding to right rear hand corner and remove right hand hood hinge.
3. Remove all electrical connections from components mounted on evaporator housing.
4. On 1977-79 models, remove thermostatic cycling switch.
5. Remove heater hose at core side of water valve, then remove retaining screws and position valve aside.
6. Remove right wheelhousing assembly.
7. Carefully disconnect remaining heater hoses from heater core.

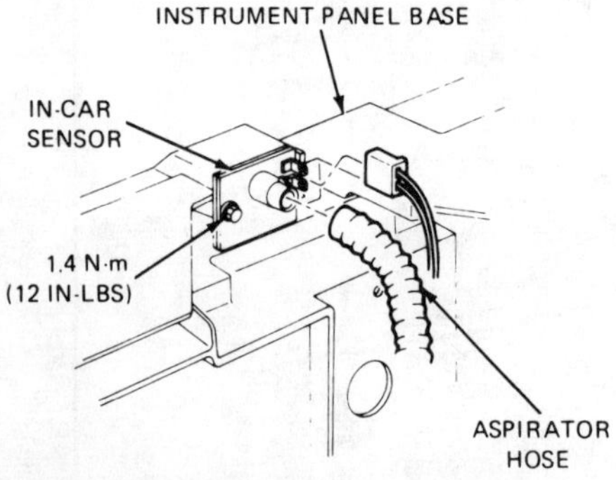

Fig. 123 In-car sensor installation. 1979-80 Eldorado & 1980 Seville

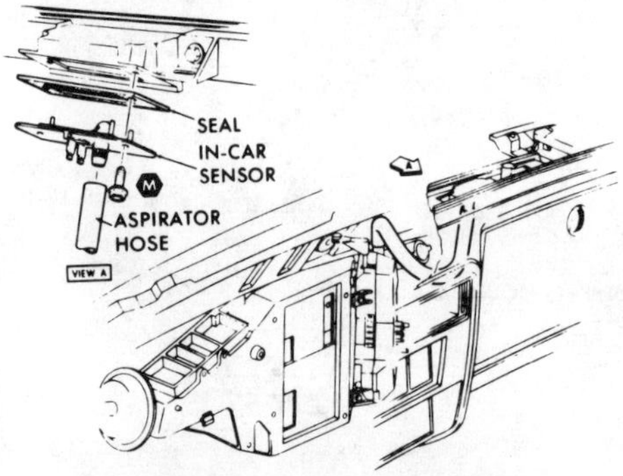

Fig. 124 In-car sensor installation. 1977-80 except Eldorado & Seville

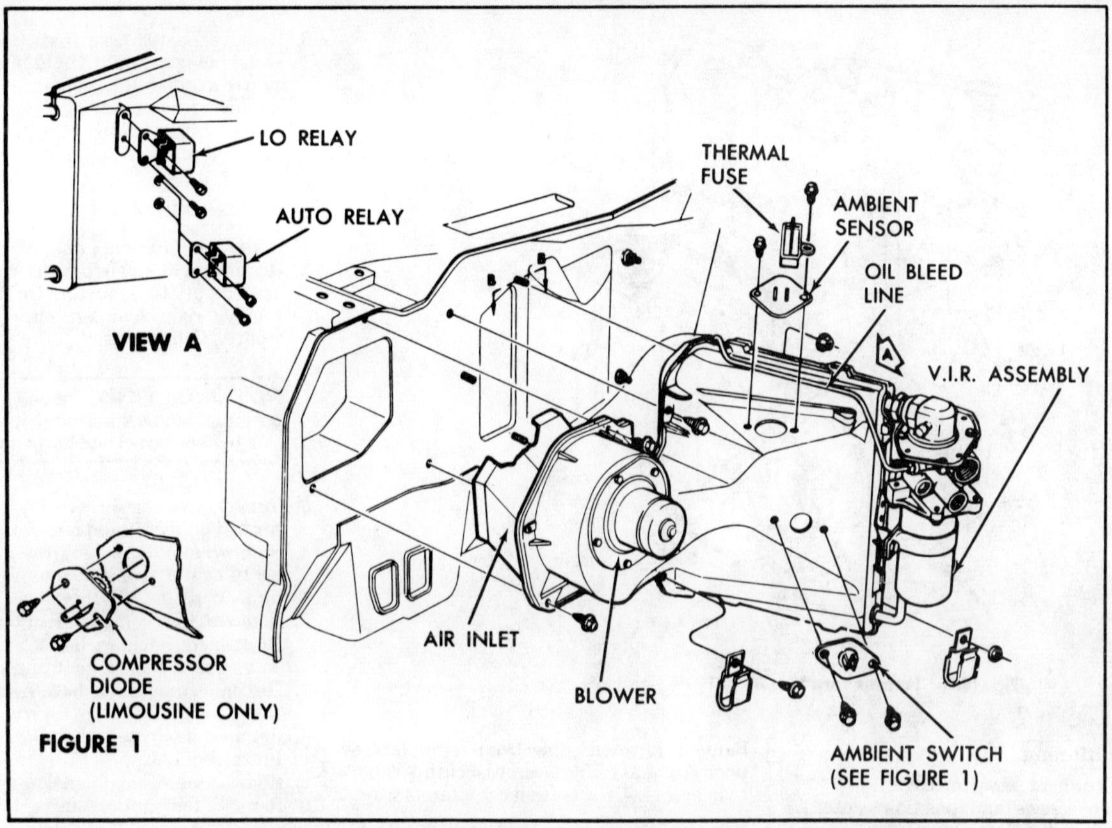

Fig. 125 Evaporator core & blower motor assembly. 1974-76 (Typical)

8. On 1974-76 models, remove V.I.R. unit as described in V.I.R. Service Section. On 1977-79 models, disconnect refrigerant lines from accumulator, then cap lines and fittings.
9. Remove evaporator housing retaining screws and nuts and remove unit from vehicle 125 and 126.
10. Remove evaporator case halves retaining screws, then separate case and remove core.
11. Reverse procedure to install.

1977-78 Eldorado

1. Disconnect battery ground cable.
2. Discharge refrigerant from system. After system is fully discharged of refrigerant, disconnect evaporator inlet and outlet pipes.
3. Disconnect connectors from thermostatic cycling switch, ambient sensor and blower relay.
4. Remove thermostatic cycling switch, then remove the right side tie strut at wheel housing and position it aside.
5. Remove the two nuts and four screws securing evaporator case to cowl, then move evaporator case away from cowl and remove the six screws holding case halves together.
6. Remove case half, then remove evaporator core, Fig. 126.
7. Reverse procedure to install.

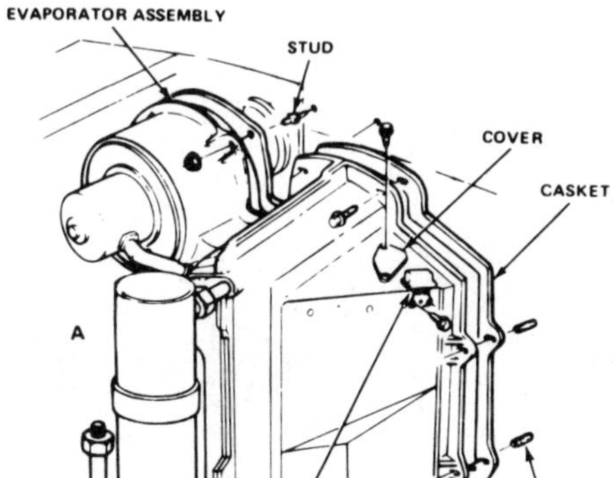

Fig. 126 Evaporator core & blower motor assembly. 1977-80 Eldorado & Seville

1977-80 Brougham & DeVille

1. Disconnect battery ground cable.
2. Discharge refrigerant from system.
3. Disconnect electrical connectors from blower motor, resistor and thermostatic switch.
4. Remove right hand windshield washer nozzle and air inlet screen.
5. Remove thermostatic switch and module cover.
6. After refrigerant system has completely discharged, disconnect evaporator inlet and outlet lines. Seal all open connections to prevent entry of dirt and moisture.
7. Remove evaporator core by lifting up and out of module. Fig. 107.
8. Reverse procedure to install.

1979-80 Eldorado & 1980 Seville

1. Disconnect battery ground cable and discharge refrigerant from system.
2. Remove electrical connectors from thermostatic cycling switch and remove the switch. On 1980 vehicles remove the electrical connector at the power module.
3. Remove insulation from connection of capillary line and evaporator inlet pipe, noting location of line on pipe so that replacement can be made in the same location.
4. Loosen clamps securing line to pipe, remove line and clean evaporator outlet pipe with a suitable solvent.
5. Disconnect refrigerant lines at accumulator and immediately plug all line connections and accumulator openings to prevent entry of dirt and moisture.
6. Remove 6 screws securing left and right evaporator case halves, then remove 2 nuts securing evaporator case left half to cowl and remove evaporator case left half by cutting case to cowl gasket at the joint between evaporator case halves, Fig. 126.
7. Disconnect accumulator and remove evaporator core.
8. Reverse procedure to install.

HEATER CORE, REPLACE

NOTE: After draining radiator and disconnecting heater hoses and battery ground cable proceed as follows:

1974-76 Models (Except Seville) & 1977-78 Eldorado

CAUTION: On vehicles equipped with an Air Cushion Restraint system, turn ignition switch to "Lock," disconnect battery ground cable and tape end, thereby deactivating system.

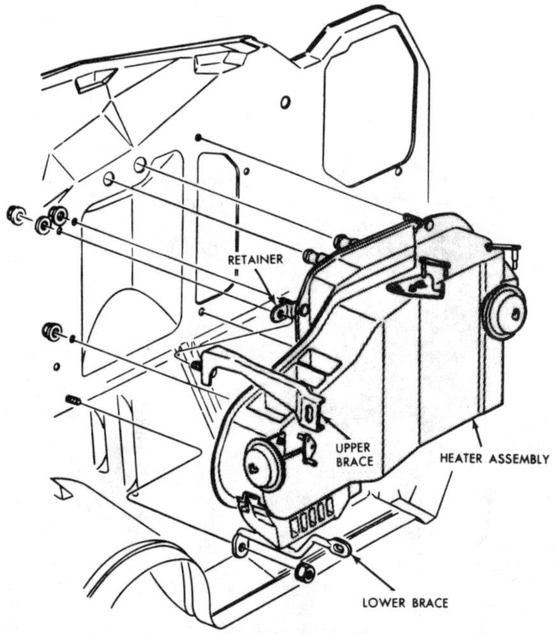

Fig. 127 Heater core less A/C. 1974 (Typical)

Less Air Conditioning

1. Remove instrument panel top cover.
2. Remove two screws and position center ventilation duct and sleeve out of way.
3. Remove vacuum hoses from diverter door and defroster door vacuum actuators.
4. Remove Bowden cable from temperature door arm and remove screw securing cable to heater case. Reposition cable out of way.
5. Remove two screws and nuts securing heater case to cowl and pull case away from cowl, Fig. 127.
6. Remove heater case from position under panel being careful to hold in upright position to avoid spilling coolant.
7. Remove two screws and retainer at each side of core securing core to case and remove core.

With Air Conditioning

1. Remove instrument panel top cover.
2. Remove right and left outlet A/C hoses and center outlet connector. On 1974 models, remove right and left outlet A/C hoses from A/C distributor.

NOTE: On 1974-76 models, remove center outlet support and connector from position between cowl and horizontal support.

3. Remove two screws securing A/C distributor to heater case and remove distributor.
4. Remove one screw securing defroster nozzle to heater case and remove nozzle from under clips and out of car.
5. Remove glove compartment liner.
6. Disconnect vacuum hoses at recirc door, water valve, control head and supply hose.
7. Disconnect aspirator hose from in-car sensor.
8. Remove instrument panel center vertical brace, Fig. 128.
9. Remove instrument panel horizontal brace.
10. Remove four nuts securing heater case to cowl on engine side of dash.
11. Remove heater case from position under panel being careful to hold case as upright

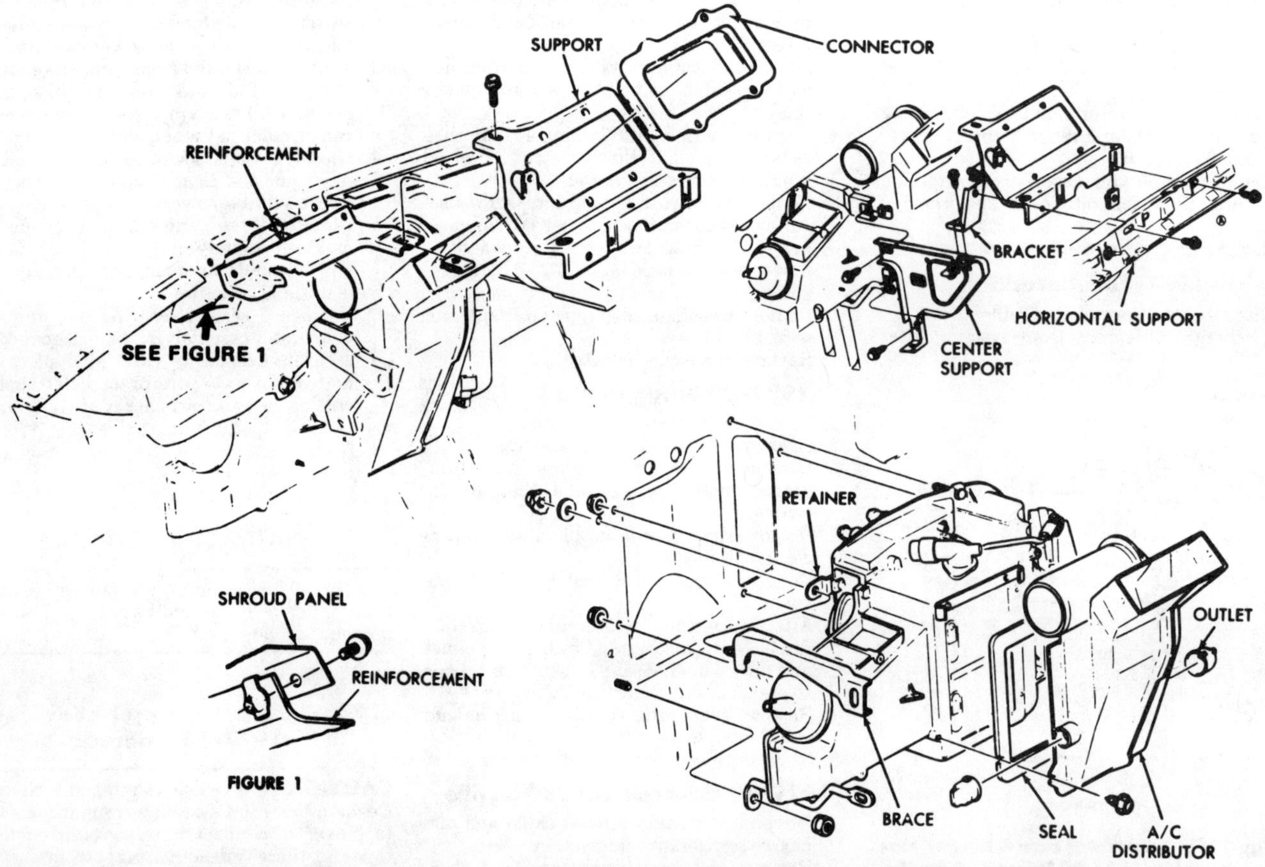

Fig. 128 Heater core with A/C. 1974-76 all, 1977-78 Eldorado & 1977-79 Seville (Typical)

as possible to avoid spilling coolant.

12. Remove rubber seal from around core water nipples.
13. Remove screw and clip from under seal.
14. Remove two screws and clip from opposite end of case and remove core.
15. Reverse procedure to install.

1976-79 Seville

1. Remove right hand wheel housing.
2. Plug heater core tubes to prevent spilling coolant into vehicle when core is removed.
3. Remove instrument panel right hand insert and applique.
4. Remove fuel injection control unit.
5. Remove radio, glove box door and glove box.
6. Remove A/C programmer.
7. Remove right hand lower instrument panel.
8. Remove screws retaining A/C distributor to heater case and remove distributor.
9. Remove clips retaining vacuum harness to heater case and disconnect vacuum hoses at heater mode door, bulkhead grommet and defroster actuator. Position vacuum harness out of way.
10. Disconnect hose from in-car sensor aspirator.
11. From under hood, remove three nuts retaining heater case to cowl, then from inside vehicle, remove two screws retaining heater case to cowl and move heater assembly from cowl.
12. Disconnect vacuum hose from A/C mode door actuator and remove heater assembly, Fig. 128.

CAUTION: Heater core pipes are very long. Use care when removing heater assembly.

13. Remove rubber seal from around heater core pipes, then remove screw and clip from beneath seal.
14. Remove screws from opposite side of core and remove core.
15. Reverse procedure to install.

1977-80 Brougham & DeVille

1. Disconnect electrical connectors from blower motor, blower resistors and thermostatic cycling switch.
2. Remove right hand windshield washer nozzle and remove right hand air inlet screen from plenum and cover plate at center of plenum opening.
3. Remove screws retaining thermostatic cycling switch and position switch aside.
4. Remove the 16 fasteners retaining cover and remove cover.
5. Remove one screw and retainer holding heater core to frame at top.
6. Place temperature door in the max. heat position and reach through temperature housing and push lower forward corner of heater core away from housing, Fig. 107, then rotate core parallel to housing. This will cause core to snap out of lower clamp. Remove heater core from temperature housing.
7. Reverse procedure to install.

1979-80 Eldorado & 1980 Seville

1. Plug heater core outlets to prevent spillage when removing heater core.
2. Remove instrument panel sound absorbers, then lower steering column.
3. Remove instrument cluster trim panel and instrument cluster as described under Instrument Cluster, Replace.
4. Remove radio front speakers.
5. Remove screws attaching manifold to heater case.
6. Remove instrument panel upper and lower attaching screws, then disconnect parkbrake release cable.
7. Disconnect instrument panel wiring harness from dash wiring assembly.
8. Disconnect right hand remote control mirror cable from instrument panel pad.
9. Disconnect speedometer cable from clip and temperature control cable from heater case.
10. Disconnect radio and A/C wiring, vacuum lines and all necessary wiring to remove instrument panel. If equipped with pulse wipers, remove wiper switch, and then disconnect wire connector from instrument cluster carrier and separate pulse wiper jumper harness from wiper switch wire connector.
11. Remove instrument panel and wiring harness assembly.
12. Remove four screws retaining defroster nozzle to cowl and one screw at heater case, then remove defroster nozzle.
13. Disconnect vacuum hoses to programmer and vacuum actuators.
14. Disconnect wire connector from programmer.
15. From engine side of dash, remove four nuts retaining heater case to cowl, Fig. 109.
16. From passenger compartment side of dash, remove one screw retaining heater case to cowl, and remove heater case.
17. Remove four screws securing heater core to case and remove heater core.
18. Reverse procedure to install.

BLOWER MOTOR, REPLACE

1974-1980

CAUTION: On vehicles equipped with an Air Cushion Restraint system, turn ignition switch to "Lock," disconnect battery ground cable and tape end, thereby deactivating system.

1. Disconnect battery ground cable.
2. On 1979-80 Eldorado and 1980 Seville, remove right side wheelhousing.
3. Remove cooling hose from nipple and blower.
4. Disconnect electrical connector at lead to motor.
5. Unfasten and remove motor, Figs. 107, 125 and 126.

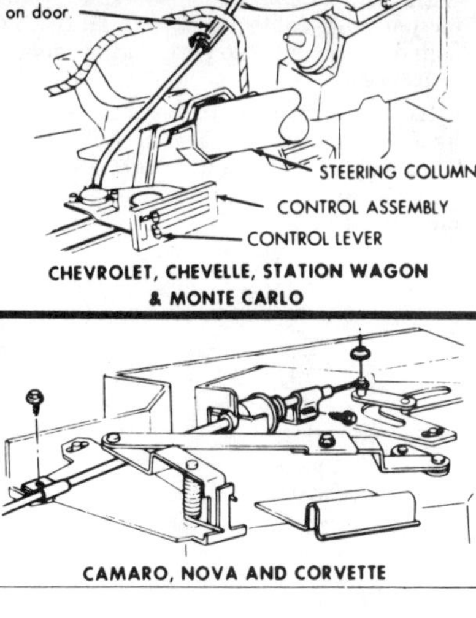

Fig. 129 Temperature control cable adjustment. 1974-76 All & 1977 Except Chevrolet Full Size

Chevrolet & GMC

TEMPERATURE DOOR CABLE ADJUSTMENT

1974-77 All Models Less A/C

1. With cables attached to control assembly, move levers to full "Off" position.
2. Attach cable wires to door lever and tighten cable bracket attaching screws.
3. Check for proper cable operation.

1977-80 Full Size; 1978-80 Malibu & Monte Carlo Less A/C

1. Move temperature lever to full "Cold" position, then rotate temperature door crank to full "Cold" position.
2. Adjust turnbuckle until cable loop aligns with pin on crank, Fig. 130.
3. Move lever to full "Hot". Door should be seated. If not, repeat adjustment procedure.

1978-80 Camaro Less A/C

1. Adjust temperature cable by moving to "Max" hot.

1974-76 Chevelle, Chevrolet & Monte Carlo With A/C

1. Connect cable securely at both ends.
2. Position control temperature lever in extreme maximum heat position.
3. Adjust turnbuckle, Fig. 129, until control temperature lever moves approximately 1/16 inch back from end of slot.
4. Cycle temperature lever to maximum cold position and return to maximum heat position. There should be approximately 1/16 inch of springback.
5. Repeat adjustment procedure, if necessary.

1974-79 Nova & 1974-78 Camaro & 1974-80 Corvette With A/C

1. Attach cable to control assembly.
2. Set temperature control lever in the "Off" position.
3. Place opposite loop of cable on selector duct temperature lever pin, Fig. 129.
4. Push cable sheath toward heater lever until valve seats and lash is eliminated from cable and control.
5. Tighten screw to secure adjustment.

1979-80 Camaro With A/C

1. Place temperature lever 1/16 inch from full "Cold" position and adjust turnbuckle until temperature door is seated.
2. Move temperature lever through full length of travel. Door should be heard seating and lever should stop slightly before end of slot.

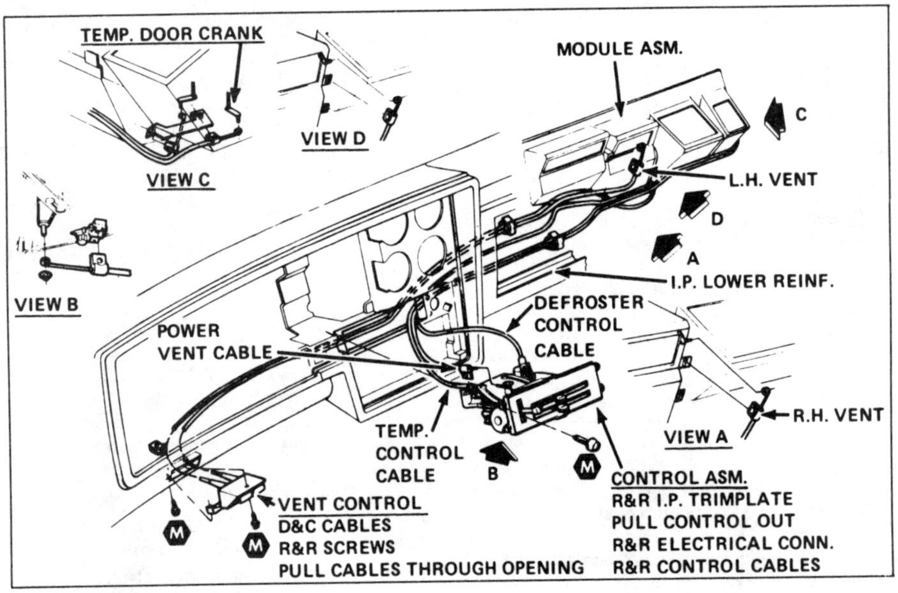

Fig. 130 Temperature control cable adjustment. 1977-80 Chevrolet Full Size & 1978-80 Malibu & Monte Carlo

3. If door is not heard seating or lever is more than ⅛ inch from end of slot, repeat adjustment procedure.

1979-80 Full Size

1. Disengage temperature cable retaining clip from bellcrank post.
2. Snap temperature door operating rod into retaining clip so that 2/5 inch of rod is exposed beyond outer edge of retainer.
3. Re-engage temperature cable retaining clip to bellcrank post.
4. Move temperature lever through full length of travel. Temperature door should be heard seating at the open and closed positions.

1980 Citation

Less A/C

The temperature cable incorporates a slider type self-adjust feature, Figs. 87 and 88. As the control head temperature lever is moved through its full range of travel, the cable clip will assume a position assuring that the temperture valve will seat in the full hot and full cold positions.

With A/C

1. Connect cable assembly to control heat, Fig. 86.
2. Place temperature lever in "Off" position.
3. Connect cable to temperature door post and push sheath toward lever until door seats and all lash is removed from cable and control head lever.
4. Tighten cable securing screw.

Chevrolet & GMC Light Duty Truck

1974-77 Sprint, 1974-80 El Camino & 1979-80 Caballero

Refer to Chevelle, Malibu and Monte Carlo for service procedures.

1974-77 G Models, 1974-80 C & K Models Less A/C

1. Disconnect battery ground cable.
2. On G Model air and defroster cables, unsnap the engine cover front latches, then remove the two cover to floorpan screws and remove engine cover. On C-K models remove glove box and door as an assembly.
3. Disconnect the respective cable clip and disconnect the cable from door.
4. Remove the cable retaining screw, then, while holding the cable with pliers, rotate the mounting tab on the cable to lengthen or shorten cable. Avoid pinching cable too tightly as damage to the cable will result.
5. Reverse procedure to install.

1978-80 G Models Less A/C

1. Disconnect battery ground cable.
2. Attach inner cable and sheath to instrument panel control.
3. With instrument panel installed, move temperature cable to "Cold" and attach loop on inner cable to temperature door on heater case.
4. Attach cable sheath to heater case, then move temperature lever to full "Heat". This will require some effort due to the force required to slide inner cable clip to its proper position.

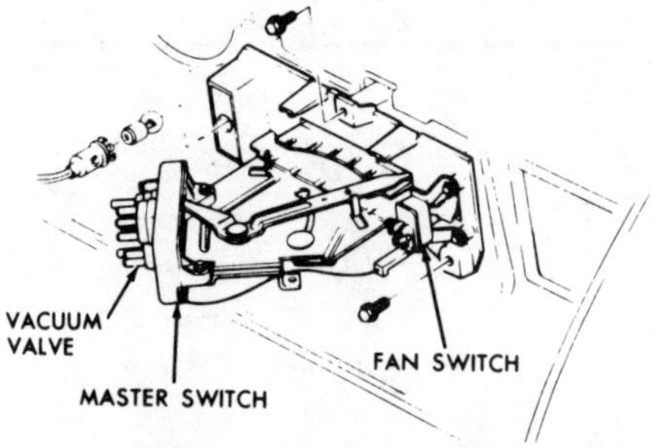

Fig. 131 Control panel installation. 1974-80 Camaro (Typical)

1974-76 G Models With A/C

1. Disconnect battery ground cable.
2. Disconnect the respective cable clip and disconnect the cable from door.
3. Remove the cable retaining screw, then, while holding the cable with pliers, rotate the mounting tab on the cable to lengthen or shorten cable. Avoid pinching cable too tightly as damage to the cable will result.
4. Reverse procedure to install.

1974-80 C & K Models With A/C

1. Remove glove box, then loosen cable attaching screw at the selector duct assembly.
2. Make sure the cable is installed in the bracket on the selector duct assembly, then place temperature lever in "Cold" position and hold while tightening cable attaching screw.

CONTROL PANEL, REPLACE

Camaro

1974-77 Less A/C & 1974-80 With A/C

1. Disconnect battery ground cable.
2. On 1974-76 models, remove ashtray assembly.
3. Remove steering column panel to instrument panel attaching screws. Carefully lower steering column panel. On 1974-78 models, the left lap cooler and control panel is also lowered with the steering column.
4. On 1979-80 models, remove gauge cluster and glove box.
5. Disconnect electrical and vacuum connections from control panel, Fig. 131. Then, remove temperature control cable set screw. Remove cable from control panel.
6. Remove control panel mounting screws and the control panel.
7. Reverse procedure to install.

1978-80 Less A/C

1. Disconnect battery negative cable.
2. Remove glove box and lower steering column trim panel.
3. Disconnect temperature cable at heater case and remove control panel retaining screws, then pull control panel forward.
4. Disconnect cables and electrical connectors from control panel.
5. Reverse procedure to install.

Chevelle, Malibu & Monte Carlo

1974-77 Less A/C

1. Disconnect battery ground cable.
2. Remove radio control knobs, instrument cluster shroud and center radio speaker.
3. Remove 3 control head assembly attaching screws, then pull control assembly rearward and disconnect cables.
4. Disconnect cables and wiring harness at blower switch and remove control assembly.
5. Reverse procedure to install.

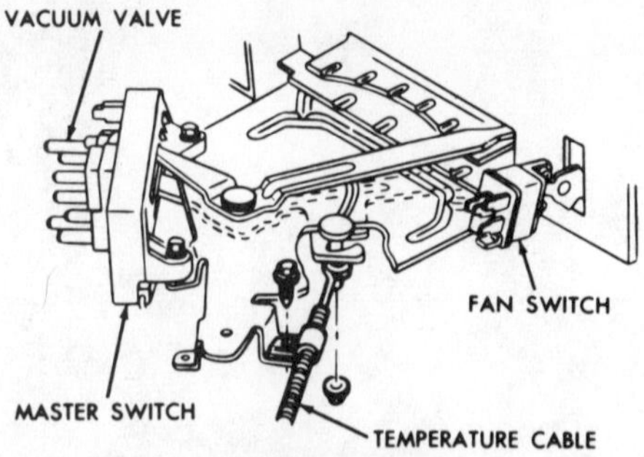

Fig. 132 Control panel installation. 1974-77 Chevelle & Monte Carlo

1978-80 Less A/C

1. Disconnect battery ground cable.
2. Remove instrument panel trim plate and pull control assembly out.
3. Disconnect electrical connectors and control cables from control assembly.
4. Reverse procedure to install.

1974-80 With A/C

1. Disconnect battery ground cable.
2. On 1974-78 models, remove radio and clock knobs.
3. Remove instrument cluster bezel mounting screws and bezel or trimplate.
4. Remove control panel mounting screws. Pull control panel rearward for access to vacuum hoses and electrical harnesses, Fig. 132 and 133.
5. Disconnect vacuum hoses, electrical connectors and Bowden cable.
6. Remove control panel.
7. Reverse procedure to install.

Chevette

1. Disconnect battery ground cable.
2. Remove instrument panel bezel attaching screws. Remove radio knobs, nuts and washers. Then, remove instrument panel bezel.
3. Remove control panel mounting screws.
4. Carefully slide control panel from instrument panel. Use caution not to damage Bowden cables, vacuum hoses and electrical connectors.
5. Disconnect vacuum hoses, electrical connectors and Bowden cables. Then, remove control panel from vehicle.
6. Reverse procedure to install.

Chevrolet

1974-77 Less A/C

1. Disconnect battery ground cable.
2. Remove cigar lighter, instrument cluster shroud and the three control panel attaching screws. Then, push control panel forward and lower from instrument panel.

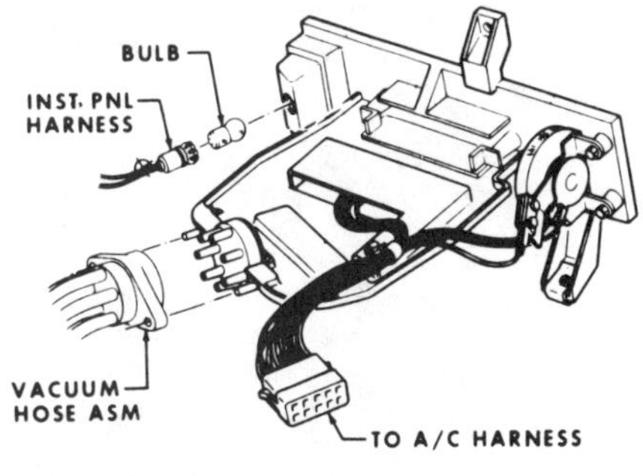

Fig. 134 Control panel installation. 1974-76 Chevrolet Full Size w/Four Season

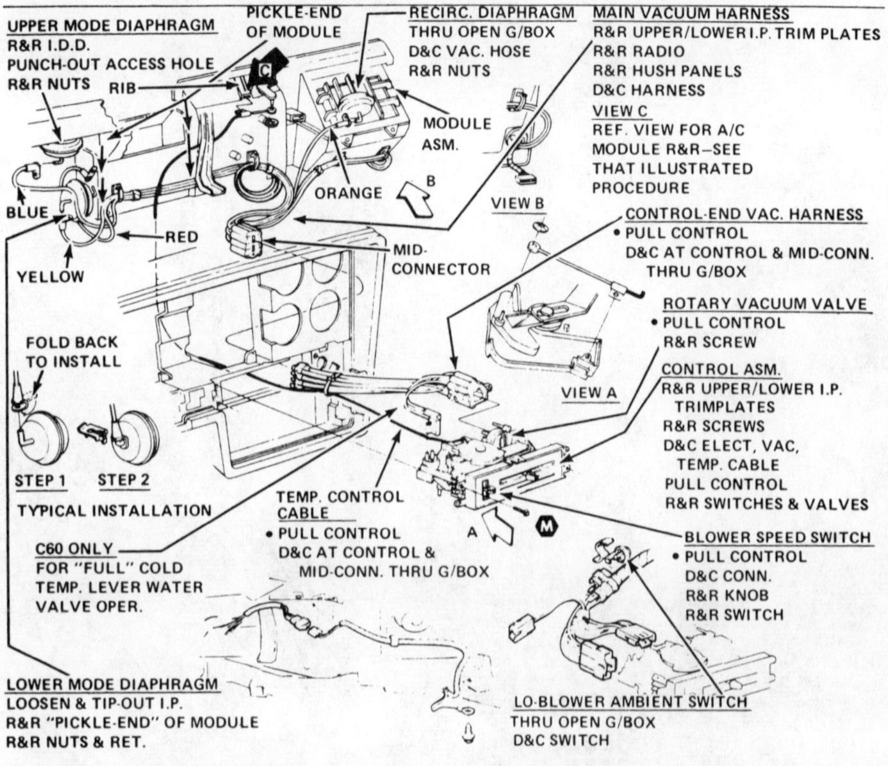

Fig. 133 Control panel installation. 1978-80 Malibu & Monte Carlo

3. Disconnect control cables and electrical connectors from control panel.
4. Reverse procedure to install.

1978-80 Less A/C

1. Disconnect battery ground cable.
2. Remove ashtray, gauge and control trim panels and glove box.
3. Disconnect temperature control cable at heater module, then remove control panel retaining screws.
4. Pull control panel out and disconnect electrical connectors and temperature control cables.
5. Reverse procedure to install.

1974-76 Four Season A/C

1. Disconnect battery ground cable.
2. Remove headlight switch knob and instrument cluster shroud.
3. Remove control panel mounting screws. Disconnect vacuum hoses and electrical connectors, Fig. 134. On 1974-76 models, disconnect Bowden cable. On all models, remove control panel from vehicle.
4. Reverse procedure to install.

1974-1976 Comfortron

1. Disconnect battery ground cable.
2. Remove cigar lighter and radio knobs.
3. Remove instrument cluster shroud. Then, remove control panel mounting screws.
4. Disconnect vacuum hoses and electrical connectors, Fig. 135.
5. Remove control from vehicle.
6. Reverse procedure to install.

1977-80 Four Season & Comfortron A/C

1. Disconnect battery ground cable.
2. Remove screws securing radio-A/C control bezel to instrument panel and remove bezel.
3. Remove 3 control head assembly attaching screws and lower assembly from instrument panel, Fig 136.
4. Disconnect wiring, cables and vacuum hoses from control assembly and remove control.
5. Reverse procedure to install.

Corvette

1. Disconnect battery ground cable.
2. Remove floor console trim plate screws and remove trim plate.
3. Remove control panel screws and lift up, pulling slightly rearward.
4. Disconnect control cables, vacuum hoses and electrical connectors from control panel, Figs. 137 and 138.
5. Remove control panel from vehicle.
6. Reverse procedure to install.

Monza

1975 Monza 2+2

1. Disconnect battery ground cable.
2. Remove clock set knob, then remove the

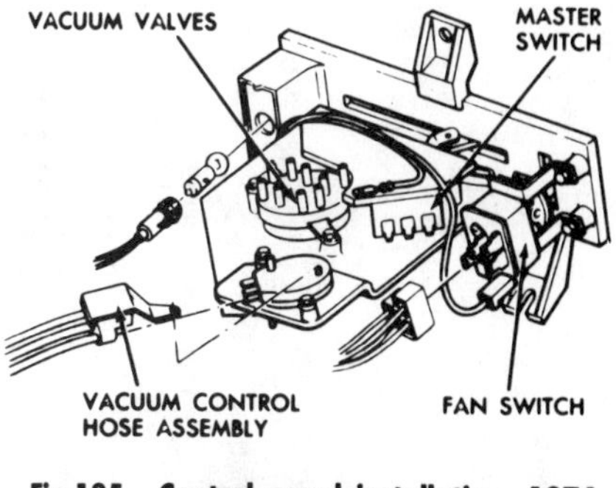

Fig.135 Control panel installation. 1974-76 Chevrolet Full Size w/Comfortron

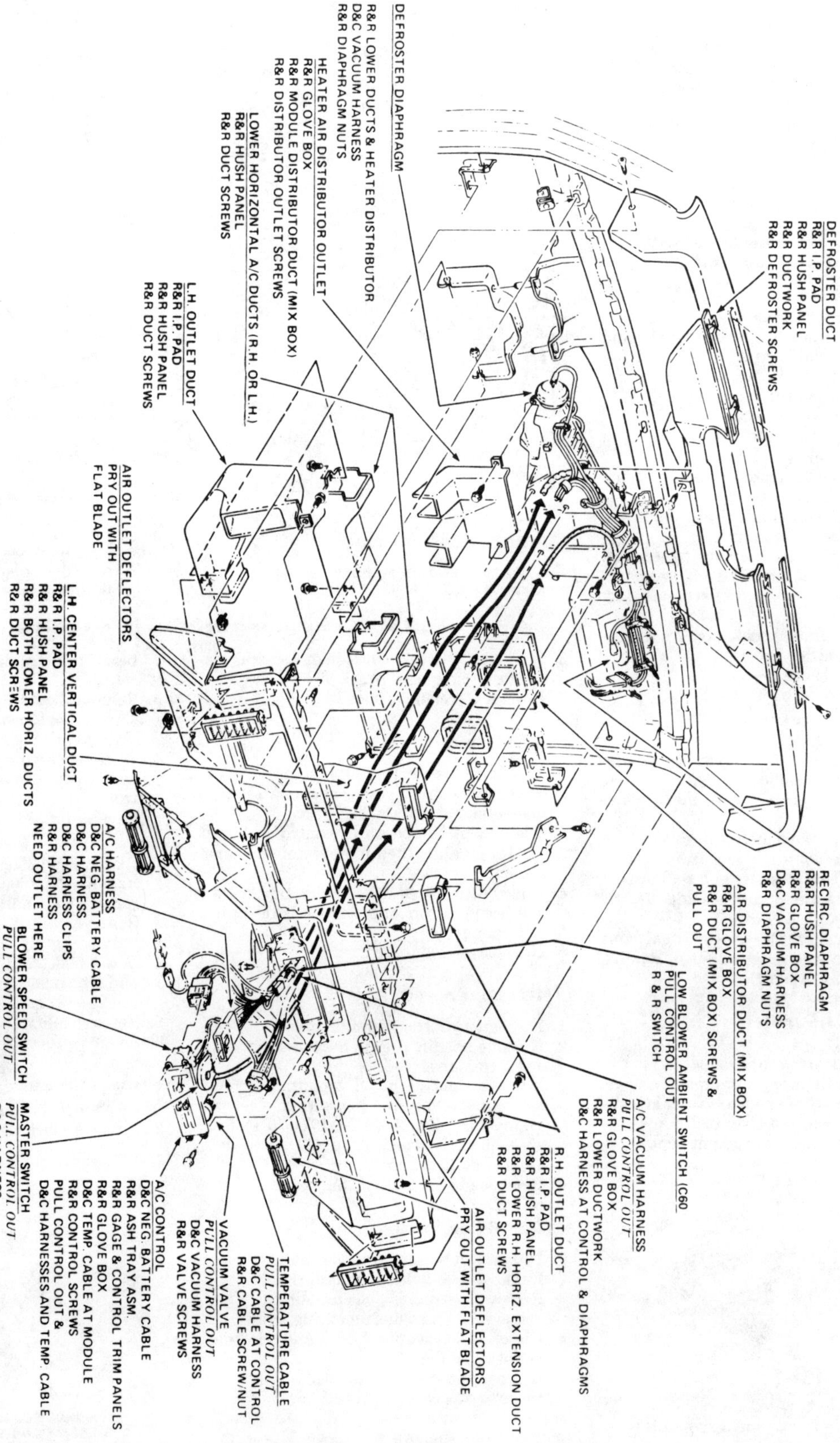

Fig. 136 Instrument panel components. 1977-80 Chevrolet & Pontiac Full Size w/A/C

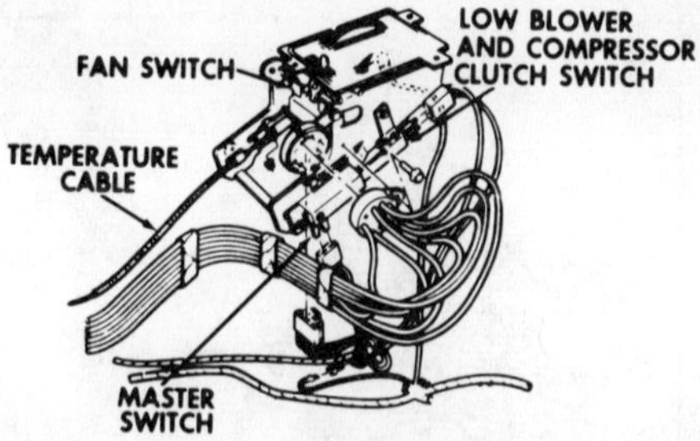

Fig. 137 Control panel installation. 1974-77 Corvette

three instrument cluster trim panel screws and remove panel.

3. Remove four screws from left side of pad, two from top of cluster opening and two below cluster opening.
4. Remove glove box, then remove the four nuts retaining right side of pad and remove instrument panel pad.
5. Remove steering column retaining nuts and lower steering column.
6. Remove one screw from inside upper column bracket, then disconnect cluster housing and pull rearward.
7. Disconnect control cables, vacuum hoses and electrical connectors, Fig. 89, from control panel and remove panel.
8. Reverse procedure to install.

1975-76 Towne Coupe & 1976-80 Monza Except Station Wagon

1. Disconnect battery ground cable.
2. Remove left and right dash outlets.
3. Remove instrument panel pad.
4. Remove steering column attaching nuts.

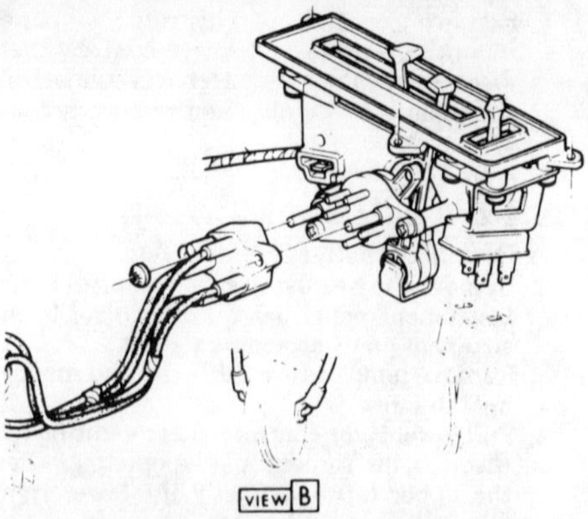

Fig. 138 Control panel installation. 1978-80 Corvette

Lower steering column assembly, resting steering wheel on seat.

5. Remove instrument panel to dash attaching screws. Place a protective cover over steering column. Then, lower instrument panel onto steering column.
6. Remove control panel mounting screws. Disconnect temperature door Bowden cable, wiring and vacuum connectors, Fig. 89, then remove control panel from vehicle.
7. Reverse procedure to install.

1977-79 Station Wagon

Refer to the 1974-77 Vega procedure.

Nova

1974

1. Disconnect the battery ground cable.
2. Remove radio.
3. Remove the control head to instrument

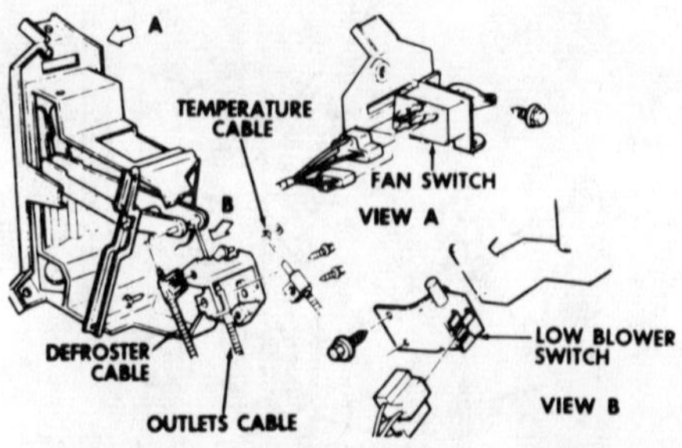

Fig. 139 Control panel installation. 1974 Nova, Apollo, Omega & Ventura

panel screws and lower the control slightly.

4. Disconnect the Bowden cables and electrical connections, Figs. 139 and 140, and then lower the control from the dash.
5. Reverse procedure to install.

1975-79

1. Disconnect battery ground cable.
2. Remove radio and control assembly retaining screws.
3. Disconnect temperature control cable from heater case and from control, Fig. 139.
4. Disconnect electrical connectors and remove control assembly.
5. Reverse procedure to install.

Vega

1974-77

1. Disconnect battery ground cable.
2. Remove intermediate duct leading to left hand outlet.
3. Remove instrument cluster bezel.

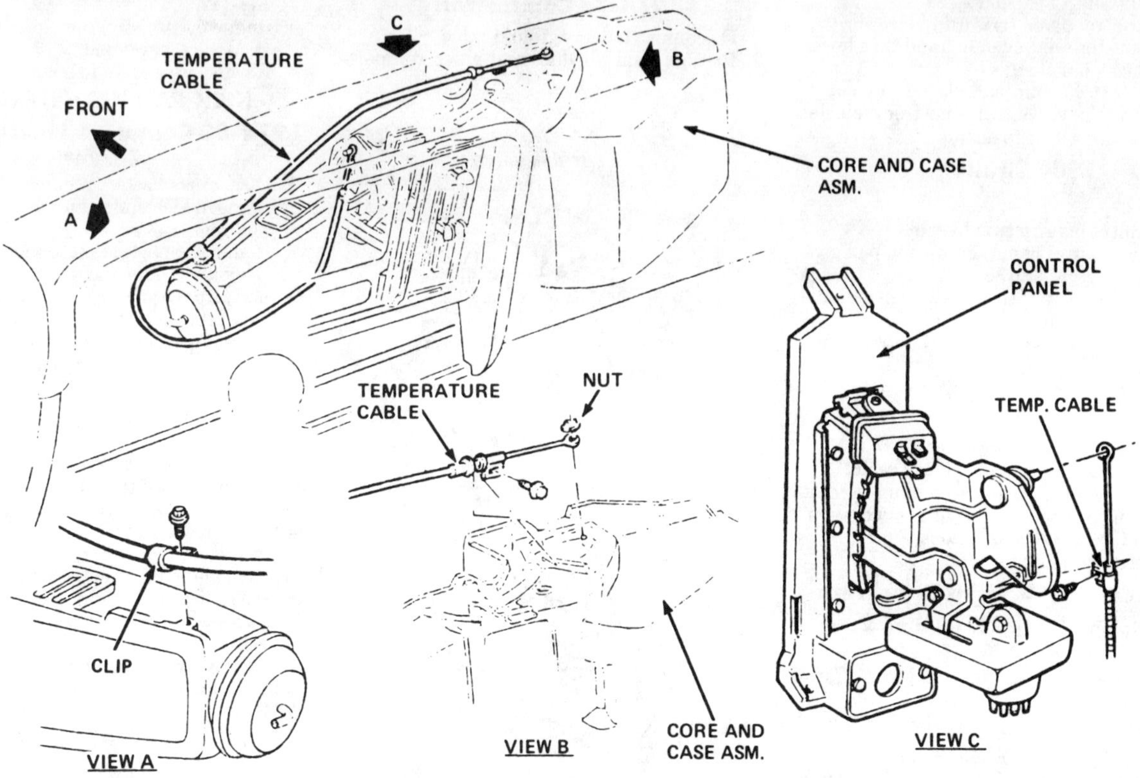

Fig. 140 Control panel installation. 1975-79 Nova, Apollo, Omega, Phoenix & Ventura

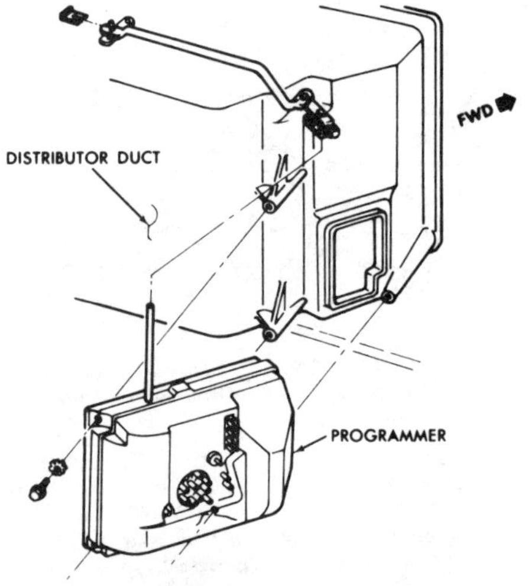

Fig. 141 Programmer installation. 1974-76 Chevrolet Full Size w/Comfortron

4. Remove ash tray and retainer.
5. Remove control panel mounting screw.
6. Remove screws securing instrument cluster to instrument panel.

NOTE: One screw above the steering column cannot be removed. The cluster will pivot on this screw sufficiently for control panel removal.

7. Remove screws securing center distributor duct to selector duct and the screws securing the duct to the instrument panel upper retainer.
8. Disconnect temperature door Bowden cable from control panel, Fig. 89.
9. Push control panel forward to clear instrument cluster reinforcement and then lower the control panel.
10. Disconnect vacuum and electrical connectors and remove control panel from vehicle.
11. Reverse procedure to install.

1980 Citation

Less A/C

1. Disconnect battery ground cable.
2. Remove control panel screws and pull control out, Figs. 87 and 88.
3. Disconnect electrical connection and cables from control.
4. Remove blower switch from control panel, if necessary.
5. Reverse procedure to install.

With A/C

1. Disconnect battery ground cable.
2. Disconnect temperature control cable.
3. Remove control panel retaining screws, then pull control out and disconnect vacuum and electrical connections, Fig. 86.
4. Reverse procedure to install.

Chevrolet & GMC Light Duty Truck

1974-77 Sprint, 1974-80 El Camino & 1978-80 Caballero

Refer to Chevelle, Malibu and Monte Carlo for service procedures.

1974-80 G Models Less A/C

1. Disconnect battery ground cable.
2. On 1974-77 models, remove ignition switch from instrument panel. On 1979-80 models remove instrument panel bezel.
3. Remove control to instrument panel mounting screws and lower control, then disconnect cables and electrical connectors from control assembly and remove assembly.
4. Reverse procedure to install.

1974-80 G Models With A/C

1. Disconnect battery ground cable.
2. Remove the headlamp switch control knob, instrument panel bezel, and control to instrument panel attaching screws.
3. Remove temperature cable clip and mounting tab screw.
4. Pull lower right control panel mounting tab through instrument panel opening, then the upper tab and finally the lower right tab.
5. Disconnect electrical and vacuum connections and remove control assembly.
6. Reverse procedure to install.

1974-80 C & K Models Less A/C

1. Disconnect the battery ground cable.
2. Remove the instrument panel bezel, then disconnect cables and electrical connectors from control panel.
3. Remove the control through the opening above the control.
4. Reverse procedure to install.

1974-80 C & K Models With A/C

1. Disconnect battery ground cable.
2. Remove radio, instrument panel bezel and control to instrument panel screws, then lower the control panel.
3. Disconnect cables and electrical and vacuum harness from control, then remove control from vehicle.
4. Reverse procedure to install.

PROGRAMMER, REPLACE

1974-76 Comfortron

1. Disconnect battery ground cable.
2. Remove cold air duct and glove compartment from vehicle.
3. Loosen link screw and remove programmer link from programmer output shaft, Fig. 141.
4. Disconnect programmer vacuum hoses and electrical connectors.
5. Remove programmer toward rear of car to avoid damage to blower motor resistor.
6. Reverse procedure to install.
7. After installation, adjust temperature door link as previously outlined.

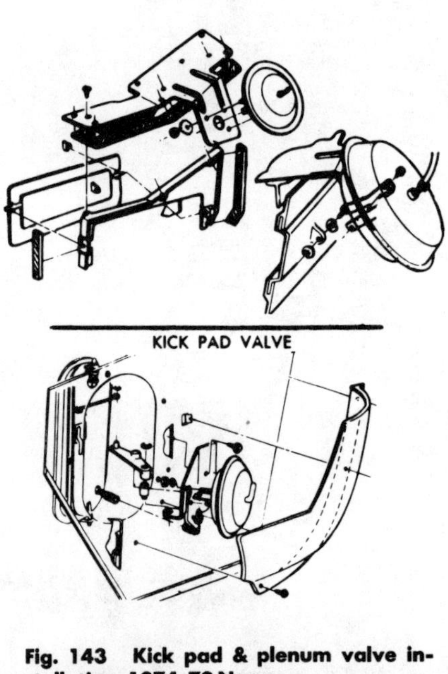

Fig. 143 Kick pad & plenum valve installation. 1974-79 Nova

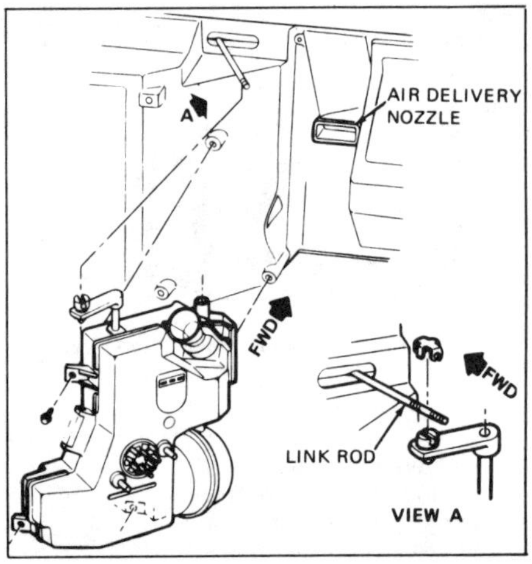

Fig. 142 Programmer installation. 1977-80 Chevrolet Full Size w/Comfortron

1977-80 Comfortron

1. Disconnect battery ground cable.
2. Remove glove box insulation tray, disconnect air mix door rod, then vacuum, electrical and sensor connectors at programmer, Fig. 142.
3. Remove programmer mounting screws then the programmer.
4. Reverse procedure to install.

KICK PAD VALVE, REPLACE

1974-80 Camaro & Nova; 1978-80 Corvette

1. Disconnect battery ground cable.
2. Remove kick and valve cover, Figs. 143, 144 and 145.
3. Remove valve bracket to kick panel screws.
4. Disconnect link and spring at plenum door and remove valve and bracket assembly.
5. Separate valve from bracket.
6. Reverse procedure to install.

1974-78 Corvette

1. Disconnect battery ground cable.
2. Remove three forward right door sill plate screws.
3. Remove right kick panel screws, Fig. 145. Then, while lifting up slightly on forward edge of sill plate, remove kick panel.
4. Disconnect right distributor duct at dash outlet end.
5. Remove valve bracket screws. Disconnect valve bracket spring, remove link from valve arm, then the valve and bracket assembly.
6. Remove push nut from valve arm. Then, separate valve from bracket.
7. Reverse procedure to install.

PLENUM VALVE, REPLACE

Camaro

1. Lift right hand end of cowl vent plastic screen from retainers.

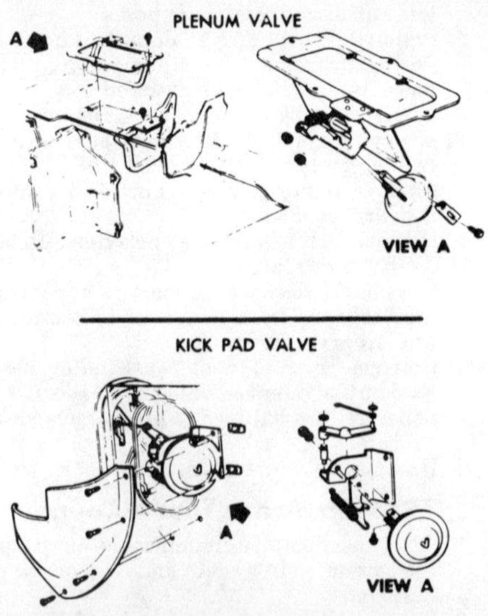

Fig. 144 Kick pad & plenum valve installation. 1974-80 Camaro

2. Remove plenum valve assembly mounting screws, Fig. 144. Disconnect vacuum hose and remove assembly from vehicle.
3. Remove plenum valve mounting screws and separate valve from bracket.

NOTE: Adjust valve to close with vacuum applied before installation.

4. Reverse procedure to install.

Chevelle & Monte Carlo

1974-77

1. Disconnect battery ground cable.
2. Lower right hand air distributor duct then remove glove box.
3. Disconnect vacuum line at diaphragm, remove diaphragm attaching screws, then disconnect actuator link and remove diaphragm.
4. Reverse procedure to install.

Chevrolet

1. Remove glove compartment.
2. Disconnect valve vacuum hose.
3. Remove valve mounting screws and pull valve rearward to disconnect clip.
4. Remove clip and the valve from vehicle.

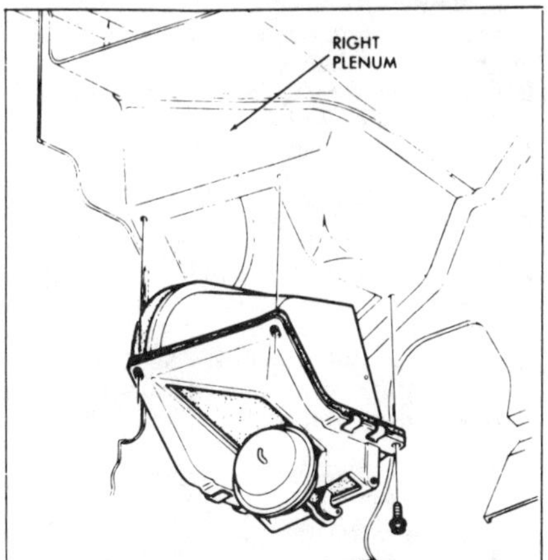

Fig. 146 Air inlet door vacuum motor installation. 1974-77 Astre, Vega & 1975-80 Monza, Skyhawk, Starfire & Sunbird

5. Reverse procedure to install.

Corvette

1974-76

1. Remove plenum screen (right end).
2. Remove plenum door bracket screws and remove door, valve and bracket as an assembly, Fig. 145.
3. Remove valve mounting screws and the valve.
4. Reverse procedure to install.

Nova

1. Remove windshield wiper arms and cowl vent grille.
2. Remove plenum valve assembly mounting screws, Fig. 143. Disconnect vacuum line, place door in closed position and lift assembly from vehicle.
3. Remove valve to bracket and link screws and the valve.
4. Reverse procedure to install.

MODE SHIFT VACUUM MOTOR, REPLACE

1974-80 Comfortron

1. Disconnect battery ground cable.
2. Remove vacuum valve retaining studs and lift vacuum valve to remove drive arm from boss on vacuum motor mechanism.
3. Remove retaining clip and power spring from motor, then disconnect vacuum hose from port on motor.
4. Remove 2 motor retaining screws, then the motor.
5. Reverse procedure to install.

AIR INLET DOOR VACUUM MOTOR, REPLACE

1974-76 Vega & 1975-80 Monza

1. Disconnect battery ground cable.
2. Remove vacuum hose clip from vacuum motor and disconnect hose.
3. Pry vacuum motor arm from door retaining clip, Fig. 146. Remove vacuum motor retaining nuts and the vacuum motor.
4. Reverse procedure to install.

1977-80 Full Size; 1978-80 Malibu & Monte Carlo

1. Remove module top cover and carefully pry door open.
2. Disconnect door linkage and remove door.
3. Reverse procedure to install.

BLOWER RELAY, REPLACE

Exc. Comfortron, Chevette, Monza & Vega

1. Disconnect battery ground cable.
2. Disconnect relay wiring harness connections.
3. Remove relay mounting screws and the relay.
4. Reverse procedure to install.

A/C & TIME DELAY RELAYS, REPLACE

1974-77 Vega & 1975-80 Monza

1. Disconnect relay electrical connector.
2. Remove relay mounting screws and the relay.
3. Reverse procedure to install.

BLOWER & COMPRESSOR CLUTCH SWITCH, REPLACE

1974-77 Vega; 1978-80 Monza

1. Remove control panel as outlined under

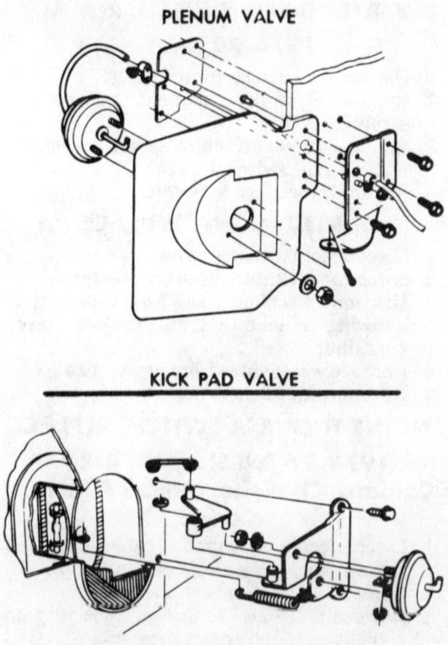

Fig. 145 Kick pad & plenum valve installation. 1974-80 Corvette

"Control Panel, Replace." It is not necessary to disconnect vacuum hoses or the Bowden cable.

2. Remove switch mounting screws and the switch from control panel.
3. Reverse procedure to install.

HI-BLOWER SWITCH, REPLACE

1974-77 Vega; 1978-79 Monza

1. Disconnect wiring connector from switch, located on the air inlet housing.
2. Remove switch mounting screws and switch.
3. Reverse procedure to install.

BLOWER RESISTOR, REPLACE

1. Disconnect battery ground cable and resistor wiring harness connector.
2. Remove resistor mounting screws and the resistor.
3. Reverse procedure to install.

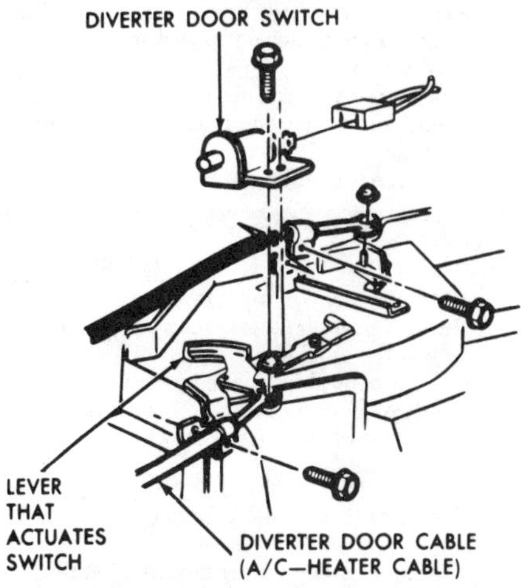

Fig. 147 Diverter door switch installation. 1974-79 Nova

DIVERTER DOOR SWITCH, REPLACE

1974-79 Nova

1. Disconnect battery ground cable.
2. Remove glove compartment and door assembly.
3. Remove screws attaching switch to selector duct and the switch, Fig. 147.
4. Reverse procedure to install.

WATER VALVE, REPLACE

1. Disconnect battery ground cable.
2. Disconnect vacuum hose from water valve.
3. Disconnect coolant hoses from water valve, allowing coolant to drain into a suitable container.
4. Remove water valve from intake manifold.
5. Reverse procedure to install.

ENGINE THERMAL SWITCH, REPLACE

1974-76 Full Size; 1974-80 Camaro, Chevelle, Malibu & Monte Carlo

1. Disconnect battery ground cable.
2. Disconnect switch electrical connector.
3. Remove switch from cylinder head.
4. Reverse procedure to install. Check engine coolant level and adjust if necessary.

BLOWER OR MASTER SWITCH, REPLACE

1975-80 Monza Towne Coupe

1. Perform steps 1 through 5 as outlined under "Control Panel, Replace" procedure.
2. Disconnect switch wiring connector.
3. Remove switch mounting screws and the switch.
4. Reverse procedure to install.

VACUUM VALVE, REPLACE

1975-76 Monza Towne Coupe

1. Perform steps 1 through 5 as outlined under "Control Panel, Replace" procedure.
2. Disconnect vacuum hose from vacuum valve.
3. Remove vacuum valve mounting screws and the vacuum valve.
4. Reverse procedure to install.

COMPRESSOR CUT-OFF SWITCH, REPLACE

1974-77 Vega & 1975-80 Monza

This switch, located on the carburetor, disengages the compressor clutch during full throttle operation on vehicles equipped with automatic transmissions.

1. Disconnect switch electrical connector.
2. Remove switch mounting screws and the switch.
3. Reverse procedure to install.

IN-CAR SENSOR, REPLACE

1974-76 Comfortron

1. Remove glove compartment.
2. Remove in-car sensor mounting screws and lower the sensor.
3. Disconnect electrical connector and aspirator hose. Then remove sensor from vehicle.
4. Reverse procedure to install.

1977-80 Comfortron

1. Remove instrument panel pad attaching screws, then remove pad.
2. Remove in-car sensor mounting screw, then disconnect wire connector and aspirator hose and remove sensor.
3. Reverse procedure to install.

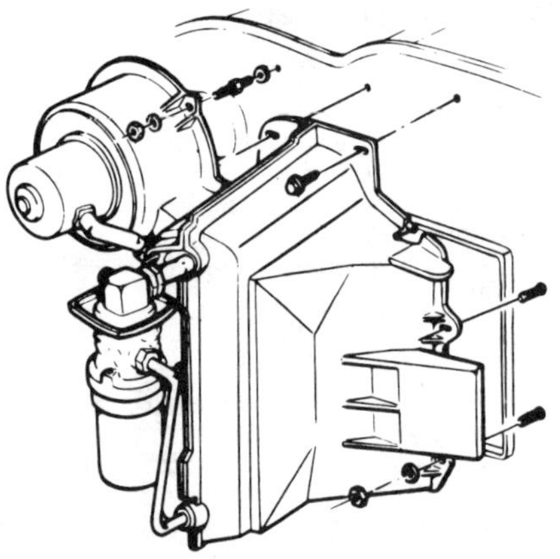

Fig. 148 Evaporator core & blower motor assembly. 1974-76 Camaro & Firebird

AMBIENT TEMPERATURE SENSOR & SWITCH, REPLACE

1974-76

1. Disconnect battery ground cable and ambient temperature switch electrical connector.
2. Remove switch mounting screws and the switch from vehicle.
3. Reverse procedure to install.

1977-80 Comfortron

1. Remove hood seal from lip of module.
2. Disconnect wire connector from sensor, then loosen leaf screen just enough so that sensor can be lifted from lip of module.
3. Reverse procedure to install.

AUTO & LO RELAYS, REPLACE

1974-76 Comfortron

1. Disconnect electrical connector from relay.
2. Remove relay mounting screws and the relay.
3. Reverse procedure to install.

EVAPORATOR CORE, REPLACE

1974-76 All Exc. Chevette, Monza & Vega

1. Disconnect battery ground cable, Figs. 92, 93, 95, 96 and 148, discharge refrigerant from system and remove air cleaner.
2. On 1974-76 Corvette, disconnect coolant recovery system reservoir and place aside.
3. On 1974-76 models, disconnect oil bleed line and core inlet and outlet lines at VIR assembly. Cap all lines.
4. On 1974-76 except Corvette, remove VIR bracket screw and place VIR assembly aside with lines still attached, then disconnect vacuum lines and remove vacuum tank and bracket.
5. Disconnect all wires and vacuum lines attached to evaporator housing, then remove evaporator cover to dash panel stud nuts and cover to evaporator case screws.
6. Carefully pull cover upward and outward and remove evaporator core.
7. Reverse procedure to install using new O-rings coated with refrigerant oil. Add three ounces of clean refrigerant oil into new core.

1975-76 Monza 2+2

1. Disconnect battery ground cable, discharge refrigerant from system, drain cooling system and disconnect heater hoses.
2. Remove glove box, right air outlet duct, deflector, instrument bezel and pad and left air outlet duct deflector and feed duct.
3. Lower steering column, then remove instrument panel assembly and remove control assembly from instrument panel.
4. Remove radio, defroster duct and center distribution duct.
5. Remove VIR assembly as described under "VIR Service" section.
6. Disconnect electrical connectors at control assembly and from plenum and disconnect lines from vacuum tank.
7. Remove cowl to heater-distributor case stud nuts and remove assembly, Fig. 99.
8. Separate case halves and remove evaporator core.
9. Reverse procedure to install.

1975-76 Monza Towne Coupe

1. Disconnect battery ground cable, discharge refrigerant from system and drain cooling system.
2. Remove VIR as described under "VIR Service" section.
3. Remove blower case retaining screws and remove blower case, Fig. 99.
4. Disconnect heater core hoses. Plug heater core tubes to avoid spilling coolant into passenger compartment.
5. Remove evaporator to dash panel cover plate screws and remove plate, then remove floor outlet duct and glove box.
6. Remove right and left hand dash outlets and instrument panel pad.
7. Remove right hand instrument panel to dash and kick pad screws, then loosen left hand instrument cluster to instrument panel screws and remove right hand instrument panel and lap cooler.

NOTE: Pull on right end of cluster to gain clearance for panel removal.

8. Disconnect hoses from unit and remove modulator duct to heater-evaporator screw.
9. Pull carpet and pad rearward to allow clearance, then pull heater-evaporator unit rearward until core tubes clear dash panel. Carefully pull to the right and disconnect temperature door Bowden cable and electrical connectors from heater-evaporator assembly.
10. Remove heater-evaporator case half screws and separate halves.
11. Remove core retaining screws and remove evaporator core.
12. Reverse procedure to install.

1974-77 Vega & 1978-79 Monza "S" Coupe & Station Wagon

1. Disconnect battery ground cable, discharge refrigerant from system and drain cooling system.
2. Disconnect all wires attached to blower-evaporator assembly and its components.
3. Disconnect evaporator inlet and outlet lines.
4. Remove accumulator retaining screws and set aside.
5. Remove attaching screws and remove blower-evaporator assembly from vehicle, Fig. 99.
6. Remove thermostatic switch from evaporator case.
7. Remove core retaining plate from rear of case, then separate case halves and re-

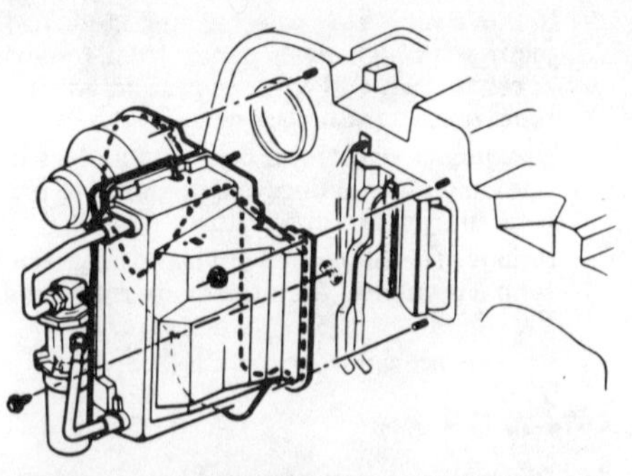

Fig. 149 Evaporator core & blower motor assembly. 1974-77 Corvette

move core.

8. Reverse procedure to install.

1976-80 Chevette

1. Disconnect battery ground cable and discharge refrigerant from system.
2. Remove heater core case to distributor assembly retaining screws and roll assembly forward and away from distributor case.
3. Disconnect thermostatic switch sensing element. Bend capillary tube away from case.
4. Disconnect temperature control cable and remove temperature door arm access cover. Position control door arm so as not to interfere with evaporator case cover removal.
5. Remove blower motor.
6. Remove sealer from around exposed portion of outlet tube, then remove retaining screws and remove upper evaporator case.
7. Remove evaporator core.
8. Reverse procedure to install.

1977 Camaro, Chevelle, Monte Carlo, 1977-79 Nova & 1977-80 Corvette

1. Disconnect battery ground cable, then purge system of refrigerant.
2. Remove air cleaner assembly, A.I.R. pipe at manifold, PCV, carburetor heat stove and engine thermal switch as necessary.
3. On 1977 Corvette, remove coolant recovery system container attaching screw and position container aside, then disconnect oil bleed and core inlet and outlet lines at VIR assembly. Cap refrigerant lines and fittings.
4. On all models except 1977 Corvette, remove accumulator bracket attaching screws and position accumulator aside with compressor and condenser lines attached. Disconnect vacuum lines and remove vacuum tank and bracket assembly.
5. Disconnect all remaining electrical and vacuum lines attached to the evaporator assembly.
6. Remove evaporator cover to dash panel stud nuts and cover to evaporator case attaching screws, Figs. 149, 150 and 151.
7. Pull cover outward and upward, using care not to damage core.
8. Remove evaporator core from case, using care not to damage oil bleed line or core.
9. Reverse procedure to install.

1977-80 Monza Except "S" Coupe & Station Wagon

1. Disconnect battery ground cable and discharge refrigerant system.
2. Remove floor outlet duct, glove box, left and right hand dash outlets and instrument panel pad.

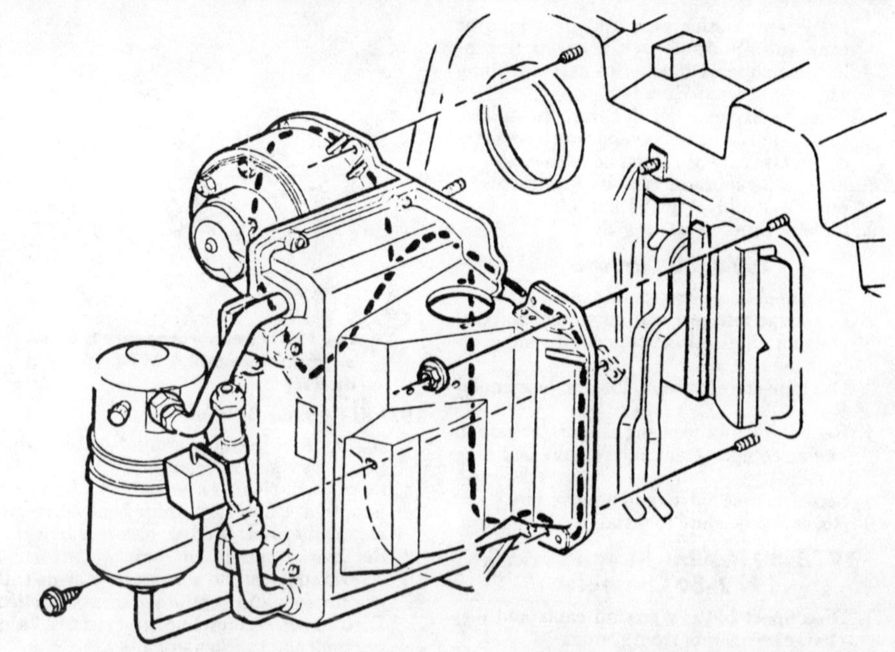

Fig. 150 Evaporator core & blower motor assembly. 1978-80 Corvette.

3. Disconnect vacuum hoses at valves on left end of heater-evaporator unit, then remove insulation tray and loosen console assembly and slide rearward.
4. Remove steering column attaching nuts and lower column until it rests on driver's seat.
5. Remove instrument panel to dash attaching screws, place protective covering over steering column, and lower instrument panel to steering column. Separate speedometer cable, radio wiring and control head connectors.
6. Remove right hand instrument panel and lap cooler as an assembly, then remove modular duct to heater-evaporator screw and remove modular duct.
7. Disconnect temperature door bowden cable and wiring, heater hoses at core tubes and 3 heater case stud nuts. Plug all lines and openings to prevent coolant spillage.
8. Disconnect evaporator core inlet and outlet lines and oil bleed line at accumulator assembly.
9. Remove the accumulator bracket attaching screws and remove the accumulator. Plug all lines and openings to prevent entry of dirt or moisture.
10. Remove the heater-evaporator case assembly from vehicle and separate heater and evaporator halves. Remove evaporator core attaching screws, then the core, Fig. 00.
11. Reverse procedure to install.

1977-78 Chevrolet

1. Disconnect battery ground cable and discharge refrigerant from the system.
2. Remove diagnostic connector to upper case attaching screws.
3. Remove connectors from blower motor, resistor, blower relay and thermostatic switch.
4. Remove wiring harness plug-in retainer from blower case shroud.
5. Remove module screen assembly to cowl attaching screws and remove screen. Then remove upper case half to lower case half cowl attaching screws. Two screws are located inside the air intake area at the case separation point.
6. Remove thermostatic switch, disconnect

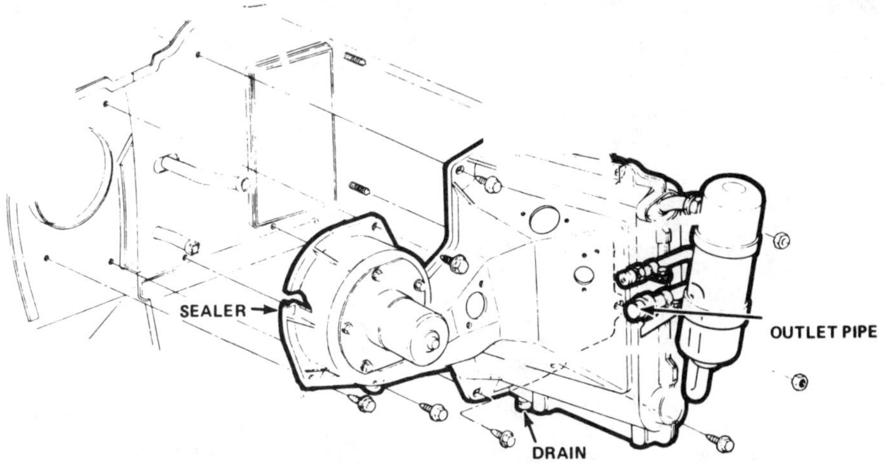

Fig. 151 Evaporator core & blower motor assembly. 1977-80 Camaro & Firebird

evaporator outlet pipe and inlet pipe at evaporator and remove the evaporator inlet pipe support bracket to case attaching screws and remove bracket.
7. Remove expansion tube from inlet line, then remove heater evaporator core case cover. Use care not to damage sealer.
8. Remove evaporator core by lifting up and out of assembly, Figs. 98 and 100.
9. Reverse procedure to install.

1978-80 Camaro

1. Disconnect battery ground cable.
2. Discharge refrigerant from system and disconnect all electrical connectors and cooling tube.
3. Disconnect refrigerant lines at accumulator.
4. Remove screws securing evaporator case to engine compartment and remove case, Fig. 151.
5. Separate case halves and remove core.
6. Reverse procedure to install.

1978-80 Malibu, Monte Carlo & 1979-80 Chevrolet

1. Disconnect battery ground cable and discharge refrigerant from system.
2. Disconnect module rubber seal and remove module screen.
3. Remove right hand windshield wiper arm.
4. Remove hi blower relay and thermostatic switch.
5. Disconnect air electrical connectors at module top, then remove module front cover, Figs. 98 and 100.
6. Remove accumulator bracket screws, then disconnect refrigerant lines at accumulator and liquid line and remove core.
7. Reverse procedure to install using new sealant around evaporator case.

1980 Citation

1. Disconnect battery ground cable and discharge refrigerant system.
2. Remove air cleaner.
3. Disconnect air conditioning electrical connections along the cowl and position wiring harness aside.
4. Remove vacuum tank.
5. Disconnect liquid line at evaporator inlet and the low pressure line at accumulator, Fig. 102.
6. Remove outer blower case to cowl screws, then the blower and case assembly.
7. Remove evaporator from vehicle.
8. Reverse procedure to install.

Chevrolet & GMC Light Duty Truck

1974-76 El Camino & Sprint

1. Disconnect battery ground cable. Fig. 96. Discharge refrigerant from system and remove air cleaner.
2. Disconnect oil bleed line and core inlet and outlet lines at VIR assembly. Cap all lines.
3. Remove VIR bracket screw and place VIR assembly aside with lines still attached then disconnect vacuum lines and remove vacuum tank and bracket.
4. Disconnect all wires and vacuum lines attached to evaporator housing, then remove evaporator cover to dash stud nuts and cover to evaporator case screws.
5. Carefully pull cover upward and outward and remove evaporator core.
6. Reverse procedure to install using new O-rings coated with refrigerant oil. Add three ounces of refrigerant oil into the new core.

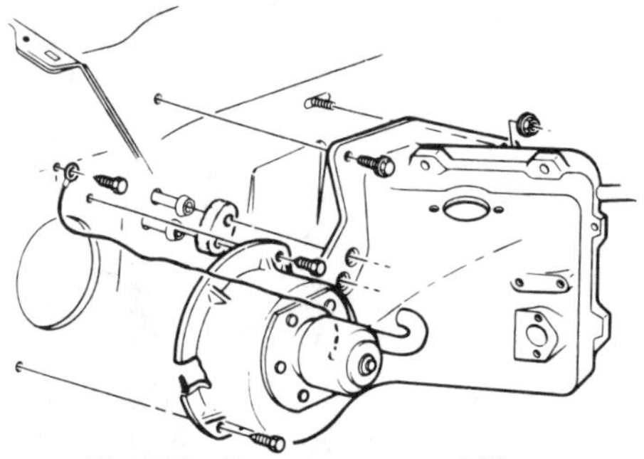

Fig. 152 Evaporator core & blower motor assembly. 1974-80 C & K models

1977 El Camino & Sprint

1. Disconnect battery ground cable, then purge system of refrigerant.
2. Remove air cleaner assembly, A.I.R. pipe at manifold, PCV, carburetor heat stove and engine thermal switch as necessary.
3. Remove accumulator bracket attaching screws and position accumulator aside with compressor and condenser lines attached. Disconnect vacuum lines and remove vacuum tank and bracket assembly.
4. Disconnect all remaining electrical and vacuum lines attached to the evaporator assembly.
5. Remove evaporator cover to dash panel stud nuts and cover to evaporator case attaching screws
6. Pull cover outward and upward, using care not to damage core.
7. Remove evaporator core from case, using care not to damage oil bleed line or core.
8. Reverse procedure to install.

1978-80 Caballero & El Camino

1. Disconnect battery ground cable and discharge refrigerant from system.
2. Disconnect module rubber seal and remove module screen.
3. Remove right hand windshield wiper arm.
4. Remove hi blower relay and thermostatic switch.
5. Disconnect air electrical connectors at module top, then remove module front cover, Fig. 98.
6. Remove accumulator bracket screws, then disconnect refrigerant lines at accumulator and liquid line and remove core.
7. Reverse procedure to install using new sealant around evaporator case.

1974-80 C & K Series

1. Disconnect battery ground cable and discharge refrigerant system.

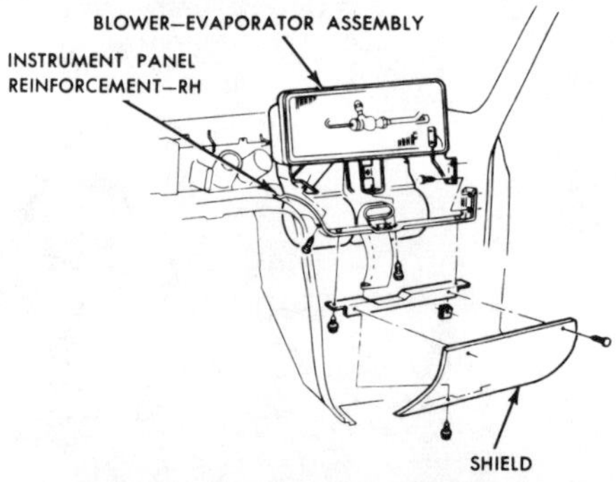

Fig. 153 Evaporator core & blower motor assembly. 1974-77 G models.

2. Remove nuts from selector duct studs projecting through dash panel, then remove cover to dash and cover to case screws and remove evaporator case cover.
3. Disconnect evaporator core inlet and outlet lines and plug all lines and openings to prevent entry of dirt or moisture.
4. Remove thermostatic switch and expansion tube assemblies, then the evaporator core, Fig. 152.
5. Reverse procedure to install.

1974-77 G Series

1. Disconnect battery ground cable and discharge refrigerant system.
2. Remove screws securing blower evaporator shield, then the shield. Remove shield bracket to instrument panel reinforcement screws and remove bracket.
3. Remove heater intermediate duct to A/C distributor duct screw then remove duct.
4. Open hood, remove lower tab mounting bolts and remove hood.
5. Loosen both steering column instrument panel reinforcement screws and remove one screw, then remove left foot cooler bracket to instrument panel reinforcement screws, disconnect outlet from duct and remove outlet and duct assembly.
6. Disconnect speedometer cable, remove instrument panel to lower reinforcement attaching screws, then rotate instrument panel up.
7. Disconnect evaporator inlet and outlet connections and plug all lines and openings to prevent entry of dirt or moisture.
8. Remove blower-evaporator support bracket to door pillar and forward engine housing screws and lower blower-evaporator assembly. Disconnect all wiring, pull drain hose through the dash panel and remove blower-evaporator as an assembly.
9. Remove expansion valve sensing bulb clamps and disconnect line from the evaporator core. Disconnect expansion valve inlet and outlet lines and remove expansion valve. Plug all lines and openings to prevent entry of dirt or moisture.
10. Remove evaporator core line clamp screws, then the core to case attaching screws and remove evaporator core assembly, Fig. 153.
11. Reverse procedure to install.

1978-80 G Series

1. Disconnect battery ground cable and discharge refrigerant system.
2. Remove coolant recovery tank and bracket, then disconnect all wiring from core case assembly.
3. Remove bracket at evaporator case, then right hand marker lamp for access.
4. On 1978-79 models, remove thermostatic switch from case and capillary tube insulation.
5. On 1978-79 models, remove thermal bulb to inlet line clamp and remove thermostatic switch.
6. Disconnect accumulator inlet and outlet lines and 2 brackets attaching accumulator to case.
7. Disconnect evaporator inlet line, remove 3 nuts and one screw attaching module to dash panel and remove core case assembly from vehicle, Fig. 154.
8. Separate case sections and remove core.
9. Reverse procedure to install.

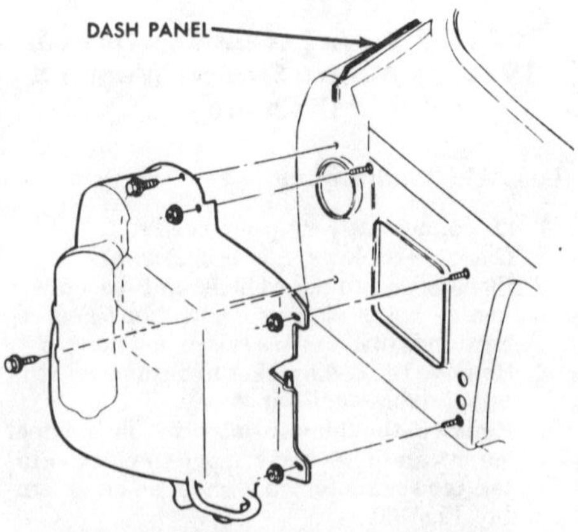

Fig. 154 Evaporator core & blower motor assembly. 1978-80 G models

Chevrolet & GMC Medium Duty Truck

1974-80 40 to 60 Series

1. Disconnect battery ground cable and discharge refrigerant system.
2. Remove air cleaner assembly.
3. Remove outlet line and disconnect inlet line. Cover all open lines and openings to prevent entry of dirt or moisture.
4. Disconnect all wiring and vacuum hoses attached to evaporator, then remove evaporator cover to fire wall stud nuts and cover to evaporator case screws.
5. Pull cover outward and upward, being careful not to damage core. On some engines it may be necessary to remove the Air Injection Reactor Manifold Assembly.
6. Remove core retaining screws and remove core.
7. Reverse procedure to install, Fig. 155.

HEATER CORE, REPLACE

1974-80 Except Chevette, Monza & Vega; 1977-80 Chevrolet Full Size; 1978-80 Chevelle, Malibu & Monte Carlo; 1980 Citation

Less Air Conditioning

1. Disconnect battery ground cable.
2. Drain radiator, disconnect heater hoses at core and plug openings to prevent spillage of water.
3. Remove nuts from air distributor duct studs on engine side of firewall, Figs. 105 and 156 through 159.
4. On 1974-76 Chevrolet and 1974-77 Chevelle and Monte Carlo, remove distributor duct retaining screw.
5. On Nova, remove glove box and door and drill out lower right hand distributor stud with a ¼" drill.
6. On Camaro, remove glove box, radio, and defroster duct to distributor duct screw.
7. On Corvette, remove right instrument panel pad, right hand dash braces, center dash console duct and floor outlet duct, radio and center dash console.
8. On all models, pull distributor duct from firewall being careful not to bend cable.
9. Disconnect cable and resistor wires and remove distributor and core.
10. On all other models, remove core assembly from duct.
11. Reverse procedure to install.

1974-76 Chevrolet Full Size

With Air Conditioning

1. Disconnect battery ground cable and compressor clutch connector.
2. Drain radiator and disconnect heater hoses from core. Plug hoses and core openings.
3. Disconnect vacuum hose from vacuum check valve and push hose grommet through firewall.
4. Remove accessible screws and stud nuts from heater and air selector duct from dash side of firewall, Fig. 156.
5. Remove two screws and three stud nuts, accessible from side of dash.
6. Remove the screws securing the fender skirt, holding skirt away from rear of wheel-well, remove the last screw.
7. Remove lap cooler assembly, glove box door and glove box.
8. Remove floor outlet duct and instrument panel pad. On 1974-76 Comfortron, disconnect aspirator and in-car sensor connector.
9. Remove distributor duct hoses and connector, then remove distributor duct from selector.
10. Remove defroster duct screw and move duct to gain access to selector and heater core. On 1974-76 Comfortron, disconnect in-line vacuum connector and line from outside air diaphragm, then disconnect electrical and vacuum connections from programmer.

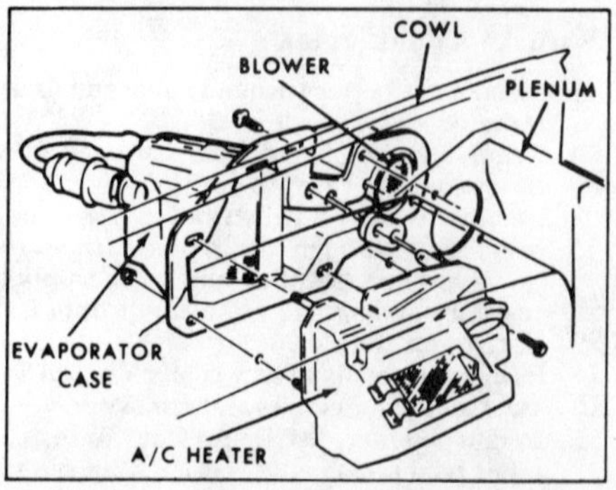

Fig. 155 Evaporator, heater core & blower motor assembly. 1974-80 40-60 Series (Typical)

11. Disconnect cables and vacuum lines from selector, then remove heater core and selector. Separate heater core from selector by removing clamp screws.
12. Reverse procedure to install.

1974-76 Chevelle & Monte Carlo

With Air Conditioning

1. Disconnect battery ground cable and drain radiator.
2. Disconnect heater hoses from core and plug hoses and core openings.
3. Remove distributor case stud nuts projecting through firewall, Fig. 157. On 1974-76, remove resistor assembly and the last stud nut through its opening.
4. Remove glove box door and glove box.
5. Remove right hand lap cooler and kick pad cover.
6. Disconnect center duct hoses and remove center duct to selector duct screws, then remove center duct.
7. Remove floor distributor duct.
8. Remove lower right hand distributor duct to dash panel screw.
9. On all models, remove air selector retaining screws, then lower selector assembly. Disconnect all wiring, vacuum lines and cables.
10. On all models, remove core and housing from selector assembly, then core from housing.
11. Reverse procedure to install.

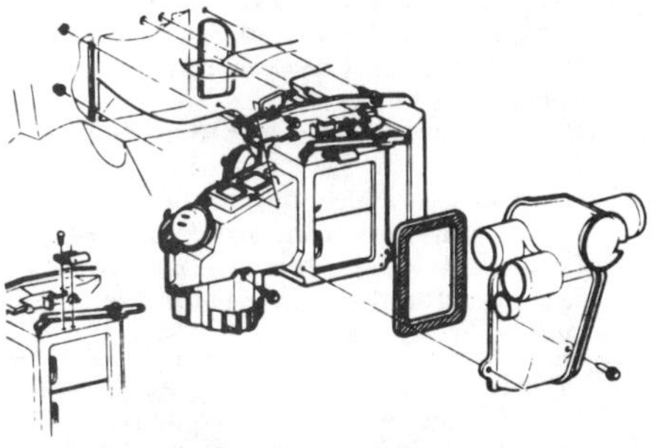

Fig. 156 Heater assembly. 1974-76 Chevrolet Full Size

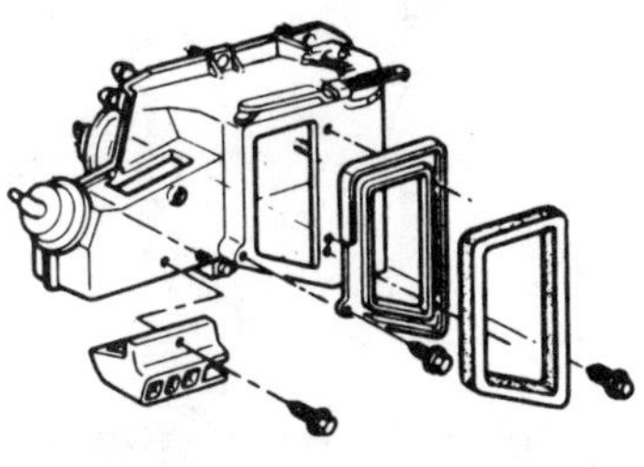

Fig. 157 Heater assembly. 1974-77 Chevelle & Monte Carlo

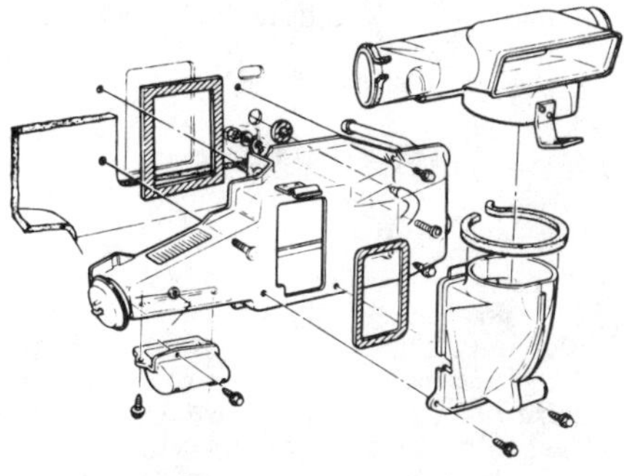

Fig. 158 Heater assembly. 1974-80 Camaro

1977 Chevelle & Monte Carlo

With Air Conditioning

1. Disconnect battery ground cable and drain radiator.
2. Remove heater hoses from core and plug hoses and core tubes to prevent spillage.
3. Remove nuts from distributor case studs projecting through dash panel, then remove resistor assembly and reach through resistor opening and remove nut from last distributor stud.
4. Remove right hand lap cooler duct to instrument panel screws and remove duct.
5. Remove center distributor duct to instrument panel and center outlet strap screws and remove center duct, then remove glove box.
6. Remove floor outlet, defroster duct to distributor duct screw and distributor duct to dash panel screw at lower right of the distributor duct, Fig. 157.
7. Pull distributor assembly rearward far enough for the distributor studs and core tubes to clear dash panel. Lower distributor and disconnect all wiring and vacuum connectors.
8. Disconnect temperature door bowden cable and remove distributor assembly from vehicle. Remove screws attaching core to distributor, then the core.
9. Reverse procedure to install.

1974-79 Nova

With Air Conditioning

1. Disconnect battery ground cable and drain radiator.
2. Disconnect heater hoses and plug hoses and core openings to prevent coolant spillage.
3. Remove accessible stud nut from air selector duct, Fig. 105.
4. On Camaro, remove right side lower rocker molding and the lower fender attaching bolts.
5. On all models, remove right hand fender skirt to fender and skirt reinforcing screws. Lower skirt to wheel and remove remaining stud nut.
6. Remove glove box door, glove box and right hand kick pad recirculating air valve.
7. Remove center and floor ducts. Remove screws securing selector duct halves together, then separate.
8. On Nova, remove selector duct right half to firewall screws and remove duct. On Camaro, remove screws securing selector duct to firewall and remove duct.
9. On all models, disconnect all wiring and cables. Scribe location of temperature camming plate on selector duct and remove plate and core.
10. Reverse procedure to install.

1974-80 Camaro

With Air Conditioning

1. Disconnect battery ground cable.
2. Drain radiator and disconnect heater hoses from core. Plug openings to prevent spillage of coolant.
3. On 1977-78 models, remove rearmost fender to skirt screws and pry skirt away from fender flange. Use a piece of wood to block skirt away from fender flange.

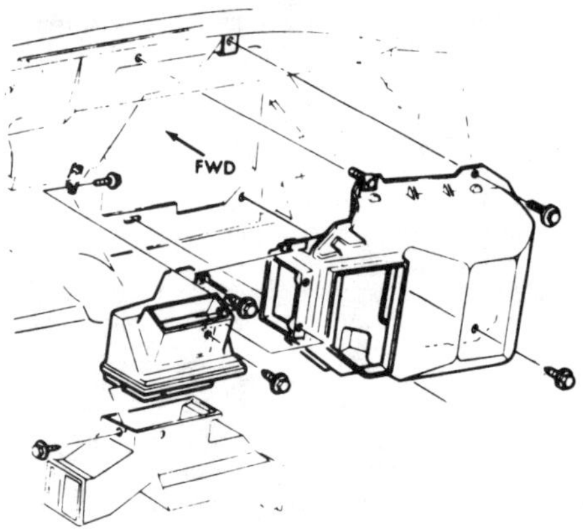

Fig. 159 Heater assembly. 1974-79 Corvette

4. Remove nuts from distributor studs on engine side of firewall, Fig. 158.
5. Remove glove box and radio.
6. Remove defroster duct to distributor duct screw. With radio removed, the defroster duct can be pulled rearward to gain clearance for distributor duct removal.
7. Carefully pull distributor from firewall and disconnect wiring and cables. Remove duct and core from vehicle.
8. Reverse procedure to install.

1974-80 Corvette

With Air Conditioning

1. Disconnect battery ground cable. Drain radiator and disconnect heater hoses from core. Plug openings to prevent spillage of coolant.
2. Remove nuts from engine side distributor duct, Fig. 159.
3. Remove right hand dash pad and center instrument cluster. Remove dash braces.
4. Disconnect right dash outlet duct from center duct. Remove screws attaching center duct to selector duct and remove center duct.
5. Remove screws attaching selector duct to firewall and pull selector rearward and to the right. Disconnect cables and wiring.
6. Remove selector duct from car. Remove temperature door camming plate from duct and remove core and housing.
7. Reverse procedure to install.

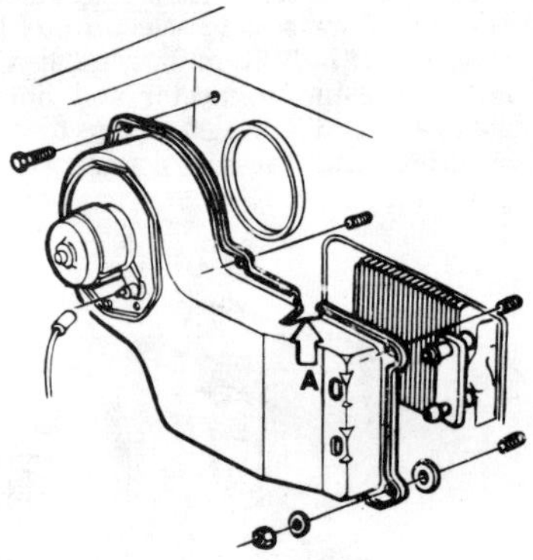

Fig. 160 Heater assembly less A/C. 1974-77 Vega except Cosworth; 1975-77 Astre; 1978-79 Monza "S" Coupe & Station Wagon, Sunbird Station Wagon

1974-77 Vega Except Cosworth & 1978-79 Monza Station Wagon & "S" Coupe

Less Air Conditioning

1. Disconnect battery ground cable.
2. Disconnect blower motor lead wire.
3. Place a pan under vehicle and disconnect heater hoses at core connections and secure ends of hoses in a raised position.
4. Remove the coil bracket to dash panel stud nut and move coil out of way.
5. Remove the blower inlet to dash panel screws and nuts and remove the blower inlet, blower motor and wheel as an assembly, Fig. 160.
6. Remove the core retainer strap screws and remove the core.
7. When replacing core, be sure the blower inlet sealer is intact.
8. Reverse procedure to install.

1975-76 Cosworth Vega

Less Air Conditioning

1. Disconnect battery ground cable.
2. Disconnect heater hoses from heater core. Plug hoses and core tubes.
3. Remove nuts and washers from heater air distributor studs on engine side of dash.
4. Remove floor outlet to air distributor duct screw and the outlet.
5. Remove air distributor to dash screws, then pull duct assembly rearward until core tubes clear dash openings. Tilt assembly downward and disconnect Bowden cables from distributor assembly.
6. Disconnect resistor wiring harness and remove air distributor assembly from vehicle.
7. Remove core retaining strap screw and the heater core.
8. Reverse procedure to install.

1974-77 Vega & 1978-79 Monza Station Wagon & "S" Coupe

With Air Conditioning

1. Disconnect battery ground cable.
2. Position a pan under heater core tubes, then disconnect heater hoses and secure in a raised position. Cap core tubes to prevent coolant spillage during removal.
3. From engine side of dash panel remove nuts from selector duct studs, Fig. 161.
4. Remove glove box and door as an assembly.
5. Remove right hand outlet to instrument panel attaching screws, then remove outlet and flexible hose.
6. Remove intermediate duct to left hand outlet.
7. Remove steering column to toe pan plastic retainer, insulation and attaching screws.
8. Remove steering column to dash panel attaching nuts, then lower column and support on driver's seat.

NOTE: Place protective tape over steering column to prevent damage to finish when lowering instrument cluster assembly.

9. Remove instrument panel bezel, then remove ash tray and retainer.
10. Remove A/C control to instrument panel attaching screws.
11. Disconnect wire connector and antenna lead from radio.
12. Remove instrument cluster to dash panel attaching screws and allow cluster assembly to rest on steering column.
13. Disconnect speedometer cable, then push A/C control forward and allow control to rest on floor, using care not to kink cable.
14. Remove center duct to selector duct attaching screws and duct to upper instrument panel retainer, Fig. 162.
15. Slide center duct to left to clear lower instrument panel to cluster tab, then slide to right to remove.
16. Remove defroster duct to selector duct attaching screws.
17. Remove remaining selector duct to dash panel attaching screws, then pull duct rearward and disconnect all electrical leads and vacuum lines.
18. Disconnect temperature door cable and remove selector duct assembly.

NOTE: Ensure all electrical leads and vacuum lines are disconnected before removing duct assembly.

19. Pry off or punch out temperature door bell crank, then remove temperature door.
20. Remove backing plate attaching screws and temperature door cable retainer.
21. Remove heater core and backing plate as an assembly.
22. Remove core straps, then remove heater core.
23. Reverse procedure to install.

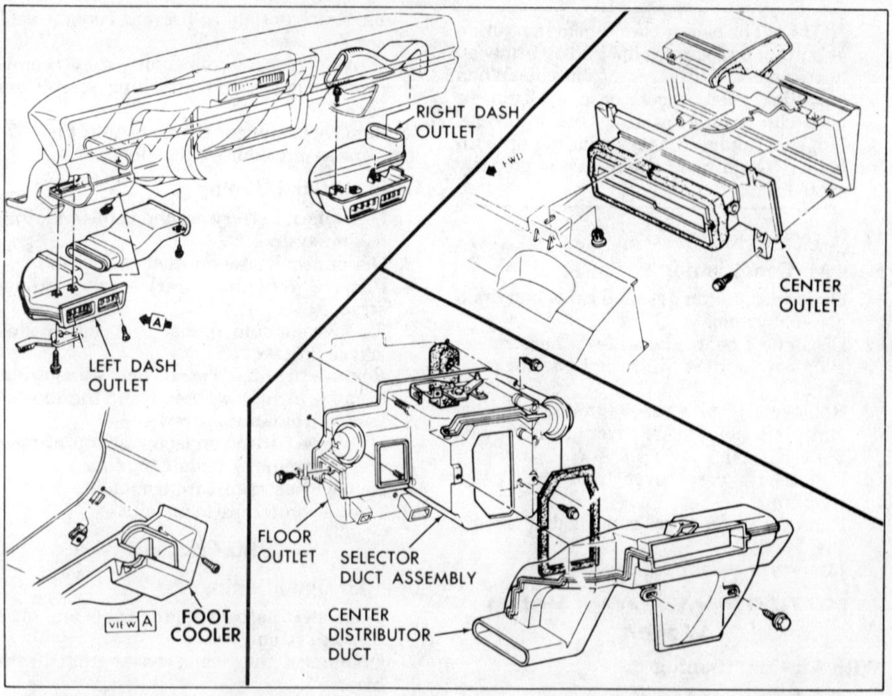

Fig. 161 Air distribution ducts and outlets on models with A/C. 1974-77 Vega & 1975-76 Astre; 1978-79 Monza Station Wagon & "S" Coupe, Sunbird Station Wagon

1975-80 Monza Except Station Wagon & "S" Coupe

Less Air Conditioning

1. Disconnect battery ground cable and disconnect blower motor electrical lead.
2. Place a pan under vehicle, then disconnect heater hoses from core and secure hoses in a raised position.
3. Remove blower inlet to dash panel screws and nuts, then remove blower inlet and blower motor and wheel assembly.
4. Remove core retaining strap screws and remove core.
5. Reverse procedure to install making certain that blower inlet sealer is intact, replace as necessary.

1975-76 Monza

With Air Conditioning

1. Disconnect battery ground cable.
2. Remove glove box and right air duct.
3. Remove instrument panel bezel and pad.
4. Remove left air duct and feed duct.
5. Lower steering column and remove instrument panel.
6. Remove control assembly, radio, defroster duct and large distribution duct.
7. Purge system of refrigerant.
8. Disconnect heater hoses at core.
9. Clean VIR exterior surface and line connections, then disconnect compressor inlet line, oil bleed line and condenser outlet line from VIR assembly.

NOTE: Cap all line and VIR ports.

10. Loosen evaporator inlet and outlet lines and remove VIR mounting clamp, then slide VIR assembly from evaporator lines.
11. Remove heater-distributor case to cowl attaching nuts and remove case, Fig. 99, then disconnect wiring harness and vacuum lines.
12. Separate heater case from distributor, then remove heater core from case.
13. Reverse procedure to install.

1976-80 Chevette

1. Disconnect battery ground cable and drain cooling system.
2. Disconnect heater hoses from core and plug openings in core.
3. On 1976-77 models, remove heater core housing to dash panel retaining screws, then remove housing from vehicle and remove heater core from housing.
4. On 1978-80 models, remove screws from core cover on engine side of dash, then remove cover and lift core out.
5. Reverse procedure to install.

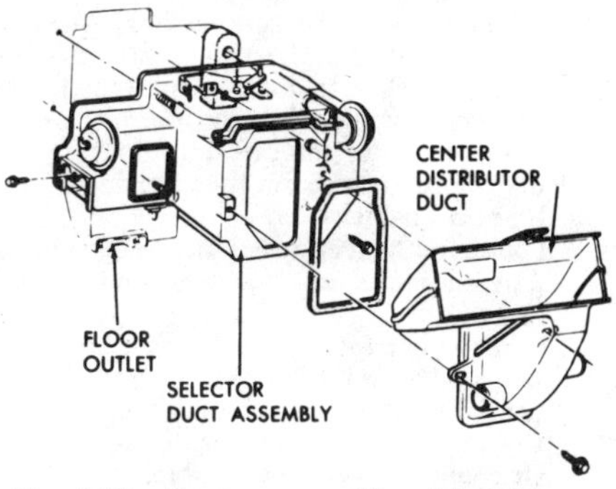

Fig. 162 Heater assembly with A/C. 1974-77 Vega & 1975-77 Astre; 1978-79 Monza & Station Wagon "S" Coupe, Sunbird Station Wagon

1977-80 Chevrolet Full Size

With Air Conditioning

1. Disconnect battery ground cable and drain cooling system.
2. Disconnect heater hoses at heater core. Cap core outlets to prevent coolant spillage.
3. Remove diagnosis connector to upper case attaching screws.
4. Disconnect wire connectors from blower motor, resistor, blower relay and thermostatic switch.
5. Remove wiring harness retainer from blower case shroud.
6. Remove module screen to cowl attaching screws and remove screen.
7. Remove upper case to lower case attaching screws.

NOTE: Two screws are located inside air intake area at case separation point.

8. Disconnect wire connector from thermostatic switch, then remove screws attaching switch to evaporator case. Carefully remove insulation and loosen two clamps enough to pull formed end of switch capillary tube from under clamps attaching tube to evaporator inlet pipe for installation.
9. Remove evaporator inlet pipe support bracket to case attaching screws and remove bracket.
10. Remove heater-evaporator core case cover, using care not to damage sealer.
11. Remove heater core to case attaching screws at top of heater core, then remove heater core, Fig. 100.

NOTE: The heater core is held in position at bottom by a spring clip. Pull up firmly on heater core to disengage from clip. When installing, position core base in alignment with clip before lowering core into case. Upper retaining bracket will line up with hole at top of core when core is properly seated.

12. Reverse procedure to install.

Less Air Conditioning

1. Disconnect battery ground cable and drain cooling system.
2. Disconnect heater hoses from heater core. Plug core outlets to prevent coolant spillage.
3. Remove attaching screws from around perimeter of heater core cover on engine side of dash panel.
4. Pull heater core cover from dash panel mounting.
5. Remove heater core from module assembly, Fig. 106.
6. Reverse procedure to install.

1977-80 Monza Except Station Wagon

With Air Conditioning

NOTE: The heater core can be removed without purging the A/C refrigerant system or removing the evaporator case half of the assembly by using the following procedure.

1. Disconnect battery ground cable.
2. Remove floor outlet duct, then remove glove box and door as an assembly.
3. Remove left and right hand dash outlets using a suitable tool.
4. Remove instrument panel pad.
5. Disconnect vacuum hoses at valves on left end of heater and evaporator unit.
6. Remove insulation tray located below instrument cluster, then loosen console and slide console rearward.
7. Remove instrument panel to dash attaching screws. Place a protective covering over steering column, then lower instrument panel to steering column. Disconnect speedometer cable, radio leads and control head connectors.
8. Remove right hand instrument panel and lap cooler as an assembly.
9. Remove modular duct to heater and evaporator case screw, then remove duct assembly.
10. Disconnect temperature door cable and wiring harness.
11. Disconnect heater hoses from heater core. Plug core tubes to prevent spillage when removing heater core.
12. Remove three heater case stud nuts located in engine compartment.
13. Remove heater core case to evaporator core case attaching screws.
14. Drive on studs to break case loose from dash panel.
15. Remove heater core case assembly.
16. Remove heater core to case attaching screws, then remove heater core.
17. Reverse procedure to install.

1978-80 Malibu & Monte Carlo

Less Air Conditioning

1. Disconnect battery ground cable and drain cooling system.
2. Disconnect heater hoses from heater core. Plug core outlets to prevent coolant spillage.
3. Disconnect wire connectors, then remove front module cover attaching screws and cover.
4. Remove heater core from module, Fig. 107.
5. Reverse procedure to install.

With Air Conditioning

1. Disconnect battery ground cable and drain cooling system.
2. Disconnect heater hoses at heater core.
3. Remove retaining bracket and ground strap.
4. Remove module rubber seal and module screen, Fig. 98.
5. Remove right hand windshield wiper arm.
6. Remove high blower relay and thermostatic switch mounting screws.
7. Disconnect wire connector at top of module, then remove module top cover.
8. Remove heater core from module.
9. Reverse procedure to install.

1980 Citation

Less Air Conditioning

1. Disconnect battery ground cable and drain cooling system.
2. Disconnect the heater hoses from heater core.
3. Remove radio noise suppression strap.
4. Remove heater core cover retaining screws and the cover, Figs. 109 and 110.
5. Remove heater core from vehicle.
6. Reverse procedure to install.

With Air Conditioning

1. Disconnect battery ground cable and drain cooling system.
2. Disconnect heater hoses from heater core.
3. Remove right side hush panel and open the glove box.
4. Remove heater duct retaining screw and the duct.
5. Remove instrument panel support bracket.
6. Remove heater case side cover retaining screws and the cover.
7. Remove heater core retaining clamps and the inlet and outlet tube support clamps.
8. Remove heater core from case.
9. Reverse procedure to install.

Chevrolet & GMC Trucks

1974 "G" Series With A/C

1. Disconnect battery ground cable, drain cooling system and discharge refrigerant system.
2. Disconnect heater hoses from core tubes and plug core openings.
3. Remove screws securing the blower-evaporator shield and the shield. Remove the shield bracket to instrument panel reinforcement screws, then the bracket.
4. Remove heater intermediate duct to A/C distributor duct screw, then the duct.
5. Remove engine cover.
6. Loosen both steering column to instrument panel reinforcement screws and remove one screw.
7. Remove left foot cooler bracket to instrument panel reinforcement screws, disconnect the outlet from the duct and remove the outlet and bracket assembly.
8. Disconnect speedometer cable.
9. Remove instrument panel to lower reinforcement attaching screws. Position the instrument panel upwards to gain access to the air distribution duct.
10. Disconnect evaporator inlet and outlet lines. Immediately cap all openings.
11. Remove blower-evaporator support bracket to door pillar and forward engine housing screws, then lower the blower-evaporator case.
12. Disconnect electrical connections and pull evaporator drain hose through dash panel.
13. Remove blower-evaporator assembly from vehicle.
14. Remove heater distributor duct to case screws and pull duct rearward from heater case retainer.
15. Remove heater core case to dash screws, then the heater case and core assembly. Tilt top of case rearward until core tubes clear dash openings.
16. Remove heater core retaining straps and the heater core.
17. Reverse procedure to install.

1974-77 El Camino & Sprint Less A/C

1. Disconnect battery ground cable.
2. Drain radiator, disconnect heater hoses at core and plug openings to prevent spillage of water.
3. Remove nuts from air distributor duct studs on engine side of firewall.
4. Remove distributor duct retaining screw.
5. Pull distributor duct from firewall being careful not to bend cable.
6. Remove core assembly from duct.
7. Reverse procedure to install.

1974-77 El Camino & Sprint With A/C

1. Disconnect ground cable and drain radiator.
2. Disconnect heater hoses from core and plug hoses and core openings.
3. Remove distributor case stud nuts projecting through firewall. On 1974-76 models, remove resistor assembly and the last stud nut through its opening.
4. Remove glove box door and glove box.
5. On all models, remove right hand lap cooler and kick pad cover.
6. Disconnect center duct hoses and remove center duct to selector duct screws, then remove center duct.
7. Remove floor distributor duct.
8. Remove lower right hand distributor duct to dash panel screw.
9. On all models, remove air selector retaining screws, then lower selector assembly. Disconnect all wiring, vacuum lines and cables.
10. On all models, remove core and housing from selector assembly, then core from housing.
11. Reverse procedure to install.

1974-77 "G" Series Less A/C

1. Disconnect battery ground cable and drain cooling system.
2. Disconnect heater hoses from core tubes and plug core openings.
3. Disconnect right hand air distributor hose from heater case and position aside.
4. Disconnect bowden cable from temperature door.
5. Remove distributor duct to heater case screws and pull duct rearward from heater case retainer.
6. Remove heater case to dash screws, then the case from vehicle, Fig. 163.
7. Remove heater core retaining strap screws and the heater core from case.
8. Reverse procedure to install.

1974-80 10-30 Series Exc. "G" Series Less A/C

1. Disconnect battery ground cable and drain cooling system.
2. Disconnect heater hoses from core tubes and plug core openings.
3. From engine compartment, remove nuts from distributor duct studs.
4. From inside vehicle, remove glove box and door assembly.
5. Disconnect cables from air-defrost and temperature doors.
6. Remove the floor outlet and the defroster duct to heater distributor duct screw.
7. Remove heater distributor to dash panel screws and pull assembly rearward, then disconnect wiring harness.
8. Remove heater distributor from vehicle, Fig. 164.
9. Remove the heater core retaining straps and the heater core.
10. Reverse procedure to install.

1974-80 10-30 Series Exc. "G" Series With A/C

1. Disconnect battery ground cable and drain radiator.
2. Disconnect heater hoses from core and plug core openings.
3. Remove glove box and door assembly.
4. Remove center duct to selector duct and the instrument panel screws, then the center lower and center upper ducts.
5. Disconnect bowden cable from temperature door.
6. From engine compartment, remove nuts from the three selector duct studs.
7. From inside vehicle, remove selector duct to dash panel screw.
8. Pull selector duct assembly rearward until the core tube clears dash panel, then lower the selector assembly to gain access to the vacuum and electrical harness.
9. Disconnect the vacuum and electrical harness and remove the selector duct assembly.
10. Remove the heater core mounting strap screws and the heater core.
11. Reverse procedure to install.

1974-80 40-60 Series With A/C

1. Disconnect battery ground cable and drain radiator.
2. Disconnect heater hoses from core and plug core openings.
3. From engine compartment, remove nuts from heater case studs.
4. Remove glove box and the outlet hoses.
5. Remove air distributor from heater case and position distributor away from case, Fig. 155.
6. Disconnect temperature door cable and remove heater case to firewall retaining screws.
7. Pull heater case from firewall and remove from vehicle.
8. Remove screws securing heater core mounting straps. Remove heater core.
9. Reverse procedure to install.

1974-80 40-60 Series, Conventional Cab Less A/C

1. Disconnect battery ground cable and drain cooling system.
2. Disconnect heater hoses from core tubes and plug core openings.
3. Remove heater and defroster duct and tubes from beneath instrument panel.

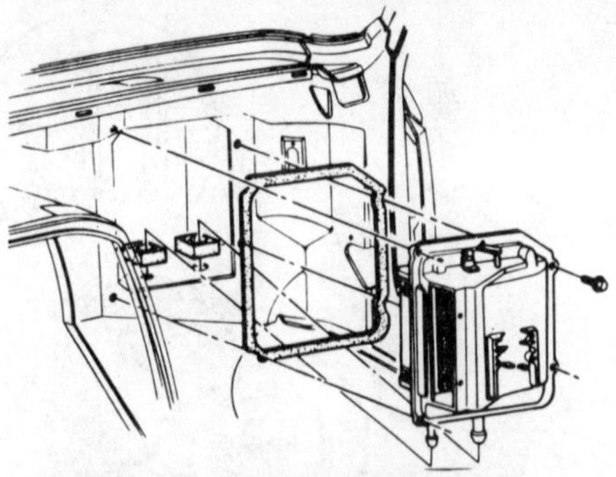

Fig. 163 Heater core less A/C. 1974-77 Chevrolet & GMC G Series

4. Open glove box door and pull downward on inside glove box door panel to clear instrument panel sheet metal.
5. Disconnect door control cable from heater assembly.
6. Remove heater to firewall retaining screws from both sides of firewall.
7. Pull heater assembly from firewall and remove from vehicle.
8. Remove core mounting clamp at core tube end and loosen clamp at opposite end.
9. Remove heater core from assembly.
10. Reverse procedure to install.

1974-80 40-60 Series, Tilt Cab Less A/C

1. Disconnect battery ground cable and drain cooling system.
2. Disconnect heater hoses from core tubes and plug core openings.
3. Separate defroster air duct tubes from air distributor.
4. Disconnect wiring from blower resistor and blower motor.
5. Remove screws securing air distributor to cab and blower duct asssembly.
6. Remove air distributor.
7. Remove screws securing heater core housing to outside air inlet duct.
8. Lower heater assembly from dash and remove from vehicle, Fig. 165.
9. Remove screws securing heater core to housing, then the heater core.
10. Reverse procedure to install.

1975-77 "G" Series With A/C

1. Remove battery and engine cover.
2. Remove blower-evaporator shield and the shield bracket.
3. Remove left floor outlet deflector and bracket.
4. Loosen steering column to instrument panel reinforcement screws and remove one screw.
5. Disconnect speedometer cable.
6. Remove instrument panel to lower reinforcement attaching screws.
7. Position instrument panel assembly rearward and disconnect radio antenna lead and electrical connectors.
8. Support instrument panel at right visor.
9. Disconnect electrical connector at brake switch.
10. Remove blower-evaporator support bracket to door pillar and forward engine housing attaching screws. Move assembly rearward for clearance.
11. Drain cooling system and disconnect heater hoses from core. Plug core openings.
12. Remove air inlet valve assembly and disconnect temperature door cable at heater case.
13. Remove heater assembly from vehicle, then the heater core from case.
14. Reverse procedure to install.

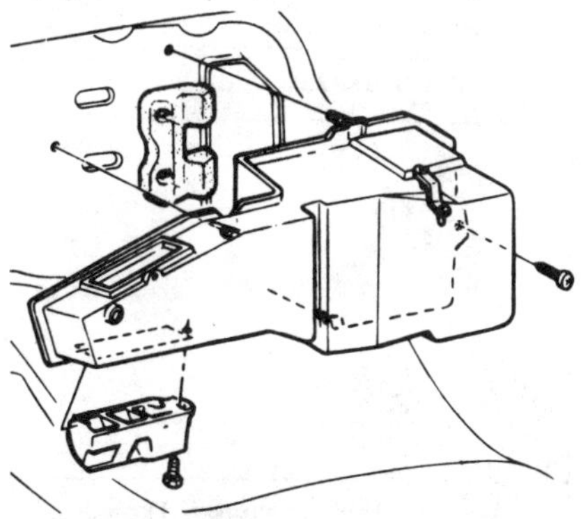

Fig. 164 Heater core less A/C. 1974-80 Chevrolet & GMC 10-30 Series except G Series

1978 "G" Series Less A/C

1. Disconnect battery ground cable.
2. Place a clean pan under the vehicle and then disconnect the heater core hoses. Plug hoses and support them in a raised position. Allow coolant from the heater core to drain into the pan.
3. Remove heater distributor duct to distributor case retaining screws and distributor duct to engine cover screw and remove duct.
4. Remove defroster to distributor case retaining screw.
5. Disconnect temperature door cable and fold cable back for access, then remove the three nuts at engine compartment side of distributor case and one screw on passenger side. Remove heater case and core as a unit. Tilt assembly rearward at top while lifting up until core tubes clear the dash openings.
6. Remove heater case and core as a unit. Tilt assembly rearward at top while lifting up until core tubes clear the dash openings.
7. Remove heater core retaining straps and remove core from heater case.
8. Reverse procedure to install.

1978-80 Caballero & El Camino Less A/C

1. Disconnect battery ground cable and drain cooling system.
2. Disconnect heater core hoses and electrical connectors.
3. Remove front module cover screws.
4. Remove heater core, Fig. 107.
5. Reverse procedure to install.

1978-80 Caballero & El Camino With A/C

1. Disconnect battery ground cable and drain cooling system.
2. Disconnect heater core hoses.
3. Remove retaining bracket and ground strap.
4. Disconnect module rubber seal and remove module screen.
5. Remove right hand windshield wiper.
6. Remove diagonal connector, hi-blower relay and thermostatic switch mounting screws.
7. Disconnect all electrical connectors at module top.
8. Remove module top cover and heater core, Fig. 98.
9. Reverse procedure to install.

1978-80 "G" Series, Less A/C

1. Disconnect battery ground cable.

2. Remove coolant recovery tank and position it aside.
3. Place a clean pan under the vehicle and then disconnect the heater core hoses. Plug hoses and support them in a raised position. Allow coolant from the heater core to drain into the pan.
4. Remove heater distributor duct to distributor case retaining screws and distributor duct to engine cover screw and remove duct.
5. Remove engine housing cover.
6. Remove instrument panel retaining screws. Remove all screws at top of instrument panel along windshield, all lower screws and right lower instrument panel support bracket at door pillar and engine housing.
7. Lower the steering column, then raise and support right side of instrument panel.
8. Remove defroster duct to distributor case retaining screw and the two screws retaining distributor to heater case.
9. Disconnect temperature door cable and fold back for access, then remove the three nuts at engine compartment side of distributor case and one screw from passenger side.
10. Remove heater case and core as a unit. Tilt assembly rearward at top while lifting up until core tubes clear dash openings.
11. Remove heater core retaining strap screws and remove core from case.
12. Reverse procedure to install.

1978-80 "G" Series With A/C

1. Disconnect battery ground cable.
2. Remove engine cover.
3. Remove steering column to instrument panel retaining bolts and lower the column.
4. Remove upper and lower instrument panel retaining screws and radio support bracket retaining screw.
5. Raise and support right side of instrument panel, then remove right lower instrument panel support bracket.
6. Remove recirculating air door vacuum actuator, then disconnect temperature cable and vacuum hoses at distributor case.
7. Remove heater distributor duct, then the two defroster duct to dash panel retaining screws below the windshield.
8. Disconnect and plug heater core hoses.
9. Remove three nuts and one screw retaining heater core case to dash panel.
10. Remove air distributor assembly from vehicle.
11. Remove gaskets to expose screws retaining sections together.
12. Remove temperature cable support bracket.
13. Remove case retaining screws, then separate the case and remove heater core.
14. Reverse procedure to install.

BLOWER MOTOR, REPLACE

1974-80 Chevelle, Malibu & Monte Carlo; 1977-80 Chevrolet; 1977-80 Camaro; 1980 Citation

1. Disconnect battery ground cable.
2. Disconnect blower lead wire at motor.
3. Remove blower to case screws and remove blower, Figs. 96, 98, 100, 102, 106, 107, 151 and 166.

1974-80 Corvette

1. Disconnect battery ground cable.
2. Remove radiator supply tank, if equipped. On 1980 models, move A/C compressor aside, remove front and right side air cleaner ducts, then move air cleaner to left side for access.

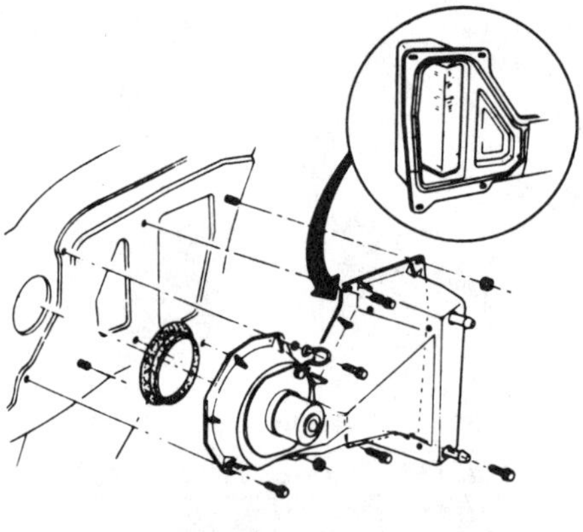

Fig. 165 Heater core & blower motor less A/C. 1974-80 Chevrolet & GMC 40-60 Series Tilt Cab

3. Disconnect blower motor wires.
4. Remove blower motor to case mounting screws and remove motor, Figs. 149, 150 and 166.

1974-76 Chevrolet; 1974-76 Camaro; 1974-79 Nova

1. Disconnect battery ground cable.
2. Disconnect blower lead and remove all hoses and wires connected to fender skirt then raise vehicle.
3. On all except Nova, remove all fender skirt retaining bolts except those retaining skirt to radiator support. On Nova, remove the eight rearmost fender skirt to fender retaining screws.
4. Pull out and down on fender skirt and wedge a 2x4 inch block of wood between skirt and fender to allow room for blower motor removal.
5. Remove blower to case retaining screws and remove blower assembly. Gently pry flange, since sealer will act as an adhesive, Figs. 93, 94, 95, 148 and 166.

NOTE: On Nova, remove blower retaining nut and separate wheel and motor before removing through fender and skirt.

6. Reverse procedure to install.

1975-80 Monza

1. Disconnect battery ground cable and disconnect blower motor electrical leads.
2. Scribe blower motor flange to case position.
3. Remove blower motor retaining screws and remove blower motor and wheel assembly, Fig. 99. Pry flange gently, if sealer acts as an adhesive.
4. Reverse removal procedure to install, aligning scribe marks made during removal.

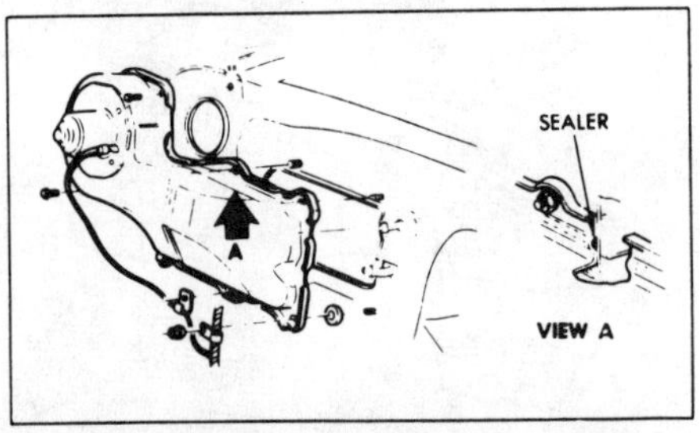

Fig. 166 Blower motor (Typical). Except Chevette, Monza, Vega, 1977-80 Chevrolet Full Size & 1978-80 Malibu & Monte Carlo

1976-80 Chevette

1. Disconnect battery ground cable.
2. Disconnect blower lead wire.
3. Remove blower motor to case attaching screws then the blower and wheel as an assembly.

NOTE: Scribe mark on blower motor and case so motor is installed in original position.

4. Remove nut and separate motor from wheel.
5. Reverse procedure to install.

NOTE: At assembly position open end of blower wheel away from blower motor. Replace sealer at blower motor flange if necessary.

Chevrolet & GMC Trucks

1974-77 El Camino & Sprint

1. Disconnect battery ground cable.
2. Disconnect blower lead wire at motor.
3. Remove blower to case screws and remove blower, Figs. 96 and 166.
4. Reverse procedure to install.

1974-77 10-30 Series Except "G" Model; 1974-80 40-60 Series Except Tilt Cab

1. Disconnect battery ground cable.
2. On some models, it is necessary to support right front of hood in raised position and remove the right hood hinge. This provides clearance for blower motor removal.
3. Disconnect blower motor lead wire at blower flange terminal.
4. Mark relationship between blower motor flange and case. If equipped, disconnect rubber cooling tube from motor.
5. Remove blower motor mounting screws and the blower motor assembly, Figs. 152 and 155. Pry the flange away from case if sealer acts as an adhesive. On some models, rotate the blower motor until flange clears hood hinge.
6. Remove nut securing blower wheel to motor shaft and separate the assembly.
7. Reverse procedure to install.

1974-77 "G" Series

1. Disconnect battery ground cable and discharge refrigerant system.
2. Remove screws securing the blower-evaporator shield and the shield. Remove the shield bracket to instrument panel reinforcement screws, then the bracket.
3. Remove heater intermediate duct to A/C distributor duct screw, then the duct.
4. Remove the engine cover.
5. Loosen both steering column instrument panel reinforcement screws and remove one screw.
6. Remove left foot cooler bracket to instrument panel reinforcement screws, disconnect the outlet from the duct and remove the outlet and bracket assembly.
7. Disconnect the speedometer cable.
8. Remove instrument panel to lower reinforcement attaching screws. Position the instrument panel upward to gain access to the air distribution duct.
9. Disconnect evaporator inlet and outlet lines, then immediately plug all open connections.
10. Remove blower-evaporator support bracket to door pillar and the forward engine housing screws, then lower the blower-evaporator assembly.

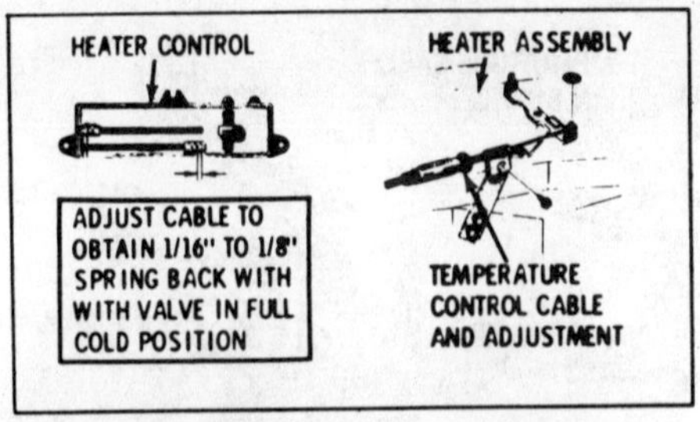

Fig. 167 Temperature control cable adjustment. (Typical). 1974-76 88, 98, 1974-77 Cutlass & 1974-80 Toronado less Tempmatic

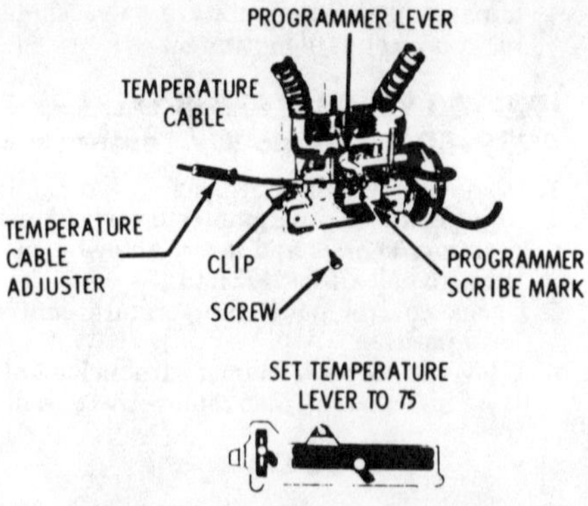

Fig. 168 Temperature control cable adjustment. 1974-76 88, 98, 1974-77 Cutlass & 1974-78 Toronado with Tempmatic

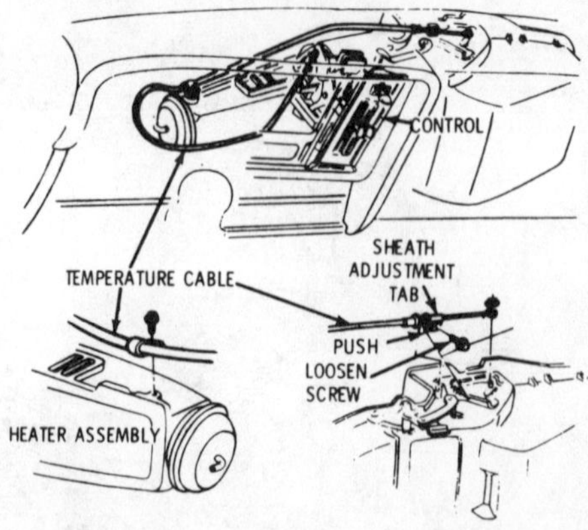

Fig. 169 Temperature control cable adjustment. 1978-79 Omega with A/C

11. Disconnect electrical connections and pull evaporator drain hose through dash panel.
12. Remove blower-evaporator assembly from vehicle, Fig. 153.
13. Separate front and rear case halves.
14. Remove blower motor retaining strap and insulator strip.
15. Remove blower motor.
16. Reverse procedure to install.

1974-80 40-60 Series, Tilt Cab

1. Perform steps 1 to 8 outlined under "Heater Core, Replace" 1974-79 40-60 Series, Tilt Cab procedure.
2. Remove screws securing blower motor to heater core housing, then the blower motor, Fig. 165.
3. Reverse procedure to install.

1978-80 10-30 Series

1. Disconnect battery ground cable.
2. On G series, remove coolant recovery tank and power antenna.
3. On C and K series, remove blower cooling tube.
4. Remove the blower motor retaining screws then remove blower motor. If necessary, pry gently on blower motor flange to remove, Figs. 152 and 154.
5. Reverse procedure to install.

1978-80 Caballero & El Camino

1. Disconnect battery ground cable.
2. On vehicles with air conditioning, remove blower motor cooling tube.
3. Disconnect blower motor electrical connectors.
4. Remove blower motor retaining screws, then remove blower motor assembly, Figs. 96 and 107.
5. Reverse procedure to install.

Oldsmobile

TEMPERATURE CONTROL CABLE ADJUSTMENT

1974-76 88, 98, 1974-77 Cutlass & 1974-80 Toronado Less Tempmatic

If properly adjusted, there must be 1/16 to ⅛ inch control lever spring-back when control lever is set at full cool position, Fig. 167. If necessary adjust temperature control cable turnbuckle.

1974-76 88, 98, 1974-77 Cutlass & 1974-78 Toronado W/ Tempmatic

1. If new cable is being installed, attach cable to control.
2. Set temperature lever to 75°.
3. Align scribe mark on programmer lever with scribe on programmer.
4. Turn adjusting nut on cable so that eyelet goes on programmer lever, Fig. 168.

1974-79 Omega

1974-77 All & 1978-79 Less A/C

1. Connect temperature control cable at control lever.
2. Place temperature control lever on cold, then move temperature door to the fully closed position.
3. Attach cable to temperature door lever and tighten cable bracket screw, then check for proper operation.

1978-79 With A/C

1. Remove glove compartment, then loosen cable adjusting screw.
2. Move temperature control lever to cold position.
3. Push slotted end of cable until door seats and lash is removed from cable, Fig. 169.

1975-80 Starfire

The temperature control cable is non-adjustable, but cable and door should be checked for freedom of operation and proper routing. When installing cable, position temperature control lever on cold and temperature door to the fully closed position.

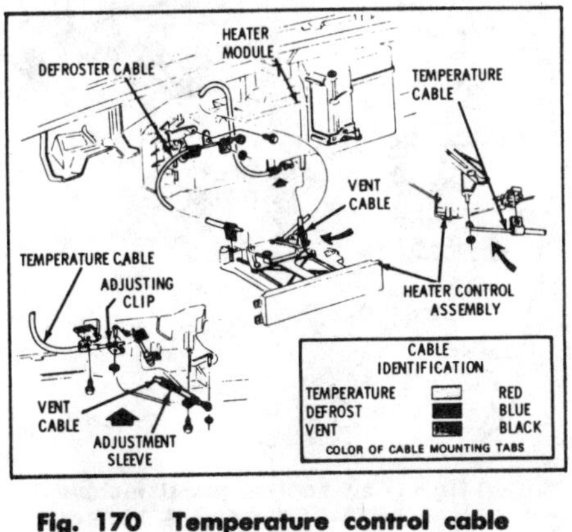

Fig. 170 Temperature control cable adjustment. 1977-80 88 & 98 less A/C

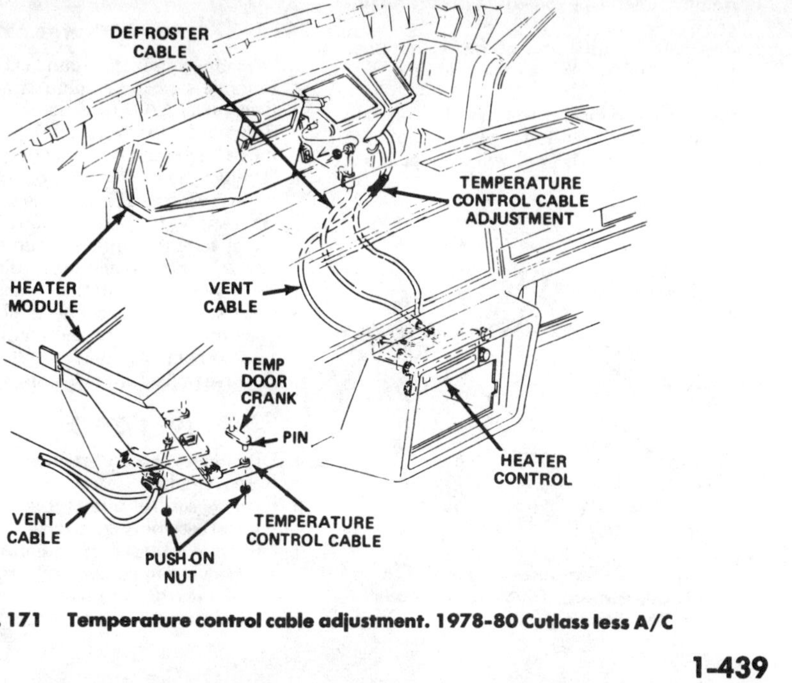

Fig. 171 Temperature control cable adjustment. 1978-80 Cutlass less A/C

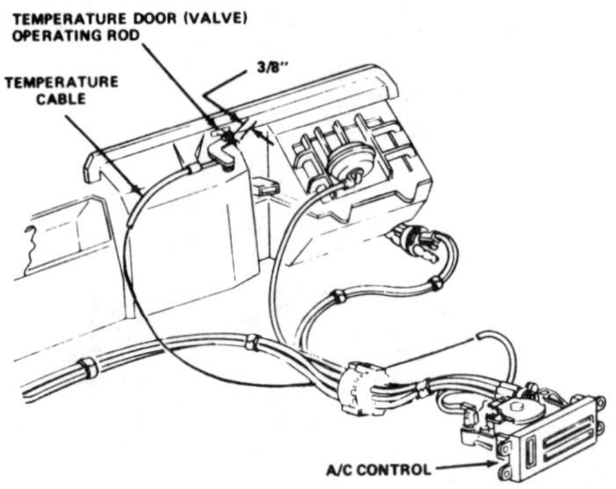

Fig. 172 Temperature control cable adjustment. 1977-80 88, 98 & 1978-80 Cutlass with A/C less Tempmatic

1977-80 88, 98 & 1978-80 Cutlass Less Tempmatic

88 & 98 Less A/C

The temperature valve cable is self-adjusting. However, an initial setting should be made to obtain correct adjustment.

1. Slide the clip to the heater end of the cable away from the end of the cable, Fig. 170.
2. Move the levers at the control to full "Hot" position. This adjusts cable for proper operation of valve.

Cutlass Less A/C

1. Move temperature lever to full "Cold" position, then rotate temperature door crank clockwise to full cold.
2. Remove glove box, reach through glove box opening and adjust cable until loop aligns with pin on crank. Install nut on pin, Fig. 171.
3. Move temperature control lever to full "Hot" position. Door should be heard hitting its seat. If not repeat procedure.

With A/C

The temperature valve operating rod is self-adjusting. However, an initial setting should be made to obtain correct adjustment.

1. Remove operating rod out of plastic retainer.
2. Position threaded end of rod on retainer with end of rod ⅜ inch past retainer, Fig. 172.
3. Snap rod into plastic retainer, then operate temperature lever full travel in hot and cold direction. The rod will ratchet in retainer and the temperature valve should hit at both ends of movement.

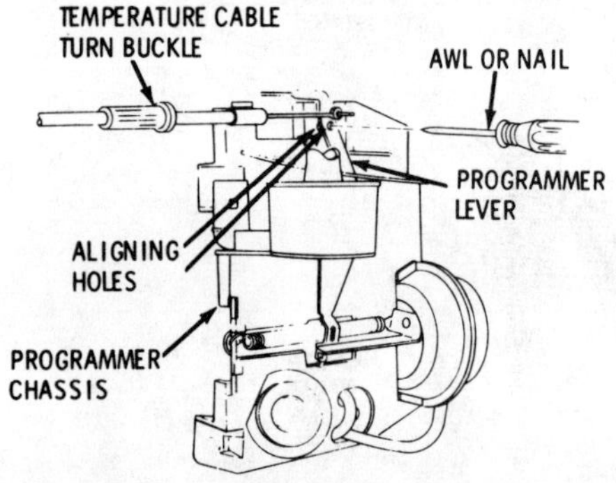

Fig. 174 Temperature control cable adjustment. 1979-80 Toronado W/Tempmatic

1977-80 88, 98, 1978-80 Cutlass & 1979-80 Toronado W/ Tempmatic

1. Remove glove compartment.
2. Align holes in programmer lever and programmer chassis and insert an awl or nail through holes, Figs. 173 and 174.
3. Check control panel temperature control lever position.
4. If lever is not at 75, adjust turnbuckle until lever indicates 75, then remove awl or nail.

1980 Omega

Less A/C

The temperature cable incorporates a slider type self-adjust feature, Figs. 87 and 88. As the control head temperature lever is moved through its full range of travel, the cable clip will assume a position assuring that the temperature valve will seat in the full hot and full cold positions.

With A/C

1. Connect cable assembly to control head, Fig. 86.
2. Place temperature lever in "Off" position.
3. Connect cable to temperature door post and push sheath toward lever until door seats and all lash is removed from cable and control head lever.
4. Tighten cable securing screw.

SLAVE DOOR ADJUSTMENT

1974-76 all, 1977 Cutlass & 1977-78 Toronado W/ Tempmatic

On Bench

1. Loosen slave door adjuster screw.
2. Apply vacuum to lower mode door diaphragm.
3. Position slave door to right against heater case and tighten link screw.

CONTROL PANEL, REPLACE

1974-77 Intermediates Exc. Omega

1. Disconnect battery ground cable.
2. Remove steering column trim screws and disconnect A/C outlet hose.
3. Remove knobs from A/C control and remove knobs and hex head nuts from radio.
4. Remove the four control panel trim retaining screws, Fig. 175, and one screw from ash tray support, then carefully pull control panel assembly from instrument panel until A/C control panel can be disconnected.
5. Remove A/C control panel retaining screws, then disconnect electrical and vacuum connectors and control cable and remove control panel.
6. Reverse procedure to install.

1974-79 Omega

1. Disconnect battery ground cable.
2. Remove radio.
3. Remove control retaining screws and lower control assembly from dash.
4. Disconnect temperature control cable and electrical connectors and remove control panel, Figs. 139 and 140.
5. Reverse procedure to install.

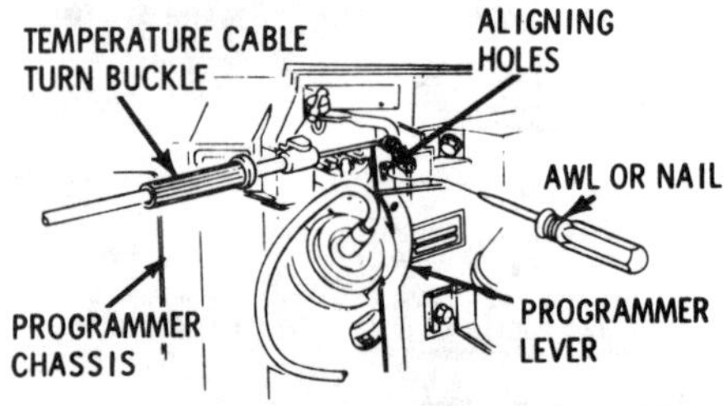

Fig. 173 Temperature control cable adjustment. 1977-80 88, 98 & 1978-80 Cutlass W/Tempmatic

1974-76 Full Size & 1977-78 Toronado

1. Disconnect battery ground cable.
2. Remove lower trim panel and disconnect temperature control cable from control panel.
3. Remove headlamp switch knob by inserting a sharp pointed tool under retainer located in notch in base of knob and pulling knob off shaft.
4. Insert a screwdriver in lower edge of applique and carefully pry applique from clips in cluster assembly.
5. Remove control panel retaining screws, Fig. 176, then pull control panel out from instrument panel and disconnect vacuum and electrical connectors and remove control panel.
6. Reverse procedure to install.

1975 Starfire

1. Disconnect battery ground cable.
2. Remove clock set knob, then remove the three instrument cluster panel screws and remove panel.
3. Remove four screws from left side of pad, two from top of cluster opening and two below cluster opening.
4. Remove glove box, then remove the four nuts retaining right side of pad and remove instrument panel pad.
5. Remove steering column retaining nuts and lower steering column.

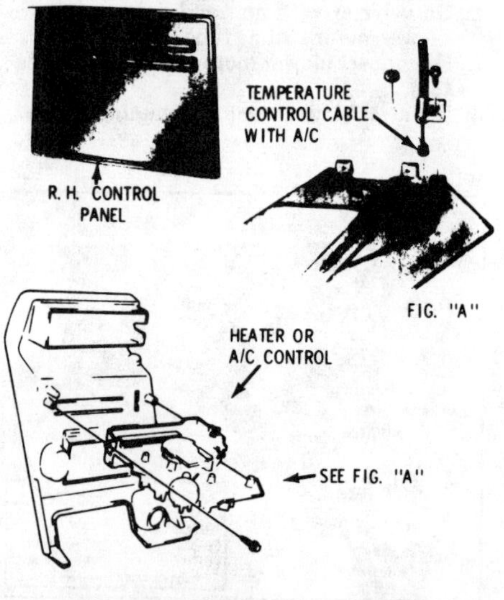

Fig. 175 Control panel installation. 1974-76 Full Size & 1977-78 Toronado

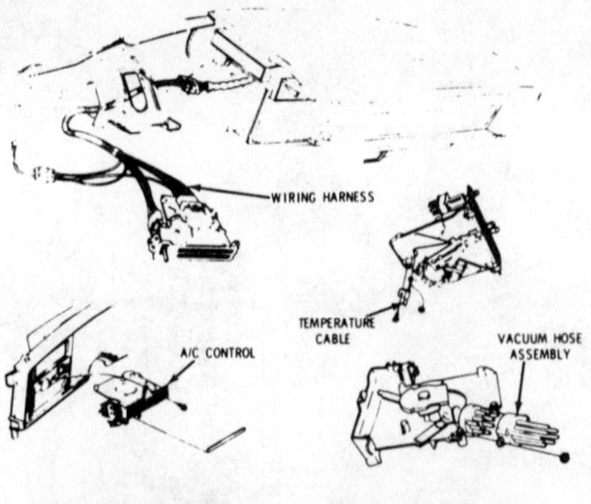

Fig. 176 Control panel installation. 1974-77 Cutlass

6. Remove one screw from inside upper column bracket, then disconnect cluster housing and pull rearward.
7. Disconnect control cables from control panel and remove panel, Fig. 89.
8. Reverse procedure to install.

1976-80 Starfire

1. Disconnect battery ground cable.
2. Remove right and left hand dash A/C outlets and remove the eleven screws from around edge of instrument panel pad.
3. Remove left A/C duct and lower outlet, then pry instrument panel pad to disengage clips and remove pad.
4. Remove steering column retaining nuts and lower steering column onto driver's seat.
5. Remove instrument panel retaining screws, then place a protective covering over steering column and lower instrument panel onto steering column.
6. Remove control panel retaining screws, Fig. 89, then disconnect control cable, electrical and vacuum connectors and remove control panel.
7. Reverse procedure to install and torque steering column retaining nuts to 25 ft. lbs.

1978-80 Cutlass

1. Disconnect battery ground cable.
2. Remove instrument panel lower trim cover.
3. Remove control mounting screws. Pull control out enough to disconnect all electrical, cable, and vacuum connections, Fig. 177.
4. Remove control.

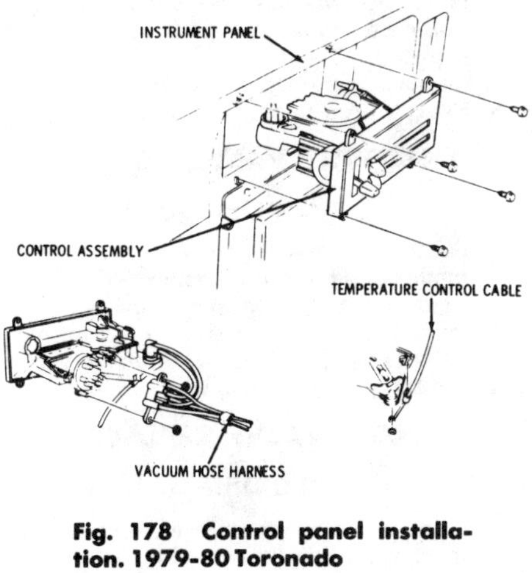

Fig. 178 Control panel installation. 1979-80 Toronado

1977-80 88 & 98

1. Remove right hand trim cover.
2. Remove control mounting screws, then pull control out just enough to disconnect vacuum and electrical connections and temperature door cable.
3. Reverse procedure to install. After installation, check cable adjustment.

1979-80 Toronado

1. Remove left hand trim cover.
2. Remove attaching screws to cluster assembly.
3. Pull control out enough to disconnect wiring, vacuum lines and temperature control cable, then remove control, Fig. 178.

PROGRAMMER, REPLACE

1974-76 Full Size & 1977-78 Toronado

1. Disconnect battery ground cable.
2. Remove right hand lower A/C duct assembly.
3. Remove temperature cable retaining screw and clip.
4. Disconnect aspirator hoses and vacuum hose.
5. Remove output retaining clip at link, Fig. 179, then remove programmer retaining screws and programmer.
6. Reverse procedure to install.

1975-78 Cutlass

1. Disconnect battery ground cable.
2. Remove glove box and disconnect heater cable from programmer.
3. Remove programmer retaining screws, then disconnect actuator rod from heater door, Fig. 179, vacuum lines and aspirator hose from programmer and remove programmer.
4. Reverse procedure to install.

1977-78 88, 98 & 1978 Cutlass

1. Disconnect battery ground cable and remove glove box.

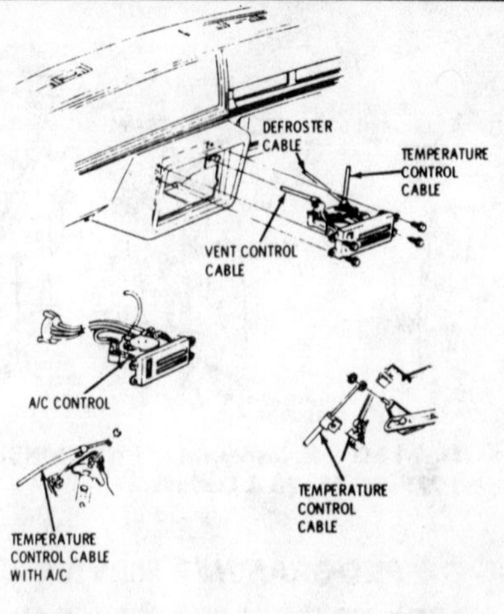

Fig. 177 Control panel installation. 1978-80 Cutlass

2. Disconnect temperature control cable and vacuum harness at programmer.
3. Disconnect in-car source hose and the temperature door link threaded rod at programmer.
4. Remove retaining screws and remove programmer, Fig. 180.
5. Reverse procedure to install.

1979-80 All

1. On full size models, disconnect battery ground cable and remove glove box. On Toronado and Cutlass remove instrument panel sound absorber.
2. Remove vacuum and aspirator hose, then temperature door rod link from crank.
3. Remove temperature cable from programmer and programmer retaining screws.
4. Remove programmer from A/C heater case, Figs. 180 and 181.
5. Reverse procedure to install.

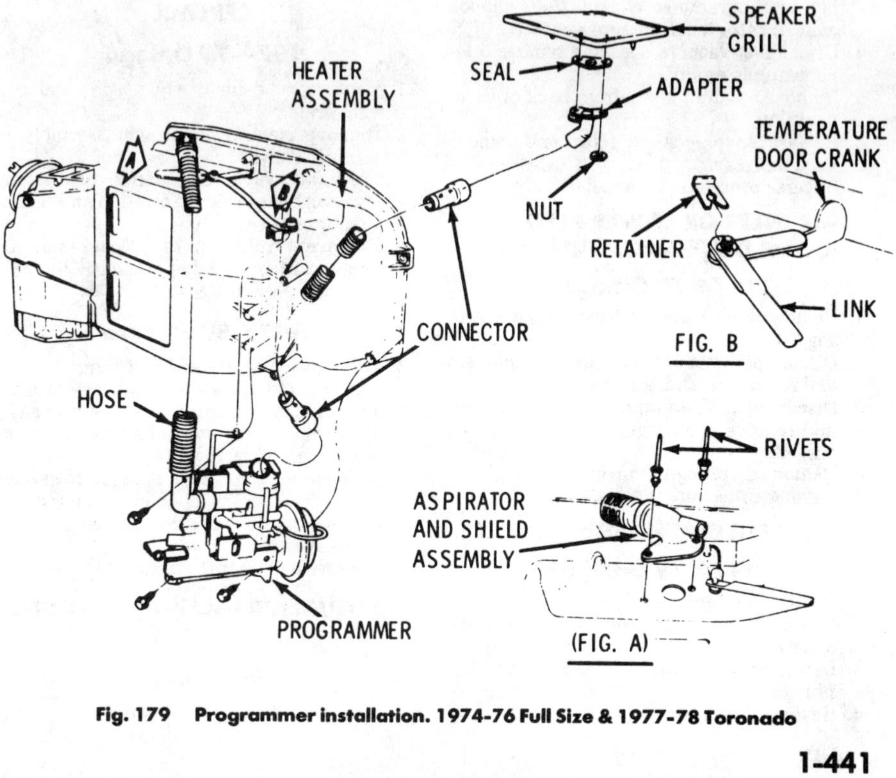

Fig. 179 Programmer installation. 1974-76 Full Size & 1977-78 Toronado

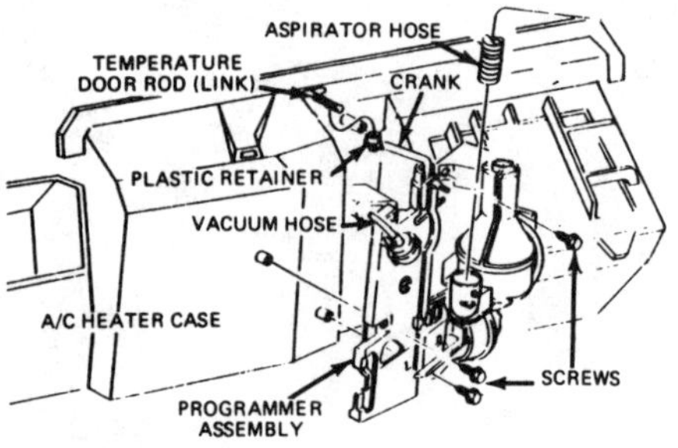

Fig. 180 Programmer installation. 1977-80 88, 98 & Cutlass

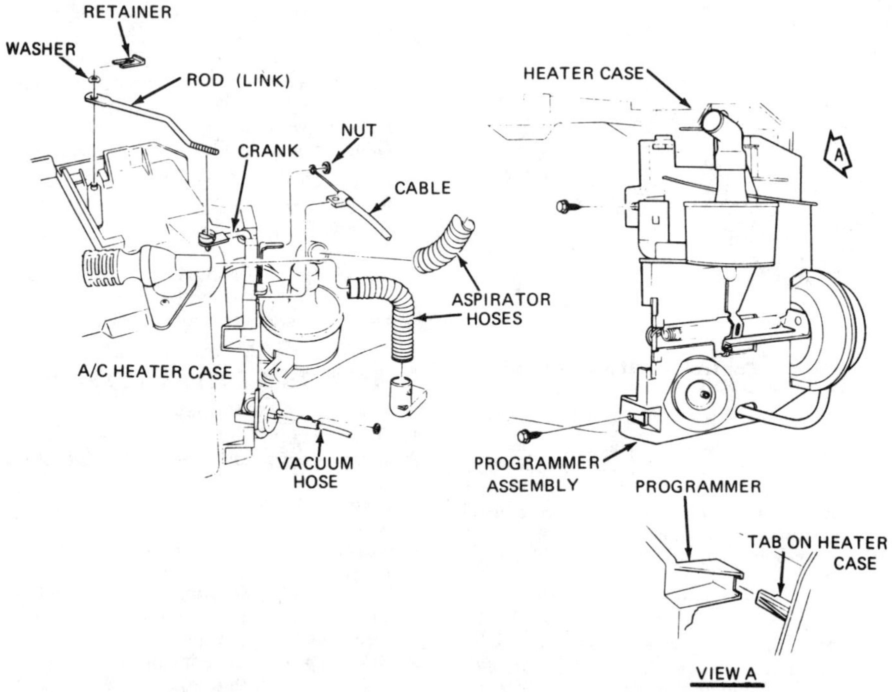

Fig. 181 Programmer installation. 1979-80 Toronado

PROGRAMMER SERVICE

1974-78 Exc. 1977-78 Full Size, 1978 Cutlass & 1979-80 Toronado

1. Remove programmer.
2. Remove screw from cover adjacent to power spring.
3. Lift cover out of latch point and remove, then disengage and remove power spring.
4. Remove screws from output shaft retainer and remove retainer.
5. Remove vacuum modulator retaining screws and remove modulator.
6. Remove nuts from vacuum motor frame and remove vacuum motor and output shaft.
7. Reverse procedure to install.

1977-80 88, 98, 1978-80 Cutlass & 1979-80 Toronado

1. Remove programmer.
2. Disconnect vacuum hose at motor, unhook power spring from programmer chassis first, then from motor. Pull motor straight out from programmer chassis, disconnect motor mechanism from programmer linkage and remove motor.
3. Remove screw holding compensator assembly to programmer chassis then unhook compensator link from programmer.
4. Disconnect vacuum hose and remove compensator assembly.
5. Remove vacuum hoses from back of vacuum relay.
6. Release snap lock on relay and remove vacuum relay.
7. Reverse procedure to install.

AIR INLET OR LOWER PLENUM DOOR DIAPHRAGM, REPLACE

1974-79 Omega

1. Remove kick panel and diaphragm cover, Fig. 182.
2. Disconnect vacuum hose from diaphragm and remove retaining screws.
3. Disconnect link and spring at recirc air door and remove diaphragm and bracket as an assembly.
4. Remove diaphragm from bracket.
5. Reverse procedure to install.

DEFROSTER DIAPHRAGM, REPLACE

1974-79 Omega

1. Remove heater outlet duct.
2. Remove actuator and link retaining screws.
3. Disconnect vacuum hose and remove diaphragm.
4. Reverse procedure to install.

1980 Omega

1. Remove defroster nozzle, then remove mode valve housing assembly.
2. Remove the two defroster diaphragm retaining screws.
3. Release actuator rod and remove diaphragm.
4. Reverse procedure to install.

PLENUM DOOR DIAPHRAGM, REPLACE

1974-79 Omega

1. Remove windshield wiper arms and cowl vent grille.
2. Remove plenum diaphragm assembly retaining screws.
3. Disconnect vacuum hose from diaphragm.
4. Position door in closed position and remove diaphragm assembly.
5. Remove diaphragm to bracket and link screws and remove diaphragm.
6. Reverse procedure to install.

1975-80 Starfire

1. Disconnect battery ground cable.
2. Disconnect vacuum hose from actuator and pry off actuator arm to door retaining clip, then remove actuator nuts and remove actuator assembly, Fig. 146.
3. Remove air inlet door housing to plenum screws and remove door and actuator, then remove hi-blower switch screws and remove switch.
4. Reverse procedure to install.

PLENUM AIR INLET DIAPHRAGM, REPLACE

1980 Omega

1. Remove windshield wipers, lower windshield molding and close out panel.
2. Remove the three screws securing plenum valve assembly, then disconnect vacuum hose and remove plenum valve.
3. Remove the two nuts securing diaphragm to plenum valve and remove diaphragm.
4. Reverse procedure to install.

RECIRCULATION DIAPHRAGM, REPLACE

1980 Omega

1. Disconnect vacuum hose, then remove max recirculation door and diaphragm assembly.
2. Remove the two nuts securing diaphragm to door assembly, them remove diaphragm.
3. Reverse procedure to install.

MODE DIAPHRAGM, REPLACE

1980 Omega

1. Disconnect vacuum hose and linkage, then remove the two nuts securing diaphragm to support.
2. Remove diaphragm assembly.
3. Reverse procedure to install.

DEFROSTER VALVE, REPLACE

1980 Omega

1. Remove defroster nozzle, then remove mode valve housing assembly.
2. Remove defroster vacuum diaphragm.
3. Push defroster valve pivot rod from mode valve housing and remove defroster valve.
4. Reverse procedure to install.

MODE VALVE, REPLACE

1980 Omega

1. Remove defroster nozzle, then remove mode valve housing assembly.

2. Remove mode valve diaphragm from adapter plate.
3. Remove adapter plate from mode valve housing.
4. Pull mode valve pivot rod from mode valve housing, then remove mode valve.
5. Reverse procedure to install.

TEMPERATURE VALVE, REPLACE

1980 Omega

1. Remove defroster nozzle, then remove mode valve housing.
2. Disconnect temperature control cable and slide valve pivot rod towards passenger side of vehicle. Remove valve.
3. Reverse procedure to install, ensuring temperature door cable is properly adjusted.

BLOWER SWITCH, REPLACE

1974-76 Cutlass

1. Remove right hand control panel trim.
2. Disconnect electrical connector and remove switch knob.
3. Remove retaining screws and remove switch.
4. Reverse procedure to install.

BLOWER, VACUUM & MASTER SWITCHES, REPLACE

1975-76 Starfire

1. Remove instrument panel as outlined under "Control Panel."
2. Disconnect electrical or vacuum connector from switch.
3. Remove retaining screws and remove switch.
4. Reverse procedure to install.

HEATER CORE, REPLACE

1974-77 Cutlass

Less Air Conditioning

1. Disconnect battery ground cable and drain radiator.
2. Disconnect heater hoses and plug hoses and core openings to prevent coolant loss.
3. Remove heater case attaching nuts and screws, disconnect control cables and remove case from dash, then core from case, Fig. 104.
4. Reverse procedure to install.

1974-79 Omega

Less Air Conditioning

1. Disconnect battery ground cable and drain radiator.
2. Disconnect heater hoses and plug hoses and core openings to prevent coolant loss.
3. Remove heater case retaining nuts from engine side of dash, Fig. 105.
4. Remove glove box and door, then pull heater case from dash.
5. Disconnect blower resistor electrical connector and control cables from case, then remove heater case from dash and core from case.
6. Reverse procedure to install.

1974-77 Cutlass

With Air Conditioning

1. Disconnect battery ground cable and drain radiator.

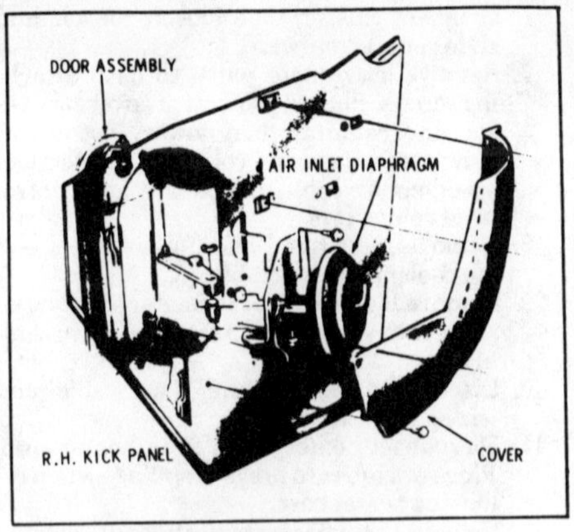

Fig. 182 Air inlet or lower plenum diaphragm installation. 1974-79 Omega, Phoenix & Ventura

2. Remove glove box and the center A/C manifold.
3. Disconnect vacuum hoses and temperature cable from heater case.
4. Disconnect heater hoses and plug hoses and core openings to prevent coolant loss.
5. Remove heater case to dash panel attaching bolts, then heater case from dash and core from case, Fig. 104.
6. Reverse procedure to install.

1974-79 Omega

With Air Conditioning

1. Disconnect battery and coolant.
2. Disconnect upper heater hose from core.
3. Remove right front fender skirt bolts and lower skirt to gain access to lower heater hose clamp. Disconnect lower hose and remove lower right hand heater core and case attaching nut.
4. Remove glove box and door.
5. Remove recirculation vacuum diaphragm at right kick panel.
6. Remove heater outlet (at bottom of heater case).
7. Remove cold air distributor duct from heater case.
8. Remove heater case extension screws and separate extension from case.
9. Disconnect cables and wiring and remove case and core assembly, Fig. 105.
10. Separate core from case.
11. Reverse procedure to install.

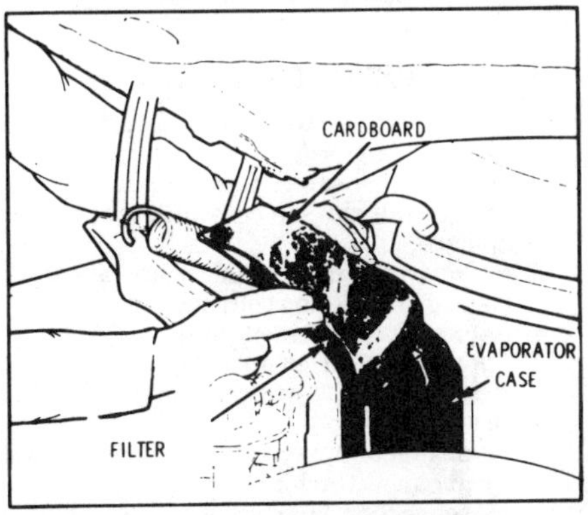

Fig. 183 Tempmatic air filter installation. 1974-75 models w/Tempmatic

1974-76 88, 98 & Toronado

CAUTION: On vehicles equipped with an Air Cushion Restraint system, turn ignition switch to "lock," disconnect battery ground cable and tape end, thereby deactivating system.

1. Disconnect battery ground cable and drain radiator.
2. Disconnect heater hoses and plug hoses and core openings to prevent coolant loss.
3. Remove heater case to dash panel attaching nuts, Fig. 103.
4. Remove instrument panel trim cover and the heater case to cowl bolts.
5. Remove lower air duct.
6. Remove instrument panel pad, then electrical connectors from glove box light and clock.
7. Remove right hand upper trim panel and on models equipped with A/C, remove manifold from heater case.
8. On all models, remove defroster duct from heater case and disconnect lower dash trim panel.
9. Disconnect temperature and defroster cables, then the vacuum hose from heater case. Remove heater case from dash, then core from case.
10. Reverse procedure to install.

1975-76 Starfire

With Air Conditioning

1. Disconnect battery ground cable.
2. Remove glove box and right air duct.
3. Remove instrument panel bezel and pad.
4. Remove left air duct and feed duct.
5. Lower steering column and remove instrument panel.
6. Remove control assembly, radio, defroster duct and large distribution duct.
7. Purge system of refrigerant.
8. Disconnect heater hoses at core.
9. Clean VIR exterior surface and line connections, then disconnect compressor inlet line, oil bleed line and condenser outlet line from VIR assembly.

NOTE: Cap all line and VIR ports.

10. Loosen evaporator inlet and outlet lines and remove VIR mounting clamp, then slide VIR assembly from evaporator lines.
11. Remove heater-distributor case to cowl attaching nuts and remove case, then disconnect wiring harness and vacuum lines.
12. Separate heater case from distributor, then remove heater core from case.
13. Reverse procedure to install.

1975-80 Starfire

Less Air Conditioning

1. Disconnect battery ground cable and disconnect blower motor electrical lead.
2. Place a pan under vehicle, then disconnect heater hoses from core and secure hoses in a raised position.

3. Remove blower inlet to dash panel screws and nuts, then remove blower inlet and blower motor and wheel assembly.
4. Remove core-retaining strap screws and remove core.
5. Reverse procedure to install making certain that blower inlet sealer is intact, replace as necessary.

1977-78 Toronado

CAUTION: On vehicles equipped with an Air Cushion Restraint system, turn ignition switch to "lock," disconnect battery ground cable and tape end, thereby deactivating system.

1. Disconnect battery ground cable and drain radiator.
2. Disconnect heater hoses and plug hoses and core openings to prevent coolant loss.
3. Remove heater case to dash panel attaching nuts, Fig. 185.
4. Remove instrument panel trim cover and the heater case to cowl bolts.
5. Remove lower air duct.
6. Remove instrument panel pad, then electrical connectors from glove box light and clock.
7. Remove right hand upper trim panel and on models equipped with A/C, remove manifold from heater case.
8. On all models, remove defroster duct from heater case and disconnect lower dash trim panel.
9. Disconnect temperature and defroster cables, then the vacuum hose from heater case. Remove heater case from dash, then core from case.
10. Reverse procedure to install.

1977-80 88 & 98

1. Disconnect battery ground cable.
2. Remove right half of hood seal from air inlet screen.
3. Remove air inlet screen.
4. Remove screws securing top of module, then disconnect electrical connectors from electrical components on top of module.
5. Remove thermostatic switch mounting screws from top of module.
6. Remove A/C diagnostic connector mounting screws and position aside.
7. Remove top of module.
8. Disconnect and remove heater core from module, Figs. 100 and 106.
9. Reverse procedure to install.

1977-80 Starfire

With Air Conditioning

NOTE: The heater core can be removed without purging the A/C refrigerant system or removing the evaporator case half of the assembly by using the following procedure:

1. Disconnect battery ground cable.
2. Remove floor outlet duct, then remove glove box and door as an assembly.
3. Remove left and right hand dash outlets using a suitable tool.
4. Remove instrument panel pad.
5. Disconnect vacuum hoses at valves on left end of heater and evaporator unit.
6. Remove insulation tray located below instrument cluster, then loosen console and slide console rearward.
7. Remove instrument panel to dash attaching screws. Place a protective covering over steering column, then lower instrument panel to steering column. Disconnect speedometer cable, radio leads and control head connectors.
8. Remove right hand instrument panel and lap cooler as an assembly.
9. Remove modular duct to heater and evaporator case screw, then remove duct assembly.
10. Disconnect temperature door cable and wiring harness.
11. Disconnect heater hoses from heater core. Plug core tubes to prevent spillage when removing heater core.
12. Remove three heater case stud nuts located in engine compartment.
13. Remove heater core case to evaporator core case attaching screws.
14. Drive on studs to break case loose from dash panel.
15. Remove heater core case assembly.
16. Remove heater core to case attaching screws, then remove heater core, Fig. 99.
17. Reverse procedure to install.

1978-80 Cutlass

Less Air Conditioning

1. Disconnect battery ground cable and drain cooling system.
2. Disconnect heater hoses from heater core.
3. Disconnect electrical connectors from heater module.
4. Remove module front cover screws and front cover, then remove heater core from module, Fig. 107.
5. Reverse procedure to install.

With Air Conditioning

1. Disconnect battery ground cable and drain cooling system.
2. Disconnect heater hoses from heater core.
3. Remove retention bracket and ground strap.
4. Remove module seal and screen, then remove right hand windshield wiper arm.
5. Remove retaining screws from hi blower relay and thermostatic switch.
6. Disconnect all electrical connectors from top of module.
7. Remove top cover from module, then remove heater core, Fig. 98.
8. Reverse procedure to install.

1979-80 Toronado

1. Disconnect battery ground cable, then drain cooling system.
2. Disconnect heater hoses at heater core, then install plugs in core outlets to prevent spillage.
3. Remove instrument panel sound absorbers, then lower steering column.
4. Remove instrument cluster.
5. Remove radio front speakers.
6. Remove three screws attaching manifold to heater case.
7. Remove four upper and three lower instrument panel attaching screws.
8. Disconnect parking brake release cable.
9. Disconnect instrument wiring harness from dash wiring assembly.
10. Disconnect right hand remote control mirror cable from instrument panel pad.
11. Disconnect speedometer cable from clip and temperature control cable at heater case.
12. Disconnect radio and A/C wiring, vacuum lines and all wiring necessary to remove instrument panel assembly. If equipped with pulse wiper, remove wiper switch and unlock wire connector from cluster carrier, then separate pulse wiper jumper harness from wiper switch wire connector.
13. Remove instrument panel and wiring harness assembly.
14. Remove defroster ducts, then disconnect lines from actuators.
15. Remove blower motor resistor.
16. From engine side of dash panel, remove three heater and A/C case retaining nuts.
17. From passenger compartment, remove screws and clip retaining heater and A/C case to dash panel.
18. Remove heater and A/C case, then remove heater core from case, Fig. 108.
19. Reverse procedure to install.

1980 Omega

Less Air Conditioning

1. Disconnect battery ground cable and drain cooling system.
2. Disconnect the heater hoses from heater core.
3. Remove radio noise suppression strap.
4. Remove heater core cover retaining screws and the cover, Figs. 109 and 110.
5. Remove heater core from vehicle.
6. Reverse procedure to install.

With Air Conditioning

1. Disconnect battery ground cable and drain cooling system.
2. Disconnect heater hoses from heater core.
3. Remove right side hush panel and open the glove box.
4. Remove heater duct retaining screw and the duct.
5. Remove instrument panel support bracket.
6. Remove heater case side cover retaining screws and the cover.
7. Remove heater core retaining clamps and the inlet and outlet tube support clamps.
8. Remove heater core from case.
9. Reverse procedure to install.

BLOWER MOTOR, REPLACE

1974 88, 98 & Toronado

CAUTION: On vehicles equipped with an Air Cushion Restraint system, turn ignition switch to "Lock," disconnect battery ground cable and tape end, thereby deactivating system.

1. Disconnect battery.
2. Remove the right front wheel.
3. Remove canister or battery.
4. Remove the three filler plates to radiator support screws.
5. Remove the filler plate to wheelhouse attaching screws. Remove filler plate.
6. Remove the blower motor attaching screws and connector, Fig. 186.
7. Remove the blower motor.
8. Reverse procedure to install.

1974-77 Cutlass

Less Air Conditioning

1. Remove the right front fender filler panel.
2. Disconnect the blower motor wiring.
3. Remove the five nuts and two screws securing the inlet assembly to dash.
4. Disengage the inlet assembly from the studs and remove from the car. The blower motor can be removed from the inlet assembly by removing the attaching screws, Fig. 186. The fan is held to the motor shaft by a nut and lock-washer.
5. Reverse procedure to install.

1975-79 Omega

Less Air Conditioning

1. Disconnect battery and detach hoses from clips on right fender skirt.
2. Raise vehicle on hoist.
3. Remove fender skirt attaching bolts except those retaining the skirt to radiator support.
4. Pull out then down on skirt and place block of wood between skirt and fender to allow clearance for blower motor removal.
5. Disconnect blower motor cooling tube and electrical connections at blower motor.
6. Remove blower motor attaching screws and remove blower motor, Fig. 186. Gently pry motor flange if sealer acts as an adhesive.
7. Reverse procedure to install.

1974-77 Cutlass

With Air Conditioning

1. Disconnect battery ground cable and blower motor feed wire.
2. Remove screws securing blower motor to dash, then the blower motor, Fig. 186.
3. Reverse procedure to install.

1975-79 Omega

With Air Conditioning

1. Disconnect battery cable.
2. Remove the fender filler plate attaching bolts and position it forward and inboard.
3. Disconnect the blower feed wire.
4. Remove the blower motor attaching screws and blower motor, Fig. 186.
5. Reverse procedure to install.

1975-76 88 & 98; 1974-78 Toronado

1. Disconnect battery ground cable.
2. Raise vehicle and remove right front wheel.
3. Referring to Fig. 187, cut along inside of bead.
4. Fold flap over to expose blower motor, Fig. 188.
5. Remove blower motor attaching screws and the blower motor through opening.

NOTE: After blower motor installation, the access hole must be sealed as follows:

(a) Drill a hole in flap and install a 5/16 inch self tapping bolt with a large flat washer, Fig. 189.
(b) Close flap and drill two holes in inner filler panel and install two 5/16 inch self tapping bolts with large flat washers, Fig. 189.
(c) With a suitable sealer, seal edge of flap. Do not use weatherstrip adhesive.

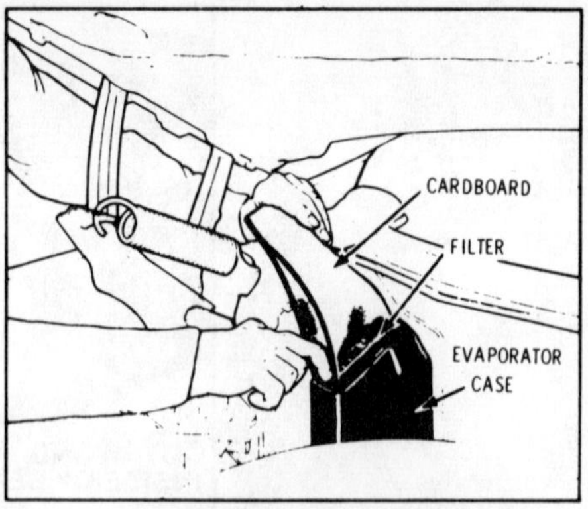

Fig. 184 Tempmatic air filter installation. 1974-75 models w/Tempmatic

1975-80 Starfire

1. Disconnect battery ground cable and disconnect blower motor electrical leads. On some models, it may be necessary to remove coolant recovery tank attaching screws and position tank aside. Do not disconnect hoses from tank.
2. Scribe blower motor flange to case position.
3. Remove blower motor retaining screws and remove blower motor and wheel assembly. Pry flange gently, if sealer acts as an adhesive.
4. Reverse removal procedure to install, aligning scribe marks made during removal.

1977-80 88 & 98

1. Disconnect battery ground cable.
2. Disconnect blower motor ground and feed wires.
3. Remove six blower motor attaching screws and blower motor, Figs. 100 and 106.
4. Reverse procedure to install.

1978-80 Cutlass

1. Disconnect battery ground cable.
2. If equipped with air conditioning, disconnect cooling tube from blower motor.
3. On all models, disconnect electrical connector from blower motor.
4. Remove blower motor retaining screws and the blower motor.
5. Reverse procedure to install.

1979-80 Toronado

1. Disconnect battery ground cable.
2. Remove right hand side fender skirt.
3. Disconnect wire connector and cooling hose from blower motor.
4. Remove five blower motor mounting screws, then remove blower motor, Fig.101.
5. Reverse procedure to install.

1980 Omega

1. Disconnect battery ground cable.
2. Disconnect blower motor electrical connections.
3. Remove blower motor attaching screws and the blower motor, Figs. 102 and 109.
4. Reverse procedure to install.

COMPRESSOR CLUTCH & MASTER SWITCH, REPLACE

NOTE: The master switch and compressor clutch switch are located on the A/C control panel.

1. Remove instrument panel bezel, and if equipped, remove clock.
2. Remove retaining switch screws and remove switch(es).
3. Reverse procedure to install.

BLOWER RESISTOR, AMBIENT SWITCH & SENSOR, REPLACE

1. On 1977-79 Omega without air conditioning, remove glove box.

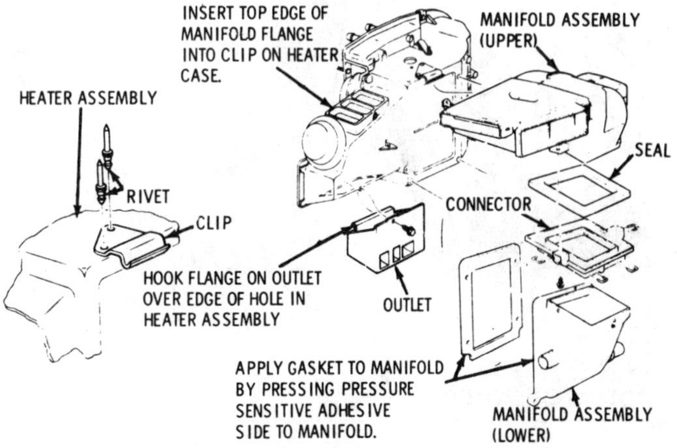

Fig. 185 Heater assembly. 1977-78 Toronado

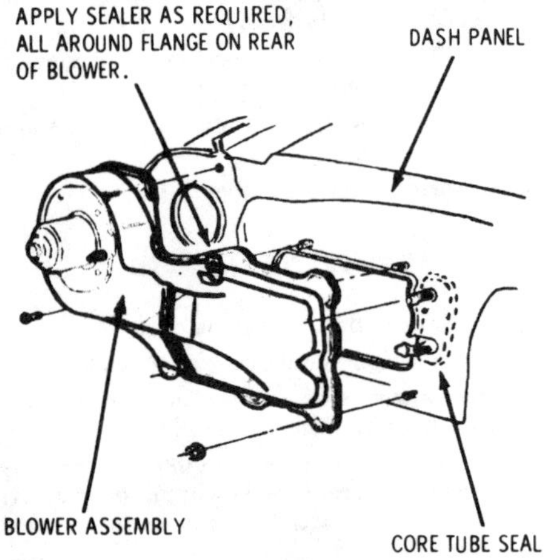

Fig. 186 Blower motor. 1974-76 88, 98; 1974-77 Cutlass; 1974-79 Omega; 1977-78 Toronado (Typical)

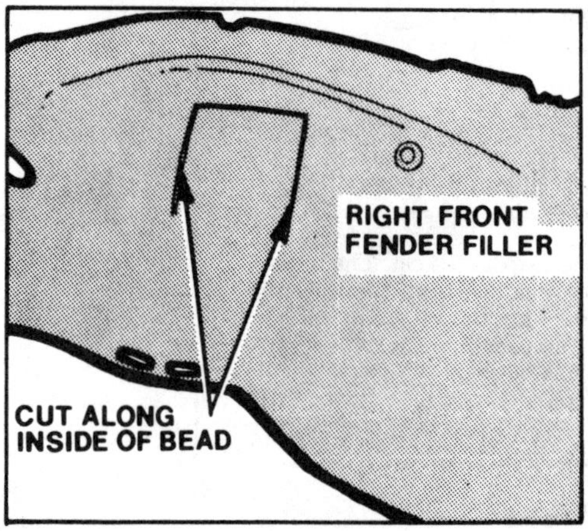

Fig. 187 Blower motor access hole location. 1975-76 88, 98; 1975-78 Toronado

2. Disconnect electrical connector.
3. Remove retaining screws and remove sensor and switch.
4. Reverse procedure to install.

TEMPMATIC AIR FILTER, REPLACE

NOTE: The air filter located in the evaporator case should be replaced every 24 months or more often under severe driving conditions.

1. Remove access doors and pull old filter from evaporator case.
2. Remove plastic wrap from new filter and insert filter and cardboard into evaporator case, Fig. 183.
3. Insert finger between cardboard fold and hold filter down while pulling cardboard out, Fig. 184.
4. Check filter to make sure it is properly seated and install access doors.

EVAPORATOR CORE, REPLACE

1974-77 Cutlass

1. Disconnect battery ground cable, discharge refrigerant from system, drain cooling system and disconnect heater hoses.
2. Remove right fender and inner filler panel, then disconnect lines from POA and expansion valve or VIR assembly. Cap all lines.
3. Disconnect harness from resistor and ambient switch, then remove evaporator to dash retaining bolts and remove evaporator assembly, Fig. 96.
4. Separate evaporator case halves and remove evaporator core.
5. Reverse procedure to install.

1978-80 Cutlass

1. Disconnect battery ground cable and discharge refrigerant system.
2. Remove module rubber seal and screen.
3. Remove right hand windshield wiper arm, diagnostic connector, hi-blower relay and the thermal switch.
4. Disconnect air electrical connectors at module top and remove module top cover.
5. Remove accumulator bracket screws and disconnect refrigerant lines at accumulator and the liquid line. Plug all lines and openings to prevent entry of dirt or moisture.
6. Remove core, Fig. 98.
7. Reverse procedure to install.

1974-76 Omega

1. Disconnect battery ground cable and discharge refrigerant from system.
2. On non-VIR equipped models, Fig. 92:
   (a) Disconnect oil bleed line and equalizer line at POA valve.
   (b) Disconnect evaporator inlet and outlet fittings at POA valve and expansion valve.
   (c) Disconnect thermo bulb from evaporator outlet line and remove expansion valve.
3. On VIR equipped models, Fig. 93, disconnect lines from VIR and remove VIR assembly, then disconnect vacuum tank.
4. Remove all screws and nuts retaining left half of evaporator case to right half and to dash, then pull left half outward and upward being careful not to damage core.
5. Remove core retaining screws and remove core.
6. Reverse procedure to install

1977-79 Omega

1. Disconnect compressor pressure switch and jump wires together, disconnect battery ground cable, then discharge refrigerant system.

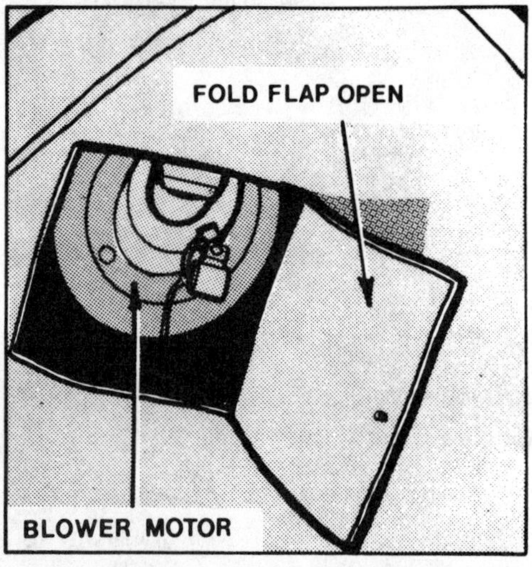

Fig. 188 Blower motor removal. 1975-76 88, 98; 1975-78 Toronado

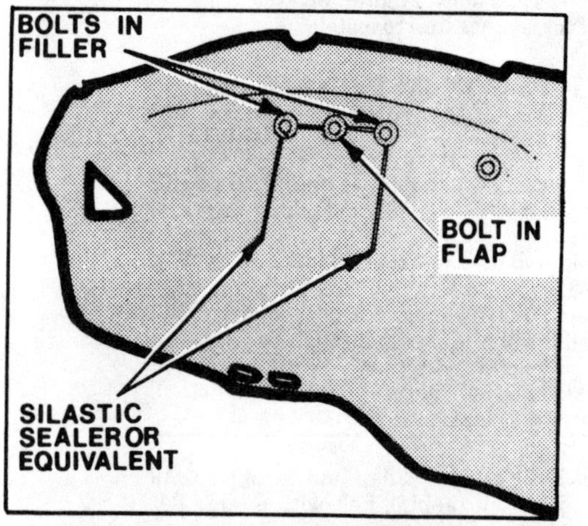

Fig. 189 Sealing blower motor access hole. 1975-76 88, 98; 1975-78 Toronado

2. Disconnect lines to accumulator, plug all lines and openings to prevent entry of dirt or moisture. Disconnect bracket to accumulator assembly.
3. Remove accumulator from evaporator, disconnect all wiring at evaporator case and remove all screws and nuts retaining left half of evaporator case to right half and also to dash.
4. Disconnect vacuum tank, pull left half of evaporator case outward and upward from core, then remove core retaining screws and core, Fig. 94.
5. Reverse procedure to install.

1974-76 Full Size & 1977-78 Toronado

1. Disconnect battery ground cable and discharge refrigerant from system.
2. Remove VIR assembly as described under "VIR SERVICE" section.
3. Remove right hand wheel and tire assembly and then remove right hand wheel housing and fender brace.
4. Remove evaporator housing retaining screws, Fig. 154.
5. Remove evaporator housing and disconnect vacuum hose, then separate housing halves and remove core.
6. Reverse procedure to install.

1977-80 88 & 98

1. Disconnect battery ground cable.
2. Disconnect wiring from blower motor, resistor, thermostatic switch and high blower relay, then remove screws from thermostatic switch and diagnostic connector. Remove thermostatic switch and diagnostic connector.
3. Discharge refrigerant from system and plug all open lines and openings.
4. Disconnect liquid line at evaporator inlet pipe, and compressor suction hose at accumulator outlet pipe.
5. Remove right end of hood seal and remove air inlet screen.
6. Remove the top 5 case to dash bolts, 9 upper to lower case screws located around flange and 2 upper to lower case screws inside air intake plenum.
7. Remove upper case by lifting straight up and off, then remove 2 accumulator pipe bracket to case screws and remove evaporator core by lifting straight up and out of case, Figs. 98 and 100.
8. Reverse procedure to install.

1975-79 Starfire

Refer to the Chevrolet Monza "Evaporator Core, Replace" procedure for evaporator core service.

1979-80 Toronado

1. Disconnect battery negative cable and discharge refrigerant system.
2. Remove blower resistor, blower relay, thermostatic switch and ground wires.
3. Remove vacuum tank, condensor line from evaporator and suction line from accumulator.
4. Remove blower wires and motor and raise car.
5. Remove 2 screws and one nut from bottom of evaporator and blower assembly.
6. Remove 2 screws and one nut from top of evaporator and blower assembly.
7. Remove evaporator and blower assembly, then the accumulator line from evaporator.
8. Remove accumulator from evaporator case, then core from case, Fig. 101.
9. Reverse procedure to install.

1980 Omega

1. Disconnect battery ground cable and discharge refrigerant system.
2. Remove air cleaner.
3. Disconnect air conditioning electrical connections along the cowl and position wiring harness aside.
4. Remove vacuum tank.
5. Disconnect liquid line at evaporator inlet and the low pressure line at accumulator, Fig. 102.
6. Remove outer blower case to cowl screws, then the blower and case assembly.
7. Remove evaporator from vehicle.
8. Reverse procedure to install.

Pontiac

CONTROL CABLES, ADJUST

NOTE: Adjustment is made after the cables are properly secured and routed.

Firebird

1974-76 All & 1977-80 With A/C

1. Refer to Figs. 190 and 191 and adjust turnbuckle until temperature door is fully seated.
2. Move temperature lever to full WARM and then back to full COLD or COOL. Temperature door should be heard hitting its seat at end of lever travel. If not, repeat above procedure.

NOTE: If control lever is over ⅛ inch away from end of slot, repeat procedure.

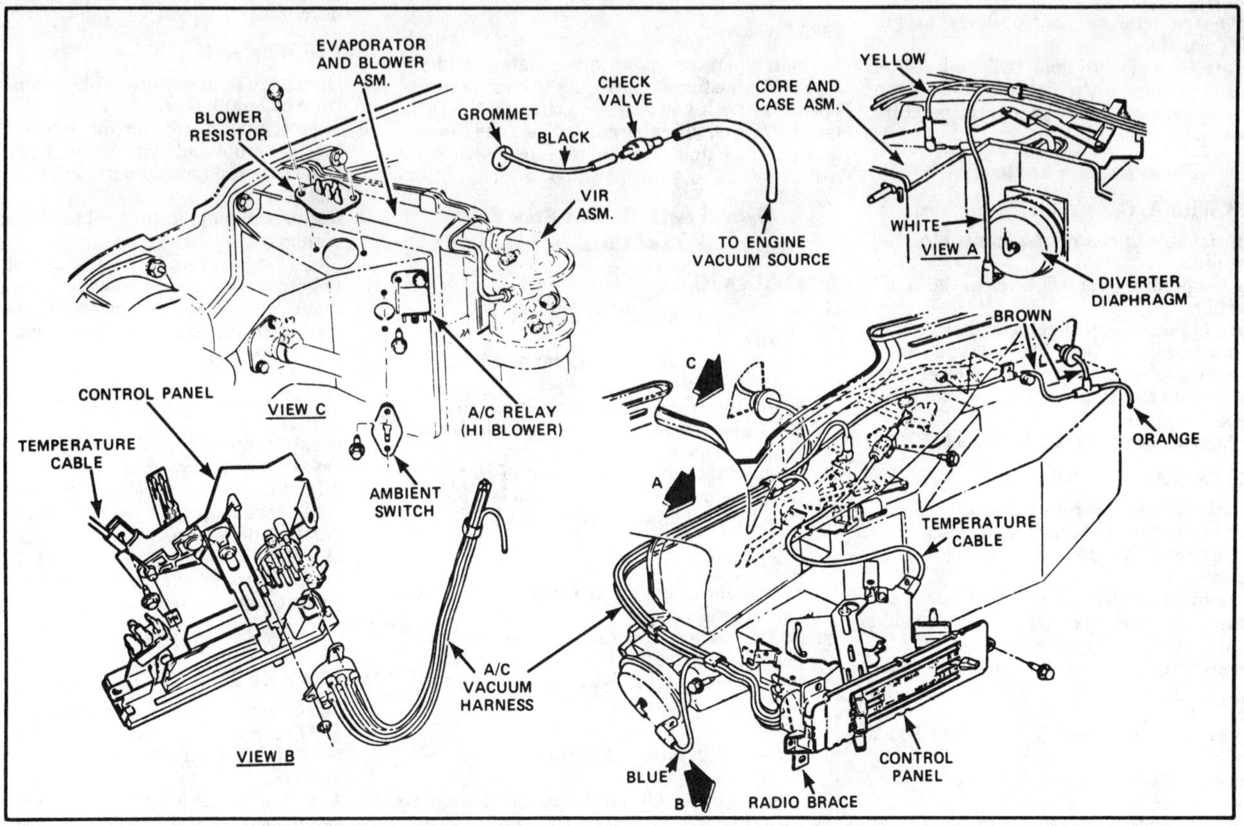

Fig. 190 Control cable & panel. 1974-75 Firebird

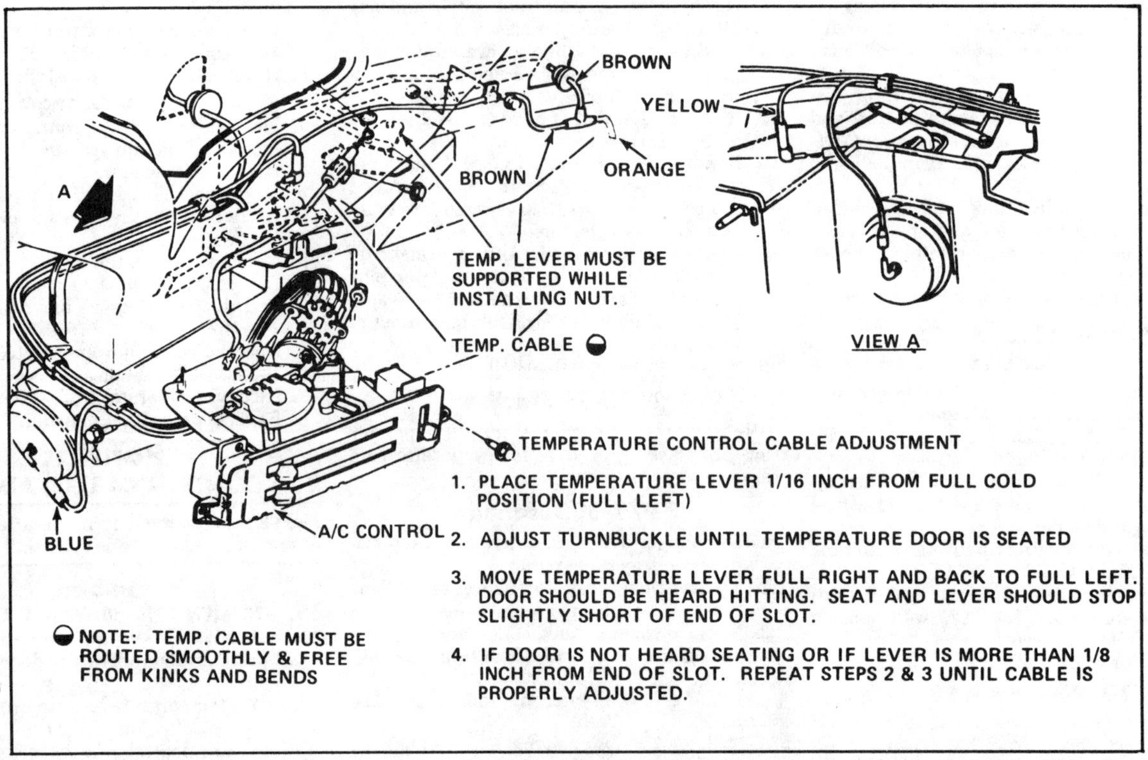

Fig. 191 Control cable & panel. 1976-80 Firebird with A/C

1977-80 Less A/C

1. Loosen nut on temperature control cable clamp.
2. Move temperature control lever to full HOT position.
3. Rotate temperature door bellcrank to full COLD position then tighten cable clamp against sheath. Ensure proper temperature door operation.

Phoenix & Ventura

1974-79 With A/C

1. Attach cable to control assembly, Figs. 139 and 140.
2. Set temperature control lever in the "Off" position.
3. Place opposite loop of cable on heater case temperature door post, Figs. 139 and 140.
4. Push cable sheath toward heater lever until valve seats and lash is eliminated from cable and control.
5. Tighten screw to secure adjustment.

1974-76 Less A/C

1. Attach cable to control assembly.
2. Move control lever to the OFF position.
3. Move heater case door to the full closed position.
4. Place opposite loop of cable on heater case temperature door lever and tighten cable attaching bracket screws.
5. Ensure proper temperature door operation.

1980 With A/C

1. Connect cable assembly to control head, Fig. 86.
2. Place temperature lever in the OFF position.
3. Connect cable to temperature door post and push sheath toward lever until door seats and all lash is removed from cable and control head lever.
4. Tighten cable securing screw.

1980 Less A/C

The temperature cable incorporates a slider-type self-adjusting feature. As the control head temperature lever is moved through its full range of travel, the cable clip will assume a position assuring that the temperature valve will seat in the full hot and full cold positions.

Grand Am, Grand Prix & LeMans

1974 With A/C

1. Place temperature control lever at full COLD.
2. Adjust turnbuckle until cam roller bottoms at end of slot in cam, Fig. 192.
3. Move temperature control lever to full HOT and then to full COLD. Cam should return to end of slot.
4. If cam does not return to end of slot or if temperature lever does not complete its full travel, repeat adjustment procedure.

1974 Less A/C

1. Position lever on top of heater case in full COLD position.
2. Adjust cable turnbuckle so that the temperature control lever will drop into the detent at the full OFF position.

1975-77 With A/C

1. Place temperature control lever at full HOT.
2. Adjust turnbuckle until temperature door seals, Fig. 192.
3. Move temperature control lever to full COLD and then to full HOT. Door must seal at both ends of temperature control lever travel. If not, repeat adjustment procedure.

1978-80 With A/C

1. Disengage temperature cable retaining clip from bellcrank post.
2. Snap temperature door rod into bellcrank retainer groove allowing 2/5 inch of rod end to be exposed beyond outer edge of retainer.
3. Reconnect temperature cable clip to bellcrank post.
4. Move temperature control lever full travel in both directions to allow for cable adjustment. Temperature door should be heard hitting in open and closed positions.

1975-80 Less A/C

1. Move temperature control lever to full COLD.
2. Rotate temperature door crank on heater case to full cold position.
3. Adjust temperature cable turnbuckle until cable wire loop lines up with pin on temperature door crank pin.
4. Move temperature lever to full HOT. If temperature door cannot be heard hitting its seat, repeat procedure.

Full Size

1974-76 With A/C

1. Place temperature control lever at full HOT. If lever does not reach end of slot in control panel, adjust turnbuckle, Figs. 177 and 178.
2. Adjust turnbuckle until roller bottoms in slot in cam on heater case and temperature lever is 1/16 to ⅛ inch from end of

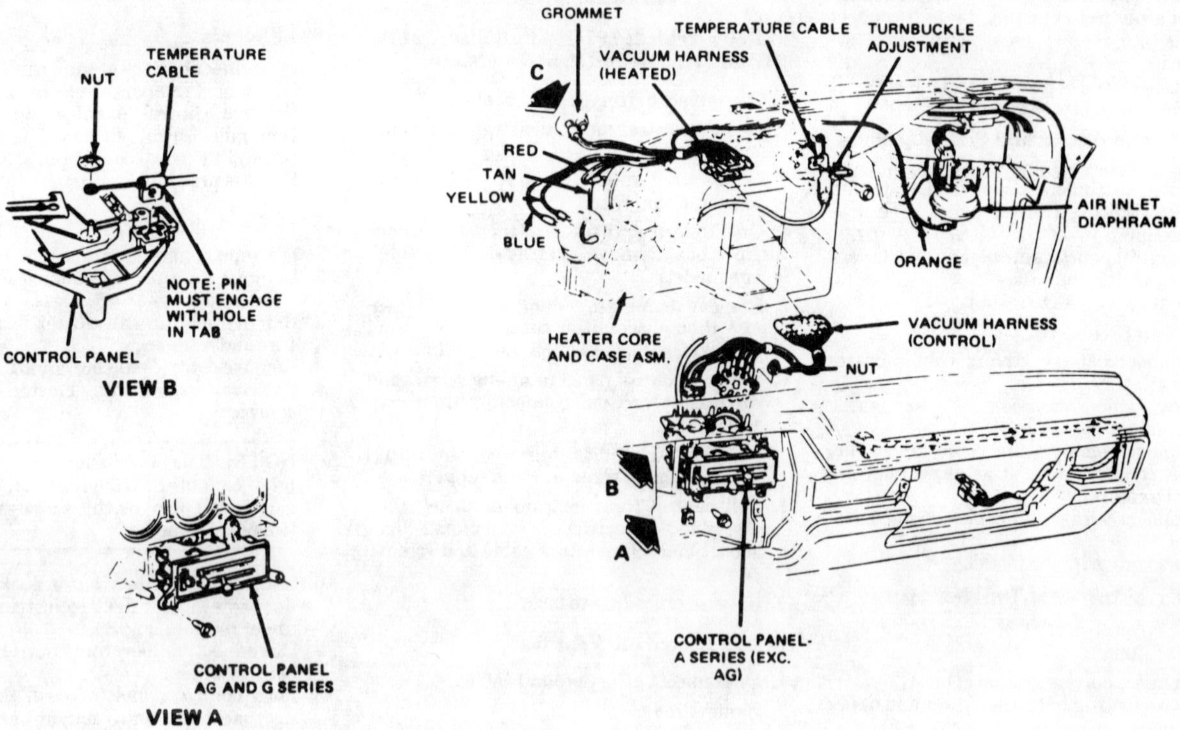

Fig. 192 Control cable & panel. 1975-77 Grand Am, Grand Prix & LeMans with A/C

slot in control panel.

3. Move temperature lever to full COLD. Lever must reach end of slot.
4. Return lever to full HOT. Lever should stop 1/16 to ⅛ inch from end of slot in control panel.
5. If temperature lever operation is not as described in steps 3 and 4, repeat adjustment procedure.

1974-76 Less A/C

Refer to 1975-80 Grand Am, Grand Prix and LeMans less A/C for temperature control cable adjustment.

1977-80 With A/C

Refer to 1978-80 Grand Am, Grand Prix and LeMans with A/C for temperature control cable adjustment.

CONTROL PANEL, REPLACE
Less A/C

1975-77 Astre

1. Disconnect battery ground cable.
2. Remove instrument panel bezel.
3. Remove control to instrument panel screws. Push control forward, then lower control enough to disconnect bowden cables and electrical leads. Ensure cables are not kinked.
4. Remove bowden cable retaining clips, then remove the cable attaching screws and disconnect the cables from the control assembly.
5. Disconnect blower switch lead and remove the control assembly.
6. Reverse procedure to install.

1976 Sunbird

1. Disconnect battery ground cable.
2. Disconnect antenna lead at the radio.
3. Remove radio control knobs and nuts.
4. Remove radio bracket to dash screws, then push radio forward and place it on the floor.
5. Remove the heater control to instrument panel screws.
6. Push control forward and lower it enough to disconnect bowden cables and electrical leads. Ensure cables are not kinked.
7. Remove bowden cable retaining clips, then remove the cable attaching screws and disconnect the cables from the control assembly.
8. Disconnect blower switch lead and remove the control assembly.
9. Reverse procedure to install.

1977-80 Sunbird

Refer to 1975-77 Astre procedurę to remove heater control panel.

1974-79 Ventura & Phoenix

1. Disconnect battery ground cable and remove radio.
2. Remove three screws securing control to instrument panel and lower control from panel.
3. Disconnect wiring and cables from control panel and remove panel.
4. Reverse procedure to install.

1980 Phoenix

1. Disconnect battery ground cable.
2. Remove control panel screws and pull control out.
3. Disconnect electrical connection and cables from control.
4. Remove blower switch from control panel, if necessary.
5. Reverse procedure to install.

1974 Firebird

1. Disconnect battery ground cable and remove instrument panel trim at steering column.
2. Remove upper instrument panel trim plate and rally gauges, if equipped.
3. Remove three control retaining screws, support bracket screw and radio.
4. Disconnect cables and wiring at control panel, then remove panel.
5. Reverse procedure to install.

1975-77 Firebird

1. Disconnect battery ground cable and remove glove box.
2. Remove upper and lower instrument panel trim.
3. Disconnect temperature cable at heater case and remove control panel retaining screws.
4. Remove screw at support bracket, then position control out from instrument panel opening.
5. Disconnect wiring and air defrost cables from control and remove control with temperature cable attached.
6. Disconnect temperature cable from control panel.
7. Reverse procedure to install.

1978-80 Firebird

1. Disconnect battery ground cable and remove glove box.
2. Remove control panel retaining screws and pull control out from instrument panel.
3. Disconnect wiring and cables from control assembly.
4. Reverse procedure to install.

1974-77 Gran Am, Grand Prix, LeMans & Full Size

1. Disconnect battery ground cable.
2. On all vehicles except Grand Am and 1975 Grand Prix, remove lower instrument panel trim plate.
3. Remove radio knobs and bezels and remove upper trim plate.
4. Remove glove box, then disconnect the three cables at the heater case. On 1977 full size models, remove temperature cable at module.

5. Remove screws and position heater control panel away from opening, then disconnect wiring and remove panel with cables attached.
6. Disconnect cables from control panel.
7. Reverse procedure to install.

1978-80 Grand Am, Grand Prix & LeMans

1. Disconnect battery ground cable.
2. Remove instrument panel trim plate and screws securing control panel to instrument panel.
3. Pull control panel from opening and disconnect all wiring and cables.
4. Reverse procedure to install.

1978-80 Full Size

1. Disconnect battery ground cable and remove ash tray.
2. Remove gauge and control trim panels, then remove glove box.
3. Disconnect temperature cable at module, remove control panel attaching screws and remove control.
4. Disconnect wiring and cables from control panel.
5. Reverse procedure to install.

With Automatic Temperature Control

1974-80 Models

1. Disconnect battery ground cable.
2. Remove instrument panel lower and upper trim plate.
3. Remove control panel retaining screws.
4. Pull control panel out from instrument panel opening, then disconnect electrical and vacuum connectors from control panel and remove control panel.
5. Reverse procedure to install.

With Custom A/C

1974-80 Firebird; 1974-80 Full Size; 1974-80 Grand Am, Grand Prix & LeMans

1. Disconnect battery ground cable.
2. Remove instrument panel trim and steering column trim.
3. On Firebird models, remove rally gauge cluster, if equipped.
4. On Firebird and full size models, remove glove box. Remove ash tray on 1977-80 full size models.
5. Remove temperature door on 1974-76 models, then disconnect control cable from heater case, Figs. 192, 193, 194 and 136.
6. Remove control panel retaining screws and disconnect vacuum connectors from control panel.
7. On Firebird models, remove rear control panel retaining screw at radio support.
8. Pull control from instrument panel, then disconnect electrical connectors, light socket and temperature cable and remove control.
9. Reverse procedure to install.

1974-79 Phoenix & Ventura

1. Disconnect battery ground cable.
2. Remove radio.
3. Remove control panel retaining screws and lower control panel slightly from instrument panel.
4. Disconnect electrical connectors and control cables from control panel and remove panel, Figs. 139 and 140.
5. Reverse procedure to install.

1980 Phoenix

1. Disconnect battery ground cable.
2. Disconnect temperature control cable.
3. Remove control panel retaining screws, then pull control out and disconnect vacuum and electrical connections, Fig. 90.
4. Reverse procedure to install.

1975-77 Astre

1. Disconnect battery ground cable.
2. Remove intermediate duct leading to left hand outlet.
3. Remove instrument cluster bezel and ash tray and retainer.
4. Remove control panel retaining screws.
5. Remove instrument cluster to panel screws.

NOTE: The screw above the steering column cannot be removed. However, the cluster will pivot on this screw to allow control panel removal.

6. Remove center distributor duct to selector duct screws and duct to instrument panel upper retainer, Fig. 89.
7. Disconnect temperature control cable from control panel.
8. Push control panel forward, then disconnect electrical and vacuum harnesses and remove control panel.
9. Reverse procedure to install.

1976-80 Sunbird Exc. 1978-79 Station Wagon

1. Disconnect battery ground cable.

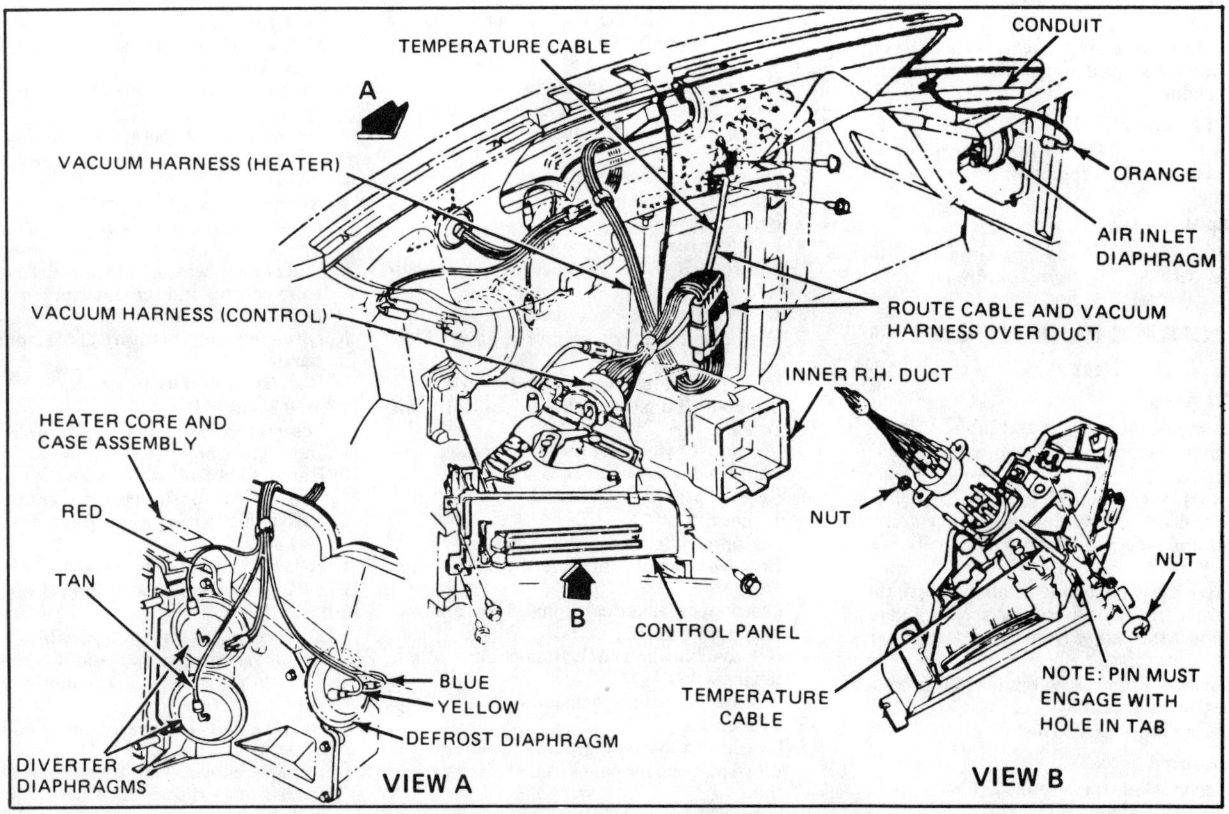

Fig. 193 Control cable & panel. 1974-77 Full Size with A/C

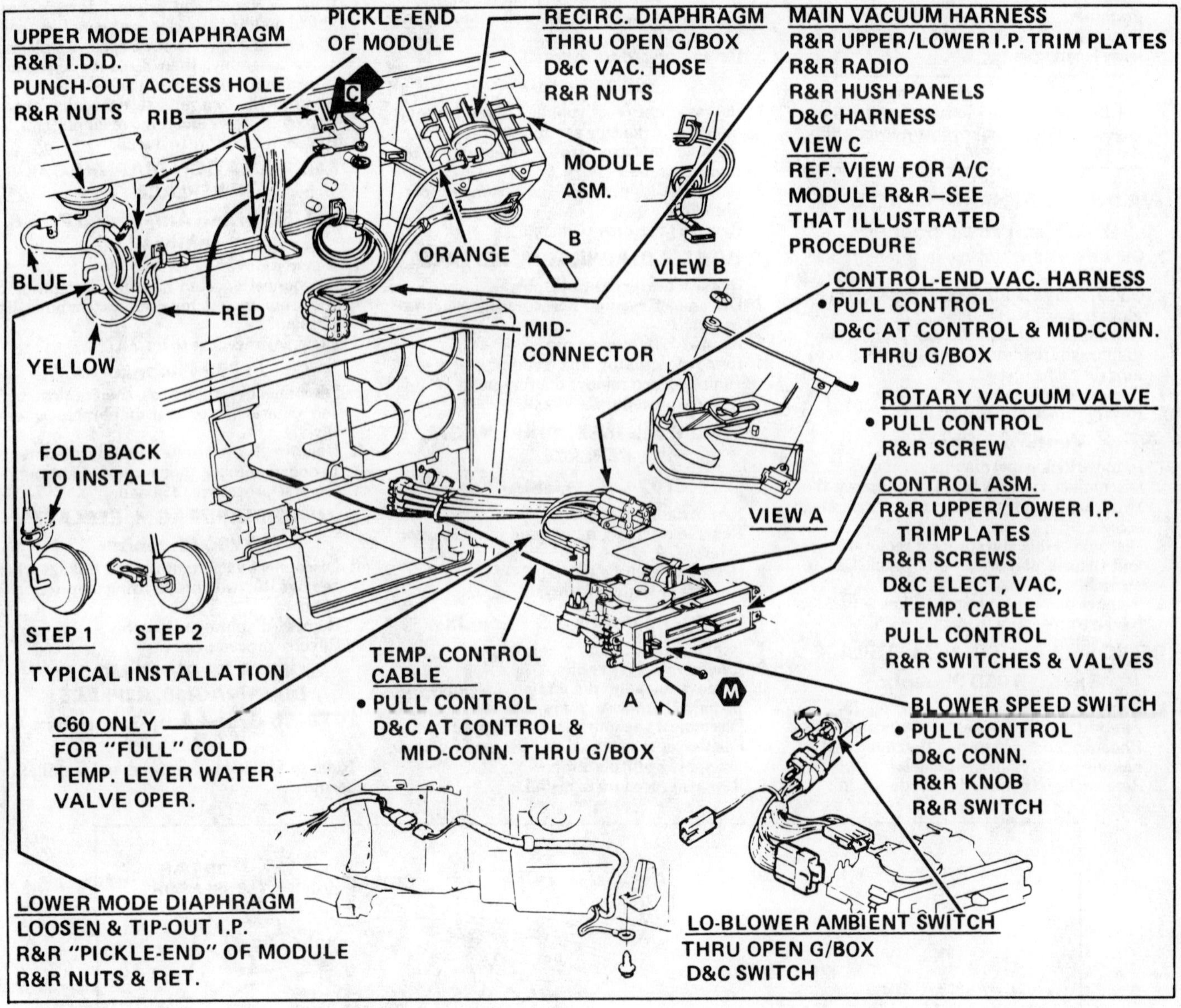

Fig. 194 Control cable & panel. 1978-80 Grand Am, Grand Prix & LeMans with A/C

2. Remove right and left hand dash outlets and instrument panel pad.
3. Remove steering column retaining nuts and rest steering wheel on driver's seat.
4. Remove instrument panel to dash retaining screws, then place a protective covering over steering column and lower instrument panel onto steering column.
5. Remove control panel to instrument panel screws, Fig. 89, then disconnect temperature control cable, electrical and vacuum connectors and remove control panel.
6. Reverse procedure to install.

1978-79 Sunbird Station Wagon

1. Disconnect battery ground cable.
2. Remove lower left hand duct, instrument cluster bezel and ash tray.
3. Remove screws securing control and cluster to instrument panel.
4. Remove cold air duct to selector duct screws and retainer.
5. Disconnect cables, vacuum hoses and wiring from control and remove control.
6. Reverse procedure to install.

PROGRAMMER, REPLACE

1974

1. Disconnect battery ground cable.
2. Remove glove box.
3. Loosen screw and remove programmer link from programmer output shaft, Fig. 195.
4. Disconnect vacuum and electrical connectors from programmer.
5. Remove programmer retaining screws, then remove programmer by carefully pulling rearward to avoid damaging heat sensor or blower motor resistor.
6. Reverse procedure to install.

1975-80

1. Remove glove box and disconnect temperature control cable from programmer.
2. Disconnect vacuum connector and in-car and ambient air source hoses from programmer, Figs. 195 through 198.
3. Disconnect temperature door link from programmer on 1974-76 models. Disconnect threaded operating rod from programmer on 1977-80 models.
4. Remove retaining screws and remove programmer.

NOTE: If original programmer is to be reinstalled, move programmer temperature control lever to full right position as viewed from front of programmer. Lift control lever locating spring out from alignment hole in programmer chassis, then move control lever to align with hole in chassis and allow locating spring to rest in control lever hole.

5. Reverse procedure to install, then set control panel temperature lever at 75. Attach mounting tab to temperature control ca-

ble to programmer chassis and adjust turnbuckle until cable loop aligns with programmer control lever and attach cable loop to lever. Move control lever to 65 to release locating spring.

CAUTION: Do not move temperature lever toward 85 while locating spring is engaged, as programmer control lever will be damaged.

AIR INLET DIAPHRAGMS, REPLACE

Exc. Ventura & Phoenix

1. On 1974-80 Firebird models, remove kick panel.
2. On 1974-80 except Firebird models, remove glove box.
3. Disconnect vacuum hoses, then remove diaphragm retaining screws and link actuator retaining screws.
4. Remove diaphragm.
5. Reverse procedure to install.

Ventura & Phoenix

1. Remove kick panel diaphragm cover.
2. Disconnect vacuum hoses and remove the two diaphragm to kick panel retaining screws.
3. Disconnect link and spring at recirc air door and remove diaphragm and bracket as an assembly, Fig. 182.
4. Remove diaphragm from bracket.
5. Reverse procedure to install.

DEFROSTER DIAPHRAGM, REPLACE

Except 1980 Phoenix

1. Remove heater case and core on 1974-77 Firebird, 1975-77 LeMans and 1975-79 Phoenix and Ventura. Remove heater module on 1977 full size models.
2. Remove heater outlet duct or distributor.
3. Remove the two retaining screws and actuator link retaining screw, if equipped.
4. Disconnect vacuum hose and remove diaphragm.
5. Reverse procedure to install.

1980 Phoenix

1. Remove defroster nozzle, then remove mode valve housing assembly.
2. Remove the two defroster diagphragm retaining screws.
3. Release actuator rod and remove diaphragm.
4. Reverse procedure to install.

DIVERTER DIAPHRAGM, REPLACE

1. Remove heater case assembly.
2. Disconnect vacuum hoses noting exact location.
3. Remove retaining screws.
4. Remove actuator link retaining screw if equipped, and remove diaphragms.
5. Reverse procedure to install.

PLENUM AIR INLET DIAPHRAGM, REPLACE

1974-80 Firebird

1. Remove cowl vent grille.
2. Remove retaining screws and link actuator screw.
3. Remove diaphragm.
4. Reverse procedure to install.

1974-79 Ventura & Phoenix

1. Remove windshield wiper arms.
2. Remove cowl vent grille.
3. Remove plenum diaphragm assembly to plenum chamber retaining screws.
4. Disconnect vacuum hose from diaphragm.
5. Position door in closed position and remove assembly by lifting out.
6. Reverse procedure to install.

1980 Phoenix

1. Remove windshield wipers, lower windshield molding and close out panel.
2. Remove the three screws securing plenum valve assembly, then disconnect vacuum hose and remove plenum valve.
3. Remove the two nuts securing diaphragm to plenum valve and remove diaphragm.
4. Reverse procedure to install.

RECIRCULATION DIAPHRAGM, REPLACE

1977-80 Grand Am, Grand Prix & LeMans

1. Remove glove box.
2. Disconnect vacuum hose from diaphragm, then remove retaining screws. Remove diaphragm.
3. Reverse procedure to install.

1980 Phoenix

1. Disconnect vacuum hose, then remove max recirculation door and diaphragm assembly.
2. Remove the two nuts securing diaphragm to door assembly, then remove diaphragm.
3. Reverse procedure to install.

MODE DIAPHRAGM, REPLACE

1980 Phoenix

1. Disconnect vacuum hose and linkage, then remove the two nuts securing diaphragm to support.
2. Remove diaphragm assembly.
3. Reverse procedure to install.

UPPER & LOWER MODE DIAPHRAGMS, REPLACE

1978-80 Grand Am, Grand Prix & LeMans

Refer to Fig. 194 for removal and installation procedure.

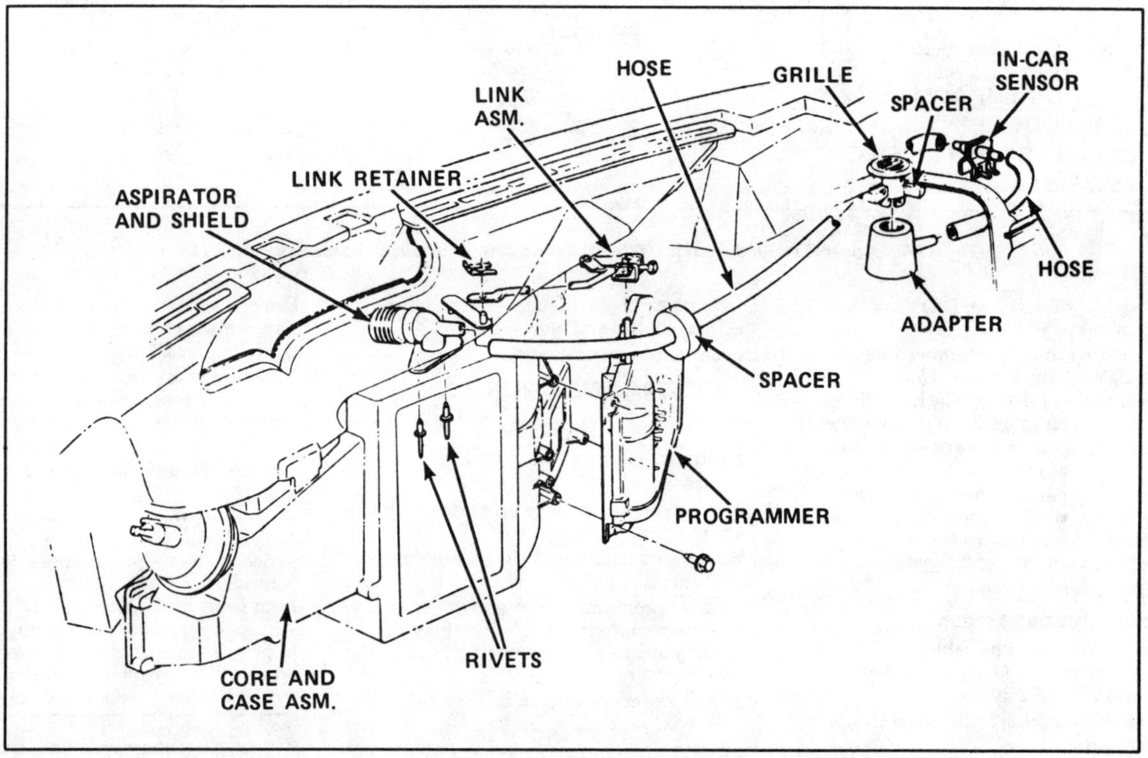

Fig. 195 Programmer assembly. 1974 Full Size

DEFROSTER VALVE, REPLACE

1980 Phoenix

1. Remove defroster nozzle, then remove mode valve housing assembly.
2. Remove defroster vacuum diaphragm.
3. Push defroster valve pivot rod from mode valve housing and remove defroster valve.
4. Reverse procedure to install.

MODE VALVE, REPLACE

1980 Phoenix

1. Remove defroster nozzle, then remove mode valve housing assembly.
2. Remove mode valve diaphragm from adapter plate.
3. Remove adapter plate from mode valve housing.
4. Pull mode valve pivot rod from mode valve housing, then remove mode valve.
5. Reverse procedure to install.

TEMPERATURE VALVE, REPLACE

1980 Phoenix

1. Remove defroster nozzle, then remove mode valve housing.
2. Disconnect temperature control cable and slide valve pivot rod towards passenger side of car. Remove valve.
3. Reverse procedure to install, ensuring temperature door cable is properly adjusted.

TEMPERATURE CONTROL CABLE, REPLACE

Exc. Astre & Sunbird, Figs. 190 Through 194

1. On 1974-76 Firebird models, remove cold air duct.
2. On all models except 1977-80 full size, remove glove box and disconnect cable at heater case.
3. Disconnect cable at control panel and remove cable. On 1974-80 full size models and 1974-80 Grand Am and Grand Prix models, it will be necessary to pull control panel out from instrument panel in order to disconnect cable.
4. Reverse procedure to install. Check operation and adjust cable if necessary.

MASTER SWITCH, REPLACE

1974-76 Firebird

1. Disconnect battery ground cable, then remove steering column trim and instrument panel trim.
2. Remove rally gauge cluster, if equipped.
3. Disconnect electrical connectors, then remove retaining screws and remove switch.
4. Reverse procedure to install.

1974 Exc. Firebird & Ventura

1. Position A/C control panel out from instrument panel opening.
2. Remove ambient switch retaining screws and remove switch.
3. Remove recirc override switch bracket retaining screw and disconnect electrical connector from switch.
4. Remove master switch board retaining screws and remove master switch board.
5. Remove control panel face plate, then lift selector lever forward and remove cam and wiper assembly.
6. Reverse procedure to install.

1975-76 Models Exc. Astre, Firebird, Sunbird & Ventura; 1977-80 All Models

1. Position A/C control panel out from instrument panel opening.
2. Remove retaining screws and remove master switch.
3. Disconnect electrical connector from switch.
4. Reverse procedure to install.

1976-80 Sunbird

1. Perform steps 1 through 5 as outlined under Sunbird "Control Panel, Replace" procedure.
2. Disconnect electrical connector from switch.
3. Remove switch retaining screws and the switch.
4. Reverse procedure to install.

BLOWER SWITCH, REPLACE

1974-76 Firebird

Perform the 1974-76 Firebird "Master Switch, Replace" procedure for blower switch replacement.

1974-76 Models Exc. Astre, Firebird, Sunbird & Ventura & 1977-80 All Models

1. Position A/C control panel out from instrument panel opening.
2. Remove retaining screws and remove switch.
3. Reverse procedure to install.

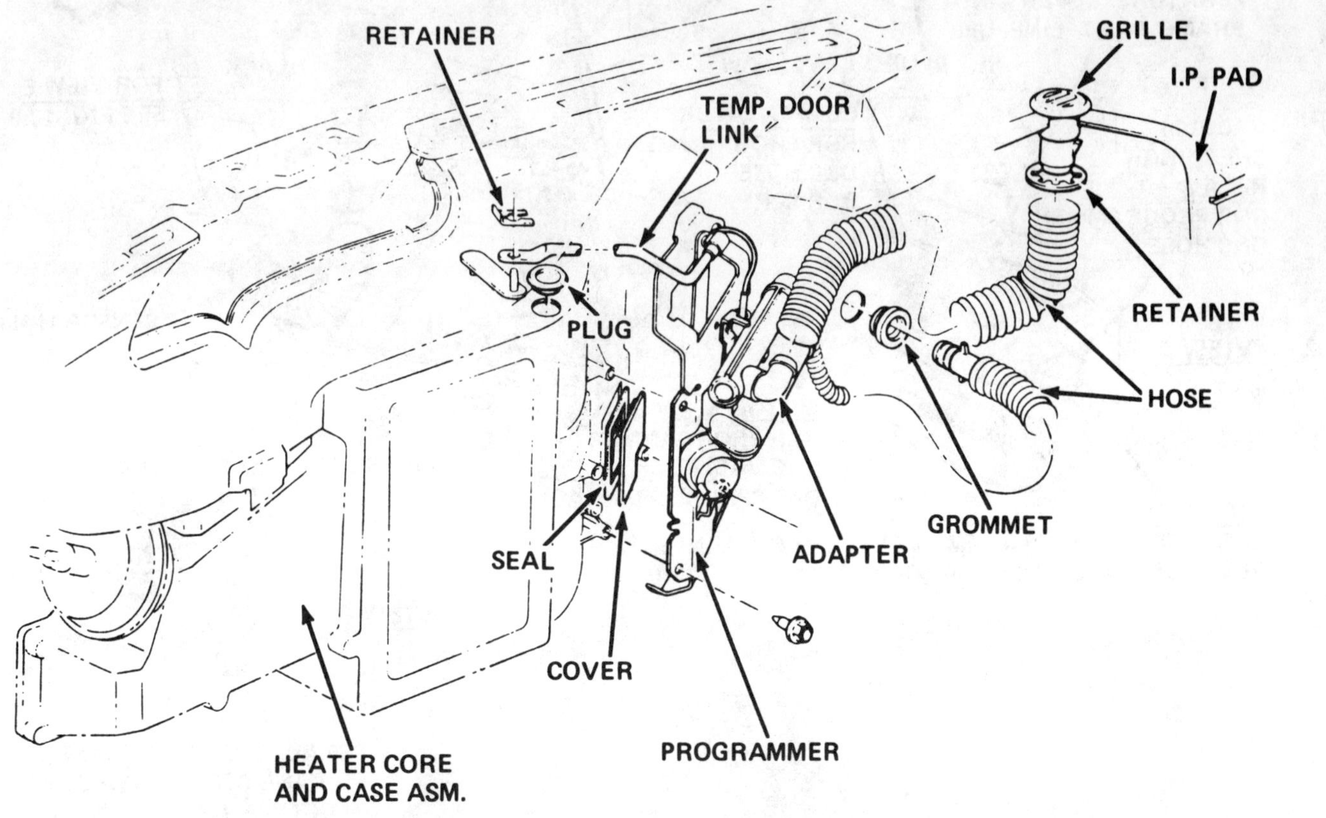

Fig. 196 Programmer assembly. 1975-76 Full Size

1975-76 Astre

1. Perform steps 1 through 4 as outlined under "Control Panel, Replace" 1975-76 Astre procedure.
2. Remove switch retaining screws and the switch.
3. Reverse procedure to install.

1976 Sunbird

1. Perform steps 1 through 5 as outlined under "Control Panel, Replace" 1976 Sunbird procedure.
2. Remove switch retaining screws and the switch.
3. Reverse procedure to install.

BLOWER RELAY, BLOWER RESISTOR & UNDERHOOD AMBIENT SWITCH, REPLACE

1. Disconnect battery ground cable.
2. Disconnect electrical connector from switch.
3. Remove retaining screw and remove switch.

A/C RELAY, REPLACE
1975-76 Astre

NOTE: This relay is mounted on the evaporator case cover.

1. Disconnect electrical connector from relay.
2. Remove relay cover retaining screws and remove relay.
3. Reverse procedure to install.

IN-CAR TEMPERATURE SWITCH, REPLACE
1974-76 Grand Am, Grand Prix, LeMans & 1974-80 Full Size

1. Remove A/C control panel.
2. Remove retaining screws and remove switch.
3. Reverse procedure to install.

1977 Grand Am, Grand Prix & LeMans

1. Remove A/C control panel, then remove insulator.
2. Remove switch retaining screws and switch.
3. Reverse procedure to install.

1978-80 Grand Am, Grand Prix & LeMans

1. Remove glove box, disconnect and remove switch.
2. Reverse procedure to install.

ATC IN-CAR SENSOR, REPLACE
1974-76 Models

1. Remove glove box.
2. Disconnect wire connector and aspirator hose from sensor.

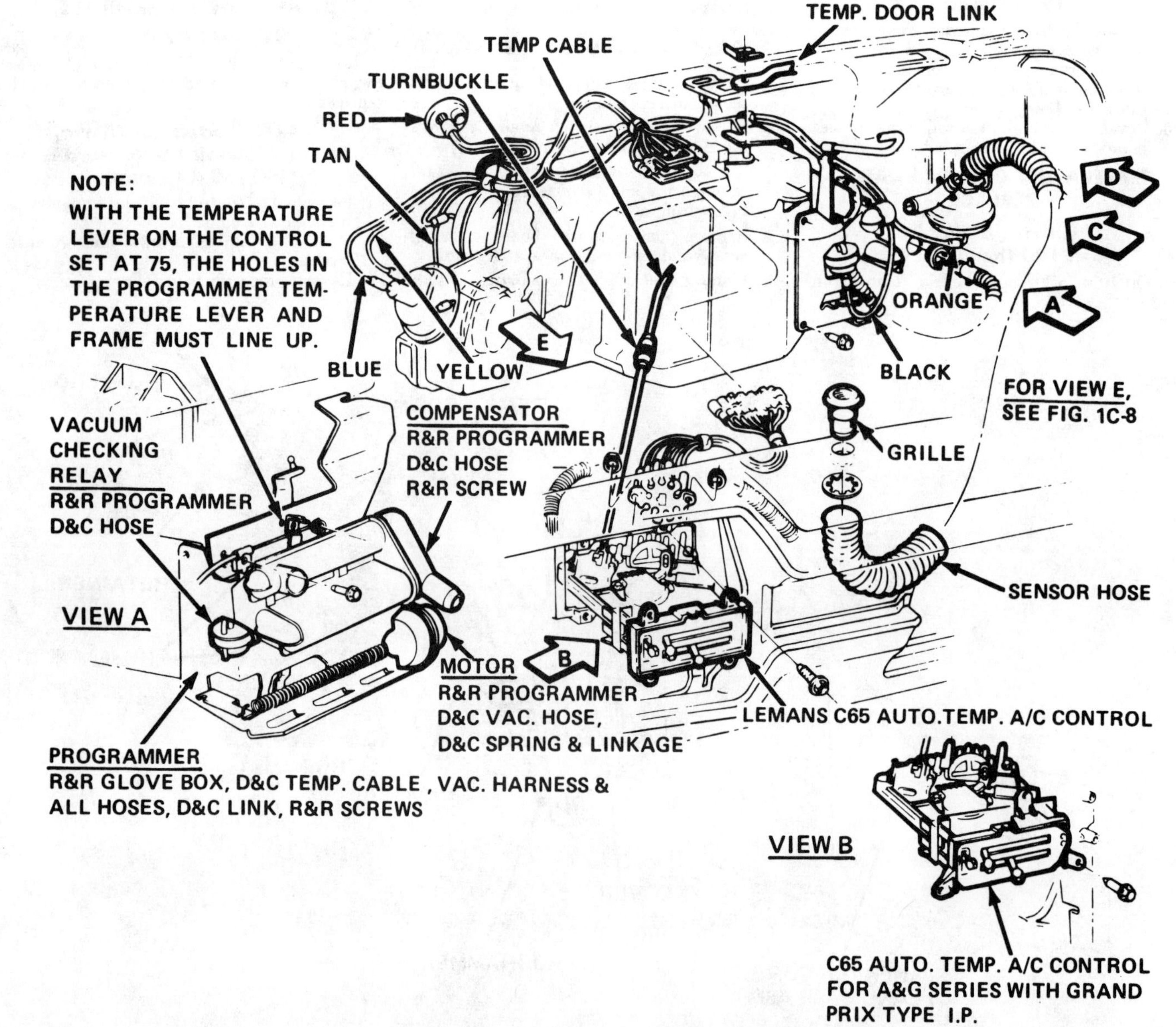

Fig. 197 Programmer assembly. 1975-77 Grand Am, Grand Prix & LeMans

3. On 1974 Full Size models, turn grille and adapter to release twist lock, then pull grille upward and remove sensor.
4. On 1974 Grand Am and Grand Prix models, release twist lock and remove sensor and grille.
5. Reverse procedure to install. Use care when routing aspirator hose to avoid pinching it behind glove box when it is installed.

1977-80 Grand Am, Grand Prix & LeMans

1. Drop glove box door by releasing side tabs.
2. Remove applique, then release snap tab.
3. Disconnect wiring and hose from sensor and remove sensor.
4. Reverse procedure to install.

1977-80 Full Size

1. Remove instrument panel pad and retaining clip from in-car sensor.
2. Disconnect wiring and hose from sensor and remove sensor.

ATC AMBIENT SENSOR & SWITCH, REPLACE

1. Remove electrical connectors and retaining screws.
2. Remove sensor and switch assembly.
3. Reverse procedure to install.

COMPRESSOR CLUTCH SWITCH, REPLACE

1974-76 Firebird

1. Disconnect battery ground cable.
2. Remove steering column lower instrument panel trim.
3. Disconnect electrical connector from clutch switch.
4. Remove retaining screw and remove switch.
5. Reverse procedure to install.

1975-76 Astre

1. Perform steps 1 through 4 as outlined under 1975-76 Astre "Control Panel, Replace" procedure.
2. Remove switch retaining screws and the switch.
3. Reverse procedure to install.

COMPRESSOR CUT-OFF SWITCH, REPLACE

1975-77 Astre & 1977-80 Sunbird

1. Disconnect electrical connector from switch.
2. Remove retaining screws and remove switch.
3. Reverse procedure to install.

HI-BLOWER SWITCH, REPLACE

1975-77 Astre

NOTE: This switch is located on the air inlet housing.

1. Disconnect electrical connector from switch.
2. Remove retaining screws and remove switch.
3. Reverse procedure to install.

THERMOSTATIC SWITCH, REPLACE

1975-77 Astre

1. Disconnect electrical connector from switch.
2. Remove switch cover retaining screws and remove switch using care to avoid damaging sensing line as it is removed from evaporator core.

NOTE: Mark location of sensing line to assure correct installation upon assembly.

3. Reverse procedure to install making sure that sensing line is making good contact with evaporator core.

TIME DELAY RELAY, REPLACE

1975-77 Astre & Sunbird

NOTE: This relay is located on the right hand fender skirt on Astre models and on the heater-evaporator case on Sunbird models.

1. Disconnect electrical connector from relay.
2. Remove retaining screws and remove switch.
3. Reverse procedure to install.

EVAPORATOR CORE, REPLACE

1974-76 Grand Am, Grand Prix, LeMans & Ventura

1. Disconnect battery ground cable and discharge refrigerant from system.
2. On Grand Am, LeMans, Grand Prix and Ventura models, remove VIR assembly as described under "VIR SERVICE" section.
3. On 1974 Ventura, disconnect oil bleed line and equalizer line at POA valve, then disconnect evaporator fittings from POA valve and expansion valve, disconnect thermo bulb from evaporator outlet and remove POA valve and expansion valve.
4. On all models disconnect all electrical connectors and vacuum lines from evaporator housing. On all except Ventura, disconnect blower motor ground wire.
5. Remove screws retaining evaporator case halves together and nuts retaining left evaporator case half to dash panel, Figs. 92, 93 and 96.
6. Carefully pry and remove left case half from dash as an assembly and remove core from case half.
7. Reverse procedure to install.

1974-76 Full Size & Firebird

1. Disconnect battery ground cable and remove VIR assembly as described under "VIR SERVICE" section.
2. Remove electrical connectors from evaporator case.
3. On Firebird models, from underneath vehicle, remove the two screws between case halves and one nut and screw retaining left half of case to dash.
4. Remove screws and nuts retaining evaporator case halves together, Figs. 95 and 148.
5. On Firebird models, remove dipstick retaining bolt and position dipstick aside.
6. Carefully pry and remove left case half and remove evaporator core.
7. Reverse procedure to install.

1977 Ventura; 1977-79 Phoenix; 1977 Grand Am, Grand Prix, LeMans & 1977-80 Firebird

1. Disconnect battery ground cable and discharge refrigerant system.
2. Disconnect all wiring from blower evaporator assembly, then disconnect blower tube.
3. Disconnect evaporator inlet and outlet lines and plug all lines and openings to prevent entry of dirt or moisture.
4. Remove screws and nuts securing blower evaporator assembly to firewall and remove assembly.
5. Remove evaporator core from assembly.
6. Reverse procedure to install.

1980 Phoenix

1. Disconnect battery ground cable and discharge refrigerant system.
2. Remove air cleaner.
3. Disconnect air conditioning electrical connections along the cowl and position wiring harness aside.
4. Remove vacuum tank.
5. Disconnect liquid line at evaporator inlet and the low pressure line at accumulator, Fig. 102.
6. Remove outer blower case to cowl screws, then the blower and case assembly.
7. Remove evaporator from vehicle.
8. Reverse procedure to install.

1977-80 Full Size

1. Disconnect battery ground cable and discharge refrigerant system.
2. Disconnect module rubber seal and remove module screen.
3. Remove right hand windshield wiper arm, diagnostic connector, if equipped, hi-blower relay and thermostatic switch screws. Unplug pressure cycling switch on 1980 models.
4. Disconnect all electrical connectors at module top and remove module top cover.
5. Remove accumulator bracket screws and disconnect refrigerant lines at accumulator and liquid line. Plug all lines and openings to prevent entry of dirt or moisture.
6. Remove evaporator core.
7. Reverse procedure to install.

1978-80 Grand Am, Grand Prix & LeMans

1. Disconnect battery ground cable and discharge refrigerant from system.
2. Disconnect module rubber seal and remove module screen.
3. Remove right hand windshield wiper arm.
4. Remove hi blower relay and thermostatic switch.
5. Disconnect air electrical connectors at module top, then remove module front cover, Figs. 98, 100 and 201.
6. Remove accumulator bracket screws, then disconnect refrigerant lines at accumulator and liquid line and remove core.
7. Reverse procedure to install using new sealant around evaporator case.

1975-77 Astre

Refer to the 1974-77 Chevrolet Vega "Evaporator Core, Replace" procedure for evaporator core service.

1976-80 Sunbird

Refer to the 1975-80 Monza Towne Coupe "Evaporator Core, Replace" procedure for evaporator core service.

HEATER CORE, REPLACE

Less Air Conditioning

Astre

Refer to the 1974-77 Chevrolet Vega "Heater Core, Replace" procedure for heater core service.

1974-76 Full Size, 1974-1977 Grand Am, Grand Prix, LeMans & 1974-80 Firebird

1. Drain radiator and disconnect heater hoses from core.

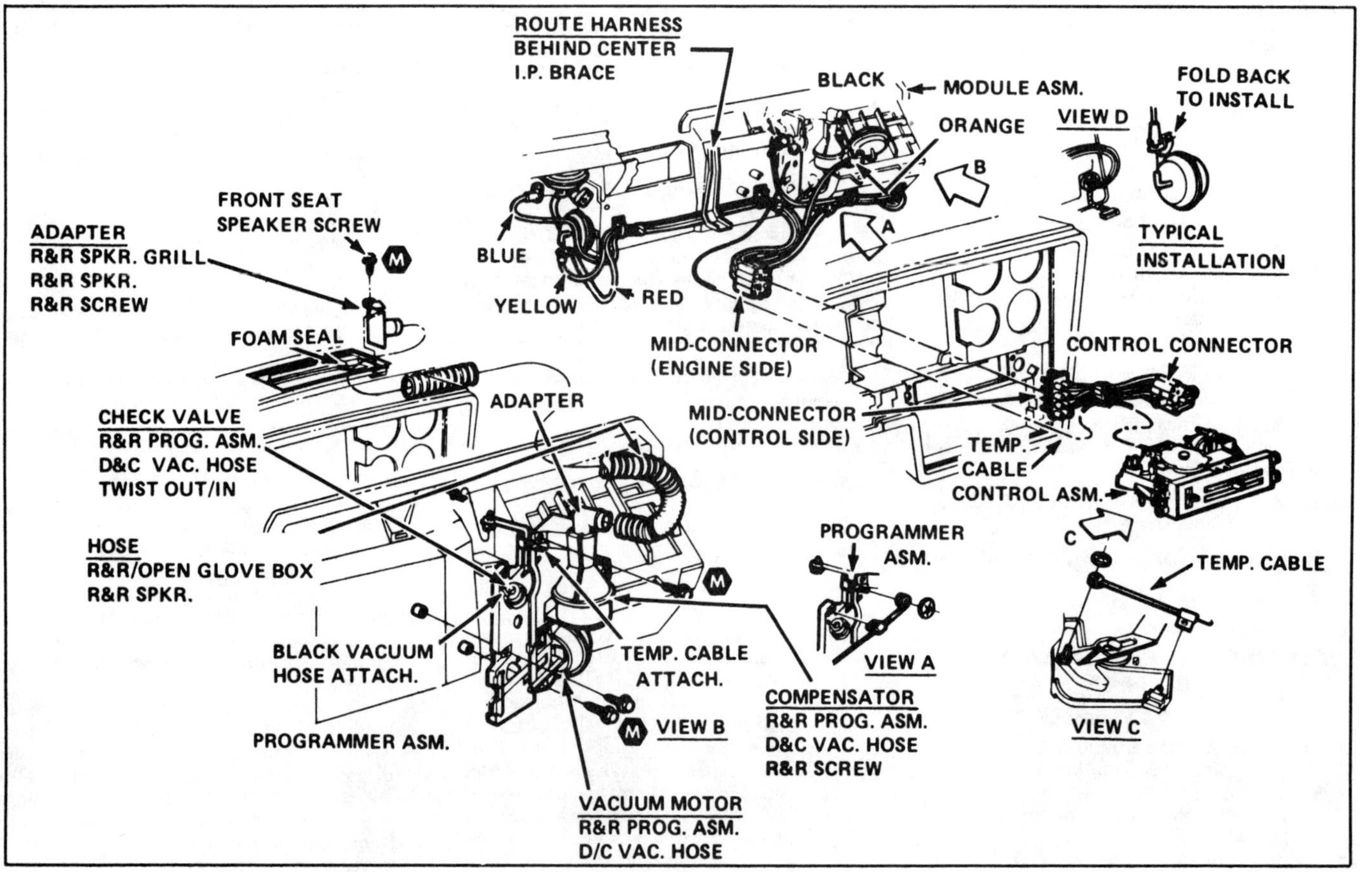

Fig. 198 Programmer assembly. 1978-80 Grand Am, Grand Prix & LeMans & 1977-80 Full Size

2. Remove retaining nuts from core case studs on engine side of dash, Figs. 158, 199 and 200.
3. Inside car, remove glove box and door on Firebird and heater outlet from case on Firebird.
4. Remove defroster duct retaining screw from heater case and pull entire heater assembly from firewall.
5. Disconnect control cables and wiring and remove assembly.
6. Remove core tube seal and core retaining strips and remove core.

1977-80 Full Size, 1978-80 Grand Am, Grand Prix & LeMans

1. Disconnect battery ground cable, drain cooling system and remove heater hoses.
2. Disconnect wiring from blower motor and on full size models, remove module front cover. On Grand Am, Grand Prix and LeMans models, remove module top cover, core bracket and ground screw.
3. Remove heater core, Figs. 106 and 107.
4. Reverse procedure to install.

1976-80 Sunbird Except 1978-79 Station Wagon

1. Disconnect battery ground cable and disconnect blower motor electrical lead.
2. Place a pan under vehicle, then disconnect heater hoses from core and secure hoses in a raised position.
3. Remove blower inlet to dash panel screws and nuts, then remove blower inlet and blower motor and wheel assembly.
4. Remove core retaining strap screws and remove core.
5. Reverse procedure to install making certain that blower inlet sealer is intact, replace as necessary.

1978-79 Sunbird Station Wagon

1. Disconnect battery ground cable.
2. Disconnect blower motor lead wire.
3. Place a pan under vehicle and disconnect heater hoses at core connections and secure ends of hoses in a raised position.
4. Remove the coil bracket to dash panel stud nut and move coil out of way.
5. Remove the blower inlet to dash panel screws and nuts and remove the blower inlet, blower motor and wheel as an assembly.
6. Remove the core retainer strap screws and remove the core, Fig. 160.
7. When replacing core, be sure the blower inlet sealer is intact.

1974 GTO, 1974-79 Ventura & Phoenix

1. Drain radiator and disconnect heater hoses from core.
2. Disconnect battery ground cable.
3. Remove retaining nuts from core case studs on engine side of dash, Figs. 80 and 81.
4. Remove glove box and door.
5. Drill out lower right hand heater case stud with ¼ inch drill from inside vehicle.
6. Pull entire heater core and case assembly from firewall.
7. Disconnect cables and wiring and remove assembly from car.
8. Remove core tube seal and core retaining strips and remove core.

1980 Phoenix

1. Disconnect battery ground cable and drain cooling system.
2. Disconnect the heater hoses from heater core.
3. Remove radio noise suppression strap.
4. Remove heater core cover retaining screws and the cover, Fig. 109.
5. Remove heater core from vehicle.
6. Reverse procedure to install.

With Air Conditioning

1974-80 Firebird

1. Drain radiator.
2. Remove glove box and door.
3. Remove cold air duct (lower right hand duct) and remove left and center lower A/C ducts.
4. Jack right front area of car and place on safety stand.

5. Remove rocker panel trim on right side and remove screws holding forward portion of rocker panel trim attaching bracket.
6. Remove three lower fender bolts at rear of fender.
7. Remove four fender to skirt bolts at rear of wheel opening.
8. Remove two fender skirt bolts near blower motor area.
9. Pry rear portion of fender out at bottom to gain access to hose clamp on lower core hose and disconnect hose.
10. Disconnect water pump to core hose at core.
11. Remove heater case retaining nuts under hood at dash.
12. Remove two heater case retaining bolts (inside car.)
13. Remove console, if equipped. If equipped with tape player, remove console with tape player intact, if equipped with tape player and no console, remove tape player.
14. Disconnect temperature cable at heater case.
15. Remove heater outlet duct.
16. Remove lower defroster duct screw at heater case.
17. Remove right kick panel.
18. Remove heater core and case, Fig. 158.
19. Disconnect vacuum hoses from heater case.
20. Remove core from case.

1974 GTO, 1974-79 Phoenix & Ventura

1. Disconnect battery and drain coolant.
2. Disconnect upper heater hose from core.
3. Remove right front fender skirt bolts and lower skirt to gain access to lower heater hose clamp. Disconnect lower heater hose clamp. Disconnect lower hose and remove lower right hand heater core and case attaching nut.
4. Remove glove box and door.
5. Remove recirculation vacuum diaphragm at right kick panel.
6. Remove heater outlet (at bottom of heater case).
7. Remove cold air distributor duct from heater case.
8. Remove heater case extension screws and separate extension from case. Fig. 105.
9. Disconnect cables and wiring and remove case and core assembly
10. Separate core from case.

1980 Phoenix

1. Disconnect battery ground cable and drain cooling system.
2. Disconnect heater hoses from heater core.
3. Remove right side hush panel and open the glove box.
4. Remove heater duct retaining screw and the duct.
5. Remove instrument panel support bracket.
6. Remove heater case side cover retaining screws and the cover.
7. Remove heater core retaining clamps and the inlet and outlet tube support clamps.
8. Remove heater core from case, Fig. 109.
9. Reverse procedure to install.

1974-76 Full Size

1. Drain radiator and disconnect heater hoses from core.
2. Remove three nuts and one screw retaining core and case to dash, Fig. 199.
3. Remove glove box and upper and lower instrument panel trim plates.
4. Remove radio.
5. Remove cold air duct and heater outlet duct.
6. Remove defroster duct to heater case screw.
7. Disconnect A/C temperature cable at heater case.
8. Disconnect vacuum hoses from diaphragms on heater case and remove core and case assembly.
9. Remove core from case (three screws).

1977-80 Full Size

1. Disconnect battery ground cable.
2. Remove right half of hood seal from air inlet screen.
3. Remove air inlet screen.
4. Remove screws securing top of module, then disconnect electrical connectors from electrical components on top of module.
5. Remove thermostatic switch mounting screws from top of module.
6. Remove A/C diagnostic connector mounting screws and position aside.
7. Remove top of module.
8. Disconnect and remove heater core from module, Fig. 106.
9. Reverse procedure to install.

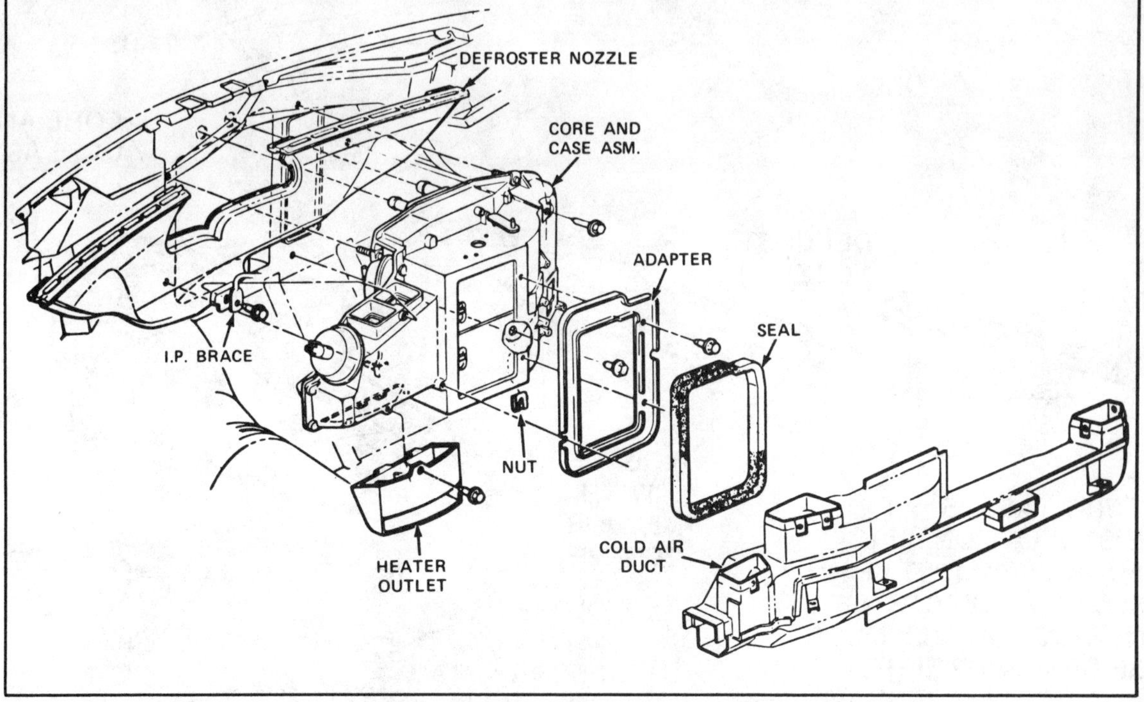

Fig. 199 Heater core & case assembly. 1974-76 Full Size

1974-77 Grand Am, Grand Prix & LeMans

1. Drain radiator and disconnect heater hose from core.
2. Remove glove box, cold air duct and heater outlet and defroster duct screw.
3. Disconnect heater core case from dash, Fig. 200. Remove blower motor resistor to gain access to case upper retaining nut.
4. Move case assembly rearward, freeing case studs from cowl and remove assembly.
5. Disconnect temperature cable and vacuum hoses from case and remove screws securing core inside of case.

1978-80 Grand Am, Grand Prix & LeMans

1. Disconnect battery ground cable.
2. Remove right hand windshield wiper arm and blade assembly.
3. Drain cooling system, then disconnect heater hoses at heater core and plug core openings.
4. Remove bracket and ground strap.
5. Remove module cover as follows:
   a. Remove seals and screens, then disconnect wire connector.
   b. Loosen and move up lower windshield reveal moulding, then remove reveal moulding cowl brackets.
   c. Tape a strip of wood below lower edge of windshield over module to prevent damage to windshield.
   d. Remove all cover attaching screws, then cut through sealing material along cowl.
   e. Carefully pry cover off from side, then lift cover away from flange of fender cowl brace.
   f. Remove heater core and seal from module, Fig. 107.

1975-76 Astre

Refer to the 1971-76 Chevrolet Vega "Heater Core, Replace" procedure for heater core service.

1975-76 Sunbird

1. Disconnect battery ground cable.
2. Remove glove box and right air duct.
3. Remove instrument panel bezel and pad.
4. Remove left air duct and feed duct.
5. Lower steering column and remove instrument panel.
6. Remove control assembly, radio, defroster duct and large distribution duct.
7. Purge system of refrigerant.
8. Disconnect heater hoses at core.
9. Clean VIR exterior surface and line connections, then disconnect compressor inlet line, oil bleed line and condenser outlet line from VIR assembly.

NOTE: Cap all lines and VIR ports.

10. Loosen evaporator inlet and outlet lines and remove VIR mounting clamp, then slide VIR assembly from evaporator lines.
11. Remove heater-distributer case to cowl attaching nuts and remove case, then disconnect wiring harness and vacuum lines.
12. Separate heater case from distributor, then remove heater core from case.

1977-80 Sunbird except 1978-79 Station Wagon

NOTE: The heater core can be removed without purging the A/C refrigerant system or removing the evaporator case half of the assembly by using the following procedure.

1. Disconnect battery ground cable.
2. Remove floor outlet duct, then remove glove box and door as an assembly.
3. Remove left and right hand dash outlets using a suitable tool.
4. Remove instrument panel pad.
5. Disconnect vacuum hoses at valves on left end of heater and evaporator unit.
6. Remove insulation tray located below instrument cluster, then loosen console and slide console rearward.
7. Remove instrument panel to dash attaching screws. Place a protective covering over steering column, then lower instrument panel to steering column. Disconnect speedometer cable, radio leads and control head connectors.
8. Remove right hand instrument panel and lap cooler as an assembly.
9. Remove modular duct to heater and evaporator case screw, then remove duct assembly.
10. Disconnect temperature door cable and wiring harness.
11. Disconnect heater hoses from heater core. Plug core tubes to prevent spillage when removing heater core.
12. Remove three heater case stud nuts located in engine compartment.
13. Remove heater core case to evaporator core case attaching screws.
14. Drive on studs to break case loose from dash panel.
15. Remove heater core case assembly.
16. Remove heater core to case attaching screws, then remove heater core.

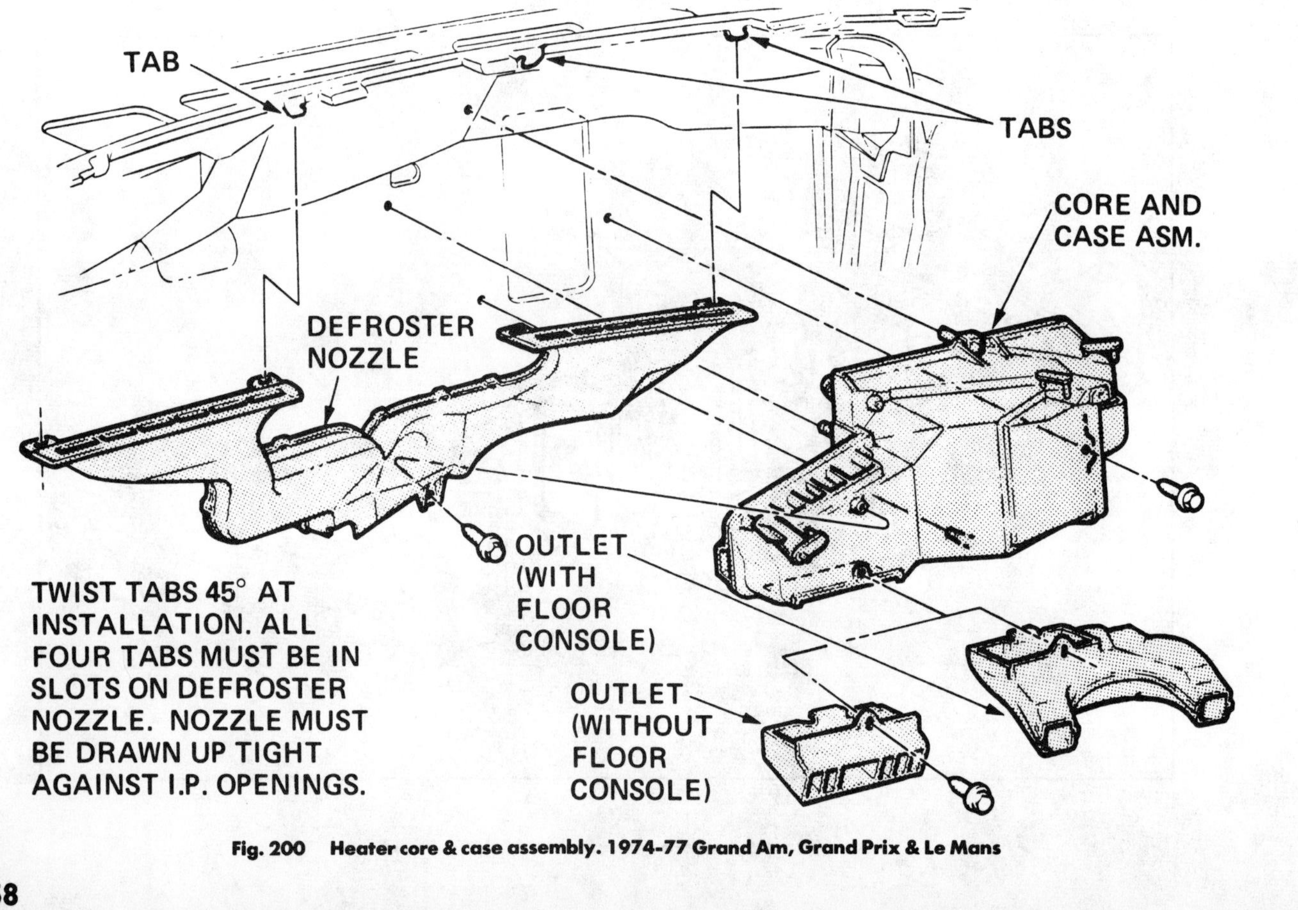

Fig. 200 Heater core & case assembly. 1974-77 Grand Am, Grand Prix & Le Mans

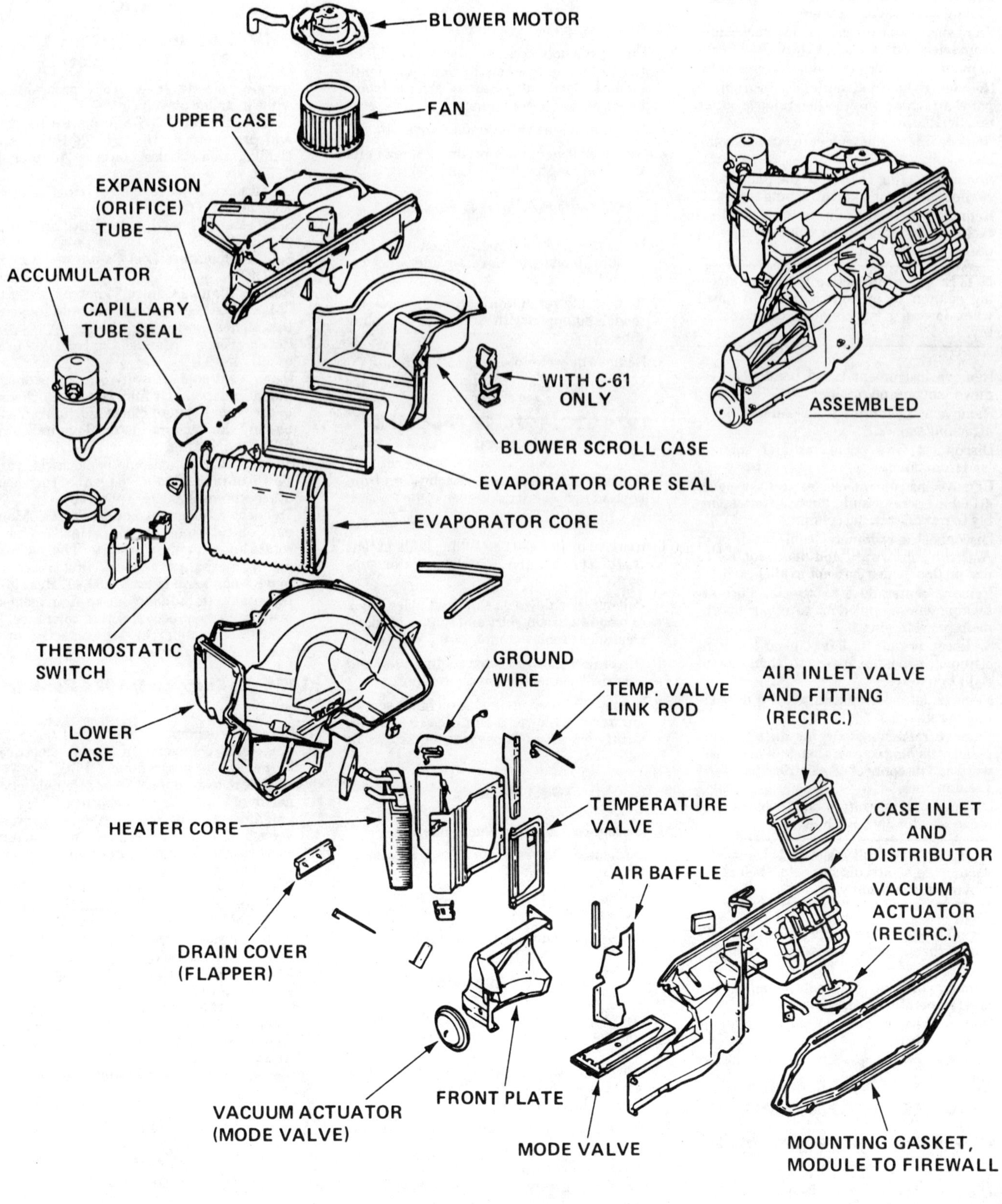

Fig. 201 Evaporator core and case assembly. 1978-80 Grand Am, Grand Prix & Le Mans

1978-79 Sunbird Station Wagon

1. Disconnect battery ground cable.
2. Position a pan under heater core tubes, then disconnect heater hoses and secure in a raised position. Cap core tubes to prevent coolant spillage during removal.
3. From engine side of dash panel remove nuts from selector duct studs, Fig. 161.
4. Remove glove box and door as an assembly.
5. Remove right hand outlet to instrument panel attaching screws, then remove outlet and flexible hose.
6. Remove intermediate duct to left hand outlet.
7. Remove steering column to toe pan plastic retainer, insulation and attaching screws.
8. Remove steering column to dash panel attaching nuts, then lower column and support on driver's seat.

NOTE: Place protective tape over steering column to prevent damage to finish when lowering instrument cluster assembly.

9. Remove instrument panel bezel, then remove ashtray and retainer.
10. Remove A/C control to instrument panel attaching screws.
11. Disconnect wire connector and antenna lead from radio.
12. Remove instrument cluster to dash panel attaching screws and allow cluster assembly to rest on steering column.
13. Disconnect speedometer cable, then push A/C control forward and allow control to rest on floor, using care not to kink cable.
14. Remove center duct to selector duct attaching screws and duct to upper instrument panel retainer.
15. Slide center duct to left to clear lower instrument panel to cluster tab, then slide to right to remove.
16. Remove defroster duct to selector duct attaching screws.
17. Remove remaining selector duct to dash panel attaching screws, then pull duct rearward and disconnect all electrical leads and vacuum lines.
18. Disconnect temperature door cable and remove selector duct assembly.

NOTE: Ensure all electrical leads and vacuum lines are disconnected before removing duct assembly.

19. Pry off or punch out temperature door bellcrank, then remove temperature door, Fig. 162.
20. Remove backing plate attaching screws and temperature door cable retainer.
21. Remove heater core and backing plate as an assembly.
22. Remove core straps, then remove heater core.

BLOWER MOTOR, REPLACE

1974-80 Grand Am, Grand Prix & LeMans

1. Disconnect blower motor feed wire.
2. On 1977 models, remove air inlet duct assembly. On 1978-80 models equipped with A/C, remove cooling tube.
3. Remove blower motor retaining screws, Figs. 111, 166.
4. Remove blower motor.

1974-76 Full Size & Firebird

1. Raise car and remove right front wheel.
2. Cut access hole approx. ¾ of the way along the outline stamped on the right hand fender skirt. Bend cut section of skirt outward for access to blower motor.
3. Disconnect blower motor feed wire.
4. Remove blower motor retaining screws and the blower motor, Figs. 95 and 166.

1977-80 Full Size & Firebird

1. Disconnect blower motor feed wire. On Firebird models, remove air inlet duct assembly.
2. Remove blower motor retaining screws. On models equipped with A/C, remove cooling tube.
3. Remove blower motor, Figs. 98 and 166.
4. Reverse procedure to install.

1974 GTO, 1974-79 Phoenix & Ventura

1. Disconnect battery and detach hoses from clips on right fender skirt.
2. Raise vehicle on hoist.
3. Remove fender skirt attaching bolts except those retaining the skirt to radiator support.
4. Pull out then down on skirt and place block of wood between skirt and fender to allow clearance for blower motor removal.
5. Disconnect blower motor cooling tube and electrical connections at blower motor.
6. Remove blower motor attaching screws and remove blower motor, Figs. 93 and 94. Gently pry motor flange if sealer acts as an adhesive.

1980 Phoenix

1. Disconnect battery ground cable.
2. Disconnect blower motor electrical connections.
3. Remove blower motor retaining screws and blower motor.
4. Reverse procedure to install.

1976-80 Sunbird

1. Disconnect battery ground cable and disconnect blower motor electrical leads. On some models, it may be necessary to remove coolant recovery tank attaching screws and position tank aside. Do not disconnect hoses from tank.
2. Scribe blower motor flange to case position.
3. Remove blower motor retaining screws and remove blower motor and wheel assembly. Pry flange gently, if sealer acts as an adhesive.
4. Reverse removal procedure to install, aligning scribe marks made during removal.

GM Auto Temperature Control Calibrations

FEEDBACK POTENTIOMETER CALIBRATION

1974-80 Buick, Cadillac & 1977-80 Chevrolet

1. Remove plastic cover from programmer while installed on vehicle.
2. Connect Tester J-23678 into wire harness and programmer. On 1974-78 Buick and Cadillac models, use Harness Adapter J-24774.
3. On all models, place control panel selector lever in "AUTO".
4. Set Tester "Manual-Automatic" switch to "Manual".
5. Set "Temperature Dial Calibrator" switch to "Off".
6. Rotate "Manual Control" knob on tester to "Max. Heat". Programmer should move to the full heat position.
7. Rotate Tester "Manual Control" knob to exactly 180 ±1.
8. Using a suitable screwdriver, slip feedback potentiometer shaft fully counterclockwise to the stop. Vacuum motor mechanism will be "in" the vacuum motor. This indicates full heat operation.
9. Slip feedback potentiometer clockwise very slowly until the first movement of the vacuum motor mechanism is observed.
10. To check adjustment, rotate Tester "Manual Control" knob to "Max. Heat". Slowly rotate knob counterclockwise. The vacuum motor mechanism should first start to move when knob is set to 180 ±1. Readjust potentiometer adjustment so first mechanism movement occurs within tolerance. If adjustment cannot be set correctly, programmer is defective.

1974-76 Chevrolet; 1974 Pontiac All

1. Remove programmer cover.
2. Plug male connector of Tester electrical harness into programmer. Plug J-23713 electrical test connector into female connector of Tester electrical harness.
3. Plug dummy plug from tester onto rotary valve and to vacuum source. Turn vacuum source on and adjust to obtain 20 inches Hg. or more.
4. Connect yellow lead of tester to positive lead and black lead of tester to negative lead of a 12 volt power supply. Voltmeter on tester should read 12 volts when in "2 Feed", "#7 Lo-Relay" and "#8 Auto-Relay" positions. If not, check voltage supply and connections.
5. Using a screwdriver turn shaft of feedback potentiometer fully counterclockwise (gear does not move). Vacuum motor should now be in full heat position (vacuum motor mechanism will not move into vacuum motor).
6. Place Manual-Automatic switch in "MANUAL" position and rotate Manual Control to "MAX.-HEAT" position.
7. Carefully adjust Manual Control to exactly "180", then turn shaft of feedback potentiometer clockwise observing for vacuum motor mechanism movement (do not watch shaft) and stop turning when movement first occurs.

8. Recheck adjustment with Manual Control, by placing Manual-Control to "MAX.-HEAT" position and observing for vacuum motor movement while slowly turning Manual Control back toward "180". Movement should first occur when Manual Control knob is exactly "180". If not, readjust feedback potentiometer so that the programmer mechanism movement occurs as described previously.

TEMPERATURE DIAL OR LEVER CALIBRATION

1974 Pontiac All

1. Connect Tester J-23678 to Automatic Temperature Control electrical harness and to programmer.

NOTE: J24774 Tester Adapter Harness is required on full size models.

2. Place control panel selector lever in "VENT" position.
3. Place Tester Manual-Automatic switch in "MANUAL" position and Temperature Dial Calibrator switch in "CAL." position.
4. Note voltmeter reading, then push COMPARE and turn temperature dial to obtain same voltmeter reading as noted previously.
5. The temperature setting on the control dial should now be at 75 for all full size models except Grand Prix and at 73 for Grand Am and Grand Prix models. If not, insert tool J-21530 into left side of temperature dial and rotate dial until the correct setting is obtained. If correct setting cannot be obtained, temperature dial is defective.

1974-76 Buick & Chevrolet

NOTE: Allow sufficient time for engine warm-up and ATC system to operate before performing calibration procedure.

1. Connect Tester J-23678 into ATC wiring harness and the programmer.
2. Place control panel selector lever in "Vent" position.
3. Set Tester "Manual-Automatic" switch to "Manual".
4. Set Tester temperature dial calibrator switch to "Cal".
5. Note voltmeter reading on Tester.
6. Press "Compare" button and note voltmeter reading.
7. With "Compare" button pressed in, rotate control panel temperature dial until voltmeter reading is the same as in Step 5.
8. Control panel temperature should be set at the same temperature as on the Tester panel. If not, use tool J-21530 to hold gear on the left side of temperature dial and slip dial to the correct setting. If temperature dial cannot be calibrated using this procedure, the dial is defective.

1977-80 Buick, Chevrolet, DeVille, Brougham & 1979-80 Pontiac Except Full Size Models

1. Remove control head and disconnect the two green wires from the temperature dial potentiometer.
2. Connect an ohmmeter to the terminals and set the temperature dial to 75 degrees.
3. If ohmmeter reading is not 36 ohms, slip temperature dial clutch until that value is noted.

1979-80 Pontiac Full Size Models

1. Disconnect harness from ambient sensor under hood and connect a jumper wire between the two harness terminals.
2. Remove glove box to gain access to A/C connectors near programmer.
3. Disconnect harness from in car sensor.
4. Disconnect 3-wire harness from programmer amplifier.
5. Connect an ohmmeter across temperature dial with leads as follows:
   a. Terminal 62 of 3-wire harness connector.
   b. Terminal 198 of in-car sensor connector.
6. If ohmmeter reading is not 36 ohms at 75 degree dial setting, adjust to attain setting.

NOTE: Ensure proper jumper wire connection as ohmmeter reading will be affected by a poor connection.

1974-76 Cadillac, 1977-79 Eldorado & Seville

1. On 1974-76 models, connect Harness Adapter J-24774 to wiring harness and programmer. Connect Tester to Harness Adapter. Then working through opening in top of glove compartment, disconnect in-car sensor electrical connector and connect wiring harness to the double connector (black wires) of the Harness Adapter.
2. On all models, set Tester "Manual-Automatic" switch to "Manual". Set "Temperature Dial Calibrator" switch to "Calibrate". Note voltmeter reading.
3. Press Tester "Calibrate" button and note voltmeter reading.
4. With "Calibrate" button pressed in, rotate control head temperature dial until voltmeter reading is the same as in Step 3 (button not pressed in).
5. On 1974-76 models, control head temperature dial should be set at 75. On all models, if setting is not correct, use Tool J-21530 to hold gear of control head temperature dial and slip the dial to the correct setting. If temperature dial cannot be calibrated using this procedure, the dial is defective.

1980 Eldorado & Seville

1. Set temperature on control panel to 75 degrees and mode selector to Auto.
2. Connect a voltmeter to the test lead from Pin W of programmer and to ground.
3. Record programmer voltage.
4. To increase temperature setting one degree, use a small screwdriver to adjust the potentiometer so that a .2 volt decrease is noted on voltmeter.

NOTE: To increase or decrease temperature setting by more than one degree, multiply the desired change by .2 volt and adjust the potentiometer so that voltage reading increases or decreases accordingly.

V. VIR SERVICE

VIR REMOVAL

NOTE: Service of the VIR assembly requires removal from vehicle. Also, whenever the unit is removed, the drier-desiccant bag must be replaced. Refer to Fig. 1 for a disassembled view of VIR unit.

1. Disconnect thermal limiter connector, discharge refrigerant system and disconnect all lines from VIR. Cap all disconnect lines.
2. Remove VIR mounting clamp and slide assembly from outlet tube, then the inlet tube.
3. Remove six mounting screws and the receiver shell. Discard drier-desiccant bag and the valve housing to receiver "O" rings.
4. Drain, measure and discard any oil found in receiver. Remove pick-up tube filter screen.

Pick-Up Tube, Replace

1. With a small screwdriver, raise tangs of pick-up tube retainer ring until ring is free of valve housing opening. Then, remove tube and discard retainer ring and "O" ring.
2. Lubricate new pick-up tube and retainer into valve housing with tool J-24182-3. Ensure tube is bottomed in opening and the tool vertically in line prior to seating tube and retaining ring.

Inlet Connector Shell, Replace

1. Remove inlet connector shell mounting screws and lift shell from valve housing. Discard shell to housing "O" ring.
2. Inspect shell and housing flanges for scratches.
3. Lubricate new shell to housing "O" ring and install in valve housing groove. Position shell on housing and install and torque mounting screws to 7-12 ft. lbs.

Expansion Valve Capsule, Replace

1. Remove inlet connector shell as outlined previously.
2. Loosen one valve capsule retaining screw three turns and remove other screw and washer.
3. Remove valve capsule with tool J-24182-1, Fig. 2. When capsule lifts free, remove tool and remaining screw and washer, then the valve capsule.
4. Remove "O" ring from expansion valve cavity with tool J-9553. Remove "O" rings from valve capsule.
5. Lubricate new "O" rings with refrigerant oil and install "O" rings in valve cavity and on valve capsule.
6. Install valve capsule into cavity and install and torque capsule retaining screws to 5-7 ft. lbs.
7. Install inlet shell connector as outlined previously.

POA Valve Capsule, Replace

1. Perform steps 1, 2 and 3 outlined under "Expansion Valve Capsule, Replace."
2. Remove valve capsule "O" ring.
3. Clean valve cavity and install new "O" ring on valve capsule.
4. Install valve capsule into housing cavity and install and torque capsule retaining screws to 5-7 ft. lbs.
5. Install inlet connector shell as outlined previously.

Moisture Indicator, Replace

1. Remove VIR as outlined previously.
2. Remove expansion valve capsule as described previously.
3. Using a pair of pliers, grasp moisture indicator at the exposed end and remove sleeve off expansion valve capsule.
4. Replace lower "O" ring of expansion valve with a new "O" ring and wipe any residue off expansion valve using a lint free cloth.
5. Place new moisture indicator on end on a clean flat and firm surface and align lower end of the expansion valve capsule with interior of sleeve. Insert end of capsule into sleeve and push down until the sleeve is fully seated in place.

NOTE: If capsule is not kept aligned with sleeve, the sleeve may become distorted enough to cause the moisture indicator material to chip.

6. Install expansion valve capsule.
7. Replace drier dessicant.
8. To replace valve housing, it is necessary to remove the expansion valve and POA valve capsules and the pick-up tube as outlined previously. Check new valve housing for dirt or foreign material, then install components removed previously.
9. Lubricate new valve housing to receiver

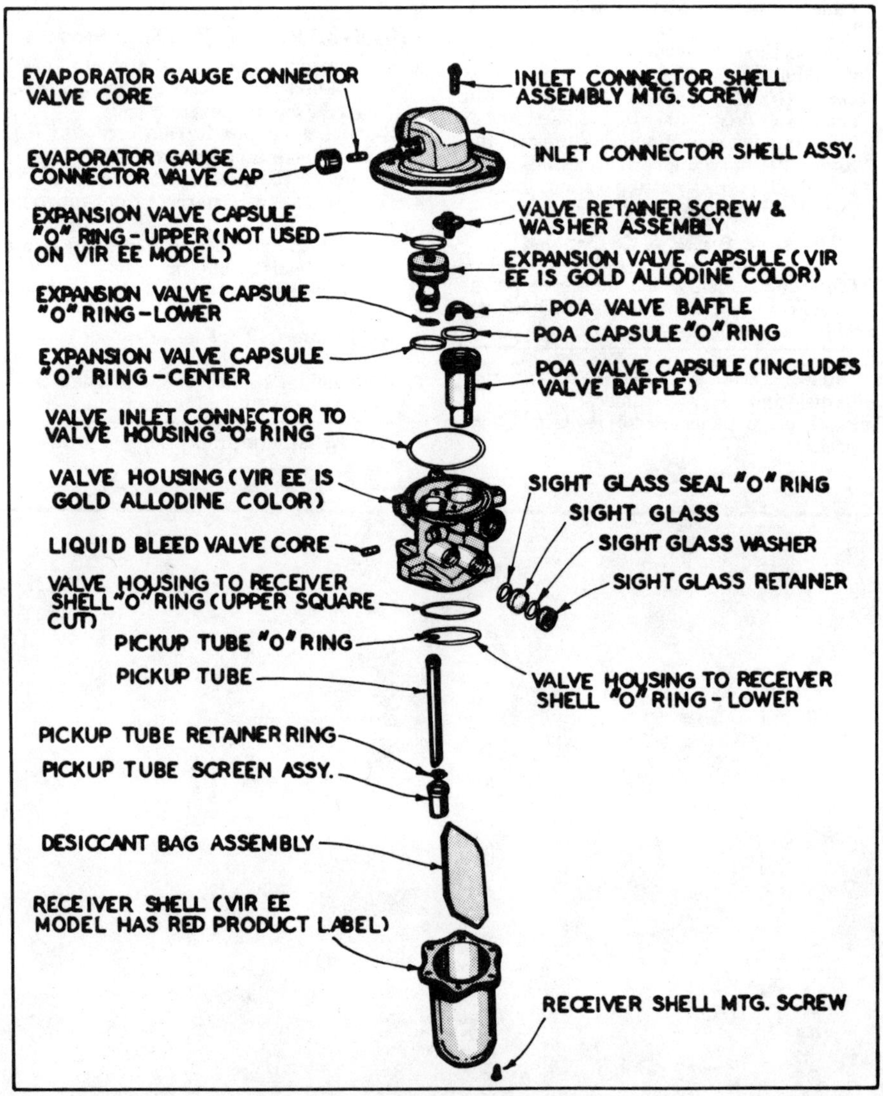

Fig. 1 **VIR exploded view**

Fig. 2 Removing expansion or POA valves

"O" ring with refrigerant oil and install ring in valve housing groove.

10. Lubricate inner upper surface of receiver with refrigerant oil and assemble filter screen to pick-up tube.
11. Add one ounce of clean refrigerant oil to receiver plus an amount of oil equal to that lost during system discharge.
12. Install new drier-desiccant bag and assemble receiver to valve housing. Torque receiver mounting screws to 7-12 ft. lbs.

VIR INSTALLATION

1. Replace all line connection "O" rings and lubricate with refrigerant oil.
2. Remove caps from disconnected lines and install VIR onto evaporator inlet line first, then the outlet line. When unit is positioned properly, install mounting clamp.
3. Torque evaporator inlet connection to 15-20 ft. lbs. and the outlet connection to 28-33 ft. lbs., Fig. 3.
4. Connect liquid bleed line and torque connection to 5-7 ft. lbs., Fig. 3.
5. Connect liquid line and torque connection to 11-13 ft. lbs., Fig. 3.
6. Connect suction line and torque connection to 28-33 ft. lbs., Fig. 3.
7. Evacuate, leak test and charge refrigerant system.
8. Connect thermal limiter connector.

Sight Glass, Replace

1. Disconnect thermal limiter connector and discharge refrigerant system.
2. Remove sight glass retaining nut with a 7/16 inch Allen wrench.
3. Place finger over sight glass, then slightly pressurize system with refrigerant to force sight glass from opening.

NOTE: Use only minumum refrigerant pressure to force sight glass from opening. Guide sight glass from opening by shifting finger pressure from side to side.

4. Remove VIR from vehicle and, using tool J-9553, remove sight glass "O" ring.
5. Lubricate new sight glass, "O" ring, thrust washer and retaining nut with refrigerant oil and install components into opening. Torque sight glass retaining nut to 20-25 inch lbs.
6. Install VIR as outlined previously.

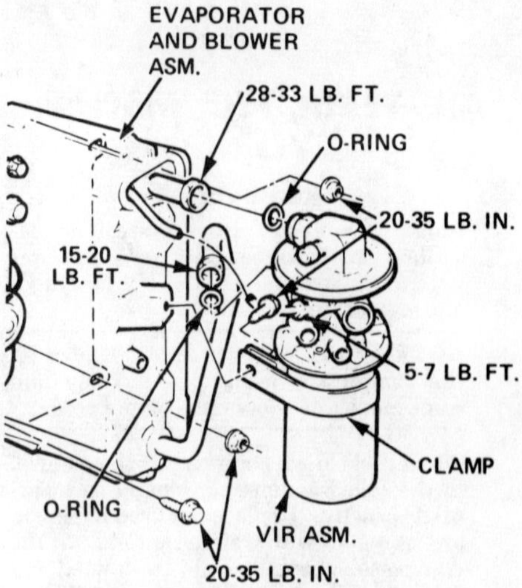

Fig. 3 VIR line connections

VI. COMPRESSOR SERVICE

AIR-TEMP, TECUMSEH & YORK

Removal

1. Isolate compressor from system as described previously, or purge system of refrigerant if compressor does not have high and low side service valves.

NOTE: These valves are not mounted on the compressor heads of Air-Temp and some models of Tecumseh compressors.

2. When replacing a compressor, the oil level in the failed compressor should be measured exactly. Replacement compressors are shipped with refrigerant oil in the crankcase. This oil should be drained and discarded. Replace with new oil in the exact amount measured in the old compressor. This keeps the amount of oil in the system constant.

NOTE: Failure to use this procedure is the most common reason for excessive oil in the crankcase. Refer to the oil level test procedure elsewhere in this manual.

3. Disconnect clutch coil wire and the refrigerant hoses. Cap open lines.
4. Loosen and remove the drive belt. Then remove mounting bolts and the compressor.

NOTE: Refer to "Unit Replacement" section for clutch service.

Installation

1. After draining oil out of the replacement compressor, install compressor to the mounting bracket. Tighten compressor mounting bolts. Fill compressor with the proper amount of refrigerant oil as previously described.
2. Remove protective caps from refrigerant hoses. Connect hoses and clutch coil wire.

CAUTION: When replacing the compressor assembly, the crankshaft should be rotated by hand several complete revolutions to clear oil accumulation from the compressor head before the clutch is energized. This is to avoid damaging the compressor reed valves.

3. Install compressor drive belt and adjust tension.
4. Evacuate, charge and operationally test the system.

Reed Plate Assembly

These compressors, Figs. 1, 2 and 3 are the reciprocating type which, after the compressor heads are removed, servicing the pistons, rods and crankshaft will pose no difficulty. Therefore, we will deal here only with reed plate assemblies.

NOTE: Although individual reeds are available and may be replaced separately on these compressors, experience has shown that it is advisable to replace the complete reed plate as an assembly. New reeds are fitted to the plate and positive action is a certainty. Too often the old plate has warped in service, and the re-use of this plate will result in failure of the new reeds and compressor.

Removal

1. Isolate compressor from system as outlined previously, or purge system of refrigerant if compressor does not have both high and low side service valves. *Service valves are not mounted to compressor heads on Air-Temp and some models of Tecumseh compressors.*
2. On models so equipped, unfasten and lift off service valves with hoses attached.

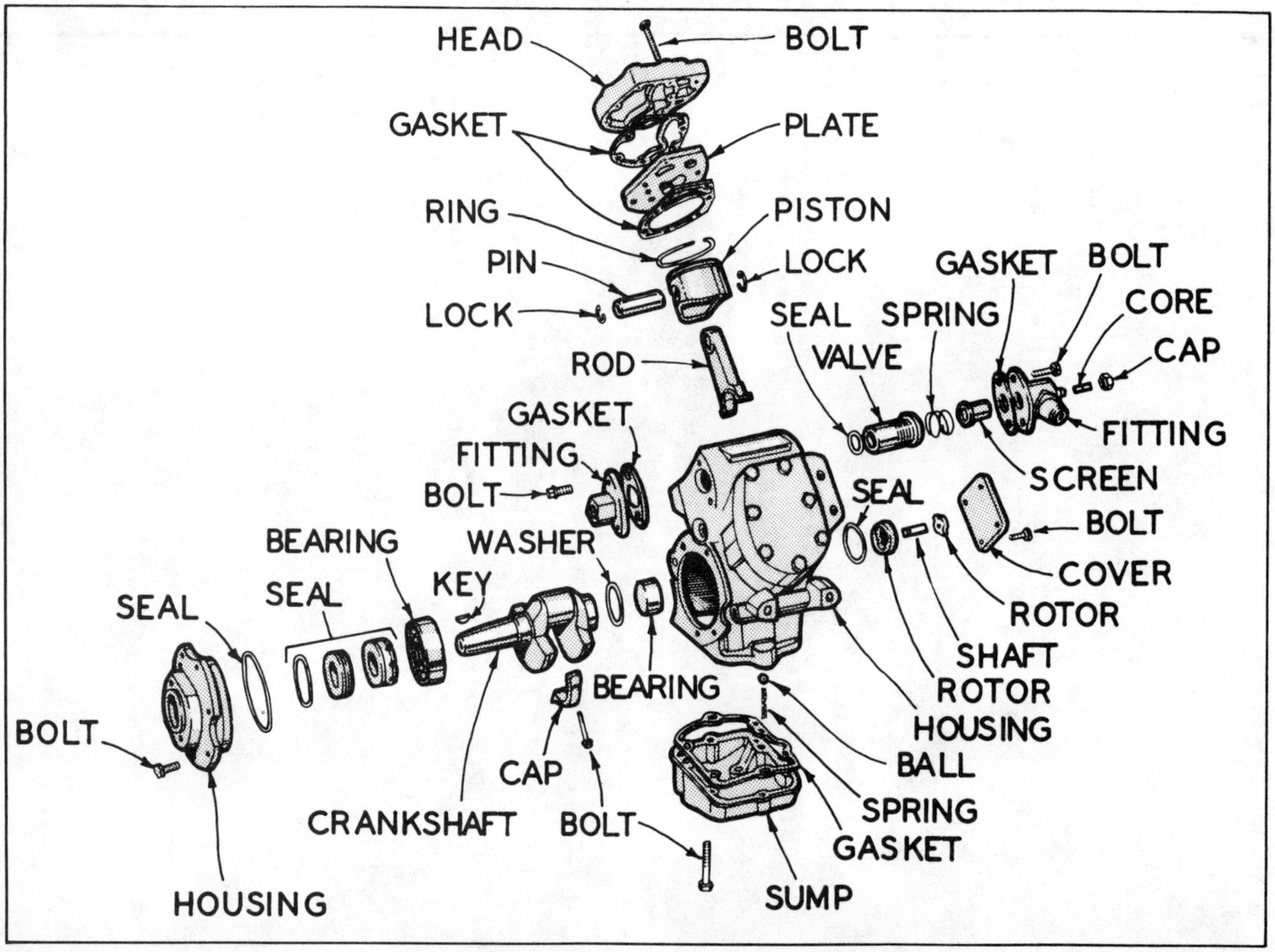

Fig. 1 Chrysler Air-Temp compressor exploded view

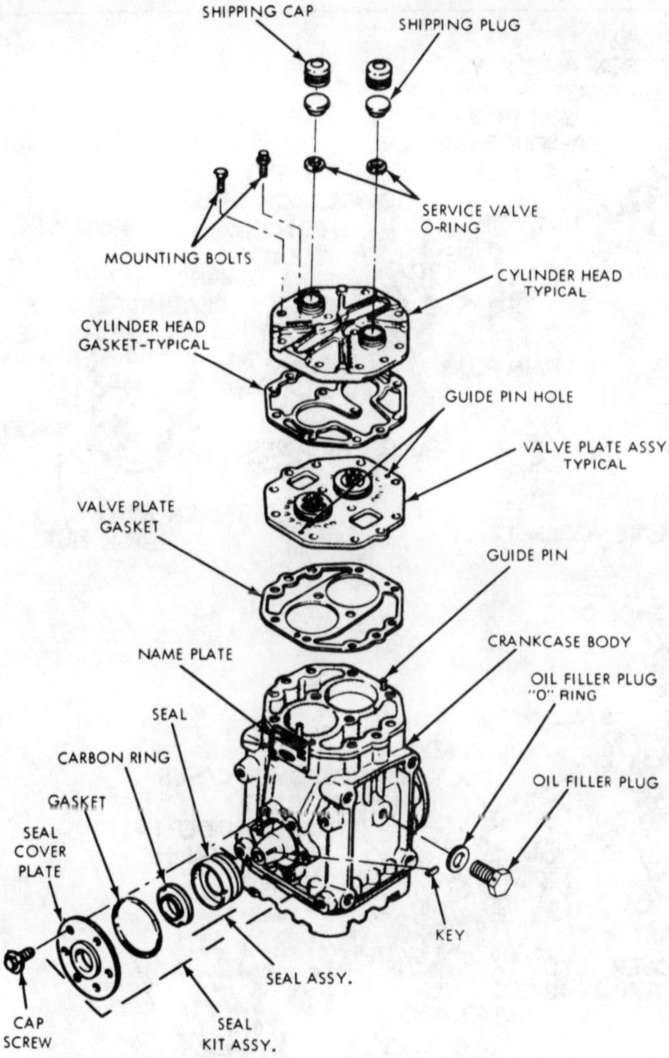

Fig. 2 York compressor exploded view

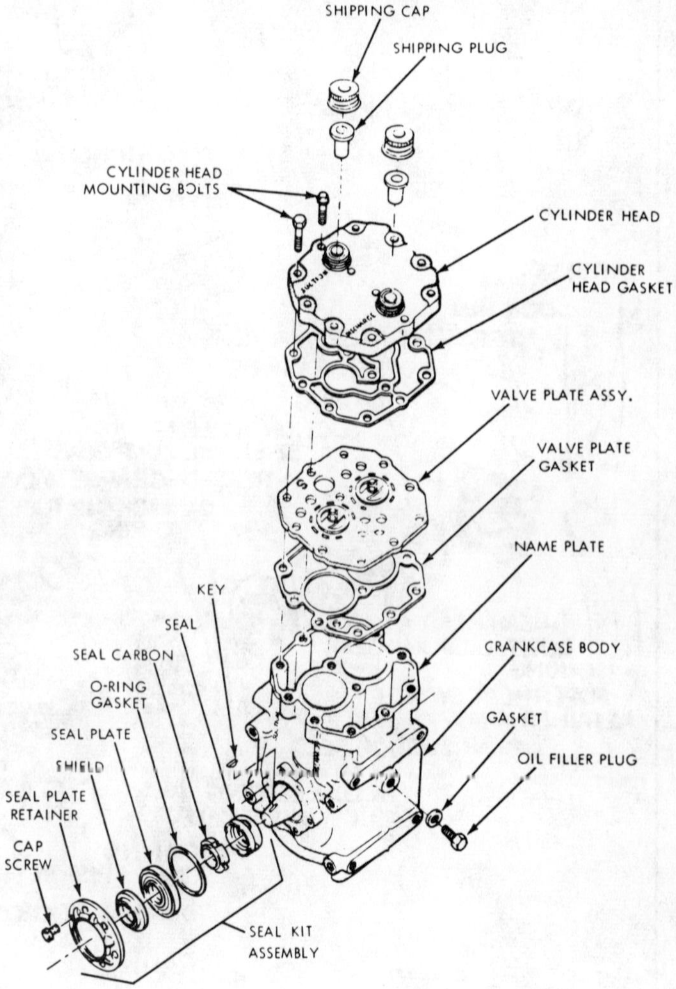

Fig. 3 Tecumseh compressor exploded view

Carefully clean all of old gaskets from mounting surface of service valves and compressor.

3. Remove compressor head bolts. Tap extruding flange on each side of head to loosen head from compressor body.
4. Lift valve plate and head from compressor body, and separate reed plate from compressor head. Clean old gasket material from compressor head and body.

NOTE: Air-Temp compressors having two heads and reed plates will require the above for each head. Several models of Air-Temp compressors with service valve mounted between the heads, separate the high and low side by a service valve mounting gasket. Failure of this gasket will give a similar indication of reed valve and/or gasket failure.

Installation

Use clean refrigeration oil on new gaskets. Use of various sealers on gaskets will result in excess sealer being picked up by the refrigerant and deposited in screens throughout system. Complete clogging of expansion valve screen can result, causing system failure.

1. Coat new gasket between reed plate and compressor liberally with refrigeration oil and install gasket on aligning pins on compressor body.
2. Install new reed plate on alignment pins or compressor body and press into place.
3. Coat new head gasket liberally with refrigeration oil and install on aligning pins of compressor body.
4. Install head on aligning pins of compressor body and press into place.
5. Install head bolts finger tight. Then tighten head bolts in a criss-cross manner, gradually and evenly, until all are tightened securely.
6. On models having service valves mounted to compressor head, coat new mounting gaskets with refrigeration oil and install on compressor head. Install bolts finger tight. Then tighten inner bolts first, gradually and evenly, then the outer bolts.
7. Follow procedure for isolating compressor, or evacuate and recharge system.
8. Adjust control for maximum performance.

FRIGIDAIRE & MOTORCRAFT

Removal

1. Discharge refrigerant from system.
2. Loosen hose connector plate retaining bolt. As the bolt is being loosened, work the high and low pressure hose fittings back and forth, breaking the seal and carefully bleeding off any remaining pressure.

NOTE: High pressure may still exist at the discharge fitting. If this pressure is released too rapidly, a considerable discharge of refrigerant and oil may result.

3. Remove bolt and hose connector plate and seal all open connections.
4. Remove compressor drive belt and disconnect compressor electrical leads.
5. Remove compressor attaching bolts and nuts, then the compressor.
6. Before compressor disassembly, drain and measure refrigerant oil remaining in compressor. Also, check oil for contamination.

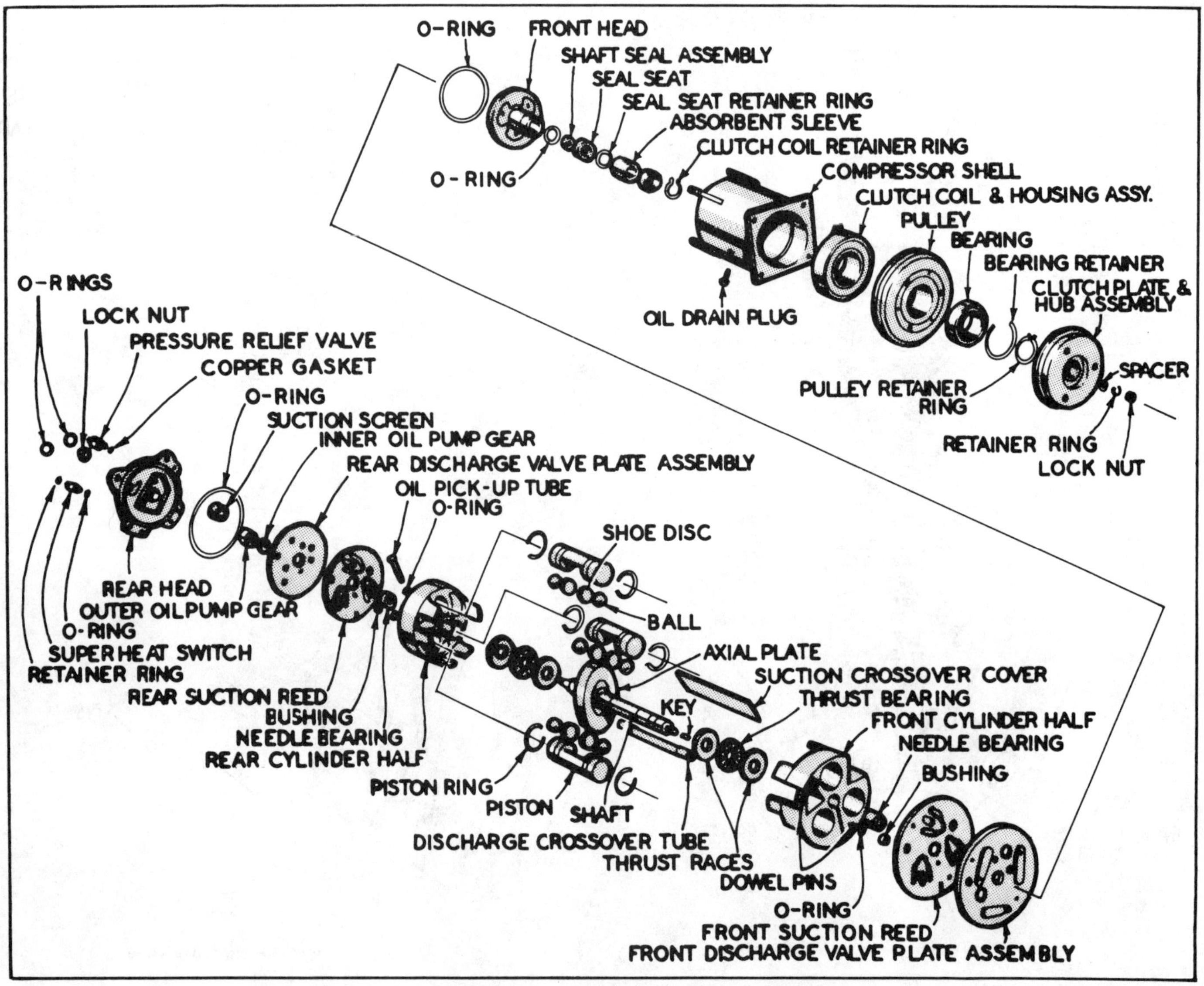

Fig. 4 Frigidaire & Motorcraft axial compressor exploded view

Refer to oil level test procedure located elsewhere in this manual.

Installation

1. Drain and discard oil shipped with new compressor.
2. Fill compressor with exact amount of new oil as measured previously.
3. Stamp refrigerant charge capacity of system on new compressor in space provided on the data plate.

NOTE: Before installing a new compressor, rotate drive shaft by hand through several revolutions to provide proper lubrication of shaft seal.

4. Position compressor on the mounting bracket and install bolts, nuts and lock washers. Bolts should be snug but not fully tight until after belt installation and adjustment.
5. Lubricate new "O" rings with refrigerant

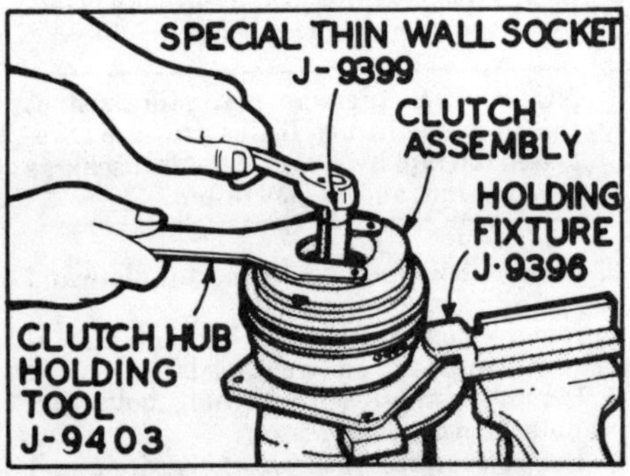

Fig. 5 Removing lock nut

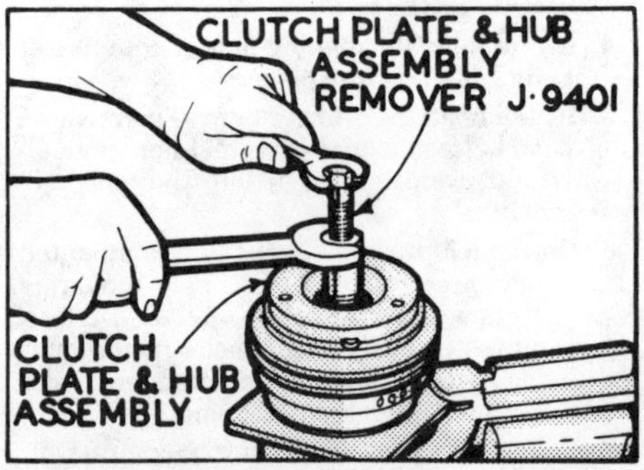

Fig. 6 Removing clutch plate and hub

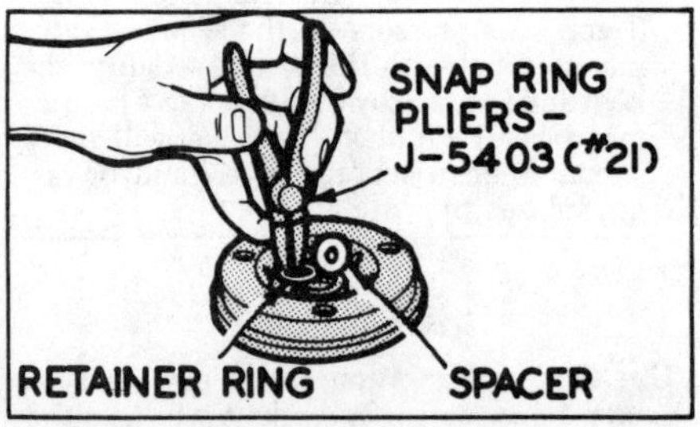

Fig. 7 Removing hub retaining ring and spacer

Fig. 8 Installing clutch plate and hub

oil and install on hose connector plate. Install plate and connect compressor electrical leads.

6. Install and adjust compressor drive belt and tighten compressor mounting bolts.
7. Evacuate, charge and operationally test the system.

NOTE: Service or repair of compressors should not be attempted unless all proper tools are at hand and a virtually spotless work area can be provided.

Axial Compressor Overhaul

When overhauling the compressor, remove only the necessary components that preliminary diagnosis indicates are in need of service. Refer to Fig. 4 for parts location and nomenclature.

Some service operations can be performed without disturbing the internal mechanism or completely removing the compressor from the car. Among them are replacement of the clutch plate and hub assembly, the pulley and bearing assembly, and pulley bearing.

The clutch coil and housing assembly may also be replaced without removing the compressor, after clutch and pulley parts have been removed. It is not necessary to disturb the shaft seal.

The shaft seal can be replaced only by removing the compressor from the car and removing the clutch plate and hub assembly to gain access to the seal.

Although certain service operations can be performed without completely removing the compressor from the car, the operations described herein are based on bench overhaul with compressor removed from car. They have been prepared in sequence in order of accessibility of the components.

NOTE: When a compressor is removed from the car for servicing, the amount of oil remaining in the compressor should be drained and measured. The measured amount of oil should be noted for future reference and then discarded. The oil should then be replaced with the same amount of new oil upon reassembly of the compressor.

Clutch Plate and Hub

Removal

1. Remove lock nut from shaft, Fig. 5.
2. Remove clutch plate and hub, Fig. 6.

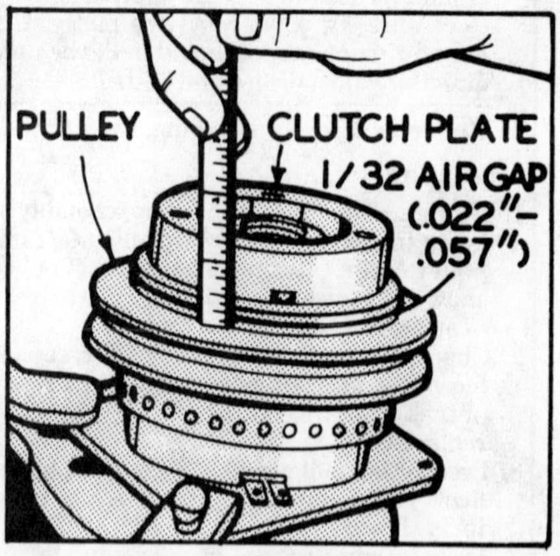

Fig. 9 Checking air gap

Then remove drive key from shaft.

3. Remove hub retainer ring and spacer, Fig. 7.

Installation

1. Install square key on shaft, allowing it to project approximately 3/16 inch out of keyway.
2. Wipe frictional surface of clutch plate and pulley clean.
3. Place clutch plate and hub assembly on shaft, aligning key with keyway in hub.

CAUTION: To avoid internal damage to compressor, do not drive or pound on hub or shaft. This could mis-position swash plate on shaft, resulting in damage to compressor.

4. Place spacer on hub, Fig. 8. Insert end of clutch plate and hub installer through spacer and thread tool onto end of shaft.
5. Hold hex portion of tool body with a wrench and tighten center screw several turns to press hub partially on shaft, Fig. 8.
6. Remove tool and check alignment of drive key with keyway in shaft. If alignment is correct, replace installer tool and continue to press hub onto shaft until there is approximately a 3/32 inch (.093) air gap between frictional surface of pulley and clutch plate.

NOTE: A zero thrust race is 3/32 inch thick and can be used as a gauge between these frictional surfaces.

7. Remove tools and install hub spacer.
8. Install hub retainer ring with flat side of ring facing spacer.
9. Install a new shaft lock nut with small diameter boss of nut against hub spacer, using thin wall socket shown in Fig. 5. Hold clutch hub with holding tool and tighten nut to 15 ft. lbs. Air gap between frictional surfaces of pulley and clutch plate, Fig. 9, should now be as specified in the *A/C Data Table*.

Compressor Pulley & Bearing

Removal

1. Remove clutch plate and hub.
2. Remove pulley retainer ring, Fig. 10.
3. Place puller pilot, Fig. 11, over end of shaft.

CAUTION: It is important that the pull

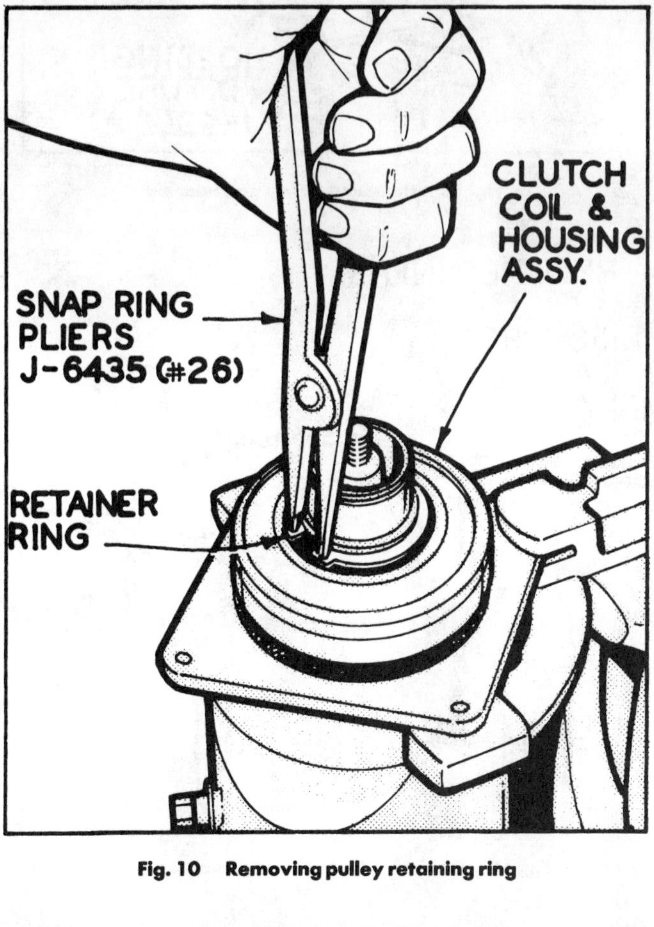

Fig. 10 Removing pulley retaining ring

er pilot be used to prevent internal damage to compressor when removing pulley. Under no circumstances should puller be used directly against drilled end of shaft.

4. Remove pulley and bearing, Fig. 11.

Installation

1. If original pulley and bearing assembly is being installed, wipe frictional surfaces of pulley clean. If frictional surfaces of pulley show any indication of damage due to overheating, pulley should be replaced.
2. Check bearing for brinnelling, excessive looseness, noise or lubricant leakage. If any of these conditions exist, bearing should be replaced.
3. Press or tap pulley and bearing assembly on neck of compressor until it seats, using tools shown in Fig. 12. Installer will apply force to inner race of bearing and prevent damage to bearing.
4. Check pulley for binding or roughness; it should rotate freely.
5. Install retainer ring.
6. Install clutch plate and hub.

Remove Internal Mechanism

1. Remove clutch plate and hub, pulley and bearing, clutch coil and housing, and shaft seal as outlined previously.
2. Invert compressor and holding fixture with front end of compressor shaft down.

NOTE: Additional oil may leak from compressor at this time. All oil must be drained into a clean container so that the total amount can be measured. A liquid measuring cup may be used for this purpose. Drained oil should then be discarded.

3. Remove rear head of compressor.
4. Wipe excess oil from teflon gasket surfaces, Fig. 13. Examine gasket surfaces and replace head should any damage be present.
5. Remove suction screen.
6. Paint an identifying mark on exposed face of inner and outer oil pump gears and remove gears.
7. Remove and discard rear head to shell O-ring.
8. Remove rear discharge valve, Fig. 14. Use two screwdrivers as shown; do not position screwdrivers between reeds and reed seats.
9. Remove rear suction reed, Fig. 15. Do not pry up on horseshoe-shaped reed valves.
10. Remove oil pick-up tube, Fig. 16. Remove O-ring from oil inlet.
11. Remove compressor from holding fixture and place it in a support block on bench.
12. Lift front head and compressor shell assembly up, leaving internal mechanism resting on support block, Fig. 17.

CAUTION: Do not tap on end of compressor shaft to remove internal mechanism. If mechanism will not slide out of compressor shell, tap on front head with a plastic hammer.

13. Rest compressor shell on its side and push front head assembly through shell, being careful not to damage teflon sealing areas on inner side of front head. Discard O-ring.

NOTE: It may be necessary to tap on outside of front head, using a plastic hammer, to overcome friction of O-ring seal between front head and compressor shell.

Fig. 11 Removing pulley and bearing

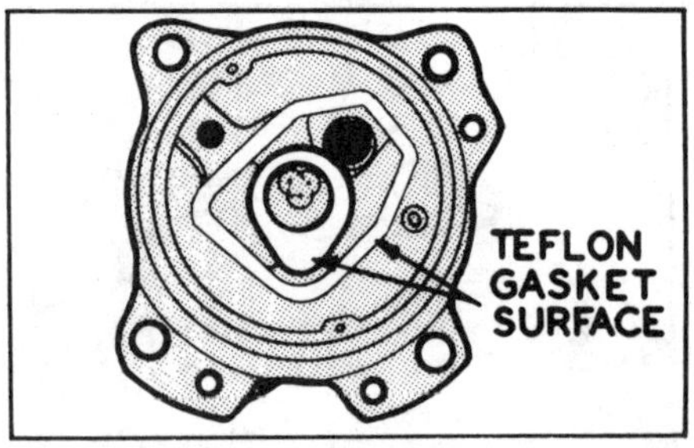

Fig. 12 Teflon gasket surfaces on rear head

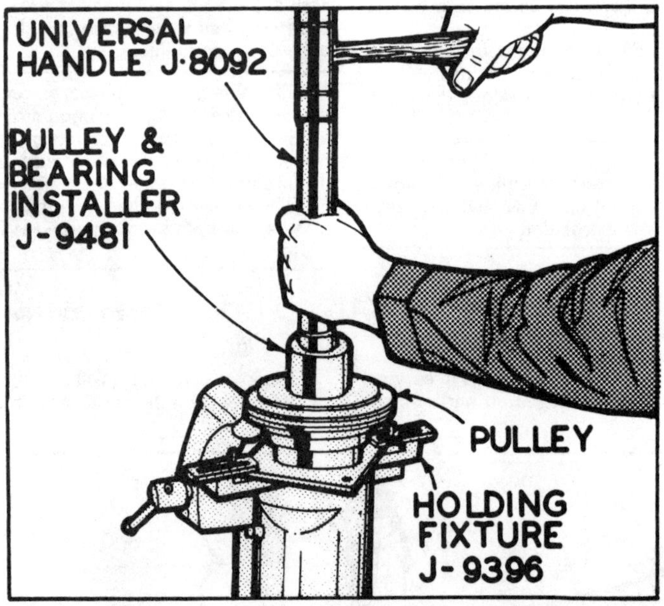

Fig. 13 Installing pulley and bearing

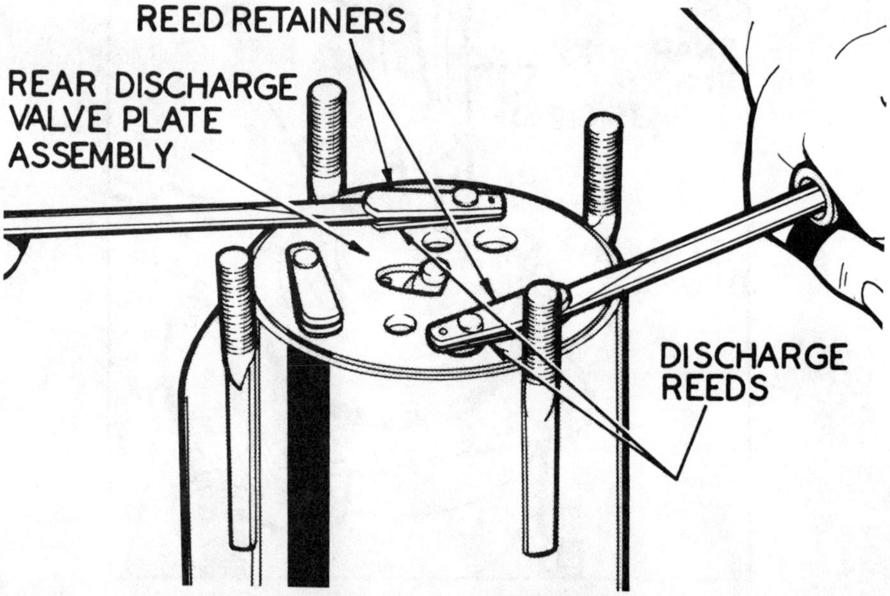

Fig. 14 Removing rear discharge valve plate

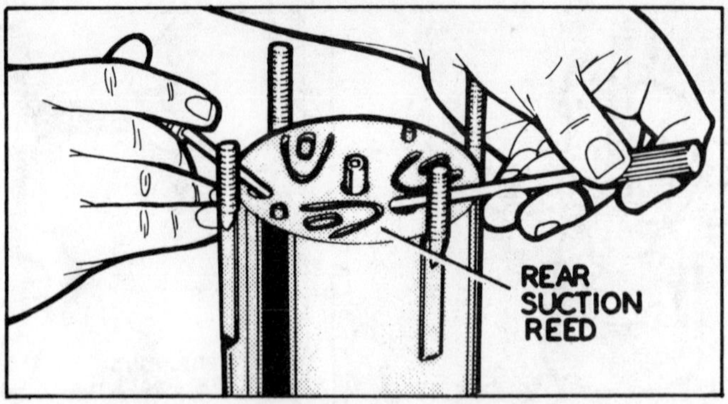

Fig. 15 Removing rear suction reed

14. Clean and examine teflon gasket surface on front head casting webs. If any damage is observed on gasket surface gasket should be replaced.
15. Remove front discharge valve assembly and front suction reed plate.
16. Remove suction cross-over cover by prying with screwdriver between cylinder casting and cover.

NOTE: Examine internal mechanism for any obvious damage. If there is major damage, due to loss of refrigerant or oil, it may be necessary to use the service internal mechanism assembly rather than replace individual parts.

Disassemble Internal Mechanism

1. Identify by a pencil mark, each piston, Fig. 17. Number piston bores in front cylinder half in like manner so that pistons can be replaced in their original locations.
2. Separate cylinder halves, Fig. 18. Make certain that discharge cross-over tube does not contact swash plate when separating cylinder halves.

CAUTION: Under no circumstances should shaft be struck at either end in an effort to separate upper and lower cylinder halves.

3. Position complete internal mechanism, rear cylinder down, on support block, and remove front cylinder half.
4. Remove and disassemble pistons, keeping all parts for each position in a separate dish or tray. Discard position shoe discs and other parts that show evidence of damage, Fig. 19.
5. Remove front thrust races and bearing; discard races, Fig. 20.
6. Remove shaft from rear cylinder half. It may be necessary to bend discharge cross-over tubes slightly in order to remove shaft.
7. Remove rear combination of thrust races and bearing from shaft; discard races.

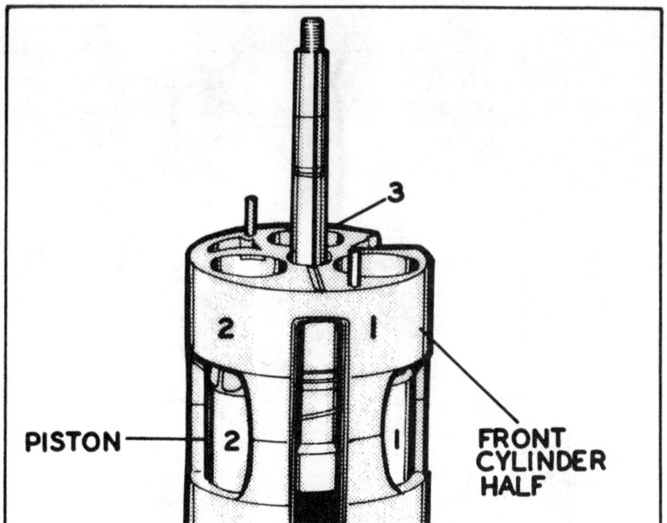

Fig. 17 Numbering of pistons and cylinder bores

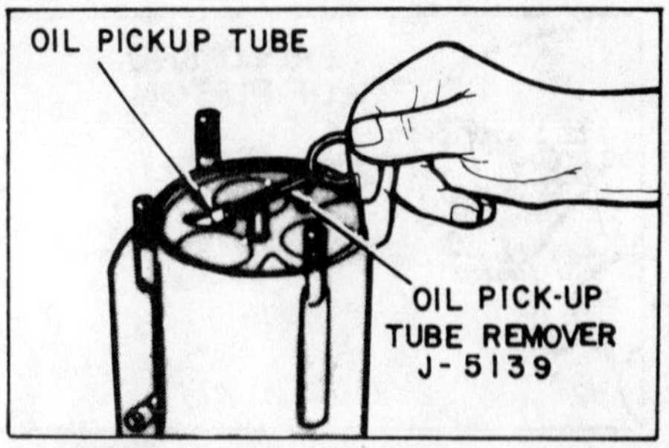

Fig. 16 Removing rear pick-up tube

8. Examine surface of swash plate and shaft. Replace as an assembly, if necessary.

NOTE: A certain amount of shoe disc wear on swash plate is normal as well as some markings indicating load of needle bearings on shaft.

9. Remove discharge cross-over tube from cylinder half, using vise grip pliers.

NOTE: This is necessary on original factory equipment only as ends of tube are swedged into cylinder halves. The discharge cross-over tube in internal mechanism assemblies that have been previously serviced have an O-ring and bushing at each end of the tube, and can be easily removed by hand.

10. Examine piston bores and needle bearings in front and rear cylinder halves. Replace front and rear cylinders if any cylinder bore is deeply scored or damaged.
11. Needle bearings may be removed if necessary by driving them out with a special thin wall socket. Insert socket in hub end (inner side) of cylinder head and drive bearing out. To install, place cylinder half on support block, hub end down, and insert bearing in end of cylinder head with bearing identification marks up. With a suitable driver, Fig. 21, drive bearing into head until tool bottoms on cylinder face.
12. Clean and air-dry parts to be re-used.

Gauging Internal Mechanism

1. Install Compressing Fixture on Holding Fixture in vise. Place front cylinder half in compressing fixture with flat side down, Fig. 22.
2. Use 4 zero thrust races and 3 zero shoe discs.
3. Install a zero thrust face, thrust bearing

Fig. 18 Separating cylinder halves

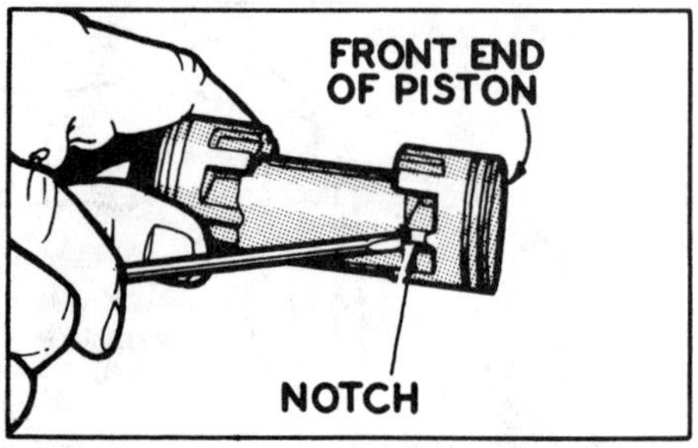

Fig. 19 Compressor piston

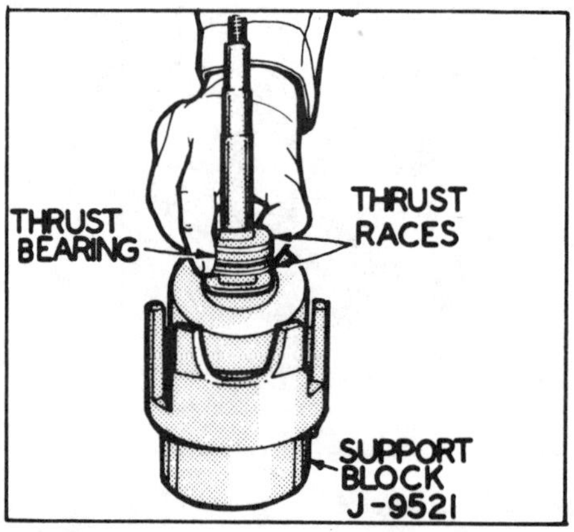

Fig. 20 Removing front thrust races and thrust bearing

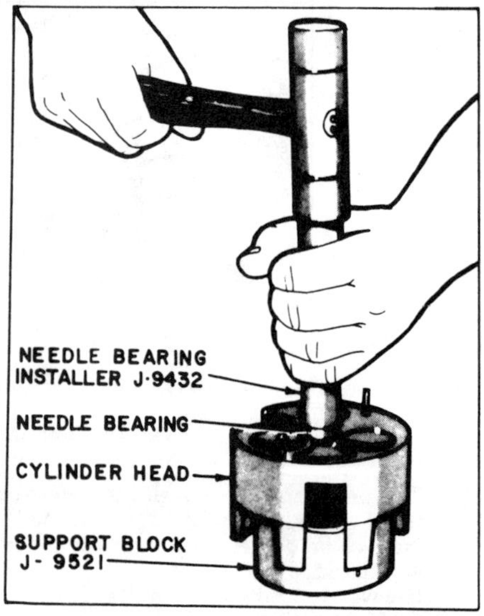

Fig. 21 Installing needle bearing

and a second zero thrust race on end of compressor shaft. Grease races and bearing with petrolatum.

4. Insert threaded end of shaft through needle bearing in front cylinder half, and allow thrust race and bearing to rest on hub of cylinder.
5. Install zero thrust race on rear end of compressor shaft so that it rests on hub of swash plate. Then install thrust plate and a second zero thrust race, Fig. 22. Grease races and bearing with petrolatum.
6. Lubricate ball pockets on No. 1 piston with refrigeration oil and place a ball in each socket.
7. Lubricate cavity of a zero shoe disc with refrigeration oil and place shoe disc over ball in front end of piston, Fig. 23. Front end of piston has an identifying notch in casting web. *Piston rings should not be installed at this time.*

NOTE: Shoe discs should not be installed on rear piston balls during this operation.

8. Rotate shaft and swash plate until high point of swash plate is over No. 1 piston cylinder bore.
9. Lift shaft assembly and hold front thrust race and bearing assembly against swash plate hub.
10. Position piston over No. 1 cylinder bore (notched end of piston on bottom and piston straddling swash plate) and lower shaft to allow piston to drop into its bore, Fig. 24.
11. Repeat Steps 6 through 10 for pistons 2 and 3.
12. Install rear cylinder half on pistons, aligning cylinder with discharge cross-over tube hole in front cylinder. Tap into place using a plastic mallet.
13. Position discharge cross-over tube holes between a pair of Compressing Fixture bolts to permit access for feeler gauge, Fig. 24.
14. Install top plate on Compressing Fixture. Tighten nuts to 15 ft. lbs., using an 0-25 ft. lb. torque wrench.
15. Measure clearance between rear ball of No. 1 piston and swash plate as follows:
    (a) Select a suitable combination of well-oiled feeler gauge leaves to fit snugly between ball and swash plate.
    (b) Attach a spring scale reading 1-ounce increments to feeler gauge. A distributor point checking scale or spring scale may be used.
    (c) Pull on spring scale to slide feeler gauge stock out from between ball and swash plate and note reading on spring scale as feeler gauge is removed, Fig. 25. Reading should be between 4 and 8 ounces.
    (d) If reading is under 4 or over 8 ounces, reduce or increase thickness of feeler gauge leaves and repeat Steps (a) through (c) until a reading of 4 to 8 ounces is obtained. Record clearance between ball and swash plate that results in a 4 to 8 ounce pull on spring scale.
16. Rotate shaft 120° and repeat Step 15 between same ball and swash plate. Record this measurement.
17. Rotate shaft 120° and again repeat Step 15 between these same parts and record measurement.
18. Select a numbered shoe disc corresponding

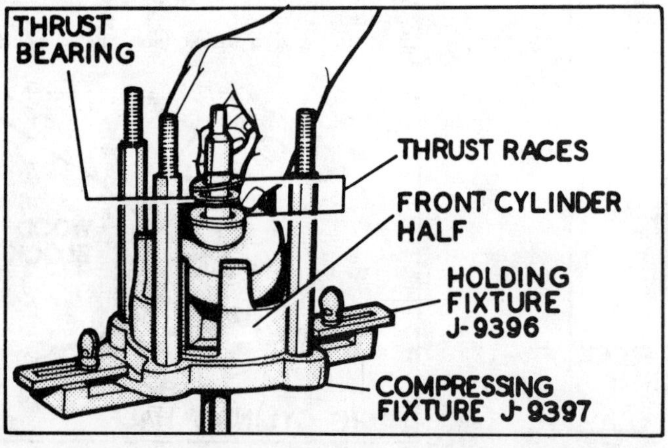

Fig. 22 Installing rear thrust races and thrust bearing

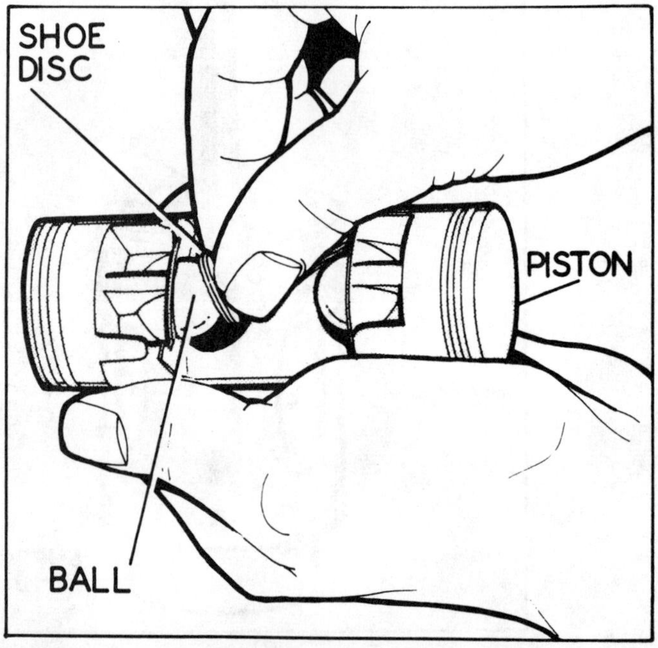

Fig. 23 Installing shoe disc

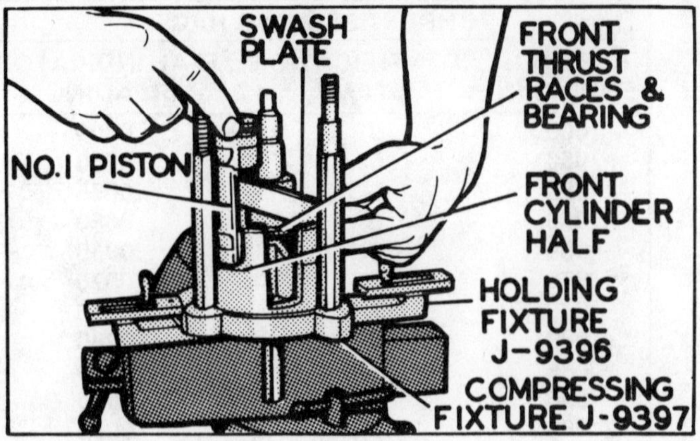

Fig. 24 Installing piston

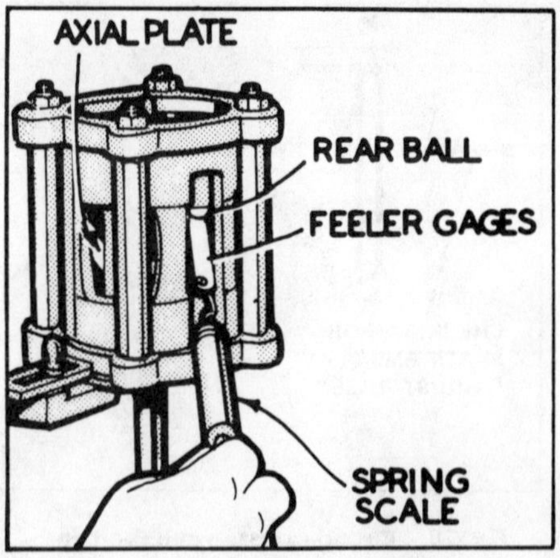

Fig. 25 Gauging rear piston ball

to minimum feeler gauge reading recorded in the 3 checks. Place shoe discs in parts dish or cups corresponding to piston No. 1 and rear ball pocket position for future reassembly.

NOTE: Shoe discs are provided in .0005 inch variations. As shown in Fig. 26, there are eleven sizes available for field servicing. All shoe discs are marked with the shoe size, which corresponds to the last three digits of the piece part number.

NOTE: Once proper selection of shoe has been made, the matched combination of disc to rear ball and spherical cavity in piston must be kept in proper relationship during disassembly after gauging operation, and during final assembly of internal mechanism.

19. Repeat in detail the same gauging operation for pistons 2 and 3.
20. Mount dial indicator on edge of Compressing Fixture, Fig. 27. Adjust dial to zero. Push front end of shaft upward and record measurement. An alternate method of selecting the proper race on compressors with teflon piston rings is to use Gage Set J-9661-01 and selecting a suitable feeler gauge leaf until the result is a 4 to 8 ounce pull on the scale between the rear thrust bearing and upper (or outer rear) thrust race, Fig. 28. If pull is just less than 4 ounces, add .0005 inch to the thickness of the feeler gauge used to measure the clearance. If the pull is just over 8 ounces, subtract .0005 inch from the thickness of the feeler gauge. Select a race TWO (2) FULL SIZES LARGER than the feeler gauge thickness.
21. On compressors without teflon piston rings, select a thrust race with a number corresponding to the amount of end play shown. On compressors with teflon piston rings, select a thrust race with a number corresponding to TWO (2) FULL SIZES LARGER than the dial indicator reading or feeler gauge measurement of the amount of end play shown.

NOTE: Fifteen thrust races are provided in increments of .0005 inch thickness and one zero gauge thickness providing a total of 16 sizes available for field service, Fig. 29. Thrust races are identified on the part by their thickness in thousands in excess of the thickness of the zero thrust race. This number also corresponds to the last three digits of the piece part number. A tolerance of .0005 inch to .0015 inch is built into thrust races to provide a running clearance between hub surfaces of swash plate and front and rear hubs of cylinder.

22. Remove Compressing Fixture top plate.
23. Separate cylinder halves while unit is in fixture. It may be necessary to use a wood block and mallet.
24. Remove rear cylinder half and remove one piston at a time from swash plate and front cylinder half. Do not lose relationship of front ball and shoe disc and rear ball. Transfer each piston, ball and shoe disc into proper part dish or cup.
25. Remove rear outer zero thrust race from shaft and install thrust race previously selected.

SHOE DISC		
PART NO. ENDING IN	IDENTIFICATION STAMP	MIN. FEELER GAGE READING
000	0	.0000
175	17-1/2	.0175
180	18	.0180
185	18-1/2	.0185
190	19	.0190
195	19-1/2	.0195
200	20	.0200
205	20-1/2	.0205
210	21	.0210
215	21-1/2	.0215
220	22	.0220

Fig. 26 Shoe disc size chart

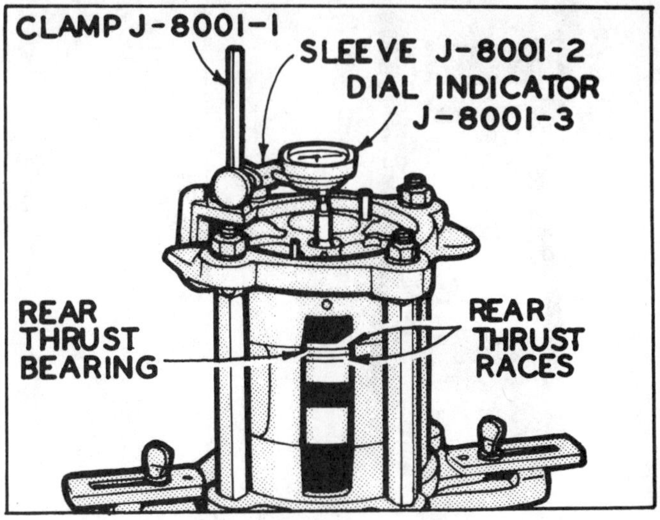

Fig. 27 Gauging rear thrust race

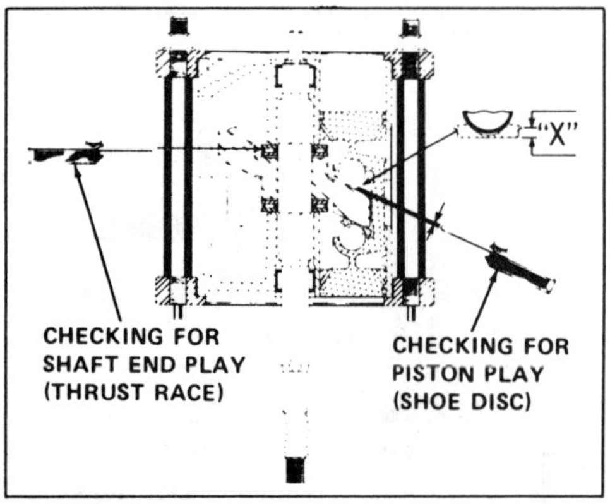

Fig. 28 Checking piston and shaft end play

THRUST BEARING RACE		
PART NO. ENDING IN	IDENTIFICATION STAMP	DIAL INDICATOR READING
000	0	.0000
050	5	.0050
055	5-1/2	.0055
060	6	.0060
065	6-1/2	.0065
070	7	.0070
075	7-1/2	.0075
080	8	.0080
085	8-1/2	.0085
090	9	.0090
095	9-1/2	.0095
100	10	.0100
105	10-1/2	.0105
110	11	.0110
115	11-1/2	.0115
120	12	.0120

Fig. 29 Thrust race size chart

Teflon Piston Ring Replacement

NOTE: The pistons on 1976-80 compressors use teflon rings. The teflon ring type piston is the same as the carbon ring type piston except for the ring groove detail. The teflon ring groove is wider, shallower and contains two ridged rings in the bottom of the groove for ring retention. The teflon ring type pistons are supplied with the teflon rings "sized" in place on the piston. However, the piston and ring assembly should be sized using piston ring gauge J-24608-1 as described further on, before installing into the compressor.

1. Remove old piston rings by carefully cutting ring with a knife or other sharp tool, holding the blade almost flat with the piston surface.

CAUTION: Use care when cutting the ring to avoid personal injury. Also, be careful not to damage piston or groove while cutting the ring.

2. Clean piston and groove with a recommended cleaning solvent and force dry with compressed clean dry air.
3. Set piston on end on a clean flat surface and place Ring Installer Guide on end of piston, then install teflon ring on Ring Installer Guide with dished or dull side down and glossy side up, Fig. 30.
4. Push the Ring Installer down over the Ring Installer Guide to install the teflon ring into the piston groove, Fig. 31. If the teflon ring is slightly mispositioned, it can be correctly positioned using a blunt-edged tool that will not damage the piston.

NOTE: Do not store the Installer Guide inside the Ring Installer, as this will cause stretching and weakening of the Ring Installer Segment Retainer O-ring. After installing the piston ring, remove the Installer Guide from the Ring Installer.

5. Lubricate piston ring area with refrigerant

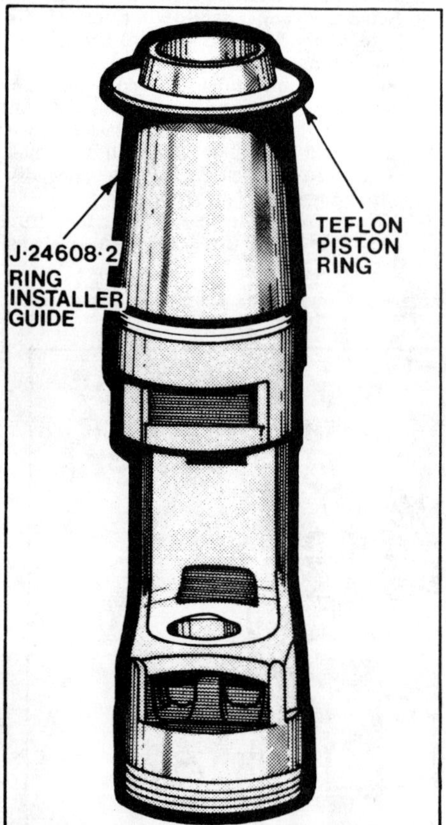

Fig. 30 Installing teflon piston ring on ring installer guide

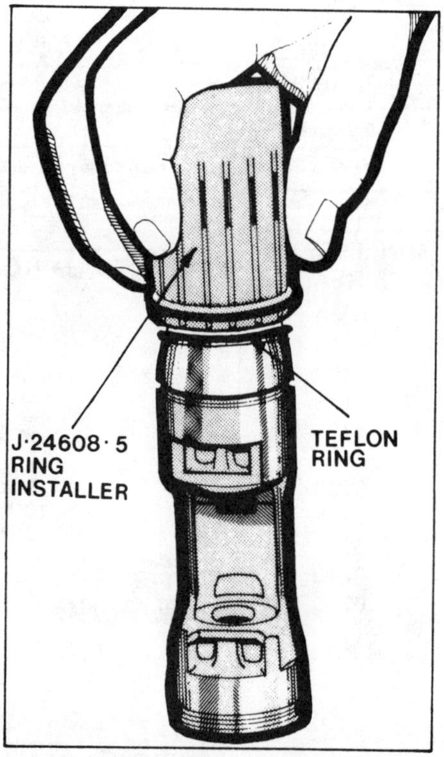

Fig. 31 Installing teflon piston ring on piston

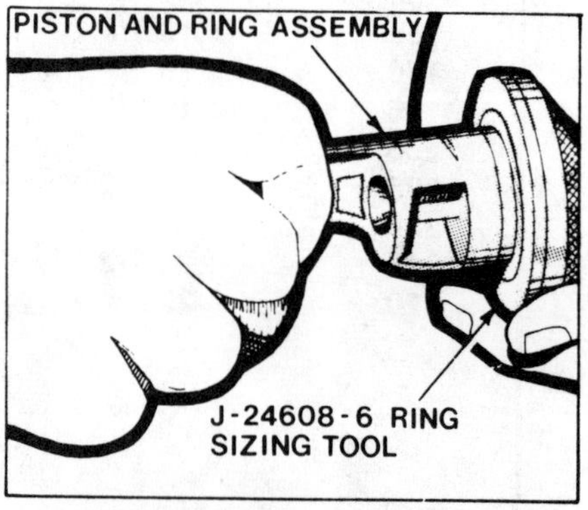

Fig. 32 Turning piston and ring assembly into ring sizing tool

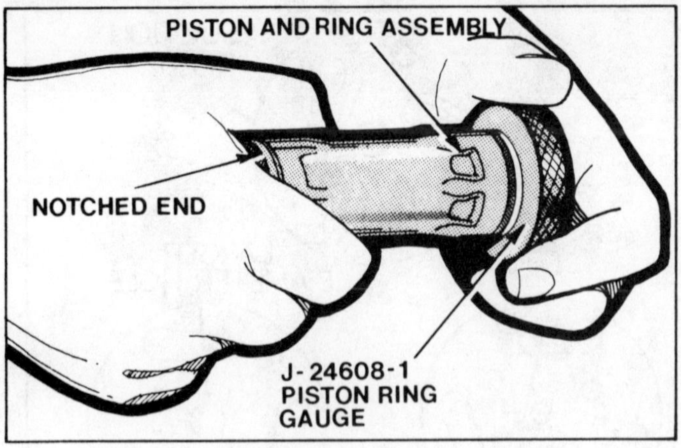

Fig. 33 Gauging piston ring

oil and rotate piston and ring assembly into Ring Sizer at a slight angle, Fig. 32. Rotate piston while pushing inward until the piston is inserted against the center stop of the Ring Sizer.

CAUTION: Do not push the piston and ring assembly into the Ring Sizer without proper positioning and rotating as described above, as the ends of the needle bearings of the tool may damage the end of the piston.

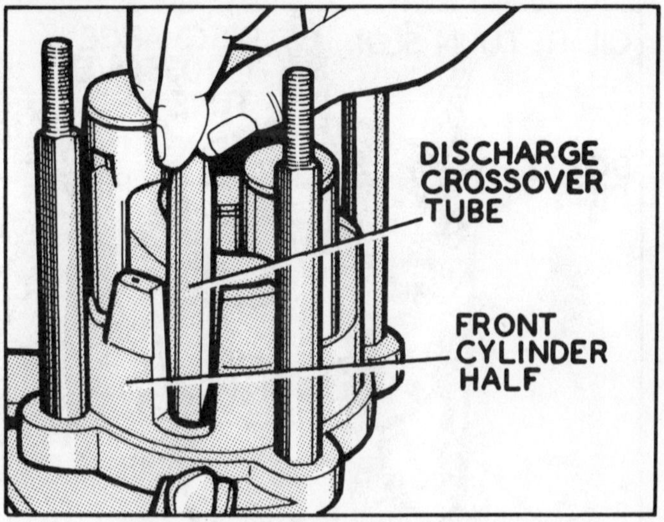

Fig. 34 Installing discharge cross-over tube

6. Rotate piston and ring assembly several complete turns, until the assembly rotates relatively free in Ring Sizer.
7. Remove piston and ring assembly and wipe end of piston and ring area with a clean cloth, then push piston and ring assembly into Ring Gage, Fig. 33. The piston should go through the Ring Gage with a force of 6 lbs. or less without lubrication. If not, repeat steps 5, 6 and 7.
8. Repeat the above procedure on opposite end of piston.

CAUTION: Do not place piston down on a dirty surface where dirt or metal chips can become imbedded in teflon ring.

9. Lubricate both ends of piston with refrigerant oil before inserting piston into cylinder bore.

CAUTION: Use care to avoid damaging teflon ring when installing piston into cylinder bore.

Internal Mechanism, Assemble 1974-75

1. Install a piston ring on each end of piston

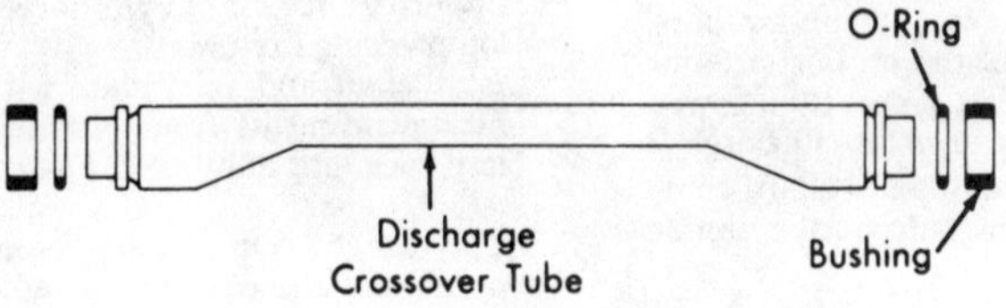

Fig. 35 Service type cross-over tube

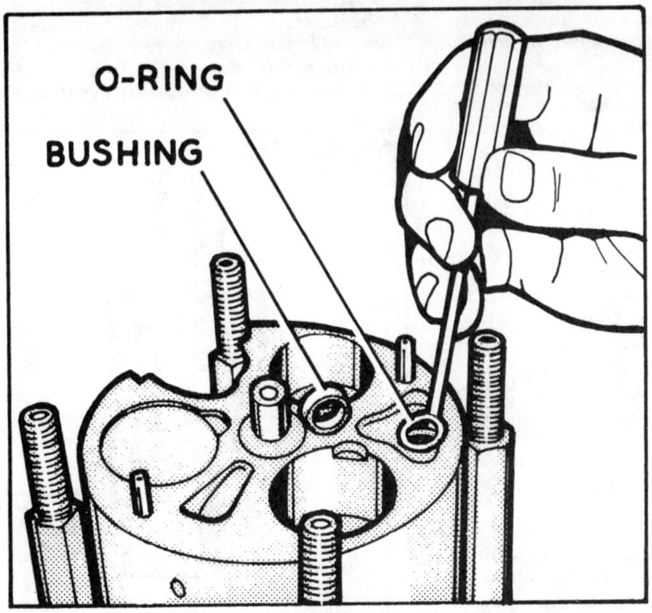

Fig. 36 Installing O-ring on cross-over tube

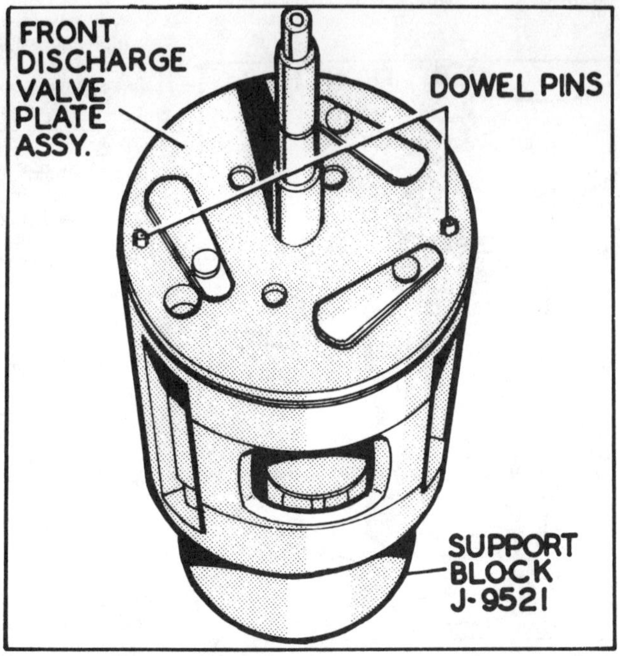

Fig. 37 Installing front suction reed

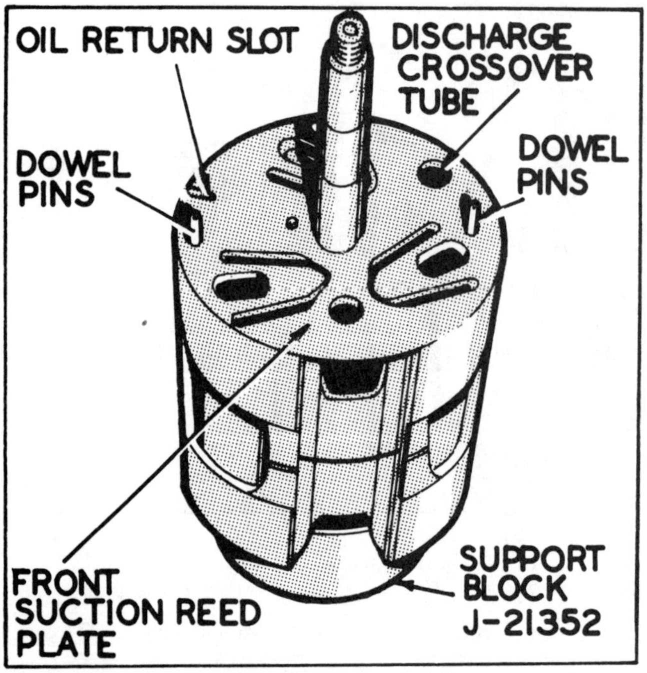

Fig. 38 Installing front discharge valve plate

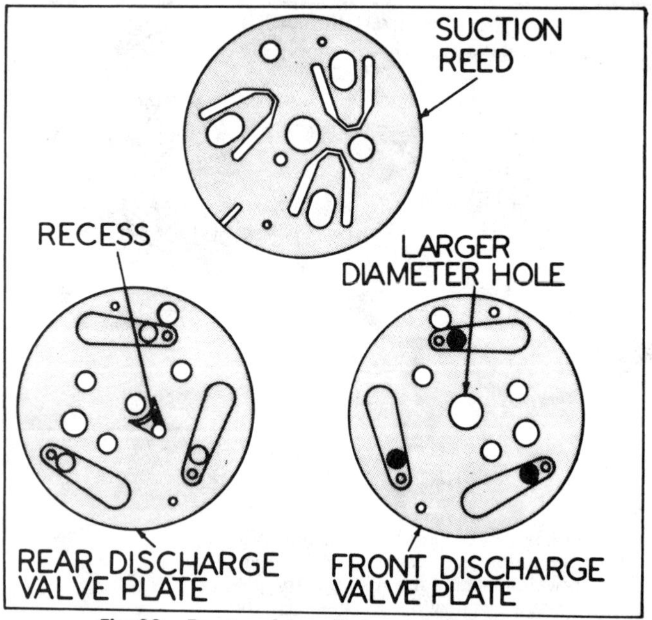

Fig. 39 Front and rear discharge valve plates

No. 1, with scraper groove toward center of piston.

2. Lubricate ball pockets of piston with refrigeration oil and place corresponding balls from parts dish or cup in each pocket.
3. Lubricate cavities of No. 1 piston shoe discs with refrigeration oil and place zero shoe disc over ball in front end of piston and numbered shoe disc over ball in rear of piston.
4. Rotate shaft and swash plate until high point of swash plate is over No. 1 piston cylinder bore. *Make certain that front thrust race and bearing assembly adhere to swash plate hub.*
5. Position piston over No. 1 cylinder bore with notched end of piston on bottom and piston straddling swash plate, then lower shaft to allow piston to drop into its bore.
6. Position piston ring gap toward shaft, compress ring and lower ring into front cylinder half.
7. Repeat Steps 1 through 6 for pistons 2 and 3.
8. Install new discharge cross-over tube in front cylinder half with bridged surface facing outboard, Fig. 34. Make certain that end of tube is properly centered in hole in front cylinder half.

NOTE: The service discharge cross-over tube is similar to the production type tube except that an O-ring and bushing is used at each end, Fig. 35. Do not install O-ring and bushing at this time.

9. Rotate shaft to position pistons in a stair-step arrangement. Position rings on each piston so that ring gaps are toward shaft, then push rings as far outboard as possible.
10. Place rear cylinder half over shaft and start pistons and rings into cylinder bores.
11. When all three pistons and rings are in their respective bores, align end of discharge cross-over tube with hole in rear cylinder half. Make certain that bridged surface of tube faces outboard for swash plate clearance.
12. When satisfied that all parts are in proper alignment, tap rear cylinder half with a mallet to seat it over locating dowel pins.
13. Remove internal mechanism from fixture and place on bench.
14. Bending suction cross-over cover slightly, start it into one end of dovetail slot in cylinder halves. Align cover with ends of cylinder faces by gently tapping end of cover with a plastic hammer.

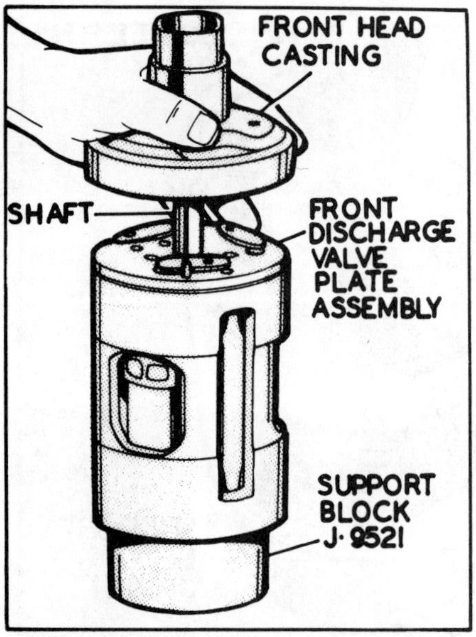

Fig. 40 Installing front head casting

Internal Mechanism, Assemble 1976-80

1. Support front half of cylinder assembly on Compressing Fixture.
2. Install shaft and axial plate with threaded end down and its front bearing race pack (zero race and numbered bearing race) installed.
3. Apply a light coat of petroleum jelly onto "numbered" shoe discs selected during gaging procedure and install all ball and shoe discs in their proper place on piston assembly.
4. Rotate axial plate until highest point is above No. 1 cylinder bore, then carefully assemble No. 1 piston, complete with ball and zero shoe disc on the front, and ball and numbered shoe disc on the rear over

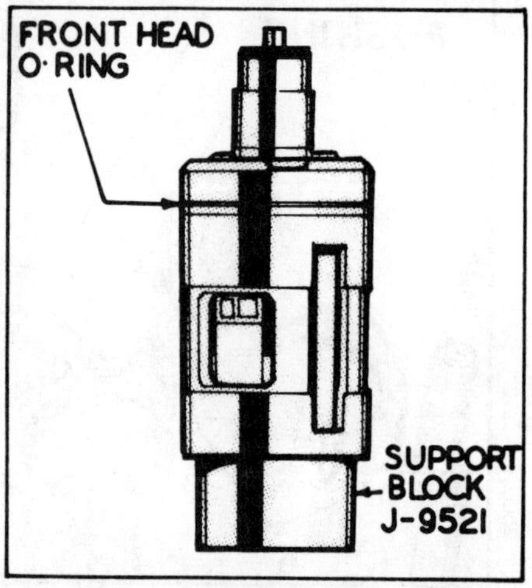

Fig. 41 Front head O-ring installed

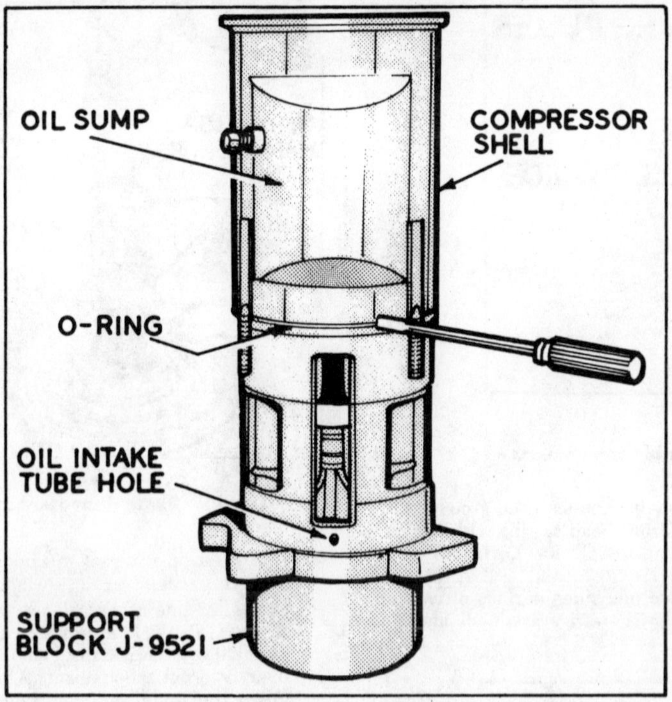

Fig. 42 Installing compressor shell

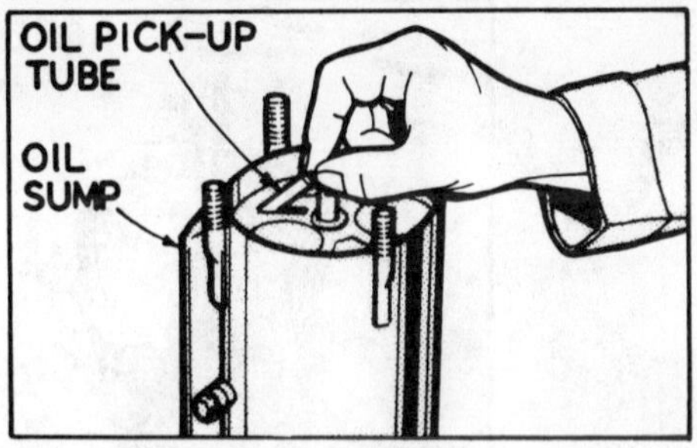

Fig. 43 Installing oil pick-up tube

axial plate. Hold front thrust bearing pack tightly against axial plate hub while lifting hub and insert piston assembly into front cylinder half, Fig. 24.

5. Repeat the above operation for the remaining cylinders.
6. Without using any O-rings or bushings, assemble one end of the new discharge cross-over tube into hole in front cylinder half, Figs. 34 and 35, assuring that flattened section of tube faces inside of compressor to allow for axial clearance.
7. Rotate shaft to position pistons in a "stair-step" arrangement, then carefully place the rear cylinder half over shaft and start pistons into cylinder bores.
8. When all three piston and ring assemblies are in their respective cylinders, align end of cross-over tube with hole in rear half of cylinder.
9. When all parts are properly aligned, seat rear half of cylinder over locating dowel pins by tapping with a wood block and mallet. If necessary, clamp cylinder in Compressing Fixture to complete drawing cylinder halves together.
10. Lubricate all moving parts with clean refrigerant oil and check for free rotation of all parts.
11. Install suction cross-over cover until flush at both ends.

Internal Mechanism, Install

1. Place internal mechanism on support block.
2. Install new O-ring and bushing on front end of discharging cross-over tube, Fig. 36. The O-ring and bushing are service parts only for internal mechanisms that have been disassembled in the field.
3. Install new dowel pins in front cylinder half, if previously removed.
4. Install front suction reed plate on front cylinder half. Align with dowel pins, suction ports, oil return slot, and discharge cross-over tube, Fig. 37.
5. Install front discharge plate assembly, aligning holes with dowel pins and proper openings in front suction reed plate, Fig. 38. Front discharge plate has a large diameter hole in center, Fig. 39.
6. Coat Teflon gasket surfaces on webs of compressor front head casting with refrigerant oil.
7. Determine exact position of front head casting in relation to dowel pins on internal mechanism. Mark position of dowel pins on sides of front head assembly and on sides of internal mechanism with a grease pencil. Carefully lower front head casting into

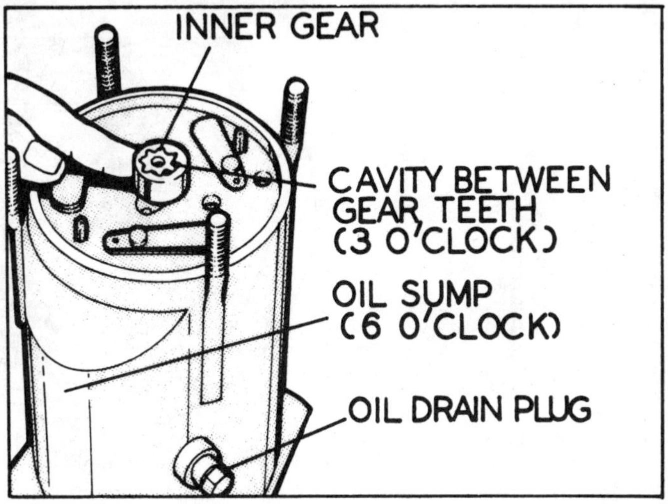

Fig. 44 Positioning oil pump gears

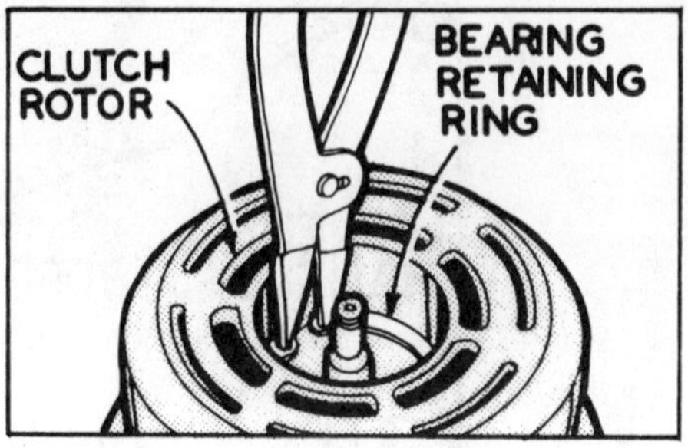

Fig. 45 Removing rotor and bearing assembly retaining ring

Fig. 46 Installing rotor and bearing assembly puller guide

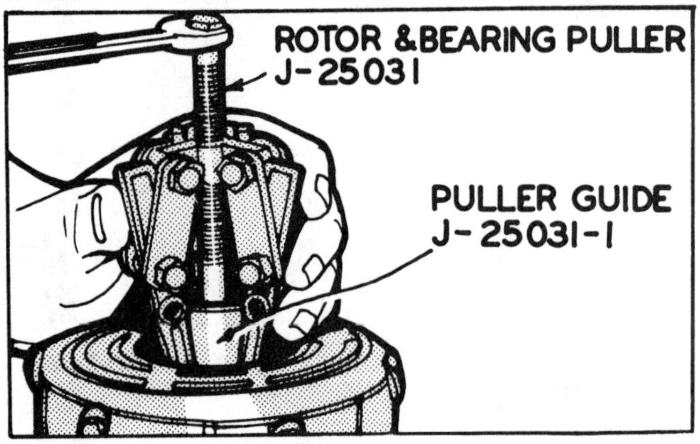

Fig. 47 Removing clutch plate and hub assembly components

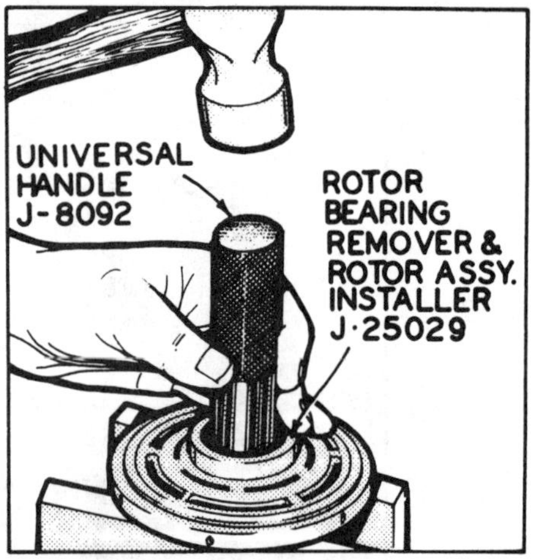

Fig. 48 Removing rotor hub bearing

position, Fig. 40, making certain that teflon sealing area and center bore of head assembly does not contact shaft as head is lowered. Do not rotate head to line up with dowel pins, as teflon sealing areas would contact reed retainers.

8. Generously lubricate angled groove at lower end of front head casting with refrigeration oil and install new O-ring in groove, Fig. 41.
9. Coat inside machined surface of compressor shell with refrigeration oil and position shell on internal mechanism, resting on O-ring seal.
10. Using flat side of a small screwdriver, gently press O-ring in around circumference of internal mechanism until shell slides down over internal mechanism. As shell slides down, line up oil sump with oil intake tube hole, Fig. 42.
11. Holding support block, invert assembly and place in Holding Fixture with front end of shaft down. Remove support block.
12. Install new dowel pins in rear cylinder half if previously removed.
13. Install new O-ring in oil pick-up tube cavity.
14. Lubricate oil pick-up tube and install in cavity, rotating compressor mechanism to align tube with hole in shell baffle, Fig. 43.
15. Install new O-ring and bushing on rear of discharge cross-over tube.
16. Install rear suction reed over dowel pins with slot toward sump.
17. Install rear discharge valve plate over dowel pins with reed retainers up.
18. Position inner oil pump gear over shaft with previously applied identification mark up.
19. Position outer oil pump gear over inner gear with previously applied identification

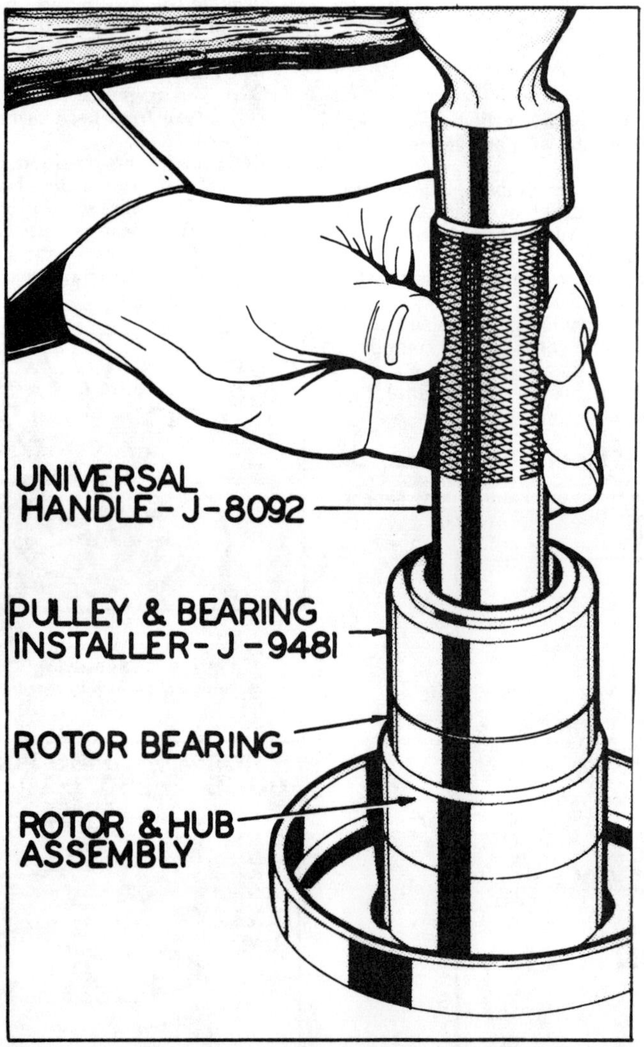

Fig. 49 Installing rotor hub bearing

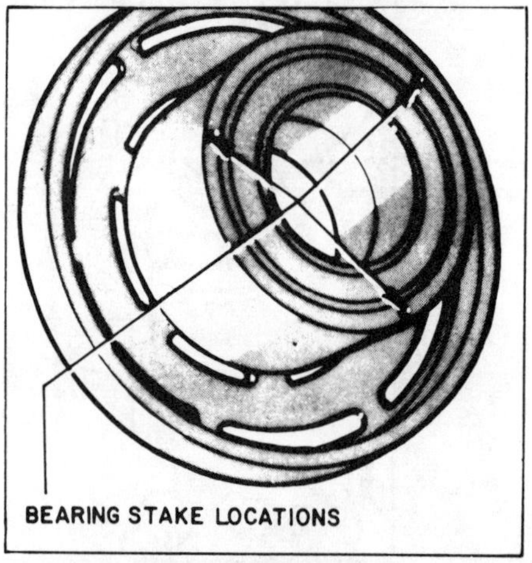

Fig. 50 Bearing stake locations

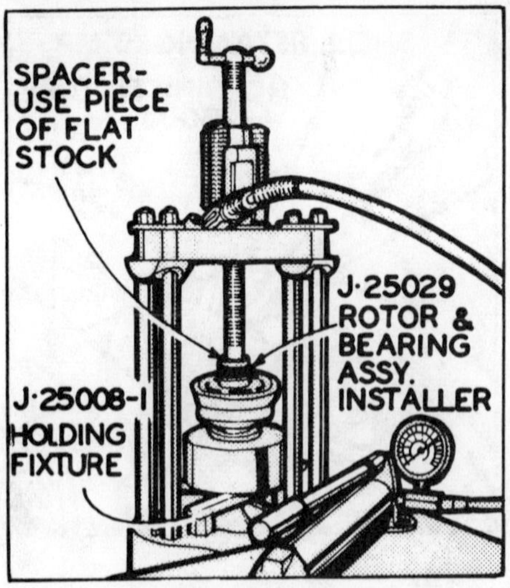

Fig. 51 Installing rotor and bearing assembly

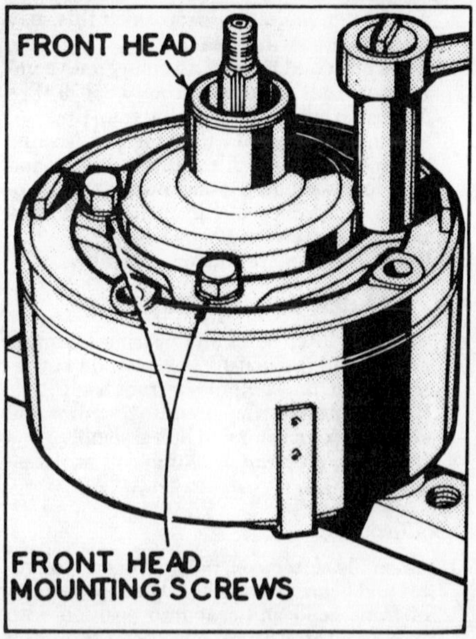

Fig. 52 Removing front head mounting screws

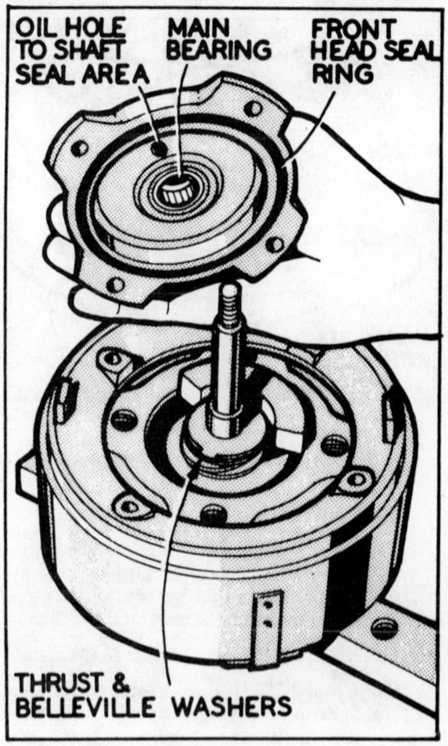

Fig. 53 Removing front head

mark up and, when facing oil sump, position outer gear so that it meshes with inner gear at the 9 o'clock position and cavity between gear teeth is at 3 o'clock position, Fig. 44.

20. Generously oil rear discharge plate around outer edge where large diameter O-ring will be placed. Oil valve reeds, pump gears and area where teflon sealing surface will contact rear discharge valve plate.
21. Lubricate new head-to-shell O-ring and install on rear discharge valve plate, in contact with shell.
22. Install suction screen in rear head casting, using care not to damage screen.
23. Coat teflon sealing surface on webs of compressor rear head casting with refrigeration oil.
24. Install rear head over studs on compressor shell. The two lower compressor mounting holes should be in alignment with compressor sump. Make certain that suction screen does not drop out of place when lowering rear head into position.

NOTE: If rear head does not slide down over dowels in internal mechanism, twist front head back and forth very slightly by hand until rear head drops over dowel pins.

25. Install nuts on threaded shell studs and tighten evenly to 20 ft. lbs., using an 0-25 ft. lb. torque wrench on 1974-75 compressor or 25 ft. lbs. using a 0-50 ft. lb. torque wrench on 1976-80 compressors.
26. Invert compressor in Holding Fixture and install shaft seal.
27. Install compressor clutch coil and housing.
28. Install compressor pulley and bearing.
29. Install compressor clutch plate and hub.
30. Add refrigeration oil as outlined elsewhere in this chapter.
31. Check for internal and external leaks.

Radial Compressor Overhaul

Clutch Rotor & Bearing

Assembly Removal

1. Remove clutch plate and hub as outlined previously.
2. Remove rotor and bearing assembly retaining snap ring with tool J-6083, Fig. 45.

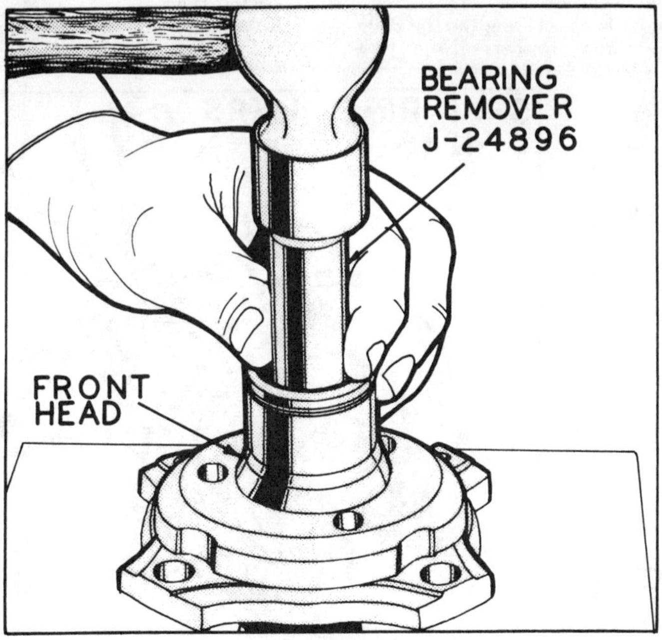

Fig. 54 Removing main bearing

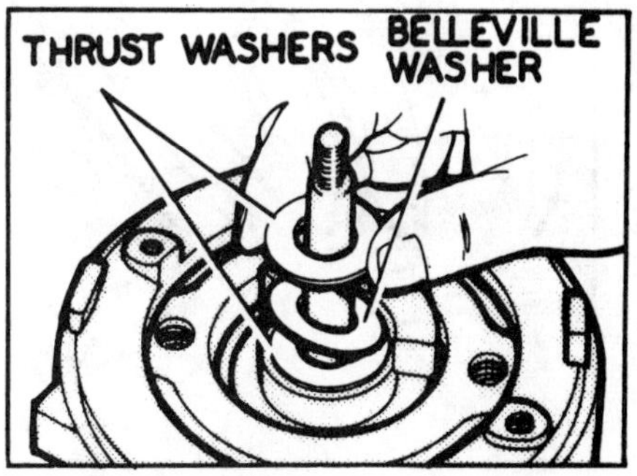

Fig. 55 Installing Belleville and thrust washers

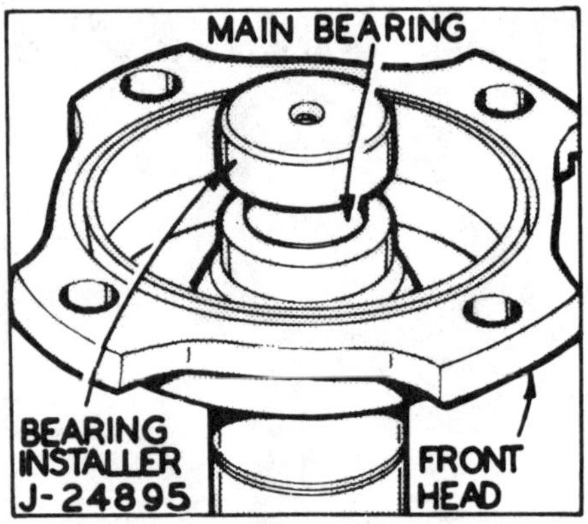

Fig. 56 Installing main bearing

Mark position of clutch coil terminals.

NOTE: If clutch rotor or bearing only are to be replaced, bend lockwashers away from pulley rim mounting screws, then remove the mounting screws and lockwashers.

---

3. Using tool J-25031, Figs. 46 and 47, remove clutch rotor and assembly parts. If the pulley rim mounting screws were removed, only the clutch rotor and bearing assembly will be removed. The clutch coil and housing is pressed onto the compressor front head and will not be removed unless the pulley rim mounting screws were not removed.

Bearing Removal

1. Remove clutch rotor and bearing as outlined previously.
2. Using tools J-8092 and J-25029, drive bearing from rotor hub, Fig. 48. It is not necessary to remove staking at rear of rotor hub to remove bearing.

Installation

1. Using tools J-9481 and J-8092, drive rotor bearing into hub. Align bearing squarely with hub bore before attempting to drive bearing into bore, Fig. 49.
2. Stake bearing, Fig. 50, however, use caution not to stake too deep since this may distort the bearing outer race.
3. Press rotor and bearing assembly onto compressor front head with tool J-25029, Fig. 51. Install assembly retaining snap ring.
4. Assemble pulley rim to rotor and bearing assembly and install mounting screws and new lockwashers. Ensure pulley rim is rotating in proper alignment, then torque mounting screws to 100 inch lbs.
5. Install clutch plate and hub assembly.

Clutch Coil & Pulley Rim

1. Remove clutch rotor and bearing assembly as outlined previously, however, do not remove from the compressor front head.
2. Remove pulley rim mounting screws and slide rim from rotor and hub assembly.
3. Replace pulley rim or clutch coil as necessary.

Installation

1. Assemble clutch coil, pulley rim and the rotor and bearing assembly. Then, place neck of front head and seat into position with tool J-25029. Ensure clutch coil terminals are positioned properly and the three protrusions on the rear of the clutch coil align with front head locator holes.
2. Install rotor and bearing assembly retaining snap ring and assemble the clutch plate and hub assembly. Check clearance between clutch plate and rotor.
3. Install and torque pulley rim mounting screws to 100 inch lbs.

Front Head & Main Bearing

Removal

1. Remove clutch rotor and bearing assembly as outlined previously, however, do not remove pulley ring retaining screws until entire assembly has been removed from the compressor front head.
2. Remove crankshaft seal parts as outlined previously.
3. Remove front head mounting bolts, Fig. 52, and the front head assembly, Fig. 53.
4. Remove front head seal ring and the Belleville and thrust washers. Note position of washers before removing.
5. With tool J-24896, drive main bearing from front head, Fig. 54.

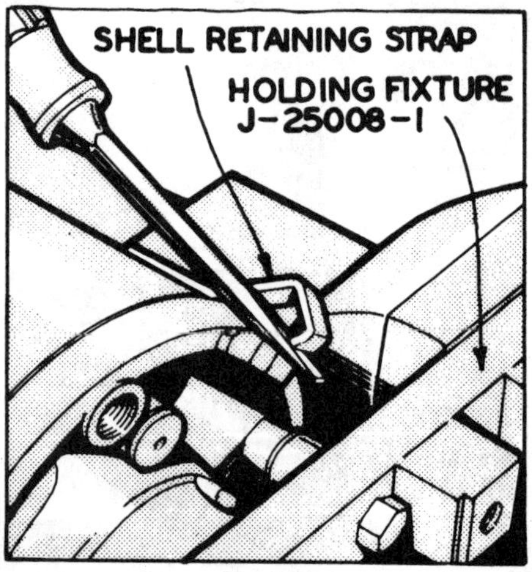

Fig. 57 Prying compressor shell retaining strap

Installation

1. Install new thrust washer on shaft with tang facing away from compressor, then place Belleville washer on shaft with the high center of washer facing away from compressor, Fig. 55. Install remaining thrust washer on shaft with tang facing toward compressor. Lubricate washers with refrigerant oil.
2. Lubricate new front head seal with refrigerant oil and install in front head seal groove.
3. Align new bearing and installer tool J-24895 with front head bearing bore and drive new bearing into head, Fig. 56.
4. Position oil hole in front head to be "Up" when installed on compressor cylinder to correspond with the "Up" position of the compressor.
5. Install and torque front head bolts to 18-22 ft. lbs.

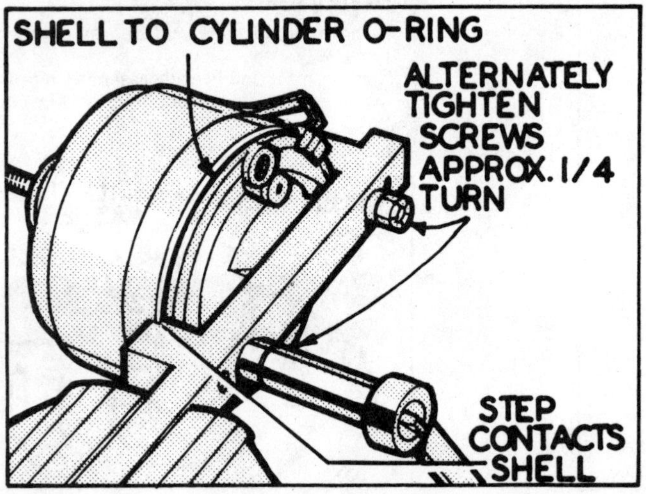

Fig. 58 Removing compressor shell

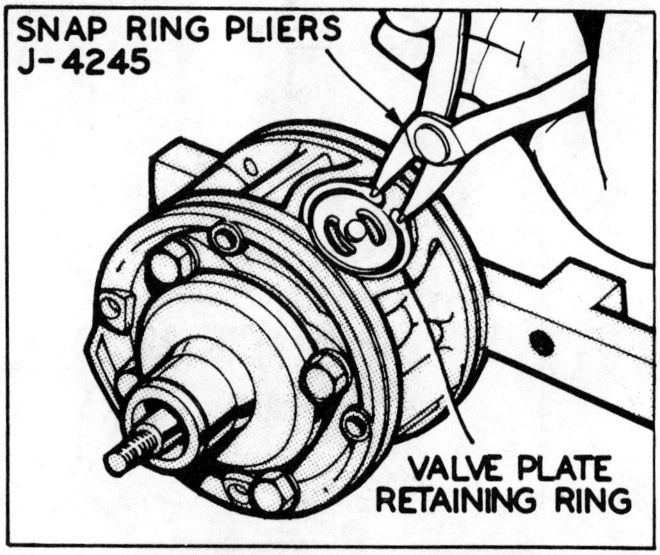

Fig. 59 Removing valve plate retaining ring

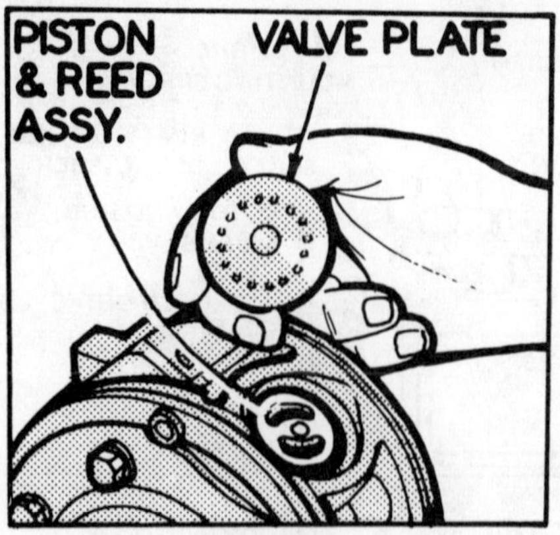

Fig. 60 Removing valve plate

6. Install new crankshaft seal.
7. Assemble clutch coil, pulley rim and the rotor and bearing assembly. Then, place neck of front head and seat into position with tool J-25029. Ensure clutch coil terminals are positioned properly and the three protrusions on the rear of the clutch coil align with the front head locator holes.
8. Install rotor and bearing assembly retaining snap ring and assemble the clutch plate and hub assembly. Check clearance between clutch plate and rotor.
9. Install and torque pulley rim mounting screws to 100 inch lbs.

Compressor Shell

Removal

1. Pry shell retaining strap from cylinder and position strap to clear cylinder when shell is removed, Fig. 57.
2. Remove compressor holding fixture and reverse the holding fixture step block protrusions engaging the compressor shell. Install medium length bolts and thread into compressor cylinder until step of fixture protrusions contact shell. Ensure step protrusions do not overlap the cylinder, but will pass both sides.
3. Alternately tighten fixture bolts approximately ¼ inch to push shell free of cylinder "O" rings, Fig. 58, then remove shell.
4. Remove valve plate retainer snap ring with tool J-4245, Fig. 59, then the compressor valve plate, Fig. 60.

Installation

1. Lubricate cylinder "O" rings with refrigerant oil and install in front and rear "O" ring grooves in cylinder.

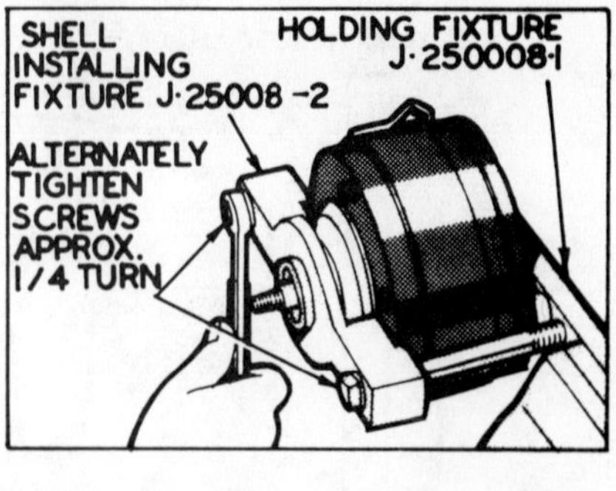

Fig. 61 Installing compressor shell

2. Place shell on cylinder and rotate retaining strap into position.
3. Install holding fixture with long bolts and plate washers, aligning step projections of fixture to contact compressor shell evenly.
4. Manually push shell close to "O" ring and tighten fixture screws finger tight. Ensure shell is aligned properly, then alternately tighten fixture bolts approximately ¼ inch to push shell over cylinder "O" rings and against shell stop flange at rear of cylinder, Fig. 61.
5. Bend shell retaining strap into position.

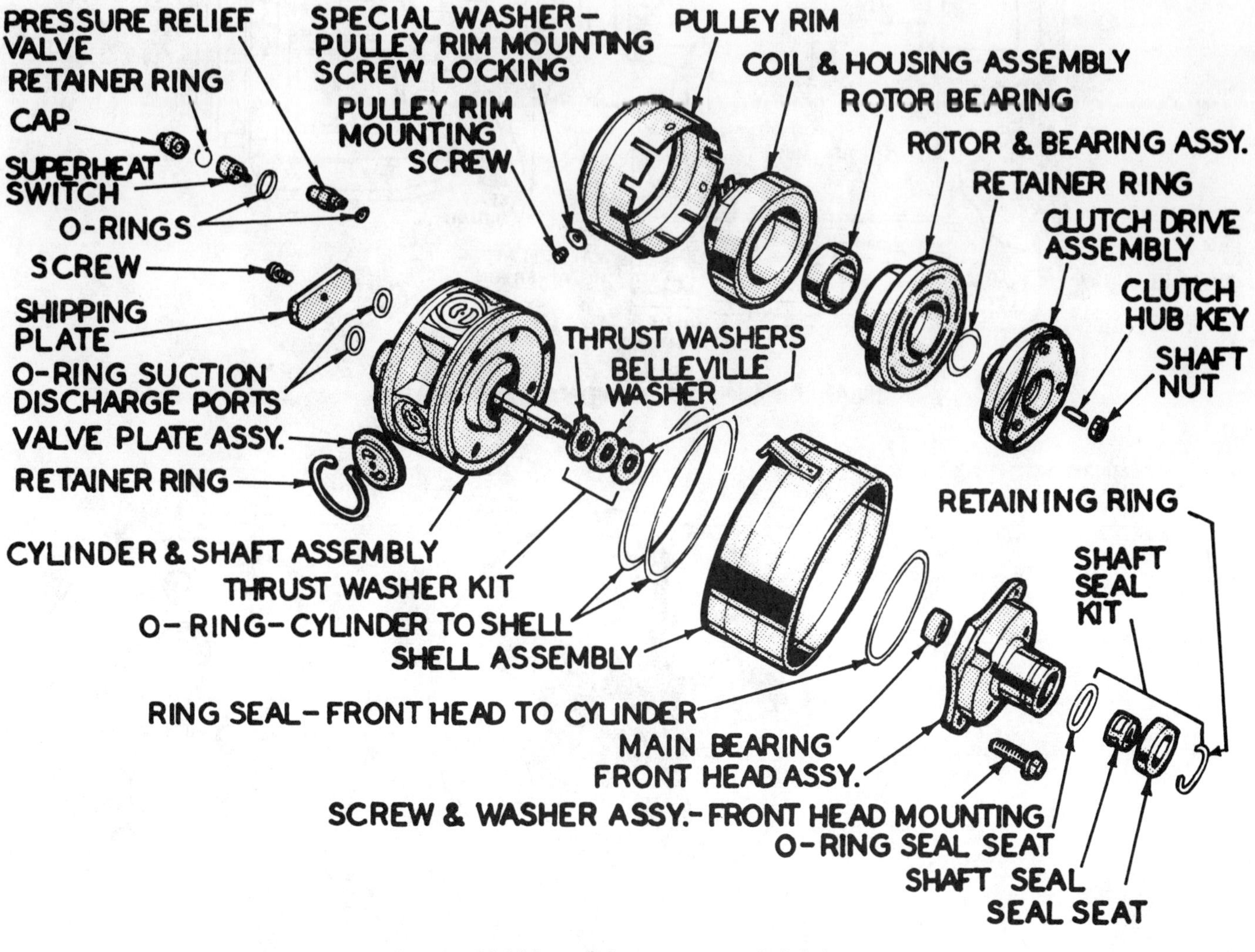

Fig. 62 Frigidaire radial compressor exploded view

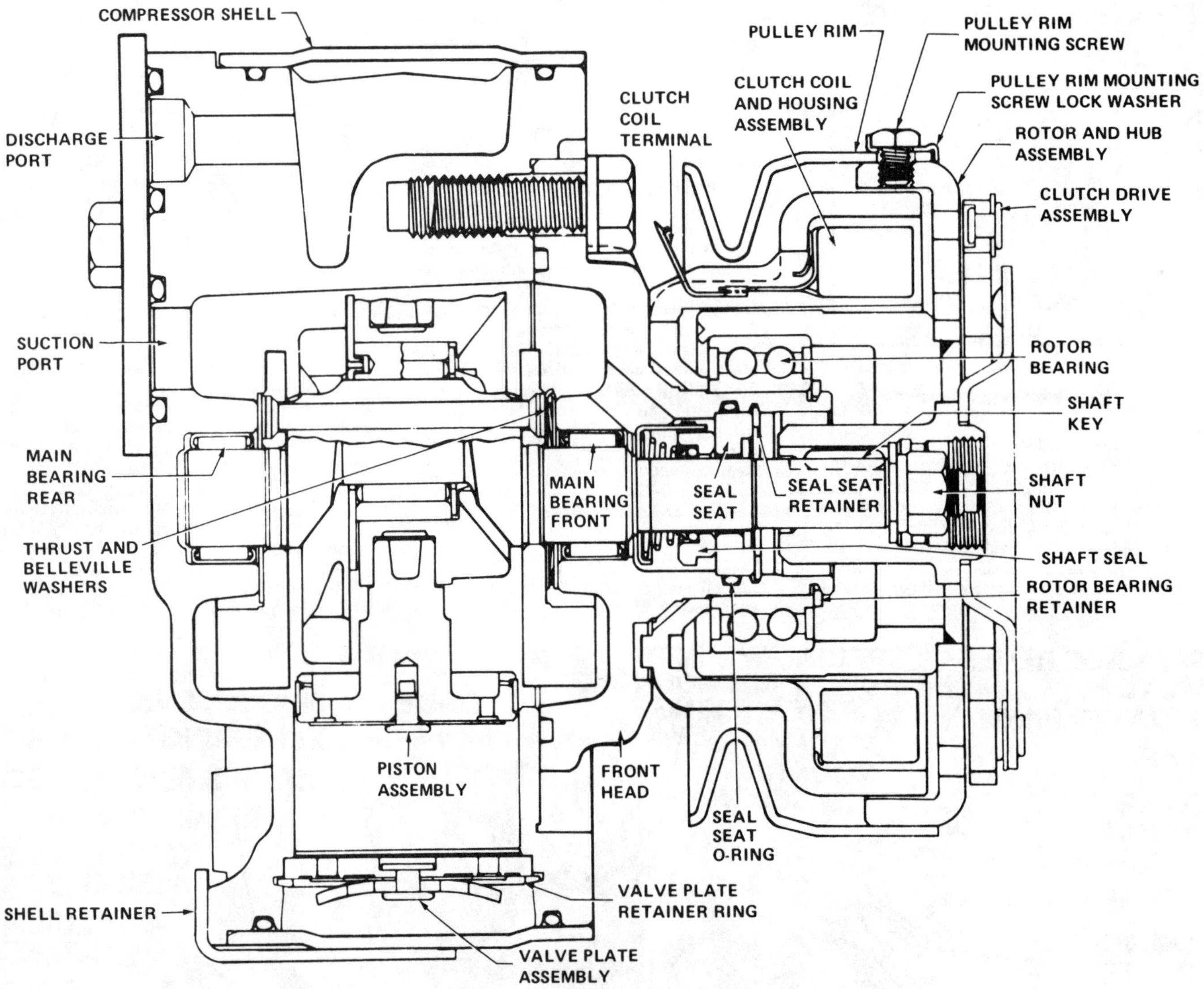

Fig. 63 Frigidaire radial compressor cross-sectional view

Nippondenso

Removal

1. Discharge refrigerant from system.
2. Remove drive belt.
3. Remove refrigerant hose manifold from compressor. Cap openings in lines and

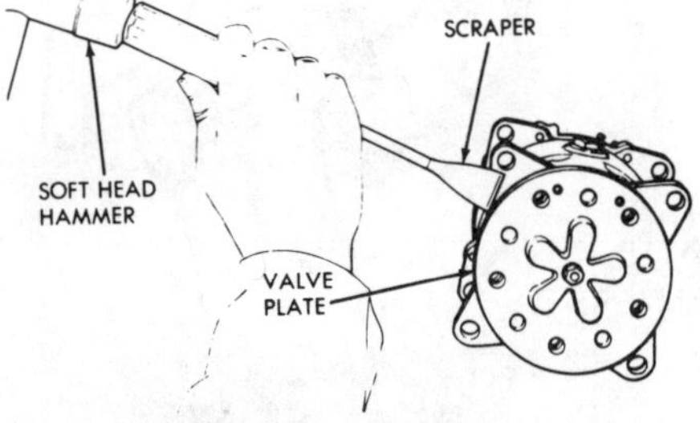

Fig. 64 Removing valve plate

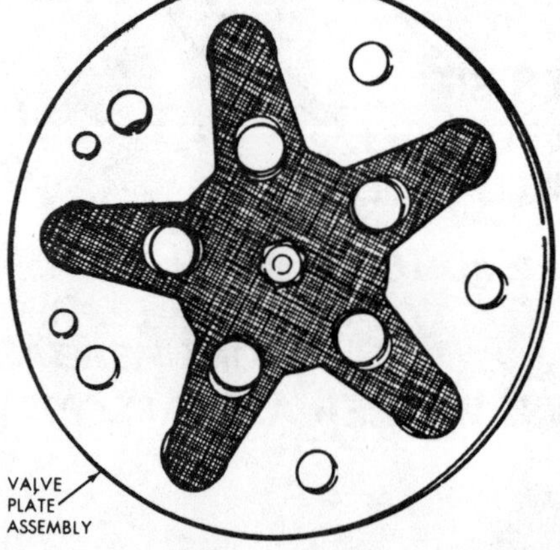

Fig. 65 Front view of valve plate assembly

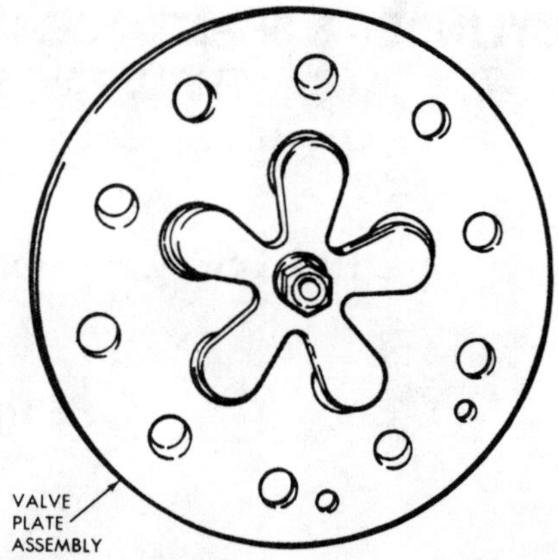

Fig. 66 Rear view of valve plate assembly

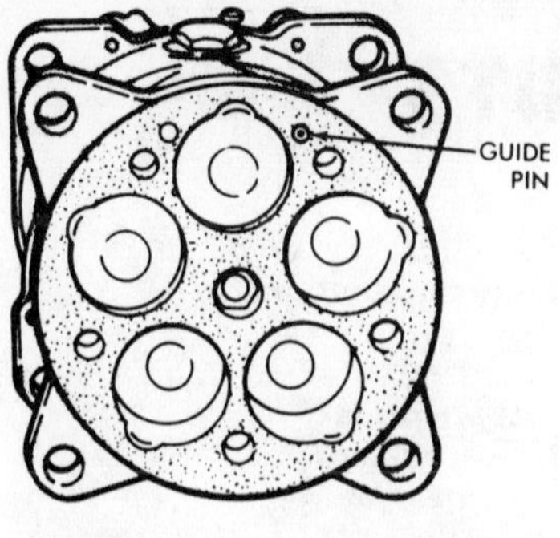

Fig. 67 Installing cylinder head gasket

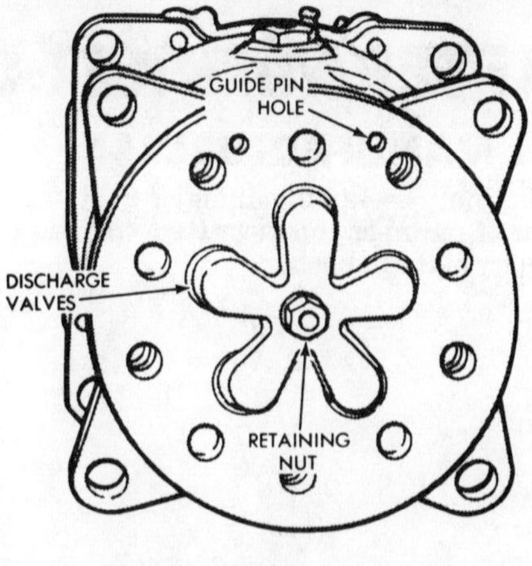

Fig. 68 Installing valve plate

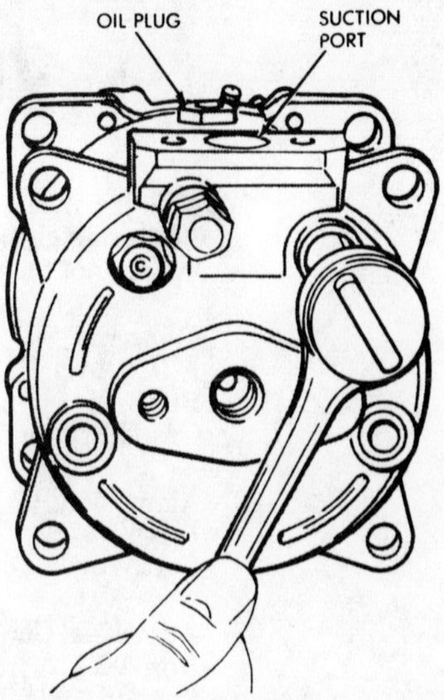

Fig. 69 Installing cylinder head

compressor to prevent entry of dirt and moisture.
4. Disconnect clutch wires at connector and disengage wire connector from compressor front brace.
5. Remove bolt at top of compressor to compressor front bracket, then remove the locknut and stud from compressor front lower inboard attachment at front bracket. If stud does not come out with locknut, install two 10 mm x 1.25 pitch nuts on stud and remove stud.
6. Remove the three lower retaining bolts and remove compressor.

Installation

1. Position compressor on mounting brackets and install the three lower retaining bolts.
2. Install the stud and locknut at front lower inboard attachment, and the bolt at top of compressor to compressor front bracket.
3. Using a new gasket lubricated with refrigerant oil, install hose manifold on compressor. Alternately torque nuts to 18-25 ft. lbs.
4. Connect clutch wires and attach connector to front brace.
5. Install and adjust drive belt.
6. Recharge system.

Sankyo & C-171

Removal

1. Discharge refrigerant from system.
2. Disconnect electrical lead from clutch.
3. Disconnect suction and discharge lines from ports. Cap lines and ports to prevent entry of dirt and moisture.
4. Remove compressor-to-bracket retaining bolts and remove compressor. It may be necessary to loosen the bracket-to-engine bolts.

Installation

1. Add the correct amount of refrigerant oil to compressor. Refer to "Oil Level, Check".
2. Install compressor and retaining bolts. Install and adjust drive belt.
3. Connect clutch wire.
4. Connect suction and discharge lines to ports, using new gaskets. Torque bolts to 170-230 inch lbs.
5. Evacuate and charge system.

Sankyo Cylinder Head & Valve Plate

Removal

NOTE: The following procedure can be performed with compressor installed in vehicle.

1. Discharge refrigerant from system.
2. Disconnect suction and discharge lines from ports. Cap lines and ports to prevent entry of dirt and moisture.
3. Remove the five bolts retaining the cylinder head. Tap cylinder head with a soft hammer to loosen the cylinder head from valve plate. If necessary, remove gasket from head or valve plate with a sharp scraper, using care not to scratch the sealing surfaces.
4. Remove valve plate from cylinder block using a soft hammer and scraper, Fig. 64.
5. Using a sharp scraper, remove gasket.
6. Inspect valve plate for missing, dimpled or bent suction and discharge valves, Figs. 65 and 66. Replace assembly if damaged. It is acceptable for suction valves to be a maximum of .010 inch above plate.
7. Using a sharp scraper, remove any gasket on head, valve plate and cylinder housing, using care to avoid scratching sealing surfaces. Make sure that cylinder head bores are free of all gasket material.

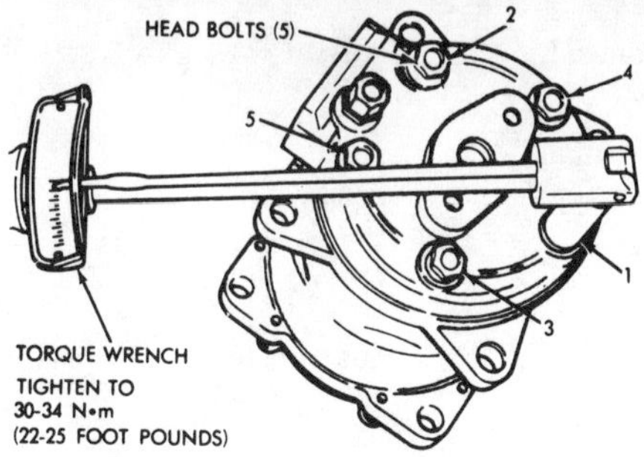

Fig. 70 Cylinder head bolts torque sequence

Installation

NOTE: The valve plate is replaceable only as an assembly complete with suction and discharge valves and gaskets.

1. Apply thin coat of refrigerant oil on surface of cylinder block to be covered by the gasket.
2. Position cylinder gasket on cylinder block so that pin in block will go through pin hole in gasket, Fig. 67.
3. Apply a thin film of refrigerant oil on top and bottom surfaces of valve plate that contact gaskets.
4. Place valve plate in position on cylinder gasket so that discharge valves (side with smaller valves, valves stops and retaining nut) are facing out and pin goes through pin hole in valve plate, Fig. 68.
5. Apply refrigerant oil on surface of cylinder head that contacts gasket and place head on gasket. The suction port should be aligned with oil plug, Fig. 69.
6. Install and lightly tighten head bolts. Torque all bolts to 22-25 ft. lbs., in sequence shown in Fig. 70.
7. Reconnect suction and discharge lines using new gaskets. Torque bolts to 170-230 inch lbs.

VII. AFTERMARKET UNITS

MANUFACTURERS***

This list includes most but not necessarily all major manufacturers of aftermarket A/C. Addresses given are those of record as this is written. Omissions, if any, are unintentional but are not the responsibility of the publisher.

Company	*Brand**	*Application***
A.R.A. Mfg. Co. Box 870 Grand Prairie, Tex. 75050	A-R-A	Most custom + universal
Alamo Mfg. Co. Box 40557 Fort Worth, Tex. 76140	AMC	Most custom + universal
Auto-Kool Corp. Box 202 Lawton, Okla. 73501	Auto-Kool	
Behr of America, Inc. 5020 Augusta Drive Fort Worth, Tex. 76106	BMW, Mercedes-Benz	Most custom + universal
Clardy Mfg. Co. Box 7068 Fort Worth, Tex. 76111	Clardy	Most custom + universal
Climatic Air Sales Co. Box 805 Dallas, Tex. 75226	Climatic Air	Most custom + universal
Coolaire Mfg. Co., Inc. 10400 S.W. 187th St. Miami, Fla. 33157	Coolaire	VW Toyota
DPD Mfg. Co., Inc. Box 18327, Sarna Sta. San Antonio, Tex. 78218	DPD	
Factory Air Conditioning Corp. 1353-55 French Pl. San Antonio, Tex. 78218	Factory Air Western Air	
Frigiking Div., Smith Jones, Inc. 10858 Harry Hines Blvd. Dallas, Tex. 75205	Datsun	Datsun Most custom + universal
Frostemp, Inc. 5643 E. Rosedale Fort Worth, Tex. 76112	Frostemp	
Heatransfer Corp. 4321 Industrial Center San Antonio, Tex. 78217	Heatransfer	VW
Kool-Temp Mfg. Co. Box 1109 Tampa, Fla. 33601	Kool-Temp	Most custom + Universal
Meier-Line Co. 17939 E. Rowland St. City of Industry, Cal. 91744	Meier-Line	VW Porsche
John E. Mitchell Co. Box 1811 Dallas, Tex. 75226	Mark IV	Most custom + universal

Nippon Denso Co. (Nippondenso of Los Angeles, Inc.) 1815 Anderson Ave. Compton, Cal. 90220	Toyota	Mark II Celica Crown Corona Corolla Hi Lux
Nordic International Co. 3701 N. 29th St. Hollywood, Fla. 33020	Nordic	
Thermo Equipment Corp. 179 Pulaski St. Newark, N.J. 07105	Duo-Therm Duo-Temp	Most custom + universal
Volkswagen Products Corp. 4900 August Dr. Fort Worth, Tex. 53207	Volksaire	VW, Porsche
W-D Manufacturing Co. 1604 N.W. 5th St. Oklahoma City, Okla. 73106	Dat-Air Toy-Air	Datsun Toyota

*Does not include "house" brands produced by the larger manufacturers for chains such as Sears, Ward, Western Auto, etc.

**Application when specified indicates that the manufacturer specializes in custom units for the make(s) listed. The phrase "most custom + universal" indicates that the manufacturer is large enough to catalog applications customized for most high-volume domestic and imported cars, the latter including Toyota and Datsun but not necessarily VW or other installations involving air cooling of the powerplant.

***Specifications can be found on page 1-494.

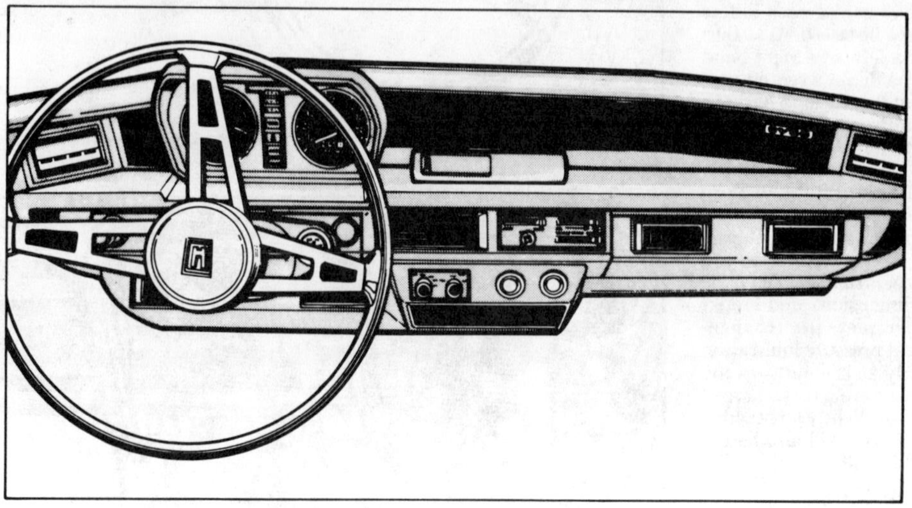

Fig. 1 ARA installation in Honda Civic CVCC

Fig. 2 Coolaire installation in Toyota Corolla

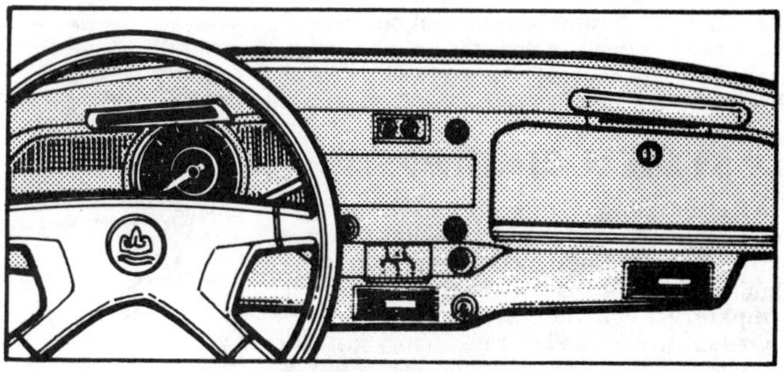

Fig. 3 Volksaire installation in Volkswagen Beetle

EXC. SANKYO SD-5 AFTERMARKET COMPRESSOR

Aftermarket compressors are most commonly supplied by Tecumseh and York. Exceptions include Meier-Line which manufactures a remarkable small high-capacity unit of its own, Nippondenso which uses an American-style reciprocating compressor built under license in Japan and the few luxury imports which use compressors designed and built in Europe. The latter are not in service in enough volume to warrant coverage, the Nippondenso unit may be easily serviced by reference to instructions for similar domestic units contained elsewhere in this book and Meier-Line volume is too low to warrant separate coverage.

The initial lubricant charge of York compressors is ⅝ pint, and dip stick depths of the initial charge are 1⅜, 2¾ and 1⅝ ins. for vertical (90-70°), inclined (50-15°) and horizontal (10-0°), respectively.

Early Tecumseh compressors, Models HA through HG, were equipped with a sight gauge and allowable operating charge in the vertical position is 5/16 in. to top of glass. When horizontal, the figures are ¼ to ⅜ up on the glass. Beginning with Model HH, the use of built-in mufflers eliminated the sight gauge and a wire dip stick must be used. Ranges on the dip stick are 1⅛ to 1½ ins. for a vertical mount and 1⅜ to 1 9/16 for horizontal. The initial factory charge was 11 oz.

The Tecumseh HA 850 compressor differs from those used in factory installations in that it will function in either a left- or a right-hand mount as viewed from the driver's seat as well as in vertical or horizontal positions on both sides. If it is mounted horizontally to the left, the internal oiling system must be changed and detailed instructions for this are contained on a red decal attached to the case. Initial factory charge of this compressor is 10 oz. and the dip stick should read a minimum of ¾ in. in any mounting position.

The still newer Tecumseh HG 850, Fig. 10, is identical in mounting dimensions and crankshaft tapers, allowing complete interchangeability, but the differential pressure lubrication system differs importantly as it eliminates the oil pump. This compressor is factory charged with 11 oz. of oil. A model of slightly greater displacement known as HG 1000 will also be encountered in aftermarket installations.

Compressor Clutch

All but the oldest aftermarket A/C systems in service are equipped with electromagnetic clutches whose servicing is described elsewhere in this book, Fig. 11. It is important in warranty work that clutch rotor bearings not be replaced in the field. Instead, replace the rotor assembly or the field coil assembly complete when one or the other is defective.

SYSTEM DIAGNOSIS

Aftermarket A/C operates on exactly the same principles as factory-installed systems and troubleshooting procedures are therefore similar except for the need to check automatic temperature control systems sometimes required on the more elaborate factory installations. You will also find the aftermarket installations more accessible and therefore easier to service.

Fig. 4 ARA installation in Mazda 808

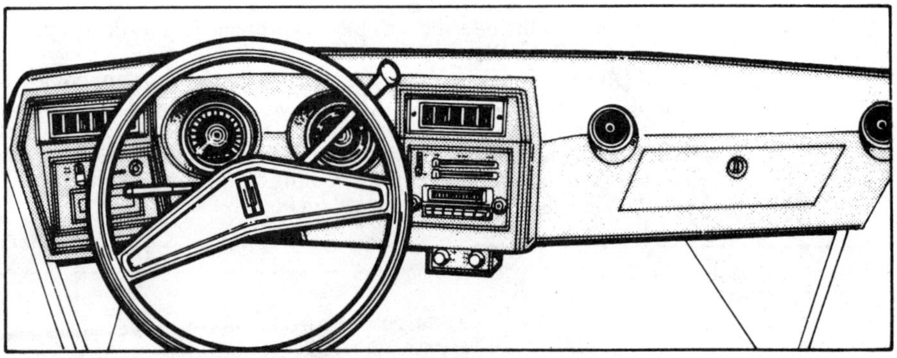

Fig. 5 Clardy installation in Oldsmobile Cutlass

Fig. 6 Clardy installation in Toyota Celica

Fig. 7 ARA installation in Chevrolet

Fig. 8 Clardy installation in Aspen & Volare

Fig. 9 Mark IV installation in Chevrolet

The most common complaint is insufficient cooling from the evaporator and an orderly search for the cause can be conveniently divided into three tests:

Test No. 1

High Side Low, Low Side High

Probable cause, defective compressor:

1. Reed valves leaking.
2. Leaking head gasket.
3. Internal leak in the compressor.
4. If cylinders or pistons are defective, replace the compressor.

High Side Low, Low Side Low

Check sight glass:

1. Bubbles in sight glass. Refrigerant is low and the cause may be a small leak.
   (a) Check for leakage and correct.
   (b) Add refrigerant until bubbles disappear and both test gauges show a normal reading.
2. No bubbles in sight glass. Gauge readings are excessively low. Possibly no liquid in sight glass.
   (a) A serious leak may be indicated.
   (b) Expansion valve screen may be clogged.
   (c) Expansion valve may be stuck closed. Replace valve.

High Side Normal, Low Side Low

Low side gauge may or may not approach vacuum while testing. Check for moisture in system.

1. Purge refrigerant from system.
2. Replace drier.
3. Pump down system and recharge.

High Side High, Low Side Normal

Low side gauge will not drop. Check for air in system.

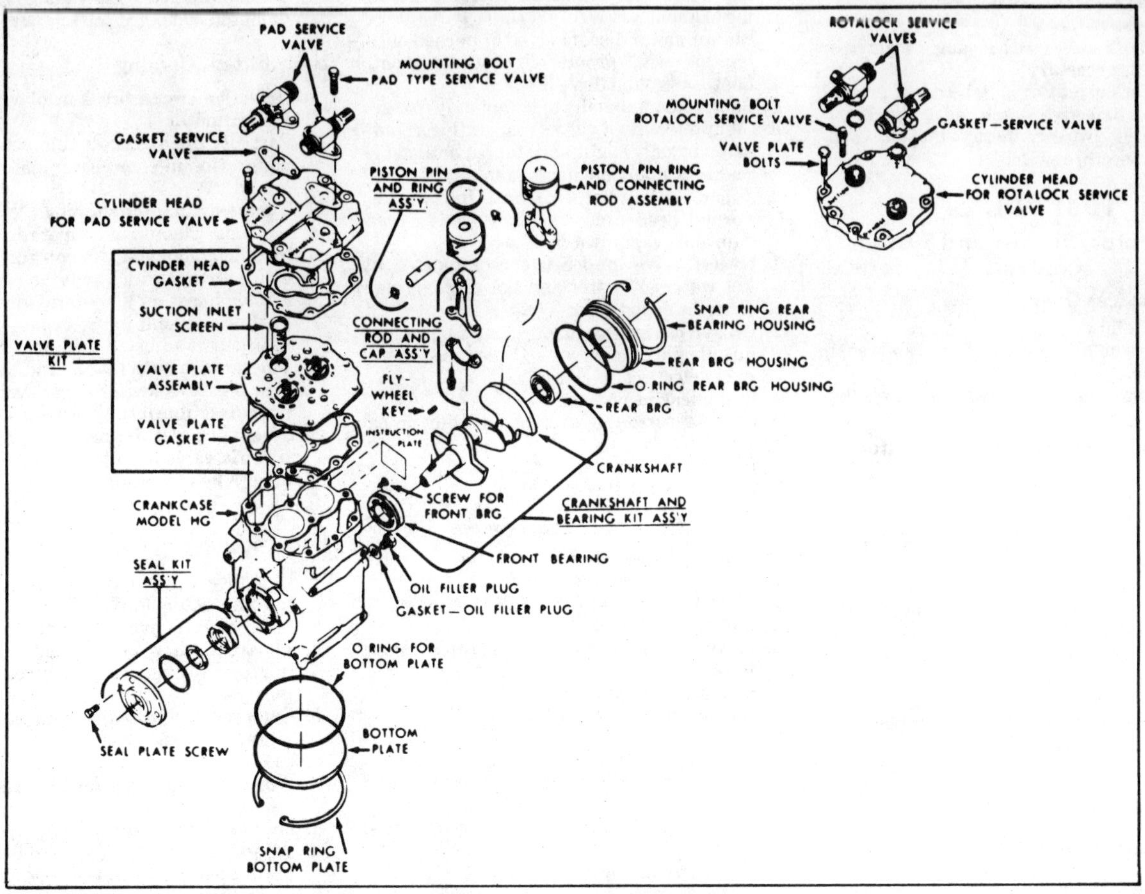

Fig. 10 Tecumseh HG850 compressor exploded view

1. Purge refrigerant from system.
2. Replace drier.
3. Pump down system and recharge.

High Side High, Low Side High

Probable causes:

1. Refrigerant overcharge.
   (a) Sight glass may show bubbles.
   (b) Bleed off refrigerant until a heavy stream of bubbles appears.
   (c) Charge with refrigerant until bubbles are eliminated from sight glass.
2. Defective expansion valve.
   (a) Check valve by R-12 method.
   (b) If valve responds to test, remove bulb from tailpipe and clean contacts.
   (c) If valve does not respond to test, replace.
3. Condenser.
   (a) May be blocked and not allow sufficient air flow. Remove bug screen and clean condenser.
   (b) Clearance between radiator and condenser must be 2 in. If not, reposition condenser.
4. Car cooling system.
   (a) Check fan belt tension and adjust, using tension gauge.
   (b) Test radiator pressure cap and system thermostat. Replace if necessary.
   (c) Check all hoses and connections.
   (d) Check for proper mixture of ethylene glycol in coolant. Flush and replace as necessary.

Test No. 2

Compressor and/or Clutch Defective

1. Check system fuse.
2. Check belt tension with a gauge. Tighten or replace as necessary.
3. Check for current at clutch coil. If present check for defective clutch coil or clutch.
4. Check relay wiring, thermostat and vacuum temperature switch.

Test No. 3

Evaporator, Blower and System Controls

Evaporator Ices Over

1. Check air flow.
2. If there is no air flow, clean coils and repair or replace air ducts.
3. If air flow is good, test system controls (below).

Evaporator Discharge Air Temperature High

1. Cars with POA or EPR valves.
   (a) Check evaporator pressure.
   (b) If pressure is too high, replace POA or EPR valve.
   (c) If pressure is normal, repair or replace heater valve, vacuum servo or heater door as required.
2. Cars with thermostat or suction throttling valve.
   (a) Adjust valve to specifications.
   (b) If valve still faulty, repair or replace it.

Blower Does Not Operate

1. If blower operates at some but not all speeds, check switch and replace if necessary.
2. Check current now at motor assembly. If current is present, replace the motor.
3. If current is not present, check for faulty switch, broken wires, blown fuse or poor connections.

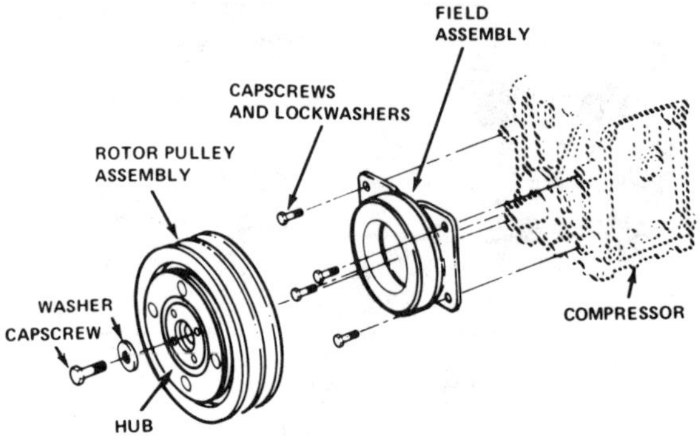

Fig. 11 Magnetic clutch. Volksaire units

Evaporator Pressure Higher or Lower Than Normal

1. On cars with POA or EPR valves, replace valve.
2. On cars with thermostat or suction throttling valve, adjust thermostat or STV to specifications. If these units do not respond, repair or replace them.

System Troubleshooting

No Cooling From System

1. Blown fuse.
2. Broken or disconnected wiring.
3. Clutch coil or solenoid burned out or disconnected.
4. Electric switch contacts in thermostat burned excessively, or a defect in the sensing element.
5. Blower motor disconnected or burned out.
6. Ignition switch ground or relay burned out.
7. Loose or broken drive belt.
8. Compressor partially or completely frozen.
9. Compressor reed valves inoperative as indicated by only slight variation of both gauge readings at any engine speed.
10. Expansion valve stuck open as indicated by normal head pressure, high suction pressure and evaporator flooding.
11. Heater valve inoperative as indicated by hot water in heater and hot discharge air from the evaporator.
12. Broken refrigerant line.
13. Fusible plug blown in A/C (not used in all) or car electrical system.
14. Major leak in system.
15. Clogged screen(s) in receiver-drier or expansion valve.
16. Plugged hose or coil.
17. Compressor shaft seal leaking.

Insufficient Cooling From System

1. Blower motor sluggish
2. Compressor clutch slipping.
3. Obstructed blower discharge passage.
4. Clogged air intake filter.
5. Clogged condenser preventing sufficient air flow.
6. Evaporator clogged.
7. Outside air vents in car open.
8. Evaporator pressure regulator, hot gas by-pass valve, suction throttling valve, or Se-lec-Trol defective or improperly adjusted as indicated by high pressure gauge reading normal and low pressure gauge reading too high.
9. Insufficient refrigerant in system.
10. Clogged screen in expansion valve as indicated by gauge pressures being normal or showing slightly increased head pressure and low suction pressure with discharge output temperature high.
11. Expansion valve thermal bulb has lost its charge as indicated by too high a low gauge reading and excessive sweating of the evaporator and suction line.
12. Clogged screen in receiver as indicated by higher than normal reading on high pressure gauge, lower than normal reading on low pressure gauge. Liquid lines and the receiver are cold to touch with possible frost.
13. Excessive moisture in system as indicated by high head pressure gauge reading.
14. Air in system as indicated by excessive head pressure and usually bubbles in sight glass.
15. Thermostat defective or improperly adjusted as indicated by low gauge reading high or clutch cycling at too high a reading.

Intermittent Cooling

1. Defective circuit breaker, blower switch or blower motor.
2. Partially open ground connection or loose connection in compressor clutch coil or solenoid.
3. Compressor clutch slipping.
4. Improper plenum door operation.
5. Insufficient vacuum for control.
6. Unit icing up which may be caused by excessive moisture in system, incorrect super heat adjustment in the expansion valve, or thermostat adjusted too low.
7. Thermostat defective as indicated by low side pressure low or excessively high and not correctible by adjustment.
8. Stuck hot gas by-pass valve or suction throttle valve as indicated by low pressure of both head and suction gauges.
9. Clogged evaporator.

Noisy System

1. Defective winding or improper connection in compressor clutch coil or solenoid.
2. Loose or excessively worn drive belt.
3. Too tight a drive belt tension.
4. Improperly aligned compressor and engine pulleys.
5. Compressor harmonic induced by engine vibration mode.
6. Noisy clutch.
7. Worn compressor inner parts or loose mounting.
8. Loose mounting panels or point on car.
9. Compressor oil level low as indicated by hot lower area of compressor case.
10. Blower fan noise caused by worn motor.
11. Idler pulley and bearing defective or misaligned.

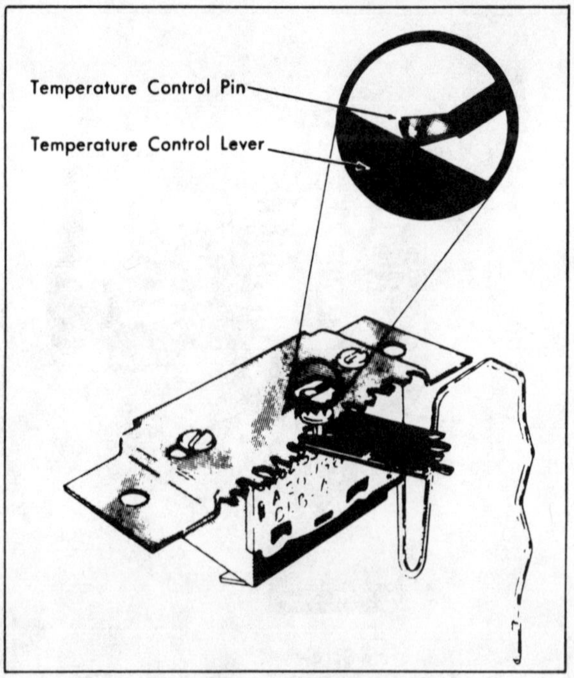

Fig. 12 Lever actuated thermostat

12. Rumbling noise caused by excessive charge in system.
13. Vibration in high pressure line, thumping noise in compressor, excessive head pressure and suction pressure, bubbles or cloudiness in sight glass—all caused by excessive charge in system.
14. Low charge in system as indicated by hissing in evaporator case at expansion valve, bubbles or cloudiness in sight glass and low head pressure.
15. Excessive moisture in the system as indicated by noisy expansion valve and low suction pressure.
16. High pressure service valve stuck shut as indicated by loud compressor knock and a high pressure gauge reading above normal.
17. Evaporator unit vibration caused by blower wheels sprung from improper handling of the unit during installation or repair.

Electrical Systems

General

The electrical system in a typical aftermarket A/C installation is simpler than the circuits found in a factory-installed design. One reason that few if any hang-ons introduce a portion of outside air into the car, depending instead entirely upon recirculation. This eliminates control devices for doors in ducts and plenum chambers, and thus there is no possibility of automatic humidity control. Neither are aftermarket A/C installations interconnected with the car heating system and thus it is not practical to introduce automatic temperature sensing systems.

Temperature controls used range from a simple lever-actuated thermostat responsive to discharge air temperature to a more sophisticated stem-operated rheostat type thermostat responsive to evaporator coil temperature. These generally are provided with a low-temperature cutout and require a 6-8° variation before they will cycle. Other electrical units in the system include a three or four speed switch for the one or two fan motors, a resistor, a circuit breaker or fuses and, of course, the electromagnetic clutch. Wiring in 12-volt system is usually #16 wire color-coded and the circuit breaker is usually rated for a 20% overload at 15 amps for 30 minutes. All driver-operated controls are housed in the evaporator case with the exception of systems requiring a trunk-mounted evaporator.

The switches and most other electrical units can be reached by separating the evaporator case from the dash mounting, using extreme care to avoid excessive force against the case itself. Hand pressure is sufficient to distort the case and cause permanent distortion of the blower wheels which are plastic in most modern units. The refrigerant lines need not be disconnected to place the evaporator case in a position on the car floor where components within can be conveniently serviced. However, use care not to force these lines.

Switches

The evaporator case almost invariably must be taken apart to reach the switches and because of the near infinite variety of shapes and mountings, it is impractical to detail this procedure. The way to open the case and remove its components will become obvious when it has been freed from its dash or console mounting.

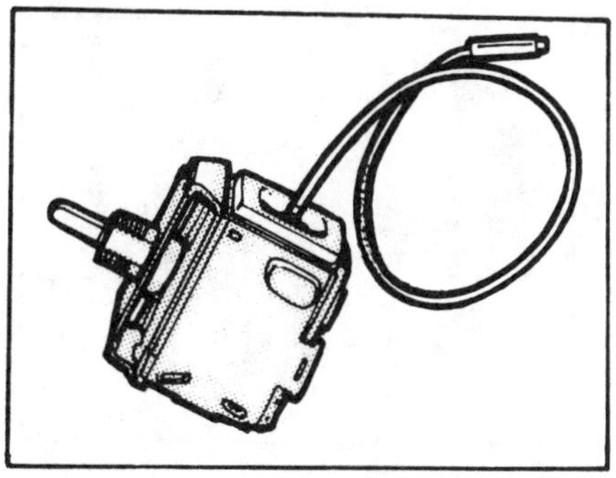

Fig. 13 Stem actuated thermostat

Switches are inexpensive and lack provisions for field adjustment. There is little you can do with them other than to clean terminals and make sure the wiring connections are secure. Replace the switch with a new one if it is malfunctioning.

Thermostat Controls

When replacing a lever-actuated thermostat, Fig. 12, it is important to position the control lever to the actuating pin correctly. Move the control lever to the extreme cold position and position the thermostat housing so that the lever and pin touch each other lightly. Then tighten the thermostat mounting screws. Care must also be taken that the sensing element is in firm, clean contact with the evaporator coil. The low temperature cutout setting of approximately 26° is made at the factory with this type of thermostat and cannot be changed.

The newer stem operated type of control, Fig. 13, contains an adjustment to change the cutout setting but because the off-on cycle is affected by any change, it is not recommended to do this in the field. The usual purpose in the adjustment is to lower discharge air temperature and the result, usually, is unsatisfactory sensitivity to temperature from an increase in the off-on cycle differential. The proper setting is around 8°.

Blower Motors

12-volt, series wound DC motors of varying sizes and shaft extension lengths and sometimes two are employed. Current draw ranges from six to 15 amps at 13.5 volts output from the car's electrical system. Theoretically, the motor(s) should last the life of the car as they utilize porous bronze bearings in contact with saturated wicks. They are impractical to repair and should be replaced in the event of failure.

Replacement blower blades may in some instances be readily mounted on the shaft extensions but be sure they are positioned exactly and that the concave side of the blades is faced toward the discharge outlet(s). It is also sometimes possible to correct armature end play by placing shims on the shaft inside the case. This is only possible, of course, when the motor housing is of the type that can be separated without destroying it. Symptom of excessive end shaft play is an audible chatter when the car is moved about violently such as on rough roads or in fast cornering.

Blower motor switches may be three- to four-position lever actuated or infinitely variable via a rheostat. In the case of the now rare trunk mounted evaporators, there is usually a push-button switch under the dash that connects via solenoids to the evaporator. None of these switches is repairable and must be replaced if defective.

Trouble and expense will be saved by testing before replacing a supposedly defective blower motor. The switch can be isolated by wiring around it to see if the motor operates. Also, poor motor operation is often caused by something as simple as loose or dirty connections, and wire from the power source that is too small and has overheated.

Compressor Clutch

Even the cheapest aftermarket A/C system now utilizes a magnetic compressor clutch that is thermostatically controlled to discharge when compressor capacity is not needed. Trouble-shooting and repair of clutches used with Tecumseh and York compressors is described elsewhere in this book.

Valves

Expansion Valve

Two different types of expansion valves will be found on aftermarket A/C systems, depending upon the system layout. The major difference between the two is the manner in which system pressure is sensed, but the basic operation is the same for both. The internal equalizer type is usually used in systems employing a short evaporator tube. In systems employing a long evaporator tube, the external equalizer type is used. This is necessary because with the long tube system, changes in compressor rotation speed can affect the pressure in the discharge portion of the evaporator without being sensed at the expansion valve. The function of an expansion valve is explained elsewhere in this book.

As is the case with factory A/C, expansion valves supplied with aftermarket kits do not require periodic service. It is important that the thermo bulb clamp remain tight as operation depends on firm contact between the bulb and the suction outlet tube. When contact is poor, excess refrigerant will flood the compressor and result in possible damage to the compressor reed valves.

Excessive wear of the expansion valve needle and seat can be caused by an inadequate charge of oil in the compressor and also by continued high operating head pressures resulting from an overcharge of refrigerant, or air in the system. Prolonged operation when the latter conditions exist admits partially condensed refrigerant at high velocity into the expansion valve. Lack of oil at the needle and seat affects the seal.

Expansion valves used with aftermarket A/C usually have superheat settings ranging from 5 to 8° and most are rated at 1½ tons refrigeration capacity.

Suction Flow Regulators

Both automatic and manual suction flow regulators, sometimes referred to as RoboTrol, Fig. 14, and SelecTrol, respectively, will be found in older installations. However, the cycling compressor clutch has replaced them in the temperature control function on practically all newer aftermarket A/C. An occasional exception will be found in a hybrid installation where a GM-type rotary compressor is used. The function of both types is to control evaporator flow to the compressor so as to obtain the coldest possible steady air temperature without allowing condensate to collect and freeze on the evaporator coil.

Although automatic valves have an adjustment screw, this adjustment is set at the factory and is normally left untouched in the field. The valve is operating normally if pressure readings on the inlet side show that it closes at 17 psi and opens at 19 psi. It will then maintain about 26 psi in the evaporator coil. In areas of high humidity such as Florida, it may be necessary to turn the adjustment screw clockwise slightly to prevent ice from forming on the evaporator coil. Correct adjustment under these conditions is difficult to achieve as a low setting will perform satisfactorily under traffic conditions but cause icing at road speeds. Conversely, a high setting may cause a drop in efficiency at all speeds. The automatic valve must be replaced if found leaking or otherwise defective.

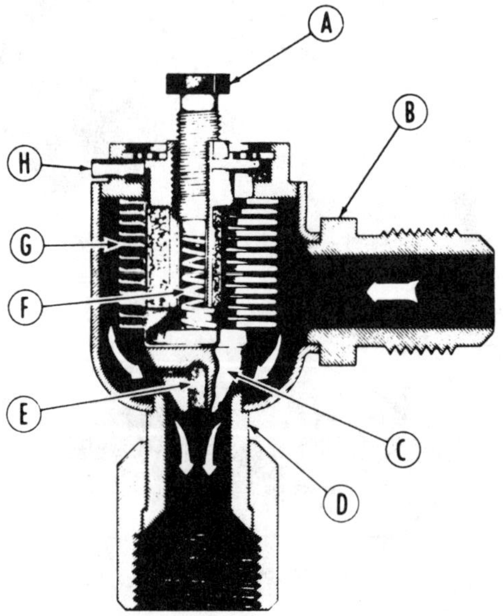

A. Adjusting Screw
B. Inlet Fitting with Backup Wrench Flat
C. Conical Valve Head
D. Outlet Fitting to Compressor
E. Oil Return By-Pass
F. Control Spring
G. Bellows
H. Spring Barrel Locking Pin

Fig. 14 RoboTrol suction flow regulator

The driver controlled type of suction flow regulator is usually located in the suction line adjacent to the evaporator coil. Turning the knob counter-clockwise (left) produces maximum refrigerant flow for coldest discharge. Turning it in the opposite direction will increase discharge air temperature appreciably.

If the control knob is binding, the sealing disc may be replaced in the field but this requires discharging the refrigerant on all but very old installations with a valve-type receiver-drier. The evaporator case must then be removed to allow access to the valve. Loosen the lock-nut and remove the adjusting screw collar (or bushing). Then push out the disc and bearing cap.

To reassemble, hold the cap small end down and insert the bearing cup open side down into its recess. Next, push the sealing disc firmly into the cap, being careful not to scrape it against the threads. Replace the guide washer and spring on the valve plugger and insert the other guide washer on the valve shoulder. Push the plunger into the body and tighten the hexagon valve cap.

To set the valve, first turn the adjusting screw against the nylon bushing until it seats and then screw the bushing into the valve cap until the valve closes tight. Insert a screw driver into the brass stem and while backing the bushing out slowly, turn the stem until it will rotate ⅞ths of a turn from stop to stop. Finally, rotate the stem counter-clockwise to the limit and insert the control knob with the indicator pointing straight up.

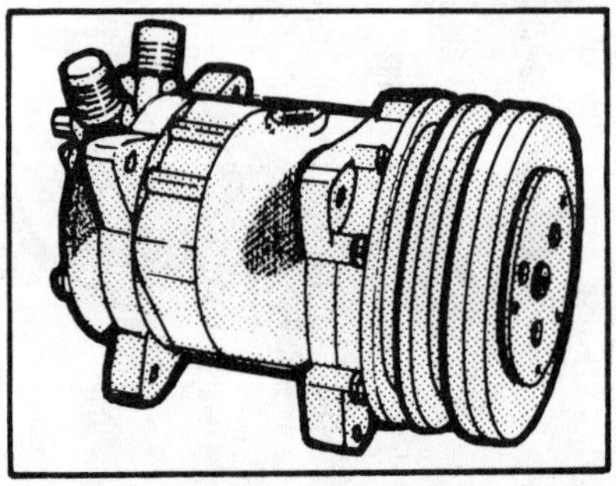

Fig. 15 Sankyo SD-5 compressor

Component Replacement

Hoses

Replacement hoses should be ordered from the manufacturer's distributor by part number to insure proper cleanliness and dryness and correct fittings. Use refrigeration oil, never sealant compounds, to seal fittings. Commercial sealants can clog the receiver-drier and the suction valve screen and cause complete failure or at the least, greatly lowered efficiency of the system. Their use will also void the warranty.

Valves

The variety and location of the various valves in aftermarket A/C systems varies too much for a detailed description of any single procedure to be profitable. As none can be repaired in the field (except for a binding driver control in a suction flow regulator as described above), the important consideration is to obtain an exact replacement. Thus, replacements should be ordered by the system maker's part number from his authorized distributor. The other main consideration is to treat these valves gently as any distortion caused by wrenches during the installation procedure will be permanent and probably cause the valve to malfunction. Cleanliness during the installation procedure is, of course, taken for granted.

Evaporator

Assuming the mounting of the evaporator assembly was performed correctly at the time of installation, replacement is simply a matter of refastening the attaching screws. In some instances, the case is mounted on brackets rather than fastening directly to the dash and these usually have provision for alignment. As the case is highly visible, it is important that alignment to the major surfaces of register be precise. Be sure to replace the grommets that protect the refrigerant hoses from abrasion and check that the drain hoses are not kinked or clogged. Check all switches for proper function.

Compressor

If the system is a custom unit, the compressor brackets will be precisely fabricated and reinstallation should pose no problems other than correct alignment of the drive pulley with the driven and idler pulleys. The easiest method of assuring alignment is to place one end of an accurate metal straight edge against the machined surface of the drive pulley and the other against the outer surface of the driven pulley on the compressor. Misalignment can't be more than 1/16 in.

First, however, it is advisable to check the drive pulley for wobble which can be caused by foreign matter present when it was initially installed. This may be done visually by first removing the ignition coil high tension lead and then have the starter actuated briefly several times. A more precise method if there is doubt is to remove the fan and hold a straight edge on the machined surface of the pulley. Then hook a spring scale at a point about a foot removed and actuate the starter as above.

Though not directly involved with compressor replacement, it's always advisable to check the clearance between the fan and the radiator which should be at least ⅜ in. Failed motor mounts can affect this clearance, as can even minor front end accident damage.

If the installation is of the universal type with the compressor mounted in other than a vertical position, it is important to reference the original position so that the replacement may be installed precisely as before. This may be difficult with the "jury-rigged" mount devised at the times of installation, and there is no easy solution that can be offered here.

Aside from misalignment, the most common cause of premature belt failure is foreign matter or burrs in any of the pulley grooves. All pulleys should be checked carefully prior to belt installation.

Belts commonly used for the compressor drive include single ½-in. "A" section, dual ⅜ or ½-in. belts and Plyflex 11MM. Tensions, as checked with a gauge, for new belts are in order 100 lbs., 70 lbs. (one belt) and 110 lbs. Tension *must* be rechecked after 20 to 30 minutes running time except the Polyflex which may be checked within 15 to 20 minutes. In order, the tensions should then be 80 lbs., 70 lbs. and 100-110 lbs. The slack after run-in is not caused by "stretch" but by the belt(s) seating in the pulleys.

Condenser

Aluminum is the material universally used for condensers furnished with aftermarket A/C kits today and thus, they do not lend to field repair. Attempts to do so without the proper equipment and skills are generally a waste of time.

Current designs for liquid-cooled cars usually have either 12 or 15 fins per inch, the latter being considered a heavy-duty type. They are usually mounted to the radiator by straps and the larger, horizontal male fitting marks the top. This must always be on top, but the same condenser will operate equally well turned around with the fittings on the opposite side. It must be placed squarely in the airstream and parallel with the radiator front at a point no more or less than ½-in. away from the radiator.

One and sometimes two condensers for air-cooled rear-engine cars such as VW Beetle are generally mounted in the luggage compartment up front and will have auxiliary electric fan(s). As kits for these cars are all of the custom variety, replacement condenser positioning will be governed by precisely fashioned brackets already in place.

Receiver-Drier

The receiver-drier in aftermarket A/C is normally positioned adjacent to the condenser in the air stream but location is unimportant as long as it is not subjected to any heat source other than the ambient temperature. It also should not be located in any closed or semi-closed area, such as under the hood to guard against overheating from absorption. Since it functions as a storage tank for the refrigerant, temperatures it is exposed to should not exceed 125° F for any length of time. Use the same precaution as observed for any refrigerant container.

The drier phase consists of a desiccant (usually silica gel) and as these materials can be fully saturated, and therefore function no longer, by moisture from the air in as little as 15 minutes, connections must be made in a minimum of time. Meanwhile, the unit must remain capped or plugged as it is received from the parts source.

No repairs are possible to either the receiver or the drier. Saturation of the drying agent soon results in moisture contamination of both the refrigerant and compressor oil and if allowed to continue, will result in acid formation and resultant deterioration of the entire system. From an owner's standpoint, the first hint of trouble is a system that will cool satisfactorily initially and then cease to cool. If turned off for 20-30 minutes, it will operate again for a short period. This is caused by the formation and subsequent melting of ice at the orifice of the expansion valve.

The serviceman can detect the cause of this

ITEM	PART NO.	DESCRIPTION
1	50110 50880	Retainer – Seal Seat – Prior to Nov. '74 Snap Ring – Seal Seat – After Nov. '74
2	56150	Seal Kit
3	50520	Oil Filler Plug O-Ring
4	50511	Oil Filler Plug
5	59610	Cylinder Head and Valve Plate Gasket Kit
6	59620 59622	Valve Plate Assembly W/Gaskets Non Antislug Prior to June '74 Valve Plate Assembly W/Gaskets Antislug After June '74
7	46330	Protective Cap (Suction Port)
8	46340	Protective Cap (Discharge Port)
9	49610 69650 69630 69610	Cylinder Head – Flare (Vertical Connections) Cylinder Head – Tube O-Ring (Vertical Connections) Cylinder Head – Rota-Lock (Vertical Connections) Cylinder Head – Rota-Lock (Horizontal Connections)
10	40471 50471	Screw – Flare-Tube O-Ring – Rota Lock Cylinder Heads (Vertical Connections) Screw – Rota Lock (Horizontal Connections)
11	40481 40485	Gasket – Aluminum – Charging Port (Flare Cyl. Head Only) Prior to Nov. '73 Gasket – Rubber O-Ring – Charging Port (Flare Cyl. Head Only) After Nov. '73
12	50590	Charging Port (Flare Cylinder Head Only)
13	40600	Charging Port Cap (Flare Cylinder Head Only)
14	50121	Shaft Key

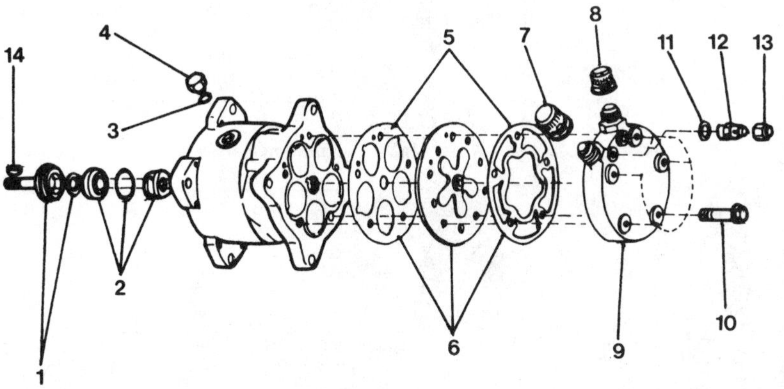

Fig. 16 Compressor field-renewal components

intermittent operation from discoloration (grayish) of the refrigerant as viewed in the sight glass and/or the same hue of discoloration of the compressor oil. Either or both systems should be purged and refilled immediately. If the neglect has been prolonged, the expansion valve should be replaced or cleaned manually. In any case, it is necessary to replace the receiver-drier.

Rarely, it is possible to find a new receiver-drier that has been shock-damaged in shipment or subsequent handling. Therefore, a check should be made of the inlet and outlet temperature of lines near the T-handle fitting at the top of the unit. If the temperature of these after the usual 15-30 minute systems checks is nearly equal, the unit is operating satisfactorily. If there is a marked difference, internal damage to the new unit is indicated and it should be replaced.

SANKYO SD-5 AFTERMARKET COMPRESSOR

The Sankyo SD-5 Figs. 15, 16 & 17, is an axial five-cylinder comprêssor. It is available in three sizes, designated SD-506, SD-507 and SD-508. The following service procedure apply to all three models. Usually, the compressor is supplied with the clutch installed. However, the compressor and clutch are available separately.

Checking Oil

A leak in any component in the air conditioning system or a ruptured hose will cause a substantial loss of refrigerant oil. After correcting the cause of leakage, check the oil in the compressor as follows:

1. Run the compressor for 10 minutes at engine idle rpm.
2. Purge all refrigerant from the compressor. Be careful not to lose any oil.
3. Position the angle gauge (tool No. 32448) across the top flat surfaces of the two front mounting ears. After centering the bubble in the level indicating portion of the angle gauge, read the inclination angle.
4. Remove the oil filler plug and position the rotor to top dead center by rotating the clutch front plate until the casting mark is visible in the center of the hole, Fig. 18.
5. If, when facing the front of the compressor, the compressor inclination is to the right, rotate the clutch front plate counterclockwise approximately 110 deg. If the compressor inclination is to the left, rotate the clutch front plate clockwise approximately 110 deg., Fig. 19.
6. Insert dipstick (tool No. 32447) to the stop position of the dipstick into the compressor.
7. Remove the dipstick and count the number of increments of oil.
8. By using the table, Fig. 20, and the previously outlined information (inclination angle of compressor and number of increments of oil), it can be determined if the oil level is correct or if oil must be added.

If the oil level is lower than that given in the table, add to specifications Suniso 5GS or Texaco Capella "E" or equivalent oil. For example, if the inclination angle is 30 deg. and the number of increments of oil is 7, add sufficient oil to bring the oil level up to 9-13 increments (preferably 13) on the dipstick.

NOTE: The increments on the dipstick do not represent any particular quantity of oil, such as cubic inches, cubic centimeters, fluid ounces, etc. Use the inclination angle, the dipstick increments and the table to determine the acceptable oil level. However, the lower numbers in the table represent 3 fl. oz. (90 cc) and the higher numbers 5 fl. oz. (150 cc).

9. When installing the oil filler plug, make sure that the sealing O-ring is not twisted and is perfectly clean, also make sure O-ring seat is free of cantamination. Torque the plug to 6-9 foot-pounds. Do not over-tighten the plug to stop a leak; remove the plug and install a new O-ring.

Truck or special applications using refrigerant hoses 15 to 25 ft. long will require the addi-

ITEM	PART NO.	DESCRIPTION
1-3-4-5-7	59800	Accessory Kit – Nut, Key, Shims, Snap Ring and Screws
2	90105	5¼" – 2-Groove – "A" – Clutch Front Plate and Rotor Pulley W/Bearings (12-Volt)
	87105	6" – 2-Groove – "A" – Clutch Front Plate and Rotor Pulley W/Bearings (12-Volt)
	99105	5¼" – 2-Groove – "A" – Clutch Front Plate and Rotor Pulley W/Bearings (24-Volt)
	96105	6" – 2-Groove – "A" – Clutch Front Plate and Rotor Pulley W/Bearings (24-Volt)
6	51040	Bearing (Set of 2)
8	89102	5¼" Field Coil, 12-Volt
	86102	6" Field Coil, 12-Volt
	99102	5¼" Field Coil, 24-Volt
	95102	6" Field Coil, 24-Volt
9	90623	5¼" – 2-Groove – "A" – 12-Volt Clutch Assembly – Complete
	87623	6" – 2-Groove – "A" – 12-Volt Clutch Assembly – Complete
	99623	5¼" – 2-Groove – "A" – 24-Volt Clutch Assembly – Complete
	96623	6" – 2-Groove – "A" – 24-Volt Clutch Assembly – Complete

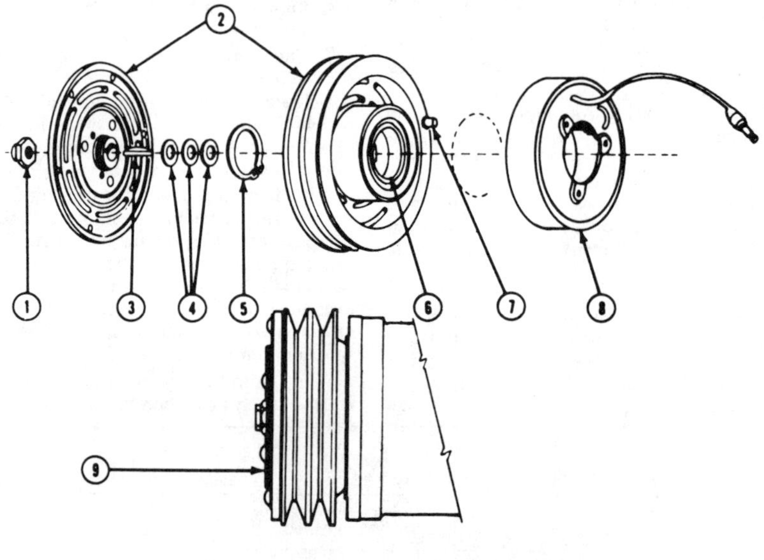

Fig. 17 Clutch components

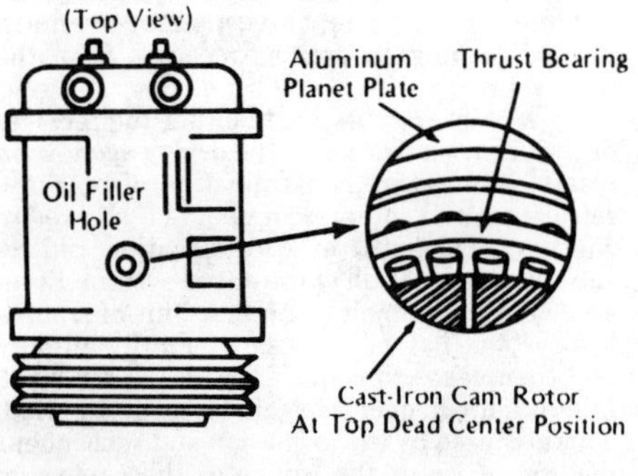

Fig. 18 Rotor positioning for oil level check

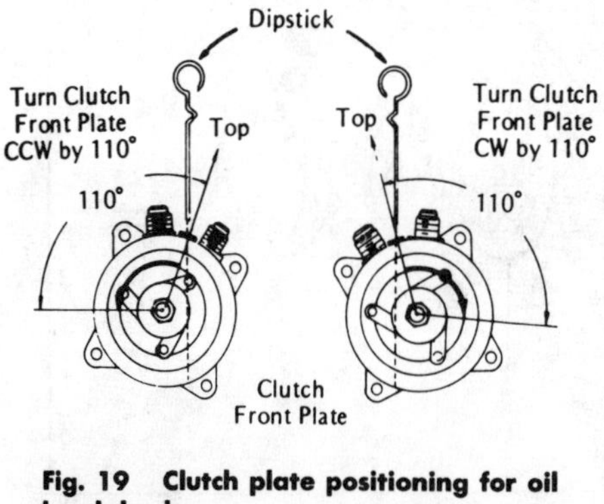

Fig. 19 Clutch plate positioning for oil level check

Inclination Angle In Degrees	Aceptable Oil Level In Increments
0	6-10
10	7-11
20	8-12
30	9-13
40	10-14
50	11-16
60	12-17

Fig. 20 Inclination angle and oil level increments chart

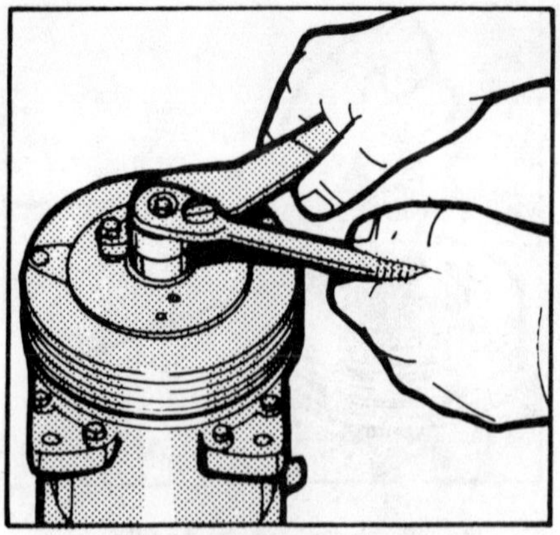

Fig. 21 Removing hexnut

Fig. 22 Removing clutch front plate

Fig. 23 Removing key and shims

dition of 5 oz. of oil to the compressor. An additional 5 oz. will be required for an additional 25-ft. increase in hose length.

NOTE: Maximum hose length must not exceed 50 ft.

Evacuation and Charging of System

The instructions contained in the manufacturer's air conditioning system installation and service manual should be followed to evacuate and charge the system and to adjust controls.

CAUTION: Sankyo compressors are designed and furnished for use with refrigerant 12 only. The use of any other refrigerant will void the warranty.

Safety Precautions

1. Always wear goggles when opening the refrigeration system. Refrigerant liquid or gas can damage the eyes permanently.
2. Never apply heat from a torch to a sealed refrigeration system. Refrigerant expands rapidly with heat and could cause an explosion.

   CAUTION: Refrigerant 12 in the presence of an open flame produces phosgene gas. It is very toxic. Never breathe it.

3. Do not use refrigerant other than Refrigerant 12.
4. Be sure gauge manifold hoses are in good condition. Never let them come in contact with the engine fan or exhaust manifold.
5. It should be noted that aluminum is not compatible with all refrigerants. Methyl chloride in particular will cause rapid deterioration of aluminum on contact.

After charging, the entire system should be leak-tested with a Halide Leak Detector.

Service and Torque Specifications

Since the majority of the compressor parts are made of aluminum alloys, care must be taken not to mar, nick or scratch machined surfaces when making repairs. Bolts should be run in until the bolt heads make contact, then tightened with a torque wrench to the specified torque.

An important factor in compressor servicing is cleanliness, and care must be exercised to prevent dirt or foreign material from entering the compressor when it is opened. All old gaskets should be replaced. All gasket surfaces should be clean and parts to be re-used should be washed in a suitable petroleum-base solvent.

TORQUE SPECIFICATIONS

Clutch retaining nut	25-30 ft. lb.
Cylinder head screws	22-25 ft. lb.
Seal seat retainer (Prior to Nov. '74)	10-13 ft. lb.
Service port w/O-ring	8-12 ft. lb.
Service port w/alum. gasket	10-14 ft. lb.
Oil filler plug	6-9 ft. lb.
Discharge (½-in.) hose fitting	18-22 ft. lb.
Suction (⅝-in.) hose fitting	22-25 ft. lb.

Shaft Seal Assembly

The shaft seal assembly is of a type that has been used by other compressor manufacturers for many years. The seals are machined to within two helium light bands of flatness. The carbon disc is lapped to the cast iron seat and the fit becomes even better as the seal assembly is run in during operation.

Replacing a seal assembly because of very slight leakage is an error that is often committed. Few, if any, mechanical seal assemblies are 100% leakproof. The rotating surfaces are separated by a very fine film of refrigerant oil.

Oil carries refrigerant and minute quantities which seep to the outside may be detected with modern ultrasensitive leak detectors which are capable of sensing leakage totaling one ounce of refrigerant over a 20-year period. Such sensitivity has a useful purpose in critical applications, but for leak-testing compressor joints or seals, a halide leak detector will pick up any leak of sufficient magnitude to require correction.

NOTE: Do not replace a seal assembly until the seal has been given an opportunity to run in and there is definite proof that replacement is really necessary.

Fig. 24 Removing shaft seal seat snap ring

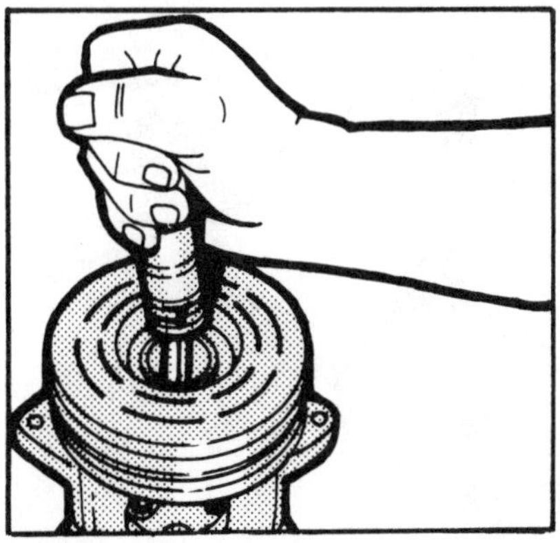

Fig. 25 Removing shaft seal seat

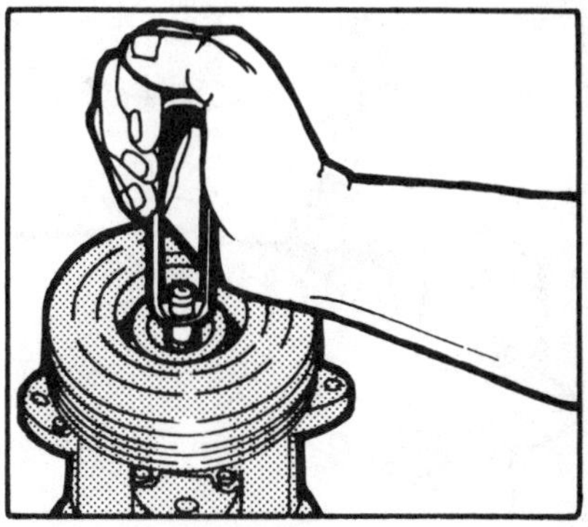

Fig. 26 Removing shaft seal

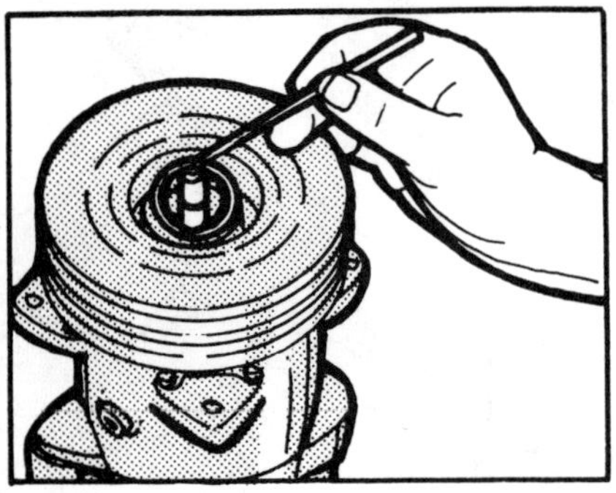

Fig. 27 Removing shaft seal seat O-ring

Fig. 28 Removing shaft seal O-ring

Fig. 29 Installing shaft seal

Maintenance Procedures

Shaft Seat Seal & Shaft Seal Replacement

1. Remove hex nut using a ¾" hex socket, Fig. 21, then remove front plate using clutch front plate puller tool No. 32416, Fig. 22.
2. Remove the key and shims from compressor shaft, Fig. 23, then the seal seat retaining snap ring using tool No. 32407, Fig. 24.
3. Remove the seal seat using seal seat tool No. 32405, Fig. 25, then the old seal using seal tool No. 32425, Fig. 26.
4. Being very careful not to damage or scratch the internal mating surfaces, remove the shaft seat seal O-ring using tool No. 32406, Fig. 27, then remove the shaft seal O-ring using same tool, Fig. 28.
5. Install the shaft seal O-ring, Fig. 28, making certain O-ring is seated properly in groove. Now install the shaft seat seal O-ring, Fig. 27.
6. With shaft seal protector tool No. 32426 placed onto shaft threads, coat shaft seal liberally with clean refrigerant oil and place seal onto the shaft seal installer tool No. 32425, Fig. 29. Slide shaft seal into place in the bore, rotating seal clockwise until seated on the provided compressor shaft flats, then remove the protector tool.

CAUTION: Do not touch the carbon ring face with your hands.

7. Coat the shaft seal seat with refrigerant oil and install it using tool No. 32405, Fig. 30.

CAUTION: Use the shaft seal seat tool to avoid installing seat upside down, then install the seal seat retaining snap ring using tool No. 32407, Fig. 31, making sure snap ring is engaged.

8. After installing key and shims of adequate thickness, align slot in hub of front plate squarely with the key then drive hub into position using tool No. 32436, Fig. 32, using the air gap gauges, tool No. 32437, make sure that the air gap is .016" to .031".

Head And Valve Plate Removal

9. Remove the screws in the head using a 13mm hex socket No. 32477, then remove the head from the valve plate by inserting the gasket removal tool, No. 32429, Fig. 33, between the valve plate and the cylinder head and tapping lightly with a soft hammer. If the head and the valve plate will not separate, hold the head and tap it lightly with a soft hammer to loosen. To remove the valve plate, insert tool No. 32429, Fig. 34, between cylinder head and valve plate.

CAUTION: Care should be taken not to damage the head surface, valve plate and reed valves.

10. Valve plate or cylinder block must be removed carefully to avoid scratching or nicking the machined sealing surfaces. Also, make sure that no gasket material remains in the cylinder bore.
11. Install the rotor assembly, Fig. 35.
12. Replace clutch rotor bearings, Figs. 36 and 37.

Head Service Ports

Removal, Repair or Replacement of the two service ports for the cylinder head can be accomplished with the compressor on the vehicle.

CAUTION: Purge the system's refrigerant charge prior to removal or repair.

Removing and Installing Valves

Remove and install valve cores using tool No. 32408.

Use a new sealing O-ring if it is necessary to change the complete service port.

Head and Valve Plate Installation

NOTE: The reed valves and valve plate are furnished only as an assembly.

The following steps should be followed in installing the assembly.

1. Apply a thin film of clean refrigerant oil on the area of the cylinder block to be covered by gaskets.
2. Place the cylinder gasket in position on the cylinder so the pin in the crankcase fits through the pin hole in the cylinder gasket.
3. Apply a very thin film of refrigerant oil on top and bottom of valve plate areas to be covered by gasket.
4. Place the valve plate in position on the cylinder gasket so the discharge valve side

Fig. 30 Installing shaft seal seat

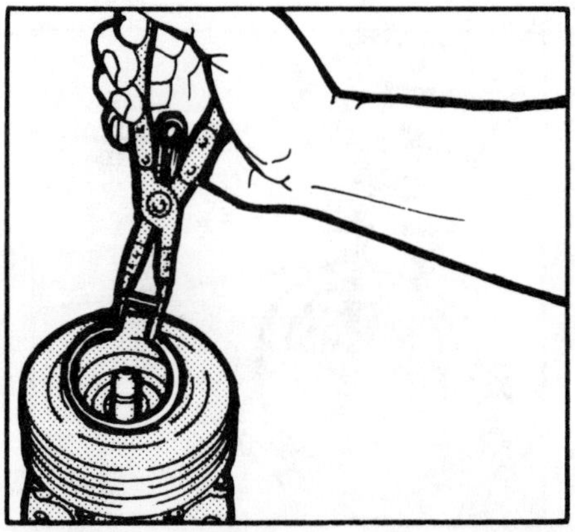

Fig. 31 Installing seal seat snap ring

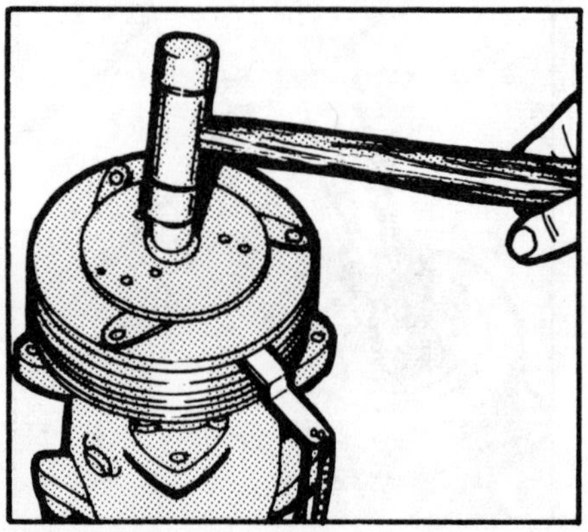

Fig. 32 Installing key and shims

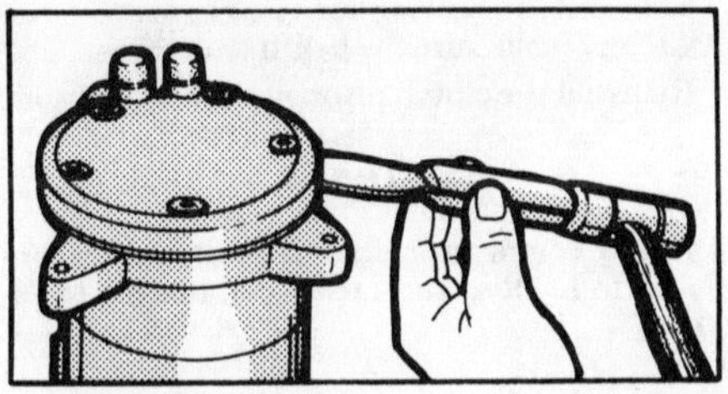

Fig. 33 Removing compressor head

(the side with the smaller valves and retainer nut) are facing up and the pin fits through the pin hole in the valve plate.

5. Place the head gasket in position on the valve plate assembly.
6. Insert the head screws and run in all screws until the head makes contact. Tighten the head screws to 22-25 ft.-lb., starting with the screw directly above the "Suc" marking on the cylinder head. Next, tighten the screw on the lower right and proceed in a pattern similar to tightening lug nuts on a wheel, alternating from one side of the head to the other.

Tracing and Locating Noise Complaints

Many underhood noise complaints can be traced to mount and drive components. Normally, if a noise occurs at one speed and not at another, the compressor can be eliminated as the source. Compressor mounting bracket bolts, loose, broken or bottomed, can be the sole cause of underhood noise.

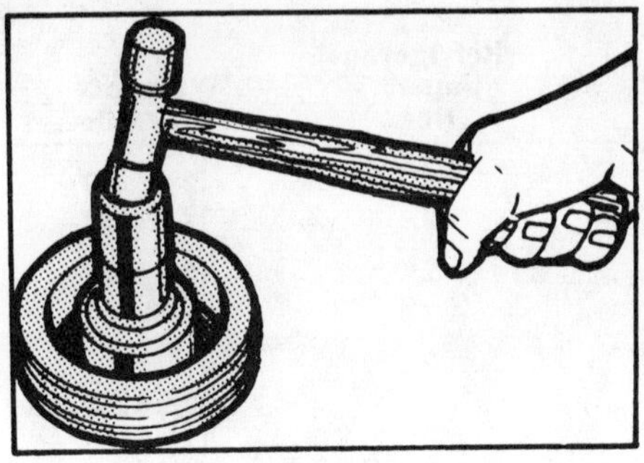

Fig. 36 Removing clutch rotor bearings

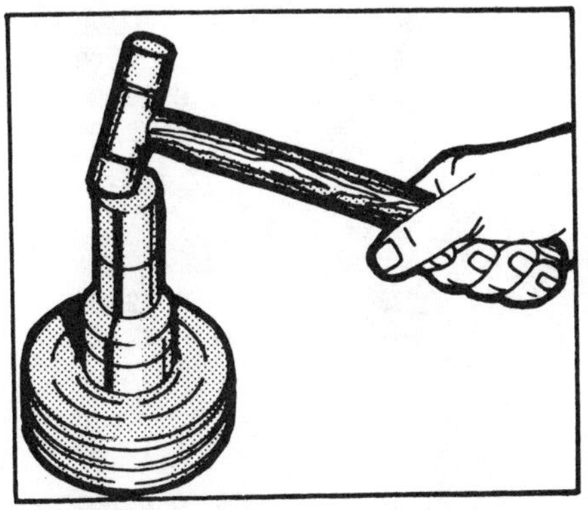

Fig. 37 Installing clutch rotor bearings

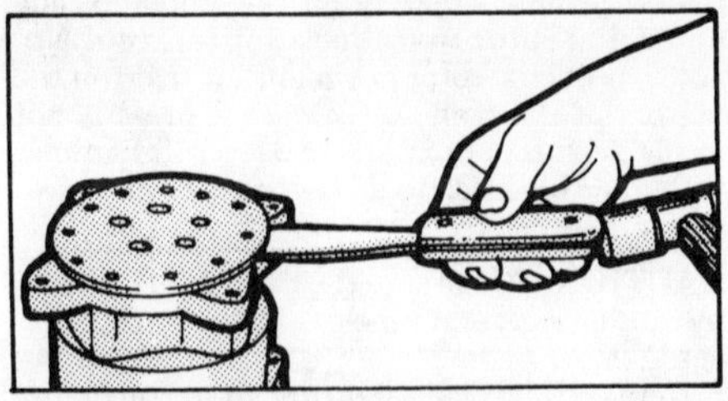

Fig. 34 Removing valve plate

Any vehicle has its critical frequencies wherein all vibrations form harmonics to generate noise. The over-the-road speed at which these critical points are found will vary with each vehicle and with each mount and drive arrangement.

Since a compressor has many moving parts, it is normal for it to generate some noise as it is operated. The refrigerant gases, as they are

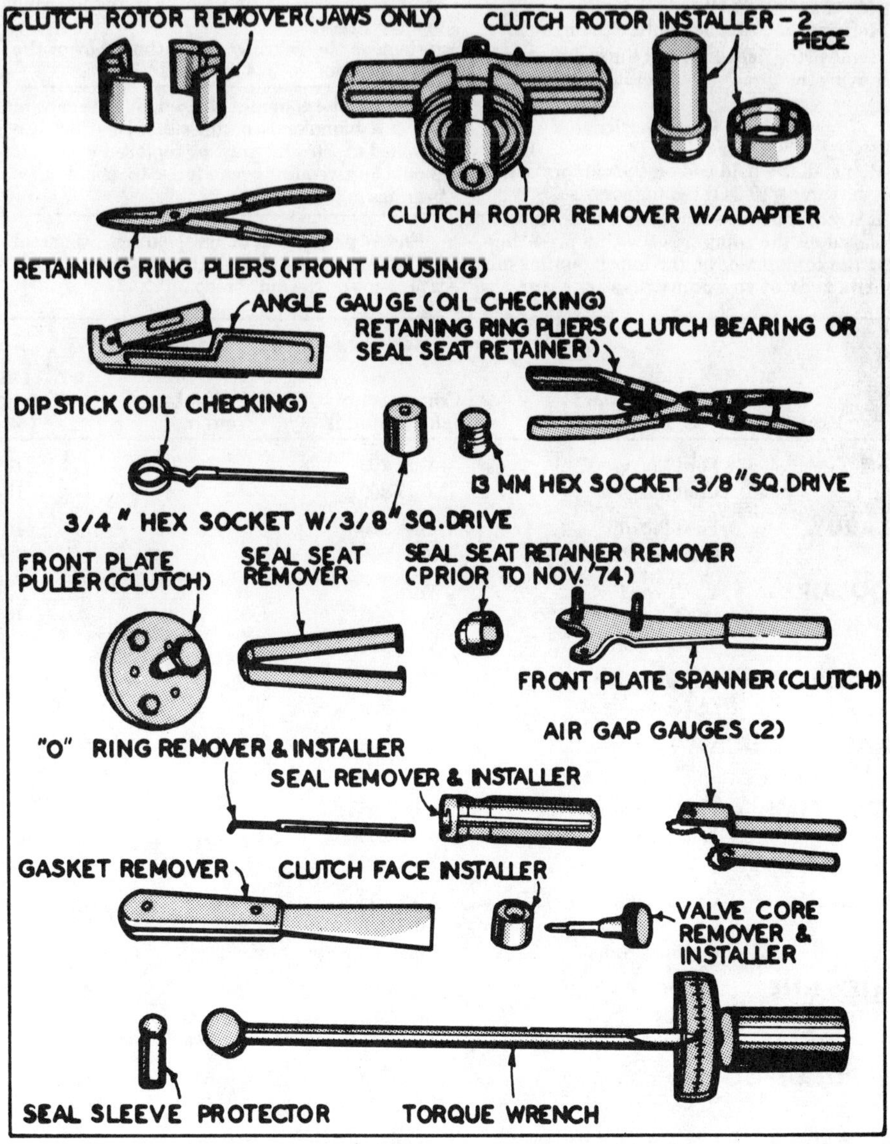

Fig. 38 Service tools

Fig. 35 Installing clutch rotor

moved by the compressor pistons, also produce noises and vibrations as a normal occurrence.

Additionally, the level of noise tolerance varies with each customer. Generally, an explanation of how the compressor functions and what it does will satisfy a customer that, since the compressor is doing work, it is natural for it to make some noise.

Clutch Removal

It will be necessary to remove the compressor from the vehicle to make repairs on any part of the clutch.

Remove the hex nut, using spanner wrench, tool No. 32409, and a ¾-in. hex socket with ratchet wrench.

Remove the front plate, using front plate puller, tool No. 32416, and the three ¼-in.-20 cap screws.

Remove the external front housing snap ring holding the rotor assembly using snap ring tool No. 32417.

Remove the external bearing snap ring using snap ring tool No. 32407.

Remove the rotor assembly. See Figure 36.

Remove the field coil from the compressor by removing the three #10-24 retaining screws.

Clutch Installation

Reinstall the field coil on the compressor using the three #10-24 retaining screws.

It is recommended that the rotor assembly be replaced on the compressor with a press. Support the compressor on the four mounting tabs on the rear of the compressor. Align the rotor assembly squarely on the front housing and using rotor installer adapter, tool No. 32435, press the rotor down until it seats on the bottom of the housing. In case a press is not available, it is permissible to use a soft hammer with tool No. 32435 to drive the rotor into position. Figure 36.

CAUTION: Make sure the rotor is not binding by turning it several times.

Reinstall internal bearing snap ring using snap ring tool No. 32407.

Reinstall front housing snap ring using snap ring tool No. 32417.

After installing key and shims of adequate thickness, align slot in hub of front plate squarely with the key and drive on, using tool No. 32436. Using the air gap gauges, tool No. 32437, make sure that the air gap is between .016 in. and .031 in., then torque the shaft nut.

Additional Service Information for Clutch

The following procedures should be followed in changing the bearings in the clutch.

After the rotor has been removed from the compressor, remove the bearing snap ring using tool No. 32407.

Remove the bearings using the arbor of the two-piece tool No. 32435. Figure 37.

NOTE: A press should be used, but the careful use of a hammer is permissible. Bearings are supplied in sets and must be replaced in sets. It should not require great force to install new bearings in the rotor.

Ensure proper seat of new bearings using tool No. 32435. Make sure that the large ring portion is pressing on the outer race.

Reinstall the bearing snap ring using tool No. 32407 and make sure the bearings are free.

Reinstall the clutch rotor on the compressor.

Belt Tension

A new compressor drive belt should be tensioned to 100 lbs., and a used-belt tension to 80-90 lbs.

Special Instructions — SD-507

The foregoing procedures apply generally to all of the Sankyo SD series, but there a few exceptions that apply to the SD-507. They are as follows:

1. When removing or replacing the clutch field coil a large snap ring, instead of three screws, will be found holding the coil in place. Use tool No. 32417 to remove and replace this snap ring.
2. Instead of the lower increments mentioned in the note following Figure 20 representing 3 fl. oz. and the higher increments representing 5 fl. oz., the lower increments for the SD-507 will represent 2.6 fl. oz. and the higher increments will represent 4.4. fl. oz.
3. When positioning the valve plate and/or gaskets, a notch will be found on the periphery of these three items. Always align the notches 180 deg. down from the oil filler plug.
4. When removing the gasket material from the cylinder head, a thorn-like projection will be found on the surface of the cylinder head. Use extreme caution to avoid nicking or marring the projection.

SPECIFICATIONS

Brand	Typical Application	Compressor Make & Model	Displ. cu. ins.	Oil Capacity (oz.)	Refrigerant Capacity (lbs.)	Warranty Mos./Miles
A-R-A	Ford Maverick	York 209	9	13	2¼	12/12,000
	Chevrolet	York 209	9	13	2½	12/12,000
CLARDY	Ford Pickup	York 210	10	10	2¾	12/12,000
	Mercury Capri	York 206	6	6½	2¼	12/12,000
COOLAIRE	VW Type I	York 206	6	13	2½	12/12,000
	VW Type III	York 206	6	13	2½	12/12,000
	Toyota Corolla	York 206	6	13	2	12/12,000
FACTORY-AIR	Toyota Corolla	York 206	6	12	1¾	12/12,000
	Chevelle	York 209	9	13	2½	12/12,000
HEATRANSFER (SEARS)	VW Type 1	York 206	6	12	2	24/24,000
KOOL-TEMP	Ford Maverick	York 206	6	12	2½	12/12,000
	Datsun 510	York 206	6	12	2½	12/12,000
	Universal	York 209	9	13	2½	12/12,000
MARK IV	Chevrolet Vega	York 209	9	13	2	12/—
	Chevrolet Pickup	York 209	9	13	2	12/—
	Toyota Corona	York 206	6	12	2	12/—
MEIER-LINE	Porsche 911	M-L 216	3½	4	2	12/12,000
	VW Type I	M-L 211	3½	4	1⅞	12/12,000
	Ford Pinto	M-L 720	3½	4	2	12/12,000
NIPPONDENSO	Toyota Corolla	N-D 2C-90	5¾	11¼	1¾	12/12,000
	Toyota Corona	N-D 2C-90	5¾	11¼	1¾	12/12,000
VOLKSAIRE	VW Type I	Tecumseh HG-500	5¼	—	1½	24/24,000
	VW Type II	York 209	9	—	3¼	24/24,000
	VW Camper	York 209	9	—	1½	24/24,000

VIII. TRUCK AND CARGO COOLERS

This section deals with truck cab air conditioning as well as body and trailer refrigeration systems.

Polarstream P-750 U

This refrigeration system operates on an entirely different principle from those systems described elsewhere in this chapter. It requires no engine or compressor and consumers no fuel in its operation. It is a pneumatically controlled liquid nitrogen system designed for undermount installation on trailers, Fig. 1. The system consists of a liquid nitrogen tank and the controls required for filling the tank and controlling the flow of liquid nitrogen into the cargo compartment. The Polarstream system is manufactured by Union Carbide Corporation. Fig. 2, shows simplified diagram of control system operation.

The liquid nitrogen tank is constructed like the familiar Thermos Bottle. The inner and outer shells are separated by insultation and a space from which all air has been evacuated. The small amount of heat that does penetrate this insulation and vacuum causes a little of the liquid to boil into vapor, or gas. Part of this gas is used to maintain pressure in the system to force the liquid out of the tank. The rest is vented through the relief valve, which limits system pressure to 22 pounds per square inch. In normal operation, a small amount of gas escapes through the relief valve almost continuously. Should the relief valve ever fail to function properly, additional safety devices prevent pressure in the tank from reaching an abnormally high level and therefore damaging the system.

The tank is filled through the filling connection and the fill valve located inside the door on the end of the tank assembly. During filling, excess gas pressure is released by opening the vent valve. The vent valve is also used to indicate when the tank is full by allowing liquid nitrogen to overflow through the open valve. The liquid level gauge indicates the amount of liquid nitrogen in the tank.

The temperature control system consists of the main control valve, the temperature controller, the liquid control valve and the spray header. When the main control (ON-OFF) valve is turned on, the temperature controller opens and causes the liquid control valve to release just enough liquid nitrogen into the cargo compartment to keep the air temperature at the set mark. The liquid nitrogen is sprayed out into the cargo space through the spray header in a pattern designed to distribute the cold nitrogen gas evenly throughout the vehicle.

SPECIFICATIONS

Total liquid nitrogen capacity	750 lbs.
Approximate system weight	
Empty	700 lbs.
Full	1,450 lbs.
Dimensions of tank assembly	
Diameter	26 inches
Overall length	90 inches
Evaporation rate $(LN)_2$ static conditions	3% of capacity per day
Normal operating pressure range	15-22 psi.
Pressure relief valve setting	22 psi.
Maximum service pressure	100 psi.
Container bursting disc rating	100 psi.
Casing vacuum bursting disc rating	18 psi.

SAFETY PRECAUTIONS

Nitrogen is a gas that makes up about four-fifths of the air we breathe. Liquid nitrogen is made by cooling air until it becomes a liquid, then removing oxygen and other gases that make up the remaining one-fifth of the air. Although nitrogen is non-flammable and non-toxic, two very important facts must be remembered when handling liquid nitrogen.

1. Liquid nitrogen is extremely cold. It boils at 320 degrees below zero, Fahrenheit (—320°F). Direct contact with the liquid or anything that has been cooled by it can freeze the skin instantly. The resulting injury is very similar to a burn.
2. Nitrogen will not support life. The air we breathe must contain a certain amount of oxygen. Breathing any atmosphere that does not contain enough oxygen can cause dizziness, unconsciousness or even death. When liquid nitrogen is released into a truck body or other enclosed space, it vaporizes into a rapidly expanding gas that

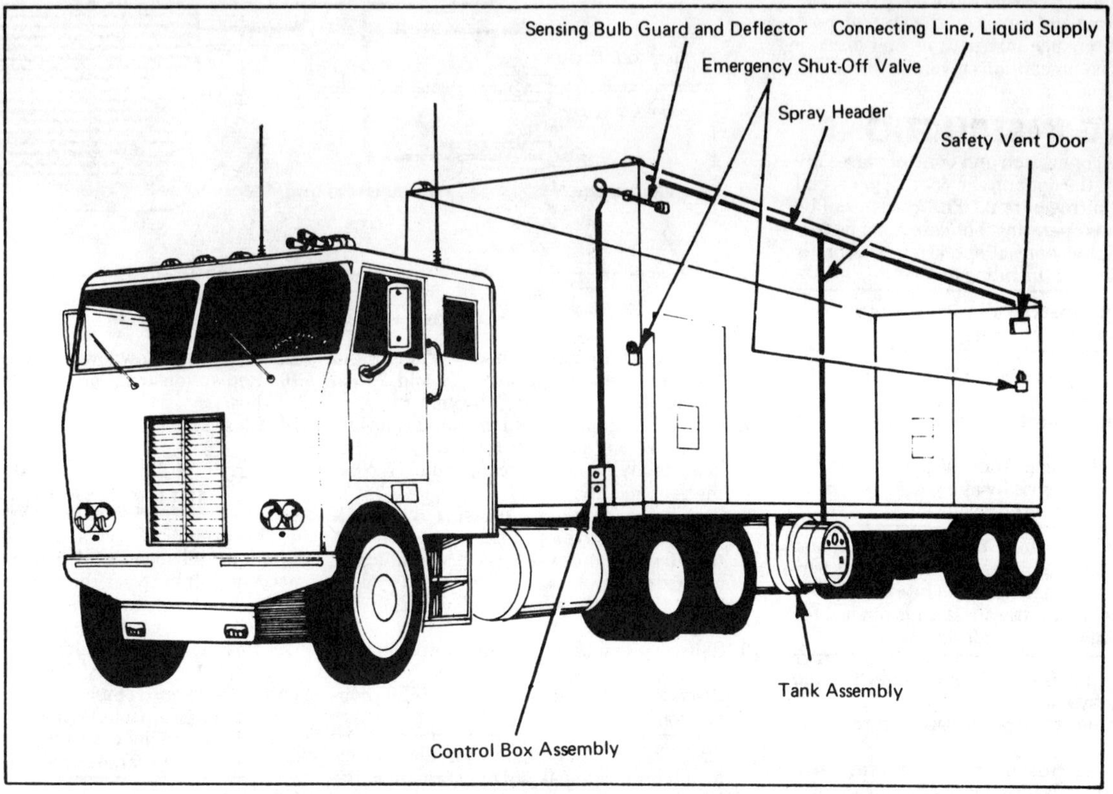

Fig. 1 Polarstream P-750U system

displaces the air from the truck and thereby lowers the oxygen concentration.

CAUTION: Nitrogen gas is odorless and colorless. It cannot be detected by human senses. High nitrogen (low oxygen) concentration in the atmosphere can cause asphyxiation completely without warning.

3. Do not enter or remain in the cargo compartment when the Polarstream system is being filled.
4. Do not enter or remain in the cargo compartment when the system is operating.
5. Never crawl between the ceiling and cargo or into "pockets" between stacks of cargo.
6. Open the doors wide and wait at least three minutes before entering the cargo space if the system has been operating for an hour or more since the last entry.
7. In vehicles equipped with partitions, open the partition fully, leave the vehicle and wait at least three minutes before entering the enclosed compartment.
8. Avoid contact with liquid nitrogen and cold piping.
9. Wear eye protection, heavy, loose-fitting gloves and long sleeves when filling the system.
10. Be careful when entering a warm, empty vehicle. High concentrations of nitrogen gas may still be present, especially toward the front of the vehicle, even after doors have been open for several minutes.

NOTE: If a paper match will burn normally in all parts of the vehicle, the atmosphere will support life.

11. Always latch or block doors fully open when working inside the vehicle.
12. Never attempt to tighten or loosen any cold fitting or to disconnect any piping until system pressure has been reduced to 0 psi. Before making any repair on the system, read the maintenance instructions and observe any special precautions given for the job to be performed.

FILLING INSTRUCTIONS

The filling connection and control valves are located inside the door on the control panel end of the liquid nitrogen tank. The system can be filled while it is operating, but care must be taken to ensure that normal operating pressure is maintained during the filling procedure.

CAUTION: Observe all safety precautions noted previously.

1. Open the door on the control housing of the tank and remove the dust cap from the fill connection. Check that connection is clean and dry.
2. Remove dust cap from fill hose connector and purge the hose by momentarily "cracking" the liquid withdrawal valve.

CAUTION: Hold fill hose firmly and point open end of hose away from all personnel. Liquid nitrogen will be ejected from this hose if withdrawal valve is opened for more than a few seconds.

3. Connect fill hose to fill connection. Tighten coupling by hand.
4. Open liquid withdrawal valve on the storage tank.
5. In quick succession, open the vent valve and the fill valve by turning handles ¼ turn clockwise.

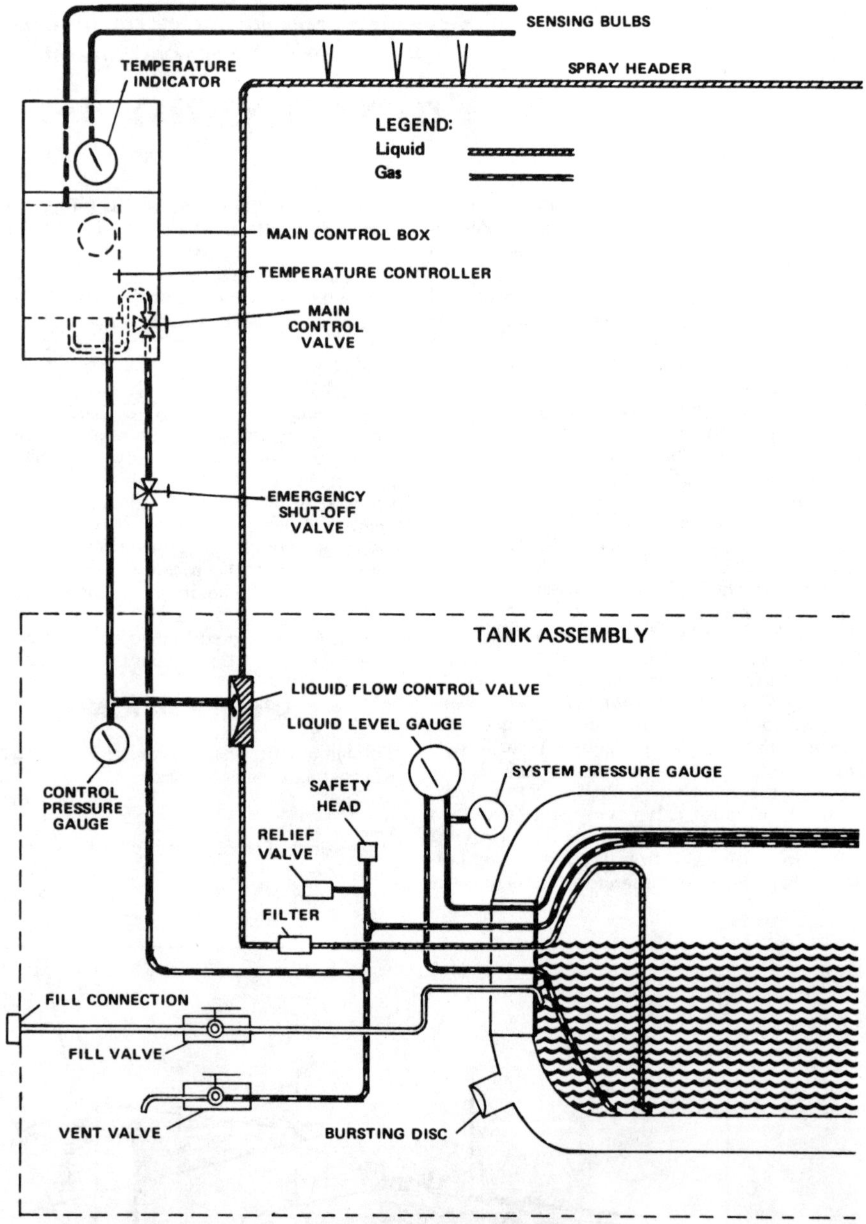

Fig. 2 Polarstream control system diagram

6. Observe container pressure gauge and maintain pressure between 15 and 20 psi during fill by "throttling" the vent valve.
7. Filling is completed when liquid sputters from the vent outlet on the left side of the housing. When this occurs, quickly close the vent and fill valves.
8. Check the container pressure gauge. If pressure is less than 15 psi, open and close the fill valve until pressure gauge indicates approximately 20 psi.
9. Close the liquid withdrawal valve on the storage tank.
10. Release pressure in fill hose by opening the hose blowdown valve.
11. Disconnect fill hose from system filling connection.

CAUTION: Be careful when disconnecting fill hose. Some liquid may remain in the hose.

12. Close hose blowdown valve on storage tank.
13. Replace dustcaps on all connections and fill hose.
14. Close and latch door.

OPERATING INSTRUCTIONS

This system is very simple to operate, once the liquid nitrogen tank had been filled, it is necessary only to set the temperature controller to the correct temperature for the cargo and turn the main control (ON-OFF) valve to the ON Position. Before unloading, turn the valve to OFF, open the doors and wait for fresh air to enter the cargo compartment before allowing any personnel inside. When unloading is finished, close the doors and turn the system back on if there is any cargo left in the compartment.

CAUTION: If there is a hissing sound from the spray header, it means that the system has not been turned off. This hissing occurs about 10 to 15 seconds before liquid nitrogen begins spraying from the headers. Turn any of the emergency shutoff valves to the OFF position and leave the compartment as quickly as possible.

NOTE: Always check out the system before starting out with a refrigerated cargo. The checkout takes only a few minutes, but it can save time and trouble by detecting problems before leaving the terminal.

1. Check that all emergency shutoff valves inside the cargo compartment are set to the ON position.
2. On the control panel of the liquid nitrogen tank, check that the liquid contents gauge indicates that the system contains enough nitrogen for the run and that the system pressure gauge indicates system pressure between 17 and 22 psi.
3. When loading is completed, close the doors and turn the main control switch to the ON position. Be sure that the temperature controller is set to the maximum temperature that will be acceptable at the end of the run.

In addition to routine operation of the control valve turning the system "Off" and "On" at unloading stops, operators should make frequent checks of the thermometer on the control box and the gauges on the tank to be sure the system is functioning normally.

1. At each stop, set main "ON-OFF" valve to the OFF position before opening the doors.
2. Open the doors wide and wait for fresh air to enter the cargo compartment before unloading. If the system has been operating for one hour or more since the previous door opening, wait at least three minutes after opening the doors before entering the cargo compartment.
3. As soon as unloading is completed, close the doors and set the "ON-OFF" valve to the ON position. If the trailer is empty, leave the valve in the OFF position.

MAINTENANCE

Introduction

Because the system is simple, it normally requires very little maintenance. Most potential problems can be avoided through careful attention to proper filling procedure and by regular inspections that will detect loose, damaged or worn parts before they can cause trouble.

This section describes the procedures that should be followed to keep the equipment in good operating condition. Also the troubleshooting and servicing procedures necessary to isolate and correct any problem that might occur.

Tools & Supplies Required

In addition to mechanic's hand tools, the following tools, special equipment and supplies are required.

High-purity dry nitrogen gas (high-pressure cylinder with pressure regulator and hose).

Pressure test adapter.

Leak test solution. (Ivory soap and water Acceptable)

Pipe compound or Teflon tape.

Routine Maintenance

Periodic and careful inspection and testing at regular intervals will often uncover small defects before they become major problems. These periodic inspections should be more detailed than the routine checks that are made every time the unit is used. All defects should be corrected as soon as they are discovered.

A detailed inspection and checkout should be performed before the beginning of the refrigeration season and at regular intervals during the operating season. All units should be inspected and tested at least twice each year. Equipment that is operated almost continuously or under unusually severe conditions should be checked out more frequently to ensure dependable operation.

Periodic Inspection and Checkout Should Include the Following:

1. Carefully inspect the entire system to detect loose, damaged or worn parts. Be sure that all bolts and other fasteners are tight. Check the operation of door latches on control box and tank housing.
2. Check that sensor bracket and capillary guard are mounted on ceiling and that sensing bulbs are properly installed and free from damage.
3. Check that spray header is securely mounted and free from damage.
4. Replace or clean the filter located in the liquid withdrawal line inside the tank housing.

   CAUTION: Vent system pressure to zero before attempting to remove this filter.

5. Check all valves and tubing connections for leaks. System pressure must be at least 18 psi during leak testing.
6. Any unit that has been out of service long enough to lose all pressure should be purged with dry nitrogen gas before it is placed in operation. (Refer to "Purging and Pressure Testing").
7. Inspect the safety vent in the rear door of the trailer. The vent door should be securely mounted and free from damage. The vent door should open and close freely without binding.
8. Perform the "Operational Checkout" procedure described below. This checkout procedure checks the operation of the controls and indicators and verifies the calibration of the temperature control system.

Operation Checkout Procedure

Perform the following procedure to check operation of the refrigeration unit.

1. Check that the liquid gauge indicates between 50 and 750 lbs. and the system pressure gauge indicates between 15 and 24 psi.
2. Turn main control valve to the ON position.
3. Turn temperature control dial to a setpoint below the temperature indicated by the thermometer. The control pressure gauge (on the tank) should indicate between 15 and 25 psi.
4. Turn temperature controller to a setpoint above the temperature indicated by the thermometer. The control pressure gauge reading should drop to zero.
5. Turn controller dial back and forth across the indicated temperature and observe dial readings at which control pressure turns on and off. These readings should be within ±5 degrees of the temperature. If they are not, the controller or the thermometer (or both) must be recalibrated.
6. Turn main control valve to the OFF position and reset temperature controller to specified setpoint.

Purging and Pressure Testing

Introduction

Systems that either have been repaired or out of service long enough to have lost pressure must be purged and pressure tested before they are filled with liquid. Purging and pressure testing should also be performed during annual preventive maintenance checkup.

During repair of a unit where piping is disconnected, moisture from the air can enter through open connections and condense in the liquid tank. Condensation can also occur in systems which have been out of service for any length of time. Before a system is filled with liquid, all traces of moisture must be removed by purging the entire system with dry nitrogen gas. Failure to remove this condensation before placing the unit in service can cause operational problems such as sticking liquid control valve, restricted liquid flow, inoperative liquid level gauge, etc.

Dry nitrogen gas necessary for this purging operation can be taken from either a high-pressure cylinder or from a liquid nitrogen storage unit equipped with a manually controlled pressure building coil.

CAUTION: Do not use compressed air, Freon, carbon dioxide or any other gases to purge or pressurize Polarstream equipment. Use of any gas other than dry nitrogen or helium will contaminate the system and cause it to malfunction.

Pressure Test Adapter

For best results, all purging and pressure testing of Polarstream systems should be accomplished by connecting the nitrogen gas source to the system fill connection. Connecting the gas supply to some other point does not check the fill connection for possible leakage.

A pressure-test adapter, Fig. 3, can be assembled from the appropriate fill hose coupling and the fittings required to permit connecting the fill coupling to the nitrogen gas hose.

Procedures

Use the following procedures to purge the system with dry nitrogen gas. All purging and pressure testing should be done through the fill connection.

CAUTION: Do not attempt to purge the system if there is liquid in the tank.

1. Connect the pressure-regulated source of dry nitrogen gas to the fill connection.

a. For Purging With a Gas From a High-Pressure Cylinder

Connect a hose from the cylinder regulator to proper pressure test adapter and install the adapter on the system fill connection. With all valves closed, open the cylinder valve and set the regulator for 50 psi.

b. For Purging With Gas From a Storage Unit

Connect the fill hose to the fill coupling. Close the storage unit main liquid withdrawal valve and open the pressure-building coil valve. This permits vapor to flow from the tank through the coil and into the fill hose.

Do not allow lines to get cold enough to form frost.

2. Open the fill and vent valves to permit nitrogen to flow through vent line for approximately 10 minutes. Then close the vent valve.
3. Turn main control valve to the ON position. Set the temperature controller to 0 deg. F. Check to be certain nitrogen gas can be heard coming out of the spray header. Continue this purging for approximately 10 minutes.
tutes.
4. Disconnect control pressure line from the tank and purge line for about five minutes.
5. After tubing is reconnected, start purging again. Watch the pressure gauge and allow system to build to 25 psi. In addition to removing any moisture that may be in the relief valve, this checks operation of the valve, which should have opened at about 22 psi.

CAUTION: If the relief valve opens at 17 psi or less, or fails to open at 25 psi, stop purging momentarily. Blow down (depressurize) the system and replace the relief valve with a new one. Do not attempt to repair the relief valve.

6. With purging now complete pressure-test the system to check for leakage. Adjust system pressure to about 15 psi. Close the fill valve and watch the system pressure gauge.
7. If system pressure decreases, there is a leak in the system. Most leaks can be detected by listening for the sound of escaping gas. Tighten loose connections or replace parts as required to stop leaks.
8. Repeat steps 6 and 7 until all audible leaks are eliminated.
9. If system pressure continues to decrease, use a leak detector solution to detect leaks.

CAUTION: Keep moisture out of the system. Do not use leak detector solution near any open fittings or connections.

10. When the purging and pressure-testing operation is completed, shut off the gas supply and disconnect it from the system. The system can now be filled with liquid nitrogen.

REPAIR AND REPLACEMENT OF PARTS

Introduction

Repair of Polarstream units generally consists of replacing the unserviceable parts. Some parts, such as the liquid control valve, the fill and vent valves, and the liquid contents gauge, can be rebuilt, but special equipment is required to repair and test these components. Rebuildng of parts is best accomplished by Authorized Polarstream Service Centers, which have the equipment and trained personnel needed for this type of work.

Before doing any repair work on this equipment, take note of any special precautions given in the instructions related to the part to be replaced In many cases, it is necessary to vent off all pressure from the liquid tank before loosening or removing parts. The procedures for depressurizing the system and for restoring normal operating pressure are as follows:

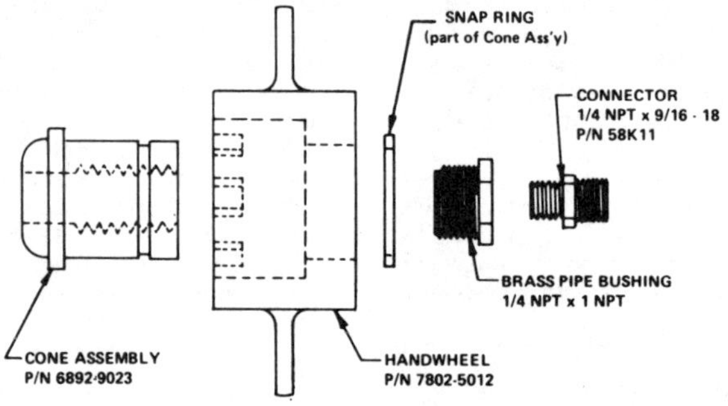

Fig. 3 Pressure test adaptor

Depressurizing The System

Before any work can be done on pressurized piping, all pressure must be vented from the system. To depressurize (blow down) the system to atmospheric pressure, open the vent valve. When the system pressure gauge drops to zero, leave the vent valve open and proceed with repair work.

Pressurizing the System

A unit that has been vented to atmospheric pressure for maintenance or that has lost pressure due to leakage or improper operating procedures must be repressurized to at least 15 psi before normal operation can be resumed.

If the system is only about three-quarters full, pressure can be restored simply by filling the tank completely full using described filling procedure. Be sure to keep system pressure above 15 psi throughout the filling operation by throttling the vent valve; do not throttle the fill valve. If the system is more than three-quarters full, use the following procedure to restore normal operating pressure.

1. Connect the unit to the bulk liquid storage unit and open the fill valve just enough to maintain system pressure of 15 to 22 psi. Do not open the vent valve or allow the relief valve to open.
2. Set temperature controller to a point at least 20 degrees below the indicated temperature.
3. Open the doors of the trailer and turn the main control valve to ON.

CAUTION: Do not enter trailer while system is on.

4. Allow the system to operate until the liquid contents gauge indicates that the tank is about three-quarters full (575 pounds or less).
5. Turn the main control valve to OFF, close the trailer doors and reset the temperature controller to the specified setpoint.
6. Refill the unit to capacity by following the proper filling procedure. Be sure to keep system pressure above 15 psi throughout the filling operation.

Pressurizing Using Helium as an Agent

A unit that is full of liquid can be pressurized very quickly with helium gas. A very small quantity of helium is required and pressurizing the system can be accomplished in only two or three minutes. The only equipment required is a cylinder of helium gas, a suitable pressure regulator, a hose and the pressure test adapter, Fig. 3, described previously.

CAUTION: Use only dry nitrogen gas or helium gas to pressurize the system. Any other gas, including compressed air, can contaminate the system and require a difficult and expensive cleaning job.

1. Use the pressure-test adapter to connect the helium supply to the fill connection of the unit. Adjust the helium pressure regulator for output pressure of about 50 psi.
2. Open the fill valve and observe the system pressure gauge. When the pressure relief valve opens, close the fill valve and watch the pressure gauge. If pressure remains in the normal operating range (15 to 25 psi), cut off the helium supply and disconnect the pressure-test adapter from fill connection.

NOTE: Some of the lines and components inside the housing of the tank assembly are accessible through the door on the end of the housing. To reach the control gauges and many of the connecting lines, it is necessary to remove the shroud section of the housing from the tank.

Remove the Tank Shroud as Follows:

1. Remove four bolts and washers that secure shroud assembly to ring assembly.
2. Slide shroud assembly off ring.
3. Reverse procedure to install the shroud onto the tank.

CAUTION: Do not open or attempt to remove the fill valve when the tank is more than one-half full. Liquid will flow from the tank through the open fill line.

Pressure Relief Valve

The relief valve should normally open at a pressure between 18 and 25 psi.

CAUTION: Do not attempt to adjust this valve. If the valve does not open at a point within the specified range, replace it with a new valve.

Safety Head

The safety head (bursting disc) on the tank vent line is not repairable. If the safety head ruptures or develops a leak, it must be replaced.

A ruptured safety head usually indicates serious trouble. Failure of the pressure relief valve to open can allow the tank pressure to go high enough to rupture the disc. After replacing the safety head, always check the operation of the relief valve and inspect the tank for possible damage.

Liquid Withdrawal Line Filter

This filter contains a sintered bronze element that traps foreign particles before they reach

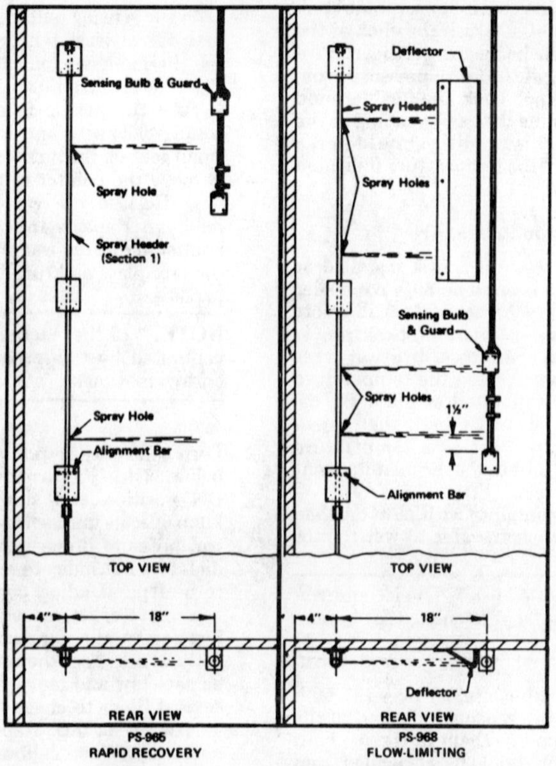

Fig. 4 Polarstream spray leaders

the liquid control valves. The filter must be inspected and cleaned periodically and replaced if clogged. A clogged filter will block or restrict the flow of liquid from the tank.

CAUTION: Do not loosen any connection on the liquid withdrawal line when the unit is pressurized. Serious injury could be caused by liquid spraying from an open line.

Filter Inspection

1. Open the vent valve and wait for all pressure to be blown down before proceeding. Leave vent valve open until work is completed.
2. Disconnect the liquid line from the filter outlet.
3. Unscrew the filter from the nipple in the end of the tank.
4. Tap the element gently and watch for any particles that might fall out.
5. Connect a pressure-regulated source of dry nitrogen gas to the outlet end of the filter. Set pressure regulator for 25 psi and purge the filter for about 30 seconds.
6. Disconnect the purge gas supply and inspect the filter. If it appears clean and gas flows freely through it, reinstall the filter in the tank. If the element is still dirty or clogged, it must be replaced.
7. Install the filter on the tank and connect the liquid line. Tighten all connections securely.
8. Pressurize the system and check for leaks at both ends of the filter.

Liquid Control Valve

The liquid control valve is a normally-closed pneumatically operated valve. The valve is fully opened when control pressure (15 to 25 psi) is applied to the valve by the temperature controller. Although field repair of the valve is not generally recommended, the valve can be disassembled by trained personnel at a Polarstream Service Center.

Fill Valve and Vent Valve

The fill valve and the vent valve are identical ball-type, manually operated valves. Teflon seals are used on the stem and on both sides of the ball to prevent leakage. A leaking fill or vent valve should be replaced. These valves can be cleaned and repaired by a Service Center which has the replacement seals and the equipment needed to pressure-test the valve for leakage after it is assembled.

NOTE: Fill and vent valves must be installed so that the end of the valve that contains the snap ring and ball vent hole is toward the tank. (The flow arrow should point away from the tank.) These valves will leak if they are installed backwards.

If the fill connection leaks excessively during filling, be sure that the connection and hose coupling are clean and undamaged. Check the seat of the fill connection for wear and scratches. Replace the connection if leakage cannot be eliminated.

Always replace missing or damaged dust caps. Do not leave the fill connection unprotected.

Liquid Contents Gauge

The liquid level gauge indicates the amount of liquid in the tank by measuring the difference between the pressure at the top of the tank and at the bottom. Because the pressure difference is very small, even when the tank is full, a very small leak in the gauge line connections or in the gauge itself can cause large errors in gauge readings.

The liquid level gauge cannot be repaired in the field. Although Service Centers can check calibration of these gauges, defective gauges must be replaced. To work on the liquid level gauge and its connecting lines, it is necessary to remove the container shroud.

Temperature Controller

Although some parts of the temperature controller can be replaced in the field, it is recommended that any repair of a controller that might affect its calibration be done by a Polarstream Service Center. Any attempt to disassemble the controller mechanism or to make adjustments other than those described in the "calibration procedures" section of these instructions can cause faulty operation of the system.

NOTE: The Partlow temperature controller used in this system vents gas continuously whenever the control pressure is turned on. The flow rate through the vent port is highest when the liquid control valve is closed (pressure off). This continuous venting should not be mistaken for leakage.

Thermometer

The proper methods for checking and adjusting the thermometer are described in the "Calibration Procedures" section of these instructions. If a thermometer is damaged or cannot be calibrated properly, the entire instrument, including the sensing bulb and capillary tube, must be replaced.

Spray Header and Sensing Bulbs

The system may be equipped with one of several types of spray headers, depending upon the requirements of a particular operation. The rapid-recovery header and the flow-limiting header are the most commonly used types. Regardless of which type has been selected, it is essential that the header be kept securely mounted and properly aligned with the temperature sensing bulbs. Align sensing bulbs with the flow-limiting and rapid-recovery spray headers, Fig. 4. Proper operation of the system depends upon the bulbs being located in an exact spray pattern, Fig. 4. If the nitrogen spray strikes the ceiling, meat rails or other obstructions, adjust the positions of the header and bulb guard to eliminate such interference. Be sure that the spray deflector is properly positioned to prevent direct spraying of liquid nitrogen on the capillary tubes. Freezing of the capillaries will cause inaccurate control and indication of cargo airspace temperature.

In addition to visual inspection to detect loose mountings and plugged holes, occasionally check the operation of the spray header by leaving all doors open and turning the system on. Stand well clear of the door while observe the spray pattern.

CAUTION: Never enter the vehicle while the system is on.

CALIBRATING PROCEDURES

Introduction

Because the refrigeration system makes use of a non-indicating temperature controller, and a separate remote-indicating dial thermometer, calibration of these components should be accomplished at the same time. These units are so sensitive that it is very difficult to calibrate them when the system is operating. The most accurate calibration can be achieved by placing the sensing bulbs in a bath of crushed ice that is slowly melting This ice bath will be at a temperature of exactly 32 deg. F. Both the controller and the thermometer are sufficiently linear that calibration at the temperature of melting ice will ensure good accuracy at any setpoint within the operating range of the controller.

In addition to the ice-bath calibration procedure, a method of checking the controller and thermometer at ambient temperature is also described in the following paragraphs. The "Quick Check" at ambient temperature should be used only as a means of checking that the controller operates at the same temperature as that indicated by the thermometer and that both are accurate.

NOTE: Operation of the temperature controllers can be monitored by observing the control pressure gauge on the tank control panel. This gauge indicates approximately the same pressure as the system pressure gauge when the controller is on (liquid control valve open) and indicates zero pressure when the controller is off (liquid valve closed). The valve in the controller also makes a clicking sound when it opens or closes.

"Quick Check" Procedure

A quick check of controller and thermometer calibration can be made at ambient temperature (when the vehicle is warm) by checking that the controller operates at the temperature indicated by the thermometer. First, set the controller to a point above the indicated temperature and turn the main control valve to the ON position. Turn the controller setpoint knob and observe the point at which the click of the controller valve can be heard or the control pressure gauge indicates control pressure is on. Rock the setpoint knob back and forth around this point to determine the exact reading at the operating point. This reading should agree within ±5 degrees of the temperature indicated by the thermometer.

Ice Bath Calibration Procedure

Whenever a new system is first installed or when any work has been done on a controller that might affect the accuracy of its calibration, or when the routine calibration check procedure discloses that system calibration is required, the thermometer and the temperature controller must be calibrated as follows: This procedure uses an ice-water bath that, when properly prepared, will act as a temperature standard of exactly 32 deg. F. (the melting point of pure fresh-water ice).

1. Fill a suitable container with pure crushed ice. Add just enough water to wet the ice completely.

   NOTE: Do not add salt. The ice and water used should be pure enough for human consumption.

2. Place the ice bath container over the bulb guard or slip both sensing bulbs out of the guard and submerge them in the ice bath container, which should be suspended from the guard. Handle bulbs and capillaries carefully. Sensing bulbs should be as close to normal mounting position as possible during calibration. Be sure that both sensing bulbs are completely covered by the ice.
3. After the sensing bulbs have been in the ice bath for at least ten minutes, check that the thermometer indicates exactly 32 deg. F. If it does not, adjust it as follows:

   Turn the plastic face plate 45 degrees counterclockwise and lift it off. Hold the small gear on the pointer shaft and carefully move the pointer in the required direction. Release the gear and recheck the reading. Repeat the adjustment until pointer indicates exactly 32 deg. Replace the face plate and turn it clockwise to lock in place.

   NOTE: If the thermometer cannot be calibrated by this procedure, replace the entire instrument.

4. Turn controller setpoint knob to a setting below 32 deg. F. Turn master control to the ON position and slowly move setpoint knob upscale until controller turns off (control pressure drops to zero). If controller dial does not indicate exactly 32 deg. F. at turn-off point, adjust controller as follows:

   Remove cover, loosen setscrew and turn hex shaft until setpoint dial indicates exactly 32 deg. F. at the turn-off point. Tighten setscrew and move dial back and forth several times to check that turnoff occurs at exactly 32 deg. Repeat adjustment as necessary. When calibration is completed, turn main control valve to the OFF position and replace controller cover.
5. Remove the ice bath container and check that capillary tubes, sensing bulbs and bulb guard are properly and securely mounted.

Roof-Mounted Truck Cab Air Conditioner

These installations function in much the same manner as a passenger car air conditioner. The main difference lies in the way most components are assembled into a single unit which is mounted on the roof of the truck cab.

FRIGIKING TC-660

The compressor is a two-cylinder, reciprocating type, with 10.35 cu. in. displacement. The unit, mounted in the engine compartment, is driven by a belt off the engine. An auxiliary power package is available if there is unsufficient space. The fixed-field magnetic clutch is thermostatically controlled.

The roof top unit contains the condenser and fan assembly, evaporator and blower assembly, receiver dryer, air handling chamber, and control panel assembly. The condenser and evaporator fan motors are series wound 12 or 24 volt, with circuit-breaker protection. The roof top unit is connected to the compressor by a ½ in. discharge hose and a ⅝ in. suction hose.

The sight glass is located on top of the receiver dryer which is mounted in the left side of the roof unit adjacent to the condenser fan. There's a viewing port in the unit cover which enables the sight glass to be observed without taking off the cover.

Controls — Operation

Two controls are mounted on an instrument panel located overhead between the air louvers. One is a thermostat control. The other is a combination unit and fan speed control switch which has four positions: OFF and three fan speeds.

To operate the unit, after the vehicle or auxiliary engine has been started, set the temperature control to the coldest position, and the fan speed to high. After the cab has been cooled, reduce the thermostat setting and fan speed to maintain the desired temperature. For rapid cooling of a hot cab, have the outside air vents closed and a window open to expel the hot air. Close the window after a few blocks. To remove smoke, open a window or vent slightly.

A combination of a low fan speed and coldest thermostat can cause icing of the evaporator, especially on a hot day. If icing occurs, turn fan to high speed, and turn thermostat off for a few minutes. Then resume cooling with a thermostat setting higher than before the frosting occurred.

To correct fogging of cab windows, turn unit on, direct air louvers toward windows until fogging clears, then turn unit off. If temperature is too cool, operate the heater at the same time.

Maintenance

Weekly

Check belts, hoses, wiring. When operating under extremely dusty or dirty conditions, check clutch for accumulations of foreign material. If necessary, disassemble clutch, blow out with compressed air.

Monthly

Add to weekly inspection procedure: Check compressor mounting bracket and drive components for cracks, or loose or missing parts. Remove unit cover, clean with compressed air and inspect condenser coil. Make sure drain hoses are in good condition. Start unit, observe sight glass for continuous presence of bubbles or foam. If refrigerant charge is low, correct leak, check compressor oil level, replace dryer, evacuate, recharge.

Yearly Spring Check-Up

Add to monthly inspection procedure: Check condenser fan mounting bolts. Remove evaporator housing cover, inspect seals and gaskets. Clean dirt from evaporator coil with vacuum cleaner. Make certain thermostat element is embedded and secure between fins of evapora-

Typical cab roof unit

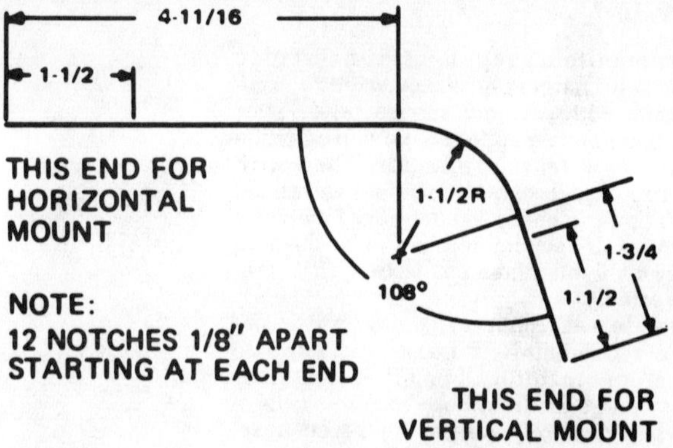

Fig. 1 Oil dipstick dimensions

tor coil. Check blower wheels for looseness, wobble, end play. Be sure evaporator blower mounting assembly bolts are secure. Turn on ignition, check fan speed control, compressor clutch operation in response to the thermostat switch position. Check unit for proper operation.

Trouble Diagnosis

Inspect entire system, as for a Spring check-up, anytime repairs of any kind are made. This takes only a few more minutes, and could prevent a major service problem later.

Performance Test

Attach manifold gauge set to compressor service valves. Before tightening gauge hoses at service valves, purge air and moisture from hoses with small amount of refrigerant from a can or tank. Close both hand valves, turn off and remove can or tank.

Mid-Seat Service Valves

Recommended mid-seat position, to dampen gauge pressure pulsations, is 1½ turns clockwise from back-seated position. Operation of service valves is explained under compressor service.

If vehicle has tilt cab, lower cab completely. Run engine at 1,000 to 1,500 rpm, turn thermostat to maximum cooling. Operate blower at high speed, insert thermometer in air stream. Make sure condenser fan and evaporator motor are operating, close doors and windows.

Immediately observe system pressures. If, after 2 to 3 minutes of operation, discharge pressure is over 350 psi or suction pressure is negative, refer to troubleshooting chart.

If conditions are normal, operate system for another 15 minutes. During this time, check for abnormal noise from clutch or compressor. After 15 minutes of operation, again check pressures, and also evaporator air temperature. Relative humidity is a factor; temperature obtained at air outlets will be lower on dry days and higher on damp days. Typical normal conditions would be as follows:

Ambient air temperature, F	70	80	90	100
Evaporator air temperature, F	35-45	40-50	45-55	50-60
Suction pressure, psi		15-30		
Discharge pressure, psi	160-200	190-240	220-270	250-300

Pressures higher or lower than those indicated above may be caused by the conditions listed in the following chart.

Condition	Pressures Discharge	Too High Suction	Pressures Discharge	Too Low Suction
Overcharge	X	X		
Air in system	X	X		
Expansion valve stuck open	X	X		
Restriction to condenser air flow (bugs, leaves, etc.)	X	X		
Condenser fan not operating properly	X	X		
Compressor malfunction (reed valve, broken rings, etc.)		X	X	
Undercharge			X	X
Moisture in system closed			X	X
Expansion valve stuck closed			X	X
Suction or liquid line restriction			X	X
Restriction to evaporator air flow			X	X
Evaporator blower not operating properly			X	X
Excessive clutch or belt slip		X	X	

Check sight glass; it should be clear. If bubbles or foam are present, check for leaks and make any necessary repairs. Check compressor oil level, install new dryer, and charge. Procedures are detailed under compressor service.

Now turn thermostat to warmer position. If unit has not already cycled., it should now begin. After making certain unit has cycled, return thermostat to coldest position, and check blower operation in all switch positions.

Next, feel hoses and components on both high and low sides of system. If vehicle has tilt cab, raise cab slightly to inspect compressor and hoses.

High Side: All hoses and components should be warm to the touch. At receiver dryer, inlet and outlet temperatures should be about the same. If there's a big difference in these two temperatures, it indicates a restriction in dryer. Heavy frost on inlet side of expansion valve may be caused by a valve defect or moisture in system.

Low Side: All lines and components should feel cold or cool to the touch.

Compressor Service

The compressor is a 2 cyl., 10.35 cu. in. displacement unit. Drive belt tension is 100-125 lb. for new belts, 80-100 lb. for old belts.

Service Valves

The service valves, mounted on top of compressor, are marked "S" for suction and "D" for discharge. Both service valves operate in same manner. The three operating positions are shown in the illustration. The mid-seated position, used when gauging system, is about 1½ turns clockwise from back-seated position.

Oil Level Check

Check oil level whenever system components have been changed, or refrigerant has been lost due to leaks. Prior to checking, run compressor 15 or 20 minutes to bring it up to operating temperature. Then stop compressor and front-seat both service valves. Slowly unscrew oil plug five full turns to allow all crankcase pressure to bleed off. With pressure relieved, remove plug.

A dipstick for checking oil level is available under Frigiking part No 36-00183, or one can be fabricated to dimensions shown in Fig. 1. Check oil level with keyway in shaft positioned to face head of compressor. This places throws of crankshaft in most favorable position for passage of dipstick. If keyway can't be seen with clutch mounted, set crankshaft throws by feel of dipstick. Oil level must be measured from lowest point in crankcase. Long leg of dipstick is used for checking oil level on horizontal mounts, short leg for vertical mounts.

Specified minimum level after operation is ⅞ in., and maximum level is 1⅛ in. Recommended oil is Texas Capilla E (viscosity 500 Saybolt), Suniso 3G, or equivalent.

After checking, tighten plug securely. Evacuate compressor and check for leaks before back-seating service valves. If plug leaks, do not overtighten. Instead, replace O-ring and, if necessary, install new plug.

Leak Test

With manifold gauge hoses connected to compressor, charging hose connected to vacuum pump, mid-seat both service valves, start pump, open both manifold hand valves, pull a vacuum for at least 10 minutes. Then, with pump running, close both manifold valves and turn off pump. Check vacuum gauge for at least 10 minutes. If vacuum does not hold, there's a leak, probably in a hose connection, that must be corrected.

To locate leak, connect charging hose to container of refrigerant R-12. Loosen charging hose connector at manifold, open R-12 container valve to purge air from hose. Tighten connection when pure refrigerant flows from hose, usually after about 5 seconds. Now break vacuum by allowing small amount of R-12 vapor (not liquid) to enter discharge side. This is done by slowly opening manifold valve.

Continue adding gas until 40 psi is indicated on both gauges. If low side gauge does not show 40 psi, system is blocked and must be repaired. With pressure equalized and holding, tap low pressure gauge with hand, record exact pressure and time.

Locate leak with electronic or halide torch leak detector. After correction, recheck. Pressure on low side gauge should hold for at least 20 minutes.

Evacuation

First, bleed system to zero pressure. Be sure this is done slowly, from low side of system, to prevent loss of oil.

Now connect charging hose to vacuum pump, and evacuate system for at least 30 minutes. Close valves, disconnect pump, and connect R-12 container. Purge charging hose, break vacuum, and charge system to about 15 psi.

Next, wait 10 minutes, remove R-12 container, and bleed system to zero pressure. Connect pump and evacuate system for at least 1 hour. If an absolute pressure gauge is available, pull vacuum until 1,000 microns is reached, or as low in the 1,000 to 2,000 micron range as possible. This deep vacuum should be pulled using a high vacuum pump, and with new or recently changed oil. Valve off pump, turn off pump, disconnect charging hose. System is now prepared for charging.

Charging

Connect charging hose to R-12 container, and purge air from hose at manifold. Let R-12 into system as a vapor (not liquid), until system will take no more. Be careful not to overcharge. An exact charge is not specified because of differences in hose lengths and other variables. Instead, system is charged until sight glass is clear.

Now close discharge, or high pressure, gauge set hand valve fully clockwise until it is tight. It is important that this valve is never open with R-12 container attached and compressor running, because compressor discharge pressure may cause the container to explode.

With discharge valve closed, start engine. Turn system controls on fully, including evaporator and condenser fans, to keep compressor running. Charge system with refrigerant until sight glass is clear. Close low side hand valve.

Vehicle cab should be cooled to lowest desired operating temperature. Check sight glass — if foam or bubbles are present, add refrigerant. Occasional bubbles, seen only for a few seconds after compressor is started, are normal and do not indicate need for additional refrigerant. Back-seat compressor service valves, disconnect manifold gauge set, and replace gauge port caps.

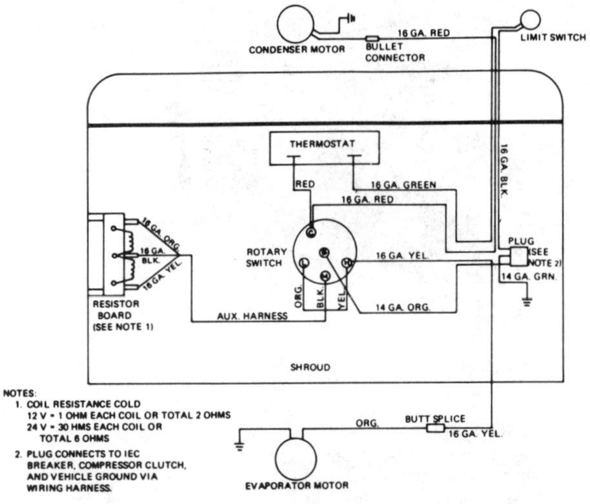

Fig. 2 Frigiking TC-660 series wiring diagram

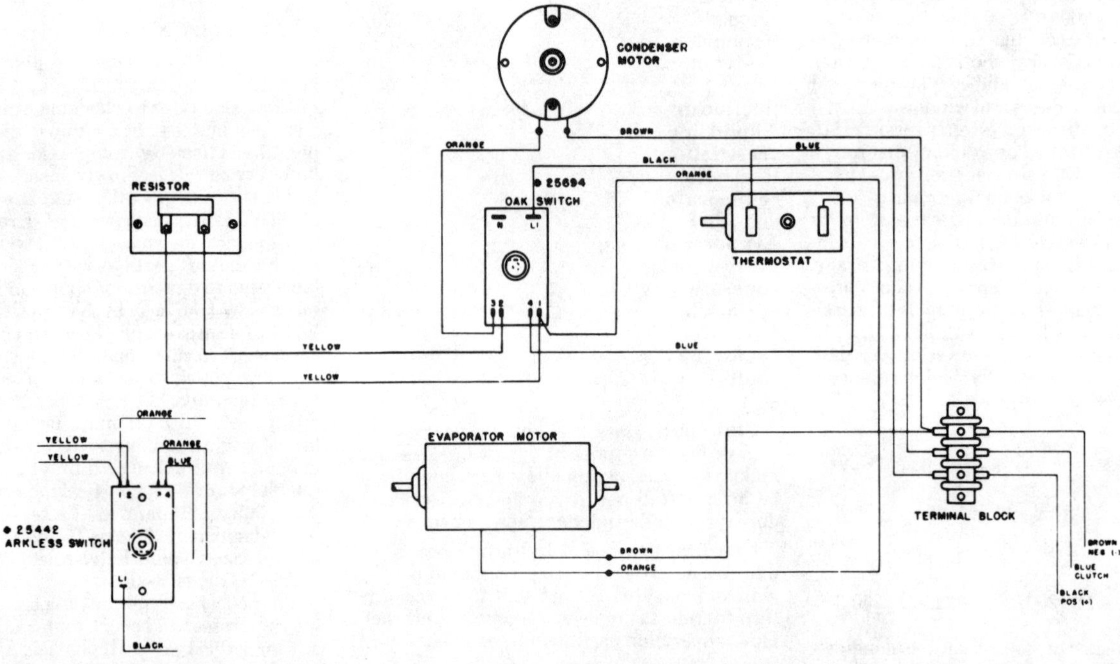

Fig. 3 Mark IV wiring diagram

Unit Removal

Roof unit can be detached from cab by removing nuts from mounting studs. With cover removed from roof unit, components are accessible for service or removal. Repair procedures are the same as for car air conditioners. A typical wiring diagram is shown in Fig. 2.

MARK IV

Mark IV cab roof installations are similar to the Frigiking. Either York or Tecumseh compressors, driven by the vehicle engine, are used. The other components are contained in the roof unit. Parts in roof unit are made accessible for service by unsnapping catches on outer cover of unit.

Checking procedures are also the same as Frigiking, with minor differences. Prior to checking for leaks, allow R-12 refrigerant gas to enter system until a pressure of 50 psi is observed on low pressure gauge. When evacuating, run vacuum pump to obtain 28 in. on low pressure gauge, add refrigerant to obtain 5 psi, then repeat evacuation to 28 in.

To charge system, first admit R-12 until system is at container pressure. With container and both manifold valves closed, run engine at about 1,500 rpm, connect 12 volt jumper wire to clutch, operate blowers. If bubbles are present in sight glass after 2 minutes of operation, open low pressure gauge manifold valve and container valve, and add charge until bubbles disappear.

Close low pressure manifold valve and turn off blowers. Bubbles should not appear in sight glass until low pressure gauge reading reaches 12 psi. Bubbles should normally appear between 12 and 8 psi. If they do not, disconnect gauge manifold center hose from container and purge small amount of refrigerant from system.

Now turn on blowers. When low pressure gauge indicates 25 to 30 psi, sight glass should not show bubbles. Turn off blowers. Bubbles should not appear at gauge reading higher than 12 psi. If they do, add R-12 refrigerant. Don't forget to remove jumper wire. A typical wiring diagram is shown in Fig. 3.

Truck Cargo Refrigeration and Heating System

Specialized refrigeration units, some of which can also be used for heating, are supplied in various applications to protect truck cargoes. Most units are mounted high on the front of the cargo container, but some are mounted underneath, at the rear or in other locations on the body. Blowers provide circulation.

The compressor is directly driven by a small engine in most applications. On some, the engine drives an alternator, and the compressor is electrically driven. Once started, the engine runs continuously, the speed varying with the cooling load. Operation is automatic. Some units have a standby electric motor which can be connected to an electrical outlet to drive the compressor when the vehicle is in the warehouse.

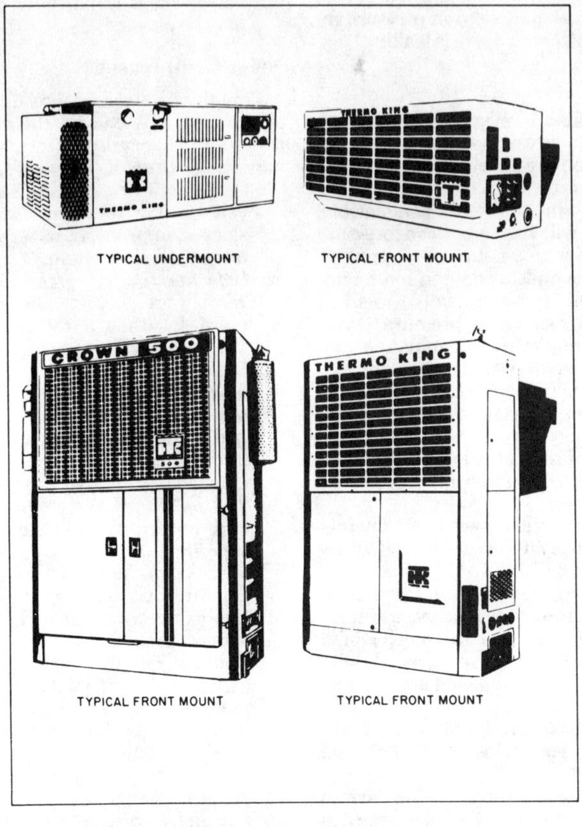

Typical refrigeration unit installations

THERMO KING MODEL MWC

The MWC is a single-unit type designed for installation in straight trucks and short trailers. The basic MWC-10 model is a front mounted refrigeration unit driven by a gasoline engine. The MWC-20 has a 3 hp electric motor added for standby use. The MWC-30 is an engine-powered model designed for both cooling and heating, and the MWC-50 is additionally equipped with a standby electric motor.

Components

The MWC is made up of two sections. The condenser section, located outside the truck body, contains the condenser coil, engine, compressor, compressor fan, control compartment and, on some models, the standby electric motor. This section is enclosed on the sides by panels, and at the top by doors.

The evaporator section, located inside the truck body, contains the evaporator coil and fan, expansion valve, dehydrator and heat exchanger. There's also a shutter which is used to close off the air funnel during defrosting, and thus prevents hot air from reaching the load.

An insulated bulkhead joins the two sections into one unit, thus simplify the installation, and preventing engine heat from entering the body.

The compressor is a model 214 Thermo King belt-driven 2 cyl. unit.

The basic engine is a wisconsin model AENLD single cylinder 8.5 hp type. Tapered roller bearings support the front and rear of the crankshaft, and the aluminum alloy piston has four rings. Valve seats have hardened inserts, and the exhaust valve has a hardened face and is equipped with a rotator. An Onan model L25 engine is used on some installations, and both engines can be run on LPG if desired.

The standby electric motor is 3 hp, 220 (or 440) volt, 3 phase, 60 cycle AC.

Controls

One-piece control panel has starter switch, ignition switch, adjustable thermostat, dial thermometer, engine hourmeter. Models 10 and 20 have manual defrost hand valve, defrost shutter control; models 30 and 50 have manual defrost switch, heat or cool switch.

Engine speed is controlled by the thermostat and idle-run controller. Controller, mounted on engine, pulls governor control arm to run engine at full governed speed when thermostat calls for refrigeration. Engine returns to idle speed when temperature matches thermostat setting.

Refrigeration-Heating System Operation

Refrigeration cycle operates basically in same manner as for car air conditioner. Refrigerant in evaporator coil absorbs heat from truck body. Heated gas in evaporator is pumped out by compressor. Accumulator, heated by engine exhaust, keeps liquid refrigerant from entering compressor. Refrigerant passes through condenser where heat is transferred to air, and then is stored in receiver tank. Heat exchanger cools refrigerant on way to evaporator, increasing efficiency.

On defrost, hot gas leaving compressor is pumped directly to drain pan heater tube to melt ice in pan before condensate from evaporator coil accumulates. Damper in evaporator is closed so warm air is not circulated over load. When coil temperature rises, heat opens defrost thermostat switch and unit returns to refrigeration cycle.

On heating cycle, compressor pumps hot gas into evaporator coil, as for defrosting, but damper remains open to circulate warm air through cargo space. In addition to heat exchanger, there's an accumulator, heated by engine exhaust system to add heat to refrigerant gas and increase heating capacity.

Three-way valve is controlled by pilot solenoid valve. Path running from defrost line through check valve to shut-off valve is used on heating cycle to pressurize receiver tank. This forces liquid refrigerant from tank up to expansion valve to increase operating pressure and add heating capacity. Second check valve in line from condenser to receiver prevents refrigerant from flowing into condenser when receiver is pressurized during heating. Small notch in seat of expansion valve permits passage of refrigerant even when valve is seated.

Trouble Diagnosis

Possible troubles, and procedures for determining their causes, are as follows:

Temperature Cannot Be Held At Required Level

Truck or trailer temperature is indication of normal or abnormal operation of unit. If load temperature cannot be brought down to thermostat setting, even when outside temperature is very high, check for following possible cause:

1. *Excessive Load:* Overload may be caused by open or leaking door, damaged gasket, break or leak in vehicle wall, wet insulation.
2. *Inadequate Air Circulation:* Load stacked too high or too tightly, blocking evaporator. Space must be left at top of load, with free flow of air under and around load.
3. *Excessive Frost On Evaporator:* Heavy frost on evaporator acts as insulator, restricts transfer of heat, reduces capacity of unit, increases running time. Abnormally low suction pressure (if unit has gauge) indicates frosted coil. Press defroster switch while unit is running.
4. *Dirt On Coils:* Dirt acts as insulator, same as frost. Keep evaporator and condenser coils clean.
5. *Incorrect Belt Tension.* Insufficient belt tension can reduce condenser and evaporator fan speeds, impairing efficiency of unit.

Suction Pressure

Suction pressure will vary with temperature in vehicle body and outside temperature. For a zero deg. body temperature, suction pressure should be about zero pounds during normal operation. At 40 deg., suction pressure should be 16 to 18 lb. If pressures are abnormal, check possible causes:

1. *Shortage Of Refrigerant:* If refrigerant charge is too low, suction pressure will be lower than normal.
2. *Excessive Frost On Evaporator:* Heavy layer of frost will cause lower than normal suction pressure. Check evaporator fan speed, obstruction in air path.
3. *Excessive Oil Circulation:* Too much refrigerant oil in system can restrict dehydrator or expansion valve orifice, cause lower than normal pressure.
4. *Faulty Expansion Valve:* Causes erratic pressure reading. If valve stays open it will cause flooding of evaporator. If it sticks closed it will cause evaporator to be starved and go into vacuum.
5. *Three-Way Valve Sticking:* If valve sticks, high pressure gas will leak into evaporator, causing high suction pressure.
6. *Restricted Refrigerant Lines:* Reduced flow of refrigerant caused by restriction in receiver outlet valve, dehydrator or expansion valve will cause lower than normal suction pressure.
7. *Expansion Valve Charge Loss:* If expansion valve capillary tube breaks and loses its charge of gas, valve will close, pressure gauge will show vacuum.
8. *Leaking Check Valve:* If check valve between defrost line and shut-off valve leaks, suction pressure will be high. To check, close shut-off valve. If suction pressure returns to normal, check valve is leaking.

Head Pressure

Head pressure also is means of diagnosing troubles. Pressure will vary with ambient temperature and suction pressure. Other factors influencing head pressure are:

1. *Air In System:* Since air is non-condensible, air in system will increase head pressure commensurate with amount of air present. To check, allow unit to come to room temperature. High pressure gauge reading more than 10 psi above pressure corresponding to temperature of ambient air indicates air in system. With compressor idle, air will collect in highest point of high side, and must be purged from discharge service valve.
2. *Restricted Air Flow:* Dirt lodged in fins of condenser coil will restrict air flow, cause high head pressure.
3. *Loose Fan Belts:* Will cause fans to operate too slowly, reduce air flow across coil, cause high head pressure.
4. *Restricted Refrigerant Lines:* Restriction in line from compressor discharge service valve, through condenser to receiver inlet valve will cause high head pressure. Be sure discharge valve is fully opened after servicing unit.
5. *Overcharge Of Refrigerant:* Overcharge will cause high head pressure. Excess must be purged.
6. *Undercharge Of Refrigerant:* Shortage of refrigerant will cause lower than normal head pressure.

Suction Line Temperature

Suction line leads from evaporator outlet to suction service valve. Line temperature is affected by same factors that affect suction pressure in system. Line will feel cool when unit is maintaining normal temperatures, and at 10 deg. line will be cold to touch. Conditions affecting temperature are as follows:

1. *Shortage Of Refrigerant:* Will result in warm suction line. Note that suction pressure will also be low, and no liquid will show in sight glass.
2. *Insufficient Refrigerant Passing Through Expansion Valve:* This will cause part of evaporator coil and suction line to be warm. Cause could be obstruction in expansion valve, valve stuck closed or clogged by ice, wax or dirt.

Liquid Line Temperature

Line leads from receiver outlet valve to evaporator inlet. During normal operation liquid line is slightly warmer than surrounding air. An extremely hot or cold line indicates abnormal condition. Some possible causes:

1. *Shortage of Refrigerant:* Shortage may cause liquid line to be hot, because gas discharged from compressor has no liquid seal at receiver outlet to hold it back.
2. *Restriction In Line:* Line restriction will cause flashing (gas will vaporize at point of restriction). Line will be cool beyond point of restriction.
3. *Clogged Dehydrator:* This will cause liquid line on outlet side of strainer or dehydrator to be cooler than line on inlet side. Line may sweat or even frost. Replace dehydrator.
4. *Receiver Outlet Valve Restriction:* If liquid line is colder at outlet side of receiver outlet valve, valve is partially closed or screen may be plugged.

Noisy Compressor

Compressor noise can result from a number of conditions. A clue to the cause can be obtained by observing other conditions which may also be present. These are as follows:

1. *Oil Leaves Crankcase:* Noise, in this case, can be due to worn pistons or cylinder sleeves, improperly set expansion valve, or loose thermostat bulb.
2. *Defective Piston Head Valve:* Leaking, worn or broken piston head valve can cause noise, insufficient refrigeration, high suction pressure, lack of compression.
3. *Defective Discharge Valve Plates:* With leaking, worn or broken discharge valve plates there can be, besides noise, a rapid rise in suction pressure on OFF cycle. Also intermittent, slight movement of engine crankshaft after unit has been shut down.
4. *Suction Line Frosted:* Expansion valve bulb not firmly attached to suction valve, superheat set too low, expansion valve stuck open, liquid refrigerant returning to compressor.
5. *Head Pressure High:* Sudden reduction of suction pressure causes oil slugging as refrigerant rapidly boils out of oil. Caused by expansion valve sticking closed from formation of ice due to moisture in system, or from wax resulting from use of poor grade of oil. If moisture is present, install new dryer. High pressure can also be caused by air or other non-condensible gases in system, or by overcharge of refrigerant.
6. *Suction Pressure High:* Noise with a high

suspection pressure and a low discharge pressure indicates a broken valve.

7. *Pressure Normal:* Loose or worn drive pulley, loose mounting bolts, worn pistons or bearings, worn or broken valves.
8. *Crankcase Overfilled With Oil:* Compressor will be noisy due to oil slugging. Can also be cause of loss of refrigeration capacity.

Compressor Test

A quick check to determine the operating efficiency of the compressor can be made as follows:

1. Install a gauge manifold.
2. With unit operating, gradually front-seat suction service valve (SSV). This will pump down compressor crankcase.
3. When compound pressure gauge indicates 20 to 25 in. of vacuum, stop unit. Pressure may rise gradually as remaining refrigerant in compressor oil evaporates.
4. Restart unit, again run to obtain 20 to 25 in. vacuum, then stop unit. It may be necessary to repeat this step several times on model 422 and 426 compressors to evaporate all refrigerant.
5. With unit stopped, note rate of pressure rise on compound gauge. Normally, pressure rise should not exceed 10 in. in a period of 3 minutes.
6. If pressure rises too rapidly, it indicates poor compressor valve plate seal. It many also indicate pilot solenoid valve leaking, or SSV not tightly front-seated.

A more thorough check of compressor's operating efficiency can be made in following manner:

1. Cover condenser coil and determine if compressor is capable of building up to 180 to 200 psi head pressure.
2. Gradually close SSV and pump down low side to obtain 20 to 25 in. vacuum.
3. Stop unit and note stability of vacuum as indicated on compound gauge. The ability to build up head pressure, and to obtain and hold 20 to 25 in. vacuum, is a good indicator of compressor efficiency.

If leaking solenoid valve is suspected, check to ensure that unit is on cooling. This can be done by checking switch setting, feeling refrigerant lines. Also be sure solenoid is not energized by incorrect wiring connection. When solenoid is energized, there's a magnetic field above coil. This can be checked easily with a screwdriver.

If solenoid valve does leak, unit will usually operate on heat. Leak will allow a greater gas flow than flow through bleed hole in three-way valve, and cause valve to shift to heat position.

Check Valve Test

A test can be made to determine whether or not check valve in shut-off valve line is sealing during refrigeration cycle. After first determining that compressor is in satisfactory condition, and pilot valve is not leaking, use following procedure:

1. Install gauge manifold.
2. Close receiver tank outlet valve. Start unit and obtain 20 to 25 in. of vacuum on compound gauge. Then stop unit and note if pressure rises. If it does, repeat pumping down operation to 20-25 in.
3. If pressure still continues to rise on gauge after shutdown, trouble is caused by check valve leaking, or leakage of three-way valve to defrost or hot gas line.
4. To determine which, close shut-off valve by turning stem clockwise. (Normal position of valve is fully open). With valve close, pump down low side as before.
5. If vacuum is now maintained, trouble is in check valve. If compound gauge pressure continues to rise, leakage to low side is probably due to leakage of three-way valve to hot gas line. Leakage in heat exchanger is also a possibility, but is unlikely.
6. Sealing of three-way valve during cooling cycle has now been checked. Next step is to check this valve during defrost or heating. (See below).

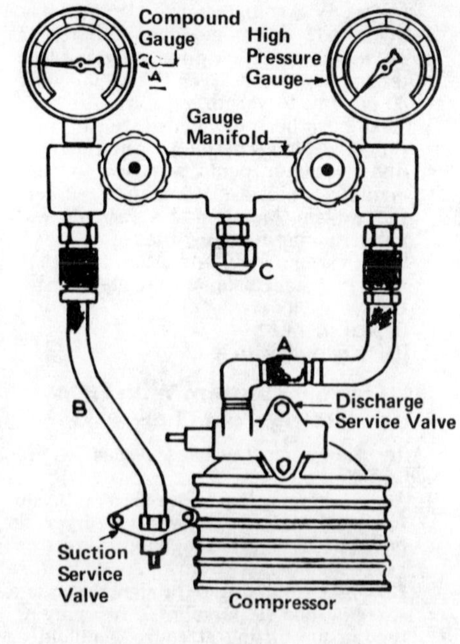

Fig. 4 Gauge manifold installation

Three-Way Valve Test

Valve test for cooling cycle is given above. Lack of performance on defrosting or heating can be due to leakage of threeway valve to condenser. One method of testing is as follows:

1. First check compressor, pilot valve, by-pass check valve to make sure they are in proper operating condition.
2. Install gauge manifold, close receiver tank outlet valve.
3. Start unit, pump down low side to obtain 15 in. on compound gauge. Stop unit, note gauge readings.
4. Energize pilot valve to shift and hold three-way valve in heating position. When three-way valve shifts, both gauges will change to indicate equal pressures. About zero lb. will be indicated on both gauges, and pressures should remain stable if valve to high side is holding the seal.
5. If gauges show a continued rise above zero, three-way valve seal to high side is leaking.

Refrigerant Checks—Service

Recommended refrigerant is R-12. Capacity of various models is shown on serial plate attached to unit. Checks, service operations require use of gauge manifold. Keep dirt out of lines.

Gauge Manifold Installation

Remove service valve cap. Make sure both service valves are in back-seated position by turning valve stems counterclockwise.

Remove caps from service port fittings. Attach high pressure gauge line finger tight to discharge service valve (DSV) service port, as shown at A, Fig. 4.

Attach compound gauge line loosely to suction service valve (SSV) service port, as shown at B.

With center fitting on manifold capped, as shown at C, open both valves on gauge manifold.

Crack open DSV to purge manifold and lines. Tighten line fittings to SSV at B, and DSV at A.

Close both valves on gauge manifold. Crack both DSV and SSV to obtain pressure readings on both low and high pressure sides of system.

Gauge Manifold Removal

Back-seat DSV. Open both gauges on gauge manifold to allow R-12 and oil to pass over to low side of system.

Back-seat SSV. Remove gauge manifold lines. Replace caps on service port fittings, plugs on gauge line hoses.

3. Some units have gauges or pressure cutouts connected to service valve port. On these, be sure to turn stem of valves ¼ to ½ turn clockwise from back-seated position to allow pressure to reach gauge or cutout.
4. Tighten valve stem packing nuts slightly to prevent leaks. Replace service valve caps. Test for leaks.

Checking Refrigerant Charge

The procedure requires bringing the unit to a stabilized condition, and then checking the sight glass. Proceed as follows:

1. Place an accurate thermometer in cargo area to indicate air temperature. Install gauge manifold.
2. Run unit until thermometer registers zero deg. Allow unit to cycle several times to stabilize conditions.
3. Compound gauge should indicate zero pounds suctions pressure. Discharge or head pressure gauge should read 150 psi. If discharge pressure is lower than this, it can be raised by covering portion of condenser coil with piece of cardboard.
4. Observe sight glass. Under above conditions refrigerant should be clear, with no bubbles. If bubbles are present, add R-12.

Adding Refrigerant

1. Install gauge manifold. Attach charging line loosely to manifold fitting. Connect other end of line to drum of R-12. Drum should be in standing position so only gas is drawn off, and should be equipped with service dryer to prevent moisture from entering system.
2. Open drum valve two turns to purge charging line. Tighten nut of charging line at manifold fitting.
3. Start unit, turn SSV clockwise two turns. Maintain 150 psi head pressure with cardboard, and zero pounds suction pressure. At two minute intervals, back-seat SSV. Operate unit for 15 minutes, then check refrigerant in sight glass. Add more R-12 if necessary.
4. When unit is fully charged, back-seat SSV, and close valve of R-12 drum.

Pumping Down

When it is necessary to open the low side of the system for any reason, or to repair the compressor, the unit is pumped down. Closing the receiver outlet valve and operating the unit will force all the refrigerant into the condenser and receiver. It is also a method of checking for a leaking three-way valve or solenoid, or leaking compressor valve plates. To check valve plates it is necessary to pump down the compressor crankcase only.

Pumping Down Low (Condenser) Side

To service low side — such as sight glass, dehydrator, expansion valve, evaporator coil, drip pan, accumulator, suction vibrasorber — proceed as follows:

1. Install gauge manifold, close receiver tank outlet valve by turning fully clockwise. Close bypass line by turning hand shut-off valve clockwise.
2. Start unit and operate until compound pressure gauge shows 15 in. vacuum. Stop unit.
3. Pressure will gradually rise as remaining R-12 evaporates. Restart unit and run until gauge shows 15 in. vacuum.
4. When system remains in vacuum, allow pressure from high side to pass to low side through gauge manifold until compound gauge shows 1 to 3 psi pressure. Low side can now be opened for service or repair.

CAUTION: Never open system when it is in a vacuum.

5. After system has been serviced, any air in system must be purged.

High Side

To service high side of system from compressor to receiver tank—parts such as three-way valve, condenser coil, check valve, discharge vibrasorber—it is necessary to pump out the charge, or to release it to the atmosphere.

Purging System

When a system has been opened for service, air has likely entered the system. The section exposed must be purged of air for efficient operation.

Purging Low Side

When unit has been pumped down to service low side, and there is refrigerant in the receiver, use following procedure:

1. Front-seat discharge service valve (DSV). Remove cap from service port.
2. Crack receiver outlet valve slightly and allow gas to escape from DSV port for a few seconds. This drives air out of low side through service port. Replace cap on service port.
3. Open receiver outlet valve slightly and build up pressure to 30 psi. Now close valve and check repaired section for leaks.
4. Back-seat all valves and replace valve caps.

Purging High Side

If high side has been opened, air must be removed with an evacuation pump.

Evacuating System

System must be evacuated with pump if high side has been opened, or unit has lost its complete charge.

1. Attach charging line from refrigerant drum to service port of SSV or to center fitting on gauge manifold.
2. Back-seat DSV, remove cap from port. Open SSV to compressor.
3. Open R-12 drum and charge system at drum pressure. Close drum valve and leak-test repaired area. Keep charge in system 30 minutes to absorb moisture. After any leaks have been eliminated, release pressure in system by partially closing DSV.
4. Run evacuator pump attached to valve manifold to pull 28 in. vacuum on refrigeration system. Maintain this vacuum, with pump running, for 20 minutes.
5. Break vacuum with R-12. Allow R-12 to remain in system 30 minutes to absorb moisture.
6. Repeat step 4.
7. Recharge unit with R-12.

Charging System With Full Refrigerant Charge

After the system has been evacuated, proceed as follows:

1. Attach charging line from refrigerant drum to inlet of a large capacity service dryer. To outlet of dryer, connect charging line from gauge manifold port or SSV port. Release R-12 gas from drum through dryer and charging line to purge air. After purging, tighten line fitting at gauge manifold or SSV.
2. Weigh drum of refrigerant and mark down for reference.
3. Open valve on R-12 drum and valve on SSV each about two turns. Drum should be in upright position. R-12 will flow into system until pressure equalizes.
4. Start unit and run until correct amount of refrigerant, as indicated on unit serial plate, has been pumped into system. Watch scale carefully during this operation to prevent undercharge or overcharge.
5. Backseat SSV, and close valve on R-12 drum. Remove charging equipment, and replace valve and port caps.

Compressor Service

Thermo King compressors are 2 and 4 cylinder reciprocating piston-type pumps, Figs. 5 and 6. The aluminum alloy pistons and replaceable cast iron sleeves are machined to close tolerances, and no rings are needed to seal the piston and cylinder wall. A rotary type refrigeration seal is used to prevent leakage around the forged steel crankshaft. The connecting rods have replaceable bearing inserts, while deep-groove ball bearings are used to support the crankshaft. Positive lubrication to the connecting rods and bearings is provided by a gear type oil pump.

Discharge valves are non-flexing ring plate design, free floating, with no pins or other retaining fasteners. A spring-loaded discharge valve cage provides pressure relief if liquid refrigerant enters the compressor body. The entire cage lifts, thus preventing breakage or deforming of the valves or reeds. The cage reseats automatically.

Model	214	422	426
Bore and stroke	2.25×1.75	2.25×1.375	2.25×1.625
No. of Cyls.	2	4	4
Displacement	13.92	21.87	25.90

First number in model number indicates number of cylinders: last two numbers indicate displacement.

New model compressors are interchangeable with early models, and both are removed and installed in the same manner. The model 426 can be used to replace the model 4A, and the 422 to replace the 4L. The only additional parts needed for the swap are a new discharge manifold and gaskets, ⅜ by 1⅞ in. capscrews, and copper gaskets.

The model 214 replaces the model 2RB. New suction tubing is required for this swap.

Compressor Lubrication

Recommended oil is Thermo King part No. 67-200. Oil capacities, with a dry compressor, are:

Compressor	Capacity	Base
2RB and 214	4 pints 8 ounces	
4A, 4L and 422	5 pints	Shallow
4A and 426	8 pints 8 ounces	Deep

On level sight glass is installed in compressor base. Oil should appear clear, with slightly bluish cast. Black oil indicates carbonization caused by air in system. Carbonized system will break down refrigeration oil, causing excessive wear of compressor parts. Brown oil indicates copper plating caused by moisture in system. Copper plating, moisture will affect valve plate reeds, break down refrigerant, affect capacity of unit. Gray or metallic oil color is indication of bearing wear or piston scoring. Metallic particles traveling with refrigerant can plug dryer and expansion valve, resulting in erratic operation. If system is contaminated, flushing with refrigerant 11 is recommended. In some cases, contamination can be corrected by changing oil, refrigerant charge and dryer.

Too much or too little oil can result from adding or removing oil under varying conditions, and can damage compressor. Before installing a replacement compressor, adjust oil level in replacement to same level as that in old compressor. This will keep same amount of oil in system.

Oil Level Check

Do not add oil to compressor before knowing that the system is definitely low. This can be determined as follows:

1. Install compound gauge at suction valve service port.
2. Install high pressure gauge at discharge valve service port.
3. Operate unit for 15 minutes or more on cooling with at least a 20 psi suction pressure and 100 psi discharge pressure. After 15 minutes, check sight glass. Oil level must be within ½ to ¼ sight glass. This method, under stable conditions, eliminates variables.

Adding Oil

To add oil, it is necessary to pump down compressor crankcase. Use following procedure:

1. Front-seat suction service valve.
2. Pump down compressor crankcase.
3. Remove filler plug, add oil. Never use engine oil.
4. Replace plug. Be sure suction service valve is opened before starting unit.

Compressor Test Procedure

The procedure for checking the operating efficiency of the compressor is detailed under Trouble Diagnosis.

Valve Plate Replacement

This operation can be done, if preferred,

without removing the compressor.

Valve Plate Removal

1. Pump down refrigerant. Front-seat suction and discharge valves.
2. Remove cap from DSV port to release pressure. Remove DSV mounting bolts, lift valve and tubing out of way.
3. Remove cylinder head bolts and washers. Tap head with soft hammer or block of wood to loosen and remove head.
4. Remove large gasket. Lift out valve plate assembly. Inspect reeds for dents or cracks. If either are found, install new parts.

Valve Plate Installation

1. Clean gasket surfaces on top of sleeves. Coat new gaskets with compressor oil, place on top of sleeves.
2. Place new valve plates on sleeves, and new gasket on valve plates.
3. Set cylinder head in position on block. Install new washers with head bolts. Torque bolts to 23 ft. lbs.
4. Place DSV in position, with new gasket under valve. Replace bolts, using new copper washers.
5. Purge air from compressor. Build up pressure in system, and check for leaks. If none are found, back-seat service valves, and open receiver outlet valve.
6. Operate unit for 10 minutes, then close receiver outlet valve and pump down unit. If vacuum holds, valve plates are not leaking. Open receiver tank outlet valve.

Compressor Removal From Unit

To remove the compressor from the unit:

1. Pump down refrigerant charge. Front-seat discharge service valve (DSV) and suction service valve (SSV).
2. Remove nuts and bolts from DSV and move it out of way.
3. Remove mounting screws from SSV.
4. Remove compressor mounting bolts, move drive belts off pulley, remove compressor from unit. Always keep service valve ports covered to prevent entry of dirt. Note oil level in sight glass.

Compressor Installation In Unit

1. Oil level should be same as that of removed compressor.
2. Loosen engine mounting bolts. Place compressor in unit. Install, but do not tighten, compressor mounting bolts.
3. Install service valves, using new gaskets. Position compressor, tighten mounting bolts.
4. Install belts, move engine to tension belts, tighten engine mounting bolts.
5. Purge air from compressor. Build up pressure in system, test for refrigerant leaks.
6. If none are found, back-seat SSV, then

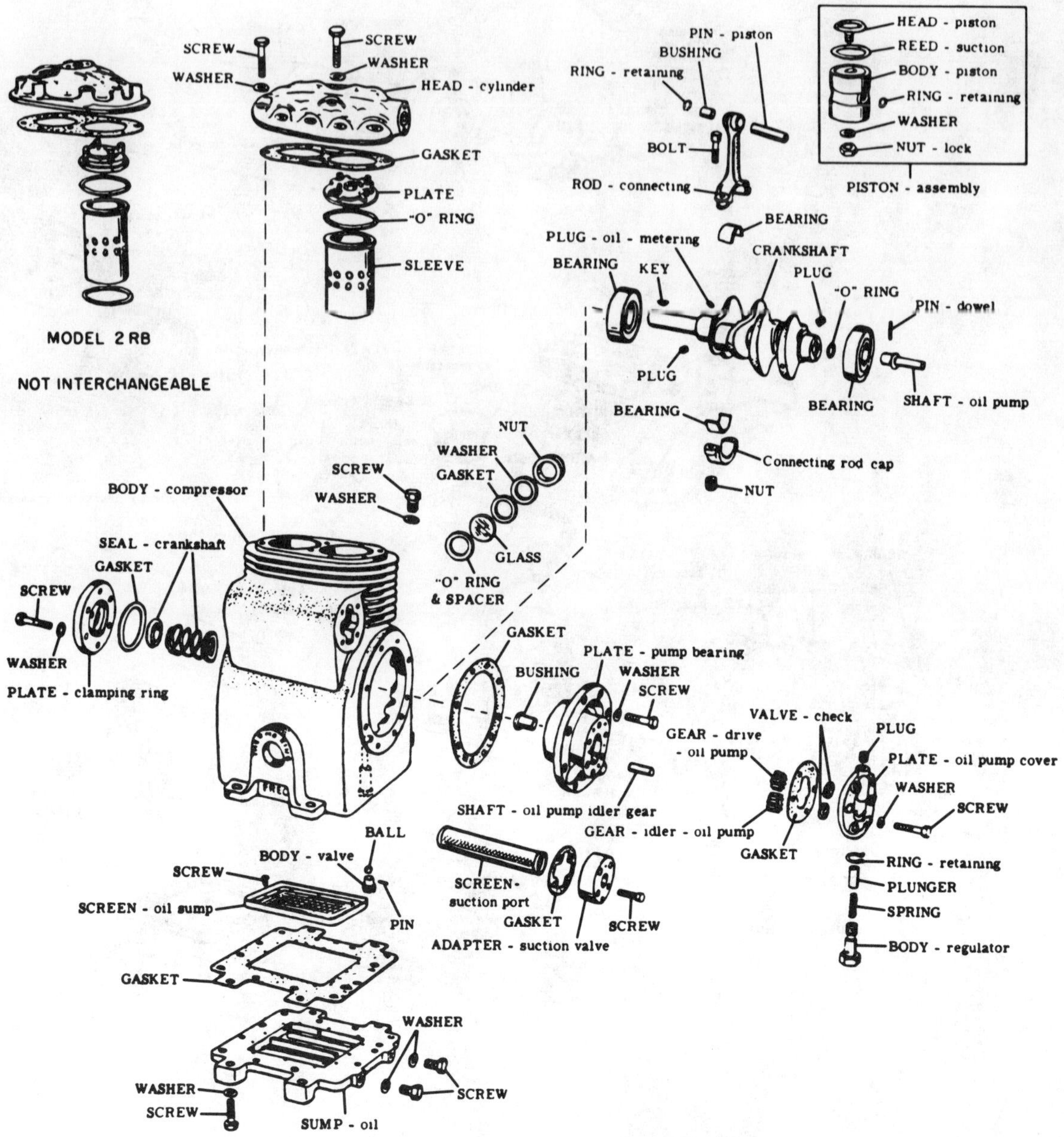

Fig. 5 Thermo King 2 cylinder compressor

close ½ turn.

7. *Back-seat DSV.* Open receiver outlet valve fully. Operate unit at least 30 minutes, then check oil level and add oil if necessary.

Compressor Overhaul-Disassembly

Following overhaul procedures cover both 2 and 4 cyl. compressors. Operations are on a 4-cyl. compressor, but 2-cyl. is identical. Clean and inspect parts during disassembly. If they are to be re-used on assembly, they should be wrapped in a clean cloth to prevent marring their surfaces. Disassembly, inspection, parts information are as follows:

1. *Oil Sump:* Drain and discard oil, remove sump and screen assembly. Inspect screen for damage, wash in kerosene or petroleum solvent. Clean oil passages to oil pump — passages are located in drain plug holes.

When replacing screen assembly on standard sump, make sure all 16 screws are properly tightened to keep dirt out of pump and bearings. With new-style deep sump, screen has heavy screen retained by four screws, screen is sealed to sump with O-ring. Old-style, standard sump cannot be used on new-

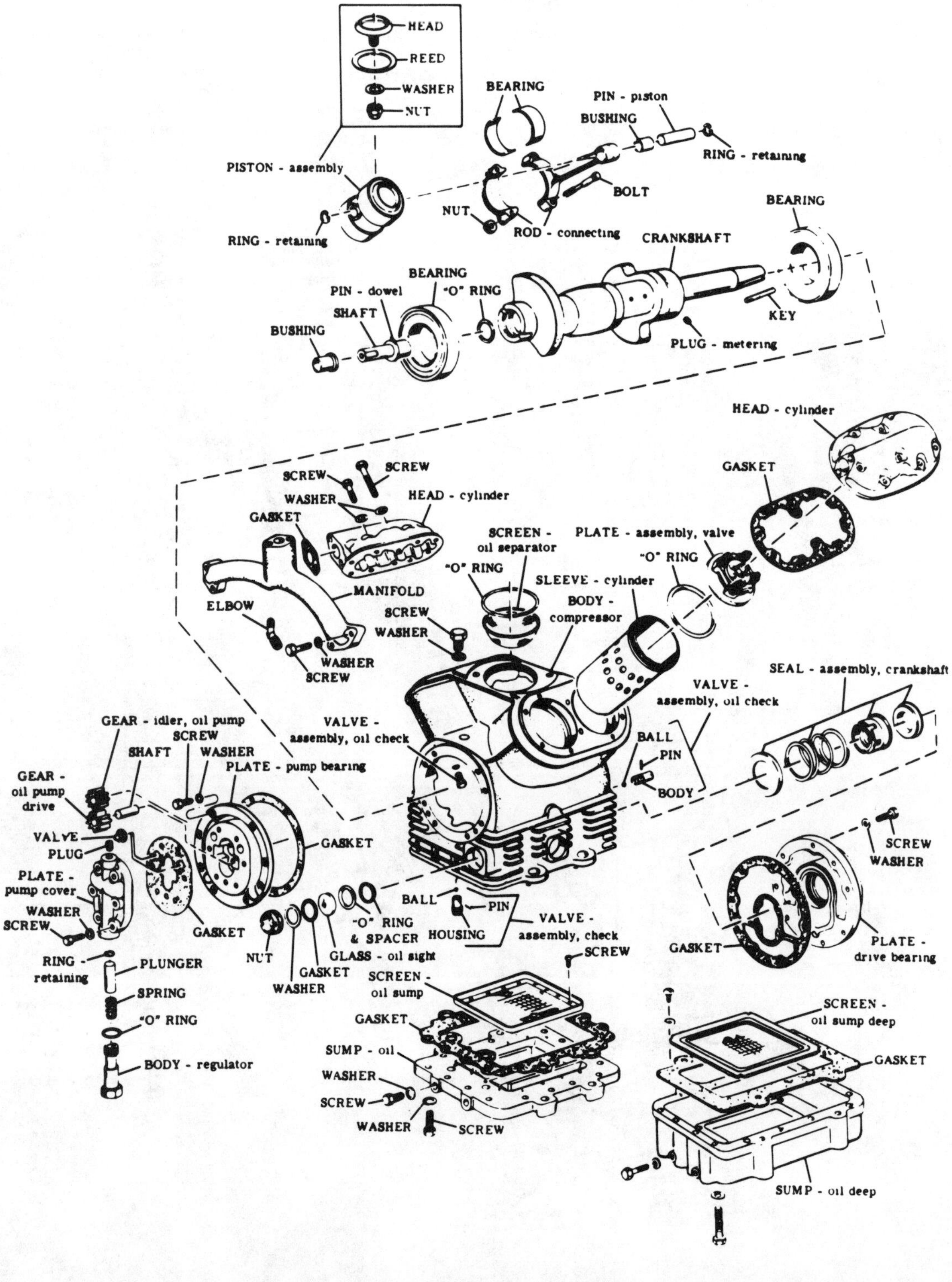

Fig. 6 Thermo King 4 cylinder compressor

style, finless body due to interference of screen frame with body.

2. *Manifold:* Remove manifold, discard gaskets.
3. *Inlet Screen:* Remove inlet port screen and gasket.
4. *Cylinder Head:* Remove head bolts and washers. Take off heads — if may be necessary to jar them loose with soft hammer or block of wood. Remove and discard large head gasket. Check heads for warpage.

Late-type heads and manifolds are constructed in three individual pieces — early-type had nine pieces. New-style heads, manifold can be adapted to 4L and 4A compressor by using 4 shorter capscrews with copper washers. Heads on 422 and 426 are the same, but manifold length is different. New style head is also used on 214 compressor.

5. *Valve Plates:* Remove valve plates, inspect for broken reeds or springs. If plate is warped, cracked, bent, has a broken spring, or if compressor does not hold a vacuum when pumped down, replace or rebuild valve plates. When repairing, use all parts in repair kit.

Early-type valve plate on 2RB, 4L, 4A uses flat gasket ring. Later type on 214, 422, 426 has fewer parts, is relieved to take O-ring. The two types are not interchangeable, but a modified valve plate is available for early models.

6. *Connecting Rods:* Use a ½ in., 12 point socket to remove nuts from rods. Crankshaft must be turned to bring counterweight to bottom to allow large end of rod to pass. Keep inserts in order so they can be installed in proper order.

Rods on all 4-cyl. compressors are interchangeable. It may be necessary to select rods which go on two outside cylinders (No. 1 and 4), because of counterweights riveted to inside of crankshaft throws. In set of four rods there should be two which will clear counterweights without reworking. Place straight edge along side of rod, and note which side of rod gives greatest clearance between face of bearing and web. Use two rods with greatest clearance to clear counterweights. If necessary, file web slightly to clear.

There are three types of rods. Fractured-type rod can be identified by the Allen screws. Do not mix different types of rods in one compressor.

7. *Piston and Sleeve:* Remove piston and sleeve as an assembly when rod cap is taken off. Keep each piston and sleeve together since they wear into matched sets. Replace pistons and sleeves if they are scored. Inspect and, if necessary, replace valve reed in top of piston.

Sleeves for 214 and 426 compressors have six ports in bottom row; 422 sleeve has 12 ports. Some early 214 models also have sleeves with 12 ports; replace these with six-port type if needed.

Piston head suction reeds were changed in thickness from .015 to .020 in. for greater reliability. Both have same part number. Always use .020 in. reeds when reassembling.

8. *Seal:* Remove seal plate by jarring it loose with hammer handle. Use screwdriver handle to push hard ring from seal plate. Seal ring is held in seal plate by O-ring on outer circumference. Replace entire seal assembly if either seal is damaged.

Remove spring and carbon ring assembly from crankshaft. Replace only as an assembly if any part is damaged.

Crankshaft seal is vented between the carbon ring, which revolves with crankshaft, and the stationary polished steel bushing which is seated in cover plate. Pressure exerted by seal spring holds two in contact. O-ring has been added to crankshaft extension on later 426, 422 models.

9. *Oil Pump Cover:* Remove pump cover plate, two check valves and gaskets. Replace cover if it is worn, cracked or warped. Be sure passage to oil pressure relief valve is cleaned.

Early 4-cyl. compressor pumps have single direction rotation, identified by arrow cast into cover. Later type rotates in either direction, is identified by oil pressure control being mounted on cover. Parts are not interchangeable. Model 426, 422 and 214 have .012 to .014 in. pump cover gasket, identified by white color. Gasket on other models is black.

Models 426, 422, 214 have sliding-piston type oil pressure relief. This can also be used on earlier models, which have spring-piston type pressure relief, but pump cover also must be installed. On late type with sliding piston, oil pressure is 45 psi, spring is 2 in. long. On early type pressure is 15 to 20 psi, spring length is 2 3/16 in.

10. *Oil Pump Gears:* Remove pump drive gear from shaft. Gears are driven by flat on shaft and flat in hole in gear. Take idler gear off shaft. If either gear shows wear, scuffing, cracks, replace both.
11. *Pump Bearing Plate:* Remove pump bearing plate bolts, install suitable tool, bump plate loose with sliding weight. Do not attempt to pry off since this can damage machined surface.
12. *Oil Pump Drive Shaft:* Drive shaft is driven through pin and slot in end of crankshaft, and sealed by O-ring in crankshaft. Remove pump drive shaft from end of crankshaft.

Drive shaft has been changed from key and keyway drive to a flat on shaft and flat in gear hole. New type can be installed on all models, but both gears must be changed when converting.

13. *Oil Return Check Valves:* Remove oil return check valves. If not removed, nylon balls will melt when case is heated to remove crankshaft.
14. *Oil Check Valves:* Remove these from bottom of compressor body. These valves permit compressor to function when rotating in either direction.
15. *Crankshaft Removal:* Removal is necessary only if shaft is bent or scored, or main bearings require replacement. Use torch to carefully heat area around seal plate. When area feels quite hot to the hand, bearing should slip out of housing. If heat is applied as shaft is pulled to rear, bearing will slip out. Do not attempt to drive bearing out of housing since this will damage bearing and housing.

Carefully remove crankshaft through rear of case. Do not spin bearings with air blast when cleaning and inspecting. To replace bearings, if necessary, use arbor press, with crankshaft properly supported. Apply pressure only to inner race — never to outer race. Crankshaft journals can be reground, if necessary, and undersize bearing inserts installed.

Long-stroke compressors have two add-on counterweights riveted to integral counterweight portion of crankshaft. When regrinding shaft, add-on weights must be marked, removed and reinstalled.

Procedure is to drill off rivet head, drive out rivet. Grind shaft to required size. Original connecting rod journal diameter is 1.372 to 1.3745 in. Replacement bearings are available in .005, .010, .020, .030 in. undersizes. Bearing-to-journal running clearance is .0008 to .0025 in. Tap holes in integral counterweights with $^5/_{16}$-10 UNC tap. Countersink add-on weights with 82 deg. tool to accept $^5/_{16}$-18 flat head screws. (This corrects original 90 deg. countersink). Reinstall weights, with screws, in proper locations. Cut off excess screw length if necessary. Stake threads with centerpunch to prevent screws from loosening.

16. *Oil Level Sight Glass:* Remove only if necessary. A spanner wrench is used to remove retainer nut.

Compressor Assembly Procedure

Use compressor oil only to oil all parts during assembly.

1. Install sight glass, if it was removed.
2. Install oil pick-up check valve assemblies in compressor body. Models 422, 426 have two in oil returns near seal, two in suction chamber adjacent to sleeves.
3. If crankshaft was removed, be sure oil holes are clear. Install oil channel plugs and metering plug.
4. Compressor crankshaft ball bearings must be dry type — do not use grease-packed automotive type. Install bearings on shaft, using arbor press with force applied to inner race. Bearing with widest race goes on drive end of shaft. Numbers on inner race face outward — they should be visible after bearing is installed.
5. Heat housing until bearing slips into place. Spin shaft to see that it rotates freely.
6. Install new O-ring in end of shaft. Install oil pump drive shaft in crankshaft. Install gasket, then set oil pump bearing plate in place by tapping it with soft hammer. Install pump drive gear on drive shaft, set pump idler gear in place. Crankshaft must spin freely.
7. Install oil pump cover gasket, two check valves, pump cover. Torque retaining bolts (with washer) to 12½ ft. lbs.
8. Assemble piston, rod and pin. Insert piston in matching sleeve. (Any new piston will fit any new sleeve).
9. Install piston-sleeve assembly in compressor. If crankshaft has counterweights, turn shaft so throw is at bottom to clear large end of rod. Torque rod cap nuts to 23 ft. lbs. Turn shaft to make sure counterweights clear rods.
10. Install valve plate O-rings on top of sleeves, and valve plates on sleeves. Set new head gasket in place. Install head with new washers. Pretighten bolts (or stud nuts) to 15 ft. lbs. Sequence is shown in illustration. Then retorque to 23 ft. lbs.
11. Install seal spring assembly, coated with compressor oil, on crankshaft. Notches in carbon member must be properly aligned with seal spring retainer.
12. Oil and install seal plate gasket. Thoroughly clean and oil seal ring. Be sure O-ring lies flat in groove and is not twisted. Install hard ring in seal plate.
13. Slide seal plate assembly into position on body, install bolts and washers and torque to 23 ft. lbs. Now make final check before installing sump — crankshaft should rotate freely.
14. Install oil sump. Gasket is mounted around screen. Install drain plugs, copper

washers in sump. Add oil to bring to same level as before disassembly. Install fill plug, suction port oil separator screen, and O-ring. Compressor is now ready for installation.

Component Service, Replace

Discharge Vibrasorber

This flexible bellows-type tube is used to prevent compressor vibration from being transmitted to refrigerant lines. Tube has wire mesh protective cover, and end caps which hold wire mesh in place and also serve as installation fittings.

To remove, release refrigerant charge. Remove DSV from compressor. Heat connections on vibrasorber until tubes can be withdrawn. To install, prepare vibrasorber and fittings for soldering. With DSV in place on compressor, set vibrasorber in place and solder joints. Use extreme care not to melt solder at wire mesh end cap attachement. Purge air, build up pressure, test for leaks.

Suction Vibrasorber

Same as discharge vibrasorber, but bellows tubing is covered with rubber to prevent frozen moisture from damaging bellows.

To remove, pump down system and equalize pressures. Melt solder on fittings to free them. When installing, wrap ends with wet cloth to prevent rubber cover from burning or loosening.

Condenser

To remove condenser, release refrigerant charge, detach front grille. Remove refrigerant lines, unfasten condenser mounting screws, and lift out of unit.

To install, position condenser in frame, and replace mounting screws. Set refrigeration lines in place and solder joints. Purge system, recharge and test for leaks.

Dehydrator

Dehydrator can be removed by pumping down system, disconnecting flare nuts at ends. End fitting is marked "IN" for proper installation.

Heat Exchanger

To remove, pump down system, detach mounting bolts that hold assembly to bulkhead. Unsolder fitting from suction header, and fittings to liquid and suction lines. Reverse procedure to install. Purge system, build up pressure, check for leaks.

Expansion Valve

This valve, controlled by sensing bulb in contact with suction line, meters liquid refrigerant into evaporator. On 10 and 20 models there is no bleed hole in valve, since valve must close off flow during off cycle. On 30 and 50 models, valve has bleed hole which has effect only on defrost and heat cycle. When three-way valve is shifted by solenoid, hot, high-pressure gas is discharged to evaporator coil and expansion valve closes, blocking refrigerant. Bleed hole allows liquid refrigerant back into system to work.

To remove valve, pump down charge, remove feeler bulb from clamp. Unfasten mounting bolts, detach valve from body. To install, reverse procedure. Feeler bulb must be in good contact with suction line or operation will be faulty. Purge, pressurize, and check for leaks.

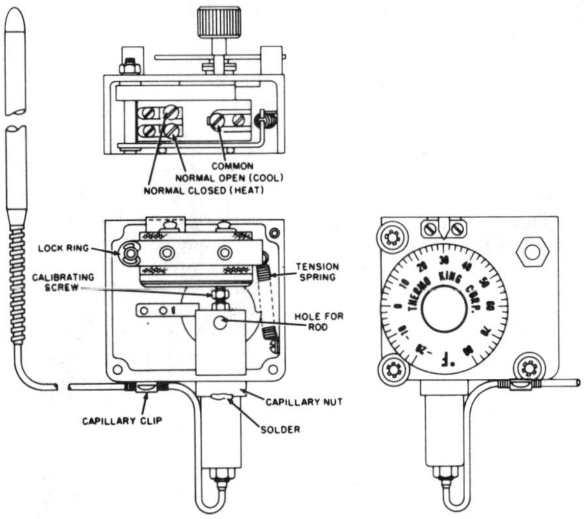

Fig. 7 Thermostat schematic

Evaporator Coil

To remove evaporator, pump down refrigerant charge, and detach defrost shutter control cable. Remove funnel, and disconnect hot gas line, liquid line and suction line. The coil can then be taken out of the housing.

Reverse the procedure to install the evaporator. Purge, build up pressure, test for leaks.

Suction Pressure Regulator Valve

Valve controls crankcase pressure at 28 to 30 psi to prevent overloading engine or electric motor during pull down. When evaporator pressure reaches pre-set pressure, valve opens fully to permit maximum pumping capacity.

Regulator is replaceable only as a unit. When replacing unit, be very careful when soldering not to overheat body of valve. Wrap body with a damp cloth, and confine heat to point of soldering.

To adjust valve, if needed, remove seal cap. Turn stem to right to increase pressure, or to left to reduce it. Replace cap.

Accumulator

Illustration shows mounting of tank adjacent to muffler, for increased efficiency on defrost or heat cycles. Tank can be replaced by usual procedure.

Suction Service Valve

To remove valve, pump down and equalize pressures. Remove screws that mount valve on adapter, unsolder valve connection to suction line.

To install valve, clean mating parts, apply flux sparingly, place valve in position. Have valve open to port and cap removed, or increase in pressure will cause solder to bubble out of joint. Heat fitting until solder is drawn in and a line of solder shows around connection. Set new gasket under valve and replace mounting screws. Purge, build up pressure, check for leaks.

Manual Defrost Valve

The valve can be removed by detaching the cap from the DSV port, releasing the refrigerant charge, and unsoldering the connections. After the tubing has been soldered to the new valve, purge, pressurize and check for leaks.

Three-Way Valve

When operating on refrigeration, hot gas from compressor enters valve and flows to condenser. Valve is held in this position by spring pressure aided by pressure of refrigerant.

On heating cycle, pilot solenoid valve opens and admits pressure from low side of compressor to operating chamber of three-way valve. This pressure and the discharge pressure move the piston to lift the valve off its seat. Hot gas line to evaporator is then open, and discharge line to condenser is shut off.

If valve sets require replacement, complete stem assembly must be installed. A special tool makes disassembling easier, and will not damage seats.

Place tool between lock ring and piston, as shown, and force ring away from piston. Piston can now be slipped off stem. Stem assembly is removed by sliding it away from piston end of valve. To reassemble, depress lock ring and slide piston into position.

Thermostat

Thermostat controls operation of unit to maintain preset temperature in vehicle body. Thermostat, when cooling is needed, switches current to solenoid which opens throttle. When demand is satisfied, thermostat switch opens, current to solenoid is interrupted, and engine speed drops to idle.

To replace thermostat power element, remove lock ring E from switch pivot, Fig. 7. Place rod in hole C, and use vise-grips to turn capillary nut clockwise to remove. Install by reversing procedure. Check thermostat calibration.

The switch can be replaced by removing lock ring E from switch pivot, unhooking tension spring D, sliding switch off switch pivot. To install new switch, first hook spring D, slide switch onto pivot, then replace ring E. Check calibration.

To recalibrate thermostat, should this ever be necessary, set it at 32 deg. Start unit and let it cycle on and off several times. Determine, using Simpson test thermometer, exact temperature at which unit shuts off. If unit stops at temperature higher than thermostat setting, as determined by test thermostat, turn calibration screw F out. If it stops at lower temperature turn screw in. Recheck after adjustment.

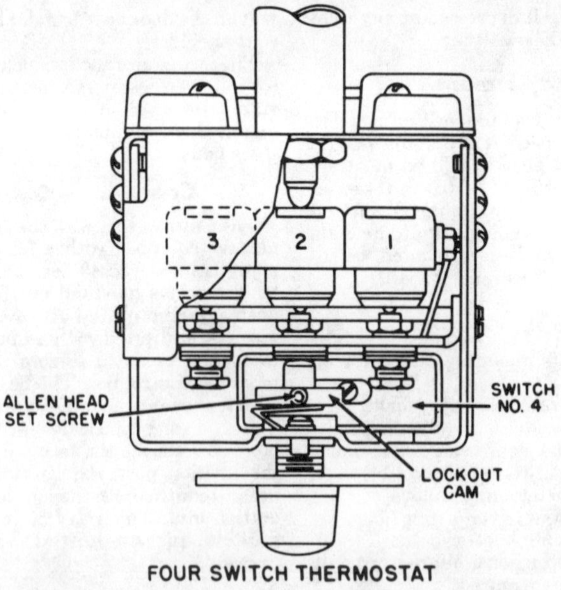

Fig. 8 Four switch thermostat

Idle-Run Controller

This mechanism works in conjunction with thermostat, and solenoid coil to run engine at idle speed during off cycle and at full speed during refrigeration cycle.

Adjust engine speed, on refrigeration cycle, to 2300 — 2400 rpm for most efficient operation. If engine speed is adjusted too low, clutch shoes will slip on drum and burn out.

To adjust speed, apply hand tachometer to crankshaft. Loosen control rod locknut and screw rod in to increase speed, or out to decrease it. Tighten locknut.

Defrost Damper

Damper prevents circulation of warm air through cargo space when unit is defrosting. When defrost relay is energized, one set of points completes circuit to damper solenoid, which pulls damper closed. When unit is off defrost, springs pull damper open.

Damper should be adjusted to close against rubber stops. Adjustment is made by screwing eyebolt into or out of plunger.

ENGINE SERVICE DATA

Make, Type	Wisconsin AENLD single-cyl. air cooled.
Oil	See lubrication chart
Fuel	Gasoline 70-75 octane, LPG optional
Valve Clearance	Intake .008 in., exhaust .016 in., cold
Breaker Point Gap	.020 in.
Ignition Timing	Set to 5 deg. BTDC. Automatic advances 15 deg.
Spark Plug Gap	.030 in.
Starting	Automotive-type electric starter
Charging	Motorola alternator. Output 35 amp at 15 volts
Engine Speed	Full 2300-2400 rpm, Low 1000 rpm

THERMO KING MODEL UWD

The UWD is a heavy duty, diesel powered, two component refrigerating and heating unit for large trailers.

Components

The condensing unit, mounted under the trailer, contains the diesel engine, 4-cyl. compressor, condenser, generator, batteries and controls for the operation of the unit. The evaporator (refrigeration) section, mounted inside the front wall of the trailer, contains the evaporator coil, fan, heat exchanger, expansion valve and air switch.

The two components are connected by three lines. A suction line returns refrigerant gas from the evaporator. A liquid line carries the liquid refrigerant to the expansion valve. A defrost and heating line carries the hot gas during the defrost operation or when the system is operating on heating.

Operation

Temperature control is fully automatic. The unit operates at high or low speeds, on either cooling or heating, as called for by the thermostat. The special thermostat will not allow the unit to switch to heating when hauling frozen loads.

Referring to the refrigeration and heating cycle diagrams, it can be seen that the system is similar to that used on the MWC model, and it operates in the same manner.

Service Operations

Troubleshooting, adjustments, repairs and other service procedures are the same as for the MWC model, except for the following items:

Pumping Down Low Side (Evaporator)

1. Close outlet valve on receiver tank.
2. Operate unit on COOL until suction pressure gauge reaches 10 in. vacuum, then stop unit.
3. Pressure will rise slowly as refrigerant oil evaporates. When pressure will remain at 2 to 3 psi, close both compressor valves. Never open system unless compound gauge indicates a slightly higher than zero pressure.

Pumping Down High Side (Condenser)

1. Close outlet valve on receiver tank.
2. Cover condenser to raise head pressure to 170 psi.
3. Operate unit until it is maintaining 10 in. vacuum.
4. Stop unit. Quickly close receiver inlet valve.

Condenser Removal

1. Pump down high side of system.
2. Remove condenser grill inside panels.
3. Unfasten clamp holding condenser return line to receiver tank.
4. Heat large end of tube reducer at condenser outlet until solder melts. Draw tube back to break connection.
5. Block up condenser coil if unit is mounted on trailer. Remove condenser mounting bolts. Bolts thread into nut strips located inside front upright of frame.
6. With condenser free of frame, heat sweated connection on elbow at condenser inlet line. As solder melts, move condenser away from frame to break connection.

Keep openings covered with masking tape during servicing. Install condenser by reversing removal procedure.

Evaporator Removal

1. Pump down the unit.
2. Disconnect defrost line at distributor, and suction line at header.
3. Remove screws attaching funnel to frame.
4. Evaporator can then be drawn out of frame.

To install evaporator reverse above procedure. Then release refrigerant charge back into system, inspect for leaks, and check for sufficient refrigerant in system.

Throttling Valve

Throttling valve controls suction pressure in compressor during operation. Pressure on springloaded bellows causes piston to move to a position that limits suction pressure to 30 psi. When evaporator pressure is less than valve setting, spring pushes valve fully open.

The valve should be lubricated by adding compressor oil through the oil cup. The pressure can be adjusted by adding or removing spacer washers behind the spring. Each washer effects a pressure change of about 2½ psi.

Accumulator

Oil metering hole in outlet tube of accumulator can be cleaned with wire inserted through plug opening in side of tank. Unit must first be pumped down.

Four-Switch Thermostat

Fourth switch on this thermostat opens No. 2 and 3 switches to prevent unit from switching to low or high speed heat when thermostat setting is 10 deg. or lower. Unit continues to run on low speed cool. Accuracy of thermostat adjustments affects fuel economy, as well as temperature control, Fig. 8.

Adjusting Sequence

1. Place sensing element of suitable test thermometer in trailer close to thermostat bulb location. Thermometer on unit lags too much for this test purpose.

2. Turn thermostat dial to zero deg. Start unit. Lower box temperature to about 30 deg.
3. Turn dial down (clockwise) until unit goes into high speed cooling.
4. Now turn dial up very slowly until unit just changes from high speed to low speed cooling. Note exact position of pointer when this occurs, but do not make any adjustments at this time.
5. Continue to turn dial up very slowly, while watching for point at which COOL indicator light goes out and HEAT indicator light comes on. This switching of lights indicates unit operating change from low-speed cool to low-speed hot. Note pointer position when lights change.
6. Temperature differential between step 4 and 5 should be 2.5 to 3 deg. If it is not, loosen locknuts on switch No. 1 and adjust this switch accordingly. Never adjust switch No. 2. Repeat steps 3, 4, 5 to confirm specified differential.
7. When unit switches from low-speed cool to low-speed heat, note exact position of pointer.
8. Continue to turn dial up very slowly, and note exact position of pointer when unit switches to high-speed heat. The temperature differential between low-speed heat and high-speed heat should also be 2.5 to 3 deg. Adjust switch No. 3 if a correction is necessary.
9. Repeat steps 7 and 8 to recheck differential between switches No. 2 and No. 3. Switches are now adjusted properly in relation to each other, but thermostat dial must be calibrated.

Thermostat Calibration

1. After above steps, start unit and allow it to cycle at least three times at 30 to 32 deg.
2. Read exact temperature on test thermometer.
3. Loosen set screw on knob on thermostat dial, and adjust dial to correspond to test thermometer reading.
4. Adjust unit thermometer to correspond with test thermometer.

No. 4 Switch Lockout Cam Setting

1. With trailer temperature above 30 deg. turn thermostat dial up.
2. Now turn dial down (clockwise) very slowly until switches No. 1, 2 and 3 can be heard to click.
3. Continue to turn dial down and note when cam actuates switch No. 4. This occurs when the fourth click is heard. At this point the dial should read 10 deg.
4. The setting of switch No. 4 can be adjusted by loosening the Allen set screw, and moving the cam in relation to the shaft. The cam must depress the switch arm at 10 deg., and keep it depressed at any temperature setting below 10 deg.

Belt Adjustment

Drive belts must be kept properly adjusted to maintain alternator speed. If alternator speed is too low, voltage and phasing will be incorrect and evaporator fans will not be driven at proper speed. With unit running on refrigeration and high speed, alternator speed should be 3600 rpm. This can be checked with hand tachometer in contact with brush end of shaft.

Drive Belt Adjustment

1. Loosen four bolts mounting alternator on bracket.
2. Loosen locknuts on adjusting eyebolts.
3. Tighten lower locknuts on adjusting bolts until belt can be depressed ½ in. under thumb pressure at center of span between alternator and water pump pulleys.
4. Tighten top locknuts on adjusting bolts.
5. Tighten alternator mounting bolts.

On units with Westinghouse alternators, only steps No. 1, 3 and 5 are required.

ENGINE SERVICE DATA

Make, Type	Mercedes Diesel OM636
Oil	See lubrication chart
Fuel	Diesel #2. Specification ASTM-D975/53T
Valve Clearance	Intake .008 in., Exhaust .006 in., cold
Fuel Timing	32 deg. BTDC
Firing Order	1-3-4-2
Starter	Automotive type
Generator	Automotive type
Cooling	8 qt. 50/50 ethylene glycol 4 lb. filler cap
Engine Speed	Full 2,200 rpm, low 1,200 rpm

THERMO KING MODEL TG

The model TG is designed for maintaining low temperatures in small delivery trucks. The three-piece system consists of an evaporator, condenser, and a compressor driven by the truck engine. On some installations the compressor is mounted underneath the body and driven by a power take-off on the transmission. On TG 20 models a standby electric motor is installed additionally to operate the system when the vehicle is not running on the road.

Components

The condenser section contains the condenser coil, fan, sight glass, receiver tank and dryer. It can be mounted behind the driver's seat on the floor board, under the truck, or bolted to the skirt.

The evaporator section includes the evaporator coil, thermostat, fan, heat exchanger, expansion valve and defrost equipment. It can be mounted at various suitable locations in the truck body.

Controls — Operation

The controls vary with the equipment. TG-10 models, equipped with a fan defrost for temperatures down to 40 deg., are controlled only by two relays mounted on the firewall, along with a dashmounted off-on switch. TG-10D models, equipped with a hot-gas defrost for temperatures down to zero deg., use a dashmounted control box. The box contains a relay which prevents accidental exhaustion of the battery during vehicle down-time. Since this relay also removes the heavy current load from the ignition post, the ignition switch can be used to control both the ignition and the refrigeration unit. This relay is also used on other models, but is mounted separately on the firewall.

Service Operations

Refrigeration and compressor troubleshooting and service operations are the same as those for the MWC model, except for the compressor drive clutch which is covered below.

Compressor Drive Clutch

The clutch is self-adjusting for wear and normally does not require service. On transmission power takeoff drives, the clutch may need a slight adjustment to eliminate drag between the shoe and the pulley when the clutch is de-energized. This adjustment is made by disassembling the clutch and adding the necessary shims to the seat at the bottom of the pulley shaft.

If the clutch fails to operate, check to make sure there's a 12 volt input to the field lead wire. If the voltage supply is satisfactory, change the clutch. Repairs are not recommended. To remove the clutch from the compressor, disconnect the lead wire. Next, remove the cap screw and washer securing the rotor pulley to the compressor shaft. Then take off the clutch field by removing the cap screw holding it against the compressor end plate.

TRANSICOLD

These transport refrigeration systems are available in various configurations for different types of installations. Models 230 and 250 are the two-section type, while models 640 and 650 are the unit "nose-mount" type. Carrier 3 cyl. compressors are used with 230 and 640 models, while 250 and 650 models have 6

THERMO KING ENGINE SPECIFICATIONS

Manufacturer	Model	Fuel	Cyls.	Bore-Stroke	Disp.	C/R	Brake P.	Oil Capac.	Oil Pres.	Cooling
Isuzu	C/201/PT	Diesel	4 OHV	3.268 × 3.622	121.6	20	32 @ 2200	18	65-95	Liquid
Mercedes	OM636	Diesel	4 OHV	2.9 × 3.9	107.8	19	24 @ 2200	6.5①	50-100	Liquid
Mercedes	OM636III	Diesel or LPG	2 OHV	2.9 × 3.9	53.9	19	13 @ 2200	9	50-100	Liquid
Waukesha	ICK	Gas. or LPG	4 L-head	2.5 × 3.125	61	5.7	16.5 @ 2200	3②	5-30	Liquid
Onan	CCK	Gas. or LPG	2 L-head③	3.25 × 3	50	5.5	13.1 @ 2400	8	20-40	Air
Onan	L25	Gas. or LPG	1 Vert.	3.25 × 3	25	7	7.5 @ 2400	4	50	Air
Wisconsin	AENLD	Gas. or LPG	1 Vert.	3 × 3.25	23	5.9	7.4 @ 2350	2.25	NA	Air

①—18 optional. ②—8 optional. ③—Horizontally opposed.

cyl. compressors. Power is supplied by either of two engines, both 4-cyl. diesels — the Perkins 108 cu. in., and the Mercedes 107.8 cu. in.

Components

Models 230 and 250 installations are in two sections. The condensing, or high side, unit is mounted under the trailer. This unit consists of the engine and accessories, generator, compressor, condenser, receiver and control panel.

The evaporator, or low side, section is available in several designs which permit it to be mounted at various locations on the trailer body. It consists of the evaporator coil, expansion valve, circulating fans and motors, and automatic defrost control.

On 640 and 650 models, the unit is mounted on the upper portion of the front wall of the trailer. In one configuration, the evaporator and condensing sections are permanently connected together by an insulating bukhead, and the bulkhead and evaporator fit into a rectangular hole in the wall of the trailer. With the other configuration, the condensing section is mounted on the front of the trailer wall, and the evaporator section behind it.

With either configuration the evaporator section contains the evaporator coil, expansion valve, heat exchanger, evaporator fan and motor, defrost pressure switch, and thermostatic switch used to terminate defrost.

The condensing section consists of the engine and accessories, generator, compressor, condenser, receiver, suction accumulator, dryer, crankcase pressure regulator (CPR) valve, and control panel.

Controls — Operation

The main control system is powered by a generator directly driven off the engine crankshaft. At the engine's lower controlled speed of 1,350 rpm, the generator produces 135 volt 60 cycle AC current on 230 and 250 models, and 135 volt 65 cycle 3 phase AC current on 640 and 650 models. At the upper controlled engine speed of 1,900 rpm, 190 volt 95 cycle current is produced on 230 and 250 models and 180 volt 95 cycle current on 640 and 650 models.

The temperature controller, or "brain," senses the temperature inside the trailer and automatically controls the cool-heating functions of the system, as well as the engine speed of the unit. The speed control solenoid, mounted on the engine, is connected by linkage to the fuel injection pump speed lever. Engine speed is maintained at either 1350 or 1900 rpm by energizing or de-energizing this solenoid.

A control valve, operated by a solenoid, controls the flow of refrigerant from the compressor to the condenser for cooling, and also to the hot gas line for heating and defrost. When the solenoid valve is de-energized the valve is in the cooling position.

A three-pole two-position relay, called the heating relay, normally has three poles open and three poles closed. This relay, when energized, stops the condenser fans, turns on the heating indicator light, turns off the cooling indicator light, and energizes the control valve solenoid to put the unit into the heating cycle. The heating relay can be energized either by the temperature controller heating contacts, or by a pole of the defrost relay.

A four-pole two position relay, called the defrost relay, normally has four poles open and three poles closed. When energized, this relay switches the circuits to put the unit into the defrost cycle. The defrost relay can be energized either by the defrost pressure switch, or by the manual switch.

A defrost pressure switch, located on the evaporator, senses the pressure difference created by the evaporator fans blowing across the evaporator coils. When the formation of ice on the coils is sufficient to cause the pressure differential to increase to a certain point the pressure switch closes. This energizes the defrost relay and starts the defrost cycle. The pressure switch starts the cycle, but it does not control the length of time it continues, or terminate the cycle.

When the defrost relay is energized, one set of contacts completes a circuit to energize the heating relay. Other contacts turn on the defrost indicator light, and also open the circuit to the heating indicator light to keep it from going on when the heating relay is energized. The defrost relay also stops the evaporator fan, and opens the circuit to the speed solenoid to put the unit into high speed.

Once energized, the defrost relay is held in a locked position by its own holding coil until defrosting is completed. This timing is determined by the defrost thermostat which is attached to the return bend on the evaporator coil and thus senses suction line temperature. The thermostat is electrically in series with the defrost relay holding circuit. When suction line temperature reaches about 65 deg. (45 deg. on 230 and 250 models) the thermostat opens the relay holding circuit and terminates the defrost cycle.

An electric heating band on the refrigerant receiver, energized only on heat and defrost cycles, increases the efficiency of these cycles. A heater passes hot gases through a coil located at the bottom of the drip pan to prevent the formation of ice during the defrost cycle.

Some units have compressor "unloaders." On 3 cyl. compressors, two cylinders are unloaded by energizing a solenoid valve. On 6 cyl. compressors, solenoid valves are located on the right and left compressor heads.

Refrigerant Checks — Service

Service valves are conventional type. If valve stem turns hard, slightly loosen stem packing.

Pumping Down System

To open, or service, the system pump most of the refrigerant into the condenser and receiver by using the following procedure.

1. Turn suction service valve (SSV) counterclockwise to back-seat it and close off gauge connection. Install pressure gauge on SSV, and open valve two turns.
2. Close valve in liquid line.
3. Operate system on refrigeration until unit builds up 10 in. vacuum, or compressor stops on low pressure switch. Electrically disconnect system to prevent accidental starting.
4. Open liquid line valve and solenoid valve momentarily until gauge shows 1 to 2 psi positive pressure. Most solenoid valves have stem for manual opening of valve.
5. Front-seat, or close, compressor discharge service valve to trap refrigerant in high side of compressor.
6. After service operation has been completed, dehydrate or purge air from line and parts exposed to atmosphere. This can be done by momentarily emitting refrigerant through parts, then reassembling. Open, or back-seat, service valves, test for leaks.

Adding Refrigerant — Partial Charge

1. Remove plug or cap from SSV.
2. Connect one end of charging line to drum of refrigerant. Position drum to draw off vapor refrigerant only.
3. Connect other end of charging line loosely to SSV gauge port, purge hose to eliminate air, tighten valve fitting. Start compressor.
4. Turn SSV off back-seated position to admit refrigerant.
5. Fill system only until lower sight glass of receiver is covered with refrigerant. Do not overfill. If a receiver is not used, admit refrigerant until continuous flashing in liquid line sight glass changes to occasional flashing. Check for non-condensibles and purge as required. Level is correct when flashing stops in liquid line.
6. Check level after unit has been operated on cooling for 30 minutes by raising discharge pressure 10 psi and observing sight glass. After unit has been correctly charged, back-seat service valve, close refrigerant drum valve, stop compressor, remove charging line, install service valve cap.

Adding Refrigerant — Full Charge

1. Dehydrate entire unit and leave in deep vacuum.
2. Connect drum of refrigerant in same manner as for partial charge.
   Note the weight of refrigerant and drum.
3. Open liquid line charging valve and allow liquid refrigerant to flow into system until specified weight of refrigerant, as observed on scales, has been added. It may be necessary, due to pressure rise in high side of system, to finish charge unit through SSV in gas form, in same manner as for partial charge.
4. Close liquid line charging valve and refrigerant drum valve. Remove charging line from valve and install valve cap.
5. Operate on high-speed cool for 20 to 30 minutes. Check compressor oil level and refrigerant level. Inspect refrigerant in moisture indicator. Liquid should be free from bubbles, and the level should be between the upper and lower sight glass in the receiver.

Compressor Service

Compressor Oil

Oil Level

Oil level should be approximately in center of compressor sight glass during operation. After 20 minutes steady operation, foaming in sight glass indicates refrigerant flooding through suction or excessive oil charge. Foaming only at start-up is normal due to refrigerant leaving oil.

Adding Oil

To add oil, pump down unit, back-seat suction service valve (SSV), and remove valve cap. Remove SSV plug and install short length of flexible tubing that reaches to bottom of a small container of compressor oil. Do not overcharge — container should hold only required amount of recommended oil. Recommended oils are: Carrier — PP36-2; Sun Oil Co. — Dual Inhibited Suniso 4G and Texas Oil Co. — Capella DI.

Momentarily crack open SSV just enough to purge air in tube through oil in container.

Now pump down unit to a vacuum. Carefully crack open SSV again to draw oil into compressor, but do not draw in air.
Order cans small enough so they can be used up when opened. Oil exposed to air more than a few hours should not be used. Oil may also be poured directly into compressor through oil filler passage in oil pump.

Removing Oil

To remove overfill, reduce crankcase pressure to 1 psi. Loosen drain plug just enough to allow oil to seep out around threads of plug. Then retighten drain plug and restart compressor.

Compressor Removal

Pump down unit. Front-seat SSV and DSV. Do not tamper with compressor terminals before relieving crankcase pressure. After pressure has been relieved, disconnect high-low connections.

Remove two mounting bolts each from SSV and DSV. Disconnect electrical wiring from compressor terminal box. Remove nuts and bolts from compressor base and mounting brackets. Slide compressor out of unit. Compressor weighs several hundred pounds; use proper equipment.

Compressor Repairs — 3 Cyl. 5F Series

Parts are covered in the order in which the compressor is disassembled.

Oil Pump

Drain oil below level of pump end bearing head, and remove bearing head. Check end play of oil pump rotor. If end play exceeds .0025 in. allowable maximum, correct by repositioning oil pump bushing in bearing head. (See below).

Turn rotor. If there is more than a slight drag, remove pump cover and check all parts.

Oil Pump Bushing

New bushing, if needed, is pressed into bearing head with small end toward pump. Locate bushing oil groove at top of bearing head.

To position oil pump bushing in bearing head, place a .001 in. shim against bushing, and between bushing and pump rotor, and install pump. Complete assembly of pump, making sure assembly is flush with cover plate surface, but does not protrude beyond bearing head surface.

Seat bushing against shim by tapping bushing with cylindrical tool. Tool can be fabricated.

Disassemble pump and remove shim. Reassemble pump and check for binding. Install bearing head, line up tang on pump rotor shaft with slot in end of crankshaft, make sure pump has proper rotation.

Refill compressor oil to proper level. Start compressor and observe oil pressure. If oil pressure does not reach 45 to 55 psi above suction pressure in 8 to 12 seconds, stop compressor and recheck pump.

Oil Return Check Valve

Valve allows oil to return from suction manifold to crankcase. Valve is normally open — closes when crankcase pressure becomes higher than suction pressure. Two disc-type check valves, located beneath partition between suction manifold and crankcase, can be removed through bottom cover or pump end of compressor. Check to see that flutter valve is not sticking, and that it seats tightly.

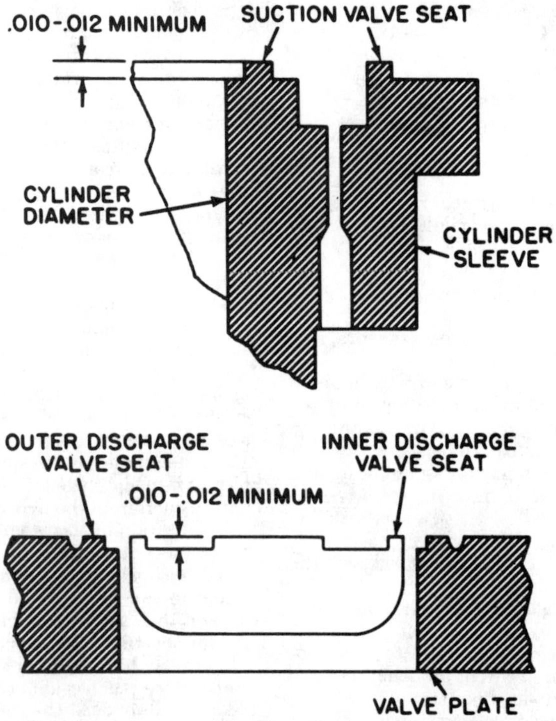

Fig. 9 Suction and discharge valve seat height

Pressure Relief Gasket

Gasket in manifold relieves refrigerant pressure from high to low side when difference in pressures reaches 400 psi. Gasket should not leak. Replace if defective, or blown from excessive pressure.

Oil Screen

Located in compressor crankcase. Screen is accessible for cleaning or insepction through hand hole cover or bottom plate.

Cylinder Heads

Remove cylinder heads. Check heads for warpage, cracking, damage to gasket surfaces.

Suction and Discharge Valve Assembly

Valve can be disassembled after removal of head. Loosen cap screws holding discharge valve seat to valve guide, and screws holding valve guide to valve plate. Now unfasten screws holding valve plate to cylinder block, remove plate from block, and guide from plate.

Inspect suction and discharge valve discs and seats for cracks or wear. On both valve discs depth of wear below face should not exceed .005 in. Height of A and B in Fig. 9, should not be less than .010 in. If it is less, replace parts. Check cylinder deck valve stops for uneven wear. If stops are worn, replace cylinder sleeves. Replace valves if cracked or worn. If seats are worn, replace complete valve plate assembly.

To reassemble, pistons must be below tops of sleeves. Place suction valve springs in valve plate recesses. Large spring coil should make full contact with bottom of recess. Set valve disc on springs, press disc into valve plate recess, slide retainer clips into place. Clips should not cover valve lifter springs or pins. Bolt valve plate to cylinder block. Remove clips.

Place discharge valve springs in valve guide spring recesses. Place disc over springs, fit inner seat in place over disc. Hand tighten bolts holding inner seat to guide. Place guide assembly on valve plate. Then tighten bolts, bend tabs on locks, and install heads.

Cylinder and Unloader Sleeves

Sleeves are removed after cylinder head, valve assembly and end bearing head have been removed. Whenever a cylinder sleeve or valve plate is replaced, use a new suction valve disc.

Move piston in mid position, position sleeve puller on top of piston. Expand puller in sleeve, turn crankshaft to force sleeve upward for removal. Disassemble unloader sleeve, pins and springs.

Break glaze when new rings are used with old sleeves. To install sleeve, move piston to top and stagger ring gaps. With a turning motion, work sleeve over pistons and rings. Seat sleeve in suction manifold partition and cylinder deck recess. Then rotate sleeve so any two suction valve springs are aligned with two unloader springs and they lie equal distances from longitudinal axis of compressor. In this position lifter pins line up with suction valve springs. Do not operate compressor with valve plate off unless sleeves are secured with cap screws.

Piston, Connecting Rod Assembly

To remove assembly, first take off cylinder head, valve plate, and bottom plate. Mark, then remove connecting rod caps. Remove piston, rod and sleeve together by pushing assembly up through cylinder deck. Keep piston in sleeve while removing.

Piston pins are push fit. Ring gap is .007 to .017 in., side clearance is .001 in. Maximum allowable wear is: Rod bearing .002 in., crankpin .003 in., cylinder bore .003 in., piston diameter

.003 in., ring end gap .030 in., ring side clearance .003 in. Do not file bearing caps.

Install piston, rod, sleeve at same time. Turn rod and install cap so chamfered sides are against radius of crankpin. Small knobs on rods and caps must be on same side of journal.

Capacity Control System

In loaded operation, oil pressure in power element increases as oil enters control circuit from oil pump. Piston is forced upward, pivoting lifter fork downward. Lifter pins drop, allowing suction valve to seat and load cylinder.

In unloaded operation, oil pressure in element drops when oil bleeds through external solenoid valve to crankcase. As oil pressure drops, piston moves downward, lifting fork pivots upward and lifts pins. This raises suction valve off its seat, unloads cylinder.

Unloader power elements can be taken off, after removal of piston and sleeve, by removing two Allen-head cap screws that hold it in position. Disassemble power element and inspect parts. After assembling, check unloader fork height.

Crankshaft

Factory-reground crankshafts and bearing sets are available in .010 and .020 in. undersizes.

Pump End Main Bearings

To disassemble, remove pump end bearing head. Maximum permissible wear is .002 in. on bearing diameter, .003 in. on journal diameter. If bearing is worn, remove bronze bearing washer, and use a puller to remove bearing.

When installing new bearing, line up oil holes. Press bearing into place with suitable equipment.

Seal End Main Bearing

With crankshaft removed, use suitable tools to remove and install bearing. Wear limits are same as for pump end bearing. When installed, edge of bearing should be 1/32 in. below surface of bronze bearing washer.

Crankshaft Seal

Oil seal in current compressors is a sleeve-type rotary bellows seal with integral seal seat. It can be used as replacement for all earlier types. Make sure crankshaft is free of burrs before installing new seal.

Compressor Repairs — 6 Cyl. 05D Series

Disassembly Operations

Remove cylinder head. Free valve plate from cylinder deck by using discharge valve hold-down cap screws as jack screws through tapped holes of valve plate after valve stops and valves have been removed, Fig. 10.

Remove bottom plate. Remove connecting rod caps, then push pistons up as far as they will go without having rings extend above cylinders. Don't bend copper tube which extends out when bottom plate is removed. If necessary, remove oil pressure relief valve.

Remove oil pump cover cap screws. This frees oil feed guide. Remove oil pump drive segment screws first, then remove bearing head. Take off snap rings to remove pump vane and related parts. Push pump rotor out of bearing head — force from bearing side and remove from opposite side. New bearing head complete with pump is available if needed.

Remove shaft seal cover plate. Be careful not to damage shaft. To remove counterweight and shaft, take out top and bottom set screws, drive out roll pin so it emerges from mass end of counterweight. Slide counterweight off shaft, and remove shaft.

If piston rings extend beyond cylinder tops, pistons and rods can be pulled out through bottom plate opening after rings are compressed.

Main bearing is precision bored in place and field replacement is not recommended. Pump end bearing is integral part of aluminum bearing head assembly. If end bearing is damaged, replace assembly.

Maximum allowable wear limits are: Main bearing journals .002 in., main bearings .002 in., rod bearings .002 in., thrust washers .025 in., piston .002 in., bore .002 in., ring gap .025 in.

New piston rings should have .005 to .013 in. end clearance, .001 to .002 in. side clearance

Assembly Operations

The assembling can be done in the following order:

Assemble oil pump in bearing head. In reassembling pump, insert pump rotor into bearing head from side opposite bearing, and with retaining ring in place on rotor. Chamfered edge of retaining ring should face in. Compress ring to fit rotor and ring into proper positions. Insert one pump vane, flat end in, in vane cylinder in bearing head, then insert plunger vane spring and guide. Insert retaining ring with ring pliers. Force spring guide down to compress plunger spring and allow retaining ring to lock in groove. Repeat for other vane.

Push piston assemblies from inside of crankcase through cylinders. Install shaft, assemble connecting rods. Chamfer on rod faces away from adjoining rod.

Install counterweight by sliding it on shaft with chamfered edge toward pump end. Position mass of weight 180 deg. opposite slot in shaft. Tap roll pin radially through hole in mass end until 3/16 in. long Allen set screw can be started. Complete setting by turning in screw. Roll pin must penetrate through weight until it meets shoulder of shaft. Longer set screw goes in narrower portion of counterweight in line with slot on shaft.

Install bearing head assembly. Bolt end cover to crankcase.

Replace shaft seal. Shaft and neoprene bellows should be lubricated with refrigerant oil. When replacing seal, pump end bearing head must be in place or seal end bearing washer may be pushed off retaining dowel. Seal may leak slightly immediately after installation, but leak should stop after short period of operation.

Assemble valve plates and cylinder heads. Before installing suction screen and bottom plate make sure shaft will turn by hand.

Compressor Capacity Control

If capacity control does not operate properly, first check solenoid valve coil located on side of cylinder head. Energize solenoid, listen for click as solenoid valve stem strikes at end of its travel. Voltage to coil must be correct.

If valve is operating, remove cylinder head. Check and clean out bleed orifice with a soft copper wire to prevent damage to orifice. Clean strainer in head. Remove any foreign material behind piston ring — the piston should slide freely in cylinder head.

Gasket must be properly installed — one bolt hole is offset slightly. Torque head bolts to 30 ft. lbs. Excessive torque may cause control piston to bind.

Component Service — Replacement

Thermal Expansion Valve

Thermal expansion valve automatically controls flow of liquid refrigerant to evaporator according to changes in superheat of refrigerant gas leaving evaporator. Valve maintains relatively constant superheat in gas leaving evaporator, regardless of suction pressure. Thus the valve has two functions — it is an automatic expansion control, and it prevents return of liquid refrigerant to compressor. Unless valve is defective, it seldom requires maintenance. Cage assembly and power assembly are matched parts, and should be replaced together.

Adjustment

Rate of refrigerant flow through valve is controlled by adjustment located inside valve. To lower temperature of evaporator, turn adjustment counterclockwise to increase flow. Turn clockwise to decrease flow and raise temperature. Do not turn adjustment more than two turns at any one time. Wait for 30 minutes for adjustment to take effect, then readjust if necessary.

Valve Removal

Pump down unit. Loosen flare nut, disconnect equalizing line from thermostatic expansion valve. Remove thermal bulb from suction line or well. Remove cap screws, lift off power assembly, remove cage assembly, use torch to heat sweated connections to valve until solder is melted. Remove valve. The valve body rarely requires removal; remove only if necessary.

Installation

Sweat valve body to line. Remove power assembly and cage assembly before brazing. Insert power and cage assemblies loosely. Two lugs on cage assembly fit into grooves in power assembly. Do not force. Purge a little refrigerant to clear lines, and tighten bolts. Fasten equalizing tube coupling nut to expansion valve. Purge a little more refrigerant to clear line. Replace thermal bulb in position.

Dryer

Removal

To replace dryer, first pump down system. Have replacement dryer ready, but do not remove seals.

Remove clamp holding dryer to bracket. Using two wrenches, uncouple dryer from flexible hoses by loosening flare connections.

Installation

Remove seals from new dryer. Work quickly to keep moisture out of system — do not make change in wet weather.

Connect inlet end to flexible hose and tighten flare connection. Connect outlet end loosely. Momentarily crack open liquid service valve on receiver to purge air, the tighten outlet connection. Install mounting clamp.

Open (back-seat) liquid receiver service valve and check for leaks. Open both compressor service valves. If system contains only purging charge (gas at refrigerant container pressure), purge, evacuate, and recharge. If system contains pumped down full charge, and dryer has been replaced to ensure a dry system, place system in operation and check for leaks at normal discharge pressure.

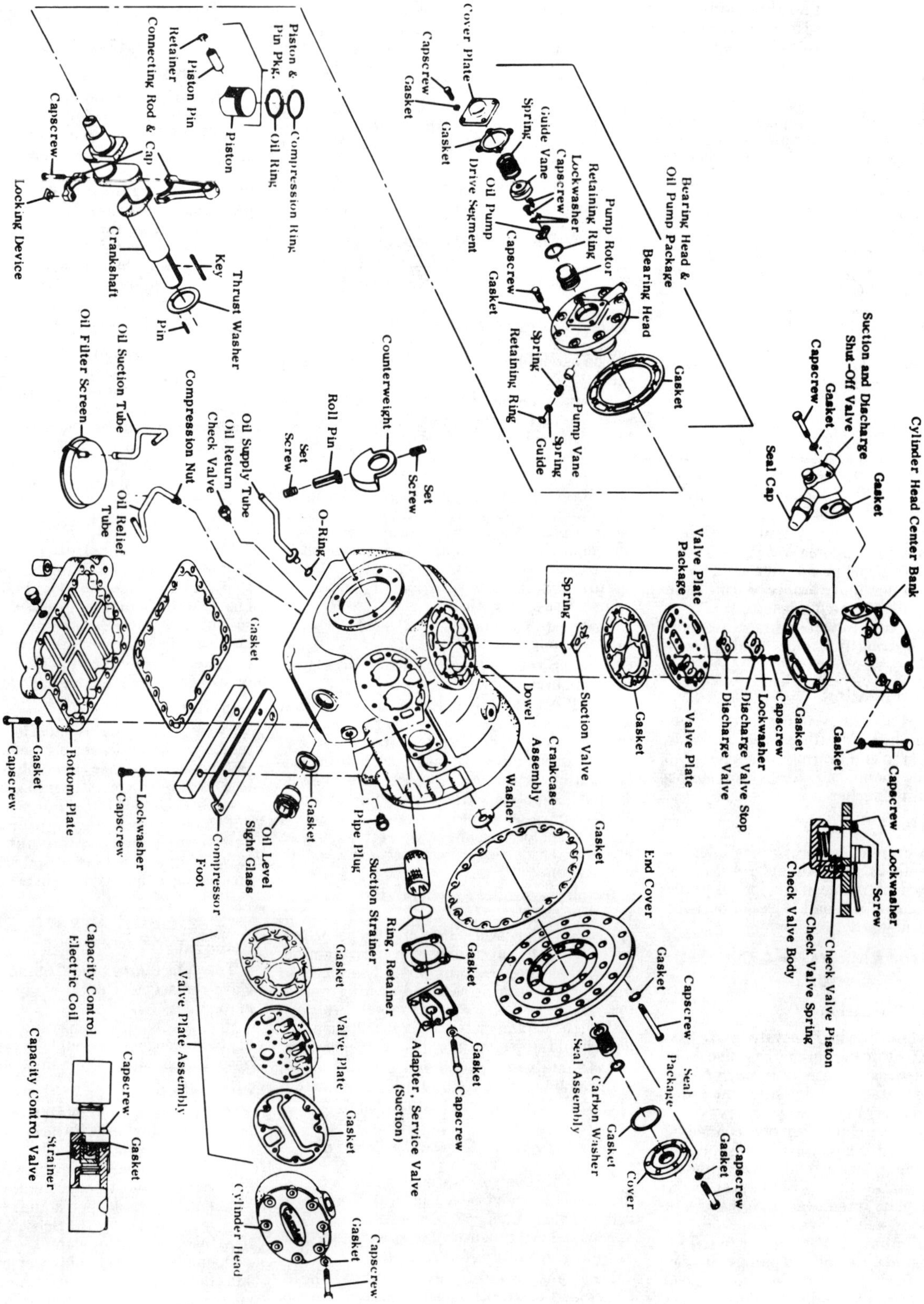

Fig. 10 Transicold 05D compressor exploded view

Crankcase Pressure Regulator (CPR) Valve

CPR valve is adjusted to maintain specified suction pressure of 15 to 17 psi. Valve is responsive to suction pressure only, and any variation of evaporator pressure has no effect on suction pressure. If evaporator pressure exceeds above setting, valve will begin to modulate to maintain desired suction pressure. When evaporator pressure falls below setting, valve opens fully.

Adjustment

To adjust CPR valve, system should be placed on heating cycle, and box temperature must be above 40 deg. This ensures evaporator pressure above 15 to 17 psi. To raise suction pressure, turn adjusting stem clockwise, or in the opposite direction to lower pressure.

Solenoid Valve

Solenoid valve located in liquid line to evaporator is actuated by temperature control switch or thermostat. Valve closes and cuts off supply of liquid refrigerant to thermal expansion valve when temperature in cargo space satisfies thermostat cutout setting. When thermostat calls for cooling, valve opens. If valve leaks in closed position, disassemble as follows:

Disassembly

Pump down system. Remove screws holding body together, carefully lift off upper part of valve (bonnet), and remvoe diaphragm. Before brazing solder-type bodies into line, bonnet and diaphragm must be removed.

Reassembly

Release a little refrigerant during reassembly. Carefully center diaphragm in body with metal buffer plate on top. Place bonnet assembly in position on body. Be sure bypass outlet enters matching hole in valve body. Install and tighten Allen-head screws. Make certain O-ring is in groove.

Changing Coils

The entire coil assembly can be lifted off enclosing tube after clamping screw has been removed at top of coil housing.

When reassembling, place parts on enclosing tube in following order:

1. Coil plate with edge up.
2. Lower coil sleeve with flange down.
3. Coil itself with lead exits at bottom.
4. Coil spring with flat edges against coil.
5. Upper coil sleeve with flange up. Sleeve projects through coil spring.
6. Coil housing Make sure coil leads do not catch over or under coil.
7. Electrical data plate. Press parts down firmly and insert clamping screw. Hold coil housing in proper position and tighten clamping screw. Replace plunger with spring end into upper end of bonnet.

Evaporator Coil Heater Replacement

Heaters are installed under evaporator coil in slots, fastened by supports, or mounted in the coil. When replacing faulty heater, be sure power is off. Use following procedure to replace the heater.

1. Disconnect heater lugs at power inlet. Disconnect all three wires where they are joined together by removing butt connector at end of coil.
2. Unfasten heater supports. On some models, evaporator coil must be removed.
3. Slide heater out of slot in coil. Check for possible causes of burnout.
4. Install new heater by reversing removal procedure. Terminal lugs may be applied and the three wires bolted together and taped, or wires may be soldered together and taped.

CONTROLLER

Temperature controller has temperature sensor in space being cooled, functions by opening or closing solenoid valve to control flow of liquid refrigerant to evaporator. Thermostat closes and competes electrical circuit to solenoid valve, thus energizing and opening valve, when temperature in cargo space rises above thermostat cut-in setting. When desired temperature is obtained, thermostat opens, de-energizing and closing solenoid valve.

Models 230, 250 — Partlow Controller

Checking Calibrating Indicated Temperature

1. Attach sensing bulb of suitable test thermometer, such as Simpson 385, to controller sensing bulb inside trailer. Run thermometer lead outside trailer.
2. Start unit, bring trailer temperature to about 34 deg., then stop unit.
3. Move red pointer to setting at least 10 deg. higher than black pointer.
4. Compare black pointer reading with test thermometer reading. They should agree within ½ deg.
5. If they do not agree, calibrate. To do this, loosen lock screw. With 3/16 in. open end wrench, turn hex shaft counterclockwise to lower temperature reading of black pointer, or in opposite direction to raise reading. Tighten lock screw.

Checking Calibrating Cooling Control Point

1. Start engine. Move red set pointer below position of black pointer to operate unit on high speed cooling.
2. Slowly move red pointer upscale. When red pointer is 3½ to 4½ deg. lower than indicated temperature of black pointer, unit should revert to low speed cooling by action of No. 1 switch. If control point is not within limits, recheck several times before assuming adjustment is necessary.
3. Continue moving red pointer slowly upscale. When red and black pointers coincide, within 1 deg., unit should go into high speed heating by action. No. 3 switch.
4. If adjustment is needed, use small ⅜ in. open end wrench to turn adjusting screw in clockwise to raise control point, or out in opposite direction to lower it.

Models 640, 650 — 12-00110 Controller

Calibration Check

Use suitable test thermometer, as on Models 230, 250.

1. Before starting unit, turn controller pointer clockwise to check calibration of No. 4 heat lock-out switch. Click should indicate it actuates when pointer is between 9 and 11 deg. It should not deactivate (there should be no second click) as pointer rotation is continued clockwise to stop pin. If recalibration is necessary, follow C steps of calibration procedure.
2. Set controller pointer at 32 deg., and start engine. Set defrost control on automatic. Cycle unit several times up and down through high and low speed heating and cooling.

After unit has stabilized, read indicator as temperature *falls*. Properly calibrated controller will give following results:

Switch	Function	RPM/Signal Light	Col.1 Temp.	Col.2 Deg. Diff.
#1	HSC to LSC	1350/cooling	38	3
#2	LSC to LSH	1350/heating	35	3
#3	LSH to HSH	1900/heating	32	

Because of temperature conditions, many times a unit will not automatically cycle into HSH. When this happens, controller pointer must be turned manually conterclockwise at

TRANSICOLD ENGINE DATA

Make, Model	Perkins Four 108	Mercedes OM 636 952
No. of Cylinders	4	4
Bore, Stroke	3.125 × 3.5	2.953 × 3.937
Compression Ratio	22 to 1	19 to 1
Horsepower	23 @ 1900	21.5 @ 1900
Operating Speed—High	1,900	1,900
Operating Speed—Low	1,350	1,350
Cooling System Capacity	12.5 qt.	12.5
Crankcase Capacity	18*	18†
Firing Order	1-3-4-2	1-3-4-2
Valve Clearance—Exh.	.012C	.010C
Valve Clearance—Int.	.012C	.008C

*22 with filter. †24 with filter.

instant unit transfers to LSH. Read controller pointer when unit transfers to HSH. This reading should be 35 deg., and not 32 deg. as when unit cycles to HSH automatically.

If controller does not perform as per Col. 1, but does perform as per Col. 2, follow A steps only of calibration procedure.

If controller does not perform as per both Col. 1 and Col. 2, follow both A and B steps of calibration procedure.

Calibration Procedure

A.

1. With unit still running, find controller pointer position where unit will cycle from LSC to LSH at thermometer reading of 35 deg.
2. Loosen two 10-32 binder-head screws on controller dial and rotate dial until 32 deg. is aligned with pointer. Tighten screws.

B.

1. Remove access cover on controller to expose switch adjustments.
2. With ⅜ in. open end wrench turn No. 1 switch adjustment in to lower, or out to raise, temperature where unit changes from HSC to LSC. Repeat with No. 3 switch adjustment to correct temperature where unit changes from LSH to HSH.

If unit will not automatically cycle into HSH, proper adustment of No. 3 switch can be made by adjusting it so that unit changes to HSH when pointer is turned to 35 deg. on dial immediately after unit changes from LSC to LSH.

C.

1. This adjustment must never precede **A** adjustments. Shut unit off and remove controller access cover
2. Turn controller pointer until set screw is visible on No. 4 heat lock-out switch actuating cam. Loosen set screw with a 5/64 in. Allen wrench and rotate cam with respect to its shaft mounting until switch actuates properly. Then tighten set screw.

IX. GLOSSARY

ABSOLUTE ZERO—Total absence of heat.

ACCUMULATOR—The accumulator acts as a refrigerant storage device. It receives vapor and some remaining liquid refrigerant and oil from the evaporator.

ACTUATOR—A device for controlling a component indirectly, instead of by hand.

AIR CONDITIONING—Control of the temperature, humidity, cleanness, and movement of air.

AIR INLET VALVE—A movable door in the plenum blower assembly that permits the selection of outside air or inside air for both the heating and air-conditioning systems.

AIR OUTLET VALVE—A movable door in the plenum blower assembly that directs air flow either into the heater core or into duct work that leads to the evaporator.

AMBIENT—Air or atmosphere that is outside of the vehicle.

AMBIENT COMPRESSOR SWITCH—Energizes or de-energizes the compressor when outside air temperatures are above or below a predetermined temperature.

AMBIENT TEMPERATURE—The temperature of surrounding air.

AMPLIFIER—Used in some automatic climate control units to boost an electrical signal from the sensors to a usable level.

ASPIRATOR—Used on some automatic temperature controlled air conditioning systems to draw a sample of passenger compartment air through the in-car sensor grille and across the in-car sensor assembly to give the sensor a more responsive temperature signal.

A.T.C.—Automatic Temperature Control. A unit which maintains a pre-set temperature in the car passenger compartment.

ATMOSPHERIC PRESSURE—The weight of air at various altitudes. The sea level pressure of air is 14.7 p.s.i. and it decreases at higher altitudes.

AUXILIARY SEAL—Prevents refrigerant oil from entering the compressor clutch.

BLEND DOOR—Controls in-car temperature by altering the path of incoming air either through or around the heater core.

BLOWER MOTOR—This component produces the air flow through the system.

BLOWER MOTOR RESISTOR—This device controls the blower motor speed as selected by the blower motor switch.

BOILING POINT—The temperature at which a liquid changes to a vapor. Water changes to steam at 212°F at sea level.

BRAZING—A high-temperature, metal-joining process satisfactory for units with relatively high internal pressures such as an air-conditioning condenser.

BTU (British Thermal Unit)—The amount of heat necessary to raise one pound of water through one degree Fahrenheit.

BY-PASS CONTROL VALVE—Same as "Hot Gas By-Pass Valve."

CHANGE OF STATE—Rearrangement of the molecular structure of matter as it changes between any two of the three physical states (solid, liquid, or vapor).

CHARGE—A specific amount of refrigerant or refrigerant oil by weight.

CHARGING—The process of placing a "charge" of refrigerant or refrigerant oil into the system.

CHARGING STATION—A unit which combines the gauge set, vacuum pump and refrigerant tank into one compact unit.

CHEMICAL INSTABILITY—An undesirable condition caused by the presence of contaminants in a refrigeration system. Refrigerant is a stable chemical by itself but in contact with contaminants may break down into harmful chemicals.

CLUTCH—A coupling which transfers torque from a driving to a driven member when desired. The compressor clutch delivers torque transmitted from the engine through a drive belt, causing the compressor drive shaft to rotate.

CLUTCH CYCLING SWITCH—Same as "Thermostatic Switch."

COLD—Total absence of heat.

COMBINATION VALVE—This valve is used on some Ford Motor Co. vehicles and combines the function of the thermal expansion valve and suction throttling valve. It regulates the liquid refrigerant flow while the interior of the vehicle and evaporator core are being cooled.

COMPRESSOR—Component of a refrigeration system that pumps refrigerant and that increases the pressure and temperature of refrigerant vapor.

CONDENSATE—Water taken from air. It forms on the exterior surface of the evaporator.

CONDENSATION—Process whereby a gas changes to a liquid.

CONDENSING TEMPERATURE—The temperature at which compressed gas in a condenser changes from a gas to a liquid.

CONDENSING PRESSURE—The head pressure read from a gauge at the high side service valve.

CONDENSOR—Component of a refrigeration system in which refrigerant vapor is changed to a liquid by the removal of heat.

CONDUCTION OF HEAT—The ability of a substance to conduct heat.

CONTAMINANTS—Anything other than refrigerant and refrigerant oil that is in a refrigeration system, such as rust, dirt, moisture, and air.

CONVECTION—The transfer of heat by the circulation of a vapor or liquid.

COOLING COIL—Same as "Evaporator."

CYCLING CLUTCH SYSTEM—A term referring to air conditioners in which conditioned air temperature is controlled by the engaging and disengaging of the compressor.

DEHYDRATE—See "Purge."

DENSITY—Weight or mass of a gas, liquid or solid.

DESICCANT—A drying agent used within air conditioning systems to absorb and hold moisture.

DIAGNOSIS—The procedure that is followed in locating the cause of malfunction.

DIAPHRAGM—A component that utilizes the pressure difference between vacuum and air to provide mechanical motion. This component is connected by a link to the unit being controlled, usually a mode, water valve, etc.

DICHLORODIFLOUROMETHANE—Same as "Refrigerant 12."

DISCHARGE—To bleed some or all refrigerant from a system by opening a valve or connection and permitting the refrigerant to escape slowly.

DISCHARGE AIR—Conditioned air as it passes through outlets and enters the passenger compartment.

DISCHARGE LINE—Connects the compressor outlet and the condenser inlet. High-pressure refrigerant vapor flows through this line.

DISCHARGE PRESSURE—Pressure of refrigerant being discharged from the compressor.

DISCHARGE SIDE—That portion of the refrigeration system under high pressure, extending from the compressor outlet to the thermostatic expansion valve.

DRIER—A device which contains the moisture absorbing agent.

DUCTS—Tubes or channels that carry and direct air flow.

ECM—Electronic Control Unit.

ENGINE IDLE COMPENSATOR—This additional aid is used on some air conditioned vehicles to prevent stalling during prolonged hot weather operation. The idle compensator is thermostatically controlled and supplies additional air to the idle mixture.

EQUALIZER LINE—Line or connection used specifically for obtaining required operation from certain control valves. Very little if any refrigerant flows though this line.

EVACUATE—To create a vacuum within a system.

EVAPORATION—Changing from a liquid to a vapor.

EVAPORATOR—Component of an air-conditioning system which conditions the air. Refrigerant liquid is changed into a vapor in this component.

EXPANSION TUBE—This component is located in the evaporator inlet of some General Motors vehicles. It controls the flow of liquid, has no moving parts and is non-adjustable.

EXPANSION VALVE—Same as "Thermostatic Expansion Valve."

FAHRENHEIT—Thermometer which uses a scale based on 32° as the freezing point of water.

FLUSH—To remove solid particles such as metal chips and dirt. Refrigerant passages are purged (flushed) with Refrigerant-12. Exterior surfaces of system components are flushed with water or an acceptable solvent.

FREEZE PROTECTION—Controlling the evaporator temperature so that moisture on its surface will not change to ice and block air flow.

GAS—A vapor which contains no particles of liquid.

GAUGE SET—One or more instruments attached to a manifold (a pipe fitting with several outlets for connecting pipes) and used for measuring pressure.

H-VALVE—A type of expansion valve used on some Chrysler Corp. vehicles which does not use a sensing tube.

HEAD PRESSURE—Same as "Discharge Pressure."

HEATER CORE—A water-air heat exchange unit which provides heat to the inside of the vehicle.

HEAT EXCHANGE—The flow of heat between two objects of different temperatures. This flow is always from the warmer to the colder of the two.

HIGH-LOAD CONDITION—Refers to those times when an air-conditioning system must operate continuously at maximum capacity to provide the cool air required, such as at high temperature and high humidity.

HIGH PRESSURE CUT-OUT SWITCH—This switch, used on some Chrysler Corp. vehicles, is wired in series in the compressor clutch circuit. It cuts off the electrical power supply to the clutch circuit when head pressure reaches 375 psi ±10 psi and holds it in the off position until the head pressure drops to about 250 psi.

HIGH-PRESSURE LINES—Lines from the compressor outlet to the thermostatic expansion valve inlet that carry high-pressure refrigerant. The two longest high-pressure lines are the "discharge" and "liquid" lines.

HIGH-PRESSURE LIQUID LINE—Same as "Liquid Line."

GLOSSARY

HIGH PRESSURE RELIEF VALVE—The function of this valve is to protect the air conditioning system from damage due to excessive pressure. This valve vents off only a small amount of refrigerant necessary to reduce system pressure and reseats itself.

HIGH-PRESSURE VAPOR LINE—Same as "Discharge Line."

HIGH SIDE—Same as "Discharge Side."

HIGH SIDE PRESSURE—Same as "Discharge Pressure."

HUMIDITY—See "Moisture."

HVAC SYSTEM—Heater Vent Air Conditioning System.

INSTABILITY—See "Chemical Instability."

LATENT HEAT—The amount of heat absorbed or released by a substance undergoing a change of state. Latent heat of vaporization is the amount of heat required to change a liquid to a vapor. Latent heat of fusion is the amount of heat required to change a liquid to a solid. Latent heat of melting is the amount of heat required to change a solid to a liquid.

LEAK DETECTOR—Equipment for locating an opening where the refrigerant may escape. The three common types of detectors are the alcohol torch, gas torch, and electronic detector.

LED—Light emitting diode.

LIQUID—A solid column of fluid without gas pockets is pure liquid.

LOW PRESSURE PROTECTION SYSTEM—This system incorporates a pressure sensitive electrical switch wired in series with the compressor clutch circuit. This switch cuts off the electrical power supply to the clutch circuit when refrigerant pressure drops below the control point of the switch.

LOW SIDE—Also the suction side (from expansion valve orifice to intake reed valve in compressor).

MICROPROCESSOR—An integrated circuit capable of receiving data, processing it and supplying results. The results are usually of an internally stored program.

MODE—A condition or setting.

MODE DOOR—Controls in-car temperature by altering the path of incoming air either through or around the heater core.

MODULE—The assembly that contains the blower, air intake, evaporator, heater core and temperature door.

MUFFLER—Device to minimize pumping sounds from the compressor.

OIL BLEED LINE—Insures positive oil return to compressor at high compressor speed and under low charge conditions.

OIL INDUCER—A tool used to add refrigerant oil into the A/C system.

OIL INJECTION CYLINDER—A cylinder containing a measured quantity of refrigerant oil added when servicing the Air-Conditioning system.

OPERATIONAL TEST—Same as "Performance Test."

ORIFICE TUBE—This tube is located in the evaporator inlet pipe, and meters the flow and reduces the pressure of the liquid refrigerant.

PERFORMANCE TEST—Taking temperature and pressure readings under specified conditions to determine if an air-conditioning system is operating satisfactorily.

PHOSGENE—A poisonous gas produced when Freon 12 is burned in an open flame.

PLENUM—The air intake chamber located on the cowl at the base of the windshield.

PLENUM BLOWER ASSEMBLY—Air passes through this assembly on its way to the evaporator. It is located on the engine side of the fire wall and contains air ducts, air valves, and a blower that permits selection of air from the outside or inside of the car and directs it either to the evaporator core or to the heater core.

POWER SERVO—Used in some automatic climate control systems to control the various program components in response to a vacuum signal from the transducer.

PRESSURE—Force per unit of area. The pressure that refrigerant exerts within the system is indicated on gauges in pounds per square inch.

PRESSURE LINE—See "Discharge Line." All refrigerant lines are under pressure.

PRESSURE SENSING LINE—Prevents the compressor suction pressure from dropping below a predetermined pressure, by opening the thermostatic expansion valve, allowing liquid refrigerant to flood through the evaporator.

PRIMARY SEAL—Prevents refrigerant from escaping around the compressor shaft.

PROGRAMMER—Used on automatic temperature controlled air conditioning systems to control blower speeds, air mixture doors and vacuum doors in a predetermined sequence.

P.S.I.G.—Pounds per square inch, gauge pressure.

PURGE—To remove moisture and air from a system or a component by flushing with a dry gas, such as nitrogen or Refrigerant-12.

RADIATION—The flow of heat through space.

GLOSSARY

RECEIVER-DEHYDRATOR—A container for storing liquid refrigerant from the condenser. A sack of desiccant in this container removes small traces of moisture that may be left in the system after purging and evacuating.

REFRIGERANT—The chemical compound used in a refrigeration system to produce the desired cooling.

REFRIGERANT OIL—Lubricates seals, gaskets and moving parts of the compressor.

REFRIGERANT-12—The refrigerant used in air-conditioning systems. It is sold under such trade names as Freon-12, Genetron-12, Isotron-12, Prestone-12, and Ucon-12.

REFRIGERATION—The removal of heat by mechanical means.

REFRIGERATION CYCLE—Complete course of operation of refrigerant back to a starting point, evidenced by temperature, pressure, and liquid-vapor changes of the refrigerant as it circulates through the system.

RELATIVE HUMIDITY—Actual moisture content of the air in relation to the total moisture that the air can hold at a given temperature. If the air contains ¾ of the moisture it can hold at its existing temperature, the relative humidity is 75%.

SCHRADER VALVE—A spring-loaded valve where a connection can be made to the refrigeration system. A gauge set can be used on a Schrader valve only with an adapter.

SCREENS—Fine mesh metal screens are located in the receiver-dehydrator, thermostatic expansion valve, and compressor. They prevent solid particles from being circulated through the system and carried to vital moving parts where they might cause damage.

SENSOR—This device is used on automatic temperature control units to sense temperature. Sensors are thermistors which vary resistance inversely with temperature changes. The change in resistance of the sensor is the basic input signal to the control system.

SIGHT GLASS—A window usually in the top of the receiver-dehydrator used to observe refrigerant flow.

SPECIFIC HEAT—The quantity of heat which is required to change one pound of a substance one degree Fahrenheit.

SUCTION LINE—Connects the evaporator outlet and the compressor inlet. Low-pressure refrigerant vapor flows through this line.

SUCTION PRESSURE—Compressor intake pressure as indicated by a gauge set.

SUCTION SIDE—That portion of the refrigeration system under low pressure, extending from the thermostatic expansion valve to the compressor inlet.

SUCTION THROTTLING VALVE—Prevents evaporator core freeze up and controls temperature of air that flows from the evaporator.

SUCTION THROTTLING VALVE P.O.A.—Prevents evaporator core freeze up, compensates for compressor speed changes and evaporator load — is unaffected by elevation above sea level.

SUPERHEAT—Any temperature above the vaporization point of a given liquid.

TEMPERATURE—Heat intensity measured in degrees Fahrenheit.

TEMPERATURE REGULATION VALVE—Same as "Hot Gas By-Pass Valve."

TE VALVE—Same as "Thermostatic Expansion Valve."

THERMOSTATIC EXPANSION VALVE—Component of a refrigeration system that controls the rate of refrigerant flow to the evaporator. It is commonly called the "TE Valve."

THERMOSTATIC SWITCH—An adjustable component used in a cycling clutch system to engage and disengage the compressor. It prevents water (condensate) from freezing on the evaporator core and controls the temperature of air that flows from the evaporator.

TORQUE—A turning force such as that required to seal a connection, usually in foot-pounds or inch pounds.

TRANSDUCER—Used in some automatic temperature control units to convert an electrical signal to a proportionate vacuum output to control the power servo.

VACUUM—Less than atmospheric pressure and expressed as inches of mercury (in. Hg.).

VACUUM MOTOR—A unit which, through the use of a diaphragm, converts a vacuum signal to mechanical action.

VACUUM PUMP—A mechanical device to evacuate a system.

VALVES-IN RECEIVER (VIR)—The VIR assembly incorporates the thermostatic expansion valve, POA suction throttling valve and receiver-dehydrator and sight glass into one unit. The construction of this unit eliminates the need for an external equalizer line between the thermostatic expansion valve and POA suction throttling valve.

VAPORIZATION POINT—The temperature at which a given liquid changes to a gas.

X

VACUUM & WIRING

DIAGRAM

SECTION

American Motors

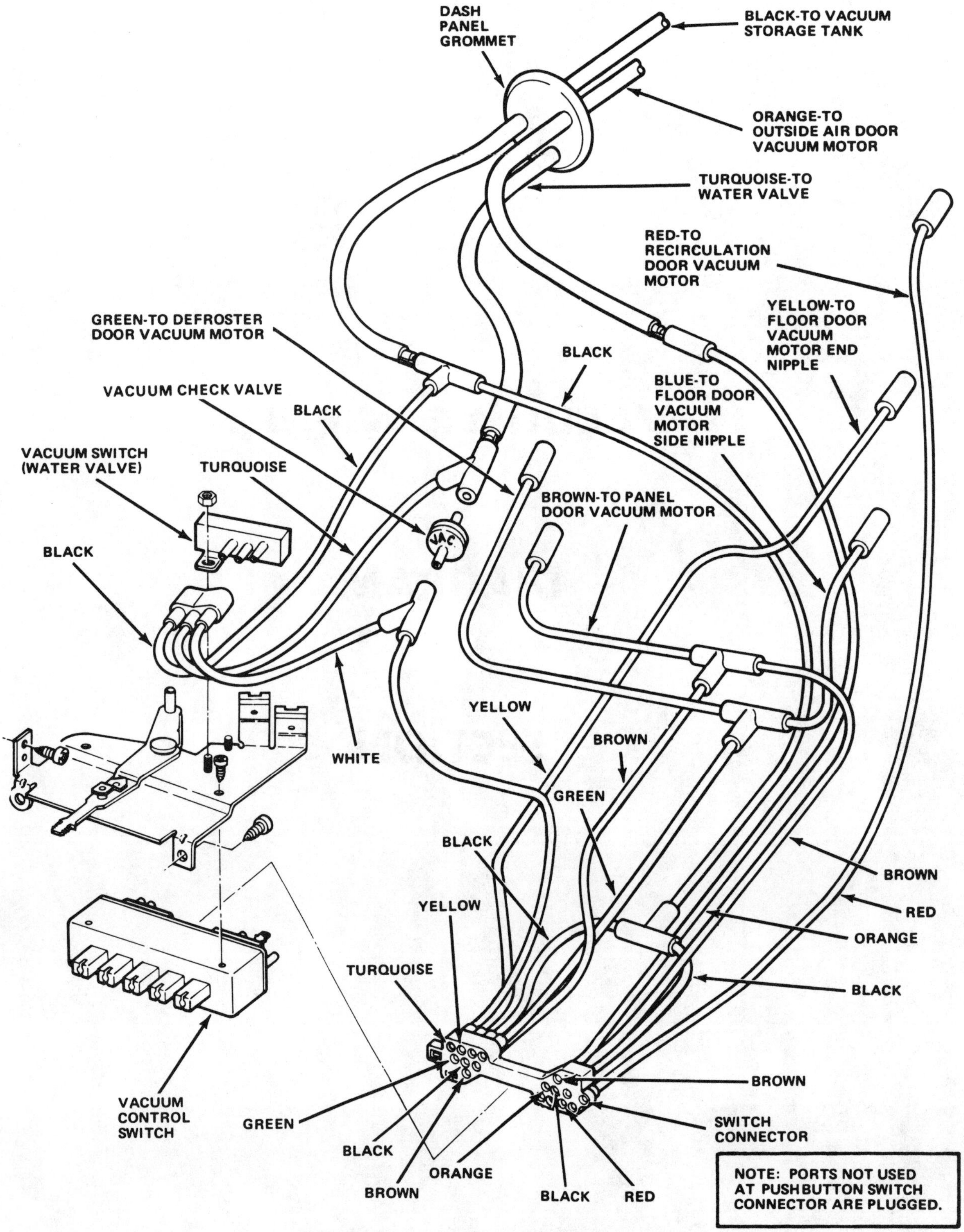

1975 AMC Pacer A/C & Heater Vacuum Circuit

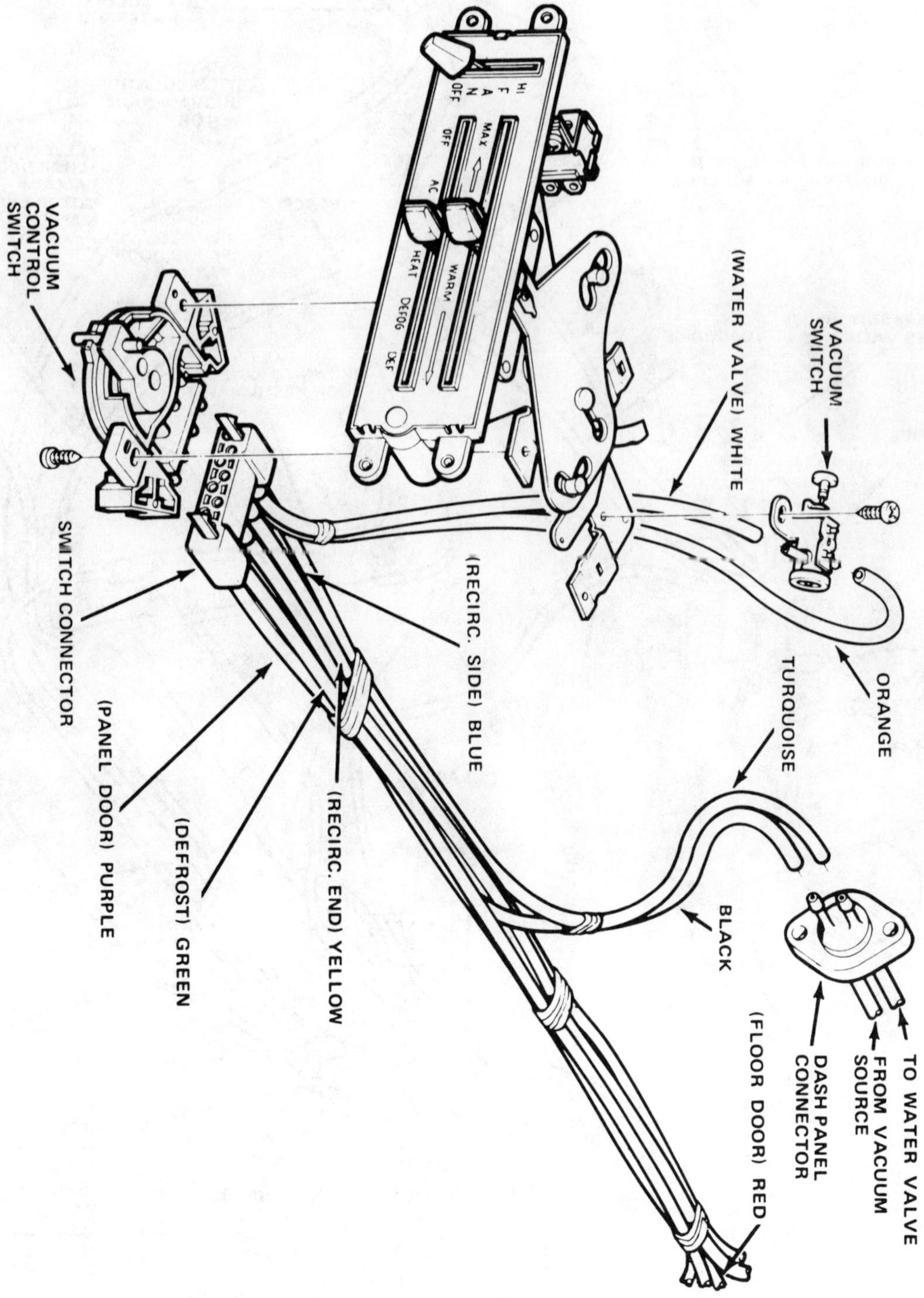

1976-78 AMC Matador A/C & Heater Vacuum Circuit

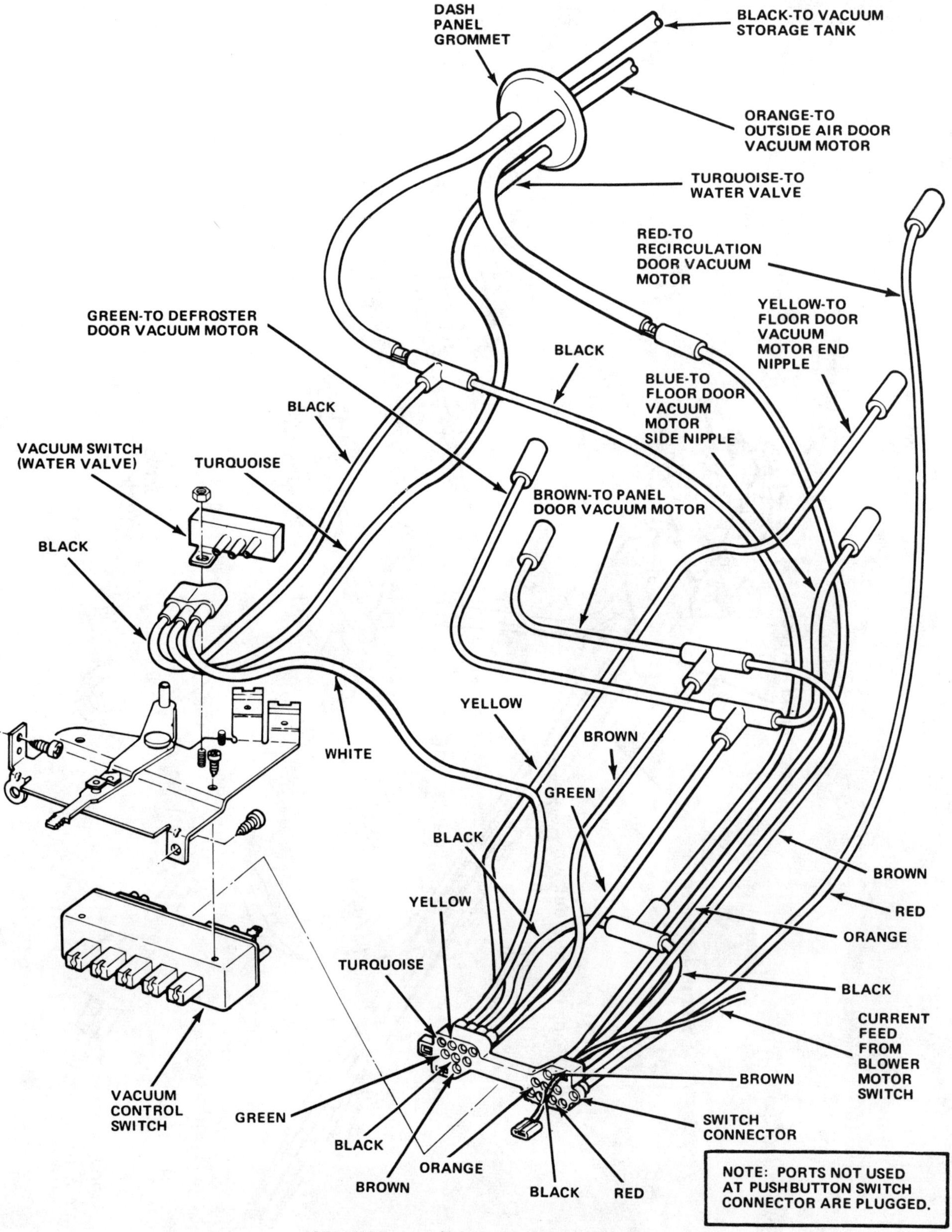

1976 AMC Pacer A/C & Heater Vacuum Circuit

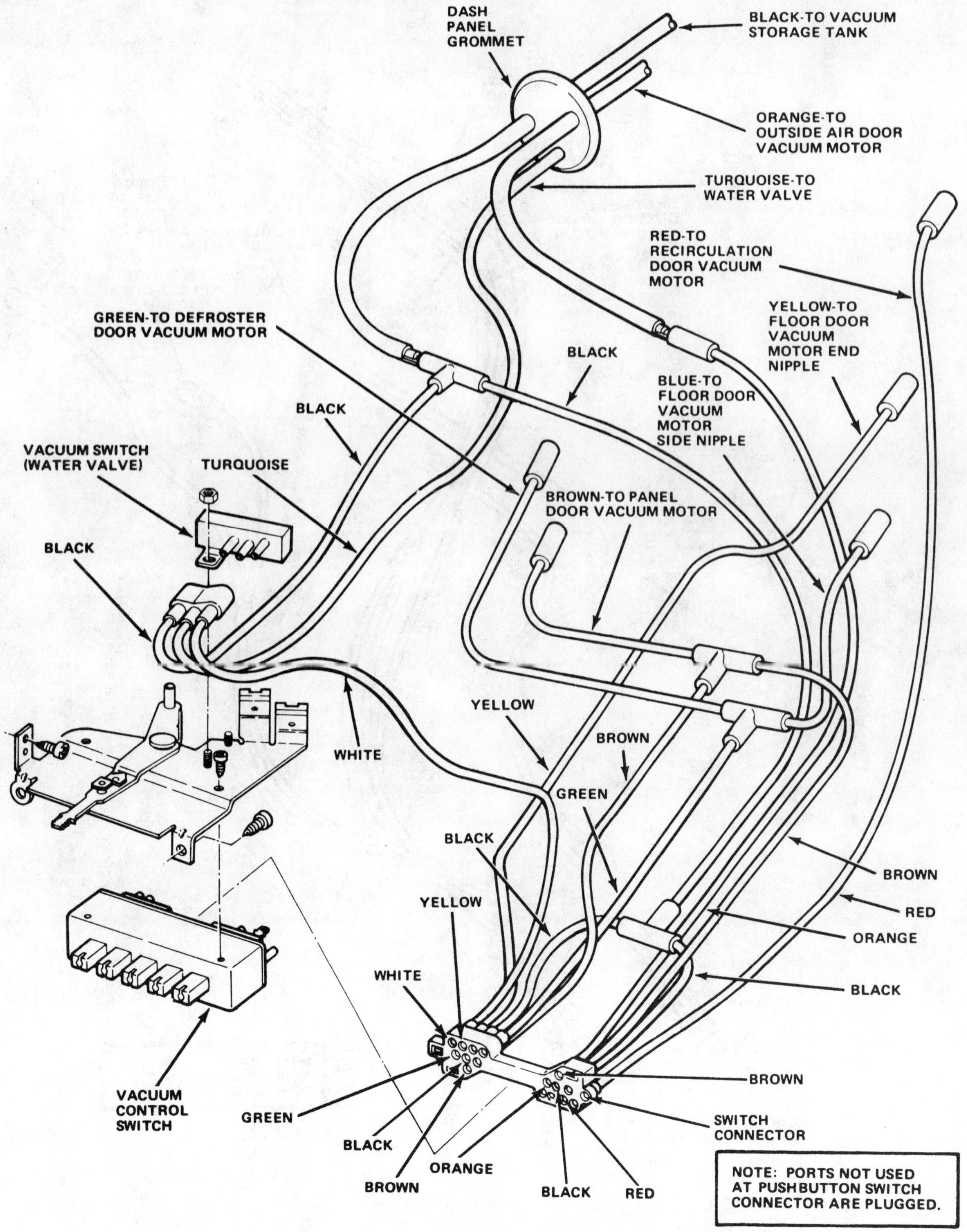

1977 AMC Pacer A/C & Heater Vacuum Circuit

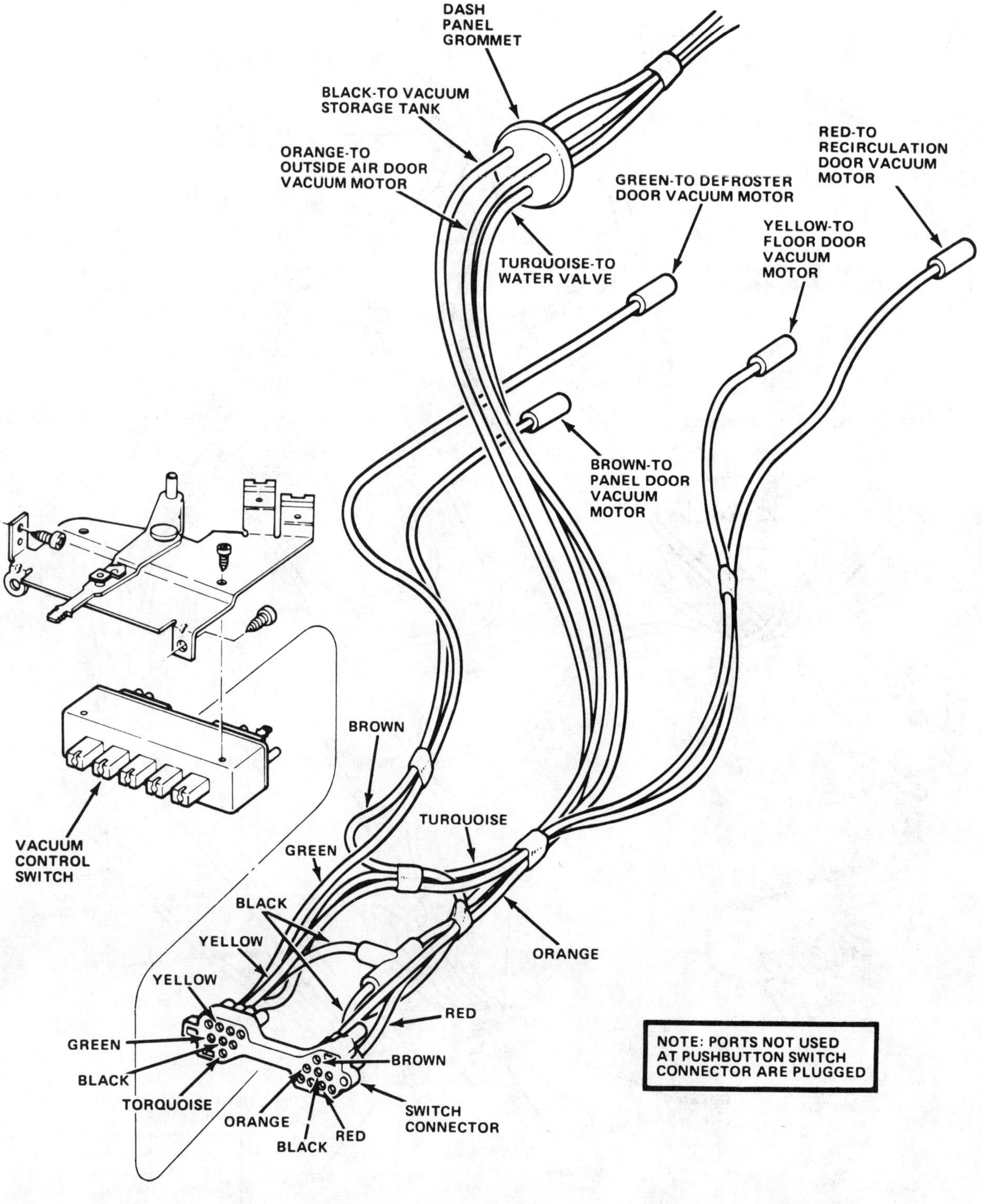

1978-80 AMC Pacer A/C & Heater Vacuum Circuit

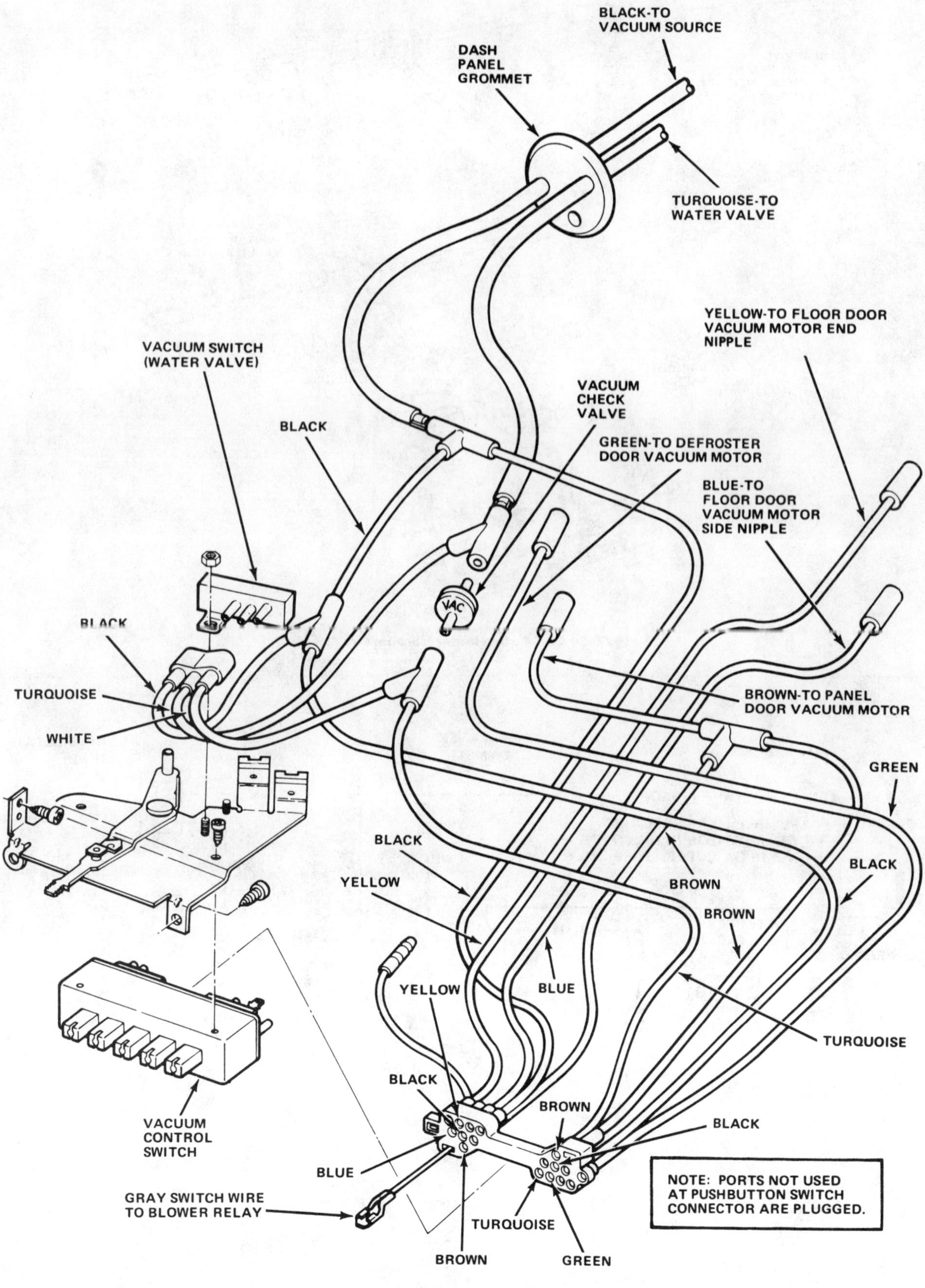

1975-76 AMC Pacer Heater Vacuum Circuit

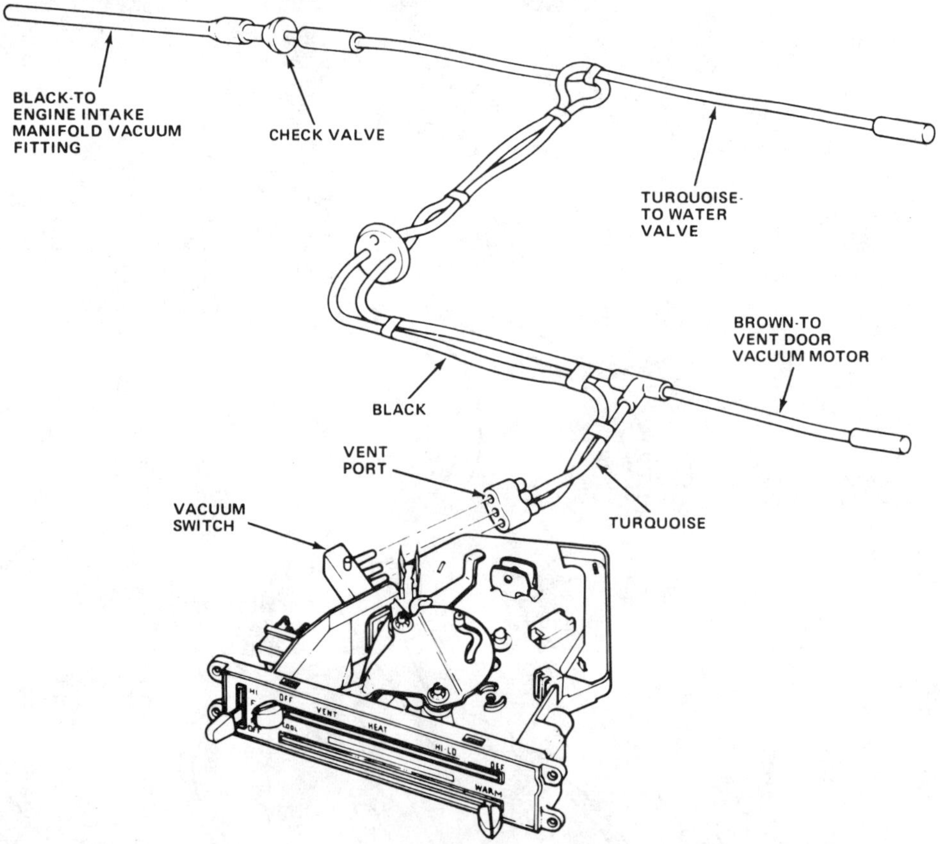

1977-80 AMC Pacer Heater Vacuum Circuit

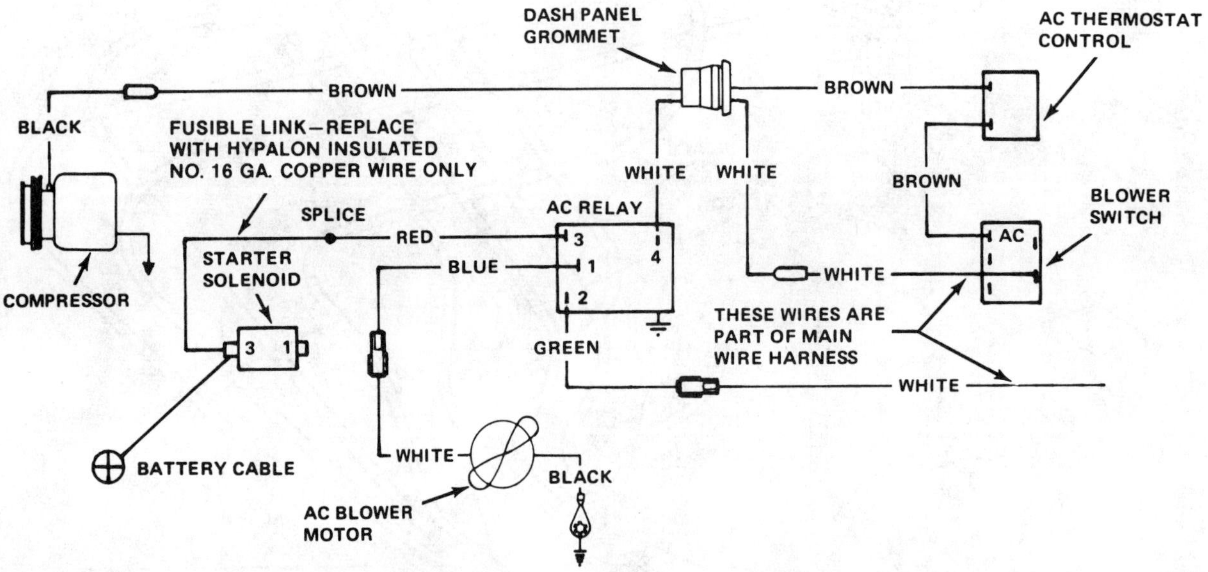

1974 AMC Javelin A/C & Heater Wiring Circuit

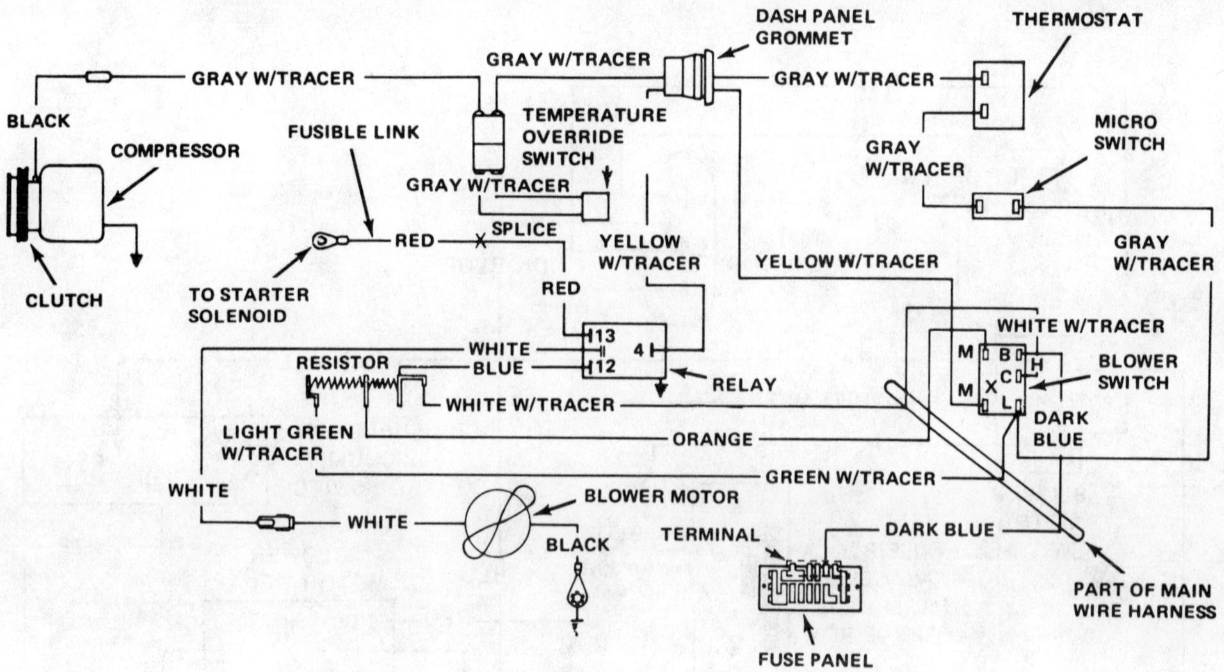

1974 AMC Ambassador; 1974-76 Matador A/C & Heater Wiring Circuit

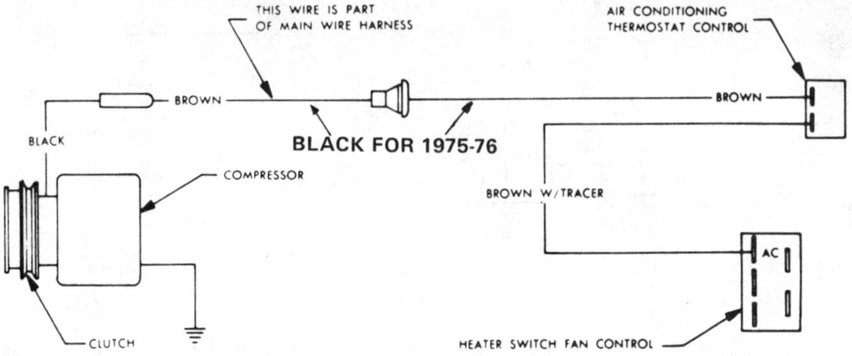

1974-76 AMC Gremlin & Hornet A/C & Heater Wiring Circuit

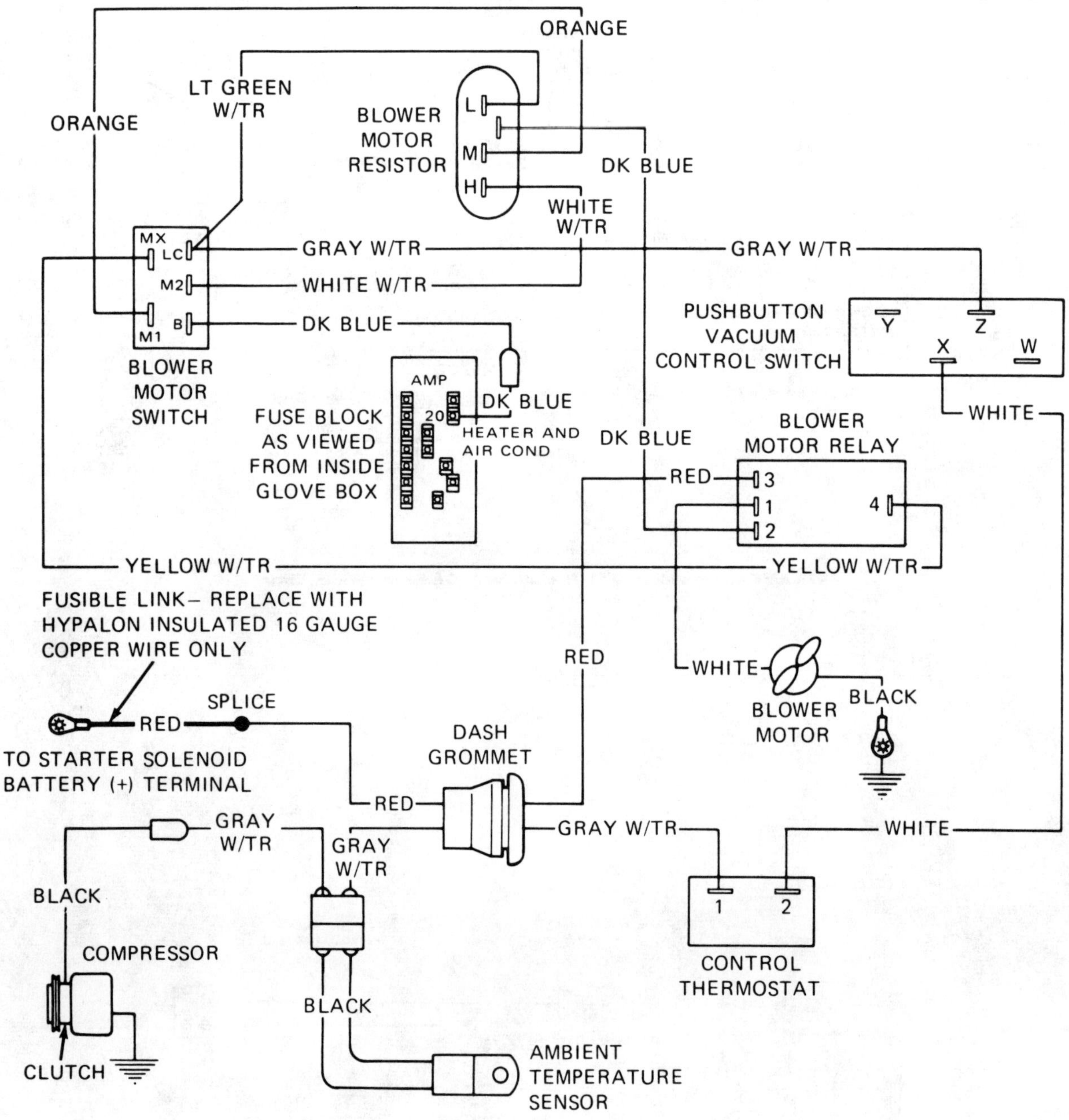

1975 AMC Pacer A/C & Heater Wiring Circuit

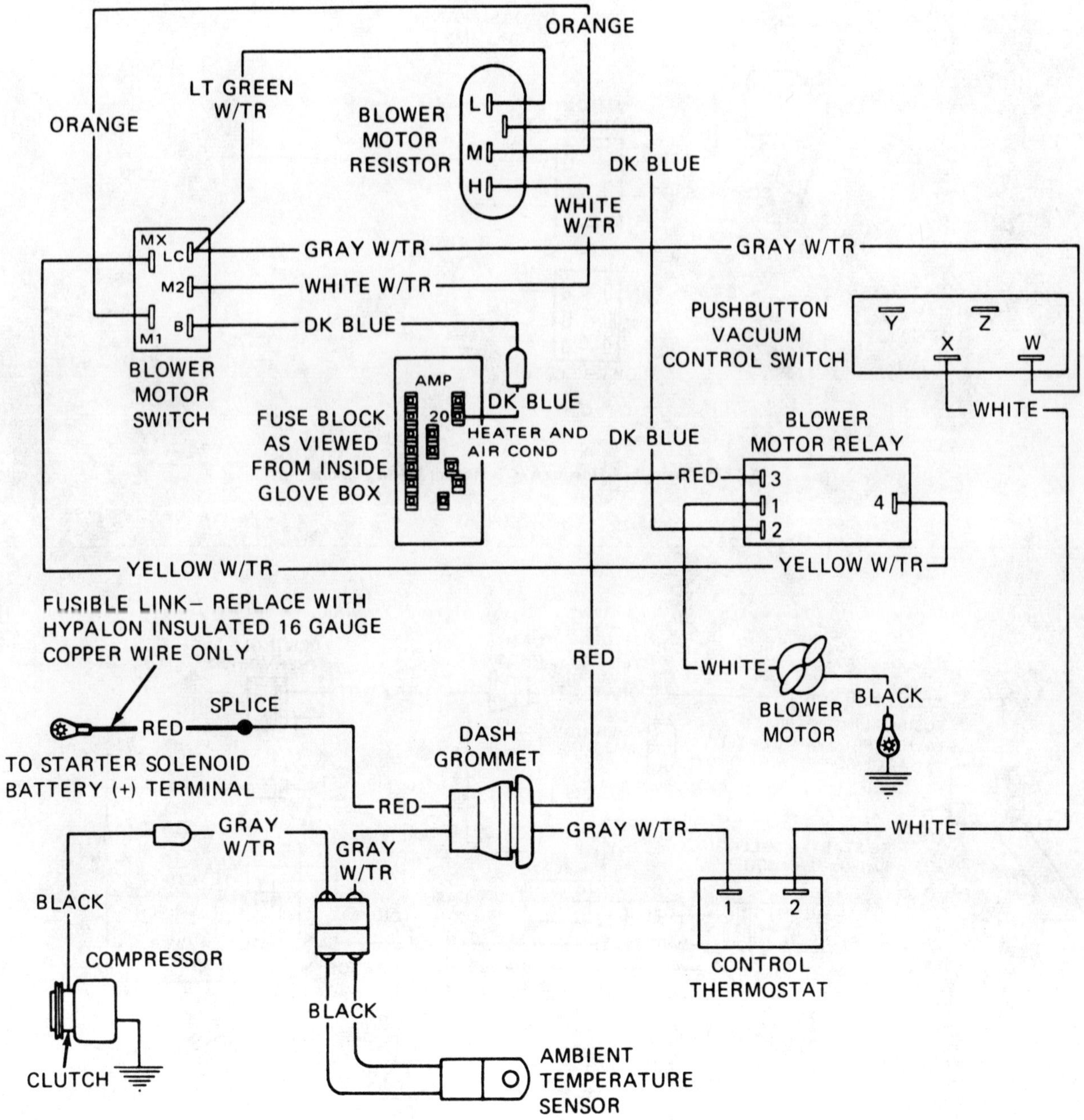

1976 AMC Pacer A/C & Heater Wiring Circuit

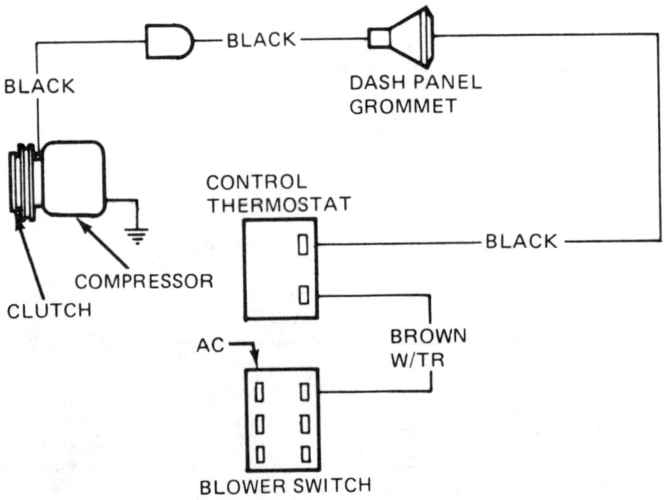

1977 AMC Gremlin & Hornet A/C & Heater Wiring Circuit

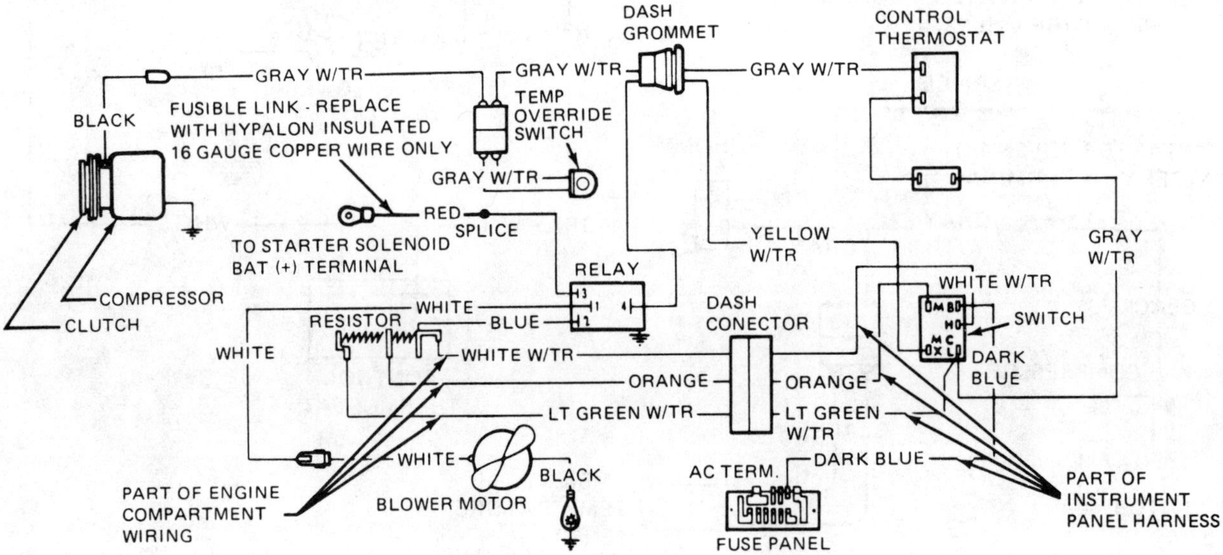

1977-78 AMC Matador A/C & Heater Wiring Circuit

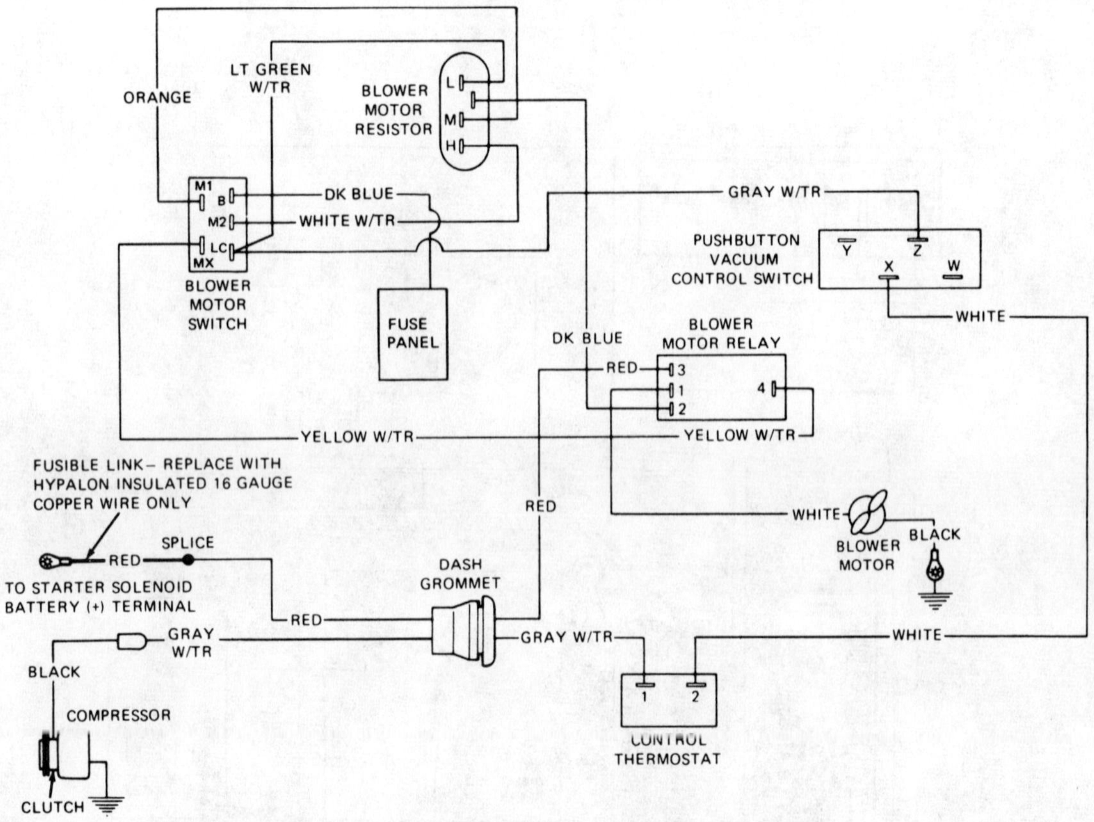

1977-80 AMC Pacer A/C & Heater Wiring Circuit

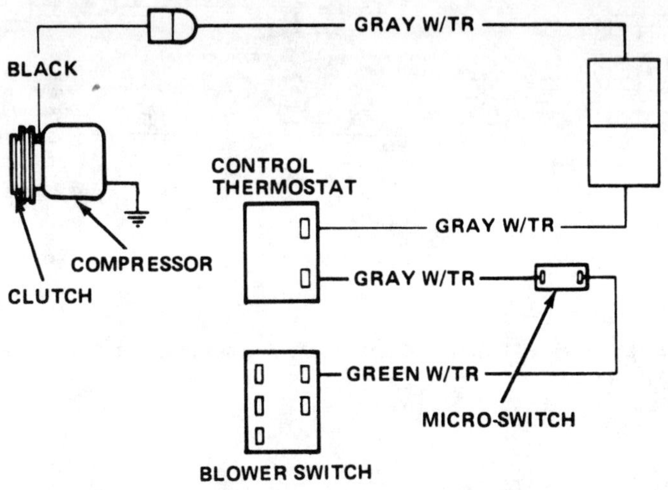

1978 AMC Gremlin; 1978-80 AMC AMX, Concord & Spirit; 1980 AMC Eagle A/C & Heater Wiring Circuit with 6-232, 258 Engine

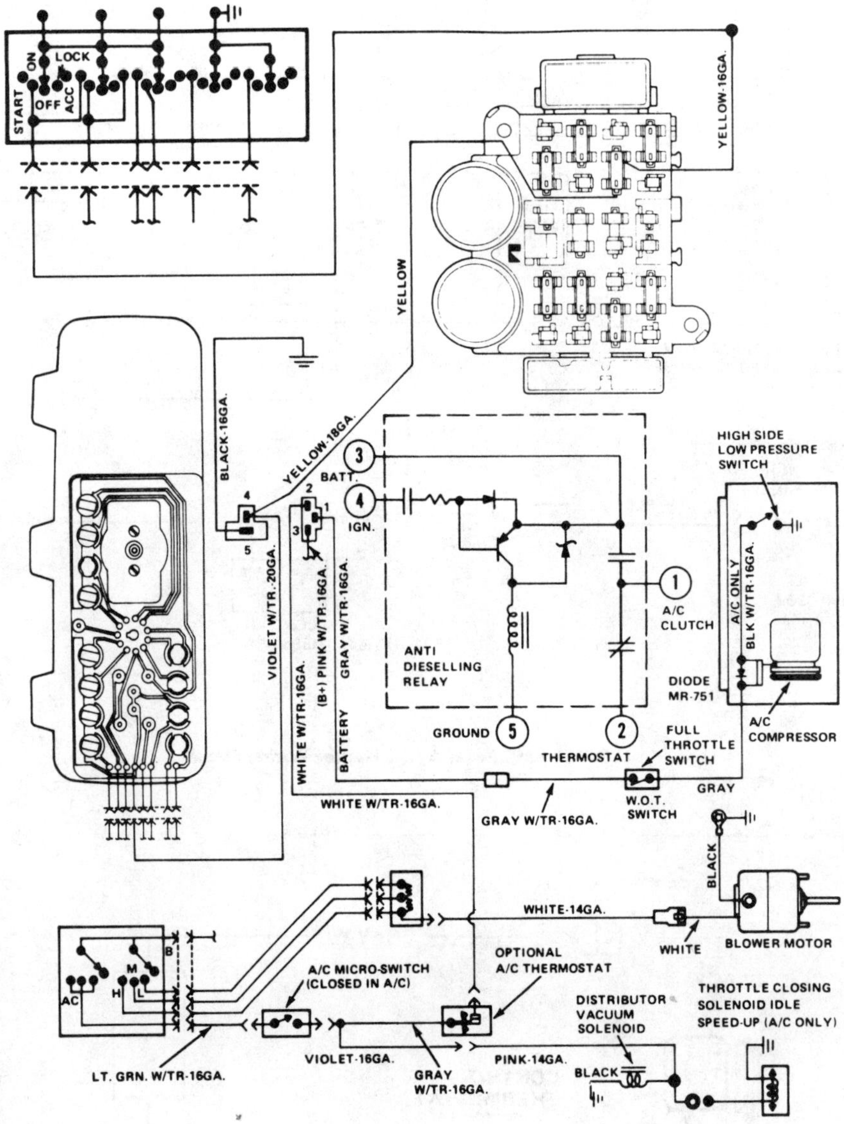

1980 AMC Concord & Spirit A/C & Heater Wiring Circuit with 4-151 Engine

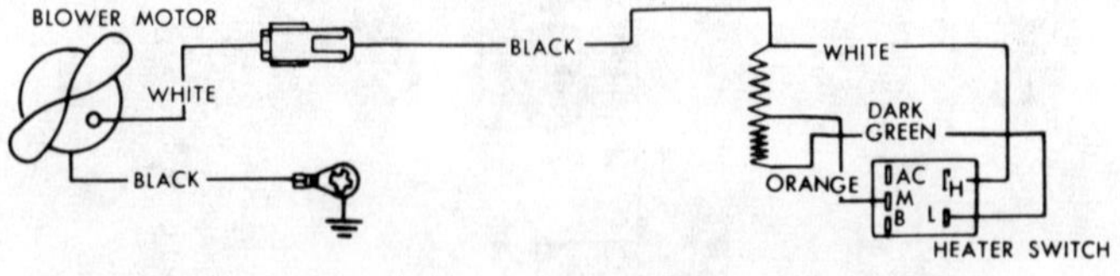

1974 AMC Gremlin & Hornet Heater Wiring Circuit

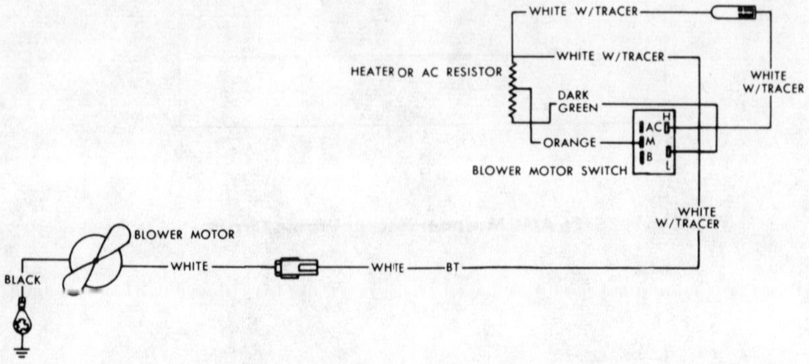

1974 AMC Javelin Heater Wiring Circuit

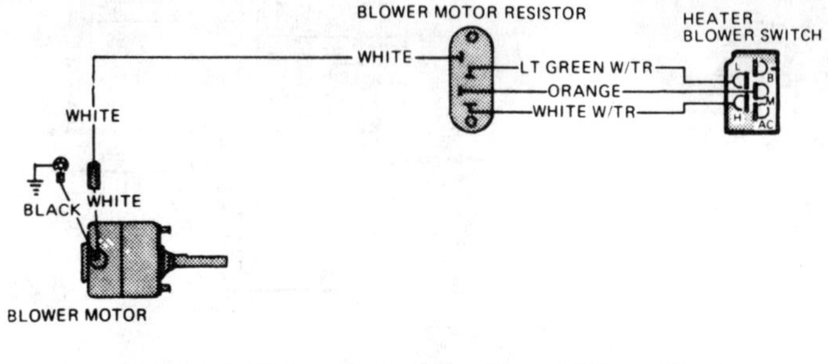

1975-76 AMC Gremlin & Hornet Heater Wiring Circuit

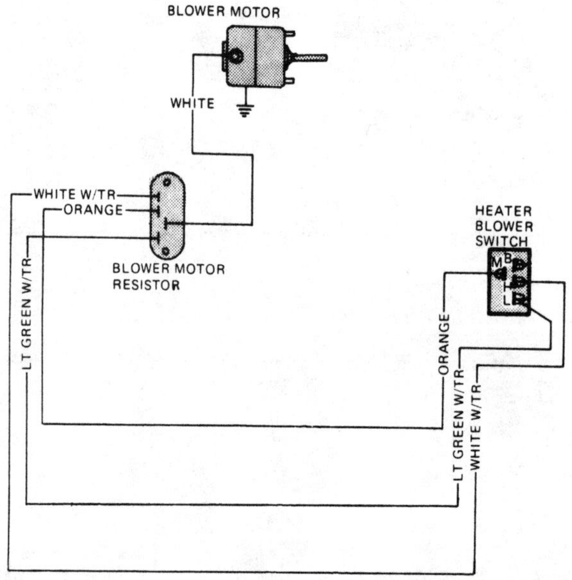

1975-76 AMC Matador Heater Wiring Circuit

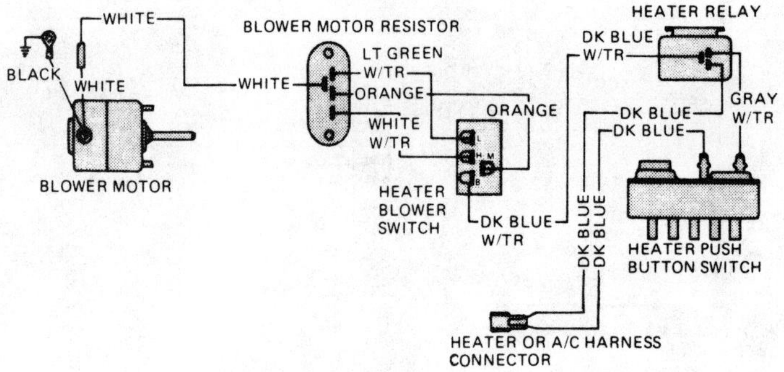

1975-76 AMC Pacer Heater Wiring Circuit

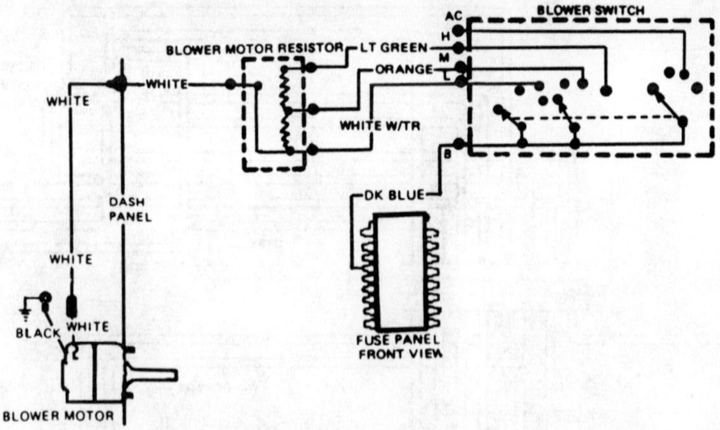

1977 AMC Hornet & 1977-78 Gremlin Heater Wiring Circuit

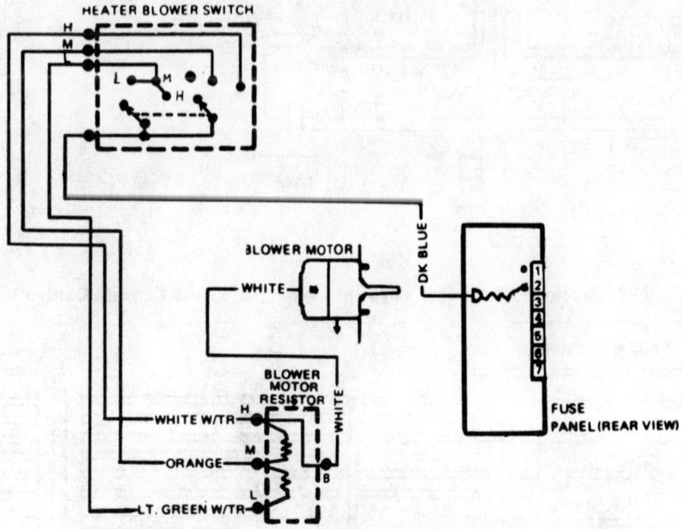

1977-78 AMC Matador Heater Wiring Circuit

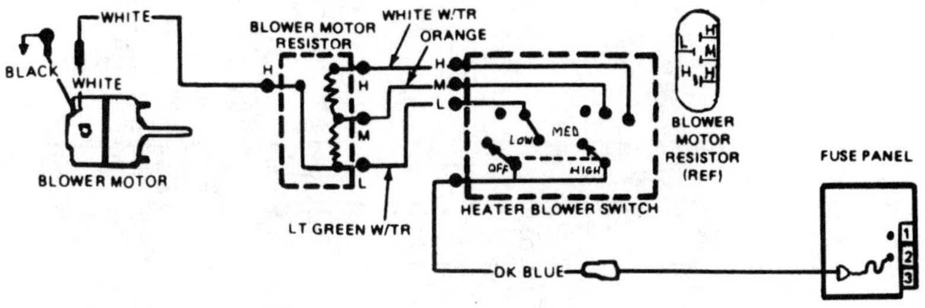

1977-79 AMC Pacer Heater Wiring Circuit

Buick — Full Size

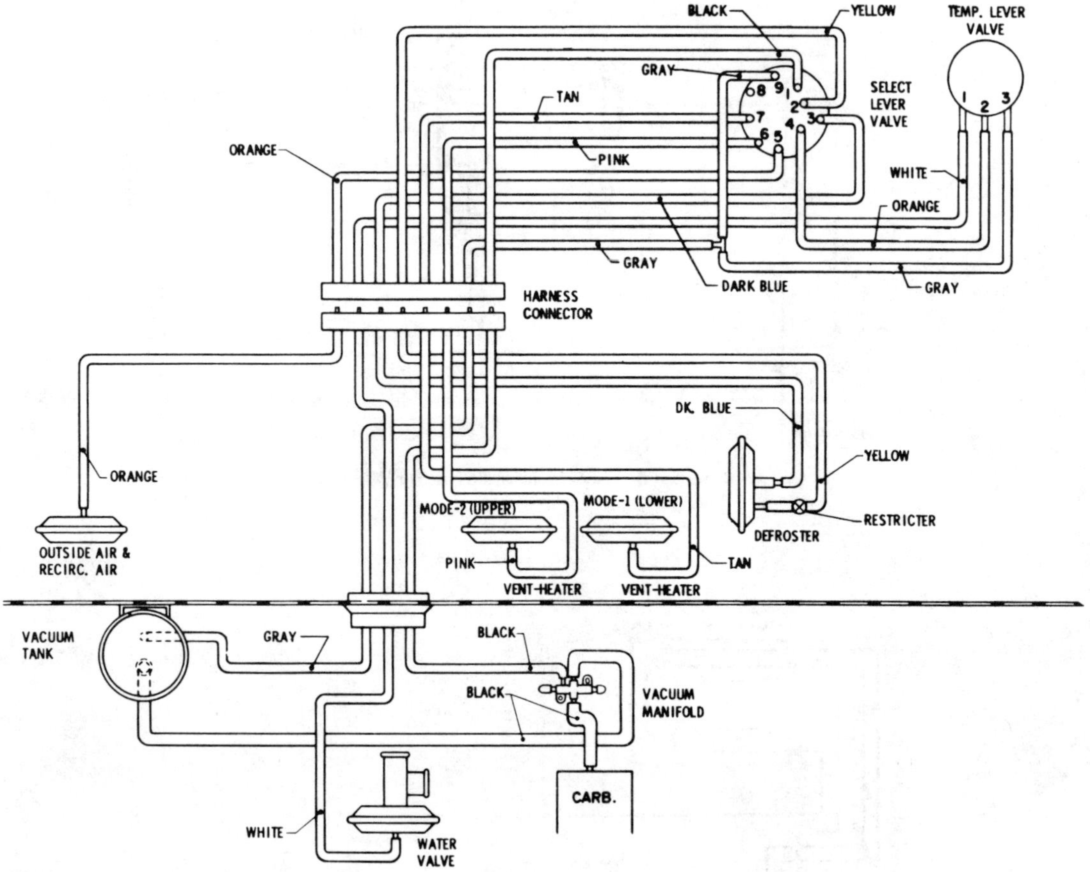

1974 Buick A/C & Heater Vacuum Circuit (Less Auto Climate Control)

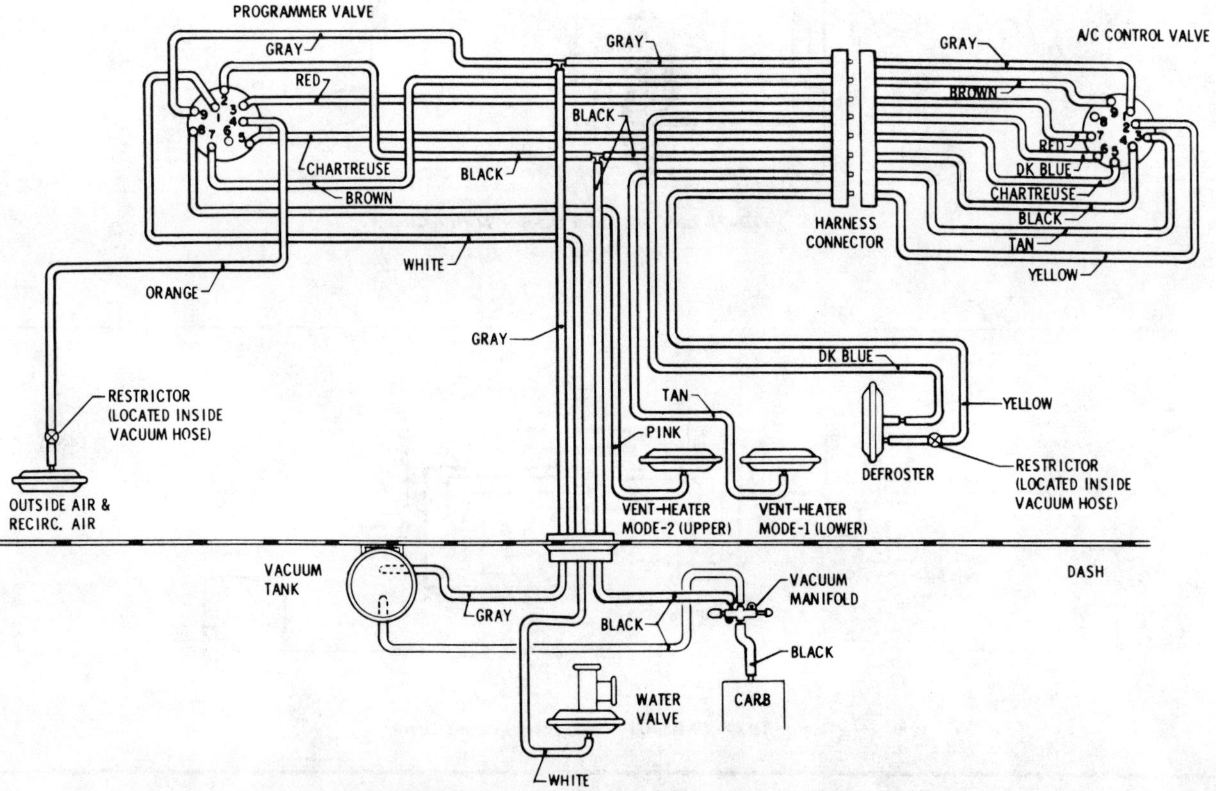

1974-75 Buick A/C & Heater Vacuum Circuit (With Auto Climate Control)

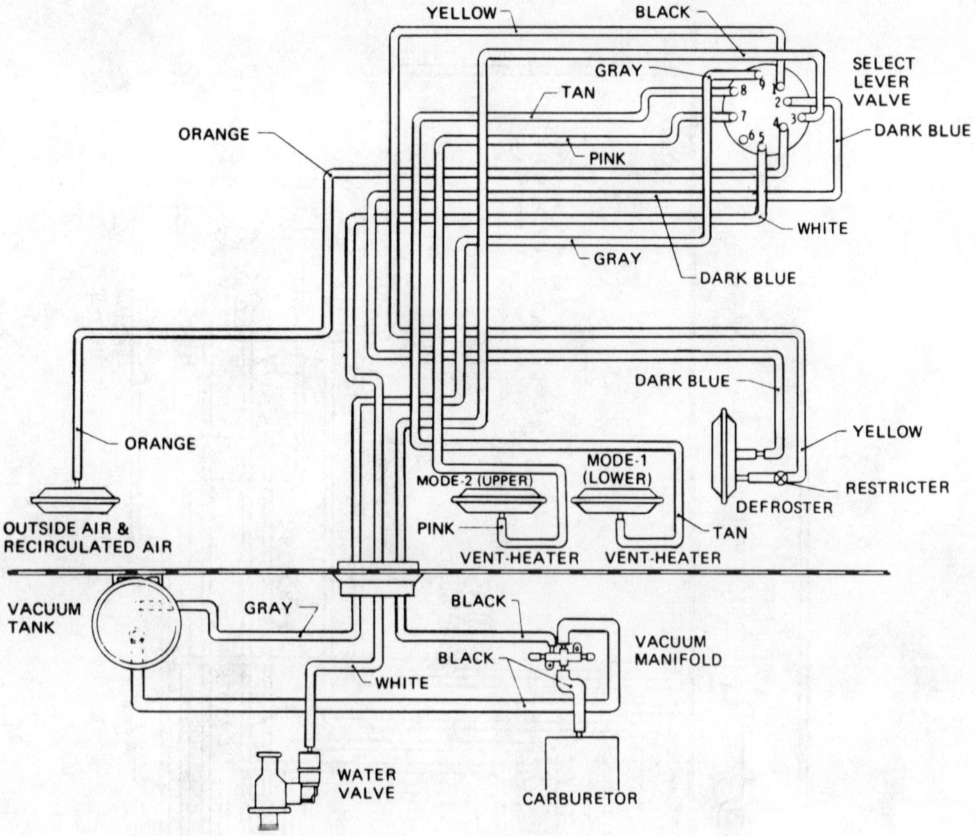

1975 Buick A/C & Heater Vacuum Circuit (Less Auto Climate Control)

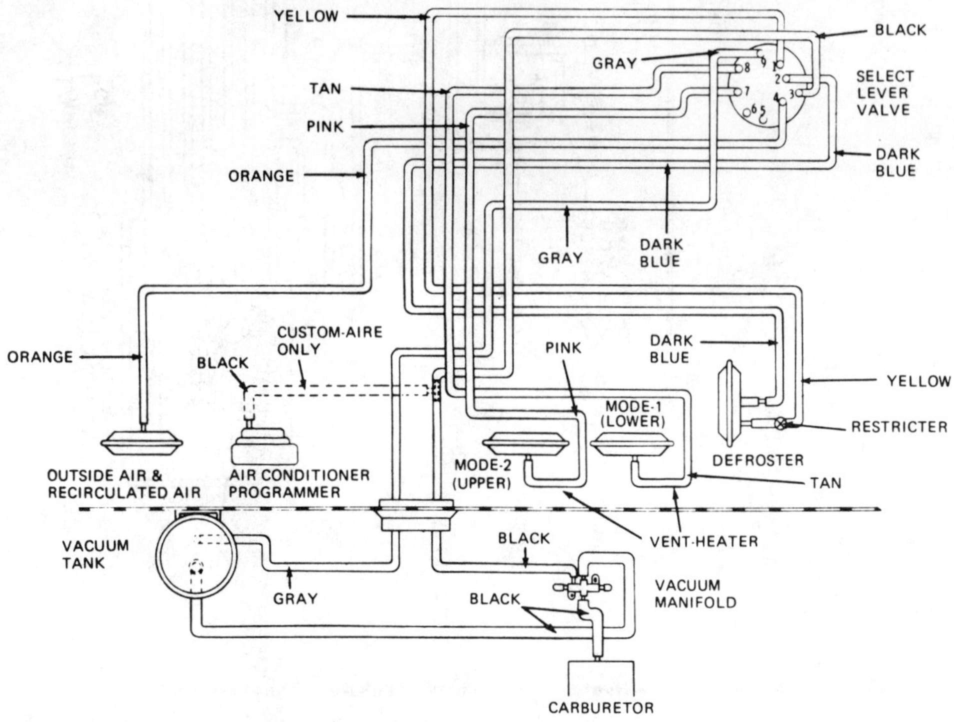

1976 Buick A/C & Heater Vacuum Circuit (With Custom Aire or Manual A/C)

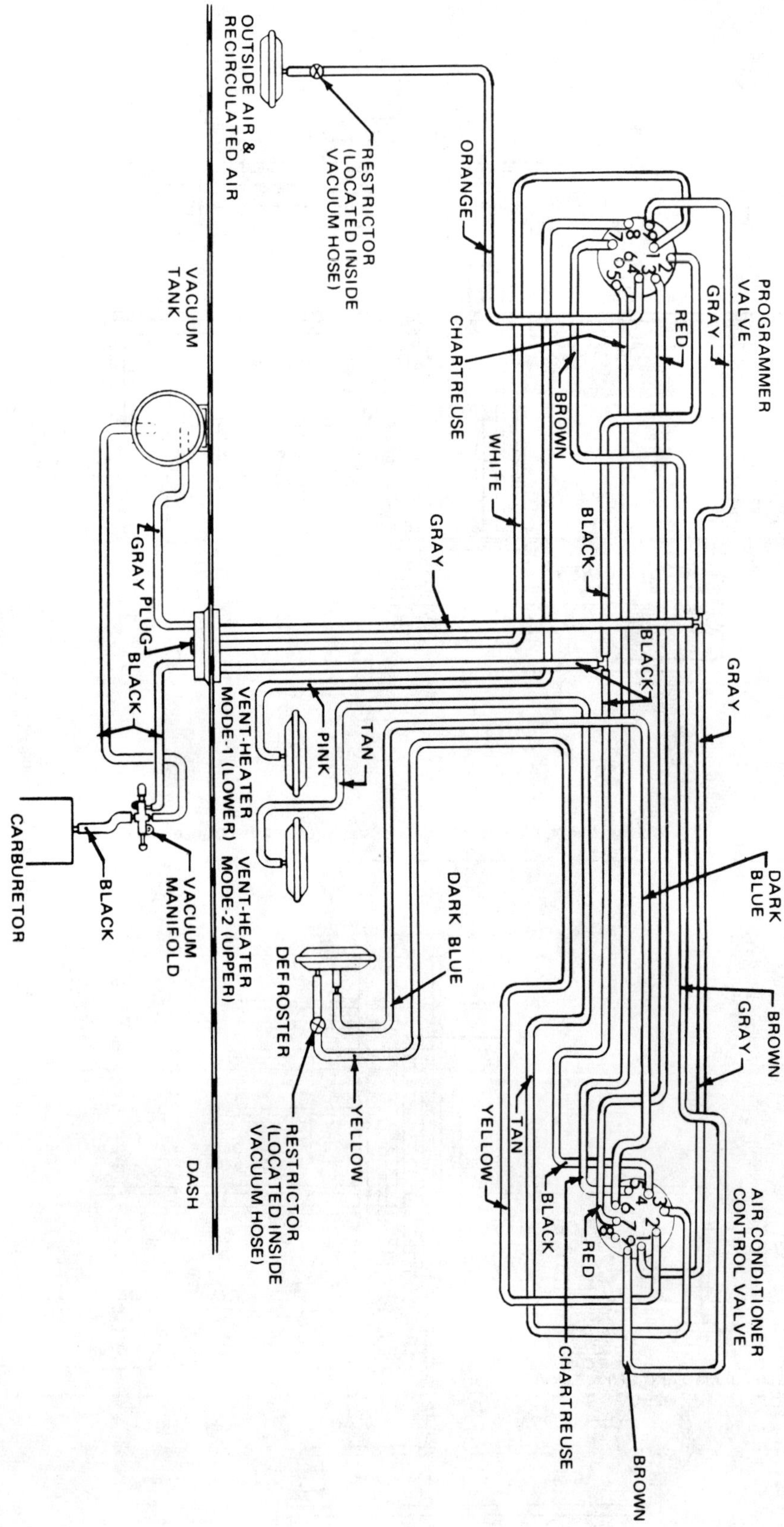

1976 Buick A/C & Heater Vacuum Circuit (With Auto Climate Control)

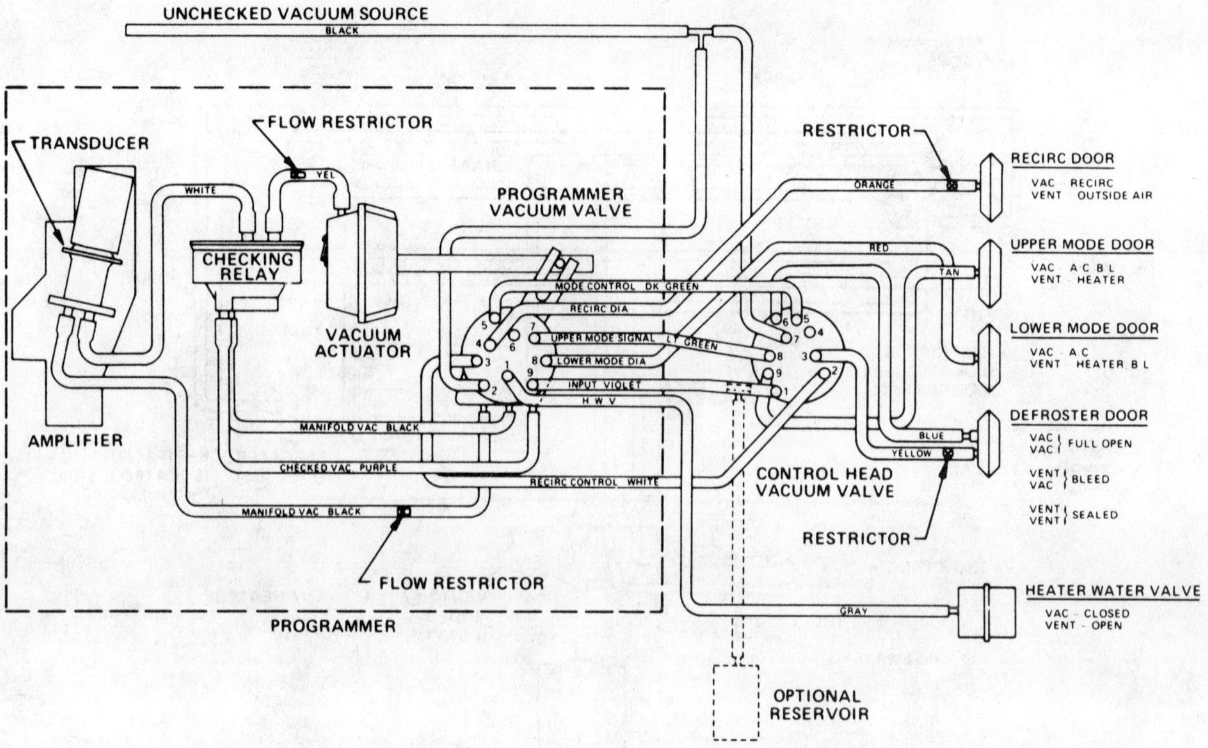

1977 Buick Electra & Riviera A/C & Heater Vacuum Circuit (With Auto Climate Control)

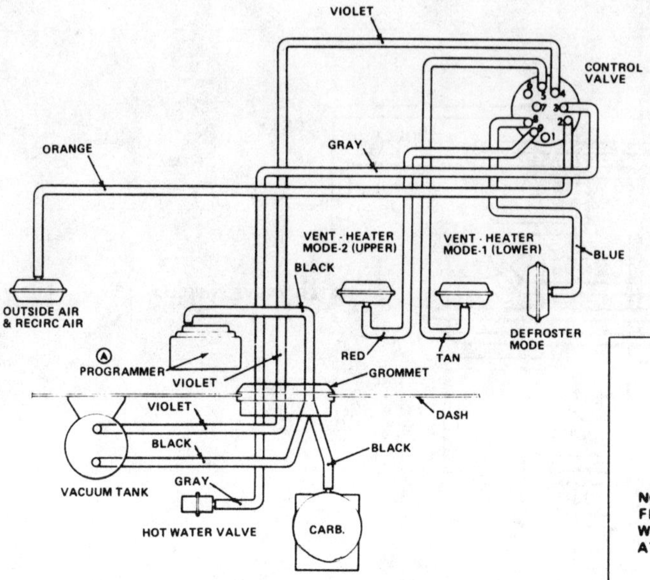

1977 Buick, LeSabre & Riviera A/C Vacuum Circuit (With Custom Aire)

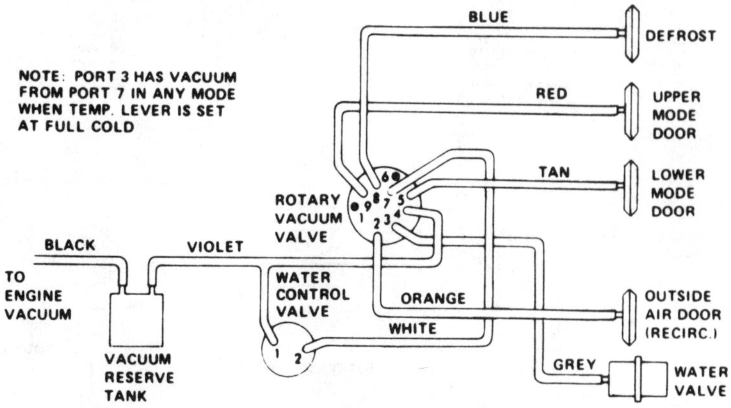

1980 Buick Electra & LeSabre A/C & Heater Vacuum Circuit (Less Auto Climate Control)

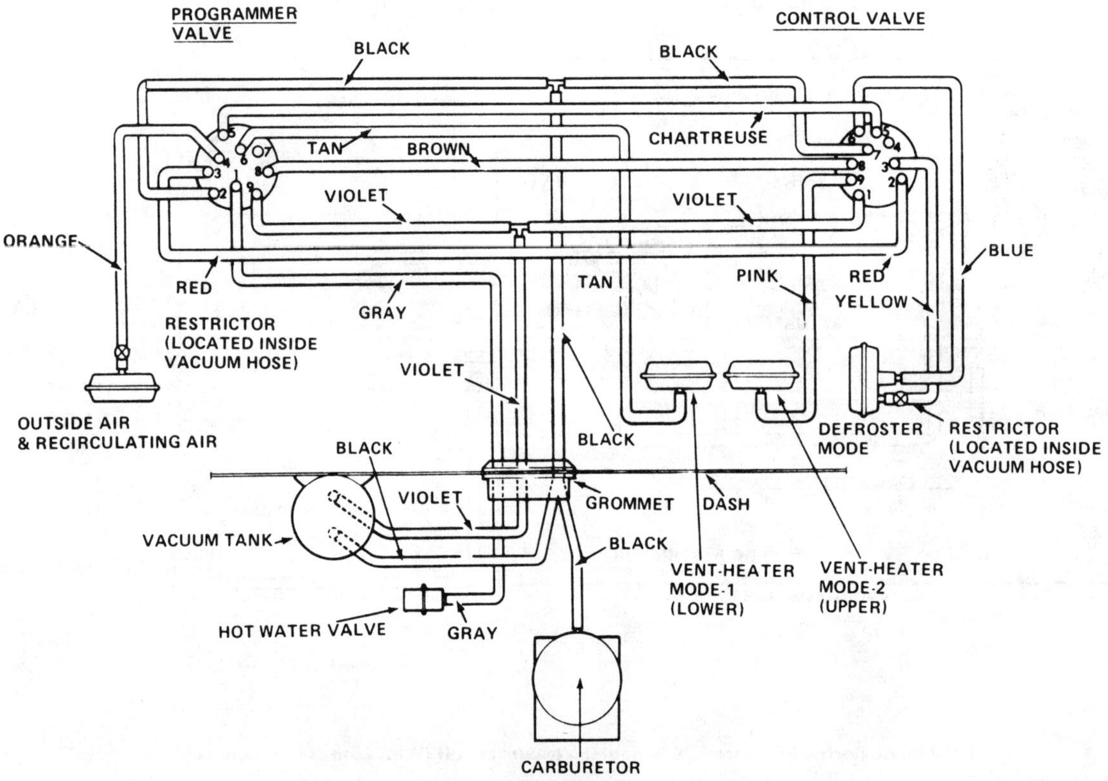

1978 Buick A/C & Heater Vacuum Circuit (With Auto Climate Control)

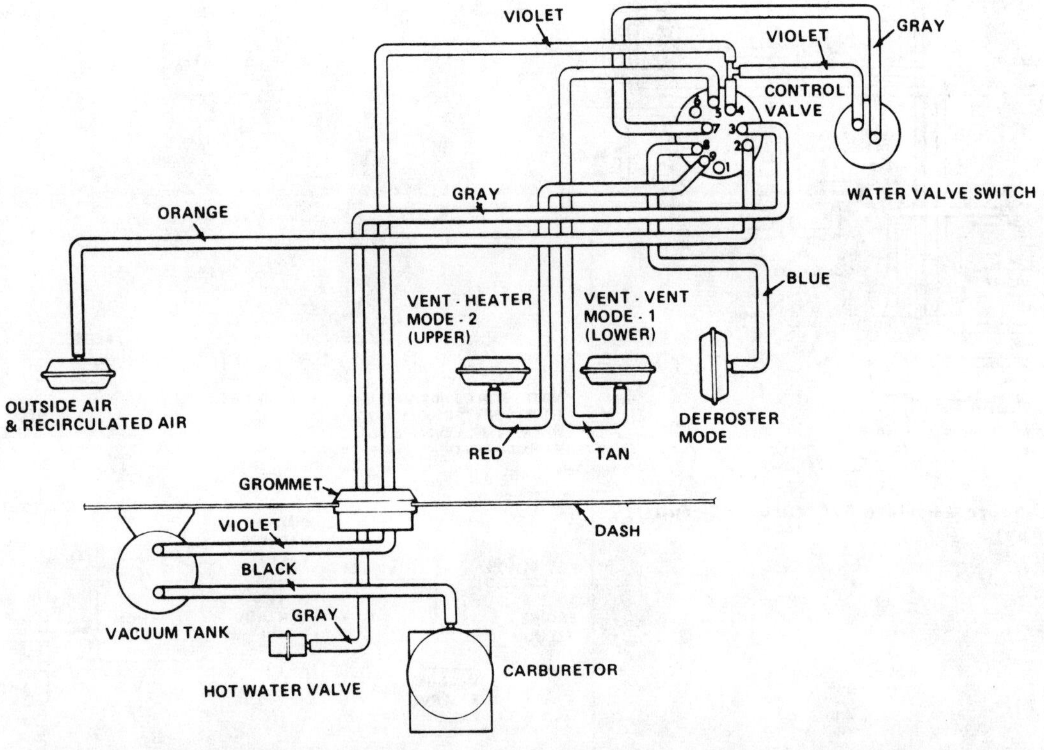

1978 Buick Riviera; 1978-79 Buick Electra & LeSabre A/C & Heater Vacuum Circuit (Less Auto Climate Control)

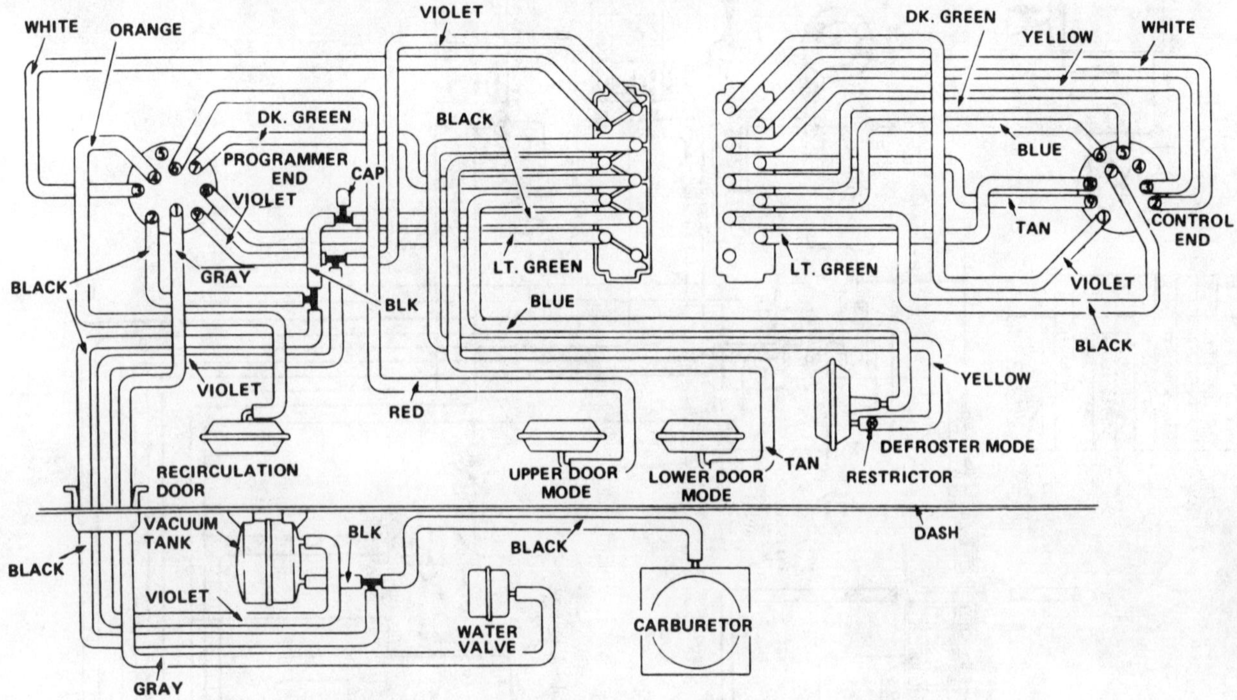

1979-80 Buick Riviera A/C & Heater Vacuum Circuit (With Auto Climate Control)

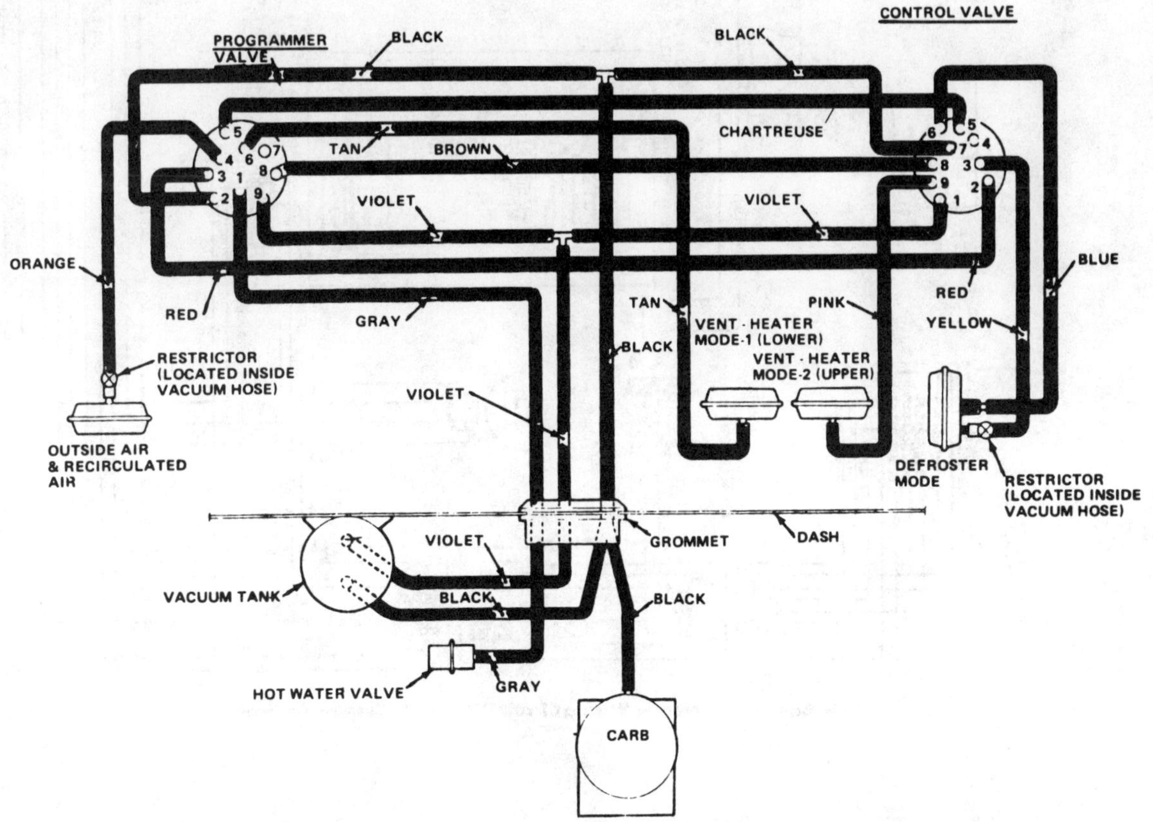

1979-80 Buick Electra & LeSabre A/C & Heater Vacuum Circuit (With Auto Climate Control)

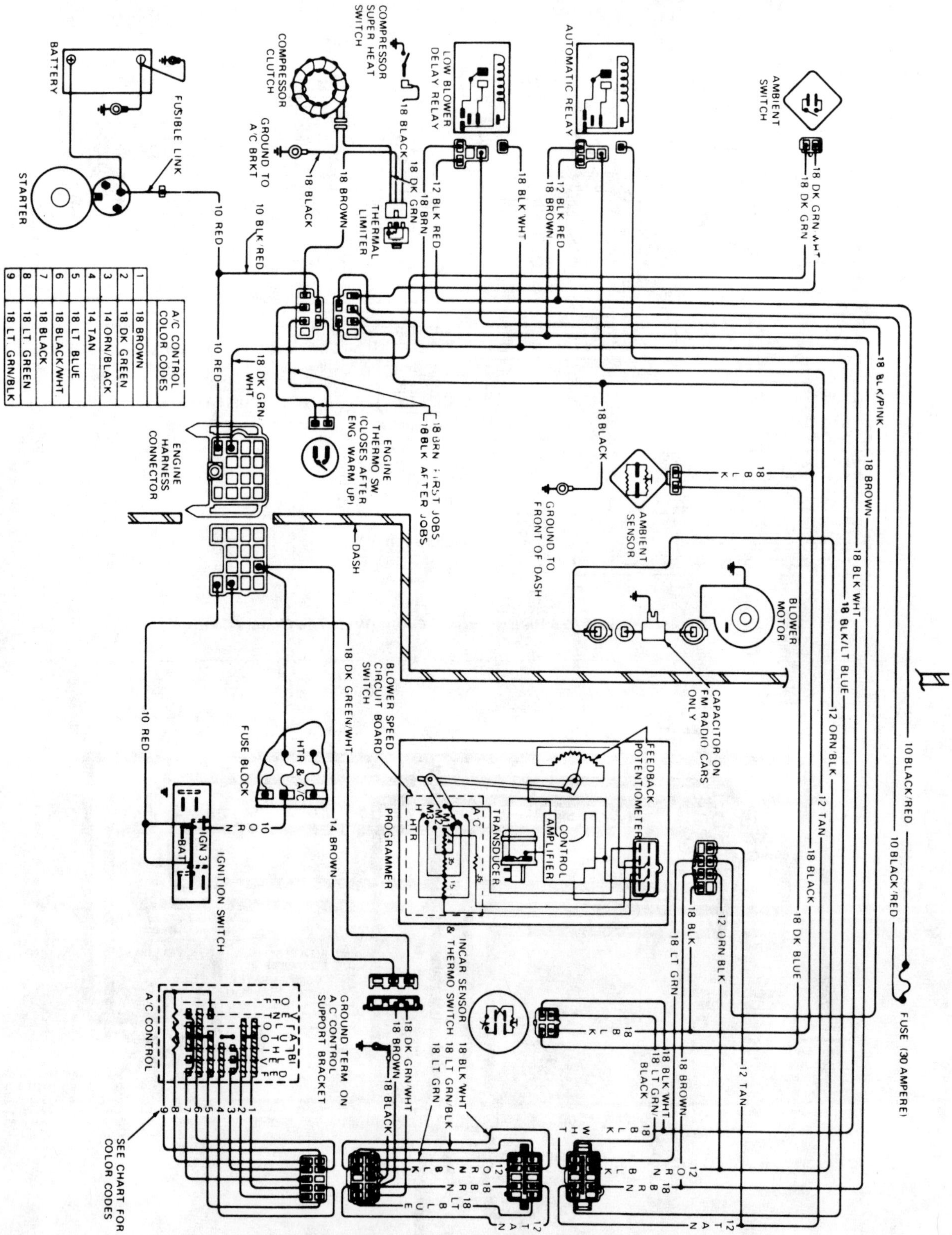

1974 Buick A/C & Heater Wiring Circuit (With Auto Climate Control)

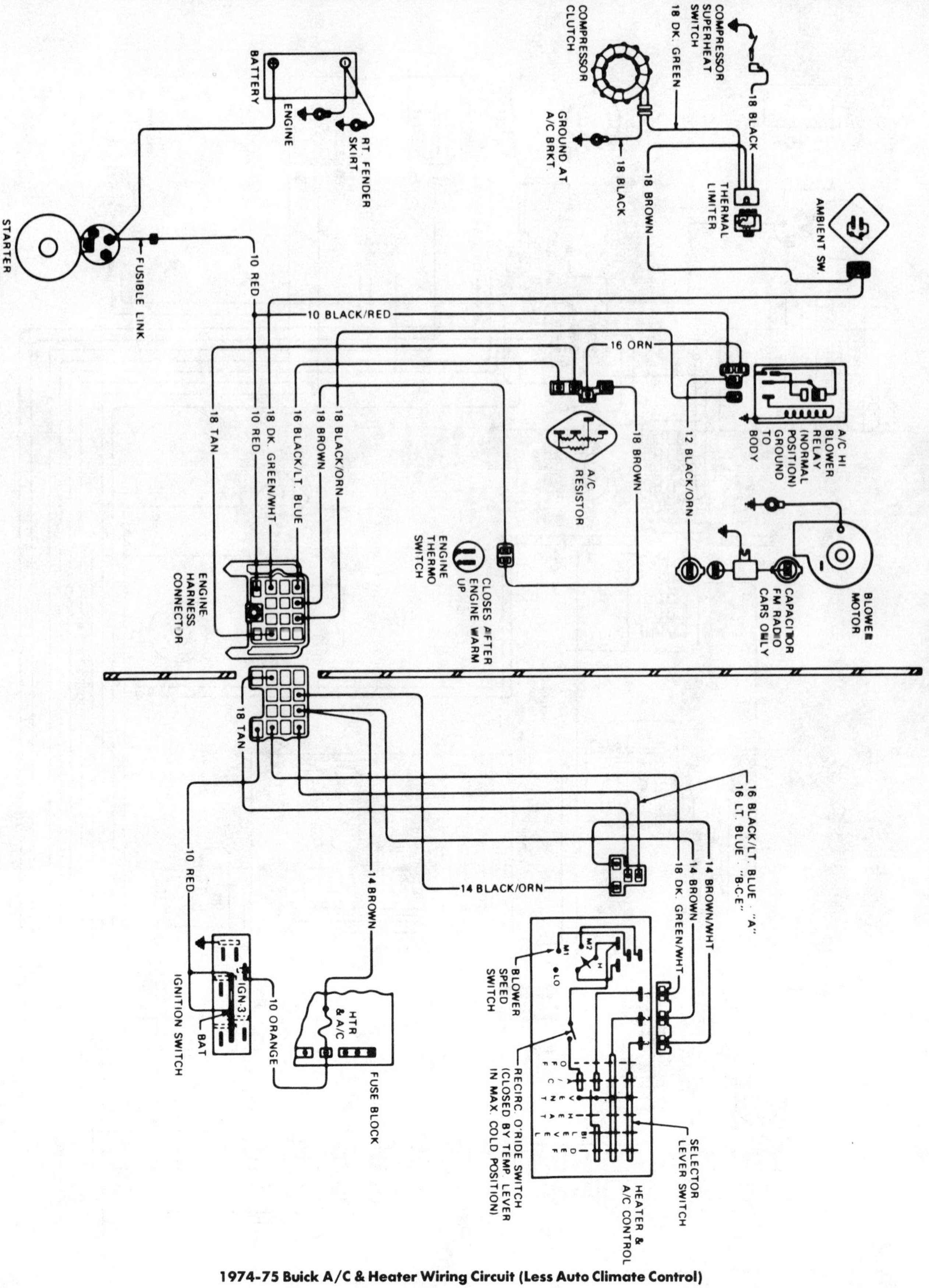

1974-75 Buick A/C & Heater Wiring Circuit (Less Auto Climate Control)

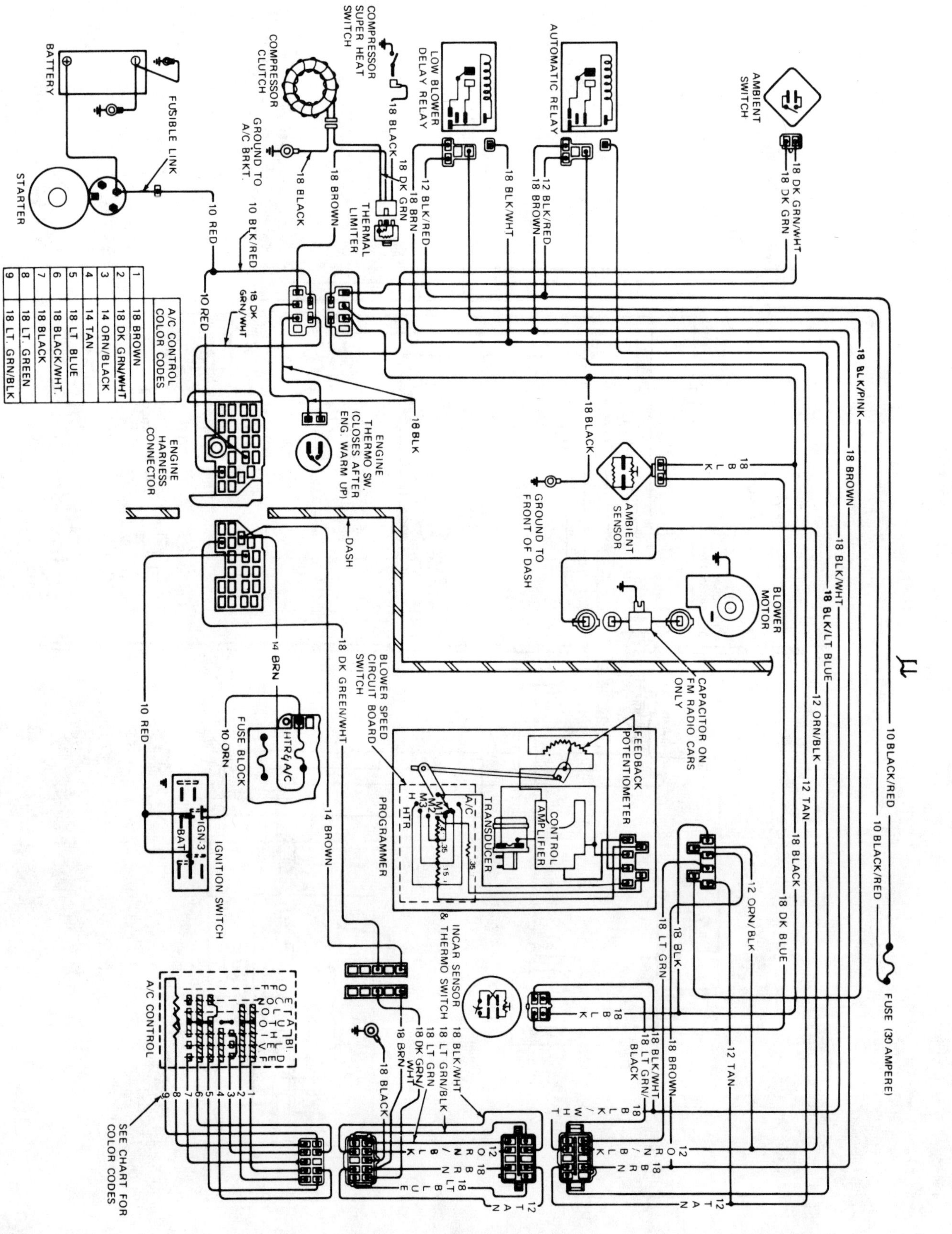

1975 Buick A/C & Heater Wiring Circuit (With Auto Climate Control)

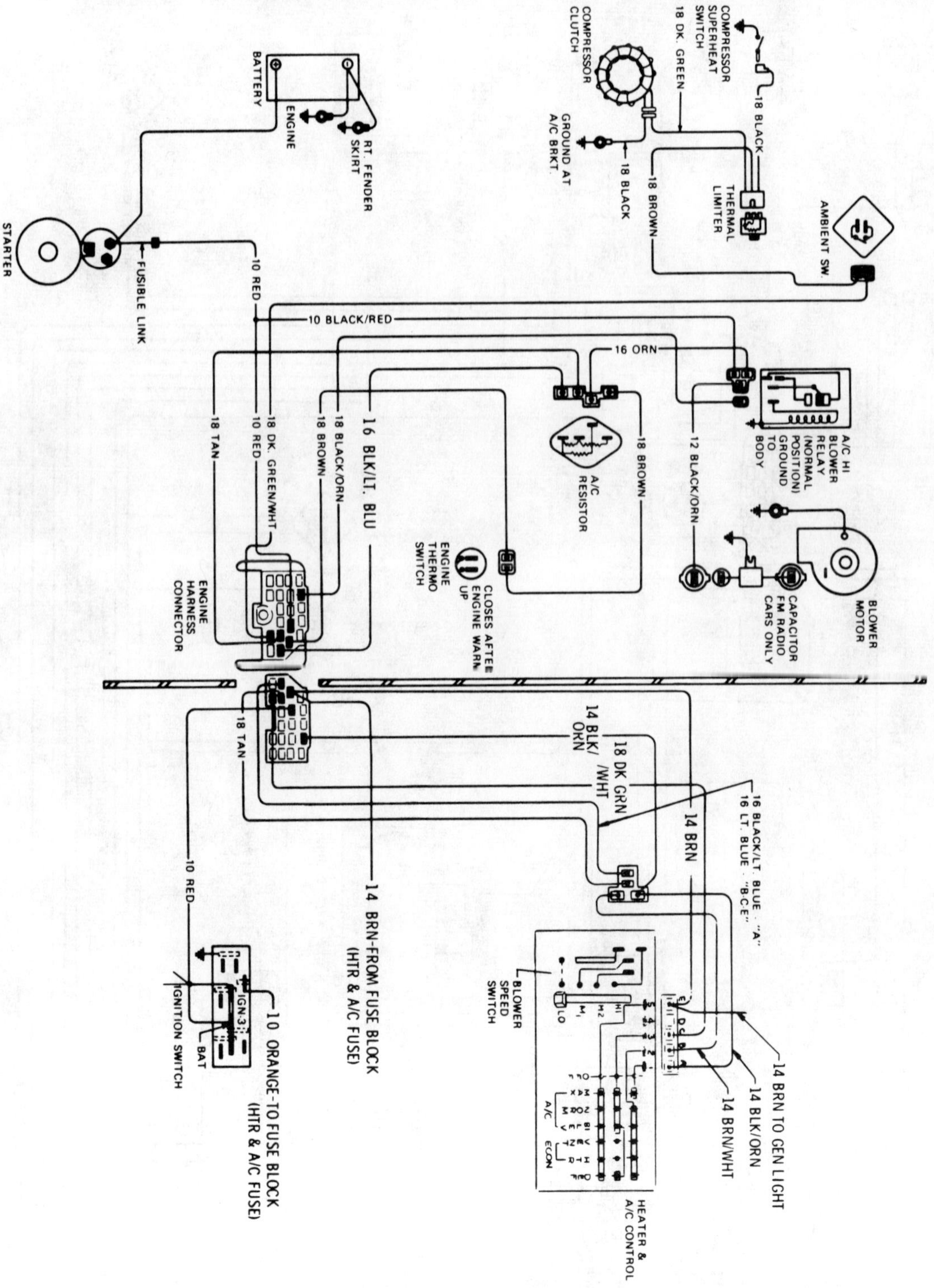

1976 Buick A/C & Heater Wiring Circuit (Less Auto Climate Control or Custom Aire)

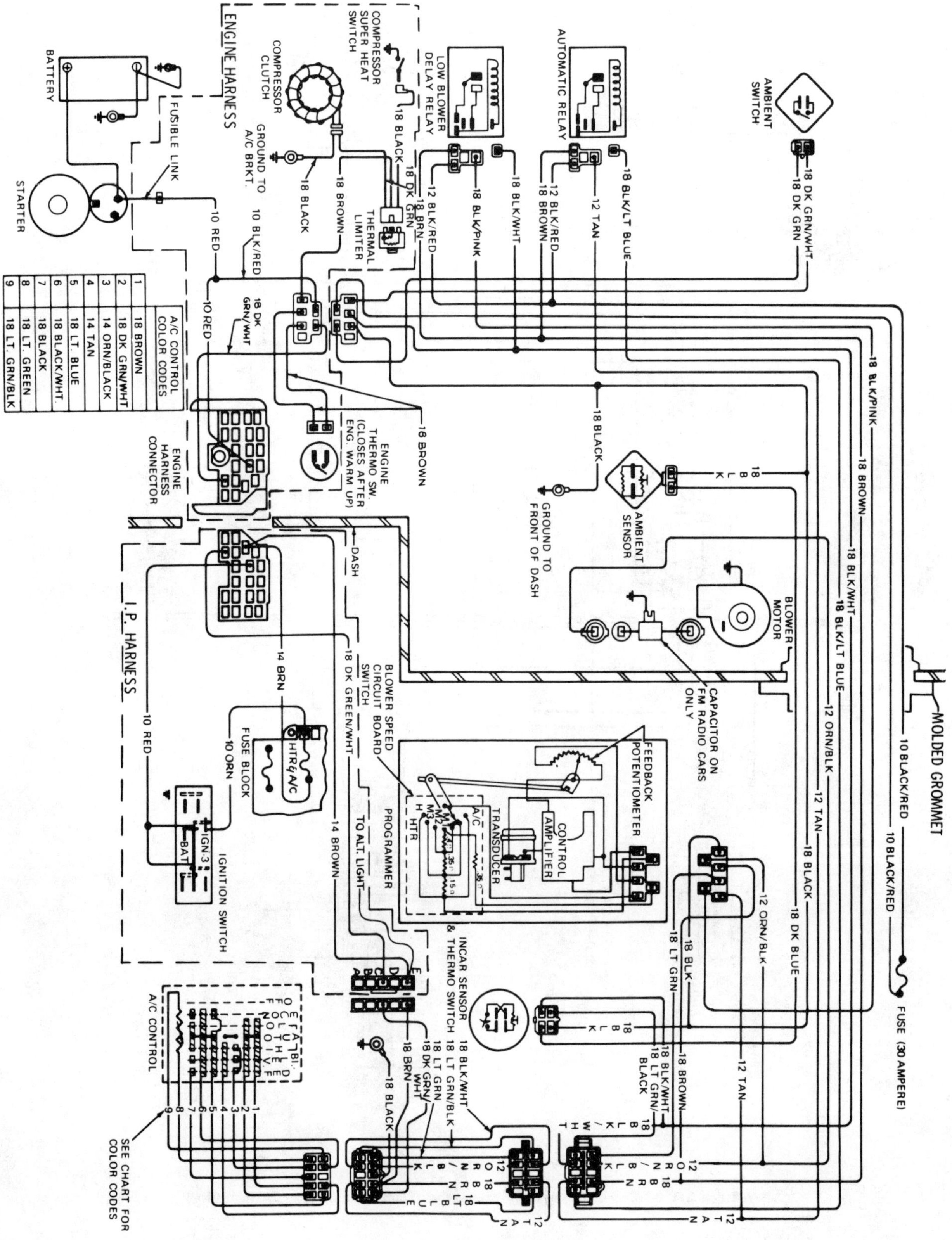

1976 Buick Except LeSabre A/C & Heater Wiring Circuit (With Auto Climate Control)

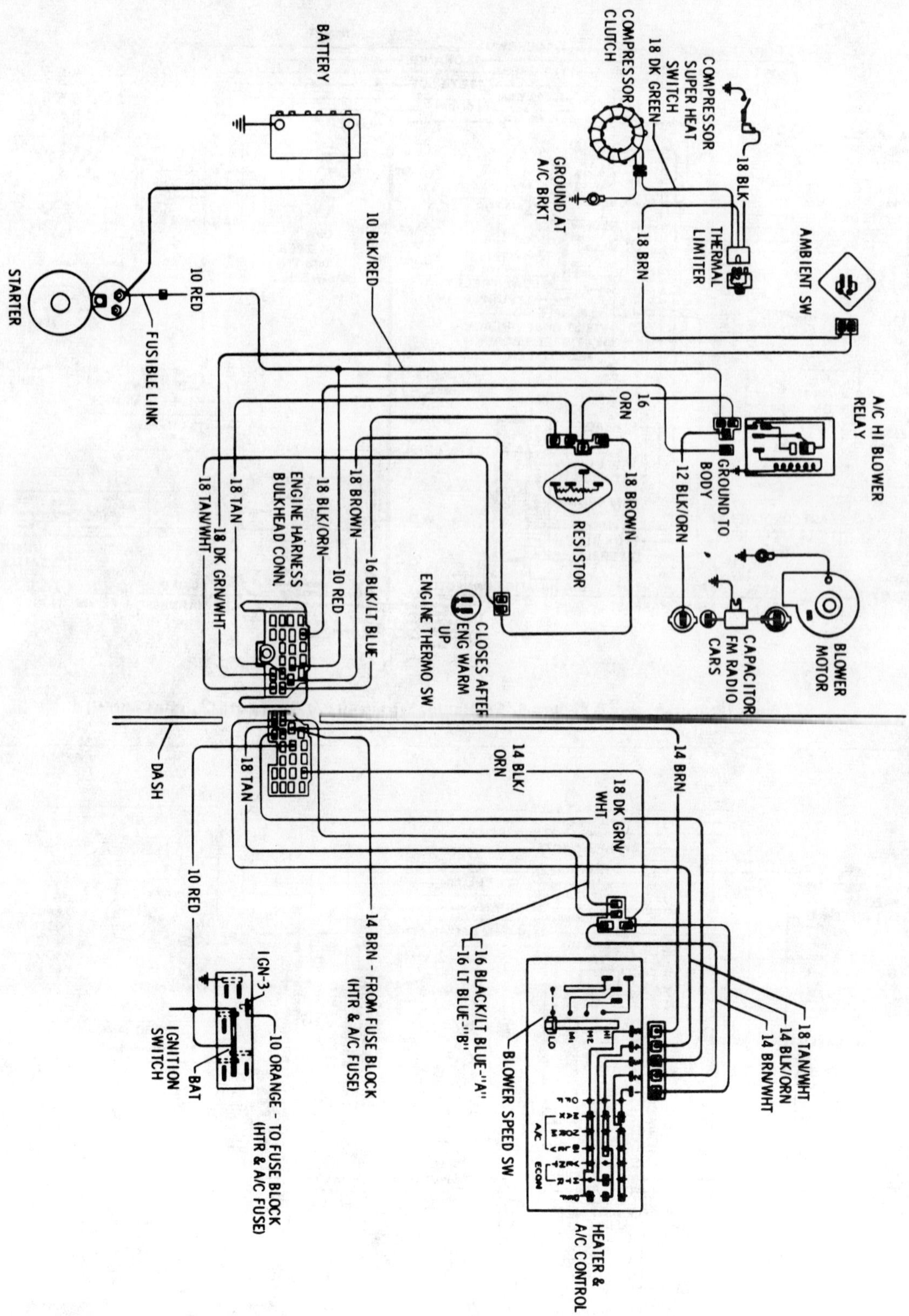

1976 Buick LeSabre A/C & Heater Wiring Circuit (With Custom Aire)

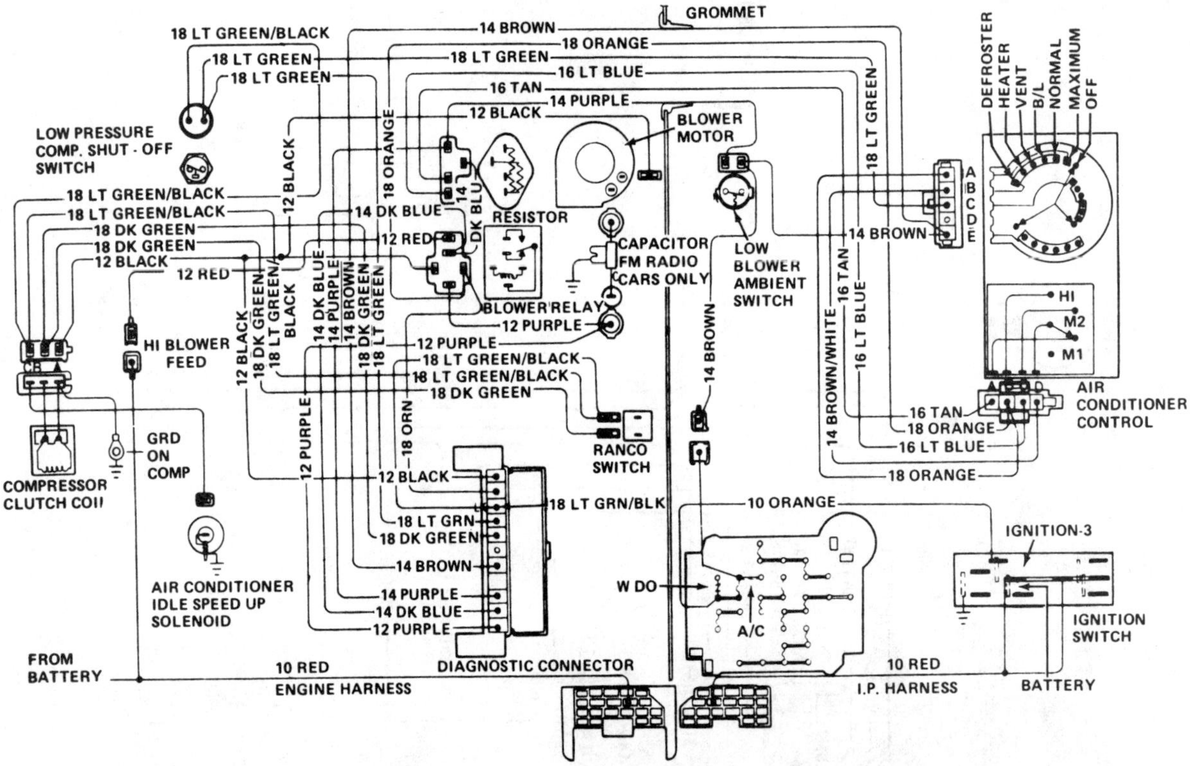

1977 Buick Electra, LeSabre & Riviera A/C & Heater Wiring Circuit (Less Auto Climate Control)

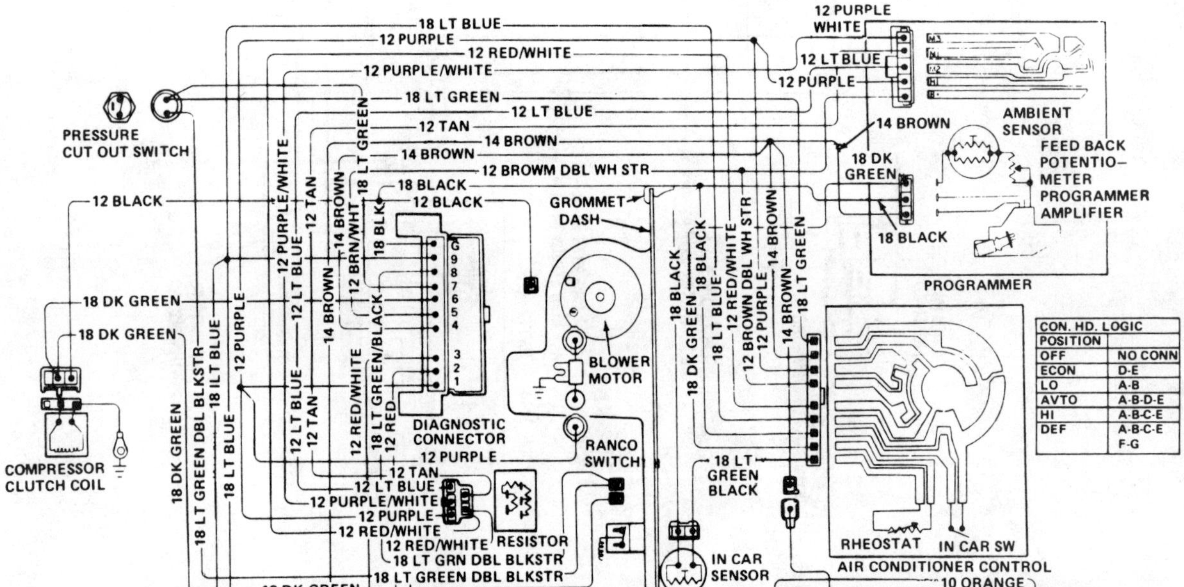

1977 Buick Electra, LeSabre & Riviera A/C & Heater Wiring Circuit (With Auto Climate Control)

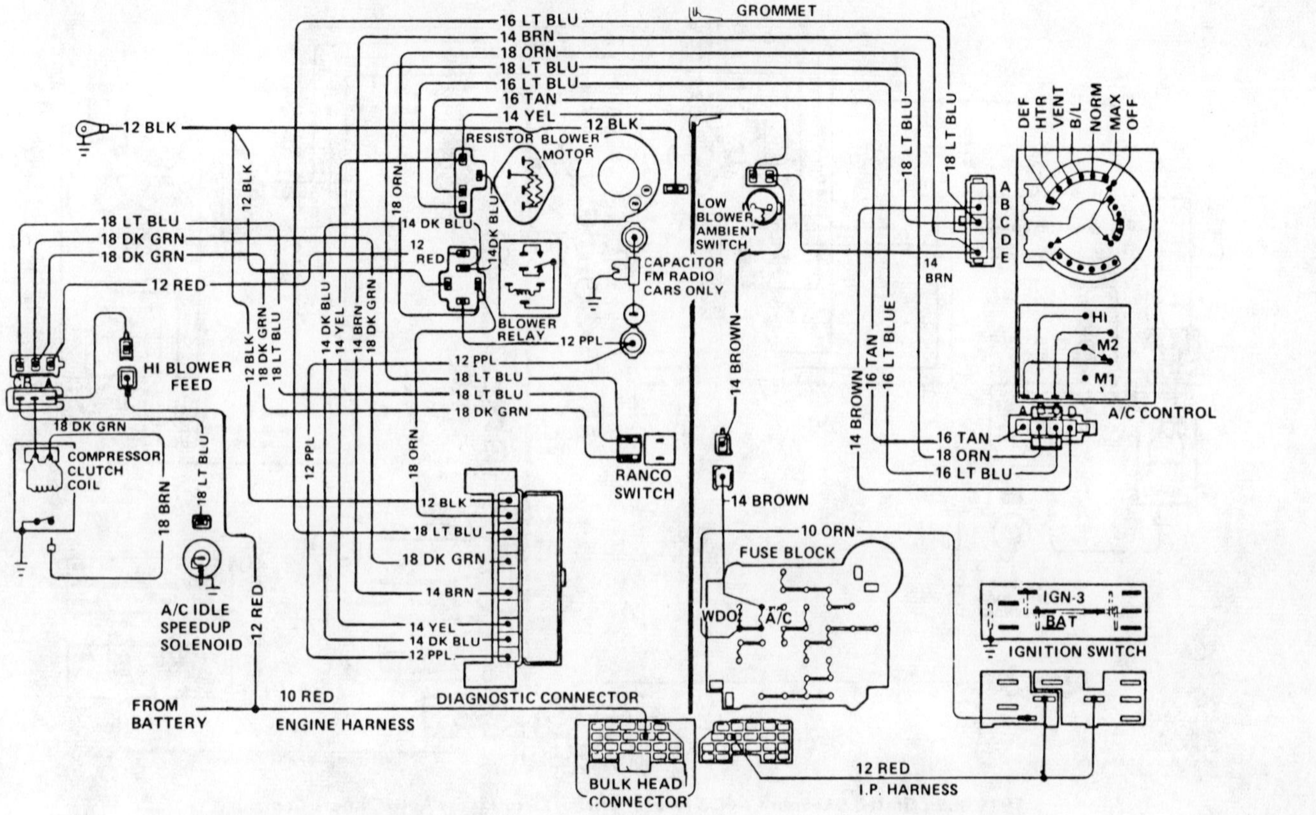

1978 Buick Electra, LeSabre & Riviera A/C & Heater Wiring Circuit (Less Auto Climate Control)

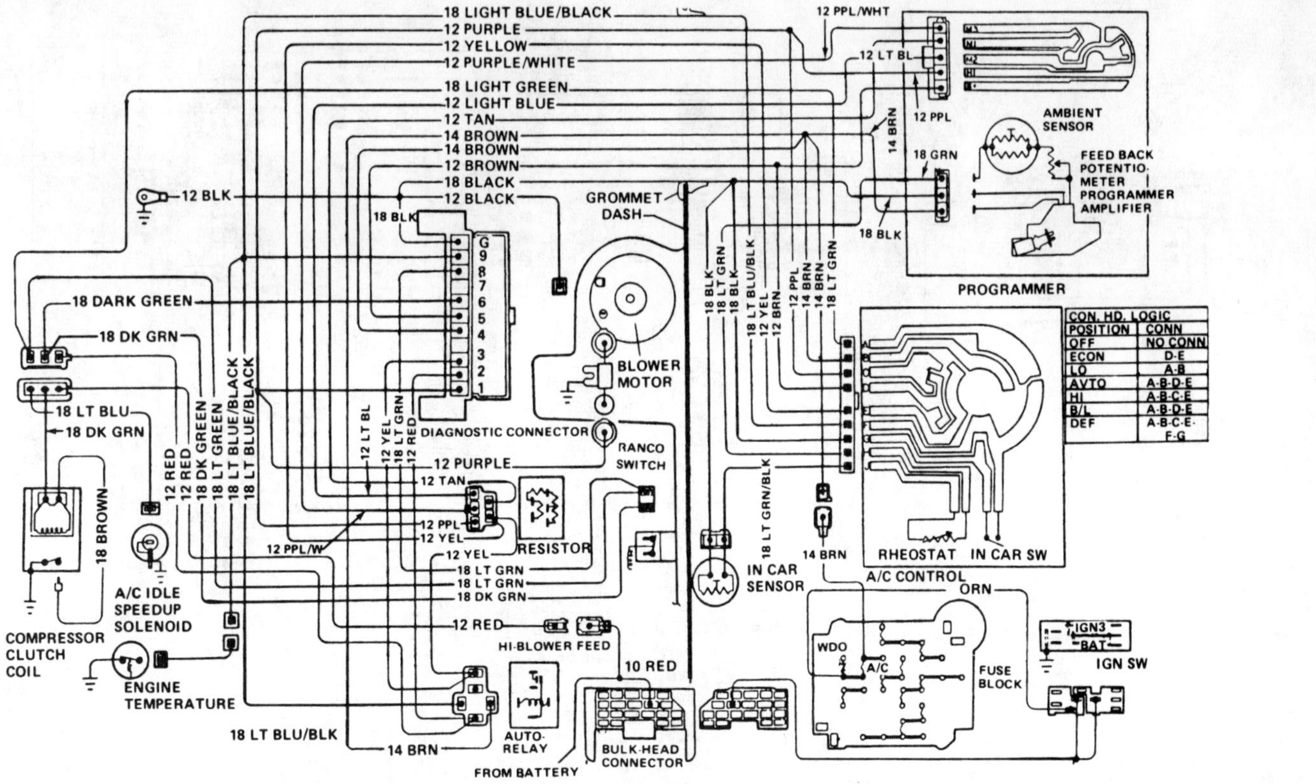

CON. HD. LOGIC	
POSITION	CONN
OFF	NO CONN
ECON	D-E
LO	A-B
AVTO	A-B-D-E
HI	A-B-C-E
B/L	A-B-D-E
DEF	A-B-C-E-F-G

1978 Buick Electra, LeSabre & Riviera A/C & Heater Wiring Circuit (With Auto Climate Control)

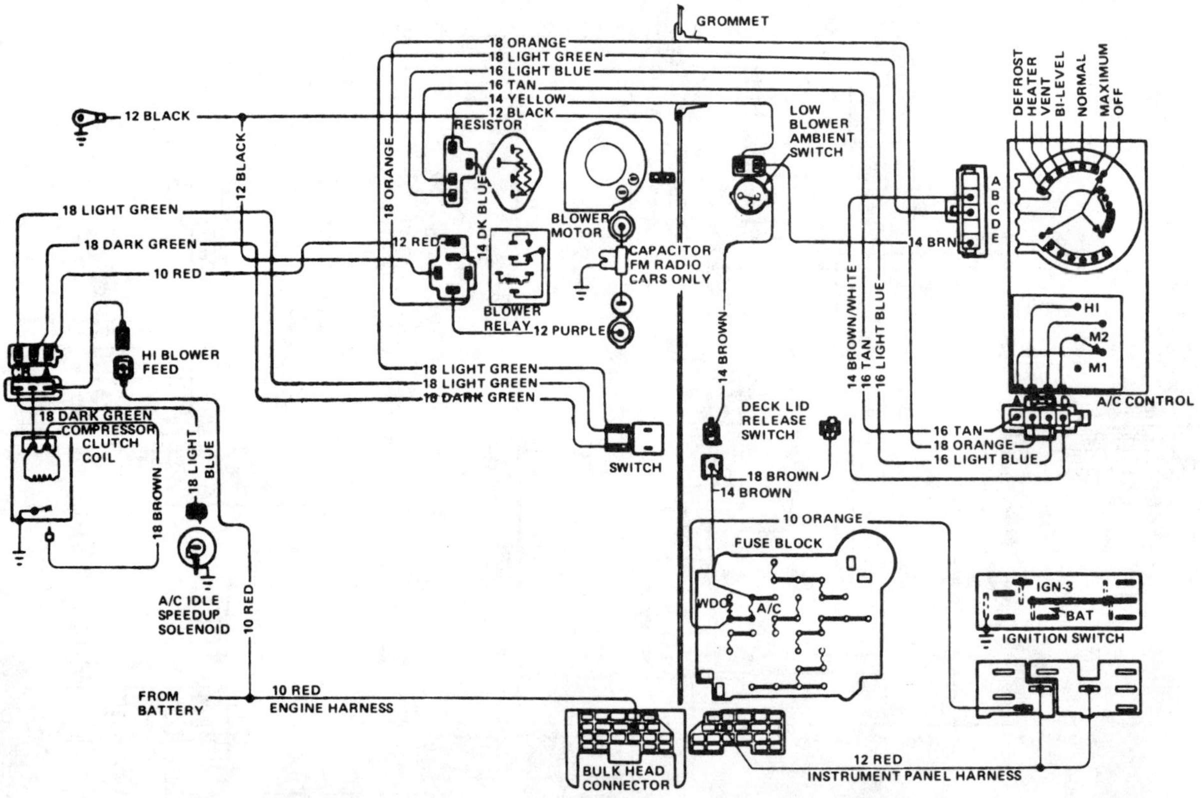

1979 Buick Electra & LeSabre A/C & Heater Wiring Circuit (Less Auto Climate Control)

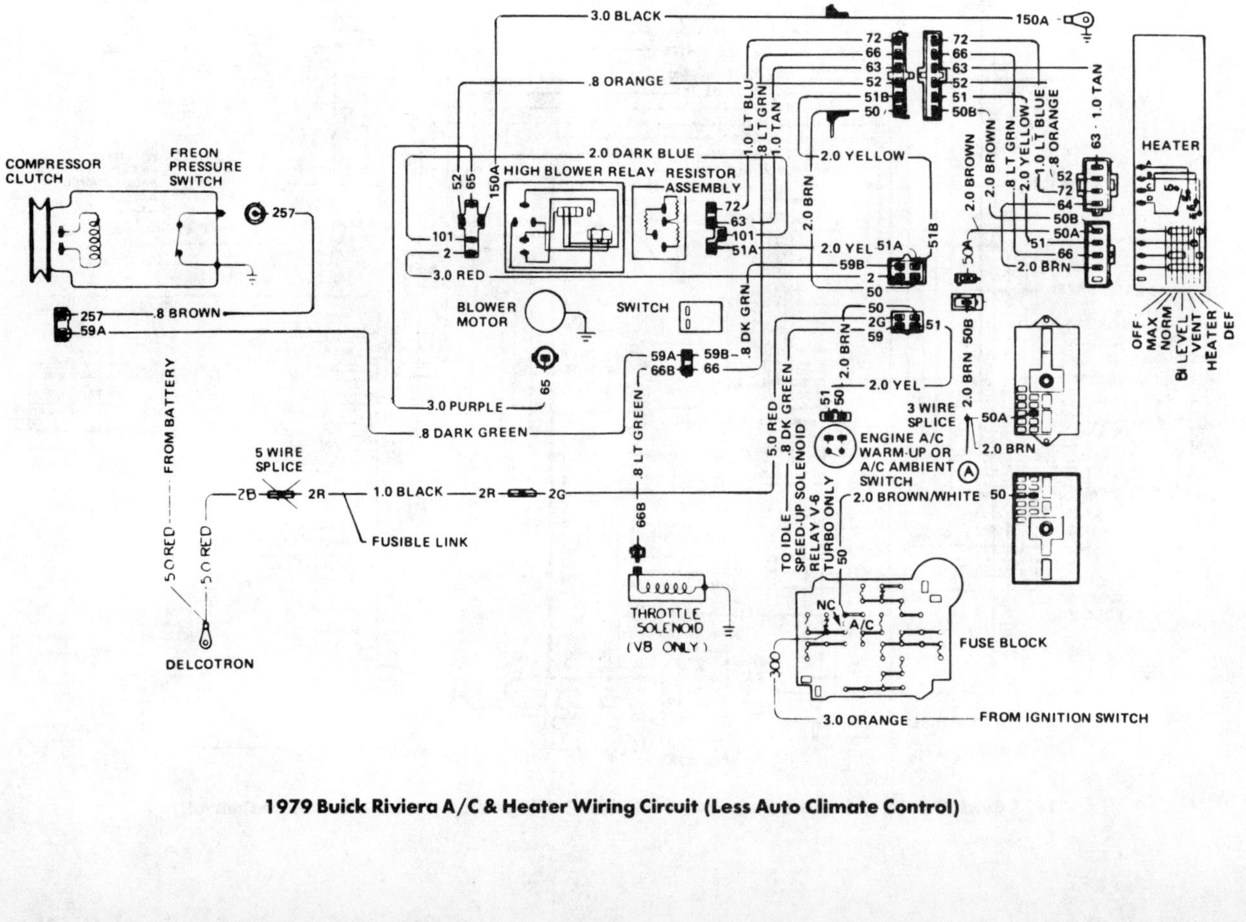

1979 Buick Riviera A/C & Heater Wiring Circuit (Less Auto Climate Control)

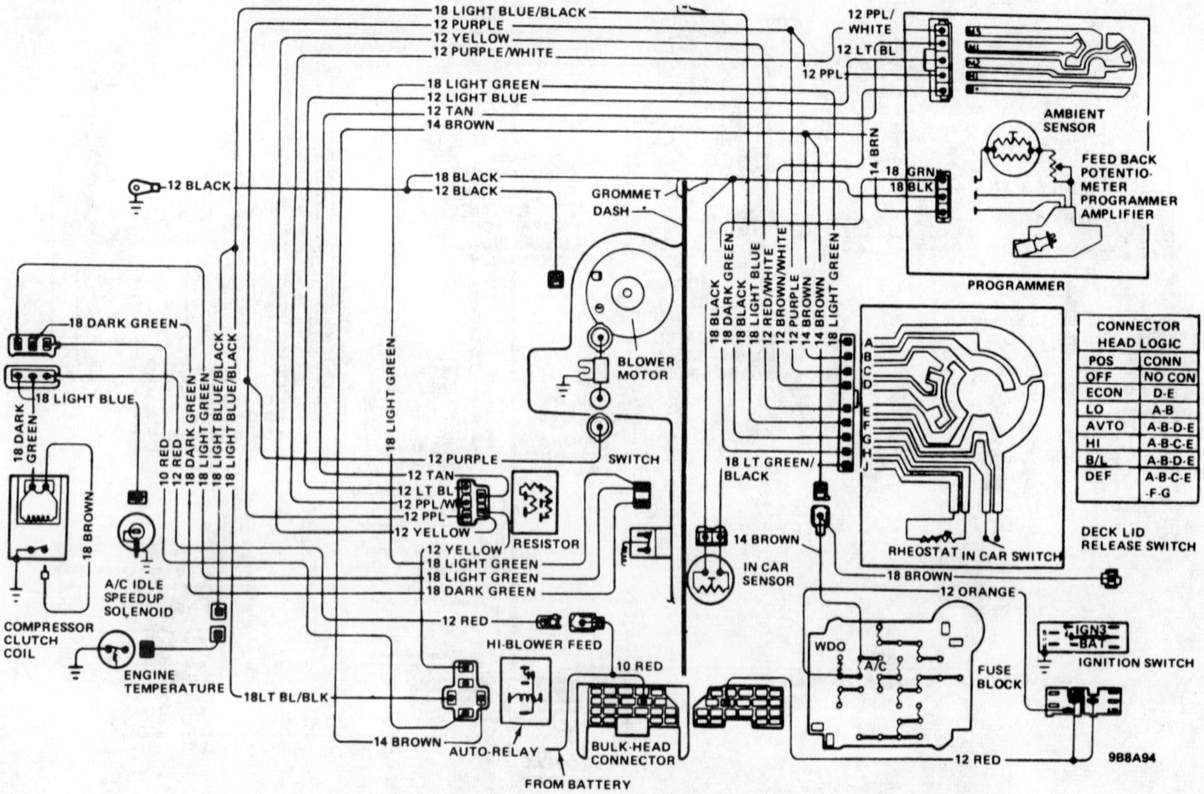

1979 Buick Electra & LeSabre A/C & Heater Wiring Circuit (With Auto Climate Control)

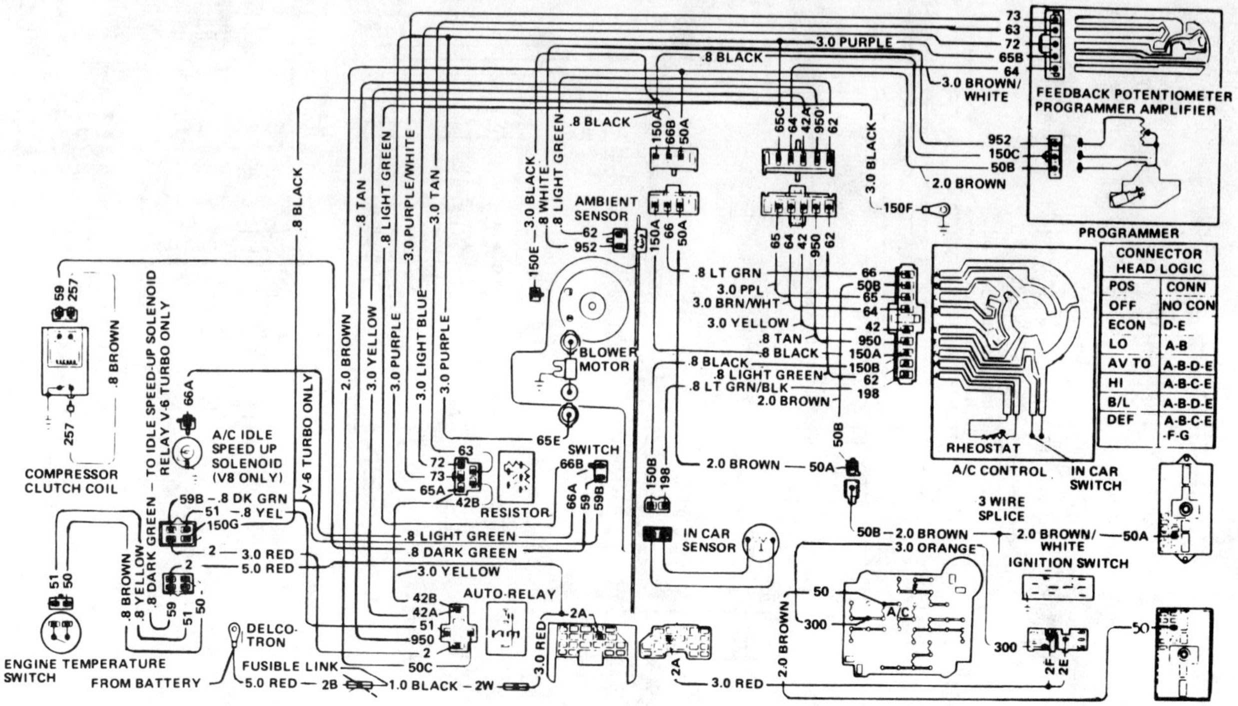

1979 Buick Riviera A/C & Heater Wiring Circuit (With Auto Climate Control)

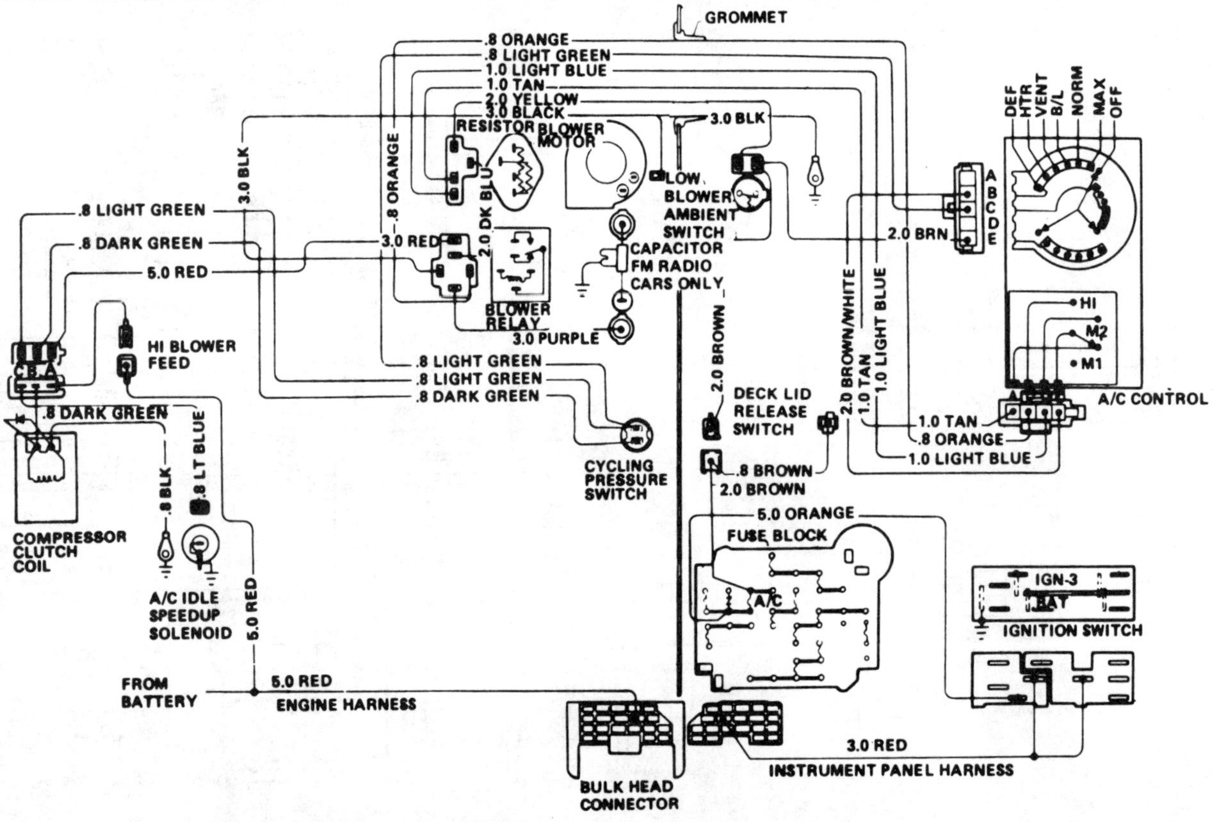

1980 Buick Electra & LeSabre A/C & Heater Wiring Circuit (With Manual A/C)

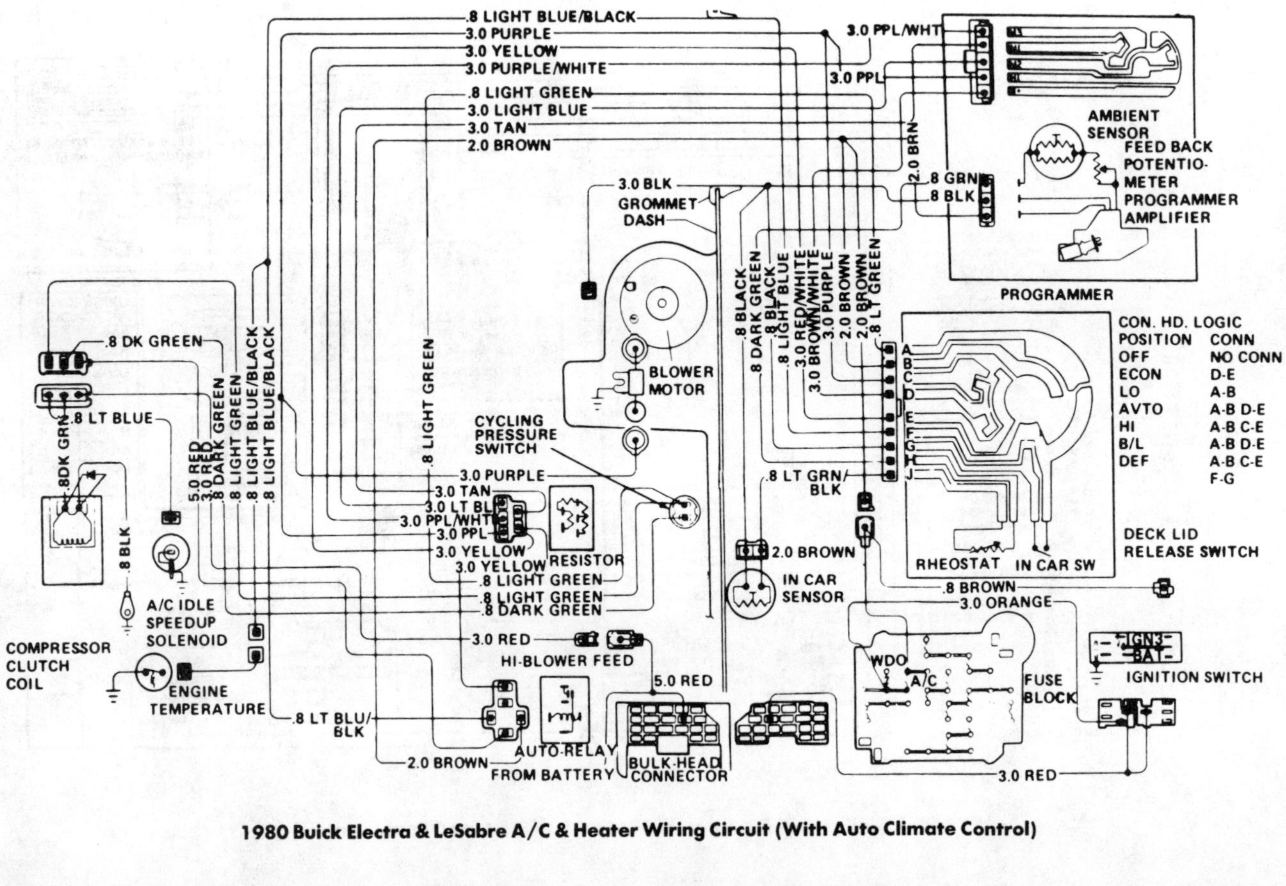

1980 Buick Electra & LeSabre A/C & Heater Wiring Circuit (With Auto Climate Control)

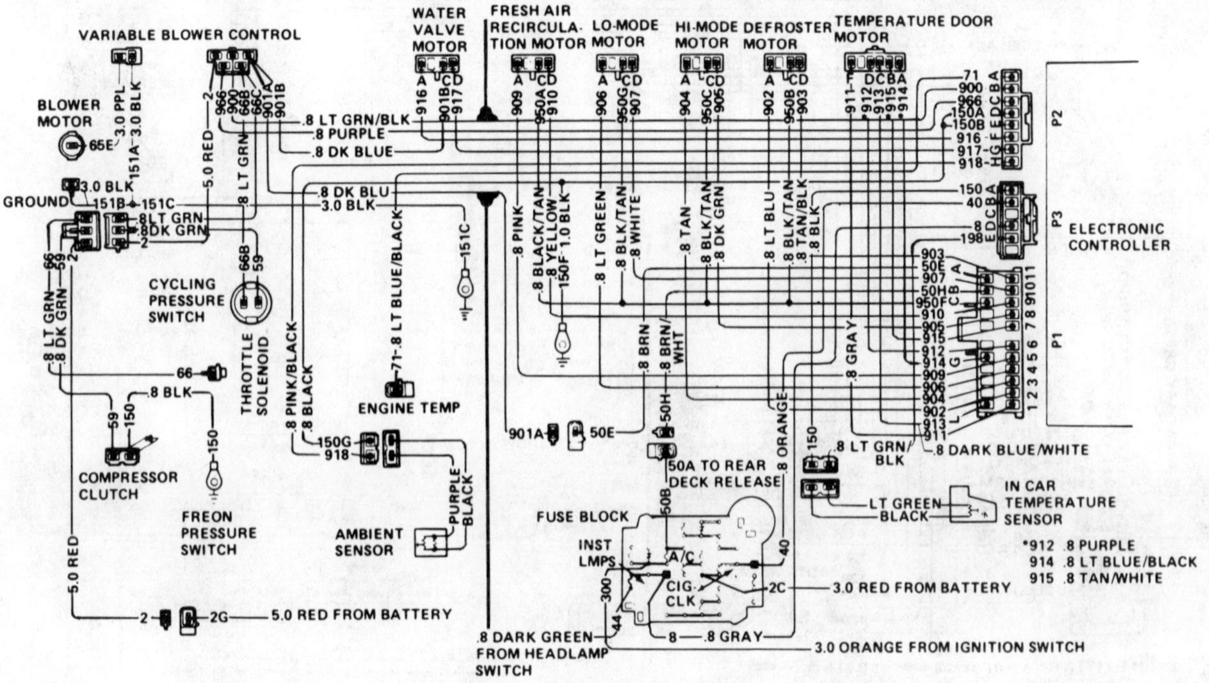

1980 Buick Electra & LeSabre A/C & Heater Wiring Circuit (With Electronic Touch Climate Control)

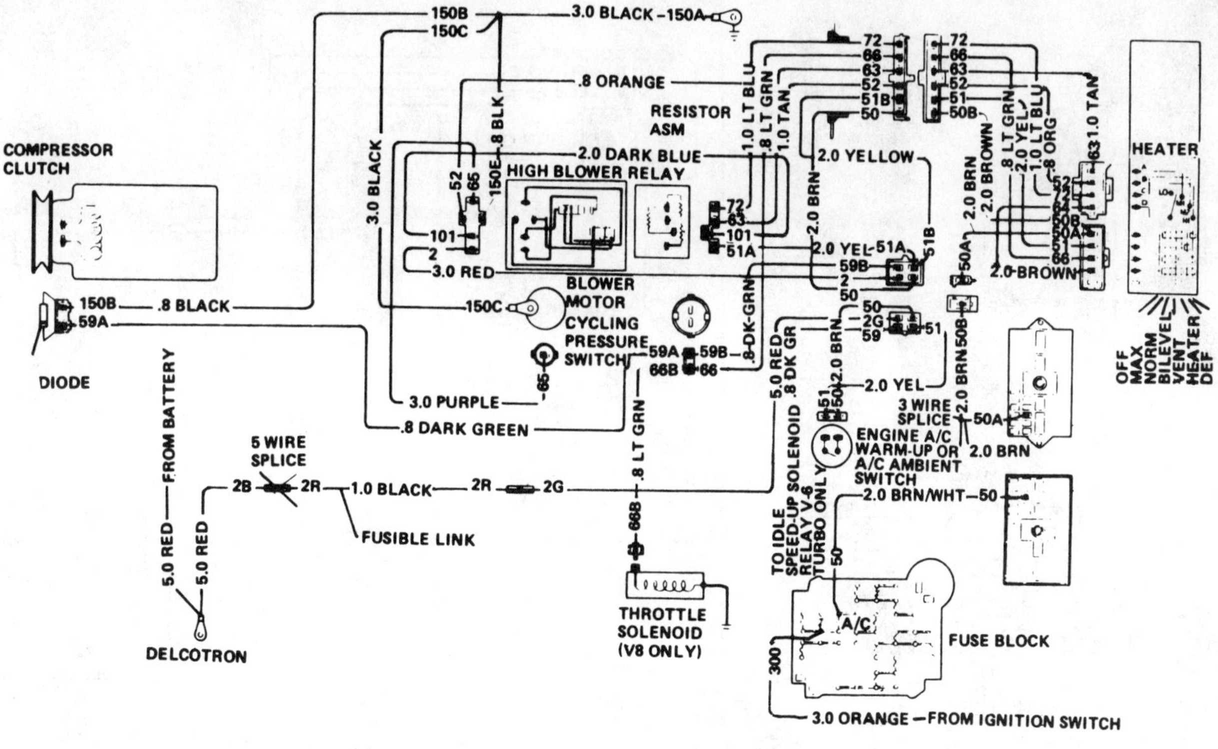

1980 Buick Riviera A/C & Heater Wiring Circuit (Less Auto Climate Control)

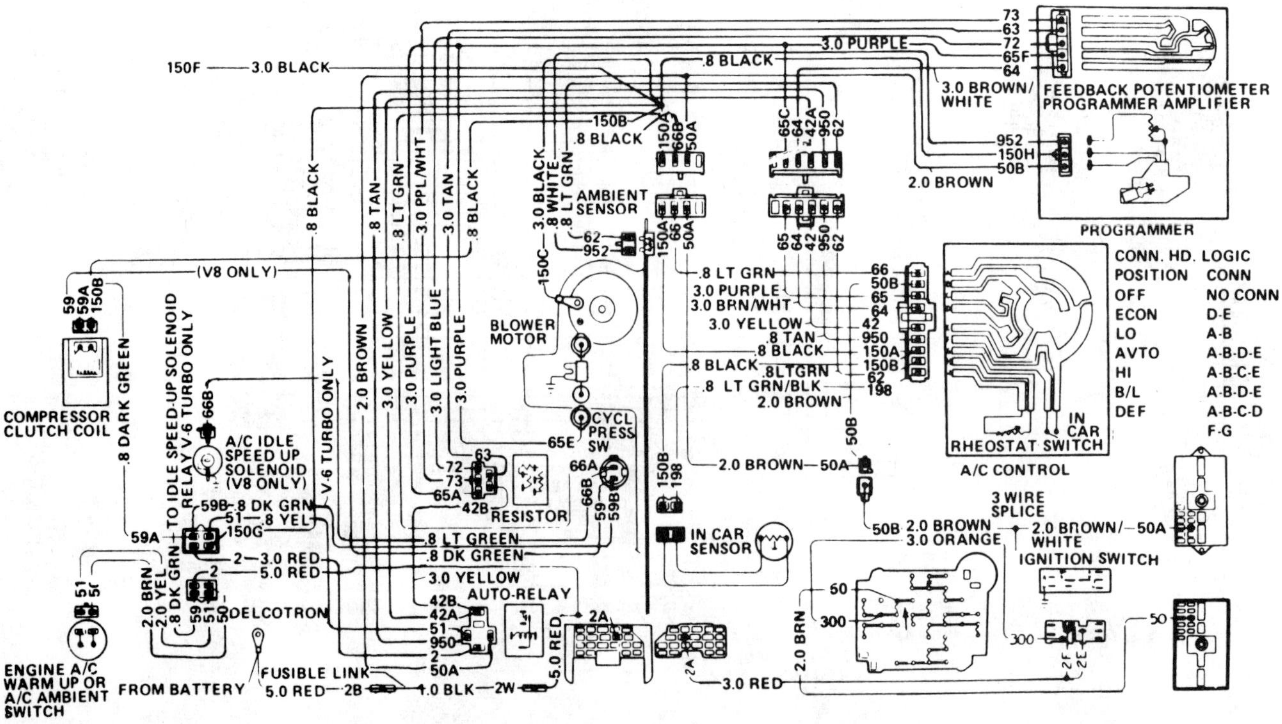

1980 Buick Riviera A/C & Heater Wiring Circuit (With Auto Climate Control)

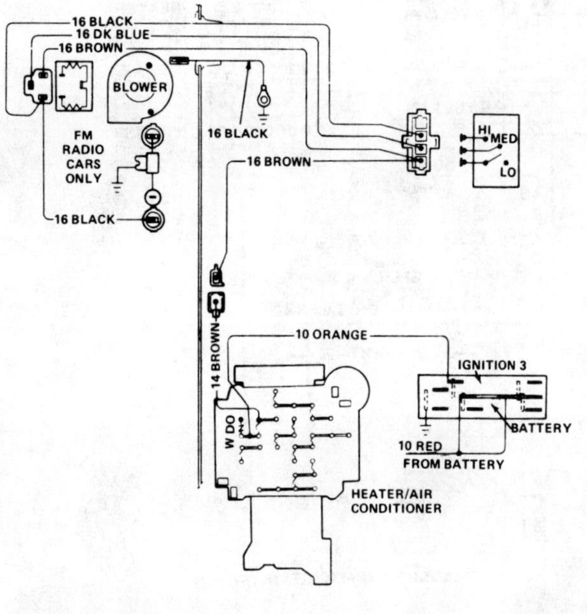

1977-79 Buick Electra, LeSabre & 1977-78 Riviera Heater Wiring Circuit

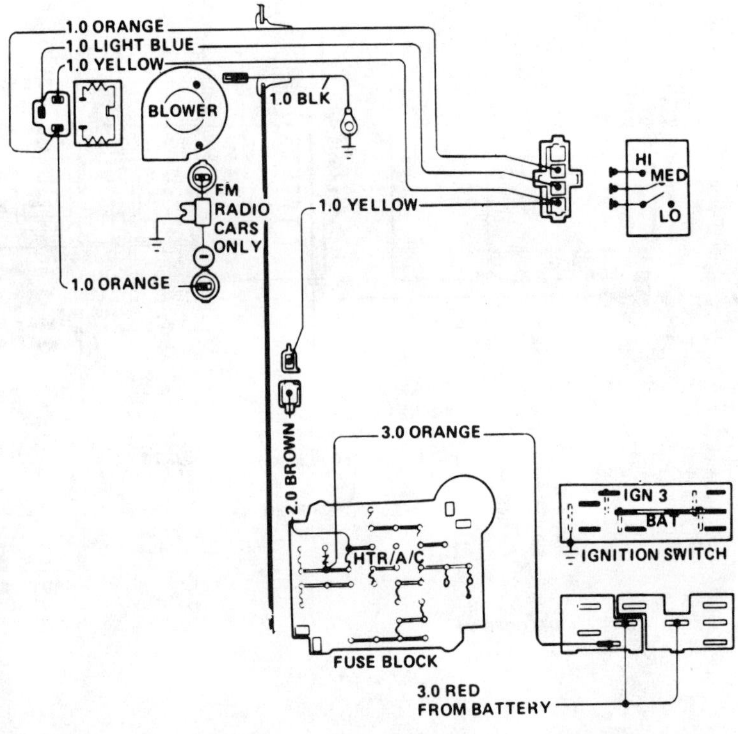

1980 Buick Electra & LeSabre Heater Wiring Circuit

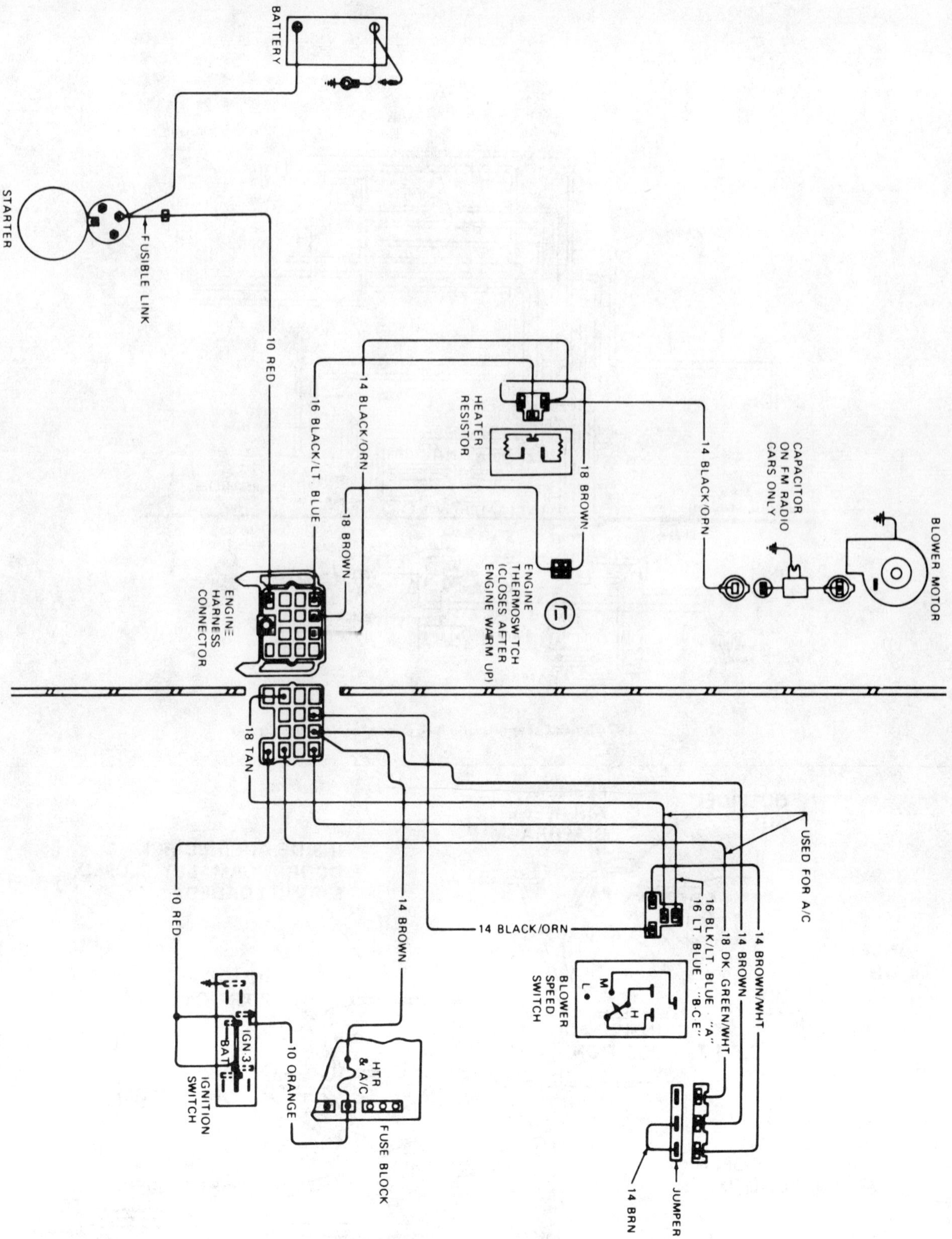

1974 Buick Heater Wiring Circuit

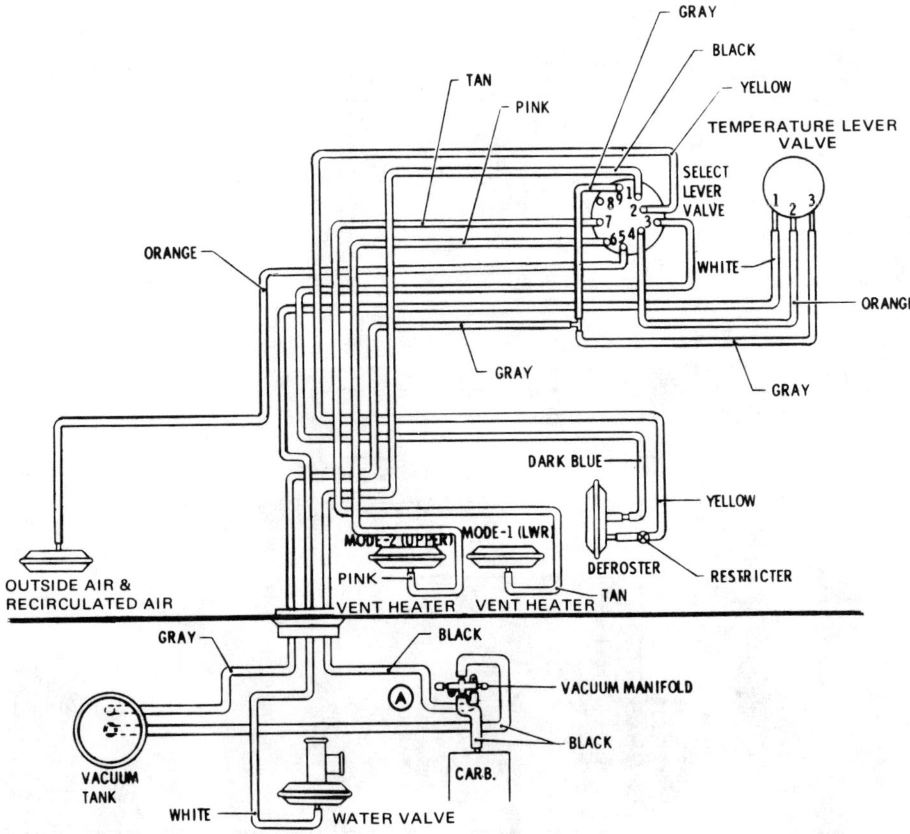

1974 Buick Except Apollo A/C & Heater Vacuum Circuit

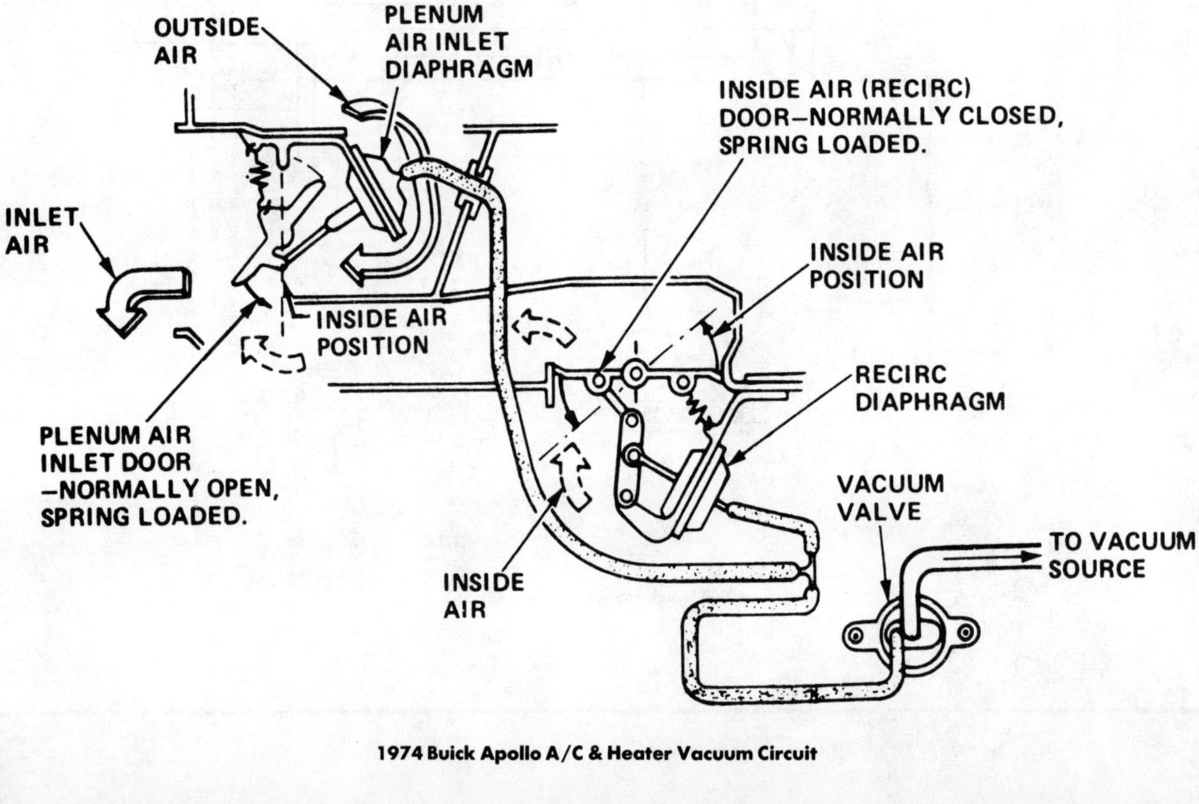

1974 Buick Apollo A/C & Heater Vacuum Circuit

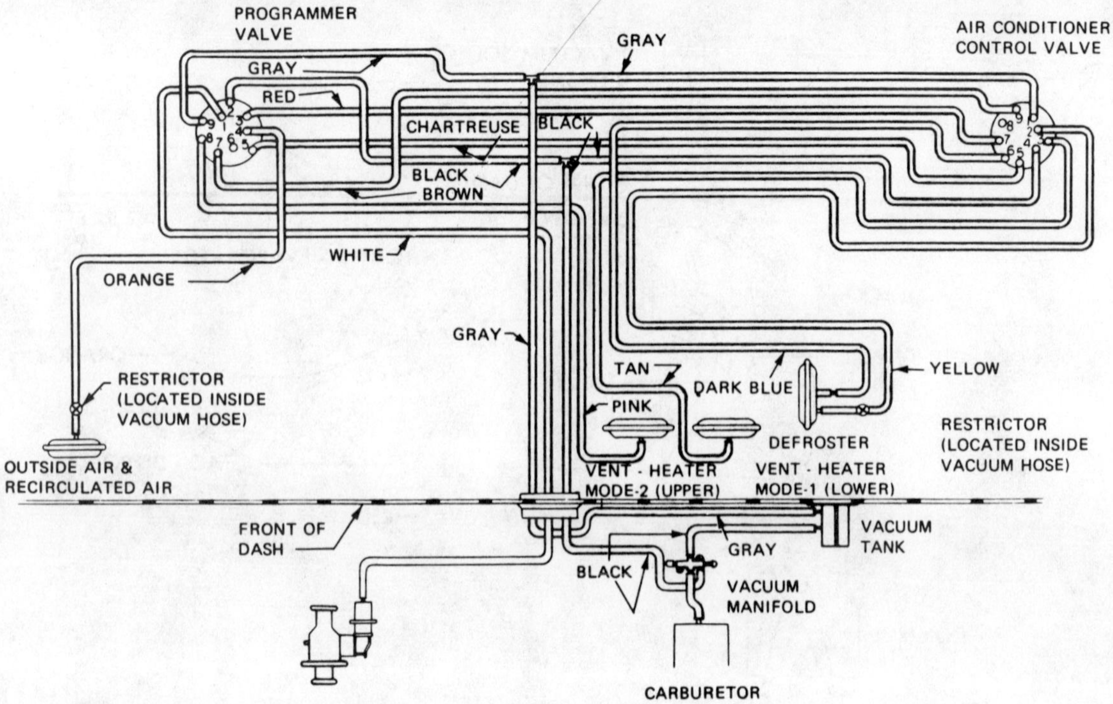

1974-75 Buick Except Apollo, Skyhawk & Skylark A/C & Heater Vacuum Circuit (With Auto Climate Control)

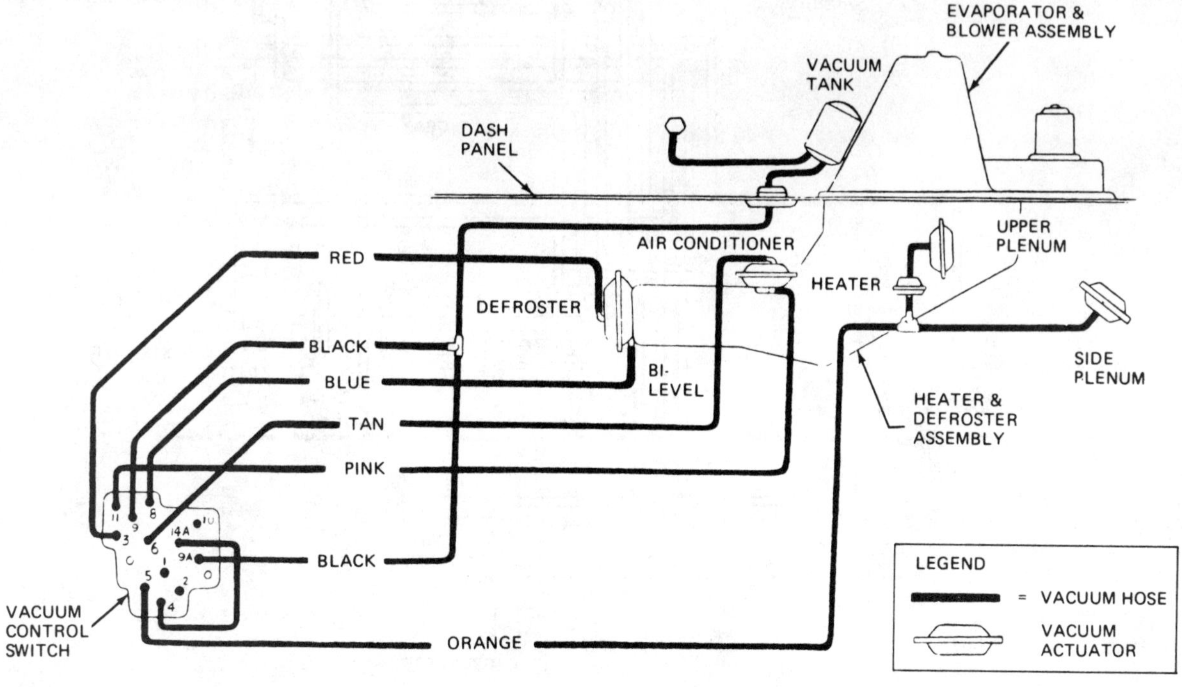

1975 Buick Apollo & Skylark A/C & Heater Vacuum Circuit

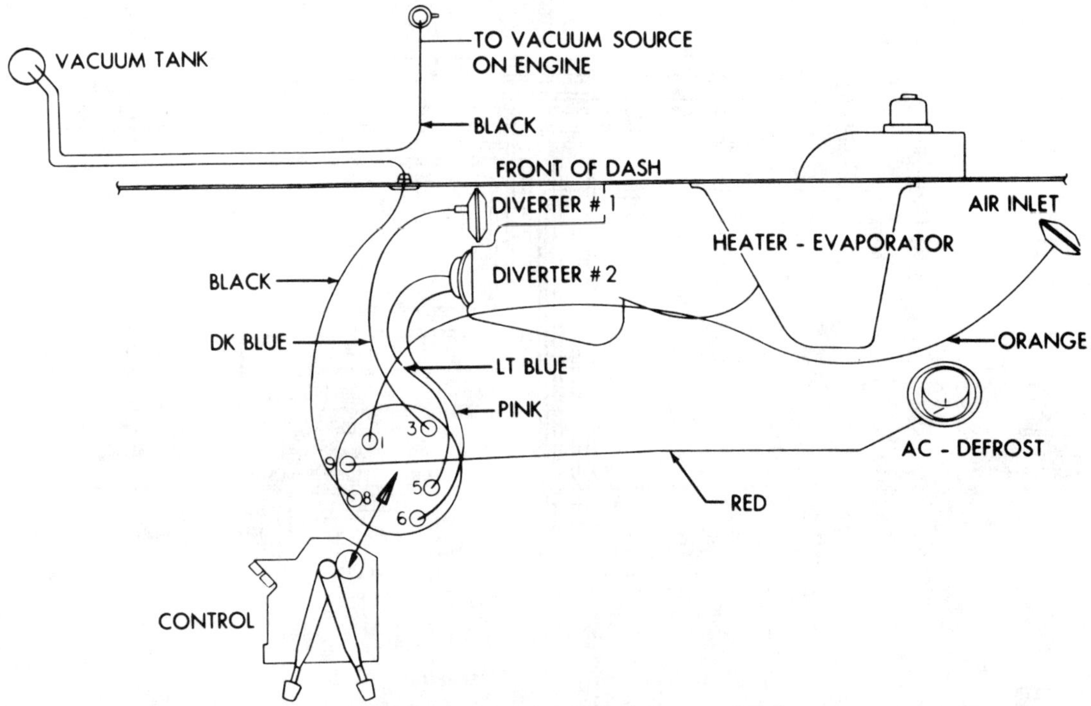

1975-76 Buick Skyhawk A/C & Heater Vacuum Circuit

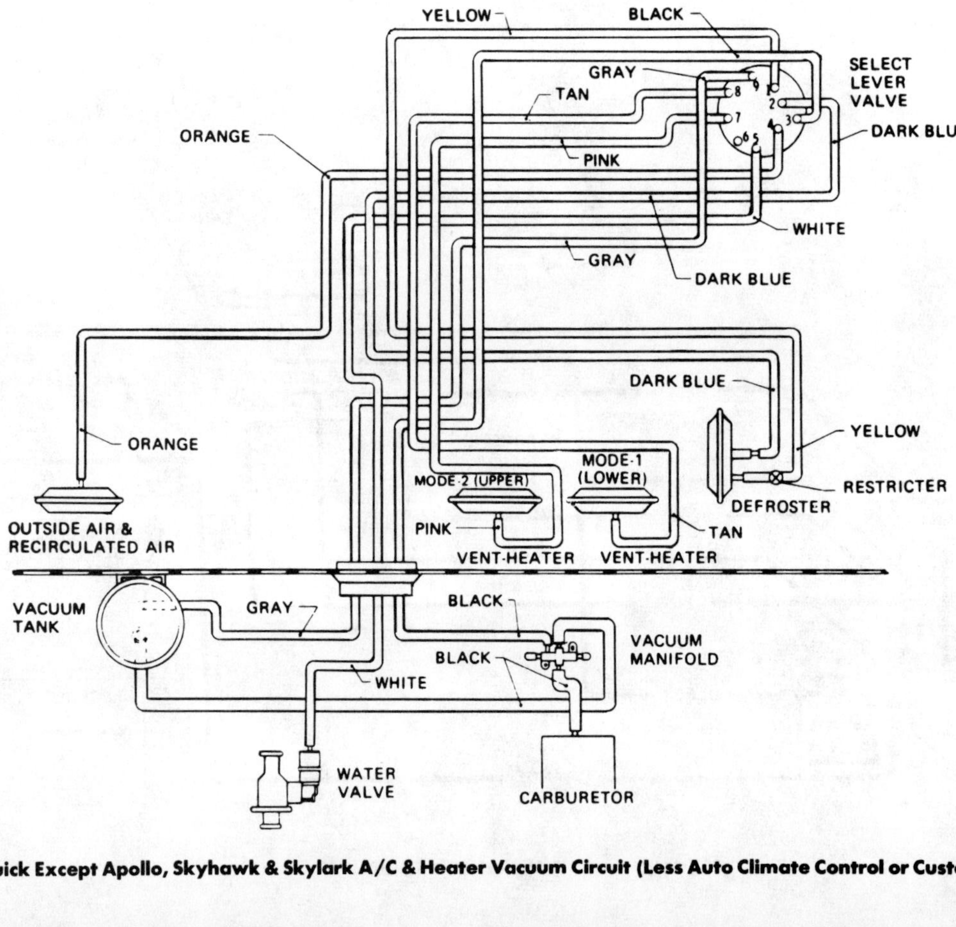

1975-76 Buick Except Apollo, Skyhawk & Skylark A/C & Heater Vacuum Circuit (Less Auto Climate Control or Custom Aire)

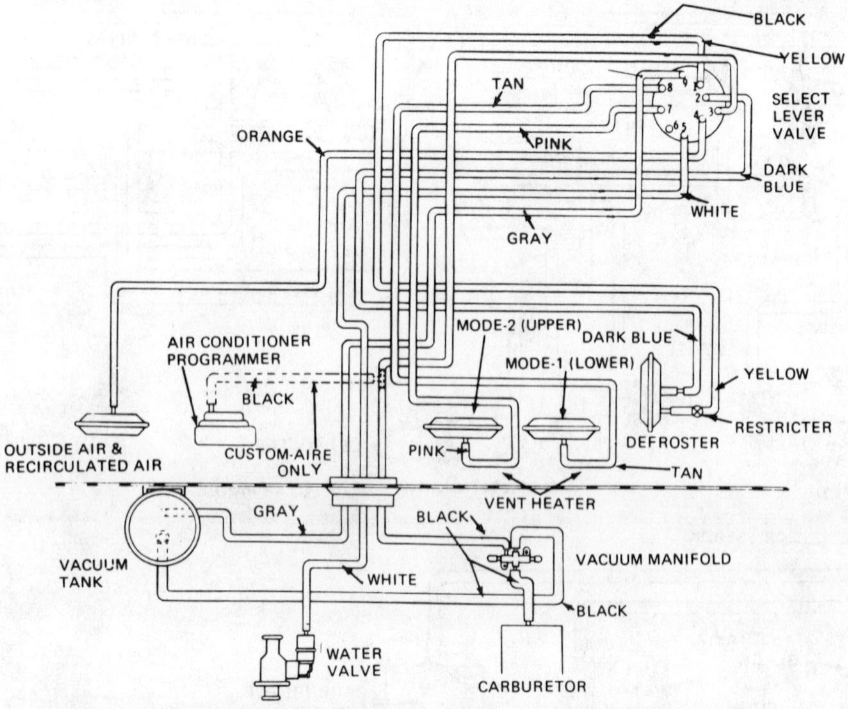

1976-77 Buick Century & Regal A/C & Heater Vacuum Circuit (With Custom Aire)

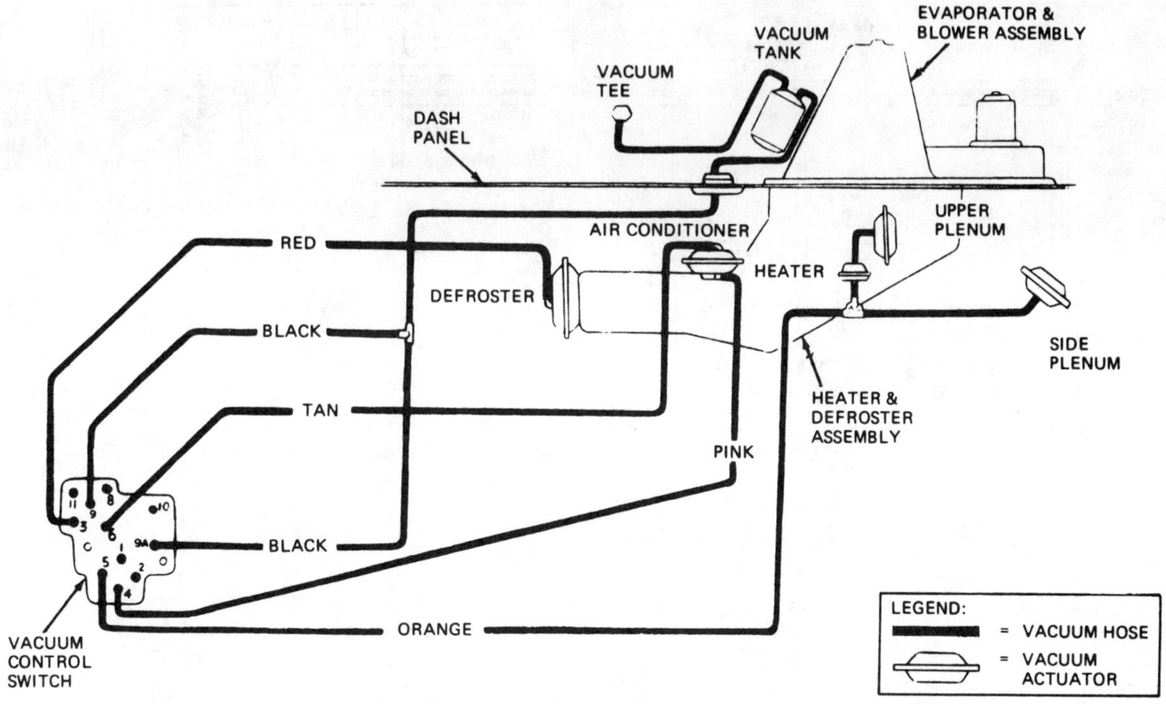

1976-79 Buick Skylark A/C & Heater Vacuum Circuit

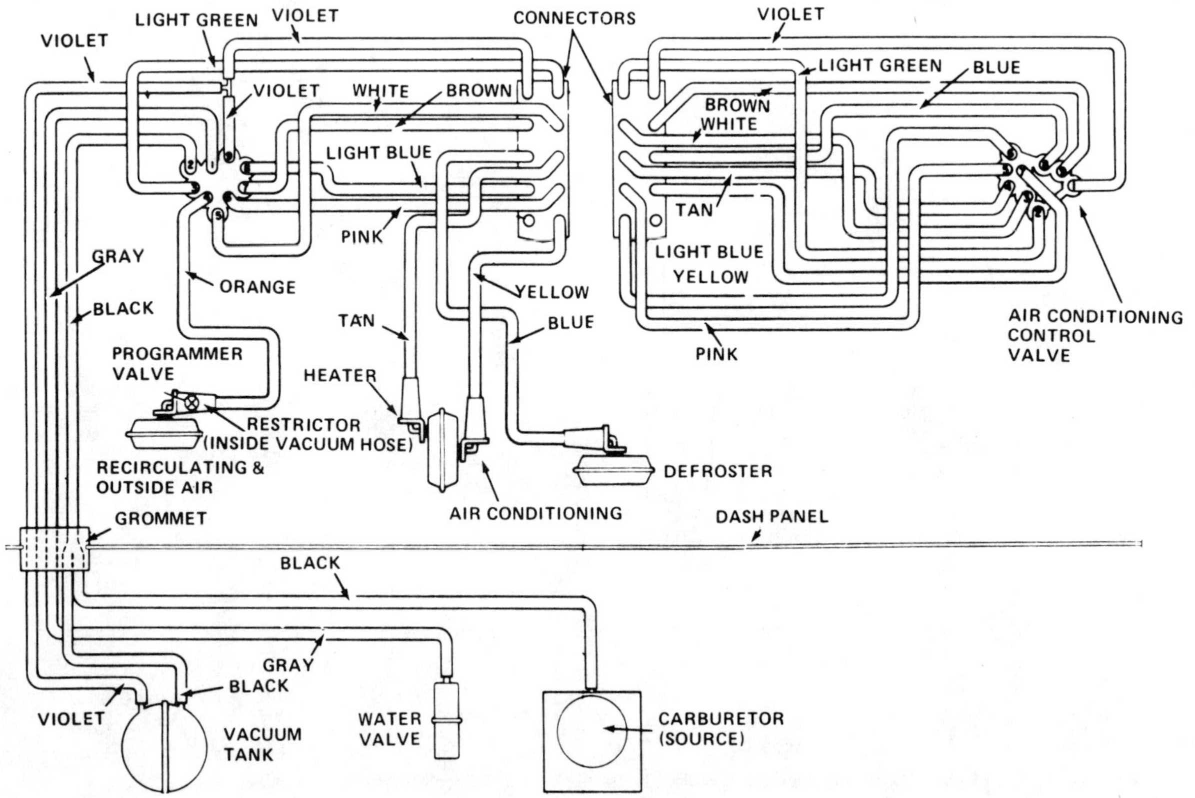

1978 Buick Century & Regal A/C & Heater Vacuum Circuit (With Auto Climate Control)

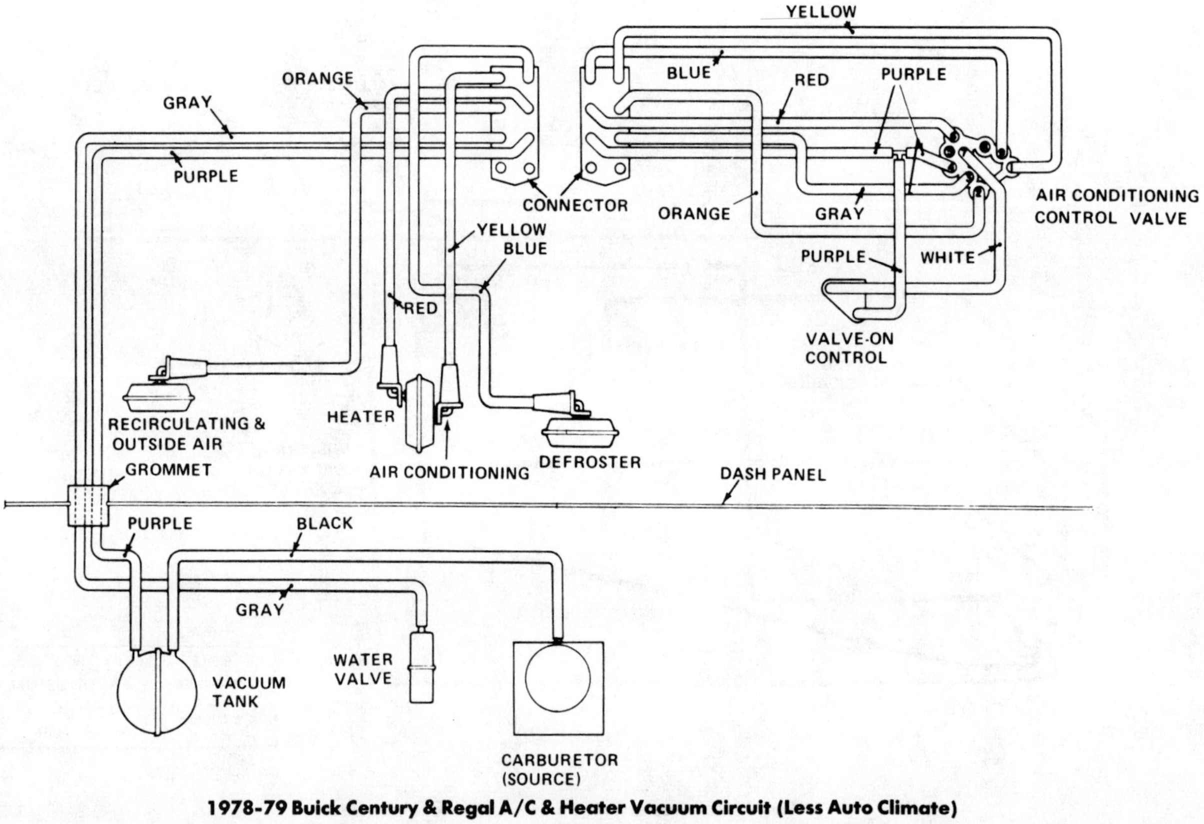

1978-79 Buick Century & Regal A/C & Heater Vacuum Circuit (Less Auto Climate)

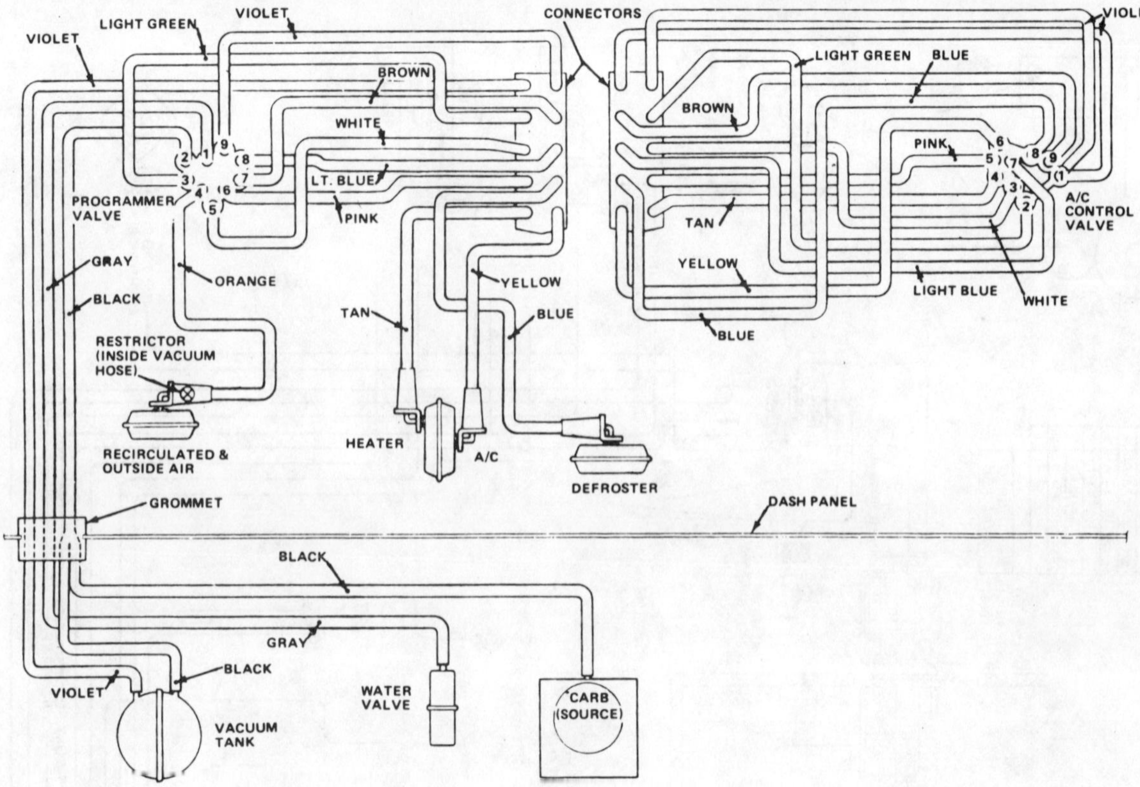

1979-80 Buick Century & Regal A/C & Heater Vacuum Circuit (With Auto Climate Control)

PORT	FUNCTION		
	A/C	BI-LEVEL	HEAT
1	VAC	VENT	VENT
6	VENT	VENT	VAC

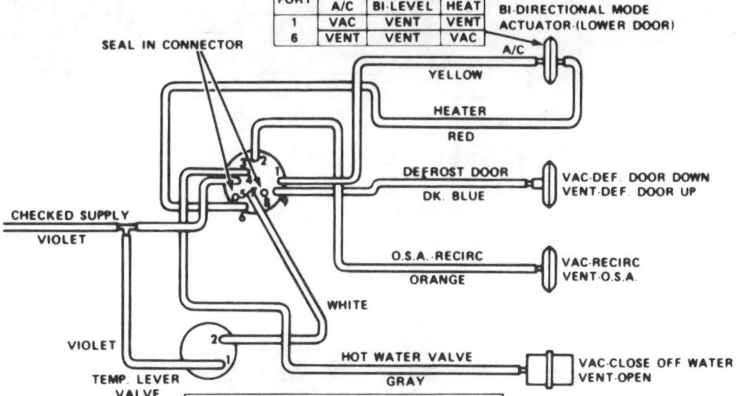

TEMPERATURE LEVER VALVE			
PORT NO.	PORT CONNECTION	COLD	HOT
1	VAC INPUT	CONN. 2	SEAL
2	W/V SIGNAL	CONN. 1	VENT

1980 Buick Century & Regal A/C & Heater Vacuum Circuit (Less Auto Climate Control)

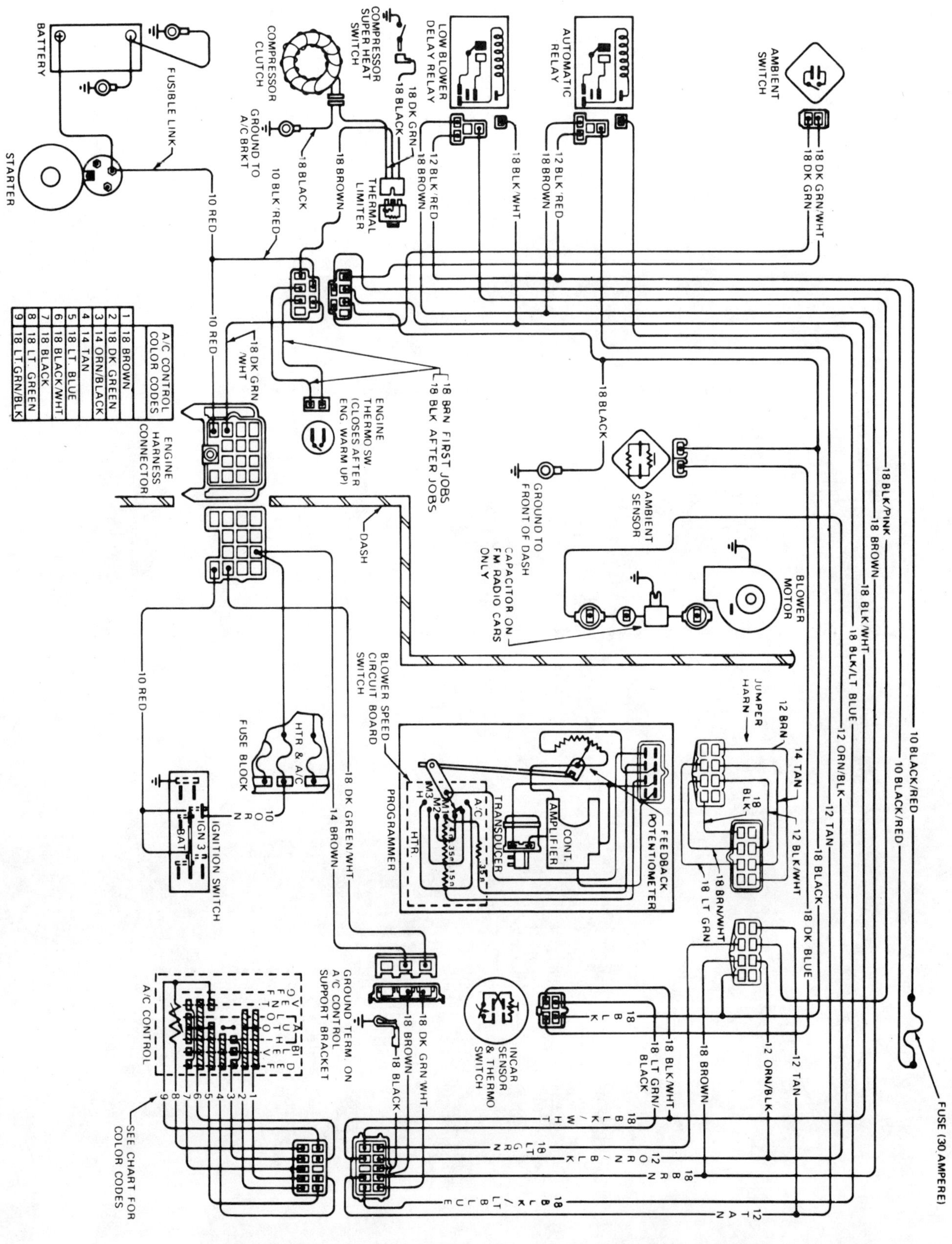

1974 Buick Except Apollo A/C & Heater Wiring Circuit (With Auto Climate Control)

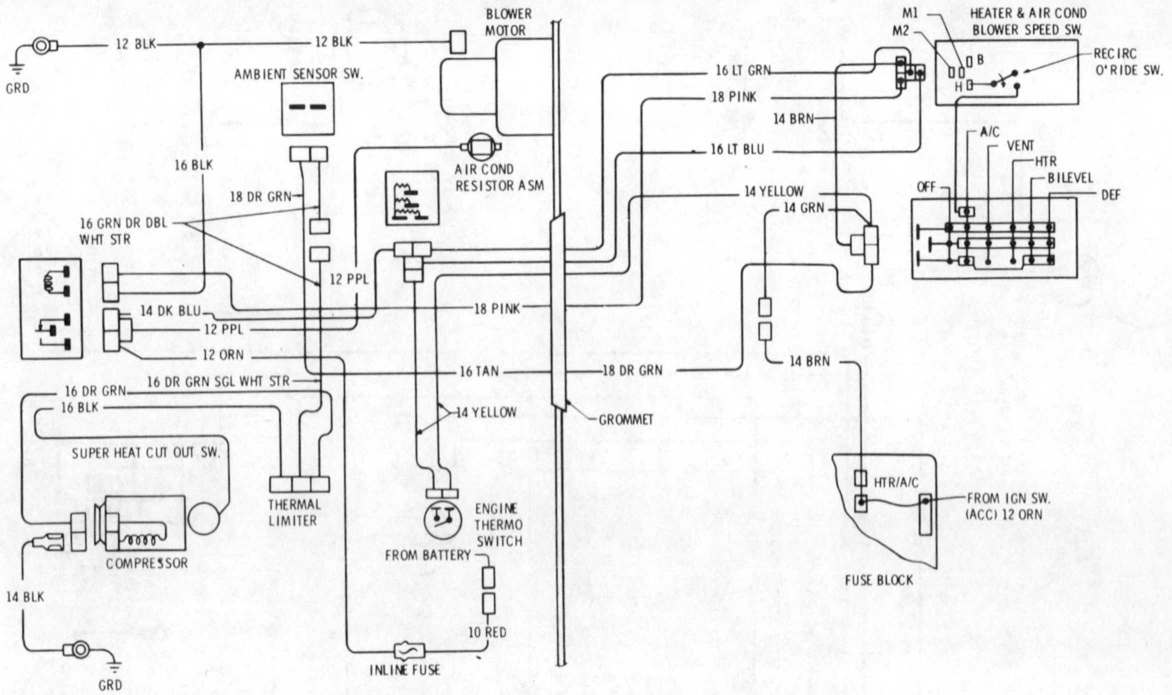

1974 Buick Apollo A/C & Heater Wiring Circuit

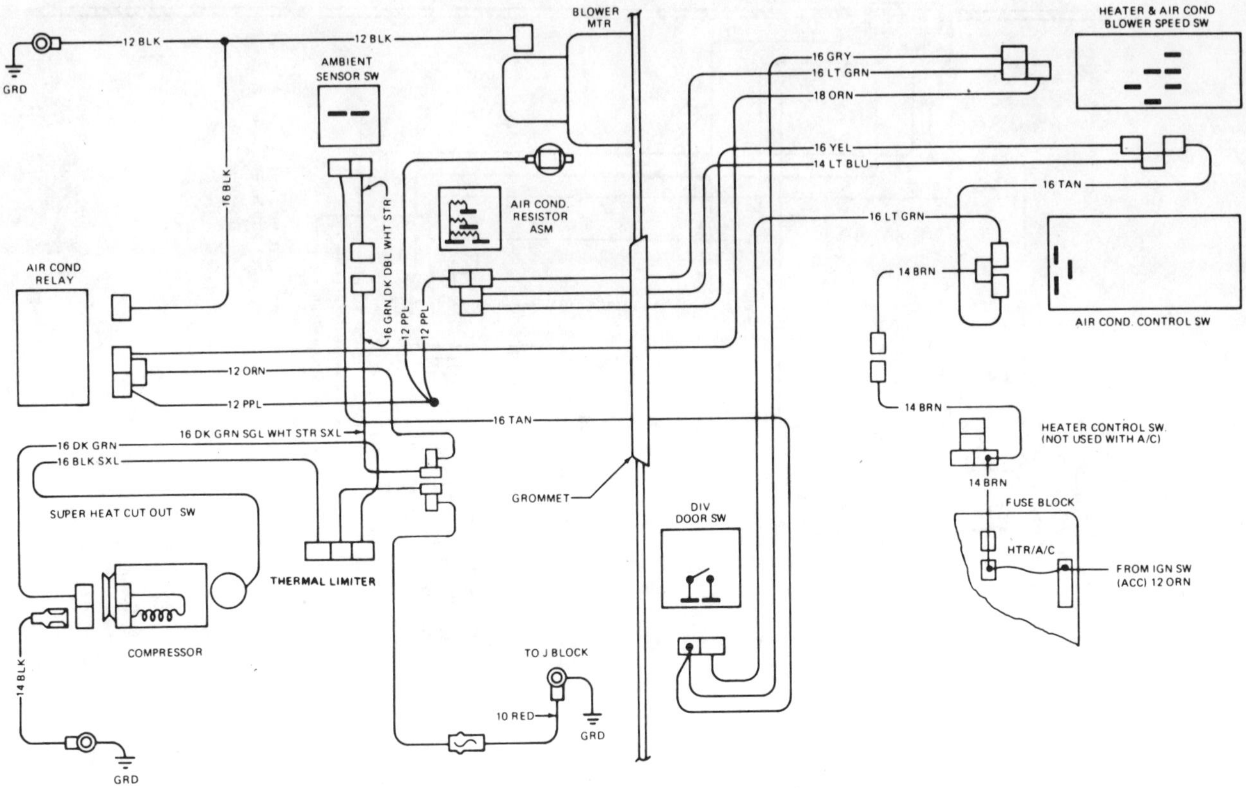

1975 Buick Apollo & Skylark A/C & Heater Wiring Circuit

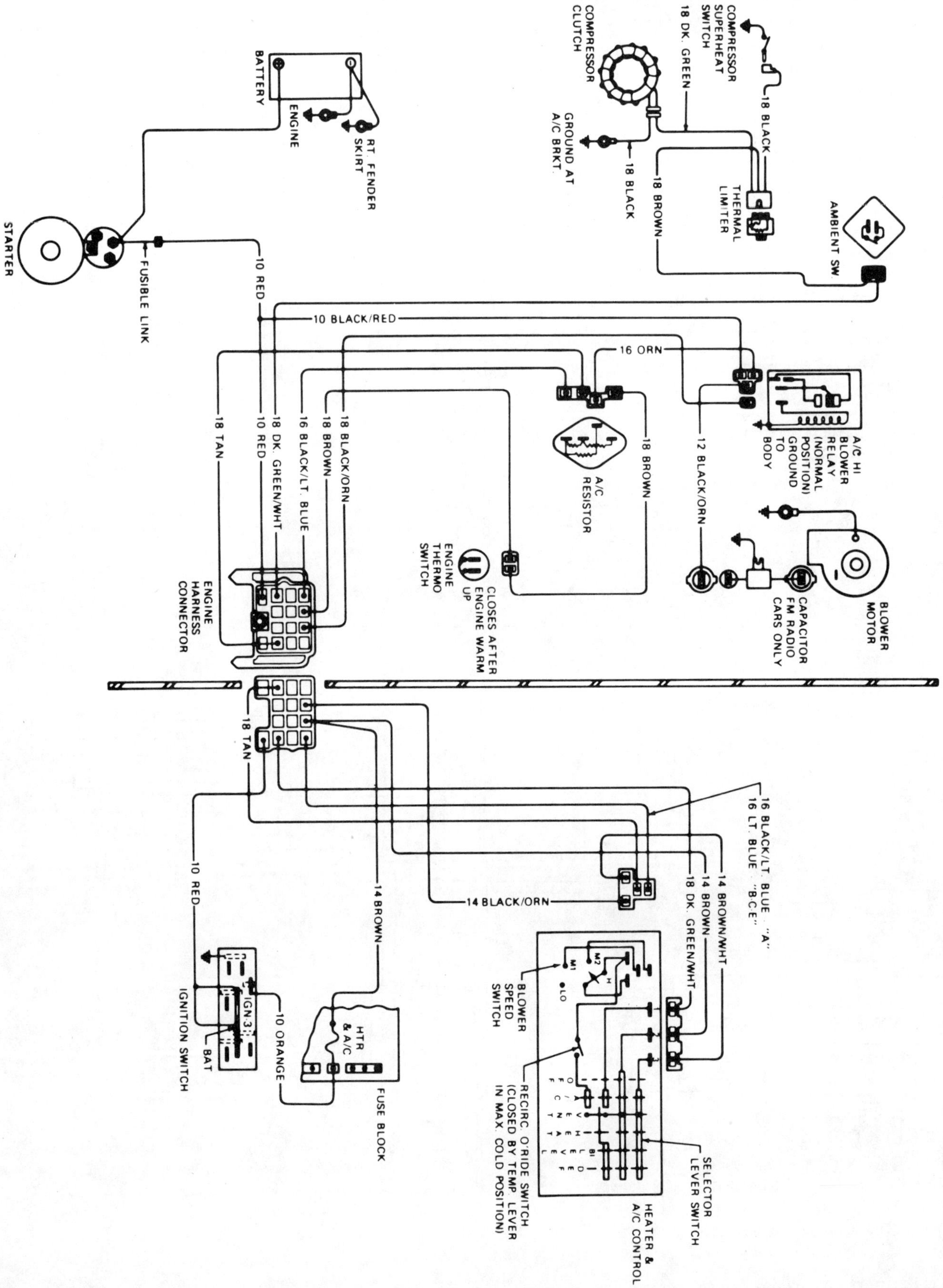

1974-75 Buick Except Apollo, Skyhawk & Skylark A/C & Heater Wiring Circuit (Less Auto Climate Control)

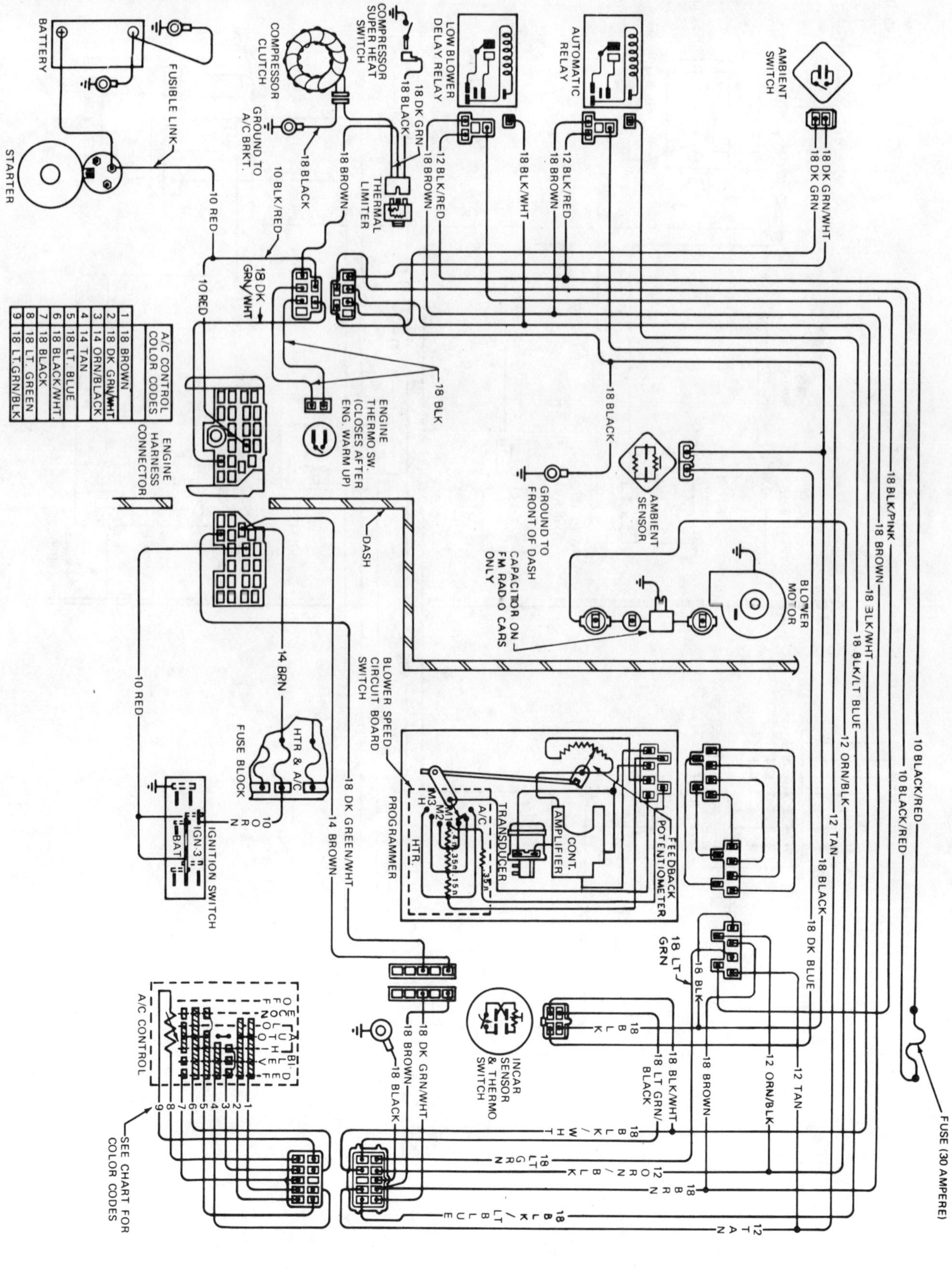

1975 Buick Except Apollo, Skyhawk & Skylark A/C & Heater Wiring Circuit (With Auto Climate Control)

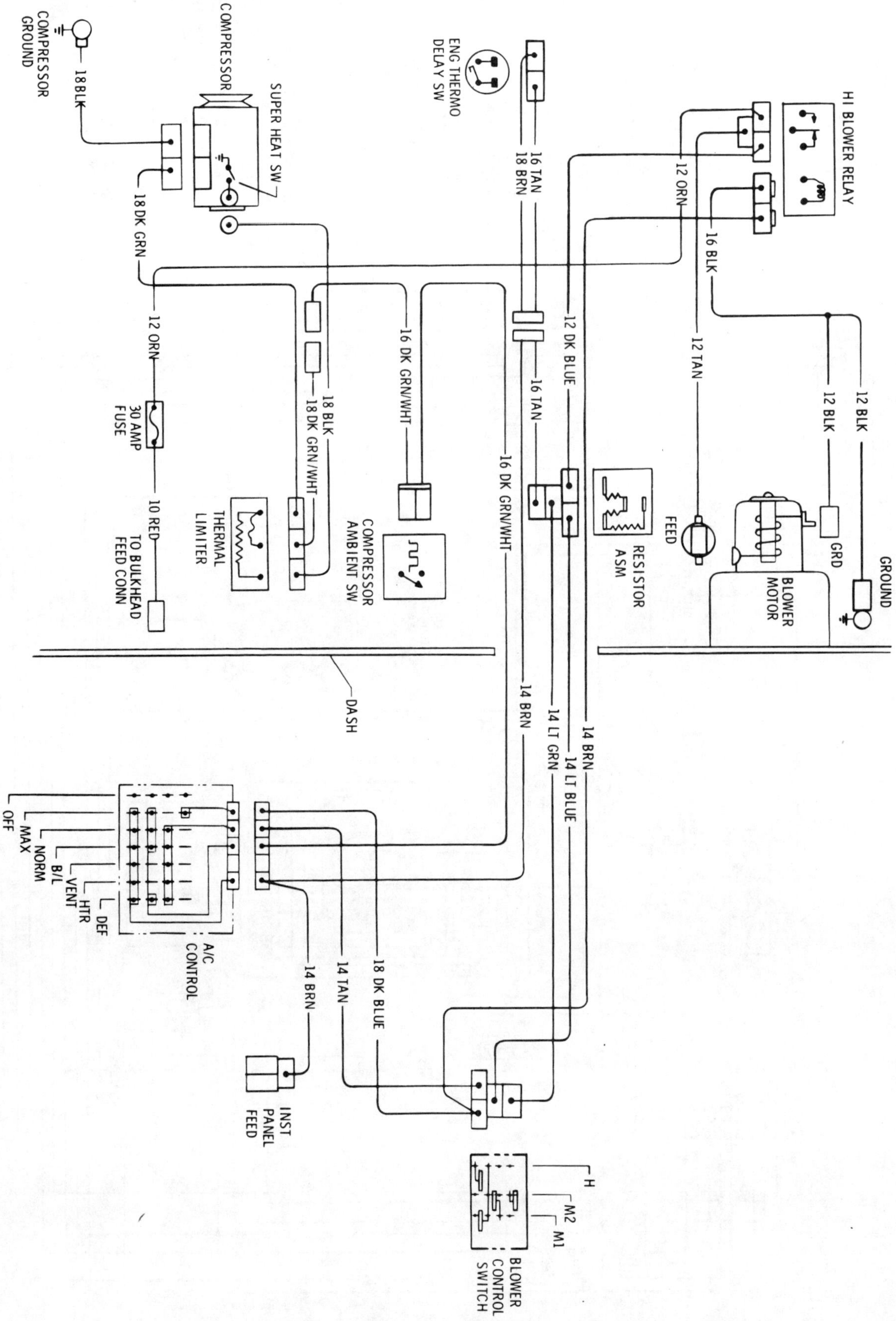

1975-76 Buick Skyhawk A/C & Heater Wiring Circuit

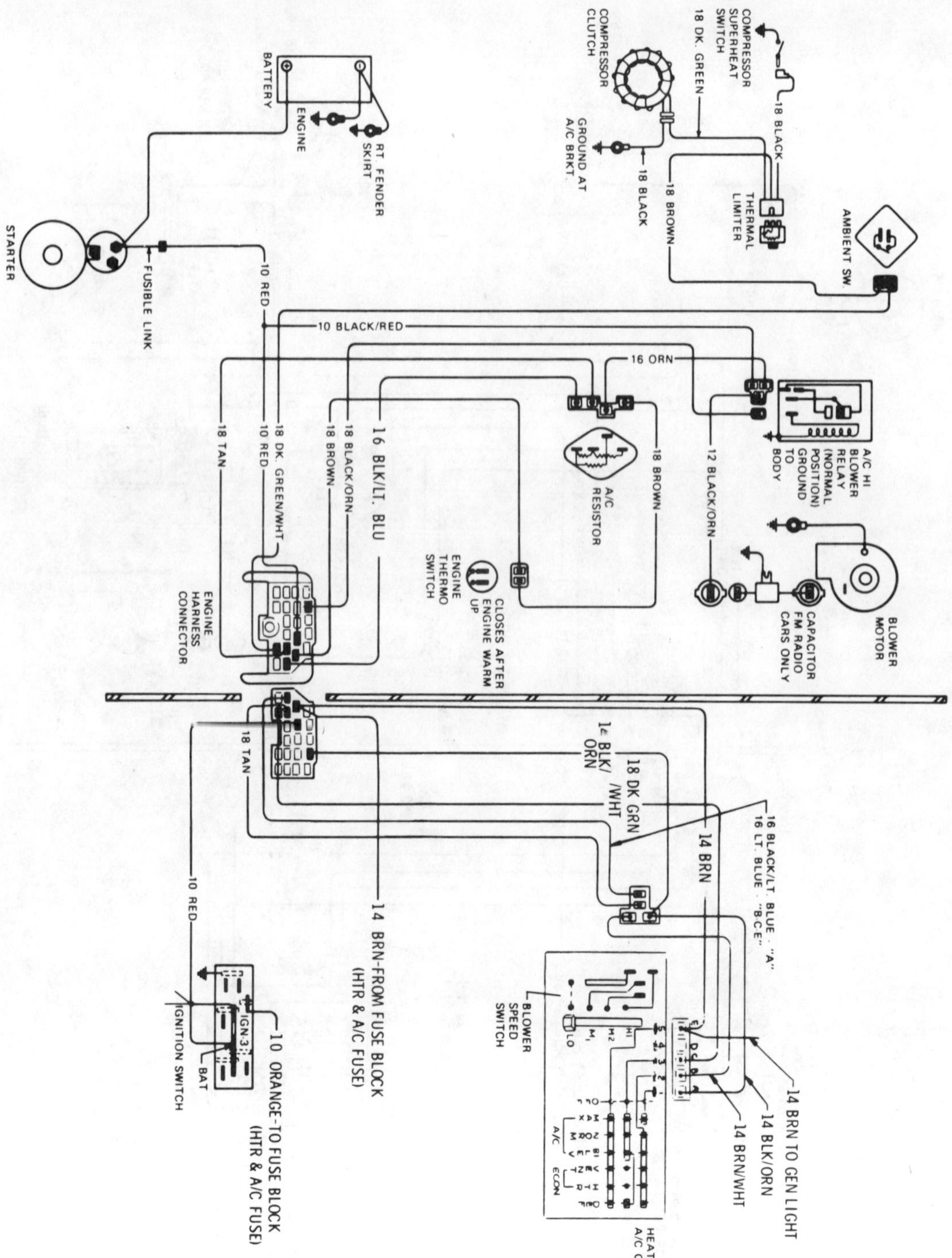

1976 Buick Century & Regal A/C & Heater Wiring Circuit (Less Custom Aire)

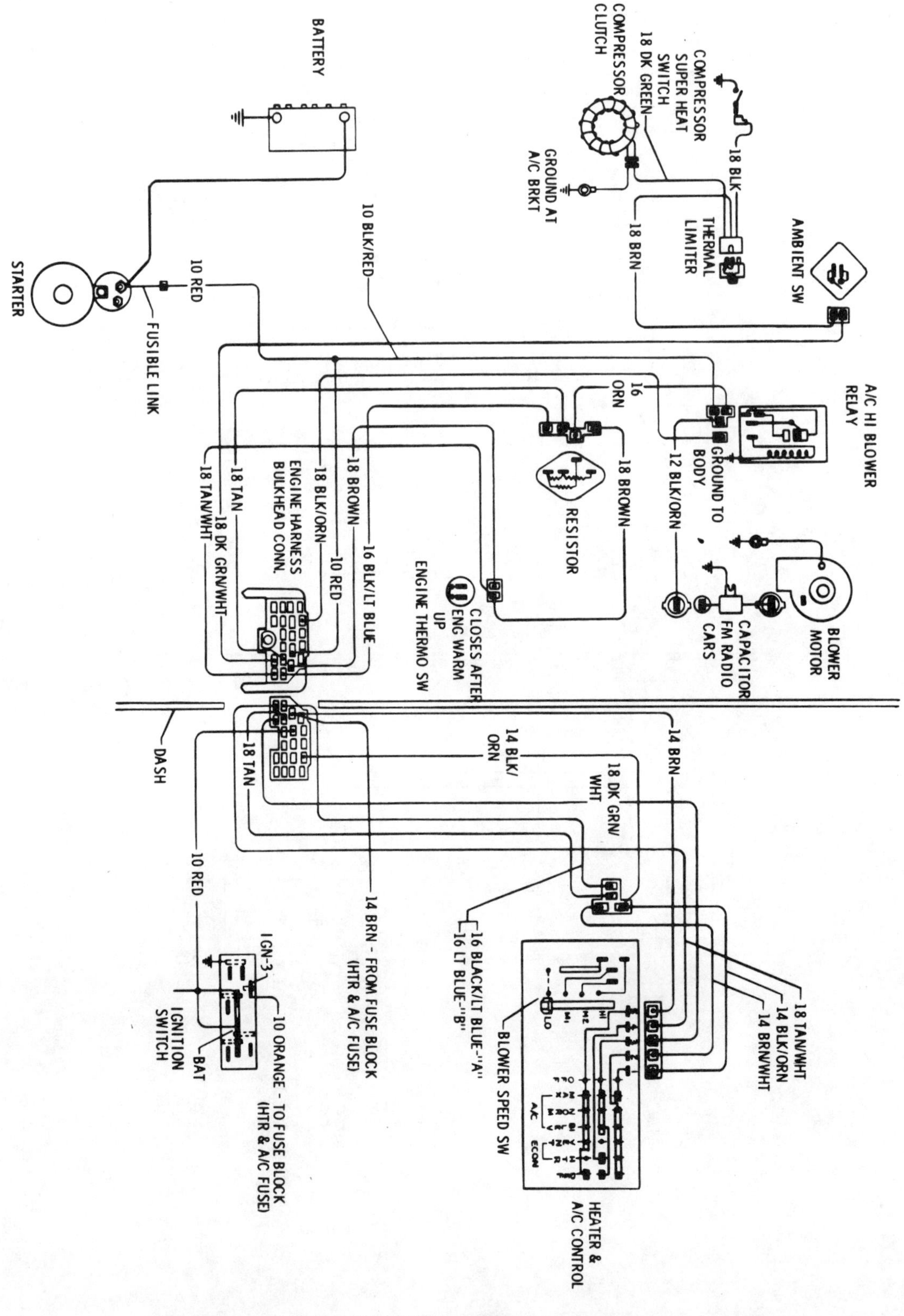

1976 Buick Century & Regal A/C & Heater Wiring Circuit (With Custom Aire)

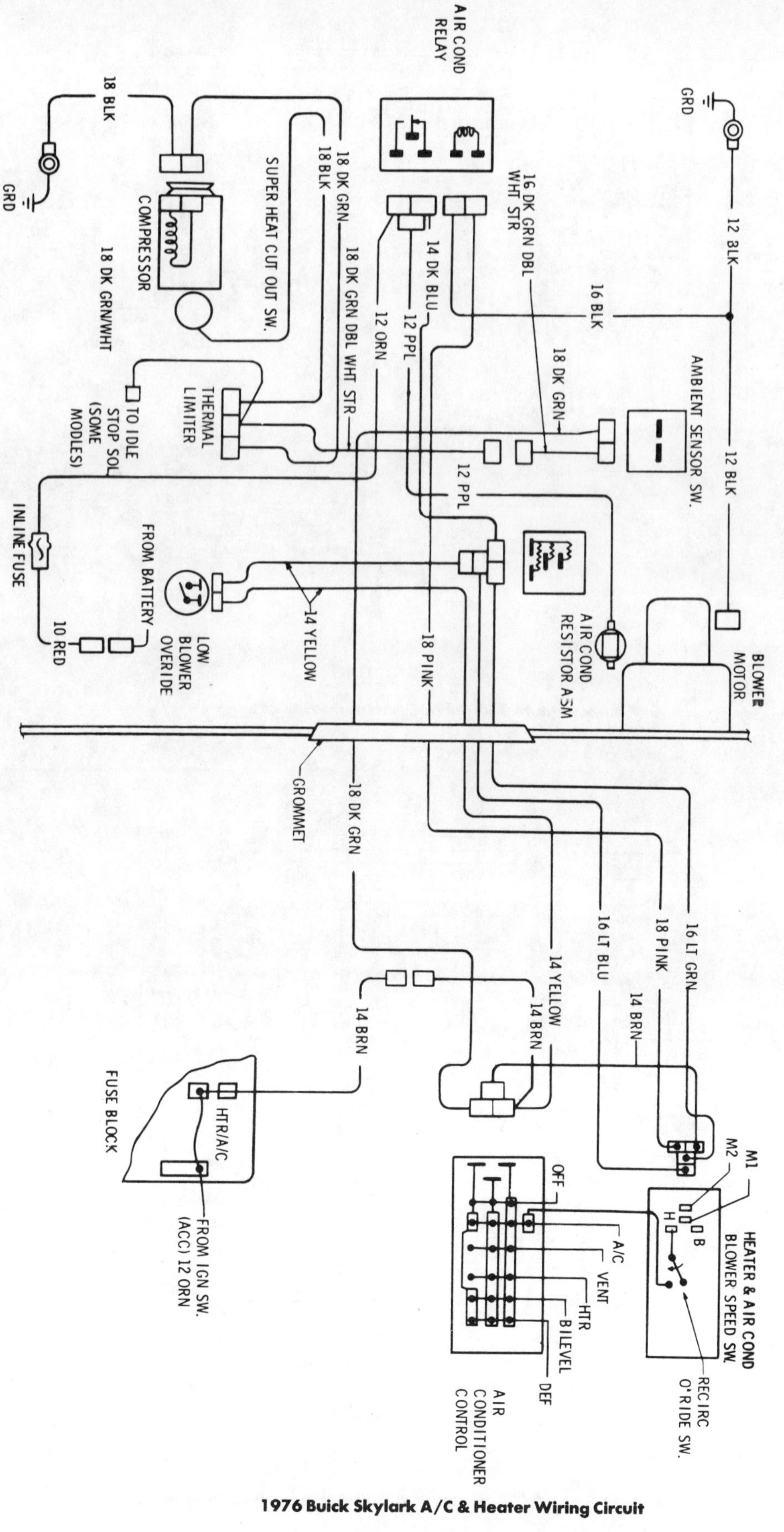

1976 Buick Skylark A/C & Heater Wiring Circuit

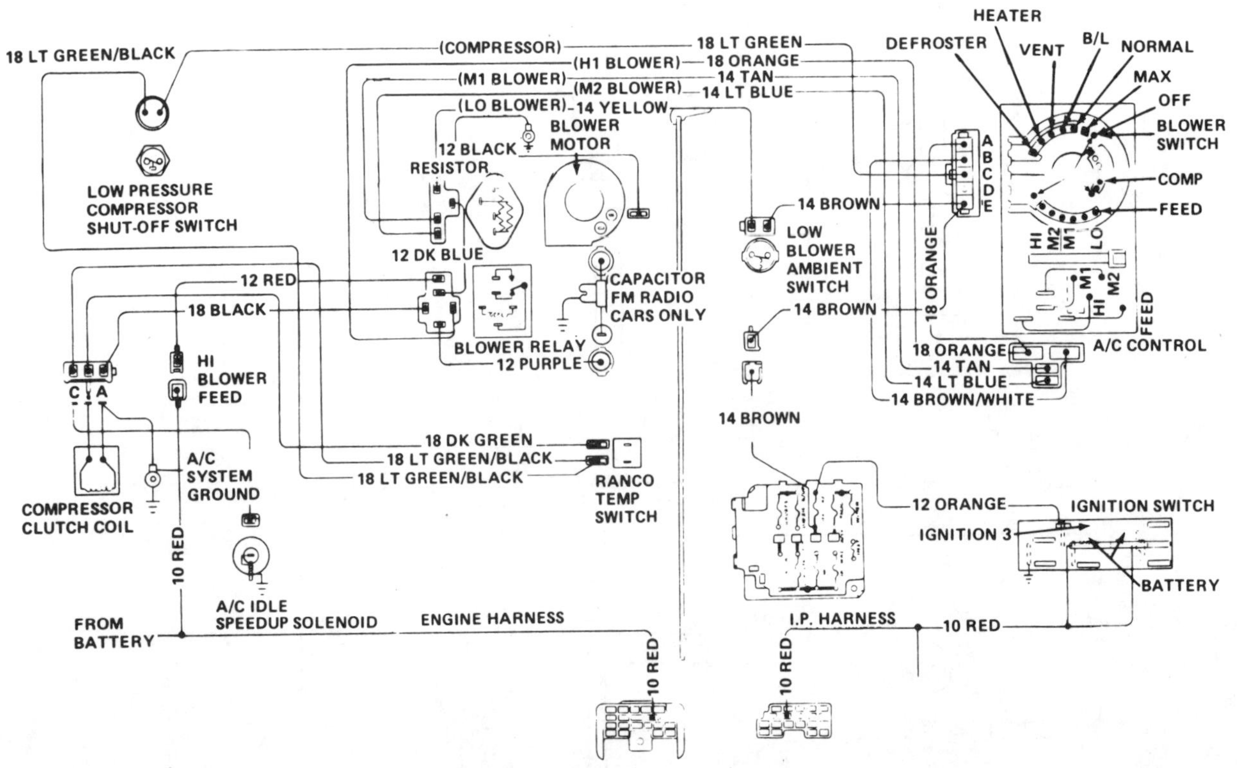

1977 Buick Century & Regal A/C & Heater Wiring Circuit

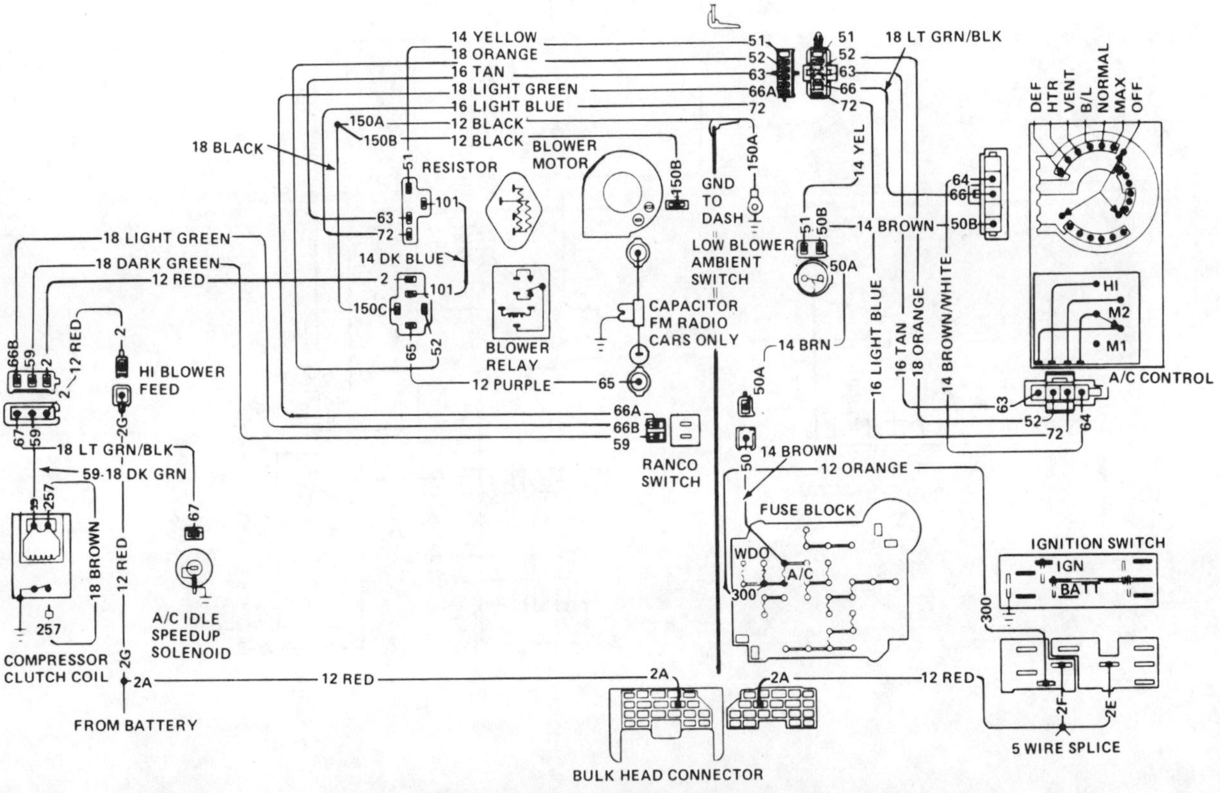

1978-79 Buick Century & Regal A/C & Heater Wiring Circuit (Less Auto Climate Control)

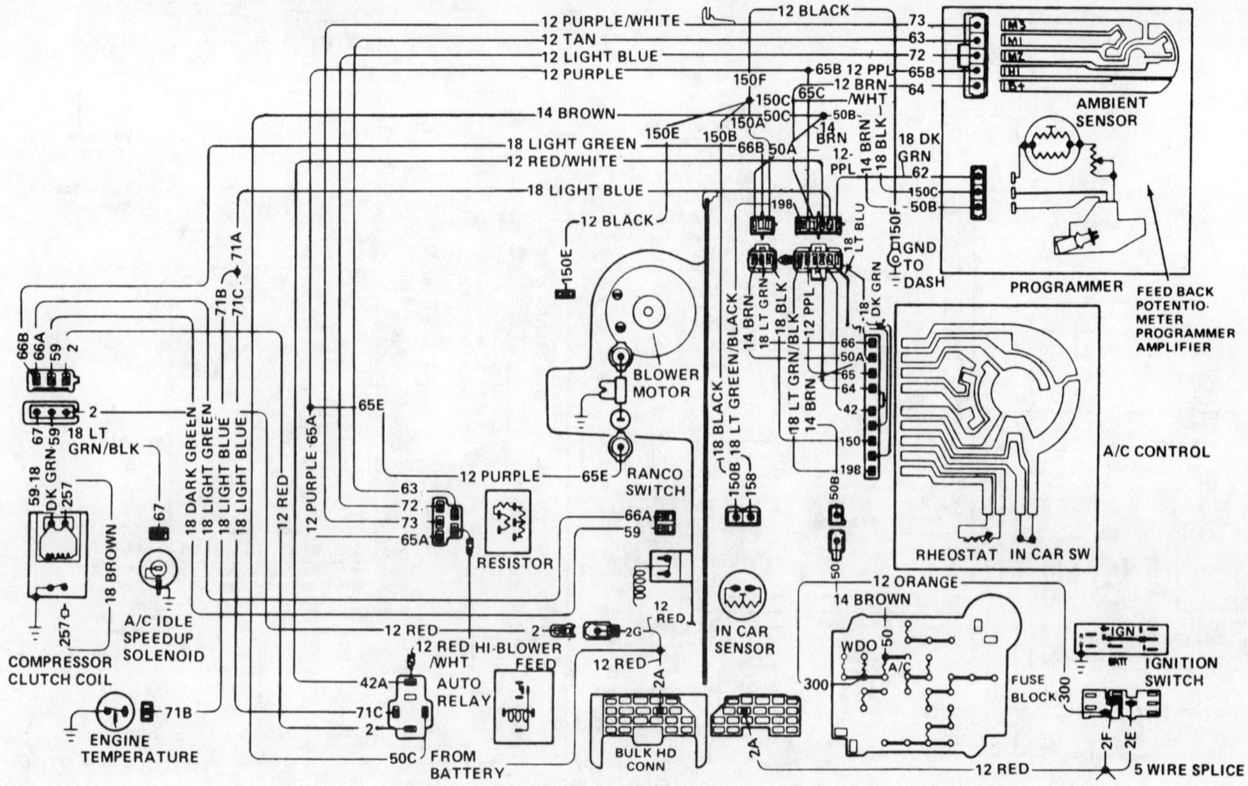

1978-79 Buick Century & Regal A/C & Heater Wiring Circuit (With Auto Climate Control)

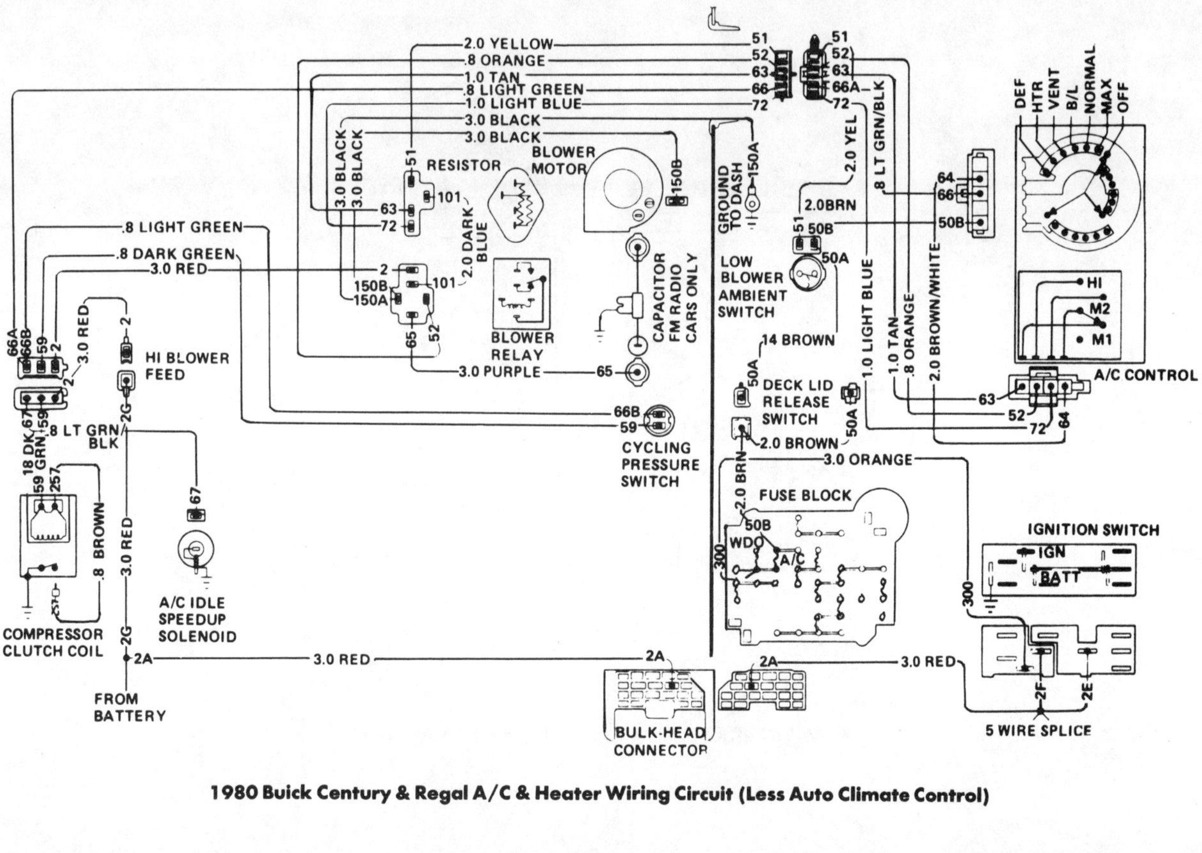

1980 Buick Century & Regal A/C & Heater Wiring Circuit (Less Auto Climate Control)

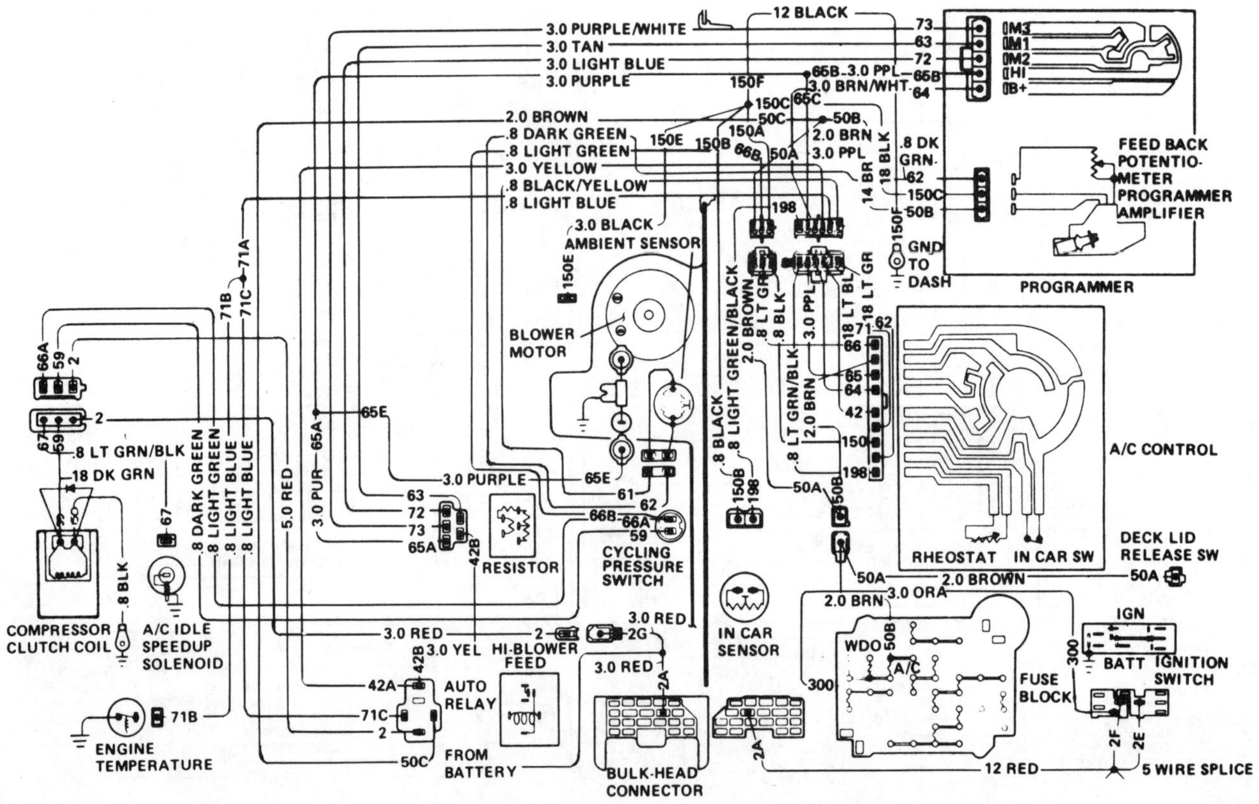

1980 Buick Century & Regal A/C & Heater Wiring Circuit (With Auto Climate Control)

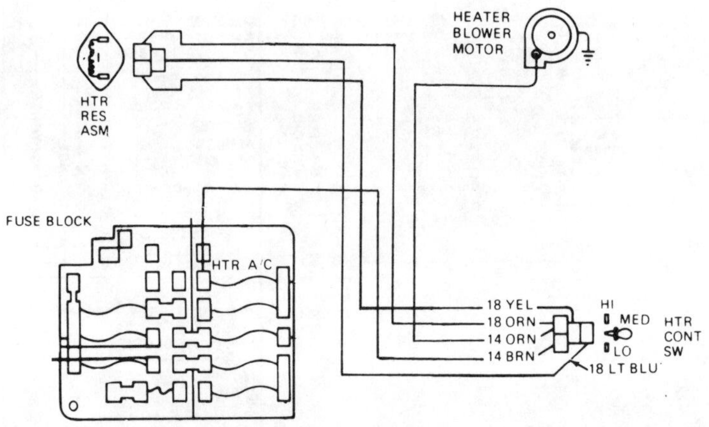

1974-76 Buick Apollo & Skylark Heater Wiring Circuit

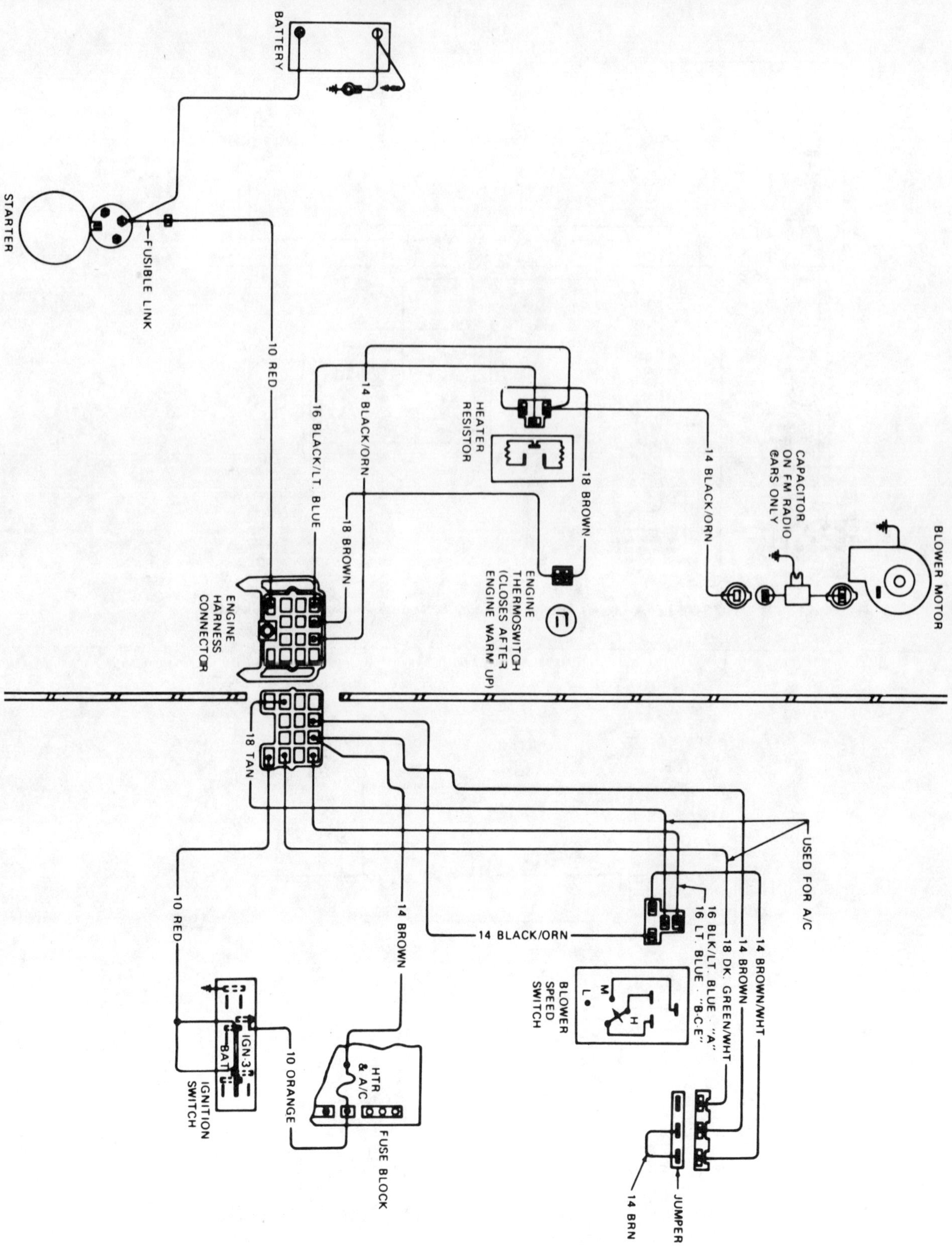

1974 Buick Except Apollo Heater Wiring Circuit

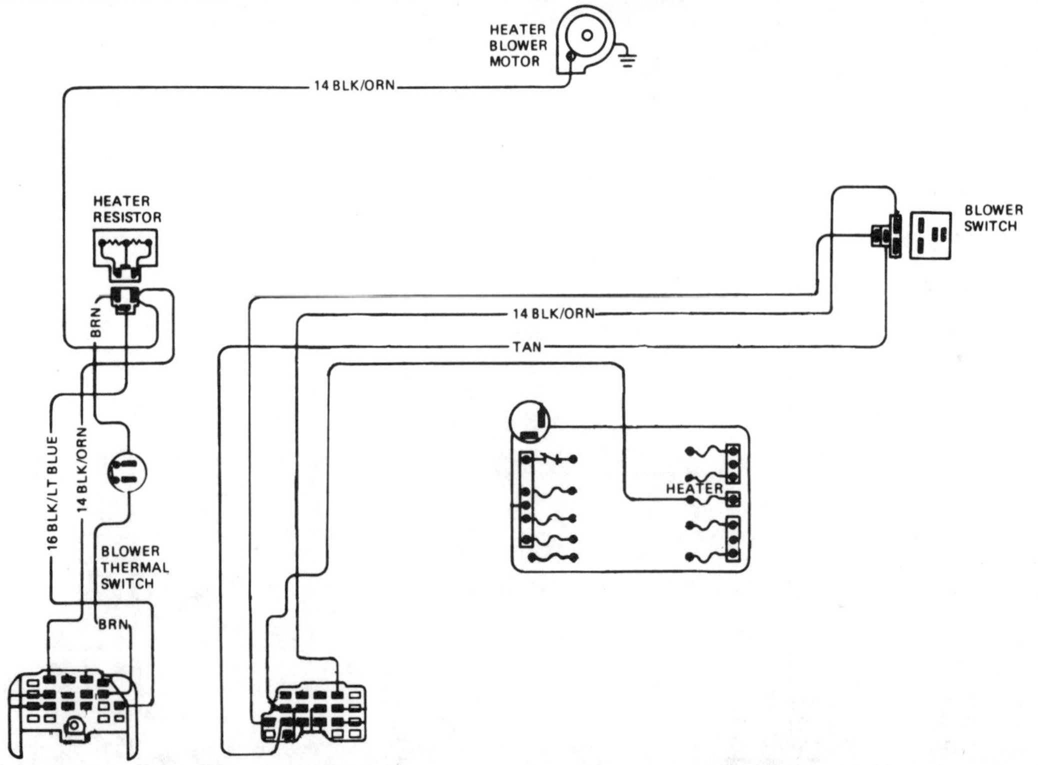

1975-76 Buick Except Apollo, Skyhawk & Skylark Heater Wiring Circuit

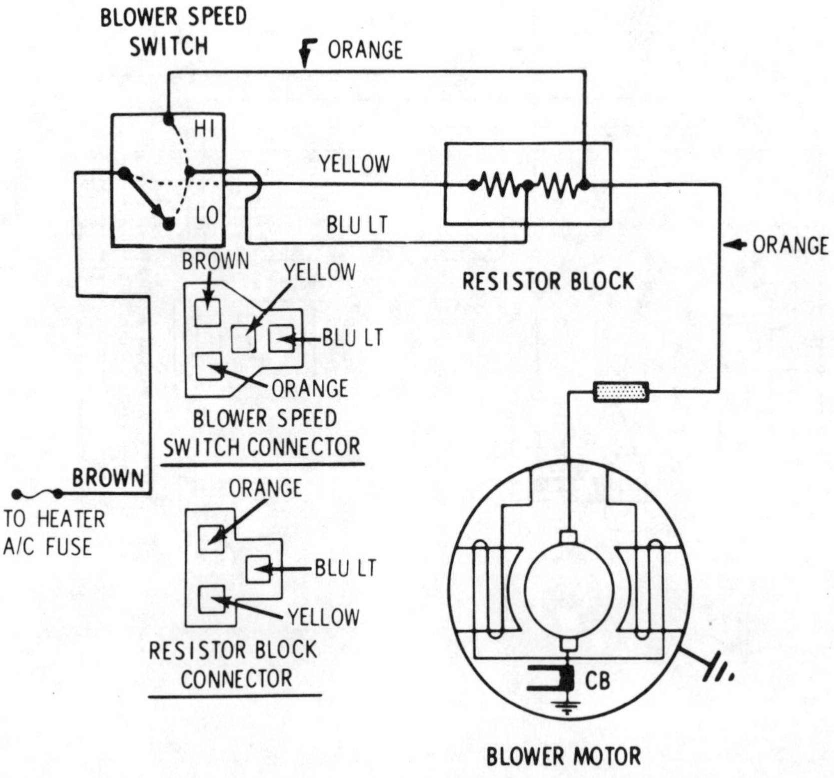

1975-76 Buick Skyhawk Heater Wiring Circuit

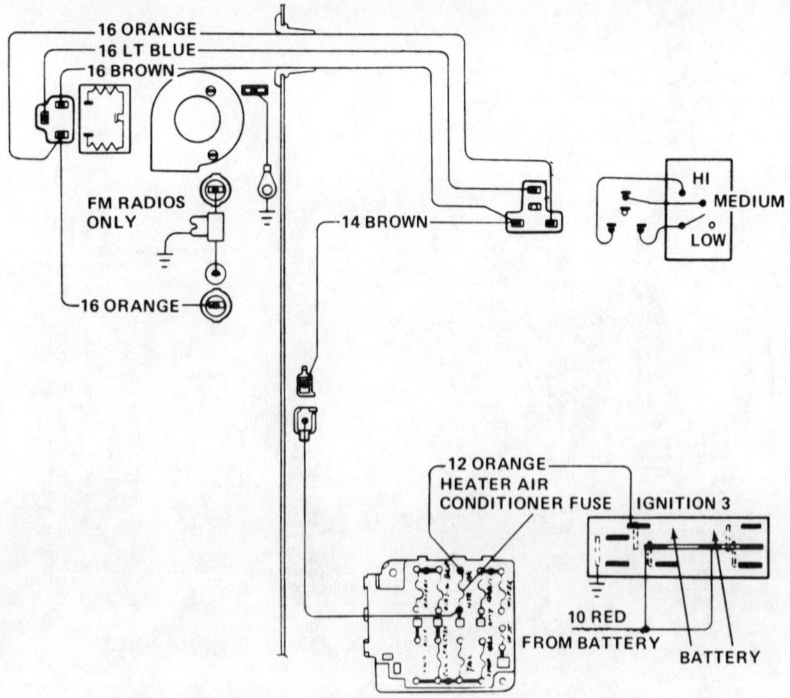

1977 Buick Century & Regal Heater Wiring Circuit

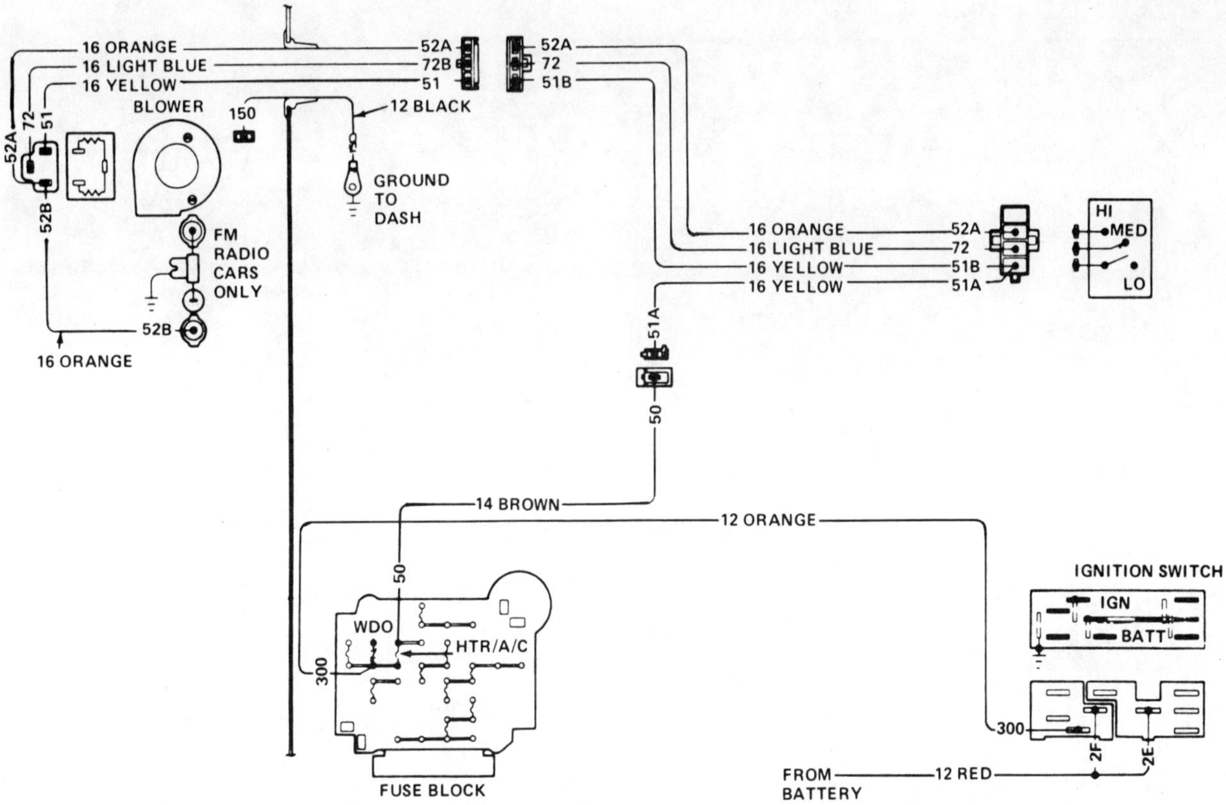

1978-80 Buick Century & Regal Heater Wiring Circuit

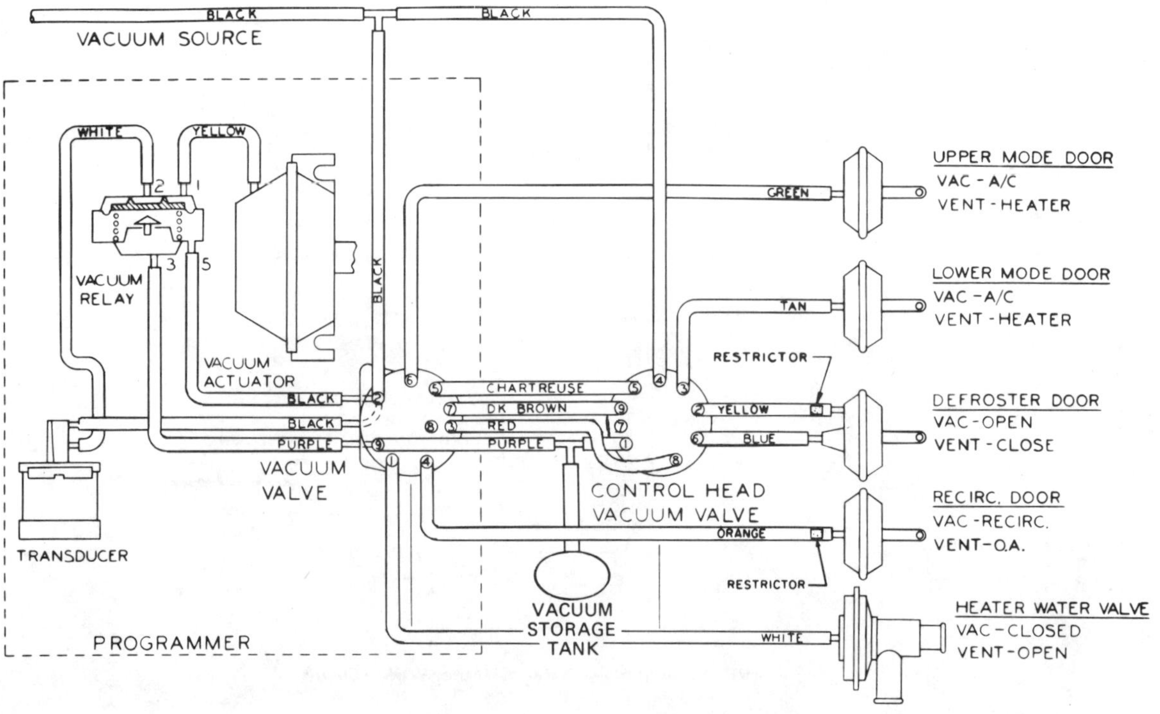

1974 Cadillac A/C & Heater Vacuum Circuit

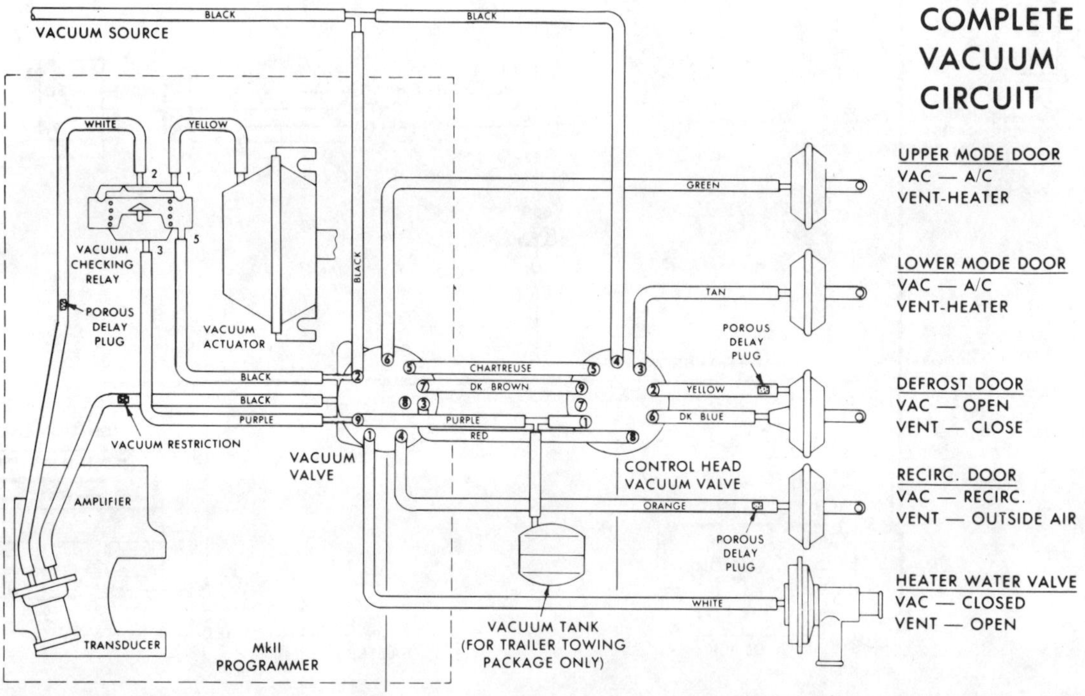

1975-76 Cadillac Except Seville A/C & Heater Vacuum Circuit

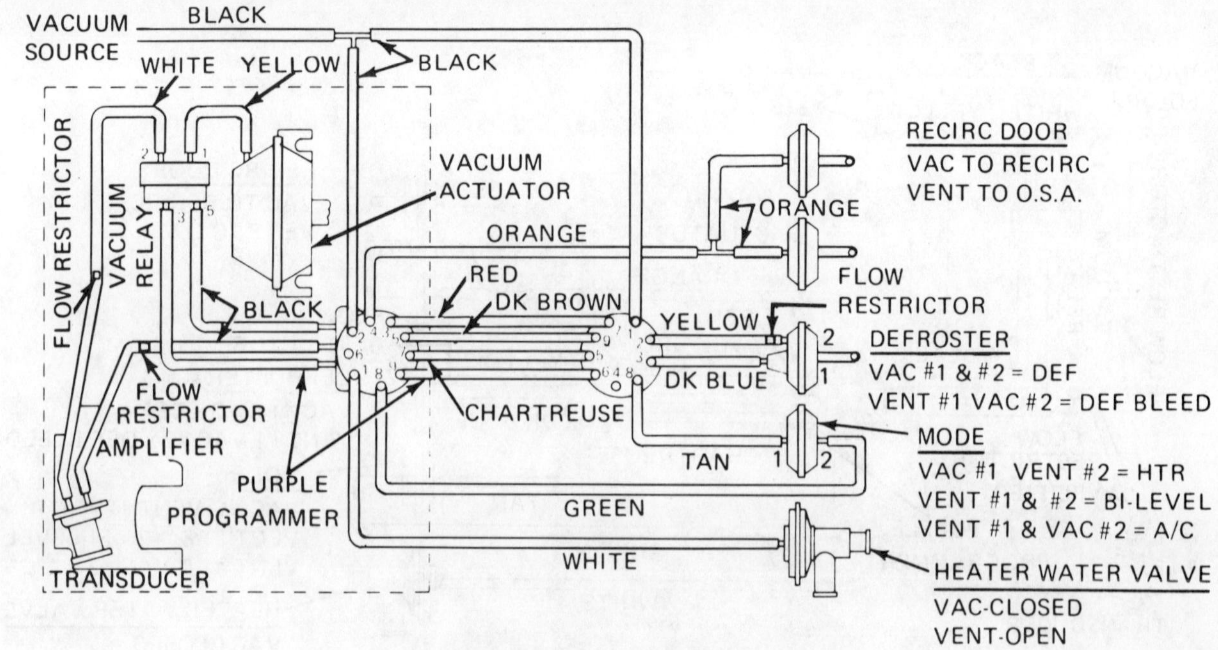

1976 Cadillac Seville A/C & Heater Vacuum Circuit

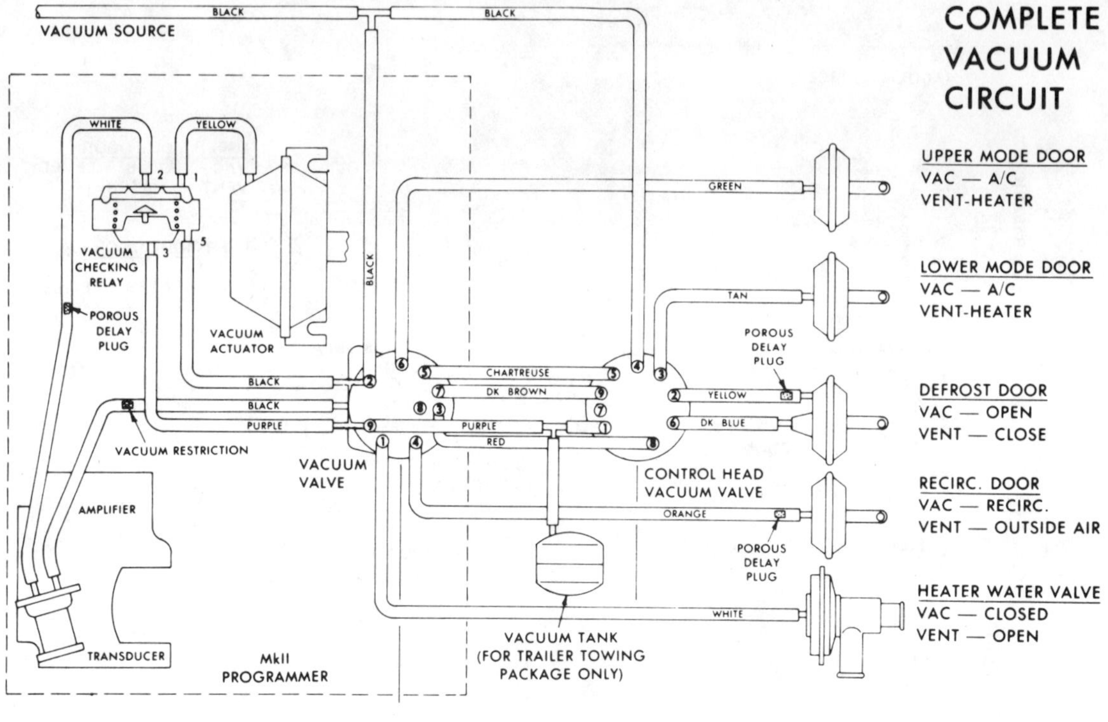

1977-78 Cadillac Eldorado A/C & Heater Vacuum Circuit

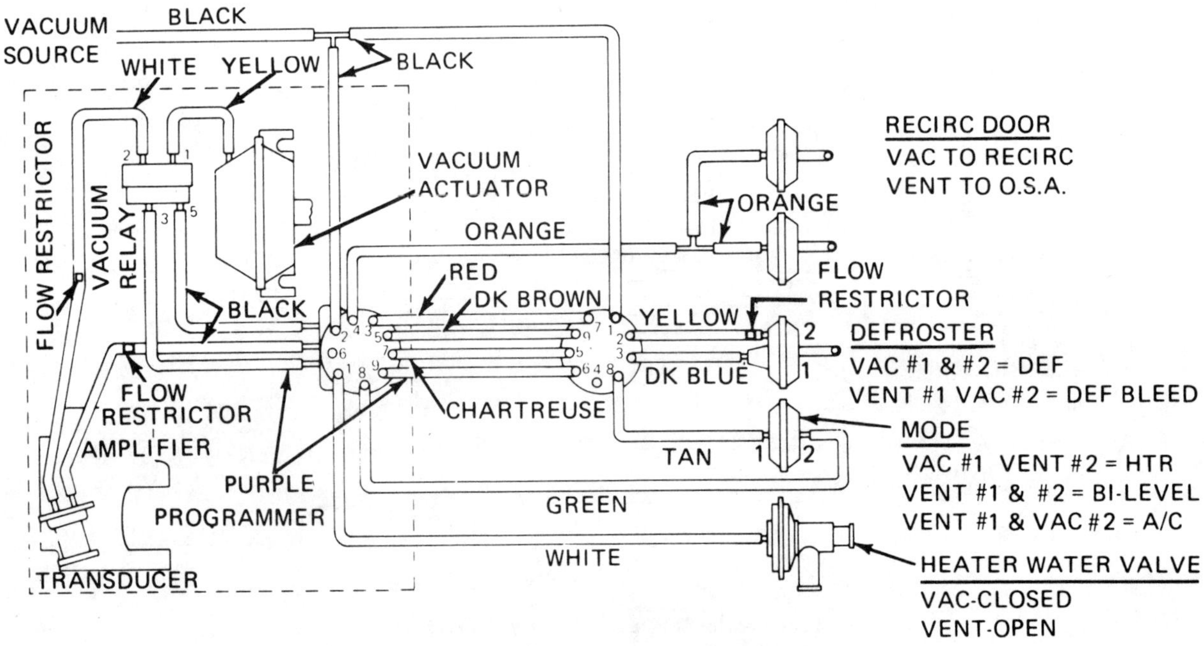

1977-79 Cadillac Seville A/C & Heater Vacuum Circuit

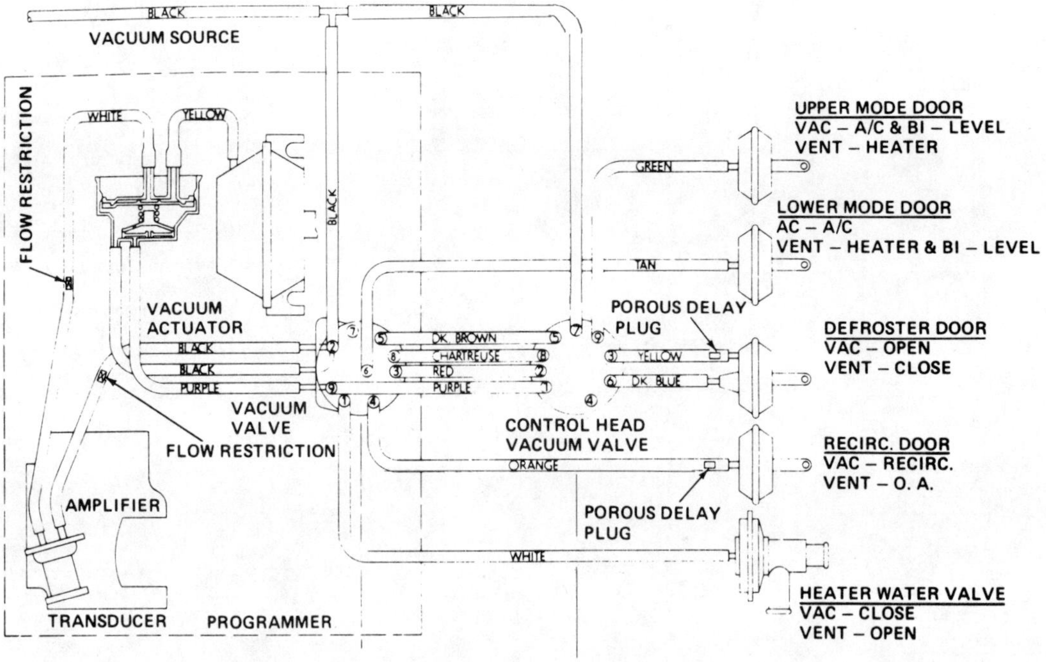

1977-80 Cadillac Except Seville & Eldorado A/C & Heater Vacuum Circuit

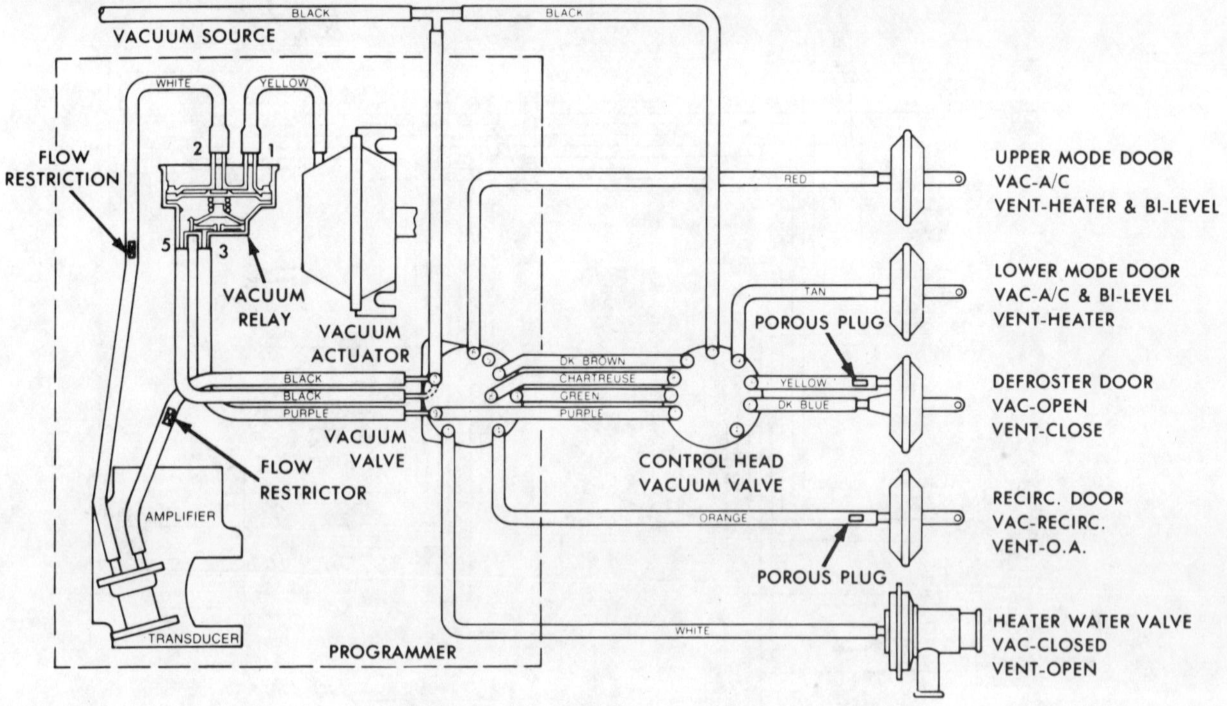

1979 Cadillac Eldorado A/C & Heater Vacuum Circuit

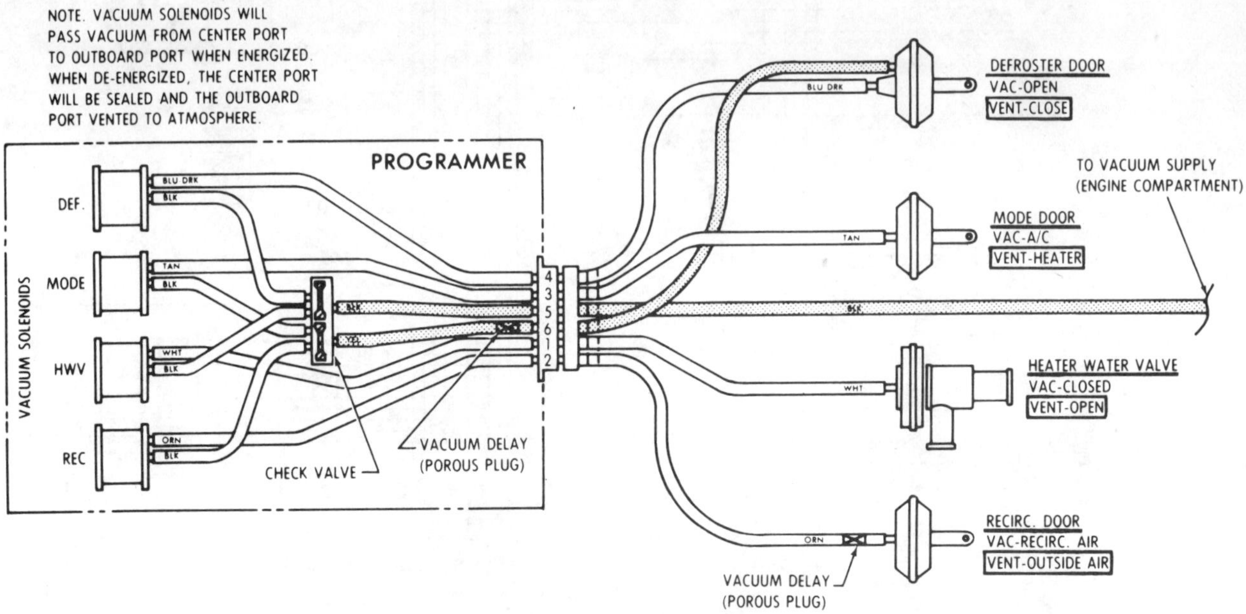

1980 Cadillac Eldorado & Seville A/C & Heater Vacuum Circuit

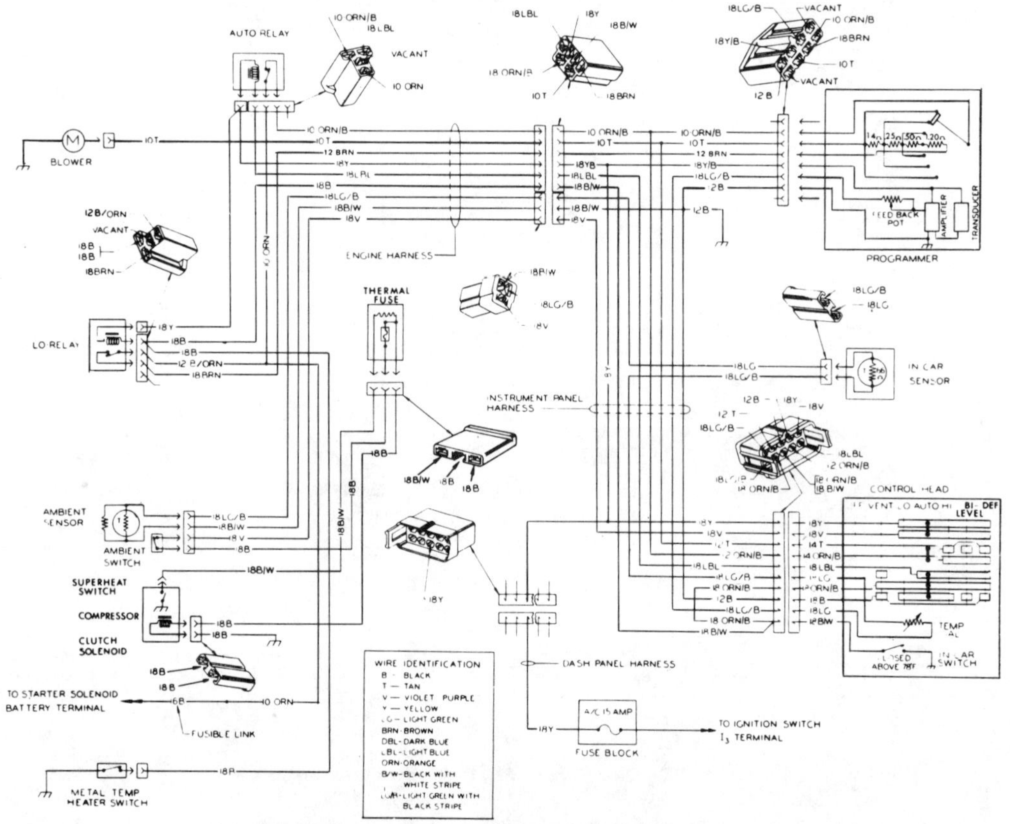

1974-75 Cadillac A/C & Heater Wiring Circuit

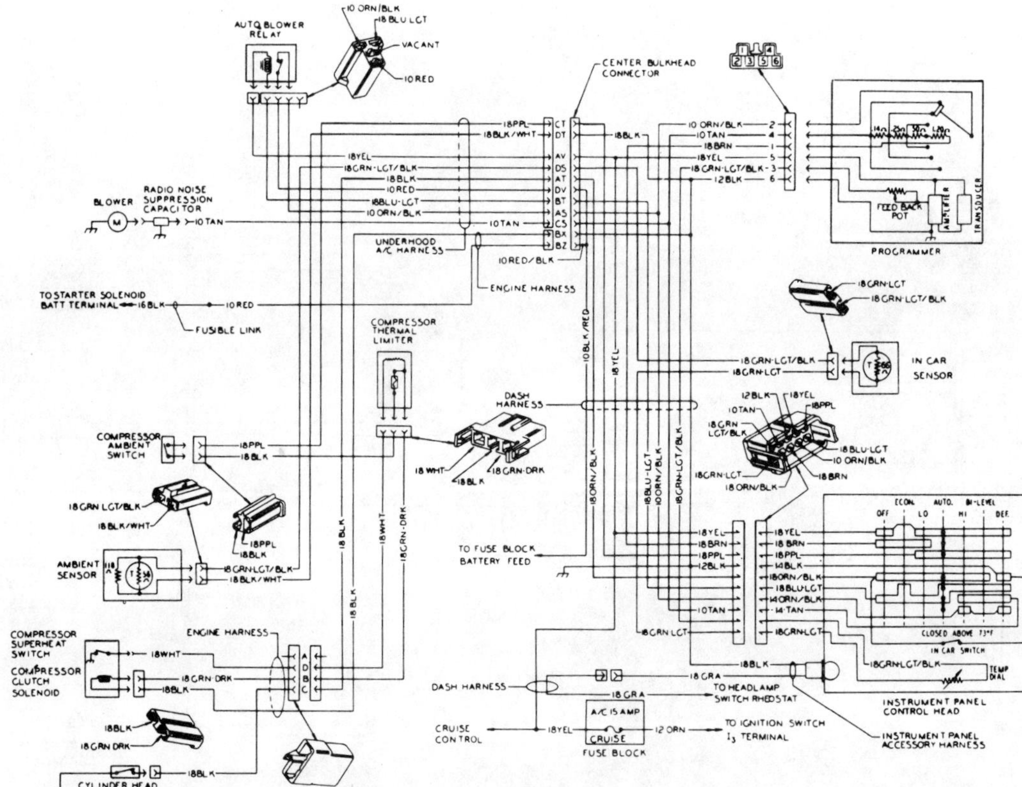

1976 Cadillac Except Seville A/C & Heater Wiring Circuit

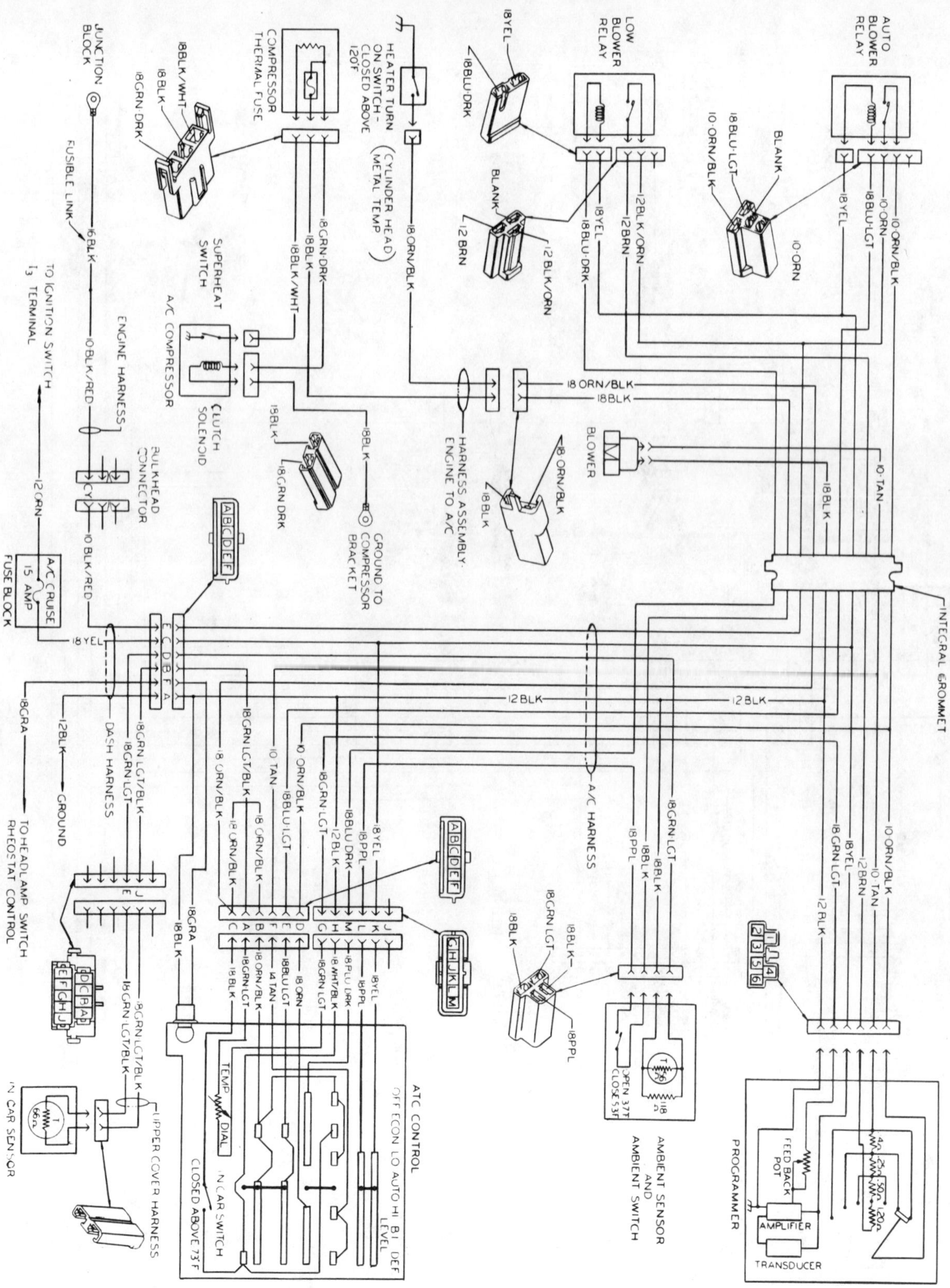

Early 1976 Cadillac Seville A/C & Heater Wiring Circuit

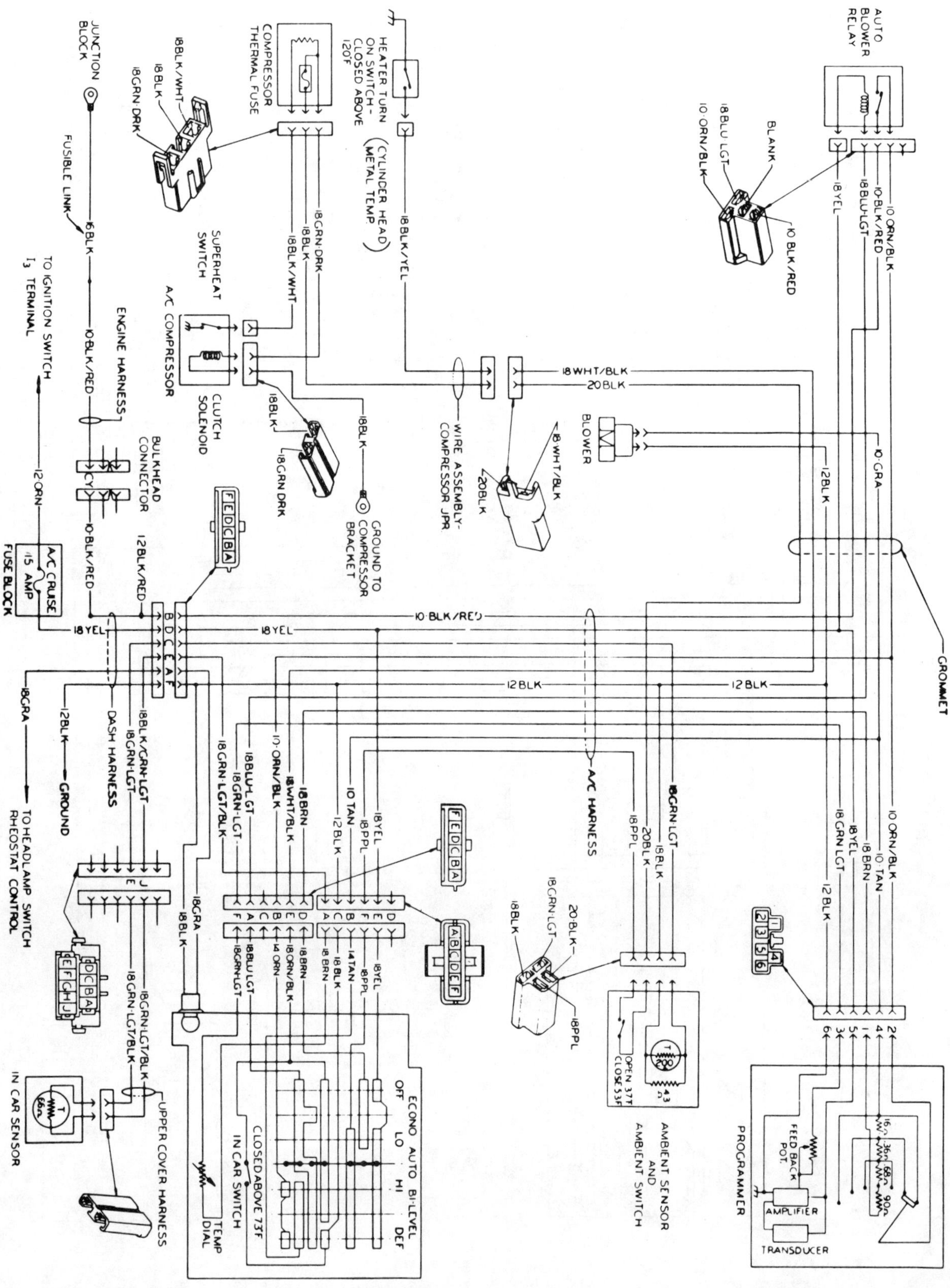

Late 1976 Cadillac Seville A/C & Heater Wiring Circuit

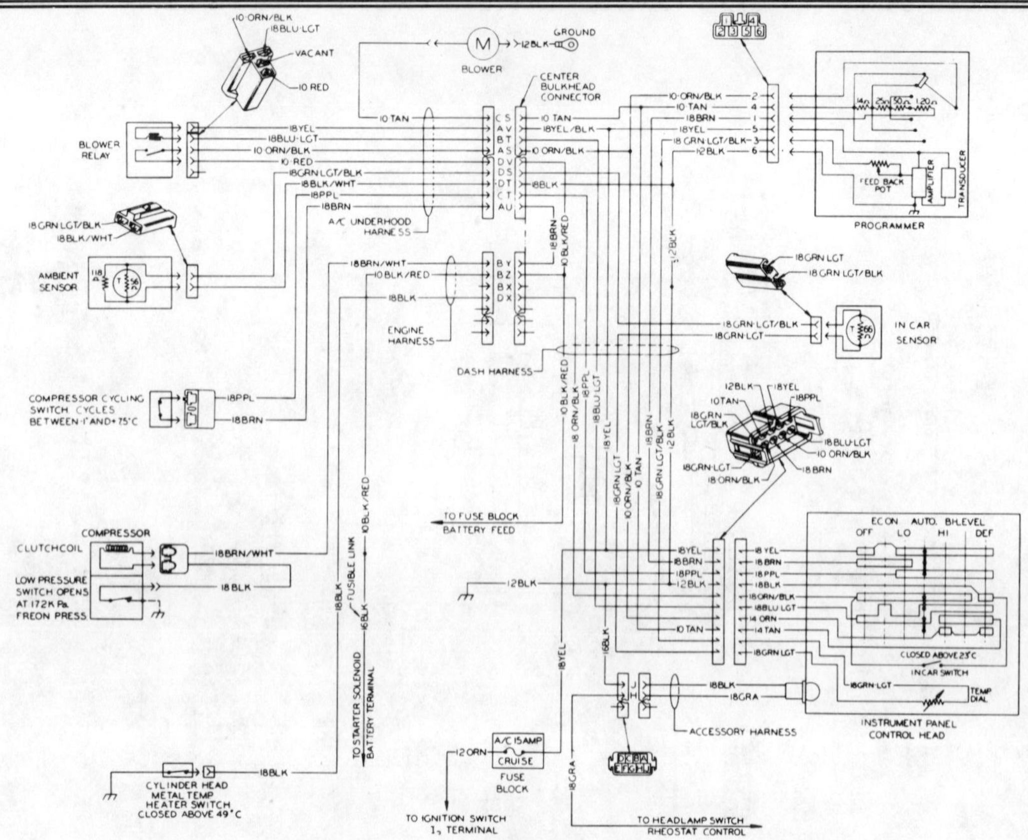

1977-78 Cadillac Eldorado A/C & Heater Wiring Circuit

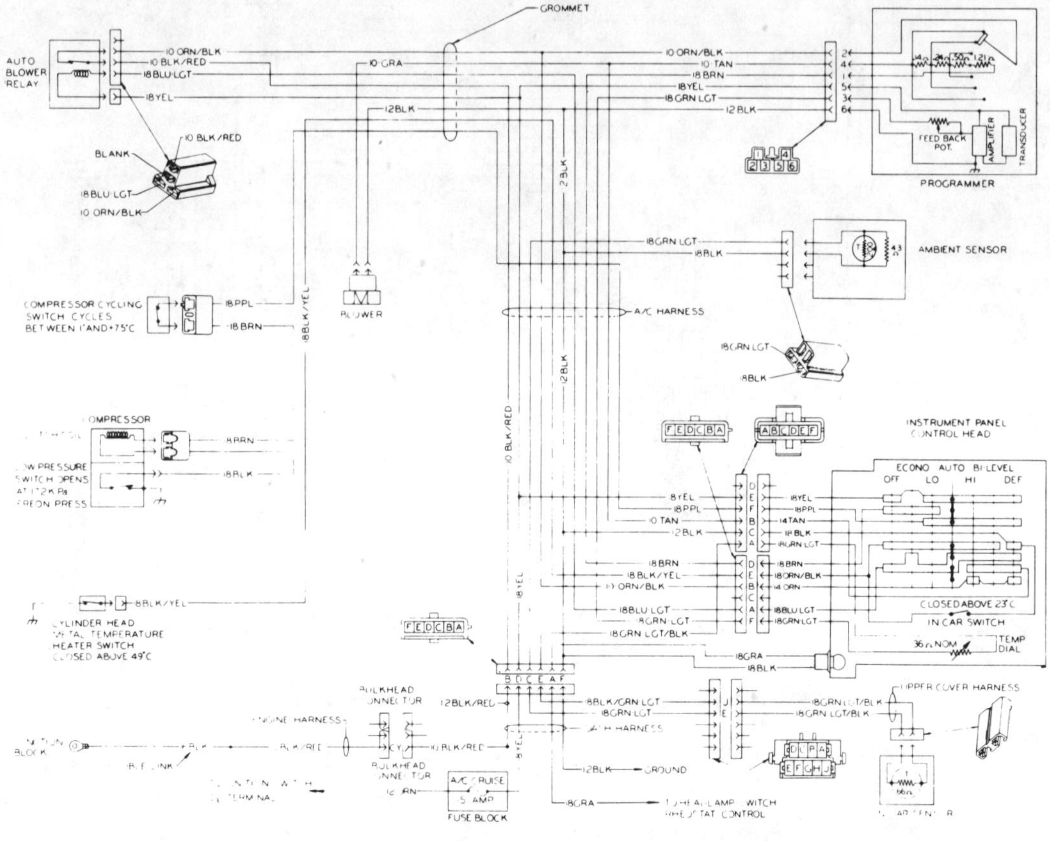

1977-78 Cadillac Seville A/C & Heater Wiring Circuit

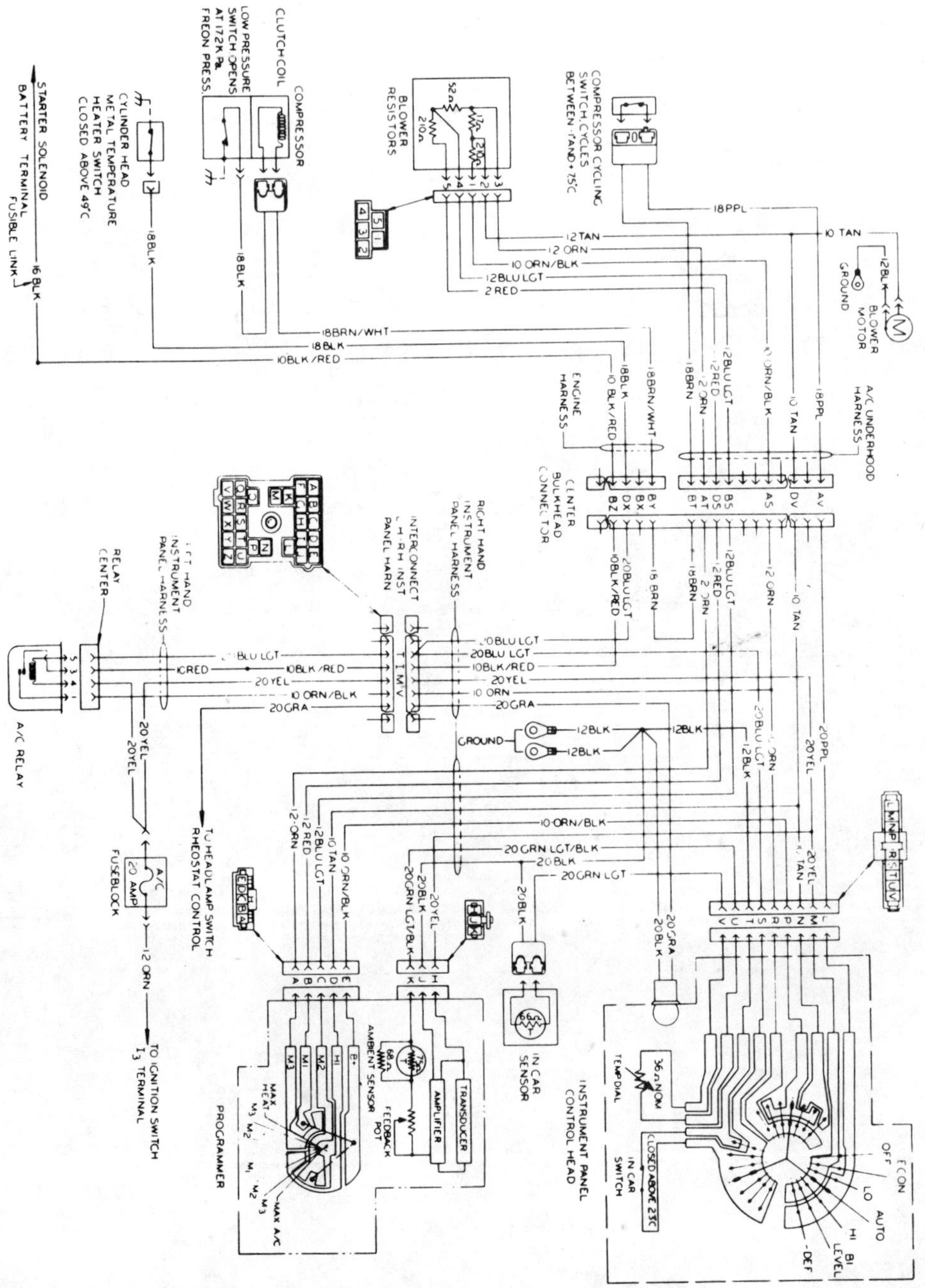

1977-80 Cadillac Except Eldorado & Seville A/C & Heater Wiring Circuit

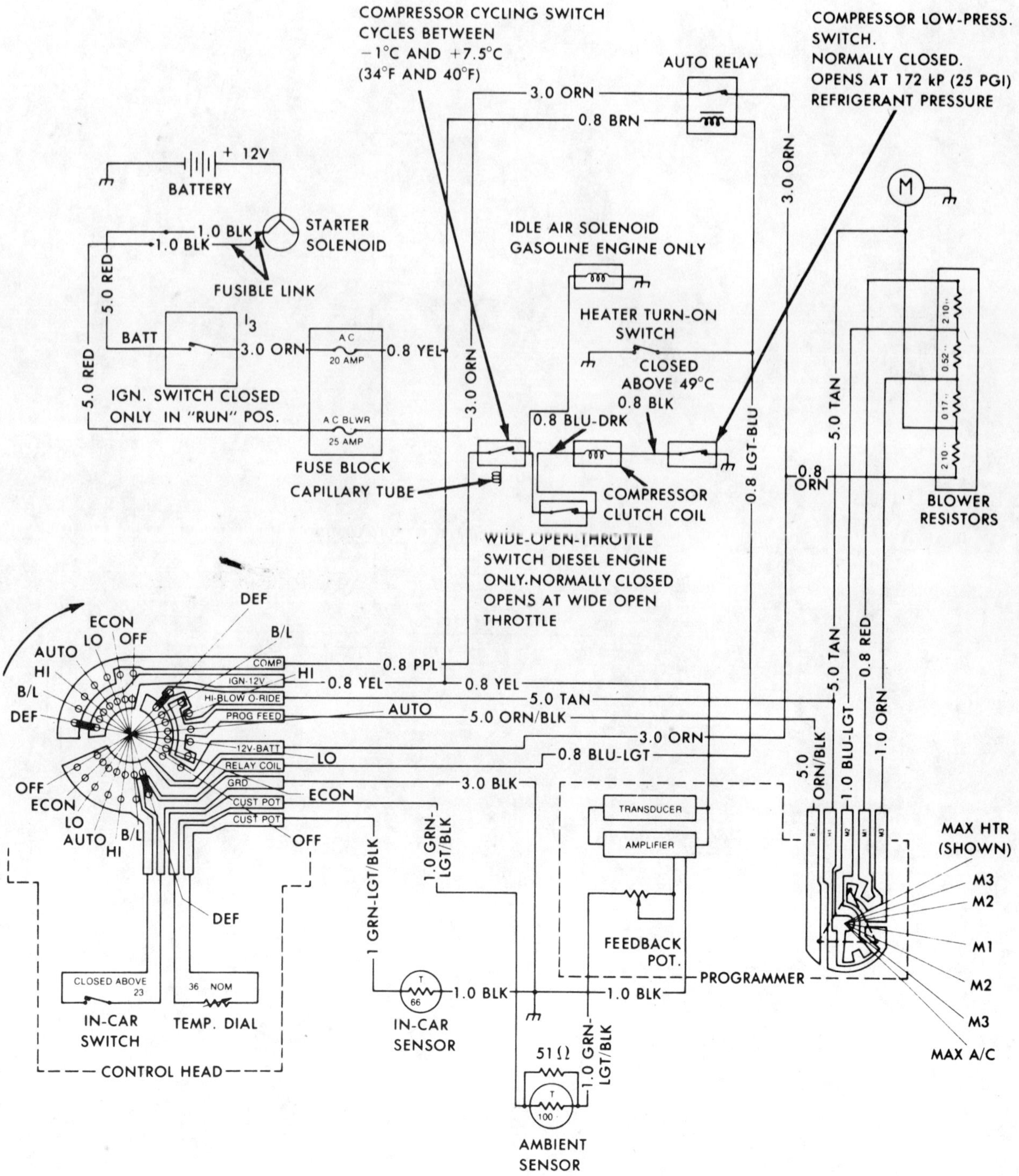

1979 Cadillac Eldorado A/C & Heater Wiring Circuit

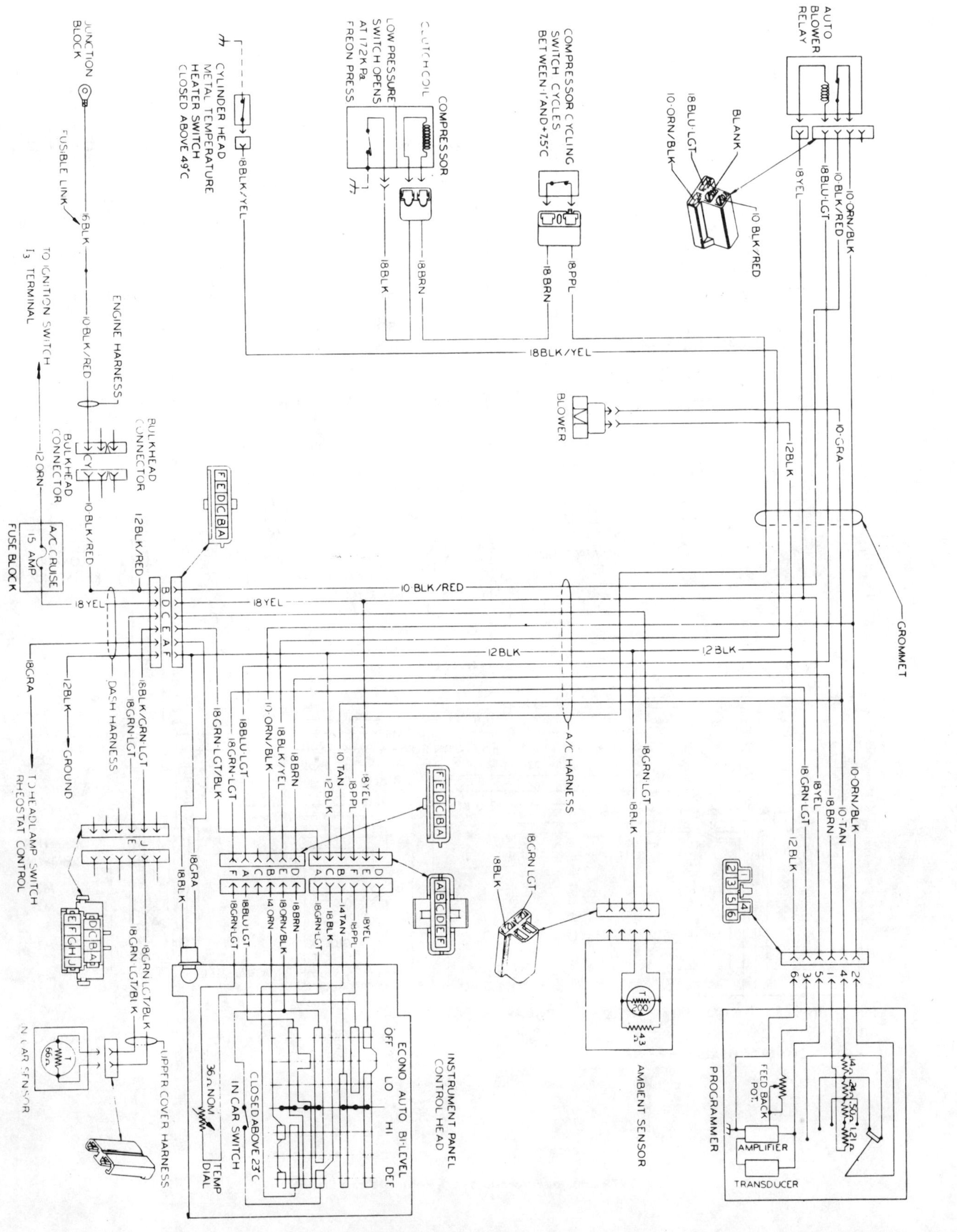

1979 Cadillac Seville A/C & Heater Wiring Circuit

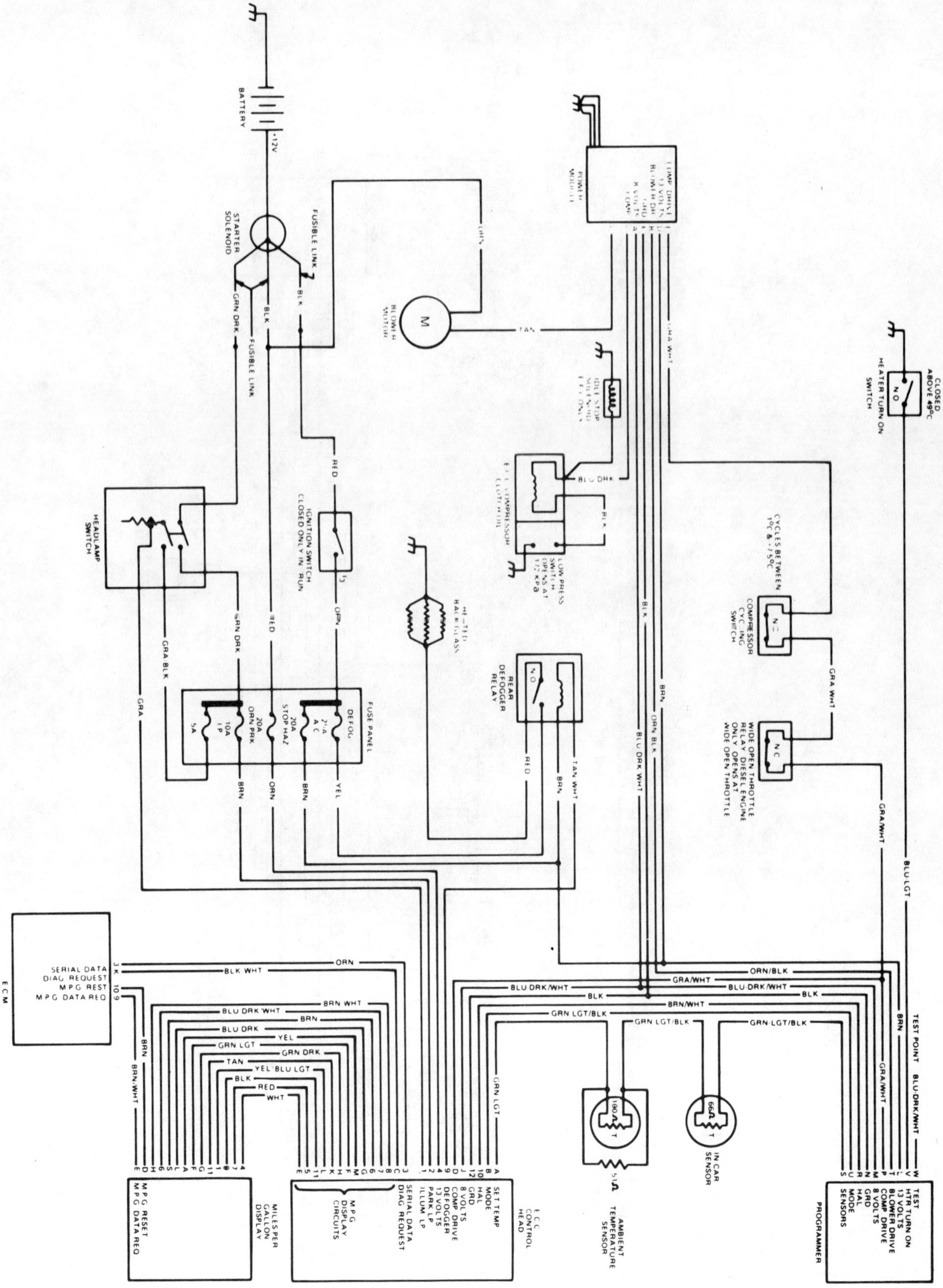

1980 Cadillac Eldorado & Seville A/C & Heater Wiring Circuit (With Electronic Climate Control, Part 1 of 2)

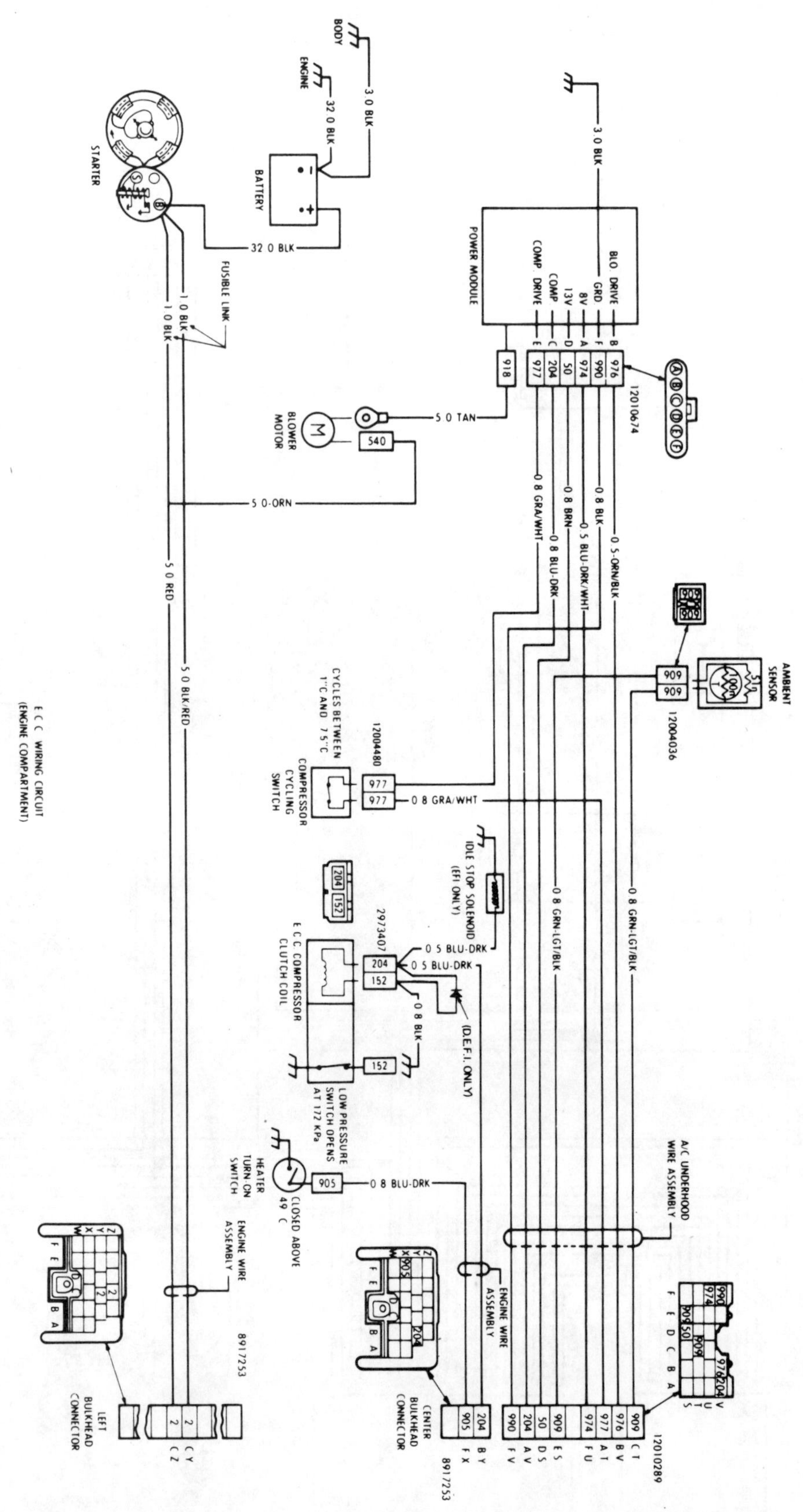

1980 Cadillac Eldorado & Seville A/C & Heater Wiring Circuit (With Electronic Climate Control, Part 2 of 2)

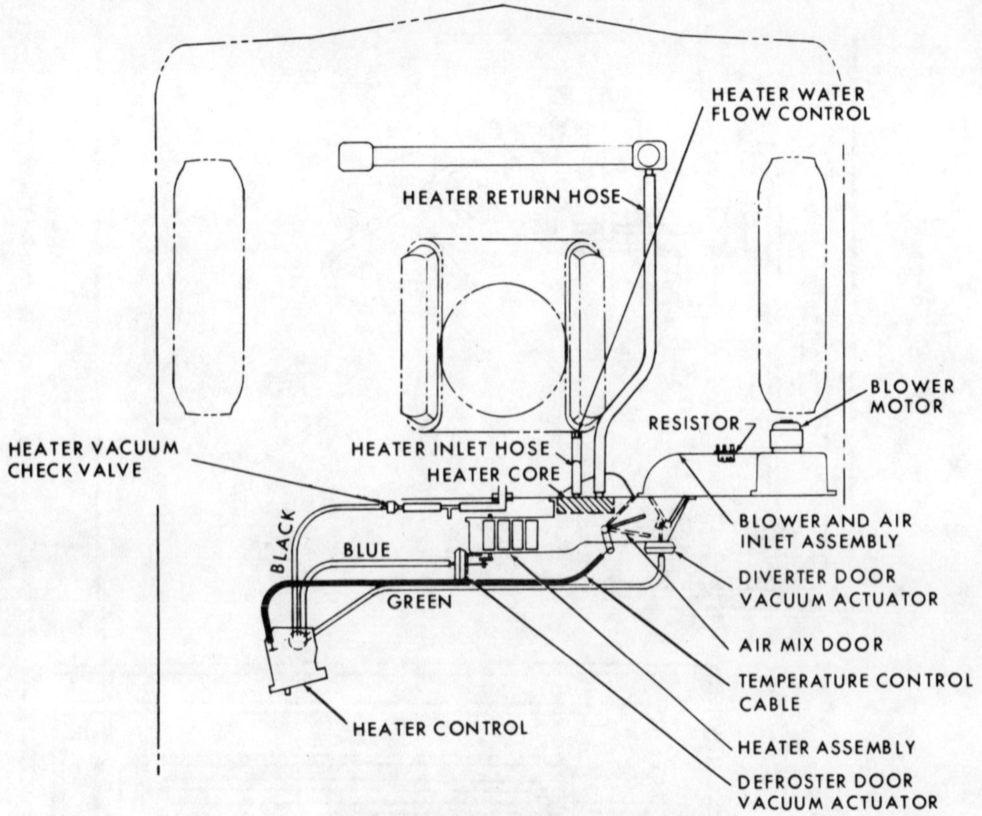

1974 Cadillac Heater Vacuum Circuit

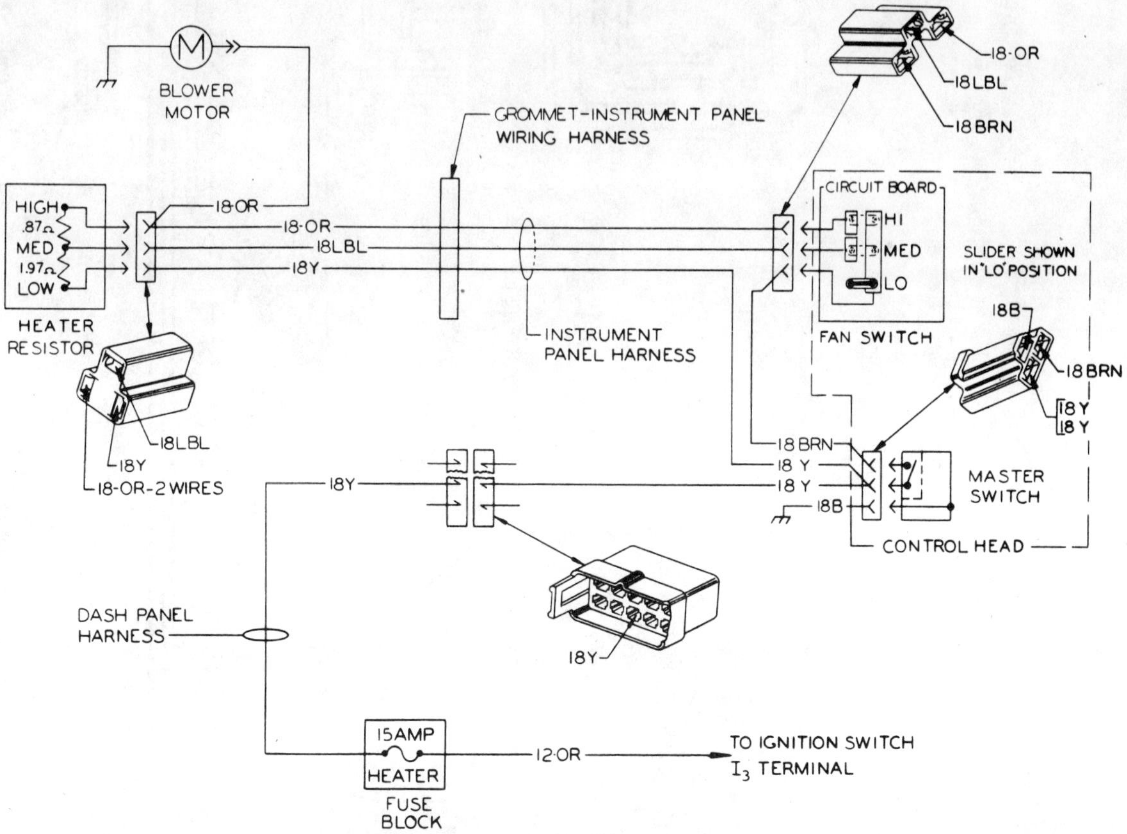

1974 Cadillac Heater Wiring Circuit

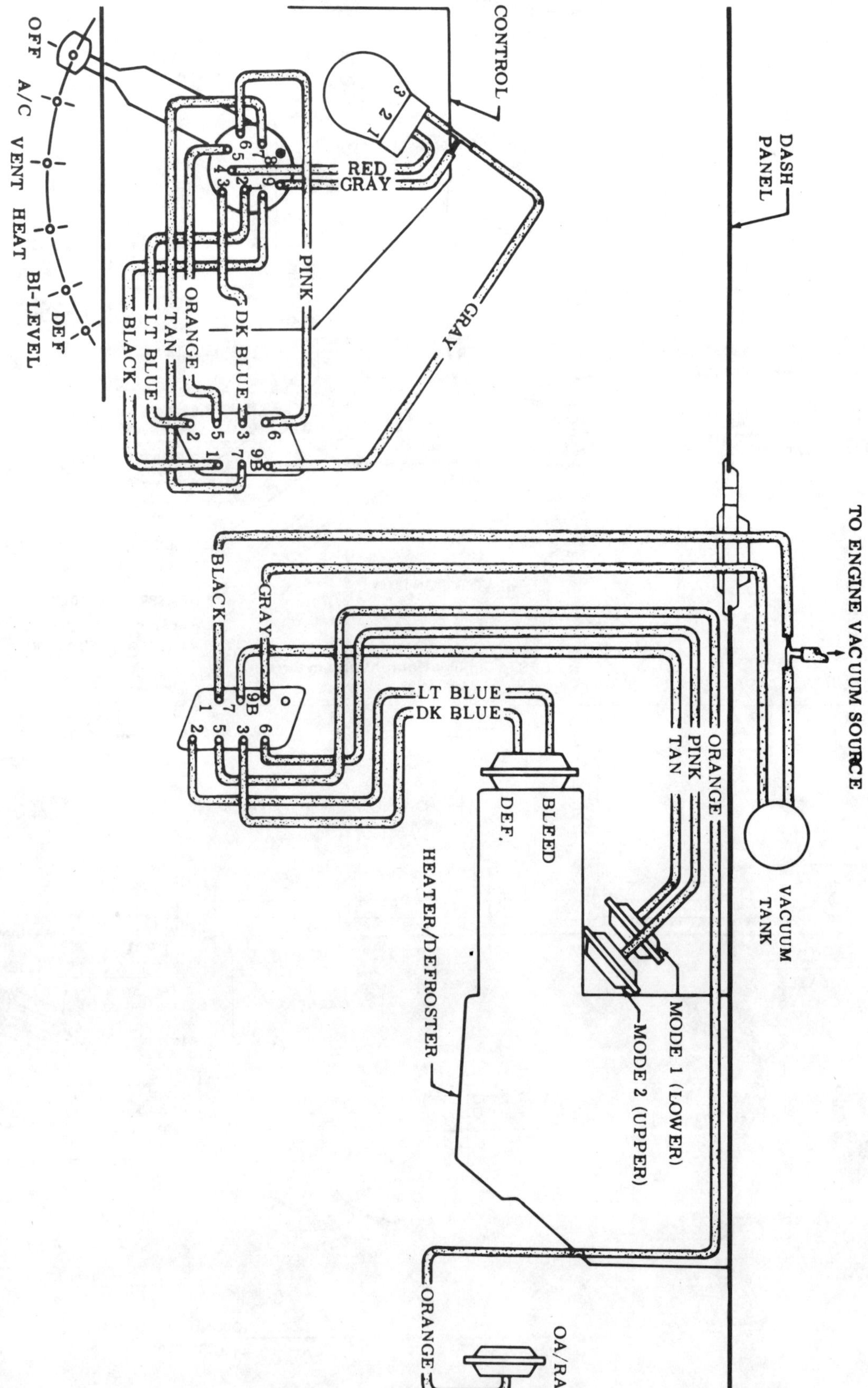

1974-76 Chevrolet Four Season A/C & Heater Vacuum Circuit

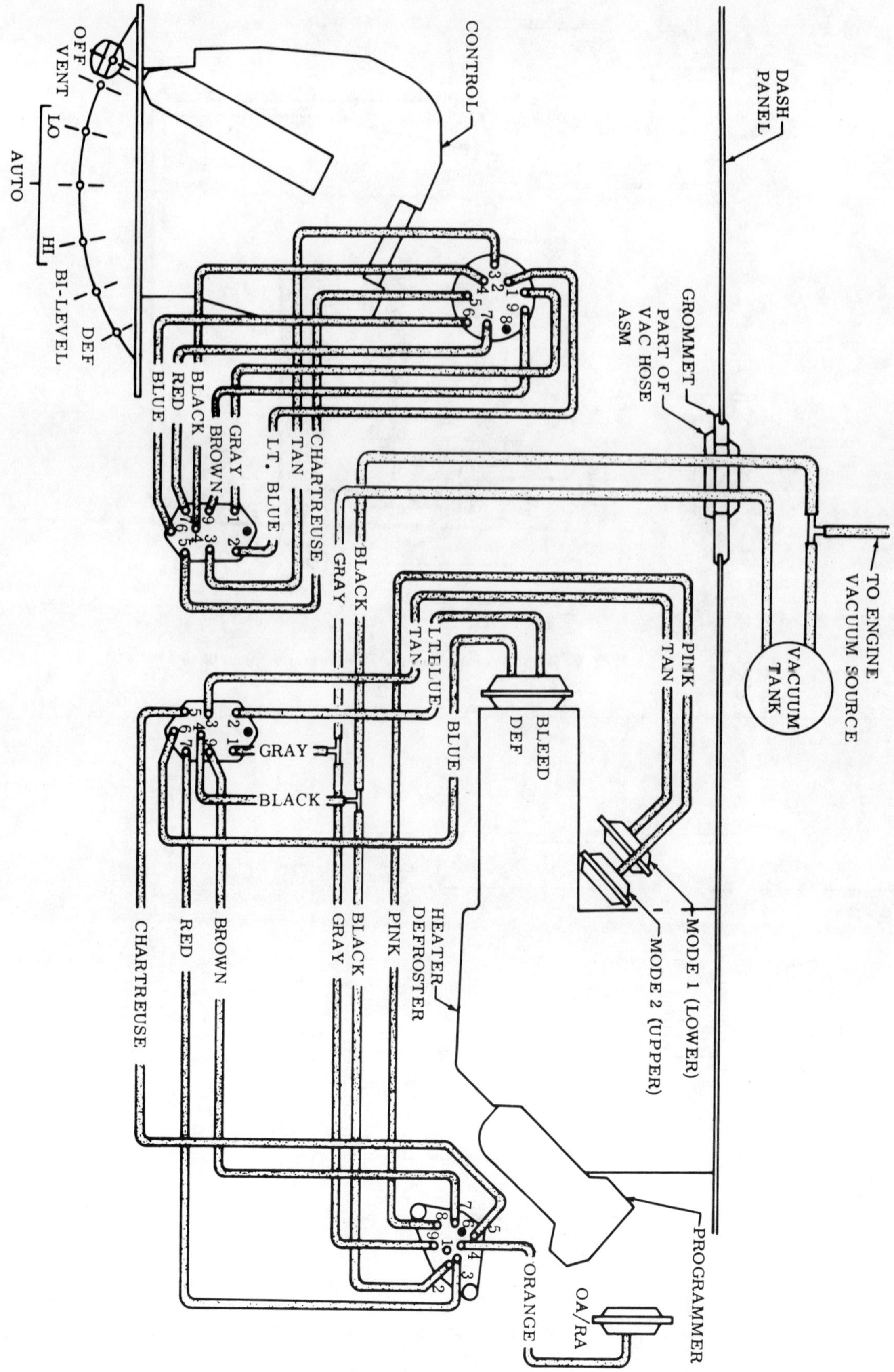

1974-76 Chevrolet Comfortron A/C & Heater Vacuum Circuit

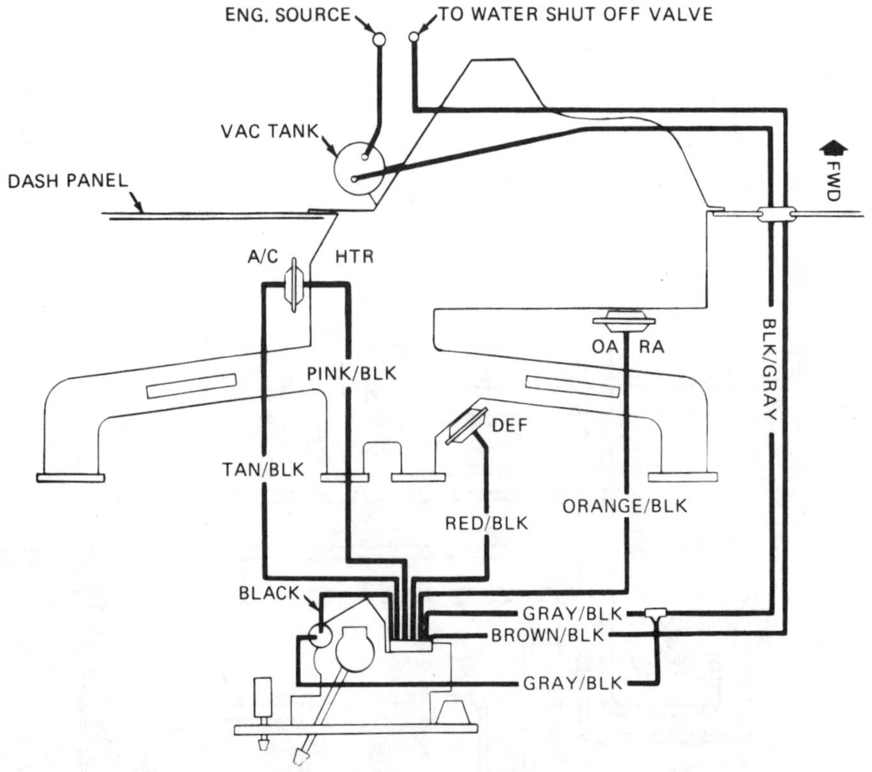

1977-80 Chevrolet Four Season A/C & Heater Vacuum Circuit

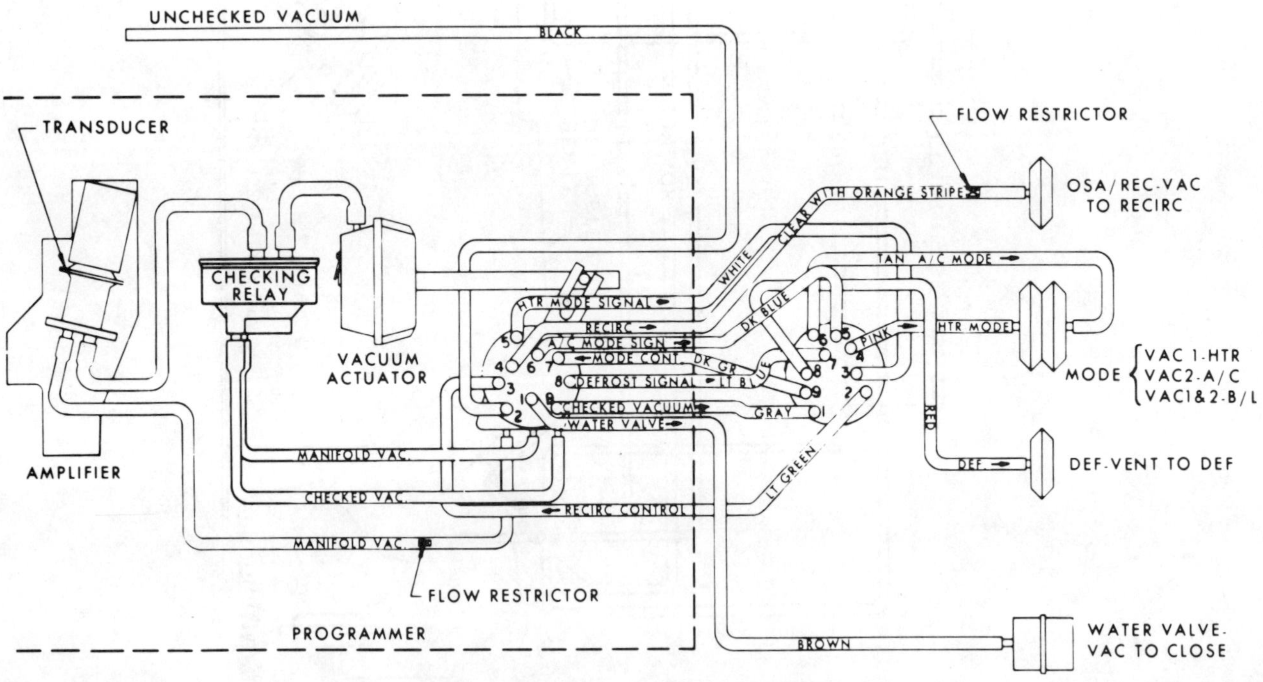

1977-80 Chevrolet Comfortron A/C & Heater Vacuum Circuit

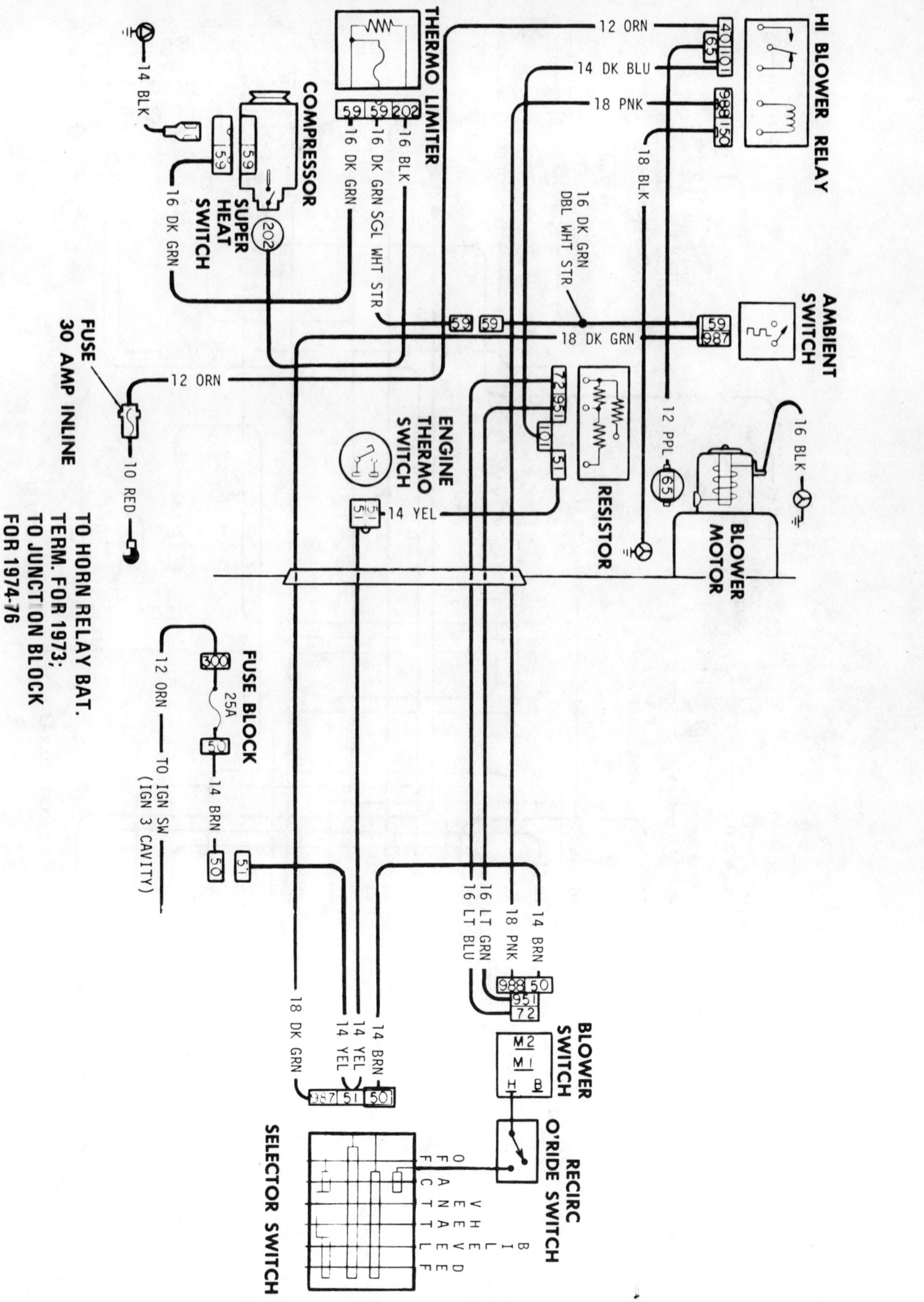

1974-76 Chevrolet Four Season A/C & Heater Wiring Circuit

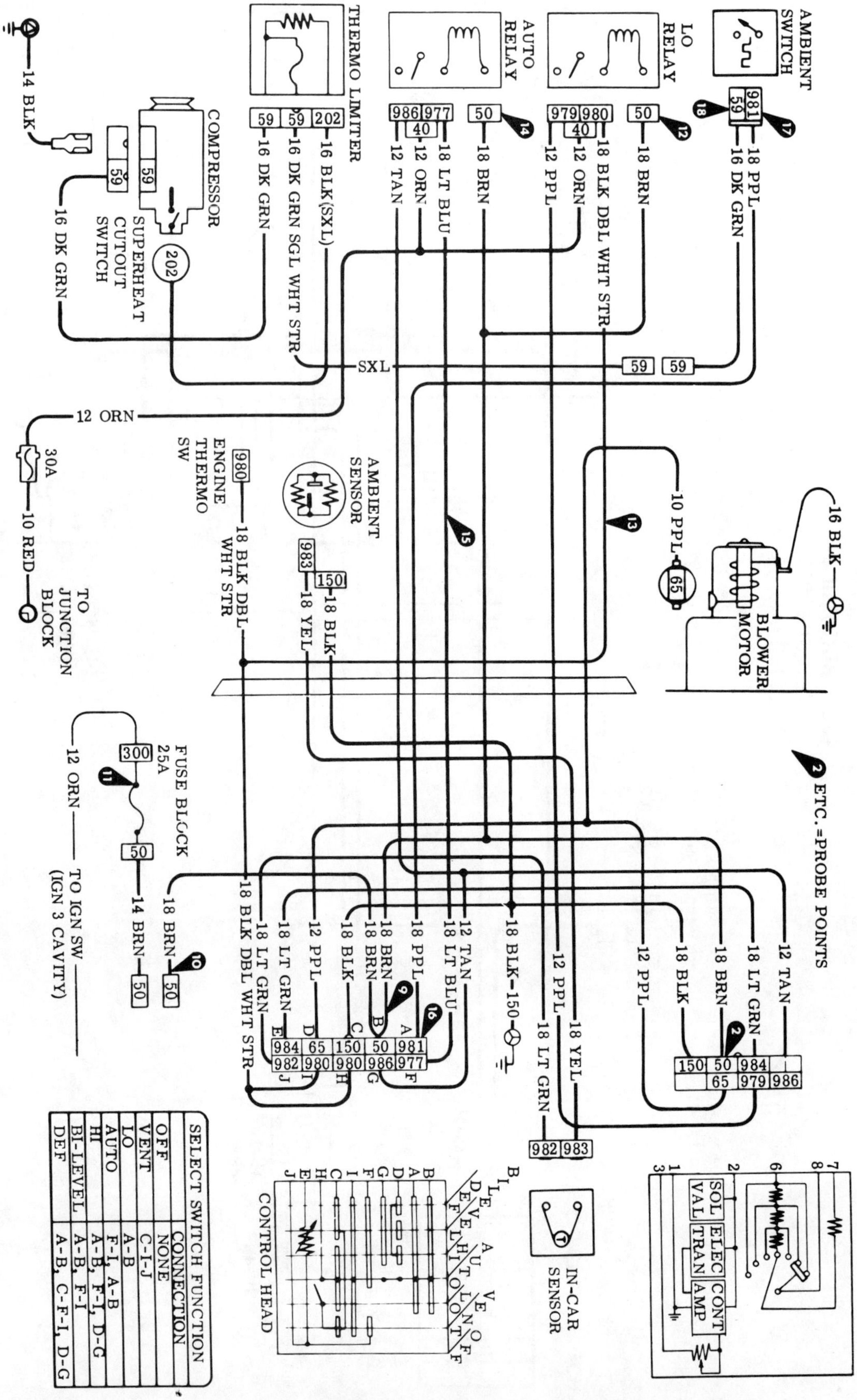

1974-76 Chevrolet Comfortron A/C & Heater Wiring Circuit

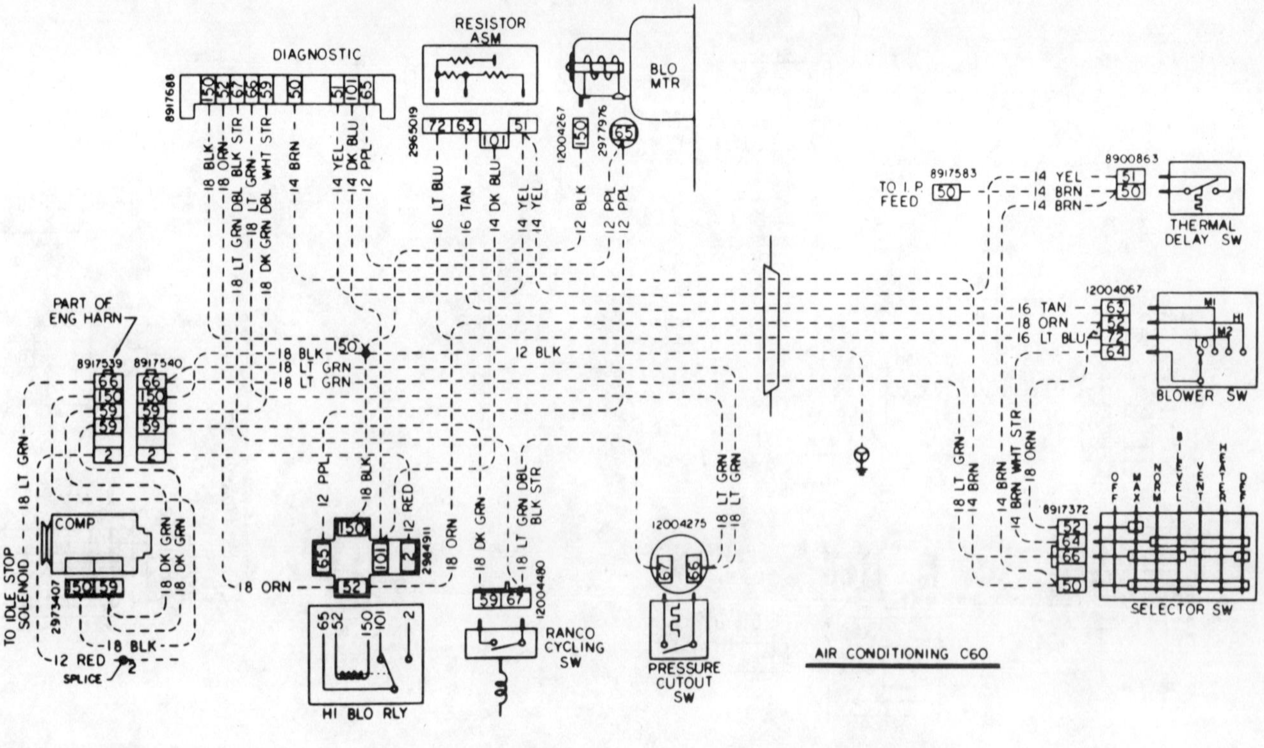

1977 Chevrolet Four Season A/C & Heater Wiring Circuit

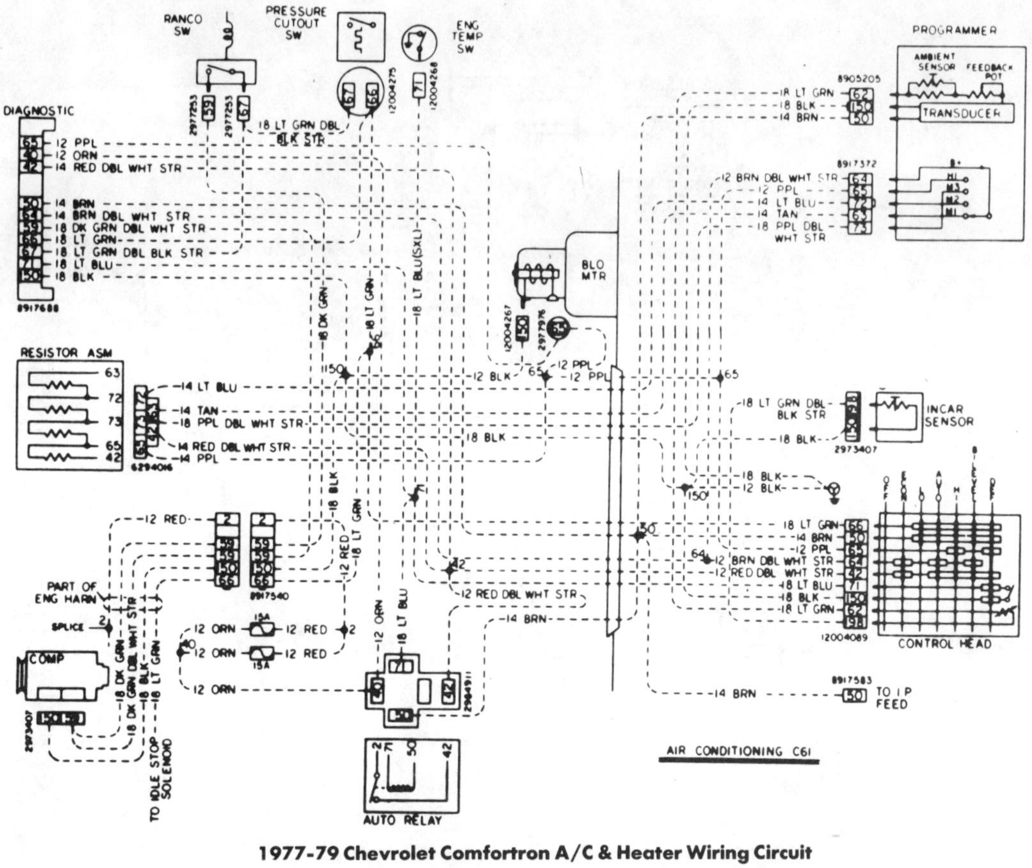

1977-79 Chevrolet Comfortron A/C & Heater Wiring Circuit

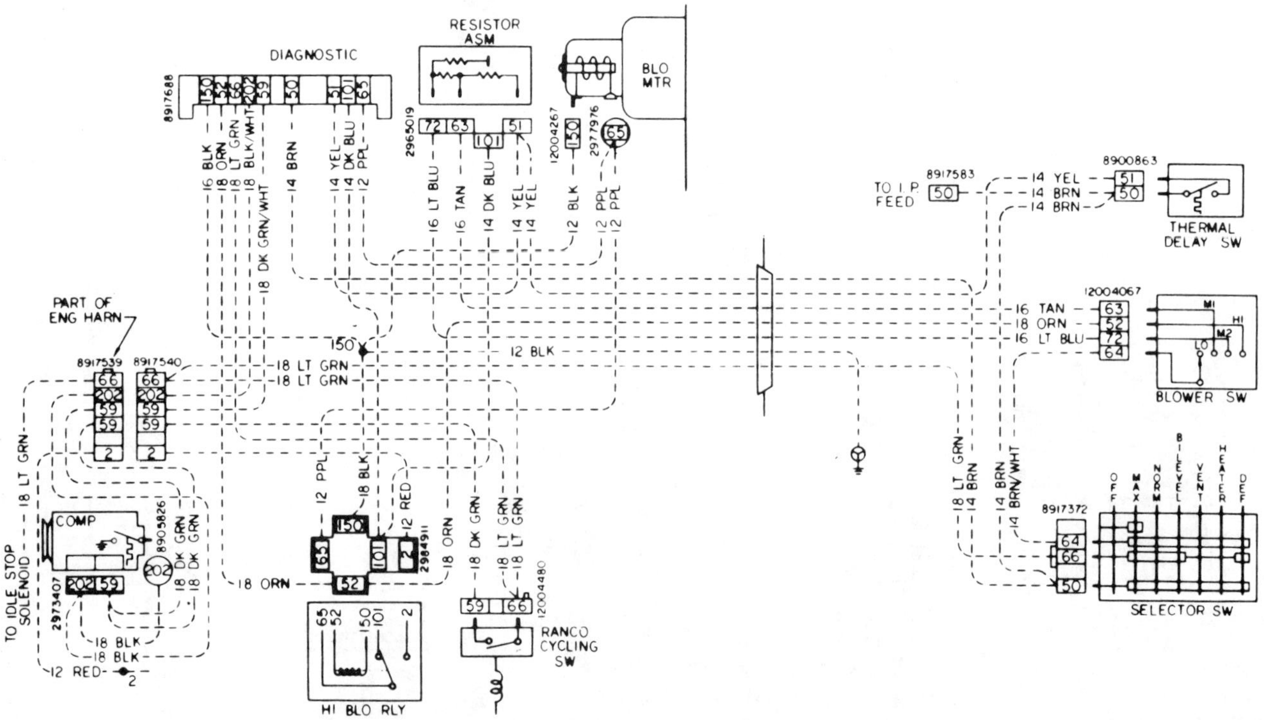

1978-79 Chevrolet Four Season A/C & Heater Wiring Circuit

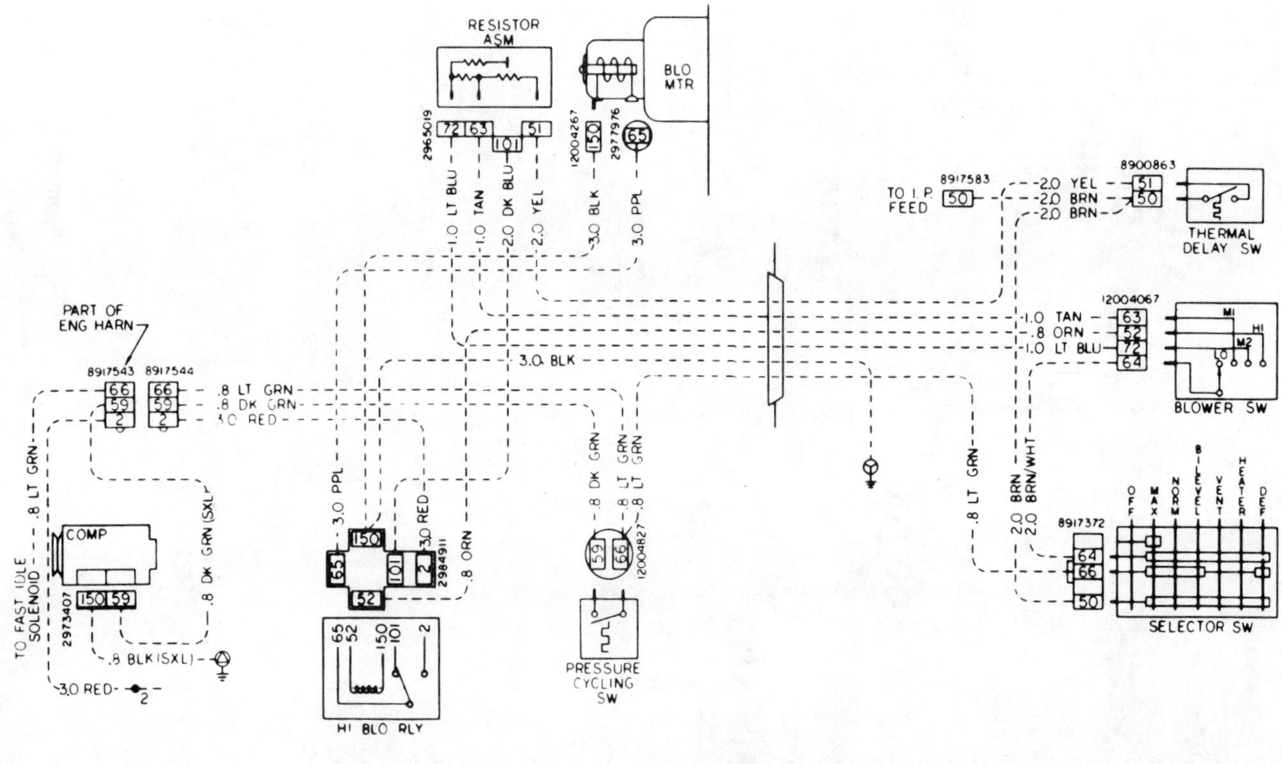

1980 Chevrolet Four Season A/C & Heater Wiring Circuit

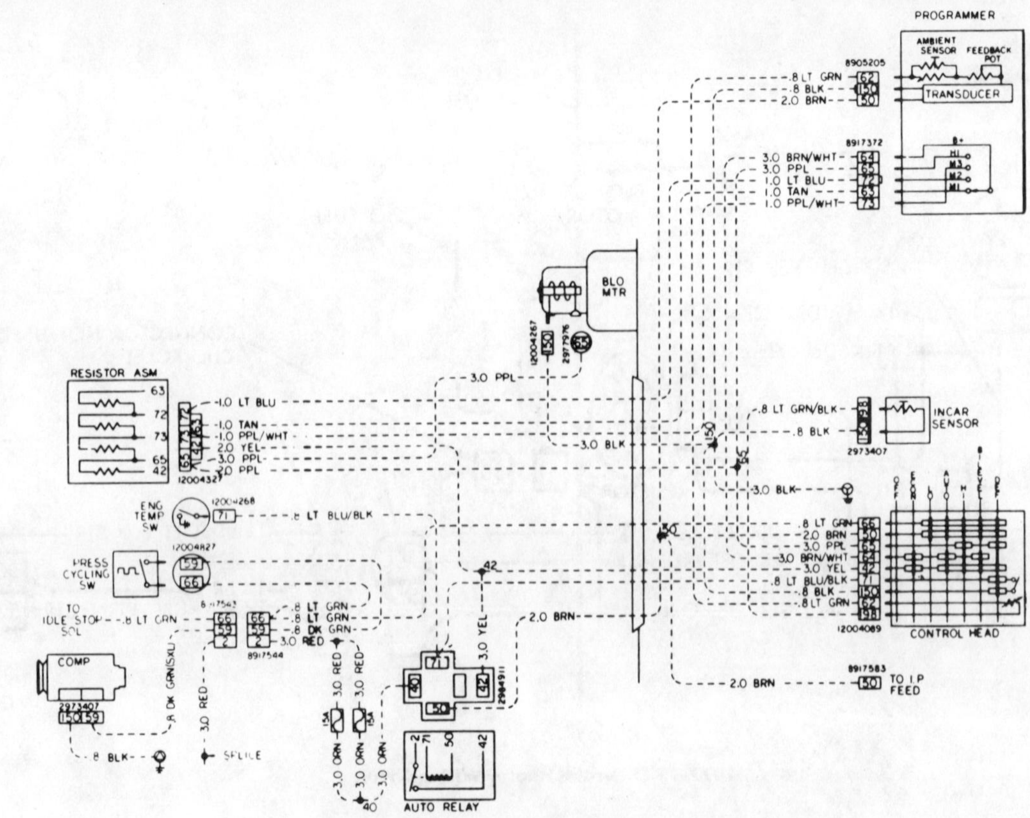

1980 Chevrolet Comfortron A/C & Heater Wiring Circuit

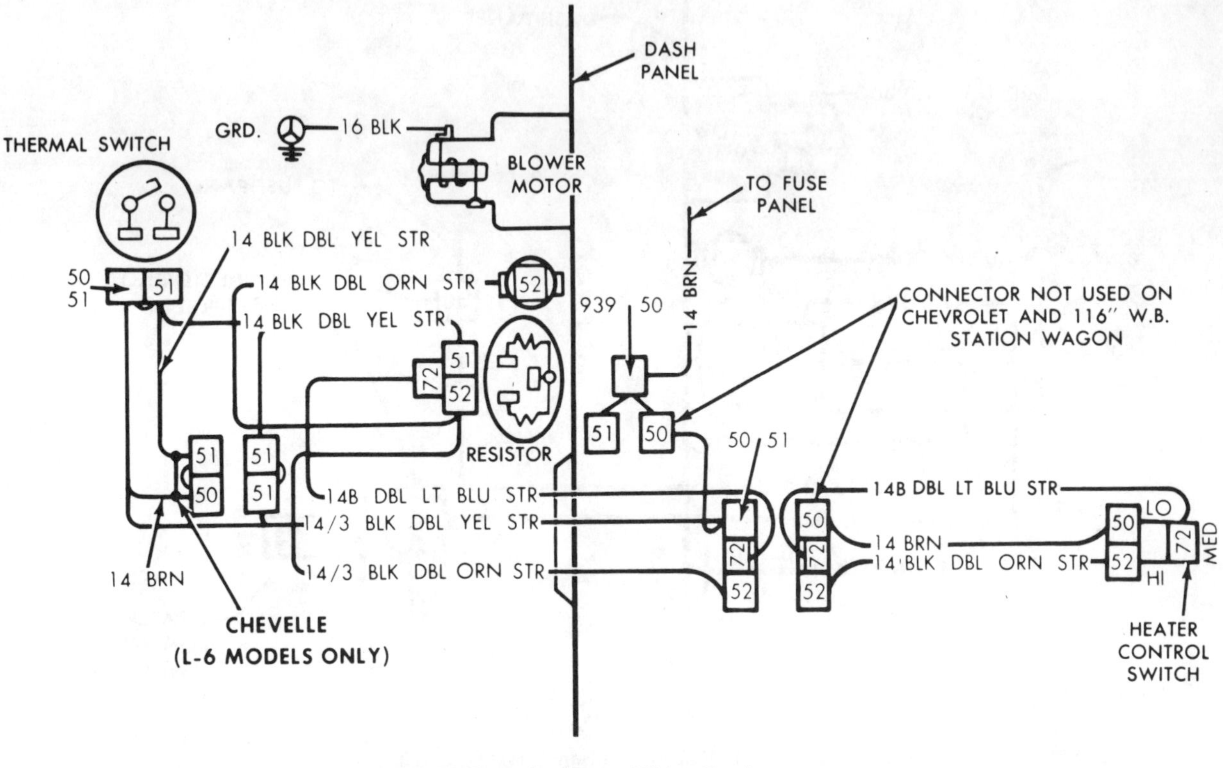

1974-76 Chevrolet Heater Wiring Circuit

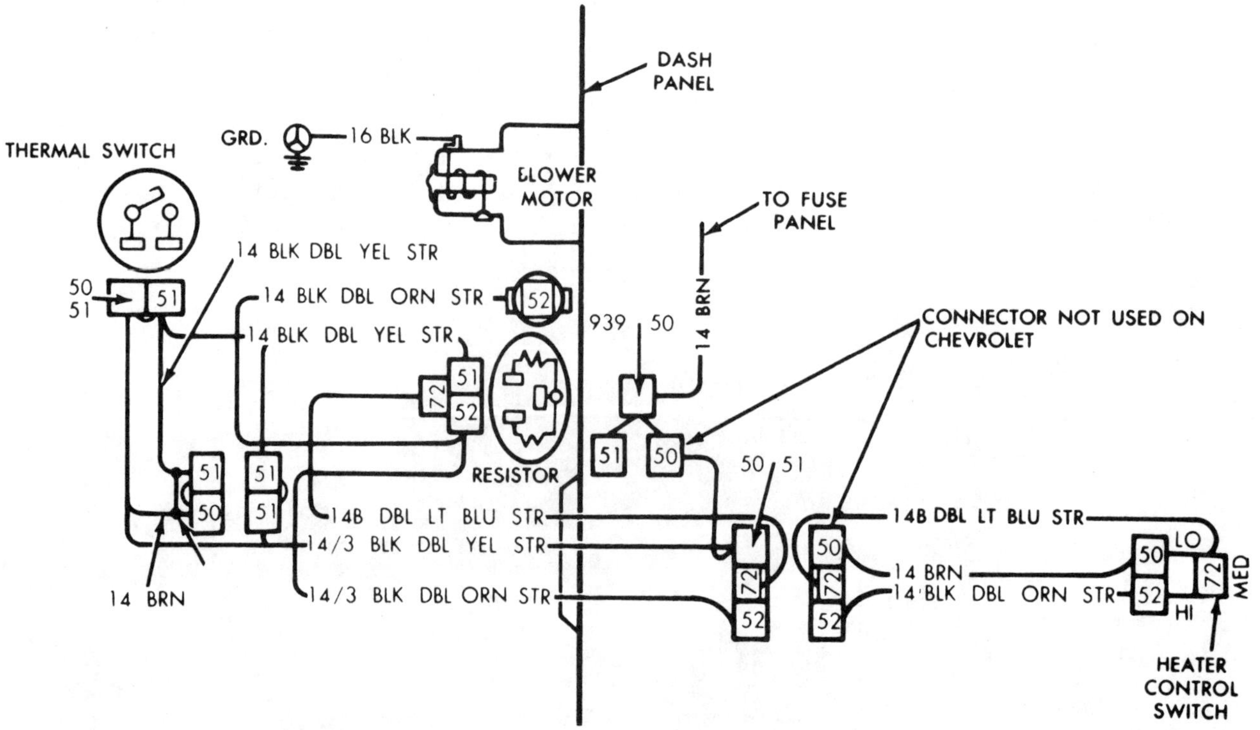

1977-79 Chevrolet Heater Wiring Circuit

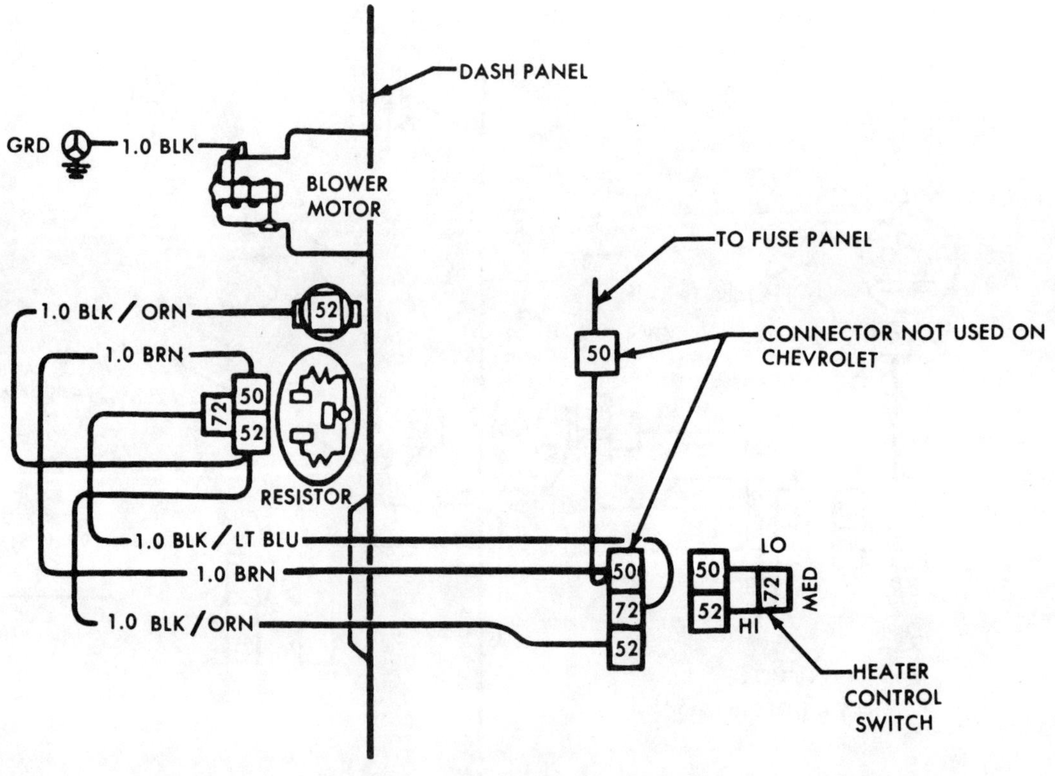

1980 Chevrolet Heater Wiring Circuit

1980
Chevrolet Citation • Buick Skylark
Oldsmobile Omega • Pontiac Phoenix

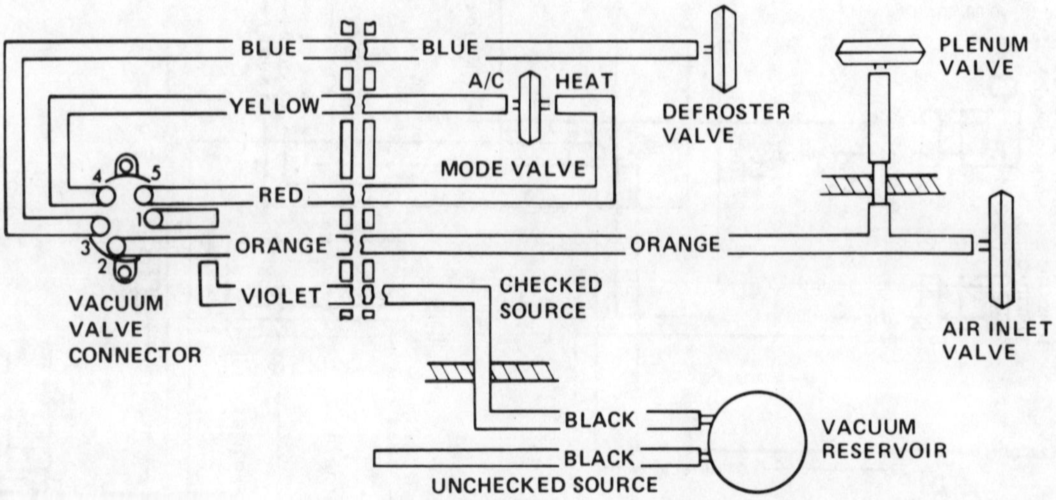

VACUUM VALVE LOGIC		MODE LEVER POSITION						
PORT	CONNECTION	OFF	MAX	NORM	BI-LEVEL	VENT	HEAT	DEFROST
1	INPUT	3,5	2,3,4	3,4	3	3,4	3,5	5
2	OSA RECIRC	VENT	VAC	VENT	VENT	VENT	VENT	VENT
3	DEFROST	VAC	VAC	VAC	VAC	VAC	VAC	VENT
4	A/C MODE	VENT	VAC	VAC	VENT	VAC	VENT	VENT
5	HEAT MODE	VAC	VENT	VENT	VENT	VENT	VAC	VAC

1980 Citation, Omega, Phoenix & Skylark A/C & Heater Vacuum Circuit

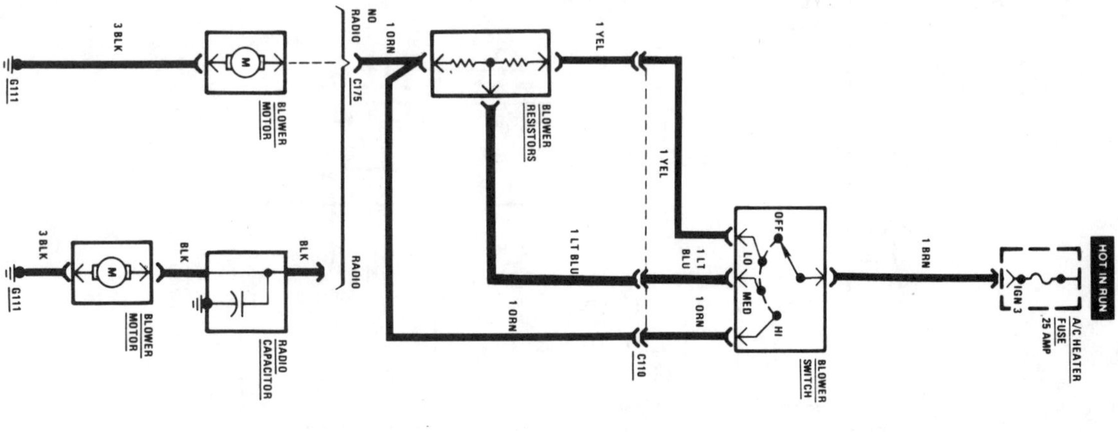

1980 Citation, Omega, Phoenix & Skylark Heater Wiring Circuit

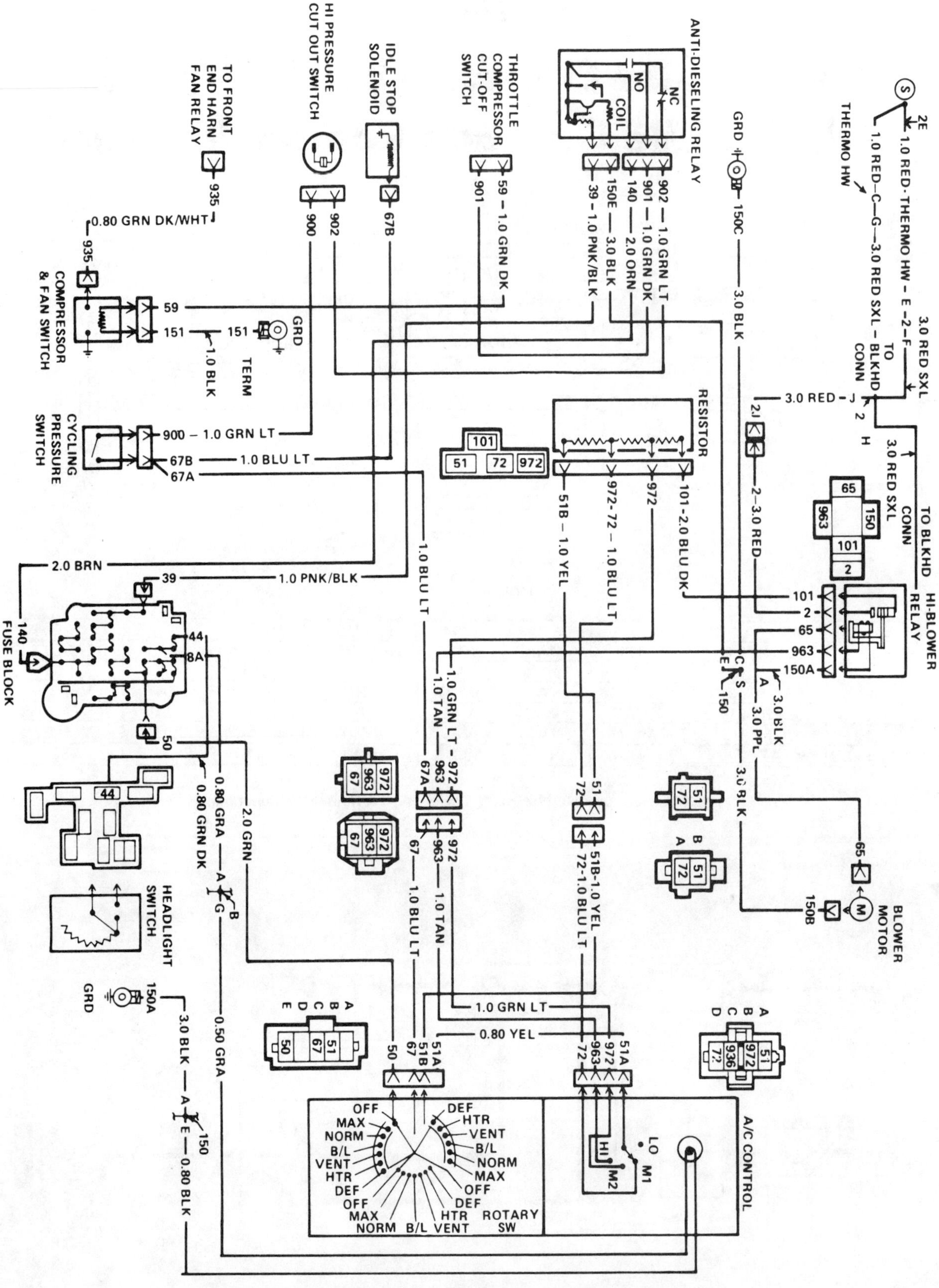

1980 Citation, Omega, Phoenix & Skylark A/C & Heater Wiring Circuit with 4-151

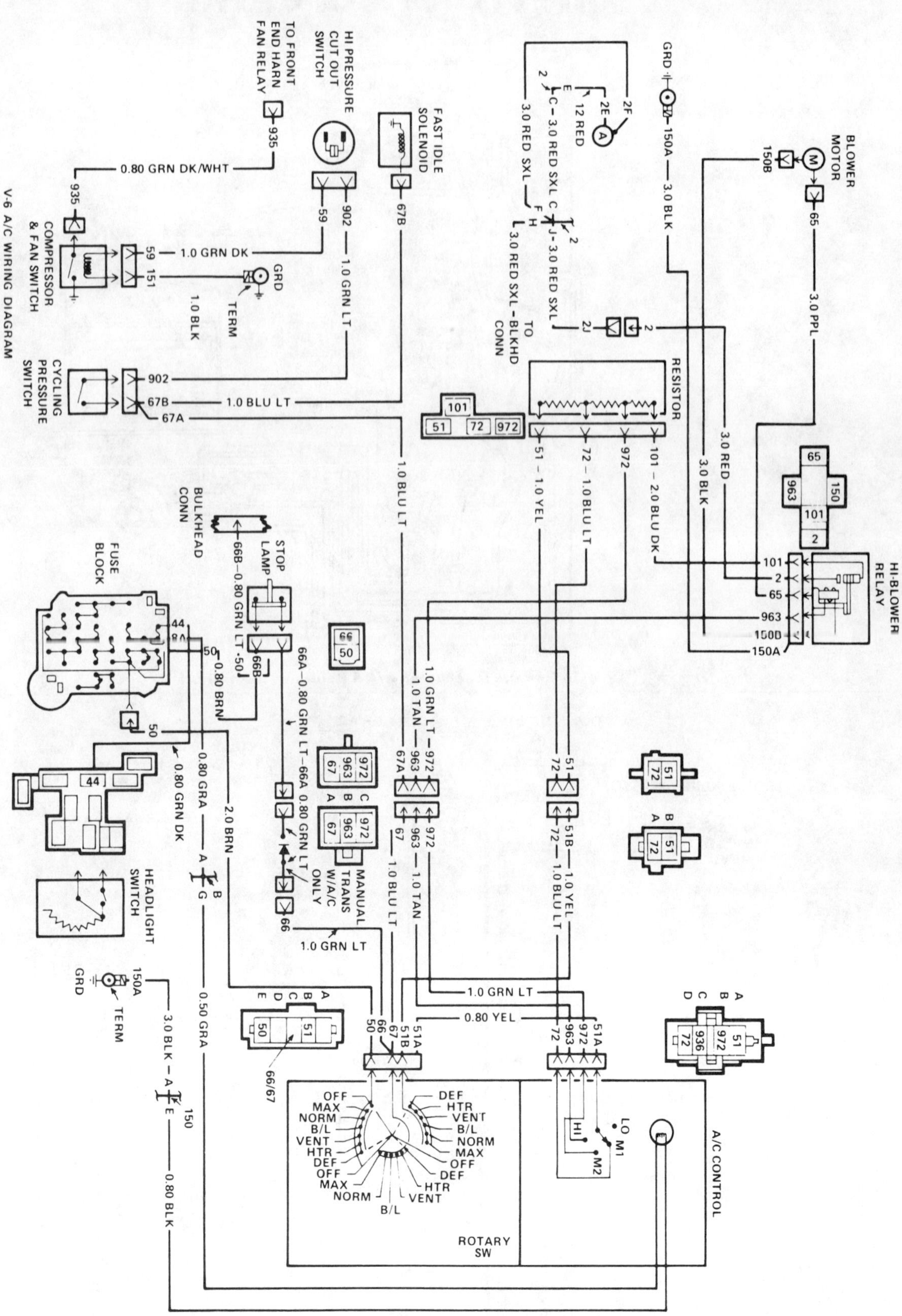

1980 Citation, Omega, Phoenix & Skylark A/C & Heater Wiring Circuit with V6-173

Chevrolet Corvette

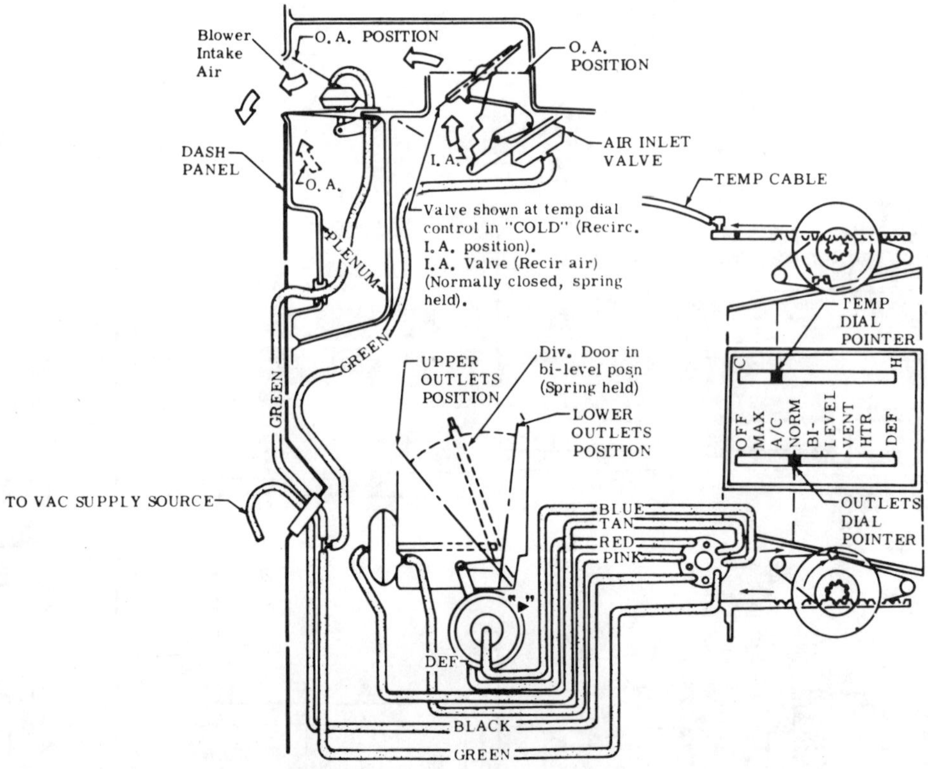

1974-76 Corvette Four Season A/C & Heater Vacuum Circuit

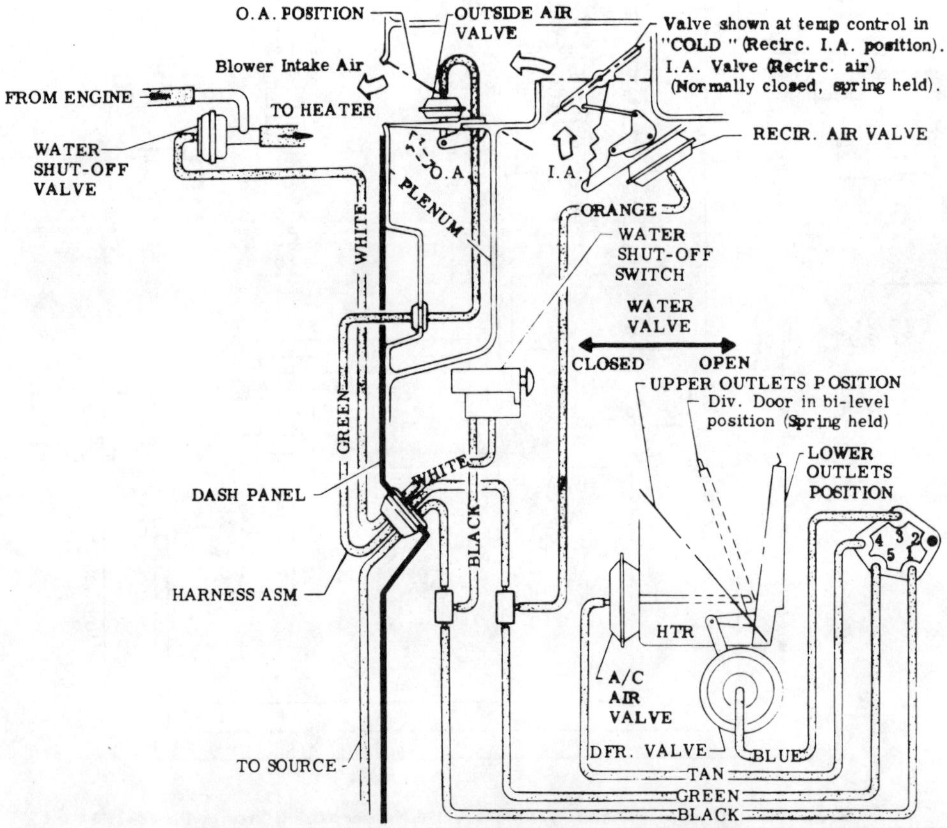

1977-80 Corvette Four Season A/C & Heater Vacuum Circuit

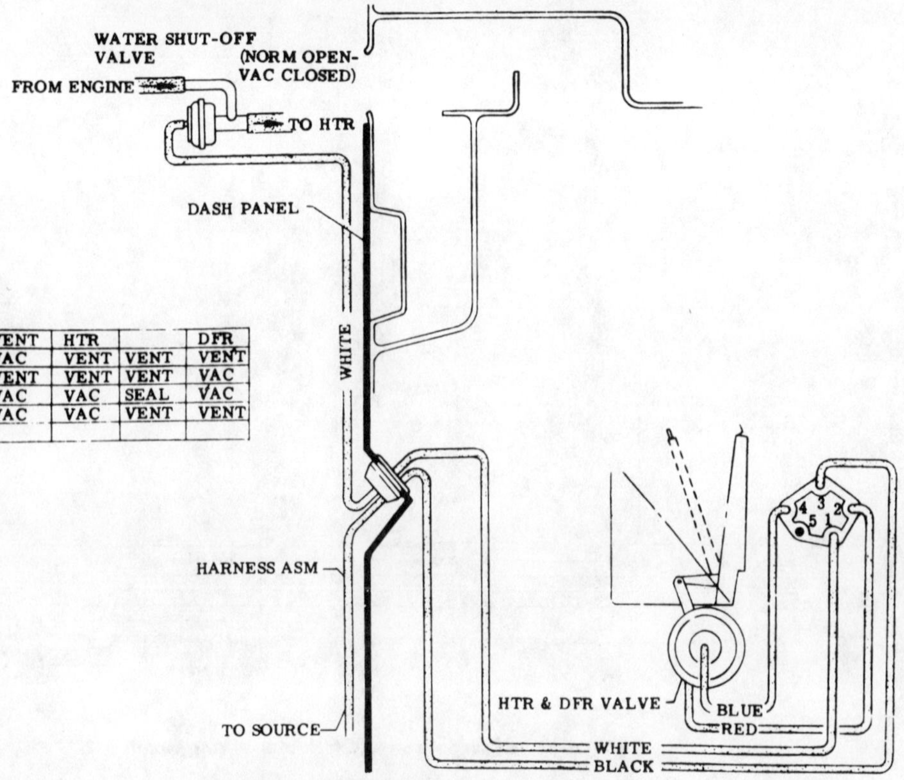

CONN	Port	VENT	HTR		DFR
WATER VLV	1	VAC	VENT	VENT	VENT
DFR	2	VENT	VENT	VENT	VAC
INPUT CK	3	VAC	VAC	SEAL	VAC
HTR	4	VAC	VAC	VENT	VENT
NOT USED	5				

1979 Corvette Heater Vacuum Circuit

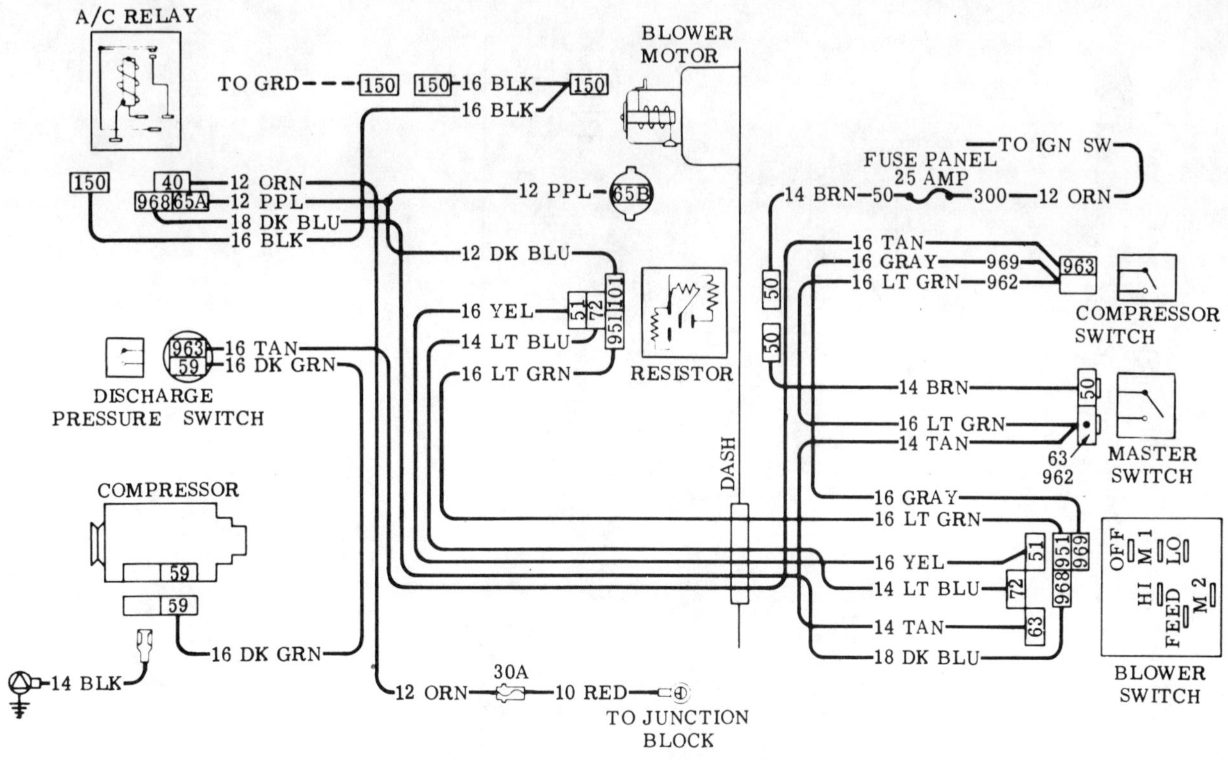

1974-76 Corvette Four Season A/C & Heater Wiring Circuit

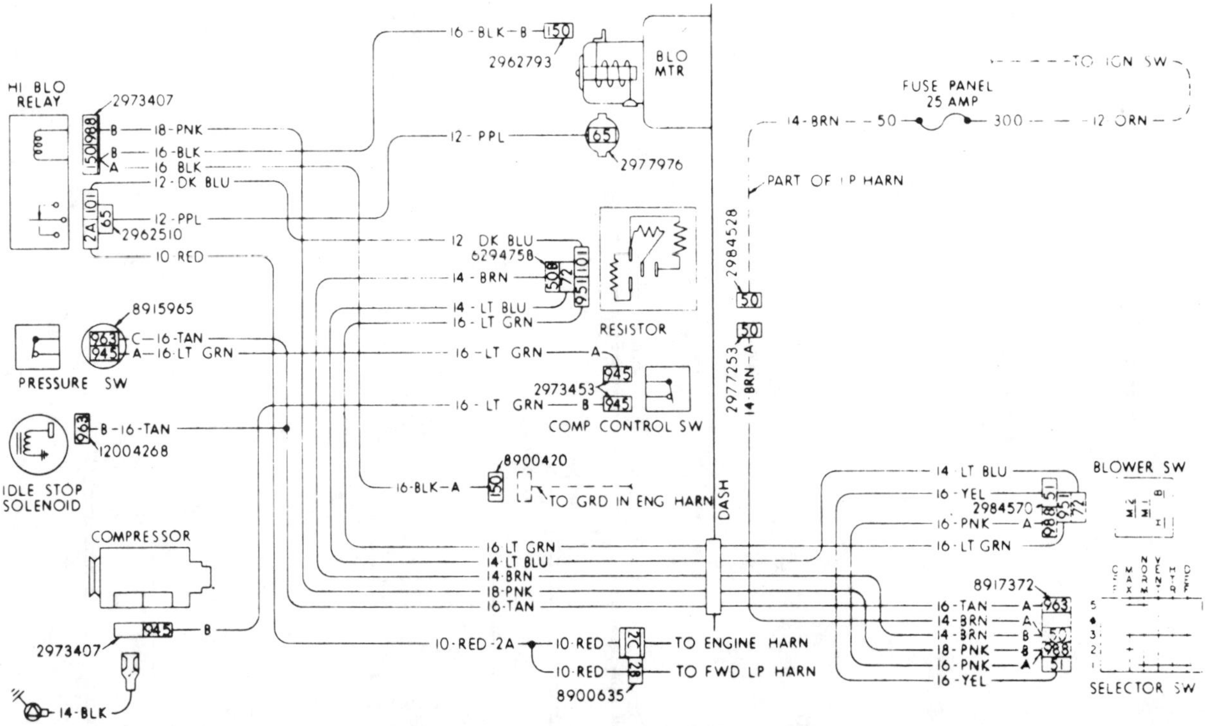

1977 Corvette Four Season A/C & Heater Wiring Circuit

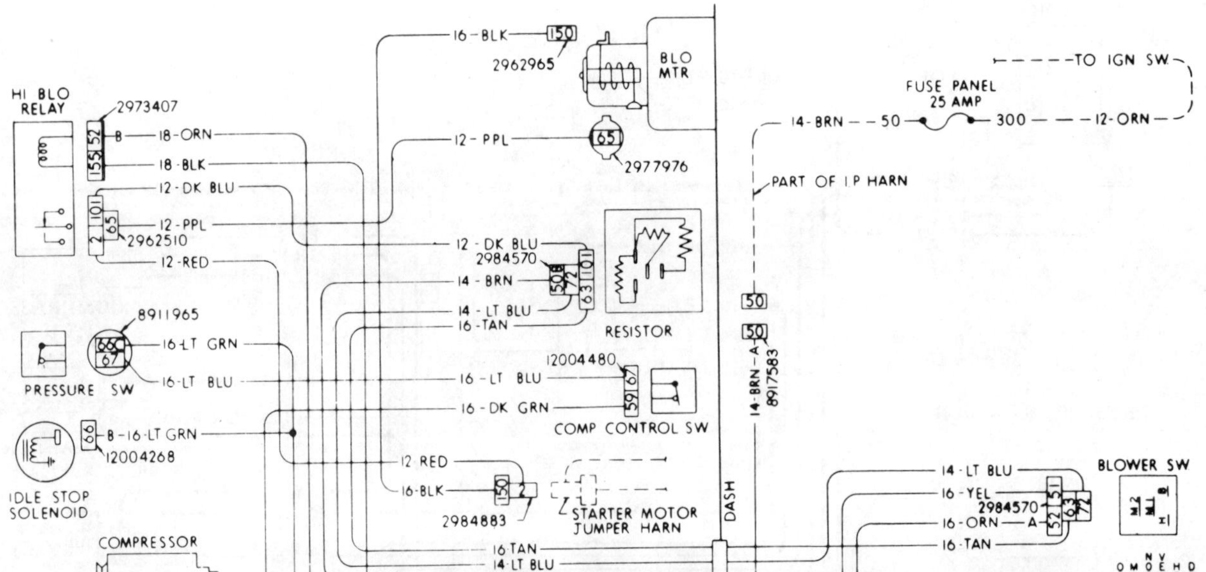

1978-80 Corvette Four Season A/C & Heater Wiring Circuit

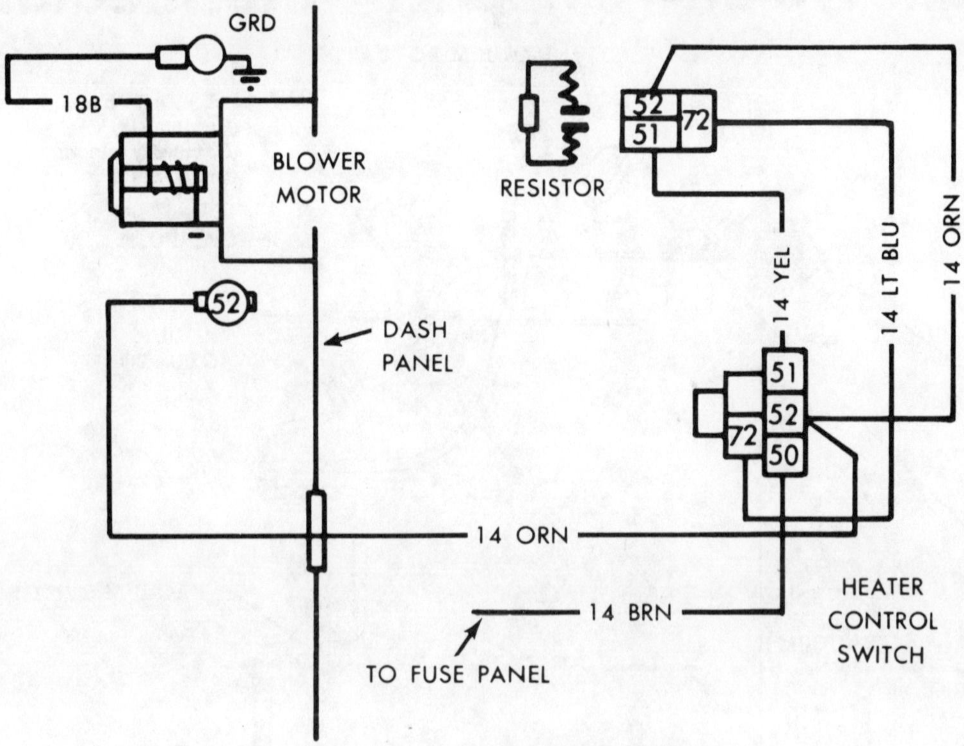

1974-76 Corvette Heater Wiring Circuit

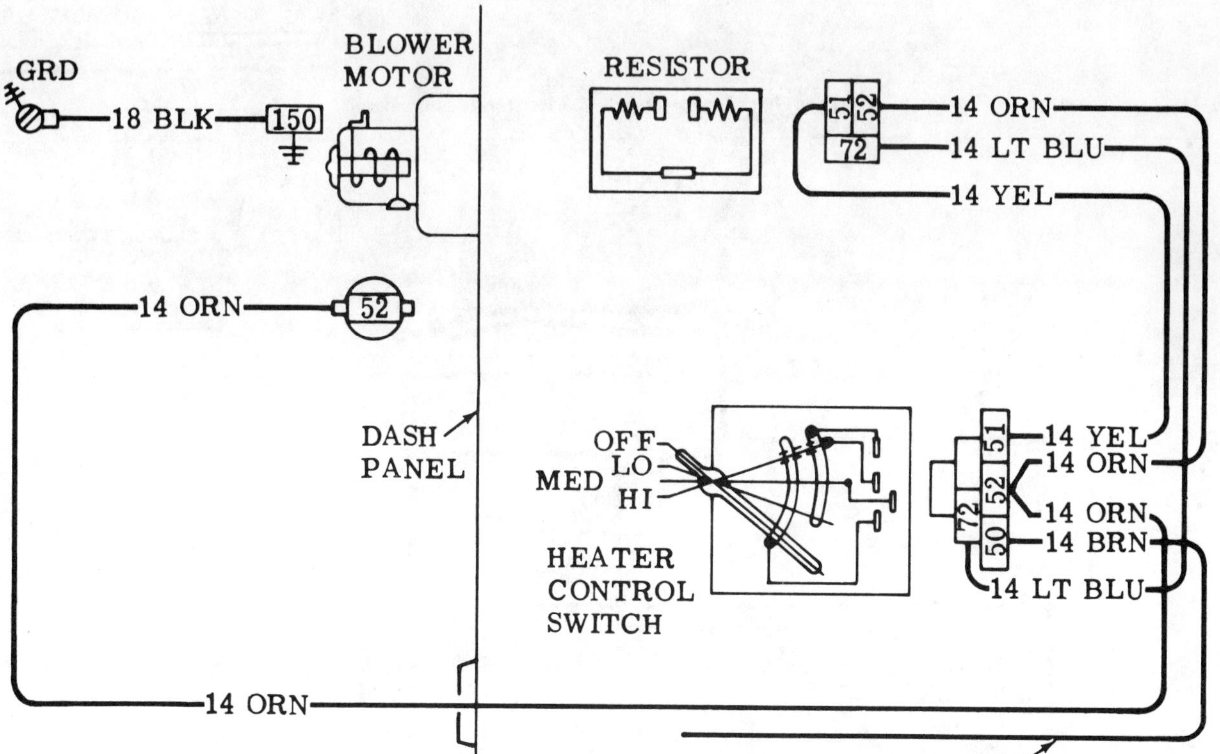

1978-79 Corvette Heater Wiring Circuit

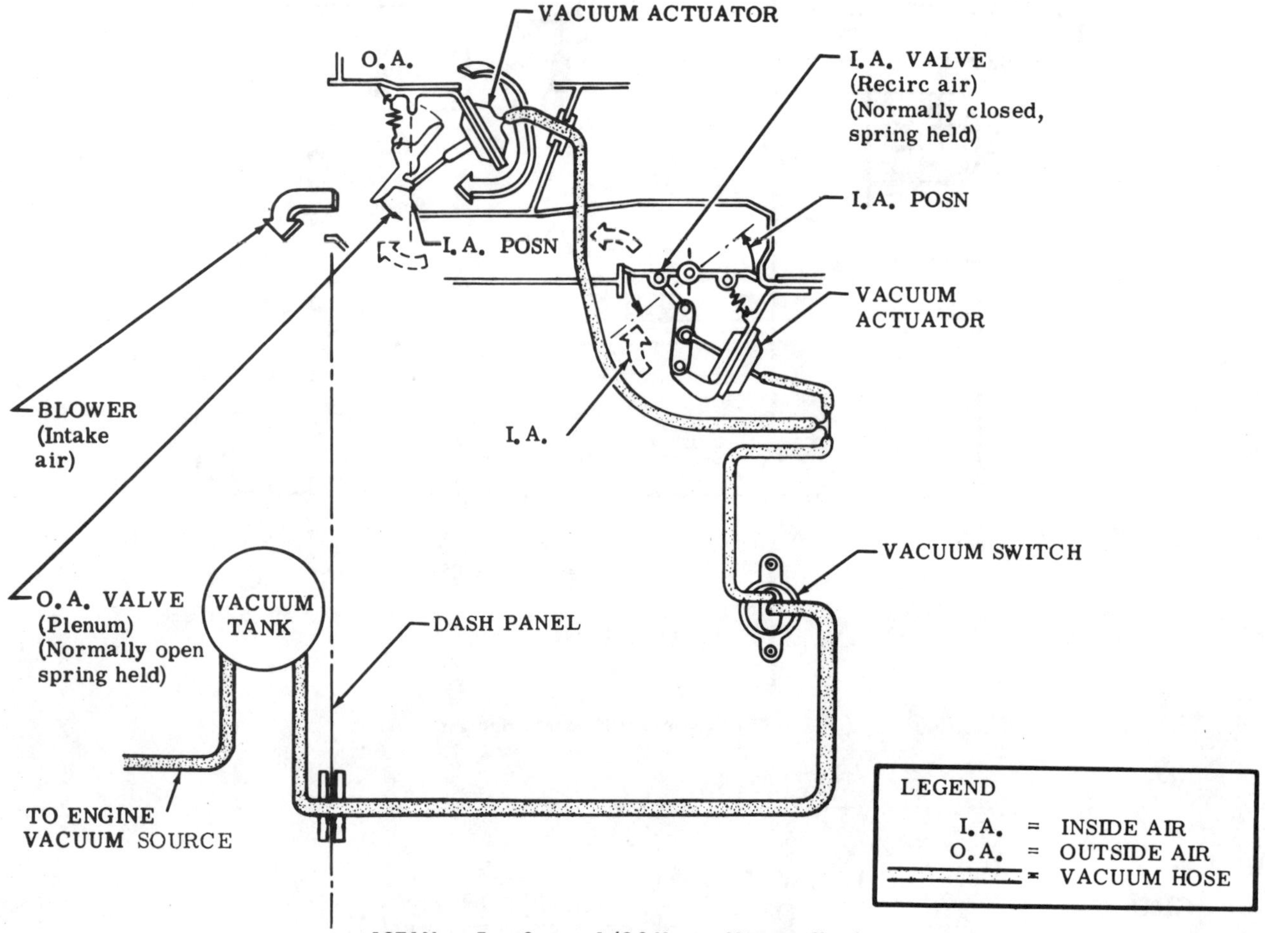

1974 Nova Four Season A/C & Heater Vacuum Circuit

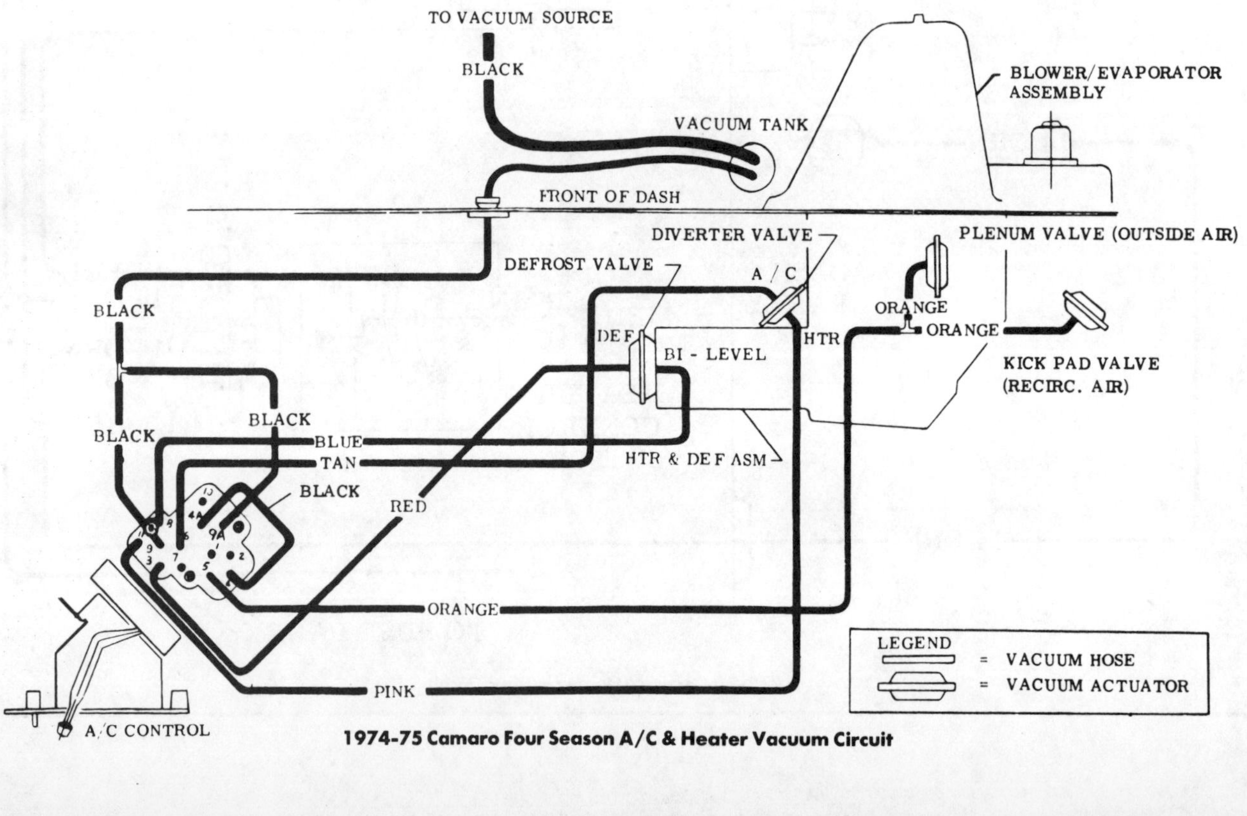

1974-75 Camaro Four Season A/C & Heater Vacuum Circuit

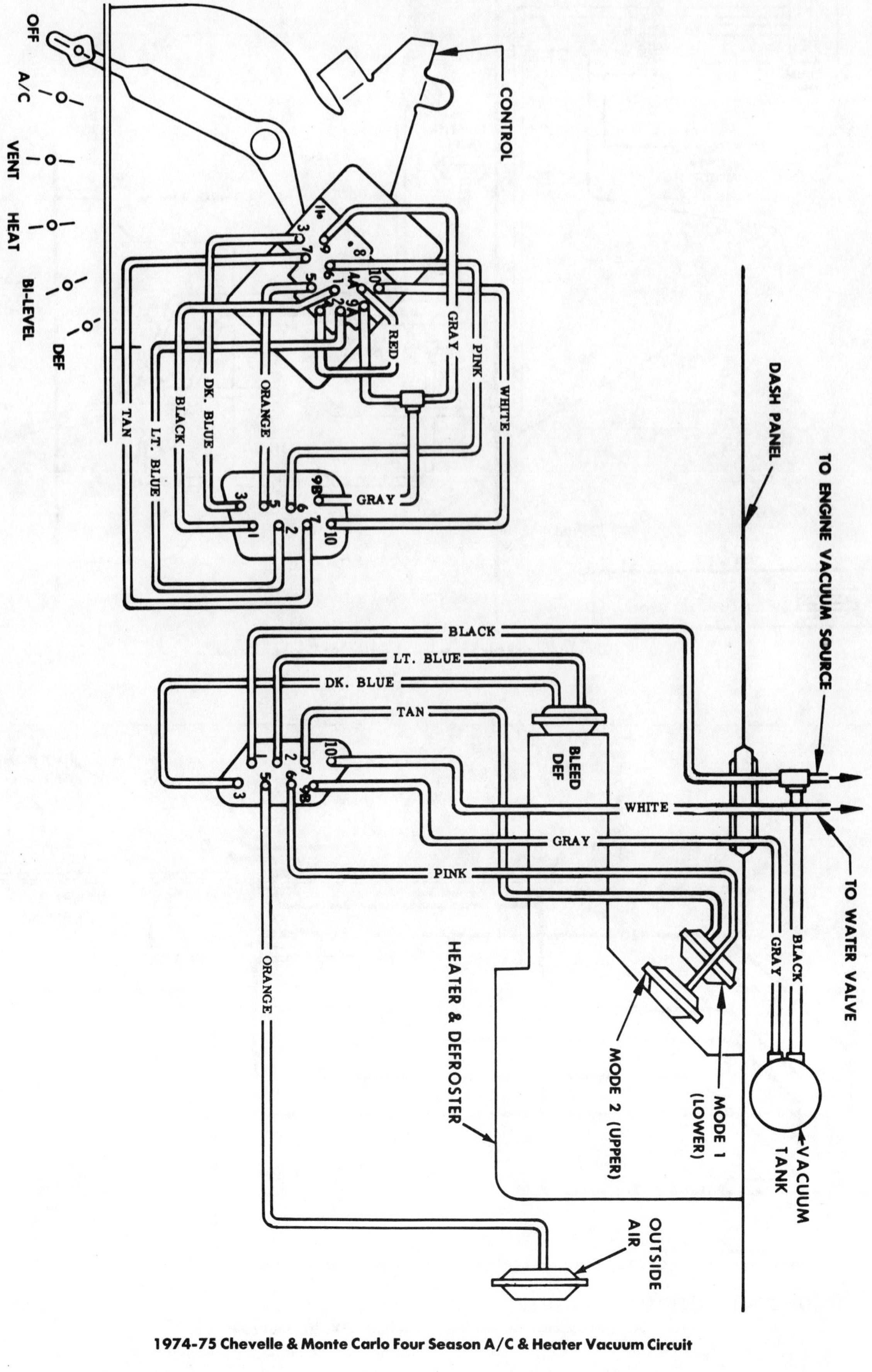

1974-75 Chevelle & Monte Carlo Four Season A/C & Heater Vacuum Circuit

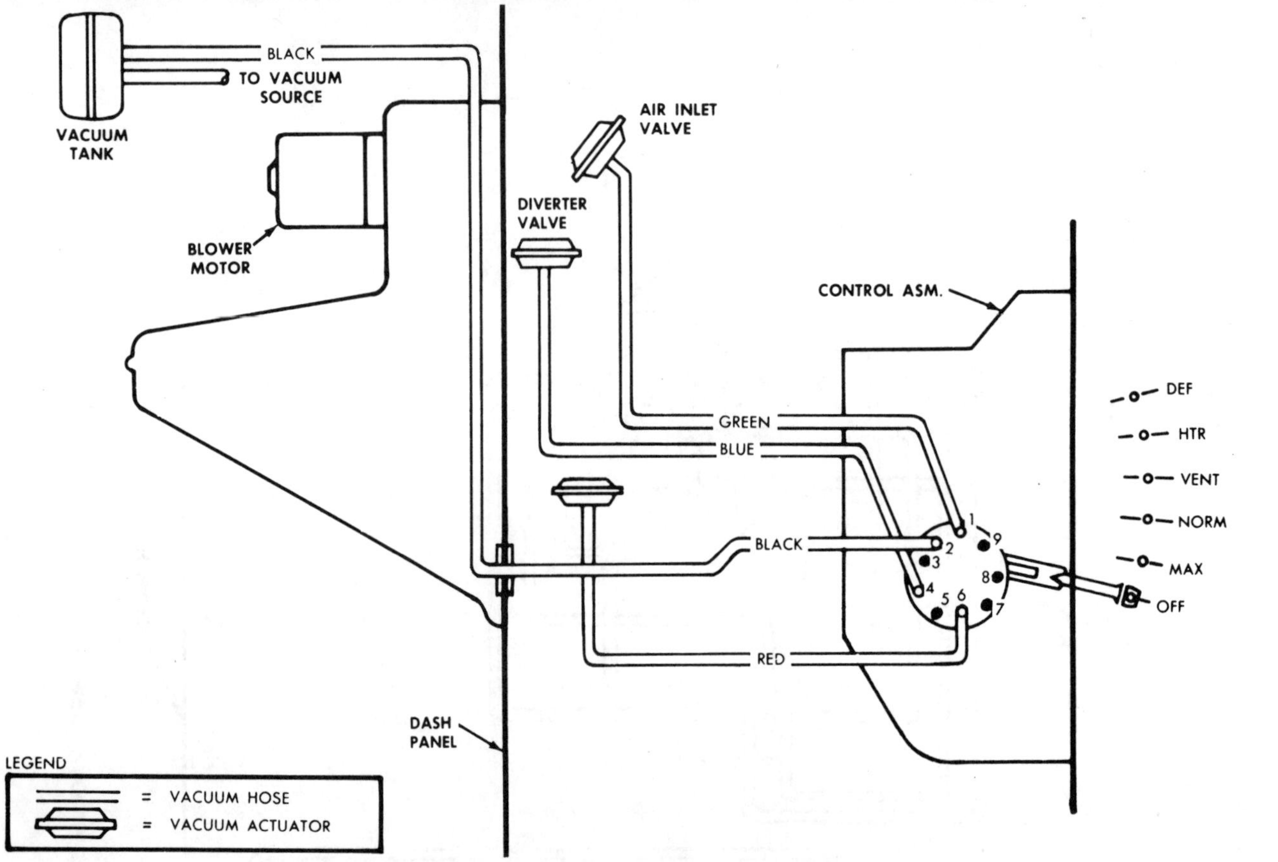

1974-76 Vega Four Season A/C & Heater Vacuum Circuit

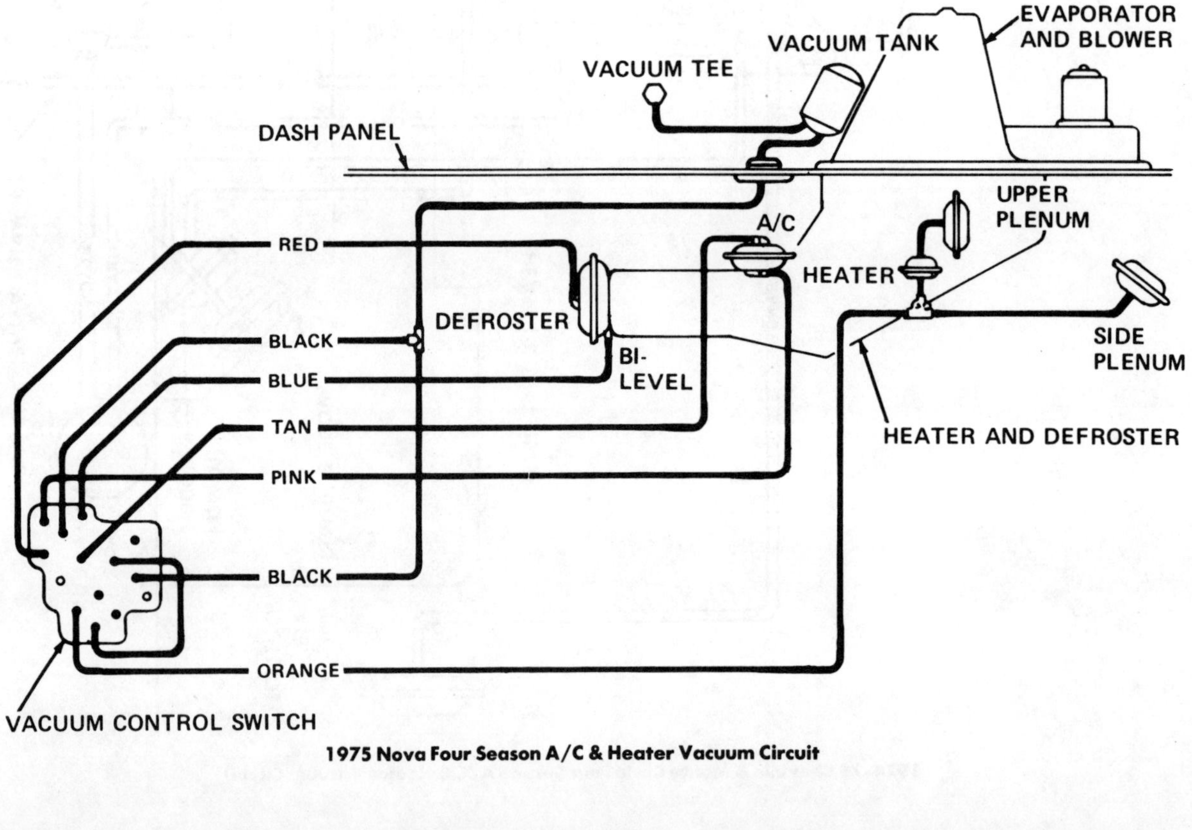

1975 Nova Four Season A/C & Heater Vacuum Circuit

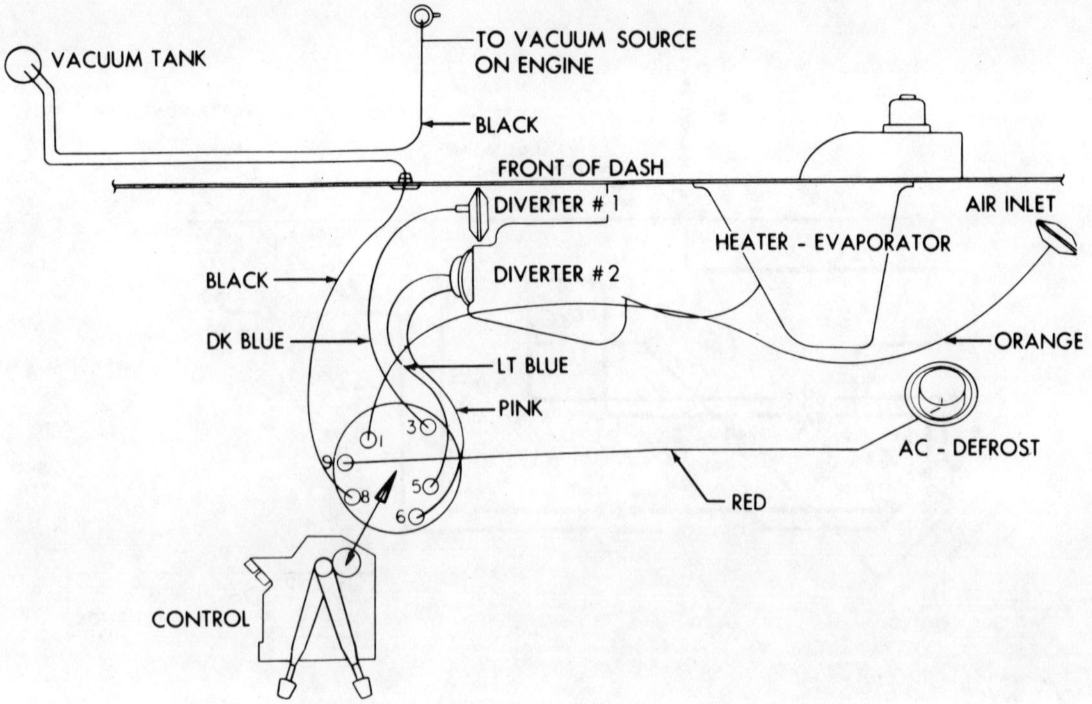

1975-77 Monza Four Season A/C & Heater Vacuum Circuit

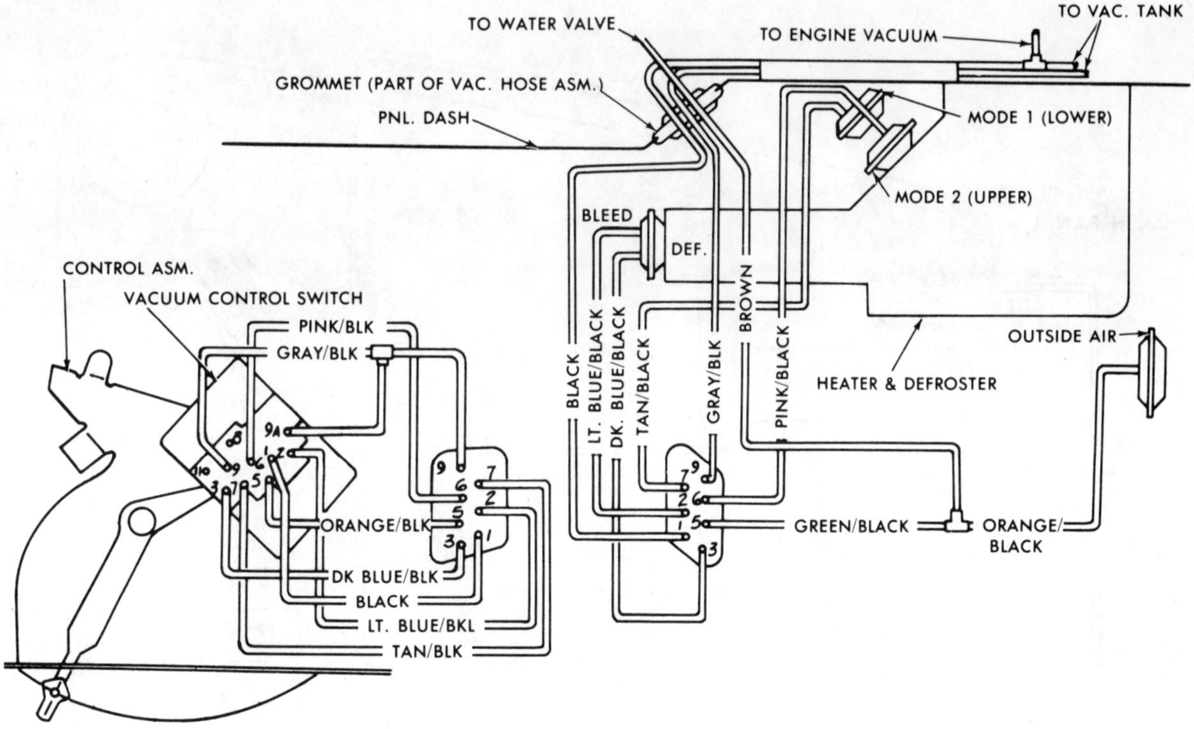

1976 Chevelle & Monte Carlo Four Season A/C & Heater Vacuum Circuit

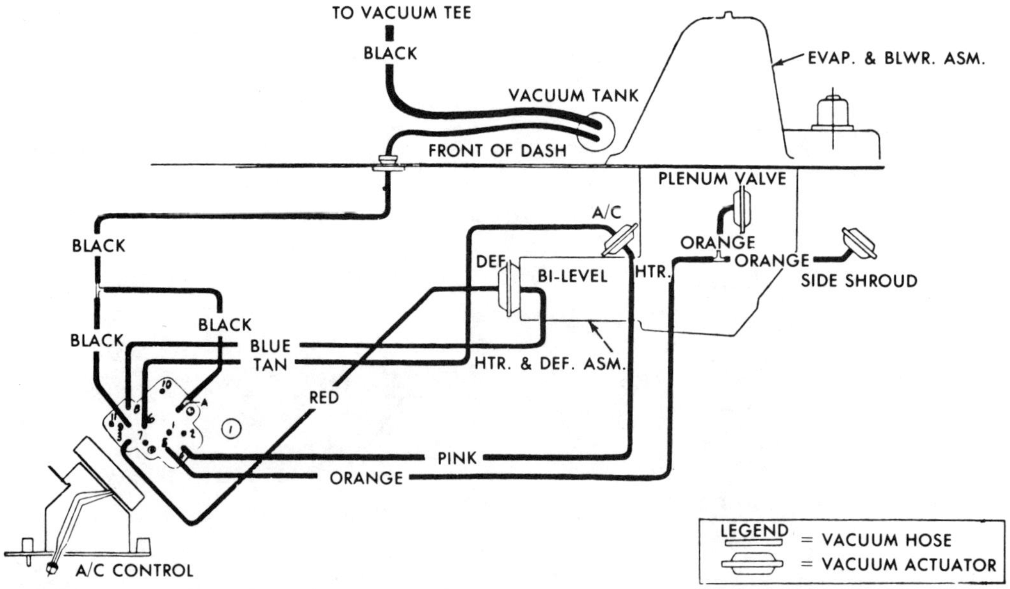

1976-79 Nova, 1976-80 Camaro & 1977 Chevelle Four Season A/C & Heater Vacuum Circuit

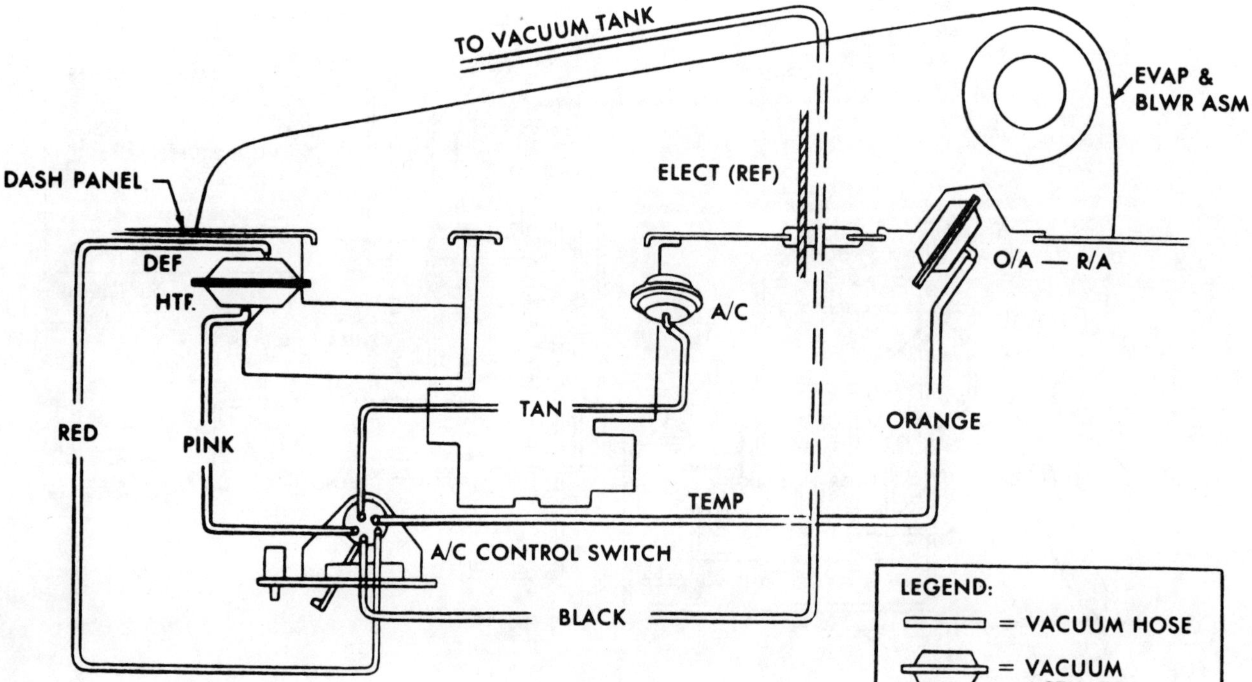

1976-80 Chevette A/C & Heater Vacuum Circuit

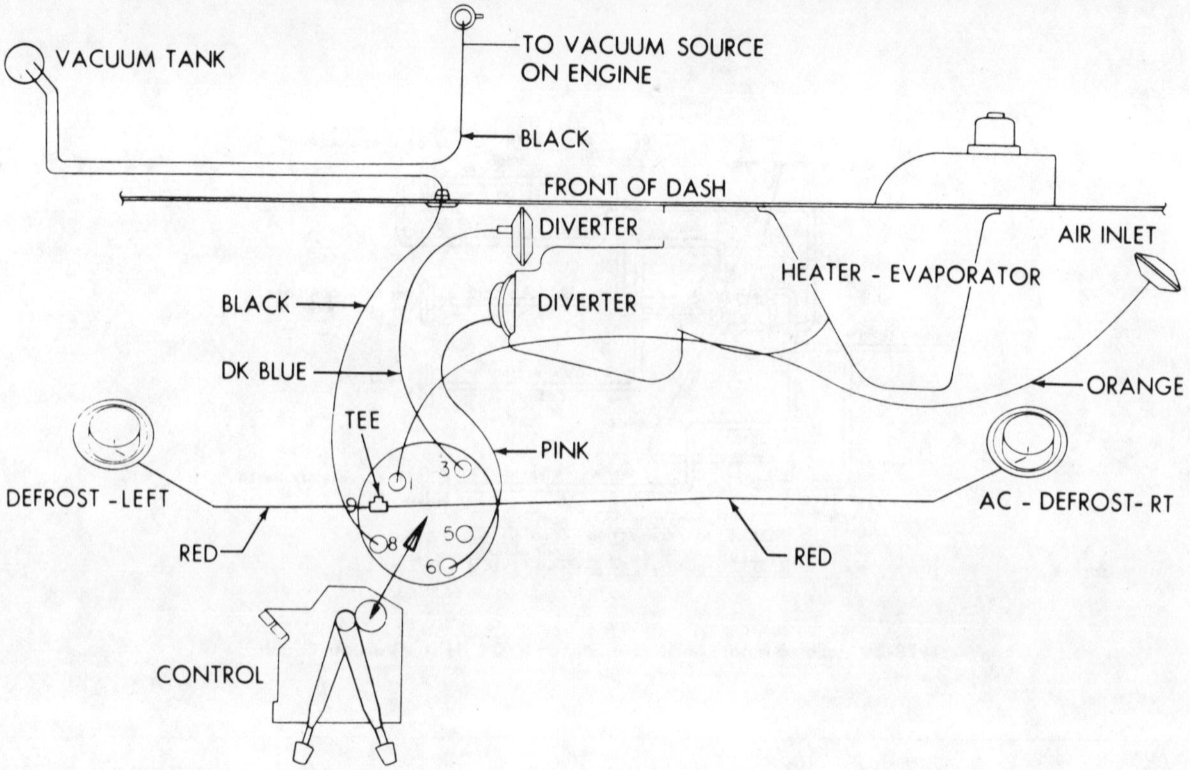

1978-79 Monza Except Station Wagon A/C & Heater Vacuum Circuit

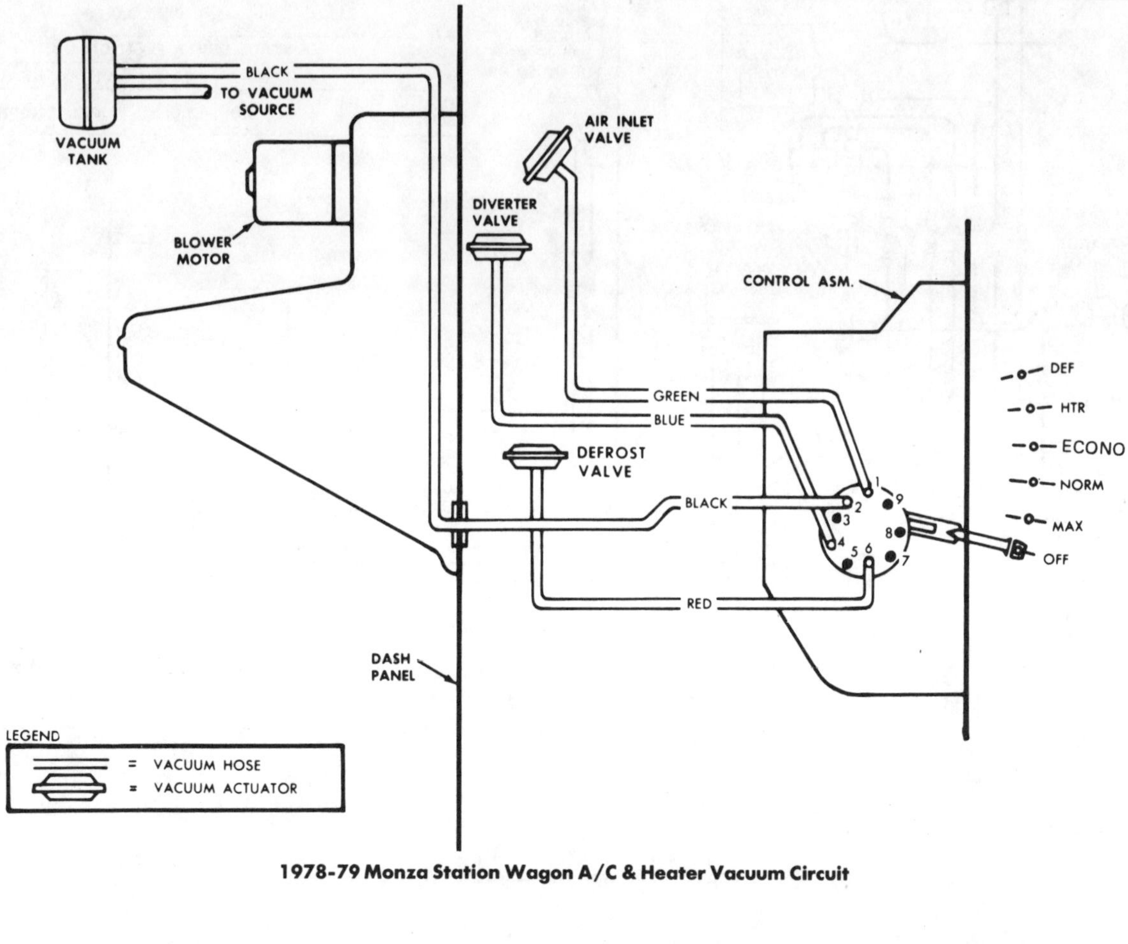

1978-79 Monza Station Wagon A/C & Heater Vacuum Circuit

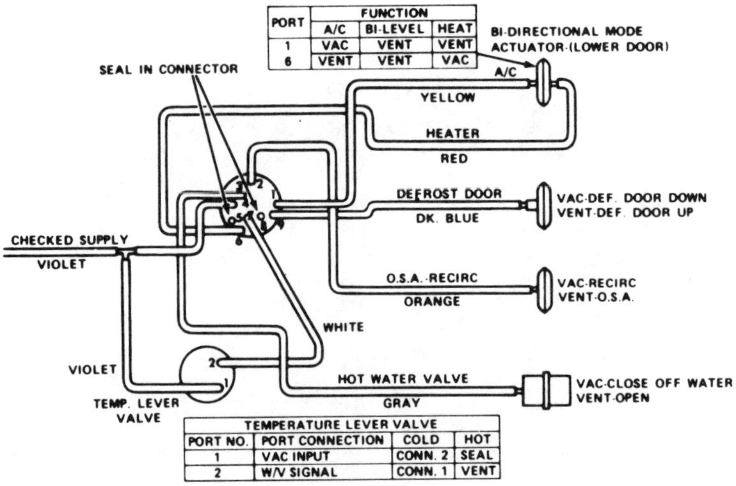

PORT	FUNCTION		
	A/C	BI-LEVEL	HEAT
1	VAC	VENT	VENT
6	VENT	VENT	VAC

TEMPERATURE LEVER VALVE			
PORT NO.	PORT CONNECTION	COLD	HOT
1	VAC INPUT	CONN. 2	SEAL
2	W/V SIGNAL	CONN. 1	VENT

1978-80 Malibu & Monte Carlo Four Season A/C & Heater Vacuum Circuit

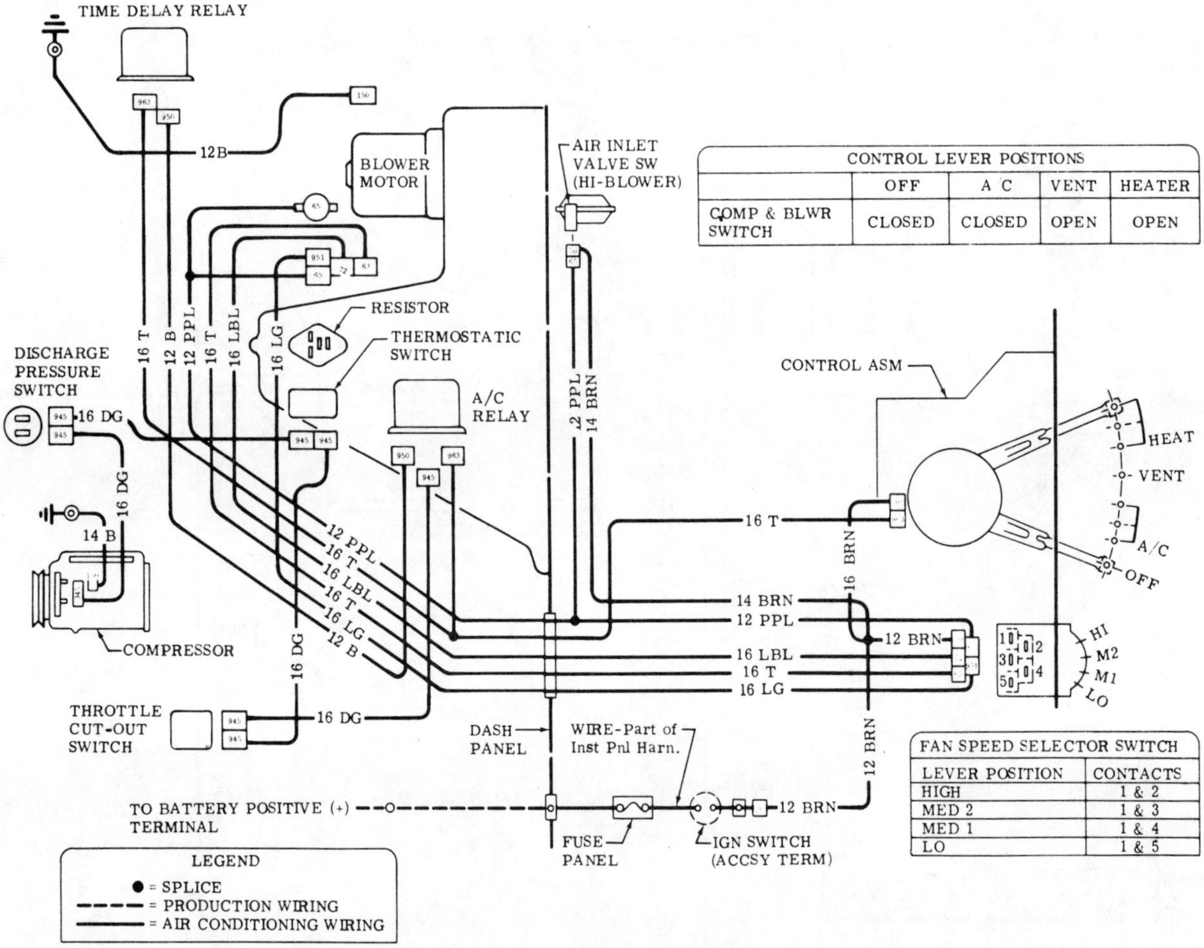

CONTROL LEVER POSITIONS				
	OFF	A C	VENT	HEATER
COMP & BLWR SWITCH	CLOSED	CLOSED	OPEN	OPEN

FAN SPEED SELECTOR SWITCH	
LEVER POSITION	CONTACTS
HIGH	1 & 2
MED 2	1 & 3
MED 1	1 & 4
LO	1 & 5

1974-76 Vega Four Season A/C & Heater Wiring Circuit

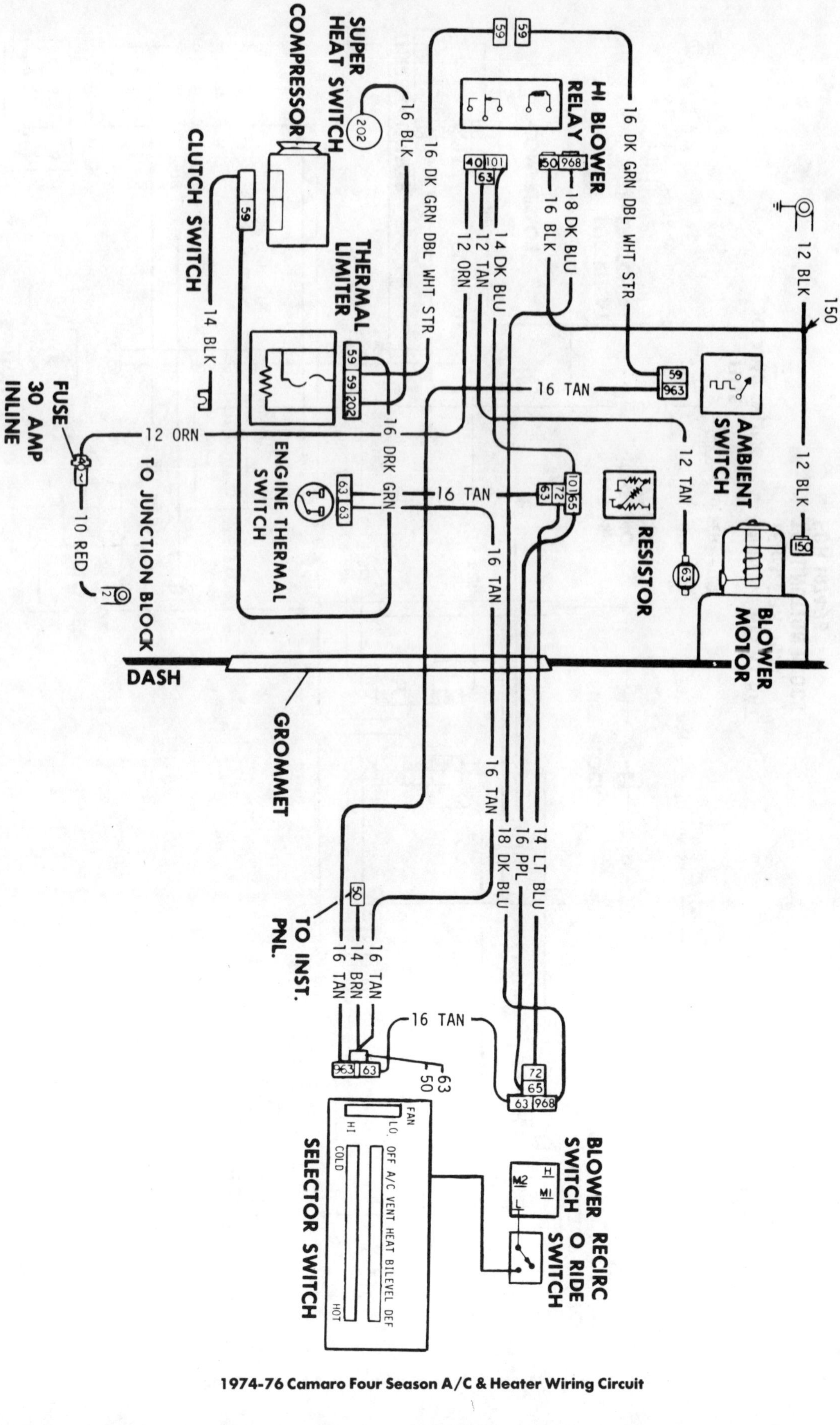

1974-76 Camaro Four Season A/C & Heater Wiring Circuit

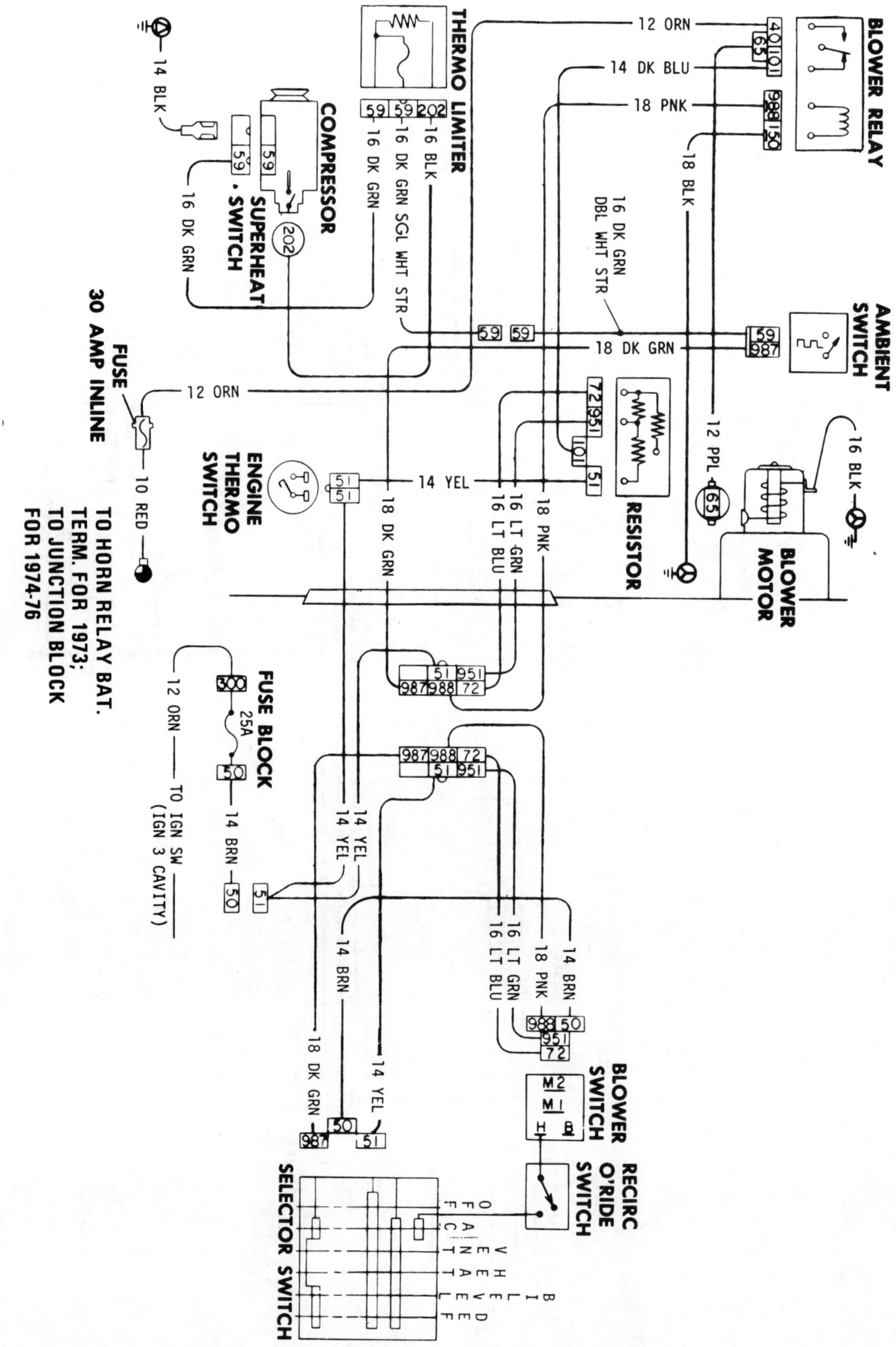

1974-76 Chevelle & Monte Carlo Four Season A/C & Heater Wiring Circuit

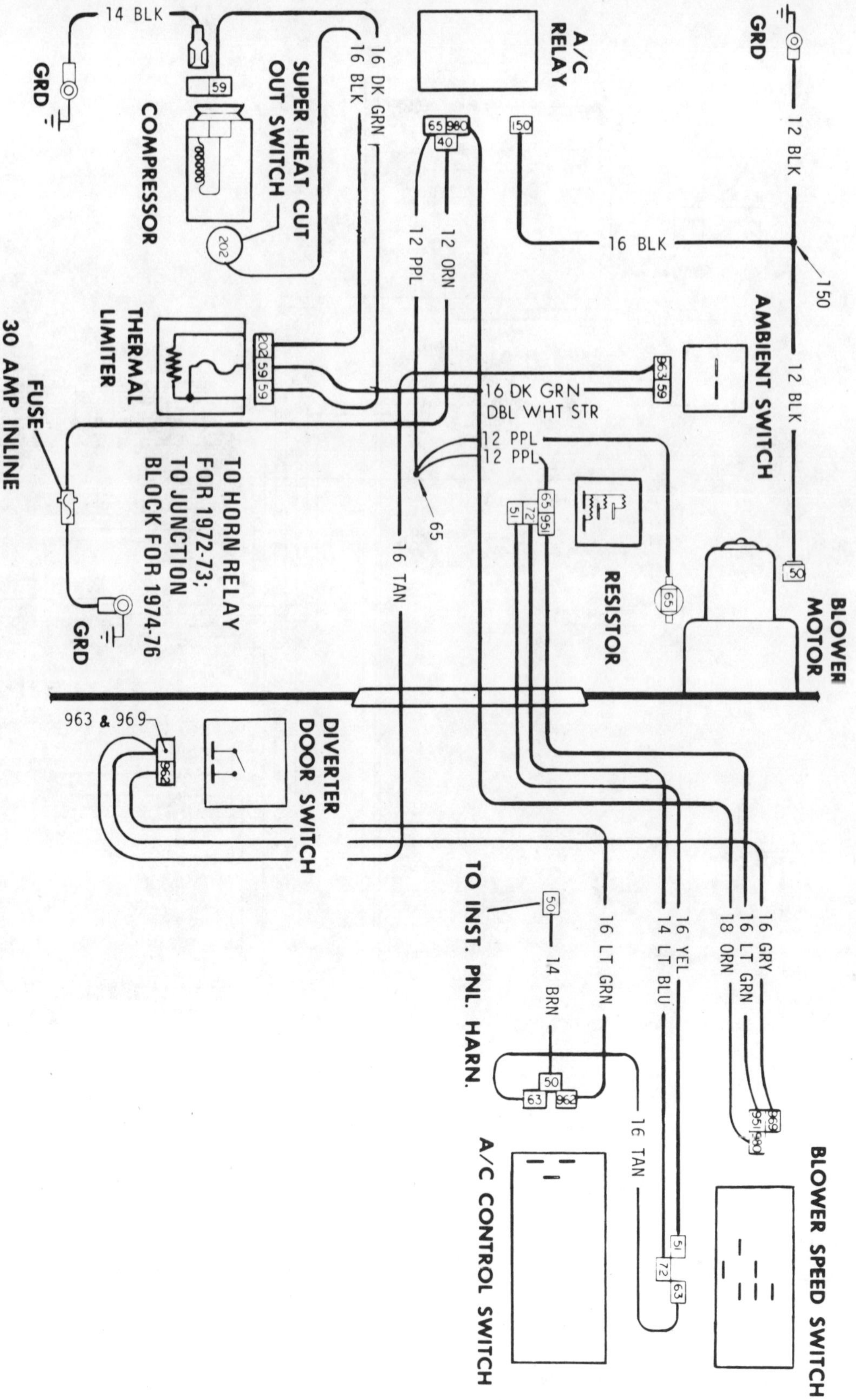

1974-76 Nova Four Season A/C & Heater Wiring Circuit

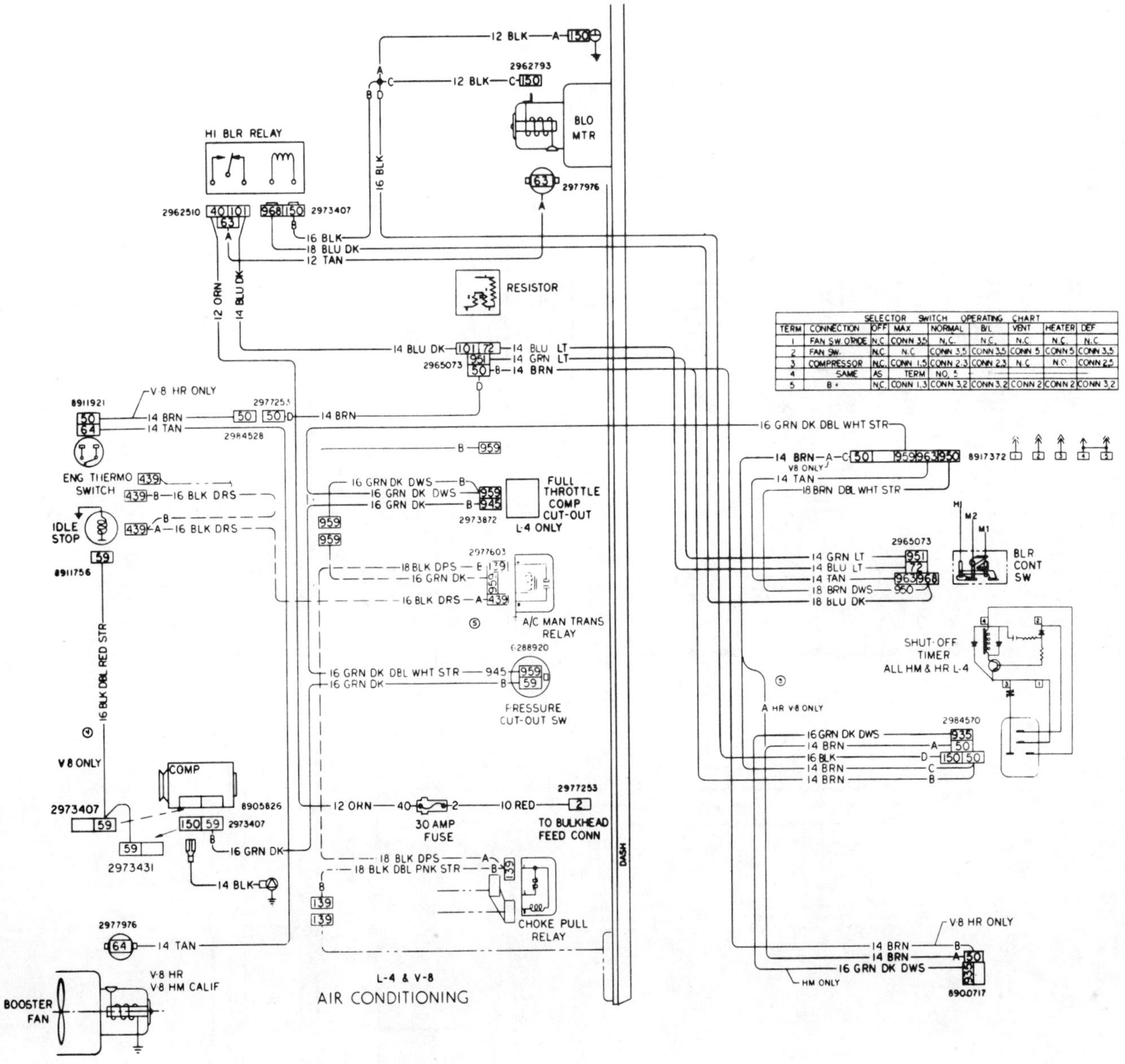

SELECTOR SWITCH OPERATING CHART								
TERM	CONNECTION	OFF	MAX	NORMAL	B/L	VENT	HEATER	DEF
1	FAN SW. O'RIDE	N.C.	CONN 3,5	N.C.	N.C.	N.C.	N.C.	N.C.
2	FAN SW.	N.C.	N.C.	CONN 3,5	CONN 3,5	CONN 5	CONN 5	CONN 3,5
3	COMPRESSOR	N.C.	CONN 1,5	CONN 2,3	CONN 2,3	N.C.	N.C.	CONN 2,5
4	SAME	AS	TERM	NO. 5				
5	B+	N.C.	CONN 1,3	CONN 3,2	CONN 3,2	CONN 2	CONN 2	CONN 3,2

1975-76 Monza Four Season A/C & Heater Wiring Circuit

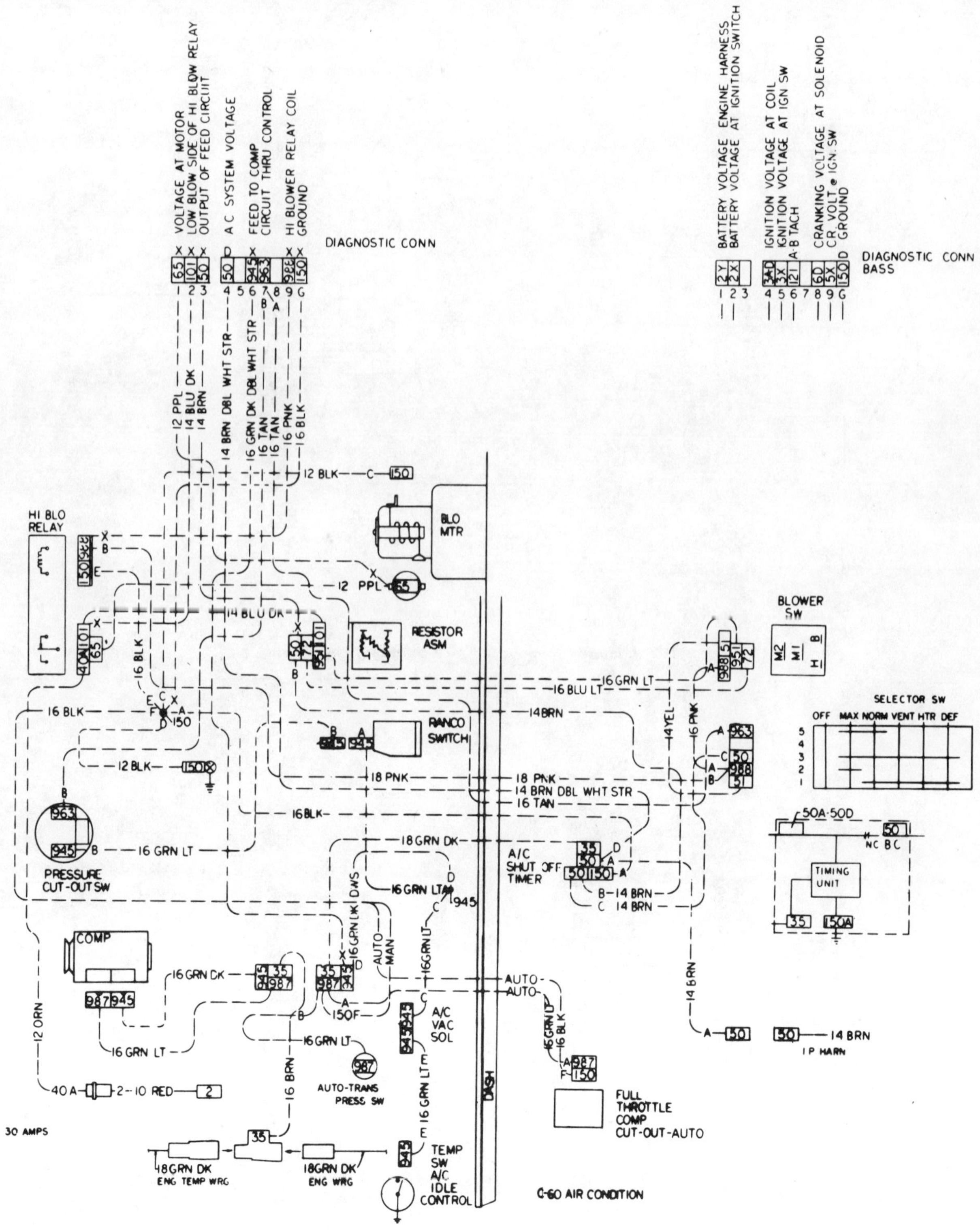

1976 Chevette Four Season A/C & Heater Wiring Circuit

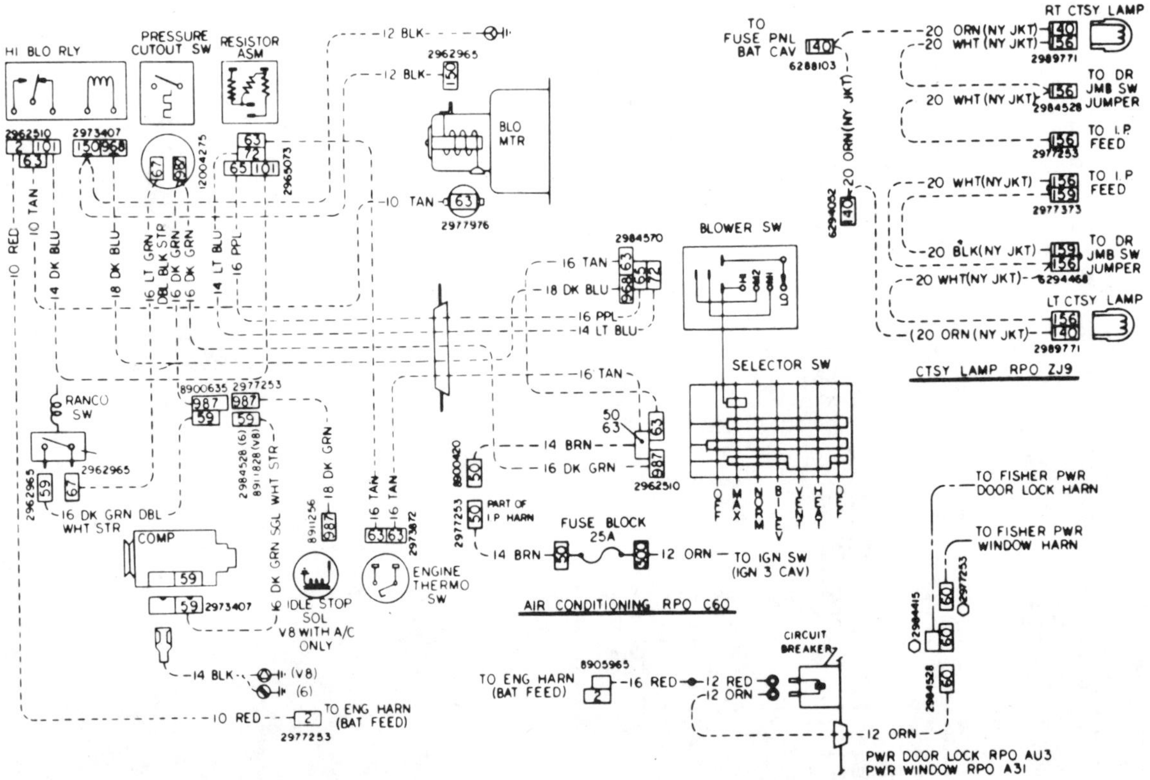

1977 Camaro Four Season A/C & Heater Wiring Circuit

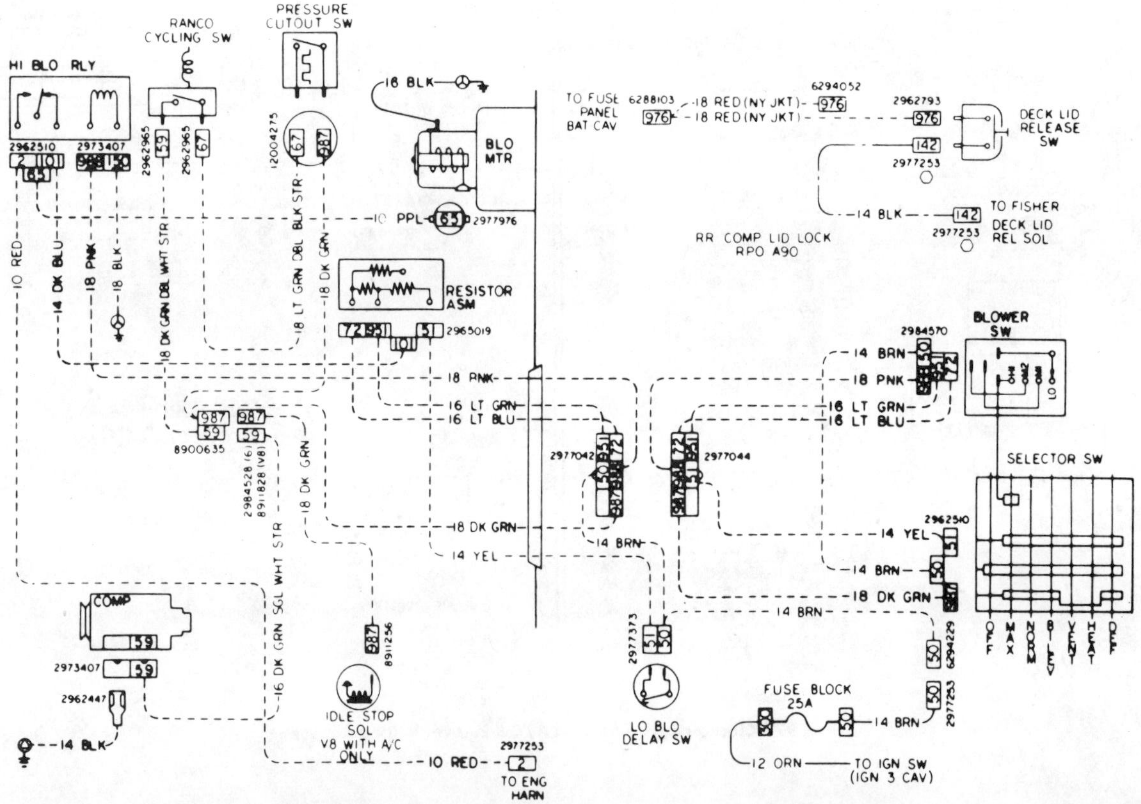

1977 Chevelle & Monte Carlo Four Season A/C & Heater Wiring Circuit

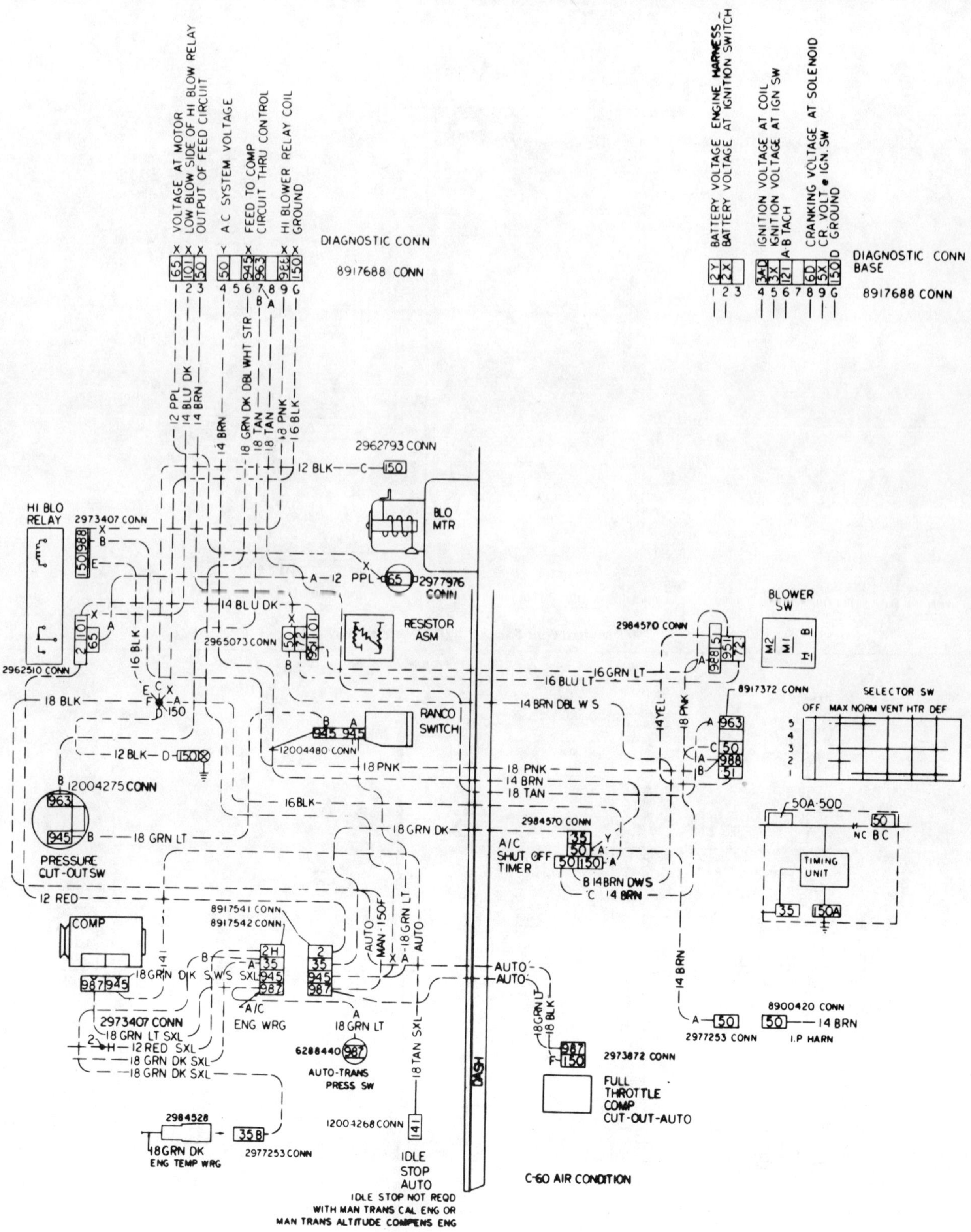

1977 Chevette Four Season A/C & Heater Wiring Circuit

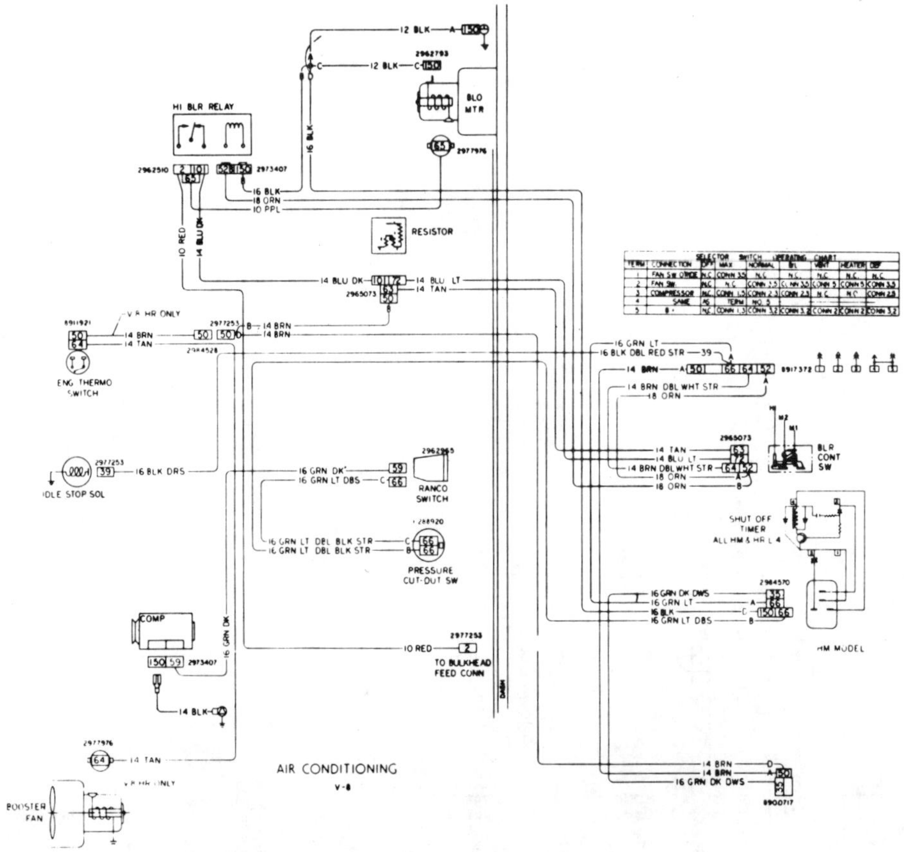

1977 Monza Four Season A/C & Heater Wiring Circuit

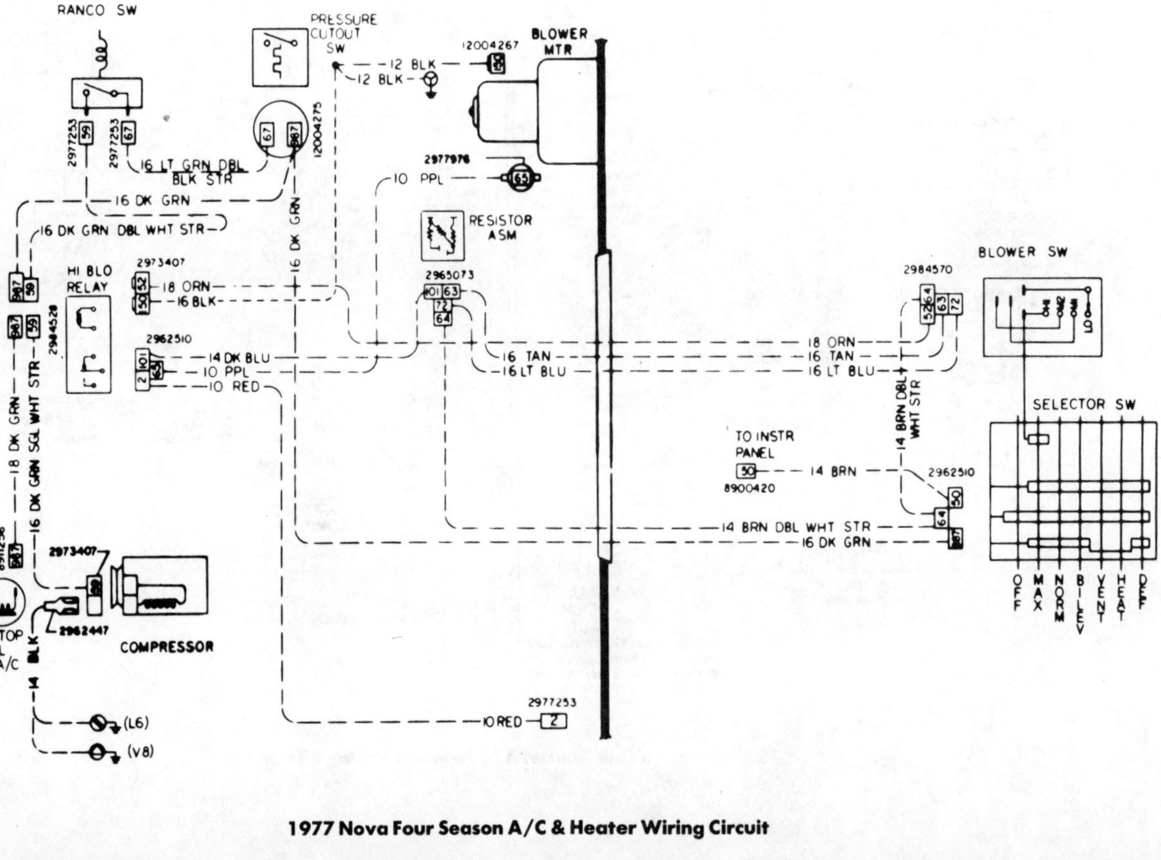

1977 Nova Four Season A/C & Heater Wiring Circuit

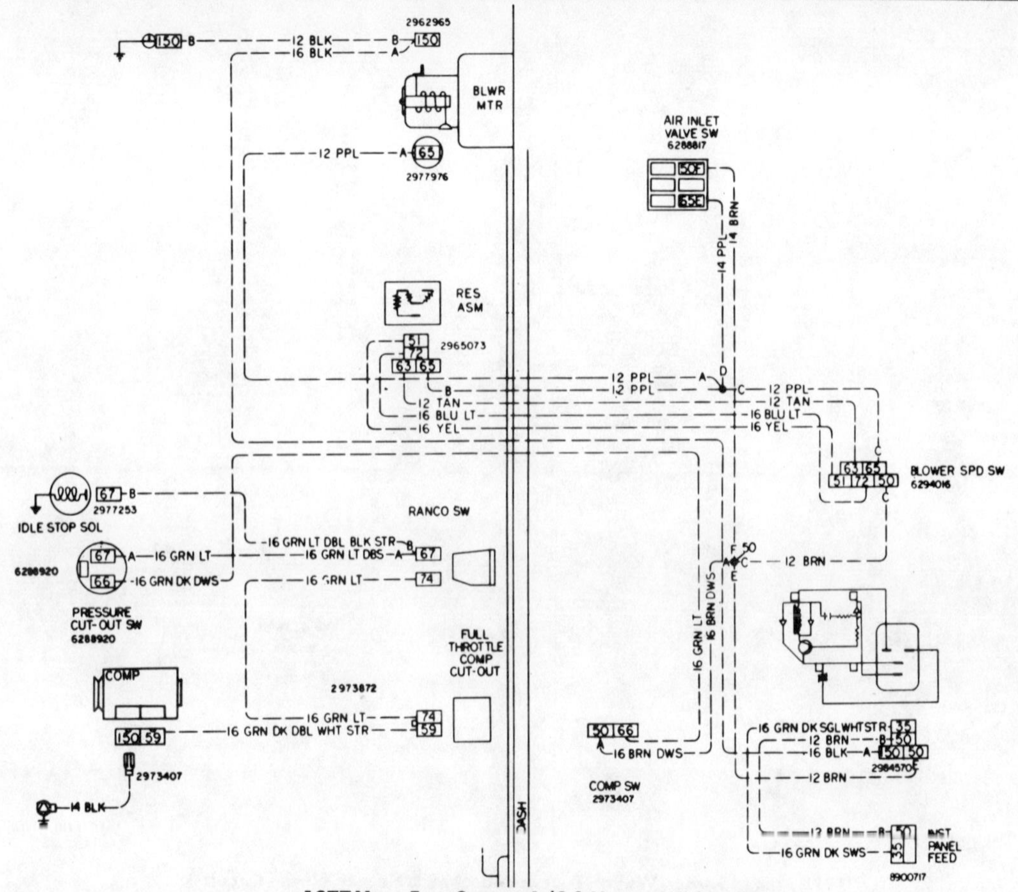

1977 Vega Four Season A/C & Heater Wiring Circuit

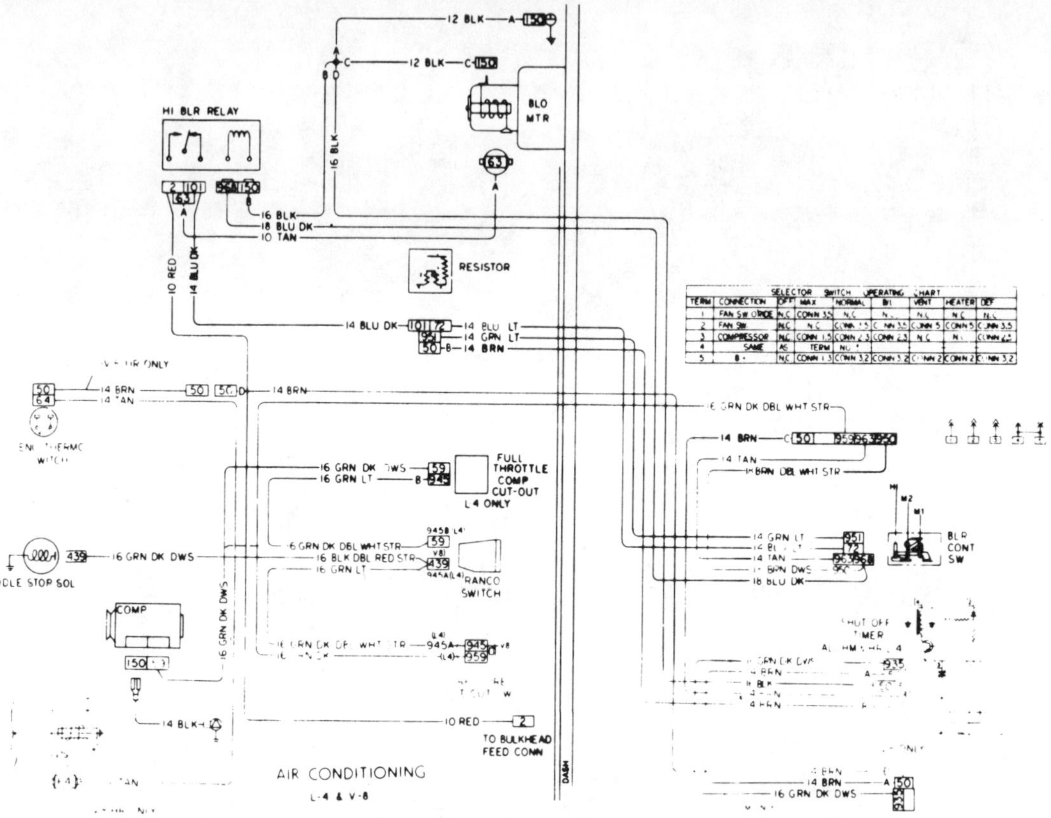

SELECTOR SWITCH OPERATING CHART								
TERM	CONNECTION	OFF	MAX	NORMAL	B/L	VENT	HEATER	DEF
1	FAN SW ORIDE	N.C	CONN 3,5	N.C	N.[illegible]	N.C	N.C	N.C
2	FAN SW.	N.C	N.C	CONN 1,5	CONN 3,5	CONN 5	CONN 5	CONN 3,5
3	COMPRESSOR	N.C.	CONN 1,5	CONN 2,3	CONN 2,3	N.C	N.[illegible]	CONN [illegible]
4	SAME	AS	TERM	NO. 1				
5	B+	N.C	CONN 1,3	CONN 3,2	CONN 3,2	CONN 2	CONN 2	CONN 3,2

1978 Monza Except Station Wagon Four Season A/C & Heater Wiring Circuit

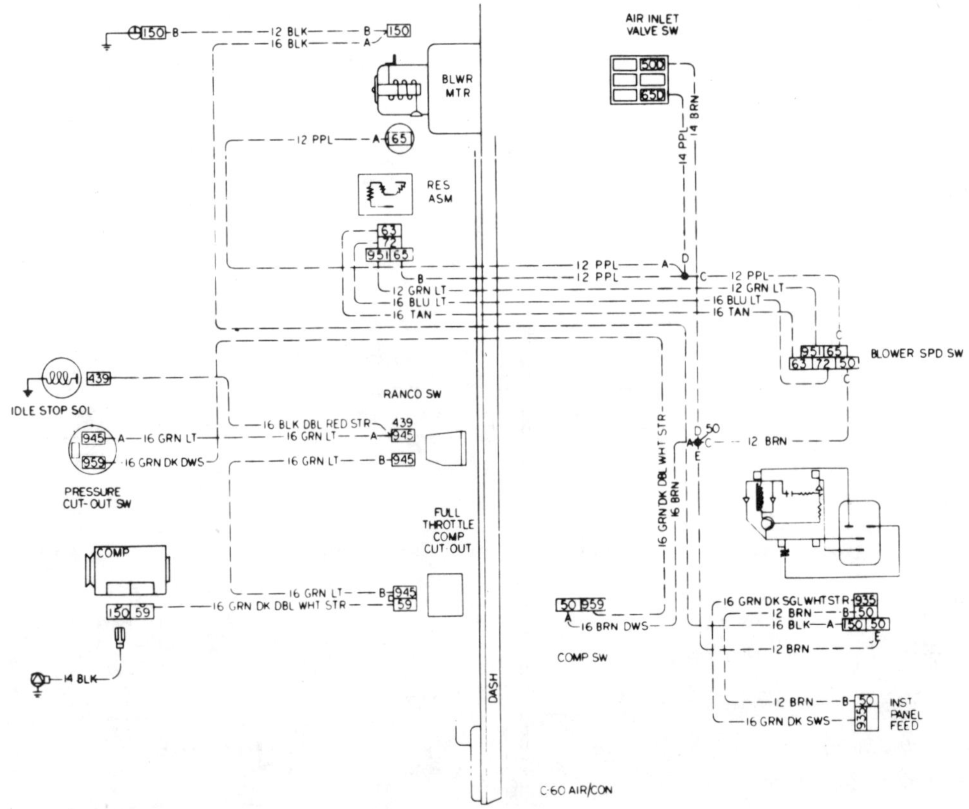

1978 Monza Station Wagon Four Season A/C & Heater Wiring Circuit

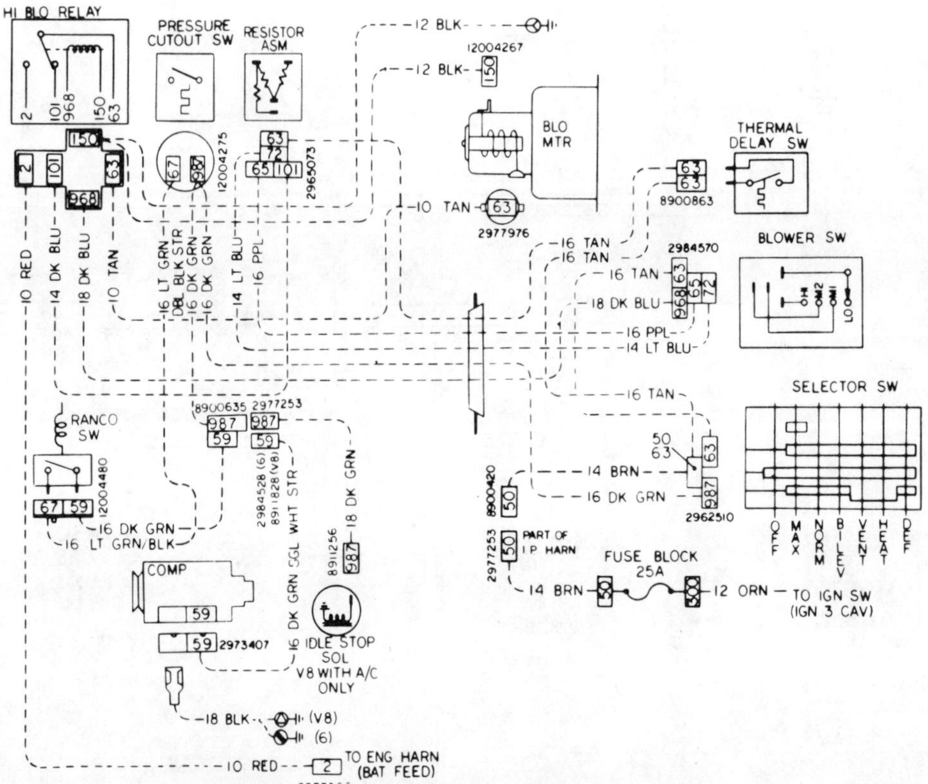

1978-79 Camaro Four Season A/C & Heater Wiring Circuit

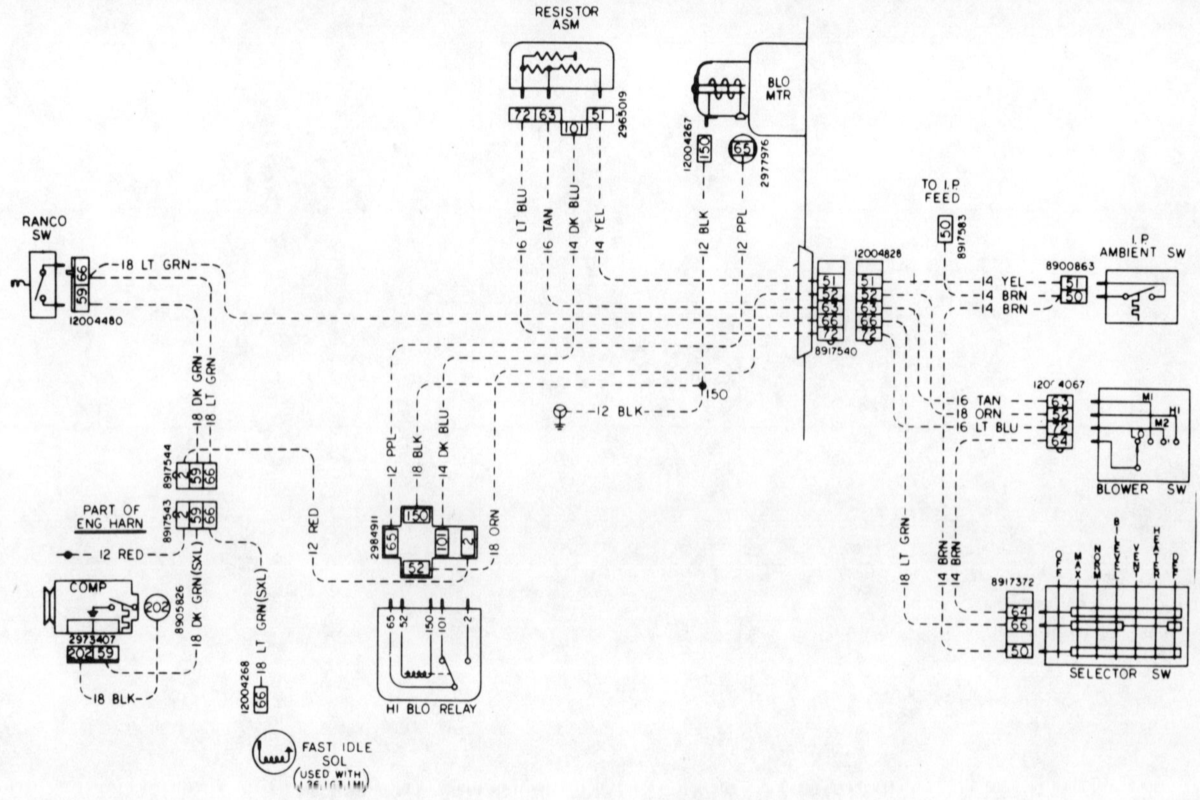

1978-79 Malibu & Monte Carlo Four Season A/C & Heater Wiring Circuit

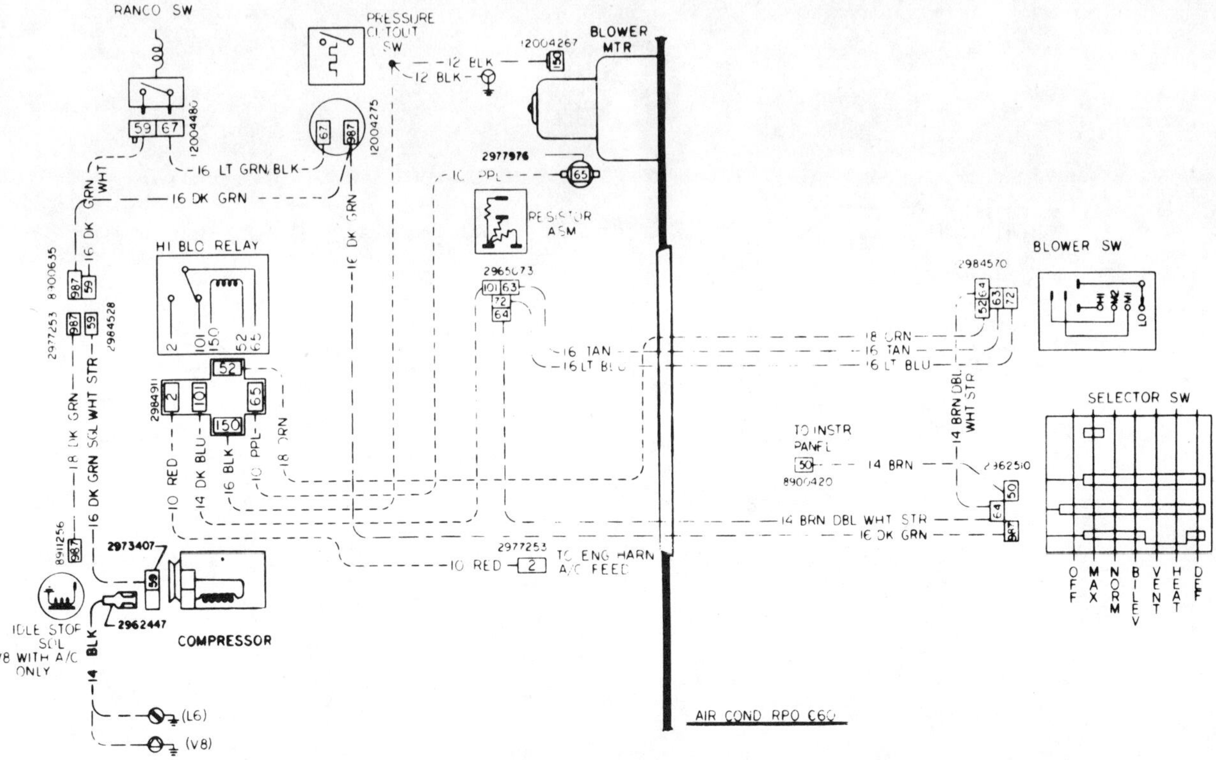

1978-79 Nova Four Season A/C & Heater Wiring Circuit

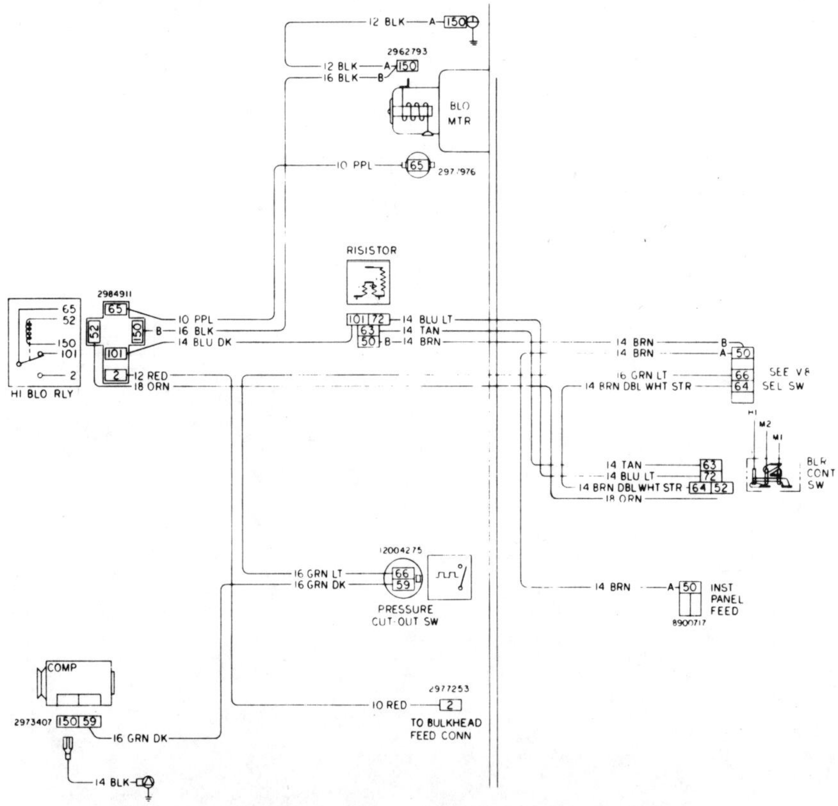

1979 Monza With 4-151 Except Station Wagon Four Season A/C & Heater Wiring Circuit

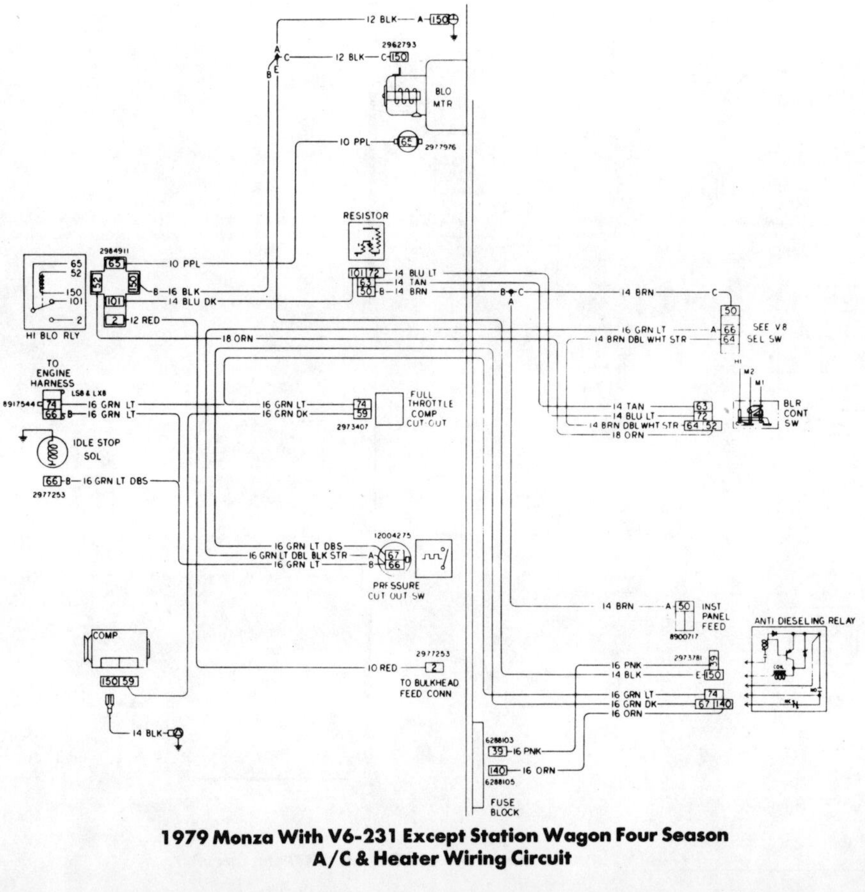

1979 Monza With V6-231 Except Station Wagon Four Season A/C & Heater Wiring Circuit

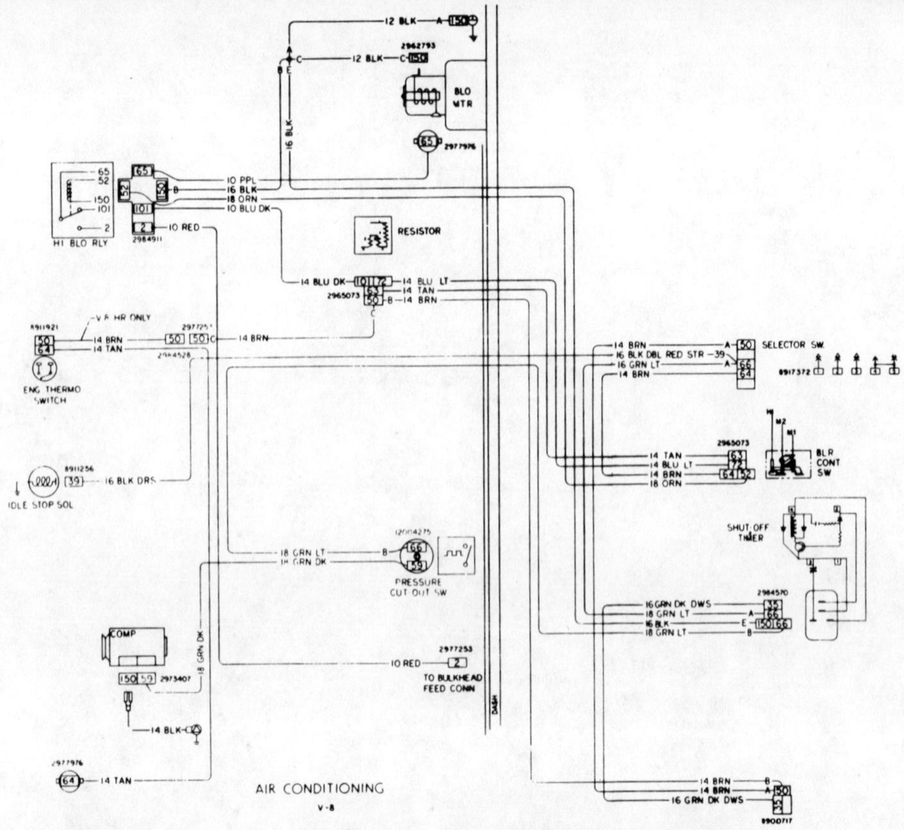

1979 Monza With V8 Except Station Wagon Four Season A/C & Heater Wiring Circuit

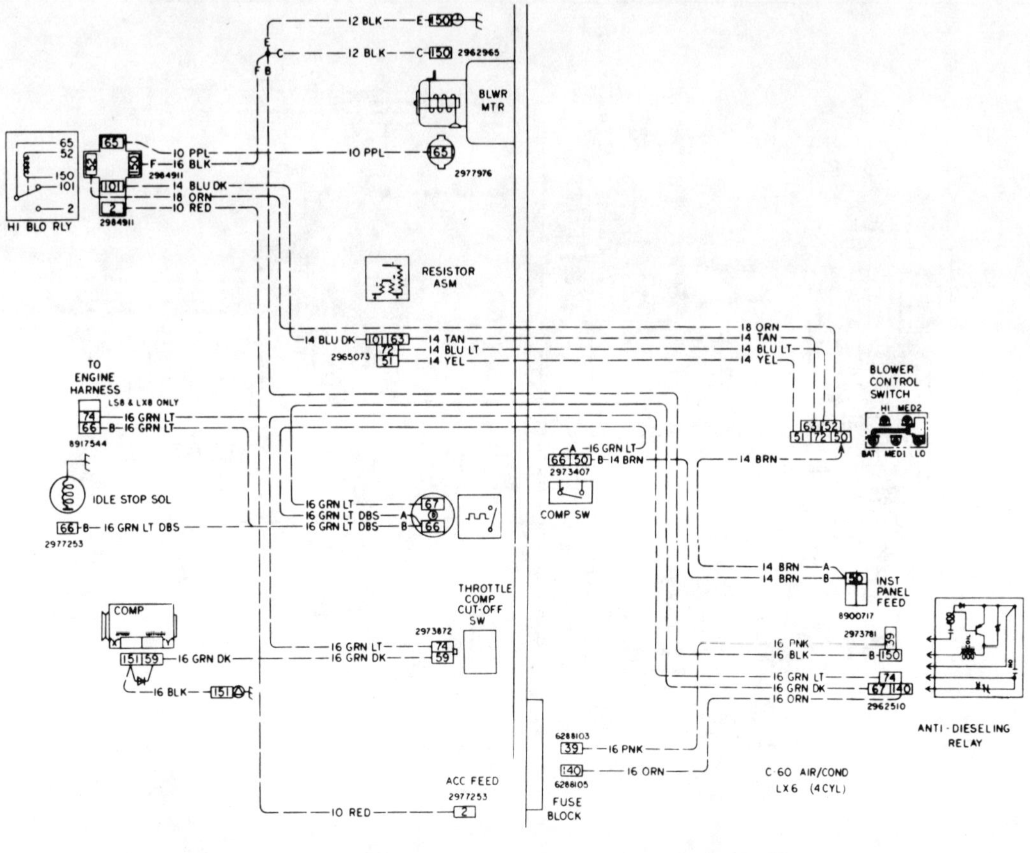

1979 Monza Station Wagon With 4-151 Four Season A/C & Heater Wiring Circuit

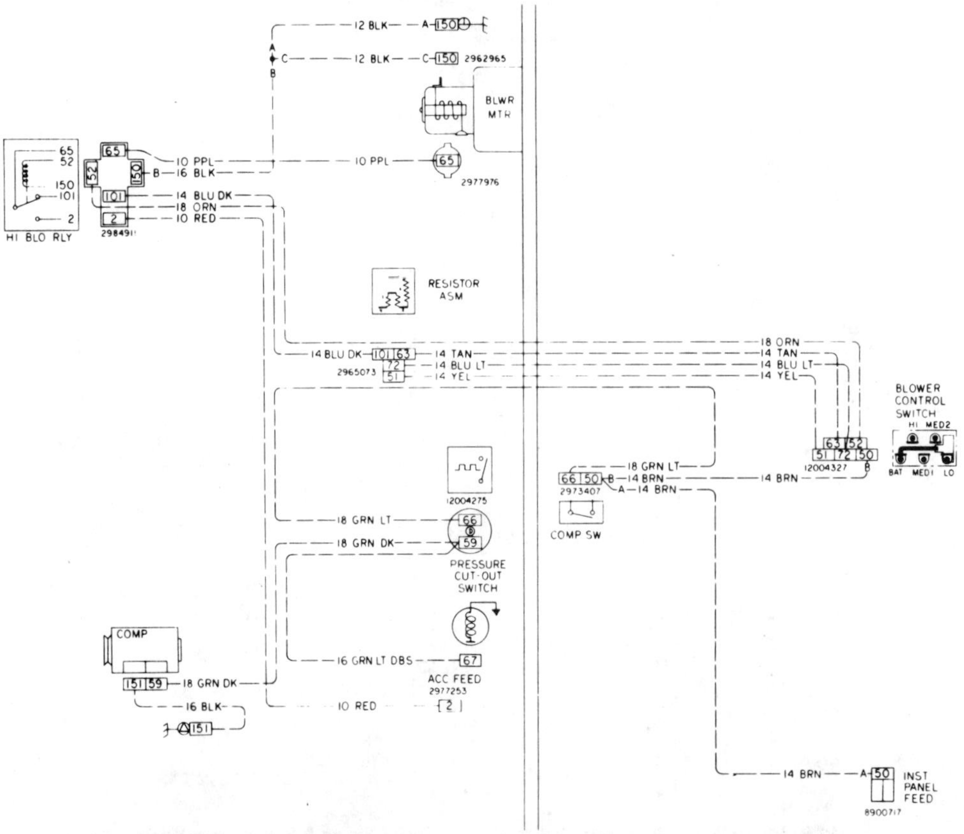

1979 Monza Station Wagon With V6-231 Four Season A/C & Heater Wiring Circuit

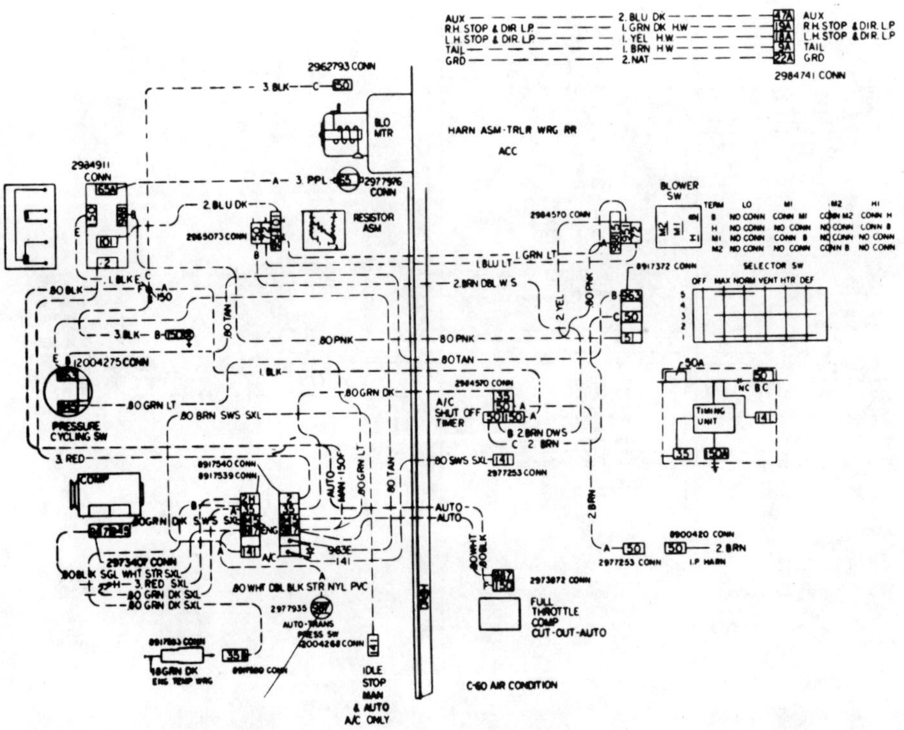

1979-80 Chevette A/C & Heater Wiring Circuit

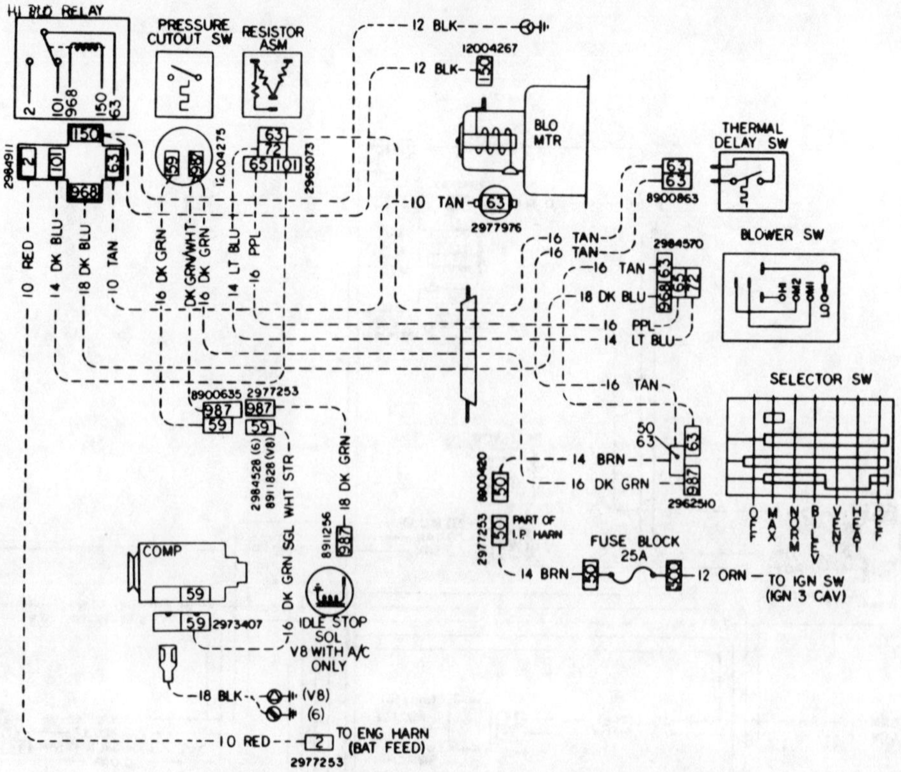

1980 Camaro Four Season A/C & Heater Wiring Circuit

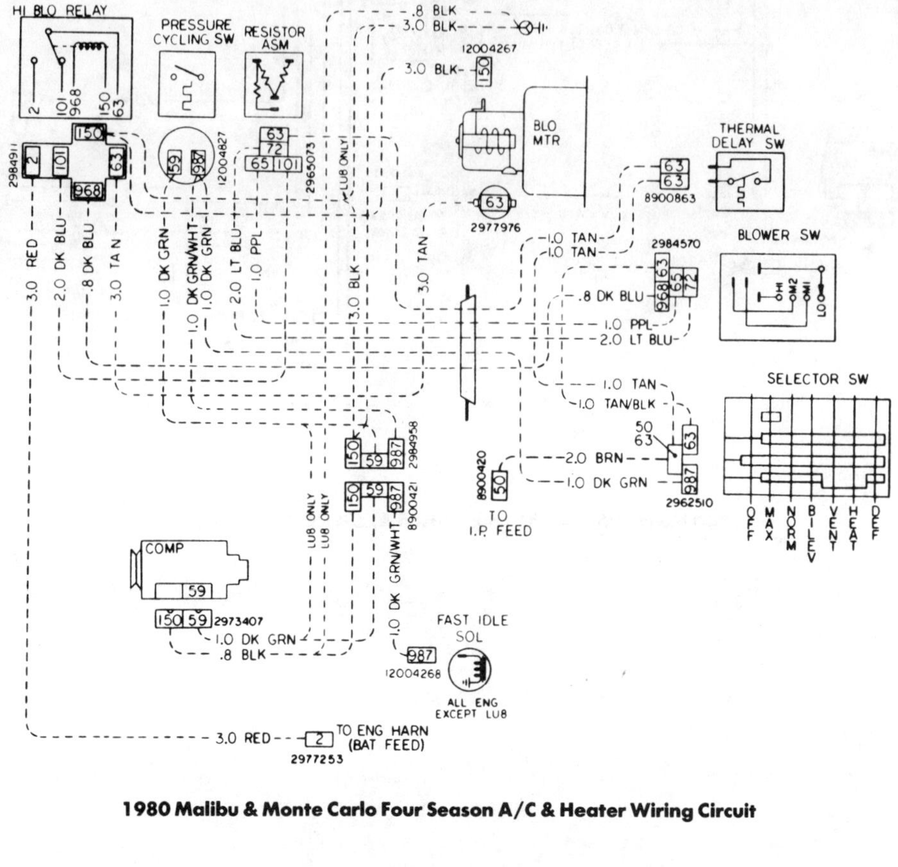

1980 Malibu & Monte Carlo Four Season A/C & Heater Wiring Circuit

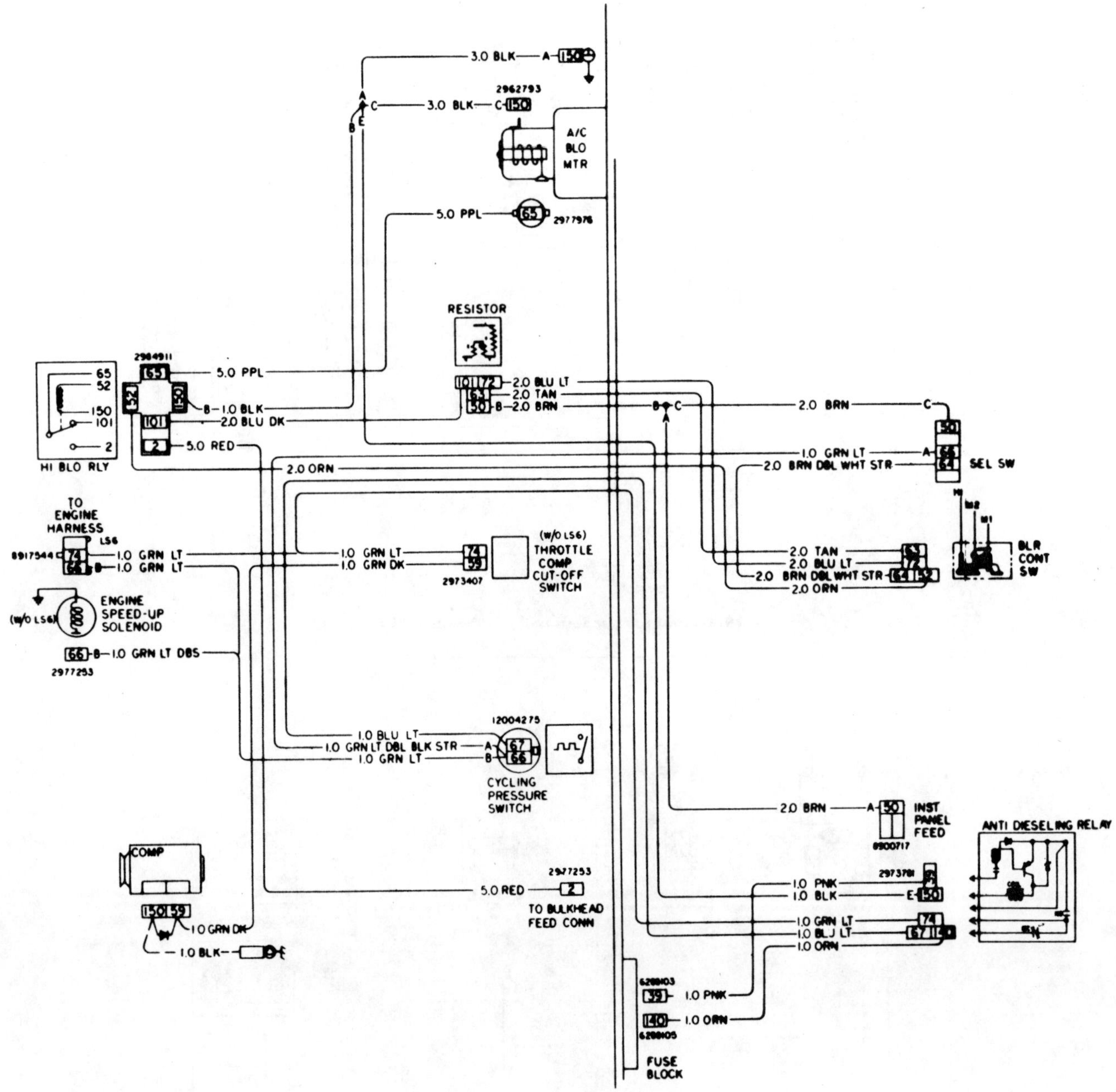

1980 Monza With 4-151 Four Season A/C & Heater Wiring Circuit

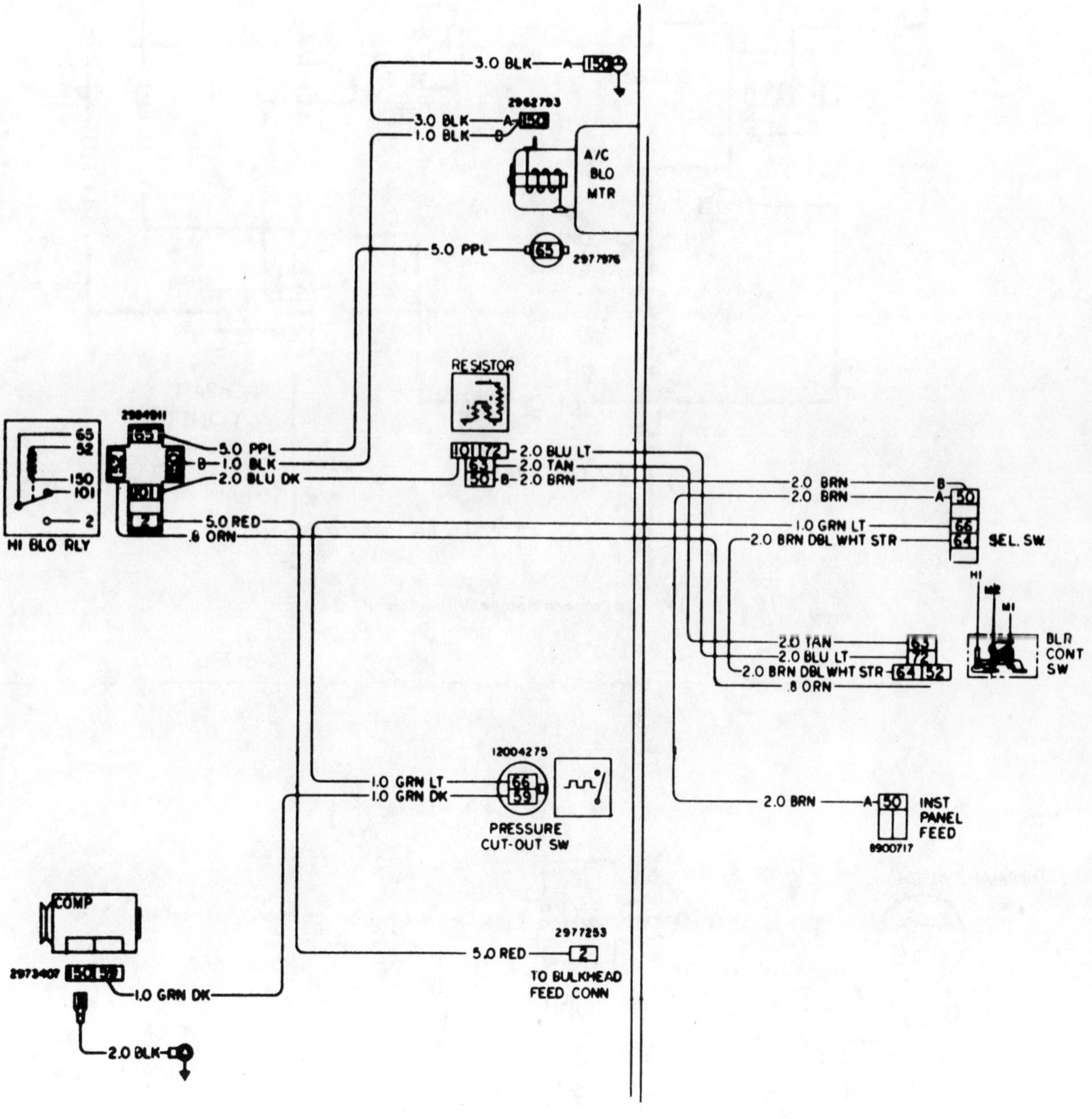

1980 Monza With V6-231 Four Season A/C & Heater Wiring Circuit

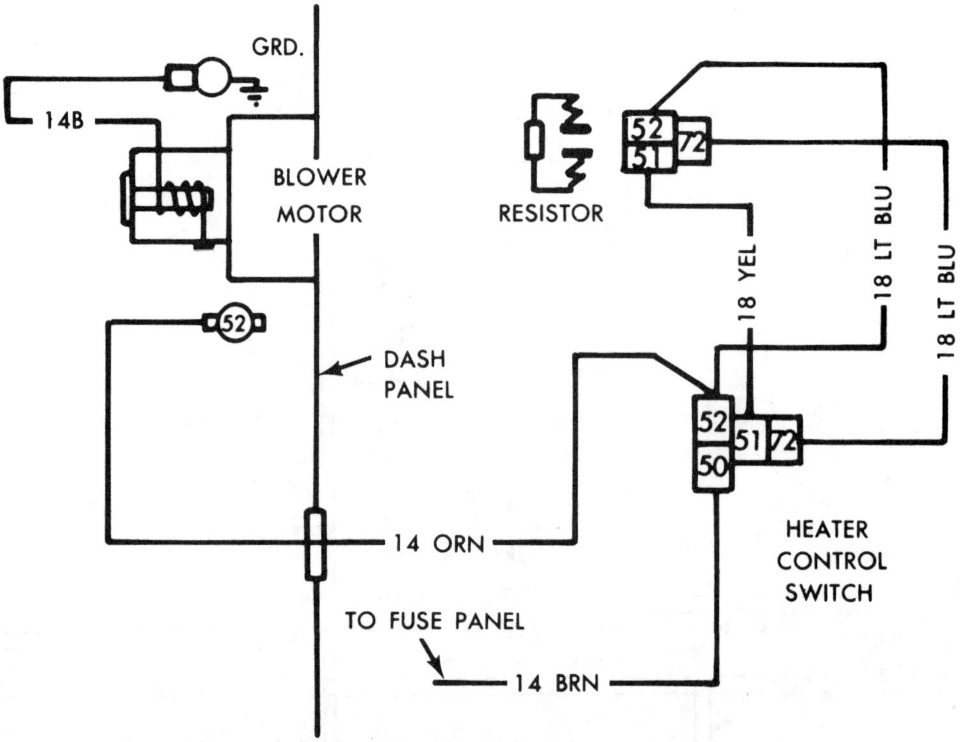

1974-76 Camaro Heater Wiring Circuit

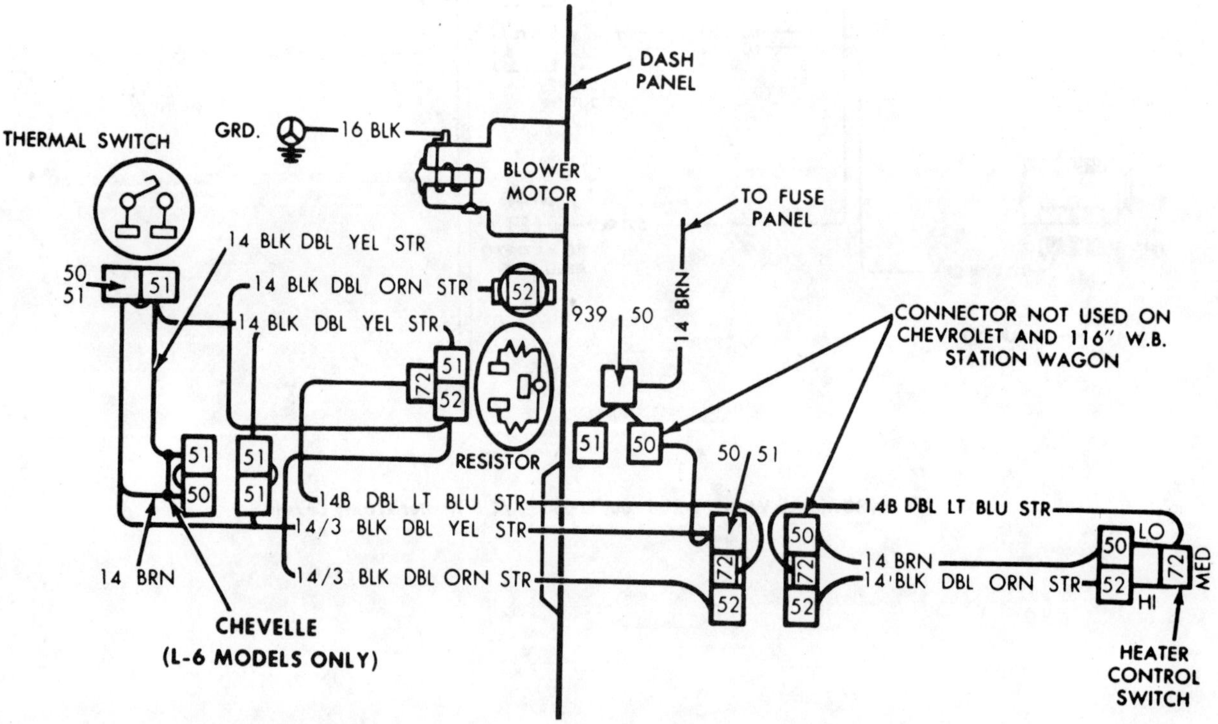

1974-77 Chevelle & Monte Carlo Heater Wiring Circuit

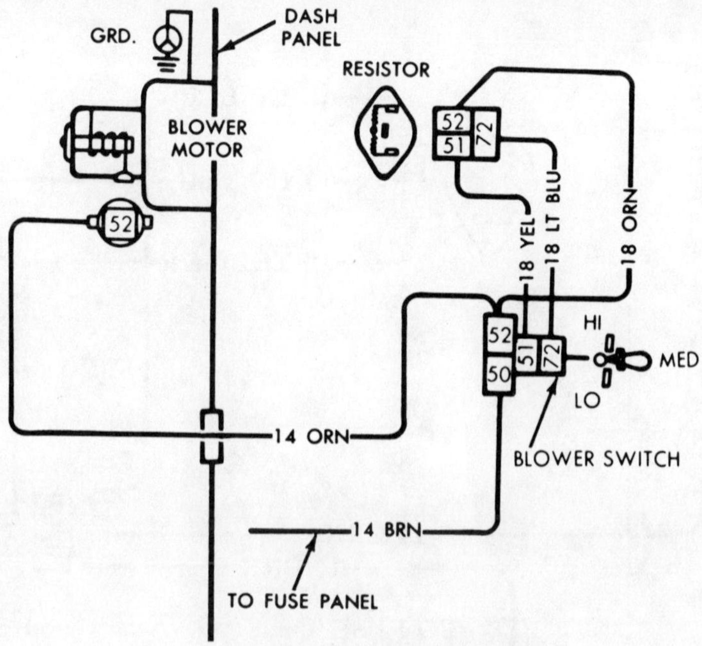

1974-77 Nova Heater Wiring Circuit

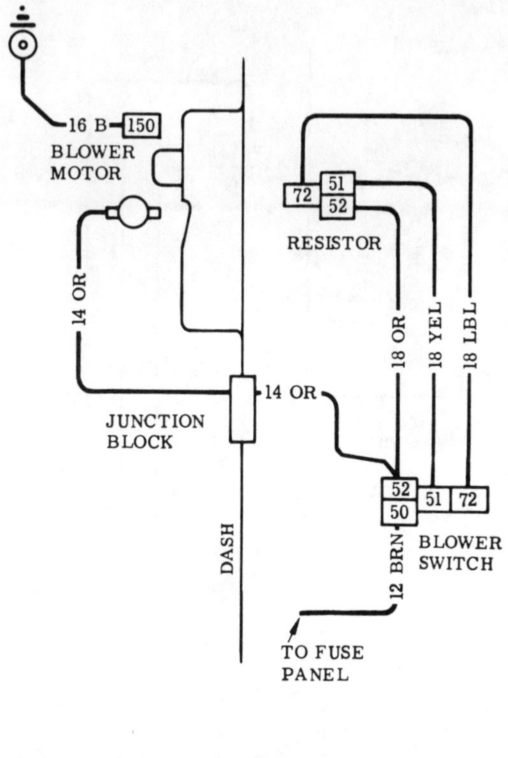

1974-77 Vega & 1975-77 Monza Heater Wiring Circuit

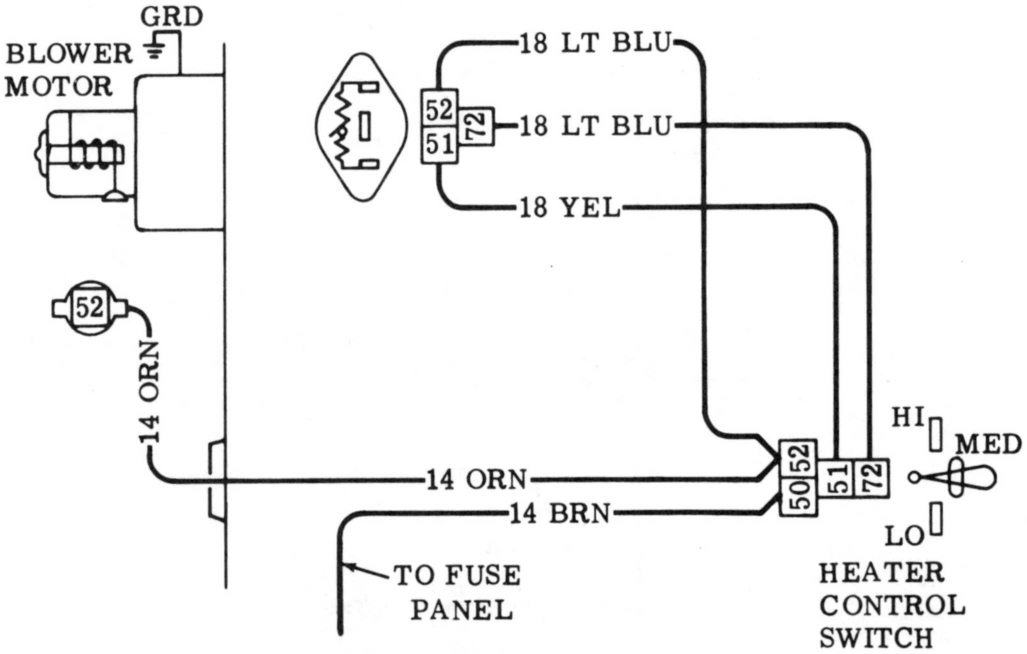

1977-80 Camaro Heater Wiring Circuit

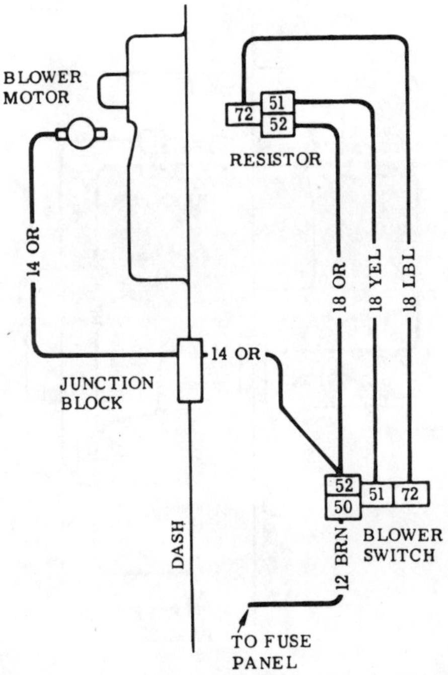

1978-79 Monza Heater Wiring Circuit

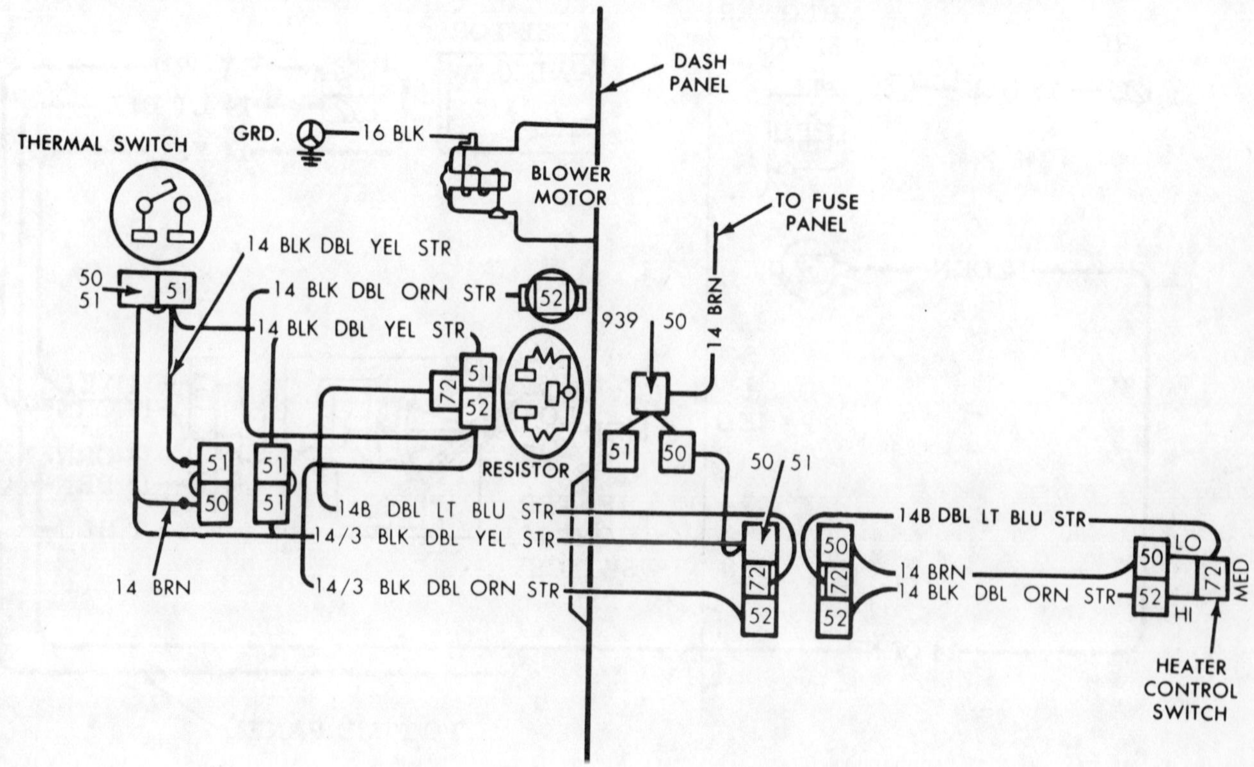

1978-79 Malibu & Monte Carlo Heater Wiring Circuit

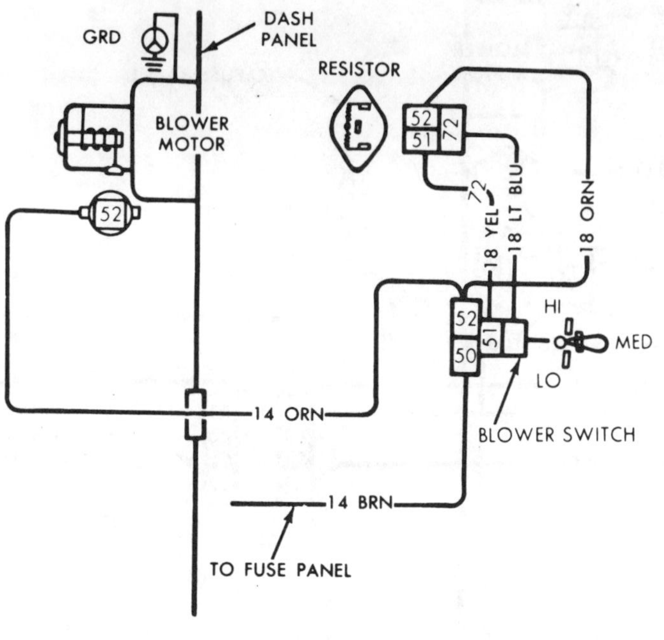

1978-79 Nova Heater Wiring Circuit

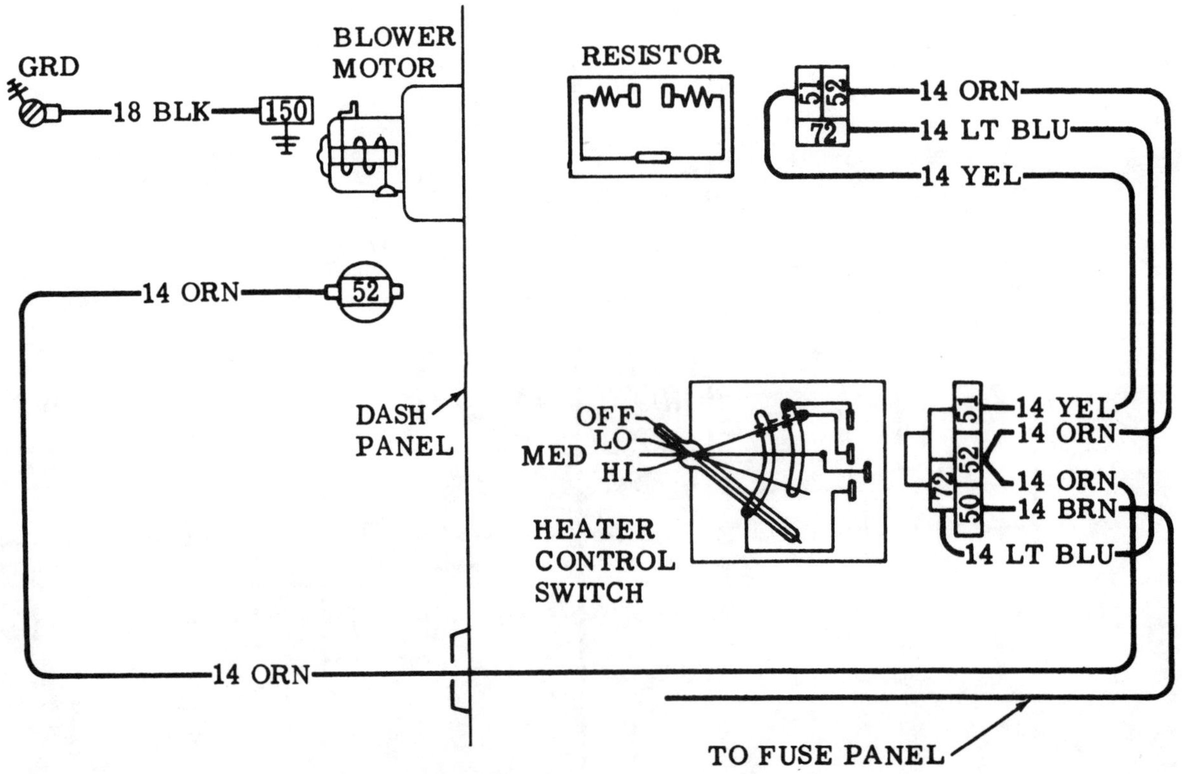

1979-80 Chevette Heater Wiring Circuit

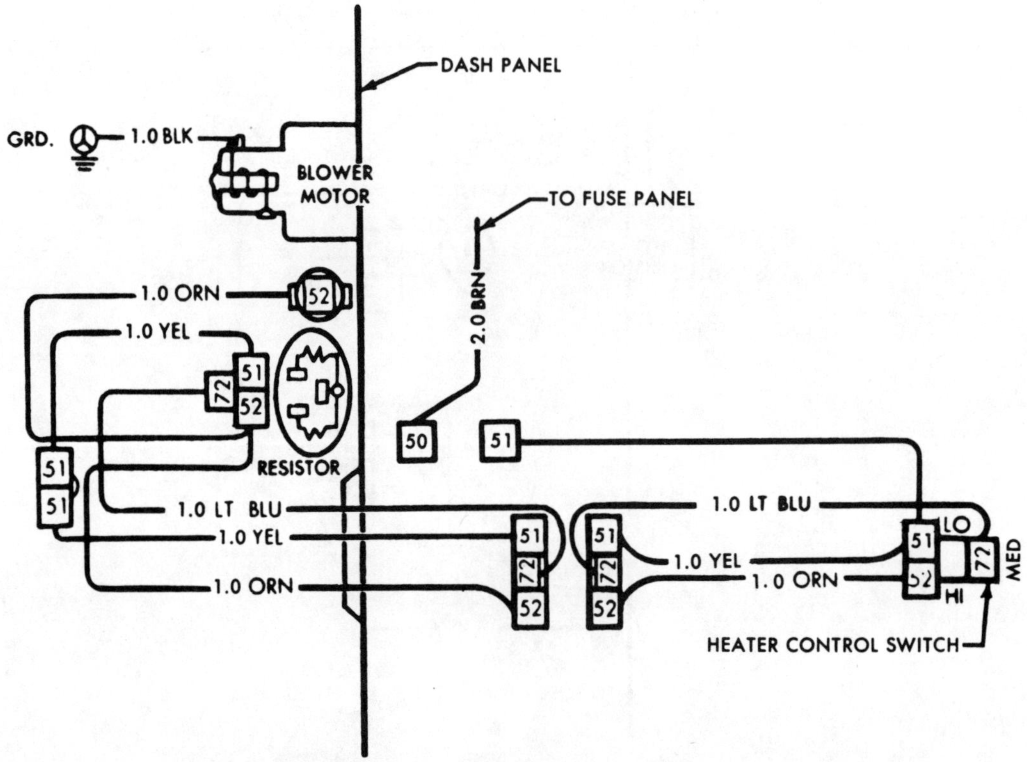

1980 Malibu & Monte Carlo Heater Wiring Circuit

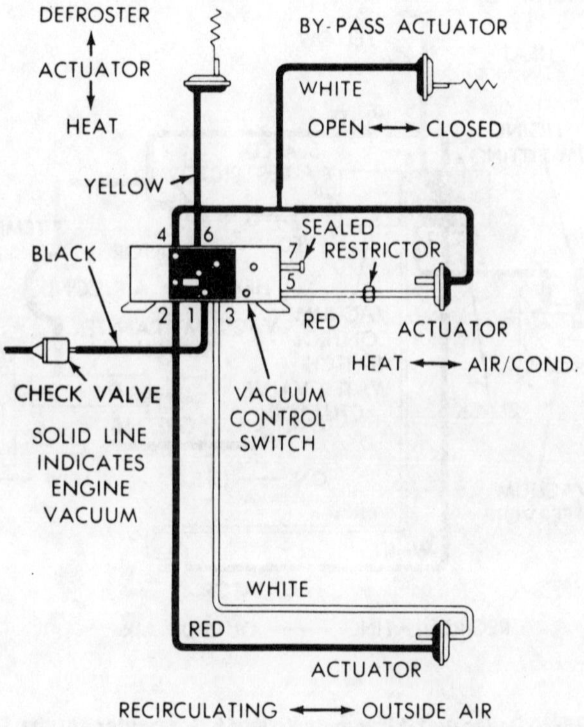

1974 Dodge Charger & Coronet; Plymouth Satellite A/C & Heater Vacuum Circuit

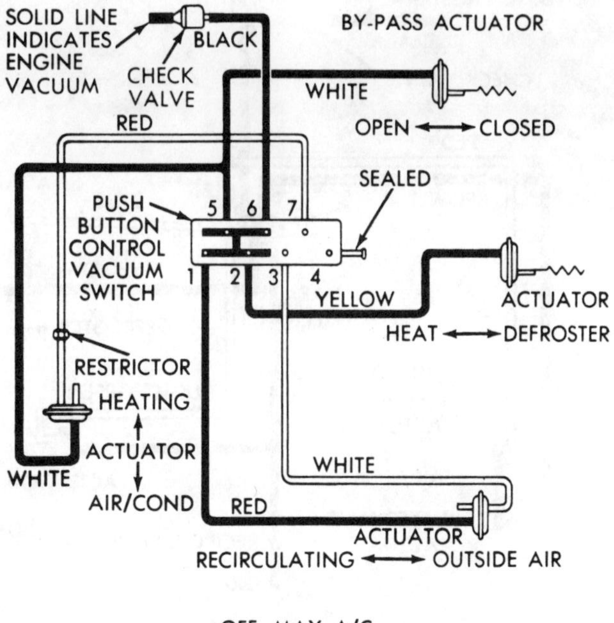

1974 Dodge Challenger & Plymouth Barracuda A/C & Heater Vacuum Circuit

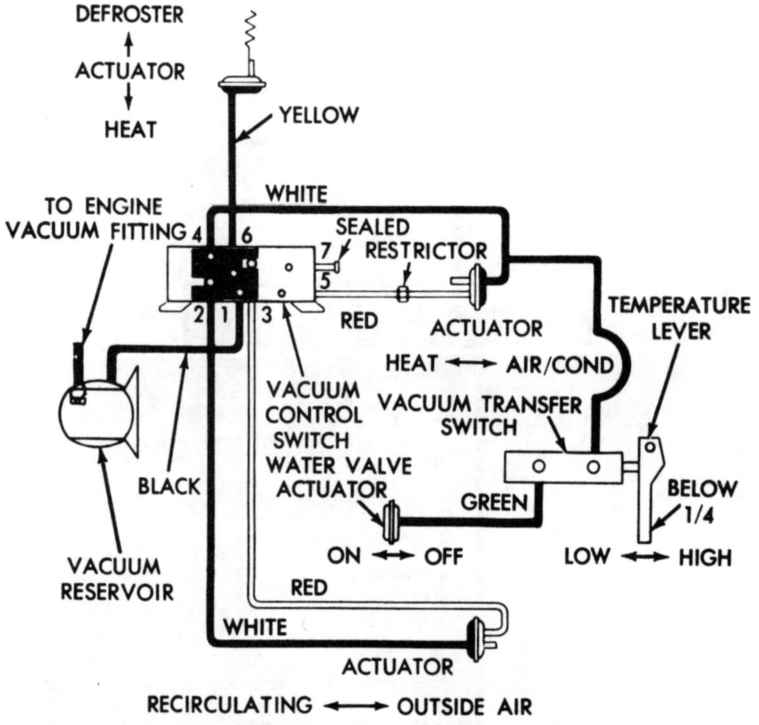

1974-76 Dodge Dart & Plymouth Valiant A/C & Heater Vacuum Circuit

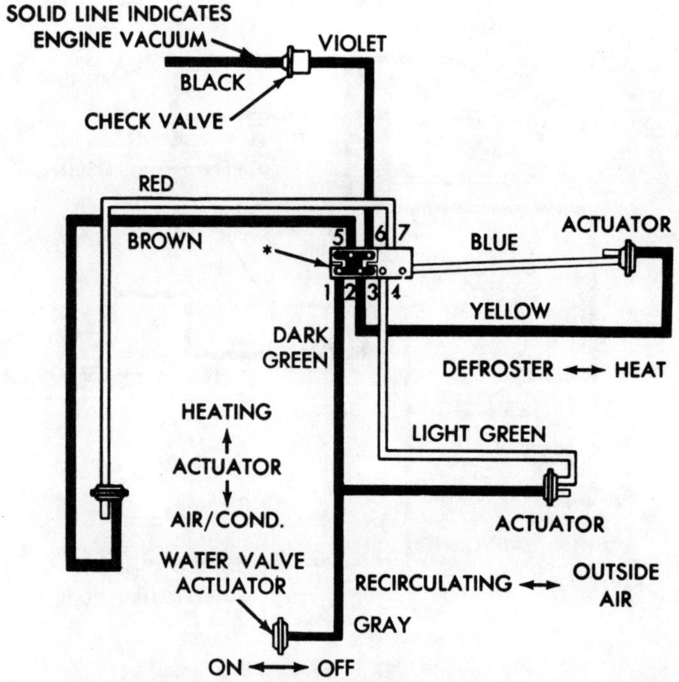

1974 Plymouth Fury; 1974-75 Imperial; 1974-76 Chrysler Newport & New Yorker, Dodge Monaco;
1975-76 Plymouth Gran Fury A/C & Heater Vacuum Circuit (Less Auto Temp)

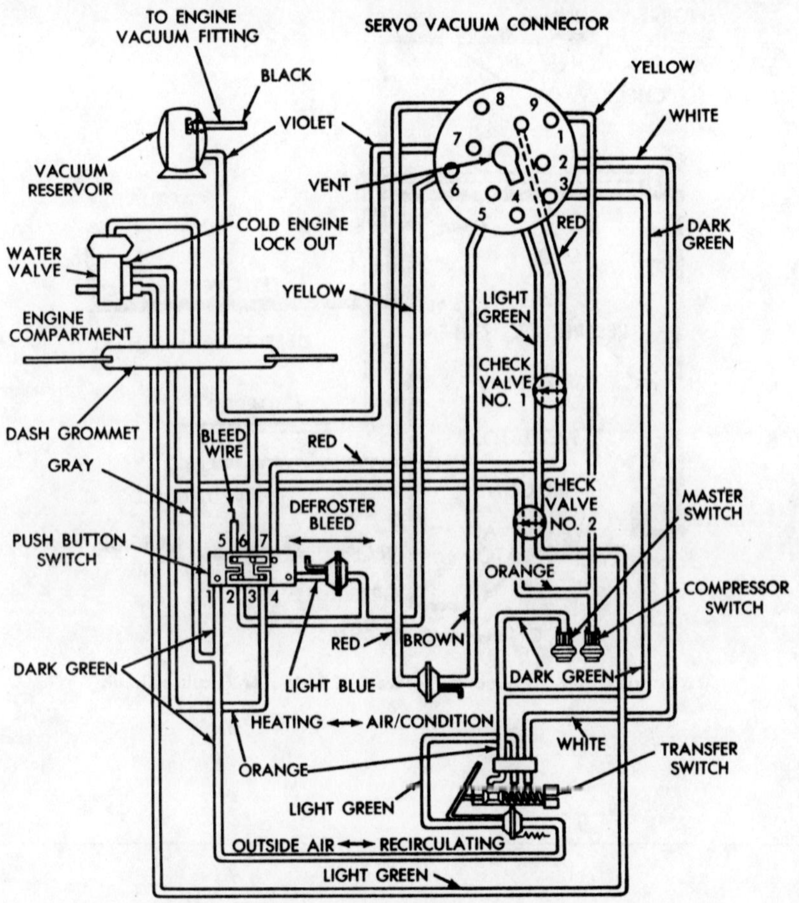

1974 Plymouth Fury; 1974-75 Imperial; 1974-76 Chrysler Newport & New Yorker, Dodge Monaco; 1975-76 Plymouth Gran Fury A/C & Heater Vacuum Circuit (With Auto Temp)

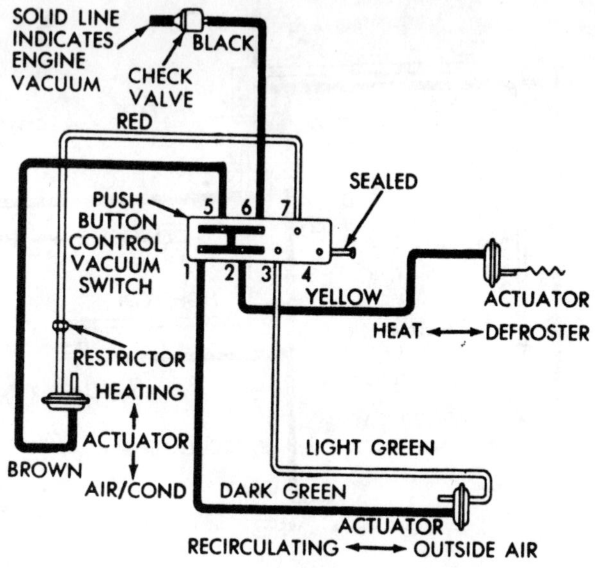

1975-76 Chrysler Cordoba; Dodge Charger & Coronet; Plymouth Fury A/C & Heater Vacuum Circuit

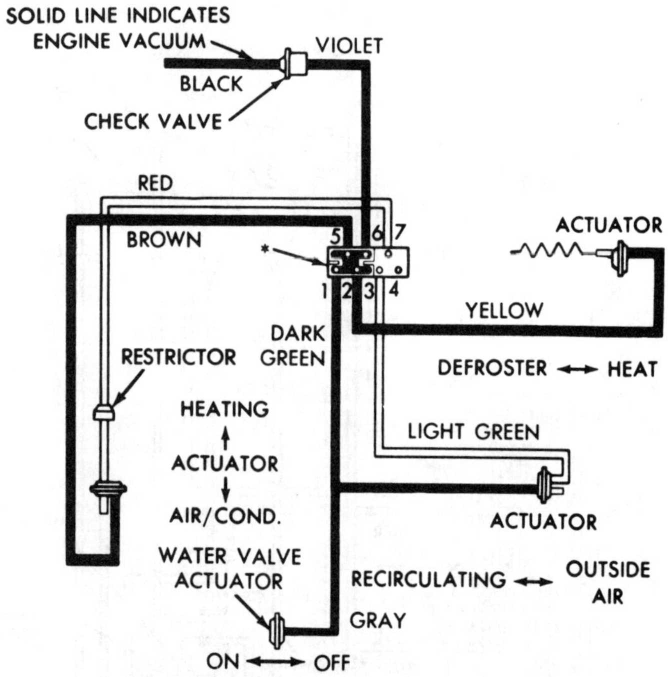

1976 Dodge Aspen & Plymouth Volare A/C & Heater Vacuum Circuit

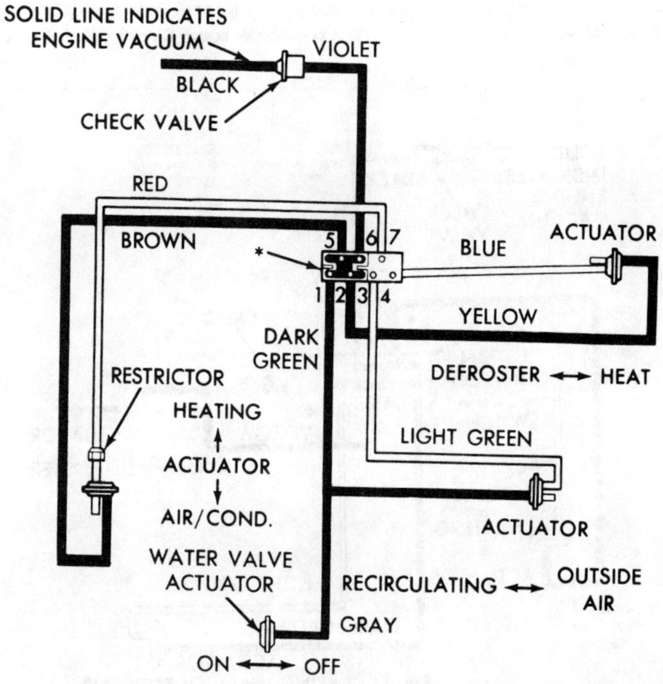

1977-78 Chrysler Newport & New Yorker; Dodge Royal Monaco; Plymouth Gran Fury A/C & Heater Vacuum Circuit (Less Auto Temp)

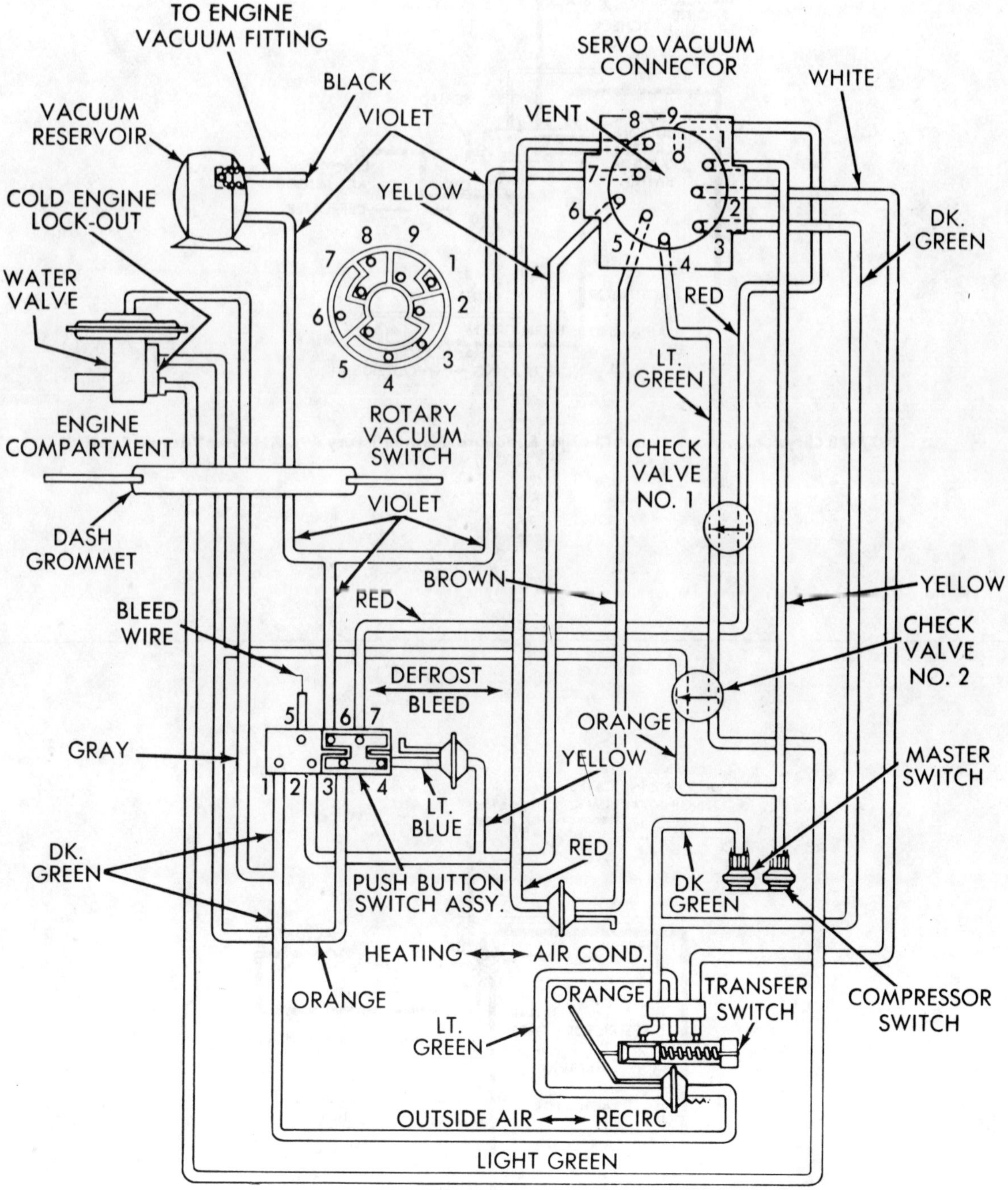

1977 Dodge Royal Monaco & Plymouth Gran Fury; 1977-78 Chrysler Newport & New Yorker A/C & Heater Vacuum Circuit (With Auto Temp)

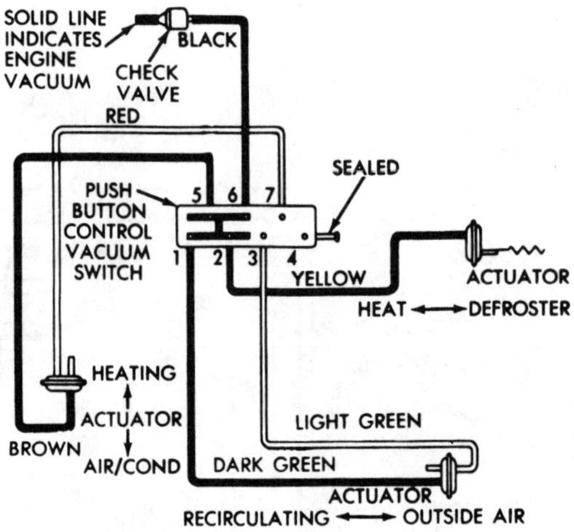

1977-78 Chrysler, Cordoba, Dodge Charger & Monaco, Plymouth Fury A/C & Heater Vacuum Circuit

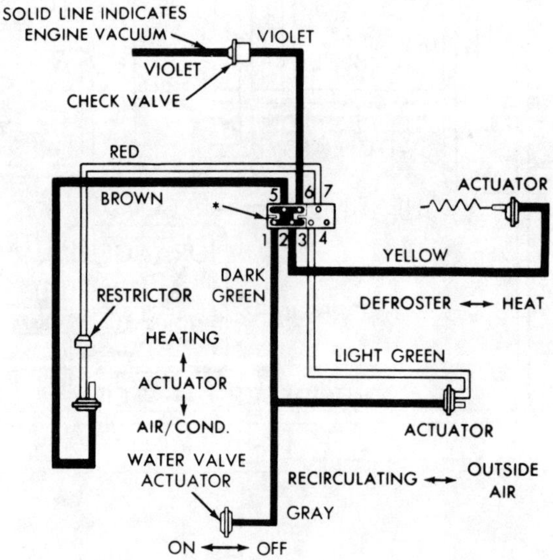

1977-79 Chrysler LeBaron; Dodge Aspen & Diplomat; Plymouth Volare A/C & Heater Vacuum Circuit

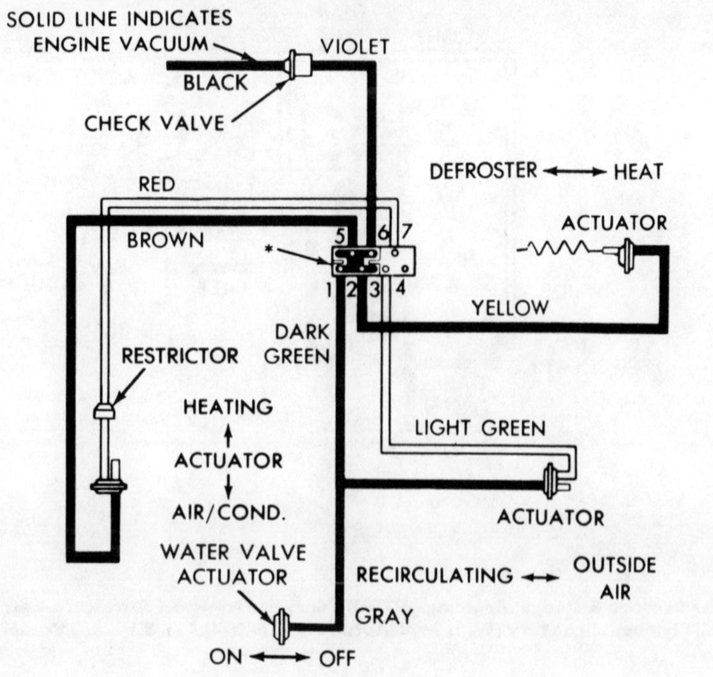

1978-80 Dodge Omni, & Plymouth Horizon A/C & Heater Vacuum Circuit

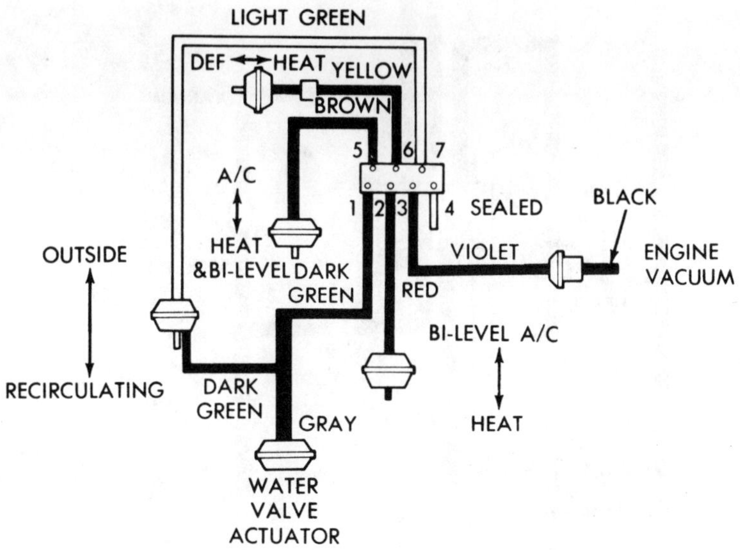

1979 Chrysler Newport & New Yorker & Dodge St. Regis (Less Semi-Auto Temp Control); 1979 Dodge Magnum (Less Semi-Auto Temp Control); 1979-80 Chrysler Cordoba (Less Semi-Auto Temp Control); 1980 Dodge Mirada (Less Semi-Auto Temp Control); 1980 Chrysler Cordoba & LeBaron, Dodge Diplomat & Mirada (With Semi-Auto Temp Control) A/C & Heater Vacuum Circuit

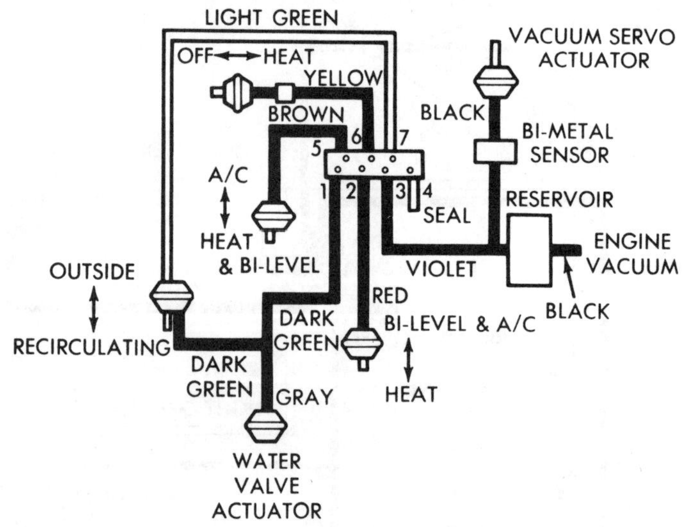

1979 Chrysler Cordoba & Dodge Magnum; 1979-80 Chrysler Newport & New Yorker, Dodge St. Regis; 1980 Plymouth Gran Fury (With Semi-Auto Temp Control) A/C & Heater Vacuum Circuit

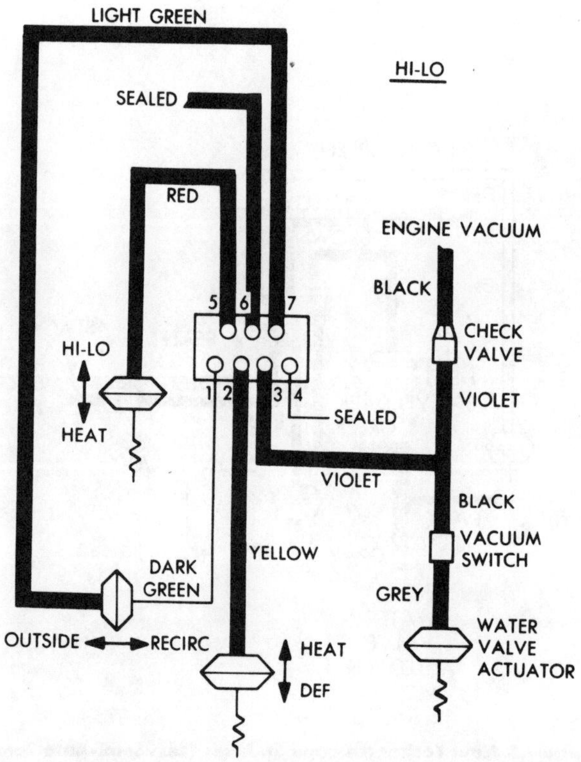

1979 Dodge Magnum; 1979-80 Chrysler Cordoba, Newport & New Yorker, Dodge St. Regis; 1980 Dodge Mirada & Plymouth Gran Fury Heater Vacuum Circuit

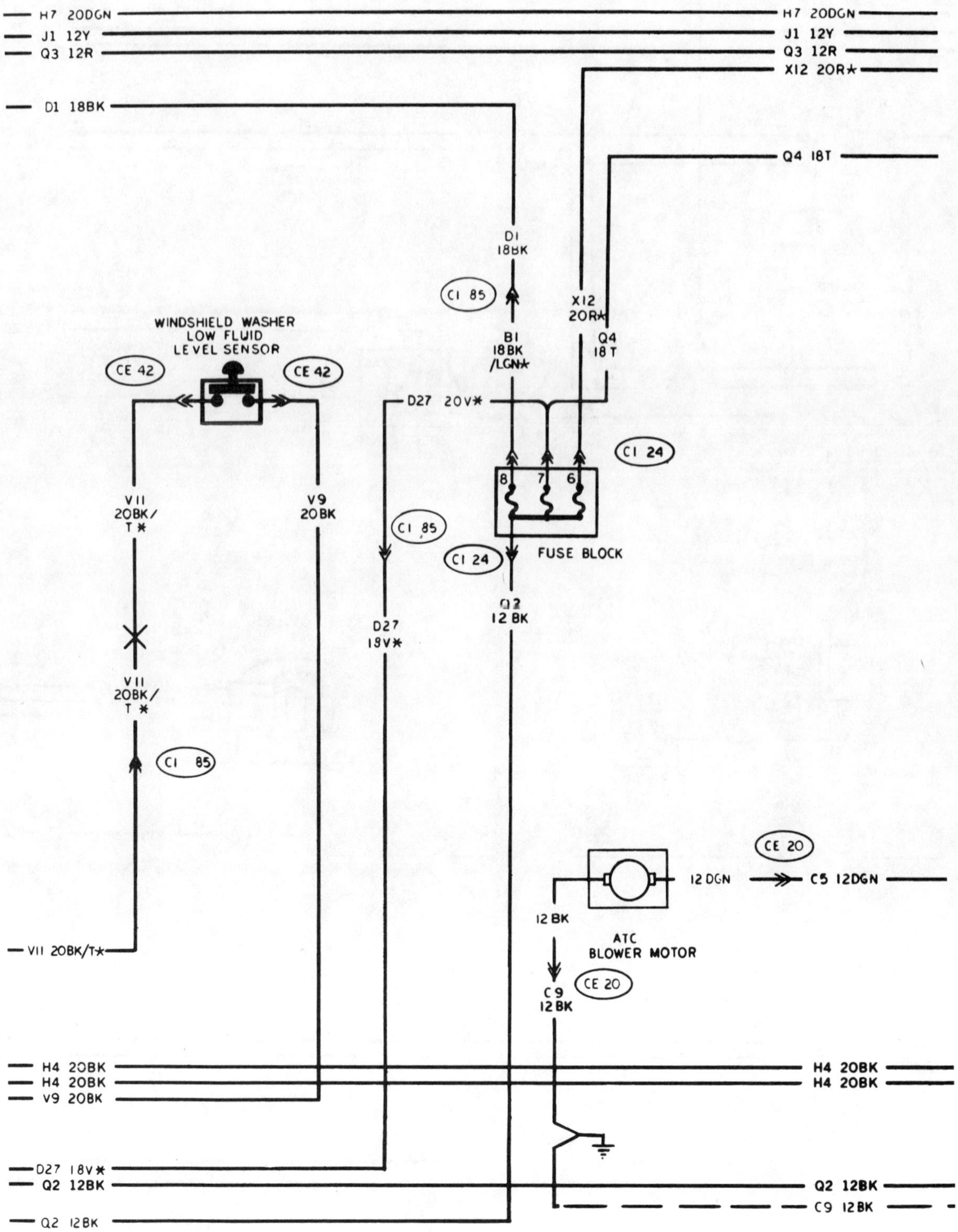

1974 Chrysler Newport & New Yorker; Dodge Monaco & Polara; Imperial; Plymouth Fury A/C & Heater Wiring Circuit Part 1 of 8

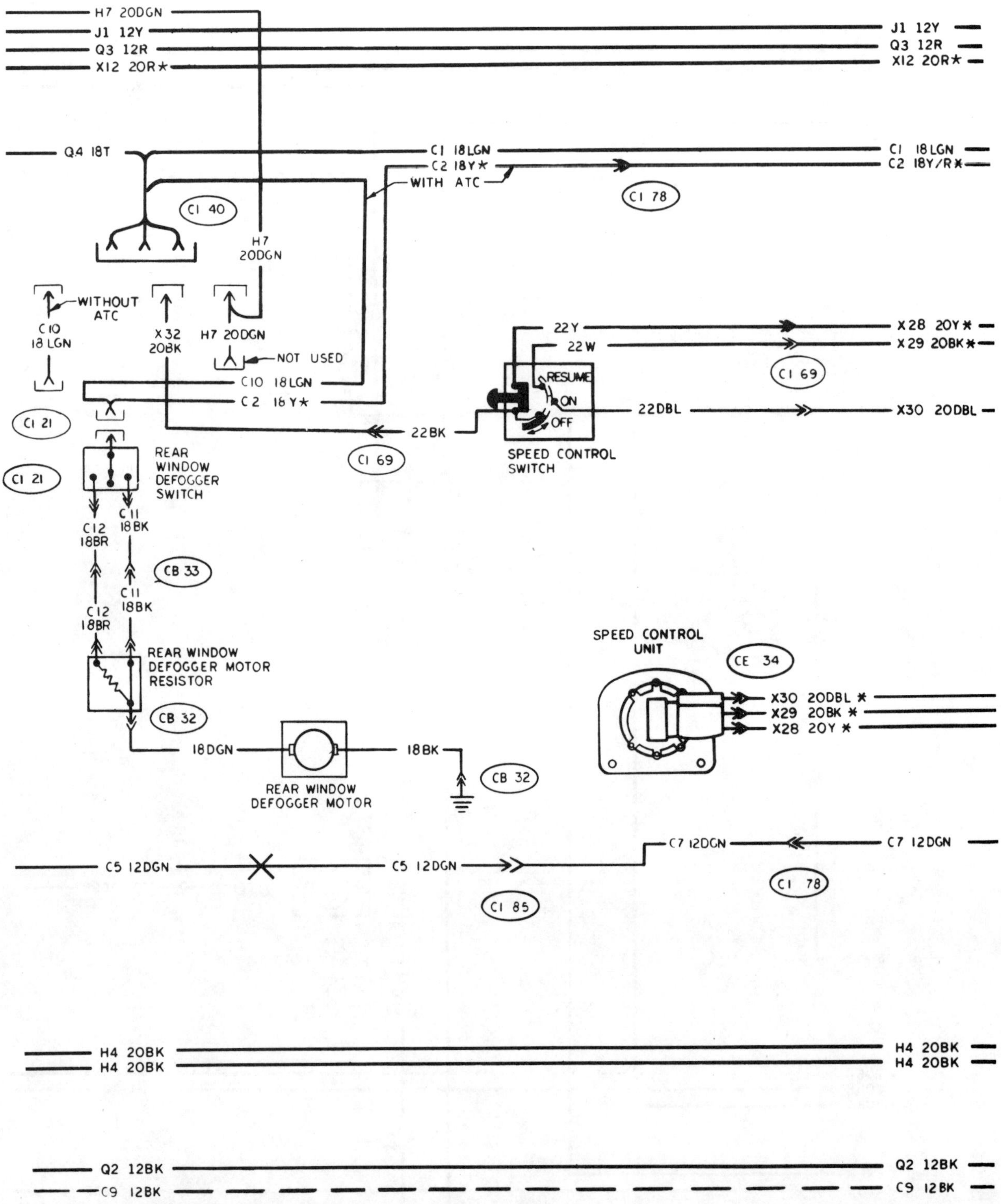

1974 Chrysler Newport & New Yorker; Dodge Monaco & Polara; Imperial; Plymouth Fury A/C & Heater Wiring Circuit Part 2 of 8

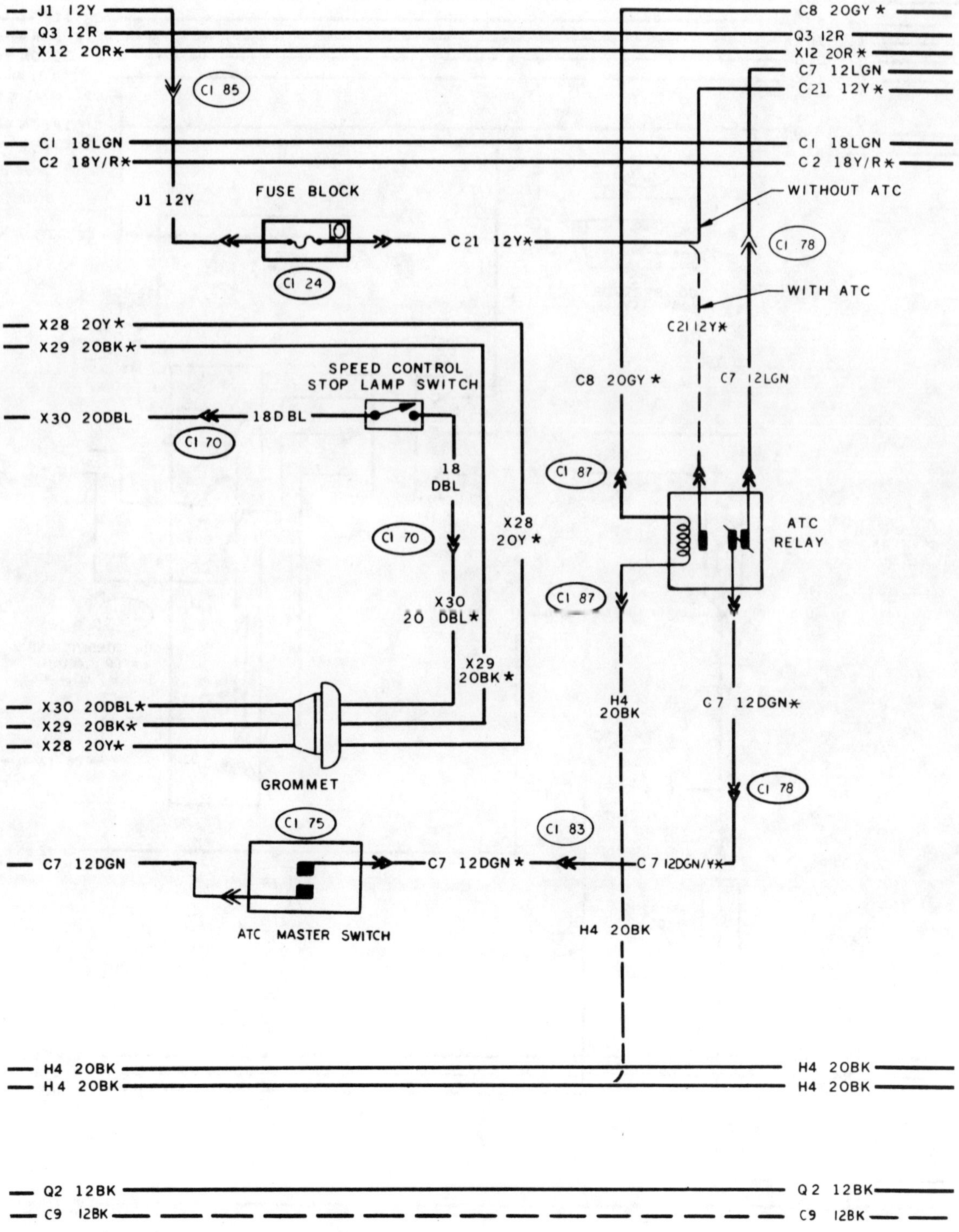

1974 Chrysler Newport & New Yorker; Dodge Monaco & Polara; Imperial; Plymouth Fury A/C & Heater Wiring Circuit Part 3 of 8

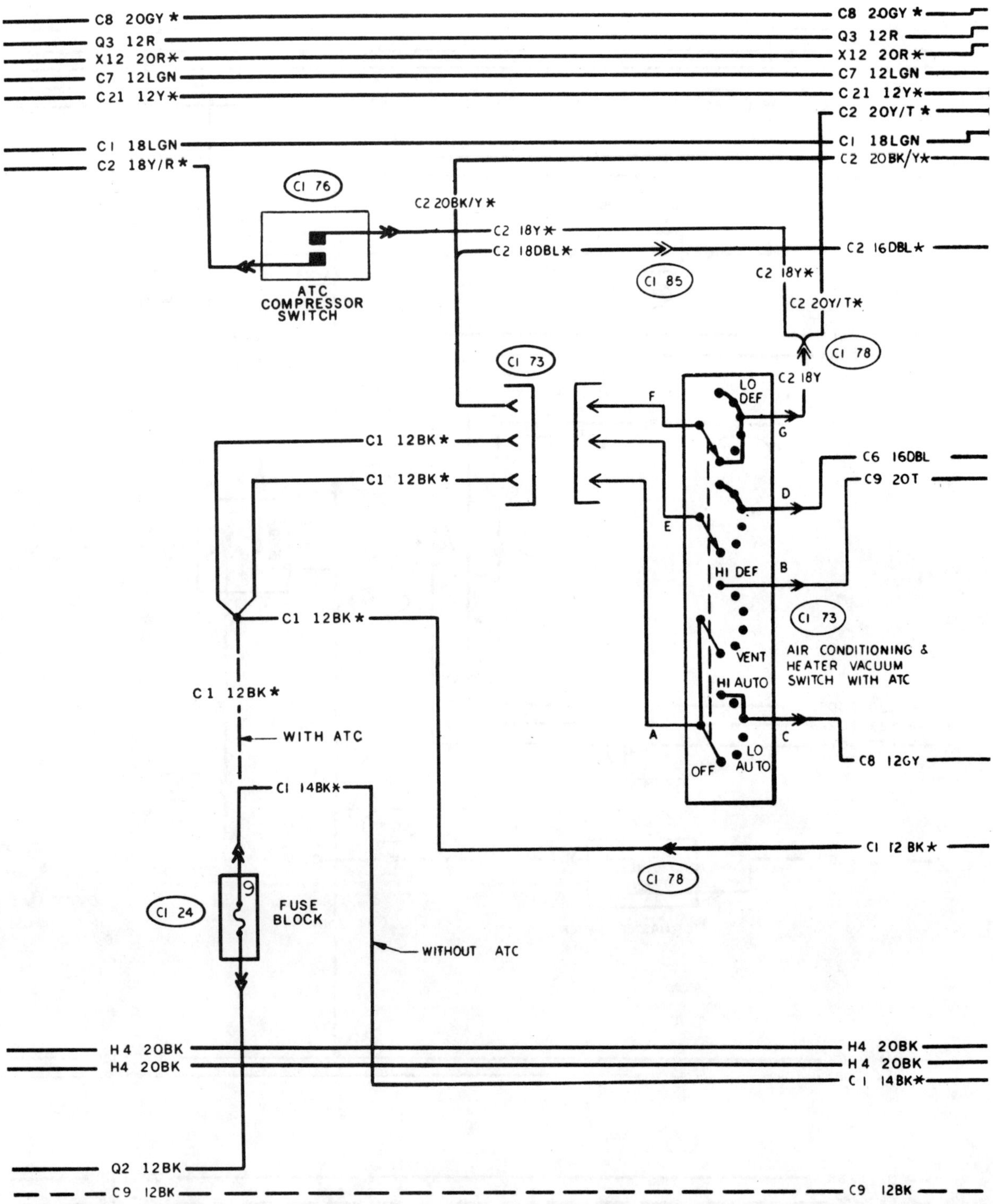

1974 Chrysler Newport & New Yorker; Dodge Monaco & Polara; Imperial; Plymouth Fury A/C & Heater Wiring Circuit Part 4 of 8

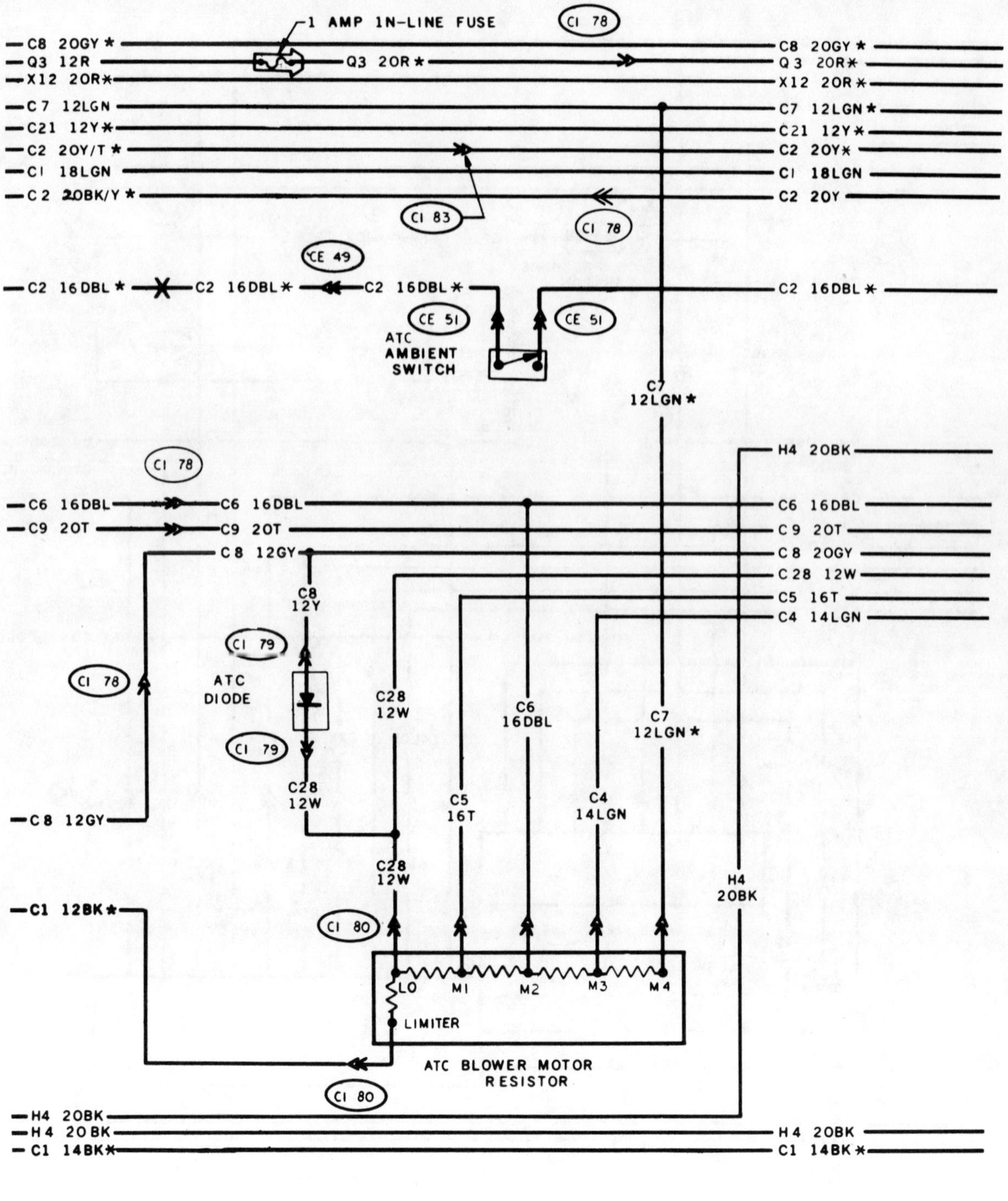

1974 Chrysler Newport & New Yorker; Dodge Monaco & Polara; Imperial; Plymouth Fury A/C & Heater Wiring Circuit Part 5 of 8

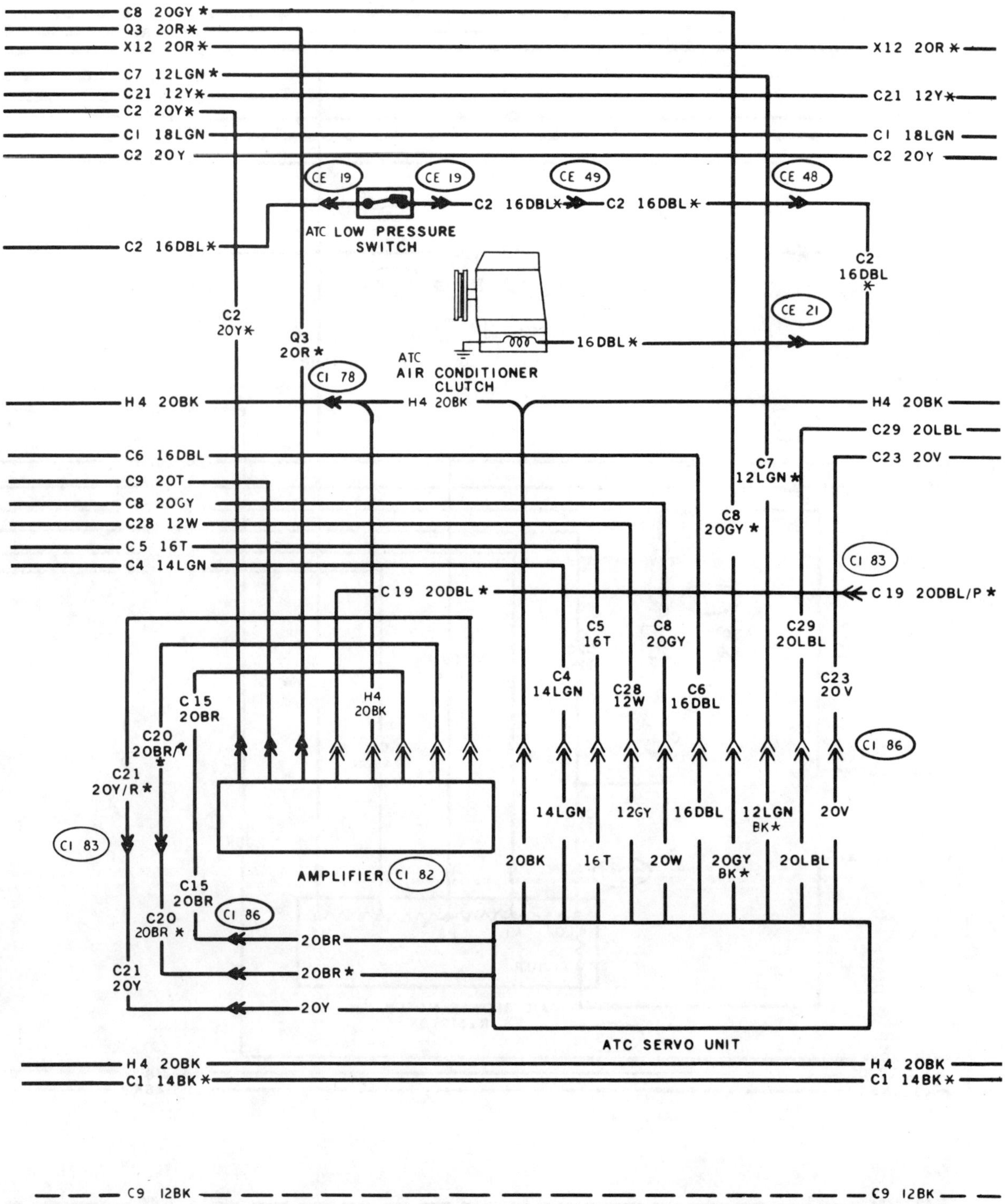

1974 Chrysler Newport & New Yorker; Dodge Monaco & Polara; Imperial; Plymouth Fury A/C & Heater Wiring Circuit Part 6 of 8

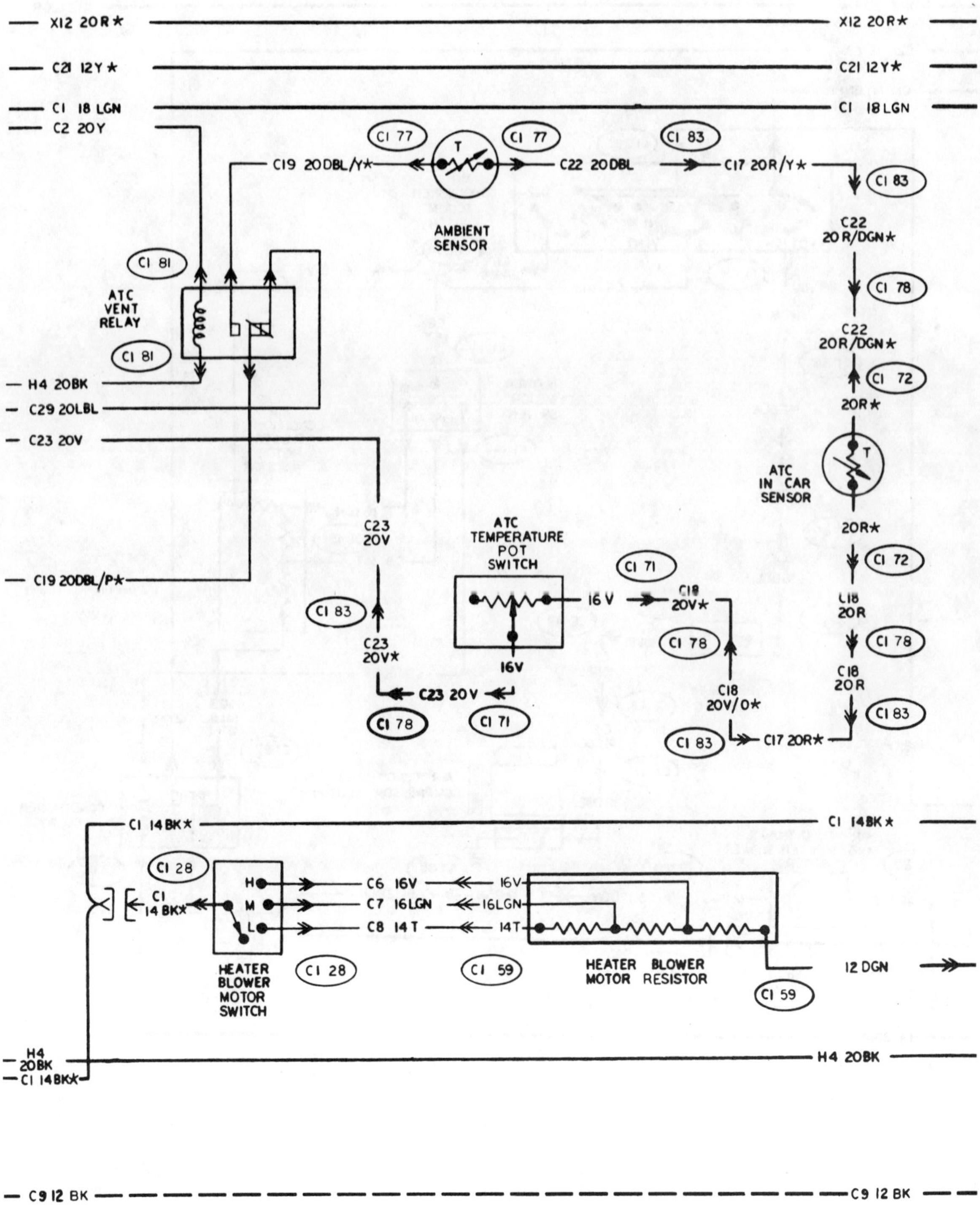

1974 Chrysler Newport & New Yorker; Dodge Monaco & Polara; Imperial; Plymouth Fury A/C & Heater Wiring Circuit Part 7 of 8

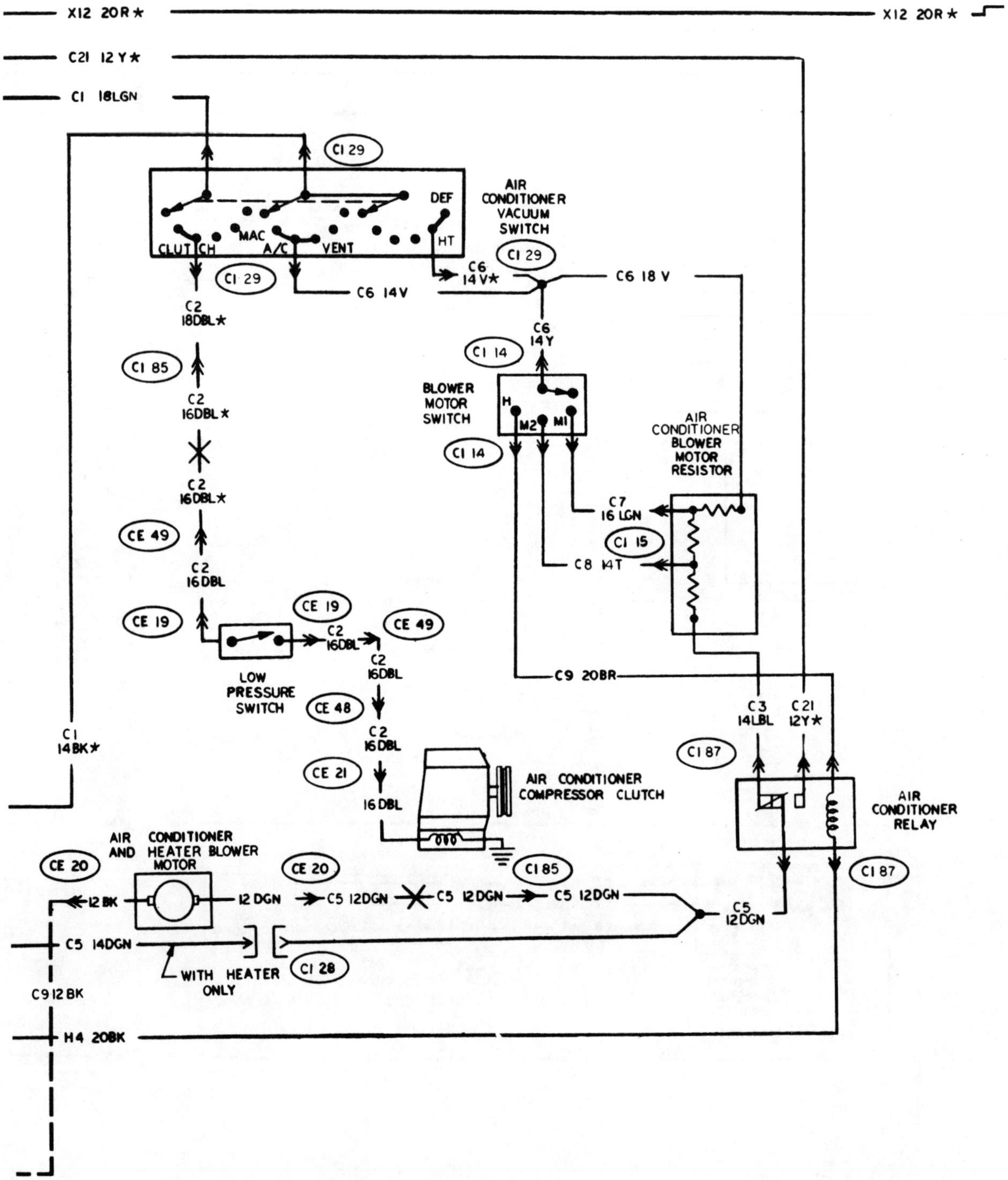

1974 Chrysler Newport & New Yorker; Dodge Monaco & Polara; Imperial; Plymouth Fury A/C & Heater Wiring Circuit Part 8 of 8

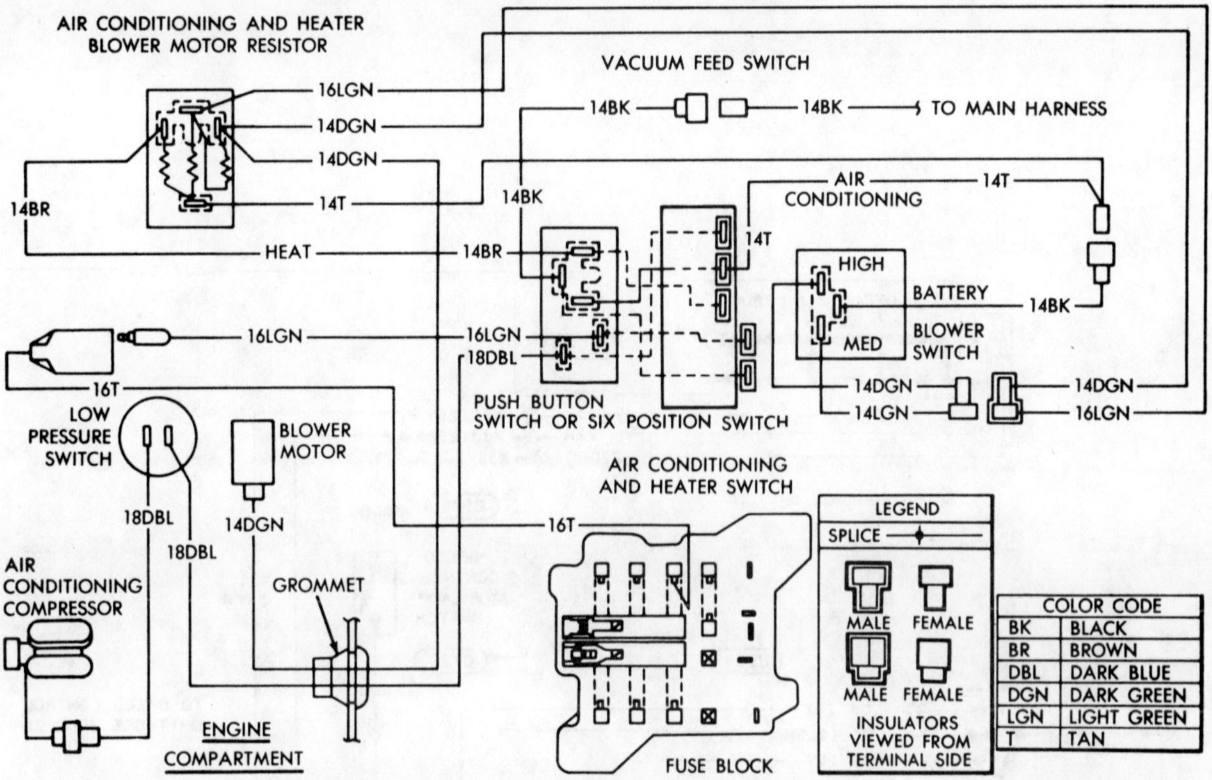

COLOR CODE	
BK	BLACK
BR	BROWN
DBL	DARK BLUE
DGN	DARK GREEN
LGN	LIGHT GREEN
T	TAN

1974 Dodge Challenger, Charger & Coronet; Plymouth Barracuda & Satellite A/C & Heater Wiring Circuit

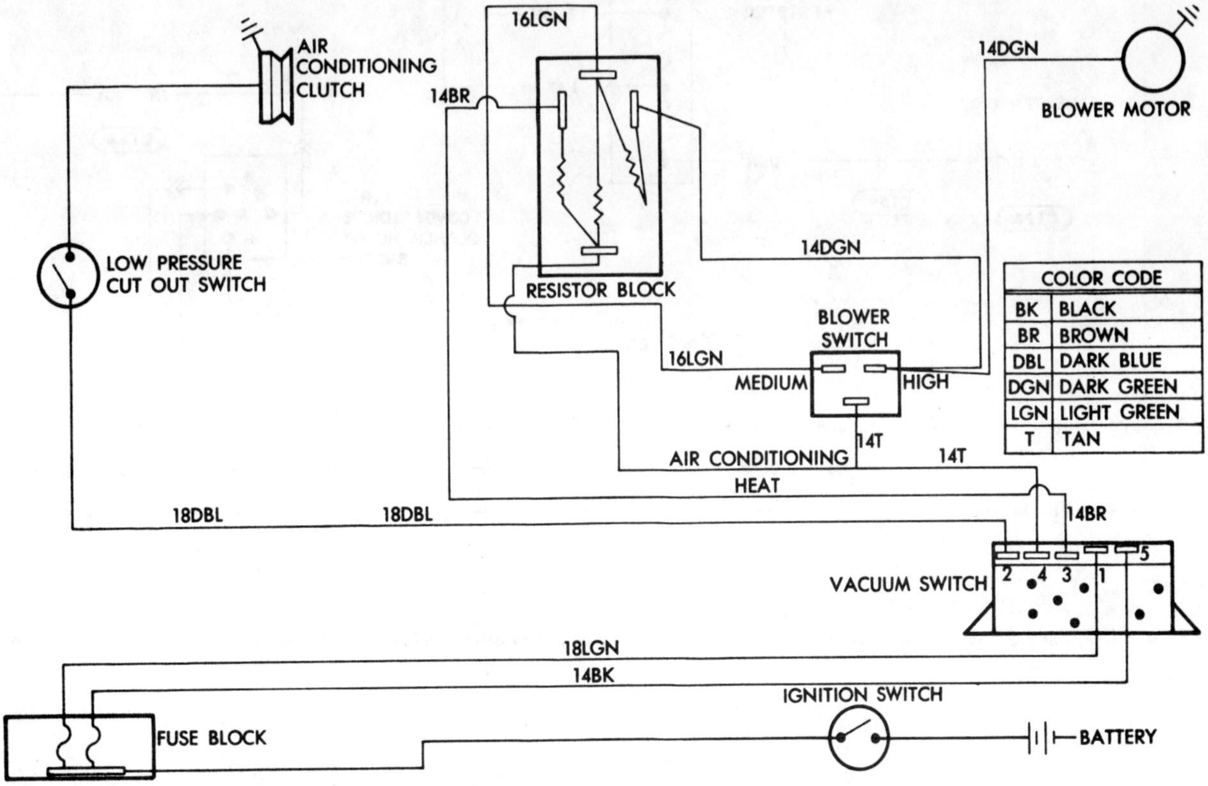

COLOR CODE	
BK	BLACK
BR	BROWN
DBL	DARK BLUE
DGN	DARK GREEN
LGN	LIGHT GREEN
T	TAN

1974-75 Dodge Dart & Plymouth Valiant A/C & Heater Wiring Circuit

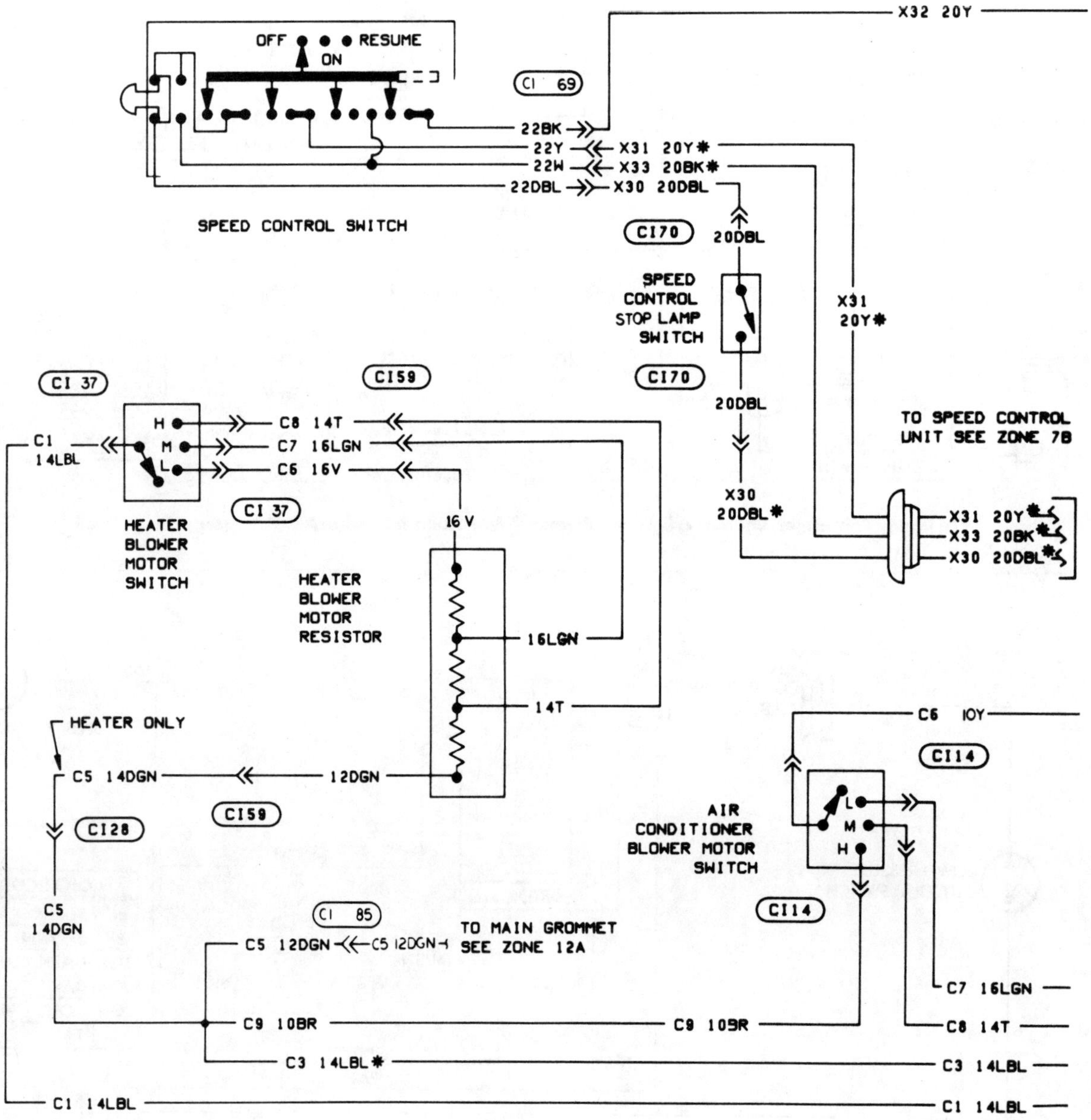

1975 Chrysler Newport & New Yorker; Dodge Monaco; Imperial; Plymouth Gran Fury A/C & Heater Wiring Circuit Part 1 of 6

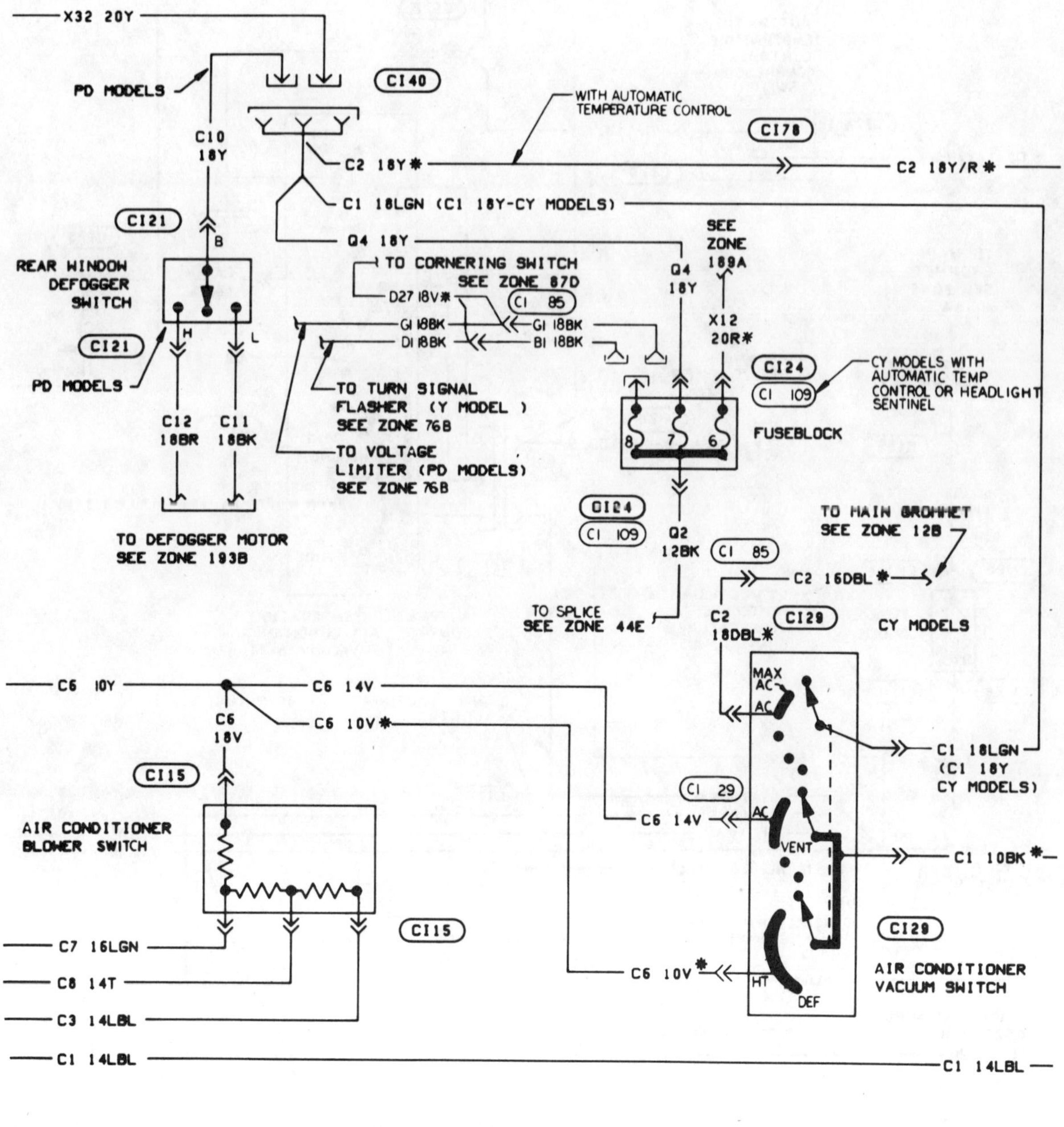

1975 Chrysler Newport & New Yorker; Dodge Monaco; Imperial; Plymouth Gran Fury A/C & Heater Wiring Circuit Part 2 of 6

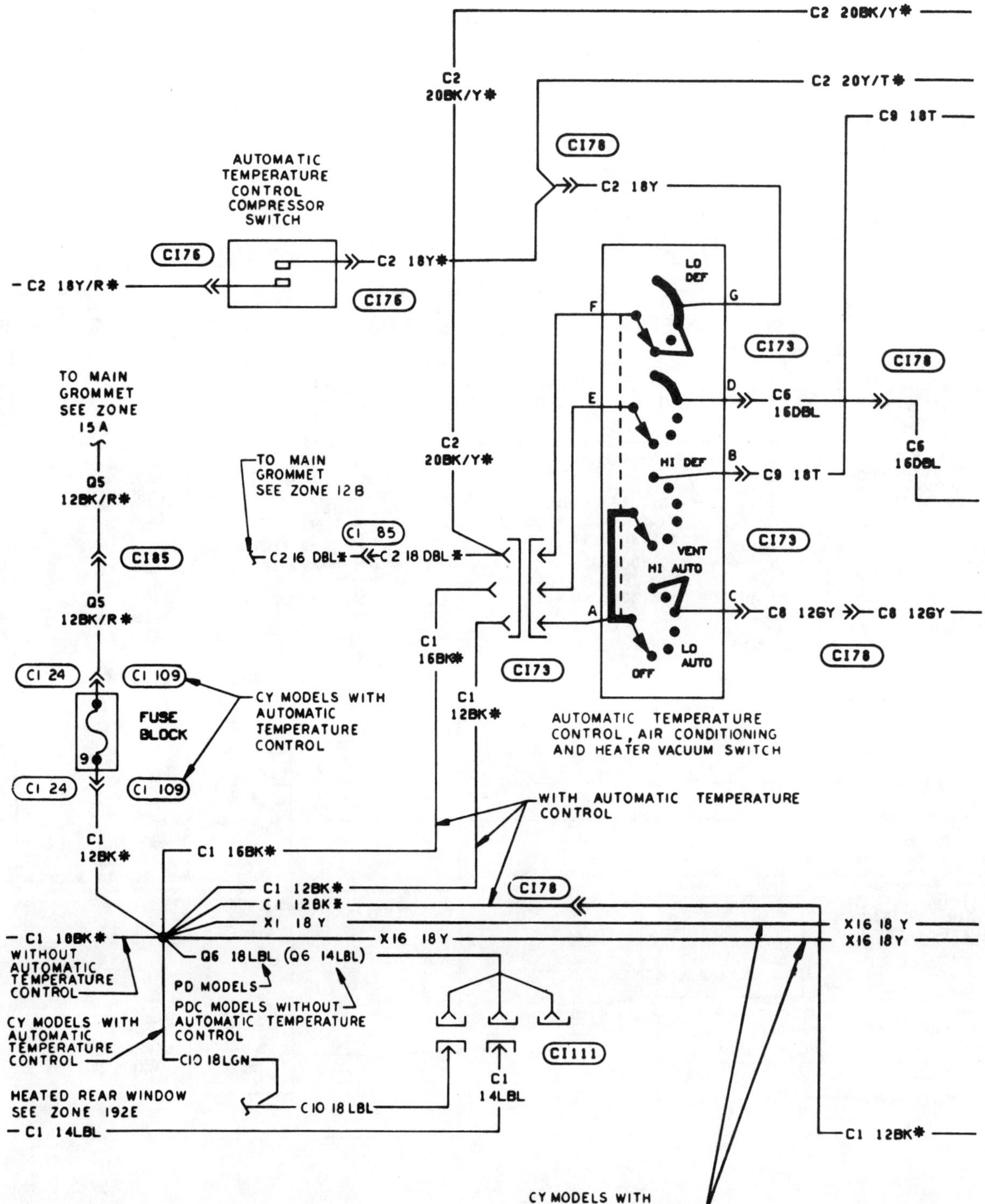

1975 Chrysler Newport & New Yorker; Dodge Monaco; Imperial; Plymouth Gran Fury A/C & Heater Wiring Circuit Part 3 of 6

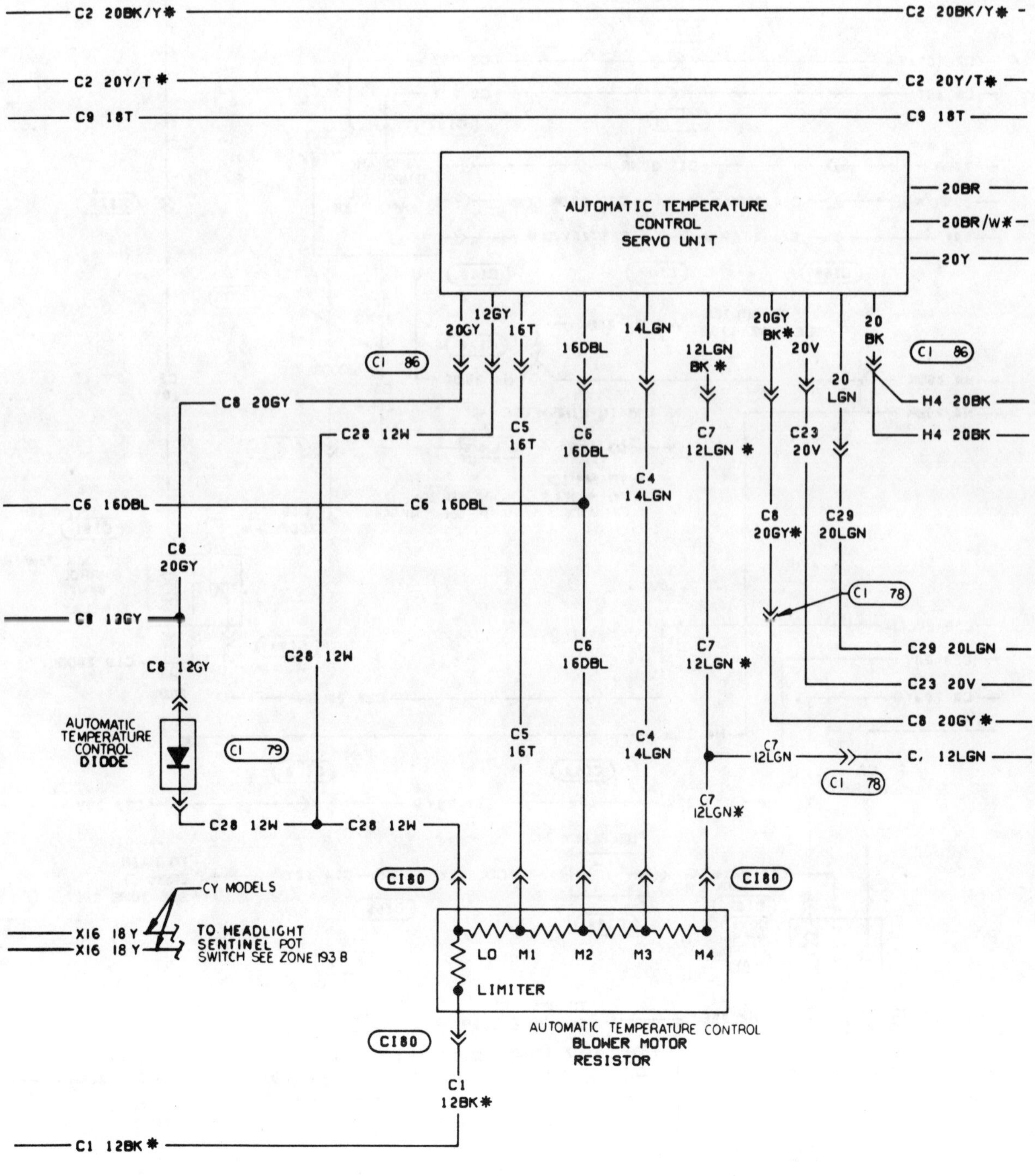

1975 Chrysler Newport & New Yorker; Dodge Monaco; Imperial; Plymouth Gran Fury A/C & Heater Wiring Circuit Part 4 of 6

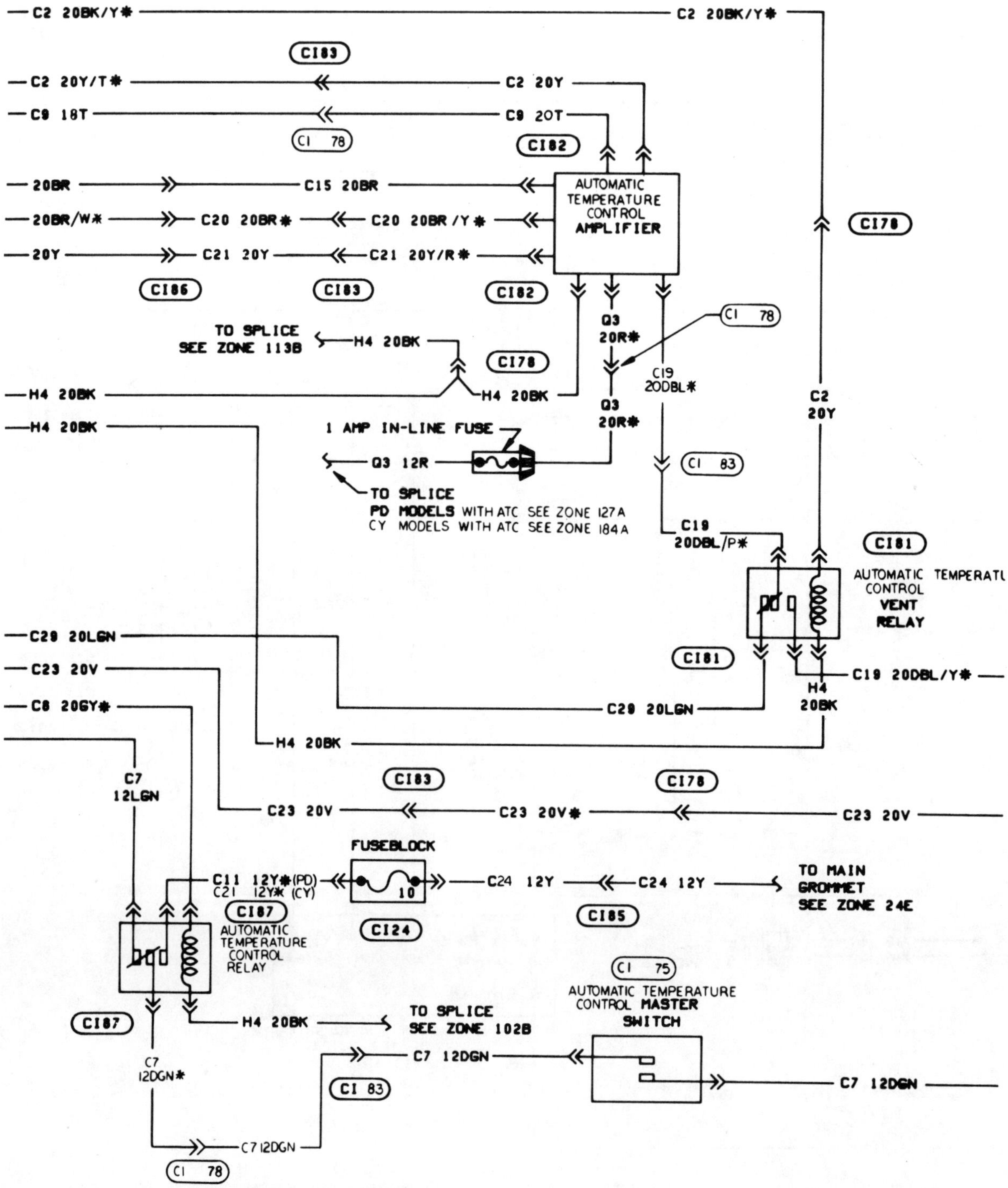

1975 Chrysler Newport & New Yorker; Dodge Monaco; Imperial; Plymouth Gran Fury A/C & Heater Wiring Circuit Part 5 of 6

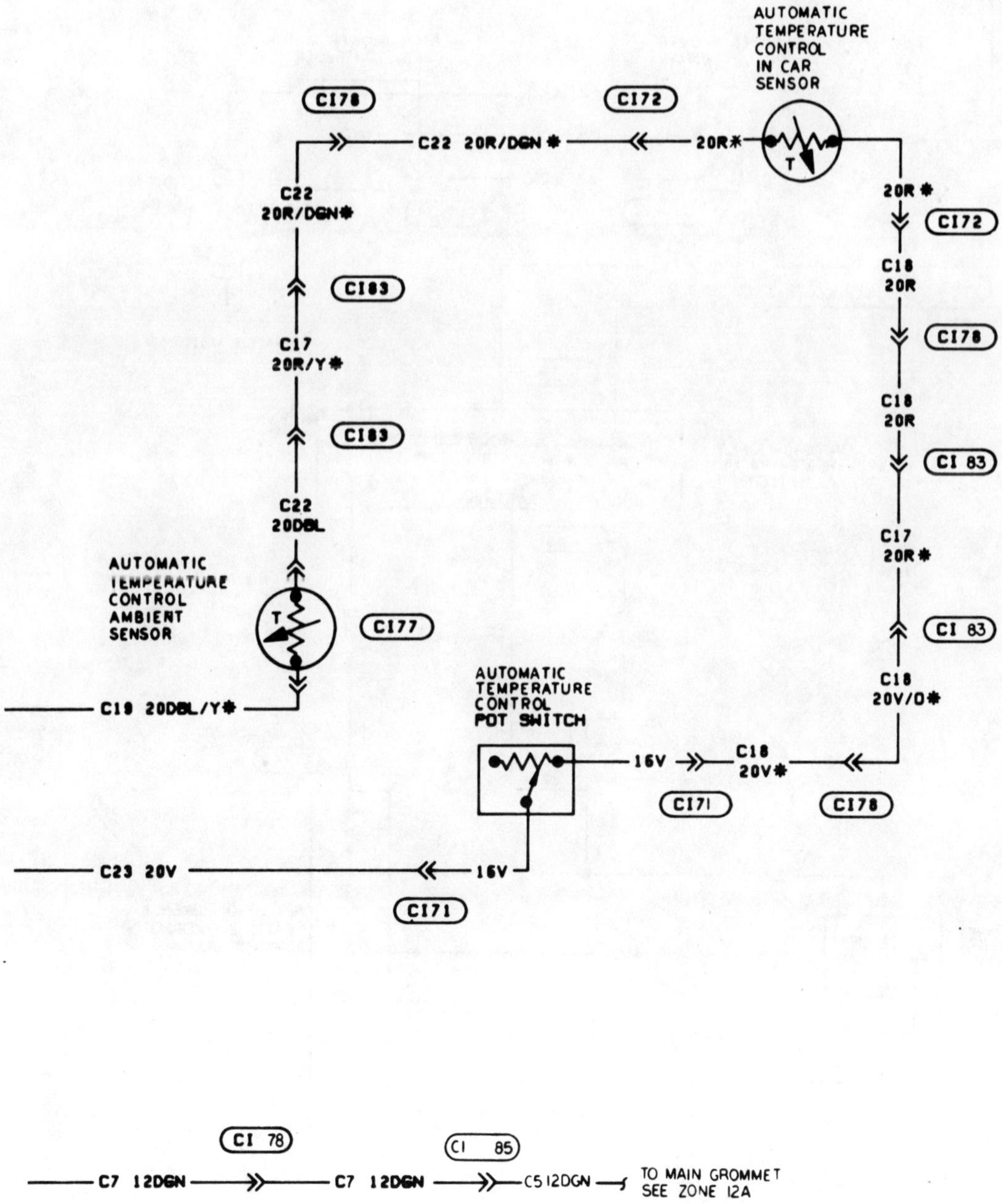

1975 Chrysler Newport & New Yorker; Dodge Monaco; Imperial; Plymouth Gran Fury A/C & Heater Wiring Circuit Part 6 of 6

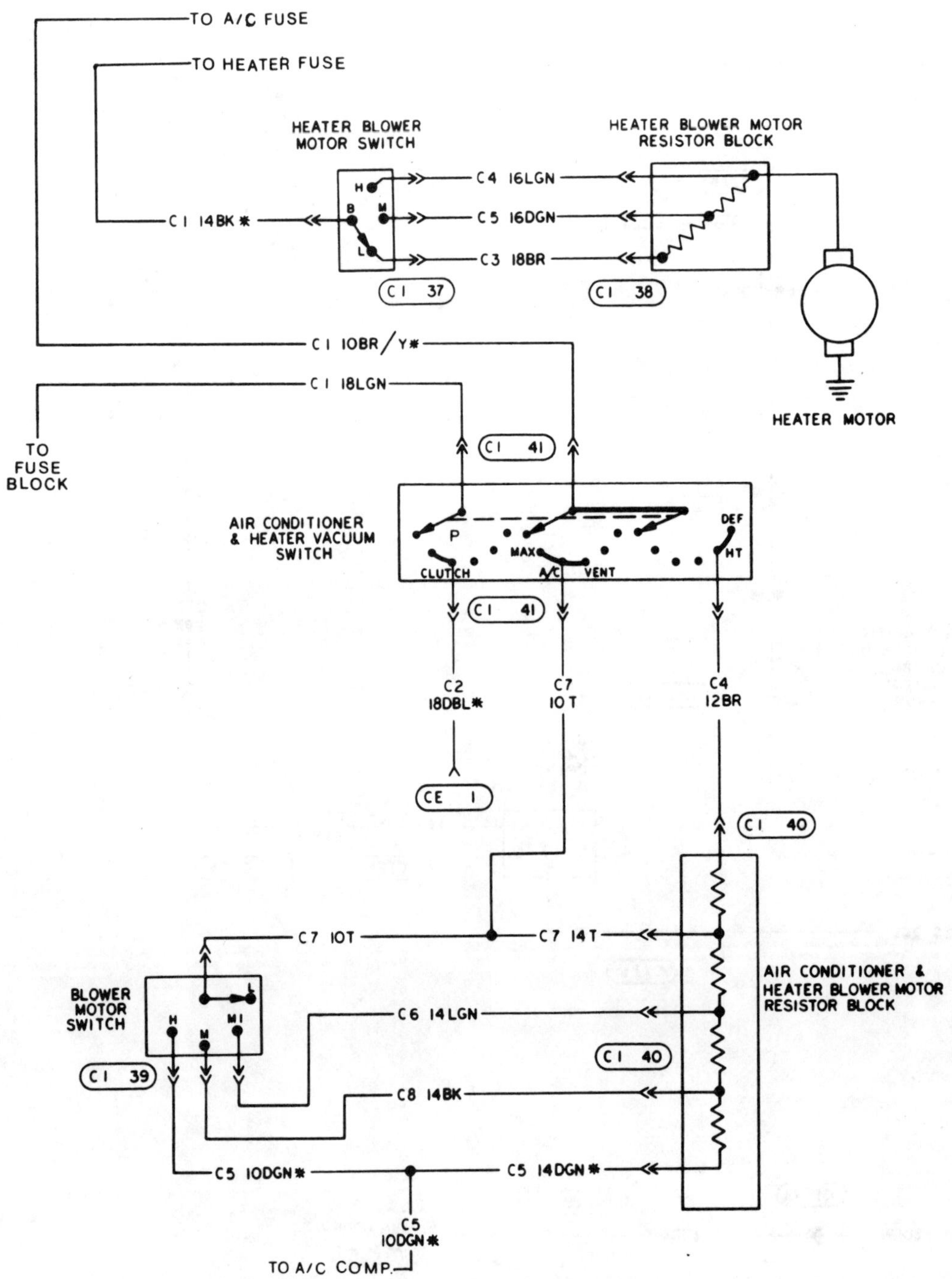

1975 Chrysler Cordoba, Dodge Charger & Coronet; Plymouth Fury A/C & Heater Wiring Circuit

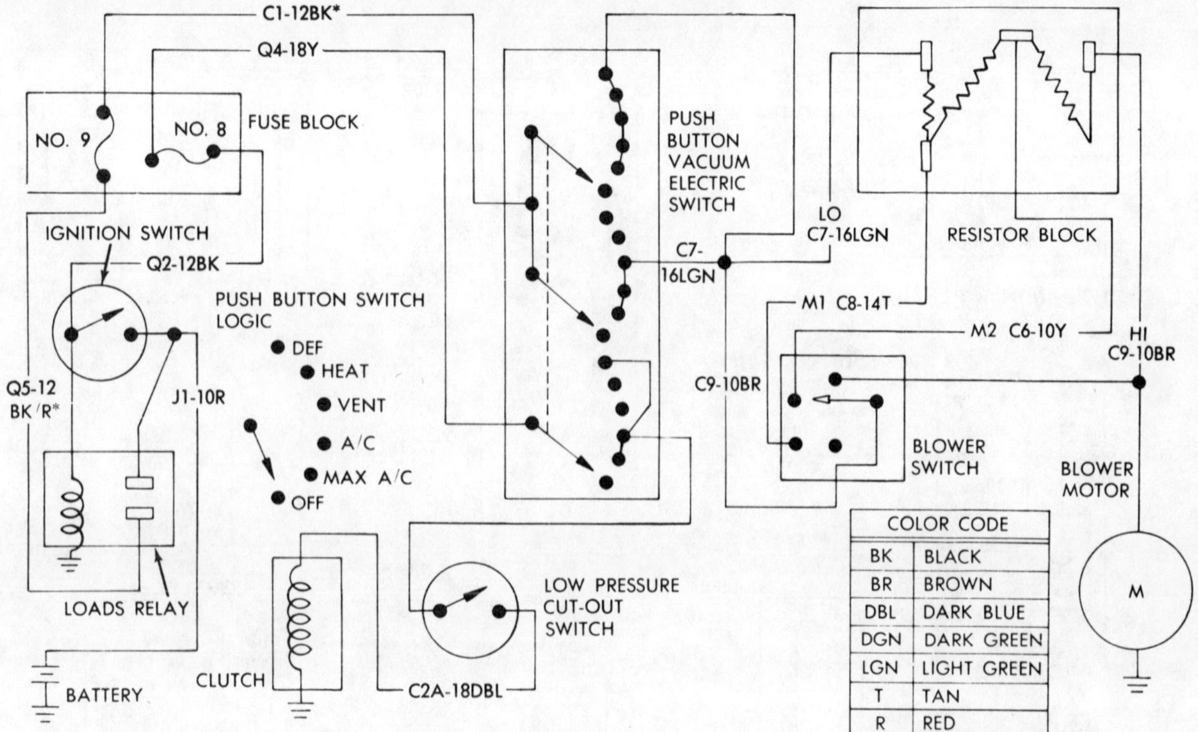

COLOR CODE	
BK	BLACK
BR	BROWN
DBL	DARK BLUE
DGN	DARK GREEN
LGN	LIGHT GREEN
T	TAN
R	RED
Y	YELLOW

1976 Chrysler Newport & New Yorker; Dodge Royal Monaco; Plymouth Gran Fury A/C & Heater Wiring Circuit (Less Auto Temp)

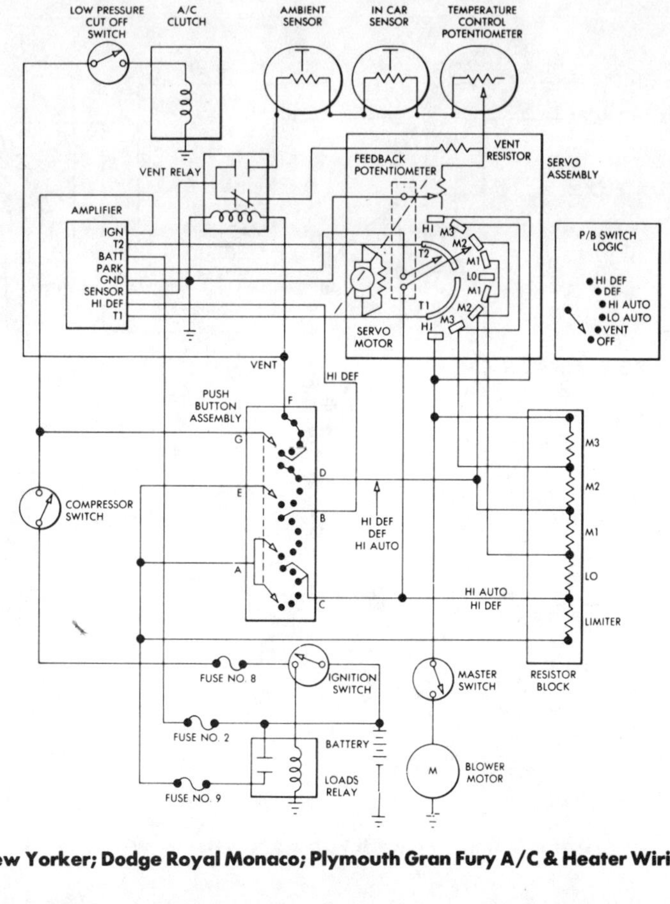

1976 Chrysler Newport & New Yorker; Dodge Royal Monaco; Plymouth Gran Fury A/C & Heater Wiring Circuit (With Auto Temp)

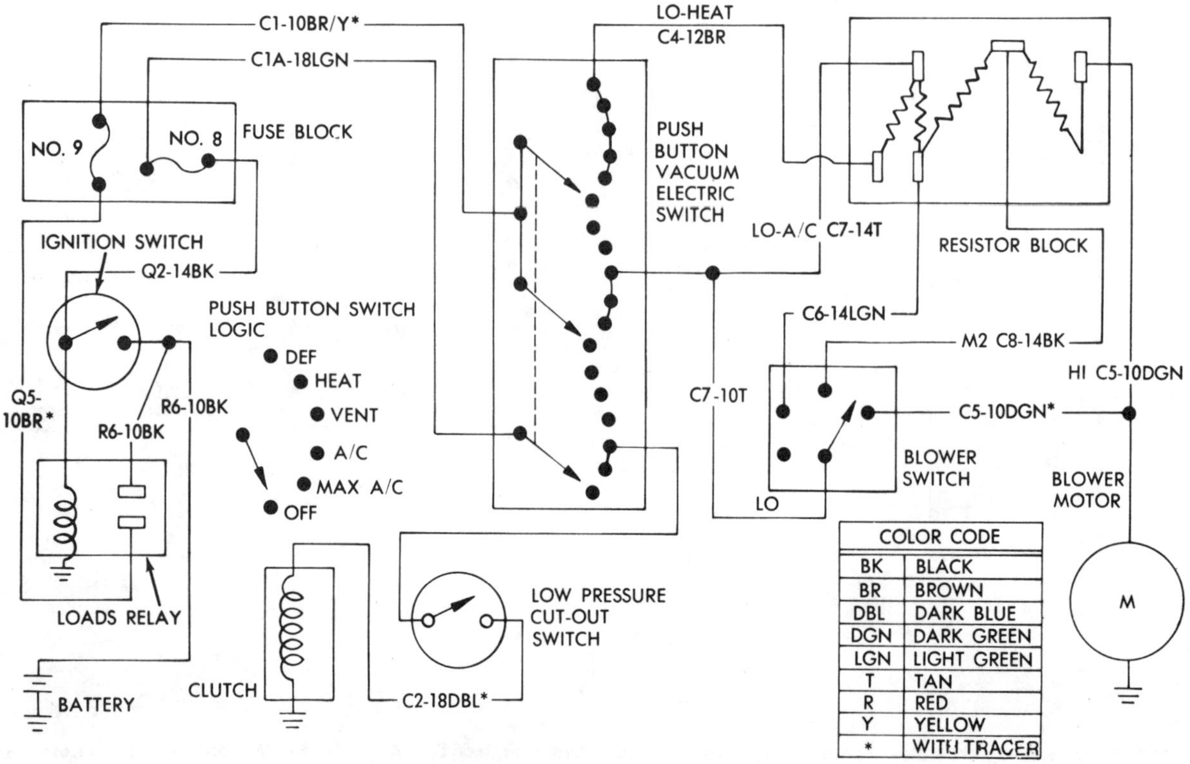

1976 Chrysler Cordoba; Dodge Charger & Coronet; Plymouth Fury A/C & Heater Wiring Circuit

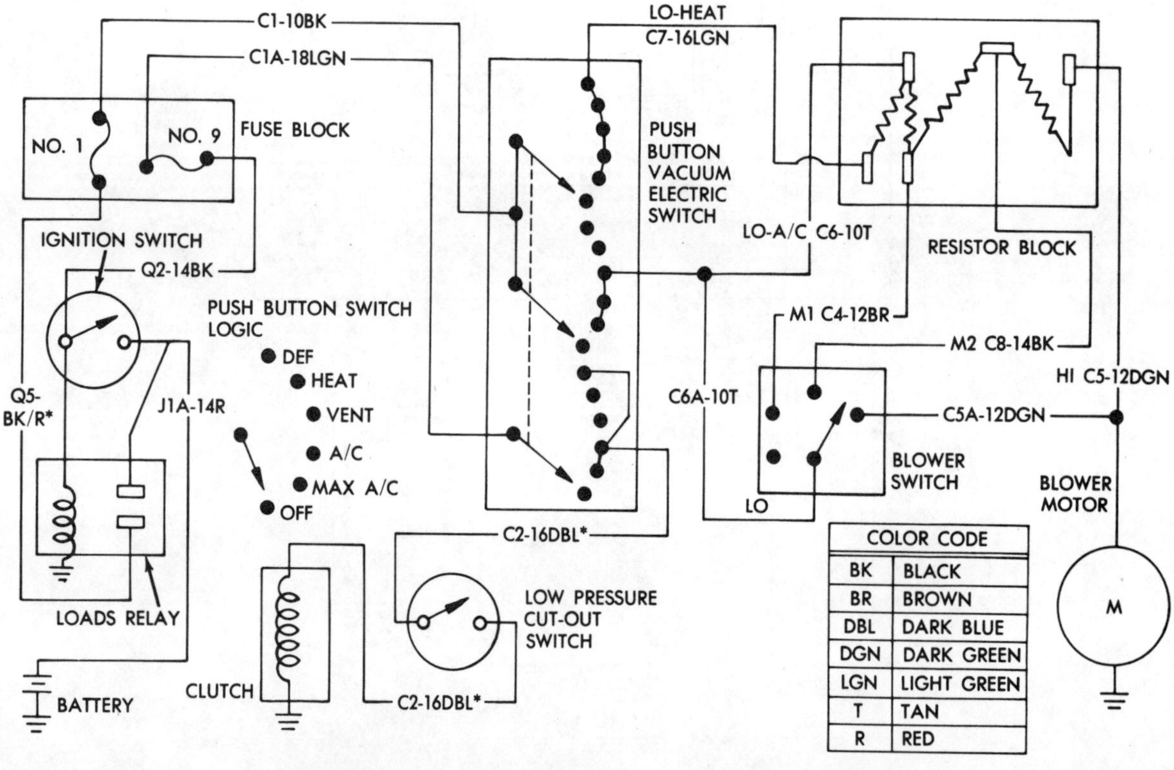

1976 Dodge Aspen & Plymouth Volare A/C & Heater Wiring Circuit

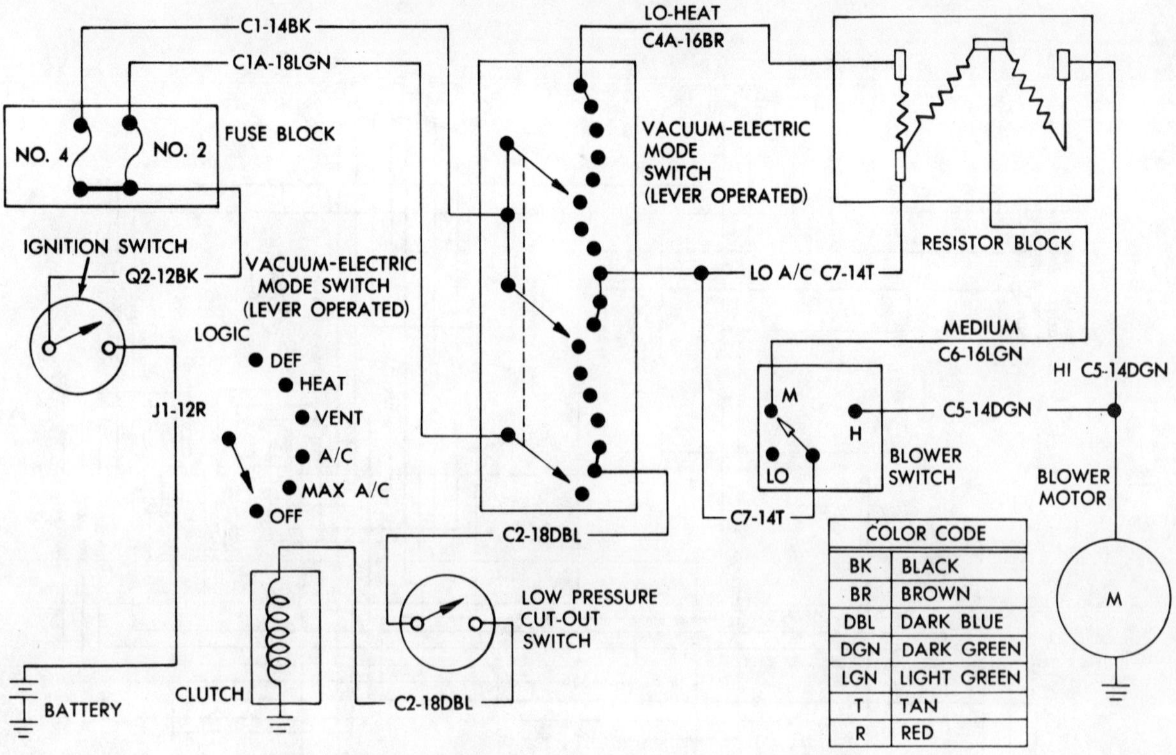

COLOR CODE	
BK	BLACK
BR	BROWN
DBL	DARK BLUE
DGN	DARK GREEN
LGN	LIGHT GREEN
T	TAN
R	RED

1976 Dodge Dart & Plymouth Valiant A/C & Heater Wiring Circuit

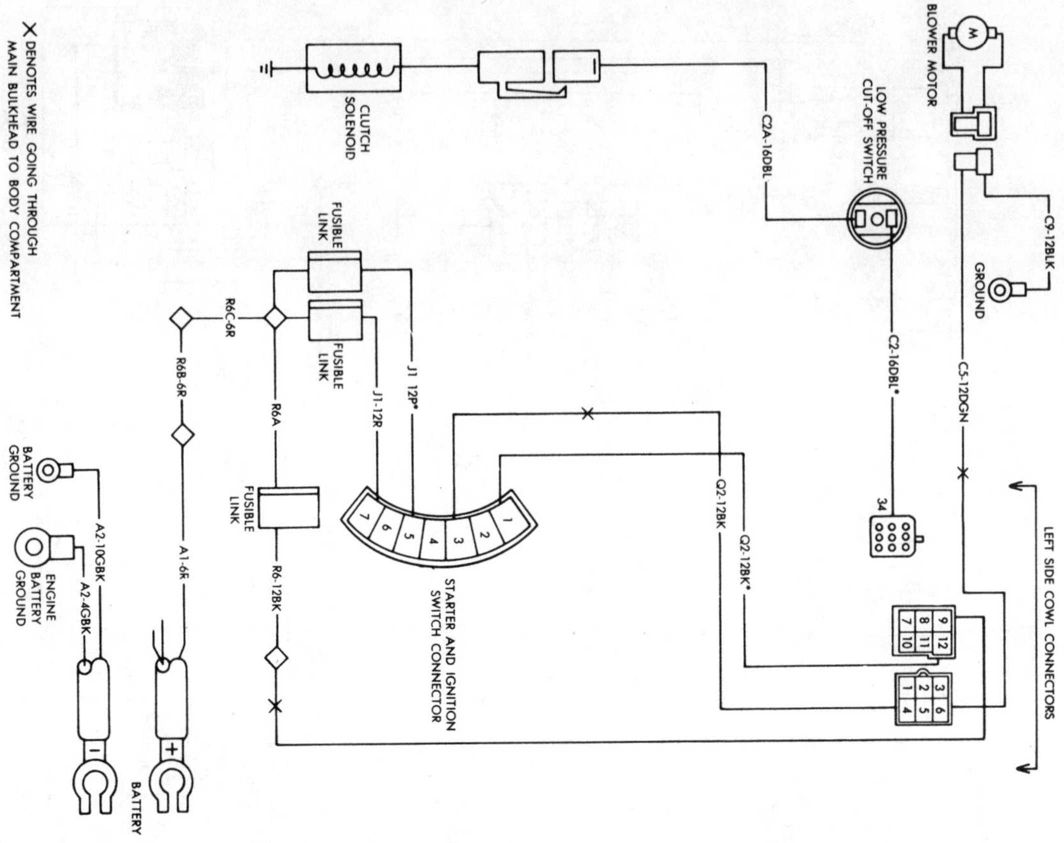

1977 Chrysler Newport & New Yorker; Dodge Royal Monaco; Plymouth Gran Fury A/C & Heater Wiring Circuit (With Auto Temp) Part 1 of 2

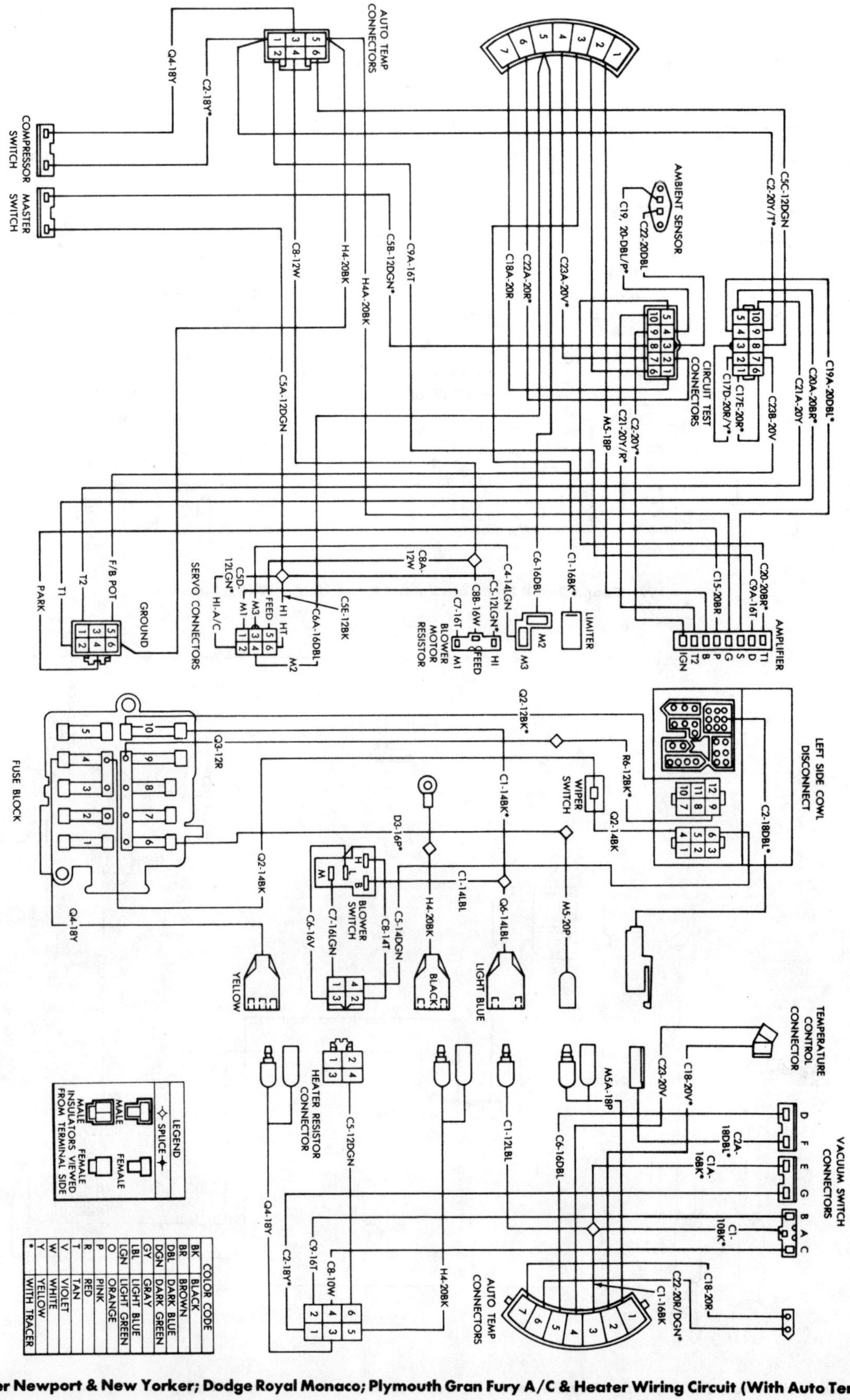

1977 Chrysler Newport & New Yorker; Dodge Royal Monaco; Plymouth Gran Fury A/C & Heater Wiring Circuit (With Auto Temp) Part 2 of 2

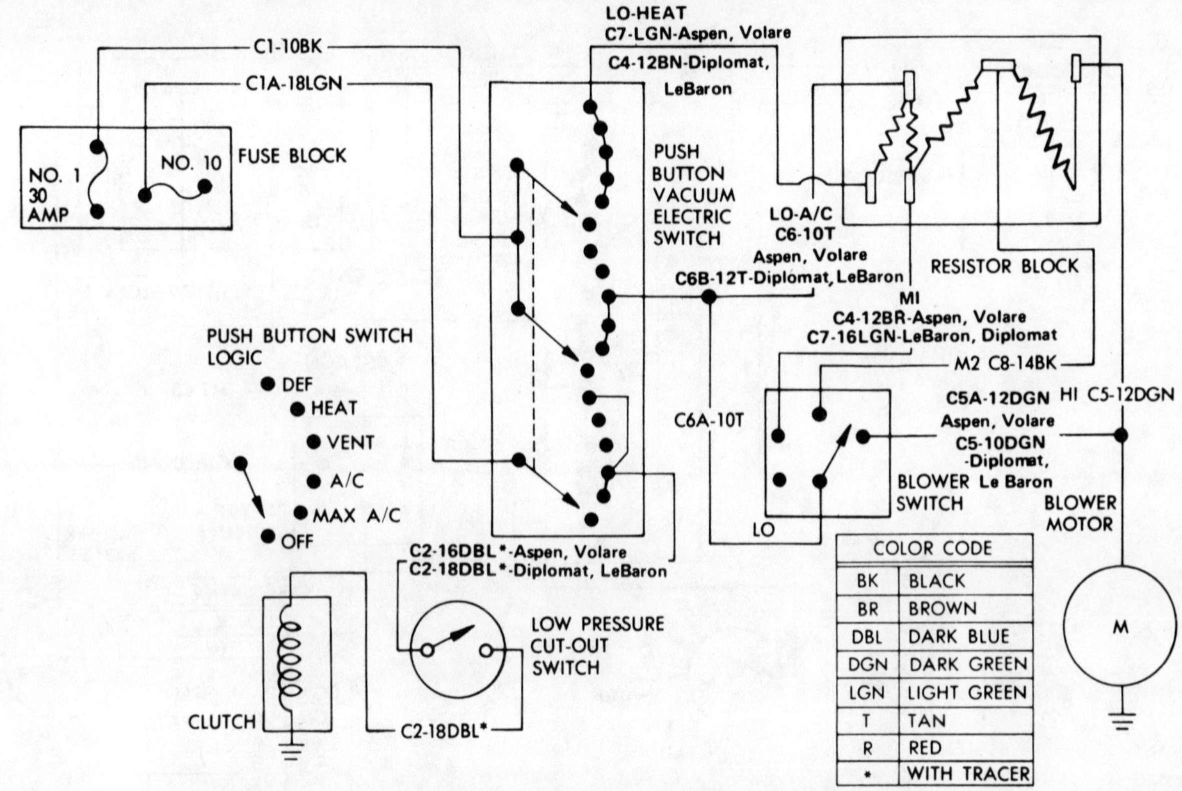

1977 Chrysler LeBaron; Dodge Aspen & Diplomat; Plymouth Volare A/C & Heater Wiring Circuit

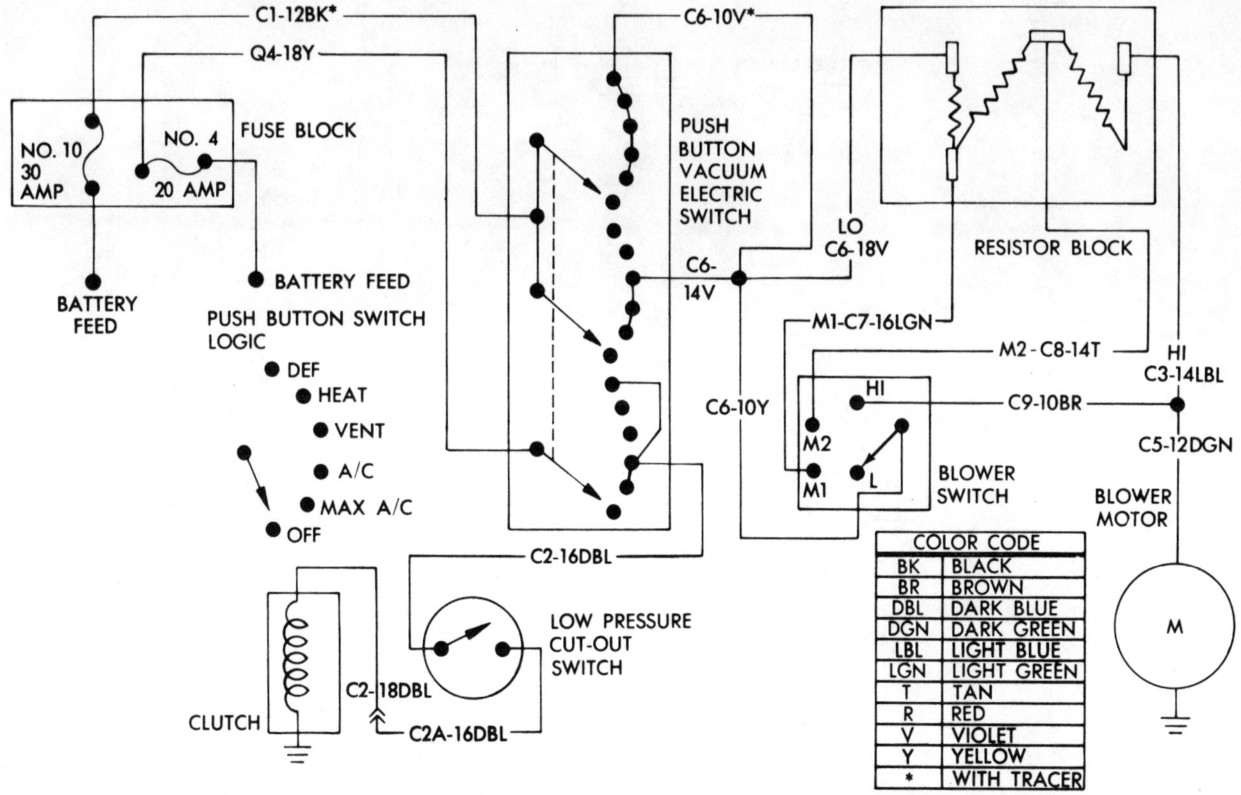

1977-78 Chrysler Newport & New Yorker; Dodge Royal Monaco; Plymouth Gran Fury A/C & Heater Wiring Circuit (Less Auto Temp)

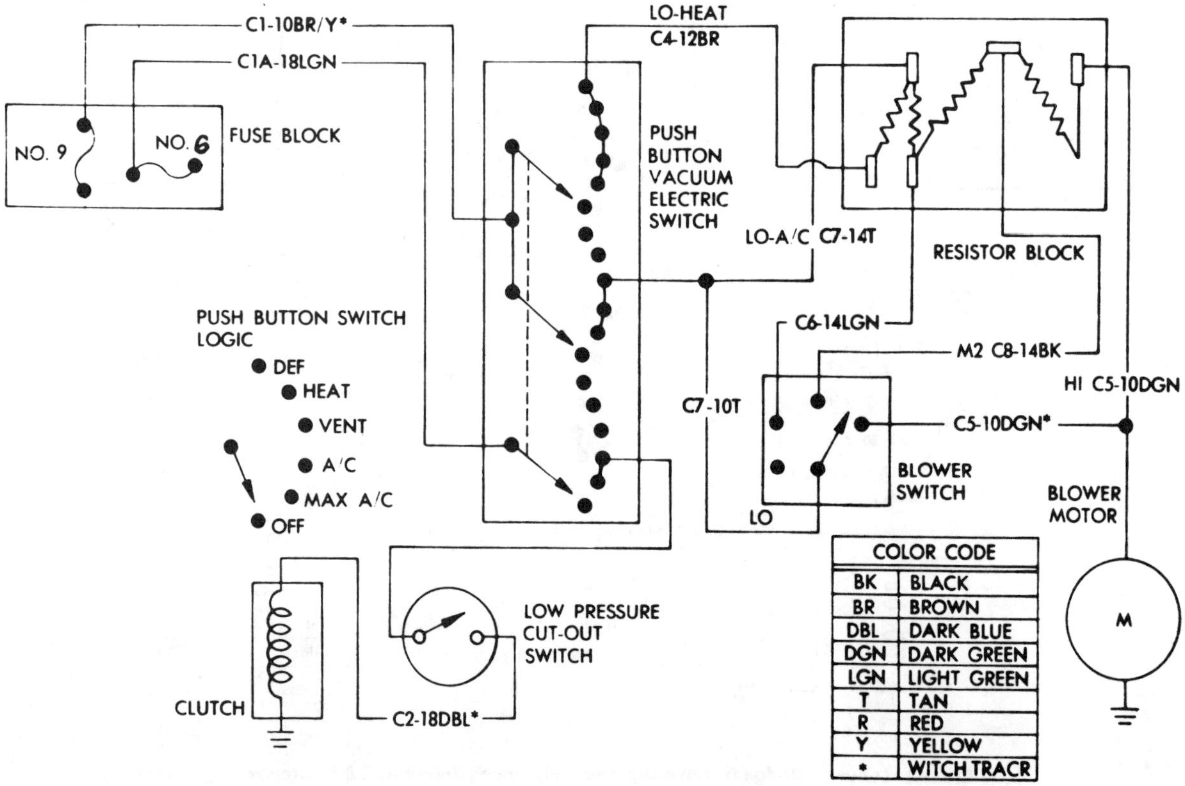

COLOR CODE	
BK	BLACK
BR	BROWN
DBL	DARK BLUE
DGN	DARK GREEN
LGN	LIGHT GREEN
T	TAN
R	RED
Y	YELLOW
*	WITCH TRACR

1977-78 Chrysler Cordoba; Dodge Charger & Monaco; Plymouth Fury A/C & Heater Wiring Circuit

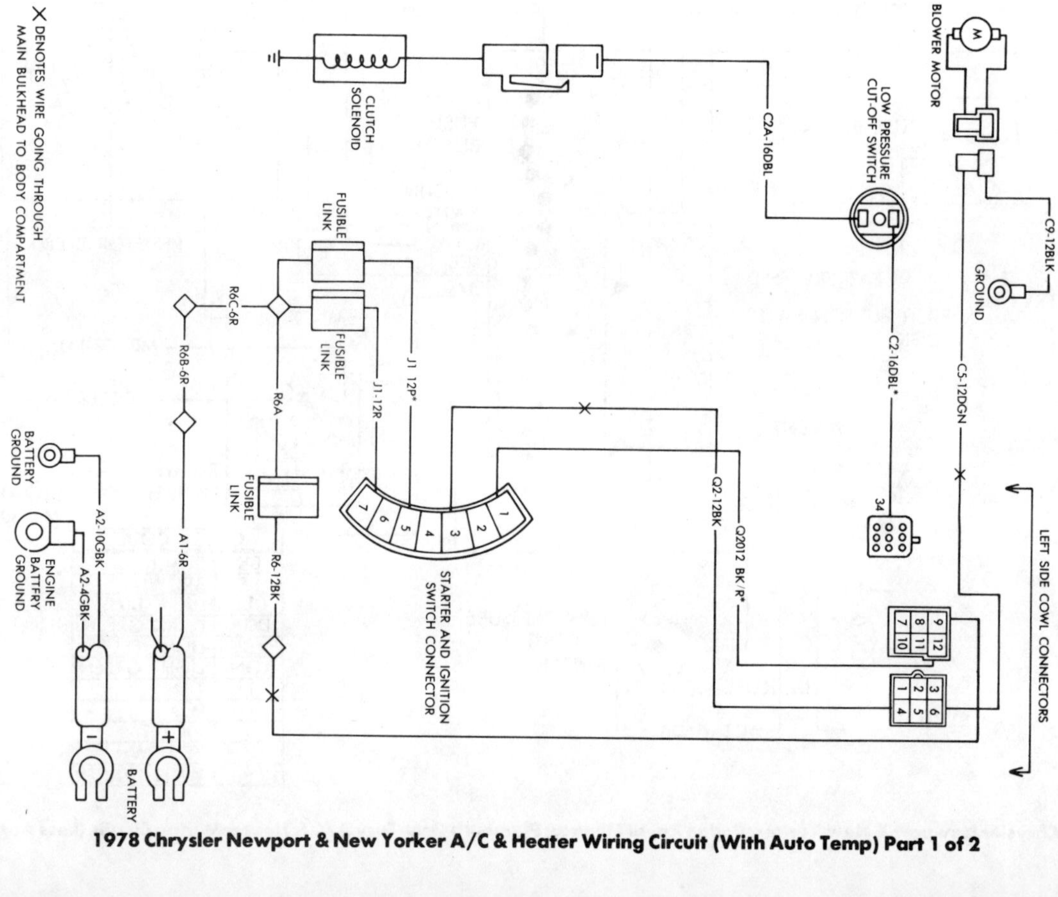

1978 Chrysler Newport & New Yorker A/C & Heater Wiring Circuit (With Auto Temp) Part 1 of 2

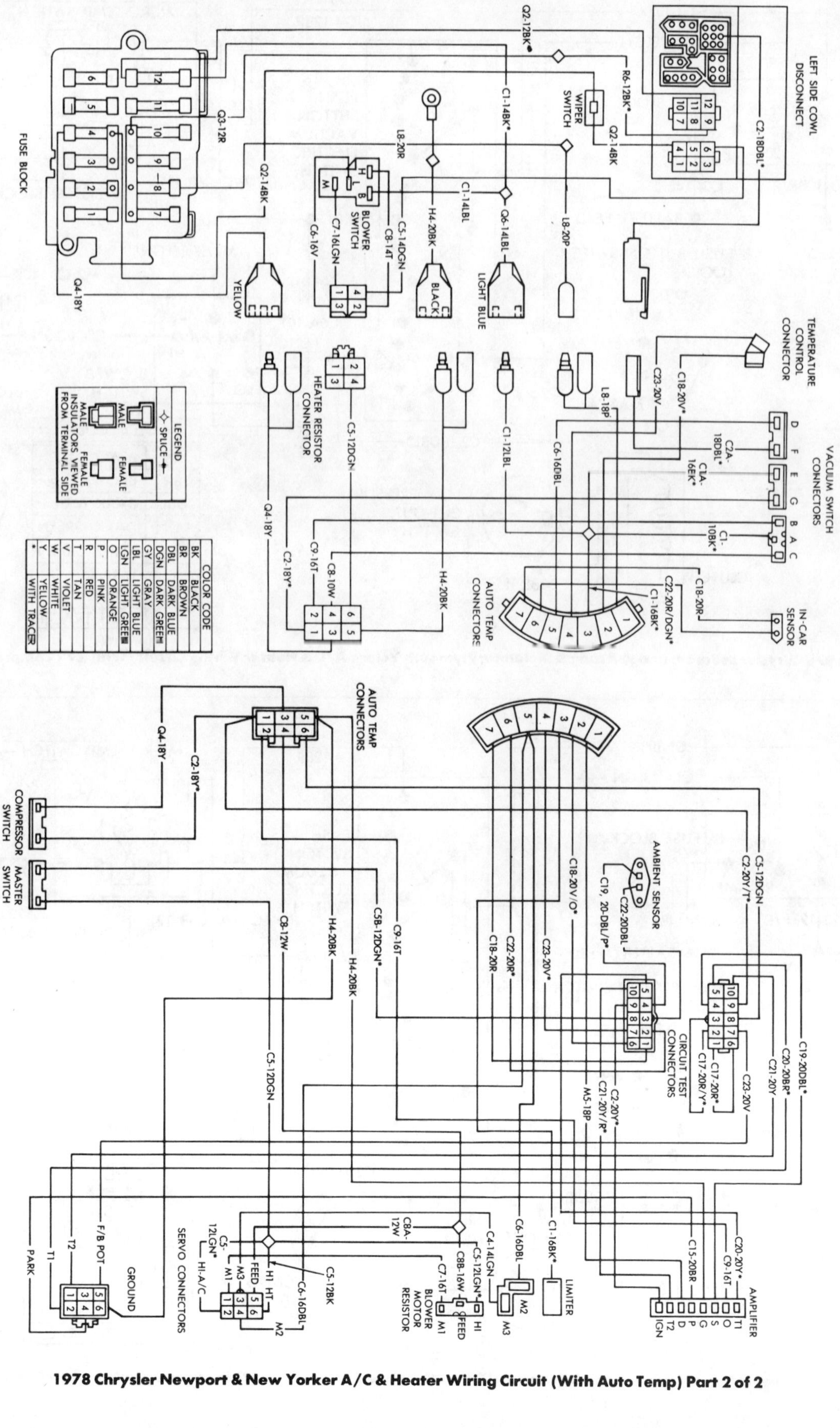

1978 Chrysler Newport & New Yorker A/C & Heater Wiring Circuit (With Auto Temp) Part 2 of 2

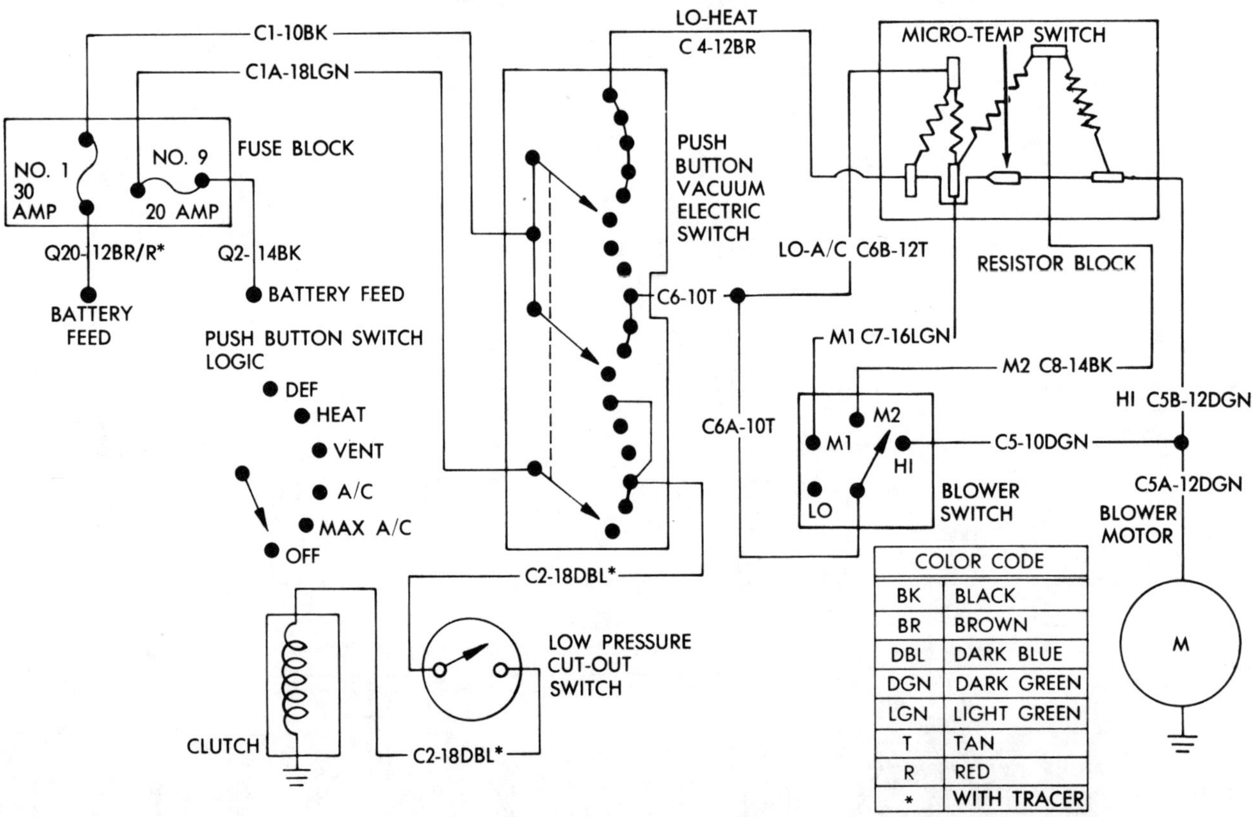

1978 Chrysler LeBaron; Dodge Aspen & Diplomat; Plymouth Volare A/C & Heater Wiring Circuit (With RV2 Compressor)

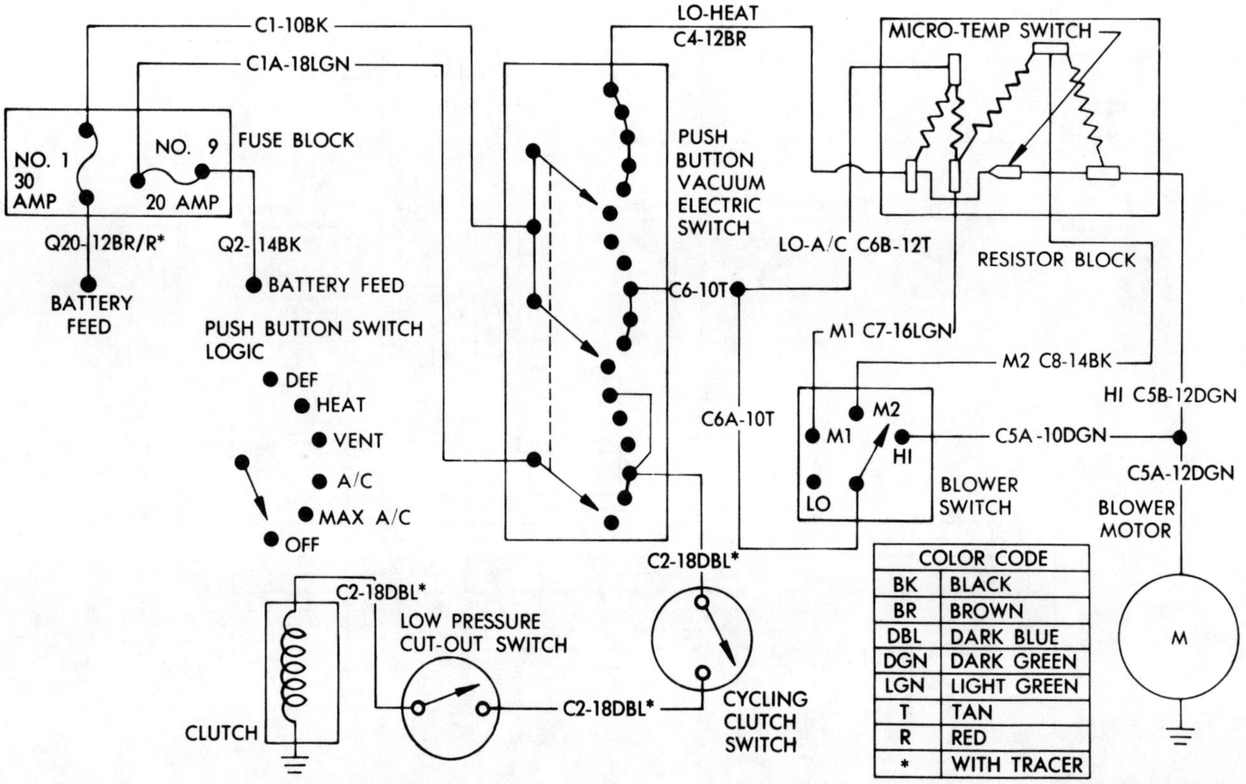

1978 Chrysler LeBaron; Dodge Aspen & Diplomat; Plymouth Volare A/C & Heater Wiring Circuit (With Sankyo Compressor)

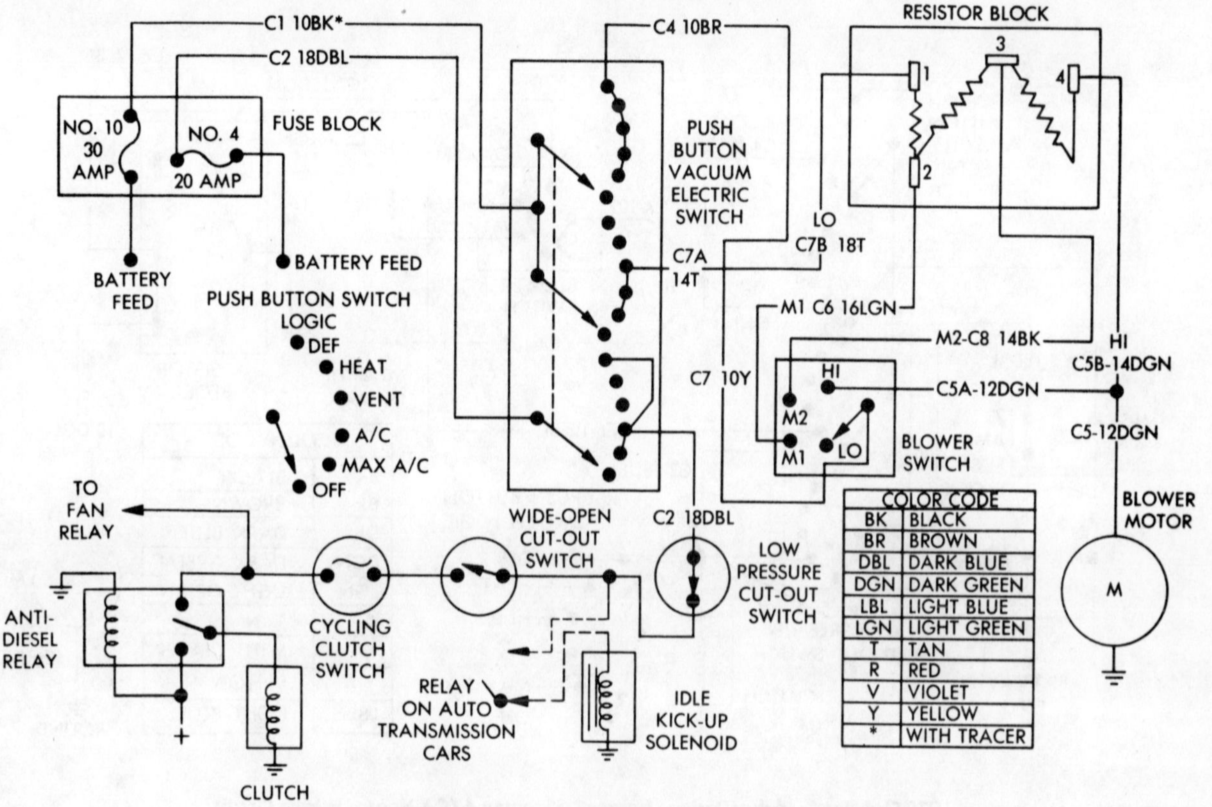

COLOR CODE	
BK	BLACK
BR	BROWN
DBL	DARK BLUE
DGN	DARK GREEN
LBL	LIGHT BLUE
LGN	LIGHT GREEN
T	TAN
R	RED
V	VIOLET
Y	YELLOW
*	WITH TRACER

1978 Dodge Omni & Plymouth Horizon A/C & Heater Wiring Circuit

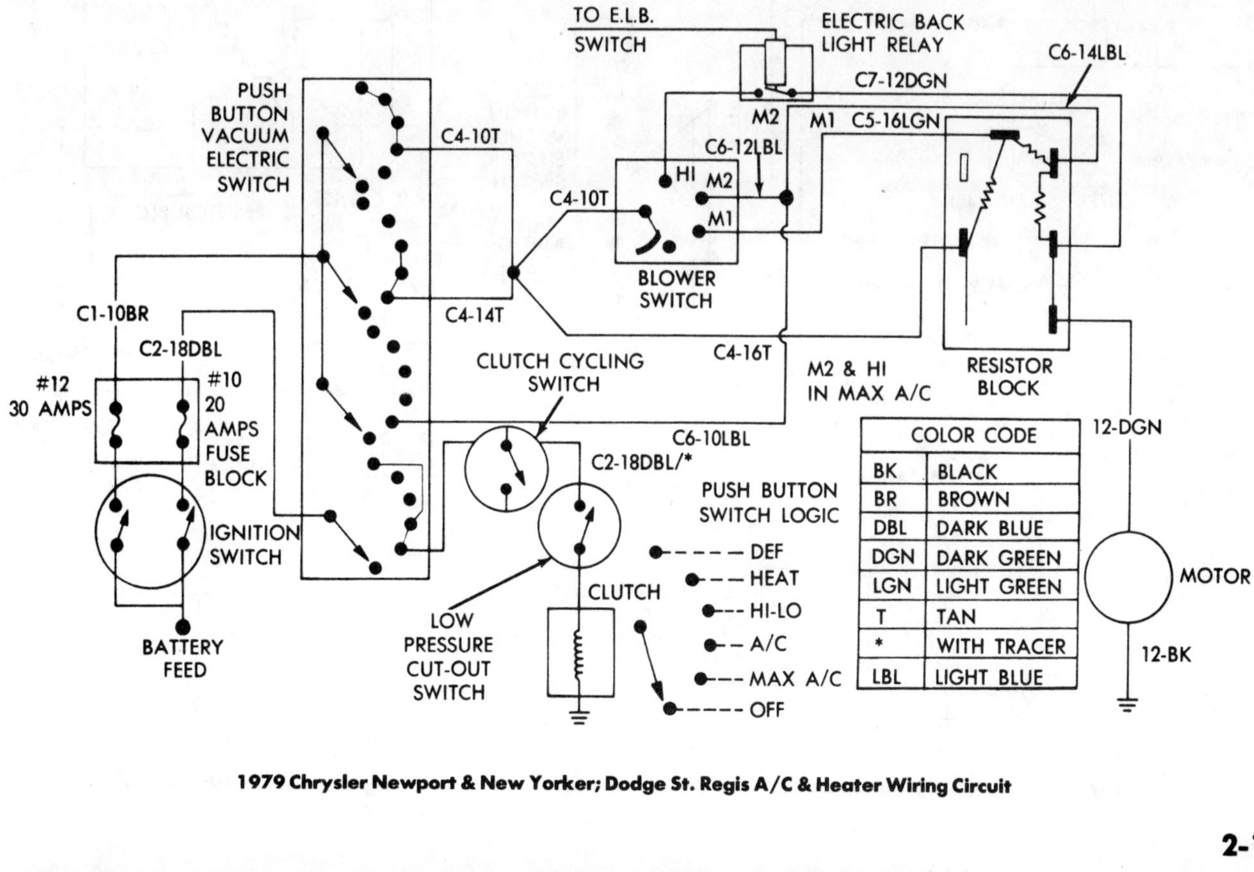

COLOR CODE	
BK	BLACK
BR	BROWN
DBL	DARK BLUE
DGN	DARK GREEN
LGN	LIGHT GREEN
T	TAN
*	WITH TRACER
LBL	LIGHT BLUE

1979 Chrysler Newport & New Yorker; Dodge St. Regis A/C & Heater Wiring Circuit

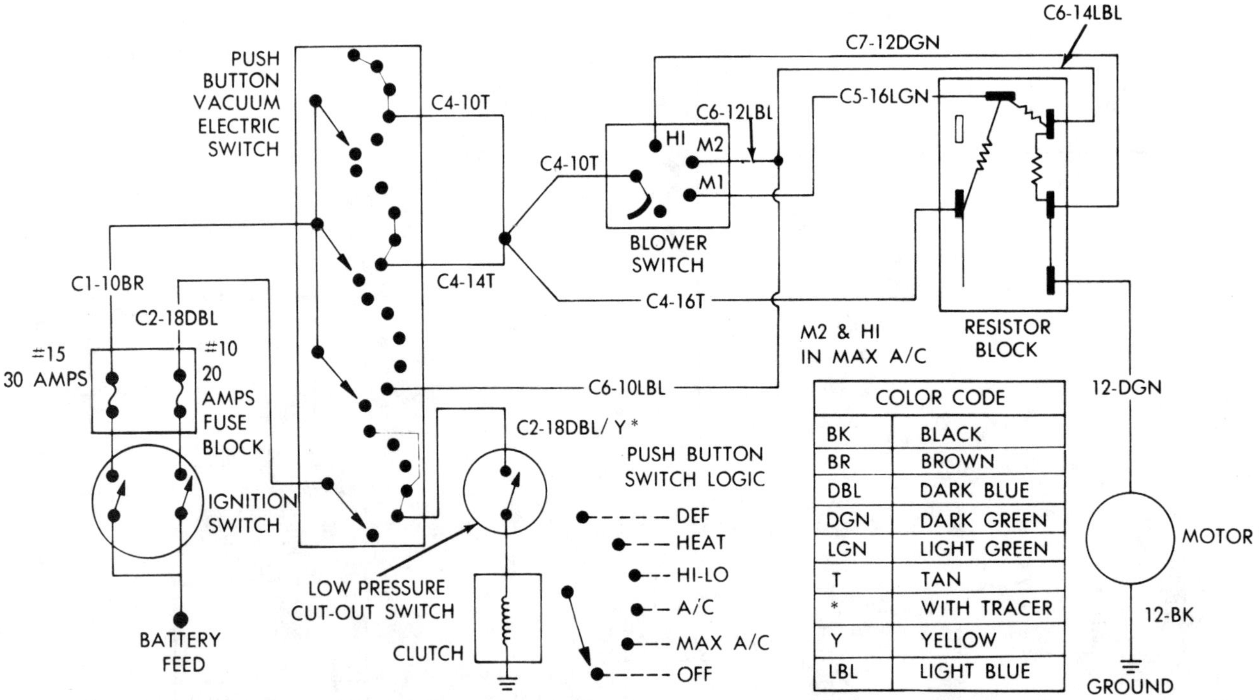

COLOR CODE	
BK	BLACK
BR	BROWN
DBL	DARK BLUE
DGN	DARK GREEN
LGN	LIGHT GREEN
T	TAN
*	WITH TRACER
Y	YELLOW
LBL	LIGHT BLUE

1979 Chrysler Cordoba; Dodge Charger & Magnum A/C & Heater Wiring Circuit

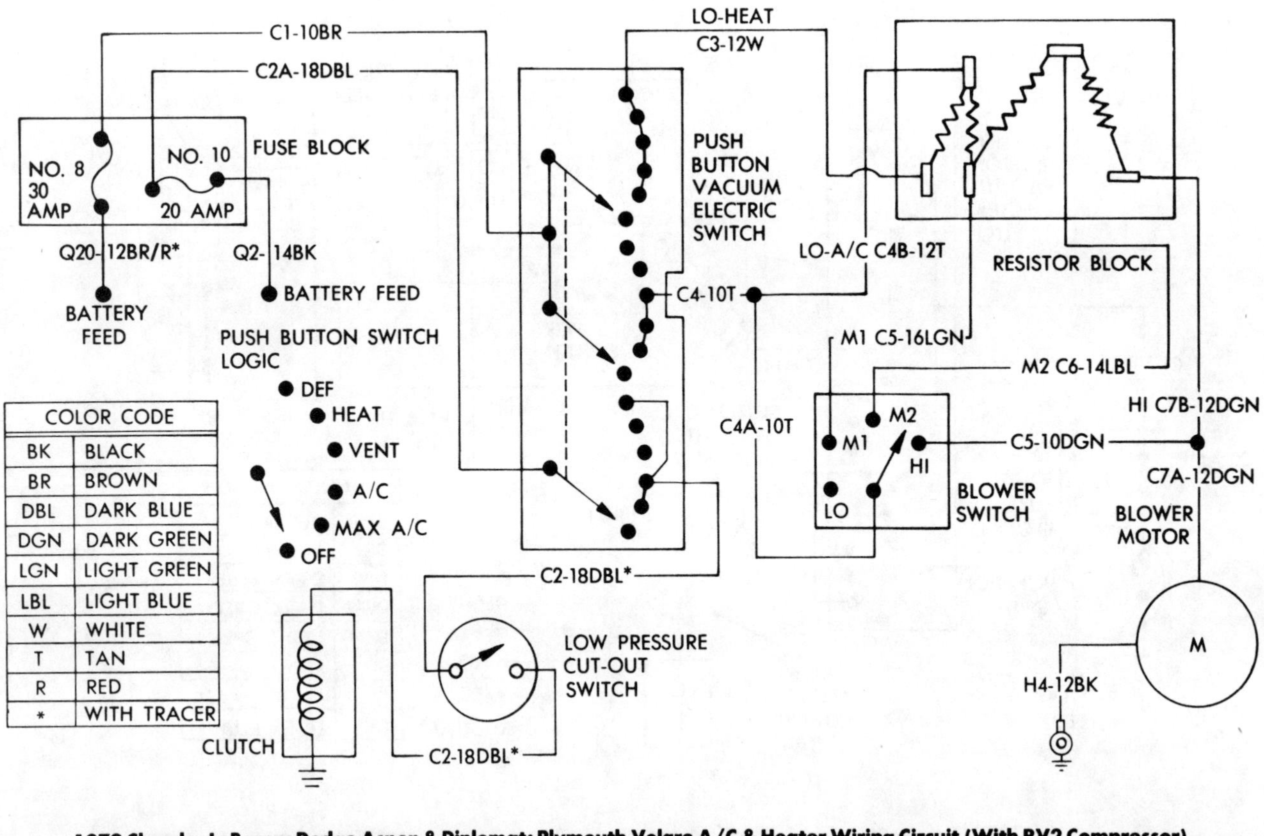

COLOR CODE	
BK	BLACK
BR	BROWN
DBL	DARK BLUE
DGN	DARK GREEN
LGN	LIGHT GREEN
LBL	LIGHT BLUE
W	WHITE
T	TAN
R	RED
*	WITH TRACER

1979 Chrysler LeBaron; Dodge Aspen & Diplomat; Plymouth Volare A/C & Heater Wiring Circuit (With RV2 Compressor)

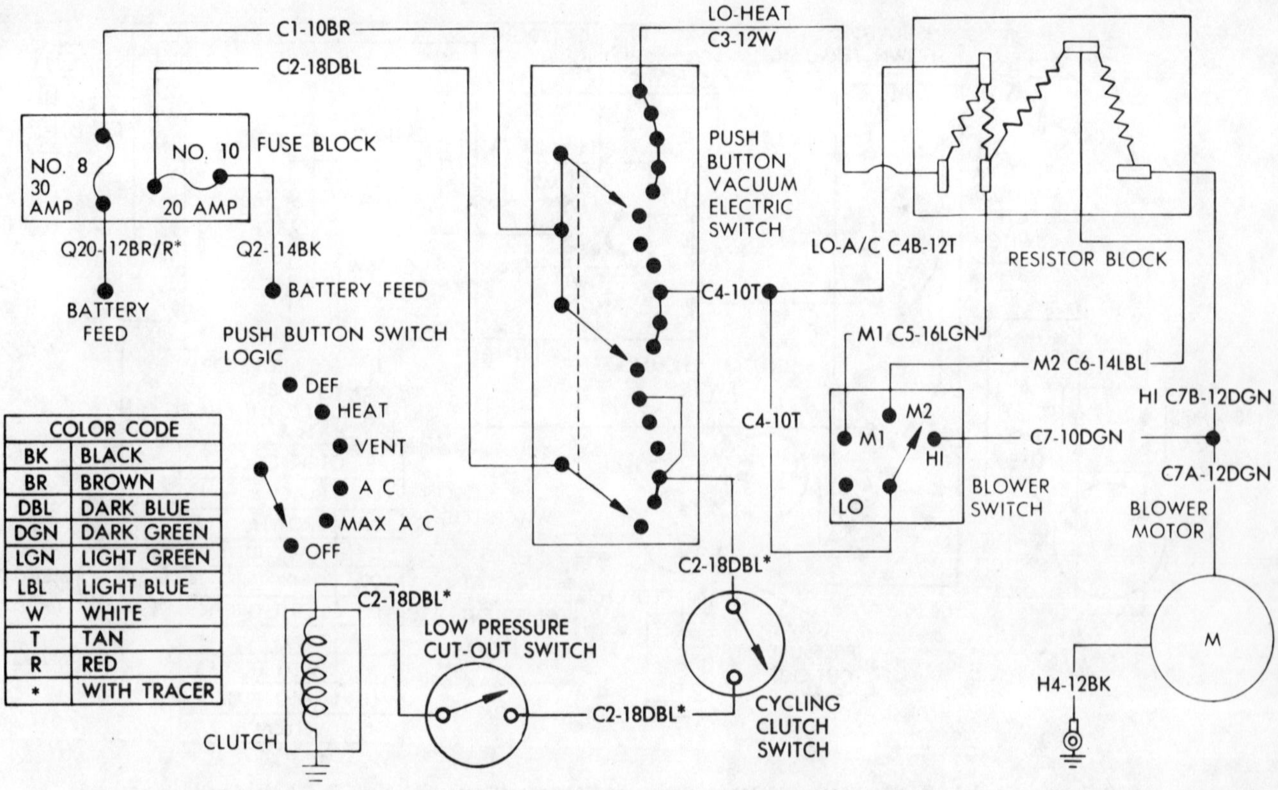

COLOR CODE	
BK	BLACK
BR	BROWN
DBL	DARK BLUE
DGN	DARK GREEN
LGN	LIGHT GREEN
LBL	LIGHT BLUE
W	WHITE
T	TAN
R	RED
*	WITH TRACER

1979 Chrysler LeBaron; Dodge Aspen & Diplomat; Plymouth Volare A/C & Heater Wiring Circuit (With C-171 or Sankyo Compressor)

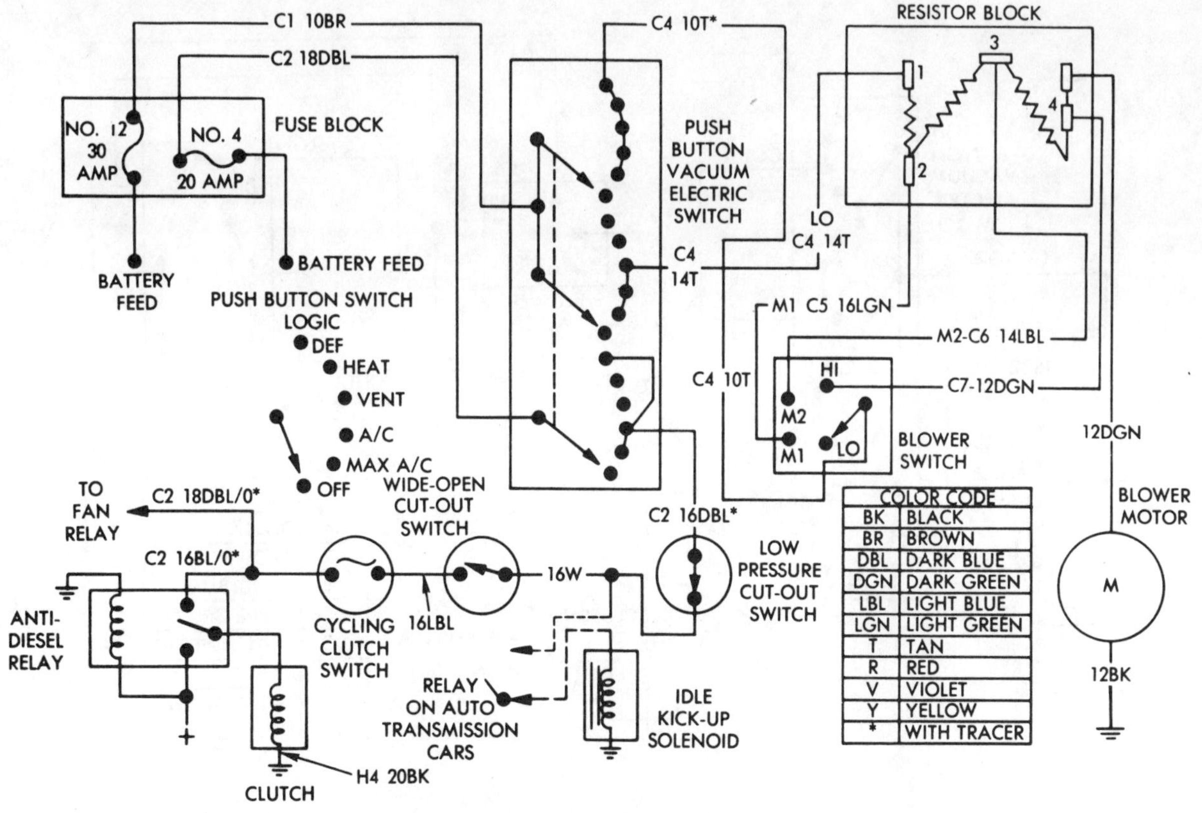

COLOR CODE	
BK	BLACK
BR	BROWN
DBL	DARK BLUE
DGN	DARK GREEN
LBL	LIGHT BLUE
LGN	LIGHT GREEN
T	TAN
R	RED
V	VIOLET
Y	YELLOW
*	WITH TRACER

1979 Dodge Omni & Plymouth Horizon A/C & Heater Wiring Circuit

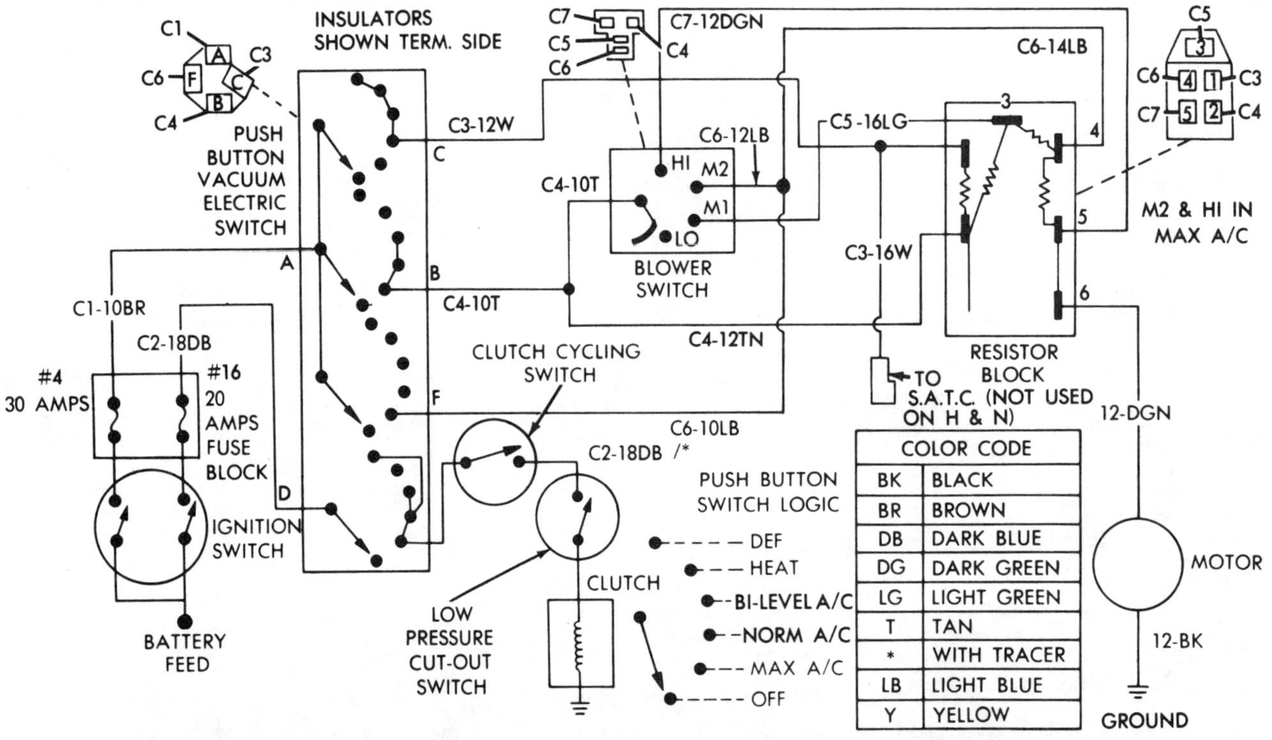

1980 Chrysler Cordoba & LeBaron; Dodge Aspen, Diplomat & Mirada; Plymouth Volare A/C & Heater Wiring Circuit

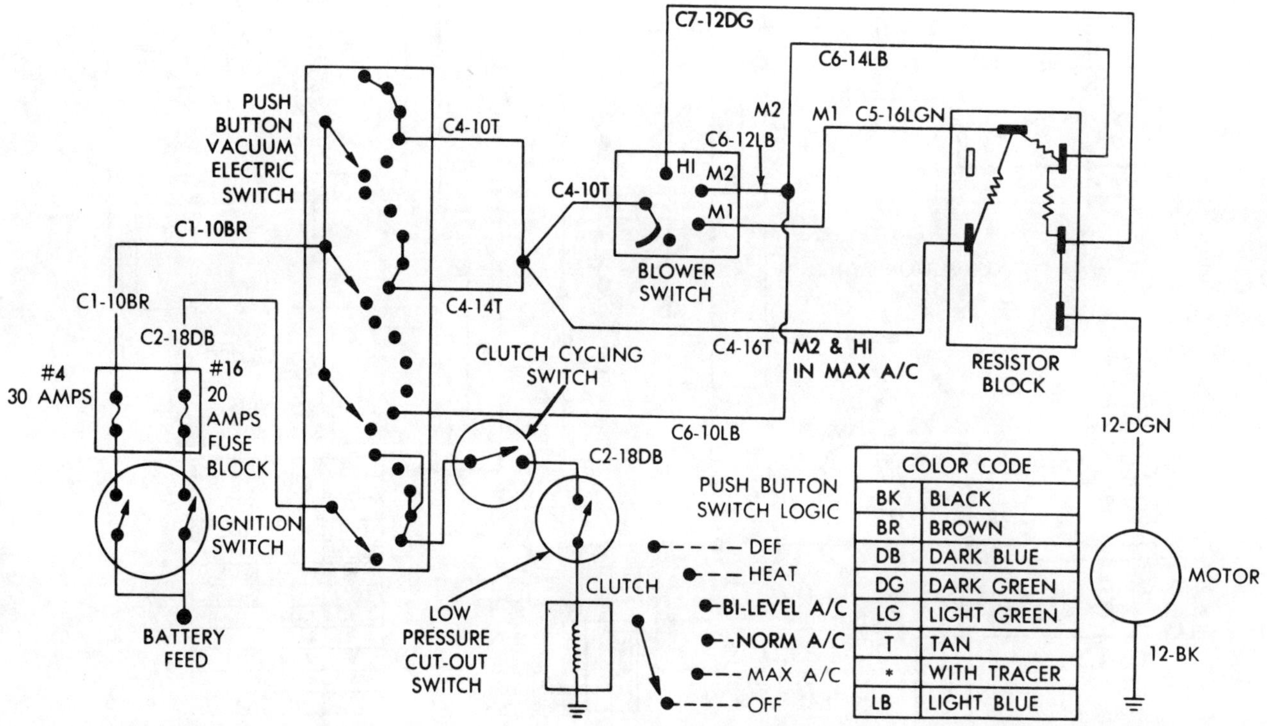

1980 Chrysler Newport & New Yorker; Dodge St. Regis; Plymouth Gran Fury A/C & Heater Wiring Circuit

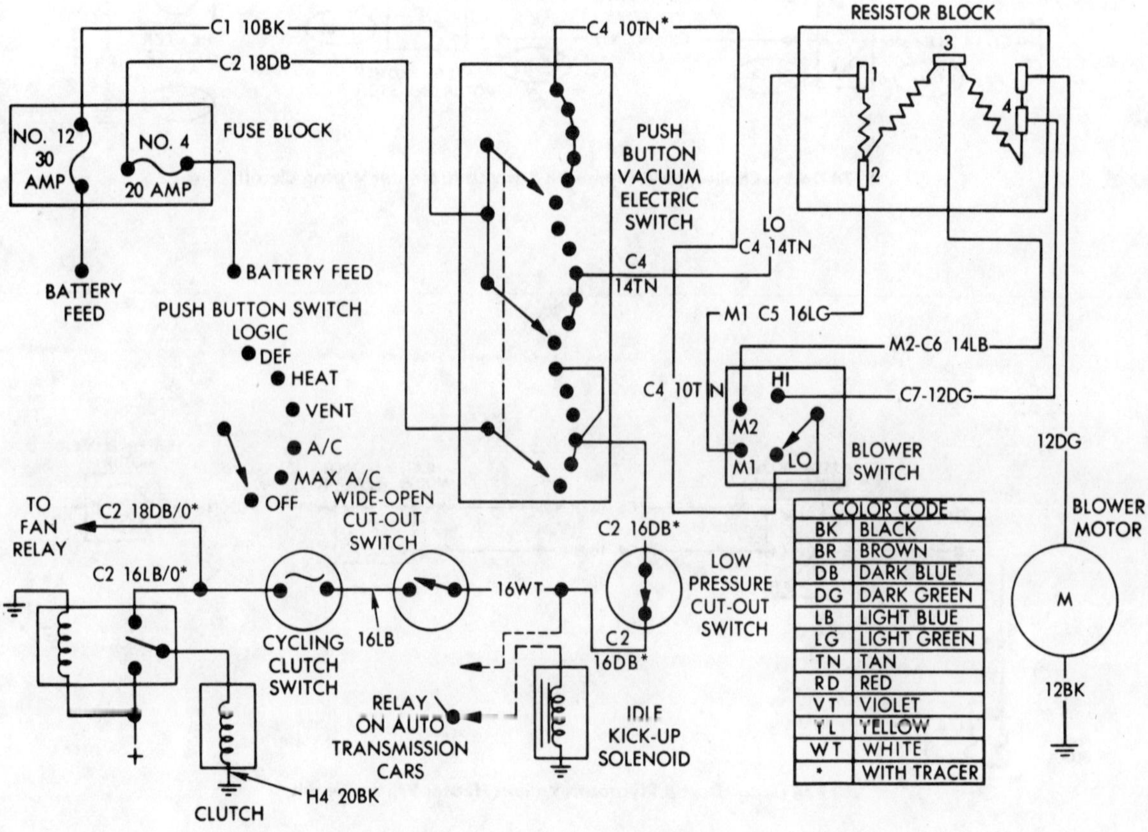

1980 Dodge Omni & Plymouth Horizon A/C & Heater Wiring Circuit

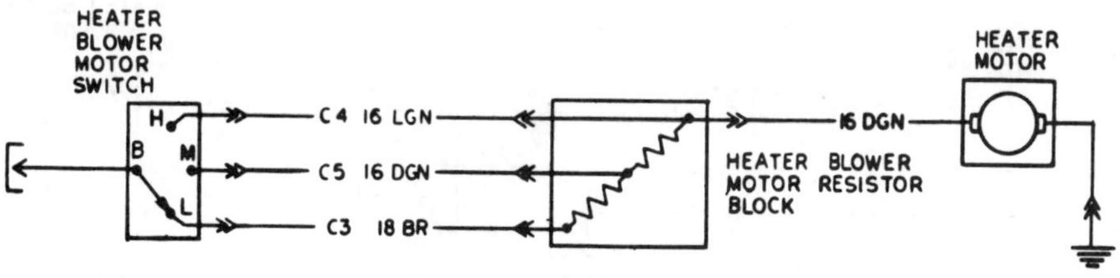

1974 Dodge Charger & Coronet; Plymouth Satellite Heater Wiring Circuit

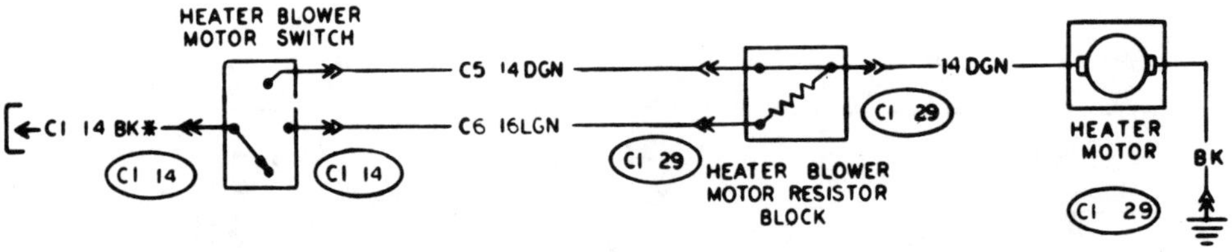

1974 Dodge Challenger & Plymouth Barracuda Heater Wiring Circuit

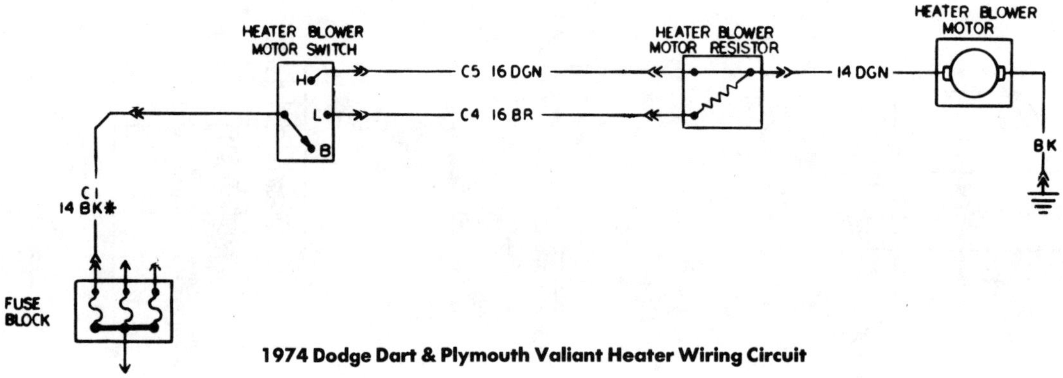

1974 Dodge Dart & Plymouth Valiant Heater Wiring Circuit

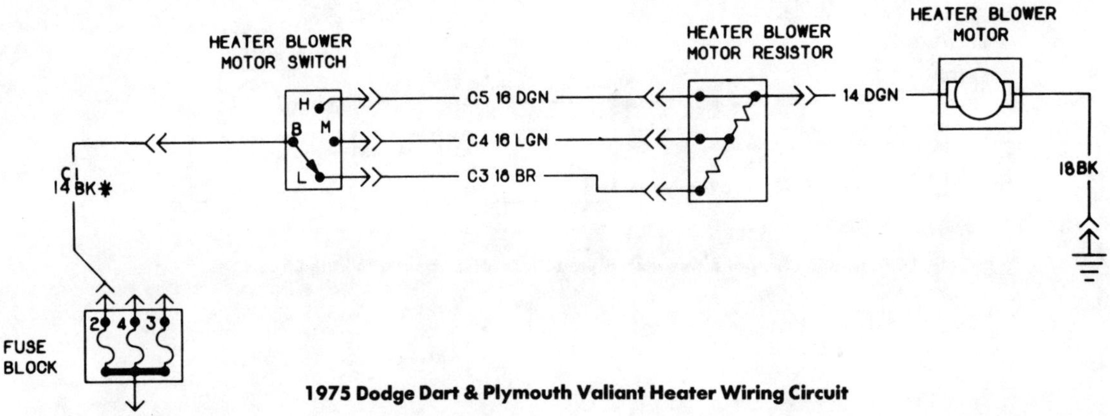

1975 Dodge Dart & Plymouth Valiant Heater Wiring Circuit

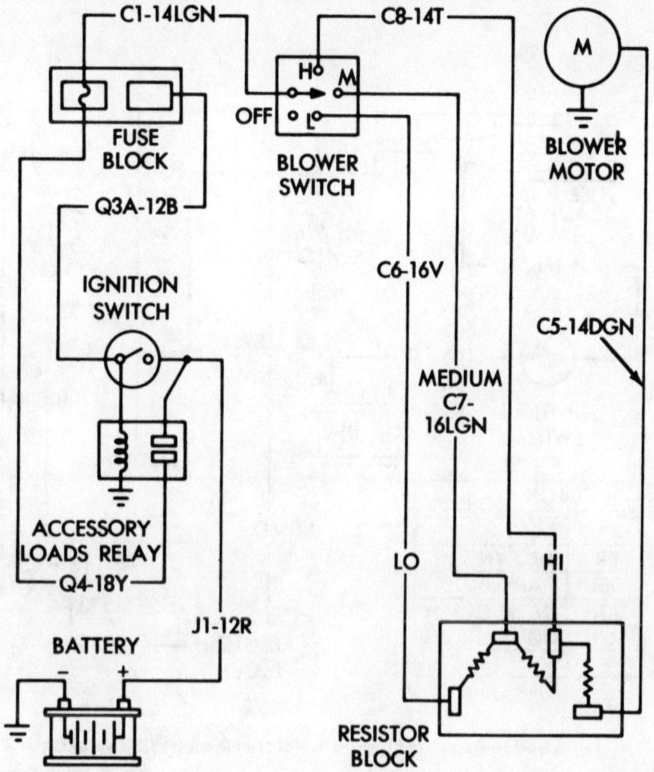

1976 Chrysler Newport & New Yorker; Dodge Monaco; Plymouth Gran Fury Heater Wiring Circuit

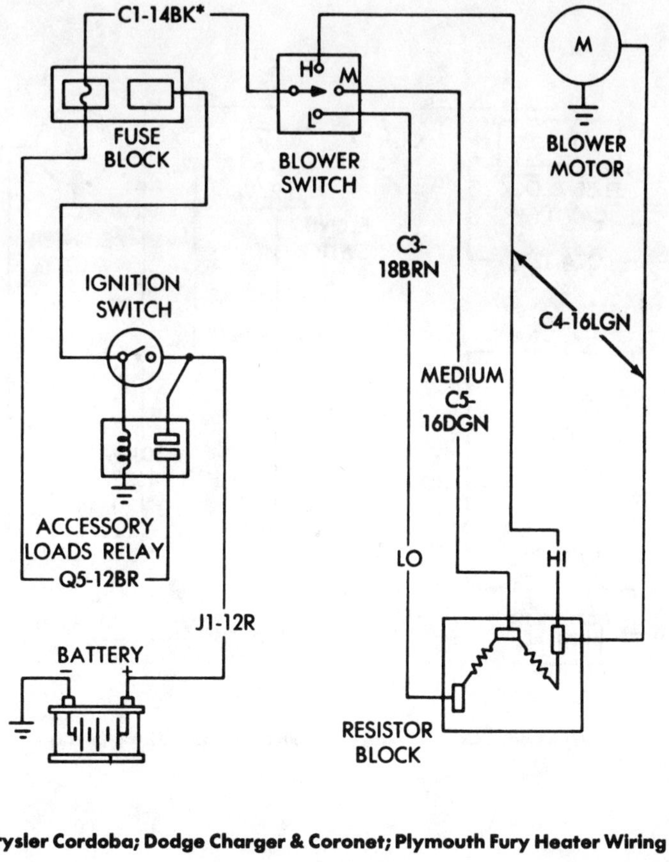

1976 Chrysler Cordoba; Dodge Charger & Coronet; Plymouth Fury Heater Wiring Circuit

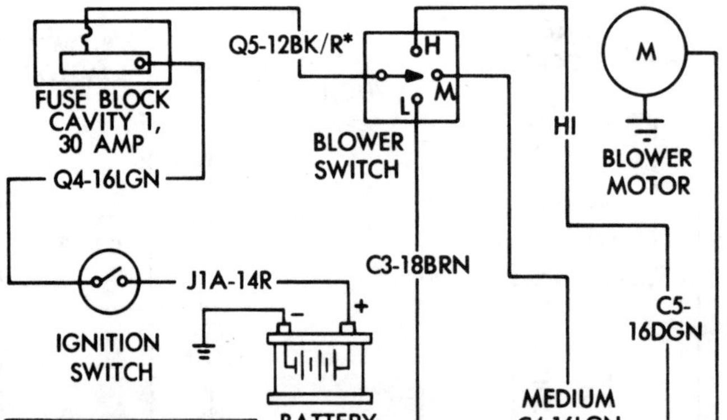

COLOR CODE	
BK	BLACK
BR	BROWN
DBL	DARK BLUE
DGN	DARK GREEN
LGN	LIGHT GREEN
*	WITH TRACER

1976 Dodge Aspen & Plymouth Volare Heater Wiring Circuit

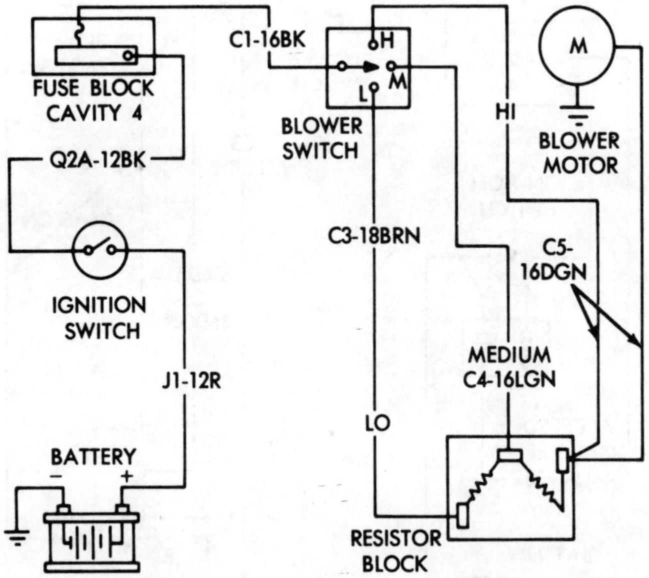

1976 Dodge Dart & Plymouth Valiant Heater Wiring Circuit

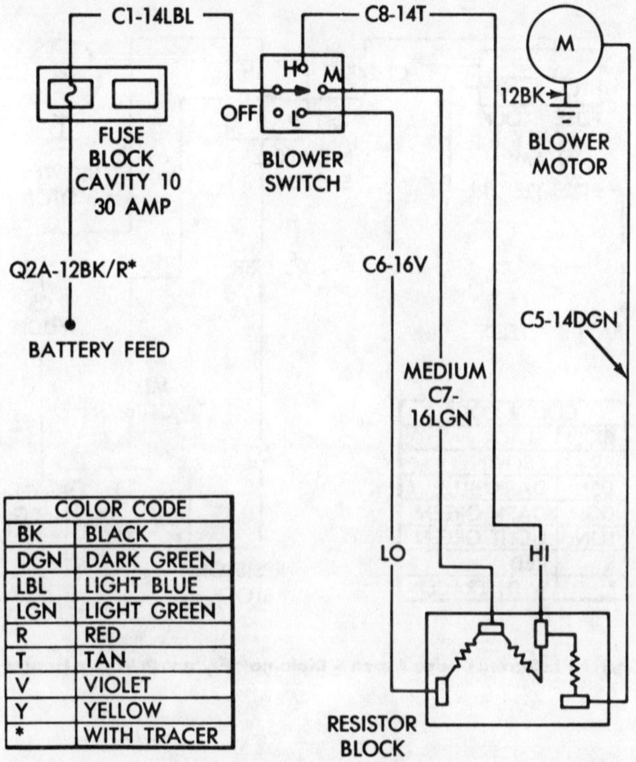

COLOR CODE	
BK	BLACK
DGN	DARK GREEN
LBL	LIGHT BLUE
LGN	LIGHT GREEN
R	RED
T	TAN
V	VIOLET
Y	YELLOW
*	WITH TRACER

1977-78 Chrysler Newport & New Yorker; Dodge Royal Monaco; Plymouth Gran Fury Heater Wiring Circuit

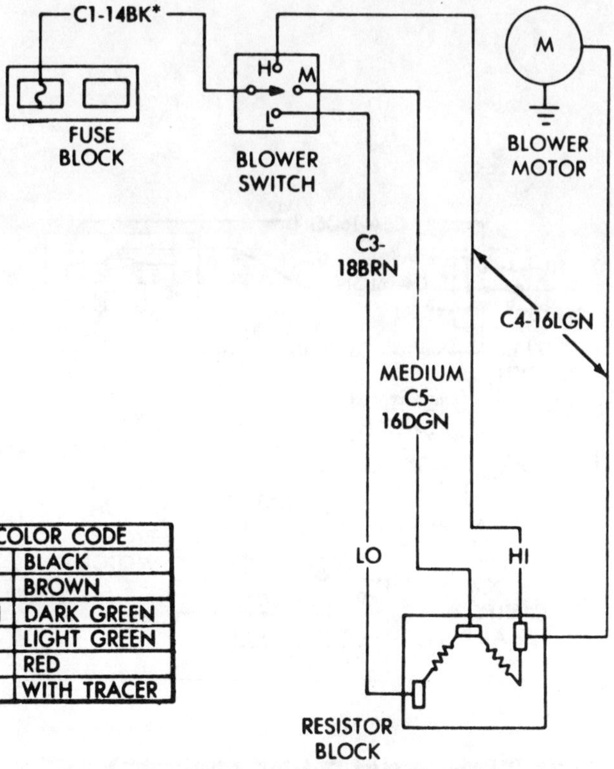

COLOR CODE	
BK	BLACK
BR	BROWN
DGN	DARK GREEN
LGN	LIGHT GREEN
R	RED
*	WITH TRACER

1977-78 Chrysler Cordoba; Dodge Charger & Monaco; Plymouth Fury Heater Wiring Circuit

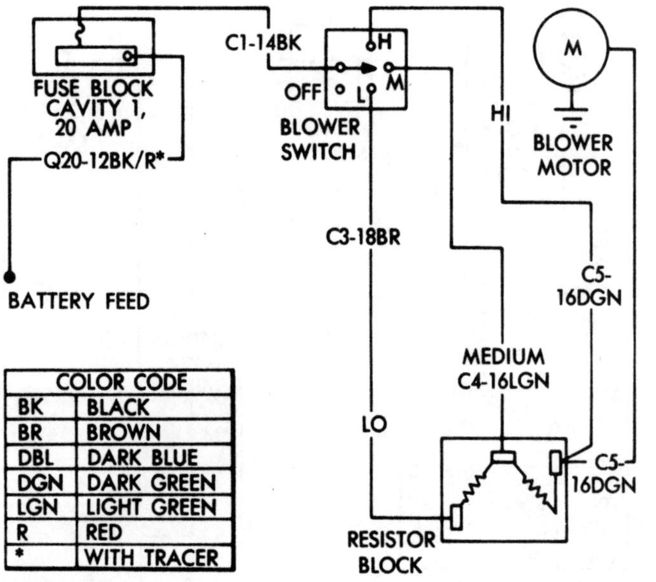

COLOR CODE	
BK	BLACK
BR	BROWN
DBL	DARK BLUE
DGN	DARK GREEN
LGN	LIGHT GREEN
R	RED
*	WITH TRACER

1977-78 Chrysler LeBaron; Dodge Aspen & Diplomat; Plymouth Volare Heater Wiring Circuit

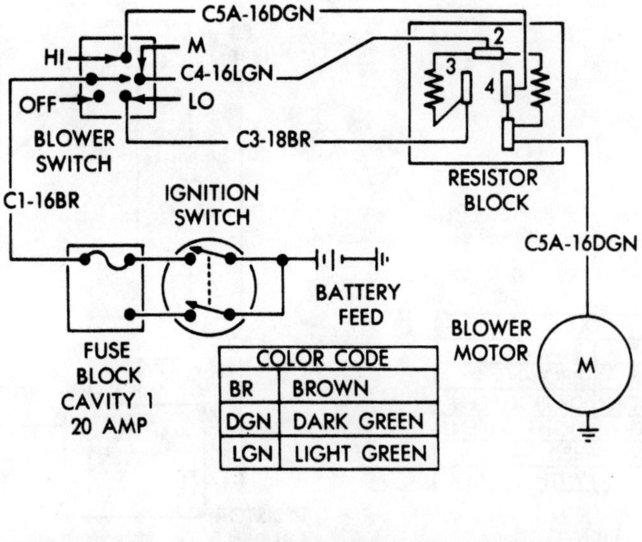

COLOR CODE	
BR	BROWN
DGN	DARK GREEN
LGN	LIGHT GREEN

1978-79 Dodge Omni & Plymouth Horizon Heater Wiring Circuit

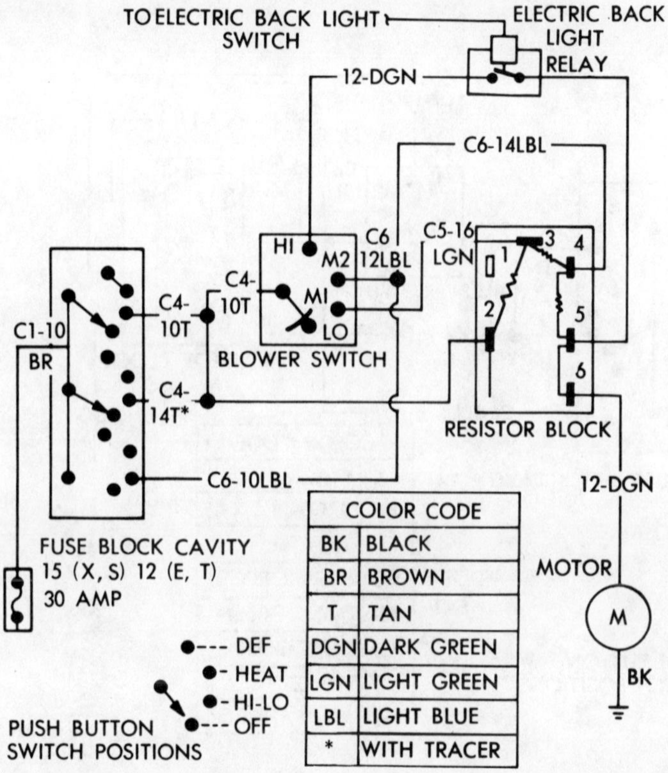

COLOR CODE	
BK	BLACK
BR	BROWN
T	TAN
DGN	DARK GREEN
LGN	LIGHT GREEN
LBL	LIGHT BLUE
*	WITH TRACER

1979 Chrysler Cordoba, Newport & New Yorker; Dodge Magnum & St. Regis Heater Wiring Circuit

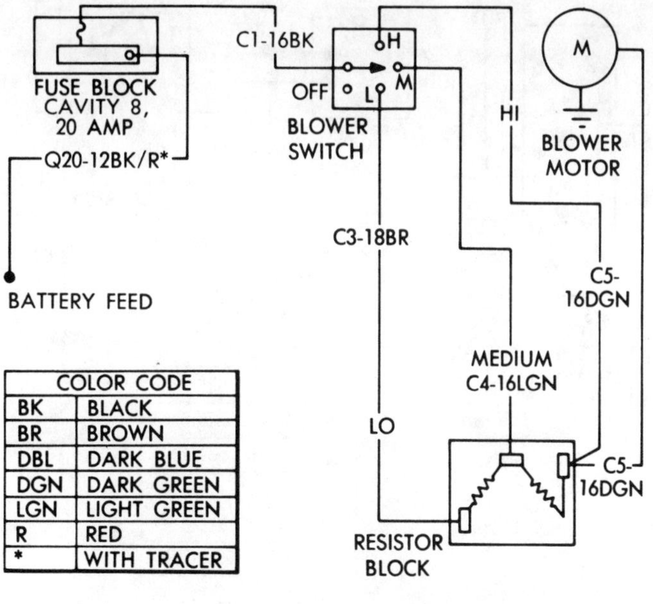

COLOR CODE	
BK	BLACK
BR	BROWN
DBL	DARK BLUE
DGN	DARK GREEN
LGN	LIGHT GREEN
R	RED
*	WITH TRACER

1979 Chrysler LeBaron; Dodge Aspen & Diplomat; Plymouth Volare Heater Wiring Circuit

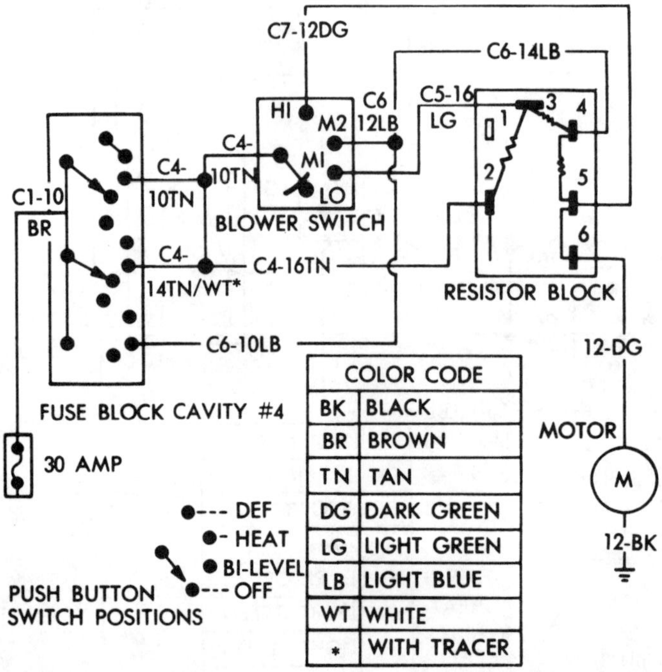

COLOR CODE	
BK	BLACK
BR	BROWN
TN	TAN
DG	DARK GREEN
LG	LIGHT GREEN
LB	LIGHT BLUE
WT	WHITE
*	WITH TRACER

1980 Chrysler Newport & New Yorker; Dodge St. Regis; Plymouth Gran Fury Heater Wiring Circuit

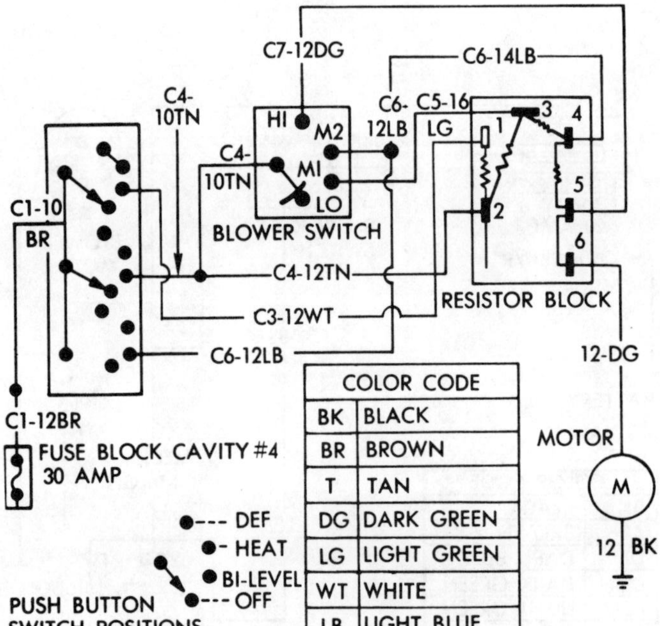

COLOR CODE	
BK	BLACK
BR	BROWN
T	TAN
DG	DARK GREEN
LG	LIGHT GREEN
WT	WHITE
LB	LIGHT BLUE
*	WITH TRACER

1980 Chrysler Cordoba & Dodge Mirada Heater Wiring Circuit

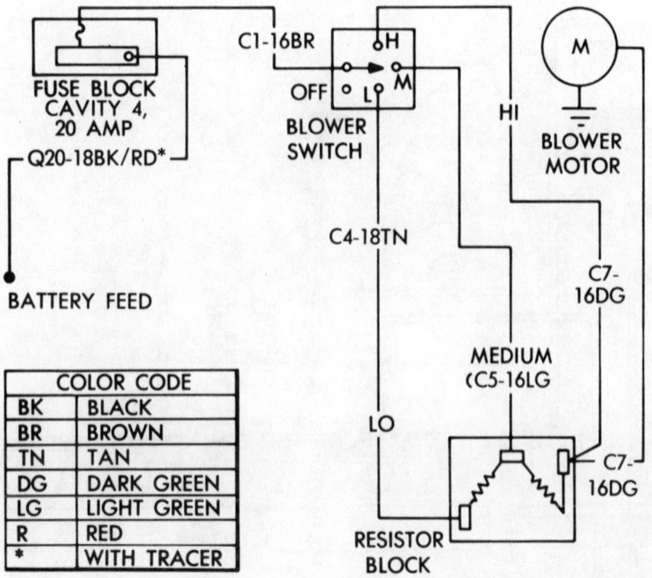

COLOR CODE	
BK	BLACK
BR	BROWN
TN	TAN
DG	DARK GREEN
LG	LIGHT GREEN
R	RED
*	WITH TRACER

1980 Chrysler LeBaron; Dodge Aspen & Diplomat; Plymouth Volare Heater Wiring Circuit

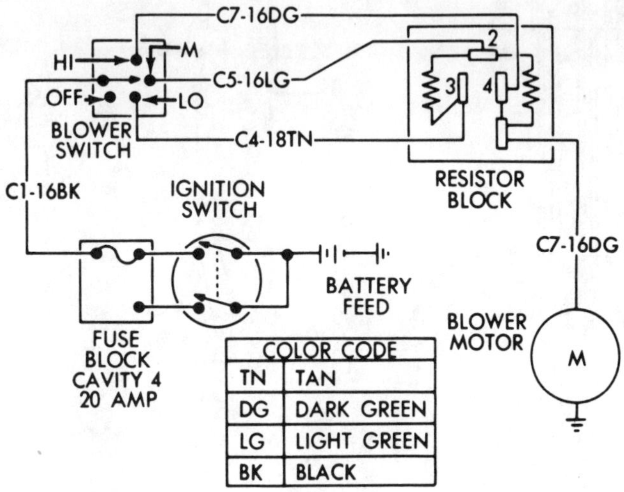

COLOR CODE	
TN	TAN
DG	DARK GREEN
LG	LIGHT GREEN
BK	BLACK

1980 Dodge Omni & Plymouth Horizon Heater Wiring Circuit

Ford Motor Company

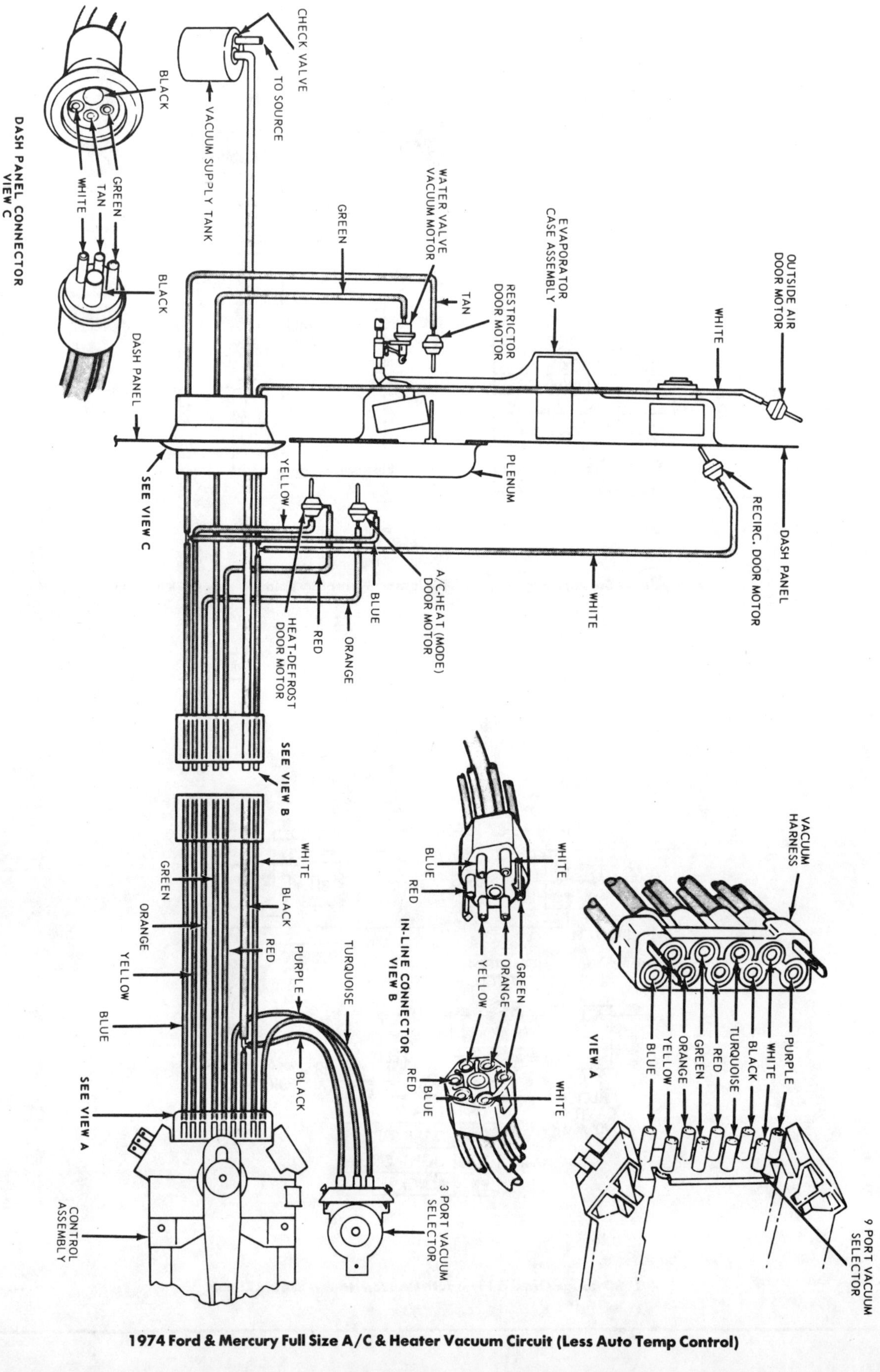

1974 Ford & Mercury Full Size A/C & Heater Vacuum Circuit (Less Auto Temp Control)

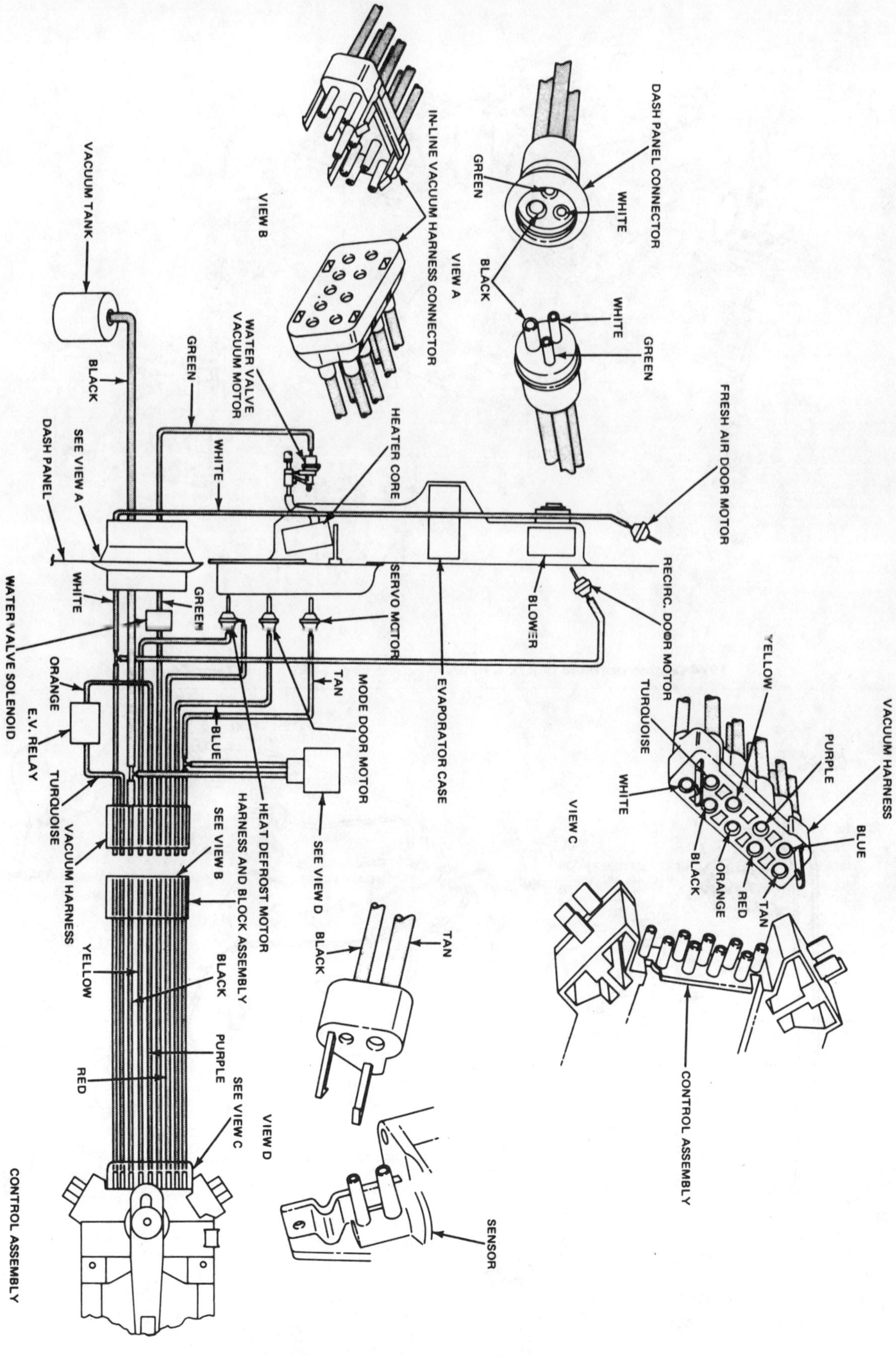

1974 Ford & Mercury Full Size A/C & Heater Vacuum Circuit (With Auto Temp Control)

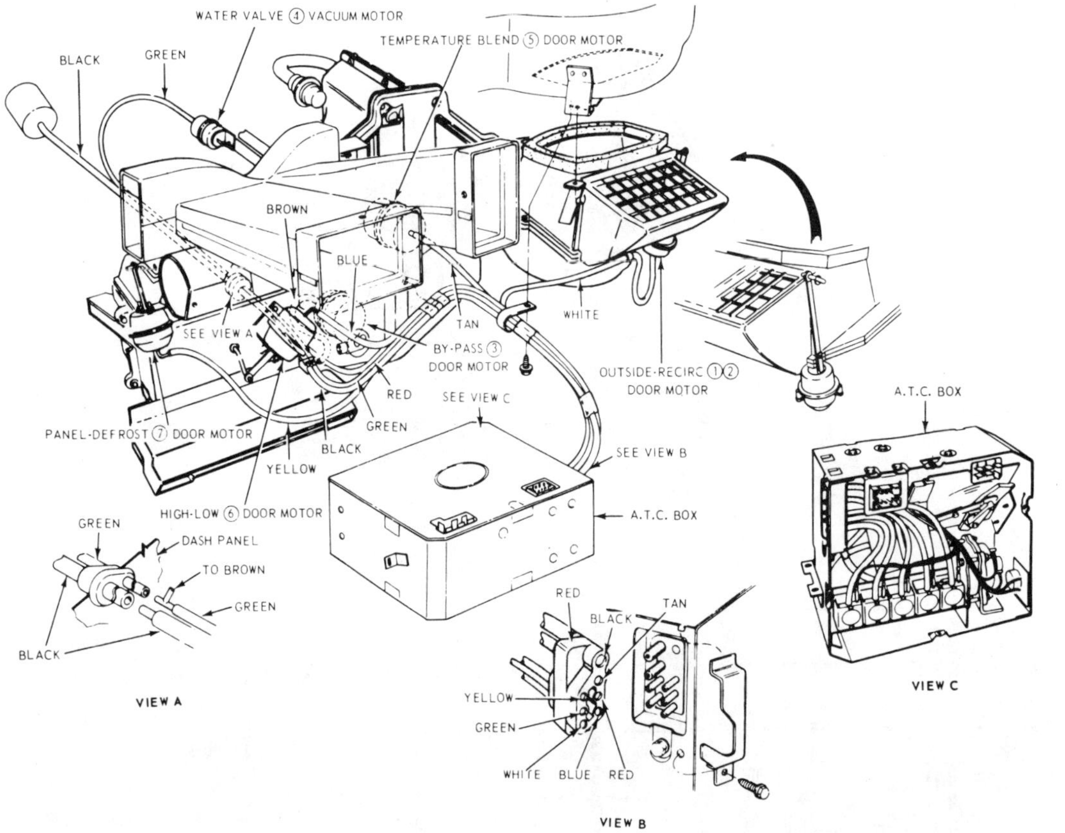

1974 Continental Mark IV A/C & Heater Vacuum Circuit (With Auto Temp Control)

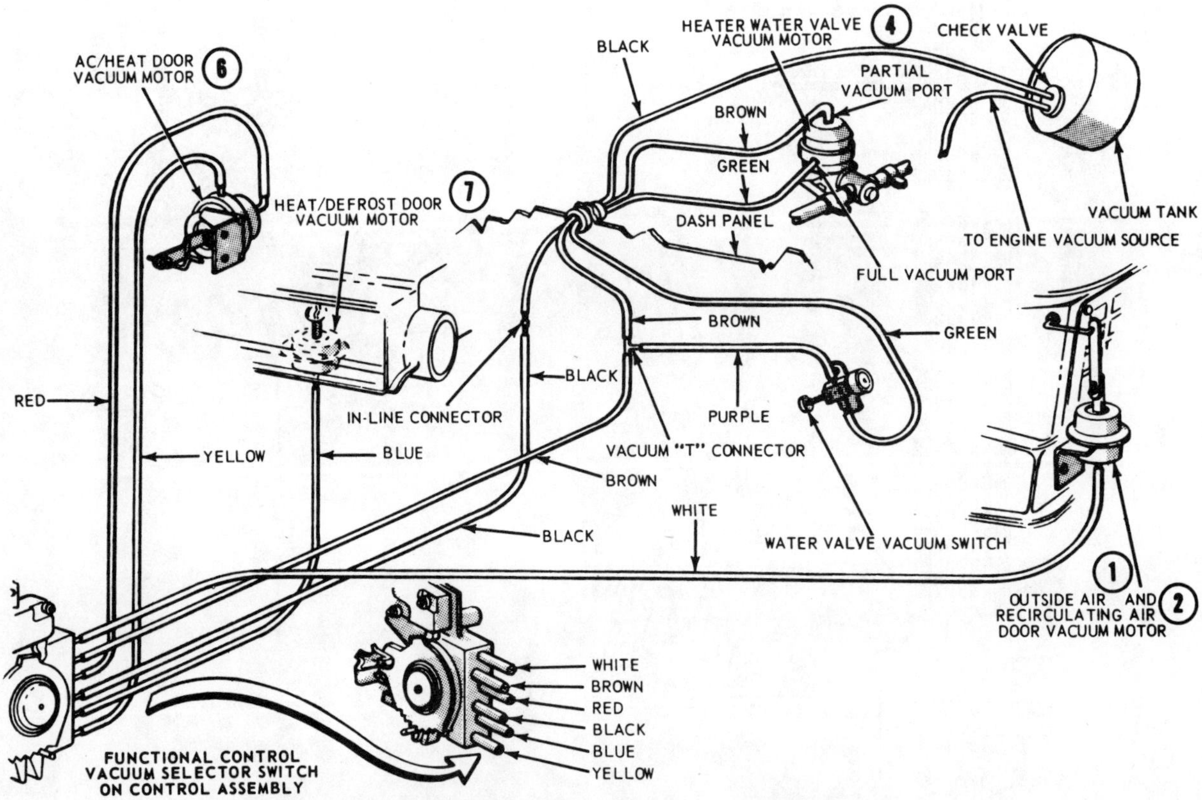

1974 Ford Maverick & Mercury Comet A/C & Heater Vacuum Circuit

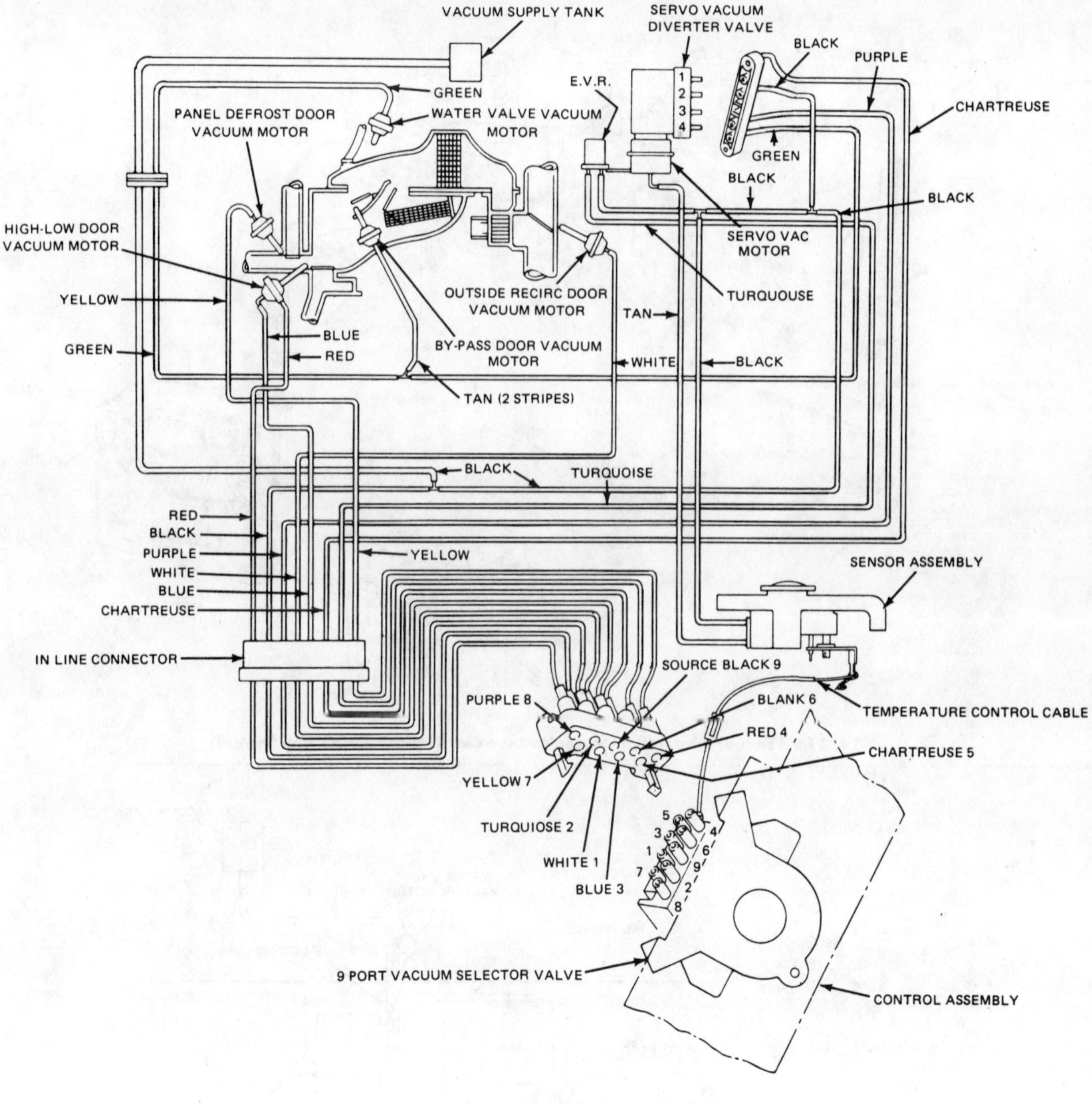

CONTROL HEAD VACUUM TEST CHART

VACUUM PORT	HOSE COLOR	VACUUM SELECTOR POSITION					FUNCTION
		OFF	VENT	AUTO	DEFOG	DEFROST	
4	RED	–	–	TO 3 AND 5	–	–	TO HIGH-LOW DOOR (6b) SIDE
5	CHARTREUSE	SEAL	SEAL	TO 3 AND 4	SEAL	SEAL	FROM SERVO DIVERTER VALVE #1 PORT
6	NOT CONNECTED	SEAL	SEAL	SEAL	SEAL	SEAL	NOT CONNECTED
3	BLUE	–	–	TO 4 AND 5	V	–	TO HIGH-LOW DOOR (6a) TOP
9	BLACK	S	S	S	S	S	SOURCE VACUUM
1	WHITE	V	–	TO 2	–	–	TO OUTSIDE-RECIRC DOOR (1)(2)
2	TURQUOISE	SEAL	SEAL	TO 1	SEAL	SEAL	FROM E. V. RELAY
7	YELLOW	–	V	TO 8	–	–	TO PANEL-DEFROST DOOR (7)
8	PURPLE	SEAL	SEAL	TO 7	SEAL	SEAL	FROM SERVO DIVERTER VALVE #3 PORT

V = VACUUM. [–] = NO VACUUM (ATMOSPHERE) S = SYSTEM SOURCE VACUUM

1974 Ford Elite & Torino; Mercury Cougar & Montego A/C & Heater Vacuum Circuit (With Auto Temp Control)

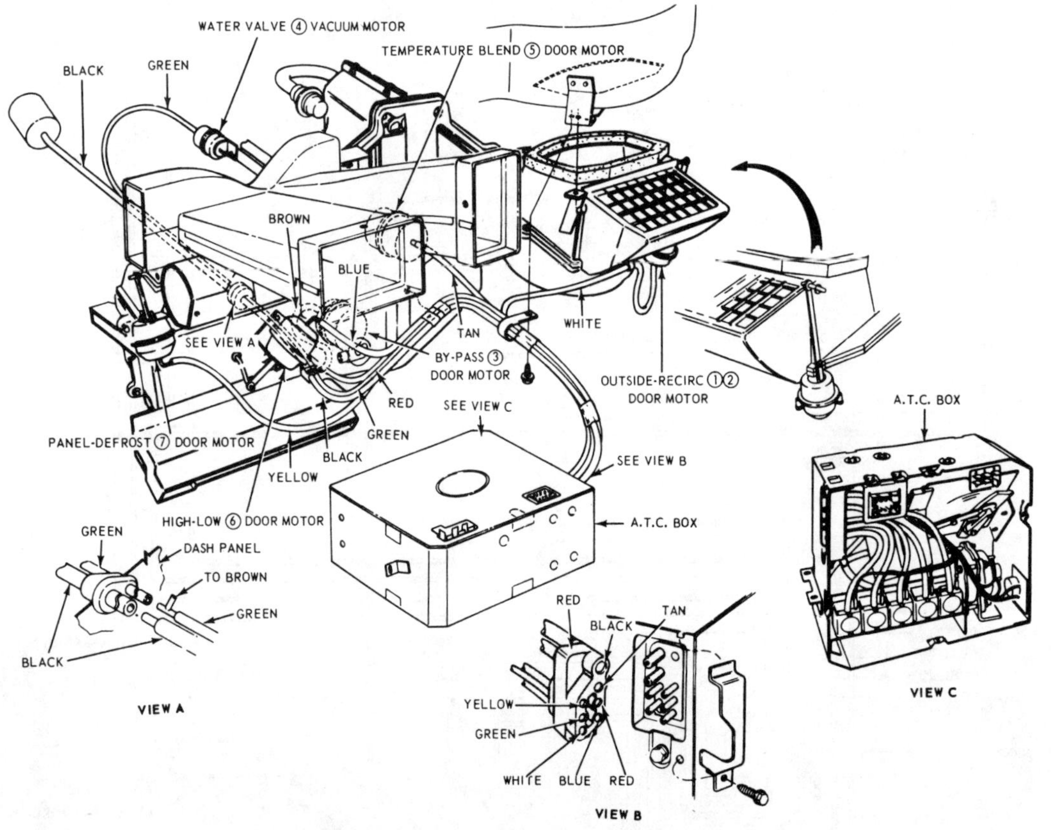

1974 Ford Thunderbird A/C & Heater Vacuum Circuit (With Auto Temp Control)

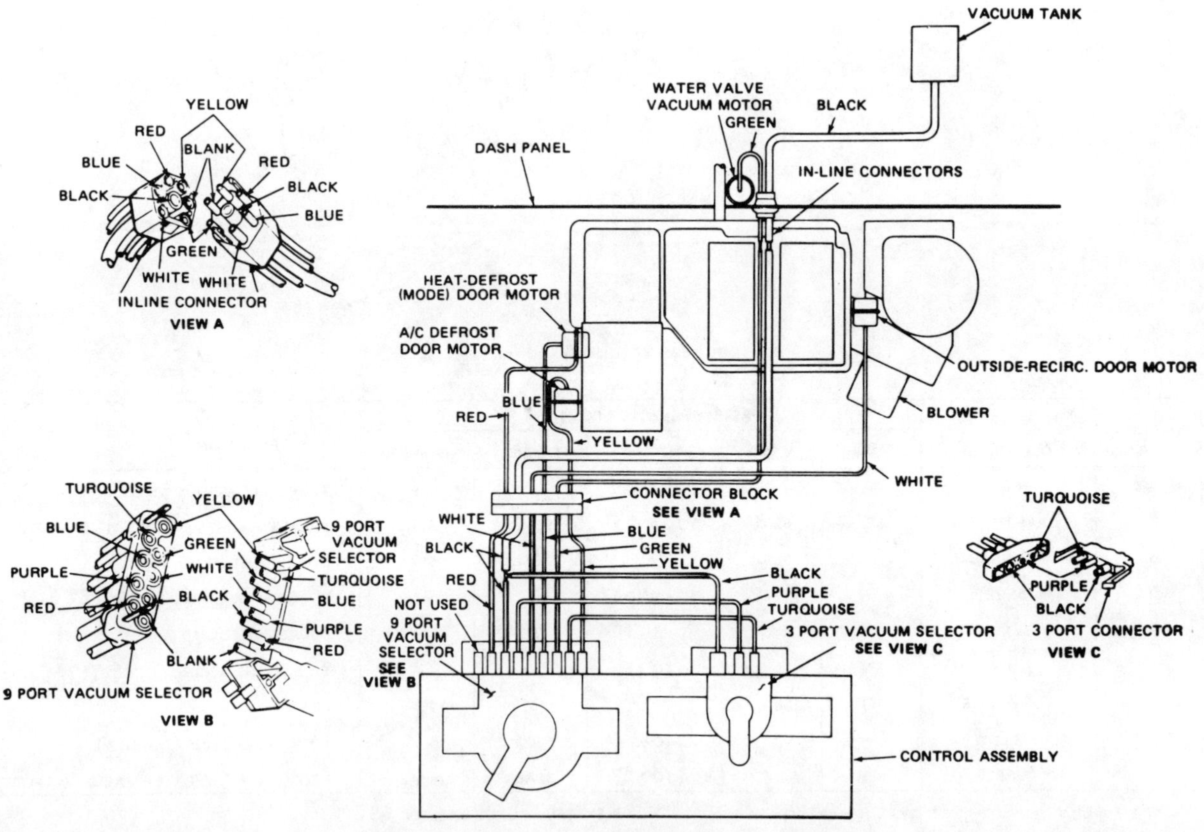

1974-76 Ford Mustang A/C & Heater Vacuum Circuit

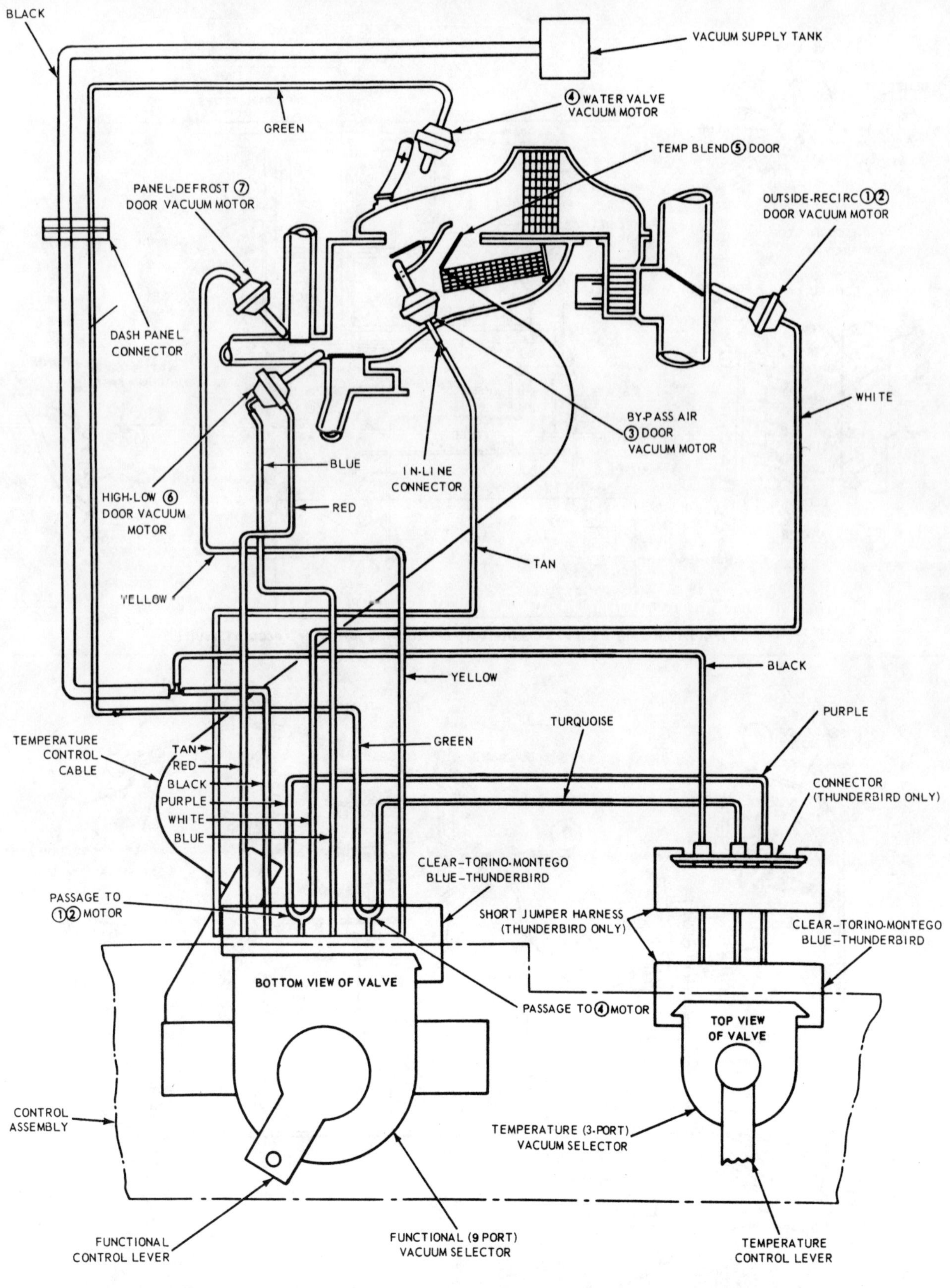

1974-76 Ford Elite & Torino; Mercury Cougar & Montego A/C & Heater Vacuum Circuit (Less Auto Temp Control)

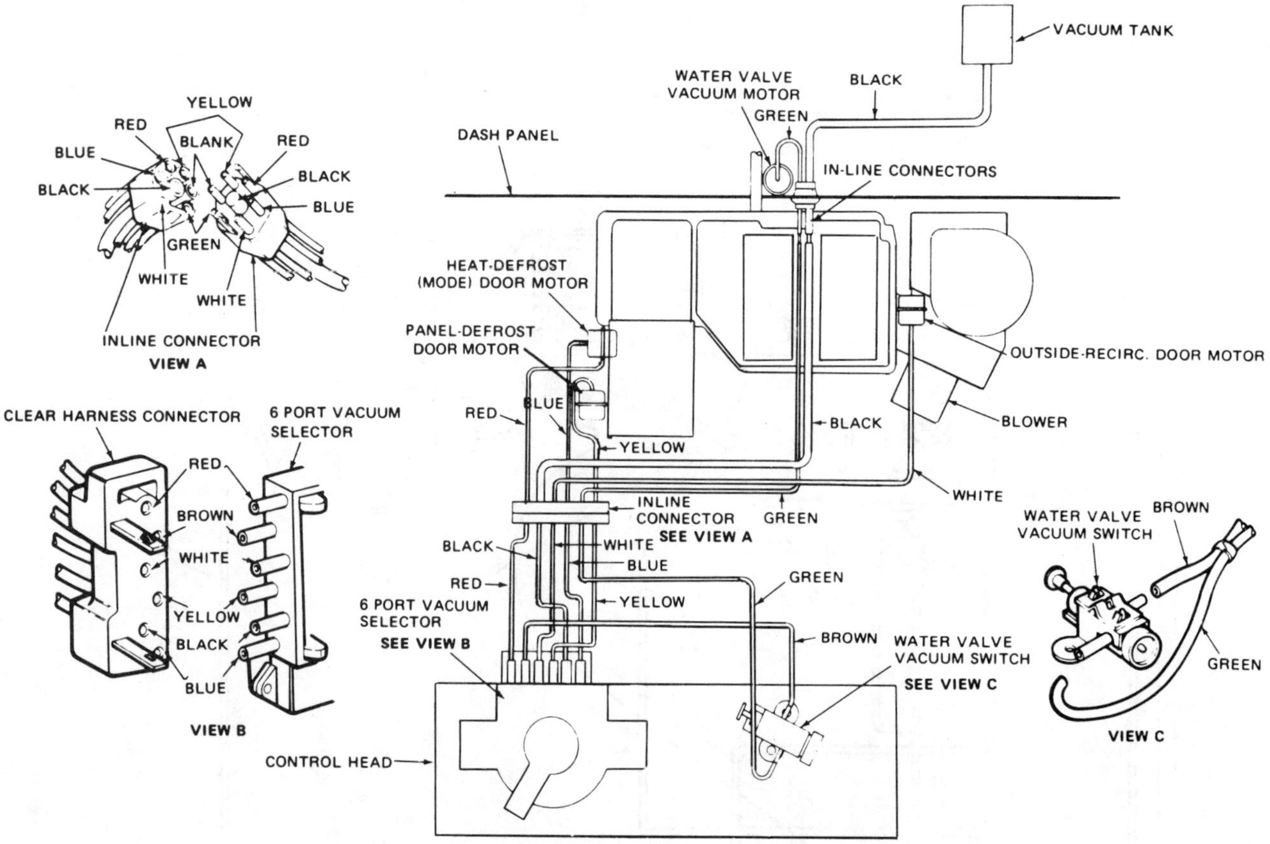

1974-76 Ford Pinto & 1975-76 Mercury Bobcat A/C & Heater Vacuum Circuit

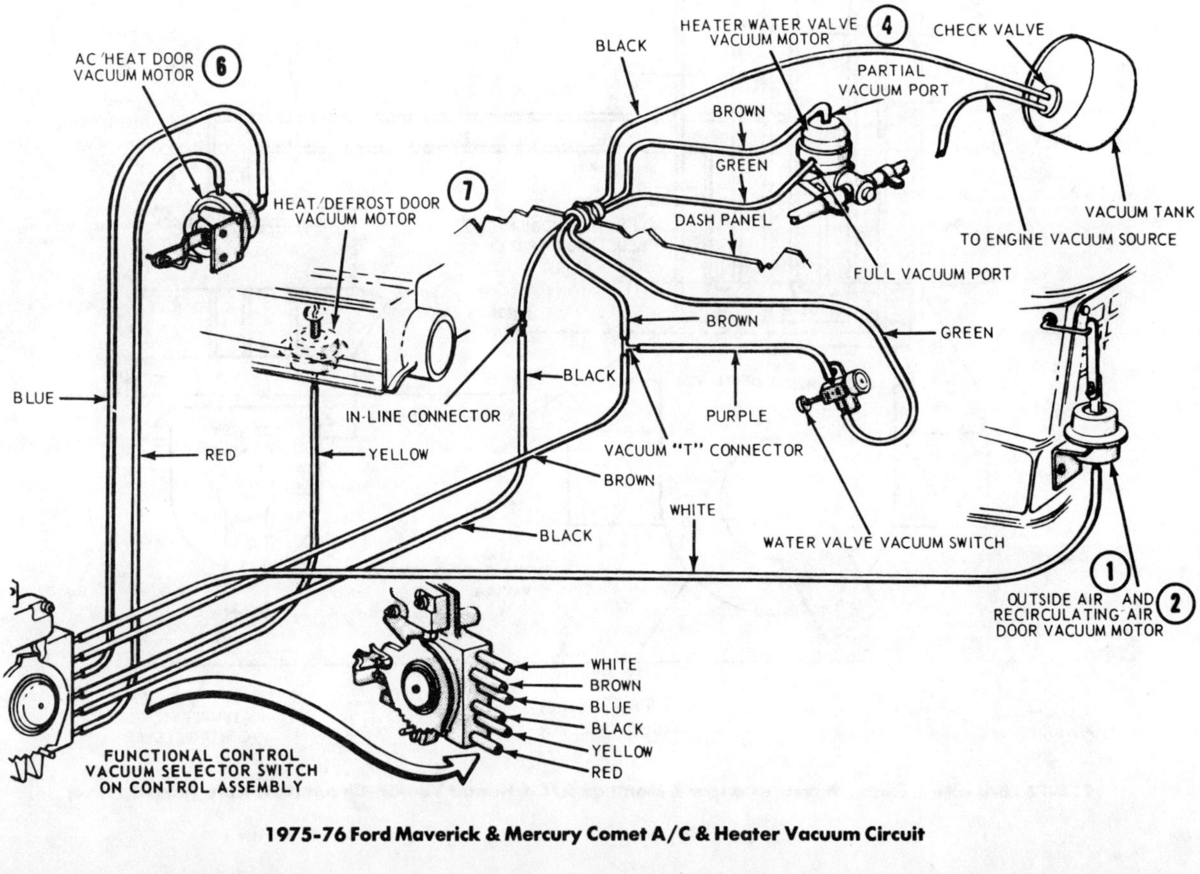

1975-76 Ford Maverick & Mercury Comet A/C & Heater Vacuum Circuit

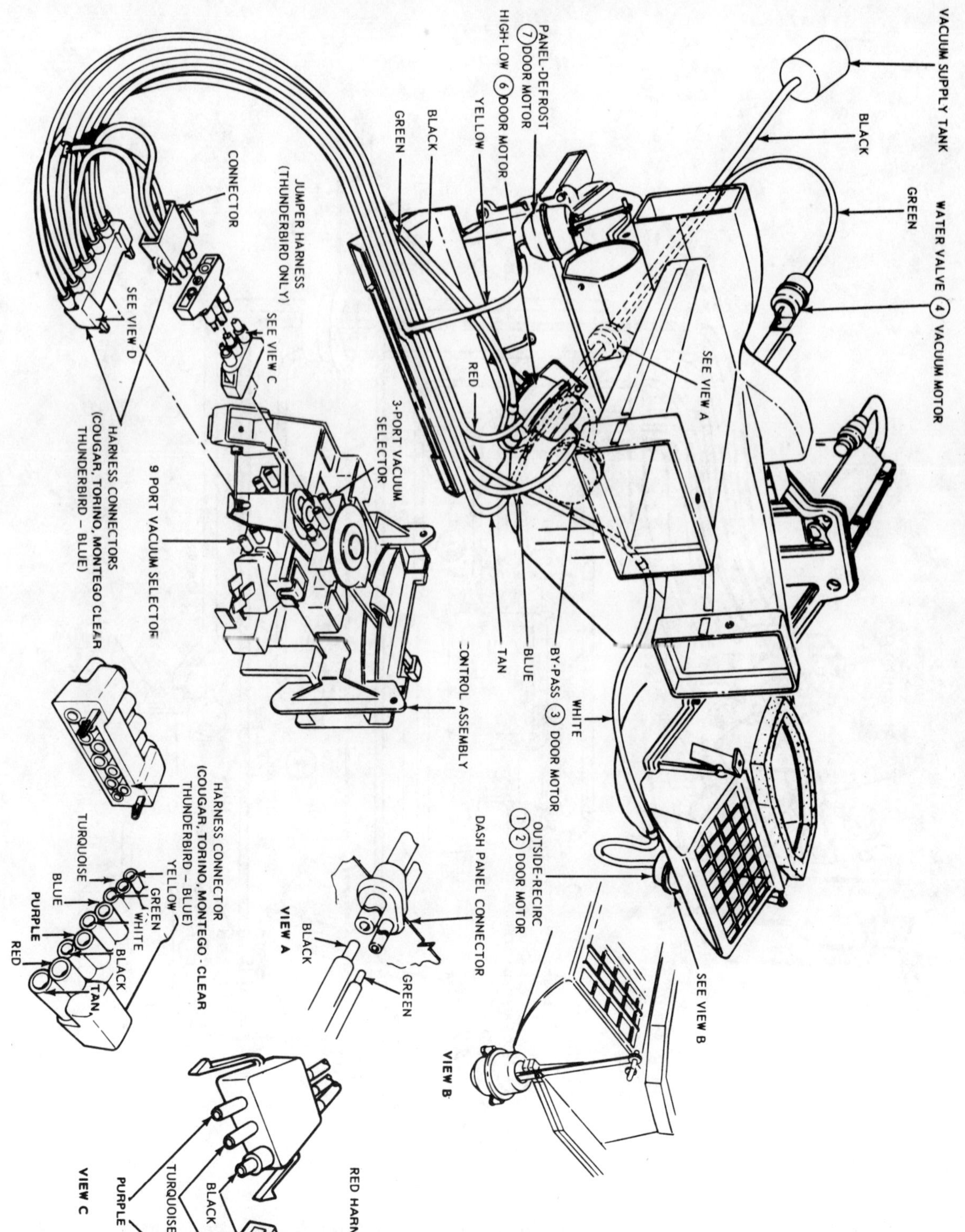

1974-76 Ford Thunderbird A/C & Heater Vacuum Circuit (Less Auto Temp Control)

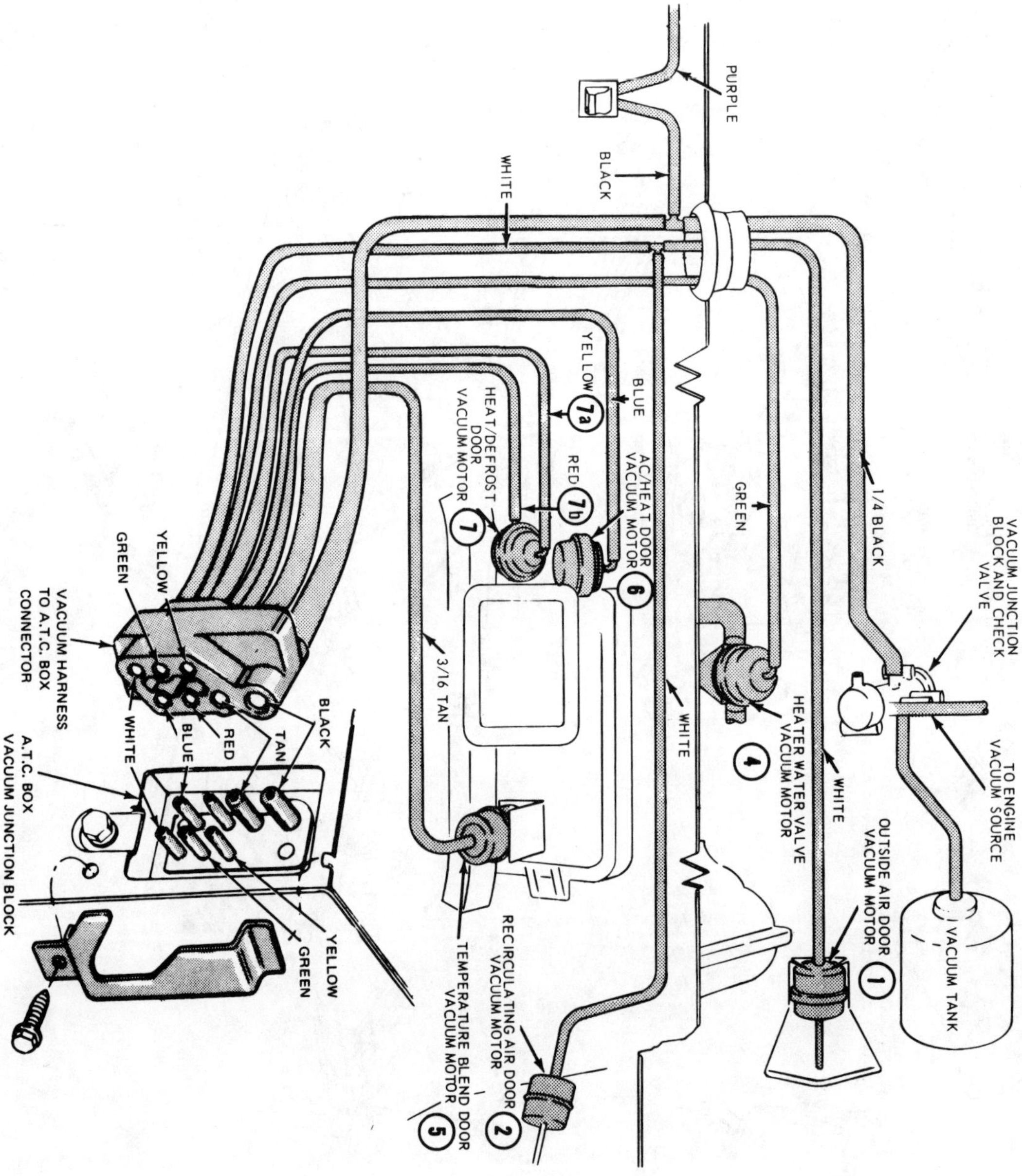

1974-76 Lincoln Continental A/C & Heater Vacuum Circuit (With Auto Temp Control)

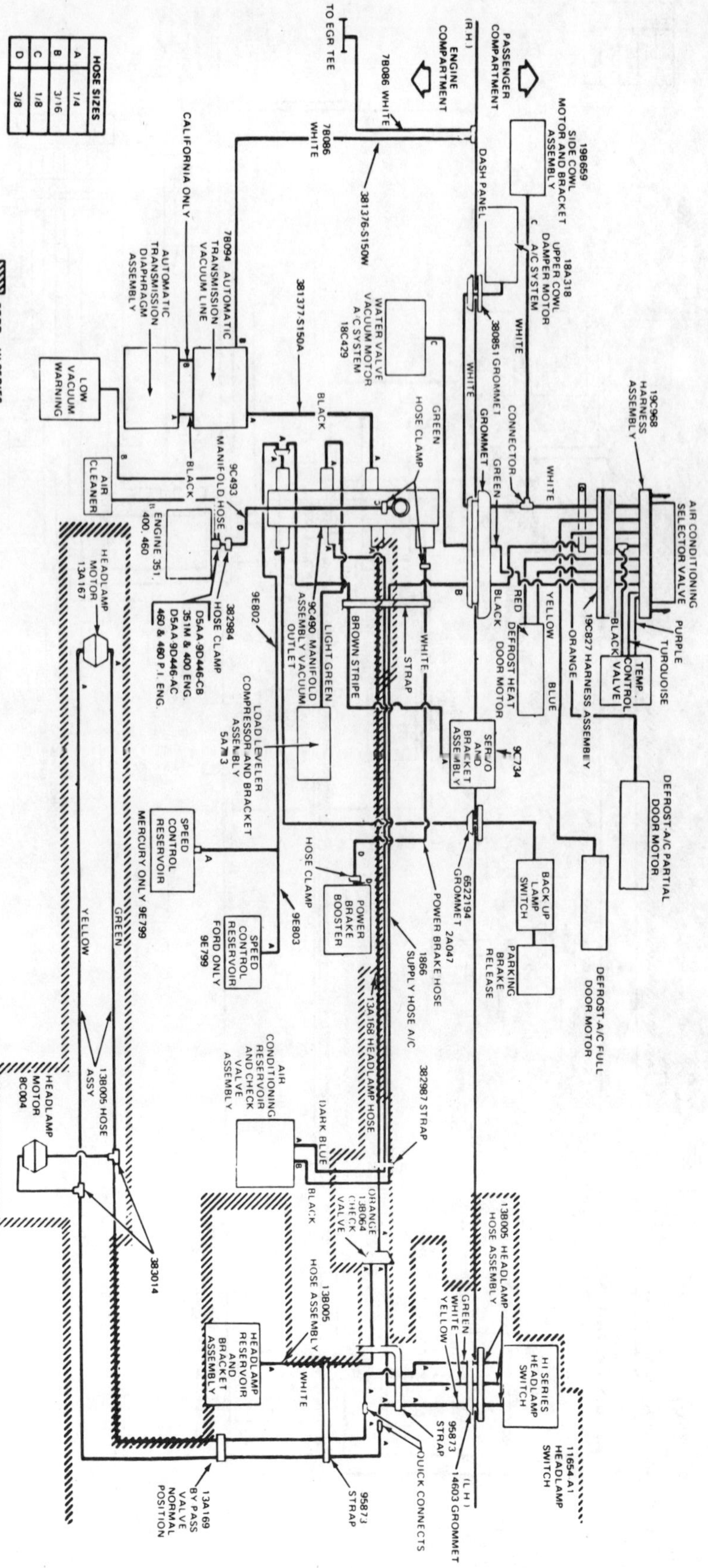

1975-76 Ford & Mercury Full Size A/C & Heater Vacuum Circuit (Less Auto Temp Control)

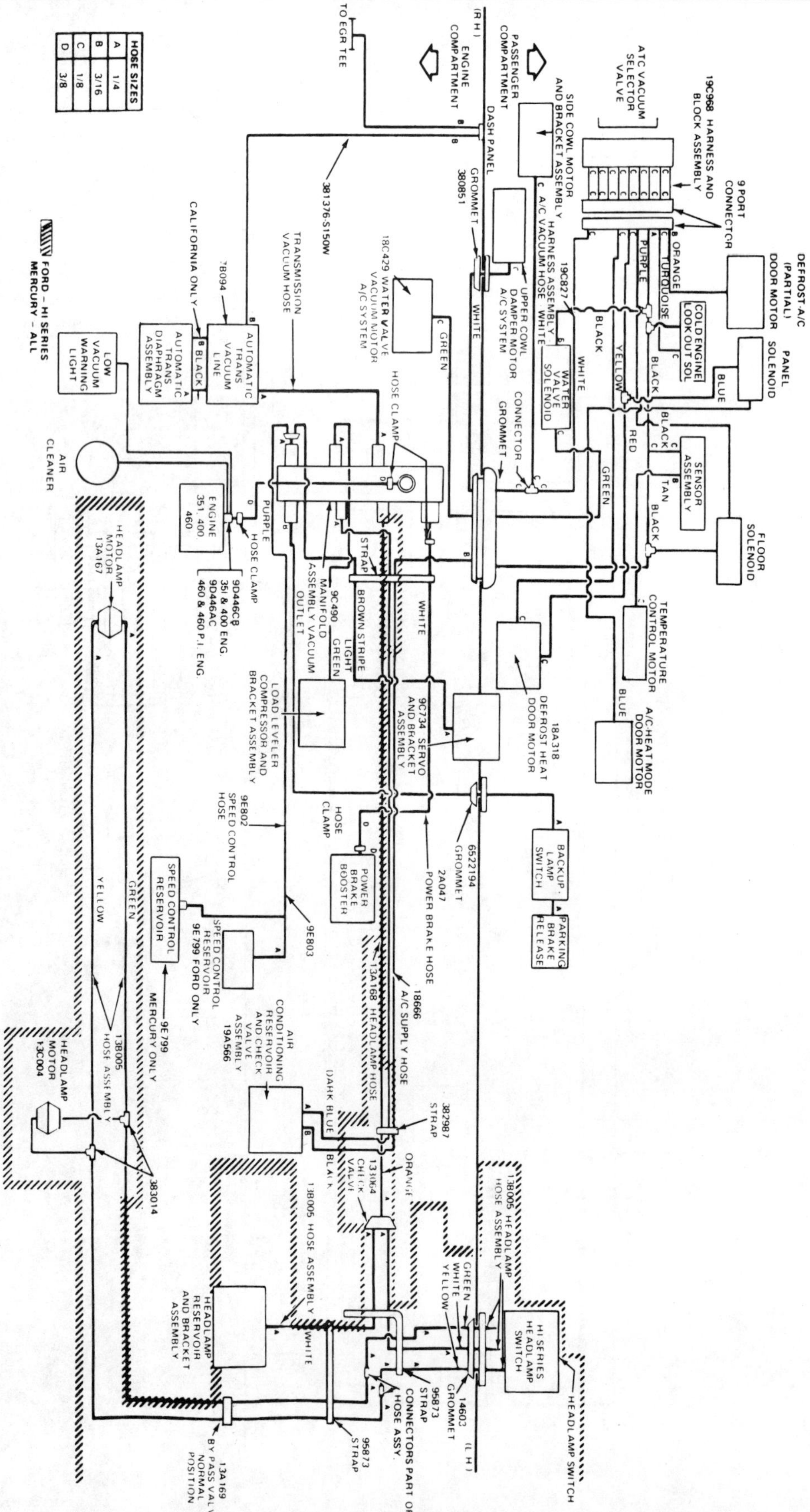

1975-76 Ford & Mercury Full Size A/C & Heater Vacuum Circuit (With Auto Temp Control)

CONTROL HEAD VACUUM TEST CHART

VACUUM PORT	HOSE COLOR	VACUUM SELECTOR POSITION					FUNCTION
		OFF	VENT	AUTO	DEFOG	DEFROST	
4	RED	V	–	TO 3 AND 5	–	–	TO HIGH-LOW DOOR (6b) SIDE
5	CHARTREUSE	SEAL	SEAL	TO 3 AND 4	SEAL	SEAL	FROM SERVO DIVERTER VALVE #1 PORT
6	NOT CONNECTED	SEAL	SEAL	SEAL	SEAL	SEAL	NOT CONNECTED
3	BLUE	V	–	TO 4 AND 5	V	–	TO HIGH-LOW DOOR (6a) TOP
9	BLACK	S	S	S	S	S	SOURCE VACUUM
1	WHITE	V	–	TO 2	–	–	TO OUTSIDE-RECIRC DOOR (1)(2)
2	TURQUOISE	SEAL	SEAL	TO 1	SEAL	SEAL	FROM E. V. RELAY
7	YELLOW	–	V	TO 8	–	–	TO PANEL-DEFROST DOOR (7)
8	PURPLE	SEAL	SEAL	TO 7	SEAL	SEAL	FROM SERVO DIVERTER VALVE #3 PORT

V = VACUUM. [–] = NO VACUUM (ATMOSPHERE) S = SYSTEM SOURCE VACUUM

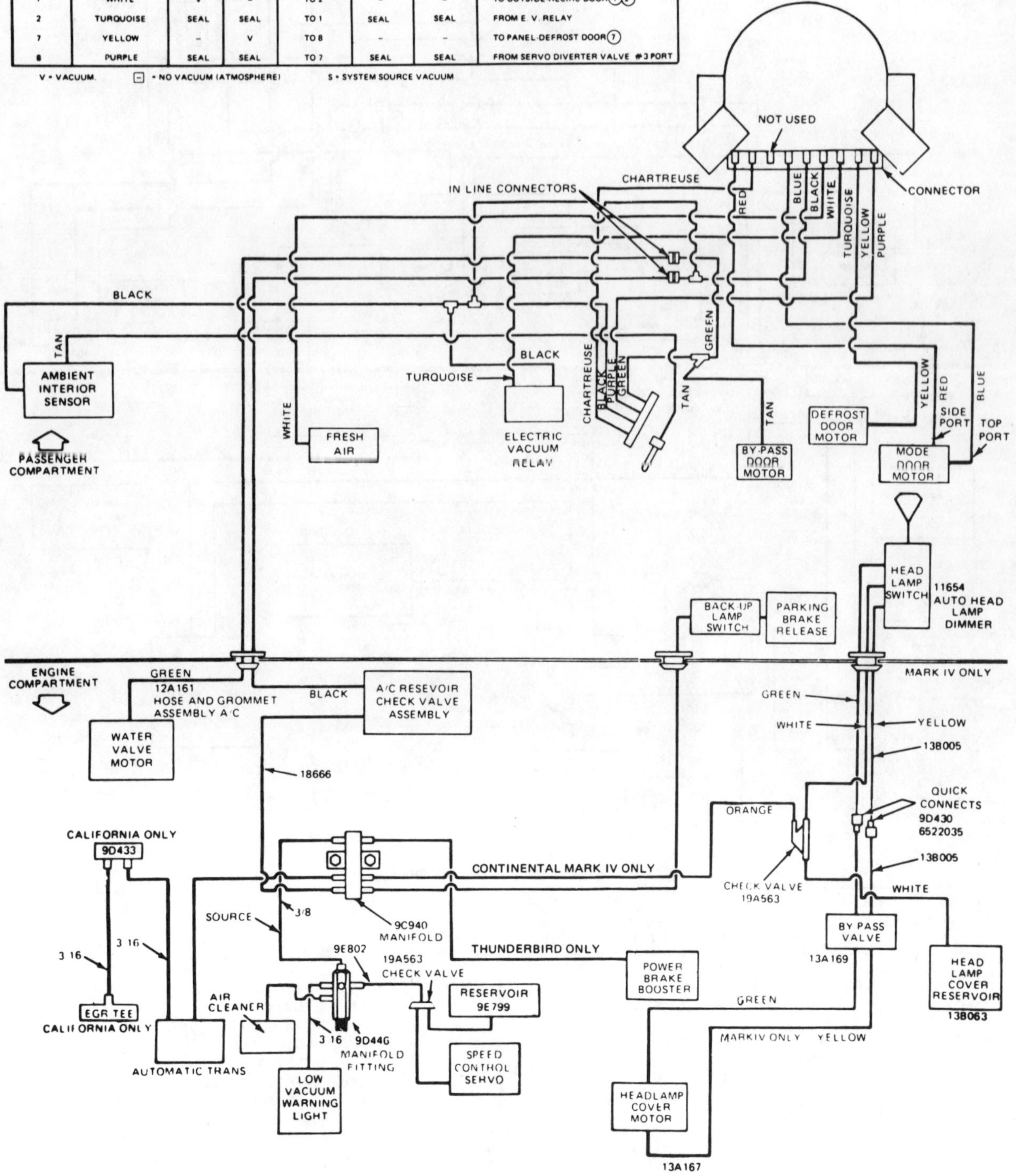

1975-76 Continental Mark IV A/C & Heater Vacuum Circuit (With Auto Temp Control)

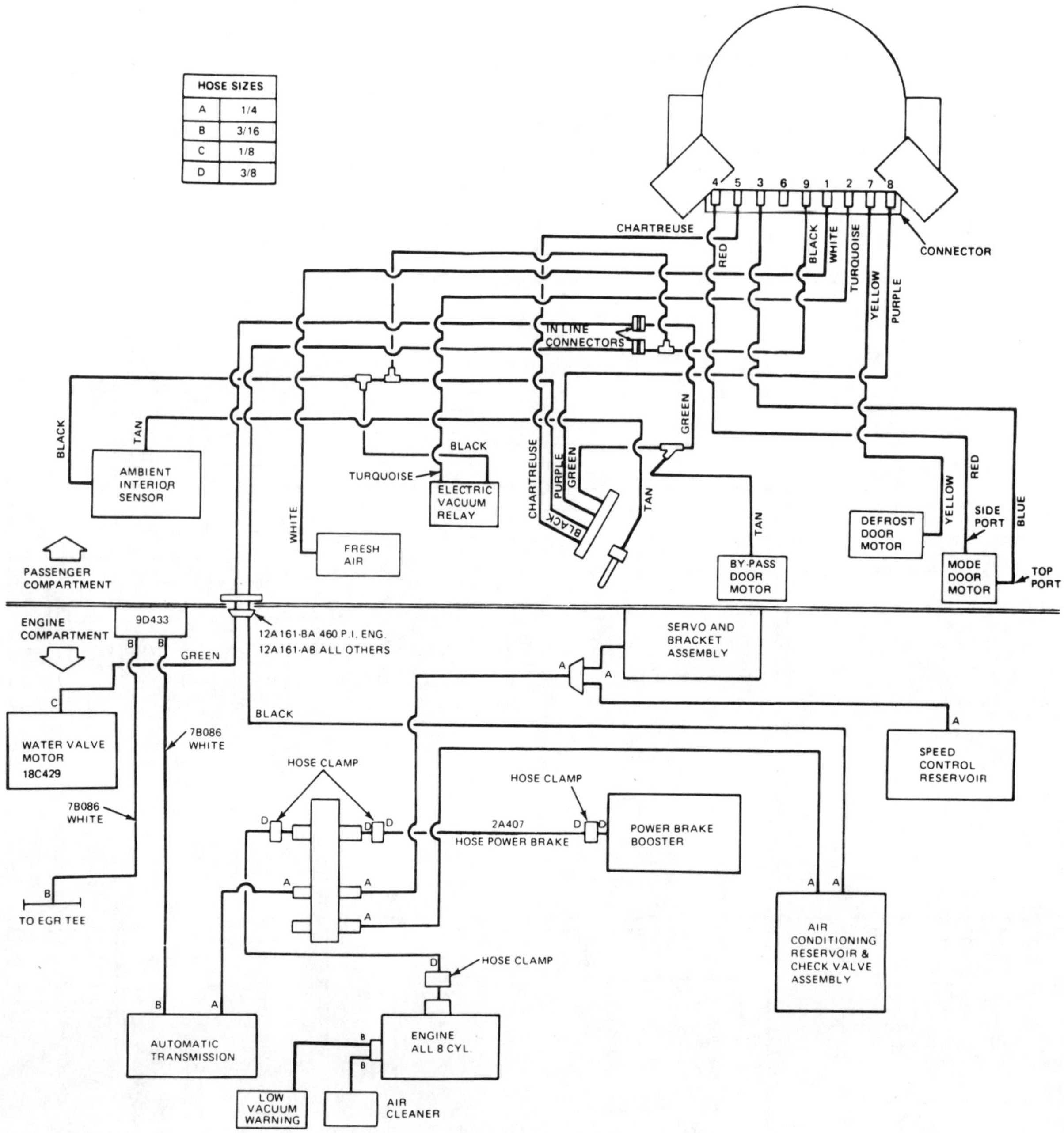

CONTROL HEAD VACUUM TEST CHART

VACUUM PORT	HOSE COLOR	VACUUM SELECTOR POSITION					FUNCTION
		OFF	VENT/EC.	AUTO	DEFOG	DEFROST	
4	RED	V	–	TO 3 AND 5	–	–	TO HIGH-LOW DOOR (6b) SIDE
5	CHARTREUSE	SEAL	SEAL	TO 3 AND 4	SEAL	SEAL	FROM SERVO DIVERTER VALVE # 1 PORT
6	NOT CONNECTED	SEAL	SEAL	SEAL	SEAL	SEAL	NOT CONNECTED
3	BLUE	V	–	TO 4 AND 5	V	–	TO HIGH-LOW DOOR (6a) TOP
9	BLACK	S	S	S	S	S	SOURCE VACUUM
1	WHITE	V	–	TO 2	–	–	TO OUTSIDE-RECIRC DOOR (1)(2)
2	TURQUOISE	SEAL	SEAL	TO 1	SEAL	SEAL	FROM E.V. RELAY
7	YELLOW	–	V	TO 8	–	–	TO PANEL-DEFROST DOOR (7)
8	PURPLE	SEAL	SEAL	TO 7	SEAL	SEAL	FROM SERVO DIVERTER VALVE # 3 PORT

V = VACUUM. ⊟ = NO VACUUM (ATMOSPHERE) S = SYSTEM SOURCE VACUUM

1975-76 Ford Elite & Torino; Mercury Cougar & Montego A/C & Heater Vacuum Circuit (With Auto Temp Control)

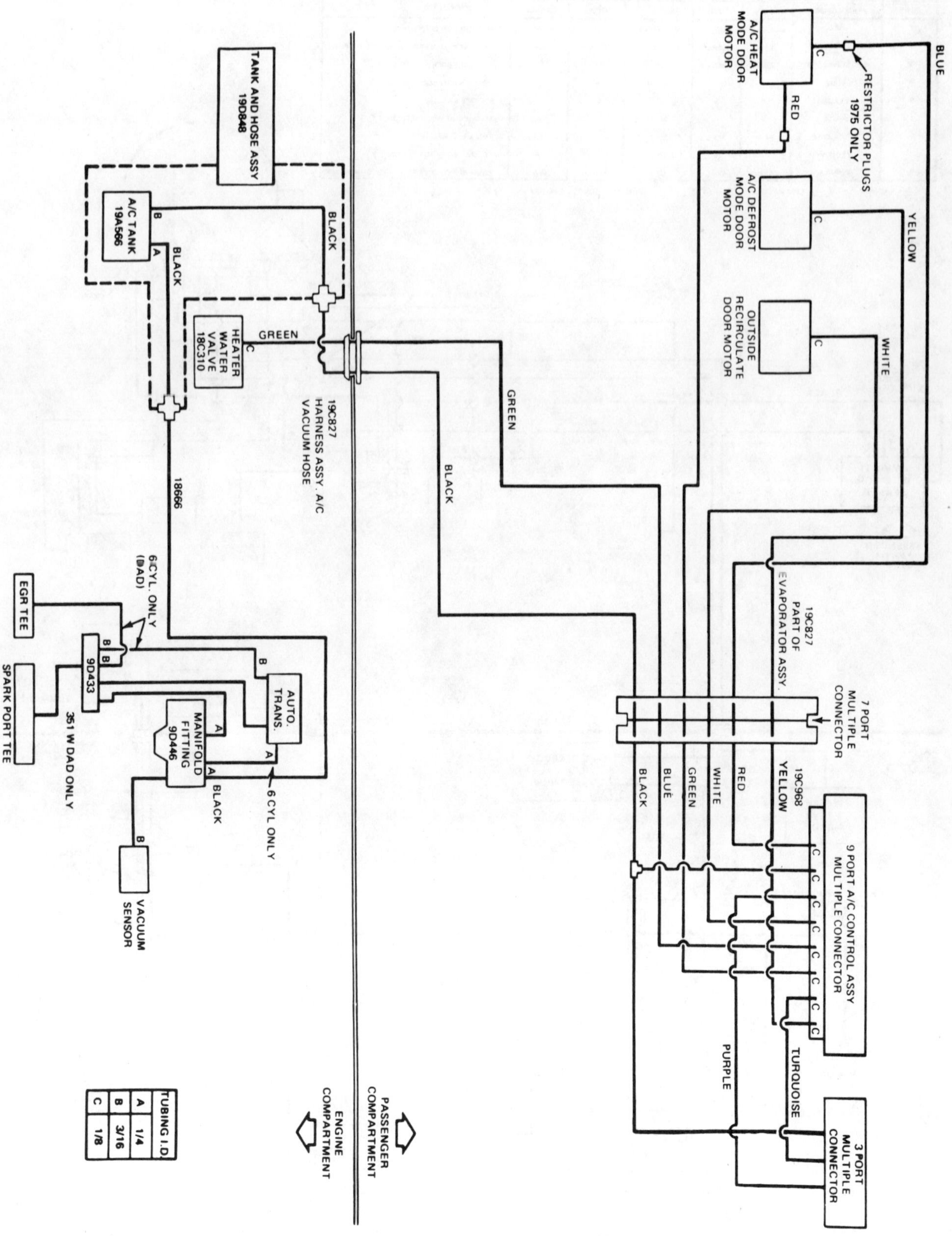

1975-76 Ford Granada & Mercury Monarch A/C & Heater Vacuum Circuit

CONTROL HEAD VACUUM TEST CHART

VACUUM PORT	HOSE COLOR	VACUUM SELECTOR POSITION					FUNCTION
		OFF	VENT	AUTO	DEFOG	DEFROST	
4	RED	V	–	TO 3 AND 5	–	–	TO HIGH-LOW DOOR (6b) SIDE
5	CHARTREUSE	SEAL	SEAL	TO 3 AND 4	SEAL	SEAL	FROM SERVO DIVERTER VALVE #1 PORT
6	NOT CONNECTED	SEAL	SEAL	SEAL	SEAL	SEAL	NOT CONNECTED
3	BLUE	V	–	TO 4 AND 5	V	–	TO HIGH-LOW DOOR (6a) TOP
9	BLACK	S	S	S	S	S	SOURCE VACUUM
1	WHITE	V	–	TO 2	–	–	TO OUTSIDE-RECIRC DOOR (1)(2)
2	TURQUOISE	SEAL	SEAL	TO 1	SEAL	SEAL	FROM E. V. RELAY
7	YELLOW	–	V	TO 8	–	–	TO PANEL-DEFROST DOOR (7)
8	PURPLE	SEAL	SEAL	TO 7	SEAL	SEAL	FROM SERVO DIVERTER VALVE #3 PORT

V = VACUUM   [–] = NO VACUUM (ATMOSPHERE)   S = SYSTEM SOURCE VACUUM

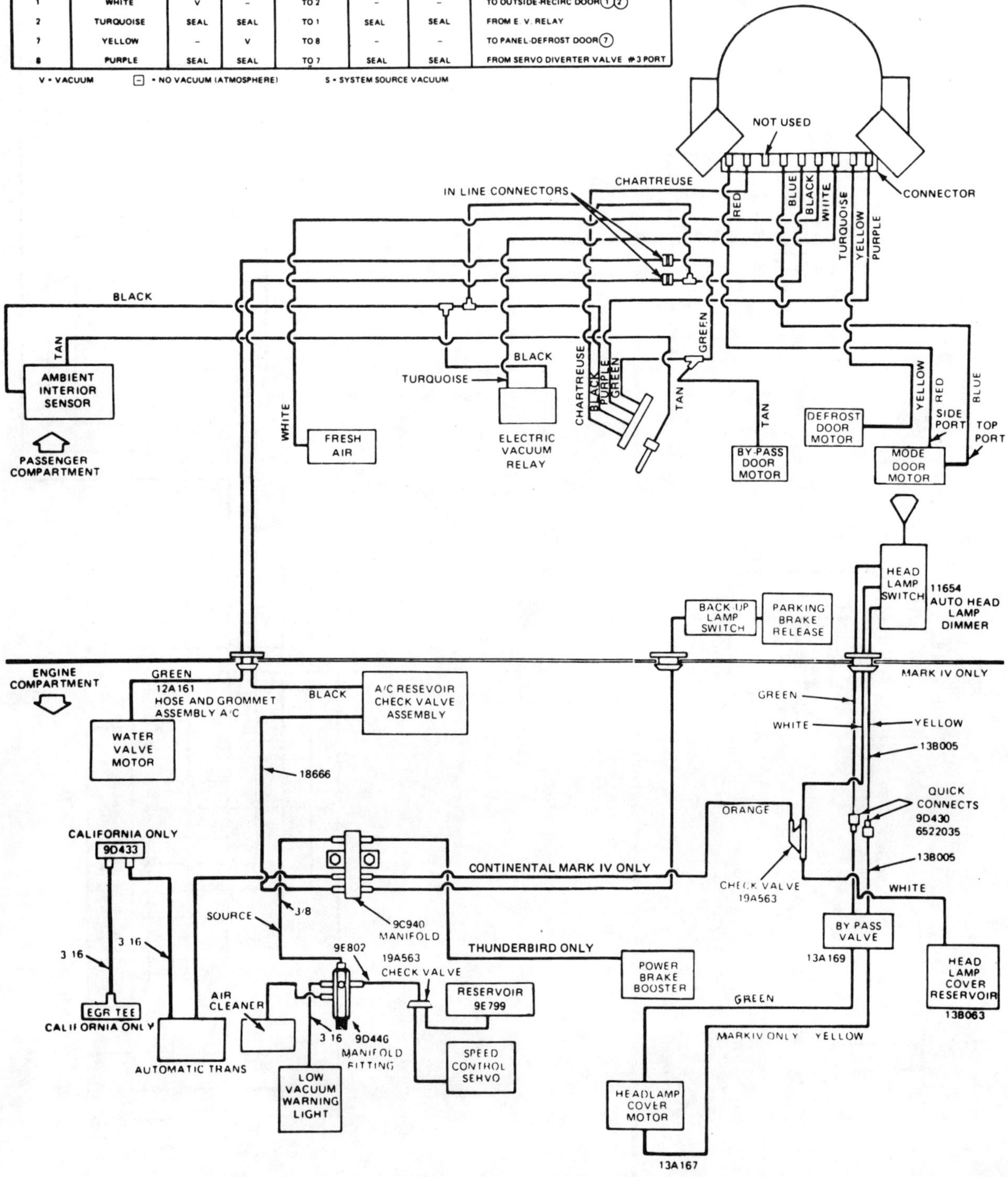

1975-76 Ford Thunderbird A/C & Heater Vacuum Circuit (With Auto Temp Control)

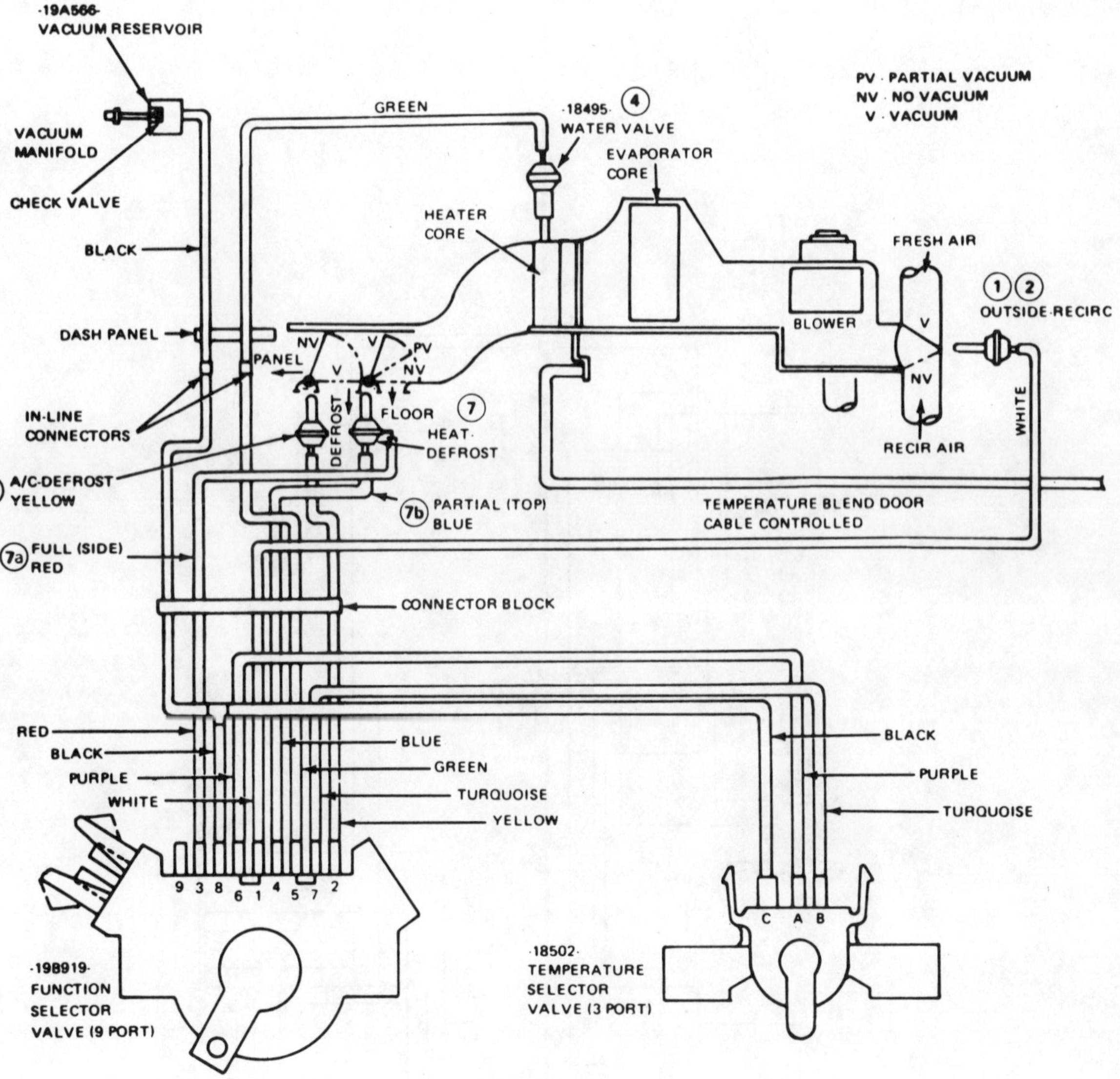

PORT	FUNCTION	FUNCTION SELECTOR VALVE LEVER POSITIONS							HOSE COLOR
		OFF	ECONOMY VENT	A/C	HI-LO	HEAT	∇ DEFOG	DEFROST	
1	Outside–Recirc.	V		To 6					White 1-2
2	A/C–Defrost		V	V	V				Yellow 6
3	Full Heat–Defrost	V				V			Red 7a
4	Partial Heat–Defrost	V			V	V	V		Blue 7b
5	To Water Valve	V	V	To 7	To 7	To 7	To 7	To 7	Green 4
6	From Temp Port A	Seal	Seal	To 1	Seal	Seal	Seal	Seal	Purple
7	From Temp Port B	Seal	Seal	To 5	To 5	To 5	To 5	To 5	Turquoise
8	Source	V	V	V	V	V	V	V	Black
9		NOT USED IN THIS SYSTEM							

PORT	FUNCTION	TEMP SELECTOR VALVE LEVER POSITIONS				HOSE COLOR
		COOL		TEMPERATURE	WARM	
C	Source	V	V	V	V	Black
A	Water Valve	V				Turquoise
B	Outside–Recirc	V	V			Purple

V - VACUUM BLANK - ATMOSPHERE

1977 Ford & Mercury Full Size A/C & Heater Vacuum Circuit (With Auto Temp Control)

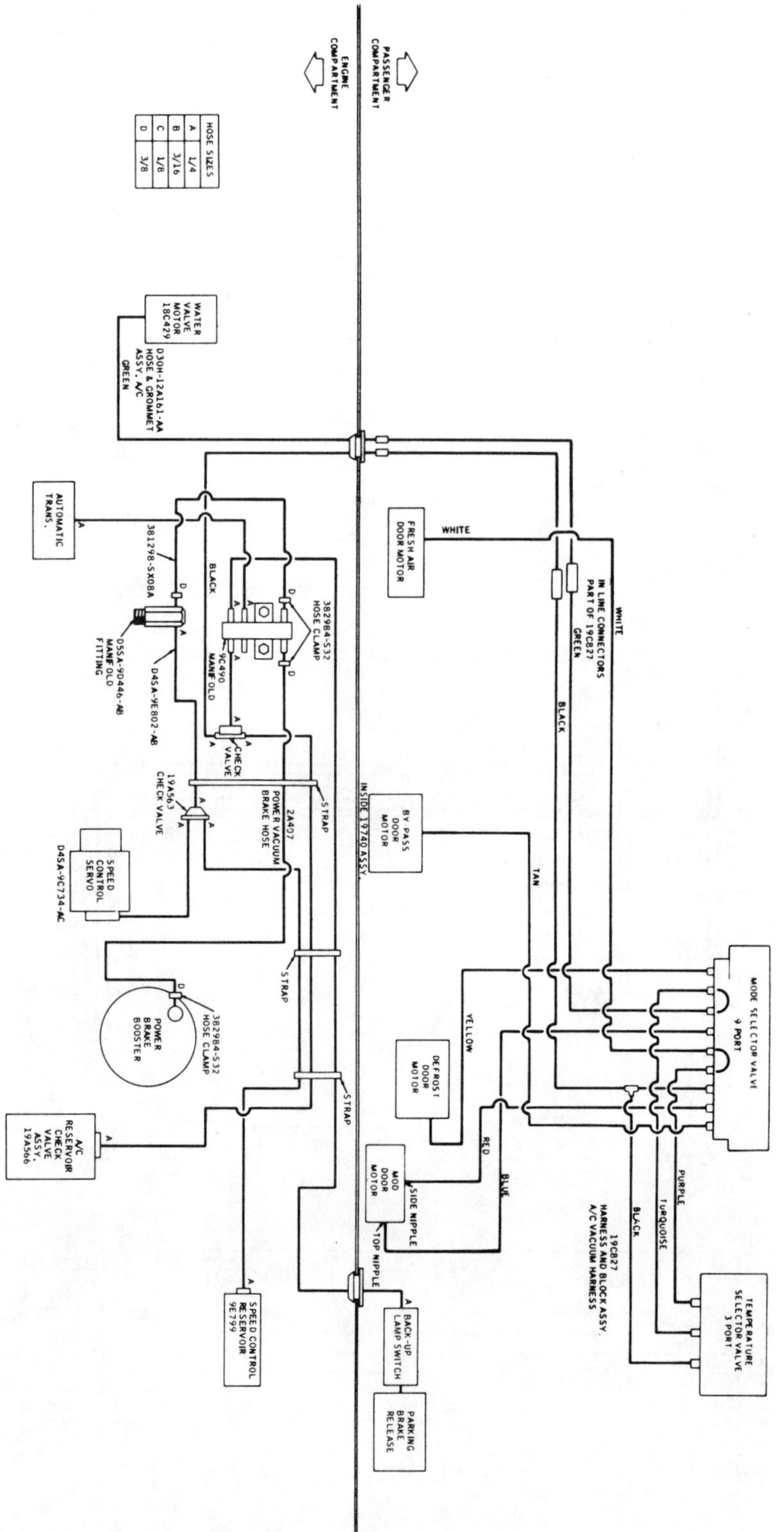

1977 Ford LTD II & Mercury Cougar A/C & Heater Vacuum Circuit (Less Auto Temp Control)

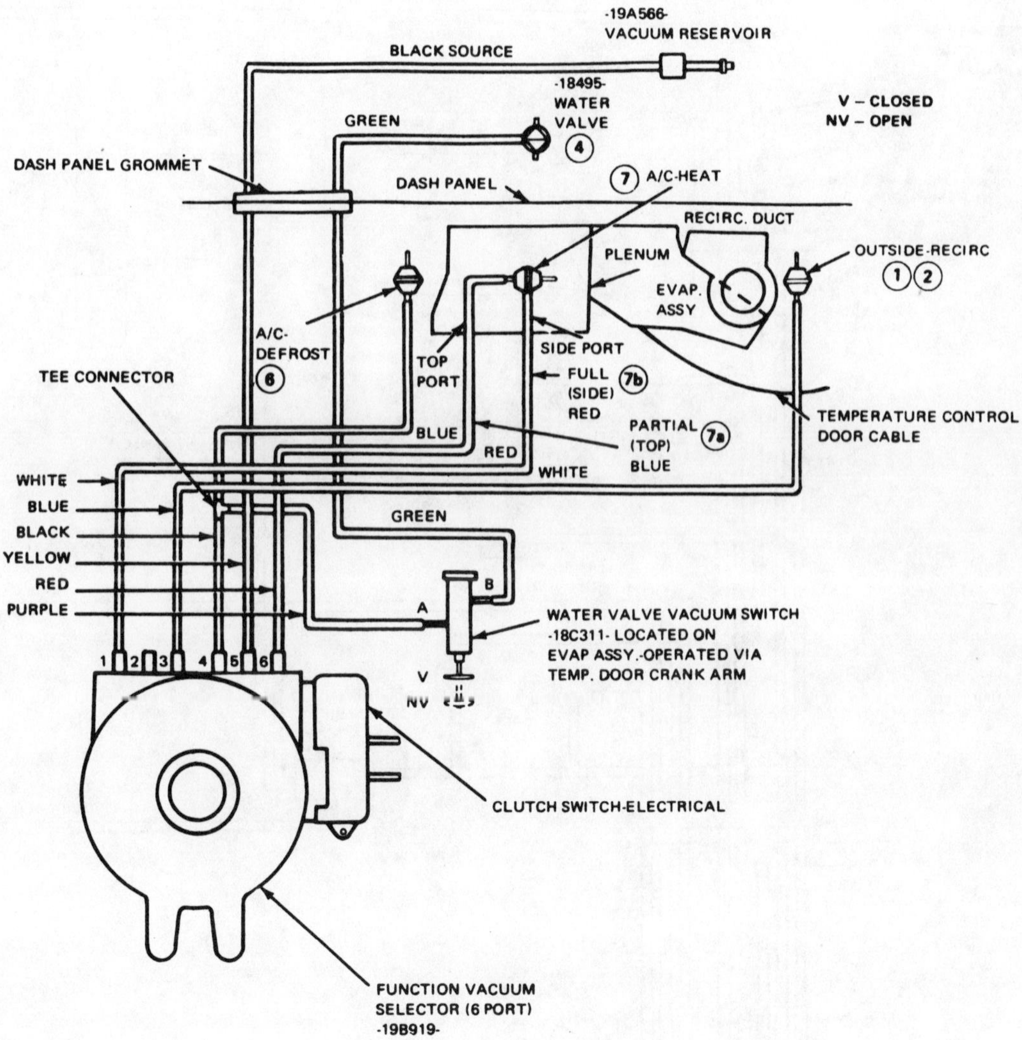

VACUUM SELECTOR CONTROL FUNCTION

PORT	COLOR	FUNCTION	MAX	FRESH	OFF	HI–HEAT	LO–HEAT	DE–FOG	DEICE
1	White	Recirculate	V		V				
2	Blank	None–Plugged	V	V					
3	Blue	Partial A/C–Heat				V	V	V	
4	Black	Source	V	V	V	V	V	V	V
5	Yellow	A/C-Defrost	V	V		V			
6	Red	Full A/C-Heat					V		
ELECTRICAL CLUTCH SWITCH			ON	ON	OFF	OFF	OFF	ON	ON

WATER VALVE VACUUM SWITCH FUNCTION

		COOL	TEMPERATURE	WARM
A	Purple	V	V	V
B	Green	V		

V = VACUUM   BLANK = NO VACUUM (ATMOSPHERE

1978 Ford Maverick & Mercury Comet A/C & Heater Vacuum Circuit

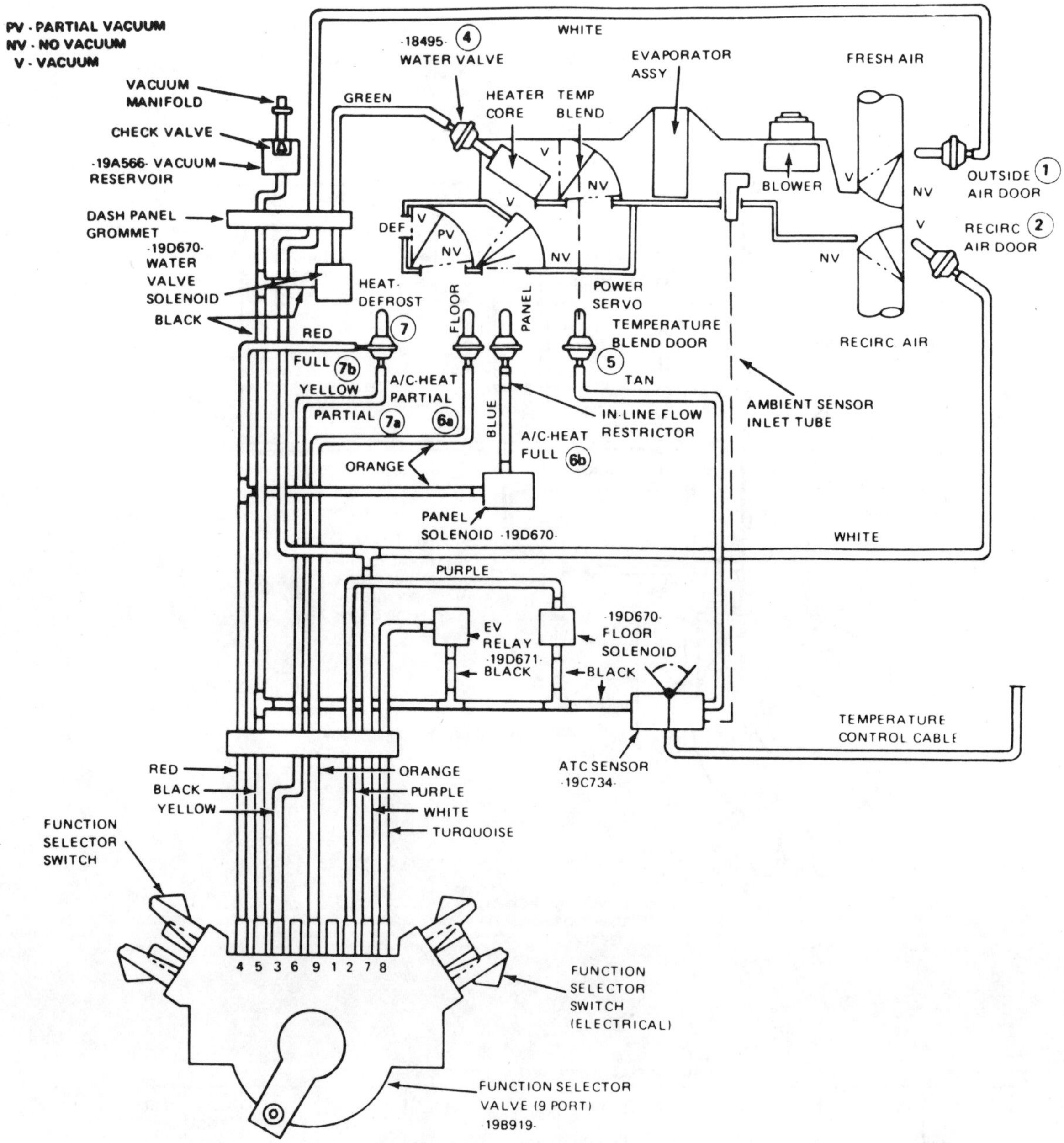

PORT	FUNCTION	FUNCTION SELECTOR VALVE POSITIONS					HOSE COLOR
		OFF	ECONOMY VENT	AUTO	DEFOG	DEFROST	
1	No Function	Seal	Seal	Seal	Seal	Seal	–
2	Floor Solenoid	Seal	Seal	To 9	Seal	Seal	Purple
3	Heat/Def Partial	V	V	V	V		Yellow 7a
4	Heat/Def Full	V	V	V			Red 7b
5	Source Vacuum	V	V	V	V	V	Black
6	No Function	Seal	Seal	Seal	Seal	Seal	–
7	Outside–Recirc.	V		To 8			White 1-2
8	EV Relay	Seal	Seal	To 7	Seal	Seal	Turquoise
9	A/C–Heat Partial		V	To 2			Orange 6a

1977 Ford Mustang A/C & Heater Vacuum Circuit

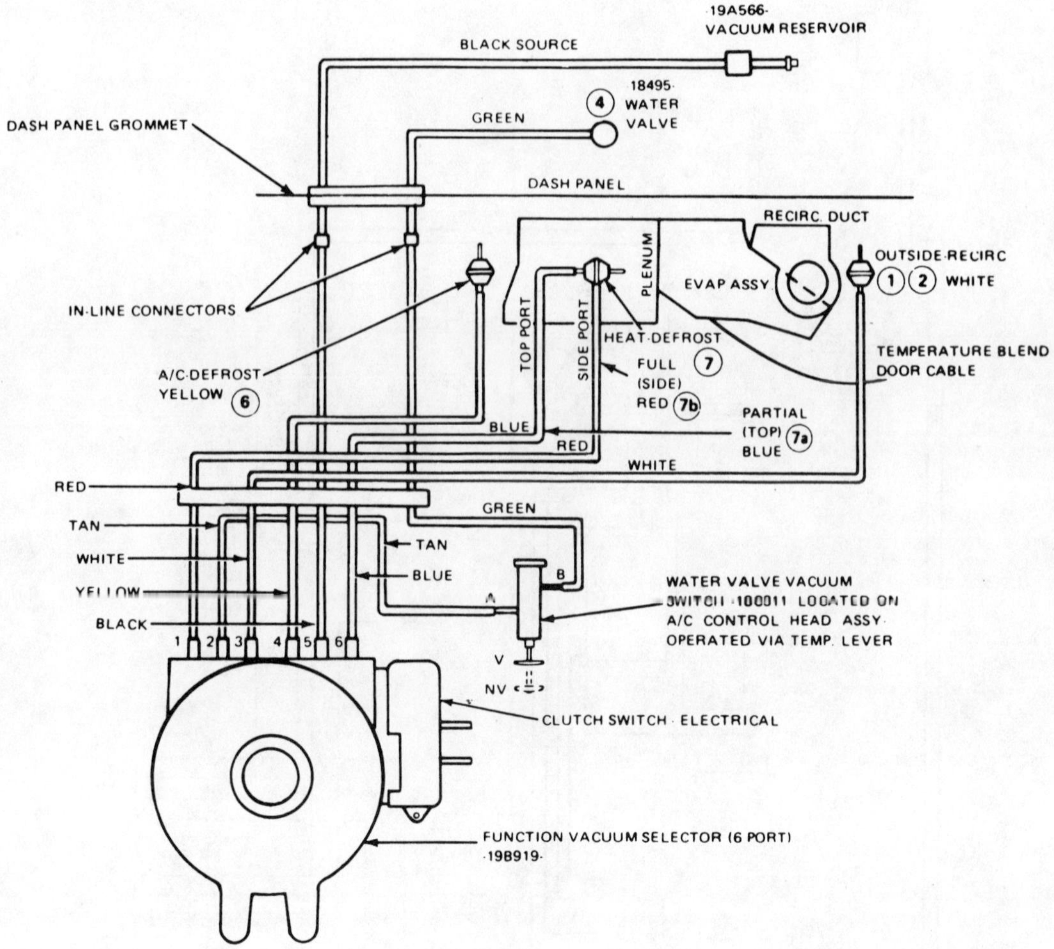

VACUUM SELECTOR CONTROL FUNCTION									
PORT	COLOR	FUNCTION	MAX	FRESH	OFF	ECONOMY VENT	HEAT	–	DEFROST
1	Red	Full Heat-Defrost					V		
2	Tan	Water Valve	V	V					
3	White	Recirculate	V		V				
4	Yellow	A/C-Defrost	V	V		V			
5	Black	Source	V	V	V	V	V	V	V
6	Blue	Partial Heat-Defrost				V	V	V	
	ELECTRICAL CLUTCH SWITCH		ON	ON	OFF	OFF	OFF	ON	ON

WATER VALVE VACUUM SWITCH FUNCTION

		COOL	TEMPERATURE	WARM
A	Tan	V	V	V
B	Green	V		

V = VACUUM BLANK = NO VACUUM (ATMOSPHERE)

1977 Ford Pinto & Mercury Bobcat A/C & Heater Vacuum Circuit

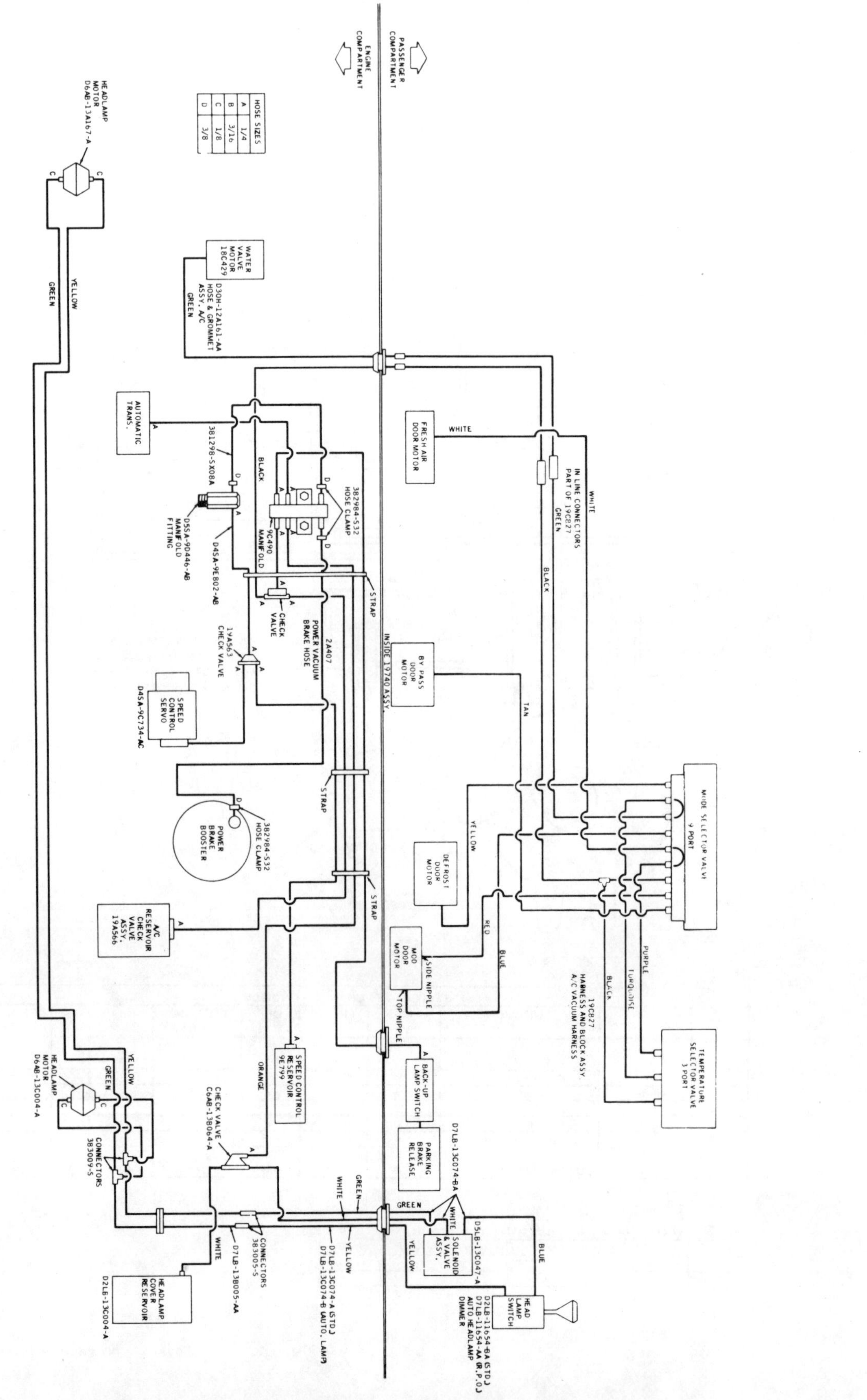

1977 Ford Thunderbird A/C & Heater Vacuum Circuit (Less Auto Temp Control)

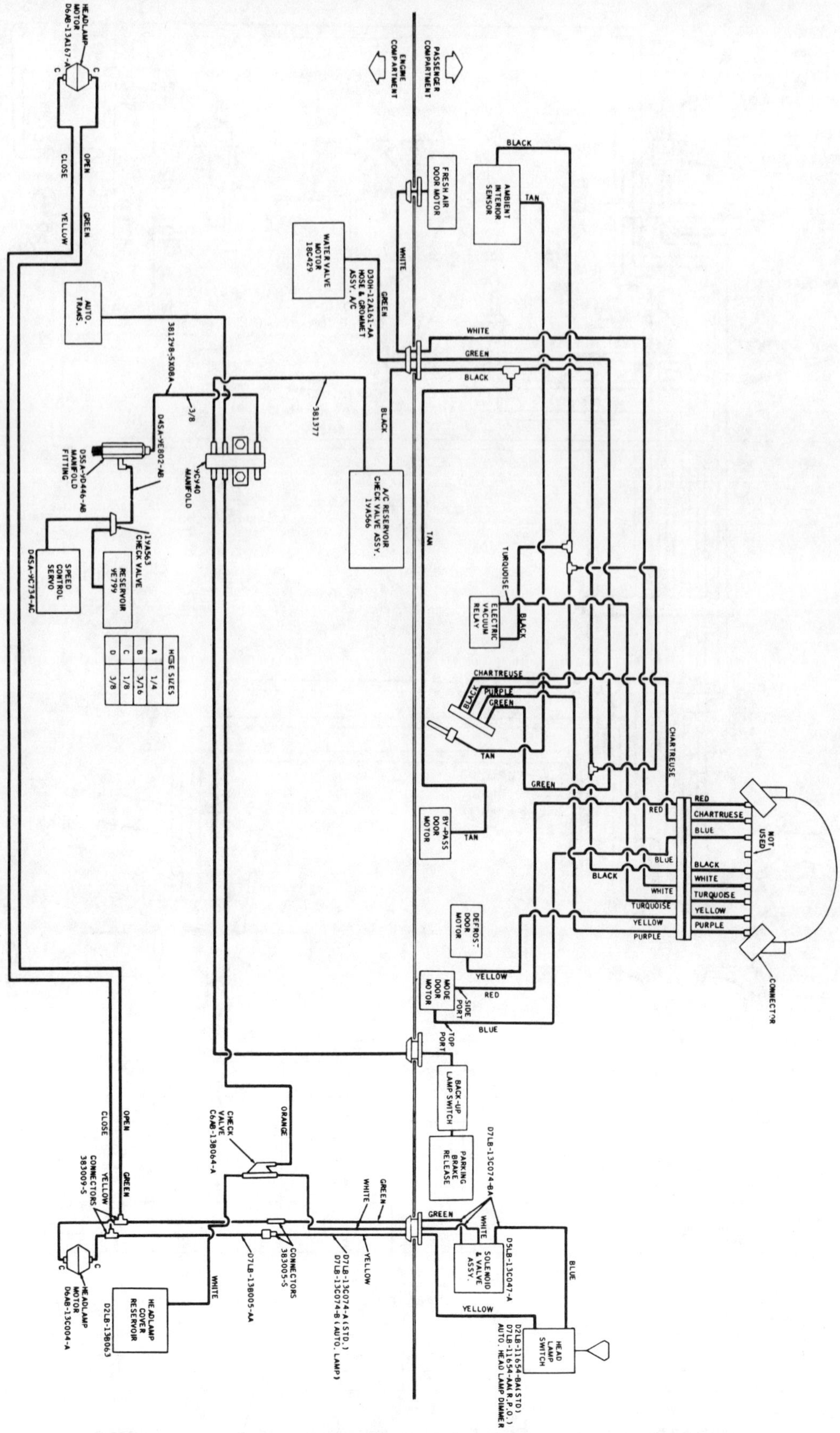

1977 Lincoln Continental A/C & Heater Vacuum Circuit (With Auto Temp Control)

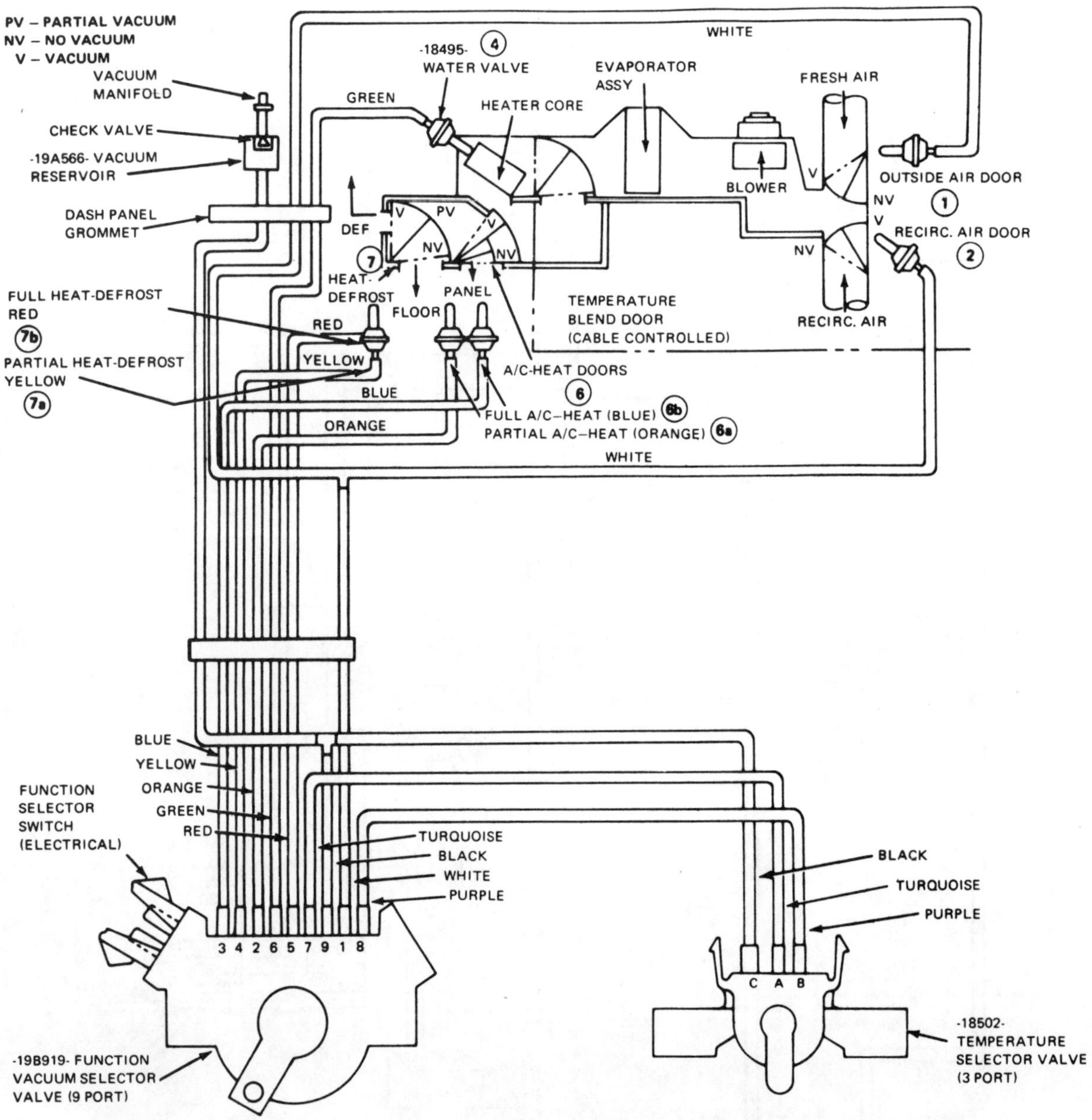

PORT	FUNCTION	FUNCTION SELECTOR VALVE POSITIONS							HOSE COLOR
		OFF	ECONOMY VENT	A/C	HI-LO	HEAT	▽ (DEFOG)	DEFROST	
1	Outside - Recirc.	V		To 8					White 1-2
2	Partial A/C-Heat		V	V	V				Orange 6a
3	Full A/C–Heat		V	V					Blue 6b
4	Partial Heat–Defrost	V	V	V	V	V	V		Yellow 7a
5	Full Heat-Defrost	V	V	V	V	V			Red 7b
6	To Engine Water Valve	V	V	To 7	To 7	To 7	To 7	To 7	Green 4
7	From Temp. Valve Port A	Seal	Seal	To 6	To 6	To 6	To 6	To 6	Turquoise
8	From Temp. Valve Port B	Seal	Seal	To 1	Seal	Seal	Seal	Seal	Purple
9	Source Vacuum	V	V	V	V	V	V	V	Black

PORT	FUNCTION	TEMPERATURE SELECTOR VALVE LEVER POSITIONS				HOSE COLOR
		COOL		TEMPERATURE	WARM	
C	Source Vacuum	V	V	V	V	Black
A	Water Valve	V				Turquoise
B	Outside–Recirc.	V	V			Purple

1977-78 Ford & Mercury Full Size A/C & Heater Vacuum Circuit (Less Auto Temp Control)

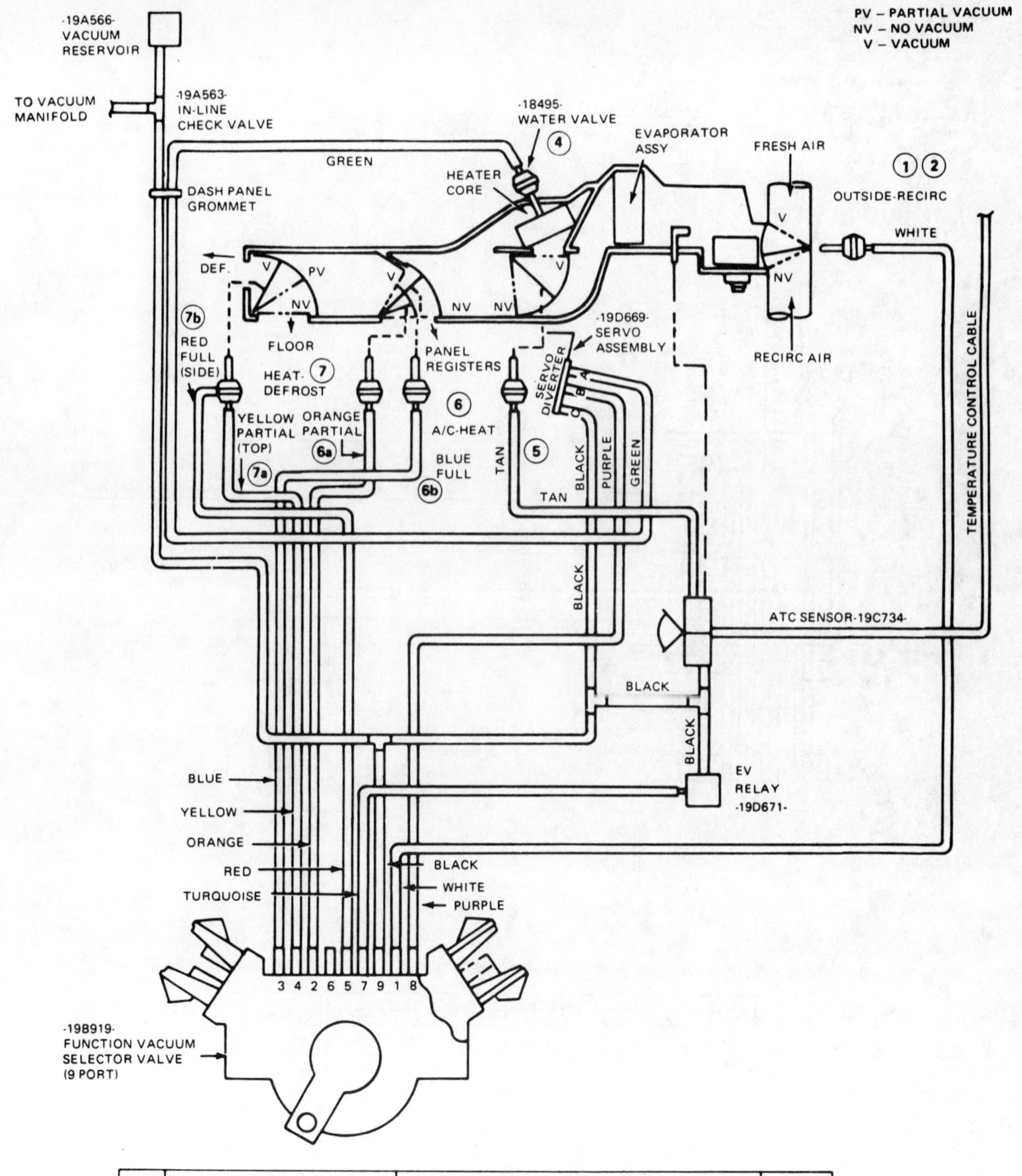

PORT	FUNCTION	FUNCTION SELECTOR VALVE LEVER POSITIONS							HOSE COLOR
		OFF	ECONOMY VENT	A/C	HI–LO	HEAT	▽ DEFOG	DEFROST	
1	Outside–Recirc.	V		To 8		To 7			White 1-2
2	Partial A/C–Heat		V	V	V				Orange 6a
3	Full A/C–Heat		V	V					Blue 6b
4	Partial Heat–Defrost	V	V	V	V	V	V		Yellow 7a
5	Full Heat–Defrost	V	V	V	V	V			Red 7b
6	No Function	Seal	Seal	Seal	Seal	Seal	Seal	Seal	
7	From EV Relay	Seal	Seal	Seal	Seal	To 1	Seal	Seal	Turquoise
8	From Servo Diverter Port B	Seal	Seal	To 1	Seal	Seal	Seal	Seal	Purple
9	Source Vacuum	V	V	V	V	V	V	V	Black

V – VACUUM BLANK – ATMOSPHERE

1977-79 Ford LTD II & Thunderbird; Continental Mark V & Mercury Cougar A/C & Heater Vacuum Circuit (With Auto Temp Control)

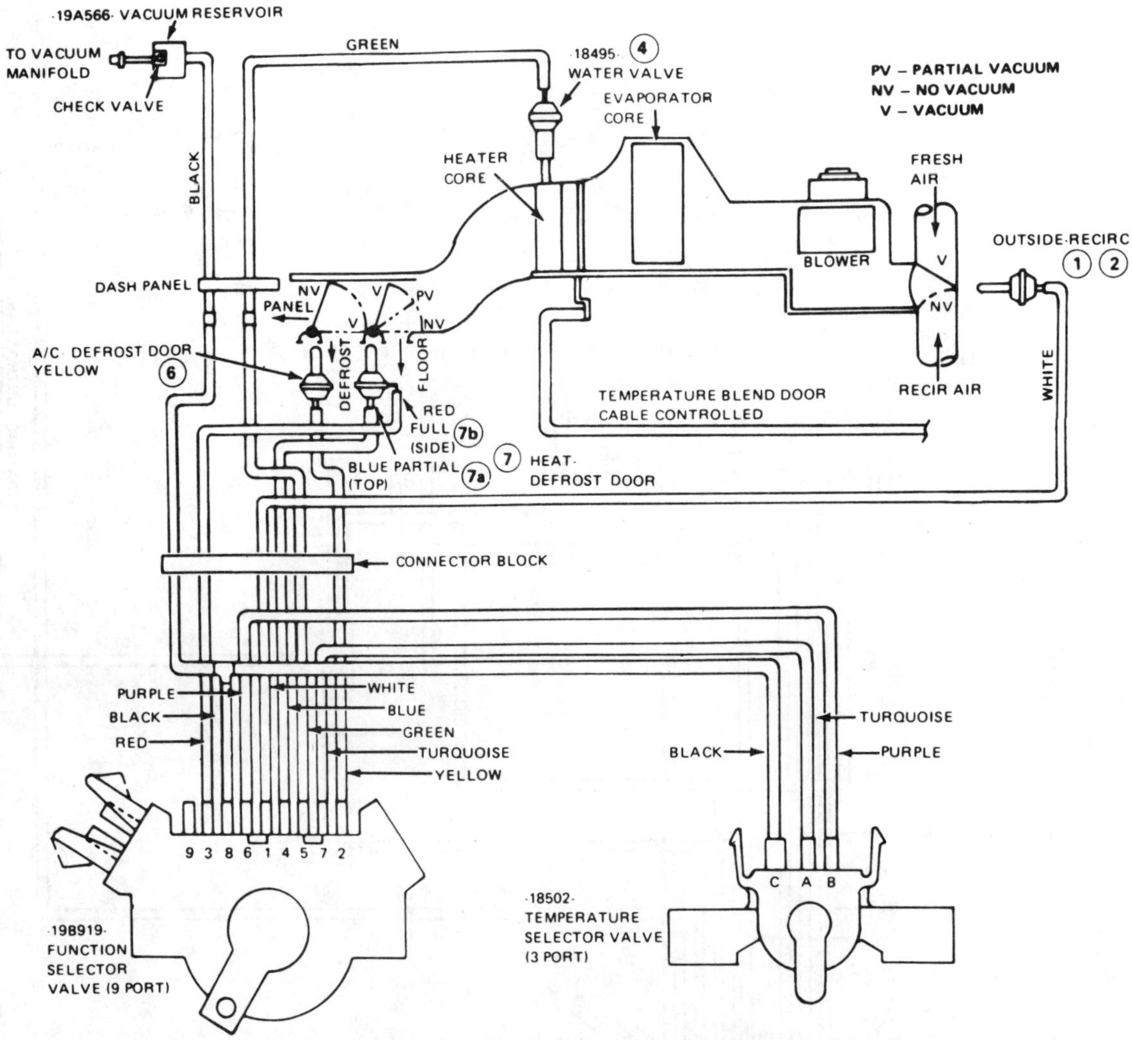

TEMP SELECTOR VALVE LEVER POSITIONS						
PORT	FUNCTION	COOL		TEMPERATURE	WARM	HOSE COLOR
C	Source	V	V	V	V	Black
A	Water Valve	V				Turquoise
B	Outside–Recirc	V	V			Purple

PORT	FUNCTION	FUNCTION SELECTOR VALVE LEVER POSITIONS							HOSE COLOR
		OFF	ECONOMY VENT	A/C	HI–LO	HEAT	▽ DEFOG	DEFROST	
1	Outside–Recirc.	V		To 6					White 1-2
2	A/C–Defrost		V	V	V				Yellow 6
3	Full Heat–Defrost	V				V			Red 7b
4	Partial Heat–Defrost	V			V	V	V		Blue 7a
5	To Water Valve	V	V	To 7	To 7	To 7	To 7	To 7	Green 4
6	From Temp Selector Port B	Seal	Seal	To 1	Seal	Seal	Seal	Seal	Purple
7	From Temp Selector Port A	Seal	Seal	To 5	To 5	To 5	To 5	To 5	Turquoise
8	Source	V	V	V	V	V	V	V	Black
9	NOT USED IN THIS SYSTEM								

V VACUUM BLANK – ATMOSPHERE

1977-80 Ford Granada & Mercury Monarch A/C & Heater Vacuum Circuit (Less Auto Temp Control)

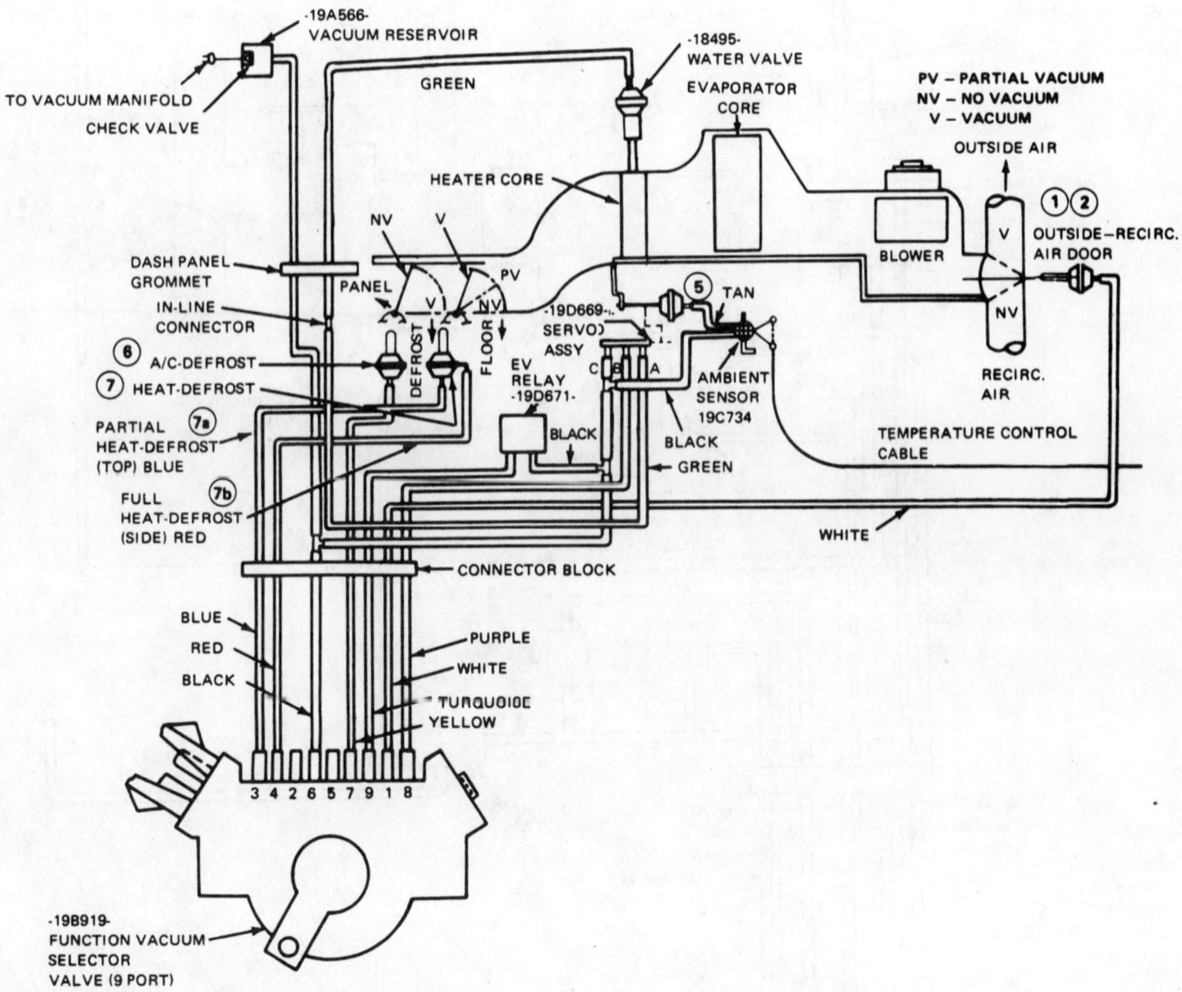

PORT	FUNCTION	FUNCTION SELECTOR VALVE LEVER POSITIONS							HOSE COLOR
		OFF	ECONOMY VENT	A/C	HI–LO	HEAT	▽ DEFOG	DEFROST	
1	Outside–Recirc.	V		To 8		To 2			White 1-2
2	From EV Relay	Seal	Seal	Seal	Seal	To 1	Seal	Seal	Turquoise
3	Partial Heat–Defrost	V			V	V	V		Blue 7a
4	Full Heat–Defrost	V				V			Red 7b
5	No Function	Seal	Seal	Seal	Seal	Seal	Seal	Seal	–
6	No Function	Seal	Seal	Seal	Seal	Seal	Seal	Seal	–
7	Defrost A/C–Defrost		V	V	V				Yellow 6
8	From Diverter Valve No. 2	Seal	Seal	To 1	Seal	Seal	Seal	Seal	Purple
9	Source	V	V	V	V	V	V	V	Black

V – VACUUM

BLANK – ATMOSPHERE

1977-80 Ford Granada, Lincoln Versailles & Mercury Monarch A/C & Heater Vacuum Circuit (With Auto Temp Control)

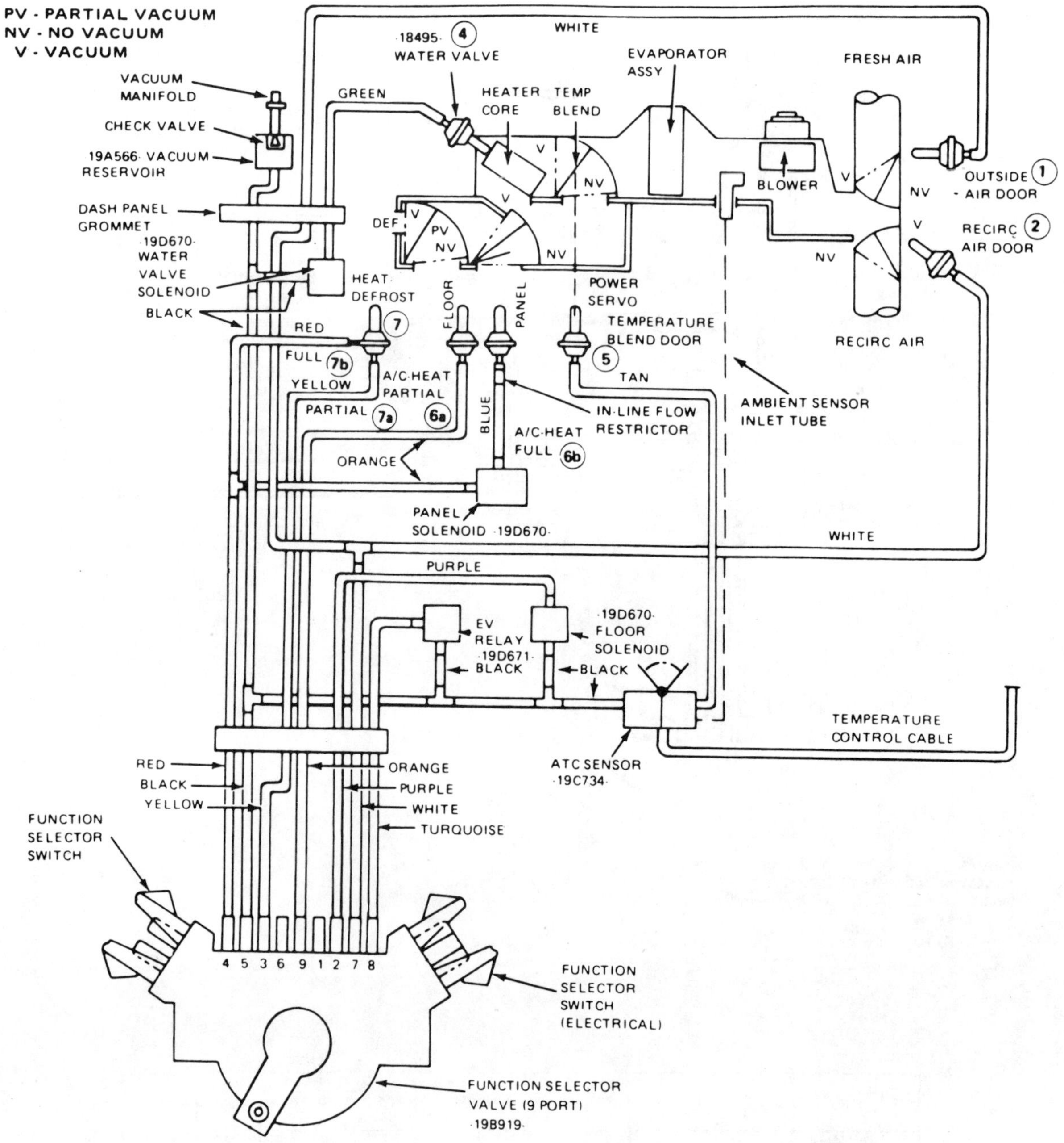

PORT	FUNCTION	FUNCTION SELECTOR VALVE POSITIONS					HOSE COLOR
		OFF	ECONOMY VENT	AUTO	DEFOG	DEFROST	
1	No Function	Seal	Seal	Seal	Seal	Seal	–
2	Floor Solenoid	Seal	Seal	To 9	Seal	Seal	Purple
3	Heat/Def Partial	V	V	V	V		Yellow 7a
4	Heat/Def Full	V	V	V			Red 7b
5	Source Vacuum	V	V	V	V	V	Black
6	No Function	Seal	Seal	Seal	Seal	Seal	–
7	Outside–Recirc.	V		To 8			White 1-2
8	EV Relay	Seal	Seal	To 7	Seal	Seal	Turquoise
9	A/C–Heat Partial		V	To 2			Orange 6a

1978 Ford Mustang & Pinto; Mercury Bobcat A/C & Heater Vacuum Circuit

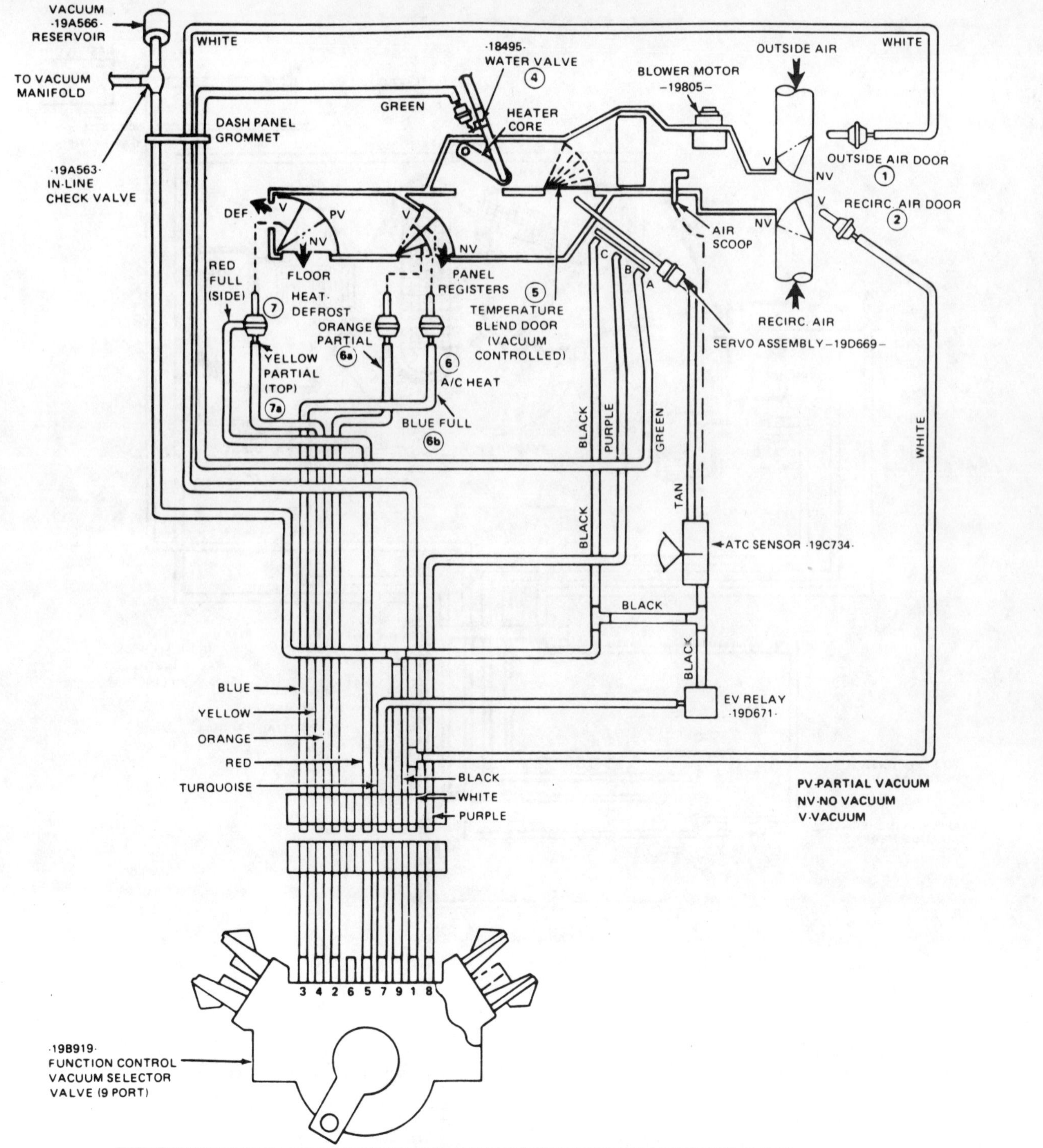

PORT	FUNCTION	FUNCTION SELECTOR VALVE LEVER POSITIONS							HOSE COLOR
		OFF	ECONOMY VENT	A/C	HI-LO	HEAT	∇ DEFOG	DEFROST	
1	Outside-Recirc	V		To 8		To 7			White 1-2
2	Partial A/C Heat		V	V	V				Orange 6a
3	Full A/C Heat		V	V					Blue 6b
4	Partial Heat-Defrost	V	V	V	V	V	V		Yellow 7a
5	Full Heat-Defrost	V	V	V	V	V			Red 7b
6	No Function	Seal	Seal	Seal	Seal	Seal	Seal	Seal	
7	From EV Relay	Seal	Seal	Seal	Seal	To 1	Seal	Seal	Turquoise
8	From Servo Diverter Valve Port B	Seal	Seal	To 1	Seal	Seal	Seal	Seal	Purple
9	Source Vacuum	V	V	V	V	V	V	V	Black

V - VACUUM

BLANK - ATMOSPHERE

1978 Ford & Mercury Full Size; 1978-79 Lincoln Continental A/C & Heater Vacuum Circuit (With Auto Temp Control)

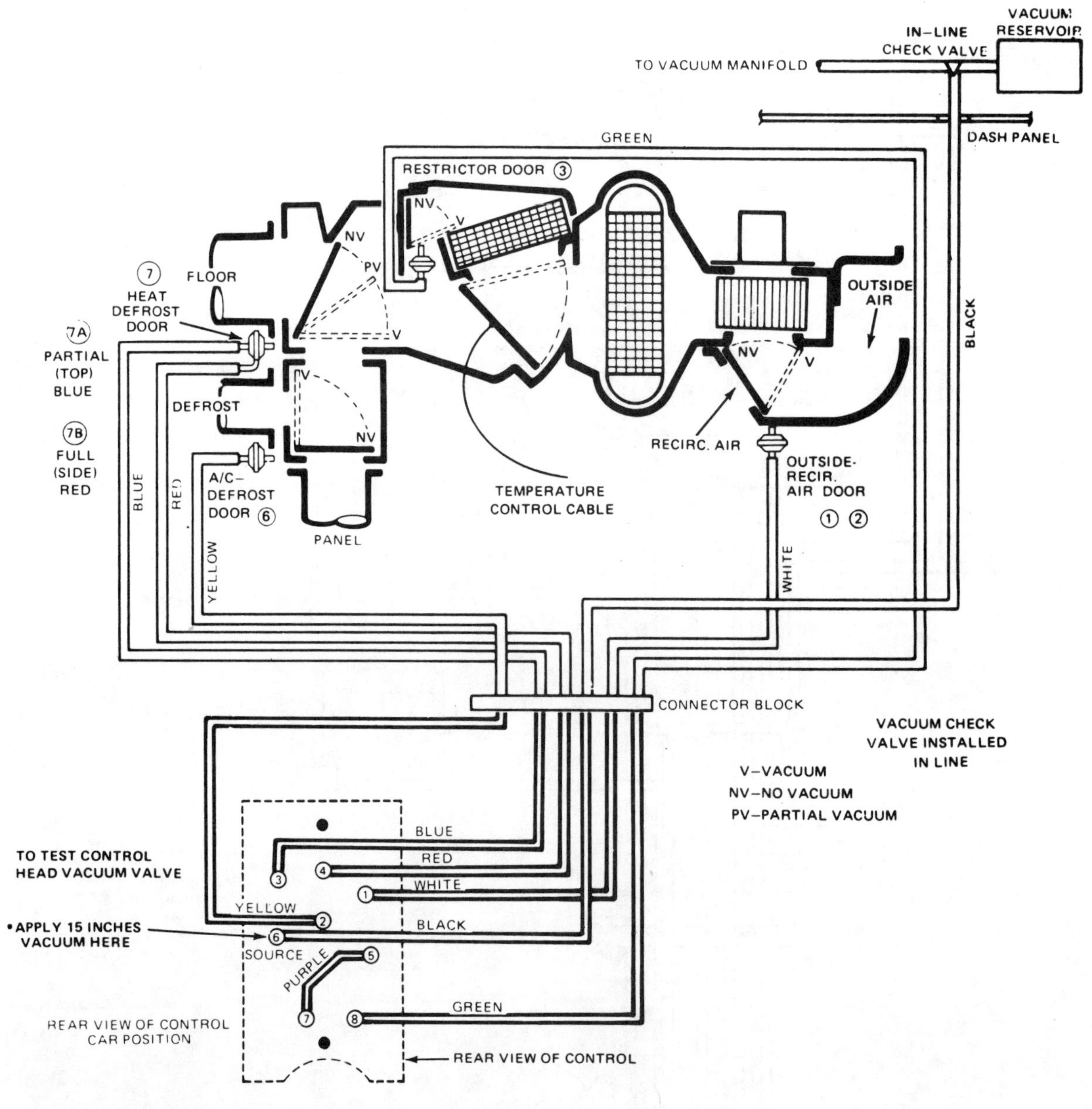

•CHECK FOR VACUUM TO OTHER PORTS PER CHARTS

•REPLACE CONTROL HEAD IF NOT PER CHARTS

FUNCTION SELECTOR LEVER		DETENT POSITIONS						
PORT	FUNCTION	OFF	MAX	NORM	VENT	HI-LO	HEAT	DEF
1	OUTSIDE-RECIRC	V	V	–	–	–	–	–
2	PANEL–DEF	V	V	V	V	V	–	–
3	PANEL–FLOOR	V	–	–	–	V	V	–
4	FULL FLOOR	V	–	–	–	–	V	–
5	TO PORT 7	–	V	V	V	–	–	–
6	SOURCE	V	V	V	V	V	V	V

TEMPERATURE LEVER POSITION			
PORT	FUNCTION	COOL	WARM
7	FROM PORT 5	EXT. CONNECTED TO PORT 5	
8	RESTRICTOR	V	
		0° 12°	42°

1978-80 Ford Fairmont & Mercury Zephyr A/C & Heater Vacuum Circuit

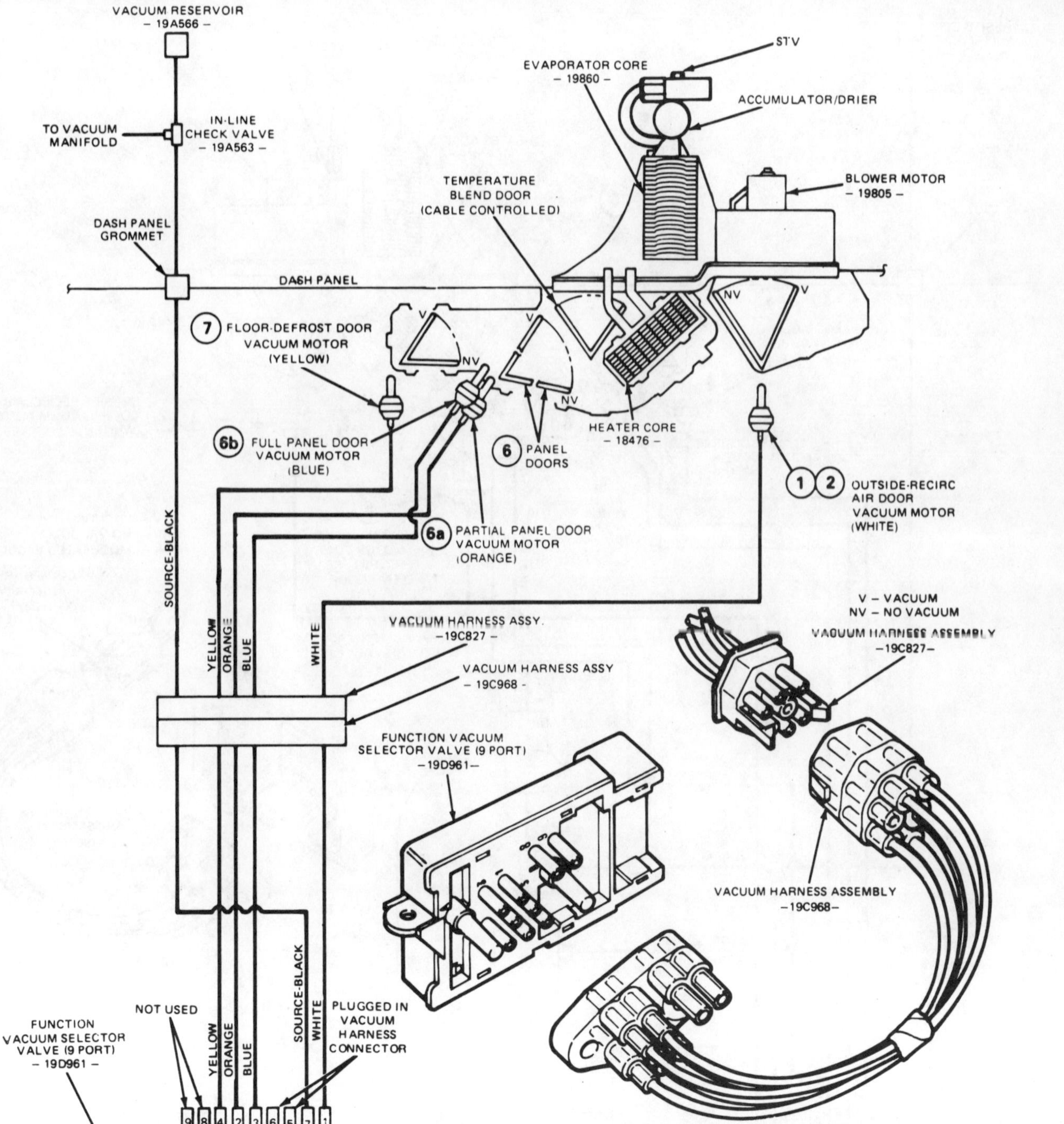

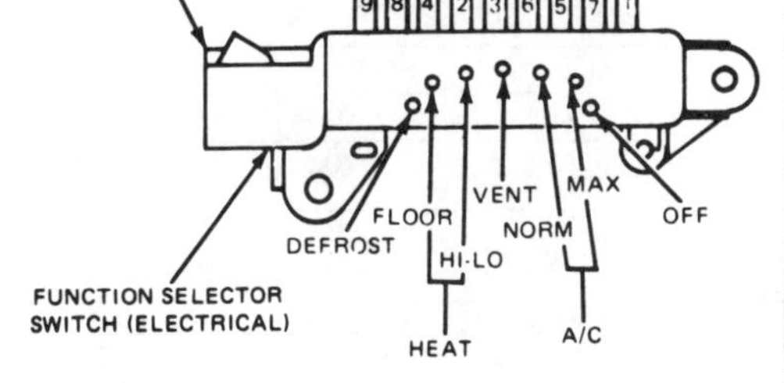

PORT	FUNCTION	FUNCTION SELECTOR LEVER POSITIONS							HOSE COLOR
		OFF	A/C MAX	A/C NORM	VENT	HEAT HI-LO	HEAT FLOOR	DEFROST	
1	OUTSIDE RECIR.	V	V						WHITE 1-2
2	PARTIAL PANEL		V	V	V	V			ORANGE 6a
3	FULL PANEL		V	V	V				BLUE 6b
4	FLOOR-DEFROST	V	V	V	V	V	V		YELLOW 7
5	NONE		TO 6	TO 6	TO 6	TO 6	TO 6	TO 6	PLUGGED
6	NONE	SEALED	TO 5	TO 5	TO 5	TO 5	TO 5	TO 5	PLUGGED
7	SOURCE	V	V	V	V	V	V	V	BLACK
8	NONE								OPEN
9	NONE								OPEN

V VACUUM BLANK ATMOSPHERE

1979-80 Ford & Mercury Full Size A/C & Heater Vacuum Circuit (Less Auto Temp Control)

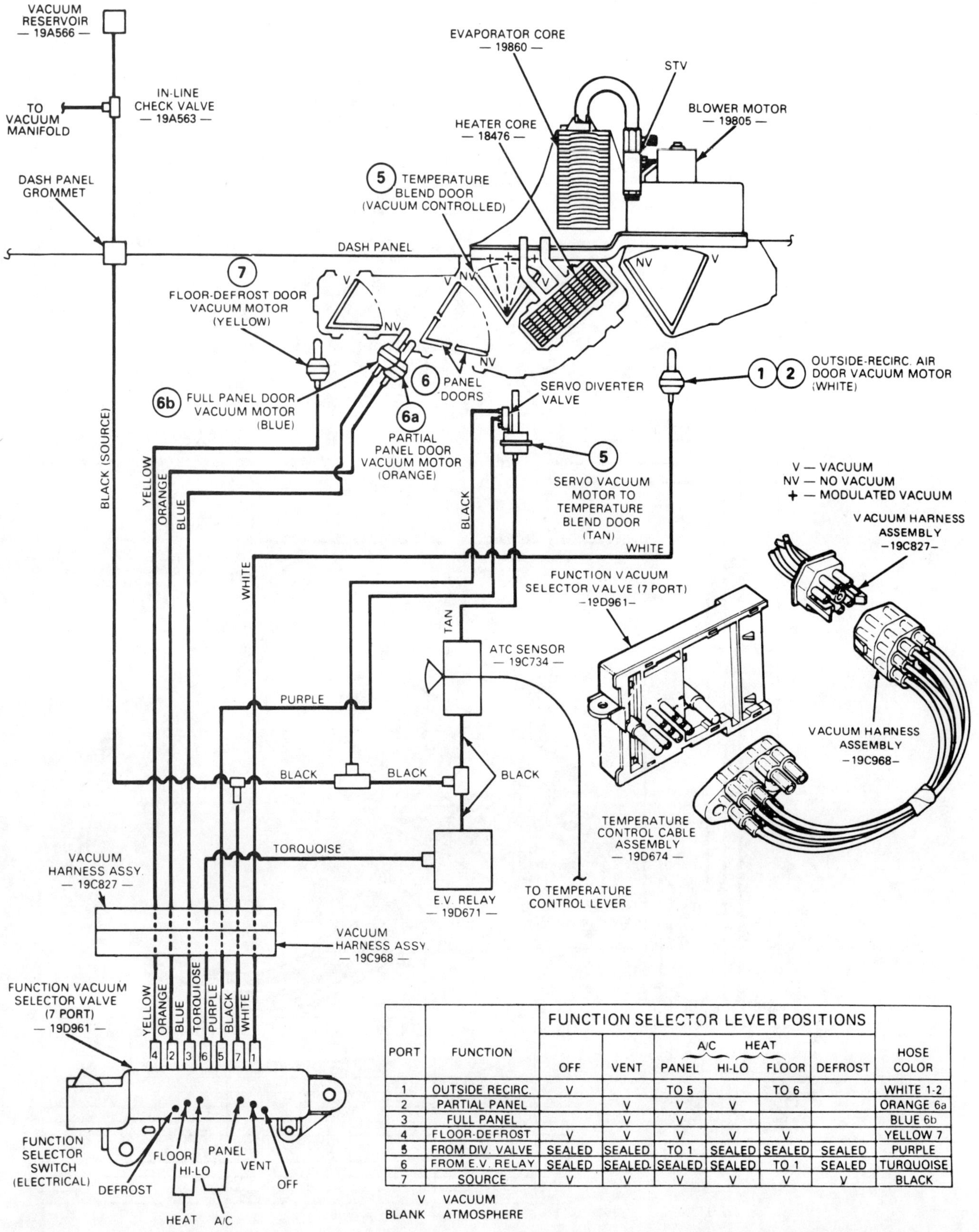

PORT	FUNCTION	FUNCTION SELECTOR LEVER POSITIONS						HOSE COLOR
				A/C	HEAT			
		OFF	VENT	PANEL	HI-LO	FLOOR	DEFROST	
1	OUTSIDE RECIRC.	V		TO 5		TO 6		WHITE 1-2
2	PARTIAL PANEL		V	V	V			ORANGE 6a
3	FULL PANEL		V	V				BLUE 6b
4	FLOOR-DEFROST	V	V	V	V	V		YELLOW 7
5	FROM DIV. VALVE	SEALED	SEALED	TO 1	SEALED	SEALED	SEALED	PURPLE
6	FROM E.V. RELAY	SEALED	SEALED	SEALED	SEALED	TO 1	SEALED	TURQUOISE
7	SOURCE	V	V	V	V	V	V	BLACK

V VACUUM
BLANK ATMOSPHERE

1979-80 Ford & Mercury Full Size; 1980 Continental Mark VI & Lincoln Continental A/C & Heater Vacuum Circuit (With Auto Temp Control)

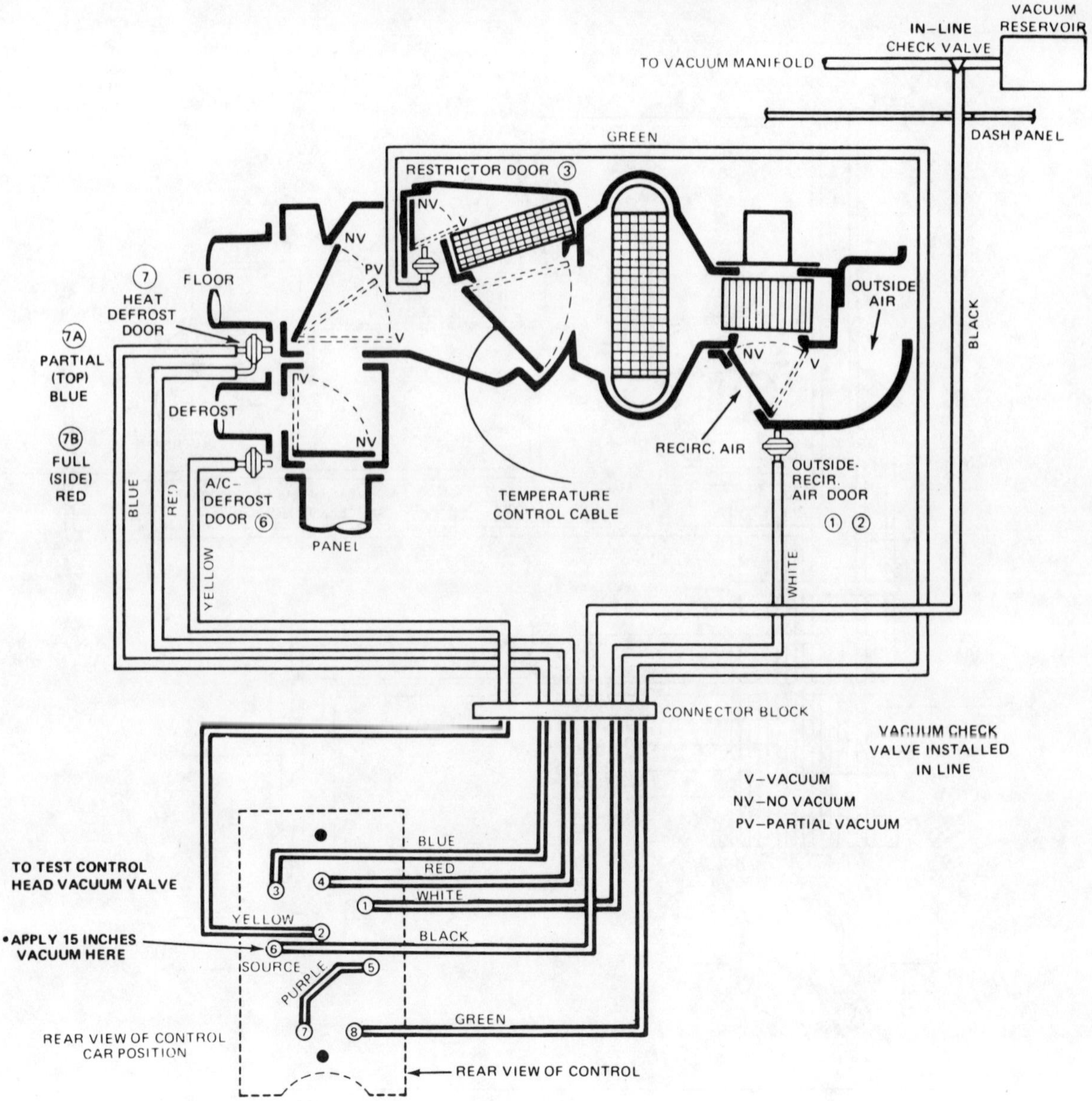

•CHECK FOR VACUUM TO OTHER PORTS PER CHARTS

•REPLACE CONTROL HEAD IF NOT PER CHARTS

FUNCTION SELECTOR LEVER		DETENT POSITIONS						
PORT	FUNCTION	OFF	MAX	NORM	VENT	HI LO	HEAT	DEF
1	OUTSIDE-RECIRC	V	V	–	–		–	–
2	PANEL - DEF	V	V	V	V	V	–	–
3	PANEL FLOOR	V	–	–	–	V	V	–
4	FULL FLOOR	V	–		–	–	V	–
5	TO PORT 7	–	V	V	V	–	–	–
6	SOURCE	V	V	V	V	V	V	V

TEMPERATURE LEVER POSITION			
PORT	FUNCTION	COOL	WARM
7	FROM PORT 5	EXT. CONNECTED TO PORT 5	
8	RESTRICTOR	V	–
		0^0 12^0	42^0

1979-80 Ford Mustang & Mercury Capri A/C & Heater Vacuum Circuit

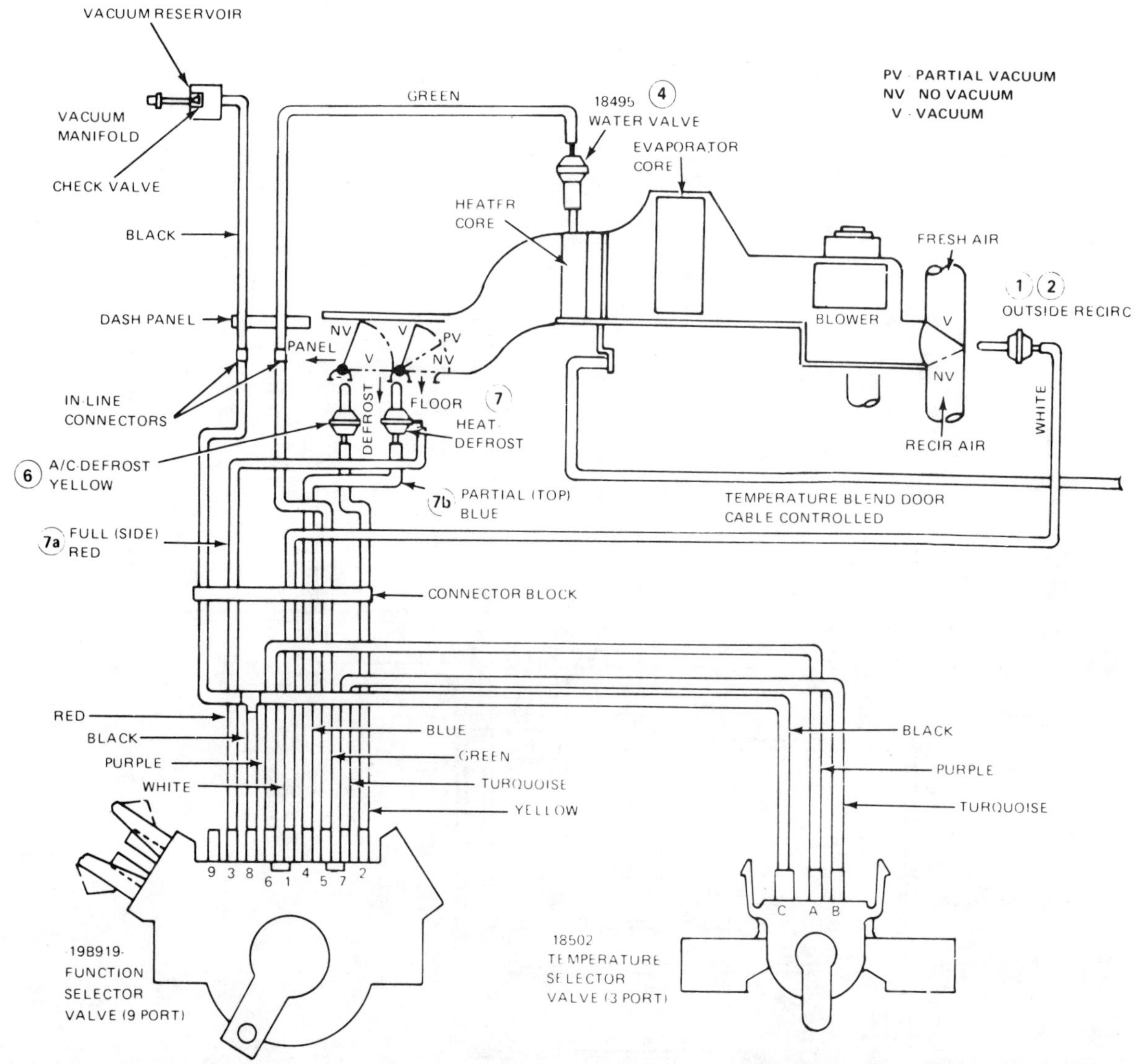

PORT	FUNCTION	FUNCTION SELECTOR VALVE LEVER POSITIONS							HOSE
		OFF	ECONOMY VENT	A/C	HI LO	HEAT	∇ DEFOG	DEFROST	COLOR
1	Outside- Recirc.	V		To 6					White 1 2
2	A/C Defrost		V	V	V				Yellow 6
3	Full Heat Defrost	V				V			Red 7a
4	Partial Heat - Defrost	V			V	V	V		Blue 7b
5	To Water Valve	V	V	To 7	To 7	To 7	To 7	To 7	Green 4
6	From Temp Port A	Seal	Seal	To 1	Seal	Seal	Seal	Seal	Purple
7	From Temp Port B	Seal	Seal	To 5	To 5	To 5	To 5	To 5	Turquoise
8	Source	V	V	V	V	V	V	V	Black
9		NOT USED IN THIS SYSTEM							

TEMPERATURE SELECTOR VALVE FUNCTION						
PORT	FUNCTION	COOL		TEMPERATURE	WARM	HOSE COLOR
C	Source	V	V	V	V	Black
A	Outside Recirc	V				Purple
B	Water Valve	V	V			Turquoise

V - VACUUM   BLANK - ATMOSPHERE

1979-80 Ford Pinto & Mercury Bobcat A/C & Heater Vacuum Circuit

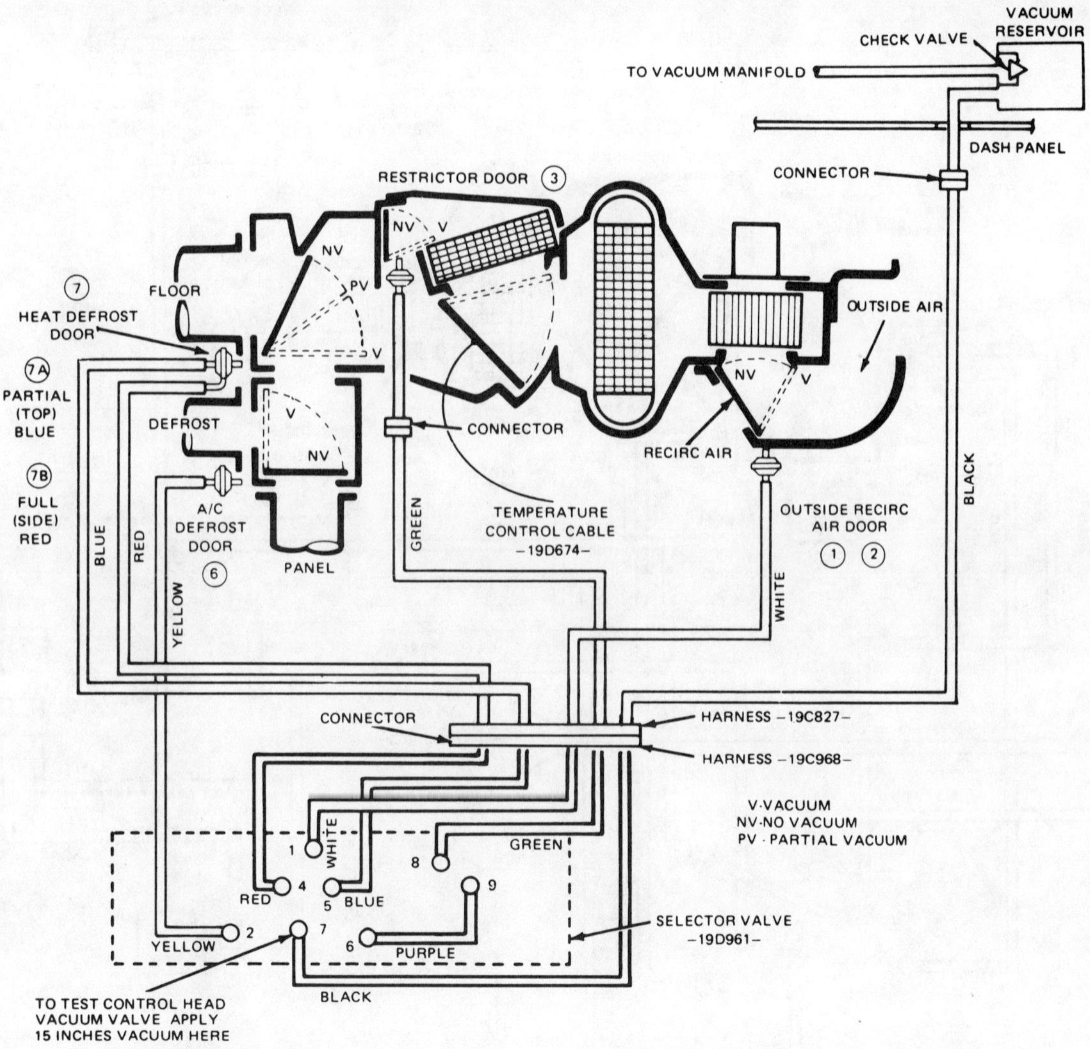

FUNCTION SELECTOR LEVER		DETENT POSITIONS						
PORT	FUNCTION	OFF	MAX	NORM	VENT	HI-LO	FLOOR	DEF
1	OUTSIDE-RECIRC	V	V	—	—	—	—	—
2	PANEL-DEF	V	V	V	V	V	—	—
3	SEALED	—	—	—	—	—	—	—
4	FULL FLOOR	V	—	—	—	—	V	—
5	PANEL-FLOOR	V	—	—	—	V	V	—
6	TO PORT 9	—	V	V	V	—	—	—
7	SOURCE	V	V	V	V	V	V	V

TEMPERATURE LEVER POSITION			
PORT	FUNCTION	COOL	WARM
9	FROM PORT 6	EXTERNALLY CONNECTED TO PORT 6	
8	RESTRICTOR	V (0°–12°)	— (12°–50°)

0° 12° 50°

V VACUUM
– ATMOSPHERE

1980 Ford Thunderbird & Mercury Cougar XR-7 A/C & Heater Vacuum Circuit (Less Auto Temp Control)

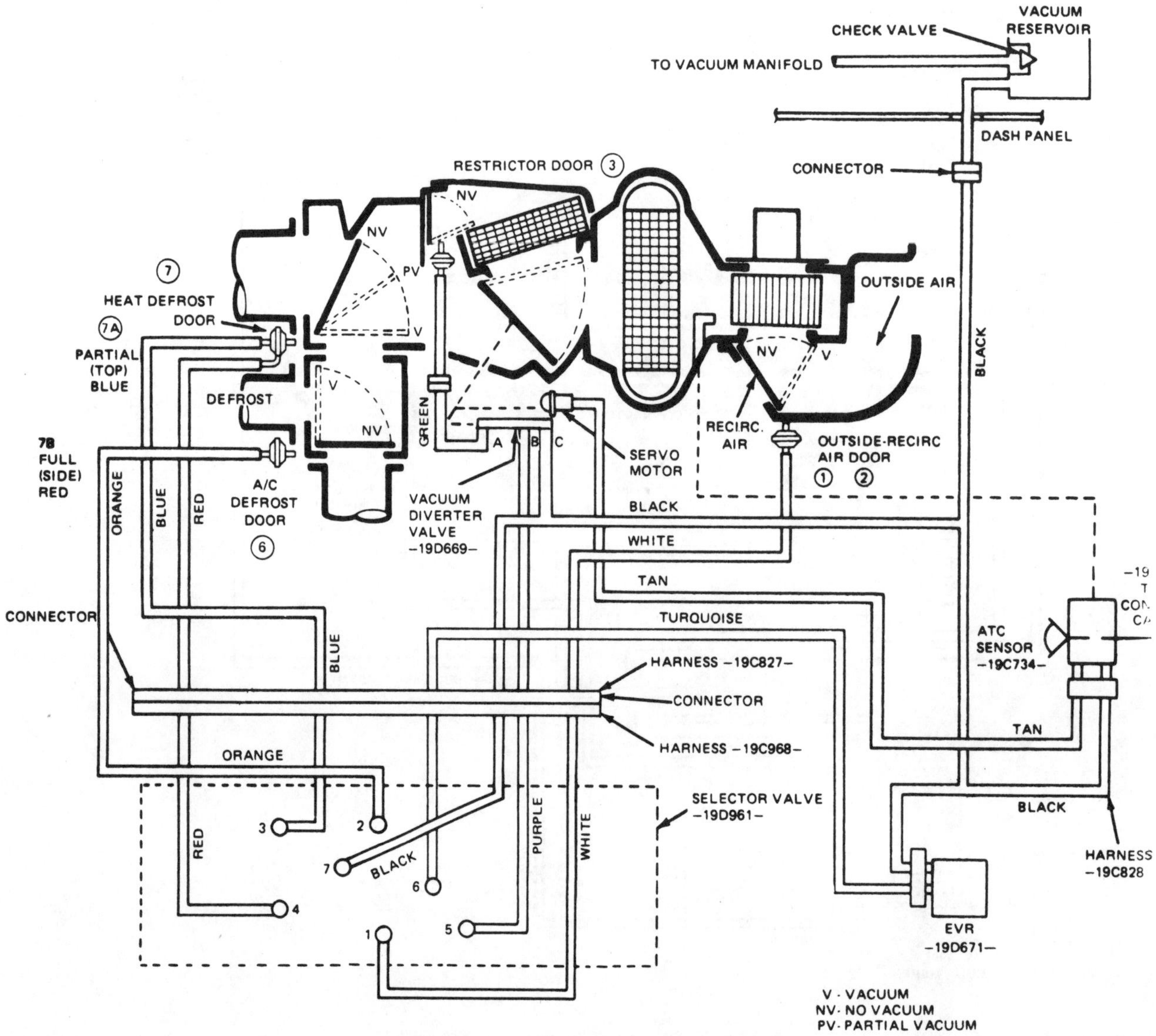

FUNCTION SELECTOR LEVER		DETENT POSITIONS					
PORT	FUNCTION	OFF	VENT	PANEL	HI-LO	FLOOR	DEF
1	OUTSIDE-RECIRC	V	—	TO-5	—	TO-6	—
2	PANEL-DEFROST	—	V	V	V	—	—
3	PANEL-FLOOR	—	—	—	V	—	—
4	FULL FLOOR	V	—	—	—	V	—
5	FROM DIVERTER	V	—	TO-1	—	V	—
6	FROM EVR	—	—	—	—	TO-1	—
7	SOURCE	V	V	V	V	V	V

V VACUUM

– ATMOSPHERE

1980 Ford Thunderbird & Mercury Cougar XR-7 A/C & Heater Vacuum Circuit (With Auto Temp Control)

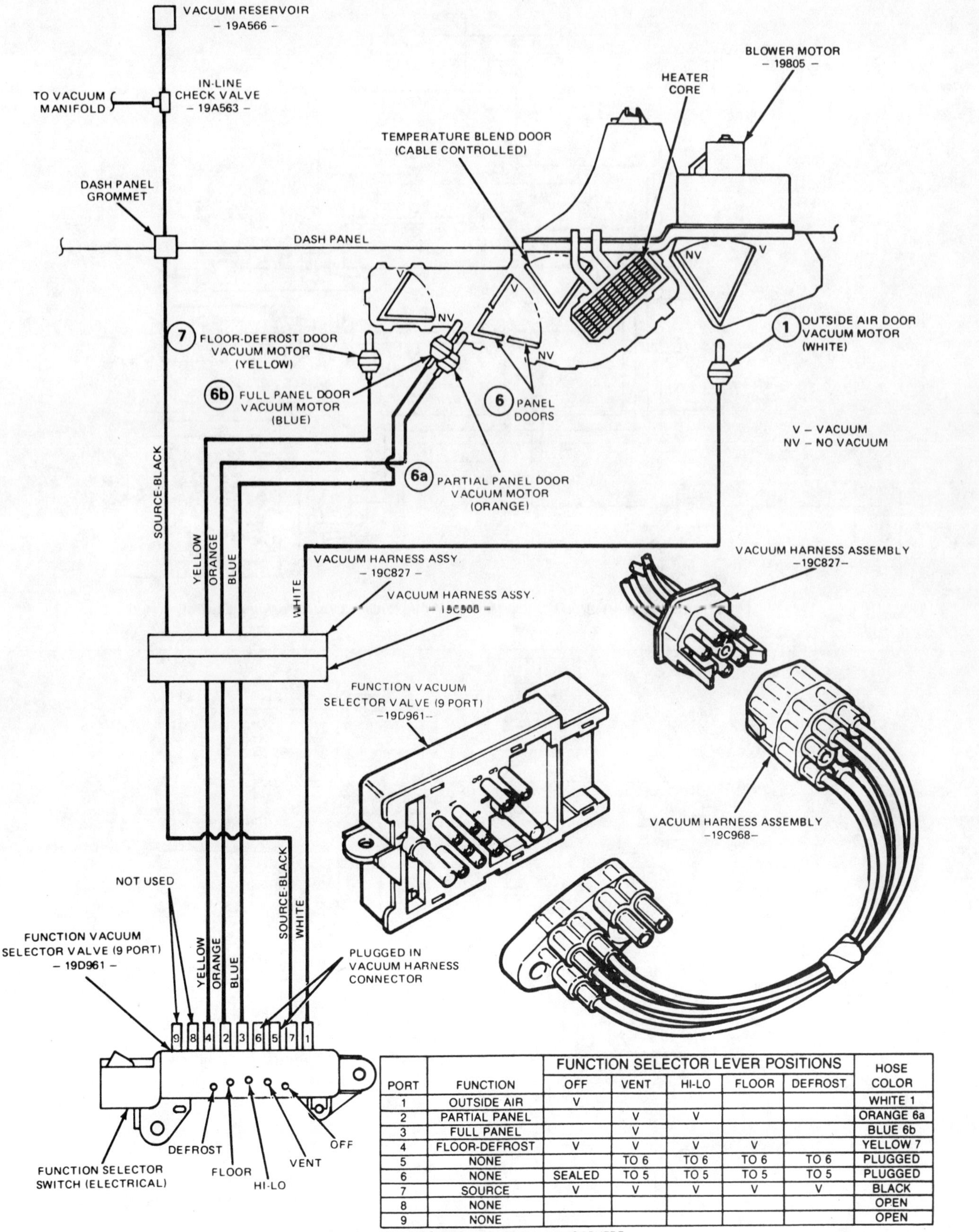

PORT	FUNCTION	FUNCTION SELECTOR LEVER POSITIONS					HOSE COLOR
		OFF	VENT	HI-LO	FLOOR	DEFROST	
1	OUTSIDE AIR	V					WHITE 1
2	PARTIAL PANEL		V	V			ORANGE 6a
3	FULL PANEL		V				BLUE 6b
4	FLOOR-DEFROST	V	V	V	V		YELLOW 7
5	NONE		TO 6	TO 6	TO 6	TO 6	PLUGGED
6	NONE	SEALED	TO 5	TO 5	TO 5	TO 5	PLUGGED
7	SOURCE	V	V	V	V	V	BLACK
8	NONE						OPEN
9	NONE						OPEN

V = VACUUM BLANK ATMOSPHERE

1979-80 Ford & Mercury Full Size Heater Vacuum Circuit

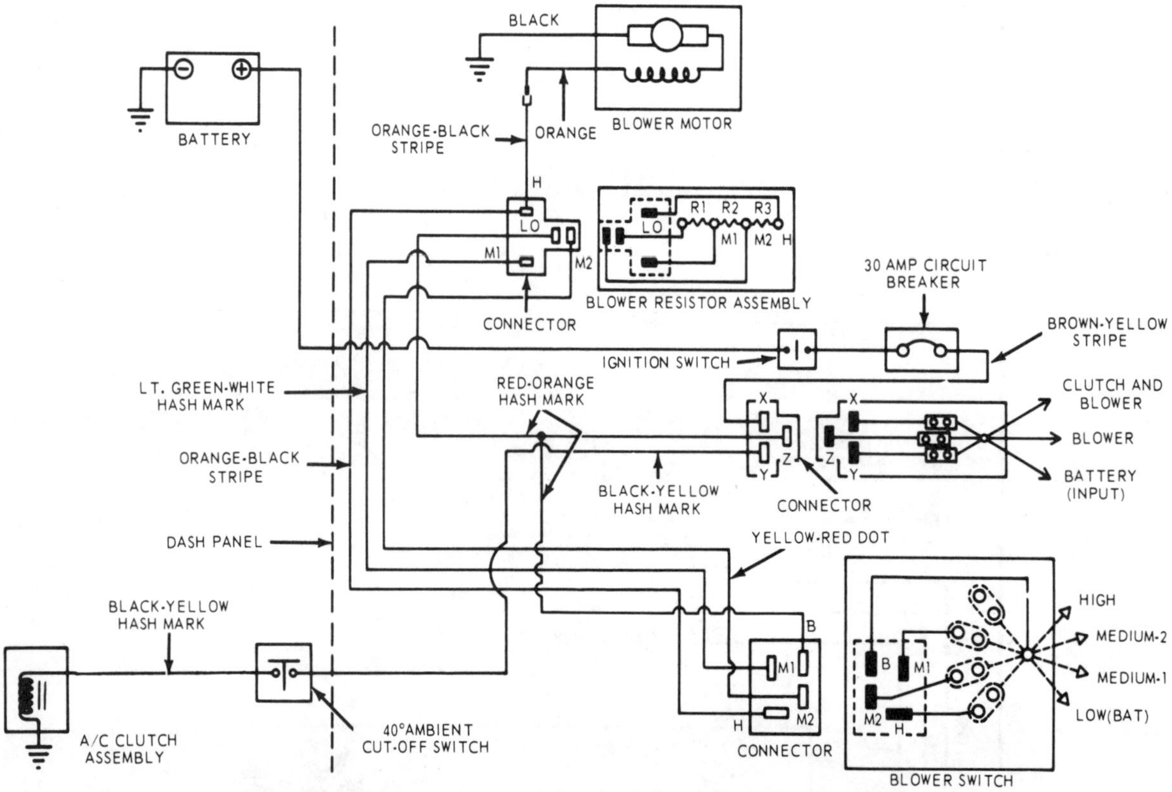

1974 Ford & Mercury Full Size A/C & Heater Wiring Circuit (Less Auto Temp Control)

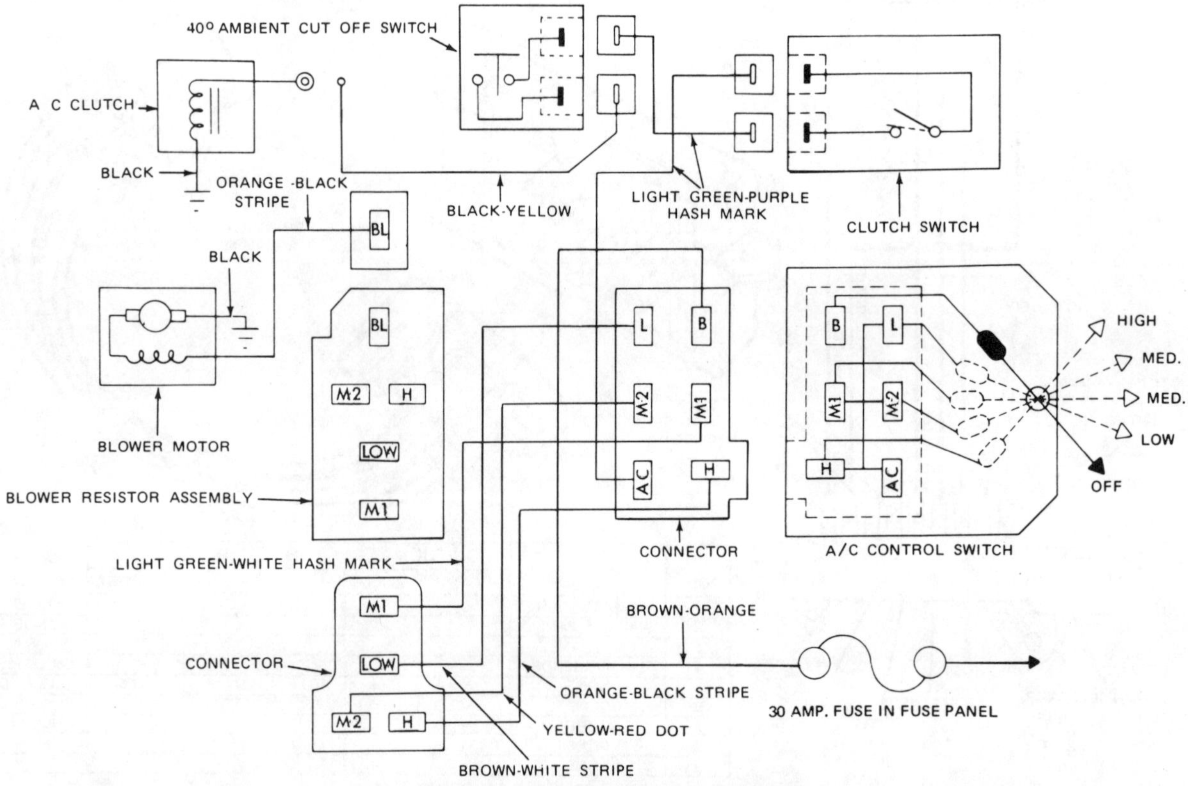

1974 Pinto A/C & Heater Wiring Circuit

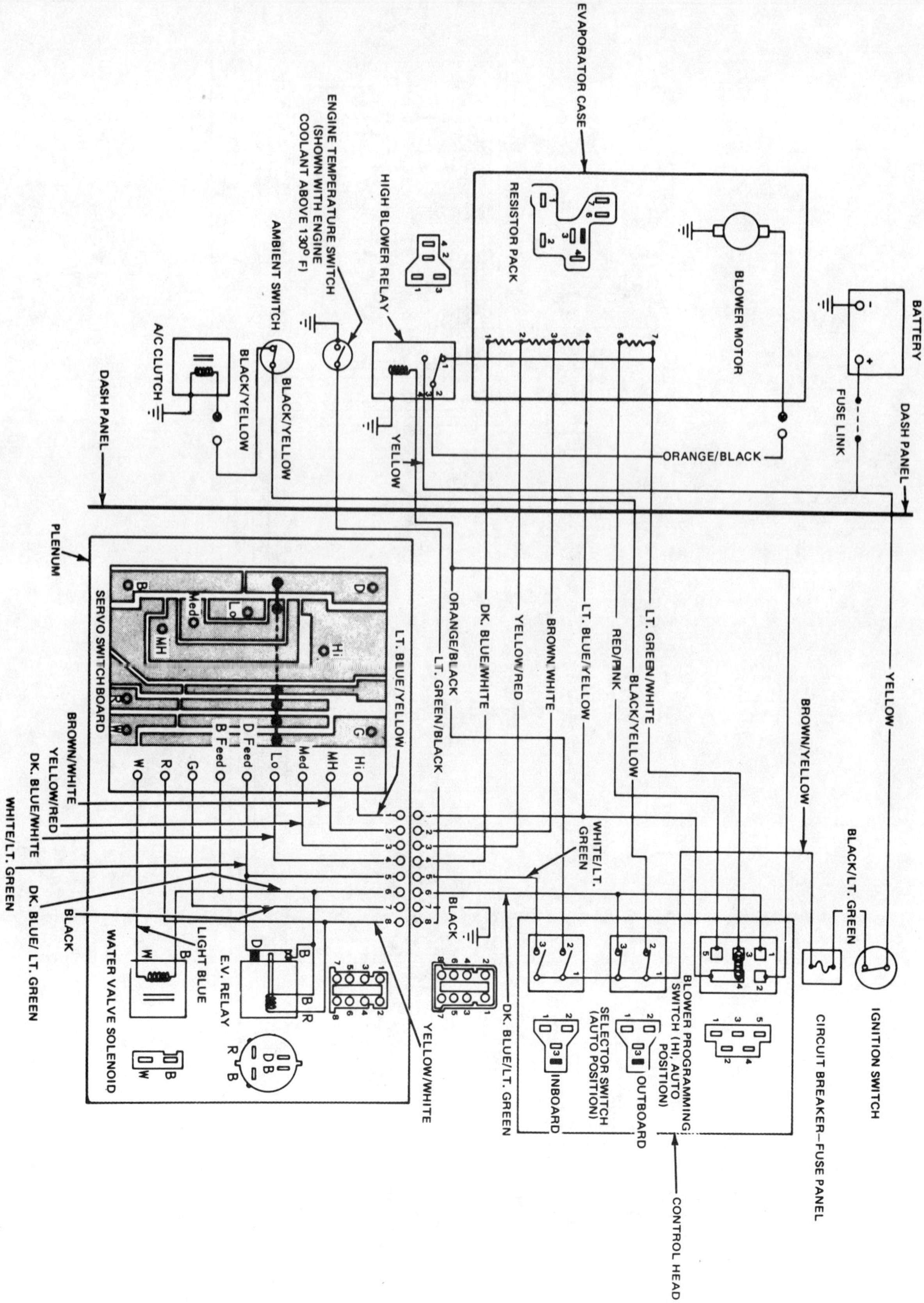

1974 Ford & Mercury Full Size A/C & Heater Wiring Circuit (With Auto Temp Control)

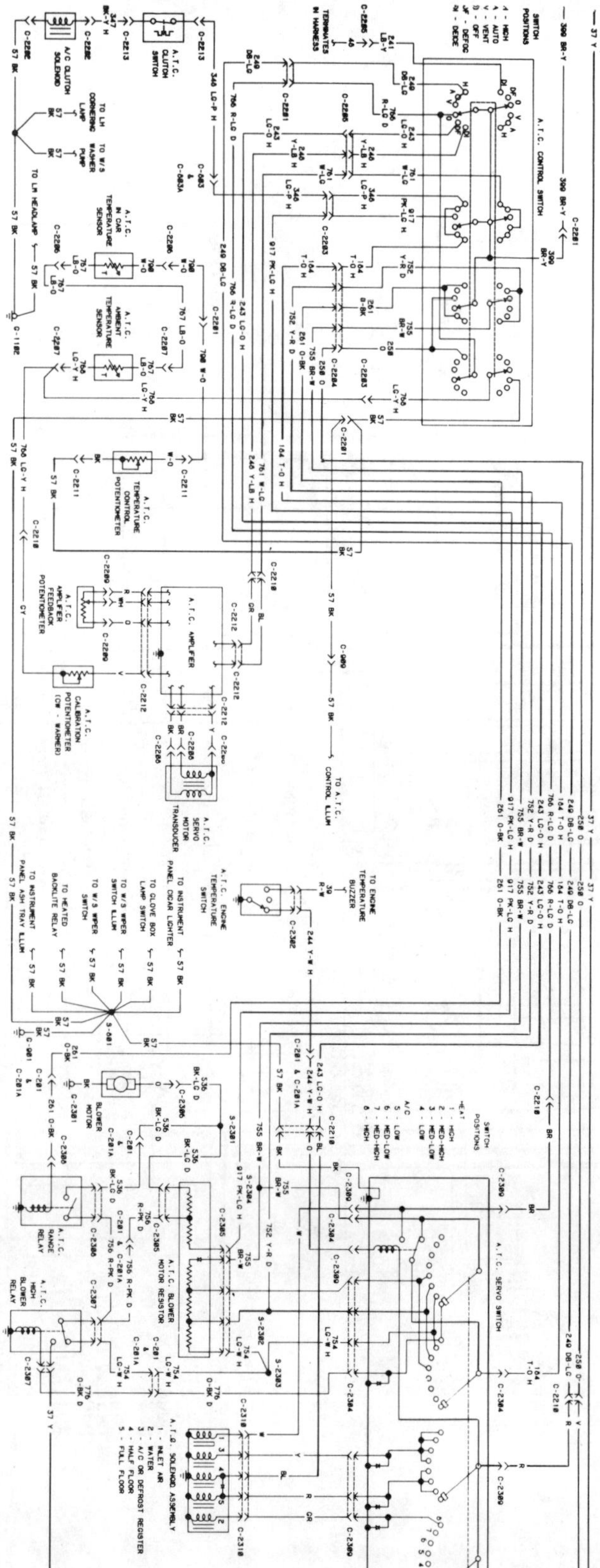

1974 Continental Mark IV A/C & Heater Wiring Circuit (With Auto Temp Control)

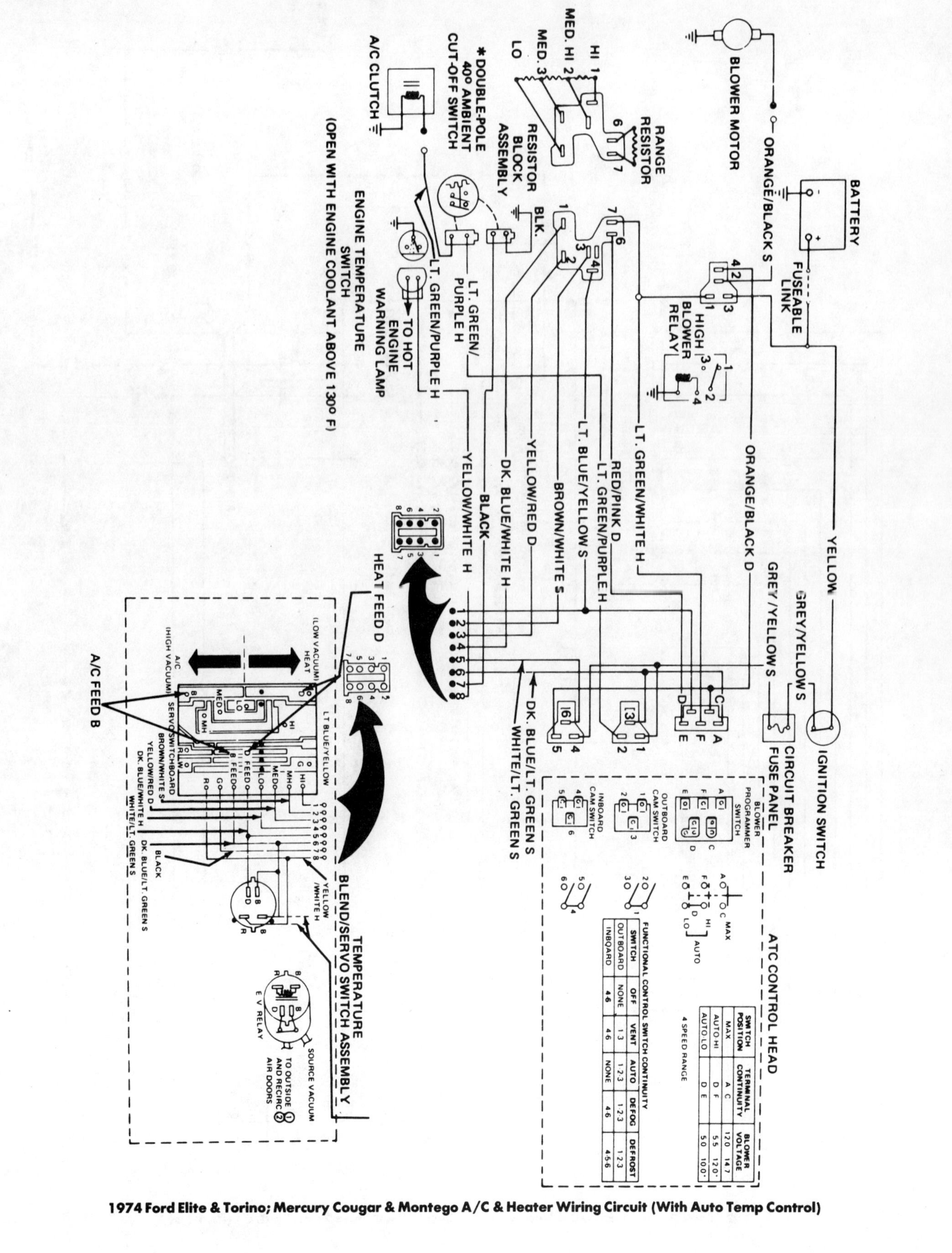

SWITCH POSITION	TERMINAL CONTINUITY	BLOWER VOLTAGE
MAX	A C	12.0 14.7
AUTO HI	D F	5.5 12.0
AUTO LO	D E	5.0 10.0

FUNCTIONAL CONTROL SWITCH CONTINUITY

SWITCH	OFF	VENT	AUTO	DEFOG	DEFROST
OUTBOARD	NONE	1-3	1-2-3	1-2-3	1-2-3
INBOARD	4-6	4-6	NONE	4-6	4-5-6

1974 Ford Elite & Torino; Mercury Cougar & Montego A/C & Heater Wiring Circuit (With Auto Temp Control)

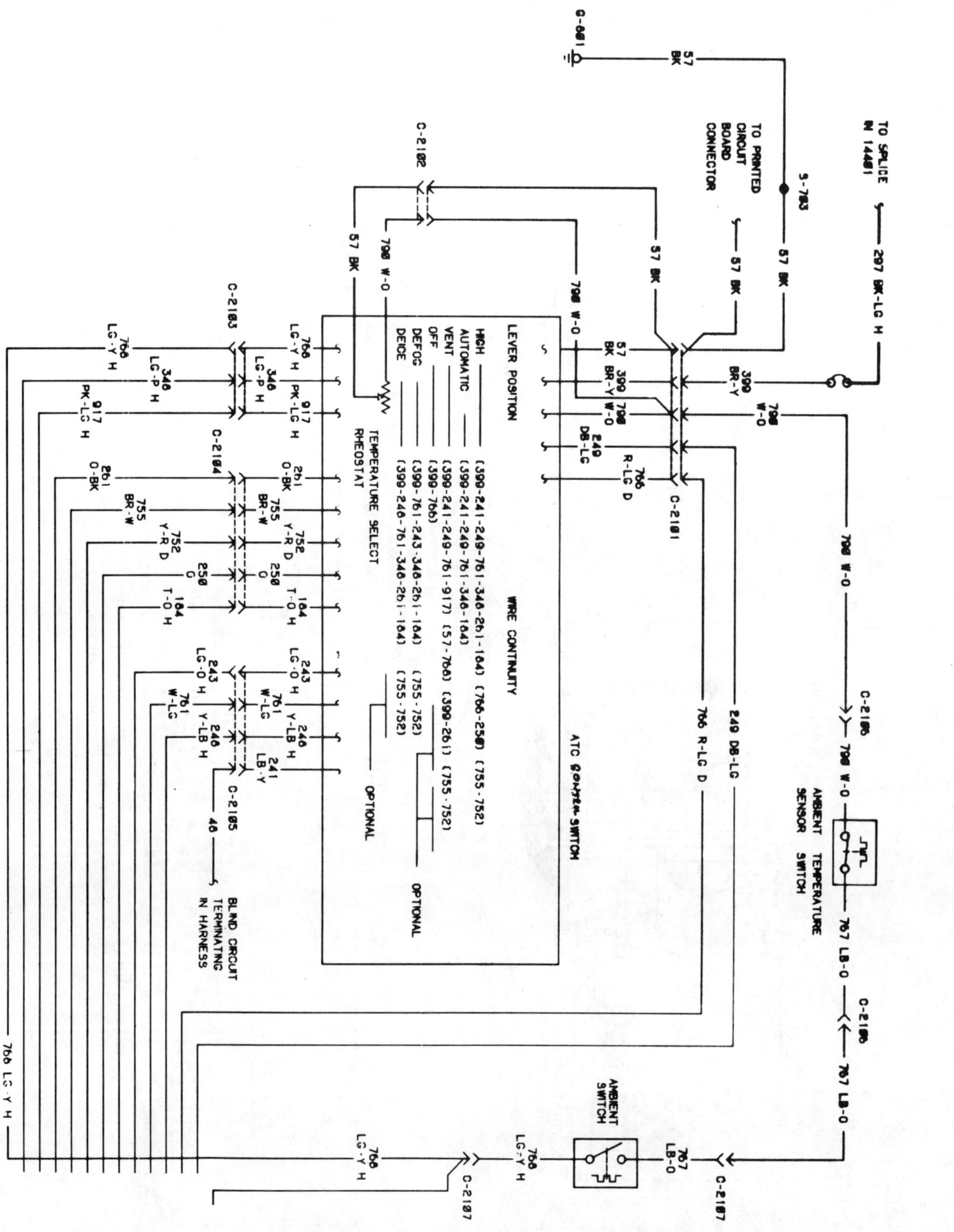

1974 Ford Thunderbird A/C & Heater Wiring Circuit (With Auto Temp Control) (Part 1 of 2)

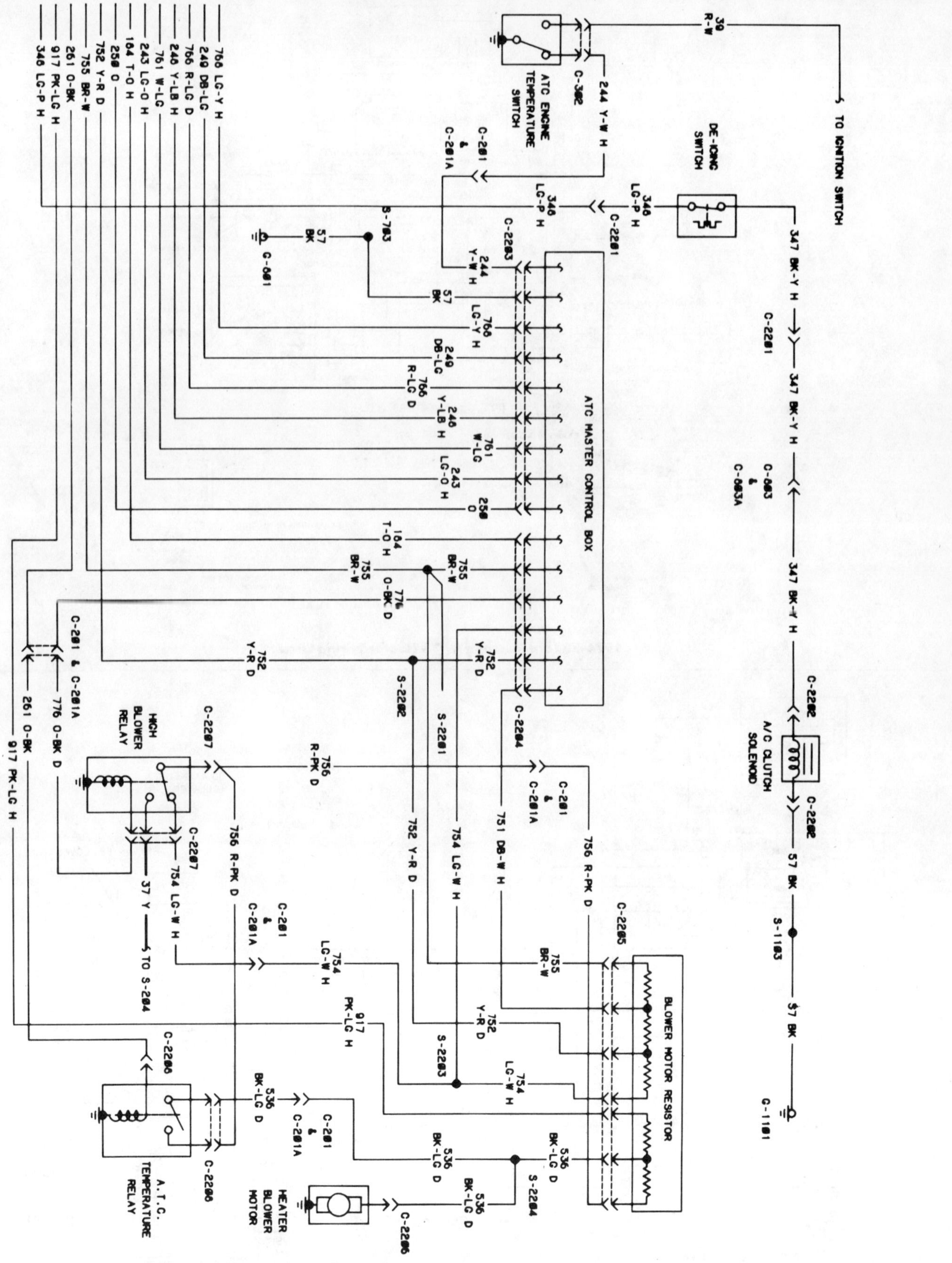

1974 Ford Thunderbird A/C & Heater Wiring Circuit (With Auto Temp Control) (Part 2 of 2)

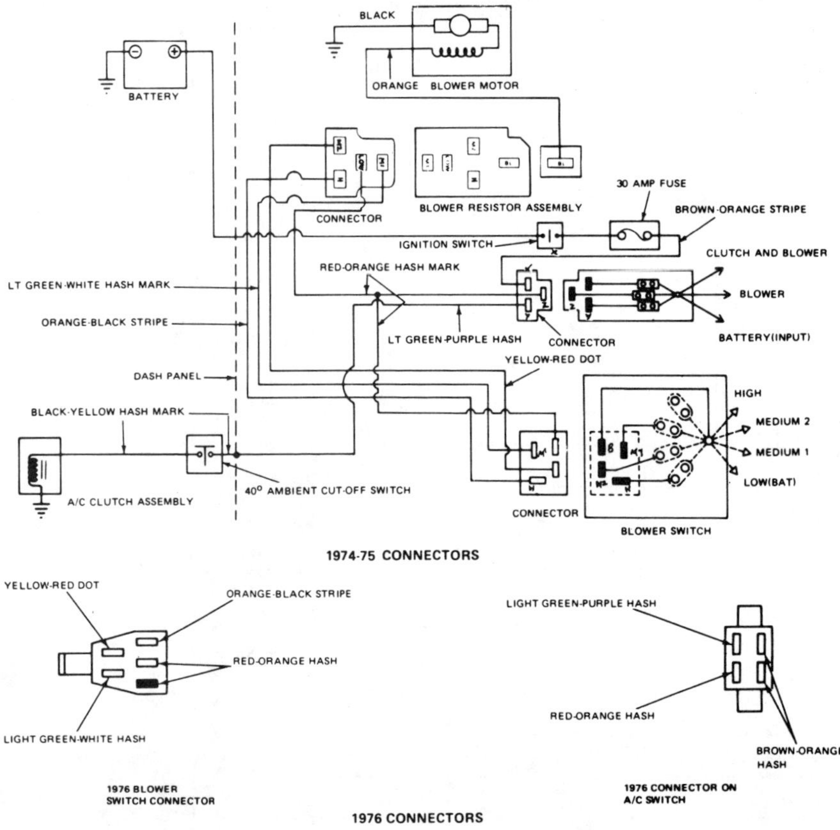

1974-76 Ford Mustang A/C & Heater Wiring Circuit

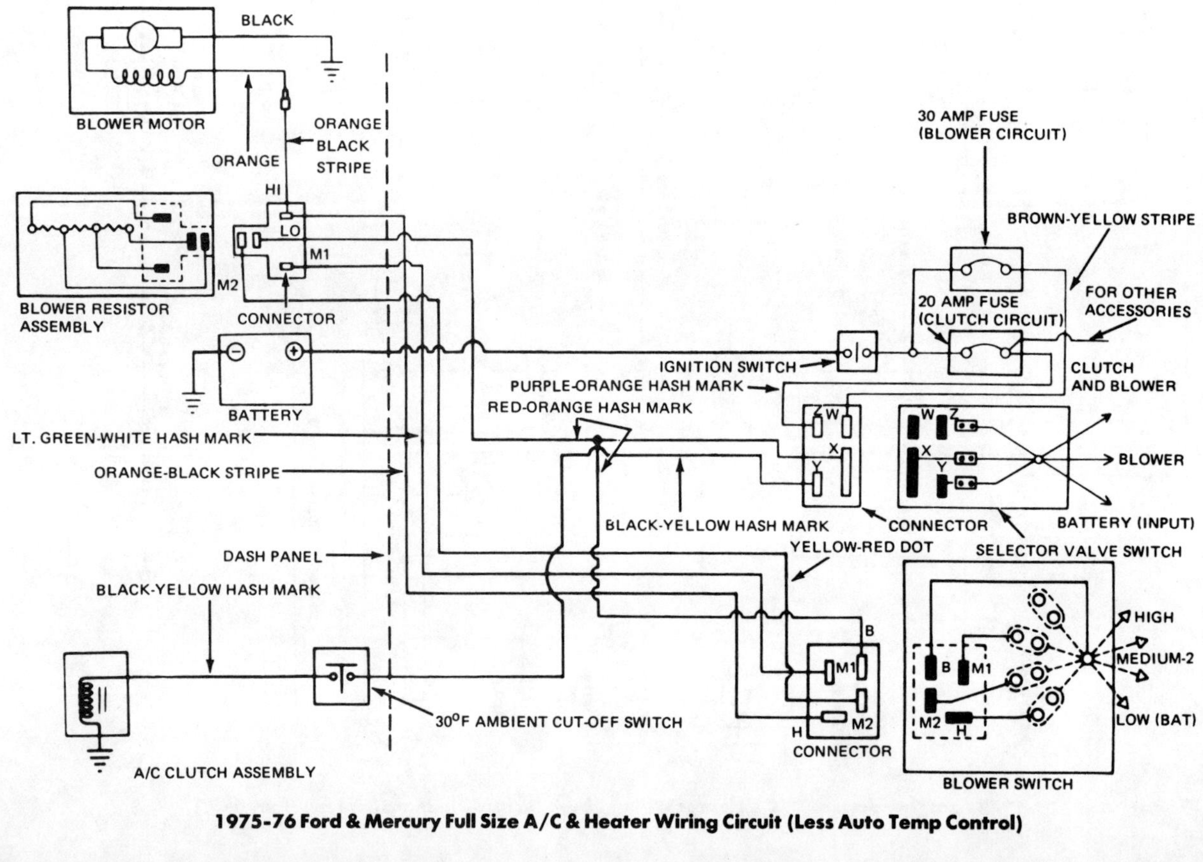

1975-76 Ford & Mercury Full Size A/C & Heater Wiring Circuit (Less Auto Temp Control)

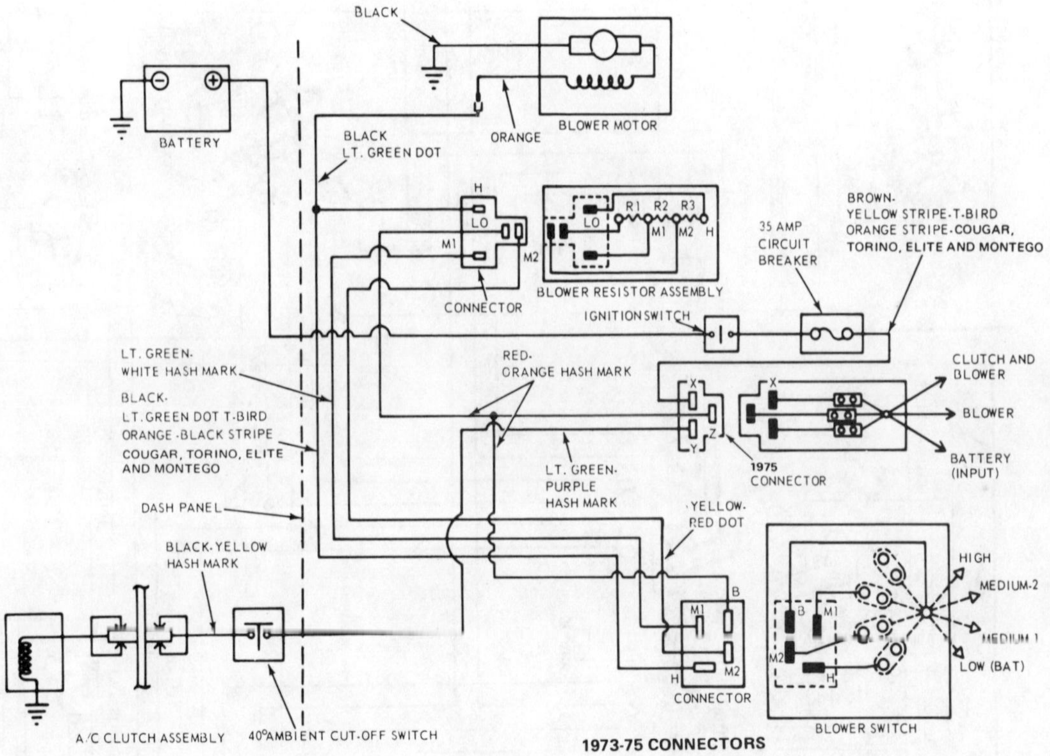

1973-75 CONNECTORS

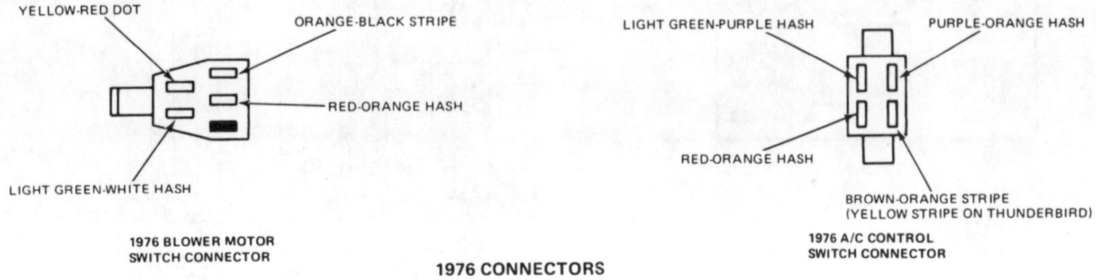

1976 CONNECTORS

1974-76 Ford Thunderbird A/C & Heater Wiring Circuit (Less Auto Temp Control)

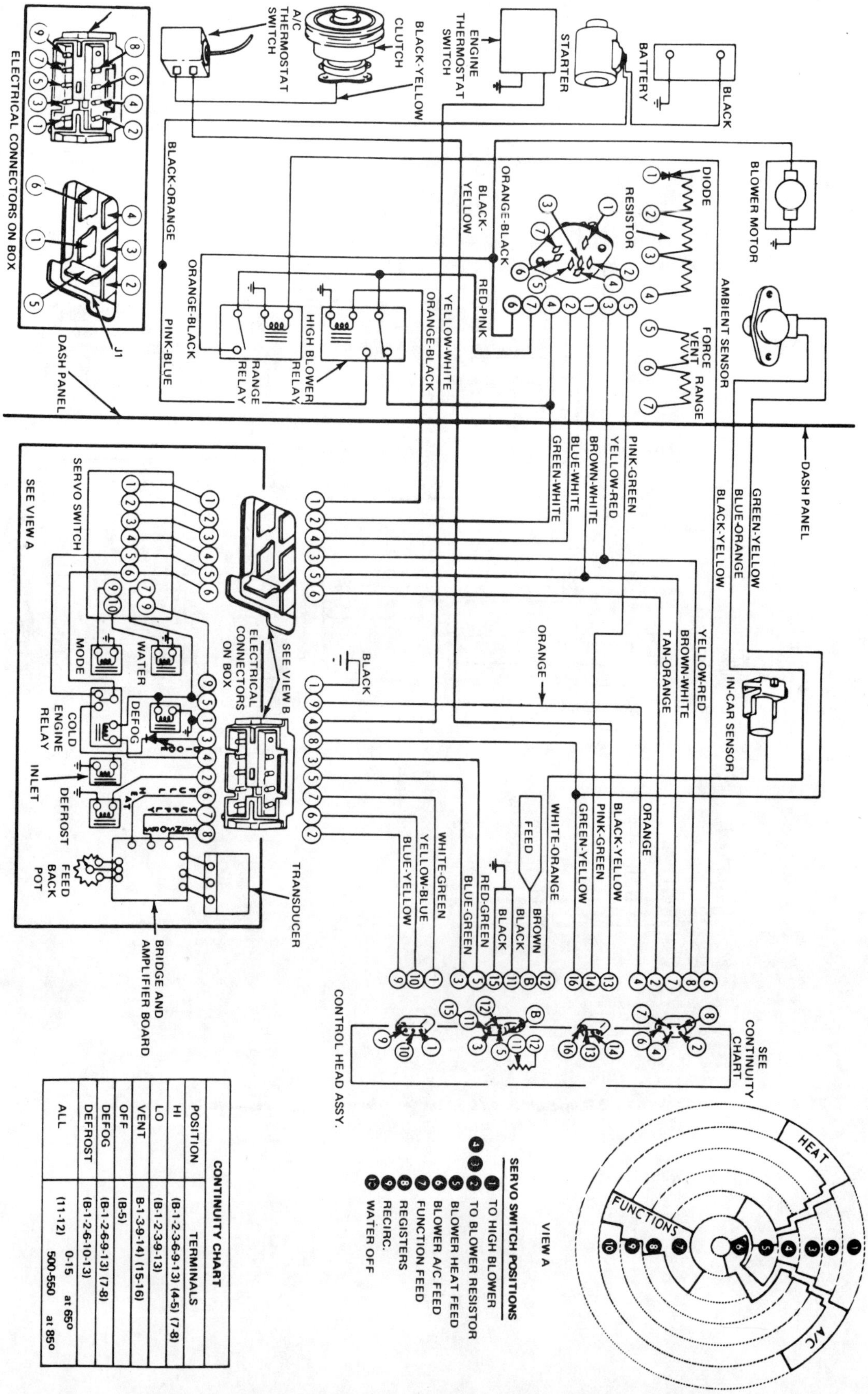

CONTINUITY CHART

POSITION	TERMINALS
HI	(B-1-2-3-6-9-13) (4-5) (7-8)
LO	(B-1-2-3-9-13)
VENT	B-1-3-9-14) (15-16)
OFF	(B-5)
DEFOG	(B-1-2-6-9-13) (7-8)
DEFROST	(B-1-2-6-10-13)
ALL	(11-12) 0-15 at 65° 500-550 at 85°

1974-76 Lincoln Continental A/C & Heater Wiring Circuit (With Auto Temp Control)

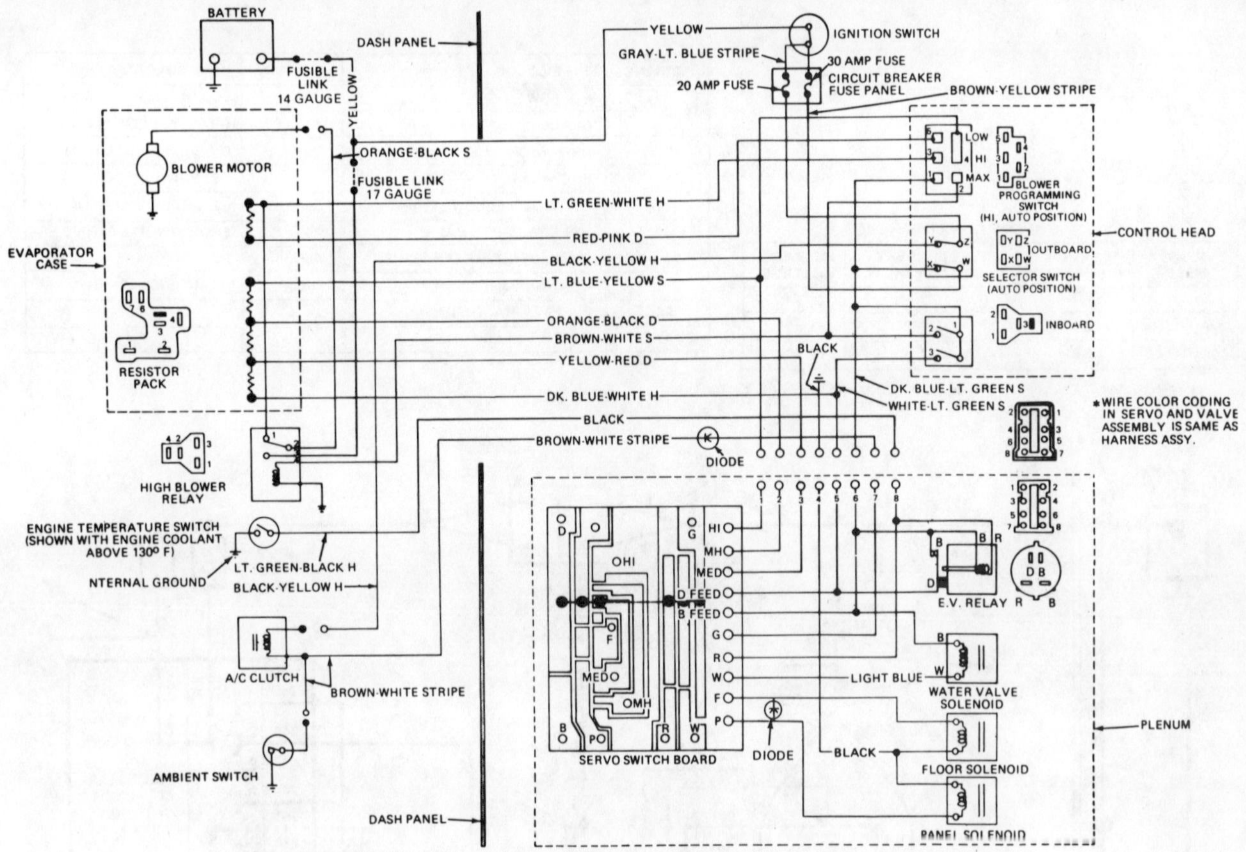

1975-76 Ford & Mercury Full Size A/C & Heater Wiring Circuit (With Auto Temp Control)

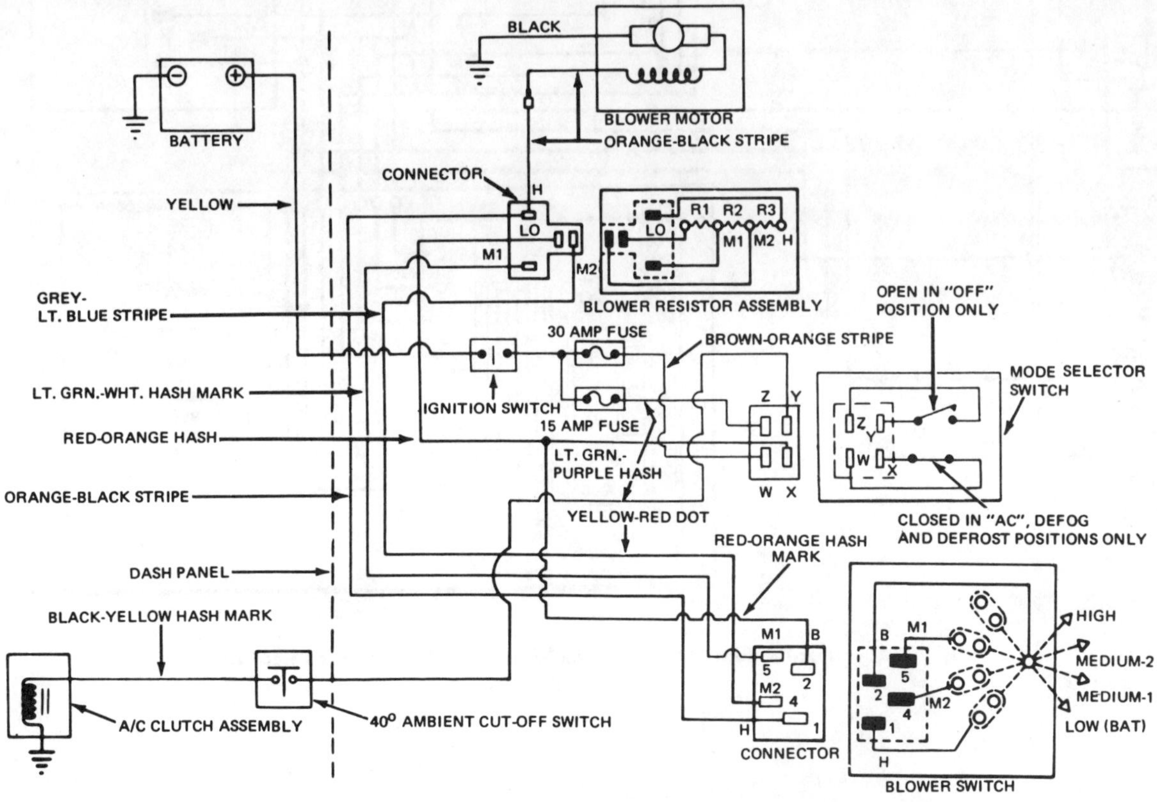

1975-76 Ford Granada & Mercury Monarch A/C & Heater Wiring Circuit

CIR. NO.	WIRE COLOR	CIR. NO.	WIRE COLOR
37	YELLOW	348	LT. GREEN/PURPLE HASH
57	BLACK	687	GRAY/YELLOW STRIPE
181	BROWN/ORANGE STRIPE	751	DK. BLUE/WHITE HASH
182	BROWN/WHITE STRIPE	752	YELLOW/RED DOT
241	LT. BLUE/YELLOW STRIPE	754	LT. GREEN/WHITE HASH
244	YELLOW/WHITE HASH	755	BROWN/WHITE STRIPE
249	DK. BLUE/LT. GREEN STRIPE	756	RED/PINK DOT
261	ORANGE/BLACK STRIPE	776	ORANGE/BLACK DOT
298	PURPLE/ORANGE HASH		
347	BLACK/YELLOW HASH		

BLOWER SWITCH

BLOWER SPEED	TERMINAL CONNECTIONS	TEST CURRENT CAPACITY (AMPS.)	BLOWER VOLTAGE
MAX.	4-1	2	12.0-14.7
AUTO HI	5-2	29	5.5-12.0 *
AUTO LO	5-3	22	5.0-10.0 *

* 4-Speed Range

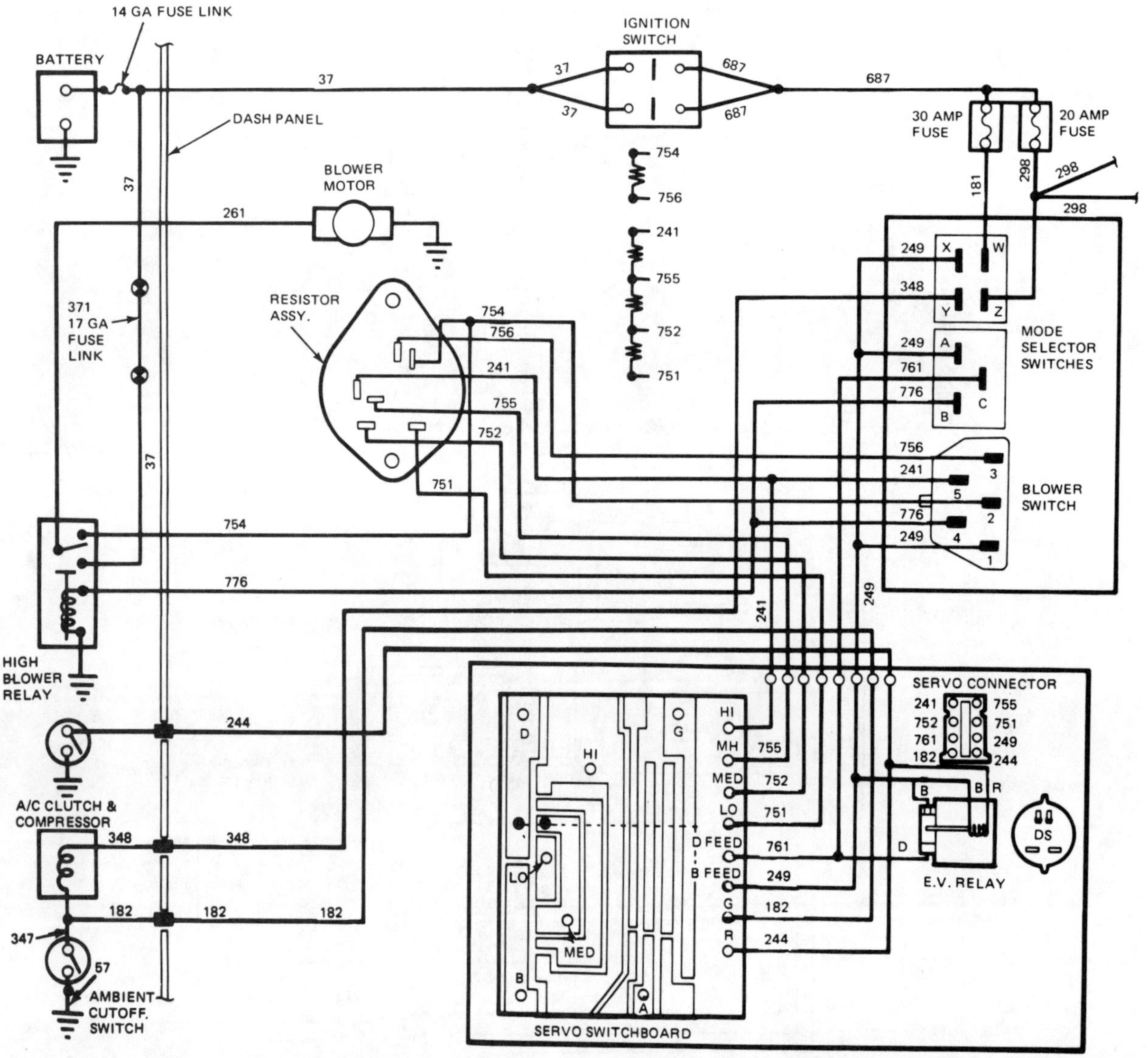

1975-76 Continental Mark IV A/C & Heater Wiring Circuit (With Auto Temp Control)

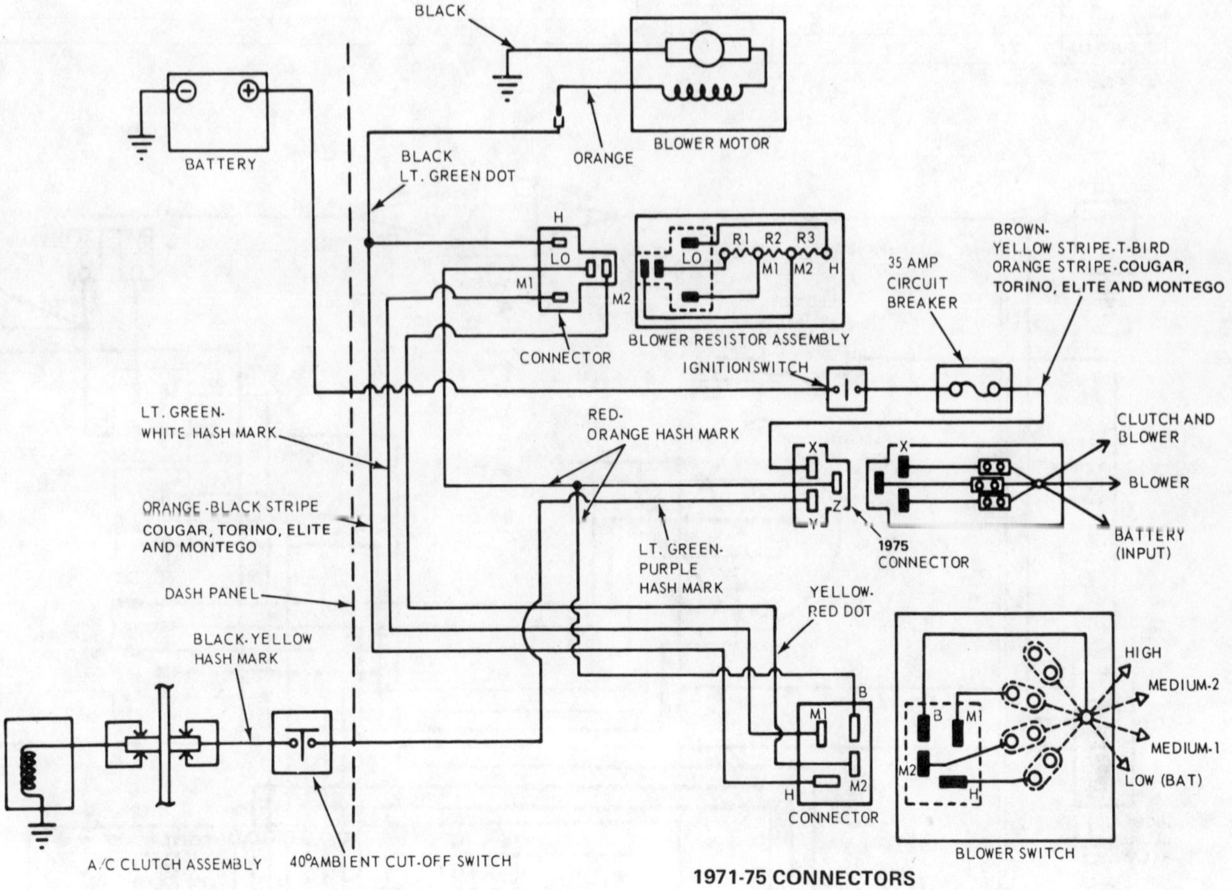

1971-75 CONNECTORS

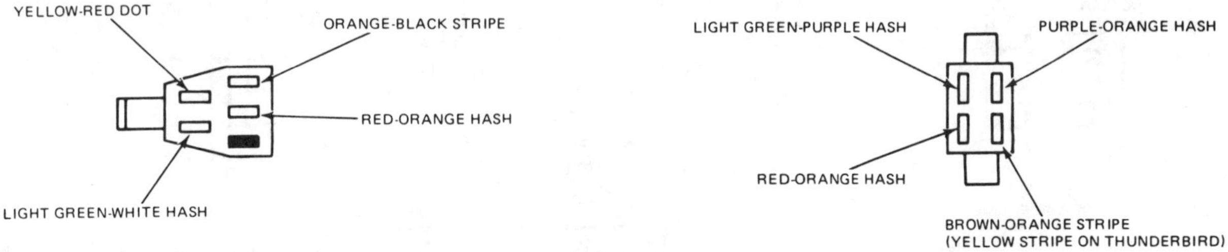

1976 BLOWER MOTOR SWITCH CONNECTOR

1976 A/C CONTROL SWITCH CONNECTOR

1976 CONNECTORS

1974-76 Ford Elite & Torino; Mercury Cougar & Montego A/C & Heater Wiring Circuit (Less Auto Temp Control)

CIR. NO.	WIRE COLOR	CIR. NO.	WIRE COLOR
37	YELLOW	348	LT. GREEN/PURPLE HASH
57	BLACK	687	GRAY/YELLOW STRIPE
181	BROWN/ORANGE STRIPE	751	DK. BLUE/WHITE HASH
182	BROWN/WHITE STRIPE	752	YELLOW/RED DOT
241	LT. BLUE/YELLOW STRIPE	754	LT. GREEN/WHITE HASH
244	YELLOW/WHITE HASH	755	BROWN/WHITE STRIPE
249	DK. BLUE/LT. GREEN STRIPE	756	RED/PINK DOT
261	ORANGE/BLACK STRIPE	776	ORANGE/BLACK DOT
298	PURPLE/ORANGE HASH		
347	BLACK/YELLOW HASH		

BLOWER SWITCH

BLOWER SPEED	TERMINAL CONNECTIONS	TEST CURRENT CAPACITY (AMPS.)	BLOWER VOLTAGE
MAX.	4-1	2	12.0-14.7
AUTO HI	5-2	29	5.5-12.0 *
AUTO LO	5-3	22	5.0-10.0 *

* 4-Speed Range

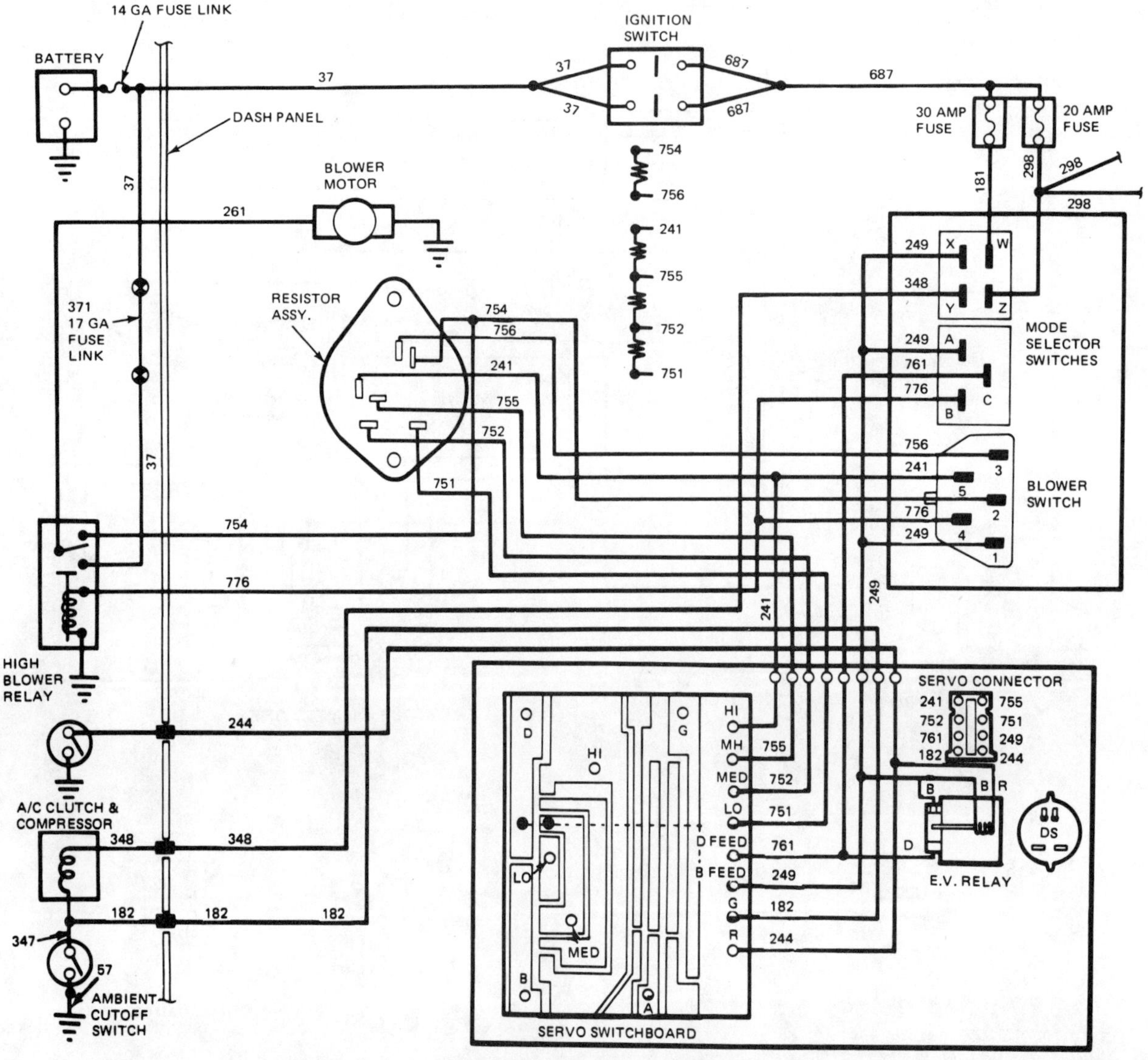

1975-76 Ford Elite & Torino; Mercury Cougar & Montego A/C & Heater Wiring Circuit (With Auto Temp Control)

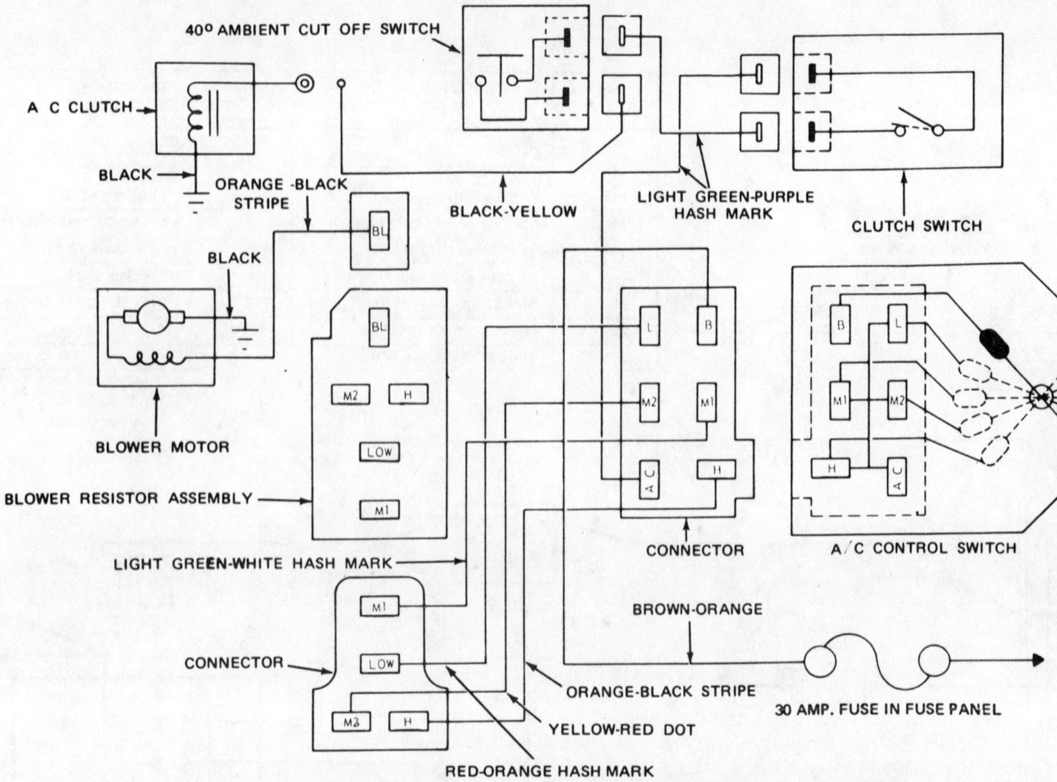

1975-76 Ford Pinto & Mercury Bobcat A/C & Heater Wiring Circuit

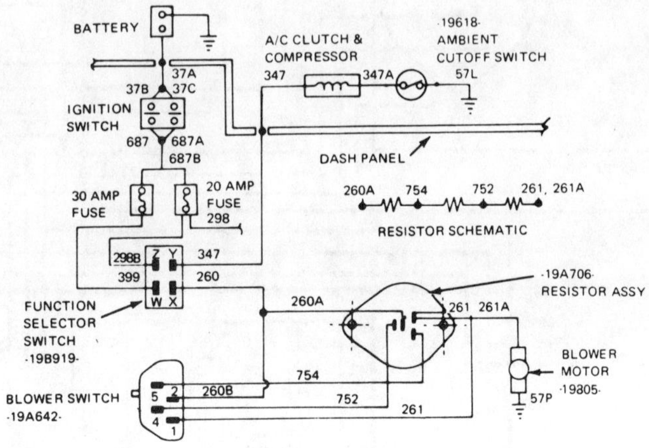

FUNCTION SELECTOR SWITCH -19B919-							
SWITCH SCHEMATIC	TERMINAL CONNECTIONS						
	OFF	ECONOMY VENT	A/C	HI–LO	HEAT	DEFOG ▽	DEFROST
X Y W Z	None	W - X	W - X Y - Z	W - X	W - X	W - X Y - Z	W - X Y - Z

BLOWER SWITCH -19A642-			
BLOWER SPEED	TERMINAL CONNECTIONS	CURRENT IN CAR (AMPS)	BLOWER VOLTAGE
Low	2	6	7.0 - 8.5
Med. Low	2 - 5	10	8.5 - 10.0
Med. High	2 - 4 - 5	17	10.0 - 12.5
High	2 - 1 - 4	28	12.5 - 14.8

CIRCUIT NO.	WIRE COLOR
57P	Black
260A	Red W/Orange Hash
260B	Red W/Orange Hash
261A	Orange W/Black Stripe
298B	Purple W/Orange Hash
347	Black W/Yellow Hash
399	Brown W/Yellow Stripe
752	Yellow W/Red Dot
754	Lt. Green W/White Hash
37A	Yellow
37B-C	Yellow
687A	Grey W/Lt. Blue Stripe
687B	Grey W/Lt. Blue Stripe

1977 Ford & Mercury Full Size A/C & Heater Wiring Circuit (Less Auto Temp Control)

CIR. NO.	WIRE COLOR	CIR. NO.	WIRE COLOR
37	YELLOW	348	LT. GREEN/PURPLE HASH
57	BLACK	687	GRAY/YELLOW STRIPE
181	BROWN/ORANGE STRIPE	751	DK. BLUE/WHITE HASH
182	BROWN/WHITE STRIPE	752	YELLOW/RED DOT
241	LT. BLUE/YELLOW STRIPE	754	LT. GREEN/WHITE HASH
244	YELLOW/WHITE HASH	755	BROWN/WHITE STRIPE
249	DK. BLUE/LT. GREEN STRIPE	756	RED/PINK DOT
261	ORANGE/BLACK STRIPE	776	ORANGE/BLACK DOT
298	PURPLE/ORANGE HASH		
347	BLACK/YELLOW HASH		

BLOWER SWITCH

BLOWER SPEED	TERMINAL CONNECTIONS	TEST CURRENT CAPACITY (AMPS.)	BLOWER VOLTAGE
MAX.	4-1	2	12.0-14.7
AUTO HI	5-2	29	5.5-12.0 *
AUTO LO	5-3	22	5.0-10.0 *

* 4-Speed Range

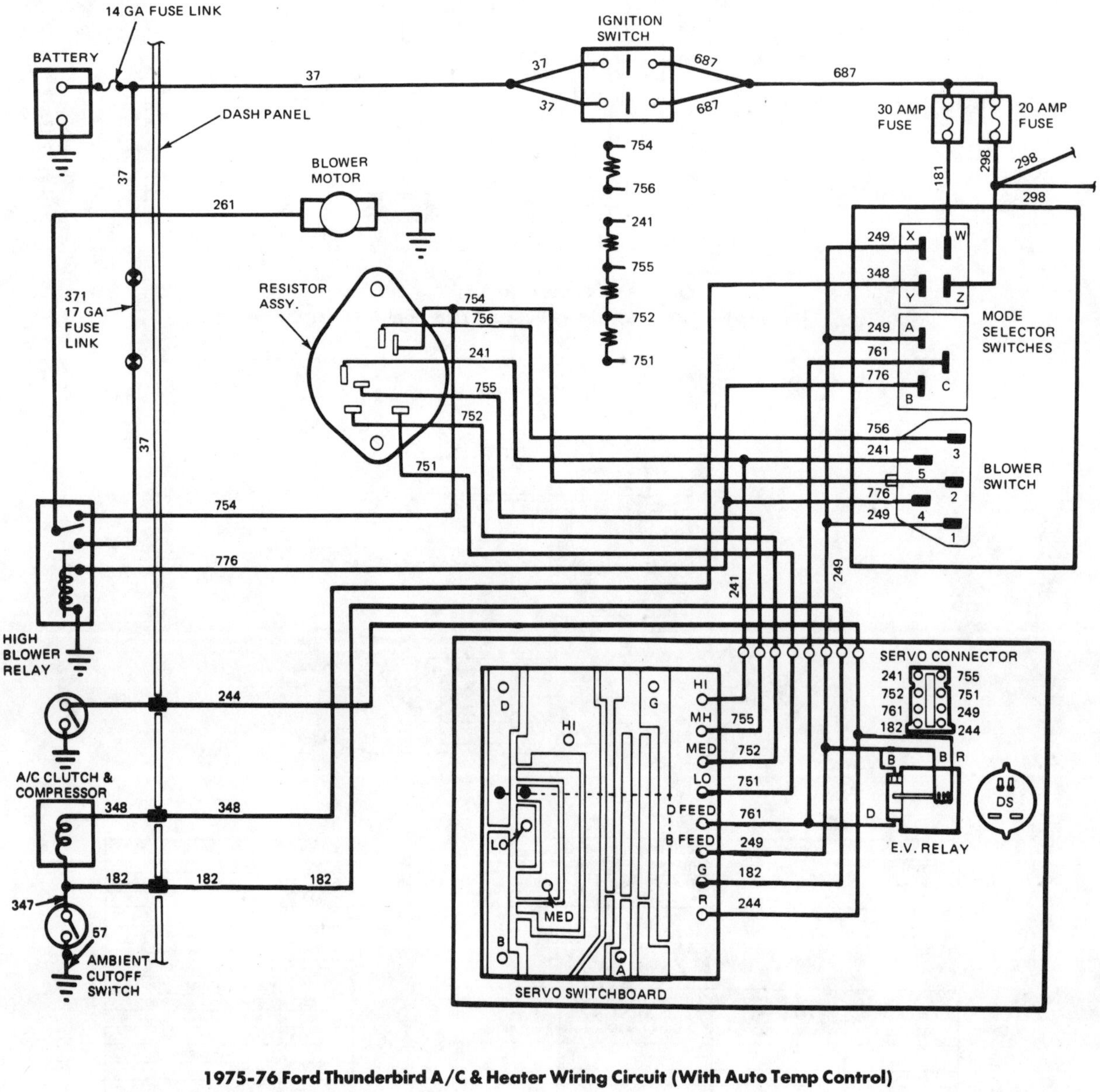

1975-76 Ford Thunderbird A/C & Heater Wiring Circuit (With Auto Temp Control)

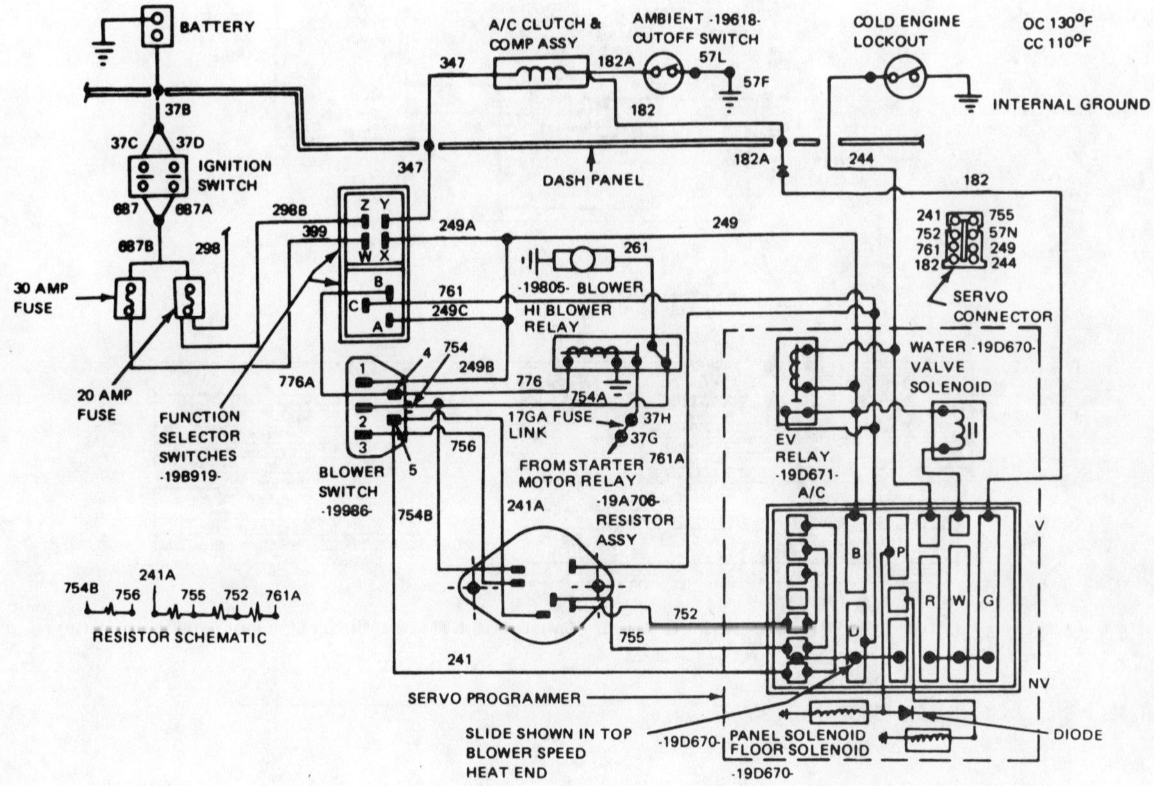

FUNCTION SELECTOR SWITCH -19B919-					
SWITCH SCHEMATIC	TERMINAL CONNECTIONS				
	OFF	ECONOMY VENT	AUTO	DEFOG	DEFROST
X Y W Z	None	W-X	W-X Y-Z	W-X Y-Z	W-X Y-Z
A C B	A-C	A-C	None	A-C	A-B-C

BLOWER SWITCH -19986-			
BLOWER SPEED	TERMINAL CONNECTIONS	CURRENT IN CAR (AMPS)	BLOWER VOLTAGE
Max	4-1	2	12.0-14.7
Auto Hi	5-2	28	5.5-12.0*
Auto Lo	5-3	22	5.0-10.0*
*4 SPEED RANGE			

CIRCUIT NO.	WIRE COLOR
37B-H	Yellow
57L-F	Black
57N	Black
182-A	Brown W/White Stripe
182-A	Brown W/White Stripe
241A	Lt. Blue W/Yellow Stripe
244	Yellow W/White Hash
249A-B-C	Dk. Blue W/Lt. Green Stripe
261	Orange W/Black Stripe
298B	Purple W/Orange Hash
347	Black W/Yellow Hash
399	Brown W/Yellow Hash
752	Yellow W/Red Dot
754A-B	Lt. Green W/White Hash
755	Brown W/White Stripe
756	Red W/Pink Dot
761-A	White W/Lt. Green Stripe
776-A	Orange W/Black Dot
37C-D	Yellow
687B	Gray W/Lt. Blue Stripe
687A	Gray W/Lt. Blue Stripe

1977 Ford & Mercury Full Size A/C & Heater Wiring Circuit (With Auto Temp Control)

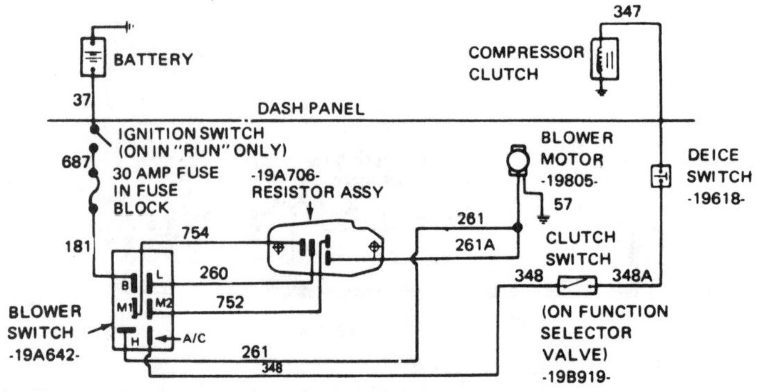

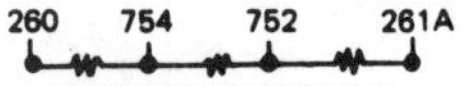

RESISTOR SCHEMATIC

BLOWER SWITCH FUNCTION -19A642-		
SWITCH POSITION	TERMINAL CONNECTIONS	CURRENT IN CAR (AMPS)
Off	Battery	–
Low	Batt - A/C-Low	3.5
Med 1	Batt - A/C-M1 - Low	6
Med 2	Batt - A/C-M2 -M1	10
High	Batt - A/C-High - M2	20

CLUTCH SWITCH FUNCTION (PART OF -19B919-)							
	A/C			HEAT		DEF.	
CONTACTS	MAX.	FRESH	OFF	HI	LO	FOG	ICE
CLOSED	X	X	–	–	–	X	X

CIRCUIT NO.	WIRE COLOR
37	Yellow
57	Black
260	Red/Orange Hash
261A	Orange/Black Stripe
348A	Lt. Green/Purple Hash
752	Yellow/Red Dot
754	Lt. Green/White Hash
687	Gray/Yellow Stripe
181	Brown/Orange Stripe
347	Black/Yellow Hash

1977 Ford Maverick & Mercury Comet A/C & Heater Wiring Circuit

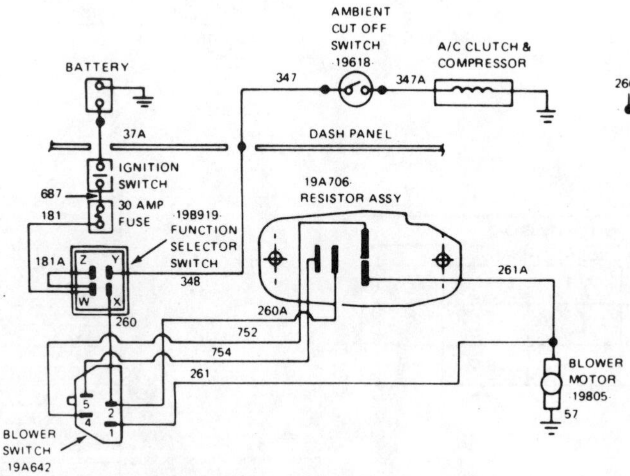

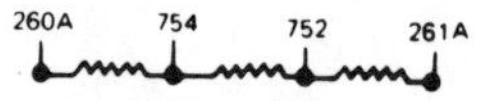

RESISTOR SCHEMATIC

FUNCTION SELECTOR SWITCH 19B919							
SWITCH SCHEMATIC	TERMINAL CONNECTIONS						
	OFF	ECONOMY VENT	A/C	HI-LO	HEAT	∇ DEFOG	DEFROST
X Y W Z	None	W X	W X Y Z	W X	W X	W X Y Z	W X Y Z

BLOWER SWITCH 19A642			
BLOWER SPEED	TERMINAL CONNECTIONS	CURRENT IN CAR (AMPS)	BLOWER VOLTAGE
Low	2	5	7.0 9.0
Med Low	2 5	8	9.0 11.5
Med High	2 4 5	15	11.5 13.0
High	2 1 4	22	13.0 14.8

CIRCUIT NO	WIRE COLOR
37A	Yellow
57	Black
181A	Brown W/Orange Stripe
260A	Red W/Orange Hash
261A	Orange W/Black Stripe
347A	Black W/Yellow Hash
348	Lt. Green W/Purple Hash
687	Gray W/Lt. Blue Hash
752	Yellow W/Red Dot
754	Lt. Green W/White Hash

1977 Ford Mustang A/C & Heater Wiring Circuit

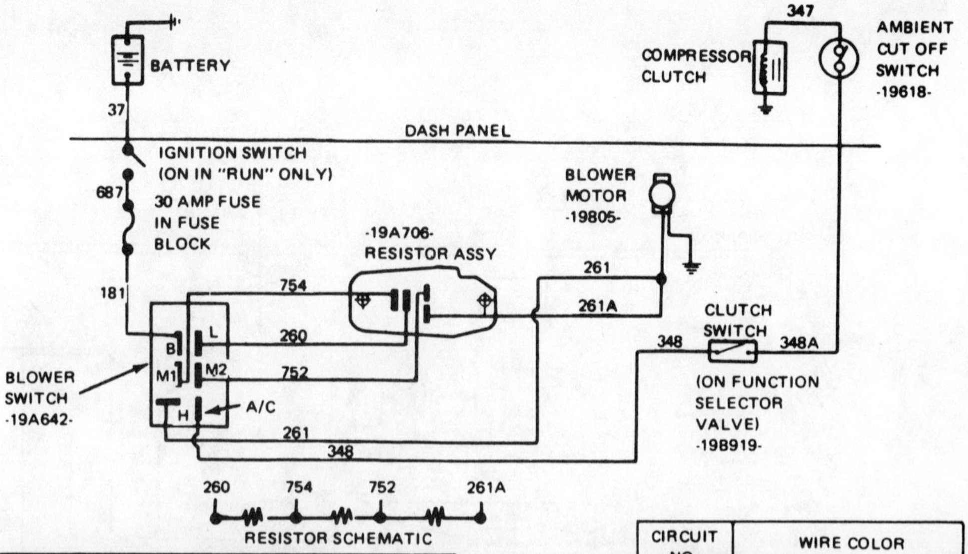

BLOWER SWITCH FUNCTION -19A642-		
SWITCH POSITION	TERMINAL CONNECTIONS	CURRENT IN CAR (AMPS)
Off	Battery	–
Low	Batt - A/C-Low	3.5
Med 1	Batt - A/C-M1 - Low	.6
Med 2	Batt - A/C-M2 -M1	10
High	Batt - A/C-High - M2	20

CIRCUIT NO.	WIRE COLOR
37	Yellow
57	Black
260	Red/Orange Hash
261A	Orange/Black Stripe
348A	Lt. Green/Purple Hash
752	Yellow/Red Dot
754	Lt. Green/White Hash
687	Gray/Yellow Stripe
181	Brown/Orange Stripe
347	Black/Yellow Hash

CLUTCH SWITCH FUNCTION (PART OF -19B919-)							
	A/C			ECONOMY		–	
CONTACTS	MAX	FRESH	OFF	VENT	HEAT	(DEFOG)	DEFROST
CLOSED	X	X	–	–	–	X	X

1977 Ford Pinto & Mercury Bobcat A/C & Heater Wiring Circuit

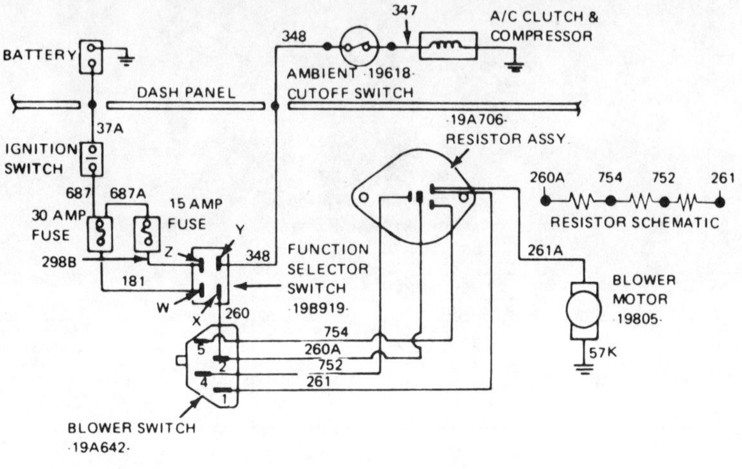

FUNCTION SELECTOR SWITCH -19B919-							
SWITCH SCHEMATIC	TERMINAL CONNECTIONS						
	OFF	ECONOMY VENT	A/C	HI-LO	HEAT	▽ DEFOG	DEFROST
X W Z Y	None	W - X	W - X Y - Z	W - X	W - X	W - X Y - Z	W - X Y - Z

CIRCUIT NO.	WIRE COLOR
37A	Yellow
57A	Black
181	Brown W/Orange Stripe
260A	Red W/Orange Hash
261A	Orange W/Black Stripe
347	Black W/Yellow Hash
348A	Lt. Green W/Purple Hash
687	Gray W/Lt. Blue Stripe
687A	Gray W Lt. Blue Stripe
752	Yellow W Red Dot
754	Lt. Green W White Hash

BLOWER SWITCH -19A642-			
BLOWER SPEED	TERMINAL CONNECTIONS	CURRENT IN CAR (AMPS)	BLOWER VOLTAGE
Low	2	8	7.0 - 9.0
Med Low	2 - 5	11	9.0 - 11.5
Med High	2 - 4 - 5	20	11.5 - 13.0
High	2 - 1 - 4	25	13.0 - 14.8

1977-78 Ford Granada & Mercury Monarch A/C & Heater Wiring Circuit (Less Auto Temp Control)

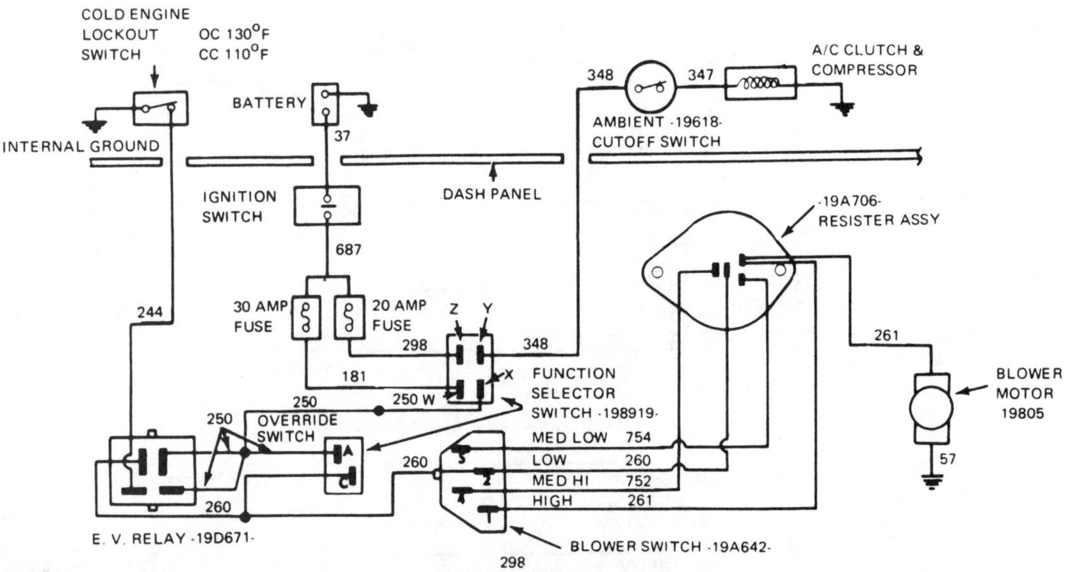

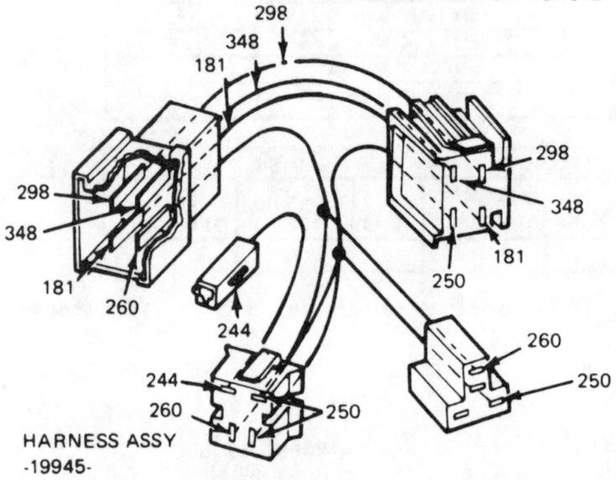

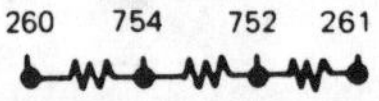

RESISTOR SCHEMATIC

FUNCTION SELECTOR AND OVERRIDE SWITCH -19B919-							
SWITCH SCHEMATIC	TERMINAL CONNECTIONS						
	OFF	ECONOMY VENT	A/C	HI-LO	HEAT	▽ DEFOG	DEFROST
X Y W Z	None	W - X	W - X Y - Z	W - X Y - Z	W - X	W - X Y - Z	W - X Y - Z
A C	A - C	A - C	A - C	A - C	None	A - C	A - C

CIRCUIT NO.	WIRE COLOR
37	Yellow
57	Black
181	Brown W/Orange Stripe
260	Red W/Orange Hash
261	Orange W/Black Stripe
347	Black W/Yellow Hash
348	Lt. Green W/Purple Hash
687	Gray W/Yellow Stripe
752	Yellow W/Red Dot
754	Lt. Green W/White Hash
244	Yellow/White Hash
298	Purple W/Orange Hash
250	Orange

BLOWER SWITCH -19A642-			
BLOWER SPEED	TERMINAL CONNECTIONS	CURRENT IN CAR (AMPS)	BLOWER VOLTAGE
Low	2	8	5
Med Low	2 - 5	11	7
Med High	2 - 4 - 5	20	10
High	2 - 1 - 4	25	14

1977 Lincoln Versailles; 1977-78 Ford Granada & Mercury Monarch A/C & Heater Wiring Circuit (With Auto Temp Control)

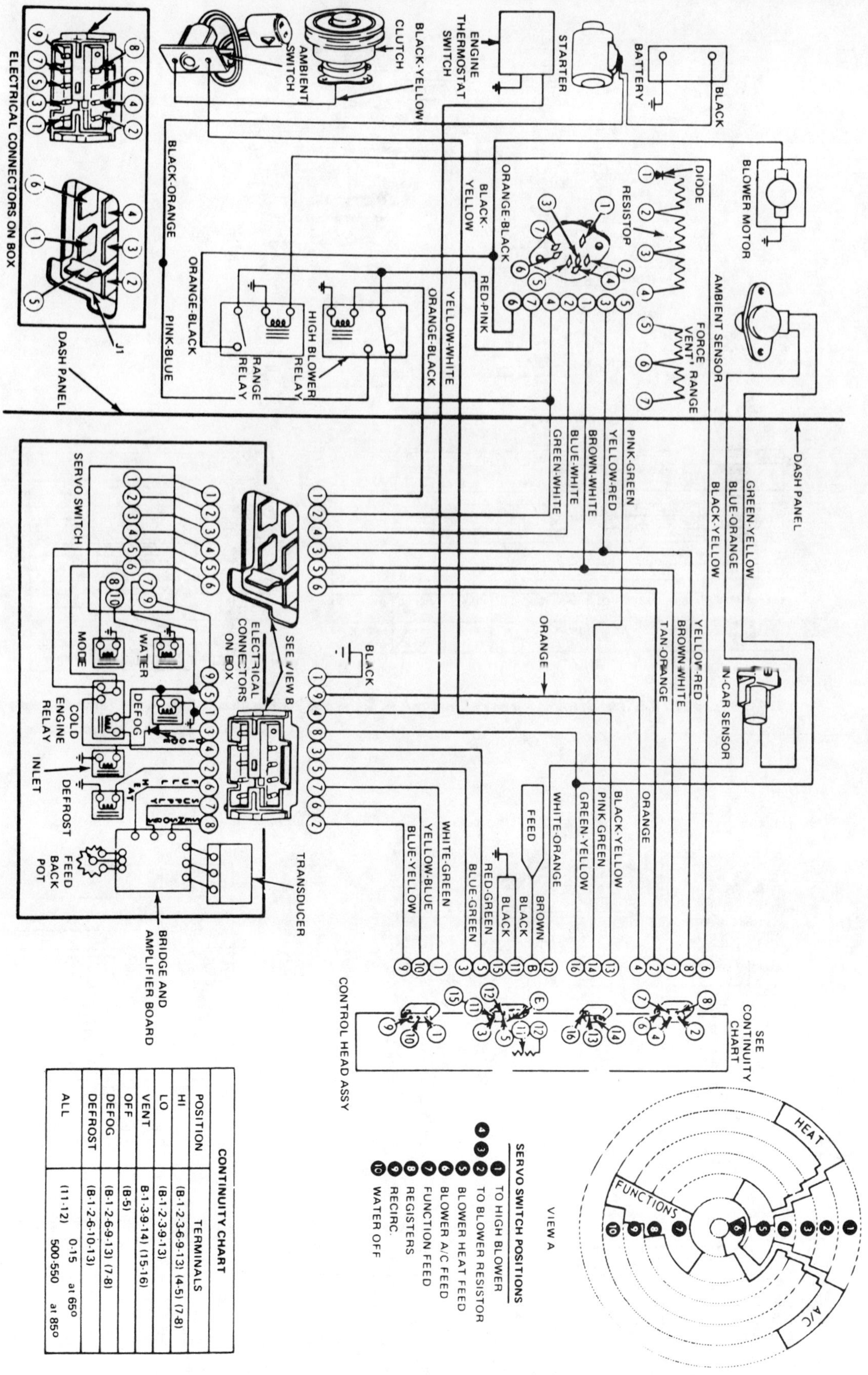

CONTINUITY CHART

POSITION	TERMINALS
HI	(B-1-2-3-6-9-13) (4-5) (7-8)
LO	(B-1-2-3-9-13)
VENT	B-1-3-9-14) (15-16)
OFF	(B-5)
DEFOG	(B-1-2-6-9-13) (7-8)
DEFROST	(B-1-2-6-10-13)
ALL	(11-12) 0-15 at 65° 500-550 at 85°

1977 Lincoln Continental A/C & Heater Wiring Circuit (With Auto Temp Control)

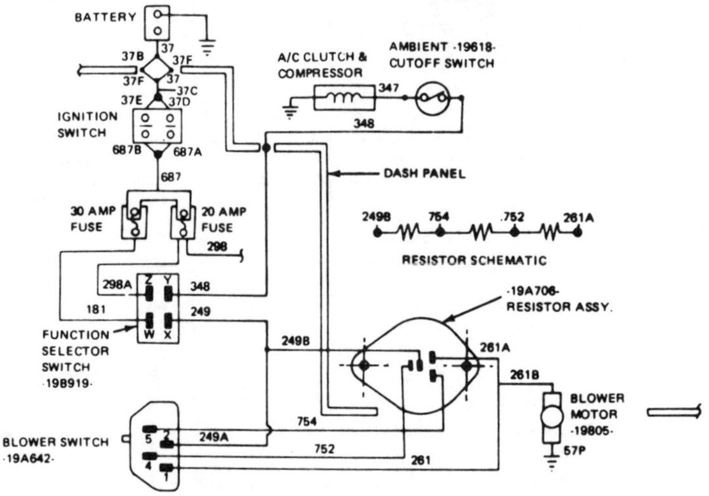

BLOWER SWITCH -19A642-			
BLOWER SPEED -	TERMINAL CONNECTIONS	CURRENT IN CAR (AMPS)	BLOWER VOLTAGE
Low.	2	6	5
Med. Low	2 - 5	8	7
Hed High	2 - 4 - 5	15	10
High	2 - 1 - 4	22	12.4

FUNCTION SELECTOR SWITCH -19B919-							
SWITCH SCHEMATIC	TERMINAL CONNECTIONS						
	OFF	ECONOMY VENT	A/C	HI-LO	HEAT	DEFOG ▽	DEFROST
X Y W Z	None	W - X	W - X Y - Z	W - X	W - X	W - X Y - Z	W - X Y - Z

CIRCUIT NO.	WIRE COLOR
347	Black/Yellow Hash
57P	Black
261A	Orange W/Black Stripe
249B	Dk. Blue W/Lt. Green Stripe
249A	Dk. Blue W/Lt. Green Stripe
261B	Orange W/Black Stripe
298A	Purple W/Orange Hash
348	Lt. Green W/Purple Hash
181	Brown W/Orange Stripe
752	Yellow W/Red Dot
754	Lt. Green W/White Hash
37F	Yellow
37C-D-E	Yellow
687	Grey W/Yellow Stripe
687A-B	Grey W/Yellow Stripe

1977-78 Ford LTD II & Thunderbird; Mercury Cougar A/C & Heater Wiring Circuit (Less Auto Temp Control)

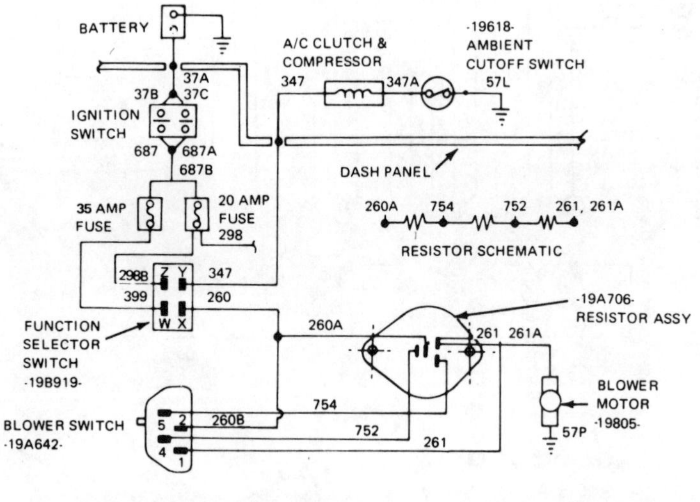

CIRCUIT NO.	WIRE COLOR
57P	Black
260A	Red W/Orange Hash
260B	Red W/Orange Hash
261A	Orange W/Black Stripe
298B	Purple W/Orange Hash
347	Black W/Yellow Hash
399	Brown W/Yellow Stripe
752	Yellow W/Red Dot
754	Lt. Green W/White Hash
37A	Yellow
37B-C	Yellow
687A	Grey W/Lt. Blue Stripe
687B	Grey W/Lt. Blue Stripe

FUNCTION SELECTOR SWITCH -19B919-							
SWITCH SCHEMATIC	TERMINAL CONNECTIONS						
	OFF	ECONOMY VENT	A/C	HI-LO	HEAT	DEFOG ▽	DEFROST
X Y W Z	None	W - X	W - X Y - Z	W - X	W - X	W - X Y - Z	W - X Y - Z

BLOWER SWITCH -19A642-			
BLOWER SPEED	TERMINAL CONNECTIONS	CURRENT IN CAR (AMPS)	BLOWER VOLTAGE
Low	2	6	7.0 - 8.5
Med. Low	2 - 5	10	8.5 - 10.0
Med. High	2 - 4 - 5	17	10.0 - 12.5
High	2 - 1 - 4	28	12.5 - 14.8

1978 Ford & Mercury Full Size A/C & Heater Wiring Circuit (Less Auto Temp Control)

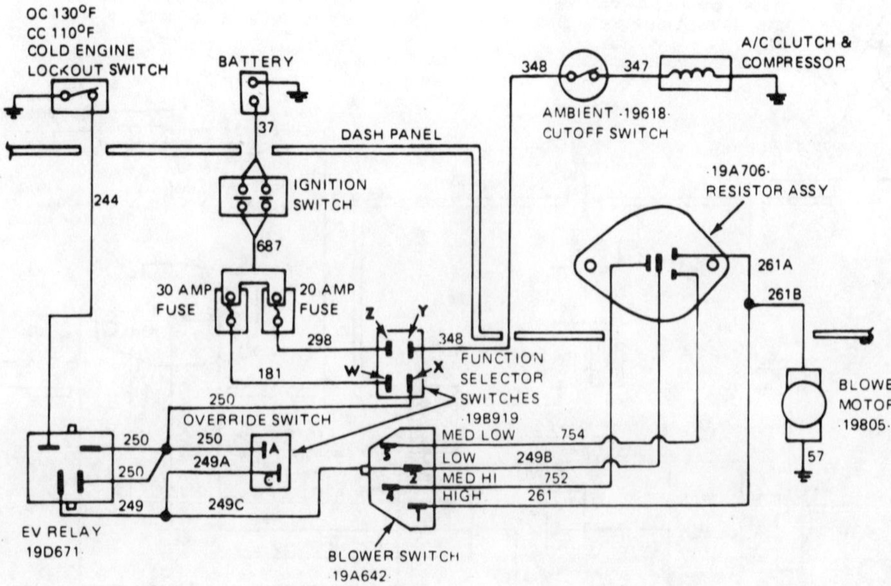

RESISTOR SCHEMATIC

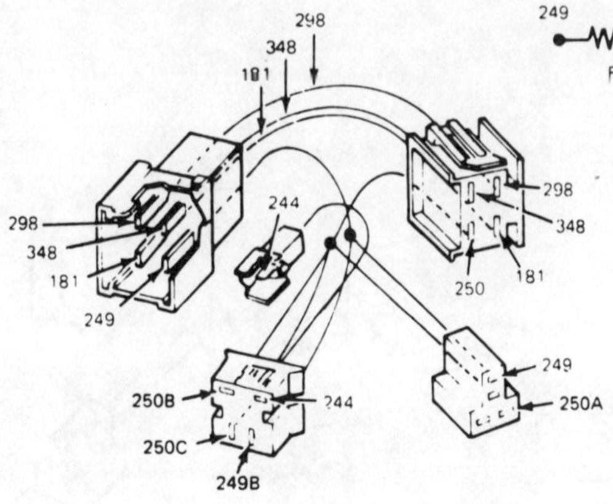

HARNESS ASSY
·19945

FUNCTION SELECTOR SWITCH ·19B919·							
SWITCH SCHEMATIC	OFF	ECONOMY VENT	A/C	HI-LO	HEAT	▽ DEFOG	DEFROST
A—C	A · C	A · C	A · C	A · C	Open	A · C	A · C
X Y W Z	None	W · X	W · X Y · Z	W · X Y · Z	W · X	W · X Y · Z	W · X Y · Z

BLOWER SWITCH ·19A642·			
BLOWER SPEED	TERMINAL CONNECTIONS	CURRENT IN CAR (AMPS)	BLOWER VOLTAGE
Low	2	6	5
Med Low	2 · 5	10	7
Med High	2 · 4 · 5	17	10
High	2 · 1 · 4	25	12.4

CIRCUIT NO.	WIRE COLOR
57	Black
347	Black W/Yellow Hash
348	Lt. Green W/Purple Hash
298	Purple W/Orange Hash
181	Brown W/Orange Stripe
752	Yellow W/Red Dot
754	Lt. Green W/White Hash
37	Yellow
687	Gray W/Lt. Blue Stripe
249	Dk. Blue W/Lt. Gr. Stripe
250	Orange
261	Orange W/Black Stripe
244	Yellow W/White Hash

1977-78 Continental Mark V; Ford LTD II & Thunderbird; Mercury Cougar A/C & Heater Wiring Circuit (With Auto Temp Control)

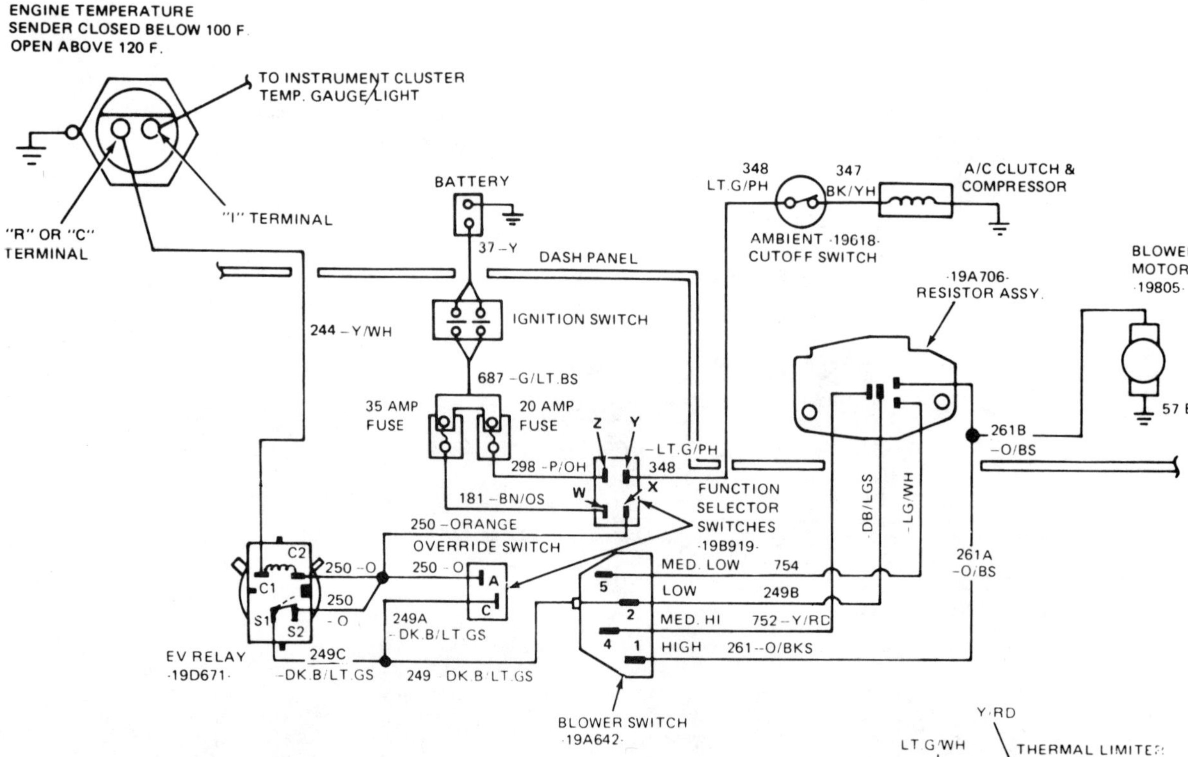

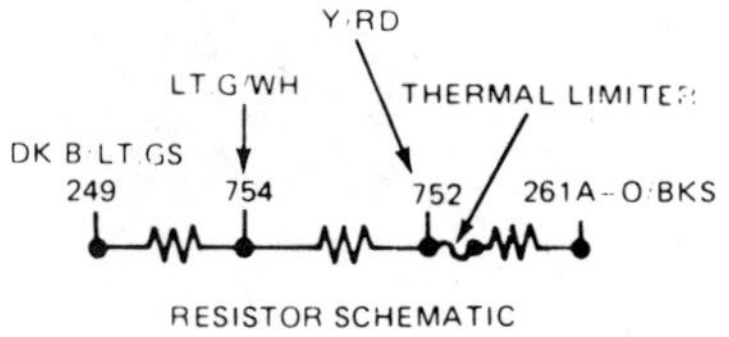

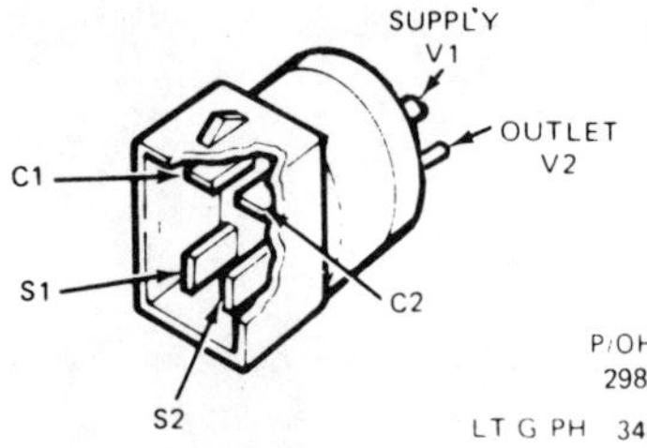

EV RELAY -19D671-

COIL TERMINALS	SWITCH TERMINALS	VACUUM SWITCH	
C1 C2	S1 S2	V1	V2
12.8 VOLTS	OPEN	OPEN	OPEN
0 VOLTS	CLOSED	SEALED	VENTED

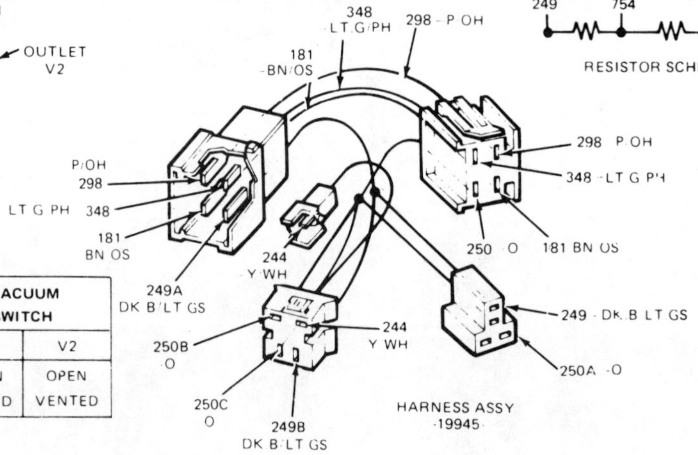

FUNCTION SELECTOR SWITCH -19B919-							
SWITCH SCHEMATIC	OFF	ECONOMY VENT	A/C	HI-LO	HEAT	▽ DEFOG	DEFROST
A — C	A - C	A - C	A - C	A - C	Open	A - C	A - C
X Y / W Z	None	W - X	W - X Y - Z	W - X Y - Z	W - X	W - X Y - Z	W - X Y - Z

BLOWER SWITCH -19A642-			
BLOWER SPEED	TERMINAL CONNECTIONS	CURRENT IN CAR (AMPS)	BLOWER VOLTAGE
Low	–	6	5
Med Low	2 - 5	10	7
Med High	2 - 4 - 5	17	10
High	2 - 1 - 4	25	12.4

1978 Ford & Mercury Full Size A/C & Heater Wiring Circuit (With Auto Temp Control)

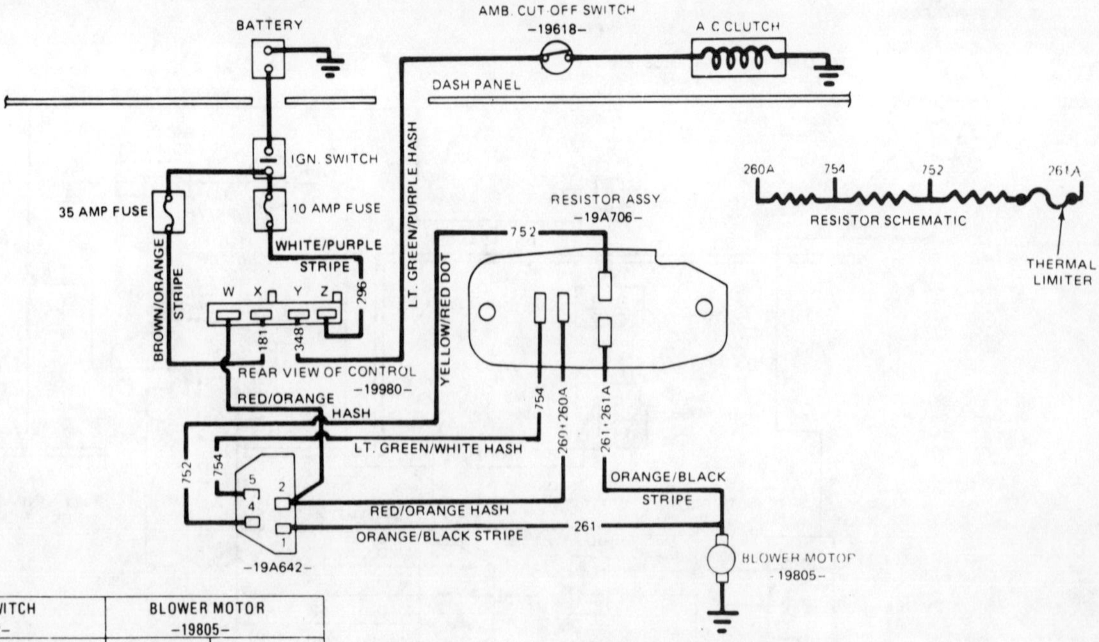

BLOWER SWITCH -19A642-		BLOWER MOTOR -19805-	
BLOWER SPEED	TERMINAL CONTINUITY CONNECTIONS	CURRENT (AMPS) IN-CAR	VOLTAGE
LO	NONE	4	3.9
MED-LO	2–5	8	6.2
MED-HI	2–4–5	I13	8.9
HI	1–2–4	23	12.8

FUNCTION SELECTOR SWITCH–19980–									
TERMINALS	AMP	TERMINAL CONNECTIONS							
		OFF	MAX	NORM	VENT	HI LO	HEAT	DEF
Y (A/C CLUTCH) Z (BATTERY)	4	NONE	Y Z	Y Z	NONE	NONE	NONE	Y Z
W (BLOWER) X (BATTERY)	35	NONE	W X	W X	W X	W X	W X	W X

1978 Ford Fairmont & Mercury Zephyr A/C & Heater Wiring Circuit

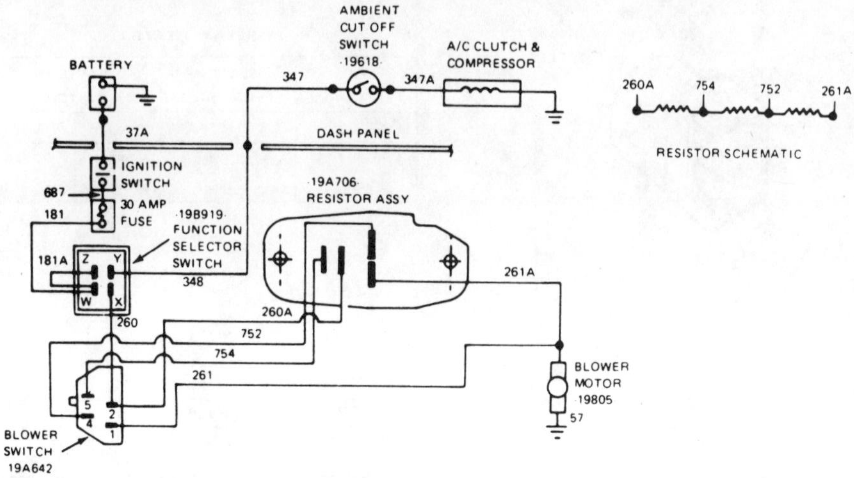

SWITCH SCHEMATIC	FUNCTION SELECTOR SWITCH 19B919 TERMINAL CONNECTIONS						
	OFF	ECONOMY VENT	A/C	HI-LO	HEAT	DEFOG	DEFROST
X Y W Z	None	W X	W X Y Z	W X	W X	W X Y Z	W X Y Z

CIRCUIT NO	WIRE COLOR
37A	Yellow
57	Black
181A	Brown W/Orange Stripe
260A	Red W/Orange Hash
261A	Orange W/Black Stripe
347A	Black W/Yellow Hash
348	Lt. Green W/Purple Hash
687	Gray W/Lt. Blue Hash
752	Yellow W/Red Dot
754	Lt. Green W/White Hash

BLOWER SWITCH 19A642			
BLOWER SPEED	TERMINAL CONNECTIONS	CURRENT IN CAR (AMPS)	BLOWER VOLTAGE
Low	2	5	7.0 9.0
Med Low	2 5	8	9.0 11.5
Med High	2 4 5	15	11.5 13.0
High	2 1 4	22	13.0 14.8

1978 Ford Mustang & Pinto; Mercury Bobcat A/C & Heater Wiring Circuit

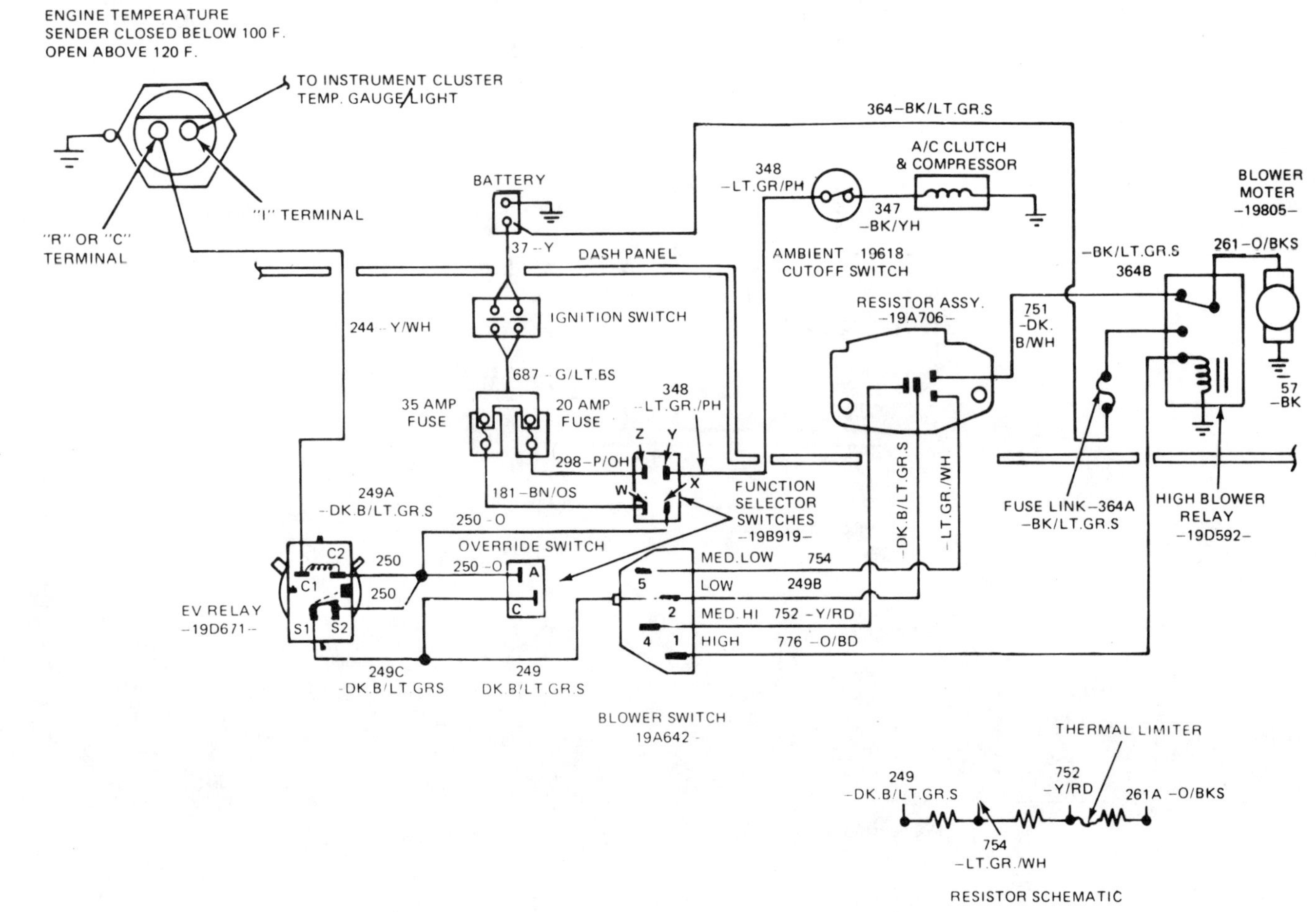

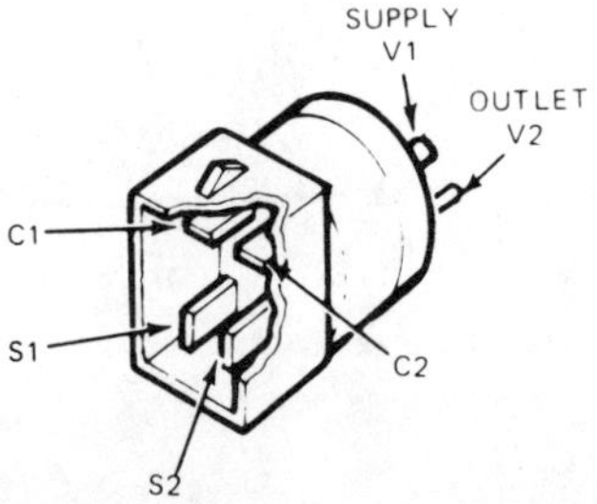

EV RELAY 19D671

COIL TERMINALS	SWITCH TERMINALS	VACUUM SWITCH	
C1 C2	S1 S2	V1	V2
12.8 VOLTS	OPEN	OPEN	OPEN
0 VOLTS	CLOSED	SEALED	VENTED

FUNCTION SELECTOR SWITCH 19B919

SWITCH SCHEMATIC	OFF	ECONOMY VENT	A/C	HI-LO	HEAT	▽ DEFOG	DEFROST
A—C	A - C	A - C	A - C	A - C	Open	A - C	A - C
X Y W Z	None	W - X	W - X Y - Z	W - X Y - Z	W - X	W - X Y - Z	W - X Y - Z

BLOWER SWITCH 19A642

BLOWER SPEED	TERMINAL CONNECTIONS	CURRENT IN CAR (AMPS)	BLOWER VOLTAGE
Low	-	6	5
Med Low	2 - 5	10	7
Med High	2 - 4 - 5	17	10
High	2 - 1 - 4	25	12.4

1978 Lincoln Continental A/C & Heater Wiring Circuit (With Auto Temp Control)

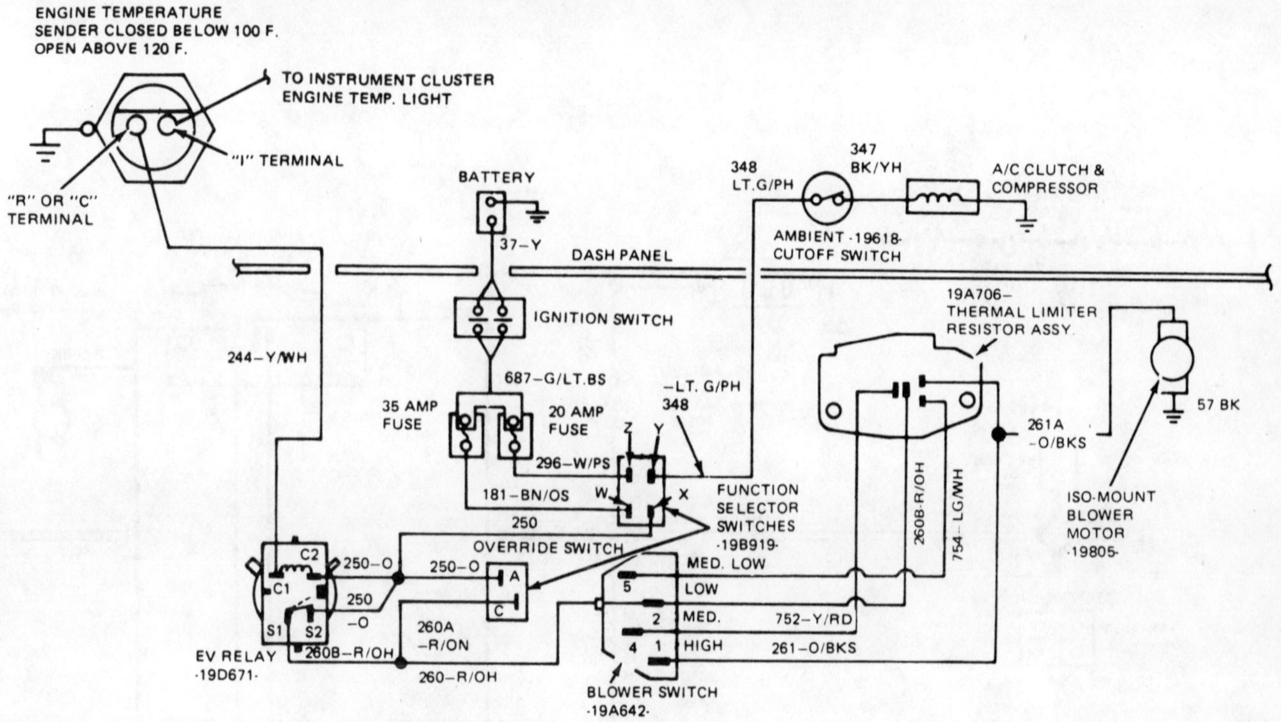

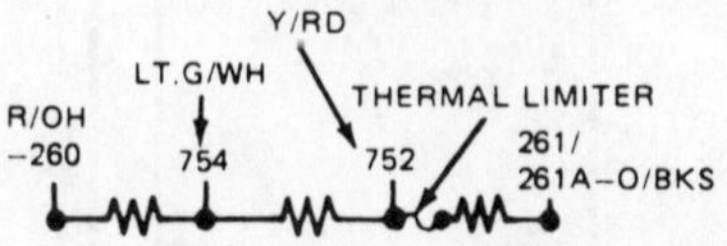

RESISTOR SCHEMATIC

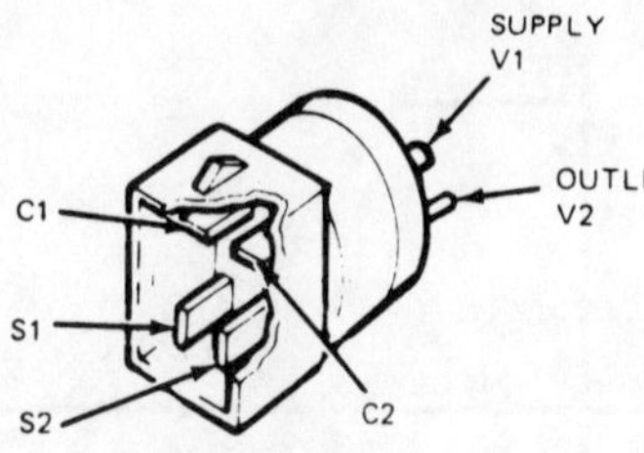

E.V. RELAY – 19D671		
COIL TERMINALS	SWITCH TERMINALS	VACUUM SWITCH	
C1 C2	S1 S2	V1	V2
12.8 VOLTS	OPEN	OPEN	OPEN
0 VOLTS	CLOSED	SEALED	VENTED

FUNCTION SELECTOR SWITCH -19B919-							
SWITCH SCHEMATIC	OFF	ECONOMY VENT	A/C	HI-LO	HEAT	▽ DEFOG	DEFROST
A—C	A - C	A - C	A - C	A - C	Open	A - C	A - C
X–Y, W–Z	None	W - X	W - X Y - Z	W - X Y - Z	W - X	W - X Y - Z	W - X Y - Z

BLOWER SWITCH -19A642-			
BLOWER SPEED	TERMINAL CONNECTIONS	CURRENT IN CAR (AMPS)	BLOWER VOLTAGE
Low	–	8	5
Med Low	2 - 5	11	7
Med High	2 - 4 - 5	20	10
High	2 - 1 - 4	25	14

1978 Lincoln Versailles A/C & Heater Wiring Circuit (With Auto Temp Control)

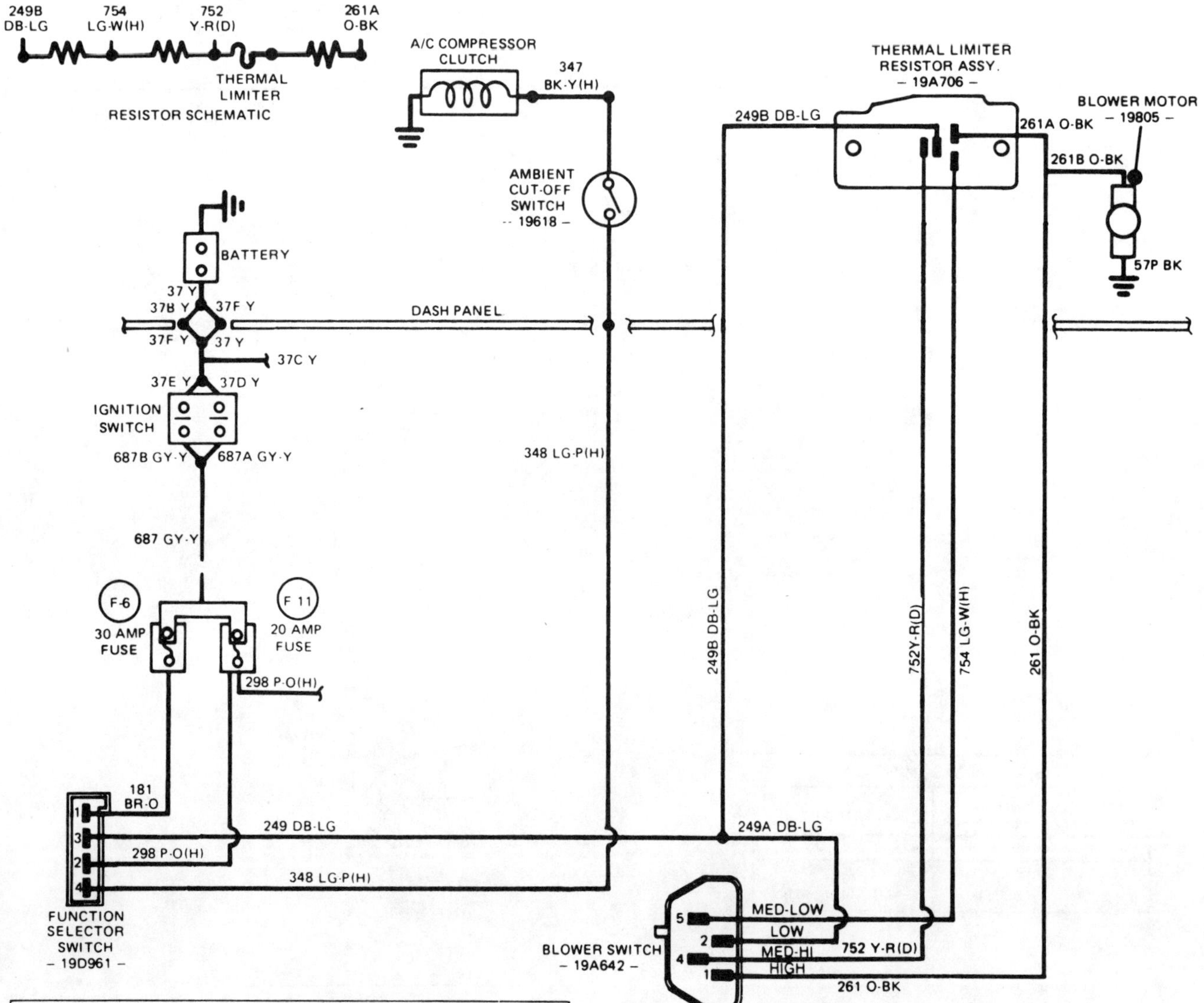

FUNCTION SELECTOR SWITCH — 19D961							
SWITCH SCHEMATIC	TERMINAL CONNECTIONS						
		A/C			HEAT		
	OFF	MAX	NORM	VENT	HI-LO	FLOOR	DEFROST
1 2 3 4	NONE	1-3 2-4	1-3 2-4	1-3	1-3	1-3	1-3 2-4

BLOWER SWITCH — 19A642			
BLOWER SPEED	TERMINAL CONNECTIONS	CURRENT IN CAR AMPS	BLOWER VOLTAGE
LOW	—	6	5
MED LOW	2-5	8	7
MED HIGH	2-4-5	15	10
HIGH	2-1-4	25	12.4

1979 Ford & Mercury Full Size A/C & Heater Wiring Circuit (Less Auto Temp Control)

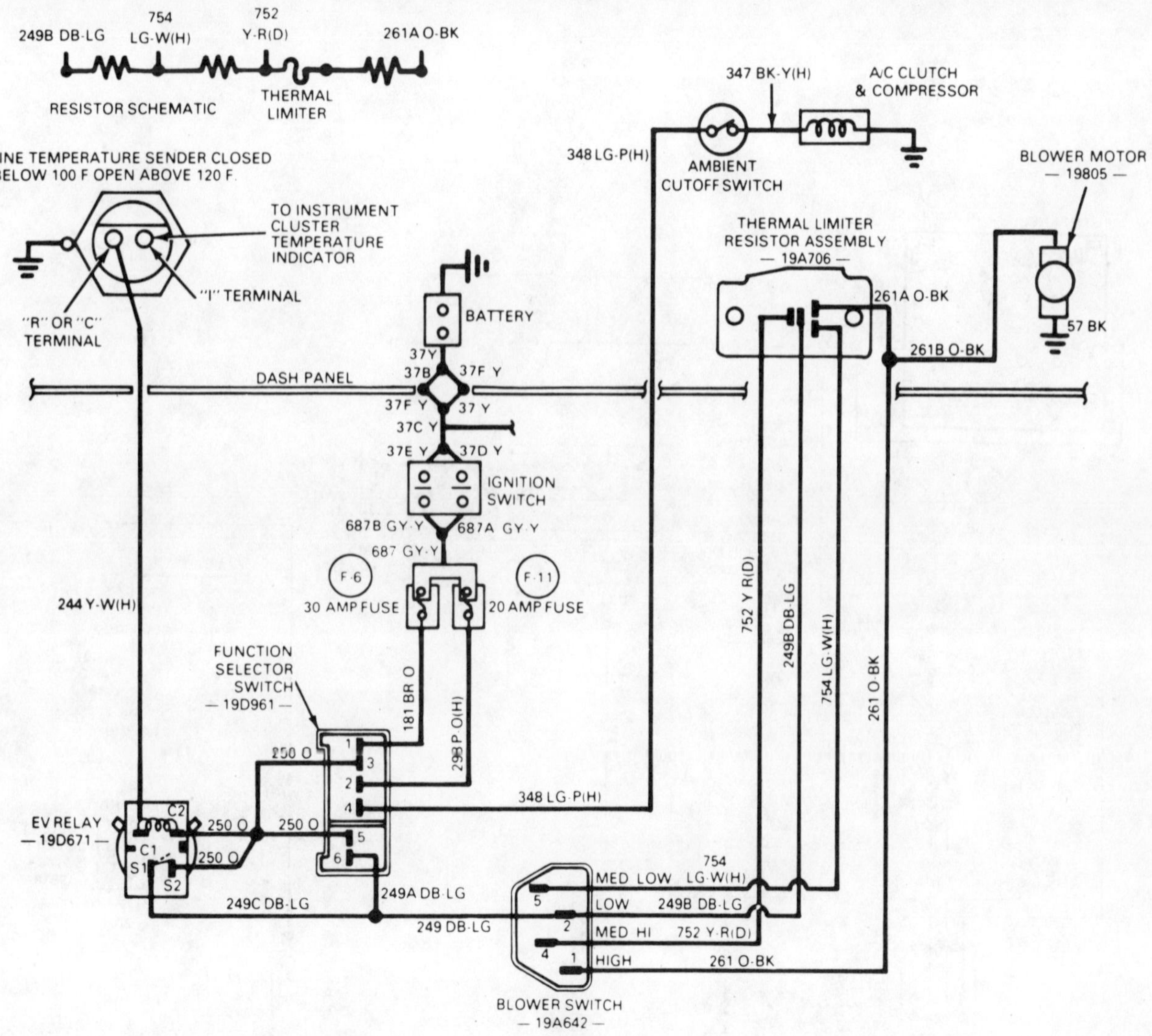

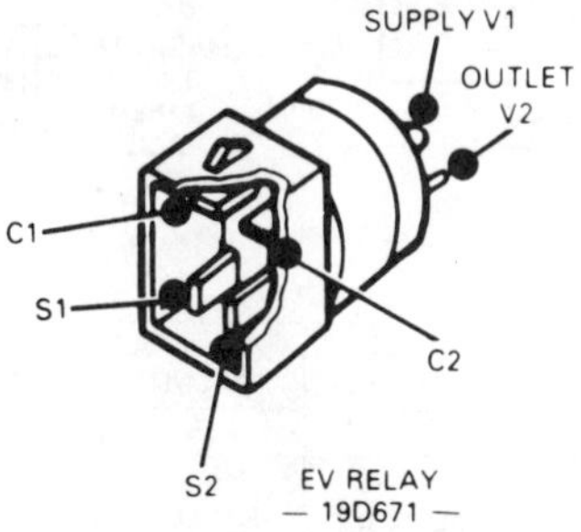

EV RELAY — 19D671 —			
COIL TERMINALS	SWITCH TERMINALS	VACUUM SWITCH	
C1 C2	S1 S2	V1	V2
12 8 VOLTS	OPEN	OPEN	OPEN
0 VOLTS	CLOSED	SEALED	VENTED

FUNCTION SELECTOR SWITCH — 19D961 —						
	TERMINAL CONNECTIONS					
			A/C		HEAT	
SWITCH SCHEMATIC	OFF	VENT	PANEL	HI-LO	FLOOR	DEFROST
1-2, 3-4	NONE	1-3	1-3 2-4	1-3 2-4	1-3	1-3 2-4
5-6	NONE	5-6	5-6	5-6	NONE	5-6

BLOWER SWITCH — 19A642 —			
BLOWER SPEED	TERMINAL CONNECTIONS	CURRENT IN CAR (AMPS)	BLOWER VOLTAGE
LOW	—	6	5
MED LOW	2 5	10	7
MED HIGH	2 4 5	17	10
HIGH	2 1 4	25	12 4

1979 Ford & Mercury Full Size A/C & Heater Wiring Circuit (With Auto Temp Control)

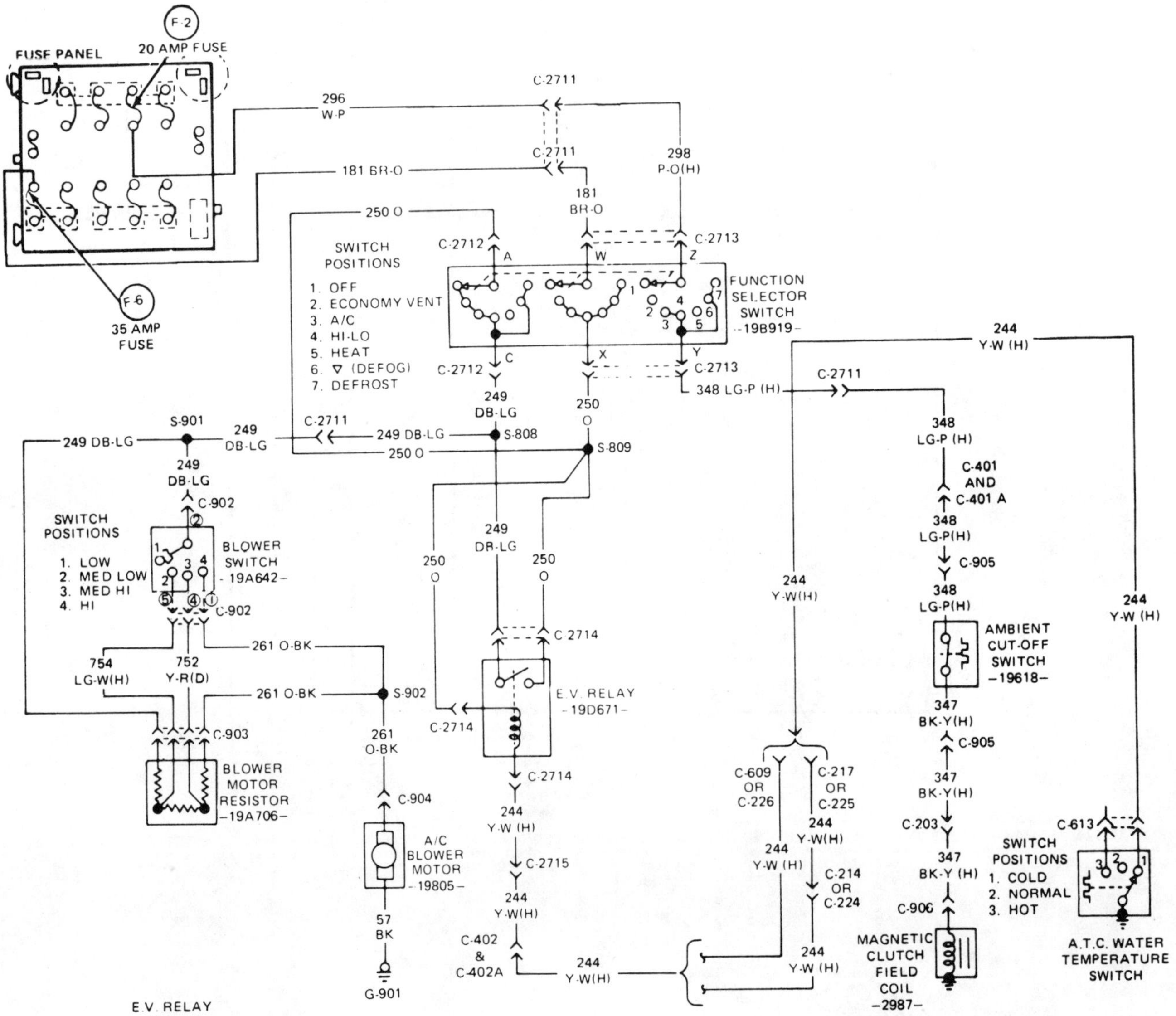

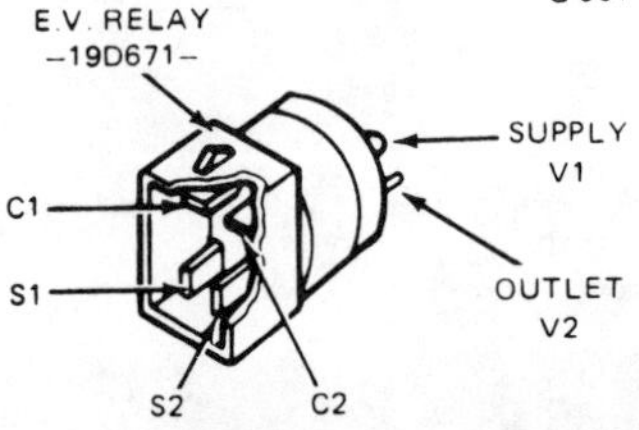

E.V. RELAY -19D671-

COIL TERMINALS	SWITCH TERMINALS	VACUUM SWITCH	
C1 C2	S1 S2	V1	V2
12.8 VOLTS	OPEN	OPEN	OPEN
0 VOLTS	CLOSED	SEALED	VENTED

BLOWER SWITCH -19A642-

BLOWER SPEED	TERMINAL CONNECTIONS	CURRENT IN CAR (AMPS)	BLOWER VOLTAGE
Low		6	5
Med. Low	2 - 5	10	7
Med. High	2 - 4 - 5	17	10
High	2 - 1 - 4	25	12.4

FUNCTION SELECTOR SWITCH -19B919-

SWITCH SCHEMATIC	OFF	ECONOMY VENT	A/C	HI–LO	HEAT	DEFOG	DEFROST
A—C	A - C	A - C	A - C	A - C	Open	A - C	A - C
X Y W Z	None	W - X	W - X Y - Z	W - X Y - Z	W - X	W - X Y - Z	W - X Y - Z

1979 Ford Fairmont & Mercury Zephyr A/C & Heater Wiring Circuit

FUNCTION SELECTOR SWITCH -19B919-						
TERMINAL CONNECTIONS						
OFF	ECONOMY VENT	A/C	HI-LO	HEAT	DEFOG ▽	DEFROST
None	(W)-(X)	(W)-(X) (Y)-(Z)	(W)-(X)	(W)-(X)	(W)-(X) (Y)-(Z)	(W)-(X) (Y)-(Z)

BLOWER SWITCH -19A642-			
BLOWER SPEED	TERMINAL CONNECTIONS	CURRENT IN CAR (AMPS)	BLOWER VOLTAGE
Low	(2)	6	7.0 - 8.5
Med. Low	(2)-(4)	10	8.5 - 10.0
Med. High	(2)-(5)	17	10.0 - 12.5
High	(2)-(1)	28	12.5 - 14.8

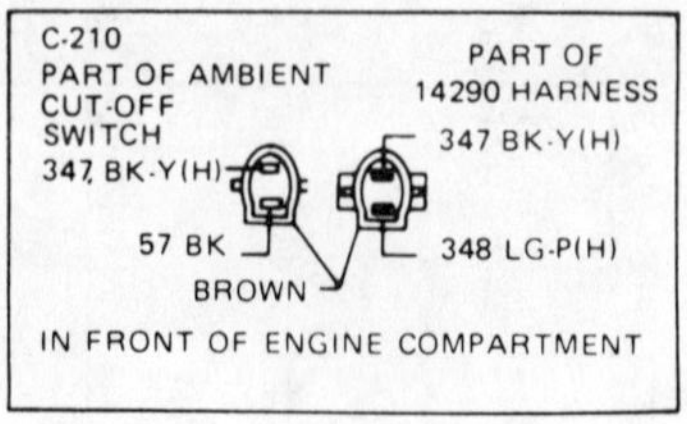

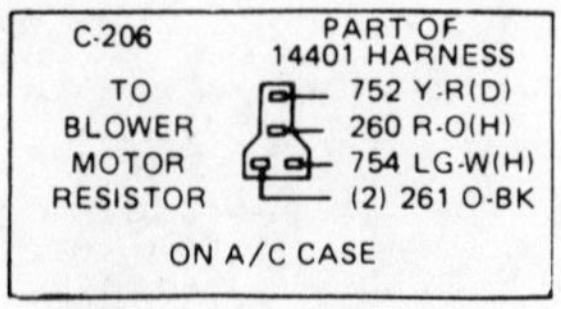

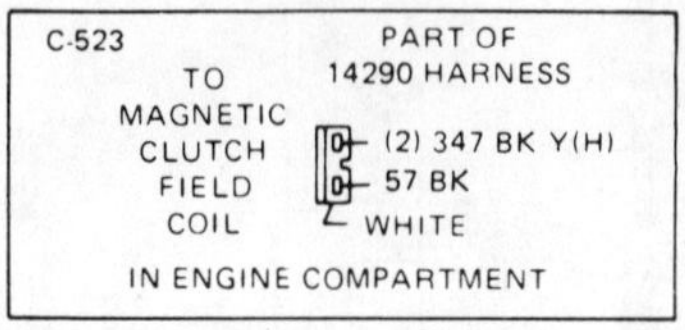

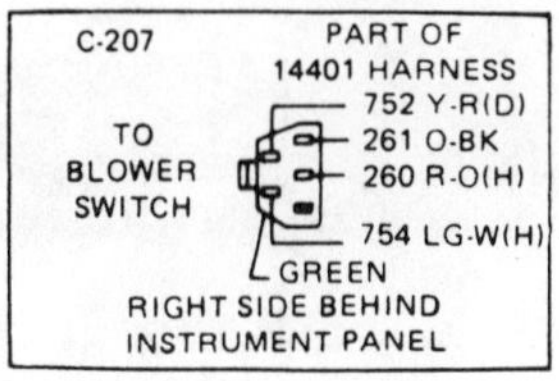

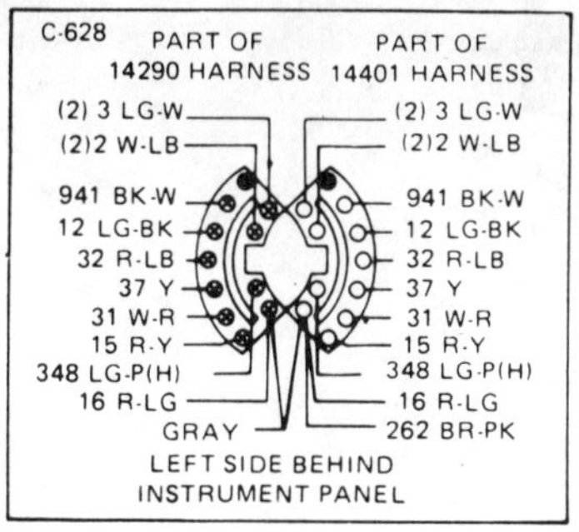

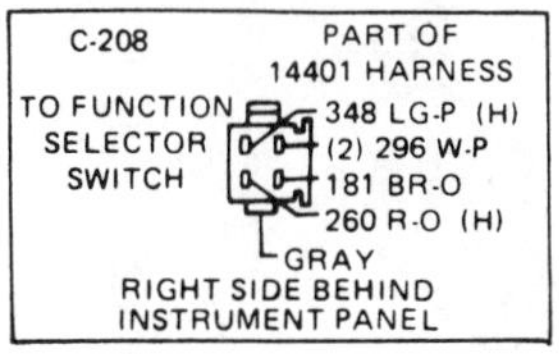

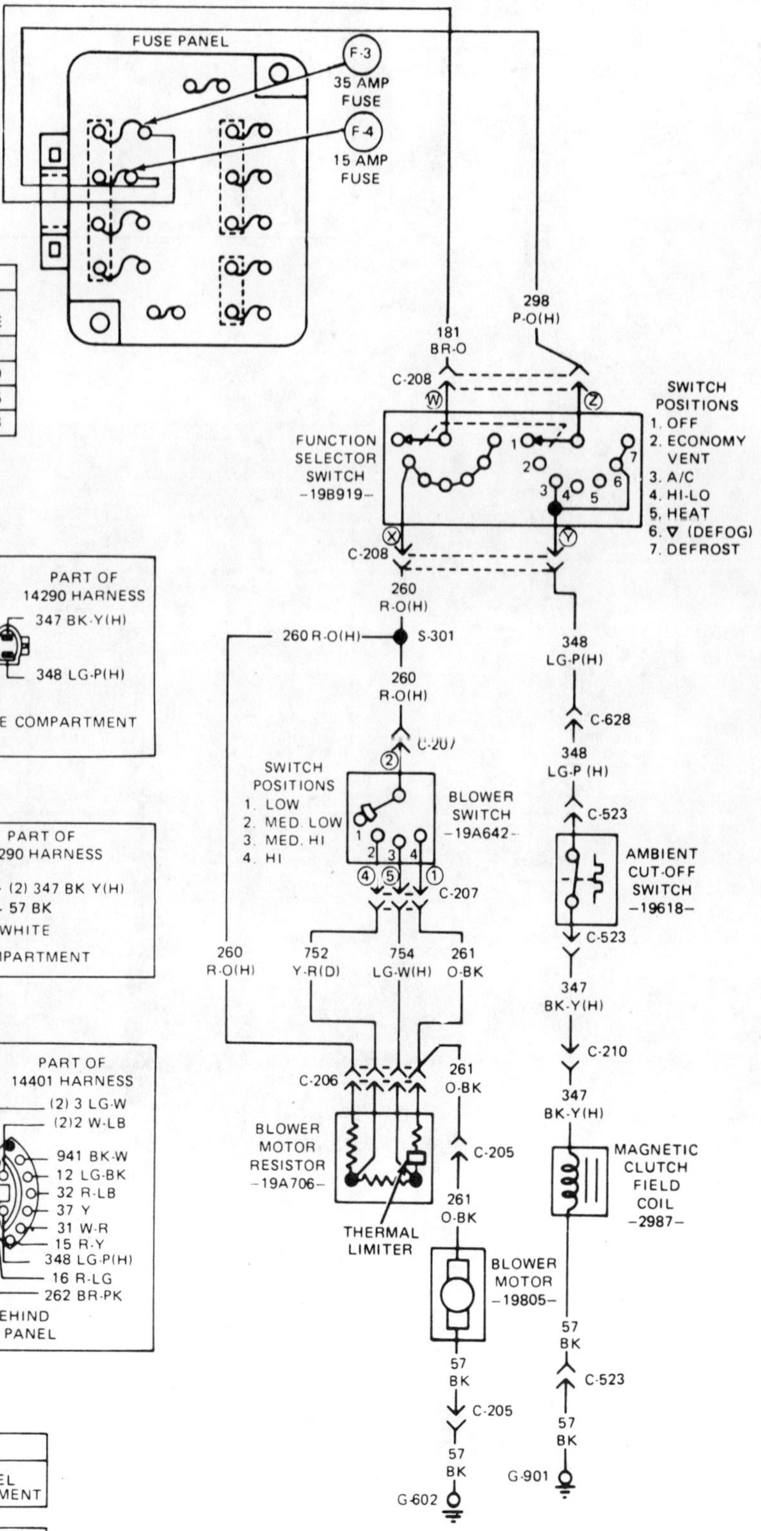

GOUND CODES
G-602 IN 14401 RIGHT SIDE OF INSTRUMENT PANEL G-901 IN 14290 RIGHT SIDE OF ENGINE COMPARTMENT

SPLICE CODE
S-301 IN 14401 NEAR A/C BLOWER SWITCH

1979 Ford Granada & Mercury Monarch A/C & Heater Wiring Circuit (Less Auto Temp Control)

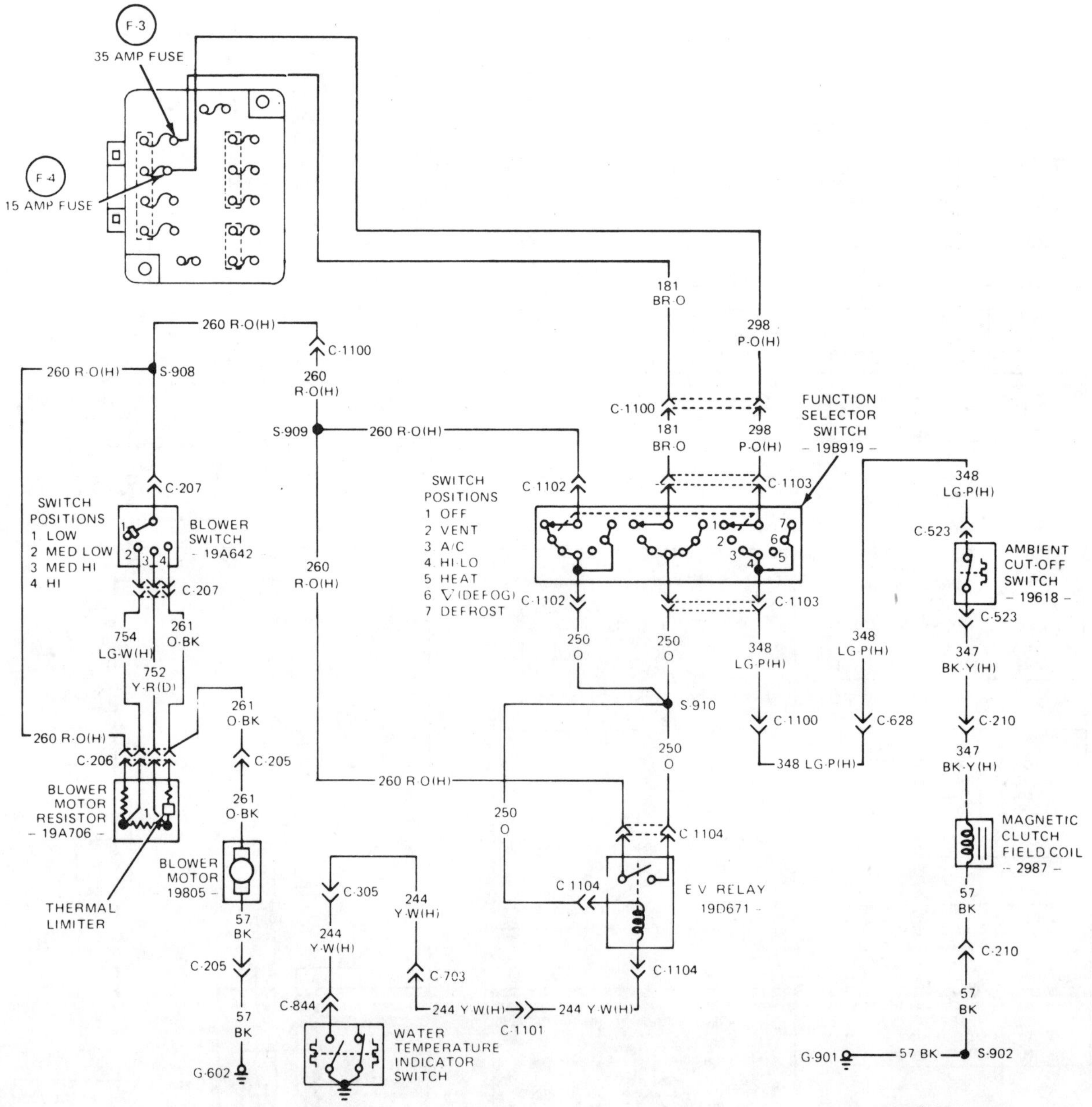

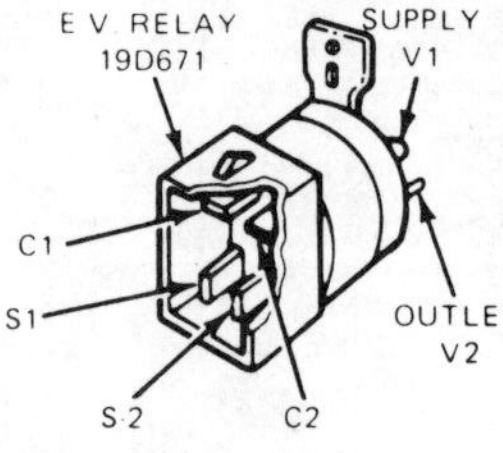

FUNCTION SELECTOR SWITCH -19B919-							
SWITCH SCHEMATIC	OFF	ECONOMY VENT	A/C	HI-LO	HEAT	DEFOG	DEFROST
A—C	A-C	A-C	A-C	A-C	OPEN	A-C	A-C
X Y W Z	NONE	W-X	W-X Y-Z	W-X Y-Z	W-X	W-X Y-Z	W-X Y-Z

BLOWER SWITCH -19A642-			
BLOWER SPEED	TERMINAL CONNECTIONS	CURRENT IN CAR (AMPS)	BLOWER VOLTAGE
LOW	②	8	5
MED LOW	② ⑤	11	7
MED HIGH	② ④	20	10
HIGH	② ①	25	14

E V RELAY -19D671-			
COIL TERMINALS	SWITCH TERMINALS	VACUUM SWITCH	
C1 C2	S1 S2	V1	V2
12.8 VOLTS	OPEN	OPEN	OPEN
0 VOLTS	CLOSED	SEALED	VENTED

1979 Ford Granada; Lincoln Versailles; Mercury Monarch A/C & Heater Wiring Circuit (With Auto Temp Control)

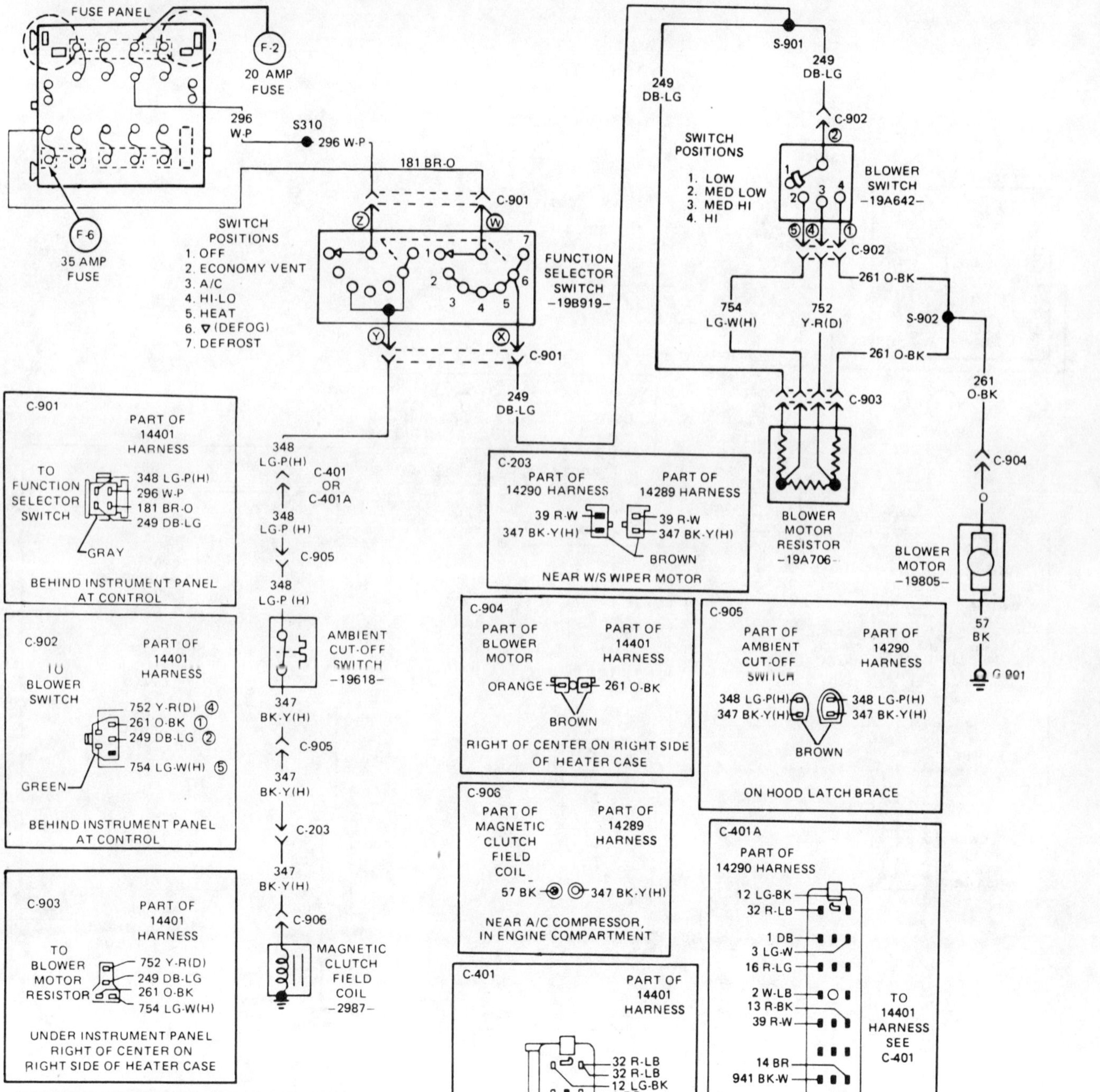

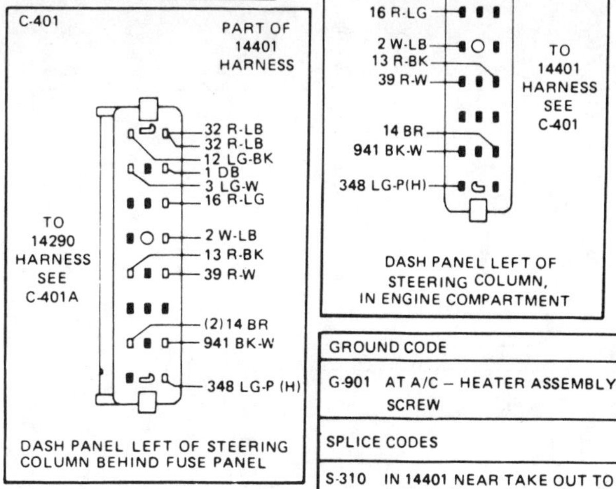

BLOWER SWITCH -19A642-			
BLOWER SPEED	TERMINAL CONNECTIONS	CURRENT IN CAR (AMPS)	BLOWER VOLTAGE
Low	②	6	5
Med. Low	②-⑤	8	7
Med. High	②-④	15	10
High	②-①	22	12.4

FUNCTION SELECTOR SWITCH -19B919-						
TERMINAL CONNECTIONS						
OFF	ECONOMY VENT	A/C	HI-LO	HEAT	DEFOG ▽	DEFROST
None	Ⓦ-Ⓧ	Ⓦ-Ⓧ Ⓨ-Ⓩ	Ⓦ-Ⓧ	Ⓦ-Ⓧ	Ⓦ-Ⓧ Ⓨ-Ⓩ	Ⓦ-Ⓧ Ⓨ-Ⓩ

GROUND CODE	
G-901	AT A/C – HEATER ASSEMBLY MOUNTING SCREW

SPLICE CODES	
S-310	IN 14401 NEAR TAKE OUT TO HEADLIGHT SWITCH
S-901	IN 14401 AT TAKE OUT TO FUNCTION SELECTOR SWITCH
S-902	IN 14401 NEAR TAKE OUT TO RADIO

1979 Ford LTD II & Thunderbird; Mercury Cougar A/C & Heater Wiring Circuit (Less Auto Temp Control)

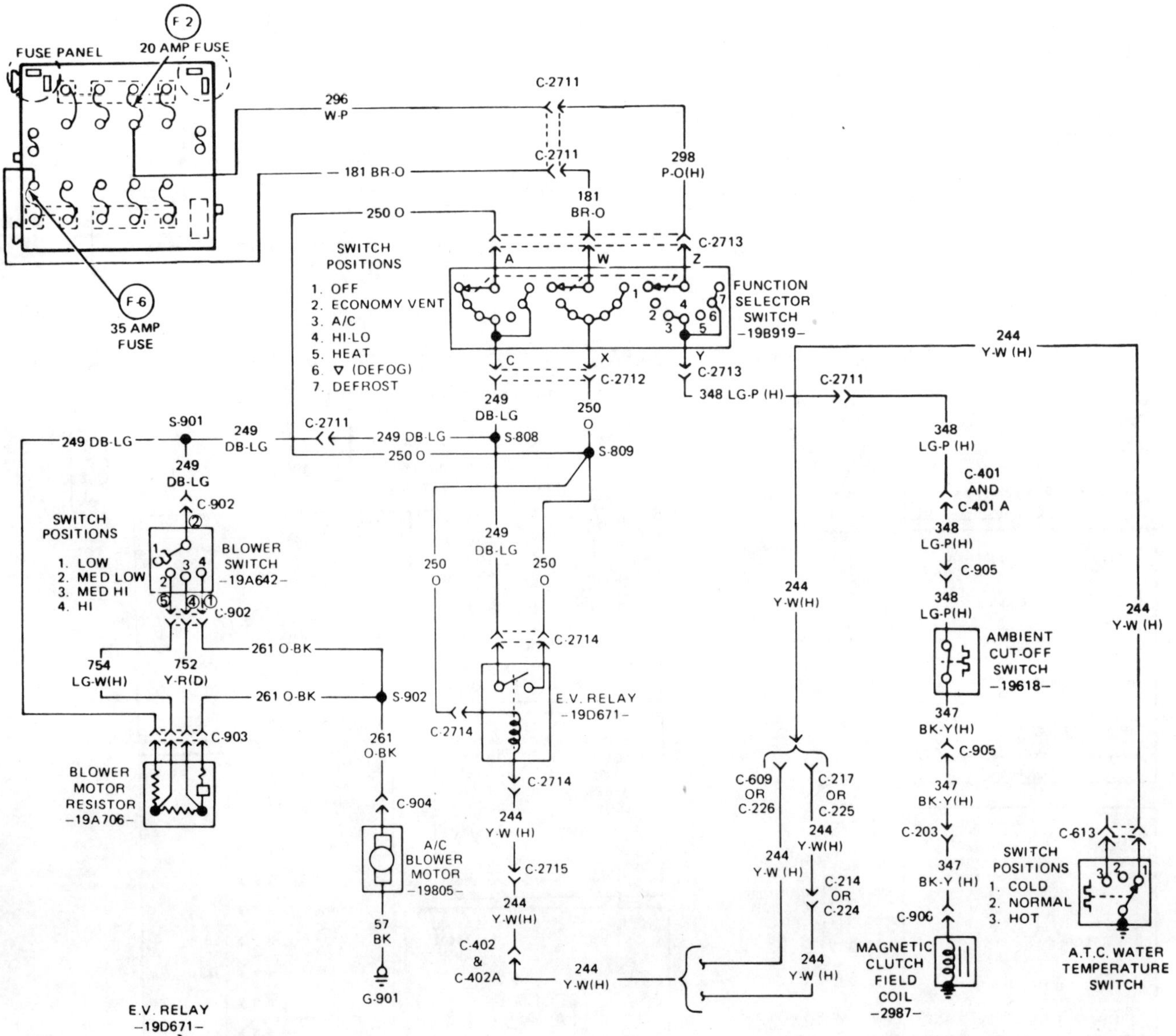

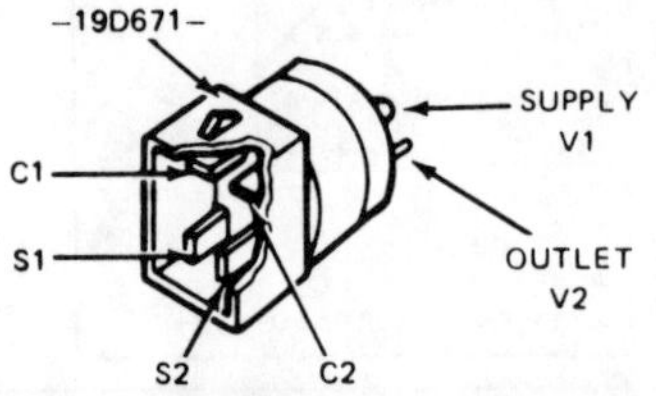

E.V. RELAY -19D671-

COIL TERMINALS	SWITCH TERMINALS	VACUUM SWITCH	
C1 C2	S1 S2	V1	V2
12.8 VOLTS	OPEN	OPEN	OPEN
0 VOLTS	CLOSED	SEALED	VENTED

BLOWER SWITCH -19A642-			
BLOWER SPEED	TERMINAL CONNECTIONS	CURRENT IN CAR (AMPS)	BLOWER VOLTAGE
Low	2	6	5
Med. Low	2 - 5	10	7
Med. High	2 - 4	17	10
High	2 - 1	25	12.4

FUNCTION SELECTOR SWITCH -19B919-							
SWITCH SCHEMATIC	OFF	ECONOMY VENT	A/C	HI-LO	HEAT	DEFOG	DEFROST
A—C	A - C	A - C	A - C	A - C	Open	A - C	A - C
X—Y (W, Z)	None	W - X	W - X Y - Z	W - X Y - Z	W - X	W - X Y - Z	W - X Y - Z

1979 Continental Mark V; Ford LTD II & Thunderbird; Mercury Cougar A/C & Heater Wiring Circuit (With Auto Temp Control)

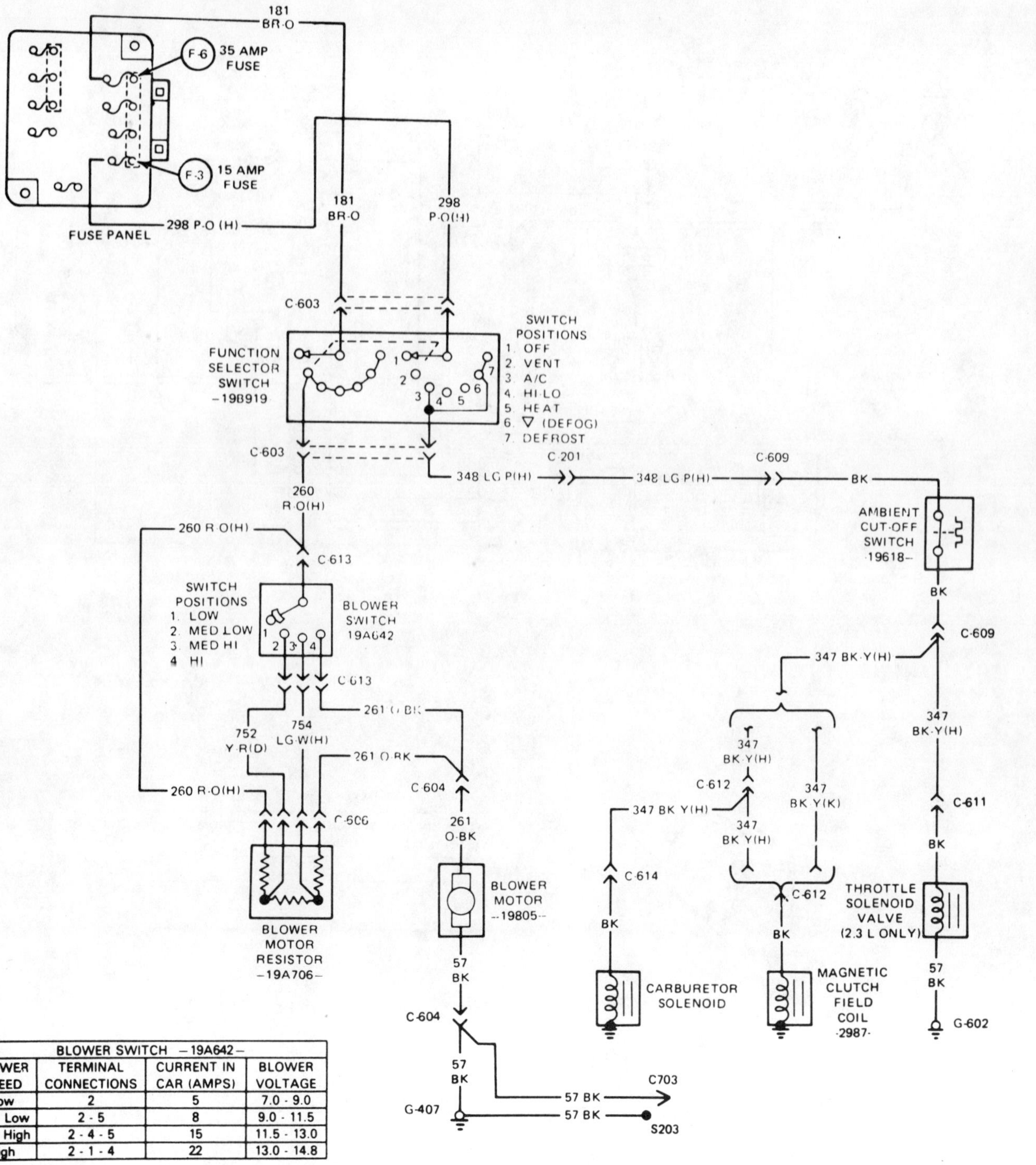

BLOWER SWITCH -19A642-			
BLOWER SPEED	TERMINAL CONNECTIONS	CURRENT IN CAR (AMPS)	BLOWER VOLTAGE
Low	2	5	7.0 - 9.0
Med Low	2 - 5	8	9.0 - 11.5
Med High	2 - 4 - 5	15	11.5 - 13.0
High	2 - 1 - 4	22	13.0 - 14.8

FUNCTION SELECTOR SWITCH -19B919-							
SWITCH SCHEMATIC	TERMINAL CONNECTIONS						
	OFF	ECONOMY VENT	A/C	HI-LO	HEAT	▽ DEFOG	DEFROST
X Y W Z	None	W - X	W - X Y - Z	W - X	W - X	W - X Y - Z	W - X Y - Z

GROUND CODES
G-407 IN 14401 HARNESS BEHIND CENTER OF INSTRUMENT PANEL G-602 AT THROTTLE SOLENOID VALVE ASSEMBLY LOCATED ON DASH PANEL NEAR HOOD LEFT HINGE
SPLICE CODES
S-203 IN 14401 HARNESS NEAR TAKE OUT TO PRINTED CIRCUIT BOARD

1979 Ford Mustang & Pinto; Mercury Bobcat & Capri A/C & Heater Wiring Circuit

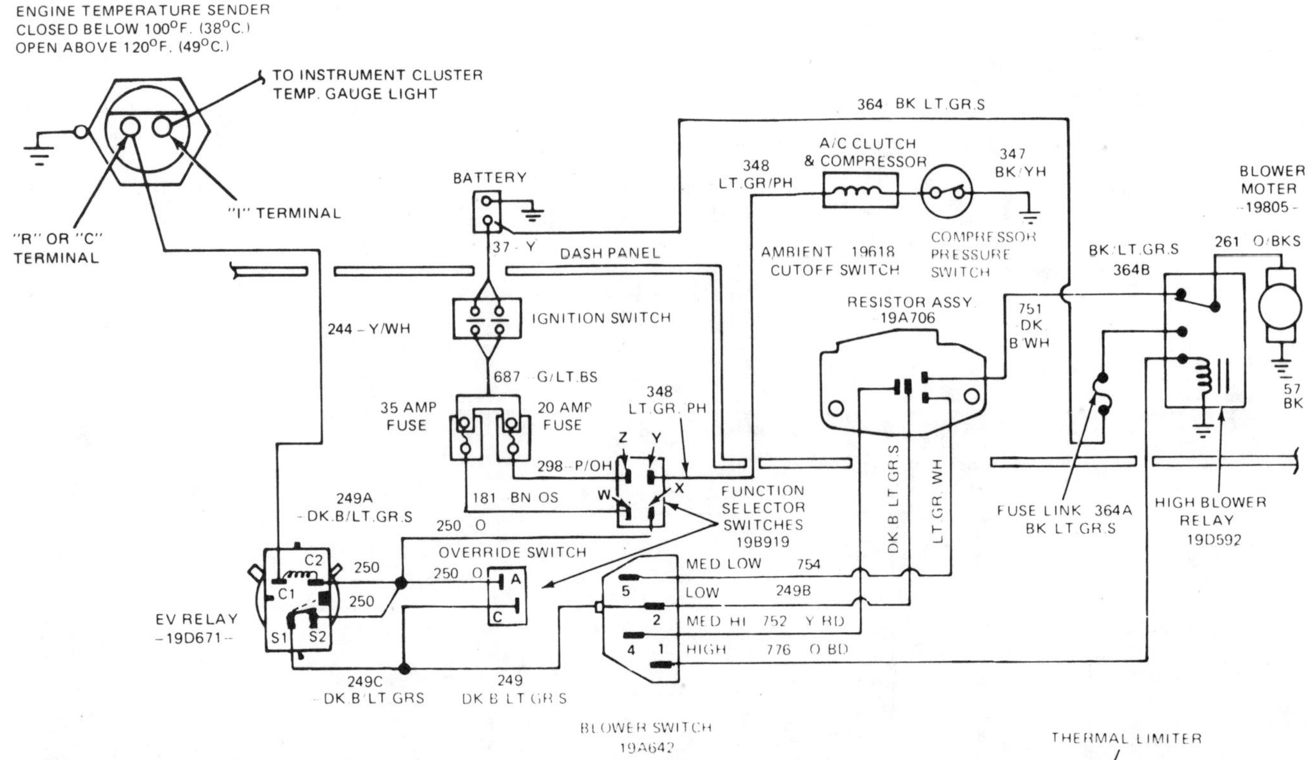

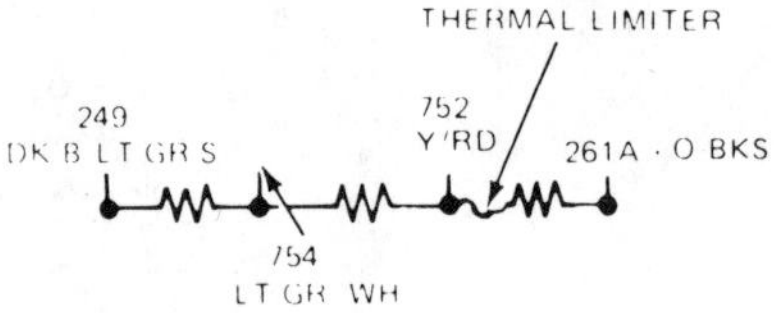

RESISTOR SCHEMATIC

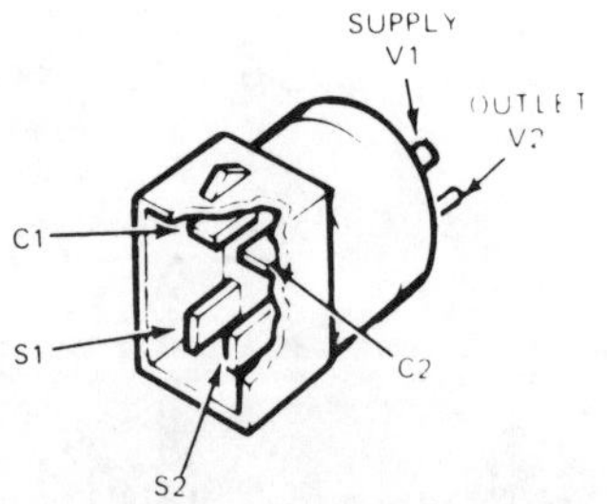

EV RELAY 19D671

COIL TERMINALS	SWITCH TERMINALS	VACUUM SWITCH	
C1 C2	S1 S2	V1	V2
12.8 VOLTS	OPEN	OPEN	OPEN
0 VOLTS	CLOSED	SEALED	VENTED

FUNCTION SELECTOR SWITCH 19B919

SWITCH SCHEMATIC	OFF	ECONOMY VENT	A/C	HI LO	HEAT	▽ DEFOG	DEFROST
A—C	A C	A C	A C	A C	Open	A C	A C
X Y W Z	None	W X	W X Y Z	W X Y Z	W X	W X Y Z	W X Y Z

BLOWER SWITCH 19A642

BLOWER SPEED	TERMINAL CONNECTIONS	CURRENT IN CAR (AMPS)	BLOWER VOLTAGE
Low	-	6	5
Med Low	2 5	10	7
Med High	2 4 5	17	10
High	2 1 4	25	12.4

1979 Lincoln Continental A/C & Heater Wiring Circuit (With Auto Temp Control)

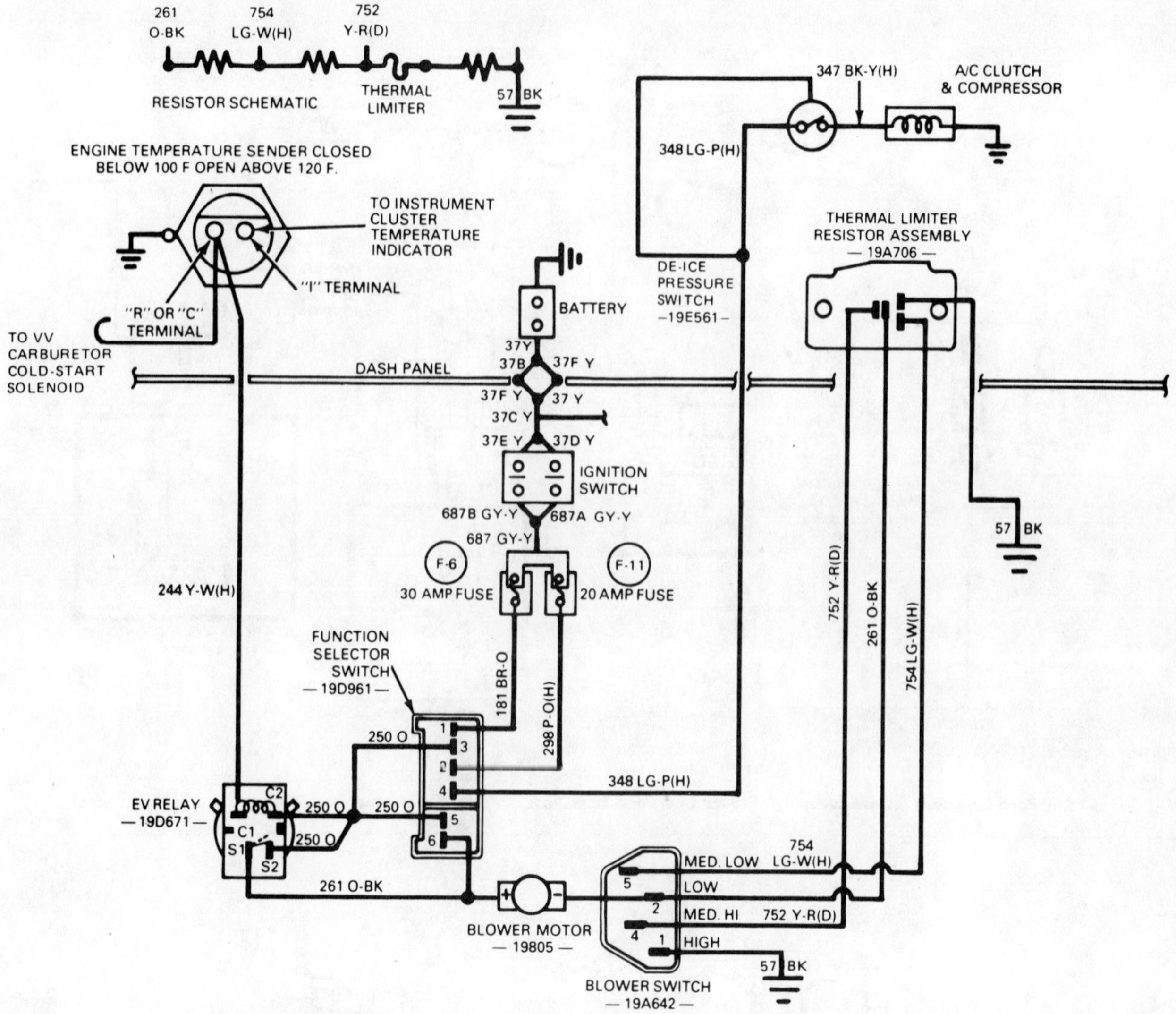

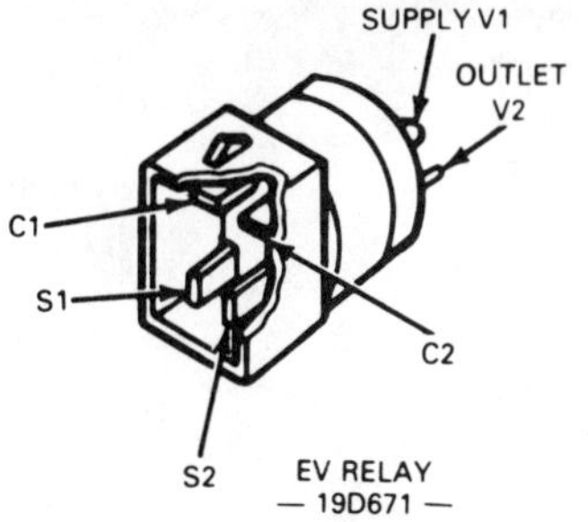

EV RELAY
— 19D671 —

EV RELAY — 19D671 —			
COIL TERMINALS	SWITCH TERMINALS	VACUUM SWITCH	
C1 C2	S1 S2	V1	V2
12.8 VOLTS	OPEN	OPEN	OPEN
0 VOLTS	CLOSED	SEALED	VENTED

FUNCTION SELECTOR SWITCH — 19D961 —						
	TERMINAL CONNECTIONS					
			A/C		HEAT	
SWITCH SCHEMATIC	OFF	VENT	PANEL	HI-LO	FLOOR	DEFROST
1 2 / 3 4	NONE	1-3	1-3 2-4	1-3 2-4	1-3	1-3 2-4
5 6	NONE	5-6	5-6	5-6	NONE	5-6

BLOWER SWITCH — 19A642 —			
BLOWER SPEED	TERMINAL CONNECTIONS	CURRENT IN CAR (AMPS)	BLOWER VOLTAGE
LOW	—	6	5
MED LOW	2-5	10	7
MED HIGH	2-4-5	17	10
HIGH	2-1-4	25	12.4

1980 Continental Mark VI; Ford & Mercury Full Size; Lincoln Continental
A/C & Heater Wiring Circuit (With Auto Temp Control)

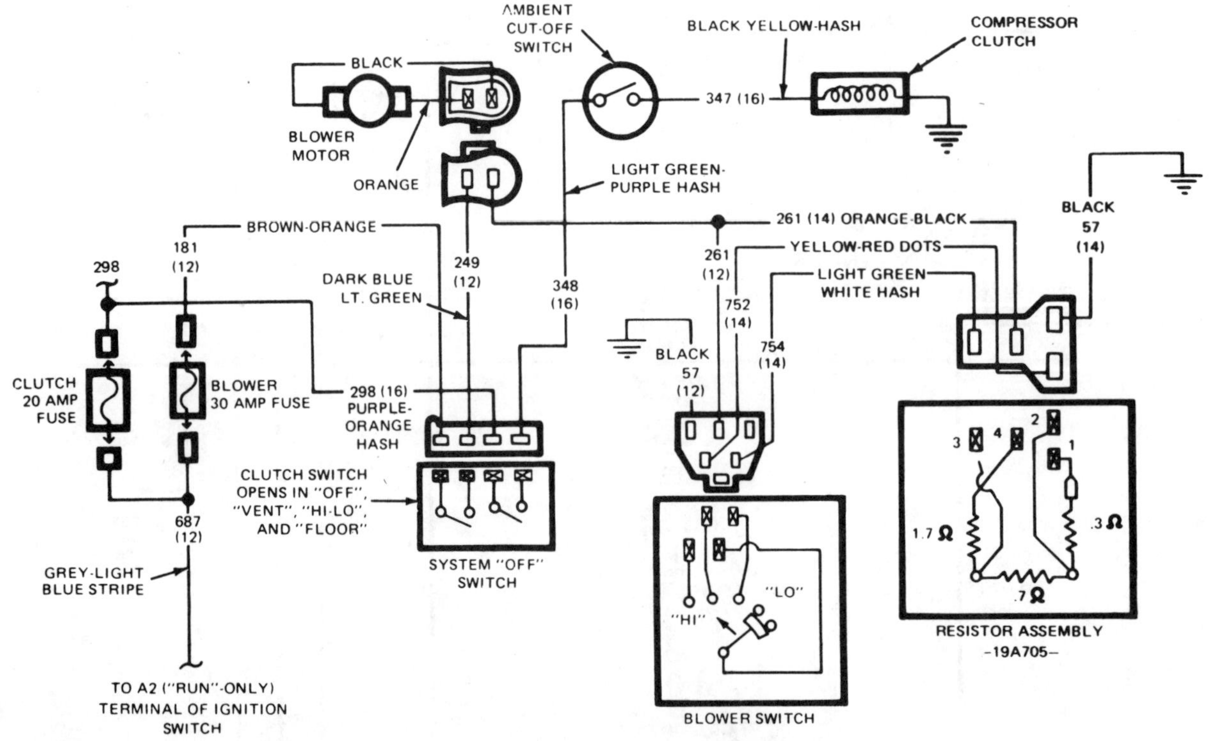

1980 Ford Fairmont & Thunderbird; Mercury Cougar & Zephyr A/C & Heater Wiring Circuit (Less Auto Temp Control)

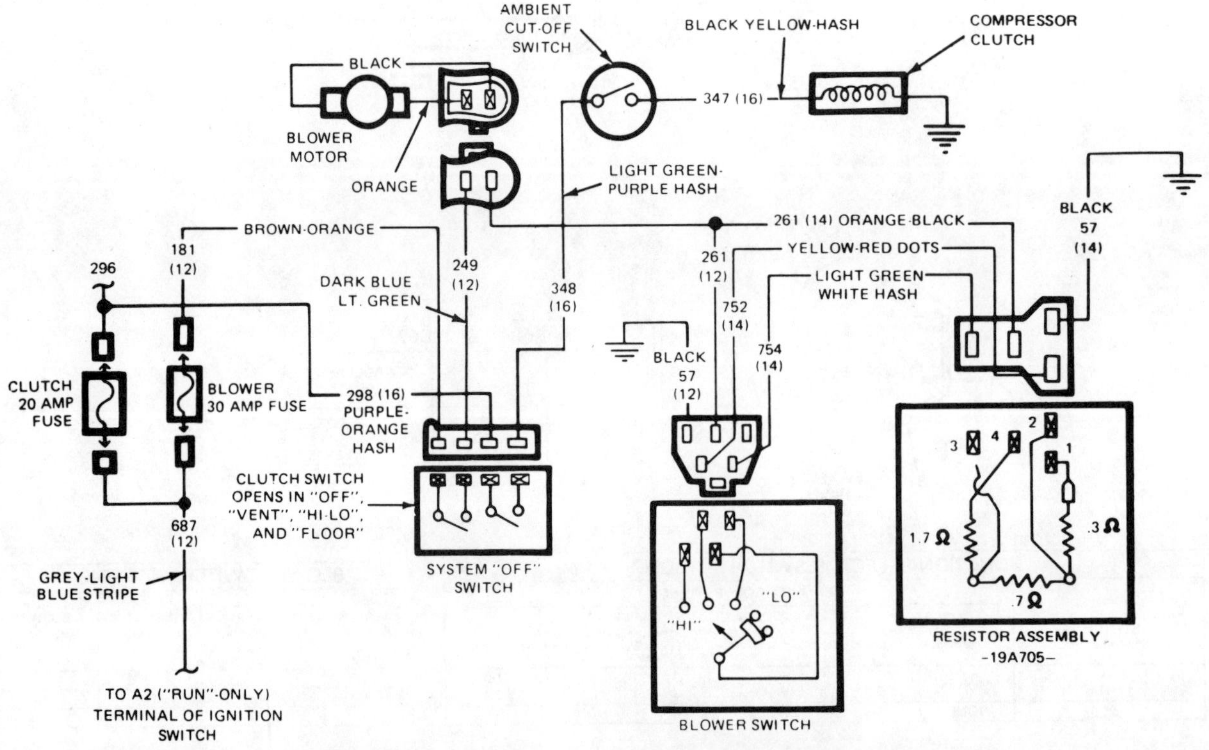

1980 Ford Mustang & Mercury Capri A/C & Heater Wiring Circuit

FUNCTION SELECTOR SWITCH -19B919-						
TERMINAL CONNECTIONS						
OFF	ECONOMY VENT	A/C	HI–LO	HEAT	DEFOG ▽	DEFROST
None	Ⓦ-Ⓧ	Ⓦ-Ⓧ Ⓨ-Ⓩ	Ⓦ-Ⓧ	Ⓦ-Ⓧ	Ⓦ-Ⓧ Ⓨ-Ⓩ	Ⓦ-Ⓧ Ⓨ-Ⓩ

BLOWER SWITCH -19A642-			
BLOWER SPEED	TERMINAL CONNECTIONS	CURRENT IN CAR (AMPS)	BLOWER VOLTAGE
Low	②	6	7.0 - 8.5
Med. Low	②④	10	8.5 - 10.0
Med. High	②⑤	17	10.0 - 12.5
High	②①	28	12.5 - 14.8

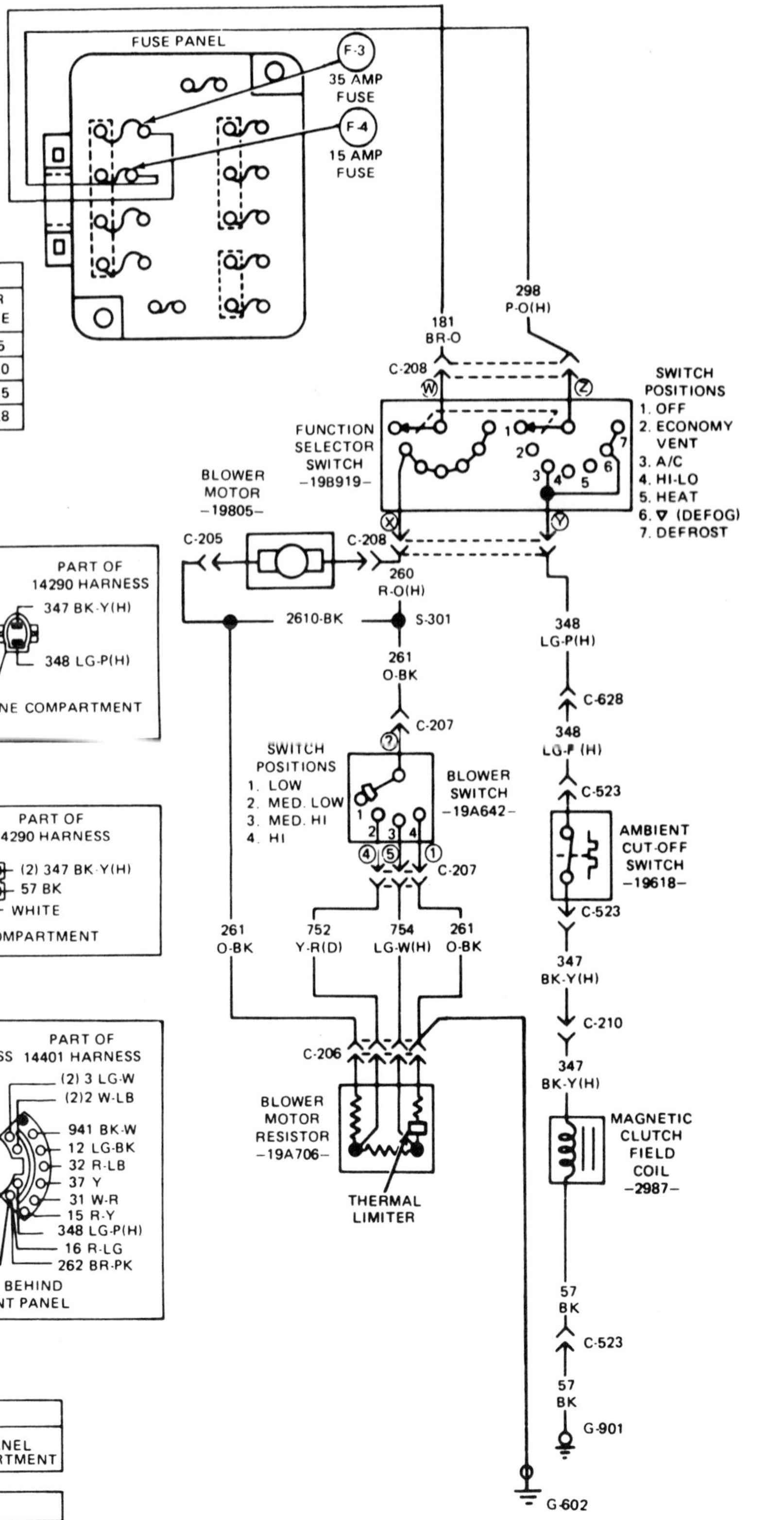

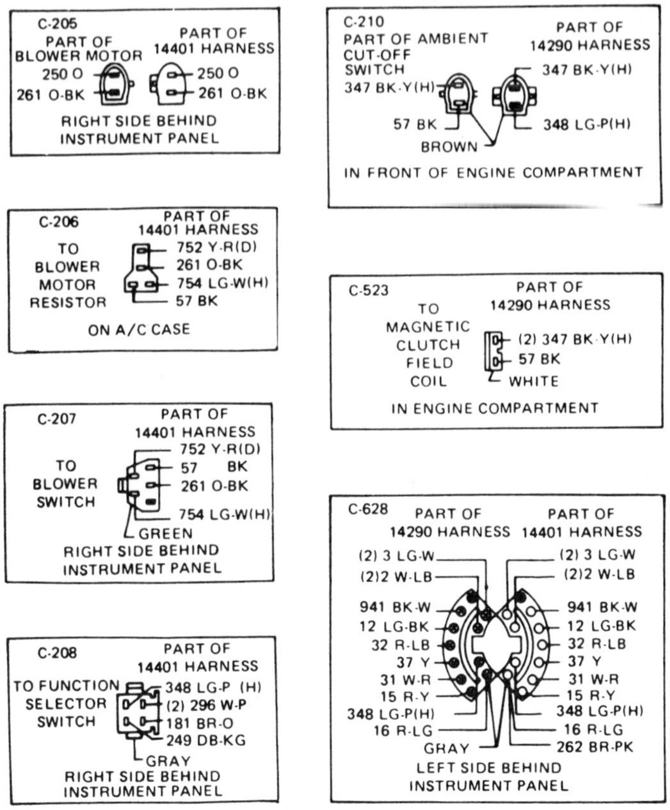

GOUND CODES
G-602 IN 14401 RIGHT SIDE OF INSTRUMENT PANEL G-901 IN 14290 RIGHT SIDE OF ENGINE COMPARTMENT

SPLICE CODE
S-301 IN 14401 NEAR A/C BLOWER SWITCH

1980 Ford Granada & Mercury Monarch A/C & Heater Wiring Circuit (Less Auto Temp Control)

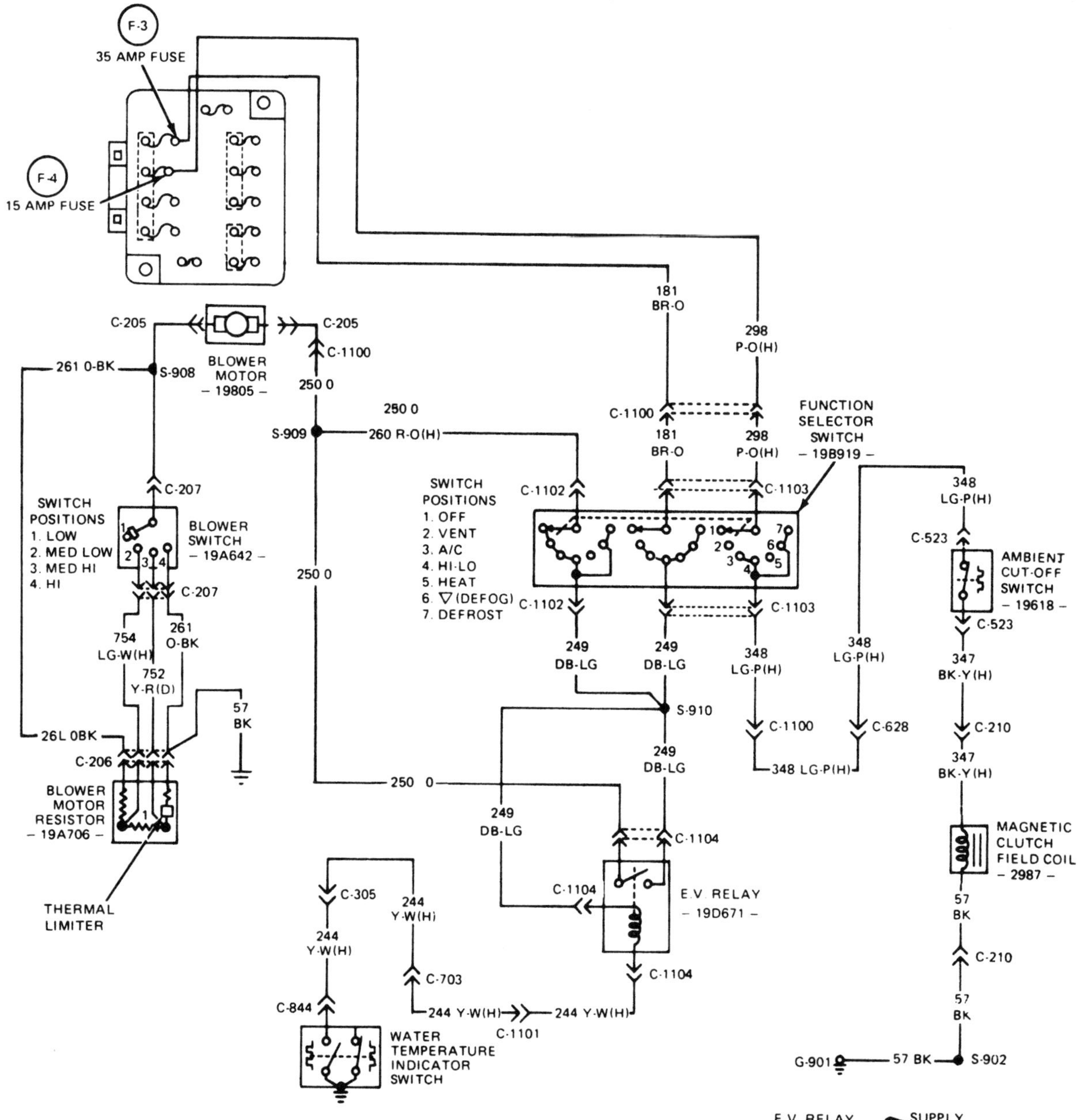

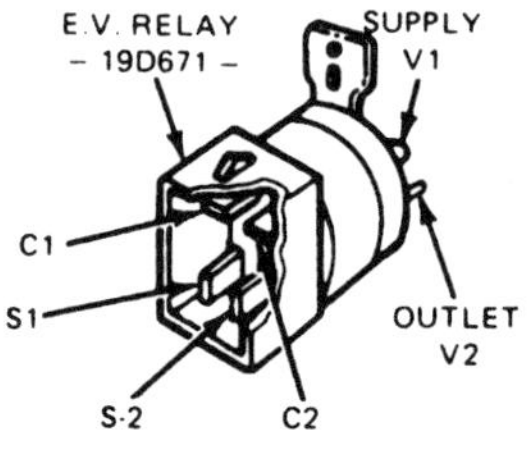

FUNCTION SELECTOR SWITCH –19B919–							
SWITCH SCHEMATIC	OFF	ECONOMY VENT	A/C	HI-LO	HEAT	DEFOG	DEFROST
A—C	A-C	A-C	A-C	A-C	OPEN	A-C	A-C
X Y W Z	NONE	W-X	W-X Y-Z	W-X Y-Z	W-X	W-X Y-Z	W-X Y-Z

BLOWER SWITCH –19A642–			
BLOWER SPEED	TERMINAL CONNECTIONS	CURRENT IN CAR (AMPS)	BLOWER VOLTAGE
LOW	②	8	5
MED LOW	② ⑤	11	7
MED HIGH	② ④	20	10
HIGH	② ①	25	14

E.V. RELAY –19D671–			
COIL TERMINALS	SWITCH TERMINALS	VACUUM SWITCH	
C1 C2	S1 S2	V1	V2
12.8 VOLTS	OPEN	OPEN	OPEN
0 VOLTS	CLOSED	SEALED	VENTED

1980 Ford Granada; Lincoln Versailles; Mercury Monarch A/C & Heater Wiring Circuit (With Auto Temp Control)

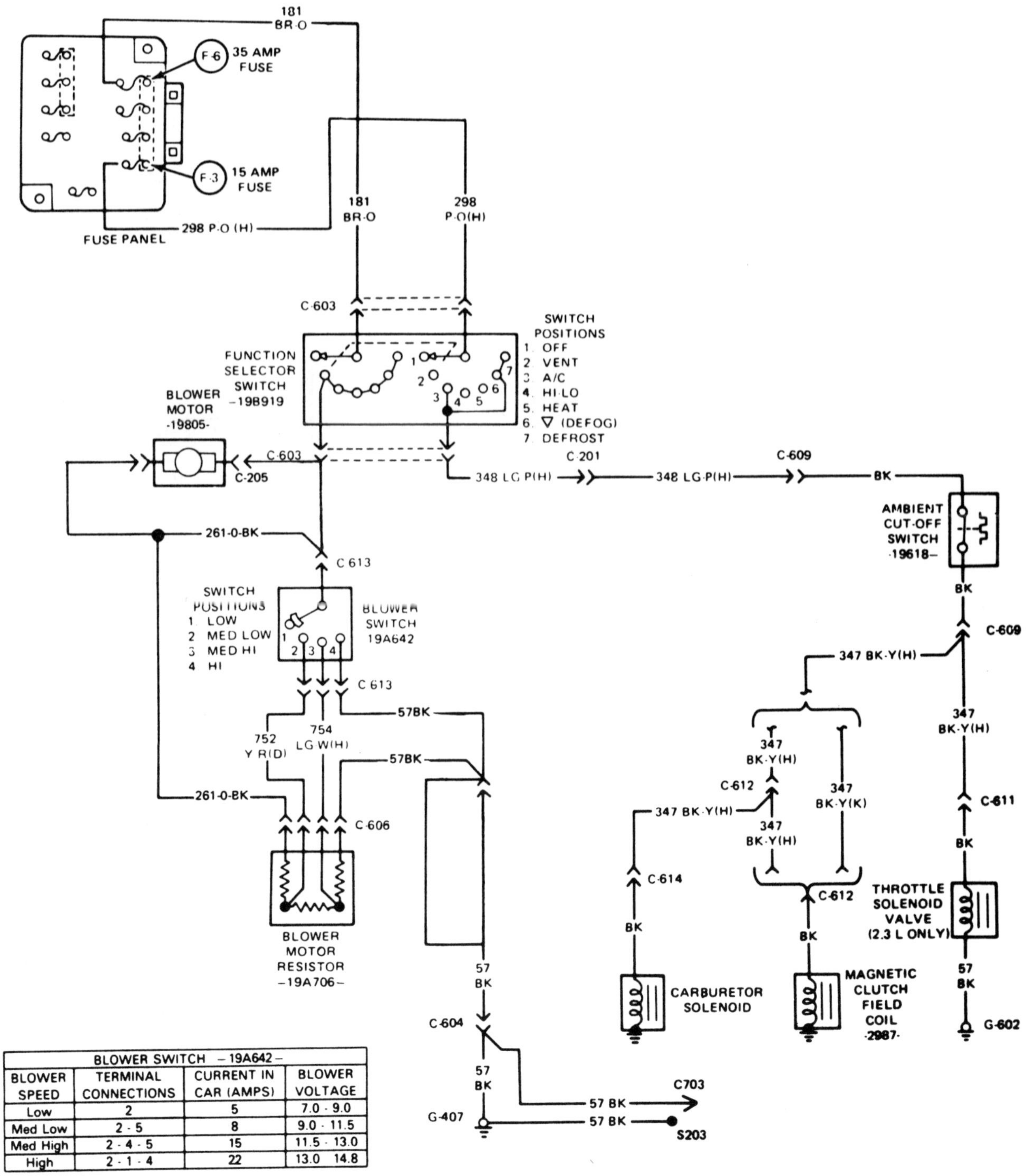

BLOWER SWITCH -19A642-			
BLOWER SPEED	TERMINAL CONNECTIONS	CURRENT IN CAR (AMPS)	BLOWER VOLTAGE
Low	2	5	7.0 - 9.0
Med Low	2 - 5	8	9.0 - 11.5
Med High	2 - 4 - 5	15	11.5 - 13.0
High	2 - 1 - 4	22	13.0 14.8

FUNCTION SELECTOR SWITCH -19B919-							
SWITCH SCHEMATIC	TERMINAL CONNECTIONS						
	OFF	ECONOMY VENT	A/C	HI-LO	HEAT	— DEFOG	DEFROST
X Y W Z	None	W - X	W - X Y - Z	W - X	W - X	W X Y - Z	W - X Y - Z

GROUND CODES
G-407 IN 14401 HARNESS BEHIND CENTER OF INSTRUMENT PANEL G-602 AT THROTTLE SOLENOID VALVE ASSEMBLY LOCATED ON DASH PANEL NEAR HOOD LEFT HINGE
SPLICE CODES
S-203 IN 14401 HARNESS NEAR TAKE OUT TO PRINTED CIRCUIT BOARD

1980 Ford Pinto & Mercury Bobcat A/C & Heater Wiring Circuit

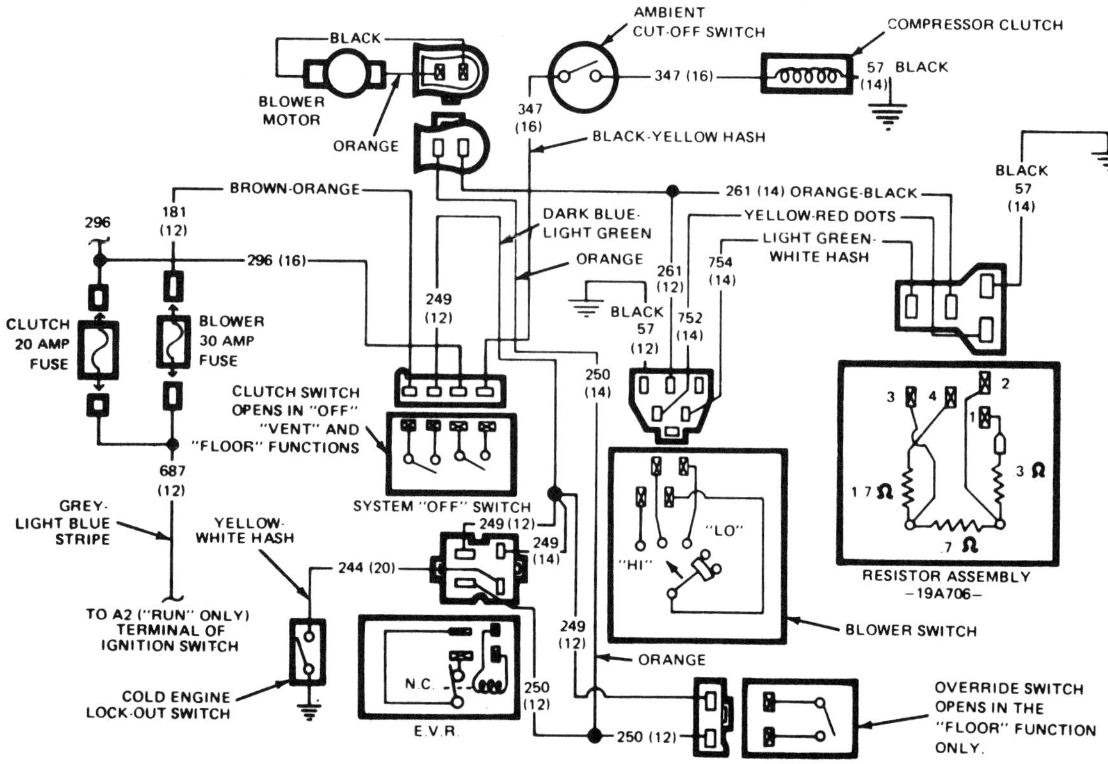

1980 Ford Thunderbird & Mercury Cougar XR-7 A/C & Heater Wiring Circuit (With Auto Temp Control)

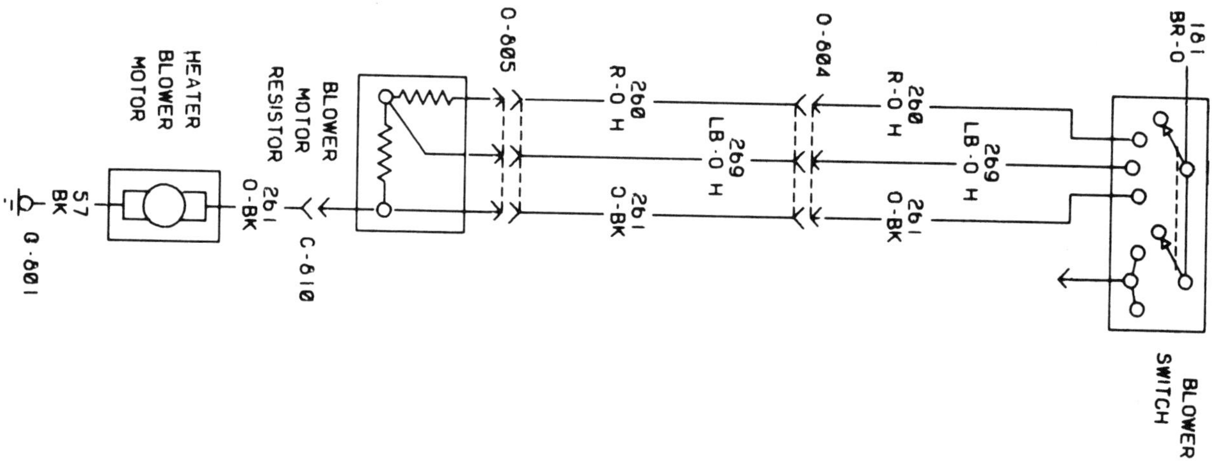

1974 Ford Maverick & Mercury Comet Heater Wiring Circuit

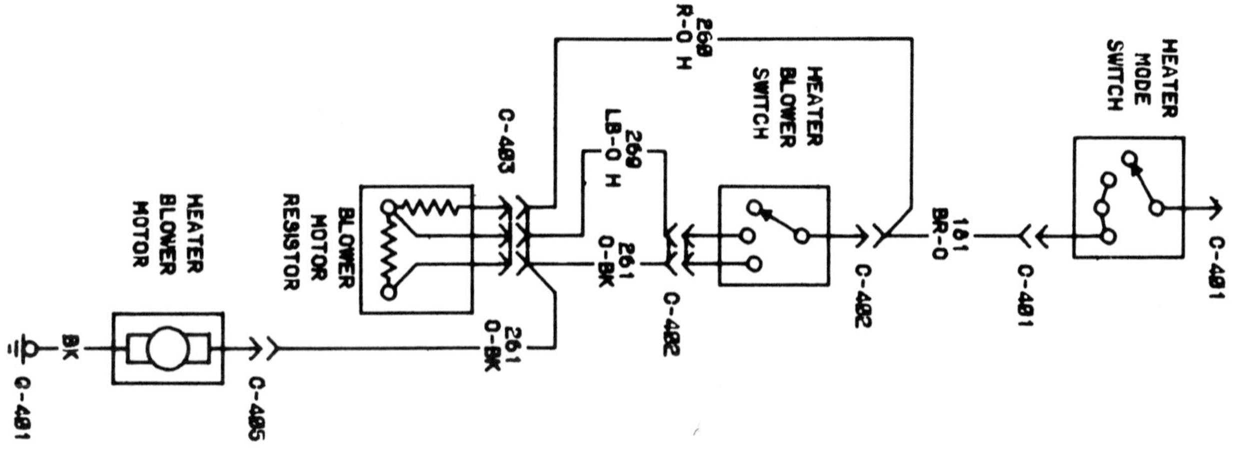

1974-76 Ford & Mercury Full Size Heater Wiring Circuit

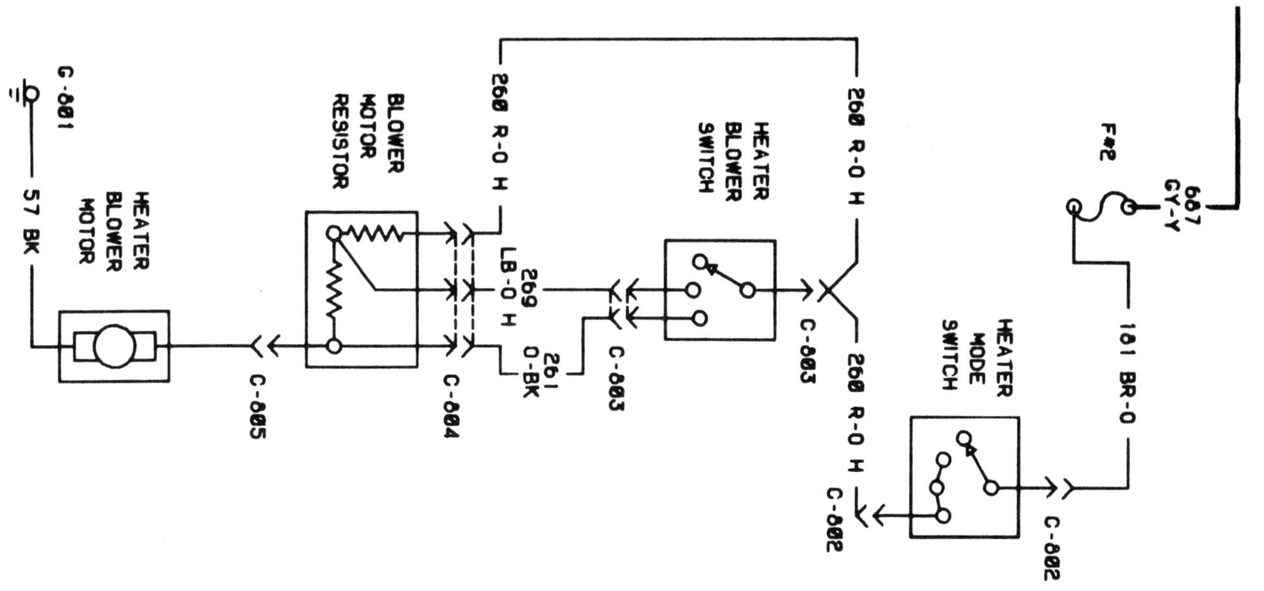

1974-76 Ford Elite & Torino; Mercury Cougar & Montego Heater Wiring Circuit

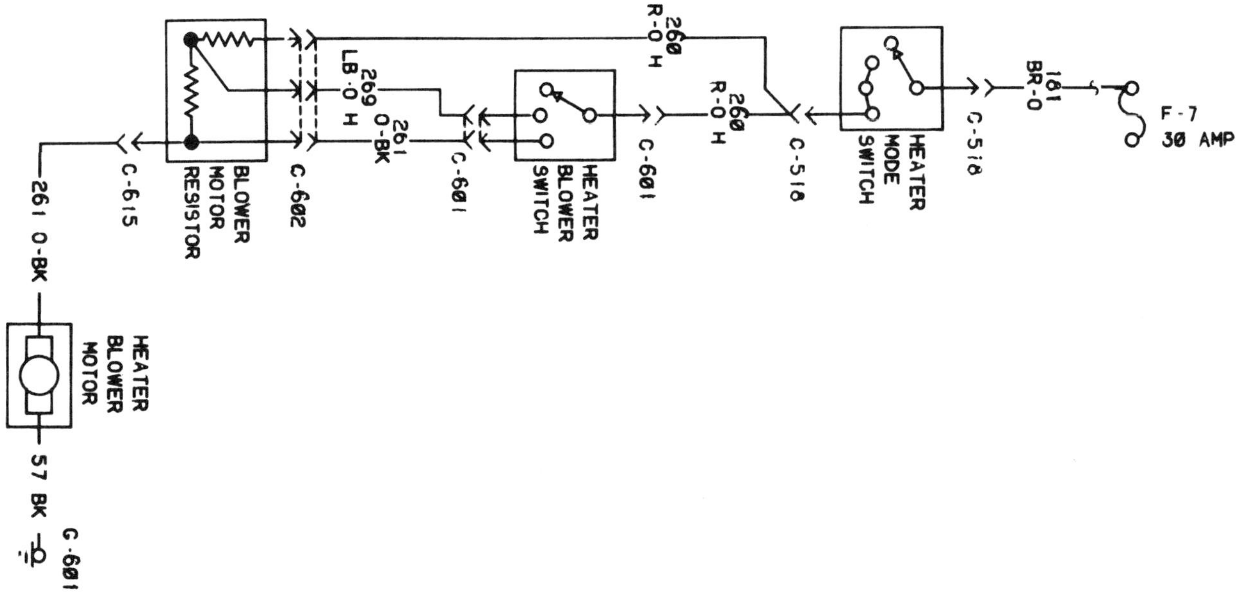

1974-76 Ford Mustang Heater Wiring Circuit

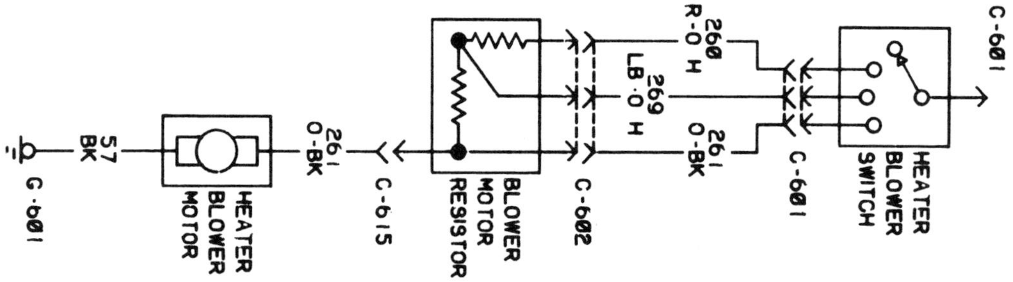

1974-76 Ford Pinto & Mercury Bobcat Heater Wiring Circuit

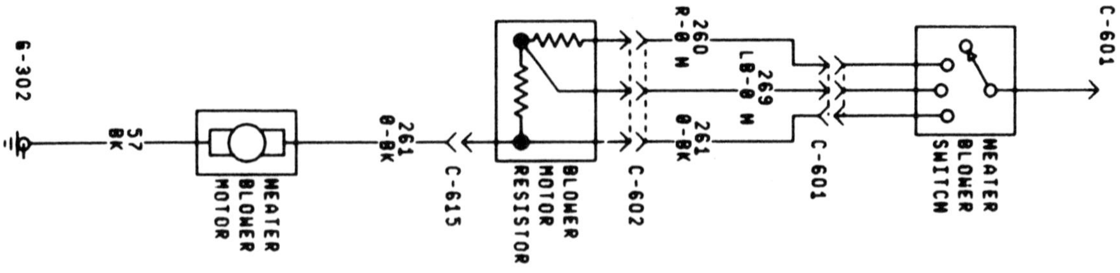

1975-76 Ford Maverick & Mercury Comet Heater Wiring Circuit

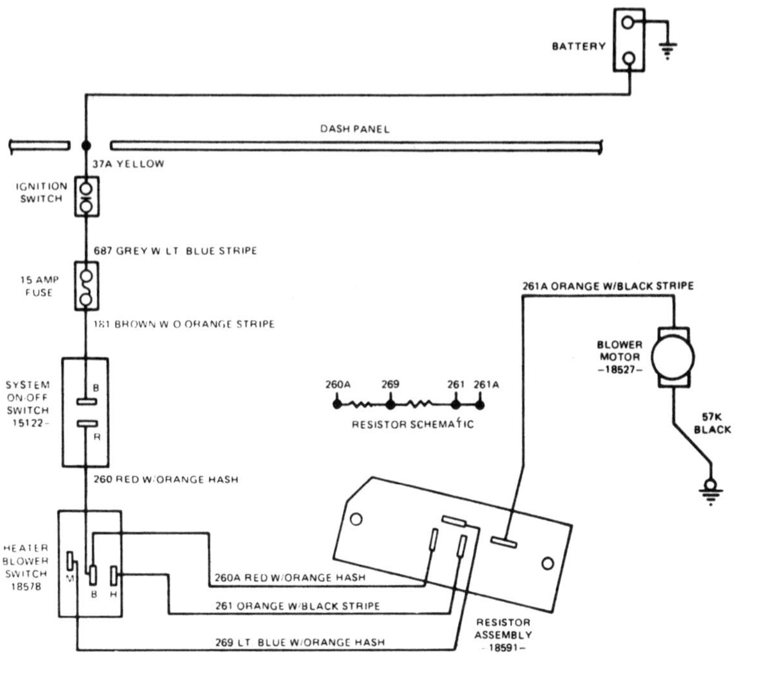

SYSTEM ON-OFF SWITCH 15122-			
SWITCH SCHEMATIC	TERMINAL CONNECTIONS		
	OFF	HEAT	DEFROST
B—R	NONE	B-R	

HEATER BLOWER SWITCH -18578-			
BLOWER SPEED	TERMINAL CONNECTIONS	CURRENT IN CAR (AMPS.)	BLOWER VOLTAGE
LOW	NONE	3.6	4.6
MEDIUM	M-B	5.3	7.4
HIGH	M-B-H	8.5	12.8

1978 Ford Fairmont & Mercury Zephyr Heater Wiring Circuit

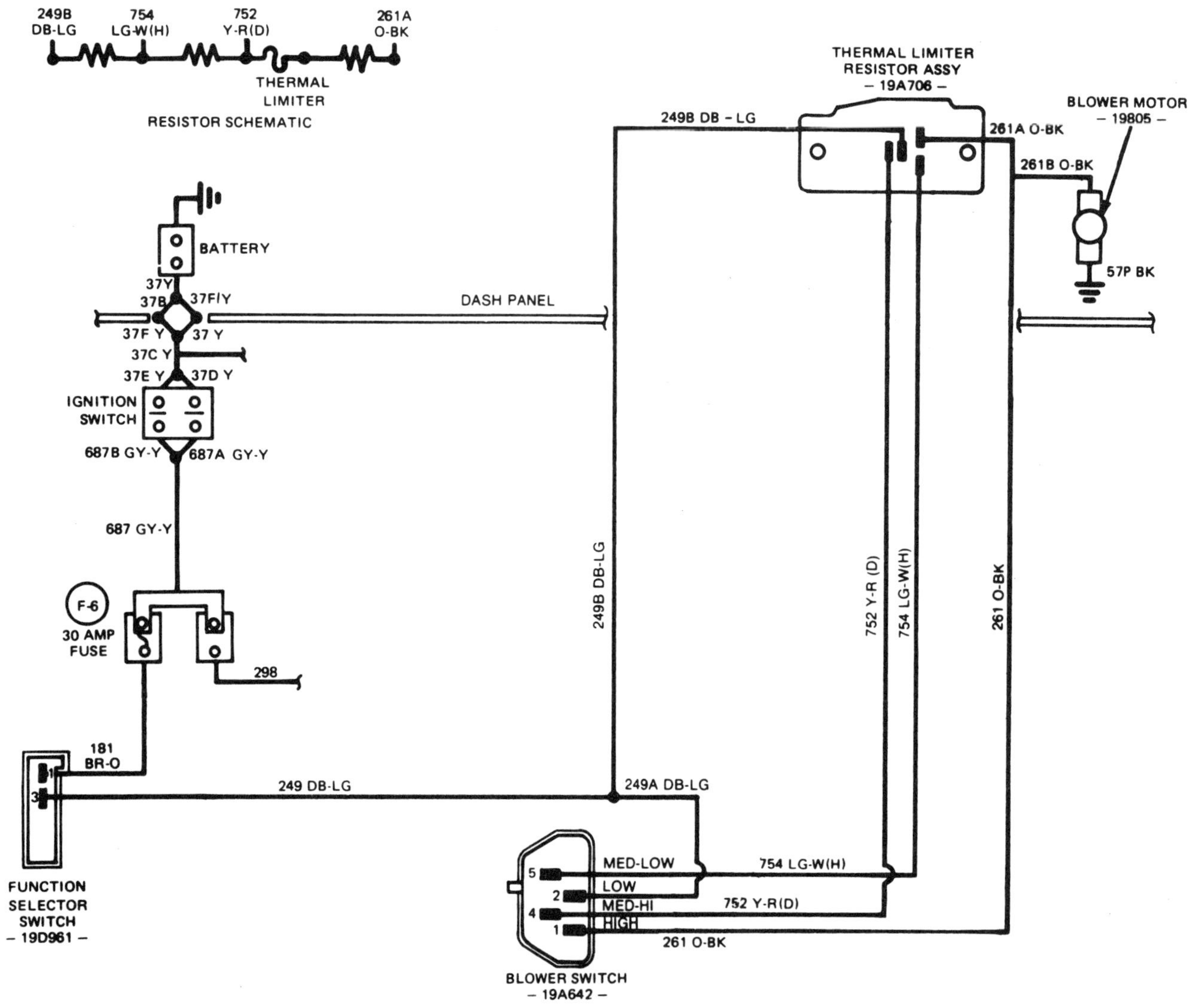

BLOWER SWITCH — 19A642			
BLOWER SPEED	TERMINAL CONNECTIONS	CURRENT IN CAR AMPS	BLOWER VOLTAGE
LOW	—	6	5
MED LOW	2-5	8	7
MED HIGH	2-4-5	15	10
HIGH	2-1-4	25	12.4

FUNCTION SELECTOR SWITCH — 19D961					
SWITCH SCHEMATIC	TERMINAL CONNECTIONS				
	OFF	VENT	HI-LO	FLOOR	DEFROST
1 3	NONE	1-3	1-3	1-3	1-3

1979-80 Ford & Mercury Full Size Heater Wiring Circuit

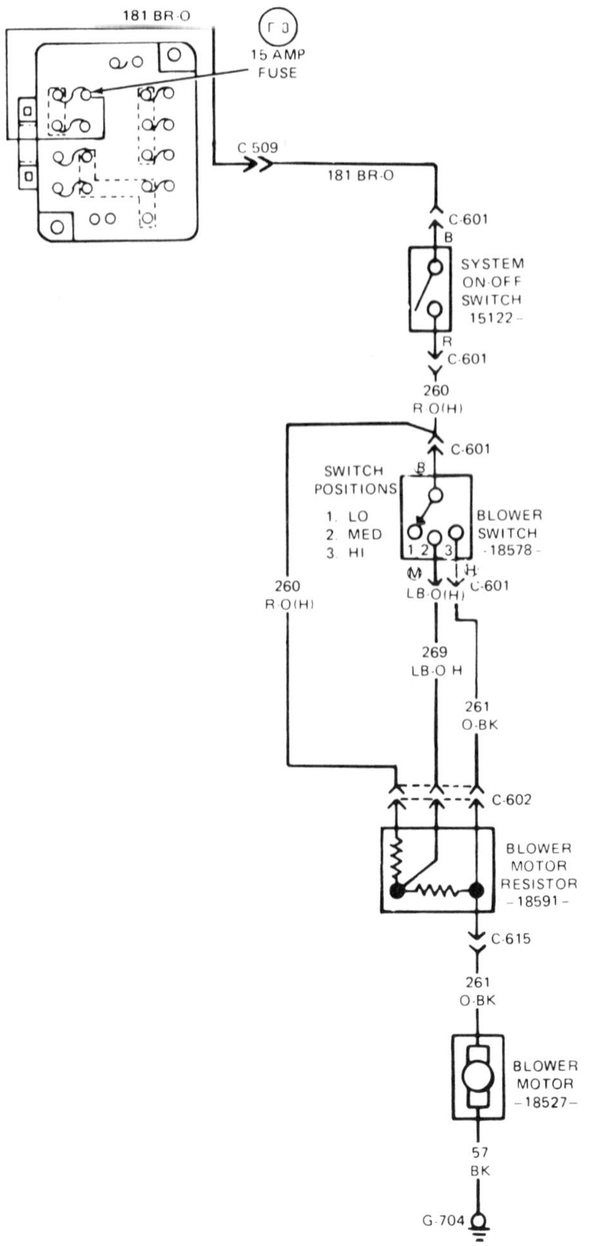

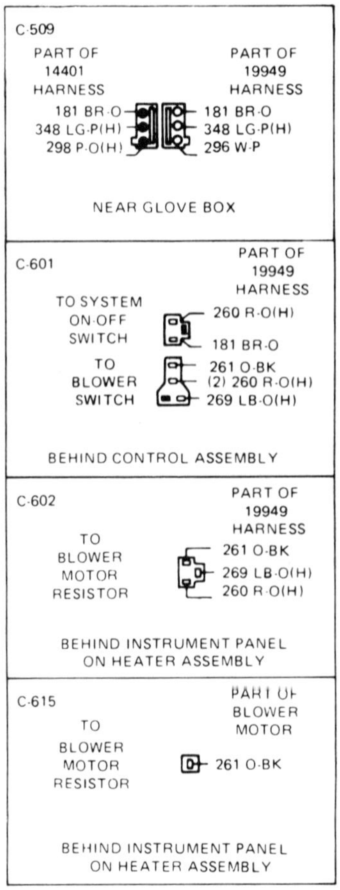

SYSTEM ON-OFF SWITCH -15122-			
SWITCH SCHEMATIC	TERMINAL CONNECTIONS		
	OFF	HEAT	DEFROST
B — R	NONE	B-R	

HEATER BLOWER SWITCH -18578-			
BLOWER SPEED	TERMINAL CONNECTIONS	CURRENT IN CAR (AMPS)	BLOWER VOLTAGE
LOW	NONE	3.6	4.6
MEDIUM	M-B	5.3	7.4
HIGH	M-B-H	8.5	12.8

GROUND CODE
G-704 PART OF BLOWER MOTOR-ON COWL SIDE PANEL

1979 Ford Fairmont & Mercury Zephyr Heater Wiring Circuit

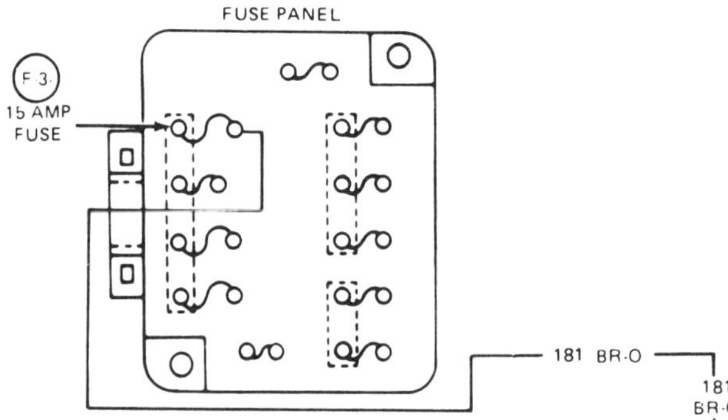

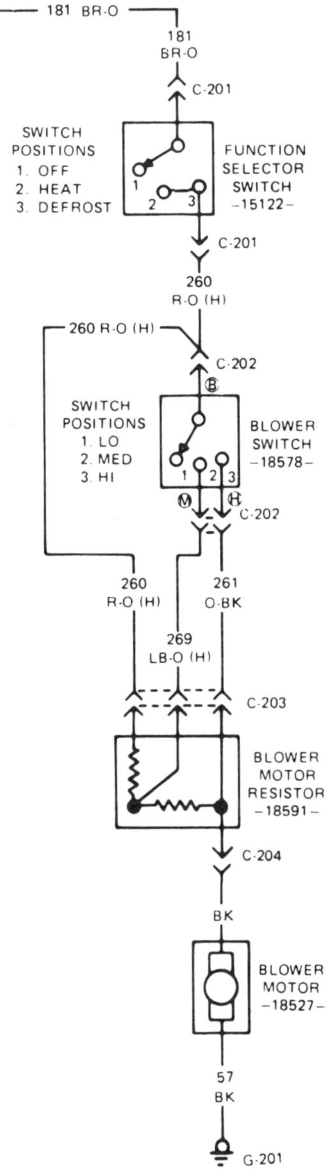

BLOWER SWITCH -18578-			
BLOWER SPEED	TERMINAL CONNECTIONS	CURRENT DRAW IN CAR (AMPS)	BLOWER VOLTAGE
LO	Ⓑ	4.0	6
MED	Ⓑ-Ⓜ	5.5	8.5
HI	Ⓑ-Ⓗ	8.0	12.8

C-201 TO FUNCTION SELECTOR SWITCH	PART OF 14401 HARNESS 260 R-O (H) 181 BR-O	RIGHT SIDE BEHIND INSTRUMENT PANEL
C-202 TO BLOWER SWITCH	PART OF 14401 HARNESS 261 O-BK (2)260 R-O(H) 269 LB-O(H)	RIGHT SIDE BEHIND INSTRUMENT PANEL
C-203 TO BLOWER MOTOR RESISTOR	PART OF 14401 HARNESS 261 O-BK 269 LB-O (H) 260 R-O (H)	ON HEATER HOUSING
C-204 TO BLOWER MOTOR RESISTOR	PART OF BLOWER MOTOR 261 O-BK	ON HEATER HOUSING

GROUND CODE
G-201 NEAR HEATER BLOWER MOTOR

1979 Ford Granada & Mercury Monarch Heater Wiring Circuit

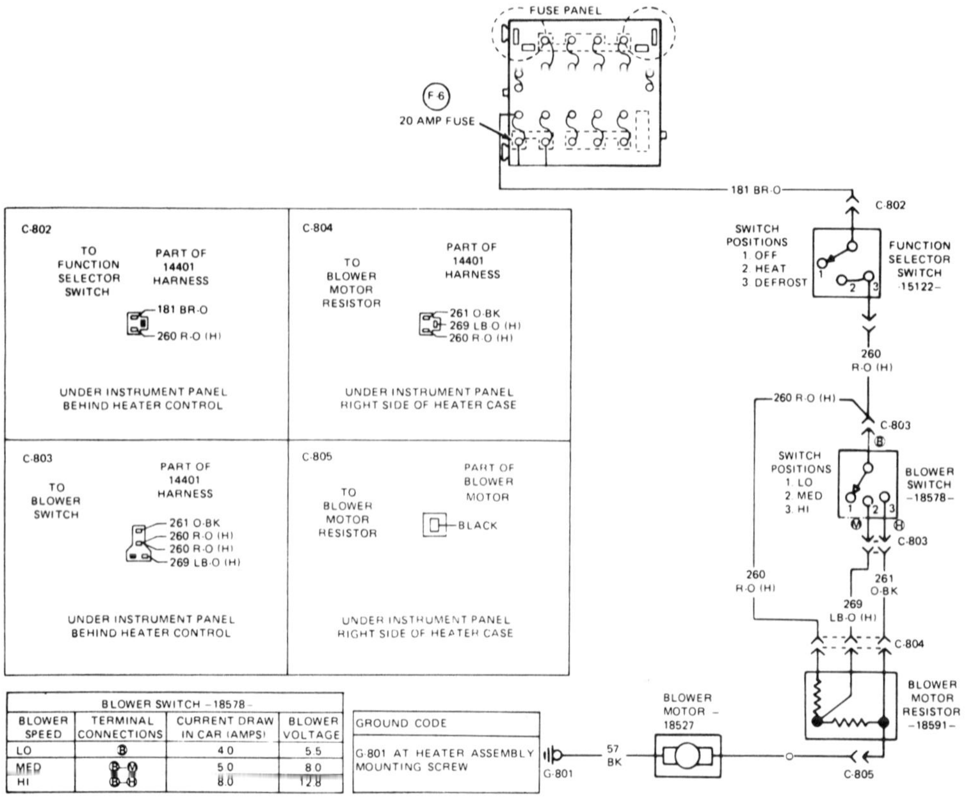

BLOWER SWITCH -18578-			
BLOWER SPEED	TERMINAL CONNECTIONS	CURRENT DRAW IN CAR (AMPS)	BLOWER VOLTAGE
LO	B	4.0	5.5
MED	B-M	5.0	8.0
HI	B-H	8.0	12.8

GROUND CODE
G-801 AT HEATER ASSEMBLY MOUNTING SCREW

1979 Ford LTD II & Thunderbird; Mercury Cougar Heater Wiring Circuit

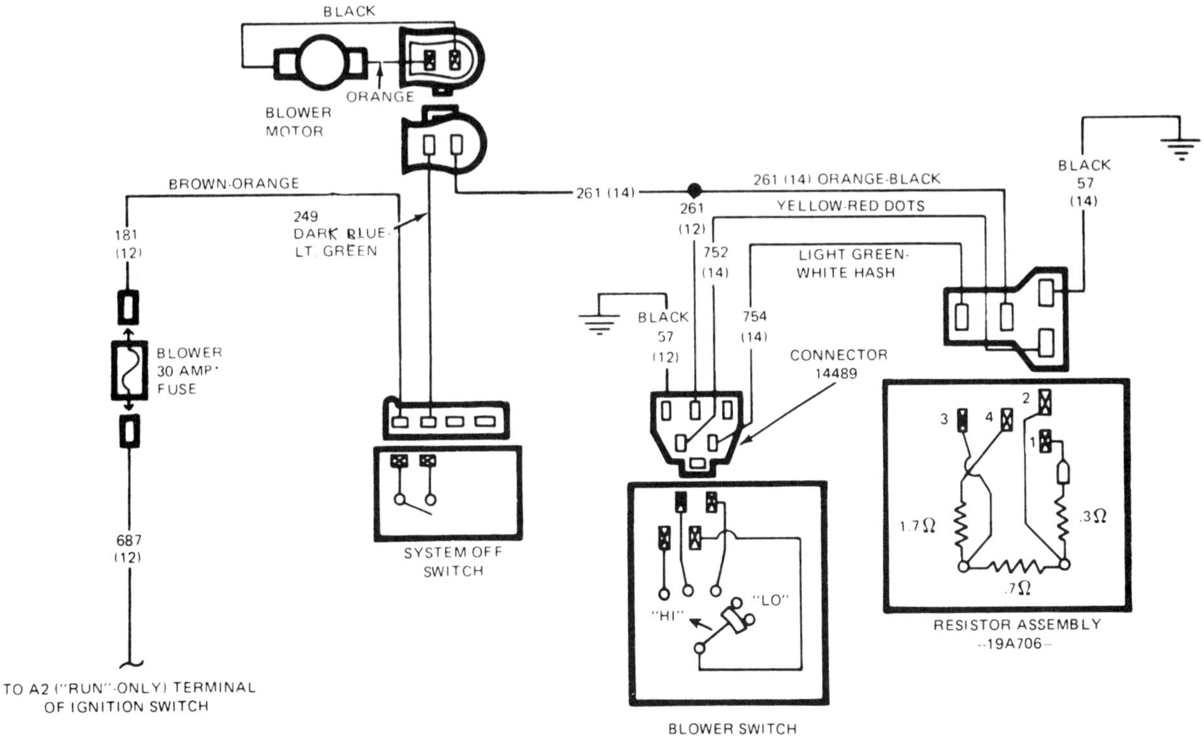

1980 Ford Thunderbird & Mercury Cougar XR-7 Heater Wiring Circuit

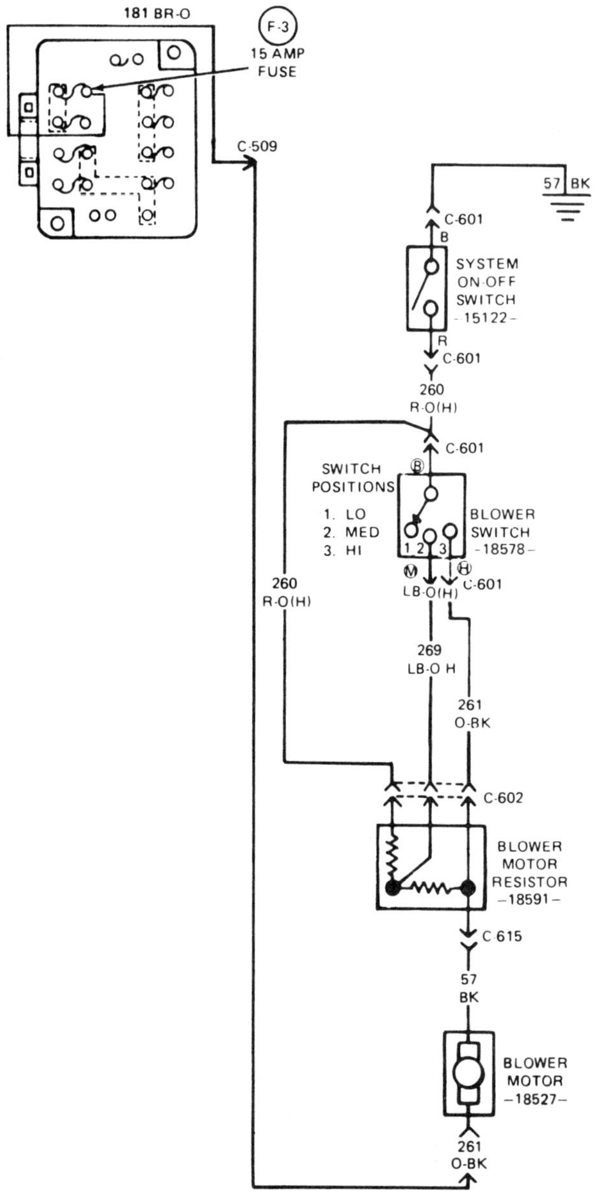

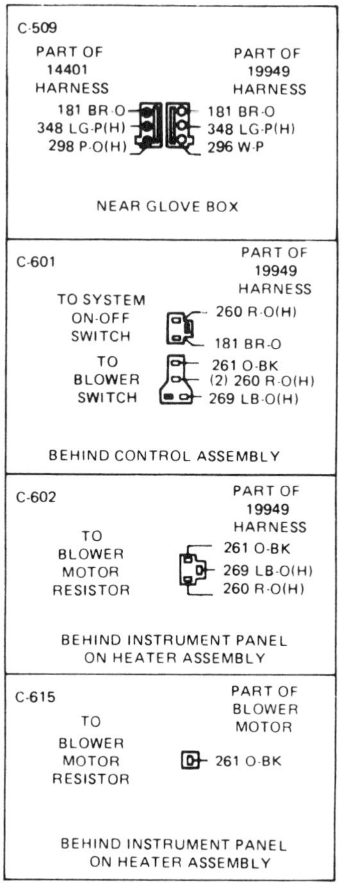

SYSTEM ON-OFF SWITCH -15122-			
SWITCH SCHEMATIC	TERMINAL CONNECTIONS		
	OFF	HEAT	DEFROST
B R	NONE	B-R	

HEATER BLOWER SWITCH -18578-			
BLOWER SPEED	TERMINAL CONNECTIONS	CURRENT IN CAR (AMPS.)	BLOWER VOLTAGE
LOW	NONE	3.6	4.6
MEDIUM	M-B	5.3	7.4
HIGH	M-B-H	8.5	12.8

GROUND CODE
G-704 PART OF BLOWER MOTOR-ON COWL SIDE PANEL

1980 Ford Fairmont & Mercury Zephyr Heater Wiring Circuit

Oldsmobile

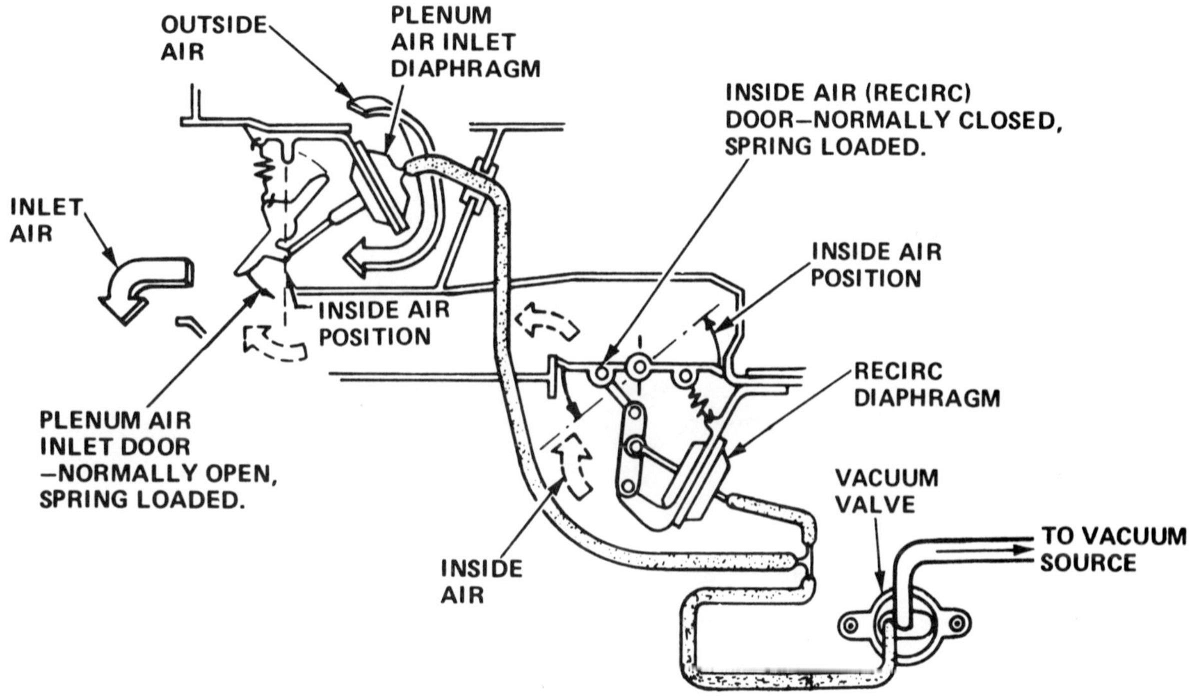

1974 Omega A/C & Heater Vacuum Circuit

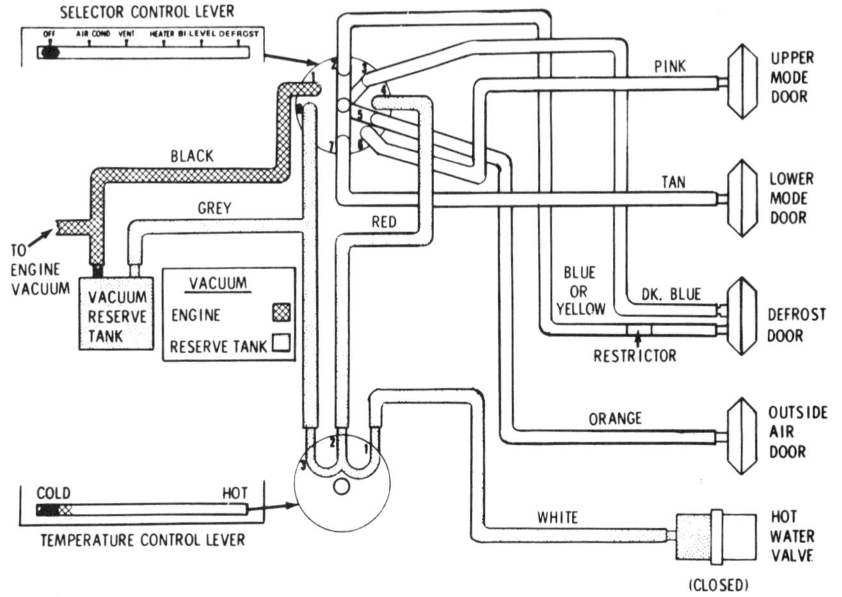

1974 Cutlass, 88, 98 & Toronado A/C & Heater Vacuum Circuit (Less Tempmatic)

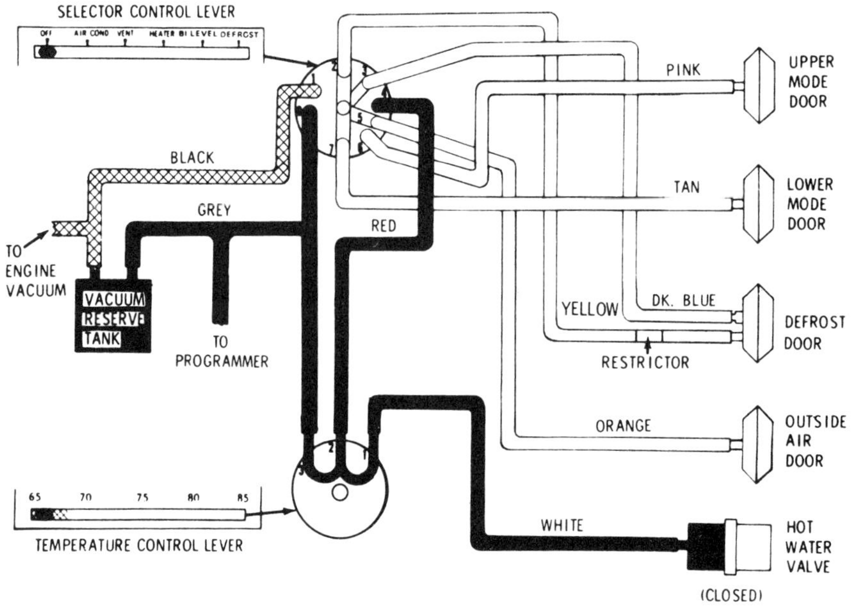

1974 88, 98 & Toronado A/C & Heater Vacuum Circuit (With Tempmatic)

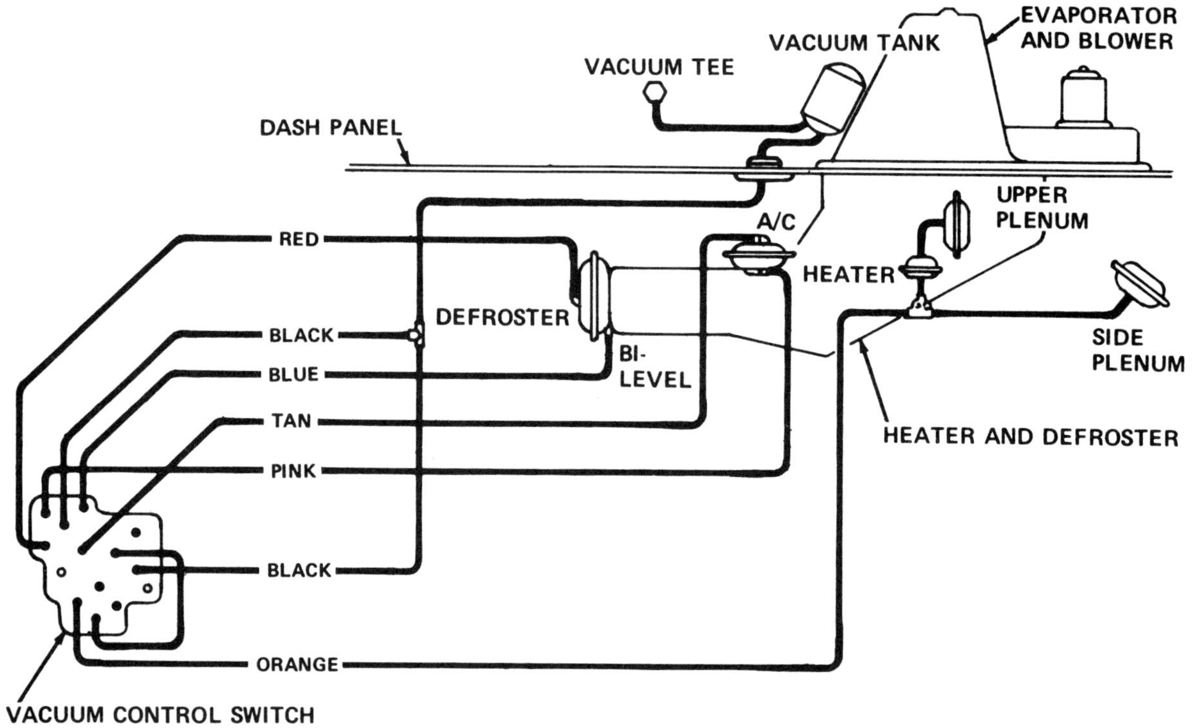

1975 Omega A/C & Heater Vacuum Circuit

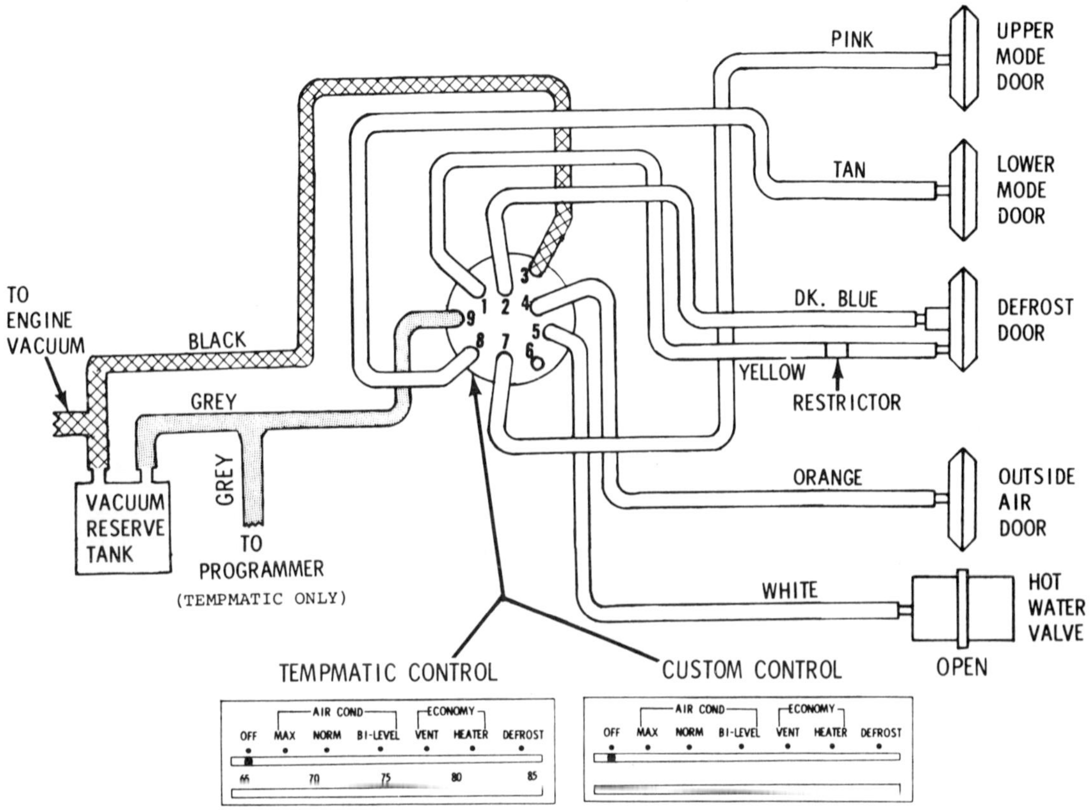

1975-76 88 & 98; 1975-78 Toronado A/C & Heater Vacuum Circuit (With Custom or Tempmatic)

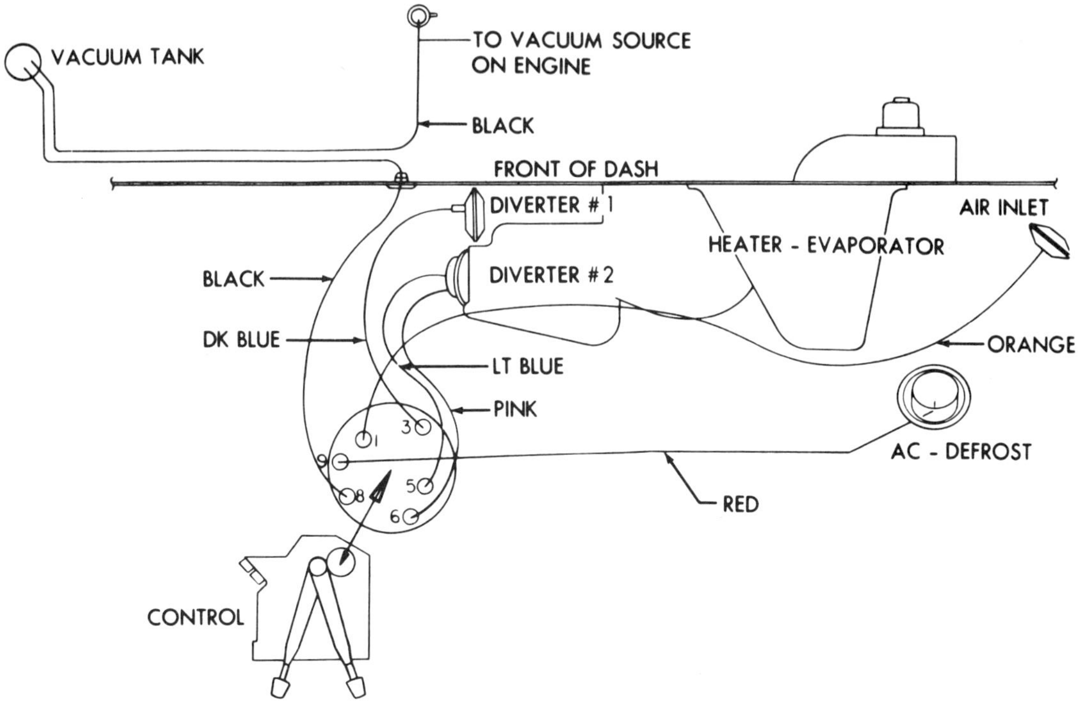

1975-77 Starfire A/C & Heater Vacuum Circuit

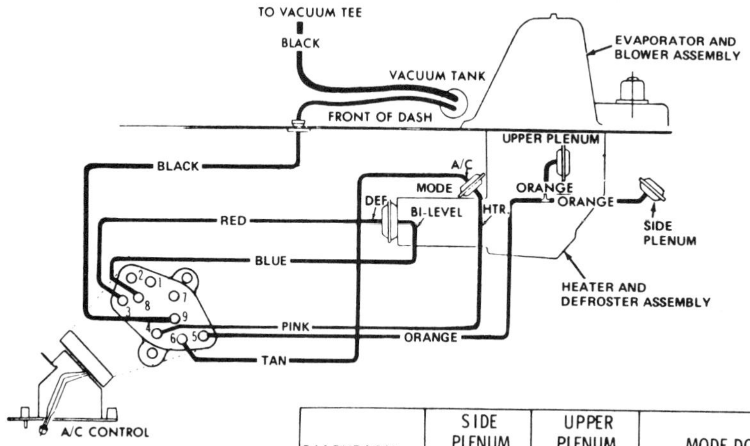

DIAPHRAGM	SIDE PLENUM ORANGE	UPPER PLENUM ORANGE	MODE DOOR TAN	MODE DOOR PINK	DEFROSTER DOOR RED	DEFROSTER DOOR BLUE
OFF	VENT	VENT	VENT	VENT	VENT	VENT
A/C MAX	VAC.	VAC.	VAC.	VENT	VENT	VENT
A/C NORM	VENT	VENT	VAC.	VENT	VENT	VENT
A/C BI-LEVEL	VENT	VENT	VENT	VENT	VENT	VENT
VENT	VENT	VENT	VENT	VENT	VAC.	VENT
HEAT	VENT	VENT	VENT	VAC.	VAC.	VENT
DEFROSTER	VENT	VENT	VENT	VAC.	VENT	VAC.

1976 Omega A/C & Heater Vacuum Circuit

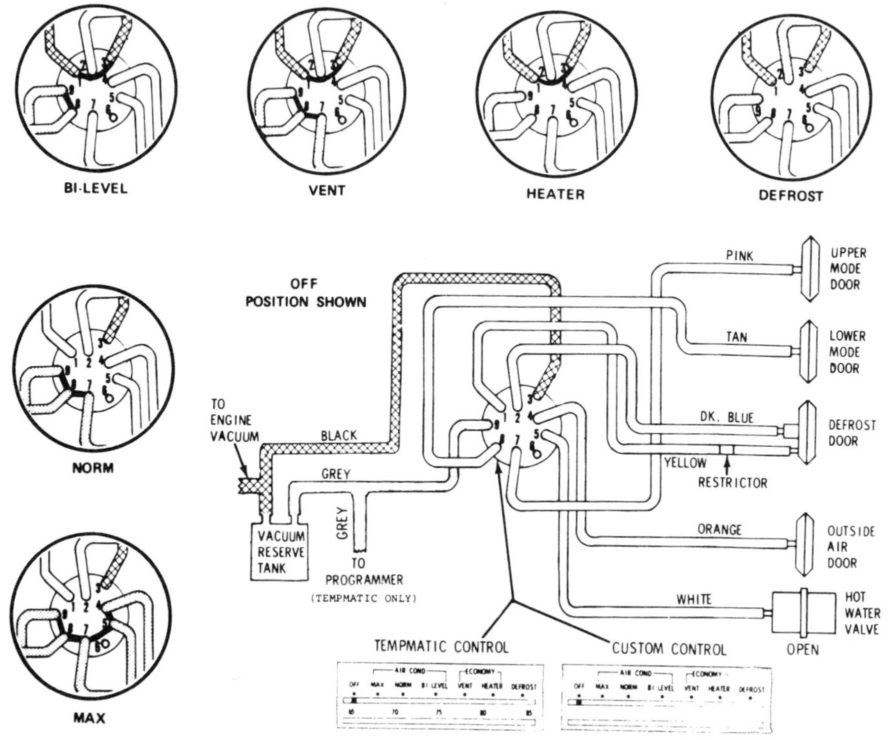

1975-77 Cutlass A/C & Heater Vacuum Circuit (With Custom or Tempmatic)

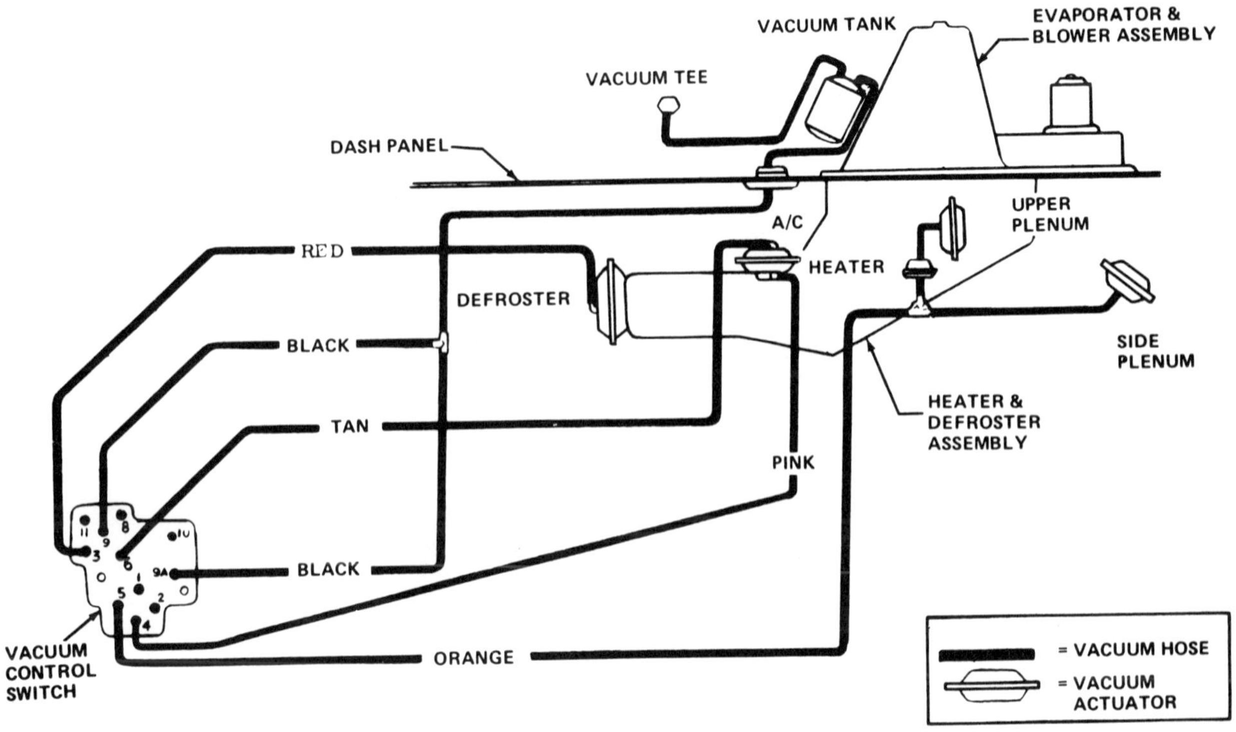

1977-79 Omega A/C & Heater Vacuum Circuit

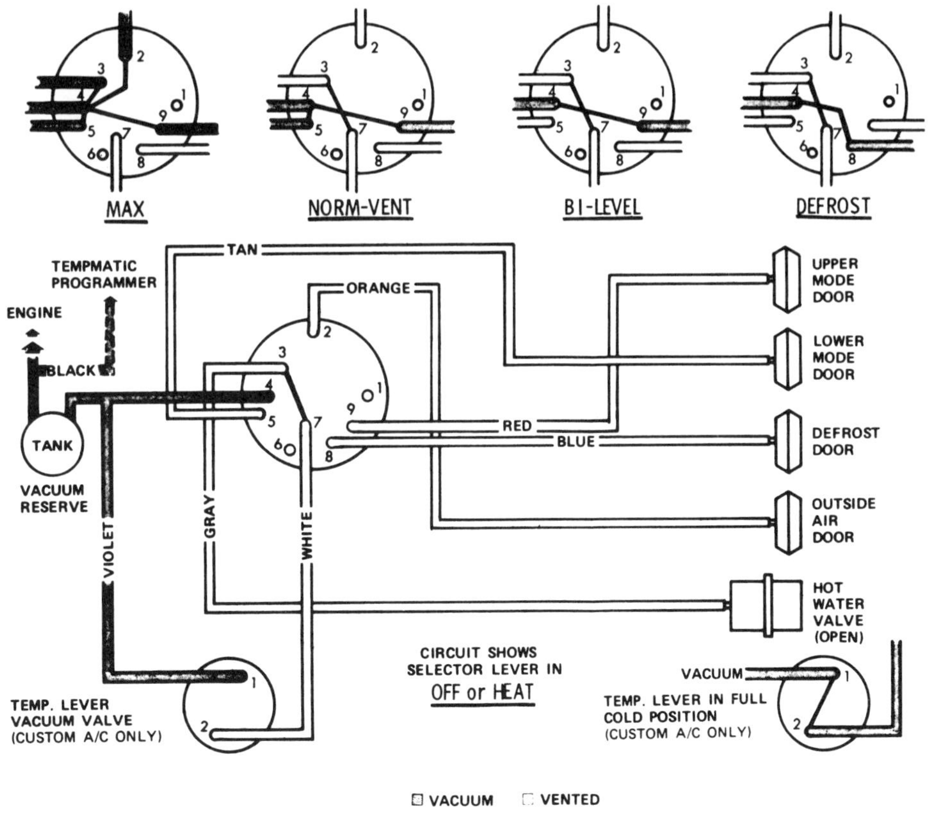

1977-79 88 & 98 A/C & Heater Vacuum Circuit (With Custom or Tempmatic)

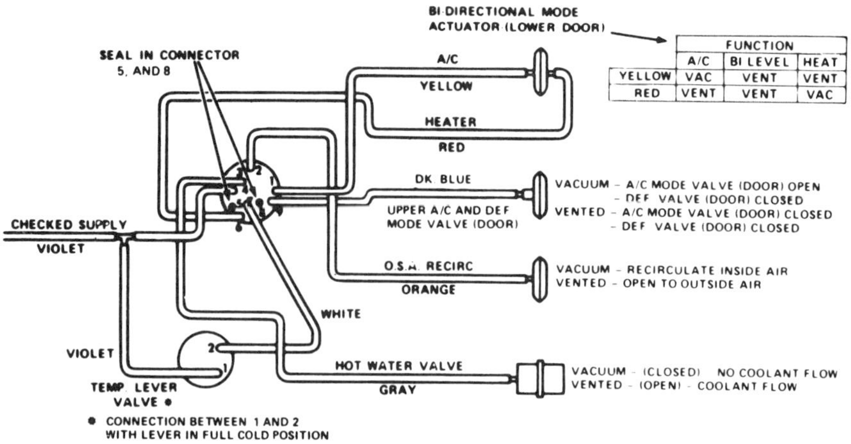

	FUNCTION		
	A/C	BI LEVEL	HEAT
YELLOW	VAC	VENT	VENT
RED	VENT	VENT	VAC

SELECT LEVER VALVE

PORT NO.	OFF	LEVER POSITION					
		MAX.	NORMAL	BI-LEVEL	VENT	HEATER	DEFROST
1	VENT	CONN. 4	CONN. 4	VENT	CONN. 4	VENT	CONN. 4
2	VENT	CONN. 4	VENT	VENT	VENT	VENT	VENT
3	CONN. 7	CONN. 4	CONN. 7	CONN. 7	CONN. 7	CONN. 7	CONN. 7
4	CONN. 6	CONN. 1,2,3,9	CONN. 1,9	CONN. 9	CONN. 1,9	CONN. 6	CONN. 1
5							
6	CONN. 4	VENT	VENT	VENT	VENT	CONN. 4	VENT
7	CONN. 3	SEAL	CONN. 3	CONN. 3	CONN. 3	CONN. 3	CONN. 3
8							
9	VENT	CONN. 4	CONN. 4	CONN. 4	CONN. 4	VENT	VENT

1978-79 Cutlass (With Custom or Tempmatic); 1980 Cutlass (Less Tempmatic) A/C & Heater Vacuum Circuit

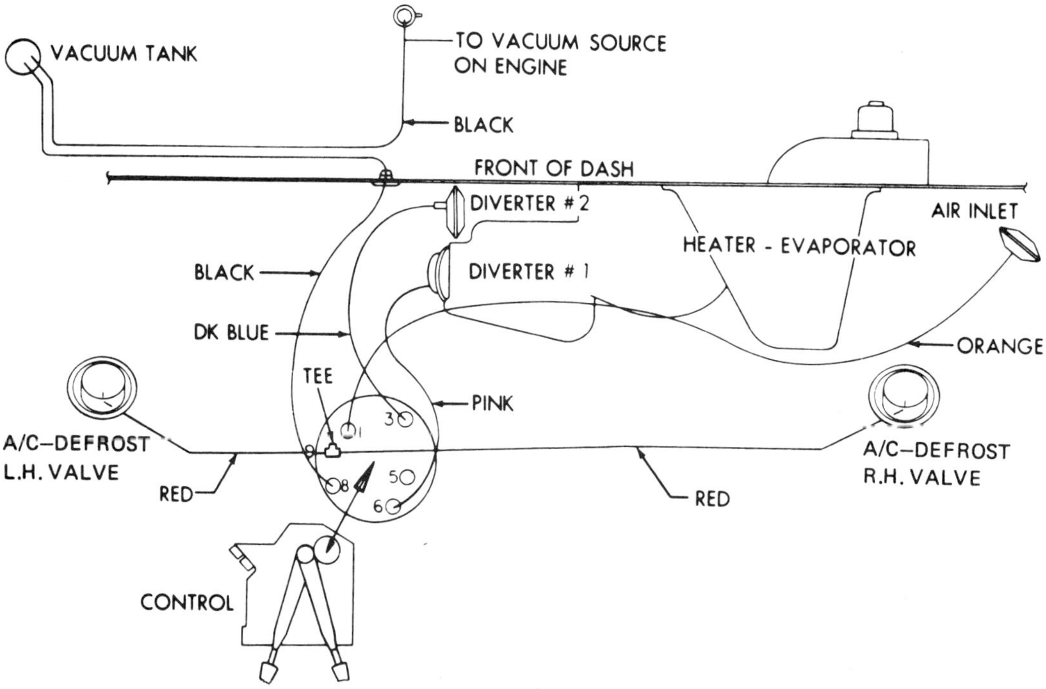

1978-80 Starfire A/C & Heater Vacuum Circuit

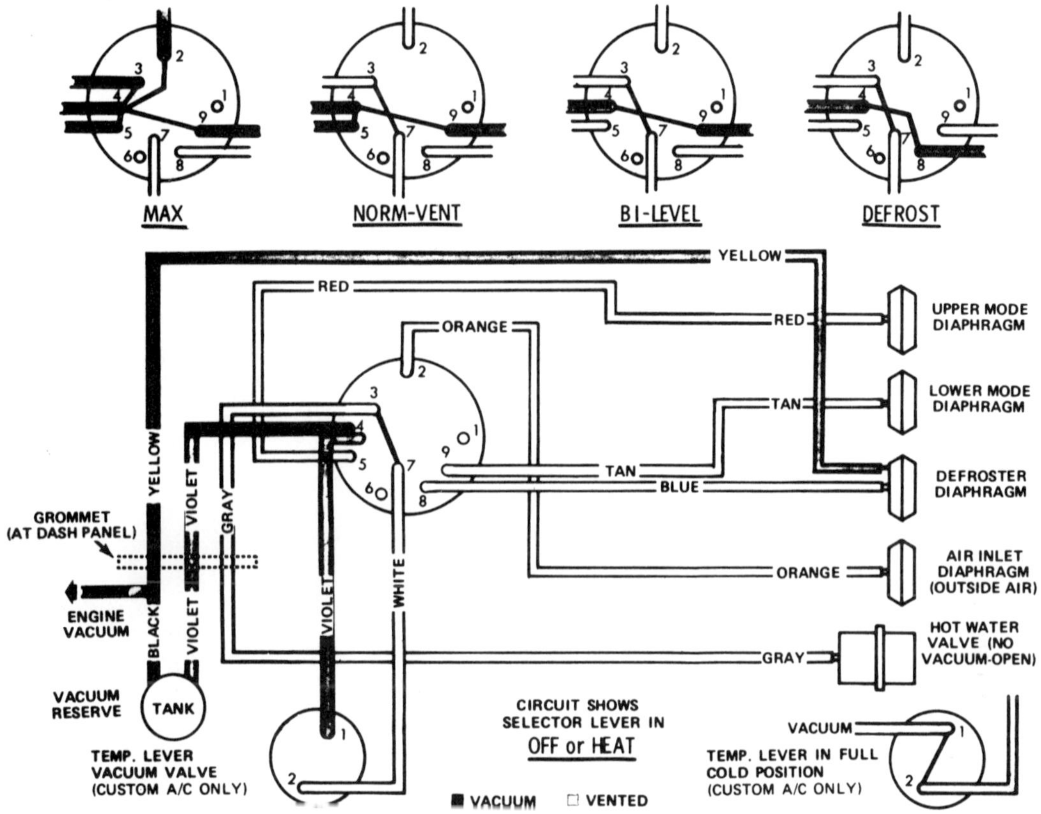

1979-80 Toronado A/C & Heater Vacuum Circuit (Less Tempmatic)

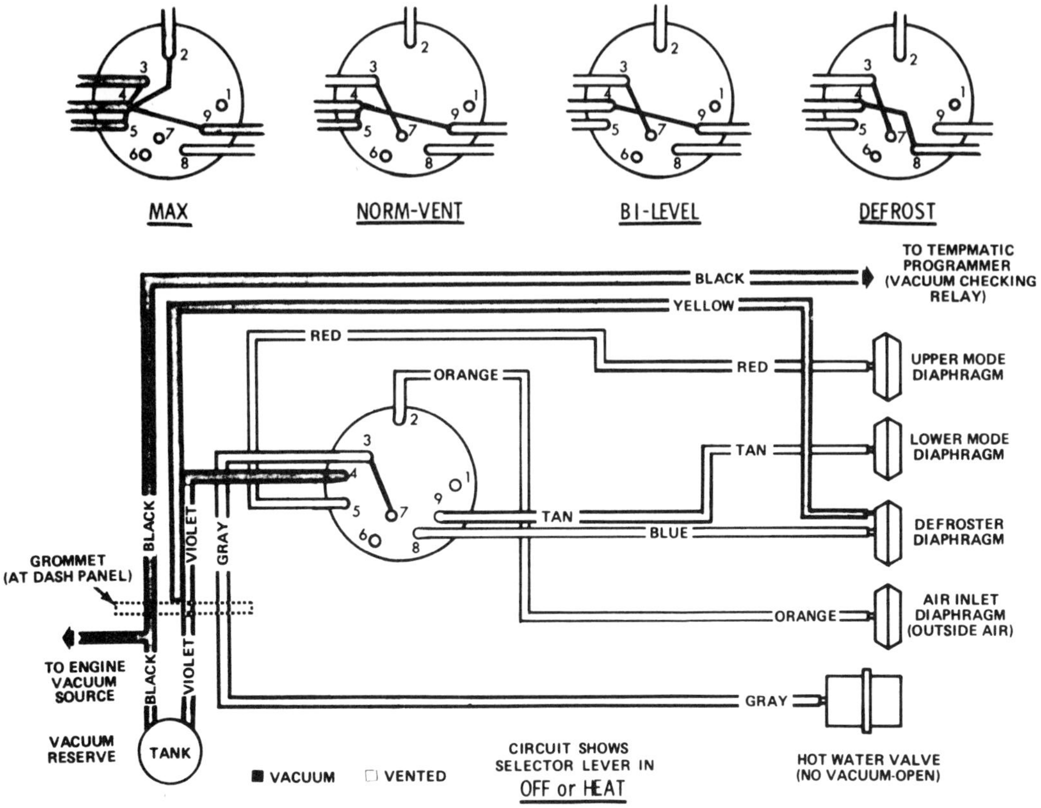

1979-80 Toronado A/C & Heater Vacuum Circuit (With Tempmatic)

	FUNCTION		
	A/C	BI-LEVEL	HEAT
YELLOW	VAC	VENT	VENT
RED	VENT	VENT	VAC

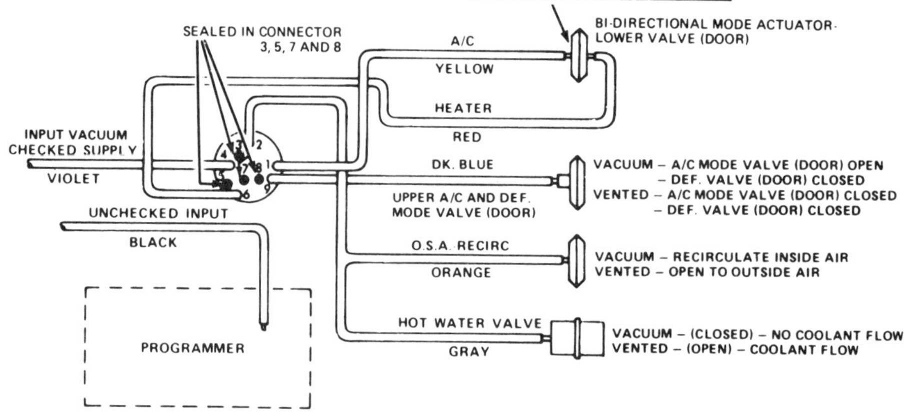

SELECT LEVER VALVE							
PORT NO.	LEVER POSITION						
	OFF	MAX.	NORMAL	BI-LEVEL	VENT	HEATER	DEFROST
1	VENT	CONN. 4	CONN. 4	VENT	CONN. 4	VENT	CONN. 4
2	VENT	CONN. 4	VENT	VENT	VENT	VENT	VENT
3							
4	CONN. 6	CONN. 1,2,3,9	CONN. 1,9	CONN. 9	CONN. 1,9	CONN. 6	CONN. 1
5							
6	CONN. 4	VENT	VENT	VENT	VENT	CONN. 4	VENT
7							
8							
9	VENT	CONN. 4	CONN. 4	CONN. 4	CONN. 4	VENT	VENT

1980 Cutlass A/C & Heater Vacuum Circuit (With Tempmatic)

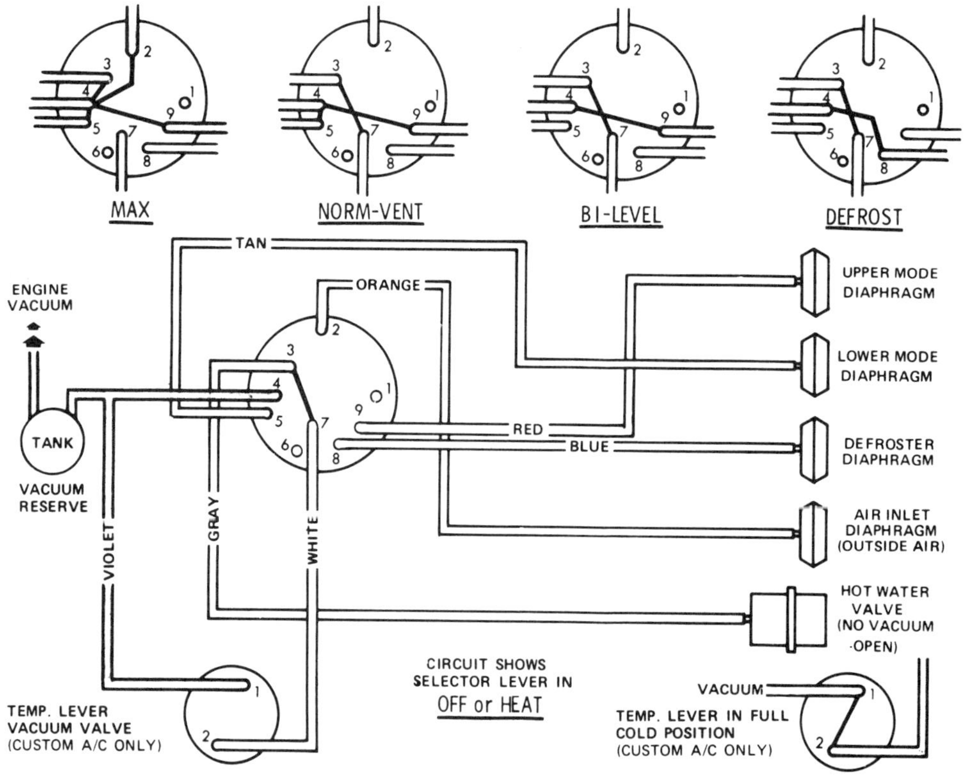

1980 88 & 98 A/C & Heater Vacuum Circuit (Less Tempmatic)

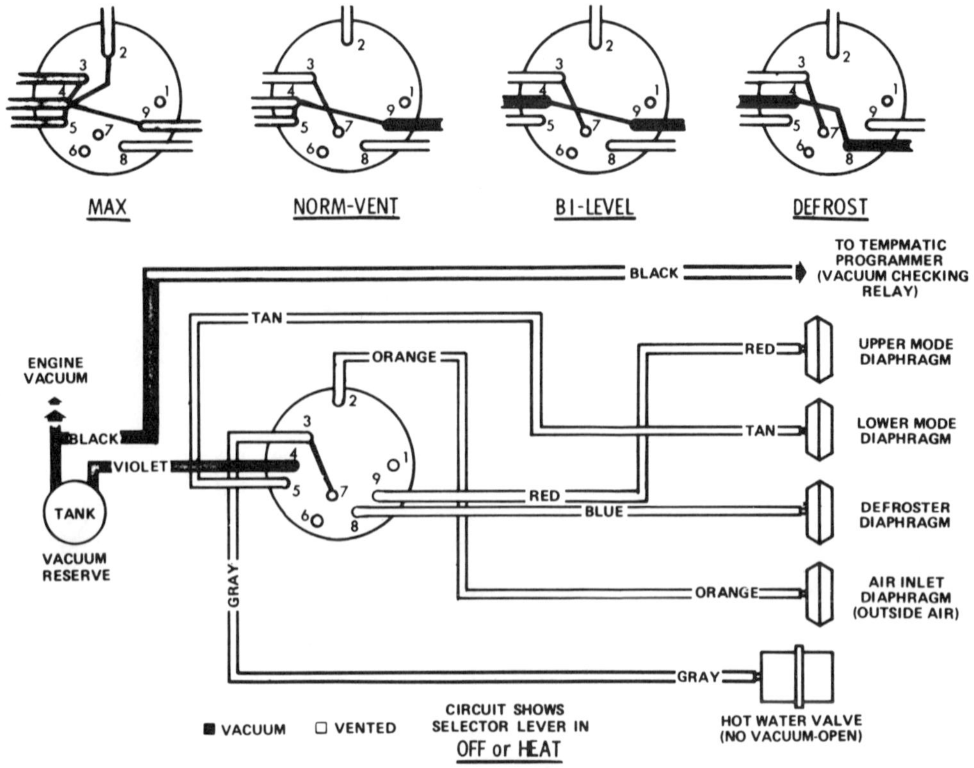

1980 88 & 98 A/C & Heater Vacuum Circuit (With Tempmatic)

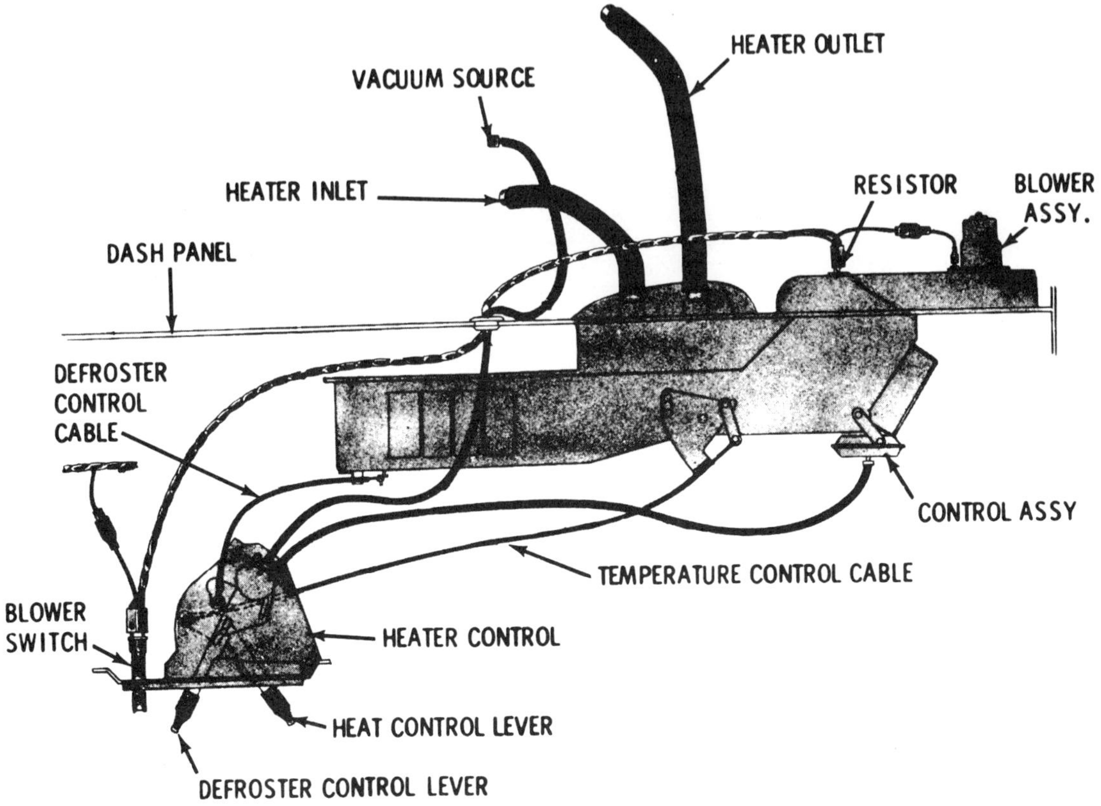

1974-76 88, 98 & Toronado Heater Vacuum Circuit

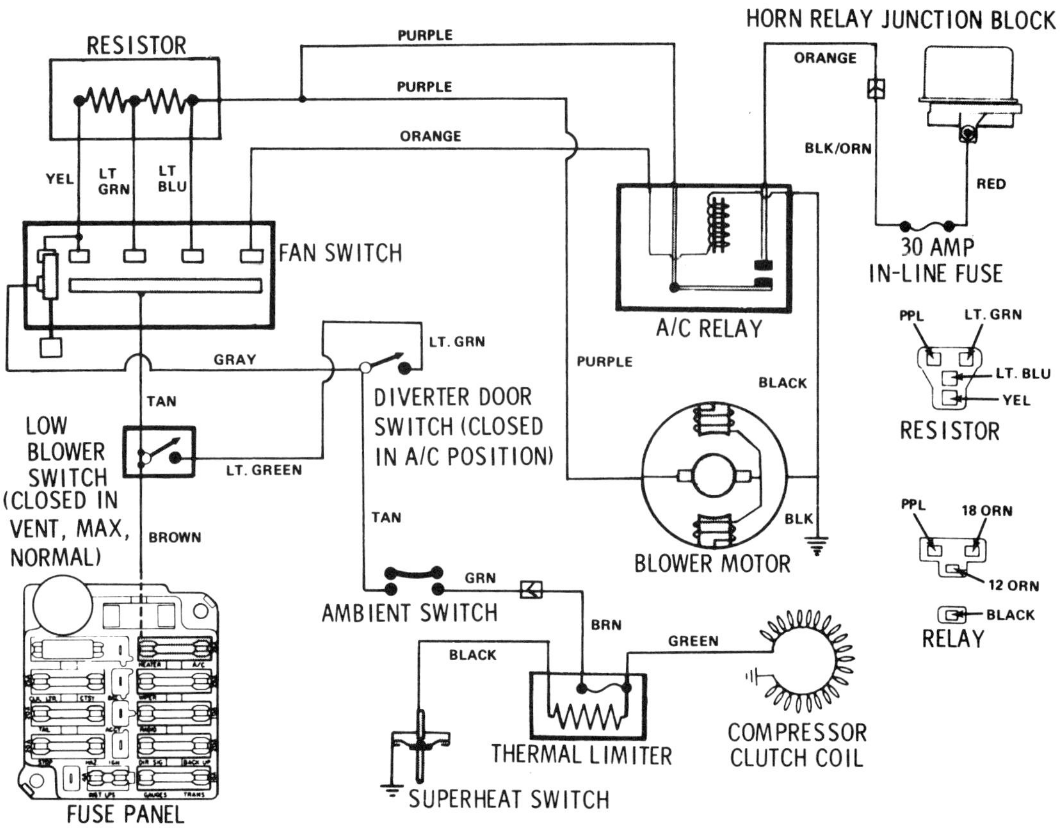

1974 Omega A/C & Heater Wiring Circuit

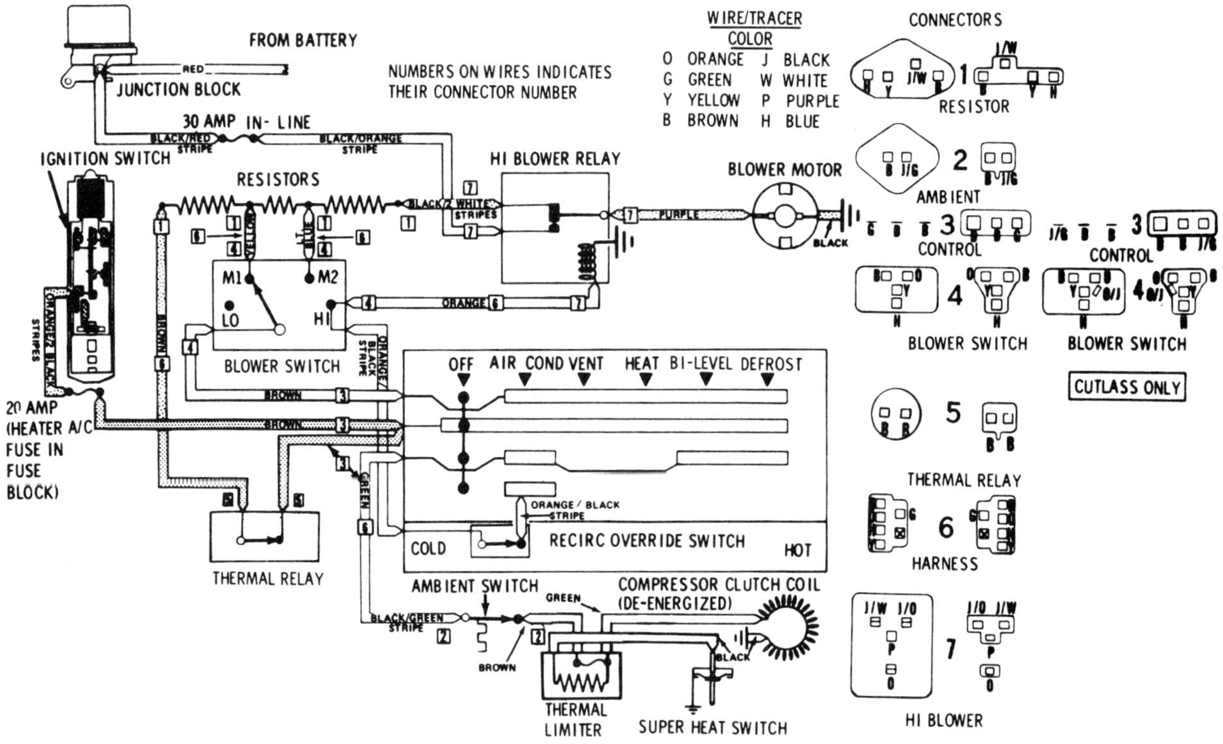

1974 Cutlass, 88, 98 & Toronado A/C & Heater Wiring Circuit (Less Tempmatic)

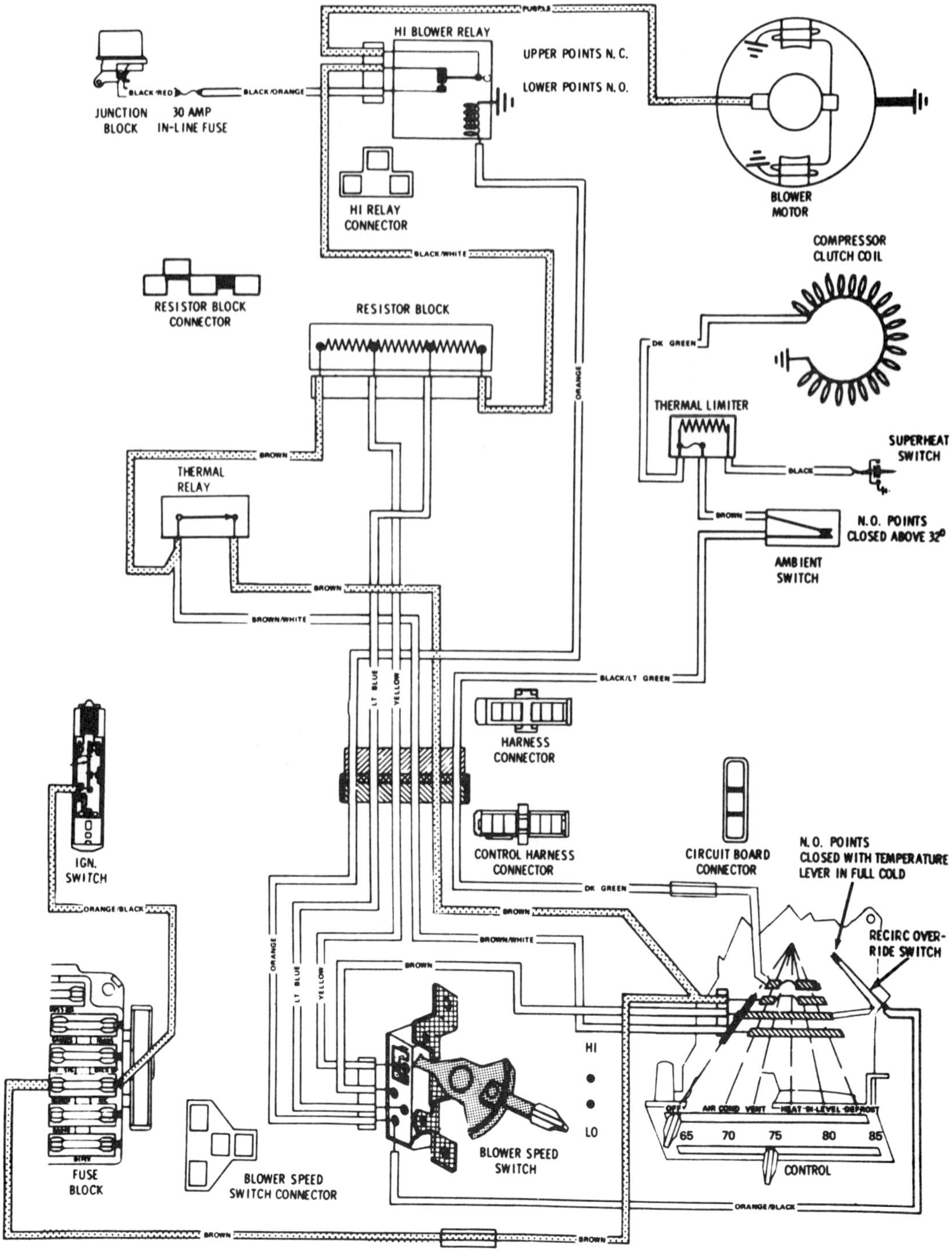

1974 88, 98 & Toronado A/C & Heater Wiring Circuit (With Tempmatic)

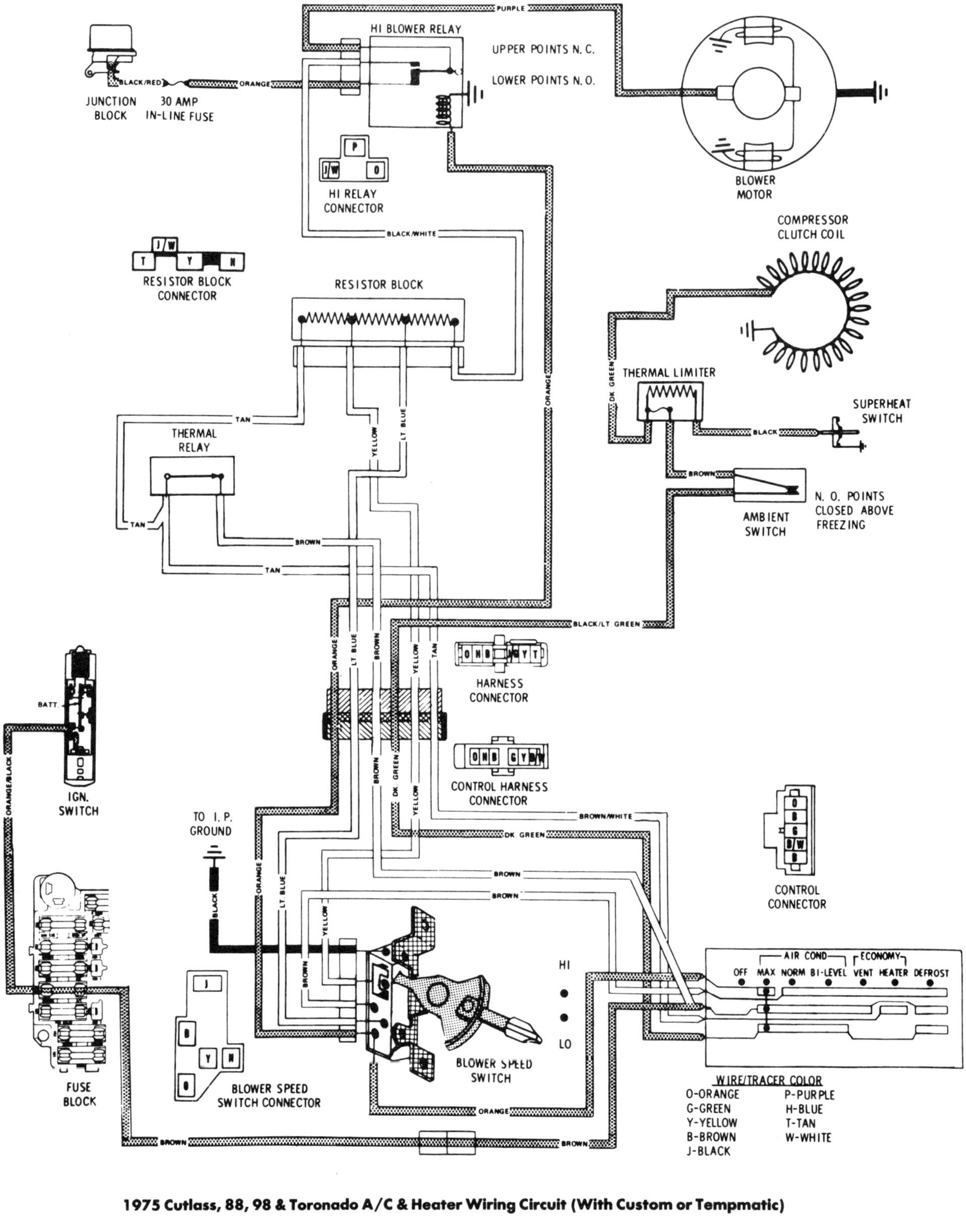

1975 Cutlass, 88, 98 & Toronado A/C & Heater Wiring Circuit (With Custom or Tempmatic)

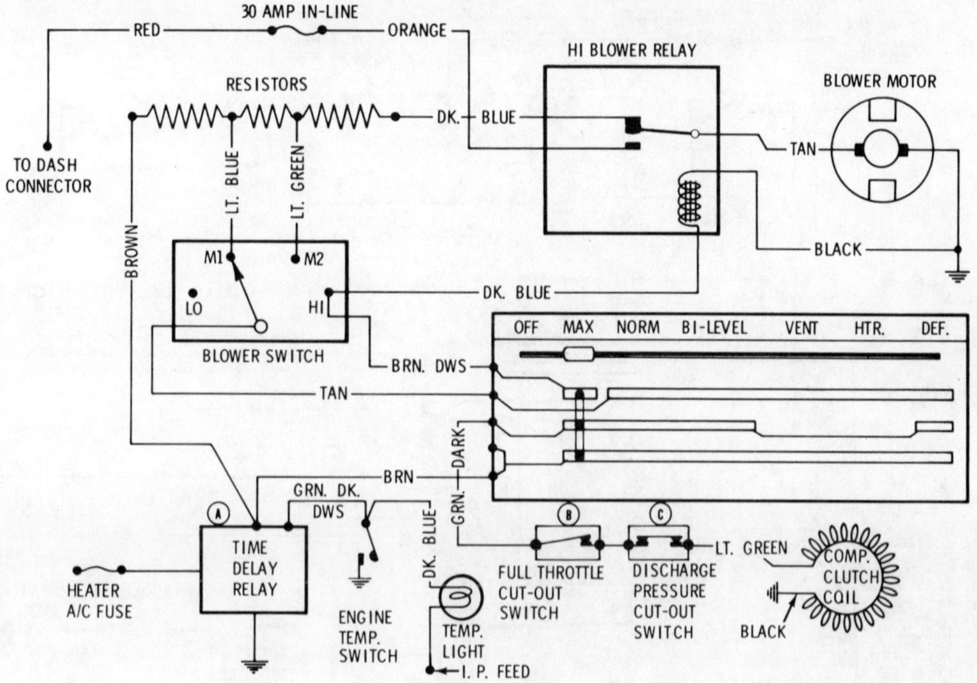

1975-76 Starfire A/C & Heater Wiring Circuit (With 4-140)

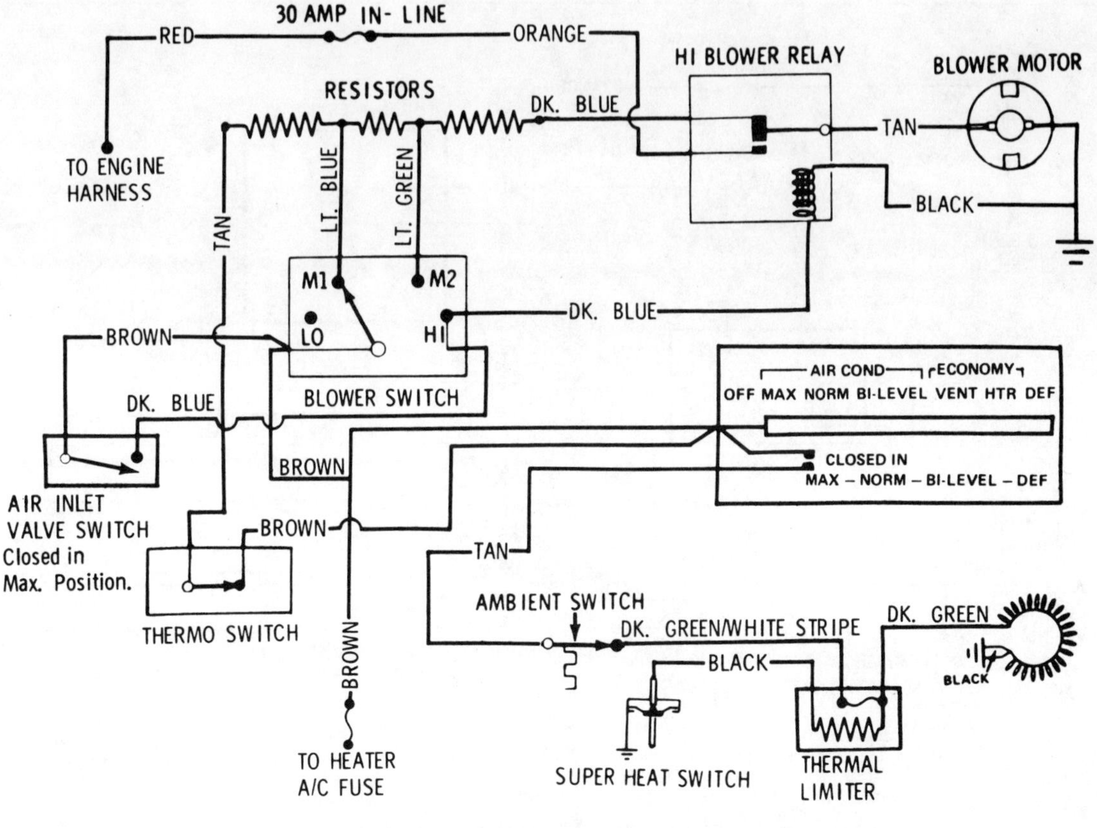

1975-76 Starfire A/C & Heater Wiring Circuit (With V6-231)

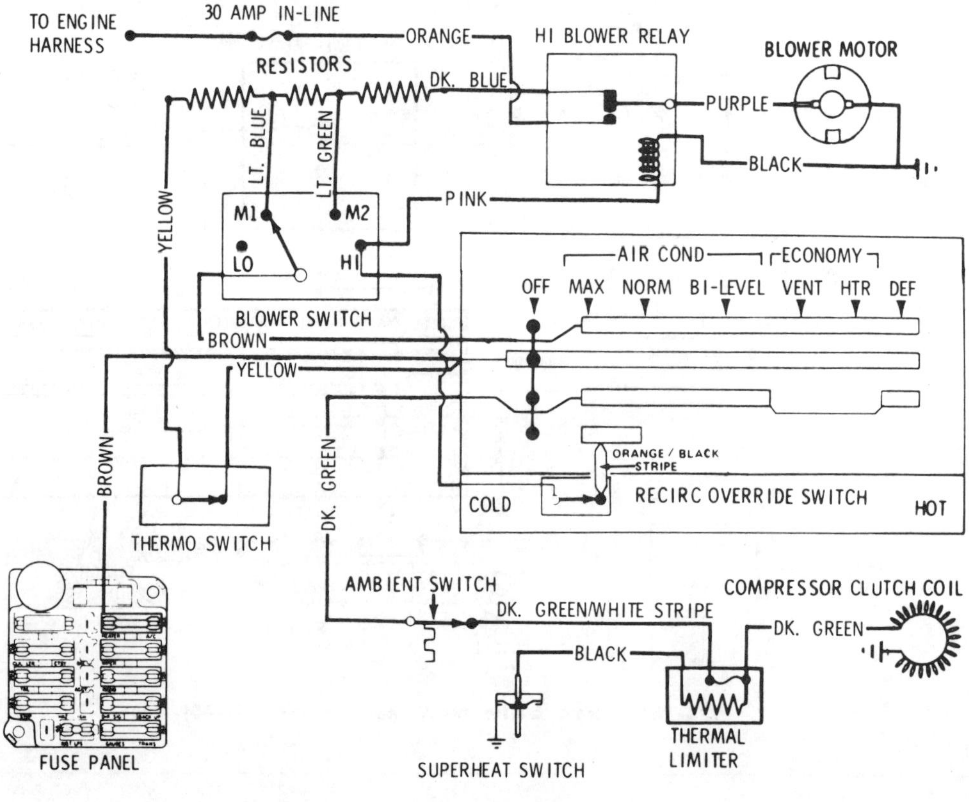

1975-76 Omega A/C & Heater Wiring Circuit

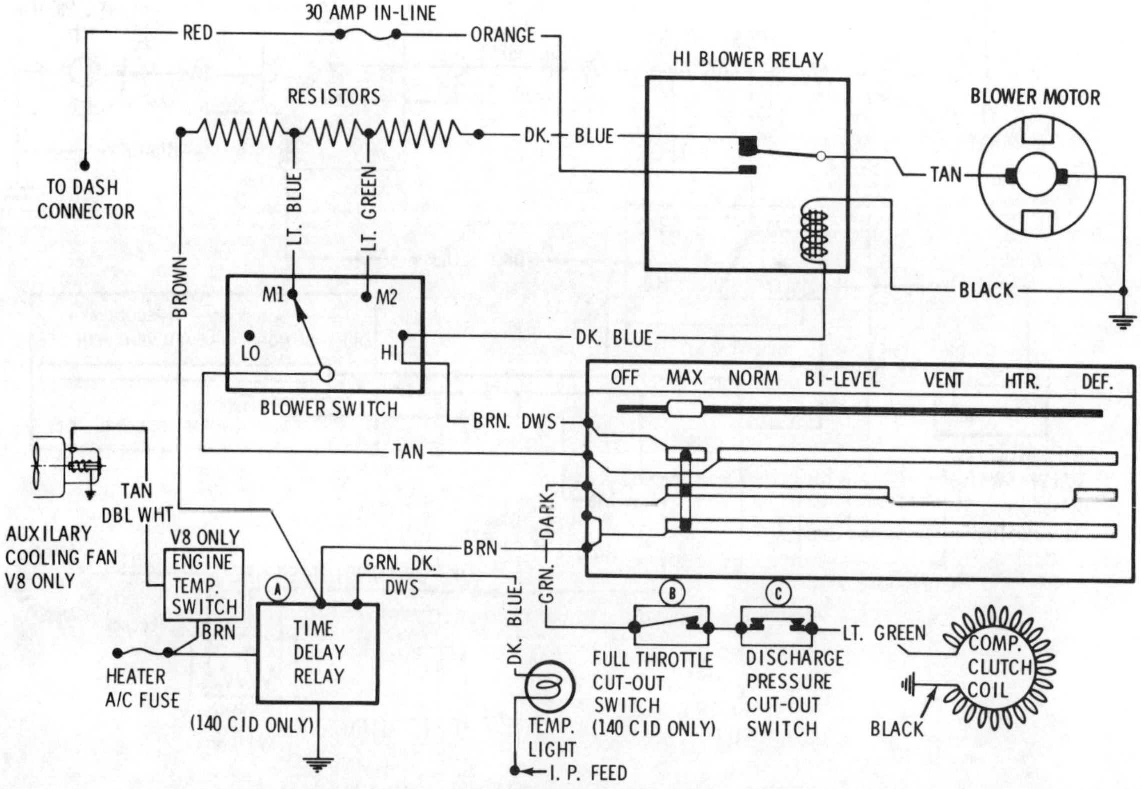

1977 Starfire A/C & Heater Wiring Circuit

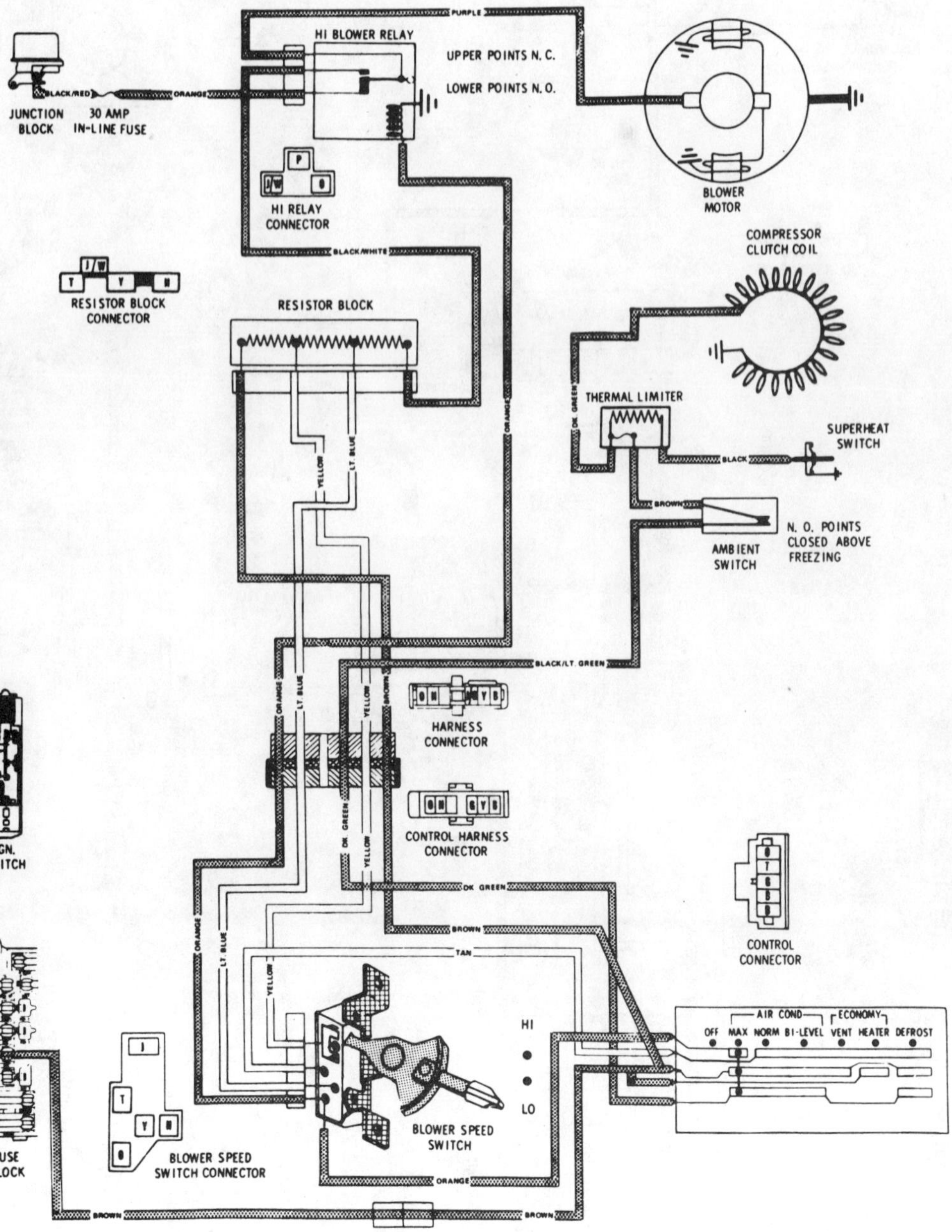

1976 Cutlass, 88, 98 & Toronado A/C & Heater Wiring Circuit (With Custom or Tempmatic)

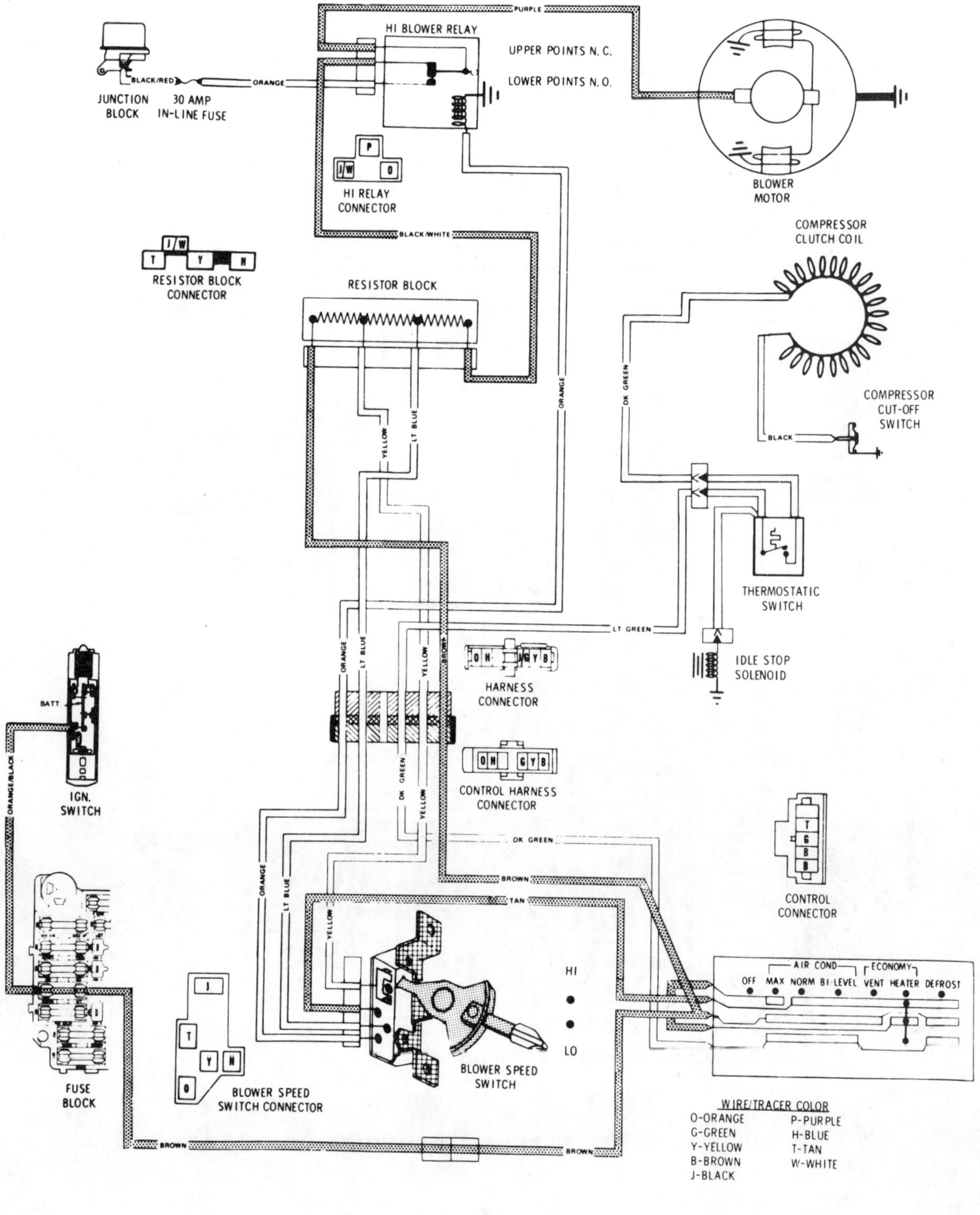

1977 Cutlass A/C & Heater Wiring Circuit (With Custom or Tempmatic)

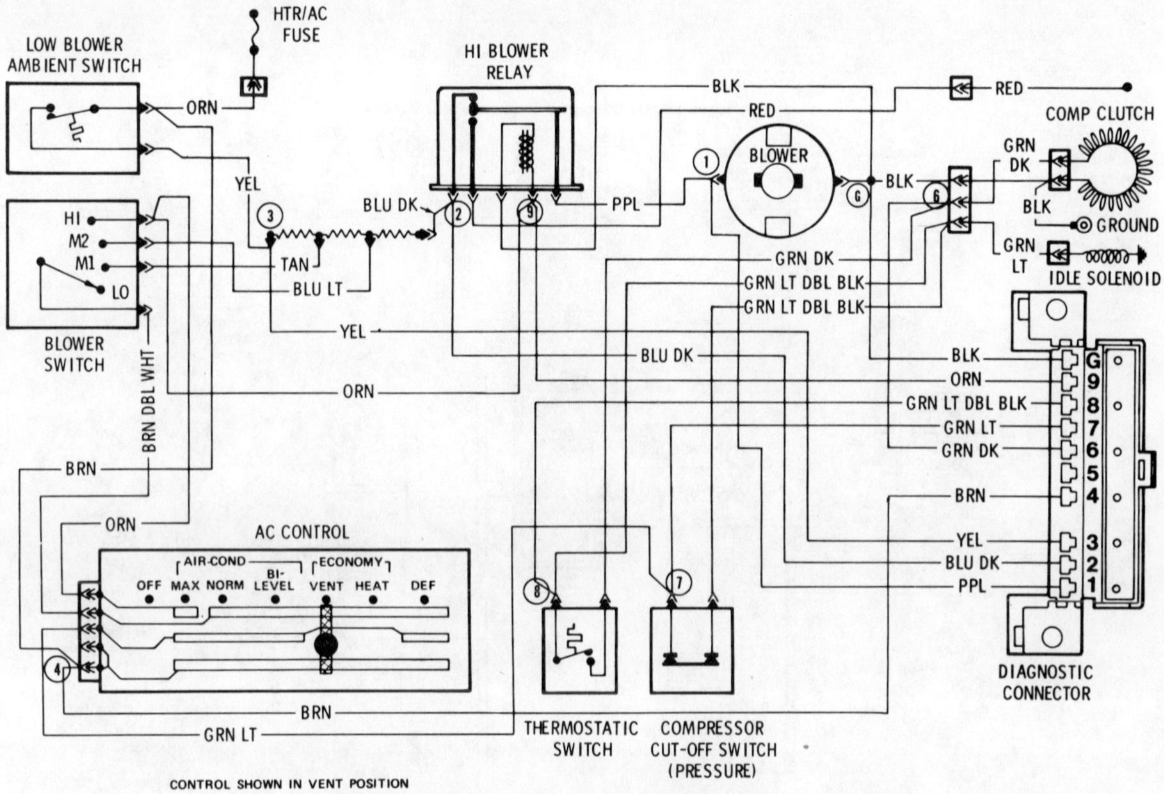

1977 88 & 98 A/C & Heater Wiring Circuit (With Custom or Tempmatic)

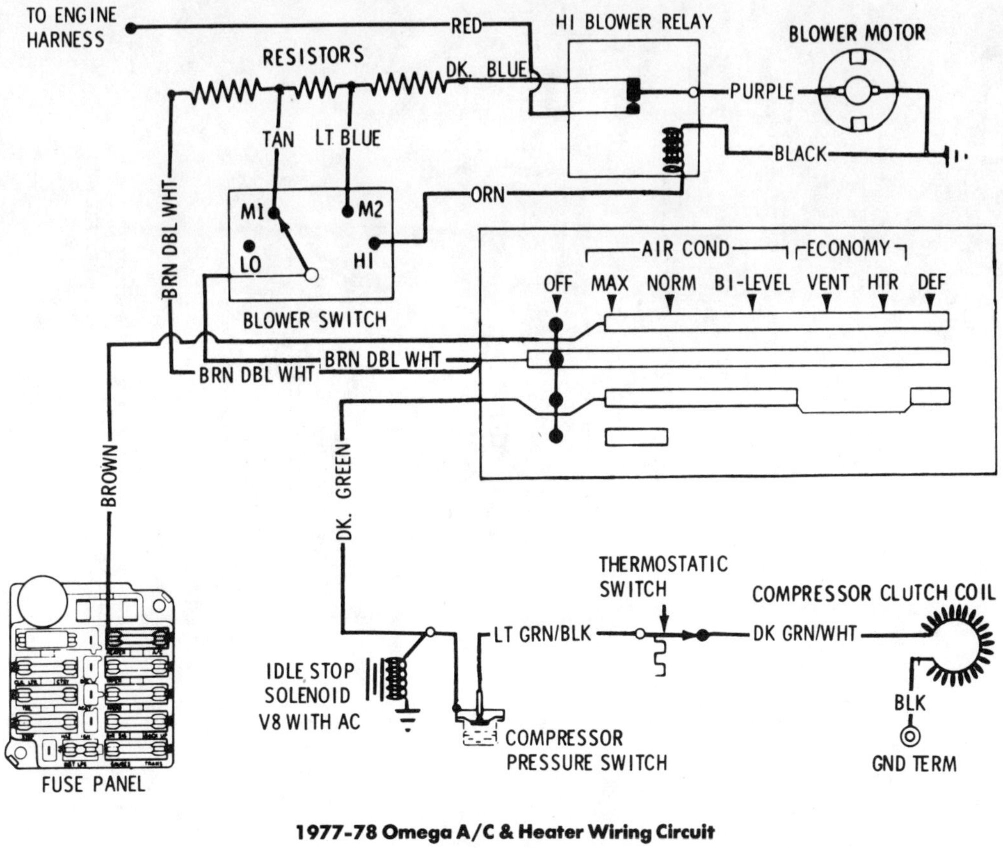

1977-78 Omega A/C & Heater Wiring Circuit

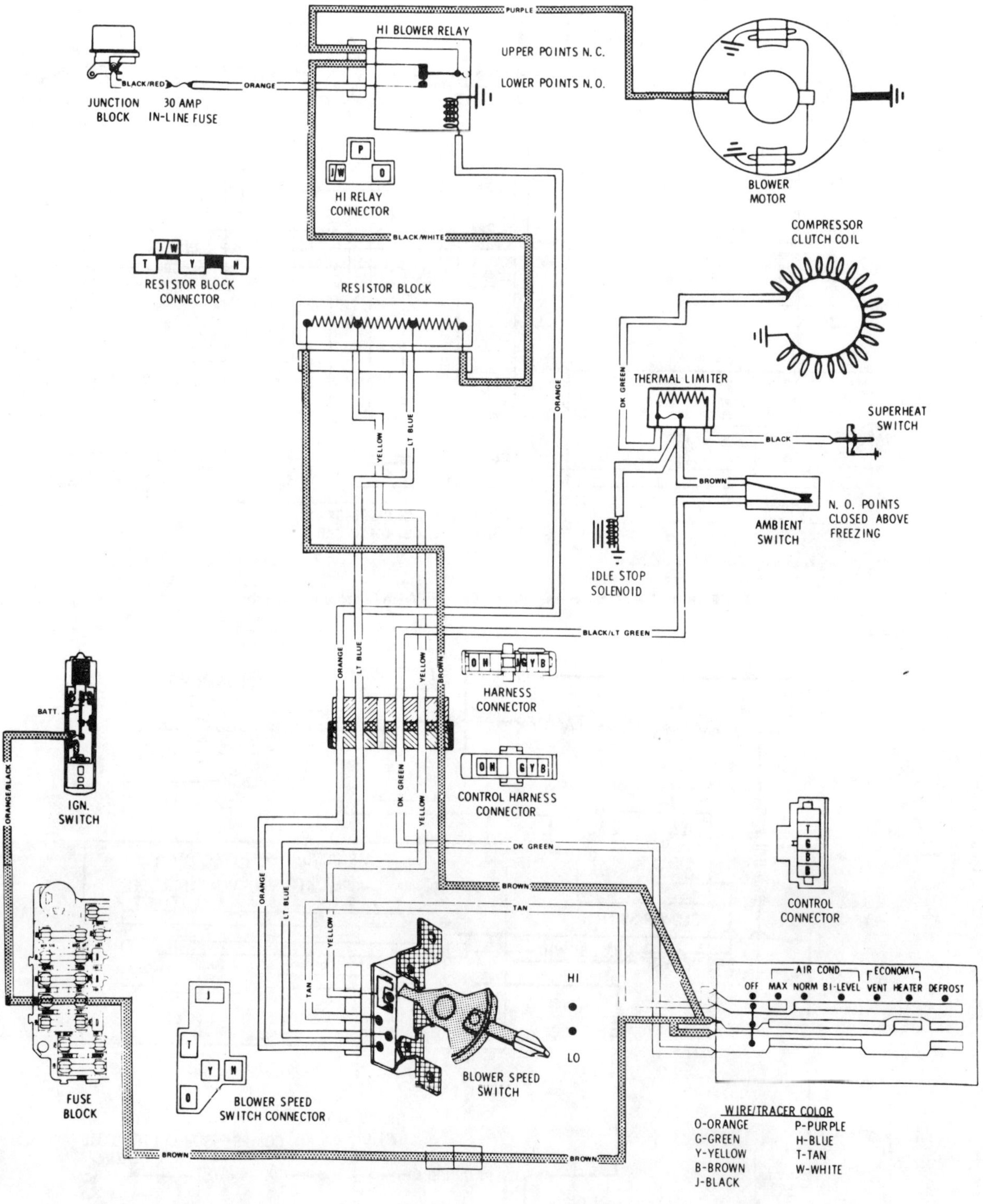

1977-78 Toronado A/C & Heater Wiring Circuit (With Custom or Tempmatic)

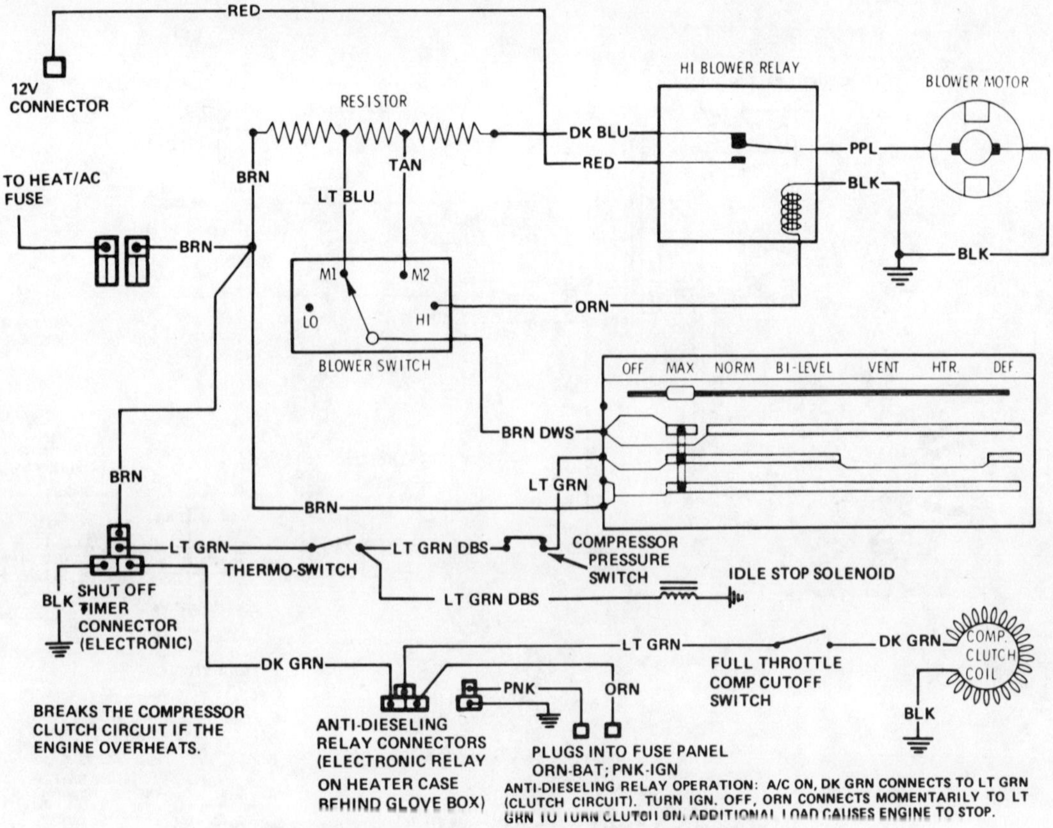

1978 Starfire A/C & Heater Wiring Circuit (With 4-151)

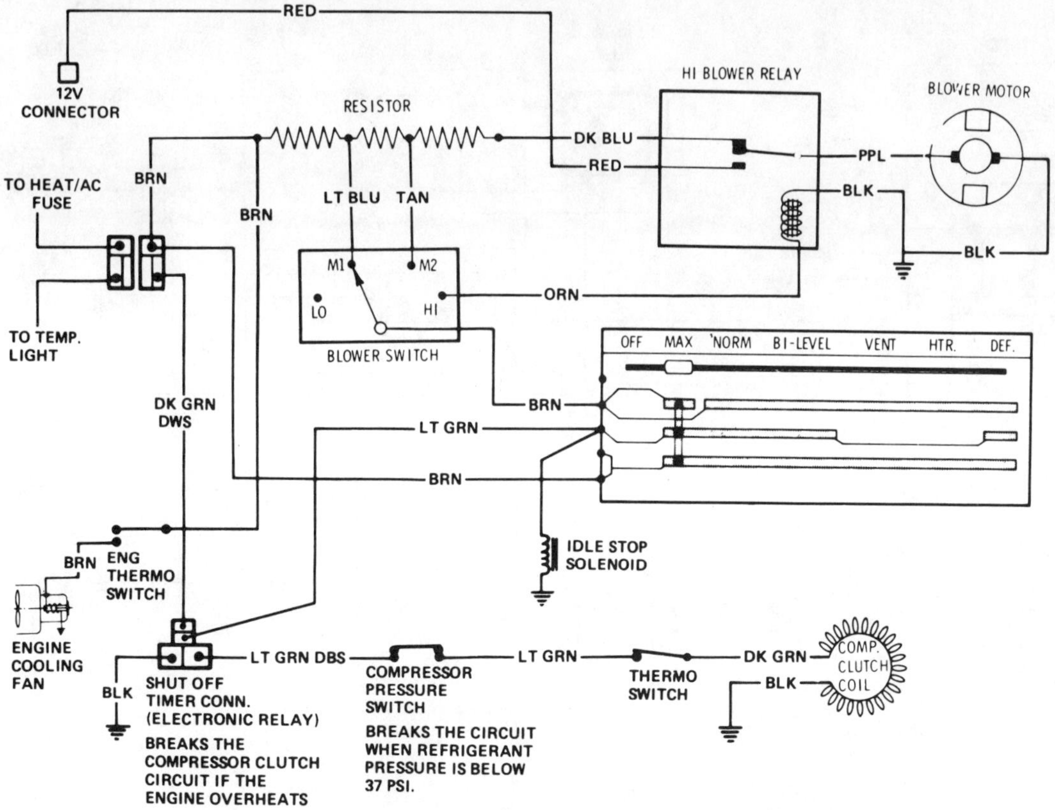

1978 Starfire A/C & Heater Wiring Circuit (With V6-231)

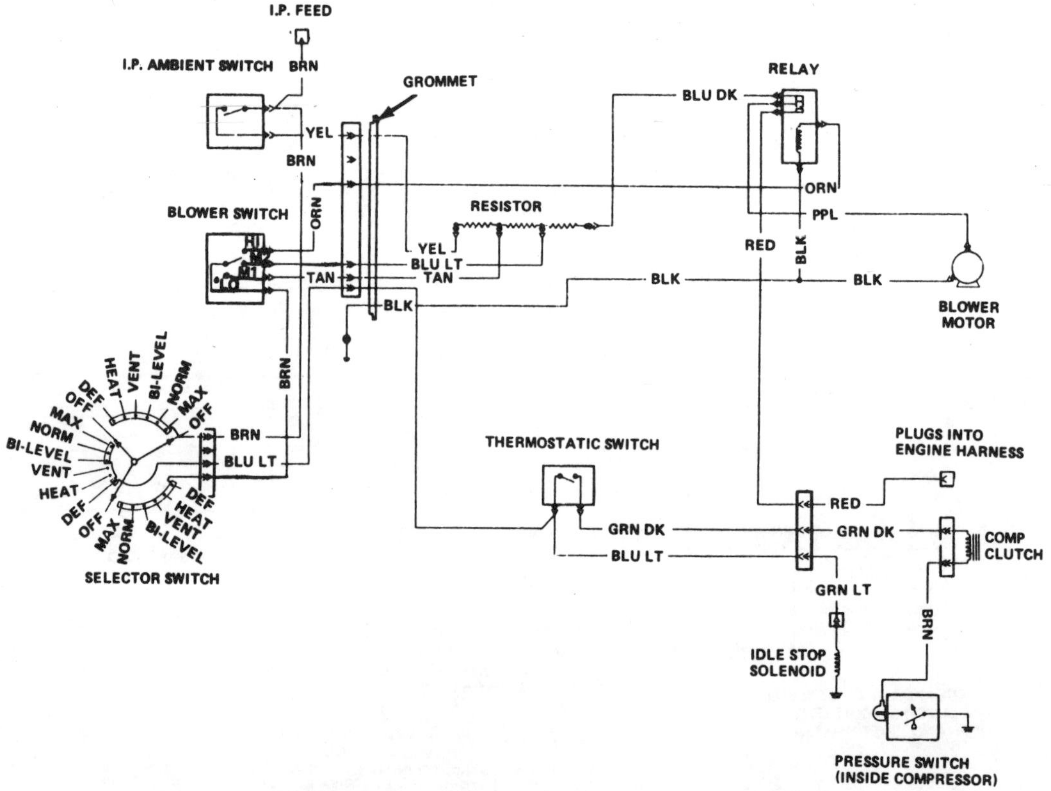

1978 Cutlass A/C & Heater Wiring Circuit (With Custom or Tempmatic)

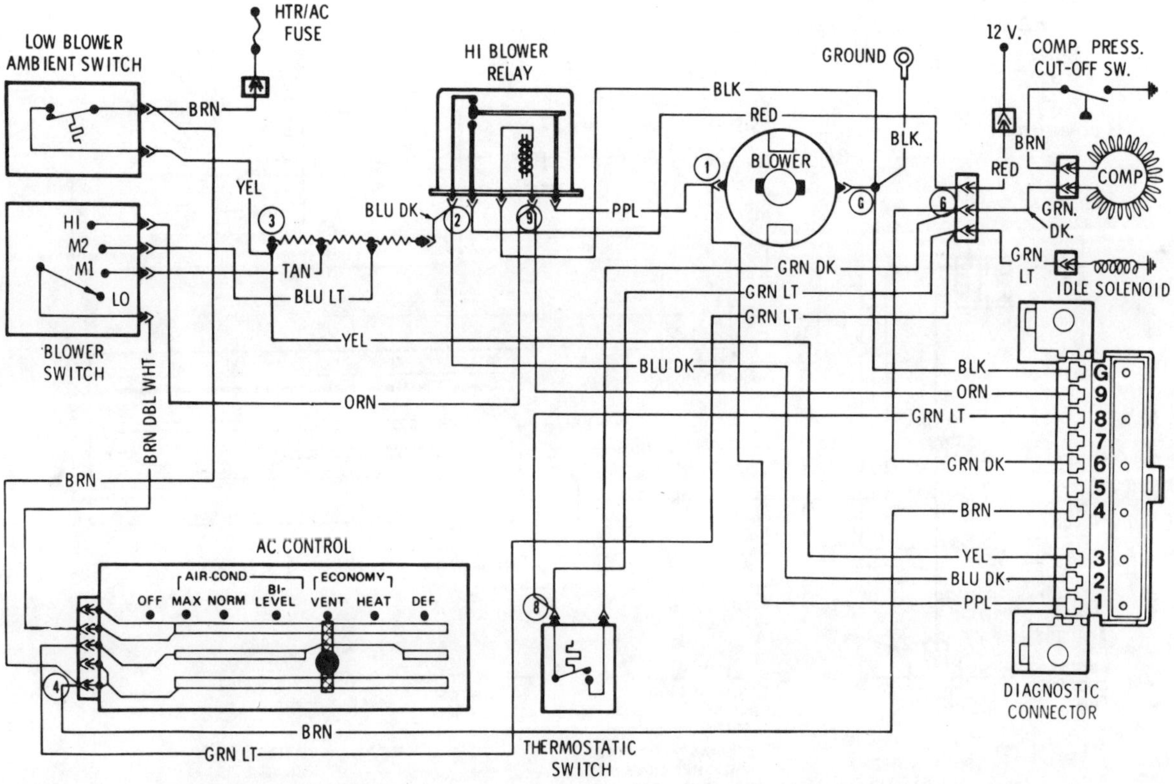

CONTROL SHOWN IN VENT POSITION

1978 88 & 98 A/C & Heater Wiring Circuit (With Custom or Tempmatic)

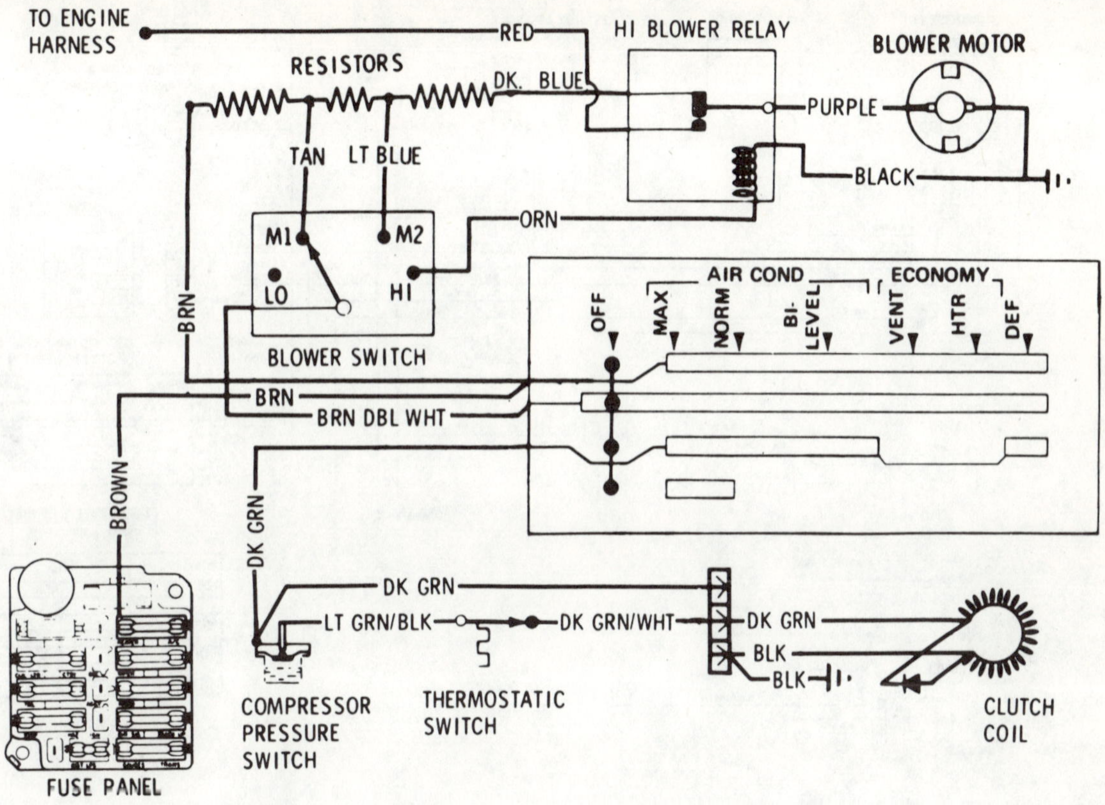

1979 Omega A/C & Heater Wiring Circuit (With V6-231)

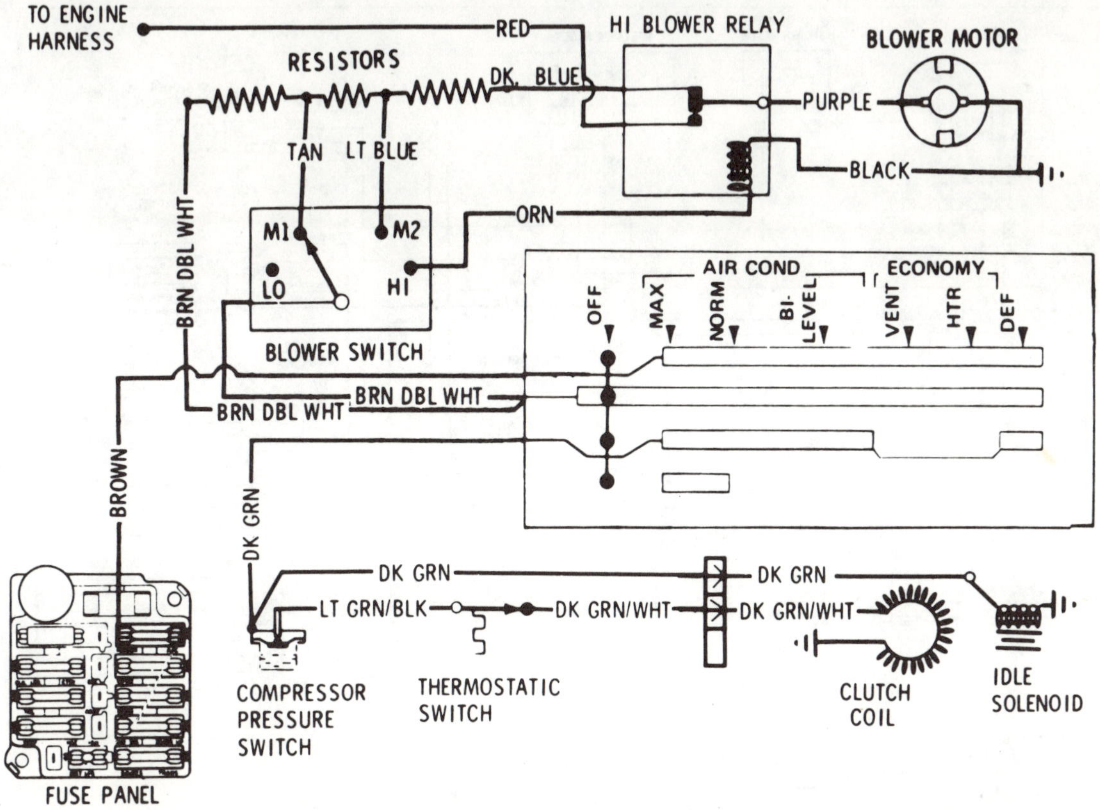

1979 Omega A/C & Heater Wiring Circuit (With V8-305 or 350)

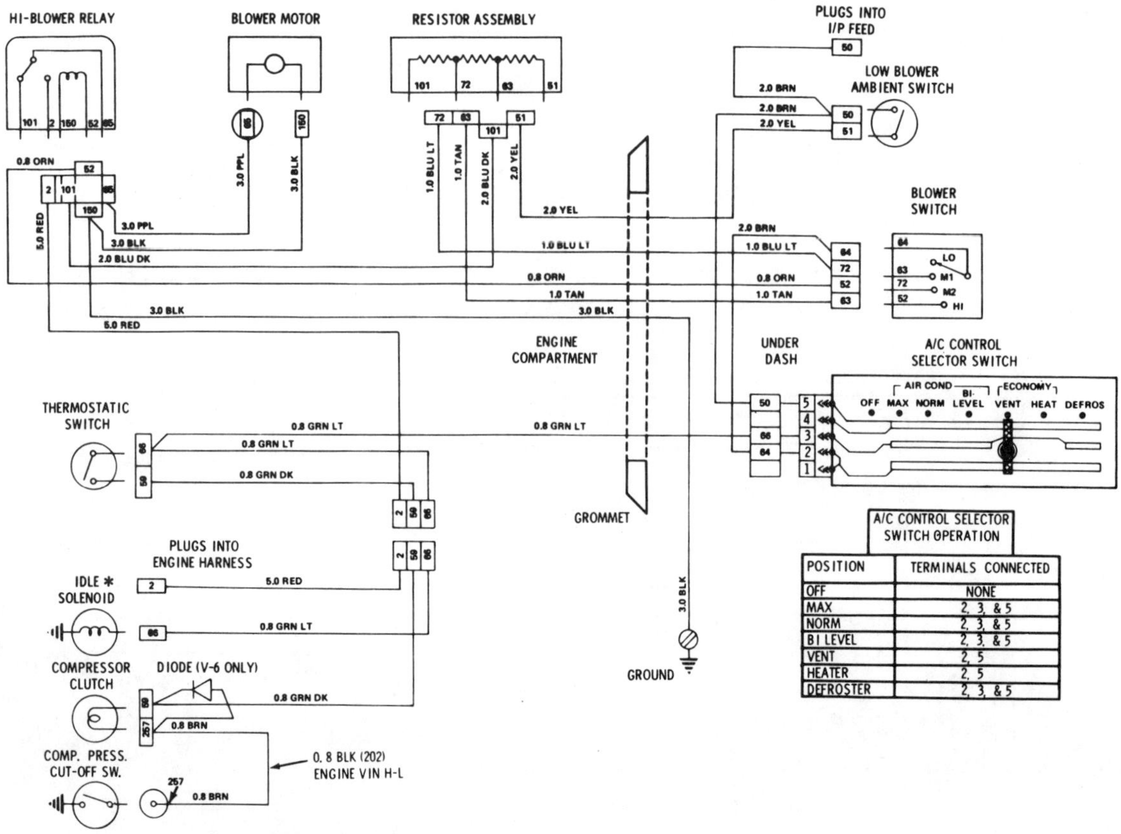

A/C CONTROL SELECTOR SWITCH OPERATION

POSITION	TERMINALS CONNECTED
OFF	NONE
MAX	2, 3, & 5
NORM	2, 3, & 5
BI LEVEL	2, 3, & 5
VENT	2, 5
HEATER	2, 5
DEFROSTER	2, 3, & 5

1979 Cutlass A/C & Heater Wiring Circuit (With Custom or Tempmatic)

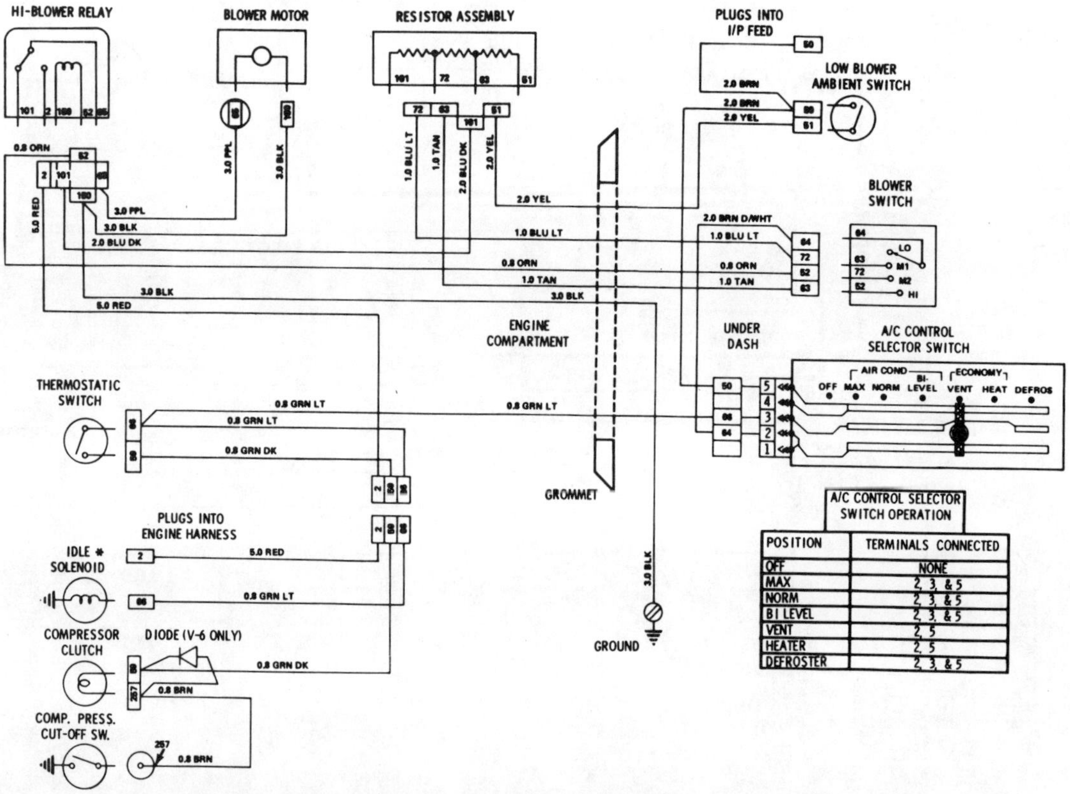

A/C CONTROL SELECTOR SWITCH OPERATION

POSITION	TERMINALS CONNECTED
OFF	NONE
MAX	2, 3, & 5
NORM	2, 3, & 5
BI LEVEL	2, 3, & 5
VENT	2, 5
HEATER	2, 5
DEFROSTER	2, 3, & 5

1979 88 & 98 A/C & Heater Wiring Circuit (With Custom or Tempmatic)

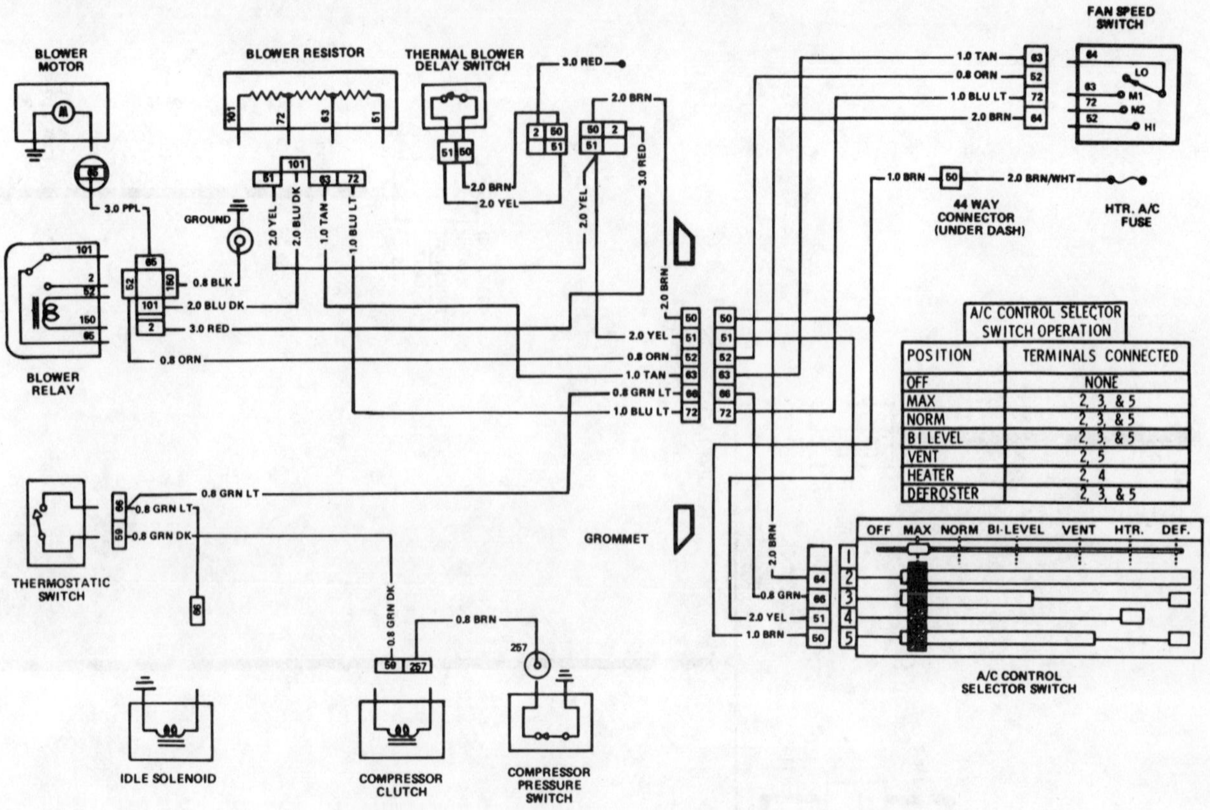

A/C CONTROL SELECTOR SWITCH OPERATION	
POSITION	TERMINALS CONNECTED
OFF	NONE
MAX	2, 3, & 5
NORM	2, 3, & 5
BI LEVEL	2, 3, & 5
VENT	2, 5
HEATER	2, 4
DEFROSTER	2, 3, & 5

1979 Toronado A/C & Heater Wiring Circuit (With Custom or Tempmatic)

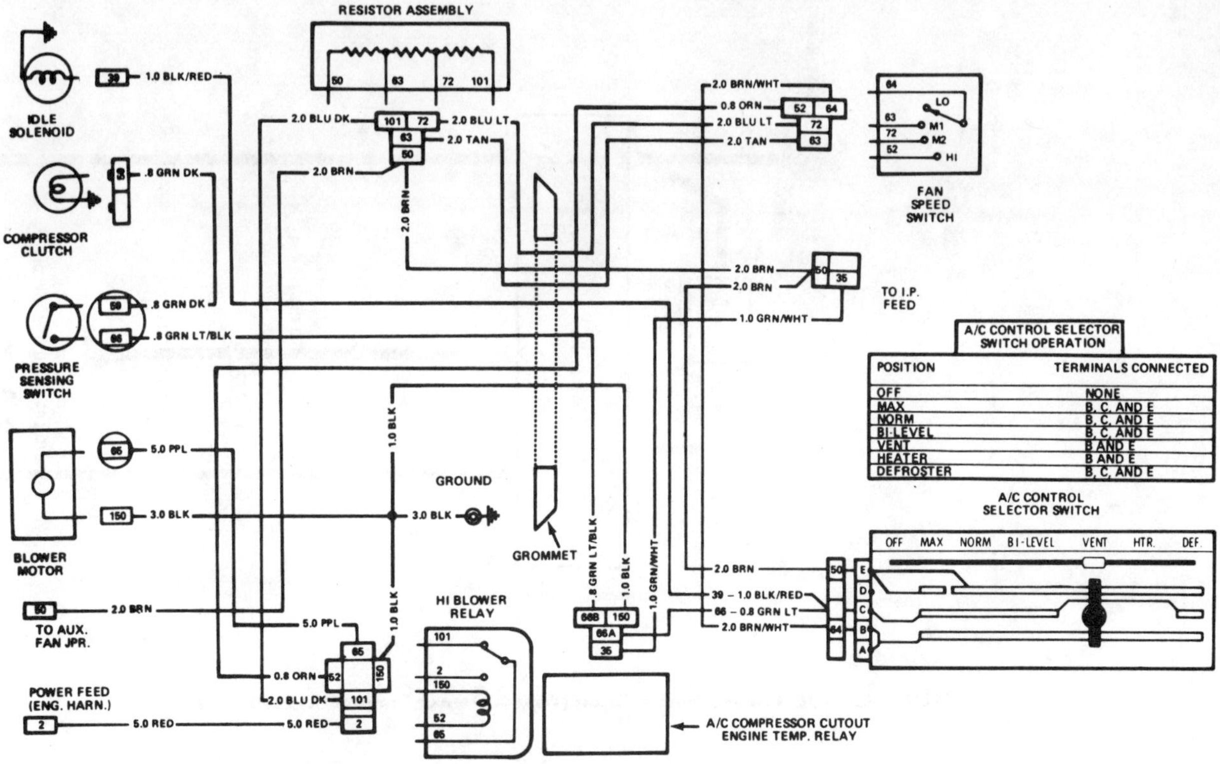

A/C CONTROL SELECTOR SWITCH OPERATION	
POSITION	TERMINALS CONNECTED
OFF	NONE
MAX	B, C, AND E
NORM	B, C, AND E
BI-LEVEL	B, C, AND E
VENT	B AND E
HEATER	B AND E
DEFROSTER	B, C, AND E

1979 Starfire A/C & Heater Wiring Circuit (With V8-305)

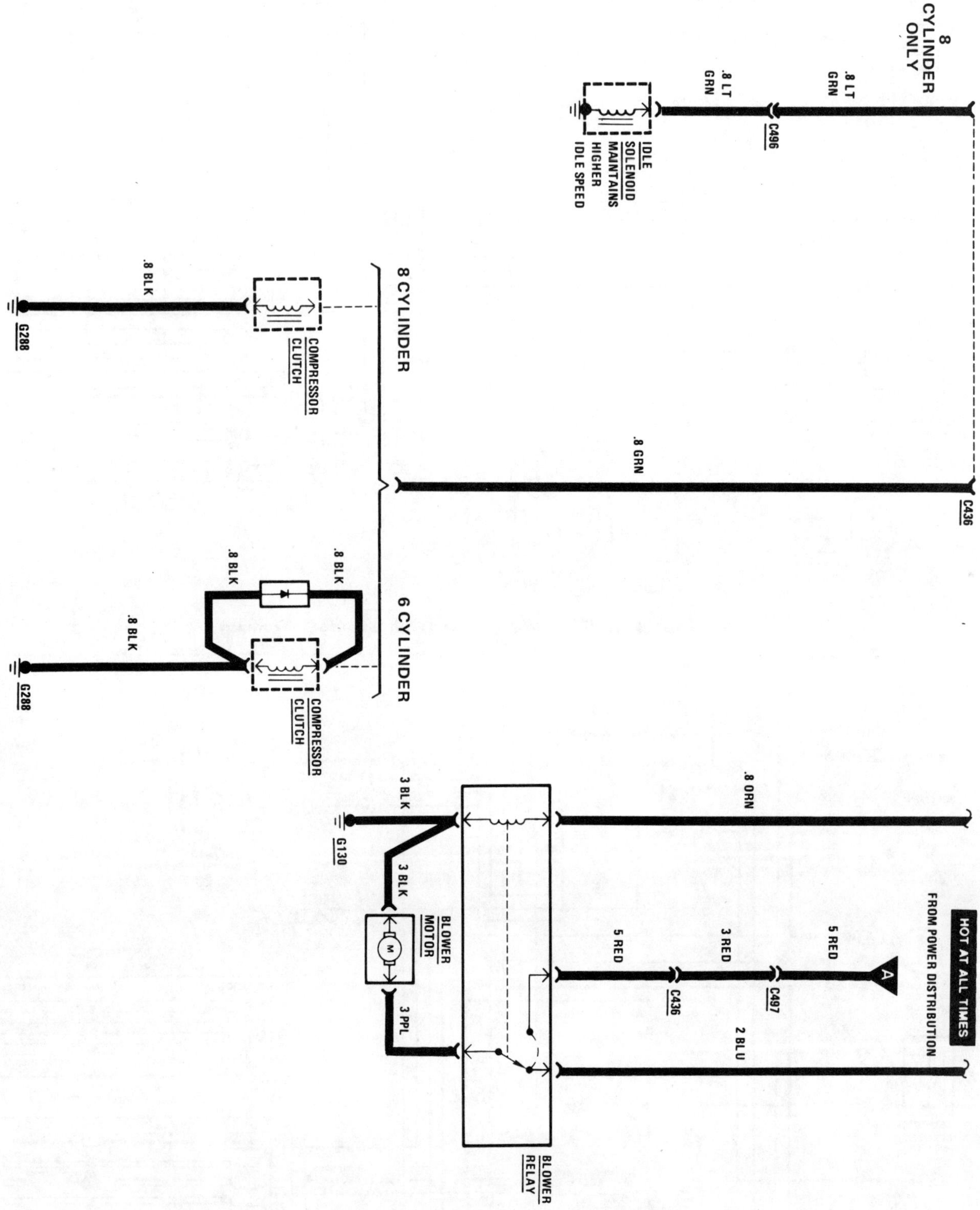

1980 Cutlass A/C & Heater Wiring Circuit (With Custom or Tempmatic) Part 1 of 2

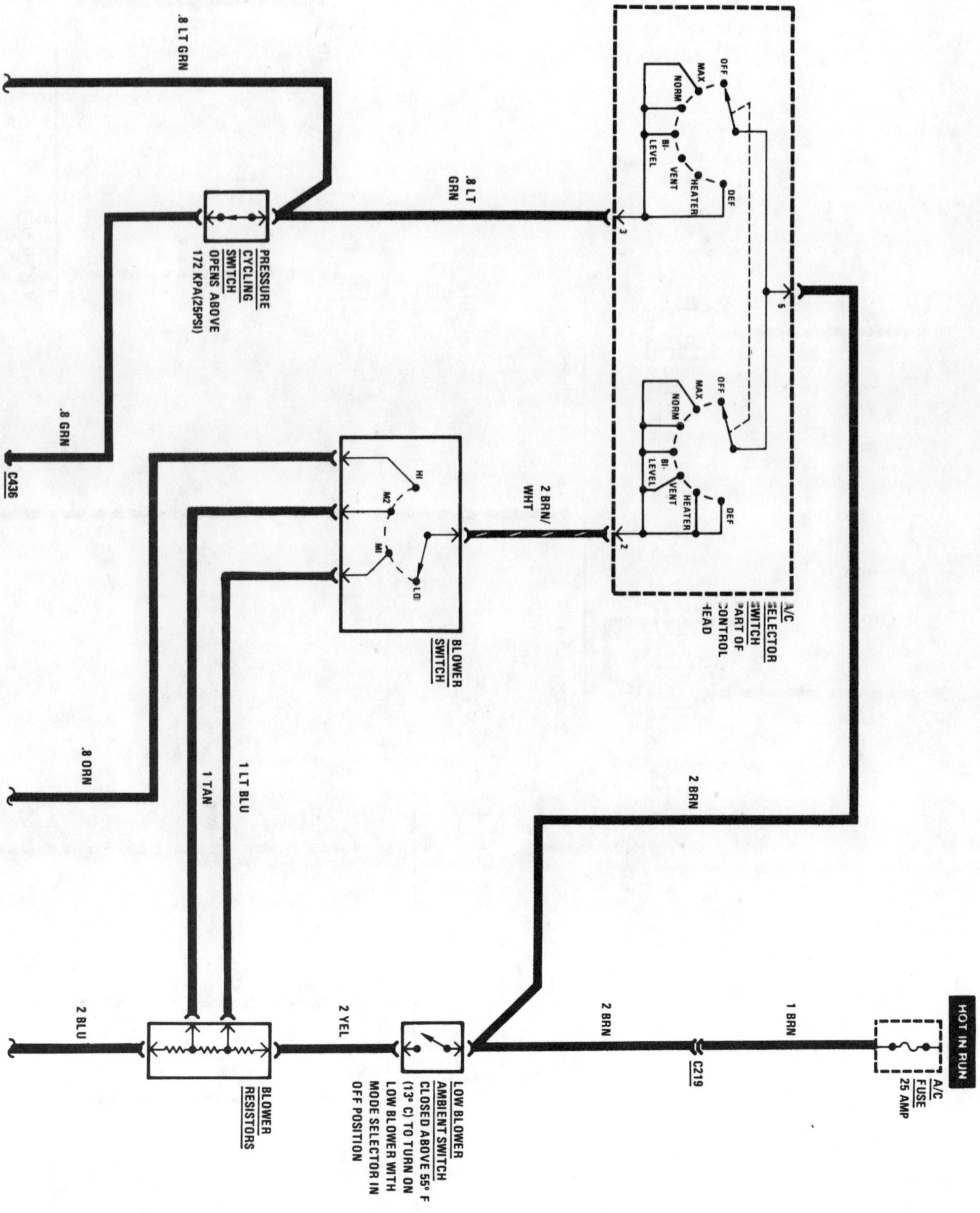

1980 Cutlass A/C & Heater Wiring Circuit (With Custom or Tempmatic) Part 2 of 2

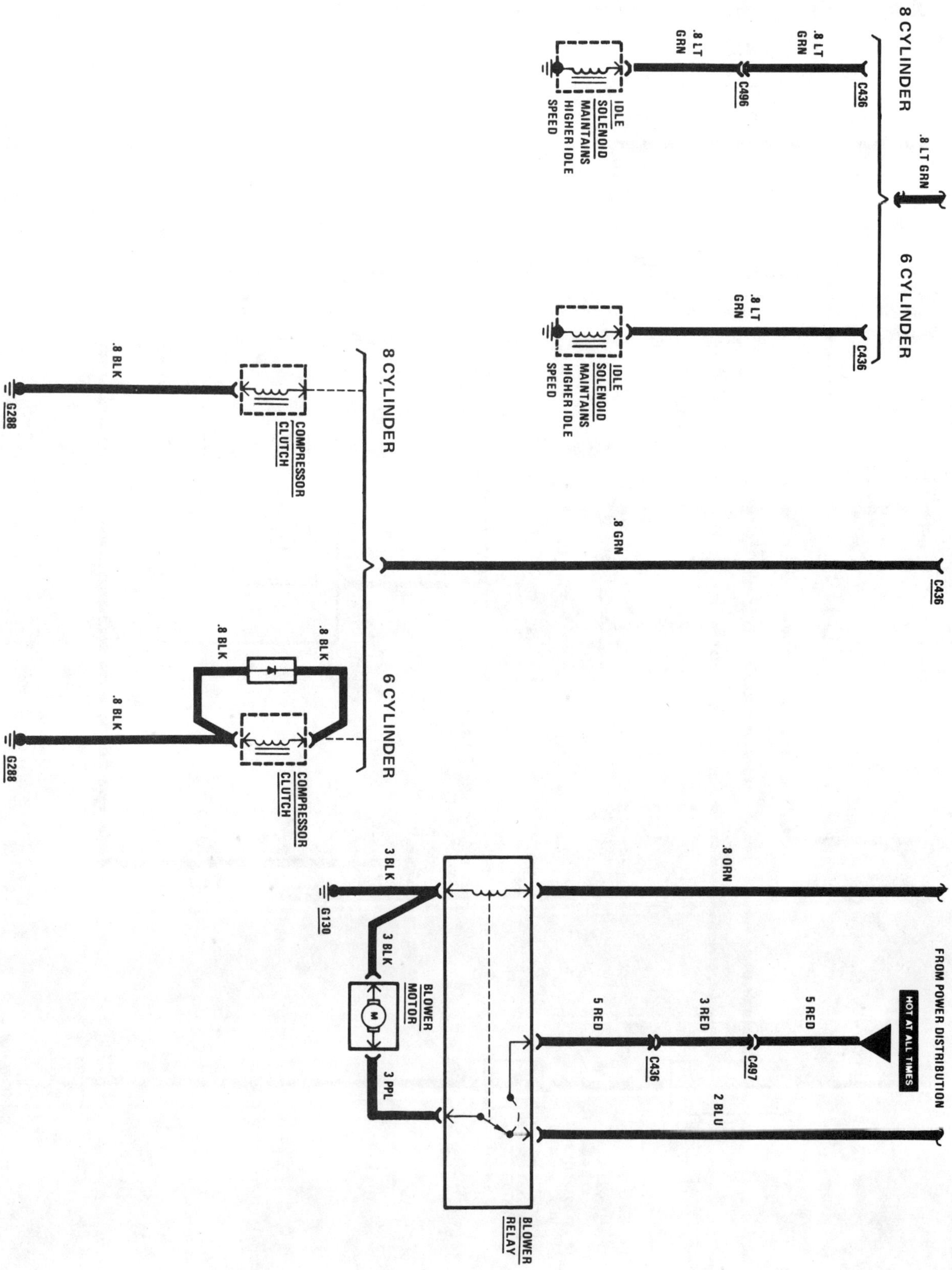

1980 88 & 98 A/C & Heater Wiring Circuit (With Custom or Tempmatic) Part 1 of 2

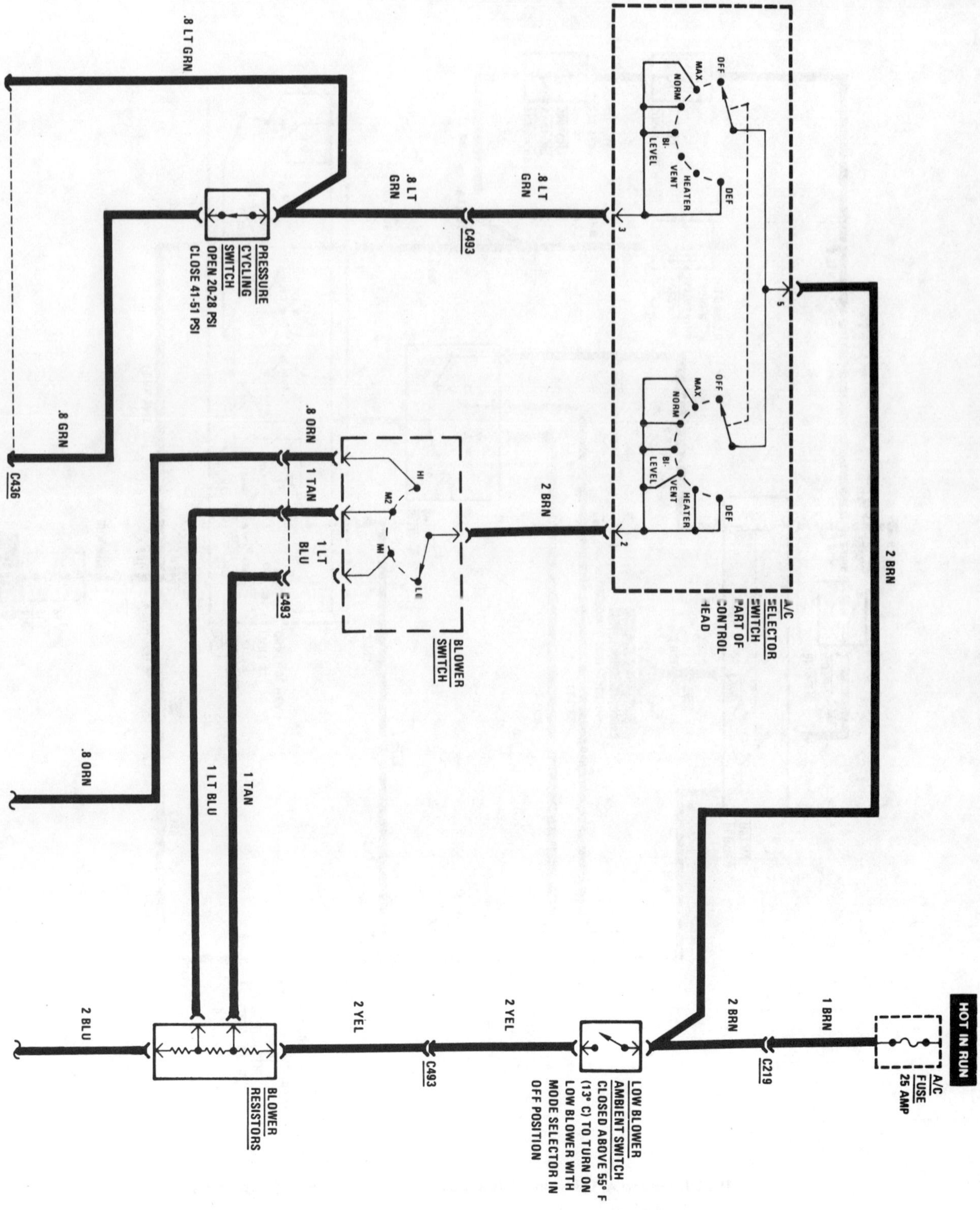

1980 88 & 98 A/C & Heater Wiring Circuit (With Custom or Tempmatic) Part 2 of 2

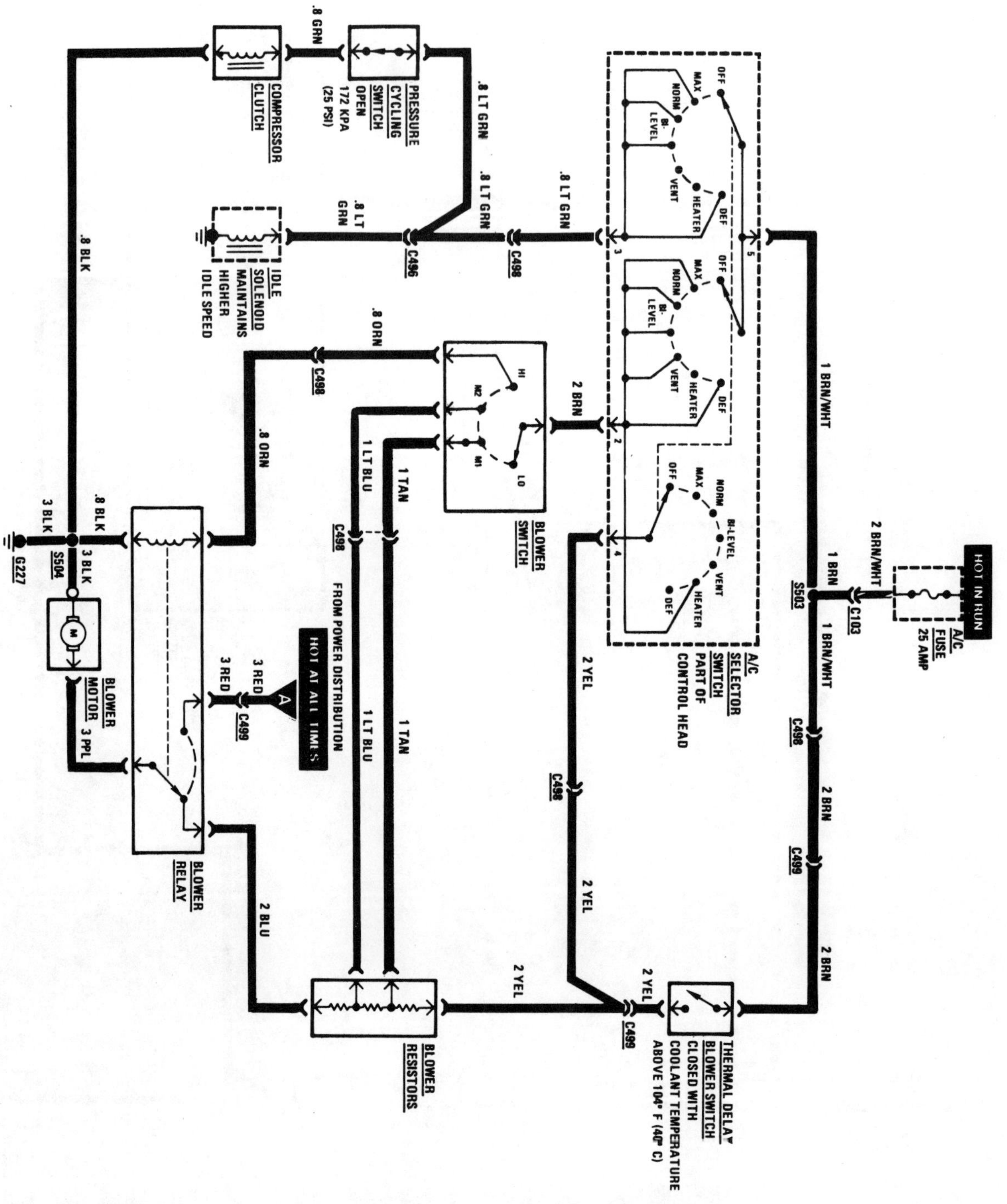

1980 Toronado A/C & Heater Wiring Circuit (With Custom or Tempmatic)

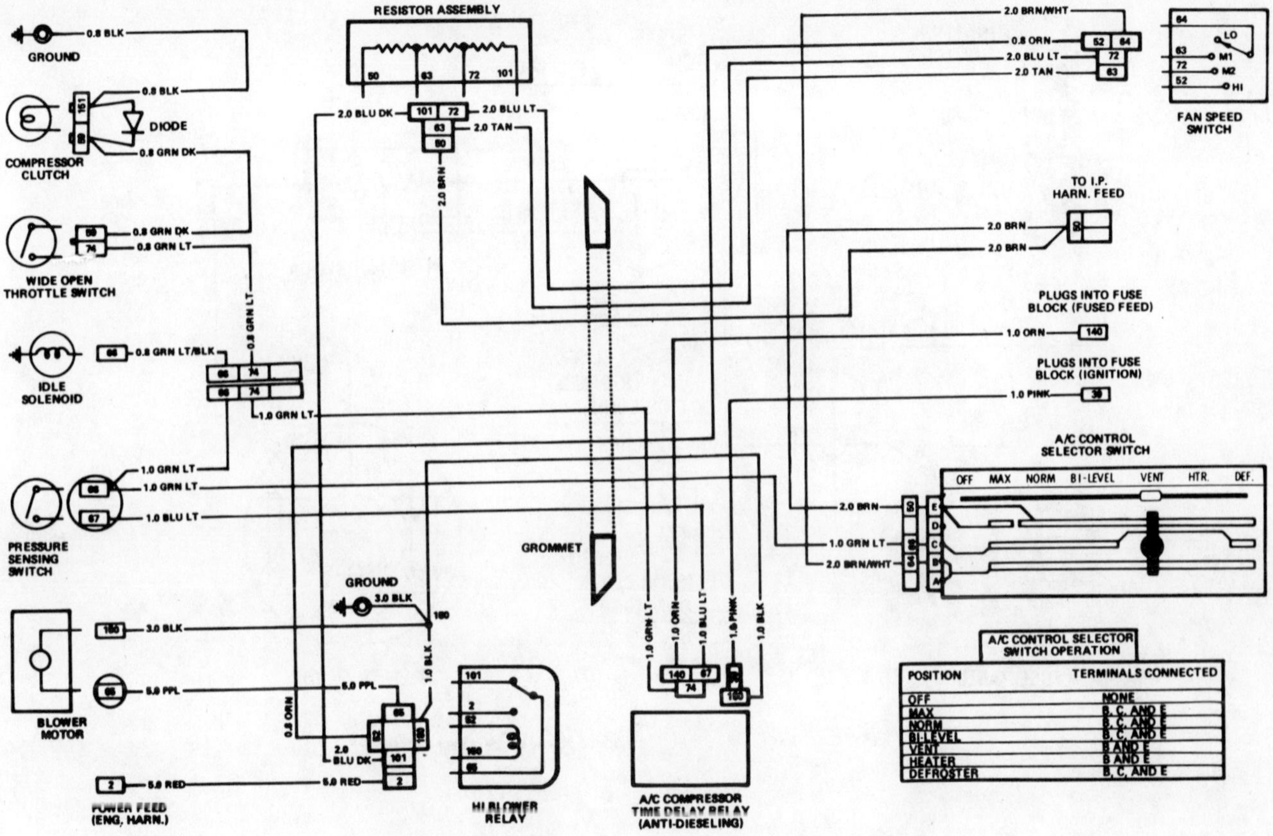

A/C CONTROL SELECTOR SWITCH OPERATION

POSITION	TERMINALS CONNECTED
OFF	NONE
MAX	B, C, AND E
NORM	B, C, AND E
BI-LEVEL	B, C, AND E
VENT	B AND E
HEATER	B AND E
DEFROSTER	B, C, AND E

1979-80 Starfire A/C & Heater Wiring Circuit (With 4-151)

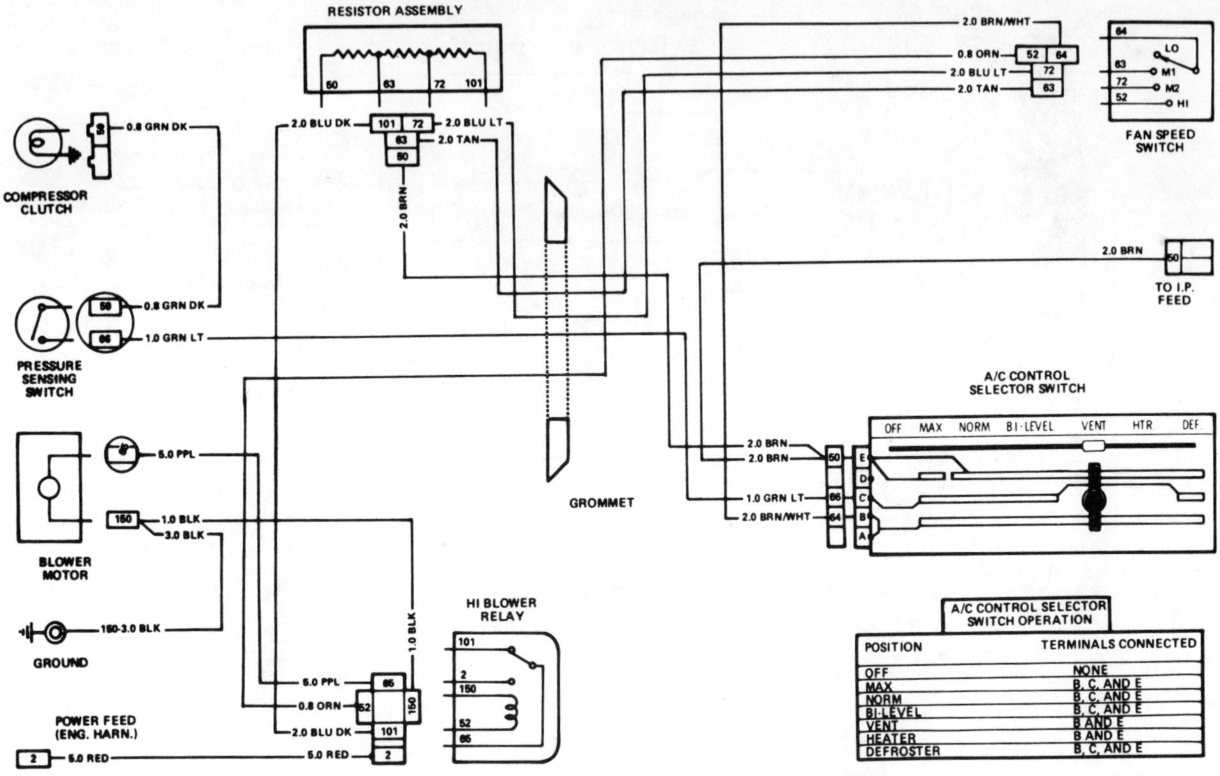

A/C CONTROL SELECTOR SWITCH OPERATION

POSITION	TERMINALS CONNECTED
OFF	NONE
MAX	B, C, AND E
NORM	B, C, AND E
BI-LEVEL	B, C, AND E
VENT	B AND E
HEATER	B AND E
DEFROSTER	B, C, AND E

1979-80 Starfire A/C & Heater Wiring Circuit (With V6-231)

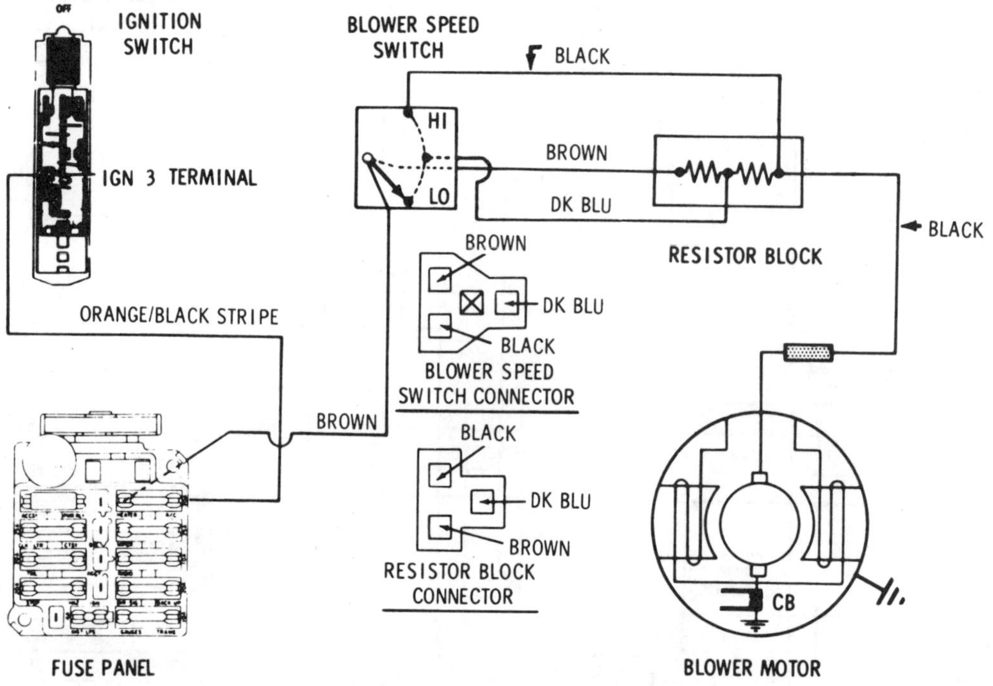

1974 Cutlass, 88, 98 & Toronado Heater Wiring Circuit

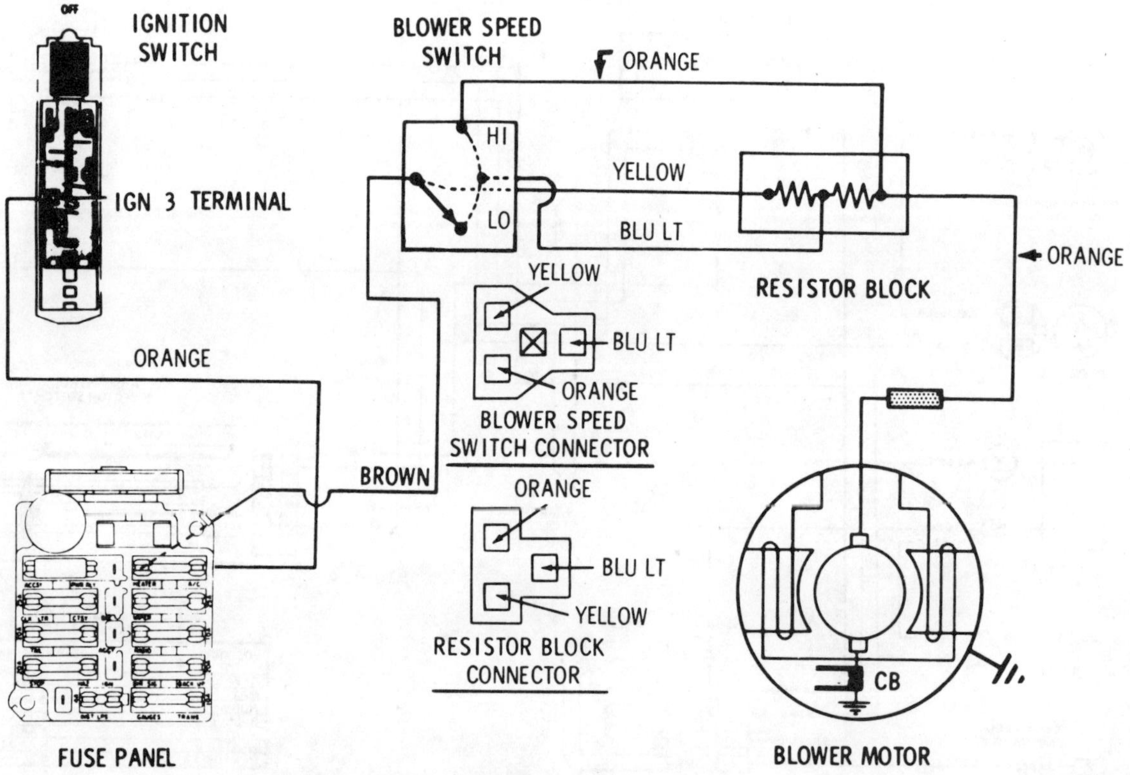

1974-75 Omega Heater Wiring Circuit

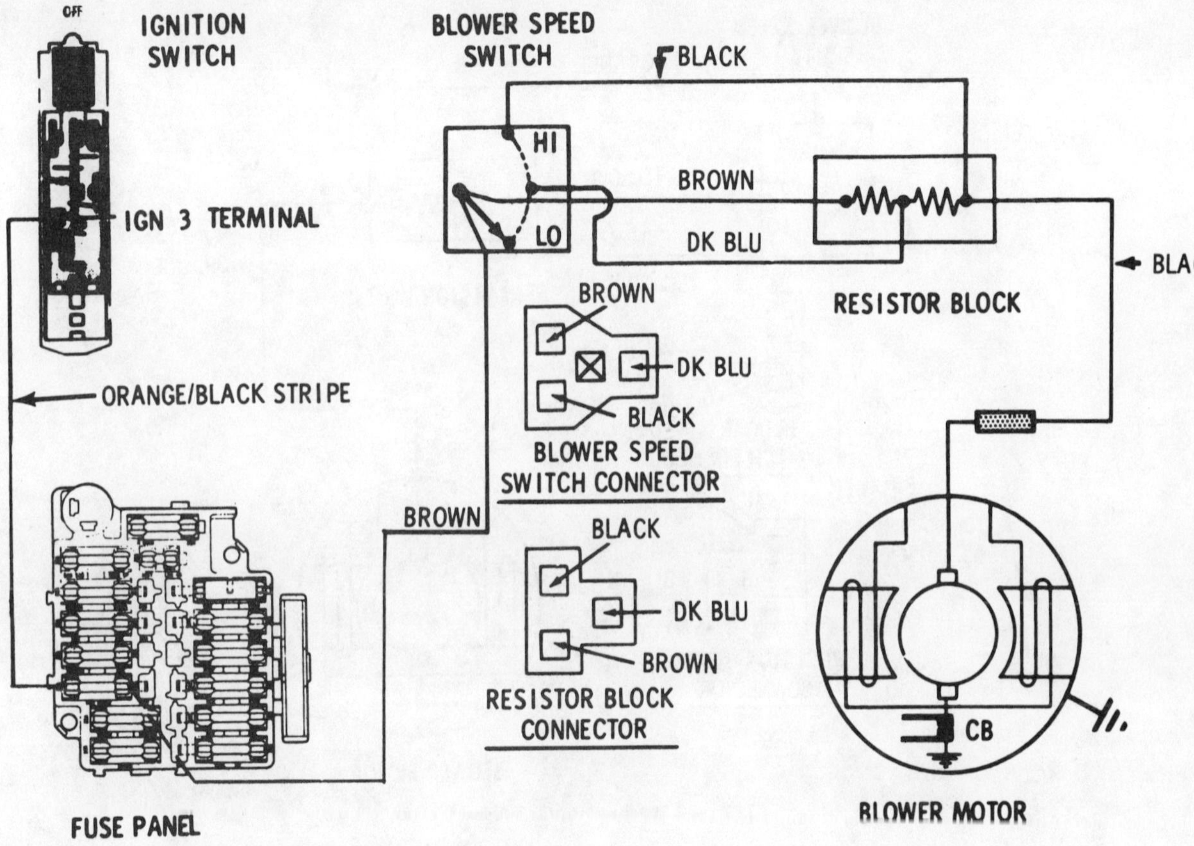

1975-76 88, 98 & Toronado Heater Wiring Circuit

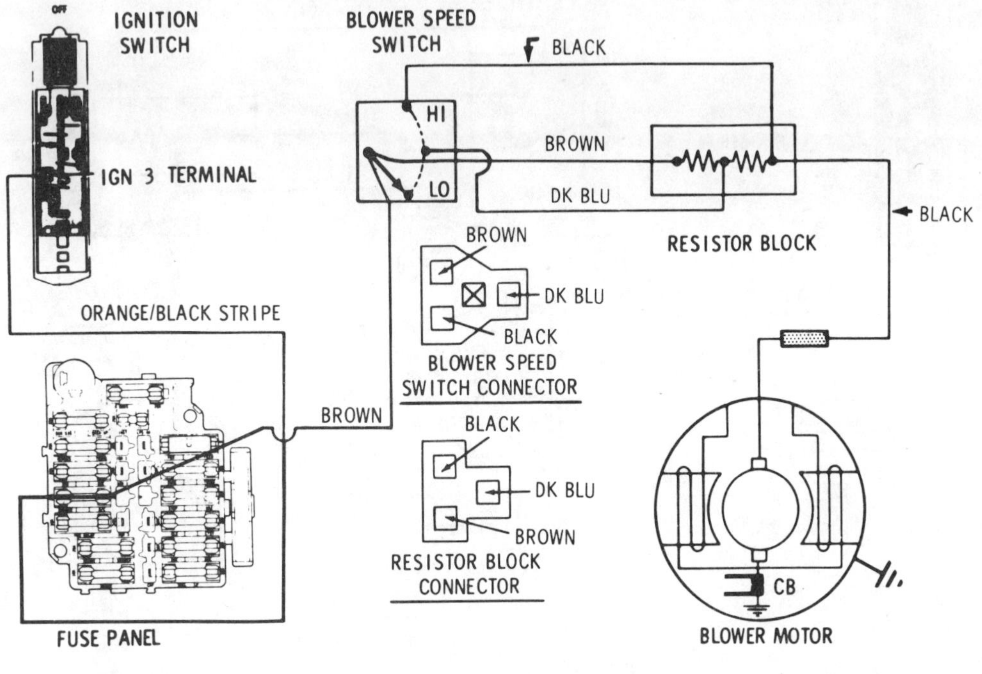

1975-77 Cutlass Heater Wiring Circuit

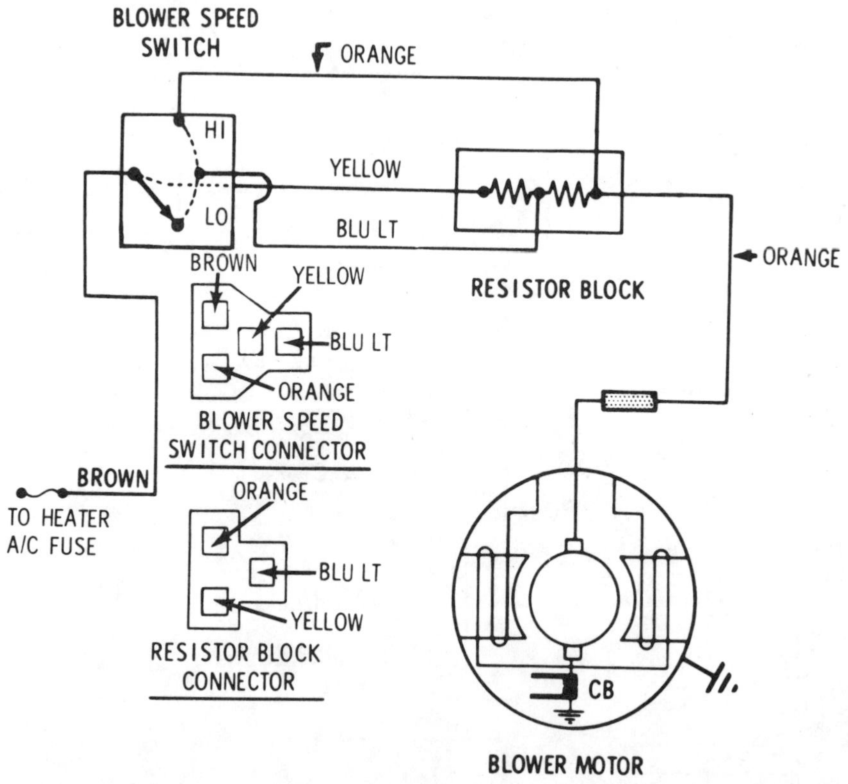

1975-80 Starfire Heater Wiring Circuit

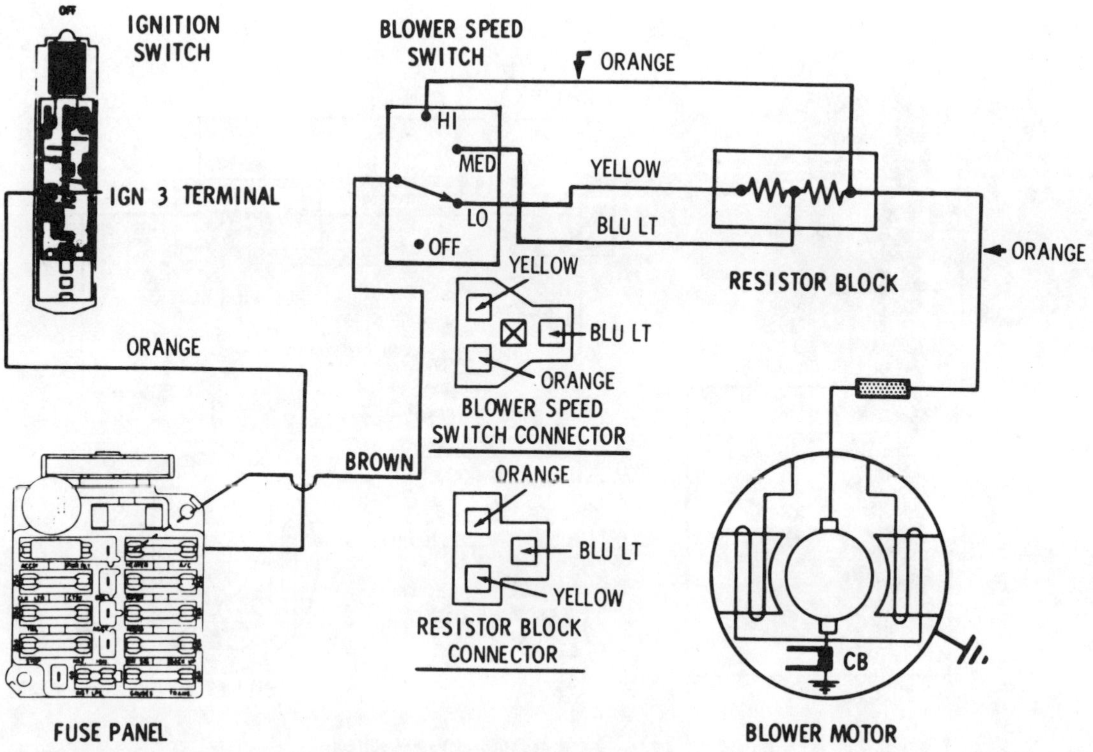

1976-79 Omega Heater Wiring Circuit

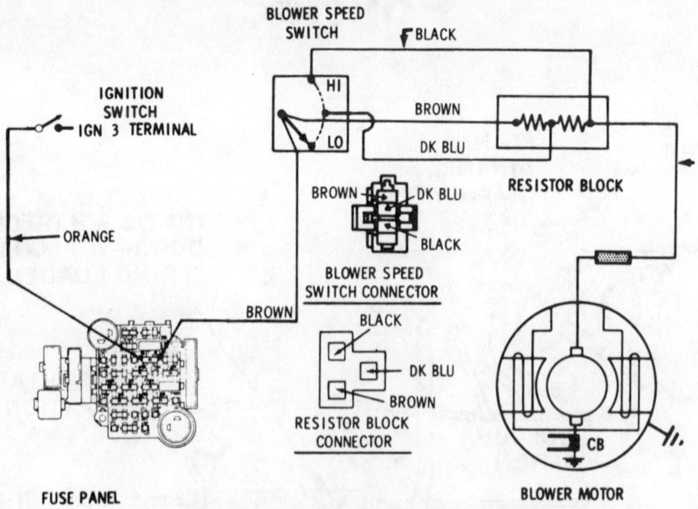

1977-78 88 & 98 Heater Wiring Circuit

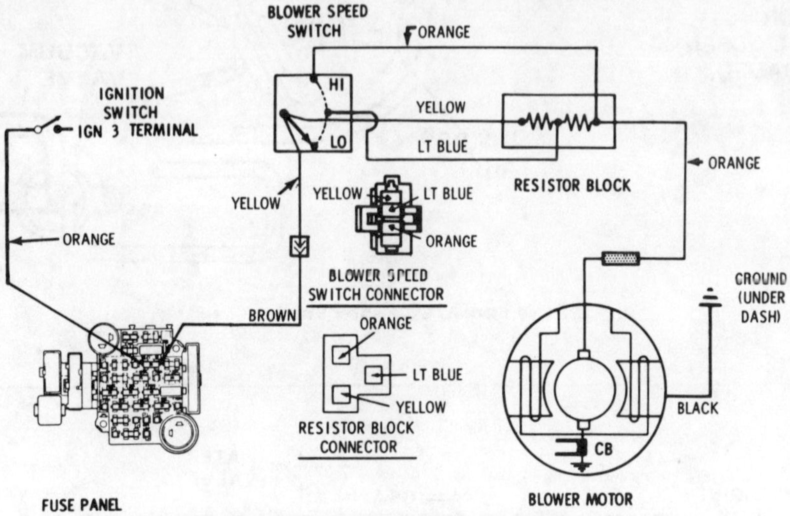

1979-80 88 & 98 Heater Wiring Circuit

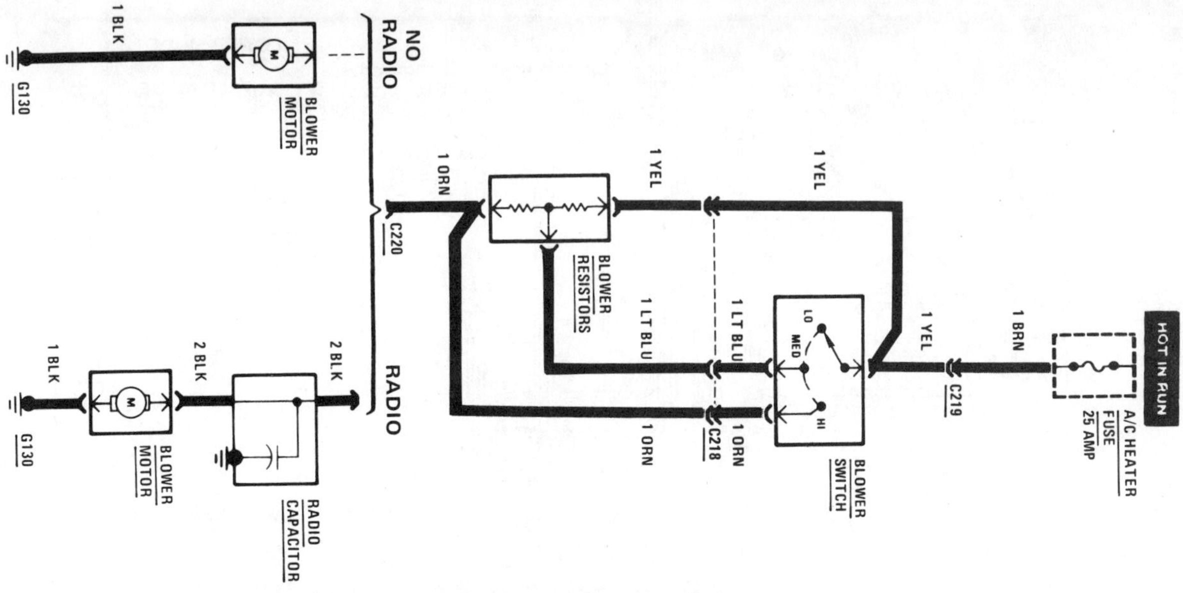

1980 Cutlass Heater Wiring Circuit

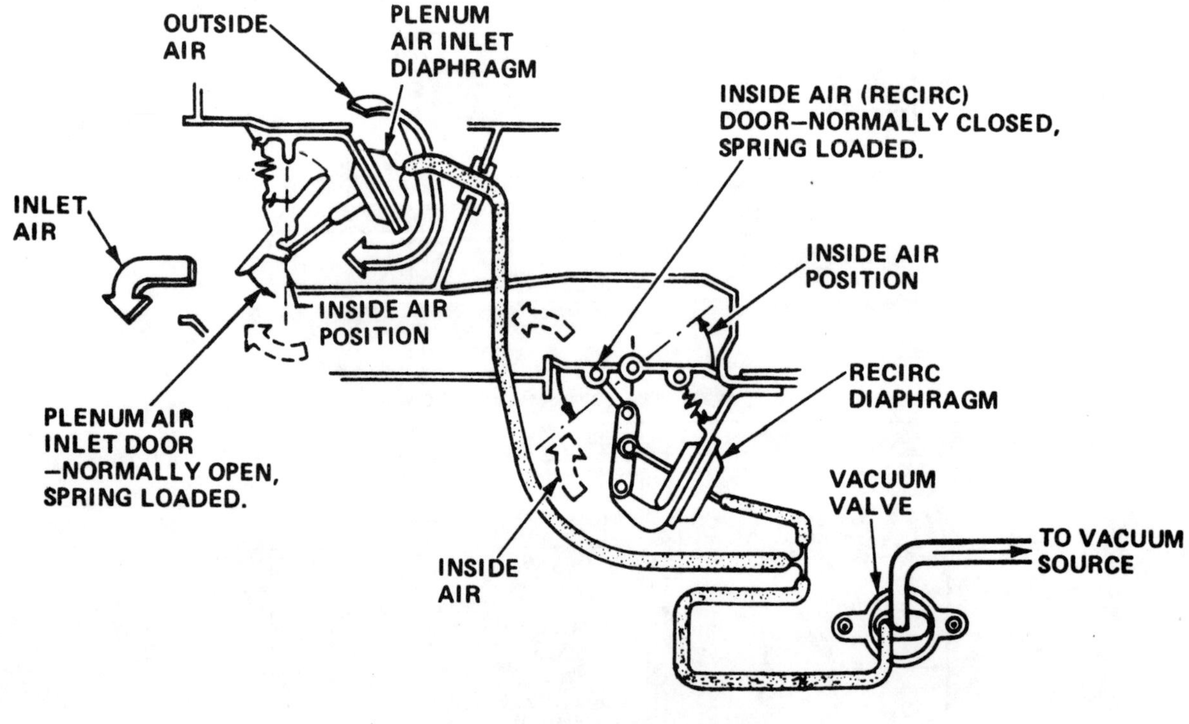

1974 Ventura A/C & Heater Vacuum Circuit

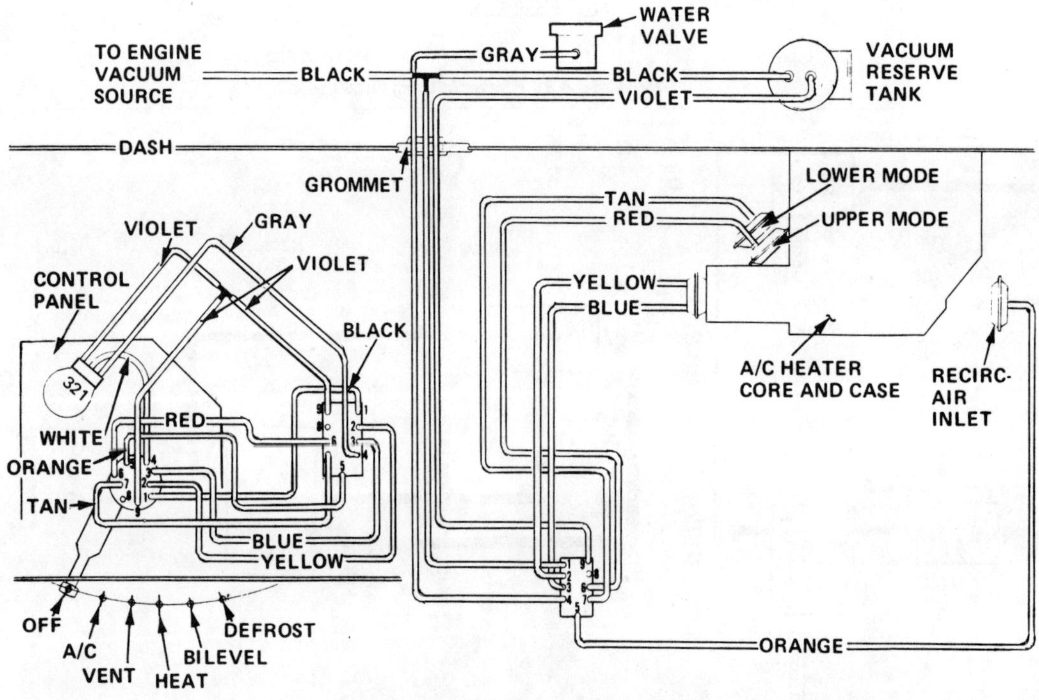

1974 Bonneville, Catalina, Grand AM, Grand Prix, Grand Ville & LeMans A/C & Heater Vacuum Circuit (Less Auto Temp Control)

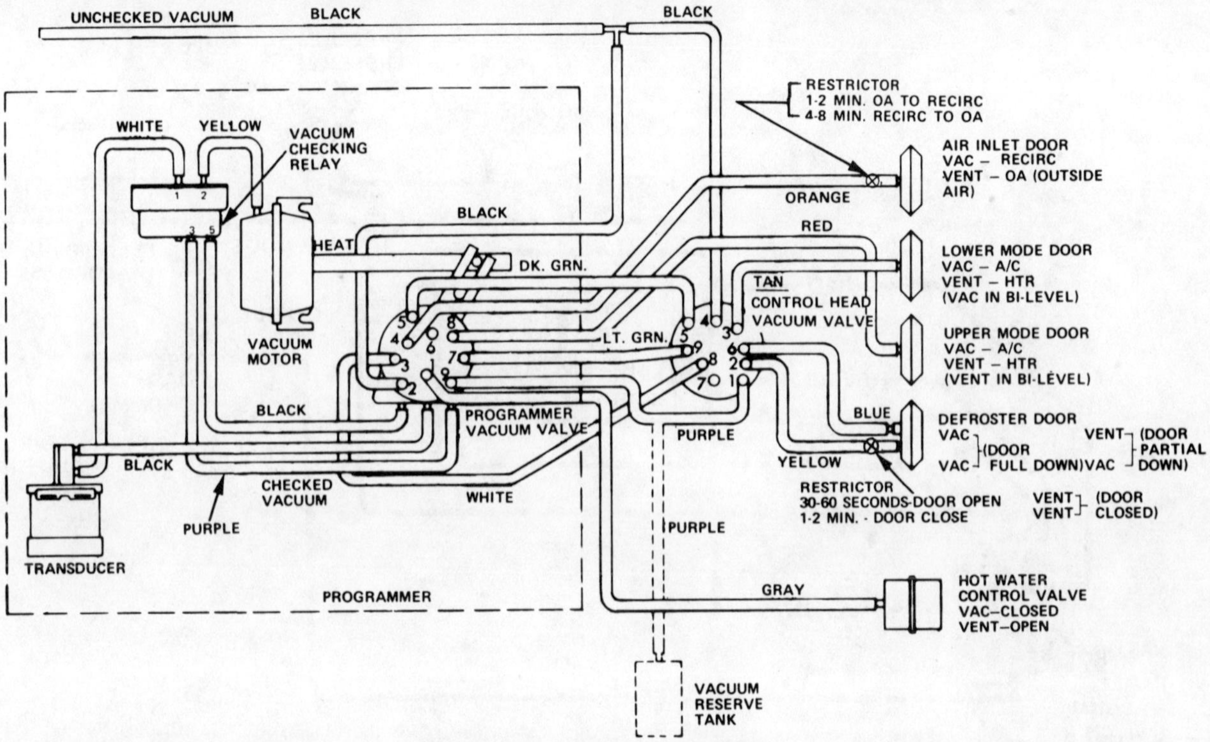

1974 Bonneville, Catalina, Grand Am, Grand Prix & Grand Ville A/C & Heater Vacuum Circuit (With Auto Temp Control)

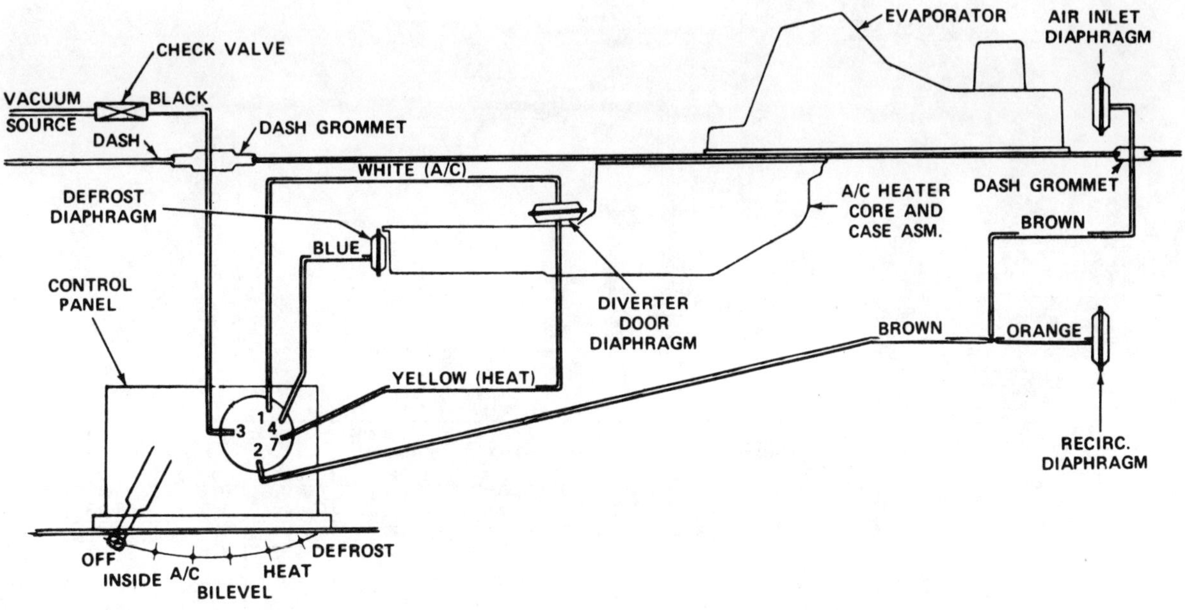

1974-80 Firebird A/C & Heater Vacuum Circuit

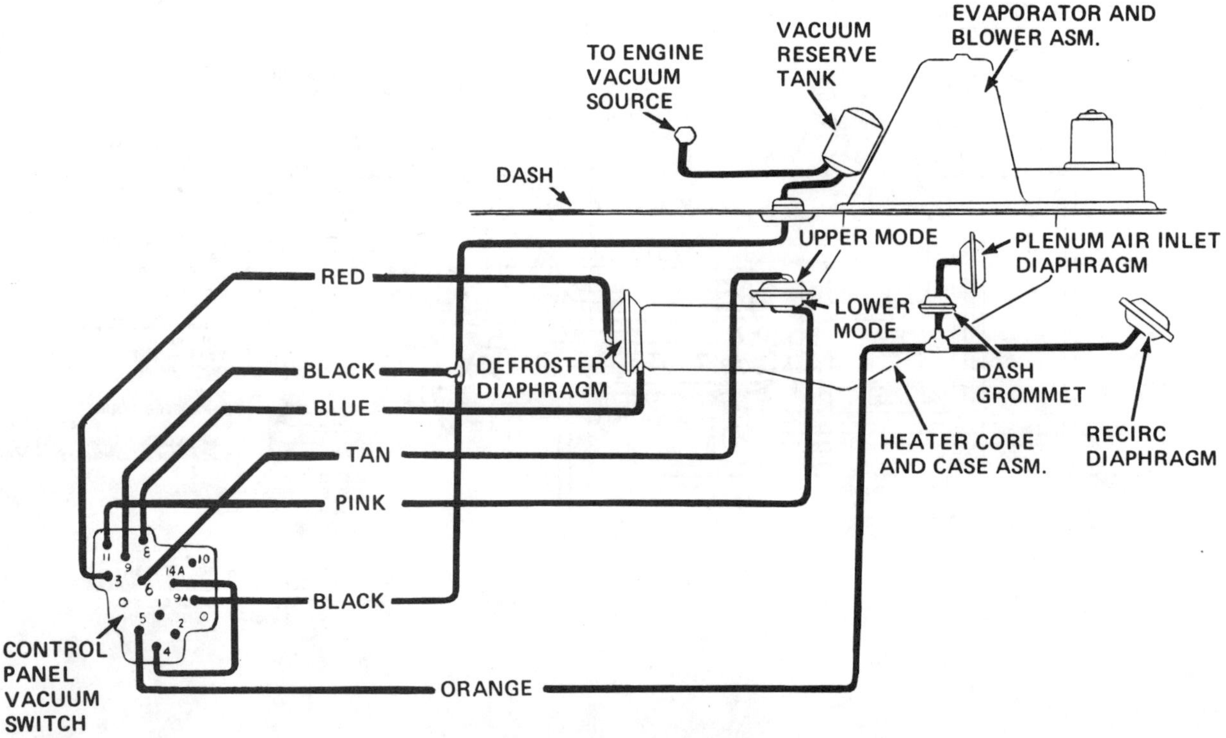

1975 Ventura A/C & Heater Vacuum Circuit

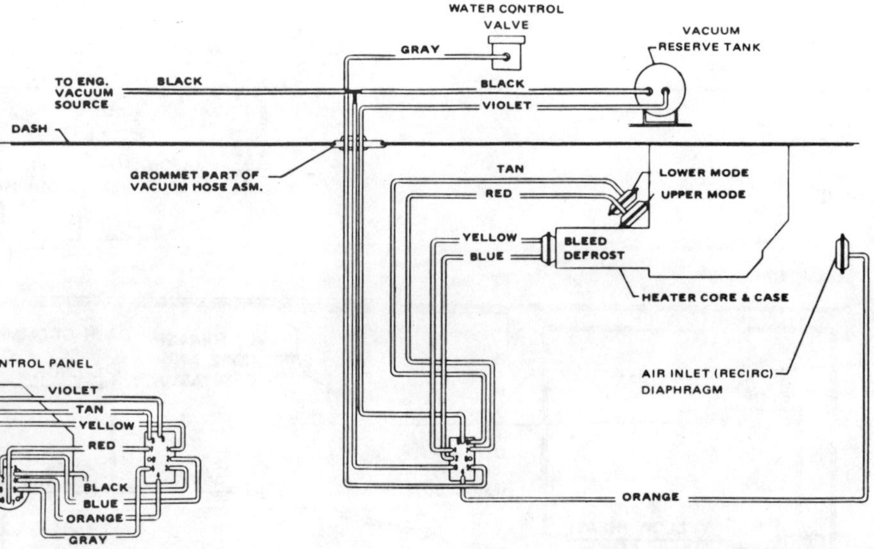

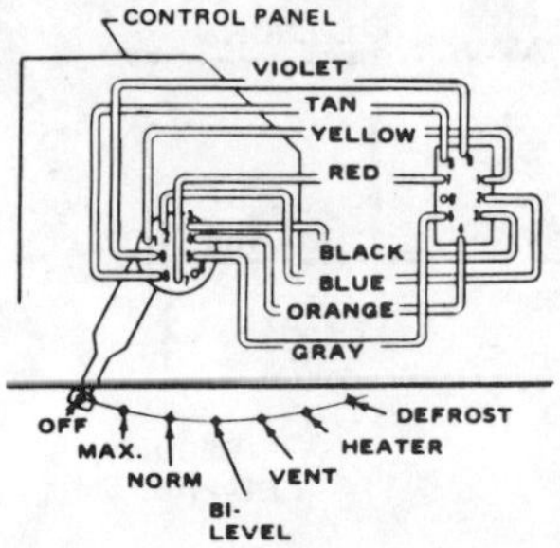

VACUUM VALVE CONNECTIONS

PORT NO.	PORT CONNECTION	SELECTOR LEVER POSITION						
		OFF	MAX	NORMAL	BI-LEVEL	VENT	HEATER	DEFROST
1	BLEED	VENT	VENT	VENT	PORT 3	PORT 3	PORT 3	PORT 3
2	DEFROST	VENT	VENT	VENT	VENT	VENT	VENT	PORT 9
3	INPUT (UNCHECKED)	SEAL	SEAL	SEAL	PORT 1	PORT 1	PORT 1	PORT 1
4	OUTSIDE AIR (RECIRC)	VENT	PORT 5,7,8,9	VENT	VENT	VENT	VENT	VENT
5	WATER CONTROL VALVE	VENT	PORT 4,7,8,9	VENT	VENT	VENT	VENT	VENT
6	(NOT USED)							
7	UPPER MODE	VENT	PORT 4,5,8,9	PORT 8,9	VENT	PORT 8,9	VENT	VENT
8	LOWER MODE	VENT	PORT 4,5,7,9	PORT 7,9	PORT 9	PORT 7,9	VENT	VENT
9	INPUT (CHECKED)	SEAL	PORT 4,5,7,8	PORT 7,8	PORT 8	PORT 7,8	SEAL	PORT 2

1975-76 Bonneville, Catalina & Grand Ville; 1975-77 Grand Am, Grand LeMans, Grand Prix & LeMans A/C & Heater Vacuum Circuit (Less Auto Temp Control)

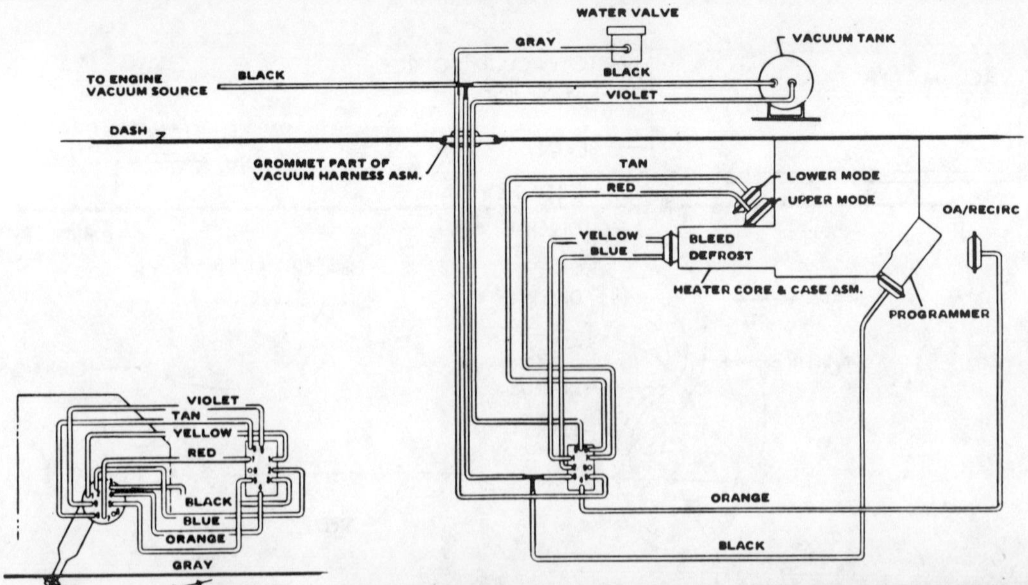

PORT NO.	PORT CONNECTION	SELECTOR LEVER POSITION						
		OFF	MAX	NORMAL	BI-LEVEL	VENT	HEATER	DEFROST
1	BLEED	VENT	VENT	VENT	PORT 3	PORT 3	PORT 3	PORT 3
2	DEFROST	VENT	VENT	VENT	VENT	VENT	VENT	PORT 9
3	INPUT (UNCHECKED)	SEAL	SEAL	SEAL	PORT 1	PORT 1	PORT 1	PORT 1
4	OUTSIDE AIR (RECIRC)	VENT	PORT 5,7,8,9	VENT	VENT	VENT	VENT	VENT
5	WATER CONT. VALVE	VENT	PORT 4,7,8,9	VENT	VENT	VENT	VENT	VENT
6	(NOT USED							
7	UPPER MODE	VENT	PORT 4,5,8,9	PORT 8,9	VENT	PORT 8,9	VENT	VENT
8	LOWER MODE	VENT	PORT 4,5,7,9	PORT 7,9	PORT 9	PORT 7,9	VENT	VENT
9	INPUT (CHECKED)	SEAL	PORT 4,5,7,8	PORT 7,8	PORT 8	PORT 7,8	SEAL	PORT 2

1975-76 Bonneville, Catalina & Grand Ville; 1975-77 Grand Am, Grand LeMans, Grand Prix & LeMans A/C & Heater Vacuum Circuit (With Auto Temp Control)

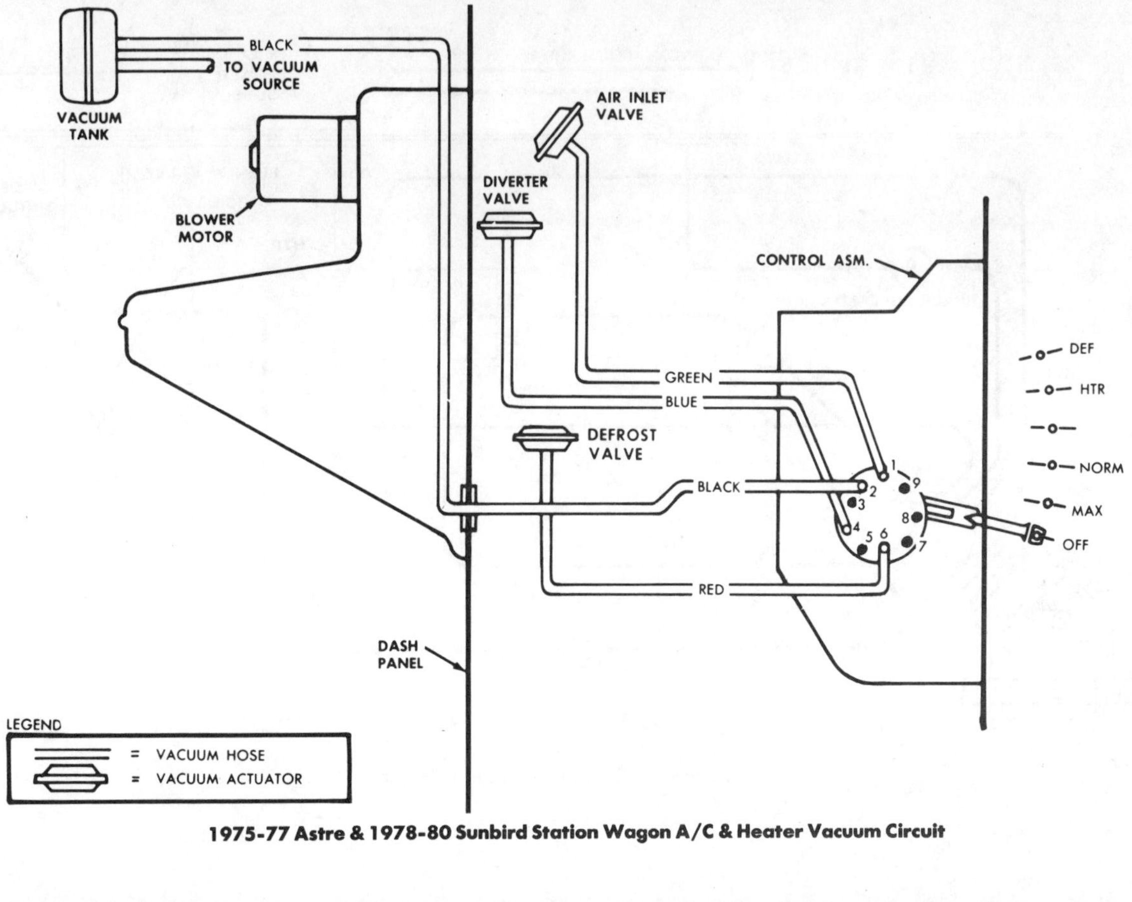

1975-77 Astre & 1978-80 Sunbird Station Wagon A/C & Heater Vacuum Circuit

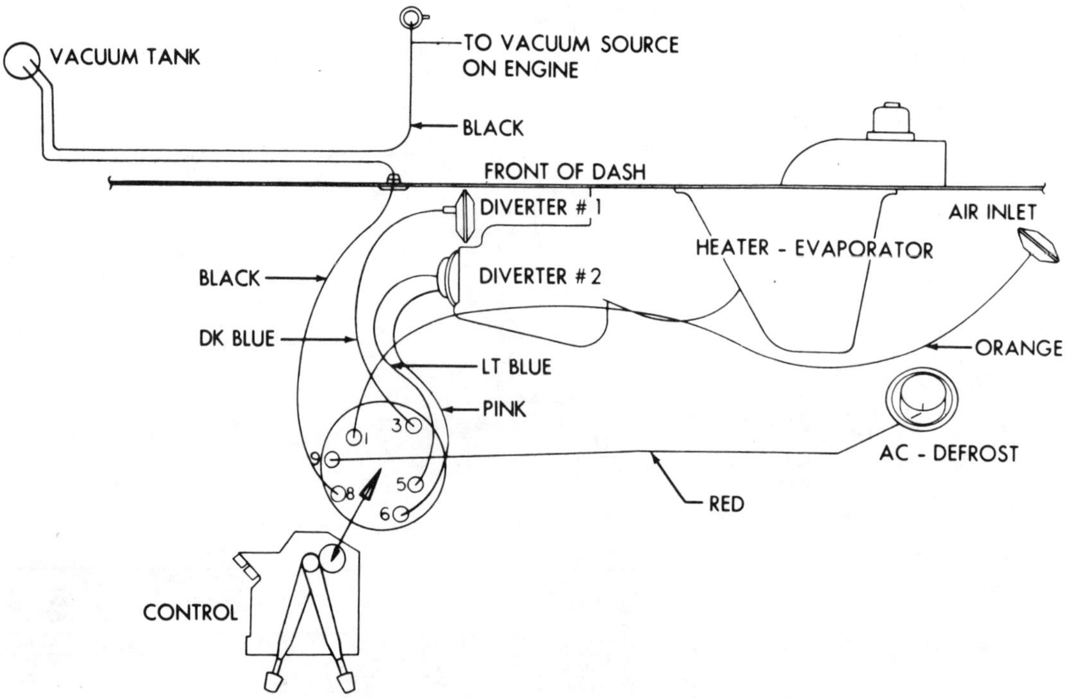

1976 Sunbird A/C & Heater Vacuum Circuit

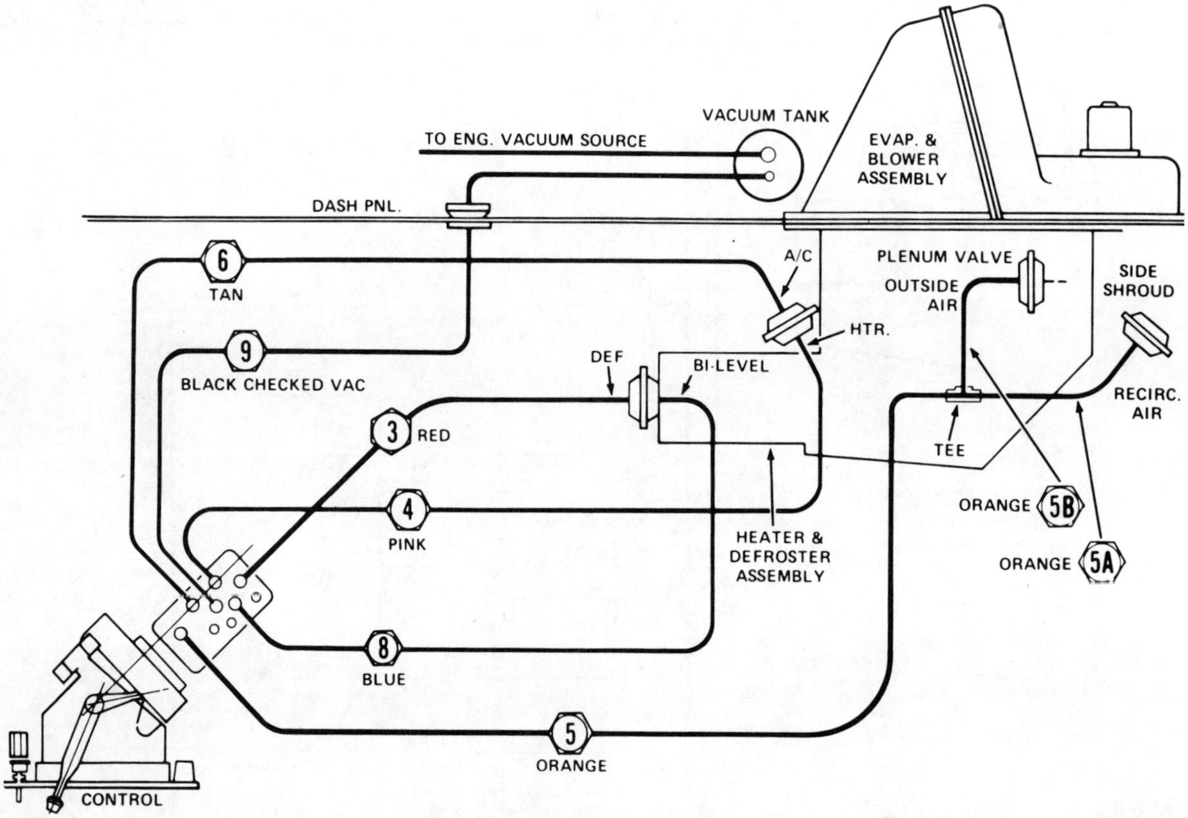

1976 Ventura A/C & Heater Vacuum Circuit

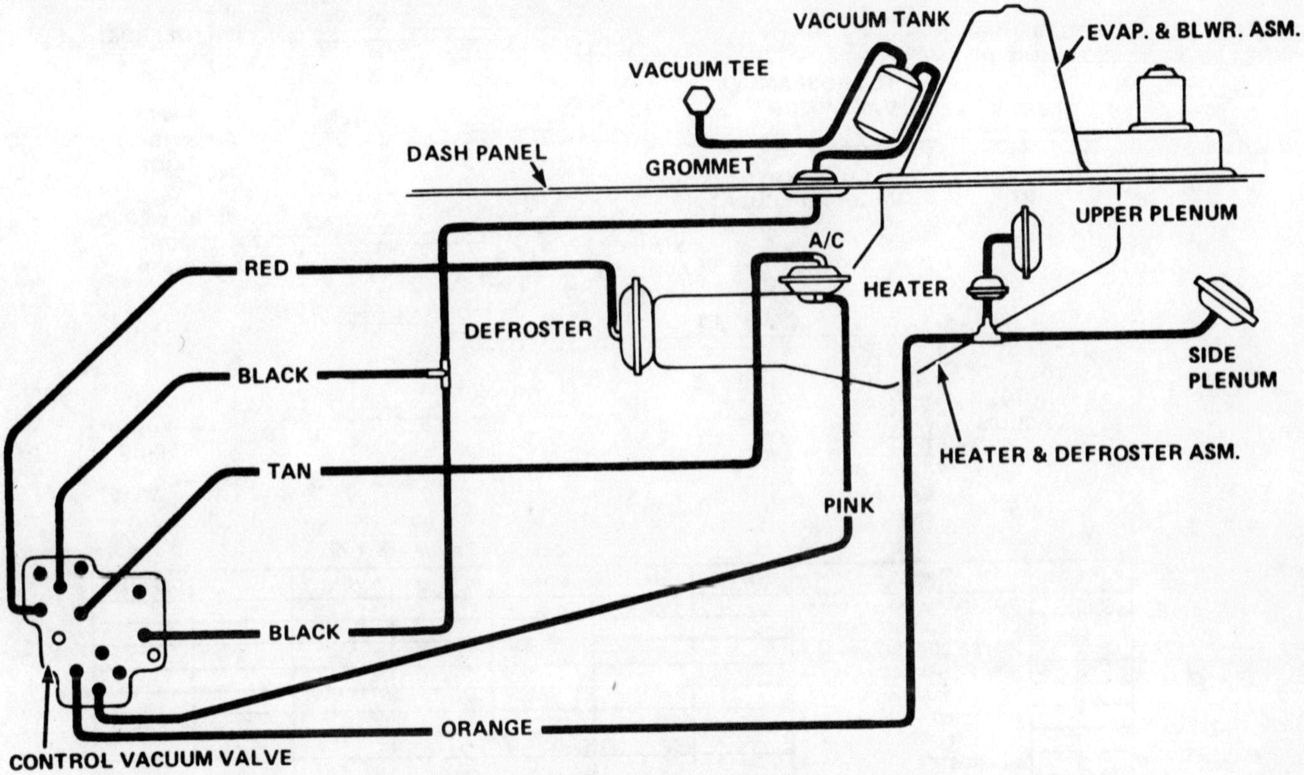

1977 Ventura & 1977-79 Phoenix A/C & Heater Vacuum Circuit

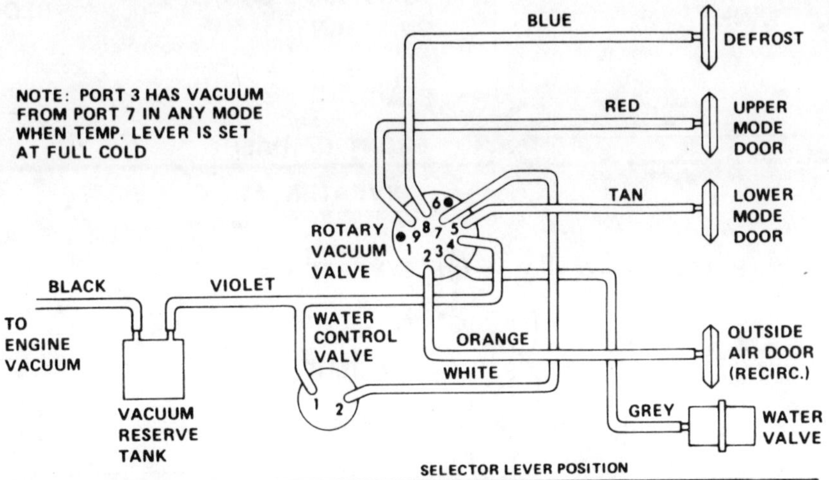

PORT NO.	PORT	SELECTOR LEVER POSITION						
		OFF	MAX	NORM	BI LEVEL	VENT	HEAT	DEFROST
1	NOT USED							
2	OUTSIDE AIR (RECIRC)							
3	WATER CONTROL VALVE							
4	INPUT-CHECKED							
5	LOWER MODE DOOR							
6	NOT USED							
7	INPUT WATER CONTROL							
8	DEFROSTER							
9	UPPER MODE DOOR							

1977-78 Bonneville, Catalina, Grand Safari & Safari A/C & Heater Vacuum Circuit (Less Auto Temp Control)

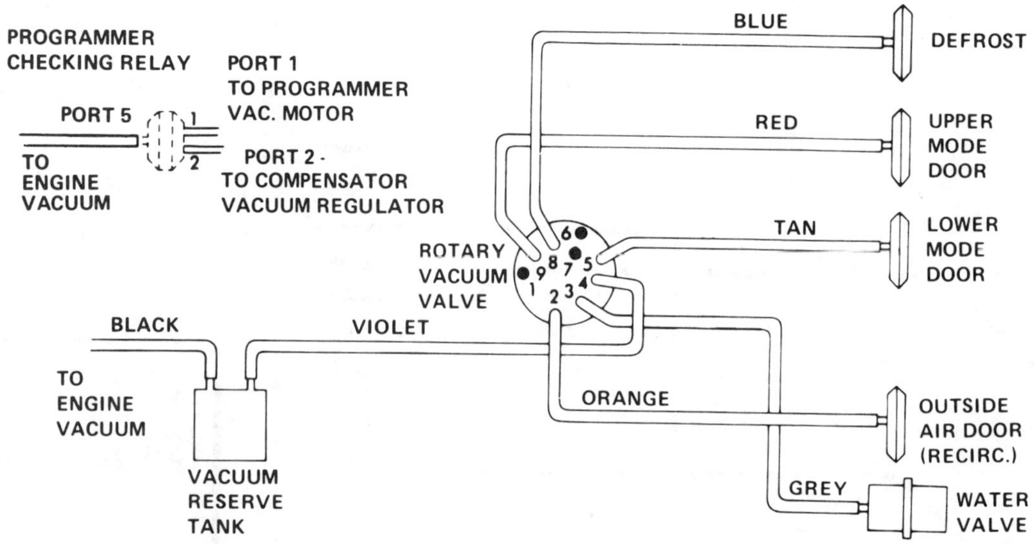

PORT NO.	PORT	SELECTOR LEVER POSITION OFF	MAX	NORM	BI LEVEL	VENT	HEAT	DEFROST
1	NOT USED							
2	OUTSIDE AIR (RECIRC)							
3	WATER CONTROL-VALVE							
4	INPUT-CHECKED							
5	LOWER MODE DOOR							
6	NOT USED							
7	NOT USED							
8	DEFROSTER							
9	UPPER MODE DOOR							

1977-78 Bonneville & Catalina A/C & Heater Vacuum Circuit (With Auto Temp Control)

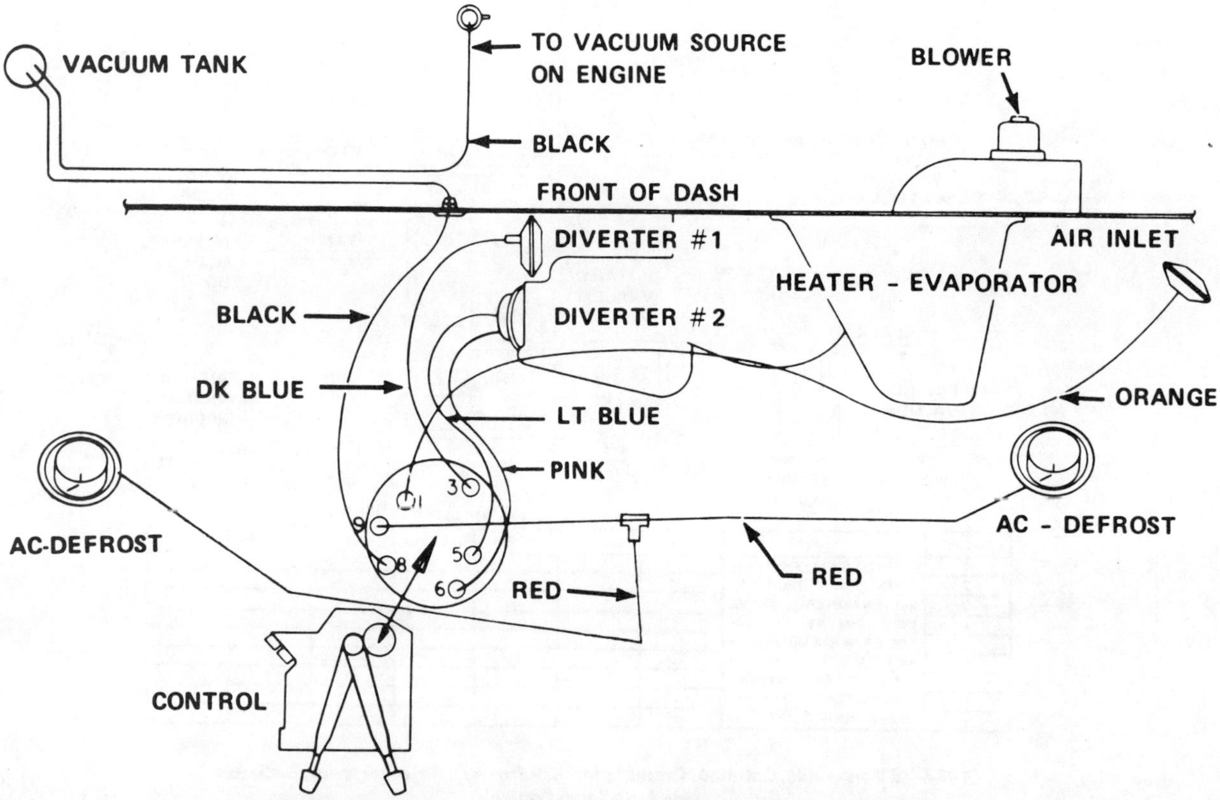

1977-80 Sunbird Except Station Wagon A/C & Heater Vacuum Circuit

PORT	FUNCTION		
	A/C	BI-LEVEL	HEAT
1	VAC	VENT	VENT
6	VENT	VENT	VAC

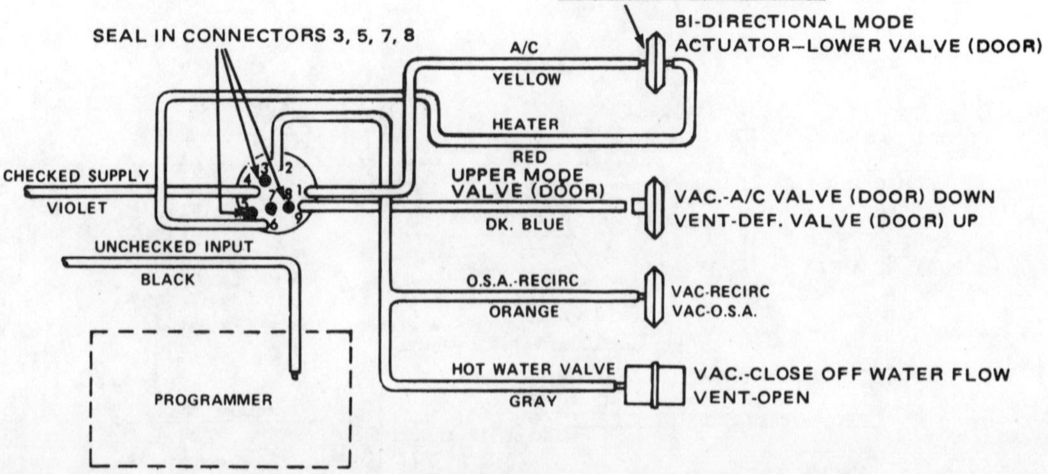

SELECT LEVER VALVE

PORT NO.	PORT CONNECTION	LEVER POSITION						
		OFF	MAX.	NORMAL	BI-LEVEL	VENT	HEATER	DEFROST
1	A/C	VENT	CONN. 4	CONN. 4	VENT	CONN. 4	VENT	CONN. 4
2	O.S.A.—RECIRC.	VENT	CONN. 4	VENT	VENT	VENT	VENT	VENT
3	NOT USED—SEAL IN CONNECTOR							
4	INPUT	CONN. 6	CONN. 1,2,3,9	CONN. 1,9	CONN. 9	CONN. 1,9	CONN. 6	CONN. 1
5	NOT USED—SEAL IN CONNECTOR							
6	HEATER	CONN. 4	VENT	VENT	VENT	VENT	CONN. 4	VENT
7	NOT USED—SEAL IN CONNECTOR							
8	NOT USED—SEAL IN CONNECTOR							
9	DEFROST	VENT	CONN. 4	CONN. 4	CONN. 4	CONN. 4	VENT	VENT

1978 Grand Am, Grand LeMans, Grand Prix & LeMans
A/C & Heater Vacuum Circuit (With Auto Temp Control)

PORT	FUNCTION		
	A/C	BI-LEVEL	HEAT
1	VAC	VENT	VENT
6	VENT	VENT	VAC

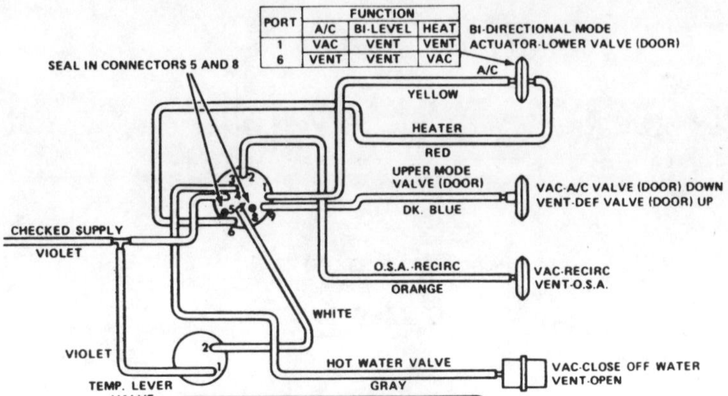

TEMPERATURE LEVER VALVE

PORT NO.	PORT CONNECTION	COLD	HOT
1	VAC INPUT	CONN. 2	SEAL
2	W/V SIGNAL	CONN. 1	VENT

SELECT LEVER VALVE

PORT NO.	PORT CONNECTION	OFF	LEVER POSITION					
			MAX.	NORMAL	BI-LEVEL	VENT	HEATER	DEFROST
1	A/C	VENT	CONN. 4	CONN. 4	VENT	CONN. 4	VENT	CONN. 4
2	O.S.A.—RECIRC.	VENT	CONN. 4	VENT	VENT	VENT	VENT	VENT
3	WATER VALVE	CONN. 7	CONN. 4	CONN. 7	CONN. 7	CONN. 7	CONN. 7	CONN. 7
4	INPUT	CONN. 6	CONN. 1,2,3,9	CONN. 1,9	CONN. 9	CONN. 1,9	CONN. 6	CONN. 1
5	NOT USED—SEAL IN CONNECTOR							
6	HEATER	CONN. 4	VENT	VENT	VENT	VENT	CONN. 4	VENT
7	W/V SIGNAL	CONN. 3	SEAL	CONN. 3	CONN. 3	CONN. 3	CONN. 3	CONN. 3
8	NOT USED—SEAL IN CONNECTOR							
9	DEFROST	VENT	CONN. 4	CONN. 4	CONN. 4	CONN. 4	VENT	VENT

1978-80 Grand Am, Grand LeMans, Grand Prix & LeMans A/C & Heater
Vacuum Circuit (Less Auto Temp Control)

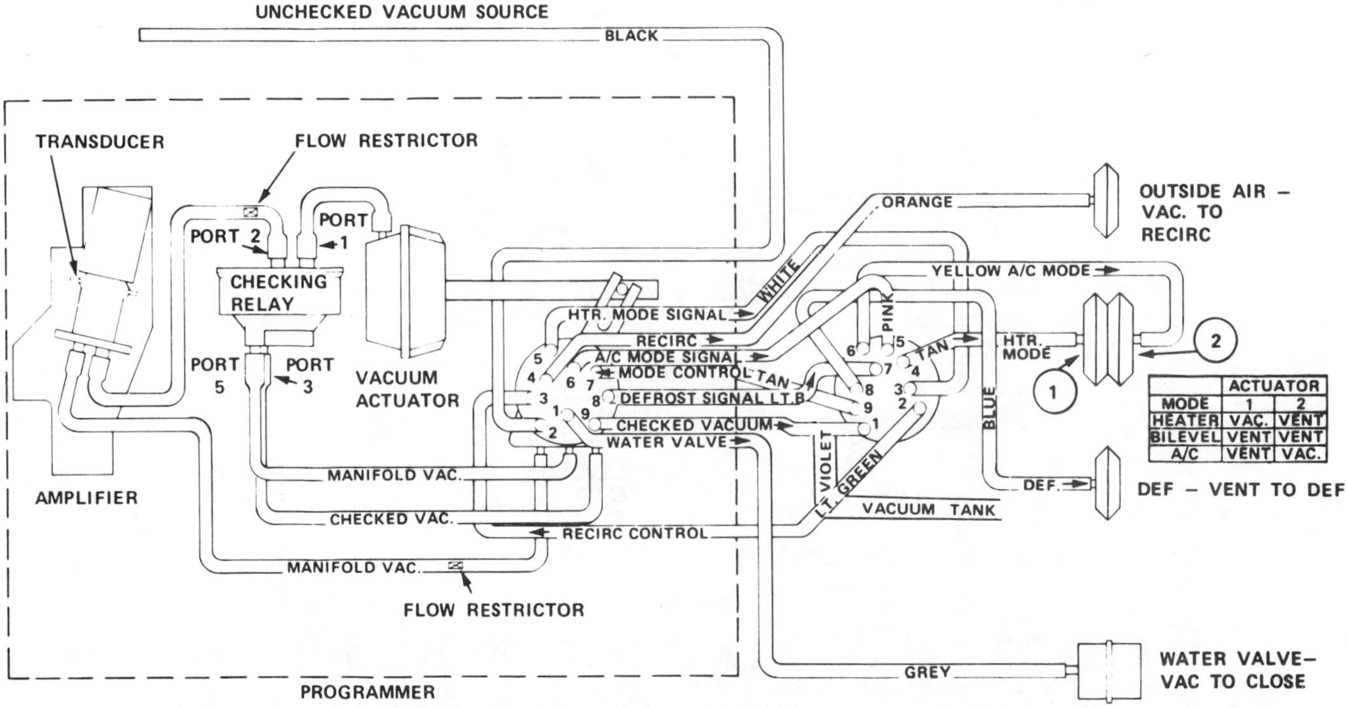

1979-80 Grand Am, Grand LeMans, Grand Prix & LeMans A/C & Heater Vacuum Circuit (With Auto Temp Control)

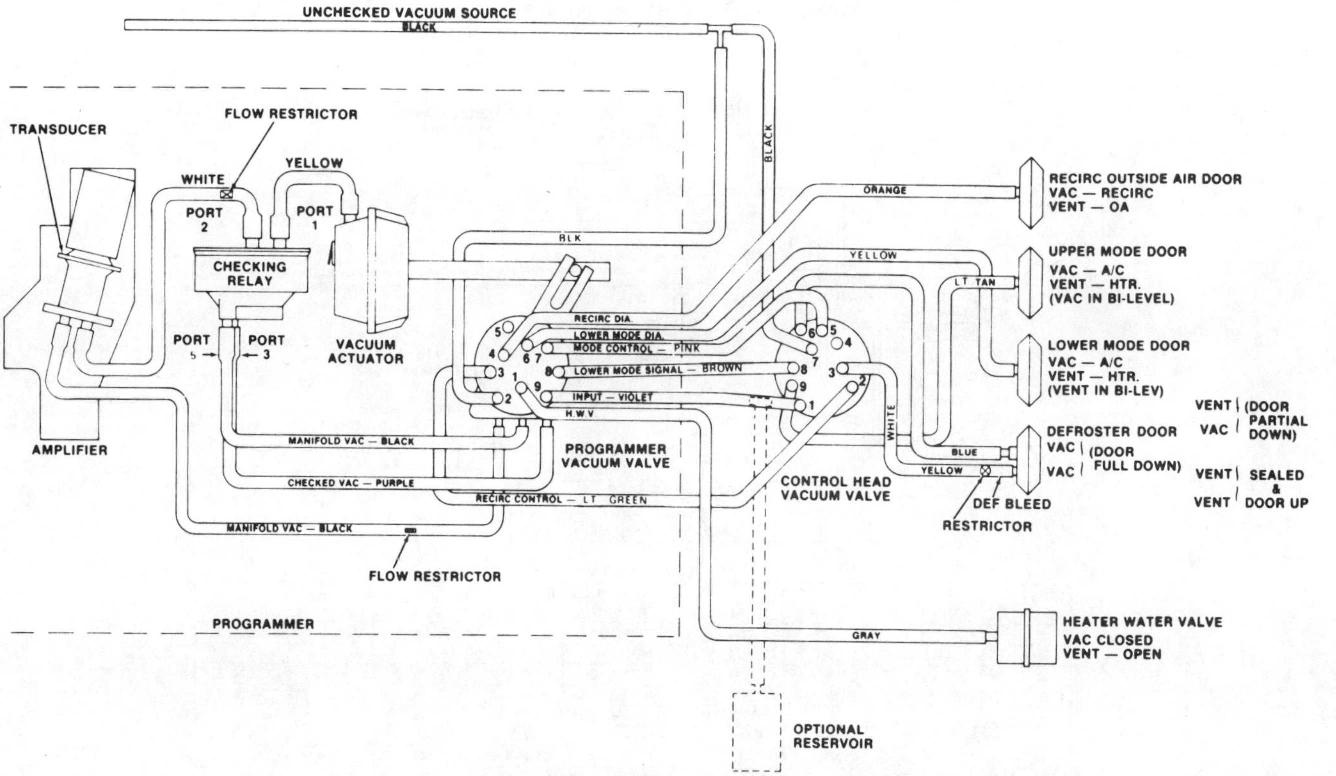

1979-80 Bonneville & Catalina A/C Heater Wiring Circuit (With Auto Temp Control)

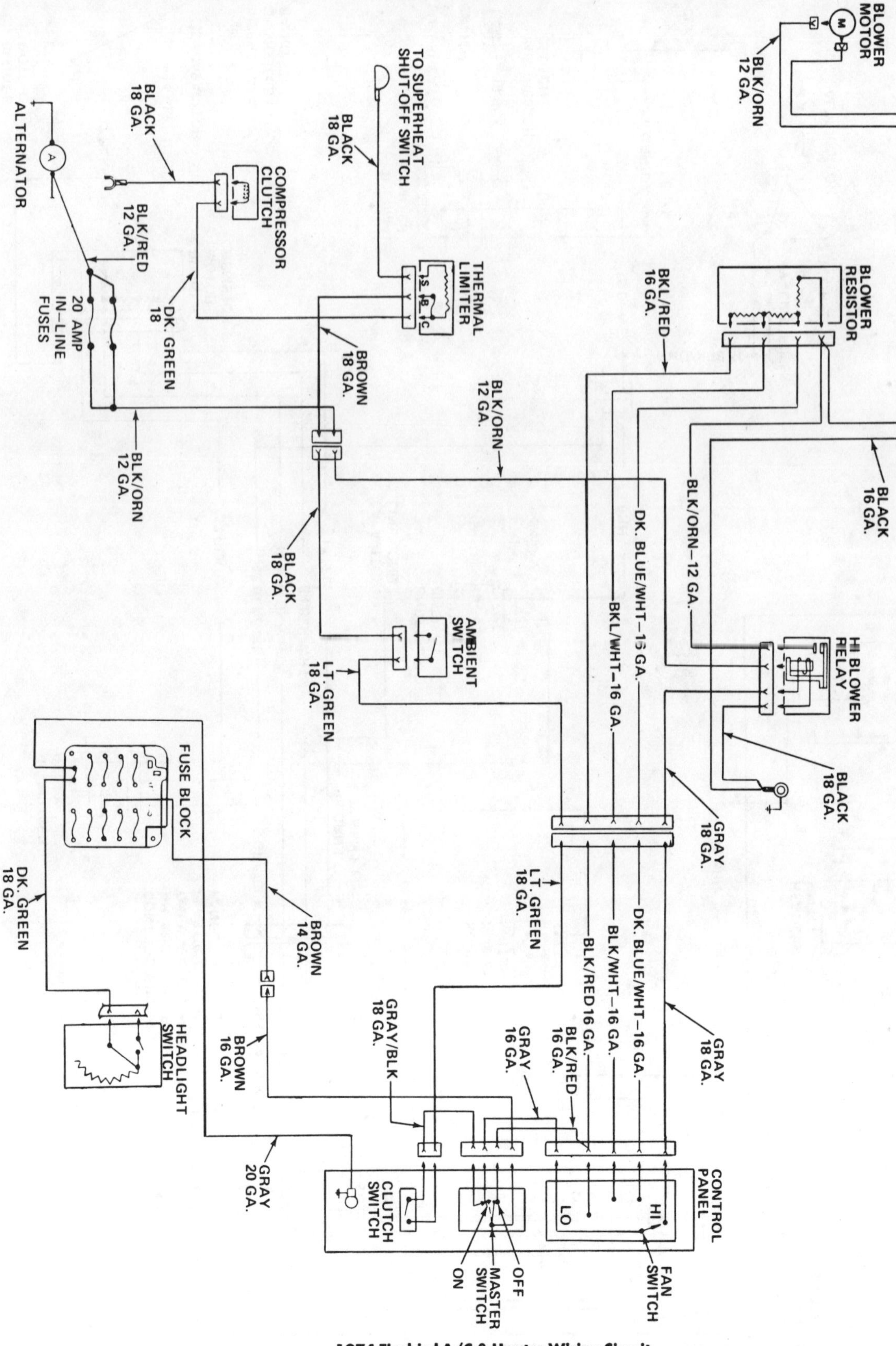

1974 Firebird A/C & Heater Wiring Circuit

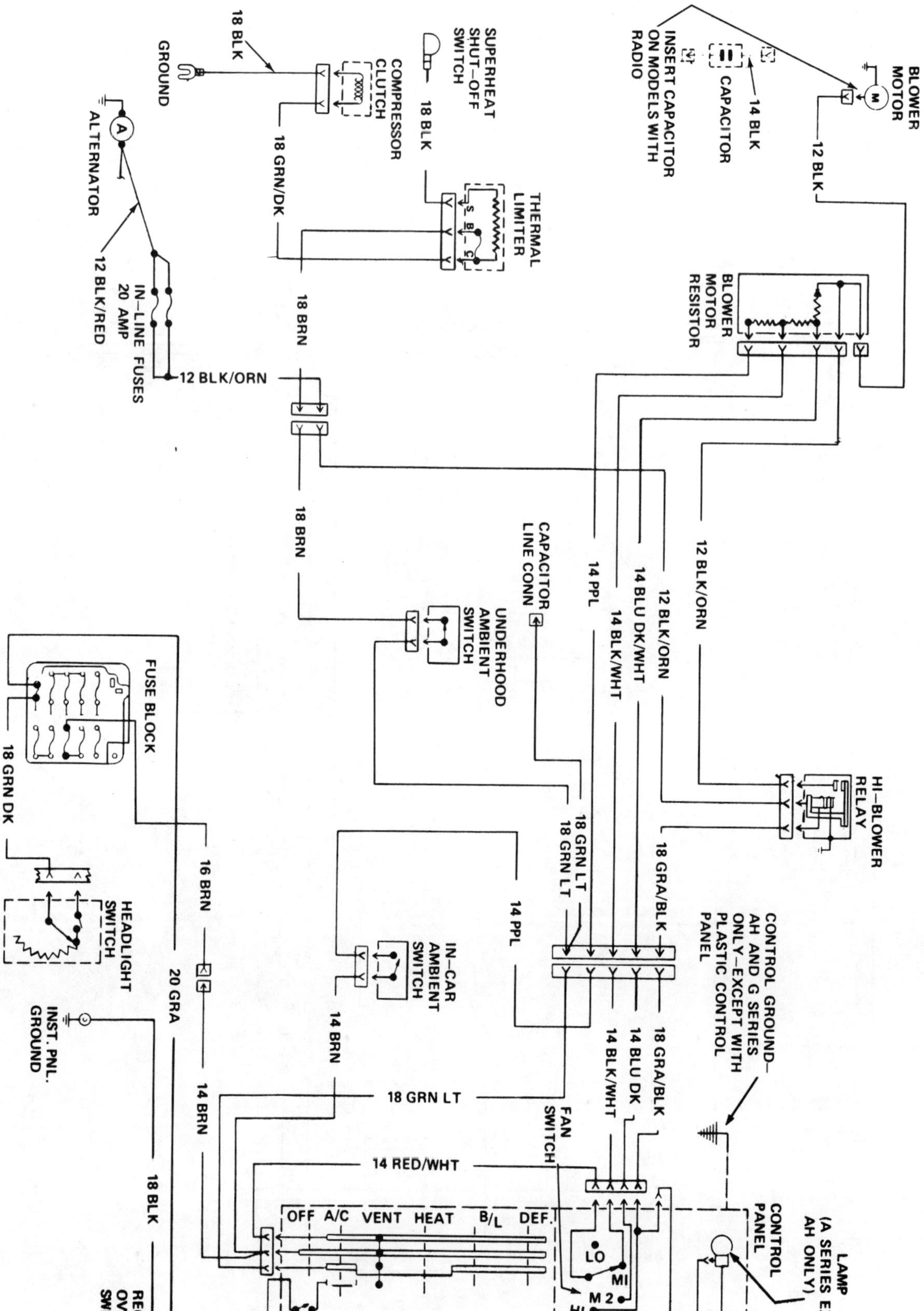

1974 Grand Am, Grand Prix & LeMans A/C & Heater Wiring Circuit (Less Auto Temp Control)

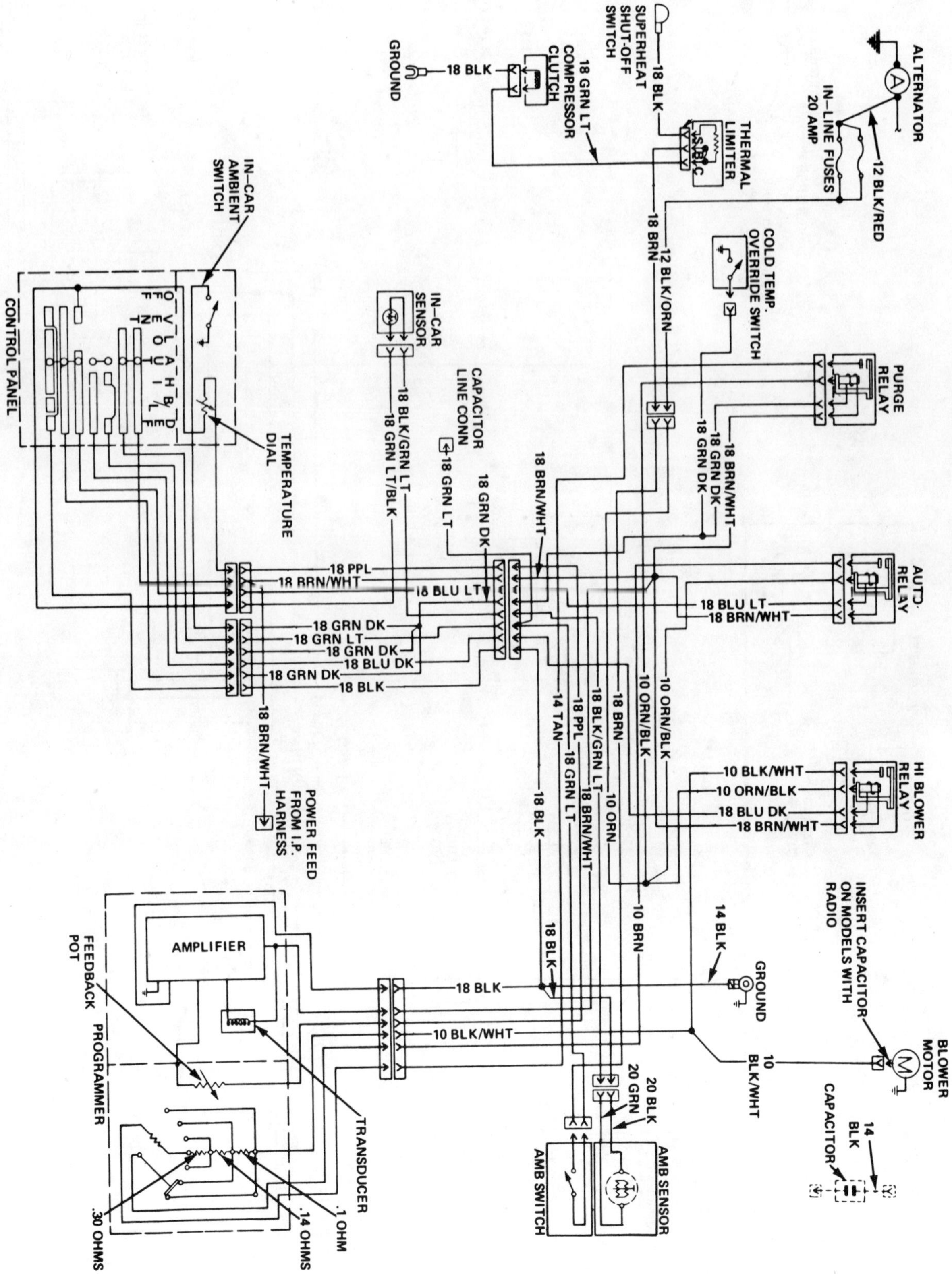

1974 Grand Am & Grand Prix A/C & Heater Wiring Circuit (With Auto Temp Control)

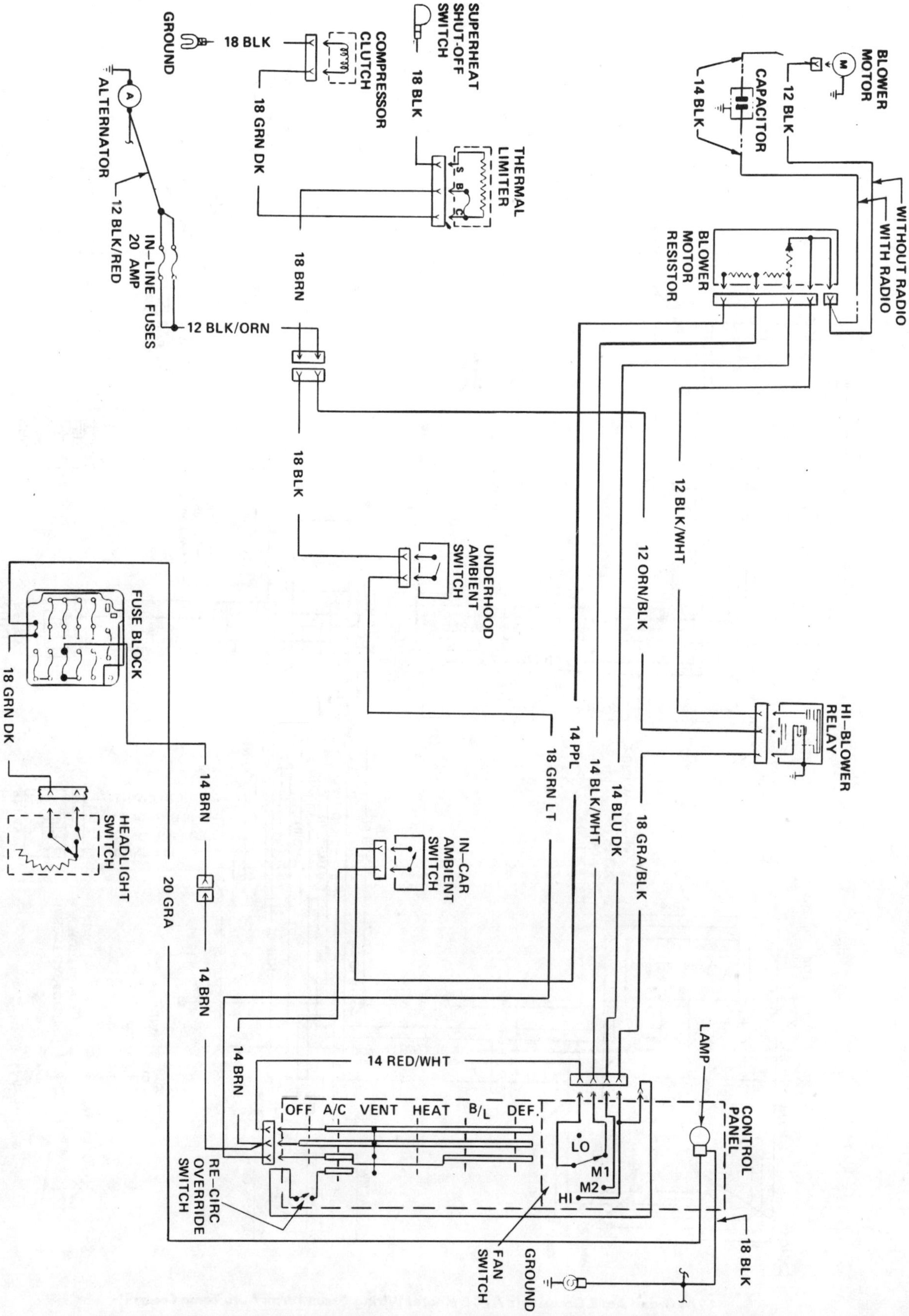

1974 Bonneville, Catalina & Grand Ville A/C & Heater Wiring Circuit (Less Auto Temp Control)

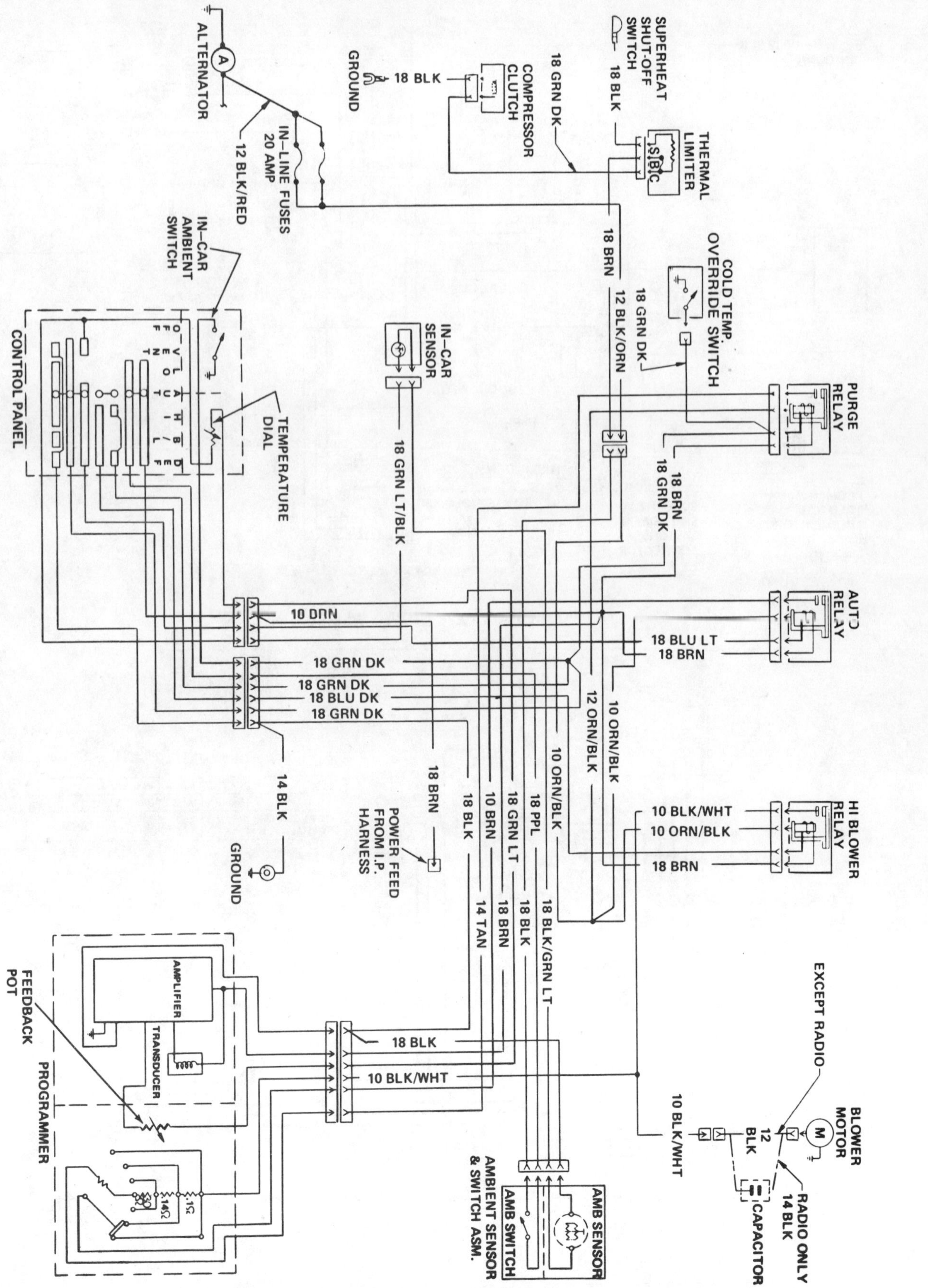

1974 Bonneville, Catalina & Grand Ville A/C & Heater Wiring Circuit (With Auto Temp Control)

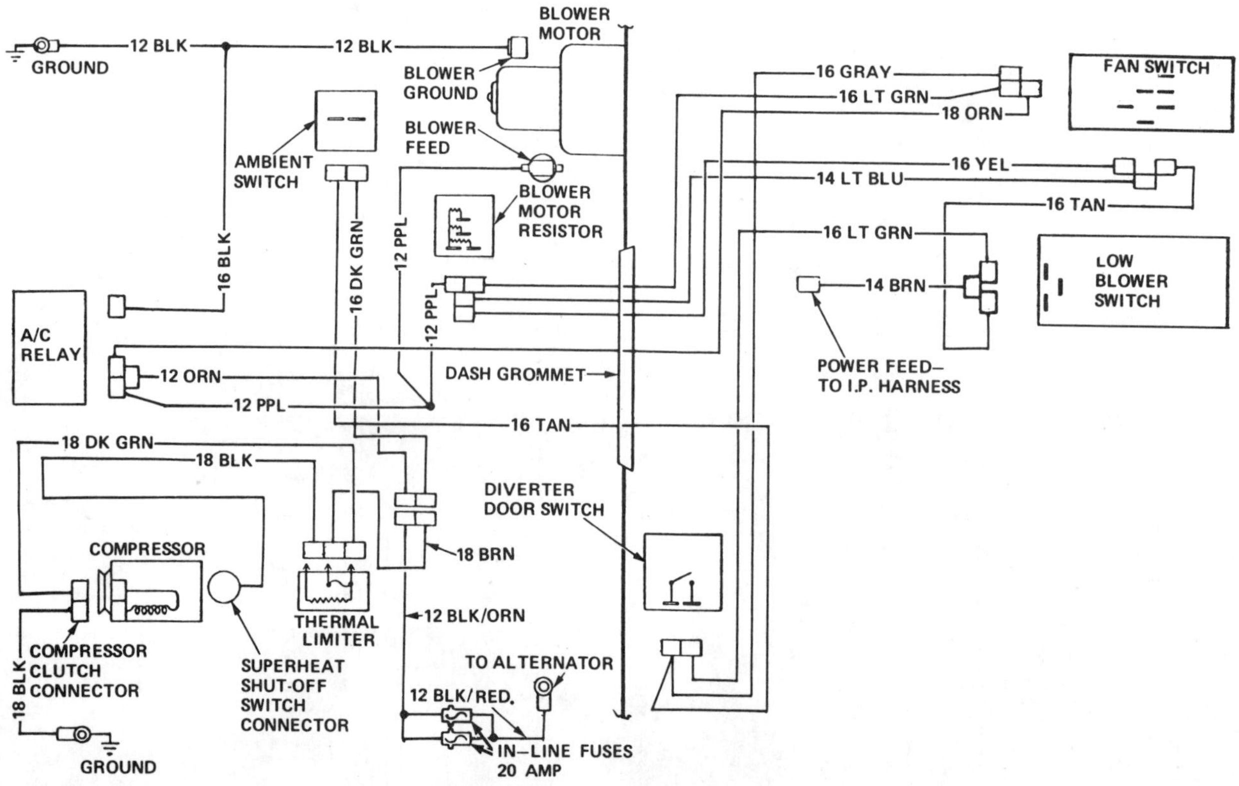

1974 Ventura A/C & Heater Wiring Circuit

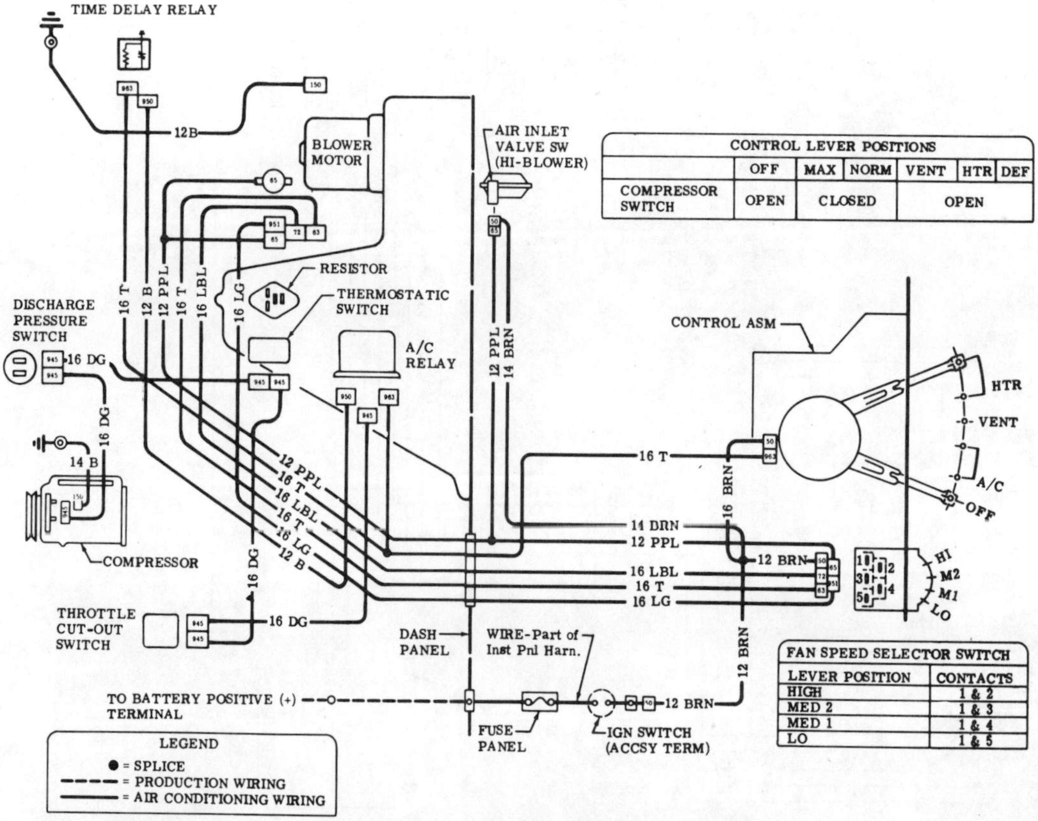

CONTROL LEVER POSITIONS	OFF	MAX	NORM	VENT	HTR	DEF
COMPRESSOR SWITCH	OPEN	CLOSED		OPEN		

FAN SPEED SELECTOR SWITCH	
LEVER POSITION	CONTACTS
HIGH	1 & 2
MED 2	1 & 3
MED 1	1 & 4
LO	1 & 5

1975 Astre A/C & Heater Wiring Circuit

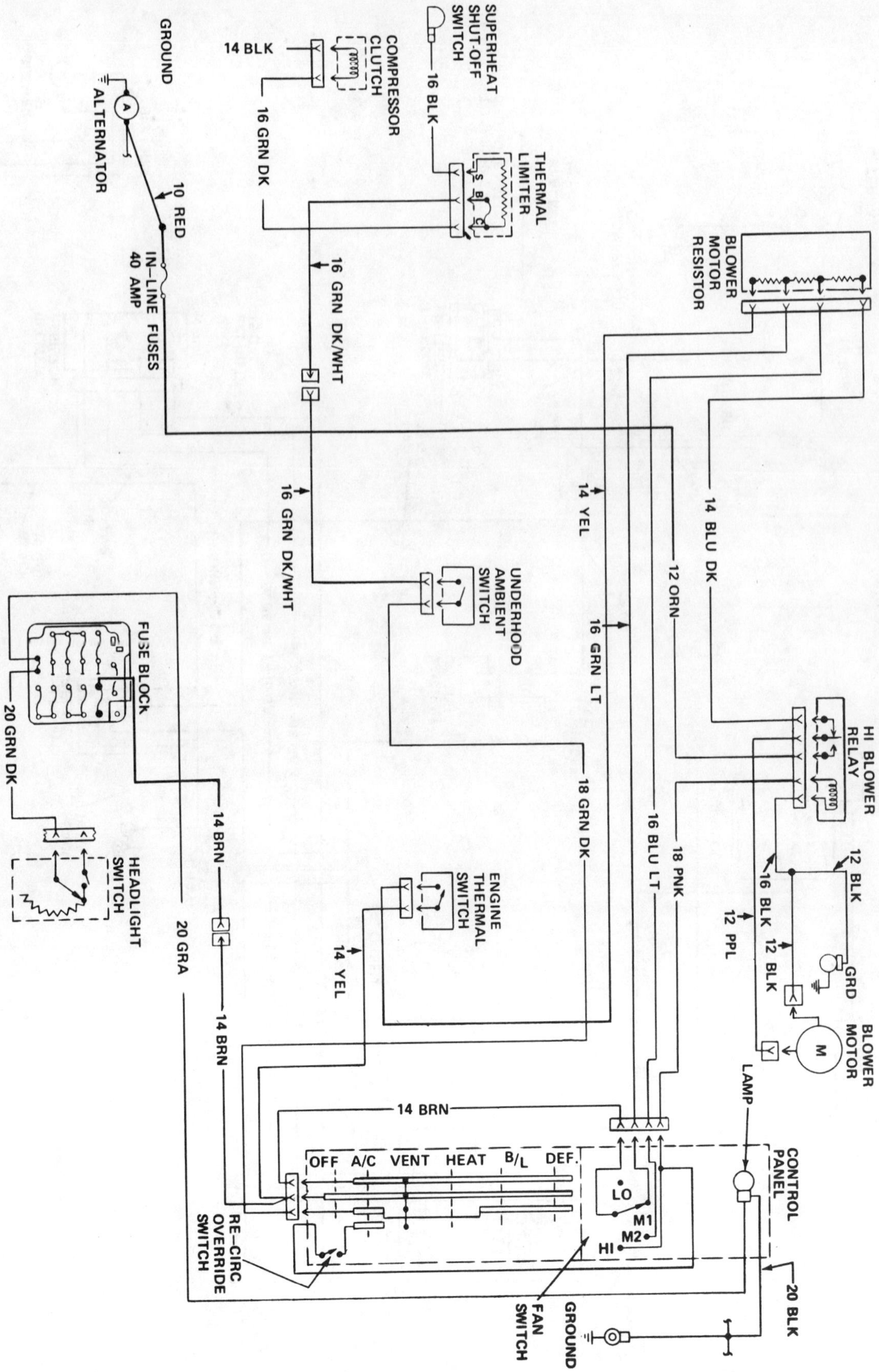

1975 Ventura A/C & Heater Wiring Circuit

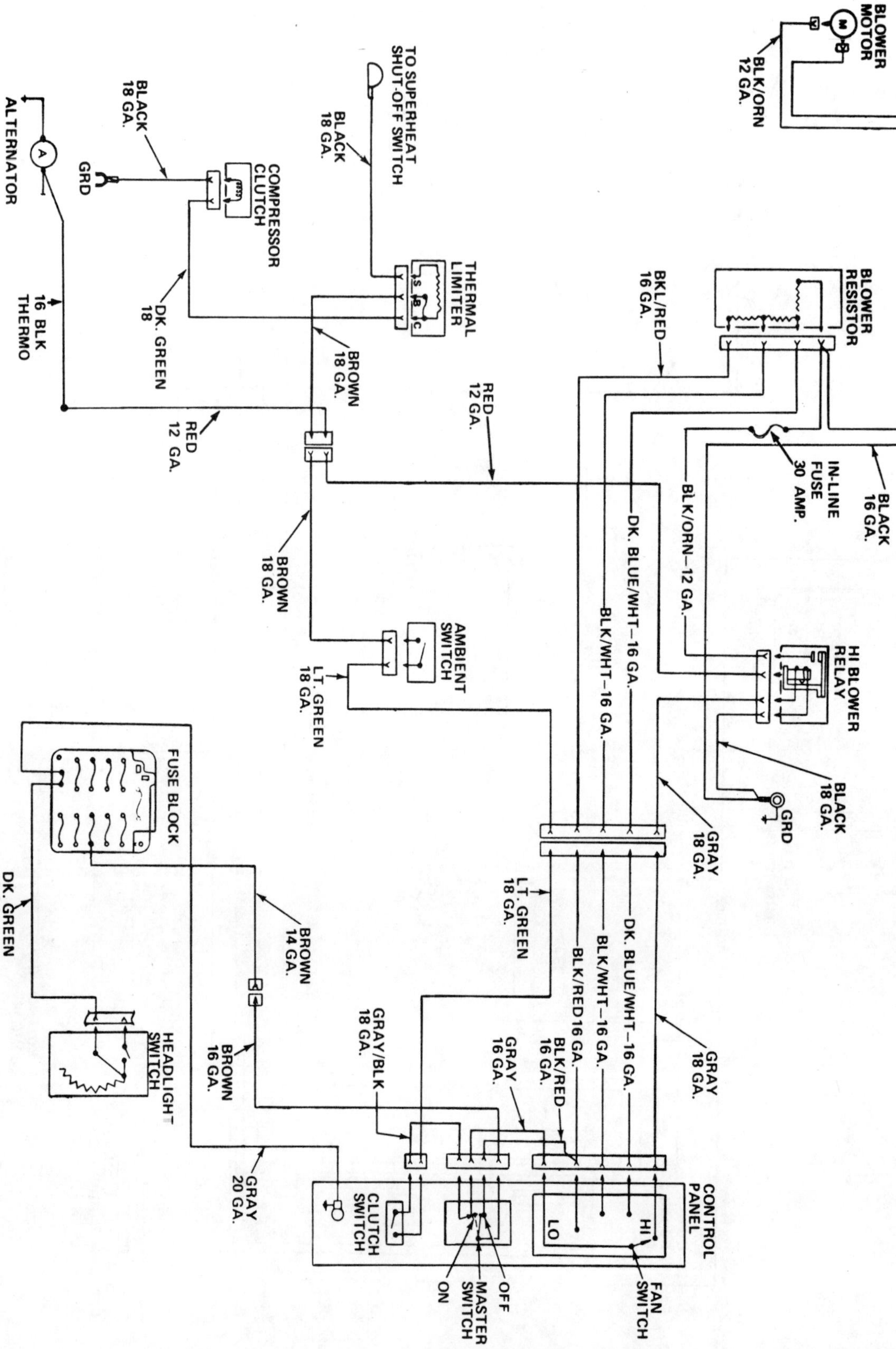

1975 Firebird A/C & Heater Wiring Circuit

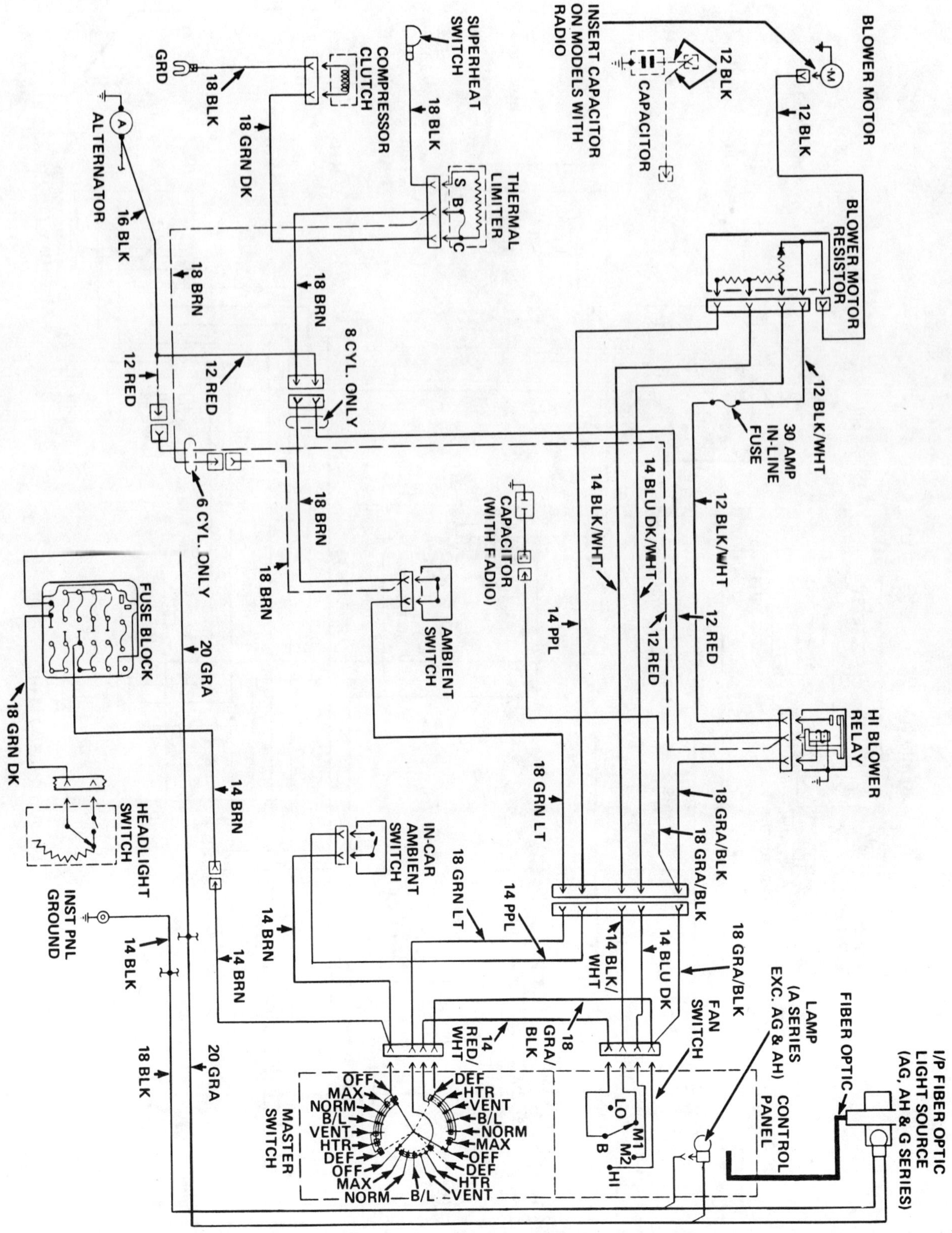

1975 Grand Am, Grand LeMans, Grand Prix & LeMans A/C & Heater Wiring Circuit (Less Auto Temp Control)

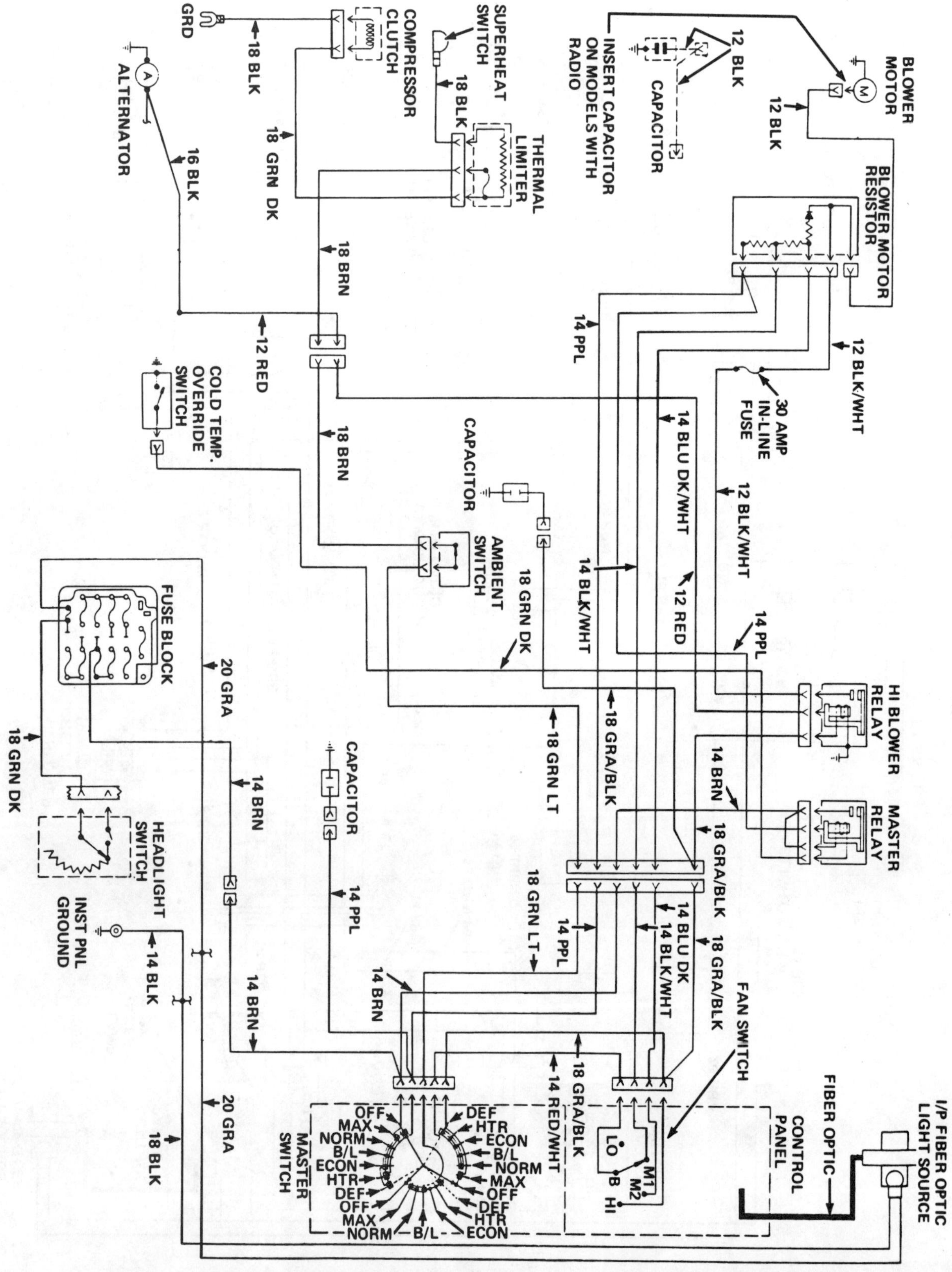

1975 Grand Am, Grand LeMans, Grand Prix & LeMans A/C & Heater Wiring Circuit (With Auto Temp Control)

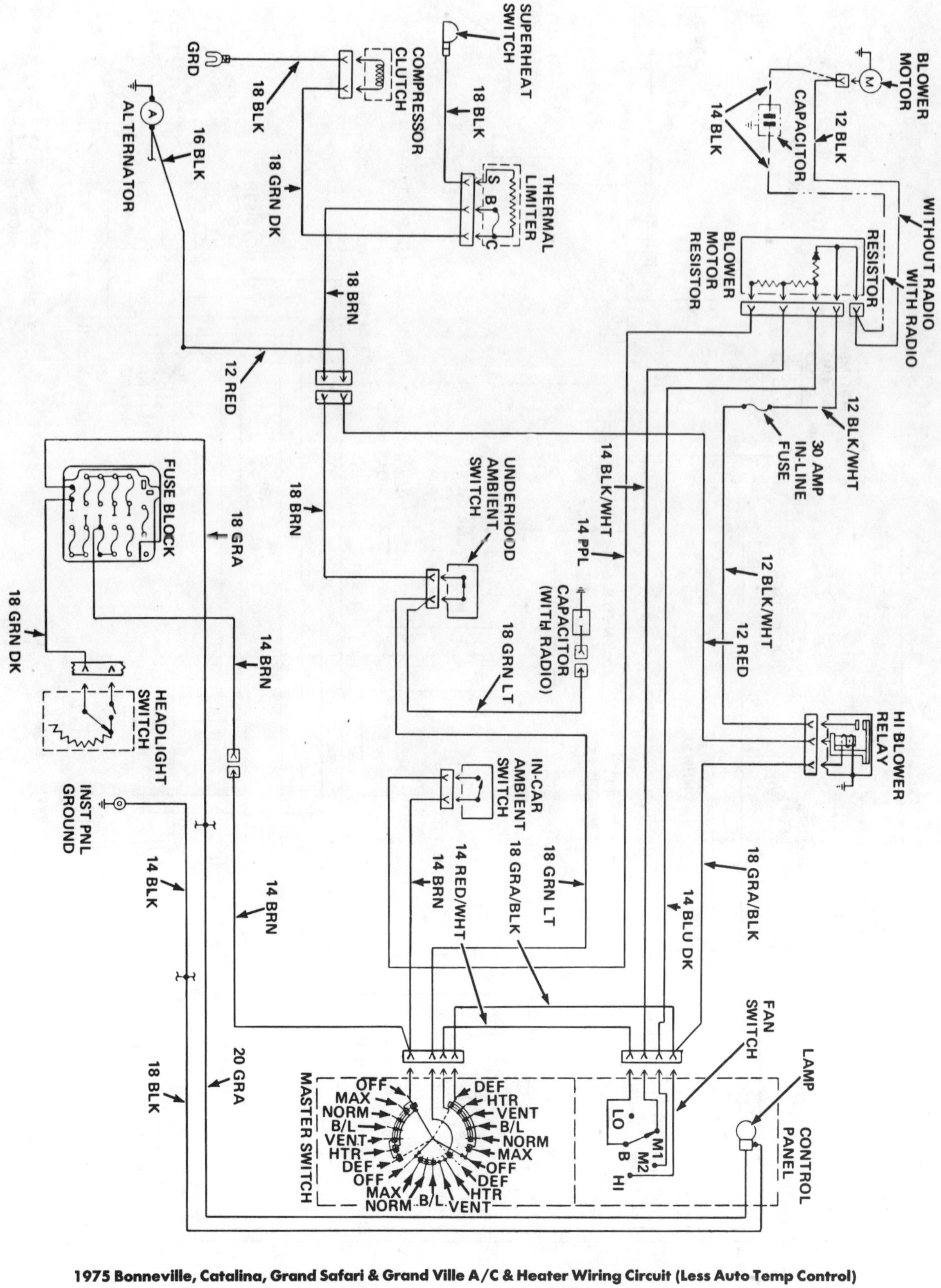

1975 Bonneville, Catalina, Grand Safari & Grand Ville A/C & Heater Wiring Circuit (Less Auto Temp Control)

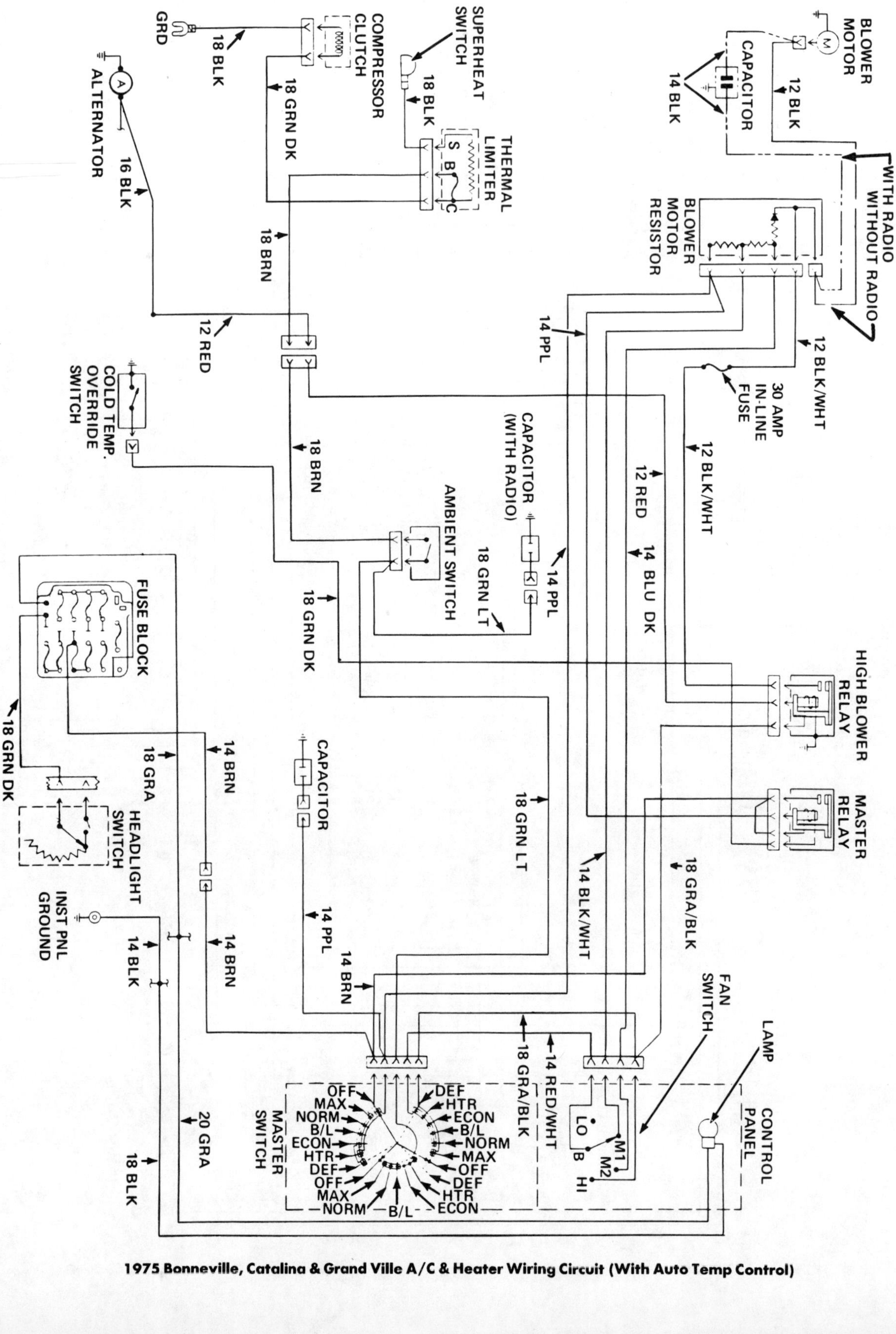

1975 Bonneville, Catalina & Grand Ville A/C & Heater Wiring Circuit (With Auto Temp Control)

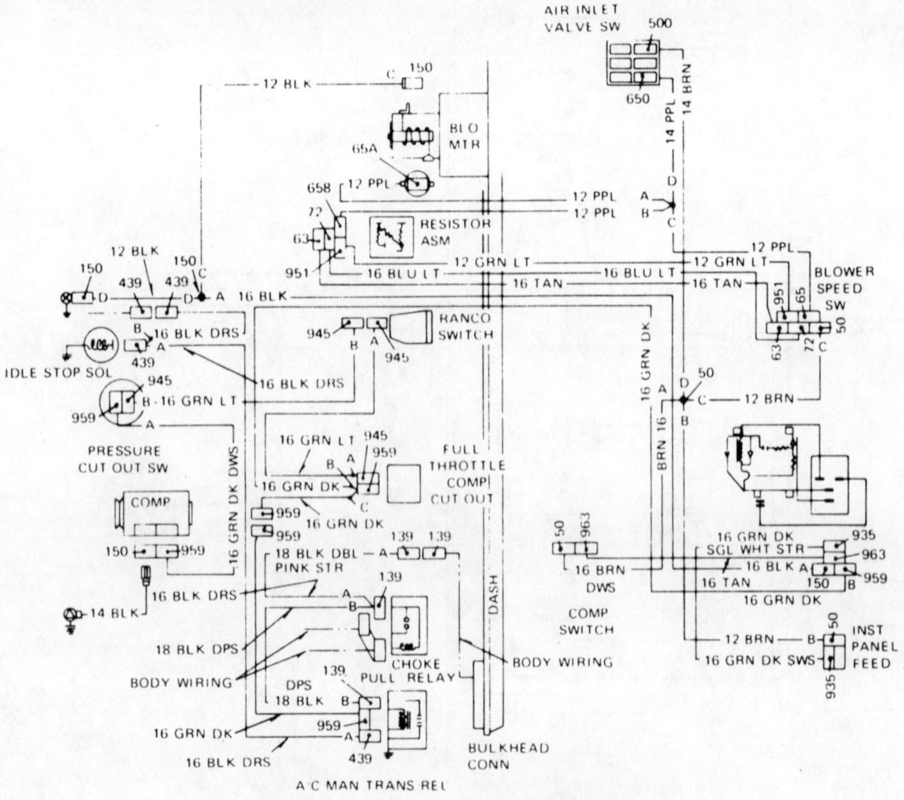

1976 Astre A/C & Heater Wiring Circuit

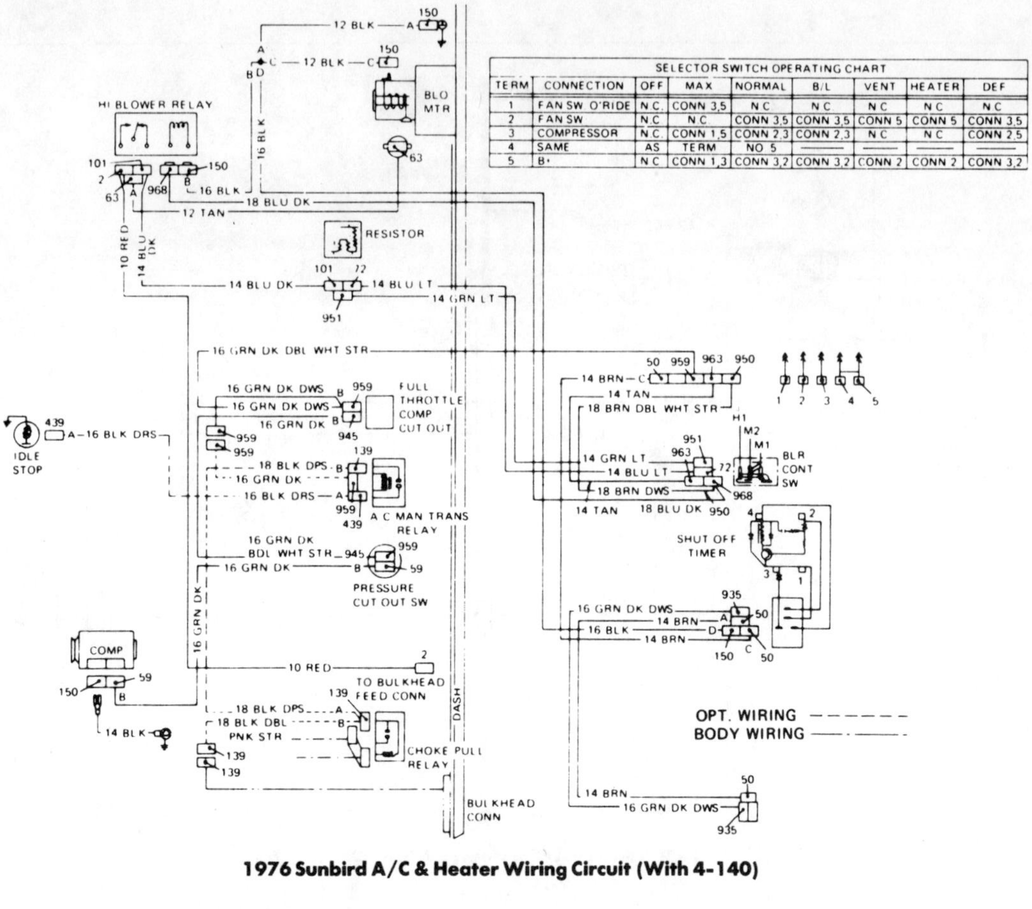

SELECTOR SWITCH OPERATING CHART								
TERM	CONNECTION	OFF	MAX	NORMAL	B/L	VENT	HEATER	DEF
1	FAN SW O'RIDE	N.C.	CONN 3,5	N C	N C	N C	N C	N C
2	FAN SW	N.C	N.C	CONN 3,5	CONN 3,5	CONN 5	CONN 5	CONN 3,5
3	COMPRESSOR	N.C.	CONN 1,5	CONN 2,3	CONN 2,3	N C	N C	CONN 2,5
4	SAME	AS	TERM	NO 5	———	———	———	———
5	B+	N C	CONN 1,3	CONN 3,2	CONN 3,2	CONN 2	CONN 2	CONN 3,2

1976 Sunbird A/C & Heater Wiring Circuit (With 4-140)

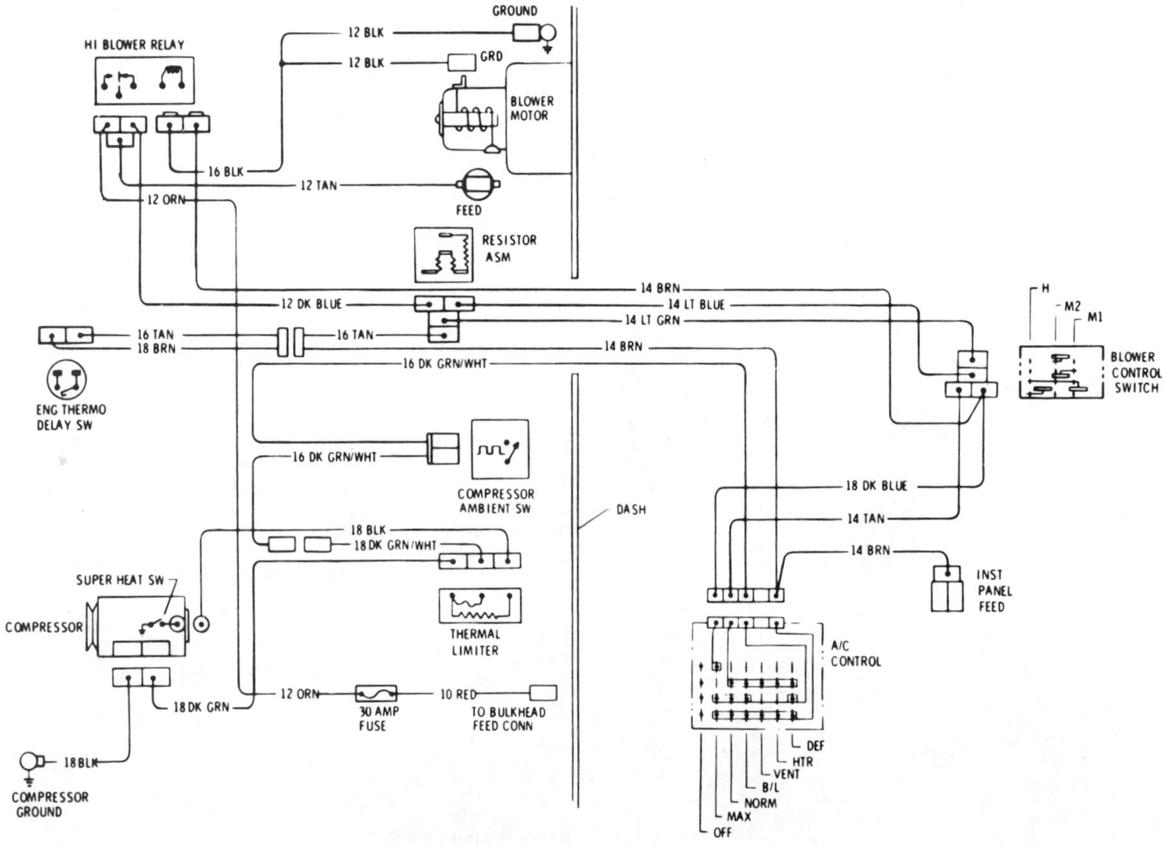

1976 Sunbird A/C & Heater Wiring Circuit (With V6-231)

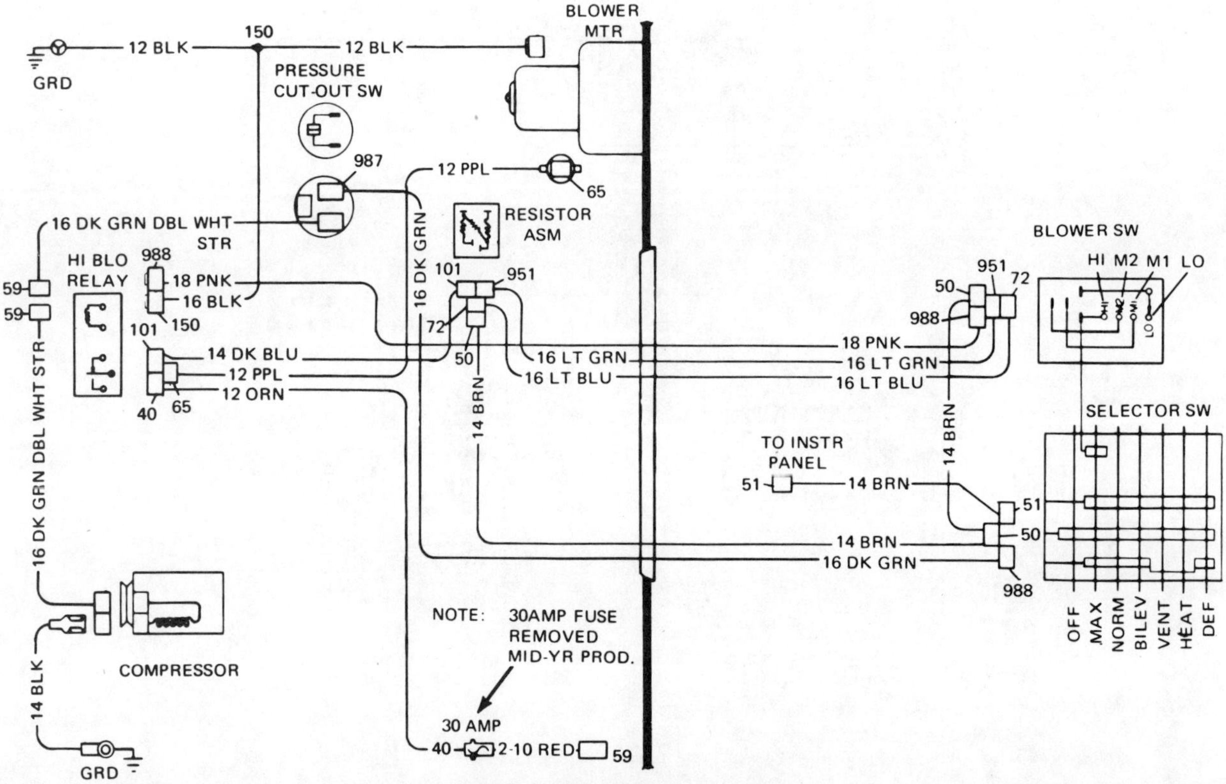

1976 Ventura A/C & Heater Wiring Circuit

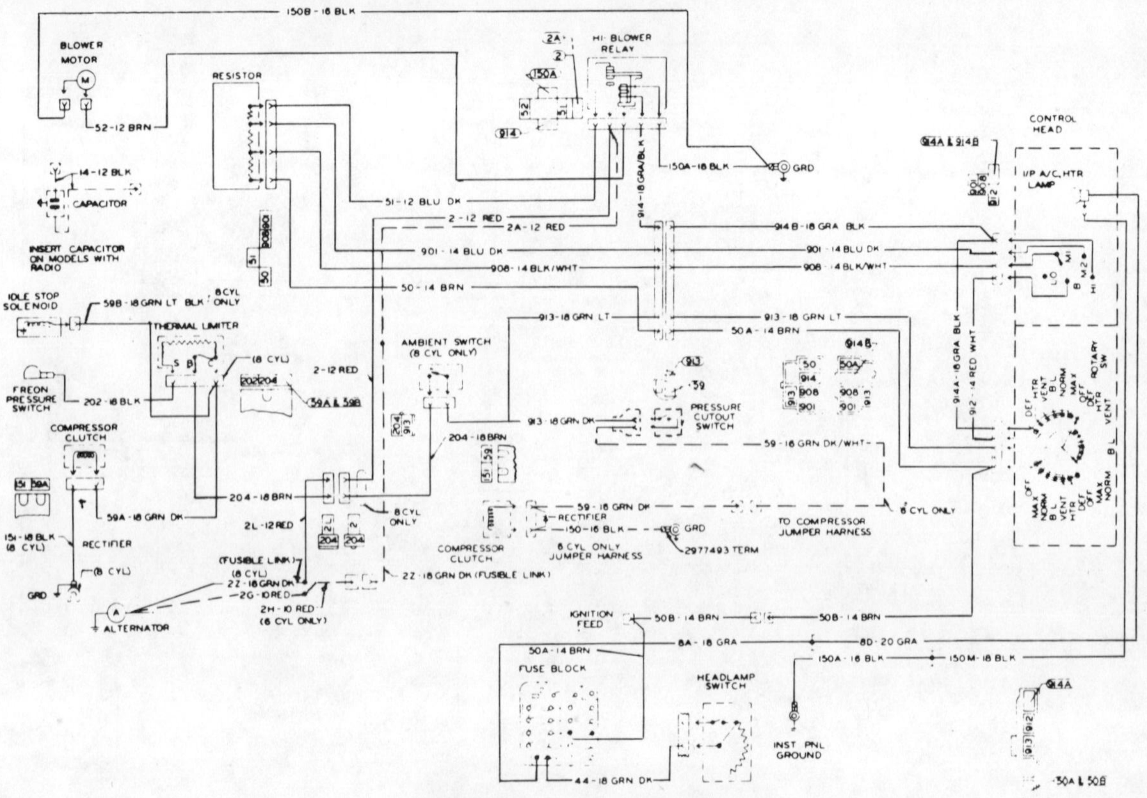

1976 Firebird A/C & Heater Wiring Circuit

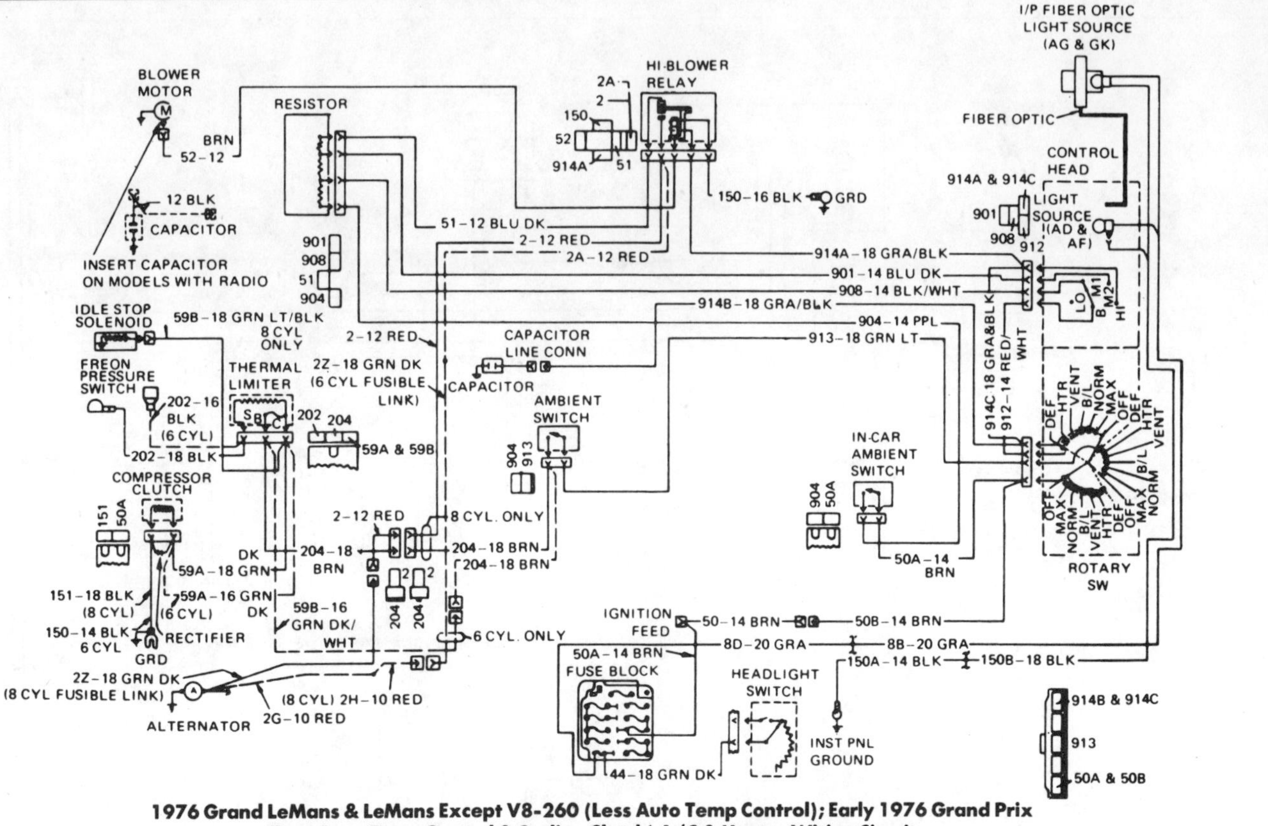

1976 Grand LeMans & LeMans Except V8-260 (Less Auto Temp Control); Early 1976 Grand Prix (Less Auto Temp Control & Cycling Clutch) A/C & Heater Wiring Circuit

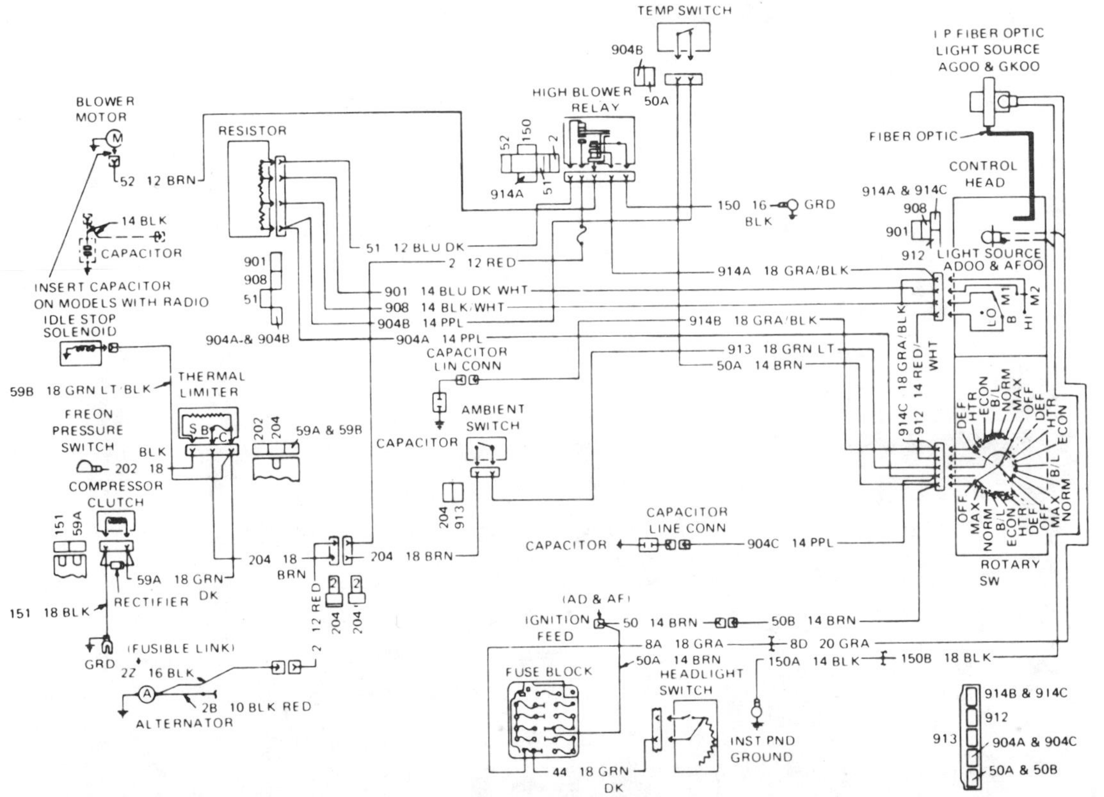

1976 Grand LeMans & LeMans Except V8-260 (With Auto Temp Control); Early 1976 Grand Prix (With Auto Temp Control, Less Cycling Clutch) A/C & Heater Wiring Circuit

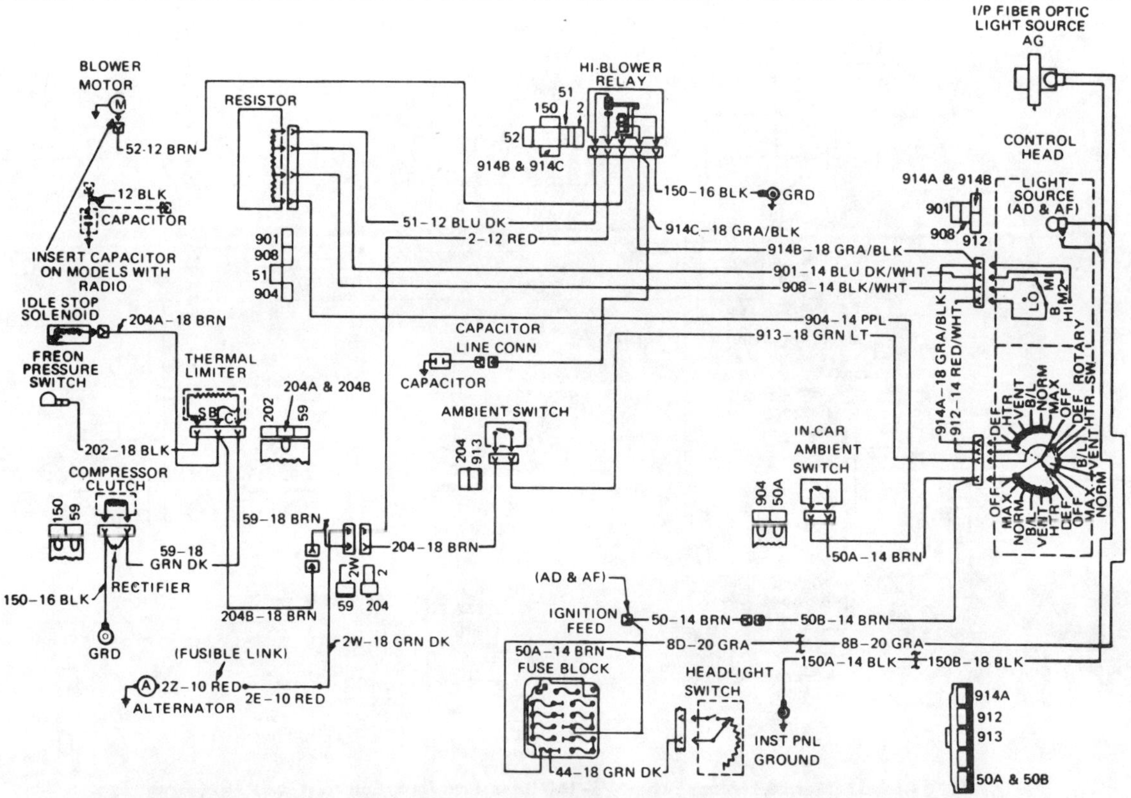

1976 Grand LeMans & LeMans With V8-260 A/C & Heater Wiring Circuit (Less Auto Temp Control)

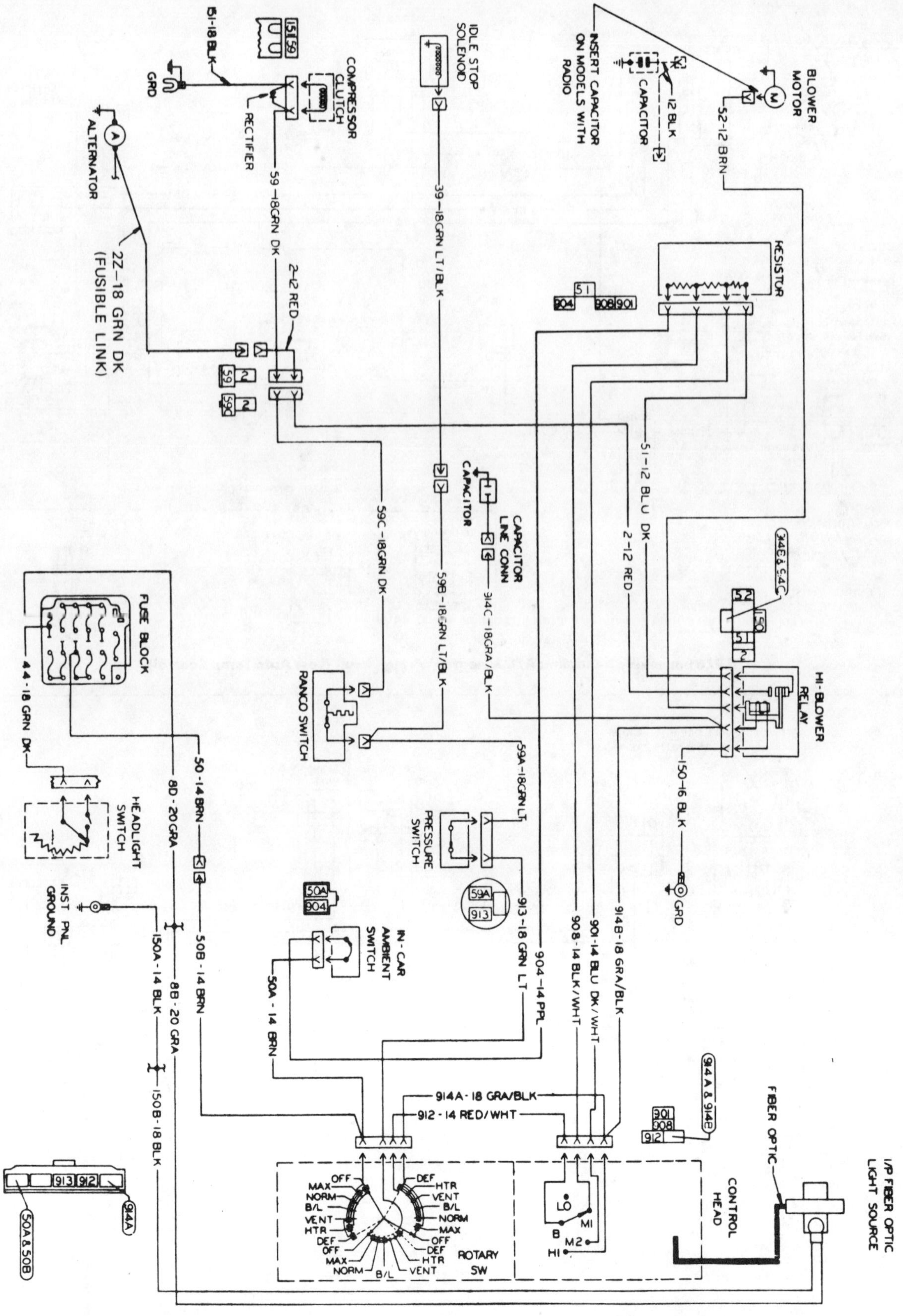

Late 1976 Grand Prix A/C & Heater Wiring Circuit (With Auto Temp Control & Cycling Clutch)

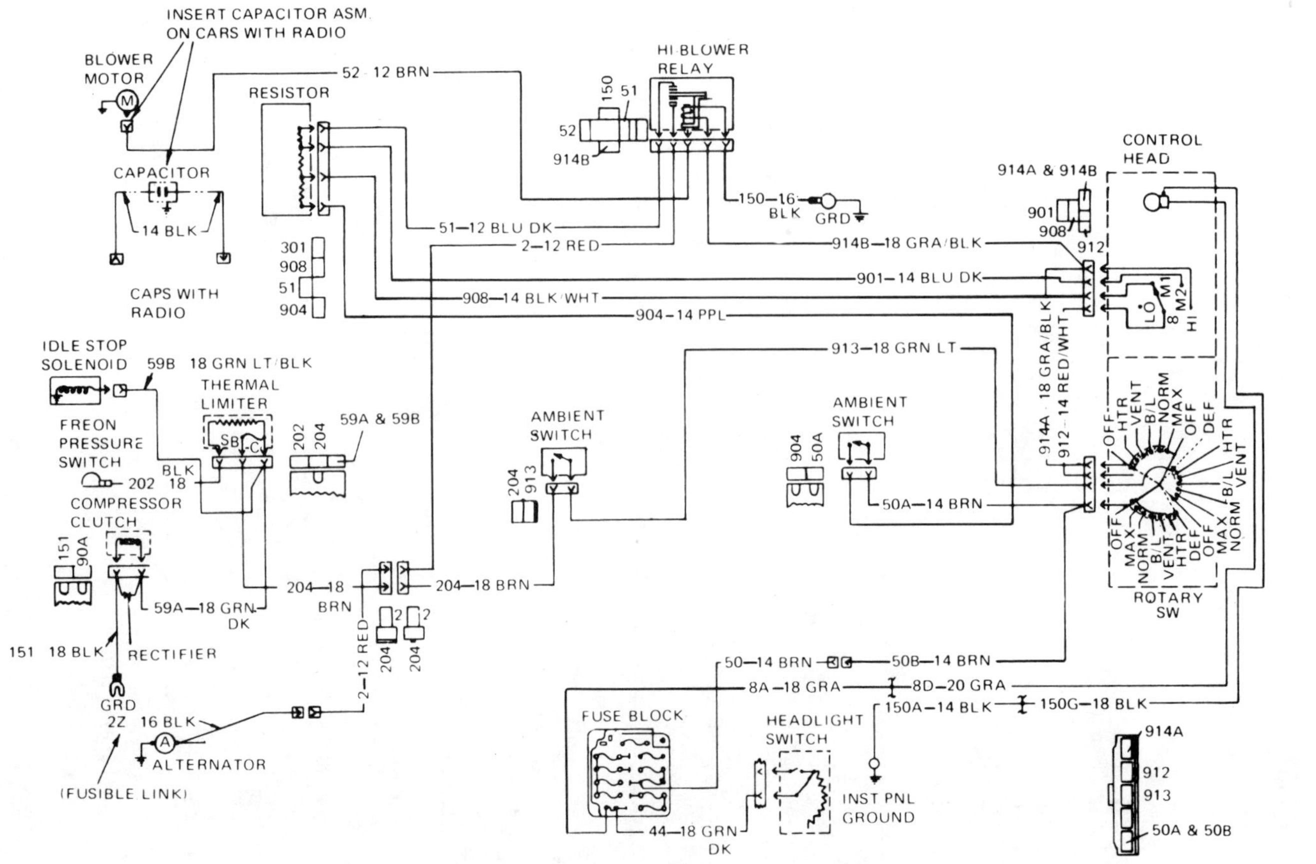

1976 Bonneville & Catalina A/C & Heater Wiring Circuit (Less Auto Temp Control)

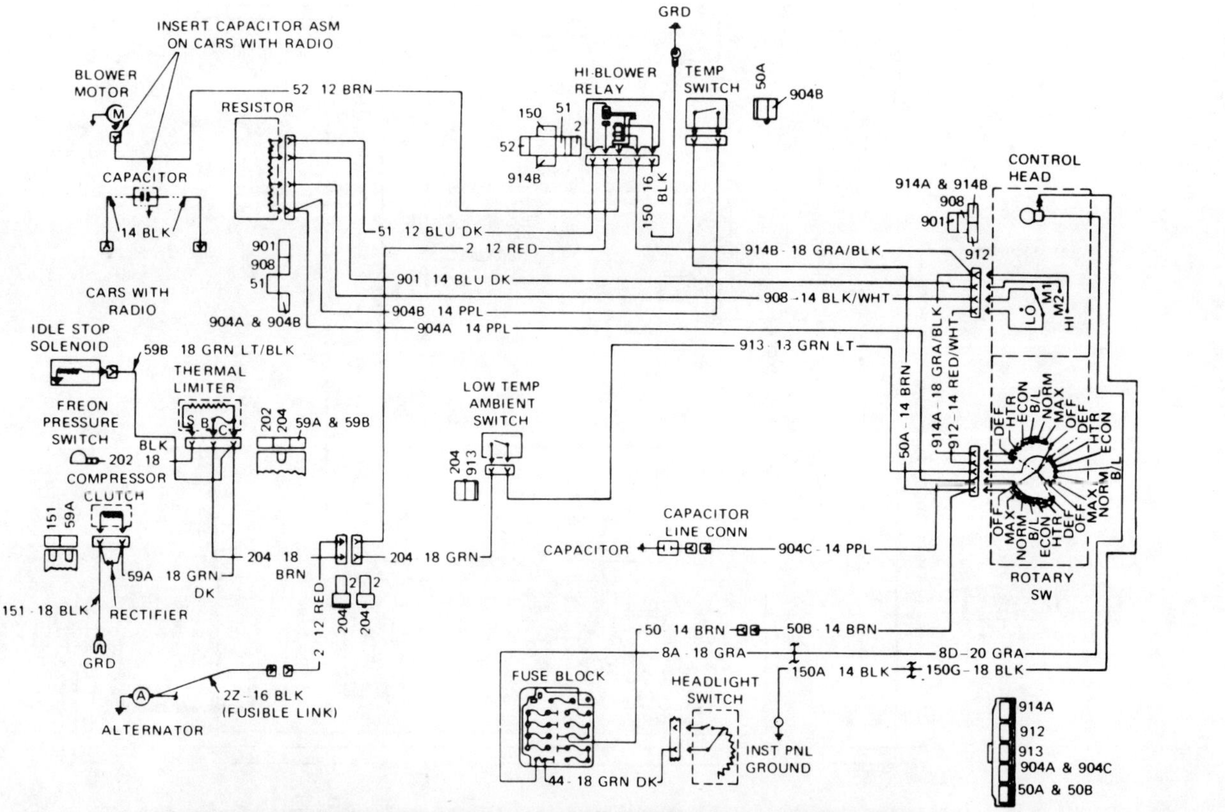

1976 Bonneville & Catalina A/C & Heater Wiring Circuit (With Auto Temp Control)

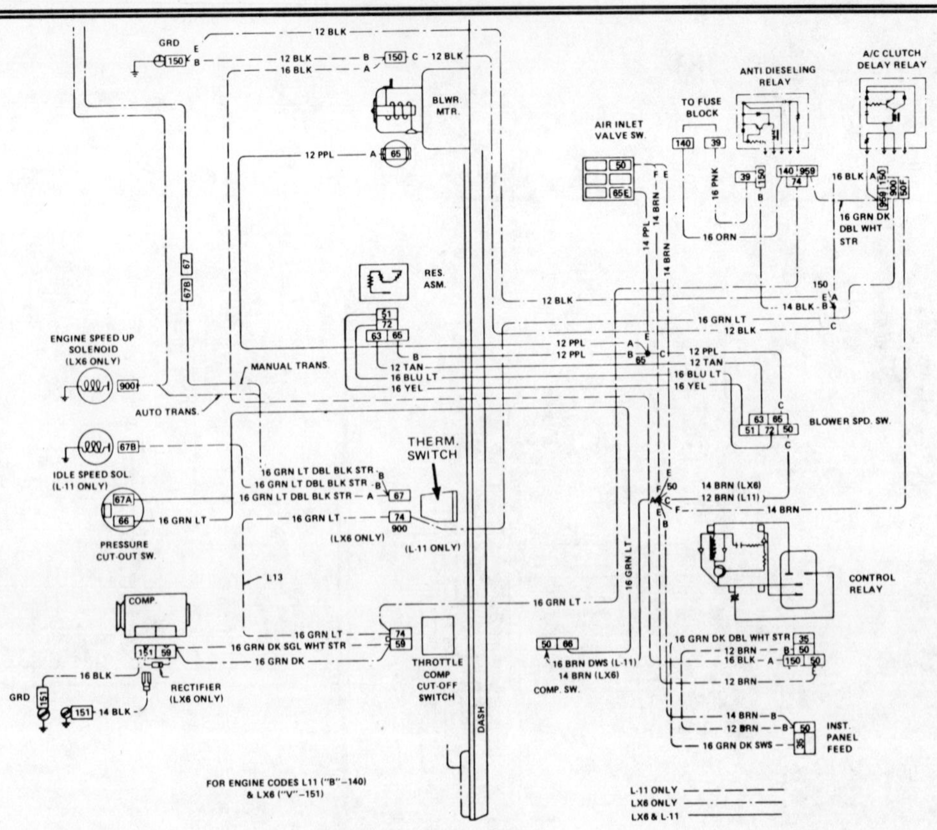

1977 Astre A/C & Heater Wiring Circuit

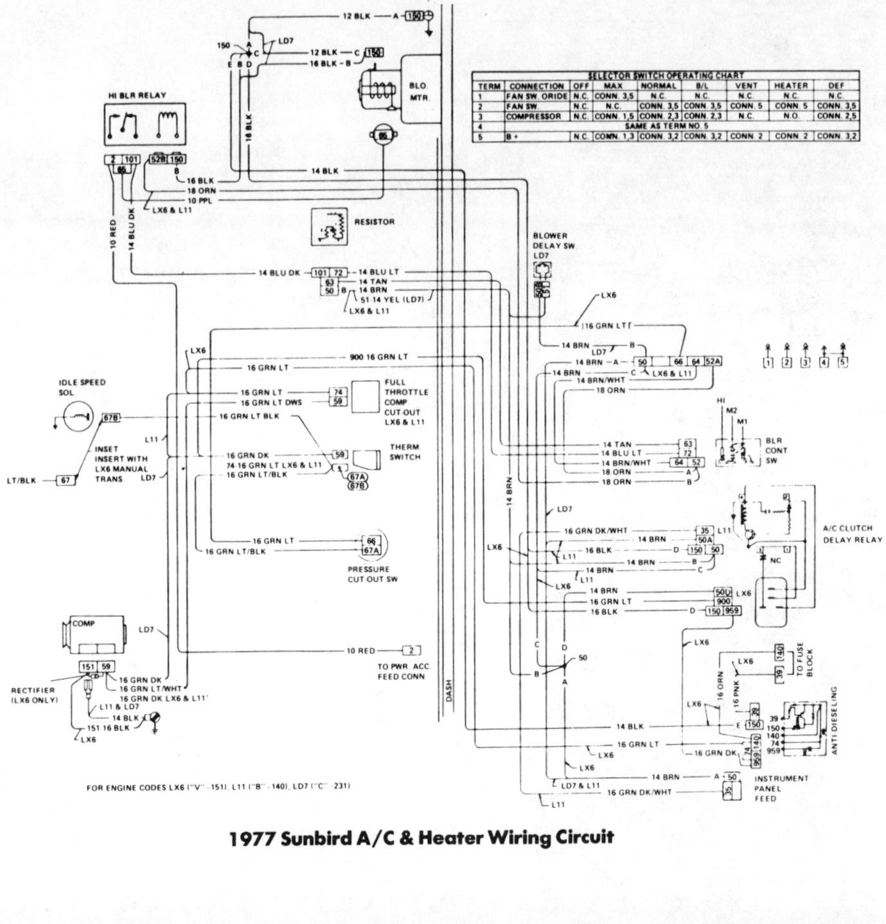

SELECTOR SWITCH OPERATING CHART								
TERM	CONNECTION	OFF	MAX	NORMAL	B/L	VENT	HEATER	DEF
1	FAN SW. ORIDE	N.C.	CONN. 3,5	N.C.	N.C.	N.C.	N.C.	N.C.
2	FAN SW.	N.C.	N.C.	CONN. 3,5	CONN. 3,5	CONN. 5	CONN. 5	CONN. 3,5
3	COMPRESSOR	N.C.	CONN. 1,5	CONN. 2,3	CONN. 2,3	N.C.	N.O.	CONN. 2,5
4	SAME AS TERM NO. 5							
5	B +	N.C.	CONN. 1,3	CONN. 3,2	CONN. 3,2	CONN. 2	CONN. 2	CONN. 3,2

1977 Sunbird A/C & Heater Wiring Circuit

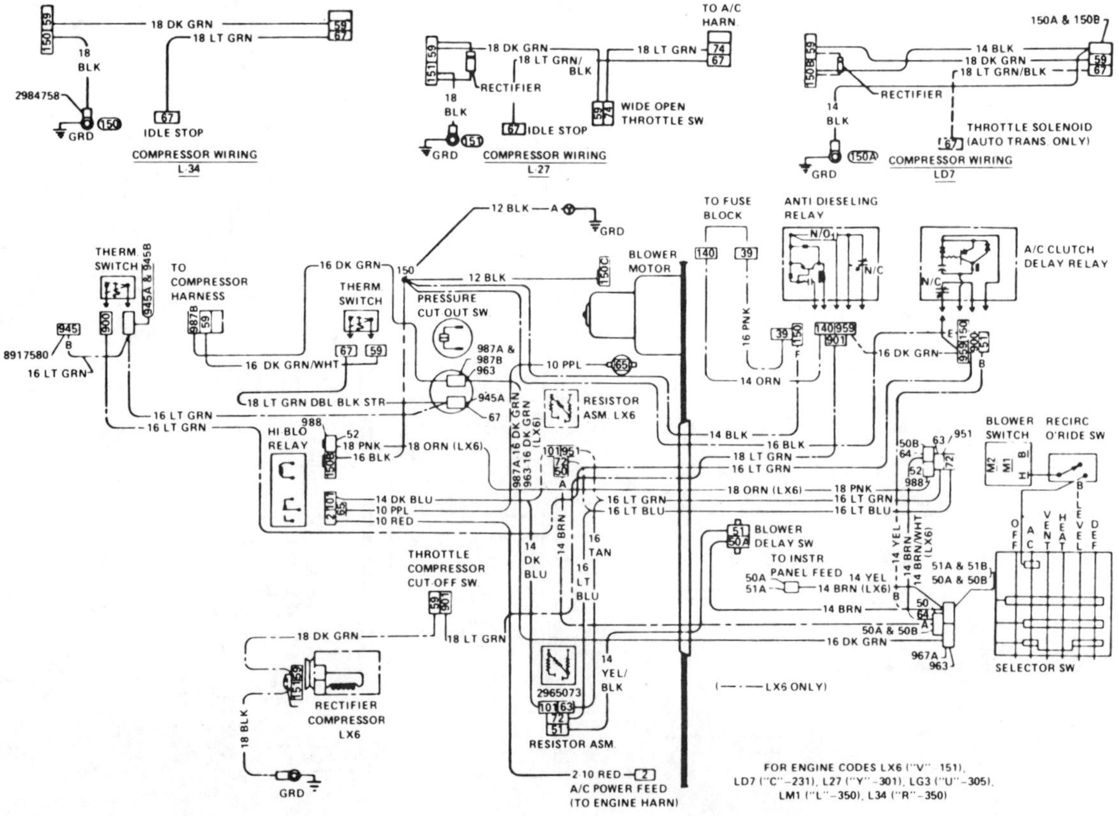

1977 Phoenix & Ventura A/C & Heater Wiring Circuit

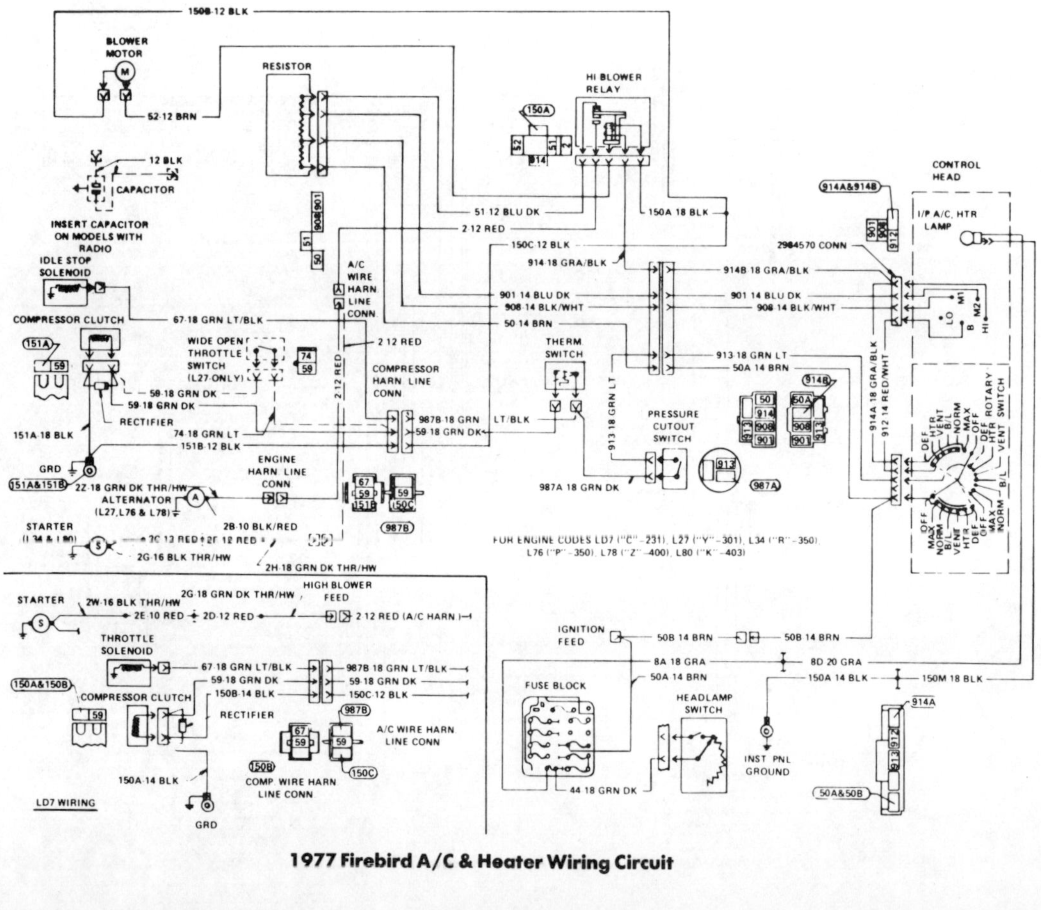

1977 Firebird A/C & Heater Wiring Circuit

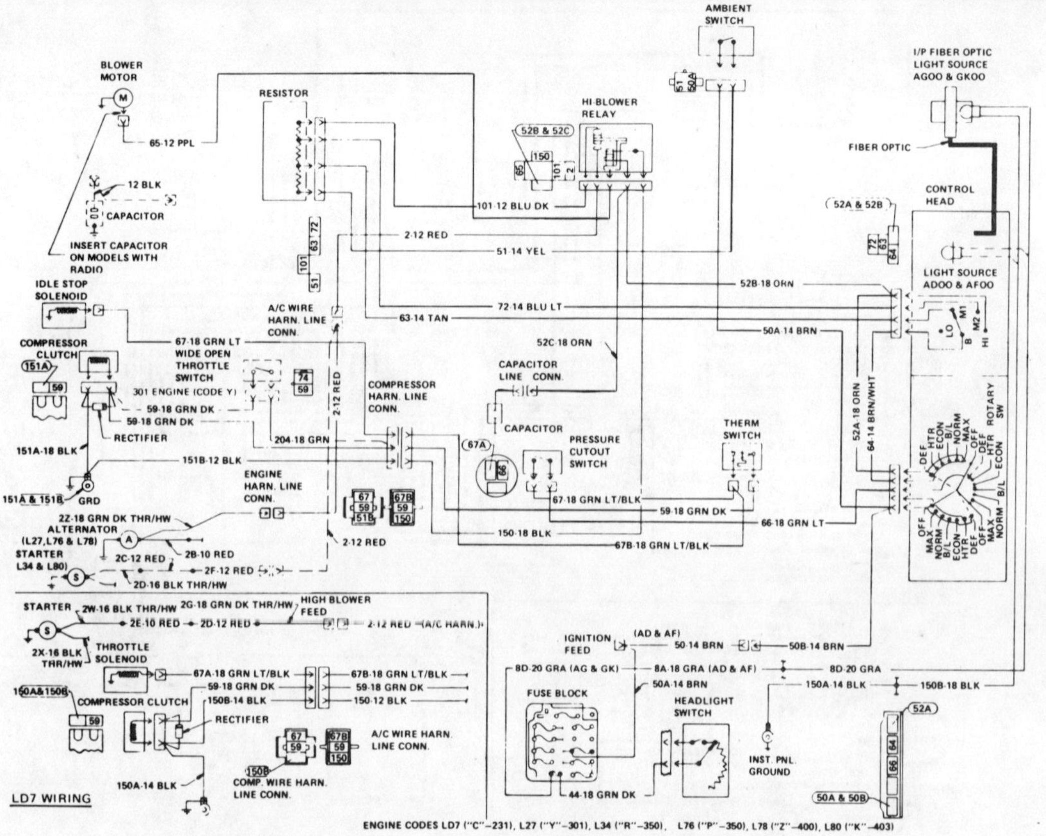

1977 Grand Am, Grand LeMans, Grand Prix & LeMans A/C & Heater Wiring Circuit (With Custom or Auto Temp Control)

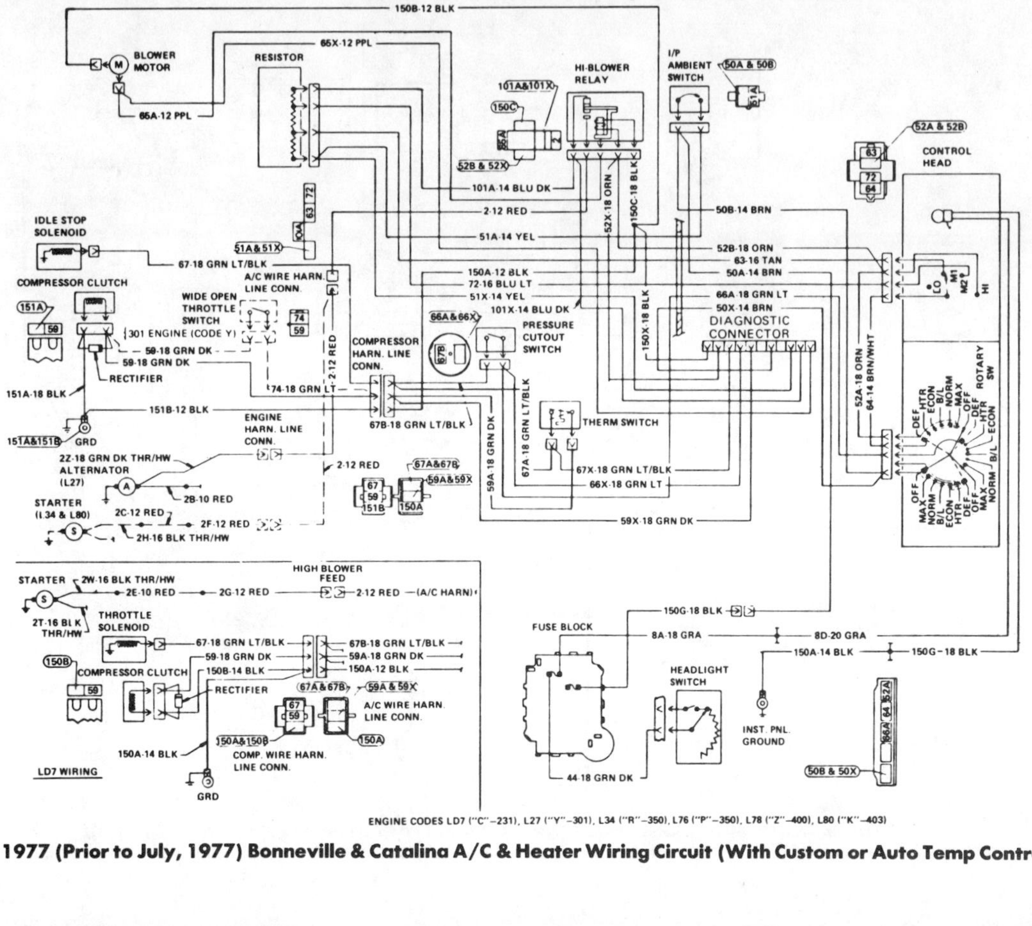

Early 1977 (Prior to July, 1977) Bonneville & Catalina A/C & Heater Wiring Circuit (With Custom or Auto Temp Control)

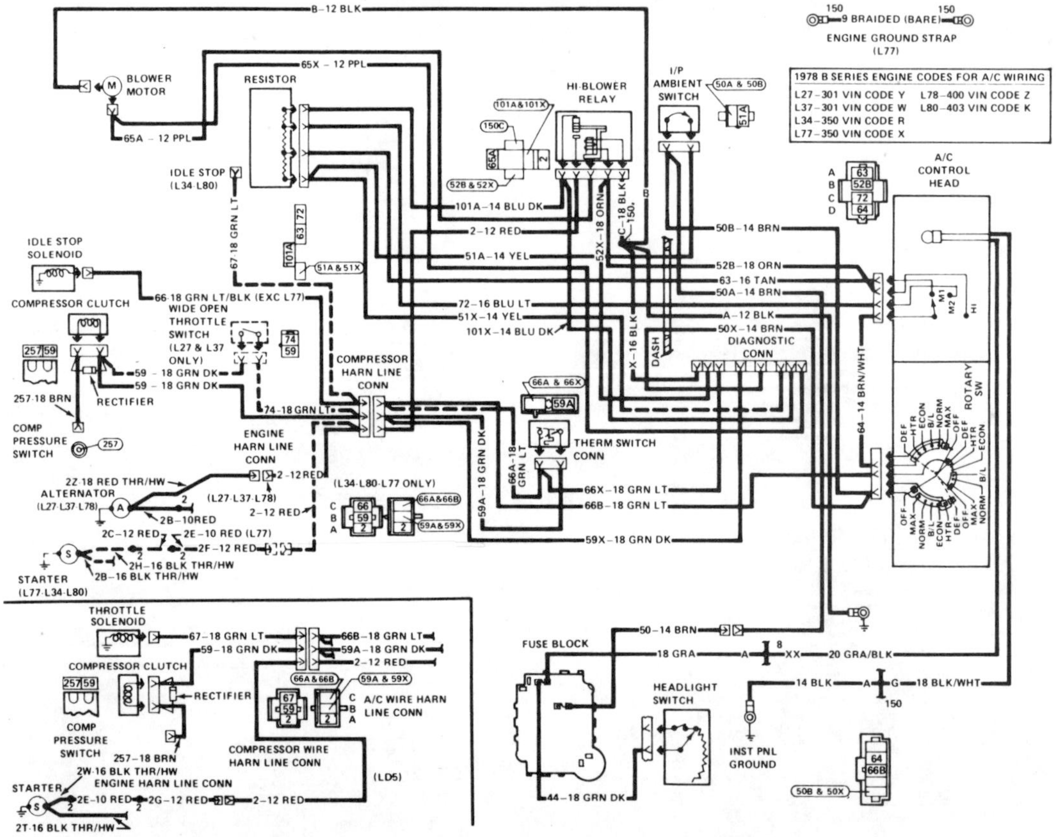

Late 1977 (Starting July 1977) & 1978 Bonneville & Catalina A/C & Heater Wiring Circuit (With Custom or Auto Temp Control)

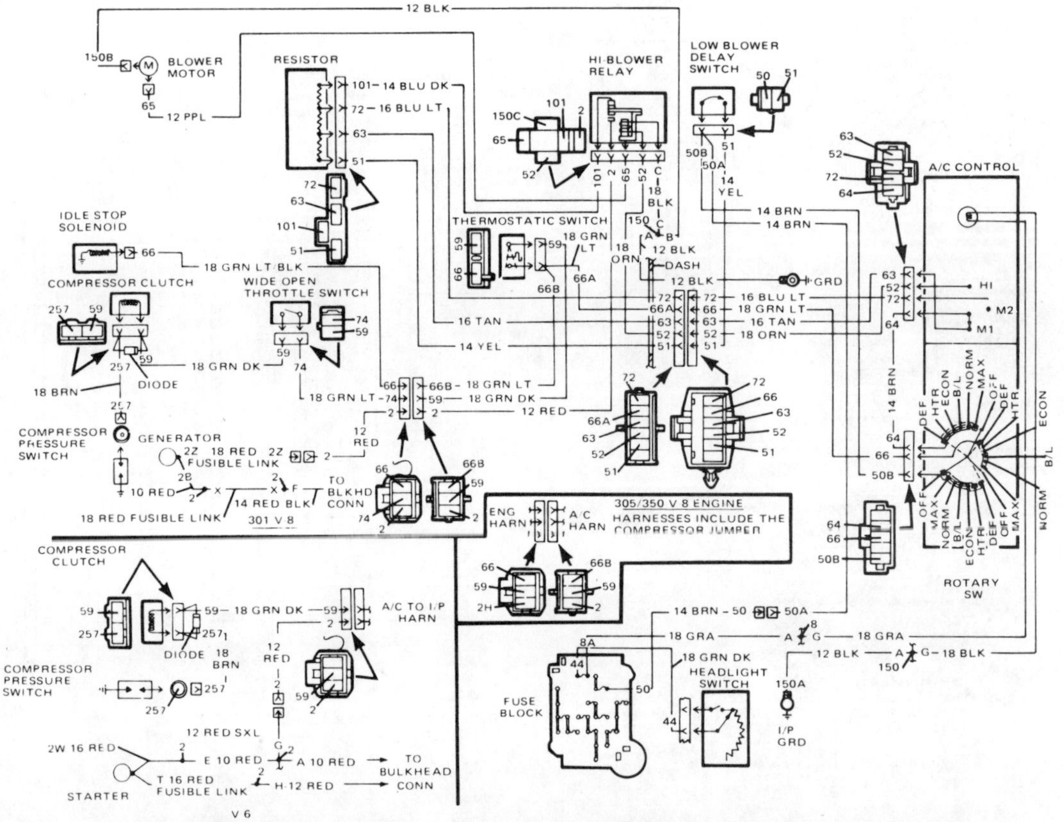

1978 Grand Am, Grand LeMans, Grand Prix & LeMans (With Custom or Auto Temp Control); 1979 Grand Am, Grand LeMans, Grand Prix & LeMans (Less Auto Temp Control) A/C & Heater Wiring Circuit

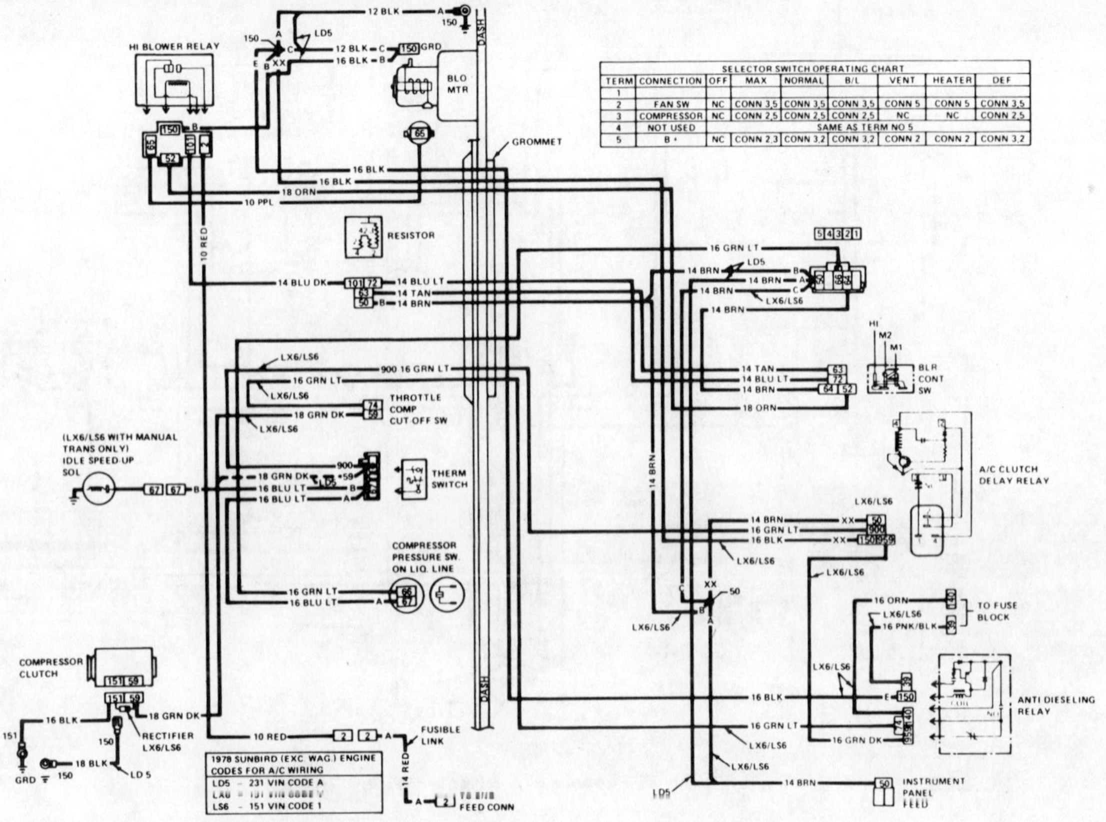

SELECTOR SWITCH OPERATING CHART								
TERM	CONNECTION	OFF	MAX	NORMAL	B/L	VENT	HEATER	DEF
1								
2	FAN SW	NC	CONN 3,5	CONN 3,5	CONN 3,5	CONN 5	CONN 5	CONN 3,5
3	COMPRESSOR	NC	CONN 2,5	CONN 2,5	CONN 2,5	NC	NC	CONN 2,5
4	NOT USED	SAME AS TERM NO 5						
5	B +	NC	CONN 2,3	CONN 3,2	CONN 3,2	CONN 2	CONN 2	CONN 3,2

1978-79 Sunbird Except Station Wagon A/C & Heater Wiring Circuit

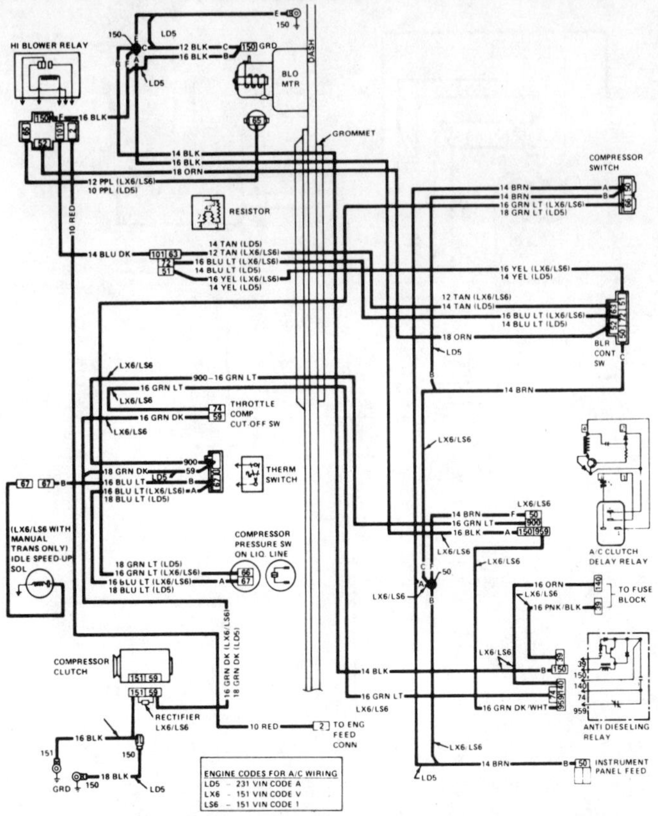

1978-79 Sunbird Station Wagon A/C & Heater Wiring Circuit

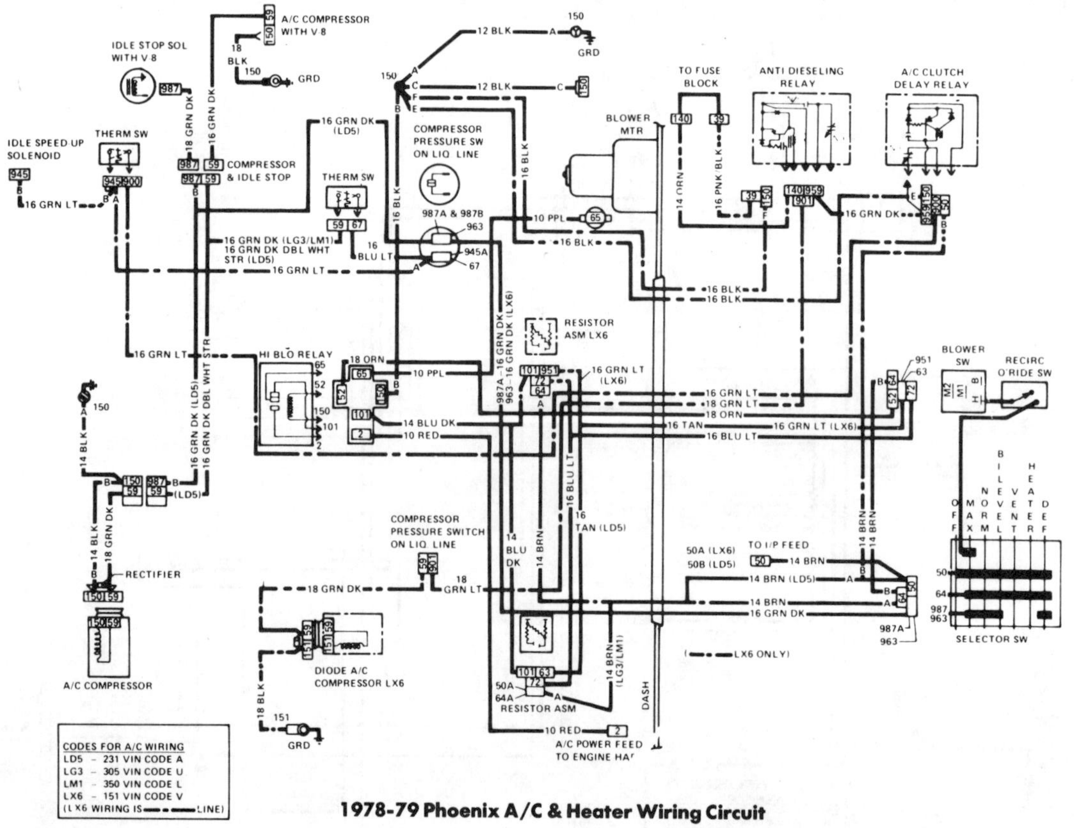

1978-79 Phoenix A/C & Heater Wiring Circuit

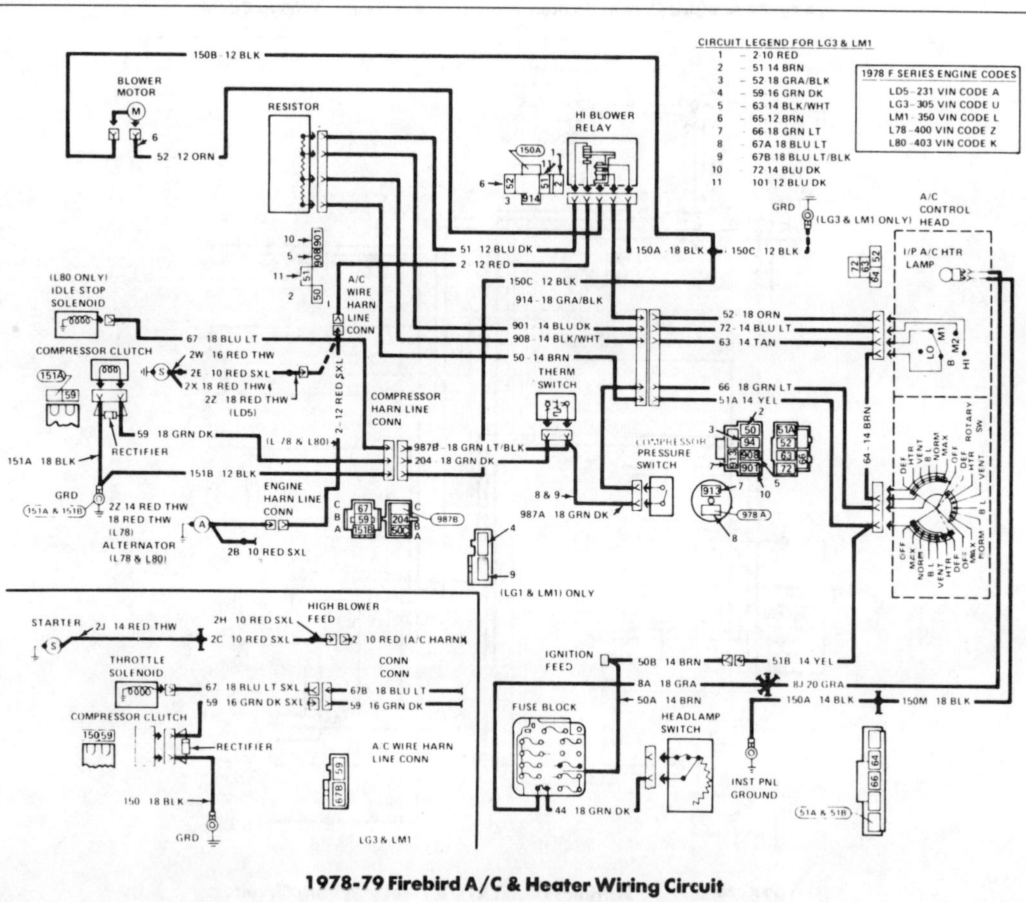

1978-79 Firebird A/C & Heater Wiring Circuit

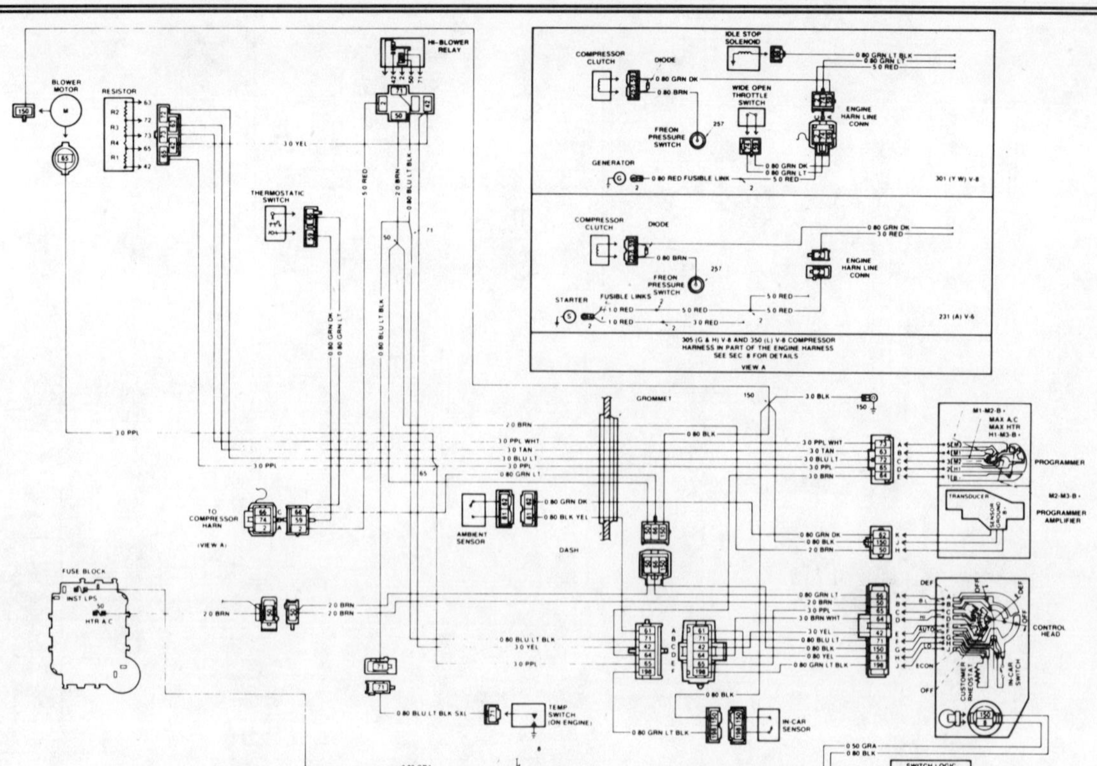

1979 Grand Am, Grand LeMans, Grand Prix & LeMans A/C & Heater Wiring Circuit (With Auto Temp Control)

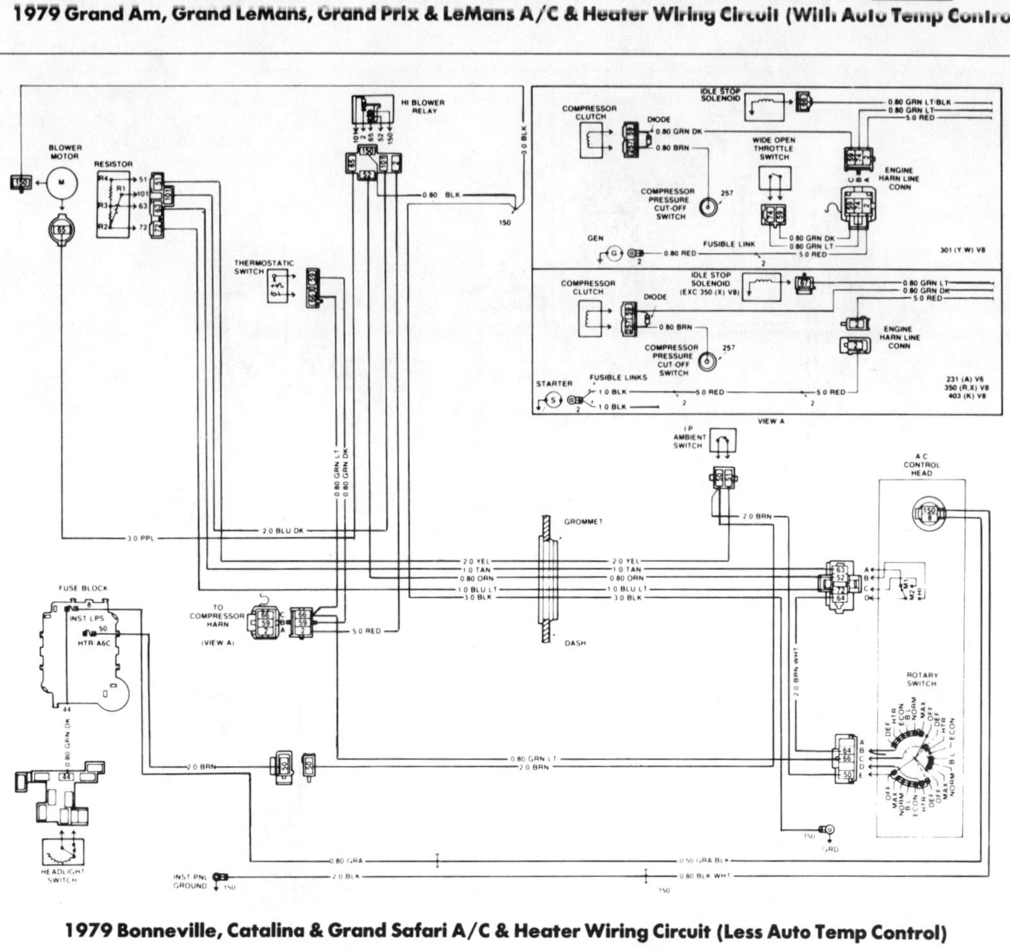

1979 Bonneville, Catalina & Grand Safari A/C & Heater Wiring Circuit (Less Auto Temp Control)

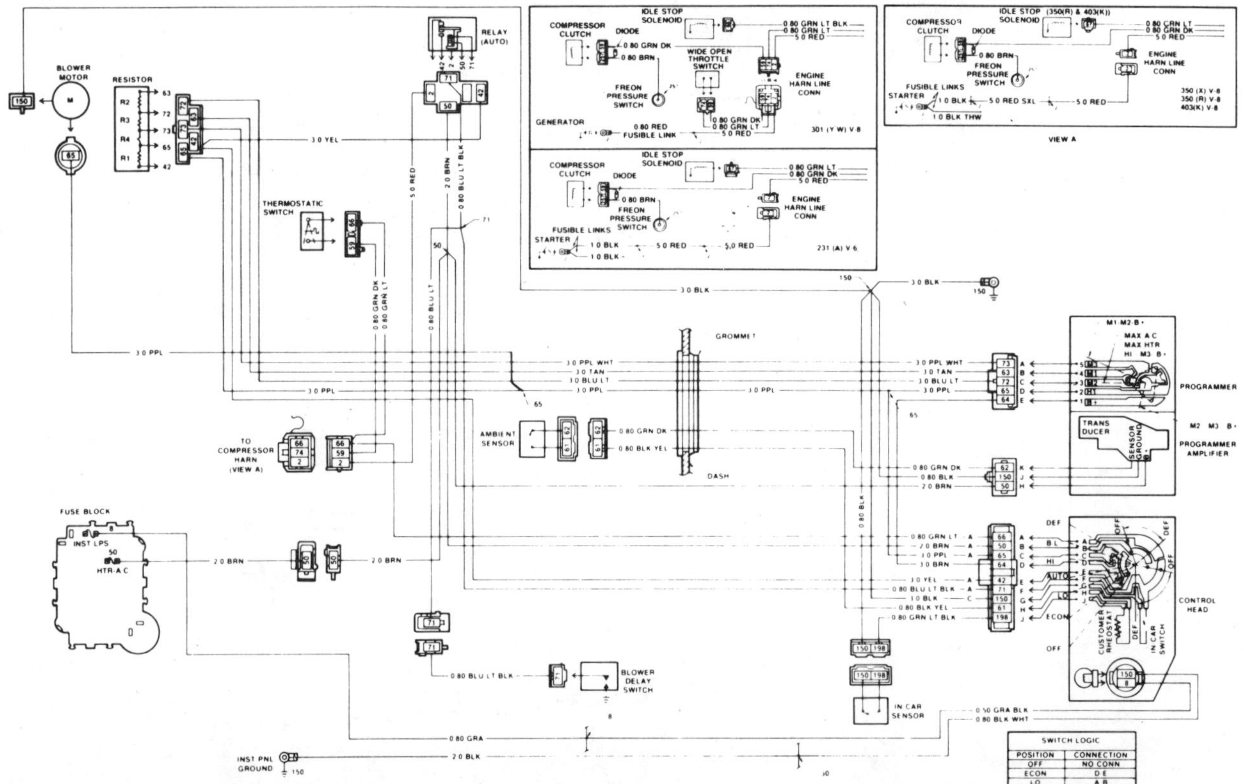

SWITCH LOGIC	
POSITION	CONNECTION
OFF	NO CONN
ECON	D E
LO	A B
AUTO	A B D E
HI	A B C E
BI LEVEL	A B D E
DEFROST	A B C E F G

1979 Bonneville & Catalina A/C & Heater Wiring Circuit (With Auto Temp Control)

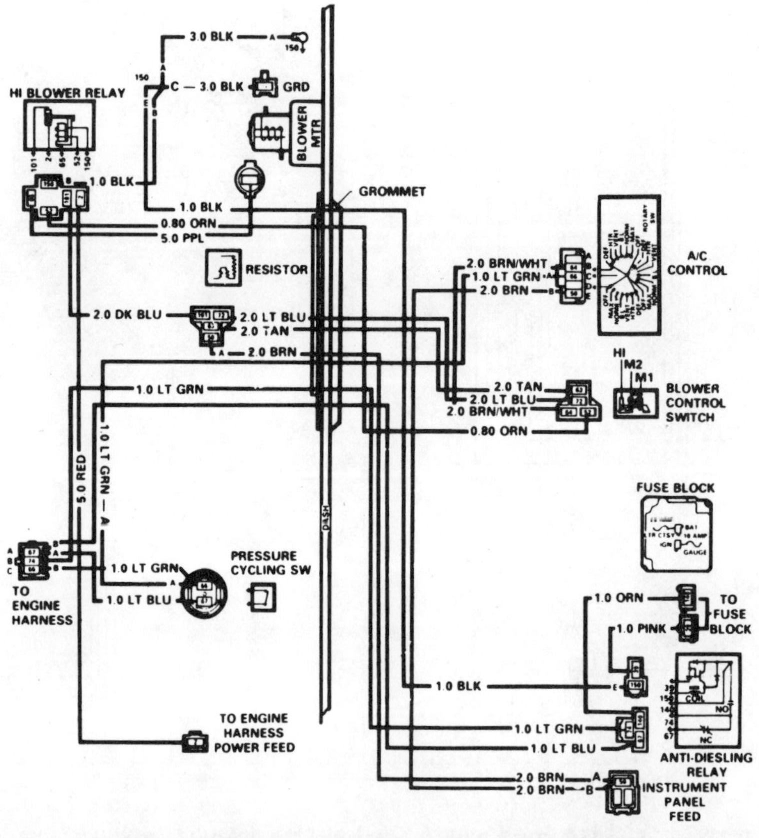

1980 Sunbird A/C & Heater Wiring Circuit (With 4-151)

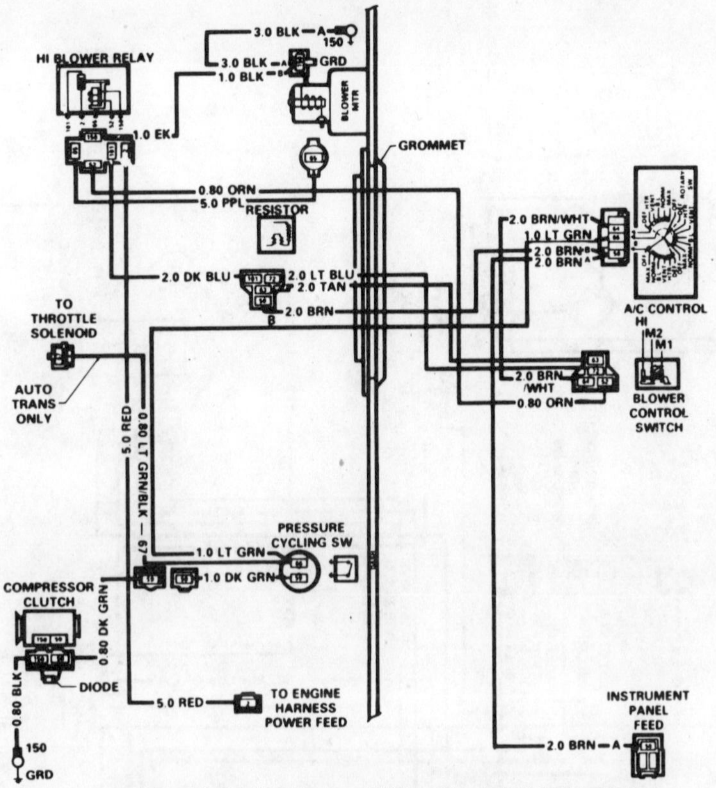

1980 Sunbird A/C & Heater Wiring Circuit (With V6-231)

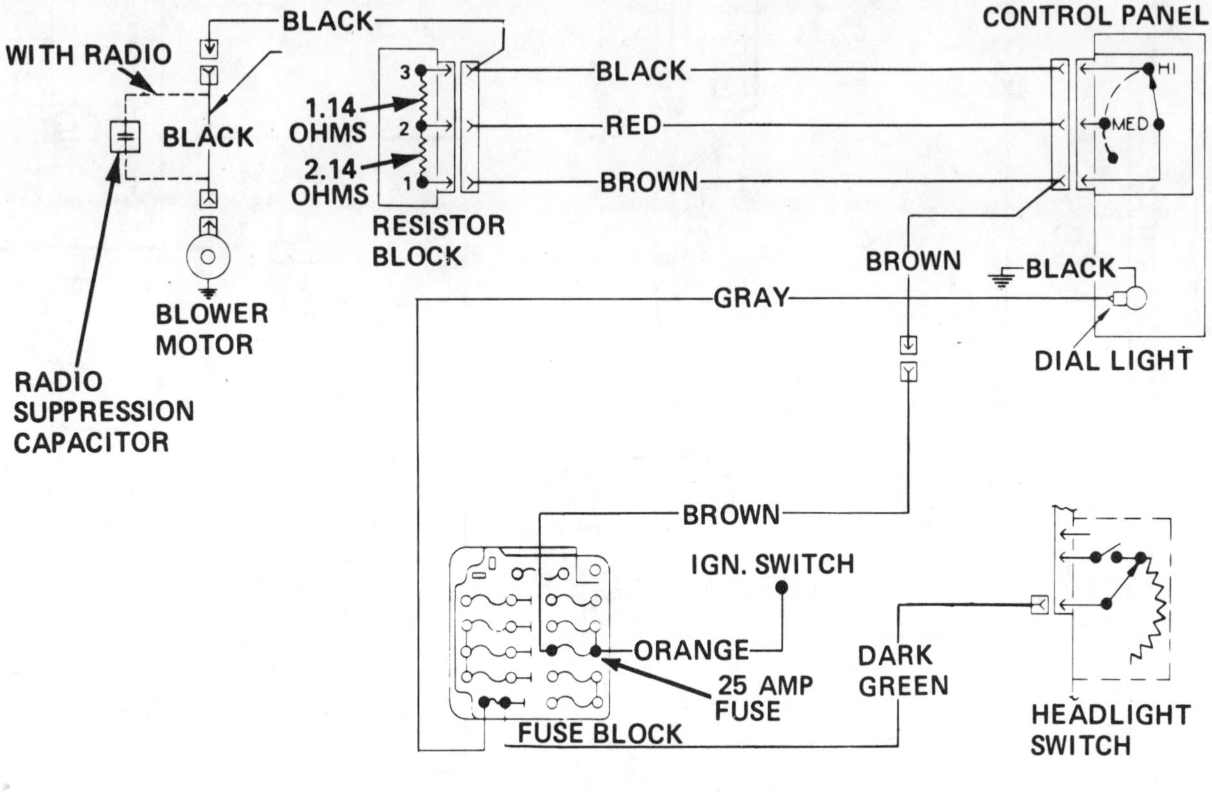

1974 Bonneville, Catalina & Grand Ville Heater Wiring Circuit

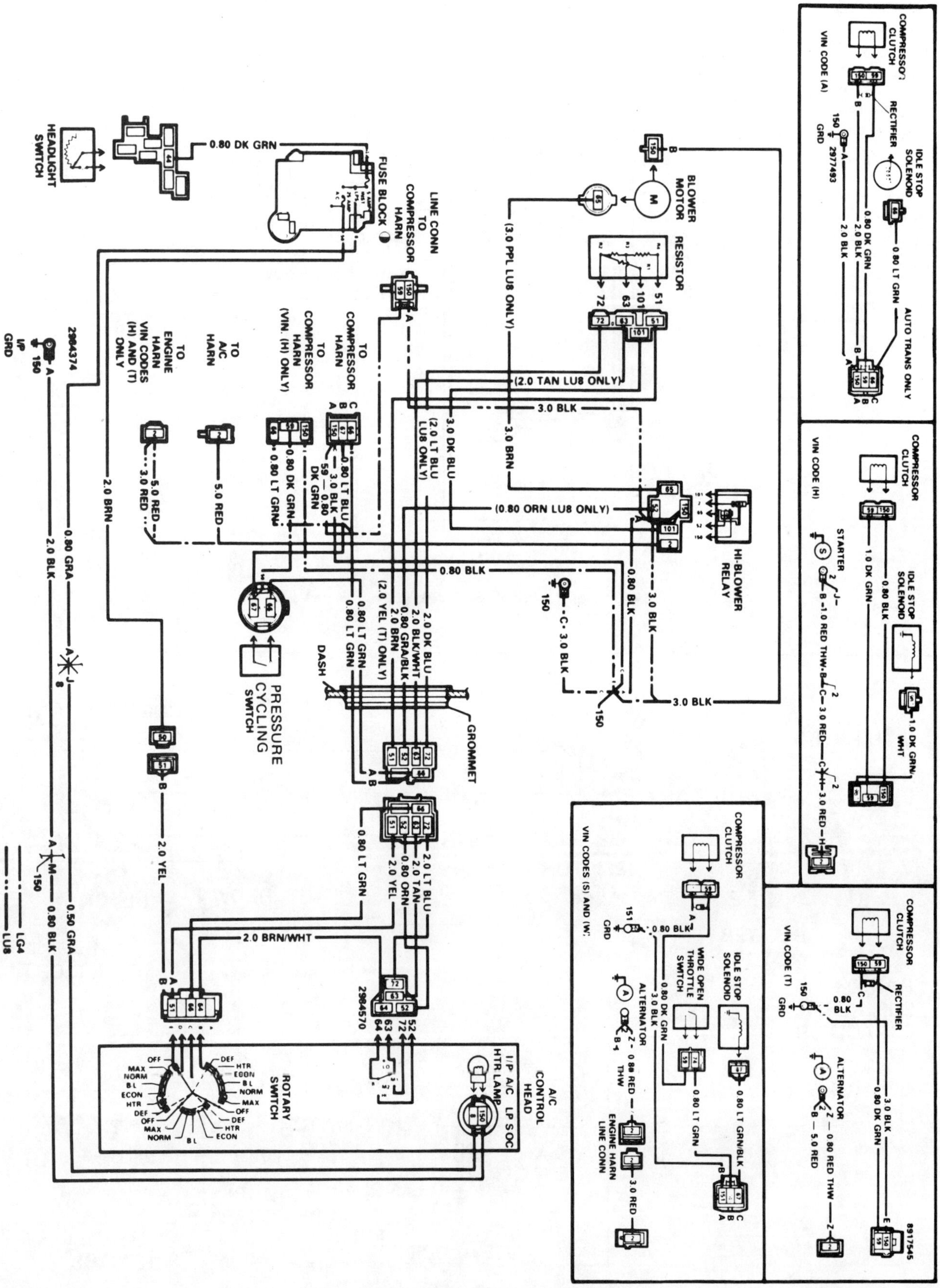

1980 Firebird A/C & Heater Wiring Circuit

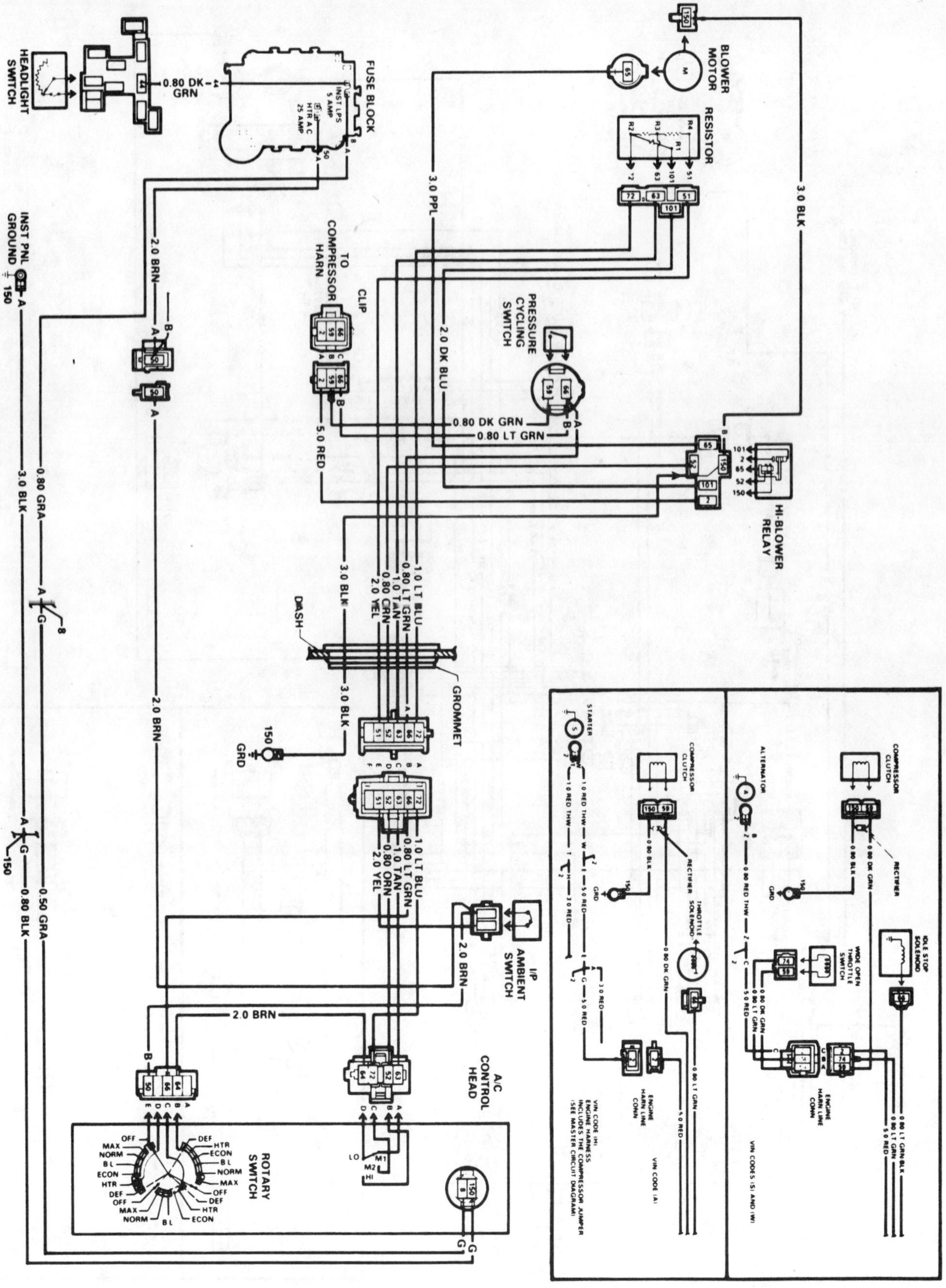

1980 Grand Am, Grand LeMans, Grand Prix & LeMans A/C & Heater Wiring Circuit (Less Auto Temp Control)

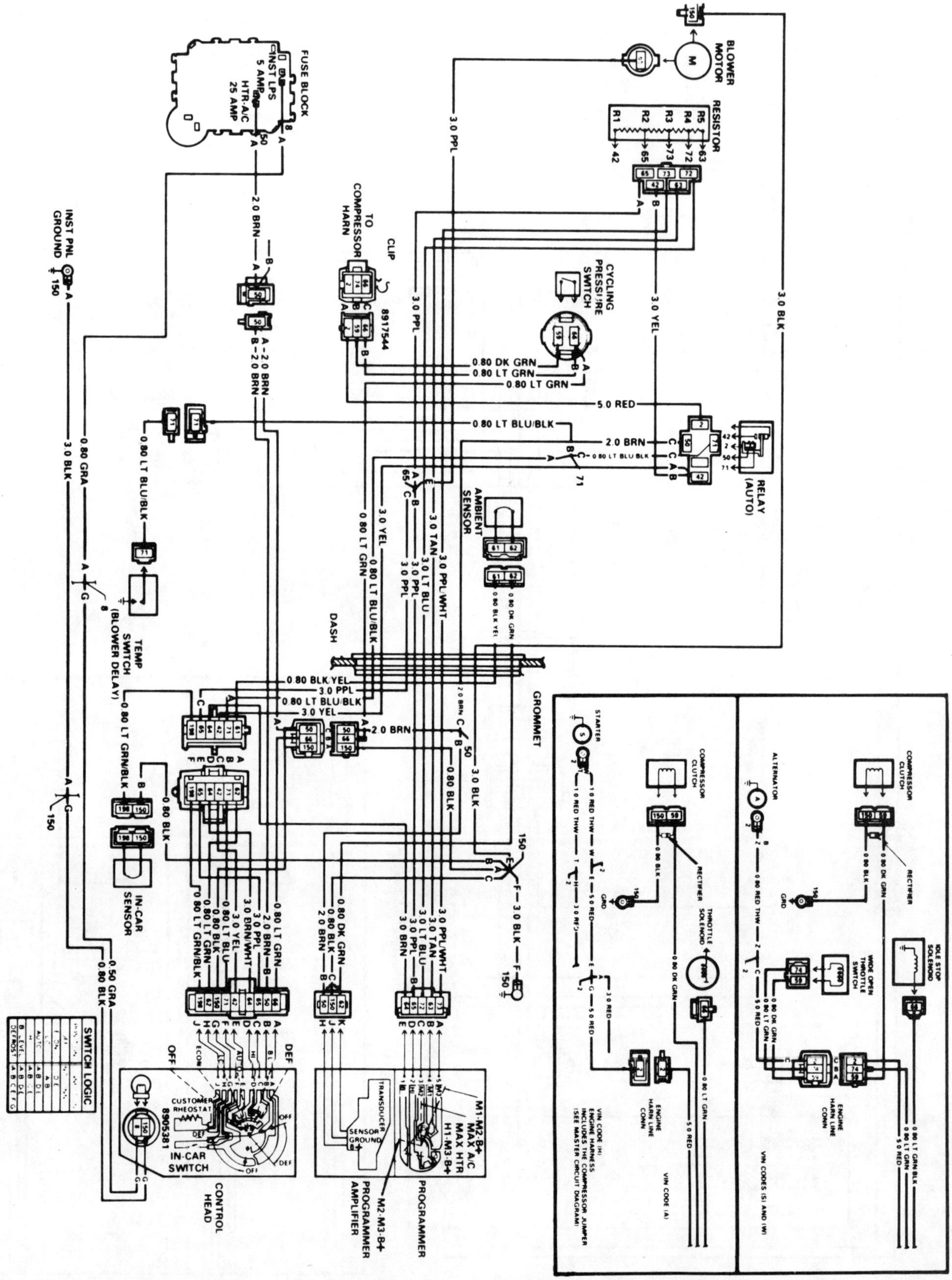

1980 Grand Am, Grand LeMans, Grand Prix & LeMans A/C & Heater Wiring Circuit (With Auto Temp Control)

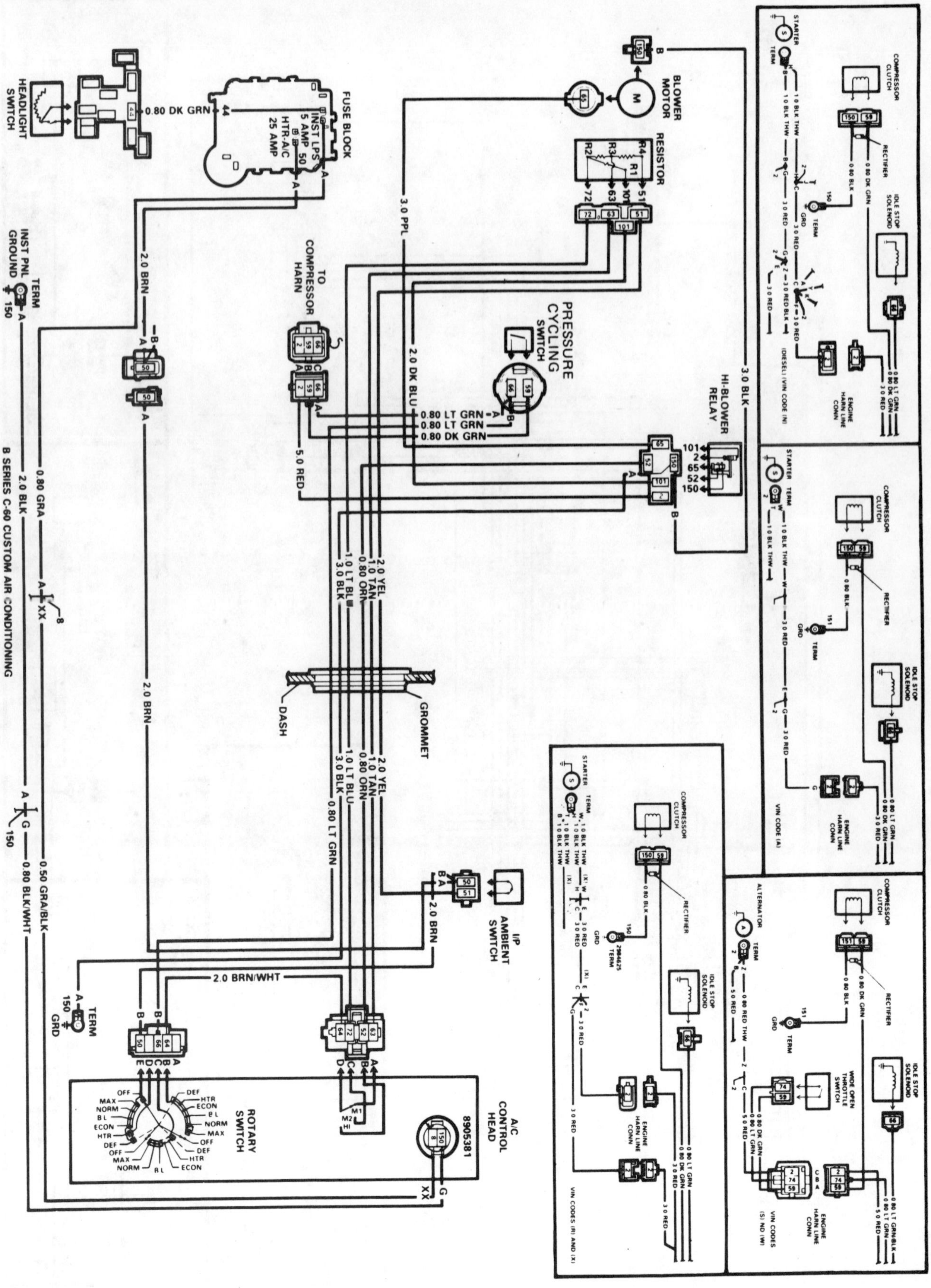

1980 Bonneville & Catalina A/C & Heater Wiring Circuit (Less Auto Temp Control)

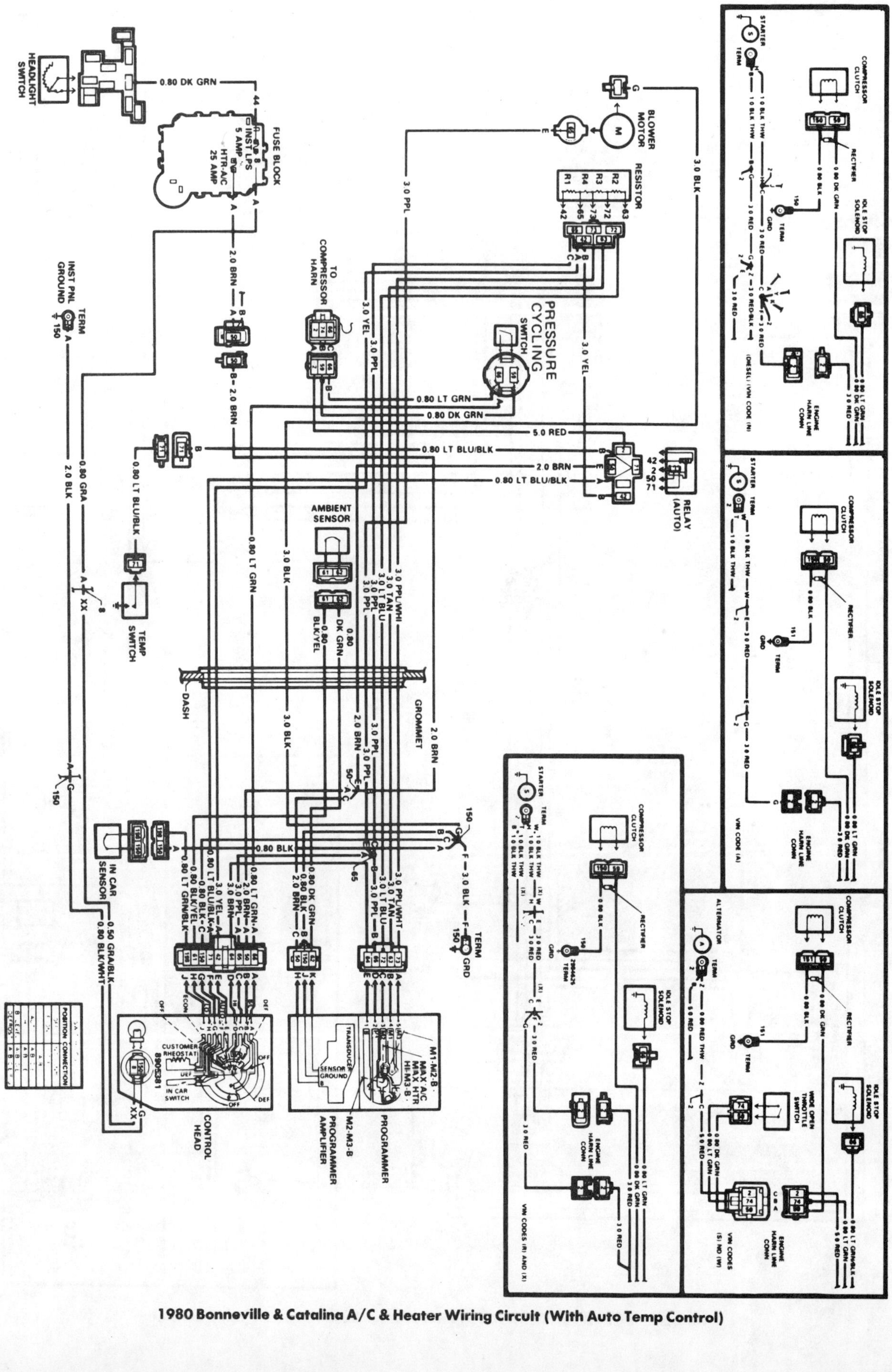

1980 Bonneville & Catalina A/C & Heater Wiring Circuit (With Auto Temp Control)

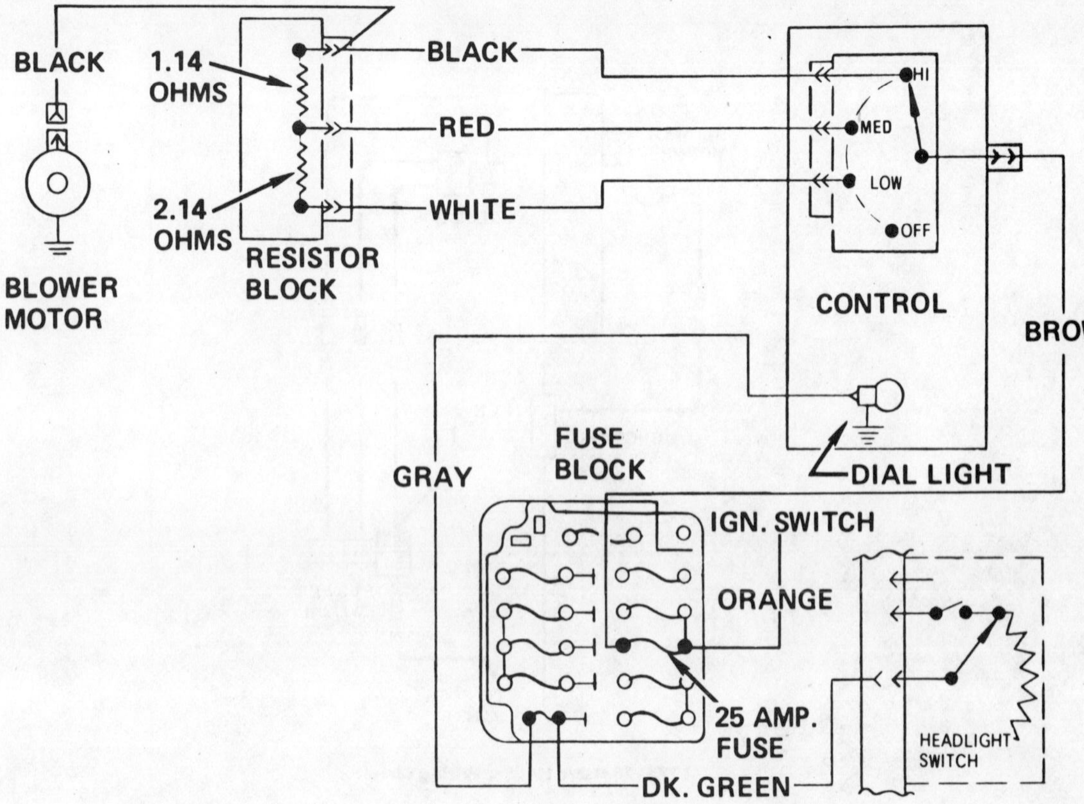

1974-76 Firebird Heater Wiring Circuit

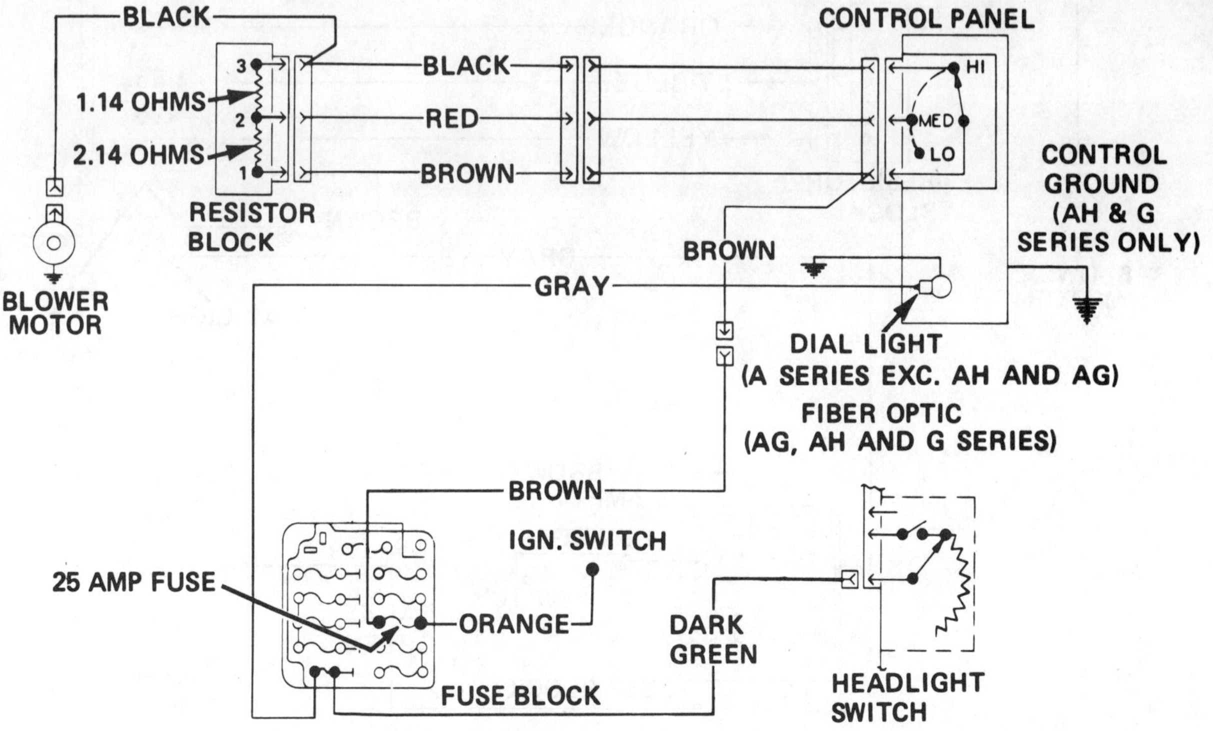

1974-76 Grand Am, Grand LeMans, Grand Prix & LeMans Heater Wiring Circuit

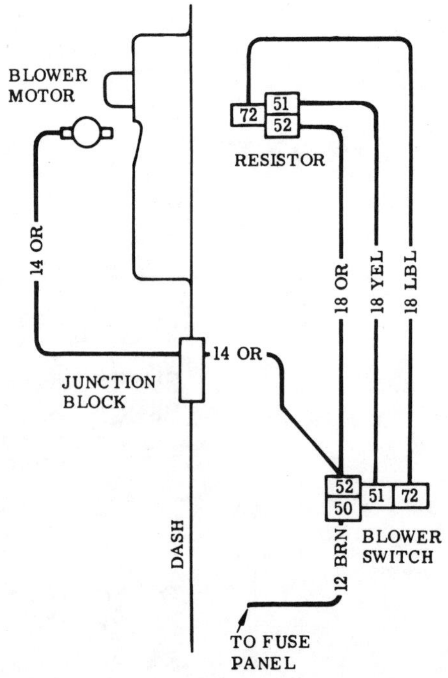

1975-76 Astre Heater Wiring Circuit

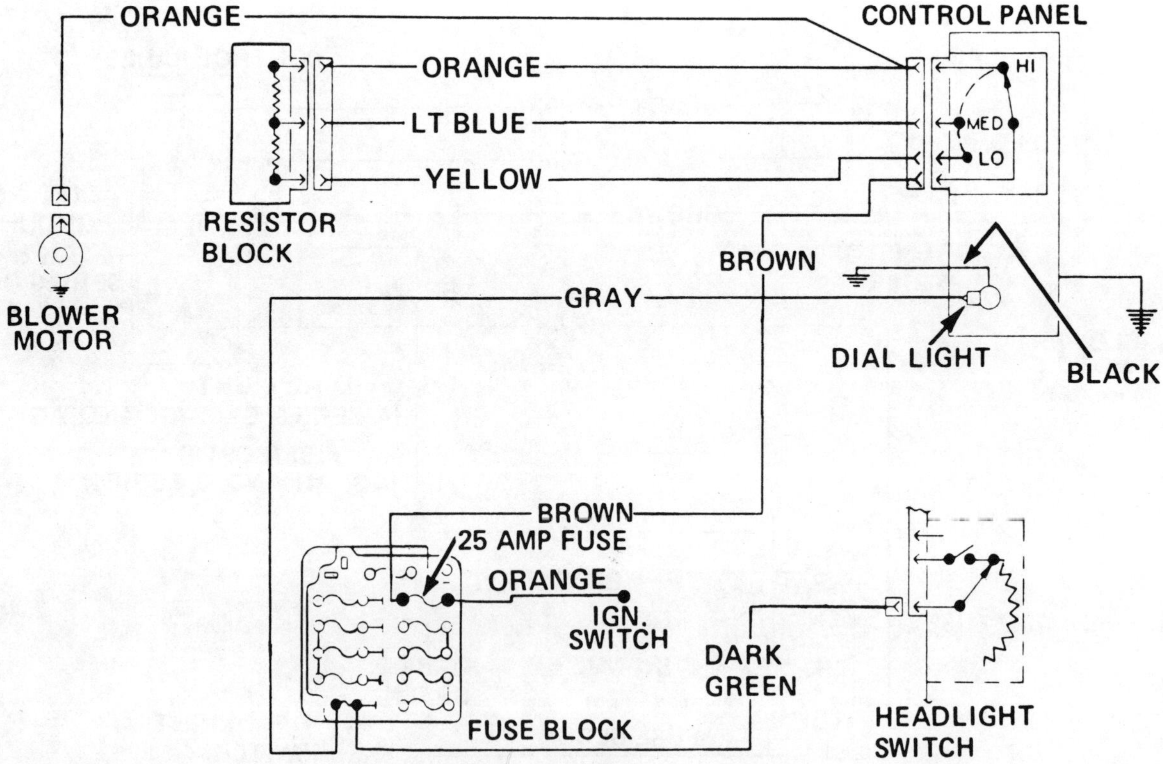

1975-76 Ventura Heater Wiring Circuit